S0-AYR-876

TX 5-653-781

INSTRUMENT ENGINEERS' HANDBOOK

Third Edition

Process Software and Digital Networks

INSTRUMENT ENGINEERS' HANDBOOK

Third Edition

Process Software and Digital Networks

Béla G. Lipták

EDITOR-IN-CHIEF

ISA–The Instrumentation, Systems, and Automation Society 

CRC PRESS

Boca Raton London New York Washington, D.C.

This reference text is published in cooperation with ISA Press, the publishing division of ISA—Instrumentation, Systems, and Automation Society. ISA is an international, nonprofit, technical organization that fosters advancement in the theory, design, manufacture, and use of sensors, instruments, computers, and systems for measurement and control in a wide variety of applications. For more information, visit www.isa.org or call (919) 549-8411.

Library of Congress Cataloging-in-Publication Data

Instrument engineers' handbook : process software and digital networks / Béla G. Lipták, editor-in-chief.—3rd ed.
p. cm.
"The third edition of the *IEH* was initially planned for three volumes ... to cover the subjects of process measurement, process control, and process software. Chilton published the first two volumes in 1995. In October 2000, CRC Press obtained the rights to publish the third volume."—Pref.
Includes bibliographical references and index.
ISBN 0-8493-1082-2 (alk. paper)
1. Process control—Handbooks, manuals, etc. 2. Measuring instruments—Handbooks, manuals, etc. I. Lipták, Béla G. II. Lipták, Béla G. Instrument engineers' handbook. Process measurement and analysis III. Lipták, Béla G. Instrument engineers' handbook. Process control.

TS156.8 .I56 2002
681′.2—dc21 2002017478

Direct all inquiries to CRC Press LLC, 2000 N.W. Corporate Blvd., Boca Raton, Florida 33431.

Visit the CRC Press Web site at www.crcpress.com

International Standard Book Number 0-8493-1082-2
Library of Congress Card Number 2002017478
Printed in the United States of America 1 2 3 4 5 6 7 8 9 0
Printed on acid-free paper

Dedicated to you, my colleagues, the instrument and control engineers, hoping that by applying the knowledge found in these pages, you will make the world a better, safer, and happier place and thereby will also advance the recognition and the respectability of the I&C profession

CONTENTS

CONTRIBUTORS

The name of the author is given at the beginning of each section. Here, all the contributors of this volume are listed in alphabetical order, giving their academic degrees and affiliations they held at the time of publication.

MIGUEL J. BAGAJEWICZ PhD, AIChe, Professor, University of Oklahoma, Norman, Oklahoma, U.S.A.

CHET S. BARTON PE, BSEE, Senior Process Automation Engineer, Jacobs Engineering, Baton Rouge, Louisiana, U.S.A.

JONAS BERGE Engineer, Smar, Singapore

PETER GRAHAM BERRIE BScE, PhD, AIChe, Marketing Communications, Endress+Hauser Process Solutions AG, Reinach, Switzerland

VIPUL A. BHAVSAR BE (I&C), Diploma in Management, Consultant–Control Systems Engineer, Canada

STUART A. BOYER PE, BSc, EE, President, Iliad Engineering, Inc., Canada

GEORGE C. BUCKBEE PE, BSChE, MSChE, Control Engineer, Top Control, Clarks Summit, Pennsylvania, U.S.A.

ERIC J. BYRES PE, Research Faculty, British Columbia Institute of Technology, Canada

DANIEL E. CAPANO President, Diversified Technical Services, Inc., Stamford, Connecticut, U.S.A.

RICHARD H. CARO BchE, MS, MBA, CMC Associates, Acton, Massachusetts, U.S.A.

HARRY L. CHEDDIE PE, BSc, Principal Engineer, Exida.com, Sarnia, Ontario, Canada

SCOTT C. CLARK BS, ChE, Project Engineer, Merck & Co., Inc., Elkton, Virginia, U.S.A.

ASGEIR DRØIVOLDSMO MS, Research Scientist, OECD Halden Reactor Project, Halden, Norway

SHIBI EMMANUEL MTech, BTech, Head of I&C, Dar Al Riyadh Consultants, Al Khobar, Saudi Arabia

HALIT EREN BSc, MEng, PhD, MBA, Senior Lecturer, Curtin University of Technology, Perth, Australia

LUDGER FÜCHTLER Dipl. Eng., Marketing Manager, Endress+Hauser Process Solutions AG, Reinach, Switzerland

BRUCE J. GEDDES PE, BSME, Framatome ANP DE&S, Charlotte, North Carolina, U.S.A.

JOHN PETER GERRY PE, BSChE, MSChE, President, ExperTune, Inc., Hubertus, Wisconsin, U.S.A.

ASISH GHOSH CE, MIEE, Independent Consultant, Wrentham, Massachusetts, U.S.A.

IAN H. GIBSON BSc, Dipl. ChE, Dipl. Inst.Tech., Principal Technical Specialist-Process & Control Systems, Fluor Australia Pty Ltd, Melbourne, Australia

HASHEM MEHRDAD HASHEMIAN MSNE, ISA Fellow, President, Analysis and Measurement Services Corp., Knoxville, Tennessee, U.S.A.

HEROLD I. HERTANU PE, MSEE, President, HLP Associates, New York, U.S.A.

KARLENE A. HOO BS, MS, PhD, AIChe, Associate Professor, Texas Tech University, Lubbock, Texas, U.S.A.

MICHAEL FRANK HORDESKI PE, BSEE, MSEE, Consultant, Jablon Computer, Atascadero, California, U.S.A.

JAMES E. JAMISON PE, BScChE, Senior Lead Instrument and Process Control Engineer, Bantrel, Inc., Calgary, Canada

KLAUS H. KORSTEN Dipl. Eng., Marketing Manager, Endress+Hauser Process Solutions AG, Reinach, Switzerland

KLAUS-PETER LINDNER Dipl. Info., New Technology Specialist, Endress+Hauser Process Solutions AG, Reinach, Switzerland

BÉLA G. LIPTÁK PE, MME, Consultant, ISA Fellow and recipient of Process Automation Hall of Fame award for 2001, Stamford, Connecticut, U.S.A.

MICHAEL N. LOUKA BS, Section Head, IFE Halden Virtual Reality Centre, Halden, Norway

M. SAM MANNAN PE, CSP, PhD, AIChe, Professor of Chemical Engineering, Texas A&M University, College Station, Texas, U.S.A.

EDWARD M. MARSZAL PE, BSChE, Principal Engineer, Exida, Columbus, Ohio, U.S.A.

GREGORY K. McMILLAN BSEngPhy, MSEE, Adjunct Professor, Washington University, St. Louis, Missouri, U.S.A.

DANIEL MIKLOVIC BSEE, MSSM, CMFgE, Vice President and Research Director, Gartner, Sammamish, Washington, U.S.A.

DOUG MORGAN BSChE, Project Engineer, Control Systems International, Irvine, California, U.S.A.

MICHAEL J. PIOVOSO PE, BSEE, MSEE, PhD, Associate Professor, Pennsylvania State University, Malvern, Pennsylvania, U.S.A.

WALLACE A. PRATT, JR. BSEE, Chief Engineer, HART Communication Foundation, Austin, Texas, U.S.A.

ALBERTO ROHR EE, Dr. Ing., Consultant, Vedrano al Lambro (MI), Italy

DERRICK KEITH ROLLINS, SR. BS, MS, PhD, AIChe, Associate Professor, Iowa State University, Ames, Iowa, U.S.A.

MICHEL RUEL PE, BScA, President, TOP Control, Inc., Hubertus, Wisconsin, U.S.A.

GURBINDER B. SINGH BE, MBA, MCSE, Consultant—Control Systems Engineer, Chicago, Illinois, U.S.A

ROBERT J. SMITH II BSEET, Electrical & Instrumentation Engineer/Information Technology Manager, Associated Professional Engineering Consultants, Inc., Cincinnati, Ohio, U.S.A.

DAVID A. STROBHAR PE, BSHFE, President, Beville Engineering, Inc., Dayton, Ohio, U.S.A.

ANGELA ELAINE SUMMERS PE, PhD, AIChe, President, SIS-TECH Solutions, LLC, Houston, Texas, U.S.A.

G. KEVIN TOTHEROW BSEE, President, Sylution Consulting, Jesup, Georgia, U.S.A.

ANN TUCK BSME, Control Systems Assistant Chief Engineer, Bechtel Corporation, Frederick, Maryland, U.S.A.

IAN VERHAPPEN PE, BScE, Engineering Associate, Syncrude Canada Ltd, Fort McMurray, Alberta, Canada

STEFANO VITTURI Dr. Eng., Researcher, CNR-LADSEB, Padova, Italy

HARRY H. WEST PE, CSP, PhD, AIChe, Adjunct Professor of Chemical Engineering, Texas A&M University, College Station, Texas, U.S.A.

PREFACE

THE MATURING OF THE I&C PROFESSION

The first volume of the *Instrument Engineers' Handbook* (*IEH*) described the devices and methods used in performing automatic industrial process measurement and analysis. The second volume of the *IEH* dealt with automatic process control devices and systems used in our various industries. This third volume of the *IEH* provides an in-depth, state-of-the-art review of all the existing and evolving digital communication and control systems. Although the transportation of digital information by buses and networks is a major topic in this volume, the total coverage of the volume includes much more. This volume also describes a variety of process control software packages, which are used in plant optimization, maintenance, and safety-related applications. A full chapter is assigned to plant design and updating, while safety and operations-related logic systems and the design of integrated workstations and control centers are also emphasized. The volume concludes with a substantial appendix, providing such practical information as bidders' lists and addresses, steam tables, materials selection for corrosive services, and much more.

It is hoped that the publication of this third volume of the *IEH* will contribute to increasing the safety and efficiency of all control systems. Although in the previous editions of the *IEH* we have advocated the use of intelligent self-monitoring and self-diagnosing instrumentation, now it seems that the time has come to take the next step and aim for unattended and self-optimizing industrial control systems. It is time to proceed from the level of self-monitoring and self-diagnosing packages to self-healing systems requiring a minimum of maintenance.

Ours is a relatively young profession. I do hope that this third volume of the *IEH* will also improve the respectability and professional standing of the instrumentation and control (I&C) profession, which is still evolving. Yet, if we compare the professional standing and the self-image of instrumentation and control engineers to those of, for example, mechanical or chemical engineers, we find ourselves at a disadvantage.

The list of disadvantages starts at the universities, which offer ME or ChE degrees, but not too many of them offer degrees in I&C engineering. Some do not even have an I&C department. Even those that do often tend to treat control as if it were a subfield of mathematics. At such universities control issues are often discussed in the "frequency domain," and control problems are analyzed by using partial differential equations and Laplace transfer functions. Under such conditions, the engineering students, when first exposed to the field of process control, often receive the wrong impression of what the I&C profession is all about.

Our engineering societies could also do a better job to improve our professional image. The main goal of such engineering societies as ASME or AIChE is to serve the professional development of their members. These societies focus on preparing scientific publications, on generating high-quality engineering standards, or on organizing courses aimed at assisting the professional advancement of their members. In contrast to that, the leadership of some I&C societies is dominated not by the users, but by the manufacturers, and focuses not on the professional advancement of their members, but on serving the commercial interests of the vendors.

The differences between the professional standings of I&C and other engineering disciplines are also visible in most operating plants, where one has no difficulty in finding a resident ME or ChE, but when one asks for the resident I&C engineer, the answer often is, "We have only instrument maintenance technicians, the vendors take care of our instrument engineering." This shows an elementary lack of understanding of the most basic requirement of good control. It is that **in order to properly control a process, one must fully understand its unique personality, and vendors can seldom, if ever, do that**.

Another observable difference is demonstrated by the bookshelves of the members of the different engineering disciplines. If one walks into the office of an ME, it is likely that one will see one or more editions of *Marks' Handbook* on the bookshelf. The same holds true for the offices of ChEs, except that there it will be *Perry's Handbook* on the bookshelves. In contrast, the bookshelves of most I&C engineers are likely to be flooded by vendors' catalogs but few professional handbooks are likely to be seen there.

FRIDAY'S NOTES

Having painted this rather dark picture, I should bring the topic of our professional standing into its proper perspective. We all know that it takes time for an engineering profession to mature, and we also know that I&C is a very young profession. I will try to demonstrate the youth of our profession by using the example of my handbook.

In 1962, at the age of 26, I became the Chief Instrument Engineer at Crawford & Russell, an engineering design firm specializing in the building of plastics plants. C&R was growing and my department also had to grow. Yet, at the age of 26 I did not dare to hire experienced people, because I did not think that I could lead and supervise older engineers. Yet the department had to grow, so I hired fresh graduates from the best engineering colleges in the country. I picked the smartest graduates and, having done so, I obtained permission from the C&R president, Sam Russell, to spend every Friday afternoon teaching them.

In a few years my department had not only some outstanding process control engineers, but C&R also saved a lot on their salaries. By the time I reached 30, I felt secure enough to stop disguising my youth. I shaved off my beard and threw away my thick-rimmed, phony eyeglasses. I no longer felt that I had to look older, but all the same, my Friday's notes still occupied a 2-ft-tall pile on the corner of my desk.

In the mid-1960s an old-fashioned Dutch gentleman named Nick Groonevelt visited my office and asked: "What are all those notes?" When I told him, he asked: "Does your profession have a handbook?" I answered with my own question: "If it did, would I be teaching from these notes?" (Actually, I was wrong in giving that answer, because Behar's *Handbook of Measurement and Control* was already available, but I did not know about it.) "So, let me publish your notes and then the instrument engineers will have a handbook," Nick proposed, and in 1968 the first edition of the *Instrument Engineers' Handbook* was published.

In 1968, the Soviet tanks, which I fought in 1956, were besieging Prague, so I decided to dedicate the three volumes of the *IEH* to the Hungarian and Czech freedom-fighters. A fellow Hungarian–American, Edward Teller, wrote the preface to the first edition, and Frank Ryan, the editor of *ISA Journal*, wrote the introduction. My co-authors included such names as Hans Baumann, Stu Jackson, Orval Lovett, Charles Mamzic, Howard Roberts, Greg Shinskey, and Ted Williams. It was an honor to work with such a team. In 1973, because of the great success of the *IEH*, I was elected to become the youngest ISA fellow ever. But the fact still remains that ours is a very young profession: when the *IEH* came out, Marks' and Perry's handbooks were in their fifth or sixth editions!

PROGRESS

The third edition of the *IEH* was initially planned for three volumes. They were to cover the subjects of process measurement, process control, and process software. Chilton published the first two volumes in 1995. The publishing process was then interrupted when Walt Disney acquired Chilton in 1996. I could do nothing but wait for work on the series to resume. In October 2000, CRC Press obtained the rights to publish the third volume.

This delay, though unfortunate, also had some positive consequences. First, CRC agreed with ISA to market the *IEH* jointly. Second, the onset of the age of digital communications made it possible for me to find the best experts in the world for every key topic in this volume. This was an important consideration because the three volumes of the *IEH* explore nearly 1000 diverse topics from anenometers to weirs and from controlling airhandlers to controlling wastewater treatment processes. Finding the best authors possible in an efficient manner would have been next to impossible before the Internet.

Now, as I start to invite co-authors for the fourth edition of this handbook, the Internet continues to be an invaluable research and communication tool. By the click of a button (liptakbela@aol.com) experts residing anywhere in the world can also contact me and offer to contribute to the IEH, thus sharing their knowledge, accumulated over a lifetime, with the international community of I&C professionals.

THE FUTURE

When Yale University invited me to teach I&C, I did not like it that my course was being offered by its chemical engineering department, because Yale did not have an independent I&C department. On the other hand, I was pleased that I was allowed to discuss control theory in the "time domain." Therefore, I used no mathematical abstractions, partial differential equations, or Laplace transfer functions. Instead, I talked about easily understandable terms like transportation lags and capacities, dead times, and time constants. In short, the course I gave was down to earth and practical. It was based on my old "Friday's notes." So, while teaching I&C in a ChE department was unfortunate, teaching I&C in the time domain, and not in the frequency domain, was a step forward.

In working with the publishers of the *IEH* over the past decades, I was also reminded of the unrecognized nature of the I&C profession. Between the various editions, I have seen the *IEH* promoted in the categories of chemical engineering, electrical engineering, and computer engineering books, but seldom in its own category. This, too, has bothered me. It just seems to be taking too long to recognize that I&C is a separate and distinct profession. When CRC agreed to a joint publication with ISA, this was a small but significant step toward gaining full recognition for our slowly maturing I&C profession.

In general, it is high time for our universities and publishers to recognize the existence of instrument engineering as a distinct and respectable profession. It is time for industrial management to understand that the availability of in-house instrument engineering know-how is essential. It is time for instrument societies to focus less on the advertising dollars from the vendors and more on helping the users by providing education and standardization. It is hoped that in the long run these steps will help in gaining the deserved respect for our slowly maturing I&C profession.

DIGITAL SYSTEMS

In the past, we first standardized on the pneumatic signal range of 3 to 15 PSIG and later on the electronic transmission and control signal range of 4–20 mA DC. As we move from the analog to the digital age, we also need a uniform worldwide standard for digital signals, which is universal. We need a fully open network, one that is *not* based on any particular vendor's standard. Yet today, there exist some 30 digital protocols, which all call themselves fieldbuses. What the International Electrotechnical Commission (IEC) did was not to issue a single standard, but simply to take the conglomeration of eight of these disparate, proprietary, non-interoperable vendors' standards and combine them into a single document (IEC 61158). This is intolerable! This is like expecting to run a plant from a control center in which eight operators might speak eight different languages.

While some progress has been made in providing an Open System Interconnect (OSI) model and while most vendors support Ethernet-TCP/IP (transmission control protocol/Internet protocol) connectivity at the business interface level, much remains to be done. With the passage of time, interoperability among the field device network front runners (Foundation Fieldbus, HART, and PROFIBUS-PA/DP) has also improved, but (because of the war for the dominance at the field level of the application layer) interoperability still remains a marketing term of blurred meaning. This is more than undesirable! This is unsafe! **The responsibility of the I&C engineering societies and of this handbook of mine is nothing less, but to work for a truly open and universal digital network standard**.

Greg Shinsky was right when he warned that smart controllers alone cannot solve the problem of dumb users. No, the problem of dumb users can only be solved by education and by placing the interests of the profession ahead of those of individual manufacturers. To achieve that goal, both the various I&C engineering societies (including ISA) and our publications, such as my handbook, have important roles to play. If we all pitch in, we can improve not only the next edition of the *IEH* and the professional atmosphere at ISA, but we can also increase the respectability and maturity of the instrument engineering profession as a whole.

CONTROL OF NON-INDUSTRIAL PROCESSES

A few years ago a group of social scientists invited me to Harvard University to talk about the control of non-industrial processes. I started the lecture by listing the information that we need about any process, before we can start designing a system to control it. Among the information needed, I mentioned the set point and the manipulated variable (the control valve), which we must have to build a control loop. As an example, I mentioned that, if we wanted to build a control loop, which would control the population of our planet, we would have to agree on both a "set point" and a "manipulated variable" for it. I am not saying that it is necessarily up to us humans to control the population of the world. What I am saying is that we have not even agreed on the desired set point or on the manipulated variable for such a loop!

Someone from the audience interrupted me at this point and asked about the control modes for such a population controller. "Should the controller be proportional and integral (as in the case of level control) or proportional and derivative (as in the case of batch neutralization)?" he asked. "One should only start thinking about control modes when the loop itself exists, not before," I responded. Therefore, humankind will first have to decide if there is a maximum limit to the total population of our planet (set point). Once there is general agreement on that, we will have to agree on the manipulated variables (on the means to be utilized to keep the population below that limit). Reaching such agreements will not be easy because the derivative settings of our political institutions (anticipation into the future) are very short (usually only 4 years) and because there is no past precedent in our culture for making decisions of such magnitude.

Controlling Evolution

It is difficult for us to be concerned about events that are likely to occur after we are dead. It is difficult, because human evolution in the past has been under the control of nature and it is hard for us to accept that we too are responsible for the future of the planet. I do not mean to suggest that we have "conquered nature" or that we are controlling our own evolution. No, that is not the case! The Creator has placed nature in the driver's seat of all evolutionary processes on this planet. Yet, He has permitted humans to change some of nature's conditions. For example, He allowed humans to minimize the discomforts caused by the weather and also allowed us to reduce the time it takes to travel, by using up some of the exhaustible resources of the planet.

Therefore, if humankind fails to come up with a set point and a manipulated variable for the population control loop, nature is likely to select and implement it for us. One can only hope that there is enough derivative (enough anticipation of future events) in our combined wisdom to prevent that from happening, because if we wait for nature to close the population control loop, the "throttling" will neither be smooth nor gradual. "So, in order to control population, we need to modify human attitudes, human culture?" asks a balding gentleman. "Yes, we can view the relationship between culture and population control as a cascade loop, where culture is the master controller which is herding a number of slave loops, one of them being population," I responded.

Controlling Cultural and Social Processes

Culture is a mostly dead-time process. It takes 10 to 20 years for a child to form his or her opinions and values. At the beginning of our individual lives, every child's mind is a blank page. Our moral and ethical standards are inscribed by our churches, our schools, by the media, and by the examples of our role models, most importantly by our parents. To understand culture as a controllable process, we must also realize that culture is the sum total of all the beliefs and convictions of all members of our society.

For a society to function smoothly, at least three generations should share the same moral and ethical values. When the prevailing culture changes faster, this can result in a difference between the moral standards of the generations. As a consequence of cultural conflicts between societies of different nations or among the generations of the same nation, "their manipulated variables will interact." These interactions can be desirable (elimination of prejudices, clarifying environmental interrelationships) or undesirable (materialism, selfishness, amoral or cynical attitudes). Whatever the nature of the changes, "de-coupling of the loops" is necessary, if society is to function smoothly and effectively. Therefore, the methods used in de-coupling the interacting control loops can also be used in controlling social or cultural processes.

Economic and Political Processes

De-coupling is also needed when the interests of various segments of society collide. As long as the de-coupling (conflict resolution) can be subordinated to the shared ethical and moral standards of society ("one nation under God," "all men are created equal," etc.), a hierarchical control system (multilayered cascade) will function properly. On the other hand, problems will arise if the set point (the moral standards of this shared cascade master) becomes fuzzy. Such fuzziness is occurring right now, because business is already "globalized" while the political, legal, educational, or other institutions are not. It is the fuzziness of this cascade master that allows perfectly moral and ethical individuals to purchase goods made by child labor in environmentally polluting plants. Such fuzziness could be eliminated by inserting a slave control loop (implemented by, say, a color-coded label on all imported goods).

I also talked about the importance of the degrees of freedom of the controlled process. The number of these degrees identifies the number of process variables that can be independently controlled (in case of a train—one, ship—two, etc.). If we try to control more variables than the number of degrees of freedom that the process has, the control loops will fight each other and the controlled process will become unstable. This discussion of degrees of freedom led to questions related to controlling such complex processes as the economy and the political system.

HERDING AND ENVELOPE CONTROLS

In multivariable processes one cannot use a single set point, but must implement either a "herding" or an "envelope" control configuration. When implementing herding control, all controlled variables are observed simultaneously and continuously, but correction is applied to only one variable at a time. The selected control variable is the one that is farthest away from where it should be. A herding loop does the same thing that a herding dog does when it is herding 1000 sheep by going after one sheep at a time (by manipulating only one variable at a time), the one that is farthest away from the desired overall goal or aim of the control system.

Envelope control is different. Here, an allowable gap (upper and lower limits) is assigned to each controlled variable. From the totality of these gaps, a multidimensional control envelope results. If all the controlled variables are inside this envelope, no corrective action is taken. If a controlled variable drifts to one of the boundaries of the envelope, a correction is initiated.

Envelope control is best suited for **controlling the economy**, because our overall economic well-being is a function of several variables of similar importance (unemployment, inflation, corporate profits, interest rates, budget deficits, etc.). In contrast, the herding control model is more suitable for political process, because in that process, one consideration is more important that all the others. This critical consideration is to guarantee that all votes have the same weight.

In **controlling the political process** all other variables should be subordinated to this one goal, and all variables should be herded in the direction of guaranteeing equal influence to all well-informed voters. In this sense, the one-party systems completely eliminate all degrees of freedom, while the two-party systems are superior, but still restrictive. This control analysis suggests that maximizing the degrees of freedom of the political process would also optimize it. If that conclusion is correct, then the public financing of the campaigns of all acceptable political candidates and the elimination of private contributions could be a step in the right direction.

NATIONALISM AND GLOBALIZATION

It was already noted that the "dead-time" of forming cultural attitudes and cultural loyalties can take decades. It is worth noting that our loyalty to "our culture" and to the traditions of our extended family (our nation) harms no one. It is a form of healthy loyalty, which should never be given up or exchanged. Yet, this loyalty should not stand in the way of developing other loyalties. A person with multiple loyalties is a richer and happier person.

If one can maintain one's 100% loyalty to one's own culture while simultaneously developing an understanding and respect for another, that person has become a richer individual, a 200% person. Understanding and accepting this will be the great test of the "globalization process." Globalization can aim for a multicultural and hence richer global society, but it can also result in a uniformly commercialized and hence poorer culture for all people. The choice is ours, but to control the globalization process, we must also understand its time constants, which in case of electronic commerce can be seconds, while in the cases of culture or ethical and moral standards can be decades or even centuries.

MATCHING OF TIME CONSTANTS

Similarly to the control of culture, the time constants of global biology and its relationship with the preservation of the species must also be understood. For example, it takes thousands of years to displace the waters in all the oceans only once. Therefore, the irreversible consequences of the pollutants that we are discharging into our receiving waters today might not fully evolve for a couple of millennia. The time constants of the processes involving atmospheric pollution and global warming are similarly long and just as poorly understood.

The time requirements of switching to nonpolluting and inexhaustible energy sources are also critical. We do not know which of the proposed inexhaustible processes will ultimately replace the fossil fuels as our new, long-term energy supply. We do not know which of a dozen proposals will eventually work. We do not know if solar energy, collected by artificial islands around the equator will fuel a hydrogen-based economy or we will "burn" the carbon dioxide in the air, use solar cells, wind turbines, or what. What we do know is that fossil fuels are exhaustible, that the disposal problems associated with nuclear waste are unsolved and that the time needed to develop an economy based on nonpolluting and inexhaustible energy sources is long. So the wisdom of process control would suggest that we had better get started!

We will only be able to adjust our actions to protect and serve the future generations when we fully understand the time constants of the cultural and physical processes of our planet. To do that, it is not only necessary to understand the basic principles of process control but it is also necessary to help the process control profession gain the kind of respect and maturity that it deserves.

The goal of the three volumes of the *Instrument Engineers' Handbook* is nothing less than that. I do hope that your verdict will be that the co-authors of these volumes have made an important contribution to increasing the respectability of the I&C profession.

Béla G. Lipták
(E-mail: Liptakbela@aol.com)

DEFINITIONS

AMPACITY The current (amperes) a conducting system can support without exceeding the temperature rating assigned to its configuration and application.

ATTENUATION Loss of communication signal strength.

BACKPLANE Physical connection between individual components and the data and power distribution buses inside a chassis.

BALUN Balanced/unbalanced. A device used for matching characteristics between a balanced and an unbalanced medium.

BANDWIDTH Data-carrying capacity, the range of frequencies available for signals. The term is also used to describe the rated throughput capacity of a given network medium or protocol.

BASEBAND A communication technique where only one carrier frequency is used to send one signal at a time. Ethernet is an example of a baseband network. Also called narrowband. Contrast to **broadband**.

BONDING The practice of creating safe, high-capacity, reliable electrical connectivity between associated metallic parts, machines, and other conductive equipment.

BROADBAND A communication technique that multiplexes multiple independent signals simultaneously, using several distinct carriers. A common term in the telecommunications industry to describe any channel with a bandwidth greater than a voice-grade channel (4 kHz). Also called wideband. Contrast to **baseband.**

CAPACITANCE The amount of charge, in coulombs, stored in a system necessary to raise the potential difference across it 1 V; represented by the SI unit farad.

DATA SERVERS A standard interface to provide data exchange between field devices and data clients.

DEMULTIPLEXING Separating of multiple input streams that were multiplexed into a common physical signal back into multiple output streams.

DEVICE DESCRIPTION A clear and unambiguous, structured text description that allows full utilization/operation of a field device by a host/master without any prior knowledge of the field device.

ETHERNET A baseband local area network specification developed by Xerox Corporation, Intel, and Digital Equipment Corporation to interconnect computer equipment using coaxial cable and transceivers.

FIELDBUS An all-digital, two-way, multidrop communications system for instruments and other plant automation equipment.

FIREWALL Router or access server, designated as a buffer between any public networks and a private network.

GROUND A conducting connection, whether intentional or accidental, between an electrical circuit or equipment and the earth, or to some conducting body that serves in place of earth. (See NFPA 70-100.)

GROUND FAULT PROTECTOR Device used to open ungrounded conductors when high currents, especially those due to line-to-ground fault currents, are encountered.

HOME RUN WIRING Wire between the cabinet where the fieldbus host or centralized control system resides and the first field junction box or device.

HUB (SHARED) Multiport repeater joining segments into a network.

IMPEDANCE Maximum voltage divided by maximum current in an alternating current circuit. Impedance is composed of resistive, inductive, and capacitive components. Like direct current circuits, the quantity of voltage divided by current is expressed in ohms.

INTERFACE (1) Shared boundary. For example, the physical connection between two systems or two devices. (2) Generally, the point of interconnection of two components, and the means by which they must exchange signals according to some hardware or software protocol.

INTEROPERABILITY A marketing term with a blurred meaning. One possible definition is the ability for like devices from different manufacturers to work together in a system and be substituted one for another without loss of functionality at the host level (HART).

LAMBDA The desired closed-loop time constant, often set to equal the loop lag time (λ).

LATENCY Latency measures the worst-case maximum time between the start of a transaction and the completion of that transaction.

LINE DRIVER Inexpensive amplifier and signal converter that conditions digital signals to ensure reliable transmissions over extended distances without the use of modems.

MANCHESTER A digital signaling technique that contains a signal transition at the center of every bit cell.

MODEM Modulator-demodulator. Device that converts digital and analog signals. At the source, a modem converts digital signals to a form suitable for transmission over analog communication facilities. At the destination, the analog signals are returned to their digital form. Modems allow data to be transmitted over voice-grade telephone lines.

MULTIPLEXING Scheme that allows multiple logical signals to be transmitted simultaneously across a single physical channel. Compare with demultiplexing.

NETWORK All media, connectors, and associated communication elements by which a given set of communicating devices are interconnected. A network may consist of several segments joined by repeaters. Networks may be joined using bridges.

PLENUM Air distribution ducting, chamber, or compartment.

PROTOCOL Formal description of a set of rules and conventions that govern how devices on a network exchange information.

RACEWAY A general term for enclosed channels, conduit, and tubing designed for holding wires, cables, or busbars.

SEGMENT The section of a network that is terminated in its characteristic impedance. Segments are linked by repeaters to form a complete network.

SERVICE Term used by NFPA-70 (NEC) to demarcate the point at which utility electrical codes published by IEEE (NESC) take over. Includes conductors and equipment that deliver electricity from utilities.

SMART FIELD DEVICE A microprocessor-based process transmitter or actuator that supports two-way communications with a host; digitizes the transducer signals; and digitally corrects its process variable values to improve system performance. The value of a smart field device lies in the quality of data it provides.

STICTION Combination of sticking and slipping when stroking a control valve.

SUBCHANNEL In broadband terminology, a frequency-based subdivision creating a separate communications channel.

SWITCHED HUB Multiport bridge joining networks into a larger network.

THROUGHPUT Throughput is the maximum number of transactions per second that can be communicated by the system.

TIMEOUT Event that occurs when one network device expects to hear from another network device within a specified period of time, but does not. The resulting timeout usually results in a retransmission of information or the dissolving of the session between the two devices.

TOPOLOGY Physical arrangement of network nodes and media within an enterprise networking structure.

ABBREVIATIONS, NOMENCLATURE, ACRONYMS, AND SYMBOLS

2D — two dimensional
3D — three dimensional

A

a — acceleration
A — (1) area; (2) ampere, symbol for basic SI unit of electric current; also amp
Å — Ångstrom (=10^{-10} m)
abs — absolute (e.g., value)
AC, ac, or a-c — alternating current
ACFM — volumetric flow at actual conditions in cubic feet per minute (=28.32 alpm)
ACL — asynchronous connection-less
ACMH — actual cubic meter per hour
ACMM — actual cubic meter per minute
ACS — analyzer control system
ACSL — advanced continuous simulation language
A/D — analog to digital
ADIS — approved for draft international standard circulation
A&E — alarm and event
AF or a-f — audio frequency
AGA3 — American Gas Association Report 3
AI — analog input
a(k) — white noise
ALARP — as low as reasonably practicable
alt — altitude
amp — ampere; also A, *q.v.*
AMPS — advanced mobile phone system or service
AMS — asset management solutions
AO — analog output
AP — access point
APC — automatic process control
APDU — application (layer) protocol data unit
API — application programming interface or absolute performance index
°API — API degrees of liquid density
APM — alternating pulse modulation
ARA — alarm response analysis
ARIMA — autoregressive integrated moving average
ARP — address resolution protocol
ASCII — American Standard Code for Information Interchange
AS-i — actuator sensor interface
ASIC — application specific integrated chips
ASK — amplitude shift keying
asym — asymmetrical; not symmetrical
atm — atmosphere (=14.7 psi)
AUI — attachment unit interface
aux — auxiliary
AWG — American wire gauge

B

b — dead time
°Ba — Balling degrees of liquid density
bar — (1) barometer; (2) unit of atmospheric pressure measurement (=100 kPa)
barg — bar gauge
bbl — barrels (=0.1589 m^3)
BCD — binary coded decimal
BCS — batch control system
°Bé — Baumé degrees of liquid density
BFW — boiler feed water
bhp or b.h.p. — braking horsepower (=746 W)
°Bk — Barkometer degrees of liquid density
blk — black (wiring code color for AC "hot" conductor)
bp or b.p. — boiling point
BPCS — basic process control system
bps — bits per second
BPSK — binary phase shift keying
Bq — becquerel, symbol for derived SI unit of radioactivity, joules per kilogram, J/kg
°Br — Brix degrees of liquid density
B2B — business to business
BTU — British thermal unit (=1054 J)
BWG — Birmingham wire gauge

C

c — (1) velocity of light in vacuum (3×10^8 m/s); (2) centi, prefix meaning 0.01

C	coulombs
°C	Celsius degrees of temperature
ca.	*circa*: about, approximately
CAC	channel access code
CAD	computer-aided design
cal	calorie (gram, =4.184 J); also g-cal
CAN	control area network or control and automation network
CATV	community antenna television (cable)
CBM	condition-based maintenance
cc	cubic centimeter (=10^{-6} m^3)
CCF	common cause failure
ccm	cubic centimeter per minute
CCR	central control room
ccs	constant current source
CCS	computer control system
cd	candela, symbol for basic SI unit of luminous intensity
CD	compact disk, compel data, or collision detector
C^D	dangerous coverage factor
CDF	cumulative distribution function
CDMA	code division multiple access
CDPD	cellular digital packet data
CEMS	continuous emissions monitoring system
CENP	combustion engineering nuclear power
CFM or cfm	cubic foot per minute (28.32 lpm)
CF/yr	cubic foot per year
Ci	curie (=3.7 × 10^{10} Bq)
CI	cast iron
CIM	computer-integrated manufacturing
CIP	computer-aided production or control and information protocol (an application layer protocol supported by DeviceNet, ControlNet, and Ethernet/IP)
CLP	closed-loop potential factor
cm	centimeter (=0.01 m)
CM	condition monitoring
CMMS	computerized maintenance management system
CMPC	constrained multivariable predictive control
CMOS	complementary metal oxide semiconductor
cmph	cubic meter per hour
CNC	computerized numerical control
CNI	ControlNet International
CO	controller output
COM	component object model
COTS	commercial off-the-shelf
cos	cosine, trigonometric function
cp or c.p.	(1) candle power; (2) circular pitch; (3) center of pressure (cp and ctp may also be used for centipoises)
cpm	cycles per minute; counts per minute
cps	(1) cycles per second (=Hz); (2) counts per second; (3) centipoises (=0.001 Pa.s)
CPS	computerized procedure system
CPU	central processing unit
CRC	cyclical redundancy check or cyclic redundancy code (an error detection coding technique based upon modulo-2 division. Sometimes misused to refer to a block check sequence type of error detection coding)
CRLF	carriage return-line feed
CRT	cathode ray tube
CS	carbon steel
CSMA/CD	carrier sense, multiple access with collision detection
CSS	central supervisory station
cSt	centistoke
CSTR	continuous-stirred tank reactor
CTDMA	concurrent time domain multiple access
cvs	comma-separated variables

D

d	(1) derivative; (2) differential as in dx/dt; (3) deci, prefix meaning 0.1; (4) depth; (5) day
D	diameter; also dia and ϕ or derivative time of a controller
DA	data access
D/A	digital to analog
DAC	device access code; digital-to-analog converter
DAE	differential algebraic equation
DAMPS	digital advanced mobile phone system or service
dB	decibels
DBPSK	differential binary phase shift keying
DC	diagnostic coverage
DC or dc	direct current
DCE	data communications equipment
DCOM	distributed COM
DCS	distributed control system
DD	data definition or dangerous component failure is detected in a leg, or a device description written in using DDL
DDC	direct digital control
DDE	dynamic data exchange
DDL	device description language (an object-oriented data modeling language currently supported by PROFIBUS, FF, and HART)
deg	degree; also °(π/180 rad)
DEMUX	demultiplexer
DES	data encryption standard
DFIR	diffused infrared
DG	directed graph
DH	data highway
DI	discrete (digital) input
dia	diameter; also D and ϕ
DIAC	dedicated inquiry access code
DIR	diffused infrared
DIS	draft international standard

DIX	Digital-Intel-Xerox (DIX is original specification that created the *de facto* Ethernet standard; IEEE 802.3 came later after Ethernet was well established)
d(k)	unmeasured disturbance
D(k)	measured disturbance
DLE	data link escape
DLL	dynamic link library
DMM	digital multimeter
DO	dissolved oxygen or discrete (digital) output
DP	decentralized periphery
d/p cell	differential pressure transmitter (a Foxboro trademark)
DPDT	double pole double throw (switch)
DQPSK	differential quadrature phase shift keying
DSL	digital subscriber line
DSR	direct screen reference
DSSS	direct sequence spread spectrum
DT	dead time (seconds or minutes)
DTE	data terminal equipment
DTM	device type manager (an active-X component for configuring an industrial network component; a DTM "plugs into" an FDT)
DU	dangerous component failure occurred in leg but is undetected
DVM	digital voltmeter

E

e	(1) error; (2) base of natural (Naperian) logarithm; (3) exponential function; also exp (–x) as in e^{-x}
E	(1) electric potential in volts; (2) scientific notation as in $1.5E–03 = 1.5 \times 10^{-3}$
E{.}	expected value operator
EAI	enterprise application integration
EAM	enterprise asset management
EBCDIC	extended binary code for information interchange
EBR	electronic batch records
EDS	electronic data sheet (DeviceNet)
E/E/PE	electrical/electronic/programmable electronic
E/E/PES	electrical/electronic/programmable electronic system
EFD	engineering flow diagram
e.g.	*exempli gratia*: for example
EHC	electrohydraulic control
EHM	equipment health management
E&I	electrical and instrumentation
e(k)	feedback error
E.L.	elastic limit
emf	(1) electromotive force (volts); (2) electromotive potential (volts)
EMI	electromagnetic interference
EMI/RFI	electromagnetic and radio-frequency interference
$e_m(k)$	process/model error
EN	European standard
EPA	enhanced performance architecture
EPC	engineering-procurement-construction (firm)
EPCM	engineering, procurement, and construction management (companies)
EQ or eq	equation
ERM	enterprise resource manufacturing
ERP	enterprise resource planning or effective radiated power
ESD	emergency shutdown (system)
ESN	electronic serial number
exp	exponential function as in exp (–at) = e^{-at}; also e

F

f	frequency; also freq
F	farad, symbol for derived SI unit of capacitance, ampere·second per volt, A·s/V
°F	Fahrenheit degrees [$t_{°C} = (t_{°F} - 32)/1.8$]
FAT	factory acceptance testing
FBAP	function block application process (FF)
FBD	function block diagram
FCC	fluidized catalytic cracker
FCOR	filtering and correlation (method)
FCS	frame check sequence
FDE	fault disconnection electronics
FDL	fieldbus data link
FDMA	frequency division multiple access
FDT	field device tool (an MS-Windows-based framework for engineering and configuration tools)
FE	final elements
FEED	front end engineering and design
FES	fixed end system
FF or F.F.	Foundation Fieldbus
FF-HSE	Foundation Fieldbus, high-speed Ethernet
FH	frequency hopping
fhp	fractional horsepower (e.g., $^1/_4$ HP motor)
FHSS	frequency hopped spread spectrum
FIFO	first-in, first-out
Fig.	figure
FISCO	Fieldbus Intrinsic Safety COncept
fl.	fluid
fl.oz.	fluid ounces (=29.57 cc)
FMEA	failure mode and effects analysis
FMS	fieldbus message specification or fieldbus messaging services/system
FNC	function byte
FO	fiber optic
FOP	fiber-optic probe
fp or f.p.	freezing point
FPM or fpm	feet per minute (=0.3048 m/min)
fps or ft/s	feet per second (=0.3048 m/s)
FRM	frequency response method
FS or fs	full scale

FSC fail safe controller
FSK frequency shift keying
FTA fault tree analysis
FTP file transfer protocol
FTS fault tolerant system

G

g acceleration due to gravity (=9.806 m/s^2)
G giga, prefix meaning 10^9 or process gain
gal gallon(s) (=3.785 liters)
GB giga-byte, 1,000,000,000 bytes
GbE gigabit Ethernet
gbps gigabits per second
G_c feedback controller transfer function
g-cal gramcalorie; also cal, *q.v.*
G_d unmeasured disturbance transfer function
G_D measured disturbance transfer function
G_D approximate feedforward transfer function model
GEOS geosynchronous Earth orbit satellites
G_{ff} feedforward controller transfer function
GHz giga-hertz
GIAC general inquiry access code
GLR generalized likelihood ratio
G-M Geiger-Mueller tube, for radiation monitoring
G_m model transfer function
G_p process transfer function
gph gallons per hour (=3.785 lph)
GPM or gpm gallons per minute (=3.785 lpm)
GPS global positioning satellite or system
gr gram
grn green (wiring code color for grounded conductor)
GSD Profibus version of an electronic data sheet
GUI graphical user interface
Gy gray, symbol for derived SI unit of absorbed dose, joules per kilogram, J/kg

H

h (1) height; (2) hour
H (1) humidity expressed as pounds of moisture per pound of dry air; (2) henry, symbol of derived SI unit of inductance, volt· second per ampere, V·s/A
HAZOP HAZard and OPerability studies
HC horizontal cross-connect
HAD historical data access
HART highway accessible remote transducer
HEC header error check
HFE human factors engineering
HFT hardware fault tolerance
hhv higher heating value
HIPPS high-integrity pressure protection system
HIPS high-integrity protection systems
HIST host interoperability support test
HMI human–machine interface
H1 field-level fieldbus; also refers to the 31.25 kbps instrinsically safe SP-50, IEC61158-2 physical layer
hor. horizontal
HP or hp horsepower (U.S. equivalent is 746 W)
H&RA hazard and risk analysis
HSE high-speed Ethernet (host-level fieldbus)
HSI human–system interface
HTML hypertext markup language
HTTP hypertext transfer protocol
HVAC heating, ventilation, and air conditioning
H/W hardware
Hz hertz, symbol for derived SI unit of frequency, one per second (1/s)

I

I integral time of a controller in units of time/repeat
IA instrument air
IAC inquiry access code
IAE integral of absolute error
ibidem in the same place
IC intermediate cross-connect
I&C instrumentation and control or information and control
ICA independent computing architecture
ICCMS inadequate core cooling monitoring system
ICMP Internet control message protocol
ID inside diameter
i.e. *id est*: that is
I&E instrument and electrical
IEH *Instrument Engineers' Handbook*
IETF Internet engineering task force
IIS Internet information server
IL instruction list
ILD instrument loop diagrams
IMC internal model control
in. inch (=25.4 mm)
in-lb inch-pound (=0.113 N × m)
I/O input/output
IP Internet protocol
I-P current to pressure conversion
IPL independent protection layer
IR infrared
IRQ interrupt request queue
IS intermediate system
ISE integral of squared error
ISM industrial, scientific, medical
ISP Internet service provider or interoperable system project
IT information technology (as in IT manager or IT department)
ITAE integral of absolute error multiplied by time
ITSE integral of squared error multiplied by time

ITT intelligent temperature transmitters
IXC interexchange carrier

J

J joule, symbol for derived SI unit of energy, heat, or work, newton-meter, N·m
JIT just-in-time manufacturing

K

k kilo, prefix meaning 1000
K kelvin, symbol for SI unit of temperature or process gain (dimensionless)
Kbs, Kbps kilo bits per second
KBs kilo bytes per second
k-cal kilogram-calories (=4184 J)
kg kilogram symbol for basic SI unit of mass
kg-m kilogram-meter (torque, =7.233 foot-pounds)
kip thousand pounds (=453.6 kg)
km kilometers
K_p proportional gain of a PID controller, q.v.
kPa kilo-pascals
kVA kilovolt-amperes
kW kilowatts
kWh kilowatt-hours (=3.6×10^6 J)

L

l liter (=0.001 m^3 = 0.2642 gallon)
L (1) length; (2) inductance, expressed in henrys
LAN local area network
LAS link active scheduler (FF)
lat latitude
lb pound (=0.4535 kg)
LCD liquid crystal display
LCM life cycle management
LCSR loop current step response
LD ladder diaphragm
LDP large display panel
LEC local exchange carrier
LED light-emitting diode
LEL lower explosive limit
LEOS low Earth orbit satellites
lim. or lim limit
lin. linear
liq. liquid
LLC logical link control
lm lumen, symbol for derived SI unit of luminous flux, candela·steradian, cd·sr
ln Naperian (natural) logarithm to base e
LNG liquefied natural gas
LOC limiting oxygen concentration
log or $\log_{10}$ logarithm to base 10; common logarithm
long. longitude
LOPA layers of protection analysis
LP liquefied petroleum or propane gas
lph liters per hour (0.2642 gph)
lpm liters per minute (0.2642 gpm)
LQG linear quadratic Gaussian
LRC longitudinal redundancy check
LSB least significant bit
LTI linear time-invariant
LVDT linear variable differential transformer
lx lux, symbol for derived SI unit of illuminance, lumen per square meter, lm/m^2

M

m (1) meter, symbol for basic SI unit of length; (2) mulli, prefix meaning 10^{-3}; (3) minute (temporal), also min
M (1) thousand (in commerce only); Mach number; (2) molecular weight; mole; (3) mega, prefix meaning 10^6
mA or ma milliamperes (=0.001 A)
MAC medium access control
MACID medium access control identifier
MAP manufacturing automation (access) protocol
MAU media access unit
MAWP maximum allowable working pressure
max maximum
MB mega-byte, 1,000,000 bytes
Mbs, mbps megabits per second
MBs mega bytes per second
MC main cross-connect
mCi or mC millicuries (=0.001 Ci)
m.c.p. mean candle power
MCP main control panel
MDBS mobile database station
MDIS mobile data intermediate system
med. medium or median
MEDS medium Earth orbit satellites
m.e.p. mean effective pressure
MES manufacturing execution system or management executive system or mobile end station
MFD mechanical flow diagram
mfg manufacturer or manufacturing
mg milligrams (=0.001 g)
MHz megahertz
mho unit of conductance, replaced by siemens, S, *q.v.*
mi miles (=1.609 km)
MI melt index
MIB management information base
micro prefix = 10^9; also μ (mu) or μm and sometimes u, as in ug or μg, both meaning microgram (=10^{-9} kg)
micron micrometer (=10^{-6} m)
MIMO multiple-input multiple-output
MIMOSA machinery information management open system alliance

min (1) minutes (temporal), also m; (2) minimum; (3) mobile identification number
MIS management information system
ml milliliters (=0.001 l = l cc)
mm millimeters or millimicrons (=0.001 m)
mmf magnetomotive force in amperes
MMI man–machine interface
MMS machine monitoring system or manufacturing message specification
MOC management of change
MODBUS a control network
MODEM modulator/demodulator
mol mole, symbol for basic SI unit for amount of substance
mol. molecules
MOON M out of N voting system
MOSFET metallic oxide semiconductor field-effect transistor
mp or m.p. melting point
MPa megapascal (10^6 Pa)
MPC model predictive control
mph miles per hour (1.609 km/h)
mps or m/s meters per second
MPS manufacturing periodic/aperiodic services
mR or mr milliroentgens (=0.001 R)
mrd millirads (=0.001 rd)
mrem milliroentgen-equivalent-man
MRP material requirement planning or manufacturing resource planning
ms milliseconds (=0.001 s)
MS Microsoft
MSA metropolitan statistical areas
MSB most significant bit
MSD most significant digit
MSDS material safety data sheet
MT measurement test
MTBF mean time between failures
MTSO mobil telephone switching offices
MTTF mean time to failure
MTTFD mean time to fail dangerously
MTTFS mean time to spurious failure
MTTR mean time to repair
MTU master terminal unit
MUX multiplexer
MVC minimum variance controller
MW megawatts (=10^6 W)

N

n (1) nano, prefix meaning 10^{-9}; (2) refractive index
N newton, symbol for derived SI unit of force, kilogram-meter per second squared, $kg \cdot m/s^2$
N_0 Avogadro's number (=6.023×10^{23} mol)
NAP network access port/point
NAT network address translation
NC numeric controller
NDIR nondispersive infrared
NDM normal disconnect mode
NDT nondestructive testing
NEC National Electrical Code
NESC National Electrical Safety Code
NEXT near end cross talk
nF nanofarad
NIC network interface card
nm nanometer (10^{-9} meter)
NRM normal response mode
NRZ nonreturn to zero (NZR refers to a digital signaling technique)
NTP network time protocol
NUT network update time

O

OD outside diameter
ODBC open database connectivity or communication
oft optical fiber thermometry
ohm unit of electrical resistance; also Ω (omega)
OJT on-the-job training
OLE object linking and embedding
OLE_DB object linking and embedding database
OPC object link embedding (OLE) for process control
or orange (typical wiring code color)
OS operator station or operating system
OSEK German for "open system interfaces for in-car electronics"
OSFP open shortest path first
OSI open system interconnect (model) or open system integration
OSI/RM open system interconnect / reference model
OT operator terminal
OTDR optical time domain reflectometers
oz ounce (=0.0283 kg)

P

p (1) pressure; (2) pico, prefix meaning 10^{-12}
Pa pascal, symbol for derived SI unit of stress and pressure, newtons per square meter, N/m^2
PA plant air
PAN personal area network
Pas pascal-second, a viscosity unit
PAS process automation system (successor to DCS)
PB proportional band of a controller in % (100%/controller gain)
PC personal computer (MS-Windows based)
PCA principal component analysis
PCCS personal computer control system

PCL	peer communication link
PCS	process control system or personal communication services
pct	percent; also %
PD	positive displacement or proportional and derivative
PDA	personal digital assistant
PDF	probability density function
PdM	predictive maintenance
PDU	protocol data unit
PE	polyethylene
PES	programmable electronic system
pf	picofarad (=10^{-12} F)
PF or p.f.	power factor
PFC	procedure functional chart
PFD	(1) process flow diagram; (2) probability of failure on demand
PFDavg	average probability of failure on demand
pH	acidity index (logarithm of hydrogen ion concentration)
PHA	process hazards analysis
pi or pl	Pouiseville, a viscosity unit
PI	proportional and integral
PID	proportional, integral, and derivative (control modes in a classic controller)
P&ID	piping (process) and instrumentation diagram (drawing)
PIMS	process information management system
PLC	programmable logic controller
PLS	physical layer signaling or projection to latent structures
PMA	physical medium attachment
PMBC	process model-based control
PMF	probability mass function
ppb	parts per billion
ppm	parts per million
PPM	pulse position modulation
PPP	point-to-point protocol
ppt	parts per trillion
precip	precipitate or precipitated
PSAT	pre-start-up acceptance test
psi or PSI	pounds per square inch (=6.894 kPa)
PSI	pre-start-up inspection
PSIA or psia	absolute pressure in pounds per square inch
PSID or psid	differential pressure in pounds per square inch
PSIG or psig	above atmospheric (gauge) pressure in pounds per square inch
PSK	phase shift keying
PSM	process safety management
PSSR	pre-start-up safety review
PSTN	public switched telephone network
PSU	post-start-up
pt	(1) point; (2) part; (3) pint (=0.4732 liter)
PT	pass token
PTB	Physikalisch-Technische Bundesanstalt
PV	process variable (measurement) or the HART primary variable
PVC	polyvinyl chloride
PVDF	polyvinylidene fluoride
PVLO	process variable low (reading or measurement)
PVHI	process variable high (reading or measurement)

Q

q	(1) rate of flow; (2) electric charge in coulombs, C
q^{-1}	backward shift operator
Q	quantity of heat in joules, J, or electric charge
°Q	Quevenne degrees of liquid density
QA	quality assurance
QAM	quadrature amplitude modulation
QoS	quality of service
QPSK	quadrature phase shift keying
qt	quart (0.9463 liter)
q.v.	*quod vide*: which see
QV	quaternary variable

R

r	radius; also rad
r^2	multiple regression coefficient
R	(1) resistance, electrical, ohms; (2) resistance, thermal, meter-kelvin per watt, m·K/W; (3) gas constant (=8.317×10^7 erg·mol^{-1}, °C^{-1}); (4) roentgen, symbol for accepted unit of exposure to x and gamma radiation (=2.58×10^{-4} C/kg)
rad	(1) radius; also r; (2) radian, symbol for SI unit of plane angle measurement or symbol for accepted SI unit of absorbed radiation dose (=0.01 Gy)
RAID	redundant array of inexpensive disks
RAM	random access memory
RASCI	responsible for, approves, supports, consults, informed
RCU	remote control unit
R&D	research and development
RDP	remote desktop protocol
Re	Reynolds number
rem	measure of absorbed radiation dose by living tissue (*r*oentgen *e*quivalent *m*an)
rev	revolution, cycle
RF or rf	radio-frequency
RFC	request for comment (an Internet protocol specification)
RFI	radio-frequency interference
RFQ	request for quotes
RH	relative humidity
RI	refractive index
RIP	routing information protocol

r(k)	set point
RMS or rms	square root of the mean of the square
RNG	rung number
ROI	return on investment
ROM	read-only memory
RPC	remote procedure call (RFC1831)
RPG	remote password generator
RPM or rpm	revolutions per minute
rps	revolutions per second
RRF	risk reduction factor
RRT	relative response time (the time required to remove 90% of the disturbance)
RS	recommended standard
RSA	rural service areas
RTD	resistance temperature detector
RTO	real-time optimization or operation
RTOS	real-time operating system
RTR	remote transmission request
RTS	ready (or request) to send
RTS/CTS	request to send/clear to send
RTU	remote terminal unit
RWS	remote work station

S

s	second, symbol for basic SI unit of time, also sec; or Laplace variable
S	siemens, symbol for derived SI unit of conductance, amperes per volt, A/V
SAP	service access point
sat.	saturated
SAT	site acceptance test or supervisory audio tone
SC	system codes
SCADA	supervisory control and data acquisition
SCCM	standard cubic centimeter per minute
SCFH	standard cubic feet per hour
SCFM	standard cubic feet per minute (airflow at 1.0 atm and 70°F)
SCM	station class mark
SCMM	standard cubic meter per minute
SCO	synchronous connection oriented
SCR	silicon controlled rectifier
SD	component in leg has failed safe and failure has been detected
SDN	send data with no acknowledge
SDS	smart distributed system
SEA	spokesman election algorithm
sec	seconds; also s
SER	sequence of event recorder
SFC	sequential function chart
SFD	system flow diagram or start of frame delimiter
SFF	safe failure fraction
SFR	spurious failure rate
SG or SpG	specific gravity; also sp.gr.
SID	system identification digit (number)
SIF	safety instrumented function
SIG	special interest group
SIL	safety integrity level
sin	sine, trigonometric function
SIS	safety instrumented system
SISO	single-input single output
SKU	stock keeping units
SLC	safety life cycle
slph	standard liters per hour
slpm	standard liters per minute
SMR	specialized mobile radio
SMTP	simple mail transfer (management) protocol
SNMP	simple network management protocol
SNR	signal-to-noise ratio
SOAP	simple object access protocol (an Internet protocol that provides a reliable stream-oriented connection for data transfer)
SOE	sequence of events
SOP	standard operating procedure
SP	set point
SPC	statistical process control
SPDT	single-pole double-pole throw (switch)
sp.gr.	specific gravity; also SG
SPRT	standard platinum resistance thermometer
sq	square; also □
SQC	statistical quality control
SQL	standard query language
sr	steradian, symbol for SI unit of solid angle measurement
SRD	send and request data with reply
SRS	safety requirements specification
SS	stainless steel
SSL	secure socket layers
SSU	Saybolt universal seconds
ST	structual text, also a fiber optic connector type
std.	standard
STEP	standard for the exchange of product model data
STP	shielded twisted pair
STR	spurious trip rates
SU	component in leg has failed safe and failure has not been detected
SV	secondary variable
S/W	software
s_y^2	sample variance of output y

T

t	(1) ton (metric, = 1000 kg); (2) time; (3) thickness
T	(1) temperature; (2) tera, prefix meaning 10^{12}; (3) period (=1/Hz, in seconds); (4) tesla, symbol for derived SI unit of magnetic flux density, webers per square meter, Wb/m^2

$T\frac{1}{2}$	half life
tan	tangent, trigonometric function
Tau	process time constant (seconds)
TCP	transmission (or transport) control protocol
TCP/IP	transmission control protocol/Internet protocol
td	process dead time (seconds)
T_d	derivative time (in seconds) of a PID controller
TDM	time division multiplexing
TDMA	time division multiple access
TDR	time domain reflectometry
TFT	thin film transistor
Ti	integral time (in seconds) of a PID controller
TI	time interval between proof tests (test interval)
TMR	triple modular redundancy
TOP	technical office protocol
TQM	total quality management
T.S.	tensile strength
TSR	terminate and stay resident
TV	tertiary variable
°Tw	Twadell degrees of liquid density

U

u	prefix = 10^{-6} when the Greek letter μ is not available
UART	universal asynchronous receiver transmitter
UBET	unbiased estimation
UCMM	unconnected message manager
UDP	user/universal data protocol (an Internet protocol with low overhead but no guarantee that communication was successful)
UEL	upper explosive limit
$u_{fb}(k)$	feedback controller output
UFD	utility flow diagram
$u_{ff}(k)$	feedforward controller output
UHF	ultrahigh frequency
UHSDS	ultrahigh-speed deluge system
u(k)	controller output
UML	universal modeling language
UPS	uninterruptible power supply
UPV	unfired pressure vessel
USB	universal serial bus
UTP	unshielded twisted pair
UUP	unshielded untwisted pair
UV	ultraviolet

V

v	velocity
v or V	volt, symbol for derived SI units of voltage, electric potential difference and electromotive force, watts per ampere, W/A
VBA	visual basic for applications
VCR	virtual communication relationship
VDU	video display unit
vert.	vertical
VFD	variable frequency drive
VFIR	very fast infrared
VHF	very high frequency
VMS	vibration monitoring system
VPN	virtual private network
VR	virtual reality
VRML	virtual reality modeling language
vs.	versus
V&V	verification and validation

W

w	(1) width; (2) mass flow rate
W	(1) watt, symbol for derived SI unit of power, joules per second, J/s; (2) weight; also wt
w.	water
WAN	wide area network
Wb	weber, symbol for derived SI unit of magnetic flux, volt·second, V·s
WG	standard (British) wire gauge
wh	white (wiring code color for AC neutral conductor)
WI	wobble index
WLAN	wireless local area network
WPAN	wireless personal area network
WS	workstation
wt	weight; also W

X

X	reactance in ohms
XML	extensible markup language
x ray	electromagnetic radiation

Y

y(k)	process output
yd	yard (=0.914 m)
yr	year

Z

Z	(1) atomic number (proton number); (2) electrical impedance (complex) expressed in ohms
zeb	zero energy band

GREEK CHARACTERS

$\eta(b)$	normalized performance index
$\eta(b+h)$	extended horizon performance index
λ	desired closed-loop time constant

$\lambda^{DU} = 1/MTTF^{DU}$	failure rate for dangerous undetected faults
$\lambda^{S} = 1/MTTF^{S}$	spurious trip rate
μm	microns
θ	process dead time (seconds or minutes)
σ^2	population variance
σ_y^2	population variance in output y
σ_{mv}^2	theoretical minimum variance
τ	process time constant (seconds or minutes)
τ_F	PV filter time constant
ψ	impulse weights

NOTES

1. Whenever the abbreviated form of a unit might lead to confusion, the abbreviation should not be used and the name should be written out in full.
2. The values of SI equivalents were rounded to three decimal places.
3. The words meter and liter are used in their accepted spelling forms instead of those in the standards, namely, metre and litre, respectively.

SOCIETIES AND ORGANIZATIONS

ACC	American Chemistry Council
ACS	American Chemical Society
AGA	American Gas Association
ANSI	American National Standards Institute
APHA	American Public Health Association
API	American Petroleum Institute
ARI	Air Conditioning and Refrigeration Institute
ASA	American Standards Association
ASCE	American Society of Civil Engineers
ASME	American Society of Mechanical Engineers
ASRE	American Society of Refrigeration Engineers
ASTM	American Society for Testing and Materials
BSI	British Standards Institution
CCITT	Consultative Committee for International Telegraphy and Telephony
CENELEC	European Committee for Electrotechnical Standardization
CII	Construction Industry Institute
CIL	Canadian Industries Limited
CNI	ControlNet International
CSA	Canadian Standards Association
DARPA	Defense Advanced Research Projects Agency
DIN	Deutsche Institut fuer Normung
DOD	Department of Defense (United States)
DOE	Department of Energy (United States)
EIA	Electronic Industries Association
EIA/TIA	Electrical Industries Alliance/Telecommunications Industries Association
EPA	Environmental Protection Agency (United States)
EPRI	Electric Power Research Institute
FCI	Fluid Control Institute
FDA	Food and Drug Administration (United States)
FF	Fieldbus Foundation
FIA	Fire Insurance Association
FM	Factory Mutual
FPA	Fire Protection Association
HCF	HART Communication Foundation
IAEI	International Association of Electrical Inspectors
ICE	Institute of Civil Engineers
ICEA	Insulated Cable Engineer's Association
IEC	International Electrotechnical Commission
IEEE	Institute of Electrical and Electronic Engineers
IETF	Internet Engineering Task Force
IPTS	International Practical Temperature Scale
IrDA or IRDA	Infrared Data Association
ISA	Instrumentation, Systems, and Automation Society (formerly Instrument Society of America)
ISO	International Standards Organization
ISTM	International Society for Testing Materials
JBF	Japan Batch Forum
KEPRI	Korean Electric Power Research Institute
LPGA	National LP-Gas Association
MCA	Manufacturing Chemists' Association
NAMUR	German standardization association for process control (Normenarbeitsgemeinschaft für Meß- und Regelungstechnik in der chemischen Industrie)
NASA	National Aeronautics and Space Administration

NBFU	National Board of Fire Underwriters
NBS	National Bureau of Standards
NEMA	National Electrical Equipment Manufacturers Association
NFPA	National Fire Protection Association
NIST	National Institute of Standards and Technology
NSC	National Safety Council
NRC	Nuclear Regulatory Commission
NSPE	National Society of Professional Engineers
ODVA	Open DeviceNet Vendors Association
OSHA	Occupational Safety and Health Administration (United States)
OTS	Office of Technical Services
PNO	Profibus User Organization
SAMA	Scientific Apparatus Manufacturers Association
TIA	Telecommunications Industries Alliance
USNRC	U.S. Nuclear Regulatory Commission
WBF	World Batch Forum

Overall Plant Design

1

1.1 AUDITING EXISTING PLANTS FOR UPGRADING 5

1.2 PROJECT MANAGEMENT AND DOCUMENTATION 14

1.3 OPERATOR TRAINING, COMMISSIONING, AND START-UP 29

1.4 FLOWSHEET SYMBOLS AND FUNCTIONAL DIAGRAMMING FOR DIGITALLY IMPLEMENTED LOOPS 42

1.5 HISTORICAL DATA STORAGE AND EVALUATION 79

1.6 INTEGRATION OF PROCESS DATA WITH MAINTENANCE SYSTEMS 91

1.1 Auditing Existing Plants for Upgrading

G. K. TOTHEROW

Manufacturing is only one part of a business. The needs of the business can change rapidly from forces outside manufacturing and the control and information systems must follow and support the short-term and long-term business goals and needs of the business. The process of auditing and upgrading control systems is primarily one of determining engineering solutions for business problems. Many companies embrace the concept of continual improvement. These companies constantly review and evolve plant systems to support their continued improvement in manufacturing.[1] However, most companies only review their automation systems in connection with a major project. There is a good reason for this behavior. It is expensive and disruptive to the plant operation to upgrade control systems.

Some people recommend auditing existing plant systems compared with "world-class" or best-practices standards. These audits have their place, but the value realized by upgrading existing systems is in achieving business needs and goals. The time to stop upgrading plant automation and information systems is when it is not a good financial decision as a way to meet the business needs. The only way to determine if the upgrade to the existing systems is a good financial decision is to audit the existing systems against the functionality needed to achieve company and plant goals. The purpose of this chapter is to provide a methodology to audit a plant for upgrading systems. A side benefit to the methodology given is that the justification for the upgrade project is written from the audit information.

There are two types of plant upgrades. First, there is what could be described as the maintenance audit to avoid obsolete components, eliminate worn-out components, or conform to new regulatory requirements to keep the plant operating. Then, there is the upgrade for process improvement or manufacturing cost savings that will show a return-on-investment from the upgrade. Both share the common theme of keeping the plant achieving business goals. The goals and the project funding are different but the prerequisites to the audit, the methodology of the audit, the integration of old and new components, and the recommendation report are the same when reviewing a plant system for upgrade.

There are five prerequisites to a meaningful consistent automation system audit:

1. Understand the company and plant goals.
2. Determine the functionality that is needed from plant-systems to achieve or contribute to those goals.
3. Establish or communicate the plant standard components and systems that can be maintained effectively with the available support personnel and spare parts.
4. Identify key processes, machinery, or areas of the plant for special attention.
5. Choose the best person to perform the audit.

This section first discusses the five prerequisites to the audits, the methodology for conducting the audits, and particular issues of integrating new technology into existing plants.

PREREQUISITES TO AUDITING

Every plant and every industry has different equipment, raw materials, and personnel. It stands to reason that every plant and industry will have a different recipe to optimize profits. State-of-the-art controls that reduce variation will not provide the same return on the investment in one process, one line, one plant, or one company, as they will in another. The same is true with respect to head count reduction, reducing maintenance costs, and increasing reliability. All these factors are very important to every manufacturer, but the degree to which they are important varies between plants, industries, company financial standing, and the general economy. For this reason it is of foremost importance to fully understand plant and company goals and audit systems against those goals.

Plant personnel, corporate experts, or outside consultants may conduct system audits and make recommendations. The degree to which they understand the plant will certainly be different so the purpose of the five prerequisites to an audit is to ensure that pertinent background information and goals of the audit are clearly understood. Specific prerequisites to an audit will need to be contracted or expanded based on the industry and the scope of the project. The prerequisites here will give the outside expert a good idea of how to meet the needs of the company and will help the plant technical person convince others to share the direction and success of the upgrade.

Goals

If this entire section were devoted to the importance of understanding company and plant goals, it would still not be enough. The importance of focusing on project goals and the complementing company goals and initiatives cannot be overstated.

Manufacturing is only one facet of a business, and it seldom drives the company; it swings to the demands of marketing, sales, design, and other business factors. Manufacturing may be asked to produce greater quantities, increase quality, decrease delivery, shave costs, reduce downtime, add grades, change packaging, or any number of other things to meet business demands. Understanding the current business goals is highly important prior to modifying manufacturing systems.

There was a time when only the highest-level people in the company knew the corporate goals. And under them only a few people knew the business and manufacturing goals of the plants. And lower still only a few clearly understood the department goals. Today, most companies have a mission statement and a policy of passing down written goals so that those below can establish supporting goals. Whether easy or difficult to attain, the company, plant, and area goals are important to auditing an existing plant because these goals and company initiatives will be the basis for justifying the audit recommendations.

It is good to be given company, plant, and specific goal information, but it must be understood that many business decisions and business information will only be disclosed on a "need to know" basis. It is also a fact of some businesses that goals change rapidly and they may not be communicated effectively. So, the first work that we do is to write a specific goal for the audit in terms of the company or plant goal to communicate effectively the basis for the audit. The audit of an existing plant for upgrading should always have a specific, stated goal. The following are some likely goals for the audit:

- Reduce variation of product
- Increase throughput
- Increase reliability
- Avoid obsolescence
- Adhere to safety or environmental regulations
- Reduce maintenance costs
- Decrease manufacturing changeover or process modification time

Functionality

The world of plant and process automation is changing very rapidly. Whereas 25 years ago an operator interface device might be a panel with a gauge and push buttons, today the operator interface device might be a wireless hand-held computer. It is far too easy and common to jump from the project goal or problem statement to looking for equipment or systems that a vendor says will solve the problem. An important intermediate step is to obtain a layperson's description written in simple language that tells what the "ideal" systems must do for the operators, maintenance mechanics, and managers to allow them to accomplish the stated goal or solve the problem. Most of the functional description should come from the users and area process managers along with an estimate of the financial payback for solving the problem or achieving the goal. The functional description and the estimated return on that functionality are often acquired by interviewing the appropriate operating and maintenance personnel.

One purpose of the written functionality description is that it breaks the components of the existing plant into digestible pieces and describes the "ideal" as established from the goals. It does so in terms of the functionality rather than component descriptions. This is necessary because the physical component often provides several functionalities. A control system may provide the controls, operator interface device, alarming, and other functionality. The system may perform near the level of the ideal functionality in one or two areas and may perform far below the ideal functionality in others. The audit recommendations could advocate replacement of the system or it might recommend add-on components to enhance the system capability in the poorly performing functional areas.

A second reason for the "ideal" functionality and the estimated financial payback for the functionality is that it will provide an estimate for the preliminary return on investment for the upgrade project. Better yet, the return for providing the functionality comes from the plant personnel who must support the project.

The functional description should address:

- Process measurement
- Final control element
- Input/output system wiring
- Control needs
- Redundancy
- Operator interface needs
- Alarm handling needs
- Historical process data needs
- Management information needs
- Production/cost/scheduling needs
- Maintenance needs
- Customer information needs

Plant Standards

Manufacturing and process facilities should establish and maintain a list of preferred components and vendors that have the functionality needed to achieve plant and project goals. Plant standards are useful to set a general direction in the components and ways a facility will try to meet company and plant goals and avoid obsolete components. Other common uses of a standard is to establish better vendor relationships, minimize spare parts, minimize decision making, minimize training costs, and ensure consistent and predictable results. That standards help in all the ways listed above needs no explanation; however, using standards to establish general direction and the period of review of standards needs further clarification.

Few plants can justify the capital financing or the production downtime to replace components across the facility when new devices are proved better to meet the needed functionality or when new industry trends and standards are established. Innovation must be integrated into the facility. The plant standard should lead in setting the direction to keep the components from becoming obsolete and promoting devices that will better satisfy the functionality desired. When should the existing standards be reviewed and modified? The answer is more often than is done for most plants. Any of the following events signal a time to review plant standards:

- The manufacturer announces a discontinuation of the product.
- An organizational or industry standards committee selects another standard.
- The plant is planning a major expansion or revision.
- A problem of service develops with established vendors or manufacturers.

Consider the example below, which shows a company that sets and changes its plant standards to establish a strategic direction.

A chemical plant built in 1961 installed Foxboro transmitters using 10–50 mA DC current signals. Once installed, the transmitters and wiring worked well and met the functionality criteria for field instrumentation and field wiring. Some time after the ISA standards committee adopted 4–20 mA DC as the standard for field signal wiring (ISA S50.1-1972) the plant decided to adopt the 4–20 mA transmitters as their new standard. The next major opportunity to install transmitters was during a major expansion and renovation project in 1975. The plant replaced 100 of the old transmitters in the area of the renovation and put them in the storeroom for spares for the rest of the plant that kept the old 10–50 mA DC transmitters.

In 1999, the plant had another major renovation project. This time the project convinced the plant to use Foundation Fieldbus for the signal wiring. The plant adopted Foundation Fieldbus as its new standard field wiring. By 2000, 30% of the plant was using new smart transmitters and Foundation Fieldbus, 65% of the plant was using transmitters with 4–20 mA DC signal wiring, and 5% of the transmitters were the old 10–50 mA DC Foxboro transmitters. The company will continue to install the Foundation Fieldbus standard with each project.

The hypothetical plant used the "plant standard" to set an appropriate direction to avoid obsolescence and integrated the new standard along with the old where both met the required functionality. The 4–20 mA DC only transmitters and signal cable will not meet necessary functionality requirements when the plant demands smart transmitters with online diagnostics.

A final word concerning standards is to guard against using the "standard" to thwart innovation that meets the functionality of its intended use and achieves company or project goals better, faster, or cheaper.

Identify Key Areas for Special Attention

There are key areas, processes, and control loops in every plant that are crucial to quality, production, or profitability. They will be referred to as key success areas. These key success areas and their impact on the operation should be noted and understood by the auditors prior to reviewing the systems for upgrading.

The purpose of this prerequisite item is to ensure that audit recommendations address any issues that might affect the process or operations at these points. These key areas may be ISO (International Standards Organization) tagged control loops, OSHA (Occupational Safety and Health Administration) regulated areas, FDA (Food and Drug Administration) certified processes, or just important areas of the plant. The audit to upgrade existing systems in the plant should specifically address the potential impact to these regulated areas. The importance of the key success areas of the plant will be well understood by the operations and management people that will be curious about how any changes may affect their operations. There may be some merit in specifically addressing these areas in the upgrade recommendations even to note that there is no effect on the process or operation at that point.

The important point to remember is that the information about these key success areas of plant operation should be communicated to those responsible for the audit. It is certain that if the potential impact on these areas is not addressed up-front, the issue will be questioned later.

Who Audits the Plant?

The last prerequisite before performing the audit is to determine who should perform the audit. There are several persons or groups that can be made responsible for the auditing so the recommendations of this section may be a little difficult to understand as we have not yet defined all of the steps and expectations of the audit. This subsection should be reread after the audit steps are reviewed if the reader disagrees with the author's opinion.

Webster's New World Dictionary[2] defines audit as follows:

> **5.** *any thorough examination and evaluation of a problem.*

The person, or group, conducting the plant review and making upgrade recommendations should have the time to dedicate to doing a thorough examination of the existing systems, experience in evaluating and making appropriate recommendations to resolve problems, and ability to write a report that will show the value of implementing the recommendations. Choosing the best person to perform the job among the several who regularly perform such tasks is as important as successful accomplishment of the goals.

Figure 1.1a is a subjective chart showing a rating of the qualifications of the persons who regularly conduct such work. The chart shows the ranking of the various people on a scale of 1 to 5, with 5 the best.

The categories on the chart are explained below.

Plant engineer—A technical person at the plant with 3 to 7 years' automation experience

Corporate engineer—A senior-level engineer who travels between various plants providing technical troubleshooting and project support

	Plant Engineer	Corporate Engineer	Vendor	Engineering Firm	Consultant
Process Knowledge	5	5	3	4	4
Plant Knowledge	5	5	3	4	4
Industry Knowledge	4	5	3	3	5
Problem Resolution	2	3	4	4	5
Time	2	3	2	5	5
Experience	3	4	3	3	5
Influence	2	3	2	5	5
Cost	4	3	5	2	1
Project Cost	3	4	2	2	5

FIG. 1.1a
Chart showing a subjective ranking of the relative strengths of persons who might perform a plant audit: 5 is best, 1 is worst.

Vendor—A technical salesperson who might be called to assess the existing systems and make recommendations

Engineering firm—An outside engineer from a firm that does detail engineering and project management

Consultant—An individual with extensive technical background and experience who is a specialist in that process industry or is an expert in the technical area of the upgrade project

Process knowledge—Knowledge of the plant process in the area of the audit

Plant knowledge—Understanding of the plant, goals, systems, and personnel

Industry knowledge—Understanding of the business, regulations, projects, and trends for that process industry

Problem resolution—Capability of determining equipment and system changes that will resolve the technical problems and add the functionality that the audit reveals

Time—The dedicated time to study the problem, determine resolution, and write reports

Experience—Generic estimate of the experience each has in this type of audit and problem resolution

Influence—Assessment of the capabilities of the person or group to have the recommendations implemented at the plant

Cost—The cost of the audit

Project cost—The degree to which the auditor will work without bias on the company's behalf to gain return-on-investment and save capital money on the project

Four of the audit prerequisites are listed to give the outside experts the detail plant knowledge that they need to do a very thorough audit. Unfortunately, there is no efficient way to transfer the knowledge and experience of the outside expert to plant personnel. Even if the plant had a technical person with the time and experience to adequately examine and evaluate systems for upgrading, that plant person may not have the political influence to be the catalyst for change that is needed to convince the plant to implement the recommendations.

THE AUDIT

The introduction to this chapter mentioned that there are two types of projects: the maintenance upgrades to avoid obsolete components, eliminate problem components, or conform to new regulatory requirements to keep the plant operating; and the capital project for process improvement or manufacturing cost savings that will show a return on investment from the upgrade. The primary differences between these types of audits are perhaps the scale of the job and the internal funding differences. Otherwise, both audits are essentially the same and share the same characteristics and steps. Either type of project will require finding the best place to replace the functionality of the old components with new. Both projects will require the new components to integrate with old components. The maintenance project to replace individual components relies more on the plant standards that set strategic directions to ensure the solutions are synchronized with the plant long-term goals.

Check Sheet for Control System Audits

- Model numbers of components
- Instrument installation
- Communication capability of electronic devices
- Physical condition of instruments, valves, controllers, and wiring
- Valve position, cycling at typical conditions
- Operating procedures, problems, suggestions
- Documentation
- Problems with regulatory control
- Multiple operator interface devices to various systems

FIG. 1.1b
List of some of the items that should be noted by observation or discussions with operators during a control system audit.

Now, we understand the company, plant, process area, and upgrade project goals. We interviewed operations and management personnel and we know the functional specifications for an ideal system that would allow operations achieve the project goals. We have estimates from operators, managers, and maintenance personnel of the time, product, rework, quality, and other tangible savings that they can achieve with functionality discussed. We understand the plant standards for various components in the audited systems. We identified the key success areas of the plant and noted their impact on profitability. We selected an auditor and gave him or her all our background material. So, now what methodology should our expert follow?

1. The first step of the audit is to review the prerequisites with the auditor. Remember that the purpose of the prerequisites is to share plant-specific knowledge and establish a functionality that would enable the plant to accomplish the goals.
2. Perform a physical audit of the systems (Figure 1.1b). Thoroughly examine the existing systems auditing functionality, models of components, and operating procedures. Take special notice of communication ports and communication capability of electronic devices. Look carefully at the physical condition of signal wiring and input/output (I/O) systems as well as the installation of the existing equipment. Note again that the systems should be audited in functional groups rather than physical components. The operator interface device includes at least three functional groups: interface to process data and control, alarm management, and historical data.
3. Define the gap that exists between the functionality needed and the functionality in the present systems. Consider submitting this gap analysis for review and approval by operations and management as an intermediate step.
4. Evaluate the upgrades needed to close or eliminate the functional gap using plant standard equipment where applicable.
5. Evaluate the modifications to the operation and maintenance practices needed with the system upgrades to achieve and sustain the project goals.
6. Make a formal recommendation of the most effective upgrade that will evolve the present systems and supply the functionality needed to achieve the goals of the project. Recommend the changes needed to operation and maintenance practices that are necessary to achieve and sustain improvements. Provide a cost estimate of the upgrade and a rough return on investment from the information gathered in the ideal functional specification. Define the operating performance goals that can be achieved by following the recommendations, and determine the measurable results that will be the success criteria for the project. The recommendation should also state the estimated length of time that the system will remain viable.
7. Audit the system performance as compared to the project goals and the agreed-upon project success criteria approximately 6 months after the upgrade project is complete.

These seven steps constitute an outline for auditing a plant for automation system upgrades. The procedure does not address what to look for or how to evaluate the selection of various components. Other sections of the *Instrument Engineers' Handbook* adequately address the selection and installation of control elements and transmitters, networking, control systems, operator interface devices, and other technical information. The last part of this section focuses on a few of the fundamental issues of the integration of new and old components in upgrading existing systems.

UPGRADING EXISTING SYSTEMS

The person performing the audit should have knowledge beyond that of the typical plant personnel of the trends, alliances, and evolving technologies that will provide a great return on investment for manufacturing and process industries. These items may not be in the functional specification. The auditor has an obligation to make the plant aware of trends, evolving technologies, and integration issues so that the plant can determine the value of an immediate investment. The rest of this section on auditing an existing plant for upgrades addresses some of the issues of integration of old and new and other items that the auditor should include in the upgrade recommendation report.

Evolution

Systems should evolve, not become extinct.

Evolution should be the plan while auditing existing systems and should definitely be a primary consideration in the evaluation phase of the system upgrade. In the 1980s and 1990s a distributed control system (DCS) meant proprietary I/O systems, controllers, data highways, operator interface devices, and process historians from a single vendor. Initial investment to purchase these systems was high, and the cost of outages and disturbance to operations installing them were even greater. Suppliers replaced components and had major new releases often because the "systems" encompassed both the hardware and software in many areas of functionality and because competition and development were very active. A result is that it is common for customers to have two "systems" from the same vendor with new operator interface devices that cannot communicate with old controllers, or new controllers that will not communicate with old operator devices. Today, many companies operate with independent "systems" from various suppliers on the same plant site.

Like others, Fisher Controls, now part of Emerson Process Management, began assisting customers with life-cycle planning for the DCS in the early 1990s. These life-cycle programs provide migration paths to keep systems current and provide notification of new products and manufacturing changes and supply of older products.[3] (Honeywell offers several options of LifeCycle Management (LCM) to help its customers integrate new technology with predictable costs.[4]) These programs help successfully avoid component obsolescence and assure that the components interface through several vintages, but they do not always lessen the cost and disruption to plant operations. There is another option to supply the evolution needed.

In the late 1990s open standards, increased functionality, and reliability of personal computers and the need for process information combined to enable, and force, vendors to create access to their systems. Today, a plant system can be defined as an arrangement of independent components connected to form a unity for the achievement of specified functionality. The components do not need to be from one vendor. Architecture of the system is very important. Hardware and physical connectivity to the proper data highway systems enable very highly flexible and upgradable functionality through upgrading software. The components can be selected because they are the best of breed, or selected on the basis of lowest cost to fill the functional requirement. With the proper architecture, the hardware and software components that comprise the process control and information system can evolve at different speeds over many years with minimal impact on operations and minimum cost.

The audit should identify existing components supporting interfaces to open systems as well as existing components from suppliers that refuse to provide open interfaces. Figure 1.1c shows a typical 1990-vintage DCS with a good, high-speed interface to other systems. Figure 1.1d shows PCs as new operator interface devices connected with redundant links to the DCS and PLCs (programmable logic controllers). Additional functionality is added to the operator interface device through software to allow retrieval of grade specifications from the specification management system and downloading of grade set points and tuning parameters to the existing control systems.

Audit the Installation and Process

Few situations are more frustrating or more futile than trying to correct process design problems with process control. Similarly, changing manufacturers or styles of instruments will hardly improve the control problems caused by instrument installation errors. Every control and system upgrade project should strive to correct the physical process and instrument installation problem of past projects. Physical problems involve piping and mechanical work that is usually expensive, and controls engineers are always under pressure to make the existing system work without modification. However, the best

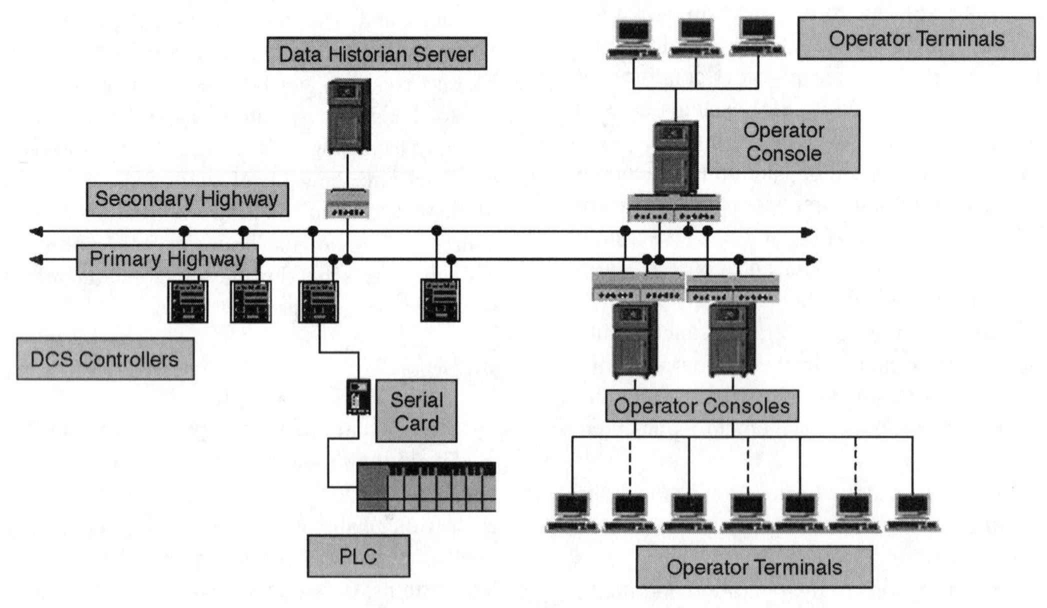

FIG. 1.1c
Typical DCS architecture circa 1990.

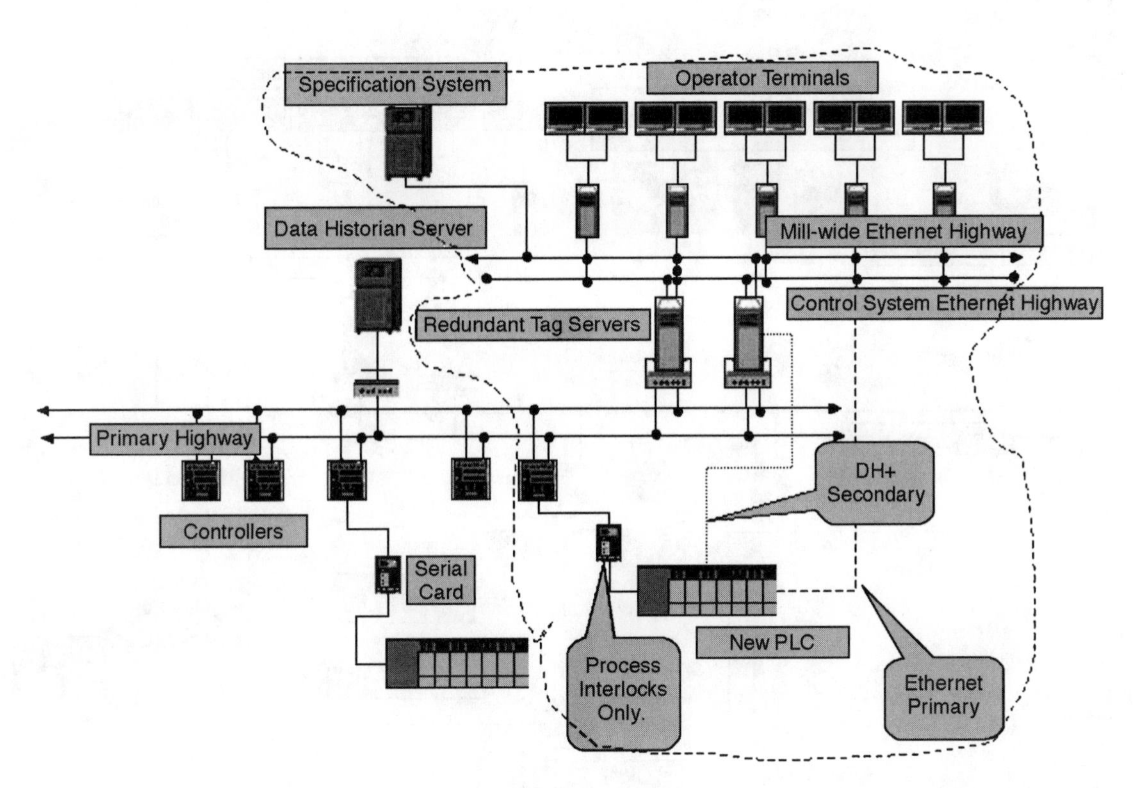

FIG. 1.1d
Illustration of the same DCS as Figure 1.1c with a new DCS controller, PLC, new operator interfaces, and supervisory controls added.

approach is to address the problems directly. Allowing operations and management to think that a new controller or new instrumentation will resolve a physical problem is as great a disservice as failing to recognize the physical problem. The audit is the time to recognize and note process and instrument installation problems.

Sometimes it is not feasible to conduct an audit to scrutinize every process and every instrument and control valve installation looking for installation problems. The process design problems are particularly difficult to see without analyzing Process and Instrument Diagrams (P&ID) and yet with existing systems this is not something that is typically part of the control system upgrade project. Scrutiny of the process and control system designs must be focused on certain areas prior to the field audit.

One way to identify process areas for extra scrutiny is through recognizing indicators of process problems from other parts of the audit. Figure 1.1e lists some of the key indicators that a control problem is more involved than just needing a new controller or upgraded system.

Indications of Physical Process Problems

- Control of process was always poor.
- Periodic or seasonal fluctuations in controllability of the process.
- Instruments and valves that work in other places are not working.
- Fast oscillations in process characteristic properties after mixing.
- Fast process transients and oscillations while in manual mode.
- Control valves that operate at extremes of their range.
- Excessive process equipment and control component failures.

FIG. 1.1e
Do not try to correct problems and disturbances introduced by poor process design and nonfunctional process equipment when the best solution is to recognize and correct the process.

Process Information and System Integration

Great process control is not enough. Process information is more valuable than control in many industries today. This is not totally without reason or justification since the information is needed for product tracking, product genealogy, offline statistical analysis, regulatory compliance, and marketing and customer relations. Process information increasingly interfaces to enterprise resource planning (ERP) systems, enterprise asset management (EAM) systems, manufacturing execution systems (MES), and data historians. Every system audit and recommendation should address these issues.

The functional specification in the prerequisites to the audit should contain a statement about the desired interfaces to other systems. Today, any audit should address this issue whether it is in the functional specification or not. The trend is clear that more process information and process system health information are desired by higher-level systems. It is also true that the process control system and the operator, or process manager as the position is often called, increasingly need access to many more systems than just a process controller. The integration between the management systems will be bidirectional where quality control persons may need to see the key information about the current product and the

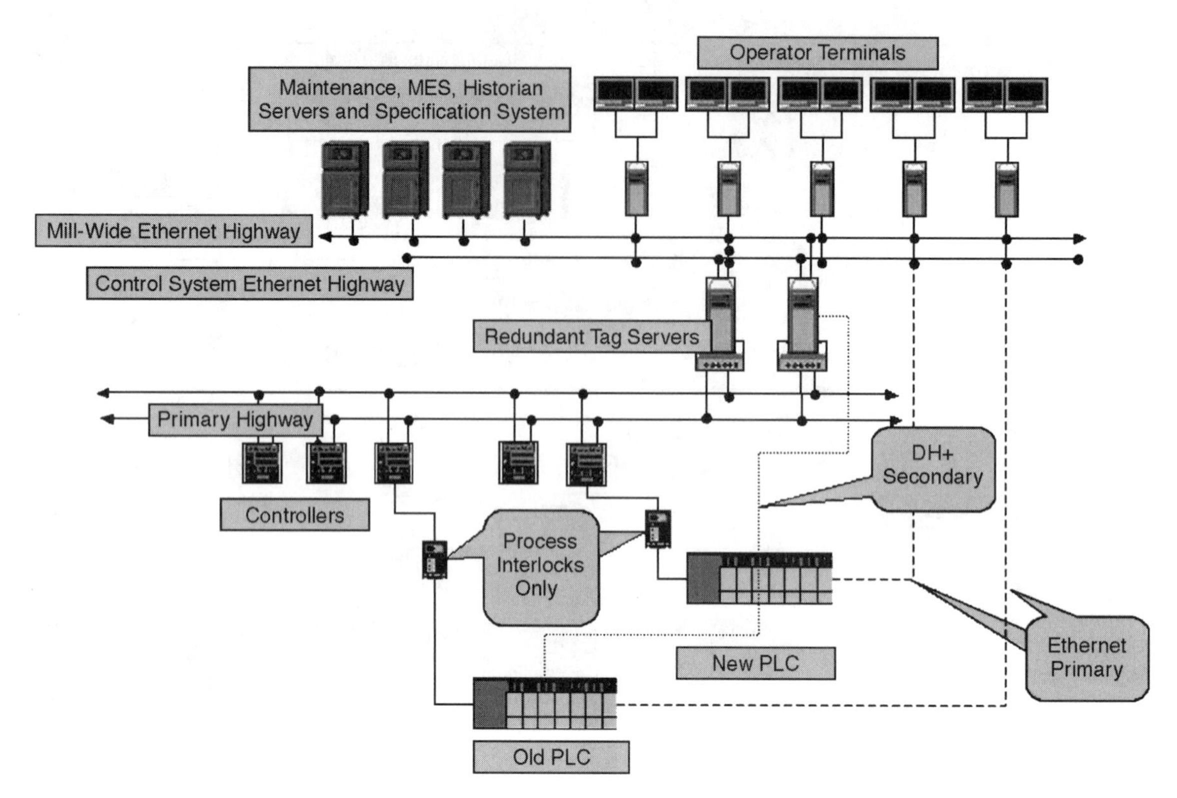

FIG. 1.1f
Illustration of the same DCS as in the previous figures with direct operator interface communication to the PLC, and the addition of interfaces to other plant information systems that need process data.

operator may need to see, approve, and download the specification for the next run in the plant. Operators, or their systems, need access to the maintenance system, e-mail, upstream and downstream processes information, historical data, time and attendance systems, possibly accounting to show real-time cost of production, and various other systems in the future. Figure 1.1f shows the connectivity for the operator interface devices to integrate to the other plant information systems.

System Diagnostics and Redundancy

System diagnostics and redundancy, like process information and system integration, is an area that the auditor may understand better than the plant personnel. So too, the auditor making recommendations for upgrading systems should include recommendations on the system diagnostics and redundancy even when the functional specification does not address the issues.

System diagnostics begins with the transmitters and final control elements in the field. Smart transmitters, valves, and a device network are the basis for a system to alert folks of process and instrument problems and provide the diagnostics to isolate the problem.

The 4–20 mA DC has been the standard for signal wiring since the 1960s but it now looks extravagant to run a pair of copper wires in a plant to every instrument for just one piece of information. Fieldbus technologies are discussed in Section 4.7. The plant may need direction and recommendations on the diagnostic and quality information available.

Regulatory control is as important as ever, but supervisory controls and coordinating plant controls are needed in most plants to make a step change in quality and productivity. Changing one regulatory controller for another is not a recipe for success. Section 1.8 of this chapter discusses hierarchical control. The auditor must consider that recommendations address virtual sensors, automatic loop tuning, statistical process control, and model-based control. Also, the control systems recommendations should make the plant aware of redundancy options, diagnostic capabilities, and automatic telephone dialing system alarms.

Redundancy of control systems, where it is required, is often considered a dreaded but necessary expense. Where applicable, the control system upgrade recommendation should address the advantages of redundancy for purely economic reasons. Systems that continue to function through a component failure avoid forced downtime, lost product, the cost of emergency maintenance support, interruptions to production scheduling, and other problems. As systems are integrated, redundancy may be needed in communication networks, interface devices, and software to ensure that the systems continue to function through a failure. The technical expert must make the plant aware of opportunities. Section 2.9 discusses system architecture for increased reliability.

References

1. Walton, M., *The Deming Management Method,* New York: Dodd, Mead & Company, 1986.
2. *Webster's New World Dictionary,* 2nd college ed., New York: World Publishing Company, 1970.
3. *Proceedings 1992 PROVOX Users Group Meeting Report,* Austin, TX: Fisher Controls Company, 1992.
4. Honeywell Web Site, 2001, http://www.iac.honeywell.com/pulp_paper/Services/InternetServiceLifeCycleMgtContent.htm.

Bibliography

McMillan, G. K., *Process Industrial Instruments and Controls Handbook,* 5th ed., New York: McGraw-Hill, 1999.

1.2 Project Management and Documentation

S. EMMANUEL

The dynamic fiery engine of a well-run project advances inexorably to its objectives on tracks laid of good documentation. And documentation had better be good, when it refers to the details of the central nervous system of a process plant. Faulty connections in the brain of the plant could result in the worst nightmares or if the project is lucky a dysfunctional plant. Good documentation is only one of the prerequisites for successful instrumentation and control (I&C) projects. The multifaceted nature of I&C projects means that the project has implications for almost every department of a plant owner's organization. Satisfying each of these stakeholders will require the best of project management skills. This section describes the structure and underlying relationships of traditional I&C documentation and its effective exploitation in conjunction with project management techniques for successfully managing an I&C project.

For the purposes of this section, the following definitions apply:

Project: A temporary endeavor undertaken to create a unique product or service
Owner: The entity with final responsibility for the complete operation of the facility
Contractor: The owner's hired representative, providing any combination of engineering, procurement and construction services

The descriptions and terminology used here are common practice in the process industry. The documentation and project management practices would be applicable as is in related industrial sectors with some changes in names of drawings and terminology.

GOOD DOCUMENTATION PRACTICES

Good documentation practices have to be enforced from the preliminary engineering phase of the project. As the project progresses, the stewardship of the engineering documents changes hands from design contractors to construction contractors or vendors to eventually reside with the owner. Thus, it is the owner's commitment to good documentation practice during each project phase that will ensure that documents always change hands in a state appropriate to that phase. Owners ensure this by enforcing review and approval cycles and maintaining a system of documentation.

An independent reviewer, representing the owner's interest, is valuable for the project. Design reviews take place during the design phases of the project when most of the documents are produced. Thoroughness of review and approval can be ensured by the use of standard checklists for each type of document and each phase of the project. These checklists are developed applying the three "C"s criteria for good documentation:

- Completeness
- Correctness
- Consistency

Completeness criteria ensure coverage of scope of the project and adequacy of the level of detail on the drawings with respect to the phase of the project. However, a project sometimes has to proceed with incomplete information and in such cases the portion of a document containing such information must be clearly marked out (shading or clouds are frequently used) with a appropriate clarification in the form of remarks or notes. The correctness criteria ensure the integrity of data on the documents with respect to I&C engineering principles, requirements of other engineering disciplines, existing plant or site conditions, and applicable standards. Consistency criteria refer to maintaining correct cross references within the project and plant documents, using specified file, border, and title block formats, using the correct drawing naming and numbering schemes, using consistent terminology and units, using drawing templates, seed files, etc. Consistency is also improved if repetition of the data is minimized, which ensures that, when the data change, only a minimum number of documents are affected.

Maintaining a system of documentation requires that the owner define and administer, as a minimum, a system of naming and revising documents and a system of tagging plant, area, equipment, and instruments. An administered system of tagging and document numbering serves to anchor the documentation system of the plant, allowing these attributes to be used as key attributes or to index into lists of equipment and documents. Although document and revision numbering conventions vary in practice, instrument tagging conventions follow ISA S5.1.[8]

In addition, particularly in large plants involving various construction and engineering agencies, it is common for owner organizations to:

- Maintain written procedures for all activities related to official documentation
- Maintain and enforce the use of standard libraries of cell or block reference files for standard equipment
- Retain control of drawing seed or template files for each type of document and database structure
- Require the use of standard installation details, materials, and use of standard coding and classification schemes

These kinds of standardization affect all phases of the project, saving time and cost and improving the quality of the construction.

PROJECT CRITERIA DOCUMENT

Capturing the requirements of a project with all its complexity is a difficult task. The traditionally quoted project objectives, that is, cost, schedule, quality, and safety, are always paramount. However, every project needs to satisfy unique requirements, which the whole project team should be aware of from the beginning so that the project ends in success. Hence, irrespective of the project phase, a project summary or criteria document helps focus efforts on the requirements of the system, indicating physical, economic, or logistical constraints that are applicable and the standards to which the system needs to comply. This could take the form of a checklist or a preformatted document with blanks to be filled in at the time of project start or estimation. Table 1.2a is a sample design criteria format for the I&C portion of a design project. On this form, design criteria typically refer to scope documents and functional specification documents to extract the system requirements. Typically, a preliminary site visit will reveal a host of constraints and new requirements, which could well be consolidated into this design criteria report. The accepted standards and practices should be listed. Standard symbols for computer-aided drafting (CAD) drawings as defined by owner and/or contractor should be documented. If the owner provides a list of preferred manufacturers and any additional practices or schematics that further clarify owner design requirements, they must also be cited here. Finally, the sizing of a project provides a measure of the complexity and uses preliminary input/output (I/O) and field instrument counts as the basis. When the design is nearing completion, this document becomes a tool to ensure that the design will meet all the functional requirements, constraints, and standards. Table 1.2a is also an indication of the many ways I&C projects can differ from one another and emphasizes the uniqueness of each project.

THE PREFERRED VENDORS/TECHNOLOGY LIST

Owners do not always have a published list; however, the owner organizations are often precommitted to a certain brand or flavor of technology for various valid reasons, such as:

- The plant is already using the brand or technology, which means spares are available, operators need not be retrained, components have site-proven reliability.
- To avoid obsolescence by buying products based on standard protocols (e.g., HART or Fieldbus) and by buying the most recent hardware and software.
- Certain vendor architectures are a better fit for certain kinds of process (e.g., batch control vs. continuous control).

Identifying the preferred vendors at the beginning of the project has several advantages:

- The vendor can be a part of the project team from the beginning, bringing in experts on the specific system who are then able to provide optimal architectures based on vendor's products to meet the plant's present and future needs.
- Design can proceed with fewer uncertainties, leading to more consistent and complete designs.
- Overall project duration is shortened because of consistent and complete designs and also because the vendor can start production earlier on the project cycle.

Disadvantages are the possibility of higher costs due to the lack of competition and the danger of excluding consideration of better and newer technologies.

I&C DOCUMENTATION SYSTEM

The system of I&C documentation has evolved over the years with each document having an inherent structure of information that is useful during specific phases of the project. For example, loop diagrams are most useful as an aid for maintenance and troubleshooting, whereas interconnection drawings are used predominantly during the construction phase. The system of documentation has also evolved to minimize repetition of data between documents and to minimize the refinements required by each iterative phase of the project. Thus, the cable block diagram aids the optimal allocation of junction boxes and cables during a FEED or basic engineering phase, the details of which would be required to be worked out only during detailed engineering in the form of layouts of cable routing and interconnection drawings. This structure of the documentation is presented as a dependency map in Figure 1.2b. Given the uniqueness of each project, this is only a suggested sequence for creation of new documents. Some of the documents may not be formal deliverables but are used as intermediate design documents only. The dependency map is useful in determining the impact of changes to a project, which is described in one of the following sections.

TABLE 1.2a
A Project Criteria Form for Instrumentation and Control Design Projects

Project No:
Plant Name:
Project Name:

1. Scope Definition Document:
(Provide reference to scope document stating specific chapters, pages and drawings as applicable)

2. Project:

Type:	☐ Revamp/Upgrade	☐ Grassroots	☐ Expansion	☐ Modifications
Scope:	☐ FEED(Prelim/Prop.)	☐ Detail Eng	☐ Proc/Const Support	☐ As-Built
Contract:	☐ Direct	☐ Sub-Contract : Contractor:		
Contract Type:	☐ Reimbursable	☐ Lump sum		

3. Deliverables

File Formats :	Drawings:	☐ Microstation	☐ AutoCad	☐ Other:Specify
	Specifications:	☐ MSWord	☐ MSExcel	☐ Other:Specify
	Database:	☐ MS Access		☐ Other:Specify
Symbology		☐ ISA SP5.1-5.4	☐ Owner's Standard	☐ Other:Specify
Templates / Seed Files:	Drawings :	☐ Owner's Standard	☐ Other:Specify	
	Specifications:	☐ Owner's Standard	☐ ISA SP20	☐ Other:Specify
	Hook up Dwgs:	☐ Owner's Standard	☐ Other:Specify	
Units:		☐ SI	☐ Imperial	
Quantity:	P&Ids: PFDs:	Logic Diag:	Graphic Pages:	
Status of Existing drawings		☐ Schematics only	☐ Prelim/Unverified	☐ Final/ As-Built

4.Client Approved /Preferred Vendor List:

5. Project Content:

5.1 Sophistication:

Control ☐Not in Scope	☐ DCS ☐ Discrete	☐ PLC ☐ Relay	☐ Special Purpose controllers(Compr.,Turbines,etc) ☐ Pneumatic.	
	Major System Vendor:			
Transmission :	☐ Field Bus	☐ Smart	☐ Conventional	☐ Pneumatic
ESD & Safety:	☐ TMR	☐ 1oo2	☐ Relay	☐ Fire Detection & Control
Software:	☐ Batch	☐ SPC	☐ Optimizing /Advanced control	

5.2 Size:

Controller I/Os:	AI	AO	DI	DO	Networked I/O	Control Loops	Monitor Loops
DCS							
PLC							
ESD							
Special Purpose Controllers							

Field Devices	Qty	Control Power	Value
Total Instruments		Instrument Air Pressure	
Local Instruments		Control Voltage/Hz	
Control Valves			
Analyzers		Power Voltage/Hz	
MOVs			
Junction Boxes			

5.3 Other:

Type of Protection: ☐ Explosion Proof ☐ Intrinsic Safety
Type of Cabling: ☐ Above Ground ☐ Underground

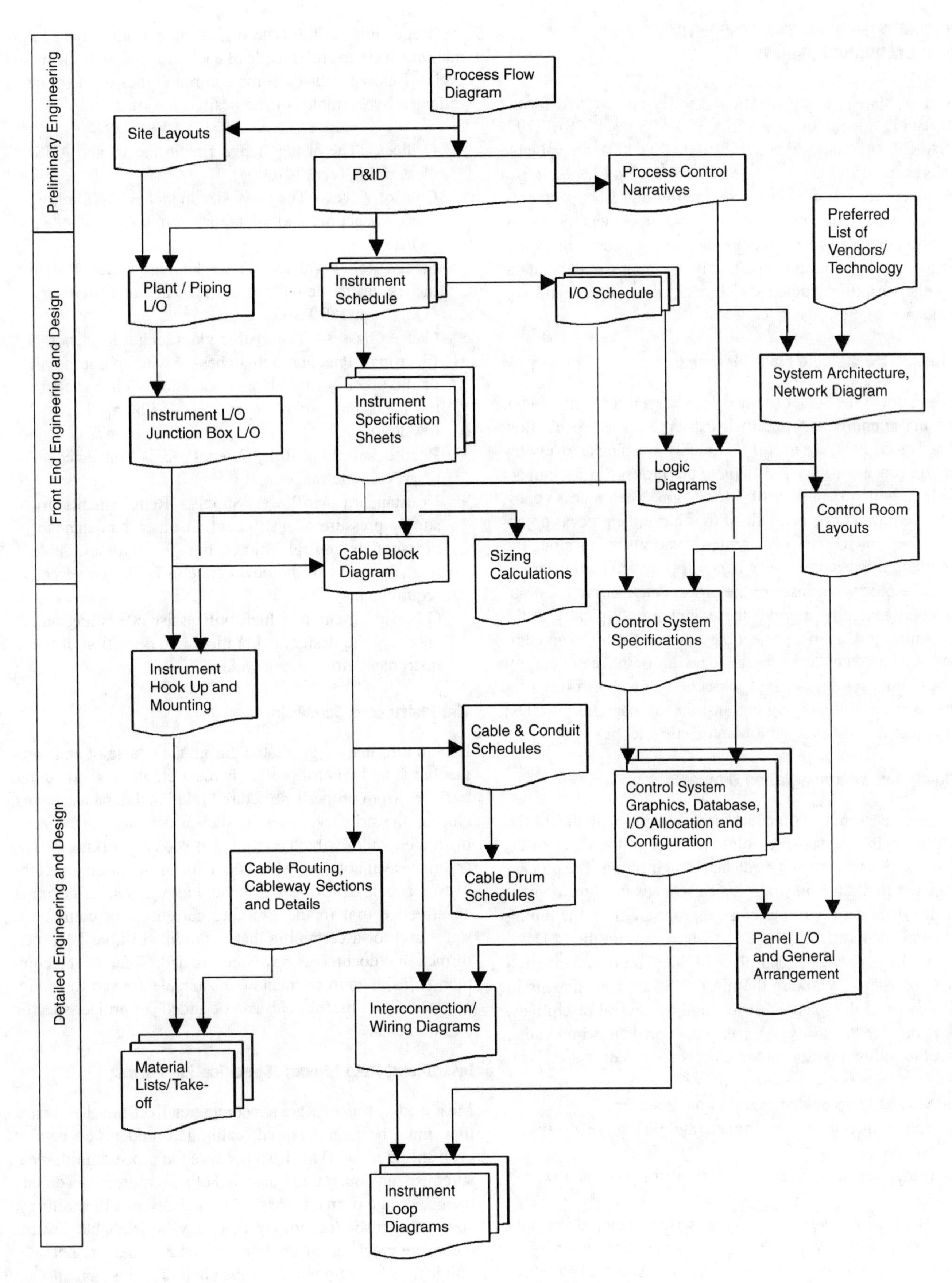

FIG. 1.2b
A document dependency map of instrumentation and control system design project.

DOCUMENTS—PURPOSE, CONTENTS, AND STANDARD FORMATS

Various standards organizations specify content and format of data in these documents. ISA S5.1 to S5.4,[8–11] ISA S20,[12] ISA RP60.4,[21] and PIP practice PIP PCEDO001[7] are valuable references. ISA S51.1[14] and *IEH*, Volume 1, Chapter 1 is a reference for terminology to be used in all I&C documents. Many owners also maintain their own documentation standards. Because project requirements vary considerably, the following description is given only to emphasize the content for the various documents with respect to purpose and significance to the project.

Process/Mechanical Flow Diagrams

These drawings depict the flow of material through a plant, the major equipment, and their capacities. The process flow diagrams (PFDs) are useful for quickly grasping an overview of the process because it depicts the process in a compact fashion without a clutter of information. The process conditions required to be maintained to achieve the process objectives are shown on these diagrams at various points along the flow and this becomes very handy for the I&C engineer.

The process licensee or the process engineer in the organization generally provides these drawings. Typically, at the preliminary stage of a project the control or instrument engineer's involvement is restricted to review of the major control loops in the system. A small subset of the symbols is used for the creation of the piping and instrument diagrams (P&IDs) to depict the instrumentation and control loops.

Piping and Instrumentation Diagrams

In the process industry the I&C design begins with the P&ID, which is also the starting point for some of the other disciplines such as piping and mechanical engineering. The process engineer in the organization typically leads the development of this document with inputs as appropriate from the piping and I&C departments. P&IDs are derived from the PFDs.

P&IDs enjoy a special place in the plant documentation system. They are among the most utilized of documents in all phases of the plant life cycle and by several disciplines. To maximize the utility of this key document, it is imperative that the following aspects are closely followed:

- Standard symbols are used throughout.
- Sufficient details of instrument and piping are always provided.
- Presentation provides for clarity of the process flow.

The facility owner typically prescribes standard symbols to be used in the P&IDs. The owner's instrumentation symbols and tagging conventions typically adhere to ISA S5.1[8] and *IEH*, Volume 1, Chapter 1. However, other systems of symbols are also used.

Depending on the type of instrument, additional information is very useful outside of each instrument circle on the P&ID. Below are listed some common types of instruments followed by examples of the desired information.

- Orifices—The orifice flange size in inches and ANSI class rating (e.g., 8″-600#).
- Control Valves—The valve size in inches, ANSI rating class, air action, and air failure action (e.g., 8″-900# AO/AFC).
- Safety Relief Valves—The inlet and outlet size in inches, orifice letter and set pressure in PSIG (e.g., 6Q8, set at 150 PSIG).
- Gauge Glasses—The visible glass length in inches.
- Electric or Pneumatic Switches—Switch point in unit of the process variable and actuation with respect to the measured variable (e.g., "H" for high, "LL" for low-low).
- Recorders or Indicators (Direct Process Connected)—Chart or scale range.
- Nonstandard Air/Power Supply—Requirements for supply pressure of instrument air other than standard pressure or control voltage not generally available (e.g., for use with shutdown systems or similar special requirements).
- Other information, which will assist in reading and checking the design, calibration, and operation of the instrumentation, may be added.

The Instrument Schedule

Design information generated during the course of an instrumentation and control project is most effectively stored and retrieved from properly structured relational database tables. One of the primary tables in such a database is the main instrument index, which is created as soon as working P&IDs are made available and which contains information on all the instruments in the project. Such an index thereafter becomes an effective tool for determining current work completion status and for ascertaining that required work has been performed and documents have been issued. Some of the common fields in an instrumentation schedule are shown in Figure 1.2c. Other useful fields can be added per project-specific requirements.

Instrument Data Sheets (Specification Forms)

Most field instruments are not commercial, off-the-shelf items; they must be manufactured, calibrated, and tested against each specification. This is so because the process fluid, measurement, and operation range and other requirements dictate the use of one of many possible combinations of metallurgy, size, and sensor technology that may be available. Instruments are precision-made, high-value items that are not economical to be maintained on the shelf. The instrument data sheets are prepared to present the basic information for the instrument requisition. When completed, they provide a

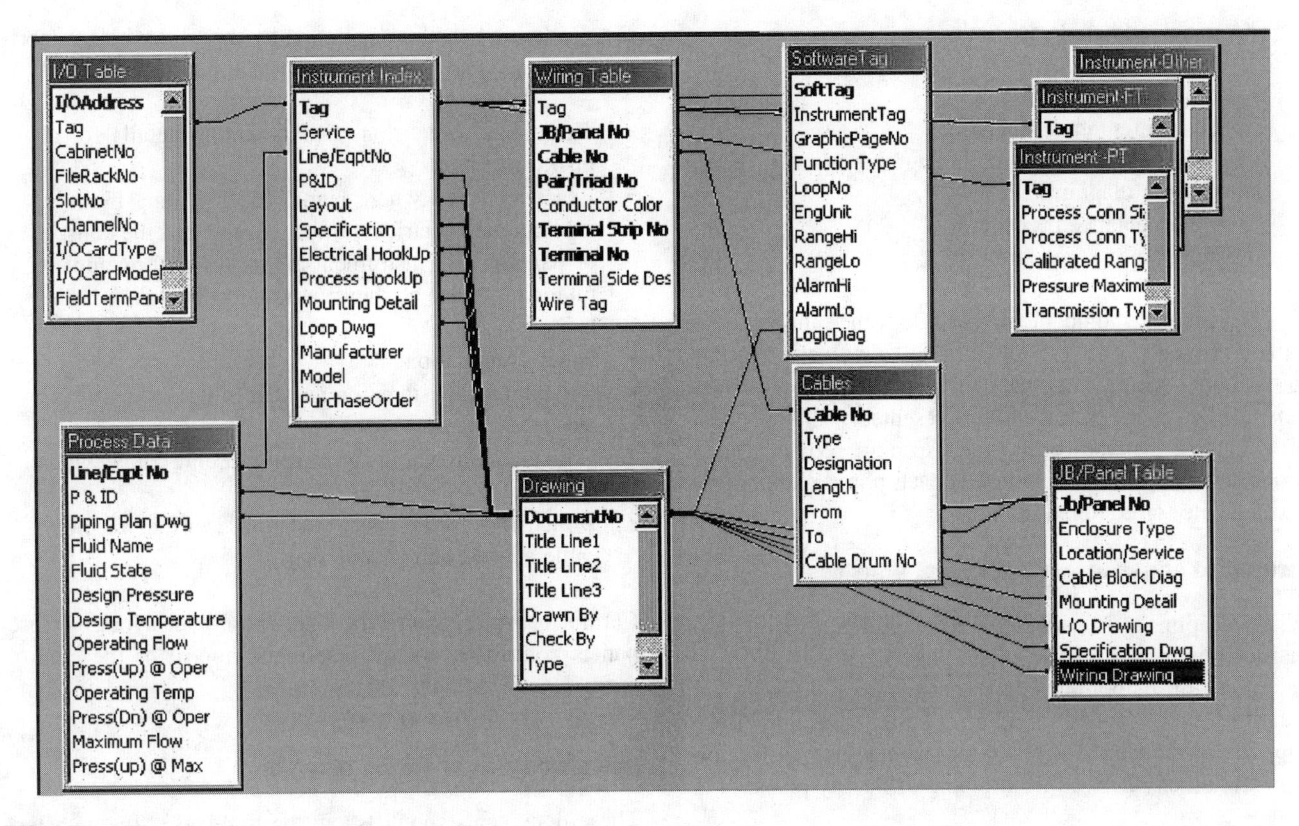

FIG. 1.2c
A database view of the object model of I&C system and documentation.

readily accessible, concise summary of information with sufficient detail to allow a vendor to select the appropriate instrument. ISA S20 or the equivalent should be used. A database is a very effective tool for maintaining large numbers of specification sheets.

System Architecture/Network Diagram

This drawing will show in block diagram format all major control system components, including routers, switches, gateways, servers, subsystems, etc. The drawing shall identify where all components are physically located (e.g., Central Control Room, Substation, Utilities Control Building, etc.). The system cable connection philosophy shall be shown. The cable/wiring details shall specify the media (i.e., fiber, coax, twisted pair, etc.), speed of communication link, protocol, and whether the link is redundant.

Control System Documentation

Control system design is a very intense phase of the project during which the owner departments, engineering contractors, and vendors have to exchange a large volume of information back and forth. Properly done, the documentation can be useful in many ways:

- Approval by the players having jurisdiction
- Check for code compliance, quality assurance, and control
- Basis for cost estimate, budget, and schedule
- Support of field installation, test, calibration, and maintenance
- Manufacturer's test records, basis for shipping-damage claims
- Basis for planning for future expansion, modification, and duplication
- Record of configuration and reference for operations in the future
- Historical records

The vendor or systems integrator produces the bulk of the documentation for a large control system following a functional specification document generated during the detail design phase of the project. The functional specification should provide sufficient details for a vendor to price the system and provide details of how each function is realized in the system. Before bid award the vendor's final offer document should detail the material and services provided. Along with the technical specification, the vendor's project delivery schedule, spare parts prices, support availability, and installation schedule are also included in a bid evaluation. On award, the vendor produces detailed documents, which shall typically comprise:

- Detailed system description, system planning information
- Vendor standard documentation for configuration, management, maintenance, and operation

- I/O, panel, and cable schedules
- Detailed panel/console drawings showing layouts and interconnections
- System configuration database
- Software documentation including source code, ladder, or logic diagrams
- Factory acceptance procedure
- Site acceptance procedure

The project should monitor and advance the status of each document from "preliminary" to "approved" and then to "as-built" with a well-specified plan agreed to by all the players. This plan is frequently represented with a responsibility matrix that indicates which specific people need or produce each type of document at each phase, the purpose, and the date for completion.

Instrument and Junction Box Layout Drawings

When locations of instruments are imposed on a facility plot plan or piping plan, the resultant cluster patterns allow for determination of optimal placement of junction boxes and identification of routing of conduits based on availability of supports, clearance and access requirements, etc. These drawings are directly derived from piping or facility plans.

Cable Block Diagrams

Cable block diagrams are created to show all the cables and the panels or instruments on which the ends terminate. These drawings are particularly useful in large projects where it is worthwhile standardizing on a small set of cable sizes and junction boxes. In an iterative manner, the process merges and splits home run cables and junction boxes until an optimal set of cables and junction boxes is reached. These drawings are also useful intermediates to the production of cable routing drawings, interconnection drawings, and cable and conduit schedules.

Control Room Layout Drawings

Control room/interface building/analyzer house layout drawings shall show in plan view the location of consoles, panels, control racks, computer racks and peripherals, logic racks, termination racks, and the position of each item of major equipment. The spacing and arrangement of cabinets should adhere to requirements for easy access for maintenance, heat dissipation, and electrical interference. Also shown are segregated pathways for cables carrying analog signals, emergency shutdown (ESD) signals, discrete signals, and power. ISA RP- 60.4 provides the detailed information for the design of control centers.

Panel Layouts and General Arrangement

In the preliminary design phase only the overall dimensions, location of instrument items, shape, graphic layout, and general layout are provided. The detailed layout shall be provided by the vendor and shall be approved by the owner before fabrication of the console or panel and shall include:

- The cutout dimensions and mounting details for all instrument items
- Exact locations where signal cabling, data highway cabling, and electrical power wiring enter the console or panel-board; instrument air supply and pneumatic tubing must also be shown as applicable to each installation
- Panel illumination
- Location and designation number of terminal strips or electrical junction boxes
- Schematic layout of pneumatic tubing runs, when applicable
- Dimensions, equipment location, wiring raceway, cable entries, and terminal strips

Careful consideration to the size, color, and layout of these panels contributes to good aesthetic appeal of the control rooms.

Interconnection or Wiring Diagrams

These documents should show the terminal strips with terminal numbers for junction boxes, field control panels, marshaling cabinets, and instrument cabinets and racks, with the appropriate identification. They should also show the connections of instrument multicore cables to the terminals, with identification by cable number and core/pair number. A table can represent the wiring for most junction boxes and marshaling panels just as well as by a conventional drawing. The advantage of a table is that it is more easily modified and maintained. However, a drawing shall be used if the wiring is complex as when multiple jumpers, resistor elements, and other devices are to be shown on the same document.

Cable Routing, Cableway Sections, and Details

The conduit/tray layout and cable routing (drawing or list) should show the general conduit/tray layout and the following cable/conduit data: identification number, approximate length, type, and routing. The routing drawing is generally imposed on a piping or facility plot plan. If the routing is underground, this drawing must accurately show all underground facilities and must be consulted and updated in the event of any change to these facilities. In addition to specifying routing of the cable, this drawing helps to estimate the wiring and cabling requirements for the plant.

Grounding System Drawings

The grounding system drawings should show grounding connections to the appropriate power supply systems as well as earth grounding locations for all instrument power systems, wiring system shield grounding, distributed control system grounding, and grounding for other appropriate instrumentation and control systems.

Instrument Loop Diagrams

Instrument loop diagrams (ILDs) are the culmination of the I&C design work as they contain the complete design information on a loop-by-loop basis. The objective is to depict the complete signal flow path from the originating device to the operator's desk accompanied by all associated information such as range, size, control, alarm settings, etc. See ISA S5.4[10] for formats and content. ILDs are extremely useful at the time of commissioning, start-up, and troubleshooting. However, ILDs repeat information that is already present in other documents. Maintenance departments in operating plants depend on ILDs and generally keep them updated. Problems arise when these changes are not always reflected in the other drawings. One solution is instrumentation documentation tools, which generate ILDs from a central database and other documents, as described in the following sections.

Logic Diagrams

Some plants maintain their process logic separate from the system-generated documentation of the logic. This serves to provide a common format of drawings irrespective of the vendor's format and the hardware used. ISA S 5.2[9] and *IEH*, Volume 1, Chapter 1 provide guidelines on format and content of binary logic. A consolidated view of the logic is also highly desirable for emergency shutdown portion of the logic, sometimes represented in the form of a cause-and-effect matrix. Logic diagrams are ideally generated from written process narratives and P&IDs and form the basis for vendors' programming of the system.

INSTRUMENT INSTALLATION CHECKOUT AND CALIBRATION/CONFIGURATION PROCEDURE

A detailed procedure should be prepared to define the responsibility for instrument installation, including complete calibration (or configuration for smart transmitters, etc.) and operational checks of all instrument loops, sequencing and interlock systems, annunciators, and shutdown systems. A checkout form should be used to record calibration and loop check sign-off/approval.

DECOMMISSIONING DOCUMENTS

Decommissioning documents are often required to show the portion to be decommissioned on a revamp job. Procedures for proper disposal or refurbishment of instruments must be provided.

DOCUMENTATION—AN INFORMATION MANAGEMENT PERSPECTIVE

The level of information technology employed on projects varies considerably; however, studies[1,2] established its effectiveness and identified the use of electronic data management as one of the prime techniques most likely to shorten project duration with no increase in total cost. The instrumentation and control portion of a project is particularly amenable to the effective utilization of information technology. Simple use of office software enables the creation and maintenance of specification sheets, schedules, and list documents. Widely available computer-aided drafting and design tools have long been part of the repertoire used by engineering companies to create drawings. However, to benefit fully from this technology it is essential to be able to link together the data resident in various traditional document formats so that data can be viewed, maintained, correlated, and shared.

Traditional documentation tools are essentially document-centric, which means information is stored with each document or file as the final atomic unit, whereas in the engineering world the atomic unit is the individual component that makes up the overall system. The components can typically be classified into various types (instruments, cables, junction boxes, I/O) with each type having a set of attributes common to that type (e.g., every instrument has a tag no., service, P&ID no., etc). Components of different types in a system are associated with each other in terms of relationships that relate some of their attributes in one type to that of another. For example, an "I/O" allocation to a particular field "instrument" is an associative type of relationship between the "I/O" and "instrument" components in the system. Other types of relationships between components are possible. For example, a cable made of several conductors represents a "whole-part" type of relationship. Similarly, the "instrument" component is a generalization of the specific type, for example, the pressure gauge. This is an example of a "general-specific" type of relationship. The components or objects also have to satisfy certain constraints and conditions with respect to the value of each of its attributes and in its relationships with other components (e.g., each conductor end can have only one termination point). Object technology refers to data structures, which enable storage of data associated with real-world components along with their relationships and rules of interaction with other components. Clearly, the object technology is better suited for storage of engineering data, which obviates the translation process from an object-centric world to a document-centric space required for the document-centric approach. See Reference 3 for a description of solutions addressing this issue.

Engineering information needs to be exchanged between the various disparate players in a project. Any data storage format needs to be sharable and usable by the whole project community. Various consortiums of companies[3] and standard bodies[4,5] have been established to achieve standardization of various aspects of this process. These standards provide the bedrock for building the next generation of software tools, which will allow information sharing across disciplines and spanning entire plant life cycles.

Underlying relationships and data structures of the components of a system are always constant irrespective of the approach a project may take to create the documents. Figure 1.2c depicts a part of the structure of the engineering data of components used in a typical I&C project. When using a database, the blocks would be implemented as database tables, whereas in the traditional documentation these would be called schedules or list documents. On a database table or list, each of the items in each of the blocks would appear as a field name or column title. Each row in the table or list represents one real-world object or entity such as an instrument, cable, junction box, or I/O channel. Ensuring that each row represents an actual object helps to keep the database structure intuitive and easily understood. Using a relational database, the relationships between these objects can be modeled easily. For example, allocating I/O to a specific field instrument would be simply implemented by providing a link field, in this case the instrument tag number in the I/O table. More intricate relationships would require to be modeled using a table itself, such as, for example, the wiring/termination table represents a relationship between the instrument, cable, and panel tables, as depicted in Figure 1.2c. This table actually represents the wiring of the cables to a termination box and hence is the database equivalent of the more traditional interconnection or wiring diagram.

The advantages of using electronic databases over paper documentation stem from the underlying advantages of database systems. That is, the ability to find, search, sort, arrange and link data in various ways allows easier maintenance, retrieval, and storage of data. Moreover, databases allow multiple access to the data, at the same time allowing concurrent engineering in large projects. The relational database systems allow data to be stored in a meaningful and intuitive manner. Also, relational databases can minimize the storage of the same data in multiple tables and thereby ensure that, if a piece of data needs to change, that change needs to be performed in just one place and is reflected in all the other views and documents generated from the same database. Advanced users of the database would also incorporate data integrity checks to ensure that the data as entered satisfy various engineering, technocommercial, and physical constraints. For example, integrity checks could ensure that a wiring database does not have electrical shorts or discontinuities, instrument ranges cover the process operating ranges, material used are compatible with the process fluid, approved vendors are used, etc. Accruing full benefits from using the database approach requires its use starting from the initial design phase of the project. The CII study[1] provides additional benefits in terms of schedule and cost reduction to a project on use of electronic media in general.

The data structure represented in Figure 1.2d also extends the previous structure to include data for individual instrument types, process engineering information,

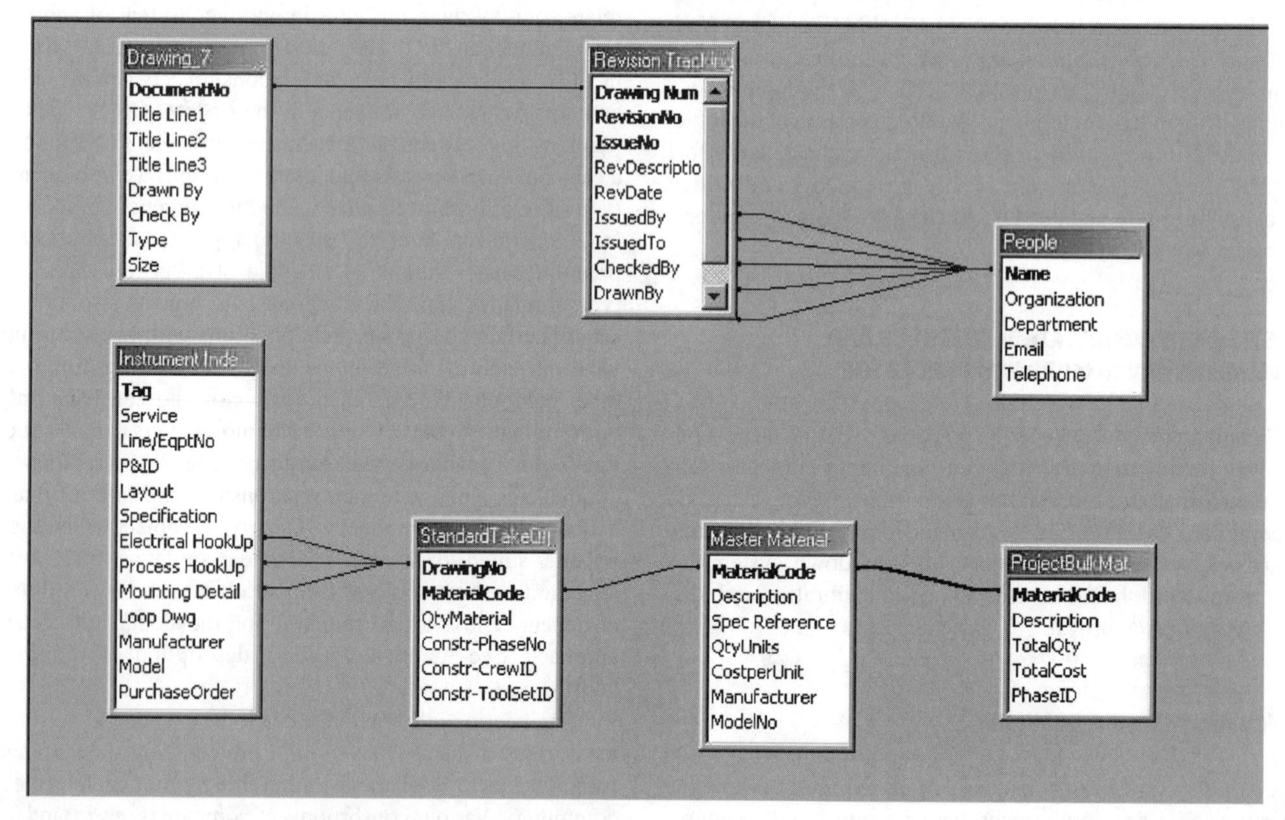

FIG. 1.2d
Extension of object model to include project organization and estimation.

material take-off data, conduits and cable trays, etc. At this level of detail, the database becomes a critical asset over the whole project and plant life cycle. This is because in addition to the engineering phase, the database could be used effectively for estimation, procurement, construction, maintenance, and demolition. This is particularly true when these databases are used with control systems, which allow data to be imported or exported to and from their internal databases and in conjunction with smart instrumentation. These systems allow the database to be always synchronized with the actual configured settings on the controllers and the field instruments. Thus, plant owners should define and maintain control over the formats and structures of these databases to be able to efficiently reuse the database between the various phases of the project.

Evidently, not all the engineering information can be effectively translated into a database. Drawings are still the main media for exchanging engineering information regarding layouts of equipment, installation and dimensional details, circuit diagrams, logic diagrams, system architecture drawings, block diagrams, flow diagrams, etc. Hence, an essential advantage of I&C design tools over a pure database approach is the integration of drawings with the database.

COMMERCIAL INSTRUMENTATION DOCUMENTATION TOOLS

Owner firms and large engineering-procurement-construction (EPC) firms are taking the initiative to migrate I&C documentation to full-fledged instrumentation knowledge management systems, with the following advantages:

- The efficiency of data access and accuracy of information translates to operational efficiency, safety, and profitability. Technicians, operators, engineers, and managers have simultaneous access to the same data presented in the appropriate form.
- Benefits are realized over the whole plant life cycle, spanning design, engineering, operation, maintenance, troubleshooting, revamps, and expansion.
- Connectivity to other systems allows sharing of data. Other applications include plant control systems, plant maintenance management systems, P&IDs, enterprise resource planning (ERP) systems, and optimization packages.
- Enhanced proficiency of handover is achieved between the different phases of the project life cycle.
- Several tasks, which otherwise would have to be done manually, are automated.
- Management of document revisions and updates is eased by maintaining an audit trail.

INtools from Intergraph Corporation, PlantSpace Instrumentation for Bentley Systems, Inc., and AutoPlant Instrumentation Workgroup from Rebis are three of the commercially available instrumentation documentation tools. Table 1.2e provides a brief comparison of current features of the three systems.

PROJECT MANAGEMENT—AN I&C PERSPECTIVE

A project is described by its distinctive characteristics as a temporary endeavor undertaken to achieve a unique product or service. The temporary characteristic of a project endows it with a time frame and a distinct beginning and an end. The uniqueness of the product or service implies that every project is different in some way or other. Project management is the application of a knowledge base, techniques, and tools to achieve project objectives. Clearly, because of the temporary and unique nature of projects, any knowledge base, techniques, or tools would not be applicable in the same way for all projects. In this section after providing an overview of projects and project management in general, the specific characteristics of I&C projects or subprojects are presented. Several textbooks and publications deal with various aspects of overall project management; a primary reference is the PMBOK guide.[6]

Because of its temporary and unique nature, each project is typically implemented in an iterative fashion. This allows the project to refine the project-specific knowledge to the degree required at each of the iterative phases. Some of these phases are often treated as subprojects, which then have the same general characteristics of the main project but phases are typically executed by different teams and at different locations from the preceding phases. As an example, a project for setting up a chemical processing plant begins with process engineering, which establishes the process and economic feasibility of the project. The next phase, sometimes called front-end engineering and design (FEED), would establish a budgetary cost and schedule and create preliminary engineering drawings. The project then advances in sequence to detailed engineering, procurement, construction, and start-up and commissioning phases. Each of these phases could be treated as a subproject, executed by different contractors at disparate locations.

Thus, the time domain may be mapped with respect to the life-cycle phases of the plant, which also encompasses a typical project in the following sequence:

- Preliminary engineering/feasibility study
- FEED or proposal engineering
- Detailed design and planning
- Procurement
- Construction
- Manufacture, assembly, configuration, and testing
- Start-up and commissioning
- Operations and maintenance
- Decommissioning

The significant players or stakeholders in an I&C project are the following:

- The owner (typically represented by a project manager)
- The user (typically represented by a plant operations or maintenance person)

TABLE 1.2e
Comparison of Some Commercial Instrumentation Documentation Tools

	INtools	*PlantSpace Instrumentation*	*AutoPlant Instrumentation Workgroup*
Software architecture	Client/server or standalone	Multiuser or stand-alone	Multiuser or stand-alone
Database type	Oracle or MS-SQL or Sybase SQL Anywhere	xBase or MS-SQL or Oracle	xBase or MS-SQL or Oracle
Modular	Yes	Yes	Yes
Native drawing format	AutoCAD, Microstation	Microstation	AutoCAD
Instrument index	Available	Available	Available
Instrument specification sheet	Available	Available	Available
Process data	Available	Available	Available
Wiring and connection tools	Available	Available	Available
Loop drawings	Available	Available	Available
Free-format forms and reports	Available	Available	Available
Flow element, control valve, relief valve, and thermowell calculation and sizing	Available	Available	
Instrument hook-up drawings	Available	Available	Available
Template drawing and symbol libraries	Available	Available	Available
Document management	Available	Available	Available
Interface to DCS systems	Available		
Interface to plant design solutions	Intergraph PDS	PlantSpace products	AutoPlant P&ID
Instrument spare parts management	Available		
Calibration history data entry and reporting	Available		
Construction work packaging	Available		
Instrument maintenance management	Available		
Support for Fieldbus Foundation-based design	Available		
Hardware requirements	166 MHz Pentium/24 MB Ram/200 MB free disk	32 MB Ram/120 MB free disk/Pentium	
Operating system	Microsoft Windows NT 3.5/95/98	Microsoft Windows NT/95/98	

INTools is a registered trademark of Intergraph Corporation, PlantSpace and MicroStation are registered trademarks of Bentley Systems Inc., AutoPlant is a registered trademark of Rebis, AutoCAD is a registered trademark of AutoDesk, Inc., Windows, Windows NT, MS-SQL Server are registered trademarks of Microsoft Corporation.

- The engineering and design organization
- The construction contractor
- The major systems vendors

The contractors and vendors have their own project organizations headed by a project manager. Very often, they are integrated in the owner–project management team by having owner representatives on the contractor's project team. The roles of these players are more or less matched with the project phases. Thus, the engineering consultant is heavily involved in the initial planning phases, whereas the construction contractor and the system vendors are active during procurement, construction, and start-up phases.

A project is a highly integrated endeavor; however, it draws its knowledge base from wide-ranging knowledge areas. The *PMBOK Guide*[6] classifies them into the following knowledge areas:

- Project integration management
- Project scope management
- Time management
- Cost management
- Quality management
- Human resources management
- Communications

- Risk management
- Procurement

In addition to the above knowledge areas, intimate knowledge of the specific context (site conditions, weather conditions, local culture, language, politics, local events, etc.) in which the project operates is essential for effective daily project management. The project perspective from each player's viewpoint differs only in the emphasis on knowledge areas on which each depends. The following highlights some aspects of the application of the knowledge areas as they pertain to I&C projects while also providing a general overview of project management. The practitioner interested in project management as a discipline in itself should consult the references listed in the bibliography to this section.

Project Integration Management

Project integration management is most effectively applied during the design phases of the project. It provides a means to ensure that the I&C specified is consistent with the designs of other disciplines, such as process, mechanical, electrical, and piping, with vendor-supplied packages and with owner's existing plant equipment and systems. The cost of correcting errors is greater the earlier they are committed in the project and the later they are found and corrected. Integration management is an important aspect, as I&C has considerable tie-ins with each of the other disciplines. To prevent design inconsistencies it is important to:

- Ensure that I&C design team has adequate data to begin its work, for example, existence of a comprehensive written scope definition, project criteria, or summary sheet and proper scheduling of I&C work to coincide with availability of information. Design engineers are required to visit the site to gauge for themselves the site conditions before they begin work.
- Ensure adequate opportunities of communication between disciplines and user representatives, for example, by weekly scheduled review meetings, encouraging the project team to flag information used by others that is under review, and encourage multidisciplinary review of final drawings. This is especially important for multi-location projects.
- Ensure comprehensive reviews of all vendor documentation and resolution of all incompatibilities preferably at the design phase or, if that is not possible, ensure that proper change control procedures are in place even during the procurement and construction phases.
- Ensure comprehensive reviews by owner's departments before construction.
- Support of design engineers should be available on an as-required basis even during the procurement and construction phases.
- By the same token, it is desirable to include persons with construction and implementation experience right from the preliminary phases of the project. Formal constructability programs as described by the "Constructability Implementation Guide" from CII[16] could augment the final project design.
- Ensure effective change control procedures are in place at all phases. Changes are inevitable in a project. Project documentation must have means to indicate changed areas in a document or to indicate information that may have to be refined at a later date.
- Ensure effective change control, which depends in part on correct estimates of the total impact of the change. The dependency map depicted in Figure 1.2b may be used as a rough guide for a project manager who has to estimate the degree of change in I&C projects. To utilize this document the project manager needs to determine the highest-level document affected by the given change. For example, if the change requires that an instrument be wired to a different junction box in the field, the highest-level document that is affected is the cable block diagram. The total change then is roughly restricted to all subsequent drawings, that is, cable routing drawings, cable schedule, interconnection drawings, ILDs, and material take-off. Once the changes are implemented on the drawings, material and construction impacts can be estimated easily.

Project Scope Management

Managing scope requires that members of the team be fully aware of the scope. Early in the project an opportunity should be provided for the I&C team to clarify all aspects of the scope. Project criteria documents as indicated in Table 1.2a are useful in ensuring that all aspects of an I&C system are accounted for. For construction and vendor organizations the instrument index, cable schedule, bill of quantities, and the I/O list provide basic project size indications and hence are used as a basis for the project scope.

Scope change is initiated typically by user departments and engineering or construction contractors as a result of late changes to project specifications, conflicts with mandatory standards, or as an impact of another concurrently executed project or subproject. A scope change in the design phase is generally recognized as significant if the change results in a change in the quantity of drawings (e.g., additional measurement points means more specification sheets and ILDs). Similarly, a vendor or construction contractor is reimbursed if the change results in additional material or service. Scope changes that do not entail an increase in deliverables are more difficult to estimate and control. However, the total impact could still be gauged the same way as described in the previous paragraph.

Time Management

Breaking up the project into various phases allows better understanding of the specific nature of each phase. These phases (i.e., design, construction, etc.) are typically executed as separate projects by specialized organizations and are further divided into activities. A complex activity is further subdivided into a group of simpler activities until each activity can be easily identified, measured for completion, and assigned to a specific team. Activities can only be performed in a certain order determined by the nature of the activity. Each activity requires a set of resources, which may include skilled manpower, material, and tools. Each activity is performed in a time frame, which is estimated based on the available set of resources. Overall project progress is an agglomerate of the progress of each activity. Determining how to measure the progress of an activity is generally mutually agreed between the owner and the contractor. This is particularly important when the progress figures are used to make partial payments to the contractor. A large project may require thousands of activities, which necessitates specialized software tools. Many commercially available software packages support this aspect of project planning and control very well. They ease the collection of data, calculation and reporting of progress, expended resources, etc. See project management Web sites[15] for a good review of the software available for this purpose.

When instrument documentation is based on a database, there is an opportunity to build a plan of execution, schedule activities, and base progress measurements using the data in the database. For example, a construction project uses the instrument index to sort out instruments by type and installation detail in a certain section of the plant (which is filtered by the location plan drawing or plant no.). Because all these installations are similar, they are scheduled as one activity, an assignment is made to a specific installation team, dates and time frames are set, and then provisions are made in the database to monitor the progress. It would be ideal if the data could be seamlessly transported between project planning and design databases. However, this level of integration is still not commercially available and is implemented only with in-house databases on smaller projects. Thomas Froese et al.[17] reports on efforts to set up a standard industrywide interface for such applications. This approach has several advantages.

- Interchange of data without manual reentry into a planning software
- Early initiation of project construction planning
- Opportunities for further automating the process, for example, segregating the material required for each activity, ensuring that the material is available by linking to a warehouse database, automatically checking for manpower or tools contention, etc.

Cost Management

More accurate cost estimates are possible now than ever before. This is as a result of greater accessibility of information in electronic format. Data from previous projects as well as current data directly from the supplier can easily be linked to databases created for estimates of the current project. Major control system vendors also provide access to application programs that help estimate cost of new systems as well as incremental expansions of existing systems. These are also being made available through their Web sites.

The advent of the World Wide Web allows purchasers to locate the cheapest provider of a product or service. Some Web sites provide searchable databases of multivendor industrial material and a means to place an order based on selections made through these sites.

Quality Management

Quality management is now an essential component of every project organization. A quality system provides a means to verify that the project is following the procedures laid down by the project organization for assuring itself and the clients that the final output will meet the requirements. During the design phase, quality systems rely on the use of checklists, formal multidisciplinary, multidepartment review of documents, formal approval cycles, maintaining logs of client comments, etc. Factory acceptance tests, site acceptance tests, incoming material inspection, and phased site inspection are required for proper quality assessment during the construction phase.

Human Resources Management

The project team selection is crucial to the success of the project. The I&C engineering discipline is truly multidisciplinary in its knowledge areas, but at the same time the final system is a specialized field. Hence, the I&C team is frequently composed of people with various engineering backgrounds. An audit of the existing team's strengths and weaknesses can be used to determine the profile of new team members required to augment it for executing a given project. A framework for such an audit would include I&C knowledge areas (e.g., control system design, field instrumentation, etc.), specific process application knowledge (e.g., olefins process, catalytic cracker process, etc.), technical skill sets (e.g., database, programming, calibration, etc.), and business skill sets (e.g., project management, construction supervision, etc.). One of the most important requirements, however, is that one should be able to apply the principles of I&C engineering appropriately to the project at hand, come up with a range of solutions, and choose the scheme that is optimal for the project. Because the I&C components are rapidly being upgraded, it is extremely important for the I&C engineer to keep current with the technological developments. This will critically determine the range of solutions that will be determined. Thus, while choosing candidates for a project, it is valuable to select people with experience on projects that have utilized the latest technology. The project must plan to procure the services of specialists even if it is on an *ad hoc* basis wherever specialized knowledge and experience are required.

Project Communications

The multifaceted nature of an I&C project makes communication between the project players a very crucial aspect of the success of the project. Collaborative tools from commercial organizations are being introduced in the market based on the Internet including ones that are specifically targeted to I&C projects.[18] These would especially be beneficial to multilocation projects.

Risk Management

Retaining control of a project in the face of unplanned events requires that the risks inherent in a project plan or design be identified and wherever appropriate a response to such possible events be formulated. In the I&C design phase risk is reduced by better documentation, specifications based on standard components, and use of standard installation practices and materials. Whenever a new technology is to be introduced, all aspects of the proposed device or systems must be studied thoroughly and an alternative backup scheme must be designed. Cost to the project in terms of money and schedule must be weighed against the probable advantages of the new technology.

Risk management also requires apportioning risk appropriately to all players concerned by adopting the correct contracting strategy. Lump-sum contracts require the contractor to bear all the risk. In reimbursable contracts, the owners bear all the risk. Wherever a contractor perceives risks during a bid stage, the risk is priced into the offer, thus raising the final price. Risk is reduced, as more project information becomes available. Hence, reimbursable contracts are more appropriate for the early project phases and lump-sum contracts for the later phases when there is a considerable amount of detailed information available.

PROCUREMENT

During the procurement phase, particular attention is required with respect to obsolescence and long-lead items for I&C project components.

Control system components are particularly prone to obsolescence. Hence, offers of those systems, which do not have the latest version of hardware and software, must be evaluated in light of the cost of earlier replacement of these components. Current obsolescence rate for larger DCS systems are of the order of 15 to 20 years after their first introduction, by which time spares and technical support thin out. Using a model introduced 5 years ago would, without considering other factors, have a hidden cost of up to 30% on account of the need to replace it earlier than the latest model. Cost of replacement of a control system also entails the cost of loss of production, which may be considerable.

Accelerating the overall project schedule requires that items with a long-lead time for procurement be identified, specified, and ordered early in the project to ensure that these items are available at site when required. Procurement lead times for I&C components are higher for:

- Larger sizes of control valves and actuators
- Nonstandard sizes, nonstandard materials
- Systems requiring extensive configuration and programming
- Larger size of the order; a large order to one manufacturer may exceed its capacity to deliver in the required time

Identification of long-lead items is frequently done at the design stage and at times these items need to be ordered by the owner as a construction team's procurement team may not be in place in time.

CONCLUSIONS

The structure of the information in traditional I&C project documentation has been delineated and its implication to project management developed with respect to a dependency map. The commercial documentation packages, which exploit these relationships to automate several design and engineering tasks, have been studied. Future trends and development in I&C documentation and project management have been indicated. A practitioner's guide to I&C project management is presented in an overall framework of project management knowledge areas.

References

1. Construction Industry Institute, "Schedule Reduction," Construction Industry Institute, Publication 41-1, April 1995.
2. Rivard, H., "A Survey on the Impact of Information Technology on the Canadian Architecture, Engineering and Construction Industry," *Electronic Journal of Electronic Technology,* Vol. 3, 1998, http://itcon.org/1998/2/.
3. Reezgui, Y. and Cooper, G., "A Proposed Open Infrastructure for Construction Project Document Sharing," *Electronic Journal of Electronic Technology,* Vol. 2, 1998, http://itcon.org/1998/2/.
4. IAI aecXML Domain Committee, http://www.aecxml.org/.
5. Object Management Group, Inc., http://www.omg.org/.
6. Project Management Institute, *Guide to Project Management Body of Knowledge—PMBOK Guide,* Newtown Square, PA: Project Management Institute, 2000.
7. Process Industry Practice, "PIP PCEDO001, Guideline for Control Systems Documentation," April 1997.
8. ISA, ISA S5.1—"Instrumentation Symbols and Identification," Research Triangle Park, NC: ISA, 1984 (R 1992).

9. ISA, ISA S5.2—"Binary Logic Diagrams for Process Operations," Research Triangle Park, NC: ISA, 1976 (R 1992).
10. ISA, ISA S5.3—"Graphic Symbols for Distributed Control/Shared Display Instrumentation, Logic and Computer System," Research Triangle Park, NC: ISA, 1983.
11. ISA, ISA S5.4—"Instrument Loop Diagrams," Research Triangle Park, NC: ISA, 1991.
12. ISA, ISA S20—"Specification Forms for Process Measurement and Control Instruments, Primary Elements and Control Valves," Research Triangle Park, NC: ISA, 1981.
13. ISA, ISA S91.01—"Identification of Emergency Shutdown Systems and Controls That Are Critical to Maintaining Safety in Process Industries," Research Triangle Park, NC: ISA, 1995.
14. ISA, ISA S51.1—"Process Instrumentation Terminology," Research Triangle Park, NC: ISA, 1979 (R 1993).
15. Project Manager, http://www.project-manager.com.
16. Construction Industry Institute, "Constructability Implementation Guide," CII Special Publication 34-1, May 1993.
17. Froese, T. et al., "Industry Foundation Classes for Project Management—A Trial Implementation," *Electronic Journal of Electronic Technology,* Vol. 2, 1999, http://itcon.org/1999/2/.
18. Entivity, Inc., http://www.automationprojectnet.com.
19. ISA, ISA-RP60.1, "Control Center Facilities," Research Triangle Park, NC: ISA, 1991.
20. ISA, ISA-RP60.2, "Control Center Design Guide and Terminology," Research Triangle Park, NC: ISA, 1995.
21. ISA, ISA-RP60.4, "Documentation for Control Centers," Research Triangle Park, NC: ISA, 1990.

Bibliography

Battikha, N.E., *Developing Guidelines for Instrumentation and Control: Implementing Standards and Verifying Work Performed,* Research Triangle Park, NC: ISA Press, 1995.

Coggan, D.A. and Albert, C.L., *Fundamentals of Industrial Control,* Research Triangle Park, NC: ISA Press, 1992.

Harris, D., *Documenting the Start-Up Process,* Research Triangle Park, NC: ISA Press, 2001.

Kerzner, H., *Project Management: A Systems Approach to Planning, Scheduling, and Controlling,* 7th ed., New York: John Wiley & Sons, 2000.

Mulley, R., *Control System Documentation: Applying Symbols and Identification,* Research Triangle Park, NC: ISA Press, 1993.

1.3 Operator Training, Commissioning, and Start-Up

G. C. BUCKBEE

Partial List of Suppliers:

Simulation: ABB Automation; Adaptive Resources; Altersys; APV; Aspen Technology; Cape Software; Echip Inc.; ExperTune; Fastech; The Foxboro Company; Honeywell Hi-Spec; SIMONS STI; Neuralware; OSI Software; Siemens Energy & Automation; Simulation Sciences, Inc.; Steeplechase Software, Inc.; Wonderware Corp; XynTech, Inc.

Start-up services: ABB Automation; ACE; Automated Control Concepts; Black & Veatch; CI Technologies; Fisher-Rosemount Systems; Honeywell Industrial Automation & Controls; HiTech; Javan-Walter; Kentrol; Matrikon Consulting; Metso Automation; Rockwell Automation; TopControl; Ultramax Corp.

Training: ABB Automation; Allen-Bradley; ESW; The Foxboro Company; Honeywell Industrial Automation & Controls; HiTech; Intellect; Metso Automation; Rockwell Automation; Rockwell Software; TopControl

Whether a control system project is perceived as a success or a failure often depends on commissioning, training, and start-up. The best control system in the world will not do well if the operators are not inclined to use it. Careful planning and execution of the commissioning, operator training, and start-up activities will reduce the short-term project risks, and help to ensure long-term success of the project.

COMMISSIONING

Commissioning is the process of placing a system into service. In a nutshell, the purpose of commissioning is to ensure that the system is fit for use. Skipping this step causes a high degree of risk to the project, the equipment, and the operating personnel.

To be properly commissioned, the system should be checked for proper construction and operation. New equipment and software should be verified. Operating procedures should be tested and fine-tuned. Process centerlines should be established or confirmed. Abnormal operations and safety procedures should also be tested. This section provides detail about the nature of these checks, giving guidance to a project manager or engineer.

Figure 1.3a provides an overview of the commissioning process. For small systems, commissioning may take only a few hours. For new plants, this could last weeks or even months.

Commissioning takes time. Usually, it is one of the last steps before system start-up. Whether or not construction was completed on time, there is usually a lot of pressure to complete the commissioning phase as quickly as possible. A clear plan and strategy for commissioning will help to define the time requirements, and also to help keep the work on track.

There are several key components of a commissioning plan:

Clear goals and objectives
Staffing
Schedule
Communications plans

Figure 1.3b presents a checklist to be used in planning the commissioning phase. Each of these items is discussed in more detail below.

Clear Goals and Objectives

The goals and objectives for the commissioning phase will be used to drive the entire process. It is important to be very clear on these at the start, because they will ultimately determine the length of the commissioning phase and the timing for the start of good production. Typical goals for commissioning are shown in Table 1.3c below. Columns to the right indicate increasing degrees of rigor for the commissioning phase.

Staffing

Staffing should include people from operations, maintenance, engineering, and management. Each of these players has one or more key roles. Specific roles that should be filled include safety, operating procedures, construction interface, engineering design, communications, leadership, maintenance systems, and technical testing. On small projects, one person may play many roles.

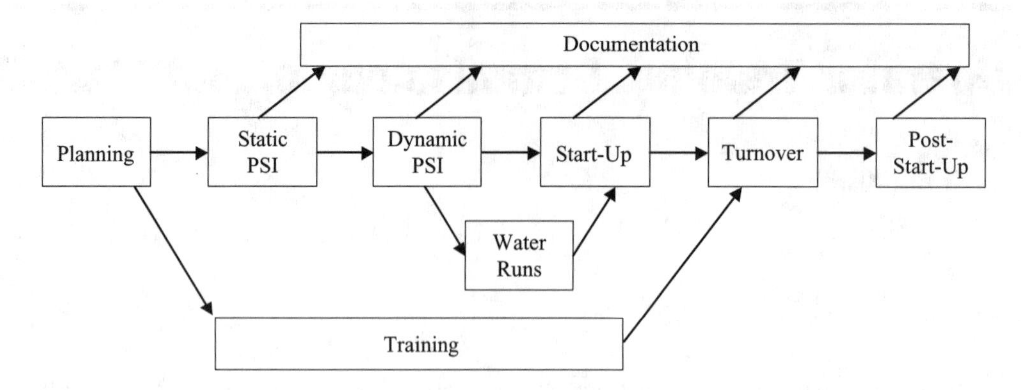

FIG. 1.3a
Overview of commissioning activities.

Item	Responsibility	Complete?
Goals & Objectives	Proj. Mgr.	15-Jan
Turn-Over Criteria	Ops. Mgr.	22-Jan
Safety Plan	Ops. Mgr.	
Decision-Making Plan	Ops. Mgr.	
Commissioning Schedule	Proj. Mgr.	
Milestone Dates	Proj. Mgr.	
Schedule Communication	Proj. Mgr.	
Staffing Plan	Proj. Mgr.	
Staffing vs. Schedule	Proj. Mgr.	
Confirm Availabilities	Ops. Mgr.	
Communicate the Plan	Ops. Mgr.	
Training Plan	Lead Operator	
Communications Plan	Ops. Mgr.	
Classroom Space Secured	Lead Operator	
Materials Developed	Lead Operator	
Commissioning Plan		
Special Test Equipment	Tech. Leads	
Logistical Issues (e.g., space and crowd control)	Tech. Leads	
Start-Up Plan	Process Engr.	
Document Sets Ordered	Tech. Leads	

FIG. 1.3b
Planning and strategy checklist.

On large projects a whole team may be required to fill a role. Although everyone bears some responsibility for the safety of themselves and those around them, it is important to assign one person to look out for interference between groups, and to identify changes in the overall situation that could lead to safety concerns. This person will be responsible for interpreting safety rules and codes, and for ensuring that proper procedures and practices are followed. The safety leader should have the ability to stop any and all work until safety issues are resolved.

An experienced operator or a process engineer usually develops operating procedures. This person helps to define and interpret operating procedures during normal and abnormal operation. Start-up, shutdown, and emergency procedures will need to be developed, modified, and reviewed.

On large projects, a construction interface may be needed to coordinate any design or construction changes that may be identified during the construction phase. This person may need to interpret design information or help to resolve procurement issues.

TABLE 1.3c
Goals for Commissioning

Commissioning Activity	*Low Degree of Rigor*	*Items to Add for Medium Degree of Rigor*	*Items to Add for High Degree of Rigor*
Control valve checks	Stroke each valve 0 and 100%	Check valves for hysteresis, stiction, and nonlinearity	Confirm design conditions for pressure drop and flow throughout range of operation
Instrument checks	Field checks for installation errors	Systematically confirm instrument spans with DCS and display configurations	Run the process under a wide range of conditions, confirming proper instrument behavior
Power supplies	Confirm power at correct voltage levels	Measure voltage sag under load conditions	Record power and current through equipment start-ups
Communications	Confirm that all nodes are communicating, both directions	Spot-checks on data being passed; confirm network loading	Check every piece of data being passed; record network activity levels
Variable speed drives	Confirm direction of rotation, min and max speeds, logic	Confirm ramp rates and full load amps	Check internal signals, cabinet heat
Production	Produce product at X rate	Produce on-spec product at X rate	Produce on-spec product at full rate, for Y hours
Alarms	Confirm audible and visual alarm devices working; confirm all critical alarms	Confirm all alarms by testing or soft jumpering signals	Test and confirm all alarms, including rate of change and deviation alarms; confirm all informational messages
Interlocks	Confirm all safety interlocks through rigorous testing	Confirm equipment protection interlocks	Confirm all interlocks through factorial testing of all possible process conditions
Human–machine interface	Check all graphics against design; confirm that all instrumentation is represented; proper tags used for all on-screen values; complete system backup	Confirm color standards, animation, and target actions	Point-by-point check of every feature; confirm historical data collection
DCS configuration	Complete system backup; confirm restore from backup; check specifics recommended by DCS vendor	Confirm unit-level operations; confirm all sequence programming through testing; controller tuning	Confirm point configurations for every point; document all changes to configuration

Engineering design contacts are needed to interpret designs, or make adjustments to the design based on commissioning results. They may also need to lead or provide input to changes in operating procedures. Engineers may also lead the commissioning work for part of the system design.

With so many activities taking place during commissioning, having clear communications is important. In the broad sense, communications includes everything from planning and update meetings to shift-to-shift notes to log books and documentation sheets to formal project updates to management. Again, everyone bears some responsibility for communications, but it is important to lay out a specific plan for how the information will be passed to all of the players. The leadership roles are needed to keep everything running smoothly. Priorities will shift. Key resources will become unavailable. The pressure to cut corners will always be there. It is the role of the leaders to set clear priorities on a daily or even hourly basis, and to help keep the effort focused, working toward the clearly defined objectives. On large projects, there may be separate commissioning leaders for each discipline. Typically, the leaders will also fill other more technical roles as well.

The role of maintenance in the commissioning should not be overlooked. After all, the maintenance crew will soon have the responsibility to resolve any problems. During commissioning, maintenance contacts will typically participate in system installation, calibrations, and testing. They will also want to confirm that the right tools and spare parts are on hand, along with the documentation to help to troubleshoot future problems.

Technical testing roles may take many forms during commissioning. Each control system may bring its own set of

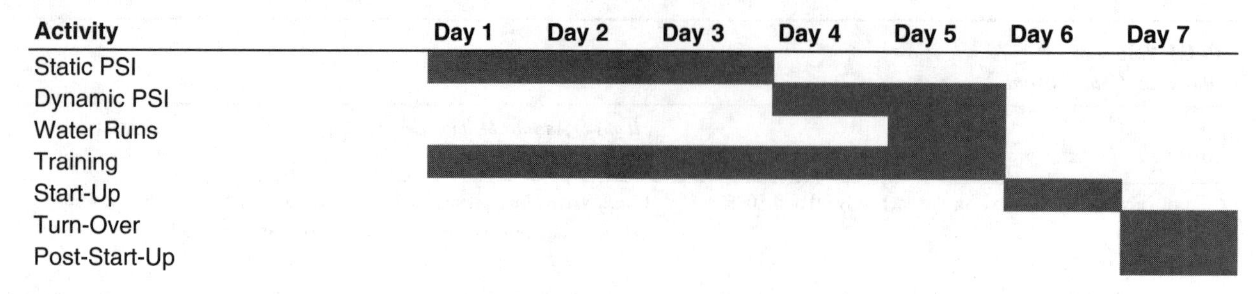

FIG. 1.3d
Sample commissioning schedule.

"experts." You may need one for the drives, one for the distributed control system (DCS), and one for the programmable logic controller (PLC). In addition, specific process expertise may be needed for the process or equipment.

Schedule

A single schedule for all commissioning and training activities will help keep plans on track. There are many interactions between commissioning steps, so it may help to use a scheduling tool that allows one to establish links between tasks. One may also want to consider linking resource allocation needs to the schedule.

To help keep everyone "on the same page," the schedule should be accessible to the entire staff. Typically, a large-format schedule is hung on the walls of a meeting room, or in a high-traffic hallway. Completed tasks are highlighted, and comments are written on the schedule. With a thumbtack, hang a string at the current date and time as a reference. Since commissioning often comes after a substantial amount of construction or installation work, the schedule is likely to be affected by any changes in the construction schedule. Be prepared to modify this schedule as the situation develops.

Figure 1.3d shows a sample commissioning schedule.

Communications

Because there are so many parallel activities during commissioning, a good communications plan is needed.

For 24-hour operations, plan on a shift-change meeting or the passing of detailed notes from each shift to the next. Nothing is more frustrating than repeating work unnecessarily. There may be several different groups passing information at the shift change, so adjust meeting schedules to accommodate the larger system. A typical 24-hour meeting schedule is shown in Figure 1.3e. Written updates should be limited, so that people can focus on the task at hand. But a daily note to "the outside world" can be very helpful in holding off unwanted "visitors."

Hang placards or signs when major milestones on the schedule are passed. For example, when power is first applied, or when part of the process goes from a dormant state to operational. This will help remind people of the safety risks that are now in play.

Of course, project documentation is a critical component of the communications plan. Be sure to designate a "Master Set" of drawings. Other sets may be used for fieldwork, but the master should be updated to reflect any changes or errors. For fieldwork, consider the use of more portable, smaller-format drawings, such as 11″ × 17″ books.

When equipment is received for installation, it should be checked. Quite often, after it is checked, a colored tag is attached, to indicate its status. This process is often referred to as "tagging." Figure 1.3f shows a sample instrument "green tag."

For instruments, the following checks should be performed prior to tagging:

- Confirmation of engineering range and engineering units
- Confirmation of materials of construction
- Confirmation of flange and fitting sizes
- Confirmation of calibration; either bench-test the instrument or confirm that the supplier has done this

With a tag affixed, the instrument is ready to be handed over to the construction crew.

Pre-Start-Up Inspection

Pre-start-up inspection (PSI) is the process of checking the actual installation against the design. Typically, PSI is completed in two phases: static PSI and dynamic PSI. During static PSI, the design is checked visually, without running the process. Figure 1.3g shows a sample static PSI check sheet.

During dynamic PSI testing, equipment is actuated and run. Motors are tested for proper rotation, valves are stroked, and power is applied to various parts of the process. Instruments, logic, and programming are checked under dynamic conditions.

Time	Meeting	Comments
7:00 AM	Shift Transition	Safety, Current Status
8:00 AM		
9:00 AM		
10:00 AM		
11:00 AM	Schedule Update	Management Team
12:00 AM		
1:00 PM	Engineering Changes	Fix New Problems
2:00 PM		
3:00 PM		
4:00 PM		
5:00 PM		
6:00 PM		
7:00 PM	Shift Transition	Safety, Current Status
8:00 PM		
9:00 PM		
10:00 PM		
11:00 PM	Schedule Update	
12:00 AM		
1:00 AM		
2:00 AM		
3:00 AM		
4:00 AM		
5:00 AM		
6:00 AM		

FIG. 1.3e
24-hour meeting schedule based on 12-hour shifts.

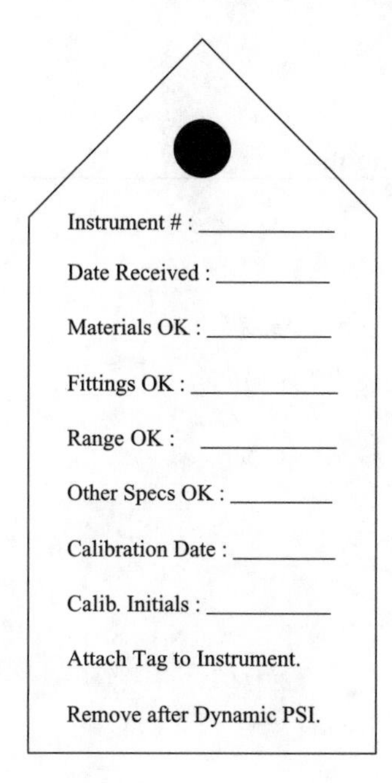

FIG. 1.3f
Sample instrument tag.

Before proceeding to dynamic PSI, make sure to communicate broadly that "things may move." For those who have been working on a dormant process for a while, it can be quite a shock if something suddenly starts up next to them.

Figure 1.3h shows a sample dynamic PSI check sheet. Use check sheets like this to make sure the proper operation of all equipment has been tested. The goal here is to find the problems now, before the process starts using expensive raw materials and energy.

Partial Operation and Water Runs

During "water runs," the process is tested with water, or other low-cost raw materials. Process equipment and controls can be tested rigorously, under conditions that approximate normal operation. Large-scale projects will often use water runs for testing, especially when raw materials are costly or if there is large potential for scrap.

Partial operation and water runs are an excellent time to ensure that controller tuning is correct. All control loops should be run in their normal mode of operation. Process and valves should be checked for hysteresis, stiction, and nonlinearity.

Of course, operating procedures should be confirmed during these test runs. Test the procedures with a variety of operators to clarify any ambiguity in the recorded procedures. A logbook or database should be used to record all problems and recommended changes. The project team will need to prioritize these fixes, and decide which ones will have to wait until after start-up.

When the turnover criteria have been met, the process is turned over to operations. At this point, operations will

Item	Date	Time	Initials	Comments
Field Check				
Location				
Orientation				
Flanges				
Seals				
Tubing				
Jumpers/Settings				
Labeled				
Terminations				
Electrical				
Power Supply				
Instrument Wiring				
Smart Communication				
Terminations				
DCS/PLC Config.				
Range High & Low				
Alarm Limits				
I/O Address				
Filtering				
PV Read				
Diagnostic Check				
Display Check				

FIG. 1.3g
Static PSI check sheet, level sensor.

Item	Date	Time	Initials	Comments
Range Check				
Zero				
50%				
100%				
Alarms				
Field Checks				
Flow Direction				
Leaks				
Controller				
Gain				
Reset				
Derivative				
Filter				
DCS/PLC Config.				
Display Check				
PV Movement				
OP Movement				
Controller Direction				

FIG. 1.3h
Dynamic PSI check sheet, control loop.

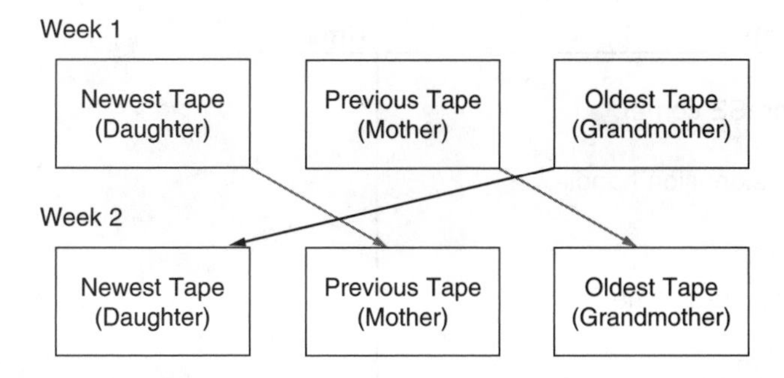

FIG. 1.3i
Backup system tape rotation.

attempt to make good product. Continued technical and management support by the project team may be required for some time during this phase of the project.

Documentation

During commissioning, countless corrections and changes are recorded. Prints have been updated, cross-references changed, and a variety of databases have been modified. A clear plan for updating documentation will help ensure that these changes are not lost.

All computer-based systems should be backed up to removable media. This includes distributed control systems (DCS), programmable logic controllers (PLCs), supervisory control and data acquisition systems (SCADA), and process data historians.

Typically, several generations of backups will be kept to allow rollback to a known, stable operation. The backups should be labeled with the system and date, and then stored in a safe location. Figure 1.3i shows how tapes or other removable media can be reused, while maintaining a rollback system. The oldest backup is overwritten, becoming the newest backup.

OPERATOR TRAINING

Operator training is probably the most important part of a project start-up. This is particularly true of process control-related projects. A well-run project, with solid control strategies, state-of-the-art equipment, and poor operator training will fail, every time.

In most plants, the operator is the pivotal decision maker. The operator runs the process, and spends more time tracking, adjusting, and managing the operation than anyone else. When designing an advanced control scheme, always remember: It is the operator who decides whether to put the controls in auto!

For operators to do their job well, they need to understand the process, the controls, and the procedures. Each of the sections below outlines the required training for each of these topics. Most control system projects come on the heels of a process improvement. Operator training will need to emphasize the changes that were made to the process. New unit operations should receive the most attention. A change to the process flow is also a critical topic.

Operators will need to understand the purpose of the process and the specific role of each piece of equipment. Specific process and equipment limitations, such as pressure limits, should be covered in detail. In many cases, excellent training on these topics is available from the equipment vendors.

The operator must learn to use the control system: not only learn, but also see value in the use of the new control system. If they do not understand it, they will not use it. Many million-dollar control systems are being run in manual because the operators do not trust the new controls. One of the most visible components of a new control system is the operator interface, sometimes called the human–machine interface (HMI). In most modern systems, this is a graphical interface, displayed on a computer screen.

Imagine that someone rearranged all the controls in your car. The gas pedal is now used for steering, the brake for acceleration, and the steering wheel is used to turn on the headlights. You would likely be quite confused and frustrated when learning to drive the car in this configuration. This is how the operator feels when the HMI is changed! Be sure that the training program allows sufficient exposure to the new HMI to overcome the fear and distrust that will naturally occur. It is a good idea to have some key operators participate in the design of the HMI screens. This will allow operators to provide input and develop a level of comfort with the new HMI. The control strategies will also need to be covered. Review P&ID, Scientific Apparatus Manufacturers Association (SAMA), or other control system drawings to explain the relationships between the various measured variables and control valves or other actuators.

Spend some time on logic, especially on process interlocks. The operator should be trained in both the purpose of the interlocks and the details of their operation. Time is precious during start-up, and a knowledgeable operator is the quickest way through it.

Operating Procedures

With any change in the control system comes some changes in the operating procedures. Operating procedures provide guidance to the operator during both normal and abnormal operation. Sometimes special procedures are required for start-up and shutdown of a system. Operators will need to be trained in the theory of operation, key sequences, and in the minutia of proper keystrokes and mouse-clicks required to use the HMI and to run the process. This is a tall order and requires well-planned training.

During normal operation, the operator will monitor and adjust the process. A solid training program will provide guidance to the operator. To monitor a steady-state operation, the operator should be familiar with expected centerlines for process operation. The operator will need to understand

Item	Comments	Time
Close All Drain Valves		
Hot Water Supply Tank	First floor, SE corner.	
Separator Drain		
Process Tank A	Bring an extension handle.	
Communications		
Call Boilerhouse x3765	Notify of start-up time.	
Establish Base Conditions		
Hot Water Tank to 50%	LIC2735	
Steam Turbine at 30%, Venting		
FIC2732 in Cascade		
TIC3675 in Auto	Setpoint should be 175°F.	
Bring Product Online		
Slowly open Product Valve	Bring a radio.	
Establish Mat Thickness	1/2 inch to start, then 1 inch	
Perform Initial Quality Check		
Basis Weight		
Caliper	1 inch + or – 0.05 inches	

FIG. 1.3j
Sample start-up checklist.

normal process variation, and be able to recognize when things are not normal.

When things are not progressing normally, the operator will likely receive many alarms from the control system. The operating procedures should provide guidance on how to respond to alarm conditions.

All alarms are not created equal. Many DCSs allow for several types of alarms for each loop. A single loop could have low-low, low, high, high-high, deviation, and rate-of-change alarms. The operator will need to understand the significance of each type of alarm and be able to determine the right course of action. (*Note:* If an alarm requires no action, maybe it should not be an alarm!) Again, it is a good idea to have experienced operators provide input on the selection of alarms, alarm limits, and alarm priorities.

During process start-up and shutdown, a specific sequence of operation must be followed. Operating outside of these sequences can result in process problems or delays. To move quickly and efficiently through these phases, the operator must be trained in the correct sequence of operation.

The start-up sequence is often intricately entwined with process interlocks. This is necessary to protect the process and the equipment. For example, a tank may need to be filled before the outlet pump can be started. If the operator tries to start the pump before the tank is filled, process safety interlocks will likely prevent the pump from running. Without proper training, the operator will waste valuable time trying to figure out why the pump is not running.

One way to minimize these mistakes is to make use of a start-up checklist. This can be generalized or very specific, depending on the complexity of the operation, the severity of mistakes, and the experience level of the operators. Figure 1.3j shows a sample start-up checklist.

During start-up training, special attention must be paid to controller modes. Each loop must be placed in the proper mode (manual, auto, or cascade). In the case of interacting loops, such as cascade or ratio controllers, this is particularly important. Figure 1.3k is an example of a job aid to ensure controllers are placed in the proper mode.

Description	Tag	S/U Mode	S/U SP	Units
Hot Water Tank Level	LIC1271	AUTO	50	% Level
Inlet Water Temperature	TIC1273	AUTO	170	°F
Inlet Steam Flow	FIC1274	CAS	2.63	M#/h
Supply Tank Pressure	PIC3701	AUTO	45	psig
Supply Tank Level	LIC3635	MAN	---	
Emulsion Flow	FIC2654	AUTO	1.22	gpm
Glue Flow	FIC2678	AUTO	3.75	gpm
Dye Addition	FIC1376	AUTO	1.05	gpm

FIG. 1.3k
Job aid for controller modes.

Item	Comments	Time
Bleed off system pressure		
Process Tank A	PIC1124, reduce SP to 5 psig	
Process Tank B	PIC1127, reduce SP to 0 psig	
Cross-feed valve open	Manually, first floor, SW corner	
Cool down		
Call Boilerhouse x3765	Notify of shut-down	
Cool down Unit 1	TIC1765, SP to 90°F	
Open spray cooling valve	FV263, open	
Wait until system is cool	TIC1765 less than 100°F	
Open drains		
Process Tank A	Manual Valve, SE corner	
Process Tank B	Manual Valve, east side	

FIG. 1.3l
Sample shutdown checklist.

During shutdown, the operator must safely return the process to a resting or idle state. Done properly, the shutdown procedure will also leave the process ready for the next start-up. In many cases, there may be several shutdown procedures, depending on the intent of the break in production. For example, there may be a brief shutdown for product changeover, or a complete shutdown and system purge for annual maintenance.

The operator should be trained in all of the shutdown sequences, and understand the intent of each variation in the procedure. As in start-up sequence training, particular attention should be paid to the controller modes and to the final positions of control valves. Figure 1.3l is a sample shutdown checklist. Note that both automated and manual operations are typically combined on such a checklist, to avoid costly mistakes.

Nothing ever goes completely as planned. Equipment fails, mistakes are made, and system software cannot be prepared for every possible twist of events. When the process is in an abnormal situation, the operator assumes the role of crisis manager. Like a pilot of a crippled airplane, the operator must return the process to a safe and stable condition, in light of all manner of constraints and problems.

Of course, planned procedures for abnormal conditions are the first line of defense. For failures that can be reasonably expected, there should be a clear set of procedures for safe recovery of the process. If we are on the airplane, we all hope and expect that the airplane pilot has been completely trained in safe recovery in the event of blown tires, engine flame-outs, and hydraulic leaks. Similarly, the operator should be trained in loss of power, loss of air, and instrument failure.

In all but the most critical of processes, it is unlikely that the effort will be made to identify *all* possible failures. The operator must learn how the control system will respond to abnormal conditions. Of particular importance are process safety interlocks and failure modes for all control valves.

Process safety interlocks are designed to protect the people, the process, and the product. Under abnormal circumstances, these interlocks respond quickly to contain the process to prevent injury or damage to the process or product. The operator must be trained in the trigger mechanisms for all process safety interlocks, and the specific consequences of these interlocks.

Most control valves can fail in one of three conditions:

1. Fail open (air to close)
2. Fail closed (air to open)
3. Fail at last position

To be effective during a loss of air or power, the operator must know how each valve will fail. With this understanding, operators will know where to focus their energy during the management of an abnormal situation.

Approach to Training

Modern training methods give great flexibility to operations training. Most training today recognizes that adults have a different learning style from children. Adults learn well by analogy and by doing. A good training program will take this into account, allowing operators to make connections to their previous knowledge, and giving them a chance to test their understanding by hands-on practice.

There are many approaches that meet the criteria above. Depending on the scope of the project, there are many ways for the training to be planned. These include direct involvement in project planning, traditional classroom training, simulation, the use of job aids, and on-the-job training.

By involving operators directly in the process design, they gain not just understanding, but also ownership in the new systems. Operators who feel ownership for the new process and controls will put their best efforts forward and find a way to make things work, even when it seems that everything will fall apart.

There are several ways to help build operator ownership:

1. Operators design and/or configure HMI graphics.
2. Operators provide input to the process and control design.
3. Operators create or input to operating procedures.
4. Operators create or develop their own training program.

Obviously, it is not possible to involve everyone. But gaining the alignment and ownership of a few key leaders among the operators will go a long way toward a successful start-up. These key leaders will slowly "leak" out information about the project, bringing people up to speed over time, rather than saving it all for one big training "event."

Traditional classroom training still has a place in operator training. It is an efficient method of passing theoretical background, rationale, and overview training. But most people can attest to the loss of interest and attention span that occurs after "too much" classroom training.

There are no fixed rules to determine how much is too much time in the classroom. But there are a number of rules of thumb that will help to make the best of this time.

- Know the audience. If their specific concerns are unknown, ask! They'll tell.
- Cover theoretical material early in the day.
- Get through the post-lunch slump by using exercises and walk-downs. Get people out of their chairs.
- Keep it interactive. Ask lots of open-ended questions. Encourage discussion between students, not just student-to-instructor.
- Keep it real. Make frequent ties back to the student's particular interests.
- Take frequent breaks. Allow time for the material to sink in; 2 hours is plenty of time to sit in a chair.

Remember that there are other, more effective ways to train people. Take a look at the methods below to get some ideas how to make the training more appealing, interesting, and effective for the students. The more interested and interactive they are, the more they will learn.

Simulation

To develop familiarity and trust in the HMI and the new controls, the operator needs focused training. Simulators for operator training have become quite popular in recent years. Simulators can provide a very close approximation of the actual operation of the process.

All simulators are not created equal. When specifying a simulator for operator training, the following items should be specified:

- Degree of fidelity of process simulation
- HMI accuracy
- Normal vs. abnormal operation
- Access to instructors

The degree of process simulation can range from simple tail-to-mouth emulation, all the way to complex, dynamic, highly accurate, first-principles process models. For operator training, consider the importance of process dynamics. If most loops are fast, and little intervention is required during transients, then the simulation can ignore the dynamics. If, however, long process lags require operators to plan their actions carefully, and make adjustments, then a more sophisticated model of process dynamics is likely required.

Troubleshooting Card

IF	AND	THEN
Low Flow Alarm	Supply Pressure Normal	Check for plug. Flush line.
	Supply Pressure Low	Check pump suction.
High Flow Alarm	Supply Pressure Normal	Check feed pump speed SP. Should be less than 800 rpm.
	Supply Pressure High	Check pressure SP. Should be less than 120 psig.

Training Card
Control Loop Start-Up

ACTION	STEPS
1. Place Level Control in Auto	A. Select "Feed Tank" display button. B. Place cursor over LT1275, click once. Auto/Manual Target appears. C. Click once on the Auto selection. Confirmation target appears. D. Click on "Yes" to confirm the loop should be in Auto.
2. Establish Temperature SP	A. Select "Unit 1 Temp" display button. B. Place cursor over TIC1776, right click once. Setpoint Entry appears. C. Type in the new value for the Setpoint. Confimation target appears. D. Click on "Yes" to confirm the loop should be in Auto.

FIG. 1.3m
Sample job aids.

The accuracy, or fidelity, of the simulation is one of the most significant factors in determining the cost of the simulation. The modeling cost rises exponentially with the accuracy. Before specifying tight modeling tolerances for process gains and dynamics, make sure they are needed.

Simulation of abnormal situations can also drive up both the realism and the cost. It is helpful to incorporate at least a few common abnormal situations in the simulator. These can be treated as "scenario training" by the operators, and will help operators think through their new skills in a "real-life" setting.

A simulator may be used to train the operators on the HMI. This way, the operators can practice keystrokes and sequences. They can also learn to navigate for information in the new control system.

With today's PC-based control systems, it is usually quite simple to provide simulated interfaces that very closely match the actual HMI that will be used. If at all possible, training should occur using the same keyboard, mouse, or touchscreen that will be installed. Quite often, this is as simple as making the HMI software and/or equipment available for a few days, and allowing the operators to test the screens at their leisure.

Students will have questions, and they will try things that nobody expected. An instructor should be available to the students, even if the simulator is designed for self-paced training. The instructor can also be used to develop and implement the abnormal scenarios, and to grade student performance.

Job aids and manuals provide the operator with a source of reference materials when much of the training has been forgotten. Job aids can be flowcharts, checklists, lookup tables, or any other reference item. They are typically used to help jog the operator's memory.

Job aids become especially valuable for complex operations, or for operations that are not performed frequently. Many plants only shut down a few times a year. It is not reasonable to expect that an operator will remember the specifics of the shutdown procedure without some help. A job aid will help guide this process. A few sample job aids are shown in Figure 1.3m. In some cases, these job aids are laminated and given to the operators-in-training. In other cases, they may be collected in a book, and kept in a convenient location in the control room.

On-the-Job Training

There is nothing like a little pressure to help keep students focused! On-the-job training, or OJT, provides the most realistic training to the operator. For small changes and enhancements to the existing control system, this can be very effective.

The operator can learn on the real system, with real problems and real consequences.

However, for large changes, the consequences can be so great that OJT is too risky. In these cases, OJT can be used to supplement the offline training, or as final qualification of operators. During OJT, it is helpful to allow access to the "experts," whether they are process engineers, experienced operators, or people at other plants who have started up a similar process. With access to the experts and time to learn and practice, OJT can be a highly effective training method.

START-UP

Starting up after major changes requires a coordinated effort. In the case of large processes, this is usually accomplished in stages; for example, each unit operation is run individually before attempting to run the whole thing.

During process start-up, clear lines of communications are essential to success. There are many parties with a stake in the outcome, and everyone must be kept "in the loop." Construction crews, operators, maintenance, engineering, and management all have a stake in the outcome, each with specific needs and desires that differ from the others.

For small projects, this is as simple as keeping a bulletin board or whiteboard updated with current status. For large projects, maintaining communications requires considerably more effort. In addition to status boards, daily or twice-daily meetings can help to facilitate the flow of information.

One special note on communications: Keep unnecessary people out of the control room! During start-up, the control room is the hub of all activity. It is an excellent source of information. But too many visitors will create chaos in the control room. Management should direct people to stay out of the control room unless absolutely required to be there. Give operators the authority to clear the control room when it gets too busy.

Two-way radios are almost indispensable during start-up. Most control systems projects these days will have a centralized control area, while the field equipment is widely distributed throughout the plant. These radios will be widely used for tasks like stroking valves, checking instrument wiring, and confirming the actual state of the process.

To avoid errors and delays, choose a high-quality set of radios, with rechargeable batteries. If it is a large project, consider the use of multiple frequencies, setting aside one channel for each part of the project team: construction, electrical, mechanical, and operating.

Start-up can be a time of chaos. Despite the best-laid plans, there will be broken equipment, installation errors, and unforeseen operating conditions. The plan will need to be adjusted. In the matter of minutes, things can change dramatically.

In times of great change, leadership and decision making are key tasks. Decision makers should never be far from the action. Among the decision makers, leadership will shift from engineering to construction, then to operations, as the start-up progresses.

As decisions are made, be sure to communicate them broadly, across the disciplines, and across any other shifts. This is especially true for schedule changes.

Field Changes

During static and dynamic PSI, problems will be uncovered. To fix these problems often requires changes to piping, wiring, and configuration. Left unmanaged, these changes can add considerably to the cost of a project. To keep these costs in line, a field change management system should be put in place.

A field change management system accomplishes the following:

Defines the change required and the reason for it
Estimates the cost and schedule impacts
Provides clear approvals for the additional work
Ensures that drawings and other documents are updated

A simple field change management system is simply a form filled out and reviewed during daily project review meetings. Sophisticated systems will use a database to identify and track all changes.

Turnover

At some point during start-up, the process is officially turned back over to the operation. At this point, the construction, engineering, and other project resources transition from a "project mode" into an "ongoing maintenance" mode. Of course, the project team is not completely off the hook, as it will need to follow up on lingering issues during the post-start-up phase. To ensure a smooth transition, it often helps to define the point of turnover more clearly. Typically, the point of turnover is defined by specific production criteria. This should include measures of both quality and quantity, or rate.

Achieving turnover is a significant milestone for most projects. It is a good time to celebrate the achievements, perhaps with an impromptu pizza party.

Post-Start-Up

Post-start-up, or PSU, is a time to resolve lingering problems, and to tie up any loose ends. The difference between a good project and an excellent project is often determined by the effort that takes place after start-up.

Documentation is the most important part of PSU. Master drawings must be updated and filed. Reference manuals should be turned over to the maintenance staff. This is particularly true for instruments and other control system components.

Continued vigilance is needed during PSU to resolve lingering equipment or process problems. This can be frustrating.

Item#	Item	Resp	Target Date
1	Replace broken handle, FV1267	John	1-Apr
2	Revise level alarm limits	Ann	6-Apr
3	Install software patch R467	Bryan	20-Apr
4	Update training manual for FIC2654	Diane	6-Apr
5	Repair Operator station furniture	John	1-May
6	Red-Line drawings into CAD	Mark	6-Apr
7	Cross-charges to Project X	George	1-May
8	Add FIC45 to Unit 2 Load Graphic	Chris	6-Apr

FIG. 1.3n
Sample PSU checklist.

After completing millions of dollars of project work, the last thing one wants to do is to chase down a 50-cent O-ring. But that O-ring may have a significant impact on the performance or the life of the instrument. No item is too small to track. Tracking these PSU items is a job for a spreadsheet or database. Figure 1.3n shows a sample PSU checklist.

There is a fine line here, of course. If PSU takes too long, routine maintenance items will begin to crop up and make their way on to the PSU list. Be sure to nip this in the bud by establishing clear responsibilities with the maintenance organization.

CONCLUSIONS

The goal of any project is successfully to install and start up a money-making process. Well-planned and well-executed operator training, commissioning, and start-up are all key elements to achieving this goal.

Making operators part of the project team and providing quality training will assure ownership and commitment from the operators.

A well-thought-out commissioning plan with input and commitment from all parties will help keep project costs down and the schedule on track by avoiding confusion, and help focus efforts on achieving production quality, reliability, and rate targets.

Methodical execution of the start-up phase of the project using simple commonsense planning, coordination, and communication can help avoid catastrophe and bring the project to a successful end.

Remember that planning a good start-up is an essential key to success for a project. Use training as a tool to build both skills and commitment to the project. With a good plan, and interested operators, the start-up will be successful.

Bibliography

Battikha, N. E., *The Management of Control Systems: Justification and Technical Auditing,* Research Triangle Park, NC: Instrument Society of America, 1992.

Bruce, A. and Langdon, K., *Essential Managers: Project Management,* New York: Dorling Kindersley Publishing, 2000.

Harris, D., *Start-Up: A Technician's Guide,* Research Triangle Park, NC: Instrument Society of America, 2000.

Harris, D., *Documenting the Start-Up Process,* Research Triangle Park, NC: Instrument Society of America, 2000.

MacKie, D., *Engineering Management of Capital Projects,* New York: McGraw-Hill, 1985.

1.4 Flowsheet Symbols and Functional Diagramming for Digitally Implemented Loops

J. E. JAMISON

The purpose of this section is to help the reader establish a uniform means of depicting and identifying mainly digital-based instruments, instrumentation systems, and functions, as well as application software functions used for measurement, monitoring, and control. It is done by presenting a designation system including graphic symbols and identification codes as well as functional diagramming techniques that were formerly known as the SAMA system. For this section, the author has tried to minimize conventional instrument symbol coverage and has maximized digital or digital-related symbols and functionality.

It must be noted that a significant part of this section is the revision work of the ISA SP5.1 subcommittee and a great deal is based on draft working documents that are utilized at the time of this writing and in which the author has been actively involved. There are other portions of this section that are based on the author's experience in industry and are not any part of the SP5.1 subcommittee proposed forthcoming revision. A DISCLAIMER TO ANY FUTURE ISA STANDARDS DOCUMENTS IS HEREBY STATED. THE READER IS CAUTIONED THAT THE DRAFT ISA DOCUMENT THAT PROVIDED MUCH OF THE INFORMATION IN THIS SECTION HAS NOT BEEN APPROVED AS OF THE TIME OF THIS WRITING. IT CANNOT BE PRESUMED TO REFLECT THE POSITION OF ISA OR ANY OTHER COMMITTEE, SOCIETY, OR GROUP. The intent is to pass along to the reader the best and latest thinking on this subject at this point in time although a number of items are contentious and are ultimately subject to change in this continuously evolving field of digital control systems and digital data buses.

SCOPE

General

The procedural needs of various users are different and these differences are recognized when they are consistent with the objectives of this standard, by providing alternative symbol and identification methods.

A limited number of examples are provided later that illustrate (with the emphasis on digital systems/loops) how to accomplish the following:

a) Design an identification system and construct an identification number.
b) Use graphic symbols to construct:
   1) Schematic diagrams of instrument devices and functions in monitoring and control loops.
   2) Function diagrams of instruments, loops, and application software functions.
   3) Schematic and ladder diagrams of electrical circuits.
c) Add information and simplify diagrams.

Examples of symbol applications are generally shown as they are applied in the oil and chemical processing industries as in the original version of this standard, but the principles shown are applicable to most other industries.

Specific applications are to be addressed in greater detail and will be forthcoming in the future planned S5.1 (now ANSI/ISA-5.01.01) series of Technical Reports dedicated to the various processing, generating, and manufacturing industries. These will include processes such as continuous and batch chemical, oil and metal refining, pulp and paper, water and waste treatment, power generation and distribution, and discrete parts manufacturing.

Application to Industries

The proposed revised ISA S5.1 (now ANSI/ISA-5.01.01) standard will be suitable for use in the above-mentioned process industries and discrete parts manufacturing that require the use of control system schematic and functional diagramming to describe the relationship to processing equipment and the functionality of measurement and control equipment.

Certain fields, such as astronomy, navigation, and medicine, use very specialized instruments that are different from conventional industrial process instruments. No specific effort was made to have the ISA standard meet the requirements of those fields. However, it is expected that in certain areas such as control functional diagrams, they will prove applicable for such specialized fields.

Application to Work Activities

The proposed revised ISA S5.1 (now ANSI/ISA-5.01.01) standard will be suitable for use whenever reference to measurement and control instrumentation, control device functions, or software applications functions is required for the purposes of symbolization and identification. Such references may be required for the following uses, as well as others:

a) Design sketches.
b) Teaching examples.
c) Technical papers, literature, and discussions.
d) Instrumentation system diagrams, loop diagrams, logic diagrams, and functional diagrams.
e) Functional descriptions.
f) Conceptual drawings: process flow diagrams (PFD) and utility flow diagrams (UFD).
g) Construction drawings: engineering flow diagrams (EFD), mechanical flow diagrams (MFD), piping and instrument diagrams (P&ID), system flow diagrams (SFD).
h) Specifications, purchase orders, manifests, and other lists.
i) Identification and tag numbering of instruments and control functions.
j) Installation, operating, and maintenance instructions, drawings, and records.

The standard is intended to provide sufficient information to enable anyone with a reasonable amount of process and instrumentation knowledge the means to understand the methods of measurement and control of the process.

The detailed knowledge of a specialist in instrumentation and control systems is not a prerequisite to understanding the standard.

Application to Classes of Instrumentation and to Instrument Functions

The symbolism and identification methods provided in the standard are applicable to all classes and types of measurement and control instruments and functions.

The methods can be used for, but are not limited to, describing and identifying:

a) Discrete instruments and their functions.
b) Shared display and control functions.
c) Distributed control functions.
d) Computer control functions.
e) Programmable logic controller display and control functions.
f) Application software display and control functions.

Extent of Loop and Functional Identification

The ISA S5.1 standard (now ANSI/ISA-5.01.01) provides identification codes and methods for the alphanumeric identification of monitoring and controlling loops, instruments, and functions. The user is free to apply additional identification by serial, equipment, unit, area, or plant number, or any other additional means required for the unique identification of a loop, instrument, or function.

A unique function identification number shall identify each instrument, its inherent functions, and each configurable function that requires or allows a user-assigned unique microprocessor or computer address required by a loop.

Extent of Symbolization

The standard provides symbol sets for the graphic depiction of a limited or total functionality for instruments and other devices, entire monitor/control loops, or control circuits. The amount of detail to be shown by the use of symbols depends on the purpose and audience for which the document is being prepared.

A sufficient number of symbols should be used to show the functionality of the instrumentation or control loop being depicted. However, it is not considered necessary to provide a symbol for each instrument device and each function within a loop.

Additional construction, fabrication, installation, and operation details of an instrument are better described in a suitable specification, data sheet, drawing, sketch, or other document intended for those requiring such details.

Inclusion of the New S5.1 Standard (now ANSI/ISA-5.01.01) in User/Owner Documents

This is a new concept in ISA standards at this point in time. Mandatory use of the standard is required by users/owners based on the following statements.

When the latest issue of the standard is included in user/owner's engineering and/or design guidelines or standards by reference and:

a) "Without exception," then the standard in its entirety shall be mandatory.
b) "With exceptions," then the parts of the standard:
   1) "Excepted to" shall be fully described and detailed.
   2) "Not excepted to" shall be mandatory.

When a previous issue of the standard is included by reference with or without exception in user/owner's engineering or design guidelines or standards, that standard in part or in its entirety shall be mandatory until such time as the user/owner's guidelines or standards are revised. When the new issue is used as a guide in the preparation of user/owner's guidelines or standards, symbols, and letter and symbol meanings different from those in the standard shall be fully described and detailed.

Symbols and the meanings of letters and symbols from previous issues of the S5.1 standard (now ANSI/ISA-5.01.01) that are different from those contained in this new issue may continue to be used provided they are fully described and detailed.

DEFINITIONS RELATED TO FLOWSHEET DIAGRAM SYMBOLOGY

General

For the purpose of understanding the ISA S5.1 standard (now ANSI/ISA-5.01.01), the following definitions and terminology apply. For a more complete treatment, see ISA-S51.1 and the ISA-S75 series of standards. Terms italicized within a definition are also defined in this clause.

Definitions

ACCESSIBLE A feature of a discrete device function or feature of an interactive shared system function or feature that can be used or seen by an operator for the purpose of performing control operations, such as set point changes, auto-manual transfer, or on-off operations.

ALARM An indicating instrument that provides a visible and/or audible indication if and when the value of a measured or initiating variable is out of limits, has changed from a safe to an unsafe condition, and/or has changed from a normal to an abnormal operating state or condition.

a) Actuation may be by binary switch or function or analog transmitter or function.
b) Indication may be by annunciator panel, flashing light, printer, buzzer, bell, horn, siren, and/or shared graphic display systems.

ANALOG A signal or device that has no discrete positions or states and changes value as its input changes value. When used in its simplest form, as in "analog signal" as opposed to "binary signal," the term denotes a continuously varying quantity.

APPLICATION SOFTWARE Software specific to a user application that is configurable and in general contains logic sequences, permissive and limit expressions, control algorithms, and other code required to control the appropriate input, output, calculations, and decisions. See also *software*.

ASSIGNABLE A system feature permitting channeling or directing of a signal from one device to another without the need for changes in wiring either by means of patching, switching, or via keyboard commands to the system.

AUTO-MANUAL STATION A manual loading station or control station that also provides switching between manual and automatic control modes of a control loop. See also *manual station*.

BALLOON An alternate term for the circular symbol used to denote and identify the purpose of an instrument or function that may contain a tag number. See preferred term *bubble*.

BEHIND THE PANEL A location that in a broad sense means "not normally accessible to an operator," such as the rear of an instrument or control panel, an enclosed instrument rack or cabinet, or an instrument rack room within an area that contains a panel.

BINARY A signal or device that has only two discrete positions/states, and when used in its simplest form as in "binary signal" as opposed to "analog signal," the term denotes an "on-off" or "high-low" state.

BOARD A freestanding structure consisting of one or more sections, cubicles, or consoles that has groups of discrete instruments mounted on it, houses the operator-process interface, and is chosen to have a unique designation. See *panel*.

BUBBLE The preferred term for the circular symbol used to denote and identify the purpose of an instrument or function that may contain a tag number. See alternate term *balloon*.

COMMUNICATION LINK A wire, cable, or transmitter network or bus system that connects dedicated microprocessor-based and computer-based systems so that they share a common database and communicate according to a rigid protocol in a hierarchical and/or peer-to-peer relationship. See also *data link*.

a) Wires or cables may be of twisted pair, coaxial, telephone, or fiber-optic construction.
b) Transmitters may be radio, telephone, or microwave devices.

COMPUTER CONTROL SYSTEM A system in which all control action takes place within a control computer, such as a main frame computer or minicomputer, which may be single or redundant.

COMPUTING DEVICE Preferred term for a device that performs one or more calculations or logic operations, or both, and transmits one or more resultant output signals. See also *computing relay*.

COMPUTING FUNCTION A hardware or software function that performs one or more calculations or logic operations, or both, and transmits one or more resultant output signals.

COMPUTING RELAY Alternate term for a device that performs one or more calculations or logic operations, or both, and transmits one or more resultant output signals. See also *computing device*.

CONFIGURABLE A term for devices or systems whose functional or communication characteristics can be selected or rearranged through setting of program switches, application software, fill-in-the-blank forms, pull-down menus, entered values or text, or other methods, other than rewiring as a means of altering the configuration.

CONTROLLER A device having an output that varies to regulate a controlled variable in a specified manner that may be a self-contained analog or digital instrument, or may be the equivalent of such an instrument in a shared-control system.

a) An automatic controller varies its output automatically in response to a direct or indirect input of a measured process variable.
b) A manual controller, or manual loading station, varies its output in response to a manual adjustment; it is not dependent on a measured process variable.
c) A controller may be an integral element of other functional elements of a control loop.

CONTROL STATION A manual loading station that also provides switching between manual and automatic control modes of a control loop. See also *auto-manual station*.

a) The operator interface of a distributed control system may be referred to as a control station.

CONTROL VALVE A device, other than a common, hand-actuated process block valve or self-actuated check valve, that directly manipulates the flow of one or more fluid process streams.

a) The designation "hand control valve" shall be limited to hand-actuated valves that when used for process throttling require identification as an instrument or control device.

CONVERTER A device that receives information as one form of an instrument signal and transmits an output signal as another form, such as a current to pneumatic signal converter.

a) An instrument, which changes a sensor's output to a standard signal, is properly designated as a transmitter and not a converter. Typically, a temperature element (TE) connects to a transmitter (TT) and not to a converter (TY).
b) A converter is sometimes referred to as a transducer, a completely general term not recommended for signal conversion.

DATA LINK A wire, cable, or transmitter network or bus system that connects field-located devices with dedicated microprocessors so that they share a common database and communicate according to a rigid protocol in a hierarchical or peer-to-peer relationship to other such devices or compatible microprocessor-based systems. See also *communication link*.

a) Wire or cable may be of twisted pair, coaxial, telephone, or fiber-optic construction.
b) Transmitters may be radio, telephone, or microwave devices.

DETECTOR A device that is used to detect the presence of something, such as flammable or toxic gases or discrete parts. See also *primary element* and *sensor*.

DEVICE A piece of instrument hardware that is designed to perform a specific action or function, such as a controller, indicator, transmitter, annunciator, control valve, etc.

DIGITAL A signal or device that generates or uses binary digit signals to represent continuous values or discrete states.

DISCRETE A term used to describe:

a) Signals that have any number of noncontinuous distinct or defined states or positions. Binary signals are a subset. See *binary*.
b) Instruments or devices that have a separate entity, such as a single case controller or recorder.

DISTRIBUTED CONTROL SYSTEM Instrumentation, input/output devices, control devices, and operator interface devices, which in addition to executing stated control and indication functions also permit transmission of control, measurement, and operating information to and from single or multiple user-specifiable locations, connected by single or multiple communication links.

FIELD INSTRUMENT An instrument that is not mounted on a panel or console or in a control room but commonly in the vicinity of its primary element or final control element. See *local instrument*.

FINAL CONTROL ELEMENT A device, such as a control valve, that directly controls the value of the manipulated variable of a control loop.

FUNCTION The purpose of, or the action performed by, a device or application software.

IDENTIFICATION The sequence of letters or digits, or both, used to designate an individual instrument, function, or loop.

INSTRUMENT A device used for direct or indirect measurement, monitoring, or control of a variable.

a) Includes primary elements, indicators, controllers, final control elements, computing devices, and electrical devices such as annunciators, switches, and pushbuttons.
b) Does not apply to an instrument's internal components or parts, such as receiver bellows or resistors.

INSTRUMENTATION A collection of instruments or functions or their application for the purpose of measuring, monitoring, or controlling, or any combination of these.

LOCAL INSTRUMENT An instrument that is not mounted on a panel or console or in a control room but commonly in the vicinity of its primary element or final control element. See *field instrument.*

LOCAL PANEL A panel that is not a central or main panel and is commonly located in the vicinity of plant subsystems or subareas.

a) The term "local panel instrument" should not be confused with "local instrument."

LOOP A combination of two or more instruments or control functions arranged so that signals pass from one to another for the purpose of measurement indication or control of a process variable.

MANUAL LOADING STATION A device or function that has a manually adjustable output, and may also have indicators, lights, or other functions, which are used to actuate or modulate one or more devices. It does not provide switching between auto-manual modes of a control loop.

MEASUREMENT The determination of the existence or magnitude of a process variable.

MONITOR A general term for an instrument or instrument system used to measure or sense the status or magnitude of one or more variables for the purpose of deriving useful information. This sometimes means an analyzer, indicator, or alarm.

MONITOR LIGHT A light that indicates which of a number of normal (but not abnormal) conditions of a system or device exists. See also *pilot light.*

PANEL A freestanding or built-in structure consisting of one or more sections, cubicles, consoles, or desks that has groups of instrument hardware mounted in it. It could house the operator–process interface and is given a unique designation.

PANEL-MOUNTED An instrument or other device that is mounted in a panel or console and is accessible for an operator's normal use.

a) A function that is normally accessible to an operator in a shared-display system is the equivalent of a discrete panel-mounted device.

PILOT LIGHT A light that indicates which of a number of normal conditions of a system or device exists. It is not an alarm light that indicates an abnormal condition. See also *monitor light.*

PRIMARY ELEMENT An external or internal instrument or system element that quantitatively converts the measured variable into a form suitable for measurement. See also *detector* and *sensor*:

a) An orifice plate is an external primary element.
b) The sensing portion of a transmitter is an internal primary element.

PROCESS Any operation or sequence of operations involving a change of energy, state, composition, dimension, or other properties that may be defined with respect to zero or some other defined initial value.

PROCESS VARIABLE Any measurable property of a process; used in this standard to apply to all variables other than instrument signals between devices in a loop.

PROGRAM A repeatable sequence of actions that defines the state of outputs as a fixed relationship to the state of inputs.

PROGRAMMABLE LOGIC CONTROLLER A controller, usually with multiple inputs and outputs, that contains an alterable program that is:

a) Typically used to control binary and/or discrete logic or sequencing functions.
b) Also used to provide continuous control functions.

RELAY A device whose function is to pass on information in an unchanged form or in some modified form, often used to mean the preferred term computing device.

a) Relay is a term applied specifically to an electric, pneumatic, or hydraulic switching device that is actuated by a signal and to functions performed by a relay.

SCAN To sample, in a predetermined manner, each of a number of variables periodically and/or intermittently.

a) A scanning device is often used to ascertain the state or value of a group of variables, and may be associated with other functions such as recording and alarming.

SENSOR A separate, or integral part, or function of a loop or an instrument that first senses the value of a process variable. It assumes a corresponding predetermined and intelligible state, and/or generates an output signal indicative of or proportional to the process variable. See also *detector* and *primary element.*

SET POINT An input variable that sets the desired value of the controlled variable manually, automatically, or by means of a program in the same units as the controlled variable.

SHARED CONTROL A feature of a control device or function that contains a number of preprogrammed algorithms that are user retrievable, configurable, and connectable. It allows user-defined control strategies or functions to be implemented and is often used to describe the control features of a distributed control system.

a) Control of multiple process variables can be implemented by sharing the capabilities of a single device of this kind.

SHARED DISPLAY The operator interface device such as video, light emitting diode, liquid crystal, or other display unit; used to display process control information from a number of sources at the command of the operator. It is often used to describe the visual features of a distributed control system.

SOFTWARE The programs, codes, procedures, algorithms, patterns, rules, and associated documentation required for the operation or maintenance of a microprocessor- or computer-based system. See also *application software*.

SOFTWARE LINK The interconnection of system components via communication networks or functions via software or keyboard instruction.

SUPERVISORY SET POINT CONTROL SYSTEM The generation of set point or other control information by a computer control system for use by shared control, shared display, or other regulatory control devices.

SWITCH A device that connects, disconnects, selects, or transfers one or more circuits and is not designated as a controller, a relay, or a control valve. As a verb, the term is also applied to the functions performed by switches.

TEST POINT A process connection to which no instrument is permanently connected, but which is intended for the temporary or intermittent connection of an instrument.

TRANSDUCER A general term for a device, such as a primary element, transmitter, relay, converter, or other device, which receives information in the form of one or more physical quantities, modifies the information or its form if required, and produces a resultant output signal.

TRANSMITTER A device that senses a process variable through the medium of a sensor or measuring element and has an output whose steady-state value varies only as a predetermined function of the process variable. The sensor can be an integral part, as in a direct connected pressure transmitter, or a separate part as in a thermocouple-actuated temperature transmitter.

IDENTIFICATION SYSTEM GUIDELINES

 From ISA Draft Standard 5.01.01, Instrumentation Symbols and Identification, Draft 4, which was not a final and approved standard at the time of publication of this handbook. To obtain an updated copy of ANSI/ISA-5.01.01, please contact ISA Standards, www.isa.org.

General

This subsection establishes an identification system for instrument loop devices and functions that is logical, unique, and consistent in application with a minimum of exceptions, special uses, or requirements. The identification system is used to identify instrumentation in text and in sketches and drawings when used with graphic symbols as described in Subsection "Graphic Symbol System Guidelines."

The identification system provides methods for identifying instrumentation required to monitor, control, and operate a processing plant, unit operation, boiler, machine, or any other system that requires measurement, indication, control, modulation, or switching of variables. Primary instrumentation, hardware and software devices and functions that measure, monitor, control, and calculate, and application software functions that require or allow user-assigned identities, shall be assigned a functional identification.

Secondary instrumentation such as hardware devices that measure and monitor, as well as level glasses, pressure gauges, and thermometers, shall be assigned only a functional identification.

Loop and functional identification shall be assigned in accordance with the guidelines in the standard or with modified guidelines based on the standard, established by the user or owner of the plant, unit, or facility in which the instrumentation is to be installed.

A unique loop identification number shall be assigned to identify each monitoring and control loop. A unique instrument identification/tag number based on the loop identification number shall be assigned for each monitoring or control loop to identify each:

a) Hardware device and integral functions.
b) Application software function, that requires or allows a user-assigned unique microprocessor or computer address.

A monitor or control loop consists of some or all of the following (as indicated):

a) Measurement of the process variable (monitor and control):
    1) Measuring element device, such as an orifice plate or thermocouple.
    2) Measurement transmitter with an integral measuring element, such as a pressure transmitter, or without an integral measuring element, such as a temperature transmitter and thermocouple.
b) Conditioning of the measurement or input signal (monitor and control):
    1) Calculating devices.
    2) Calculating functions.
c) Monitoring of the process variable (monitor):
    1) Indicating or recording device.
    2) Application software display function.
d) Controlling of the process variable (control):
    1) Indicating or recording control device.
    2) Application software control function.
e) Conditioning of the controller or output signal (control):
    1) Calculating devices.
    2) Calculating functions.

f) Modulation of the manipulated variable (control):
   1) Control valve modulation or on-off action.
   2) Manipulation of another control loop set point.
   3) Limiting another control loop output signal.

Secondary instrumentation shall be assigned instrument identification/tag numbers or other forms of identification in accordance with the guidelines established in the ISA standard or with modified guidelines based on the standard established by the user/owner of the plant, unit, or facility in which the instrumentation is to be installed.

Examples of instrument identification systems will be found in a future series of S5.1 (now ANSI/ISA-5.01.01) Technical Reports.

Instrument Index

Loop identification numbers and instrument identification/tag numbers shall be recorded in an instrument index, which shall be maintained for the life of the facility for the recording and control of all documents and records pertaining to the loops and their instrumentation and functions.

An instrument index shall contain references to all instrumentation data required by owner or government regulatory agency management-of-change requirements. It should contain, as a minimum, for each loop:

a) Loop identification number.
b) Service description.
c) Instrument identification/tag numbers.
d) Piping and instrument drawing numbers.
e) Instrument data sheet numbers.
f) Location plan numbers.
g) Installation detail numbers.

Guideline Modifications

These guidelines may be modified to suit the requirements of:

a) Existing user-designed identification and numbering schemes that are not included in this standard.
b) Computer databases used for record keeping.
c) Microprocessor-based monitoring or control systems.

When modified guidelines are adopted, they shall be fully described and detailed in the user/owner's engineering or design standards.

Multipoint, Multifunction, and Multivariable Devices

Input and output devices and functions that are components of a multipoint device shall have a tag suffix that delineates between the different components.

Multifunction devices that receive a single input signal, send out two or more output signals, or perform two or more functions may be assigned readout/passive or output/active function multifunction [U] and shall have a loop number assigned according to the measured/initiating variable.

Multivariable devices that receive two or more input signals, transmit one or more output signals, and have been assigned measured/initiating variable multivariable [U] shall have:

a) Each different input shall be assigned its own loop identification number and each output indicating, recording, switching, alarming, etc. device and function that is actuated solely by a single variable shall be assigned instrument/tag numbers that identify them as part of these loops.
b) Each device and function, indicating, recording, switching, alarming, etc., that is actuated by more than one of the multivariables shall be assigned instrument/tag numbers that identify them as part of the multivariable loop.

Loops that perform two or more functions from a single measured/initiating variable may have:

a) Each function assigned a unique instrument/tag number and shown on diagrams as multiple tangent bubbles for the integral functions and multiple individual bubbles for the non-integral functions.
b) One readout/passive and/or output/active function designated by succeeding letter [U], for the integral functions and multiple individual bubbles for the non-integral functions, and, if necessary, a note or comment defining the integral functions.

Systems Identification

Instrumentation is often assembled into systems for various reasons including ease of purchase, ease of application, compatibility, etc. These systems may need to be identified on drawings and in text.

Some of the more common instrumentation systems and the system codes for identifying them are:

ACS	Analyzer Control System
CCS	Computer Control System
CEMS	Continuous Emissions Monitoring System
DCS	Distributed Control System
MMS	Machine Monitoring System
PCCS	Personal Computer Control System
PLC	Programmable Logic Controller
SIS	Safety Instrumented System
VMS	Vibration Monitoring System

Suffixes may be added to the instrumentation system codes [SC] when required:

a) [SC] 1, [SC] 2, etc., when more than one system is used in a complex.
b) [SC]-M, [SC]-L, when main and local systems are used in a unit.
c) [SC]-[unit identifier].

Measured/Initiating Variable

10- P - *01 A	- Loop Identification Number
10	- Optional Loop Number Prefix
-	- Optional Punctuation
P	- Measured/Initiating Variable
-	- Optional Punctuation
*01	- Loop Number
A	- Optional Loop Number Suffix

First Letters

10- P D - *01 A	- Loop Identification Number
10	- Optional Loop Number Prefix
-	- Optional Punctuation
P D	First Letters
P	- Measured/Initiating Variable
D	- Variable Modifier
-	Optional Punctuation
*01	- Loop Number
A	- Optional Loop Number Suffix

FIG. 1.4a
Typical loop identification number. (© Copyright 2000 ISA—The Instrumentation, Systems, and Automation Society. All rights reserved. Used with permission. From ISA Draft Standard 5.01.01, Instrumentation Symbols and Identification, Draft 4, which was not a final and approved standard at the time of publication of this handbook. To obtain an updated copy of ANSI/ISA-5.01.01, please contact ISA Standards, www.isa.org.)

10- P D A L - *01 A- A- 1	- Loop Identification Number
10	- Optional Loop Number Prefix
-	- Optional Punctuation
P *01 A	- Loop Number, Measured Variable
P D *01 A	- Loop Number, First Letters
-	- Optional Punctuation
*01	- Loop Number
A	- Optional Loop Number Suffix
P D A L	- Functional Identification Letters
P D	- First Letters
P	- Measured/Initiating Variable
D	- Variable Modifier
A L	- Succeeding Letters
A	- Function Identifier
L	- Function Modifier
-	- Optional Punctuation
A	- Tag Number Suffix
-	- Optional Punctuation
1	- Tag Number Suffix

FIG. 1.4b
Typical instrument identification/tag number. (See statement of permission following Figure 1.4a.)

Loop Identification Number

A loop identification number is a unique combination of letters and numbers that is assigned to each monitoring and control loop in a facility to identify the process or machine variable that is being measured for monitoring or control (Figure 1.4a).

Loop identification numbers are assigned:

a) Numerals in parallel, serial, or parallel/serial sequences.
b) Letters or letter combinations selected from Table 1.4c Identification Letters (Column 1, Measured/Initiating Variables and Column 2, Variable Modifiers).

Loop identification number numerals shall be assigned to loop variables letters according to one of the following sequencing methods:

a) Parallel: duplicated numerical sequences for each loop variable letter or letter combination.
b) Serial: the same numerical sequence regardless of loop variable letter or letter combination.
c) Parallel/Serial: parallel sequences for selected loop variable letters or letter combinations and a serial sequence for the remainder.

Loop number numerical sequences are normally three or more digits, -*01, -*001, -*0001, etc.:

a) Where -* can be any digit from 0 to 9.
b) Coded digits related to drawing numbers or equipment numbers.
c) *00, *000, *0000, etc. are not used.

Gaps may be left in any sequence to allow for the addition of future loops. (See Tables 1.4c, d, e, and f for various combinations of allowable instrumentation identification/tag numbers.)

Typical Instrument Identification/Tag Number

See Figure 1.4b.

IDENTIFICATION LETTER TABLES

TABLE 1.4c
Identification Letters

	First Letters		Succeeding Letters		
	Column 1	Column 2	Column 3	Column 4	Column 5
	Measured/Initiating Variable	Variable Modifier	Readout/Passive Function	Output/Active Function	Function Modifier
A	Analysis		Alarm		
B	Burner, combustion		User's choice	User's choice	User's choice
C	User's choice			Control	Close
D	User's choice	Differential, deviation			Deviation
E	Voltage		Sensor, primary element		
F	Flow, flow rate	Ratio			
G	User's choice		Glass, gauge, viewing device		
H	Hand				High
I	Current		Indicate		
J	Power		Scan		
K	Time, schedule	Time rate of change		Control station	
L	Level		Light		Low
M	User's choice				Middle, intermediate
N	User's choice		User's choice	User's choice	User's choice
O	User's choice		Orifice, restriction		Open
P	Pressure		Point (test connection)		
Q	Quantity	Integrate, totalize	Integrate, totalize		
R	Radiation		Record		
S	Speed, frequency	Safety		Switch	
T	Temperature			Transmit	
U	Multivariable		Multifunction	Multifunction	
V	Vibration, mechanical analysis			Valve, damper, louver	
W	Weight, force		Well		
X	Unclassified	X-axis	Unclassified	Unclassified	Unclassified
Y	Event, state, presence	Y-axis		Auxiliary devices	
Z	Position, dimension	Z-axis		Driver, actuator, unclassified final control element	

Source: See statement of permission following Figure 1.4a.

General

This clause provides in tabular form the alphabetic building blocks of the Instrument and Function Identification System in a concise, easily referenced manner.

Table 1.4c, Identification Letters, defines and explains the individual letter designators to be used as loop and functional identifiers in accordance with the guidelines of Subsection "Identification System Guidelines."

The letters in Table 1.4c shall have the mandatory meanings as given in the table except:

a) The user shall assign a variable name to the user's choice letters in Column 1 and a function name to the user's choice letters in Columns 3 through 5 when such letters are used.
b) The user may assign meanings to the blanks in Columns 2 to 5 if needed.

Table 1.4d, Allowable Loop Identification Letters, provides the allowable loop identification letters and combinations according to the loop identification number construction schemes.

The letters and combinations shall have the mandatory meanings as given in the table except:

a) The user shall assign a variable name to the user's choice letters in the First Letter column.

Tables 1.4e and 1.4f, Allowable Function Identification Letter Combinations, provide allowable combinations of function identifying letters.

The letter combinations shall have the meanings given in the table, except the user:

a) Shall assign a variable and/or function to user's choice letters if used.
b) May assign a meaning to blanks if needed.
c) Cells marked N/A are combinations that shall be not allowed.

GRAPHIC SYMBOL SYSTEM GUIDELINES

 From ISA Draft Standard 5.01.01, Instrumentation Symbols and Identification, Draft 4, which was not a final and approved standard at the time of publication of this handbook. To obtain an updated copy of ANSI/ISA-5.01.01, please contact ISA Standards, www.isa.org.

General

The future revised ISA standard S5.1 (now ANSI/ISA-5.01.01) establishes a graphic symbol system and functional identification for depicting instrument loop devices and functions, application software functions, and the interconnections between them that is logical, unique, and consistent in application with a minimum of exceptions, special uses, or requirements.

The graphic symbol system shall be used to depict instrumentation in text and in sketches and drawings. When used with identification letters and numbers as described in the subsection "Identification System Guidelines," it shall identify the functionality of each device and function shown.

The graphic symbol system provides methods for schematic loop diagramming, functional diagramming, and electrical schematic diagramming of any process or system that requires measurement, indication, control, modulation, or switching of variables.

Primary instruments are hardware and software devices and functions that measure, monitor, control, or calculate (i.e., transmitters, controllers, calculating devices, control valves, etc.). Secondary instruments are hardware devices that measure and monitor, such as sight flow indicators, level and pressure gauges, thermometers, etc.

Primary and secondary instrumentation shall be depicted in accordance with the guidelines established in the ISA standard.

Specific industrial application examples of the graphic symbol system will be found in a future series of S5.1 (now ANSI/ISA-5.01.01) Technical Reports. Sketches that are not all inclusive of acceptable methods of depicting instrumentation are included in the following text to illustrate the intent of the standard. However, the individual symbols and their meanings are to be mandatory in the future, imminent standard.

Guideline Modifications

These guidelines may be modified to suit the requirements of existing user-designed graphic symbols that are not included in this standard. When modified symbols are adopted, they shall be fully described and detailed in the user/owner's engineering or design standards.

Instrument Line Symbols

See Table 1.4g. Symbols represent:

a) Instrument and device connections at process measurement points.
b) Connections to instrument power supplies.
c) Signals between measurement and control instruments and functions.

Lines shall be:

a) Fine in relation to process equipment and piping lines.
b) As short as possible consistent with clarity.

Measurement and Control Devices and/or Function Symbols

See Table 1.4h.

Symbols represent discrete devices that perform continuous and/or on–off functions that do not share control or display functions for:

a) Measurement (transmitters, primary elements).
b) Indication (indicators, annunciators).
c) Control (controllers, control valves, switches, solenoids).

Limited operator accessibility (set point changes, control mode transfers, etc.) and unlimited engineer or technician accessibility through location and enclosure methods are shown.

TABLE 1.4d
Allowable Loop Identification Letter Schemes

First Letters	Measured/Initiating Variable	Scheme 1	Scheme 2	Scheme 3	Scheme 4	Scheme 5	Scheme 6	Scheme 7(1)		Scheme 8(1)		Scheme 9(1)	
								Parallel	Serial	Parallel	Serial	Parallel	Serial
		Parallel Meas./Init. Var.	Parallel Meas./Init. Var. w/Var. Mod.	Parallel First Letters	Serial Meas./Init. Var.	Serial Meas./Init. Var. w.Var. Mod.	Serial First Letters	Measured/Initiating Variable		Measured/Initiating w/Variable Modifier		First Letters	
A	Analysis	A-*01	A-*01	A-*01	A-*01	A-*01	A-*01	A-*01		A-*01		A-*01	
B	Burner combustion	B-*01	B-*01	B-*01	B-*02	B-*02	B-*02		B-*01		B-*01		B-*01
C	User's choice	C-*01	C-*01	C-*01	C-*03	C-*03	C-*03		C-*02		C-*02		C-*02
D	User's choice	D-*01	D-*01	D-*01	D-*04	D-*04	D-*04		D-*03		D-*03		D-*03
E	Voltage	E-*01	E-*01	E-*01	E-*05	E-*05	E-*05		E-*04		E-*04		E-*04
F	Flow, flow rate		F-*01	F-*01		F-*06	F-*06			F-*01		F-*01	
FF	Flow ratio	F-*01	FF-*02		F-*06	FF-*07		F-*01		FF-*02			
FQ	Flow total		FQ-*03	FQ-*01		FQ-*08	FQ-*07			FQ-*03		FQ-*01	
G	User's choice	G-*01	G-*01	G-*01	G-*07	G-*09	G-*08		G-*05		G-*05		G-*05
H	Hand	H-*01	H-*01	H-*01	H-*08	H-*10	H-*09		H-*06		H-*06		H-*06
I	Current	I-*01	I-*01	I-*01	I-*09	I-*11	I-*10		I-*07		I-*07		I-*07
J	Power	J-*01	J-*01	J-*01	J-*10	J-*12	J-*11		J-*08		J-*08		J-*08
K	Time	K-*01	K-*01	K-*01	K-*11	K-*13	K-*12		K-*09		K-*09		K-*09
L	Level	L-*01	L-*01	L-*01	L-*12	L-*14	L-*13	L-*01		L-*01		L-*01	
M	User's choice	M-*01	M-*01	M-*01	M-*13	M-*15	M-*14		M-*10		M-*10		M-*10
N	User's choice	N-*01	N-*01	N-*01	N-*14	N-*16	N-*15		N-*11		N-*11		N-*11
O	User's choice	O-*01	O-*01	O-*01	O-*15	O-*17	O-*16		O-*12		O-*12		O-*12
P	Pressure		P-*01			P-*18				P-*01		P-*01	
PF	Pressure ratio	P-*01	PF-*02	P-*01	P-*16	PF-*19	P-*17	P-*01		PF-*02			
PK	Pressure schedule		PK-*03			PK-*20				PK-*03		PK-*03	
PD	Pressure difference		PD-*04	PD-*01		PD-*21	PD-*18			PD-*04			
Q	Quantity	Q-*01	Q-*01	Q-*01	Q-*017	Q-*22	Q-*19		Q-*13		Q-*13		Q-*13

R	Radiation	R-*01	R-*01	R-*01	R-*018	R-*23	R-*20		R-*14		R-*14		R-*14
S	Speed	S-*01	S-*01	S-*01	S-*019	S-*24	S-*21		S-*15		S-*15		S-*15
T	Temperature		T-*01			T-*25				T-*01		T-*01	
TF	Temperature ratio	T-*01	TF-*02	T-*01	T-*20	TF-*26	T-*22	T-*01		TF-*02			
TK	Temperature schedule		TK-*03			TK-*27				TK-*03		TD-*01	
TD	Temperature difference		TD-*04	TD-*01		TD-*28	TD-*23			TD-*04			
U	Multivariable	U-*01	U-*01	U-*01	U-*21	U-*29	U-*24		U-*16		U-*16		U-*16
V	Vibration, machine analysis	V-*01	V-*01	V-*01	V-*22	V-*30	V-*25		V-*17		V-*17		V-*17
W	Weight, force		W-*01			W-*31					W-*18		W-*18
WD	Weight difference		WD-*02			WD-*32					WD-*19		WD-*19
WF	Weight ratio	W-*01	WF-*03	W-*01	W-*23	WF-*33	W-*26		W-*18		WF-*20		WF-*20
WK	Weight loss (gain)		WK-*04			WK-*34					WK-*21		WK-*21
WQ	Weight total		WQ-*05			WQ-*35					WQ-*22		WQ-*22
X	Unclassified	X-*01	X-*01	X-*01	X-*24	X-*36	X-*27		X-*19		X-*23		X-*23
Y	Event, state, presence	Y-*01	Y-*01	Y-*01	Y-*25	Y-*37	Y-*28		Y-*20		Y-*24		Y-*24
Z	Position, dimension		Z-*01	Z-*01		Z-*38	Z-*29				Z-*25		Z-*25
ZX	Position, X-axis		ZX-*02	ZX-*01		ZX-*39	ZX-*30				ZX-*26		ZX-*26
ZY	Position, Y-axis		ZY-*03	ZY-*01		ZY-*40	ZY-*31				ZY-*27		ZY-*27
ZZ	Position, Z-axis	Z-*01	ZZ-*04	ZZ-*01	Z-*26	ZZ-*41	ZZ-*32		Z-*21		ZZ-*28		ZZ-*28
ZD	Gauge deviation		ZD-*01	ZD-*01		ZD-*42	ZD-*33				ZD-*29		ZD-*29
ZDX	Gauge X-axis deviation		ZDX-*02	ZDX-*01		ZDX-*43	ZDX-*34				ZDX-*30		ZDX-*30
ZDY	Gauge Y-axis deviation		ZDY-*03	ZDY-*01		ZDY-*44	ZDY-*35				ZDY-*31		ZDY-*31
ZDZ	Gauge Z-axis deviation		ZDZ-*04	ZDZ-*01		ZDZ-*45	ZDZ-*36				ZDZ-*32		ZDZ-*32

Source: See statement of permission following Figure 1.4a.
Note (1): Assignment shown is one of many possibilities.

TABLE 1.4e
Allowable Readout/Passive Function Identification Letter Combinations

		A(1)						B	E	G	I	L	N	O	P	Q	R	W	X
		Absolute Alarms			Deviation Alarms														
First Letters	Measured/Initiating Variable	H	M	L	D	DH	DL	User's Choice	Sensor, Primary Element	Gauge, Glass (2)	Indicate	Light	User's Choice	Orifice Restrict	Point (Test Conn.)	Integrate Totalize	Record	Well	Unclassified
A	Analysis	AAH	AAM	AAL	AAD	AADH	AADL		AE	N/A	AI			N/A	AP	N/A	AR	N/A	
B	Burner combustion	BAH	BAM	BAL	BAD	BADH	BADL		BE	BG	BI	BL		N/A	N/A	N/A	BR	N/A	
C	User's choice	CAH	CAM	CAL	CAD	CADH	CADL		CE	CG	CI	CL					CR		
D	User's choice	DAH	DAM	DAL	DAD	DADH	DADL		DE	DG	DI	DL					DR		
E	Voltage	EAH	EAM	EAL	EAD	EADL	EADL		EE	EG	EI	EL		N/A	EP	N/A	ER	N/A	
F	Flow, flow rate	FAH	FAM	FAL	FAD	FADH	FADL		FE	FG	FI	FL		FO	FP	FQ	FR	N/A	
FF	Flow ratio	FFAH	FFAM	FFAL	FFAD	FFADH	FFADL		FE	N/A	FFI	N/A		N/A	N/A	N/A	FFR	N/A	
FQ	Flow total	FQAH	FQAM	FQAL	FQAD	FQADH	FQADL		N/A	N/A	FQI	N/A		N/A	N/A	N/A	FQR	N/A	
G	User's choice	GAH	GAM	GAL	EAD	GADL	GADL				GI						GR		
H	Hand	N/A	N/A	N/A	N/A	N/A	N/A		N/A	N/A	HI	N/A		N/A	N/A	N/A	HR	N/A	
I	Current	IAH	IAH	IAL	IAD	IADH	IADL		IE	N/A	II	IL		N/A	IP	N/A	IR	N/A	
J	Power	JAH	JAM	JAL	JAD	JADH	JADL		JE	N/A	JI	JL		N/A	JP	JQ	JR	N/A	
K	Time	N/A	N/A	N/A	N/A	N/A	N/A		N/A	N/A	KI	KL		N/A	N/A	KQ	KR	N/A	
L	Level	LAH	LAM	LAL	LAD	LADH	LADL		LE	LG	LI	LL		N/A	LP	N/A	LR	N/A	
M	User's choice	MAH	MAM	MAL	MAD	MADL	MADL				MI						MR		
N	User's choice	NAH	NAM	NAL	NAD	NADL	NADL				NI						NR		
O	User's choice	OAH	OAM	OAL	OAD	OADL	OADL				OI						OR		
P	Pressure	PAH	PAM	PAL	PAD	PADH	PADL		PE	PG	PI	PL		N/A	PP	N/A	PR	N/A	
PD	Pressure differential	PDAH	PDAM	PDAL	PDAD	PDADH	PDADL		PDE	PDG	PDI	PDL		N/A	PDP	N/A	PDR	N/A	
PF	Pressure ratio	PFAH	PFAM	PFAL	PFAD	PFADH	PFADL		N/A		PFI	N/A		N/A	N/A	N/A	PFR	N/A	
PK	Pressure schedule	PKAH	PKAM	PKAL	PKAD	PKADH	PKADL		N/A		PKI	PKL		N/A	N/A	N/A	PKR	N/A	
Q	Quantity	QAH	QAM	QAL	QAD	QADH	QADL		N/A		QI	QL		N/A	N/A	N/A	QR	N/A	
R	Radiation	RAH	RAM	RAL	RAD	RADH	RADL		RE	RG	RI	RL		N/A	RP	RQ	RR	N/A	
S	Speed	SAH	SAM	SAL	SAD	SADH	SADL		SE	SG	SI	N/A		N/A	SP	N/A	SR	N/A	
T	Temperature	TAH	TAM	TAL	TAD	TADH	TADL		TE	TG	TI	TL		N/A	TP	N/A	TR	TW	
TD	Temperature differential	TDAH	TDAM	TDAL	TDAD	TDADH	TDADL		TE	TDG	TDI	TDL		N/A	N/A	N/A	TDR	N/A	

TF	Temperature ratio	TFAH	TFAM	TFAL	TFAD	TFADH	TFADL	N/A	N/A	TFI	N/A	N/A	N/A	N/A	TFR	N/A
TK	Temperature schedule	TKAH	TKAM	TKAL	TKAD	TKADH	TKADL	N/A	N/A	TKI	TKL	N/A	N/A	N/A	TKR	N/A
U	Multivariable	N/A	N/A	N/A	N/A	N/A	N/A	N/A	N/A	N/A	N/A	N/A	N/A	N/A	N/A	N/A
V	Vibration, machine analysis	VAH	N/A	VAL	VAD	VADH	VADL	VE	VG	VI	N/A	N/A	VP	N/A	VR	N/A
W	Weight, force	WAH	WAM	WAL	WAD	WAD	WADL	WE	N/A	WI	WL	N/A	N/A	N/A	WR	N/A
WD	Weight difference	WDAH	WDAM	WDAL	WDAD	WDAD	WDADL	WE	N/A	WDI	WDL	N/A	N/A	N/A	WDR	N/A
WF	Weight ratio	WFAH	WFAM	WFAL	WFAD	WFAD	WFADL	WE	N/A	WFI	N/A	N/A	N/A	N/A	WFR	N/A
WK	Weight loss (gain)	WKAH	WKAM	WKAL	WKAD	WKAD	WKADL	N/A	N/A	WKI	WKL	N/A	N/A	N/A	WKR	N/A
WQ	Weight total	WQAH	WQAM	WQAL	WQAD	WQAD	WQADL	N/A	N/A	WQI	WQL	N/A	N/A	N/A	WQR	N/A
X	Unclassified	XAH	XAM	XAL	XAD	XAD	XADL	XE	XG	XI	XL	N/A	N/A	N/A	XR	N/A
Y	Event, state, presence	YSAH	N/A	YAL	N/A	N/A	N/A	N/A	YG	YI	YL	N/A	N/A	N/A	YR	N/A
Z	Position, dimension	ZAH	ZAM	ZAL	ZAD	ZADH	ZADL	ZE	ZG	ZI	ZL	N/A	N/A	N/A	ZR	N/A
ZX	Position, X-axis	ZXAH	ZXAM	ZXAL	ZXAD	ZXADH	ZXADL	ZXE	ZXG	ZXI	ZXL	N/A	N/A	N/A	ZXR	N/A
ZY	Position, Y-axis	ZYAH	ZYAM	ZYAL	ZYAD	ZYADH	ZYADL	ZYE	ZYG	ZYI	ZYL	N/A	N/A	N/A	ZYR	N/A
ZZ	Position, Z-axis	ZZAH	ZZAM	ZZAL	ZZAD	ZZADH	ZZADL	ZZE	ZZG	ZZI	ZZL	N/A	N/A	N/A	ZZR	N/A
ZD	Gauge deviation	ZDAH	ZDAM	ZDAL	ZDAD	ZDADH	ZDXADL	ZDE	ZDG	ZDI	N/A	N/A	N/A	N/A	ZDR	N/A
ZDX	Gauge X-axis deviation	ZDXAH	ZDXAM	ZDXAL	ZDXAD	ZDXADH	ZDXADL	ZDXE	ZDXG	ZDXI	N/A	N/A	N/A	N/A	ZDXR	N/A
ZDY	Gauge Y-axis deviation	ZDYAH	ZDYAM	ZDYAL	ZDYAD	ZDYADH	ZDYADL	ZDYE	ZDYG	ZDYI	N/A	N/A	N/A	N/A	ZDYR	N/A
ZDZ	Gauge Z-axis deviation	ZDZAH	ZDZAM	ZDZAL	ZDZAD	ZDZADH	ZDZADL	ZDZE	ZDZG	ZDZI	N/A	N/A	N/A	N/A	ZDZR	N/A

Source: See statement of permission following Figure 1.4a.

N/A = not allowed.

Note (1): Alarm combinations are given with Function Modifiers for deviation from set point and absolute values. Adding [H] or [L] forms low–low and high–high alarm Functional Identifications.

Note (2): Readout/Passive Function [G] (glass, gauge) is shown for local direct connected devices, such as flow sight glasses, level glasses, pressure gauges, and thermometers, and also for weigh scales and position indicators. These devices provide a simple view of a process condition. The Readout/Passive Function [I] (indicate) may continue to be used in facilities where it is currently used.

TABLE 1.4f
Allowable Output/Active Function Identification Letter Combinations

		C				K	S			T			U	V	X	Y	Z
		Controller					Switch			Transmitter							
First Letters	Measured/ Initiating Variable	C(1)(2)	IC(3)	RC(3)	CV(4)	Control Station	H	M	L	T	IT	RT	Multi-functiion	Valve Damper Louver	Unclassified	Compute Convert Relay	Actuator Drive
A	Analysis	AC	AIC	ARC	N/A	AK	ASH	ASM	ASL	AT	AIT	ART	AU	AV	AX	AY	
B	Burner, combustion	BC	BIC	BRC	N/A	BK	BSH	BSM	BSL	BT	BIT	BRT	BU	BV	BX	BY	BZ
C	User's choice	CC	CIC	CRC		CK	CSH	CSM	CSL	CT	CIT	CRT	CU	CV	CX	CY	
D	User's choice	DC	DIC	DRC		DK	DSH	DSM	DSL	DT	DIT	DRT	DU	DV	DX	DY	
E	Voltage	EC	EIC	ERC	N/A	EK	ESH	ESM	ESL	ET	EIT	ERT	EU	N/A	EX	EY	EZ
F	Flow, flow rate	FC	FIC	FRC	FCV	FK	FSH	FSM	FSL	FT	FIT	FRT	FU	FV	FX	FY	
FF	Flow ratio	FFC	FFIC	FFRC	N/A	FFK	FFSH	FFSM	FFSL	N/A	N/A	N/A	N/A	N/A	FFX	FFY	
FQ	Flow total	FQC	FQIC	FQRC	FQCV	FQK	FQSH	FQSM	FQSL	FQT	FQIT	FQRT	N/A	FQV	FQX	FQY	
G	User's choice	GC	GIC	GRC		GK	GSH	GSM	GSL	GT	GIT	GRT	GU	GV	GX	GY	
H	Hand	HC	HIC	N/A	HCV	N/A	N/A	N/A	N/A	N/A	N/A	N/A	N/A	HV	HX	HY	
I	Current	IC	IIC	IRC	N/A	IK	ISH	ISM	ISL	IT	IIT	IRT	IU	N/A	IX	IY	IZ
J	Power	JC	JIC	JRC	N/A	JK	JSH	JSM	JSL	JT	JIT	JRT	JU	N/A	JX	JY	JZ
K	Time	KC	KIC	KRC	N/A	N/A	KSH	KSM	KSL	N/A	N/A	N/A	N/A	N/A	KX	KY	
L	Level	LC	LIC	LRC	LCV	LK	LSH	LSM	LSL	LT	LIT	LRT	LU	LV	LX	LY	
M	User's choice	MC	MIC	MRC		MK	MSH	MSM	MSL	MT	MIT	MRT	MU	MV	MX	MY	
N	User's choice	NC	NIC	NRC		NK	NSH	NSM	NSL	NT	NIT	NRT	NU	NV	NX	NY	
O	User's choice	OC	OIC	ORC		OK	OSH	OSM	OSL	OT	OIT	ORT	OU	OV	OX	OY	
P	Pressure	PC	PIC	PRC	PCV	PK	PSH	PSM	PSL	PT	PIT	PRT	PU	PV	PX	PY	
PD	Pressure difference	PDC	PDIC	PDRC	PDCV	PDK	PDSH	PDSM	PDSL	PDT	PDIT	PDRT	PDU	PDV	PDX	PDY	
PF	Pressure ratio	PFC	PFIC	PFRC	N/A	PFK	PFSH	PFSM	PFSL	N/A	N/A	N/A	N/A	N/A	PFX	PFY	
PK	Pressure schedule	PKC	PKIC	PKRC	N/A	PKADH	PKSH	PKSM	PKSL	N/A	N/A	N/A	N/A	N/A	PKX	PKY	
Q	Quantity	QC	QIC	QRC	QCV	QADH	QSH	QSM	QSL	QT	QIT	QRT	QU	N/A	QX	QY	
R	Radiation	RC	RIC	RRC	N/A	RADH	RSH	RSM	RSL	RT	RIT	RRT	RU	RV	RX	RY	

S	Speed	SC	SIC	SRC	SCV	SADH	SSH	SSM	SSL	ST	SIT	SRT	SU	SV	SX	SY	
T	Temperature	TC	TIC	TRC	TCV	TADH	TSH	TSM	TSL	TT	TIT	TRT	TU	TV	TX	TY	
TD	Temperature differential	TDC	TDIC	TDRC	N/A	TDADH	TDSH	TDSM	TDSL	TDT	TDIT	TDRT	TDU	TDV	TDX	TDY	
TF	Temperature ratio	TFC	TFIC	TFRC	N/A	TFADH	TFSH	TFSM	TFSL	N/A	N/A	N/A	N/A	N/A	TFX	TFY	
TK	Temperature schedule	TKC	TKIC	TKRC	N/A	TKADH	TKSH	TKSM	TKSL	N/A	N/A	N/A	N/A	N/A	TKX	TKY	
U	Multivariable	UC	UIC	URC	N/A	N/A	USH	USM	USL	UT	N/A	N/A	N/A	N/A	UX	UY	
V	Vibration, machine analysis	VC	VIC	VRC	N/A	VADH	VSH	VSM	VSL	VT	VIT	VRT	VU	VV	VX	VY	
W	Weight, force	WC	WIC	WRC	WCV	WAD	WSH	WSM	WSL	WT	WIT	WRT	WU	WV	WX	WY	
WD	Weight difference	WDC	WDIC	WDRC	N/A	WDAD	WDSH	WDSM	WDSL	WDT	WDIT	WDRT	WDU	N/A	WDX	WDY	
WF	Weight ratio	WFC	WFIC	WFRC	N/A	WFAD	WFSH	WFSM	WFSL	N/A	N/A	N/A	N/A	N/A	WFX	WFY	
WK	Weight loss (gain)	WKC	WKIC	WKRC	N/K	WKAD	WKSH	WKSM	WKSL	N/A	N/A	N/A	N/A	N/A	WKX	WKY	
WQ	Weight total	WQC	WQIC	WQRC	N/A	WQAD	WQSH	WQSM	WQSL	N/A	N/A	N/A	N/A	N/A	WQX	WQY	
X	Unclassified	XC	XIC	XRC	N/A	XAD	XSH	XSM	XSL	XT	XIT	XRT	XU	XV	XX	XY	XZ
Y	Event, state, presence	YC	YIC	YRC	N/A	N/A	YSH	YSM	YSL	YT	YIT	YRT	YU	N/A	YX	YY	YZ
Z	Position, dimension	ZC	ZIC	ZRC	N/A	ZADH	ZSH	ZSM	ZSL	ZT	ZIT	ZRT	ZU	ZV	ZX	ZY	ZZ
ZX	Position, X-axis	ZXC	ZXIC	ZXRC	N/A	ZXADH	ZXSH	ZXSM	ZXSL	ZXT	ZXIT	ZXRT	N/A	ZXV	ZXX	ZXY	ZXZ
ZY	Position, Y-axis	ZYC	ZYIC	ZYRC	N/A	ZYADH	ZYSH	ZYSM	ZYSL	ZYT	ZYIT	ZYRT	N/A	ZYV	ZYX	ZYY	ZYZ
ZZ	Position, Z-axis	ZZC	ZZIC	ZZRC	N/A	ZZADH	ZZSH	ZZSM	ZZSL	ZZT	ZZIT	ZZRT	N/A	ZZV	ZZX	ZZY	ZZZ
ZD	Gauge deviation	ZDC	ZDIC	ZDRC	N/A	ZDADH	ZDSH	ZDSM	ZDSL	ZDT	ZDIT	ZDRT	N/A	ZDV	ZDX	ZDY	ZDZ
ZDX	Gauge X-axis deviation	ZDXC	ZDXIC	ZDXRC	N/A	ZDXADH	ZDXSH	ZDXSM	ZDXSL	ZDXT	ZDXIT	ZDXRT	N/A	ZDXV	ZDXX	ZDXY	ZDXZ
ZDY	Gauge Y-axis deviation	ZDYC	ZDYIC	ZDYRC	N/A	ZDYK	ZDYSH	ZDYSM	ZDYSL	ZDYT	ZDYIT	ZDYRT	N/A	ZDYV	ZDYX	ZDYY	ZDYZ
ZDZ	Gauge Z-axis deviation	ZDZC	ZDZIC	ZDZRC	N/A	ZDZK	ZDZSH	ZDZSM	ZDZSL	ZDZT	ZDZIT	ZDZRT	N/A	ZDZV	ZDZX	ZDZY	ZDZZ

Source: See statement of permission following Figure 1.4a.

N/A = not allowed.

Note (1): The combinations in the [C] column do not have operator visible indication of measured variable, set point, or output signal, when used with discrete hardware single case instruments.

Note (2): The combinations in the [C] column may also be used for a controller function configured in a shared or distributed control system.

Note (3): The combinations in the [IC] and [RC] columns indicate the order to be followed in forming the Functional Identification of a controller device or function that also provides indication or recording.

Note (4): The combinations in the [CV] column indicate the order to be followed in forming the Functional Identification for self-actuated control values.

TABLE 1.4g
Instrument Line Symbols (proposed for the next revision of ISA S5.1, now ANSI/ISA-5.01.01 at the time of this writing)

No.	Symbol	Application
01		1. Instrument impulse line from process 2. Instrument impulse line from equipment 3. Analyzer sample line from process 4. Functional instrument diagram signal lines
02	ST	1. Heat [cool] traced instrument impulse line from process 2. Heat [cool] traced instrument impulse line from equipment 3. Heat [cool] traced analyzer sample line from process 4. Type of tracing may be indicated as ET = electrical, RT = refrigerated, ST = steam, etc.
03		1. Generic instrument impulse line connected to process line 2. Generic instrument impulse line connected to equipment
04		1. Heat [cool] traced generic instrument impulse line connected to process line 2. Heat [cool] traced generic instrument impulse line connected to equipment 3. Process line or equipment may or may not be traced
05		1. Heat [cool] traced instrument connected to process impulse line 2. Instrument impulse line may or may not be traced
06		1. Flanged instrument connection to process line 2. Flanged instrument connection to equipment
07		1. Threaded instrument connection to process line 2. Threaded instrument connection to equipment
08		1. Socket welded instrument connection to process line 2. Socket welded instrument connection to equipment
09		1. Welded instrument connection to process line 2. Welded instrument connection to equipment
10	AS	1. Instrument air supply 2. Indicate supply pressure as required, e.g., AS-60psig, AS-400kPa, etc. 3. IA [instrument air] or PA [plant air] may be used for AS 4. Use as required
11	ES	1. Instrument electric power supply 2. Indicate voltage and type as required, e.g., ES-24VDC, ES-120VAC, etc. 3. Use as required
12		1. Undefined signal 2. Use for PFDs 3. Use for discussions or diagrams where type of signal, pneumatic or electronic, is not of concern
13		1. Pneumatic signal
14		1. Electric signal 2. Electronic signal 3. Functional instrument diagram signal lines
15		1. Hydraulic signal
16		1. Filled thermal element capillary tube
17		1. Guided electromagnetic signal 2. Fiber-optic cable 3. Guided sonic signal
18		1. Unguided electromagnetic signal 2. Unguided sonic signal 3. Alternate radio communication link; see symbol 22

TABLE 1.4g Continued
Instrument Line Symbols (proposed for the next revision of ISA S5.1, now ANSI/ISA-5.01.01 at the time of this writing)

No.	Symbol	Application
19	—o——o—	1. Communication link or system bus, between devices and functions of a microprocessor-based system 2. System internal software link
20	—●——●—	1. Shared communication link or bus (not system bus) between two or more independent microprocessor-based systems 2. Shared data link from/between field-located microprocessor-based devices or functions
21	--o-----o-·-	1. Dedicated communication link or bus (not system bus) between two or more independent microprocessor-based systems 2. Dedicated data link from a field-located microprocessor-based device and/or function
22	N N	1. Dedicated radio communications link (not systems bus) between radio transmitting and receiving devices or systems 2. Unguided radio signal 3. Alternate unguided electromagnetic signal; see symbol 18
23	—◉——◉—	1. Mechanical link or connection
24		1. Signal connector 2. Drawing-to-drawing signal connector 3. Internal signal connector used to avoid long signal lines
25		1. Signal connector 2. Internal signal connector used to avoid long signal lines 3. Drawing-to-drawing signal connector

Source: See statement of permission following Figure 1.4a.

Table 1.4i covers analog, digital, and/or discrete shared control devices and/or functions for continuous control, indication, calculation, etc. that are microprocessor based and configurable. They communicate with each other and share control or display functions in applications like distributed control and programmable logic systems.

Limited operator accessibility (set point changes, control mode transfers, etc.) and unlimited engineer accessibility is through local- or wide-area communications networks, keyboards, and video displays as shown.

Table 1.4j deals with analog, digital, or discrete control devices or functions for on–off or binary control, indication, calculation, etc. that are microprocessor based and configurable. They communicate with each other and share control or display in distributed control and programmable logic systems.

Limited operator accessibility (set point changes, control mode transfers, etc.) and unlimited engineer accessibility is through local- and/or wide-area communications networks, keyboards, and video displays as also shown.

The devices and functions in Table 1.4k include process plant computer-implemented regulatory and/or advanced control analog/digital/discrete control and indication functions that are main frame computer or mini-computer based.

Limited operator accessibility (set point changes, control mode transfers, etc.), and unlimited engineer accessibility is through local and/or wide area communications networks, keyboards and video displays are also shown.

Multipoint, Multifunction, and Multivariable Devices and Loops

Multipoint devices are indicators or recorders that may be single or multivariable and receive input signals from two or more primary elements or transmitters.

Multifunction devices are controllers or switches that receive input signals from two or more primary elements or transmitters and control two or more manipulated variables.

Multivariable devices are indicators, recorders, or controllers that receive input signals from two or more primary elements or transmitters and control one manipulated variable. Single-variable or multivariable multipoint indicators and recorders for two or three points shall be drawn with bubbles either:

a) Tangent to each other in the same order, left to right, as the pen or pointer assignments.

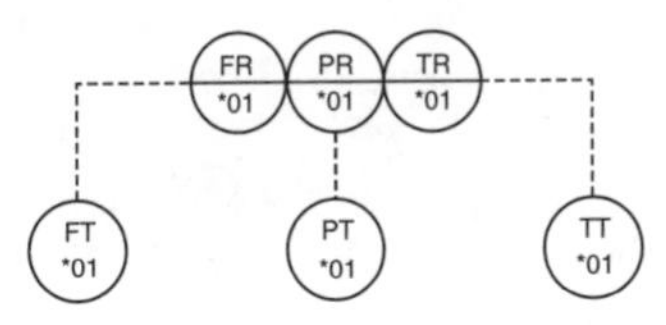

b) Separate from each other, with pen or pointer number indicated preferably in upper right or left quadrant and

TABLE 1.4h

Discrete Devices and/or Functions (proposed for the next revision of ISA S5.1, now ANSI/ISA-5.01.01 at the time of this writing)

No.	*Symbol*	*Location and Accessibility*
01		1. Field or locally mounted 2. Not panel or cabinet mounted 3. Normally accessible to an operator
02		1. Central or main control room 2. Front of main panel mounted 3. Normally accessible to an operator
03		1. Central or main control room 2. Rear of main panel mounted 3. Not normally accessible to an operator
04		1. Secondary or local control room 2. Field or local control panel 3. Front of secondary or local panel mounted 4. Normally accessible to an operator
05		1. Secondary or local control room 2. Field or local control panel 3. Rear of secondary or local panel or cabinet mounted 4. Not normally accessible to an operator
06		1. Signal processor identifier located in upper right or left quadrant of symbols above 2. Signal processor identifier attached to symbols where affected signals are connected

Source: See statement of permission following Figure 1.4a.

a note defining instrument or device indicated in preferably lower right or left quadrant.

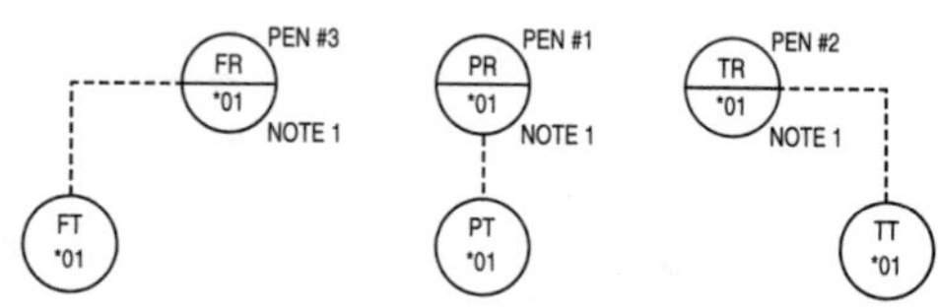

Multipoint indicators and recorders for four or more points may be drawn with bubbles separate from each other, with point number indicated by adding a suffix to the tag numbers:

a) Single variable:

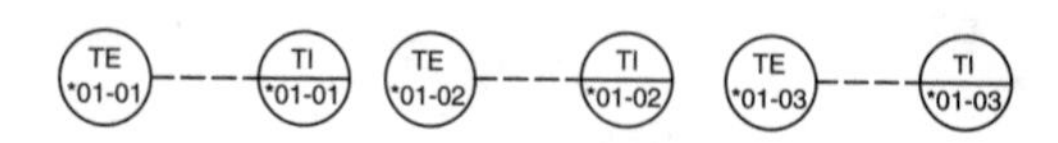

b) Multivariable:

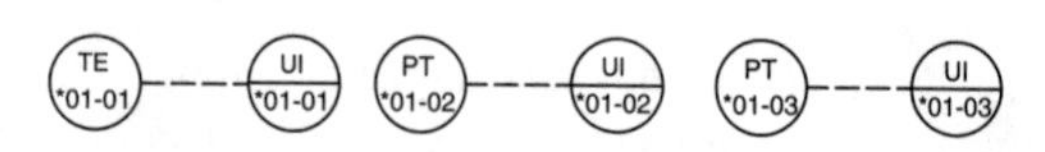

TABLE 1.4i

Shared Continuous Devices and/or Functions (proposed for the next revision of ISA S5.1, now ANSI/ISA-5.01.01, at the time of this writing)

No.	*Symbol*	*Location and Accessibility*
01		1. Dedicated single function device 2. Field or locally mounted 3. Not panel or cabinet mounted 4. Normally accessible to an operator at device
02		1. Central or main console 2. Visible on video display 3. Normally accessible to an operator at console
03		1. Central or main console 2. Not visible on video display 3. Not normally accessible to an operator at console
04		1. Secondary or local console 2. Field or local control panel 3. Visible on video display 4. Normally accessible to an operator at console
05		1. Secondary or local console 2. Field or local control panel 3. Not visible on video display 4. Not normally accessible to an operator at console
06		1. Mathematical function located in upper right or left quadrant of symbols above 2. Mathematical function attached to symbols where affected signals are connected

Source: See statement of permission following Figure 1.4a.

Multivariable controllers may be drawn with bubbles for each measured variable input and for the output to the final control element; measured variable indicators may be

a) Shown:

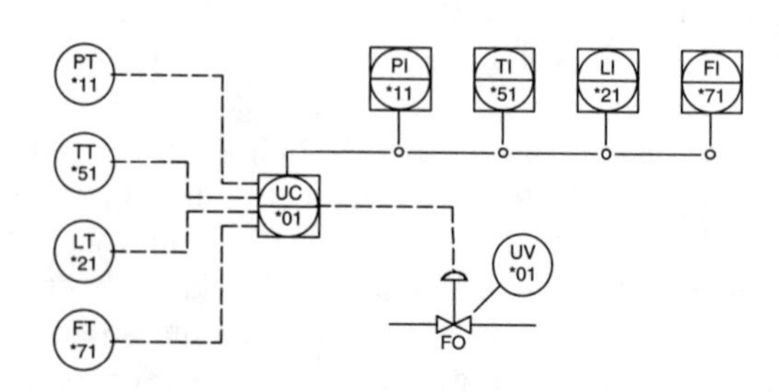

TABLE 1.4j

Shared On–Off Devices and/or Functions (proposed for the next revision of ISA S5.1, now ANSI/ISA-5.01.01, at the time of this writing)

No.	Symbol	Location and Accessibility
01		1. Field or locally mounted 2. Not panel or cabinet mounted 3. Normally accessible to an operator at device
02		1. Central or main console 2. Visible on video display 3. Normally accessible to an operator at console
03		1. Central or main console 2. Not visible on video display 3. Not normally accessible to an operator at console
04		1. Secondary or local console 2. Field or local control panel 3. Visible on video display 4. Accessible to an operator at console
05		1. Secondary or local console 2. Field or local control panel 3. Not visible on video display 4. Not normally accessible to an operator at console
06		1. Mathematical function located in upper right or left quadrant of symbols above 2. Mathematical function attached to symbols where affected signals are connected

Source: See statement of permission following Figure 1.4a.

b) Assumed:

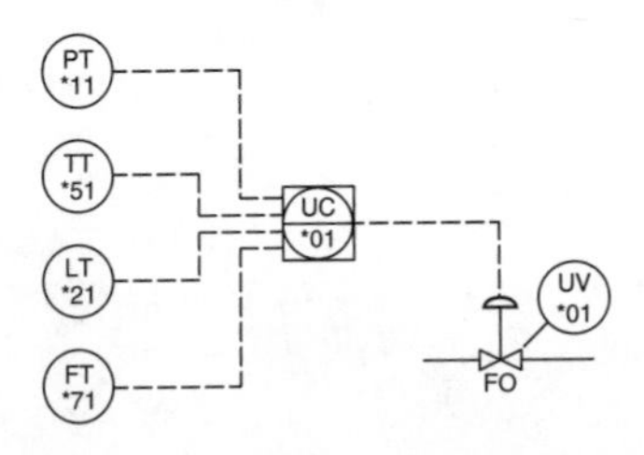

Multifunction controllers shall be drawn with bubbles for each measured variable input and output to final control elements; measured variable indicators may be

a) Shown:

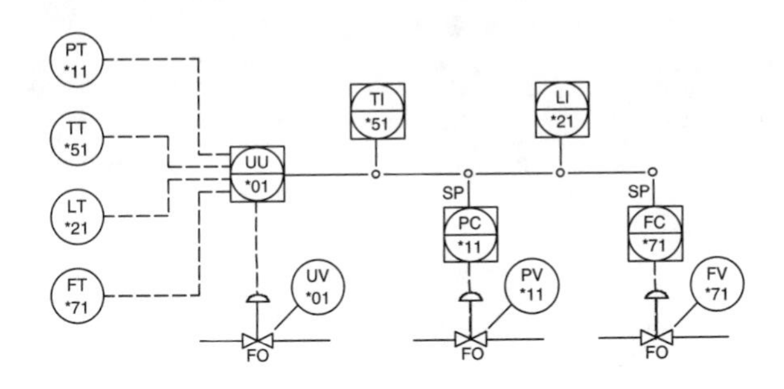

TABLE 1.4k

Computer Devices and/or Functions (proposed for the next revision of ISA S5.1, now ANSI/ISA-5.01.01, at the time of this writing)

No.	Symbol	Location and Accessibility
01		1. Undefined location 2. Undefined visibility 3. Undefined accessibility
02		1. Central or main computer 2. Visible on video display 3. Normally accessible to an operator at console or computer terminal
03		1. Central or main computer 2. Not visible on video display 3. Not normally accessible to an operator at console or computer terminal
04		1. Secondary or local computer 2. Visible on video display 3. Normally accessible to an operator at console or computer terminal
05		1. Secondary or local computer 2. Not visible on video display 3. Not normally accessible to an operator at console or computer terminal

Source: See statement of permission following Figure 1.4a.

b) Assumed:

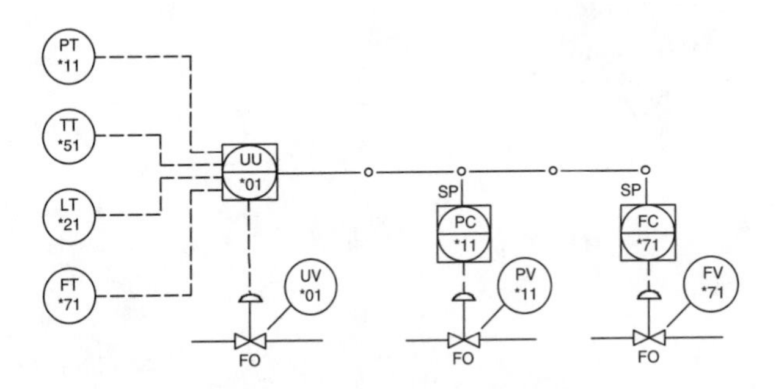

Fieldbus Devices, Loops, and Networks

Comments and Exceptions (Including Non-ISA Industrial Practice) Instrument and control systems staff working at engineering, procurement, and construction management (EPCM) companies have had to improvise on P&ID symbols for fieldbus devices, loops, segments, and networks throughout the late 1990s and early 2000 years. This has happened while waiting for the draft standard work outlined in this section to be discussed and approved as the latest revision to ISA S5.1 (now ANSI/ISA-5.01.01). (For specific details on fieldbus technologies, refer to later chapters and sections in this volume.)

Certain techniques and shortcuts used by several EPCM companies and how they have handled fieldbus symbology will be mentioned in this subsection. A few companies have generated their P&IDs using the proposed Instrument Line Symbol No. 20 (see Table 1.4g) as the shared data links or Foundation Fieldbus (FF) segments between field located microprocessor-based devices as well as FF host systems. In this way, it is implicit that the devices connected together by that symbol are fieldbus devices and do not need any further symbology or identification on the P&IDs. This symbol has also been used for other fieldbuses such as PROFIBUS-PA, PROFIBUS-DP, AS-i Bus, and DeviceNet (Figure 1.4l and 1.4m).

Another symbol used by the EPCM companies for fieldbus has been the Instrument Link Symbol No. 19 (see Table 1.4g), which is the current existing symbol (ANSI/ISA S5.1-1984 (R1992)) and normally the one used for data links and DCS data highways. This has been done at times when the EPCM company client/owner had specific, custom P&ID symbology standards and was reluctant to change a worldwide standard to a new symbol such as No. 20. Once again, any field devices such as transmitters and control valves that are connected together by Symbol No. 19 are now known to be fieldbus-type devices. The disadvantage is the P&IDs must be studied carefully to determine real communication link, DCS data highway, system bus, or internal software link applications from the fieldbus applications.

Another EPCM company used the conventional analog electronic signal (Symbol No. 14 in Table 1.4g) at the urging of its client, but added the suffix "FB" to each fieldbus device bubble on the P&IDs. Once again, this was not a standardized approach and led to ambiguity and misunderstanding. It is highly recommended that the proposed draft revision Instrument Line Symbol No. 20 (which we hope will be approved by the time this volume is released) be used for all types of fieldbus segments and networks.

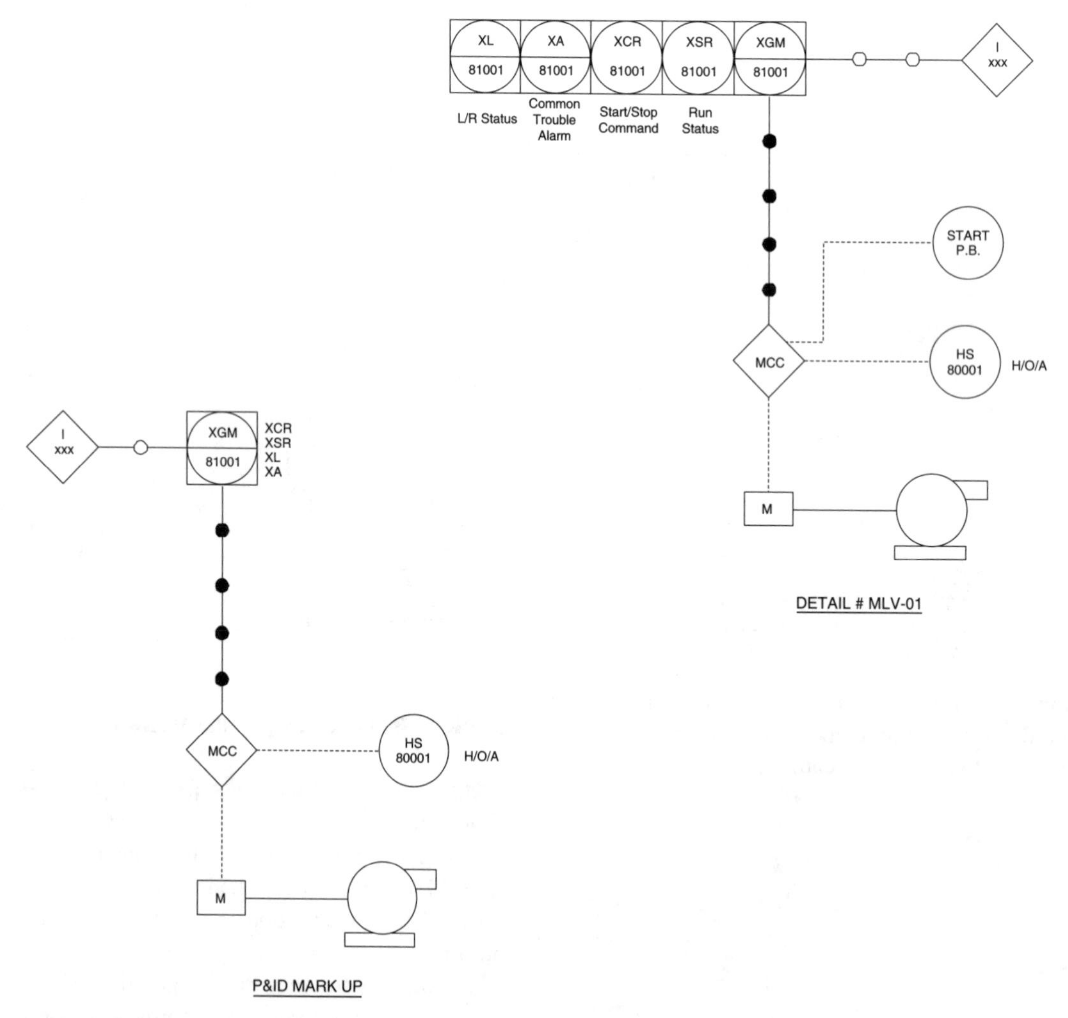

FIG. 1.4l
Low voltage motor control on DeviceNet (detail and P&ID mark-up).

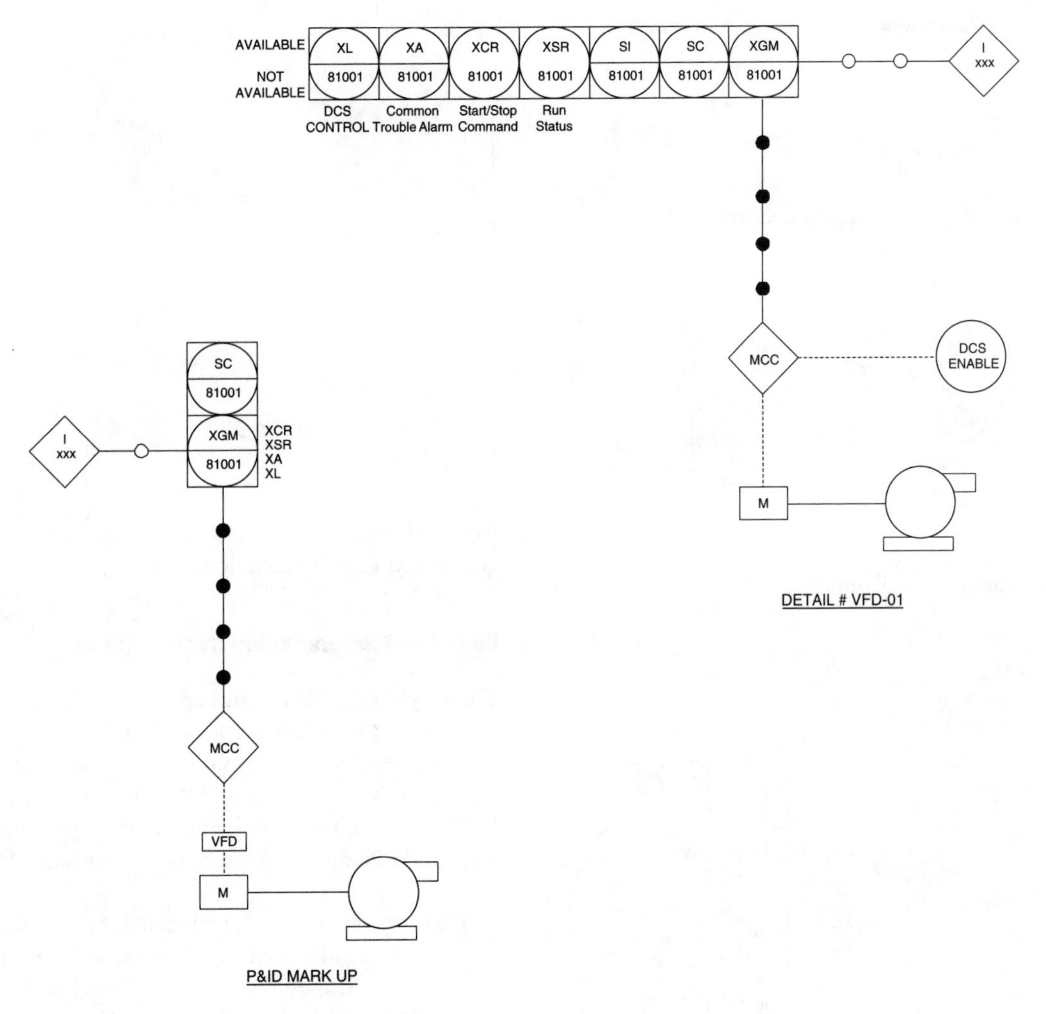

FIG. 1.4m
VFD motor control on DeviceNet (detail and P&ID mark-up).

Fieldbus P&ID Examples: DeviceNet Figures 1.4l and 1.4m are the practical methods of one EPCM company of establishing a P&ID detail and mark-up for a low voltage motor control plus a VFD motor control implemented with DeviceNet as the fieldbus. It should be pointed out that these figures do not completely conform to the ISA S5.1 (now ANSI/ISA-5.01.01) proposed standard and are a compromise born out of necessity.

FUNCTIONAL DIAGRAMMING FOR DIGITAL SYSTEMS (ex-SAMA)

 From ISA Draft Standard 5.01.01, Instrumentation Symbols and Identification, Draft 4, which was not a final and approved standard at the time of publication of this handbook. To obtain an updated copy of ANSI/ISA-5.01.01, please contact ISA Standards, www.isa.org.

Instrument and Control Systems Functional Diagramming

Symbol tables are given for use in preparing instrument and control loop functional diagrams, which are not normally shown on PFDs and P&IDs. Functional instrument and loop diagrams used to depict functionally monitoring and control loops in functional instrument diagrams, functional logic diagrams, application software diagrams, sketches, and text shall be prepared from:

a) Instrument line symbols
b) Instrument functional diagramming symbols
c) Mathematical function block symbols

Equivalent Loop, Functional Instrument, and Electrical Diagrams*

a) Loop schematic:

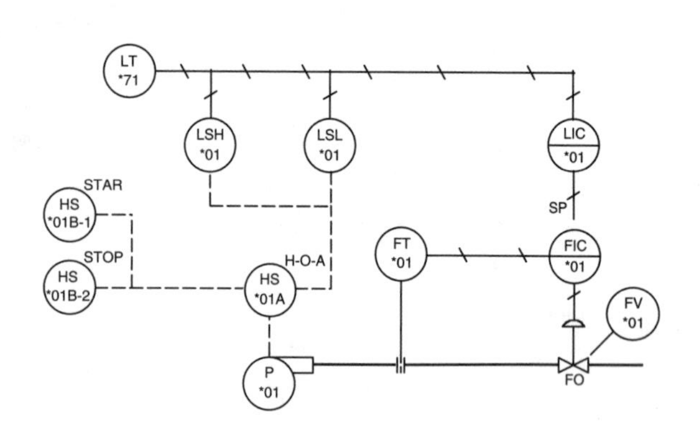

b) Functional instrument diagram:

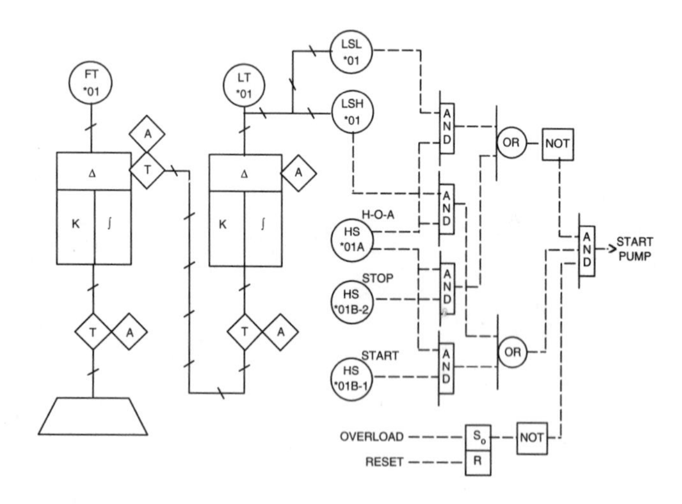

* See statement of permission following Figure 1.4a.

c) Electrical schematic diagram:

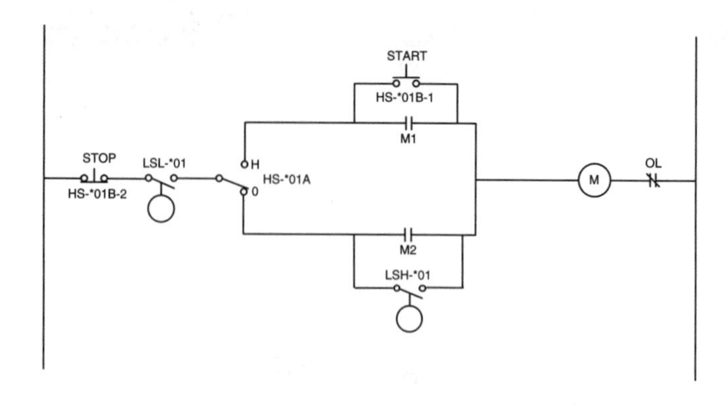

Note: No equivalent electrical schematic exists for the process control instrumentation.

Functional Diagramming Symbol Tables**

These symbols (Table 1.4n) are not normally used on P&IDs but are used to diagram control systems at the hardware and function level for configuration and other purposes.

These symbols (Table 1.4o) are not normally used on P&IDs but are used to diagram control systems at the hardware and function level for configuration and other purposes:

> Binary logic switching and memory functions in analog or sequential control schemes. In Truth Tables and Graphs: Logic One ("1") is "true" and Logic Zero ("0") is "false."

**Symbols extracted by ISA, with permission, from Scientific Apparatus Manufacturers' Association SAMA Standard RC22-11-1966 (R1973) Functional Diagramming of Instrument and Control Systems, which is no longer supported by SAMA.

TABLE 1.4n
Functional Diagramming Symbols—Instrument and Mathematical Functions (proposed for the next revision of ISA S5.1, now ANSI/ISA-5.01.01, at the time of this writing)

No.	Symbol	Description
01	○	1. Measuring device 2. Input device 3. Readout device 4. Output device 5. Symbols from Tables 1.4h to 1.4k may be used
02	□	1. Automatic controller 2. Single mode controller 3. Discrete device driver 4. Insert function symbols, as required to define controller algorithm, from Table 1.4p 5. Use for vertical diagramming

TABLE 1.4n Continued

Instrument and Mathematical Functions (proposed for the next revision of ISA S5.1, now ANSI/ISA-5.01.01, at the time of this writing)

No.	Symbol	Description
03		1. Automatic controller 2. Two-mode controller 3. Insert function symbols, as required to define controller algorithm, from Table 1.4p 4. Use for vertical diagramming
04		1. Automatic controller 2. Three-mode controller 3. Insert function symbols, as required to define controller algorithm, from Table 1.4p 4. Use for vertical diagramming
05		1. Automatic signal processor 2. Insert function symbol from Table 1.4p 3. Use for vertical diagramming
06		1. Automatic controller 2. Two-mode controller 3. Insert function symbols, as required to define controller algorithm, from Table 1.4p 4. Use for horizontal diagramming
07		1. Automatic controller 2. Two-mode controller 3. Insert function symbols, as required to define controller algorithm, from Table 1.4p 4. Use for horizontal diagramming
08		1. Automatic signal processor 2. Insert function symbol from Table 1.4p 3. Use for horizontal diagramming 4. May be rotated 90°
09		1. Final control element 2. Control valve 3. Insert function symbol identifier from Table 1.4p (No. 14)
10		1. Final control element with positioner 2. Control valve with positioner 3. Insert function symbol identifier from Table 1.4p (No. 14)
11	*	1. Manual signal processor 2. (*) = A, adjustable signal generator (*) = T, signal transfer
12	A T	1. Manual auto station

Source: See statement of permission following Figure 1.4a.

TABLE 1.4o

Functional Diagramming Symbols—Binary Logic, Memory, and Time Functions (proposed for the next revision of ISA S5.1, now ANSI/ISA-5.01.01, at the time of this writing)

No.	Symbol/Truth Table	Definition/Graph

01

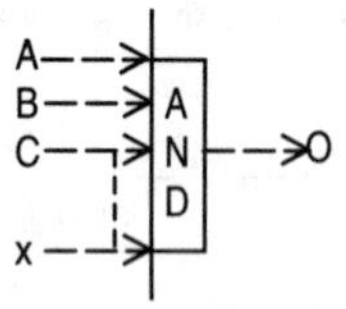

1. AND gate
2. Output true only if all inputs are true

	A	B	C	x	O
1	0	0	0	0	0
2	1	0	0	0	0
3	0	1	0	0	0
4	0	0	1	0	0
5	0	0	0	1	0
6	1	1	0	0	0
7	1	0	1	0	0
8	1	0	0	1	0
9	0	1	1	0	0
10	0	1	0	1	0
11	0	0	1	1	0
12	1	1	1	0	0
13	1	1	0	1	0
14	1	0	1	1	0
15	0	1	1	1	0
16	1	1	1	1	1

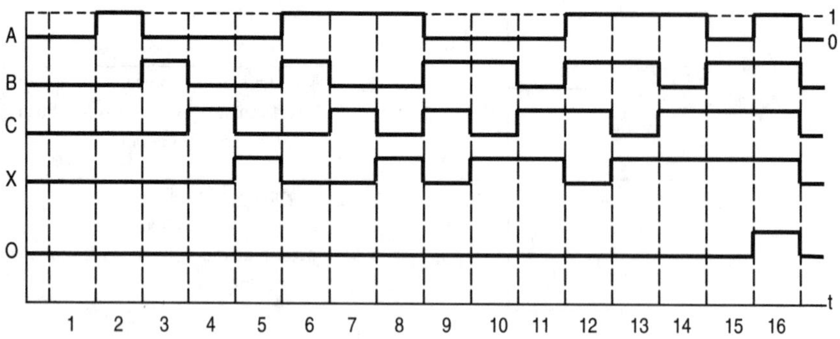

02

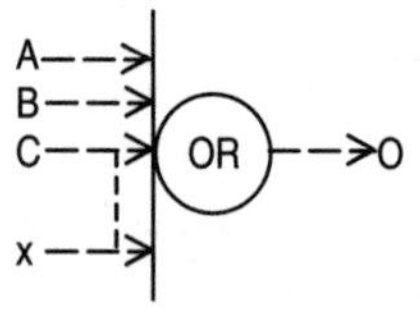

1. OR gate
2. Output true if any input true

	A	B	C	x	O
1	0	0	0	0	0
2	1	0	0	0	1
3	0	1	0	0	1
4	0	0	1	0	1
5	0	0	0	1	1
6	1	1	0	0	0
7	1	0	1	0	0
8	1	0	0	1	0
9	0	1	1	0	0
10	0	1	0	1	0
11	0	0	1	1	0
12	1	1	1	0	0
13	1	1	0	1	0
14	1	0	1	1	0
15	0	1	1	1	0
16	1	1	1	1	0

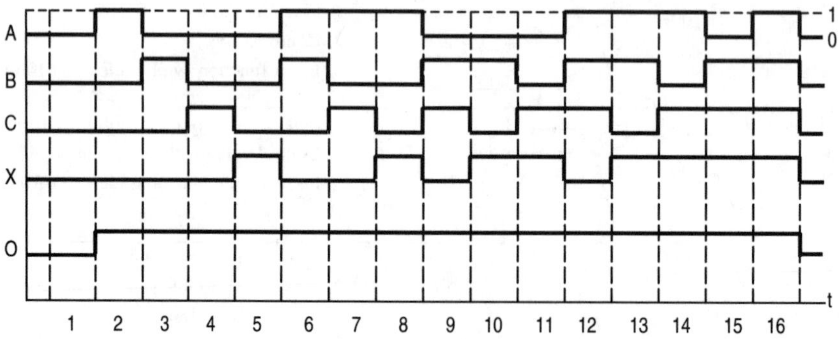

TABLE 1.4o Continued

Functional Diagramming Symbols—Binary Logic, Memory, and Time Functions (proposed for the next revision of ISA S5.1, now ANSI/ISA-5.01.01, at the time of this writing)

No.	Symbol/Truth Table	Definition/Graph

03

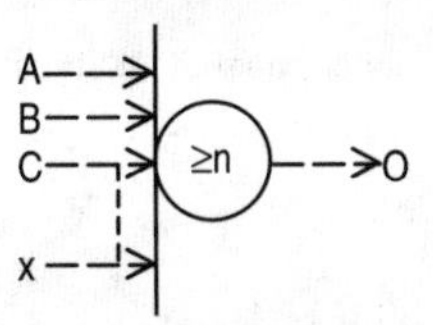

1. Qualified OR gate with greater than or equal to qualifications
2. Output equals "1" if number of inputs equal to "1" are greater than or equal to "n" inputs
3. Truth table and graph are for "n" equal 2

	A	B	C	x	O
1	0	0	0	0	0
2	1	0	0	0	0
3	0	1	0	0	0
4	0	0	1	0	0
5	0	0	0	1	0
6	1	1	0	0	1
7	1	0	1	0	1
8	1	0	0	1	1
9	0	1	1	0	1
10	0	1	0	1	1
11	0	0	1	1	1
12	1	1	1	0	1
13	1	1	0	1	1
14	1	0	1	1	1
15	0	1	1	1	1
16	1	1	1	1	1

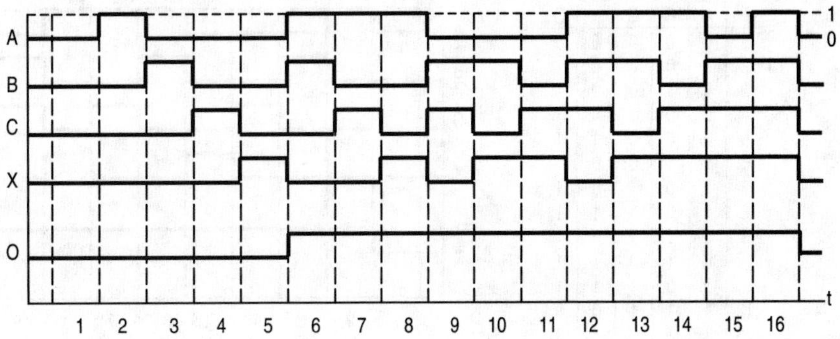

04

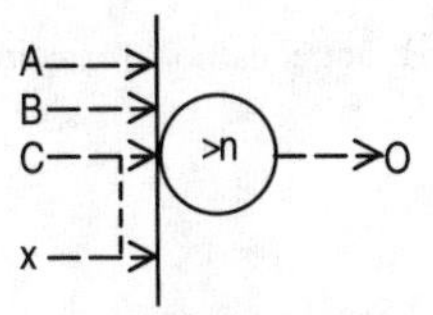

1. Qualified OR gate with greater than qualifications
2. Output equals "1" if number of inputs equal to "1" are greater but not equal to "n" inputs
3. Truth table and graph are for "n" equal 2

	A	B	C	x	O
1	0	0	0	0	0
2	1	0	0	0	0
3	0	1	0	0	0
4	0	0	1	0	0
5	0	0	0	1	0
6	1	1	0	0	1
7	1	0	1	0	1
8	1	0	0	1	1
9	0	1	1	0	1
10	0	1	0	1	1
11	0	0	1	1	1
12	1	1	1	0	1
13	1	1	0	1	1
14	1	0	1	1	1
15	0	1	1	1	1
16	1	1	1	1	1

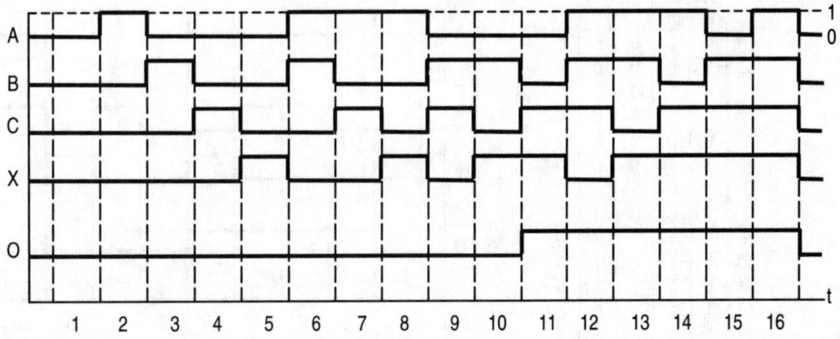

TABLE 1.4o Continued

Functional Diagramming Symbols—Binary Logic, Memory, and Time Functions (proposed for the next revision of ISA S5.1, now ANSI/ISA-5.01.01, at the time of this writing)

No.	Symbol/Truth Table	Definition/Graph

05

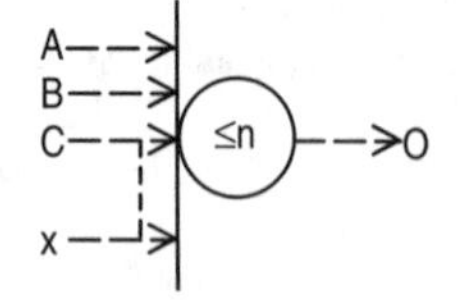

1. Qualified OR gate with less than or equal to qualifications
2. Output equals "1" if number of inputs equal to "1" are less than or equal to "n" inputs
3. Truth table and graph are for "n" equal 2

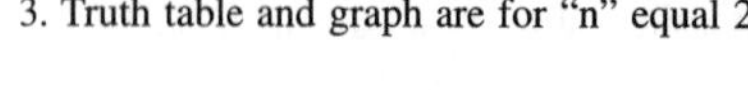

	A	B	C	x	O
1	0	0	0	0	0
2	1	0	0	0	1
3	0	1	0	0	1
4	0	0	1	0	1
5	0	0	0	1	1
6	1	1	0	0	1
7	1	0	1	0	1
8	1	0	0	1	1
9	0	1	1	0	1
10	0	1	0	1	1
11	0	0	1	1	1
12	1	1	1	0	0
13	1	1	0	1	0
14	1	0	1	1	0
15	0	1	1	1	0
16	1	1	1	1	0

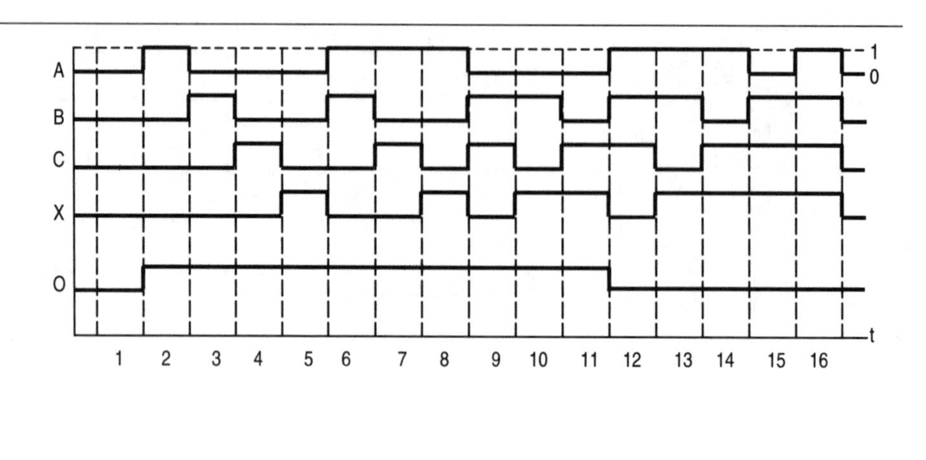

06

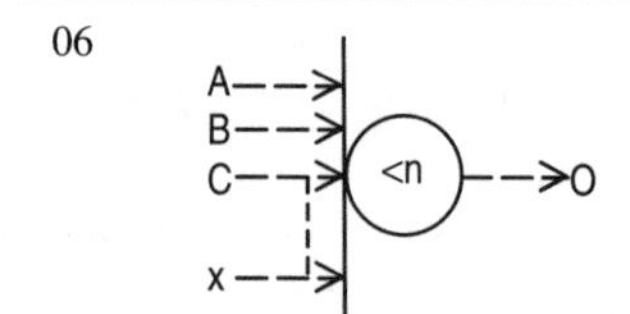

1. Qualified OR gate with less than qualifications
2. Output equals "1" if number of inputs equal to "1" are less but not equal to "n" inputs
3. Truth table and graph are for "n" equal 2

	A	B	C	x	O
1	0	0	0	0	0
2	1	0	0	0	1
3	0	1	0	0	1
4	0	0	1	0	1
5	0	0	0	1	1
6	1	1	0	0	0
7	1	0	1	0	0
8	1	0	0	1	0
9	0	1	1	0	0
10	0	1	0	1	0
11	0	0	1	1	0
12	1	1	1	0	0
13	1	1	0	1	0
14	1	0	1	1	0
15	0	1	1	1	0
16	1	1	1	1	0

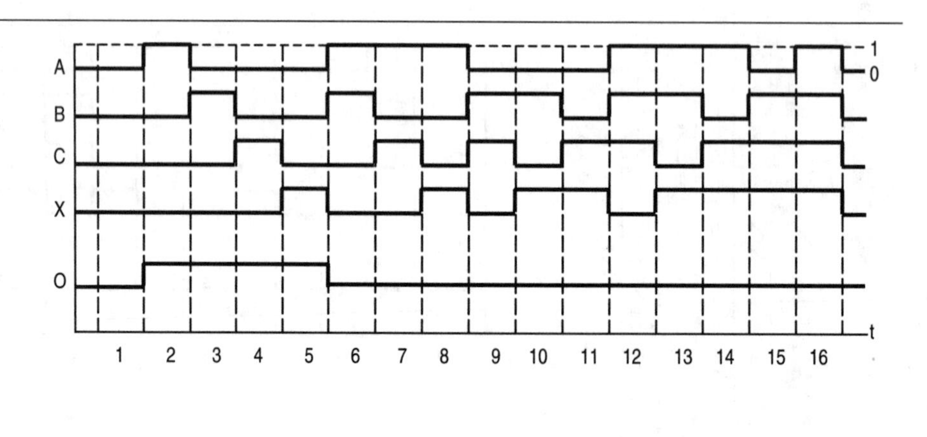

TABLE 1.4o Continued

Functional Diagramming Symbols—Binary Logic, Memory, and Time Functions (proposed for the next revision of ISA S5.1, now ANSI/ISA-5.01.01, at the time of this writing)

No.	Symbol/Truth Table	Definition/Graph
07	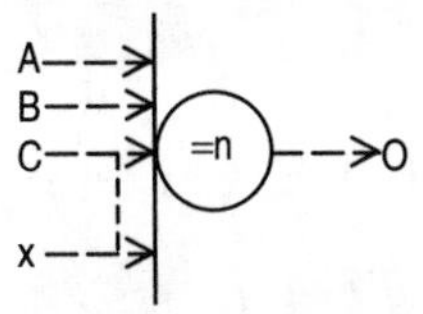 A, B, C, x → (=n) → O	1. Qualified OR gate with equal to qualifications 2. Output equals "1" if inputs equal to "1" are equal to "n" inputs 3. Truth table and graph are for "n" equal 2

	A	B	C	x	O
1	0	0	0	0	0
2	1	0	0	0	0
3	0	1	0	0	0
4	0	0	1	0	0
5	0	0	0	1	0
6	1	1	0	0	1
7	1	0	1	0	1
8	1	0	0	1	1
9	0	1	1	0	1
10	0	1	0	1	1
11	0	0	1	1	1
12	1	1	1	0	0
13	1	1	0	1	0
14	1	0	1	1	0
15	0	1	1	1	0
16	1	1	1	1	0

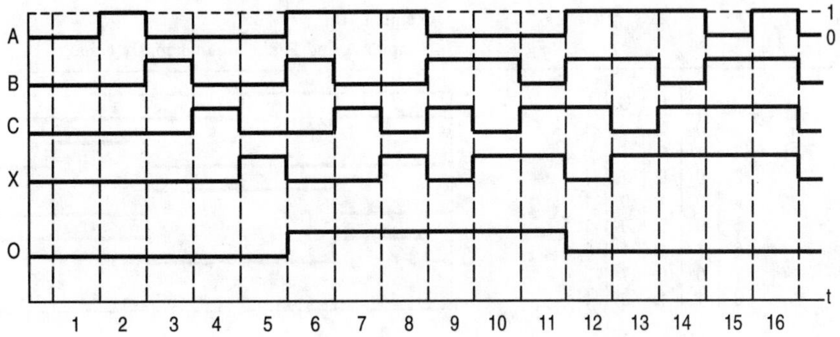

No.	Symbol/Truth Table	Definition/Graph
08	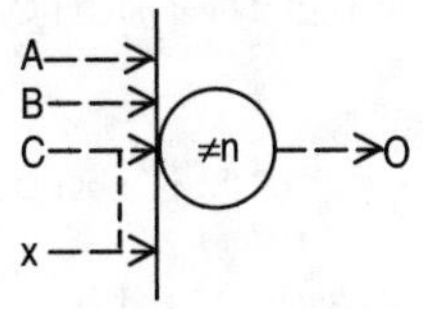 A, B, C, x → (≠n) → O	1. Qualified OR gate with not equal to qualifications 2. Output equals "1" if inputs equal to "1" are not equal to "n" inputs 3. Truth table and graph are for "n" equal 2

	A	B	C	x	O
1	0	0	0	0	0
2	1	0	0	0	1
3	0	1	0	0	1
4	0	0	1	0	1
5	0	0	0	1	1
6	1	1	0	0	0
7	1	0	1	0	0
8	1	0	0	1	0
9	0	1	1	0	0
10	0	1	0	1	0
11	0	0	1	1	0
12	1	1	1	0	1
13	1	1	0	1	1
14	1	0	1	1	1
15	0	1	1	1	1
16	1	1	1	1	1

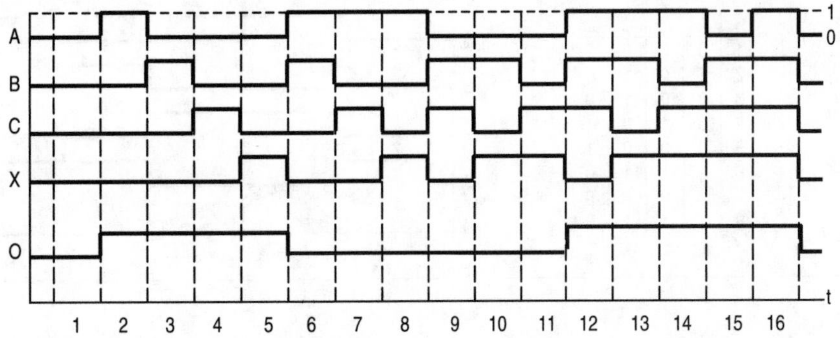

No.	Symbol/Truth Table	Definition/Graph
09	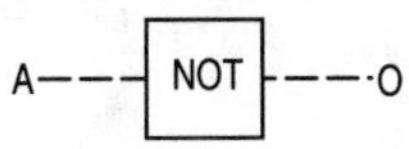 A → NOT → O	1. NOT gate 2. Output false if input true 3. Output true if input false

A	O
1	0
0	1

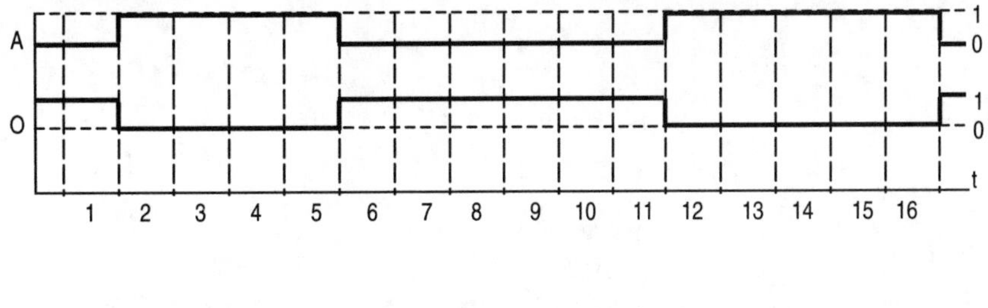

TABLE 1.4o Continued

Functional Diagramming Symbols—Binary Logic, Memory, and Time Functions (proposed for the next revision of ISA S5.1, now ANSI/ISA-5.01.01, at the time of this writing)

No.	Symbol/Truth Table	Definition/Graph
10	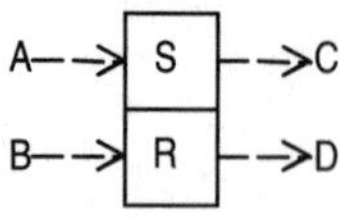 A—> S —>C B—> R —>D	1. Basic memory 2. Outputs C and D are always opposite 3. If input A equals "1," then output C equals "1" and D equals "0" 4. If input A changes to "0," output C remains "1" until input B equals "1," then C equals "1" and D equals "0" 5. If input B equals "1," then output D equals "1" and C equals "0" 6. If input B changes to "0," output D remains "1" until input A equals "1," then D equals "1" and C equals "0" 7. If inputs A and B are simultaneously equal to "1," then outputs C and D change state

	A	B	C	D
1	0	0	0	1
2	1	0	1	0
3	0	0	1	0
4	0	1	0	1
5	0	0	0	1
6	1	1	1	0
7	0	0	1	0
8	1	1	0	1

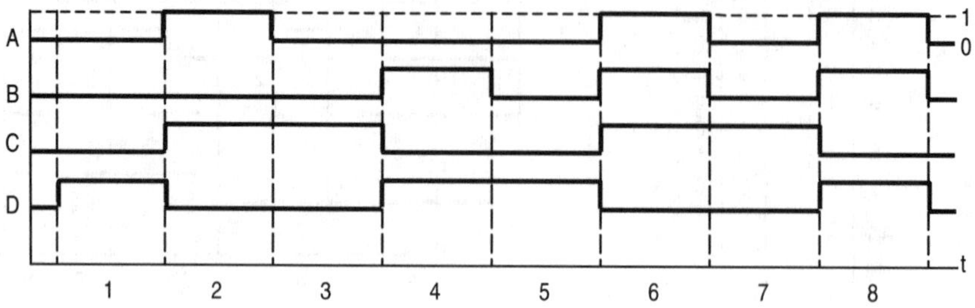

No.	Symbol/Truth Table	Definition/Graph
11	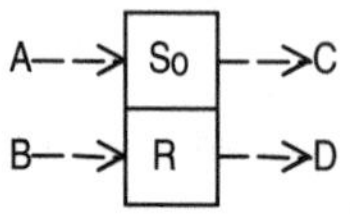 A—> So —>C B—> R —>D	1. Set dominant memory ("S_o Dominant") 2. Outputs C and D are always opposite 3. If input A equals "1," then output C equals "1" and D equals "0" 4. If input A changes to "0," output C remains "1" until input B equals "1," then output C equals "1" and D equals "0" 5. If input B equals "1," then output D equals "1" and C equals "0" 6. If input B changes to "0," output D remains "1" until input A equals "1," then output D equals "1" and C equals "0" 7. If inputs A and B are simultaneously equal to "1," then output C equals "1" and D equals "0"

	A	B	C	D
1	0	0	0	1
2	1	0	1	0
3	0	0	1	0
4	0	1	0	1
5	0	0	0	1
6	1	1	1	0
7	0	0	1	0
8	1	1	1	0

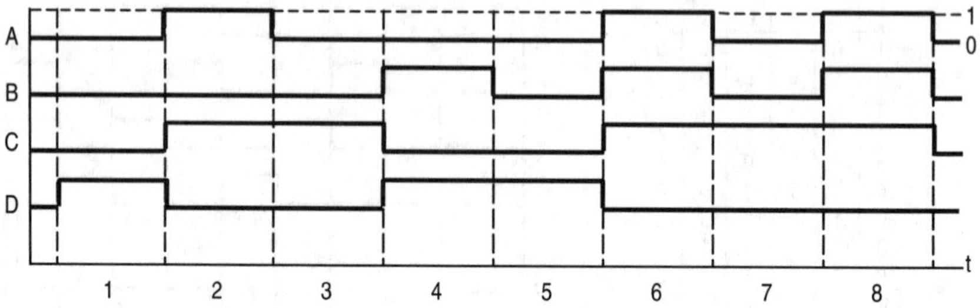

TABLE 1.4o Continued

Functional Diagramming Symbols—Binary Logic, Memory, and Time Functions (proposed for the next revision of ISA S5.1, now ANSI/ISA-5.01.01, at the time of this writing)

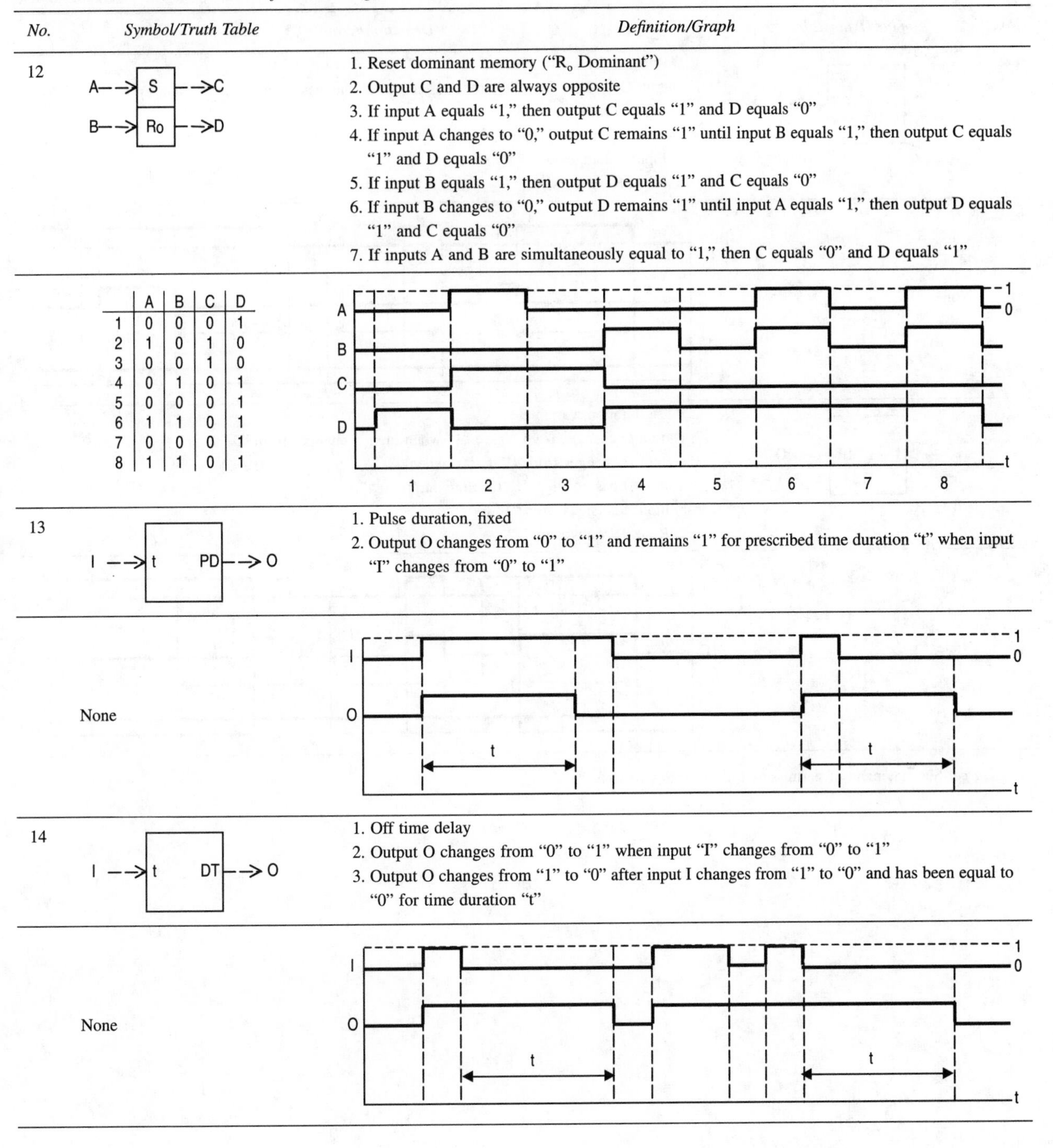

No. | *Symbol/Truth Table* | *Definition/Graph*

12

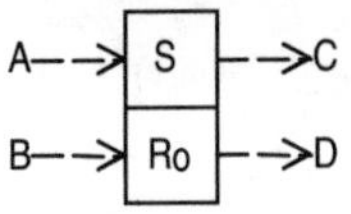

1. Reset dominant memory ("R_o Dominant")
2. Output C and D are always opposite
3. If input A equals "1," then output C equals "1" and D equals "0"
4. If input A changes to "0," output C remains "1" until input B equals "1," then output C equals "1" and D equals "0"
5. If input B equals "1," then output D equals "1" and C equals "0"
6. If input B changes to "0," output D remains "1" until input A equals "1," then output D equals "1" and C equals "0"
7. If inputs A and B are simultaneously equal to "1," then C equals "0" and D equals "1"

	A	B	C	D
1	0	0	0	1
2	1	0	1	0
3	0	0	1	0
4	0	1	0	1
5	0	0	0	1
6	1	1	0	1
7	0	0	0	1
8	1	1	0	1

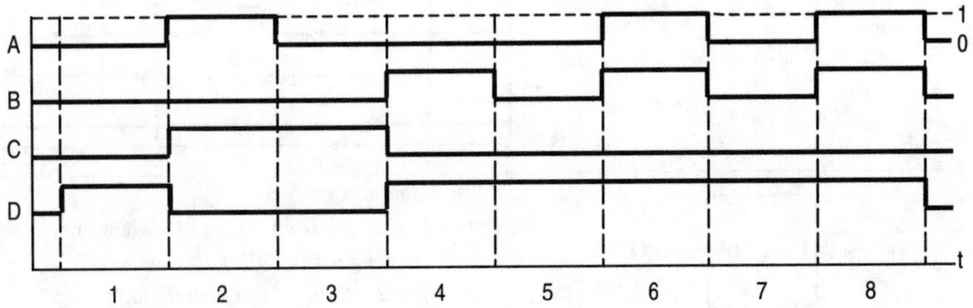

13

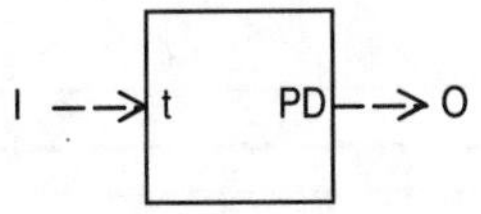

1. Pulse duration, fixed
2. Output O changes from "0" to "1" and remains "1" for prescribed time duration "t" when input "I" changes from "0" to "1"

None

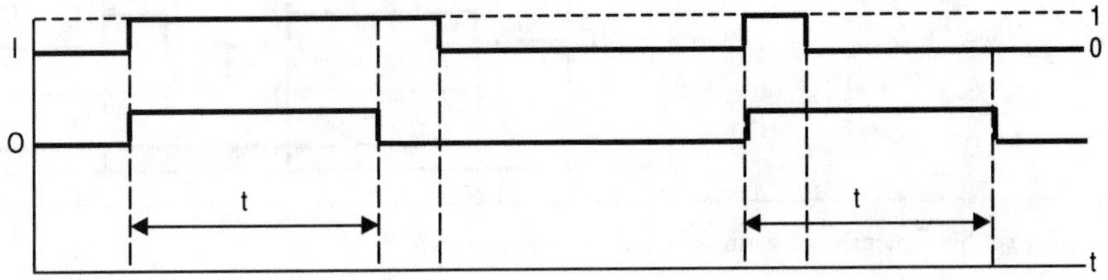

14

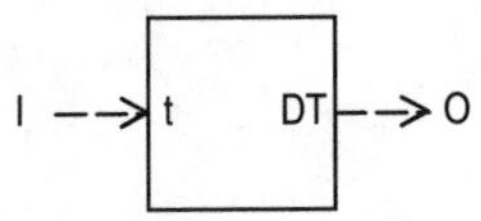

1. Off time delay
2. Output O changes from "0" to "1" when input "I" changes from "0" to "1"
3. Output O changes from "1" to "0" after input I changes from "1" to "0" and has been equal to "0" for time duration "t"

None

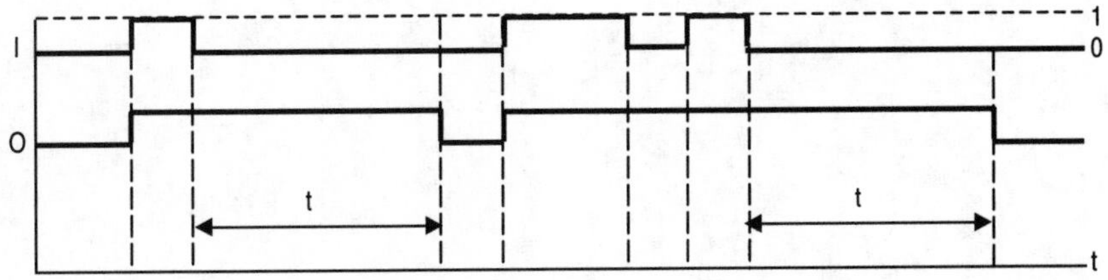

TABLE 1.4o Continued

Functional Diagramming Symbols—Binary Logic, Memory, and Time Functions (proposed for the next revision of ISA S5.1, now ANSI/ISA-5.01.01, at the time of this writing)

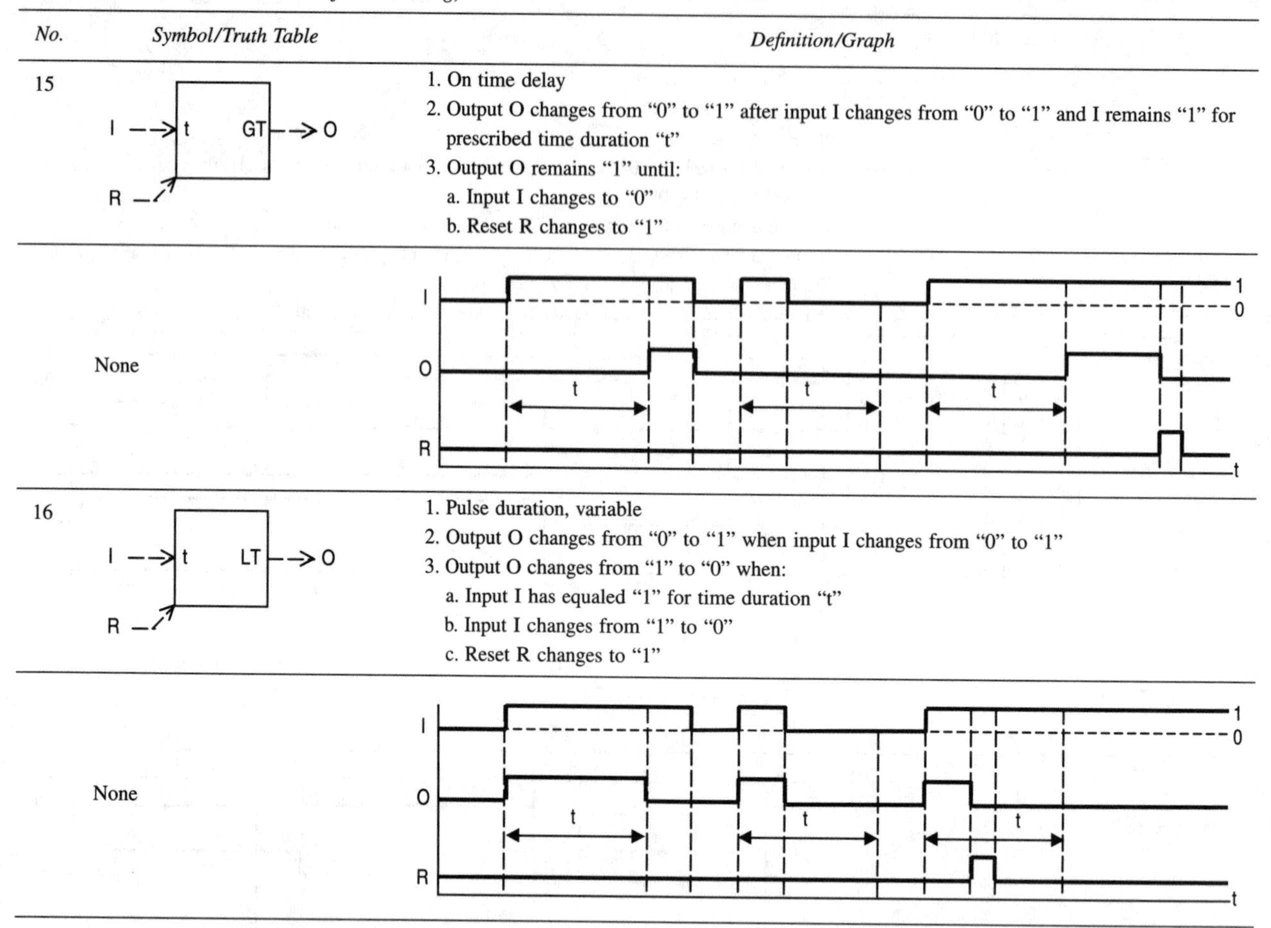

No.	Symbol/Truth Table	Definition/Graph
15	I —> t GT —> O; R	1. On time delay 2. Output O changes from "0" to "1" after input I changes from "0" to "1" and I remains "1" for prescribed time duration "t" 3. Output O remains "1" until: a. Input I changes to "0" b. Reset R changes to "1"
	None	I, O, R, t, 1, 0
16	I —> t LT —> O; R	1. Pulse duration, variable 2. Output O changes from "0" to "1" when input I changes from "0" to "1" 3. Output O changes from "1" to "0" when: a. Input I has equaled "1" for time duration "t" b. Input I changes from "1" to "0" c. Reset R changes to "1"
	None	I, O, R, t, 1, 0

Source: See statement of permission following Figure 1.4a.

TABLE 1.4p
Mathematical Function Block Symbols (proposed for the next revision of ISA S5.1, now ANSI/ISA-5.01.01, at the time of this writing)

No.	Symbol/Function	Equation/Graph	Definition
01	Σ Summation	$M = X_1 + X_2 \ldots + X_n$ 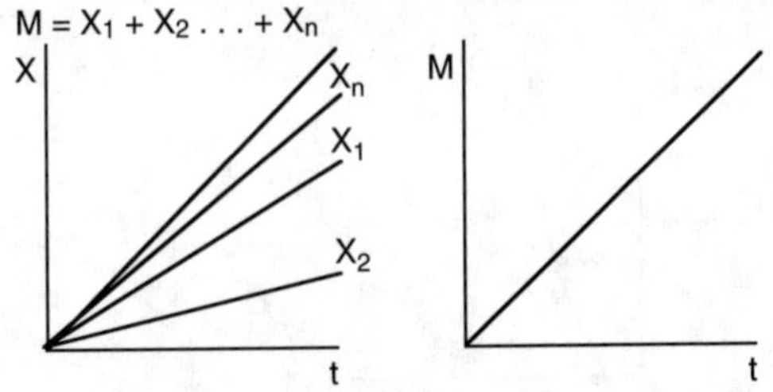	1. Output equals algebraic sum of inputs
02	Σ/n Average	$M = X_1 + X_2 \ldots + X_n/n$ 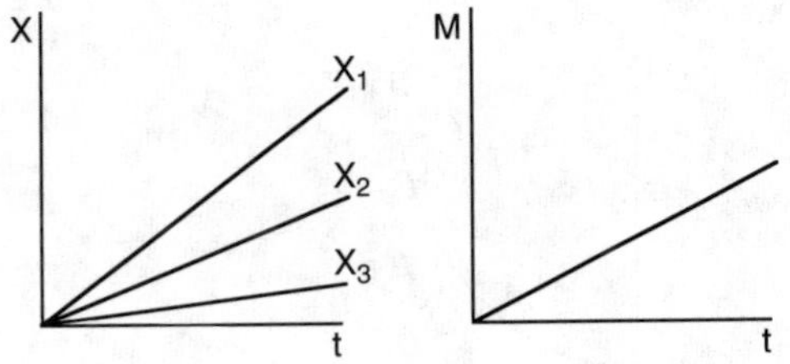	1. Output equals algebraic sum of inputs divided by number of inputs
03	Δ Difference	$M = X_1 - X_2$ 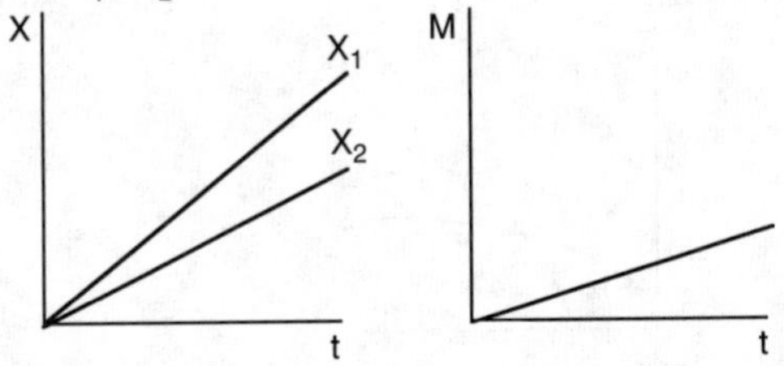	1. Output equals difference of two inputs
04	X Multiplication	$M = X_1 X_2$ 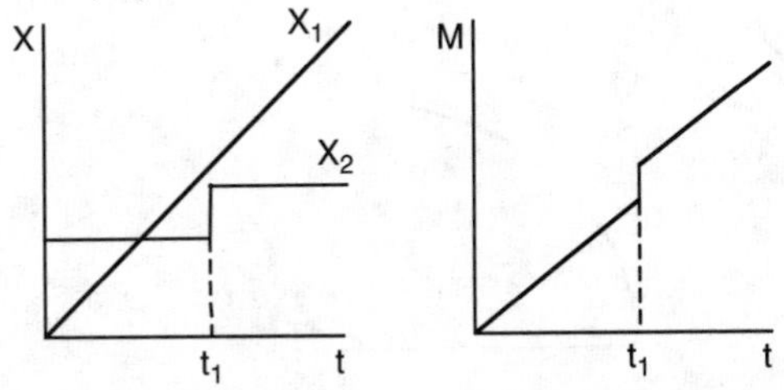	1. Output equals product of two inputs
05	÷ Division	$M = X_1/X_2$ 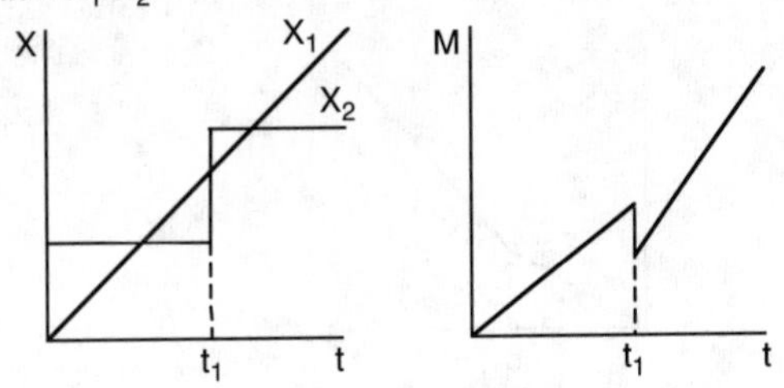	1. Output equals quotient of two inputs

TABLE 1.4p Continued

Mathematical Function Block Symbols (proposed for the next revision of ISA S5.1, now ANSI/ISA-5.01.01, at the time of this writing)

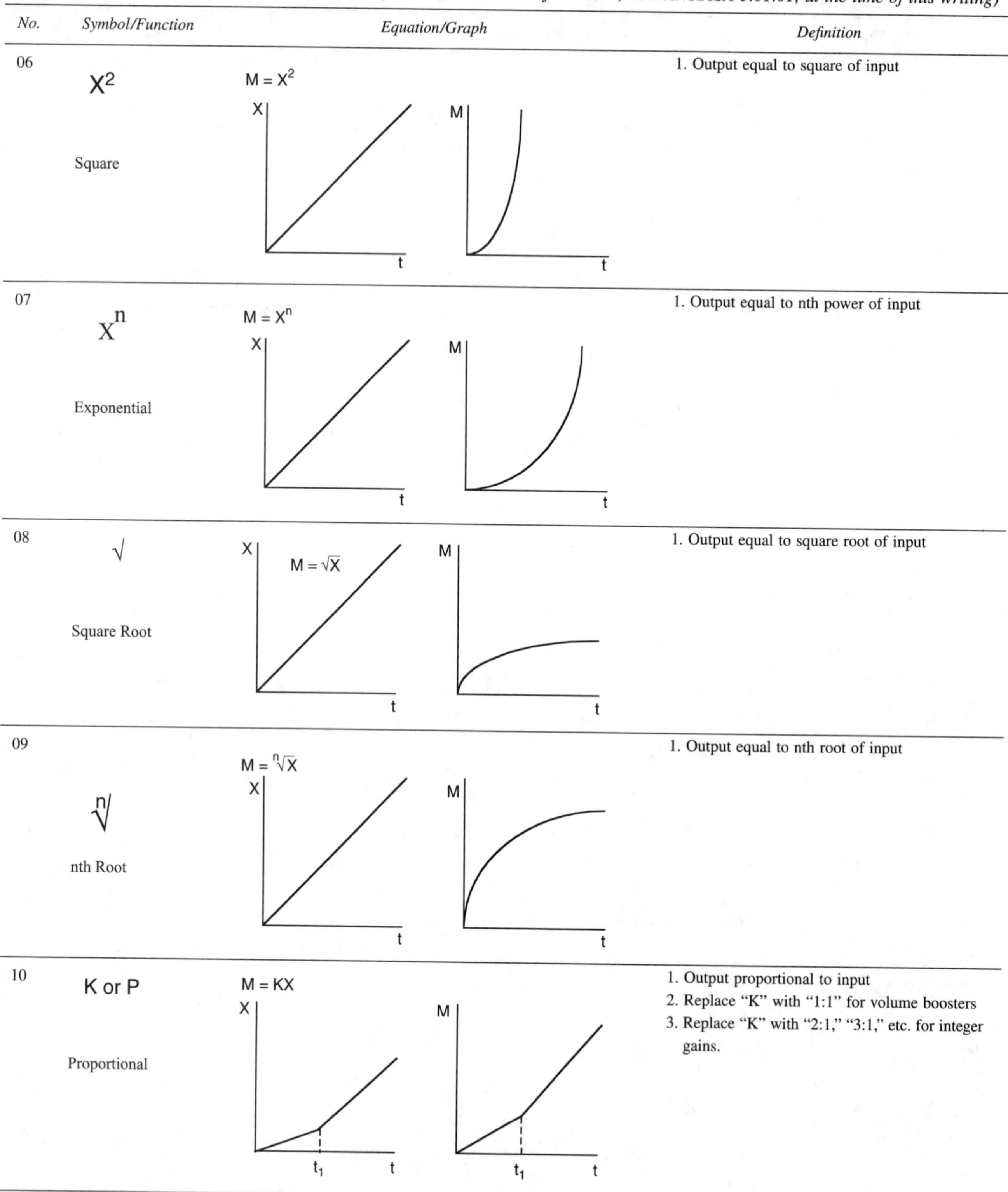

No.	*Symbol/Function*	*Equation/Graph*	*Definition*
06	X^2 Square	$M = X^2$ X, t; M, t	1. Output equal to square of input
07	X^n Exponential	$M = X^n$ X, t; M, t	1. Output equal to nth power of input
08	$\sqrt{}$ Square Root	$M = \sqrt{X}$ X, t; M, t	1. Output equal to square root of input
09	$\sqrt[n]{}$ nth Root	$M = \sqrt[n]{X}$ X, t; M, t	1. Output equal to nth root of input
10	K or P Proportional	$M = KX$ X, t_1, t; M, t_1, t	1. Output proportional to input 2. Replace "K" with "1:1" for volume boosters 3. Replace "K" with "2:1," "3:1," etc. for integer gains.

TABLE 1.4p Continued

Mathematical Function Block Symbols (proposed for the next revision of ISA S5.1, now ANSI/ISA-5.01.01, at the time of this writing)

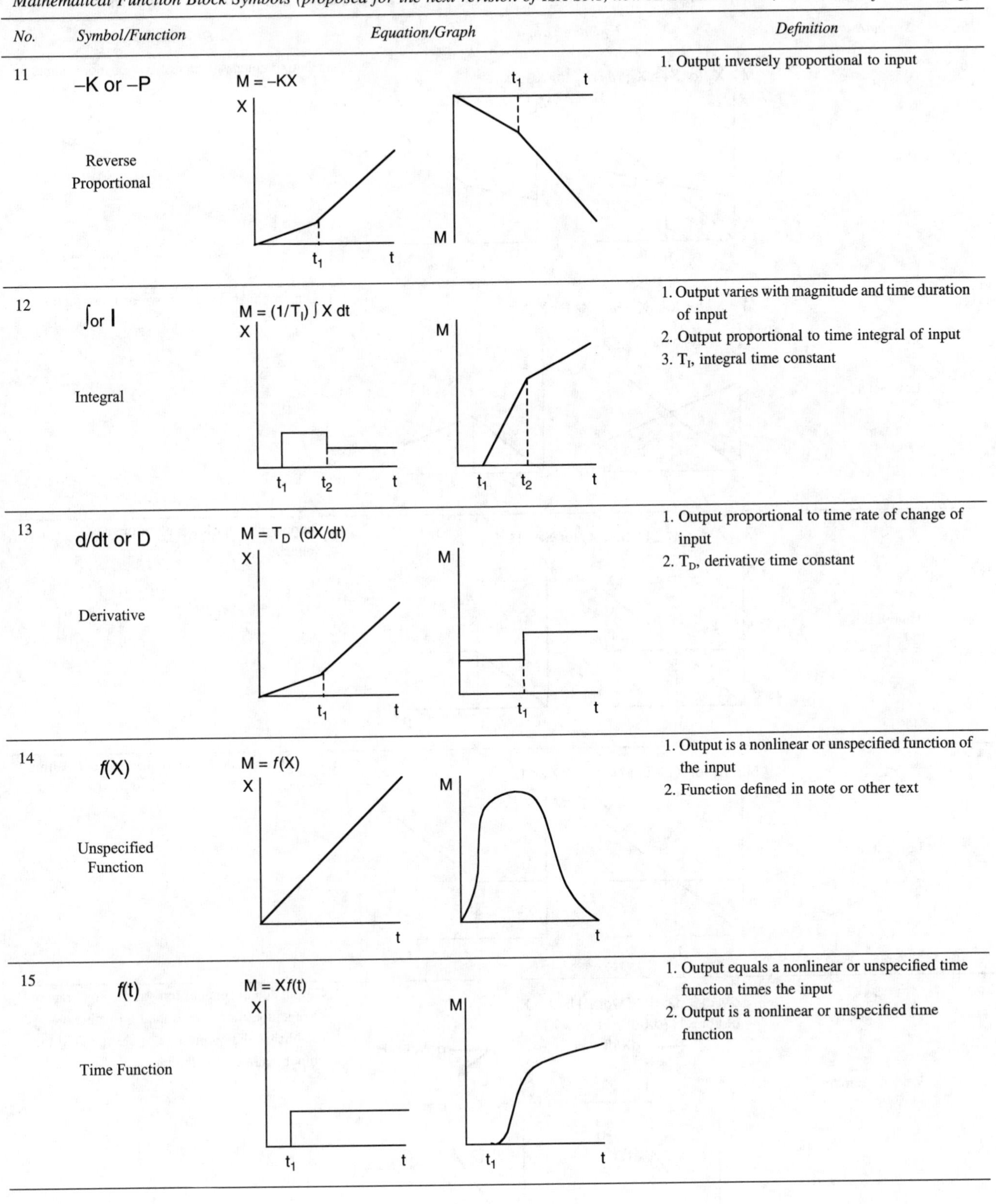

No.	Symbol/Function	Equation/Graph	Definition
11	–K or –P Reverse Proportional	M = –KX	1. Output inversely proportional to input
12	∫ or I Integral	M = (1/T$_I$) ∫ X dt	1. Output varies with magnitude and time duration of input 2. Output proportional to time integral of input 3. T$_I$, integral time constant
13	d/dt or D Derivative	M = T$_D$ (dX/dt)	1. Output proportional to time rate of change of input 2. T$_D$, derivative time constant
14	*f*(X) Unspecified Function	M = *f*(X)	1. Output is a nonlinear or unspecified function of the input 2. Function defined in note or other text
15	*f*(t) Time Function	M = X*f*(t)	1. Output equals a nonlinear or unspecified time function times the input 2. Output is a nonlinear or unspecified time function

TABLE 1.4p Continued

Mathematical Function Block Symbols (proposed for the next revision of ISA S5.1, now ANSI/ISA-5.01.01, at the time of this writing)

No.	*Symbol/Function*	*Equation/Graph*	*Definition*
16	$>$ High Select	$M = X_1$ for $X_1 \geq X_2$, $M = X_2$ for $X_1 \leq X_2$ 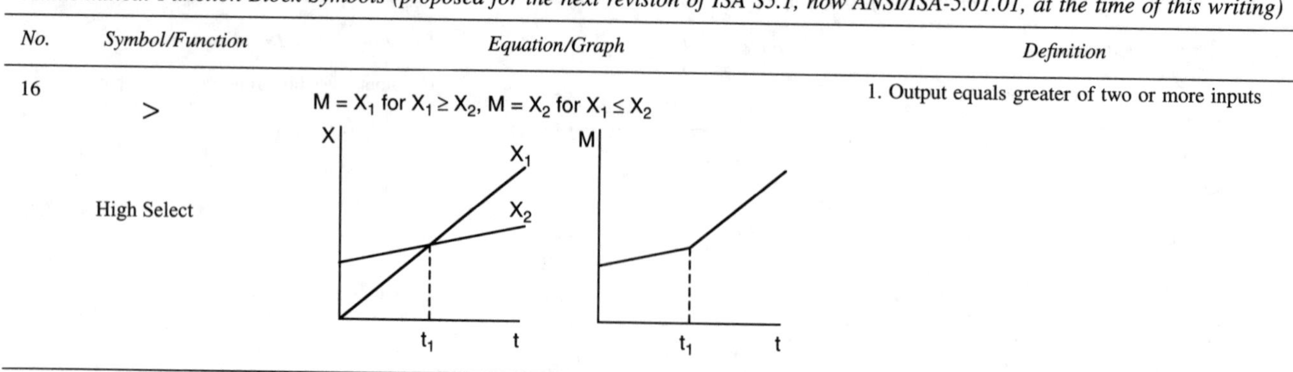	1. Output equals greater of two or more inputs
17 	$<$ Low Select 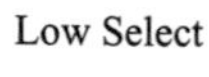	$M = X_1$ for $X_1 \leq X_2$, $M = X_2$ for $X_1 \geq X_2$ 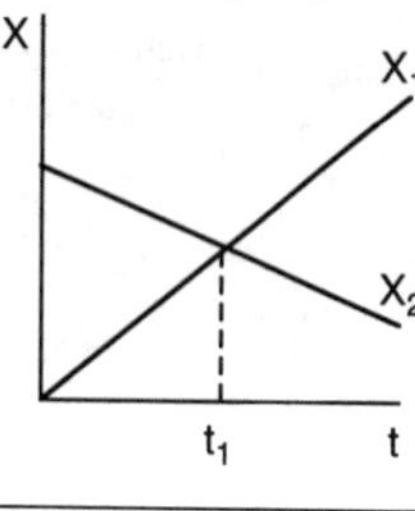	1. Output equals lesser of two or more inputs
18	$\ngtr$ High Limit	$M = X_1$ for $X_1 \leq H$, $M = X_2$ for $X_1 \geq H$ 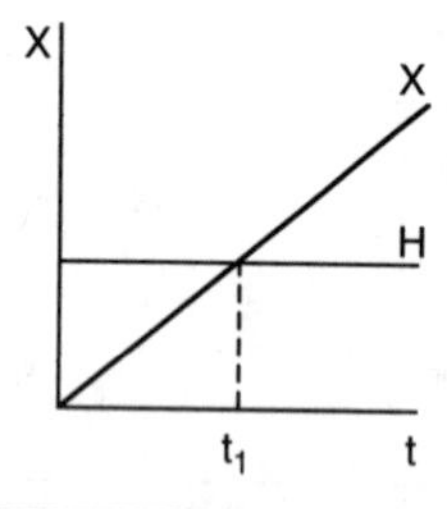	1. Output equals the lower of the input or high limit values
19	$\nless$ Low Limit	$M = X_1$ for $X_1 \geq L$, $M = X_2$ for $X_1 \leq L$ 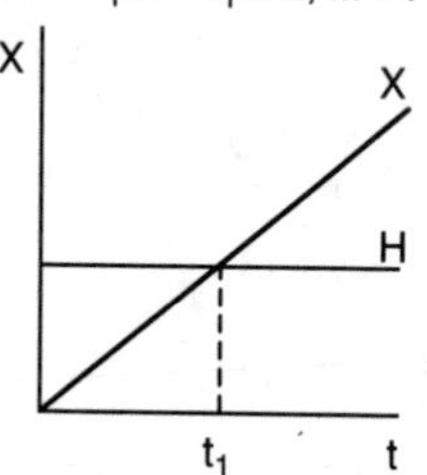 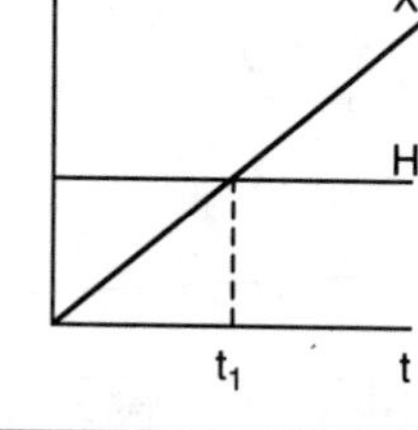	1. Output equals the higher of the input or low limit values 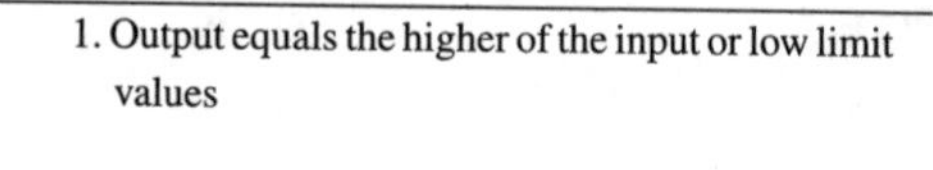
20	$\forall$ Velocity Limiter 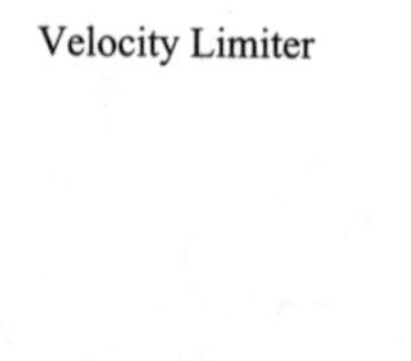	$dM/dt = dX/dt$ ($dX/dt \leq H$, $M = X$) $dM/dt = H$ ($dX/dt \geq H$, $M \neq X$) 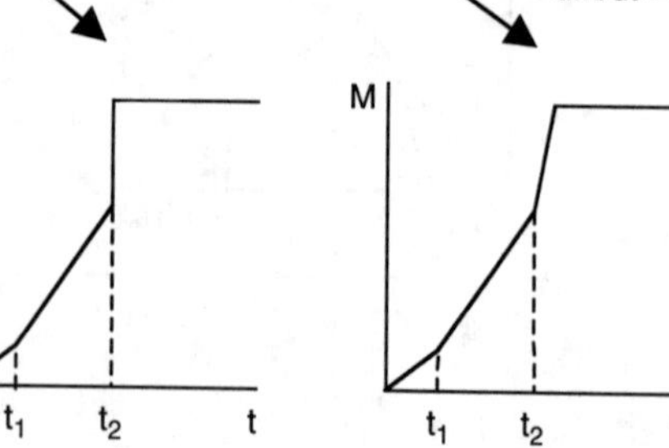 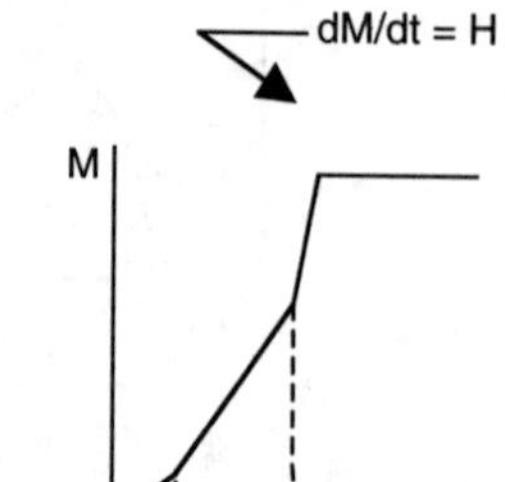	1. Output equals input as long as the input rate of change does not exceed the limit value that establishes the output rate of change until the output again equals the input

TABLE 1.4p Continued
Mathematical Function Block Symbols (proposed for the next revision of ISA S5.1, now ANSI/ISA-5.01.01, at the time of this writing)

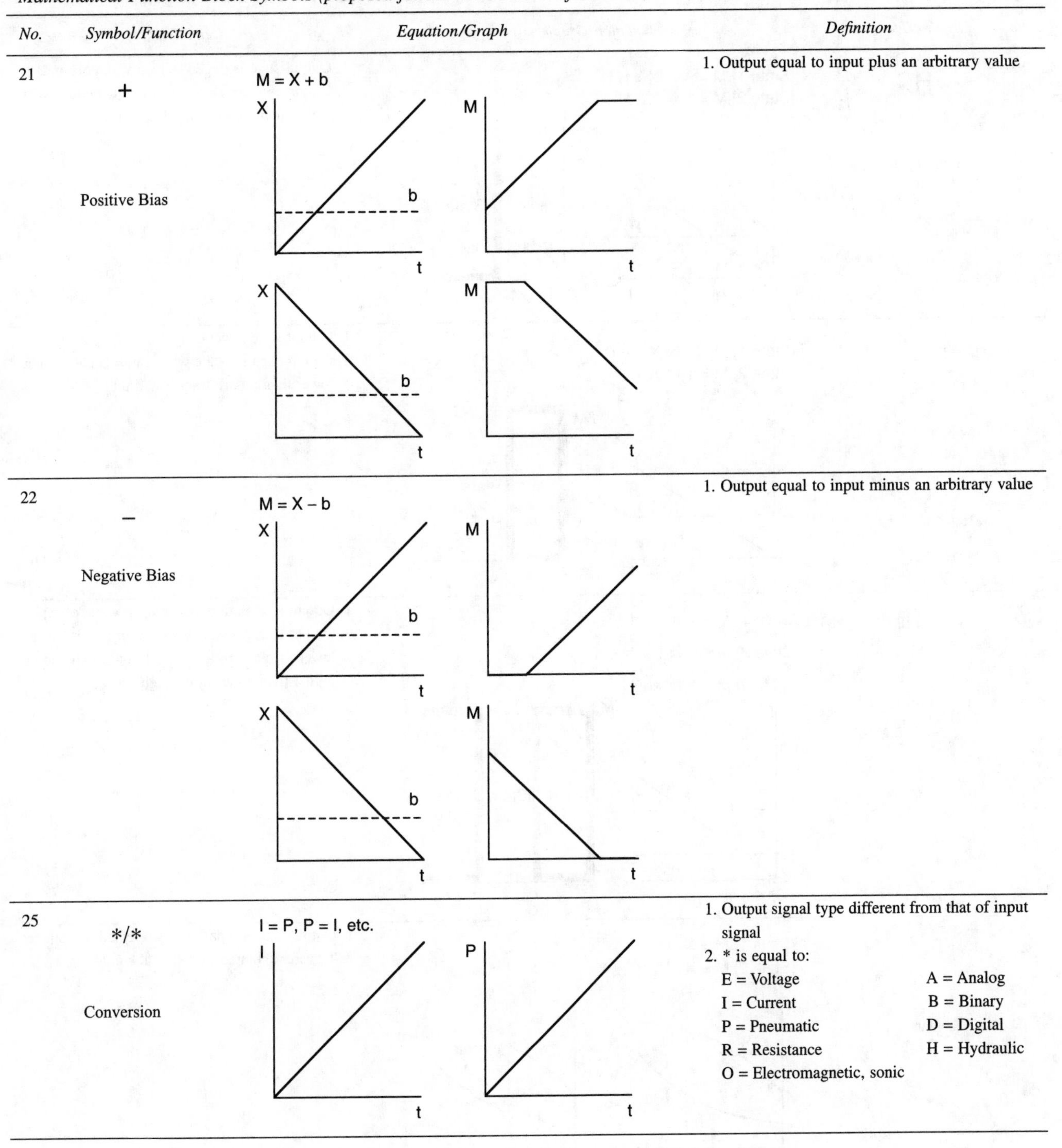

No.	*Symbol/Function*	*Equation/Graph*	*Definition*
21	+ Positive Bias	M = X + b	1. Output equal to input plus an arbitrary value
22	– Negative Bias	M = X – b	1. Output equal to input minus an arbitrary value
25	*/* Conversion	I = P, P = I, etc.	1. Output signal type different from that of input signal 2. * is equal to: E = Voltage A = Analog I = Current B = Binary P = Pneumatic D = Digital R = Resistance H = Hydraulic O = Electromagnetic, sonic

TABLE 1.4p Continued

Mathematical Function Block Symbols (proposed for the next revision of ISA S5.1, now ANSI/ISA-5.01.01, at the time of this writing)

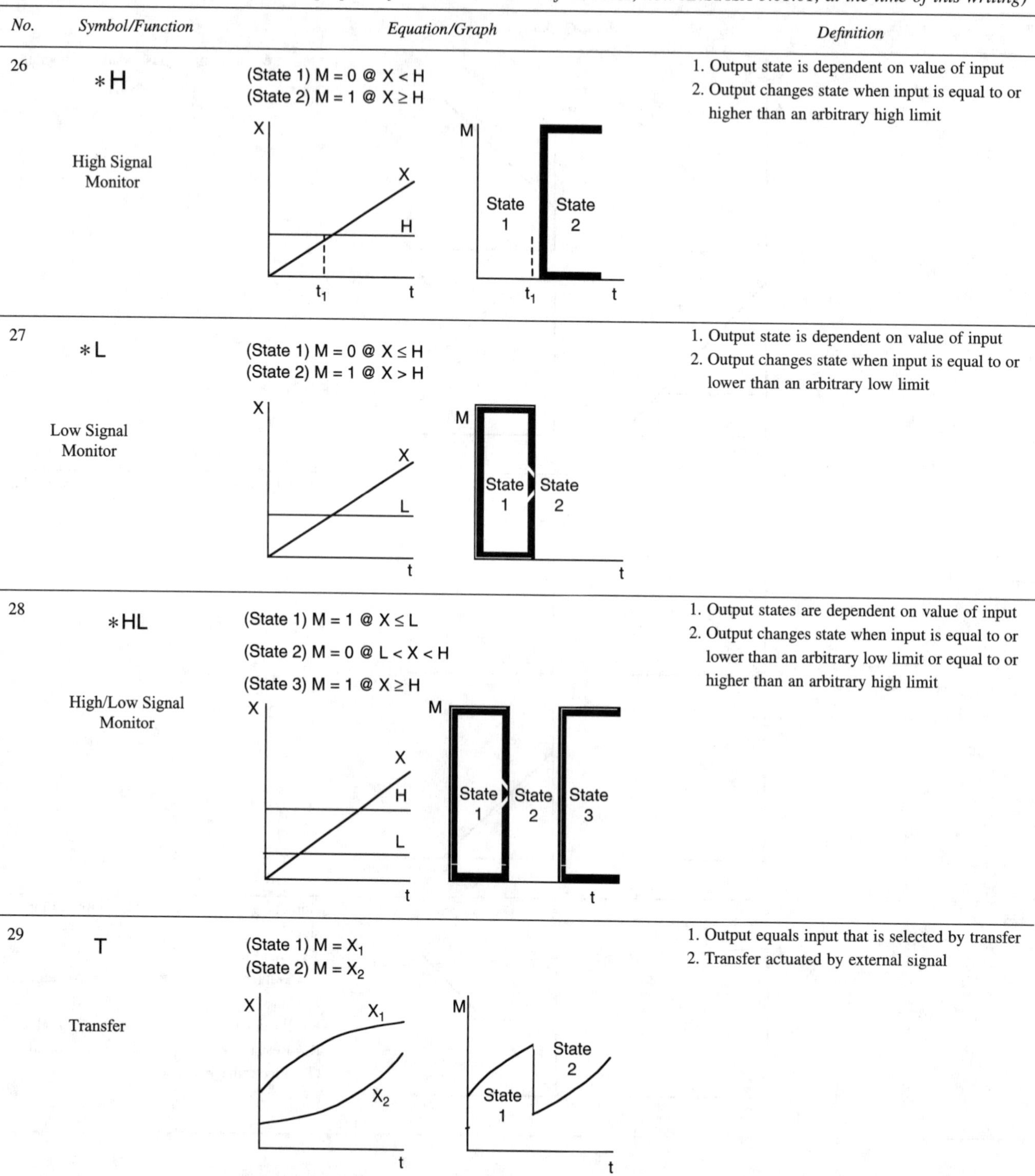

No.	*Symbol/Function*	*Equation/Graph*	*Definition*
26	∗H High Signal Monitor	(State 1) $M = 0$ @ $X < H$ (State 2) $M = 1$ @ $X \geq H$	1. Output state is dependent on value of input 2. Output changes state when input is equal to or higher than an arbitrary high limit
27	∗L Low Signal Monitor	(State 1) $M = 0$ @ $X \leq H$ (State 2) $M = 1$ @ $X > H$	1. Output state is dependent on value of input 2. Output changes state when input is equal to or lower than an arbitrary low limit
28	∗HL High/Low Signal Monitor	(State 1) $M = 1$ @ $X \leq L$ (State 2) $M = 0$ @ $L < X < H$ (State 3) $M = 1$ @ $X \geq H$	1. Output states are dependent on value of input 2. Output changes state when input is equal to or lower than an arbitrary low limit or equal to or higher than an arbitrary high limit
29	T Transfer	(State 1) $M = X_1$ (State 2) $M = X_2$	1. Output equals input that is selected by transfer 2. Transfer actuated by external signal

Source: See statement of permission following Figure 1.4a.

1.5 Historical Data Storage and Evaluation

G. C. BUCKBEE

Partial List of Suppliers: ABB Instrumentation; Canary Labs; Crystal Reports; Fluke; The Foxboro Company; Honeywell; National Instruments; Oil Systems; Omega Engineering, Inc.; SAS; Sequentia; Siemens; Squirrel; Trendview Recorders; Toshiba International; Wonderware; Yokogawa; Zontec

Accurate and reliable production and process data are essential to continuous process improvement. Today's computer-based control systems are capable of generating huge volumes of data. Simply capturing all that data, however, will not solve the problem. To be useful, it must be the right data, readily accessible, in the right format. The design of such an information system requires an understanding of both the process and the business environment. This chapter considers the design and development of a data historian system for industrial processes. These systems may be known as real-time information systems, data historians, or process information systems.

CLARIFYING THE PURPOSE OF THE DATA SYSTEM

Figure 1.5a shows the basic functions of a data historian. Fundamentally, data is collected, historized, then analyzed. Reports are developed and decisions are made. By improving understanding of the process, the data historian plays a critical role in process improvement. A real-time information system must meet the needs of many users. Operators want reliable short-term process information. Managers want reliable and accurate production accounting. Engineers typically want both short-term and long-term data with sophisticated analysis tools. The information technology department wants something that will integrate well with other systems, while requiring minimal maintenance. To avoid major disappointment at project completion, it is absolutely critical to describe clearly the scope and functionality of the system at the beginning.

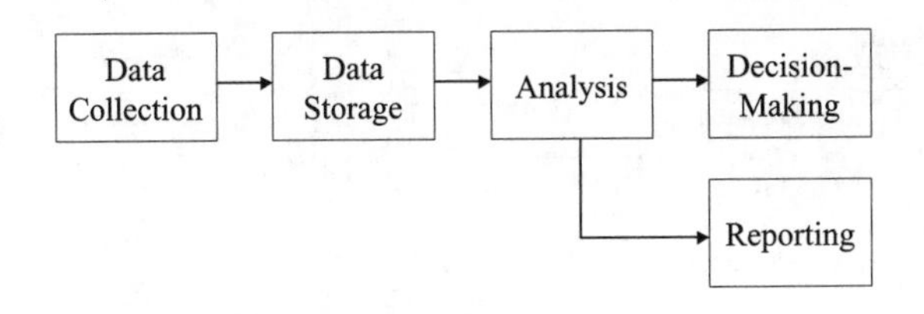

FIG. 1.5a
Basic functions of a data collection system.

Fundamentally, the information system exists to support decision making for each of these different users. Consider each the following questions for each piece of data:

1. Is this the right data? Based on the user's needs, make sure that all relevant data are being captured. Are the raw data or filtered values desired? Should online samples be collected, just the lab data, or both? Is the right point in the process being measured?
2. Does the data come from the most reliable and accurate source?
3. Can the data be readily accessed, at the location where the decision is made? If the data are not accessible, the data are of no use to anybody. If the operator needs the data, then the data should be available on the plant floor. In team-based environments, the information should be available in team rooms, meeting rooms, or wherever the team gathers. Engineers should be able to place the data onto the same computer system as their analysis tools.
4. Are the data displayed in the right format for a decision to be made? As the saying goes, a picture is worth a thousand words. Trends are common, easy to use, and extremely valuable. For production data, cold, hard numbers in an easy-to-read report are usually desired.

With so many different users, gaining a clear definition of the scope can be a challenge. Although most people can easily visualize scope in the form of equipment, it is more important to define the system functionality first. By functionality, we mean defining all of the expected capabilities of the system. This starts out as a general description, and evolves into a very detailed description of the system.

Requirements from each of the system end users must be collected, prioritized, and combined. A simple tool for clarifying the system needs is the IS/IS NOT chart, shown in Table 1.5b. The entries in this chart will vary greatly from plant to plant, or even from department to department. Clarifying this on paper and gaining management alignment early will help to ensure a successful project. The following sections provide

TABLE 1.5b
IS/IS NOT Analysis for a Sample Mill

IS	IS NOT
Process data collection	Maintenance data collection
Production data collection	Financial
Daily reporting	Monthly reporting
Short-term (minutes, hours, days)	Long-term (months, years)
Trending, reporting	X-Y plotting, statistics
Read-only from DCS	Download to DCS
Continuous data	Batch data

some suggestions for system scope. Of course, there is also hardware and infrastructure scope to be considered.

Interactions and Integration with Other Systems

If the data historian is expected to communicate with other electronic systems, the nature of that communication must be spelled out. To clarify the communications between the two systems, one should answer the following questions:

1. What information is going to be passed?
2. What is the trigger for passing the data? (time, an event, etc.)
3. What is the form of the data? (scalar number, records, databases, etc.)
4. Where does the "Master data" reside?

When several different information systems are involved, it is helpful to use a data flow diagram to understand the interaction between systems. Figure 1.5c shows a sample data flow diagram. For the information system to be successfully integrated with operations, consider the following issues:

1. How will the data be used? This is the critical issue. Will the operator use the data from a control chart to make adjustments to the process? Will operators have the authority to do so? Do they have the right training to do it? Will the production data be used to make daily decisions?
2. How can it be ensured that the data will be entered only once? Multiple points of data entry will lead inevitably to discrepancies and confusion, not to mention effort lost by having to enter the same data twice.
3. How can the manual data entry be minimized? Let's face it … manual data entry is a rather boring chore, and people make mistakes when they are bored. Can the data be collected directly from the control system or transferred electronically from lab equipment?

Integration with Maintenance

Section 1.6 provides a great deal of detail about the integration of maintenance functions with data historians. Only a small summary is presented here. Some process data are particularly valuable for maintenance operations. For example, hours of use of large equipment are often needed to plan maintenance schedules. To meet the needs of maintenance, the following should be considered:

1. How is maintenance work planned? How is it scheduled? Is there an existing electronic scheduling system?
2. Is the right process data being measured for maintenance needs? For example, are oil temperatures, or vibration levels, being measured and recorded online? If so, this can be very useful data for maintenance.
3. How will the maintenance personnel gain access to the data? Will they have physical access to a computer or terminal in their workspace? Or will they work from printed reports?

Integration with Management

The management team is another key customer of the data system. In fact, many information systems are justified based on the value of the data to the management team. To meet the needs of the management team, consider the following:

1. What types of routine reports are needed? Daily and monthly reports are the most commonly requested.

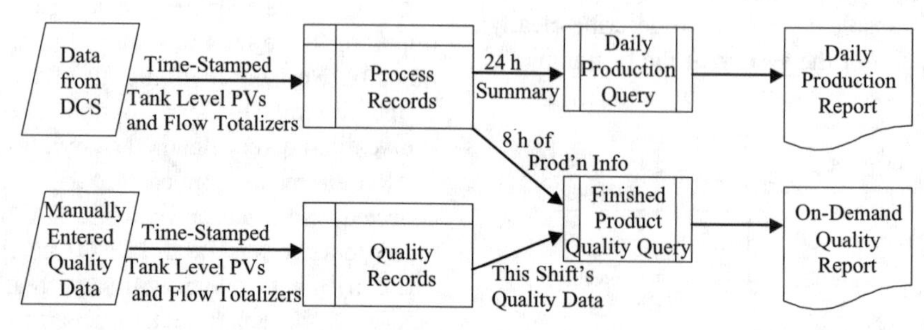

FIG. 1.5c
Sample data flow diagram.

2. Which items should be tracked? Typically, the top priorities are production rate data, then material and energy usage. Environmental data have also become increasingly important in recent years.
3. What types of analysis tools are needed? Does the management require a particular type of analysis tool, such as control charts?

Beware of trying to provide all the analysis tools for everyone's needs. This will drive up the cost of the project considerably, and many of these tools will never be used. A preferred approach is to make the data available in an open format, then to let each analysis tool be justified individually. Using this approach, the process engineer may use a different set of tools than the shift manager, but they will both be looking at the same data.

DATA COLLECTION

The first part of the information system is the data collection. The data must be retrieved from the control system. With the proliferation of computer-based control systems, there are myriad ways to gather data. Specific network connection issues are reviewed in Chapter 4. Data to be collected generally fall into two categories: continuous and event data. The type of process operation will often determine the data collection needs.

In a continuous process, materials are supplied and transformed continuously into finished product. Examples include pulp and paper mills, oil refineries, utilities, and many other large-scale processes. Data collection is needed to track production, raw material and energy consumption, and for problem solving.

In a batch process, fixed procedures, or recipes, are used to produce finished products. Ingredients are added, and processing operations are performed, in a specific sequence. Data collection needs include keeping historical records of batch times, operations, ingredient amounts, and lot numbers.

Continuous data are collected to keep a record of the process. Most often, data are collected regularly, at a given frequency. At a minimum, the process variable (PV) of each loop is recorded. Some systems allow the collection of controller output and set point as well. Continuous data collection is most often used for continuous processes.

In older systems, data are typically collected once per minute. But newer systems allow fast collection, as fast as once per second, or even faster.

Older DCS (distributed control systems) systems have little memory available for data storage. To conserve storage space, DCS systems will rotate out the data as the data age. It is typical to keep up to a week's worth of fast data, then to compress the data by taking averages. This way, a long history of averages can be kept, even though some resolution is lost.

Short-term continuous data are of great interest to the operator. Good resolution and good trending tools are critical for these analyses. The long-term averages are typically of interest to management, and are used to maintain production records and to evaluate costs.

Event Data

Any process event can also trigger the collection of data. For example, the lifting of a safety relief valve may trigger the collection of information about vessel pressure, level, and contents. In fact, each alarm can be considered an event.

In batch systems, the collection of event data takes on even more significance. As each significant step of the batch process passes, records of material use, time, energy use, and processing steps are recorded. These data can be used later to reconstruct the manufacture of each batch of product.

Event data are often collected on a triggered, or interrupt basis, rather than at a fixed collection frequency.

For anything more complicated than simple logging, collecting event data requires advance planning to define the event record. Each piece of data is placed into a field in a data record. Figure 1.5d represents a typical data record.

One type of event that is very important to track is operator activity. By tracking the actions of the operator, it is possible to reconstruct events after the fact. Typically, the following

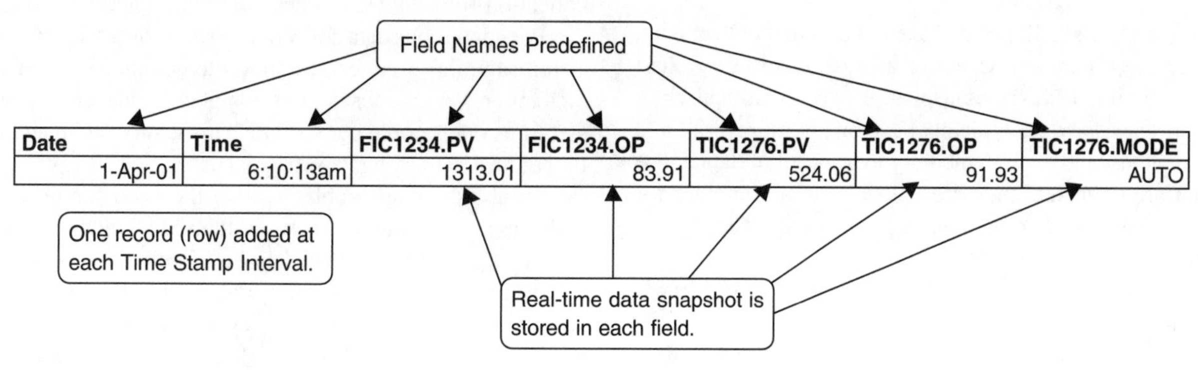

FIG. 1.5d
Sample data record.

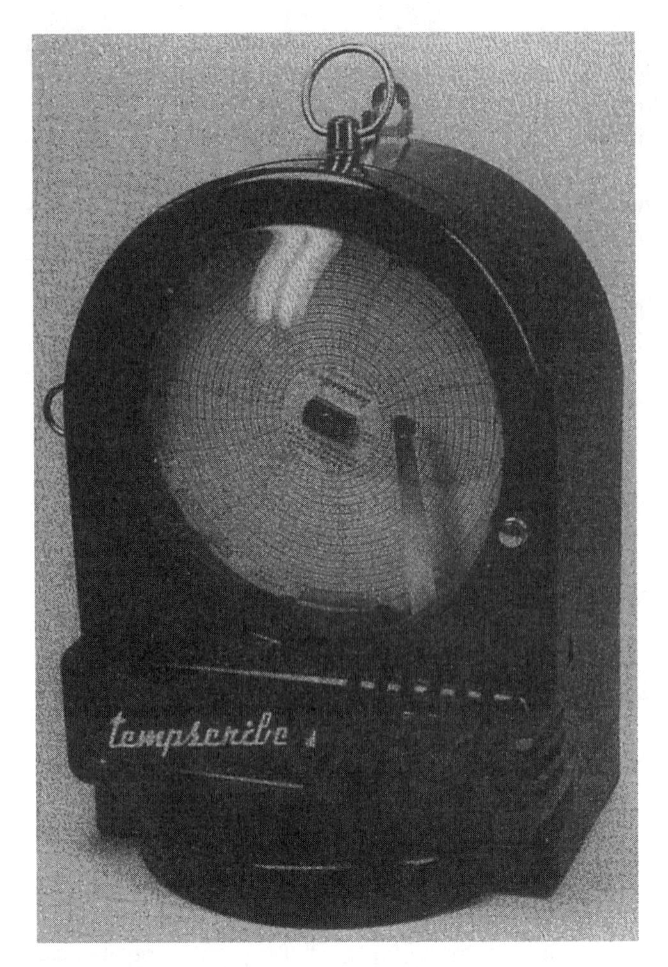

FIG. 1.5e
Chart recorder.

activities are logged by DCS systems:

1. Operator changes, such as set point changes, control output changes, or controller mode changes.
2. Reactions to alarms, such as acknowledgments.

A good activity log will identify the change that was made, the date and time of the change, and the control station from which the change was made.

Data Loggers

Data loggers are small, stand-alone devices that collect data. They are the digital equivalent of a brush recorder or chart recorder. Figure 1.5e shows a picture of a traditional chart recorder. These devices typically have one or more 4–20 mA input channels, and enough memory to log a lot of data. Once the data are collected, they are uploaded to a computer for analysis. The typical components of a data logger are shown in Figure 1.5f.

Multichannel models allow the collection of up to 30 channels simultaneously.

Most data loggers can be battery powered, allowing them to be used in the field. If the logger is to be used in extreme conditions, check to be sure that it can handle the temperature, moisture, and other environmental conditions. If data are to be collected over many hours or days, be sure to check the system life on one set of batteries, or inquire about alternate power supply arrangements.

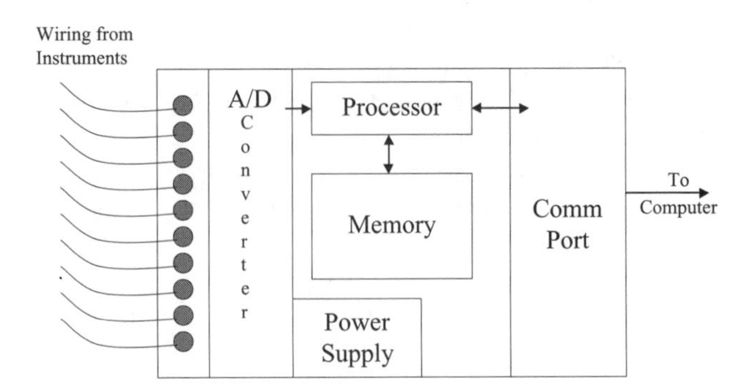

FIG. 1.5f
Data logger.

Choose a system that is capable of collecting data quickly enough to satisfy the needs. For long-term data collection in a remote location, 1 min or even 1 h per sample may be enough. But if the data logger is being used for troubleshooting a problem, much faster data collection may be necessary, on the order of 1 ms per sample.

Some systems are designed to be idle, until a trigger signal is received. This feature is very helpful for troubleshooting an intermittent problem. Set up the trigger signal to detect the first sign of a problem, then collect data.

Once data are collected, the data must be uploaded to a computer for analysis. Most data loggers support a serial link connection to a PC for this purpose.

Data loggers typically do not have on-board analysis functions. However, some newer systems include a laptop computer that is capable of a great deal of analysis.

Data Collection Frequencies

As obvious as it sounds, data must be collected quickly enough to be useful. But oversampling of data simply wastes storage space. So how is the required sampling time determined? Shannon's sampling theorem states that we must sample at least twice as fast as the signal that we hope to reconstruct. Figure 1.5g shows the compromise between data collection frequency and data storage and network loading.

For continuous data collection systems on a DCS platform, 1-min sampling frequencies are quite common. Newer DCS systems are capable at sampling at higher frequencies, typically as fast as one sample per second. This is more than adequate for production tracking, raw material usage, and energy tracking.

Troubleshooting problems drive the need for faster data collection. For example, new systems that monitor process dynamics may require that data be sampled at 1 s or faster. When troubleshooting, we are often trying to determine the sequence of events. Because process and safety interlocks happen fairly quickly (milliseconds), sampling will have to be much faster to satisfy these needs. To troubleshoot these very fast problems, a separate, dedicated data logger is often used.

For troubleshooting event-type data, it is not the collection frequency, but the precision of the time stamp that is of concern. Some DCS systems, such as Foxboro I/A, time-stamp the data at the input/output (I/O) level, providing excellent resolution. Others pass the data along at a given collection frequency, only to be time-stamped when it reaches the historian device.

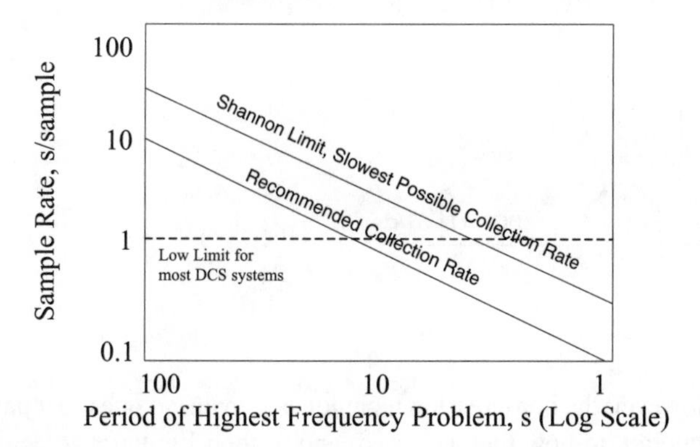

FIG. 1.5g
Sample rate vs. problem frequency.

ARCHITECTURE OF A DATA HISTORIAN SYSTEM

The data historian often provides the only link between the control system and the plantwide computer network. The system must be designed to accommodate the needs of both. The typical architecture of a data historian system is shown in Figure 1.5h.

In a modern data historian system, most signals are collected in the DCS or PLC (programmable logic controller), and data storage is relatively inexpensive. Communications is the bottleneck. To improve system capacity and capability, the communications will have to be optimized.

For starters, we will need to look at internal communications in the DCS or PLC. Some DCS systems will pass data along internal networks at a given scan frequency, typically 1 s at best. Others will use a deadband and exception-reporting mechanism to reduce internal communications load. Figure 1.5i shows how the deadband algorithm works to reduce data storage requirements.

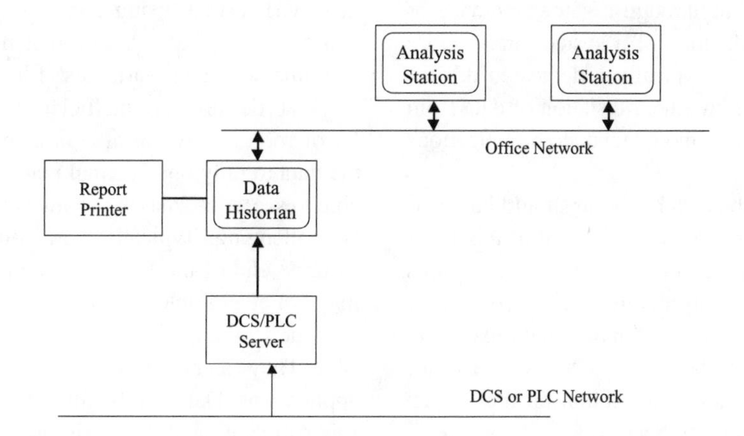

FIG. 1.5h
Typical architecture.

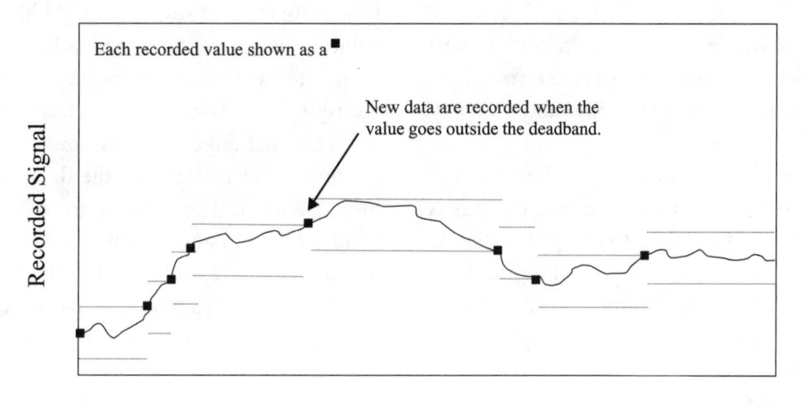

FIG. 1.5i
Diagram of deadband reporting.

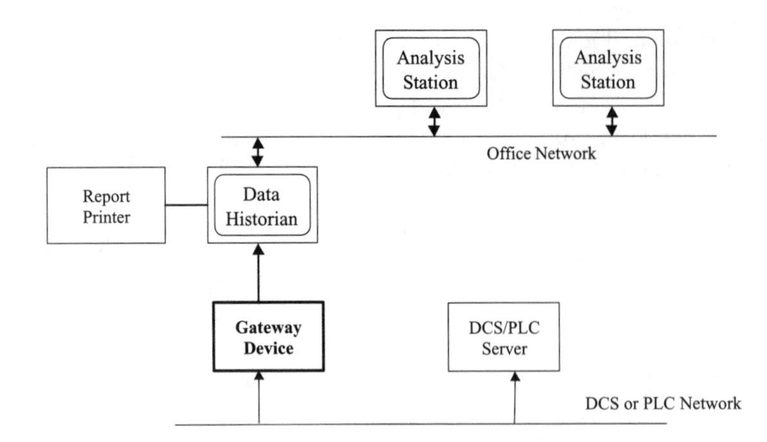

FIG. 1.5j
Diagram of a gateway device.

In the past DCS systems used proprietary networks. This changed dramatically in the 1990s, as open network systems made their way into the DCS world. At this writing, PLCs remain proprietary at the processor and I/O level, although the trend is toward more open systems, using industry standards.

Communications with the data historian may be via the proprietary network, or may be through a gateway device of some sort. Figure 1.5j is a diagram of a system architecture with a gateway device. This is typically a dedicated device, sold by the DCS vendor. It allows the translation of data from the proprietary network into a more open communications standard.

Data that have reached the data historian should be available on an open network. The *de facto* standard at this layer is Ethernet for the physical communications. Communications across this network may use any number of proprietary or industry-standard protocols, such as dynamic data exchange (DDE), object linking and embedding (OLE), or OLE for process control (OPC). For more information on these open communications protocols, refer to Section 4.9.

DATA STORAGE

Data must be stored somewhere, usually on a server hard disk. There are many considerations, including the format of data storage, the ease of retrieval and exporting, and a variety of security issues.

Because large amounts of data are being collected, it is easy to consume disk storage space. Traditional databases and simple flat-file techniques are simple to use, but consume a large amount of space. In the past, this was a problem, because of the cost of storage media. Many varieties of data compression techniques were developed and employed to reduce the cost of storage.

As the cost of storage media has dropped dramatically in recent years, the cost of storage has become quite small. Unless one is archiving thousands of data points at rates greater than once per second, storage costs should be only a minor concern. However, the compression techniques remain, so one must pay attention to how they are used, and to their limitations.

Flat files represent the simplest form of data storage. Data are stored in rows and columns. Each row represents a set of data from one instant in time. Each column contains data from a specific tag or parameter. The data are typically stored in ASCII format, using commas or tabs to delimit between columns. One that uses commas may be known as a .csv, or "comma-separated variables" file.

Flat files are very inefficient for storage, as they take up a lot of room. They are also limited in accuracy. If one stores the data to only one decimal place, that is all one will ever be able to see. Also, making changes to which data will be stored is challenging. Typically, new columns are added on to the extreme end of the file. Deleting columns is even trickier, or may not be possible, depending on the application being used.

The strength of flat files is that they are extremely accessible. They can be opened by word processing and spreadsheet applications. Data can be imported into almost any data analysis software package. A simple program written in Basic or C can easily manipulate the data. Data can be easily edited, copied, and pasted. Figure 1.5k shows an example of a flat file.

Some vendors store data in a simple compressed or binary format. In these cases, data might be reduced to a 12- or 16-bit value to be stored. This typically results in a dramatic reduction, often ten times or more, in storage requirements. With the right extraction tool, data retrieval is also quite fast.

The challenge with compressed storage is that it is not as easy to view or extract the data. A more sophisticated tool must be used. These tools are often made available by the vendor of the historian software. When purchasing a historian that uses compressed files, be sure to understand export capabilities of the system. Otherwise, one may be locked into using that vendor's tools for analysis.

The relational database has become a more common method for data storage in recent years. A relational database organizes the data in a series of tables. By establishing relationships between these tables, a flexible and very powerful database can be developed. Changes, additions, and deletions

```
Date,Time,FIC1234.PV,FIC1234.OP,TIC1276.PV,TIC1276.OP,TIC1276.MODE
1-Apr-01,6:10:13am,1313.01,83.91,524.06,91.93,AUTO
1-Apr-01,6:11:13am,1313.27,83.52,524.87,91.81,AUTO
1-Apr-01,6:12:13am,1313.39,83.13,524.62,91.67,AUTO
1-Apr-01,6:13:13am,1312.87,84.06,524.67,90.31,AUTO
1-Apr-01,6:14:13am,1312.31,83.87,523.06,91.11,MAN
1-Apr-01,6:15:13am,1311.76,84.11,522.17,91.11,MAN
1-Apr-01,6:16:13am,1310.02,84.22,523.44,91.11,MAN
```

FIG. 1.5k
Sample flat file.

Table :Batch Records	Table :Products	Table :Operators
Batch No:	Product ID:	Tech ID:
Start Time:	Description:	Last Name:
Product ID:	Min Density:	First Name:
Density:	Max Density:	Date of Hire:
Weight, lbs.:	Target Density:	Operating Team:
Lead Tech ID:		

FIG. 1.5l
Relational database design for a simple batch system.

of data can also be accomplished. In addition, many of these systems will automatically track security information, such as who made the changes and when.

Relational databases carry a lot more overhead than do simple flat files or binary files. For this reason, they may not be capable of high data throughput, or they may require faster hardware, or more sophisticated networks than a simpler file format. When designing a relational database system, be sure to request performance data from the vendor. Performance data should include an assessment of the rate at which data can be stored and retrieved.

Relational databases are especially powerful where event-type data are to be stored. For example, batch operations often require recording of specific data from each step in the process. Each step can be recorded in the database. Later, these records can be processed using the powerful relational tools, such as combining and filtering. Figure 1.5l shows a relational database design for a simple batch system.

Some data historian vendors use proprietary algorithms for data storage. These techniques are often optimized for the specific system. Depending on the approach used, these formats may also provide substantial reductions in storage requirements.

The caution, of course, is that one must be able to retrieve the data. As in the compressed, or binary files, it is not desirable to be limited to the data analysis tools provided by a single vendor. So be sure to find out about export capabilities and formats.

It is also important to understand if the algorithm used causes a loss in resolution of the data. For example, some systems use a simple deadband to minimize the data collection frequency. A new data point is only recorded if the value changes by more than a certain deadband, for example, ±1%. Because most operation is typically at steady state, data storage is greatly reduced. One must be aware, however, that the data retrieved may have been collected at a variety of sampling intervals. This may limit the ability to perform certain types of analysis on the data.

The swinging-door algorithm is a modification of the deadband method. In this approach, the previous two data points are used to develop a straight-line extrapolation of the next data point. A deadband around this line is used. As long as the value continues to track this trend line, within the deadband, no more data are collected. Once a new point is collected, a new trend line is calculated, and the process starts over. Figure 1.5m shows how the swinging door algorithm works.

Where to Store Data

Given the open capabilities of today's DCS and PLC systems, there are many options for the location of the stored historical data. Data can be stored within the DCS, in the human–machine interface (HMI) system or on a separate box. When making this decision, keep in mind the following.

Use industry-standard computers and hard drives for data storage. The cost of the hardware for the data storage will be much less than the DCS vendor's proprietary platform, and there will be greater flexibility. Redundancy and sophisticated security are available in most readily available industrial servers.

Consider the use of two networks. One will be for the control system, and the other will be for data retrieval and analysis. This separates the network loading, and eliminates the risk of a "power user" analyzing last year's data and bringing down the control network. A third network, typically the largest, is the plantwide local-area network (LAN).

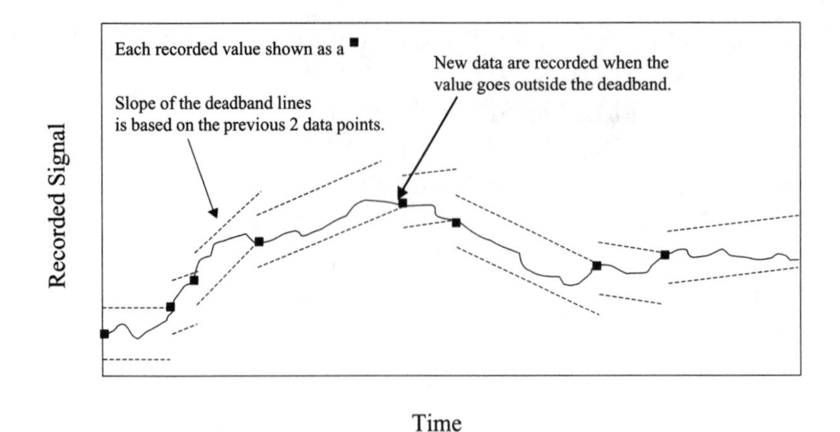

FIG. 1.5m
Swinging door data storage algorithm.

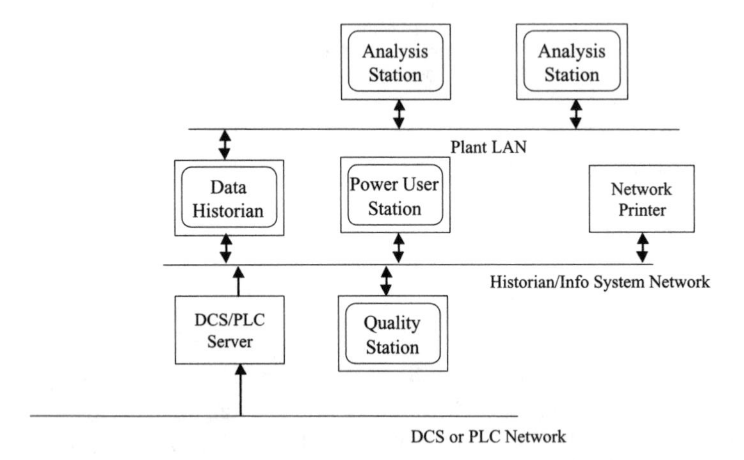

FIG. 1.5n
Diagram of recommended network arrangement.

Figure 1.5n shows a recommended network arrangement. The idea here is to minimize the network traffic on each LAN. At a minimum, keep data historian traffic off the control LAN as much as possible: Protect the basic function of the control network, which is to control the process. For systems connected to a plant or corporate LAN, coordinate efforts with the information technology (IT) group.

Data Compression

Some of the algorithms mentioned above are used for data compression. That is, data compression reduces the amount of data storage required, thereby reducing network load and hard disk storage requirements.

Meta-Data

Meta-data can be defined as "data about the data." For example, many DCS systems include meta-data about the quality of a process variable signal. The signal may be "out of range," "good," "manually substituted," or "questionable."

Capturing the meta-data is more of a challenge than simply collecting the number. Moreover, most analysis tools are not yet sophisticated enough to associate the meta-data to the raw values for analysis.

Most newer control systems will make signal quality data available to a data historian.

The Cost of Data Storage

There was a time when data storage was expensive. In fact, on older DCS systems, this may still be true. In these cases, one must be careful about what to store, and for how long. As the cost of electronics dropped in the 1990s, the cost of electronic storage plummeted dramatically. At this time, the cost of storage is ridiculously small. As an example, take the cost of storing one PV, collected continuously every second for 1 year.

$$2 \text{ bytes} \times 60 \text{ s/min} \times 60 \text{ min/h} \times 24 \text{ h} \times 365 \text{ days}$$
$$= 6 \text{ MB/year}$$

In 2001, a 20-GB drive, or 20,000 MB, cost about $400. So the cost of each year's worth of PV storage was roughly

$400 * 6 MB/20,000 MB = $0.12! And that was without any compression! So in most cases, cost of data storage is only a minor factor when designing a system.

HARDWARE SELECTION

Typically, data historians will bridge multiple networks. On one side, they collect data from the control system network, and on the other they make the data available to the plant LAN. This arrangement will minimize traffic on the control system network, improving capacity and stability of the control system.

Designing a computer network has become a fairly complicated subject. Refer to Chapter 4 for a more complete treatment of networks.

Data are typically stored on a hard drive system. This technology changes rapidly, and hard disks of greater and greater capacity are available each year. Data historian systems are usually designed to have the largest available hard drive. The incremental cost is small, and there will be room to collect data at higher frequencies and for longer time periods with a larger hard drive.

Hard drives, of course, are mechanical devices with moving parts. Although hard drive reliability has improved greatly, failures are still quite common over a 5- or 10-year life. Redundant systems, such as RAID arrays, are used to minimize the risk of lost data due to hard drive failure. There are many ways to implement these redundancy strategies. Discuss this with a knowledgeable computer consultant before attempting this.

Some RAID arrays use a parity-checking scheme, which is shown in Figure 1.5o. If any single drive fails, the parity-check drive can be used to reconstruct the lost data. The parity drive totals the bits in each column from each of the other drives. If the total is even, it records a zero. If the total is odd, it records a 1. Lost data are reconstructed by totaling the remaining drives, and checking the odd or even result against the parity.

To analyze the data, a computer station that is capable of accessing the data and running the software analysis tools is necessary. In most cases, the lowest-cost option will be to run the analysis software on the computers that already exist on the plant LAN. This has the added advantage of making the data and the completed analysis available via standard business applications, such as e-mail and word processing.

If the existing computers are not capable of handling the analysis, or if a dedicated station is desired, then it will be necessary to purchase additional stations. These stations should have a high-speed network card to gather the data from the historian. The amount of raw processing power required depends on the amount of data that will be analyzed. Most analysis stations will work well with low- to mid-range processors. A high-powered process may be required when analyzing very long-term data (years), or when doing complex analysis, such as correlation studies or frequency analysis.

Analysis stations should also have a simple way of exporting the data to other PCs. This can be directly via LAN connection, or via floppy, removable hard drive (or ZIP drive), or via a CD-writer. The choice of media depends upon what type of readers are available to the majority of downstream data users.

In smaller applications, one may be tempted to locate the analysis software directly on the same machine that is used for data collection and storage. This is not recommended. Frequent user access to the machine may cause operating system failures or machine re-boots. Data may be lost while the machine is recovering.

Backup Media

The backup media should be chosen based on the following criteria: durability, ease of use, and cost.

Durable media will last for many years. It may be surprising that CDs have an expected life span of only 10 years, and they are one of the better media available. And one should be sure to select a supplier that will be around for a while. One is not likely to need to restore from a backup for several years. If a disaster occurs, and it is necessary to restore the data, it is important to be sure that the device driver will work with the latest operating system. Backup media should also

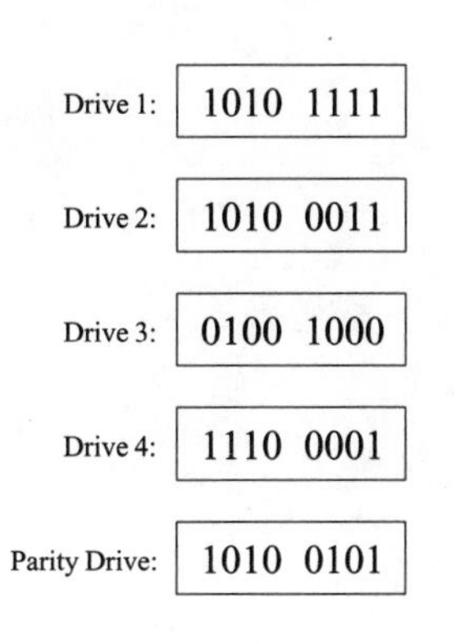

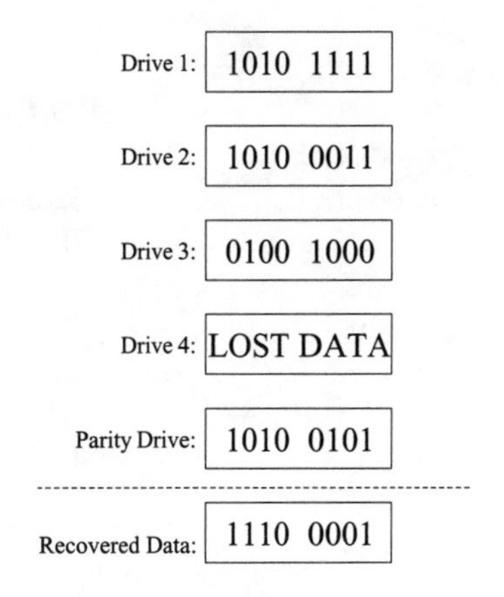

FIG. 1.5o
Diagram of parity-checking scheme.

be easy to use for both backup and recovery. There should be software backup tools that guide the process. In some of the better systems available today, the backup process can be automatically triggered, based on time and day. This way, one simply pops in the new media sometime before the next backup.

Also, be sure that the media can be easily labeled. If it is not labeled, it will not be possible to find the right version for recovery.

When selecting media for backup, the cost of media is almost as important as the cost of the drive hardware. Remember that one will be filling up the media on a daily, weekly, or monthly basis. A lot of media will be used over the life of the system.

ANALYSIS AND EVALUATION

The most basic of analysis tools is summarization. Using tools such as totalization or averaging, one can convert material usage rates and production rates into more meaningful numbers, such as tons produced per day, or tons consumed per month.

Because many DCS and PLC systems are equally capable of totalizing and averaging the numbers, a strategic choice about where to perform the calculation is necessary. In most cases it is best to perform the calculation as far down in the system as possible. This ensures consistency and also maintains the data in the event of loss of communication to the data historian.

When establishing any sort of summarization, be sure to document the basis for the calculation. Of particular importance is the time frame chosen for daily averaging. Some plants use midnight-to-midnight, some choose shift-start to shift-start times (i.e., 7 A.M. to 7 A.M.) to report production. Make sure that the data historian matches the established business practices. Managers love reports! It provides some insight into the process, and more importantly, it provides information for direct accountability, particularly for production information. In many cases, managers' bonuses are based directly on the production numbers that appear on their reports. So expect a lot of interest from management when planning the reporting system. Similarly, be sure to document the basis of any calculations, perhaps in a user's manual. Generally speaking, the daily production report is the heart of many reporting systems. This report will show daily production totals, and may also include usage of raw materials and energy. The most sophisticated of these reports will include pricing and profit data obtained through a link to another business tracking system.

Monthly reports simply totalize and collate the information that was contained in the daily reports. Depending on the needs of management, quarterly or annual reports may also be required. Figure 1.5p shows a typical daily production report.

Daily Production Report
XYZ Widgets, Inc.

Start Date	Start Time	Stop Date	Stop Time
1-Apr-01	7:00:00 AM	2-Apr-01	7:00:00am

Production

Total Tons :	527.3	
Good Tons :	506.3	Good Tons: 96.02%
Product A :	126.6	
Product B :	269.5	
Product C :	110.2	
Scrap :	21.0	

Raw Materials

Ingredient A:	433.1 tons
Ingredient B:	66.1 tons
Ingredient C:	32.3 tons
Gas:	1276.3 MCF

Time Analysis

Product A:	5:21
Product B:	12:17
Product C:	4:41
ChangeOver:	0:41
Maintenance:	1:00

FIG. 1.5p
Sample production report.

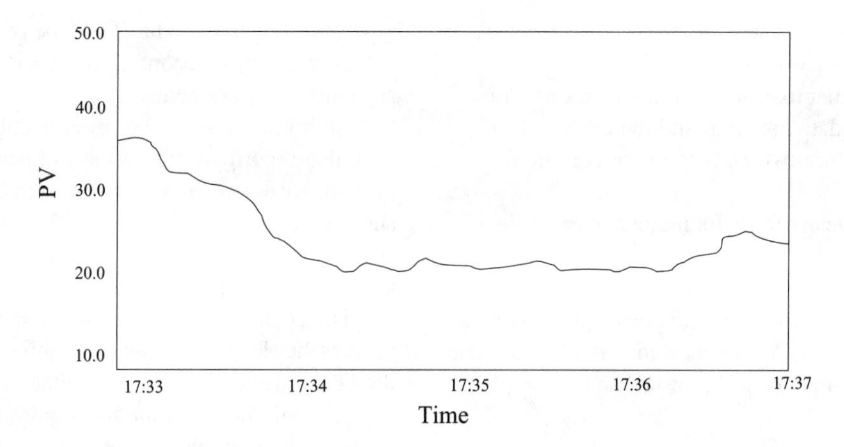

FIG. 1.5q
Diagram of a simple trend.

When designing the report, it is best to try to provide the data in a format that is directly usable. Try to avoid providing a report so that the user can enter the data into a spreadsheet. If users are going to do this every day, week, or month, why not design the system to provide the spreadsheet directly?

As reporting systems have become more sophisticated, it is now possible to add trends or other graphical information to the report.

Trends, of course, are a basic analysis tool included with most DCS and HMI systems. But they are usually limited to short-term analysis.

For data that stretches over months or even years, most DCS and HMI-based trending packages cannot handle it. So a trending package that is designed for the data historian will most likely be required. Figure 1.5q shows a simple trend.

Advanced trending packages will allow the user to adjust the timescale and the displayed range of data. This is commonly called "zoom-in" and "zoom-out" capability.

Even the most well conceived analysis system will not meet the needs of all users. "Power users" will always come up with some new analysis that they simply must have to solve a process or business problem. Although it is tempting to modify the design of the data historian or analysis tools, it is usually prudent (and cost-effective) simply to make the data available through export to a common data format.

The fact is that 99% of the users will not need the special feature that the power users wants. By exporting the data, power users can then apply whatever tools they wish. The responsibility for justifying and maintaining the specialized analysis tool falls back to them.

A simple export tool will simply dump out the data in a standard ASCII, CSV, or Excel file. More sophisticated export tools may allow one to filter or sort the data first, before exporting. Statistical quality control requires special trending and/or alarming tools. For more detail on the use of statistical process control tools, see *IEH*, Volume 2, Section 1.20 "Statistical Process Control."

To assess the relationship between variables, correlation tools are required. A variety of these tools such as JMP and SAS have been around for many years.

Data Filtering and Editing

Data are not perfect. Sensors fail. Wires fall off. Instruments drift out of calibration. Unexpected disturbances occur. To complete analysis of imperfect data, it is desirable to have data filtering and editing tools.

Filtering tools allow data selection based on a variety of criteria. The most obvious will allow selection based on a time window. But it may be desirable to be able to select data using more complex criteria, such as whether or not the point was in alarm, or whether a start-up was in progress, or whether it was a Sunday. A good set of filtering tools can be used to find patterns in the data quickly, and to prove or disprove hypotheses about plant problems.

Data editing tools allow the user to modify data that was collected automatically. This can be used to fill in gaps, or to correct for known instrument drift, for example. In the most sophisticated of systems, that database will keep track of all edits that were performed on the data.

SYSTEM TESTING

When starting up a data historian, plan on extensive system testing. One will want to confirm that the right data are being recorded, at the right frequency, and stored in the right place. Also, it is important to test the backup and restore system. Waiting until after a catastrophe to find out that all the valuable data were never backed up is not desirable.

SUPPORT OF THE DATA HISTORIAN SYSTEM

The biggest support issue will be to decide who owns the maintenance of the system. In most plants, the data historian falls right on the border between the instrument and control department and the IT department. The input of both groups is important to establish a successful maintenance plan.

Be sure to discuss issues such as:

- Where will the hardware be housed?
- Who will complete the backups?

- Who will be called in if there are hardware problems?
- Who will handle software problems?
- How to work together to handle communications problems between the data historian and the DCS or PLC. *Remember*: There are two ends to every communications problem!
- Where to find budgetary funds for maintenance, service, and upgrades.

There are many ways to structure the agreement between the various parties successfully. What is most important at this stage is to discuss the issues openly, and to establish a basic plan.

Security

Depending on the policies of a company, data security may be a very big concern. In many plants, information such as production rates, downtime, and material usage are considered proprietary information.

To use the data to the fullest extent, the historian should be resident on the plantwide or corporate network. However, the larger the connected network, the greater the concerns about unauthorized access to critical business information.

The security system for a data historian can range from simplistic to very complex. It is best to consult with the IT department to ensure that an appropriate security system is selected. The goal is to maintain the security of the data, while making authorized access as simple as possible. The best systems will allow integration of data historian security with the existing plant network security.

A more detailed treatment of security issues can be found in Section 2.7, "Network Security."

Backup, Archive, and Retrieval

Be sure to establish a backup system. It would be a shame to collect all of this valuable data, then lose the data because of a hard drive crash. Backups are typically made to some type of removable media. The type of media chosen depends on several factors: economics, expected life of the data, and regulatory requirements.

Archives back up the process data. A complete backup will also record configuration information for the operating system, hardware, and data historian configuration. Archives should be made daily or weekly, and a complete backup should be done when the system is installed, and at least annually thereafter.

The economics of removable media seem to change each year, as the electronics industry continues to evolve. Check with the IT department or a PC consultant to evaluate these options.

A word about regulatory requirements: If the data must be kept on file to meet government regulations, be careful in the choice of media. Most electronic media (even CD-ROMs) have an expected life of only 10 years. Even if the media lasts, will the media be capable of replay in 10 years, on the next generation of computers? Unfortunately, paper and microfilm remain the media of choice for storage that must last more than 10 years.

Of course, a backup is useless if it cannot be retrieved. The retrieval system should allow data to be restored fully. Depending on the requirements, one may want to restore the data to the primary data historian, or one may want to restore the data to another off-line system. Be sure to inquire about the capabilities.

Bibliography

Eckes, G., *General Electric's Six-Sigma Revolution,* New York: John Wiley & Sons, 2000.

Hitz, P., *Historical Data in Today's Plant Environment,* Research Triangle Park, NC: ISA Press, 1998.

Johnson, J. H., "Build Your Own Low-Cost Data Acquisition and Display Devices," 1993.

Lipták, B., *Optimization of Unit Operations,* Radnor, PA: Chilton, 1987.

Liu, D. H. F., "Statistical Process Control," in *Process Control (Instrument Engineer's Handbook),* 3rd ed., Radnor, PA: Chilton, 1995, pp. 138–143.

1.6 Integration of Process Data with Maintenance Systems

G. K. TOTHEROW

Businesses are using every advantage and innovation to produce their products at a lower cost. Manufacturing initiatives like just-in-time manufacturing (JIT), total quality management (TQM), lean manufacturing, and agile manufacturing are implemented by every major producer. Quality programs and control and automation projects are used to reduce operating expenses while general cost savings programs, outsourcing, and other initiatives are used to drive down fixed costs. Process control and maintenance management are two areas of manufacturing that are seeing incredible technical advances and system acceptance because of the return on investment they provide on the manufacturing process. E-business and enterprise resource planning (ERP) systems exploded on the scene in the late 1990s, making businesses see the potential of integrating data from the plant floor all the way up to the business systems. This has had an effect on every other plant system. Many people see synergy and further savings to the plant by interfacing systems that developed separately for years. The purpose of this section is to discuss the benefits and the methods of integrating the process information with the maintenance systems.

The section will describe plant floor automation that sets the stage for electronic data transfer, the maintenance function, computerized maintenance management system (CMMS), preventive maintenance and condition monitoring systems, the information that helps operations and maintenance, and the integration to the CMMS.

PLANT FLOOR SYSTEMS

The Honeywell TDC 2000 was the first commercial distributed control system (DCS) when it was introduced in 1975.[1] Thus began the large-scale computerization of the plant floor. The concept behind the distributed control systems was to use microprocessor technology for implementing control, but unlike a computer control system, control loops were distributed among many smaller controllers each with its own central processor units. The controllers were linked together along with operator consoles and engineering stations through a communication highway. In succeeding years the DCS continuously expanded functionality, access to process data, and interfaces to numerous other systems. New open system instrument highways, device highways, local-area networks (LANs), and wide-area networks (WANs) expand the DCS structure and the complexity of the plant floor systems.

Process and manufacturing plants are evolving into complex systems of electronic controllers and communication networks that have a plethora of information available to assist and sometimes confound plant personal. Smart transmitters and smart valves communicate on device highways that carry the process data, calibration information, transmitter diagnostics, and other information. Special instrumentation indicates the health of equipment and process systems. Expert systems infer process and system problems that are not evident to operations. Control systems produce data and alarms and even system diagnostics to assist with problem detection and correction. The leading edge of proven industrial systems is well beyond producing health alarms and systems that continue to operate through component failure—the systems now must help the plant personnel respond to the alarms more efficiently.

Plant systems are becoming more open and better at assimilating process information for the user from a variety of controllers and input/output (I/O) devices. The proprietary DCS of the past is being replaced with open platforms. Human–machine interface (HMI) and plant historian devices are selected at many plants specifically to integrate process data from several different systems and older legacy systems. Wonderware's InTouch, Rockwell's RSView32, Emerson's Intellution, or any of a host of other HMI products can gather process information from virtually any control or I/O device to present the information to the operator. Their interface capabilities and ability to bridge between dedicated control highways and general plant data highways make them ideal to act as the interface system for all real-time data to other plant systems (Figure 1.6a). Plant historian systems like the Aspen Tech Info21, the OSI PI, and the Wonderware Industrial Sequel server are also used to blend data from several different systems and may be used as the mechanism to interface process data to ERP systems and the CMMS.

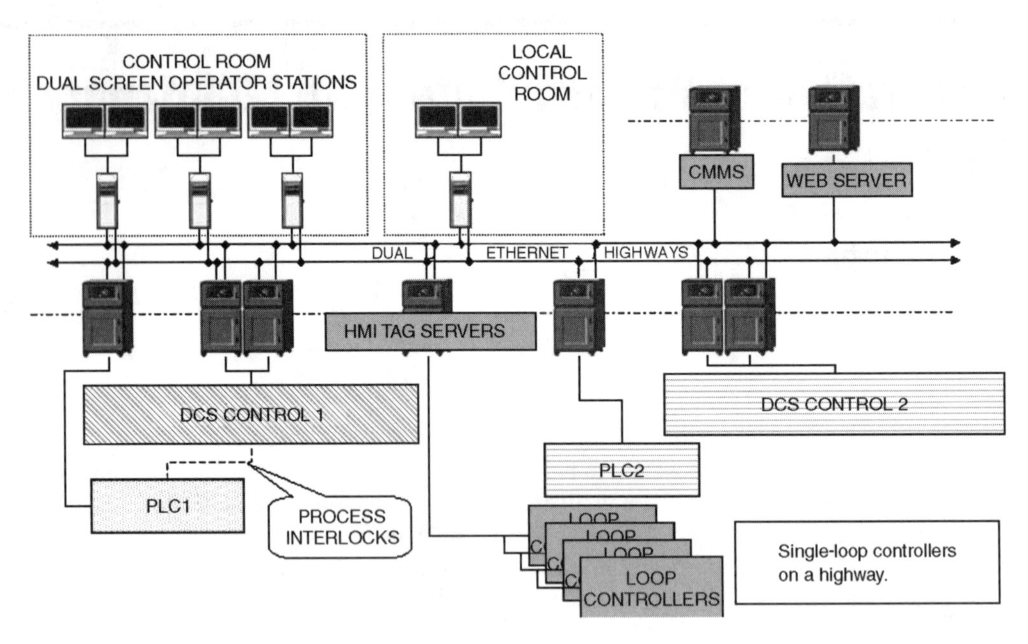

FIG. 1.6a
Independent HMIs interfacing to proprietary controls.

MAINTENANCE

Maintenance of manufacturing and process equipment is a large part of the cost of operation in many industries. In his book *Uptime*, John Dixon Campbell[2] shows the estimated costs of maintenance as a percent of the total cost of production for several industries. The costs of all of the industries varied from mining at 20 to 50% and fabrication and assembly with estimates of 3 to 5%. Terry Wireman,[3] in his book *Computerized Maintenance Management Systems*, estimates maintenance costs of process industries at 15 to 40% of the total cost of production. Wireman goes on to explain that given the profit margins in most of the industries, $1.00 saved in maintenance is as much as $3.00 in new sales.

Companies have a compelling reason to try simultaneously to increase the effectiveness of maintenance while lowering the cost of asset management. The only way to increase the effectiveness of the maintenance is to concentrate on performing only the tasks that must be done to keep assets operating at peak performance.

Wireman characterizes maintenance as having five primary functions: maintenance of existing equipment, equipment inspections and service, equipment installation, maintenance storekeeping, and craft administration.[3] Practices and systems are evolving in manufacturing and process industries to reduce the cost of maintenance, increase the production capacity, and increase the service years of assets. The study of the maintenance function is becoming increasingly popular with numerous excellent books available on the subject and dozens of consultants teaching maintenance management systems and techniques. These initiatives mirror the manufacturing initiatives and strategies implemented at the corporate level. The purpose of the maintenance strategies is to help determine what is important to define as a maintenance task, to whom to assign the task, and identification of the key indicators of maintenance performance.

Maintenance is a business. Performing the five functions of maintenance described by Wireman using the best maintenance methods and practices requires loads of documentation, data, and data analysis. It is possible to operate the business of maintenance manually, but like balancing the books of a large financial institution without computers, it is not feasible. It is the function of the CMMS to automate the business of maintenance and store the records and data for tracking, analysis, reporting, and predicting.

COMPUTERIZED MAINTENANCE MANAGEMENT SYSTEM

The software that automates the business of maintenance is a CMMS or, somewhat interchangeably, an enterprise asset management (EAM) system. Some experts will state that an EAM is not the same as a CMMS because it is multisite or integrated to other enterprise-level systems.[4] This section will use the term CMMS because it is more descriptive of the software, it has been used far longer, and it does not necessarily exclude multisite systems or integration to other systems.

CMMSs have not been in general use for very long. The third edition of the *Maintenance Engineering Handbook* in 1997 never discussed computerization of maintenance. In 1977, flow-of-work requests were paper work tickets that were used to schedule and document work.[5] Yet, the CMMS

TABLE 1.6b
Some of the Many Condition Monitoring Systems That Can Be Used in a Predictive Maintenance Program

Detection System	*Failure*	*Equipment*
Vibration and lube analysis	Bearing loss	Rotating machinery
Spectrographic and ferrographic analysis	Lubrication	Rotating machinery
Time resistance tests	Insulation	Electrical
Thermography	Discharge	Electrical
Corrosion meters, thickness checks	Corrosion	Tanks, vessels
Process system—heat transfer calculations	Fouling	Heat transfer equipment
Valve performance testing	Stiction, hysteresis	Valves
Transmitter diagnostics	Transmitter, element	Process instrumentation
Control system diagnostics	Component, communication	Control system

has been around long enough that some plants are installing their second or even third system. It was, and still is, very common for companies to design their CMMSs to duplicate their standard practices and procedures of maintenance. It was, and still is, very common for a CMMS alone to fail to improve the effectiveness of maintenance. Nevertheless, several new and exciting things are changing the CMMS to dramatically improve its effectiveness. Computers and the software that comprise a CMMS are lower in cost and higher in functionality, making the CMMS available to smaller organizations. ARC was predicted worldwide shipments of CMMS software and service was $1.31 billion in 2001 rising to $1.61 billion by 2005.[6] Maintenance consultants are convincing industry to implement best maintenance practices along with installation of the CMMS rather than automating ineffective strategies. And last, the CMMS is being integrated into plant and company systems to further automate the tasks of maintenance.

CONDITION MONITORING AND PREDICTIVE MAINTENANCE

Predictive maintenance (PdM) is a maintenance function that uses monitoring equipment as indicators of the health of assets to detect signs of wear that lead to failure of a component.[7] Condition monitoring (CM) is very similar, and both CM and PdM are used interchangeably when they describe the systems. If there is a discernible difference between the two, it is that CM describes the instruments and systems that monitor the equipment condition and PdM describes the function of scheduling maintenance based on the predicted future degradation and failure. This section will use PdM to describe the maintenance function and CM to describe instrumentation and systems.

CM systems are most commonly used in industry (see Table 1.6b) in the following areas:

- Vibration analysis
- Oil analysis
- Thermography
- Corrosion monitoring
- Process parameter trending
- Instrument/control element monitoring

CM systems have a function different from the CMMS and have developed separately in the workplace; each has its own merits and users.[8] PdM based on equipment conditions has been around for a long time and has been used a great deal with certain types of equipment—transformers and rotating equipment such as motors, generators, and turbines. If preventative maintenance can be described as "fix it before it breaks," then PdM can be described as "fix it when it really needs to be fixed."[9] PdM requires interpretation of the data to determine what caused change in the data and when the piece of equipment might fail. A highly trained technician is needed to interpret the data to determine causes and predict equipment failures. However, expert systems in the PdM system interpreting the online data can generate a work request in the CMMS for the technician to review the problem. This is one of the major benefits of integrating the CM system of online instruments with the CMMS.

OPERATION AND MAINTENANCE NEEDS

Operating personnel use control systems to manipulate equipment to produce a product. Maintenance is responsible for keeping the control systems and equipment functioning. The control systems and the equipment itself provide valuable

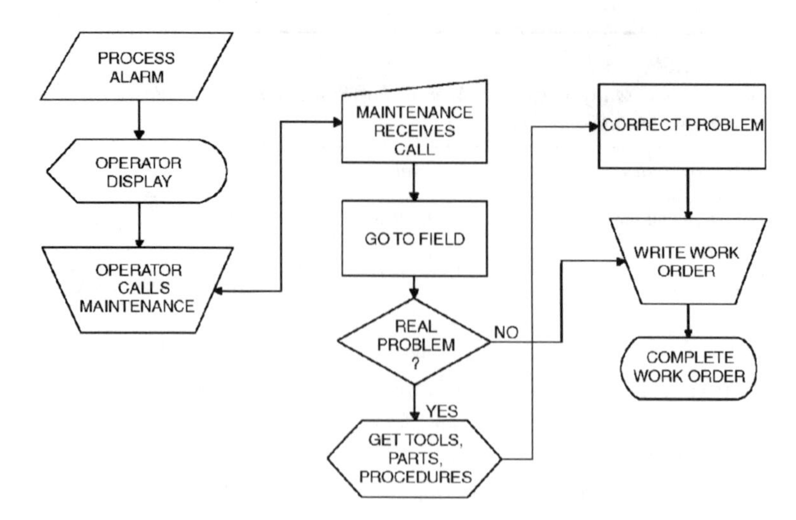

FIG. 1.6c
Response to a system alarm.

information to both operations and maintenance. The CMMS is used to help maintenance work more effectively, but operating personnel initiate many of the maintenance work requests and need to know when work will be performed on process equipment. Where, and to whom, the information is displayed is more commonly a function of the system rather than a logical decision of whom and in what form the information should be presented. Today, both the CMMS and the control systems have evolved to be able to transfer information like the business systems.

Often, process and manufacturing plants have complex systems of electronic controllers and communication networks. Good engineers build into these systems diagnostic information and alarming to indicate the health of the systems and the process. Process operators need system alarms when a fault causes a problem that requires a change to the process. What about an alarm that the redundant control system data highway has failed? This system fault does not require a change to the process operation. The system fault is only an alarm event to prompt the operator to call maintenance to correct the problem. Today's systems put operating personnel, with limited technical knowledge of system alarms, as the only method of alerting maintenance of equipment problems. This creates inefficiencies, delays, and unnecessary trips to the field as shown in Figure 1.6c. As the plant floor technologies become even more complex and redundant, it is likely that even more systems will need to notify different groups within maintenance of problems that do not immediately affect the process operation. A better situation is created when the control system equipment alarms are integrated into the CMMS to produce an automatic emergency work order with complete documentation. In fact, it is completely feasible to connect the process control systems to the CMMS to have a process alarm generate a work order, prepare a work execution package, assign the maintenance mechanic, and notify the mechanic by personal digital assistant (PDA). This ensures that the correct person is called, that the person has the best information available on the system and the problem, and that the whole incident is captured by the CMMS so that the historical and financial information is kept in the records. Alarms can be written in the control system for network failures, communication failures, controller errors, equipment monitoring instrumentation, and transmitter and valve problems with the intent purpose of generating a work order and capturing the problems. The flowchart of this process is shown in Figure 1.6d.

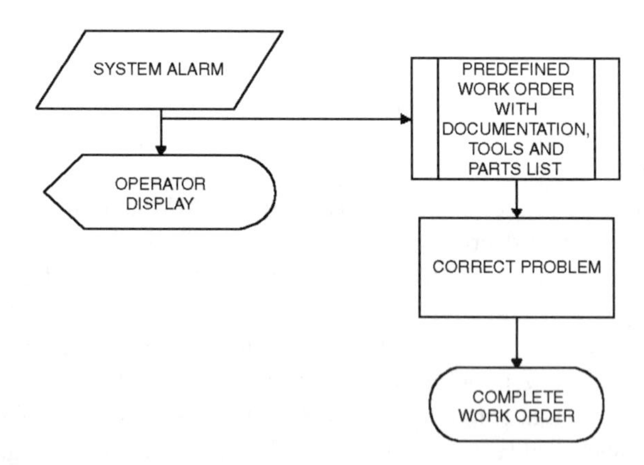

FIG. 1.6d
System alarm that is integrated into the CMMS.

Integration of systems implies bidirectional information flow. Operating personnel need periodic access to the CMMS from their HMI to enter a work request or to determine if maintenance corrected a process or equipment problem. Equipment histories and other information from the CMMS improve the ability of operating personnel to perform their jobs efficiently. Similarly, maintenance needs information that is in the process systems such as motor stop/starts, equipment operating hours, operation outside of design specification, temperature alarms, statistical process control (SPC) alarms, vibration and temperature alarms, and many more. Control system HMIs are sometimes installed in maintenance shops specifically for maintenance technicians to see system diagnostics. Certainly, this can save a little time in diagnosing the problem if the maintenance person is in the shop at the time of the problem. But maintenance works most effectively when working from a planned work order and capturing all information and history of the work in the CMMS. Cost savings from better, more efficient asset management is driving this integration of the systems.

CMMS INTEGRATION

Business systems like accounting, payroll, and purchasing were computerized long before other plant functions. When the maintenance storeroom and other functions became automated with the CMMS, there was obvious overlap with information in the other business systems. There was a great need for electronic data exchange interfaces between the CMMS and plant financial systems long before the plant floor was automated with computers. It is best to review the interfaces between the CMMS and financials before discussing the integration with other systems.

The CMMS is a business system that is used to order, purchase, and receive materials and services, track work-hours, charge time and material against plant equipment and projects, track costs, plan work orders, and numerous other functions. Most of this information must be reconciled with other company business systems such as accounting and the general ledger, payroll, time and attendance, production scheduling, and even customer management systems. This was accomplished manually by taking reports from one system and periodically typing the information in the other. Many companies wrote code between the dissimilar systems to interface data between them on a point-by-point basis to increase the efficiency and accuracy of information needed by both systems. Figure 1.6e shows an extremely complicated point-to-point integration model for exchanging information between the systems.

The point-to-point integration of systems has problems and limitations in large systems such as "breaking" when the software of either system is modified and high costs associated with maintaining and modifying the custom code. Some companies decided that the cure is buying one system that can perform all of the business functions in a company from marketing, to production scheduling, to maintenance management. This system is called an enterprise resource planning (ERP) system. ERP solutions met with limited success and some spectacular failures in the late 1990s; yet, they are still the first choice solutions for some companies based more on the tight integration of the various packages. However, many companies cannot afford the extravagant cost of an ERP and need to use a best-of-breed approach to selecting systems.

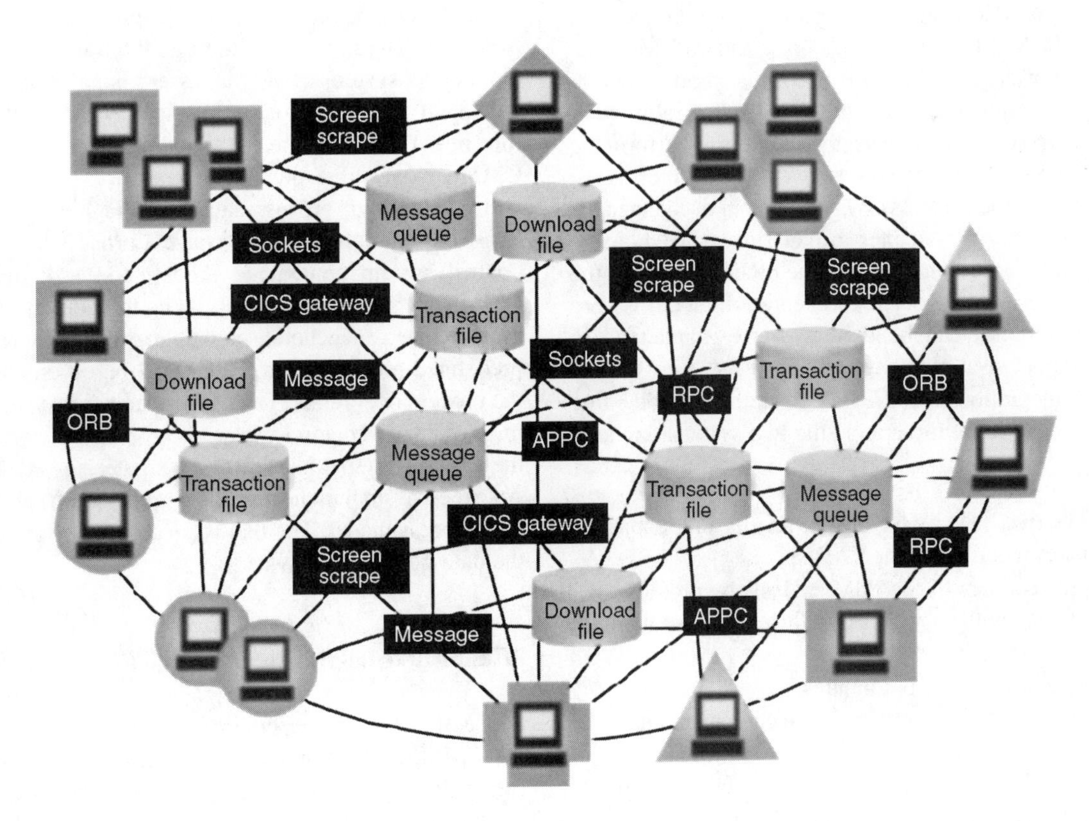

FIG. 1.6e
A complicated point-to-point interface between systems. (Courtesy of Wonderware Corporation.)

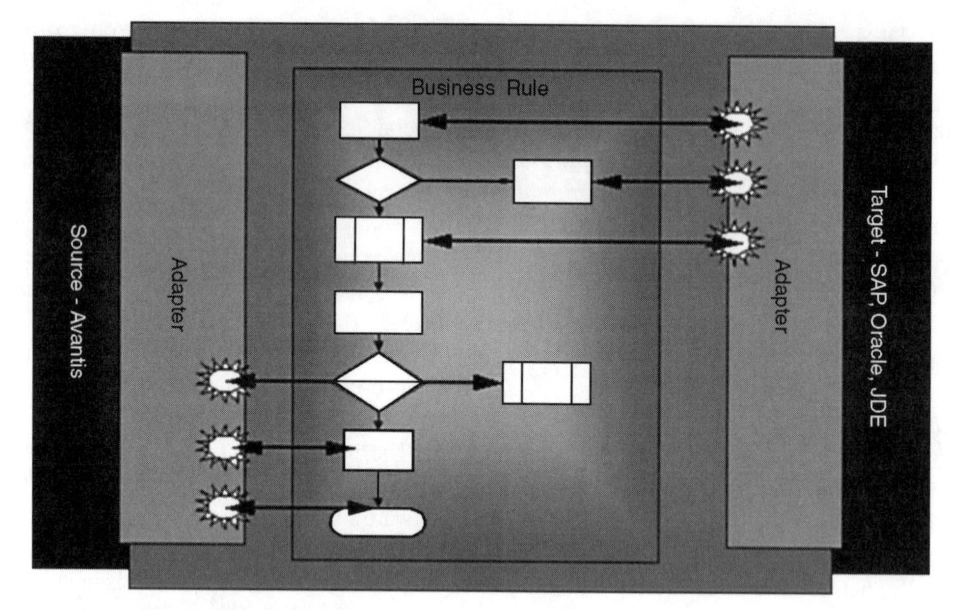

FIG. 1.6f
The Wonderware adapter for interfacing the CMMS with any business system. (Courtesy of Wonderware Corporation.)

These companies still want tight integration of systems. Vendors of best-of-breed systems are moving to fill that need using configurable adapter software to integrate data between systems. Figure 1.6f shows the enterprise application integration (EAI) of Wonderware for interfacing its CMMS with any business system.[10]

The EAI uses an adapter to expose objects in systems, such as a CMMS or ERP, using industry-standard XML. The XML version of the object can then be understood by one or more other systems. The EAI will have configurable business-rule-based software between the adapters to perform the linking of the objects in the two systems.[11] The adapters serve to isolate each of the systems. When a system is changed, or even replaced, only the adapter to that system needs to be changed with no additional work to the other systems or the business middleware. This type of integration model is very important because it can be used to make it practical to interface data between many systems.

CMMS integration with the PdM systems can add value to both. The benefits of integrating the two systems include:

- More effective PdM by having the results and recommendations recorded in the CMMS
- Improved accuracy of the PdM analysis by providing the technician with easy access to the equipment work history
- Identification of repetitive failures
- More effective communication of the equipment health throughout the organization
- Automatically generating PdM work orders from alarms of potential equipment failure from the expert system in the CM system
- More effective use of highly trained maintenance technicians by assisting with troubleshooting, documentation, and lowering the number of healthy checkups on equipment

CM systems intersect with the process data systems in two ways. The first way is through the transmitter and valve diagnostic systems like the asset management solutions (AMS) of Emerson Process Management. AMS looks at the transmitter and valve health signals and can interface with a CMMS to generate predefined responses. The second way CM is used with process data is to diagnose process problems. It is straightforward to have a process alarm from the control system generate a predefined work order in the CMMS. A more difficult task is the integration of an expert system that can generate a whole range of messages and predefined procedures to the CMMS based on anomalies in the process operations. The use of the expert system to determine, read, and diagnose process problems is very similar to the expert systems needed to automate any PdM function. An expert system in the process world could be an SPC alarm or a process model. The real issue comes back to interfacing the data between the systems.

INTEGRATION TECHNIQUES AND BUSINESS ALLIANCES

It does not take an expert in predictive analysis to see that the CMMS, CM systems, and process data systems of the future will be capable of being tightly integrated. The integration between these varied systems is evolving but is still in its infancy right now. What are some the integration techniques one would expect to see employed to interface these systems to each other?

- MIMOSA—The Machinery Information Management Open System Alliance group is comprised of CM vendors, CMMS vendors, end users, machinery insurers, control system vendors, and system integrators. A major focus of the alliance is to provide recommendations and conventions to unify machinery data from the various plant systems to provide companies with the ability to create comprehensive machine information management systems.[12]
- ODBC (Open Database Connectivity) Drivers—The ODBC connectivity of many of the systems provides the ability to exchange data or it allows a data warehouse to be used between two systems.
- OPC (OLE for Process Controls)—A data exchange format defined by a set of standard interfaces based on Microsoft OLE/COM technology.
- Web Content—Dynamic HTML or Web portals may be employed to allow access to information from the various systems. Although not really exchanging the information, the Web page can integrate the information for the viewer.
- XML Adapters—Extensible Markup Language adapters, like the one shown in Figure 1.6f, can be used to provide the interface between the systems.

Business alliances are forming quickly among the various process data, CMMS, and CM system companies to help the companies integrate their products even before industry groups have worked out the open standards between all the vendors. It should be noted that nearly all the vendors have alliances or an integration strategy. The list of integrated systems and alliances below is not complete but gives an idea of the integration efforts by some of the major suppliers.

- Foxboro has fully integrated the MAXIMO CMMS product and Invensys' Avantis CMMS with the Foxboro I/A Series DCS.
- The Honeywell Equipment Health Management (EHM) system is integrated into both MAXIMO and SAP maintenance to generate work orders automatically.
- Wonderware has fully integrated its InTouch HMI software with Avantis CMMS software.
- Emerson Process Management AMS CM system is interfaced with the MAXIMO CMMS to create a work order when a valve or instrument needs calibration.
- Datastream, a CMMS company, and PREDICT/DLI, a provider of diagnostic tools and services for PdM tasks, formed a strategic alliance agreement to integrate the detection of machine wear with the scheduling of maintenance and cost analysis.

SUMMARY

Businesses must use every advantage to produce their products at a lower cost. Maintenance is the largest controllable expenditure in a plant. Computerized maintenance management systems are necessary to manage the business of maintenance. The CMMS has been interfaced to various business systems for years to automate the maintenance business. The plant floor systems are electronic and open to enable data transfer. Operation and maintenance functions can be further automated and made more efficient through the integration of process data and maintenance data from the CMMS. Interfaces between process data and maintenance data are available and improving. Business alliances are showing that the vendors see a future where there will be tight integration of their products with other systems to enable plants to operate more efficiently.

References

1. Freeley, J. et al., "The Rise of Distributed Control," *Control Magazine*, December 1999.
2. Campbell, J. D., *Uptime: Strategies for Excellence in Maintenance Management*, New York: Productivity Press, 1995.
3. Wireman, T., *Computerized Maintenance Management Systems*, 2nd ed., New York: Industrial Press, 1994.
4. Wireman, T., "What's EAM Anyway," *Engineer's Digest*, December 1998; Web site Indus International, www.indusinternational.com.
5. Higgins, L. R. and Morrow, L. C., Eds., *Maintenance Engineering Handbook*, 3rd ed., New York: McGraw-Hill, 1977.
6. ARC Report, April 2001.
7. Wireman, T., *Developing Performance Indicators for Managing Maintenance*, New York: Industrial Press, 1998.
8. Wetzel, R., *Maintenance Technology Magazine*, December 1999.
9. Lofall, D. and Mereckis, T., "Integrating CMMS with Predictive Maintenance Software," www.predict-dli.com.
10. Wonderware Corporation, "The Avantis E-Business Integration Strategy Making the Connection," White paper.
11. Wonderware Corporation, "The Avantis E-Business Integration Strategy Making the Connection," White paper.
12. MIMOSA Web page, www.mimosa.org.

1.7 Applications, Standards, and Products for Grounding and Shielding

D. MORGAN

Partial List of Suppliers:

Grounding Accessories, Equipment, and Services: Cpeitalia Spa; ElecSys, Inc.; Hydrocom; SPGS; Lyncole; Micro-Ray Electronics, Inc.; Newson Gate Ltd; Panduit; PowerEdge Technologies, LLC; Priority Cable; Sankosha Engineering PTE, LTD; Senior Industries; Shore Acres Electrical (SAE), Inc.; Silver Lock; Superior Grounding Systems

Shielding Products: APM Advanced Performance Materials; Die-Cut Products Company; Dixel Electronics Ltd; EAC Shielding; Hi-Pac Technologies, Inc.; Holland Shielding; MAJR Products Corp.; Omega Shielding; RFI Shielding Limited; Shield Resources Group; TBA Electro Conductive Products; The MμShield Company; Thermshield LLC; Tecknit; Texas Instruments

Electrostatic Discharge Products: CPS Custom Products and Services, Inc.; Legge Systems; Micro-Ray Electronics, Inc.; NRD Static Control; Richmond Technology; Robust Data Comm; Static Specialists Co., Inc.; ZMAR Technology; ESD Systems.com

Computer Flooring (ZSRGs): Computer Floors, Inc.; Computer Power Technology; Harris Floor Tech., Inc.; Pugliese Interiors Systems, Inc.; Tate

To the instrument engineer, grounding and shielding probably do not generate the interest enjoyed by more academic and less practical pursuits. It would seem preferable if grounding and shielding could be reduced to a few simple rules. Certainly, continually revisiting an issue on which safety is so dependent would have its drawbacks. In practice, an engineer with a few years of experience is likely to run across not only a good deal of variability in grounding and shielding application, but also a good deal of confusion. In fact, all too often it is simply accepted that there is a trade-off between safety and noise reduction. Getting it right seems more of an art than a science. However, when one starts with the laws of physics and tacks on knowledge of codes and standards, one will be pleasantly surprised to find complete compatibility between what sometimes seem conflicting goals.

GROUNDING- AND SHIELDING-RELATED STANDARDS

Electrical/electronic standards come in two basic forms: those designed for field engineer guidance and those aimed at establishing standards for equipment vendors. Standards-producing bodies are located worldwide. Some standards bodies, like the National Fire Protection Agency (NFPA) and the National Electrical Manufacturers Association (NEMA), are national in scope. Others, like the International Electrotechnical Commission (IEC), are not aligned with a single country. Usually, there is no problem with adopting a standard not specifically targeted for a given country. For example, in 1999, NEMA published the results of a project to analyze NFPA 70—"National Electrical Code" (NEC) and IEC 60364—"Electrical Installations in Buildings." "Electrical Installation Requirements: A Global Perspective" is a 90-page report that concluded NEC code compliance leads to compliance with IEC 60364 Section 131. It also concluded that unlike NEC, IEC 60364 is not intended for direct use by designers, installers, or verification bodies, only as a guide to develop national wiring principles.

Where safety is an issue, federal, state, and local authorities adopt certain standards as law. The standards themselves are not necessarily considered law without legislation that specifically adopts them. Therefore, it is important to identify which body has jurisdiction over an installation and which standards it enforces. The exception is Part S of the Occupational Safety and Health Administration (OSHA) law, 29 Code of Federal Regulations 1910. Organizations that issue standards and develop codes related to shielding and grounding include the following:

National Fire Protection Agency (NFPA) (1 Battery Park, P.O. Box 9101, Quincy, MA 02269-9101. 1-800-344-

3555. www.nfpa.org): Publishes the U.S. "National Electrical Code" and approximately 300 other fire safety–related documents. Also sanctions a handbook reference published by McGraw-Hill. The code itself is primarily text based and uses highly specific wording. The handbook supplements the code with detailed discussions and useful diagrams, figures, and photographs. NFPA also provides a cost-effective Web-based subscription service. NEC code covers nonutility electrical installation topics applicable to many facilities. The same rules apply to households, refineries, and steel mills. Most municipalities, as well as OSHA, adopt parts of NEC code as law.

Institute of Electrical and Electronics Engineers (IEEE) (Operations Center, 445 Hoes Lane, Piscataway, NJ 08855-1331. 732-981-0060. www.ieee.org): IEEE is a leading international authority in technical areas ranging from computer engineering, biomedical technology and telecommunications, to electric power, aerospace, and consumer electronics. Formed of ten worldwide regional groups; IEEE publishes the "National Electrical Safety Code" (NESC) and the associated handbook. The NESC is what utilities in the United States look to for guidance. IEEE publications cover areas of interest to both field engineers and vendors. Of particular interest is the IEEE "color" book series. IEEE color books are loaded with useful figures and examples. IEEE standards frequently reference NEC code.

International Electrotechnical Commission (IEC) (3 rue de Varembe, P.O. Box 131, CH–1211 Geneva 20, Switzerland. +41 22 919 02 11. www.iec.ch.com): Founded in 1906, IEC is a world organization that publishes international "consensus" standards for all electrical, electronic, and related technologies. More than 50 countries are represented in standards development. IEC publications cover areas of interest to both field engineers and vendors. The IEC mandate is to publish international consensus standards to remove technical barriers to trade, to develop new markets, and to fuel economic growth. IEC standards represent the core of the World Trade Organization Agreement on Technical Barriers to Trade. Although the focus of IEC is on commerce, the field engineer can still benefit from a familiarity with its publications.

CSA International (Canadian Standards Association) (178 Rexdale Boulevard, Toronto, Ontario, M9W 1R3, Canada. 416-747-4000. www.csa.ca): CSA is an independent Canadian standards publishing organization that publishes the Canadian Electrical Code.

International Association of Electrical Inspectors (IAEI) (901 Waterfall Way, Dallas, TX 75080-7702. 972-235-1455. www.iaei.com): Publishes the *Soares Book on Grounding*. IAEI publishes material targeted at field engineers. NFPA offers the *Soares Book* as part of its standards collection.

National Electrical Manufacturers Association (NEMA) (1300 North 17th Street, Rosslyn, VA 22209. 703-841-3200. www.nema.org): NEMA sets standards that electrical equipment vendors follow to provide "certified" options to field engineers. Most engineers are familiar with NEMA ratings for enclosures.

Electronic Industries Alliance/Telecommunications Industries Alliance (EIA/TIA) (2500 Wilson Boulevard, Arlington, VA 22201. 703-907-7500. www.eia.org): EIA/TIA publishes standards primarily dealing with communications equipment and cabling. EIA/TIA standards are the primary focus of telecommunications and network engineers. Still, EIA/TIA standards are becoming more and more relevant to instrument engineering as instrument buses and control system connectivity evolve.

Insulated Cable Engineers Association (ICEA) (P.O. Box 440, South Yarmouth, MA 02664. 508-394-4424. www.icea.net): ICEA is a professional organization dedicated to developing cable standards for electric power, control, and telecommunications industries. The most important issue at this point involves upgrading the standard for high-speed Ethernet cabling. The most recent adopted standard is for Category 6 cable designed for gigabit Ethernet.

Standards, codes, and U.S. laws relevant to grounding and shielding from the above organizations can be found in Table 1.7a. Obviously, the amount of available material is a bit daunting. Still, one can appreciate the critical importance of familiarity with at least a few such as NEC, NESC, OSHA, and IEEE color books.

POWER GROUNDING BASICS

Power grounding practices for electrical safety, static electricity management, shielding, and noise reduction are interrelated and do not conflict with one another. Grounding practices for safety and signal integrity purposes are rooted firmly in the fields of electrostatics, magnetism, and radiation. Grounding has many purposes including:

1. *Providing a low impedance path for fault currents back to over current protection devices.* When a fault to ground occurs, it is desirable to interrupt and isolate the offending circuit and equipment. An overcurrent protection device needs to see a much larger current than it does during normal operation to perform this function. When all equipment is low impedance bonded back to the grounding electrode via the equipment grounding conductor, current due to line faults to equipment or ground will cause momentarily high currents back at the overcurrent protection device.
2. *Reducing shock hazards to personnel.* When a circuit faults to conductive equipment, the equipment becomes part of the circuit. If the equipment is grounded, the current flows at "zero" voltage and is therefore not a shock hazard. When all equipment is low impedance bonded back to the overcurrent device, such faults should not persist and should be immediately obvious.
3. *Providing fire protection.* If a low impedance path back to circuit overcurrent protection is provided, fault currents will not radiate heat energy and can readily be isolated. Because 100 A flowing in 0.1 Ω produces 1 kW of heat, if the overcurrent device is designed to trip at 120 A, current will continue flowing until the conductor melts, possibly causing a fire.

TABLE 1.7a
Grounding- and Shielding-Related Standards

Code Body	*Standard Number–Year*	*Description*
NFPA	70-1999	National Electric Code (NEC)
	1999	NEC Handbook
	77-2000	Recommended Practice on Static Electricity
	79-1997	Electrical Standard for Industrial Machinery
	780-1992	Standard for the Installation of Lightning Protection Systems
UL	96A	Lightning Protection Systems
OSHA	29 CFR 1910-303	Subpart S, Electrical—Design Safety Standards for Electrical Systems—General Requirements
	29 CFR 1910.304	Subpart S, Wiring Design and Protection
	29 CFR 1910.305	Subpart S, Wiring Methods, Components, and Equipment for General Use
	29 CFR 1910.307	Subpart S, Hazardous Locations
U.S. Department of Commerce/NIST	FIPS Pub 195	Federal Information Processing Standards Publication, Federal Building Grounding and Bonding Requirements for Telecommunications
IEEE	C2-1997	National Electrical Safety Code (NESC)
	1997	NESC Handbook
	1943-1997	NESC Interpretations Collection
	141-1993(R1999)	"Red Book," Recommended Practice for Electric Power Distribution for Industrial Plants
	142-1991	"Green Book," Recommended Practice for Grounding Industrial and Commercial Power Systems
	241-1990(R1997)	"Gray Book," Recommended Practice for Electric Power Systems in Commercial Buildings
	242-1986(R1991)	"Buff Book," Recommended Practice for Protection and Coordination of Industrial and Commercial Power Systems
	399-1997	"Brown Book," Recommended Practice for Industrial and Commercial Power Systems Analysis
	446-1995(R2000)	"Orange Book," Recommended Practice for Emergency and Standby Power Systems for Industrial and Commercial Applications
	493-1997	"Gold Book," Recommended Practice for the Design of Reliable Industrial and Commercial Power Systems
	602-1996	"White Book," Recommended Practice for Electric Systems in Health Care Facilities
	739-1995	"Bronze Book," Recommended Practice for Energy Management in Commercial and Industrial Facilities
	902-1998	"Yellow Book," Guide for Maintenance, Operation, and Safety of Industrial and Commercial Power Systems
	1015-1997	"Blue Book," Recommended Practice for Applying Low-Voltage Circuit Breakers Used in Industrial and Commercial Power Systems
	1100-1999	"Emerald Book," Recommended Practice for Powering and Grounding Sensitive Electronic Equipment
IAEI	1996	*Soares Book on Grounding*
CSA	C22.1-98-1998	Canadian Electrical Code
IEC	60364-2000	Electrical Installation of Buildings, Parts 1–7
	60485-1974	Digital Electronic D.C. Voltmeters and D.C. Electronic-to-Digital Converters
	60204-2000	Safety of Machinery—Electrical Equipment of Machines
	60621-2-1987	Electrical Installations for Outdoor Sites under Heavy Conditions (including open-cast mines and quarries), Part 2
	61200-704-1996	Electrical Installation Guide—Part 704—Construction and Demolition Site Installations
	60297-5-102-2001	Mechanical Structures for Electronic Equipment—Dimensions of Mechanical Structures of the 482.6 mm (19 in) Series—Part 5-102: Subracks and Association Plug-In Units—Electromagnetic Shielding Provision

TABLE 1.7a Continued
Grounding- and Shielding-Related Standards

Code Body	Standard Number–Year	Description
	60512-9-1992	Electromechanical Components for Electronic Equipment; Basic Testing Procedures and Measuring Methods—Part 9: Miscellaneous Test
	60512-23-3-2000	Electromechanical Components for Electronic Equipment; Basic Testing Procedures and Measuring Methods—Part 23-3: Test 23c: Shielding Effectiveness of Connectors and Accessories
	61000--2000	Electromagnetic Compatibility (Shielding)
	61312-2	Protection against Lightning Electromagnetic Impulse (LEMP)—Part 2: Shielding of Structures, Bonding Inside Structures, and Earthing
	61587-2-1999	Mechanical Structures for Electronic Equipment—Tests for IEC 60917 and IEC 60297—Part 3: Electromagnetic Shielding Performance Tests for Cabinets, Racks, and Subracks
	60297-5-102-2001	Mechanical Structures for Electronic Equipment—Dimensions of
	60950-1999	Safety of Information Technology Equipment
NEMA/IEA	NEMA WC55-1992/ICEA S-82-552	Instrumentation Cables and T.C. Wire
	NEMA WC57-1995/ICEA S-73-532	Standard for Control Cables
	NEMA WC74-2000/ICEA S-93-639	Shielded Power Cables 5,000—46,000 V
	ANSI C37.32-1996	The Grounding System That Works—Steel Conduit (EMT, IMC, and Galvanized Rigid)
ICEA	S-80-576-2000	Category 1 & 2 Individually Unshielded Twisted Pair Indoor Cables for Use in Communications Wiring Systems
	S-90-661-2000	Category 3, 5, & 5e Individually Unshielded Twisted Pair Indoor Cable for Use in General Purpose and LAN Communication Wiring Systems
	P-60-573	Guide for Tapes, Braids, Wraps & Serving Specifications
TIA/EIA	568A-1995 (and 568)	Commercial Building Telecommunications Wiring Standard, North America
	TSB-39-1991	Additional Cable Specifications for Unshielded Twisted Pair Cable
	TSB-40A-1994	Additional Transmission Specifications for Unshielded Twisted Pair Connecting Hardware
	TSB-53-1992	Additional Specifications for Shielded Twisted Pair (STP) Connecting Hardware
	TSB-67	Transmission Performance Specifications for Field Testing of UTP Cabling Systems
	607-1994	Commercial Building Grounding and Bonding Requirements for Telecommunications
	TSB-96	Level II-E Test Equipment, Field Certification of Installed Category 5 Channels for Use with 10000 Base-T
	6EIA/IS-647-1996	Requirements for the Control of Electromagnetic Interference Emissions and Susceptibility Characteristics of Equipment Intended to Operate in Severe Electromagnetic Environments

4. *Providing protection from lightning.* If lightning were to enter a facility and thereafter experience high impedance to ground, extremely high voltages and flashover between structures and electronics could result. It is very important to provide a low impedance path to ground for lightning to take in every facility.
5. *Limiting voltages.* Voltages must remain within limits that do not cause breakdown of insulation or otherwise lead to undesirable current paths or lethal potential differences. It is simple to limit voltages in power distribution systems by grounding one of the conductors.
6. *Managing static electricity.* Static electricity can result in surprisingly high voltages, which may lead to spark, explosions, and damage to fine electronics. Grounding and bonding can be used to eliminate charge buildup and potential differences between equipment.
7. *Controlling electrical noise.* Because the potential of the earth varies, current will flow in conductive connections between different locations in a facility. These currents can become superimposed on electronic communications equipment if they are not shorted to ground. Similarly, transients from sources such as lightning can be greatly limited with proper grounding. In short, stable, low impedance grounding is necessary to attain effective shielding of low-level circuits, to provide a stable reference for making voltage measurements, and to establish a solid base for the rejection of unwanted common mode signals.

GROUNDING, BONDING, AND OVERLOAD PROTECTION

One of the fundamental purposes of safety grounding and bonding is to provide low impedance current paths when normal ungrounded conductors fault to ground. If a fault current travels through high impedance, the overload protection device cannot distinguish the fault condition from normal operation. Furthermore, the impedance can radiate heat. When low impedance paths are provided via grounding and bonding, sudden high currents will trip the overload device causing the faulted circuit to be quickly isolated by the overcurrent device (breaker). For example, from the relationship V = IR, resistance in a 120-V, 100-A load is 1.2 Ω. Were the ungrounded conductor in this circuit to fault, the fault circuit would require much lower resistance for the current to rise much above 100 A and trip the breaker. In fact, *10 to 15 times* more current is required for quick action. This suggests a need for less than 0.12-Ω resistance in the grounding and bonding system. Figure 1.7b demonstrates this principle schematically. Here the ground resistance meets code requirements of less than 25 Ω. The circuit breaker only senses a 6-A increase in current.

NEC code strictly prohibits relying on earth as an equipment grounding conductor. This is because earth's resistance and voltage can vary not only from location to location and depending on depth of an electrode, but also from time to time in the same location. This phenomenon has been documented in detail.[1]

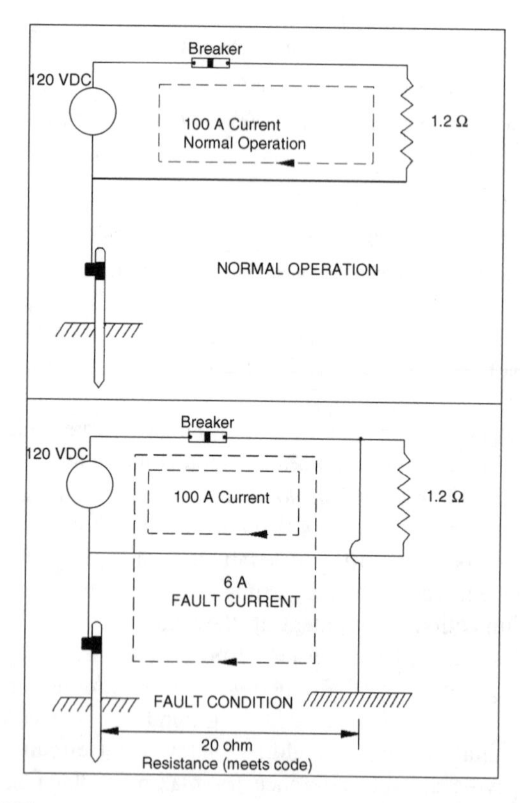

FIG. 1.7b
Development of a fault current in earth.

GROUNDING ELECTRODE RESISTANCE

Much research has focused on identifying techniques for achieving low resistance grounding rods. Many problems encountered with communications reliability can be traced to high grounding electrode resistance. The standard method of measuring soil resistively is known as the four-probe Werner method.

Lightning, in particular, places stringent requirements on grounding electrode resistance. Radio tower and telecommunications hub installations are particularly affected. Often expensive boards and the like can be destroyed leading to long downtimes. The lower the resistance of the electrode system (and the better the bonding), the more likely all of the lightning energy will be dissipated in the ground. Lightning energy has been shown to dissipate *horizontally* across the ground. Thus, horizontal electrodes have been shown to be preferable to vertical ones. In the past, salt solutions were injected around grounding rods to decrease electrode resistance. Maintenance and environmental issues have led to a decrease in popularity of this technique. More and more, conductive concrete backfill is used effectively to increase electrode surface area, soil contact, and grounding system capacitance. It has also been shown that long, deep electrodes can greatly lower resistance. Deep electrodes also offer a stabler ground over time.

POWER GROUNDING DEFINITIONS

Accepting that the NFPA NEC provides the most in-depth, authoritative, and widely accepted set of guidelines on grounding, it is important to understand the specific terms employed in the publication. Some of these terms include:

BONDING (BONDED) Bonding is the practice of providing electrical continuity among all conducting **equipment** in a facility. It has implications for overcurrent protection as well as static electricity management.

BONDING JUMPERS The conductors used to bond equipment together and to certain conductors.

BRANCH CIRCUITS Circuits after a final overcurrent protection device.

CIRCUIT BREAKER Device used to open circuits manually or automatically based on a maximum current trip point.

EFFECTIVELY GROUNDED Intentionally connected to earth through a low impedance conductor of sufficient capacity. Also called *solidly grounded.*

EQUIPMENT General term for fittings, devices, appliances, fixtures, and apparatus used in electrical installations.

EQUIPMENT GROUNDING CONDUCTOR The conductor used to connect non-current-carrying conductive equipment to the grounded conductor, grounding electrode, or both at the service equipment and at the source of a separately derived system if desired.

GROUND An intentional or accidental connection between electrical circuits or equipment and earth. In marine and aviation applications ground refers to a substitute for earth.

GROUNDED Connected to earth or its substitute.

GROUNDED CONDUCTOR A circuit conductor that is intentionally grounded. Also called a neutral in AC applications. NEC considers there to be only one grounded conductor in a facility.

GROUNDING CONDUCTOR A conductor used to connect equipment to the grounded conductor and hence the grounding electrode.

GROUNDING ELECTRODE The primary ground point, located at the service entrance, in a facility. Although there can only be one grounding electrode in a facility, local grounding electrodes are permitted.

GROUNDING ELECTRODE CONDUCTOR Conductor used to connect grounding electrode to equipment grounding conductor and the grounded conductor.

LISTED Certified for its purpose by an organization whose authority is accepted by enforcement bodies with appropriate jurisdiction.

PLENUM Air duct.

RACEWAY General term for enclosed equipment that houses wires, cables, or bus bars.

SEPARATELY DERIVED SYSTEM A wiring system whose power is derived from a battery, generator, or transformer not directly connected electrically to supply conductors originating in another system. In instrumentation, a UPS (uninterruptible power supply) or an isolation transformer usually demarcates a separately derived system.

SERVICE The conductors and equipment for delivering electric energy from the serving utility to the wiring system of a facility.

NEC ARTICLE 250

NEC Article 250 is dedicated to grounding. Although the code is specific in its requirements, it also helps to consult the NEC handbook for examples, diagrams, and photographs of appropriate applications. The material is broken into ten sections:

A. **General:** Establishes the reasons for grounding, namely, proper overcurrent protection operation, limiting voltage surges due to lightning and contact with higher voltages, and stabilization of voltage to earth during normal operation. Establishes that all conductive materials that may come in contact with electrical equipment must be bonded to the grounding electrode system. Requires establishment of a continuous and properly sized fault current path other than the earth.

B. **Circuit and System Grounding:** Establishes which circuits should be grounded. To date, the only circuits that should *not* be grounded include cranes in explosive dust environments, health-care facilities, and electrolytic cells. Also establishes the proper technique for grounding at the service and service entrances.

C. **Grounding Electrode System and Grounding Electrode Conductor:** Sets forth appropriate designs and locations for grounding electrode systems. Lists equipment to be bonded together as part of the system including water pipes, building steel, ground rings, gas pipes, and electrodes. Grounding rods must be more than 8 ft in length. Also establishes that the grounding electrode conductor must be aluminum, copper, or copper-clad aluminum. Relates size of service entrance conductors to grounding electrode conductors (NEC Table 250-66). Also establishes requirements for connecting the grounding electrode to the grounding conductor. Methods include exothermic welding, listed lugs, listed pressure connectors, etc.

D. **Enclosure, Raceway, and Service Cable Grounding:** Critical section that requires grounding and bonding of all such equipment. Not only does this have important safety implications, but it also forms the basis for good instrumentation wiring design.

E. **Bonding:** Details requirements for equipment bonding to ensure electrical continuity and capacity back to overcurrent protection device. Bonding must be done using listed equipment. For example, conduit is bonded to cable tray and junction boxes using special, listed conduit-to-cable tray clips and bushings. Conduit must also be of the listed type to ensure low impedance, durable continuity at segment joints. Bond junctions must be free of any insulting materials such as paint.

F. **Equipment Grounding and Equipment Grounding Conductors:** Establishes the identification methods for equipment grounding conductors (bare, clear, green insulation, or green with a yellow stripe). Also lists specific non-current-carrying equipment that *must* be connected to the equipment grounding conductor. Such equipment includes elevators, cranes, switchboards, water pumps, lighting fixtures, and skid-mounted equipment.

G. **Methods of Equipment Grounding:** Details various methods and equipment to be used in grounding and bonding conductive equipment.

H. **Direct-Current Systems:** Requires grounding for DC systems of 50 to 300 V. Makes an exception for systems with ground fault detection equipment in industrial, professionally maintained settings.

I. **Instruments, Meters, and Relays:** Requires grounding of the secondary of power supplies with primary connections of 300 V or more. Requires grounding of cases of instrument power supplies, meters, and relays.

J. **Grounding of Systems and Circuits of 1 kV and over.**

GROUNDING EXAMPLES

Two of the articles in Section B of NEC Article 250 (250-24 and 250-30), as summarized on p. 103, prescribe specific requirements for grounding at service entrances and separately derived systems. It is difficult to form a picture of an approved installation without the aid of examples from the handbook or other sources. Three instructive examples are given below.

Service Entrance

With a service entrance, power is taken directly to a panel board equipped with overcurrent protection devices for distribution directly to branch circuits or to downstream separately derived systems. NEC requires installation of a grounding electrode system with 25-Ω resistance to ground maximum or two ground electrodes near the distribution panel. The code and the handbook also make recommendations for measuring ground rod resistance. The grounding electrode is used to establish neutral reference to ground as well as to establish the ground reference for the equipment grounding conductor. Proper connection of lightning protection equipment to the grounding electrode is also an important consideration.

Once the equipment ground is established, NEC requires bonding all conductive equipment and machinery to it. This is achieved in two ways. First, all metallic panel equipment and conduit is bonded to the equipment grounding conductor. Conduit penetration bonding is accomplished with special bonding bushings and wedges that provide connections for the equipment grounding conductor. From here, all raceway and cable tray used are listed and properly connected. Raceway/cable/enclosure interfaces require use of listed bonding jumpers and connectors when other connection means are not approved.

Second, an equipment grounding conductor, "green wire," is run with all circuits (except where provided by the code) and to all conductive equipment, machinery, and structural steel. It is important to survey a facility to ensure there are no discontinuities in the equipment grounding system. Equipment grounding conductor connection to structural steel should be done explicitly at one point. Conductive conduit and cable tray can only be assumed to provide equipment grounding conductor functionality where it is bonded using appropriate, listed equipment. Where nonconductive flexible conduit is used to mate conductive conduit to equipment such as pumps and valves, the green wire is used to provide the continuity that the flexible conduit interrupted. It is therefore important to ensure that the conductive conduit run is properly bonded as well. This may involve making a connection with a listed connector to the equipment grounding conductor or to properly bonded cable tray. It should be noted, however, that only neutral blocks in the service panel should be bonded to the panel. All other neutral blocks in subsequent subpanels should be insulated from their panel (250-24). Many of these details are summarized in Figure 1.7c.

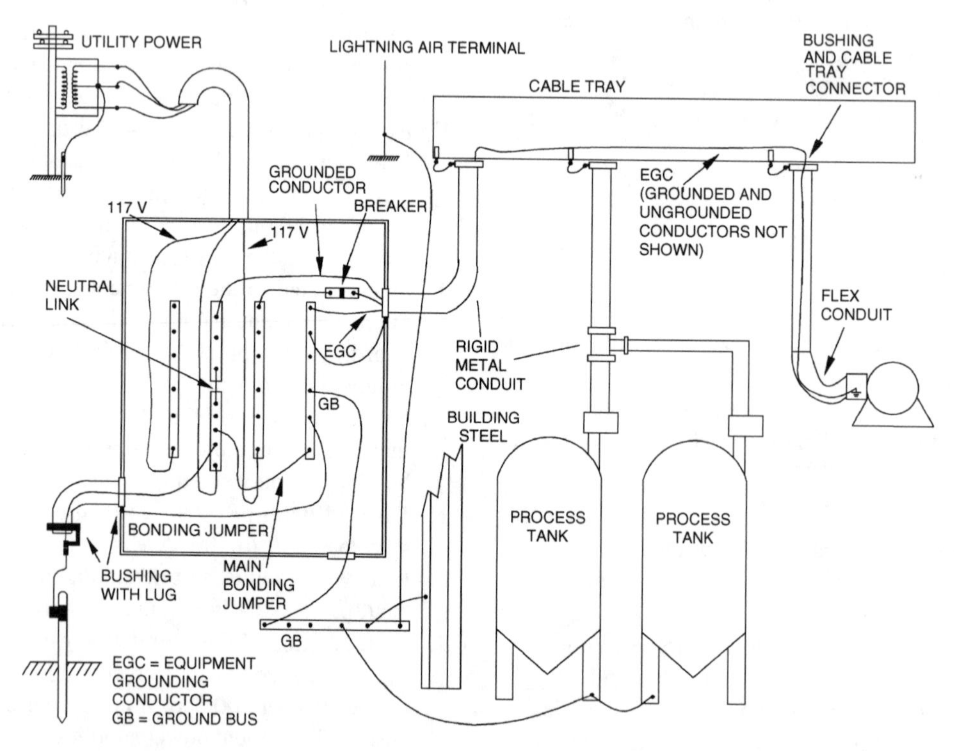

FIG. 1.7c
Service entrance grounding and bonding of equipment.

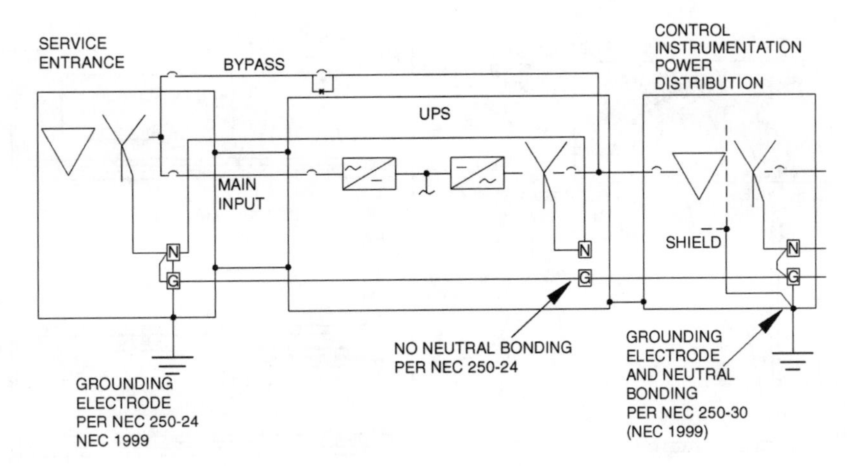

FIG. 1.7d
Grounding of separately derived intrumentation and control power. (See IEEE 142-1991 for other options.)

Separately Derived Instrumentation Power System

A key safety function of power grounding is to provide a low impedance path for returning ground faults back to overcurrent protection devices. Fortunately, this safety function can be maintained while still providing common mode noise attenuation. To provide isolation and common mode noise reduction, instrumentation power networks warrant use of an isolation transformer. NEC defines such a system as "separately derived."

Separately derived systems derived from transformers require that the primary power source be properly grounded at the service entrance. The secondary of the transformer must be regrounded as well. Grounding of the secondary of a transformer always involves a three-step process. First, one sizes the bonding jumper between the center tap or neutral of the secondary and the transformer case. Next, one sizes the grounding electrode conductor using NEC Table 250-66. Finally, one selects the grounding electrode.

NEC requires that a single grounding electrode have less than 25 Ω of resistance or that two be used (250-56). Instrumentation systems have more stringent requirements. For example, an intrinsic safety ground must have 1 Ω or less resistance to ground. The IEEE Green Book recommends ground resistances of 2 to 5 Ω for industrial systems.

Equipment grounding conductor and equipment bonding concepts demonstrated in the previous example all still hold. Figure 1.7d demonstrates a preferred method of grounding an instrumentation power system consisting of the following:

1. Single UPS module
2. Nonisolated UPS bypass
3. Distribution panel with Delta-Wye isolation transformer

Here the UPS is not separately derived source because its neutral is bonded to the grounding conductor at the service entrance. Keeping in mind that subpanels should not be bonded to neutrals, the neutral in the UPS is isolated from its equipment grounding conductor. Because the isolation transformer is a separately derived system, its neutral is bonded to the grounding conductor. Keep in mind that *the* grounding conductor refers to the main one at the service entrance. Also note that NEC 250-30(3) allows one to use *the* grounding electrode in place of a *local* grounding electrode at the transformer. NEC 250-30(2) requires, in either case, that the grounded conductor of the transformer be connected to *the* grounding electrode. It also requires that the grounded conductor of the derived system and the equipment grounding conductor be bonded. Thus, the service and separately derived systems will always be bonded.

In this arrangement, the service entrance power can be supplied at 208 or 480 V. A 480-V input/480-V output UPS is recommended to reduce the cost of feeder wire and to reduce voltage drop. Stepping down to 120 V is accomplished in the isolation transformer.

Note that for this configuration, noise elimination effectiveness is dependent on how close the transformer is located to the loads. One of the benefits of placing the transformer after the UPS is that the UPS does not have to be located close to the loads to attenuate noise properly.

Single-Point Grounding of Power Supplies

It is desirable to "isolate" control system and instrumentation power from other power sources in a facility. Figure 1.7d demonstrates the preferred method of obtaining this goal while meeting code requirements. Note that all equipment is grounded to a single point. What Figure 1.7d does not show is how this single-point grounding concept is carried over to the numerous points where grounding is required in a typical control system and instrumentation network. Although safety may seem to be less of a concern at the low voltages normally encountered, single-point grounding has implications for noise elimination as well. For one, if connections are made haphazardly along a very long ground bus, potential differences develop at the various points due to bus resistance.

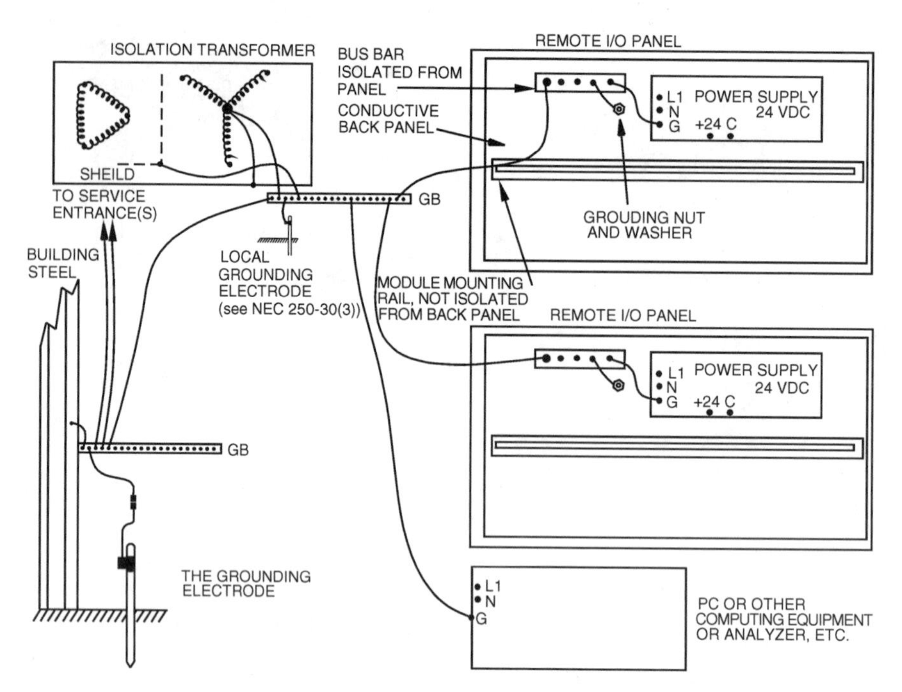

FIG. 1.7e
Single-point grounding of power supplies, computing equipment, and signal cabinetry.

These potential differences would then lead to current loops. Although it is not possible to connect every ground to a single point, Figure 1.7e suggests an approximate method that is widely accepted. Note that only equipment grounding conductors are shown. Power conductors, conduit, cable tray, and the bonding thereof are not shown. The principles put forth in Figure 1.7d still apply.

Although the input/output (I/O) modules are not shown, it is assumed that they are powered by the power supply or by a grounded neutral plus ungrounded conductor circuit derived from the transformer. Discrete signal neutral wiring is not shown because the neutral can come from a variety of sources. It is further assumed the I/O modules provide connections for cable shields at grounded terminals adjacent to the signal terminals and that all signals are grounded either at the modules (analog) or via their neutral conductors (discrete). This implies that analog modules provide grounding via their connection to the mounting rail. Grounding of signals at the I/O module is the most common method of signal grounding. Still, not all I/O modules conform to this approach.

THE UNGROUNDED SYSTEM

NEC code allows for use of ungrounded systems in certain situations. Usually, the idea is to eliminate unexpected equipment outages. In such facilities, a single fault to ground will not cause overload protection to operate. Ground fault indicators are thus necessary to alert maintenance staff when and where a fault has occurred so it can be cleared. Otherwise, the purpose of not grounding, i.e., avoidance of unexpected outages, would not be met once a second fault causes unexpected tripping of overcurrent protection. Ground fault detectors are basically current switches placed around ground conductors or around all the phases in a power line. Although it seems simple enough, it is often very difficult to design a ground fault detection network that makes locating the fault a trivial matter. With a grounded system, fault location is obvious. One should evaluate other techniques for increasing availability, such as redundancy, prior to deciding on an ungrounded system.

No matter what the reason for not grounding, one is still required to detect ground faults. This is usually accomplished using what NEC and IEEE call “relays.” Relays are current switches that detect:

1. Unbalanced current flow in an AC circuit (the faulted conductor experiences more flow than the other) or
2. Current flow in the grounding electrode system.

RESISTANCE GROUNDING

The magnetic field component of very high fault currents can produce mechanical stresses beyond what some machinery is designed to handle. Motor and generator frames can be torn apart by a magnetic field associated with a fault current. Grounding through a carefully sized resistance counteracts this problem. Ground fault and ground resistance monitoring is employed in such systems as well.

SHIELDING THEORY

The fundamental basis for shielding and grounding in instrumentation comes from electrostatics. The fundamental basis of electrostatics is charge. Charge, like matter and energy, is conserved. Charge is given the symbol Q and is measured in

TABLE 1.7f
Dielectric Constants of Common Materials

Material	*Dielectric Constant (k)*
Vacuum	1
Air	1.00054
Water	78
Paper	3.5
Porcelain	6.5
Fused quartz	3.8
Pyrex glass	4.5
Polyethylene	2.3
Amber	2.7
Polystyrene	2.6
Teflon	2.1
Transformer oil	4.5
Titanium dioxide	100

coulombs (C). It takes 6.28×10^{18} electrons to produce 1 C of charge. 1 C of charge flowing past a point results in 1 A (ampere) of current. Benjamin Franklin originally perceived charge as coming in two flavors: that which attracted a glass rod rubbed with silk (negative) and that which repelled it (positive). Since charge is capable of doing work, like rotating a glass rod, it must produce a force:

$$f = \frac{Q_1 Q_2}{r^2 k} \qquad 1.7(1)$$

The parameter r is the distance between the charges. This is known as Coulomb's law. The parameter k, known as the dielectric constant, is necessary because the force between two charges is reduced when they are embedded in an insulating medium other than a vacuum. See Table 1.7f for examples of some common material dielectric constants. All dielectrics also have a maximum voltage gradient (V/r) that they can withstand before breaking down and flashing over.

The ability to perform "work at a distance" is a property of vector fields like gravity. The strength of an electric field, E, produced by a charge in a dielectric medium is

$$E = \frac{Q_1}{kr^2} \qquad 1.7(2)$$

Similarly, the force between two charges can be written in terms of the electric field:

$$f = Q_2 E \qquad 1.7(3)$$

The above can be restated to say work is required to move a charge from one point to another in an electric field. Consider the situation where a point charge (Q_p) is moved from infinity to a charged sphere (Q_s). The work done can be calculated from

$$W = \int_{\infty}^{r} \frac{Q_s Q_p}{kr^2} dr = \frac{Q_s Q_p}{kr} \qquad 1.7(4)$$

or

$$W = \int_0^Q \frac{Q_p}{C} dQ_p = \frac{1}{2}\frac{Q_p^2}{C} = \frac{1}{2}CV^2 \qquad 1.7(5)$$

A scalar measure of the amount of work required per unit charge is known as the volt. Volts measure work in joules per unit charge; 1 volt is 1 joule/coulomb.

$$\frac{W}{Q} = V = \frac{Q}{kr} \qquad 1.7(6)$$

This means charge and voltage are proportional.

$$Q = (C)V \qquad 1.7(7)$$

This C is known as capacitance (1 farad = 1 coulomb/volt) because it describes the capacity of a sphere to carry a charge at a given voltage. Simply stated, charge and voltage are proportional and capacitance is the slope. It should be noted that capacitance depends only on dielectric constant and geometry.

$$C = \frac{Q}{V} = Q\left(\frac{kr}{Q}\right) = kr \qquad 1.7(8)$$

When dealing with flat plate capacitors, capacitance is related to area (A) and distance of separation (d):

$$C = \frac{Q}{V} = \frac{kA}{d} \qquad 1.7(9)$$

It is more correct to consider a vacuum having a finite capacivity such that k is the ratio of the capacitance of a medium to that in free space. The so-called permittivity of a vacuum is fixed at 8.854×10^{-12} C^2/Nm^2. All other materials increase this value but increase capacitance.

What can be concluded is that capacitance is a property of two surfaces with equal yet opposite charges. The charges are capable of doing work or being worked upon. Every line of force that originates from one surface terminates on the other. Therefore, a capacitive system, called a capacitor, can store energy. Usually, this is looked upon as an ability to separate charges. As long as the charges are separate, energy is stored. When they are brought together, energy is released. We can also conclude that more energy can be stored when a higher-dielectric-constant material is placed between capacitor surfaces. Usually conductors are used to guide the emerging electric field.

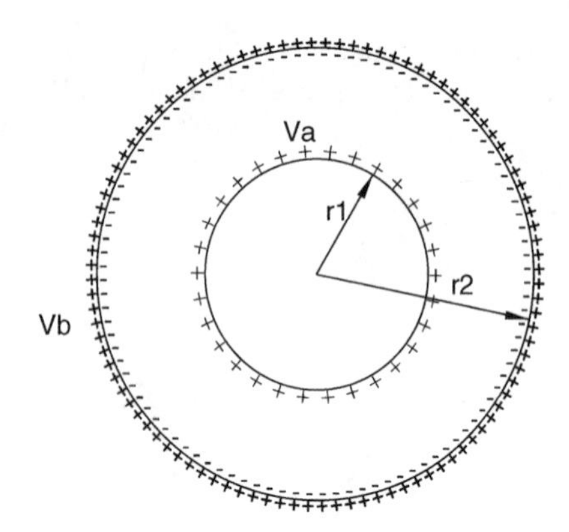

FIG. 1.7g
Electric fields on a charged sphere surrounded by a conducting shell.

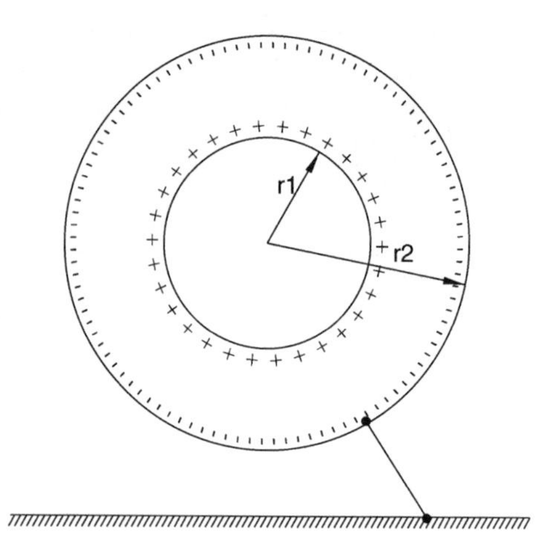

FIG. 1.7h
No electric field surrounds a charged sphere shielded by an earthed shell.

All this becomes very important when trying to understand the physical basis for electronic shielding. To begin, consider Figure 1.7g, which details what happens when a charge is placed on a sphere surrounded by a conducting shell. Benjamin Franklin was the first to study such a system in 1755. He concluded that the electric field flux due to the charge on the sphere could not penetrate the shell. Because it is conserved, charge-induced flux can only stop and start at a charge that negates it or creates it. Therefore, there must be a charge of –Q on the inside of the shell and a charge of +Q on its outer surface. The potential difference between the two surfaces, given that the charge (Q) on each surface is the exact opposite of the other, is given by

$$V = Va - Vb = \frac{Q}{k}\left(\frac{1}{r_1} - \frac{1}{r_2}\right) \qquad 1.7(10)$$

The capacitance of system can be calculated from

$$C = \frac{Q}{V} = \frac{kr_1r_2}{r_2 - r_1} \qquad 1.7(11)$$

Although the system has a net charge of zero, each surface can be highly charged. Conversely, voltage on either conductor is undefined. The charge and geometry of the system fix the voltage difference. Capacitance has an analogous meaning in the ideal gas law. If voltage is considered to be analogous to pressure, then capacitance is analogous to the temperature-corrected volume a gas occupies; see Equation 1.7(6).

$$n = \left(\frac{V}{RT}\right)P \qquad 1.7(12)$$

Connecting the shell to an infinite conducting plane, e.g., earth, will cause current briefly to flow as charge is directed away. As a result, the situation shown in Figure 1.7h is

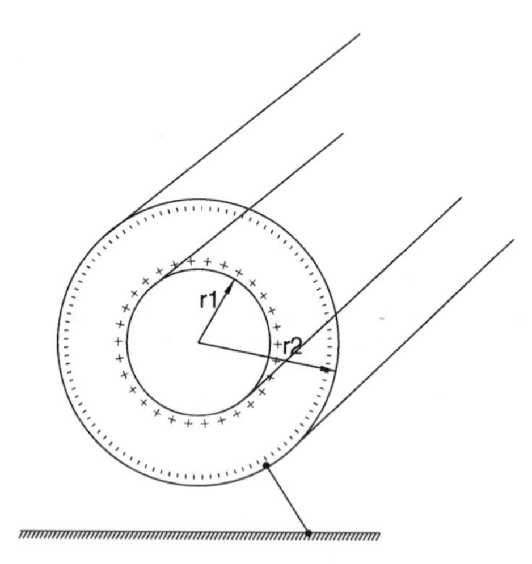

FIG. 1.7i
Distributions of charges in a shielded wire.

obtained. The inner sphere is charged +Q and the entire outer sphere is induced with a charge –Q. At this point, there is no field outside of the shell.

A similar treatment for a long cylindrical conductor is given in Figure 1.7i. Unlike spherical conductors where the voltage is considered undefined, the voltage at a wire is considered infinite. The voltage between two wires is related to their charge, dielectric permittivity, and geometry by

$$V_D = -\frac{Q}{2\pi\varepsilon}\ln\frac{r_1}{r_2} \qquad 1.7(13)$$

The capacitance, or ratio of charge to voltage, is given by

$$C = \frac{Q}{V_D} = \frac{2\pi\varepsilon}{\ln\frac{r_1}{r_2}} \qquad 1.7(14)$$

As an example, consider a system of three conductors with charges $Q_1 = 0$, Q_2, and Q_3. Because of the behavior just discussed, each conductor senses a voltage due the others:

$$V_1 = \frac{Q_2}{4\pi r_{21}} + \frac{Q_3}{4\pi r_{31}} \quad \text{(etc.)} \qquad \mathbf{1.7(15)}$$

$$V_2 = \frac{Q_1}{4\pi r_{12}} + \frac{Q_3}{4\pi r_{32}} \qquad \mathbf{1.7(16)}$$

$$V_3 = \frac{Q_1}{4\pi r_{13}} + \frac{Q_2}{4\pi r_{23}} \qquad \mathbf{1.7(17)}$$

Here one can begin to appreciate the importance of mutual capacitance between conductors. One can take the analysis further[2] by applying Green's theorem to arrive at values for what is known as *mutual elastance*:

$$s_{12} = \frac{V_2}{Q_1} \quad \text{and} \quad s_2 = \frac{V_1}{Q_2} \qquad \mathbf{1.7(18)}$$

Somewhat like the inverse of capacitance, elastance is dependent only on the geometry of the system. One can invert Equations 1.7(15 to 17) to arrive at a set of equations for mutual capacitances (c_{ij}):

$$\begin{aligned} V_1 &= c_{11}Q_1 + c_{12}Q_2 + c_{13}Q_3 \\ V_2 &= c_{21}Q_1 + c_{22}Q_2 + c_{32}Q_3 \\ V_3 &= c_{13}V_1 + c_{23}V_2 + c_{33}V_3 \end{aligned} \qquad \mathbf{1.7(19)}$$

These relationships mean that induced voltages must result in charge movements. Charge movements, in turn, lead to current flow. When the shield in Figure 1.7j is broken, the spheres experience mutual capacitance and *leak* current flows in the capacitance (Figure 1.7k).

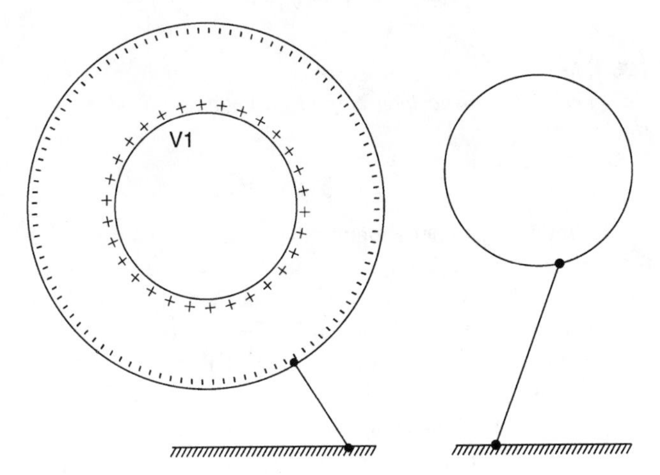

FIG. 1.7j
Absence of capacitance between a shielded and nonshielded sphere.

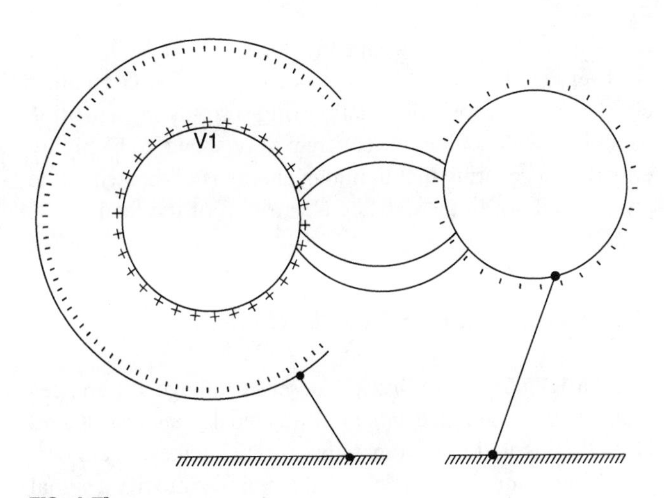

FIG. 1.7k
Capacitive current flow due to poor shielding.

The ratio of the voltage difference to the current flowing in a capacitance is known as *reactance*:

$$X_{c_{11}} = \frac{1}{\theta c_{11}} \quad \text{and} \quad X_{c_{12}} = \frac{1}{\theta c_{12}} \qquad \mathbf{1.7(20)}$$

If one considers the inherent time constants of the electric field, one can begin to see why high-frequency electrical energy reduces energy storage capacity. This is the tie that binds high frequency and static electrical behavior.

The basis for shielding is simply the fact that the potential on a properly designed shielded conductor cannot influence the charge on another conductor outside of the shield, or vice versa. Shielding is never perfect and cable shielding is rated in terms of its *leakage capacity* per foot. It is usually on the order of picofarads.

It is highly instrumental to consider some of the uses for capacitors. Capacitors are used in systems that:[3]

1. Reduce voltage fluctuations in power supplies
2. Transmit pulsed signals
3. Generate or detect radio-frequency oscillation radiation
4. Produce electronic time delays

The astute engineer will be quick to realize that the fundamental abilities of capacitance end up being more of a liability than an advantage in certain situations.

LIGHTNING

Lightning events are analogous to capacitor discharging. Usually, clouds have a negative charge with respect to the earth. When the voltage difference exceeds the dielectric strength of air (0.8 kV/mm), breakdown occurs and a lightning flash is observed. The voltage at the point of the strike is close to zero but is driven to millions of volts by electrons seeking to dissipate. Instead of dissipating downward, a "skin" effect is observed and the dissipation occurs radially. Grounding electrodes in the form of a shallow ring surrounding a facility, backfilled with conductive concrete have shown to be highly effective in ensuring that lightning energy is dissipated in the ground and not in the delicate electronics of the facility.[4]

ELECTROSTATIC INSTRUMENT SHIELDING

The basic purpose of I/O is to sense voltage differences relative to a reference potential created by an instrument. Usually, this will involve passing a signal across a gain created by an amplifier. Single-ended amplifiers modify a signal relative to the same reference potential on both their input and output. Charge and voltage measurements are usually handled with a single-ended amplifier. With single-ended amplifiers, it is important to limit non-signal-related current flowing unbalanced in the zero signal reference conductor. Such current flow is by definition normal mode pickup. When a single-ended amplifier and its power supply are completely suspended in shield without touching it, capacitive coupling at different voltages will occur. The result will be current flow back to the inputs of the amplifier (feedback) (Figure 1.7l).

Grounding the signal zero reference and floating the shield does not help. The potential difference of the earth between conductor 3 and the shield capacitance of the shield at the opposite end causes a current to flow in the input side of conductor 3 (Figure 1.7m). Only by directly referencing the signal common to the shield can one effectively eliminate capacitive feedback to the amplifier (Figure 1.7n). This leads to an important rule of shielding: *An electrostatic shield should be connected to the zero reference potential of a signal.*

Because a shield conductor will normally be run in parallel with conductors referenced to potentials other than its zero reference potential, current will flow in the mutual capacitances created by the shield and its environment. However, these currents will not flow in the signal conductors if the shield is properly grounded.

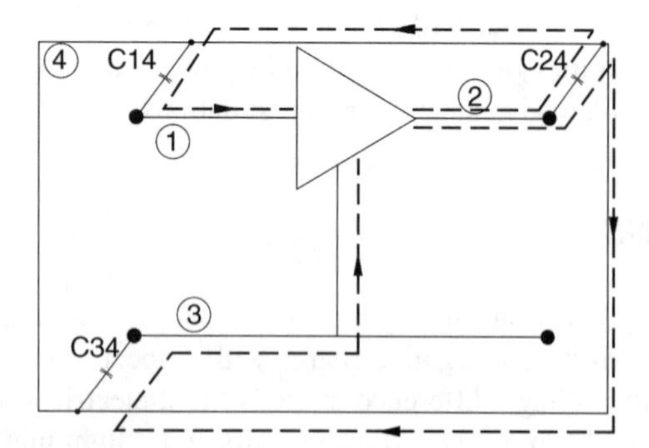

FIG. 1.7l
Feedback currents for an amplifier with an ungrounded shield.

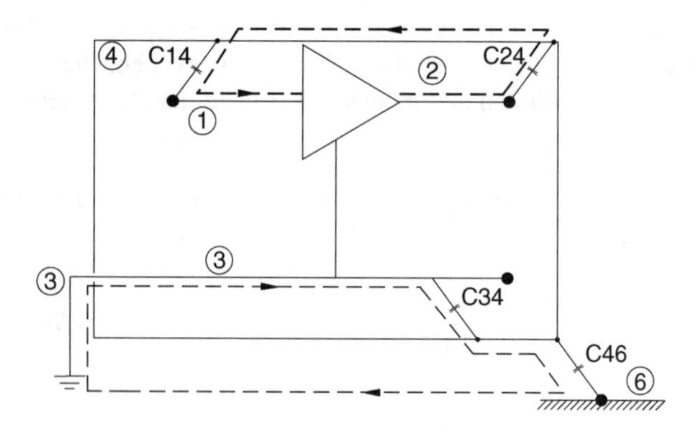

FIG. 1.7m
Feedback currents when common conductor is grounded, but shield is not.

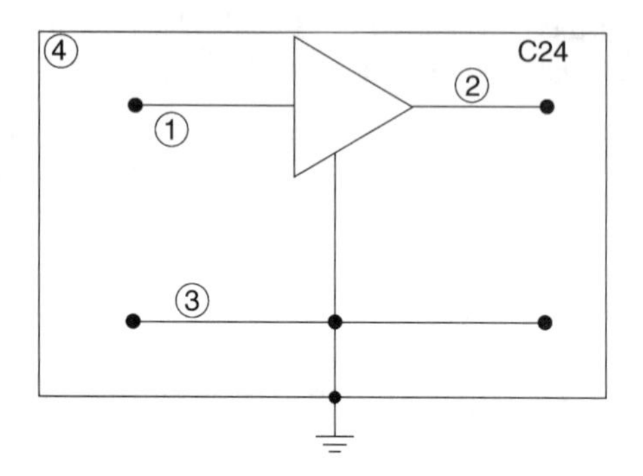

FIG. 1.7n
Properly grounded amplifier shield and common conductor.

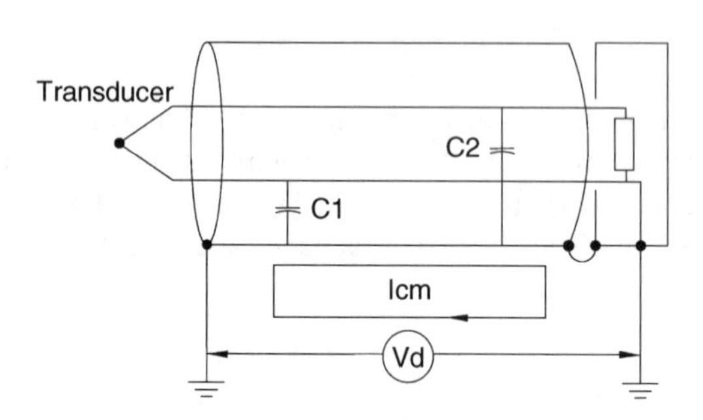

FIG. 1.7o
Common mode ground loop current in multiply grounded signal shield.

Examining another situation that should be avoided leads to a second rule. In Figure 1.7o, the shield is earthed twice. The potential difference of the earth at the two points leads to a current in the shield and earth. The effect is an increase in *common mode pickup* due to current flowing in capacitances 1 and 2. Common mode pickup refers to pickup that is equal on both signal conductors. As common mode pickup increases, more and more of it can appear in the differential signal (i.e., as normal mode noise). The ability to hinder common mode pickup from turning into normal mode noise is known as the *common mode rejection* capability of the measuring equipment. Common mode rejection is never perfect. It can be concluded that *the shield conductor should be connected to the zero signal reference potential only once, preferably at the signal common earth connection*. Often the measuring equipment, or I/O, grounds the signal from a transmitter. Shields in such systems are usually attached to terminals adjacent to the signal terminal on such I/O modules. Keeping shields and signal lines in close proximity also has important implications for limiting magnetic interference.

DIFFERENTIAL AMPLIFIERS AND ISOLATORS

In many cases, one is presented with a situation where there are two earthed, zero reference conductors at different points in a measurement system. A common example involves self-powered transmitters that are interfaced with digital control system I/O. Grounded thermocouple, strain gauge, and data link applications also call for use of differential amplifiers. Often, I/O is designed to reference a signal common to earth. Meanwhile, a self-powered transmitter provides an output referenced to the ground of its power supply. Clearly, any potential difference between the I/O zero reference and the transmitter power supply reference appears as common mode noise that can lead to normal mode pickup.

In such situations, it is advisable to utilize differential rather than single-ended I/O. When differential I/O is not an option, then an external differential amplifier, also called an *isolator*, can be implemented in the signal lines. In keeping with earlier analysis, each side of the isolator should be independently shielded. The shields should be earthed at their respective signal grounds.

Differential amplifiers and isolators couple the signals via many different techniques. Light-emitting diodes and photomultipliers, radio-frequency transmission, transformers, and even direct ohmic connections are all used. The transformer approach requires some sort of modulation and demodulation. Therefore, bandwidth and complexity can be a problem. Direct connections do not limit bandwidth, but they also do not limit the inherent common mode signal. Figure 1.7p shows a basic application of an isolator or differential amplifier. For a more in-depth treatment see Reference 5.

The internal shield of the amplifier is called a guard shield. The guard principle requires that the amplifier guard be driven

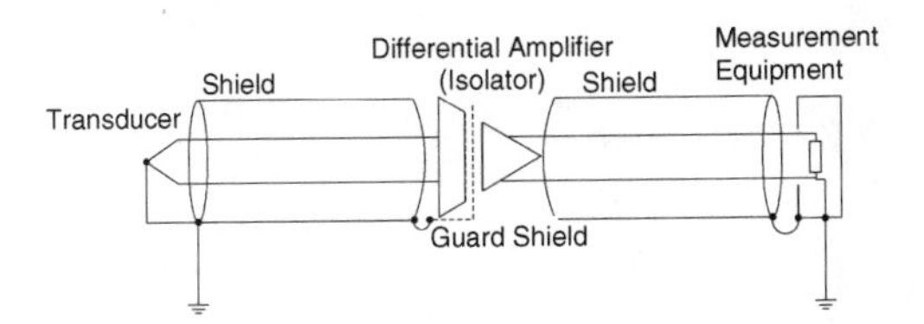

FIG. 1.7p
Application of a differential amplifier (isolator).

at the common-mode voltage appearing at the amplifier inputs. The most effective way to do this is as follows:

1. Connect amplifier guard to the signal cable shield and make sure that the cable shield is insulated from the chassis ground and from any extension of the system ground.
2. Connect the ground of the signal source to the signal cable shield. In this way, the voltage of the amplifier guard and signal cable shield is stabilized with respect to signal source.
3. Connect the signal cable shield and its tap to the signal source and the signal ground, which should be as close as possible to the signal source. This will limit the maximum common-mode voltage. The signal pair must not be connected to ground at any other point.
4. Connect the amplifier chassis, equipment enclosure, the low side of the amplifier output, and output cable shield to the system ground.

INSTRUMENT POWER TRANSFORMER SHIELDING

Power supply transformers, when used with resistance-bridge instruments, should use three shields. The primary coil shield should be connected to the grounding electrode system. The inner shield of the secondary should be connected to the output common of the power supply, which becomes the zero-signal reference conductor eventually. The outer shield on the secondary should encompass the secondary coil as well as also be connected to the zero signal reference.

FLOATING VS. GROUNDED INSTRUMENTS

The question whether to "float" or ground signals often arise. General advice can be given, but it should not be considered definitive. There are ways of getting around most problems one might encounter if the advice is ignored:

1. Operating differential equipment with one ungrounded side should be avoided. The ungrounded side will find its own value and that value could pose operational or safety problems.
2. Although a floating signal to a differential amplifier can tolerate a single fault, it is difficult to know when that fault has occurred. While the fault condition persists, performance is neither guaranteed nor defined. A second fault or even someone receiving a shock is required to detect this situation.
3. Because thermocouples are often in electrical contact with the device or structure of which the temperature is being measured, and with an intentional ground at the sensor, differential amplifiers with very good common mode rejection are required to discriminate effectively against noise in the structure or grounding system.[6]

4. Grounded thermocouples should not be used with single-ended amplifiers or I/O.
5. Ungrounded thermocouples should not be used with differential I/O.

ISOLATION TRANSFORMERS

It is reasonable to try to "isolate" instrumentation system power from the other "dirty" power in a facility using what has become known as an *isolation transformer.* It is important to understand what is actually suggested by the term *isolation* and not to take the concept too far. Serious code violations and legal problems can result from misguided application of isolation transformers.

All power in a facility should ultimately be grounded at a single point. Isolation transformers allow regrounding the secondary to a new point on the facility grounding electrode system. Still, they should be bonded to the facility grounding electrode system as well. This will provide excellent protection against common mode pickup from dirty power. Not bonding the secondary of an isolation transformer to a point on the grounding electrode system is a code violation. Floating the secondary is even more serious in most situations. In short, an isolation transformer should be installed just like any other distribution transformer over which the facility has responsibility; see Figure 1.7d.

As a further enhancement, instrument isolation transformers are usually equipped with copper or aluminum shields between the primary and secondary coils. This practice is effective in limiting common-mode current pickup. One purpose of the shield is to prevent interwinding short circuits. Another is to divert fault currents to the equipment grounding system to allow the overcurrent device associated with a circuit served by the transformer to operate correctly. Along the same lines, another function is to protect against lightning damage, as the lightning will flow readily to ground via a grounded shield rather than through one's instrumentation. The shield should be solidly connected to the grounding electrode system via as short a lead as possible to minimize lead inductance. Long leads can also serve as lightning rods. Grounding the shield to a grounding rod not grounded to the equipment ground is a dangerous code violation.

Above 100 kHz, transformer shields lose effectiveness in terms of reducing common mode noise. This problem is dealt with by equipping transformers with integral passive LC filters.

POWER SUPPLY SHIELDING

Proper use of a floating power supply to power instrumentation loops varies depending on the type of signal generator and whether differential or single-ended instrumentation is being employed. With resistance bridge signal generators, it is important to ground the bridge and shield(s) appropriately if either differential or single-ended equipment is used. The various shielding issues are beyond the scope of this article. See Reference 7 for a more detailed discussion.

DIGITAL COMMUNICATIONS SHIELDING

Even though shielding concepts are firmly rooted in physics, there is a high degree of variability in the application of high-speed digital communications cable shielding. For example, most of the world's electronic 100Base-T networks use unshielded twisted pair wiring (UTP). One would think that the situation is a result of shielded twisted pair (STP) wiring failing to perform adequately. Actually, STP wiring performs better. However, given the self-healing mechanisms of the TCP/IP (transmission control protocol/Internet protocol), the typical information technologist would hardly be cognizant of a problem. An instrument engineer may not enjoy the same amount of leeway. Roughly 20 digital instrumentation buses are used worldwide. Although they differ in their requirements for shielding, it can be concluded that shielding will improve performance (see Table 1.7q).

MAGNETIC FIELD INFLUENCES

The purpose of the shielding measures described above is to manage electrostatic influences. Magnetic influences are not addressed by these techniques. Only time-varying magnetic influence is a concern. Fortunately, they are easy to manage via control of "loop areas." The effect of a time-varying magnetic field on a circuit is a function of the loop area it exposes to the field (Lenz's law). Larger loop areas can couple more "flux linkages" as they float by. Conversely, current flow loop areas should be kept small to limit field generation. When building steel alone is used for a ground return path, very large loop areas can result. Twisting of signal wire also effectively limits loop area. It also creates opposing currents from the same field. Obviously, signal conductor pairs should reside in the same cable.

It is also important to consider how shields are terminated at the zero reference potential. If a shield is directed a significant distance from the signal conductors it once ran with, a large loop area can result. Even "pigtailing" of shields can pose a problem as it creates a half turn of inductance. Pigtails can also radiate noise in an enclosure. The preferred methods either employ so-called "back-shell" connectors or utilize I/O that provides a terminal for shield grounding.

EMI AND RF SHIELDING

A singly grounded shield does not impede the flow of the magnetic component of electromagnetic interference (EMI) and radio-frequency (RF) energy. On the other hand, grounding an instrument shield more than once violates electrostatic

TABLE 1.7q
Digital Network Cable Specifications

Network Name	Associated Standards	Cable
PROFIBUS DP/PA	EN 50170/DIN 19245 part 3 (DP), part 4(PA), IEC 1158-2(PA)	UTP
INTERBUS-S	DIN 19258, EN50.254	UTP
ControlNet		COAX
DeviceNet	ISO 11898 and 11519	UTP, COAX
Remote I/O		UTP
ARCNET	ANSI/ATA 878.1	UTP
AS-I		UTP
Foundation Fieldbus—H1	ISA SP50/IED 61158	UTP
Foundation Fieldbus—HSE	IEEE 802.3u, TCP and UDP	UTP
IEC/ISA SP50 Fieldbus	IEC 1158/ANSI 850	UTP
Seriplex		4 wire shielded
WorldFTP	IEC 1158-2	UTP
SDS	ISO 11989	UTP
CANopen		UTP
Ethernet	IEEE 802.3	UTP
Modbus Plus		UTP
Modbus	EN 1434-3, IEC 870-5	UTP
Genius Bus		STP

shielding rules. Still, one can make use of a second shield to some degree to eliminate time-varying interference. The most common way to minimize RF pickup is to use conduit, cable tray, and metallic enclosures throughout a system. Power safety requirements for bonding and multiple grounding of such equipment make it ideal for shielding signal lines from RF influences.

Quite often, shielding is also applied to cabinets as well as whole rooms. A wide range of equipment is available to help with this task. Raised computer floors formed with electrically bonded stringers (also called zero signal reference grids, or ZSRGs) are an important example. Special gaskets and waveguide ventilation covers are also prominent. Other forms of ZSRGs include building structural elements such as Q-Decking and rebar. Proper bonding of noncontinuous grids is an important consideration as well. Shielded rooms often need to be single-point grounded with respect to the outside world to isolate them from low-frequency currents. When this is done, surge protectors are required for bridging isolation gaps during lightning strikes.

SHIELDED CABLE

There are two basic types of cable shielding, woven braid and aluminum–Mylar lapped foil. Braided shields give only 85% coverage of the signal line. They are adequate when coupled with signal conductors that are twisted at least ten times per foot (33 twists per meter). Braid-shielded cable leakage capacity is about 0.1 pF/ft (0.3 pF/m). Thus, there will probably be noise at microvolt levels.

Lapped foil shielding is usually equipped with a drain wire for use in connecting to the zero reference potential. The drain wire should be well bonded to the conductive layer of the foil and *external to* the shield interior. This type of shielding provides 100% coverage of the signal line, reducing the leakage capacity to 0.01 pF/ft (0.03 pF/m). To be most effective, the wires should have six twists per foot (20 twists per meter). The lapped foil is coated with an insulating layer to prevent accidental grounding. Care should be taken in bending such cable to prevent the foil from opening at kinks.

INTRINSIC SAFETY GROUNDING AND SHIELDING

The two basic types of intrinsic safety barriers, Zener Diodes and galvanic isolators, utilize different grounding techniques. Zener diode barriers require a high-integrity intrinsic safety ground with no more than 1-Ω resistance to ground. Galvanic isolators function like differential amplifiers coupled magnetically. Because of the nonelectrical coupling, and a high level of isolation between the primary and secondary windings, the ground connection is not necessary with galvanic barriers.

Intrinsic safety poses a problem where shielding is concerned. Barriers either provide no shield connections or the possible connections would ground the shield. Furthermore, conductors that run between nonintrinsically safe and intrinsically safe areas without appropriate barrier isolation can act as conduits of unsafe energy levels. Barrier manufacturers do not recommend using shielding in such installations. Where no shielding is used, it is often better to use digital transmitters and communications as the nature of such communications means that they either work or do not work. The result is either the data that the transmitter intended or a detectable error. With analog systems, problems are subtler and more difficult to detect.

THE STATIC ELECTRICITY PROBLEM

Static electricity not only presents a problem for those who handle fine electronics frequently, but it is also a serious safety issue. In the presence of flammable vapors, a spark from one's finger could mean an explosion. In an electronics-manufacturing environment, it could mean scrapping five or ten $2000 PCI cards a day. One is apt to experience static electricity discharge every time one removes clothes from a clothes dryer. In fact, the very word *electricity* is derived from the Greek word for a substance that when rubbed, produced shocks—amber or *electron*. The conclusion to draw is that static electricity is everywhere, and everywhere it is, it is a potential problem.

Nonconductive materials are the problem. They can easily acquire a charge via agitated contact with another substance. They either gain or lose electrons. Large ice crystals falling down and contacting small ones going up in clouds are the source of lightning. All phases of nonconductive material, including gases, are susceptible to gaining a charge from agitated contact with another substance.

NFPA 77, *Recommended Practice on Static Electricity*, 2000 edition, is the definitive source for information on how static electricity dangers can be mitigated. It is loaded with simple recommendations that can mean saving a life. For example, one technique to avoid static buildup in an agitated tank is to make sure that the agitator is sufficiently covered while in operation to minimize splashing.[8] It also makes sense to let liquids "relax" after being agitated and before other materials, especially solids and low-flash-point liquids, are combined with them. Also, whenever a liquid transfer takes place via a nonconductive route (e.g., air or rubber hose) prior to the transfer, the holding and receiving vessels should be ohmically connected.

In general, the key to static electricity management is to employ proper grounding and bonding. Discharges occur only when there is a charge separation between two objects that come into close contact. If those objects are allowed to discharge in a safe manner, no problem results. Following NEC code for other reasons should ensure most of the bases related to grounding and bonding are covered. An extra, recommended precaution is to familiarize oneself with the wide variety of products that exist for controlling static electricity. They range from indicating bonding clamps for liquid transfers to conductive smocks for electronics shop workers.

PRODUCTS

There are many essential accessories required for proper grounding applications that would garner an instrument engineer's interest. They include conductive concrete, bushings, connectors, grounding electrodes, ground wire, and resistance meters. Also important are the various products for managing static electricity discharge including special bonding jumper clamps that indicate whether they are properly connected and static charge meters. From the field engineer's perspective, shielding products include braid and Mylar shielded, twisted pair cable, and EMI shielded enclosures, gaskets, and ventilation devices. Electronics manufacturers will be interested in products that manage static electricity conductive clothing, wrist straps, heel straps, bins, conductive coatings, gaskets, coatings, mats, and other flooring.

Surge protectors are also popular and are required on large substations. Surge protectors come in several forms. The purpose of all types is to direct current on a conductor to ground during a large surge but to isolate it from ground during normal operation. Surge protectors are a form of "after the fact" protection in the sense that good grounding practice can eliminate surge issues in most installations. Specialized ground fault detector/interrupters are also used often. Usually, these devices detect ground faults in installations that apply complicated code rules to operate ungrounded.

CONCLUSION

However uninteresting grounding and shielding may seem to the casual observer, the instrument engineer is faced with a host of technical issues. Fortunately, there are sources to turn to for specific guidance on grounding. Although the primary emphasis of such guidance is safety, it does not in any way conflict with quality signal practices. The engineer can draw on simple rules for establishing shielding practice as well. Some general rules to follow include:

1. Never use the signal cable shield as a signal conductor, and never splice a low-level circuit.
2. The signal cable shield must be maintained at a fixed potential with respect to the circuit being protected.
3. The minimum signal interconnection must be a pair of uniform, twisted wires, and all return current paths must be confined to the same signal cable.
4. Low-level signal cables should be terminated with short, untwisted lengths of wire, which expose a minimum area to inductive pickup.
5. Reduce exposed circuit area by connecting all signal pairs to adjacent pins in a connector.

6. Cable shields must be carried through the connector on pins adjacent to the signal pairs.
7. Use extra pins in the connector as a shield around signal pairs by shorting pins together at both ends and by connecting to signal cable shield.
8. Separate low-level circuits from noisy circuits and power cables by maximum physical distance of up to 3 ft (0.9 m) and definitely not less than 1 ft (0.3 m).
9. Cross low-level circuits and noisy circuits at right angles and at maximum practical distance.
10. Use individual twisted shielded pairs for each transducer. Thermocouple transducers may be used with a common shield when the physical layout allows multiple pair extension leads.
11. Unused shielded connector in a low-level signal cable should be single-end grounded with the shield grounded at the opposite end.
12. High standards of workmanship must be rigidly enforced at all times.

References

1. Conroy, M. D. and Richard, P. G., "Deep Earth Grounding vs. Shallow Earth Grounding," in *Proceedings of the Power Quality Conference,* 1993.
2. Morrison, R., *Grounding and Shielding Techniques in Instrumentation,* New York: John Wiley & Sons, 1977, pp. 17–20.
3. Halliday, D. and Resnick, R., *Physics, Part Two,* New York: John Wiley & Sons, 1978, p. 652.
4. "Conductive Concrete Electrodes: A New Grounding Technology," *Electrical Line,* July, pp. 22–23, 1999.
5. Morrison, R., *Grounding and Shielding Techniques in Instrumentation,* New York: John Wiley & Sons, 1977, pp. 49–66.
6. Lovoula, V. J., "Preventing Noise in Grounded Thermocouple Measurements," *Instruments and Control Systems,* January, 1980.
7. Morrison, R., *Grounding and Shielding Techniques in Instrumentation,* New York: John Wiley & Sons, 1977, pp. 73–75.
8. National Fire Protection Association, NFPA Standard 77, 2000, p. 23.

Bibliography

Bouvwens, A. J., *Digital Instrumentation,* New York: McGraw-Hill, 1984.

Conroy, M. D. and Richard, P. G., "Deep Earth Grounding vs. Shallow Earth Grounding," *Proceedings of the Power Quality Conference,* 1993.

Denny, H., "Grounding for the Control of EMI," Virginia DWCI, 1983.

Halliday, D. and Resnick, R., *Physics, Part Two,* New York: John Wiley & Sons, 1978.

Klipec, B., "Reducing Electrical Noise in Instrument Circuits," *IEEE Transactions on Industry and General Applications,* 1967.

Lightning Protection Systems, Pub. No. 96A, Northbrook, IL: Underwriters' Laboratories.

Lovoula, V. J., "Preventing Noise in Grounded Thermocouple Measurements," *Instruments and Control Systems,* January, 1980.

McPartland, J. F. and McPartland, B. J., *National Electrical Code Handbook,* New York: McGraw-Hill, 1999.

Morrison, R., *Grounding and Shielding Techniques in Instrumentation,* New York: John Wiley & Sons, 1977.

Morrison, R. and Lewis, W. H., *Grounding and Shielding in Facilities,* New York: John Wiley & Sons, 1990.

"National Electrical Code 1999" (NFPA 70), Quincy, MA: NFPA, 1998.

"Recommended Practice for Grounding of Industrial and Commercial Power Systems," IEEE Standard 142-1992, *IEEE Green Book,* New York: IEEE, 1991.

"Recommended Practice for Powering and Grounding of Sensitive Electronic Equipment," *IEEE Emerald Book,* Standard 1100-1992, New York: IEEE, 1992.

Robinson, M. D., *Grounding and Lightning Protection,* Publ. EL-5036, Palo Alto, CA: Electric Power Research Institute, 1987.

Simmons, J. P., *Soares Book on Grounding,* 7th ed., Dallas, TX: IAEI, 1999.

1.8 Concepts of Hierarchical Control

H. I. HERTANU

Concepts of hierarchical control are based on functionality and hardware architecture. Functionality and hardware/software interplay is an iterative process that helps configure a control and management system for a given enterprise.

The concept of hierarchical control should be viewed in terms of functionality and hardware/software *architecture*. The term functionality applies to a well-defined task structure that meets the challenges of a modern integrated industrial enterprise operation, whereas the term *architecture* encompasses a variety of hardware/software building blocks that change more rapidly with the advancement of microelectronics, optoelectronics, software, and digital technology.

FUNCTIONALITY

The most commonly used image of the overall task structure that illustrates the functional hierarchy of an industrial enterprise, i.e., process plant, manufacturing plant, power plant, energy distribution network is the *pyramid* (Figure 1.8a).

The pyramid functional hierarchy is divided in several sections from the base upward. Each section operates on a set of input data received from other sections and generates a set of output data to other sections. Each section is provided with its own monitoring. Historically, each section was assigned a level number: level 1A, dedicated controllers; level 1B, direct digital control (DDC); level 3, supervisory control, etc. Today more generic functional names are in order such as functional sections, leaving their respective structures to depend on the technological progress.

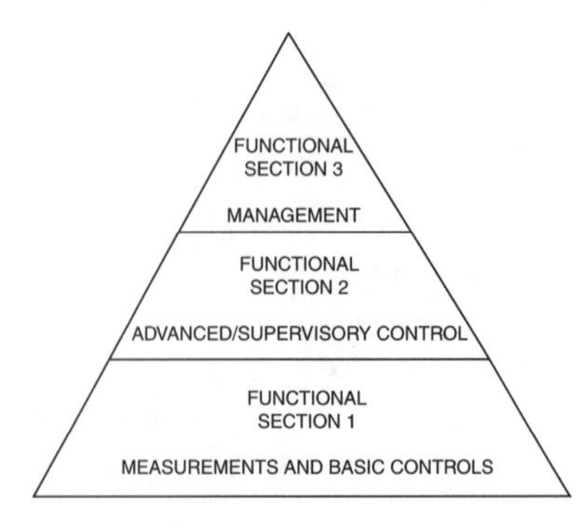

FIG. 1.8a
Functional hierarchy.

Measurements and Basic Controls (Functional Section 1)

This section includes the *fundamentals* of the overall instrumentation and control strategy of any industrial entity. Primarily, all the process variables and final control elements that provide the totality of any control system input and output data are part of this section and constitute the keystone of the global functionality of a control system in terms of accuracy and timeliness of response.

The basic control functions are also an integral part of the *fundamentals*. A variety of independent control algorithms operate over sets of inputs and/or outputs providing the following:

- Continuous/regulatory controls, i.e., PID (proportional, integral, derivative) algorithms interconnected in various combinations as required by a designed control strategy
- Sequential/interlock controls designed to perform step by step in real-time process or manufacturing operations called by their nature as normal operations, start-up/shutdown, safety, emergency operations
- Alarm processing and control based on a preestablished priority of response
- Real-time data collection
- Equipment monitoring
- Control system integrity monitoring, controlling malfunctions

Advanced/Supervisory Controls (Functional Section 2)

This section includes complex algorithms that are designed to address the overall control strategy of a particular industrial entity.

Control by objectives, that is, quantitative and qualitative targets assigned to the production of goods, environmental protection, energy consumption and conservation, productivity, and efficiency, represents a modern and global approach to advanced controls.

The advanced control algorithms applied selectively to achieve the objectives outlined above are as follows:

- Envelope control, which throttles a set of manipulated variables based on a matrix of constraints imposed on a set of process variables and process set points to control a process and meet predetermined objectives.

- Model base controls, which are of different vintages, i.e., internal model control (IMC), model predictive control (MPC), process model base control (PMBC), nonlinear control, most often implemented by means of neural networks, and, last but not least important, an artificial intelligence base model or a model developed using an expert system.
- Optimization control, which is today included in one of the algorithms mentioned above.
- Maintenance control, which ensures the correct operation of all equipment and control system elements by monitoring the real-time parameters needed for scheduling preventive maintenance.
- Advanced controls, which rely on the following two support functions:
  - Material and energy balance monitoring
  - Monitoring and archiving key production and operational parameters

Management (Functional Section 3)

In the past decade it has become necessary to gradually restructure and integrate various functions coexisting within an enterprise, with the controls as defined in Functional Sections 1 and 2, above. The management function complexity consists of constant interplay among marketing, financing, human resources, production, and inventory tasks to generate and optimize profit while supplying the market with quality goods in a timely fashion.

The management function operates on a variety of data. The decision-making process—a key management prerogative—evolves from analyzing data and drawing conclusions that are translated and returned to the overall management and control system as target data and set points.

The management function relies on the concept of data (information) organization and processing. An entire scientific domain is being developed, dedicated to data (information) organization and processing known as IT (information technology), informatics, or by the French term *informatique*.

The fundamental structure used in data (information) organization and processing is the database. A database stores and retrieves information. However, a state-of-the-art database stores information and also interrelationships among pieces of information stored. Furthermore, the database must be capable of retrieving information on the basis of an interrelationship that was not perceived when the information was stored.

The database concept has created a management structure that helps to integrate marketing/sales, financing, production, inventories, human resources, cost control, and administration. This structure is ERP (enterprise resource planning). A predecessor to ERP is MRP1 (material requirement planning), which later evolved into MRP2. Initially through MRP1 it was possible to plan interrelated requirements, which involves splitting the bill of materials for the end products on the basis of sales forecasts or real orders and determining production targets and set points. MRP1 provided a continuous calculation of net material requirements. MRP2, which developed from MRP1, takes into consideration fluctuation in forecast data by including simulation of a master production schedule, thus creating long-term control. A more general feature of MRP2 is its extension to integrate all the management tasks mentioned above and illustrated in Figure 1.8b.

ERP is the most modern model for integration of all management tasks by creating an enterprise unified relational database on which various algorithms and models (planning, marketing/sales, financing, cost control, quality control, and profitability) would operate. ERP would provide information to generate a research and development plan, and an investment plan for the enterprise.

ERP will generate information for planning and monitoring the enterprise preparedness and prevention of major crisis such as explosions, environmental disasters, floods, and fire.

The application of artificial intelligence to enterprise management and advanced controls also will be supported by ERP.

Using ERP to create an overall enterprise management function is a monumental undertaking. ERP *is not an off-the-shelf application package*. It is a concept plus a recommended set of tools, which to fit to the operation of a particular industrial enterprise requires operational changes within the enterprise and customization of various ERP tools.

HARDWARE ARCHITECTURE

Input/Output Systems

Distributed input/output (I/O) is the most important development in control system hardware in the past 10 years. A specialized data highway connecting all types of field devices of a control system has finally emerged. Although there is no universally accepted standard communication protocol and its supporting standardized hardware in place, a few implementations do coexist, such as Foundation Fieldbus, Hart, AS-i bus, PROFIBUS-DP, and high-speed Ethernet.

Controllers

These are specialized microprocessor-based devices capable of executing all the control tasks included in Functional Section 1. These devices operate in real time and could be structured to provide self-tuning, redundancy control, and communication. In this category one should add specialized controllers such as PLCs (programmable logic controllers), RCUs (remote control units), NCs (numeric controllers),

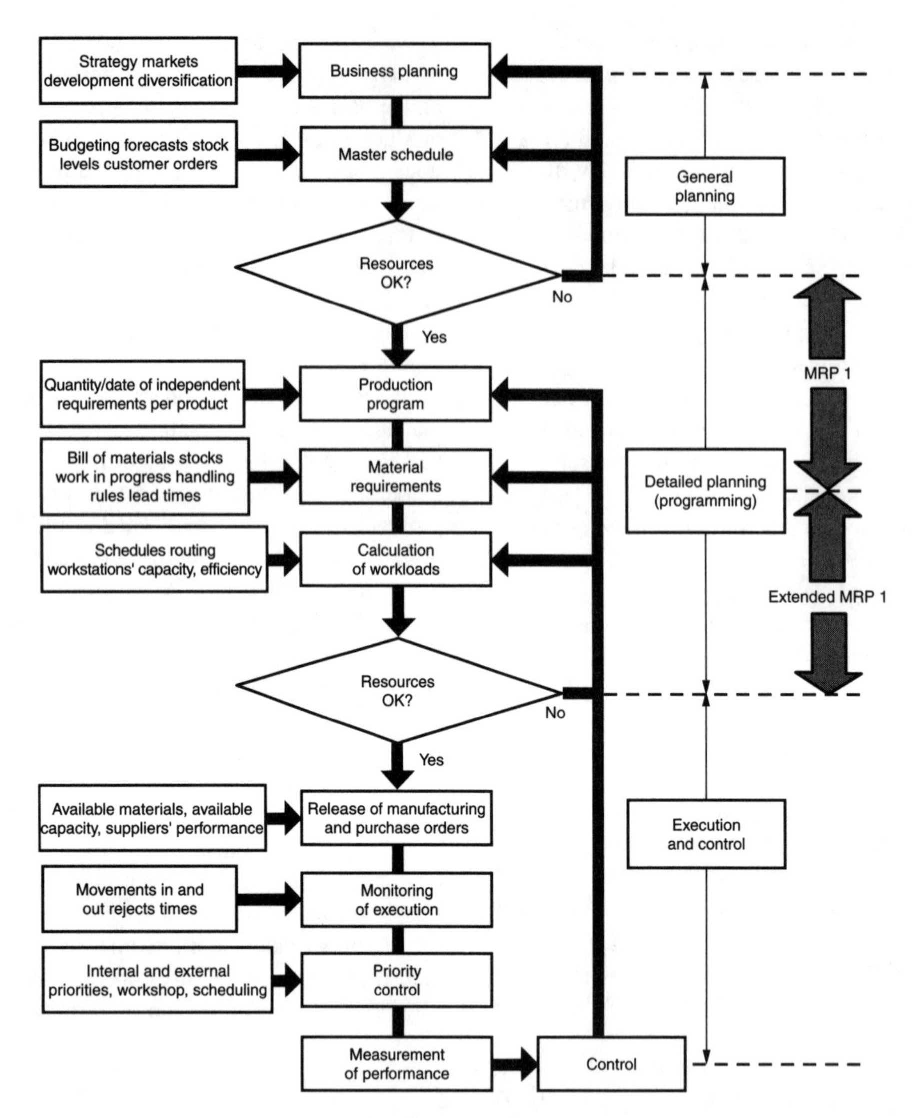

FIG. 1.8b

Flowchart for MRP1 and MRP2 (extended MRP1). (From Waldner, J.B., Principles of Computer-Integrated Manufacturing, New York: John Wiley & Sons, 1992. With permission.)

process analyzers, weighing systems, laboratory instruments, etc.

Workstations

Workstations are created to provide the interface between human operators and the control and management system to which these devices are connected. Workstation hardware includes video monitors, keyboards, a mouse, trackballs, video cameras, barcode readers, and a fully equipped computer that services all these operator I/O devices. The workstations provide a complete operator-friendly environment, which consists of a broad range of functions associated with operating and managing a given industrial enterprise. Well-designed and flexible interactive video displays, input commands, and a hierarchy of protection passwords have become well-accepted operation tools.

Communications

The key hardware elements of control and management systems exchange information in real time, via different types of communication networks. These networks are constructed to provide horizontal communication to ensure data transfer within the domain of each functional section, as defined above. Furthermore, vertical communication is also provided for data transfer between the domains of Functional Sections 1, 2, and 3. The communication technology applied at present to control and management systems is not standardized. One group of manufacturers and users tends to support a communication network standard (hardware, protocol, and software), while another group promotes a different standard.

However, for a while several communication standards will coexist. Selecting and using them would influence the selection and use of a particular control and management system, which fits best an industrial enterprise.

It is helpful to mention some of the communication buzzwords that are used in connection with control and management systems.

Data highway is a generic term adopted by manufacturers, which developed proprietary communication networks, designed to service mainly proprietary control devices.

Foundation Fieldbus is a *de facto* standard for a communication network dedicated to field instruments and control devices.

Hart is a *de facto* standard for a communication network dedicated also to field instruments. Hart is a precursor standard to Foundation Fieldbus.

Ethernet is a well-established communication standard born at Xerox Corp. in the late 1970s and adopted as a standard by the IEEE in 1988 as IEEE 802.3. Over the years Ethernet improved in terms of speed (now 100 Mbaud, in the near future 1 Gbaud) and message integrity in spite of its nondeterministic protocol.

Internet and *intranet* represent essentially the same communication technology. The Internet became a worldwide communication standard allowing access to information needed by all fields of human interests. The intranet limits access to users of a particular enterprise, provides specialized information domains, and enhances security of information transactions. Internet/intranet navigation is assured by browsers such as Microsoft Explorer, Netscape Navigator, etc.

Architectural Concepts

The fundamental concept in today's system architecture is the distribution of functions executed by a distribution of hardware key elements. These distributions (functional and hardware) operate as an integrated system due to links to communication networks that ensure fast, accurate, and secure exchange of information, in real time.

Several other architectural concepts should be considered in the process of engineering control and management systems:

- Availability, that is, the inclusion in the system structure of adequate redundancy, which would ensure uninterrupted system operation
- Interoperability, that is, having all system points of access configured alike, therefore allowing the usage of identical procedures for sending and retrieving system information
- The open system concept, meaning the structuring of a system by using components made by different manufacturers
- Flexibility in terms of expansion in size and/or functionality

STRUCTURAL CONFIGURATION

Structuring a system is based on an in-depth knowledge of the enterprise operation in its entirety.

In principle one should adopt a bottom-up configuration, which determines the type, quantity, and distribution of all the instruments, control devices, and any other primary equipment that needs to be connected to the system. Then, in accordance with various functional requirements, one configures the communications requirements and the measurement and control hardware as part of Functional Section 1. Next using similar criteria we configure successively Functional Sections 2 and 3.

A block diagram of a bottom-up system structural configuration is shown in Figure 1.8c. For clarity, the diagram shows only one hardware/software block of a kind. An example of a commercially available system is shown in Figure 1.8d.

HARDWARE/SOFTWARE INTERPLAY

Engineering a management and control system is in itself a complex and an iterative process where the interplay between functionality and hardware/software should ultimately produce an optimal structural configuration meeting given requirements. Familiarity with the IT domain and specifically with its application to control and management of industrial enterprises is a must. Hundreds of minisystems, large systems, microprocessor-based devices, workstations, communications equipment, and software packages are offered to potential users. Making sense of this myriad of products is daunting. Here are some IT considerations and trends.

Hardware

Each and every device included in a control and management system, such as controllers, workstations, and servers, contains a classical computer architecture consisting of microprocessors, memories, an internal communication bus, communications ports, external storage devices, monitors, printers, keyboards, etc. An operating system makes the hardware execute predetermined instructions. This commonality in the hardware structure has reduced system design complexity and has enhanced its operability and maintainability. Trends in hardware are toward higher-speed microprocessors, reliability of all components, and cost reductions. Furthermore, the utilization of commercially available equipment such as PCs vs. hardware dedicated solely to industrial control is an issue that is still being debated.

The introduction and proliferation of "intelligent instruments," "intelligent control devices," and "intelligent I/O subsystems" impact the control system hardware design, whereas more processing power is dedicated to communications, and some measurement/control tasks are executed at the instrument and control device level.

Operating Systems

The following operating systems are being used in control and management systems: Windows NT, UNIX, LINUX, and various proprietary operating systems. NT is widely used

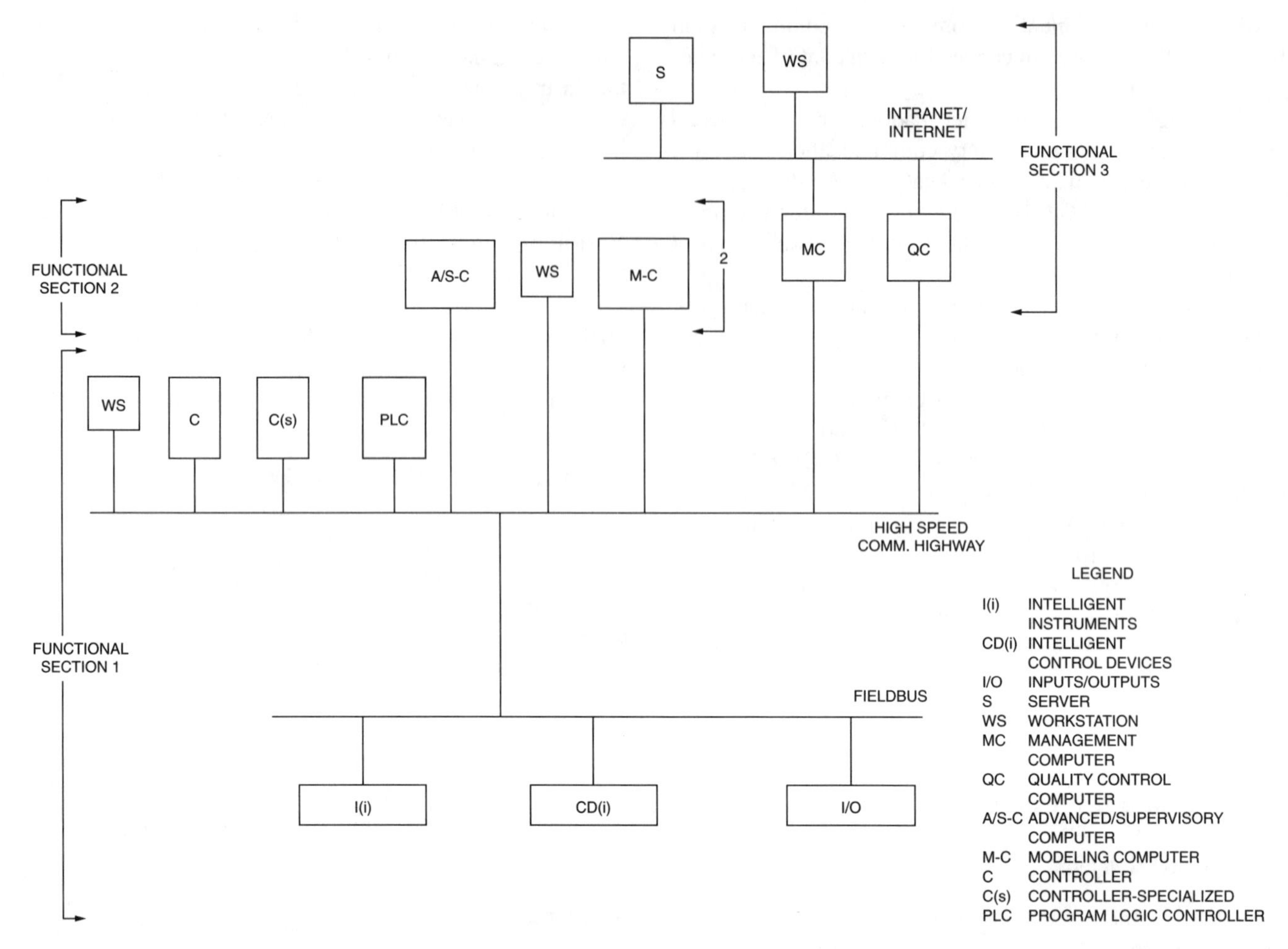

FIG. 1.8c
Hardware architecture—conceptual block diagram.

and firmly established. NT supports all applications included in Functional Sections 2 and 3 well. There are some concerns regarding NT supporting real time process control (Functional Section 1). Propriety operating systems are used more frequently in the process control domain.

UNIX, including its proprietary extensions, is used in large and very large systems supporting all functional sections. UNIX has a good stability and security track record and has been in operation for over 30 years.

LINUX is a rather new operating system with a limited application base. However, its acceptability as a viable alternative to NT remains to be proved in the future.

Communication Protocols

Ethernet high-speed protocol, TCP/IP protocol, and other proprietary or *de facto* protocols coexist today in industrial control and management systems. It would be difficult to predict which of these protocols, if any, will be universally adopted.

Application Software

The application software encompasses a vast domain of programs, mostly proprietary, which perform all the tasks that are included in Functional Sections 1, 2, and 3. The following is a list of important programs dedicated to perform control and management functions:

- I/O processing
- Integrated control tool set for total control solutions including logic/sequential control and continuous control domains
- Graphic and display tools for generating interactive process control and management displays
- System-wide information management
- System communication networks management
- Production control and information management tools, such as:
  Analyzer management software
  Batch software
  Historian
  Process optimizer

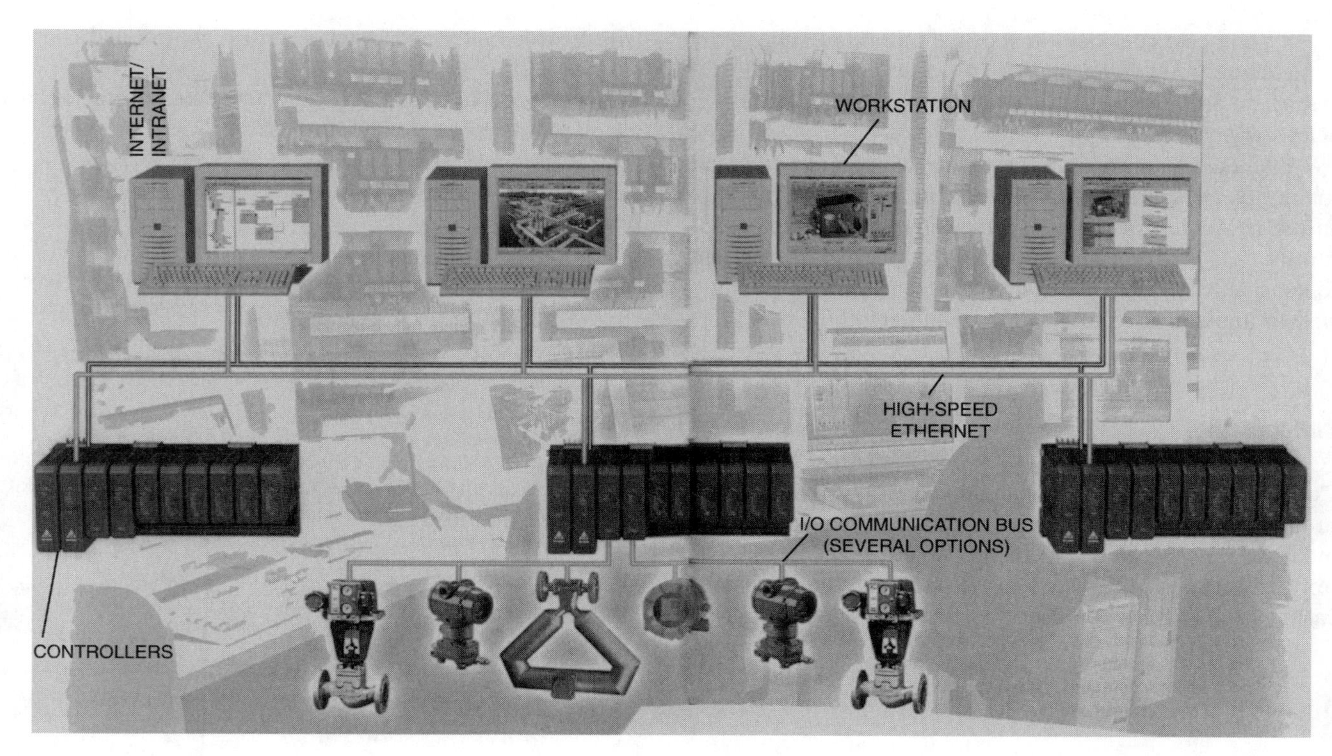

FIG. 1.8d
Control and management system. (Courtesy of Control Associates, Inc.)

Production modeling tools
Statistical process control
Spreadsheet
- Software development tools for system configuration and/or programming of custom solutions

SYSTEM HIERARCHY INTERPLAY

The hierarchy interplay is between functionality defined for Functional Sections 1, 2, and 3, the application software which will implement a given function such as batch control, alarm processing, operator–system interface, and the system hardware and software best suited to execute it. This is an iterative process, which takes place until a system hierarchy is conceived.

In this iterative process, additional criteria should be considered:

The size of the control and management system (small, medium, large)
The geographical spread of the enterprise
Expansion/modernization of an existing enterprise
The overall system response time and also very specific response time of real-time critical tasks such as I/O processing

Reviewing commercially available hardware and software packages dedicated to enterprise control and management and trying to match them with functional requirements is a complex undertaking. The better the match, the fewer the surprises that may occur during implementation and operation.

This iterative process helps to better define various functions once one has a better understanding of the application software and system software and hardware available. The conclusions of this process may determine if a single source control and management system supplier would be acceptable or if a *hybrid* system may be more suitable.

SYSTEM SPECIFICATION, SELECTION, DESIGN, AND IMPLEMENTATION

These are well-established engineering tasks, which are well served if a control hierarchy is conceived following the concepts described in this chapter. It helps in the generation of a functional specification, which comes closer to the available technology while translating the enterprise goals to feasible functions.

The selection process of system offers/proposals is helped by making a more objective selection and by avoiding "shoe-in" systems that claim to fit all applications. Increasingly, hybrid systems are the answer for medium and large systems. The hybrid systems include specialized hardware/software packages supplied by different manufacturers integrated into a control and management hierarchy best suited for a particular enterprise.

System cost considerations of the initial investment, the expenditure outlay over the life of the system, and the ROI (return on investment) of this undertaking are also part of the selection process.

The design and the system implementation depend mainly on personnel competence, experience, and training. In the iteration process, where the interplay between functionality and hardware and software takes place, it is important to consider if one has the resources (technical and managerial) to implement a particular system hierarchy.

CONCLUSIONS

Concepts of hierarchical control provide directions to follow in the application of IT to total control and management of industrial enterprises. The demands for better and better control and management of enterprises become more sophisticated to respond to global market pressures. IT rapid progress will provide better and faster means to control and conduct business.

The concepts of hierarchical control are adaptable tools, which will keep pace with progress.

Bibliography

Feeley, J., Merritt, R., Ogden, T., Studebaker, P., and Waterbury, B., "Process Control from 1990," *Control Magazine,* www.controlmag.com, December 1999.

Hebert, D., "Where's the Support?" *Control Magazine,* November 2000, p. 47.

Hertanu, H., "The Cost of Distributed Control," *Control Magazine,* January 1995.

O'Brien, L., "DCS Grows Up," *Control Magazine,* January 2000, p. 73.

Studebaker, P., "All for One—The Top 10 Trends in Process Control," *Control Magazine,* www.controlmag.com, April 1998.

Waldner, J.-B., *Principles of Computer-Integrated Manufacturing,* New York: John Wiley & Sons, 1992.

Waterbury, B., "Distributed Intelligence Goes to Work," *Control Magazine,* August 2000, p. 41.

1.9 Analog and Discrete Input/Output, Costs and Signal Processing

H. EREN

Typical I/O Products:

Digital I/O
32-bit digital I/O, up to 80 MB/s
48-bit, isolated D/O

Data Acquisition (DAQ) Boards
5 MS/s, 12-bit, 16 analog inputs
1.25 MS/s, 12-bit, 16-64 analog inputs
333 kHz, 16-bit, 16 analog inputs
204 kS/s, 16-bit 2 in/2 out, DSP based
204 kS/s, 16-bit 2 in/2 out, host based

Counter Timers
32-bit, counter timers, up to 80 MHz

Analog Inputs
500 kS/s, 12-bit, 16 analog inputs
24-bit, thermocouple inputs
20 MS/s, 8-bit, 2 analog inputs

Digitizers
60 S/s, 24-bit, 8 channel logger

Typical Data Acquisition Board (DAQ):

Analog Input
Analog inputs: 12-bit, 2 or 4 inputs
Maximum sampling: 5 MS/s
Minimum sampling: 1 kS/s
Input range: ±42 V, ±20 V, ±10 V, ±5 V, ±1 V, ±500 mV
Input coupling: AC or DC
Over voltage: ±42 V
Relative accuracy: ±0.5 LSB
Data transfer: DMA, interrupts, programmed I/O
Dynamic range: 200 mV to 10 V, 75 dB; 20 to 50 V, 70 dB
Amplifiers: ±200 nA bias current; ±100 pA offset current; 1 MΩ input impedance, CMRR 60 dB (typical)
Bandwidth: 5 MHz, 500 mV to 50 V

Analog Output
No. of channels: 2
Resolution: 16-bits, 1 in 65,536
Maximum update: 4 MS/s, one channel; 2.5 MS/s two channels
Relative accuracy: ±4 LSB
Voltage output: ±10 V
Output impedance: 50 Ω ± 5%
Protection: Short to ground
Settling time: 300 ns to ±0.01%
Slew rate: 300 V/μs
Noise: 1 mV_{RMS}, DC-5 MHz

Digital I/O
No. of channels: 8 input, 8 output
Compatibility: 5 V/TTL
Data transfer: programmed I/O
Input low voltage: 0 V to 0.8 V
Input high voltage: 2 V to 5 V

Timing
Base clock: 20 MHz and 100 kHz
Clock accuracy: ±0.01%
Gate pulse duration: 10 ns, edge-detect
Data transfer: Programmed I/O

Trigger
Analog trigger: Internal (full scale); external ±10 V; 8-bits resolution, bandwidth 5 MHz;
Digital trigger: 2-triggers; start stop trigger, gate, clock; 10 ns min pulse width; 5 V/TTL; external trigger impedance 10 kΩ

Other Characteristics
Warm-up time: 15 min
Calibration interval: 1 year
Temp. coefficient: ±0.0° ppm/°C
Operating temp.: 0 to 45°C
Storage temp.: –20 to 70°C
Relative humidity: 10 to 90%
Bus interface: Master, slave
I/O connector: 68-pin, male SCSI
Power requirements: 4.65 to 5.25 VDC, 3 A
Physical dimensions: 31 × 11 cm
Price: about U.S.$2000

Partial List of Suppliers:

Access I/O Products, San Diego, CA; Acqutek Corp., Salt Lake City, UT; Advantech Automation Corp., Cincinnati, OH; Advanced Circuit Technology, An Amphenol Co., Nashua, NH; Advanced Data Capture Corp., Concord, MA; Agilent Technologies, Inc., Palo Alto, CA; Allen-Bradley, A Rockwell Automation Brand, Duluth, GA; Amtech, Hudson, WI; Analog Devices, Inc., Norwood, MA; Analog Interfaces, Inc., Alliance, OH; Analogic Corp., Peabody, MA; Arrow Electronics, Inc., Melville, NY; Axiom Technology Inc., Chino, CA; ABB Automation, Inc.—TOTALFLOW® Products Group, Houston, TX; ADAC Corp. (American Data Acquisition Corp.), Woburn, MA; APT Instruments, Litchfield, IL; ATI Industrial Automation, Apex, NC; Beckhoff Automation LLC, Minneapolis, MN; Bruel & Kjaer, Norcross, GA; Bulova Technologies, Inc., Lancaster, PA; Burr-Brown Corp., Tucson, AZ; Celerity Systems, Inc., Cupertino, CA; Chase Scientific Co., Aptos, CA; ComputerBoards, Inc., Middleboro, MA; Data Entry Systems, Huntsville, AL; Data Flow Systems, Inc., Melbourne, FL; Dataforth, A Burr-Brown Co., Tucson, AZ; Davis Instruments, LLC, Baltimore, MD; Daytronic Corporation, Dayton, OH; Devar, Inc., Bridgeport, CT; Diamond Systems Corp., Palo Alto, CA; Digital Equipment Corp., Maynard, MA; Digital Interface Systems, Inc., Youngstown, OH; Dranetz Technologies, Inc., Edison, NJ; DSP Technology, Inc., Fremont, CA; Echotek Corp., Huntsville, AL; Entrelec, Inc., Irving, TX; Global American, Inc., Hudson, NH; Harris Corp., Controls Div., Melbourne, FL; Industrial Computer Systems, Greensboro, NC; Intelligent Instrumentation, Inc., Tucson, AZ; ICS Advent, San Diego, CA; Jenzano, Inc., Port Orange, FL; Jon Goldman Associates, Orange, CA; Kaye Instruments, Inc., Bedford, MA; Keithley Instruments, Inc., Cleveland, OH; KSE Corp., Pottstown, PA; LPTek Corp., Westbury, NY; Measurement Systems International, Inc. (MSI), Seattle, WA; Microstar Laboratories, Inc., Bellevue, WA; MSC Industrial Supply Co., Melville, NY; National Instruments Corp., Austin, TX; NEWCO, Inc., Florence, SC; Phillips Electric, Inc., Cleveland, OH; PC Instruments, Inc., Akron, OH; PCS Computer An Advantech Sub., Cincinnati, OH; Quatech, Inc., Akron, OH; Racal Instruments, Inc., Irvine, CA; Real Time Devices, Inc., State College, PA; RC Electronics, Inc., Santa Barbara, CA; Schneider Electric, North Andover, MA; Signatec, Inc., Corona, CA; Solartron Mobrey, Houston,

TX; Sonix, Inc., Springfield, VA; Symmetric Research, Kirkland, WA; Synetcom Digital, Inc., Torrance, CA; Texas Electronics, Inc., Dallas, TX; Trig-Tek, Inc., Anaheim, CA; TEAC America, Inc., Montebello, CA; TEK Industries, Inc., Manchester, CT; TRW, Inc., Cleveland, OH; Ultraview Corp., Orinda, CA; Veda Systems, Inc., California, MD; Vesta Technology, Inc., Wheat Ridge, CO; Wang Laboratories, Inc., Billerica, MA; Warren-Knight Instrument Co., Philadelphia, PA; WaveEdge Technologies, Burke, VA; Westinghouse Electric Corp., Process Control Div., Pittsburgh, PA; Worldnet Computers, Inc., Aliso Viejo, CA; WAV, Inc., West Chicago, IL

The majority of process equipment such as sensors, transducers, and final control elements are analog devices, and they generate or operate on analog signals. In the fields of process control, automatic control, instrumentation, communications, etc., the signals generally can be processed in three different ways: (1) by analog techniques that directly deal with analog signals, (2) by converting the analog signals into digital or digital signals to analog and implementing the systems by using digital instruments, and (3) by dealing with the signals purely in digital form as digital-to-digital input and output.

In modern times when computers and microprocessor-based equipment are heavily used in processes, the analog signals must be converted to digital values. The key feature of modern instruments is the conversion from analog signals into digital signals and vice versa by means of analog-to-digital converters (A/D or ADC) and digital-to-analog converters (D/A or DAC), respectively. This section begins by explaining the basic concepts related to these conversions and the circuits used to carry them out. An important advantage of digital techniques is that they can be managed and communicated relatively easily and reliably in comparison to the analog methods. Also, in addition to A/D and D/A conversions, other indispensable operations (such as sampling or signal conditioning) are required to make full use of modern instruments and instrumentation systems. The whole process of carrying out all these operations is referred to as the data-acquisition systems. In this section, the basic functions of these systems and the theoretical foundations of their operation are presented.

In many scientific and engineering applications and in process control and instrumentation, the proper design of devices to carry out monitoring and controlling is important. In this respect two types of devices can be found:

1. **Monitoring devices** to display and record the values of process variables
2. **Control devices** to act on processes to control their variables

Figure 1.9a illustrates a typical monitoring and controlling process. The *sensors* are used to measure the values of the variables. The *display* and the *recorder* instruments are utilized for visually observing the values of the process variables and for storing these values. The *controller* instruments carry out the control functions, determining the necessary orders of operation for the *actuators*.

The sensors used for measuring the process variables normally produce analog voltage or current signals. These signals can be dealt in two ways: (1) using analog instruments that deal directly with analog signals, and (2) converting the analog signals into digital forms and implementing them on digital instruments. A typical arrangement of a digital control system is given in Figure 1.9b.

Digital systems have many advantages over their analog counterparts, particularly if the system is complex. This is because digital systems are based on digital processors, which provide powerful capabilities for signal processing and handling simply by means of software or firmware implementation. Another important advantage of digital systems is that once a signal is converted to digital form, it can be managed by any kind of digital computer, and this provides a wide range of possibilities for signal communication, storage, and visual display. A typical microprocessor-based digital system configuration is given in Figure 1.9c. In this section, analog signal processing will not be dealt with in detail; however, further information can be obtained (e.g., Garrett, 1987) from the references in the bibliography given at the end of this section.

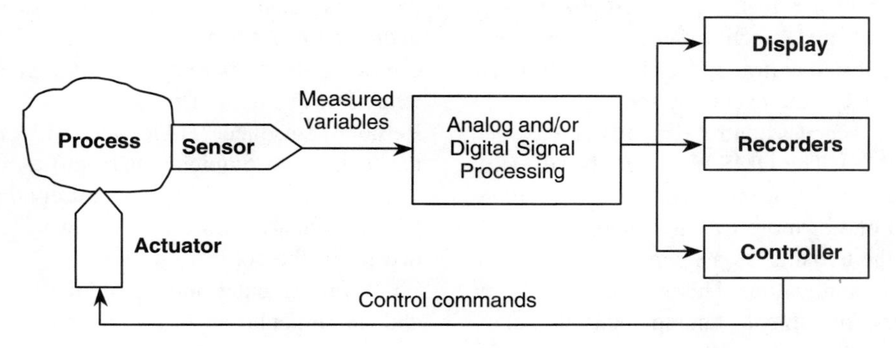

FIG. 1.9a
Monitoring and controlling of a process.

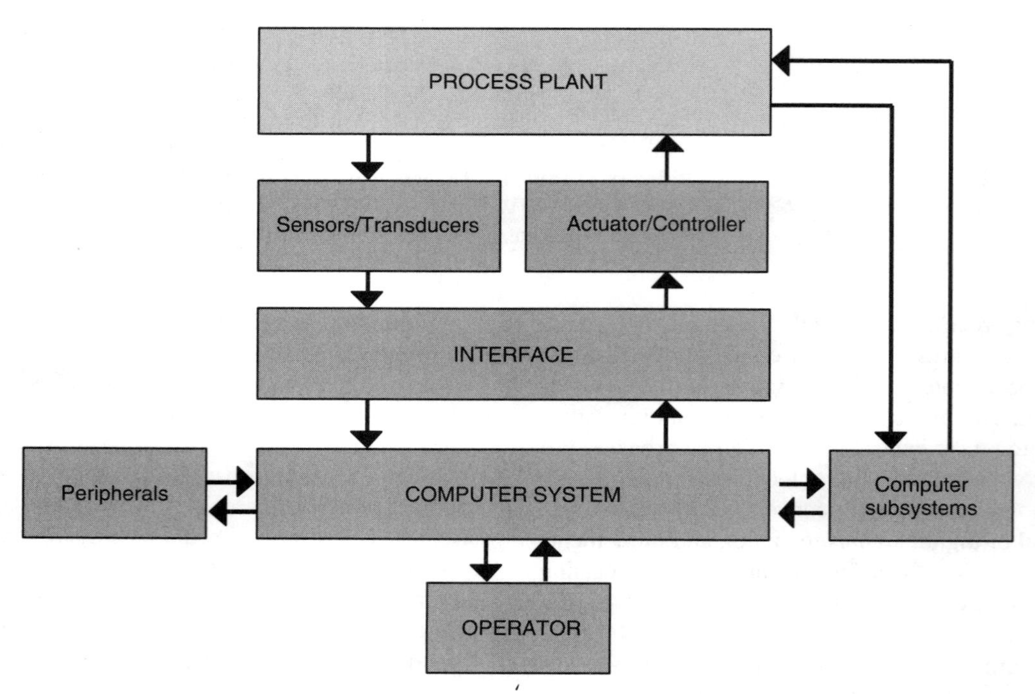

FIG. 1.9b
A typical digital process control system.

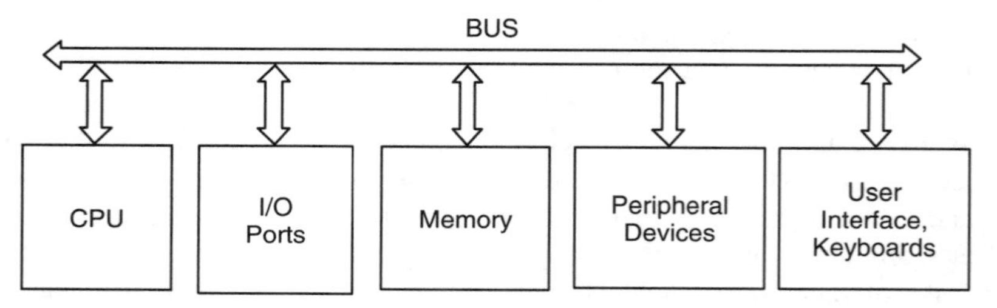

FIG. 1.9c
A basic microcontroller architecture.

In the majority of applications, processes are monitored continuously by the data acquired from the sensors, transducers, and other measuring systems that are operating on the plant and factory floor. The advances in sensor technology provide a vast array of measuring devices that can be used in a process. In all applications, reliable and effective measuring of the process variables is essential so that further decisions can be made concerning the operations of the overall system.

Essentially, modern process control depends on the open measuring system (OMS). The OMS has a number of components, such as the data acquisition, data processing, graphical presentation, and one or more transit systems. Technology and topology of the transit systems depend on its age of design and the specific application. This means that a complete variety of sensors, instruments, analog and digital components, controllers, interfaces, and bus systems can be expected to exist on the plant floor, accumulated over time.

The OMS field devices in use today can be grouped in three categories: (1) units with traditional analog and discrete input/output (I/O), (2) hybrid analog and digital devices, and (3) purely digital devices. The first type of device is usually connected to a controller by means of a dedicated 4–20 mA analog current loop. The second type can be used with both analog and digital communication systems. For example, a protocol called Highway Addressable Remote Transducer (HART) uses digital communication signal technology implemented onto a conventional 4–20 mA analog signal. The third type, purely digital devices, usually requires custom interface hardware and custom software drivers in the control systems.

A fundamental and essential part of digital instruments and instrumentation is the conversion from analog signals into digital signals, and vice versa. These concepts will be considered next.

A/D AND D/A SIGNAL CONVERSIONS

The conversion of an analog signal into the digital signal equivalent consists of representing the values taken by the analog signal in a set of determined sequences of n bits. The conversion is carried out in two steps: first the signal is quantized and later it is coded. *Quantizing* means representing the continuous values of the analog signal using a set of discrete values, and *coding* means representing these discrete values by bit sequences. The number of bits of these sequences determines the number of possible values of the conversion: 2^n for n bits.

An electronic circuit called an analog-to-digital converter (ADC) carries out the quantization and coding processes. The ADC operates with signals of determined amplitude, for example, between –V and V volts. Another electronic circuit called a digital-to-analog converter (DAC) carries out the conversion from digital signals into analog signals. This device converts digital codes of n bits into a signal of 2^n discrete levels of voltages or currents.

D/A Converters

Digital words of binary information require DACs for the conversion of the binary words into appropriately proportional analog voltages. The typical D/A converter has three major elements—(1) a resistive or capacitive network, (2) switching circuitry, operational amplifiers, and latches, and (3) a reference voltage. Also, an important requirement is the means to switch the reference voltage from positive to negative for achieving a bidirectional analog signal at the output.

Figure 1.9d shows a typical n-bit DAC. The set of latches holds the binary number, which is to be converted to an analog voltage signal. The output of each latch controls a transistor switch, which is associated with a determined resistor in the resistor network. The voltage reference, which is connected to the resistor network, controls the range of the output voltage. The operational amplifier works as an adder circuit. The operation of the D/A converter is as follows:

1. Switches set the digital word, i.e., switch is closed for logic ONE and switch is grounded for logic ZERO.
2. The weighted resistors determine the amount of voltage V_{OA} at the output.

For each of the resistors in the network, when its corresponding switch is on, the operational amplifier adds a voltage to V_{out}, which is inversely proportional to the value of the resistor and directly proportional to the reference voltage and the feedback resistor (R). For example, for the 2R resistor the operational amplifier adds the following value to V_{out}:

$$V_{out} = V_{ref}\frac{R}{2R} = \frac{V_{ref}}{2} \qquad 1.9(1)$$

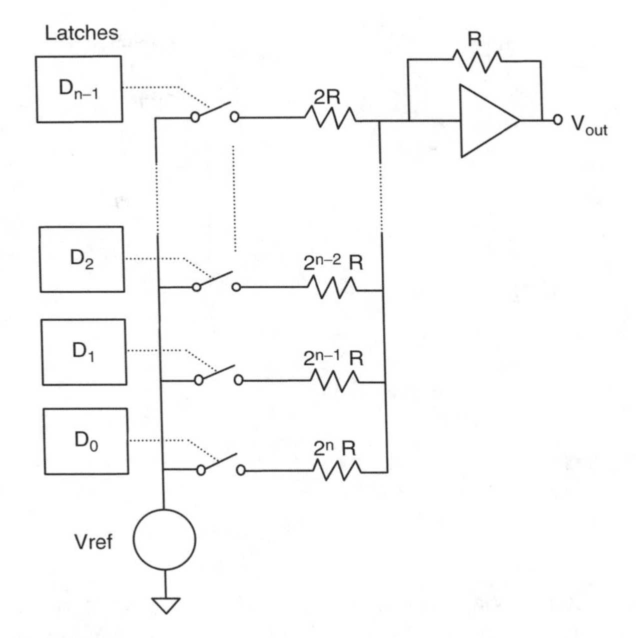

FIG. 1.9d
Basic DAC design.

The total value of V_{out} can be calculated by the following formula:

$$V_{out} = V_{ref}\left(\frac{D_{n-1}}{2} + \frac{D_{n-2}}{4} + \cdots + \frac{D_1}{2^{n-1}} + \frac{D_0}{2^n}\right) \qquad 1.9(2)$$

where D_{n-1}, D_{n-2}, etc. represent the values of the bits of the binary number to be converted. In this formula each active bit contributes to V_{out} with a value proportional to its weight in the binary number, and thus the output voltage generated by the DAC is proportional to the value of the input binary number.

Weighted Current D/A Converter The weighted current DAC generates a weighted current for each bit containing a logic ONE. This kind of DAC is useful when an analog current is to be implemented directly, although if a voltage is needed, the output voltage is simply the voltage across a load resistor. The circuit diagram for the weighted current DAC is shown in Figure 1.9e.

When the switches are closed, in the nonzero case, Kirchoff's current law states that (provided the currents are all traveling in the same direction) the current leaving the junction is the addition of all the currents entering the junction.

One of the advantages for this type of DAC is that, for unipolar analog outputs, it is easy to implement constant-current drive circuits (using weighted resistors to set the value of each current source). However, limitations exist especially in accuracy. There will be leakage currents in the zero condition

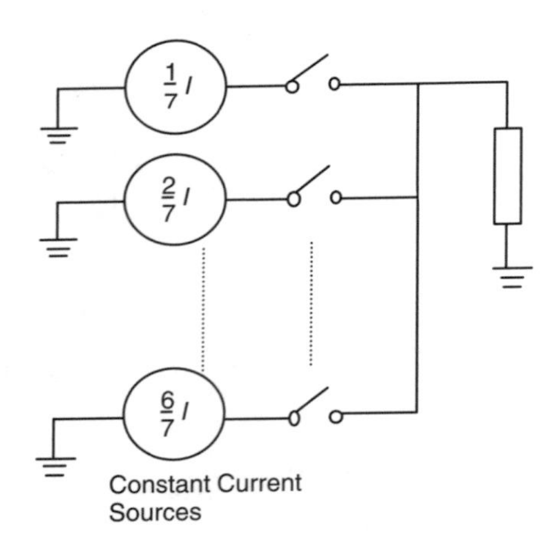

FIG. 1.9e
A 3-bit weighted current DAC.

(all switches open), resulting particularly at higher temperatures. This type of D/A conversion is suitable for low to medium accuracy and can be used when a high rate of conversion is necessary.

A/D Converters

ADCs and DACs provide the interface between the analog world (the real world) and the world of digital computation and data processing. Many of the applications for A/D interfaces are quite evident as we use them in our daily lives. Temperature controllers, computers and related peripherals, electronic fuel injection, and CD music systems are just a few examples of the typical applications that require A/D interfaces.

The analog signals converted to digital form can arise mainly from sensors (or transducers) measuring a myriad of different real-world sources including temperatures, motions, sounds, electrical signals, etc. When these signals are converted to digital forms signal, the signal can then be processed using, for example, digital filters or algorithms. The processed signals can then be converted back to analog forms to drive devices requiring an analog input such as mechanical loads, heaters, or speakers. The ADCs and DACs therefore provide the vital link between the real world and the world of digital signal processing.

There are many types of ADCs in practical use because of the diverse range of applications for ADCs. Direct ADCs determine equivalent digital values from a reference analog signal fed into the input. Several types of ADCs are as follows.

Counter Ramp ADCs Counter ramp converters are one of the simplest implementations of direct ADCs. However, the trade-off for simplicity is speed, and these types of converters are the slowest ADCs available. For an analog input, $2^n - 1$ processing steps are required to convert the input to an n-bit digital word. Figure 1.9f shows the block diagram for the counter ramp ADC. The conversion process of the counter ramp ADC is as follows:

1. The reset pulse resets the counter to zero thus driving the D/A decoder output V_{OA} to 0 V.
2. The clock starts and the counter begins counting, driving the D/A decoder such that the output V_{OA} is proportional to the counter.
3. V_{OA} is compared against the analog reference voltage V_{IA} and when $V_{OA} > V_{IA}$; then the comparator output changes state such that the clock is switched off and the counter stops. The parallel digital word in the counter is the quantized version of the analog input voltage.

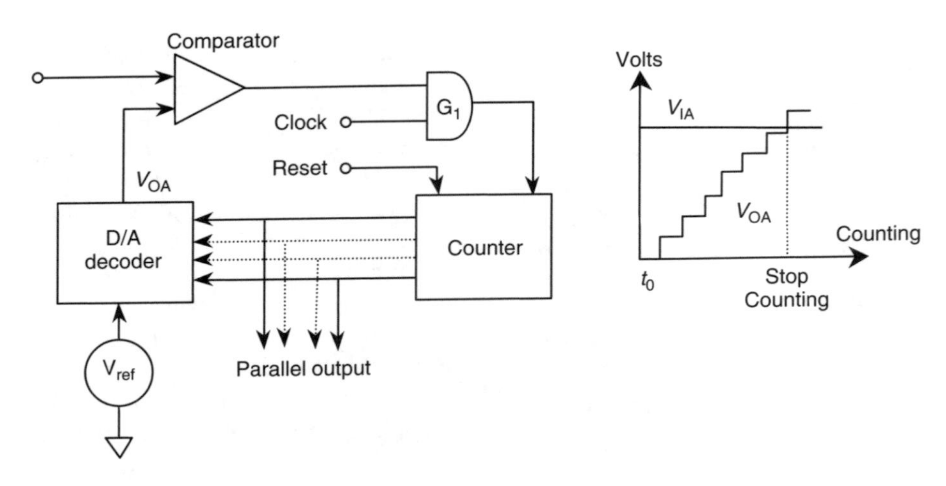

FIG. 1.9f
Counter ramp ADC.

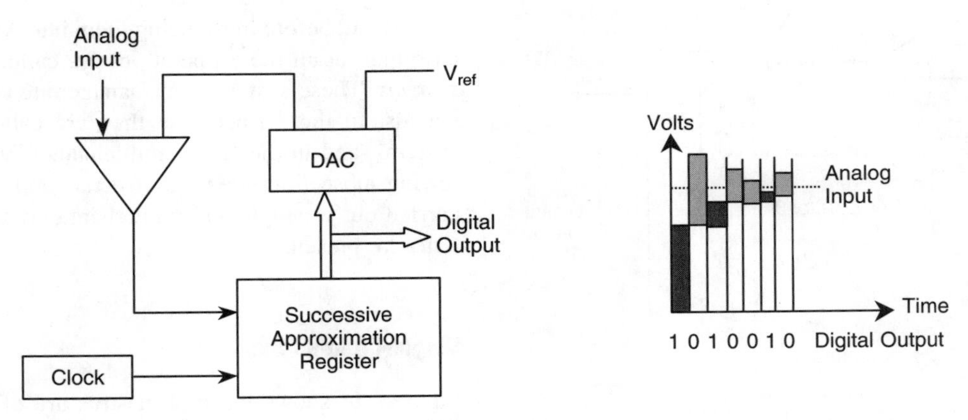

FIG. 1.9g
(Left) Scheme of a successive approximation ADC. (Right) Output bit generation in the successive approximation process.

It is clear that the input voltage V_{IA} must be constant for the whole process for a successful conversion to take place. This is achieved by a *sample and hold* operation, where the analog signal is sampled and "held" until the ADC has completed the conversion process. But the problem with this for the counter ramp converter is that for high sampling speeds, the counter and the clock may have to operate extremely fast, especially if the analog signal is highly varying. Therefore, the counting speeds can prohibit this type of ADC from being used for high-speed converters.

An improvement to the counter ramp ADC as discussed above is the tracking converter, which essentially has the same structure, but uses an up–down counter. The counter would count up or down depending on the direction of the change in magnitude of the analog input. For example, if the analog input voltage decreased, the output of the comparator would sense the change and send a message to the counter to count down. Similarly, if the input voltage increases, the counter would count up.

Successive Approximation ADC (Serial) Figure 1.9g (left) shows the basic design of a successive approximation ADC. The analog input to this system is successively compared with the voltage generated by a DAC. The digital input to the DAC, which is stored in the successive approximation register (SAR), is adjusted according to the result of each comparison. If the converter is of n bits, the conversion process requires n comparisons, and the result of the conversion is the final value stored in the SAR.

The conversion process for this ADC is as follows:

1. The programmer begins the conversion process by trying a logic ONE into the most significant bit.
2. The output of the decoder is compared against the input analog signal. If the analog signal is smaller than the decoder output V_{OA}, then the logic ONE is deleted and replaced with a ZERO. If the analog signal is larger than V_{OA}, then the logic ONE remains.
3. The SAR then goes to the next bit and tries logic ONE for that bit and the process of comparison is repeated until the signal has been converted.

Figure 1.9g (right) shows the conversion process of a value of an analog input. Only the calculation of the five most significant bits of the digital output is shown. The first comparison is carried out with the binary number 10…00, which is stored in the SAR. Then as the analog input is greater than the voltage generated by the DAC with this number, the most significant bit of the SAR is held in "1." Once the first step is finished, a new comparison is carried out. Then the following bit (the second) of the SAR is set to "1." At that moment the binary number in the SAR is 1100…00. Consequently, the value of the analog input is greater than the voltage generated by the DAC, so the second bit of the output is set to "0." This process continues in the same way until the n bits of the output are calculated.

Again, the analog input is sampled and held at the comparator. This means that the conversion sample rate is limited by the fact that each bit must be tested before the conversion of the word is completed.

The speed of the conversion is much greater than that of the counter ramp, but the trade-off for speed is complexity, especially with the programmer logic, which has the job of taking the D/A decoder through different steps of the successive approximation process.

Flash ADCs (Parallel) Flash or parallel ADCs are the fastest but most complicated ADCs available. The flash converter, like the successive approximation converter, works by comparing the input analog signal with a reference signal, but unlike the previous converter, the flash converter has as many comparators as there are representation levels except for one. For example, an 8-bit converter would have 256 representation levels and thus 255 comparators. Therefore, there are $2^n - 1$ comparators for an n-bit converter. This is shown in the block diagram of the flash converter in Figure 1.9h.

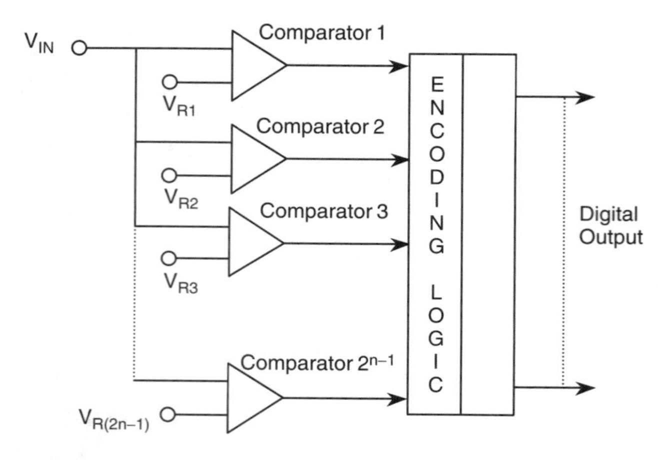

FIG. 1.9h
Flash ADC.

The input analog voltage is connected to the input of every comparator so that a comparison can be made with each of the reference voltages representing the quantization levels. The outputs of the comparators then drive the encoding logic to generate the equivalent digital output. The conversion rate for flash ADCs is very fast because only one step is required for complete conversion, and depending on the propagation time of the encoding logic, a digital output can be generated almost in real time.

The main disadvantage of the flash converter is, of course, the number of comparators required to implement large-scale converters. For every additional binary bit of resolution added to the ADC, the number of comparators required is essentially doubled. For example, 255 comparators are required for an 8-bit converter. If 4 more bits were added, 4095 comparators would be required for a 12-bit converter. Along with the number of comparators an equal number of reference voltages is also required. For a 12-bit converter, 4095 reference voltages are also needed. It must also be noted that the input capacitance increases linearly with the number of comparators implemented; thus, input impedance becomes larger and it is impractical to incorporate input signal buffers.

DATA-ACQUISITION SYSTEMS

In addition to the A/D conversions, the conversion of analog signals into digital signals requires further operations (such as the amplification and filtering of the analog signals provided by sensors or the sampling of these signals) to be carried out, before introducing them into ADCs. The systems carrying out all these operations are called *data-acquisition systems*. These systems can manage one or multiple analog signals. In the former case they are called single-channel systems, and in the latter, multichannel systems. In the following subsections the basic structure and the main functions carried out in single and multichannel data-acquisition systems are presented.

Single-Channel Systems

Figure 1.9i shows the typical structure of a single-channel data-acquisition system. The signals generated by sensors are not suitable for ADCs. Generally, they are of low amplitude and are mixed with noise, which perturbs the quality of the measurement provided by the sensor. So these signals must be transformed into an appropriate form. This process is called *signal conditioning*. In low-cost applications, the output generated by signal-conditioning circuits can be directly introduced into an ADC. However, this approach is not valid for most applications, when higher precision is required. In these cases, the conditioned signal must be sampled before the A/D conversion. In the following paragraphs the basic concepts of signal conditioning and the circuits to carry out the sampling operation are described.

Analog Signal Conditioning In a great number of applications, the adaptation between the analog signal generated by a sensor and the type of input demanded by an ADC requires two basic operations: amplification and filtering.

Amplification is used to adapt the amplitude range of the signal generated by a sensor to the amplitude range of the ADC receiving the signal. This is because the voltage ranges generated by sensors are much smaller than those managed by ADCs, which are normally 0 to 5 V or 0 to 10 V. In addition, the input of an ADC is unipolar, that is, one of the terminals of the ADC is connected to the reference terminal. However, if the signal provided by the sensor is differential, the amplifier must convert this signal into unipolar form. In this case a differential amplifier must be used.

Filtering is used to modify the frequency spectrum of the signal generated by a sensor. There are two basic reasons for using filters in data-acquisition systems: to reduce any type of interference, for example, that produced by the electric

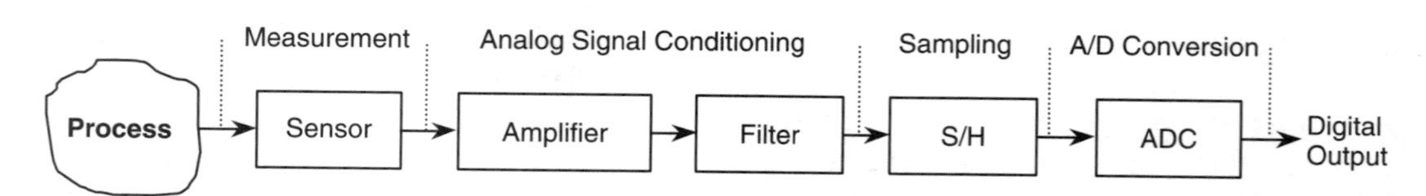

FIG. 1.9i
Basic structure and functions of a single-channel data-acquisition system.

network (50 Hz); and to diminish the bandwidth of the signals to reduce their noise and avoid the aliasing phenomena.

Sample-and-Hold Circuits An ADC requires a finite time, called conversion time t_c, to achieve conversion operations. The input voltage change during the conversion process introduces an undesired uncertainty into the generated output. Full conversion accuracy is only obtained if this uncertainty is kept below the converter resolution. This means that

$$\left(\frac{dV}{dt}\right)_{max} \leq \frac{FS}{2^n t_c} \qquad 1.9(3)$$

where n is the number of bits of the converter, t_c the conversion time, and FS the full-scale value. This formula indicates that even in the case of slow evolution signals, full conversion accuracy cannot be obtained. To overcome this difficulty, a sample-and-hold (S/H) circuit is utilized. The S/H circuit acquires the value of the input analog signal (*sample*) and holds it throughout the conversion process. Thus, a fixed level of voltage is provided to the converter during the conversion process and the uncertainty provoked by the variation of the input voltage is eliminated.

Figure 1.9j shows the basic structure of an S/H circuit. It is made up of a capacitor, a switch with its corresponding control circuit, and two amplifiers to adapt the input and output impedances of the system. The control circuit sends commands SAMPLE and HOLD alternately to switch S_1. This switch closes when it receives the SAMPLE command. Then capacitor C_h charges up (or down) until it reaches the level of the input signal, and thus a sample is taken. Immediately after, the control circuit generates a HOLD command, which causes switch S_1 to open. Then the voltage sample is held in capacitor C_h until a new sampling operation begins.

Multichannel Systems

When there are several channels in a data-acquisition, frequently all the channels share one or more resources of the system. In this case, a switch to assign the common resource alternately to each channel is required. The function carried out by this switch is called *multiplexing*. Two approaches exist to achieve the multiplexing function: analog and digital multiplexing.

Analog Multiplexing Analog multiplexing means that the multiplexing function is carried out on analog signals. Figure 1.9k shows the most common configuration of an

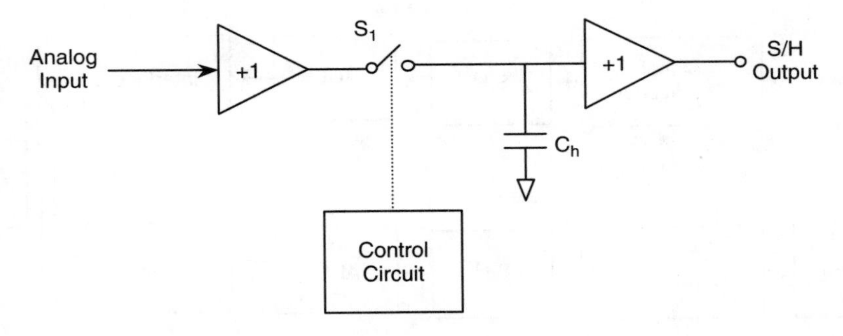

FIG. 1.9j
Basic structure of a sample-and-hold circuit.

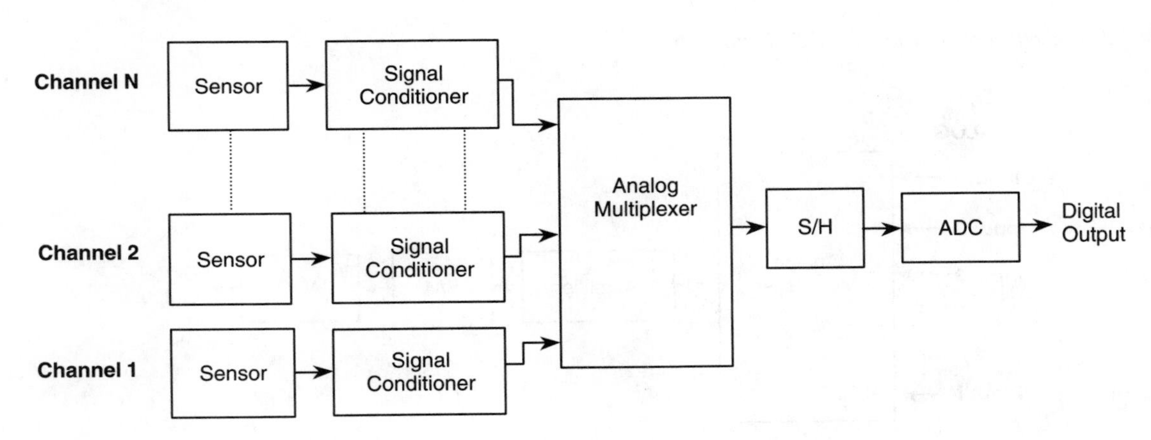

FIG. 1.9k
Multichannel data-acquisition system with analog multiplexing.

analog-multiplexed data-acquisition system. The multiplexer alternately connects one of the channels of the system to an S/H circuit. This circuit samples the signal in the channel and holds it, so that the ADC can carry out the conversion. The S/H circuit enables the multiplexer to switch to another channel if needed while the ADC is still carrying out the conversion.

Digital Multiplexing When there are many acquisition channels, both analog multiplexer and shared resources must be very fast, even if the signals in the channels are slow. In this case an alternative configuration to that represented in Figure 1.9l can be used. This alternative configuration is based on using one ADC for each channel of the acquisition system. The outputs of the converters are connected to a common bus by using interface devices. The bus is shared with the digital processor used to process the digital outputs of the converters. Using the bus, the digital processor can alternately access the outputs generated by each ADC, thus carrying out the *digital multiplexing* of the signals.

Data-Acquisition Boards

A data-acquisition board is a plug-in board, which allows the treatment of a set of analog signals by a computer. So these boards are the key elements to connect a computer to a process to measure or control it. Data-acquisition boards normally operate on conditioned signals, that is, signals that have already been filtered and amplified. However, some special boards can deal with the signals directly generated by sensors. These boards include the necessary signal-conditioning circuits on board.

The number of analog inputs normally managed by a data-acquisition board is 8 or 16. These inputs are treated by a system like that represented in Figure 1.9m. This system is made up of an analog multiplexer, an amplifier with programmable gain, a sample-and-hold circuit, and an ADC. All the analog input channels share all the common elements of

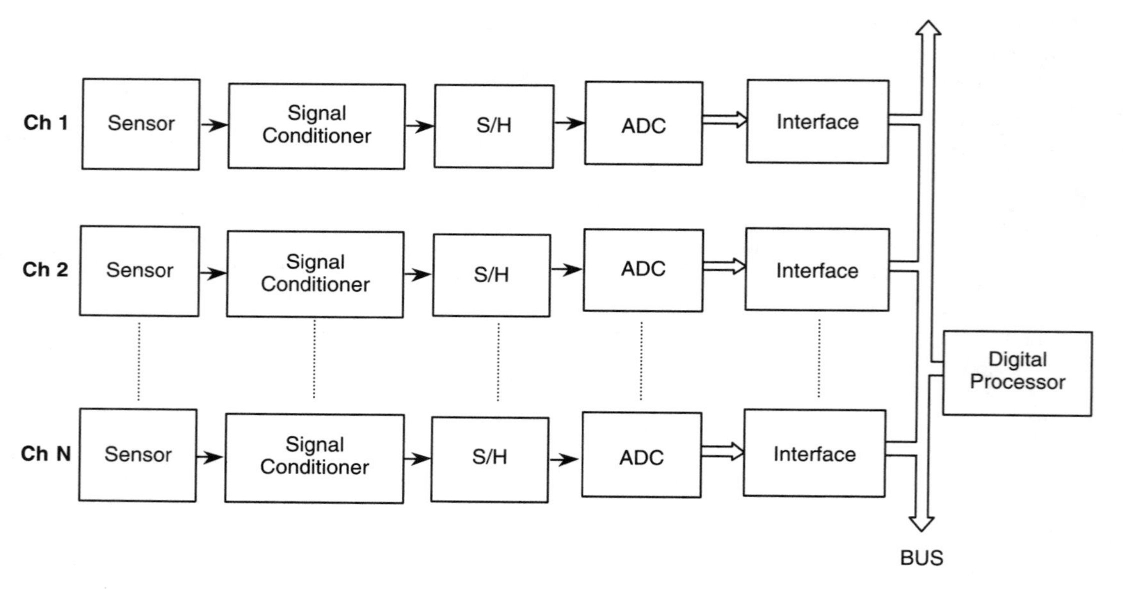

FIG. 1.9l
Multichannel data-acquisition system with digital multiplexing.

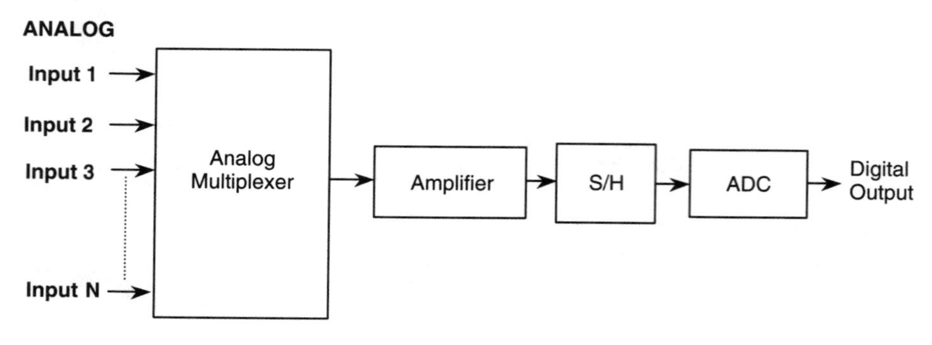

FIG. 1.9m
Input system of a data-acquisition board.

this system. Some data-acquisition boards also offer analog outputs to carry out control operations. When available, the normal number of analog outputs is two.

In addition to analog inputs and outputs, data-acquisition boards also supply digital I/O lines and counters. Digital lines are used for process control, communication, with peripheral devices, etc. Counters are used for applications such as counting the number of times an event occurs, or generating a time base.

The summary of the features offered by a typical data-acquisition board is the following:

Analog inputs: 8 or 16
Resolution: 12 or 16 bits
Sampling rate: Between 30,000 and 330,000 samples/second
Analog outputs: 2
Digital I/O: 8 or 16
Counters: 1 or 2

However, many other boards with more inputs, higher resolution, and faster scanning speeds can be found in the enormous and highly competitive market for data-acquisition systems.

DIGITAL TO DIGITAL I/O

One of the major advantages of A/D conversion is that digital information can be managed and communicated much more easily and reliably than analog information. This is a very important characteristic, because instrument communication is a fundamental fact in laboratories, industrial environments, and in the context of complex devices or machines.

Digital communications are carried out using physical communication links, normally referred to as *buses*. A bus is a set of wires with a set of defined physical properties (such as the maximum length allowed for the wires), mechanical definitions (for example the definition of the connectors used to connect devices to the bus), and communication protocols that allow handshaking (dialogue) among the devices connected to the bus. A crucial aspect of digital communications is the standardization of the communication buses. Once an standard is defined, any manufacturer can develop an instrument for that standard, and this instrument will be able to communicate with any other instrument developed for the same standard. A widespread and general-purpose standard for communicating digital instruments is the RS-232-C. It is used in any kind of environment (industrial, laboratory, etc.). For communication of laboratory instruments, specific buses have been developed such as the GPIB bus. For industrial communications other types of buses, known as *fieldbuses*, have been defined. The following subsections provide a brief overview of these communication buses.

Distributed Systems and Networks

Digital systems are made from groups of instruments, controllers, microprocessors, and computers that are connected to form networks by the use of local and global bus systems. The hardware architectures of the digital control systems can be configured in centralized, decentralized, hierarchical, or distributed configurations by a number of different network topology as exemplified in Figure 1.9n.

The information flow between the nodes, individual instruments, and the computers are regulated by protocols. According to the IEEE, a networking protocol is "a set of conventions or rules that must be adhered to by both communicating parties to ensure that information being exchanged between two parties is received and interpreted correctly." The protocol defines the following:

1. Network topology supported: star, ring, bus (tree), etc.
2. ISO reference model layers implemented: physical, data link, network, transport, session, presentation, and application
3. Data communication modes: simplex, half-duplex, or duplex
4. Signal types: digital or analog
5. Data transmission modes: synchronous or asynchronous, etc.
6. Data rate supported: several bps (bits per second) to several Gbps
7. Transmission medium supported: twisted pair, coaxial cable, optical, microwave, etc.
8. Medium-access control methods: CSMA/CD (carrier sense multiple access with collision detection), control token, etc.
9. Data format: mainly based on data transmission modes and individual protocol specifications
10. Type, order of messages that are to be exchanged
11. Error detection methods: parity, block sum check, CRC (cyclic redundancy check)
12. Error control methods: echo checking, ARQ (automatic repeat request), sequence number I, etc.
13. Flow control methods: X-ON/X-OFF, window mechanisms, sequence number II, etc.

ISO reference model layer layout for the protocols is shown in Table 1.9o. The reference model has seven layers, each of which is an independent functional unit and each of which uses functions from the layer below it and provides functions for the layer above it. The lowest three layers are network-dependent layers, the highest three layers are network-independent layers (application-oriented), and the middle layer (transport layer) is the interface between the two.

RS-232-C

RS-232-C is designed to carry out point-to-point communications between two devices, data terminal equipment (DTE) and data communication equipment (DCE). The DTE is normally a computer and the DCE, a peripheral, which can be any kind of digital instrument.

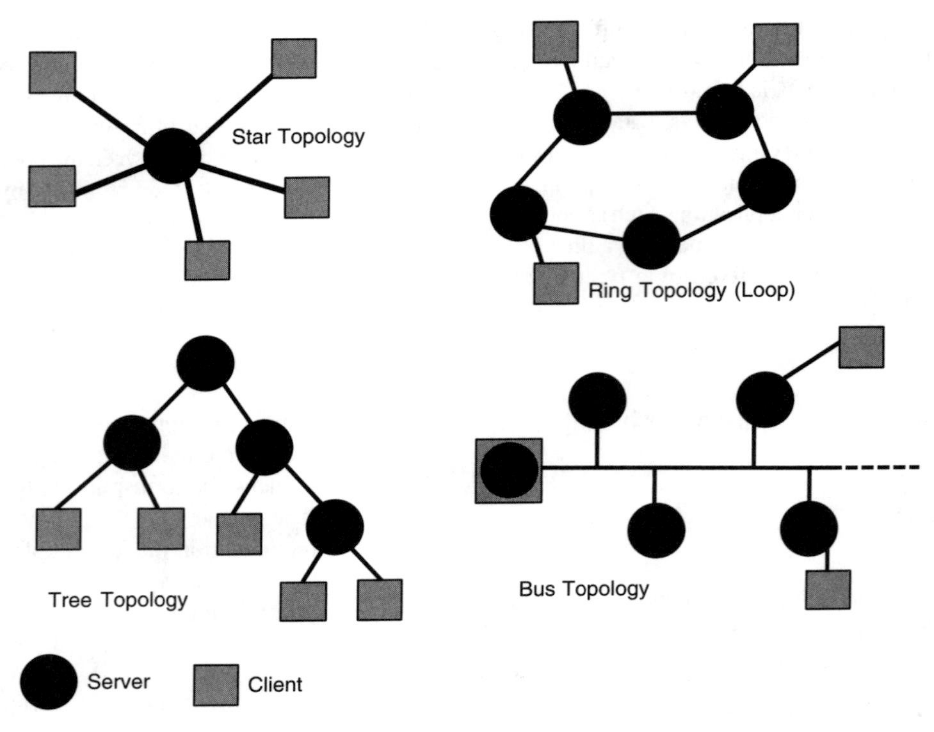

FIG. 1.9n
Examples of network topology.

TABLE 1.9o
ISO Reference Model Protocol

1	*2 Layer*	*3 Application*	*4 Protocols*
1	Physical	Electrical, mechanical, process; functional control of data circuits	ISO/IEEE 802.4, phase coherent Carrier, Broadband 10 Mbs, etc.
2	Link	Transmission of data in LAN; establish, maintain and release data links, error, and flow	IEEE 802.4 Token Bus, IEEE 802.2 Type 1 connections
3	Network	Routing, segmenting, blocking, flow control, error recovery, addressing, and relaying	ISO DIS 8473, Network services, ISO DAD 8073 (IS)
4	Transport	Data transfer, multiplexing, movement of data in network elements, mapping	ISO Transport, Class 4. ISO8073 (IS)
5	Session	Communication and transaction management, synchronization, administration of control	ISO Session Kernel. ISO 8237 (IS)
6	Presentation	Transformation of information such as file transfer; data interpretation, format, and code transformation	Null/MAP transfer. ISO 8823 (DP)
7	Application	Common application service elements (CASE); manufacturing message services (MMS); network management	ISO 8650/2 (DP), RS-511, ISO 8571 (DP), IEEE802.1

The standard defines electrical, mechanical, and functional characteristics. The *electrical* characteristics include parameters such as voltage levels and cable impedance. The *mechanical* section describes the number of pins of the connectors and the function assigned to each pin. Although the connector itself is not specified, in practice DB-25 or DB-9 connectors are almost always used. The *functional* description defines the functions of the different electrical signals to be used.

Two important limitations of RS-232-C are the low transfer rates (typically 20 kb/s) and the maximum length of the wires (around 16 m). To overcome these limitations, the new standards RS449 (mechanical) and RS423 (electrical) have been defined. These new standards are upward compatible with RS-232-C and can operate at data rates up to 10 Mb/s and at distances of up to 1200 m. However, changing to a new standard is a long and costly process, so the penetration of this new standard in the market is, as yet, very limited.

Despite its limitations, RS-232-C communications are widely used, especially in low-cost installations. However, if more-sophisticated communications are required, more specialized buses, such as the GPIB or fieldbuses must be used.

The GPIB (IEEE 488)

The general purpose interface bus (GPIB) is oriented to permit communication among digital instruments (oscilloscopes, digital multimeters, etc.), computers, and peripheral devices (such as printers), which are normally found in the context of a laboratory. Hewlett Packard developed the first version of the GPIB bus in the 1960s. This version was called Hewlett Packard Interface Bus (HPIB). In addition to instrumentation equipment, Hewlett Packard used this bus to connect a wide variety of peripherals, including printers and magnetic-storage devices, to its computers. This bus quickly gained great popularity, and in 1978 the IEEE drafted the standard IEEE 488, which defined the mechanical, electrical, and functional characteristics of the bus. From then on, the bus became known as the general-purpose interface bus (GPIB). The first version of this standard was the IEEE 488.1. In 1987, the IEEE published an extension to the original standard. This new version, which was called IEEE 488.2, defined some common commands to be managed by all GPIB instruments, command syntax, and the data structures used by the instruments when communicating with the IEEE 488 bus.

The main characteristics of the GPIB bus are the following:

- The maximum number of devices connected to the bus is 15. One device must operate as the bus controller.
- Devices can be linked using a linear or a star configuration, or a combination of both. The maximum length of the cable connecting the devices is 20 m, and the average separation between devices must not exceed 2 m. However, bus extenders and expanders are available to overcome these system limitations.
- The bus has 16 lines, 8 for data and 8 for control.
- The message transfer is asynchronous, and is controlled using handshaking lines.
- The maximum data transfer rate over the bus is 1 MB/s; however, the current throughput obtained from the system is much lower than this maximum rate.

Figure 1.9p shows the structure of the GPIB bus. Lines DI01 to DI08 are used for transferring data (or commands) between instruments. The other eight lines are for control: three for handshaking to coordinate data transfer and five for other bus control functions.

Any device connected to the GPIB bus may send or receive data through the bus. The device sending data is called a *talker*, and that receiving data, a *listener*. In addition, at least one device must act as a bus controller. This device is called a *controller*. A typical bus controller is a computer (a PC, for example), which operates both as a talker and a listener. Examples of devices that only act as talkers are digital voltmeters and counters. Examples of devices that are only listeners are x-y plotters and waveform generators. Magnetic tape drives and programmable spectrum analyzers are both talkers and listeners.

The crucial advantage of the GPIB bus is that it is widely supported. This bus has been on the market for a long time,

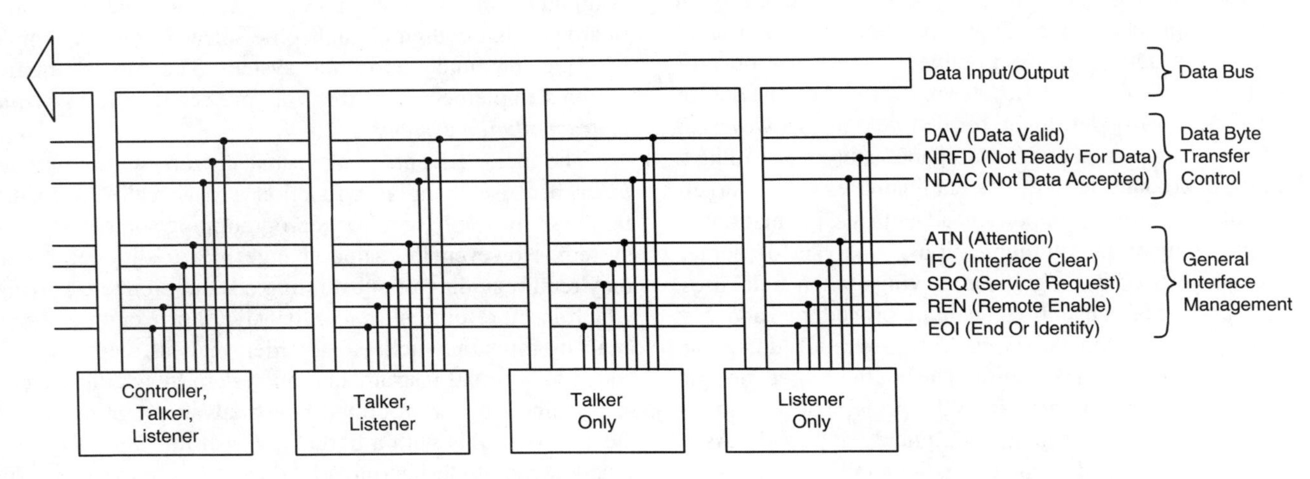

FIG. 1.9p
General GPIB structure.

and GPIB controller boards for the most popular computers (including PC compatibles, Macintosh, and Sun workstations) can be found. There is also plenty of software to support GPIB systems, available for the most-used operating systems in the market, including DOS, Windows, OS/2, Mac, and UNIX. In addition, most manufacturers of high-end instrumentation systems support the GPIB interface. So there are thousands of instruments available with GPIB interfaces.

VXIbus

Virtual instrumentation systems can be developed at reasonable prices using PCs and data acquisition boards. However, when very high performance and compact size are required, the best choice is VXIbus. VXI stands for VME eXtensions for Instrumentation. The VME standard is a definition of a backplane bus designed to develop industrial computer-based control systems. VME is widespread in industrial environments, and there are thousands of products for this bus. The VXI standard defines some modifications on the VME standard to adapt it to instrumentation applications.

VXIbus modules have no front panels. The only way to control them and make measurements with them is through a virtual instrument running on a system controller, which is a processor connected to the bus. Most VXIbus modules are now delivered with their corresponding virtual instrument software. An outstanding characteristic of VXIbus is its performance. The modules connected to a VXIbus system can exchange data much faster than, for example, a set of stand-alone instruments connected via a GPIB bus.

The Fieldbuses

Fieldbuses are designed to communicate digital devices in industrial environments. Sensors, actuators, frequency converters (to control electrical motors), position controllers, PLCs, and industrial computers are examples of the wide variety of devices that can be found in an industrial plant and that must communicate with each other. A fieldbus is an interconnection system that allows the communication of all these types of devices using a defined communication protocol. To connect a device to a determined fieldbus, the device must be specifically developed for that fieldbus. However, at present there are a great number of manufacturers providing a full range of products for the different fieldbuses in the market.

Fieldbuses provide important advantages in relation to traditional industrial communications. Analysis of actual fieldbus projects shows significant savings when fieldbuses are used instead of conventional cabling procedures, and this is true even for small installations. Costs are reduced in plant design, installation, and operation. During the project design stage, fully integrated, user-friendly project design tools allow field devices to be integrated, quickly and easily. As a result of this centralized engineering, the number of potential sources of error is substantially reduced. In the installation and cabling of the field devices the cost advantage is particularly clear. Instead of thick bundles of cables, just one two-wire bus cable is laid. In addition to saving time and material, a further benefit is that the likelihood of wiring errors is drastically reduced.

In operation, the benefits of fieldbus solutions are seen primarily in troubleshooting and maintenance. Errors can be more easily detected, analyzed, and localized, and the stocking of spares is simplified. When subsequent plant upgrades arise, the fieldbus is the better choice, too. Instead of complex recabling or utilization of expensive cable reserves, the existing bus is simply extended to the new field device.

During the 1990s, a significant number of fieldbus definitions were achieved. However, each bus definition presents particular characteristics and, consequently, some buses are more suitable for determined installations than others. For example, the AS-I bus (1993) is oriented to low-level communications, connecting the sensors and actuators of a plant with its automation system. Other buses are devoted to managing higher-level communications, such as PROFIBUS (1994/1995), which allows communication among PLCs, drive units, I/O modules, and industrial computers. A particular type of fieldbus is the CAN bus (1995), which was specifically developed for the automotive industry. The CAN bus is used to communicate and control the great number of digital devices, which can be found in a vehicle: ABS breaks, air bags, electric windows, etc. Other important examples of fieldbuses on the market are INTERBUS-S (1984), DeviceNet (1994) and the Foundation Fieldbus H1 (1995).

Virtual Instruments

Traditional instrumentation systems are made up of multiple stand-alone instruments, which are interconnected to carry out a determined measurement or control operation. Currently, however, the functionality of all these stand-alone instruments can alternatively be implemented by using a computer (for example, a PC), a plug-in data-acquisition board for that computer, and some software routines implementing the functions of the system. The instrumentation systems implemented in this way are referred to as *virtual instrumentation systems.*

The major advantage of virtual instrumentation is *flexibility,* because changing a function of a virtual instrumentation system simply requires reprogramming some part of the system. However, the same change in a traditional system may require adding or substituting a stand-alone instrument, which is more difficult and expensive. For example, a new analysis function, such as a Fourier analysis, can be easily added to a virtual instrumentation system by adding the corresponding software routing to the analysis program. Nevertheless, to do this with a traditional instrumentation system, a new stand-alone instrument (a spectrum analyzer in this case) would have to be added to the system.

Virtual instrumentation systems also offer advantages in displaying and storing information. Computer displays can show more colors, and allow users to change quickly the format of displaying the data received by a virtual instrument. Virtual displays can be programmed to resemble familiar instrument panels, including buttons and dials. Computers also have more mass storage than stand-alone instruments and, consequently, virtual instruments offer more flexibility in storing measurements.

Software for Virtual Instrumentation To develop the software of a virtual instrumentation system a programming language or a special software package must be utilized. However, the option of using a traditional programming language (C, for example) can generate several problems. One of them is that it may be very difficult to program the graphical elements of a virtual instrumentation system using C. Another disadvantage is that it is not an easy language to learn, and the engineer designing the virtual instrument should also be an experienced programmer.

Today, a more utilized option is the Microsoft Visual Basic programming language, which runs under the Windows operating system. Visual Basic has become quite popular in the development of virtual instrumentation systems, because it combines the simplicity of the Basic programming language with the graphical capabilities of Windows. Another important reason for its popularity is that it is an open language, meaning that it is easy for third-party vendors to develop products that engineers can use with their Visual Basic programs. Several companies now sell Dynamic Link Libraries (DLLs), which add custom controls, displays, and analysis functions to Visual Basic. The controls and displays mimic similar functions found on stand-alone instruments, such as toggle switches, slide controls, meters, and light-emitting diodes (LEDs). By combining various controls and displays, any virtual instrument can easily be programmed.

A third option for developing virtual instrumentation systems is to use a software package specifically designed for the development of these systems. The crucial advantage of these packages is that it is not necessary to be a Windows programmer to use them. Several packages of this type exist, but only one of them has reached great diffusion, the LabVIEW. National Instruments developed this package.

LabVIEW was designed to be extendible, so new functional modules can be added to a LabVIEW program. For example, a manufacturer of an interface card may provide the user with a LabVIEW driver, which appears as a virtual instrument representing the card and its functionality in the LabVIEW environment. LabVIEW modules can also be written using general-purpose languages, such as C or C++. These languages provide great flexibility to program functions that perform complex numerical operations with the data received by a virtual instrument.

THEORY OF SIGNAL ACQUISITION

This section presents the theoretical foundations for the representation of analog information in digital format. The transformation between the two types of information is carried out during A/D conversions and can be analyzed considering three different processes: sampling, quantization, and codification. The following subsections briefly describe these processes.

The Sampling Process

An analog signal is first sampled at discrete time intervals to obtain a sequence of samples that are usually spaced uniformly in time. This is achieved ideally by multiplying a periodic train of Dirac delta functions spaced T_s s apart (where T_s represents the sampling period and its reciprocal $f_s = 1/T_s$ is the sampling rate) with the (arbitrary) analog signal. The signal obtained from this process of continuous-time uniform sampling is referred to as the ideal sampled signal (ideal because it is realistically impossible to implement a Dirac delta function). Because of the duality of the Fourier transform, by sampling a signal in the time domain (multiplication), the signal becomes periodic in the frequency domain (convolution of delta-train and baseband spectrum).

Therefore, if the analog signal is to be completely reconstructed in the D/A conversion process, then the sampling rate must be such that it is equal to or greater than twice the bandwidth of the message signal. This is known as the Nyquist criterion. Failure to either limit bandwidth or increase sampling rate such that the Nyquist criterion is satisfied may introduce an effect called aliasing, where the spectrum at a sample frequency multiple-overlaps with its adjacent spectrum.

To eliminate the effects of aliasing, two corrective measures can be implemented:

1. Sample at higher than the Nyquist rate
2. Pass presampled signal through a low-pass filter to band-limit the baseband signal such that the bandwidth of the signal is less than the sampling rate

By using a continuous signal as input, the sampling operation obtains a signal, which is discrete in the time domain and continuous in the amplitude domain. The sampling of a signal consists of multiplying itself by a train of impulses separated by a time T, which is called the *sampling period*. The sampled signal is also a train of impulses, modulated in amplitude, which is a signal discrete in the time domain, but continuous in the amplitude domain. Figure 1.9q shows the sampling process of a signal X(t) (Figure 1.9q(a)), using a train of impulses P(t) (Figure 1.9q(b)), to obtain the sampled signal $X_P(t)$ (Figure 1.9q(c)).

One fundamental requirement of A/D conversions is that the digital signal obtained in a conversion process must be

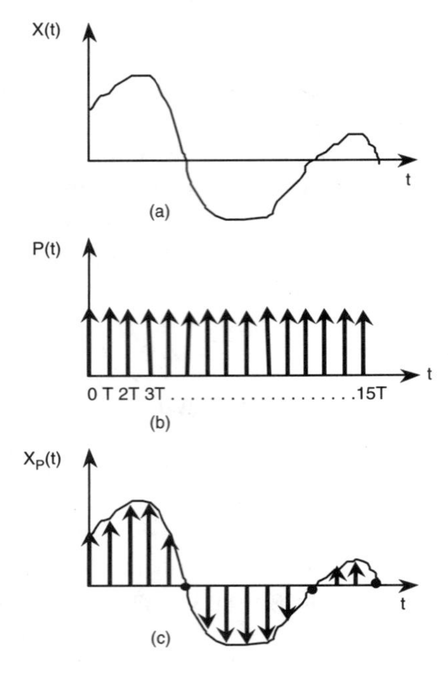

FIG. 1.9q
(a) Representation of an analog signal. (b) Train of impulses to sample the analog signal. (c) The sampled signal, which is a train of impulses, modulated in amplitude.

fully representative of the original analog signal. This means that the analog signal is completely recoverable using the values of the digital signal. To guarantee the complete recoverability of the digital signal, the sampling period used to carry out the conversion must fulfill the postulates of the *sampling theorem*. This theorem, which is of crucial importance in the theory of signal acquisition, states that "under certain conditions a continuous-time signal can be completely represented by and recoverable from knowledge of its instantaneous values of samples equally spaced in time." The conditions to be fulfilled are the following:

1. The original analog signal must be band-limited. This means that the Fourier analysis of the signal must determine that the frequency content of the signal ranges from 0 to a maximum value of f_M.
2. The sampling frequency ($f_s = 1/T$) must not be lower than twice the highest frequency present in the signal. That is: $f_s > 2f_M$.

Figure 1.9r shows the phenomenon that occurs when the sampling frequency (f_s) is less than twice the highest frequency (f_M) of the signal. This phenomenon, which is called *aliasing*, provokes overlapping of frequency patterns of the sampled signal. When aliasing occurs, it is impossible to recover the original signal.

In the real world it is very difficult to find band-limited signals, because sensors produce signals with an infinite number of frequency components. So the ideal conditions for the sampling theorem are not fulfilled. Therefore, there is no possibility of processing real signals without error. However, by carrying out an appropriate treatment of signals, this error can be minimized. One important action, which can be achieved, is the conversion of the real signal into a band-limited signal before the sampling process. A low-pass filter, which causes the information contained in the higher frequencies to be lost, carries this out. However, as long as this information is of little importance, the complete process is valid while f_s are greater than twice f_M.

Quantization

Amplitude quantization is defined as the process by which a signal with a continuous range of amplitudes is transformed into a signal with a discrete amplitudes taken from a finite set of possible amplitudes. The quantization process used in practice is usually memoryless. That is, the quantizer does not depend on previous sample values to predict the next value. Although this is not an optimal solution, it is commonly used in practice.

The precision of the quantizer is limited by the number of discrete amplitude levels available for the quantizer to approximate the continuous signal amplitudes. These levels are referred to as representation or reconstruction levels and the spacing between two representation levels is called a step size.

Uniform quantizers are a family of quantizers that have evenly spaced representation levels. Two types of uniform quantizers are midtread and midrise quantizers. These are shown in Figure 1.9s.

Despite that uniform quantizers are more commonly used, depending on the application, there are circumstances that make it more convenient to use non-uniform quantizers. For example, the ranges of amplitudes covered by voice signals, from passages of loud talk contrasted to passages of weak talk is on the order of 1000 to 1. For this reason, a non-uniform quantizer with the feature that the step size increases as amplitude increases. Therefore, infrequently occurring loud talk is catered to, but weak passages of talk are more favored with smaller step sizes.

The use of quantization introduces an error between the continuous amplitude input and the quantized output, which is referred to as quantization noise. Quantization noise is both nonlinear and signal dependent; thus, quantization noise cannot be alleviated by simple means. However, provided that the step size is sufficiently small, it is reasonable to assume that the quantization error is a uniformly distributed random variable, and, therefore, the effect of quantization noise is similar to that of thermal noise.

By modeling the quantization noise as a uniformly distributed random variable, it is possible to ascertain a signal-to-noise ratio (SNR) for the quantizer. As stated earlier, the precision of the quantizer is dependent on the number of representation levels available. This also corresponds to the number of bits per sample R (the number of representation

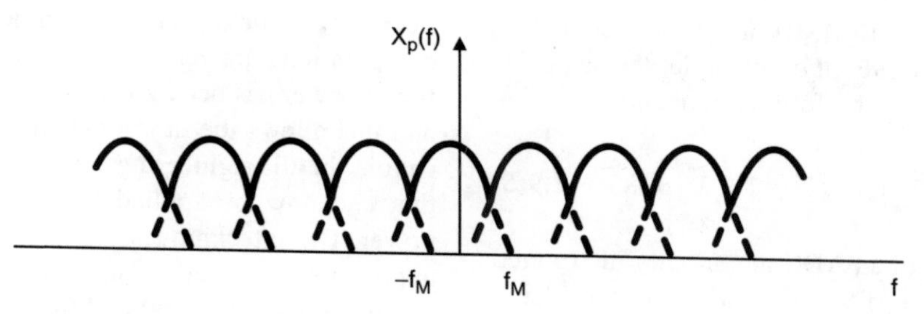

FIG. 1.9r
Representation of the aliasing phenomena.

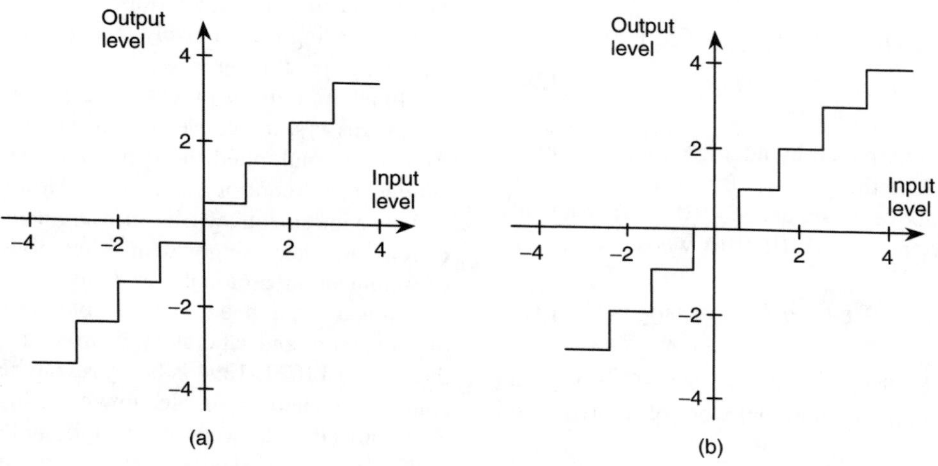

FIG. 1.9s
Types of uniform quantizers—(a) midrise and (b) midtread.

levels $L = 2^R$) The SNR is exponentially proportional to the number of bits per sample R, so increasing the number of bits exponentially increases the SNR. However, increasing the number of bits also increases the bandwidth so there is a trade-off.

The type of quantization is known as *uniform quantization*, and its main characteristic is that the absolute quantization error stays constant in the whole quantification range. As a result, the relative error is variable, and it is maximum for the minimum amplitudes of the original analog signal. Some applications require the relative quantization error to be held constant. In this case another type of quantization, called *non-uniform quantization*, is used. Non-uniform quantization is based on varying the size of the quantization interval proportionally to the amplitude of the analog input signal.

Coding

Coding is the process of representing the finite output states of a quantizer using sequences of n bits: a different sequence of bits is assigned to each output state. For n bits 2^n different sequences can be generated and, consequently, 2^n output states can be coded. The codes formed using bits are normally known as *binary codes* and they can be of two types: unipolar and bipolar.

Unipolar Codes Unipolar codes are used to represent unipolar quantities, that is, quantities with a predefined sign (positive or negative). The most common unipolar codes are *natural binary, BCD* (binary-coded decimal), and *Gray*. To illustrate the coding process, the natural binary code is described below.

In the natural binary code, each bit of a number represents a successive power of 2 according to the position that the bit occupies in the number. Given a number of n digits (represented by $D_{n-1}, D_{n-2}, \ldots D_1$ and D_0), the value of the number is determined by the following expression:

$$D_{n-1}D_{n-2}\ldots D_1D_0 = D_{n-1}\times 2^{n-1} + D_{n-2}\times 2^{n-1} + \cdots + D_1\times 2^1 + D_0\times 2^0 \qquad 1.9(4)$$

D_{n-1} is the most significant bit (MSB) of the number, and D_0 is the least significant bit (LSB). For example, the sequence of bits "1010101" represents the following quantity:

$$1010101 = 1 \times 2^6 + 1 \times 2^4 + 1 \times 2^2 + 1 \times 2^0 = 85 \qquad \textbf{1.9(5)}$$

However, the output code of an ADC is normally interpreted as a fraction of its full-scale (FS) value, which is considered the unit. So the binary codes provided by an ADC in its output must be interpreted as fractional numbers, which is equivalent to dividing the value of each bit by 2^n. Taking this into account, the value of a sequence of n bits is

$$D_{n-1}D_{n-2}\ldots D_1D_0 = D_{n-1} \times 2^{-1} + D_{n-2} \times 2^{-2} + \cdots + D_1 \times 2^{1-n} + D_0 \times 2^{-n} \qquad \textbf{1.9(6)}$$

The bit sequence is interpreted by adding an implicit binary point at the left end of the number. By using this form of representation, the above bit sequence ("1010101") actually represents the binary number 0.1010101, whose value is

$$1 \times 2^{-1} + 1 \times 2^{-3} + 1 \times 2^{-5} + 1 \times 2^{-7} = 0.6640625 \qquad \textbf{1.9(7)}$$

In this way, if the FS voltage value of a DAC is V_{FS}, the voltage V_o corresponding to an output word of n bits will be

$$V_o = V_{FE} \sum_{i=0}^{i=n-1} (D_i/2^{n-i}) \qquad \textbf{1.9(8)}$$

For example, for a DAC of 8 bits and a FS value of 10 V, the output word "11010000" represents the following voltage value:

$$V_o = 10 \times (2^{-1} + 2^{-2} + 2^{-4}) = 8.125 \text{ V} \qquad \textbf{1.9(9)}$$

Bipolar Codes Bipolar codes are used to represent bipolar quantities, that is, quantities that can be either positive or negative. The most common bipolar codes are two's complement, one's complement, sing-magnitude, and offset binary. These codes are used when in an A/D conversion process is necessary to preserve the positive–negative excursion of the original analog signal.

CONCLUSIONS AND COMMENTS

The utilization of digital processors and computers in the development of instrumentation systems has improved the capabilities of these systems extraordinarily. Computers provide flexibility in the design of instrumentation systems and great computational power to process signals. Computer computational power has been growing constantly since its invention, and this allows the development of more challenging applications in all engineering fields such as the machine vision that can now be applied thanks to the existing computer power. These techniques are beginning to be widely used in applications devoted to industry (e.g., to detect defects in manufactured products) and in the control of manufacturing lines.

Real-time monitoring of very complex devices (the reader can imagine, for example, the complexity of a strip mill in a steel factory) may require the processing of hundreds of signals in small time slots. However, this can only be achieved using the relatively recent technique of interfacing and signal processing.

In addition to the processing capacities provided by digital processors, the conversion of analog signals into digital format has multiplied the communication abilities of measurement and control instruments. However, the communication requirements of new measurement and control applications are increasingly demanding. Because of this, new communication protocols, systems, and devices are being researched and defined. For example, for the communication of computers and laboratory instruments, the utilization of the standard IEEE 1394 is being researched. The IEEE 1394 connection enables simple, low-cost, high-bandwidth data communication between computers and a wide variety of digital electronic devices. This communication standard, whose transfer rates reach 400 Mb/s, can overcome the insufficient communication speed of the GPIB bus for some demanding applications. Currently, a new protocol, called IICP (Instrument and Industrial Control Protocol), is being defined to connect instruments, control devices, and computers using the IEEE 1394 standard.

BIBLIOGRAPHY

Caristi, A. *IEEE-488: General Purpose Instrumentation Bus Manual,* New York: Academic Press, 1989, 253 pp.

Demler, M. J., *High Speed Analog-to-Digital Conversion,* New York: Academic Press, 1991.

Garrett, P. H., *Computer Interface Engineering for Real-Time Systems—A Model Based Approach,* Englewood Cliffs, NJ: Prentice-Hall, 1987.

Georgopoulos, C. J., *Interface Fundamentals in Microprocessor-Controlled Systems,* Dordrecht, the Netherlands: D. Reidel, 1985.

Haykin, S., *Communication Systems 3/E,* New York: John Wiley & Sons, 1994.

Hoeschelle, D. F., *Analog-to-Digital and Digital-to-Analog Conversion Techniques,* New York: John Wiley & Sons, 1994.

Jain, R., Werth, J., and Browne, J. C., *Input/Output in Parallel and Distributed Computer Systems,* Boston: Kluwer Academic Publishers, 1996.

Johnson, G., *LabVIEW Graphical Programming: Practical Applications in Instrumentation and Control,* New York: McGraw-Hill, 1997, 625 pp.

Kaiser, V. A. and Lipták, B. G., "DCS-I/O Hardware and Setpoint Stations," in *Instrumentation Engineers' Handbook–Process Control,* 3rd ed., Liptak, B. G., Ed., Radnor, PA: Chilton, 1995.

Putnam, F., "Internet-Based Data Acquisition and Control," *Sensors,* No. 11, 1999.

Romanchik, D., "Choose the Right Bus for Electronic Test Systems," *Quality,* No. 3, 1997.

Soclof, S., *Design and Application of Analog Integrated Circuits,* Englewood Cliffs, NJ: Prentice-Hall, 1991.

Tompkins, W. J. and Webster, J. G., Eds., *Interfacing Sensors to the IBM PC,* Englewood Cliffs, NJ: Prentice-Hall, 1989, 447 pp.

Van de Plassche, R., *Integrated Analog-to-Digital and Digital-to-Analog Converters,* Dordrecht, the Netherlands: Kluwer Academic Publishers, 1994.

1.10 Estimating the Cost of Control System Packages

G. C. BUCKBEE

Cost estimates are used for several purposes. For management, cost estimates are a tool to assess the economic viability of a project. For financial personnel, they are used to forecast cash flow and corporate profits. For project managers, estimates are a tool for accountability. A good cost estimate will help all these people perform well in their jobs.

As control systems have moved from hardware-intensive systems to software-intensive systems, cost estimating has changed dramatically. Today, as much as half of the system costs may be for software licenses. Clarifying the control system objectives is the primary tool for controlling software costs.

In the past, proprietary hardware, software, and networks limited the engineer's options for system design and cost management. Today, the use of open system architectures allows much greater flexibility to mix and match control system components.

The improved flexibility does have a price, however. The engineer must take on the role of system integrator, understanding allowable combinations of hardware, software, and networks—all this while keeping in mind that one vendor's "industry standard" is not the same as another's.

SUPPLIERS

In lieu of a list of suppliers, it is suggested that the reader consult with an experienced engineer, project manager, or project cost engineer. The reader should look for someone with specific experience on projects similar to the one the reader wants to estimate.

DESIRED ACCURACY OF THE ESTIMATE

Whenever starting to develop an estimate, one should first understand the desired level of accuracy for this estimate. Typically, an order-of-magnitude estimate will be used to assess the feasibility of a project. A more detailed estimate will be used when gaining management approval for a project. Finally, a highly detailed estimate would be used to manage the project execution on a daily basis.

The effort required increases exponentially with the accuracy desired. Figure 1.10a shows the relationship between

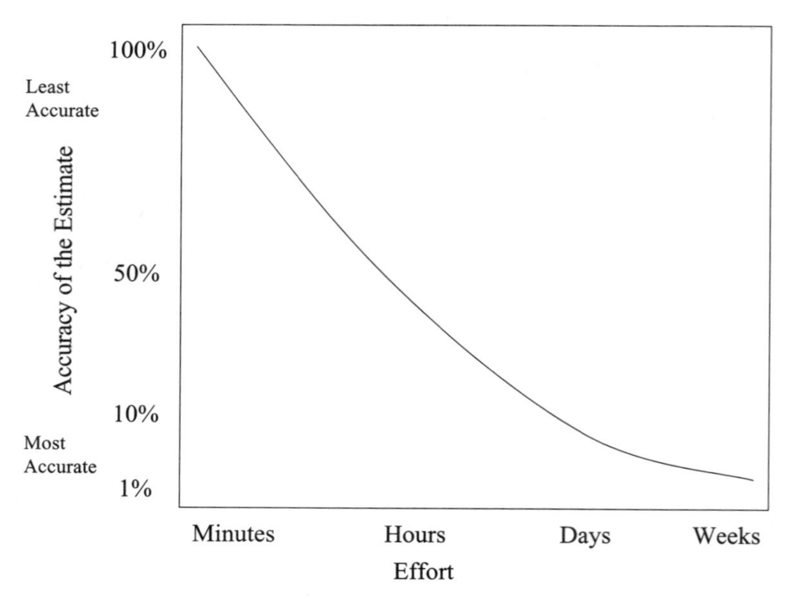

FIG. 1.10a
Estimation effort vs. accuracy.

effort and accuracy. If the desired accuracy is unclear, try to determine how the estimate will be used, so that the appropriate level of effort can be expended on it.

CLARIFY SCOPE AND OBJECTIVES OF THIS CONTROL SYSTEM

Before starting the estimate, try to determine the expected system scope and the true purpose of this new addition with as much clarity as possible. The best way to control cost is to maintain tight control of the project scope and objectives. If the scope is unknown when develop the estimate is developed, then the estimate probably is not of much value.

ESTIMATING TECHNIQUES

There are many techniques to develop cost estimates. Typically, the more accurate the estimate, the more effort is required. The difference between a simple estimate and a detailed estimate could be 100 times, or even 1000 times more effort. Be sure to know how much accuracy is necessary when developing the estimate.

One of the simplest estimating techniques is to mirror a recently completed project. Look around for a recently completed project that is roughly the same scope and complexity. In a large company, a similar project may exist within the company. If not, it may be necessary ask some control system suppliers to provide references to some of their customers who have recently completed similar projects.

If it is not possible to find a project that is close, it may be necessary to try a scaling technique for estimating. If the previous project was of roughly the same level of complexity, then the cost can be scaled by the ratio of input/output (I/O) count in the new project divided by the I/O count in the old project.

In both the mirroring approach and the scaling technique, basic assumptions should be checked. For control system projects, estimating labor can be quite tricky. Understanding of whether or not labor for engineering, design, installation, configuration, and start-up support must be estimated is essential.

Detailed bogey analysis can be used when a very accurate estimate is needed. For this technique, the project scope must be listed in great detail. Then, for each item, a "bogey manual," or estimating book, should be consulted. Many companies have their own estimating book. Contact the engineering or project management group to see if it has one.

For the most accurate cost estimate, get quotes from the vendors for everything. This, of course, takes a great deal of time, but will provide a very detailed estimate, with a high degree of accuracy. But this can be difficult to do early in the project, because there is probably insufficient information about the project scope to request a quote.

Of course, going for the most detailed estimate at the start is not required. Starting with a simple estimate and then adding more and more detail can be an effective method to maximize the return on the investment of time.

Figure 1.10b shows some sample project estimates. As the project detail becomes clearer, the estimate becomes more and more refined. Some of the initial assumptions may change, and costs may shift between categories. Total project cost will likely rise more detail is added to the estimate. To offset this effect, use a larger contingency number when

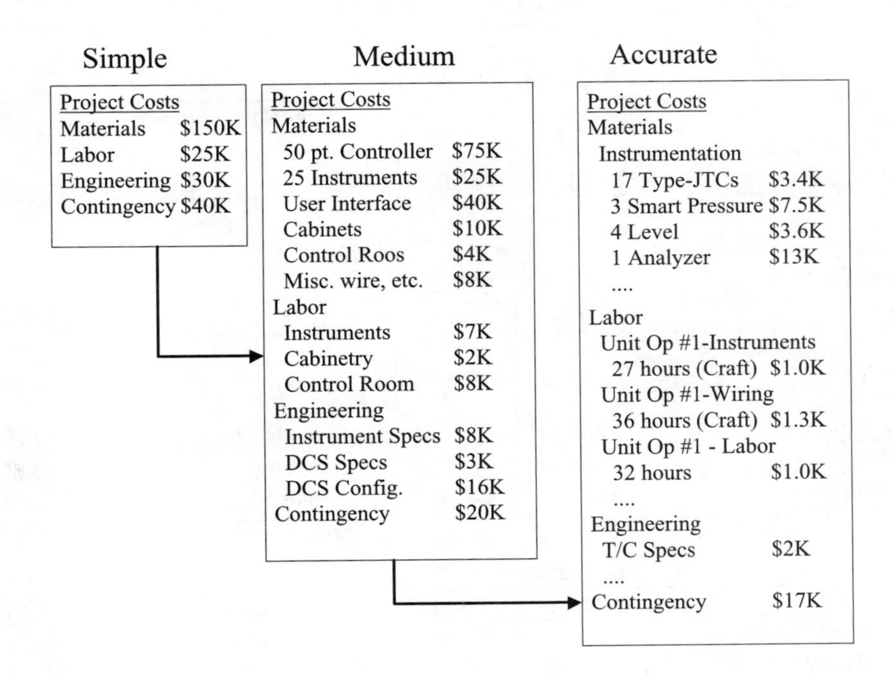

FIG. 1.10b
Improving estimate accuracy.

working with a rough estimate, then reduce the contingency as the estimate detail develops.

SOFTWARE TOOLS

For all but the most complex estimates, a simple spreadsheet is all that is needed. Fancy financial tools are not typically required and rarely work the way expected. But it is necessary to be organized. Developing a spreadsheet to estimate project costs will help one write everything down in an organized fashion, and will provide a basis for future discussions. With an electronic version of the estimate, one can quickly evaluate the effects of changes to the project. Table 1.10c shows a sample spreadsheet for estimating project costs.

CONTROLLER COSTS

The costs for controllers will depend, of course, on the type of controller used. Table 1.10d provides some estimates for controller costs.

TABLE 1.10c
Spreadsheet for DCS Project Cost Estimate

Materials

Instrumentation — **Sub-Total $27,311**

No.	Item	Cost per	Total	Source/Assumptions
17	Type J Thermocouples, modified	$195	$3,315	Quote from XYZ Distributor
3	Smart Pressure Transmitters	$2,475	$7,425	Quote from ABC Company
4	Level Sensors	$924	$3,696	Similar to project X
1	Analyzer	$12,875	$12,875	Competitive bid

User Interface — **Sub-Total $26,941**

No.	Item	Cost per	Total	Source/Assumptions
3	Computers	$1,774	$5,322	800 MHz, 128M RAM, 80GBHD
3	Touch-Screen Monitors	$795	$2,385	Quote from DCS vendor
1	Color Laser Printer	$1,879	$1,879	Can we do better?
3	Software User Licenses	$5,785	$17,355	Discuss corporate discount

I/O — **Sub-Total $6,375**

No.	Item	Cost per	Total	Source/Assumptions
5	T/C Input cards, 4 channels each	$1,275	$6,375	Vendor price list

Materials

Instrumentation — **Sub-Total $27,311**

No.	Item	Cost per	Total	Source/Assumptions	Capital	Expense
17	Type J Thermocouples, modified	$195	$3,315	Quote from XYZ Distributor	$3,315	
3	Smart Pressure Transmitters	$2,475	$7,425	Quote from ABC Company	$7,425	
4	Level Sensors	$924	$3,696	Similar to project X	$3,696	
1	Analyzer	$12,875	$12,875	Competitive bid	$12,875	

Labor — **Sub-Total $11,025**

No.	Item	Cost per	Total	Source/Assumptions	Capital	Expense
8	Days Site-Clearance	$480	$3,840	T&M contractor		$3,840
3	Days Wiring Install	$795	$2,385	T&M contractor	$2,385	
4	Days Production Support	$1,200	$4,800	Required by Op. Dept.		$4,800

TABLE 1.10d
Controller Costs

Controller Type	*Cost Range*
Single-loop controller, pneumatic	$500–$3000
Single-loop controller, electronic	$700–$4000
Multiloop controller, electronic	$6000–$30,000, between $500 and $1000 per loop, including I/O

In many distributed control systems (DCS), it is not possible to assign a cost directly to a control loop. This is because a control loop is simply a program running on a processor. In these cases, determining the number and type of processors may be a better way to estimate the cost. It is also necessary to be careful about the physical location of those controllers that can be more and more remotely mounted. Thus, more controllers (processors) may be needed than the basic specifications of the manufacturer. Redundancy typically doubles the controller costs. Some manufacturer offer a redundancy N:1 so there is one backup for two, three, or four controllers, reducing the cost of redundancy. Reliability of control system components is now proved to be excellent for most major manufacturers. Redundancy is often overkill for most applications. To justify redundancy, verify loss in production in the history of the plant for shutdowns or failures not involving human error. Do not forget that many other things can shut down a system, such as fire, broken cables, loss of power, voltage surge, human error, etc.

Always estimate for at least 10% more loops than exist in the original design. This allows room to make changes either during the project or in the future and will be a small additional cost that will save many headaches later.

OPERATOR STATION COSTS

Estimating the cost of operator stations has become very challenging. In the past, an operator station was a bundled set of hardware and software, typically sold by the DCS manufacturer at a fixed price. But today, the hardware is typically a personal computer, and a large portion of the costs is in the software. In this section, we will discuss hardware cost estimation. A section on software costs follows.

Operator station hardware is typically a fairly new PC, loaded up with the latest complement of RAM memory, and drives. A 17-in. monitor is the minimum size typically seen. This basic configuration can be roughly estimated at between $2000 and $3000. In fact, although PCs have changed dramatically over the past 10 years, the cost of a typical desktop system has not changed much at all. You just get more for your money!

Of course, adding options and features drives up the cost. The most common options for operator stations are the addition of larger monitors; 21-in. monitors are quite common. Add $500 to $1000 per station if larger monitors are desired. Be sure to note that larger furniture may be needed to accommodate larger monitor sizes.

Another common option is a touch-screen. Plan on an additional $1000 per touch screen.

Furniture and cabinetry can have a large impact on operator station costs. At the low end, in a well-protected control room, it may be acceptable to use industrial-grade office furniture, at a cost of $200 per operator station. Custom-fitted enclosures will cost as much as $3000 per station. Do not forget chairs for the operators; plan on $300 to $1000 for these chairs. (This is a good place to splurge—a happy operator will make a project run better in the end.)

If a station will be located on the operating floor, it may be advisable to enclose it in a NEMA-rated cabinet, such as those manufactured by Icebox. Plan on adding $2000 to $10,000 for each station, depending on options.

INSTRUMENTATION

Instrument costs can vary from under $100 for a simple thermocouple, to several thousand dollars for a smart transmitter. Analytical instruments, such as gas analyzers, can cost $100,000 or even more.

To estimate instrument costs accurately, develop a list of the instruments required, and specify as much information as possible before asking for quotes. At a minimum, specify the instrument type, range, type of output signal, and service. If it is possible to be more specific about such things as flange sizes and materials of construction, then a more accurate estimate will result.

CONTROL VALVES

Control valves can cost from a few thousand dollars to tens of thousands of dollars. The primary factors determining control valve cost are the size and type of valve. Figures 1.10e through 1.10i provide approximate cost figures for various valve types, assuming simple actuators on the valves.

As with instruments, listing the valves required, along with their basic specifications will lead to the best estimate. This list can be provided to a valve supplier for a quote. Be sure to discuss a discount from list price, especially if many valves are being ordered.

Most vendors will request detailed information to price a valve accurately. Time invested to gather good data will strongly reflect on loop performance and valve life.

The minimum information needed on the specification is the valve size, type, and service, as well as specification whether positioners on the valves are needed.

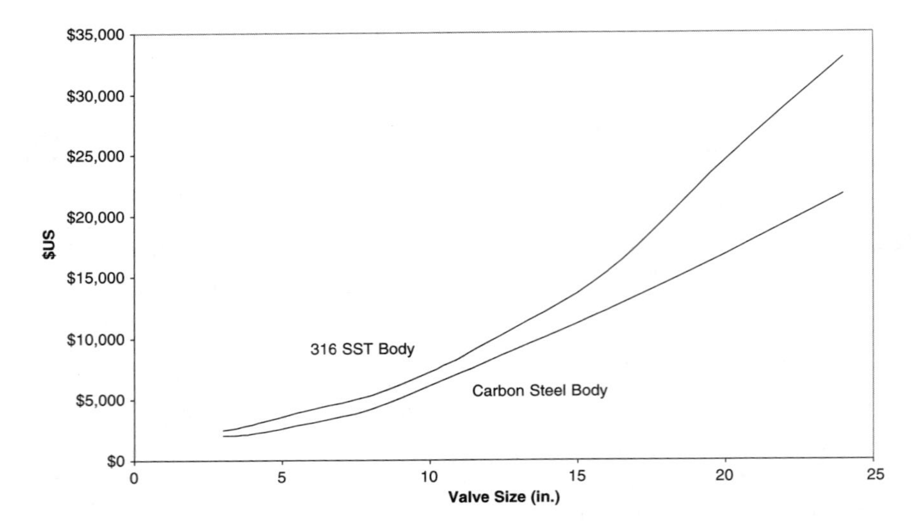

FIG. 1.10e
Cost of butterfly valves.

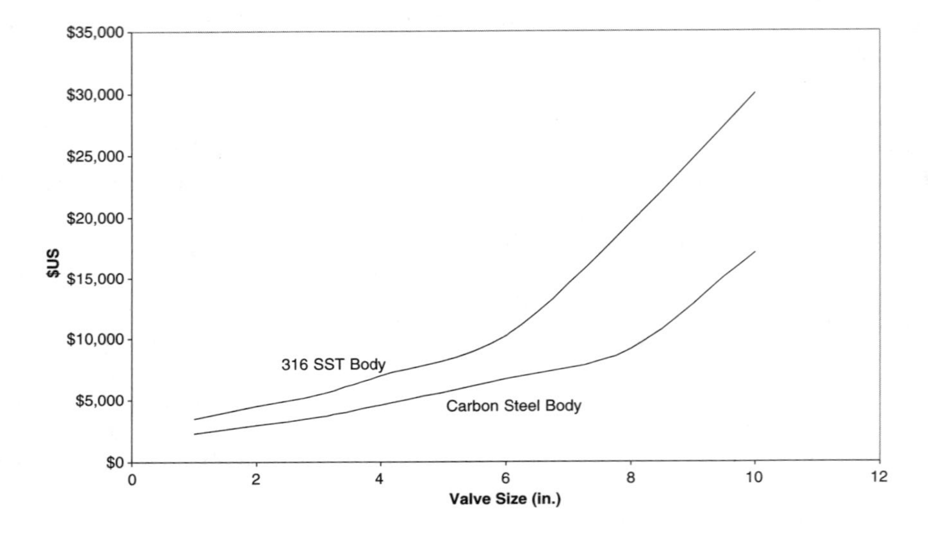

FIG. 1.10f
Cost of globe valves.

As a negotiating tool, it is a good idea to ask the supplier to provide quotes by line item, and also to provide a quote for everything purchased as a group.

MOTORS AND DRIVES

A general rule of thumb for motors is roughly $100 per horsepower. This estimate works well in the range of 10 to 300 HP, for a basic, no-frills motor. Of course, specialty motors, such as servo motors, DC motors, or linear motors, will cost more.

The cost of drive components varies widely, depending on the size, quality, durability, and accuracy of the drive. As with PID controllers, stand-alone single-axis systems, or large, integrated, multiaxis drives may be considered. The cost of the drive power supply will vary most directly with motor size.

SOFTWARE COSTS

Software is absolutely the most confusing and challenging item to estimate in a control system project. There are so many options.

If one has not chosen a control system provider, then the task will be even more difficult. Each vendor, it seems, takes a different approach to software packaging and pricing. For some, it is bundled in the hardware cost; for others everything must be purchased separately. This makes it nearly impossible to estimate software costs until the field of vendors has been narrowed.

The best approach will be to revisit the project objectives and scope definition. Look specifically at each piece of desired functionality, and see if software must be added for each. For example, is the statistical analysis package really

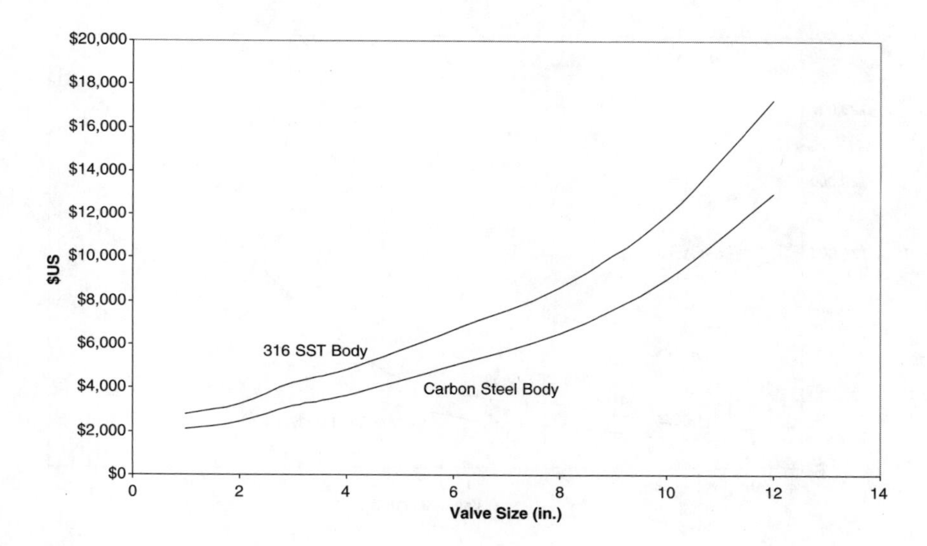

FIG. 1.10g
Cost of plug valves.

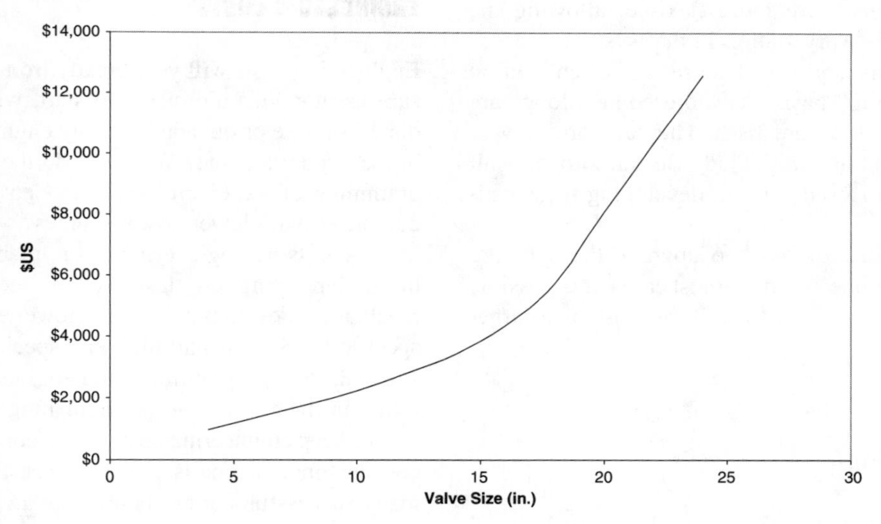

FIG. 1.10h
Cost of gate valves—carbon steel body.

necessary, or is the need simply for control charts? Look at each piece of software quoted by the vendor, and ask which of the project objectives will be satisfied by it.

Work closely with the vendors while estimating this part of the project. The options are changing every day.

A word about software licensing. Contrary to popular belief, most software is not "purchased." It is licensed. In most cases, making copies or modifying the software is prohibited. The details of the licensing agreements are enough to drive anyone crazy. (Has anyone ever really read the license agreement that pops up when installing any Microsoft software?)

The cost of producing copies of software is ridiculously small. CDs can be printed for pennies apiece. The printed manuals typically cost far more to produce than the software media. The vendor's cost comes largely from development and testing effort, which can be quite large. Every vendor has a different model for determining how to recapture the development expense. Pricing remains largely market driven. This is especially true for small-volume specialty software, sold as an add-on to a larger control system. Remember that market-driven means "negotiable." Be sure to ask about alternative pricing on software.

When estimating the software costs, knowledge of some specifics about the licensing terms is necessary. Some software is licensed on a "per seat" basis, meaning that one must pay for one copy for each station where it will be installed.

Other software is licensed for a certain "number of concurrent users." With five concurrent users, the software could be installed on ten stations, but would only be able to be used on any five of the stations at a time.

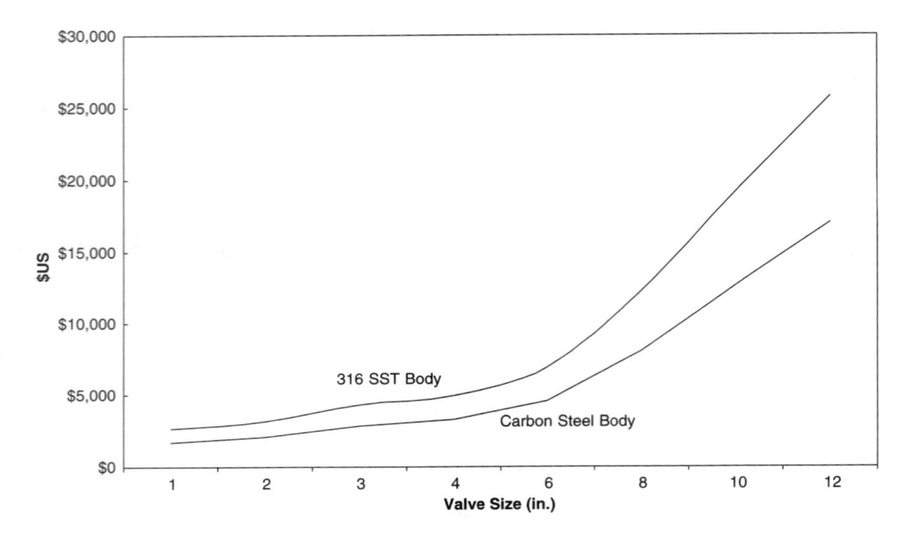

FIG. 1.10i
Cost of conventional ball valves.

"Per system" licenses are more flexible, allowing the software to be used on every station in that system.

For control systems, software licenses are often sold on a "per tag" or "per loop" basis. As more control loops are added, the cost of the software rises. The software may or may not be incrementally upgraded. Be sure to consult with the control system vendor when developing these estimates.

Also, make sure that the cost to upgrade the software license in the future is understood. In most cases, it is possible to apply any current payments toward the cost of a larger license at a later date.

MAINTENANCE COSTS

Annual software maintenance fees are typically 15% of the original licensing price, although the range varies somewhat. The first year's maintenance fee is usually included in the up-front pricing.

Be sure to ask the vendor about these costs. Also, be sure that what is included is clear. Obtaining software updates, bug fixes, and updated documentation, at a minimum, is probably necessary.

Hardware and equipment maintenance programs are also available from the vendor, in a wide variety of formats, ranging from 24-hour phone support to on-site maintenance. Remember to negotiate these costs up-front, before the equipment purchase, to preserve leverage in the negotiation.

A good hardware maintenance agreement will extend the warranty on hardware, and will often include discounted pricing for upgrades to newer hardware.

The 24-hour phone support is almost a base requirement for most control systems. Ascertain whether this is a fee-per-call or fee-per-year type of service.

ENGINEERING COSTS

Engineering costs will vary greatly from project to project. Be sure assumptions are clear. To start, will the engineering be done in-house or outside? Are any engineering costs included in the vendor's quote? Who will do the configuration or programming of the control system? Can some of the work be completed with lower-cost resources?

For outside engineering costs, plan on $30 to $150 per hour, depending on the skill set required. Drafting and mechanical design will be at the lower end of costs. Industry-specific DCS skills and high-end specialists are at the opposite end. PLC programming and engineering can typically be found in the $50 to $90 per hour range.

To keep engineering costs down, consider the use of lower-cost personnel for parts of the project. For example, there are many successful stories about using a computer-literate operator to design, develop, and configure the human–machine interface (HMI) screens. Also, lower-cost designers or administrative personnel can be used to key-punch tag configuration data into the control system or instrument database.

TRAINING AND START-UP COSTS

Some amount of training will be required for operators, electrical and mechanical maintenance personnel, engineers, and management. Costs for a trainer and equipment can range from $200 to $3000 per day. To estimate these costs, start with the number of people to be trained, and the number of days of training per person. A typical class will have from 6 to 20 students in it, depending on the material and the level of interaction required.

Simulator costs can range from several thousand dollars to millions. It is critical to be clear on the scope and intent of the simulator work to contain costs.

INSTALLATION COSTS

Labor costs vary widely, based on the local labor market. Check with engineering contractors or union halls for the prevailing labor rates in the area.

Also, in developing the estimate, keep in mind that some work, such as demolition and site clearance, can be completed by relatively low-skilled laborers, while other work, such as wiring terminations or instrument installation, required ... or higher-skilled tradesman. For the best estimate, organize con- costs by type, and assign each a different labor rate. F ... This struction work, include some overhead for supervi ... sure to can range from 5 to 30% of the direct labor cost construc- discuss the allocation of supervision costs with tion contractor before the estimate is finalize

CONTROL ROOM INCIDENTAL COSTS

...cts I have managed,
In the dozens of control syste ...e to update or upgrade
almost every one included so ...m a fresh coat of paint to
the control room. This can r ...ese costs should be included
a completely redesigned r ...timate project cost, and they
in the estimate. They ... the operator happy, which will
will go a long way ...ful in the end.
help the project ... building codes and regulations. In
Watch out ...n significant building renovations are
the United S ...s with Disabilities Act requires certain
made, th ...the area accessible to those with disabili-
revisio ...oom may sometimes fall under these guide-
ties ...nanges can easily double the cost of the work.
li ...nderstand the trigger points for these regulations.
...with a local civil engineer for details.

...ES

Do not forget about the taxman. Sales taxes and property taxes are the primary concerns. Depending on state and local laws, one may or may not need to pay sales tax on the purchase of new equipment and software. Also, whether or not the equipment is capitalized will play a role in future tax liabilities.

Most equipment, such as control systems, instruments, and valves, are capitalized, making them subject to special tax treatment. Services may or may not be capitalized, depending on the circumstance. Some software, also, may be capitalized, but may be on a different depreciation schedule.

When preparing a budget, it is often helpful to list capital and expense items in separate columns, to make it clear how the finances will be impacted. The sample budget in Figure 1.10c shows how this can be done.

Tax laws are often complicated, and subject to interpretation. Be sure to talk with the company accounting or capital management people to make sure that the right amount for taxes has been included.

WORKING WITH VENDORS

Vendors are, of course, the best source of pricing information for their products. By working closely with vendors, it is possible to develop a very accurate estimate of the costs. In fact, by encouraging vendors to help develop the estimate, they will build some ownership for the budget and for the success of the project.

The trick, of course, is not to "let the fox into the hen house." That is, if the vendor knows in advance what the top-end budget is, then the project will probably come in at or near that number. Every time.

When the vendor is closely involved, be sure that unneeded scope is not added. Be sure that to maintain control over what will be supplied. Watch out for add-ons, such as added software packages, extra operator stations, and overpriced maintenance programs.

CONTINGENCY COSTS

With all of the details filled in on the estimate, it is almost time to go forward to management. But have we remembered everything?

It is typical to build some contingency monies into each estimate. This could range from 5 to 25% of the estimate, depending on the degree of confidence one has in the project scope and in one's estimating techniques. Figure 1.10j represents the relationship between degree of certainty and amount of project contingency. It should be obvious that the advice of an experienced person will allow reduction of the contingency requirements.

If any physical or construction work is to be performed, be sure to leave some money in contingency. There are almost always unexpected surprises behind that wall being torn down.

To reduce the contingency funds in a project, several techniques can be used. Each of these techniques will increase the degree of certainty and accuracy of the estimate, reducing the need for contingency money.

Engineering studies will improve the accuracy of the estimate by investigating those areas with the most uncertainty. Although this requires some up-front spending, it is usually money that would be spent anyway.

Benchmark comparisons with other projects will also build the degree of confidence and the accuracy of the estimate. Spending more time looking for a similar project will help. Find out how much contingency money was included, and how that money was spent.

Vendor quotes for parts of the project will also lock in some of the costs in the estimate. If time will be spend doing this, focus on the large pieces of capital hardware and software. These will typically have the largest impact on the estimate.

Labor estimates can be improved by doing field walk-downs with construction planners or supervisors. They will be able to point out the problem areas, and may also have

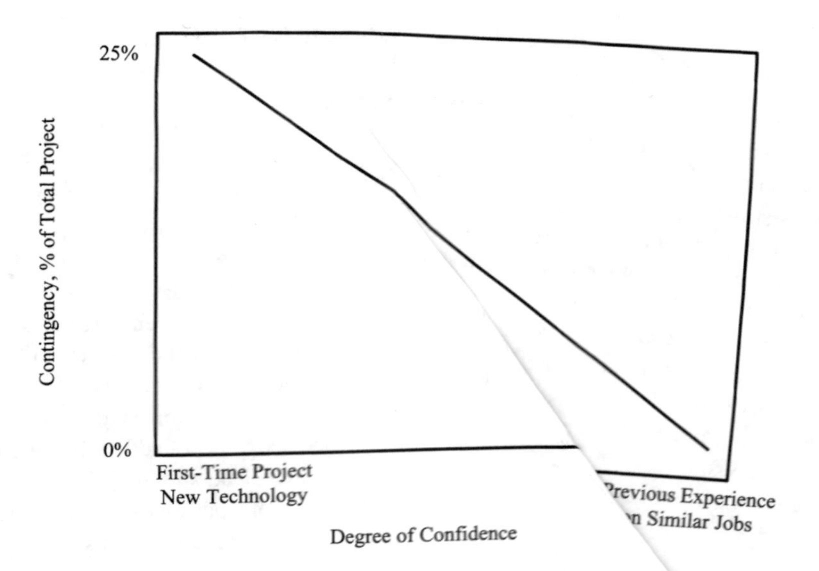

FIG. 1.10j
Project contingency must rise as confidence in data drops.

some ideas to cut costs. If construction personnel can be involved early, the labor estimate will be more accurate.

ESTIMATING VS. BIDDING

One way to develop an estimate is to place the project up for bid. This way, the bidders provide the estimate. If the job is bid as a fixed-price contract, then one can be fairly sure the estimate is accurate.

Including at least two competing bidders keeps the process honest. Quite often, a competitive bidding process will result in a lower overall estimate than an in-house estimate. This is because there is a shift in the ownership of risk, in the labor pool to be used, and in the incentives.

A typical process for bidding is to develop a request for proposal (RFP). This contains all the requirements for the project. The RFP is offered to several bidders, and they are given a timetable to respond. It is very important to provide a forum for these bidders to ask questions, to make sure that they understand the scope that is needed.

After bidder questions are resolved, some time is allowed for the bidders to finalize their bids. Vendor selection should happen fairly quickly after final bids are submitted. Figure 1.10k shows a typical bid process.

The trick to successful bidding is to be specific enough to obtain what is truly wanted, while being general enough to allow creativity and flexibility on the part of the bidders. The more flexibility they have, the lower the cost will be.

To bid a project, be sure to start early. You will need to start the conceptual and preliminary design stages earlier, so that you have enough detail to develop the RFP.

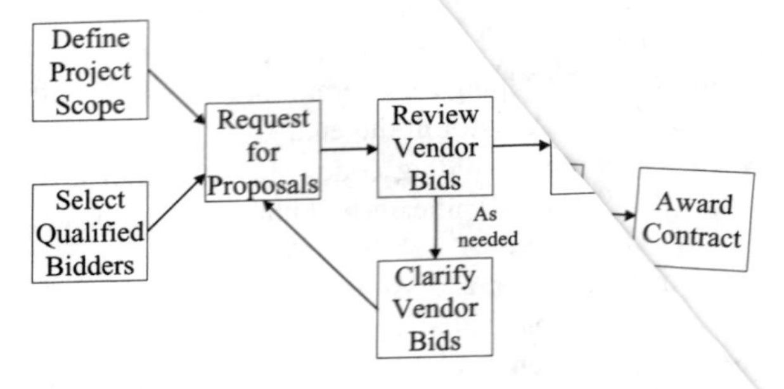

FIG. 1.10k
The bidding process.

SUBMITTING THE BUDGET

When the time comes to submit the project budget or estimate, it is important to communicate not just the final dollar figure, but also some of the key project parameters. Clearly communicate items such as the scope boundaries, the largest potential risks, key contingency items, and the degree of accuracy that went into developing the estimate. Management will use this information to plan its own budgets.

Bibliography

Battikha, N. E., *The Management of Control Systems: Justification and Technical Auditing,* Research Triangle Park, NC: Instrument Society of America, 1992.

Jones, C. and Jones, T. C., *Estimating Software Costs,* New York: McGraw-Hill, 1998.

Peters, M. and Timmerhaus, K., *Plant Design and Economics for Chemical Engineers,* New York: McGraw-Hill, 1980.

Designing a Safe Plant

2

2.5 PROCESS SAFETY MANAGEMENT 182

2.6 REDUNDANT OR VOTING SYSTEMS FOR INCREASED RELIABILITY 192

2.7 NETWORK SECURITY 198

2.8 SAFETY INSTRUMENTED SYSTEMS: DESIGN, ANALYSIS, AND OPERATION 209

2.10
INTELLIGENT ALARM MANAGEMENT 252

2.11
SAFETY INSTRUMENTATION AND JUSTIFICATION OF ITS COST 268

2.12 INTERNATIONAL SAFETY STANDARDS AND CERTIFICATION (ANSI/ISA-S84, IEC 61511/61508/62061, ISO 13849) 278

2.1 Hazardous Area Classification

E. M. MARSZAL

This section describes hazardous area classifications and presents an overview of how area classifications are selected. Hazardous (classified) areas are those areas where arcing, heat, or misoperation caused by electrical devices could initiate a fire or explosion. When an area is classified, a number of different techniques can be used to ensure that electrical equipment in that area will not cause a hazardous condition. These techniques include intrinsic safety, purging, and explosion-proof enclosure, which are described in other sections of this handbook.

Hazardous area classification, and design of compliant equipment, is not only good engineering practice; it is often given the weight of the law. In the United States, the Occupational Safety and Health Administration (OSHA) requires hazardous area classification and design compliance in regulation 29 CFR 1926.449 for construction and in regulation 29 CFR 1910.307 for occupational safety and health. In addition, many local governments incorporate the National Electrical Code (NEC) standard that defines the requirements of area classification into local legislation, by reference. More information about compliance with electrical safety regulations can be obtained by contacting OSHA at the following address:

U.S. Department of Labor
Occupational Safety and Health Administration
200 Constitution Avenue, N.W.
Washington, D.C. 20210

The process of selecting an area classification based on the conditions of a room containing electrical equipment is a straightforward exercise. The classification process consists of identifying the flammable materials that exist, or are stored in the facility, determining the flammable characteristics of those materials, determining how those flammable materials might come into contact with an electrical device, and then selecting the appropriate standard and lookup table to determine the extents and characteristics of the classified area. The characteristics of a classified area are given by its Class, Division, and Group, if the traditional method of classification is used, or by a Class, Zone, and Group if the newer zone classification method is used.

The proces s of area classification should not be taken lightly. Improper use of general-purpose electrical equipment in an area that contains flammable materials might lead to a fire or explosion. On the other hand, use of classified area equipment and wiring techniques is much more costly and labor intensive than general-purpose equipment, and thus should be avoided if possible.

HAZARDOUS AREA CLASSIFICATION OVERVIEW

An area classification defines the expected environment in which an electrical or electronic device will reside in terms of the presence and type of flammable gases or vapors, flammable liquids, combustible dust, or ignitable fibers or flyings. The environment of electrical or electronic devices is of concern because the potential energy stored and consumed by these devices is often great enough to ignite flammable mixtures. The determination of area classification should be performed for each individual room, section, or area of a process. A classification will rarely extend to an entire plant.

Hazardous area classifications are performed by following standards, in an attempt to match characteristics of an area with equipment that is designed to safely operate in that atmosphere. This section focuses on classes described in NFPA Standard 70—The National Electrical Code. The National Electrical Code allows two methods for hazardous area classification. The first is the more traditional Class–Division–Group scheme (Division Classification System). The second is the Class I–Zone method (Zone Classification System), which was incorporated into the 1999 edition of the code.

DIVISION CLASSIFICATION SYSTEM

The Division Classification System is the original classification system defined by NFPA in the National Electrical Code, and is used extensively in the United States. The Division Classification System requires that hazardous areas be identified with a class, division, and group. The class identifies whether the flammable material that may be present is a gas, dust, or fiber. The division defines the likelihood that the flammable material will be present by defining how it is used. The group categorizes the physical properties of the material.

The class defines the type of flammable material that is present. The class selected will be either 1, 2, or 3, and is shown with the corresponding Roman numeral. Class I locations are those where flammable gases may be present in flammable concentrations. Class II locations are those where a combustible dust is present. Class III indicates a location where the presence of easily ignitable fibers and flyings are present, but not in the form of an air/fiber suspension that is ignitable (air/fiber suspensions should be considered combustible dust).

Divisions

The division generally defines the probability of flammable materials being present and how the flammable material is used in that area. Division 1 areas have the following characteristics.

- Flammable concentrations of gases and vapors can exist for long periods of time.
- Flammable concentrations occur because of repair or maintenance activities or leakage from process equipment.
- Failure or misoperation of process equipment might release flammable mixtures and simultaneous failure of electrical equipment can cause ignition.

In addition to the general conditions that define Division 1, flammable dusts mixtures of Class II Type E that are electrically conductive and present in hazardous quantities are also classified as Division 1. Essentially, Division 1 indicates a continuous to frequent presence of a flammable mixture. Presence of flammable mixtures may occur during routine operations.

Division 2 areas have the following characteristics, in general.

- Flammable gases and volatile flammable liquids are used, but normally contained in a closed system. Release will only occur through accidental breach of the system or abnormal operations.
- Accumulation of flammable materials is normally prevented by positive mechanical ventilation, but may occur if ventilation fails.
- The area is adjacent to a Division 1 area and flammable material might be transferred unless prevented by positive means such as clean-air pressure ventilation with safeguards against ventilation failure.

In addition to the general conditions defining Division 2, Class II locations where combustible dust can accumulate on or near electrical equipment and can impede safe dissipation of heat from electrical equipment or may be ignited from abnormal operation of electrical equipment or its failure are also classified as Division 2. In essence, Division 2 indicates that presence of flammable mixtures is infrequent to rare. Presence of flammable mixtures in a Division 2 location will occur only as a result of process accidents or abnormal process operation.

For Class III locations, Division 1 is reserved for areas where the process that creates the fibers and flyings occurs and Division 2 is assigned to areas where the materials are stored, but not processed.

TABLE 2.1a
Division Classification Group Assignment for Groups B, C, and D

Group	*MESG Range*	*MIC Ratio Range*
Group B	≤0.45 mm	≤0.40
Group C	>0.45 mm but ≤0.75 mm	>0.40 but ≤0.8
Group D	>0.75 mm	>0.8

Groups

Classes are further broken down into groups that assign categories of chemical and physical properties to the hazardous material. Guidance for determining the group of a substance can be found in "Recommended Practice for the Classification of Flammable Liquids, Gases, or Vapors and of Hazardous (Classified) Locations for Electrical Installations in Chemical Process Areas" (NFPA 497) and "Guide to Fire Hazard Properties of Flammable Liquids, Gases, and Volatile Solids" (NFPA 325).

Class I is subdivided into four groups: A, B, C, and D. Group A is composed of only acetylene. Group B, C, and D materials are flammable gases or vapors produced by the evaporation of flammable or combustible liquids that may burn or explode when combined with air and ignited. Groups B, C, and D are differentiated by experimentally determined values of Maximum Experimental Safe Gap (MESG) and Minimum Igniting Current Ratio (MIC Ratio). MESG and MIC Ratio ranges corresponding to groups B, C, and D are shown in Table 2.1a.

Group B is typified by hydrogen gas, Group C is typified by ethylene gas, and Group D is typified by propane gas. Experimental values of MESG and MIC Ratio for a variety of chemicals along with corresponding Division Classification System division and group assignments can be found in NFPA 497.

Class II is subdivided into three groups: E, F, and G. Group E includes atmospheres that contain metal dusts, or other dusts whose properties (e.g., particle size, abrasiveness, and conductivity) present similar hazards. Group F includes atmospheres contain carbon-based dusts with more than 8% volatiles entrapped in the dust. This group also includes carbon-based dusts that have been "sensitized" so that they present an explosion hazard. Group F includes such dusts as coal, coke, charcoal, and graphite. Group G contains all Class II materials that are not included in Groups E or F, including flour, grain, and wood. More information on Class II group determination can be found in NFPA 499, "Recommended Practice for Classification of Combustible Dusts and of Hazardous (Classified) Locations for Electrical Installations in Chemical Process Areas."

Class III is not subdivided into groups.

ZONE CLASSIFICATION SYSTEM

The zone classification system was introduced in the 1999 edition of the National Electrical Code as an alternative to the division classification system that is more in line with

TABLE 2.1b
Standards for Selecting Hazardous Area Zones

Standard	*Title*
IEC 79-10	Electrical Apparatus for Explosive Gas Atmospheres, Classification of Hazardous Areas
API RP 505	Classification of Locations for Electrical Installations at Petroleum Facilities Classified as Class I, Zone 0, Zone 1, or Zone 2
ISA S12.24.01	Electrical Apparatus for Explosive Gas Atmospheres, Classifications of Hazardous (Classified) Locations
IP 15 (The Institute of Petroleum, London)	Code of Safe Practice in the Petroleum Industry, Part 15: Area Classification Code for Petroleum Installations

IEC standards on hazardous area classification. This system separates hazardous areas into three zones: Zone 0, Zone 1, and Zone 2. Industry and application specific guidance for selecting zones can be found in standards listed in Table 2.1b.

Zones

The zone classification system separates hazardous areas into three zones: Zone 0, Zone 1, and Zone 2. A Zone 0 area indicates that there will be flammable mixtures of gaseous vapor present continuously, or for long periods of time. Zone 1 designation of an area indicates that flammable mixtures of gas or vapor are likely to exist under normal conditions. Flammable mixtures may occur as the result of repair or maintenance operations or from minor leakage from sources such as valve and pump packing. Flammable mixtures in Zone 1 may also occur as the result of equipment breakdown or misoperation of the process that can also cause failure of electrical equipment such that sources of ignition are created. Zone 1 classification can also be the result of contiguous location to a Zone 0 area where positive separation between environments are not provided or properly safeguarded.

Zone 2 areas are those where flammable mixtures are unlikely during normal operation and will quickly disperse if they do occur. Areas where flammable materials are processed but kept within closed systems are generally considered Zone 2. Zone 2 includes areas where accumulation of flammable gases is prevented by ventilation and areas that are adjacent to Zone 1 areas but are not positively separated by safeguarded means.

Groups

The Zone Classification System separates hazardous gases and vapors into two groups, I and II. Group II gases and vapors are further subdivided into three subgroups, IIA, IIB, and IIC. Group I is reserved for describing "firedamp," which is a mixture of gases whose principal component is methane, which is typically found in mines. The National Electrical Code does not apply to mine installations, so Group I is not used.

TABLE 2.1c
Zone Classification System Group Assignment for Groups IIA, IIB, and IIC

Group	*MESG Range*	*MIC Ratio Range*
Group IIC	≤0.5 mm	≤0.45
Group IIB	>0.5 mm but ≤0.9 mm	>0.45 but ≤0.8
Group IIA	>0.9 mm	>0.8

TABLE 2.1d
Comparison of Zone and Division System Groups

Zone System Group	*Division System Group*
IIA	Group D
IIB	Group C
IIC	Groups A and B

The Group II subgroups separate compounds into three subgroups in a fashion that is similar to the Division Classification System, using MESG and MIC to classify compounds. The classification proceeds according to Table 2.1c.

Group IIA includes such materials as methane and propane, Group IIB includes materials such as ethylene, and Group IIC includes materials such as hydrogen and acetylene. Zone classification groups can be roughly compared to division classification groups using Table 2.1d.

KEY FACTORS CONTRIBUTING TO CLASSIFICATION

If a hazardous area is to be classified, selection of the class and group can be a trivial exercise: selecting class and looking up the group in table 2-1 of NFPA 497 for gases and vapors and table 2-5 of NFPA 499 for dusts. The complexity and confusion in the area of classification often lies in the determination of the likelihood of a flammable material being present.

The classification process takes into account the fact that all sources of hazards (e.g., gas, vapor, dust) have different ignition properties and will produce different blast effects if an explosion occurs. Some of the key factors that influence the selection of an area classification are listed and described below.

FLASH POINT The temperature at which a material is capable of volatilizing a sufficient quantity of vapor to flash when exposed to an ignition source.

AUTO IGNITION TEMPERATURE The temperature at which a material will spontaneously ignite when exposed to air without requiring an external ignition source.

EXPLOSIVE LIMITS The limits of concentration of a mixture of flammable material in air, measured as percent by volume at room temperature, where the mixture can be ignited. Limits are given in terms of both Lower Flammability Limit (LFL) and Upper Flammability Limit (UFL).

VAPOR SPECIFIC GRAVITY Ratio of the weight of volume of pure vapor the weight of an equal volume of dry air.

LEAK SOURCES Locations in a process or container holding flammable materials where the flammable material could escape, such as flanges, screwed connections, and valve packing.

VENTILATION In cases where the release of flammable material is slow, such as through minor leaks in valve packing, accumulation of a flammable mixture can be accomplished by ventilation of the area, either through natural means or forced mechanical ventilation.

GUIDANCE FOR APPLICATIONS

The task of classification should be performed with great care and be based on a complete understanding of electrical usage and experience with the process under control, including the characteristics of the hazardous materials and potential leak sources. Experience in various industries and particular services has often been compiled into specific rules for defining the extents of hazardous (classified) areas. These classification procedures can be found as separate articles in the National Electrical Code, as shown in Table 2.1e, and in industry group standards, as shown in Table 2.1f.

TABLE 2.1e
National Electrical Code Articles Describing Area Classification

Article	Title
511	Commercial Garages, Repair and Storage
513	Aircraft Hangers
514	Gasoline Dispensing and Services Stations
515	Bulk Storage Plants
516	Spray Application, Dipping and Coating Processes
517	Health Care Facilities
518	Places of Assembly
520	Theaters, Audience Areas of Motion Pictures and Television Studios and Similar Locations
525	Carnivals, Circuses, Fairs, and Similar Events
530	Motion Picture and Television Studios and Similar Locations
540	Motion Picture Projectors
547	Agricultural Buildings
550	Mobile Homes, Manufactured Homes, and Mobile Home Parks
551	Recreational Vehicles and Recreational Vehicle Parks
553	Floating Buildings
555	Marinas and Boatyards

TABLE 2.1f
Separate Standards Describing Area Classification

Services	Standard	Title
Flammable and Combustible Liquids	NFPA 30	Flammable and Combustible Liquids Code
Spray Applications	NFPA 33	Standard for Spray Applications Using Flammable or Combustible Liquids
Dipping and Coating	NFPA 34	Standard for Dipping and Coating Processes Using Flammable or Combustible Liquids
Organic Coating Manufacturing	NFPA 35	Standard for the Manufacture of Organic Coatings
Solvent Extraction	NFPA 36	Standard for Solvent Extraction Plants
Laboratories	NFPA 45	Standard for Fire Protection for Laboratories Using Chemicals
Gaseous Hydrogen Systems	NFPA 50-A	Standard for Gaseous Hydrogen Systems at Consumer Sites
Liquefied Hydrogen Systems	NFPA 50-B	Standard for Liquefied Hydrogen Systems at Consumer Sites
LPG (Liquefied Petroleum Gas)	NFPA 58	Liquefied Petroleum Gas Code
LPG at Utility Plants	NFPA 59	Standard for Storage and Handling of Liquefied Petroleum Gas at Utility Plants
LNG (Liquefied Natural Gas)	NFPA 59A	Standard for the Production, Storage, and Handling of Liquefied Natural Gas
Chemical Processes	NFPA 497	Recommended Practice for the Classification of Flammable Liquids, Gases, or Vapors, and of Hazardous (Classified) Locations for Electrical Installations in Chemical Process Areas
Chemical Processes with Dusts	NFPA 499	Recommended Practice for the Classification of Combustible Dusts and of Hazardous (Classified) Locations for Electrical Installations in Chemical Process Areas
Wastewater Treatment	NFPA 820	Standard for Fire Protection in Wastewater Treatment and Collection Facilities
Petroleum Refining	ANSI/API RP 500	Recommended Practice for Classification of Locations for Electrical Installations at Petroleum Facilities Classified as Class I, Division 1 and 2
Flammable Dusts	ANSI/ISA S12.10	Area Classification in Hazardous (Classified) Dust Locations
Offshore Production Platforms	ANSI/API RP 14F	Recommended Practice for Design and Installation of Electrical Systems for Offshore Production Platforms

Bibliography

NFPA 70, "National Electrical Code," Batterymarch, MA: National Fire Protection Association, 1999.

NFPA 497, "Recommended Practice for the Classification of Flammable Liquids, Gases, or Vapors, and of Hazardous (Classified) Locations for Electrical Installations in Chemical Process Areas," Batterymarch, MA: National Fire Protection Association, 1997.

NFPA 499, "Recommended Practice for the Classification of Combustible Dusts and of Hazardous (Classified) Locations for Electrical Installations in Chemical Process Areas," Batterymarch, MA: National Fire Protection Association, 1997.

29 CFR 1910.307, "Hazardous (Classified) Locations," Washington, D.C.: U.S. Department of Labor, Occupational Safety and Health Administration, 1981.

2.2 Intrinsic Safety Rules for Fieldbus Installations

J. BERGE

Intrinsically safe installation becomes simpler and more economical if the correct equipment is used:

- Isolated safety barriers
- Safety barriers with built-in repeaters
- Safety barriers with built-in terminators
- Devices with low power consumption
- A typical entity concept barrier for fieldbus such as Smar SB302 costs U.S.$400
- A typical FISCO model barrier for fieldbus such as Smar DF47 costs U.S.$450
- A typical repeater for fieldbus such as Smar RP302 costs U.S.$300

Foundation™ Fieldbus and PROFIBUS PA are based on the IEC 61158-2 standard that provides communication and power on the same pair of wires. Essentially, there is not a great difference between a regular fieldbus network and an intrinsically safe network. Basic topology remains the same with terminators in each end (see Section 4.2). For an intrinsically safe installation, a safety barrier is used in place of the regular power supply impedance. Other differences include fewer devices per wire. When this wire is run in a hazardous area, intrinsic safety is one of the better protection methods. Fieldbus was specially developed to enable it to be used in intrinsically safe systems. For example, field instruments can be designed in such a way that they do not inject power onto the bus, which is one of the requirements for intrinsic safety.

Installation of instruments is made easy by fieldbus. Foundation Fieldbus H1 was specially created to enable reuse of existing cables, and therefore it does not have as many special requirements for knowledge and tools as do other networks. Make sure to follow local regulations and carefully read the manuals and follow the manufacturer's instructions before installing any device.

INTRINSIC SAFETY

The concept of intrinsic safety is the same for fieldbus as it is for conventional installation, i.e., the basic rules for limiting the energy to a level below what is required for ignition are not changed. In an intrinsically safe system power is only supplied from a single source, through the safety barrier. All the field instruments are passive in the sense that they only consume power (Figure 2.2a).

Fieldbus devices operate on a voltage anywhere between 9 and 32 VDC from the bus. Because the concept of intrinsic safety is to limit power, the bus usually operates at the lower end of this range. Separately powered devices are also possible, but even so these devices are forced to draw at least 10 mA quiescent current from the network to be able to manipulate the current during communication without injecting power on the bus. The fieldbus connection is electrically almost identical for all devices including analog and discrete performing input as well as output functions. Therefore, a fieldbus-based system requires only a single barrier type. The main distinguishing difference between field devices from different suppliers is their power consumption. It is important to select devices with low power consumption to enable as many devices as possible to be connected to each barrier because the power available to an intrinsically safe bus is limited. The barrier may be a zener barrier or an intrinsically safe galvanic isolator. Intrinsically safe fieldbus devices are current sinking and do not provide power to the network. The main limiting factor to the number of devices

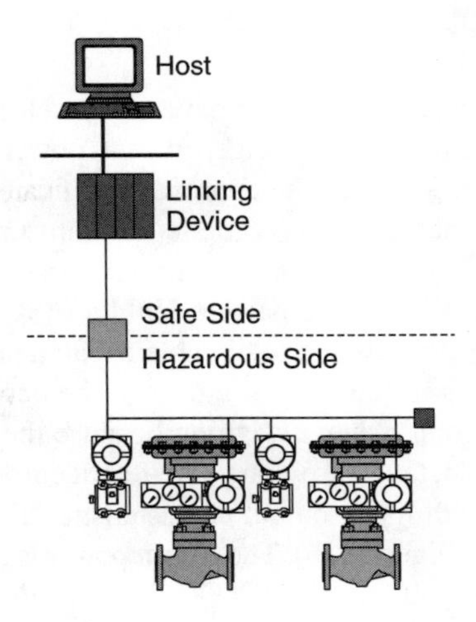

FIG. 2.2a
The safety barrier limits the amount of energy on the network in the hazardous area.

on the intrinsically safe network segment is the power consumption pushing the limit well below 32 devices. However, barriers may be multidropped and still result in 16 devices per interface port, which is important to keep system cost down. It is a good idea to look for the lowest possible device power consumption because it minimizes the number of barriers required and thereby reduces the system complexity and cost.

The fieldbus network terminators at each end of the bus contain a capacitor and, therefore, to be used in intrinsically safe installations, must also be certified intrinsically safe. Because power is scarce, the terminator should be totally passive without current consumption in order not to reduce cable length or device count. IEC 61158-2 barriers are different from barriers for other kinds of networks. Therefore, make sure to use the correct type of barrier.

For intrinsic safety design and installation, follow local regulations and codes of practice and make sure to read the respective user manuals. Connect only intrinsically safe certified devices in the hazardous area.

A basic rule of intrinsic safety is that there can only be a single source of power; therefore, safety barriers cannot be redundant, i.e., only one safety barrier can be connected to each hazardous area segment. Typically, the barriers are mounted in the safe area because it is much cheaper and simpler. Sometimes distance or other limitations force the barriers to be mounted in the hazardous area, in which case flameproof enclosures with flameproof seals are needed. When nonisolated safety barriers are used, special care should be taken to follow the instructions for grounding in the product manual. The traditional entity concept and the newer FISCO model are two schemes for supplying intrinsically safe power to the network. FISCO provides more power thereby enabling more devices and longer cable than the entity concept.

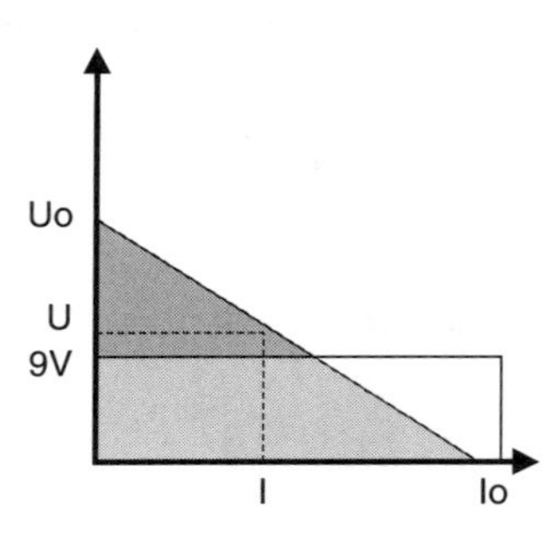

FIG. 2.2b
Linear characteristics of entity concept barrier.

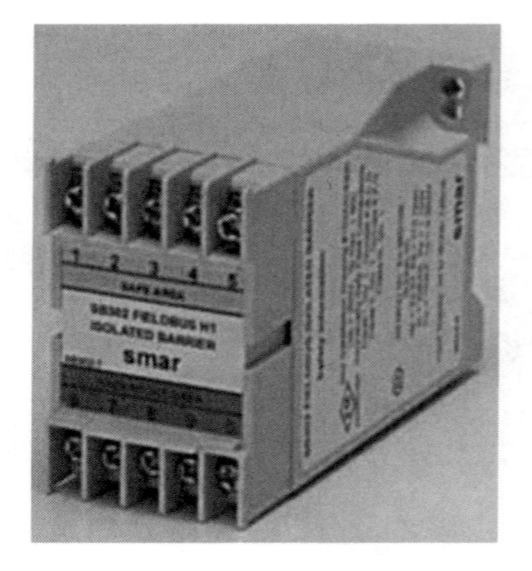

FIG. 2.2c
Isolated repeating entity concept barrier. (Courtesy of Smar.)

ENTITY CONCEPT

Matching devices and barrier have to be selected based on the entity parameters for voltage, current, power, capacitance, and inductance stated in the approval certificate for the components. It is necessary to compile the entity parameters of all devices to match the barrier because several devices are multidropped off a single barrier. Unlike FISCO, in the traditional entity concept the total cable capacitance and inductance for the hazardous side segment of the network must be taken into account when designing because the cable capacitance and inductance are considered concentrated.

Barriers with linear output characteristics are used for the entity concept (Figure 2.2b). The output power is approximately 1.2 W or some 60 mA at 11 VDC for Exia IIC. Only a few devices can be connected to each linear barrier due to this current limit. Similarly, the cable length is limited by the low voltage output as only a small voltage drop can be accommodated.

A field device can only handle a limited voltage, power, and current. Similarly, the total amount of inductance and capacitance allowed on the bus is limited. A barrier should be selected on a worst-case basis, i.e., it must have a voltage, current, and power output lower than what is permitted by the lowest corresponding entity parameter in the field devices (Figure 2.2c).

The total external capacitance and inductance of all the devices connected to the safe side plus the network cable must be within the limits of the barrier. The cable data sheet is typically the best place to obtain the cable parameters. The inductance value for the cable will very often exceed the value allowed for the barrier. However, it is common to compare the inductance/resistance (L/R) ratio allowed by the barrier to that of the cable instead of the inductance value itself. It is often found that the limiting factor for distance in intrinsically safe installations based on the entity concept is the cable capacitance. Cables without screen have a lower capacitance than those with screen.

An easy way to evaluate the network segment is to make a table of the entity parameters for all components on the network segment (Table 2.2d).

The maximum allowed cable capacitance is the balance after all the device capacitances have been subtracted from the capacitance allowed by the barrier. The maximum allowed

TABLE 2.2d
Entity Concept Example

Tag	Ui (Vmax), V	Ii (Imax), mA	Pi (Pmax), W	Ci, nF	Li, μH	L/R, μH/Ω	Iq, mA
PT-456	24	250	1.5	5	8		12
LT-789	24	250	1.5	5	8		12
PCV-456	24	250	1.5	5	8		12
LCV-789	24	250	1.5	5	8		12
Terminator	24	250	1.5	0	0		0
Wire				100	275	25	0
	min	min	min	sum	sum		sum
	Uo (Voc)	Io (Isc)	Po (Pm)	Co (Ca)	Lo (La)	Lo/Ro (L/R)	
Barrier requirement	24	250	1.5	120	307	25	48
Selected barrier	21.4	200	1.1	154	300	30.7	60

cable distance should be calculated based on the maximum allowed cable capacitance.

Example Entity parameters for typical devices and a cable are inserted (see Table 2.2d). The cable in this example is 500 m long and has 200 nF/km capacitance and a 25 μH/Ω L/R ratio, which are typical values.

The output voltage, current, and power allowed for the devices in this example exceed the barrier output, the current provided by the barrier is sufficient, and the capacitance allowed by the barrier is also sufficient. Although the inductance allowed is too low, a typical barrier has a permissible L/R ratio of 30.7 μH/Ω, which can still safely be used for this application.

Based on the barrier and cable in the example above the maximum cable length due to capacitance can be calculated to 670 m by subtracting the device capacitance from the barrier limit:

$$\frac{154 - 4 \times 5}{200} = 0.670 \qquad \textbf{2.2(1)}$$

FAULT DISCONNECTION

Fault disconnection electronics (FDE) is a circuit on the bus connection in the device that limits the current drawn from the network in case of a device circuitry malfunction. This prevents the entire network from being short-circuited by the failure of the electronics in a single device.

FIELDBUS INTRINSICALLY SAFE CONCEPT

FISCO was developed by PTB (Physikalisch-Technische Bundesanstalt) to simplify and maximize the utilization of networking in hazardous areas. Unlike the entity concept, in the FISCO model the cable capacitance and inductance are not considered concentrated or unprotected as long as the cable parameters are within given limits. FISCO-type barriers have no specified permitted capacitance or inductance for the same reason.

FISCO barriers have an output characteristic that is trapezoidal (Figure 2.2e), typically providing 1.8 W of output power for Exia IIC. This greater power in turn enables more devices to be connected as compared to a traditional entity barrier (Figure 2.2f).

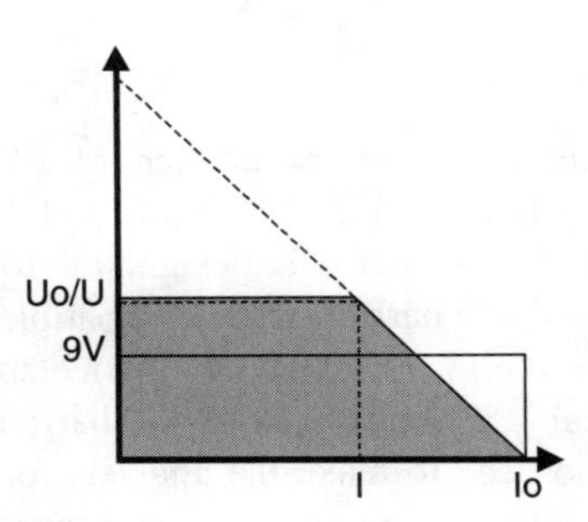

FIG. 2.2e
Trapezoidal characteristics of a FISCO barrier.

FIG. 2.2f
Isolated repeating FISCO barrier. (Courtesy of Smar.)

TABLE 2.2g
FISCO Cable Restrictions

R (loop)	15–150 Ω/km	(4.6–46 Ω/1000 ft)
L	400–1000 μH/km	(120–300 μH/1000 ft)
C	80–200 nF/km	(24–60 nF/1000 ft)

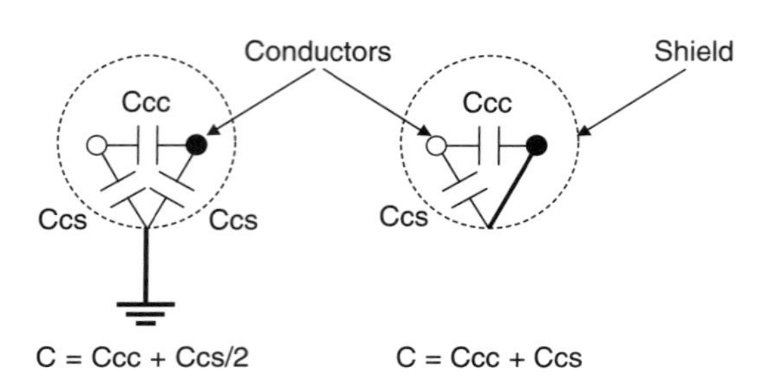

FIG. 2.2h
Computing equivalent capacitance of the cable.

Example The number of devices that can be connected to a FISCO barrier with a 100 mA-output current capacity based on a typical device power consumption of 12 mA is eight, simply calculated as:

$$\frac{100}{12} = 8 \qquad \textbf{2.2(2)}$$

The capacitance and inductance of FISCO certified devices are negligible (less than 5 nF and 10 μH, respectively). FISCO field devices are designed to tolerate the high power from a FISCO barrier. To be compatible with a typical FISCO barrier, the Pi (Pmax) of the device should be larger than the typical 1.8 W provided by the barrier.

There is no need to make the analysis for the cable and devices in the network for inductance and capacitance. Cables with parameters within the ranges specified in Table 2.2g can be used in FISCO installations for lengths up to 1 km with a maximum spur length of 30 m.

The connection of the cable shield affects the effective cable capacitance that should be calculated using the equations in Figure 2.2h. If the conductors are isolated from an earthed shield, the capacitance is lower than if the shield is connected to one of the signal conductors at the barrier.

It is necessary to select a barrier that has voltage, current, and power output lower than what is permitted for the field device with the worst-case parameter.

Example Parameters for eight typical devices are listed Table 2.2i.

The worst-case field device is higher than the output voltage, current, and power of the barrier in this example, and the current provided is sufficient. Thus, the suggested barrier can be used.

TABLE 2.2i
FISCO Example

Tag	*Ui (Vmax), V*	*Ii (Imax), mA*	*Pi (Pmax), W*	*Iq, mA*
FT-012	24	250	2	12
TT-789	24	250	2	12
LT-456	24	250	2	12
PT-123	24	250	2	12
FCV-012	24	250	2	12
TCV-789	24	250	2	12
LCV-456	24	250	2	12
PCV-123	24	250	2	12
Terminator	24	250	2	0
	min	min	min	sum
	Uo (Voc)	*Io (Isc)*	*Po (Pm)*	
Barrier requirement	24	250	2	96
Selected barrier	15	190	1.8	100

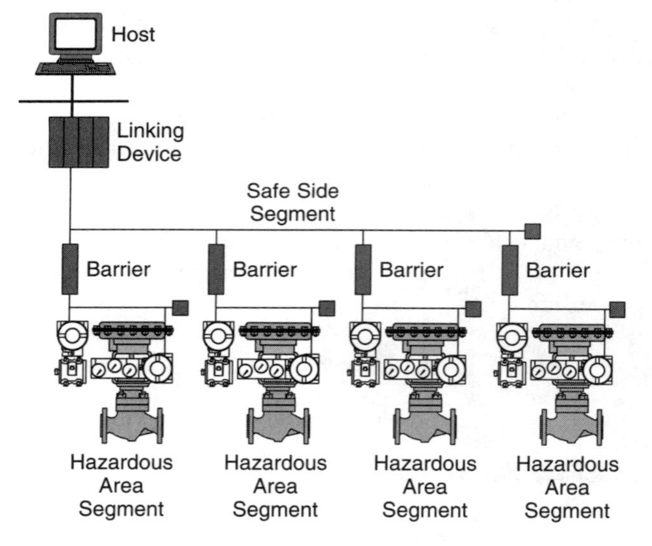

FIG. 2.2j
Network formed by four intrinsically safe and one regular segment.

The maximum cable length is calculated based on voltage drop as for a regular non-intrinsically safe installation (see Section 4.2) because there is no cable capacitance or inductance limitation as long as the specified cable parameters are followed.

For an intrinsically safe installation with eight devices using a safety barrier with 13.8 V output, the maximum distance is calculated as 1.1 km considering only the voltage drop. However, other aspects set the limit for Exia IIC as

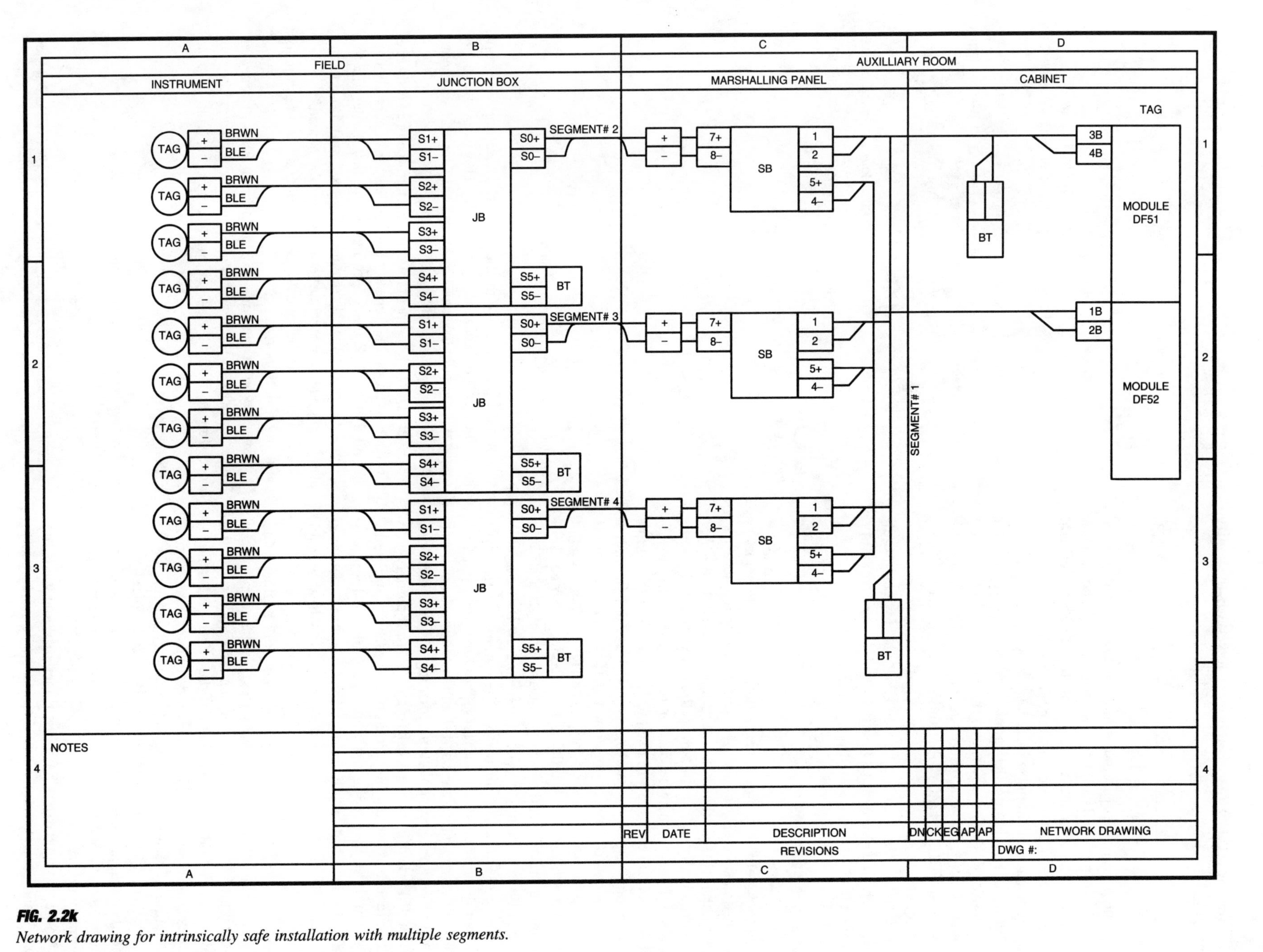

FIG. 2.2k
Network drawing for intrinsically safe installation with multiple segments.

1 km for the FISCO model and even lower for the entity concept.

$$\frac{13.8-9}{12\times 10^{3}\times 8\times 2\times 22} = 1.1 \qquad 2.2(3)$$

If the power supply output voltage is lower, or the device power consumption is higher, the distance will be shorter. Conversely, the opposite also applies. It is therefore critical for both intrinsically safe and regular installations that the device current consumption is as low as possible.

When dimensioning the number of devices per barrier the worst-case device FDE current should be taken into account. That is, leave a margin as wide as the device with the highest FDE current so that this device can fail without affecting others.

It is important that the barrier and field devices are all FISCO certified. Non-FISCO barriers and devices cannot be used in FISCO-style installations.

REPEATING BARRIERS

A fieldbus network can consist either of a single segment or of several segments joined by repeaters to form a network. Every network segment has a terminator in each end. For non-intrinsically safe installations the network usually only has a single segment, whereas for intrinsically safe installations several segments are used to form a network linking the segments together using repeaters. It would be rather uneconomical to connect only one barrier per host communication port because each safety barrier, in particular a linear barrier, connects only a few devices. Some field instruments have a current consumption so high that only two can be connected to each linear barrier. It is possible to run one safe area segment from the communication port to several multidropped barriers provided that barriers with built-in repeaters are used. Each repeating barrier has a hazardous area segment, thus forming a larger network with a full 16 devices per network (Figure 2.2j). If barriers with built-in repeaters are used, external repeaters are eliminated.

A single network diagram replaces multiple-loop diagrams for installation documentation. A typical network diagram for an intrinsically safe installation is shown in Figure 2.2k.

There is a limit to the length of a network segment due to the signal attenuation along the wires. A repeater refreshes the timing of the signal received in one end and boosts the level as the signal is regenerated in the other end using a fieldbus controller chip. Thus, the repeater overcomes signal attenuation by restoring the original symmetry and amplitude of the signal. The repeater is bidirectional and typically galvanically isolated. Repeaters may thus also be used to connect intrinsically safe segments together to form a larger network with many devices.

Nonrepeating barriers have low impedance that loads the network and also attenuates and distorts the communication signal as it passes through the barrier. Therefore, only one such barrier per network is possible. However, by using repeaters, the network can be divided into several segments, each with one barrier (Figure 2.2l).

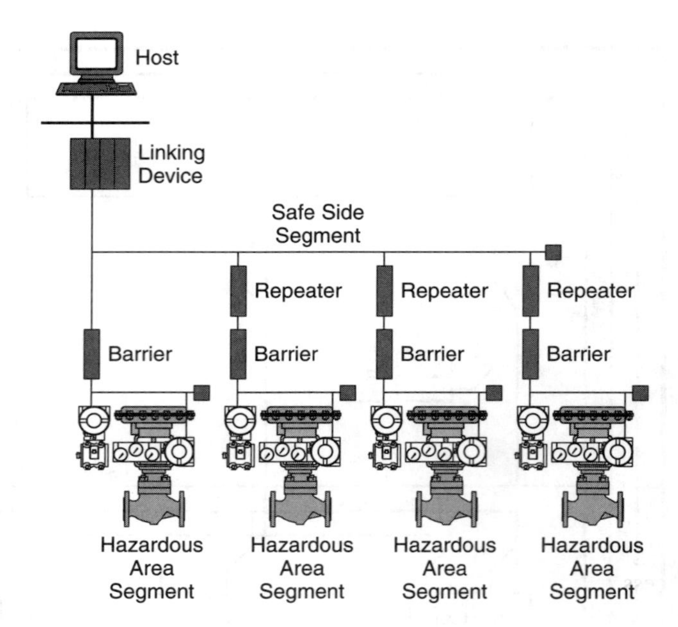

FIG. 2.2l
Repeaters with separate barriers.

FIG. 2.2m
Fieldbus repeater. (Courtesy of Smar.)

Every network segment must have a terminator in each end. However, typically the repeater already has terminators built in and all that need be done is to make sure that the terminators are enabled or disabled as required (Figure 2.2m). Power must also be supplied on separate terminals for the repeater.

It is helpful if the repeater is isolated and has a terminator built in on both ports.

Every type of network uses a different kind of repeater. Make sure to use a repeater designed with characteristics according to IEC 61158-2 and not some other kind of bus.

Bibliography

Fieldbus Foundation, "31.25 kbit/s Intrinsically Safe Systems, Application Guide AG-163," 1996.

Johannesmeyer, U., PTB report W-53e, "Investigations into the Intrinsic Safety of Field Bus Systems," Braunschweig: PTB, 1994.

PROFIBUS, "Technical Guideline, PROFIBUS-PA User and Installation Guideline," version 1.1, order No. 2.092, September 1996.

2.3 Purging ... Inerting Systems

E. L. S... (1993) H. H. WEST, M. S. MANNAN (2002)

...s piping and process vessels is nec-
...centrations of oxygen, thereby reduc-
...sion or toxic exposure hazard. Purging
...erm addition of an inert gas, such as
...oxide, steam, flue gas, etc., to a piping
...or chemical process vessel. The piping, pro-
...ank previously may have contained flammable
...es. Purging then renders the vapor space (also
... space or headspace) not ignitable for a specific
..., such as during a maintenance outage period.
...ntrast, inerting (sometimes called blanketing) is the
...ng-term maintenance of an inert atmosphere within a vapor space of a container or chemical process vessel during operation.

Purging and inerting can also be used to rid process units or piping of toxic materials.

SAFE OXYGEN LEVELS

The goal of purging and inerting techniques is to reduce the confined space oxygen concentration below the limiting oxygen concentration (LOC). The LOC is the concentration of oxygen below which a fire or explosion cannot occur for a specific chemical in the vapor space of the vessel. The LOC for common flammable gases and vapors, using nitrogen or carbon dioxide as diluent, are listed in Table 2.3a. A safety margin must be maintained between the LOC and a normal working concentration in a system. Conservative purge control practice uses at least two percentage points below the LOC. For example, if the LOC of ethanol using nitrogen as diluent is 10.5% by volume, the purge control point would be 8.5%. Some companies suggest even larger purge control margins of safety.

TABLE 2.3a
Limiting Oxygen Concentration in Air Using Nitrogen or Carbon Dioxide

	LOC Using N_2	*LOC Using CO_2*
Natural gas	12%	14.5%
Ethanol	10.5%	13%
Hydrogen	5%	5.2%
Acetone	4%	6%

SAFE TOXIC CHEMICAL LEVELS

The same techniques described here for purging or inerting flammable vapors can be extended to toxic chemical vapors. However, the concept of LOC must be replaced by a toxic vapor concentration considered safe, which is dependent on the specific chemical involved. Some highly toxic chemicals, such as nerve gases and chemical weapon agents, require part per billion purge control. These highly toxic chemicals require extraordinary purge control procedures and equipment.

TYPES OF PURGE SYSTEMS

There are several methods that can be used to achieve purge or inerting goals. The various techniques of purging and inerting are detailed here.

Siphon Purging

Siphon purging involves filling a closed container to be purged with water or some other available liquid product, followed by introducing the purge gas (such as nitrogen or carbon dioxide) into the vapor space as the liquid is drained. Required purge gas volume equals the volume of the vessel. The rate of purge gas flow corresponds to the volumetric rate of liquid discharge.

Consideration of the potential for evaporation during siphon purging is required for volatile liquids. Siphon purging may not be appropriate if the liquid is above its flash point due to evaporation into the vapor or ullage space of the container.

Vacuum Purging

Vacuum purging is one of the most common vessel inerting procedures, provided the vessel is designed for the maximum vacuum pressure that can be developed by the vacuum pump.

There are three basic steps in a vacuum purge:

1. Drawing a vacuum on the vessel
2. Relieving the vacuum with an inert gas to atmospheric pressure
3. Repeating steps 1 and 2 until the desired oxygen concentration is reached

The amount of purge gas required depends on the number of vacuum evacuations needed to develop the desired oxygen concentration. By using the ideal gas law, the volume of purge gas required is

$$V_p = ZV(P_h - P_l)/P_h$$

where

V_p = volume of the purge gas required
Z = number of purge cycles
V = volume of the vessel vapor space
P_h = highest pressure, normally atmospheric
P_l = lowest pressure

Similarly, the number of purge cycles can be estimated by

$$Z = \ln(C_l/C_h)/\ln(P_l/P_h) \qquad \textbf{2.3(1)}$$

where

C_h = concentration original
C_l = concentration final

Pressure Cycle Purging

Piping or closed containers may be purged by adding an inert gas under pressure. Then, after the gas has sufficiently mixed throughout the vapor space, it can be vented to atmosphere.

More that one pressure cycle may be necessary to attain the desired LOC level. The same equations developed for vacuum purging can be used for pressure purging.

Sweep-through Purging

Sweep-through purging involves introducing the purge gas into a vessel at one opening and withdrawing the mixed purge gas and vapors at another opening with the exit gases vented to the atmosphere or to an air pollution control device. This method depends on sweeping out a residual flammable vapor. The quantity of purge gas required depends upon the physical arrangement. Sweep-through purging is commonly used when the vessel is not ready for pressure or vacuum cycling.

The volume of purge gas required is estimated by the following equation:

$$V_p = V/K[\ln(C_l/C_h)] \qquad \textbf{2.3(2)}$$

where

V_p = volume of the purge gas required
K = mixing factor
V = volume of the vessel vapor space
C_h = concentration final
C_l = concentration original

The mixing factor is detailed in NFPA 96,[6] with the default value of 0.25 used for most situations.

Fixed-Rate Continu

Fixed-rate purging inv
into a vessel at constant r
of a mixture of original va
the vessel or ullage headspac
purge is typically controlled b
rate considering potential tempera

The maximum temperature cha
tanks operating at or near atmospheric
the sudden cooling from a barometric
summer thunderstorm. The flow rate of in
to prevent the vessel from falling significan
spheric pressure is determined by the rule of th

two standard cubic feet per hour (2 SCFH) of in
per square foot of tank surface, including both she
roof surfaces

Figure 2.3b presents a simple method of flow control for a continuous introduction of purge gas.

Variable-Rate Continuous Purging

Variable-rate (sometimes called demand) purging involves feeding purge gas into the piping or closed container at a rate that is a function of demand. Demand is essentially based on maintaining a certain identified pressure within the vessel that is slightly above the surrounding atmosphere, typically 1 or 2 in. of water pressure in an atmospheric tank.

Variable-rate purging offers an obvious advantage over continuous applications because the inert gas is supplied only when it is actually needed, thereby reducing the total quantity

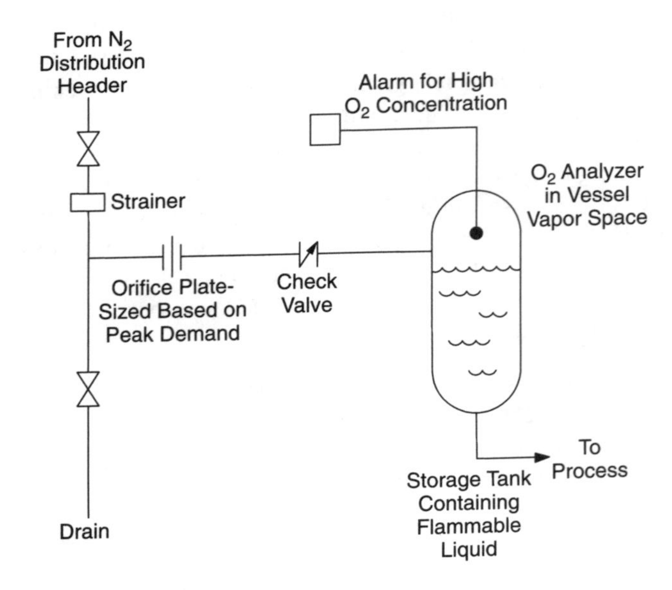

FIG. 2.3b
Simple method of flow control for continuous introduction of purge gas.

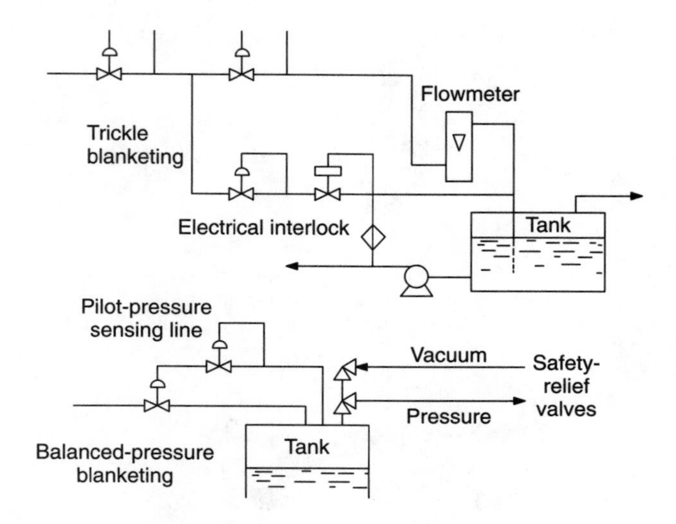

FIG. 2.3c
Methods for variable-rate purging of closed containers and tanks: trickle blanketing and balanced-pressure blanketing.

of inert gas requirements and product losses and avoiding disposal problems. This advantage is dependent on flow control devices actually at very low pressure differentials that are sometimes difficult to maintain and are sometimes questionable in their accuracy. For variable-rate applications, a purge control system is required to maintain a pressure in the vapor space above the liquid. This system should include an automatic inert gas addition valve or backpressure regulator to control the oxygen concentration below the LOC.

The control system should feature an analyzer to monitor the oxygen concentration continuously and allow inert gas to enter the space to maintain the oxygen concentration at safe levels within a reasonable margin of safety. An increase in the concentration above the safe point should initiate an alarm and shut down the operation. When oxygen concentration cannot be continuously monitored, the purge control system should be designed to operate at no more than 50% of the LOC and checked or inspected on a regular schedule.

Figure 2.3c depicts two methods that can be used for variable-rate purging of closed containers and tanks: trickle blanketing and balanced pressure blanketing.

PURGE FLOW REGULATORS*

Applications: Low flow regulation for air bubblers, for purge protection of instruments, for purging electrical housings in explosion-proof areas, and for purging the optical windows of smokestack analyzers

Purge Fluids: Air, nitrogen, and liquids

Operating Pressure: Up to 450 PSIG (3 MPa)

Operating Temperaure: For glass tube up to 200°F (93°C)

Ranges: From 0.01 cc/min for liquids and from 0.5 cc/min and higher for 1/4 in. (6 mm) glass tube rotameter can handle 0.05 to 0.5 GPM (0.1 pm) of water or 0.2 to 2 SCFM (0.3 to 3 cmph) of air

Inaccuracy: Generally 2 to 5% of range (laboratory units are more accurate)

Costs: A 150 mm glass-tube unit with 1/8 in. (3 mm) threaded connection, 316 stainless steel, and 16-turn high-precision valve is $260; the same with aluminum frame and standard valve is $100. Adding a differential pressure regulator of brass or aluminum construction costs about $150 (of stainless steel, about $500). For highly corrosive services, all-Teflon, all-PPFA and all-CTFA units are available which, when provided with valves, cost $550 with 1/4 in. (6 mm) and $1300 with 3/4 in. (19 mm) connetions

Partial List of Suppliers: Aaborg Instruments & Controls, Inc.; Blue White Industries; Brooks Instrument, Div. of Rosemount; Fischer & Porter Co.; Fisher Science; Flowmetrics, Inc.; ICC Federated, Inc.; Ketema, Inc. Schutte and Koerting Div.; Key Instruments; King Instrument Co.; Krone America, Inc.; Matheson Gas Products, Inc.; Omega Engineering, Inc.; Porter Instrument Co., Inc.; Scott Specialty Gasews; Wallace & Tiernan, Inc.

Purge flows are low flow rates of either gases or liquids. They usually serve to protect pressure taps from contacting hot or corrosive process fluids or from plugging. They can also protect electrical devices from becoming ignition sources by maintaining a positive inert gas pressure inside their housings or to protect the cleanliness of the optics of analyzers through purging.

The low flow rates of the purge media can be detected by a variety of devices including capillary, miniature orifice, metering pump, positive displacement, thermal, and variable-area-type sensors. Capillary flow elements are ideal for the measurement of low flow rates and are frequently combined with thermal flowmeters to provide a high-precision, high-rangeability, but also higher-cost flow regulator. Integral orifices can be used on both gas and liquid flow measurement, whereas positive displacement meters are most often used to detect the flow of liquids. The purge flow of liquids can be both measured and controlled by metering pumps. In addition, Volume 2 of the *Instrument Engineers' Handbook*, third edition, includes a complete section devoted to flow regulators.

There is only one type of purge flow regulator that has not been covered in other parts of this reference set: the rotameter-type purge meter. This is the least expensive and most widely used purge meter, and for these reasons it is described in the separate section that follows.

* This material was prepared by E. L. Szonntagh as Section 2.18 of Volume 1 of the *Instrument Engineers' Handbook* and is reprinted from there.

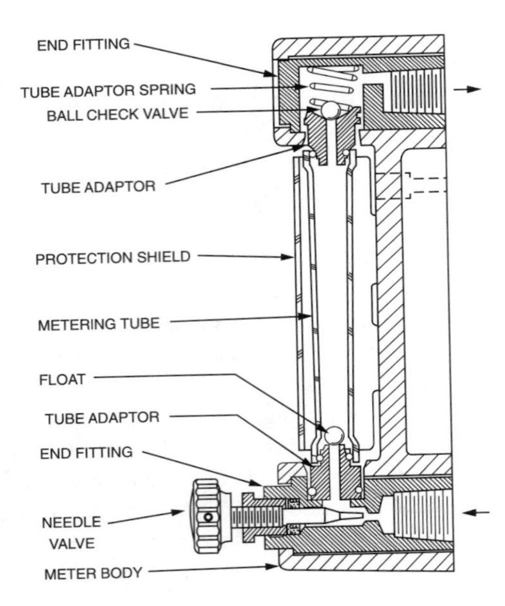

FIG. 2.3d
Purge rotameter with integral needle valve.

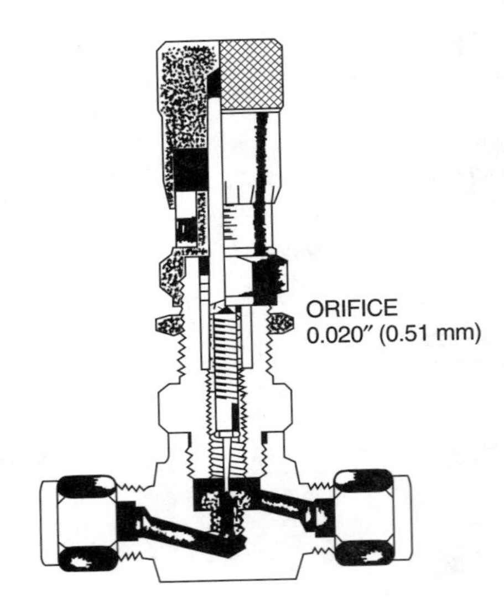

FIG. 2.3f
Fine-adjustment needle valve with vernier scale. (Courtesy of Swagelok Co.)

TABLE 2.3e
Gas Properties under the Standard Conditions of 29.92 in. of Mercury and 70°F (760 mm of Mercury and 21°C)

Gas	Density (lb/ft^3)	μ Viscosity (micropois)	Specific Gravity
Air	0.0749	181.87	1.000
Argon	0.1034	225.95	1.380
Helium	0.0103	193.9	0.138
Hydrogen	0.0052	88.41	0.0695
Nitrogen	0.0752	175.85	0.968
Oxygen	0.0828	203.47	1.105
Carbon dioxide	0.1143	146.87	1.526

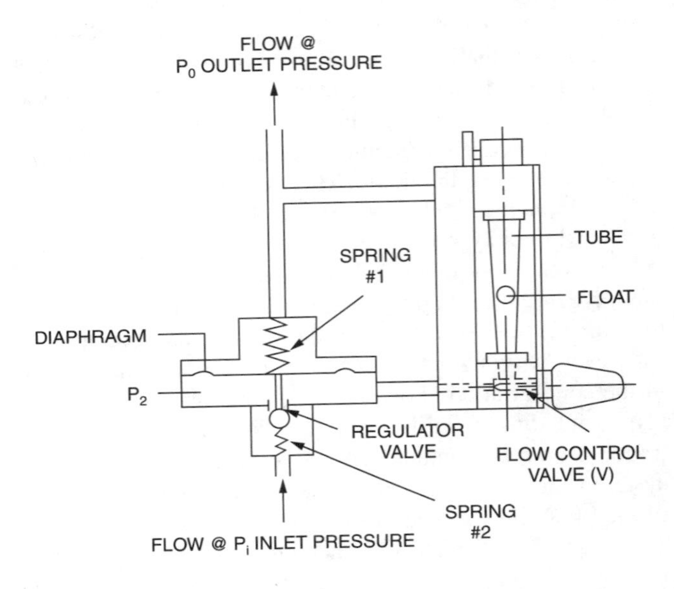

FIG. 2.3g
Purge flow regulator consisting of a glass tube rotameter, an inlet needle valve, and a differential pressure regulator. (Courtesy of Krone America, Inc.)

Purge Rotameters

These are perhaps the most widely used flowmeters and certainly are the most widely used form of the rotameter. These meters take many forms, all of which are inexpensive, and are intended for low flow measurement. Most purge meters are selected to handle inert gases or liquids at low flow rates where these fluids are used as a purge; therefore, accuracy is not critical. Repeatability is normally the required performance characteristic. Purge meters are available with optional needle control valves. Figure 2.3d shows a typical purge-type rotameter with integral needle control valve.

The metering needle valves are usually multiple-turn units provided with long stems. The opening around their needle-shaped plugs can approach capillary dimensions. The flow rate through these devices is a function not only of the opening of the valve and the pressure differential across it, but also of both the density and the viscosity of the purge media. Table 2.3e provides information on the density and viscosity of a number of purge gases. Figures 2.3f shows a high-precision needle valve, which is provided with a vernier-type scale that allows a more accurate setting of the valve opening. The dual scale increases the precision and reproducibility of setting by subdividing the smallest reading of the first scale onto the second.

When the purge flowmeter is combined with a differential pressure regulator, it becomes a self-contained flow controller (Figure 2.3g). By adjusting springs 1 and 2 for a constant

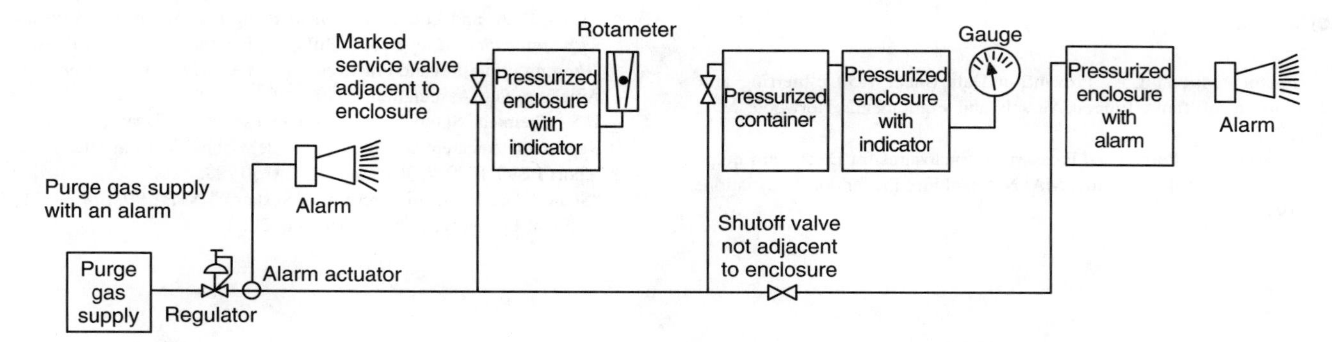

FIG. 2.3h
Example 1: Shows indicators may be used if purge gas supply has an alarm and the shutoff valve is adjacent to the enclosure.

pressure difference of about 60 to 80 in. of water (150 to 200 cm of water) this constant pressure drop ($P_2 - P_o$) is maintained across the flow control valve (V) and the purge flow is thereby fixed. Figure 2.3g describes a configuration in which the outlet pressure (P_o) is constant and the inlet pressure P_i is variable. Units are also available for bubbler and purge applications where the inlet pressure P_i is constant and the outlet P_o is variable. In that case the constant pressure drop across the valve (V) is maintained to equal ($P_i - P_2$). Purge flow controllers on gas service are usually provided with a range of 0.2 to 2 SCFH (6 to 60 slph), an accuracy of 5% of full scale over a range of 10:1, a pressure rating of 150 to 300 PSIG (1 to 2 MPa), and a maximum temperature limit of 212 to 572°F (100 to 300°C).

Purged and Pressurized Enclosures

National Fire Prevention Association (NFPA) Standard 496 on Purged and Pressurized Enclosures for Electrical Equipment[2] provides specific requirements for purging electrical equipment. Special requirements for process instrumentation, such as process analyzers, are also included.

Three levels of inert gas pressurization of electrical equipment are defined based upon electrical hazard classification:

1. Type Z Pressurization (for Division 2)
2. Type Y Pressurization (for Division 1)
3. Type X Pressurization (for special condition Division 1)

These three pressurization concepts are detailed throughout the NFPA 496 standard. An example of purge gas configuration systems is shown in Figure 2.3h.

Safety Considerations

Purge gases from an enclosed vessel being purged must be vented to a safe location, because incidents have occurred were nitrogen purges have caused asphyxiation.[5]

Many workers have been accidentally exposed to inert gases. If a person enters an atmosphere of nitrogen, he or she can lose consciousness without any warning symptoms in as little as 20 s; death can follow in 3 to 4 min.

Whenever nitrogen or any other asphyxiating gases are used for inerting or purging, strict adherence is required for confined space entry procedures, personal protective equipment, and training.

Purge System Design Considerations

Although nitrogen and carbon dioxide are the most common inert purge gases, steam is sometimes used. If so, steam must be supplied at a rate sufficient to maintain the vessel temperature greater than 160°F. Care must be taken to avoid condensation by cooling, which could collapse the closed vessel by implosion. This 160°F temperature limit is based on experience and is considered sufficient to prevent condensation of diluent steam.

Possible sources of nitrogen purge gases include commercially available high-pressure cylinders, on-site air separation plants that remove oxygen from air and recover nitrogen by liquefaction or pressure swing adsorption, and cryogenic tanks containing nitrogen and an associated vaporizer. Carbon dioxide may also be available from commercial pressurized cylinders or from flue gas.

Cross connections between the inert purge gas source and any other system must not be allowed.

The following factors should be considered in the design of any system to reduce oxygen concentrations to safe levels:

- Required reduction in oxygen concentration
- Variations in process, process temperature, process pressure, and chemicals
- Purged gas supply source and equipment installation
- Compatibility of the purged gas with the process
- Operating controls
- Maintenance, inspection, and testing
- Leakage of purged gas to surrounding areas
- Use of breathing apparatus by personnel

References

1. Factory Mutual Loss Prevention Data Sheet 7-59, "Inerting and Purging of Tanks Process Vessels and Equipment," Norwood, MA, 1977.
2. "Standard on Purged and Pressurized Enclosures for Electrical Equipment," NFPA 496, Boston, MA: National Fire Prevention Association, 1998.
3. Crowl, D. A. and Louvar, J., *Chemical Process Safety Fundamentals with Applications,* Englewood Cliffs, NJ: Prentice-Hall, 1990, 195–200.
4. "Purging Principles and Practices," report XK0775, Washington, D.C.: American Gas Association, 1990.
5. U.S. Chemical Safety and Hazard Investigations Board, "Summary Report on Accident at Union Carbide Hahnville Louisiana Plant," report PB99-159972, Washington, D.C., 1998.
6. "Standard on Explosion Prevention Systems," NFPA 96, Boston, MA: National Fire Prevention Association, 2000.

2.4 High-Integrity Pressure Protection Systems

A. E. SUMMERS

Features:	Instrumented systems that are typically designed to meet safety integrity level 3 per ANSI/ISA 84.01-1996 and IEC 61511
Purpose:	To mitigate identified overpressure scenarios
HIPPS-Related Codes, Standards, and Recommended Practices:	American Society of Mechanical Engineers (ASME), "Boiler and Pressure Vessel code, Section VIII—Pressure Vessels," United Engineering Center, New York, NY
	American Petroleum Institute (API), RP 521, "Guide for Pressure Relieving and Depressuring Systems," Washington, D.C.
	Instrumentation, Systems, and Automation Society (ISA), ANSI/ISA S84.01-1996, "Application of Safety Instrumented Systems (SIS) for the Process Industry," Research Triangle Park, NC
	International Electrotechnical Commission (IEC), IEC 61508, "Functional Safety of Electrical/Electronic/Programmable Electronic Safety Related Systems," Geneva, Switzerland
	International Electrotechnical Commission (IEC), IEC 61511, "Functional Safety: Safety Instrumented Systems for the Process Sector," Geneva, Switzerland

In the process industry, an important safety consideration is the prevention of loss of containment due to vessel or pipeline overpressure situations. Loss of containment can result in impact to human life and the environment, when flammable, explosive, hazardous, or toxic chemicals are released to the atmosphere. Loss of containment can also result in economic impact due to production unit replacement/repair costs and production losses.

Industry standards from the American Petroleum Institute (API) and American Society of Mechanical Engineers (ASME) provide criteria for the design and protection of vessels and pipelines from rupture or damage caused by excess pressure. In conventional designs, pressure relief devices, such as pressure relief or safety valves, are used as the primary means of pressure protection. The design of each pressure relief device is based on the assessment of overpressure scenarios, such as typically experienced with the total loss of cooling or power supply.

Conventional pressure relief system design, including relief header and flare sizing, does not examine the reduction in potential loading due to hazard mitigation provided by operator response to alarms or to the initiation of instrumented systems, including basic process control systems (BPCS) or safety instrumented systems (SIS). In fact, until 1996, the ASME codes mandated the use of pressure relief devices for protection of pressure vessels.

However, in some applications, the use of pressure relief devices is impractical. Typical cases include:

- Chemical reactions so fast the pressure propagation rate could result in loss of containment prior to the relief device opening; examples are "hot spots," decompositions, and internal detonation/fires
- Chemical reactions so fast the lowest possible relieving rate yields impractically large vent areas
- Exothermic reactions occurring at uncontrollable rates, resulting in a very high propagation rate for the process pressure (the pressure propagation rate for these reactions is often poorly understood)
- Plugging, polymerization, or deposition formed during normal operation, which have historically partially or completely blocked pressure relief devices
- Reactive process chemicals relieved into lateral headers with polymerization and thus plugging, rendering the relief device useless
- Multiphase venting, where actual vent rate is difficult to predict
- Pressure relief device installation creates additional hazards, due to its vent location

In such applications, the installation of the pressure relief device provides minimum risk reduction. Consequently, other methods of preventing overpressure must be utilized to achieve measurable risk reduction.

Adding to the complexity, in many countries around the world, there is increased pressure from community and regulatory authorities to reduce venting and combustion of gases. In these countries, it is now unacceptable to flare large volumes of gas. The need to balance safety requirements and environmental requirements has resulted in increased focus on using an alternative approach to pressure protection.

Fortunately, API 521 and Code Case 2211 of ASME Section VIII, Division 1 and 2, provide an alternative to pressure relief devices—the use of an instrumented system to protect against overpressure. When used, this instrumented system must meet or exceed the protection provided by the pressure relief device. These instrumented systems are SIS, because their failure can result in the release of hazardous chemicals or the creation of unsafe working conditions. As SISs, they must be designed according to the U.S. standard ANSI/ISA 84.01-1996 or the international standard IEC 61511. The risk typically involved with overpressure protection results in the need for high SIS integrity; therefore, these systems are often called high-integrity pressure protection systems (HIPPS) or high-integrity protection shutdowns (HIPS).

CODE REQUIREMENTS

Until August 1996, ASME required the use of pressure relief devices for pressure vessels designed in accordance with Section VIII, Division 1, para UG-125(a) Section VIII, Division 2, para, AR-100. The approval of ASME Code Case 2211 in August 1996 changed this position by defining the conditions for which overpressure protection may be provided by a SIS instead of a pressure relief device.

The new ruling is designed to enhance the overall safety and environmental performance of a facility by utilizing the most appropriate engineered option for pressure protection. Although no specific performance criteria is included in Code Case 2211, the use of HIPPS must result in an installation as safe or safer than the conventional design. The overpressure protection can be provided by an SIS in lieu of a pressure relief device under the following conditions.

1. The vessel is not exclusively in air, water, or steam service.
2. The decision to utilize overpressure protection of a vessel by system design is the responsibility of the user. The manufacturer is responsible only for verifying that the user has specified overpressure protection by system design, and for listing Code Case 2211 on the Data Report.
3. The user must ensure the maximum allowable working pressure (MAWP) of the vessel is higher than the highest pressure that can *reasonably* be achieved by the system.
4. A quantitative or qualitative risk analysis of the proposed system must be made, addressing credible overpressure scenarios, and demonstrating the proposed system is independent of the potential causes for overpressure, is as reliable as the pressure relief device it replaces, and is capable of completely mitigating the overpressure event.
5. The analysis conducted for (3) and (4) must be documented.

RECOMMENDED PRACTICES

API recommends a number of practices for addressing pressure relieving and depressuring systems in the petroleum production industry. For example, API 521 describes flare system design methods that require assessing the relief load based on credible overpressure scenarios for relief valve sizing and on the simultaneous venting of all affected vessels for main flare header sizing. The fourth edition of API 521 allows taking credit for a favorable response of the instrumented systems. Despite API 521 permitting design alternatives, API 521 Part 2.2 recommends the use of HIPPS only when the use of pressure relief devices is *impractical*.

STANDARDS

The international standard, IEC 61508, "Functional Safety of Electrical/Electronic/Programmable Electronic Safety Related Systems," establishes a framework for the design of instrumented systems that are used to mitigate safety-related risks. The U.S. standard, ANSI/ISA 84.01-1996, "Application of Safety Instrumented Systems (SIS) for the Process Industry," and the international standard, IEC 51611, "Functional Safety: Safety Instrumented Systems for the Process Sector," are intended to address the application of SISs in the process industries.

The objective of these standards is to define the assessment, design, validation, and documentation requirements for SISs. Although these design standards are not prescriptive in nature, the design processes mandated by these standards cover all aspects of design, including risk assessment, conceptual design, detailed design, operation, maintenance, and testing. Because HIPPS is a type of SIS, the requirements of these standards, as pertaining to each specific HIPPS application, must be investigated and applied thoroughly.

The SIS standards are performance based, with the safety integrity level (SIL) as the primary performance measurement. The SIL must be assigned by the user based on the risk reduction necessary to achieve the user's risk tolerance. It is the user's responsibility to ensure consistent and appropriate SIL assignments by establishing a risk management philosophy and

TABLE 2.4a
Safety Integrity Levels

SIL IEC 61508	*SIL ANSI/ISA S84*	*PFD*
1	1	0.1 to 0.01
2	2	0.01 to 0.001
3	3	0.001 to 0.0001
4	Not applicable	0.0001 to 0.00001

risk tolerance. The risk reduction provided by the HIPPS is equivalent to the probability of failure on demand attributable to all of the HIPPS devices from the sensor through the logic solver and final elements. The relationship between the SIL and probability to fail on demand (PFD) is shown in Table 2.4a.

The SIL establishes a minimum required performance for the HIPPS. The SIL is affected by the following:

1. Device integrity determined by documented and supportable failure rates
2. Redundancy and voting using multiple devices to ensure fault tolerance
3. Functional testing at specific intervals to determine that the device can achieve the fail-safe condition
4. Diagnostic coverage using automatic or online methods to detect device failure
5. Other common causes including those related to the device, design, systematic faults, installation, and human error

Because the criteria used to establish the SIL affect the entire life cycle of the HIPPS, the SIL forms the cornerstone of the HIPPS design.

HIPPS JUSTIFICATION

A decision tree can be utilized to facilitate the justification for HIPPS in the process industry. Figure 2.4b is a simplified decision tree showing the key steps in assessing and designing a HIPPS.

The successful implementation of HIPPS requires examination of applicable regulations and standards, including local and insurer codes that may mandate the use of pressure relief devices. From ASME Code Case 2211, the vessel cannot be exclusively in air, water, or steam service. This exclusion is intended to prevent building utility systems (e.g., residential boilers) from being installed without pressure relief devices. API 521 recommends the use of HIPPS only when the use of a pressure relief device is impractical. It is the user's responsibility to establish the definition of "impractical." Any applicable regulation or standard should be reviewed during early project front-end loading to ensure that the HIPPS approach is acceptable.

ASME Code Case 2211 requires a qualitative or quantitative risk analysis of the potential overpressure scenarios. This hazard or risk analysis is initiated when the process design is sufficiently complete to allow identification of the causes and magnitude of potential overpressure events. Typically, this means the following information is available:

- Process flow diagrams
- Mass and energy balances
- Equipment sizing and design requirements
- Piping and instrumentation diagrams (P&IDs)

Sometimes, the design team wants to wait until the design is essentially complete before examining the pressure relief requirements and the HIPPS design. Unfortunately, this is the most expensive point to be making instrumentation modifications. For example, when design is declared as complete, piping and instrumentation diagrams (P&IDs) have been finished, specified equipment MAWP are fixed, long-delivery items have been ordered, and field installation drawings are nearing completion. This is *not* the time to determine that a HIPPS must be installed. Installing a HIPPS at this point would require redundancy in field inputs and outputs, a redundant logic solver, and online testing facilities, all of which have significant impact on project documentation, schedules, and budget.

The hazard analysis should follow a structured, systematic approach, using a multidisciplinary team consisting of representatives from process engineering, research and development, operations, health and safety, instrumentation and electrical, and maintenance. Typical hazard analysis approaches include "what-if" analysis, "what-if"/checklist analysis, hazard and operability study (HAZOP), failure modes, effects, and criticality analysis (FMECA), fault tree analysis (FTA), or event tree analysis (ETA).

The hazard analysis examines operating (e.g., start-up, shutdown, and normal operation) and upset conditions that result in overpressure. The "Causes of Overpressure" provided in API Recommended Practice 521 should be reviewed to ensure completeness of the hazard analysis. For example, the hazard analysis should examine the following initiating causes for overpressure events:

- Loss of utilities, such as electric power, steam, water, etc.
- Runaway reactions
- Fire exposure
- Operating errors
- Maintenance errors
- Block outlet
- Equipment failures
- Instrumentation malfunctions

The hazard analysis should document the propagation of each potential overpressure event from the initiating cause to the final consequence (also referred to as the "overpressure scenario").

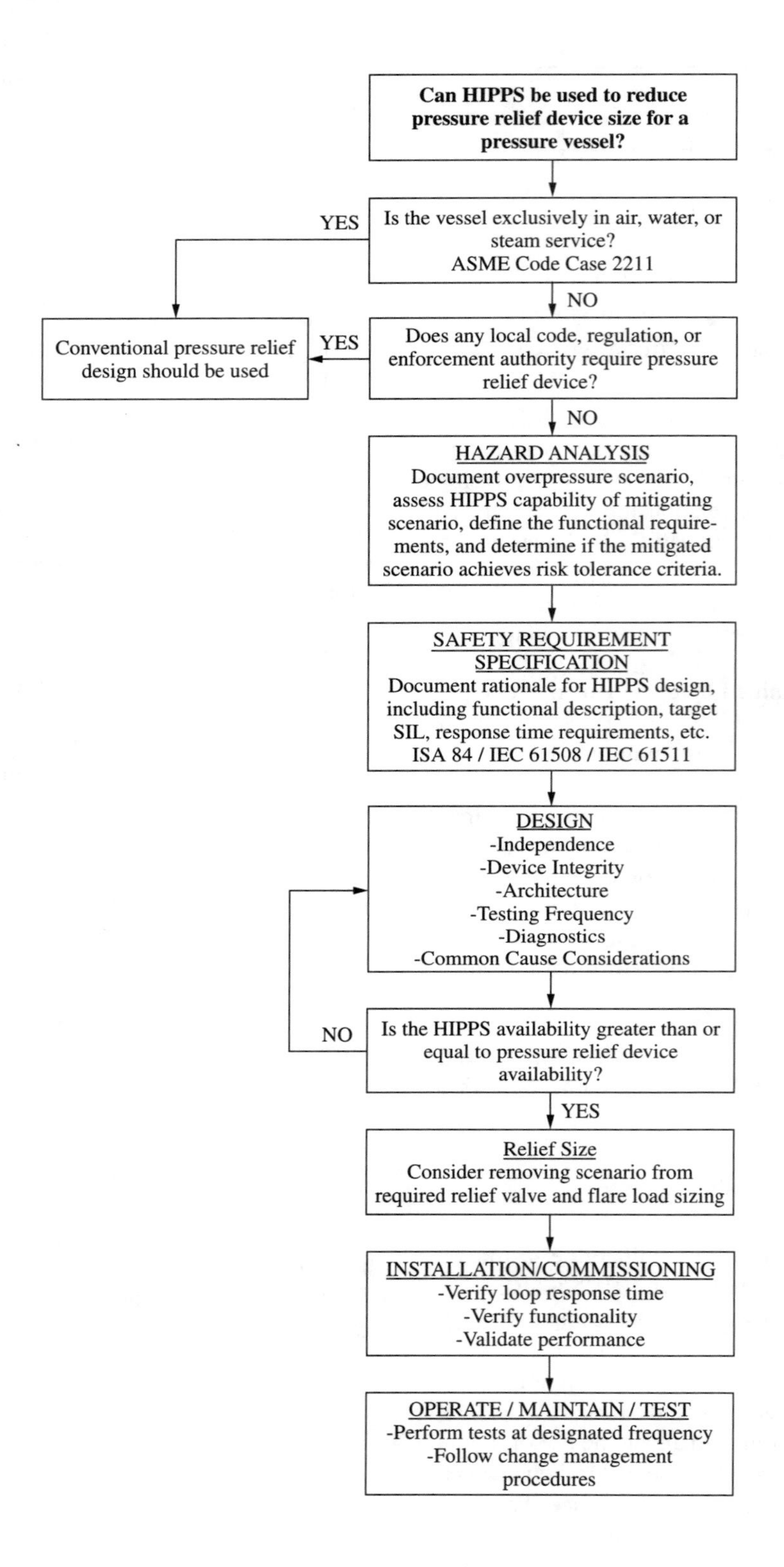

FIG. 2.4b
Simplified design tree.

ASME Code Case 2211 requires that the user ensure that the MAWP per Section VIII, Division I, para UG-98 of the vessel is greater than the highest pressure that can reasonably be achieved by the system. "Reasonably be achieved" is not defined in Code Case 2211. However, many users define "reasonably be achieved" utilizing documented risk tolerance criteria. The risk of each overpressure scenario is evaluated in terms of frequency and consequence. During the hazard analysis, the mitigated frequency of each overpressure scenario is determined by assessing the initiating cause frequency and risk reduction provided by any independent protection layers, such as HIPPS. If the risk, as defined by the mitigated frequency and consequence, achieves or is below the risk tolerance criteria, the scenario is considered for removal from the relief device and flare loading calculations.

When it is determined that an overpressure scenario achieves the risk tolerance criteria, the assessment team must ensure that the documentation adequately describes the justification for the team's decision prior to removing the scenario from the sizing calculations for the pressure relief device or the flare system. Careful consideration must be given when removing a scenario, because if the assumptions used to justify the removal are incorrect, the vessel may be left unprotected or underprotected, resulting in loss of containment and a possible hazardous situation.

SAFETY REQUIREMENT SPECIFICATION

A safety requirement specification (SRS) must be developed to address each overpressure scenario that will be mitigated using HIPPS. The SRS describes how and under what conditions the SIS will mitigate each overpressure scenario, including a functional logic description with trip set points and device fail-safe state. Only those scenarios that can be successfully mitigated by the SIS can be *considered* for removal from the pressure relief and flare loading calculations. For example, in hydrocarbon applications, the fire case scenario often cannot be removed from the sizing calculations due to the inability of HIPPS to mitigate the cause of overpressure.

When specifying the process performance of HIPPS, the process dynamics must be evaluated to ensure that the HIPPS response time is fast enough to prevent overpressure of the vessel. The response time must be evaluated by considering the time it takes to sense that there is an unacceptable process condition; the scan rate and data processing time of the logic solver; and initiation of the final element. For general process industry applications, HIPPS valves are typically specified to have closure times of less than 5 s. However, the actual required closure must be determined for each installation. The valve specification must include acceptable leakage rate, since this affects downstream pressures and relief loading. The valve specification must also ensure that the actuator provides sufficient driving force to close the final element under the worst-case, upset pressure condition.

In addition to the safety functional requirements, the SRS also includes documentation of the safety integrity requirements, including the SIL and anticipated testing frequency. At a minimum, the target SIL for the HIPPS should be equivalent to the performance of a pressure relief device. Reliability information for a single-valve relief system is provided in "Guidelines for Process Equipment Reliability Data" by the Center for Chemical Process Safety. The data in Table 2.4c indicate that for spring-operated pressure relief devices, the minimum target SIL should be SIL 3, which is equivalent to a probability to fail on demand in the range of 1E-03 to 1E-04. When using a pilot-operated pressure relief device, the minimum target SIL should be SIL 2, which is equivalent to a PFD in the range of 1E-02 to 1E-03. Because of the range of PFD, many users choose to design HIPPS at a target SIL 3, regardless of the pressure relief device type.

TABLE 2.4c
Pressure Relief Device Failure to Open on Demand

Pressure Relief Device Type	*Failure to Open on Demand*		
	Lower	*Mean*	*Upper*
Spring operated	7.9E-06	2.12E-04	7.98E-04
Pilot operated	9.32E-06	4.15E-03	1.82E-02

The SRS must also specify exactly how the HIPPS will be configured to achieve the target SIL. The high availability requirements for HIPPS drive the choices made concerning device integrity, diversity, redundancy, voting, common cause concerns, diagnostic requirements, and testing frequency.

DEVICE INTEGRITY AND ARCHITECTURE

It is important to recognize that the HIPPS includes all devices required to reach the desired fail-safe condition for the process. The HIPPS includes the entire instrument loop from the field sensor through the logic solver to the final elements, along with other devices required for successful SIS functioning, such as SIS user interfaces, communications, and power supplies. For example, if the final elements are air-to-move valves and the safe action requires valve closure, instrument air availability must be considered when determining the overall HIPPS availability. Because all devices used in HIPPS contribute to the potential probability of failure on demand for the HIPPS, the structure of the instrumented loop must be defined and evaluated as a system to verify that the entire loop meets SIL requirements. A brief discussion of SIS devices follows.

Process Sensors

The process variables (PV) commonly measured in HIPPS are pressure, temperature, and flow. Traditionally, these variables were monitored using discrete switches as the input sensor to the SISs. Switches worked well for three reasons: (1) most trip conditions are discrete events, i.e., a high pressure, high temperature, or low flow; (2) relay systems and early programmable logic controllers (PLCs) processed discrete signal much more easily than analog signals; and (3) switches were usually less expensive than analog transmitters.

The evolution of PES (programmable electronic system) technology has made it easy to use analog PV inputs. The use of transmitters to measure these variables is now preferred over the use of switches. Switches only give a change in output when they are activated and can "stick" or experience some other failure mode that is revealed only when the switch

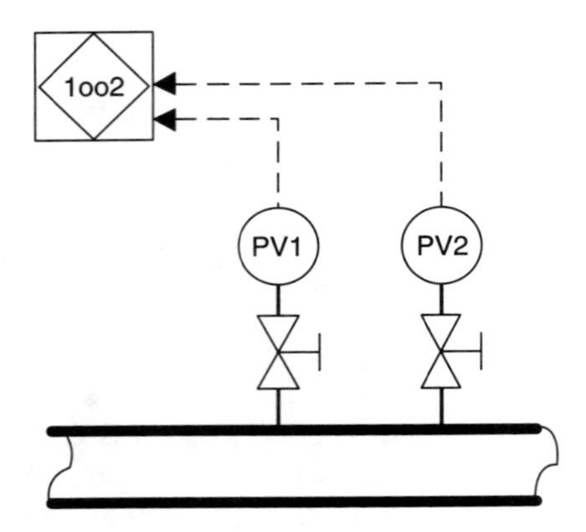

FIG. 2.4d
Installation illustration of 1oo2 field input devices.

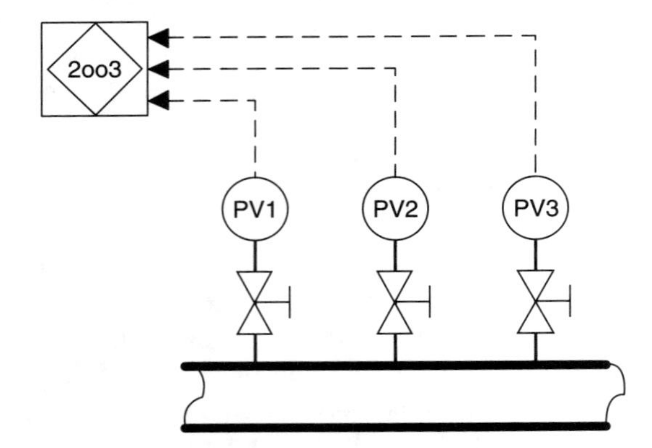

FIG. 2.4e
Installation illustration of 2oo3 field input devices.

is tested or a demand is placed on it. Transmitters can be continuously monitored and the operability of the transmitters readily observed. A single transmitter providing multiple levels of trip/alarm functions (i.e., low, high, and high-high level) can replace multiple switches. With transmitter redundancy employed, out-of-range or deviation alarming can be implemented to ensure a high level of availability.

Most HIPPS applications require 1oo2 or 2oo3 transmitters on all field inputs. Figures 2.4d and 2.4e provide illustrations of typical installations. The use of redundant inputs enables the system designer to incorporate diagnostics into the HIPPS, which significantly reduces the probability to fail on demand for the field inputs. Separate process connections are also recommended to decrease common cause faults, such as plugged process taps. Utilizing diversity in the process variable measurement, where practical, is also recommended to reduce common cause failures and consequently the PFD.

Logic Solver

The logic solver hardware must be designed to meet the assigned SIL. Since many HIPPS are designated as SIL 3, the logic solver is specified to be compliant with SIL 3 performance requirements, as provided in IEC 61508. The logic solver can be relays, solid state, or PES. If a PES is used, the selected PES should provide a high level of self-diagnostics and fault tolerance. Redundancy in the I/O modules and main processor is necessary to meet fault tolerance requirements. The outputs must also be configured as deenergize to trip to ensure the HIPPS works under loss of power.

ANSI/ISA 84.01-1996, IEC 61508, and IEC 61511 require that the safety logic be independent of the basic process control system logic. Adequate independence of the safety logic reduces the probability that a loss of the basic process control system hardware will result in the loss of HIPPS functioning. From a software standpoint, independence also reduces the possibility that inadvertent changes to the HIPPS safety functionality could occur during modification of basic process control functions.

Final Elements

The majority of HIPPS utilize dual devices in a 1oo2 configuration. The final elements are typically either (1) relays in the motor control circuit for shutdown of motor operated valves, compressors, or pumps or (2) fail-safe valves opened or closed using solenoids in the instrument air supply.

Figures 2.4f, 2.4g, and 2.4h provide illustrations of typical installations when fail-safe valves are used as the final elements. At least one of the valves must be a dedicated shutdown valve. The second valve can be a control valve, but it must be configured fail-safe, have no minimum stops, and its actuation must be controlled by the HIPPS logic solver. The system designer should also examine the initiating cause for the various scenarios to be mitigated using the HIPPS. If the initiating cause for the overpressure scenario is the failure of the control valve, the system designer should strongly consider providing the redundant isolation using two block valves rather than using the control valve and a block valve.

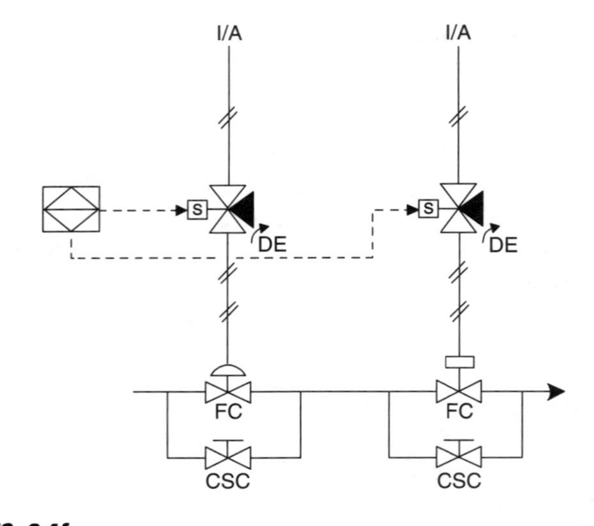

FIG. 2.4f
Installation illustration for final elements showing 1oo2 valves and 1oo1 solenoids.

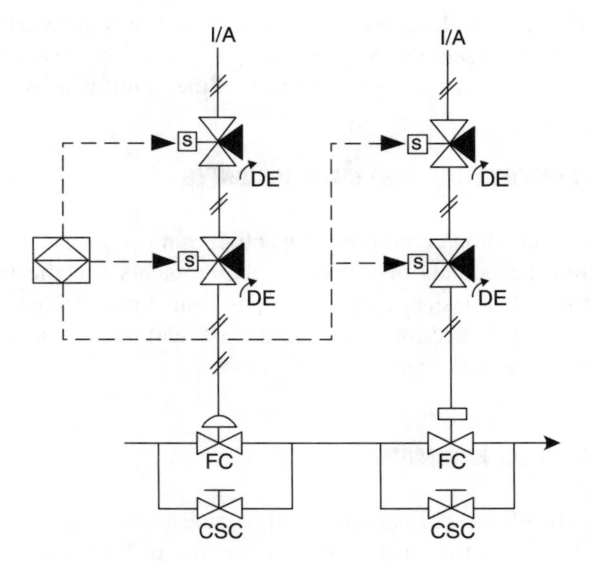

FIG. 2.4g
Installation illustration for final elements showing 1oo2 valves and 1oo2 solenoids.

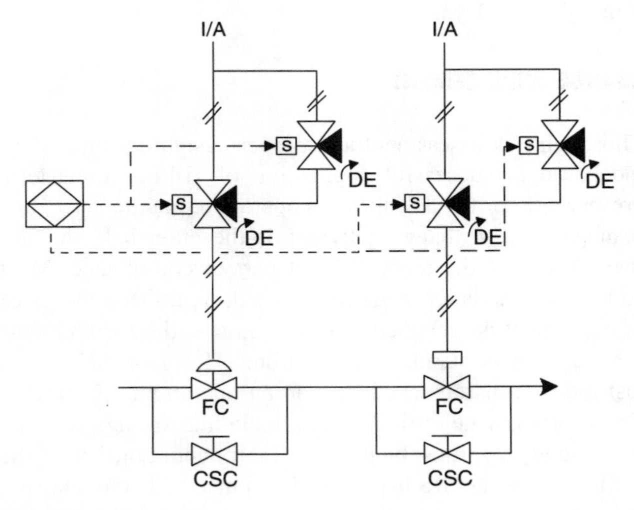

FIG. 2.4h
Installation illustration for final elements showing 1oo2 valves and 2oo2 solenoids.

Solenoid-operated valves (solenoids) configured as de-energize to trip are used to actuate the fail-safe valves. Solenoids can be configured 1oo1 or 1oo2, but spurious closure of the valves due to solenoid coil burnout can cause process disruptions, loss of production, and downtime. The solenoids can also be configured as 2oo2 to reduce spurious trips, as long as adequate testing is performed to uncover stuck valves or plugged vent ports. The solenoid should be mounted as close to the valve actuator as possible to decrease the required transfer volume for valve actuation. The exhaust ports should be as large as possible to increase speed of valve response.

DIAGNOSTICS

Diagnostic capability should be designed into HIPPS. The ability to detect failures of devices online significantly improves the availability of the HIPPS. For example, the use of signal comparison on analog inputs allows annunciation of transmitter failures to the control room. To support the claimed risk reduction associated with diagnostics, operation procedures must require that these alarms be responded to promptly with a work order for repair within the mean time to repair specified in the safety requirements specification. Maintenance procedures must also place high priority on repair of HIPPS devices.

TESTING FREQUENCY

If all failures were self-revealing, there would be no need to test safety system devices. Shutdown valves that do not close completely, solenoid valves that are stuck in position, and pressure switches with stuck closed contacts are all examples of covert, dangerous failures. If safety system devices are not tested, dangerous failures reveal themselves when a process demand occurs, often resulting in the unsafe event that the safety system was designed to prevent. Testing is performed for one reason, and one reason only, to uncover failures.

The appropriate testing of HIPPS is key to ensure that the availability requirements are satisfied. Architecture, redundancy, and device integrity have a significant effect on the probability to fail on demand and therefore testing frequency requirements. To determine the required testing frequency, quantitative risk assessment is the accepted approach by most users. In general, all HIPPS components require a testing frequency in the range of 3 to 12 months. Online and offline testing provisions should be provided to permit each device to be completely function-tested. Any required bypasses must be managed through a change management process with appropriate access security.

Whatever the testing frequency, it is essential that the testing is performed throughout the HIPPS system life. Any changes in the testing frequency must be validated by quantitative methods to ensure that the availability is not lowered to an unacceptable level.

COMMON CAUSE FAILURES

A common cause failure (CCF) occurs when a single failure results in the failure of multiple devices. ASME Code Case 2211 requires that sufficient independence be demonstrated to ensure reliability of the HIPPS performance. To minimize CCFs, the initiating causes of each scenario identified during the hazard analysis should be examined. Then, the HIPPS hardware and software should be designed to function independently from these initiating causes. For example, if a control transmitter is listed as an initiating cause to the scenario, the control transmitter cannot be the sole means for detecting the potential incident. At least one additional transmitter will be required for the HIPPS.

Once independence of the HIPPS devices is demonstrated, CCFs related to the design must be examined. The following are often cited as examples of CCFs:

- Miscalibration of sensors
- Fabrication flaws
- Pluggage of common process taps for redundant sensors
- Incorrect maintenance
- Improper bypassing
- Environmental stress on the field device
- Process fluid or contaminant prevents valve closure

The most critical failure is that the SRS is incorrect at the beginning of the design process and the HIPPS cannot effectively detect or prevent the potential incident. Improper system specification can compromise the entire HIPPS.

Industrial standards and corporate engineering guidelines and standards can be utilized to reduce the potential for CCF. The proposed or installed HIPPS design can be compared with these standards. Deviation from the standards can be corrected through design revision or documented to justify why this specific application has different requirements.

Checklists can also be used to reduce potential CCFs. A checklist analysis will identify specific hazards, deviations from standards, design deficiencies, and potential incidents through comparison of the design to known expectations, which have been expressed as checklist questions.

In some cases, it may be necessary to consider the impact of potential CCFs when verifying whether the HIPPS can achieve the target SIL. In such cases, the potential CCFs will need to be considered in the quantitative performance evaluation.

"AS SAFE OR SAFER" VERIFICATION

The HIPPS must provide an installation that is as safe or safer than the pressure relief device that it replaces. For documentation of the "as safe or safer" and compliance with the target SIL, the design of any HIPPS should be quantitatively verified to ensure it meets the required availability. Quantitative verification of SIL for HIPPS is the generally accepted approach for most users of HIPPS. A guidance report by ISA (expected final in 2002), ISA TR84.02, recommends use of one of the following methods for SIL verification:

1. Markov Models
2. Fault Tree Analysis
3. Simplified Methods

Any of these techniques can be utilized to determine whether the design meets the required SIL. If it does not meet the required SIL, the design must be modified until it does.

IMPLEMENTATION AND COMMISSIONING

Implementation/commissioning activities must be performed within the bounds of the safety requirements specification and detailed design. Any deviations from these documents must be evaluated for impact on the SIL and on any assumptions made with regard to performance.

OPERATION AND MAINTENANCE

The HIPPS must be operated, maintained, and tested throughout the life of the plant. The high integrity of HIPPS is often achieved through the use of frequent testing. Once the required testing frequency is determined for a particular HIPPS design, the testing must be performed at that frequency. If the SIL verification calculation states that the testing is to occur at a 6-month interval, it must be done at 6 months, not 1 year.

CHANGE MANAGEMENT

Thorough risk assessment and proper design constitute half the battle to successful application of HIPPS. Long-term preservation of the SIL through operation, maintenance, and management of change activities is the other half and, for many users, is the most difficult part of compliance. Most codes and standards focus solely on design. Once the piece of equipment is "certified" for compliance, the requirements for the code or standard are fulfilled. However, SIL is not just a design parameter. It is also an operational parameter. The choices made during design, including voting, diagnostics, and testing, must be preserved throughout the life of the facility. Once the SIS is designed and installed, and a testing frequency is chosen, the SIL is fixed and can only be changed by modification of one of the major design parameters. Consequently, the HIPPS SIL serves as a "management of change" checkpoint.

ADVANTAGES AND DISADVANTAGES OF HIPPS

It is poor safety practice to install and rely on pressure relief devices in services where the sizing of the device is poorly understood or known to be inadequate due to chemical reactions, multiphase fluids, or plugging. In these applications, alternatives, such as HIPPS, should be examined to ensure mitigation of overpressure events.

Industry is increasingly moving toward utilizing HIPPS to reduce flare loading and prevent the environmental impact of pressure venting. They are becoming the option of choice to help alleviate the need to replace major portions of the flare system in existing facilities when adding new equipment or units. If the header and flare system must be enlarged, significant downtime is incurred for all of the units that discharge to that header. The capital and installation cost associated with HIPPS is attractive when compared with the downtime or equipment cost of flare modification. Another benefit is that the process unit will not flare as much as a process unit designed for full flare loading. In some areas of the world, this is becoming important as regulatory agencies place greater restrictions on flaring of process gases.

The main disadvantage of HIPPS is the careful documentation, design, operation, maintenance, and testing to ensure compliance with standards. Specific regulatory and enforcement jurisdiction requirements must be determined. In some instances, approval of local authorities is required. Regulatory and standards requirements must be understood by all parties, including facility management and instrumentation and electrical, operations, and maintenance personnel.

Any justification for HIPPS must be thoroughly documented through a hazard analysis, which identifies all potential overpressure scenarios and demonstrates that the HIPPS can adequately address each scenario. The ability of the HIPPS to address overpressure adequately is limited by the knowledge and skill applied in the identification and definition of overpressure scenarios.

HIPPS systems are more complex, requiring the successful functioning of multiple devices to achieve the performance of a single pressure relief device. The user must verify that HIPPS will work from a process standpoint and that the HIPPS design results in an installation as safe as or safer than a conventional design. The effectiveness of the system is highly dependent on the field design, device testing, and maintenance program. Consequently, the user must understand the importance of application-specific design aspects, as well as the associated costs of the intensive testing and maintenance program whenever a HIPPS is utilized. When a pressure relief device is not installed or is undersized based on conventional design, the HIPPS becomes the "last line of defense," whose failure potentially results in vessel rupture.

Finally, there is no "approved" rubber stamp in any regulation or standard for the use of HIPPS for reduction in the size of relief devices and associated flare system for pressure vessels or pipelines. Substantial cautionary statements are made in the standards and recommended practices, concerning the use of HIPPS. No matter what documentation is created, the user still has the responsibility to provide a safe and environmentally friendly operation.

Bibliography

"Guidelines for Process Equipment Reliability Data," New York: Center for Chemical Process Safety of the American Institute of Chemical Engineers, 1989.

"Safety Instrumented Systems (SIS)—Safety Integrity Level (SIL) Evaluation Techniques," ISA dTR84.0.02, Draft, Version 4, March 1998.

Summers, A. E., "Techniques for Assigning a Target Safety Integrity Level," *ISA Transactions,* Volume 37, pp. 95–104, 1998.

Summers, A. E., "Using Instrumented Systems for Overpressure Protection," *Chemical Engineering Progress,* November 1999.

Summers, A. E., "S84—The Standard for Safety Instrumented Systems," *Chemical Engineering,* December 2000.

Summers, A. E. and Raney, G., "High Integrity Protection Systems and Pressure Relief Systems," IMECE Conference, American Society of Mechanical Engineers, Nashville, TN, November 1999.

Summers, A. E., Ford, K., and Raney, G., "Estimation and Evaluation of Common Cause Failures," 1999 Loss Prevention Symposium, American Institute of Chemical Engineers Spring Meeting, Houston, TX, March 1999.

Windhorst, J. C. A., "Over-pressure Protection by Means of a Design System Rather Than Pressure Relief Devices," CCPS International Conference and Workshop on Risk Analysis in Process Safety, American Institute of Chemical Engineers, Atlanta, GA, October 1998.

2.5 Process Safety Management

H. H. WEST AND M. S. MANNAN

Process safety management (PSM) is generally defined as the application of management systems to the identification, understanding, and control of chemical process hazards to prevent accidental releases.

During the past 20 years, a number of chemical or related incidents in the petrochemical industry have adversely affected surrounding communities. A few of these incidents, such as the vapor cloud explosion in Flixborough, U.K. in 1974,[12] the LPG explosion in Mexico City in 1984,[12] the toxic material release in Bhopal India in 1984,[12] and the fire and radiation release in Chernobyl, U.S.S.R.,[12] were reported worldwide. Both governmental agencies and trade organizations responded by developing standards and regulations to improve process safety. The American Petroleum Institute (API)[10] and the American Chemistry Council (ACC)[15] started to work with their members to develop PSM organizational guidelines. The Clean Air Act Amendments of 1990 required both the Occupational Safety and Health Administration (OSHA) and the U.S. Environmental Protection Agency (EPA) to develop standards to protect workers and the public and environment, respectively. OSHA fulfilled its mandate by promulgating the PSM standard.[8] The EPA[6] also added PSM requirements to the Risk Management Program in 40 CFR Part 68.

The PSM regulation covers chemical and petroleum processing plants that contain one of the listed toxic materials or flammable gases and liquids above a threshold quantity. The threshold quantity for flammable liquids is, for example, 10,000 lb or the equivalent of about 20 barrels of oil. Although a first glance would suggest that only those portions of the chemical process unit containing the listed toxic or flammable fluids would be subject to the OSHA PSM rule, actually any adjacent unit that can ultimately cause a catastrophic release of the toxic or flammable fluids will also be covered by PSM. In addition to the vast majority of chemical manufacturing facilities and petroleum processing plants, facilities such as water chlorination and ammonia refrigeration may also be covered.

The OSHA PSM rule contains a few exemptions, such as gasoline retail facilities, oil and gas drilling operations, and pipelines. Unattended facilities are also exempt from PSM regulation.

ELEMENTS OF PROCESS SAFETY MANAGEMENT

Essentially, API Recommended Practice 750, the ACC Responsible Care program, the 1990 Clean Air Act Amendments, and OSHA 1910.119 define a structured PSM system, which requires substantial documentation of current chemical processing industry practice. Table 2.5a presents the American Institute of Chemical Engineers'[4] definitions of the primary elements of PSM, which are slightly different from the OSHA definition. The 14 elements of the OSHA PSM program include:

1. Employee participation
2. Current process safety information
3. Process hazards analysis and documentation
4. Management of change procedure
5. Operating procedures
6. Training with documentation
7. Mechanical integrity, including scheduled process equipment tests/inspections
8. Formal pre-start-up safety reviews
9. Emergency response plans
10. Investigation of incidents
11. Periodic compliance audits
12. Contractor issues
13. Safe work practices
14. Trade secrets

Although the process documentation is formally defined in the section entitled "Process Safety Information," each of the other PSM elements contains additional documentation requirements. Compliance with each PSM element, therefore, adds additional documentation as a practical necessity.

Figure 2.5b presents a graphical representation of these program elements, emphasizing the critical role of process information for the other elements. Documentation of each of the above program elements is essential to compliance. A brief description of these key safety program elements follows, with emphasis on the implications for instrument and control engineers.

TABLE 2.5a
AIChE Elements of Process Safety Management

1. Accountability: Objectives and Goals
   - Continuity of Operations
   - Continuity of Systems (resources and funding)
   - Continuity of Organizations
   - Company Expectations (vision or master plan)
   - Quality Process
   - Control of Exception
   - Alternative Methods (performance vs. specification)
   - Management Accessibility
   - Communications
2. Process Knowledge and Documentation
   - Process Definition and Design Criteria
   - Process and Equipment Design
   - Company Memory (management information)
   - Documentation of Risk Management Decisions
   - Protective Systems
   - Normal and Upset Conditions
   - Chemical and Occupational Health Hazards
3. Capital Project Review and Design Procedures (for new or existing plants, expansions, and acquisitions)
   - Appropriation Request Procedures
   - Risk Assessment for Investment Purposes
   - Hazards Review (including worst credible cases)
   - Siting (relative to risk management)
   - Plot Plan
   - Process Design and Review Procedures
   - Project Management Procedures
4. Process Risk Management
   - Hazard Identification
   - Risk Assessment of Existing Operations
   - Reduction of Risk
   - Residual Risk Management (in-plant emergency response and mitigation)
   - Process Management during Emergencies
   - Encouraging Client and Supplier Companies to Adopt Similar Risk Management Practices
   - Selection of Businesses with Acceptable Risks
5. Management of Change
   - Change of Technology
   - Change of Facility
   - Organizational Changes That May Affect Process Safety
   - Variance Procedures
   - Temporary Changes
   - Permanent Changes
6. Process and Equipment Integrity
   - Reliability Engineering
   - Materials of Construction
   - Fabrication and Inspection Procedures
   - Installation Procedures
   - Preventive Maintenance
   - Process, Hardware, and Systems Inspections and Testing (pre-start-up safety review)
   - Maintenance Procedures
   - Alarm and Instrument Management
   - Demolition Procedures
7. Human Factors
   - Human Error Assessment
   - Operator/Process and Equipment Interfaces
   - Administrative Controls vs. Hardware
8. Training and Performance
   - Definition of Skills and Knowledge
   - Training Programs (e.g., new employees, contractors, technical employees)
   - Design of Operating and Maintenance Procedures
   - Initial Qualification Assessment
   - Ongoing Performance and Refresher Training
   - Instructor Program
   - Records Management
9. Incident Investigation
   - Major Incidents
   - Near-Miss Reporting
   - Follow-Up and Resolution
   - Communication
   - Incident Recording
   - Third-Party Participation as Needed
10. Standards, Codes, and Laws
    - Internal Standards, Guidelines, and Practices (past history, flexible performance standards, amendments, and upgrades)
    - External Standards, Guidelines, and Practices
11. Audits and Corrective Actions
    - Process Safety Audits and Compliance Reviews
    - Resolutions and Close-Out Procedures
12. Enhancement of Process Safety Knowledge
    - Internal and External Research
    - Improved Predictive Systems
    - Process Safety Reference Library

EMPLOYEE PARTICIPATION

This element of the regulation requires developing a written plan of action regarding employee participation; consulting with employees and their representatives on the conduct and development of other elements of process safety management required under the regulation; providing to employees and their representatives access to process hazard analyses and to all other information required to be developed under this regulation.

FIG. 2.5b
Process safety management program elements.

PROCESS SAFETY INFORMATION

To begin any formal process safety review, accurate process information must be supplied to the specialists charged with the review. Hence, the PSM regulations require maintaining the following types of process information:

- Basic Chemical Data Sheets, such as MSDS indicating chemical hazards
- Process Flow Diagrams
- Process Equipment Specification Sheets
- Safety Protective System Design Basis, including interlocks, detection/suppression systems, ventilations systems, and pressure/vacuum relief systems
- Plot Plans with Electrical Area Hazard Classification
- Piping and Instrumentation Drawings (P&ID)
- Maximum and minimum safe operating limits (temperature, pressure, flow, etc.) for each process equipment and major piping segment must be documented
- An evaluation of the consequences of deviations from these safe operating limits
- Design codes and standards employed
- Material and energy balances

The most important consideration is the requirement that these process information resources be kept up to date.

Control system designers must supply information to the operator in a format that will permit continual update to reflect changes. The control system operator has the duty to keep the safety information up-to-date, including P&IDs, temporary device jumper lists, etc.

Note that any operations beyond these defined safe operating limits is considered a process change and must be formally authorized. Hence the "easy to modify" feature of modern control systems must be managed properly to avoid safety problems.

PROCESS HAZARD ANALYSIS

Several process hazard analysis (PHA)[4] (also commonly called process safety analysis or system safety analysis techniques) procedures are available:

- Fault Tree Analysis
- Failure Mode and Effects Analysis (FMEA)
- HAZard and OPerability (HAZOP) studies
- Safety system checklists
- SAFE charts
- What-If studies
- Checklist Analysis
- DiGraph Analysis

OSHA has not specified any single approved method, but has taken a flexible approach where owners must select the most appropriate method of analysis for their facility. Table 2.5c provides an example of a set of What-If questions for an instrumentation PHA.

Formal documentation of the design basis of safety systems began in the aircraft and nuclear industries. Fault tree analysis as well as failure mode and effects analysis provided a format for documenting the decisions used to design safety response systems (as well as other portions of the entire plant) in these high-risk industries. These techniques eventually formed the tenets of system safety analysis, which is more rigorously defined in safety management guidelines, such as U.S. military specification 882.[7]

The first attempts to apply the above system safety analysis concepts to the chemical processing industry occurred in conjunction with the liquefied natural gas (LNG) industry during the late 1960s and early 1970s. The potential widespread public damage of a major LNG release prompted requirements for indepth safety reviews by government regulators. In response to major chemical accidents in the 1970s and 1980s, additional chemical safety procedures were developed. The HAZOP procedure formalized a checklist approach to safety audit studies. The offshore petroleum industry introduced the API Recommended Practice 14C Safe Chart,[11] documenting the safety protective systems. The United Kingdom Health and Safety Executive[2] has promulgated chemical safety documentation requirements specifically aimed at modern computer-controlled safety protective systems. The EPA[13] also requires formal hazard analysis and documentation for selected extremely hazardous chemicals under the Superfund Reauthorization Act (SARA) Title III enactments.

TABLE 2.5c
What-If Questions for Instrumentation PHA

Question	*Comment*
Have instruments critical to process safety been identified and listed with an explanation of their safety function and alarm set points?	
Has the process safety function of instrumentation been considered integrally with the process control function throughout plant design?	
Is every significant instrument or control device backed up by an independent instrument or control that operates in an entirely different manner? In critical processes, are these first two methods of control backed up by a third, ultimate safety shutdown?	
What would be the effect of a faulty sensor transmitter, indicator, alarm, or recorder? How would the failure be detected?	
If all instruments fail simultaneously, is the collective operation still fail-safe? Are partial failures also fail-safe (e.g., one instrument power bus remaining energized while others fail)?	
How is the computer control system configured? Are there backups for all hardware components (computers, displays, input/output modules, programmable logic controllers, data highways, etc.)?	
How is computer control software written and debugged? If there is a software error, is the backup computer also likely to fail as a result of the same error? Should extremely critical shutdown interlocks be hardwired instead?	
Is there a computer with outputs to process devices? If so, is computer failure detection implemented? Can any output or group of outputs from the computer cause a hazard?	
Where sequence controllers are used, is there an automatic check, together with alarms, at key steps after the controller has called for a change? Is there a check, together with alarms, at key steps before the next sequence changes?	
What are the consequences of operator intervention in computer-controlled sequences?	
Does the control system verify that operator inputs are within an acceptable range (e.g., if the operator makes a typographical error, will the control system attempt to supply 1000 lb of catalyst to a reactor that normally requires only 100 lb)?	
Uninterruptible power supply (UPS) for supporting the process control computer? Is it periodically tested under load? Does the UPS also support critical devices that may need to be actuated or does it only support information and alarm functions?	
Does the operator–machine interface incorporate good human factors principles?	
Is adequate information about normal and upset process conditions displayed in the control room?	
Is the information displayed in ways the operators understand?	
Is any misleading information displayed, or is any display itself misleading?	
Is it obvious to operators when an instrument is failed or bypassed?	
Do separate displays present information consistently?	
What kinds of calculations must operators perform, and how are they checked?	
Are operators provided with enough information to diagnose an upset when an alarm sounds?	
Are operators overwhelmed by the number of alarms associated with an upset or emergency? Should an alarm prioritization system be implemented? Can operators easily tell what failure/alarm started the upset?	
Is there a first alarm or critical alarm panel?	
Are the displays adequately visible from all relevant working positions?	
Do the displays provide adequate feedback on operator actions?	
Do control panel layouts reflect the functional aspects of the process or equipment?	
Are related displays and controls grouped together?	
Does the control arrangements logically follow the normal sequence of operation?	
Are all controls accessible and easy to distinguish?	
Are the controls easy to use?	
Do any controls violate strong populational stereotypes (e.g., color, direction of movement)?	
Are any process variables difficult to control with existing equipment?	

TABLE 2.5c Continued
What-If Questions for Instrumentation PHA

Question	*Comment*
How many manual adjustments must an operator perform during normal and emergency operations?	
When adjacent controls (e.g., valves, switches) have a similar appearance, what are the consequences if the incorrect control is used?	
Are redundant signal or communication lines physically separated (i.e., run in separate cable trays, one run aboveground and another underground)?	
Are signal cables shielded or segregated from power cables (i.e., to avoid electromagnetic interferences and false signals)?	
Are there control loops in the process which are not connected into the computer control system? How do operators monitor and control from the control room?	
Are automatic controls ever used in manual mode? How do operators ensure safe operation while in manual mode?	
What emergency valves and controls can operators not reach quickly and safely while wearing appropriate protective clothing?	
What procedures have been established for testing and proving instrument functions and verifying their alarm set points are correct? How often is testing performed?	
Are the means provided for testing and maintaining primary elements of alarm and interlock instrumentation without shutting down the process?	
Are controls deliberately disabled during any phase of operation? How are alarm set points and computer software protected from unauthorized changes?	
What happens when such an instrument is not available?	
Are instrument sensing lines adequately purged or heat-traced to avoid plugging?	
What are the effects of atmospheric humidity and temperature extremes on instrumentation? What are the effects of process emissions? Are there any sources of water—water lines, sewer lines, sprinklers, roof drains—that could drip into or spray onto sensitive control room equipment?	
Is the system completely free of instruments containing fluids that would react with process materials?	
What is being done to verify that instrument packages are properly installed, grounded, and designed for the environment and area electrical classification? Is instrument grounding coordinated with cathodic protection for pipes, tanks, and structures?	
Are the instruments and controls provided on vendor-supplied equipment packages compatible and consistent with existing systems and operator experience? How are these instruments and controls integrated into the overall system?	

The HAZOP Concept

The HAZOP process is based on the principle that a team approach to hazard analysis will identify more problems than when individuals working separately combine results. The HAZOP team is made up of individuals with varying backgrounds and expertise. The expertise is brought together during HAZOP sessions and, through a collective brainstorming effort that stimulates creativity and new ideas, a thorough review of the process under consideration is made.

The HAZOP team focuses on specific portions of the process called "nodes." Generally, these nodes are identified from the P&ID of the process before the study begins. A process parameter is identified, for example, flow, and an intention is created for the node under consideration. Then a series of guidewords is combined with the parameter "flow" to create a deviation. For example, the guideword "no" is combined with the parameter flow to give the deviation "no flow." The team then focuses on listing all the credible causes of a "no flow" deviation beginning with the cause that can result in the worst possible consequence the team can think of at the time. Once the causes are recorded, the team lists the consequences, safeguards, and any recommendations deemed appropriate. The process is repeated for the next deviation and so on until completion of the node. The team then moves on to the next node and repeats the process.

Fault Tree Analysis

Fault tree analysis is the most widely used method in quantitative risk analysis. It is a deductive procedure for determining the various combinations of hardware and software failures, including human errors that could result in the occurrence of specified undesired events (referred to as top events) at the system level. A deductive analysis begins with a general conclusion, then attempts to determine the specific causes of

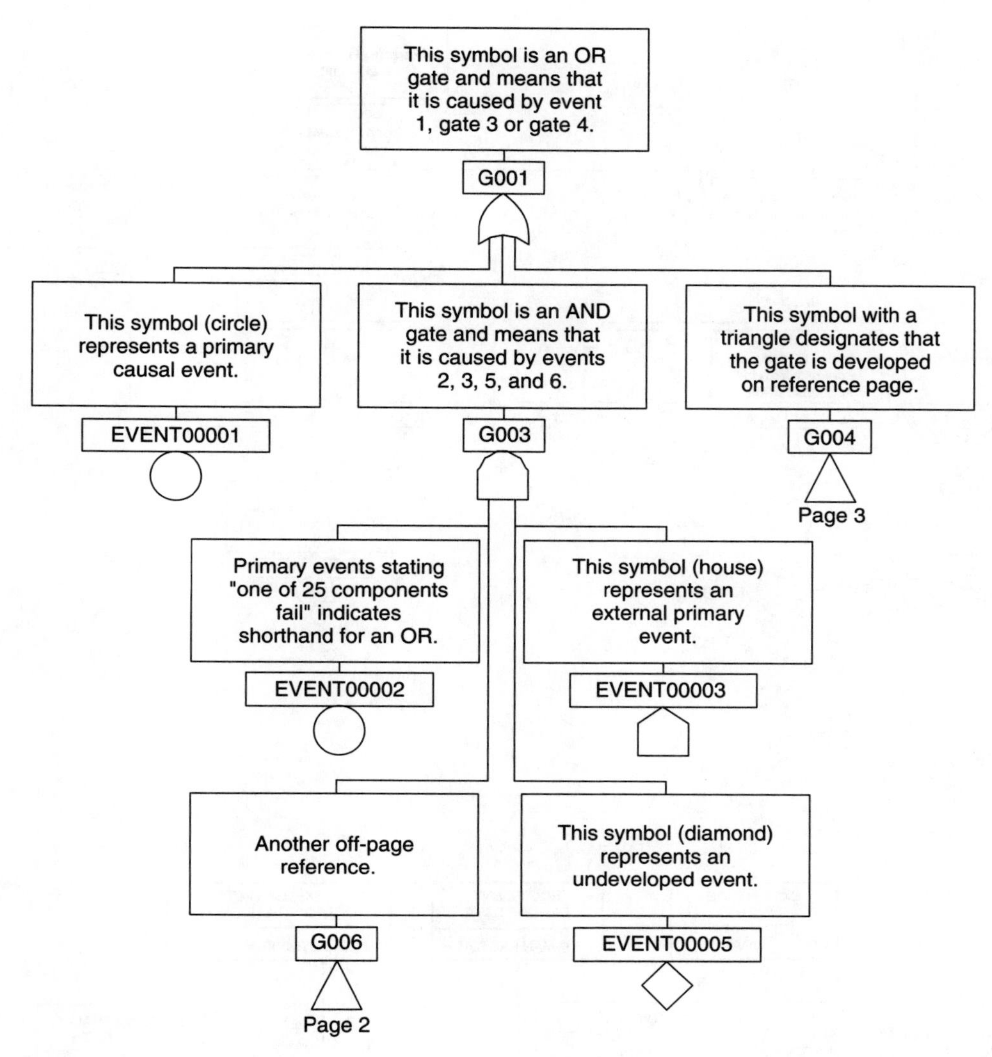

FIG. 2.5d
Fault tree symbol legend.

this conclusion. This is often described as a "top-down" approach.

The main purpose of fault tree analysis is to evaluate the probability of the top event using analytical or statistical methods. These calculations involve system quantitative reliability and maintainability information, such as failure probability, failure rate, or repair rate. Fault tree analysis can provide useful information concerning the likelihood of a failure and the means by which such a failure could occur. Efforts to improve system safety and reliability can be focused and refined using the results of the fault tree analysis.

A gate is used to describe the relationship between the input and output events in a fault tree. Fault trees can have several different kinds of gates. The kinds of gates typically used in fault tree construction are described and shown in Figure 2.5d. Figure 2.5e presents an example of a fault tree, using various types of gates and associated probability data for each respective intermediate gate.

Chemical Process Quantitative Risk Assessment

Since there is no absolute theoretical foundation for defining the "realistic" or "credible" accident, a broad range of emergency response and protective control systems has been used throughout the chemical industry.

The current state of the art in safety assessment procedures for both existing and new chemical processing plants is conceptually illustrated in the block diagram of Figure 2.5f. This decision process uses quantitative methods to define accident size and frequency, which is then combined with the corresponding damage potential to arrive at the estimated risk. Risk reduction techniques are shown in the procedural feedback loop. The double border blocks in Figure 2.5d show the role of the ESD (emergency shutdown) and hazard control systems in the overall risk assessment procedure.

The American Institute of Chemical Engineers has published several books on process hazard analyses.[4]

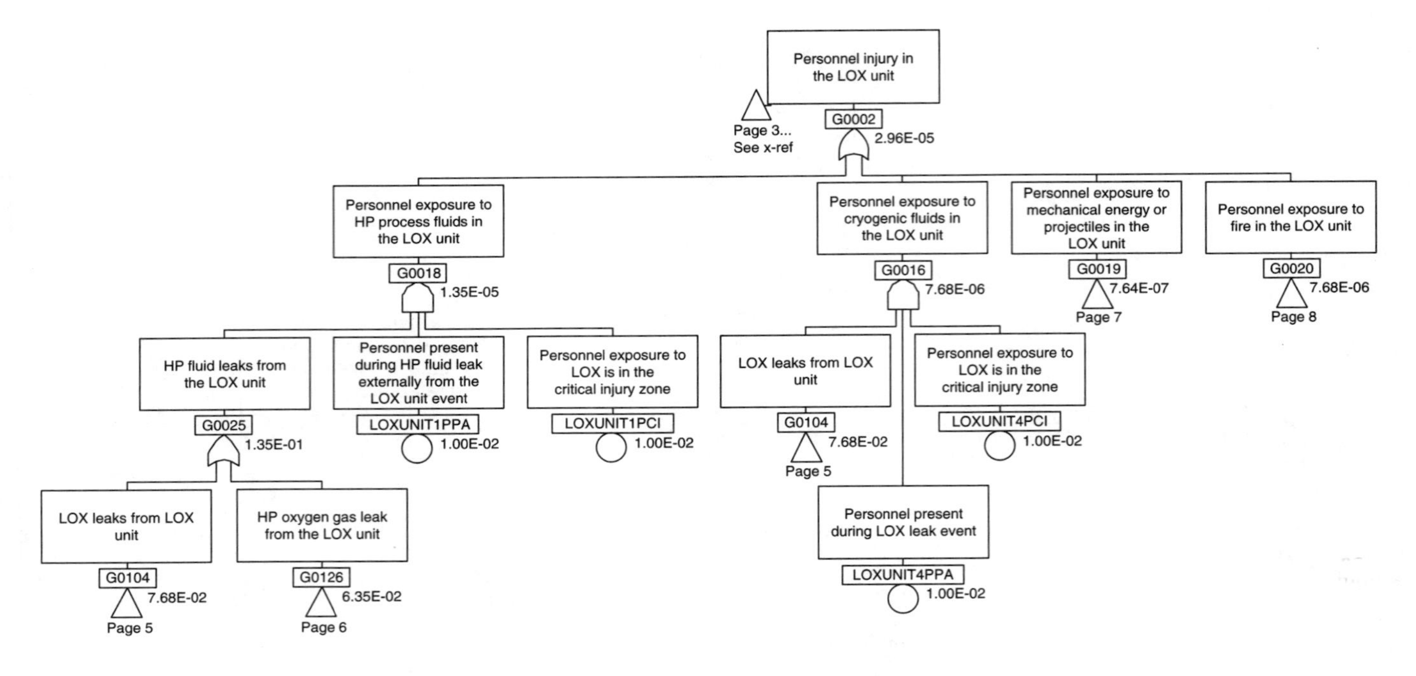

FIG. 2.5e
Fault tree example.

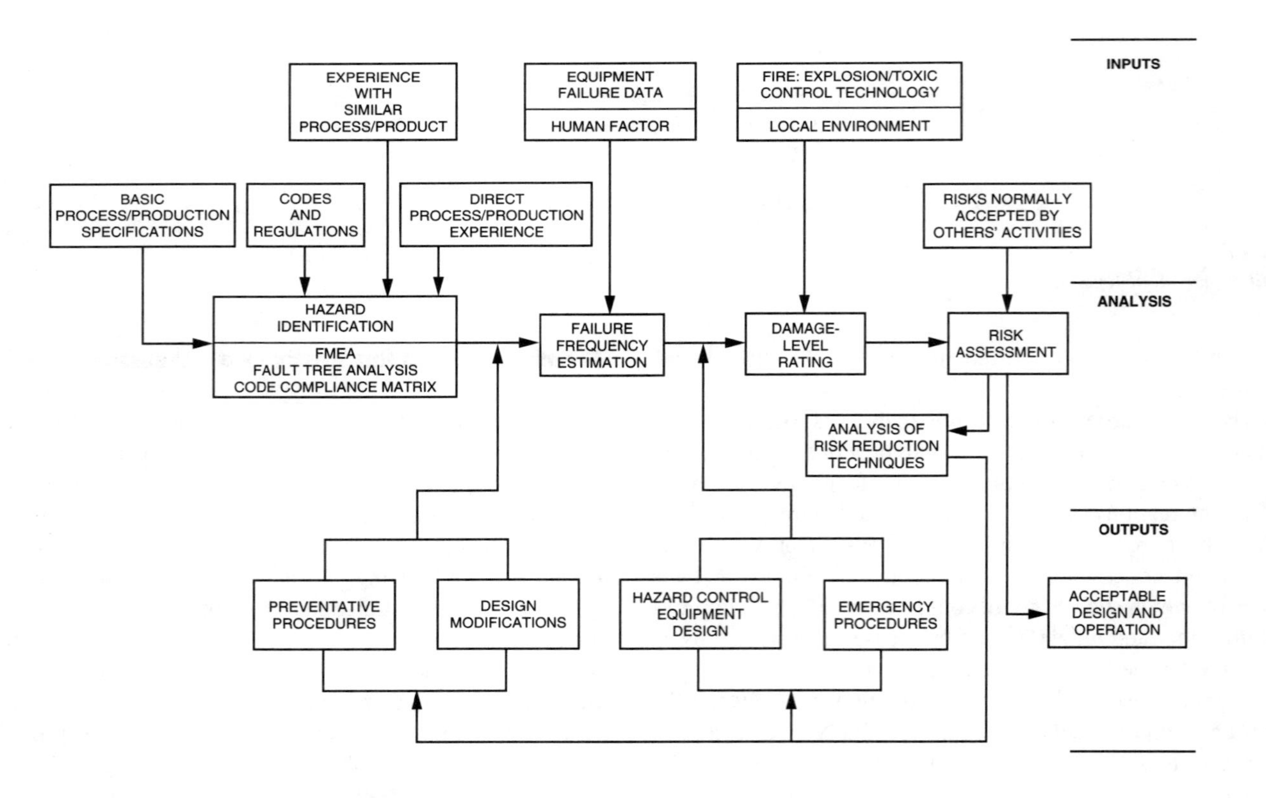

FIG. 2.5f
Elements of system safety analysis.

Failure Analysis of Safety Protective Systems

The performance capability and reliability of the safety protective system plays a vital role in both the process failure frequency and the quantified damage estimate.

To calculate the reliability of the modern computer protective systems, the failure modes and corresponding failure frequencies of each major component part must be obtained. A Failure Mode and Effects Analysis (FMEA) or fault tree study of the protective control system must begin with a clear definition of all of the components. This study can highlight potential design flaws as well as quantify the system reliability. Complicating the analysis is the various chemical plant operating modes; start-up, shutdown, maintenance lockout, normal design production rates, partial system availability, etc.

New failure modes of the new computer input/output (I/O) units, particularly in comparison with the former relay logic panel technology, are cause for some concern. Electromechanical relays fail predominantly in one way, so that designers can develop fail-to-safety strategies with confidence. The newer solid-state relays have several failure modes, which may lead to fail-to-danger situations.

Software failure estimates are rarely considered, relying on the system field test to reveal all covert software problems. This assumption results in an overly optimistic system reliability. More detail regarding analysis of SISs is contained in Section 2.4, "High-Integrity Pressure Protection Systems."

Location of Control Room

To accommodate pneumatic or hydraulic instrumentation, control rooms were traditionally located near the chemical process equipment. With computer control and digital data transmission technology, the control room can be located remotely. This would place most of the plant operations personnel out of the range of a chemical fire or explosion.

With recent OSHA interpretations regarding "providing a safe work place" extending to chemical plant control rooms, remote location and/or blast wall design concepts may be necessary. Analysis supporting the design and location of control rooms is now required.

MANAGEMENT OF CHANGE

The PSM standard requires a formal safety analysis of every process change, except replacement-in-kind or standard maintenance items. The results of this analysis must be disseminated to all affected groups (operations, training, maintenance, process engineering, contractors, etc.) before operations resume.

The purpose of the management of change (MOC) procedure is to ensure that safety, loss prevention, environmental and engineering evaluations are provided at the planning stages of process changes.

A formal authorization procedure, typically tied to the process work order system, is needed to guarantee effective safety review of appropriate process change requests. The definition of "process change" includes all modifications to equipment and procedures (excluding replacement-in-kind). Process changes can range from installation of new or additional equipment, to installation of different gasket or seal materials, to *instrument set point adjustments outside of allowable limits*. Therefore, careful definition of the allowable limits of operational control must be documented.

Version control of operating software is, accordingly, mandated by the PSM requirements. The flexibility of modern control systems to allow operator override or local logic changes must be administratively controlled.

WRITTEN OPERATING PROCEDURES

Surprisingly, prior to promulgation of the OSHA PSM rule, many chemical plants did not have up-to-date written operating procedures. Although no specific level of detail has been required by OSHA, the target should be the trained relief operator who needs a quick refresher, if needed. The operating procedures manual must include various phases of operation, such as start-up, shutdown, operation with various equipment out-of-service, etc.

If a procedure is not in the operating procedures manual, the operator should not attempt the unauthorized procedure. The operating procedures must be available prior to system start-up and must be kept current. Online tutorials are more frequently used to maintain these up-to-date criteria.

TRAINING WITH DOCUMENTATION

As with any new technology, training of operations and maintenance staff is critical. Certification of training is becoming more prevalent, particularly with the trend of government requirements for annual updated training certifications. Now, even experienced chemical operators must periodically refresh their knowledge of the operating procedures and the training of new operators must be documented. Simulators and formal tests are not required, but sufficient information to explain the method of training is needed. Utilization of the up-to-date operating manuals in the process-specific aspects of training satisfies several process safety objectives. Documentation of the training is required.

Note also that the MOC section required training each operator before a process change, either equipment or procedural, is implemented.

MECHANICAL INTEGRITY

Design code compliance is important. If a process unit is not compliant with the applicable design code, then an analysis of its capability to contain the hazardous chemical inventory properly is required. In some cases, the safety protective system is critical to mechanical integrity.

The recommended maintenance schedule and procedures for all process equipment, including safety protective systems, must be provided. Maintenance records, including test and inspection results, must be available for audit.

INCIDENT ANALYSIS AND REPORTS

In the emotion-charged atmosphere immediately following a significant accident, accurate information of all types is important to the emergency response team, management, and all other interested parties. The subsequent accident analysis effort must begin with accurate process safety data to reconstruct the scenario leading to the incident. In addition to the process safety documentation, the incident analysis team must interview witnesses and other individuals with special relevant knowledge, document the consequences of the accident, and secure all related information (such as maintenance records, environmental reports, medical data, etc.).

The regulations require investigation and analysis of "near-miss" events that may have catastrophic potential. Many definitions of a near-miss event include the operation of safety critical or safety protective instrumentation and control systems. This is based on the presumption that an incident would have occurred if the safety critical control system were inoperative. Again, careful definition of safety critical control systems must be made, since this may trigger an investigation procedure.

CONTRACTOR ISSUES

The interface between employee and contractor responsibilities is specifically required along with a documentation trail to prove compliance. The PSM regulation identifies the responsibilities of the employer regarding contractors involved in maintenance, repair, turnaround, major renovation, or specialty work, on or near covered processes. The host employer is required to consider safety records in selecting contractors, inform contractors of potential process hazards, explain the emergency action plan of the facility, develop safe work practices for contractors in process areas, periodically evaluate contractor safety performance, and maintain an injury/illness log for contractors working in process areas. In addition, the contract employer is required to train its employees in safe work practices and document that training, assure that employees know about potential process hazards and the host employer's emergency action plan, assure that employees follow safety rules of facility, advise the host employer of hazards that the contract work itself poses and of hazards that are identified by contract employees.

In the contractor paragraph, OSHA has used a belt and suspender approach. Both the host employer and contract employer have specific responsibilities that they must fulfill. The need for flexibility, quick turnarounds, and specialized services is the main reason process plants are contracting out increasingly significant portion of their daily work, particularly maintenance work to contractors.

EMERGENCY RESPONSE PLAN

The OSHA 29 CFR 1910.38 rules,[9] "Employee Emergency Plan and Fire Prevention Plan," which applies to chemical processing and storage facilities, have been in force for more than 10 years. Now it must be coordinated with the PSM requirements. EPA regulations, particularly the community right-to-know rules contained in the Emergency Planning and Community Right to Know Act (EPCRA) or SARA Title III legislation,[10] also require written emergency response procedures. Other federal agencies[11] and some state agencies have instituted safety rules, which include elements of emergency preparedness.

SAFE WORK PRACTICES

Although the OSHA PSM regulation specifically targets hot work permit procedures, the need to coordinate all safe work practices has expanded the interpretation of this section of the PSM rule to include other recently revised or enacted OSHA rules, such as:

Lockout/Tagout (OSHA 1910.147)
Confined Space Entry (OSHA 1910.146)
Personal Protective Equipment (OSHA 1910 subpart I)
Hazard Communication (OSHA 1910.1200)

The important concept is the *integration* of these safe work practices into the PSM operating procedures manual, training requirements, etc. For example, the *specific* valve configuration to conduct a particular piping break may require documentation in the operating procedures manual (for the operators responsibilities) and additional documentation in the maintenance procedures (for the maintenance mechanics) as well as reference to general plant lockout/tagout, personal protective equipment, and hazard communication procedures. Obviously, a disjointed or inconsistent set of procedures is not compliant with the OSHA regulations.

TRADE SECRETS

Similar to the trade secret provisions of the hazard communication regulation, the PSM regulation also requires information to be available to employees from the process hazard analyses and other documents required by the regulation. The regulation permits employers to enter into confidentiality agreements to prevent disclosure of trade secrets.

ISSUES IN PROTECTIVE SYSTEM TECHNOLOGY

The following issues are highlighted to point out areas of current diversity of opinion.

Separate and Distinct Systems

The issue of separation of protective systems from the process measurement and control system is an area of potential controversy. For harmless chemicals and relatively mild processing conditions, a single unified system may be optimal. However, the most dangerous conditions and chemicals warrant separate systems. Between these extremes lie many potential combinations.

For example, a float device may be used to measure the liquid level continuously. This same sensor may trigger logic in the controller for a high alarm or even a high-high shutdown activation. Yet, other designs may insist on using a separate high-level sensor for alarm and another for shutdown (with the process measurement as a confirming backup).

API Recommended Practice 14-C requires separate and distinct safety protective systems. It goes even further by recommending *two* independent safety protective devices for each identified failure mode.

However, the OSHA PSM rule does not have any specific safety protective system requirements, simply suggesting good engineering practice.

Minimum Operational Safety Protective System Defined

A recent application of quantitative risk assessment procedures is to define the minimum acceptable online safety protective system. If sufficient portions of the safety system are in the shop for maintenance or are awaiting replacement parts, then the process may not proceed. The calculated risk would, therefore, exceed corporate targets. Hence, the process must not continue. This is a relatively new concept that has significant implications for the entire chemical industry.

Insurability

Prior to the government regulations on safety analysis, some safety managers claimed that increased safety analysis would assist the underwriter in justifying lower insurance premiums. In comparison to the insurance discount that accrues with cars containing air bag restraints, it is argued that highly protected chemical plants also deserve discounts.

If the discount is granted, based upon more capable protective systems, then it may also follow that insurance is not in force when an accident occurs during a time when the safety system is non-operable.

Environmental Impact

The concern for environmental damage minimization has prompted the EPA to delve deeper within chemical plant operations. Through the regulatory permit application procedure, the capabilities of the protective systems are examined for pollution prevention and minimization in the event of a chemical process spill.

References

1. West, H. H. and Brown, L. E., "Analyze Fire Protection Systems," *Hydrocarbon Processing,* August 1977.
2. "Programmable Electronic Systems in Safety Applications," London: U.K. Health and Safety Executive, 1987.
3. "Guide to Hazard and Operability Studies," London: Chemical Industries Health and Safety Council, 1976.
4. "Guidelines for Hazard Evaluations," New York: Center for Chemical Process Safety, American Institute of Chemical Engineers, 1985.
5. "Guidelines for Quantitative Risk Assessment," New York: Center for Chemical Process Safety, American Institute of Chemical Engineers, 1985.
6. U.S. Code of Federal Regulations, Environmental Protection Agency, Title 40 U.S. Code of Federal Regulations Part 68, Washington, D.C.: U.S. Government Printing Office, 2000.
7. Military Specification 882, "System Safety Analysis," 68, Washington, D.C.: U.S. Government Printing Office, 1996.
8. U.S. Code of Federal Regulations, Title 29, Sections 1910.119, Washington, D.C.: U.S. Government Printing Office; and *Federal Register,* February 24, 1992.
9. Clean Air Act Amendments of 1990, Report 101-952, Conference Report to Accompany Senate Bill S.1630, October 26, 1990.
10. API recommended practice 750, "Management of Process Hazards," Washington, D.C.: American Petroleum Institute, January 1990.
11. API recommended practice 14C, "Analysis, Design, Installation and Testing of Basis Surface Safety Systems for Offshore Production Platforms," 4th ed., Washington, D.C.: American Petroleum Institute, January 1986.
12. "Guidelines for Safe Operations and Maintenance," New York: Center for Chemical Process Safety, American Institute of Chemical Engineers, 1995.
13. Title 40 U.S. Code of Federal Regulations Parts 355, 370 and 372, Emergency Planning and Community Right to Know Rules, Washington, D.C.: U.S. Government Printing Office, 2000.
14. New Jersey Toxic Catastrophic Prevention Act, Trenton, NJ: New Jersey Department of Environmental Protection, 2000.
15. American Chemistry Council, "Responsible Care," Washington, D.C., 2001.

2.6 Redundant or Voting Systems for Increased Reliability

I. H. GIBSON

As the old adage has it, a chain is no stronger than its weakest link. This is demonstrably true in control systems design, where statistical theory is used to predict the behavior of complex systems. When a single sensor/logic/effector chain has inadequate reliability, redundant systems can be brought into play to reduce the probability of failure on demand or to reduce the probability of false tripping or to yield both functions.

There are several approaches available, and the correct choice of method depends on the relative sensitivity of the systems to the two failure modes, in economic and human-factor risk, and the cost and practicality of proof testing.

I/O REDUNDANCY AND VOTING

When assembling a high-reliability system with current generation equipment, the logic solver is the most reliable component of the chain, as it will normally be provided with a high level of internal diagnostics, and redundant input/output (I/O). The vast majority of faults in logic systems come from faults in functional specification, where the equipment performs in accordance with the designer's stated requirement, but the concept itself is faulty.

The field sensor(s) will be, for preference, analog devices rather than switches, as the latter, by their very nature, provide little facility for online fault detection.

The least reliable components are normally the effectors, the final control elements. IEC 61508 usefully divides safety-related systems into two categories; demand-type systems, where proof testing is practical at a frequency significantly higher than the anticipated frequency of demand, and continuous-type systems where this is not so.

Demand-type systems are then rated to PFD_{ave}—the average probability of failure on demand over the period between successive proof tests. It should be noted that the PFD_{ave} tends to deteriorate between successive proof tests, as no proof test can enable the return of the system to its original reliability rating, because it can detect only some fraction of hidden faults. Continuous rated systems are more difficult to build to equivalent levels of functionality, because of the difficulty involved in demonstrating the absence of concealed faults. A typical example would be automotive ABS braking systems, where the demand rate can be several times per minute. This is also a reason process control systems are not rated as safety integrity level 1 (SIL 1) or higher without specific certification.

Redundancy is the term used for multiple devices performing the same function. *Voting* is the term used to describe how the information from redundant devices is applied. One can have redundant devices without voting, or with manual voting if an operator can check the output of the redundant devices and decide which output is to perform the required function. A simple example is the traditional temperature measurement installation, where a bimetallic thermometer is installed in one thermowell, and a duplex thermocouple element is installed in a second well, with one element connected to an alarm/shutdown system, and the other to the process control system. Here the thermometer can provide local operator indication, and the two thermocouple elements offer redundant information to the control room operator. This system also demonstrates one of the weaknesses of redundant systems—common mode failure. Both of the duplex elements are identical, and are closely coupled mechanically and electrically. Drift in calibration (such as from manganese migration between sheaths and elements), phase change hysteresis, or straight mechanical damage can produce similar effects in both components. A slip during maintenance can easily take out both measurements at once and it is impossible to repair one element without taking the other out of service.

NooM TERMINOLOGY

N out of M redundancy (commonly written as NooM, e.g., 2oo3) is a method to increase reliability. The terminology denotes the number of coincident detected trip signals required for the system to trip. The term NooMD is often used to denote a system with inherent diagnostic capability, which can detect and take action on some defined percentage of otherwise unrevealed failures.

Table 2.6a shows the effect of adding parallel, identical quality information to a logic system in the absence of common mode effects. In the equations in Table 2.6a, TI denotes the time interval between proof tests, MTTR is the mean time to repair, the $\lambda^{DU} = 1/MTTF^{DU}$ is the failure rate for dangerous undetected faults, and $\lambda^{S} = 1/MTTF^{S}$ is the spurious trip rate.

TABLE 2.6a
Redundancy and Probability of Failure

Redundancy	Average Probability of Failure on Demand	Probability of False Tripping
1oo1	$\lambda^{DU} \times \frac{TI}{2}$	λ^{S}
1oo2	$\frac{(\lambda^{DU} \times TI)^2}{3}$	$2\lambda^{S}$
2oo2	$\lambda^{DU} \times TI$	$2(\lambda^{S})^2 \times MTTR$
2oo3	$(\lambda^{DU} \times TI)^2$	$6(\lambda^{S})^2 \times MTTR$
1oo3	$\frac{(\lambda^{DU} \times TI)^3}{4}$	$3\lambda^{S}$

As far as practical, redundant measurements should utilize differing technologies, both hardware and software, to minimize common mode effects, but this needs to be balanced against the innate probabilities of failure of the equipment. There is no point in using different technology if one of the devices is appreciably less reliable than the other.

Common-mode failures and systematic faults add separate terms to both probabilities, limiting the effect of redundancy in reducing the effects of individual failure.

The probability of failure on demand is the sum of the probabilities of failure of each series component in the system. This means that the chain is always poorer than the least reliable component.

The discussion above is deliberately simplified. More precise methods take into account common-mode failures, the probability of further failures during the repair period, and systematic errors. The effect of online diagnostic cover is also important in many cases.

Several approaches to these calculations can be found in the ISA TR84.0.02 series, and in IEC 61508 Part 4. Computer programs that analyze complex systems are readily available from several sources, such as exida.com and Honeywell. In all cases, however, the most difficult part of the calculation is obtaining accurate failure rate data for the components. Some indicative figures are to be found in ISA TR84.0.02 Part 1, which were derived from the records of several major chemical plants. The variation between these sets is appreciable. Other data can be found in OREDA, which covers most of the oil platforms in the North Sea. In all cases, it must be borne in mind that these reflect previous generations of equipment, and all manufacturers have been working toward higher reliability.

With newly designed equipment, historical data cannot be relied on and analytical techniques such as FMEDA (Failure Modes, Effects and Diagnostic Analysis) are applied by the manufacturers to provide estimates of the frequencies of the various types of potential failures.

A published example of such data for the Siemens-Moore Critical Transmitter is shown in Tables 2.6b and 2.6c below.

TABLE 2.6b
Estimated Failure Rates for Siemens Moore 345 Transmitter

Type	Output Response	Failures per 10^9 hours
DD—Dangerous detected	Short/fail overrange (FO) output ≥ 21 mA	17.5
	Open/fail underrange (FU) output ≤ 3.6 mA	162.7
	Failure (3.7 mA)	295.4
	Fail-safe reaction (FO, FU, or 3.7 mA)	475.6
SD—Safe detected	Fail-safe reaction to 3.7 mA	174.1
SU—Safe undetected	Output normal	86.5
DU—Dangerous undetected	Bad output (readback outside 2% of calculated value)	29.2
AU—Diagnostic annunciation failure	None	94.2

TABLE 2.6c
Alternative Format for Table 2.6b Data

Parameter	Value	Remarks
MTTF	147.7 years	Assume constant failure rate, where $MTTF = 1/(\lambda^{DD} + \lambda^{DU} + \lambda^{SD} + \lambda^{AU})$
MTTFD	226 years	Assume constant failure rate, where $MTTFD = 1/(\lambda^{DD} + \lambda^{DU})$
C^D	94.2%	$C^D = \lambda^{DD}/(\lambda^{DD} + \lambda^{DU})$

MTTF: mean time to failure; MTTFD: mean time to fail dangerously; C^D: dangerous coverage factor.

These are for an ambient of 40°C, and are in units of "failures per 10^9 hours." As Siemens-Moore notes, rates may vary considerably from site to site depending on chemical, environmental, and mechanical stress factors.

These figures are indicative of the upper end of current transmitter technology.

VOTING AS A CONTROL STRATEGY

Voting of analog information can be useful in control as well as in safety systems. Many modern control systems carry a quality bit from each input through their calculations, and this can be used to switch calculations on failure of input.

One of the more useful tools is the use of Median Signal Select. The median of an odd number of signals is the value that has an equal number of signals larger and smaller than the designated one. Thus, if we consider three transmitters measuring a common variable, with measurements 25, 26, and 30%, the median signal is 26%. The median is less sensitive to gross errors than the mean or average, and will allow a system to continue operating in reasonable control if any one device fails either high or low. The median of three signals can be computed by selecting the highest of the outputs from three low signal selects. Thus, for signals A, B, and C, the

$$\text{MEDIAN}(A, B, C) = \text{MAX}(X, Y, Z),$$

where X = MIN(A, B), Y = MIN(B, C), and Z = MIN(C, A).

The same result can be obtained from the lowest of three high selects. See Figure 2.6d.

For an even number of inputs, greater than three, the median is taken as the average of the two most central points.

Care must be taken in using median signal selection with a derivative action controller, as the switching between transmitters must cause a step change in the rate of change of the measurement.

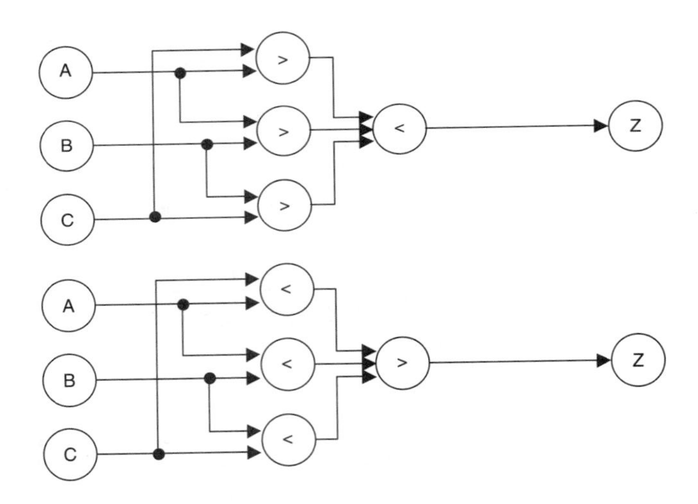

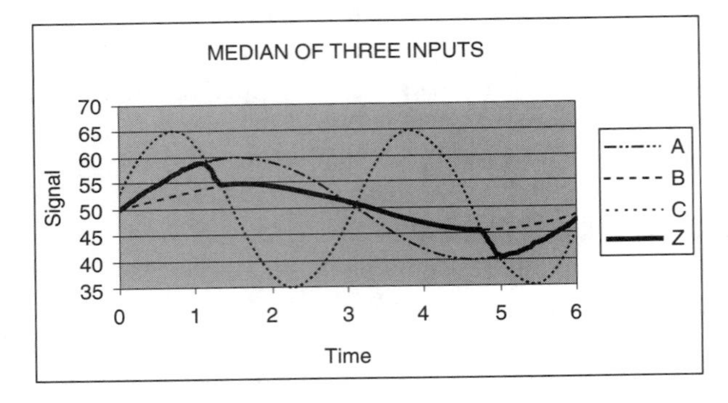

FIG. 2.6d
Median signal select.

With three identically ranged transmitters, one can extract redundant data to determine high alarm, low alarm, and control measurement. The median signal becomes the control measurement, the highest good-quality signal gives the high alarm, and the lowest good-quality signal gives the low alarm. "Good quality" is normally taken as between 0 and 100%, but this may be narrowed considerably by using other information, such as being less than 3 standard deviations away from the historical mean of all signals, or extracted from the on-board diagnostic messages from a "smart" transmitter.

NAMUR DIAGNOSTIC LEVELS

The German "NAMUR Empfehlung" NE-43 defines a set of diagnostic signal levels that can be handled within an analog 4–20 mA system. These are given in Table 2.6e.

It should be noted that this requires the receiving device to be able to interpret the full 0 to 22 mA signal range, and not clip either top or bottom. Many PLC (programmable logic controller) systems have A/D converters that will not read above 20 mA, and some will not read below 4 mA.

With the growing application of digital signal transmission techniques, ranging from HART, which superimposes a digital data train on a 4–20 mA carrier without changing the average value of the 4–20 mA, through to PROFIBUS PE or Foundation Fieldbus (FF), which can provide a mass of diagnostic data from the field equipment to a properly configured control system, the practicality of monitoring both the process and the input and output control devices for developing problems is now apparent and can offer advantages in availability by, in effect, reducing the proof-test interval to an individual measurement/transmit cycle.

This is only useful if the information is actually used, and the application of "Asset Management Systems" in association with information extracted from the process control measurements is a growing field.

TABLE 2.6e
NAMUR Transmitter Signal Ranges for Nominal 4–20 mA Signals

Condition	*Current Range (mA)*
Short circuit or transducer failed high	Output ≥ 21.0
Overrange	20.5 ≥ Output > 21.0
Maximum scale	20.0
Minimum scale	4.0
Underrange	4.0 > Output ≥ 3.8
Transducer detected failure (low)	3.8 > Output > 3.6 (3.7 Nominal)
Open circuit or transducer failed low	3.6 ≥ Output

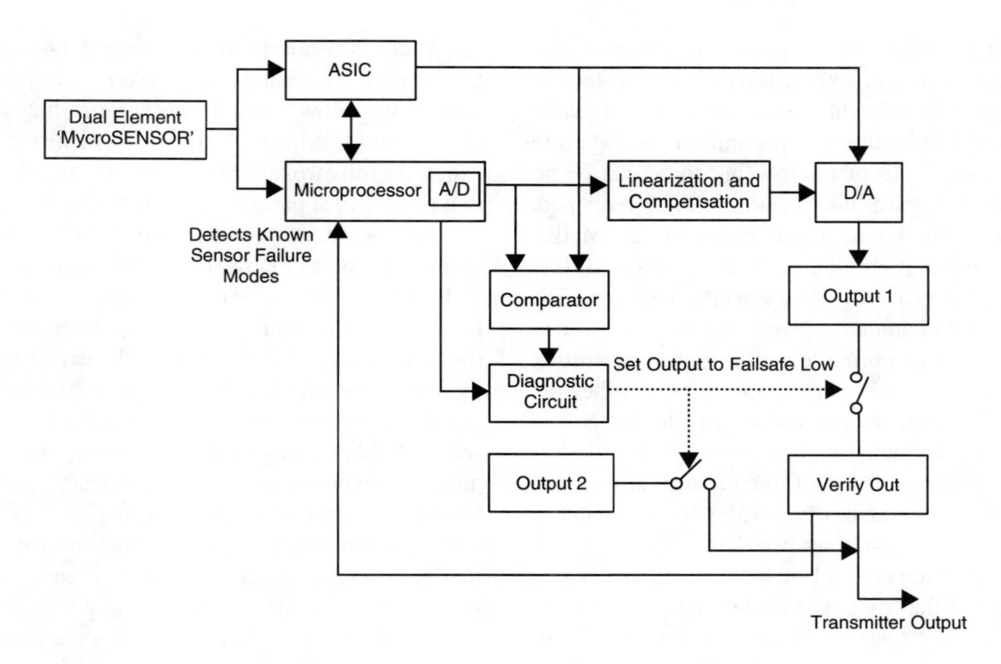

FIG. 2.6f
Siemens-Moore 345 XTC Critical Transmitter block diagram. Redrawn from the Siemens-Moore 345 Instruction Manual.

A recent development in process measurement technology has been the appearance of SIL 2 rated pressure and differential pressure transmitters from at least two suppliers (Siemens-Moore XTC series and ABB 600T series). SIL 2 signifies that the equipment offers a $MTTF^{DU}$ better than 100 years. Such a level of integrity would normally require use of two transmitters with data comparison.

Both the XTC and 600T apply redundant measurement systems with onboard diagnostic techniques to compare the (duplicated) process sensor measurements through the internal signal conversion/temperature compensation and the actual output current, and provide diagnostic messages of detected problems for an installed cost higher but comparable with older designs. These designs do suffer from a possibility of common-mode failure if the process connection becomes plugged, and both sensors see a constant value, but an analysis of expected process noise can in some cases detect this fault. The suppliers both claim SIL 3 ($MTTF^{DU}$ greater than 1000 years) if two such transmitters are fitted in a redundant installation with suitable voting; this would normally require three good-quality transmitters at higher installed cost; see Figure 2.6f.

The effector or final control element is traditionally the most expensive item in an individual loop. Not only are process valves large, high-pressure equipment, but they are exposed to the flow of the process fluid, which may contain erosive, corrosive, or gummy components to reduce the reliability of the device.

For a shutdown system to be testable, the effector must be tested. Unless the valve is closed under flowing conditions, there can be no certainty that it will operate when required. Even then, most block valves cannot be given better than SIL 1 rating, although at least one manufacturer (Mokveld) claims SIL 3 on the basis of long experience and documented in-service reliability on certain of its designs.

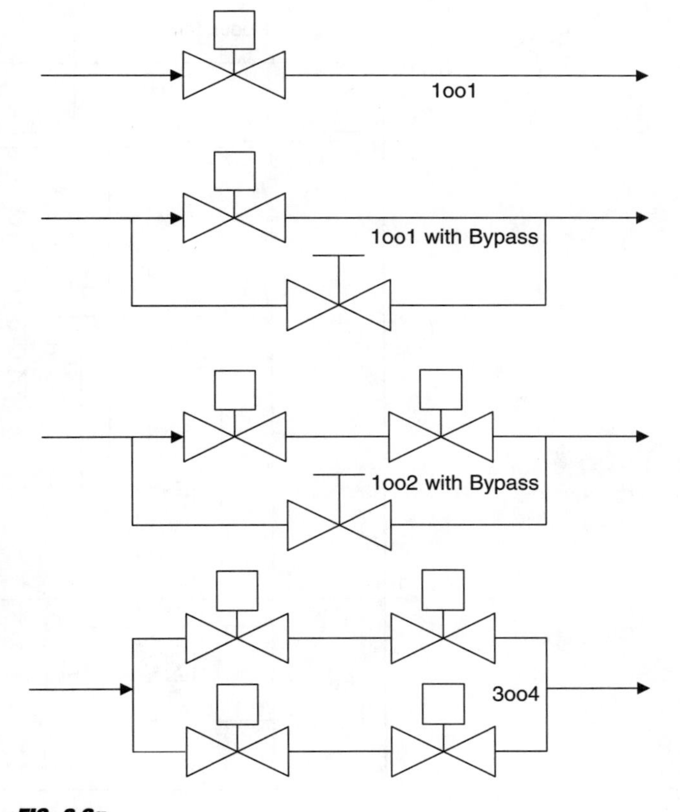

FIG. 2.6g
Valve redundancy patterns.

Because modern plants are commonly required to stay on line for several years, on-stream testing is necessary to validate the availability of the shutdown system. It is possible in some cases to trip a shutdown valve and reopen it before the upstream process is out of bounds, but there can be no certainty that once the valve has closed, it can be reopened. This is a 1oo1 system. Some manufacturers (e.g., Metso/Neles, Emerson) offer partial-stroke testing systems that allow the valve to be stroked partway, while measuring the pressure/torque/time characteristic on the actuator. Such diagnostic information can provide warning of deterioration of the valve and actuator but cannot determine whether the trim will give tight shutoff. The partial-stroke limiting equipment can also be a potential source of error.

If on-stream testing is essential, then a normally closed manual bypass can be provided, fitted with position switches to demonstrate that it is closed in service. This is the 1oo1 with bypass, which gives a lower level of security, as the bypass may be left open and the trip system inhibited. For a higher level of security, a 1oo2 system can be used, with two shutdown valves in series. This can give a SIL 3 rating; again, a test bypass can be fitted at the expense of possible maloperation.

There is a possibility of using 3oo4 voting, but this is unlikely to be economic. In this format, any one valve closing cannot stop flow through the system, but any three valves closing must stop the flow. This permits each valve to be tested on full stroke without loss of safety coverage. (Either two of a vertical pair will also shut off flow.) See Figure 2.6g.

Redundant design solenoid valve assemblies are also commercially available (e.g., Norgren/Herion, ASCO/Sistech). These can use redundant output signals from the PLC to allow online testing of the trip solenoids without taking the safety system into bypass, at the expense of considerably increased complexity. An example is shown as Figure 2.6h.

The common practice of having a control valve with a series shutdown valve offers a low-cost redundancy, but care must be taken to ensure independence of trip and control functions. In many cases the cause of a problem can be a stuck control valve, in which case the trip facility will be ineffective. Care needs to be taken when using double-acting piston actuators, as merely tripping the air supply to the positioner may be ineffective for an extended period. Instead, the trip signal should vent one side of the cylinder and apply supply pressure to the other.

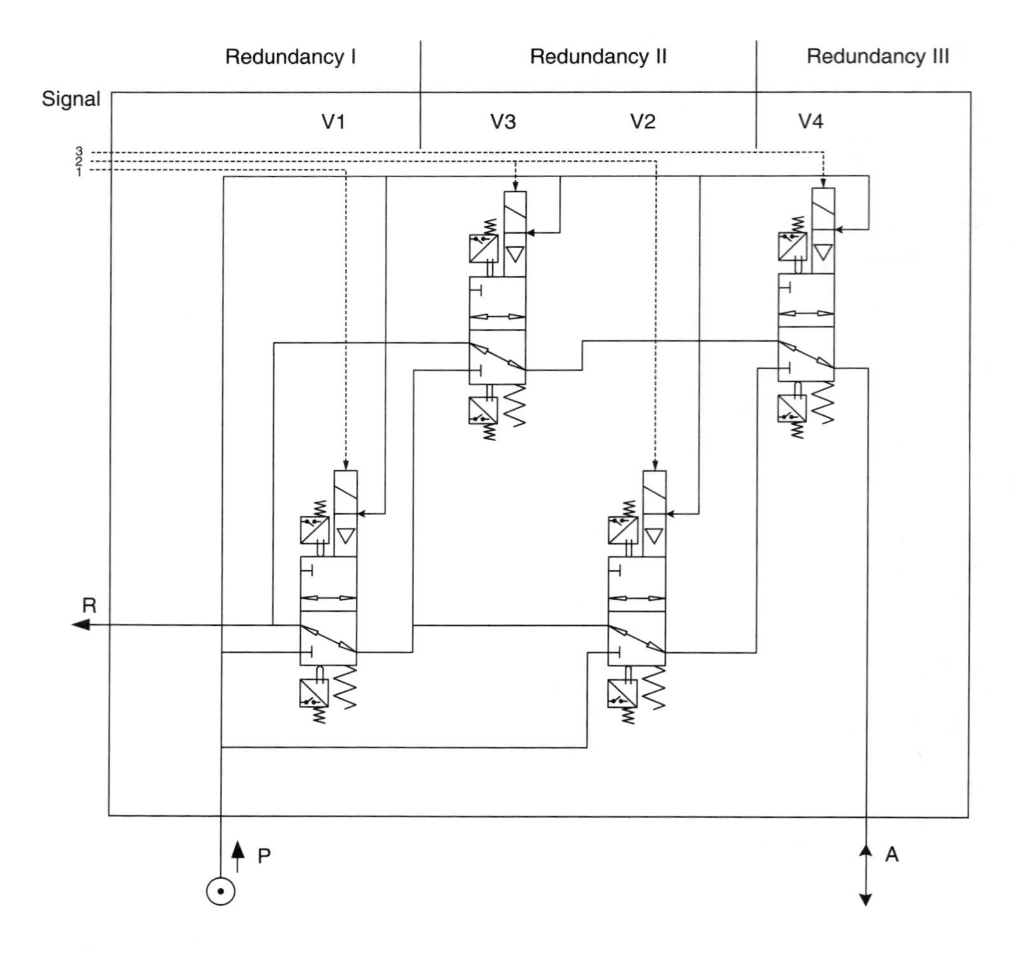

FIG. 2.6h
Schematic representation of a Herion 2oo3 voting solenoid valve. Internal Piloted Herion "2oo3" solenoid valve with switching position monitoring. V1, V4, and (V2/V3) can be independently energized (triplicated output) or from a single source.

It is good practice to trip the control signal to manual/closed whenever the shutdown is activated, but no credit can be taken for this in a SIL calculation, as the entire distributed control system (DCS) system is normally rated below SIL 1.

Control valves have traditionally been provided with single-block isolation and a globe valve bypass of comparable capacity to the control valve. This offers poor functionality in many cases; if the process is a high pressure one, then upstream and downstream double-block and bleed isolation may be mandated for access to the control valve, and the globe valve will often be unsuitable for throttling the process flow for an extended period while the control valve is removed for service. A parallel-redundant control valve with single-block isolation (or even upstream-only isolation) may be an acceptable alternative, to carry operation through to the next plant shutdown.

The concept of "Independent Protection Levels" is discussed elsewhere. This takes the concept of redundancy to a higher level, where the same risk is mitigated by the use of separate systems.

2.7 Network Security

M. F. HORDESKI

The design of a security system starts with an assessment of potential threats based on the order of their significance. The significance of threats should include an analysis of the operating environment to be protected.

Damage to a control system may be caused by malicious or nonmalicious activities. In most cases, the security measures applied to minimize the potential of malicious damage will prevent or to some extent minimize the potential of nonmalicious damage. This does not include hazards like fire or flooding. Attacks on information may be performed through legitimate or illegitimate hardware interfaces attached to the network or to the communication system used by this network. Limiting access to legitimate interfaces through physical security measures increases the security of information but may not completely solve the problem.

The problem of an attack by a legitimate user cannot be solved completely since every legitimate user has some privileges in the system. Users need these privileges to perform their jobs. Their activity can be monitored to prevent access to unauthorized information, but it is impossible to stop users who have legitimate access to information from memorizing it and then passing it to people who have no such access rights. The likelihood of this type of information theft can be minimized by the appropriate personnel policies enforced by the system owners. The design of the security system needs to consider cost effectiveness and user acceptability.

Another problem involves information passing the border of the area controlled by the information owner and being entrusted to a second party. The information is subject to misuse by this party or intruders operating as a result of weak security policies. As shown in Table 2.7a, types of misuse of information include information disclosure where the attacker reads the information, information modification where the attacker changes the information, and service denial where the attacker delays or prevents the information from being delivered.

TABLE 2.7a
Types of Misuse of Information

Information disclosure	Attacker reads the information
Information modification	Attacker changes the information
Service denial	Attacker delays or prevents the information from being delivered

Encryption can be used to prevent information disclosure. Information modification can be detected by encryption techniques combined with redundancy. Service denial attacks can be detected with encryption combined either with time stops or message numbering.

The problem of identification exists in any communication that takes place between two parties. An intruder may pretend to be a legitimate user or a user may pretend to be some other user.

Security should not be based on the secrecy of the design. Secrecy of the design may be required for some operations, but it should not be the only protection of the system.

The rule of sufficient and necessary privilege is important. All users should have sufficient privileges to perform their tasks but only those privileges that are absolutely necessary to perform the required functions.

Access to every part of the system should be monitored. Parts of the system should share as little as possible to minimize the potential of information leaks. But, when necessary, privilege partition execution rules should require the cooperation of the different parts.

The security of the computers on the network also depends on the communications channel. The physical security of the channel is important as well as the logical security of communications over this channel.

PHYSICAL SECURITY

Physical security cannot be supplied by the system supplier and thus becomes the responsibility of the owner. In most cases a local-area network (LAN) is controlled by one party and the consistent implementation of physical security will be based on rules defined by the network owner.

Physical security measures apply to all components of the network: workstations, communication devices, and wiring. As shown in Table 2.7b, physical security measures include site access control, hardware security, organization policies, and hazard protection.

TABLE 2.7b
Physical Security Measures

Site access control
Hardware security
Organization policies
Hazard protection

Physical security can be enforced over the whole installation where the owner has administrative and physical control of the whole network and communication system. In this private network, a legitimate user may be an attacker.

Physical security cannot be fully enforced when the communications system is under control of some other party. Or the communication media may be accessible in the case of communication through a satellite.

Site Access Control

Site access control deals with the management of physical entry to the location where the LAN is found. Control of this entry is accomplished through a variety of techniques. Control points are used to check and admit traffic. Other entrances to the site, such as windows, loading docks, vents, and other potential passages, must be blocked. The control points are monitored by guards or controlled by automatic badge readers, fingerprint checkers, or other automated methods. Picture badges may be required to enter the site.

Updated plans of the layout of LANs (cables, communication equipment, data processing equipment) and site layout plans should be available to all security personnel. Entry and exit logs for each entrance to the site should include name of person, company, time of entry, and time of exit. When there is more than one entrance to the site, the logs should be compared for matching entries on a regular basis, usually once a day. Escorts for persons without appropriate clearances should be provided. The goal of these measures is to ensure that only those individuals authorized to enter a zone are allowed to do so. All others will be turned away.

Hardware Security

This deals mostly with the management of the control facility and its assets. The main goal is to organize and manage the assets in such a way that will minimize the compromising of any information associated with them.

This includes limiting the number of personnel having access to network devices. Network equipment can be locked in separate rooms with access either by key or magnetic card. Locks on the network devices can also limit access by key or magnetic card.

Terminal screens should face walls without windows; this also applies to printers or plotters. The printout should not be able to be seen through the windows. Computer printouts containing sensitive information should be shredded. Magnetic storage media should be completely erased prior to moving them to a lower-security area of the system.

Cables and wires should be installed in such a way that uncontrolled access to them is difficult. Routing open cables through maintenance closets leaves them open to tampering. They can be enclosed in conduit or closed wire racks to minimize the ability to access the media. The cables could be integrated into the floors or ceilings of an office as it is being built and encased in solid conduits. Cable routes can also be alarmed to indicate if tampering is being attempted.

Hazard protection is accomplished through the design of the facility containing the network and its components. The facility must protect the equipment from water damage and power surge. Some of this may be accomplished through the application of adequate building codes and measures. If water damage is a concern, the equipment should be certified for the water conditions expected. Applications in food processing, for example, could be subject to water spraying.

SECURITY POLICIES

The organization must construct policies that will provide the desired level of security. The documentation must provide a clear definition of security policies and security-related responsibilities of the personnel. There also needs to be a clear distinction between sensitive and nonsensitive information.

There should be guidelines on the reporting of security violations, password, badge, and magnetic strip card protection. Records of all security violations should be maintained and backups of all information essential for continuous operations are needed.

A security training program is needed for new employees with update sessions for all personnel. Background checking for personnel is expensive but provides a level of security that may not be obtained in any other way.

Collect badges, cards, and delete access rights for all persons dismissed or moved to other jobs not requiring these rights. Any user should not have more privileges than absolutely necessary to perform the job. For sensitive operations, there can be a separation of duties where two or more persons have to cooperate to perform an operation.

ENCRYPT TO PROTECT NETWORK DATA

A major security concern is the ease with which data can be intercepted and then read. It is not that difficult to tap into a network and eavesdrop as packets go by. One solution is to encrypt network data so that it is unreadable by anyone who does not have the key required to unlock encryption.

Encryption involves transforming normal text into a garbled, unintelligible format. A cipher is used to transform the standard text, called plaintext, to a scrambled form, called ciphertext. Encryption, which involves encipherment, is the process of transforming the plaintext to ciphertext. Decryption or decipherment is the opposite process of transforming the ciphertext to plaintext.

The potency of a cipher depends on the secrecy of the enciphering algorithm or the secrecy of its parameters. The security of a cipher should not rely on the secrecy of the algorithm. Modern technology allows the algorithm implementation to be in the form of an integrated circuit that is resistant to reverse engineering attacks.

The algorithm should involve secret parameters called keys. In most applications except military or secret government communications, the algorithms are public knowledge and the security of the cipher depends on the secrecy of the keys. The encryption key can be public and the decryption key known only to its owner.

When the encipherment and decipherment use the same key, the cipher is called symmetric. In symmetric ciphers, the key is known only to the sender and the receiver. The message enciphered under this key identifies the sender as an author of the message. When the encipherment and decipherment keys are different, the cipher is called asymmetric. Asymmetric ciphers are known as public key ciphers since there is no reason to keep both keys secret.

Public Key Encryption Systems

In public key encryption systems, both the originator and the recipient must have the same key to decrypt a file. One half of the key is made public and published in a directory. The other half of the key remains private. Without this portion of the key, encrypted data cannot be decrypted. The size of the key helps determine the difficulty involved in breaking encryption scheme. In public key systems, the fact that a message is encrypted with the public key does not confirm any information about the message originator since the encryption key is public knowledge.

Although the size of the key is significant, even more significant for public key encryption systems is authentication. This is the process that determines if the parties exchanging keys are authorized to do so and are not intruders.

The authentication process usually involves computer-to-computer communications checking of a public key that could contain 512 bits. Once the authentication process is complete, then the two computers can exchange encrypted data securely.

A well-known example of a public key encryption scheme is the Data Encryption Standard (DES) developed by IBM and the National Bureau of Standards (now the National Institute of Standards and Technology, NIST). This standard has a key length of only 56 bits and some schemes use double-length DES keys to increase the effective key length to over 100 bits and gain greater security. In this approach, the first half of the key is used for encrypting the result, while the second half of the key is used for decrypting the result. Finally, the result is then reencrypted using the first half key again.

Private Key Encryption

A private key encryption system uses a computer as a key distribution center to pass out keys to both sender and receiver. The sending computer sends a set of credentials consisting of user name, time, and date. This acts as a ticket, which is encrypted and then transmitted to the receiving computer where the same key is applied to decrypt the message. An example of this type of encryption system is the Kerberos system developed at MIT. A private key encryption scheme tends to be much faster than a public key system.

Message Authentication Code

If the sender and receiver use a public key or private key encryption system, the message may be transmitted successfully, although it might have been garbled during transmission. The message authentication code is a process that involves appending a cyptographic checksum on the end of each message. The sending and receiving computers must know the cyptographic key so that both can correctly compute the checksum.

GOVERNMENT NETWORK SECURITY

The U.S. government is mandated by the Computer Security Act to have a level of security equal to the National Security Agency Orange Book C2 level. This level requires a security system for each user with computer files automatically overwritten by random 1s and 0s when they are deleted. All users are required to have login IDs and passwords.

The act encourages the use of devices such as electronic keys to improve password security. All user actions must be audited as a part of an audit trail. C2 security requires that hardware be protected through logical hardware locks and keys. C2 requires network interface cards with DES encryption chips. Separate RAM that is not addressable by the computer, as well as other security programs, must be used for storage of encryption.

CALLBACK MODEMS

Callback modems can be effective for preventing intruders from logging onto the network remotely. These dial-back systems identify the user and prevent unauthorized external access to the network. When a remote user attempts to log on to the network, the security device requests an access code and then hangs up. If the code is valid, the device calls back the user at a prearranged number.

IDENTITY VERIFICATION

Verification of a user's identity is essential to physical security by controlling the physical access to hardware. When this type of access is granted, it controls a user's access to the system resources.

Identity verification of users can be based on personal characteristics or on tokens in their possession. Methods based on personal characteristics include fingerprints, hand geometry, retinal patterns, voiceprints, and signatures. Methods involving tokens include magnetic strip cards, smart tokens, keys, and passwords.

The methods based on personal characteristics are newer and require sophisticated methods of measurement so they tend to be more expensive. Methods based on personal characteristics can also have higher margins of error because the measured characteristics depend on the physical and mental condition of the individual. These methods are often more difficult for users to accept, because direct contact of the sensor with parts of the human body may be required.

In these methods, measurements of the user's characteristics are taken and transformed into a representation of the user's identity, which is stored by the system and used in future identification attempts. Tokens can be stolen, but falsification of the personal characteristic is difficult and in many cases impossible.

The characteristics measured depend on the method. In identity verification based on a fingerprint, the configuration of characteristic lines such as loops, arches, whorls, and ending and splitting lines is considered. In methods based on signature, static and dynamic characteristics are measured. Static characteristics reflect the shape of the signature, while dynamic ones reflect the speed and pressure applied to the pen.

In some applications, the time of a single user identification may be important. The time required for identification among the various methods may vary, but most of them are in the range of seconds. The error margin in the different methods ranges from a few percent to almost 0% for methods based on retinal patterns.

Because the methods used for personal characteristic verification involve comparison of the measurement with a threshold value stored in the system, two types of errors can exist:

1. A legitimate user can be rejected.
2. An illegitimate user can be accepted.

A lowering of threshold values increases the number of errors of the first type, and an increase of the threshold values increases the number of the second types of errors.

The acceptance of illegitimate users is a security violation, but the rejection of legitimate users creates a nuisance problem and encourages users to shortcut the security system.

OPERATING SYSTEM SECURITY

Security management at the operating system can use four basic levels: login/password, user (trustee), directory/folder, and file (Table 2.7c).

TABLE 2.7c
Security Levels at the Operating System

Security Levels
Login/password
User (trustee)
Directory/folder
File

These levels act as four doors through which a user must pass to reach the data contained in files on the network. If users lack the proper keys at any one of the four doors, they are effectively barred from the data.

Because network operating systems allow network users to share applications and files across the network, they usually have sophisticated file protection methods that are tied directly to the security features of the network. The combination of these rights determines which directories, files, and applications each user can access on the network.

Security starts with user accounts and passwords, which allow only authorized persons to log in to the network. Passwords are unique to individual users and are usually encrypted so that not even the network administrator can see what they are.

Once logged in to the file server, a user's access to files may be limited by trustee right assignments. Trustee security is established through basic rights like Read, Write, Open, Create, and Delete. A combination of these rights is needed to enable users to perform particular functions. For example, to run an application, users need at least Read and Open rights to the program files of the application. The Read right allows the user to view an existing, open file. Users need the Open right to open the file.

One can also assign trustee rights to users in groups to control access to an application or directory. For example, if several employees may need to access those files containing reports, one can create a group and give that group the necessary rights to the applications used to modify the report files. The rights allowed in a rights mask can be coupled with trustee rights to give an added layer of security. The mask determines what rights any user may exercise and overrides individual users' trustee rights.

File attributes (or flags) are one way of preventing accidental changes. A file may be flagged Shareable or Nonshareable, Read/Write, or Read/Only.

A file can also be flagged Execute Only or Hidden. A hidden flag hides the files from a directory listing. Execute Only allows applications to be run or deleted, but not copied or renamed.

An operating system like NetWare can store security files such as group security, director security, and file security in a special-purpose database. In NetWare this is called the bindery. Each server maintains its own bindery containing the security information for all the users and groups defined for that server. The access and security are file server based.

The user account must have the necessary rights to access the application.

Each entry in the bindery is composed of an object and its associated properties. An object can be a user name, a group name, the name of a resource (such as a print queue), or any other entity (logical or physical) that has a name. The operating system assigns a unique ID number to each object. The properties of an object describe the information known about the object.

The properties for a user object would include the user's password, trustee rights, and a list of groups to which the user belongs. The properties of a group would consist of a list of the group members and the group trustee rights. A property for a resource could be a list of users authorized to use the resource.

LOGIN AND PASSWORD SECURITY

Login and password security are closely connected. To log in to a file server, a person must correctly enter a valid user name which is unique to each user on the network, followed by the correct password for that user account. If the person enters either one or both of these items incorrectly, the person is denied access to the file server.

Login security determines how secure this login process is whereas password security is a subset of login security. It affects how secure the password part of the login process is. Login security depends on account restrictions, which include password restrictions and time restrictions that determine when a user can log in. Station restrictions may also be used that limit a user to logging in only at specific workstations.

Account Restrictions

Several login-related restrictions can be placed on a user's account. The ability to invalidate an account temporarily without deleting it altogether is useful when it has been determined that an intruder has been trying to log in using that account. One can also set an expiration date on any user account and limit the number of concurrent logins a user can perform. If the maximum number of connections is set to one, the user can log in on only one workstation at a time.

Password Restrictions

Passwords are a critical component of the overall security system. To preserve the confidentiality of passwords, they are not allowed to appear on the screen, either when a user is logging in or when the user changes the password. For maximum security, every user account including GUEST accounts should be required to have a password.

Passwords should be short enough to remember, yet unusual and difficult to guess. To be secure, passwords should be at least eight characters long. Allowing passwords shorter than eight characters increases the chances of someone cracking a password and illegally entering the network. The shorter the password, the easier it is to guess and test. The maximum number of characters one can type in the logon box will make it harder for hackers, but also harder for users to log on.

Place restrictions on all user-defined passwords. Do not use words for passwords that contain the birth date, middle name, and/or spouse's and children's names, because this makes it easier for hackers to obtain access to the password. Also, do not use any common word that can be found in a dictionary. A common hacker technique is to run lists of common words such as names of famous places or people. Use more complicated rather than simple passwords.

The later Windows NT versions have a service pack with features that improve the security of the network passwords. It requires the user passwords to contain characters from at least three of the following four groups:

1. English uppercase letters (A to Z)
2. English lowercase letters (a to z)
3. Arabic numerals (0 to 9)
4. Punctuation or other special characters

Many Internet sites require one to enter a username and password to access information. Users should never use their Windows NT logon password to sign on to a Web site. There is the chance that someone could intercept the password and try it to log onto the system.

NT Login and Registry Security

There is a Microsoft utility called TweakUL. This utility stores the username and password values in the Windows Registry in an unencrypted format. Hackers can access the registry.

If the system is set up for automatic logon, one can still protect the login information by editing the permissions associated with the segment of the registry where the information is automatically stored.

First, open the Windows NT registry editor and choose Permissions from the security dropdown menu. Then, change the default permission for the Everyone group from Read to Special Access. Next, deselect the Query Value check box. Now, when someone other than an administrator tries to access the key, it will appear grayed out.

Windows NT can clear the username on the logon screen. NT by default displays the name of the last person who logged onto the system. This can be a security threat, because a hacker may guess a user's password based on the account name or the login environment. One can clear the previous username by adding a new string value called

DontDisplayLastUserName

Security is improved if there are only READ capabilities on the registry keys. The default setting is WRITE on some of the most important keys. These WRITE privileges allow a user to gain access to the server. This can be prevented by changing the permissions on these keys to READ. Use the registry editor to change the security/permissions on these keys.

When a user starts an NT workstation, it automatically creates some shared resources. These include a share to the root directory of the hard drive that remains invisible during browsing. To prevent this automatic resource sharing one can edit the NT registry settings.

Windows NT holds data in swap files and temp files. To speed access and improve performance, pages of data are swapped in and out as needed. When the system shuts down, the swap file stays on the hard drive. An intruder could view data stored in the swap file. The NT system can be configured to clear the swap file every time the system shuts down.

This can be done by using the NT registry editor. Go to HKEY_LOCAL_MACHINE and add

ClearPageFileAtShutdown

This will clear all pages except those that are still active during shutdown.

One of the common paths a hacker uses to gain access to a network is through services running on the server. These include the ClipBook Server, Plug-&-Play, and Net Logon. Consider shutting down those that are not needed.

Allowing users to select and change their own passwords is more user-friendly than assigning preset passwords that users cannot change. Assigned passwords often have no meaning to users and may be awkward to remember, especially if changed frequently. Users often resort to writing down frequently changed, hard-to-remember passwords, and posting them near their workstations, which defeats the whole purpose of a password.

Periodic Password Changes

Passwords may expire after a certain period of time. When a user's password expires, the user receives a message during the next login attempt indicating that the password has expired. Users can log in with the old password if a certain number of grace logins are allowed. If a user exceeds the maximum number of grace logins without changing the password, the user will be locked out of the system until the user's account is cleared.

The length of time between forced password changes depends on the type of network. Typically, network users should change their passwords every 30 to 60 days.

Users should be required to come up with a new, unique password every time their old one expires. Users are sometimes tempted to alternate between the same two or three passwords so they can remember them more easily, but this practice compromises password security.

Another technique invalidates the password on a specific date. If the date an employee is leaving is known, one can set the password to expire on the employee's last workday.

Users may select passwords that mirror their name or the name of their spouses, children, or pets. Requiring unique new passwords every 30 or 60 days will soon exhaust these names or nicknames for spouses, pets, and children and make it more difficult for people trying to log on illegally.

Another password-related problem found in many large companies is the yellow stick-on note with the network password written on it attached to the top of a monitor or side of a computer.

Some users place their network passwords in batch files to speed up the login process. These batch files can be examined by a proficient intruder who may use the TYPE command to learn the password. One solution to this potential problem is to restrict users to a particular workstation.

Time restrictions can be used to limit the hours of the day when users can access the network. Users can also be prevented from logging in on certain days of the week or logins could be restricted on Saturday and Sunday.

Station restrictions allow a user to log in only from a specific workstation on the network. If the workstation the user is trying to log in from does not match those specified in the station restriction, the user will not be allowed to complete the login process.

Systemwide Login Restrictions

Some login restrictions can be set up so they apply universally across the network (Table 2.7d). You can set a common expiration date for all accounts as well as a systemwide minimum password length. Periodic changes on all user passwords can be required along with login time restrictions. The number of days between password changes can affect security as well as the number of grace logins.

Illegal attempts to enter a network using a particular login can indicate a concerted attempt to break into the network or a forgetful user. One can set a limit for the number of unacceptable login attempts. Network login removal should be part of each employee's checkout procedure.

Dynamic Passwords

Dynamic passwords are a way of discouraging intruders from successfully logging onto the network. Each time users log in, they have a new password. To accomplish this, users have a handheld remote password generator (RPG) device and special software must be run on the network.

TABLE 2.7d
Login Restrictions

A common expiration date for all accounts
A systemwide minimum password length
Periodic changes on all user passwords
The number of days between password changes
The number of grace logins
Login time restrictions

A user logs on and is greeted with a challenge number. The user retypes this number along with a personal identification number into the RPG. This device then generates a one-time-only new password. Each RPG is electronically linked to an employee's personal identification number and cannot be used by anyone else. Outside intrusion techniques that can compromise information and security on a network include network packet sniffers, password attacks, IP (Internet protocol) spoofing, middle-man attacks, denial-of-service attacks, and application layer attacks.

Packet Sniffers

Networked computers communicate serially by sending one data segment after another. The large data segments are broken into smaller pieces, which are the network packets.

When the network packets are in clear text, the packets are not encrypted and can be processed and read by a packet sniffer. This could be implemented in either software or hardware. Packet sniffers are designed for network fault analysis. They can capture the packets off the network and perform processing for data analysis. The packet sniffer software can capture the packets sent over a physical network wire. There are freeware and shareware packet sniffers that allow any user to get into a system without any knowledge of the inner workings of the system.

A packet sniffer could be used to get information that is on the network, including user account names and the passwords used. Other packet sniffer information includes the layout and operation of the network, including what computers run specific operations, how many computers are on the network, and which computers have access to others.

Password and Middleman Attacks

Password attacks mean that there are repeated attempts to identify a user account or password. These attacks occur from brute-force, repeated attempts, which are usually automated.

If attackers gain access to the network, they could modify the routing tables for the network. Then, all network packets can be routed to the attacker before they are transmitted to their final destination. This allows the attacker to monitor all network traffic and become what is called a middleman.

A middleman attack can come from someone who is working for the Internet service provider (ISP). The attacker could gain access to the network packets transferred between the network and any other network using packet sniffers and routing or transport protocol software.

These attacks could result in the theft of information, denial of services, corruption of transmitted data, introduction of new or false information into network sessions, interruption of control, and the degradation of network performance.

Denial-of-Service Attacks

Denial-of-service attacks attempt to make a service unavailable. This can be done by exhausting or slamming a server on the network. Denial-of-service attacks may use Internet protocols, such as transmission control protocol (TCP) and the Internet control message protocol (ICMP). These attacks probe a weakness in the system. Flooding the network with undesired network packets is one technique that is used. Another technique is to plant false information about the status of network resources.

Application and Trojan Horse Attacks

Application attacks use some weakness in the software that is used on servers. The attacker will try to gain access with the permission of the account running the application.

Trojan horse program attacks use a program that the attacker substitutes for another program. The substitute program can furnish the functions of the normal program, but it also can monitor login attempts and seize user account and password information.

A Trojan horse program may display a screen, banner, or prompt that looks like a valid login sequence. The program will process the information that the user types in and e-mail it to the attacker.

Disk Lockout

Some network operating systems have a local disk lockout feature, which restricts users from accessing their local disk drives while in certain applications. This can prevent them from copying confidential information.

A related technique is to use diskless workstations so that intruders have no drive to use for copying. An adjunct measure is making key program files executable only so that they cannot be copied.

PROTECTION FROM VIRUSES

In recent years, computer viruses have been publicized as one of the most serious threats known to networks. Some of this threat is real and some is the result of media sensationalism. However, no one can ignore the possibility of virus infection in the network. It is important to recognize virus symptoms and know what corrective actions are available. One should also know how to prevent the spread of viruses on the network.

Recognizing a Virus

A computer virus is a small program that propagates within a computer system by generally attaching to other software and reproducing a part of itself. Many variations exist, known by equally offensive names such as bombs, worms, and Trojan horses. The distinctions between each variation are not as important as knowing that they all spread by infecting user programs. Just as a biological virus spreads by infecting its host and then being passed to other hosts, a single infected program on a network can quickly take over the network as the virus spreads to other programs. Most viruses are self-replicating

TABLE 2.7e
Typical Virus Effects

Increasing system response times
Displays of strange messages
Modifying or destroying data
Physical damage of hardware
Network asleep and does not respond

sections of computer code that hide in computer programs. They attach themselves to other programs and ride along with these programs when they are copied.

They can be attached to a disk given by a friend or colleague. They can be transmitted over phone lines when programs are downloaded or e-mails are acknowledged.

Viruses have also been known to infest commercial software. Some companies have had to recall thousands of copies of programs that had mysterious messages flashing across users' screens. Most software companies now use virus detection programs to check all new software.

Some virus types begin working immediately after infection; others are dormant until some predetermined condition or date triggers the destructive mechanism. Typical virus effects are shown in Table 2.7e.

One warning sign of virus infection is a noticeable slowdown in system response time that gets worse over time. Another is a program that takes up more and more RAM with every succeeding execution.

Another sign of a virus is abnormal screen displays, such as cascading characters, screens filled with snow, ominous messages appearing at random, and files mysteriously disappearing or listed on the screen as not found.

Worms are a type of virus program that try to take advantage of any weaknesses in an operating system to replicate themselves. They will try to consume all data and cause the system to crash.

A Trojan horse is a type of virus program that appears innocent but contains hidden instructions to do destructive things. A Trojan horse program can contain all kinds of unwelcome guests including time bombs that are set to destroy data and other programs on a specific date when triggered by the computer's own clock.

A typical virus like the Joshi strain may alter the partition table program with its own code. The virus then occupies a portion of the partition table, and when the system is booted up the virus reserves a section of RAM for itself. The virus copies itself over to this area of RAM and begins to perform its tasks. These include any attempts by the system to access disk drives, view, or modify the partition table for self-preservation. It now resides in RAM and deflects any attempts to reboot the system.

While the virus lives in a computer, it tries to infect any disk that is accessed from the disk drive. When a certain date occurs, the Joshi virus announces itself and demands that the user wish the creator (Joshi) a happy birthday.

Antivirus Utilities

It is best to use a combination of virus detection and removal utilities, like Antitoxin or Disinfectant, with preventive programs such as Vaccine or GateKeeper. Most of these are available from online services or through users' groups. Do not expect them to provide 100% protection.

There are also browser-based services by application service providers or software companies like Norton. These check for new virus programs on a daily basis.

Various techniques are used against these uninvited software guests. Some virus protection software depends on data integrity. They may check the file length and let the user know if any parameters have changed since they were last inspected. To compare files, some programs use a checksum or cyclic redundancy check. A checksum is a unique number derived by running files through a data compression program such as ARC or ZIP. The checksum is almost certain to differ between two files that are not exactly identical.

Scanning programs look for strings of code that are associated with certain viruses. They alert the user when a potential virus is detected.

Another approach is terminate-and-stay-resident (TSR) virus detection programs. These programs reside in computer memory until a new file is loaded. It then scans the file. This approach is faster than file-server programs, but it uses up RAM. A typical antivirus program may be able to handle hundreds of viruses, but new viruses are always being hatched. Most software vendors provide updates to handle new viruses.

Most antivirus programs detect viruses and also remove them from infected files, boot sectors, and partition tables. Some programs also immunize files to protect them from virus infection. The immunized files can then detect and eliminate viruses that attack them.

The network can be protected against viruses in the following ways:

1. Do not allow users to upload or download executable files on the network.
2. Provide write-protect status to executable files and use write-protect tabs on disks.
3. Keep virus-scanning software on the network to check all new programs.
4. Have a regular backup program so that the network can recover should a virus attack.
5. Limit user rights.
6. Store and protect original program disks.
7. Have all users log on and off.
8. Check new software on a stand-alone system before installing it on a network.
9. Do Internet searching on stand-alone computers.
10. Use a firewall.

PREVENTIVE/CORRECTIVE MEASURES

Preventing unauthorized access is the first step toward controlling what enters the network. Only legitimate users should be allowed on the network. Enforce tight password security and require users to change passwords frequently.

Enable the security features provided with the network operating system so that users have access to data on a need-to-use basis. Flag all executable files as Read-Only to prevent them from being written to.

Keep a log of all accesses or attempted accesses to the network. The log should record at least the time, date, and node address of each access. This information can help pinpoint when and where any illegal break-ins occur. An intruder detection/lockout option in the operating system can also help in this area.

A common way for a virus to enter a computer system is through a contaminated floppy disk. Always use a boot diskette that is known to be safe.

To avoid acquiring contaminated software, software disks should be in original packaging. Pirated programs could harbor infection. Public domain, shareware, or freeware program suppliers should obtain the software directly from the authors and check all disks for viruses before distributing them. If it is uncertain where a program came from, contact the author and compare the creation data and file size with the version in question. Do not use any program that does not provide a contact address or phone number. Software of unknown origin is the most likely to contain viruses.

Thoroughly test all new software on an isolated system before uploading it to the network. There have been cases where commercial software, including antivirus programs, contained viruses.

Make working copies of all original disks and write-protect the originals. Store them in a safe place so there are always uninfected masters available if virus contamination should occur.

Regularly check system programs for unusual behavior, look for unexplained size changes in executable files, a change in the number of hidden files, files with unusual creation dates and times, unusually large or small files, and files with unconventional names or extensions.

Remove any suspicious programs with a utility that completely overwrites the disk space formerly occupied by the deleted files. DOS type DEL or ERASE commands do not actually remove files from the disk until the space is needed for other files.

Make regular backup copies of program and data files. If a virus does destroy data, that data can be restored from a backup copy. However, some viruses lie dormant for a while, so it is possible for them to be copied onto the backup disks.

Encrypt data whenever possible, especially when the data are sent over telephone lines or other transmission media. This security measure discourages unauthorized tampering with data during transmission.

INTERNET ACCESS

Levels of security should be implemented for applications with Internet access. This will restrict the levels of access that are available to remote users. This form of security is available with NT, Internet Information Server (IIS) security, SSL (Secure Socket Layers), digital certificates, and encryption. Protected data access via the Internet includes read-only access of process graphics using standard NT and IIS security.

In Windows NT–based systems, general security is supported through the operating system and IIS. This allows one to enable and disable access to files and directories. SSL is a protocol developed by Netscape that allows a secure transaction between standard browsers and Web servers on the Internet. Both the Netscape Navigator and Internet Explorer browsers support SSL, which uses the public key method of encryption.

DIGITAL CERTIFICATES

Public keys can be authenticated with digital certificates. A digital certificate involves a digital identification that is used with encryption to secure the authenticity of the parties involved in the transaction. Digital certificates provide a type of identification or passport that allows users to navigate across multiple networks. First, the client (user) initiates the connection; then the server sends back a digital ID. The client responds by sending the server its digital ID. Next, the client sends the secure key to the server to initiate the secure session. Users retain their access levels and credentials (digital IDs) for each network.

Digital certificates provide the ability to identify users, access levels, and authorized functions in multiple networks. When a user's levels are defined and a digital certificate is issued, the user's transactions can be audited.

SECURING THE NETWORK WITH FIREWALLS

Network security can include firewalls and proxy servers. Proxy servers and firewalls act as traffic cops. They regulate who enters and exits the network.

A firewall monitors the traffic crossing network perimeters and sets restrictions according to the security policy. Perimeter routers can also be used. These can be located at a network boundary. Firewalls commonly separate internal and external networks. The firewall server acts as the gateway for all communications between trusted networks and untrusted and unknown networks.

The least secure network is the outermost perimeter network of the system. This network is the easiest area to access, making it the most frequently attacked. In some cases, the least secure network will be the Internet. Then, the firewall server will go between the Web server and the secure network.

A firewall protects networked computers from intrusion that could compromise operations, resulting in data corruption or denial of services. Firewalls are implemented as hardware devices or software programs running on a host computer. Internet connectivity may also be provided by this computer. The firewall device or computer provides an interface between the private network it is intended to protect and the public network that is exposed to intruders.

Because this firewall interface resides at a junction point between the two networks, it becomes a gateway or interconnection between them. The original firewalls were simple linking routers that segmented the network into distinct physical subnetworks. This limited the damage that could spread from one subnet to another like a firedoor or firewall.

Today's firewall checks the traffic routed between two networks. If this traffic meets certain criteria, it is routed between the networks. If it does not meet these criteria, it is stopped. The firewall filters both inbound and outbound traffic. It allows one to manage public access to private networked resources such as control applications. It can be used to log all attempts to enter the private network and can trigger alarms when hostile or unauthorized entry is attempted.

Firewalls can filter packets based on their source and destination addresses and port numbers. This is known as address filtering. Firewalls can also filter distinct types of network traffic. This is known as protocol filtering because the decision to forward or reject traffic depends on the protocol used. Firewalls can also filter traffic based on the packet characteristics.

Firewalls cannot prevent individual users with modems from dialing into or out of the network and bypassing the firewall. Because employee mischief cannot be controlled by firewalls, policies involving the use and misuse of passwords and user accounts must be enforced.

One group that was accused of breaking into information systems run by ATT, British Telecommunications, GTE, MCI World Com, Southwestern Bell, and Sprint did not use any high-tech methods. It used industrial espionage including dumpster pilfering and tricking employees into revealing sensitive information.

Dumpster pilfering, which is sometimes called garbology, involves looking through company trash. Firewalls cannot be effective against these techniques, but firewall software may be used to isolate the network interface from the remote access interface.

IP Spoofing

IP spoofing occurs when an attacker outside the network pretends to be a trusted computer on the system by using an IP address that is within the range of IP addresses for the network. The attacker may also use an authorized external IP address to provide access to specific resources on the network.

An IP-spoofing attack is limited to the injection of data or commands into an existing stream of data that is passed between a client and server application or a peer-to-peer network connection. For bidirectional communication, the attacker must change all routing tables to point to the spoofed IP address.

An attacker may also emulate one of the internal users in the organization. This could include e-mail messages that appear to be authentic. These attacks are easier when an attacker has a user account and password, but they are still possible by combining spoofing with a knowledge of messaging protocols.

Many firewalls examine the source IP addresses of packets to check if they are proper. A firewall may be programmed to allow traffic through if it comes from a trusted host. An intruder or cracker may try to gain entry by spoofing the source IP address of packets sent to the firewall. If the firewall thinks that the packets originated from a trusted host, it might let them through unless other criteria were not met. The intruder would need to know the rule base of the firewall to attack this type of weakness.

One technique against IP spoofing is to use a virtual private network (VPN) protocol such as IPSec. This involves encryption of the data in the packet as well as the source address. The VPN software or firmware decrypts the packet and the source address and performs a checksum. If either the data or the source address has been changed, this is likely to be detected and the packet will be dropped. The intruder would need access to the encryption keys actually to penetrate the firewall.

Firewall Problems

Firewalls bring some problems of their own. Information security involves constraints on the users that are not always popular. Firewalls restrict access to certain services. Users need to log in and out, to memorize different passwords, and not to write them down on a note on the computer screen. Firewalls can also result in traffic bottlenecks. They concentrate security at one spot, providing a single point of failure.

Firewalls can protect private LANs from hostile intrusion from the Internet. Many LANs are now connected to the Internet where Internet connectivity would otherwise have been too great a risk.

Firewalls allow network administrators to offer access to specific types of Internet services to selected LAN users. This selectivity is an important part of any information management program. It involves protecting information assets and controlling access rights. Privileges are granted according to job description and needs.

Types of Firewalls

There are four basic types of firewalls:

1. Packet filters
2. Circuit level gateways
3. Application level gateways
4. Stateful multilayer inspection firewalls

Packet Filters

Packet filtering firewalls work at the network level of the OSI model, or the IP layer of the TCP/IP protocol stack (Table 2.7f). They are usually part of a router, which is a device that receives packets from one network and forwards them to another network.

In a packet filtering network, each packet is compared with a set of criteria before it is forwarded (Table 2.7g). Depending on the packet and the criteria, the firewall can drop the packet, forward it, or send a message to the originator. Rules can include source and destination IP address, source and destination port number, and protocol used.

Packet filtering firewalls are characterized by their low cost. Most routers support packet filtering. Even if other firewalls are used, implementing packet filtering at the router level provides an initial degree of security at a low network layer. This type of firewall only works at the network layer, however, and does not support more complex rule-based models.

TABLE 2.7f
OSI and TCP/IP Software Layers

OSI Model	TCP/IP Model
7 Application	
6 Presentation	5 Application
5 Session	
4 Transport	4 Transport Control Protocol (TCP) User Datagram Protocol (UDP)
3 Network	3 Internet Protocol (IP)
2 Data Link	2 Data Link
1 Physical	1 Physical

TABLE 2.7g
Firewalls

Packet filtering network	Traffic is filtered based on rules including IP address, packet type, and port number; unknown traffic is only allowed to (IP) level 3 of the network stack
Circuit-level gateway	Traffic is filtered based on session rules, such as when a session is initiated by a recognized computer; unknown traffic is only allowed to (TCP) level 4 of the network stack
Application-level gateway	Traffic is filtered based on application rules for a specific application or protocol; unknown traffic is allowed up to the application level at the top of the stack
Stateful multilayer inspection firewall	Traffic is filtered at the top three levels based on application, session, and packet filtering rules; unknown traffic is allowed up to (IP) level 3 of the network stack

Network address translation (NAT) routers offer the advantages of packet filtering firewalls and can also hide the IP addresses of computers behind the firewall. They also offer a level of circuit-based filtering.

Circuit-level gateways work at the session layer of the OSI model, or the TCP layer of TCP/IP. They monitor TCP handshaking between packets to determine if a requested session is legitimate. Information passed to a remote computer through a circuit-level gateway appears to have originated from the gateway. This is useful for hiding information about protected networks.

Circuit-level gateways are relatively inexpensive and have the advantage of hiding information about the private network they protect. However, they do not filter individual packets.

Application-level gateways, which are also called proxies, are similar to circuit-level gateways except that they are application specific. They filter packets at the application layer of the OSI model. The incoming or outgoing packets cannot access services for which there is no proxy.

Because they examine packets at the application layer, they can filter application specific commands. This cannot be done with either packet filtering firewalls or circuit-level firewalls because they do not know the application-level information.

Application-level gateways can also be used to log user activity and logins. They offer a high level of security, but they can have a serious impact on network performance. This is due to the use of context switches that slow down network access. They are not transparent to end users and require manual configuration for each client computer.

Stateful multilayer inspection firewalls couple several aspects of other types of firewalls. They filter packets at the network layer, determine if session packets are legitimate, and evaluate packets at the application layer. They allow a direct connection between client and host, which reduces any problems caused by the lack of transparency of application-level gateways.

They use algorithms to recognize and process application-layer data rather than running application-specific proxies. Stateful multilayer inspection firewalls offer a high level of security, good performance, and transparency to end users. They are expensive and because of their complexity may be potentially less secure than simpler types of firewalls if not administered properly.

Bibliography

Firewalls Q and A, www.vicomsoft.com/knowledge/reference/firewalls, December 13, 2000, 1–14.

Fortier, P. J., Ed., *Handbook of LAN Technology,* 2nd ed., New York: McGraw-Hill, 1992.

Liebing, E. and Neff, K., *NetWare Server Troubleshooting and Maintenance Handbook,* New York: McGraw-Hill, 1990.

Schatt, S., *Understanding Network Management: Strategies and Solutions,* New York: Windcrest/McGraw-Hill, 1993.

2.8 Safety Instrumented Systems: Design, Analysis, and Operation

H. L. CHEDDIE

Vendors of Programmable Electronic Systems (PES):

ABB, http://www.abb.com/: ABB manufacturers several lines of dual and TMR (triple modular redundancy) safety programmable logic controllers (PLCs); GE Fanuc, http://www.gefanuc.com: GE Fanuc has a special configuration of its Series 90–70 PLC approved for safety applications; Hima, http://www.hima.com/: Hima is a manufacturer of a line of safety PLCs; Honeywell, http://www.iac.honeywell.com/: Honeywell manufactures a redundant fail-safe controller (FSC) that is used in high-integrity safety-critical applications; ICS Triplex, http://www.icstriplex.com/: ICS Triplex manufactures Regent and Trusted, two lines of safety PLCs based on TMR technology; Pilz, http://www.pilz.com/: Pilz manufactures a line of safety PLCs targeted primarily at the machine control industry; Rockwell Automation, http://www.automation.rockwell.com/: Rockwell Automation and HIMA partner in the development and production of customized safety-specific controllers and software for safety-critical machinery; Siemens-Moore, http://www.smpa.siemens.com/: Siemens-Moore manufactures a line of safety PLCs targeted for the process and machine control industries; Triconex, http://www.Triconex.com/: Manufacturer of a TMR safety PLC; Yokogawa Industrial Safety Systems, http://www.yis.nl/: YIS supplies the PROSAFE family of safety PLCs and high-integrity solid state safety systems.

Firms Providing Consulting Services for Safety Instrumented Systems:

ACM Consulting, Calgary, AL, Canada, http://www.acm.ab.ca/: ACM is an engineering/consulting company that supplies turnkey automation projects in SCADA, control, and safety applications; AE Solutions, http://www.aesolns.com/: AE Solutions, based in Greenville, SC, is an industrial automation and systems integration company, offering a comprehensive suite of process automation products and services to provide complete systems integration; Exida.com, http://www.exida.com: Exida.com offers a full suite of products and services for automation system safety and reliability. Its products and services are used by many industries including oil and gas, petrochemical, chemical, nuclear, and power; Honeywell Safety Management Systems (HSMS), http://europe.iac.honeywell.com/products/fsc/: HSMS specializes in developing and implementing safety solutions for the process industries; Silvertech International, http://www.silvertech.co.uk/: Silvertech International, with headquarters in the United Kingdom, is an engineering firm providing design, engineering, installation, and maintenance services for high-availability and high-reliability control and safety systems; SIS-Tech, http://www.sis-tech.com/: SIS-Tech Solutions offers specialized services and industry knowledge aimed at safety instrumented system (SIS) evaluation and design; Premier Consulting Services (PCS), http://www.triconex.com/premier_consultants/: PCS provides services in risk analysis, evaluation, and design of safety/emergency shutdown and other safety and critical control–related applications.

WHAT IS A SAFETY INSTRUMENTED SYSTEM?

In the field of process instrumentation and control, instrumentation engineers are mainly involved with two types of systems that may not be, but usually are, independent or separate from each other. These systems are commonly termed basic process control systems (BPCSs) and safety instrumented systems (SISs).

The exact definitions of BPCS and SIS vary. Definitions from ANSI/ISA-84.01 standard[1] are as follows:

> **Basic Process Control System (BPCS):** A system that responds to input signals from the equipment under control and/or from an operator and generates output signals, causing the equipment under control to operate in the desired manner. Some examples include control of an exothermic reaction, anti-surge control of a compressor, and fuel/air controls in fired heaters. Also referred to as a Process Control System.

> **Safety Instrumented System (SIS):** System composed of sensors, logic solvers, and final control elements for the purpose of taking the process to a safe state when predetermined conditions are violated. Other terms commonly used include Emergency Shutdown System (ESD, ESS), Safety Shutdown System (SSD), and Safety Interlock System.

Definitions from IEC 61511[4] are as follows:

> **Basic Process Control System (BPCS):** System that responds to input signals from the process, its associated equipment, other programmable systems and/or an operator and generates output signals causing the process and its associated equipment to operate in the desired manner but which does not perform any safety instrumented functions with a claimed SIL ≥ 1.

> **Safety Instrumented System (SIS):** Implementation of one or more safety instrumented functions. An SIS is composed of any combination of sensor(s), logic solver(s), and final elements(s).

Although the various definitions may change and can be subject to different interpretations the basic objectives of the systems are consistent. The BPCS is intended and designed to maintain a safe, operable and profitable operation, and SIS are safety systems used on hazardous processes to minimize the risk of injury to personnel, harm to the environment, or damage to plant equipment. It is one of several protection layers in industrial plants to prevent or mitigate the consequences of hazardous events. Other mitigation layers can be relief devices, containment vessels, or emergency response.

There are several fundamental differences between BPCSs and SISs, notably:

- SISs are passive systems whereas BPCSs are active systems, i.e., faults or failures in BPCS systems tend to reveal themselves immediately to the process operator, whereas many SIS failures may lead to an unannunciated, unprotected, unsafe, and dangerous state.
- The design of SISs may be subject to numerous national and international standards as the result of regulations and legislation.

The BPCS should be designed to reduce the need for the SIS to operate because operation of the SIS has in many instances severe economic and other safety implications, i.e., for many processes the unsafest period is during start-up. Because the BPCS is also intended to maintain a safe operation, it can and should also be regarded as a safety system, and should be designed to provide as much safety as possible. For many hazardous processes it is not always possible for the BPCS to be able to provide the level of safety that is required. In these instances a separate and independent system is usually required to perform the enhanced safety requirements.

Although this section of the *IEH* handbook deals specifically with the design of SIS, it has to be emphasized that safe process design is the first line of defense for safety, and BPCS the second. Many of the techniques that apply to the design of the SIS are also applicable to the BPCS. Although a separate SIS may be required to provide enhanced safety functions, this should not diminish the safety requirements of the process and BPCS.

There are numerous standards, guidelines, articles, and textbooks available dealing with the design of SIS (refer to reference list at the end of the section). The full implementation of an SIS includes many steps involving several disciplines, i.e., safety engineering, process engineers, instrument engineers, etc. This section of the *IEH* is not intended to provide all information and data required to design SIS. Rather, it is intended to introduce the reader to the concepts, and also to list the steps and provide general direction and guidelines for the implementation of SISs. For each step, references where more detailed information can be obtained are cited. Also, for some steps, additional details and issues are discussed to supplement the referenced texts. It should be noted that the steps listed in this section are guidelines only and deviate somewhat from those listed in current and draft standards. Examples of such deviations are in justifying SIS and ensuring that each life cycle step is adequately completed before proceeding to the next.

MAIN COMPONENTS ASSOCIATED WITH AN SIS

An SIS in reality is a system that carries out certain specific safety functions commonly termed safety instrumented functions (SIFs). A very simple SIF can be, "If the level in vessel A exceeds a specified value then close valve X." A single SIS can consist of one or several SIFs, and SIF may consist

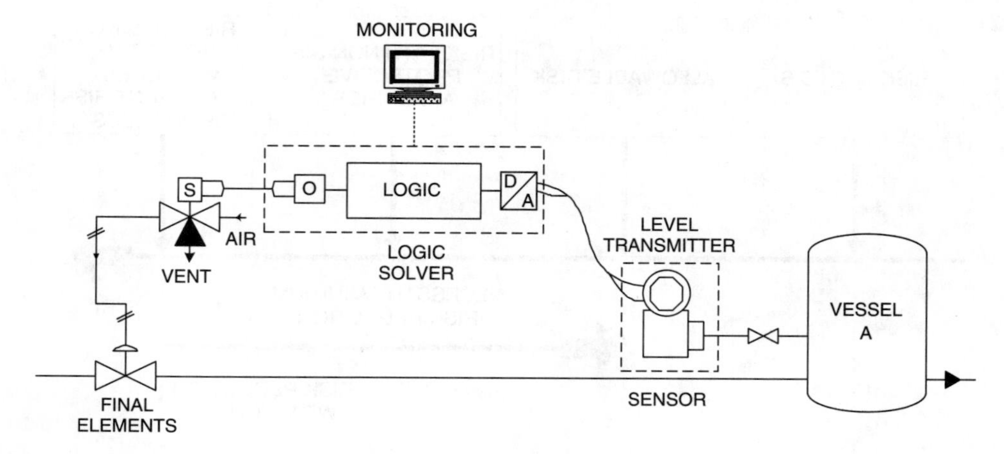

FIG. 2.8a
Components of an independent SIS.

of single, multiple, or multiple groups of sensors and actuators. Also, the same sensor(s) or group of sensor(s) and final element(s) may be part of many SIFs.[15]

Figure 2.8a shows the components of an independent SIS that may be required to provide the above safety function. In reviewing the figure it is possible to list all the components that may make up the SIS in the following categories—sensor, logic solver, final elements, monitoring device, and other devices.

Sensor
- Process block valve
- Leg lines
- Tracing and insulation of leg lines
- Transmitter
- Transmitter housing and heating/cooling
- Transmitter manifold assembly

Logic solver
- Backplane
- Power supplies
- Processor
- Input cards
- Output cards
- Communication cards

Final elements
- Actuator or solenoid valve
- Trip valve

Monitoring device
- Monitor
- Processor

Other devices
- Safety barriers
- Power supplies
- Interdevice cables
- Terminals
- Air supplies
- Junction boxes
- Cabinets

An SIS is not just a single sensor, logic solver, and trip valve. Each component listed above has an impact on the overall performance and reliability of the SIS and must be taken into consideration in the analysis of each SIF.

RISK REDUCTION

By installing an SIS and providing certain specific SIF we are actually reducing the risk of a hazardous or unwanted event occurring. Risk is a function of probability and consequence. In most instances, the SIS reduces the probability of the unwanted event to reduce the overall risk. In the example shown in Figure 2.8a, the SIS reduces the probability or likelihood of the liquid in the vessel overflowing.

Figure 2.8b shows the overall risk reduction achieved by all safety-related systems and other risk reduction facilities.[4]

There is an inherent risk of every process, which if unacceptable would require preventive or mitigation measures that may include SIS and non-SIS means. The non-SIS protective measures are capable of reducing the consequences or the likelihood of the hazardous events. A relief valve reduces the likelihood of an unacceptably high pressure building up in a vessel or pipe, and a containment dyke reduces the consequences of a spill occurring if a tank filled with a hazardous liquid ruptures or overflows.

The objective of an SIS is to provide additional risk reduction to reduce the risk to an acceptable level, and for SIS this is usually accomplished by reducing the likelihood of the event occurring.

Figure 2.8b indicates that risk falls into two regions: acceptable (below allowable risk pointer) and nonacceptable (above allowable risk pointer). There is a principle called ALARP (as low as reasonably practicable) in which there are

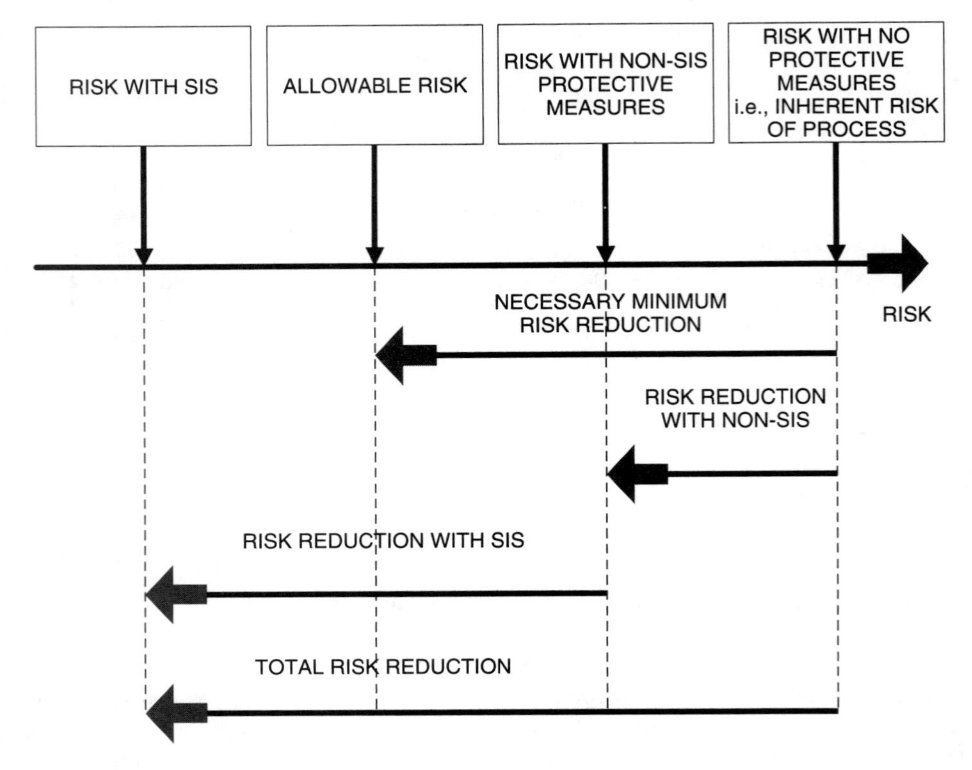

FIG. 2.8b
General risk reduction model.

three regions:

1. Unacceptable
2. Tolerable
3. Generally acceptable

The ALARP principle states that risk in the tolerable region is acceptable if it is impractical or too costly for further risk reduction to occur, i.e., further risk reduction is "beyond management control." In this case we accept the risk. Figure 2.8c summarizes this concept. In the figure:

Point A represents the inherent risk of the process with no mitigation.
Point B shows the risk reduction with a non-SIS system that reduces the likelihood of the event, e.g., relief valve.
Point C shows the risk reduction with a non-SIS system that reduces the consequences of the event, e.g., a containment vessel or dyke.
Point D shows the risk reduction with an SIS providing a level of risk reduction in which the final risk is in the tolerable region.
Point E shows the risk reduction with an SIS providing a level of risk reduction in which the final risk is in the acceptable region. This would be the case where it can be justified to install an SIS to reduce the risk to this level.

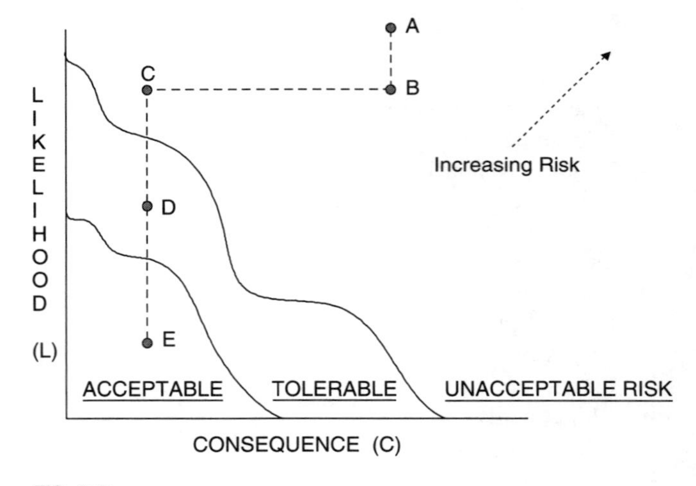

FIG. 2.8c
Risk reduction concept with SIS and non-SIS systems.

PHASES OF SIS OVERALL IMPLEMENTATION

Listed below are six phases associated with the full implementation of SIS.

Phase 0 Develop corporate or site-specific SIS design guidelines and management system.

Before SIS can be designed for any plant, general SIS design guidelines should be in place for that plant. The guideline would then form the basis for the various design and implementation phases. The guidelines can be a corporate standard, a site-specific internal engineering practice or a national/international standard developed by ISA or IEC. Many plants may find it impractical to use the national/international standards directly because of site preferences, federal, state, provincial, or local regulations. They would therefore prepare their own, or amend the national/international standards to suit their specific requirements. Some standards, e.g., NFPA 8501/2 for Burner Management Systems, or API RP 14C for offshore production platforms, tend to be followed with little or no changes to their requirements.

The design guidelines usually include the requirements for all activities involved in specifying, designing, installing, commissioning, validating, startup, operation, testing, repairing, modifying, assessing, and auditing SIS. Once the guidelines are complete, a management system has to be in place to ensure that the requirements of the guidelines are being followed. The design guidelines specify what has to be done, and how. The management system determines whether it is being done, and if so, whether it is accomplishing what it is supposed to accomplish.

Once phase 0 is complete, the actual design, installation, and operation of each individual SIS can be subdivided into the five phases that are listed below.

Phase 1 Identify how much risk reduction is required for the process. By identifying the hazards associated with a process we are able to determine how much risk reduction is required.

Phase 2 Specify, justify, design, and install the SIS to achieve risk reduction. The SIS is installed to satisfy the availability requirements based on the risk reduction required. The higher the risk reduction, the higher the availability requirements of the SIS. Each SIF proposed for the risk reduction needs to be justified. As part of this phase the SIS designer should demonstrate that the design satisfies the risk reduction requirements.

Phase 3 Operate, maintain, and test the system to sustain risk reduction. Operations and maintenance personnel have to operate and maintain the system to ensure that the integrity requirements of the SIS are being maintained.

Phase 4 Modify system as required. All systems will require modification at some time. The same steps as for a new installation have to be followed.

Phase 5 Determine that risk reduction is being achieved. After the SIS has been in operation we need to determine whether the risk reduction is being achieved, and also whether the original assumptions used in the design are correct.

Note: Phases 1 to 4 are also referred in some documents[3,4] as Analysis (phase 1), Realization (phase 2), and Operation (phases 3 and 4) phases.

TABLE 2.8d
Phase 0 Activities

Phase	*Item*	*Activity*
0. Develop SIS design guidelines and management system	0.1	Define safety life cycle (SLC) activities
	0.2	Identify and list responsibilities for each SLC activity
	0.3	Prepare guideline for the implementation of all phases
	0.4	Establish management system for SIS and incorporate into overall safety planning and procedures manual for corporation or site

In phase 0, the overall objective is to have an SIS design guideline and management system in place. The management system ensures that the requirements of the design guidelines are being followed and are accomplishing the original requirements. Phase 0 is usually the most difficult and poorly implemented phase; the end result is an ineffective SIS. This phase is usually part of the standards development for an organization. Usually, this is the responsibility of a team dedicated to the task with input from the associated disciplines.

To complete this phase the four activities listed in Table 2.8d need to be completed. Key activities associated with each of the implementation phases are noted in Table 2.8e.

TABLE 2.8e
Phase 1 to 5 Activities

Phase	*Item*	*Activity*
1. Identify how much risk reduction is required for the process	1.1	Process design review
	1.2	Hazard and risk analysis
	1.3	Identify SIF and SIL requirements
2. Design and install the SIS to achieve risk reduction	2.1	Prepare SRS
	2.2	Define SIS architecture and design concept
	2.3	Justify the need for each SIF
	2.4	SIS detailed engineering
	2.5	Install, commission, and complete pre-start-up acceptance tests
3. Operate, maintain, and test the system to ensure long-term risk reduction	3.1	Complete pre-start-up safety reviews and start-up
	3.2	Operate and maintain
4. Modify system as required	4.1	Modify or decommission
5. Determine that risk reduction is being achieved	5.1	Assessment of overall SIS performance

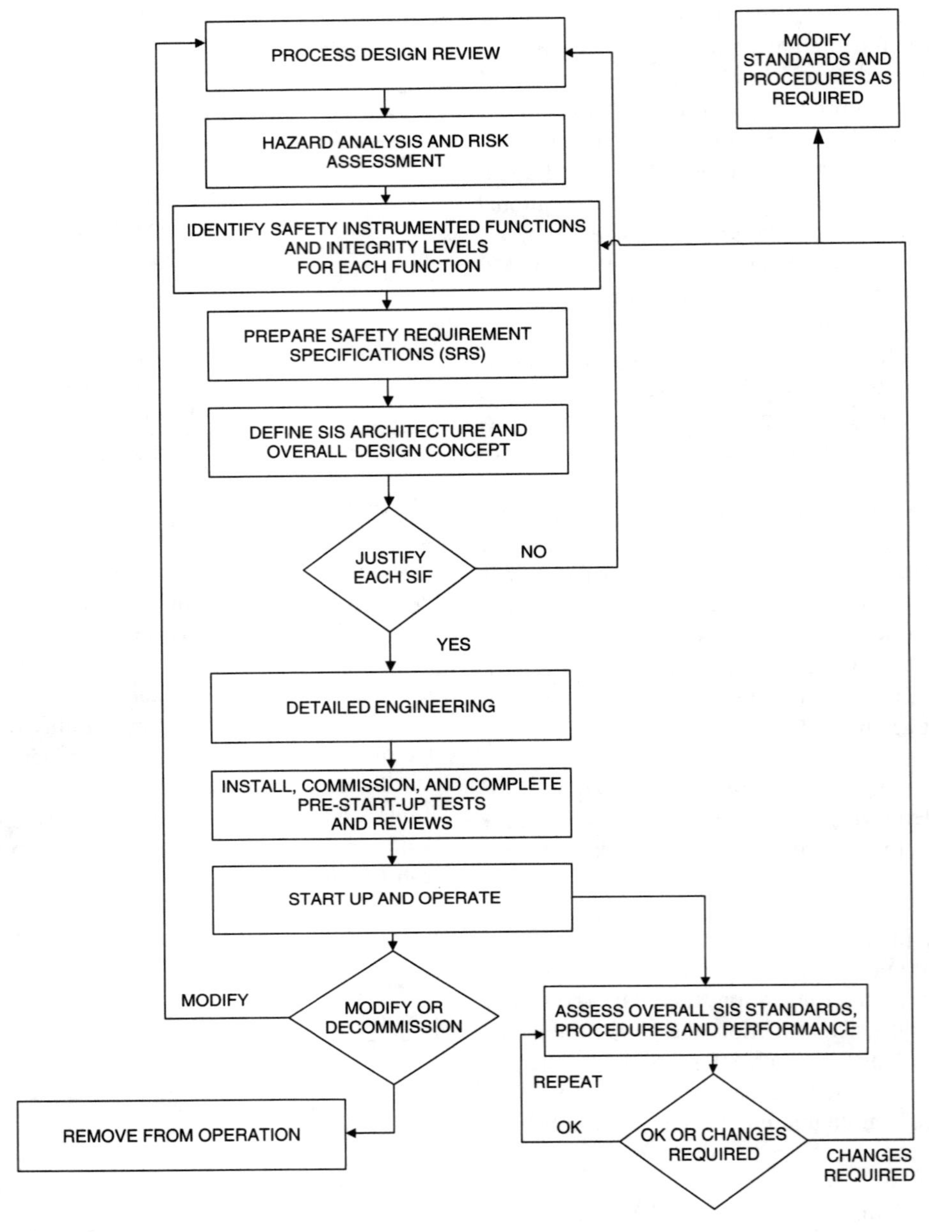

FIG. 2.8f
Typical safety life cycle (SLC).

ITEM 0.1: SAFETY LIFE CYCLE ACTIVITIES

Figure 2.8f is a typical safety life cycle (SLC) that lists the essential steps in the overall implementation of SIS. The SLC is intended to list key activities involved in the full implementation of an SIS from conception to final decommissioning, i.e., from cradle to grave. Each organization needs to develop a model that can be easily incorporated into its existing organization structure and overall safety planning procedures. It is important that a model is not force-fitted into an organization. The SLC for every organization should be unique, and for every project the SLC steps will also vary based on its complexity and requirements.

ITEM 0.2: RESPONSIBILITIES MATRIX

The completion of the SLC activities is usually the responsibility of a team comprising individuals from several disciplines with varying responsibilities. A RASCI chart forms a responsibility matrix and is a very effective tool used to identify individuals and their responsibilities as members of

TABLE 2.8g
RASCI Chart

Item/Responsibility	*Plant/Project Management*	*Loss Control/Safety Engineering*	*Process Engineering*	*Process Control Systems Engineering*	*Operations*	*Maintenance*
1.1 Process design review[a]	A	I	R	I	I	I
1.2 Process hazard and risk analysis[b]	A	R	C	I	I	I
1.3 Identify SIF and SIL requirements	A	R	S	C	I	I
2.1 Prepare SRS	A	I	S	R	I	I
2.2 Define SIS architecture and design concept	A	I	I	R	I	I
2.3 Justify the need for each SIF	A	C	S	R	I	I
2.4 SIS detailed engineering	A	I	I	R	I	I
2.5 Install commission complete pre-start-up acceptance tests SIS	R	I	I	A	I	I
3.1 Pre-start-up safety review and start-up	A	I	S	S	R	S
3.2 Operate and maintain	A	I	I	I	R	R
4.1 Modify or decommission	A	I	R	I	I	I
5.1 Assessment[c]	A	I	I	I	I	I

[a]All members of the team need to review the process for sufficient understanding in order to participate fully in the hazard and risk analysis.
[b]In the United States, OSHA requires that one person on the team must have experience and knowledge specific to the process being evaluated.[16]
[c]The assessment is an ongoing activity, and should be completed by a special and independent team established by the plant management.

R = Responsible: performs tasks; has overall responsibility for completion of tasks associated with this item; A = Approves: determines that task is completed and meets standards and/or gives authorization to continue to the next step; S = Supports: provides resources enabling completion of task; C = Consults: provides advice or expertise; I = Informed: is notified that a task is in progress and/or completed; participates as needed.

the team. The acronym RASCI stands for **r**esponsible for, **a**pproves, **s**upports, **c**onsults, and **i**nformed.

The RASCI chart (Table 2.8g) is used to document the team members and their roles—who actually performs the work for each phase and step, who approves the work, who provides the resources, who the consultants are, and who has to be kept informed.

The chart defines the responsibilities for the design, implementation, and operational phases. The design guidelines and standards should be in place at this stage.

It should be also be noted that for each step, one group has been assigned the responsibility for ensuring that the standards have been met for that step, and gives authorization to continue to the next step.

ITEM 0.3: SIS DESIGN GUIDELINE PREPARATION

SIS designs have a major impact on personnel safety, environmental protection, and protection of equipment and property. As such, the guidelines are not just basic design and engineering documents, but they also reflect the values and culture of an organization.

The guidelines need to be detailed yet simple enough to enable any group within or outside the company to be able to engineer the SIS.

It should include the data, procedures, and requirements for all phases. Typical contents should include the following:

- Scope of document: Defines what is included in document
- Safety life cycle model: Details of the model that is relevant to the site
- Responsibilities: RASCI chart according to Table 2.8g
- Hazard and risk assessment techniques: Describes standards and procedures for hazard analysis, risk analysis, and procedures to identify safety instrumented functions (SIF) and safety integrity level (SIL)
- Safety requirements specification (SRS): Defines techniques, content, and format for completing the SRS
- SIS conceptual design: How to select and define SIS technology, architecture, and components for each SIF
- Detailed design requirements: Establishes design principles, guidelines, and techniques for designing SIS
- SIS documentation requirements: Type and quality of documentation required for detailed engineering

- Installation, commissioning, and start-up of SIS systems: Typical procedures and methods need to be documented
- SIS operation and maintenance: Overall operation and maintenance philosophy for SIS needs to be established
- Management of change procedures: Procedures for carrying out changes need to be established
- SIS assessment and performance monitoring: Defines the terms of reference for an independent assessment team
- References: Terms, acronyms, and reference documentation

ITEM 0.4: IMPORTANCE OF ESTABLISHING A MANAGEMENT SYSTEM FOR SIS

No matter how well a system is designed and installed, it will not accomplish its objectives unless adequate and effective management systems, procedures, and practices are in place for success and are rigorously enforced.

SISs are only one aspect of safety within an organization. Examples of other areas are personnel protection equipment, incident investigations, equipment tagging/lockout, and safe work permits. All these areas including SIS should fall under the umbrella of process safety management (PSM) and be integrated and well coordinated as part of the overall health and safety planning. This will ensure that the systems and resources in place for managing overall safety will also apply to SIS.

For a design guideline to be effective it should also be fully integrated into the overall safety-planning manual for the corporation or site. It is best if it is not a stand-alone or sectionalized into many documents.

Successful SIS operation is therefore more than just providing and installing the SIS hardware. It also involves the full implementation and total respect for all policies and procedures.

Each step of the safety life cycle should be subject to a review process to verify that the desired expectations and results of that step were achieved before proceeding to the next step. We should not wait until the SIS is in operation to determine that the requirements of some steps were not adequately fulfilled. Any activity or set of activities requires the necessary checks for effectiveness and possible improvements.

After the SIS has been in operation, assessments should be carried out on an ongoing basis to determine whether the risk reduction is being accomplished, and also whether the original assumptions used in the design were correct.

Some of the management issues that need to be addressed with respect to SIS therefore are as follows:

- Is the SIS design, installation, operation, and maintenance part of an overall safety plan?
- Have all recommendations from the hazard analysis been implemented?
- Is the SIS design guideline being followed?
- Has each step in the SLC been completed as required?
- Has the SIS been designed and installed to satisfy the requirements of the safety requirements specification?
- After installation, is the SIS actually providing the risk reduction required? Are data and assumptions used in the design correct? Is functional safety being achieved?
- Are operating, maintenance, and change request procedures adequate, and are they being followed?
- How is the safety integrity level being sustained on a long-term basis?
- Are ISO quality systems and procedures being used through the whole SLC?
- Are personnel involved in SLC activities competent to carry out their tasks?
- Is the documentation adequate, accurate, and accessible?

ITEM 1.1: PROCESS DESIGN REVIEW

Overall Objective

To ensure that all members of the team involved in the SLC steps have the necessary information and are familiar with the process or proposed changes so as to be able to fully participate in the subsequent SLC steps.

Key Inputs Required

- Process description including chemistry
- Process flow diagrams
- Process piping and instrumentation diagrams

Outputs Expected

Understanding of the process or process changes by team members.

Additional References

None

Additional Details and Issues

The conceptual design of the process has to be fully completed before this step is scheduled.

ITEM 1.2: HAZARD AND RISK ANALYSIS

Overall Objective

To identify, evaluate, and determine methods to control the hazards associated with the process. OSHA[16] states that the process hazard analysis shall address:

- The hazards of the process
- The identification of any previous incident which had a likely potential for catastrophic consequences in the workplace
- Engineering and administrative controls applicable to the hazards and their interrelationships such as appropriate application of detection methodologies to provide early warning of releases (acceptable detection methods might include process monitoring and control instrumentation with alarms, and detection hardware such as hydrocarbon sensors)
- Consequences of failure of engineering and administrative controls
- Facility siting
- Human factors and
- A qualitative evaluation of a range of the possible safety and health effects of failure of controls on employees in the workplace

Key Inputs Required

- Process design, layout, staffing arrangements
- Material data sheets
- Area classification drawings
- Process description including chemistry
- Process flow diagrams
- Process piping and instrumentation diagrams

Outputs Expected

Identification of all relevant hazards, events causing hazards, and protection layers to prevent hazards, including recommended SIFs.

Additional References

References 4, 8, 10, and 16

Additional Details and Issues

OSHA[16] recommends the following methodologies as appropriate to determine and evaluate the hazards of the process:

(i) What-If [Works well for stable processes with little changes]
(ii) Checklist [Works well for stable processes with little changes]
(iii) What-If/Checklist [Used effectively as the first step in any analysis; also used for processes that are well defined]
(iv) Hazard and Operability Study (HAZOP) [Very effective but also time consuming, and requires a large amount of resources]
(v) Failure Mode and Effects Analysis (FMEA) [Used when the process contains a large amount of systems and mechanical equipment]
(vi) Fault Tree Analysis [Systematic top-down approach using graphical representations; allows the frequency of occurrence of events to be quantitatively evaluated]
(vii) An appropriate equivalent methodology

ITEM 1.3: IDENTIFY SAFETY INSTRUMENTED FUNCTIONS (SIF) AND SAFETY INTEGRITY LEVEL (SIL) REQUIREMENTS

Overall Objective

To determine the SIFs required to achieve the necessary risk reduction and SIL requirements for each SIF.

Key Inputs Required

- Process safety information
- Process hazard analysis report including list of recommended SIF
- P&ID (process and instrumentation diagram)
- Process descriptions

Outputs Expected

A description of the required safety instrumented functions and associated safety integrity requirements

Additional References

References 4, 8, and 12

Additional Details and Issues

Definitions[1,4]

PROBABILITY OF FAILURE ON DEMAND (PFD) A value that indicates the probability of a system failing to respond to a demand. The average probability of a system failing to respond to a demand is referred to as PFDavg.

AVAILABILITY 1 - PFDavg.

SAFETY INSTRUMENTED FUNCTION (SIF) A function to be implemented by an SIS that is intended to achieve or maintain a safe state for the process, with respect to a specific hazardous event.

SAFETY INTEGRITY The probability of an SIS satisfactorily performing the required safety functions under all stated conditions within a stated period of time.

SAFETY INTEGRITY LEVEL (SIL) One of the four possible discrete safety integrity levels (SIL 1, SIL 2, SIL 3, SIL 4) of SISs for the process industries. SIL 4 has the highest level of safety integrity; SIL 1 has the lowest. SILs are defined in terms of PFD according to Table 2.8h.

TABLE 2.8h
Safety Integrity Levels: Probability of Failure on Demand

SIL	Average Probability of Failure on Demand (PFD_{avg})	Risk Reduction (RRF)
4	<0.0001	>10,000
3	0.001–0.0001	1000–10,000
2	0.01–0.001	100–1000
1	0.1–0.01	10–100

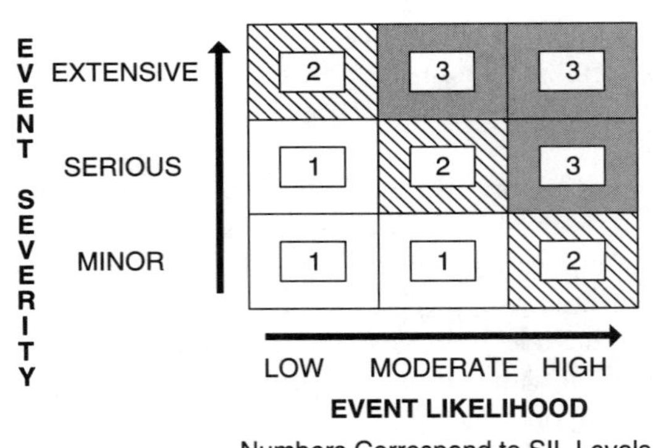

FIG. 2.8i
Risk matrix—SIL Level 3.

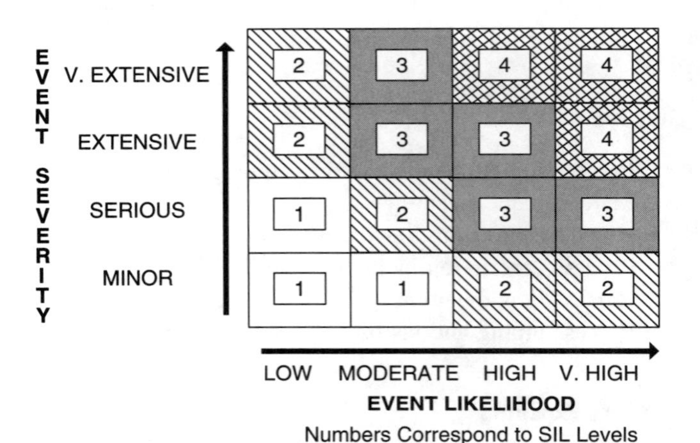

FIG. 2.8j
Risk matrix—SIL Level 4.

General Comments

- Depending on the nature and complexity of the process it is sometimes possible to combine or integrate this step with step 1.2 so that both steps can be completed at the same time.
- The SIL assessments should include personnel safety, plant equipment, and the environment. By doing so we usually end up with a higher SIL than if personnel safety alone is considered.
- It is important to recognize that, although the RRF (risk reduction factor) for, say, SIL 1 vary from 10 to 100, we have to assume for SIL selection that the actual reduction may be 10 rather than 100.

Techniques for SIL Selection Technique used depends on training and experience of personnel. Frequently used techniques are as follows:

Hazard or Risk Matrix[4] The hazard or risk matrix (Figures 2.8i and 2.8j) is a simple-to-use qualitative method, and because of its simplicity it is probably the most widely used technique in the process industries. It shows the risk reduction (SIL level) based on the likelihood and severity of a hazardous event. Figure 2.8i can be used for plants with a max SIL level of 3 and Figure 2.8j with a maximum of 4. The main disadvantage of this method is that because it is highly qualitative, it tends to lead to higher than warranted SIL levels. Reference 4 provides additional details and definitions for severity and likelihood. The risk matrix and risk graph have to be calibrated to satisfy the actual requirements of an organization.

Risk Graph Refer to Figure 2.8k. The risk graph[4] uses four parameters to make a SIL selection. They are consequence of risk (C), occupancy (F), probability of avoiding the hazard (P), and demand rate (W). Consequence represents the average number of fatalities that are likely to result from a hazard when the area is occupied. Occupancy is a measure of the amount of time that the area is occupied. The probability of avoiding the hazard will depend on the methods that are available for personnel to know that a hazard exists and also the means for escaping from the hazard. The demand rate is the probability of the initiating event occurring.

Layer of Protection Analysis (LOPA) For the safety matrix and the risk graph both the frequency and consequence of the unwanted risk event are estimated using very generalized guidelines. LOPA is a technique that helps estimate the frequency of the unwanted event more accurately. It requires examination of the system to determine all the layers of protection that exist, and to place a numerical value on the frequency of occurrence or probability that they will fail. The frequency of occurrence, taking into consideration all layers, can then be calculated using probability equations. [Refer to Section 2.9, Equation 2.9(8) for calculations.] The final frequency/probability value can be used as part of the risk graph or risk matrix or, if corporate tolerable risk guidelines are available, the probability values can be used as part of those guidelines.

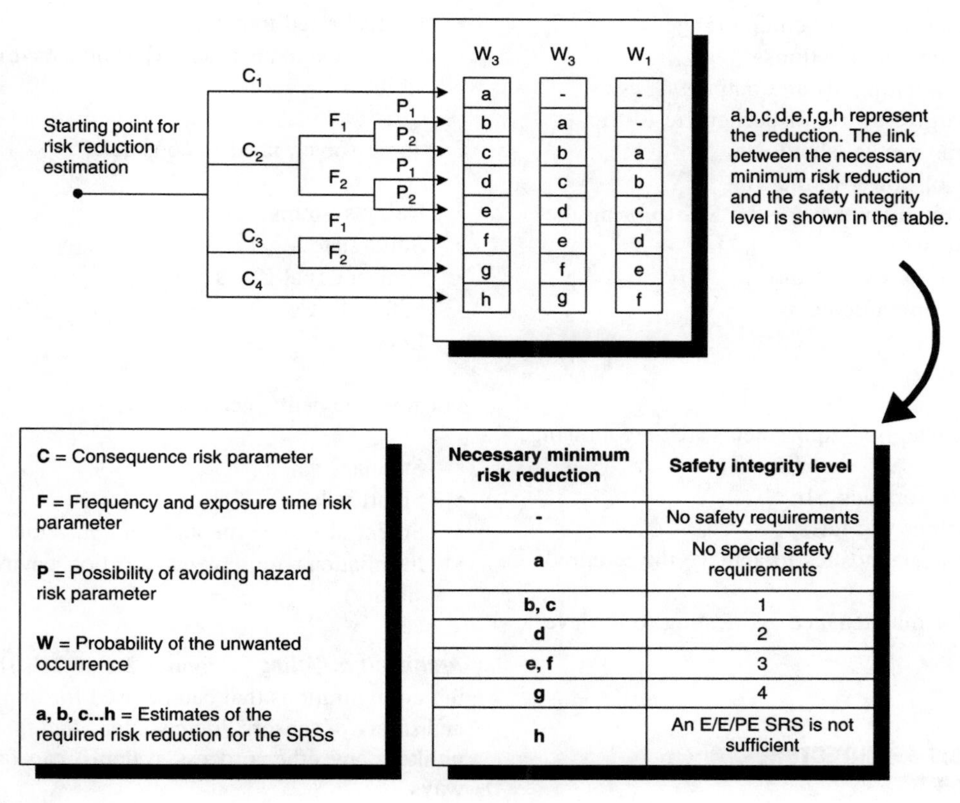

FIG. 2.8k
A risk graph (determination of SIL).

Example: Figure 2.8a showed an SIS for a high liquid level shutdown. The layers of protection without the SIS could be an independent high-level alarm. For this system we have:

1. BPCS, which automatically control the level. The frequency in which this system fails so as to cause a high level was calculated to be 0.02/year.
2. An independent high-level alarm. The PFD_{avg} is 0.1.

From Equation 2.9(8) the frequency at which a high level will occur is 0.02 ∗ 0.1/year.

Reference 13 provides an excellent description of the use and application of this technique.

ITEM 2.1: PREPARE SAFETY REQUIREMENTS SPECIFICATIONS

Overall Objective

To produce a collection of documents that contains the detail requirements of the safety instrumented functions that have to be performed by the SISs. This includes the safety functional requirements (what the system does) and the safety integrity requirements (how well it does it) of the system.

Key Inputs Required

- PHA (process hazards analysis) report
- List of SIF
- Required SIL for each SIF
- Logic requirements
- Other relevant data, e.g., timing, maintenance, bypass

Outputs Expected

Safety requirements specifications, hardware and software

Additional References

References 1, 4, and 7

Additional Details and Issues

The main safety functional requirements are as follows:[1]

- Definition of the safe state
- Process inputs and their trip points

- Process parameter normal operating range
- Process outputs and their actions
- Relationship between inputs and outputs
- Selection of energize-to-trip or deenergize-to-trip
- Consideration for manual shutdown
- Actions on loss of power to the SIS
- Response time requirements for the SIS to bring the process to a safe state
- Response actions for overt fault
- Operator interface requirements
- Reset functions
- Logic requirements

The main *safety* integrity requirements are the following:[1]

- The required SIL for each SIF
- Acceptable spurious trip rates
- Requirements for diagnostics to achieve the required SIL
- Requirements for maintenance and testing to achieve the required SIL

ITEM 2.2: DEFINE SIS ARCHITECTURE AND DESIGN CONCEPT

Overall Objective

To define the requirements for SIS technology, architecture, and SIS components to ensure that SIL is met for each SIF. Verification that the systems proposed will satisfy the SIL requirements.

Key Input Required

- SRS

Outputs Expected

- Selected SIS architecture, technology, and hardware selection for sensors, logic solvers, and final elements
- List of components and configuration for each SIF
- SIL verification for each SIF

Additional References

References 1, 4, 6, and 7

Additional Details and Issues

For sensor selection, need to consider:

- Switches
- Conventional transmitters
- Smart transmitters
- Critical rated transmitters
- Installation requirements (location, environment, accessibility)

For logic solver, need to consider:

- Relay systems
- Solid-state logic
- Conventional PLCs
- Safety PLCs
- Location

For final elements, need to consider:

- Standard valve
- Smart valve
- Solenoid valve type and configuration
- Installation requirements (location, environment, accessibility)

Architecture/Voting Figures 2.8l to 2.8s show different voting configurations that can be used for improved availability and safety of sensors, logic solvers, and final elements. SISs, unlike many other process systems, can fail in two defined ways:

1. Fail-safe, creating a nuisance trip of unit (impacts process availability)
2. Fail-to-danger, thus losing protection provided by SIS (impacts process safety)

The architecture selected must therefore not only satisfy the safety requirements of the process, but also production and operational issues. Frequent failures of the SIS that cause excessive shutdown and start-up cycles will not only affect production, but also create hazardous start-up states. Depending on the voting configuration used in an SIS, we can increase

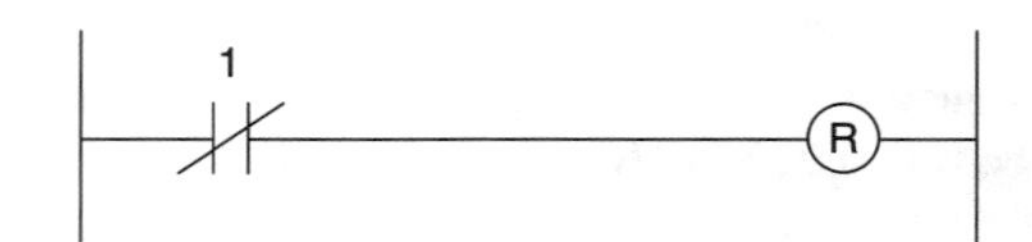

FIG. 2.8l
1oo1 configuration.

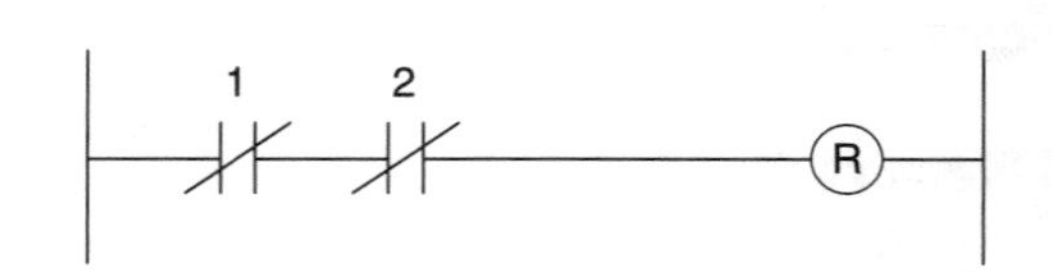

FIG. 2.8m
1oo2 configuration.

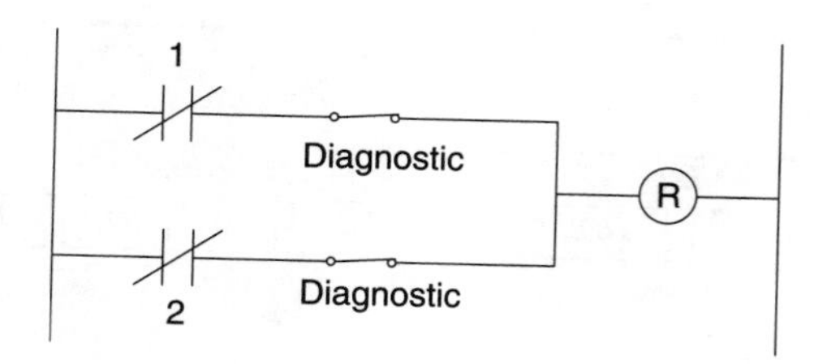

FIG. 2.8n
1oo2D configuration. Because of its extensive diagnostics a 1oo2D system performs as a 1oo2 system for safety, and as a 2oo2 for availability.

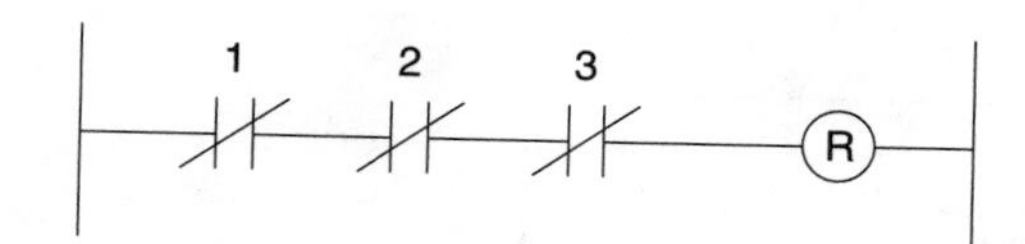

FIG. 2.8o
1oo3 configuration.

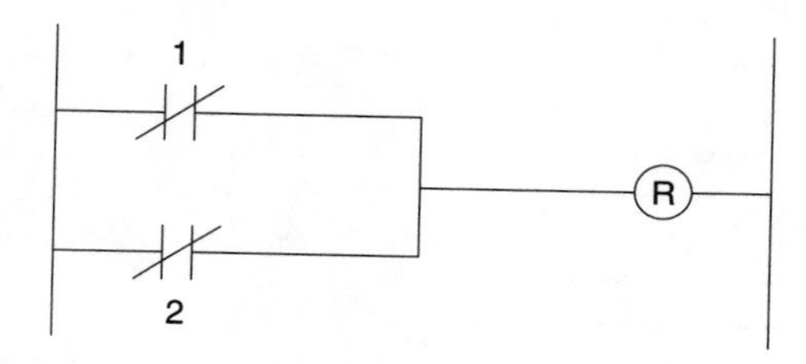

FIG. 2.8p
2oo2 configuration.

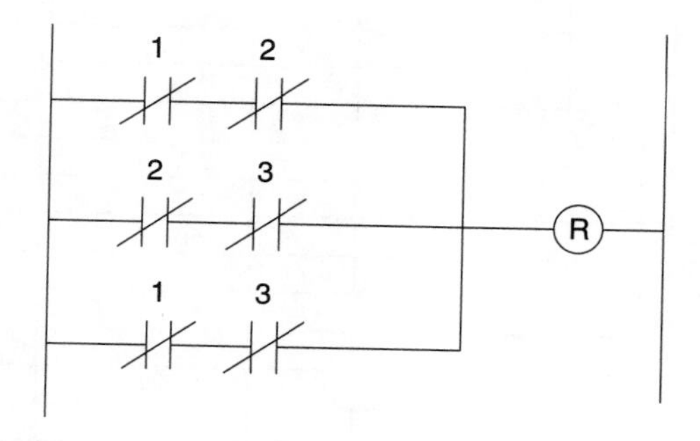

FIG. 2.8q
2oo3 configuration.

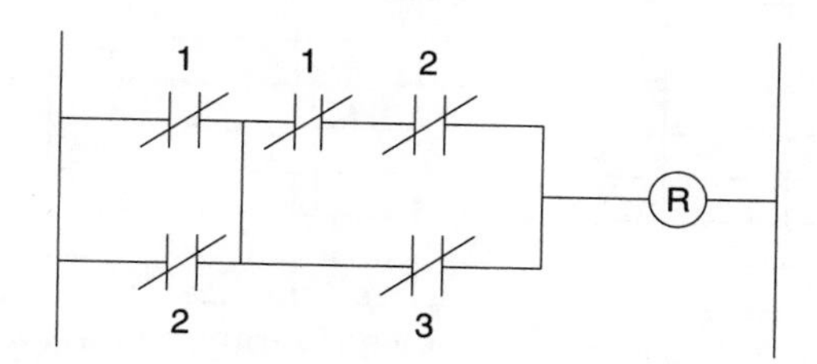

FIG. 2.8r
2oo3 alternative configuration.

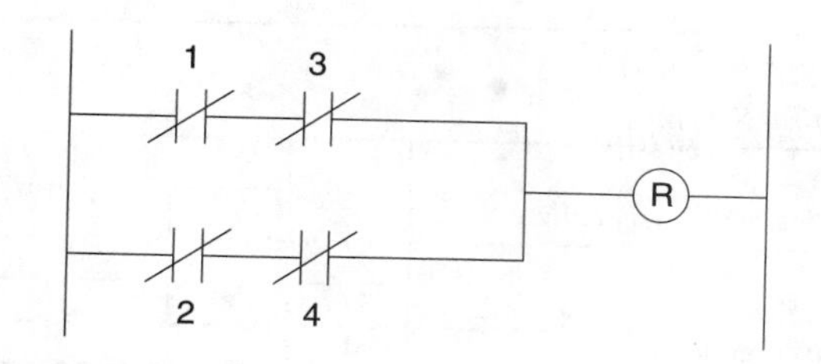

FIG. 2.8s
2oo4 configuration.

or decrease the availability and/or safety of the process. Table 2.8t lists some common voting schemes as applied to sensors, logic solvers, and actuators. The table shows the various conditions under which we will fail safe or fail to danger based on the following types of failure of leg components.

SD = Component in leg has failed safe and failure detected
SU = Component in leg has failed safe and failure not detected
DD = Component in leg has failed dangerously and failure detected
DU = Component in leg has failed dangerously and failure not detected

SIL Verification:

Techniques for SIL Verification[2]

Use of simplified equations The simplified equations (see Tables 2.8u and 2.8v) can be used for relatively simple systems. The answers that are obtained are reasonably accurate. More complex systems involving multiple groups of sensors and actuators with different voting usually require specialized software[14] to complete the analysis accurately. In many instances the designer will use the simplified equations for quick verification and use specialized software for final verification and documentation.

Fault tree analysis Fault tree analyses calculate maximum failure rate. These rates can be used because the results will be conservative.

Markov models This method is very complex and requires an expert analyst or specialized software.[14]

Safety and Availability Calculations for SIS The key safety and availability calculations are the average probability of failure on demand (PFD_{avg}) and spurious trip rates (STR). The actual calculation can be quite complex and involved depending on the complexity of the SIF and the calculation method. Most end users would carry out a simple "order of

TABLE 2.8t
Voting Schemes

Voting and System Failure State	Legs	Leg1				Leg2				Leg3			
		SD	SU	DD	DU	SD	SU	DD	DU	SD	SU	DD	DU
1oo1–Fail safe	1		x										
	1	x											
1oo1–Fail to danger	1			x									
	1				x	NOT APPLICABLE							
1oo1D–Fail safe	1			x									
	1	x											
	1		x										
1oo1D–Fail to danger	1				x								
1oo2–Fail safe	2	x											
	2					x							
	2		x										
	2						x						
1oo2–Fail to danger	2			x				x					
	2			x					x				
	2				x			x					
	2				x				x				
1oo2D–Fail safe	2			x				x					
(See note)	2	x				x							
	2		x				x			NOT APPLICABLE			
1oo2D–Fail to danger	2				x				x				
	2			x					x				
	2				x			x					
	2				x	x							
	2	x							x				
2oo2–Fail safe	2	x				x							
	2	x					x						
	2		x			x							
	2		x				x						
2oo2–Fail to danger	2			x									
	2				x								
2oo3–Fail safe	3	x				x							
	3	x					x						
	3	x								x			
	3	x									x		
	3		x			x							
	3		x				x						
	3		x							x			
	3		x								x		
	3					x				x			
	3					x					x		
	3						x			x			
	3						x				x		
2oo3–Fail to danger	3			x				x					
	3			x					x				
	3			x								x	
	3			x									x
	3				x			x					
	3				x				x				
	3				x							x	
	3				x								x
	3							x				x	
	3							x					x
	3								x			x	
	3								x				x

Examples: A 1oo1 system will fail to danger if a dangerous detected failure (DD) or a dangerous undetected failure (DU) of the component occurs. A 2oo3 system will fail to danger if a dangerous detected failure (DD) in leg 2 and leg 3 occurs (1 of 12 conditions for fail to danger).

Note: A 1oo2D can be configured to trip (fail safe) if an SU failure occurs in either leg.

TABLE 2.8u
Diagnostic Coverage, Common Cause, and Failure Rate Equations

Equation	Equation No.
$\lambda = \lambda^{S} + \lambda^{D}$	2.8(1)
$\lambda^{S} = \lambda^{SU} + \lambda^{SD}$	2.8(2)
$\lambda^{D} = \lambda^{DU} + \lambda^{DD}$	2.8(3)
$\lambda^{SD} = C^{S} * \lambda^{S}$	2.8(4)
$\lambda^{SU} = (1 - C^{S}) * \lambda^{S}$	2.8(5)
$\lambda^{DD} = C^{D} * \lambda^{D}$	2.8(6)
$\lambda^{DU} = (1 - C^{D}) * \lambda^{D}$	2.8(7)
$\lambda^{N} = (1 - \beta)\lambda$	2.8(8)
$\lambda^{C} = \beta\lambda$	2.8(9)

magnitude" calculation to look at the general impact of various voting configurations prior to the detailed and more complex calculations. Listed below are some simplified equations for PFD_{avg} and STR calculations. Refer to Reference 2 for more details.

Terminology, Symbols, and Equations

SFR — spurious failure rate
PFD_{avg} — average probability of failure on demand
RRF — risk reduction factor: RRF = $1/PFD_{avg}$
MTTFS — mean time to spurious failure (SAFE failure)
MTTFD — mean time to dangerous failure
β — beta, common cause factor for redundant configurations
C^{D} — diagnostic coverage for dangerous failures
C^{S} — diagnostic coverage for safe failures
λ — lambda, used to denote the failure rate of a module or a portion of a module; normally expressed in failures per million (10^6) hours
λ^{C} — used to denote the common cause failure rate for a module or portion of a module
λ^{D} — used to denote the dangerous failure rate for a module or portion of a module; these failures can result in a dangerous condition
λ^{S} — used to denote the safe or false trip failure rate for a module or portion of a module; these failures typically result in a safe shutdown or false trip
λ^{DD} — the dangerous detected failure rate for a module or portion of a module; detected dangerous failures are failures that can be detected by online diagnostics
λ^{DU} — used to denote the dangerous undetected (covert) failure rate for a module or portion of a module; dangerous undetected failures are failures that cannot be detected by online diagnostics
λ^{SD} — safe detected failure rate for a module or portion of a module
λ^{SU} — safe undetected failure rate for a module or portion of a module
λ^{N} — normal failure rate
λ^{C} — common cause failure rate

Table 2.8v provides some simplified equations and additional details for various voting configurations.

The PFD_{avg} for a safety system (SIS) is the sum of the PFD_{avg} for the sensors (S), logic (L), final elements (FE), and other devices (OD), i.e.,

$$PFD_{SIS} = \sum PFD_{S} + \sum PFD_{L} + \sum PFD_{FE} + \sum PFD_{OD}$$

TABLE 2.8v
Simplified Equations

Voting	Probability of Failure on Demand (PFD_{avg})	Spurious Failure Rate (SFR)	Typical Degraded Mode(s)	Channels Required for Dangerous Failures	Channels Required for Trip
1oo1	$\lambda^{DU} \times \frac{TI}{2}$	λ^{S}	$1 \Rightarrow 0$	1	1
1oo2	$\frac{(\lambda^{DU})^2 \times TI^2}{3}$	$2\lambda^{S}$	$2 \Rightarrow 0$	2	1
1oo2D	$\frac{(\lambda^{DU})^2 \times TI^2}{3}$	$(\lambda^{S})^2 \times MTTR$	$2 \Rightarrow 1 \Rightarrow 0$	2	1
1oo3	$\frac{(\lambda^{DU})^3 \times (TI^3)}{4}$	$3\lambda^{S}$	$3 \Rightarrow 0$	3	1
2oo2	$\lambda^{DU} \times TI$	$(\lambda^{S})^2 \times MTTR$	$2 \Rightarrow 1 \Rightarrow 0$	1	2
2oo3	$(\lambda^{DU})^2 \times TI^{2tbc}$	$6(\lambda^{S})^2 \times MTTR$	$3 \Rightarrow 2 \Rightarrow 0$	2	2
2oo4	$(\lambda^{DU})^3 \times (TI)^3$	$12(\lambda^{S})^3 \times MTTR^2$	$4 \Rightarrow 2 \Rightarrow 0$	2	2

Note: The 1oo2D system operates as 1oo2 for safety, and as a 2oo2 for availability.

The overall MTTFS for the SIS being evaluated is obtained as follows:

$$STR_{SIS} = \sum STR_S + \sum STR_L + \sum STR_{FE} + \sum STR_{OD}$$

$$MTTFS = \frac{1}{STR_{SIS}}$$

Example Use of simplified equations: Consider the system shown in Figure 2.8a. After completing the hazard analysis, SIF determination, and SIL selection, it was agreed that either a high level or a high pressure in the vessel must close the trip valve and a SIL 2 system was required. A diagram of the system based on the initial design concept is shown in Figure 2.8w.

Tables 2.8x, 2.8y, and 2.8z list the PFD_{avg} and MTTFS results for various cases using the equations in Table 2.8v for 1oo1 and 1oo2 voting.

Example For Case 1, the PFD_{avg} for the 1oo2 level and pressure transmitter is

$$= \frac{\lambda_{PT}^{DU} \times \lambda_{LT}^{DU} \times (TI)^2}{3}$$

$$= \frac{2 \times 10^{-6} \times 3.6 \times 10^{-6} \times (8760)^2}{3}$$

$$= 1.84E\text{-}04$$

Case 3 (see p. 225) with a PFD_{avg} of 6.08E-03 would satisfy our requirements.

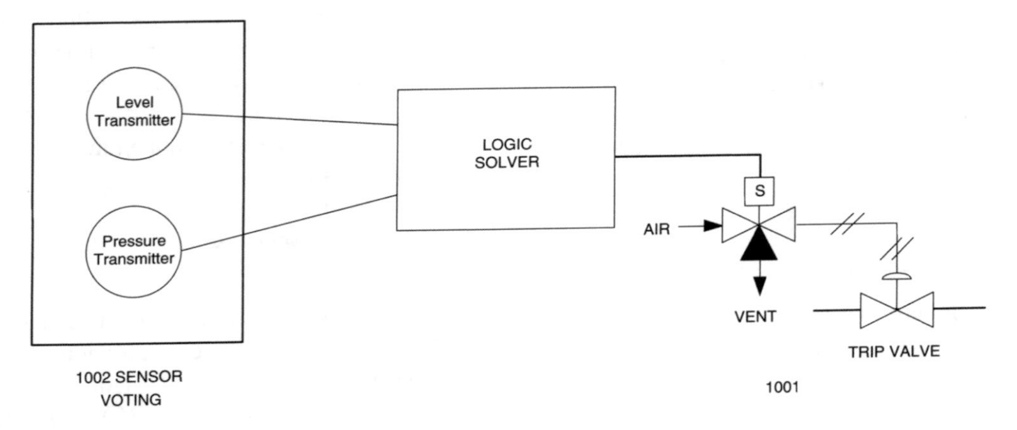

FIG. 2.8w
Calculation example.

TABLE 2.8x
Case 1—1oo2 Voting Level and Pressure Transmitter, DCS, Solenoid, and Valve

Device	λ^S (/year)	λ^D ($/10^6$ h)	C^D	TI (h)	λ^{DU} ($/10^6$ h)	PFD_{avg}	STR (/year)
Pressure transmitter	0.02	4.00E-06	0.5	8760	2.00E-06		
Level transmitter	0.03	6.00E-06	0.4	8760	3.60E-06		
1oo2 voting level and pressure transmitter						1.84E-04	0.05
Logic solver	0.01	1.00E-06	0.9	8760	1.00E-07	4.38E-04	0.01
Solenoid valve	0.05	1.00E-05	0	8760	1.00E-05	4.38E-02	0.05
Main trip valve	0.02	1.20E-06	0	8760	1.20E-06	5.26E-03	0.02
Auxiliaries	0.05	5.00E-07	0	8760	5.00E-07	2.19E-03	0.05
				PFD_{avg}		5.19E-02	
				RRF		19.27965	
				SIL		1	
				STR(SIS)			0.18
				$MTTFS^{spurious}$			5.55 yr

TABLE 2.8y
Case 2—Same as Case 1 except 1oo2 Voting for Solenoid Valve

Device	λ^S (/year)	λ^D (/10^6 h)	C^D	TI (h)	λ^{DU} (/106 h)	PFD_{avg}	STR (/year)
Pressure transmitter	0.02	4.00E-06	0.5	8760	2.00E-06		
Level transmitter	0.03	6.00E-06	0.4	8760	3.60E-06		
1oo2 voting level and pressure transmitter						1.84E-04	0.05
Logic solver	0.01	1.00E-06	0.9	8760	1.00E-07	4.38E-04	0.1
Solenoid valve	0.05	1.00E-05	0	8760	1.00E-05	2.56E-03	0.1
Main trip valve	0.02	1.20E-06	0	8760	1.20E-06	5.26E-03	0.02
Auxiliaries	0.05	5.00E-07	0	8760	5.00E-07	2.19E-03	0.05
				PFD_{avg}		1.06E-02	
				RRF		94.10799	
				SIL		1	
				STR(SIS)			0.23
				MTTFS			4.35 yr

TABLE 2.8z
Case 3—Same as Case 2 except Test Frequency for Solenoid and Trip Valve Changed to Every 6 Months

Device	λ^S (/year)	λ^D (/10^6 h)	C^D	TI (h)	λ^{DU} (/10^6 h)	PFD_{avg}	STR (/year)
Pressure transmitter	0.02	4.00E-06	0.5	8760	2.00E-06		
Level transmitter	0.03	6.00E-06	0.4	8760	3.60E-06		
1oo2 voting level and pressure transmitter						1.84E-04	0.05
Logic solver	0.01	1.00E-06	0.9	8760	1.00E-07	4.38E-04	0.01
Solenoid valve	0.05	1.00E-05	0	8760	1.00E-05	6.39E-04	0.1
Main trip valve	0.02	1.20E-06	0	8760	1.20E-06	2.636E-03	0.02
Auxiliaries	0.05	5.00E-07	0	8760	5.00E-07	2.19E-03	0.05
				PFD_{avg}		6.08E-03	
				RRF		164.4831	
				SIL		2	
				STR(SIS)			0.23
				MTTFS			4.35 yr

ITEM 2.3: JUSTIFY THE NEED FOR EACH SIF

Overall Objective

To justify the need for each SIF.

Key Inputs Required

- List of SIF and SIL
- SIS architecture and conceptual design for each SIF
- Cost impact of incidents

Outputs Expected

- Costs associated with each SIF
- Justification for each SIF

Additional References

References 6, 7, 11, and 12

Additional Details and Issues

- SISs are installed mainly for personal, environmental, and equipment protection.

- Assigning impact criteria for personal protection is a very complex issue, so the risk guidelines for personal protection tend to be followed without the need for additional justification.
- For environmental incidents and equipment damage, it is possible to determine the cost impact, and based on the cost we should be able to justify the need for and the integrity of the SIS being proposed.
- The SIS can only be truly justified after its architecture and conceptual design are completed. This is the only time when the cost of each SIF can be adequately determined.

Justification Example A chemical plant located close to a main river continuously discharges service water into the river. Based on an agreement with the local environmental protection authority, if the pH and/or the conductivity of the service water exceeds certain limits, the water has to be diverted to a holding pond.

Existing System The pH and the conductivity of the service water effluent at each operating unit in the plant are monitored by the unit operating personnel and if limits are exceeded an off-site operator is immediately dispatched to divert the water manually to the holding pond.

- The demand rate is every 6 months (2/year) (based on operations data).
- For the existing system there is only one layer of protection, i.e., operator's manual intervention. The PFD for this layer is 0.1.
- The frequency of occurrence of a discharge based on the single layer of protection is therefore:

$$2 * 0.1 = 0.2/\text{year}$$

Cost of Discharge The total cost to the company if a discharge occurs is estimated to be $50,000:

Annual cost of a discharge = $50,000 * 0.2 = $10,000

Proposed System An automatic interlock system is proposed to accomplish the above. The interlock system is to consist of pH and conductivity transmitters, trip relays, relay logic, and an electrically operated motorized valve (mov), which closes a gate to divert flow to the holding pond. The complete system has to be installed at an off-site location. The mov has a backup air supply and it closes automatically if a power failure occurs. Because of the remote location and need for redundancy, the cost is higher than normally expected.

A SIL 2 system was recommended for this application. Some general design data for the interlock system are as follows:

Life of system	15 years
Initial design and installation cost of SIS	$80,000
Interest rate	5%
Initial design and installation cost/year over 15 years	$7,700*
Maintenance, operating costs/year	$3,500
Total annual cost	$11,200

Since the annual cost of the SIS exceeds the benefits, from a monetary point of view the system cannot be justified. One may argue that the expenditure can be justified because the system will reduce environmental incidents and enhance the public image of the plant or corporation. This then becomes a management decision.

ITEM 2.4: SIS DETAILED ENGINEERING

Overall Objective

To complete the detailed engineering of the SIS in conformance with the SRS and the conceptual design. To provide the documentation, configurations, and programming to enable the system to be purchased, installed, tested, commissioned, and maintained.

Key Inputs Required

- Safety requirement specification
- SIS architecture and conceptual design
- Software requirements

Outputs Expected

- Detailed engineering drawings
- Documentation
- Programming
- Pre-start-up acceptance test procedures
- Pre-start-up safety review procedures
- Procedures and schedules that are required to operate, periodically test, and maintain the system

Additional References

References 1, 4, and 7

* This is the cost of borrowing $80,000, 15 years, 5% interest rate.

Additional Details and Issues

Key Design Requirements

- Separate BPCS and SIS (wiring, field devices, and logic solver).
- All common components designed to highest SIL for all SIF.
- Requirements for maintenance and testing must be provided as part of the design.
- The operator's interface must not impact in any way the SIS configuration or application software.
- Proof testing facilities must be provided if required.

Documentation Requirements The following documentation should be part of the final engineering package:

- SRS
- SIL verification calculations
- Instrument and other equipment lists
- Instrument specification sheets
- Loop diagrams
- Electrical schematics
- Logic descriptions
- Location drawings
- Junction box and cabinet connection diagrams
- Power panel schedules
- Tubing diagrams
- Spare parts list
- Vendor manuals and specifications
- Checkout procedures
- Pre-start-up acceptance tests (PSAT) procedures
- Pre-start-up safety review (PSSR) procedures
- A block diagram of the overall system
- Application software and description

ITEM 2.5: INSTALL, COMMISSION, AND COMPLETE PRE-START-UP ACCEPTANCE TESTS

Overall Objectives

- Complete factory acceptance for logic solver prior to shipment
- To install the SIS according to design documents, commission, calibrate, and test the SIS components
- To verify that the system has been installed as per the detail designs requirements and installation documentation, and that it performs according to the SRS

Key Inputs Required

- Safety requirements specification
- Factory acceptance test procedures
- Pre-start-up acceptance test procedures
- Installation drawings
- PES program code and descriptions
- Commissioning procedures

Outputs Expected

- Factory acceptance report
- Commissioning report
- PSAT report
- The SIS is ready for the PSSR

Additional References

References 1, 4, and 7

Additional Details and Issues

Factory Acceptance Tests The factory acceptance tests should confirm that:

- The logic solver and associated input/output (I/O) satisfies the SRS before the system is shipped from factory or staging area.

Commissioning Activities The commissioning activities should confirm that:[1]

- All equipment and wiring are properly installed.
- Power sources have been properly connected, are within specifications, and are operational.
- All instruments have been properly calibrated.
- All trip settings are correct.
- Field devices are operational.
- Logic solver and I/O are operational.
- Grounding has been properly connected.
- No physical damage is present.
- The interfaces to other systems and peripherals are operational.

Pre-start-up Acceptance Test (PSAT) A PSAT provides a full functional test of the SIS to show conformance with the SRS. The PSAT activities should confirm that:[1]

- SIS communicates with the BPCS, hardwired annunciators, and other systems.
- Sensors, logic, computations, and final control elements perform in accordance with the SRS.
- Safety devices are tripped at the set points as defined in the SRS.
- The proper shutdown sequence is activated.
- The SIS provides the proper annunciation and proper operation display.
- All computations are accurate.

- General resets, bypasses, and bypass reset functions operate correctly.
- Manual shutdown systems operate correctly.
- Complete online testing procedure ensures that procedure is correct and workable.
- SIS documentation is consistent with actual installation and operating procedures.

ITEM 3.1: PRE-START-UP SAFETY REVIEW (PSSR) AND START-UP

Overall Objective

To ensure that the necessary safety reviews are completed and the complete safety system has been properly tested before start-up.

Key Inputs Required

- Completion of all pre-start-up acceptance tests
- Procedure for completing PSSRs

Output Expected

- Results and full documentation of all pre-start-up safety reviews
- After the PSSR is completed satisfactorily, the SIS may be placed in operation

Additional References

References 1, 4, and 7

Additional Details and Issues

Prior to start-up the SIS has to be fully operational and on line. All bypasses and forces are to be removed. The PSSR should include the following SIS activities:

- Verification that the SIS was constructed, installed, and tested in accordance with the SRS.
- Safety, operating, maintenance, management of change (MOC), and emergency procedures pertaining to the SIS are in place and are adequate.
- Employee training has been completed and includes appropriate information about the SIS.
- All bypass functions including forces and disabled alarms have been returned to their normal positions.
- All process isolation valves are set according to the process start-up requirements and procedures.
- All test materials (e.g., fluids) have been removed.

ITEM 3.2: OPERATE AND MAINTAIN

Overall Objective

The purpose of this step is to ensure that adequate training has been provided to the operating and maintenance personnel and that adequate maintenance and operating procedures are in place, as well as to ensure that the requirements of the SRS and the functional safety of the SIS are sustained throughout its operational life.

Key Inputs Required

- Completion and documentation of PSSR activities
- Operating procedures
- Maintenance procedures
- Function test procedures and schedule
- Operator and maintenance personnel training

Output Expected

Correct operation of SIS.

Additional References

References 1, 4, and 7

Additional Details and Issues

- Bypassing may be required for periodic maintenance. For hazardous processes, adequate controls and procedures must be in place to maintain safety when systems are bypassed.
- Periodic testing of system is essential to sustain the as designed SIL level.
- Operation and maintenance procedures shall be developed and tested to ensure that a safe state will be maintained during abnormal operating conditions or while maintenance is being carried out.
- Operators shall be trained on the function and operation of the SIS.
- Maintenance personnel shall be trained as required to sustain full functional performance of the SIS (hardware and software) to its targeted integrity.

ITEM 4.1: MODIFY OR DECOMMISSION

Overall Objective

The purpose of this step is to ensure that for any modification to the SIS the MOC procedures are followed, approvals have been obtained to carry out the changes, the proper safety reviews are completed, and the documentation has been updated to reflect the changes.

Key Inputs Required

- Approved change request
- Proposed modifications to the SRS relating to the changes
- MOC procedure detailing steps to be followed for change

Outputs Expected

- Results of safety analysis
- Modified installation drawings
- PES program code changes
- SIS modified testing procedures
- Approval for field modification
- Approval for SIS decommissioning if required

Additional References

References 1, 4, and 7

Additional Details and Issues

For SIS modification we have to ensure that:

- A hazard analysis and risk assessment is completed.
- Prior to carrying out any modification to a safety instrumented system, procedures for authorizing and controlling changes are in place.
- Modification activity shall not begin without proper authorization.
- Modification shall be performed with qualified personnel who have been properly trained.

For SIS decommissioning we have to ensure that:

- There will be no impact on areas of plant still in operation.
- Prior to carrying out any decommissioning of an SIS, procedures for authorizing and controlling changes shall be in place.
- The procedures shall include a clear method of identifying and requesting the work to be done and identifying the hazards.

ITEM 5.1: ASSESSMENT OF OVERALL SIS PERFORMANCE

Overall Objective

After the SIS has been in operation, an investigation needs to be conducted to determine whether the functional safety is being achieved, i.e., is the risk reduction being achieved, and are the original assumptions used in the design correct? The aim is also to determine whether standards and procedures are adequate and whether there is compliance with requirements.

Key Inputs Required

- Assessment plan and procedures
- Results of hazard analysis
- Safety requirements specification
- List of standards and procedures
- Unit and equipment performance data

Outputs Expected

- SIS Functional Safety Assessment Report with follow-up recommendations

Additional References

References 4 and 12

Additional Details and Issues

The assessment(s) should be headed by a person who was not directly involved in the safety life cycle activities for the particular SIS. The team should also include technical and operations personnel with the expertise needed for the evaluations, and should be able to obtain past operational and performance data for the SIS and unit.

Items to be reviewed as part of the assessment should be:

- Comparing the actual demand rate for each SIF vs. the assumptions made during the hazard analysis
- Comparing the actual failure rate for devices vs. the values used in the SIL verification calculations
- Reviewing the safety, operating, maintenance, and emergency procedures pertaining to the SIS in place and whether they are being followed
- Reviewing the employee training that has been completed
- Reviewing information and documentation relating to the SIS that has been provided to the maintenance and operating personnel

References

1. ANSI/ISA 84.01-1996 "Application of Safety Instrumented Systems for the Process Industries," Instrument Society of America S84.01 Standard, Research Triangle Park, NC: Instrument Society of America, February 1996.
2. ISA-TR84.0.02, Part 1: Introduction; Part 2: Simplified Equations; Part 3: FTA; Part 4: Markov; Part 5: Markov Logic Solver, "Safety Instrumented Systems (SIS)—Safety Integrity Level (SIL) Evaluation Techniques," Instrument Society of America TR84.0.02 Technical Report, Research Triangle Park, NC: Instrument Society of America, 1998.
3. IEC 61508, Parts 1–7 IEC Publication 61508-1999, "Functional Safety of Electrical/Electronic/Programmable Electronic Safety-Related Systems," Geneva, Switzerland: International Electrotechnical Commission, 1998.
4. IEC 61511, Parts 1–3 "Draft Standard—Functional Safety: Safety Instrumented Systems for the Process Industry Sector," 1999.
5. Smith, D. J., *Reliability, Maintainability, and Risk,* Oxford: Butterworth-Heinemann, 1973.
6. Goble, W. M., *Control Systems Safety Evaluation and Reliability,* 2nd ed., Research Triangle Park, NC: Instrument Society of America, 1998.
7. Gruhn, P. and Cheddie, H. L., *Safety Shutdown Systems: Design, Analysis, and Justification,* Research Triangle Park, NC: Instrument Society of America, 1998.
8. *Guidelines for Safety Automation of Chemical Processes,* New York: Center for Chemical Process Safety, American Institute of Chemical Engineers, 1993.

9. *Guidelines for Process Equipment Reliability Data with Data Tables,* New York: Center for Chemical Process Safety, American Institute of Chemical Engineers, 1989.
10. *Guidelines for Chemical Process Quantitative Risk Analysis,* New York: Center for Chemical Process Safety, American Institute of Chemical Engineers, 1989.
11. Beckman, L. V., "Safety Performance vs. Cost Analysis of Redundant Architectures Used in Safety Systems," Proceedings of ISA/96 Conference and Exhibit, Research Triangle Park, NC: Instrument Society of America, 1996.
12. "CFSE Study Guide," available from Exida.com, Sellersville, PA.
13. Marszal, E. M., "Decrease Safety System Costs by Considering Existing Layers of Protection," paper available as a free download from exida.com web site.
14. "SILVer, Safety Integrity Level Verification tool," available from Exida.com, Sellersville, PA.
15. "On-line Web Training," available from Exida.com, Sellersville, PA.
16. Occupational Safety and Health Administration (OSHA). Standard Title: Process Safety Management of Highly Hazardous Chemicals, 29 CFR 1926.64.

2.9 Reliability Engineering Concepts

H. L. CHEDDIE

Software and Consulting Services for Reliability:

Barringer & Associates, Inc., Humble, TX, USA, http://www.barringer1.com/. Consultants for solving reliability problems in industries by using engineering, manufacturing, and business expertise.

Exida.com, http://www.exida.com. Exida.com offers a full suite of products and services for automation system safety and reliability. Their products and services are used by many industries including oil and gas, petrochemical, chemical, nuclear, and power.

Life Cycle Engineering, SC, USA, http://www.lce.com/. Life Cycle Engineering specializes in maintenance engineering, maintenance management, and computer maintenance services.

Meridium, Roanoke, VA, USA, http://www.meridium.com/. The Meridium Enterprise Reliability Management System (ERMS) was developed to provide an advanced software package capable of tracking, evaluating, and improving plant reliability.

Relex Software Corp., Greensburg, PA, USA, http://www.relexsoftware.com/. Relex Software specializes in software analysis tools, training courses, and seminars in all aspects of reliability engineering.

Reliability Center, Inc., Hopewell, VA, USA, http://www.reliability.com/. Training and consulting services to help clients achieve continuous reliability improvement in manufacturing processes.

ReliaSoft Corp., Tucson, AZ, USA, http://www.ReliaSoft.com/. ReliaSoft Corporation provides software and services that combine the latest theoretical advances with practical tools for the practitioner in reliability engineering. ReliaSoft also provides a support site (http://www.Weibull.com/) for Weibull software. This site is also a source of information, discussion, and tools for reliability engineering.

SoHaR Inc., Beverly Hills, CA, USA, http://www.sohar.com/. SoHaR (an acronym for Software and Hardware Reliability) provides research and development of dependable computing for critical applications.

IMPORTANCE OF RELIABILITY FOR INSTRUMENT ENGINEERS

Instrumentation and control systems in various industries are becoming more and more complex. Not only are these systems expected to operate for longer periods without causing any loss to production, but they also have to ensure that the processes they control will operate in a safe and environmentally responsible manner. As the complexity of the systems increases, we need to understand clearly what impact the additional complexities have on the overall system reliability.

This section introduces the topic of reliability engineering so that the basic concepts can be better understood with as little complex mathematics as possible. The examples relate as much as possible to instrumentation to make the subject matter more meaningful to instrument engineers. It is hoped this section will spur a deeper interest in this interesting and worthwhile topic.

The field of reliability engineering has grown in importance since World War II, and it is now realized that all engineering professions require a solid foundation of basic reliability concepts. This section of the handbook addresses the basic concepts required by instrument engineers to complete some of their essential tasks. Some of the important items that require understanding by instrument engineers are (1) how is reliability measured, (2) failure distribution, and (3) how to calculate failure rates. These techniques require a basic knowledge of probability and statistics. This section therefore focuses on the following main topics:

- Introduction to probability and statistics
- Block diagram analysis

- Failure distributions
- Failure rate calculations

Most instrument engineers would have completed some preliminary course in basic statistics, but not necessarily in reliability theory, which is heavily based on statistics and probability. There are many misconceptions by instrument personnel when discussing the performance and reliability of devices, e.g., mean time between failure (MTBF) is often confused with expected life, the misconception being that if a device has an MTBF of, say, 10 years, then the expected life is 10 years. This section hopes to clarify some of these misconceptions. There are numerous excellent textbooks available on the subject of reliability and statistics. This section is very basic and is not intended to cover the full body of knowledge required by instrument engineers in the field of reliability engineering. One of the reasons for including this section in the handbook is to highlight and stress the importance of this field to encourage further reading. A list of the main texts dealing with reliability engineering is included at the end of this section.

The U.S. military and the automotive industries have been the pioneers in developing techniques and standards for system reliability and improvement. These techniques have been widely accepted by the manufacturing industries, and we are now gradually seeing the recognition of their importance by process industries.

Most individuals would think of items as reliable if they carry out their required functions when required. For example, a toaster may be considered unreliable if it periodically burns the toast. This could be a very subjective or biased assessment. One definition of reliability currently used in military standards is "the **probability** that a device will operate **successfully** for a specified **period of time**, and under specified conditions when used in the manner, and for the **purposes**, intended." Keywords in the definition of reliability are *probability, success, time*, and *purpose*. In reviewing the four keywords, probability indicates that there is always a likelihood that an item will fail; success indicates that we need to define the criteria for success clearly; the period of time has to be specified (in process plants this can be the time between a major turnaround, or the expected life of the plant); finally, the item must be used only for the purpose for which it was designed. Clear definitions of criteria for success, time period, and conditions for operation (e.g., physical, chemical, and human environments) must therefore accompany any use of this definition.

Another definition of reliability can be related to the science of carrying out predictive and preventive maintenance programs to maximize equipment performance and minimize life cycle costs.

Reliability is a characteristic inherent in design of systems. Systems that are inherently unreliable because of poor design cannot be expected to perform adequately. We have to recognize the importance of analyzing systems up-front as part of the design effort to attain acceptable reliability.

PROBABILITY, STATISTICS, AND BLOCK DIAGRAM ANALYSIS

The following are some of the numerous techniques for carrying out reliability analysis of systems:

- Block diagram analysis
- Fault tree analysis
- Markov analysis
- Failure modes and effects analysis (FMEA)

Block diagrams are reviewed in this section by virtue of their simplicity and practicality.

The equations for analyzing block diagrams can be easily derived from the laws of probability. We review the laws that apply to reliability and then show how the equations are derived.

The probability of an event is a number between 0 and 1, which indicates the likelihood of occurrence of the event. Zero is the null probability; i.e., the event cannot happen, whereas one is the certainty probability that the event will definitely happen.

Population vs. Sample. A population consists of the entire group of objects about which information is wanted. A sample is part of the population. The differences between probability and statistics are summarized in Figure 2.9a. Probability and statistics are related to each other in that they ask opposite questions.

With probability, we generally know how a process works and want to predict what the various outcomes of the process will be, e.g., a manufacturer that produces numerous identical devices would quite likely know the failure distribution and failure rates for the devices. With these data an end user can

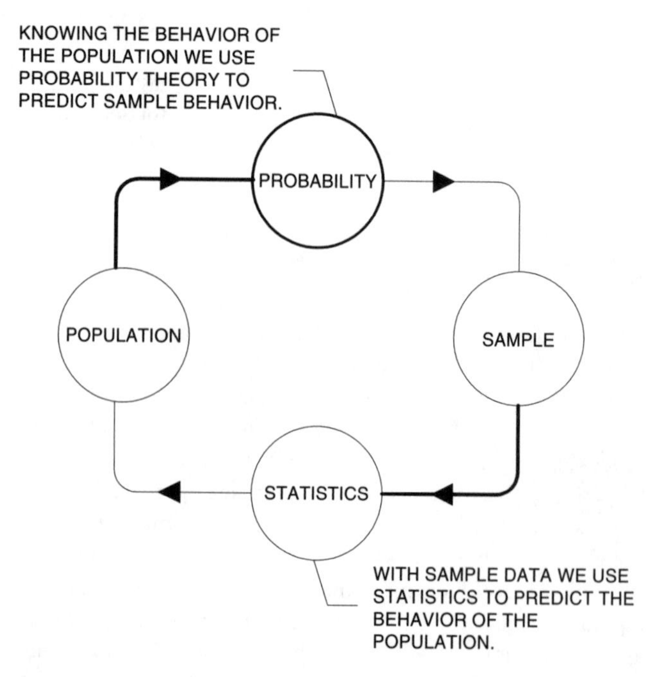

FIG. 2.9a
Relationship between probability and statistics.

determine the probability of failure of individual or group of devices. With statistics, we are able to observe various outcomes of the process, and want to know how the process works, e.g., if we do not have the failure data for a population of devices we can monitor the performance of individual devices or a group of devices and then predict the behavior of a population.

Knowledge of probability therefore helps determine the behavior of individual devices, and statistics helps to determine the behavior of a population.

PROBABILITY LAWS

There are four laws of probability that are very applicable to reliability. Before discussing these laws in any detail it is important to review some definitions.

An **experiment** is a well-defined process with observable outcomes, e.g., 100 electrical motors were put on test. The number of motors still working at the end of specific time periods can be the outcome of the experiment.

If an experiment is carried out, all possible outcomes of the experiment are called the **sample space** (S). *Example*: If five switches are taken from a lot and tested to determine if they are good or bad, the possible outcomes of the test can be that we will observe 0 to 5 defective switches. The sample space is therefore:

$$S: (0,1,2,3,4,5)$$

An **event** (E) is a collection or subset of outcomes from the sample space. An event could have zero, one, or more than one outcome. In the above example possible events can be

$$(0)$$
$$(0,1,2)$$

An event could have just one outcome and hence it is often called a **simple event**. An event with more than one outcome is called a **compound event**. The classical definition of probability says that the probability of an event P(E) is number of outcomes in E/number of outcomes in S.

Complementary events, often denoted by P(E′), consist of all outcomes in S which are not in E.

Events A and B are said to be **mutually exclusive** if both cannot occur at the same time.

Event A is said to be **independent** of event B if the probability of the occurrence of A is the same regardless of whether or not B has occurred.

The concept of complementary, mutually exclusive, and independent events as applied to instrumentation can be reviewed by looking at the various components of the pressure control loop shown in Figure 2.9b:

Complementary events: A component in the loop, e.g., pressure transmitter, will be in either an operational or failed state (based on our definition of "operational" and "failed").

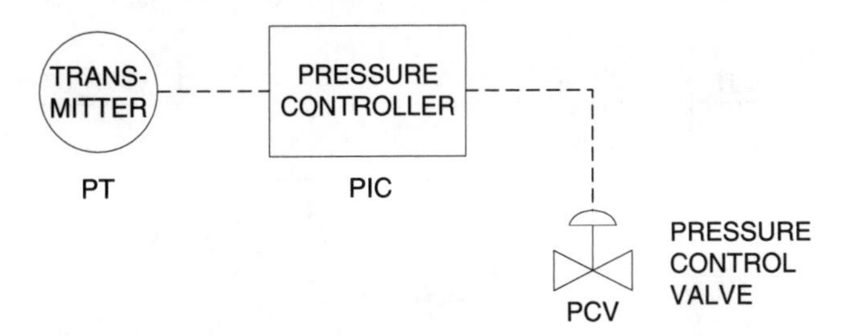

FIG. 2.9b
Loop to demonstrate complementary, mutually exclusive, and independent events.

Mutually exclusive: Failure of the transmitter, controller, or valve is considered not mutually exclusive in that the failures can occur simultaneously.

Independence: Failure of the transmitter, controller, or valve is considered independent in that any one device failure does not impact on the failure of the other two devices. A failure of the transmitter should not impact on the valve failure.

PROBABILITY LAWS APPLIED TO RELIABILITY

Law 1

- If P(A) is the probability that an event A will occur, then the probability that A will not occur is 1 – P(A).
- In reliability terms if the reliability of a device is R_A, then its unreliability U_A is $1 - R_A$ (Figure 2.9c).

Law 2

- If A and B are two events that are not mutually exclusive, then the probability that either A or B occurs is

$$P(A \text{ or } B) = P(A) + P(B) - P(A \text{ and } B) \qquad 2.9(1)$$

- *Note*: If A and B are mutually exclusive, then

$$P(A \text{ or } B) = P(A) + P(B) \qquad 2.9(2)$$

Series System Application In a series system shown in Figure 2.9d, the devices are connected in such a manner that if any one of the devices fails, the entire system fails. The

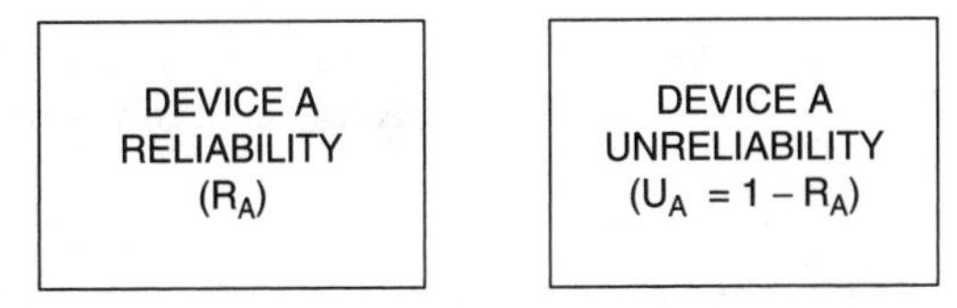

FIG. 2.9c
Application of probability Law 1.

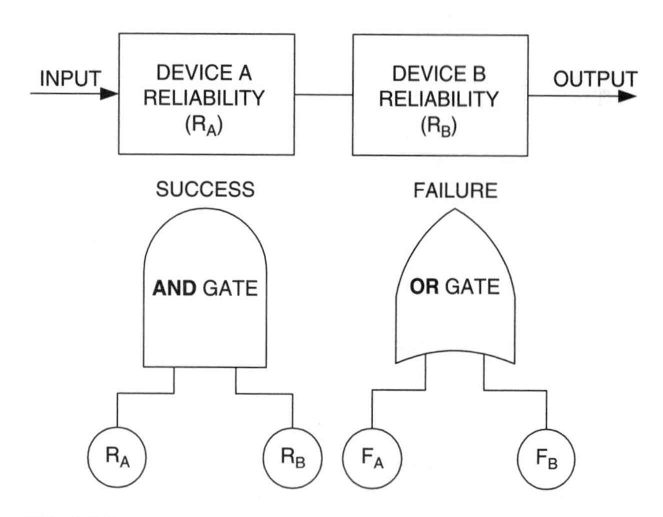

FIG. 2.9d
Series system with equivalent AND/OR gate.

assumption is made that the device failures are independent and nonmutually exclusive.

For the system to *fail*, device A or B has to fail.
The probability of A failing is $(1 - R_A)$.
The probability of B failing is $(1 - R_B)$.
From Law 2, Equation 2.9(1), the probability of A or B failing is

$$(1 - R_A) + (1 - R_B) - [(1 - R_A)(1 - R_B)]$$
$$= 1 - R_A R_B$$
$$= \text{Unreliability of System}$$

The reliability of the system would therefore be $1 - (1 - R_A R_B) = R_A \times R_B$.

The system reliability (R_s) for n devices connected in series is

$$R_s = R_1 \times R_2 \times \cdots R_n \qquad 2.9(3)$$

where R_n is the reliability of the nth component.

Law 3

- If A and B are two independent events, then

$$P(A \text{ and } B) = P(A) \times P(B) \qquad 2.9(4)$$

Parallel Systems Application In a parallel system, shown in Figure 2.9e, the system fails only when all of the components fail.

The component failures are independent, and not mutually exclusive.
For the system to fail, A and B have to fail.

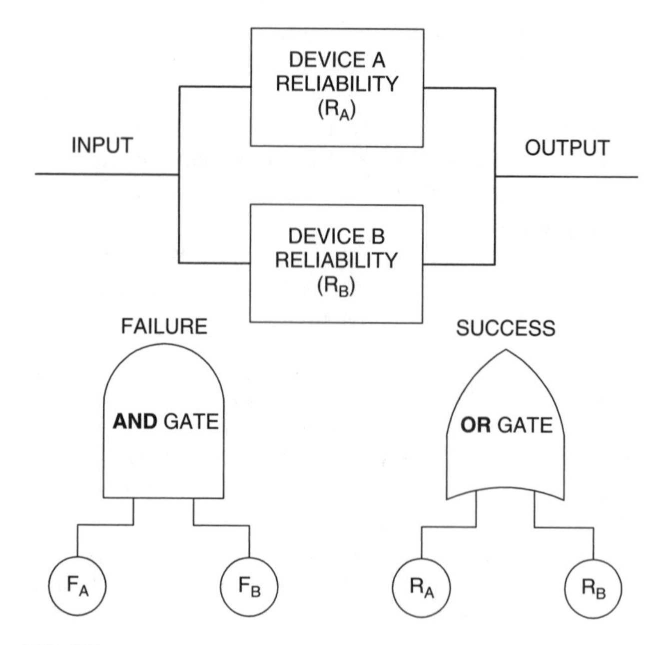

FIG. 2.9e
Parallel system with equivalent AND/OR gate.

The probability of A failing is $(1 - R_A)$.
The probability of B failing is $(1 - R_B)$.
From Law 3, the probability of A and B failing is

$$(1 - R_A)(1 - R_B) = 1 - (R_A + R_B - R_A R_B)$$

The reliability of the system would therefore be

$$R_A + R_B - R_A R_B$$

The system reliability for n devices connected in parallel is

$$R_S = 1 - [(1 - R_1) \times (1 - R_2) \times \cdots (1 - R_n)] \qquad 2.9(5)$$

where R_n is the reliability of the nth component.

Example Figure 2.9b shows a control system consisting of a pressure transmitter, pressure controller, and control valve. If the probability of failure over a 10-year period is 0.25 for the transmitter, 0.05 for the controller, and 0.3 for the valve, what is the reliability of the complete system for the next 10 years?

Solution For the system to be successful, the transmitter, the controller, and the valve have to be successful.

Probability of success for transmitter is $(1 - 0.25) = 0.75$

Probability of success for controller is $(1 - 0.05) = 0.95$

Probability of success for valve is $(1 - 0.3) = 0.7$

System reliability is therefore $0.75 * 0.95 * 0.7 = 0.499$

Law 4

Conditional Probabilities: A conditional probability is the probability of occurrence of one event given that a second event has occurred. It is indicated as P(B|A). The vertical line, |, means "given that." The conditional probability that event B will occur, given that event A has occurred, is

$$P(B|A) = P(A \text{ and } B) \div Pr(A)$$

hence

$$P(A \text{ and } B) = P(A) \times P(B|A) \quad \textbf{2.9(6)}$$

If P(B|A) = P(B), we can conclude that A and B are independent; hence, Law 3 applies.

Example A box has 50 switches, 5 of which are defective. Two switches are to be selected from the box for a job. What is the probability that both switches selected will be defective?

Solution This is an application of the conditional probability theory stated above.

Let

$$P(A) = \text{Probability that 1st switch selected is bad}$$
$$P(A) = 5 \div 50$$

Let

$$P(B|A) = \text{Probability that 2nd switch selected is bad given that A is bad}$$
$$P(B|A) = 4 \div 49$$

Therefore

$$P(A \text{ and } B) = (5 \div 50) \times (4 \div 49) = 0.00816$$

Bayes Theorem Law 4 forms the basis for Bayes theorem, which is extremely useful for solving complex block diagrams that cannot be easily broken down using the series and parallel block structure shown in Figures 2.9d and 2.9e.

If a complex system consists of several components and component A is selected as the key component. (The key component is usually a component that improves the overall reliability of the system, but the system still operates if it fails.)[6]

Bayes theorem may be stated as follows:[6]

> The reliability of a complex system is equal to the reliability of the system given that the key component is good ($R_s|A_{good}$) times the reliability of the key component (R_A), plus the reliability of the system given that the key component is bad ($R_s|A_{bad}$) times the unreliability of the key component (U_A):

$$R_S = (R_S|A_{GOOD})R_A + (R_S|A_{BAD})U_A \quad \textbf{2.9(7)}$$

Example of the Use of Bayes Theorem First consider the system shown in Figure 2.9f. The system consists of pump A and valve B connected in parallel with pump C and valve D. Each pump and its associated valve is connected in series. The block diagram for this system is shown in Figure 2.9g. The reliability for each pump and associated valve is [see Equation 2.9(3)]:

$$0.9 * 0.8 = 0.72$$

The reliability of the system would be [see Equation 2.9(5)]:

$$R_S = 1-(1-0.72)(1-0.72)$$
$$= 0.9216$$

The reliability of the system can be improved by adding an extra valve (valve E) as shown in Figure 2.9h. This should improve the overall reliability. Figure 2.9i illustrates a block diagram for the system shown in Figure 2.9h. Bayes theorem

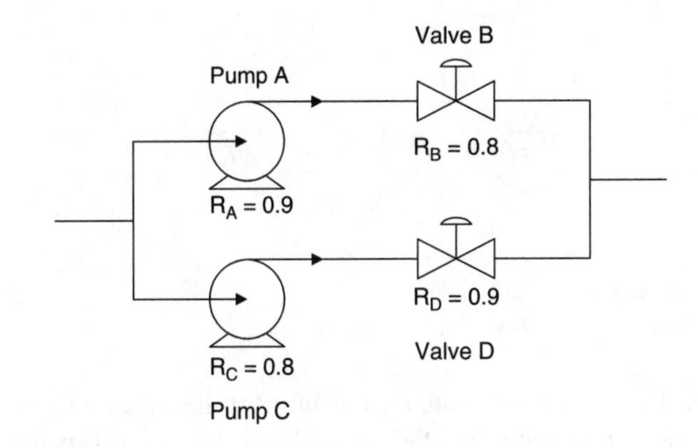

FIG. 2.9f
Pump/valve parallel configuration.

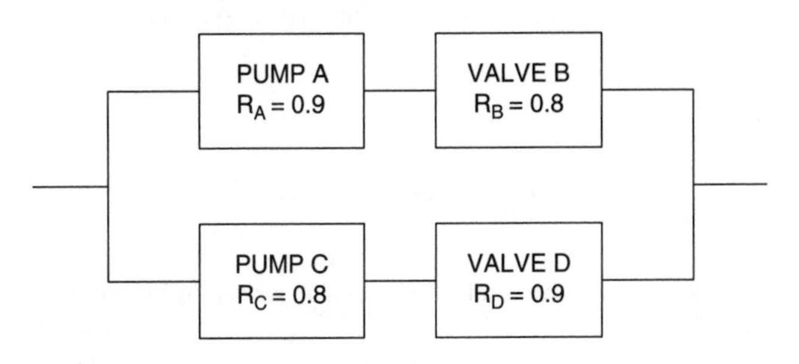

FIG. 2.9g
Block diagram for Figure 2.9f.

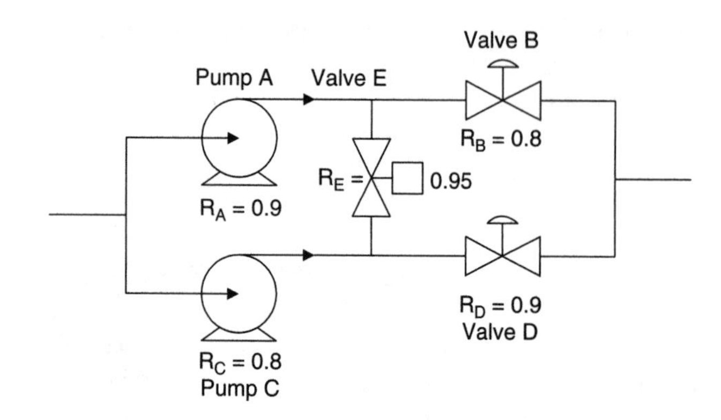

FIG. 2.9h
Component E added to improve overall reliability.

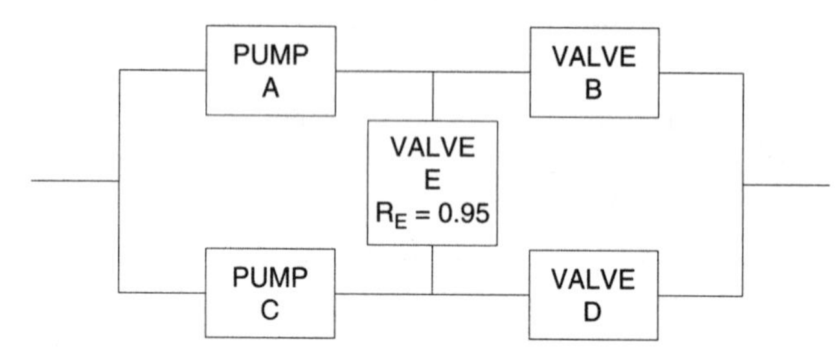

FIG. 2.9i
Block diagram for Figure 2.9h.

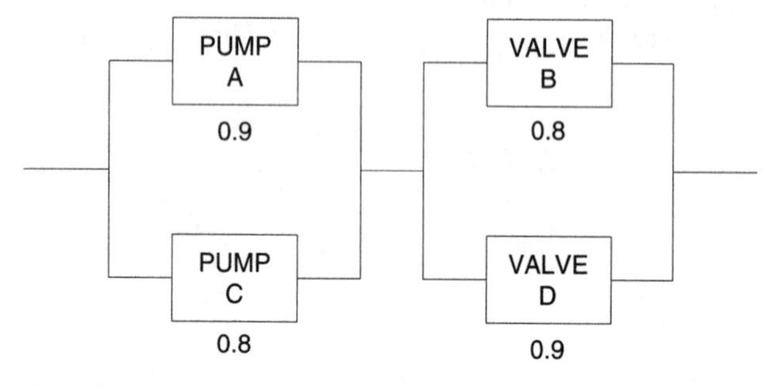

FIG. 2.9j
Block diagram—key component good.

will be used to calculate the reliability of the enhanced system. Figure 2.9j shows the system block diagram if E is the key component, and if E is good. Therefore,

$$(R_S|E_{GOOD}) = [1-(0.1)(0.2)]^2 = 0.9604$$

Therefore,

$$(R_S|E_{GOOD})R_E = 0.9604 \times 0.95 = 0.91238$$

The system block diagram if E is bad is shown in Figure 2.9k. Therefore,

$$(R_S|E_{BAD}) = 1-[(1-0.72)(1-0.72)] = 0.9216$$

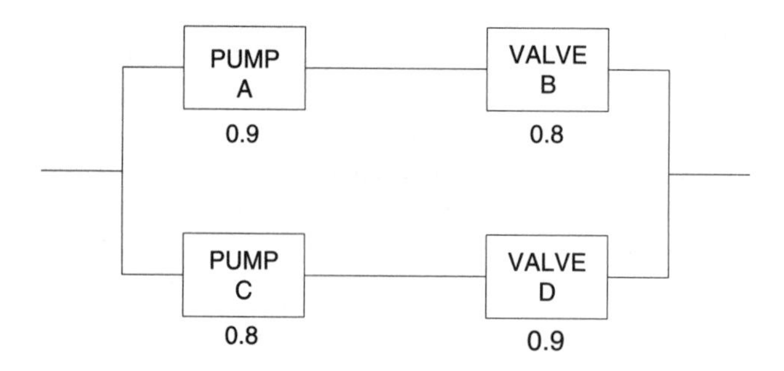

FIG. 2.9k
Block diagram—key component bad.

Therefore,

$$(R_S|E_{BAD})U_E = 0.9216 \times 0.05 = 0.04608$$
$$R_S = 0.91238 + 0.04608 = 0.95846$$

By adding component E the overall system reliability has increased from 0.9216 to 0.95846.

Permutations A permutation is an arrangement of a given set of objects, where the order of the arrangement is important. For example, the three objects a, b, and c can be permutated in six ways as:

abc, acb, bac, bca, cab, cba

The number of permutations of n objects taken r at a time:

$${}_nP_r = \frac{n!}{(n-r)!}$$

where

$$n! = n \times (n-1) \times (n-2) \cdots \times 1$$

for example,

$$4! = 4 \times 3 \times 2 \times 1 = 24$$

Combinations A combination is the number of ways r objects can be selected from n objects without regard to order. For example, cab and cba are two different permutations, but they are one and the same combination. The number of combinations of n distinct objects taken r at a time is

$$\binom{n}{r} = \frac{n!}{r!(n-r)!} \qquad \text{for } r \leq n$$

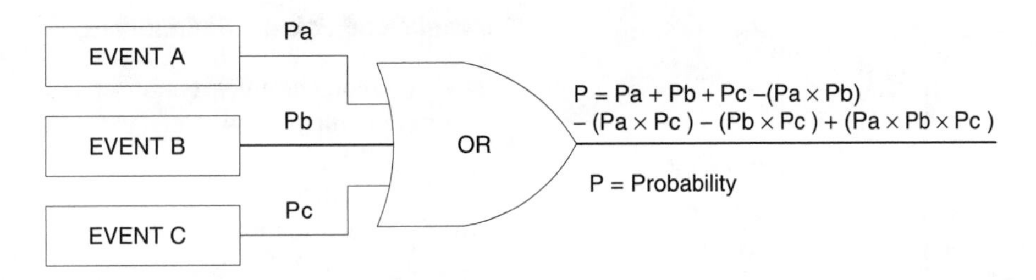

FIG. 2.9l
OR logic—probability.

FREQUENCY AND PROBABILITY CALCULATIONS FOR VARIOUS COMBINATIONS OF EVENTS

Figures 2.9l through 2.9q summarize the probability and frequency values for various events, based on the logic diagrams shown.

For Fig. 2.9l if events A, B, and C are mutually exclusive, then

$$P(\text{A or B or C}) = P(A) + P(B) + P(C)$$

APPLICATION EXAMPLE FOR LOGIC DIAGRAMS

The logic and formula shown in Figure 2.9p are widely used in analyzing various layers of protection for safety systems (see Section 2.8). An initiating event occurring at a certain frequency has the potential to create a hazardous event if several other independent layers of protection failed to operate as required.

Figure 2.9r shows a situation in which there are four layers of protection. For a hazardous event to occur, event A has to occur and all four layers of protection have to fail. The frequency (F) at which a hazardous event will occur will be

$$F = F_A \times P_A \times P_B \times P_C \times P_D \qquad \mathbf{2.9(8)}$$

where P_A, P_B, P_C, and P_D are the probabilities that the individual layer will fail to operate when required.

DISCRETE DISTRIBUTIONS AND APPLICATIONS

Introduction

In reliability analysis we are interested in counting or measuring variables that take on values at random. These variables are termed *random variables*, and are defined as variables whose value depends on the outcome of a random experiment. The word *random* is used to indicate that the value of the variable is determined by chance. Quite often uppercase letters are used to indicate random variables and lowercase letters are used to indicate the actual value. There are two types of random variables, discrete and continuous. Examples:

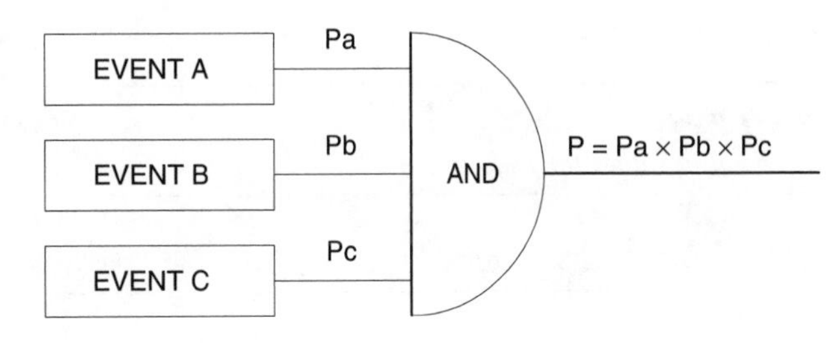

FIG. 2.9m
AND logic—probability.

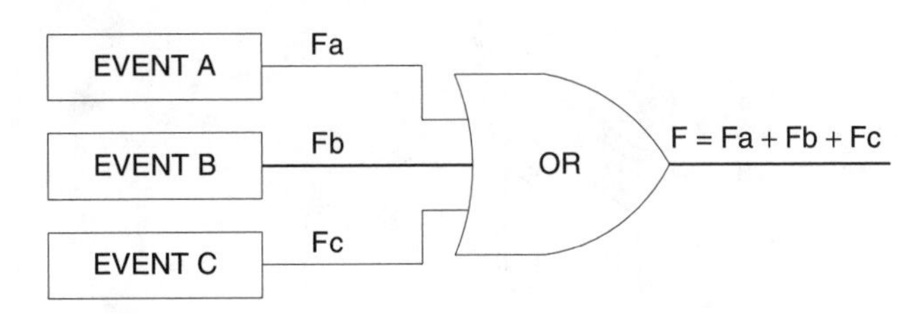

FIG. 2.9n
OR logic—frequency.

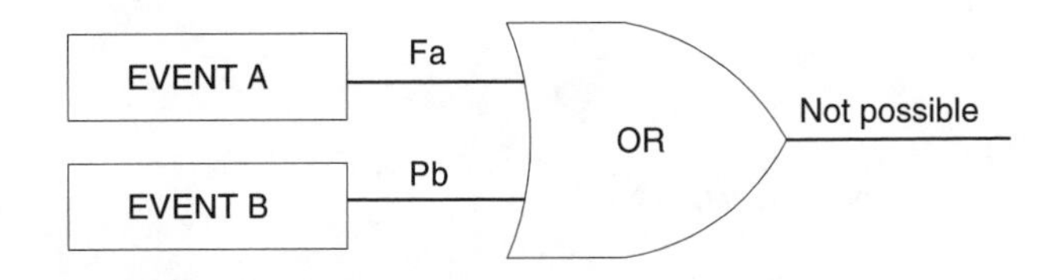

FIG. 2.9o
OR logic—frequency/probability.

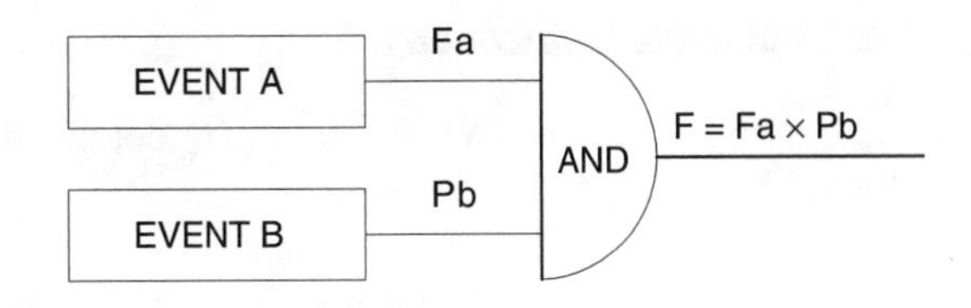

FIG. 2.9p
AND logic—frequency/probability.

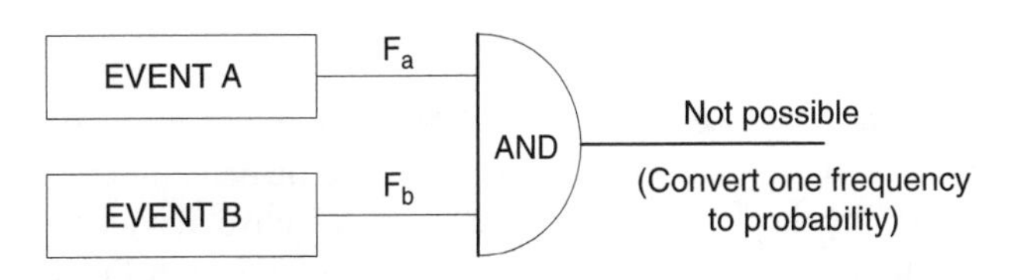

FIG. 2.9q
AND logic—frequency/time.

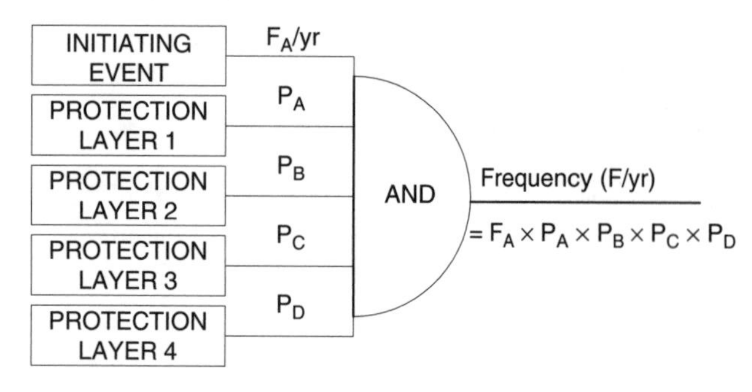

FIG. 2.9r
Layer of protection analysis.

TABLE 2.9s
Probability Values for Defects

x	0	1	2	3	4	5
p(x)	0.59	0.33	0.073	0.008	0.0004	0.00001

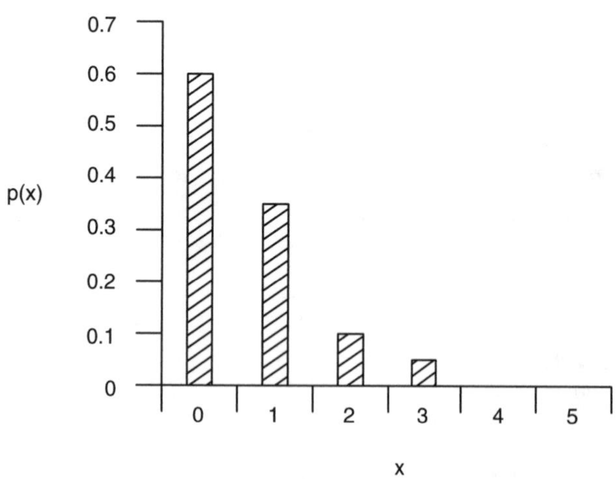

FIG. 2.9t
Probability mass function for defects.

Discrete: Number of failures/month of a set of devices
Continuous: Life of a device in years

Properties of Discrete Distributions

For discrete random variable (X), function p(x) defines the probability of X = x, also

$$p(x) \geq 0 \qquad \text{for all values of x}$$

$$\sum_x p(x) = 1$$

Example Five devices are taken from a box in which 10% of the total number of devices is known to be defective. The probability p(x) of getting 0,1,2,...5 defective devices is shown in Table 2.9s. A plot of p(x) vs. x is shown in Figure 2.9t. This plot is commonly known as the probability mass function (PMF).

Binomial and Poisson Distributions

Some common and important discrete distributions encountered in reliability analysis are:

1. Binomial distribution
2. Poisson distribution

Binomial Distribution This distribution is the model used when describing random variables that represent the number of successes (x) out of n independent trials, when each trial can result in success with probability P, and failure with probability (1 – P), i.e.,

$$P(x) = \binom{n}{x} P^x (1-P)^{n-x}, \quad x = 0,1,\ldots.n. \qquad \mathbf{2.9(9)}$$

where

$$\binom{n}{x} = \frac{n!}{x!(n-x)!}$$

and

$$n! = n(n-1)(n-2)\ldots$$

In the previous example the probability p(x) of getting one defective item from the box in which 10% is known to be defective would be

$$\binom{5}{1}(0.1)^1(0.9)^4 = 0.328$$

Application of Binomial Distribution Consider a 2oo3 system, Figure 2.9u, i.e., the system is okay if at least two of the three systems are functioning.

R = Reliability = Probability of Survival

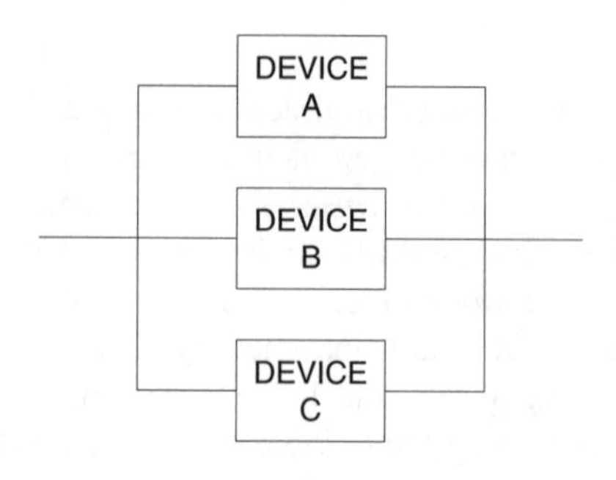

FIG. 2.9u
2oo3 redundancy.

Based on the binomial equation, the probability of two systems functioning (surviving) is

$$\binom{3}{2}R^2(1-R)^{(3-2)}$$
$$= 3R^2(1-R)$$
$$= 3R^2 - 3R^3$$

The probability of three systems functioning (surviving) is

$$\binom{3}{3}R^3(1-R)^0 = R^3$$

$$\therefore R_s = \text{Probability of 2 surviving} + \text{Probability of 3 surviving}$$
$$= 3R^2 - 3R^3 + R^3$$
$$= 3R^2 - 2R^3$$

Poisson Distribution This distribution is used to determine the probability of occurrences of a specific number of events within some period when the mean number of occurrences of the event in the period is known, i.e.,

$$P(x) = \frac{e^{-\mu}(\mu)^x}{x!}, \quad x = 0,1,2\ldots \qquad \textbf{2.9(10)}$$

where μ = mean number of occurrences of an event in a fixed interval of time.

Example of Poisson Distribution The average number of motor failures experienced per month in a plant is three. What is the probability that for a particular month no more than one failure will occur? The probability of no more than one failure is

$$P(0) + P(1) = \frac{e^{-3}(3)^0}{0!} + \frac{e^{-3}(3)^1}{1!} = 0.199$$

Application of Poisson Distribution: Standby Systems If the failure rate of the main and standby systems, shown in

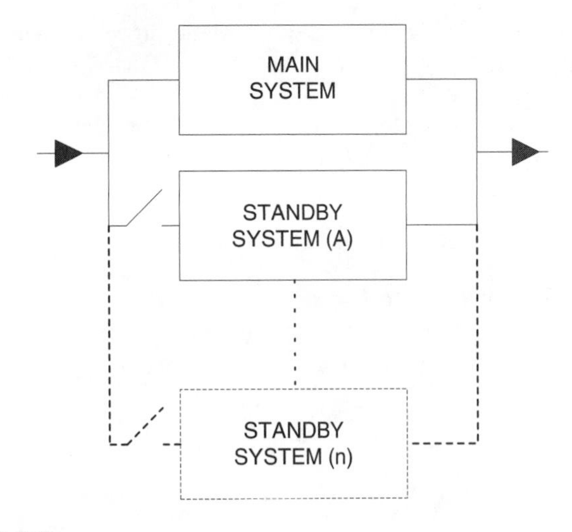

FIG. 2.9v
Standby system configuration.

Figure 2.9v is λ, then the number of failures in time t would be λt. In the above standby systems the redundant components will become operational if the main system has failed. The switching device to detect and transfer to the redundant system is assumed to be 100% reliable.

If the life distribution of the components is exponential with failure rate λ, the reliability of the standby system with r identical, independent components is given by[7]

$$R(t) = e^{-\lambda t}\sum_{j=0}^{r-1}\frac{(\lambda t)^j}{j!} \qquad \textbf{2.9(11)}$$

λt = average number of failures in time t

Example of Standby Systems The processors for a PLC system controlling an operating unit are connected in a standby mode as shown in Figure 2.9w. The operating unit will have to be shut down if both processors fail. If the failure time for a processor is exponentially distributed with $\lambda = 50 \times 10^{-6}$/h.

- What is the probability that a unit shut down will not be caused by processor failures in 6000 hours?
- Assume the switching mechanisms have reliability = 1.0.

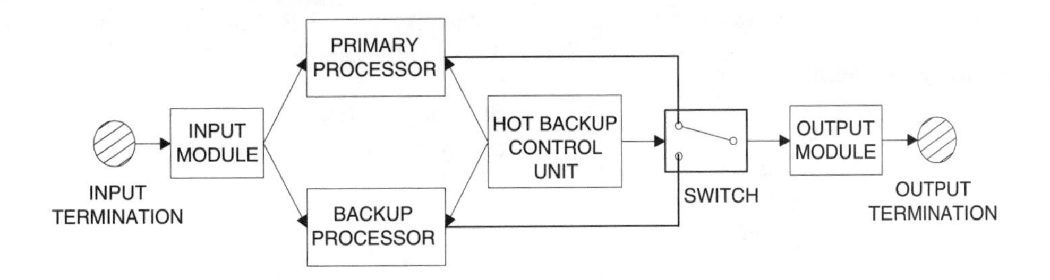

FIG. 2.9w
PLC system with redundancy processes.

- How does this compare with a single processor with no backup?

$$R(t) = e^{-\lambda t}\sum_{J=0}^{r-1}\frac{(\lambda t)^J}{J!}$$

$$\lambda = 50 \times 10^{-6}/h, \quad t = 6000\ h, \quad r = 2$$

Solve for J = 0, to J = (r − 1) = 1

$$\therefore R(t) = e^{-0.00005\times 6000}\left[\frac{(0.3)^0}{0!}\right]+\left[\frac{(0.3)^1}{1!}\right]$$

$$= e^{-0.3}(1+0.3)$$

$$= 0.963$$

For a single processor with no backup:

$$R(t) = e^{\lambda t}$$

$$= e^{-0.3}$$

$$= 0.7408$$

r Out of n Systems For a system with n identical, independent components in parallel, the system will work if at least r out of n components are working. If R is the reliability of the components for time t, then the reliability of the system is[7]

$$R_s = \sum_{k=r}^{n}\binom{n}{k}R^k(1-R)^{(n-k)} \qquad \textbf{2.9(12)}$$

For a three out of four system, n = 4, r = 3.

For a 3oo4 system, the system reliability would be

$$R_s = \binom{4}{3}R^3(1-R)^1+\binom{4}{4}R^4(1-R)^0$$

$$= 4R^3 - 3R^4$$

CONTINUOUS DISTRIBUTIONS AND APPLICATIONS

In Figure 2.9t we showed the probability of discrete random variables. For continuous random variables, the probability of any specific value occurring would be zero. A different technique therefore has to be adopted in creating probability plots for continuous random variables. This involves producing frequency distributions.

Frequency and Probability Distributions

A frequency distribution is constructed from data by arranging values into classes and representing the frequency of occurrence of any class by the height of the bar. The frequency of occurrence corresponds to probability.

Example Table 2.9x shows the number of failures recorded on an annual basis for a particular item. The frequency distribution is shown in Figure 2.9y.

TABLE 2.9x
Failure Data

Time (year)	*Failures Recorded at End of Period*
1	75
2	55
3	42
4	31
5	24
6	18
7	14
8	10
9	7
10	5

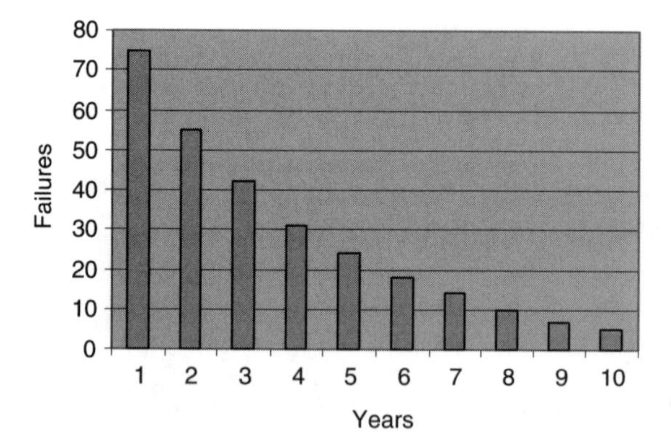

FIG. 2.9y
Frequency distribution.

A probability distribution or probability density function (PDF) is the proper statistical term for a frequency distribution constructed from an infinitely large set of values where the class size (time value) is infinitesimally small. It is a mathematical model of the frequency distribution. This distribution is commonly denoted mathematically as f(t).

For the frequency distribution shown in Figure 2.9y, the PDF will resemble Figure 2.9z. The probability of failure from time t to (t + Δt) would be represented by the shaded area in the figure. The failure rate is the number of failures from t to (t + Δt) divided by the number that have survived to time t.

It is sometimes possible to create a mathematical model of the distribution. By knowing the model, it is easy to predict the possible behavior of individual items from the population.

Reliability in Terms of PDF Based on our definition of reliability, i.e., probability of success for t > T, and referring to Fig. 2.9aa,

$$R(t) = \int_T^{\infty} f(t)\,dt$$

$$U(t) = \int_0^T f(t)\,dt$$

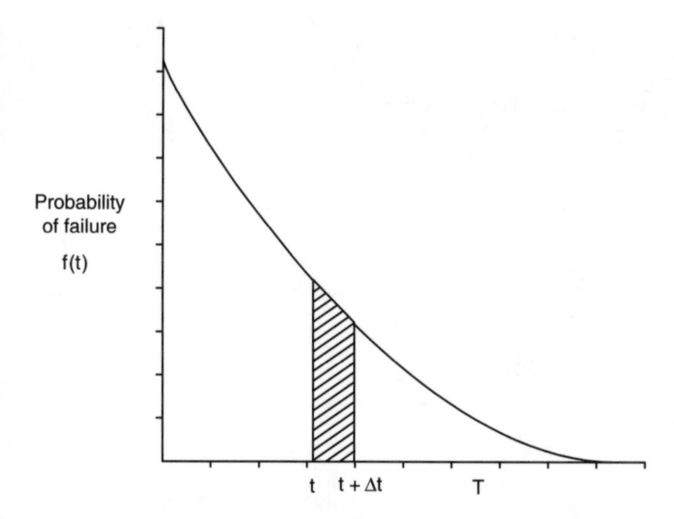

FIG. 2.9z
PDF from frequency distribution.

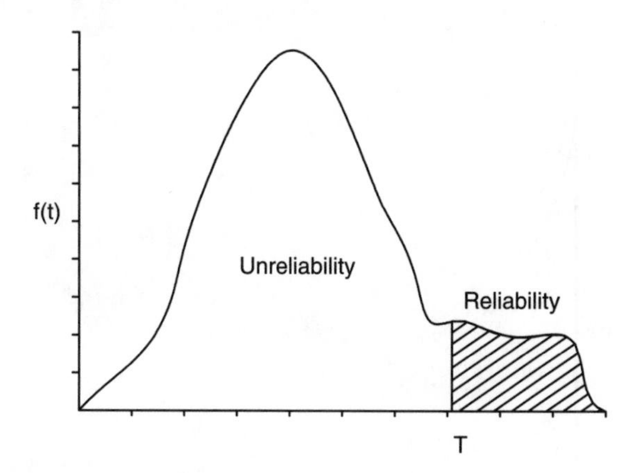

FIG. 2.9aa
Probability distribution function.

U(t) is commonly known as the cumulative distribution function (CDF).

R(t) + U(t) = 1, and it can be proved that[4] if h(t) is the hazard function or failure function,

$$h(t) = \frac{f(t)}{R(t)} \qquad 2.9(13)$$

Normal, Exponential, and Weibull Distributions

Some common and important continuous distributions encountered in reliability analysis are

1. Normal distribution
2. Exponential
3. Weibull

Summaries of the distributions are as follows.

The Normal Distribution The normal distribution is one of the most important distributions; its PDF is bell shaped. The distribution is identified by specifying two values, its mean and variance. The mean is located at the peak of the distribution, and the variance determines its shape, whether it would be spread out or have all the area concentrated at its peak.

The PDF for the normal distribution is

$$f(x) = \frac{1}{\sigma\sqrt{2\pi}} e^{-(1/2)[(x-\mu)/\sigma]^2} \qquad 2.9(14)$$

where μ is the mean and σ^2 is the variance.

To determine the probability associated with the normal distribution, we calculate the area under the curve. For example, Figure 2.9bb shows a distribution for a random variable x with a mean $\mu = 20$ and $\sigma = 5$. To determine the probability that $x \geq 20$ we have to calculate the area under the curve for $x \geq 20$. Because the distribution is symmetrical, the probability = 0.5. To calculate the probability for different values of x, we use a table developed for a special normal distribution called the "standard normal distribution." This distribution has a mean $\mu = 0$, and $\sigma = 1$.

Any normal distribution can be shifted and rescaled to the standard normal distribution using the Z transformation, which is defined as

$$Z = \frac{x - \mu}{\sigma} \qquad 2.9(15)$$

Z measures how far x is from μ in terms of σ.

Once a value of Z is computed, the proportion of values that lie on either side of the value can be found by consulting a Z table (refer to Tables 2.9jj and 2.9kk).

Example of Normal Distribution What is the reliability of a device at 'x' = 10,000 hours if, based on field data, the mean (μ) = 15,000 hours and σ = 3,000 hours?

$$Z = \frac{10000 - 15000}{3000} = -1.67$$

From Table 2.9jj the area to the left of Z = 0.0475

$$\therefore \text{Reliability} = 1 - 0.0475 = 0.9525$$

The Exponential Distribution This distribution is used to determine the reliability of systems with constant failure rate where f(t) is the life distribution (PDF), R(t) is the reliability at t, and h(t) is the failure rate. It is a single-parameter distribution and is commonly expressed in terms of its mean, θ, and the inverse of its mean, λ. The PDF for this distribution is

$$f(t) = \frac{1}{\theta} e^{-t/\theta}$$
$$= \lambda e^{-\lambda t}, \quad t \geq 0 \qquad 2.9(16)$$

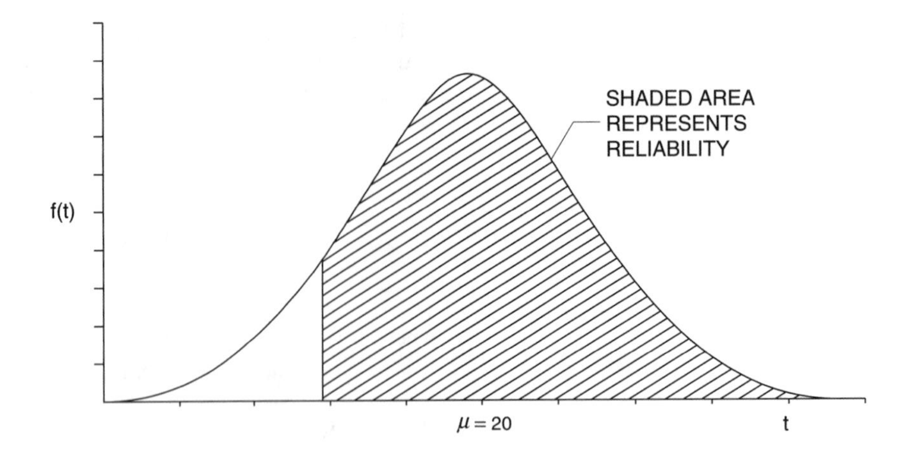

FIG. 2.9bb
Normal distribution.

where

θ = the distribution mean

$\lambda = \frac{1}{\theta}$ = the failure rate

The exponential PDF is shown in Figure 2.9cc.

The reliability function R(t) is

$$\begin{aligned} R(t) &= \int_t^{\infty} \lambda e^{-\lambda t} dt \\ &= e^{\lambda t}, \quad t \geq 0 \end{aligned} \qquad \mathbf{2.9(17)}$$

The result is shown in Figure 2.9dd.

The hazard function or failure rate is

$$h(t) = \frac{f(t)}{R(t)} = \frac{\lambda e^{-\lambda t}}{e^{-\lambda t}} = \lambda$$

The hazard function or failure rate is constant. This constant failure rate is unique to the exponential distribution, and it means that the probability of failure in a specific time interval is the same regardless of the starting point of that time interval. Also, if we know that the device has survived to time t, then the probability of failure from time t to (t + Δt) is the same as if the unit has just been put into operation from time 0 to Δt.

If an item follows the exponential distribution, the probability of failure in the interval t = 0 to t = 10 hours is the same as the probability of failure in the interval t = 1000 to t = 1010 hours.

Example A device has a mean time between failure of 80 hours. Given that the item survived to 200 hours, what is the probability of survival until t = 300 hours?

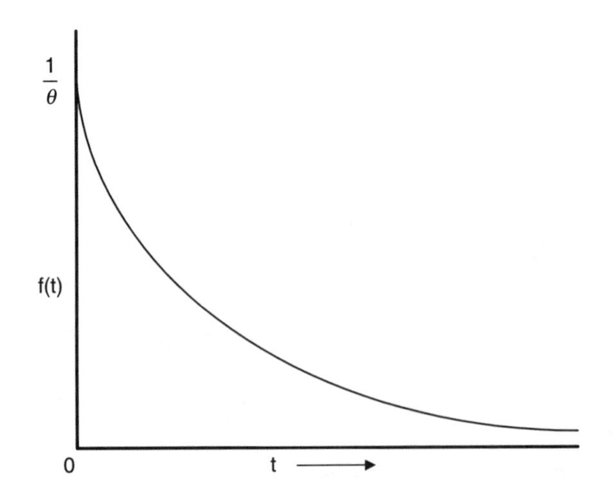

FIG. 2.9cc
PDF for exponential distribution.

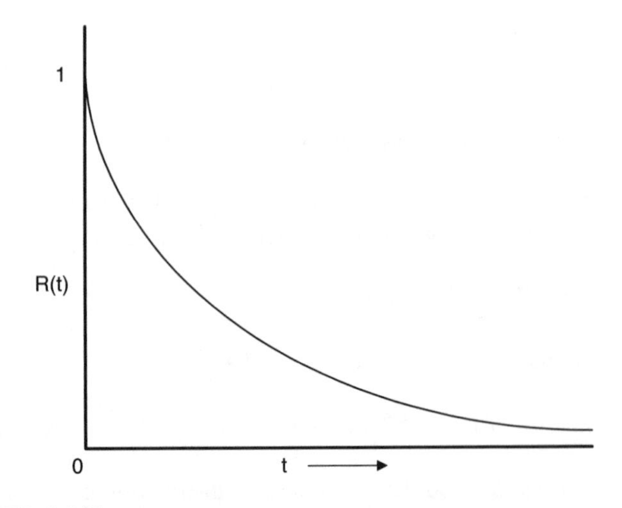

FIG. 2.9dd
The exponential reliability.

Solution The probability of survival until t = 300 hours given survival until t = 200 hours is

$$\lambda = \frac{1}{80} = 0.0125$$

$$\frac{R(300)}{R(200)} = \frac{e^{-300\times0.0125}}{e^{-200\times0.0125}}$$
$$= 0.2865$$

also

$$R(100) = e^{-100\times0.0125}$$
$$= 0.2865$$

The Weibull Distribution The Weibull distribution is the most versatile of all the reliability distributions. Depending on the values of its parameters, it can approximate the normal, the exponential, and many other distributions.

If parts fail according to a Weibull distribution, the probability that any single part will fail at a particular time, t, is represented by the CDF F(t):

$$F(t) = 1 - e^{-\alpha t^{\beta}} \quad \textbf{2.9(18)}$$

where α is called the location parameter and β is called the shape parameter. If we knew α and β, then we could insert the values into the above formula and calculate the probability of failure at any time, t. This is called a two-parameter Weibull distribution.

A three-parameter distribution is also available, when minimum life is taken into consideration. What is also very useful about the shape parameters are

For $\beta < 1$ Failure rate decreasing
$\beta = 1$ Failure rate constant (exponential distribution)
$\beta > 1$ Failure rate increasing
$\beta \geq 3.5$ Shape resembles the normal distribution

Other Weibull functions:

Reliability function $$R(t) = e^{-\alpha t^{\beta}} \quad \textbf{2.9(19)}$$

Probability density function (PDF) $$f(t) = \alpha\beta t^{\beta-1} e^{-\alpha t^{\beta}} \quad \textbf{2.9(20)}$$

Hazard rate function $$h(t) = \alpha\beta t^{\beta-1} \quad \textbf{2.9(21)}$$

The PDF for the Weibull distribution is not a single distribution, but the distributions vary for various values of β. As β increases, the distribution becomes more bell shaped, and as β decreases, the distribution looks more exponential.

Many methods exist for estimating the Weibull distribution parameters from data. The more common methods use probability plotting, hazard plotting, or maximum likelihood estimation.[8] Weibull analysis of failure data has also been simplified due to the availability of software to carry out the analysis and to determine the parameter values.[2]

Figure 2.9ee shows the typical failure rate distribution for a mechanical and electrical system. This is commonly known as the bathtub curve.

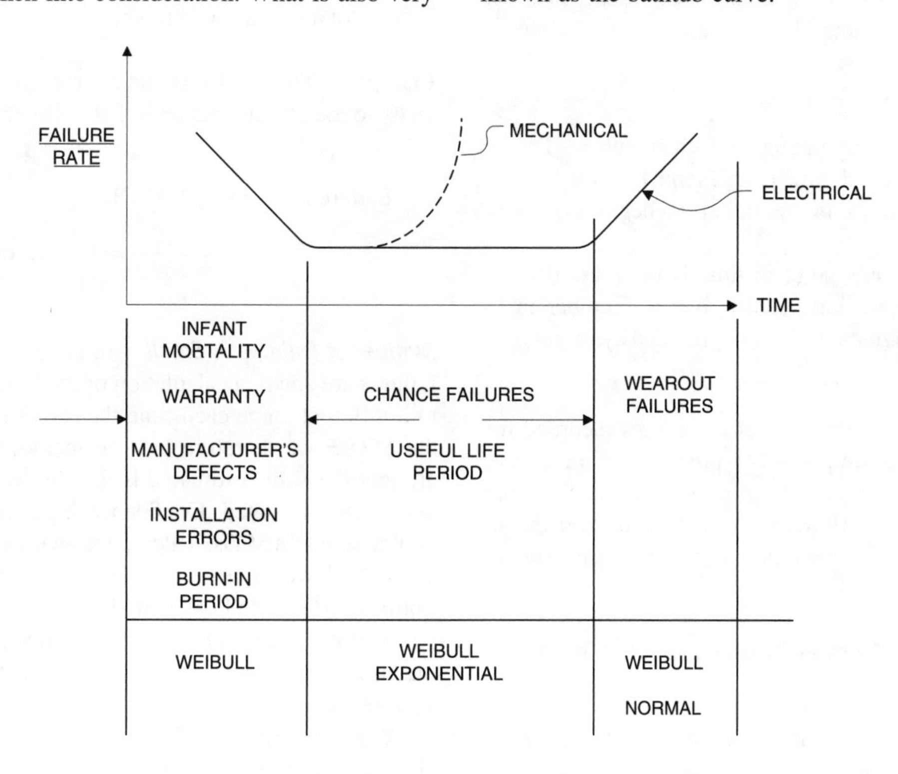

FIG. 2.9ee
Bathtub curve.

RELIABILITY MEASURES

Availability

Earlier we defined reliability as the probability that an item will perform its intended function for a *specified interval* under specified conditions. Availability is commonly defined as the probability that an item will operate at *any point in time* under specified conditions, or as the fraction of the total operating time during which the system is online and performing as specified. It is therefore a function of the failure and the repair rate.

Coombs and Ireson[1] define three types of availability: inherent, achieved, and operational. *Inherent availability* is entirely related to product design and does not take into consideration any scheduled or preventive maintenance downtime. *Achieved availability* includes preventive and corrective maintenance downtime. *Operational availability* also takes into consideration the time when the system is functionally acceptable but inactive

Failure Rate

Failure rate at any point in time is defined as the proportion that will fail in the next unit in time divided by the number of units that have survived up to that point in time.

Example Consider a control room consisting of several mimic/graphics panels with a total of 480 small lamps. The number of lamp failures experienced on a monthly basis from the original installation of 480 lamps is shown in Table 2.9ff. The data cover a period of 48 months. Column 3 lists the number of failures, Column 4 the failure rate, and Column 5 the reliability.

Data

The **failure rate** at beginning of each month = (The failures recorded during each month) divided by (number of lamps in operation at beginning of month)

The **reliability** at any point in time is the portion of the overall population that survives = (number in operation at beginning of period) divided by original population

Figure 2.9gg shows the number of failures recorded for every time period, Figure 2.9hh is a plot of the failure rate, and Figure 2.9ii is the reliability plot. From these plots one can easily analyze the performance of the lamps over the 48 months and make decisions relating to their future use or replacement.

Mean Time between Failures (MTBF) and Mean Time to Failure (MTTF)

For a population of devices, the MTBF and MTTF represent the total cumulative time of the population divided by the number of failures. MTBF and MTTF therefore represent the average time between failures. MTBF is used for items that are reparable, and the MTTF is used for nonreparable items.

MTBF/Failure Rate Calculations Based on Field Data

Instrument engineers are usually able to document the number of failures of a specific set of devices for a specific period. Usually at the end of the period only a few of the devices would have failed. This type of analysis is termed *time truncated*, and it is possible to estimate the failure rate from the data. If the number of failures and the time intervals are large, then the estimated value will be very close to the true value. If the number of failures is small, then the estimated value could be significantly different from the true value. Statistical techniques have been developed to determine our confidence that the estimated value will be equal to the actual value in cases where the number of failures is small.

We will review the two situations, i.e.,

1. Large number of failures
2. Small number of failures

Large Number of Failures If the number of failures is large, i.e., ≥10, the MTBF can be calculated from the formula.

$$\text{MTBF} = \frac{\text{T}}{\text{V}}$$

where
T = total test time = number of devices * test time
V = total number of failures

Example The total test time for a group of devices was 7640 hours. During this period 16 failures were experienced.

$$\begin{aligned}\text{Failure rate } (\lambda) &= 1/\text{MTBF}\\ &= \frac{16}{7640} = 0.0021 \text{ failures/hour}\end{aligned}$$

Number of Failures Is Small In the case where numbers of failures are small, a calculation of the MTBF will likely lead to a different value each time the test is repeated. In stating the MTBF we therefore have to include how confident we are that the value estimated is the true value, i.e., the confidence limit is stated. Confidence intervals can be single or double sided. In MTBF calculations we are usually concerned about the lower limit of the MTBF; hence, lower confidence limits are stated and calculated.

A 90% lower single-sided confidence limit therefore means that our estimated value will contain the true value 90% of the time.

If θ is the true MTBF and T is the total test time, statisticians have demonstrated that $2\text{T}/\theta$ has a χ^2 distribution. This is a special theoretical distribution developed for statistical

TABLE 2.9ff
Lamp Failures

Time (month)	*No. in Operation at Beginning of Period*	*Failures Recorded at End of Period*	*Failure Rate at Beginning of Month*	*Reliability*
1	480	8	0.0167	1.000
2	472	6	0.0127	0.983
3	466	3	0.0064	0.971
4	463	3	0.0065	0.965
5	460	1	0.0022	0.958
6	459	3	0.0065	0.956
7	456	1	0.0022	0.950
8	455	4	0.0088	0.948
9	451	1	0.0022	0.940
10	450	2	0.0044	0.938
11	448	2	0.0045	0.933
12	446	1	0.0022	0.929
13	445	1	0.0022	0.927
14	444	3	0.0068	0.925
15	441	2	0.0045	0.919
16	439	3	0.0068	0.915
17	436	3	0.0069	0.908
18	433	2	0.0046	0.902
19	431	1	0.0023	0.898
20	430	1	0.0023	0.896
21	429	1	0.0023	0.894
22	428	0	0.0000	0.892
23	428	1	0.0023	0.892
24	427	0	0.0000	0.890
25	427	2	0.0047	0.890
26	425	0	0.0000	0.885
27	425	1	0.0024	0.885
28	424	2	0.0047	0.883
29	422	1	0.0024	0.879
30	421	2	0.0048	0.877
31	419	1	0.0024	0.873
32	418	0	0.0000	0.871
33	418	1	0.0024	0.871
34	417	0	0.0000	0.869
35	417	2	0.0048	0.869
36	415	1	0.0024	0.865
37	414	2	0.0048	0.863
38	412	3	0.0073	0.858
39	409	1	0.0024	0.852
40	408	4	0.0098	0.850
41	404	3	0.0074	0.842

(Continued)

TABLE 2.9ff Continued
Lamp Failures

Time (month)	No. in Operation at Beginning of Period	Failures Recorded at End of Period	Failure Rate at Beginning of Month	Reliability
42	401	4	0.0100	0.835
43	397	5	0.0126	0.827
44	392	4	0.0102	0.817
45	388	4	0.0103	0.808
46	384	3	0.0078	0.800
47	381	5	0.0131	0.794
48	376	5	0.0133	0.783

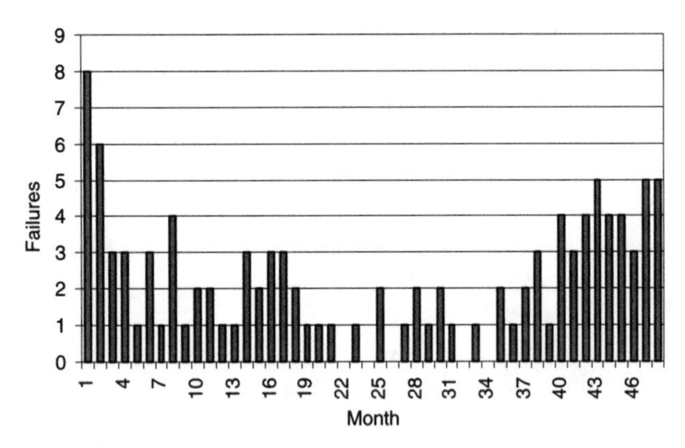

FIG. 2.9gg
Histogram.

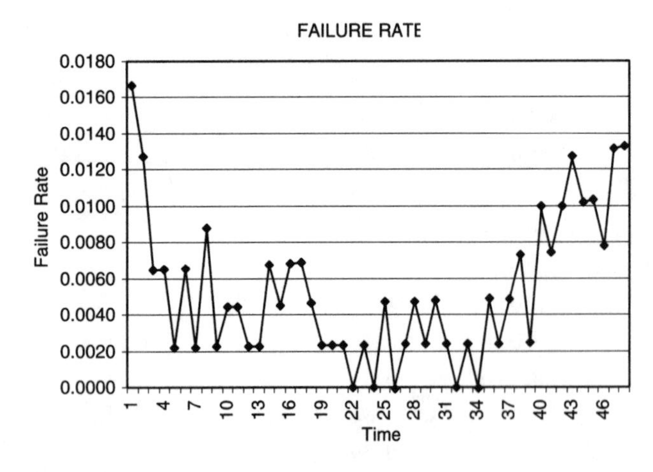

FIG. 2.9hh
Failure data.

analysis. Percentile values of this distribution, similar to the Z distribution, are shown in Table 2.9ll. For time-truncated tests and lower confidence limits, the applicable equation is[3,9]

$$\frac{2T}{\chi^2_{(\alpha,2r+2)}} = \theta \qquad \textbf{2.9(22)}$$

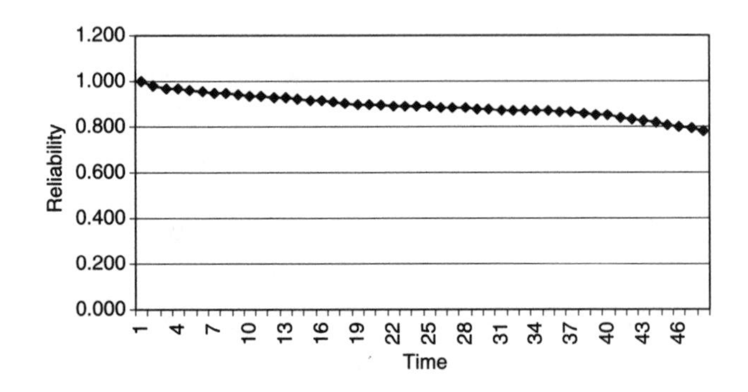

FIG. 2.9ii
Reliability.

where
T = total test time
α = (1 − confidence)
r = number of failures
ν = degrees of freedom = (2r + 2)

Example Over a period of 5 years 24 transmitters were evaluated. During this period only two failures were experienced. Calculate the failure rate (λ), assuming a single-sided lower confidence limit of 70%.

Solution

$$\begin{aligned} \text{The total test time T} &= 24 \times 5 \text{ years} \\ &= 120 \text{ years} \\ \text{Number of failures} &= 2, \quad r = 2 \\ \alpha &= (1-0.7) = 0.3 \\ \nu &= (2r+2) = 6 \end{aligned}$$

From Table 2.9ll, χ^2 for $\nu = 6$ and $\alpha = 0.3$ is 7.231

$$\theta = \frac{2 \times 120}{7.231} = 33.19 \text{ years}$$

$$\lambda = \frac{1}{\theta} = 0.03\text{f/year}$$

TABLE 2.9jj

Z Table: Negative Values

z	.00	.01	.02	.03	.04	.05	.06	.07	.08	.09
−3.80	.0001	.0001	.0001	.0001	.0001	.0001	.0001	.0001	.0001	.0001
−3.70	.0001	.0001	.0001	.0001	.0001	.0001	.0001	.0001	.0001	.0001
−3.60	.0002	.0002	.0001	.0001	.0001	.0001	.0001	.0001	.0001	.0001
−3.50	.0002	.0002	.0002	.0002	.0002	.0002	.0002	.0002	.0002	.0002
−3.40	.0003	.0003	.0003	.0003	.0003	.0003	.0003	.0003	.0003	.0002
−3.30	.0005	.0005	.0005	.0004	.0004	.0004	.0004	.0004	.0004	.0003
−3.20	.0007	.0007	.0006	.0006	.0006	.0006	.0006	.0005	.0005	.0005
−3.10	.0010	.0009	.0009	.0009	.0008	.0008	.0008	.0008	.0007	.0007
−3.00	.0013	.0013	.0013	.0012	.0012	.0011	.0011	.0011	.0010	.0010
−2.90	.0019	.0018	.0018	.0017	.0016	.0016	.0015	.0015	.0014	.0014
−2.80	.0026	.0025	.0024	.0023	.0023	.0022	.0021	.0021	.0020	.0019
−2.70	.0035	.0034	.0033	.0032	.0031	.0030	.0029	.0028	.0027	.0026
−2.60	.0047	.0045	.0044	.0043	.0041	.0040	.0039	.0038	.0037	.0036
−2.50	.0062	.0060	.0059	.0057	.0055	.0054	.0052	.0051	.0049	.0048
−2.40	.0082	.0080	.0078	.0075	.0073	.0071	.0069	.0068	.0066	.0064
−2.30	.0107	.0104	.0102	.0099	.0096	.0094	.0091	.0089	.0087	.0084
−2.20	.0139	.0136	.0132	.0129	.0125	.0122	.0119	.0116	.0113	.0110
−2.10	.0179	.0174	.0170	.0166	.0162	.0158	.0154	.0150	.0146	.0143
−2.00	.0228	.0222	.0217	.0212	.0207	.0202	.0197	.0192	.0188	.0183
−1.90	.0287	.0281	.0274	.0268	.0262	.0256	.0250	.0244	.0239	.0233
−1.80	.0359	.0351	.0344	.0336	.0329	.0322	.0314	.0307	.0301	.0294
−1.70	.0446	.0436	.0427	.0418	.0409	.0401	.0392	.0384	.0375	.0367
−1.60	.0548	.0537	.0526	.0516	.0505	.0495	.0485	.0475	.0465	.0455
−1.50	.0668	.0655	.0643	.0630	.0618	.0606	.0594	.0582	.0571	.0559
−1.40	.0808	.0793	.0778	.0764	.0749	.0735	.0721	.0708	.0694	.0681
−1.30	.0968	.0951	.0934	.0918	.0901	.0885	.0869	.0853	.0838	.0823
−1.20	.1151	.1131	.1112	.1093	.1075	.1056	.1038	.1020	.1003	.0985
−1.10	.1357	.1335	.1314	.1292	.1271	.1251	.1230	.1210	.1190	.1170
−1.00	.1587	.1562	.1539	.1515	.1492	.1469	.1446	.1423	.1401	.1379
−0.90	.1841	.1814	.1788	.1762	.1736	.1711	.1685	.1660	.1635	.1611
−0.80	.2119	.2090	.2061	.2033	.2005	.1977	.1949	.1922	.1894	.1867
−0.70	.2420	.2389	.2358	.2327	.2296	.2266	.2236	.2206	.2177	.2148
−0.60	.2743	.2709	.2676	.2643	.2611	.2578	.2546	.2514	.2483	.2451
−0.50	.3085	.3050	.3015	.2981	.2946	.2912	.2877	.2843	.2810	.2776
−0.40	.3446	.3409	.3372	.3336	.3300	.3264	.3228	.3192	.3156	.3121
−0.30	.3821	.3783	.3745	.3707	.3669	.3632	.3594	.3557	.3520	.3483
−0.20	.4207	.4168	.4129	.4090	.4052	.4013	.3974	.3936	.3897	.3859
−0.10	.4602	.4562	.4522	.4483	.4443	.4404	.4364	.4325	.4286	.4247
0.00	.5000	.4960	.4920	.4880	.4840	.4801	.4761	.4721	.4681	.4641

Body of table gives area under Z curve to the left of **z**. *Example:* P[Z < −2.63] = .0043.

TABLE 2.9kk
Z Table: Positive Values

z	.00	.01	.02	.03	.04	.05	.06	.07	.08	.09
0.00	.5000	.5040	.5080	.5120	.5160	.5199	.5239	.5279	.5319	.5359
0.10	.5398	.5438	.5478	.5517	.5557	.5596	.5636	.5675	.5714	.5753
0.20	.5793	.5832	.5871	.5910	.5948	.5987	.6026	.6064	.6103	.6141
0.30	.6179	.6217	.6255	.6293	.6331	.6368	.6406	.6443	.6480	.6517
0.40	.6554	.6591	.6628	.6664	.6700	.6736	.6772	.6808	.6844	.6879
0.50	.6915	.6950	.6985	.7019	.7054	.7088	.7123	.7157	.7190	.7224
0.60	.7257	.7291	.7324	.7357	.7389	.7422	.7454	.7486	.7517	.7549
0.70	.7580	.7611	.7642	.7673	.7704	.7734	.7764	.7794	.7823	.7852
0.80	.7881	.7910	.7939	.7967	.7995	.8023	.8051	.8078	.8106	.8133
0.90	.8159	.8186	.8212	.8238	.8264	.8289	.8315	.8340	.8365	.8389
1.00	.8413	.8438	.8461	.8485	.8508	.8531	.8554	.8577	.8599	.8621
1.10	.8643	.8665	.8686	.8708	.8729	.8749	.8770	.8790	.8810	.8830
1.20	.8849	.8869	.8888	.8907	.8925	.8944	.8962	.8980	.8997	.9015
1.30	.9032	.9049	.9066	.9082	.9099	.9115	.9131	.9147	.9162	.9177
1.40	.9192	.9207	.9222	.9236	.9251	.9265	.9279	.9292	.9306	.9319
1.50	.9332	.9345	.9357	.9370	.9382	.9394	.9406	.9418	.9429	.9441
1.60	.9452	.9463	.9474	.9484	.9495	.9505	.9515	.9525	.9535	.9545
1.70	.9554	.9564	.9573	.9582	.9591	.9599	.9608	.9616	.9625	.9633
1.80	.9641	.9649	.9656	.9664	.9671	.9678	.9686	.9693	.9699	.9706
1.90	.9713	.9719	.9726	.9732	.9738	.9744	.9750	.9756	.9761	.9767
2.00	.9772	.9778	.9783	.9788	.9793	.9798	.9803	.9808	.9812	.9817
2.10	.9821	.9826	.9830	.9834	.9838	.9842	.9846	.9850	.9854	.9857
2.20	.9861	.9864	.9868	.9871	.9875	.9878	.9881	.9884	.9887	.9890
2.30	.9893	.9896	.9898	.9901	.9904	.9906	.9909	.9911	.9913	.9916
2.40	.9918	.9920	.9922	.9925	.9927	.9929	.9931	.9932	.9934	.9936
2.50	.9938	.9940	.9941	.9943	.9945	.9946	.9948	.9949	.9951	.9952
2.60	.9953	.9955	.9956	.9957	.9959	.9960	.9961	.9962	.9963	.9964
2.70	.9965	.9966	.9967	.9968	.9969	.9970	.9971	.9972	.9973	.9974
2.80	.9974	.9975	.9976	.9977	.9977	.9978	.9979	.9979	.9980	.9981
2.90	.9981	.9982	.9982	.9983	.9984	.9984	.9985	.9985	.9986	.9986
3.00	.9987	.9987	.9987	.9988	.9988	.9989	.9989	.9989	.9990	.9990
3.10	.9990	.9991	.9991	.9991	.9992	.9992	.9992	.9992	.9993	.9993
3.20	.9993	.9993	.9994	.9994	.9994	.9994	.9994	.9995	.9995	.9995
3.30	.9995	.9995	.9995	.9996	.9996	.9996	.9996	.9996	.9996	.9997
3.40	.9997	.9997	.9997	.9997	.9997	.9997	.9997	.9997	.9997	.9998
3.50	.9998	.9998	.9998	.9998	.9998	.9998	.9998	.9998	.9998	.9998
3.60	.9998	.9998	.9999	.9999	.9999	.9999	.9999	.9999	.9999	.9999
3.70	.9999	.9999	.9999	.9999	.9999	.9999	.9999	.9999	.9999	.9999
3.80	.9999	.9999	.9999	.9999	.9999	.9999	.9999	.9999	.9999	.9999

Body of table gives area under Z curve to the left of **z**. *Example*: P[Z < 1.16] = .8770.

TABLE 2.9II
χ^2 Distribution Tables

	α									
v	0.5	0.45	0.4	0.35	0.3	0.25	0.2	0.15	0.1	0.05
1	0.455	0.571	0.708	0.873	1.074	1.323	1.642	2.072	2.706	3.841
2	1.386	1.597	1.833	2.100	2.408	2.773	3.219	3.794	4.605	5.991
3	2.366	2.643	2.946	3.283	3.665	4.108	4.642	5.317	6.251	7.815
4	3.357	3.687	4.045	4.438	4.878	5.385	5.989	6.745	7.779	9.488
5	4.351	4.728	5.132	5.573	6.064	6.626	7.289	8.115	9.236	11.070
6	5.348	5.765	6.211	6.695	7.231	7.841	8.558	9.446	10.645	12.592
7	6.346	6.800	7.283	7.806	8.383	9.037	9.803	10.748	12.017	14.067
8	7.344	7.833	8.351	8.909	9.524	10.219	11.030	12.027	13.362	15.507
9	8.343	8.863	9.414	10.006	10.656	11.389	12.242	13.288	14.684	16.919
10	9.342	9.892	10.473	11.097	11.781	12.549	13.442	14.534	15.987	18.307
11	10.341	10.920	11.530	12.184	12.899	13.701	14.631	15.767	17.275	19.675
12	11.340	11.946	12.584	13.266	14.011	14.845	15.812	16.989	18.549	21.026
13	12.340	12.972	13.636	14.345	15.119	15.984	16.985	18.202	19.812	22.362
14	13.339	13.996	14.685	15.421	16.222	17.117	18.151	19.406	21.064	23.685
15	14.339	15.020	15.733	16.494	17.322	18.245	19.311	20.603	22.307	24.996
16	15.338	16.042	16.780	17.565	18.418	19.369	20.465	21.793	23.542	26.296
17	16.338	17.065	17.824	18.633	19.511	20.489	21.615	22.977	24.769	27.587
18	17.338	18.086	18.868	19.699	20.601	21.605	22.760	24.155	25.989	28.869
19	18.338	19.107	19.910	20.764	21.689	22.718	23.900	25.329	27.204	30.144
20	19.337	20.127	20.951	21.826	22.775	23.828	25.038	26.498	28.412	31.410
21	20.337	21.147	21.992	22.888	23.858	24.935	26.171	27.662	29.615	32.671
22	21.337	22.166	23.031	23.947	24.939	26.039	27.301	28.822	30.813	33.924
23	22.337	23.185	24.069	25.006	26.018	27.141	28.429	29.979	32.007	35.172
24	23.337	24.204	25.106	26.063	27.096	28.241	29.553	31.132	33.196	36.415
25	24.337	25.222	26.143	27.118	28.172	29.339	30.675	32.282	34.382	37.652
26	25.336	26.240	27.179	28.173	29.246	30.435	31.795	33.429	35.563	38.885
27	26.336	27.257	28.214	29.227	30.319	31.528	32.912	34.574	36.741	40.113
28	27.336	28.274	29.249	30.279	31.391	32.620	34.027	35.715	37.916	41.337
29	28.336	29.291	30.283	31.331	32.461	33.711	35.139	36.854	39.087	42.557
30	29.336	30.307	31.316	32.382	33.530	34.800	36.250	37.990	40.256	43.773
31	30.336	31.323	32.349	33.431	34.598	35.887	37.359	39.124	41.422	44.985
32	31.336	32.339	33.381	34.480	35.665	36.973	38.466	40.256	42.585	46.194
33	32.336	33.355	34.413	35.529	36.731	38.058	39.572	41.386	43.745	+7.400
34	33.336	34.371	35.444	36.576	37.795	39.141	40.676	42.514	44.903	48.602
35	34.336	35.386	36.475	37.623	38.859	40.223	41.778	43.640	46.059	49.802
36	35.336	36.401	37.505	38.669	39.922	41.304	42.879	44.764	47.212	50.998
37	36.336	37.416	38.535	39.715	40.984	42.383	43.978	45.886	48.363	52.192
38	37.335	38.430	39.564	40.760	42.045	43.462	45.076	47.007	49.513	53.384
39	38.335	39.445	40.593	41.804	43.105	44.539	46.173	48.126	50.660	54.572
40	39.335	40.459	41.622	42.848	44.165	45.616	47.269	49.244	51.805	55.758
50	49.335	50.592	51.892	53.258	54.723	56.334	58.164	60.346	63.167	67.505

Values in the table are critical values of the χ^2 random variable for right-hand tails of the indicated areas, α.

References

1. Coombs, C. F., Jr. and Ireson, W. G., *Handbook of Reliability Engineering and Management,* New York: McGraw-Hill, 1996.
2. Dodson, B., *Weibull Analysis* (with software), Milwaukee: ASQC Quality Press, 1995.
3. Dovich, R. A., *Reliability Statistics,* Milwaukee: ASQC Quality Press, 1990.
4. Kapur, K. C. and Lamberson, L. R., *Reliability in Engineering Design,* New York: John Wiley & Sons, 1977.
5. Kececioglu, D., *Reliability Engineering Handbook,* Vol. 1, Englewood Cliffs, NJ: Prentice-Hall, 1991.
6. Kececioglu, D., *Reliability Engineering Handbook,* Vol. 2, Englewood Cliffs, NJ: Prentice-Hall, 1991.
7. Krishnamoorthi, K. S., *Reliability Methods for Engineers,* Milwaukee: ASQC Quality Press, 1992.
8. Nelson, W., *Applied Life Data Analysis,* New York: John Wiley & Sons, 1982.
9. Smith, D. J., *Reliability Engineering,* London: Pitman, 1972.

Bibliography

Aggarwal, K., *Reliability Engineering,* Boston: Kluwer Academic, 1993.

Amendola, A., *Common Cause Failure Analysis in Probabilistic Safety Assessment,* Boston: Kluwer Academic, 1989.

ANSI/IEEE, *Reliability Data for Pumps and Drives, Valve Actuators, and Valves,* New York: John Wiley & Sons, 1986.

Ascher, H. and Feingold, H., *Repairable Systems Reliability: Modeling, Inference, Misconceptions and Their Causes,* New York: Marcel Dekker, 1984.

Barlow, R. E., Clarotti, C. A., and Spizzichino, F., *Reliability and Decision Making,* London: Chapman & Hall, 1993, 370 pp.

Blanchard, B. S., Verma, D., and Peterson, E. L., *Maintainability: A Key to Effective Serviceability and Maintenance Management,* New York: John Wiley & Sons, 1995.

Bralla, J. G., *Design for Excellence,* New York: McGraw-Hill, 1996.

Brombacher, A. C., *Reliability by Design: CAE Techniques for Electronic Components and Systems,* New York: John Wiley & Sons, 1992.

Cluley, J. C., *Reliability in Instrumentation and Control,* Boston: Butterworth-Heinemann, 1993.

Hoyland, A. and Rausand, M., *System Reliability Theory: Models and Statistical Methods,* New York: John Wiley & Sons, 1994.

Kececioglu, D., *Reliability and Life Testing Handbook,* Vols. 1 and 2, Englewood Cliffs, NJ: Prentice-Hall, 1993.

Knezevic, J., *Reliability, Maintainability and Supportability: A Probabilistic Approach,* New York: McGraw-Hill, 1993.

Krishnaiah, P. R. and Rao, C. R., *Quality Control and Reliability,* Amsterdam: Elsevier/North-Holland, 1988.

Misra, K. B., *Reliability Analysis and Prediction: A Methodology Oriented Treatment,* New York: Elsevier, 1992.

Modarres, M., *What Every Engineer Should Know about Reliability and Risk Analysis,* New York: Marcel Dekker, 1993.

Nash, F. R., *Estimating Device Reliability: Assessment of Credibility,* Boston: Kluwer Academic, 1993.

O'Connor, P. D. T., *Reliability Engineering,* Washington, D.C.: Hemisphere, 1988.

Patton, J. D., Jr., *Maintainability and Maintenance Management,* Research Triangle Park, NC: Instrument Society of America, 1980.

Ramakumar, R., *Reliability Engineering: Fundamentals and Applications,* Englewood Cliffs, NJ: Prentice-Hall, 1993.

Rao, S. S., *Reliability-Based Design,* New York: McGraw-Hill, 1992.

Sarma, V., Singh, M., and Viswanadham, N., *Reliability of Computer and Control Systems,* New York: Elsevier Science, 1987.

Smith, A. M., *Reliability-Centered Maintenance,* New York: McGraw-Hill, 1993.

Ushakov, I., *Handbook of Reliability Engineering,* New York: John Wiley & Sons, 1994.

Villemeur, A., *Reliability, Availability, Maintainability and Safety Assessment,* Vols. 1 and 2, Chichester, U.K.: John Wiley & Sons, 1992.

Von Alven, W. H., "Reliability Engineering," prepared by the engineers and statistical staff of ARINC Research Corporation, 1964.

Zacks, S., *Introduction to Reliability Analysis: Probability Models and Statistics Methods,* New York: Springer-Verlag, 1992.

Military Standards Used in Reliability Work

MIL-HDBK-H 108 Sampling Procedures and Tables for Life and Reliability Testing (Based on Exponential Distribution)

MIL-HDBK-189 Reliability Growth Management

MIL-HDBK-217F Reliability Prediction of Electronic Equipment

MIL-HDBK-251 Reliability/Design Thermal Applications

MIL-HDBK-263 A Electrostatic Discharge Control Handbook for Protection of Electrical and Electronic Parts, Assemblies and Equipment

MIL-STD-337 Design to Cost

MIL-HDBK-338 Electronic Reliability Design Handbook

MIL-HDBK-344 Environmental Stress Screening of Electronic Equipment

MIL-STD-415 Test Provision for Electronic Systems and Associated Equipment

MIL-STD-446 Environmental Requirements for Electric Parts

MIL-STD-690C Failure Rate Sampling Plans and Procedures

MIL-STD-721C Definition of Terms for Reliability and Maintainability

MIL-STD-756B Reliability Modeling and Prediction

MIL-HDBK-781 Reliability Test Methods, Plans and Environments for Engineering Development, Qualification and Production

MIL-STD-781D Reliability Design Qualification and Production Acceptance Tests: Exponential/ Distribution

MIL-STD-785B Reliability Program for Systems and Equipment, Development and Production

MIL-STD-790E Reliability Assurance Program for Electronic Parts Specifications

MIL-STD-810 Environmental Test Methods

Technical reports Sampling procedures and tables for life testing based on the Weibull distribution. OASD installation and logistics, Washington:

- TR 3 Mean life criterion
- TR 4 Hazard rate criterion
- TR 6 Reliable life criterion

MIL-STD 1556 Government/Industry Data Exchange Program Contractor Participation Requirements

MIL-STD-1629A Procedures for Performing a Failure Mode, Effects, and Criticality Analysis

MIL-STD-1686B Electrostatic Discharge Control Program for Protection of Electrical and Electronic Parts, Assemblies and Equipment

MIL-STD-2074 Failure Classification for Reliability Testing

MIL-STD-2155 Failure Reporting, Analysis and Corrective Action System (FRACAS)

MIL-STD-2164 Environment Stress Screening Process for Electronic Equipment

MIL-STD-470B Maintainability Program Requirements for Systems and Equipment

MIL-STD-471A Maintainability Verification/Demonstration/Evaluation

MIL-HDBK 109 Statistical Procedures for Determining Validity of Suppliers Attributed Inspection

MIL-HDBK-472 Maintainability Prediction

DOD-HDBK-791 Maintainability Design Techniques

MIL-STD-2165A Testability Programs for Electronic Systems and Equipment

DOD-STD-1701 Hardware Diagnostic Test System Requirements

2.10 Intelligent Alarm Management

D. A. STROBHAR

Alarms have a unique niche in the realm of instrumentation and control. They are the only element whose sole purpose is to alter the operators' behavior. PID controllers react to some of the data the instrumentation collects and attempts to modify the process variable. Indicators and other instrumentation are used by the operators to assess aspects of the process they are controlling. Some of the instrumentation is present to collect data for historical or analytic purposes by numerous other groups (engineers, managers, accounting, maintenance personnel, etc.), but alarms are specific, they are for the operator, they are there to prompt an operator to act.

The advent of modern control systems has created a situation where the function of the alarm system is lost due to the alarm system itself. In the "dark ages" of hardwired alarm panels, each alarm had to be thought out and justified. Each new alarm was added at a cost, and perhaps at the expense of another alarm if the panels were full. But distributed control systems have little cost for adding alarms. A key to the console, some minor configuration knowledge, and a little time and an alarm can be added. The result has been a figurative explosion of alarms for process operators resulting in literal explosions at processing facilities. Critical alarms cannot be seen among the tidal wave of alarms that flood the board during even minor plant upsets.

Should we go back to using hardwired panels? An option, but hardwired alarm panels are not flawless. The accident at Three Mile Island had as a contributing cause a critical alarm located on a back panel out of sight of the operators (all of the primary panel space had been utilized). A better option is to use a little intelligence: intelligence in the selection of alarms, intelligence in understanding what alarms should and should not be used for, intelligence in presenting the alarm information, and intelligence in realizing that people have inherent characteristics and limitations in how they can process alarm information.

This section hopes to increase the intelligence of alarm system designers, instructing them that:

1. The focus of the alarm system needs to be on the operators.
2. Operator behavior is constrained by inherent characteristics and limitations of the human information processing system.
3. Alarms are only a portion of what is needed to facilitate behavior.
4. Alarm management requires understanding the capabilities and use of the control system's capabilities.

ALARM MANAGEMENT COSTS

From a project prospective, alarm management has two distinct aspects: (1) alarm system design and (2) design implementation. Many vendors offer both services as a package, while other firms specialize in one or the other. Regardless, the design effort should not require more than 3 work-hours per control loop on the unit (control loop defined as analog output). For example, a unit with 100 control loops should require approximately 300 work-hours for the design phase. The implementation of the design should require 6 to 9 work-hours per loop, or two to three times the effort of the design portion. Implementation includes the reconfiguration of the DCS (distributed control system) database, configuration of new displays, and training/management of change reviews for those affected. The level of effort described is what it *should* require, although many facilities have spent orders of magnitude greater effort in alarm management with little success.

ALARM MANAGEMENT PROBLEM

The most common problem with alarm systems today is the flood of alarms that occur during an upset. A simple example can be seen in Figure 2.10a, with each bar showing the number of alarm actuations in a 5-min increment of time. A slide valve failure on a fluid catalytic cracking (FCC) unit has resulted in a wave of alarms hitting the board operator. The distributed control system (DCS) had been configured so that the only viable means to receive alarm information was through a listing of alarms by time of actuation. With that type of presentation, the upper threshold for human signal detection will limit the number of alarms that can be detected (not processed), which is 15 signals per minute.[1] The limit for processing alarms is 5 alarms per minute.[2] These two limits are shown on the figure and show the alarms that were beyond what a human can likely process, which explains why the operator said they "abandoned" the alarm

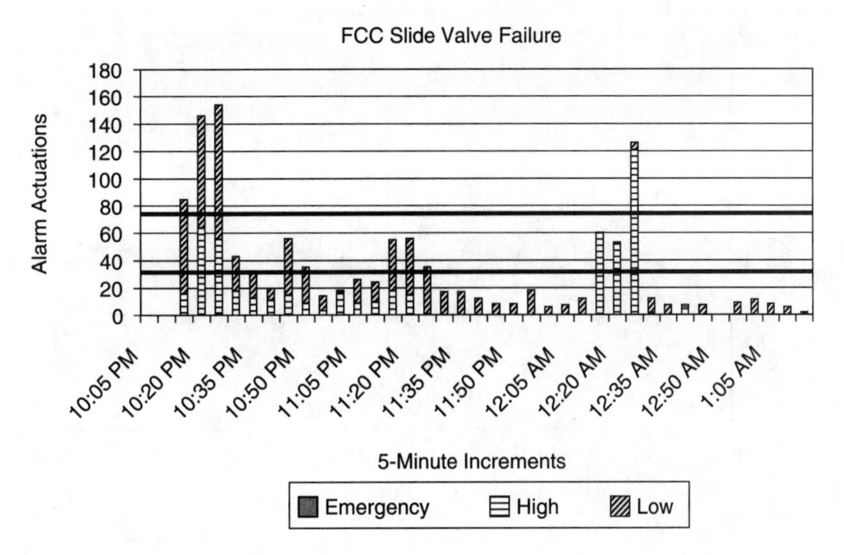

FIG. 2.10a
Typical alarm flood with limits of detection.

system during the course of this upset. During a time when the alarm system was needed most, it was of the least value.

Several factors contribute to the large number of alarms with modern control systems. First, the ease with which alarms can be added (as mentioned earlier) enables alarms to be added with minimal thought and analysis. Second, each alarm by itself might seem needed, and only appear redundant when the system is viewed as a whole (which is not very often). Third, alarms become the correction for any problem, attempting to compensate for bad system design, training, or operator selection. Fourth, alarms are used for purposes beyond their intended function of prompting behavior, becoming auditing tools for process safety management (PSM) or engineering activities. Fifth, no overriding alarm philosophy exists, resulting in the application of as many ways to alarm a process unit as there are board operators, engineers, and supervisors.

Almost any unit on distributed control has had some problem with alarm flooding; below are several incidents at plants across North America illustrating some of the problems.

1. Loss of Steam and Instrument Air—The utility area of a major refinery had one of the seven boilers trip. The resulting alarm flood obscured the alarm for boiler feedwater pump trip. Eventually, all the boilers were lost and the refinery shut down with no steam and no instrument air.
2. Loss of Instrument Air—A plant was shutting down due to the loss of raw water to the plant. The alarm indicating that an instrument air dryer had switched improperly was missed in the flood of alarms. The problem was detected when control valves for the shutdown were not responding due to a loss of instrument air.
3. Refinery Power Failure—In the midst of a refinery power failure, the coker/sulfur recovery unit (SRU) board operator had to abandon the alarm system due to the flood of alarms. A loss of seal leg on the SRU reactor was missed, as was the resulting H_2S release into the unit. A catastrophe was averted only because the outside operator had gone into fresh air at the beginning of the upset, perhaps knowing the potential for the board to miss a key alarm.
4. FCC Surge Drum Overfill—In the course of a power failure, the high-level alarm on the FCC surge drum was missed in the flood of alarms. The surge drum filled and filled the flare line with liquid, preventing the other units from being able to use the flare to relieve pressure.

Alarm and information overload are neither new nor novel to human control of complex systems. Other industries have encountered the problem before and have found a solution. The solution was the origin of the study of human factors as a scientific discipline.

KNOWLEDGE OF HUMAN FACTORS AS A SOLUTION

Process plants are not the first system to face information overload. In fact, the birth of human factors engineering as a discipline was in part a result of such a situation.[3] Information overload has proved to be the initiating problem for application of human factors technology to a variety of industries.

Aircraft underwent a rapid increase in capability with the advent of World War II. The increase in aircraft speed began its exponential climb with advances in fighter technology during the war, as seen in Figure 2.10b. With the increase in

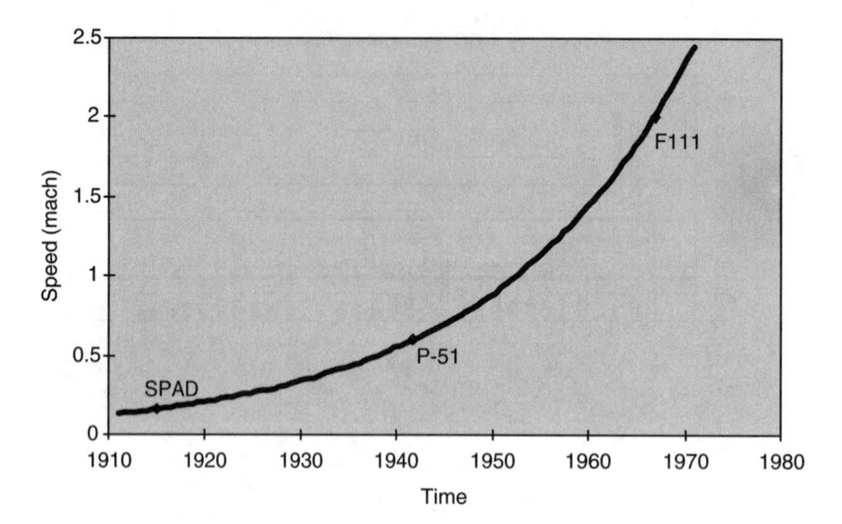

FIG. 2.10b
Increase in aircraft speed/capability.

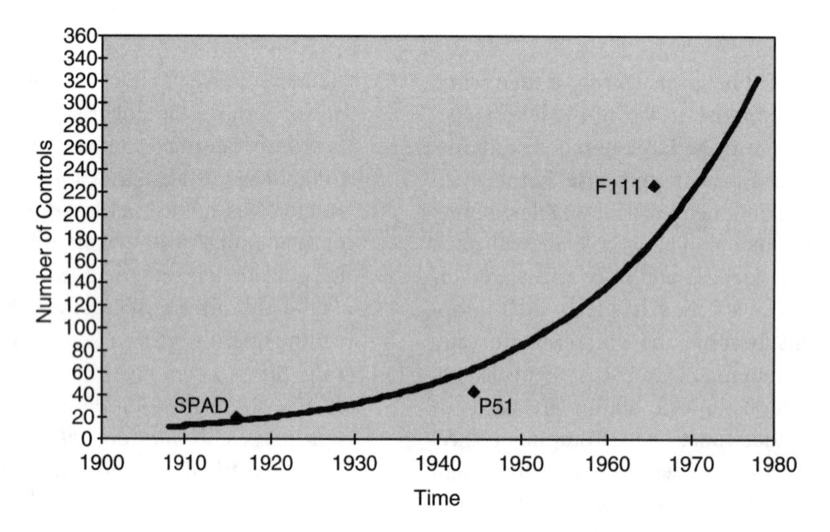

FIG. 2.10c
Increase in aircraft control/switches.

speed came a similar increase in "operator error"-related problems. The Army Air Corp (precursor of the Air Force) blamed the aircraft manufacturer for building poor aircraft, who in turn blamed the Army for selecting poor pilots. Some people speculated that the limits of human control of aircraft had been reached, that people could only handle so much of an airplane. (This sounds similar to the number of control loops per operator argument presented for span of control limits.)

A group of psychologists began to investigate the problem. What they found was that as the capability of the aircraft began to exponentially increase, so did the instrumentation in the cockpit, as seen in Figure 2.10c.[4] The pilots were being flooded with information. The solution for the aerospace industry was the identification of the critical instruments that are required to fly an airplane (largely still the same today), and placement of them in a standard configuration in each cockpit.[5] The goal was to separate the information needed to fly the plane from all the other information on weapon systems, auxiliary systems, etc. The result was the beginnings of human factors engineering, understanding how people use and process information in complex systems.

The same phenomena that the aerospace industry observed in the 1940s, the U.S. nuclear power industry saw in the 1970s. General Public Utilities operated three nuclear power plants. Figure 2.10d shows the number of hardwired alarms in each of the plants control rooms. The 1969 reactor (680 MW) had about 300 hardwired alarms. The 1974 (820 MW) reactor had about 600 hardwired alarms. A 1978 reactor (840 MW), a sister plant to the one built in 1974, had 1200 hardwired alarms. A doubling of alarms was occurring with each new plant, despite modest increase in reactor size. The 1978 reactor is Three Mile Island Unit 2, the site of the 1979 accident that triggered the beginning of the use of human factors technology in the nuclear power industry.

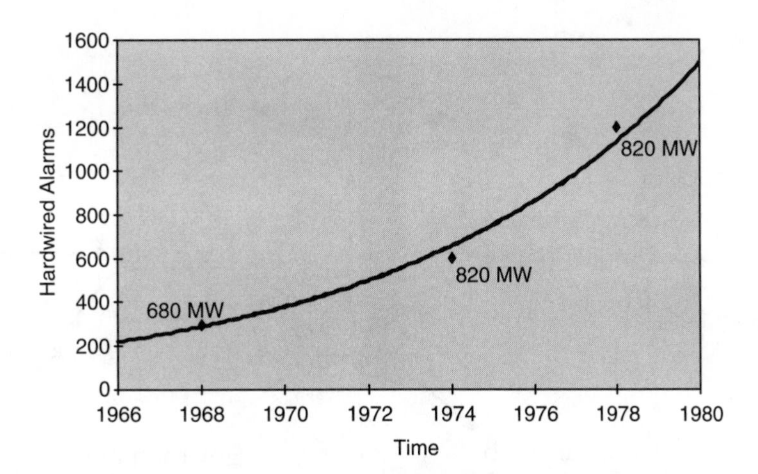

FIG. 2.10d
Increase in nuclear power plant alarms.

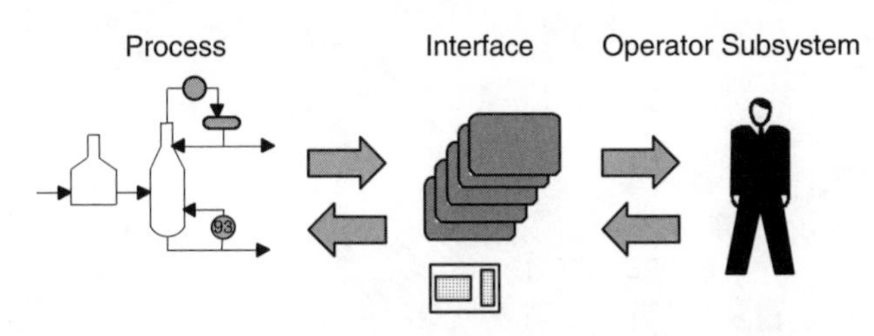

FIG. 2.10e
Model of operator–process interaction.

Intelligent alarm management involves changing the way we look at alarm systems. Figure 2.10e shows a simple model of operator–process interaction, where information on the process is sensed and presented to the operator via displays, who can effect changes on the process through those or other displays. Good alarm systems are the result of starting with the right side of the model, focusing first on the operator (what do they need to do, what decisions do they need to make, what information do they need) and then focusing on the process (what is being sensed). Poor alarm systems are the result of focusing on the process, taking an "if it is sensed, it will be alarmed" philosophy.

HUMAN FACTORS VARIABLES

Research since the 1940s has identified the variables that influence human performance. Figure 2.10f shows a number of the major variables. The major variables and their design elements include the following.

- Interface—The design of the workstation and the alarm/display system
- System—The dynamic characteristics of the system to be controlled and the level of automation (need for human intervention)
- Environment—The environments in which the performance occurs, including light, noise, and stress
- Organization—The expectations, crew dynamics, levels of authority, and motivation set forth by management
- Individual—The inherent attributes and level of skill/knowledge acquired through training

An important aspect of understanding human performance is to understand the interactive nature of the variables. Improvements in one variable can compensate for problems in another or necessitate changes in a different variable. An example is the design of the interface that interacts with the capabilities of the individual. The interface for an automated teller machine (ATM), with little presumed skill/knowledge on the part of the user, will be dramatically different from the cockpit of a fighter aircraft, in which the user has undergone extensive training. The multivariable nature of human performance provides much of the flexibility in system design that enables humans to accomplish all that we can accomplish, and also makes design/analysis of human–machine systems difficult. Human performance is not a one-dimensional problem.

The interactive and multivariable nature of human performance can often mask alarm problems. In some cases, increased staffing can be used to compensate for poor alarm system design. Extra operators work the board to handle the alarm floods. Unfortunately, a similar phenomenon is seen where alarms are used to compensate for poor training or operating practices. One plant had an operator lose level in a treated water tank for 8 hours, eventually resulting in the shutdown of the entire refinery. The solution for an operator not doing his job was to have yet another alarm (they had already been alerted to the problem once). The greatest irony in the alarm problem is the tendency to add alarms to compensate for an operator missing an alarm as a result of an alarm flood. Such was the case at a nitric acid plant, where after missing high, high-high, and hardwired high-level alarms in the flood, yet another hardwired alarm was added.

Alarms are one aspect of human–machine system performance. For optimal system performance, all the variables

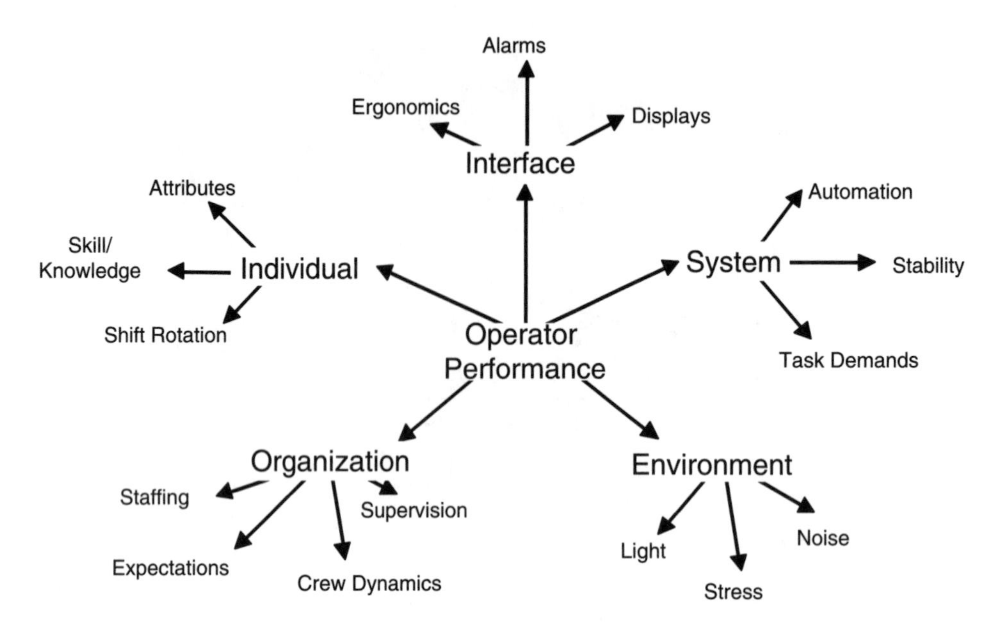

FIG. 2.10f
Human factors variables.

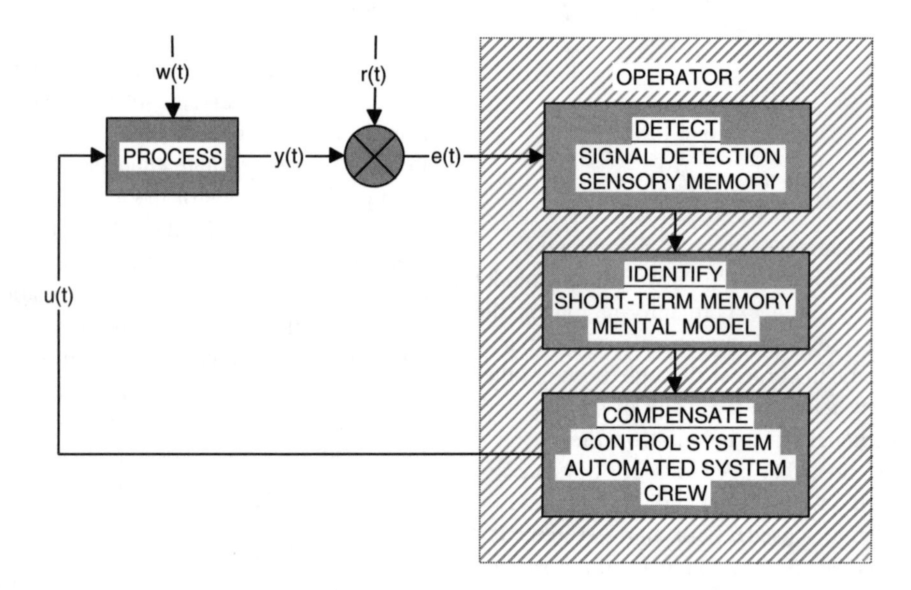

FIG. 2.10g
Model of operator–process performance.

need to be properly designed into the system. Alarms should not be substitutes for poor training or operating discipline. But a poor alarm system need not be tolerated and compensated for by staffing with extra personnel.

ALARMS AND HUMAN PERFORMANCE

Intelligent alarm management requires an understanding of the role of alarms in prompting operator behavior and the inherent characteristics and limitations of how humans process information. A variation on the classic feedback control model of human performance is seen in Figure 2.10g. In the figure, when a process undergoes a disturbance [w(t)], this causes the output of the process [y(t)] to change. If the output varies from where is should be [r(t)], then an error value is generated [e(t)]. A correction [u(t)] for the error occurs only after several things happen: the operator must (1) detect that a problem has occurred (error value large enough), (2) identify the nature of the problem, and (3) compensate for the problem.[6]

Detection

The operator first must be aware that something is wrong. Although it seems difficult to believe that this would be possible with "too many alarms," it is in fact not uncommon.

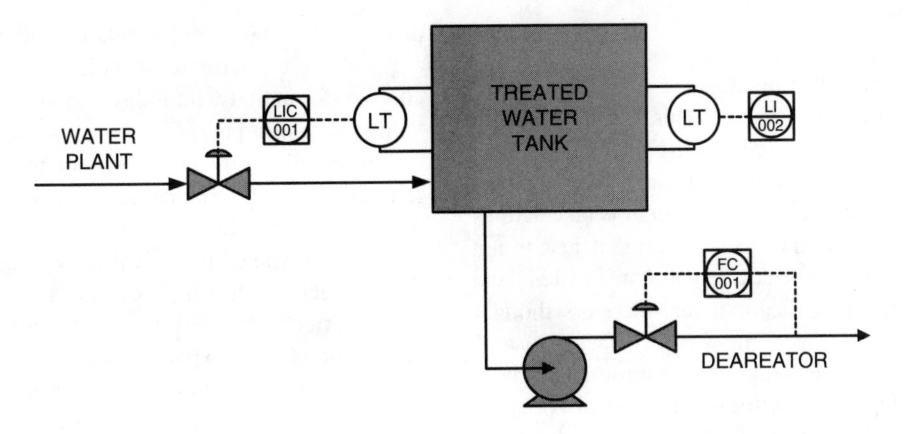

FIG. 2.10h
Case of "operator error."

Several of the earlier examples had operators who were oblivious to one of the problems they were facing. The boiler operator knew they had lost a boiler, they just failed to detect the loss of a BFW pump. An operator on a hydrotreating unit went 20 min with no flow through one pass of a fired heater. Unfortunately, upon detecting the problem, they opened the valve and promptly burned down the heater. How did they fail to detect no flow through one pass of a heater? Part of the problem was an alarm summary screen that was continually filled with alarms. One more alarm just faded into the background with all the other alarms.

Our ability to detect a signal is influenced by several factors: signal detection theory, expectancy, and our sensory memory system. Intelligent alarm management requires an understanding of each of these variables and their contribution to the operator being able to perform successfully the first step in responding to a process disturbance.

Signal detection theory holds that the ability to detect a signal is a function of the degree of difference from the background noise and the motivation of the individual.[7] The signal needs to be conspicuous, different in strength or characteristics from the background "noise," if it is to be detected. The system designer needs to ensure not only that the signal strength is appropriately strong (either bold, attention getting, larger, louder), but also that there be minimal background noise (no alarms on the screen for days, no other display elements colored red or yellow that are similar to the alarm). The tube rupture incident is in part due to a signal (the alarm) not sufficiently conspicuous from the background (one more alarm on a page of alarms). The motivation of the individual changes with the perceived risk of the situation, hence the value in conveying quickly the importance or priority of the alarm. However, if all alarms are "critical" or alarms frequently "cry wolf," then little change in motivation occurs with each alarm.

In the absence of a conspicuous signal, our expectations of what we think will or should occur influences our ability to detect the signal. Too often we fail to detect a problem because we have expectations for what we think is occurring. The basic design of the system associated with the treated water incident mentioned earlier is shown in Figure 2.10h. The operator had first received a high-level alarm from the frozen controller in the treated water tank, and raised the controller's set point. The frozen controller closed the valve that let treated water into the tank, resulting in a constant decrease in level. Shortly thereafter the operator received a low-level alarm on the redundant transmitter and again raised the set point. The operator said they did not remember receiving a low-level alarm. Given that they took the same action to both alarms, it is postulated that they thought the low-level alarm was actually a high-level alarm. Why? Because the redundant transmitters usually track and they were expecting the second alarm. The low-level alarm was not particularly conspicuous, differing from the high-level alarm only by showing a PVLO instead of a PVHI in a string of text on an alarm summary screen.

The conspicuousness of a signal is limited by how the human information processing system functions. The first stage in our information processing system is sensory memory, which occurs before conscious processing of information. Colors and patterns are partially processed prior to conscious processing. If the color or pattern has a unique meaning (e.g., red means bad, octagon means stop), then information can be transferred to the operator without conscious processing. If the color or pattern does not have a unique meaning, then additional conscious processing and associated resources will be needed.[8]

At the heart of one of the major deficiencies in alarm system design is the failure to utilize the power of sensory memory. This results when there is no unique color for alarm information. Red is variously used for warning, pumps stopped, valves closed, flames in heaters, high-pressure hydrogen or steam, etc. Detecting a red alarm in a graphic display filled with red symbols and red text is significantly harder than in one where only alarms are red. Alarms should have a unique color associated with them in the alarm and display system. This will likely require use of other coding techniques to encode information often associated with alarm colors. The natural association with red for stopped/closed will conflict with the need to keep red unique for indication of an alarm condition.

Identification

Once operators detect that a disturbance exists, they must identify the nature of the disturbance. This is likely the area of the greatest number of failures in the operator–process system. Many major process incidents are the result of improper identification of the problem; operators take actions appropriate for what they think is occurring; it just is not what is actually occurring. The accident at Three Mile Island is a case of improper identification; the operators thought they were in an overcooling event, when in fact a loss of coolant was occurring. Their actions were appropriate if overcooling was occurring, but disastrous for a loss of coolant.

At a large crude oil unit, the operators had a shutdown system with a separate hardwired alarm panel (so the alarms would not get lost in the DCS system). The alarm tiles were color-coded: white for problems with the system, red for pre-trip conditions, and green for tripped (unit safe). An upset ensued and the shutdown system tripped the heater. The operator observed the panel, focused on the red tiles (ignoring the green), and thought they should bypass the fuel gas chop valves to prevent the heater from tripping. The head operator and shift supervisor concurred with the assessment. The field operator, blown off the heater deck (otherwise unhurt) upon bypassing the valve and introducing fuel gas into a hot heater with no flames, was the first to realize they had misdiagnosed the problem. How can three senior operators have fuel gas put into a flameless, hot heater? They misidentified the problem: what they thought was going on, was not.

Two major features of how we, as humans, process information influence our ability to identify properly the source of the disturbance: short-term memory and our internal mental models of the system. Short-term memory is the site of our conscious processing. Mental models are our internal summations and predictors of how systems are to function. Our design of alarm systems needs to account for both.

The human short-term memory system is capacity limited. The limit is seven, plus or minus two, chunks of information.[9] Think of short-term memory as a set of receptacles or tanks; incoming information is stored for use in conscious processing. At about seven chunks, either new information must be excluded or old information discarded. This is part of the problem in presenting alarm information via a listing of alarms by time of actuation. At about seven alarms, the short-term memory system is going to become overloaded, and seven alarms is not very many with today's DCS.

The key to overcoming the short-term memory limitation is to utilize the fact that the limit is based upon "chunks," not bits, of data. The more data that can be processed as an entity, a "chunk," the more data that an individual can process. That was the advantage in old control rooms with banks of hardwired annunciator panels along the top. If each alarm panel corresponded to a system or subsystem, then every alarm in a major upset need not be processed. The operator could identify those panels that were lit up and to what degree, thereby inferring the location and severity of the problem. The hardwired panels facilitated chunking. Should we go back to hardwired panels? No, other means exist to chunk information with today's systems that are superior to hardwired alarm panels.

Information from our short-term memory system is transferred to our long-term memory system to diagnose the cause of the problem. High-level problem solving often involves the use of a type of analogical reasoning, called recognition-primed decision making.[10] Complex patterns are matched to our experiences to derive a course of action. If the alarm information forms no pattern (again, the presentation by time of actuation), then this type of advanced decision making cannot occur. Chunking and pattern formation is essential for proper identification of the cause of a process disturbance.

The final aspect of identification is for operators to access their mental model of the process. The highly interactive nature of process plants usually guarantees that a change in one part of the process will generate changes in another. Rarely is an alarm by itself sufficient to identify the nature of the process disturbance. Often the disturbance is detected downstream of the problem and the operator's knowledge of the process is needed to identify the exact problem. Therefore, the alarm system needs not only to alert the operator to the problem, but also to help the operator find the associated information needed to assess the alarm in the full context of system performance.

Compensation

The final, and critical, piece in the process is for the operator to compensate for the disturbance. This can be done through the control system, through some automatic system, or through field personnel. The ability for the operator to effect a change is, of course, the ultimate objective of the alarm function; therefore, the alarm system needs to aid the operator in locating the necessary controls to alter the process. A major flaw in many alarm management efforts is to overlook this aspect of the role of alarms in total system performance. Many plants view the function of alarms to alert or warn the operator, neglecting the need for them to effect some change on the system. The alarm system needs to complete the process, not only helping the operator detect and identify the problem, but taking the operator to a display that enables the operator to do something about it, to compensate for the disturbance. Alarms must be integrated with the display system for optimum human–machine system performance.

APPLIED ALARM MANAGEMENT

Creating an effective alarm system that aids the operator in the detection, identification, and compensation of process disturbances is a three-step process. First, the objectives and characteristics of the alarm system need to be defined. Many poor alarm systems are the result of assumptions: assumptions regarding what alarms are for, assumptions regarding what alarms are important, assumptions regarding how the DCS "must" present alarms. Second, the alarm system needs to be evaluated relative to the objectives. Unfortunately, this process is often undercut by an emotional attachment to a unit or part of a unit (big money-maker or high risk) that results in "exceptions" to the rules. Third, the resulting alarms need to be integrated into the display system. Rethinking how displays are structured and formatted is often required.

Defining System Requirements

Prior to selecting and prioritizing individual alarms, the role of the alarm system must be defined and articulated. Everyone "knows" why alarms exist in the system; unfortunately, frequently everyone has a different explanation for why this is so. Everyone "knows" the capabilities of the DCS for alarm processing; unfortunately, that usually means how it is being used now. It is surprising the number of control engineers who confuse the manner that has been chosen to present alarms from how the system is capable of presenting alarms. That is why it is important to ensure that everyone agrees and understands what the alarm system is for and how the capabilities of the DCS can be used prior to evaluating individual alarm systems. Although every attempt should be made to understand how the DCS is used, typically during the alarm selection process the opportunities of how to use those capabilities become fully realized, often necessitating a revisit to how the DCS will be used.

DCS Alarm Characteristics

Although there is a similarity in how most DCS systems handle alarms, they are not identical. It is important to understand the capabilities of the DCS systems for processing and presenting alarm information. In particular, the manner that is currently used or even used as an example in the product manuals probably is not the only way, or even the best way, to use the features of the system. Although this is not an in-depth analysis of each of the major DCS systems, there are some characteristics that should be understood and articulated by the system designer prior to attempting alarm system design or configuration. Several questions should be answered as a prelude to any alarm management effort.

What is an alarm?
What are the possible conditions for alarming and when will they be used?
How many priorities are available for alarms and do they have any automatic characteristics (e.g., color, sound, need to acknowledge, etc.)?
Are alarms grouped in any fashion?
How are alarms shown to the operators?
Does the system provide for creating logical combinations of alarms ("and"s/ "or"s)?

Although it may seem merely a matter of semantics, what is an "alarm" needs to be resolved immediately. The DCS likely counts any point that has exceeded an alarm limit to be an alarm. That is a hardware definition. Intelligent alarm management is user-focused, so an alarm must be audible and require acknowledgment. Alarms require the operator's attention, which may not occur if the "alarm" makes no sound and does not require acknowledgment by the operator.

Most systems have similar capabilities for alarming analog values: high, low, deviation from set point (if a controller), rate-of-change, output (if a controller). It is digital points that often create some confusion for alarming. Digital points, such as a level or pressure switch, are binary in nature. Alarming for them can typically be when they are in an abnormal state, change of state, or a mismatch between the desired and actual states. The latter requires some feedback on both the demand (open/closed, on/off) and the actual. Where possible, avoid the use of switches for alarming. Analog-based alarms are easier to test and detect failures than are discrete indications.

The priority levels that are to be used for the system need to be established. The Moore DCS has the capability for a thousand alarm priorities, probably far more than is needed. Compare this to the Honeywell DCS that has three alarm priorities that have an audible tone. Foxboro has five priorities, whereas Bailey now has 12 (up from 8 in a previous version). The criteria for assigning an alarm to each priority need to be clearly established, reflecting risk and time required to respond in a high alarm situation. More than three levels of alarms are both difficult to select and convey the difference to the operator. Because priorities are to convey the speed required in a high alarm situation, a three-level system works the best, essentially reflecting fast–faster–fastest response requirements.

Use of the priorities will interact with how to group or manage the alarms. For example, take a plant where a system may need to be turned off periodically while the system is down for regeneration, etc. A "high priority" alarm may be needed for one point in that system all the time, and another only when the system is in normal operation. The latter alarm may become a nuisance during a regeneration mode of operation. Two of the DCS priorities may be needed to cover this high priority in the Foxboro system, while the Honeywell system could use its contact cutout feature to address the point used for regeneration.

The Fisher DCS has a very flexible and complex alarm priority scheme. Other systems have an easy-to-grasp concept that an alarm condition is assigned a priority that has certain characteristics. The Fisher DCS has sets of alarm characteristics (color, flash, need to be acknowledged, position on screen) that can be assigned to a point for different operating modes. This "characteristic set" approximates alarm priorities, but is configurable for each section of the plant. The same point can have multiple ways in which the alarm will appear for a variety of operating modes. This enables complex alarm management strategies to be developed, but renders meaningless the question, "what priority is that alarm?" Unfortunately, most plant personnel have their concept of what an alarm priority means. For Fisher systems, it is imperative either to provide some translation tool or to ensure users understand the capabilities of the system prior to any alarm management effort.

Most systems have some means to group alarms or operating tags into different areas. Honeywell groups alarms into "units," which in early systems was an essential part of alarm management. Fisher systems have the points assigned to plant process areas (PPAs), which can become a powerful part of the alarm management strategy. Bailey has alarm groups, also more critical in earlier systems for alarm management than in their current system. The system designer should understand the impact, benefit, and effort required to change any grouping scheme.

The presentation of alarms also varies with the DCS systems, both audibly and visually. Audible characteristics need to be understood, such as options for alarm tones (pitch, continuous vs. momentary), and how those options are generated (are they inherently tied to priority, or is it user selectable?). Visual presentation is usually through three primary means: (1) an alarm summary screen showing a listing of points in alarm, (2) a keypad or alarm status boxes showing alarm status for groups of alarms, and (3) on-screen alarm status of individual alarms. The alarm list is useful for one or two alarms at a time, but is quickly rendered useless in a major upset. As such, the system should be designed not to rely on such a method of alarm presentation. The keypad/status works well, providing the selection of alarms (1) has eliminated nuisance alarms, (2) has removed alarms that stay "in" for long periods, and (3) shows a priority appropriate to the potential hazard. However, if the keypad has several LEDs (light-emitting diodes), how will they be mapped to priorities? For example, Honeywell uses three priorities, yet has two colors on the keypad (red and yellow). Which priority matches which keypad color, and is it consistent with color-coding on the graphics? What can be shown on a screen for an individual alarm (condition, priority, acknowledgment status) should be understood, both what is suggested by the manufacturer as well as other options. One plant chose to convey the alarm condition by color on the screen (red = PV HI, yellow = PV LO) rather than priority. Detection of the critical alarm on a screen of multiple alarms would not be as fast in that situation as one in which color had been used to convey priority.

Most systems today have some provision for creating logical combinations of alarms. This should be defined before the alarm management effort (the limitations on such logic). For example, the Foxboro DCS has pattern alarms that allow logical combinations of up to 19 tags. Honeywell allows each tag to have the alarming inhibited based upon the status of another tag. How to use these features will be discussed in more detail later.

Alarm Philosophy

Establishing an alarm philosophy or set of alarm guidelines is essential to successful alarm management. The creation of the guidelines is almost as important as the guidelines themselves, as the process forces plant personnel to think through the function and use of the plant alarm system. The philosophy needs to be adopted and enforced by senior management as the "ground rules" for alarm selection and presentation. A typical alarm philosophy should expound the basic objectives of the alarm system and their implications for the particular plant. Some of the objectives and implications are provided below.

Overall Alarm System Purpose and Function

Objective 1: Alarms are not a substitute for an operator's routine surveillance of unit operation.

Implication

- Process changes that should be caught by operators during their normal monitoring of the process, and pose no safety issues, shall not be alarmed.
- The alarm system should be an aid for the operator, not a replacement.
- Operators are expected to investigate alarms that occur, accessing the appropriate graphic.
- The normal and expected shall not be alarmed (e.g., time for samples to be taken).

Objective 2: The purpose of an alarm is to prompt a unique operator action.

Implication

- The action required in response to each alarm can be specified.
- There shall be no alarms for which there is no operator action.
- All alarms are important and should be acted upon as soon as possible, regardless of priority.
- There shall not be multiple alarms that prompt the same action. Redundant instrumentation due to shutdown systems will either (1) not be alarmed, (2) use logic to prevent multiple alarms, or (3) have alarms on deviation between the primary (alarmed) variable and other instruments. Common alarms shall be created for multiple alarms on different variables that require the same response (e.g., firebox temperatures, equipment vibration, shutdown first-out).
- Alarms shall not be used to convey status information (e.g., lab results in, steps in regen cycle, pump on/off, alarms returning to normal).
- Alarms shall not be used to convey information needed by individuals other than the operators (e.g., engineers, maintenance).
- There should be no alarms "in" for long periods during steady-state operation (the alarm screen should be dark).

Alarm Selection and Prioritization

Objective 3: Alarm set points will not be altered without proper management of change.

Implication

- Operators will not be able to change alarm set points without management of change.
- Inhibited and disabled alarms will be reviewed at the beginning of each shift.
- Alarm set points and conditions must be selected that are applicable for anticipated changes in rates and operating conditions (times of year).
- Alarms for which a fixed set point cannot be established will require preapproval of a range of changes.

Objective 4: Alarm priorities are to indicate the required speed of operator response in a *high* alarm situation.

Implication

- Priorities will reflect the *immediate and proximate* consequence of inaction as follows:
  1. **Level 1:** Immediate operator action (drop everything). Endangerment of personnel, catastrophic equipment failure/environmental impact, unit shutdown, or shutdown of other units imminent.
  2. **Level 2:** Rapid operator action required. Unit shutdown possible. Partial shutdown has occurred. Emergency priority alarm possible.
  3. **Level 3:** Prompt operator action required. High-priority alarm possible. Off-spec or production loss imminent.
  4. **Status:** No operator action required, but status or time stamp information required.

- Alarms related to (1) safety shutdown system activation and/or (2) violation of a process safety limit will not have an automatic (*de facto*) priority.
- The relationship of the number of alarms in each level is No. of Level 3 alarms > No. of Level 2 alarms > No. of Level 1 alarms.

Alarm Presentation

Objective 5: Alarms shall have an audible tone and require acknowledgment to facilitate detection by the operator.

Implication

- Alarms that do not make a sound are not considered alarms and are not covered by these guidelines.
- Alarm priority should be reflected in tonal qualities (volume, pitch).
- Alarms on adjacent consoles shall be easily distinguishable.

Objective 6: Alarms, their condition (high, low), state (unacknowledged or acknowledged), and priorities should stand out from points not in alarm on graphic displays.

Implication

- The alarm condition will be indicated on the graphic displays.
- A unique color will be used to indicate a point is in alarm and its priority.
- Unacknowledged alarms shall blink or stand out until acknowledged.

Objective 7: Alarms relating to personnel safety, environmental, and process problems will be easily distinguishable from each other.

Implication

- Each class of alarm will be visually segregated from the others.

Objective 8: Alarms will be presented to facilitate processing during high alarm events.

Implication

- Multiple alarms in an area will be "chunked" to enable higher-order processing by the operator.
- "Chunked" alarms will be tied to graphics from which control changes can be made.
- Alarms will be combined into single system alarms and presented on a graphic so that they can be processed as a system alarm.

The alarm philosophy is then applied to an individual unit in a two-step process. The first step is to determine what will be alarmed. The second step is to determine how the alarm information will be presented and integrated into the display system. If the first step is not done correctly, the second step will be impossible to do well.

Alarm Selection

Alarm selection is application of the alarm philosophy. The best approach to date is the use of a team of operators (at least two board-qualified), DCS specialists, a facilitator, and process engineers to review each point in the system for the need to alarm. Using the P&ID or material flow sheets is a good starting point to ensure that all points are reviewed. The effort needs to be well documented to ensure system integrity. For each point that is deemed necessary for alarming, the possible causes, operator actions, and consequences of inaction need to be documented. Once this has been determined, the priority can be set. A database preloaded with the DCS information has been found to be the best method to capture this information. If a point is determined not to need alarming,

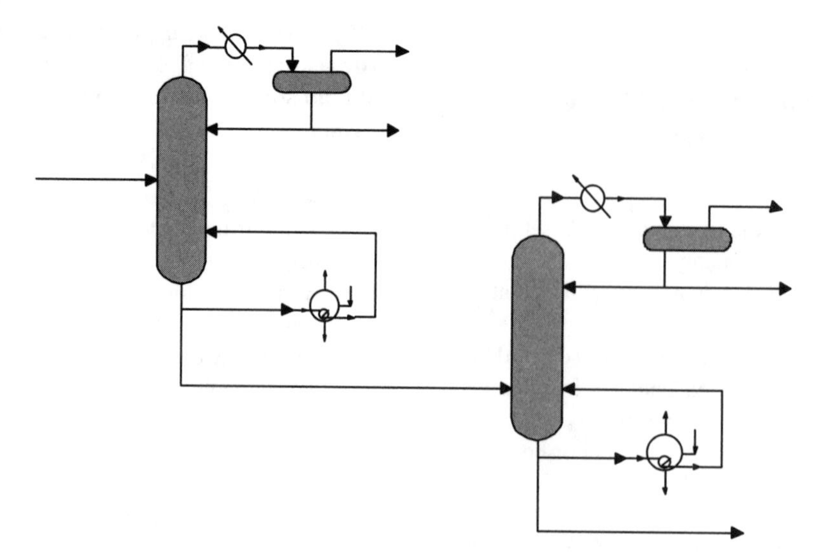

FIG. 2.10i
Alarm/system interactions.

then the basis for that should be specified, such as redundant with another alarm, no operator action, or no new action from another alarm. The process should take about 1 day for every 20 controllers and associated indicators in a unit. For example, a unit with 100 controllers should be able to be evaluated in about 1 week.

The highly coupled nature of most process plants needs to be understood when selecting alarms. Very few variables are totally independent of other variables in the process. As such, redundancy is built into the alarm system by alarms on other, interactive variables. For example, consider the process in Figure 2.10i. If the reflux flow controller were to fail open in the upstream tower, then the operator might have alarms on (1) the high flow, (2) low overhead temperature in the upstream tower, (2) low temperature in column bottoms, and (4) high pressure in the downstream overhead. Undue concern over the operator seeing the high reflux flow is not warranted, as there are other variables that would alert the operator to a problem. Similarly, a variation in one variable may not even need to be alarmed, because of the same interactive effect. If high reflux flow were not alarmed, would the operator fail to be alerted? Not if some of the other variables previously described had alarms. Alarms need to be evaluated in the context of the entire system, not on a point-by-point basis. Almost any alarm by itself can be justified; it is only when viewed as a system that the redundancy and over-alarming becomes apparent.

Specific and proximate causes for each alarm should be clearly defined and documented. For example, a low-flow alarm might be needed to alert the operator to (1) control loop malfunction, (2) pump malfunction, (3) loss of upstream level, or (4) downstream restriction. Although a power failure might also cause this alarm by shutting down the pump, the purpose of the alarm is not to alert the operator to a power failure, so it should not be listed as a cause. If the cause of the alarm is not having properly responded to a previous alarm (e.g., level in upstream vessel), then it should be identified as such.

The responses specified need to be unique and action verbs. If the response to a high alarm is the same as the response to a high-high alarm, then the action is not unique. Why are two alarms needed to tell the operator to do the same thing? If the response is, "because this is really important," then adjust the priority. Being "important" could apply to any alarm and justify the flood of alarms that occurs in an upset. Because the purpose of an alarm is to prompt action and action is observable, the specified response should be an action verb. "Watch the point more closely" is not an objective action. This could be used to justify seven alarms on the same point, as it was at one location. The responses should be "start/stop," "open/close," "isolate," "adjust," etc. Trying to convey knowledge, as in it would be "nice to know," is an endless process and results in over-alarming. If multiple alarms have the same action, such as temperatures on tube skins or in reactors, consider combining them into a single alarm. This is particularly true if the identification process will require the operator to assess the other indicators for confirmation and/or patterns, rendering any one alarm meaningless.

One of the key pieces of information in the data collected is the consequence of inaction. One of the errors often made, however, is to ascribe the final consequence of inaction, when it needs to be the immediate and proximate consequence of inaction. For example, one set of operators said that the consequence of missing a high-level alarm on an overhead accumulator was a trip of the downstream gas compressor. When asked if the compressor had a knockout drum with a high-level alarm, they said, "Of course." The real consequence of missing the accumulator high-level alarm would be a high-level alarm on the compressor knockout.

The consequence of inaction should reflect the priority level assigned. If the consequence is off-spec product, it should not be the highest priority alarm. Similarly, if the consequence is fire or explosion, then it should not be the lowest priority level. If the consequence is truly catastrophic, a question should be asked: Why is human intervention, with its inherent error rates, being used as the final level of safeguarding?

It is possible and quite common to fill out the alarm information and have the operators set the wrong alarm priority. Overprioritizing alarms is often the result of the need to overcome past alarm practices (e.g., "It needs to be a higher priority or we will lose it in the flood"), an emotional attachment to a piece of equipment (e.g., "All the compressor alarms need to be the highest priority because it's the heart of the unit"), or compensating for unusual DCS configuration (e.g., "The lowest priority levels don't have an audible tone"). Care must be taken to ensure that priorities are set based on the consequence of inaction and not a rank ordering of alarm importance within a particular system. An error frequently made is to assign a point in a low criticality system a Level 1 priority because it is the most serious event in that system, but poses far less risk (consequence of inaction) than any of the Level 2 alarms in a more critical system. The priority should not be system specific, but convey to the operator the required speed of response for his or her span of control.

Alarm management efforts have been hampered by other process safety management activities. At one location, any variable that had been identified as a process safety limit was automatically the highest alarm priority. Yet many of the limits pose little short-term risk. For example, a safety limit on a tube poses little danger if exceeded only slightly for a short time, compared with a slight exceedence of a pressure separating high-temperature hydrocarbon and oxygen. Process safety limits are not alarms and should not dictate alarm requirements. Arbitrarily assigning priority and alarms based upon process safety limits will make the plant less safe, not safer.

Several rules of thumb exist for determining if the alarm selection process was successful. The total number of alarms for a continuous process should be a 2.5:1 ratio of alarms to controllers. For example, a unit with 100 controllers should have about 250 alarms. These alarms should have a priority distribution of the following

- Level 1—10%
- Level 2—35%
- Level 3—55%

These rules of thumb are not theoretically based. They result from an analysis of numerous alarm management projects led by the author. They need to be understood as rules of thumb. They are simply averages from a variety of plants. Although deviation from these targets may be acceptable, such a deviation should "raise a flag" and cause the team to re-review the alarms that have been set. In addition, the averages were calculated based upon a high and low alarm on the same point counting as two alarms. Although individuals have argued that opposite alarms on the same point should count as one (since they are mutually exclusive), the ratio for the number of alarms was based on two alarms on the same point counting as two and this needs to be adhered to when using the ratio to determine the expected number of alarms.

ADVANCED ALARM FILTERING

Most state-of-the-art control systems provide some means to filter, suppress, or alter alarms based upon plant operating conditions or modes. Although this is seen by many as the "magic bullet" to slay alarm problems, it is rarely used effectively, if at all. Advanced alarm filtering has the potential to improve alarm management greatly, but it first requires an understanding of the baseline alarm condition. If the baseline alarms have not been properly selected, then its use of alarming filter logic in the DCS will be impossible or of little effect.

Once the alarms have been selected, that set of alarms becomes the candidate for applying some of the alarm filtering capability. Different modes of operation will have become apparent in the alarm selection process as will the potential for numerous alarms to have common causes. These modes and conditions should be combined with a list of general plant failure modes and the base set of alarms evaluated for each mode/condition. For each mode, is the alarm needed (alarms to maintain product specification are likely to be of low interest in a power failure), will it likely be a nuisance, is it redundant with another alarm? Opportunities to change alarm priority or to suppress alarms with changing plant modes can quickly be ascertained, forming a subset of the baseline unit alarms.

ALARM PRESENTATION

How do the alarms selected present themselves to the operator? The means of presentation needs to be consistent with the overall goal of conveying information that can be used to take an action. Both the auditory and visual characteristics of the system need to provide information, often before little is known by the operator about the individual alarm.

The auditory nature of the alarm needs to reflect the priority of the alarm. The tone for the alarm should convey the degree of urgency of the priority. A high pitch, fast warble alarm is going to convey more information than a low pitch, slow warble alarm. The tones should be unique for each console in a multiple console environment. The volume likewise should reflect the degree of severity. The greater the volume above ambient, the higher the alarm priority. It is suggested that 2, 4, and 6 dBA above ambient be used for the three alarm priorities.

The visual characteristics of the alarms are the first point in which the alarm system design begins to interact with the display system design. Alarm colors should be unique and also convey information. Our population stereotypes associate red with warning and yellow with caution. These colors should be used exclusively for alarms, which may force some changes in other aspects of the interface design.

The primary means for presenting alarms on many systems is through a listing of alarms that have actuated, often called an alarm summary screen. It is strongly recommended that the system not rely on this method of alarm presentation as the primary alarm interface for the operator. A sequential listing of alarms places high demands on the operator's information processing system, almost guaranteeing that alarms will be missed in a high-alarm situation. The sequential listing allows for little pattern recognition to occur and requires that the operators read the descriptor to understand the nature of the alarm. Use of other methods to provide alarm information to the operator should be the goal of the system designer, methods that aggregate the potentially large number of individual alarms into more meaningful, higher-order alarms.

Aggregation

Critical to intelligent alarm management is aggregation of the alarm information. Human short-term memory limitations demand that alarms be aggregated if the operator is ever to process them. An operator in one alarm management effort captured the need to aggregate alarm information. After the team removed 3000 alarms from the then current 4000, an operator on the team commented, "I understand why 1000 alarms is better than 4000 alarms, but am I really going to notice the difference in a flood of 1000 from a flood of 4000? Or is a flood a flood?" He was absolutely correct; reducing the number of alarms to a minimum is a necessary, but not sufficient condition, for creation of an effective alarm system.

Most DCSs have some means to aggregate alarms. Honeywell has the PRIMOD to group alarm information. Fisher allows alarming to be done by the PPA (plant process area). Bailey has event bars and ADPs (alarm display panels) to show the status of alarms in major areas of the plant. Foxboro has the "pattern alarm" feature in which the alarm status of 19 points can be aggregated. The goal is not just to use the feature, but to use it intelligently.

Alarm aggregation will ultimately create an alarm hierarchy. Operators' span of control needs to be delineated into the systems and subsystems that match their mental model, around which the alarms will be aggregated. An example is shown in Table 2.10j, in which a unit that has undergone an analysis is shown with the alarms per system. Each system is subsequently broken into subsystems, as in Table 2.10k. In this instance, the operator is controlling four units with similar operation, so the subsystems were common to the four plants. The systems and subsystems should reflect an increasing decomposition of the process. Aggregation of alarms, i.e., status by system and subsystem, allows the operator to quickly assess where alarms exist and their importance with minimum demands on the information processing system.

	Alarms			
System	Emergency	High	Low	Total
Treating	16	32	42	90
Reformer	21	73	149	243
Isomer 1	1	30	22	53
Isomer 2	2	31	25	58
Utilities	3	19	7	29
Total	43 (9%)	185 (39%)	245 (52%)	473
Expected	44 (10%)	154 (35%)	242 (55%)	440

Table 2.10j
Example of alarm distribution.

System	Subsystem			
Treating	Feed	Heaters	Reactor	Comp/Sep
Reformer	Feed	Heaters	Reactor	Comp/Sep
Isomer 1	Feed	Heaters	Reactor	Comp/Sep
Isomer 2	Feed	Heaters	Reactor	Comp/Sep
Utilities	Air	Water	Steam	DCS

Table 2.10k
Example of systems/subsystems.

Display Integration

Alarm aggregation by itself only aids through the identification step in the process; a fully functioning system requires that the alarms be integrated with the operating display. The systems and subsystems for the alarms need to be mapped to the display system structure. There should be system-level displays and subsystem-level displays. The alarms aggregated around the systems and subsystems should point operators to the displays from which they can effect a change on the process.

By using the previous example, the alarm structure defined can be mapped to displays and DCS options. Displays can be created for each system and subsystem. The systems and subsystems shown in Table 2.10k could become the rows/columns on a keypad graphic display. A flashing light on the appropriate area shows an operator the priority and location of the alarm. The aggregation enables a large volume of alarms to be processed at a higher level within the limits of short-term memory. Each area can become a target to the display from which an action can be effected. The only item missing is operational information to help identify the nature of the disturbance.

An example of an integrated alarm display is shown in Figure 2.10l for a hypothetical crude unit. A temperature profile is shown on the top portion of the display, supporting

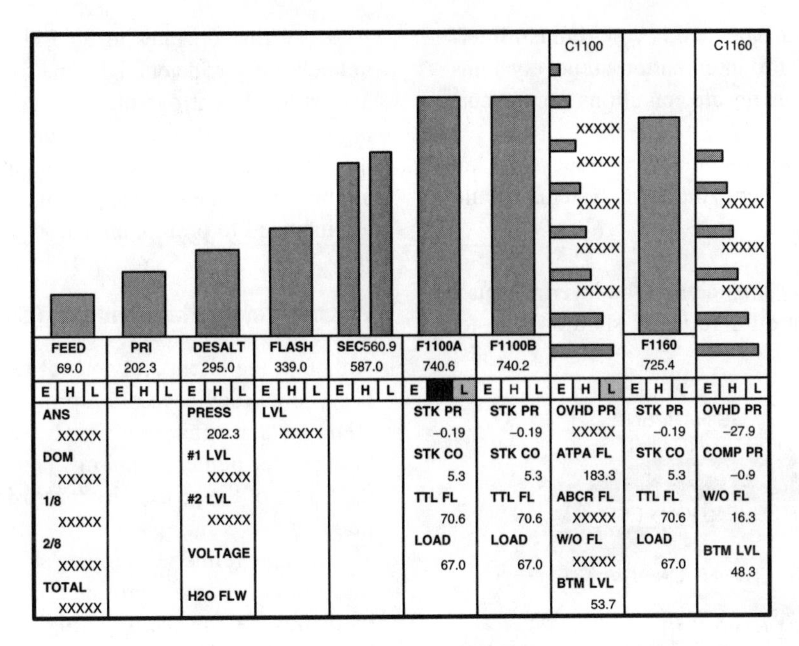

FIG. 2.10l
Example of integrated alarm/display overview.

pattern recognition. Key operating values are shown in the bottom portion of the display. Most important for alarm management are the alarm status boxes across the middle, showing the priority and number of alarms in each section of the operator's span of control. From this one display the operator knows where problems exist, even if "the flood of 1000" were to occur. Related data are present to help identify the problem. Targets to system-level displays enable compensation to be quickly accessed. Operators now have an alarm system that alerts them to a problem, conveys the urgency, helps them identify the disturbance, and directs them to a display from which a change can be effected.

Most systems have provisions to link either an alarm on the alarm summary screen to a particular display or have a "go to alarm" feature that calls up the display with the highest priority alarm. Although this method would seem to be superior to creating the information hierarchy and mapping alarms to displays, it is vastly inferior. First, it enables a poorly organized and structured display system to have the illusion that alarms and displays are integrated. Second, it presupposes (1) that operators can discern from the single alarm what display they would like to go to or (2) that operators will always want to go to the display in which the point in alarm resides. As we have discussed, the decision-making process, including what display will be of most use, is limited by and requires processing of large amounts of information as groups or chunks. If operators can discern the nature of the problem prior to accessing the graphic, they may decide that a display without the alarm point is the superior display to access. For example, problems downstream of a reactor may best be resolved by adjusting the reactor first, dealing with the downstream separation issues later. A "go to" feature would likely take the operator to the downstream display. Far superior to accessing a display from an alarm summary or with a "go to" button, creation of an integrated alarm and display hierarchy will enable operators to process the alarm information and decide what display they want to use.

ISSUES IN ALARM MANAGEMENT

Changes in the process industry over the past 10 years show the interaction between alarm management and other aspects of plant operation and safeguarding. In particular, many of the programs to enhance plant safety have degraded the effectiveness of the plant alarm systems. Some of the issues and their impact are discussed in the following sections.

Layered Protection

Too often alarm system designers fail to view alarms in the overall context of risk reduction. Alarms are one layer in preventing a plant from going into a region of unsafe operation. Given that alarms are to prompt human intervention, they should not be relied on exclusively for this purpose and their relation to the other layers needs to be understood. Figure 2.10m shows the systems that are to prevent a plant from going into an unsafe region.

1. Control System—This should compensate for minor disturbances, adjusting the process as needed.
2. Alarm System—This alerts the operator to problems beyond the capability of the control system to enable human intervention.

3. Relief Systems—If changes are beyond human intervention or too fast for intervention, relief systems divert energy to prevent unsafe conditions from occurring.
4. Shutdown Systems—If the plant continues into an unsafe region, independent shutdown systems should stop operation.

Alarms are a critical component in the layering, but they should not be a substitute for relief or shutdown systems.

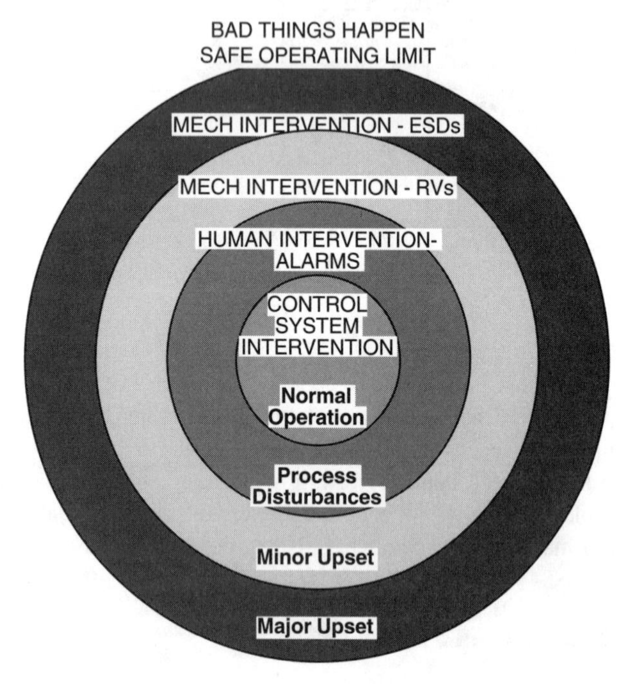

FIG. 2.10m
Role of alarms in layered protection.

A vessel that operates at 35 PSI and relieves at 38 PSI is unlikely to provide enough time for operator intervention via alarms to prevent overpressure. Alarming high pressure at a high priority level would only cause frustration and diminish the overall effectiveness of the alarm system. After-the-fact alarms violate the basic tenant of alarm management—alarms are intended to prompt an action.

Process Safety Management (PSM)

Although the intent and general results of PSM efforts have been good, they have often proven counterproductive in the arena of alarm management. Most process hazard analyses have created numerous alarms. The net effect may be to reduce the safety of the process by increasing the potential for alarm flooding.

In many refineries, process safety limits or safe operating limits have become sacrosanct. Discussing changing the limits or not making them the highest priority level is seen as heresy, no matter how inane or unachievable the limits are. Consider the system in Figure 2.10n. The exchanger had design limits on both tube and shell, so safe operating limits were set on the temperature indicators for product inlet and cooling water outlet. The result was an emergency priority alarm on cooling water return at 250°F. If this cooler actually had a return temperature of 250°F, it would (1) be an emergency, (2) no longer be a cooler but a boiler, and (3) be the result of major upsets elsewhere in the system (since the upstream tower operated at 180°F). There were no alarms on the temperature of the product out of the cooler, despite the material going to a storage tank. The alarming around this exchanger was absurd. Several meetings were required to reevaluate what was a safe operating limit and how it would be handled for exchangers. The result was to alarm the temperature of the product to storage, as the desire to prevent

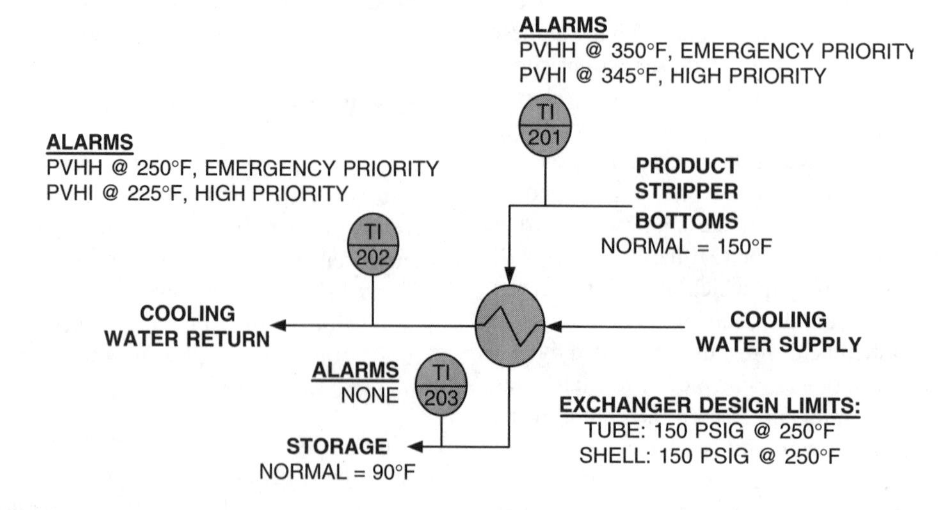

FIG. 2.10n
Example of "poor" alarming.

sending material over 210°F to storage (possible flashing of any water in the storage tank) was the most constraining limit.

SAFETY SHUTDOWN SYSTEMS

Like PSM, safety shutdown systems have in some cases increased rather than decreased risk. Triple redundant safety systems can generate triple the number of alarms. The multiple indicators can often be alarmed in the DCS and an alarm returned from the shutdown system. Proper alarming for a shutdown system requires an understanding of the information we want the operator to know about the safety system.

Prior to defining the alarming, it should be remembered that the goal of an independent shutdown system is to prevent the process from becoming unsafe, *independent* of operator action. If the shutdown system is functioning, then the safety of the plant should in no way be affected by any alarming surrounding the shutdown system. The shutdown system should prevent entering an unsafe mode of operation, regardless of any alarms given to the board operator. If no prealarms existed, then the operators might not like being surprised at a unit shutdown, but it should not cause the plant to be unsafe.

Only four general types of alarms should be associated with the shutdown system. First, there should be one alarm that alerts the operator that a particular process variable has moved/is moving in a direction that if uncorrected could lead to an automatic shutdown (the pre-trip alarm). This can be one variable tied into the shutdown system or a redundant indication of the same variable. Second, there should be one alarm that indicates that the shutdown system has activated and shut down the process or some part of the process. The operator's reference/goal state has changed and the operator now needs to initiate a new course of action. The individual causes of the shutdown are less important at this point than the shutdown itself and can be relegated to a "first-out" display. Third, there should be one alarm indicating any fault or potential for malfunction of the shutdown system. The usual response involves contacting the maintenance department, thereby allowing a common trouble for system protection. Fourth, there should be one alarm indicating the shutdown system has been bypassed or is inoperative. Again, the cause of the loss of shutdown system is now secondary to the loss of the safety protection it provides. All the other alarms from the shutdown system can become status alarms on special purpose pages, such as first-out, start-up, troubleshooting, testing, etc. But what if the point used for pre-trip fails and the operator is not warned of a trip? While annoying, it should not reduce the safety of the process if the shutdown system is properly designed.

TECHNOLOGICAL ADVANCES

New advances in technology should not alter alarm objectives, but facilitate them. New advances are being seen in equipment monitoring. How should these be alarmed? Although the advances are far more sophisticated than simple vibration monitoring, the objective is the same: alert the operator to potential equipment failure before it happens. However, in both the simple and complex case, the operator response will likely be the same: call maintenance/inspection and or take the piece of equipment off line. Focus on operator action, not the sophistication that generates the action.

References

1. Lanzetta, T., Denger, W., Cuarm, J., and Berch, D., "Effects of task type and stimulus heterogeneity on the event rate function in sustained attention," *Human Factors,* 29(6), 625–633.
2. Stanton, N., Ed., *Human Factors in Alarm Design,* London: Taylor & Francis, 1994, p. 35.
3. Meister, D., *The History of Human Factors and Ergonomics,* New York: Lawrence Erlbaum, 1999, pp. 29, 152.
4. Weiner, E. L. and Nagel, D. C., Eds., *Human Factors in Aviation,* San Diego: Academic Press, 1988, pp. 495–503.
5. Jones, R., Milton, J., and Fitis, P., Eye Fixations of Aircraft Pilots: Frequency, Duration, and Sequence of Fixations during Routine First Flight, USAF, AFTR 5975, 1949.
6. Sheridan, T. B., *Man-Machine Systems: Information, Control, and Decision Models of Human Performance,* Cambridge, MA: MIT Press, 1974, pp. 177–178.
7. Van Cott, H. P. and Kinkade, R. G., Eds., "Human Engineering Guide to Equipment Design," Washington, D.C., U.S. Government Printing Office, 1972, pp. 151–155.
8. Bourne, L. E., Jr., Dominowski, R. L., and Loftus, E. F., *Cognitive Processes,* Englewood Cliffs, NJ: Prentice-Hall, 1979, pp. 32–41.
9. Miller, G. A., "The Magical Number Seven, Plus or Minus Two: Some Limits on Our Capacity for Processing Information," *Psychological Review,* Volume 65, 1956, pp. 81–97.
10. Zsambok, C. and Klein, G., Eds., *Naturalistic Decision Making,* Mahwah, NJ: Lawrence Erlbaum, 1997, pp. 285–292.

Bibliography

Endsley, M. R. and Garland, D. J., *Situation Awareness Analysis and Measurement,* New York: Lawrence Erlbaum, 2000.

Goodstein, L. P., Andersen, H. B., and Olsen, S. E., *Tasks, Errors and Mental Models,* New York: Taylor & Francis, 1988.

Klein, G., *Naturalistic Decision Making: Implications for Design,* Dayton, OH: CSERIAC Program Office, Wright-Patterson AFB.

Salvendy, G., *Handbook of Human Factors and Ergonomics,* 2nd ed., New York: John Wiley & Sons, 1997.

Vicente, K. J., *Cognitive Work Analysis: Toward Safe, Productive, and Healthy Computer-Based Work,* New York: Lawrence Erlbaum, 1999.

Wickens, C. D., *Engineering Psychology and Human Performance,* 2nd ed., New York: HarperCollins, 1992, 266–286.

2.11 Safety Instrumentation and Justification of Its Cost

H. M. HASHEMIAN

Industrial processes such as power plants, chemical processes, and manufacturing plants depend on temperature, pressure, level, and other instrumentation not only to control and operate the process, but also, and more importantly, to ensure that the process is operating safely. Normally, for safety purposes, the output signal from a process instrumentation channel is compared with a set point that is established based on the process design limits. If the set point is exceeded, then alarms are initiated for operator action or the plant is automatically shut down to avoid an accident. In some cases, more than one instrument is used to measure the same process parameter to provide redundancy and added protection in case of failure of the primary instrument. In some processes such as nuclear power plants, up to four redundant instruments are used to measure the same process parameter. For example, the steam generator level is measured with four differential pressure sensors. Often, the same type of sensor and signal conditioning equipment are used for these redundant measurements. However, there are plants in which redundant instruments are selected from different designs and different manufacturers to provide diversity as well as redundancy and minimize the potential for a common mode failure.

When redundant instruments are used, then a logic circuit is often implemented in the process instrumentation system to help in determining the course of action that should be taken when a process limit is exceeded. For example, when there are three redundant instruments, the process is automatically shutdown if two of the three redundant instruments agree that the safe limit is exceeded.

Because industrial processes are inherently noisy due to vibration, process parameter fluctuations, interferences, and other causes, electronic or mechanical filters are sometimes used to dampen the noise and avoid false alarms and inadvertent plant trips. The selection of the filters must be done carefully to ensure that the filter does not attenuate any transient information that may indicate a sudden or significant change in a process parameter requiring action to protect the safety of the plant.

This section describes the characteristics and costs of typical instrumentation that is used in industrial processes to protect the plant and ensure the safety of both plant personnel and the general public. Improved safety instrumentation has received special attention in recent decades due to major industrial accidents that have caused loss of life and property. Although major industrial accidents have fortunately been few, a great deal of effort and money are spent to prevent these accidents. New plants are now equipped with digital instrumentation and advanced sensors that take advantage of 21st-century technology and maximize safety. Also, old plants are upgrading their safety instrumentation systems and retrofitting them with new equipment, including digital instrumentation to improve safety and overcome equipment degradation due to aging. Furthermore, new monitoring, testing, and surveillance activities have been implemented in some process industries to monitor the health of instrumentation and identify instruments that have anomalous performance. Some of these new monitoring, testing, and surveillance techniques are described in this section and additional information is given in Reference 1.

SAFETY INSTRUMENTATION CHANNEL

To protect the safety of a process, accurate and timely measurements of key process parameters are usually made on a continuous basis. The information is provided to a hardware logic module or software algorithm to determine if and when action must be taken to ensure safety.

To measure a process parameter, an array of instruments is typically employed. This array is often referred to as an "instrument channel." Figure 2.11a shows typical components of an instrument channel in a typical industrial process. The channel consists of a sensor such as a resistance temperature detector (RTD) or a thermocouple to measure temperature, or a pressure sensor to measure the absolute or differential pressure for indication of pressure, level, or flow. The connection between the sensor and the process is referred to as the "process-to-sensor interface"; an example is a thermowell in the case of temperature sensors, and sensing lines or impulse lines in the case of pressure sensors. Figures 2.11b and 2.11c show examples of how a thermowell or a sensing line is typically used in an industrial installation.

As will be seen later, it is important for process safety to verify that temperature sensors are properly installed in their thermowell and that sensing lines, which bring the pressure information from the process to the sensor, are not obstructed. Any appreciable air gap in the interface between

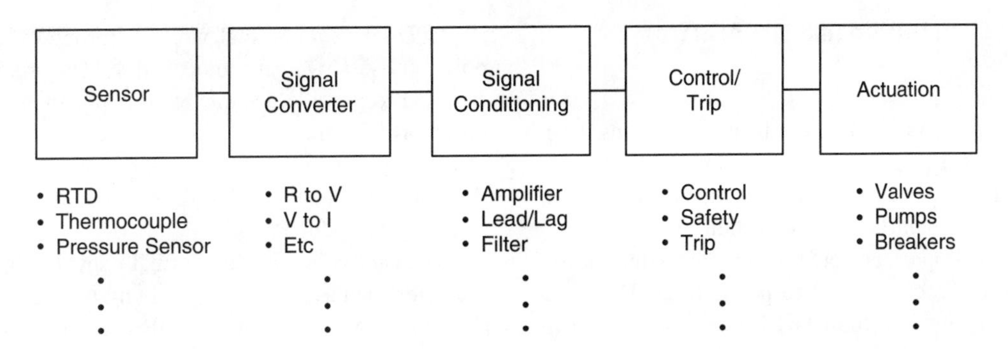

FIG. 2.11a
Block diagram of a typical instrument channel.

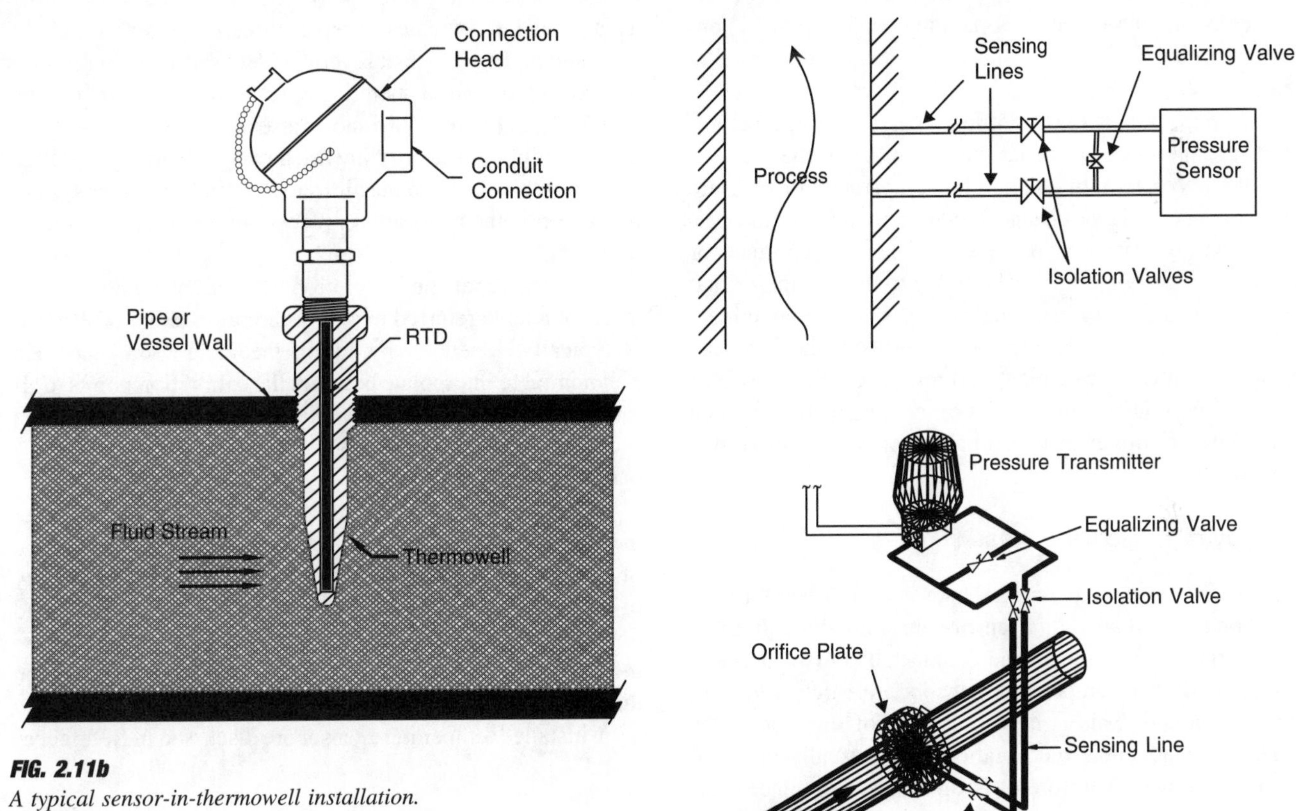

FIG. 2.11b
A typical sensor-in-thermowell installation.

FIG. 2.11c
Two views of typical pressure sensing lines.

the temperature sensor and its thermowell or clogging or obstruction of pressure sensing lines can cause measurement delays with a potential to affect the safety of the process.

As shown in Figure 2.11a, the output of the sensor is typically connected to a signal conversion or a signal conditioning unit such as a Wheatstone bridge, which converts the resistance of an RTD or a strain gauge to a measurable voltage proportional to temperature or pressure, or a current-to-voltage converter, which may be used in a pressure instrumentation channel. There are sometimes amplifiers in the instrument channel to increase the signal level and improve the signal-to-noise ratio, and electronic filters to dampen any extraneous noise. The next stage in a typical safety instrumentation channel is a logic circuit to compare the process parameter against a safety set point and ensure that the process parameter is within a safe limit. If the process parameter is found to exceed the safety set point, then actuation systems, such as relays, are used to initiate an alarm, trip the channel, or trip the plant. Normally, plant trips are not initiated unless redundant instrument channels indicate that a safe limit is exceeded.

PERFORMANCE CHARACTERISTICS OF SAFETY INSTRUMENTATION

The safety of a process is monitored on an ongoing basis during operation using instrumentation that measures the key process parameters. As such, the performance of the sensors or instrumentation system is very important to safety.

The two most important performance characteristics of instrumentation are accuracy and response time. The accuracy of an instrument is established by calibration and its response time is established by an appropriate response time testing technique. Normally, instruments are calibrated before they are installed in a process and then periodically calibrated after installation in the process. Recently, methods have been developed to verify the calibration of instruments remotely while the process is operating. Referred to as "on-line calibration verification," these methods are discussed in Reference 2.

Response time is important in processes where a parameter can experience a sudden and significant change. For example, in a reactor or a boiler, the temperature or pressure can experience a large change suddenly under certain operating conditions. In these processes, it may be important for the process sensors to respond quickly to the change so that the plant operators or the plant safety system can initiate timely action to mitigate any consequences of the anomaly. Because accuracy (calibration) and response time are two of the most important characteristics of instrumentation, a more detailed description is provided below for each of these two characteristics.

Accuracy of Temperature Sensors

The accuracy of a temperature sensor is established through its calibration. Temperature sensors are normally calibrated in a laboratory environment and then installed in the process. After installation, temperature sensors are rarely removed and recalibrated. This is especially true of thermocouples because thermocouple recalibration is not usually practical or cost-effective. Therefore, it is often best to replace thermocouples if and when they lose their calibration. More information on thermocouple calibration is provided in Reference 3.

RTDs are different from thermocouples in that RTDs can easily be recalibrated. In fact, it is often better to recalibrate an RTD than to replace it. This is because RTDs usually undergo a type of curing while in the process. They mature beyond the infant mortality stage and reach a stable stage in which they normally perform well for a long period of time (e.g., 10 to 20 years).

Industrial temperature sensors are often calibrated in an ice bath, water bath, oil bath, sand bath, or a furnace. For calibrations up to about 400°C, the ice bath, water bath, and oil bath are all that is needed. In these media, temperature sensors, especially RTDs, can be calibrated to very high accuracies (e.g., 0.1°C). Better accuracies can be achieved for RTDs with additional work or through the use of fixed point cells. However, industrial RTDs are not normally required to provide accuracies of better than 0.1°C, nor can they normally maintain a better accuracy for a long period of time. As for thermocouples, the best calibration accuracy that can reasonably be achieved and maintained is about 0.5°C up to about 400°C.

The calibration process normally involves a standard thermometer such as a standard platinum resistance thermometer (SPRT), or a standard Type S thermocouple to measure the temperature of the calibration bath. Also, it is through the standard thermometer that calibration traceability to a national standard is typically established for a temperature sensor.

The bath temperature is measured with the standard thermometer as well as the temperature sensor under calibration. This procedure is typically repeated in an ice bath (at about 0°C) and in the oil bath at several widely spaced temperatures that cover the desired operating range of the sensor (e.g., 100, 200, 300°C). Figure 2.11d shows an example of a temperature sensor calibration facility in which two oil baths and one ice bath are shown with connections to a central computer system that controls the calibration process and records the calibration data.

The results of the calibration are normally presented in terms of a table referred to as a "calibration table." This table is typically generated by fitting the calibration data to a polynomial to interpolate between the calibration points. Calibration tables can be extrapolated beyond the calibration points to a limited degree. A reasonable limit is about 20% based on research results.[4]

In addition to calibration of sensors, the accuracy of industrial temperature measurements depends on installation of the sensors and other factors. As such, the overall accuracy of a temperature instrumentation channel is evaluated based not only on the accuracy of its initial calibration, but also on the effect of installation and process operating conditions on the accuracy. These effects and how to verify the accuracy of an installed temperature sensor are discussed in References 1 and 4.

Accuracy of Pressure Sensors

Like temperature sensors, the accuracy of pressure sensors depends on how well the sensor is calibrated. The procedure for calibration of pressure sensors is more straightforward than temperature sensors. Pressure sensor calibration involves actual adjustments to sensor zero and span potentiometers as opposed to the data recording and data fitting that is required for temperature sensors.

Typically, a constant-pressure source and a pressure standard are used to calibrate pressure sensors. Depending on the sensor, its application, and its range, a number of precise pressures are applied to the sensor while its output is adjusted to indicate the correct reading. Typically, pressure sensor calibration is performed at five points covering 0, 25, 50, 75, and 100% of the span. Sometimes the calibration data are

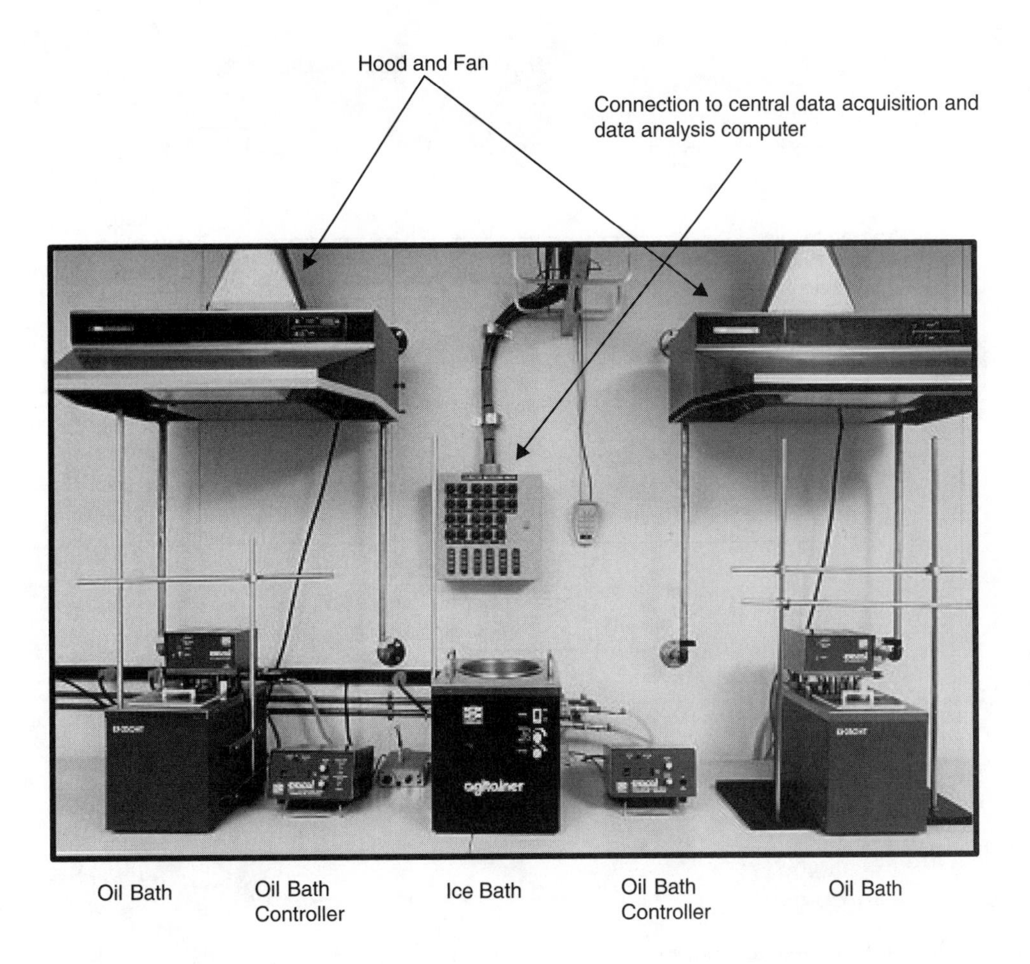

FIG. 2.11d
Photograph of a precision temperature sensor calibration facility.

recorded with both increasing and decreasing pressure inputs, and the resulting outputs are averaged to account for the hysterisis effect.

Conventionally, deadweight testers have been used in most industrial processes for the calibration of pressure sensors. Although deadweight testers are still in common use, recent calibration equipment has become available to streamline the calibration process. Referred to as "automated calibration systems," the new equipment enables the user to calibrate a pressure sensor in a fraction of the time that is normally required with conventional equipment. The new equipment incorporates computer-aided data acquisition, data analysis, automated report preparation, trending, and control of the test equipment. Figure 2.11e shows a simplified block diagram of an automated pressure sensor calibration system.

A problem with calibration of some pressure sensors is that sensors are sometimes calibrated when the plant or process is shut down or at less than normal operating conditions. That is, the effect of temperature on the calibration is not often accounted for. Furthermore, the effect of static pressure is not accounted for in calibrating differential pressure sensors. These effects may be important and can cause the calibration of a pressure sensor to be different during shutdown as compared to normal operating conditions.

RESPONSE TIME TESTING

The response time of process instrumentation channels can be very important to safety. In transient conditions, such as sudden changes in temperature or pressure, process instrumentation must respond quickly to notify the operators or initiate automatic safety actions. In some processes, instruments must respond within a fraction of a second to prevent the process from falling into an unsafe condition.

Typically, the response time of the sensor in an instrument channel is the most important component of the channel response time. This is because the electronics of an instrument channel are usually fast and are often located in mild environments of the plant. As such, the electronics typically suffer less calibration drift and less response time degradation than sensors that are normally located in or near the process in a harsh environment. In the past, the common wisdom was that the response time of sensors is a relatively constant variable, which does not change with the age of the sensor. This point has now been rejected as research results and test data gathered over the last two decades have shown that sensor response time can degrade significantly. In the case of temperature sensors, response time degradation occurs not only as a result of the sensor itself, but also from changes in

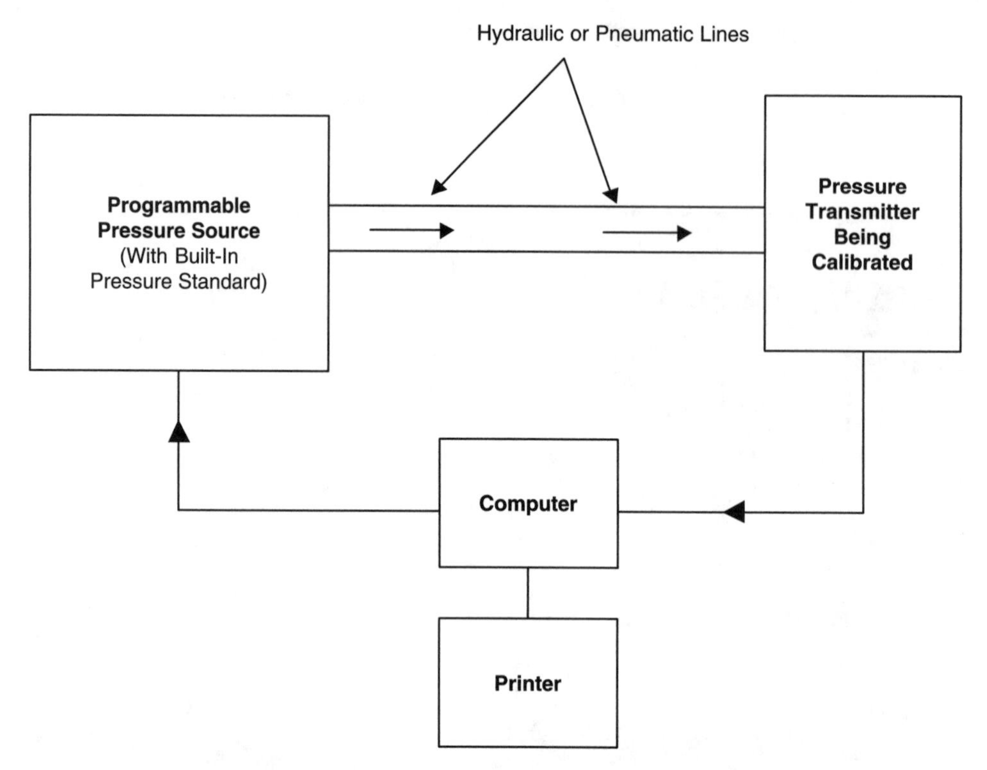

FIG. 2.11e
Block diagram of an automated pressure sensor calibration system.

sensor–thermowell interface. As for pressure sensors, response time degradations occur mostly because of blockages and voids in the sensing lines although pressure sensors themselves can also suffer response time degradation.

Because sensor response time degradation can occur, it is important to make measurements from time to time to verify that the response time of a sensor has not increased to an unacceptable level. This may be accomplished in a number of ways such as the two described below.

Response Time Testing of Temperature Sensors

In a laboratory environment, temperature sensor response time testing is performed by imposing the sensor to a step change in input. Typically, a method called the "plunge test" is used. The plunge test involves a sudden immersion of the sensor into an environment with a different temperature than the sensor. For a typical temperature sensor, the "plunge test" is performed in room-temperature water flowing at 1 m/s. This procedure is prescribed by the American Society for Testing and Material (ASTM) in a standard known as ASTM Standard E 644.[5] Figure 2.11f illustrates the test setup for plunge test of a temperature sensor.

The plunge test is useful for measurement of response time of temperature sensors in a laboratory environment. After the sensor is installed in a process, its response time can only be measured by *in situ* testing at process operating conditions. If the sensor is removed from its thermowell in the process and plunge-tested in a laboratory, the response time results could be much different from the actual in-service response time of the sensor when it is installed in the process. This is due to the effect of process conditions such as flow and temperature on response time. More importantly, the response time of a temperature sensor depends on its installation in its thermowell if a thermowell is used.

The above arguments stimulated the development of a new technique for *in situ* response time testing of temperature sensors. The method is referred to as the loop current step response (LCSR) technique. It is based on sending an electric current through the sensor leads to cause Joule heating in the sensing element of the sensor. Based on an analysis of the sensor output during the Joule heating or after the heating is stopped, the response time of the sensor is identified. This method is well established and approved by the U.S. Nuclear Regulatory Commission for the measurement of response time of temperature sensors in the safety systems of nuclear power plants. The details are presented in Reference 6.

Response Time Testing of Pressure Sensors

For pressure sensors, the response time is typically measured using a ramp pressure input. The ramp test signal is applied to the sensor under test and simultaneously to a fast-response reference sensor as shown in Figure 2.11g. The asymptotic

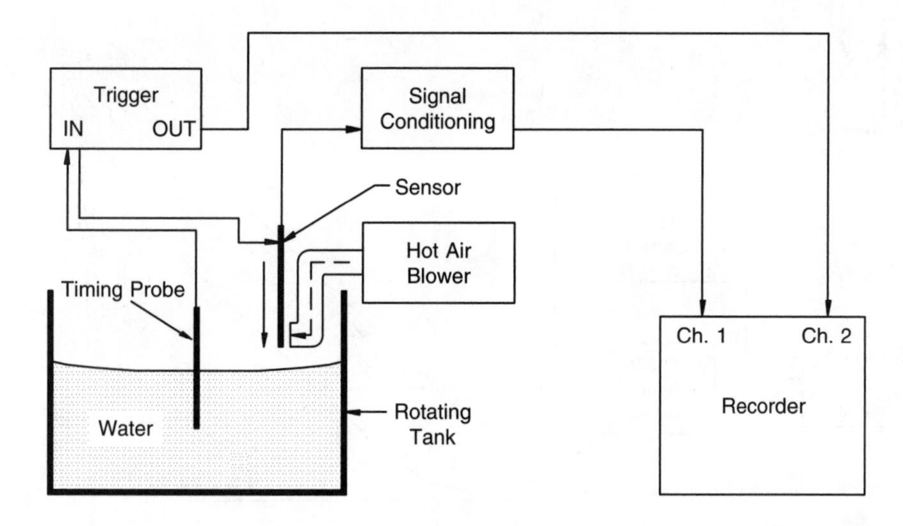

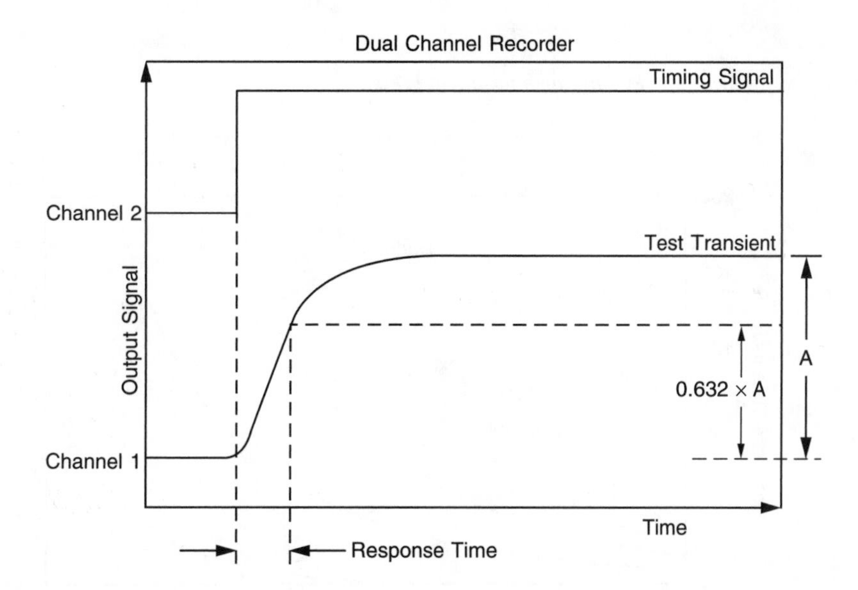

FIG. 2.11f
Illustration of equipment setup for laboratory measurement of response time of temperature sensors.

delay between the two sensors is then measured as the response time of the pressure sensor as shown in Figure 2.11g. The details are presented in Reference 7.

Unlike temperature sensors, the response time of pressure sensors does not depend on the process operating conditions. Nevertheless, a new method was developed to measure the response time of pressure sensors as installed in an operating process. Referred to as "noise analysis," the method is based on analyzing the process fluctuations that often exist at the output of process sensors while the plant is operating. The details are presented in Reference 8. The advantage of the noise analysis technique is that its results provide the response time of not only the pressure sensor, but also the sensing lines that bring the pressure information from the process to the sensor. This is important because any significant clogging or voids in the pressure sensing line can cause a major delay in measurement of transient pressure signals with a potential effect on the safety of the process. Figure 2.11h shows how the response time of representative pressure sensors may increase with sensing line blockages as they reduce the sensing line diameter.

DESIGN, DEVELOPMENT, AND TESTING REQUIREMENTS

Table 2.11i provides examples of some of the key requirements for safety instrumentation. The list is divided into initial requirements and recurring requirements as described below. Initial requirements are those to be met in producing the instruments and recurring requirements are those to be met in maintaining the instruments.

Initial Requirements

Safety-grade instrumentation shall be designed by qualified and experienced electrical and mechanical engineers or equipment designers and constructed by trained personnel

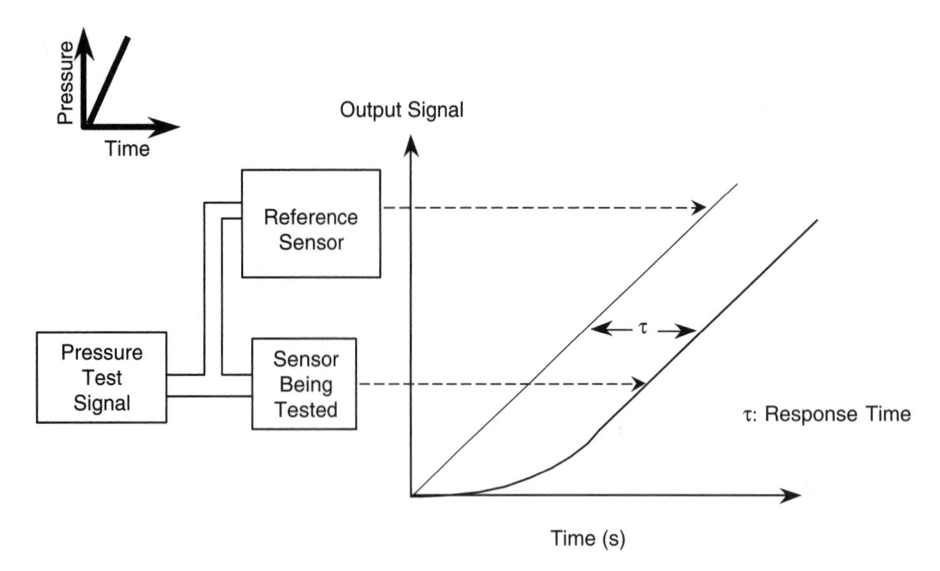

FIG. 2.11g
Illustration of equipment arrangement for response time testing of pressure sensors.

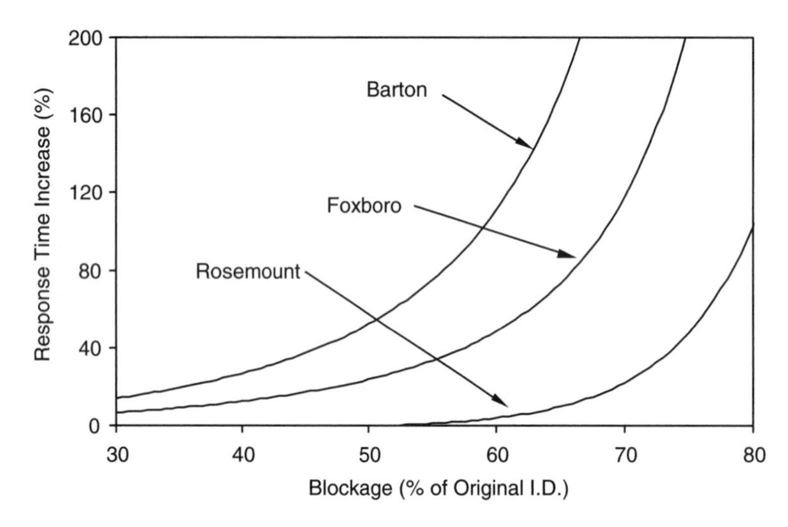

FIG. 2.11h
Effect of blockages on dynamic response time of representative pressure transmitters.

using documented procedures. High-quality material shall be used in construction of the instruments to help provide the desired performance, reliability, and a long life. The first prototype should be aged by such methods as accelerated aging techniques and subsequently tested as necessary to establish the qualified life of the instrument and verify that the instrument is sufficiently reliable for the intended service.

Upon construction, each instrument shall be tested individually under an approved quality assurance (QA) program and the test results should be documented. These tests should include calibration, response time measurements, and functionality tests as applicable. The tests must be performed using appropriate standard equipment with valid calibration traceable to a recognized national standard.

In addition to qualification tests and factory acceptance measures mentioned above, safety-grade instrumentation shall be tested after installation in the process to ensure that no damage or degradation has occurred during the installation. Examples of measurements that may be made include loop resistance measurements, insulation resistance measurements, continuity tests, capacitance measurements, cable testing, response time measurements, calibration checks, etc.

Recurring Requirements

While installed in the process, safety-grade instrumentation shall be tested periodically to verify continued reliability and compliance with the expected performance requirements.

TABLE 2.11i
Example of General Requirements for Safety-Grade Instrumentation

Initial Requirements
Good design and construction
High-quality construction material
Good calibration (accuracy): achieved by (1) precision laboratory calibration of each instrument prior to installation into the process and (2) appropriate steps taken during installation and use to minimize errors
High stability in calibration so that the sensor maintains its accuracy for a long period of time
Fast dynamic response time: (1) achieved by designing the system for fast response and (2) verified by laboratory tests of the prototype sensor
Reliability and longevity
Good insulation resistance, robust cabling, and high-quality connectors
Recurring Requirements
Periodic calibration on the bench or calibration checks and calibration verifications *in situ* while the instrument remains installed in the process operating conditions or at shutdown conditions as appropriate
Periodic response time measurements using *in situ* response time testing techniques as much as possible
Verify the condition of the process-to-sensor interface (thermowell, sensing line, etc.)
Testing of cables and connectors using *in situ* cable testing technology
Insulation resistance and capacitance measurements
Other electrical measurements (e.g., TDR)

A number of online monitoring techniques have been developed over the last decade or so for *in situ* testing of performance of process instruments while they remain installed in a process. A number of these techniques have been mentioned in this section and are found in the references that are provided at the end of this section.

Recently, in addition to instrument performance verification, tests are performed to verify the condition of cables and cable insulation materials. For example, the time domain reflectometry (TDR) technique is now in common use for testing cables in a variety of applications. This test is based on sending a signal through the cable and measuring the signal reflection. Based on the characteristics of the reflected signal, one can identify changes in the cable characteristics and verify the condition of the conductor and insulator of the cable. By using computer-aided testing, these types of measurements can be performed precisely, accurately, and effectively. Figure 2.11j shows typical results of a TDR testing of cables leading from instrument cabinets in the control room of a plant to an RTD in the field.

The frequency of testing of safety instrumentation depends on the instrument, its performance characteristics, and its importance to process safety. Some instruments may require testing on a monthly or quarterly basis and others would only need testing once every operating cycle or longer. Performance trending is an effective means for determining objective testing periods. Recently, reliability engineering techniques have been used in developing risk-based maintenance programs. These programs are used to determine the importance of instruments to plant safety and to arrive at objective maintenance schedules.

SAFETY INSTRUMENTATION COSTS

The cost of a safety instrumentation channel depends on the cost of the sensors, the cost for the instrument channel, and maintenance costs. These cost components, their typical prices, and the justification of their costs are discussed below.

Cost of Sensors

Because of high accuracy, fast dynamic response, and high reliability requirements, the instrumentation that is used to ensure the safety of a process could be much more expensive than conventional instrumentation. For example, a general-purpose RTD for a typical industrial temperature measurement application would normally cost no more than about $300.00 for a good-quality sensor. (All prices in this chapter

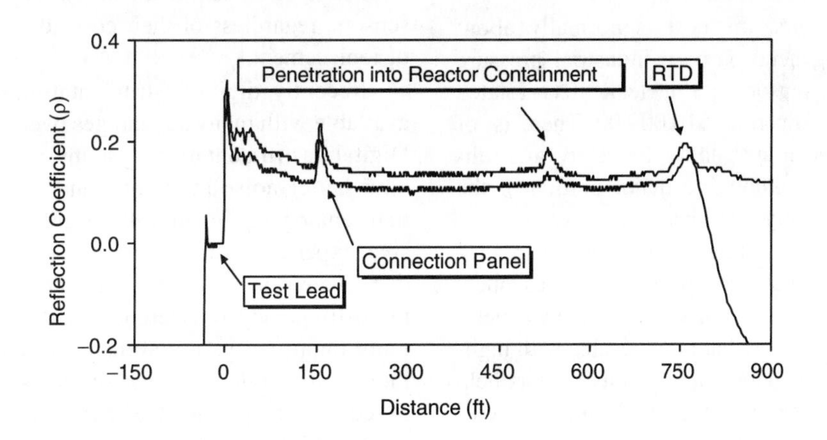

FIG. 2.11j
Example of a TDR trace for an RTD circuit.

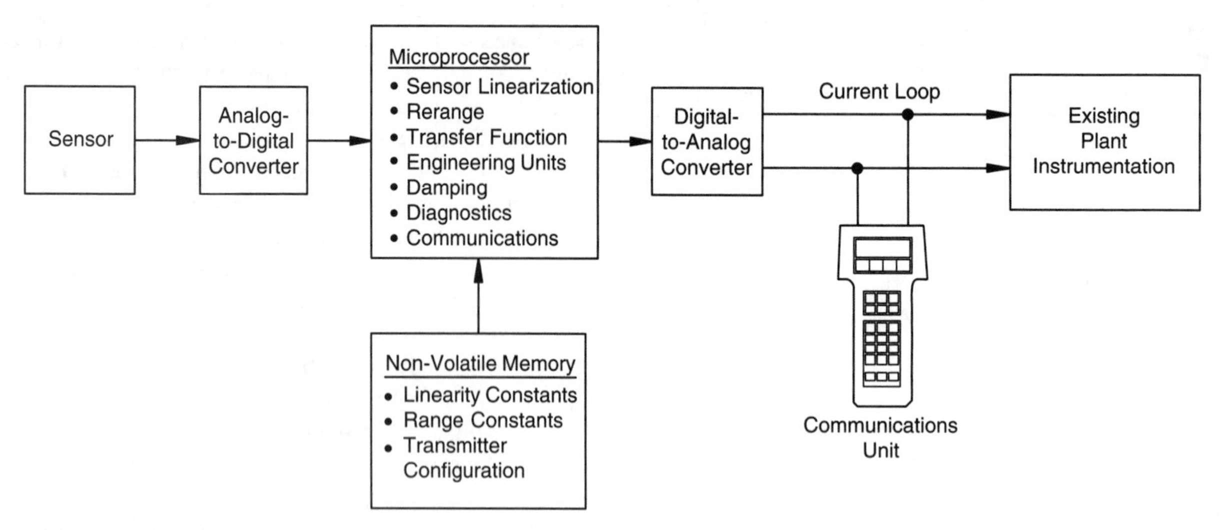

FIG. 2.11k
Typical components of a smart sensor.

are in U.S. dollars in the year 2002.) This compares with about $3000.00 for a safety-grade temperature sensor with its own calibration table, response time test results, etc. In nuclear power plants where safety is vital, the cost of plant protection instrumentation is typically very high. For example, a set of 40 RTDs and their matching thermowells for a typical safety-related installation can cost as much as $500,000.00. This cost usually includes a specific calibration table for each RTD, response time test certificates, QA documentation, nuclear qualification test results, sensor installation services, post installation testing, etc. A major portion of the high cost is for nuclear safety–related qualification, but the quality of the sensors is also superior. Reference 4 describes a research project in which commercial-grade temperature sensors were compared with nuclear-grade temperature sensors. The sensors in the research project were aged under typical process operating conditions and then tested to determine their performance after aging. The results illustrated that the nuclear grade sensors were at least twice as good as the commercial-grade sensors in terms of performance.

As for pressure sensors, the cost is normally about $1000.00 or less for a general-purpose industrial pressure sensor. The same type of sensor for a nuclear safety–related application could cost as much as $10,000.00. There is, of course, a difference between the quality, longevity, and reliability of the two sensors, although a major portion of the extra cost is for nuclear safety qualification.

Pressure sensors for safety-related applications in processes other than nuclear power plants would not cost as much, although a higher cost would be justified for the safety benefit, longevity, and good performance that is achievable with high-grade sensors. In addition to safety, high-quality sensors help with plant efficiency and economy. For example, the lower cost of maintenance of high-quality sensors over the life of the plant can easily compensate for the extra cost of the sensor.

Recently, smart pressure sensors have become available that provide superior performance at very reasonable costs. These sensors are becoming very popular in many industries, including the nuclear power industry where smart sensors are being qualified by some utilities for nuclear safety–related applications. Figure 2.11k shows typical components of a smart sensor.

Cost of Instrument Channel

The overall cost of a safety grade instrument channel could range up to about $200,000.00, including not only the sensor, but also the cables, electronics, design, and the labor to install and test the channel. As such, the cost of the sensor is typically a small portion of the cost of the total instrument channel.

Because the sensor is the most important component of an instrument channel and is usually the one that resides in the harsh environment in the field, its high cost is readily justified. Because replacing a sensor is typically very expensive, it is often important to use high-quality and reliable sensors, regardless of their cost, to minimize the frequency of replacement.

Recently, digital instrumentation systems have become available with many advantages over the analog equipment. Digital instrumentation systems are relatively drift free, have better noise immunity, and are much more versatile than analog equipment. They are, however, more expensive, experience subtle failures, and are susceptible to error due to environmental effects such as electromagnetic and radio-frequency interference (EMI/RFI). In addition, and more importantly, digital instrumentation is prone to common mode failures because of the use of the same software in redundant equipment. The fear of common mode failures is the chief reason digital instrumentation is subject to stringent licensing requirements for use in process safety systems. Furthermore, obsolescence is more of a concern with digital equipment than analog equipment. That is, rapid advances in digital technologies could make it difficult for users to purchase spare parts and obtain support for digital equipment as they aged in a process.

Maintenance Costs

Although the cost of components in an instrument channel can add to a large sum, it is really the maintenance costs that can overshadow the cost of the components. For example, a safety system instrument channel must be calibrated periodically, response time tested, maintained, and sometimes upgraded or replaced during the life of the plant. The cost of these activities over the life of the plant can exceed the original cost of the equipment itself. Therefore, any extra investment in the purchase of high-quality instruments is often justified by the savings that would result in the cost of maintenance, performance verification, or replacement.

In the past, instrument maintenance and performance verification were typically performed manually, one channel at a time. Today, automated equipment and techniques have been developed for predictive maintenance of instruments and for instrument performance verification that can test several channels at one time. Some of the techniques, such as online calibration verification methods and *in situ* response time testing techniques, were mentioned earlier. The cost of test equipment and implementation of these techniques have a range of $100,000.00 to $500,000.00, depending on the scope of the implementation. The cost is justified considering the time that it would take to perform the tests manually on one instrument at a time. Because this new test equipment often provides for testing of multiple instruments simultaneously and reduces the test time, it can help reduce the plant maintenance costs and outage time.

References

1. Hashemian, H. M. et al., "Advanced Instrumentation and Maintenance Technologies for Nuclear Power Plants," U.S. Nuclear Regulatory Commission, NUREG/CR-5501, August 1998.
2. Hashemian, H. M., "On-Line Testing of Calibration of Process Instrumentation Channels in Nuclear Power Plants," U.S. Nuclear Regulatory Commission, NUREG/CR-6343, November 1995.
3. Hashemian, H. M., "New Technology for Remote Testing of Response Time of Installed Thermocouples," U.S. Air Force, Arnold Engineering Development Center, Report Number AEDC-TR-91-26, Volume 1—Background and General Details, January 1992.
4. Hashemian, H. M. et al., "Aging of Nuclear Plant Resistance Temperature Detectors," U.S. Nuclear Regulatory Commission, Report Number NUREG/CR-5560, June 1990.
5. ANSI/ASTM E 644-78, Standard Methods for Testing Industrial Resistance Thermometers, Philadelphia: American Society for Testing and Materials, 1978.
6. Hashemian, H. M. and Petersen, K. M., "Loop Current Step Response Method for in-Place Measurement of Response Time of Installed RTDs and Thermocouples," in Seventh International Symposium on Temperature, Vol. 6, Toronto, Canada: American Institute of Physics, May 1992, 1151–1156.
7. Hashemian, H. M. et al., "Effect of Aging on Response Time of Nuclear Plant Pressure Sensors," U.S. Nuclear Regulatory Commission, NUREG/CR-5383, June 1989.
8. Hashemian, H. M., "Long Term Performance and Aging Characteristics of Nuclear Plant Pressure Transmitters," U.S. Nuclear Regulatory Commission, NUREG/CR-5851, March 1993.

2.12 International Safety Standards and Certification (ANSI/ISA-S84, IEC 61508/61511/62061, ISO 13849)

I. H. GIBSON

There are three major international standards bodies: the International Electrotechnical Commission (IEC/CEI), the International Organization for Standardization (ISO), and the International Telecommunication Union (ITU). The first two of these are associations of national standards bodies; the ITU is a United Nations organization. Although the ITU produces some standards relevant to instrument engineering practice, this section concentrates on safety systems.

THE INTERNATIONAL ELECTROTECHNICAL COMMISSION

The IEC (sometimes referenced by its French initials as CEI) was founded in 1906, and is the world organization that prepares and publishes international standards for all electrical, electronic, and related technologies. The IEC was founded as a result of a resolution passed at the International Electrical Congress held in St. Louis in 1904. The membership consists of more than 50 participating countries, including all of the world's major trading nations and a growing number of industrializing countries. The IEC Web site is at http://www.iec.ch.

THE INTERNATIONAL ORGANIZATION FOR STANDARDIZATION

The ISO, established in 1947, is a worldwide federation of national standards bodies from some 130 countries, one from each country. ISO is a nongovernmental organization, with a mission to promote the development of standardization and related activities in the world with a view to facilitating the international exchange of goods and services, and to developing cooperation in the spheres of intellectual, scientific, technological, and economic activity. It developed from an earlier organization, the International Standards Association with the acronym ISA, which can lead to some confusion with the instrument society with the same initials—the "ISA Nozzle," as an example, was originally defined by the international standards body.

The work of the ISO results in international agreements that are published as international standards. These are then reissued by the member organizations as national standards; the cooperating members are required to make no more than minimal editorial changes to the internationally agreed text. Its Web site is at http://www.iso.ch.

In some fields where the charters of IEC and ISO overlap, joint ISO/IEC documents are issued.

ASSOCIATED NATIONAL STANDARDS BODIES

In the United States, the national standards body, which is a member of both IEC and ISO, is the American National Standards Institute (ANSI). The U.S. practice in generating standards differs from most other countries, in that standards are commonly generated by technical societies and then accepted as ANSI standards. Other countries commonly generate their national standards within the national standards bodies. This leads to some difficulties in the handling of international standards. Under the charters of both ISO and IEC, the national member organizations are obligated to issue the international standards as national standards with minimal editorial changes; yet the form of most international standards tends to be more mandatory than is acceptable within the U.S. system, and sometimes agreed international standards are accepted by the responsible technical society but not accepted by ANSI.

Over the years this has led to standards such as ISO 5167 (the international standard for orifice flow measurement) being accepted by ASME (ASME MFC-3M) but not by ANSI, which held to ANSI/API 2530 for commercial and technical reasons.

THE BASIC INTERNATIONAL SAFETY STANDARD—IEC 61508

IEC 61508, "Functional Safety of Electrical/Electronic/Programmable Electronic Safety-Related Systems," was developed during the 1990s to provide a generic format within which "application sector" standards might be developed.

At the same time, it was required to be a directly applicable document for use where such application sector documents had not yet been agreed upon. In spite of the title, which might appear to limit its application to electrical/electronic/programmable electronic (E/E/PE) systems, it recognizes that in most situations safety is achieved by a number of protective systems, which rely on many technologies (for example, mechanical, hydraulic, pneumatic, and E/E/PE systems). It suggests that the framework may be used for the development of safety-related systems based on other technologies.

During the development of IEC 61508, which was the subject of very considerable objections that caused a 5-year hiatus in final formulation, the Instrument Society of America (ISA) S84 Committee (now ISA, the Instrumentation, Systems and Automation Society) generated its own standard as a "process sector" implementation. This is now ANSI/ISA S84.01. The final step to the United States withdrawing its objections to the publishing of IEC 61508 was the insertion of a clause in each part of IEC 61508, which notes that:

> In the USA and Canada, until the proposed process sector implementation of IEC 61508 is published as an international standard in USA and Canada, existing national process safety standards based on IEC 61508 (i.e., ANSI/ISA S84.01-1996) can be applied to the process sector instead of IEC 61508.

IEC 61508 is in seven parts:

- Part 1: General requirements
- Part 2: Requirements for electrical/electronic/programmable electronic safety-related systems
- Part 3: Software requirements
- Part 4: Definitions and abbreviations of terms
- Part 5: Examples of methods for the determination of safety integrity levels
- Part 6: Guidelines on the application of Parts 2 and 3
- Part 7: Overview of techniques and measures

It also includes by reference:

- ISO/IEC Guide 51:1990—"Guidelines for the inclusion of safety aspects in standards"
- IEC-Guide 104:1997—"Guide to the drafting of safety standards and the role of Committees with safety pilot functions and safety group functions"
- IEC 60050(191):1990—International Electrotechnical Vocabulary (IEV) Chapter 191 "Dependability and quality of service"
- IEC 60050(351):1975—International Electrotechnical Vocabulary (IEV) Chapter 351 "Automatic Control"

Notable in IEC 61508 is the "Full Safety Life Cycle" philosophy, which requires consideration and documentation from conceptual engineering through to final decommissioning, and the requirement that all persons involved with the system at all stages in its life shall be of demonstrated competence to perform their functions (Figure 2.12a).

THE INTERNATIONAL PROCESS SECTOR SAFETY STANDARD—IEC 61511

The process sector implementation of IEC 61508 began development in 1998, with a drafting committee drawing heavily on the ISA S84 experience. This is intended to be a three part document, and the first drafts of all parts have been issued for comment by the voting members of IEC—each of the 50 national committees is entitled to comment, and approximately a third of these will provide clause-by-clause scrutiny. The second drafts of Parts 1 and 3 closed for comment in early 2001, and it is anticipated that the final drafts for ratification should be issued during 2002. Part 2 will be somewhat later.

As it stands, IEC 61511 defines its relationship with IEC 61508 by claiming application engineering and maintenance, while leaving the design and construction of the hardware and system software to the safety equipment manufacturers under IEC 61511 (Figure 2.12b and 2.12c).

THE INTERNATIONAL MACHINERY SECTOR SAFETY STANDARD—IEC 62061

The machinery sector application of IEC 61508 has been under development for several years, and reached its second committee vote in early 2001. It is anticipated that this will reach its final form in 2002. This is designed specifically to cover the use of programmable electronic systems in personnel guarding functions on mechanisms. The approach used is more directed to high integrity than requiring reliability in combination with integrity. Thus, a 1oo3 system may be called for where 61511 might use 2oo3.

THE ANSI/ISA PROCESS SECTOR SAFETY STANDARD—ANSI/ISA S84.01

As indicated above, ANSI/ISA S84.01, although claiming to be a process sector implementation of IEC 61508, was actually produced prior to the final editorial changes to the IEC document. It is not, in itself, complete, as it does not provide enough information to perform the calculations that it requires. These are incorporated in the five-part ISA TR84.00.02, which describes four approaches of graduated accuracy and complexity to determining the safety integrity level (SIL) of a safety instrumented system (SIS), and the Markov technique, which is required for the evaluation of logic systems. This was published in early 2001. Both the standard and the technical report can be obtained through the ISA at http://www.isa.org.

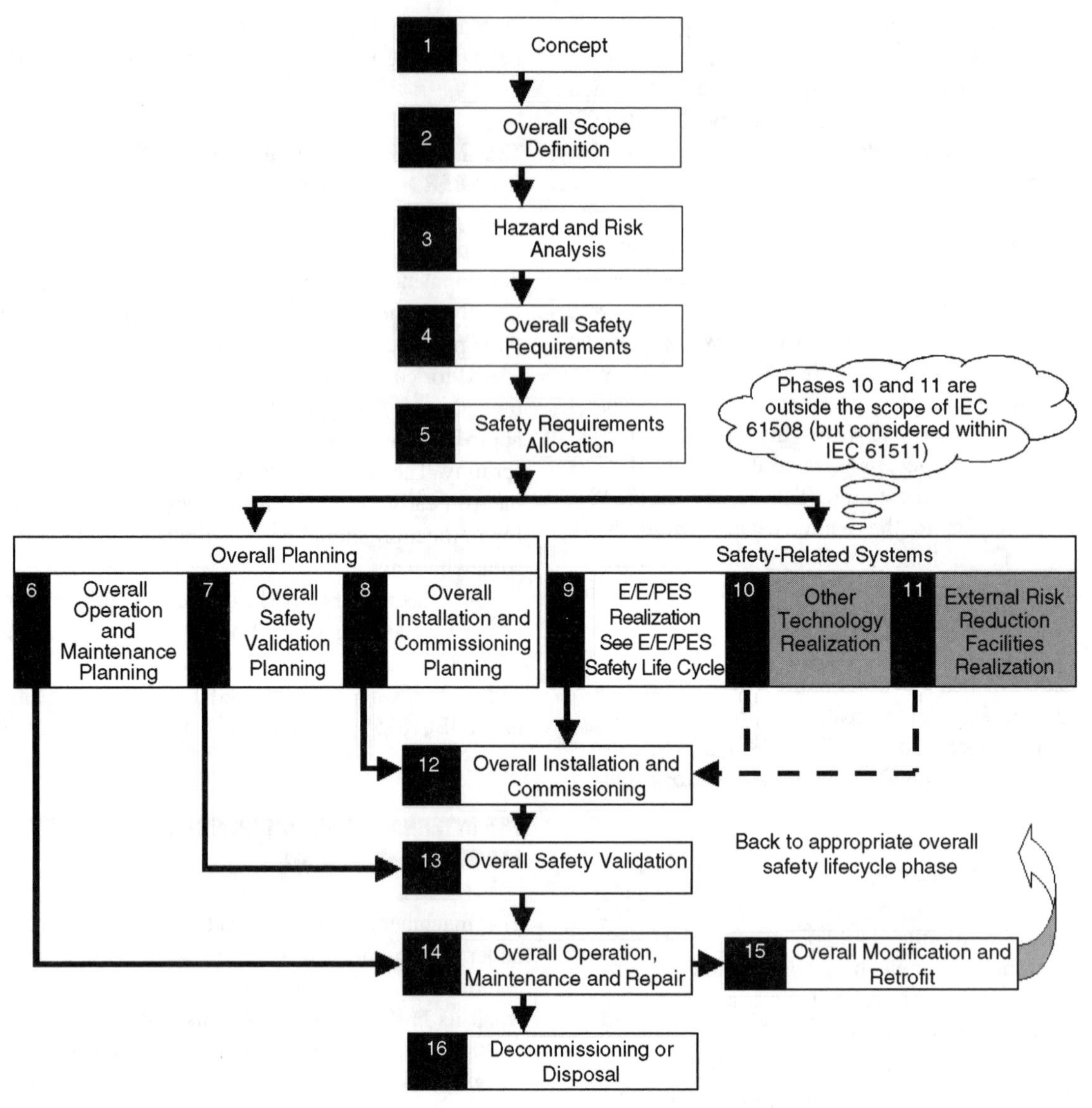

FIG. 2.12a
The IEC 61508 safety life cycle map.

A very useful third document exists, generated by the Process Industry Practices group (a consortium of major petroleum, petrochemical, and process plant design/construct organizations), which is intended as a generic, clause-by-clause extension of ANSI/ISA S84.01 to yield a specification suitable for commercial application. This is *PCESS001—Safety Instrumented Systems Guidelines*, which can be obtained from http://www.pip.org. This practice provides information to supplement Sections 5, 6, and 7 and Appendix B of S84.01 regarding the specification and design of SIS. Installation, maintenance, and testing issues are addressed only to the extent required for incorporation during SIS design. The information in this practice is presented in the form of additional examples and design considerations that should be considered by SIS designers. This practice is not a stand-alone document and should be used in conjunction with S84.01. To facilitate cross-referencing, section numbers and titles in the practice are followed by corresponding S84.01 section numbers enclosed in brackets ({ }).

Safety Integrity Level

The SIL, like pH, is a logarithmic representation. It can be represented by SIL = $\log_{10}$ (MTBF), where MTBF is the mean time between failures (years). Thus, SIL 3 signifies a mean time between failures in the range 1000 to 10,000 years. ANSI/ISA S84.01 directs its attention to the calculation of SIL of a particular SIS; the actual requirement specification of an SIL to reduce the risk of a perceived hazard to an "acceptable" level is not addressed.

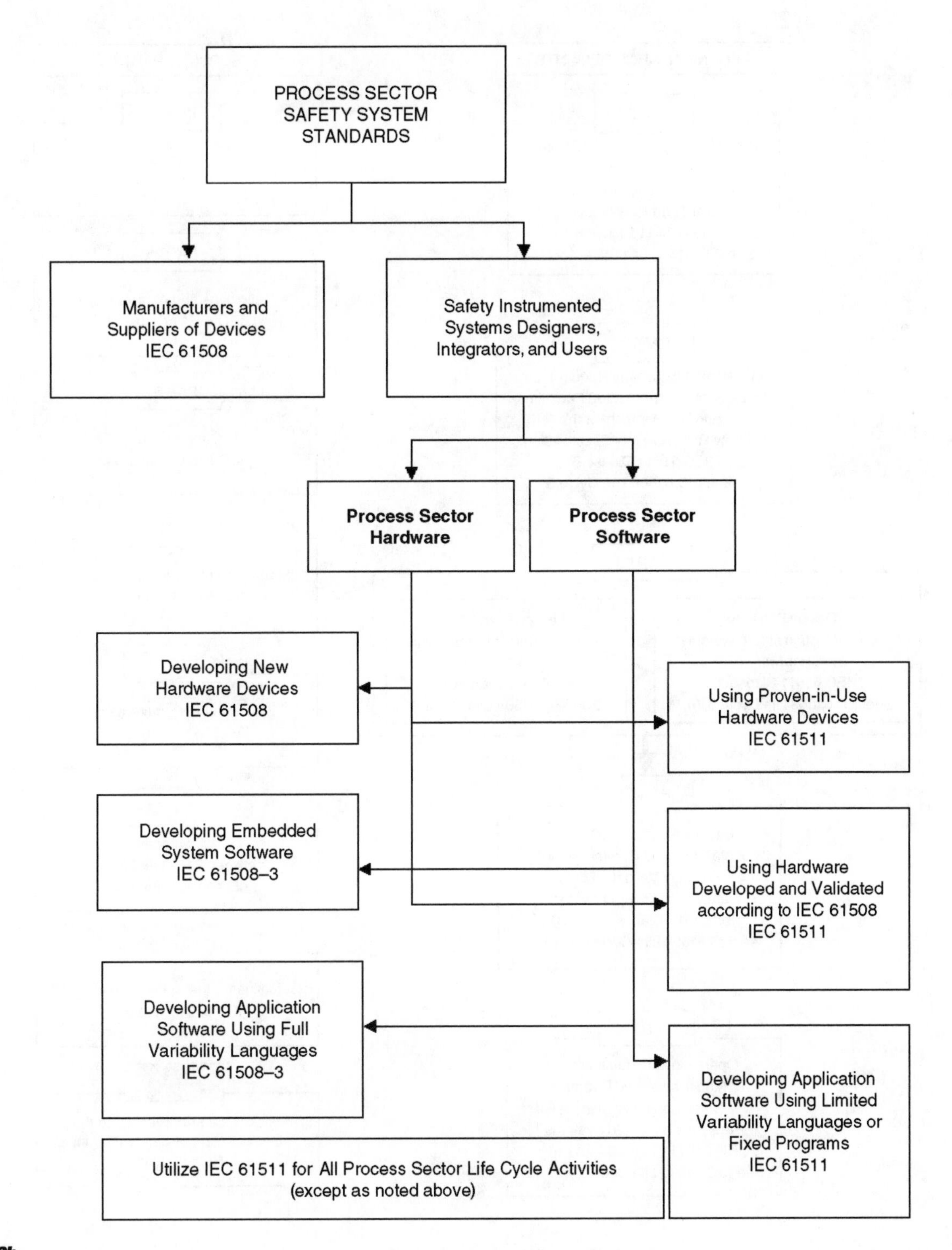

FIG. 2.12b
Relationship between users of IEC 61508 and IEC 61511.

THE EQUIPMENT HARDWARE AND SOFTWARE CERTIFICATION PROCESS

For some years prior to the development of IEC 61508, equipment for safety-related purposes was certified in Germany by the Technische Überwachung Verein (TÜV) Rheinland against German standards. These gave the AK 1 through 6 ratings, which can be roughly converted to SIL by dividing by 2 (i.e., AK 6 is roughly equivalent to SIL 3). TÜV now tests and certifies equipment against IEC 61508, and the U.S. Factory Mutual laboratories are also offering this service. It must be noted that assembling a chain of SIL 3 certified equipment will not necessarily yield a SIL 3 system, as the SIL of the assemblage must be less than that of the lowest-rated component.

The SIL rating of the components of an SIS are normally estimated from the failure rates of the individual components

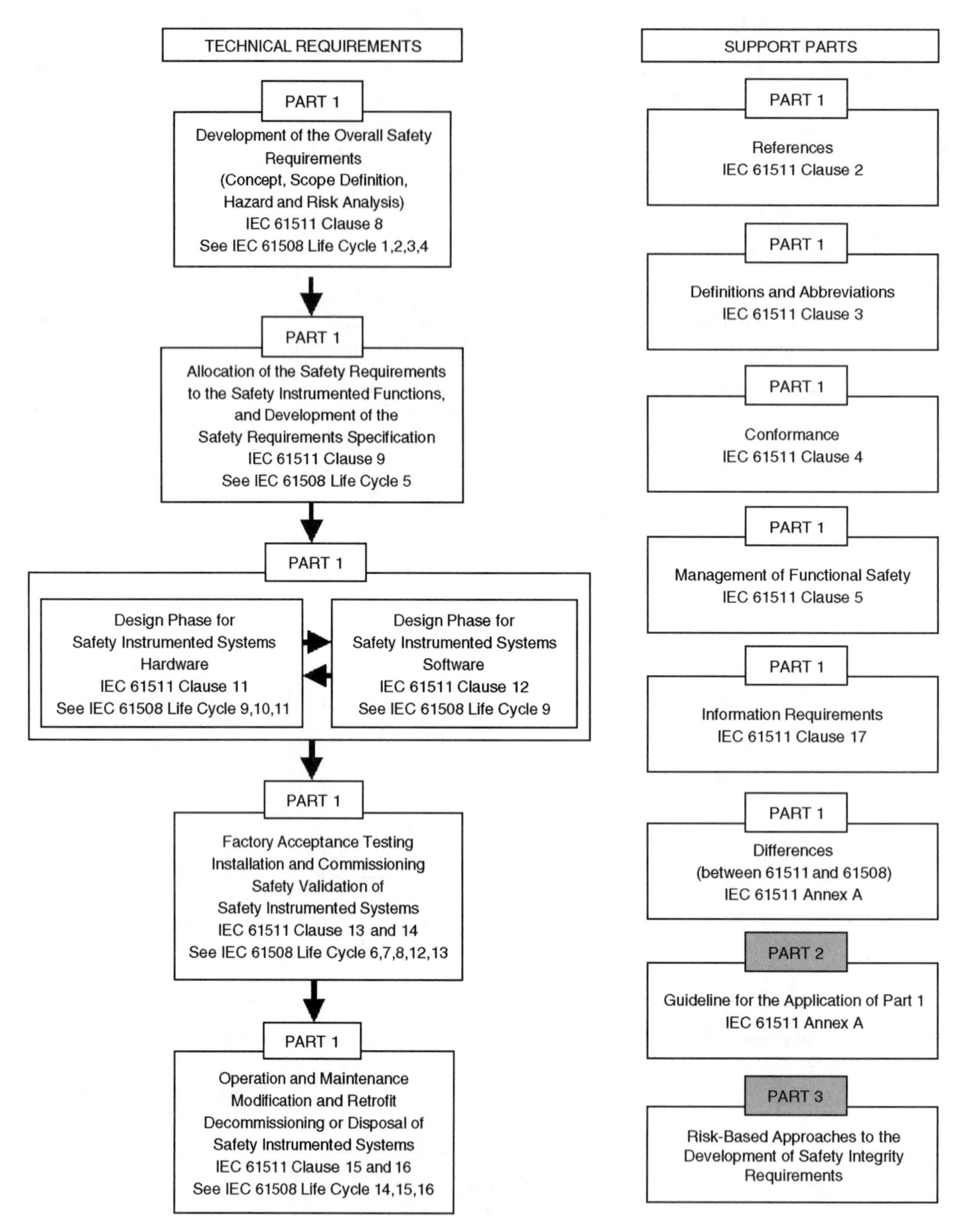

FIG. 2.12c
The IEC 61511 technical framework, compared with IEC 61508.

of the devices, as it is unusual to be able to obtain statistically valid failure rate data on current-generation instruments, particularly when the required rate is in the order of 1 in 200 to 500 years. SIL 3 does not mean that the equipment will not fail but rather that, over a sample of units, the average frequency of failure will be low.

By the time enough equipment is in service to give valid failure data, the next generation is on the market. It is also unlikely that equipment will have a uniform failure rate over the life of the equipment—the "bathtub curve" was once common, with infant mortality followed by a long, low failure period, and then an old-age period of increasing rate. Most modern equipment has a low infant mortality, and by the time old age sets in, there is little interest in the data. There are a few companies that acquire and publish failure rates on their high-end equipment, such as Mokvelt. This company claims

SIL 3 on some of its valve designs on the basis of "proven in service," with up to 50 years history on a large in-service inventory and careful spare parts tracking.

It can be argued that although hardware may fail in service, software cannot. Unfortunately, whereas software will perform the specified functions reliably, the specification of those functions is very frequently in error. The functional specifications for SISs have been found to be the cause of some 45% of the more spectacular system "failures" where the consequences have resulted in detailed investigations.

To minimize the capability of programmers to induce errors, the application languages for safety-related system software should be configured rather than programmed. The system software itself must be subjected to rigorous inspection, and any facilities that are indeterminate in cycle time avoided.

ISO STANDARDS ON SAFETY-RELATED SUBJECTS

A large number of ISO standards have relevance to instrument engineering; the ISO 13849 series is even more generic than IEC 61508, as it includes all technologies.

ISO 13849-1:1999 Safety of Machinery—Safety-Related Parts of Control Systems—Part 1: General Principles for Design

This part of ISO 13849 provides safety requirements and guidance on the principles for the design of safety-related parts of control systems. For these parts, it specifies categories and describes the characteristics of their safety functions, including programmable systems for all machinery and for related protective devices.

This part of ISO 13849 applies to all safety-related parts of control systems, regardless of the type of energy used, e.g., electrical, hydraulic, pneumatic, mechanical. It does not specify which safety functions and which categories shall be used in a particular case. This part of ISO 13849 applies to all machinery applications for professional and nonprofessional use. Where appropriate, it can also be applied to the safety-related parts of control systems used in other technical applications.

Note: See ISO/TR 12100-1:1992, 3.11.

Control Center, Workstation, and Logic Design

3

3.4 MANUFACTURING PLATFORMS AND WORKSTATIONS 323

3.5 WORKSTATION HOSTS: DESIGN CONCEPTS AND CLASSIFICATION 327

3.6 INTEGRATION OF DCS, PLC, HMI, AND SCADA SYSTEMS 334

3.7 INTEGRATION WITH RTUs, MULTIPLEXERS, FIELDBUSES, AND DATA HIGHWAYS 341

3.8 HYBRID SYSTEM WITH DISCRETE AND ANALOG CAPABILITY 351

3.9 SCADA—SUPERVISORY CONTROL AND DATA ACQUISITION 357

3.1 Operator Interface Evolution

G. K. TOTHEROW

The devices and instruments operators use to control process equipment have changed dramatically since the beginning of the digital control system age. Control rooms, which more closely resemble a NASA mission control center than the bench boards and instrument panels of the past, often have digital displays with images of the control stations, push buttons, and chart recorders used 50 years ago. Further, the digital displays are sometimes less intuitive than the bench board controls, and often are installed in a control room in a way that lacks detailed planning to support the operator's entire job. Notwithstanding these issues, the digital display human–machine interface (HMI) has allowed a single operator to monitor and control thousands of process control loops from a single station. The latest advances in HMI technology using inexpensive hardware, thin-clients, handheld computers, and cell phone technology have freed the operator from confinement to a particular location for control and information systems. Understanding the evolution of the operator interface and the changing role of the operator in a plant allows suppliers to build a better HMI product. It also allows plant engineers to pick the best technology to design systems that make the operator more effective and efficient.

The modern HMI is near the evolutional convergence of two different development paths. One development path is that of the HMI for the distributed control system (DCS). The other path is the development of the HMI for the programmable logic controller (PLC) market. This makes a chronological discussion of HMI evolution somewhat difficult without confusing the reader about which HMI is being discussed. Therefore, the author will attempt to describe the evolution of DCS and PLC HMIs separately until the two converge, and then the discussion will apply to both systems.

BASIC FUNCTIONS OF THE CONTROL SYSTEM HUMAN–MACHINE INTERFACE

Discussions about the HMI invariably become complicated because the same term is used to describe the function as well as the physical devices. Operators have always had HMIs, with some of the oldest being the faceplate of the controller or the handle of a valve. Although it can refer to any type of interface device, the term HMI usually refers to the display, computer, and software that serve as the operator's interface to a controller or control system. An analogy is the word *taxi*. A taxi is an automobile, boat, coach, or rickshaw that serves a specific function. Use the same automobile in another function and it is not a taxi. This section uses the term HMI to describe the computer hardware and software that allows interaction between the operator and the controls. Further clarification will be provided as necessary to describe specific HMI hardware, software, or a specific type of HMI. Table 3.1a is a quick reference of the specific HMI terms used in this section.

Many engineers use the HMI to perform a plethora of functions. Some of these functions are an excellent use of a computer and HMI software. But are they good uses of the primary operator interface to the control system? Should the operator's HMI also be a plant computer, or a management report writer, or a process historian, or a plant data server? Does the engineer need a taxi or just an automobile? One simple rule is that if the auxiliary function can impair or impede the performance of the device to serve as the operator's

TABLE 3.1a
Definitions of HMI Terminology

Term	*Definition*
HMI	Human–Machine Interface: Computer hardware and software that allows operators to interface with the control system
CRT HMI	Cathode Ray Tube HMI: The term is specifically used where there is a transition from hand stations to the HMI
DCS HMI	Proprietary HMI from DCS vendors that is optimized for process control
GUI	Graphical User Interface: General computer term for software that allows graphical interaction with the computer
Open HMI	HMI built on PC hardware and Windows operating system for general use; not specific to any hardware or industry
Operator's window	Open HMI that incorporates information systems and controls
PLC HMI	Proprietary HMI from PLC vendors that is optimized for discrete control; often connected as a remote PLC rack

HMI, then it should not be on the operator's HMI hardware and a second computer should be installed.

The operator interface to the plant process, whether bench board devices or HMI, has three primary functions. The first is to provide visualization of process parameters and methods with which to control the process. Second, the operator interface is to provide alarms and indications to the operator that the process is out of control or the system has failed. And third, the operator interface should provide a method, such as a trend chart, to allow the operator to understand where the process is going and how fast.

Visualization and Control

The first responsibility of the HMI is to provide the operator with a view of what the process is doing and a way to control the process. The panel board was the operator's window to the process through the various control stations and the analog instruments before the digital age of controls. The controllers themselves were mechanical devices that mounted to the bench board or the control panel. Most of the controller was hidden from view, and the part of the controller that was above the bench board for the operator to see and use was the controller faceplate. This faceplate displayed the process variable and provided access to change the mode, set point, or output to the loop. The controllers were mounted on the panel in logical clusters based on the process group. Figure 3.1b shows a power boiler control panel with pneumatic controllers to the right and a few electronic trend units and DCS hand stations on the left.

The HMI provides the same functionality to monitor and control the process in a newer, highly customized, and constantly evolving package. The HMI has made it possible to increase the number of control loops and process information presented to the operator many times higher than could be mounted in a control panel. Figure 3.1c shows the controller hand stations implemented on the HMI.

FIG. 3.1b
Power boiler control panel circa 1984; 19 hand stations are in the bottom center of the panel; 4 are digital stations for a Bailey Net90 DCS, the other 15 are pneumatic controllers. (Photograph courtesy of Jim Mahoney.)

FIG. 3.1c
Group display of hand station faceplates on a Bailey Net90 HMI circa 1985. (Photograph provided by Jim Mahoney.)

Process Alarming

The second responsibility of the operator's HMI is generation, annunciation, and manipulation of process and system alarms. Generation of the alarm was once the sole function of a switch in the field to determine if a process parameter such as a pressure or temperature exceeded specifications. Annunciation of the alarm was the responsibility of an annunciator panel. The annunciator panel itself was a backlit light box with translucent window panels with the alarm written on the panel. The discrete alarms were wired from switches to the annunciator to provide a visual and audible alarm for serious process upsets or failures. The annunciator allowed the operator to manipulate the alarm by silencing the audible portion or acknowledging the alarm. Alarming process parameters with an annunciator panel was expensive because of the field installation of switches and wiring.

The DCS incorporated process alarms into the control system and displayed them to the operator on the HMI. The DCS could get the alarm condition from a switch, generate the alarm condition in the controller from a transmitter, and display it to the HMI, or the HMI could generate the alarm based on the process variable input to the controller. Process switches were not needed to provide process alarms. The DCS made process alarms cheap and easy to implement; engineers and managers made the DCS process alarms a nuisance to operators.

Management of process alarms through silencing, acknowledging, and summarizing is now a primary function of the HMI. Advanced alarm management capability became a distinguishing feature between DCS vendors.

Trending

Operators have long depended on pen recorders that tell not only what the process is doing but also what the process did in the past. Analog pen recorders could be set to record the process at variable rates to make the chart show the process trend for a certain number of previous hours. The resolution of the recorder was inversely proportional to the amount of time the recorder could capture on paper. Getting the proper balance of time and resolution on the recorder was very important to help the operator understand the direction and rate of change of the process. The chart paper was changed as necessary and kept on file to enable process engineers and managers to look at the history of the control parameter.

The digital display HMI immediately had provisions for keeping a history of the information received and the capability to provide the operator with trending. The trending provided by the HMI was very similar to the chart recorders—the data were collected at a rate that would balance resolution with the total trend time. However, unlike the chart recorder, the operator could not change the paper once the graph was full to store the history of the process data. Many DCS companies had a difficult time building a good long-term process historian and decided relatively early that long-term historical data would not be an HMI function. Most of the DCS companies abdicated the long-term historical systems to others, choosing instead to provide interface devices to partners. This decision is still reflected in many HMI products where short-term history and trending is an integral function of the HMI, but long-term history is not.

DCS CONSOLES

The first companies that offered the digital display HMI were the DCS companies beginning with Honeywell in 1975 with the TDC 2000.[1] This was followed closely by several other major manufacturers of control products. Like many companies, Savannah Electric and Power Company bought a Bailey Net90 DCS in 1983 to replace pneumatic boiler controls. And like many others, the company merely replaced pneumatic controllers on the bench board with rows of digital hand stations connected to the DCS. Figure 3.1d shows digital hand stations for a Bailey Net90 mounted in a panel beside pneumatic controllers. The HMI "operator interface unit" displaying process graphics and graphical representations of the hand station faceplates sat beside the operator bench board. The logic of the transition was that the operators were accustomed to controlling the process by controller faceplates and process graphics were new and would take time. The expense of the HMI and the total control room rebuild was also a factor in a phased approach to introducing the HMI.

FIG. 3.1d
Four Bailey Net90 DCS hand stations mounted beside two pneumatic controllers circa 1984. (Photograph courtesy of Jim Mahoney.)

The introduction of a graphical user interface (GUI) display as the operator HMI was a technology leap that was destined to change the way many industries operated, yet many operators, and some engineers, felt more comfortable operating the process from a digital representation of their own older process controller faceplate than from a graphic of the process. The faceplate was familiar to the operators, provided all of the detail about the controller and the control loop, and was easily implemented by engineers. The problem with using the faceplate from the HMI was that the operator often had a much better intuitive feel for finding the right controller and responding quickly to upsets with the bench board controls as compared to navigating through various displays to find the right controller. Custom process graphics, or mimic displays, showing representations of pumps, motors, vessels, agitators, heaters, and other process equipment along with good overview displays and standards for navigation made controlling large processes from the HMI display much more intuitive for operators.

At Savannah Electric, the HMI offered more information than the hand stations and the operator could sit in one position to control and monitor the boiler and auxiliaries. Eventually, the operators used the HMI as their primary control panel.

Proprietary Functionality

1975 to 1985 The various HMI offerings of the early DCS companies used proprietary software and proprietary hardware and, like other systems of that generation, hardware and software were inseparable. Nowhere was this more evident than in the HMI keyboards. The keyboards contained rows and rows of special fixed keys to start, stop, open, and close the selected devices and acknowledge alarms and silence horns. The keyboards also had configurable keys for display navigation and special functions. Finally, the HMI included function keys that changed function with each display to help make the operator more efficient. DCS vendors had different ideas and standards for their individual HMIs, but all approached the operator interface as the total solution to replace the bench board. The DCS HMI included cabinets with room for mounting push buttons or telephones, wedge-shape cabinets to mount between HMI consoles to curve the console, matching tabletops for writing surfaces, and even chairs and stools.

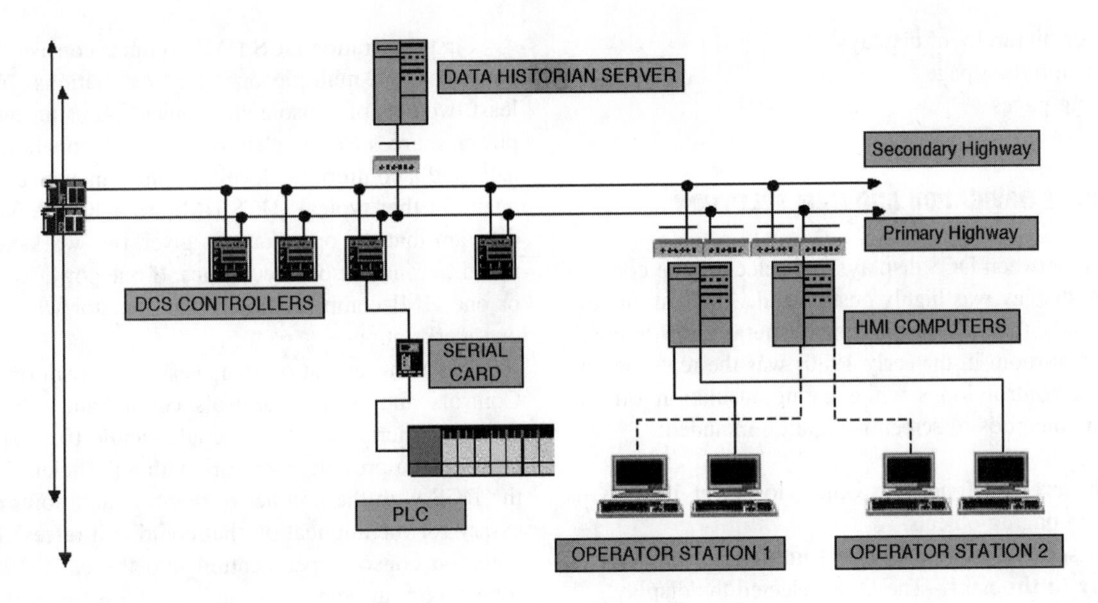

FIG. 3.1e
Typical DCS circa 1990.

1986 to 1995 The second generation of these proprietary HMI products showed increased integration with other components in the "system." Most DCS HMI products incorporated distributed alarming, where any console alarm is updated to all others; console redundancy, where functionality is automatically transferred on a failure; and a common configuration database for controllers and consoles. One price for this tight integration of proprietary systems was that by the second or third generation of DCS offerings communication problems existed between vintages. The first-generation HMI could not always communicate with second- or third-generation controllers and third-generation HMI could not always communicate with first- or second-generation controllers.

The proprietary HMI communicated only with the DCS controllers. A company that employed PLCs to do motor logic had to interface the PLC to the DCS at the controller level to have the DCS HMI stop and start the motor. Figure 3.1e shows a typical DCS interface to PLCs and historians. This added greatly to the cost of the system and added unnecessary load on the DCS highway just to give the operator access to the motor controls through the HMI.

1996 to 2001 The latest generation of HMI from most DCS vendors is based on the Microsoft Windows™ operating system and uses as much standard hardware as possible to keep costs lower. Proprietary may not be the correct terminology for most of the DCS HMI products built on the Microsoft Windows operating systems, but they are very often highly optimized with their own proprietary DCS controllers and less capable with other controllers. These pseudo-proprietary HMI products are configured through their own software to achieve the high integration. An example of this is the Siemens-Moore Product APACS system console built on the Wonderware FactorySuite products.

Organization of Information and Standards

The early DCS HMI was built specifically for a single control system. Controller faceplates were standard graphical objects that looked and functioned like the bench board hand stations from the same controls company. Popular early displays were the faceplate, group, and area.[1] The faceplate was the graphical representation of the hand station with all of the process information and loop control. The group display is a display showing four to eight faceplates in a little less detail. The area display is an overview of many process variables to quickly see if the processes are stable.

Mimic displays show graphical illustrations of the process with process information. Good mimic displays and good overview pages that show only the most pertinent information about large process areas made the operators more efficient. The HMI was so flexible and easy to change that companies quickly realized that they needed to adopt at least basic standards for building these custom displays to assure consistency between engineers and projects. Suddenly, arguments arise regarding whether a red motor indicates that the motor is energized or that the motor is stopped. Does a green valve indicate open, closed, or deenergized? The standards that most companies adopted included at a minimum object size, conditional colors, process colors, screen navigation, and screen density guidelines. Organizations such as the ISA issued standards for the development of the cathode ray tube (CRT) displays that include the items mentioned and many more.[2]

The standard displays for most companies installing a DCS included:

- Process graphic displays
- Faceplates
- System diagnostic graphics
- Controller tuning pages

- A list or hierarchy of displays
- Alarm summary page
- Trending pages

SCREEN/PAGE NAVIGATION AND ITEM SELECTION

Navigating between DCS displays and selecting the controllers on the display was highly customized with custom keyboards to make the operator efficient. Generally, the "mouse" in the control room in the early 1980s was the real one that shorted-out control loops while eating insulation off the cables. The methods of screen navigation included:

- **Touch screen**—Touch the object to select it. This included paging objects.
- **Direct screen reference (DSR) numbers from one display to the next**—The DSR selected the display.
- **Direct screen reference (DSR) numbers from the display list**—Each display on the display list was assigned a DSR number. Selecting that DSR from the display list would take the operator to the display.
- **Configured forward and backward display keys**—Each display could have configured forward and backward displays to navigate logically.
- **Show detail keys**—Devices selected could bring up a detailed display of the object or controller.
- **Defined function keys**—Console keys that take the configured display.
- **Trackballs**—The trackball was mounted in the custom keyboard and manipulated a pointing device on the display.

Later revisions of the DCS HMI saw more vendors using a mouse or trackball and less use of custom keyboards. Navigation and making control system changes with an on-screen pointing device vs. the older custom keyboards presented an interesting change for some operators. The above-average operators memorized display keys and DSRs for many loops, and it was not uncommon to see an operator "outrun" the update time of the HMI by selecting the display, entering the known DSR, and hitting the start or stop custom key from the custom keyboard. Those operators are forced to wait for the HMI to refresh from each graphic change using an on-screen pointing device. However, this problem with the best operators was more than offset by the increased speed of average or below-average operators and operator errors from trying to move too fast.

Redundancy

Many people were very worried about operating potentially dangerous processes through a computer when the DCS first arrived on the scene. DCS vendors worked hard on the technical issues of redundancy and on selling the concept of the HMI to industries that liked the hand stations.

First-generation DCS HMI products consisted of a minicomputer and multiple operator workstations. Installing at least two sets of console electronics and using two different power sources for the electronics and the operator terminals achieved a comfortable level of redundancy. See Figure 3.1e again for the "typical" DCS HMI arrangement. Notice in the diagram that the operator was given two workstations from two different console electronics. If one power source failed, or one HMI computer failed, the operator would lose only one workstation.

The first-generation consoles from vendors like Fisher Controls and Bailey Controls could handle 2500 points. Second-generation consoles could handle 10,000 points. The HMI had to provide operators with an efficient interface to the DCS with the number of displays and point count necessary for the application that could still refresh quickly. A common console specification into the early 1990s called for a maximum screen update time of 5 seconds and a refresh rate of 3 seconds. The HMI was, and still is, a very large part of the selection criteria for a distributed control system for many customers.

THE OPEN HMI

History

1980s The development of the PLC HMI was very different from that of the DCS HMI. The HMI products supplied by the PLC companies were industrial HMI field devices replacing push buttons. In general, the PLC companies concentrated their energy on making hardware and often used other companies to write configuration and HMI software. Shortly after the DCS was introduced, several small companies recognized a market niche and wrote HMI software for personal computer hardware aimed primarily at the PLC market. The open HMI was never an integral part of a proprietary control system.

Microsoft released the Windows 3.1 operating system for the PC in the 1980s, and with each successive release it was more apparent that Windows was the platform of choice for the open HMI.

1990 to 1997 The PLC became a very powerful logic controller capable of proportional integral derivative (PID) control and processor redundancy. The low price of the PLC and open HMI made them a good choice in the process industry over most of the "mini-DCS" platforms that were introduced at the time. Technology increases and low cost continued to bring the PLC and open HMI into larger service in the process industries, and functionality increased accordingly. The open HMI continued increasing in functionality compared to the DCS HMI, although the major DCS players scoffed at suggestions of an HMI on a PC in the early 1990s.

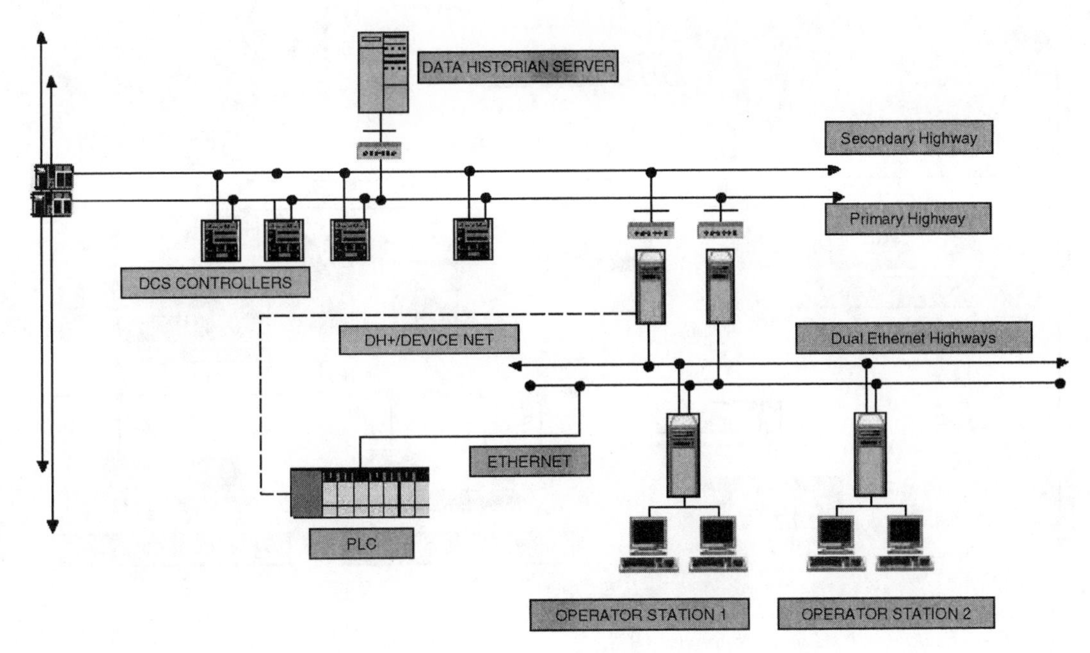

FIG. 3.1f
Illustration of a DCS using open HMIs. Note that the PLC has redundant communication directly to the HMI.

Microsoft released Windows NT 3.5 in 1994, Windows NT 3.51 in 1995, and Windows NT 4.0 in 1996.[3] Most of the open HMI companies migrated their software to the current releases from Microsoft. PLC companies released open HMI software products at this time to capitalize on the burgeoning market for HMI products.

Major DCS vendors stopped scoffing at the suggestion of utilizing PC hardware and the Windows operating system.

1997 to 2001 The open HMI began to look even more impressive for process systems when several of the leading HMI companies introduced software supporting client/server architecture and distributed alarming. The open HMI companies also began the drive to integrate their HMI with a suite of software products to provide better information to the entire corporation. The proprietary DCS HMI remained more functional for the DCS than the open HMI, although most DCS users never fully utilized the advanced functionality. The open HMI now duplicated most of the functionality of the DCS HMI, was less expensive, and could interface directly to other control systems. The open HMI could act as a common, integrated HMI for all control systems at a plant. Figure 3.1f shows the open HMI accessing data from multiple systems.

The open HMI began to replace the DCS proprietary HMI products at many process sites.

Architecture

The open HMI began as a single operator interface to a single PLC controller. The HMI software contained the tag database and the graphical displays. In a system of multiple HMI machines only one machine ran the controller input/output (I/O) driver. All the HMI machines initiated communication conversations with the I/O server to interface to the controller. This was a fine arrangement for a small system but cumbersome, with many controllers and HMI machines. The answer was a client/server architecture and remote tag access in the HMI.

The client/server architecture for the open HMI was introduced in the mid- to late 1990s. The server machine connected to the controller device or system and contained a tag database. Multiple HMI machines could act as clients to the server accessing the tags in the server. The client/server architecture improved the ability of the open HMI to interface to multiple controllers and greatly opened the door for the HMI to become the operator's window to information from any source.

The next major development in HMI architectural design is to make the operator HMI a thin-client. A thin-client workstation is a machine that displays the operator HMI displays without HMI application software on the workstation. The advantage of this architecture is the decreased cost of system management, increased system maintainability and security, and longer component life cycle. There are two types of thin-client systems.

One thin-client displays an HMI application session that is running in a Microsoft Terminal Server. The client box establishes an independent session of the HMI software on a terminal server box through an Ethernet TCP/IP (transmission control protocol/Internet protocol) network. This HMI architecture, shown in Figure 3.1g, enables every thin-client to have a full-functioned HMI that is very fast using either Microsoft Remote Desktop Protocol (RDP) or Citrix Independent Computing Architecture (ICA) protocol. This is an excellent way to use wireless handheld computers.

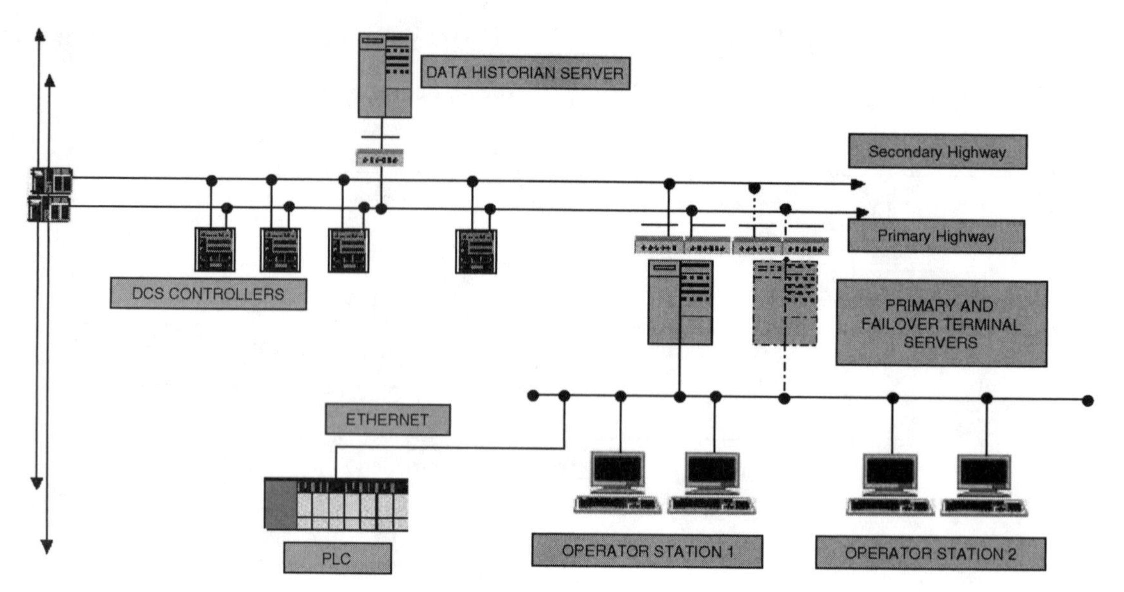

FIG. 3.1g
The architecture of thin-client open HMI on a traditional DCS.

Citrix MetaFrame software further enhances the architecture by allowing server farms and server load balancing.

The thin-client and terminal server is very similar to the older operator consoles that were made of dumb terminals from a mainframe computer. The more common thin-client in use today is one that uses an Internet browser as the client software to view the operator HMI. The advantage of the Internet browser thin-client over the terminal server is that networks which include telephone cable and modems, and PCs with browser software are typically already in place or readily available. The disadvantages of the Web browser HMI are the speed of the update and the current limitations of graphic-intensive pages in the Web server. Most of the popular commercial HMI software packages are thin-client capable with terminal services, Web servers, or both. These are relatively new technologies for the HMI software so a thorough understanding of the functionality of the systems is prudent.

The architecture for the future may be the object server architecture. This architecture is very similar to the client/server arrangement except that all control information will be built into an object server. A control object will be a control loop point and all of its functionality, including a graphic of the valve and instruments, the process variable, the set point, the alarms, and every other characteristic of the loop. The object will be available to every other program such as the HMI client to be dropped on a display, in a historian, or a Web page.

Organization of Information and Standards

Open HMI software is extremely versatile—providing a cost-effective operator HMI to a stand-alone machine or a networked series of HMIs for a system of networked controllers. The DCS vendors supplied a framework of basic functionality, symbols, control stations, diagnostics, and other application guides tailored to large process applications for their HMI. The PLC vendors supplied a framework of basic functionality for their proprietary single machine HMIs. The open HMI software companies followed the Microsoft business model of supplying shrink-wrapped software for their customers to build into a functional HMI for their particular installation. The organization of process information in the open HMI and the configuration standards fell entirely to system integrators and engineers—with mixed results.

Information on the ergonomics of human factors such as position of the chair, screen glare, and others is abundant. There is less information readily available on specific guidelines to help an engineer organize the information and build an effective HMI. The ISA has a document—ISA-TR77.60.04-1996 for fossil fuel plants—that provides some guidelines that are common to all process industries. The problem is that many companies implementing medium to small projects never dedicate resources to human-factors engineering and establishing HMI standards.

Documentation

Documentation is defined in this section to include installation, configuration, backup, and change control documentation. All documentation is a weak point in the open HMI when compared to the proprietary DCS HMI. The first weakness is that the hardware, software, network, and configuration are individual components from different companies. The engineers who install and configure the open HMI must provide the documentation for the HMI system. The open HMI software companies provide software documentation, configuration training, and configuration examples and tips to their users. The software company cannot know the system hardware or the configuration application, leaving the end user and engineer responsible for documentation of the system.

The second area of documentation weakness for the open HMI is the configuration backup and change control documentation. The DCS prescribed database backup procedures and most of the configuration systems automatically documented database changes. The configuration backup of a PC file like an HMI application is intuitively easy but change control documentation is not. Third-party software companies recognized this market niche in the late 1990s and filled this void for the open HMI companies with software that limits access to the HMI or PLC configuration and maintains a log of the changes that are made to the software packages.

Fig. 3.1h
Control room with pneumatic controllers, alarm annunciators, and push buttons circa 1984. (Photograph provided by Jim Mahoney.)

Plant Networking

One of the major technologies that fueled the HMI evolution is the increase in plant information system networks that can tie process control systems together. The DCS can be credited with introducing the plant communication network in many process companies. The network was easy to buy and easy to install because the DCS vendor sold all the network components and prescribed every phase of the installation. The network was part of the DCS. Many process industries began installing TCP/IP network systems in the 1990s for various communication needs. This networking of commodity items in process plants was a key factor in implementing large systems of open HMI operator stations.

CONTROL ROOMS

A review of the evolution of the operator's HMI cannot be complete without a discussion of the operator's function and the control room. The operator's job has always been more than merely to operate the process. The operator makes visual inspections of equipment, performs or coordinates chemical testing, and performs shutdown, lockout, and tagging of equipment. The operator implements manufacturing schedule changes and generates production reports and log sheets that describe operating conditions and show how the product was made. The operator reports productivity, dispatches support personnel, and coordinates problem resolution and maintenance activities. The operator makes thousands of individual decisions every day in reaction to material, system, and process equipment problems. The process operator is the conductor of the orchestra of people and systems that produce the product. The control system HMI must fit well into the operator's world.

The control rooms that predate the CRT HMI had individual process controller devices for valves, vanes, dampers, and other process control equipment that allowed operators to manipulate the controller output and set point. Push buttons and lights were the operator interfaces for discrete relay logic. Individual instrument gauges, thermometers, strip chart recorders, and sight glasses monitored process variables in the control room and in the field. Annunciator panels provided alarming and first-out indications.

Extensive study went into the ergonomic design and arrangement of the operator panels of control devices, annunciator systems, telephones, and the operator's desk in the control room. Engineers designed the control panels for functionality during routine and emergency situations as well as start-up and shutdowns. Operators easily adapted the controls to their own needs by attaching notes to the instruments, and using markers to indicate "sweet spots" of control or danger areas beside the controller. In a glance, the operator could usually survey the control panel and easily recognize and respond to problems (Figure 3.1h).

It can be said that the modern control room is the house that DCS built. Some control rooms are wonderfully designed and entirely practical with well-designed HMIs. Unfortunately, some companies never duplicate the ergonomic functionality of the old control room when they install HMIs. Many issues can make the control room less functional to the operator after installing the HMI. One of the most common is poor design of the HMI displays, navigation, and alarm handling for the various operating conditions. Steady-state management of the process is far different from operating through process upsets, a system start-up, or even operating isolated equipment during maintenance shutdowns. Another issue that makes the control room less functional with an HMI is making operators move between the HMI and other control room interface systems to do their job. The third item that may deteriorate the ergonomics of the control room is changing the operator's job function after installing the HMI.

The problems that make the control room less functional are all avoidable with good human-factors engineering. Section 3.3 provides a discussion of upgrading the control room, covering methods for avoiding many of the problems listed above.

2000 AND BEYOND

Process control, the HMI, and the operator's job function will continue to evolve. Major DCS companies are building their HMI console from open HMI software. Many process companies are replacing older proprietary DCS HMI consoles with open HMI software configured to provide most of the DCS functionality. The DCS HMI products are doing everything that the open HMI software companies are doing to make their products cost-effective. The open HMI software companies are increasing product functionality and integration into other software packages but still leaving all application services to others. These HMI systems have nearly converged in their development, so the open HMI has lost most of its price advantage.

The physical choices for providing the operator with a window into the process will continue to increase and will include HMI stations distributed around the plant, consolidated control rooms, personal HMI devices such as handheld computers or cell phones, and off-site HMIs. The limiting factor to making the operator more efficient is not likely to be the ability to interface to the processes but the process itself. Process industries like pulp and paper, foundries, utilities, chemical production, and others are very capital intensive. Very expensive process and process control equipment will not be replaced without a great return on investment, so existing equipment will continue to be utilized for years with new systems integrating with existing components and plant support systems.

Three initiatives that will have a large impact on the process operator HMI through 2010 will be integration of data from any system into the HMI, widely distributed and mobile operator windows, and the remote-from-site HMI. These technologies can integrate into existing plant systems, utilize existing legacy systems, and they have the capacity to fundamentally change the way process systems are operated in the future.

PROCESS MANAGERS NOT OPERATORS

Operators are controlling and monitoring an ever-increasing number of process loops because of the capabilities of the modern HMI. Simultaneously with the increased process control responsibilities, many operators are also asked to perform tasks once accomplished by forepeople such as coordinating product transitions, making shutdown and maintenance decisions about equipment, and monitoring product quality. Many control rooms are an eclectic mess of HMI consoles, various information consoles, written logs and books, video monitors, chalkboards, and written notes because of the roles of the operator. The HMI must become a picture frame for the operator with a collage of control system objects, informational objects, and financial objects from many different systems arranged sensibly. The HMI will be the operator's window into the process, the production and maintenance schedules, and likely the customer's order.

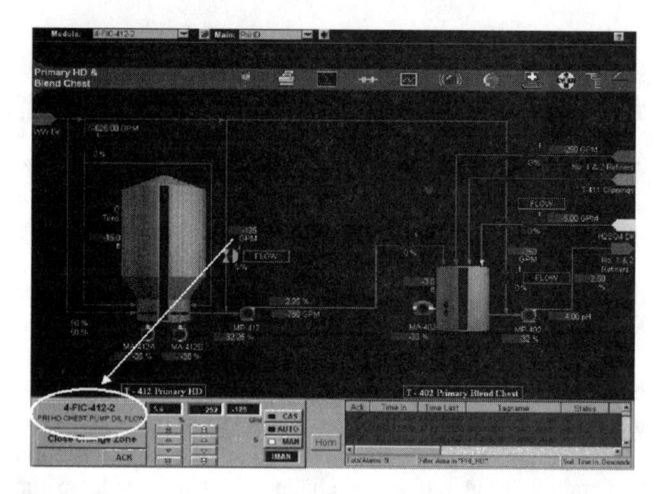

FIG. 3.1i
HMI graphic using the Orion CyberBAR display style guide. Selecting the flow on the graphic populates the change zone for operator entry. The CyberBAR near the top of the display is selected to navigate to information pages.

The HMI will require increasing organization of the display of information to convey the real-time data, lab test data, historical information, interfaces to maintenance, tracking, and integration of video cameras. Pierce Rumph of Orion CEM (Atlanta) has a hobby of studying the operator tasks and the HMI. Orion CEM, part of Emerson Process Management, is a leader in developing an HMI style guide that helps pulp and paper operators become process managers.[4] The Orion CyberBAR style guide is a navigation bar that separates information above the bar from process interaction below the bar. The bar moves up or down depending on the information contained on the display. The CyberBAR is a navigational tool that uses icons for page navigation and alarm notification. The CyberBAR also serves as an information bar calling up items such as diagnostics, trends, and standard operating procedures (SOPs). Figure 3.1i shows a display using the CyberBAR style guide. The CyberBAR near the top of the display shows navigation to information pages. The arrow shows where the operator has selected a flow loop on the display and the change zone is populated and ready to accept control inputs. Orion is also a leader in the use of audible messages from the HMI. Voice messages that notify the operator of alarms and console activated sequences and audible feedback for motor starting and stopping keep the operator engaged with the process.

Unfortunately, navigational aids and intelligent audible signals are not sufficient for the process manager trying to manage a major crisis in a process area of thousands of points. Dynamic alarm management and intelligent alarming methods to enable better response to emergencies are necessary if the operator's process area continues to increase. Process alarms must be written smart to take advantage of alarm hierarchies to auto-acknowledge minor alarms when a higher-level alarm for the same process group is active. Also, the operator's window must offer alarm disabling, silencing, and dynamic alarm level changes for various modes of operation for large highly sophisticated process industries.

Distributed and Mobile Control of the Process

Is a control room the best place for the operator to control the process? Or, should the operator move about the process area?

The control room is a relatively new invention in most manufacturing industries. Prior to the DCS many processes were run from local panel boards. Coordination of the process control with the panel boards was very difficult and the central control room increased the plant control quality by allowing for better coordination. The control systems are advanced enough that the control room is not needed to coordinate the control.

HMI technology is advancing very rapidly toward inexpensive operator stations that can be distributed throughout a process area so that an operator can easily walk around the process and still be very close to an HMI. The operator can also use wireless handheld computer technology or even cell phone technology interfaced to the HMI software to be able to check on equipment or perform testing while still monitoring the process.

Remote Operation of the Plant

If the main operator is in a central control room 100 yards from the process today, why could the operator not be 100 miles from the process? Why not 1000 miles?

Control rooms of the past had to be located within the propagation limits of the signal wiring and process lines without extraordinary expenses. Even DCS networks and Ethernet networks have practical physical limitations on the distance between a controller and the HMI.

Thin-client HMIs and the Internet now enable an operator, or a backup "expert" operator, to be located anywhere in the world. The few technical limitations associated with remote operation of the plant today are dissolving and future discussions about the location of operators will be philosophical.

RESEARCH

HMI is the subject of many research projects in private industry, governments, and academia. The work in virtual imaging is fascinating—but will it be relevant in process control? The work by NASA on changing HMI functionality associated with modes of operation[5] could have industrial applications one day, but audio-based interaction between humans and machines might provide the greatest benefits for the industrial HMI.

Sounds can be heard from any direction (except in the plant) and auditory messages can be sent in parallel with visual signals or conventional process control. Auditory messages can be spoken and received in hands-free operations providing unmatched flexibility by any other I/O method.[6] Engineers can implement auditory HMI in small phases to augment conventional control. The technology of audio HMI is only beginning, and like most technology it will be tested and refined in commercial applications long before it is introduced into industry.

THE FUTURE

The biggest gains in operational success associated with the HMI do not lie in the technology, but in the management and customization of the HMI for the particular industry, process, and individual operator. Distributed HMI devices throughout the plant, personal HMI devices, ergonomically designed central control rooms, and perhaps auditory control can all be used with great success. The companies that implement technology without a plan will probably achieve less than the companies that have a formal plan for operating the plant and provide the right HMI tools to make it happen.

Increased priority on display design, information presentation, and operator interaction will provide a good return on investment. The process industries have more similarities in their needs than differences. This creates opportunities for tool kits from outside organizations and open HMI companies to increase the efficiency and standards of HMI displays.

Jim Mahoney, a project specialist with ABB Automation, has over 20 years experience working with operator control system interfaces. He has migrated control rooms from pneumatic to digital, from first-generation HMI to the newest PC-based HMI. In a conversation about operator interfaces Jim said, "The key to a good console is well-laid-out graphics that properly mimic the process, and ease in getting around the multitude of screens. A state-of-the-art console with poor graphics does not help operators do their job and can actually be a major detriment." Few experienced people will disagree with Jim, yet few realize that an HMI with mediocre displays helps make mediocre operators and an HMI with great displays helps make great operators.

Last, remember the three fundamental functions of the HMI software: visualization and control of the process, alarming and alarm management, and historical trending. These functions were lumped together in software in the 1980s because of the available technology to produce an HMI at that time. A future generation of HMI could see each of these functions performed by different "best of breed" software and presented through the operator's window.

References

1. Freely, J., Merrit, K., Ogden, T., Studebaker, P., and Waterbury, B., "The Rise of Distributed Control," *Control Magazine,* December 1999.
2. ISA-TR77.60.04-1996, "Fossil Fuel Power Plant Human-Machine Interface CRT Displays," approved May 24, 1996.
3. Business Week Online, February 22, 1999.
4. Rumph, P., "Improved Information Interfacing Turns Operators into Process Managers," *Pulp & Paper,* August 1998.
5. Degani, A., Shafto, M., and Kirlik, A., "Modes in Human-Machine Systems: Review, Classification, and Application," Human Factors Research and Technology Web page, NASA, http://human-factors.arc.nasa.gov, 2001.
6. Stephanidis, C. and Sfyrakis, M., "Current Trends in Man–Machine Interfaces: Potential Impact on People with Special Needs," in P. R.W. Roe, Ed., *Telecommunications for All,* Brussels: ECSC-EC-EAEC, 1995. (Also available in Spanish.)

3.2 Virtual Reality Tools for Testing Control Room Concepts

A. DRØIVOLDSMO M. N. LOUKA

Cost: Desktop-based computer systems with adequate three-dimensional graphics acceleration range from $2.5K to $5K. Projection-based computer systems range from $10K to over $1M, depending on the display technology. In addition there is the cost of software, which typically ranges from $100 to $50K, depending, among other things, on its capabilities for creating detailed models efficiently and for supporting effective and accurate testing. See the subsection on VR Software Requirements for Design and Testing for further guidance.

Partial List of Suppliers: Alias|Wavefront; Auto desk; BARCO Projection Systems; Bentley Engineering, Cadcentre; Fakespace Systems; InFocus; Multigen-Paradigm; Opticore; Panoram Technologies; ParallelGraphics; Parametric Technology; REALAX; Sense8; SGI; Sony; Stereographics; Sun Microsystems; Unigraphics Solutions; Virtock Technologies

Recognized review models for the design process of control rooms[4,10,13] have a similar general approach to the handling of outputs from the human–system interface (HSI) design verification and validation (V&V). They all recommend an iterative process where deviating verification results should be used as inputs to the next stage or phase in the design process. The problem with this approach is that there is a shortage of tools supporting HSI V&V in the early phases of control center design.

Virtual reality (VR) technology has proved to be a promising, powerful, and cost-effective tool in control room design work. It enables designers to spend more time evaluating creative new ideas, helping them to identify and eliminate potential problems early in the design process, and lowers the threshold for end users to participate more actively in the design process.

This section describes a human centered generic design process for a control room and the practical use of virtual reality as a tool for testing control room concepts.

VIRTUAL REALITY

VR technology enables users to immerse themselves in an artificial environment simulated by a computer, with the ability to navigate through the environment and interact with objects in it.

Two key words in the field of VR are presence and immersion. VR immerses the user in a simulation so that the user has a sense of being present in the virtual environment. The degree of immersion depends primarily on the computer hardware used, whereas presence is achieved if the virtual environment causes the user to suspend disbelief and accept the computer-generated experience as real.

A computer system that generates an artificial world that tricks the user into feeling part of it would not be a complete definition of VR because it lacks the vital component of interactivity. VR is a three-dimensional (3D) user interface in which the user can perform actions and experience their consequences. It is a multidimensional real-time simulation rather than a linear animation with predefined camera movement. This is what distinguishes VR from recorded, computer-generated images used in movies and on television and from real-time computer animation where the user is a passive viewer. Unlike pre-rendered images of a 3D environment, the user of a VR system can usually move around freely in a virtual environment.

In the context of control room design and testing, VR technology enables a designer to interactively construct a virtual control room, view the room from the inside as if it were a physical environment, and to simulate moving around in the room, checking lines of sight, legibility of labels, clearance, etc. This flexibility is highly suited to human-centered iterative design processes.

VR SOFTWARE REQUIREMENTS FOR DESIGN AND TESTING

To evaluate whether an off-the-shelf VR software package can be used effectively for control room design and testing, it is important to investigate what facilities the software provides.

Software packages designed for product or architectural design typically provide most, if not all, of the features that are desirable. In some cases, additional features required can be implemented by a programmer, but often a little ingenuity in the design of special 3D objects that can be added to a model and be moved around at will is all that is required. Examples of such special objects are semitransparent view cones, semitransparent spheres for viewing ideal distances, virtual rulers and protractors.

The key feature required of a VR software package for control room design and testing is that the user can interactively add 3D objects to a model and move them around. How accurately and efficiently this can be done depends on the capabilties of the software.

Other useful features include:

- The ability for objects to snap to an invisible user-defined grid when positioned in a model.
- The ability to group objects together so that they can be treated as a single aggregate object.
- The ability to link objects to each other, e.g., link a computer display to a desk so that the computer display can be moved independently of the desk but moves with the desk if the desk is moved.
- The ability to constrain objects so that they can only be moved in a predefined manner (e.g., rotated around a specific axis or moved in a specific plane) or can be locked in place to prevent accidental repositioning.

The manner in which a user interacts with the 3D model depends on the selected VR hardware and software. In most cases, a desktop computer with 3D graphics acceleration and a VR software package are all that is necessary. This form of VR is often called desktop or fishtank VR. The user utilizes a 2D pointing device such as a mouse to select and manipulate objects, choose menu options, etc.

Desktop VR is an effective and relatively inexpensive option. The desktop VR experience can be enhanced by projecting the image on the computer display onto a large screen so that several users can work together in a meeting room, with one user controlling the computer. Some 3D graphics cards and projectors support stereoscopic projection, enabling users to see the model with an enhanced sense of depth. Stereoscopic projection typically requires that users wear special glasses. Some users find that stereoscopic viewing (which is also possible using a desktop display) enhances their spatial understanding; however, such display systems should only be used for brief viewing sessions (e.g., up to half an hour) to prevent nausea and headaches.

For a greater sense of immersion, multiscreen projection systems or head-mounted displays can be used to place the viewer inside the model. Although the hardware required to do this efficiently, with motion tracking and 3D input devices such as gloves, is relatively costly, immersive VR can be useful for realistic final walkthroughs of a virtual environments, to experience the environment from the perspective of the end user in as realistic a manner as possible. The purchase of an immersive VR system purely for control room design and testing is generally difficult to justify from a cost-effectiveness point of view; however, many large organizations have such facilities, which can be accessed and utilized for limited periods.

The VR software package used may provide multiuser networking features, which enable geographically spread users to view and manipulate a shared 3D model simultaneously. This feature is potentially useful in situations where it is inconvenient for a design team to meet physically. A simpler form of distributed collaboration can be achieved by distributing a copy of the 3D model files by e-mail or from a Web site to project participants for comment. The master copy can then be updated based on feedback received. In either case, the distributed users will usually need to install software that can be used to view the 3D model.

CREATING 3D OBJECTS

Model rooms and objects to position in them can be constructed using a 3D modeling package, a 3D scanning system, or a combination of both. For a control room design project, a library of rooms and objects should be prepared for use with the VR software.

Typically, a 3D computer-aided design (CAD) package is used to create 3D objects. Most CAD packages are capable of exporting 3D object data in a format understood by VR software systems. The most common format used is ISO VRML97[7]—the Virtual Reality Modeling Language. However, there are fundamental differences between 3D CAD models and 3D models intended for interactive use, which usually need to be taken into account when preparing objects created in a CAD package for use in a VR system. In particular, CAD models emphasize the physical properties of solid objects, whereas VR models emphasize the visual appearance of surfaces. When models are exported from a CAD system, curved surfaces are usually converted into polygons that approximate the original curve. Often the number of polygons generated is too large to be used efficiently as a VR model but optimization software is available that can be used to reduce the number of polygons without significantly degrading the appearance of the object. In some cases, such optimization is built into the export routines of the CAD software or the import routines of the VR system.

Although CAD systems can be used to create accurate models, most systems do not support techniques for modeling surfaces that are used by VR systems to enhance significantly the visual appearance of objects. In particular, CAD systems do not normally have any support for applying images (textures) to the surfaces of objects to "fake" detail. Textures can be applied to objects created in CAD systems by importing the geometry into a 3D modeling package that supports textures and then exporting in an interactive 3D format supported by the VR system.

Unless CAD models of objects already exist, which can be used to create VR objects, it is usually more efficient to use a 3D modeling package dedicated to the creation of models for real-time 3D graphics or animation. Such packages typically provide support for modeling at a polygon level, enabling the creation of highly efficient and attractive 3D objects. It is particularly important that objects are optimal in cases where conventional instruments are modeled in large numbers because the total number of polygons that a computer needs to render can quickly become inefficiently large, significantly affecting the usability of the system due to performance considerations.

3D scanning systems can be used to digitize an existing object or environment. The scanning process usually consists of a sensing phase and a reconstruction phase. The sensing phase captures raw data and the reconstruction phase processes the collected data to turn it into a 3D CAD model. Common scanning techniques include tracking, photogrammetry, and laser scanning.

Tracking systems work by positioning a probe on an object and telling a computer to record the position of the probe. After a sufficient number of positions have been sampled, reconstruction software can convert the points into a computer model of the object. This technique would not be appropriate for scanning in a whole room with a large number of flexible pipes. It is mainly applied to scanning individual objects.

Photogrammetry techniques process one or more photographs to create a 3D model. There are a number of different approaches to this, some of which are automated while others require human intervention.

Laser scanners are typically positioned in a room where they rapidly sample the distance between the laser device itself and the surfaces of objects in the room to create a point cloud, which is then processed to create a 3D model. To scan an entire room effectively, it is usually necessary to takes samples with the scanning device placed in a number of different locations in the room.

Photogrammetry and laser scanning techniques can be used both to scan individual objects and entire environments.

In general, the most efficient manner in which to construct a 3D model of a control room for design purposes is for a human to create models using a CAD or 3D modeling package.

USING VR IN THE CONTROL ROOM DESIGN PROCESS

VR should be viewed as a support tool during the design work, i.e., during the preliminary and detailed design, when we consider development of a control room and/or control room upgrading. To ensure usability, human factors and operational experience input must be included at an early stage in the system design. A design process can be described in many different ways, but the description in Figure 3.2a is believed to cover what selected standards[3,4,8,10,11,13] recommend for control rooms. For continuity in the development and best result, the use of a VR model should start as early in the design process as possible. Development should be an integrated part of all analysis and design phases described below.

Projects differ in their nature, and based on the job that has to be performed, the design team needs to be gathered

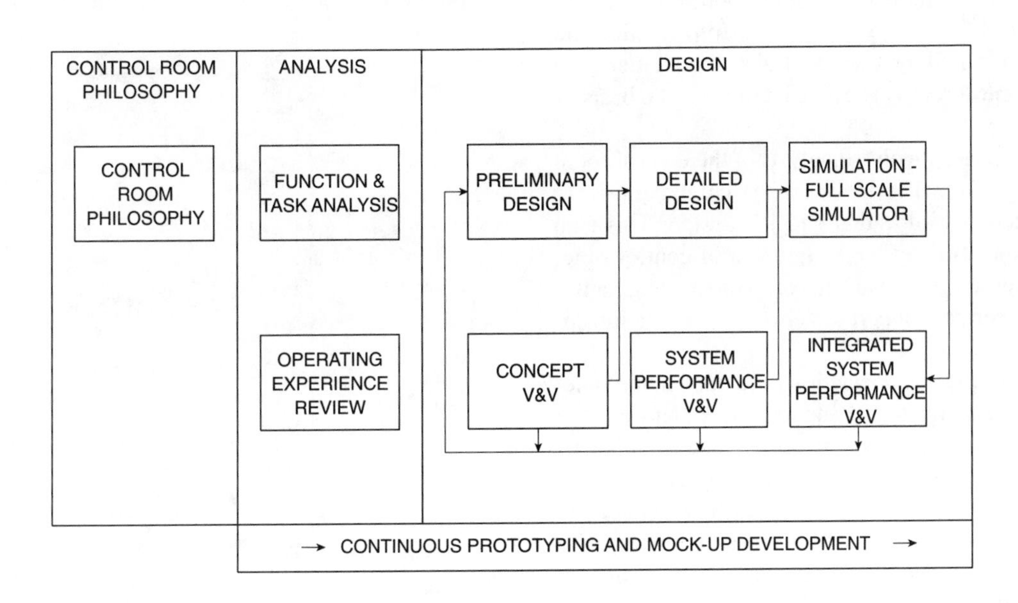

FIG. 3.2a
Outline of the human-centered design process, describing different stages of the design. The use of VR should start early in the analysis phase and continue through the whole design process.

to fulfill the needs of the project. The design team will often encompass the following professions: design engineers, system engineers, operations staff, and human factors staff. Composing the design team, one should have in mind the availability of all team members through the whole design process.

VR is a powerful tool for visualization, and good use of VR will allow the multidisciplinary design team to communicate and assess design ideas more effectively. All members of the design team possess different expertise and tend to follow different perspectives to identify an optimal solution.

The real experts on how to operate the plant are the operators who will have their daily work in the control room. The importance of good communication between the operators and the other members of the design team cannot be overestimated. For refurbishment of existing systems it is recommended that existing operators should be taken into the design team, while for new systems, both experienced and future users should participate.

The operators are not trained to express their design solutions with traditional CAD tools. Using VR tools enables operators to be part of the design process by simpler means. Good VR support software allows the operator to design a control room by clicking and dragging 3D objects, such as instrument panels, large screens, video display units (VDUs), workplace furniture, and other control room equipment. This way the cross-disciplinary team can work on a common reference to the design, which facilitates the experienced end users contribution to early fault detection and recovery.

The design process begins with a basic concept or idea, e.g., a design philosophy. The design philosophy should address all functions of the control room, i.e., all activity that is expected to take place within this environment, such as the roles of the operators, staffing size and responsibilities, control room functions, control room layout, instrumentation and control (I&C), and HSI process control needs. Experience from existing control rooms is expected to form the basis of the philosophy.

The analysis phase involves analysis of the control room functional and performance requirements culminating in a preliminary functional allocation and job design. This step usually requires analysis of the assignment of control functions to personnel (e.g., manual control), systems (e.g., automatic control), operator support systems, or combinations of the above.[2,3,10]

On the basis of the documentation developed in the functional analysis, the design team conducts a task analysis. The purpose of this analysis is to identify the detailed components of human tasks and their characteristic measures.[2,4,13] The task analysis helps to define the HSI requirements for accomplishing operator tasks. The analysis forms the basis for a number of decisions regarding whether combinations of anticipated equipment can fulfill the system performance requirements. At this stage of development, the control room exists as a concept and is described as a conceptual layout in the VR model. A set of requirements defined by the preceding analysis also exists.

The preliminary design phase covers the development of the conceptual design, including the drawing of the initial control room layout, furnishing designs, displays, and controls necessary to satisfy the needs identified in the preceding analysis phase.

Different approaches have to be taken in the conceptual design depending on the starting point of the design work. Building the control room from scratch gives the opportunity to utilize a full user-centered design strategy.

New Control Room

Building the conceptual design for a new control room uses the operator as the focal point for the new design. The purpose of this phase is not detailed design, but ensuring that all necessary equipment can be placed into the control room in a way that satisfies the functional requirements.

Starting with an empty space, the control room can be built around the operator in accordance with the requirements now specified. Figures 3.2c, 3.2d, and 3.2e show examples of how the VR model starts out as an empty space and how new components are added from predefined libraries supported by the equipment vendors (Figure 3.2b). In this approach, the conceptual design is not done in CAD, but the drawings can be built using VR software and later converted to CAD.

This example describes the development of a conceptual design for a nuclear power plant control room. In the nuclear power plant, there are normally three or four people located in the central control room. This example uses a shift supervisor who has the overall responsibility for all activities related to the control room, a reactor operator responsible for controlling the reactor, and a turbine operator who controls the turbines and generation of electricity.

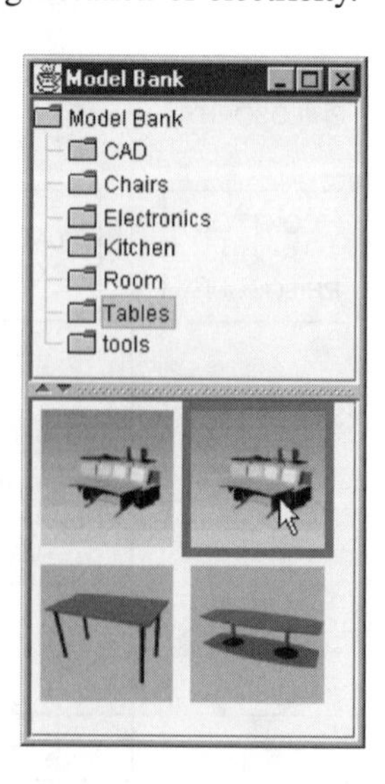

FIG. 3.2b
Library of components. A model bank or library of control room components, such as tables, chairs, VDUs, and control room equipment is an agile tool for starting the design.

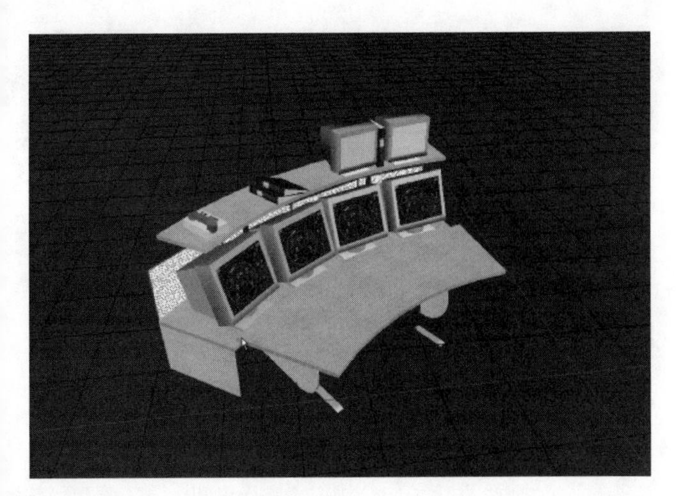

FIG. 3.2c
First one operator control station is moved into the design space from the library.

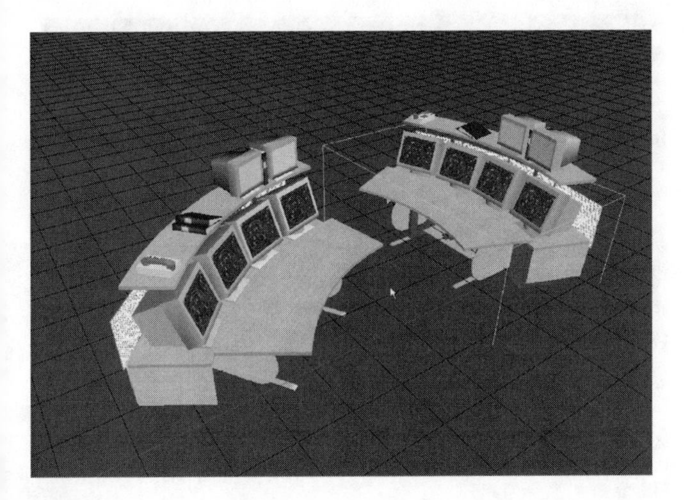

FIG. 3.2d
The second operator station is added and rotated to an approximate position.

After moving the initial three workstations into the model, the sight lines from the shift supervisor workplace (the nearest of the three workstations shown in Figure 3.2e) can be visualized and used for guidance in the further location of large screen monitors or control panels (Figures 3.2c, 3.2d, and 3.2e). The design team proceeds by adding objects to the control room and arranging them to construct a preferred layout.

Analyses and design team inputs can be considered within the context of operational crew information exchange, in order to support rapid changes in the design.

Refurbished Control Room

In this situation, the physical layout of the existing room is the limiting factor, which has to be taken into consideration. As described above, a number of techniques can be used to

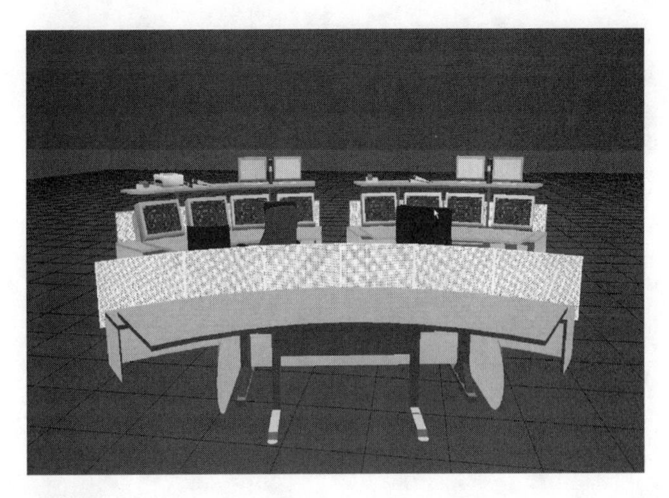

FIG. 3.2e
View from the shift supervisor's desk, at an early design phase.

convert the control room parts not under revision into a format usable for VR software packages. As shown in Figures 3.2f, 3.2g, and 3.2h, the new equipment is then fitted into the model of the existing control room. If the final V&V requires simulation (integration with a simulator) and interactivity with the model, the designer should consider a full remodeling of the instrumentation kept from the old control room. On the other hand, if simulation is not required, textures from the old control room can be used to represent the instrumentation. In the example shown in Figures 3.2f, 3.2g, and 3.2h, the walls and stand-up panels are converted from CAD drawings and the instrumentation is represented by photographs of the old panels.

V&V of the Preliminary Design

The proposed design should be evaluated on the basis of specifications, and against the applicable functional requirements, design requirements, and design criteria.[3] When a VR model has been developed during the conceptual design phase, this will replace many of the drawings and drafts. The requirements from the analysis can then be evaluated against the VR model.

V&V of the functional design assesses the functions and relationships of personnel and automated equipment in controlling plant processes, and checks the allocation of the task responsibilities of control room operators. The VR model is used instead of physical and paper mockups for scenarios with walkthrough and talk-through techniques. Figure 3.2i shows an aerial view of a control room conceptual design suitable for V&V at this stage of development. Using a conference room with a good wall projection of the model makes a convenient scene for a team working where whole crews of representative end users can participate in the audit. In the VR model, manikins can represent the crew members and indicate their positions stepping through test scenarios.

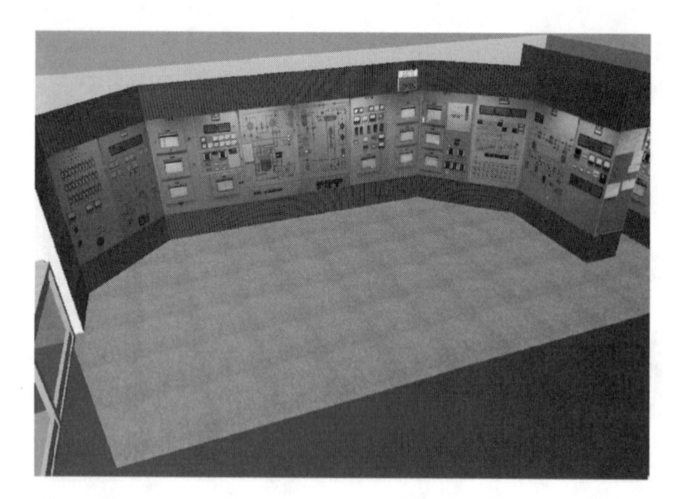

FIG. 3.2f
The starting point is an existing control room with fixed equipment intended to be used in the upgraded control room.

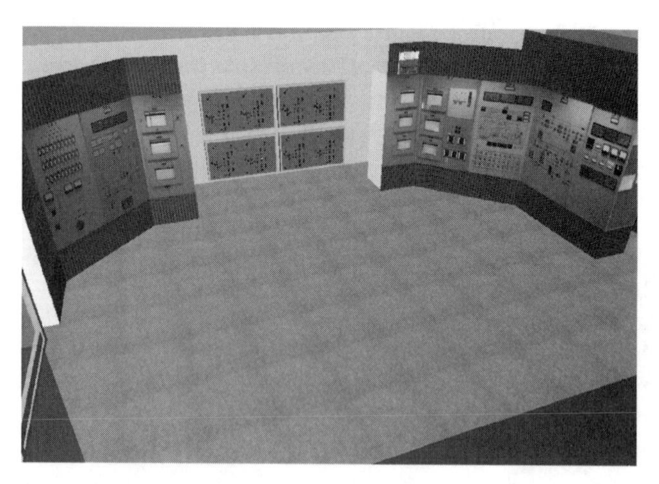

FIG. 3.2g
Large screen monitors added into the model.

FIG. 3.2h
Operator control stations added into the old control room.

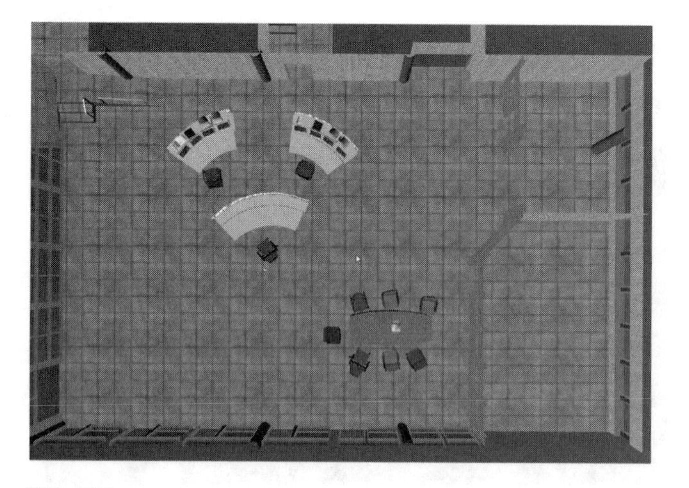

FIG. 3.2i
VR model suitable for the conceptual design standards compliance audit.

The functional review is not done on a very detailed process or I&C level; this is left for the detailed design phase. Emphasis should be put on reviewing and verifying that the functional requirements have been accommodated by design ideas and technologies that are feasible, acceptable, and in compliance with all applicable guidelines, standards, and policies. This step should be viewed as a major milestone that enables the subsequent detailed design to proceed with a minimum risk of major functional revisions and changes.[4,5]

The detailed design phase results in a detailed design specification necessary for the construction or procurement of the control room, its content, operational interfaces, and environmental facilities.

Refinement of the VR model from earlier phases takes place with the aim of producing a final decision on the layout.[6] In this phase, the control panel layout will be specified and designed. If the control room has conventional control panels, the modeling of these panels should be done in VR. If the intention is to use the VR model for the final validation (see simulation and final validation below), modeling should be done with full functionality for instruments intended used in validation scenarios. A good library of components and software support for creating control panel layouts and making connections to a simulator is a necessity for doing this work (Figures 3.2j and 3.2k).

The final control room layout can be verified against guidelines using the VR model.[6–9,15] Workstation configuration, control room configuration, environment, panel layout, and panel labeling are categories of guidelines suitable for this technique. However, the techniques for utilization of VR technology in guideline verification should be handled with care. Satisfactory verification of guidelines can be achieved only if the VR model and software provides good support for the evaluator, i.e., specially developed software tools must be present to help the verification process.[12] Tools for measurement of distance and angles like the tools shown in Figures 3.2l

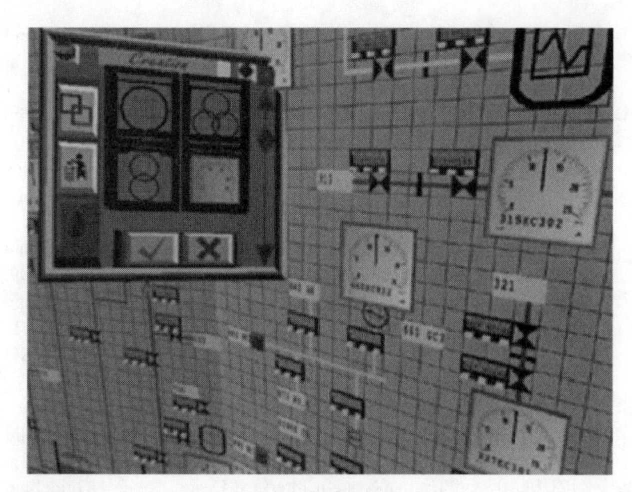

FIG. 3.2j
Building control panels using a VR design tool with a library of instruments.

FIG. 3.2l
Verification of labeling text size guidelines in a VR model. The length of the line drawn from the manikin's head to the control panel (2.62 m) is displayed in the upper right part of the figure.

FIG. 3.2k
The process of arrangement and layout decisions.

FIG. 3.2m
Verification of a stand-up panel with a tool for displaying the maximum allowed viewing angles.

and 3.2m are examples of such tools speeding and making the verification more reliable.

The level of detail in the modeling determines whether or not the specific panel layout design, label content, label lettering and design can be verified using the VR model. Textures are usually not sufficient for this purpose. To be able to judge panel labels, the text size and layout must be correctly modeled in detail or paper specifications should be used in combination with the model.

By using specifications from the lighting manufacturers, software can be used to prototype, analyze, and present how a workstation or an object will appear when specific materials and lighting conditions are used in the VR model. Levels of illumination can the read from the calculated illumination model and verified against the guidelines for different work areas in the control room.

The main point of continuous development of the VR model through the whole design process is to detect and correct problems as early as possible. In line with this principle the guidelines verification should not be left to the last stage. Using the model for verification of guidelines should be an iterative process and should take place as early as feasible. Output from the verification should go back into the revised design specifications. Making the most of the experienced plant operators and human factors staff in the design team is important in this process.[1,4,14]

The final design should be documented in a design description that includes the requirements for verification that the "as-built" design is the design resulting from the design process V&V evaluations.[10] In other words, all documents, snapshots, and comments from use of the VR model should be a part of the final design documentation.

Further, the VR model can be used in guideline verification of the final design. If the control room is a refurbished conventional control room, the VR model can also be connected to a simulator and used as replacement of conventional control boards. This way, validation scenarios and training can be run with the control room crew members performing their operations in the VR model. However, most industries do not use or require a full-scale simulator of their control rooms, and the VR model will therefore in most cases be used only for walkthrough and talk-through of validation scenarios.

On the other hand, if the control room is a new and fully computerized control room, the gain from using a VR model for full-scale simulation is more limited. If the new control room is built using standard computer equipment, most of the validation can be done from the VDUs that will comprise the final system.

CONCLUSION

Current VR technology offers the means for a complete approach for the design of control rooms. Using VR as a tool for testing control room concepts has several advantages. The greatest advantage is perhaps that the technique supports V&V in early phases of the design process, contributing to early fault detection and recovery, thus saving time and money in the later design phases. Although current standards for the design process do not take advantage of this technique, future standards, it is hoped, will integrate and recommend the use of VR in refurbishment and conceptual design of new control rooms.

References

1. Collier, S.G. and Green, M., "Verification and Validation of Human Factors Issues in Control Room Design and Upgrades (HWR-598)," Halden, Norway: OECD Halden Reactor Project, 1999.
2. Green, M., Collier, S.G., Sebok, A., Morisseu, D., Seim, L.A., and Skriver, J., "A Method for Reviewing Human Factors in Control Centre Design," Oslo: Norwegian Petroleum Directorate, 2000.
3. IEC, "Nuclear Power Plants—Main Control Room—Verification and Validation of Design," International Standard 1771 (1995-12), Geneva: International Electrotechnical Commission, 1995.
4. ISO, "Ergonomic Design of Control Centres—Part 1: Principles for the Design of Control Centres," ISO 11064-1:2000, Brussels: ISO, 2000.
5. ISO, "Ergonomic Design of Control Centres—Part 2: Principles for the Arrangement of Control Suites," ISO 11064-2:2000, Brussels: CEN, Management Centre, 2000.
6. ISO, "Ergonomic Design of Control Centres—Part 3: Control Room Layout," ISO 11064-3:1999, Brussels: CEN, Management Centre, 1999.
7. ISO, "Information Technology—Computer Graphics and Image Processing—The Virtual Reality Modeling Language—Part 1: Functional Specification and UTF-8 Encoding First Edition," ISO 14772-1:1997, Brussels: CEN, Management Centre, 1997.
8. NUREG-0700, "Human-System Interface Design Review Guideline," NUREG-0700, Rev. 1, Vol. 1, Washington, D.C.: The NRC Public Document Room, 1996.
9. NUREG-0700, "Human-System Interface Design Review Guideline: Review Software and User's Guide," NUREG-0700, Rev. 1, Vol. 3, Washington, D.C.: The NRC Public Document Room, 1996.
10. NUREG-0711, "Human Factors Engineering Program Review Model," NUREG-0711, Washington, D.C.: The NRC Public Document Room, 1994.
11. Stubler, W.F., O'Hara, J.M., Higgins, J.C., and Kramer, J., "Human Systems Interface and Plant Modernization Process: Technical Basis and Human Factors Guidance," NUREG/CR-6637, Washington, D.C.: U.S. Nuclear Regulatory Commission, Office of Nuclear Regulatory Research, 2000.
12. Drøivoldsmo, A., Nystad, E., and Helgar, S., "Virtual Reality Verification of Workplace Design Guidelines for the Process Plant Control Room," HWR-633, Halden, Norway: OECD Halden Reactor Project, 2001.
13. IEC, "Design for Control Rooms of Nuclear Power Plants," International Standard 964 (1989-03), Geneva: International Electrotechnical Commission, 1989.
14. Louka, M., Holmstrøm, C., and Øwre, F., "Human Factors Engineering and Control Room Design Using a Virtual Reality Based Tool for Design and Testing," IFE/HR/E-98/031, Halden, Norway: OECD Halden Reactor Project, 1998.
15. ANSI/HFS, "American National Standard for Human Factors Engineering of Visual Display Terminal Workstations," ANSI/HFS 100-1998, Santa Monica, CA: Human Factors Society, 1988.

3.3 Upgrading the Control Room

B. J. GEDDES

Cost: Workstations range from $10K to $25K, depending on the number of flat panel displays, video display units, computers, and input devices. Where new screens replace analog instruments in a conventional panel, cost is typically $5K per screen. These costs do not include hardware/software integration. See Section 1.10 for further information on estimating control system costs.

Partial List of Suppliers: ABB; Barco Projection Systems; Clarity Visual Systems; Emerson Process Management; Framatome ANP; General Electric; Invensys; Raytheon ELCON Digital Display Group; Rockwell Automation; Siemens Energy & Automation; Westinghouse Electric Company

As process plants age, owners must consider a variety of asset life cycle management issues, including instrument and control (I&C) system obsolescence. Plants constructed prior to the early 1970s were predominantly supplied with analog instruments and controls, either electronic or pneumatic.

It is likely that the vendors who supplied these original analog systems no longer provide parts or services to keep them running. As computer chips and information systems became available through the advent of the information age, instrument and control system vendors found innovative ways to apply this digital technology across their product lines. Systems and components are now smaller, more reliable, and fault tolerant, and they can perform multiple functions that were previously spread across multiple, discrete components.

Today, process engineers are presented with a variety of digital components and systems ranging from transmitters, switches, relays, breakers, controllers, recorders, indicators, video displays, distributed control systems, and fieldbuses and networks to connect them all together. The control system vendor may supply any or all of these new components and systems as replacements for the original equipment.

Digital information and control technology offers a variety of entry points in the analog process plant. The entire plant can be upgraded at once in one outage, or the owner and designer can choose to upgrade one system or component at a time. The technology is flexible enough to consider any upgrade path, which almost always has an effect on the control room.

The designer has two basic choices:

1. Building a New Control Room—If the plant is still operating, but has excessive cost or obsolescence problems in the I&C system or in the control room, then a new control center (building) may be economically justified. A new control center will be built as close as possible to the old control room or control center, so operations and maintenance tasks that require proximity to plant equipment can be maintained. This option also has the advantage of allowing construction to proceed while the plant is operating.
2. Upgrading the Control Room as Required (Phased Approach)—If the plant can still meet its business goals safely and effectively, and obsolescence is not yet a significant safety, reliability, or cost issue, then the designer can select an upgrade path for the plant on a system-by-system basis that accounts for time-to-obsolescence. System upgrades can be performed while the plant is in production, or during routine maintenance outages. Over time, the control room will have the same basic layout as that supplied during original construction, but the panels and furniture will be configured with modern technology where it is needed. Original indications and controls that are not obsolete and can still be easily maintained may remain as-is, but they will be reintegrated in a hybrid analog/digital design with modern human machine interfaces.

Designers must consider the impact on the control room when applying new technology in any part of the plant. Even a seemingly innocuous plant change can impact the control room. For example, a digital, fieldbus-compatible process transmitter may offer a local upgrade path for a given plant system. But this transmitter now supplies much more information than the original 4–20 mA DC transmitter, and it should be applied in the context of a total plant architecture that will impact the control room as it evolves.

It is critical to see beyond the immediate needs of the plant. Plant owners will naturally tend to apply funding and resources to the I&C systems that cause the most trouble and stop until the next acute need arises. Acute system problems usually have a significant impact on plant safety or production, and the designer owes a more proactive and comprehensive approach to the owners.

BASELINE EVALUATION

The goal of the baseline evaluation is to ensure the upgraded control room still functions well within the original plant. The functional allocations between humans and machines should be well understood and documented. Operations and maintenance tasks should be understood and documented, and information zones should be mapped and correlated with these tasks. More importantly, the baseline will also ensure that the control room operating staff's sense of how the plant is designed and operated will be maintained with the modified or new control room.

The designer must first understand how and why the original control room is designed the way it is, then understand the basic operating philosophy of the control room staff. The operating philosophy can be defined as the set of rules and procedures matched with the set of roles and responsibilities of operating staff resources that are applied during all modes of operation, including normal, abnormal, and emergency operations. Operating staff (control room supervisors, control room operators, and system operators) roles and responsibilities in an existing plant are based on the functional allocations between humans and machines, which then determine the tasks assigned to humans in all phases of plant operations.

The baseline evaluation should begin with an evaluation of the basic design of the existing plant information and control systems by reviewing loop diagrams, schematics, wiring diagrams, and panel layout/cutout drawings. Figure 3.3a represents an architectural perspective for the analog I&C systems and components in a typical older process plant. Each set of systems and components can be categorized by its basic function.

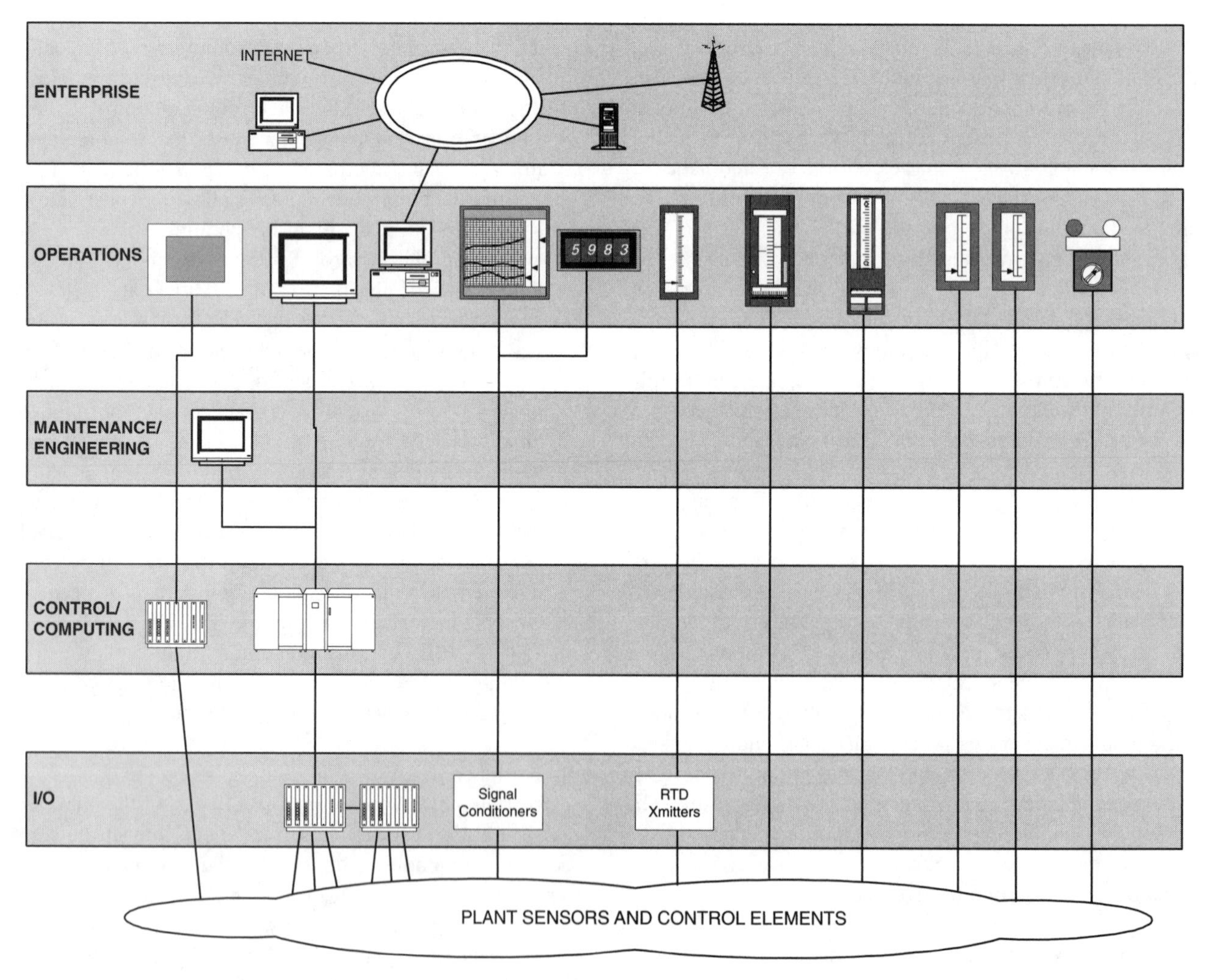

FIG. 3.3a
Analog I&C architecture.

The cloud at the bottom of Figure 3.3a comprises the basic sensors such as pressure, temperature, flow, level, and position switches, various measurement elements, and analog process transmitters. It also represents final control elements such as motor control centers, motor-operated valves, air-operated valves, dampers, and machines. This population of sensors and control elements usually numbers in the thousands for large process plants.

The input/output (I/O) layer just above the cloud at the bottom of Figure 3.3a represents the basic I/O systems that may be used by what few digital platforms are employed, such as a mainframe plant process computer.

The mainframe computer, in the control/computing layer, typically runs a real-time database program for those plant variables that need to be stored and retrieved in group displays or reports, or used as input data for complex performance calculations. The number of data points assigned to the plant process computer may number in the thousands, but its role is limited to information-only applications.

The maintenance layer represents any interfaces made available to maintenance and engineering. The only interfaces that are available to maintenance and engineering in the analog plant are confined to the one or two computers that may be used in specialized applications.

The operations layer represents the devices used by the plant operator, and it can be seen that the majority of the devices in Figure 3.3a are in this layer. These devices include trend recorders, digital indicators, vertical and horizontal meters, a variety of hand switches, single loop controllers, lamps, and annunciator light boxes. By inspection, most of the analog plant sensors in Figure 3.3a terminate directly on these devices used by the operators. Any maintenance required on these analog loops will usually take the affected devices out of service and result in a negative impact on the operation of the plant due to disruptions and loss of information and control.

Finally, the enterprise layer illustrates the basic plant or company enterprise where information technology (IT) systems are deployed. Operators may have a company PC at their disposal in the control room for retrieving enterprise information such as drawings or accounting data, or sending and receiving e-mails. This PC is only connected to the office local-area network (LAN) and is not used directly to operate or maintain the plant.

Changing any component in the lower layers of Figure 3.3a can impact the operations layer, and vice versa. For example, the plant may need to upgrade a set of plant sensors due to obsolescence. The signals from the old sensors were processed by the mainframe computer. The new sensors require a new software code to properly acquire and convert their signals to the engineering units required by the operator. The new software cannot run on the old mainframe computer, so a new computing platform is needed in the form of a server, which comes with a client software package for use on a PC or workstation in the control room. The server can also be integrated with a DCS (distributed control system). Now we have a complete system upgrade that affects every functional layer in Figure 3.3a, with significant control room impact.

FIG. 3.3b
Analog control room.

ORIGINAL DESIGN PERSPECTIVE

Figure 3.3b shows an older control room, typical of power plants and other process plants designed and built between the late 1960s and the early 1980s. The majority of the indications and controls are from original construction and are almost all analog instruments. Some plant systems have been upgraded since original construction, but only on an as-needed, system-specific basis. No serious consideration has been given yet to a total control room modernization effort.

The operator interfaces are typically located on "stand-up" control panels and are used by control room operators during normal and emergency operations. Desks are provided for senior control room operators and control room supervisors, with telephone and radio equipment and terminals connected to a mainframe plant process computer. The panels are laid out by major plant functions such as electrical systems, cooling water systems, major machine controls (e.g., turbine/generators or compressors), reactor controls, and emergency systems.

On each panel, indications and controls are grouped by specific plant systems or functions. For example, Figure 3.3c shows a feedwater control system panel from the same control room shown in Figure 3.3b. Annunciator windows are on the top plane of the panel. Feedwater heater level indications are grouped on the left-hand side of the vertical plane. Steam generator temperature and level indications are grouped on the right-hand side. Across the bottom planes are indications and controls associated with the control elements for feed pump turbines, condensate pumps, and various valve position indications and controllers. Groups of instruments and controls are bordered or separated by a thick black demarcation line called "zone banding."

FIG. 3.3c
Analog control panel.

Each group of instruments and controls can be associated with a set of operator tasks for each mode of plant operation (start-up, normal, shutdown, and emergency operations). Written procedures describe operator actions required for responses to various alarms or during various controlled transients. Operators are trained in a full-scale simulator for various operating modes and equipment failure scenarios, and they are trained to perform periodic visual scans, or surveillances, of each group of indications and controls either close up or from a "total control room" perspective. The baseline evaluation should capture all of these operator tasks and their associations with each group of indications and controls.

INFORMATION ZONES AND TASK ANALYSIS

Perhaps the most significant design aspect of an older, analog control room is that all plant indications and controls are in discrete locations on the panels. Most plant variables are only readable from one instrument in a "fixed position" on a panel, and plant system variables are grouped in "information zones" where operators are trained to look for patterns in the indications to detect, diagnose, and maintain or mitigate various plant conditions. The information zone human–machine interface (HMI) design concept is a very strong paradigm for operators, and should not be significantly modified or corrupted unless a new control room is built, complete with a new task analysis, operating procedures, and operator retraining and qualification program. In the phased approach, information zones are kept intact as each zone is upgraded over time.

Each information zone is associated with a set of operator tasks, which should be studied and evaluated using simulators or mockups. Tasks will vary with modes of plant operation. Most operator tasks in a continuous process plant will be associated with surveillance of the process systems. During routine surveillances, an operator will visually scan the control panels looking for patterns in the indications and alarms that indicate plant systems are operating within specifications. Some manual manipulations of multichannel meters or displays may be required.

In an upset mode, such as a sensor or equipment failure, or a trip of a major plant machine or the plant itself, an operator's task loading will shift dramatically, and will be focused on key plant variables and controls. Alarms will be prioritized by the operator per procedures and training plans, and the event will be diagnosed and mitigated through the cognitive strengths of the human operator. The HMI will be designed to accommodate this shift in task loading with an emphasis on decision making.

Once the task evaluation is captured, operating experience should be evaluated. Operating logs and event reports should be reviewed for positive or negative trends in plant or human performance. Operators should be interviewed to see if there are any HMI issues that need to be improved or corrected. Are there any recurring situations where operators are tasked with too many simultaneous inputs requiring concurrent actions? How do operators prioritize their actions under these scenarios? Are there any operator tasks where there is a deficiency in available information, such as resolution, timeliness, or readability? Are there any indications and controls that are too difficult to read or understand, or are simply not available under certain conditions where they are required for an operator task?

If an operating experience review indicates deficiencies, event scenario walkthroughs should be conducted in a full-scale simulator if one is available. If a full-scale simulator is not available, a full-scale mockup of the control room should be built. At first, a mockup of the old control room might seem like a waste of time, but it can support an engineering analysis of the strengths and weaknesses of the old control room without disturbing plant operations, and it can support experimentation and evaluation of new design concepts as they are integrated in the original control room if the phased approach is used.

Figure 3.3d shows a 1/8 scale mockup of the same control system panel shown in Figure 3.3c. It was made by reviewing photographs and layout and cutout drawings of the panel, then making a scale template of each plane on a personal computer using Visio™, a commercial graphics software package. Objects for each type of control panel device (recorders, indicators, controllers, etc.) were made and stored in a Visio stencil, then applied to the scale template using a "drag and drop" technique. Each panel was printed on a color printer or plotter, spray-glued to a piece of cardboard, then cut out and assembled using a hot glue gun. The total cost of the software and materials for this mockup was less than $100.

The engineering and technician time invested in building the object library, or stencil, and preparing this first panel was over 100 hours, but subsequent panels take less time and the library is now available for finishing a mockup of the entire control room. This mockup was prepared at full scale in the computer, then printed at 1/8 scale for this version.

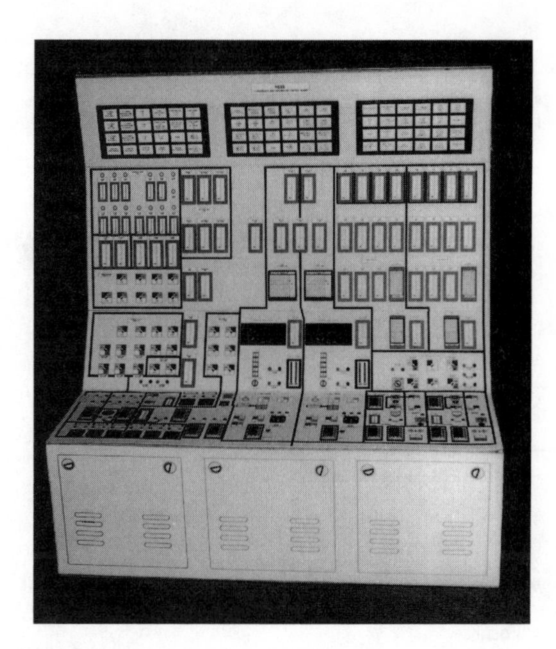

FIG. 3.3d
Panel mockup.

The same Visio files can be printed at full scale on an engineering plotter, and applied to a full-scale plywood mockup, which is useful for walkthroughs with operators.

Finally, the baseline evaluation should examine all of the operating and maintenance costs associated with the original I&C systems in the plant. If an older, analog plant is expected to have a long operating life, many if not all of the I&C systems will need to be replaced. These replacement activities will be necessary to maintain continued plant operation, to take advantage of current technology, to improve plant availability, and to facilitate further reductions in operating and maintenance costs.

The most significant economic driver is usually obsolescence, which can be changed from a liability to an asset in considering control room upgrades. When obsolescence is considered in aggregate, a strong economic case for an integrated, modern control room can be made. A system should be considered obsolete when any of the following criteria are met.

1. The system will not meet the operational demands required of it during all phases of plant operation, including normal, abnormal, offline, online, and emergency situations.
2. The parts manufacturer has formally announced that it will no longer directly support the system or component, or will require a significant increase in component cost to continue support.
3. The system or component will not interface with new components that are required in other systems. It is virtually assured that if the operational life of a process plant is extended, all system components will require some form of major component replacement.
4. The system or component cannot support plant operations that are desired by the business plans.

Once the baseline evaluation is completed, the plant owner and the designer will have an accurate assessment of the functional, performance, and economic state of the plant I&C systems and, in particular, the control room. Goals and objectives can be clearly described, and upgrade options can be estimated and undergo a total cost–benefit analysis.

BUILDING A NEW CONTROL ROOM

Building a new control room all at once allows the full potential of modern I&C systems to be immediately realized in an older process plant. A new control room in this context means the complete retirement and replacement of the original analog panels, I&C systems and HMI components with new digital systems, using a DCS or PLC (programmable logic controller) platform. The replacement systems may be installed in the original room, or may be installed in a new building or otherwise spare plant location.

If the plant can no longer meet its business goals safely and effectively because of control system degradation and obsolescence, it probably needs a major overhaul, and an entirely new control room may be in order. If a plant is being overhauled for a major capital upgrade (new reactors, boilers, compressors, turbine generators, or other major machinery), it will probably include a new process or machine control system. The designer can engineer a new control house and demolish or retire the old one in place, or gut and rebuild the existing control room. The new control room should be implemented with a workstation/console arrangement designed to maximize safety, quality, reliability, and productivity with a heavy emphasis on human engineering.

The decision to build an entirely new control room all at once has profound implications for the entire process plant. The designer can demolish the old control room by removing all components and structures, including panels and furniture, leaving floor and ceiling slabs and four bare walls, with field cables pulled back or left coiled near their penetrations.

Using this approach, an analog control room is replaced with a fundamentally new, "glass control room," or one that consists of video display units (VDU) located on operator workstations or consoles and can also include one or more large displays. The control system cabinets will typically be located in a computer room or cable spreading room adjacent, above or below the new control room, with data communications interfaces serving the consoles and workstations.

The ISO 11064 standards (Parts 1 through 8) offer a sound methodology in the design of a new control room. Their emphasis is on the design of a new control room in the context of a new plant, but they also describe design processes that can be used in designing a new control room as a retrofit in an existing facility.

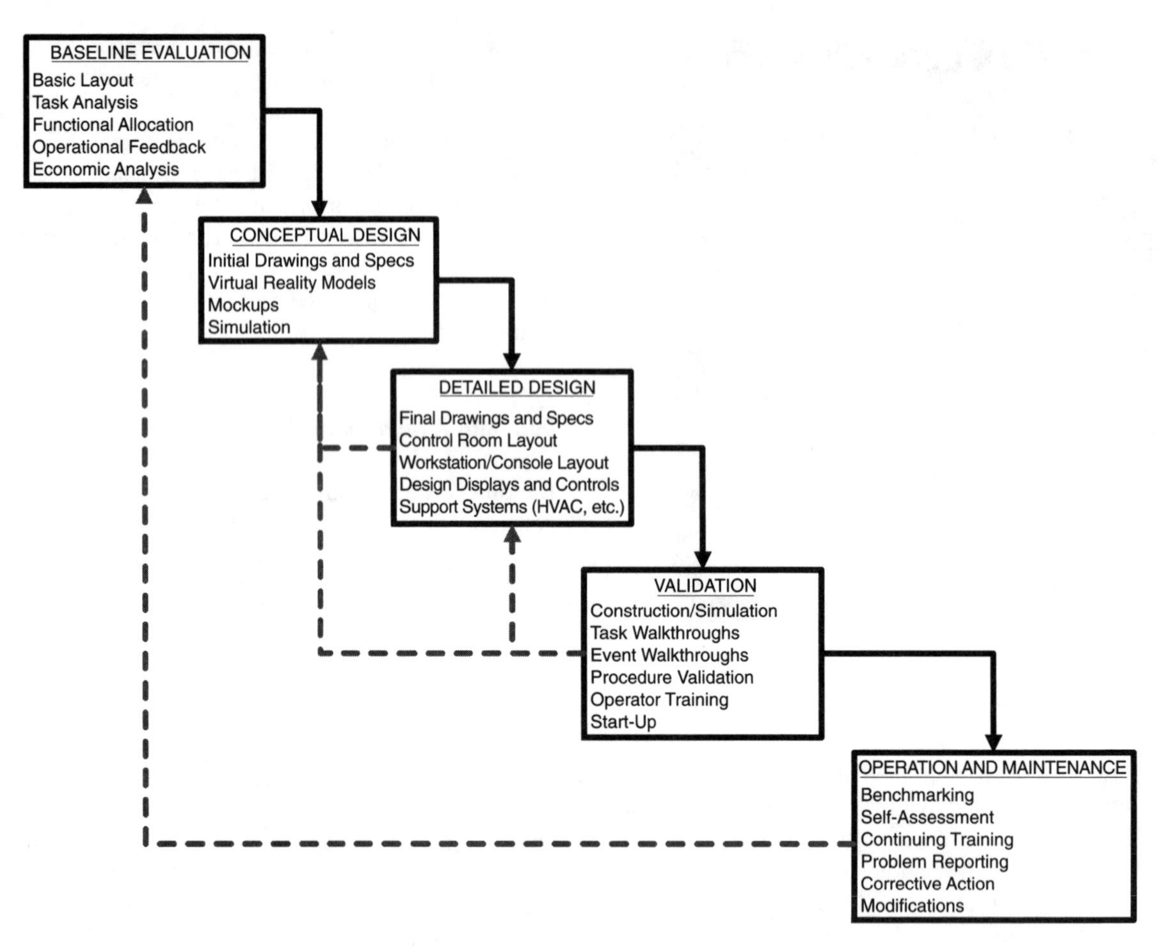

FIG. 3.3e
General design process for new control rooms.

Using a standard design methodology, as shown in Figure 3.3e, will improve safety, quality, and reliability when applying modern digital system technology. The roles and responsibilities of operators and machines can be modified and tasks and functions reallocated to achieve a higher performance standard. An increased reliance on automation can significantly alter the role of operators by reducing their task to vigilance over a constant process, while also greatly expanding their role during system upsets to one of corrective action through cognition, diagnosis, and appropriate manual responses. If a new control room is under consideration, particular attention has to be paid to ensuring that any changes in roles and responsibilities of machines and operators from the old control room to the new one continue to meet or exceed the goals and objectives outlined in the baseline evaluation. Without a careful design process, unintended functions can be created with potentially negative consequences.

Basic ergonomic principles should be applied in the design of control rooms. Ergonomic design principles include:

- Use a human-centered design approach. Design for basic physical limitations, but also emphasize human cognitive strengths such as perceptual, problem-solving, and decision-making abilities.
- Improve design through iteration. Repeat design evaluations until functional and performance goals are achieved.
- Start with the baseline evaluation performed on the original control room, and revise through each design iteration.
- Design error-tolerant systems. Allow for human error, and provide interlocks, alarms, and overrides where appropriate.
- Ensure user participation. The most knowledgeable asset on the design team can be the control room operator. Add system and process engineers, ergonomists, architects, and industrial designers to form an interdisciplinary design team.

TECHNOLOGY ADVANCES

Technology advancements in recent years have enabled a state-of-the-art approach to control room design. Compact workstations, large display panels, soft controls, fixed and selectable displays, advanced alarm systems, and computerized procedure systems greatly enhance safety, reliability, and efficiency.

A modern control room includes compact workstations, a safety console for plants that require one, and furniture and rest areas. A large display panel (LDP) can also be provided. Key features that support the ability of the operating crew to maintain efficient and safe plant operation include:

- Full-function workstations supporting direct plant control and monitoring by one or more system operators
- An identical workstation supporting normal monitoring and crew coordination functions of a control room supervisor and serving as a backup to the operator workstations
- An LDP providing overall plant operational and safety assessment
- A safety console (for those industries with safety requirements) providing control capability for all safety-related components independent of the workstations supporting safe plant shutdown even in the event of complete workstation failure

Advantages of this control room layout, shown in Figure 3.3f, include enhanced communication between operators, operational facilities for all expected crew members, good visibility of the LDP, ease of accommodating design and job allocation changes, and convenient access and egress routes. Layouts will vary depending on the number of systems in the process plant that are monitored and controlled in a main control room. If a site has multiple process plants that are independent islands, but still controlled from one central control room, then the control room layout may employ multiple sets of compact workstations, depending on operator and supervisory tasking.

The LDP is a wall-mounted overview display, which includes static and dynamic elements. High-resolution, tiled projection systems are available today with seamless borders between tiles. The fixed display section of the LDP provides continuous, parallel display of key alarm, component, system, and parameter information. This complements the workstation HMI with a spatially dedicated graphical depiction of the plant. A variable display section allows operators to display pertinent information selectively to support crew coordination. A sample LDP design is shown in Figure 3.3g.

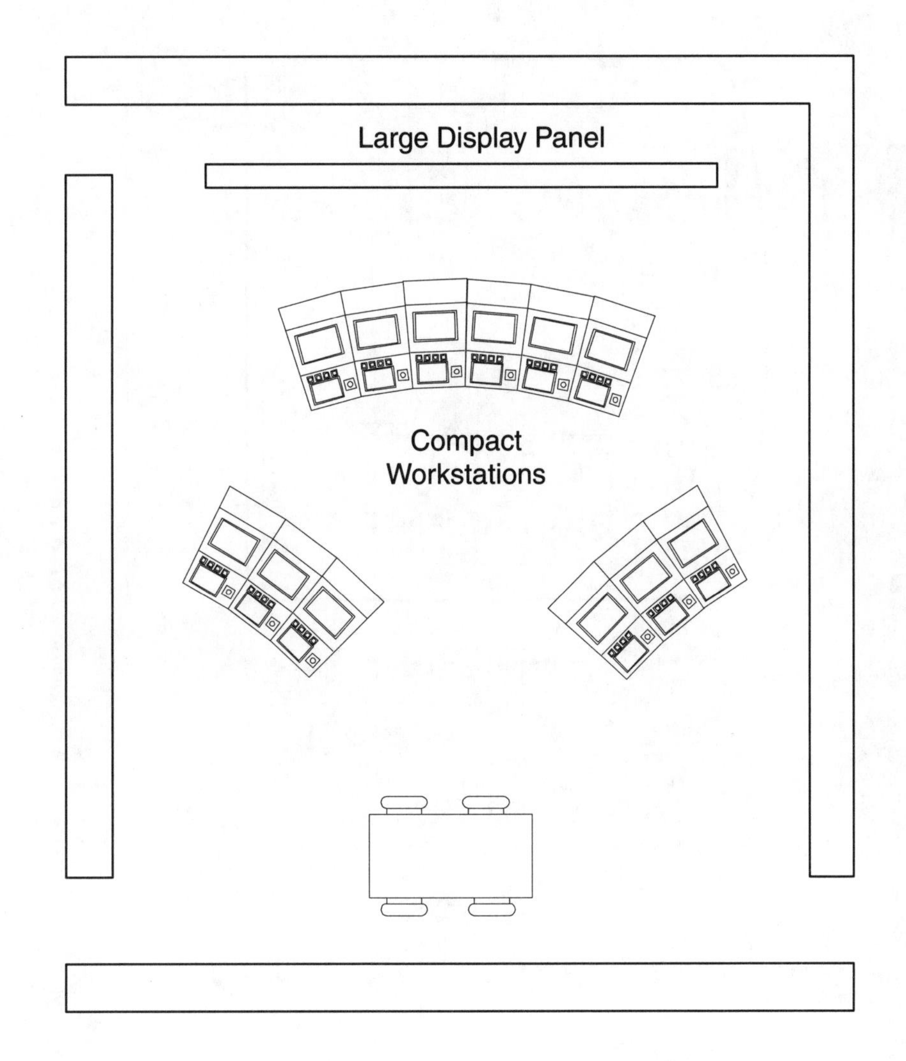

FIG. 3.3f
Sample control room layout.

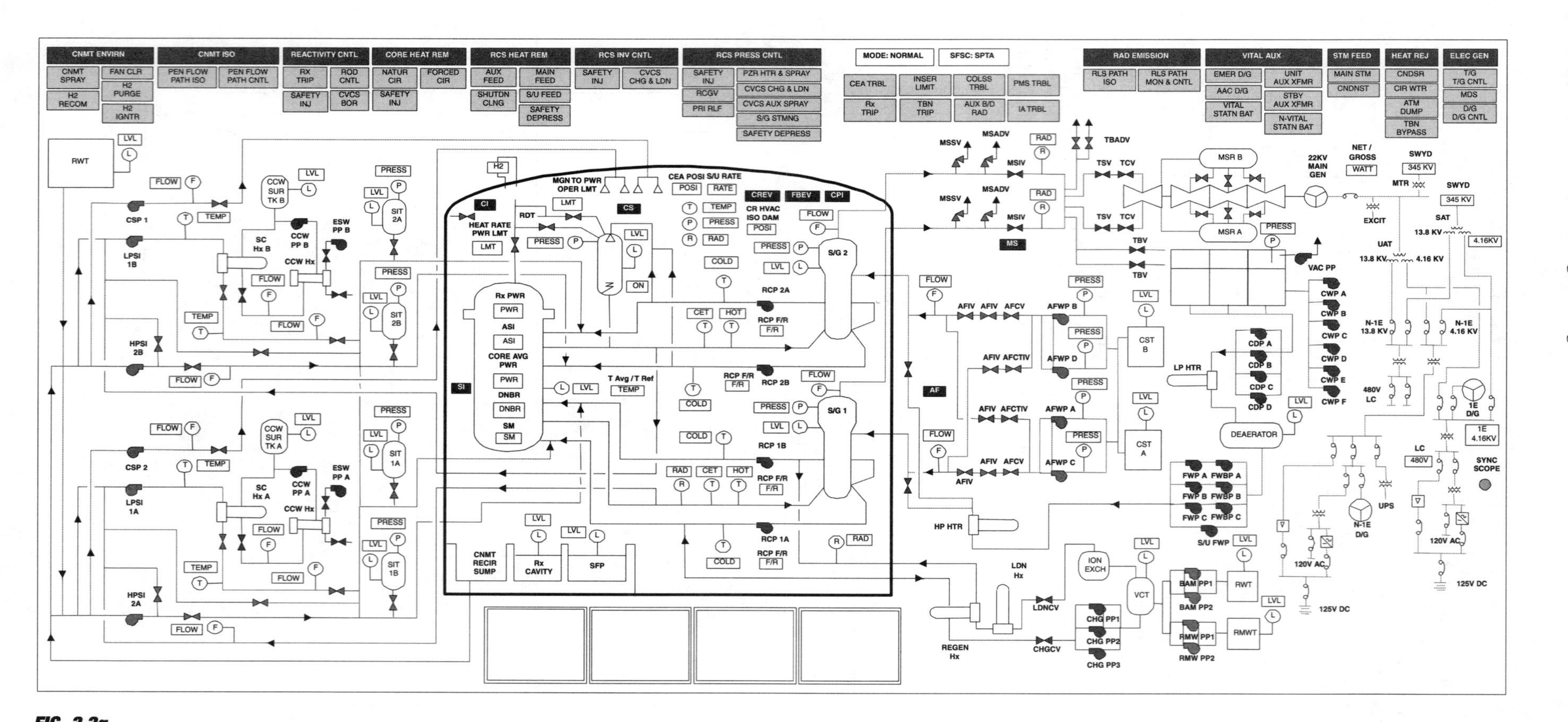

FIG. 3.3g
Large display panel. (Courtesy of Westinghouse–CENP & KEPRI.)

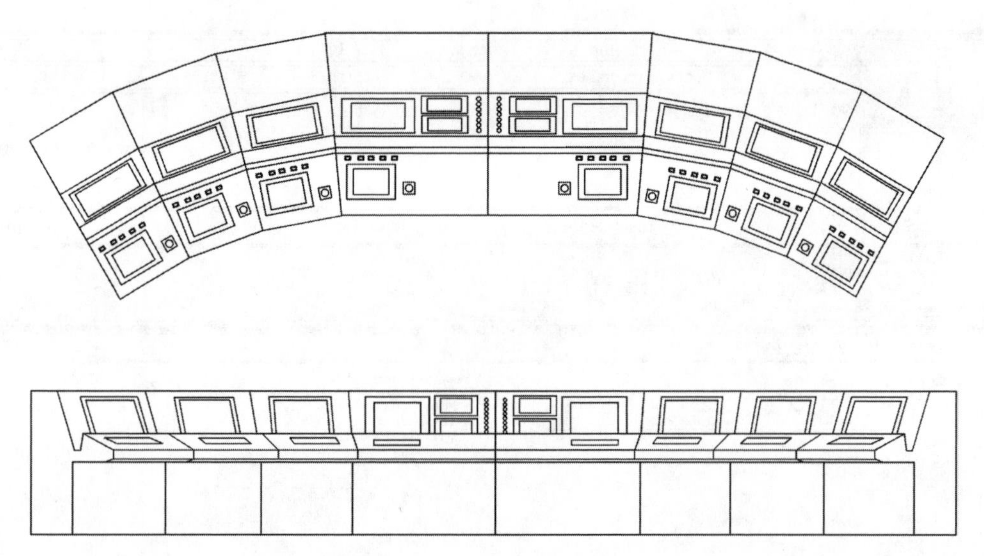

FIG. 3.3h
Typical workstation design. (Courtesy of Westinghouse–CENP & KEPRI.)

Each main control room workstation provides devices for access to all information and controls necessary for one person to monitor and control all processes associated with plant operation and safety. This includes both safety and nonsafety systems. A sample workstation layout is illustrated in Figure 3.3h. Each workstation contains the following:

- One alarm VDU with trackball user interface
- Multiple VDUs supporting process monitoring or electronic procedures with trackball user interface
- Multiple flat panel displays used as soft controllers for process and component control; each works in conjunction with one VDU, using a touch-sensitive user interface
- Dedicated, diverse push buttons for manual safety system actuation
- Laydown area for logs, drawings, backup paper procedures, etc.

The major advantages of the compact workstation approach are its (1) operational and design flexibility, (2) compactness and simplicity, (3) ability to accommodate changes cost-effectively, and (4) provision of an enhanced integrated environment for a computerized procedure system (CPS) and operator aids.

Workstations should allow simultaneous access to plant information through selectable displays on multiple VDUs per workstation. A wide variety of display formats should support system mimics, major plant functions or conditions, technical data sheets, trends and graphical information, and application program access. All are designed to support specific operator functions. Multiple methods should be provided for convenient access to the display set, including navigational access through menus, direct access through format chaining from other displays (or alarms and procedures), and a dedicated mechanism such as function buttons or voice entry. A major function of the VDU displays is to provide a soft control link allowing the operator quickly to select a component or process control on the soft controllers directly from display pages.

"Soft controls" should utilize flat panel displays to emulate the physical switches and manual/auto stations that populate conventional plant control panels. Use of software-based control allows a standard interface device to assume the role of numerous physical devices. This has the advantage of allowing operator access to all plant controls from a single workstation, design flexibility, the ability to accommodate changes easily, and simplification of hardware procurement and maintenance. Various design concepts can support control of multiple safety and nonsafety divisions from the same workstation device, while maintaining the single-failure-proof reliability of conventional channel separation and independence to support safety system requirements such as those called out in IEEE-603 (1998) "Standard Criteria for Safety Systems for Nuclear Power Generating Stations" and ISA 84.01 (1996) "Application of Safety Instrumented Systems for the Process Industries."

An advanced alarm system can be provided to improve the annunciation process, by incorporating methodologies that:

- Reduce the total number of alarms with which the operator must cope
- Distinguish between true process deviations and sensor failures
- Minimize the occurrence of "nuisance" alarms
- Prioritize the relative importance of alarms so the operator can focus on the most critical alarm conditions first while deferring less critical alarm conditions
- Determine the impact of alarms on plant operations and distinguish these from lower-level system alarms

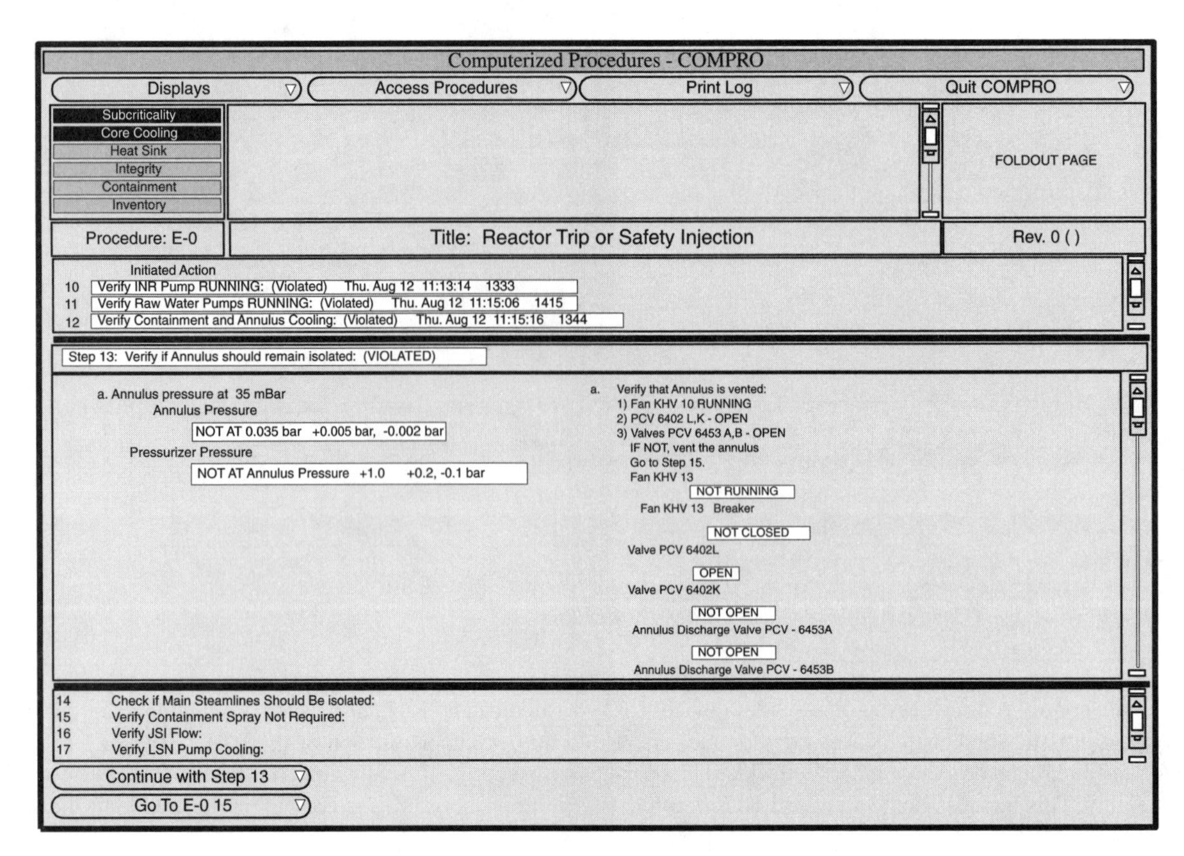

FIG. 3.3i
Computerized procedure system. (Courtesy of Westinghouse–CENP.)

Highest priority alarms, such as those for critical functions, are presented in fixed locations on the LDP. All alarms are presented in list form on a dedicated workstation VDU as well as through relevant locations in the VDU display hierarchy.

COMPUTERIZED PROCEDURE SYSTEMS

Computerized procedure systems (CPSs) are another significant technology development in modern control rooms, which take advantage of state-of-the-art technology to provide benefits in the human–machine and control room integration areas (Figure 3.3i).

The purposes of an online, data-driven, cathode ray tube (CRT)-based CPS are the following:

- To guide the user step by step through the procedures by monitoring the appropriate plant data, by processing the data and by identifying the recommended course of action.
- To provide the parallel information necessary to allow the user to assess other plant conditions that may require attention, such as, for example, notes, cautions, and foldout page items.

The computer monitors and evaluates large amounts of information quickly and efficiently. Procedure information is online and is updated essentially continuously. Hence, the operator becomes more vigilant because a large amount of procedurally required information is immediately available. The operator's mental loading is reduced because all required procedural information is displayed, including, for example, the status of the current high level step.

The human user's primary role in a CPS environment is to monitor the progression through the plant procedures while maintaining a clear picture of plant state, to take control actions on the control board when they are required and to watch for unsafe plant conditions. The user should retain both authority and responsibility for plant operation. A CPS is user paced, that is, the system should not advance to a procedure step, a note, caution, or foldout page item, or a procedure unless instructed to do so by the user.

A CPS can operate on a two-CRT workstation such that the required procedural information always appears on the first screen and supporting or supplementary procedural information appears on the second screen. Access to the supporting information should be through pull-down menus located along the top of the main screen.

The benefits of a CPS include:

- The computer and the operator complement each other for a more accurate implementation of the procedures, resulting in enhanced situation assessment by the operator.
- The system simultaneously monitors multiple plant parameters.
- The system brings all procedural information to one location.
- The system provides detailed record-keeping capability of the procedure execution.

PHASED APPROACH

A phased approach to upgrading the control room may be economically attractive. In this approach, it is important to recognize whatever design and operating constraints may be in place. The control room to be upgraded is assumed to be in service, and should continue to meet the business goals set by the owner.

The entire control room can be upgraded in phases, over time, according to whatever drivers are selected. The scope and schedule of each phase is optimized when it is consistent with the overall cost–benefit analysis of the entire upgrade, which may take years. If obsolescence issues are not acute, but are slowly accumulating, and the plant is running safely and meeting business objectives, then upgrades can be applied to individual systems or groups of systems while the plant is running or during routine maintenance outages.

If a phased approach is used, then a long-range strategic I&C modernization plan should be developed before any system or component upgrades can significantly impact the control room. Failure to do so will not have any impact on a particular system upgrade, but in the overall scheme, it can increase the total cost of the finished control room. The long-range plan will provide a conceptual design for the finished plant and control room, and it will provide a scope, schedule, and budget for each phase so that they are scheduled in the most cost-beneficial sequence. Each phase will account for the set of systems and components that should be upgraded, taking into account time-to-obsolescence and grouping of systems and components so duplication or rework on later phases is avoided. Obsolescence can be successfully applied in arguing for a comprehensive upgrade plan. Not all the original analog I&C systems will become acutely obsolete at the same time.

In the original analog control room, a unique treatment of indications and controls for each plant system was the norm. Panels were typically designed to support collections of meters, recorders, switches, and controllers grouped by plant system. It is important to avoid unique upgrades for each plant system. If unique digital solutions are applied over time, the result will be a muddled set of DCS, PLC, and other operator interfaces in the control room. All the operational goals of the plant upgrades will not only go unfulfilled but will worsen when realized in total.

Figure 3.3j illustrates the modernized plant. The functional layers are still the same, but there is a significant change in the technology applied in each layer. Figure 3.3j can be considered a hybrid digital/analog plant. The left side of this figure shows digital systems; the right side shows the original analog components. As more plant systems are upgraded, more of Figure 3.3j becomes digital. Hardware and software diversity may be applied for safety systems in each layer where required by regulatory agencies.

The key point in Figure 3.3j is that new I&C systems are expandable horizontally in this perspective, using a building-block approach that applies DCS or PLC technology linked by network communications technology, such as Ethernet, fieldbus, or any other standard that is supported by the manufacturer. This point is essential for implementing a cost-effective phased control room upgrade, where systems are expanded until obsolescence issues are resolved and modernization goals are realized.

The I/O layer employs subracks that can be field-mounted or mounted in cabinets or panels in the control room. The location of the new subracks depends on a number of factors, such as communication constraints (distance, bandwidth, number of allowable nodes) and the impact on cable costs. To optimize cable costs, the best place to mount I/O subracks is within reach of the field cables that were originally terminated on or near the old analog devices.

The control/computing layer employs DCS or PLC processors or a combination of both. Equipment selection should allow data archiving and computational servers to be added to this functional layer and integrated into the whole control and computing platform system. Hardware and software integration can be expensive and time-consuming, so it should be consistent with one set of standards.

Note that original control room meters can still be used with a new DCS/PLC platform using the new I/O subsystems. This design feature is useful when an entire process system is upgraded, where video display units cannot replace every system switch, light, meter, or annunciator. For example, a small-scale DCS can replace a turbine/generator electrohydraulic control (EHC) system with one or two main VDUs in the place of the original operator interface and still allow connections to original devices in other locations. If all original EHC interface devices in this example are scheduled for replacement with VDUs, the potential exists for other plant systems to be caught up in the EHC upgrade simply because the new VDUs take up more panel space than the original meters and switches.

If any special applications are required, such as a data archiving agent or any plant performance calculation packages, plan ahead and select a DCS/PLC hardware and software platform that can support the special application hardware and software. Pay close attention to database integration issues, because a central database is the strength of a DCS. If an original plant data acquisition system can be connected

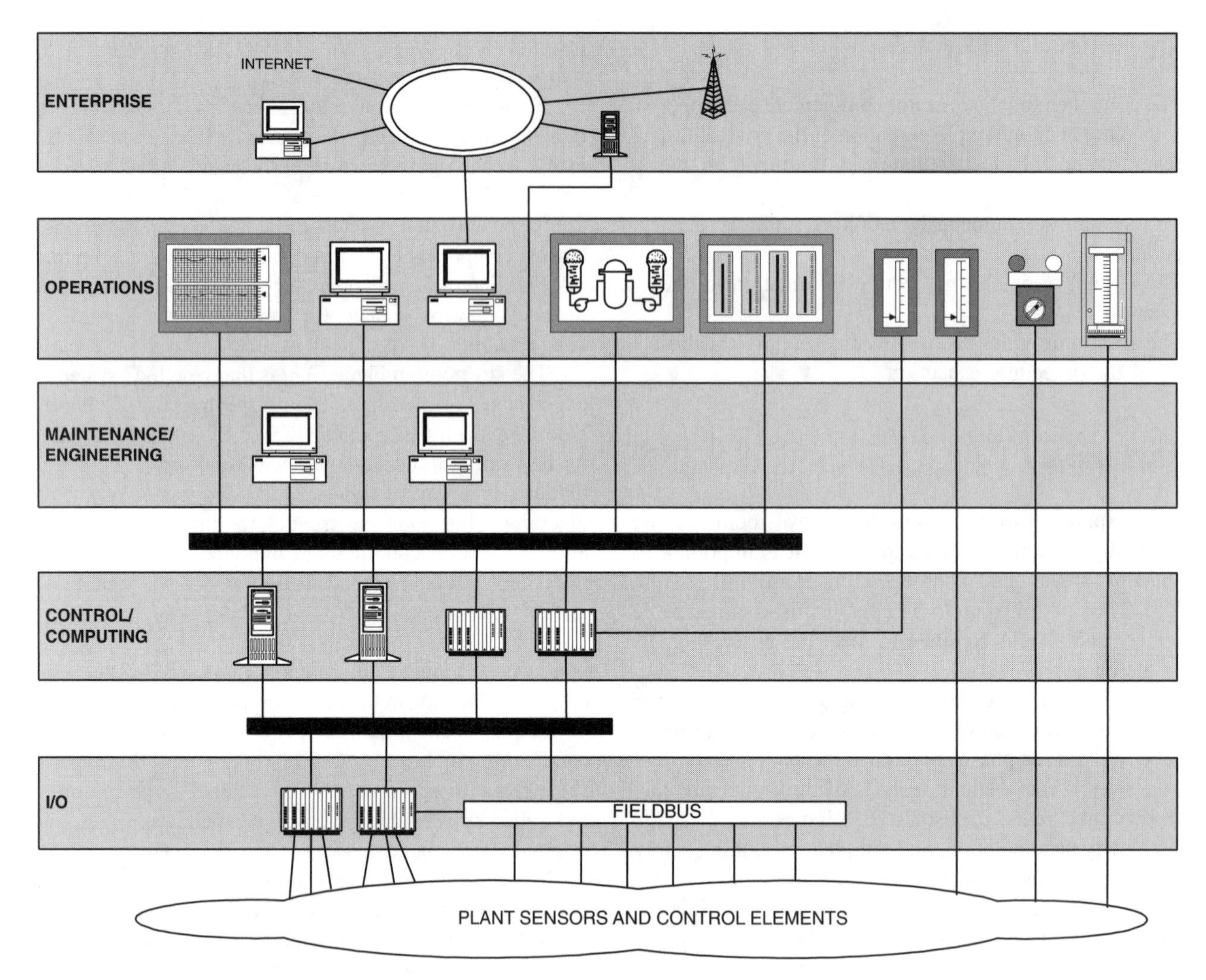

FIG. 3.3j
Hybrid I&C architecture.

to a new DCS and the data integrated into the central database, then it will be possible to display and archive the data seamlessly, transparent to the operator, without a new front end for those data points.

Network communications is utilized between all layers. The network topology is optimized for cost and connectivity options, and the first phase of the modernization effort should be carefully designed so that future phases are accounted for with at least 100% spare connectivity. Spare connectivity should support copper and fiber media options. When pulling new fiber-optic cables, select termination locations that support the long-range plan, and pull as many fibers as possible in each cable run (limited by budget or raceway fill).

The operations layer in Figure 3.3j is the most constrained layer in the phased upgrade approach. This layer is constrained by the physical layout and fundamental design of the existing control room. For example, the breakfront panels shown in Figure 3.3b are designed for a standing operator, and indications and controls are laid out by groups associated with major plant system functions. HMI resources such as flat panel displays have to be designed so that the basic information zones of the control panels can remain intact. If a particular information zone contains vertical meters and trend recorders, then flat panel displays should be sized to fill that zone (with redundancy if possible), and the DCS or PLC system should be programmed with a display page that carries the same information with the same readability of the replaced components.

Figure 3.3k is a good example of some information zones on the control room panel shown in Figure 3.3c. Shown here are four zones. The zone on top is a group of feedwater heater level indicators. The middle-left zone is a pair of heater drain tank level indicators while the middle-right zone consists of three condenser vacuum indicators. The bottom zone is a set of indicating hand switches that indicate and control valve position. All of these instruments are connected to obsolete transmitters in the field, which are being upgraded using fieldbus technology. New controllers

FIG. 3.3k
Feedwater heater indicators and controls.

are being installed, and an integrated display can be connected to the system for use by operators in the control room. The original indicators in Figure 3.3k can be maintained as-is using analog and digital I/O points connected to the controller I/O subrack, but this upgrade provides a point of entry into the control room for a modern HMI interface, which is inevitable given the I&C obsolescence issues in the plant. The feedwater heater system is a good pilot project for this technology, where its information zone remains intact using flat panel displays.

Note the size of the meters in Figure 3.3k. The character size, color, luminosity, and arrangement are designed to allow readability from 15 ft away, where the control room supervisor sits. This is a firm constraint for the system, because operator tasking is designed to allow one control room supervisor to stand back and take in all of the control panel indications simultaneously from a central location in the control room, while an operator can stand close to these indicators and manipulate them manually.

The information zone "map" for these instruments is shown in Figure 3.3l. The information is of relatively low density. These parameters are used by control room operators during normal operations, including start-up and shutdown. The operators task during normal operations is to monitor these variables during routine watch-standing operations. Their task during alarm conditions is to diagnose the cause of the alarm and verify automatic control actions have taken place and/or to take manual action as required by written

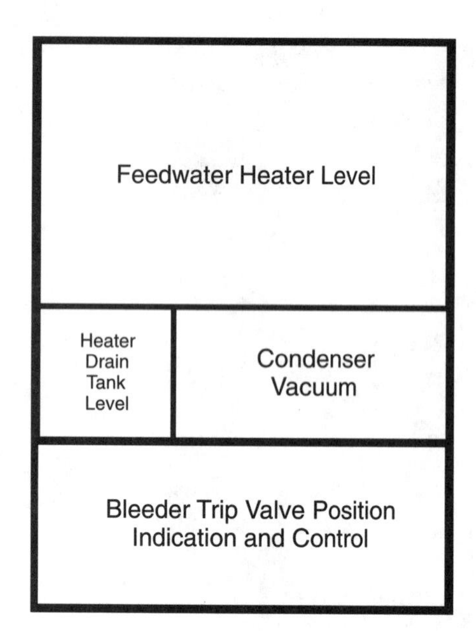

FIG. 3.3l
Feedwater heater information zone.

procedures, so that the system parameters are returned within normal operating limits.

Figure 3.3m shows the same information zone in a modern display example, which can also be applied in a mockup (any scale). This figure shows two flat panel displays, designed for installation in the place of the original instruments. The concept is to program each flat panel with a "default page" that arranges the same variables in a similar fashion to the original arrangement, with font sizes as large as or larger than the characters on the original meters. Because this is a redundant design, each flat panel can be set to display any other page in the display hierarchy.

This design maintains the same readability and operator tasking as the original design, while offering powerful new interfaces to the operator. The control room supervisor can continue to stand back and take in all indications during normal and emergency operations, while the control room operator can manually page through the various display pages available on the flat panel using a touch or trackball interface. Also, applying DCS technology for this interface supports additional flat panels or VDUs in other parts of the control room, such as an auxiliary panel or on the control room supervisor's desk. Although the default page is designed to be simple and readable consistent with the original panel design and operator tasking, the system also supports a full range of object-oriented graphic displays, event logs, archives, alarm management, and testing provisions.

Maintaining the original information zone design concept is only the start of any control room upgrade effort. Improvements in safety, reliability, readability, and a reduction in the potential for operator error should also be goals of the project. The use of a digital DCS or PLC platform enables the realization of these additional goals. Variations in colors, fonts, reverse video, and flashing objects are available with digital

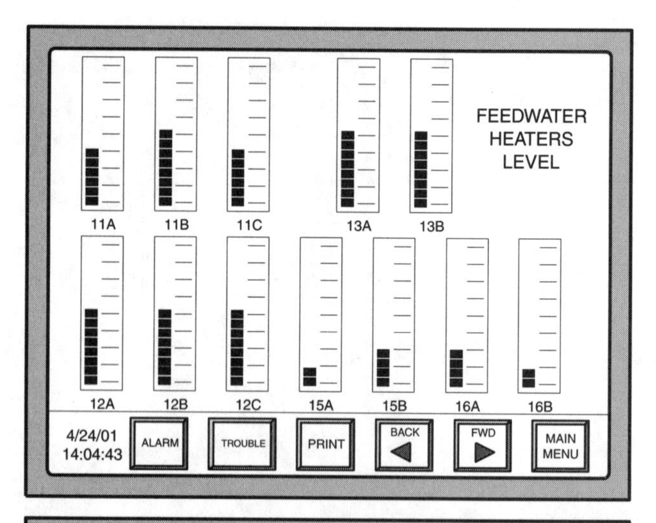

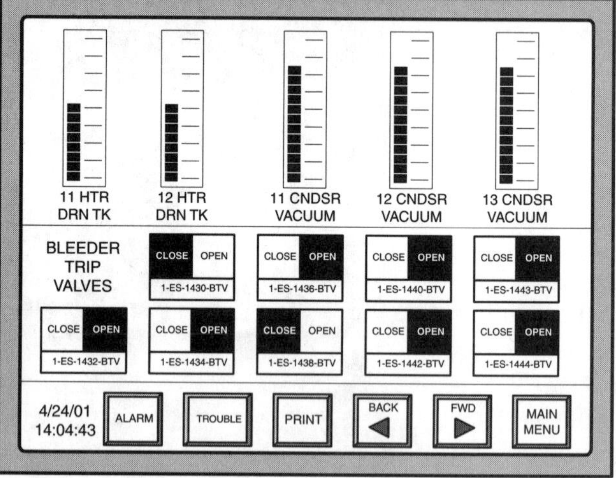

FIG. 3.3m
Upgraded feedwater heater displays.

technology. New systems can detect and alarm on a failed channel (out of range high or low). Also, technology developments such as automatic alarm prioritization and computerized procedures may be implemented in a new system such as this one.

The feedwater heater displays in the above example are a starting point for the entire panel in Figure 3.3c. As more channels of information are added to the DCS/PLC system, more displays can be added to the same panel, or larger displays can be installed with more information available per display.

PLANNING AHEAD

The first phase in a phased control room upgrade project should prove that the overall conceptual design for the control room can be implemented safely and effectively. Once it is proven that the upgraded system has a positive impact on control room operations with improved reliability and cost-effectiveness, then the system can be expanded to all systems in the control room. The pilot approach allows training of engineers, technicians, and operators to commence without a massive undertaking, and valuable lessons learned can be applied to all disciplines on future phases (design, procurement, start-up, and operations and maintenance). The key is to purchase a digital I&C platform that can be applied on a smaller-scale project then expanded into the rest of the I&C systems in the plant.

HMI STANDARDS

It is essential that the long-range plan account for the impact on the operator, and that an HMI standard be applied consistently from start to finish. Flat panels, video display units, workstations, and consoles should be uniformly designed, and software should be standardized so that all system indications and controls are seamless and transparent in a common interface style. If an operator has to move to a specific piece of HMI hardware for each plant system or point-and-click on different icons for each plant system, then the opportunity to improve safety and reliability while reducing operating costs is diminished. The only exception to this rule is when diverse indications and actuations are required, such as for manual safety system actuations (e.g., reactor trip).

HMI attributes such as color codes, abbreviations, fonts, graphic elements, display sizes, and readability should all be consistent. A human-factors engineering (HFE) program should develop standards and procedures that will be coupled with an operator and maintenance technician training program, consistent with the long-range plan.

While the goals of a control room upgrade are to eliminate obsolescence, reduce maintenance and operating costs, and improve safety and reliability, the basic control room panel and furniture layouts and operating and maintenance procedures will be constrained throughout the phased approach. However, the control room panels and furniture can still be altered in a local sense. It is acceptable to introduce different HMI methods in clearly distinct contexts. For example, it is acceptable to add modern soft controls techniques into conventional control rooms for distinct operations such as water treatment. As operators become familiar with the new HMI, the designer can be confident they will accept transitioning to the modern HMI for more critical systems. At any point in this transition process, it is important to ensure that operators are within the same HMI context (conventional or soft) for any particular plant system.

Figure 3.3n illustrates design concepts for panels, furniture, and a large display that can be evaluated for use in any analog control room. Panels can be modified and still mix manual controls, analog meters, and flat panel displays. Furniture can be modified to mount flat panels on arms that can be manipulated to suit operator tasks. For example, if an operator needs a line of sight to the control panels during routine operations, then the flat panel can swing out of the way. If an alarm requires some diagnosis, then the operator can swing the flat panel back into view and perform a specific task, or use it in combination with the large display panel in a supervisory role.

FIG. 3.3n
Demonstration control room. (Courtesy of Westinghouse Electric Company.)

Gross design problems in the control room should not be applied as constraints in the phased approach. For example, if a critical process variable is indicated on one panel and controlled on another in a way that does not allow the operator to observe and control the process simultaneously within acceptable HFE practice, even if the operator has learned to work around the problem, then the control and indication should be colocated in the new interface. This is a good example of how a VDU in each information zone can resolve old design issues.

MAINTAINABILITY

Improved maintainability of the plant and control systems in particular should be a significant goal of a control room upgrade. Information and control systems should be purchased with diagnostic aids and functions that support the ability to store short-term, local data logs and the ability to off-load data to archival systems. In addition to normal process information, there should be pages of information related to process events (warnings and alarms), diagnostic events (system and component faults), and maintenance information. This information should enable an operator to detect and mitigate an event immediately, and call the maintenance department to investigate, troubleshoot, and perform corrective actions if necessary. Corrective actions should be more timely and effective on the control system itself as well as on plant equipment.

Networks allow multiple video displays in and around the control room. When an analog plant was constrained to field- or panel-mounted indications and controls, the only data available to the maintenance or engineering department were real-time data in the control room, or sparse, historical data by way of manual operator logs. Intermittent problems were usually very difficult to repeat, diagnose, and correct. A control room upgrade should fill this void in the maintenance and engineering departments. Displays should be set aside for these departments in or adjacent to the control room, with secure access. Control system access should be controlled first by physical security, then by password security.

Maintenance and engineering displays should provide enough data to understand a plant process event or a control system event. Digital inputs and outputs should be captured in a sequence of events (SOE) log. Maintenance and engineering displays should also allow users to navigate all available information and settings on the system, including the ability to change settings, start and stop programs or tasks, and run system diagnostics in an offline mode. Modes should be controlled by password and/or key switch inputs to the system.

Displays on the main control panels should be limited to information and controls that are related only to normal and emergency plant processes. Physical security should limit access to operator displays to qualified plant operators only.

Bibliography

Harmon, D. and Shin, Y. C., "Joint Design Development of the KNGR Man-Machine Interface," Taejon, Korea: KEPRI & Combustion Engineering, 1998.

IEEE 1289, "IEEE Guide for the Application of Human Factors Engineering in the Design of Computer-Based Monitoring and Control Displays for Nuclear Power Generating Stations," Piscataway, NJ: IEEE Press, 1998.

IEEE-603, "Standard Criteria for Safety Systems for Nuclear Power Generating Stations," Piscataway, NJ: IEEE Press, 1998.

ISA 84.01, "Application of Safety Instrumented Systems for the Process Industries," Research Triangle Park, NC: Instrument Society of America, 1996.

ISO 11064-1, "Ergonomic Design of Control Centres, Part 1: Principles for the Design of Control Centres," Geneva, Switzerland: International Standards Organization, 2000.

ISO 11064-2, "Ergonomic Design of Control Centres, Part 2: Principles for the Arrangement of Control Suites," Geneva, Switzerland: International Standards Organization, 2000.

ISO 11064-3, "Ergonomic Design of Control Centres, Part 3: Control Room Layout," Geneva, Switzerland: International Standards Organization, 1999.

Ivergard, T., *Handbook of Control Room Design and Ergonomics,* London: Taylor & Francis, 1989.

Lipner, M. H., "Operational Benefits of an Advanced Computerized Procedures System," Pittsburgh, PA: Westinghouse Electric Company, 2001.

Modern Power Station Practice, 3rd ed., Vol. F, *Control & Instrumentation,* Oxford: British Electricity International, Pergamon Press, 1991.

NUREG/CR-6633, "Advanced Information Systems Design: Technical Basis and Human Factors Review Guidance," USNRC, March 2000.

NUREG/CR-6634, "Computer-Based Procedure Systems: Technical Basis and Human Factors Review Guidance," USNRC, March 2000.

NUREG/CR-6635, "Soft Controls: Technical Basis and Human Factors Review Guidance," USNRC, March 2000.

NUREG/CR-6636, "Maintainability of Digital Systems: Technical Basis and Human Factors Review Guidance," USNRC, March 2000.

NUREG/CR-6637, "Human Systems Interface and Plant Modernization Process: Technical Basis and Human Factors Review Guidance," USNRC, March 2000.

NUREG/CR-6684, "Advanced Alarm Systems: Revision of Guidance and Its Technical Basis," USNRC, November 2000.

NUREG/CR-6691, "The Effects of Alarm Display, Processing and Availability on Crew Performance," USNRC, November 2000.

Technical Report 1001066, "Human Factors Guidance for Digital I&C Systems and Hybrid Control Rooms: Scoping and Planning Study," Palo Alto, CA: Electric Power Research Institute (EPRI), November 2000.

3.4 Manufacturing Platforms and Workstations

R. J. SMITH II

Partial List of Suppliers: Debian; Microsoft Corp.; Sun Microsystems, Inc.; Momentum Computer, Inc.; Dell Computer Corp.

Manufacturing control systems have evolved over the years, from programmable logic controllers (PLC), to distributed control system (DCS), to the latest hybrid systems (DCS/PLC). For further information on PLC, DCS, DCS/PLC, see Sections 6.5, 7.7, and 7.14 of Volume 2 (Process Control) of the IEH. The first PLCs were designed to replace relay logic in large panels installed on the plant floor with smaller electromechanical versions. Throughout the years, control systems have been integrated into enterprise systems, controlling product flow from the plant floor to the customer. Although these systems vary in size from a stand-alone PLC to systems that are interconnected throughout the plant and around the world, they all have one thing in common: an operating system (OS). These operating systems can run on field devices, PLCs, DCSs, and operator workstations. They need to be operational all year around with zero downtime. For this very reason, the OS needs to be proven able to run the control application without faults.

Manufacturing platforms come in many shapes and sizes: from a manufacturing resource planning (MRP) system integrated into a plant control system to embedded operating systems in a proprietary OS running many types of devices. For most manufacturing systems on the plant floor, a real-time operating systems (RTOS) for control is not necessary, but these systems must provide deterministic operations. Real-time control comes from the interaction among the hardware, system, and application layers. All three layers must work together to allow the control functions to operate without any delay despite the actions of the OS. OSs that are real-time have installation requirements that far exceed the typical office installation requirements. Many control system manufacturers specify which OS their application runs on, and they will provide what is required. The trend for manufacturing control systems leads us onto the plant floor and out of the control room, which leaves a decentralized control system that is connected via a network that can be managed from around the world with human–machine interfaces.

COMPUTER SYSTEMS

A computer system consists of a three-layer hierarchy: hardware, system software (OS), and application software (control software) (Figure 3.4a). The interaction between each layer in a manufacturing platform is important. A manufacturing platform is the combination of computer systems and field devices. There are different types of computer systems for a manufacturing platform, all with OSs (system software). A few well-known OSs are UNIX, Linux, and Microsoft (MS) Windows NT/2000. For further information, please see *IEH*, Volume 2 (*Process Control*), Section 7. This is the reason most control system software vendors require their software to be installed on certain hardware with specific system software. Until the mid-1990s many control system vendors supplied the hardware, system software, and control software. These proprietary systems were designed to be extremely stable. In the last few years there has be a turn from fully proprietary systems to more open systems. These open systems use commercially available hardware and software allowing the design of a manufacturing platform to be flexible and very stable.

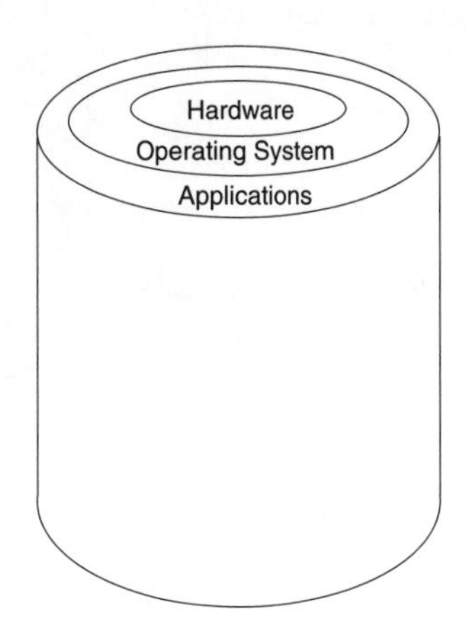

FIG. 3.4a
Computer hierarchy.

A note on "open systems": Be sure to "partner" with vendors who can offer a wide range of products and can assure they are all able to work together.

The choice of a computer system is determined by the control system application. The requirements for the hardware will vary, but will consist of a personal computer (PC) or a single board computer (SBC). The PC is the solution for a workstation or a stand-alone control system, whereas the SBC is normally integrated into a piece of equipment. There are two different types of processor that can be used: complex instruction set computer (CISC) and reduced instruction set computer (RISC). The main difference between CISC and RISC is the pool of available program instructions for each. The CISC can accomplish tasks with fewer processor steps than an RISC, but an RISC can run the task faster. Both UNIX and Windows NT/2000 can run on either type of processor.

Although UNIX and its flavors (for example, Linux) have been around for years, MS Windows NT/2000 seems to be the *de facto* solution for many open systems. However, this has not diminished the role of UNIX as an OS in manufacturing. Many vendors still have UNIX versions of their software.

UNIX and MS Windows NT/2000 both have a similar architecture (Figures 3.4b and 3.4c). As the figures show, both have a user environment and a operating environment. The user environment consists of a user, graphical user interface (GUI), application, commands, shell for the UNIX systems, and a subsystem for the MS Windows NT/2000 system. The GUI enables the user to interact with the system via a keyboard, mouse, touch screen, etc. The user and GUI serve as a one of the points of input that the system receives. For both the shell (UNIX) and subsystem (MS Windows NT/2000), they are the interface for the users or applications that are not native to that particular OS (for example, a command typed on the command line). A command line input is not a GUI execution, so the shell or subsystem must interrupt and execute. The application section can interact with the GUI, shell, or subsystem, and commands within the user environment. This allows ease of interaction from the user's standpoint. The command section interrupts and executes operations received from the shell or subsystem and application. The operating environment consists of the kernel and hardware layers. The kernel is the commander of the user environment and the hardware. Acting as a hub, it allocates time and memory to programs and handles file storage and communications with the hardware. The hardware layer is the physical component of the system; it includes processor, memory, hard drives, etc.

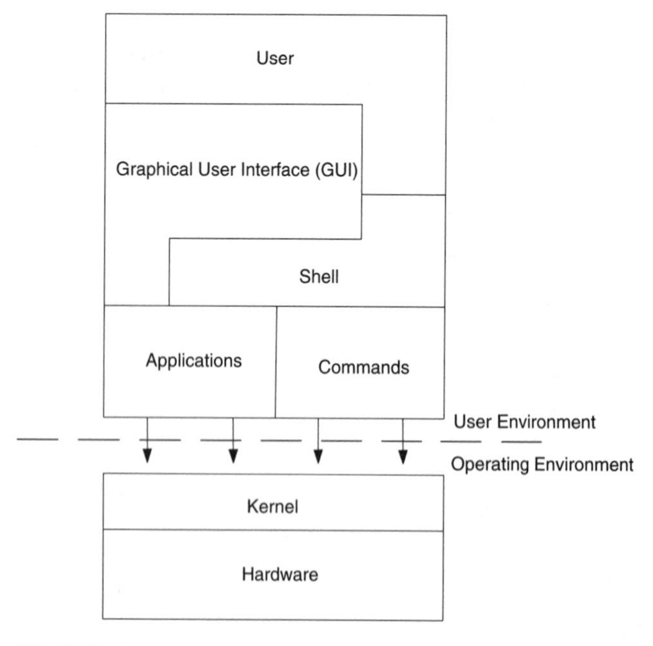

FIG. 3.4b
Basic UNIX architecture.

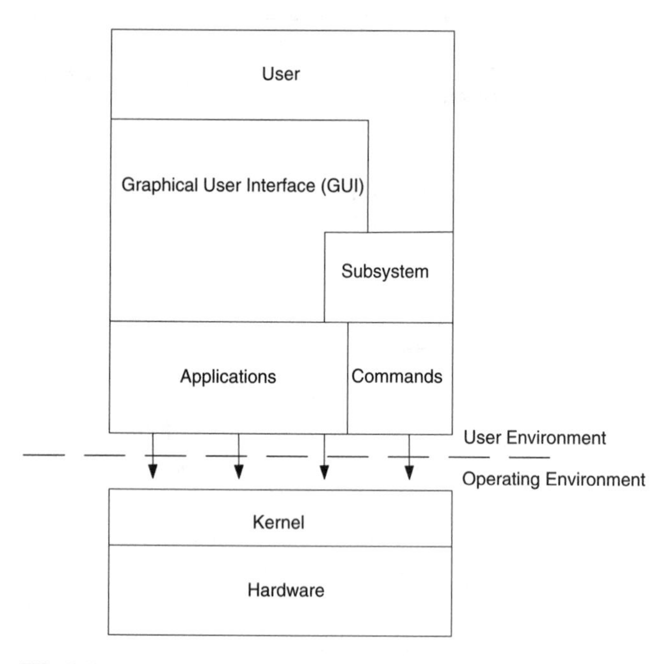

FIG. 3.4c
Basic MS Windows NT architecture.

RELIABILITY

OS reliability is the degree to which the system meets the specifications of its functions, even when subjected to unexpected conditions. Although unexpected conditions are not welcome, they do occur. Many OS vendors will provide statistics on the reliability of their OS, but this does not change the fact that a computer system will fail; the question is how soon. To prevent an OS failure, install only the software that is required on the system, stay up to date on all bug fixes, and check the hardware compatibility list for the OS. Reliability is a major concern when running a control system. The instrumentation, control modules, workstations, etc. must be dependable. There are many ways to make a control system reliable, with redundancy in the field to the control network. The proprietary and open systems do in fact have these features. The proprietary systems are designed to incorporate this throughout the control system from the operator's workstation to the input and output (I/O) points. In open systems, fault

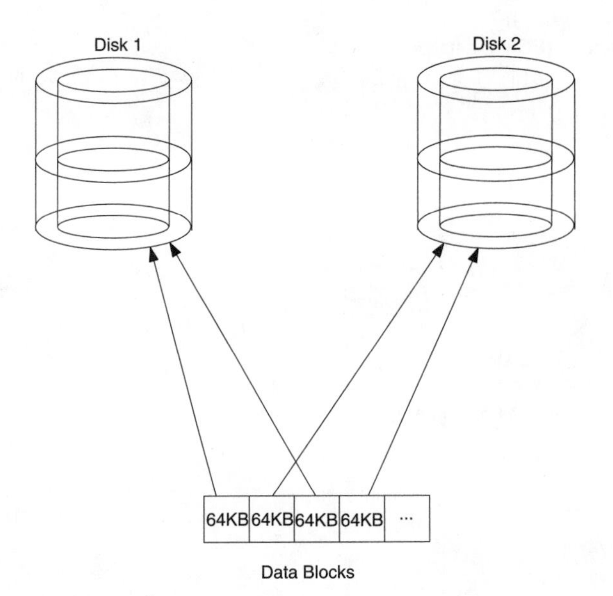

FIG. 3.4d
Level 0 disk stripping.

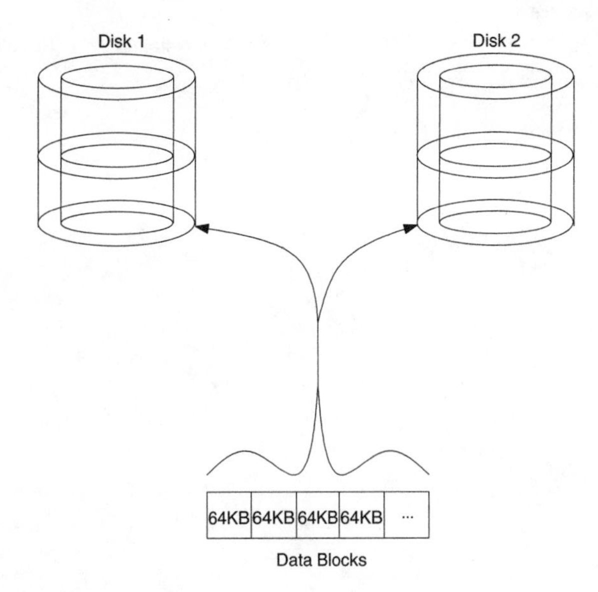

FIG. 3.4e
Level 1 disk mirroring.

tolerance is seen in the proprietary hardware or data network from the vendor. The operator workstations in an open system are usually a PC, with little fault tolerances for hardware failure. There are many options to increase the reliability of the workstations from a hardware failure. These options can be in the form of multiple processors, dual power supplies, multiple network interface cards (NIC), or different levels of redundant array of inexpensive disks (RAID). The first three are self-explanatory: two of something is better than one. RAID is available in three different levels: 0, 1, and 5. The first level, Level 0, is disk stripping. Disk stripping offers little in terms of fault tolerance because if one disk fails the data cannot be recovered, as the data are spread across two or more disks (Figure 3.4d). Level 1 disk mirroring duplicates the data onto two different disks. The two disks act as one set of data; if one of the disks fails, the other takes over until the failed disk is replaced (Figure 3.4e). Level 5 disk stripping with parity is by far the best of the three. Disk stripping with parity uses three disks, writing the strips of data with a parity bit across three different disks (Figure 3.4f). This allows the data to be regenerated if one of the disks fails. Fault tolerance and system backups (for example, tape) will keep downtime to a minimum.

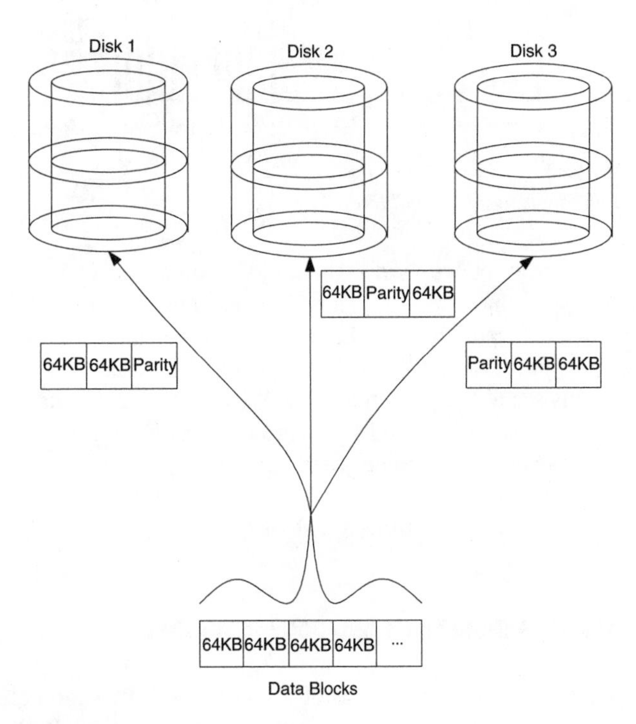

FIG. 3.4f
Level 5 disk strip with parity.

SCALABILITY

Scalability determines how well a system can grow in complexity. For example, a control system with 100 I/O points may perform adequately, but when multiple areas and additional I/O points are added the system might fail to meet response time requirements. Control systems built on today's standards with Ethernet communications including fieldbus, PROFIBUS, and ControlNet can take full advantage of the different methods of exchanging information, which makes OSs scalable. For further information, see *IEH*, Volume 2 (*Process Control*), Section 7.16. The different methods used are dynamic data exchange (DDE), distributed component object model (DCOM), OLE for process control (OPC), and Active X.

Although both UNIX and Windows NT/2000 are considered scalable, the approach in implementing them is different. UNIX was designed to be a mainframe operating system that could support many users accessing application/file services.

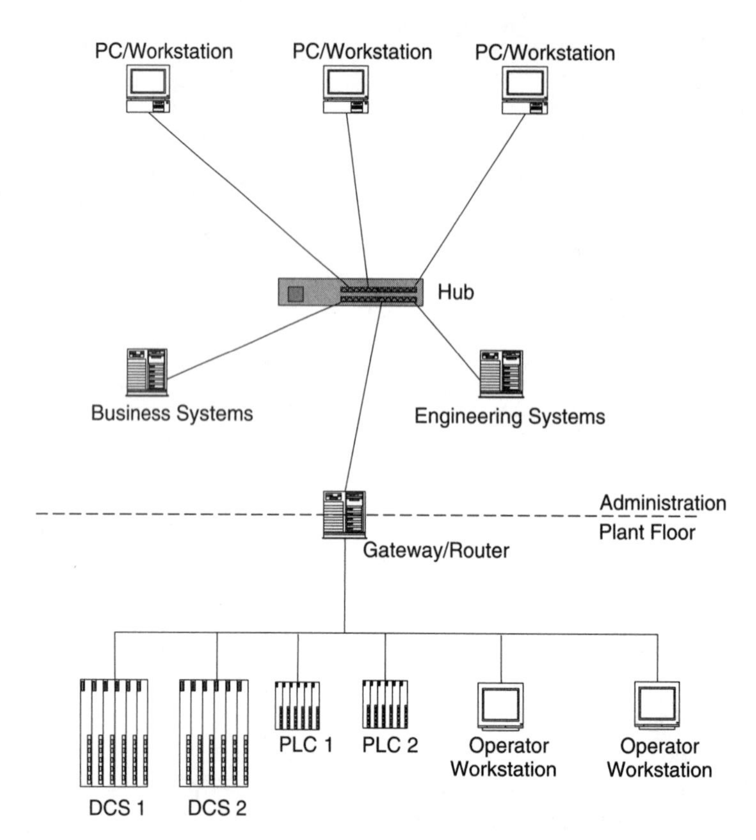

FIG. 3.4g
Manufacturing platform hierarchy.

Since the introduction of high-end PCs, some UNIX operating systems have been used for client operating systems. This allows many of the mainframe features like multiple/load sharing processors to be used on the client system and enables a highly scalable environment. The Windows NT/2000 shares the same features as UNIX for hardware and filing uses, but still lacks application robustness. Although scalability should be considered in workstation operating systems, the focus should be on the controllers and I/O scalability.

SYSTEM MANAGEMENT AND ADMINISTRATION

System management consists of a logical implementation of one or more computer systems within an organization. UNIX and Windows NT/2000 have a hierarchy of user rights, system resources (storage devices, printers, etc.), and system policies. They must be designed to provide secure and functional data exchange between the plant floor and the top-level systems. Each level from the plant floor to accounting should be isolated such that the activity in one will not affect the other, by means of gateways, routers, etc. (Figure 3.4g). The design of the plant control system to interface the business system must be implemented in such a way that the integrity of the plant operation is not hindered.

UNIX and Window NT/2000 both have a similar hierarchy of user rights, systems resources, and system policies. Although both have a domain style of shared resources across different department servers, they are implemented differently. A domain is a connected group of servers and workstations, for example, TCP/IP (transfer control protocol/Internet protocol). These domains serve as a central area where resources are shared within the group. Both UNIX and Windows NT/2000 can cross domains, which allows use of resources in other domains. Crossing domains must be implemented carefully.

Bibliography

Hennessy, J. L. and Patterson, D. A., *Computer Organization and Design*, 2nd ed., San Francisco: Morgan Kaufmann, 1998.

Hoskins, J., *IBM AS/400: A Business Perspective*, 5th ed., New York: John Wiley & Sons, 1994.

Hoskins, J. and Davies, D., *Exploring IBM RS/6000 Computers*, Gulf Breeze: Maximum Press, 1999.

Hordeski, M. F., *Computer Integrated Manufacturing*, Blue Ridge: Tab Books, 1988.

Hughes, T. A., *Programmable Controllers*, 3rd ed., Research Triangle Park, NC: The Instrumentation, Systems, and Automation Society, 2001.

Rosen, K. H., Rosinski, R. R., and Farber, J. M., "UNIX System V Release 4: An Introduction," Berkeley, CA: Osborne McGraw-Hill, 1990.

Shah, J., *VAX/VMS: Concepts and Facilities*, New York: McGraw-Hill, 1991.

Tabak, D., *Advanced Microprocessors*, 2nd ed., New York: McGraw-Hill, 1995.

Yarashus, D., *MSCE Training Guide; Windows NT 4 Exams*, Indianapolis, IN: New Riders Publishing, 1998.

3.5 Workstation Hosts: Design Concepts and Classification

G. B. SINGH

Cost:	U.S.$8000 for a full-fledged workstation server without any software loaded
List of Suppliers:	ABB Automation; Bailey Controls; Cegelec; Fisher Rosemount Systems; Foxboro Systems; GE Fanuc; Honeywell, Inc.; Westinghouse Corporation
OEMs:	Packard Bell; Hewlett Packard; IBM; Digital; Sun

CLASSIFICATION OF WORKSTATIONS

Operator stations can be divided into various categories, based on the hardware architecture and on the functions they perform.

Hardware Architecture

Those workstations classified based on the hardware architecture can be further divided into two categories: (1) diskless workstations or dumb terminals and (2) intelligent or stand-alone workstations.

Diskless Workstations This type of workstation does not have a hard disk. Its major task is to act as a dumb interface between the application program and the end user. This type of workstation was popular in the days of mainframe computers, where the main processor, which had enormous processing power, was centrally located, and the terminals were used as an interface for the end users.

Diskless workstations are defined by *webopedia* as: A workstation or PC on a local-area network (LAN) that does not have its own disk. Instead, it stores files on a network file server. Diskless workstations can reduce the overall cost of a LAN because one large-capacity disk drive is usually less expensive than several low-capacity drives. In addition, diskless workstations can simplify backups and security because all files are in one place—on the file server. Also, accessing data from a large, remote file server is often faster than accessing data from a small, local storage device. One disadvantage of diskless workstations, however, is that they are useless if the network fails (Figure 3.5a).

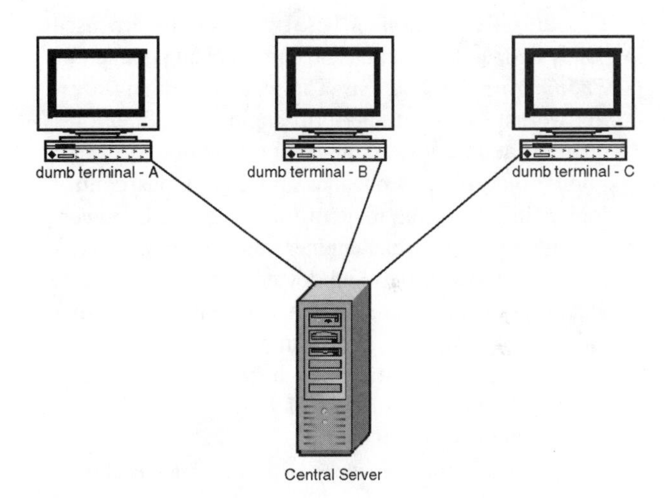

FIG. 3.5a
Diskless workstations.

Intelligent Workstations Intelligent or stand-alone workstations are fully equipped with hard disks and application software. These operator stations may also have databases and may perform various critical operation and control functions. The advantage is that intelligence is distributed across the network, and each workstation can do its piece of a task, which helps reduce processing overhead on a single computer. Also, if one of the operator stations is down, only the functionality related to that workstation is affected and all other tasks remain unaffected. The disadvantage of this type of workstation is that it requires additional hardware and software. The licenses of each software (application or database) have to be purchased for each operator station wherever the particular software is loaded. Certain processes become redundant and are duplicated over various workstations. Client/server architecture is based on this philosophy (Figure 3.5b).

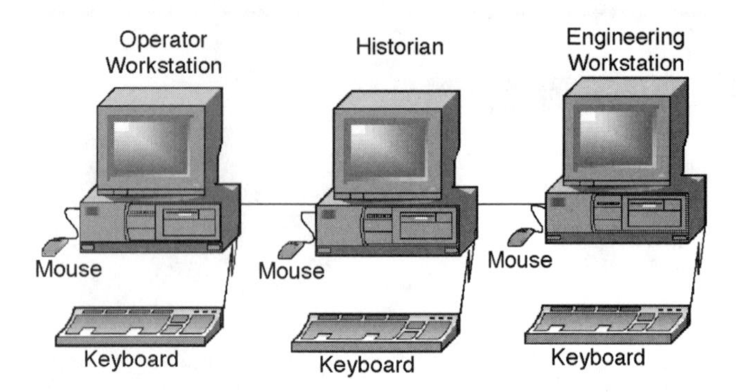

FIG. 3.5b
Intelligent workstations.

Function

The second category of classification is based on the functions workstations perform; in this grouping, workstations are known by the task to which they are primarily assigned:

- *Operator Workstations*: These workstations are used solely to operate, monitor, and control plant parameters.
- *Engineering Workstations*: The main task of engineering workstations is to perform engineering functions: creating new loops; adding various input and output points, modifying sequential and continuous control logic, simulating the logic offline, database engineering, and preparing such engineering documentation as input/output (I/O) lists and device summaries, etc.
- *Historian Workstations*: These workstations are specifically reserved for performing plantwide acquisition, storage, and retrieval of historical process and system information such as alarms, events, operator actions, and system diagnostics.
- *Application Workstations*: A hybrid of all the above-mentioned functions is performed at a single station, known as an application workstation. The workstation may be capable of database management, historical data management, plant and process control, third-party interface functions, etc. The application workstation performs operations such as online configuration and provides a broad range of configuration capabilities, including database, display, system definition, control strategy configuration, and system status monitoring.
- *X-Terminals/Application Processors/Gateways*: These workstations are dedicated servers. They are preassigned to perform a predefined task and normally they act as a black box for the users. Their wide functionality may include conversion of protocols to enable connection of various types of networks (token ring, token bus, Ethernet, etc.), DDE (dynamic data exchange), and OLE (object linking and embedding) servers, and X-terminals.
- *Gateways*: A gateway is a special host that interfaces with two or more distinct networks, to provide for extension of the address cover of each. Some gateway hosts provide protocol translation between the nets they connect. There are two fundamental types of gateways: a gateway that has the capability of performing routing functions when interconnected networks have a consistent message format, and a gateway that must translate services between interconnected networks when hosts use dissimilar message formats. In this case, the internetwork can only support services that can be mapped between the two dissimilar networks. End-to-end flow control is lost with this type of gateway.
- *Portable Workstations*: These are mobile workstations with a special wireless network interface card, which allows operation of the plant from anywhere in the plant (based on the distribution of the antennas). In contrast to the fixed type of workstation, these stations provide the convenience of remote access to ease operation and maintenance. This type of workstation is very useful in plant start-ups and troubleshooting. However, portable workstations are still not very popular in industrial mission-critical applications.

Photographs of diskless workstation–based DCS and intelligent workstation–based DCS are shown in Figures 3.5c and 3.5d, respectively.

FIG. 3.5c
A diskless workstation–based DCS.

FIG. 3.5d
An intelligent workstation–based DCS.

HARDWARE COMPONENTS OF WORKSTATIONS

Any workstation may have the following components:

Central processing unit (CPU)
Workstation display (with/without touchscreen)
Alphanumeric keyboard
Annunciator and annunciator/numeric keyboards (Figure 3.5e)
Mouse
Trackball
Permanent storage device (e.g., a hard disk)
Temporary storage device (e.g., a floppy drive, CD ROM drive, or tape drive)
RAID (redundant array of inexpensive disks) for backups
RAM

Several of these hardware components affect the performance of any workstation:

CPU: The central processing unit is the main workhorse of the workstation. Today, most CPUs are 32 bit. The higher the number of bits, the higher the processing power. RISC processors and 64-bit processors are becoming increasingly popular in workstation applications.

RAM: This is a very critical component of any operator station. Because all operator stations will have a lot of graphic capabilities and software, it is critical that the size of the RAM be adequate to achieve a faster refresh rate for the graphics. Generally, the accepted refresh rate for switching between one graphic to another is 1 s or less, and apart from various software tricks that can be used to reduce this time, the size of the RAM plays a very important role.

Monitor: The display resolution of the monitor is very important for the aesthetics of a workstation. CRTs (cathode ray tubes) have been the mainstay of computer technology. CRT monitors are measured diagonally from the outer corners of the glass tube as in 14, 17, 19, 20, or 21 in., but the viewable image size on a CRT is smaller because the corners of the glass have a radius, so they are not available as viewable area. Active matrix TFT (thin-film transistor) LCDs (liquid crystal displays) or flat panel LCD monitors are becoming more

FIG. 3.5e
A workstation with an alarm annunciator panel and keyboard.

FIG. 3.5f
Typical dual-screen workstation. (Courtesy of Foxboro Systems.)

popular in applications where CRTs have traditionally been used, thanks to their improved color and brightness, better appearance, higher resolution, lesser power consumption, and longer life. Stray magnetic fields generated by motors, high-power machinery, welders, buses, and transformers are common on the plant floor and can result in screen noise and degraded color purity, or can cause severe geometric distortions in CRT images. Flat panel monitors are not affected by magnetic fields, and hence TFT screens are becoming more and more popular. Figure 3.5f shows a dual-screen workstation.

The RAM size of the video card also plays a crucial role. The optimum size must be selected based on the graphics and their orientation.

Workstations may also consist of special utilities, such as touchscreen utilities, speech synthesizer, etc. They must be correctly selected for the application, and additional software utilities must be matched to these applications.

SOFTWARE FEATURES OF WORKSTATIONS

Most of the time the operator station acts as a gateway between the user and the plant. Thus, it must be robust enough to sustain crude handling by operators and be failproof to avoid catastrophe due to errors committed unknowingly.

There are a number of standard applications on any operator station:

- Provision of equipment information for the workstation and its associated I/O devices, buses, and printers
- Capability for change actions directed to the associated equipment
- Processing of station alarm conditions and messages and maintenance of the system date and time
- Database management, storage, retrieval, and manipulation of system data files
- Maintenance of a history of values for process-related measurements, which have been configured for retention by the historian
- Subsetting of the overall operating environment according to the type of user process engineers, process operators, and software engineers have access to

specialized functions and databases suited to their specific requirements and authorizations
- Dynamic and interactive process graphics

Following are the key features available in a Windows NT-based/Windows 2000/X-Windows/UNIX-based workstation:

- Operating system: The operating system (NT/UNIX/Linux/OS2, etc.) is the basic platform of operation of the workstation over which all other software is ported.
- System software: System management function software (data acquisition, control, and management) performs the real-time data collection, management, and storage of the process control messages and data, database management software, and its interface with the historian software.
- Graphical user interface: The graphical window interface to software applications, which is also known as display manager software, determines the user-friendliness and the aesthetics of the workstation.
- Application programming interface (API): API, used for interface with third-party devices, defines the capability of performing control and data acquisition functions for direct connection to a variety of I/O devices.
- Database management system: Database management systems such as DB/2, Informix, Access, Oracle, etc. are the workhorses where the raw data are stored. These database management systems provide front-end software such as SQL, Access, Developer 2000, and Visual Basic for manipulating the data.

Optional packages may include:

Historian package for viewing the history of various parameters
Sequence of event recording function
Trend display and troubleshooting functions
Special engineering software such as performance monitoring systems, efficiency calculators, maintenance management software, etc.

SELECTION OF CORRECT PLATFORM FOR A WORKSTATION

With the advent of various operating systems such as Windows 95, Windows 98, Windows NT, Windows 2000, Linux, UNIX, OS/2, etc., the question arises: Which is the best?

NT was groomed with the use of the latest software technology. NT can be considered a *de facto* open operating system because it is a widely used operating system and supports a wide variety of products; however, it is still a proprietary operating system, and most of the internal source code related to the system registry is not released to the public. This makes it cumbersome for administrators to secure NT positively because of the obscurity built into the operating system.

UNIX is an old but highly tested operating system. UNIX is an open operating system in which the source code is available and reviewable. A skilled team of administrators can secure UNIX systems more positively than NT.

Each of the operating systems has advantages and disadvantages. For example, a UNIX-based system comes with a very high level of security provisions, but it requires a great deal of effort and a special skill set to develop and maintain such a system. At the same time a system on Windows NT may not have the robustness of UNIX but may be easier to handle. UNIX has an advantage at the high end (enterprise-level automation) whereas for a small to medium level of enterprise automation this difference is negligible.

However, all these issues are very much debatable. Each operating system has certain advantages and disadvantages over others; the right question a system designer must ask is what is right for a particular system.

The following are questions that may help any control system engineer decide the ideal choice:

Is a highly open system or a closed system required?
Need the system have a very high level of security provision?
Is the special type of skilled labor available to operate the system that is sought?
Is user-friendliness built into the system?
What support is available for a particular system in the marketplace?
What are the third-party interfaces supported by the system?
What is the proven track record of the system?
Is the system upgradable to the next version?
What is the history of the product under consideration?
Does the vendor support upgrading or will it be necessary to change to a new system altogether?

Technical superiority is not the only consideration in the choice of an operating system. Personal, political, economic, and environmental factors also play important roles in the decision-making process of platform selection for a particular system. A particular operating system may be ideal technically, but too expensive to afford.

Often a particular feature may be absolutely essential to one company, important to another, trivial to a third, and an unacceptable impediment to a fourth.

Hardware and the interactions between hardware and software can be significant factors in both quantifiable and anecdotal results. Variations in software, even software that shares the same name and version number, can play a significant role in both quantifiable and anecdotal results.

Variations in the personnel creating, administering, and maintaining the computer system (the human factor) can be a significant factor in results. Many times considerable emphasis is placed on the selection of the platform, but how well the applications are designed and interwoven with the platform is equally important. Many times the platform may be very stable but the application program is unstable.

COMPARING VARIOUS OPERATING PLATFORMS

A brief comparison of the various attributes of each type of operating system follows (Table 3.5g).

Cost

For any buying decision, the bottom line always is determining the cheapest solution available. However, choosing an operating system and evaluating its cost is a different ballgame, because one should consider not only the initial investment cost but also the maintenance cost over the long run. Normally, the major cost categories are hardware costs and software costs, but often there are hidden costs, such as costs of after-sales support agreements, upgrades/service packs, hardware upgrades, profits lost for every hour of downtime, and last, but not least, costs for systems engineers and supporting staff.

It is very important that when the costs of systems are compared the features that are standard in one system must be compared with those in the other systems. It should be determined if these features are also available as a standard package or if they must be purchased at additional cost. NT is often chosen for budget reasons because many customers are not willing to pay for the more expensive hardware required by most commercial versions of UNIX.

Reliability

Reliability is generally considered important by end users. Even if one operating system offers more functionality, is more scalable, and offers greater ease of system management, it would be of no use when a server processing real-time mission-critical process application is plagued by frequent crashes resulting in costly downtimes.

In addition, an operating system may be extremely reliable at one kind of task and extremely unreliable at another. UNIX has been operating in the server environment for more than 30 years and now has proved its mettle as far as reliability is concerned. However, Windows NT and Windows 2000 are relative newcomers in the networking environment and are still trying to prove their credentials. Many system integrators and manufacturers have shifted from OS/2 or UNIX to Windows NT over the past few years but still UNIX is considered a more reliable system than NT. Only the coming years will establish the reliability of Windows NT as compared with UNIX.

TABLE 3.5g
Comparison of Architectural Features of Windows NT Server with SCO UNIX

Architecture	*Windows NT*	*SCO UNIX*
Multiuser operating system	Yes	Yes
Preemptive multitasking	Yes	Yes
Support to intel processor	Yes	Yes
Symmetric multiprocessing	Yes	Yes
Asymmetric multiprocessing	No	Yes
Paged virtual memory	Yes	Yes
Maximum number of user connections	Unlimited	Unlimited
Single login to network	Yes	No
Dynamic loading of services	Yes	Yes
Memory protection	Yes	Yes
Audit alerts	Yes	No
Structured exception handling	Yes	No
Installable file system	Yes	Yes
Hardware abstraction layer	Yes	No
Microkernel-based architecture	Yes	No

Manageability and Administration

This is a very important issue for system engineers who influence the buying decision of a system. For a distributed control system, manageability and administration are very important, and it is generally recognized that the Windows-based system provides easier manageability and administration because of its graphical user interface (GUI). NT has long enjoyed an intuitive user interface for managing single systems, largely benefiting from the exceptional familiarity of the Windows look-and-feel adopted by the NT GUI. For the administration of a control system, the administration functions are frequently much simplified by providing special administration programs with a user-friendly GUI. Such functions as backup, creating users, sharing resources, defining priorities in the execution of multiple tasks, defining access, and security are a few of the core management and administrative functions.

Scalability

Scalability is a measure of the ability of an operating system to *scale* from use on a very small personal computer to a very large network of computers. The existence of scalability allows for the smooth growth of a computer system: using the same operating system on the entire installed system base.

If the operating system is truly scalable, it is possible (but not always achievable) to reduce administrative, maintenance, and training costs by standardizing on a single operating system. For control system users, scalability is of prime importance because the ever-expanding nature of any business always demands the system to grow larger and larger. If this topic is discussed in the context of the commercial systems, then the issue of multiprocessing capability would arise. However, for a control system manufacturer, scalability is not only the power of the operating system to support multiple processors but also relates to the capability of the hardware that is used in the system, the network topology adopted, the media used for communication, etc. Both Windows NT– and UNIX-based systems are highly scalable. Another viewpoint of scalability is the capability of a system to support physical memory and processors. UNIX systems can, in general, support more physical memory and processors than can NT. This, in turn, leads to higher performance even though both systems can be supported on the same microprocessors.

Security

The *security* of a system is a very debatable subject. No system can claim perfect security. Normally, for a control system security is designed with the following aspects in mind:

1. Security features to avoid corruption or loss of data
2. Security features to avoid misuse (intentional or by operator error)

Regarding the first aspect, the major culprits are computer viruses, worms, or Trojan horses that may enter the system and corrupt the data of the system. To prevent this, *never* allow floppy disks or tape drives from unknown sources to be used in the system. Also, the system must be guarded with strong antivirus software, which must be upgraded regularly. Apart from this, the user rights for editing, copying, etc. must only be accessible to the system engineer. It is generally believed that UNIX-based systems are less prone to virus attacks as compared with Windows NT–based systems.

Regarding the second aspect, system programmers must strive to make the system failproof by anticipating the various combinations that a naive operator may perform on the system. Ideally, all the critical functions must be password-protected or at least they should not be activated by a single key press. There should always be an acknowledgment to each task. These issues may not be directly related to the operating system; however, they are definitely linked indirectly via the built-in security features of each operating system. For example, the Windows NT system has a built-in security feature such that, unless files are allowed to be shared, by default the files remain not shared.

Error Handling

Systems are prone to failures; therefore, some disaster management plan must always be in place. Many of the operating systems provide for annunciation of the failure of their hardware; however, if there are problems in the software/services they also need to be annunciated to the administrator. What is important for a control system engineer is that all the hardware- and software-related alarms and messages appear in the system alarms. A separate group of alarms called system alarms must always be created, and this group must be different from the process alarms. Once errors are reported, it is important to know how to recover from these errors, and here lies the real test of a system engineer. Error recoveries are predominantly governed by the hardware architecture that has been selected for the system and how the system administration tasks are done. If regular backups of the latest changes are kept, then it may be easier to restore the system to good health. Such hardware architecture as disk mirroring and RAID, normal standby configuration, etc. may be very helpful in handling the errors. Both Windows NT– and UNIX-based systems support various error recovery tools very well.

Integration of Software and Hardware

This is the crux of a computer system. A computer system consists of both hardware and software. A key recurring question is how well the software and hardware are integrated. The choice here is between close integration of hardware and software vs. the availability of a wide number of choices of vendors, which forces purchase managers to collect the different parts from different parties and then ask system engineers to synchronize the system.

It takes more than a GUI to make a computer easy to use. It takes tight integration between software and hardware.

> The Mac is still more elegant and stylish, still more tightly integrated, with better links between software and hardware, because a single company makes both the computer and operating system.—*The Wall Street Journal*

Both Windows NT and UNIX support a large variety of hardware; however, performance may vary with each type of hardware used. Also, the same application can run quite differently on different platforms, even though the platforms have the same speeds and feeds on performance. This is because optimization levels differ, and sometimes hardware vendors assist software vendors in tuning applications for their particular environments.

The ideal solution is to have a total solution from a single system integrator; for example, buying branded hardware and software from SUN, IBM, HP, or the equivalent would be a better choice than collecting an operating system from one source, a workstation from another source, and other applications from a third source.

Openness

Distributed control systems are the central supervisory and monitoring system for the plant, and there may be many third-party interfaces required to connect with the system. Thus, openness is crucial to the system. Both UNIX and NT are highly open systems. The NT system is considered more open because of the possibility of connecting it to various subsystems available commercially. Normally, these interfaces are a major cause of downtime of the system and the compatability of these interfaces must always be checked before considering them. It must be kept in mind that many times these interfaces require special hardware (gateways, bridges, etc.) and software. Integration and management of these additional gateways is often very cumbersome. If these interfaces are redundant, then the overall redundancy of the system must be reviewed after considering the redundancy of these interfaces in tandem with system redundancy.

Another measure of connectivity is the ability to run programs from other operating systems in emulation. Emulated software runs more slowly than native software, but allows for easy trading of data and use of obscure programs available on a limited number of platforms. Hardware emulation can attain the same speed as the actual system being emulated. It also has the possible advantage of sharing some computer resources (sharing hard drives, sharing monitors, etc.) and perhaps even the ability to copy and paste between systems or other levels of direct sharing, as well as saving desk space by requiring only one computer setup to run multiple systems.

CONCLUSIONS

The bottom line is that there is little to distinguish between UNIX and Windows NT. This choice is much more dependent on the environment in which the operating system has to run, the industry and the application to be served, the technical support and programming staff, the client environment, and the existing infrastructure. Eventually many of the glaring technical differences between these operating systems will become non-issues because of the industry trend toward better integration of these two operating systems, with products such as OpenNT.

Bibliography

"Microsoft Windows NT from a UNIX Point of View," white paper, Business Systems Technology Series, Seattle: Microsoft Corporation, 1995.

Norton, P., *Inside the PC,* Indianapolis, IN: Sams Publishing, 1996.

"Windows NT Server vs. UNIX," Business-Critical Application Systems, 1996.

"Windows NT Server Directory Services," Microsoft Windows NT Server Product Group, March, 1996.

GLOSSARY

CLIENT/SERVER ARCHITECTURE An architecture where processing is shared between client and server applications. The server accepts requests from the client, processes the requests, and sends information back to the client.

DATA A representation of facts, concepts, or instructions in a formalized manner suitable for communication, interpretation, or processing by people or automatic means; any representation, such as characters, to which meaning might be assigned.

DATA ACQUISITION The process of identifying, isolating, and gathering source data to be processed by some central agency into a usable form.

DRIVER A series of instructions the computer follows to reformat data for transfer to and from a particular network device. Software drivers standardize the data format between different kinds of devices. Multiplexing of messages through a driver, for example, is made possible by supervisor synchronizing commands, or primitives, that permit several sender processes to send messages to a single-receiver process port associated with the driver.

EMULATION The use of programming techniques and special machine features to permit a computing system to execute programs written for another system.

END USER A person, process, program, device, or system that functions as an information source or destination and/or employs a user–application network for the purpose of data processing and information exchange.

ETHERNET A baseband local-area network specification developed jointly by Xerox Corporation, Intel, and Digital Equipment Corporation to interconnect computer equipment using coaxial cable and transceivers.

HOST (1) In a network, a computer that primarily provides services such as computation, database access, special programs, or programming languages; the primary or controlling computer in a multiple-computer installation. (2) An abstraction of an operating environment wherein a set of processes interacts with a supervisor. The supervisor contains system processes and manages the operating environment, which includes input/output devices, directories, and file systems. A host is a convenient boundary for containing specific resources needed by other hosts. A host is virtual, and several hosts may reside on the same computer.

INTERFACE (1) A shared boundary. For example, the physical connection between two systems or two devices. (2) Generally, the point of interconnection of two components, and the means by which they must exchange signals according to some hardware or software protocol.

SERVER A processor that provides a specific service to the network, for example, a routing server, which connects nodes and networks of like architectures; a gateway server, which connects nodes and networks of different architectures by performing protocol conversions; and a terminal server, printer server, and file server, which provide an interface between compatible peripheral devices on a local-area network.

TERMINAL (1) A point at which information can enter or leave a communication network. (2) An input/output device to receive or send source data in an environment associated with the job to be performed, capable of transmitting entries to and obtaining output from the system of which it is a part.

TOPOLOGY (1) The physical arrangement of nodes and links on a network; description of the geometric arrangement of the nodes and links that make up a network, as determined by their physical connections. (2) The possible logical connections between nodes on a network, indicating which pairs of nodes are able to communicate, whether or not they have a direct physical connection. Examples of network topologies are bus, ring, star, and tree.

3.6 Integration of DCS, PLC, HMI, and SCADA Systems

D. MIKLOVIC

Costs:	Parallel links with PLCs (programmable logic controllers) or serial lines with computers can be made for as little as \$2000. For costs associated with PLCs and PCs, see Sections 3.4 and 3.1, respectively. For estimating an integrated DCS (distributed control system), see Section 3.5. For the integration cost alone, the cost is about 25% of the total hardware and control software cost.
Suppliers:	Integration is provided by systems integrators and equipment suppliers and is dependent on the specific style of integration taken. See related sections for suppliers of DCS, PLC, HMI, or SCADA applications. Architecture and engineering firms with staging facilities for assembling, integrating, and testing multivendor systems prior to final installation include AMEC; Bechtel; Brown and Root; Fluor Daniel; M. W. Kellogg; and Stone and Webster.

Some aspects of DCS system integration are discussed in other sections of this chapter. Therefore, the reader is advised to also refer to Section 4.17 in connection with DCS integration with supervisory computers, to Section 4.2 for DCS integration with PLCs, and to Sections 4.3 and 4.7 in connection with fieldbuses, multiplexers, and input/output (I/O). The first step in integrating DCSs with PLCs and/or computers is to define what is meant by "integration" for the particular project. In some cases, integration refers to a vertical communication for information exchange only. In other cases, integration includes horizontal linkages between the DCS and PLC or computer sharing control responsibility as peers in the system.[1] Some might regard the exchange of data on a few variables as integration. Others would argue that integration is only achieved when common variable names, addressing, and functions exist, all sharing a unified operator interface, with data and information in the system universally accessible throughout. The truth is that integration, like beauty, is in the eye of the beholder. What suffices as integration for one user may not even approach it for another.[2] Therefore, the contents of this section are more like a menu of system integration tools and techniques as opposed to a rigid standard. In many cases a collection of integration tools can be selected and combined. Many of the alternatives are combinational rather than either/or options.

HUMAN–MACHINE INTERFACE INTEGRATION

One of the key reasons to integrate DCSs with PLCs and computers is to obtain a superior human–machine interface (HMI). PLCs generally do not have an embedded HMI. The PLC systems have processing capability and excellent I/O systems for digital information. However, the typical HMI for a PLC-based system installed prior to 1997 is a bench board with numerous push-button switches and indicator lights.[3] Since that time, a variety of HMI technology has been deployed. Refer to Section 3.1 for additional details on HMI types.

The DCS, on the other hand, has an excellent color graphics display system (Section 3.1) in the form of an operator's station. To take advantage of this robust and user-friendly interface it is necessary to have the PLC and DCS share information. The PLC must provide the status of controlled devices to the DCS, and the DCS must provide the PLC with control signals that will start or stop a particular motor or group of motors, or open and close valves. Integration can inform the operator when a requested action is inhibited and can advise the operator what is preventing the action from occurring. For example, if a motor start is requested but it is inhibited because a limit switch on a safety device is not actuated, rather than just not starting the motor, a well-integrated system might describe the nature of the problem to the operator by changing the running light color from red (stopped) to yellow, instead of to green (the color associated with running). If the systems are very well integrated, the operator may be able to use the DCS HMI to query the PLC to determine which limit switches are preventing the requested action from taking place.[2]

The DCS can also be interfaced with a variety of computer systems for advanced control, data analysis, or other information processing–related activity. Although most computers already have cathode ray tube (CRT)-based interfaces, some are text-based and use a cryptic (to the uninitiated) series of commands or instructions. By having a simple function key or touch screen button as a means for starting a computer program for advanced control, or even a plantwide electronic mail system, operators can use a familiar and well-understood interface.

SEQUENCING

One of the most common reasons for providing a hybrid DCS/PLC system is to take advantage of the superior sequencing/interlocking capability of the PLC, as well as its low cost and highly reliable digital I/O.[2] As is noted in Section 3.10, PLCs have been optimized for sequential control. The relay ladder style of programming is well suited for the task of interlocking a number of devices. For this reason, many users elect to use the DCS for analog loop control, HMI, and basic storage, while using the PLC for sequential digital control.[2] This presents the need for integration between the devices. Often pumps or stirrers are not started until certain temperatures or levels are reached. These heating, filling, or other continuous throttling operations are monitored and controlled by the DCS as continuous control loops. The type of information exchanged between the PLC and the DCS is generally digital in nature, indicating on/off or open/closed status.

SUPERVISORY CONTROL

Although PLCs have superior sequencing capabilities, computers are better suited for implementing advanced control. The primary means of programming in a DCS system is a block-oriented process control language. Often, below this block language there is only an assembly language program that interfaces with the microprocessor used in the DCS. The use of an external computer can be more convenient than programming in assembly language or slowing the DCS processing down utilizing massive block control programs in the DCS. Because supervisory control is related to product changes, process optimization, or other "less than real time" functions, a general-purpose computer will usually suffice (Section 5.8). These devices are programmed in third-generation languages such as FORTRAN or C, or with fourth-generation programming languages that resemble the English language. A typical division between the tasks assigned to a general-purpose computer and tasks delegated to a DCS are shown in Table 3.6a. Because the types of control implemented in the computer vs. the DCS can vary or overlap, the information exchange required is also subject to great variation. Often complete process histories must be passed from the DCS to the computer if extensive model-based control is to be implemented (Section 5.7). The volume of information and the frequency and speed of transfer are critical considerations in the selection of the proper integration and communication techniques.

TABLE 3.6a
Task Distribution between DCS and Computer

	General-Purpose Computer	*DCS*
Speed of requests	Slow to fast	Fast to very fast
Types of control	Supervisory, optimizing, advanced control, data historian	Regulatory, advanced
Example	Product change, grade change, energy optimization, value optimization, SQC/SPC[a] set point change	Basic functions, i.e., flow, temperature, level, pressure, speed, production optimization

[a]SQC = statistical quality control; SPC = statistical process control.

DCS–PLC INTEGRATION

Beyond the control requirements, there often exists a need to move information from a PLC to a DCS because the DCS has a better storage mechanism for production data. The history modules of many DCS systems are merely ruggedized fixed magnetic disks. Rather than storing production-related information in the valuable PLC RAM, the information is passed to the DCS for storage and totalization. The typical types of information exchanged include actual running times for equipment, cycles or actuations for linear devices, such as some valve-positioners and solenoid devices, and so on.

Another category of information is the PLC program itself. Sometimes as a result of a product change, a new operating program must be downloaded into the PLC, because different sequencing is required for making the new product.[4] Rather than having a maintenance person load a new program into the PLC, the DCS can store a number of programs, and when a product change is initiated, the DCS can download the new program into the PLC. This takes advantage of the magnetic storage media in the DCS and its capability to synchronize the downloading step with the analog changes implemented in the process.

The advantages of tying PLCs to DCS systems are numerous. Improved HMIs for the PLC often result in improved performance and higher productivity.[3] Where advanced interfaces, which can explain why actions may be inhibited, are provided, the resulting productivity improvements through reduced downtime can be substantial. Also, the use of a computer or DCS to store alternate programs corresponding to different products or product grades can greatly simplify PLC logic and can often reduce the size of the PLC needed to perform a task.[4] Feeding production-related information to the DCS also frees the PLC from performing tasks that it cannot perform in an optimized manner. This also reduces the initial capital investment and simplifies the programming of the PLC. By integrating the

DCS with the PLC, the best features of each device can be used. The PLC can be used for sequential logic and to provide inexpensive and robust I/O. The DCS can be used to provide a better HMI and has the ability to handle analog I/O, PID (proportional, integral, derivative) control, and mid-term storage.

DCS–COMPUTER INTEGRATION

Often the historical storage capability of DCS, while better than that of a PLC, is still too limited. Many historical databases in DCS systems have a capability of storing only 32 to 72 hours of basic second-by-second information. If longer-term storage is desired, it is often accomplished through the use of a computer-based process information management system (PIMS) or data historians. These applications are fully developed in Section 5.8. In brief, these systems collect snapshots of the process typically at 1 second to several minute intervals (the once-a-minute data capture frequency is the most common). Online storage for these variables is usually provided for periods ranging from several months to as long as 5 years, although 1 or 2 years is most common. The PIMSs also have tools for analysis of such volumes of data. These include trend displays, which change scale easily, basic SQC (statistical quality control) charts, data extraction or export to advanced statistical packages, and event logging functions. These systems have prodigious appetites for information. It is not uncommon for data involving up to 10,000 points (variables) to be collected every minute. This volume of data is typical for a larger facility with several DCS systems feeding a single PIMS.

In addition to the need to move information generated by the DCS, there is also the need to move into the PIMS information that has been generated by the PLC and passed to the DCS for intermediate storage. Typical information types in this category include running times, production statistics, process measurements, and set points. The optimum blend of hardware and minimum complexity in each of the individual devices is obtained when computers are also used. The greater benefits are achieved, however, by using the computer to provide features that would be difficult if not impossible to implement in a DCS or PLC. The process analysis capabilities of a PIMS system can save a plant from thousands to millions of dollars annually. The ability to detect such phenomena as seasonal variations and long-term trends; and to couple price information to actual usage, has proved invaluable. In one paper manufacturing facility a PIMS was used to take prices from a business system and flow rates for chemicals used in paper manufacturing from a DCS system and produce a graph scaled in dollars per ton of product. Operators could adjust the usage of chemicals (within quality guidelines) to minimize the unit cost of product. This simple tool has been documented as saving over $250,000 annually.[5]

PLC–COMPUTER INTEGRATION APPROACHES

Integration via Direct I/O

The first method developed for integrating PLCs with DCSs and computers, and also computers and DCSs (for high-speed applications), was to provide direct I/O connections to each of the systems. Figure 3.6b illustrates such a PLC-to-DCS connection. In essence, this configuration creates a high-speed parallel connection that operates at the I/O scan speed of the connected devices. Although the response is certainly in real time, this method of connection has fallen into disfavor as high-speed network-based interfaces have become available.[4] Nevertheless, it still is used, particularly where the amount of information transferred is small. It should be obvious that such a wiring scheme, using one or two wires per digital I/O point, is very cable intensive. In addition, it is usually desirable to include a circuit from each side that serves as a watchdog.

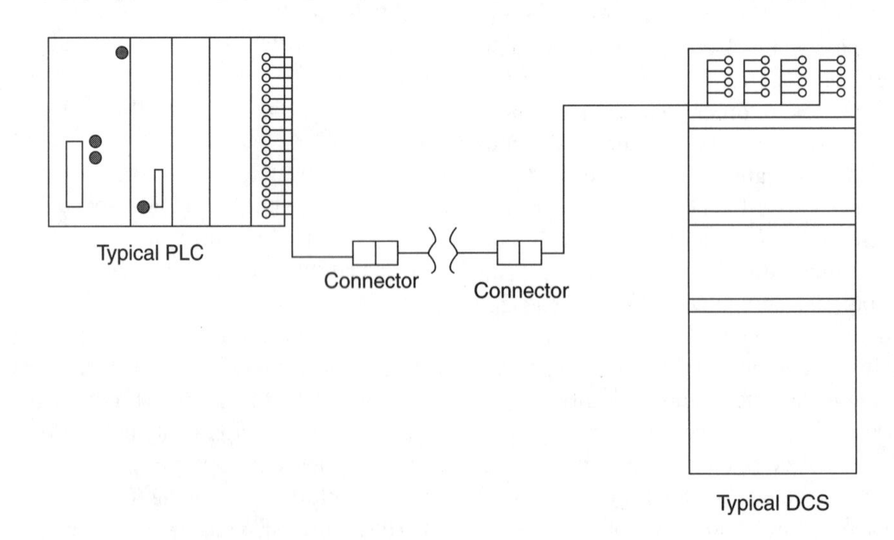

FIG. 3.6b
A direct connection between PLC and DCS systems using parallel I/O wiring.

This watchdog line is a pulse generator that turns on and off on a fixed time basis. Each side (DCS and PLC) monitors the line from the other device. If a circuit fails to oscillate at the proper frequency, the "listener" assumes that the other device has failed, alerts the operator, and follows the preprogrammed failure routine.[4] System integrators like to include connectors like those illustrated in Figure 3.6b because they ease the temporary assembly for testing and troubleshooting, as well as the subsequent disassembly for shipment and reassembly at the final destination. Although such connections greatly improve these aspects of the integration effort, it should be noted that they also introduce additional connections that are potential failure points. In process industries with free chloride or sulfide ions in the atmosphere, these connectors must be of high quality and gastight construction, or problems may arise that are extremely difficult to diagnose. Finally, because this type of connection uses so many individual wires, accurate and detailed documentation is essential. Every wire must be labeled as to its origin and destination, and all DCS loop sheets and PLC logic diagrams should be cross-referenced to facilitate troubleshooting and maintenance. Poor documentation has led to more downtime with this style of connection than have actual failures of the wires and associated connectors.

Serial Linkages

Serial data links are far more common between DCSs and computers, especially where very high-speed communication is not essential to proper performance. Some PLC-to-DCS communications are also well supported by serial links. Where the PLC serves only to provide information to the DCS for purposes of archival storage or where minimal HMI-related data are transferred, serial links are the appropriate choice. A serial link transfers the data as a string of pulses. Figure 3.6c shows how a string of pulses can represent, in this case, eight separate digital states.[5] Note that additional bits are required for synchronization and error detection. Typical transmission speeds are from 1200 bits per second (bps) to approximately 57,000 bps for direct connected serial links. Since 1999 an additional multidrop serial communications protocol, Universal Serial Bus (USB), has emerged, particularly utilized on PCs. Except for PCs connected to process control systems, it has seen little deployment on field devices, outside the laboratory. USB currently supports speeds of up to 12 Mbs.

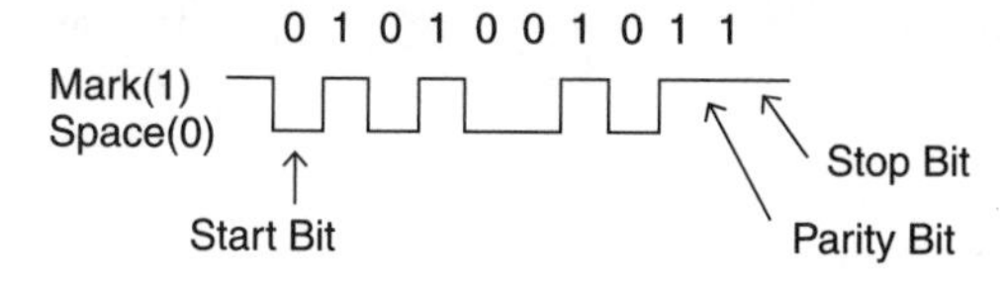

FIG. 3.6c
Serial links transfer the data as a string of pulses. In this case the eight pulses can represent eight states of equipment operation. Similar encoding is used in LANs.

The challenge with serial communications is efficiency. With a 70% data-to-overhead ratio, this represents 840 data to 13,640 bps. The major limitation of serial data communication is that the information transmitted has no specific meaning. Unlike in the parallel wiring scheme (see Figure 3.6b), there is no physical attribute to tie the information to a particular meaning. A protocol must be agreed to by both parties (PLC and DCS) defining the meaning of information transmitted across the serial link. Usually a message is prefaced with identifiers specifying the block of data to follow, and then the actual data.[6] This drops the efficiency of this transfer mechanism even further. Also, the agreement on protocols requires the cooperation of the vendors, which is not always forthcoming as the suppliers are in competition for an overlapping market.

Finally, serial data communication is much more difficult to troubleshoot. There are multiple coding schemes, data rates, and even a variety of error correction schemes. If all are not perfectly matched, communication cannot take place.[5] As with the parallel wiring discussed previously, good documentation is essential. Because there is generally a single cable with 4 to 12 wires between the devices, the physical labeling is less important as these are often documented by the equipment providers and follow either a *de facto* or *de rigueur* standard. The meaning of the data communicated, the orders, variable names, addresses, and other pertinent information must be thoroughly documented, however. Because there are so many variations, the buyer must be careful to specify in advance what standards apply in terms of bit rate, type of parity, and number of bits per character. Although serial communications represents a significant savings in both wire and I/O module hardware at both sides, it is a more difficult interface to troubleshoot and maintain. The classic tools of the electrician, such as a voltmeter or oscilloscope, are useless for troubleshooting serial links. A special-purpose data analyzer is required. In addition, because the data transfer usually operates outside the control block programs, more specialized programming skills are required to use the various data analyzers.

Network Linkages

The most efficient and practical means of linking systems today involves the use of networking technology. From the established and venerable X.25 standards, used primarily to connect to legacy systems, installed prior to 1994, to fieldbuses, and Internet-based TCP/IP (transmission control protocol/Internet protocol) networks over Ethernet LANs (local-area networks), most modern control equipment today has some form of high-speed data communication capability.

X.25

Because control systems have lives often exceeding 15 years, interfaces to older, legacy systems are still a periodic challenge facing control engineers today. X.25 is one solution for high-speed connection to these older implementations. Between computers and DCS systems, the limitations of the serial interface have been overcome by the adoption of a technology previously suited to wide-area networking, X.25. The original intent of the X.25 standard was primarily to support the data traffic from terminal to host in a wide-area scenario. Support for minicomputer to minicomputer, or minicomputer to host traffic, can be obtained by making one of the devices look like a terminal to the other system. An international standards–setting body, the CCITT, through its Study Group VII, developed what has become known as Recommendation X.25, or simply X.25. This is a standard for data terminal equipment (DTE) devices to conduct multiple (if required) communication sessions with other DTE devices over a network. It utilizes a packet switching arrangement. Generally, DTE devices do not talk to each other directly. DTE devices talk to data communications equipment (DCE) devices that talk to each other. Because X.25 became the dominant mechanism for wide-area communications, virtually every manufacturer of a minicomputer made an X.25 communications package available.[6] This high degree of availability led to the growth of X.25 as an accepted means for DCSs to interface to generic or general-purpose plantwide minicomputer systems up through the mid-1990s.

It is important to remember that, basically, X.25 serves to link devices through the equivalent of a network layer. It does provide a virtual circuit at this level (the network) and minimizes the need for extensive transport-type services. Figure 3.6d shows the services that X.25 provides, mapped against the ISO-OSI model. Thus, to provide data access, common higher-level protocols are required. To say that two devices have X.25 interfaces implies only that they can be interconnected electrically through appropriate DCE devices and can establish a virtual circuit to pass packets between the systems. Actual communication, that is, data exchange, is not implied by the existence of X.25 connections. As with serial communications, additional specification is required; however, the basic data transfer options are rigidly controlled by X.25. The final advantage of an X.25 connection is that data link speeds of up to several hundred thousand bits per second can be supported.

LAN

Today, the most common integration tool for tying both PLCs and computers to DCS systems is a network-to-network link.[5] Many PLC manufacturers support either proprietary or standard minicomputer network protocols such as TCP/IP over an Ethernet or IEEE 802.3 physical layer. See Section 4.4 for a discussion on networking protocols including Ethernet, Token Ring, and the IEEE 802 standards. TCP/IP is shorthand for a stack of protocols using either Ethernet or wide-area networks, with a collection of upper-layer services. The TCP/IP suite of protocols serves as the mechanism for the Internet, which grew out of the old ARPAnet, now replaced by a dedicated Defense Data Network.

TCP/IP has emerged as the *de facto* network protocol for computer-to-computer networking, due to the popularity of the Internet. Figure 3.6e shows how a PIMS has been tied together, and with its various components using TCP/IP to communicate. Note that the DCS or PLC devices can tie directly to the TCP/IP network if a TCP/IP interface is available, as is usually the case for the DCS, or through a simple gateway device such as a PC, in the case of most PLCs. The typical LAN speed is 10 million bps, thousands of times faster than serial communications. In some cases 100 Mbs data rates are also supported. These high data rates can support true control interaction over the network, instead of just simple information exchange. Also, most LAN technologies support basic messaging services and file transfer capabilities and have generally well-defined protocols.[6] Because these networks exist outside the realm of the control systems, troubleshooting tools and support staff are often already available.

Fieldbus

Fieldbus protocols are an additional network-to-network communications mechanism. Specific fieldbus protocols and standards are covered in Section 4.7. In fieldbus-based connections the devices to be connected, DCS, PLC, PC, or SCADA system, must both support the chosen protocol or an intermediate device; usually a general-purpose computer such as a PC must be used to act as a router. Although there are strengths to each of the common fieldbus protocols, most were designed to support I/O and field device communication among themselves and with supervisory systems. Using fieldbus protocols as a general-purpose communications path is feasible, but data volumes should be balanced against response time and the needs of the control scheme should be determined first. In general, only higher-speed fieldbus technology should be utilized if the data requirements exceed 8 Kbytes/second (KBs).

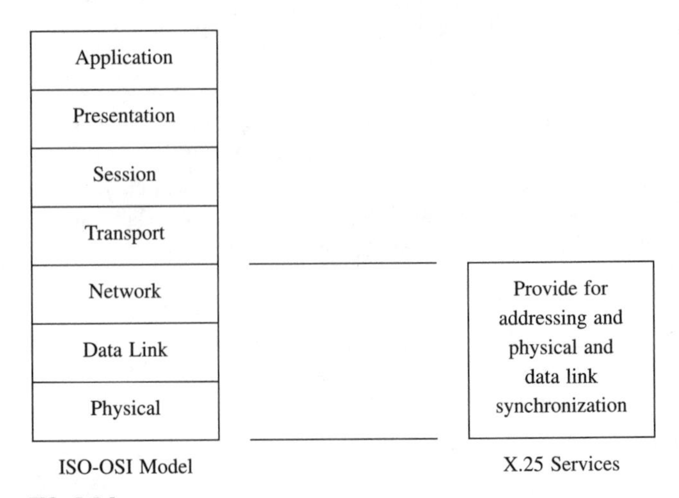

FIG. 3.6d
X.25 is a standard for connecting legacy systems without LAN support while providing high-speed data transfer using wide-area network technology.

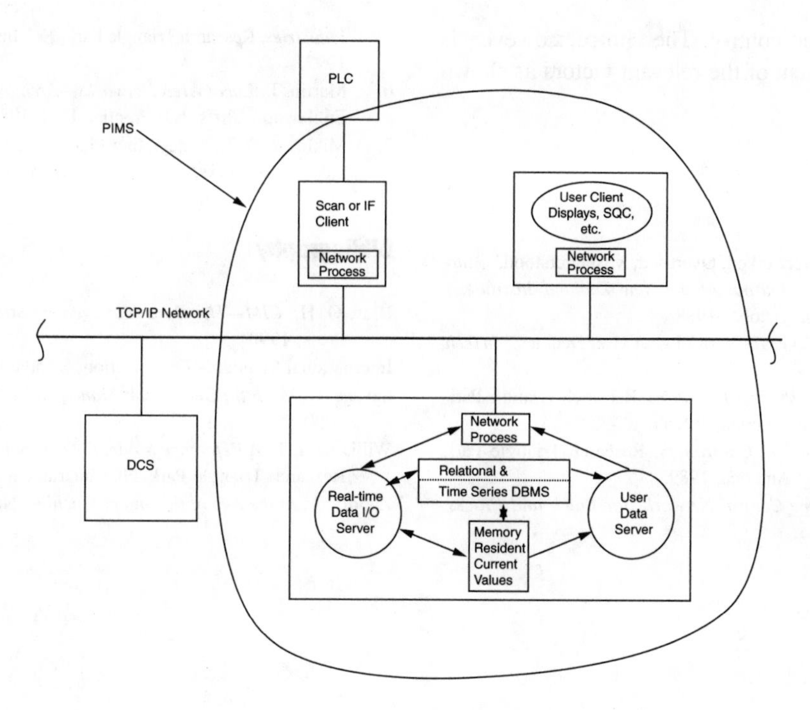

FIG. 3.6e
A typical PIMS with its components using the TCP/IP protocol to interface to both a PLC and a DCS.

TABLE 3.6f
Comparison Factors for PLC/DCS/Computer Integration

Interface Type	*Hardware Cost*	*Software Cost*	*Distance between Devices*	*Speed*	*Reliability*	*Maintainability*
Parallel	Moderate to high	Low to moderate	Low, typically less than 25 ft (8 ms)	Can be as high as 500 kbs	Low to moderate	Easily maintained by plant electricians
Normal Serial	Low	Low to moderate	50 ft (16 ms) to miles (km) for fiber-optic links	Typically less than 56 kbps	Moderate to high	Requires protocol analyzer
USB	Low	Low	5 ft (1.5 ms)	4–112 Mbs	Moderate to high	Requires special analyzers
X.25	Medium to high	Medium to high	Unlimited for practical purposes	Typical less than 1 Mbs	Low to moderate	Requires protocol analyzer
Ethernet TCP/IP	Moderate	Moderate	See IEEE 802.3 standards	Either 10 or 100 Mbs	High if industrial grade components are used	Requires network protocol analyzer
Firewire	Low	Low	15 ft	100 Mbs+	Moderate to high	Moderate to high
Fieldbus	Low to moderate once infrastructure for support in place	Moderate because fieldbus protocols are not designed for high-data volume movement	See relevant fieldbus standards but generally more than sufficient for plant-based integration efforts	See relevant fieldbus standards	Generally very high	Generally very high

SUMMARY

"System integration" is a nebulous term. It can mean just about anything the integrator desires. It is up to the system owner to define the level of integration expected and to justify the cost of integration. The current trend is away from simple parallel connection and toward LAN-to-LAN connections using standards-based open protocols such as TCP/IP. The high speeds associated with these LAN-to-LAN connections support not only massive information transfer, but also, in many cases, actual shared control. The choice, however, is dependent on an assessment of the relevant factors as shown in Table 3.6f.

REFERENCES

1. Miklovic, D. T., "CIM-Overworked, Overused, Misunderstood," *Automation and Control—The Journal of the New Zealand Institute of Measurement and Control,* October 1988.
2. Miklovic, D. and Ipock, C. G., "DCS or PLC: A User Decides," *InTech,* January 1991.
3. Bernard, J. W., *CIM in the Process Industries,* Research Triangle Park, NC: Instrument Society of America, 1989.
4. Hughes, T. A., *Programmable Controllers,* Research Triangle Park, NC: Instrument Society of America, 1989.
5. Miklovic, D. T., *Real Time Control Networks for Batch and Process Industries,* Research Triangle Park, NC: Instrument Society of America, 1992.
6. Martin, J., *Local Area Networks—Architectures and Implementations,* Englewood Cliffs, NJ: Prentice-Hall, 1989.
7. Miklovic, D. T., "Integrating Plant-Wide Databases and MAP Communication Networks," in *Proceedings of the ISA Conference,* October 1986.

Bibliography

Bray, O. H., *CIM—The Data Management Strategy,* Dearborn, MI: Digital Press, 1990.

International Standards Organization, Standard 7498, 1984.

Savage, C. M., *Fifth Generation Management,* Dearborn, MI: Digital Press, 1990.

Williams, T. J., *A Reference Model for Computer Integrated Manufacturing,* Research Triangle Park, NC: Instrument Society of America, 1989.

Zuboff, S., *In the Age of the Smart Machine,* New York: Basic Books, 1988.

3.7 Integration with RTUs, Multiplexers, Fieldbuses, and Data Highways

S. A. BOYER

This section discusses the integration of signals among parts of a DCS (distributed control system), between a DCS and field devices, such as transmitters and valves, and between a DCS and other high-level computing and auxiliary equipment. Because integration is about communication, some discussion about principles of communication leads off the section. Multiplexers, and their counterparts demultiplexers, have been used as elements of DCS since DCSs have existed, and the concept and operating description of them in this application will be developed. Remote terminal units (RTUs) or remote input/outputs (I/O) will be considered as these are the "controllers" that help put the "distributed" in DCS. Data highways allow communication among various parts of the DCS as well as between the DCS and other computer equipment using digital signals. Fieldbuses are so called because they have been developed to enable field devices to communicate among themselves and also with DCSs using a bus topology. Because both data highways and fieldbuses depend on fairly complex data communication features, some small amount of detail is provided about the concept of protocols and the physical aspects of data communication.

BACKGROUND

Integration Means Communication

A DCS surrounded by various field devices, computers, and auxiliary equipment will simply be a collection of parts unless they can pass information back and forth by communicating.

Concept of Same Medium, Same Time, Same Language

There are three conditions that have to be met in order for two individuals to communicate. These conditions must be met whether the individuals are people, machines, or a mixture of both (Figure 3.7a).

1. Same Medium
2. Same Time
3. Same Language

FIG. 3.7a
Criteria for successful communication.

First, the individuals must be using the same **medium**. If the speaker is using compressed air waves (talking), the listener must be prepared to detect compressed air waves (hearing) and not optical pulses (seeing), chemical signals (tasting), or tactile messages (feeling). Second, the **time** must be correct so that the individuals are in the proper mode. They cannot communicate if both are listening at the same time or if both are speaking at the same time. (There are exceptions to this condition, both in people and in machines.) Third, both the speaker and the listener must be using the same **language**. Language is defined in the discussion of protocol later in this section and in other sections of the *Instrument Engineers' Handbook*.

Mediums commonly used by process control systems include instrument air in pneumatic tubing, electrical current in copper wire, light pulses in optical fiber, and electromagnetic waves in the form of radio signals. For most of this discussion, the medium will be copper wire, but many of the factors discussed apply to the other media as well.

Legacy systems, with one transmitter connecting to one controller connecting to one valve, used the same medium throughout the entire system. If pneumatic pressure was the medium, it tended to be the medium throughout the entire loop. The transmitter interpreted the process parameter in terms of pneumatic pressure and sent a pressure signal to the controller. The controller received that controlled variable as a pneumatic signal, applied an algorithm to the signal using mechanical action to develop a manipulated variable, and sent that manipulated variable to the valve as a pressure signal in its output section. The communication in these systems was entirely "point to point." Communication distances were quite limited, but often the controller was located out in the process. Often the facility was operated by people who spent most of their time walking around, looking at the process, and making adjustments as they went.

The "time" factor was not very critical because parts that did the "transmitting" always transmitted, and parts that did the "receiving" always listened. Separate "media," in the form of pneumatic lines, connected each communication path. "Language" was pretty simple. As long as both the transmitter and the receiver understood that 3 psi meant the same thing, they were using the same language. Calibrating the instruments was a way of ensuring that the language being used was the same.

This approach works well for small facilities that can be enclosed in climate-controlled buildings. Larger processes result in too much time loss as the operator spends more and more time walking from one controller to the next and less time making control changes.

The next generation of control systems still used one dedicated controller per loop, but now had the process measurement translated into an electrical signal. The controller solved the algorithm electrically, and sent an electrical signal to the valve. This medium change allowed the controller to be located farther away from the process. Like the pneumatic systems, dedicated media from the sensors to the controller and from the controller to the valve meant the time aspect was still straightforward. The language aspect for the electrical signal standardized over time to 4–20 mA, but notice now that there is a medium/language translation required at the valve where an electrical signal at 4 to 20 mA must be changed to a pneumatic signal at 0 to 18 psi for the valve actuator. As plants became larger, the CCRs (central control rooms) filled with controllers, alarm annunciators, and process variable monitors. The operators became surrounded by equipment. Most of the time, the operators did not need to look at the information that was presented to them. A controller could be set once at the beginning of the shift, and be forgotten until some change occurred. Individual alarms on annunciator matrixes went off so infrequently that "alarm test" buttons had to be added to ensure that the alarm lights were still operational. Process variables did not have to be watched continuously—once per hour was usually enough. It was time to shift away from the concept of dedicated display to shared display. When the sharing of media began, the time aspect started to be a consideration.

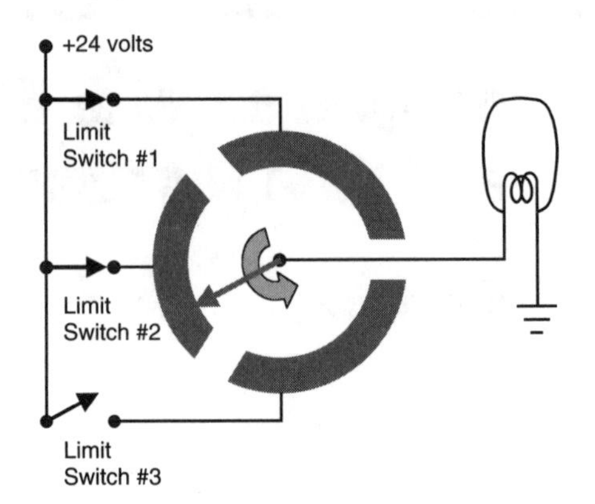

FIG. 3.7b
Simple rotary multiplexer.

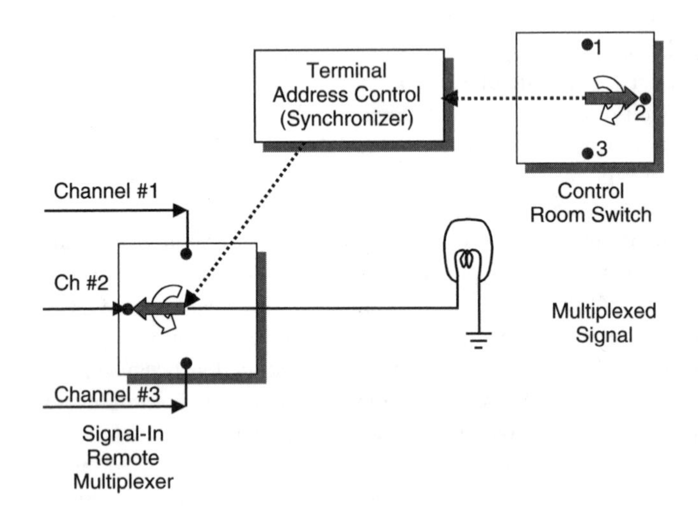

FIG. 3.7c
CCR switch synchronized with multiplexer.

MULTIPLEXERS

Manual Multiplexers

The first attempts to share displays involved wiring the signals from several individual process measurement devices to the terminals of a rotary switch. The rotor of the switch is connected to the human–machine interface (HMI), which may be as simple as a lightbulb that shines if the process device is a closed limit switch and does not shine if the limit switch is open (Figure 3.7b). With this device, an operator could check the status of three motors using only one readout device in the CCR. The space taken up on the wall or console in the CCR will be reduced by using this concept. There is a limit to the number of field devices that can be connected to the switch, but using one switch for 10 or 15 motors is practical.

This human operator-powered switch can also be used to select an analog signal from among many and output the signal to the input of an HMI such as an ammeter, calibrated to read 0 to 100%.

Multiplexing is the name given to this technique of moving signals from multiple sources over a single pair of wires. A basic multiplexer consists of terminals, to which two or more input signals are connected, a switching mechanism or circuit that can route any one of the input signals to a common output, and a driver that operates the switching mechanism.

The resulting signal will not be intelligible unless there is some decoding procedure at the other end of the pair of wires. In the early cases, the human operator could relate the switch position to some process sensor. This is what limited the number of switch positions that were practical.

As the application of multiplexers grew, the advantages of locating them out in the field became evident. Because multiple signals could now be brought to the CCR using only one twisted pair of wires, the wiring cost of the installation could drop. Remote location of the multiplexer meant that the rotor could not be turned by hand. Using a normal motor to turn it could cause errors. It was necessary to have a "terminal address control" between the switch arm in the control room and the rotor arm of the multiplexer in the field (Figure 3.7c).

Very simple multiplexer systems require that the control room operator move the control room switch to the required position to read one of the sensors. The multiplexer rotor is slaved to this switch, perhaps using a stepping motor, so that it will go to the same position that the operator selected. This is called manual remote multiplexing.

Auto-Scan Multiplexing

It is not difficult to extend this concept to have a timer drive the CCR switch so that each of the inputs is scanned regularly, and auto-scan multiplexers constitute the majority of multiplexers. Some of these multiplexers will be equipped with manual overrides so that operators can select a channel that they wish to concentrate on. As the multiplexers became larger, more sophisticated methods of relating the switch position to the multiplexed signal became necessary. The CCR switch would identify which field device was being sampled; the multiplexed signal would indicate the measurement.

Demultiplexers

Consider now that the multiplexed signals from the field will not be interpreted by the operator at a single readout, but will be diverted to a series of readouts located in the CCR. Referring to Figure 3.7d, if the multiplexed signal is input to the switch in the CCR and the multiple terminals of the switch are wired to the inputs of the readouts, we have a multiplexer in the field synchronized to a demultiplexer in the CCR.

This new device means that it is now possible to use that same twisted pair of wires for connecting a large number of field devices to the same number of devices in the CCR. It is even possible to locate multiple sensors in the field to multiple controllers in the CCR—all over one twisted pair of wires. And if we can do that, we can also connect multiple controllers in the CCR to multiple valves in the field—all over one other twisted pair of wires. Notice that the medium and the language are not much changed, but that the time has become more complex. The synchronizer must work perfectly or the information will not get through.

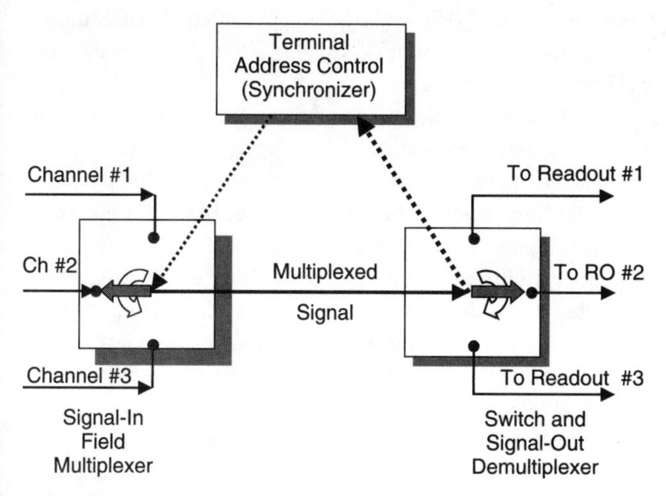

FIG. 3.7d
Demultiplexer synchronized with multiplexer.

Signal Conditioning

As one can imagine, the signals coming out of each of the demultiplexer terminals would not be a smooth signal, but a series of pulses. As the rotor moves from one terminal to the next, it passes a point where it is not connected to either, and zero volts appears on the rotor. This is one situation where a form of output conditioning can be used to improve the signal. A sample-and-hold circuit on each demultiplexer output terminal will remember what the last signal was and will continue to hold the output constant until a new signal is received from the demultiplexer rotor. The sample-and-hold circuit can be visualized as a high-input impedance amplifier whose output is switched on during the sample period and switched off during the hold period. This amplifier charges a capacitor, which remembers the value and which drives an output amplifier. Field effect transistors are used as the switches in this application. See Figure 3.7e for the final result.

Multipoint thermocouple indicators provide an example of other kinds of signal conditioning. Millivolt signals from each thermocouple junction are switched into circuitry that performs reference junction temperature compensation and linearization and generates a more powerful driving signal for the indicator. It is less expensive to buy and maintain one input signal conditioner for the multiplexer output than it would be to buy and maintain one for each thermocouple; often the multiplexing will be done before this type of signal conditioning.

Until now, we have considered only analog signals traveling down the twisted pair, and discrete signals have been treated the same as analog. With some consideration for timing, digital signals can be treated the same way. A digital signal is a string of discrete pulses, each of which, represents a value. (See "Analog-to-Digital, Conversion," p. 347.) Now we are starting to move to conditions where both time and language have become critical. For this discussion, it is important to know that all the digital signal must be included for it to be meaningful. This means that if the longest digital signal is 8 ms long, then each terminal must be connected to the rotor for at least 8 ms (Figure 3.7f). In fact, the terminal address must be added to the 8-ms signal, so this 8 ms minimum may extend to 12 or 15 ms. The lower part of Figure 3.7f shows a multiplexed signal with terminal addresses added. The address could be added to the left of the data or it could be added to the right. The language defines such things as this. Multiplexing digital signals is extremely common. Digital computers use the concept to share one CPU (central processing unit) among millions of registers, and fieldbus and data highways use it for data communication.

If many analog signals that must be digitized are located close together, it may make economic sense to multiplex

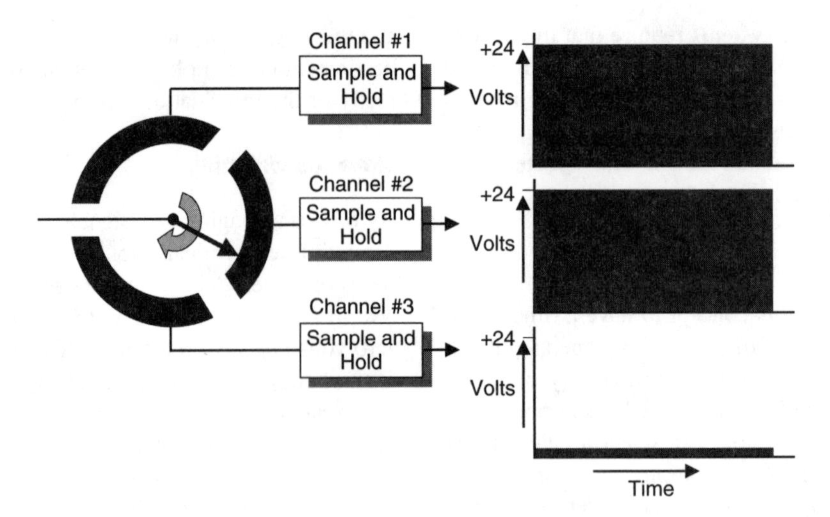

FIG. 3.7e
Demultiplexer with output sample-and-hold (smoothing) conditioning.

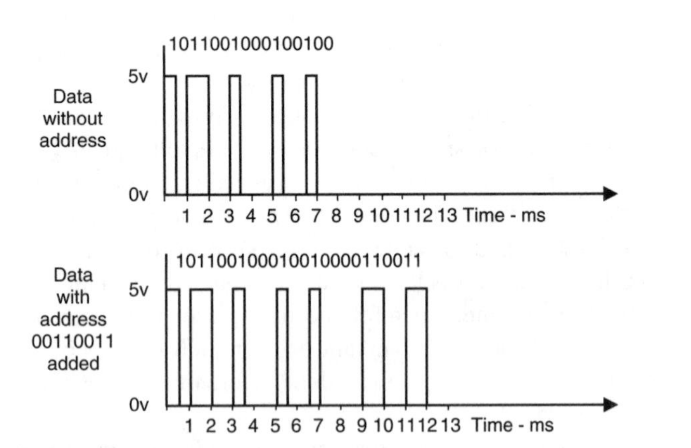

FIG. 3.7f
Digital signal without and with address.

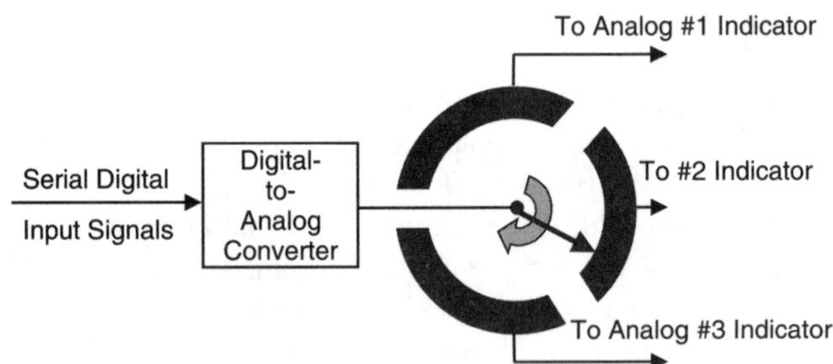

FIG. 3.7g
Series of digital signals converted to analog signals prior to demultiplexing.

them and then share the A/D converter, which can be thought of as a more complex signal conditioner. Similarly, if many analog devices that need to be driven are close to a digital signal that has the driving information, digital-to-analog (D/A) conversion is normally done before the demultiplexing, as shown in Figure 3.7g. As one can imagine, the importance of timing becomes extremely critical as the speed of these signals increases.

Multiplexers—Mechanical or Solid State?

Most of this discussion has dealt with multiplexers that appear to have rotating mechanical parts. The rotating switch concept was useful to discuss principles, but it is not used extensively in industrial multiplexers. Electromechanical multiplexers, often in the form of mercury wetted reed relays actuated by rotating magnets, are used. Advantages include very high crosstalk isolation because of their extremely high ratio of open circuit to closed circuit resistance, in the order of 10^{14}, and their accuracy, which is a function of their extremely low contact resistance. Disadvantages include low speed, with mechanical multiplexers operating in the low to mid-hundreds of points/second. Mechanical vibration issues have been a concern with this type, but experience with building them has allowed most issues to be addressed and vibration levels of 0.75 g should not be an issue. DCS multiplexers for analog process signals generally use electromechanical multiplexers.

Solid-state multiplexers do not have the speed or vibration limitations, but early models had a low enough open circuit resistance that channel to channel crosstalk was a major problem. Metallic oxide semiconductor field-effect transistors (MOSFETs) are now widely used as the switches for most solid-state multiplexers, and these multiplexers are satisfactory for multiplexing digital signals. Cost of solid-state multiplexers, since they are based on large-scale integrated circuits, is lower than mechanical multiplexers. Power consumption is lower and size is smaller than for mechanical units.

Another major advantage of solid-state multiplexers is that channels may be selected in other than a sequential order. Because the switching is electronic, the address of the next channel may be programmed in whichever order is required, and in fact this order may change from cycle to cycle as necessary.

Supercommutation and Subcommutation

A scanning system is not limited to sampling all signals at the same rate. Consider a 100-point multiplexer with a scan period of 2 s, or a scan rate of 50 points/s. If one signal were to be connected to every tenth terminal, it (the supercommutated channel) would be scanned at ten times the scan rate, or 500 points/s. The penalty for this increased speed is that it uses up more of the multiplexer capacity. This technique is often used in DCS systems when faster apparent scan rates are required.

Subcommutated channels are channels that share by means of another level of multiplexing a single terminal of the primary multiplexer. The subcommutated channels are sampled at a lower rate than the channels connected directly to the primary multiplexer. If, for example, ten subcommutated channels occupied one primary multiplexer channel, each of the ten would be sampled one tenth as often as the basic scan cycle.

Multiplexer Selection Criteria

Characteristics to consider when selecting multiplexer equipment for an application include:

1. Accuracy, or the degree of signal alteration through the multiplexer
2. Size, or number of channels switched
3. Input signal requirements such as voltage range and polarity, source impedance, and common mode rejection
4. Channel crosstalk
5. Driving signal feedthrough
6. Sampling rate
7. Physical and power requirements
8. Whether one (single-ended), two (differential), or three (including shield) lines are switched for each multiplexer point

Multiplexers and DCSs

In one form or another, multiplexers have been associated with computer-based control systems for as long as these control systems have existed. When DCSs developed from earlier direct digital control (DDC), multiplexers continued to be an integral part. In the same way as DDC shared one CPU among all of the process inputs and outputs, DCS shared several CPUs, often distributed at rack rooms throughout the facility among groups of inputs and outputs. The way that each input connects to the control logic, which is realized by an algorithm running in one of those CPUs, is by first being processed through a multiplexer and arranged into a serial queue of bits. The way that the CPU sends its signal to the many valves and other actuators that are attached to it is to feed a serial signal through an A/D converter, then through a demultiplexer, which will break the signal into parts and send each part to the proper actuator.

Frequency Multiplexing

Until now, all of our discussions have been about time division multiplexing. Frequency multiplexing is another technique for sharing a single communication path among several signals. Each user is assigned a part of the overall bandwidth available on the path. Instead of taking a sample of each input signal as in time multiplexing, each signal continuously modulates its allocated frequency band. All bands are continuously added and the result is transmitted to the receiver. At the receiver, band-pass filters separate the signal into the assigned bands, and demodulate each band to finally decode the messages. This is a costly method of multiplexing and is seldom, if ever, used to interface directly with field devices. It is enjoying a comeback in high-density data work, but does not appear to show much promise for data rates below 1 Mb/s. It is currently in use for Ethernet links over radio, is used for SCADA systems only at the high end, and is not used integrally on DCSs.

RTUS (REMOTE TERMINAL UNITS) OR CONTROLLERS

Why RTUs?

If signals from the process are concentrated at a location remote from the CCR, a remote multiplexer will allow them to be gathered on a single wire pair and sent to the CCR for input into the controllers and for monitoring through the HMI. If this concentration of input signals is also a concentration of control signals that must be delivered to the process, there is an opportunity to do more than just gather inputs and distribute outputs. The reason that DCSs are so called is because the control system is distributed. Not only is it possible to have the logic separate from the HMI, but it is also possible to have the logic (controllers) spread throughout the facility. It would be technically possible to have a completely distributed network of individual controllers, each complete with its own stand-alone power supply, I/O conditioning, logic and memory, all tied together with a link back to the CCR where the operator could monitor and control all these controllers. That, in fact, is the direction fieldbus is leading. But at the time that DCSs were developed, the communication technology had not progressed to the point where this was feasible. Instead, what was done was to build clusters of controllers, each capable of operating on a nearly stand-alone basis (usually with no operator HMI), and to use communication links to connect them to each other and to the HMI in the CCR (Figure 3.7h). These clusters resemble the RTUs of SCADA technology, but are seldom called RTUs when they are part of a DCS. More often

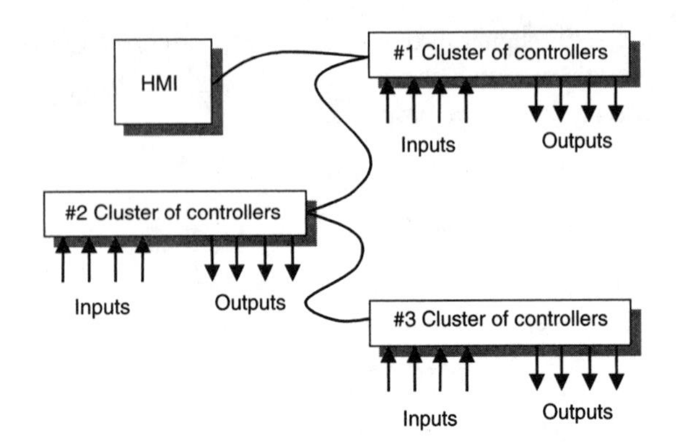

FIG. 3.7h
Distributed controllers.

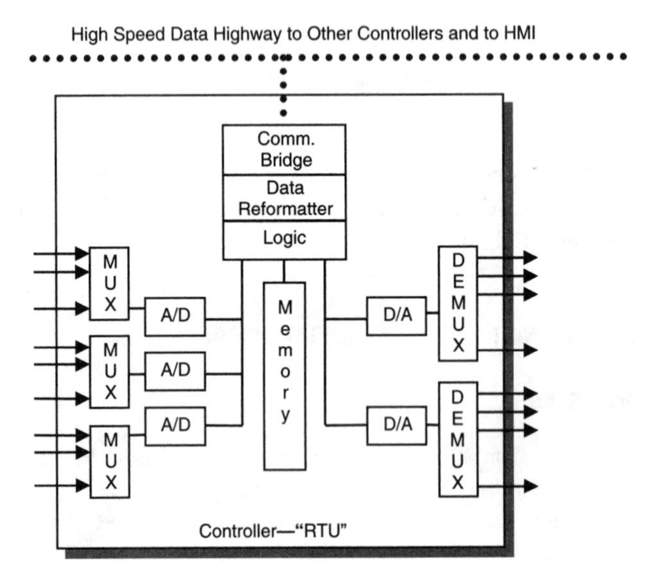

FIG. 3.7i
Controller block diagram.

in DCS discussion, they will be referred to as controller racks, controller stations, or just "controllers." To distinguish them from the controller of a single-process loop, they will be referred to in the rest of this discussion as Controllers (with a capital C).

Normally, these Controllers consist of conditioning circuitry to handle inputs from many field sensors and conditioning circuitry to handle outputs to many field devices (Figure 3.7i). They will also be equipped with multiplexers to receive signals from this input conditioning circuitry and demultiplexers to send outputs to the output conditioning circuitry. They will include memory in which is stored algorithm instructions, provision for programming nonstandard algorithms, and memory for storing constants like set points, alarm points, and so on. They will be equipped with logic for solving the algorithms and for synchronizing the entire operation. Of particular interest in this section of the *Instrument Engineers' Handbook*, these Controllers have provision for communicating with other such Controllers, with the HMI in the CCR, with archiving memory located in a secure area, and with computers that provide high-level advanced control functions. Because the communication that ties together these various elements happens at different clock speeds, the Controller must also have memory that acts as a "buffer" to store data at one data rate until they can be clocked out at the other data rate.

Analog-to-Digital Conversion

Earlier, it was noted that 4 to 20 mA is considered to be the standard for electrical analog process signals. This means that if some process parameter like pressure is to be read, the range of interest will be considered to be 0 to 100% and the signal from the pressure transmitter will be 4 mA for 0% and 20 mA for 100%. Intermediate values will vary in a linear manner.

At the Controller input processor, this 4 to 20 mA will be dropped across a resistor to produce a 1 to 5 V signal. For this example, assume that 3.500 V results. In the upper left corner of Figure 3.7j, the 3.500 V is fed to the first conversion stage, which tries to subtract one half of full scale, or 2.500 V. It can, so the most significant bit is tripped to a "1," and the remainder, 1.000, is sent to the next stage. The second conversion stage tries to subtract one half of the MSB, or 1.250 V, but because 1.250 is larger than 1.000, it cannot. The second most significant bit is not tripped to a "1." It is left as a "0." The remainder, still 1.000 V, is sent to the next stage, and so on. In this case, the analog value 3.500 V will be converted

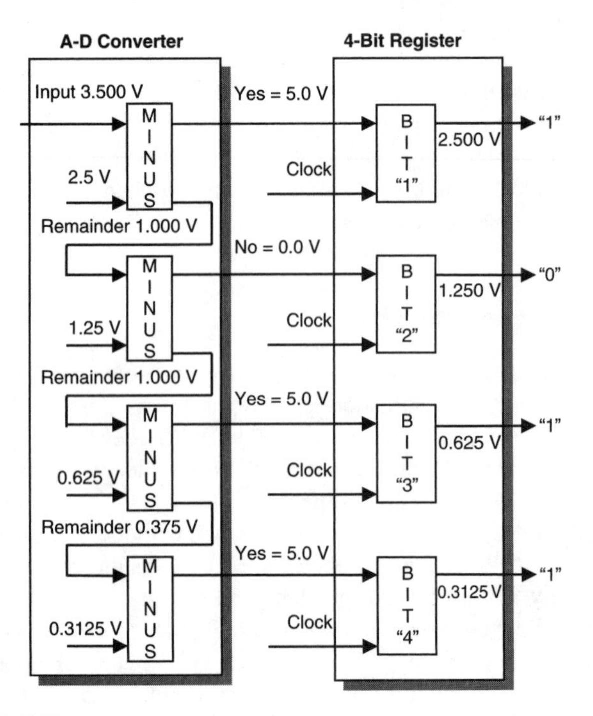

FIG. 3.7j
A/D conversion.

to a four binary digit word, 1011. Converting this back to analog yields:

$$1 \times 2.500\text{ V} = 2.500\text{ V}$$
$$+0 \times 1.250\text{ V} = 0.000\text{ V}$$
$$+1 \times 0.625\text{ V} = 0.625\text{ V}$$
$$+1 \times 0.312\text{ V} = \underline{0.312\text{ V}}$$
$$\text{Total } 1011 = 3.437\text{ V}$$

Note that this is not exactly 3.500 V, but with a 4-bit converter, we cannot expect too much. Resolution increases with the number of bits:

1 in 2 from 1-bit conversion
1 in 4 from 2-bit conversion
1 in 8 from 3-bit conversion
1 in 16 from 4-bit conversion, and so on

Most industrial A/D converters provide about 14 bits, which gives a resolution of 1 in 16,000, or 0.006%.

Logic Functions within the Controller

With the signal from each field device converted to a digital signal, the Controller, which is really a digital computer, can now operate on these signals. It can linearize the inputs, take square roots, add, subtract, etc. Certain functions that are commonly used in Controllers will be available as preprogrammed routines. Provision for programming less common functions may reside in the Controller. For example, the solution for the AGA 3 formula, to calculate fluid flow based on pressure drop across an orifice plate, will be available in most Controllers. The algorithm for calculating standard deviation in a series of inputs will probably not.

A very important function of the Controller is to communicate with the CCR. In the same way that the multiplexer connected to each field device, the HMI section in the CCR will scan each Controller to ask for updates or to provide instructions. The Controller must be able to recognize that someone is talking to it, must be able to determine if the message has arrived with no errors, and must be able to format a reply, including the information that was requested.

Digital-to-Analog Conversion

After all the logic has been completed, an instruction must be sent to the valve, motor, or switch. Since the logic has been effected in digital form and the actuator is not digital, another conversion must occur. D/A conversion happens in much the same way as does A/D conversion. For each "1" in the digital string, the corresponding voltage is added. For each "0," the voltage is not added.

DATA HIGHWAYS

A DCS is a network of digital computers, arranged to optimize reliability, cost, and operability. The fact that we refer to it as a system implies communication among the component parts, which we have agreed to call Controllers for this discussion. Think of the data highways as being serial data pathways that integrate these Controllers into a system, as shown in Figure 3.7h. These highways are internal to the DCS, essentially tying together the Controllers and HMI into a DCS. But once we are able to tie together these different digital computers, it is no great leap to connect other digital computers outside the DCS to allow the DCS to communicate with application computers, long-term storage memory, and corporate local-area networks.

Some intra-DCS communication is by proprietary methods. That means that the company that made the DCS developed a communication system for its own use and did not follow a generally available standard. In many cases, the company started with a standard and modified it to meet special conditions imposed by the needs of the company. Because similar communication needs existed in addition to DCSs, nonproprietary industry standards were developed, and DCS manufacturers are shifting away from the proprietary forms. They are finding that the reduced cost of outsourcing the standard plus the increased wealth of equipment that can access a standard highway offsets the benefits of ownership.

FIELDBUS

Recognition that large amounts of digital information could be moved reliably and quickly among Controllers in the DCS triggered the concept that digital communication could be extended to include the field devices themselves. Since the 1980s, process transmitters, which in many cases were small digital computers, were converting their output signal to a 4 to 20 mA analog form to send it to the input conditioner of a Controller where it was changed back to a digital signal. The fact that it was an analog signal meant also that only one signal could be sent on the wire at a time. It seemed a fairly straightforward step to extend the digital data highway concept all the way from the field sensors in to the Controllers and back out to the field actuators. This would allow not only the process measurement information to be brought into the DCS, but also additional data relating to equipment health and ambient conditions. Calibration checks could be made without going to the field. Range changes could be imposed on the transmitters from the control room. Multiple sensors could use the same pair of twisted wires to communicate their information to the DCS. It might even be possible to connect a transmitter directly to a valve and locate the controller (lowercase c) in one or the other field device.

It was recognized that the ambient conditions that this equipment would have to work in would be more severe than the air-conditioned buildings that housed DCS Controllers, but the transmitters with their signal-enhancing digital computers were already working in such conditions. It was known that the cables carrying the signals to and from the field devices from and to the Controllers would be passing through potentially electrically noisy areas of the factory, but it is axiomatic that digital signals are more noise immune than are analog signals. It did not seem that it should take long to define the standard that would be used to move the data among the field devices and the Controllers. In fact, it took so long to develop an industrywide standard that many equipment manufacturers developed and built their own systems and were marketing equipment built to proprietary standards before the industry standard was completed.

Fieldbus is the term given to the digital communication method for enabling field devices to talk to each other and to talk to the Controllers of DCSs. There are many fieldbuses. We can expect to see a very significant consolidation of these standards into one or two dominant ones in the next several years.

The reader is referred to Section 4 for details about the development of fieldbus and comparison of the various flavors of fieldbus.

INTEGRATION

Up until now, this section has considered some of the basic laws of communication, has dealt with the operation and application of multiplexers and demultiplexers, has discussed the concept of tying together clusters of Controllers and HMIs to form a DCS, and has mentioned the concept of a digital communication link to field devices. We now consider the factors that must be addressed to actually accomplish an integration of these various parts of the system.

The Basic Problem of Integration

We are trying to pass information that is gathered continuously in real time by a transmitter, converted into a digital signal, sent over a pair of twisted copper wires to the input conditioner of a controller, and stored in the memory of that controller. Then the information will be read from that memory, bundled with many other pieces of information, and sent over an optical fiber cable to the CCR. At this location, it will be translated from its optical form back to electrical form, stored for a short time in the memory of the HMI, be recovered from that memory, and interpreted into alphabetic characters that are presented onto a cathode ray tube or liquid crystal screen so the operator can understand them. Signals from the operator, either in the form of digital signals from his keyboard or in the form of digital signals that indicate the position of a mouse, must now be sent to a register in the HMI, be interpreted by driver programs, and be sent back along a similar route to the controller, perhaps to adjust a set point in the remote controller. The controller must know that one particular algorithm, of the many it has available to it, must be accessed and solved, using the arithmetic powers of the controller, the transmitter measurement, and the set point from the HMI. After the solution is reached, it must be stored in a register, recovered, converted from a digital to an analog signal, demultiplexed to the proper dedicated wire pair that goes to the correct valve, remembered until another value is calculated, and fed through a device to translate the analog electric signal to an analog pneumatic signal, which will move air through a steel tube to force open the valve by the proper amount. Several thousand other transmitters and valves are clamoring for their share of attention at the same time as all of this is going on.

If this sounds fairly complicated, I have been successful in sketching the complexity of what we are trying to do. The reader will notice that several media have been noted in the above paragraph. Twisted pairs of wires, optical fiber, and pneumatic tubing were all mentioned. The inference throughout the discussion was that timing is critically important to share the media among all of the potential users that are also trying to use these limited media resources. And finally, there were many languages mentioned, from analog electrical signals, to digital electrical signals, to mouse position signals, to pneumatic signals for opening and closing valves. When one considers that a company different from the one that made the controller probably made the transmitter, the valve was manufactured by a third, and the mouse by a fourth it is evident that an additional level of complexity is involved. How is it possible to integrate such a complex jumble of variables?

ISO-OSI Model

For a long time, the only way to integrate such a diverse group of parts was for a manufacturer to consider each step of the communication path, limit the types of inputs and outputs to some small, reasonable number, write communication drivers for each and every translation that was needed, build the translation hardware if media changes were needed, and try to recover the development costs. As recognition of the need for some standardization of these steps grew, standards-writing groups undertook to develop an industry standard or group of standards that could guide manufacturers in the development of such a system.

ISO is the International Organization for Standards, headquartered in Geneva, Switzerland. It and the IEC (International Electrotechnical Commission) are technical networks of national standards organizations, responsible for issuing world standards that deal with nearly everything. ISO has defined a model called the OSI (Open Systems Interconnection), and it describes all the things that have to be considered and solved for a system like the integrated

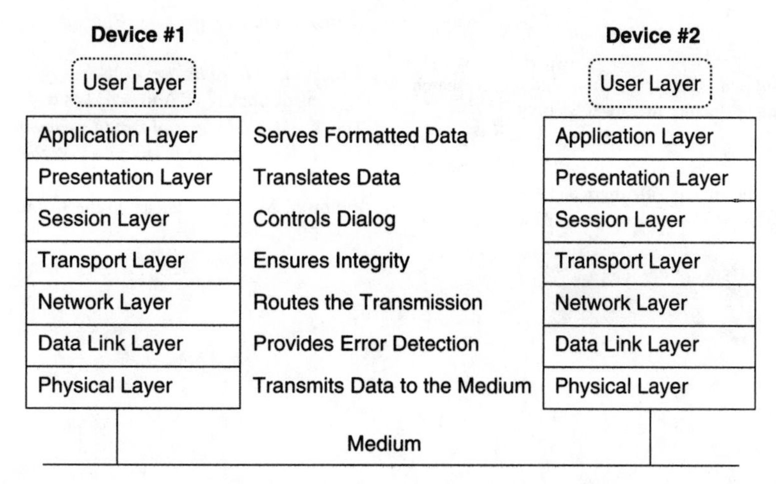

FIG. 3.7k
Open systems interconnection model—the communication stack.

control system to work. It is interesting to note that the OSI model is not limited to control systems and in fact was not even built especially for them. It in fact was built within ICS 35, which deals with information technology for business machines and was built to define any system that supposes to move data from place to place. Figure 3.7k shows the seven-layer model, with a user layer shown above each device stack. It is important to know that the seven layers describe functionality. They do not describe individual pieces of hardware or software, as will be shown in the following simplified discussion. The reason that this discussion is simplified is because more detailed discussions of OSI can be found in Chapter 4.

Starting at the bottom of Figure 3.7k, the medium, actually outside the communication stack, must be the same for both devices that we want to integrate. The first, or physical layer, describes the medium, such as a twisted pair, defines the voltage levels, connectors, and so on and provides signal modulation/demodulation.

The second, or data link layer, describes how the transmitting device will act. If it must wait until no other device is talking, the data link defines what this procedure will be. Error detecting codes, such as parity, are added by the transmitting device and are interpreted and removed at the receiving device at this layer. Note that the two layers just described provide the functions of a *modem*.

If there are only two devices connected together, the network layer is not needed and the application layer from Device 1 can connect through its data link layer/physical layer, through the medium to Device 2, and up through its physical/data link layer to its application layer. This application layer receives information from a user layer above it, formats it according to defined rules, and sends it on its way. The application layer of each device speaks the same language, and in fact cannot tell that it is not talking directly to its counterpart in the other device.

If more than two devices are connected together by the medium, the network layer defines the functionality of the addresses that are necessary for the information to reach the proper receiving device. Obviously, the more complex the communications path becomes, the more effort is required to move data successfully from one device to another.

How Does OSI Help?

OSI provides a pattern that defines the functions that must be performed to move data from one device to another, irrespective of how complex that path may be. Given that a pattern exists, standards have been written to define the steps that must be taken to implement these functions. Individual software programs have been written to match these standards and to perform the functions. The same procedure of evaluating data flow and translation must be undertaken; however, off-the-shelf drivers should now be available to translate most of the different languages. Similarly, off-the-shelf hardware devices, like Ethernet modems to realize OSI levels 1 and 2, are available inexpensively to perform most of the physical translations.

The Challenge for the Engineer

An engineer specifying a DCS and all the equipment that must be integrated with it has a responsibility to know that medium, time, and language must be considered for each two different pieces of equipment that the system will include. The engineer should plan to do extensive testing from each device to every other connected device, to ensure that they will communicate. The schedule should anticipate that equipment that has not been integrated on previous projects might not integrate as well as the engineer has been told that it will.

Bibliography

Boyer, S. A., *Supervisory Control and Data Acquisition,* 2nd ed., Research Triangle Park, NC: Instrument Society of America, 1999.

Byres, E., "Shoot-out at the Ethernet Corral," *InTech,* February 2001.

Herb, S., *Understanding Distributed Processor Systems for Control,* Research Triangle Park, NC: Instrument Society of America, 1999.

Horowitz, P. and Hill, W., *The Art of Electronics,* New York: Cambridge University Press, 1980.

Shinskey, F. G., *Process Control Systems,* New York: McGraw-Hill, 1967.

Strock, O. J., *Introduction to Telemetry,* Research Triangle Park, NC: Instrument Society of America, 1987.

Thompson, L., *Industrial Data Communications, Fundamentals and Applications,* Research Triangle Park, NC: Instrument Society of America, 1991.

Zielinski, M., "Let's Clear up the Confusion…," *Worldbus Journal,* April 2001.

3.8 Hybrid Systems with Discrete and Analog Capability

J. BERGE

- The system architecture shall be networked as far as possible.
- Both conventional and networked I/O must be supported.
- Both regulatory and logic programming languages must be supported.

When the programmable logic controller (PLC) and the distributed control system (DCS) were introduced in the early 1970s, there were several differences between them, although the basic architecture was the same. The historical importance of DCS and PLC in process and factory automation cannot be overstated. Initially, PLC only had discrete input/output (I/O) and, as the name implies, only performed fast logic control in application, mainly replacing hardwired relays and timers in manufacturing automation. DCS were predominantly analog, performing regulatory proportional, integral, derivative (PID)-type control taking the place of single-loop controllers in process automation, while being too slow for logic. PLCs did not have the redundancy and I/O module diagnostics found in a DCS. The DCS comes packaged with operator consoles tailored for the controller, whereas PLCs are the controllers alone, which have to be integrated with third-party process visualization using the appropriate driver. These differences have now been erased and the line between DCS and PLC has blurred. In the early 1990s, fieldbus technologies were introduced, which enabled a new networked architecture, where I/O is accessed over a bus and control can be done in the field. A new category of systems has evolved from the success of DCS and PLC, known as hybrid systems, which also take the concepts of distribution further. A hybrid system combines a mixture of legacy and modern technology and architecture for maximum versatility and benefits:

- Hybrid architecture: centralized and networked
- Hybrid I/O: conventional and fieldbus
- Hybrid signals: analog and discrete
- Hybrid control: regulatory, discrete, and batch

HYBRID ARCHITECTURE: CENTRALIZED AND NETWORKED CONTROL

The DCS and PLC architecture was characterized by I/O subsystems linked to a centralized controller via a proprietary remote I/O network and controllers networked to the consoles via a proprietary controller network. There may also have been a proprietary plantwide network. Field instruments usually had no communication or at best some slow proprietary communication. The complex hardware and multiple levels of networking resulted in a rather bulky and costly system (Figure 3.8a). Modern hybrid systems simplify the system architecture by migrating most of the loops into the field. Using H1 and HSE Foundation™ Fieldbus in the hybrid system reduces the networking to just two levels, where control is done in the devices in the field or possibly in a central shared controller.

In old hardwired systems, there was no communication between the host and field instruments from different manufacturers. The view from the operator console was therefore restricted to the level of the controllers or possibly the I/O subsystem in a good DCS. There was no way to obtain detailed information about the health of sensors and actuators or to configure them from the console. In the traditional architecture "system" and the "instruments" were two isolated islands. The Foundation Fieldbus technology tightly integrates system components, making field instruments an integral part of a homogeneous system. The operator workstations are referred to as the "host" of the system (Figure 3.8b).

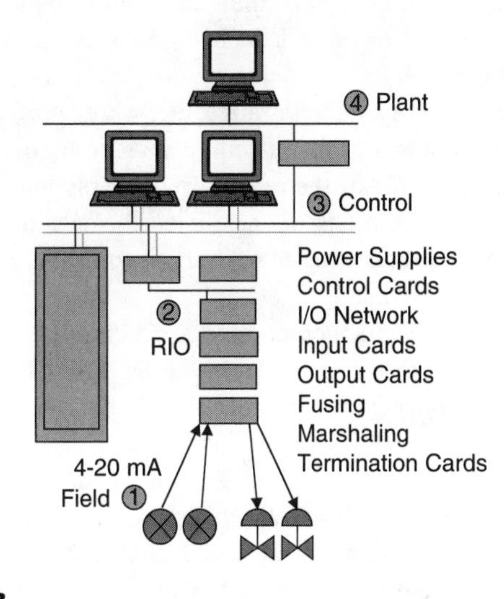

FIG. 3.8a
DCS and PLC architecture has complex hardware and networking.

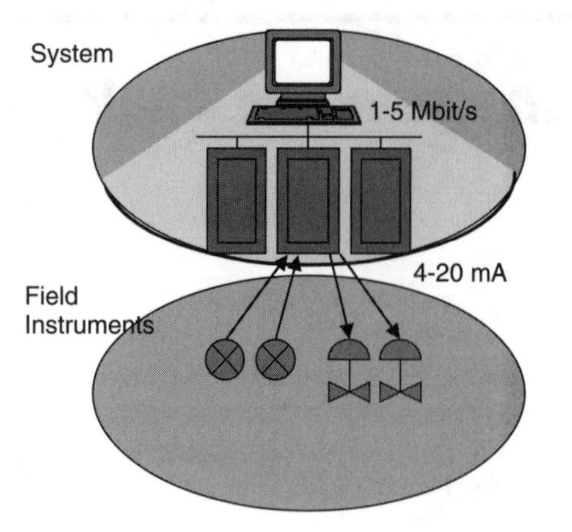

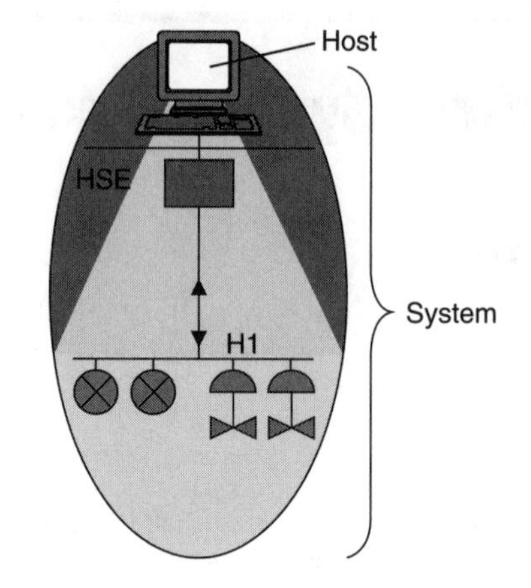

FIG. 3.8b
DCS (left) has limited view compared to Fieldbus architecture (right).

The DCS and PLC have a centralized architecture where controls are performed in one or a few controllers shared by anywhere from 30 to as many as 100 loops. Such architecture has single points of failures affecting many loops and is therefore very vulnerable. When control is done in a central controller, redundancy is a must to achieve acceptable availability, which drives cost up. This form of centralized control is also possible in a hybrid system, but it is important to use redundant controllers. Redundant controllers should be mounted in separate back planes in different panels some distance apart to minimize the number of common points of failure for the primary and secondary units. When physically separated the redundant units will not be subjected to the same stress, such as radio interference or power surges, as would be the case if the back plane were shared.

For a DCS, even a small expansion frequently became very expensive because it often required the purchase of an additional controller and other costly hardware. Too often, expensive engineering change charges were also incurred. When purchasing the DCS, users paid quite a bit extra for spare capacity to avoid this problem. The limitation of the centralized architecture is that controller resources are drained as devices are added. As instruments were connected the controller was loaded, reducing the performance. Limits of memory capacity, address range, etc. were approached. In a hybrid system this problem can be avoided by distributing as much as possible of the regulatory controls into the field while discrete control is often still performed in a central controller. This makes the hybrid architecture very scalable and easy to expand.

A hybrid system incorporates a decentralized network architecture that has evolved from the DCS and PLC architectures. The concept of distributed control has been taken further such that the devices in the field connected to the H1 Fieldbus perform PID and other functions. Because the instruments on the field-level network communicate peer to peer, sophisticated control strategies can be built involving many devices and measurements. Thus, in a hybrid system it is possible to perform control anywhere: in the transmitter, central controller, or valve positioner. When the Foundation Fieldbus function block programming language is also used in the central controller, a homogeneous system, where the same language is used throughout, is achieved. One application can be used to configure every part of the system, transmitters, positioners, and controller, all in the same language (Figure 3.8c).

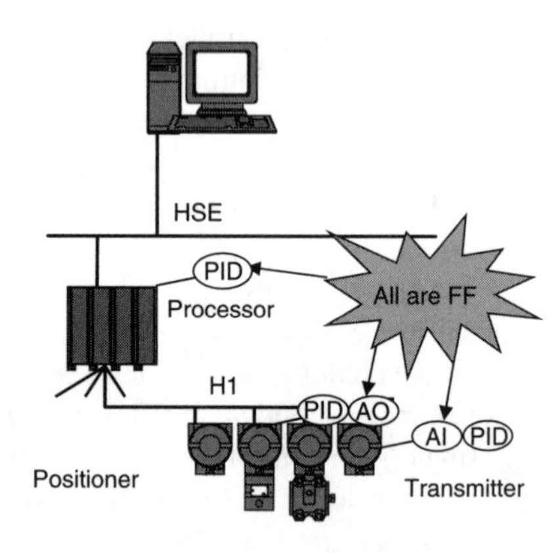

FIG. 3.8c
In a hybrid system, control can be performed anywhere using the same language.

Because the devices in the field perform most of the regulatory control tasks, the need for expensive central controllers is drastically reduced, lowering the system cost. Because no single device controls multiple loops, the fault tolerance of the system becomes higher, usually referred to as single-loop integrity, meaning that the failure of any device only affects one loop. The decentralized architecture is more scalable, making expansion simpler and cheaper. In a distributed architecture adding devices means adding more resources, such as processors and memory, making the system more powerful; therefore, adding loops has a lesser impact on the performance.

A linking device on the Foundation HSE host-level network connects to Foundation H1 field-level networks, masters the communication between the controlling field device, and gives the operator access to modes, set points,

FIG. 3.8d
Foundation HSE-H1 Fieldbus linking device. (Courtesy of Smar.)

tuning parameters, etc. (Figure 3.8d). Through the linking device a wide range of function blocks can be configured in field devices.

Fieldbus instruments increase the information available in the system by a factor of at least 200:1. Therefore, a high bandwidth host-level network is necessary to benefit fully from fieldbus. The proprietary remote I/O and controller networks in the DCS and PLC are not fast enough. Modern systems therefore have host-level networks based on the Ethernet platform. For the system to be well integrated and homogeneous, the host-level network should be an open standard and of the same protocol family as the field-level network, for example, using Foundation H1 Fieldbus at the field-level and Foundation HSE Fieldbus at the host level. This way the application and user layers of the protocol stacks are the same; only the media differ. This ensures instrument data do not become inaccessible and eliminates the need for parameter mapping. Whereas the Foundation H1 Fieldbus makes transmitters, positioners, etc. interoperable, the Foundation HSE Fieldbus makes host devices and subsystems interoperable.

The Foundation HSE network has the advantage that the protocol includes application and user layer, making Ethernet interoperable, not just connectable. It includes a standard mechanism for complete network and device redundancy over and above simple Ethernet media redundancy. An HSE device can consist of a redundant pair, e.g., a primary and secondary linking device, each with redundant ports. Switch-over in case of failure is automatic and bumpless.

HYBRID I/O: CONVENTIONAL AND NETWORKING

Although controllers and operator consoles were networked, the legacy DCS and PLC had centralized hardwired architecture where the sensors and actuators in the field are wired one-by-one to an I/O subsystem using individual signal wires. The signal wire carried only the basic analog or discrete process input variable or output value, i.e., 0 to 100% analog or on/off discrete. A hybrid system supports this type of traditional hardwired I/O for conventional devices as well as a modern distributed networking architecture for fieldbus devices. Because of the advantages of fieldbus, this technology should be used as much as possible, using conventional means only for existing equipment that cannot be replaced or for devices not yet available in a fieldbus version. A hybrid system makes it easy to integrate the modern with the existing.

A hybrid system connects to sensors and actuators in the field through discrete and analog I/O modules on a local extended back plane or as remote I/O. Remote I/O may be installed close to the sensors and actuators networked back to the controller, thereby eliminating long wire runs for the conventional I/O, with subsequent savings further reducing overall system cost. Field devices are hardwired to the modules on a point-to-point basis. The I/O modules are typically packaged with as many as 16 channels per module. Different modules exist for discrete and analog input, and output plus special modules for pulse input and temperature input. I/O subsystems tend to be complex and costly but in a hybrid system conventional I/O is very much reduced compared to DCS or PLC. A hybrid system mixes conventional I/O with fieldbus. The conventional I/O should be made available in the configuration tool as standard Foundation Fieldbus I/O block just like fieldbus instruments so that the control strategy can be configured in a homogeneous environment. Fieldbus and conventional controls integrate seamlessly.

Another alternative for a hybrid system is to use field-mounted converters for the conventional I/O that multidrop on the fieldbus network just like other Foundation Fieldbus devices (Figure 3.8e). A centralized I/O subsystem is then not required. Fieldbus converters interface directly to the linking device.

The capability of hybrid systems to deal with both fieldbus and conventional I/O makes them ideal for integration with existing systems and for reinstrumentation of old plants. This makes a smooth transition to fieldbus possible.

Field-level networks networking is the characteristic that most obviously distinguishes a networked part of the hybrid system architecture from the conventional. In a modern system field-level networking is the primary means to connect sensors and actuators. Because the remote configuration, maintenance, and diagnostics capabilities of networked devices bring many

FIG. 3.8e
Field-mounted fieldbus converter. (Courtesy of Smar.)

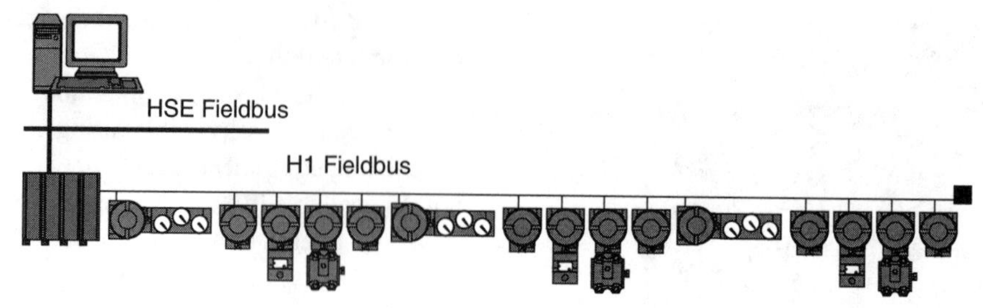

FIG. 3.8f
Multiple devices mixed on one network.

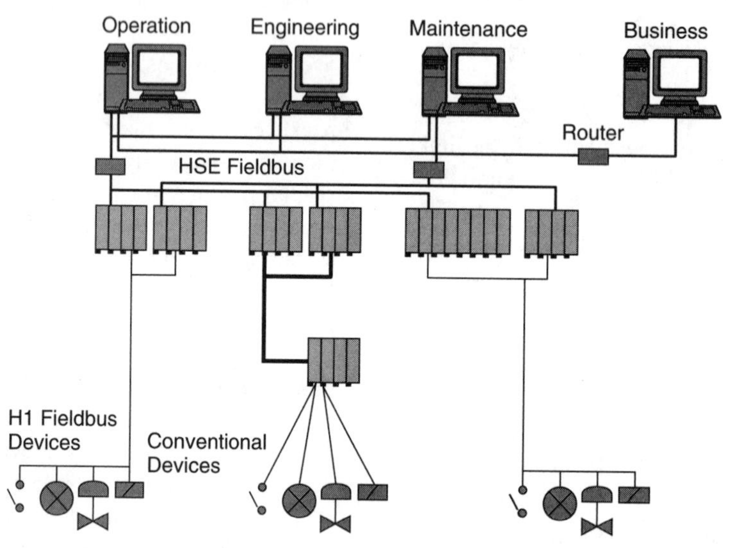

FIG. 3.8g
Redundant linking device and host-level network.

operational benefits, this should be implemented as far as possible. Networked I/O reduces wiring because as many as 16 devices can be multidropped on one fieldbus network (Figure 3.8f). These devices can be a mix of transmitters, valve positioners, on/off valves, etc. Because devices are completely digital the I/O subsystem and remote I/O networking level is eliminated. The fieldbus devices usually perform the regulatory control, drastically reducing the need for central controllers and thereby making the system cheaper and less complex than in the past; see Section 4.11.

For many legacy protocols, interfacing was as simple as adding an interface module to the I/O subsystem of a DCS and mapping the variables into the controller, because only a limited number of process parameters were communicated in the past. Fieldbus provides more configuration and diagnostics than can realistically be mapped into a database. A hybrid system must therefore be built on the fieldbus architecture including linking devices and a Foundation HSE Fieldbus host-level network incorporating dedicated software for fieldbus device, network, and control strategy configuration.

A linking device typically has some four H1 Fieldbus communication ports and connects to the HSE Fieldbus network. Devices of different types from different manufacturers can be mixed freely on a field-level network to accommodate the needs of the plant. The linking device provides the power for the field instruments all integrated as a single compact device. The linking device acts as the primary link master taking on the role as link active scheduler (LAS), i.e., it manages the communication on the different H1 networks. Redundancy is achieved by connecting a primary and secondary linking device in parallel to the same field-level network but different host-level networks (Figure 3.8g). This provides two complete and independent data paths to the operator.

The linking device ensures plug-'n'-play installation by automatically detecting and assigning addresses to devices connected to the network eliminating address duplication. Detected devices appear in the live list (Figure 3.8h).

Net 1

Tag	Id	Address
LIY-123	0003020005:SMAR-FI302:800404	0x15
LIT-123	0003020001:SMAR-LD302:801137	0x18
TIT-123	0003020002:SMAR-TT302:800640	0x19
DIO	0003020009:SMAR-FB700:800176	0x20
Bridge	0003020007:SMAR-DFI302:901967	0x10

FIG. 3.8h
Devices are automatically detected, identified, and assigned an address.

Another protocol widely used in automation is Modbus, which is found in a number of types of equipment, such as PLCs, remote terminal units (RTUs), weighing scales, flow computers, etc. A hybrid system typically has a built-in Modbus communication port; if not, an additional gateway module can usually be added. This allows data from Modbus devices contained in the registers to be used just like regular fieldbus I/O as part of the control strategy and to be displayed to the operators.

HYBRID SIGNALS: ANALOG AND DISCRETE I/O

A control system includes a mix of analog and discrete sensors and actuators. Therefore, the controller needs both analog and discrete I/O from the field. In the past the hardwired conventional analog I/O was handled by a DCS and the discrete I/O was left to PLCs. A hybrid control system can handle both analog and discrete I/O in both conventional and networked form under one roof. Mixing analog and discrete on a fieldbus network is easy because all devices appear electrically identical on the network and are wired the same way. For conventional hardware signals, the I/O subsystem has to be fitted with different types of modules to handle the various kinds of signals. The I/O subsystem can be either local, i.e., in the same backplane as the processor module, or remote in a separate back plane connected to the processor via a remote I/O network. Remote I/O makes it possible to distribute racks with I/O modules at various locations in the field close to the sensors and actuators, thus reducing hardwiring.

Discrete "binary" devices include proximity switches and solenoid valves used in logic and interlocks. Although some discrete devices such as on/off valves, electrical actuators, and converters exist in fieldbus versions, the discrete devices are still often of the conventional hardwired type. The I/O subsystem is then fitted with modules of different types to meet the requirement of the plant. Typical types include AC input, DC input, transistor output, relay output, and triac output. DC input usually comes with the option of either an NPN or a PNP sensor type. The input modules are usually available for different nominal input voltage levels. Transistor outputs are used for DC loads and triacs for AC loads. Relays can be used for high current and voltage of both AC and DC, but have limited speed and lifetime.

Analog "scalar" devices include transmitters and control valve positioners used in continuous regulatory control and monitoring. Most analog devices exist in fieldbus versions, and therefore analog devices are now usually of the networked type. If, however, conventional devices are used, the I/O subsystem is fitted with modules of different types to meet the requirements of the plant. Common types include analog input, analog output, direct temperature sensor input, and pulse input. The analog I/O modules usually accept 4 to 20 mA and 0 to 10 V, etc. Pulse input modules are capable of measuring frequencies up to 100 kHz, e.g., from turbine or vortex flow meters. The best way to measure temperature is to use field-mounted fieldbus temperature transmitters. However, a hybrid system also supports direct input of thermocouple and RTD (resistance temperature detector) sensors linearized and compensated according to various standards mixed at the same time.

HYBRID CONTROL: REGULATORY AND DISCRETE LANGUAGES

In the past, a DCS handled the regulatory controls using a control strategy configured in a function block diagram language. The PLC handled the logic using a control strategy configured in a relay ladder logic language. These languages were different for every manufacturer. A hybrid system supports both the function block diagram and ladder diagram languages. Moreover, these languages now follow standards, eliminating the need for proprietary languages. In a hybrid system the most suitable language for each control task can be chosen and yet be integrated with the others.

Logic used for interlocks and permissives is typically configured using the IEC 61131-3 ladder diagram language (Figure 3.8i). Because the discrete I/O used in the logic typically comes from conventional discrete I/O, the logic is usually executed in a central shared controller, such as a linking device processor. Whenever a shared central controller is used, it is a good idea to have a redundant configuration to ensure availability. A significant part of the logic in the past was used in conjunction with the regulatory controls to handle shutdown in case of failure and reset windup protection of valves that had reached end of travel, etc. These functions are built into the Foundation Fieldbus function block language, and therefore the amount of logic required is drastically reduced. The ladder diagram language is the most widely used language for logic since it is graphical. Typically, the ladder diagram functions supported include various contact inputs, output coils, numerical, selection, arithmetic, comparison, counters, and timers; see Section 3.12.

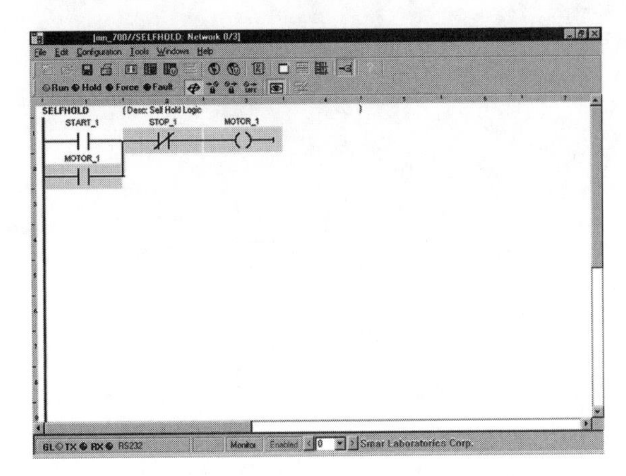

FIG. 3.8i
IEC 61131-3 online monitoring in configuration tool.

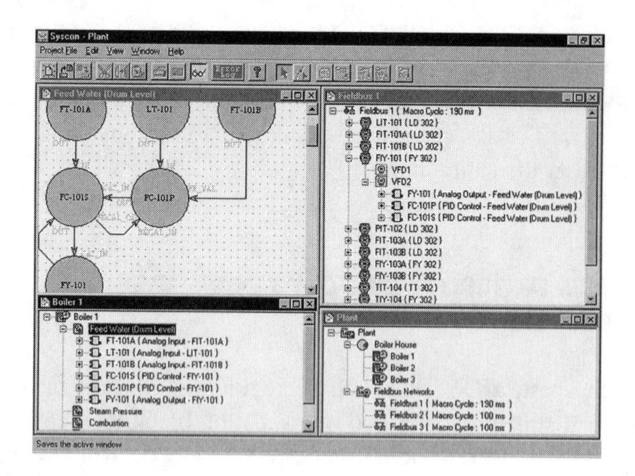

FIG. 3.8j
Foundation Fieldbus H1 and HSE network, device, and control strategy configuration.

Regulatory control of temperature, pressure, level, flow, etc. is typically configured using the Foundation Fieldbus function block diagram language (Figure 3.8j). In a hybrid system, most of the regulatory control devices are fieldbus based and the controls are executed in these, but function blocks may also reside in a central shared controller such as the linking device. The Foundation language includes blocks for input, output, controls, and arithmetic including calculation, selection, alarms, etc.

Bibliography

Berge, J., *Fieldbuses for Process Control and Engineering, Operation and Maintenance,* Research Triangle Park, N.C.: Instrumention, Systems, and Automation Society, 2002.

IEC 1131-3, "Programmable controllers—Part 3: *Programming Languages,*" IEC, 1993.

3.9 SCADA—Supervisory Control and Data Acquisition

S. A. BOYER

Partial List of Suppliers: quindar.com; abb.com; foxboro.com; honeywell.com; neles.com; tessernet.com; emersonprocess.com; modicon.com; controlmicrosystems.com; pickford.com; demanddataservices.com; autosoln.com; datap.ca

BACKGROUND

SCADA (supervisory control and data acquisition) is a control technology that has been developed for controlling geographically large processes. Pipelines, oil and gas production facilities, electrical generation and transmission systems, and irrigation systems are examples of processes that can be operated using this technology.

Extending the Operator's Reach

With geographically smaller processes, it is practical for the human operator to visit each of the process sensors to monitor the health of the process. In addition, it is effective, if the process be small enough, for the operator to go to each valve or motor that must be controlled and to make any necessary adjustments.

Soon it was noticed that even for small plants, the operator spent too much time walking and not enough time operating. It became necessary to extend the reach of the operator. Control systems were developed, first using pneumatic signaling, and later using electrical signals, to bring the information from the process to the operator and to take the operator's instructions to the valves and motors in the process.

Processes that were large, but not too large, could be wired in the same manner as were smaller plants. Individual cables to devices, and home run cables, often buried in the same way as are telephone cables, were used to communicate from operator to field. This type of control is not SCADA. It is direct control with the controller continuously communicating with the device. Direct analog and direct digital control uses this concept very effectively over distances up to a couple of kilometers.

The cost of a pair of wires connecting each sensor and actuator to operator interfaces at the central control room became higher as the distances increased, often costing ten times as much as the field device itself. As electronics became less costly, multiplexers (see Section 3.7) were used to allow signals from many devices to travel over one pair of wires.

The operation of geographically large processes still requires that sensors be monitored and valves and motors be controlled. Before the development of SCADA, telephones would be used to enable a field supervisor to instruct operators which information to gather and which actuators to adjust. The operator often had to drive, and sometimes to fly, from location to location in order to do so.

During the 1960s, improvements in communication technology were combined with improvements in human–machine interface (HMI). This resulted in a system that greatly extended the reach of the operator. With these technologies integrated properly, the operator or supervisor could sit in a comfortable central location and monitor the flow of oil through a pipeline located 15 miles away—or 1500 miles away.

Concept of Supervisory Control

That technology was still not SCADA. Early systems were simple telemetry, with communication in one direction, from field to control room. They still required that an operator be dispatched to the field site to adjust the pressure or to turn a motor on or off. The problem was that, although the communication system could reach a long distance, there was nothing available at the remote site that was capable of understanding and remembering any instructions. A latching relay could remember the simple instruction "Close," but translating that instruction from the operator, through the communication system, to the relay took the development of inexpensive programmable devices. When microcomputers became available, it was possible to build a machine that could understand the instruction from the radio, remember the instruction, compare the instruction to the current situation, and drive the relay to the proper position. Now the operator, sitting in a control room, could supervise the process by monitoring the positions of switches and sending instructions to something at the site to open and close relays. That "something," which later became known as an RTU (remote terminal unit), was doing the actual control because it was hardwired to the relay. It could also be hardwired to status switches on the field devices

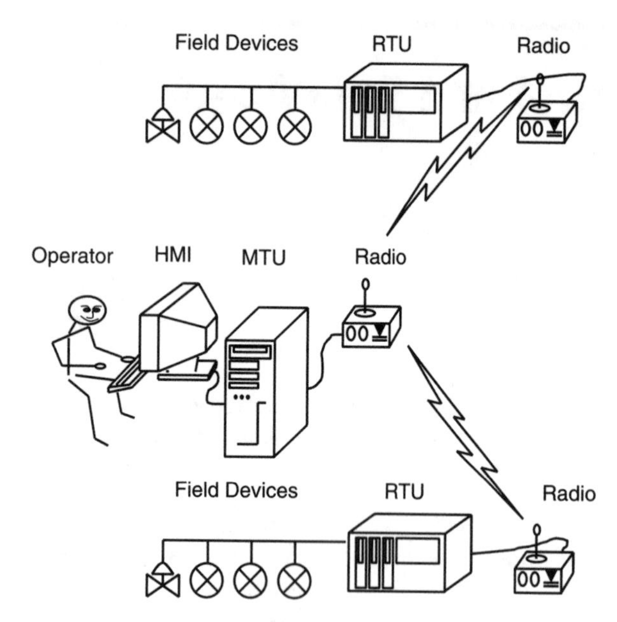

FIG. 3.9a
Simple SCADA layout.

to provide feedback to the operator that the device had moved to the required position, allowing the operator to supervise the operation of the RTU control.

Even early RTUs were capable of analog control. At prices of U.S. $80,000, RTUs in the late 1960s were equipped with drivers, or special programs, that could translate orders from the RTU to process controllers, adjusting their set points. Bringing back the process value to the RTU was simple and again completed the link to allow the operator to supervise the RTU. Supervisory control, in this context, can be thought of as instructions issued intermittently and carried out continuously.

Ultrahigh-frequency radio was used as the communication medium for these early systems. With a base radio near the operator, and smaller radios near each of the process equipment concentrations, a communication system could be built to cover very large areas for a much lower price than could be done with buried cables.

Figure 3.9a shows the layout of a simple SCADA system to illustrate the relative locations of the various important elements. Later in this section, more detail of this layout and other architectures will be considered. The operator views and controls the process through an HMI that is connected to a master terminal unit (MTU). The MTU, which in some industries is called the host computer or host station, connects through the communication system to RTUs at locations of concentrated process data and/or control points. Each RTU connects to process measurement or process actuator devices. Note that if data were moving only from the process device to the operator, the system would be a data telemetry system or a monitoring system. The fact that SCADA systems move data both to and from the operator requires that the communication system be able to operate in both directions.

BASE TECHNOLOGIES

SCADA is a technology that comprises many other technologies. Examples shown in Figure 3.9a include process measurement, data communication, logic, HMI, and actuator operation.

Process Measurement

In order for an operator to tell what is happening in a far-flung process, instruments to measure the process parameters must be installed in the process and connected to the RTU. Almost none of these instruments is exclusive to SCADA systems, and the reader should refer to other sections of the *Instrument Engineers' Handbook* for details of how process measurement instruments work and the limitations and strengths of each measurement technology.

The important consideration about measurement equipment used for SCADA is the method or format of the output. Almost all process measurement instruments were originally designed for visual output. Later, output methods were added to allow them to interface with electronic systems. There are three common families of output method used for SCADA. The first and simplest is the discrete output. Switches and relays provide discrete output. The second is analog output, with the process measurement transmitter outputting a voltage or current proportional to the process measurement. The third, and most complex, is serial output. The measurement device outputs a digital signal that describes one or more parameters of interest.

Discrete outputs can be described by considering the status indication of an electrical motor contactor as an example. When the contactor is in the "on" position, a visual flag shows red. When the contactor is "off," the flag shows green. This output method is fine for operator looking at the contactor, but another method is needed to tell a local annunciator, or an RTU, the position of the contactor (Figure 3.9b). An electrical switch is added to the contactor in such a way that the switch will be open in one of the contactor positions and be closed in the other position. Because there are only two positions that the contactor can be in, a discrete signal like a switch can unambiguously describe the contactor position. Notice that many people call such signals "digital inputs" or "digital outputs," and while it is arguable that they

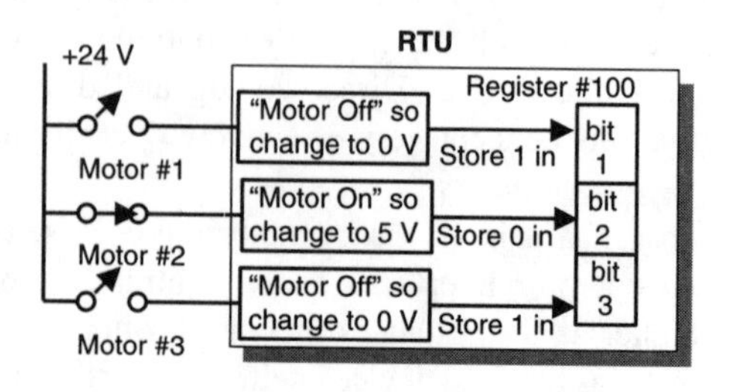

FIG. 3.9b
Discrete inputs to RTU.

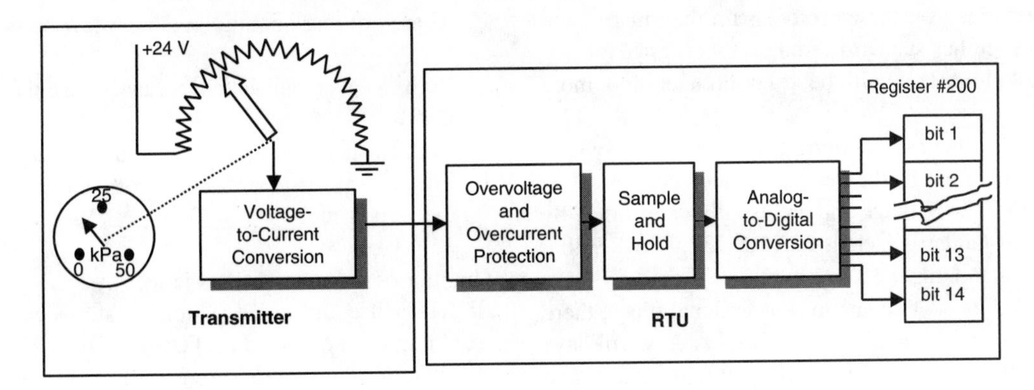

FIG. 3.9c
Analog input to RTU.

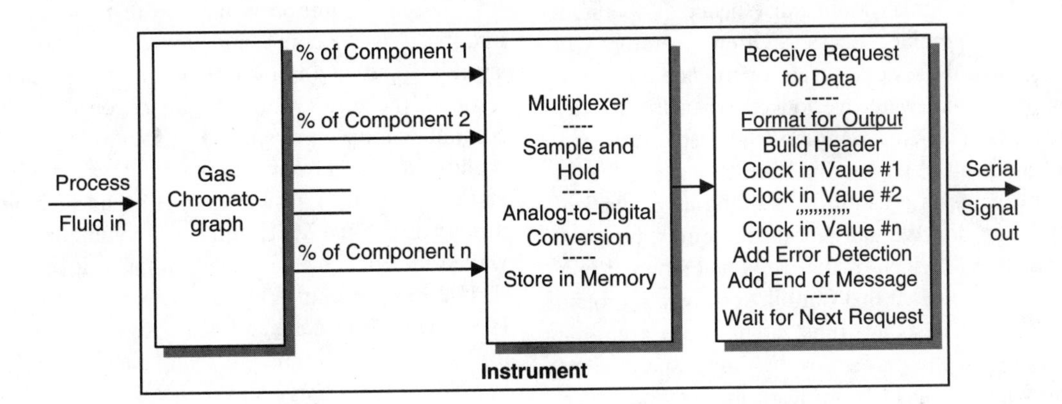

FIG. 3.9d
Instrument with serial output to RTU.

are 1-bit digital signals, the term *discrete* instead of *digital* better differentiates them from serial digital signals, which are discussed later in this section. The short form description "DI" or "DO" applies whichever term is used.

The voltage level that is applied to the switch is determined by the power supply provided for the RTU. The switch is a passive device and the RTU supplies the electrical power to the circuit. Because the RTU is a solid-state device, a low voltage of 5 to 24 V is usually used. Because the RTU input circuitry can be designed to be high impedance, very little current is needed and so voltage drop will be insignificant for these signals. Most, perhaps all, discrete signals into an RTU are direct current.

Analog outputs from a transmitter can be described by considering pressure measurement (Figure 3.9c). Most process pressure measurement is done by circular visual pressure gauges, whose needles are driven by the change of shape of a Bourdon tube. With the advance of automated systems, additions to the Bourdon tube to translate the physical movement of the tube into an accurate, repeatable pneumatic or electrical signal resulted in "pressure transmitters" whose output could be interpreted by the receiver or by the RTU. Similar modifications of temperature, level, flow, and other instruments were made to allow remote reading of instruments measuring these parameters.

The power to transmitters is usually, but not always, supplied by the RTU. Accordingly, the voltage levels will be in the 5 to 24 V range and will be direct current. Because the impedance of the transmitter is lower than the impedance of a switch, higher voltages are more common, and voltage drop in the circuit is significant. The transmitter usually outputs a signal whose current, varying between 4 and 20 mA, represents a range of 0 to 100% of the process parameter. It is obvious from a comparison of Figures 3.9b and 3.9c, that the generation and formatting of an analog signal is much more complex than for a discrete signal.

Serial outputs from a field sensor were developed when several parameters were available from the sensor. An example is the multiple outputs, one for each measured component, from a gas chromatograph (Figure 3.9d). Proprietary serial codes in early equipment made it difficult for the RTU to understand what the serial signal meant, and "drivers" that could translate the signal to the RTU became a selling point for RTUs. Gradually, *de facto* serial standards emerged, and many manufacturers used these protocols for their instrument serial signals. Fieldbus standard serial protocols are expected to provide solutions that will be even broader and more general.

The power for these serial signals is sometimes provided by the RTU, sometimes by the device. Voltage levels are low, usually in the 5 to 24 V range and are usually determined by communication standards such as EIA RS 232, EIA RS 485, or the fieldbus standard.

In addition to these three main sensor signal types, there are many others and designers of SCADA systems will have to ensure that the RTU they buy will be able to understand them or they will have to provide translation drivers. Examples are voltage pulse outputs, thermocouple millivolt signals, and pulse width signals.

When designing any remote control system, of which SCADA is one, the designer should enlist the aid of operators of the process or of similar processes. Often, operators will learn tricks to make their work easier, or to enable them to measure conditions that were overlooked when the facility was originally designed. An example is the case of an operator who had the habit of putting his hand on each of the oil well production pipes in a manifold building as he walked down the 30-ft corridor. Without any further effort, he knew at the end of his walk which of the wells had stopped producing. A warm pipe meant that hot oil from below ground was flowing. SCADA designers must either be very familiar with the operation of the process or they must spend some time working with the operators to learn the methods operators use to keep the process going. In this case, the operator did not realize that he was doing this, until the designer asked him why he put his hand on each pipe. A similar temperature measurement function was designed into the SCADA system that was installed at this facility.

Data Communication

Another look at Figure 3.9a will show that data communication is at the center of the SCADA system. This long-distance communication, and the fact that it imposes the need for supervisory control rather than direct control, is the element that characterizes SCADA as different from other process control systems.

The communication system that was modified to serve early SCADA systems was a relatively simple technology with a low data rate. As discussed later in this section, this is not necessarily true now.

Ultrahigh-frequency radios, operating in the frequency range of 300 to 1000 MHz, meet the distance requirements of most SCADA systems. They have essentially line-of-sight paths, which provide about 20 miles of range over level terrain. Frequency modulation, rather than amplitude modulation, is usually used because of its greater immunity to noise. Modulation data rates are surprisingly low, often in the range of 300 to 1200 bits per second (bps).

There are three conditions that allow communication:

1. Both speaker and listener must speak the same language.
2. Both speaker and listener must use the same medium.
3. Speaker and listener cannot both speak and listen at the same time.

The first of these conditions is met by having the MTU and all RTUs use the same communications protocol and by providing both MTU and RTUs with identical configuration information. Protocol is discussed later in this section.

The condition that they both use the same medium will be met if all of the units use radio, at the same frequency, and with the same modulation technique, to communicate the information that they have to exchange.

The last condition is the one that forced early systems into master/slave communication. If the MTU were to ask if any RTU had anything to report and more than one tried to respond, the result would be interference that would make communication impossible. The solution that worked was to prohibit any RTU from talking until it was asked a question—and requiring it to answer even if it had nothing to say. The result was that the MTU initiated a communication to RTU 1, waited until RTU 1 answered, initiated a communication to RTU 2, waited until RTU 2 answered, and so on. When all RTUs had been asked, and had answered, the MTU returned to RTU 1 and continued to scan.

Notice that this system requires each RTU to respond in order to trigger the MTU to move on to the next RTU. What happens if one of the RTUs cannot respond? Generally, a rule is established that the MTU will wait for each RTU for a small amount of time, and then will try again to talk to the same RTU. If necessary, it will do this three times before deciding that the RTU is not going to answer. It will then continue its scan and will try the noncommunicating RTU one more time on the next scan. If it still gets no answer, it will store a "communication alarm" in its memory and will tell the operator that the RTU is not working. Depending on the way the MTU is programmed, it may not try to talk to the RTU again until the operator tells it that the RTU has been fixed.

This system requires that each of the RTUs have its own name so that it will know if the MTU is talking to it. That is one more thing that must be handled by the protocol. Notice that the MTU does not need a name in master/slave communication, because any RTU can talk only to the MTU.

As systems became larger, it was noticed that the MTU could not do any communicating while it was waiting for an answer from an RTU or while that RTU was responding. To get around that inefficiency, a second radio frequency was added to the communication system. Figure 3.9e shows how this works. Now the MTU would transmit on frequency 1 (F1) and receive on frequency 2 (F2). All RTUs would receive on F1 and transmit on F2. Note that only one RTU can transmit at a time; otherwise the interference from more than

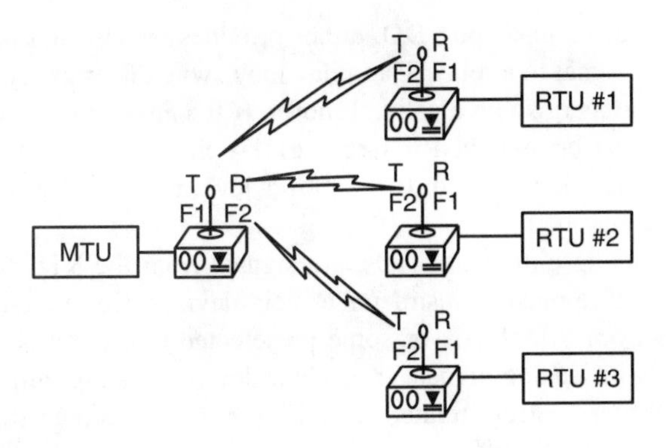

FIG. 3.9e
Master/slave communication.

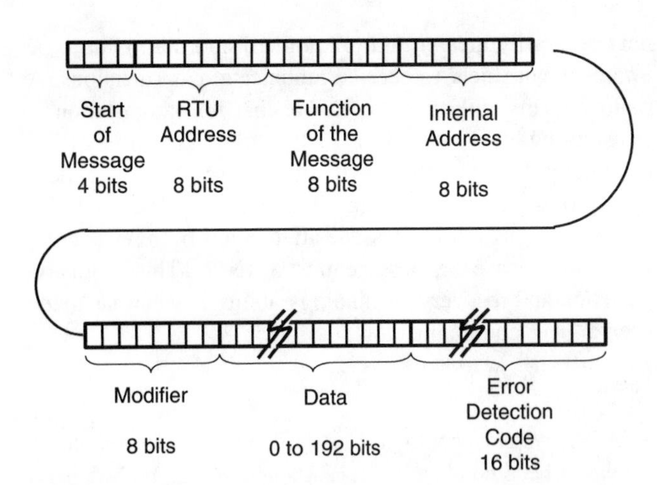

FIG. 3.9f
Simple protocol elements.

one RTU would make both signals unintelligible. Note also that none of the RTUs can hear any RTU and that the MTU can hear one RTU while the MTU is transmitting to another. Allowing the MTU to talk to the second RTU while it was still listening to the first RTU resulted in an increase in efficiency of the communication. It increased the scan rate of the system, which is another way of saying that it decreased the polling time or polling period.

A system that allows communication in only one direction is called "simplex." A system that allows communication in both directions, but only one direction at a time, is called "half duplex." A system that allows communication in both directions at the same time is called "duplex." The paragraph above describes the MTU operating with duplex communication.

A slave station cannot talk until it is told to. But as soon as RTU 1 hears a message directed to it, it will analyze the message for communication errors, store the message, formulate a response, turn on its transmitter, wait for the transmitter to stabilize, modulate the transmitter carrier to send the message to the MTU, and turn off its transmitter.

When the MTU hears the response from RTU 1, it will analyze the message for communication errors, store the message, and send a message to RTU 2. It continues to scan each RTU in the system, sometimes providing command information, but more often gathering information from the RTU. Note that the RTU cannot send even the most urgent alarm to the MTU until the MTU initiates communication by sending a message to it.

These messages are designed to be efficient. They must follow a very strict set of language rules, so that the receiver will understand exactly what the sender means by each bit that is sent. This set of rules is called the "protocol." There are hundreds of different protocols, and it is possible for an MTU to communicate to two different RTUs using two different protocols—but both RTU and MTU must know which protocol to use for each conversation. Most SCADA systems use only one protocol for all their RTUs.

Figure 3.9f is a block diagram of a simple protocol. The bit on the upper left of the diagram is the first to be modulated. First, a series of 4 bits is sent to allow the clock in the RTUs to synchronize with the clock in the MTU. Next, the address of the intended receiver is sent. An address of 8 bits will enable a maximum of 256 RTUs to be addressed. At this time, all RTUs except the one whose address matches this address will ignore the rest of the message. Next comes a description of the function of the message that is being sent, which might be a request for a group of accumulator register data, a command to turn off all the described motors, or a request for the status of the described status switches. Following that is the internal address of the first RTU register to which the message is directed. The modifier describes the number of registers after the one named, and so describes the total of the registers of interest. It may also describe the number of data words that are included in the message. Next comes a section of variable length that contains data that are being sent. It may be "0" bits long if the MTU is just asking for information, or it may contain as many as 192 bits if the MTU is issuing a command. Finally, an error-checking code, which the RTU will use to ensure that the message contains no errors, will be added to the end of the message.

Notice that while all the bits in this message are necessary, less than three quarters of them are process data. Note also that at a data rate of 1200 bps this message took less than a fifth of a second to send and the response will take about the same time. When one considers this plus the fact that the RTU transmitter settling time, which is the time that the radio transmitter takes to stabilize after it has been turned on, may be as much as one half a second, one can see that the "communication overhead" can exceed the communication time. For this reason, doubling the modulation rate will not double the communication rate.

Peer-to-peer, an alternative to master/slave, communication that allows any station to initiate communication has been developed. This system allows an RTU to send a message whenever any important parameter, such as a fire alarm, changes. Systems of hundreds of RTUs may use this method without filling the airwaves with useless communication, but each RTU must be checked occasionally to ensure that it is still operative. With peer-to-peer communication, silent does not necessarily mean healthy. Another important point to be aware of with peer-to-peer communication is collisions. A radio modem will check to ensure that no other station is transmitting before it turns on its transmitter, but if any RTU can talk at any time, there will still be cases of two or more trying to talk at the same time. Provision must be made for each to recognize if its message attempt has been unsuccessful, in which case the message must be resent. This "collision detection and recovery" technology ability is what has made peer-to-peer communication possible.

Logic

SCADA systems could not be built until electronic logic packages became affordable. RTUs and MTUs both depend on programmable devices reliably repeating fairly complex activities.

The functionality of many of today's RTUs is so common that all their programming is written in firmware and most RTUs are produced as special-purpose packages. All that is required to set up a new remote station is to connect the RTU to a radio/modem, plug it into a power supply, connect the field devices, and do some very simple configuration. In spite of this apparent simplicity, there is still a lot of logic required for it to function. Checking for communication errors, understanding what the MTU is asking for, temporarily storing switch and transmitter values, formatting instructions to field actuators, formatting messages to be sent to the MTU—all these require some level of logic.

MTUs, on the other hand, are usually much more flexible and their hardware exists as general-purpose computers. The competitive advantage of an MTU exists in its software, including its operating system, which may be proprietary, and the special application programs that form its functionality. Industry-related applications will be discussed later in this section, but several applications common to most SCADA systems will be discussed here.

As with RTUs, the MTU must be able to handle communication functions, but since the MTU is responsible for talking to many RTUs, the logic of MTU communication is more complex. The scheduling of RTU scans is one of the communication functions for which the MTU is responsible.

MTUs need to be able to present information in ways that humans can understand. The logic for this HMI will not be discussed here because it is covered extensively in Section 3.1, but it is extensive, requiring database management, scheduling, and flexibility. Finally, the configuration of the field, including the recognition of where the status of each field device is stored in each RTU, the units of measurement of each field device, and other such details, requires the application of logic at a very high level.

Field Device Actuators

RTUs send data to field devices using the same modes as they receive data. Discrete outputs, analog outputs, and serial outputs represent more than 95% of such signals.

A discrete output, DO, either provides or does not provide a signal to a relay. That relay may switch instrument air or may energize an electrical motor. If the amount of power that must be switched is large, a relay in the output of the RTU may be provided. If the power level is smaller, a transistor may be used.

Analog outputs, or AOs, are signals from the RTU that are used to make adjustments to field devices over a continuum from 0 to 100% of some preselected range. As is the case with analog signals from field devices, analog outputs usually are current related, and the normally accepted standard is 4 to 20 mV.

Serial outputs are a more recent method for signaling from an RTU to a field device. Normally, the same protocol is used for an RTU to send information to a device as is used for the device to send information to the RTU. Sometimes the device will have a proprietary protocol, but more often, a standard protocol, either supported by a standards agency or copied from a very popular, established format, will be used.

RELIABILITY

SCADA systems are distributed over very large areas. The elements of the system, consisting of RTUs and their associated field devices, are often located in places that are difficult to access. Communication systems that tie these widely separated parts together are subject to noise, equipment breakdown with long mean time to repair, and high preventative maintenance costs. Electrical power systems for remote locations, because they usually have relatively few customers, are not designed to be as robust as are locations in high-population areas. The combination of communication and electrical problems results in a level of reliability for SCADA systems that is lower than the reliability of most process control systems.

There is a quality of "chicken and egg" to this reliability. Because early SCADA systems were acknowledged to be low in reliability, SCADA-system design procedures included techniques to minimize the effects of loss of communication or loss of electrical power at the remote site. And, because these mitigation techniques worked well, less effort was expended on solving the reliability issues.

Communication

Communication noise can be thought of as any electrical or electromagnetic effect on the system that interferes with the data being communicated. Lightning and solar storms are two sources that come to mind, but in addition, the interference from other radios, intermittent power supply failures, changes in radio path attenuation, and surges from electrical equipment on the same circuit as the communication circuit will introduce noise into the system. One SCADA design maxim is to plan for the system response *when* the communication failure happens—not *if* it happens. Murphy's law says that the failure will happen at the most critical time of operation.

This combination should indicate that one does not rely on the MTU being able to communicate to the field during times of process upset. In fact, systems responsible for personnel safety, environmental release prevention, or major equipment protection should not depend on the communication system working. SCADA designers have known this for decades, and build subsystems at the local site that operate to bring the process to a safe condition even when the SCADA communication system is down. This is not different from the modern process control design technique of keeping the safety instrumented system (SIS) separate from the basic process control system; refer to Section 2.11 for more information on SIS.

An exception to this principle is the detection of and response to a pipeline leak. Most pipeline leak detection systems measure the amount of fluid going into a pipe and subtract the amount of fluid going out of the pipe. If more goes in than comes out, there is a leak. Factors like fluid compressibility and pipe stretching based on temperature and pressure can be built into the model. To measure the inlet and outlet volumes in a timely manner, a SCADA system is needed. After the leak is detected, pumps along the line must quickly be turned off and block valves must quickly be closed to isolate the leak and limit the loss of fluid. Again, this can only be accomplished by using a SCADA system. Recognizing the exposure that exists when SCADA communication fails, it is obvious that backup systems should be put in place to act when the primary leak detection system is out of service. Examples of these backups could include local "rate of change of pressure" sensors at pump stations and block valve locations, complete with hardwired local logic to shut off the pump and close the block valves.

Many SCADA systems have a central hub through which all data communications must pass. Providing redundancy in this hub, often by means of a "hot standby" radio link, is effective in increasing the reliability of the system for a small increase in capital cost. A hot standby radio continuously monitors the operation of the primary radio and assumes the responsibility when the primary fails.

Uninterruptible Power Supplies

Failure of electrical power systems will affect more than the communication system. It will result in failure of electrically powered regulatory control systems, SIS, and SCADA systems.

Mitigation of utility power failure is addressed by including a UPS (uninterruptible power supply) in the site equipment. UPSs store electrical energy in rechargeable batteries and release this energy when the utility power system fails. For power users that can operate on direct current, and most electronic equipment fits into this category, the UPS can be as simple as a battery that is kept charged from the utility power system and that continuously delivers DC power to the load. For loads that require alternating current, an electronic alternator fed from the battery must supply power at the frequency needed by the load. The added complexity of the alternator and the switching associated with it reduce the reliability of AC UPSs.

Several layers of UPS may be required at a site. First, the clock and memory of the RTU must be supplied continuously with power, but this is not different from any computer. A small, dedicated battery with a life of several years is used for this. Next, supply to critical sensors and RTU logic, sufficient to maintain the site until maintenance can restore normal power, should be made available. For equipment measuring government royalty volumes of gas or oil, there is often a requirement for 35 days of RTU memory and UPS power at a SCADA site. Third, supply to the RTU radio can usually be much shorter. Remember that the system has been designed to operate safely with the communication system down.

TOPOLOGY, ARCHITECTURE, AND NEW COMMUNICATION METHODS

Figure 3.9g illustrates several of the topologies that data communication can build on. On the left, a simplified version of the communication system from Figure 3.9a shows that the equipment is arranged with the RTUs surrounding a MTU, which forms the "star." This layout is the basis of the master/slave system, discussed earlier. One can see that any RTU can communicate with the MTU, but cannot talk to any other RTU.

The star topology is also used by cellular telephone technology. Cell phones are starting to be used as the communications basis for SCADA systems in areas where the cell infrastructure already exists. SCADA does not require continuous exchange of data, and the data that it must exchange can be concentrated to short packets. The method is called cellular digital packet data (CDPD). It is not available on all cellular telephone systems, but where it is, it can be a very effective communication solution. One of the benefits that can be taken is that the address can be an Internet protocol (IP) address instead of a telephone number. Transmission control

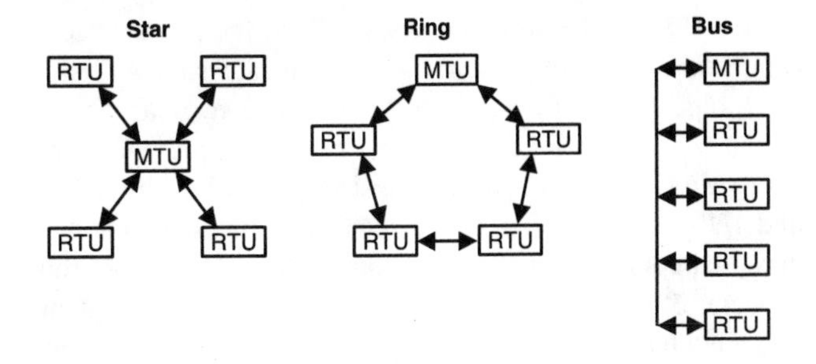

FIG. 3.9g
Data communication topologies.

protocol/Internet protocol (TCP/IP) can be used to move the packets across the cellular system. With proper attention to security, these IP-addressed RTUs can be accessed through the Internet. CDPD compresses and packages digital data and fits it into the pauses between words in normal telephone conversations, allowing large amounts of data to be moved to and from MTUs for only a very small cost. This communication method is relatively new for SCADA, but will be used more and more in the future. "Report by exception," the principle of RTU reporting only when some condition has changed, is used in a peer-to-peer mode to further reduce communication costs. Most cellular telephone companies charge their data customers on the basis of the number of bits of data that have been transmitted, so design effort to concentrate this data can pay for itself very quickly in such a system. Report by exception, by making the scan unnecessary, reduces the number of bits needed to pass along the information that really counts. An RTU reports only when it has something critical to say.

Another star topology technology that is gaining popularity is geosynchronous satellite communication. In the same way as RTU hardware prices have dropped from tens of thousands of dollars to mere hundreds of dollars, the price of fixed Earth stations that transmit to and receive from satellites has dropped to thousands of dollars. Communication systems that use satellites are not limited to well-populated areas with extensive communication infrastructure, but can be installed in nearly any location on Earth. In addition to cost reductions for capital equipment, lowering operating costs for data transfer are becoming very affordable.

The center part of Figure 3.9g shows a "ring" configuration, or daisy chain, in which any station can talk to any adjacent station. Peer-to-peer communication, installed with the ability to "store and forward" a message, allows a message to be passed from the MTU through one or more RTUs to an RTU that is outside the direct radio range of the MTU. This system is often used in modern SCADA systems to reach one or more distant RTUs. Pipelines are an example of an industry where the geographical layout of the process encourages this type of communication. Note that the confirmation waiting time of the MTU increases significantly for each extra hop in the communication link. Increasing the number of hops of the communication also results in increasing the likelihood of communication errors, and this is a factor to consider in the layout of a system.

Combining the store and forward and the repeater concept, some SCADA systems actually locate sub-MTUs at these repeater locations. A sub-MTU is one that can autonomously scan the RTUs within its range and provide some instructions and longer-term data storage, but which itself is under the control of a central MTU. A pipeline that has pumping/compression stations along its length may use this concept. If these stations also receive or sell pipeline product, they may require local administration that can justify or benefit from a local MTU. When this happens, the infrastructure at the sub-MTU will probably have increased to the point that a high-speed communication system, capable of supporting voice and high-speed data, will be needed. The signals from the MTU to the sub-MTUs will be able to be included in this high-speed data system.

The right layout in Figure 3.9g shows a "bus" configuration. Each station can send messages to, and/or receive messages from, any other station. This topology has been used infrequently in the past for SCADA, but is becoming more popular as communication system sophistication increases. Communication systems that use optical fiber cable instead of radio are likely to use a bus topology.

COMMON APPLICATIONS

Scanning

Nearly all SCADA systems have one MTU and multiple RTUs. This implies that the MTU must communicate with one RTU to instruct or query it, then move on to the next, and so on. This application is called scanning. Even for systems that have RTUs communicating on a report-by-exception basis, the MTU should keep track of which RTUs it has not heard from and scan them perhaps once a day to ensure that they are still operational.

Monitoring

For operators to control a process, it is necessary for them first to monitor the process to determine if any changes are needed. The process measurement equipment discussed earlier provides the base signals. The communication system moves the data from the process to the location of the MTU. Now the MTU must arrange the data and feed them to an HMI to present them in a manner that is logical and clear to the operator. Most of the signals that are monitored are discrete. That means that they indicate that a valve is open or closed, or a motor is on or off. Some monitored signals tell the operator the temperature of the process or the number of amps flowing through a conductor. These are analog signals.

Alarms

Signals attached to certain critical process functions require the rapid response of the operator to prevent the process from going out of control. Such signals are handled by the RTU in the same way as are monitored points. When such a signal is received by the MTU, it is recognized as different and triggers special responses within the MTU and HMI. Depending on its criticality, an alarm signal may cause colored lights to flash on the HMI screen or may set off horns or bells to attract the attention of the operator. These extra responses usually continue until the operator acknowledges the condition, after which the horn will silence and the flashing light will settle to a solid red or yellow color. Only after the signal has been acknowledged and the condition has returned to normal will the HMI return to its standard colors.

During certain phases of the process, it is expected that many points will be out of limits, and in order to lower the demands on operator attention, certain of the alarms will be masked. For example, if a power failure exists at a remote facility, an alarm should attract the operator's attention to this. It is not necessary to also warn the operator that the electric heat tracing and air compressor have stopped working.

Most alarms are logged and time-stamped with the time of the incident, time of acknowledgment, and time of return to normal. The logging is usually printed on a dedicated alarm printer as well as being logged into the MTU memory.

One alarm function that is common to most SCADA systems now is the ability to phone or send an e-mail message to selected people if alarms are not acknowledged. Lists of such people, dependent on the nature of the alarm, should be configurable in the system.

Commands

All SCADA systems provide applications to assist the operator, through the HMI, to issue commands to the field actuators attached to RTUs. These commands may be discrete, such as opening or closing valves or turning motors on or off. They may be analog, such as increasing or decreasing controller set points.

Certain control instructions are critical enough that they may justify an additional security step. They may require that the RTU receives the control message, repeats it back to the MTU, then waits until the MTU confirms that it was received correctly. This is called checkback.

Data Gathering

Many SCADA systems, in addition to providing information about the present condition of the process, are also responsible for gathering custody transfer information. If the process is an electrical transmission system, the amount of power transferred will be gathered. If the process is an irrigation system, the amount of water delivered to the customer will be needed. If the system is associated with an oil or gas production facility, the oil and gas will be measured to establish royalties. This information is often what pays the bills and its accurate measurement and "paper trail" requirements may affect the design aspects of the SCADA system such as precision of measurement, storage time of RTU memory, and frequency of scans.

Report Generation

Most SCADA systems, because they are dedicated to the monitoring and control of simple processes, are provided with report capability for the several types of reports that are needed on a regular basis by the operation. Resource production applications usually will have to submit environmental reports to some government authority. Distribution applications will need to generate invoices to send to customers. Nearly all applications will need to produce maintenance or operations reports on a daily basis to advise the staff about issues that need to be attended to soon.

SCADA applications that need more complex reporting should be provided with a free-form report generation capability.

Archiving

Archiving refers to the storage of data for later retrieval. SCADA systems gather many types of data. They may include custody transfer measurements, alarms, communication failure records, and the like. For legal and operational reasons, provision to access this information must exist on the system. Archives stored in some form of database provide this capability.

Several companies have bundled the above noted applications with HMI presentation programs. Since these bundles provide the basic software needed for many SCADA installations, they are marketed as SCADA packages. Note that buying one of these is not buying a SCADA system. In addition to this software package, the field devices, RTUs, communication system, and MTU will also have to be bought.

INDUSTRY-SPECIFIC APPLICATIONS

Oil and Gas Production

Metering the volumes of oil and gas produced by each well, in order to account for royalties to the owner of the resource, is one of the most important applications. In fact, this application determines the minimum security needs of many SCADA systems.

Monitoring and archiving environmental releases from flare stacks and liquid meters is often an operational requirement that can be handled by SCADA.

An almost unlimited number of operational alarms, to alert the operator that pressures or levels are rising too high, temperatures are becoming too low or too high, flow rates are out of limits, the facility is on fire or has a general power failure, are available as part of the alarm application.

In the later stages of oil production, most oil fields require the fluid to be pumped from the reservoir to the surface. While the pumps are operating, oil is brought to the surface. There is an entire technology, based on SCADA, called "well head monitoring," that alerts the operator to failures of the pump, failures of the rod connecting the pump to the surface, and failures of the electric motor that activates the pump. Learning immediately about these pump failures, rather than waiting until an operator makes the rounds the next day, can easily justify the installation of a SCADA system. Turning the pumps on and off without driving to each of them is an application. If the production facility is in an area where the electric utility offers special rates for "low rate times" or for "interruptible power contracts," the pumps may be turned on only during a low electric power rate period of each day, or they may be turned off quickly in response to a request by the electric utility to qualify for an interruptible power rate.

During secondary recovery, with the reservoir artificially pressurized by injecting water, SCADA systems may be used to monitor flows and pressures of water and oil to provide information to a model that tracks the real-time pressure within the reservoir. One of the purposes of a model like this is to ensure that reservoir pressure does not rise high enough to fracture and permanently damage the reservoir.

During tertiary recovery, valuable solvents may be injected to dissolve hydrocarbons that remain in the reservoir. Pressures that are too high will damage the reservoir; pressures that are too low will prevent some solvents from working properly. Because these solvents will later be recovered, it is necessary to keep track of how much was injected so that royalties do not have to be paid on them when they are produced.

Natural gas producers have applications that monitor sales made by their marketing departments, selectively turn on specific wells based on their deliverability, and automatically adjust compressor speeds to optimize production based on these sales.

Pipelines

Pipelines share the need for providing custody transfer measurements of the product that they carry. In addition to the custody transfer, these measurements may also be used to provide leak detection on the line. This application is usually based on subtracting the volume out from the volume in. Short simple pipelines may use an algorithm as simple as this; however, longer pipelines with high vapor pressure will need more complex models that may consider the elevation of parts of the line, pressure at each station, flow rate, temperature, and coefficient of expansion of the steel pipe.

When a pipeline leak is discovered, block valves are closed to isolate the line into small sections to limit the amount of fluid that may leak out and pumps/compressors are shut off to stop pressurizing the line. Fairly complex applications are needed to ensure that these drastic measures do not add to the risk of environmental and equipment damage. Starting a pipeline after an emergency shutdown requires that people attend each site to confirm that it is ready to start, but after this confirmation, the coordination of the start is usually handled by the SCADA system in another application.

Pipelines longer than several miles will probably include compressor or pump stations. These are usually considered to be part of the pipeline and their operation is included as part of the pipeline system. Automatic control of the compressors within the station will probably be done by equipment located at the station, but supervisory control, in the form of setting the pressure required at each station, will be through the SCADA system. Each station will be monitored for actual or impending failure, and maintenance will be scheduled based on information gathered by the system.

Electric Generation and Transmission

The generation of electricity by a utility requires that energy be released to turn generators. Most of the generators that are operated will provide power into the grid. To ensure that a new large load will not exceed the capacity of the generators, some "spinning reserve" generators are also kept running. Profits are improved if utilities can keep the spinning reserve low and still respond quickly to changes in the load their customers demand. Complex models that consider levels of water behind each available dam, cost of coal or gas that must be burned to produce a unit of electricity, amount of air pollution associated with each boiler, remaining time to next generator maintenance shutdown, and myriad other factors provide recommendations in real time to the system operator. Only by using SCADA can the data be gathered, collated, and presented to the operator quickly enough. SCADA also enables the operator to select generators to be turned on or off in response to measured and forecast loads.

Monitoring rainfall in real time provides information to forecast short-term reservoir level changes that may influence the choices of power source.

Transmission of electric power over great distances depends on application programs that measure current flowing through the lines and calculate line temperature, expansion, and sag. Monitoring ambient temperature and wind to fine-tune the line temperature calculations allows for optimal use of the facilities. Often, there is more than one route to move power from a source generator to a load center. Application programs to calculate the route that will result in the least amount of losses are used to guide the operator. Control of switches to isolate line sections or to connect sections to the network is normally a part of the SCADA system.

Irrigation

Irrigation systems store and distribute water for agricultural purposes. The measurement of water to each customer for billing purposes is the most obvious of the applications. Prorationing, or limiting the total amount to be shared, is a method of ensuring equity among customers. As water levels measured behind the irrigation company's dams increase and decrease in response to rainfall, the amount available to each customer will also increase and decrease, and application programs to provide this equitable distribution are used by many irrigation companies.

Pumps and valves are opened and closed based on programs to deliver preselected amounts of water to customers.

Newer Industry Applications

The city-state of Singapore operates thousands of apartment buildings, each of which has elevators. A SCADA system monitors the operation of each of these and can dispatch maintenance crews as soon as a failure is noted by the system and alarmed to the central operator.

Large open pit mines have installed GPS (global positioning satellite) systems in their 200-ton trucks to monitor truck location, schedule refueling, and optimize the loading/unloading of the trucks.

Long haul and delivery trucks and taxis have been equipped with GPS to enable the dispatcher to monitor their locations. This allows the nearest vehicle to be dispatched to the next customer, and allows a written instruction, precluding misinterpretation, to be sent to the vehicle operator.

OPTIMIZING

Dead Time

In a SCADA communication system, dead time is the time when useful data are not being transmitted or received. The reason that it is important is that large dead times mean slow scans.

Dead time includes the time after an MTU has sent a message to an RTU and is waiting for a response. It is often measured as a percentage of the total time. Keeping the waiting time short is the obvious way to reduce this source of dead time. Eliminating "store and forward" stations is one way to do this, but may not always be possible. Another way is to select a protocol that allows a long message length. More information will be relayed on each message, and the ratio of waiting to talking will drop. Be aware that short messages work best in areas of high ambient noise. A third way to reduce dead time is to select radios that stabilize very quickly after they are turned on.

Always question the need to reduce the scan time. SCADA systems have historically had scan times in the order of 15 to 30 min. The philosophy of SCADA is based on scan times of this order, and while most people would prefer to see these times reduced, they have difficulty writing down a dollar justification for shortening them.

Communication Methods

In new systems, MTUs are nearly always connected to corporate local-area networks (LANs). When this is done, access to data gathered by the company can be made available to technical and managerial people for research and monitoring of the business. Security of the data must be considered when this is done, but SCADA systems have been able to assign security levels since the early days.

New SCADA applications that operate over relatively small geographic areas are using spread spectrum radio. In some areas, spectrum congestion is high enough that the government regulatory agencies cannot allocate additional licenses for conventional radios. One of the advantages of spread spectrum radios is that they operate at a very low power output and therefore do not need to be licensed.

Pipeline companies often install optical fiber cable in the pipeline right of way. In some countries, security of service is the reason; more often, cost is the determining factor. The company already has the right of way and cable is not difficult to install. When a system is completely serviced by optical fiber with its high-speed communication, there is some philosophical question if it is still a SCADA system or has now become a large distributed control system. Most people still consider it to be a SCADA system. Before building an optical fiber communication system by putting a cable in the same trench as the pipe, one should consider that the pipe does not just lie there. As pressure and temperature changes cause it to lengthen and shorten, the pipe will move around in the ground, perhaps crushing the cable. Several such failures have been reported.

Communication between the MTU and RTUs through Internet connection was mentioned earlier, but is worth reiterating. In addition to the communication aspects, this concept offers a new potential for outsourcing. Companies are contracting to provide MTU service for new and existing RTU sites. These services include data gathering, alarm handling, archiving, report generation, and control capability. Some companies include the RTU rental and communication costs in their flat rate per RTU charges; some do not. Communication may be by satellite or cellular phone. This method of central control will change the way that SCADA serves industry.

Bibliography

Boyer, S. A., *Supervisory Control and Data Acquisition,* 2nd ed., Research Triangle Park, NC: Instrument Society of America, 1999.

Herb, S., *Understanding Distributed Processor Systems for Control,* Research Triangle Park, NC: Instrument Society of America, 1999.

Horowitz, P. and Hill, W., *The Art of Electronics,* New York: Cambridge University Press, 1980.

Strock, O. J., *Introduction to Telemetry,* Research Triangle Park, NC: Instrument Society of America, 1987.

Thompson, L., *Industrial Data Communications, Fundamentals and Applications,* Research Triangle Park, NC: Instrument Society of America, 1991.

3.10 PLC Programming

V. A. BHAVSAR

Partial List of Suppliers: Most PLC manufacturers do the PLC programming for their clients or they have systems integrators who can do the PLC programming for them. Following are a few major PLC manufacturers: ABB; Allen-Bradley; GE Fanuc; Honeywell; Klockner–Moeller; Mitsubishi; Modicon; Omron; Siemens; Triconex.

Programming Costs: PLC programming cost depends on the complexity and the length of the program. It could cost anywhere between U.S. $40 and $80 per hour for programming. Programming cost could be 50 to 100% or more of the hardware cost depending on the size and complexity of the PLC system.

This section covers the basics of PLC (programmable logic controller) programming. It provides readers with enough information about PLC system hardware and programming techniques to enable them to write simple programs and understand complex programs. It starts with a basic understanding of PLC, followed by an introduction to major hardware components, operation of PLC system, and various programming languages available. The main focus of this section is on ladder logic, the most widely used PLC language. The rest of the section gives details about basic ladder logic instructions, programming devices used to write ladder logic, ladder logic structure, how to develop ladder logic and program a PLC, programming considerations, and documentation. The section concludes with a typical ladder logic program example and recent developments in PLC programming.

PLC is an acronym for programmable logic controller. This device is also known by other names such as programmable controller or logic controller. Basically, the PLC is an electronic device that was invented to replace the electromechanical relay logic circuits used for machine automation in the automobile industry. It was first introduced in the late 1960s to a major U.S. car manufacturer by Bedford Associates as Modular Digital Controller (MODICON).

Today, PLCs are used in many industries including machining, packaging, material handling, automated assembly, and countless others. Usually, PLC manufacturers provide their customers with various applications of their products. Other useful sources to learn more about existing and new PLC applications are technical journals, magazines, and papers published on control and automation.

PLC SYSTEM HARDWARE AND OPERATION*

A PLC system consists of the following major components:

- Rack
- Power supply module
- CPU (processor/memory)
- Input/Output (I/O) modules
- Programming device

Figure 3.10a shows a typical modular PLC system.

The PLC is a sequential device, i.e., it performs one task after another. Figure 3.10b shows how a conventional PLC system operates. A PLC performs three major tasks in the following order:

1. Task 1 Read inputs—The PLC checks status of its inputs to see if they are on or off and updates its memory with their current value.
2. Task 2 Program execution—The PLC executes program instructions one by one sequentially and stores the results of program execution in the memory for use later in task 3.
3. Task 3 Write outputs—The PLC updates status of its outputs based on the results of program execution stored in task 2.

After the PLC executes task 3 it returns to execute task 1 again.

* Refer to *IEH*, Volume 2, *Process Control*, Chapter 6.

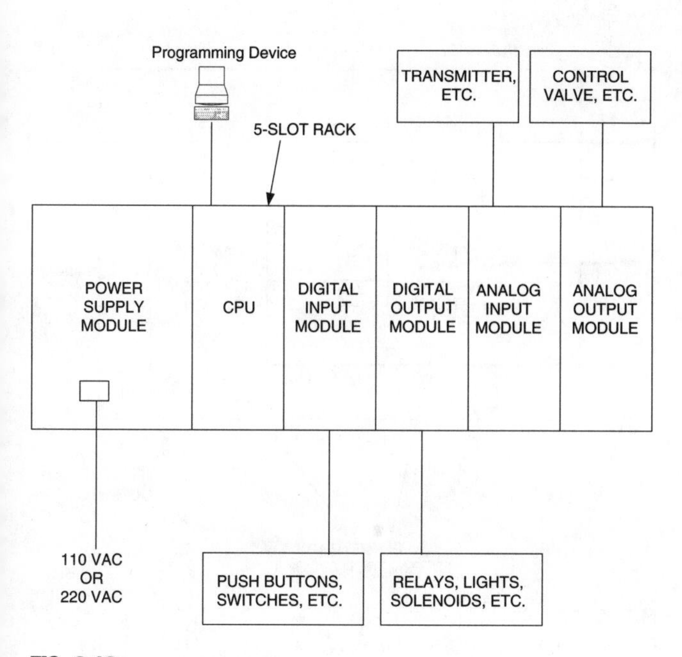

FIG. 3.10a
Modular PLC system.

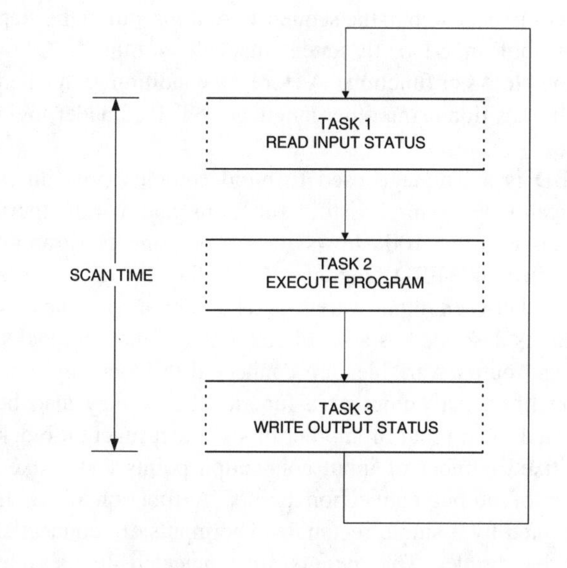

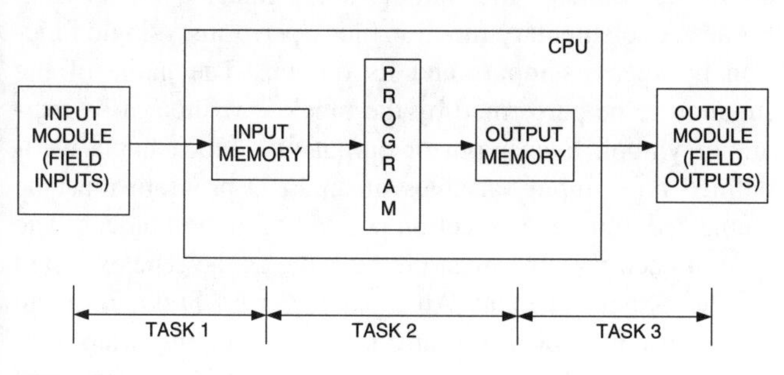

FIG. 3.10b
PLC operation.

The total time taken by the PLC to perform these three tasks is called the PLC scan time. Scan time depends on the CPU clock speed, user program length, and the number of I/Os. Typically, scan time is in milliseconds. The smaller the scan time, the faster the updates of the I/O and the program execution. As a general guide, the PLC scan time should be less than half the time it takes the fastest changing input signal to change in the system.

PLC PROGRAMMING LANGUAGES

There are two forms of PLC programming languages available:

Text—Instruction List (IL), Structured Text (ST)
Graphic—Sequential Function Charts (SFC), Function Block Diagrams (FBD), Ladder Logic

Different PLCs support one or more of the above languages for programming. These languages have their advantages and limitations, and they complement one another to provide programmers with more programming power. A brief explanation of these languages is given below, followed by a detailed explanation of ladder logic.

IL is a low-level language. It is mainly used for smaller applications or for optimizing parts of an application. IL is one of the languages for description of the actions within the steps and conditions attached to the transitions of the SFC language (this is discussed in more detail later). Instructions always relate to the current result (or IL register value). The operator indicates the operation that must be performed on the current value of IL register and the operand. The result of the operation is stored again in the IL register. An IL program is a list of instructions. Each instruction must begin on a new line, and must contain an operator and, if necessary, for the specific operation, one or more operands, separated with commas (,). A label followed by a colon (:) may precede the instruction. If a comment is attached to the instruction, it must be the last component of the line. Comments always begin with (* and end with *). Empty lines may be entered between instructions. Comments may be put on empty lines. Below are examples of instruction lines:

Label	*Operator*	*Operand*	*Comments*
Start:	LD	IN1	(* start push button *)
	AND	MD1	(* mode is manual *)
	ST	Q2	(* start motor *)

ST is a high-level structured language designed for automation processes. This language is mainly used to implement complex procedures that cannot be easily expressed with graphic languages. ST is one of the languages for the description of the actions within the steps and conditions attached to the transitions of the SFC language (this is discussed in more detail later). An ST program is a list of ST statements. Each statement ends with a semicolon separator. Comments may be freely inserted into the text. A comment must begin with (* and ends with *). These are basic types of ST statements:

- Assignment statement (variable := expression;)
- Subprogram or function call
- Function block call
- Selection statements (IF, THEN, ELSE, CASE ...)
- Iteration statements (FOR, WHILE, REPEAT ...)
- Control statements (RETURN, EXIT ...)
- Special statements for links with other languages such as SFC

Below are examples of ST lines:

```
(* input1 : start push button, output1 : start motor, mode:
system mode, man = 0, auto = 1 *)
if (mode) then
    return;
end_if;
(* start motor in manual mode*)
if (input1) then
output1 := true;
end_if;
```

SFC is a language used to describe sequential operations graphically. The process is represented as a set of well-defined steps, linked by transitions. A sequence of steps is called a task, and many such tasks make up the whole process. Figure 3.10c shows basic components (graphic symbols) of the SFC language: steps and initial steps, transitions, and links between steps and transitions. An SFC program is activated by the system when the application starts. A run time step that is active is highlighted. When program execution starts, the initial step becomes the current step by default.

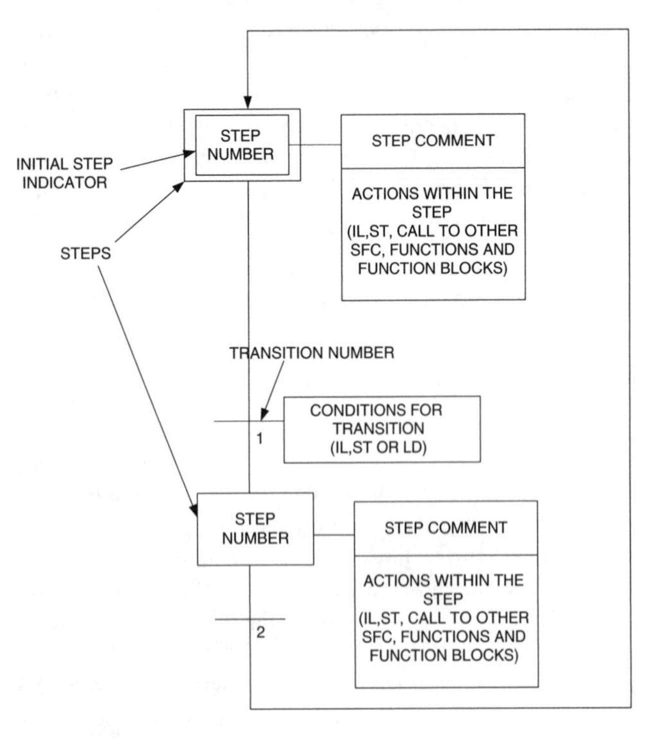

FIG. 3.10c
Basic components of SFC language.

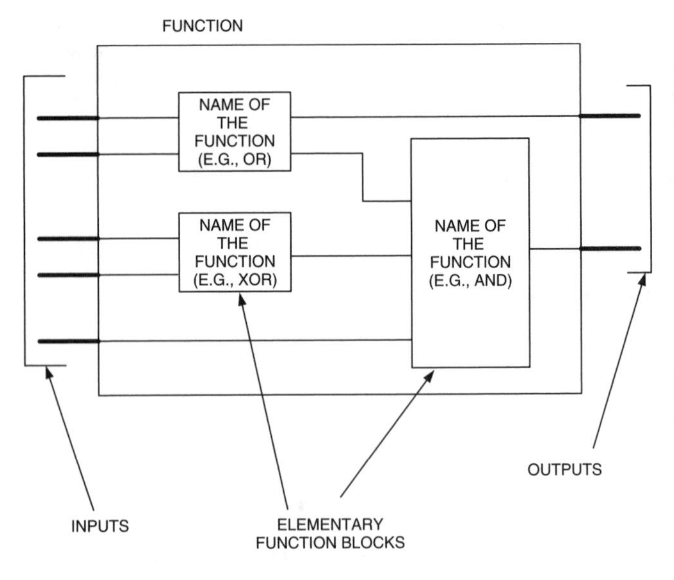

FIG. 3.10d
Basic components of FBD language.

When the conditions for transition become true, the sequence advances to next step in the sequence. Actions within the step can be Boolean, ST or IL statements, calls to other SFCs or function blocks or functions. A Boolean condition is attached to each transition using other languages ST, IL, Ladder logic (LD).

FBD is a language used to build complex procedures graphically by taking existing functions and wiring them together. Figure 3.10d shows basic components (graphic symbols) of the FBD language. FBD diagram describes a function between input variables and output variables. A function is described as a set of elementary function blocks. Input and output variables are connected to blocks by connection lines. An output of a function block may also be connected to an input of another block. Each function block has a fixed number of input connection points and a fixed number of output connection points. A function block is represented by a single rectangle. The inputs are connected on its left border. The outputs are connected on its right border. An elementary function block performs a single function between its inputs and its outputs. The name of the function to be performed by the block is written in its rectangle symbol. Each input or output of a block has a well-defined type. Input variables of an FBD program must be connected to input connection points of function blocks. The type of each variable must be the same as the type expected for the associated input. An input for an FBD diagram can be a constant expression, any internal or input variable, or an output variable. Output variables of an FBD program must be connected to output connection points of function blocks. The type of each variable must be the same as the type expected for the associated block output. An output for an FBD diagram can be any internal or output variable, or the name of the program (for subprograms only). When an output is the name of the currently edited subprogram, it represents the assignment of the return value for the subprogram (returned to the calling program).

Ladder logic is one of the most popular and widely used programming languages by electricians and programmers because it emulates the old relay-based ladder logic structure. For this reason, the rest of this section discusses ladder logic programming in greater detail.

LADDER LOGIC STRUCTURE

As ladder logic is an extension of relay logic, it makes sense to see how ladder logic and relay logic structures are analogous to each other. Also, a basic understanding of Boolean logic will help in writing ladder logic because many times a control scheme is first developed using Boolean logic and then converted into ladder logic in PLC. Figure 3.10e shows three different representations of a simple logic for starting/stopping a motor.

In relay logic, the lines L1 and L2 represent the power applied to the relay circuit. The start push button is normally an open type and the stop push button is normally a closed type. When the start push button is pressed, the contact of the start push button closes, the current flows through the start and stop push button contacts to the relay coil, and energizes the relay coil. The contact of the relay coil is closed and provides an alternative current path to the relay coil, thus keeping it energized. When the stop push button is pressed, the contact of the stop push button opens, the current path to the relay is opened, the relay is deenergized, and the relay contact opens. The motor runs as long as the motor relay stays energized.

In Boolean logic, the start push button and motor relay contact are represented as inputs to the OR logic block, the output of the OR logic block and the stop push button are represented as inputs to the AND logic block, and the motor relay coil is represented as output of the AND logic block. The normally closed stop push button is shown by inverting the stop input (logical 1). When the start push button is pressed (logical 1), the output of OR block becomes logical 1. With both inputs of the AND block logical 1, the output of the AND block becomes logical 1 and turns the motor relay on. The status of motor relay is fed back to the OR block to keep its output on. The motor relay turns off as soon as the stop button is pressed.

In PLC ladder logic, two vertical lines represent virtual power lines, and the actual electrical current is replaced by logical path. A horizontal line represents a logical rung. Inputs are shown near the left vertical line and outputs are shown near the right vertical line on a rung. When the inputs have logical 1 or TRUE state, the logical path to the outputs completes, and the outputs receive a logical 1 or TRUE state. In this example, the start and stop push button contacts are shown as inputs that represent the physical push buttons connected to a PLC input module. Inputs have their PLC memory addresses where their current values are stored (here they are 001 and 002). As the push buttons are operated, the PLC

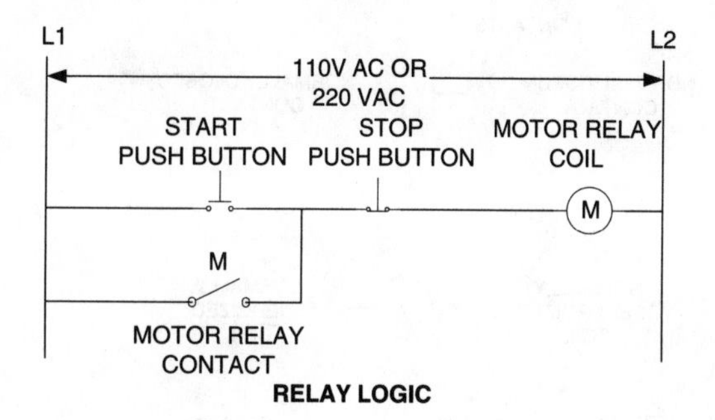

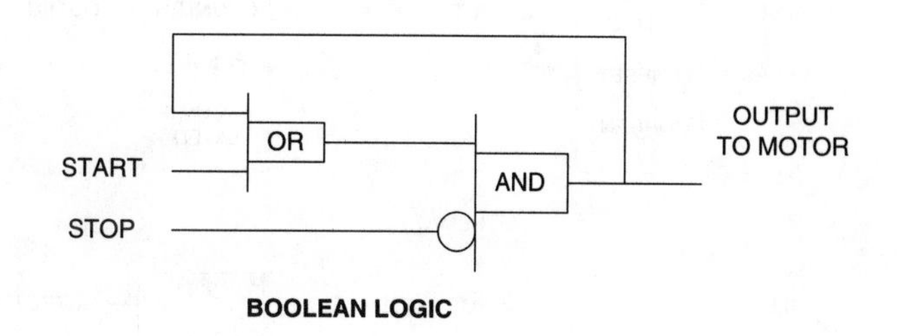

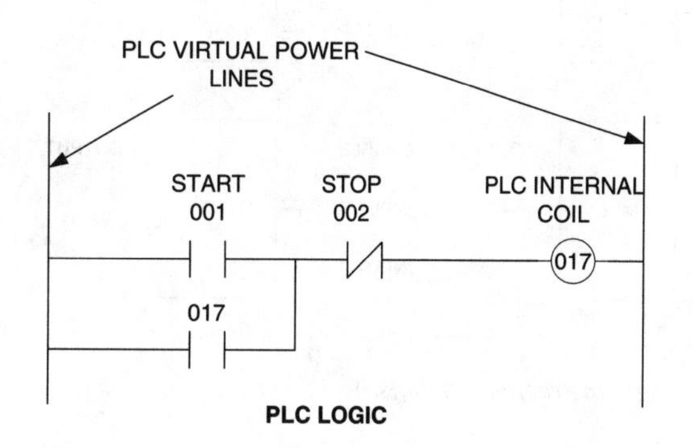

FIG. 3.10e
Three types of logic representation.

memory is updated with their current status. An open contact represents a logical 0 or FALSE, and a closed contact represents a logical 1 or TRUE in PLC memory. In this example relay coil is shown as internal coil (PLC memory address 017) that represents a physical relay connected to a PLC output module. As the status of the internal coil changes based on the push button status, the PLC memory is updated with the current status of the internal coil. An energized internal coil represents logical 1 and a deenergized internal coil represents logical 0 in PLC memory. Physical relay coil becomes energized when the PLC internal coil is energized (logical 1) and deenergized when the PLC internal coil is deenergized (logical 0). The internal coil has a large number of virtual contacts limited only by the PLC memory capacity.

LADDER LOGIC PROGRAMMING BASIC INSTRUCTIONS

Figures 3.10f and 3.10g show graphical symbols of ladder logic basic instructions discussed below.

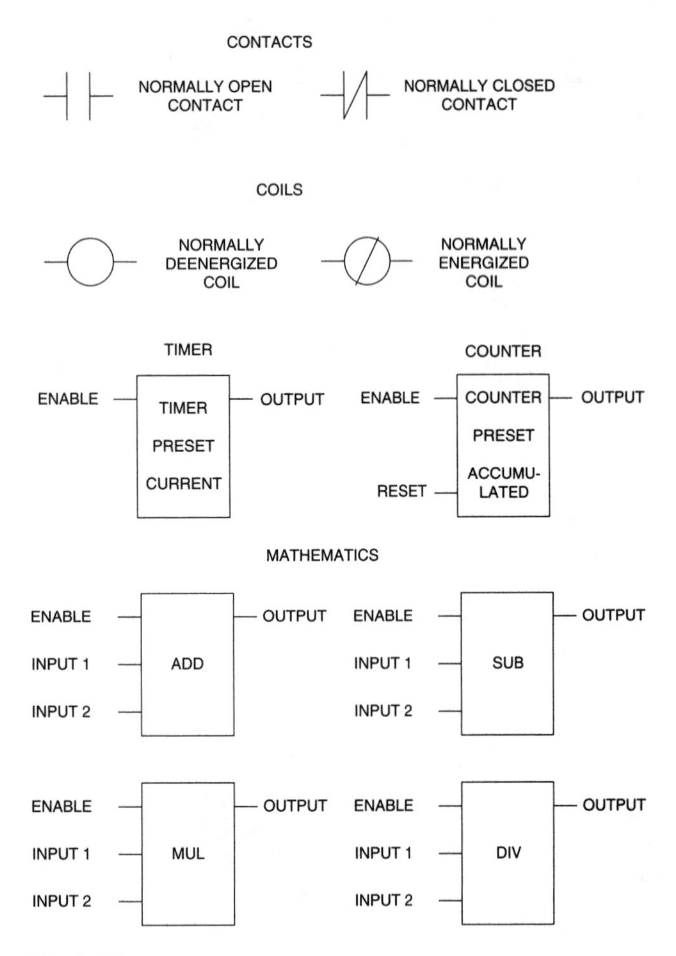

FIG. 3.10f
PLC ladder logic instruction symbols.

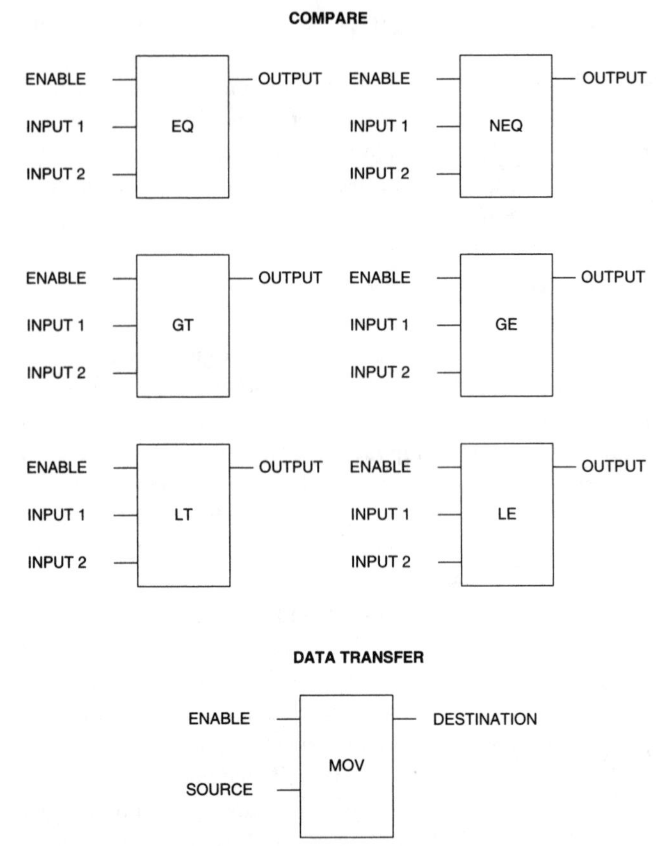

FIG. 3.10g
PLC ladder logic instruction symbols.

CONTACT—This is an input instruction. It can be used to represent an external digital input or a contact of internal (soft) relay. There are two basic types of contacts:

1. Normally open (NO) contact is used to represent an input signal that is normally off and will become on when operated, e.g., a physically connected push button with normally open contact. For a NO contact, logical 0 represents an OFF or FALSE condition and logical 1 represents an ON or TRUE condition.
2. Normally closed (NC) contact is used to represent an input signal that is normally on and will become off when operated, e.g., a normally closed contact of an internal (soft) relay coil. For an NC contact, logical 0 represents an ON or TRUE condition and logical 1 represents an OFF or FALSE condition.

COIL—This is an output instruction. It can be used to represent an external digital output or an internal (soft) relay. There are two basic types of coils:

1. Normally deenergized coil is used to represent an output signal that is normally off or deenergized and will become on or energized when all inputs on a rung preceding this coil are TRUE.
2. Normally energized coil is used to represent an output signal that is normally on or energized and will become off or deenergized when all inputs on a rung preceding this coil are TRUE.

TIMER—This is a timing instruction. A timer is used to delay an event or to time an event. Generally, a timer instruction has enable input, preset time value, elapsed time value, and an output signal. Preset and elapsed times are stored in PLC memory registers. There are two basic types of timers. Figure 3.10h shows a timing diagram for these timers.

1. An on-delay timer is used to delay turning on an output. When the enable condition becomes true, the timer starts timing, its current time value starts to go up, and after its current time reaches the preset time value, the output of the timer is turned on. The output stays on as long as the enable input signal is on. As soon as the enable signal becomes false or logical 0, the output goes off and the timer current value becomes 0. There are other variations of timers that have a reset input connection to reset the timer current value.

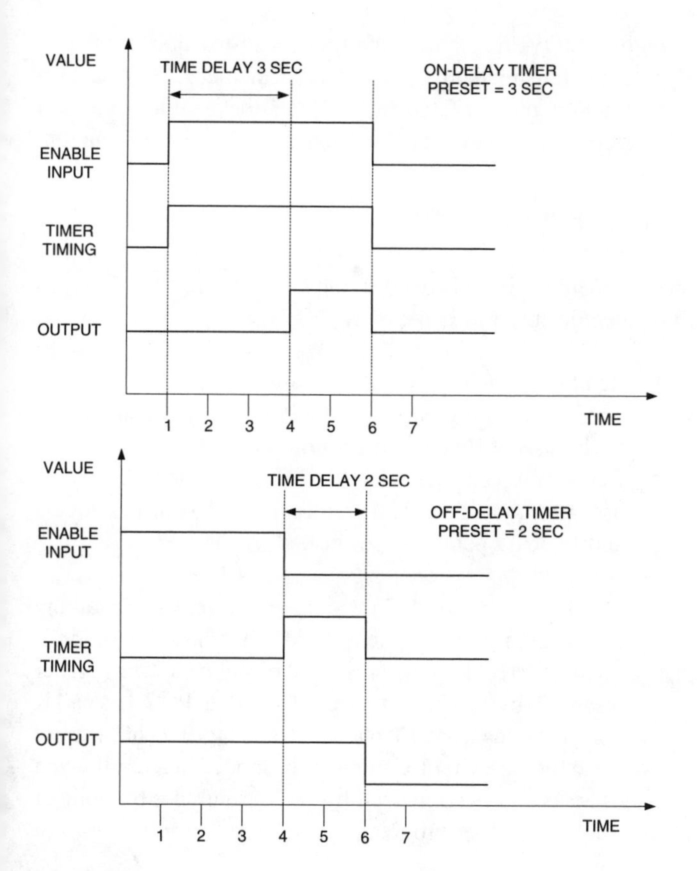

FIG. 3.10h
Timing diagram for PLC timer instruction.

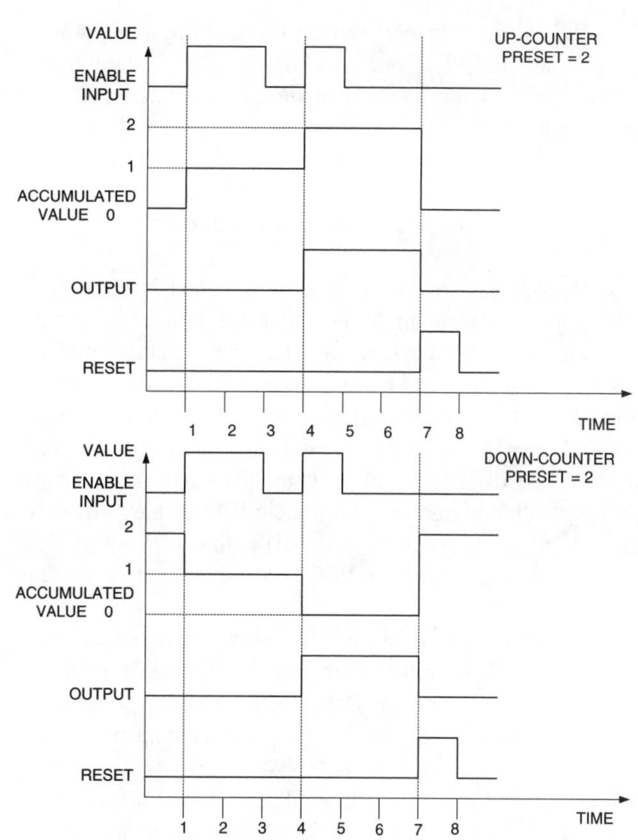

FIG. 3.10i
Timing diagram for PLC counter instruction.

2. An off-delay timer is used to delay turning off an output. As soon as the enable condition becomes true, the output of the timer turns on. When the enable condition becomes false or logical 0, the timer starts timing, its current time value starts to go up. Timer output stays on till its current time reaches the preset time value. As soon as the timer current time reaches the preset time, its output goes off, the timer stops timing, and its current value stops increasing. When enable input becomes true the next time, the timer current value is reset to 0.

COUNTER—This is a counting instruction. A counter is used to count the number of occurrences of an event. Generally, a counter instruction has an enable input, a preset count value, an accumulated count value, a reset input, and an output signal. Preset and accumulated counts are stored in PLC memory registers. There are two basic types of counters. Figure 3.10i shows a timing diagram for these counters.

1. An up-counter is used to count up. When the enable condition becomes true, i.e., transitions from logical 0 to logical 1, the counter accumulated value is incremented by 1. Every time this transition takes place, the counter accumulated value increases by 1. When the accumulated value reaches the preset value, the counter output turns on. The output stays on until the counter is reset.
2. A down-counter is used to count down. The counter accumulated value is set at the preset value. When the enable condition becomes true, i.e., transitions from logical 0 to logical 1, the counter accumulated value is decremented by 1. Every time this transition takes place, the counter accumulated value decreases by 1. When the accumulated value reaches 0, the counter output turns on. The output stays on until the counter is reset.

MATHEMATICS—Math instructions are used for data manipulations. Generally, a math instruction has an enable input, two data inputs, and one output. Data inputs are register locations or constants and output is a register location where the result of the math operation is stored. There are four basic types of math instructions.

1. ADD is used to perform addition. When the enable condition becomes true, input 2 data are added to input 1 data and the result is stored in output.
2. SUB is used to perform subtraction. When the enable condition becomes true, input 2 data are subtracted from input 1 data and the result is stored in output.

3. MUL is used to perform multiplication. When the enable condition becomes true, input 1 data are multiplied with input 2 data and the result is stored in output.
4. DIV is used to perform division. When the enable condition becomes true, input 1 data are divided by input 2 data and the result is stored in output.

Although most advanced PLCs support real math, some PLCs only support integer math, i.e., math performed on integer numbers. Hence, when division on integers is performed, the result is an integer and a loose remainder (fractional part of the result). To deal with this, first multiply the number being divided with 10 for one decimal digit accuracy and then perform the division in PLC. Thus, the result is an integer number with one implied decimal digit. To display this number on the operator interface (HMI) with one decimal digit, read the integer value from PLC and display it after dividing it by 10 in the HMI.

COMPARE—Compare instructions are used for comparing data. Generally, a compare instruction has an enable input, two data inputs, and one output. Data inputs are register locations or constants. There are six basic types of compare instructions.

1. EQ is used to check if input 1 and input 2 are equal. When the enable condition becomes true, input 1 data are compared with input 2 data to see if they are equal and, if so, the output of the EQ instruction becomes on.
2. NEQ is used to check if input 1 and input 2 are not equal. When the enable condition becomes true, input 1 data are compared with input 2 data to see if they are not equal and, if so, the output of the NEQ instruction becomes on.
3. GT is used to check if input 1 is greater than input 2. When the enable condition becomes true, input 1 data are compared with input 2 data to see if input 1 is greater than input 2 and, if so, the output of the GT instruction becomes on.
4. GE is used to check if input 1 is greater than or equal to input 2. When the enable condition becomes true, input 1 data are compared with input 2 data to see if input 1 is greater than or equal to input 2 and, if so, the output of the GE instruction becomes on.
5. LT is used to check if input 1 is less than input 2. When the enable condition becomes true, input 1 data are compared with input 2 data to see if input 1 is less than input 2 and, if so, the output of the LT instruction becomes on.
6. LE is used to check if input 1 is less than or equal to input 2. When the enable condition becomes true, input 1 data are compared with input 2 data to see if input 1 is less than or equal to input 2 and, if so, the output of the LE instruction becomes on.

DATA TRANSFER—This instruction is used for moving data. It has an enable input, a source input, and a destination output. MOVE is a basic data transfer instruction. MOVE is used to transfer data from a source memory location to a destination memory location. When the enable condition becomes true, source data are transferred to the destination.

PLC MEMORY STRUCTURE

PLC memory is embedded in the CPU. There are two types of memory used in CPU.

1. Read-only memory (ROM) is used to store the firmware or operating system of the PLC. It is not available to the user for data or user program storage.
2. Random Access Memory (RAM) is used to store the user program, data, and results of various arithmetic and logical operations performed by the CPU.

When we refer to PLC memory structure, we are talking about the user memory, i.e., RAM. PLC user memory is specified in bytes; 1 byte represents 8 bits of data. One register represents 2 bytes or 16 bits of data. If a PLC has a 1K memory, it means it has 1056 bytes or 528 registers of memory available for the user. PLC memory is divided in the following major areas and is referenced by register number/bit number in a register by user program.

1. User program memory—This memory stores user ladder logic or any other form of user program.
2. Input status memory—This memory stores the status of inputs physically connected to the PLC. Each digital field input takes 1 bit of input memory. Each analog field input takes 8 or 12 or 16 bits of input memory depending on the resolution of analog input module.
3. Output status memory—This memory stores the status of outputs physically connected to the PLC. Each digital field output takes 1 bit of output memory. Each analog field output takes 8 or 12 or 16 bits of output memory depending on the resolution of analog output module.
4. Timer status memory—This memory stores the status, preset, and elapsed time values of timers in the PLC.
5. Counter status memory—This memory stores the status, preset, and accumulated values of counters in the PLC.
6. Numeric data memory—This memory stores numeric data in PLC. It is used to store data for manipulation by the user program and the results of data manipulation.

LADDER LOGIC PROGRAMMING DEVICES

A programming device is required to write, edit, and monitor user programs, to download them into the PLC through the programming port on the CPU, to monitor and force the I/O status. It is also used for debugging the PLC program and to view PLC diagnostics. There are handheld programmers and personal computer–based programmers available for PLC programming.

A handheld programmer is a dedicated programming device manufactured and supplied by a PLC manufacturer for its PLCs. It has a keypad and an LCD (liquid crystal display). One can connect the programmer to the PLC using the programming cable that comes with programmer and do online programming or make program changes. One cannot store programs in this type of programmer.

A PC-based programmer needs the following to write PLC ladder logic.

1. PC with DOS or Windows operating system
2. Serial port
3. PLC programming software supplied by the PLC manufacturer
4. Programming cable supplied by the PLC manufacturer

PC-based programmers are very popular for the following reasons.

1. Offline programming can be done and programs can be stored on hard disk.
2. Program documentation becomes easier.
3. A large screen makes it easier to see large ladder rungs.
4. Program printouts can be made easily for permanent records.
5. Some programmers even allow logic simulation by a executing program in the PC as if it were being executed in the PLC.

Today, laptops are used in industrial environments for online programming and program changes.

PLC PROGRAMMING CONSIDERATIONS

Following are some of the major considerations when writing any PLC program:

1. Program scan time—This is a very important factor in any PLC program. As explained earlier, scan time is the time taken by the PLC to complete its three main tasks—read input status, execute PLC program, write output status. PLC repeats these three tasks continuously. If the total scan time of a program is not less than half the time at which the fastest input changes in the system, there is a possibility that the PLC will not be able to detect a change in that input and will fail to respond to the change. There are several ways to reduce the scan time in a PLC system.

- Use a faster CPU.
- Use an interrupt routine.
- Make the program compact by logic optimization (See *IEH*, Volume 2, Chapter 6) and simplification.
- Organize the program such that for elements in parallel, the one most likely to be true is highest on the ladder and so on, and for elements in series, the one least likely to be true is leftmost on the ladder and so on. This way, when one of the parallel elements is found true, the rest of the parallel elements are not required to be solved and program execution progresses to the next rung. The same is true for elements in series.

2. Program scan direction—It is a very important in PLC programming to know how a PLC program scan is performed. Some PLCs, e.g., Allen-Bradley PLCs, scan programs left to right (called rung scanning), whereas others, e.g., Modicon PLCs, scan programs top to bottom (called column scanning). Both methods are appropriate; however, a programmer should be aware of the scan method used by the PLC because it has an impact on whether a coil is energized or deenergized in the same scan or the next scan after the input conditions become true.

3. Fail-safe programming—A program should be written in such a way that it always puts the process it is controlling in safe mode when any of the critical system components fails. This is called fail-safe programming. In the system, first define the fail-safe conditions and then determine what outputs are required to be turned off or turned on to achieve those conditions. There are many components in a PLC system, such as I/O modules, power supply module, cables, etc. Programs should be written such that if there is a loss of input signal due to power loss or cabling problems or failure of input module itself, then the outputs will go to a state that is safe.

4. Emergency shutdown—There should be a hardwired relay, called a master control relay (MCR), and a normally closed (NC) emergency stop button wired in series in the system. The MCR is normally kept energized through the emergency stop button by the PLC system power. Power to the physical outputs is supplied through the contacts of this MCR. In case of an emergency, the emergency stop button is activated and the power to all the outputs is removed.

5. PLC fault handling and diagnostics—A program should also have the ability to use various diagnostics available in the CPU and to generate alarms for any fault condition in the system or failure of any hardware that might create unsafe operating conditions. There are status registers in the PLC that can be used to generate alarms or messages to help diagnose the problems and for prompt corrective actions.

PLC PROGRAM DOCUMENTATION

This is a very important aspect of PLC programming that is often overlooked or given less importance. These days programming software is available with good program documentation capabilities. Good program documentation has the following advantages:

- Helps in program debugging
- Helps program users understand the logic
- Provides online information about field connections, thus in many cases eliminates the need to have the wiring diagrams handy in troubleshooting

Program documentation involves the following main areas:

- I/O names and descriptors—These are the short, user-defined names given to field inputs and outputs with a more detailed description of the I/O signals. The name is usually a shortened field tag name and the descriptor is a short explanation about the signal.
- Internal memory names and descriptors—These are the short, user-defined names given to internal coils and registers with a more detailed description of their function. The name is usually a shortened identifier and the descriptor is a short explanation about the function.
- Program comments—These are inserted between the logic rungs in the program to explain the purpose of a block of logic and how that piece of logic works.

Program documentation is a continuous process during PLC programming. The commonly used steps for program documentation are:

- Define all the I/O names and descriptors.
- Define all the internal memory names and descriptors.
- Write comments before writing a piece of logic.
- Keep adding/editing comments and other descriptors/names as the program is written.

There are utilities available in programming software to import/export names, descriptors, and comments from/to word processors like Excel spreadsheets. Use of Excel spreadsheets in creating PLC documentation saves a lot of time.

A PLC program documentation printout consists of:

- List of I/O names and descriptors
- List of internal memory names and descriptors
- Cross-reference list for I/Os and internal memory
- Usage list of I/Os and internal memory
- Total PLC memory used and available
- PLC ladder logic listing
- PLC system configuration

PLC HARDWARE CONFIGURATION

PLC programming also involves hardware configuration. Hardware configuration lets the CPU know what components make up the whole PLC system and their addresses. The PLC has to be configured first, before it can be programmed with logic. PLC hardware configuration involves the following:

1. Select type of rack (chassis) used—5- or 10-slot, etc.
2. Select type of card that is installed in each slot— digital I/O or analog I/O, etc.
3. CPU configuration for memory allocation of various types of user memory
4. CPU configuration for communication ports—programming port, etc.
5. Address allocation for physical I/Os connected to I/O modules

PLC programming software is used to develop and download PLC configuration into the CPU.

A typical PLC system hardware and CPU configuration for GE Fanuc PLC is discussed later in this section.

PLC LADDER PROGRAM STRUCTURE

Typical GE Fanuc PLC

Figure 3.10j shows a typical GE Fanuc program structure. A program is contained in the folder and can be divided in various logic blocks. When a programmer starts to write a program, first the programmer creates a folder with an appropriate name, e.g., TEST. There is a main program block where the programmer writes the ladder logic and also makes calls to other logic blocks. These other logic blocks are declared in the main logic block and are called subroutines. In this example, the main program block is divided in three sections or subroutines, i.e., ANALOG, ALARMS, and OUTPUT, and makes calls to these subroutines. These subroutines contain specific logic as explained below.

- The ANALOG subroutine handles analog input signals and converts the raw analog value into engineering units for alarming purpose. This value in engineering units can also be displayed on any HMI.
- The ALARMS subroutine generates alarms with deadband.
- The OUTPUT subroutine generates physical outputs for annunciator alarms.

This block structure makes it easier to understand and troubleshoot the program.

Table 3.10k shows various types of memory identifiers for PLC addressing.

Table 3.10l shows a typical variable declaration table.

Figure 3.10m shows a typical PLC system hardware and CPU configuration.

Figure 3.10n shows a typical PLC logic for analog input scaling.

```
                    I D E N T I F I E R   T A B L E

          IDENTIFIER     IDENTIFIER TYPE           IDENTIFIER DESCRIPTION
          ----------     ---------------     --------------------------------
          ANALOG         SUBROUTINE               ANALOG INPUT SCALING LOGIC
          ALARMS         SUBROUTINE               ALARMS WITH DEADBAND LOGIC
          OUTPUT         SUBROUTINE               ANNUNCIATOR OUTPUT LOGIC
          TEST           PROGRAM NAME             TEST PROGRAM

 |[        BLOCK DECLARATIONS           ]

                +-------+
        SUBR  1 |ANALOG |         LANG: LD  (* ANALOG INPUT SCALING LOGIC *)
                +-------+

                +-------+
        SUBR  2 |ALARMS |         LANG: LD  (* ALARMS WITH DEADBAND LOGIC *)
                +-------+

                +-------+
        SUBR  3 |OUTPUT |         LANG: LD  (* ANNUNCIATOR OUTPUT LOGIC   *)
                +-------+

 |[      START OF PROGRAM LOGIC         ]
 |
 | << RUNG 4  STEP #0001 >>
 |
 |ALW_ON
 |%S0007  +-------------+
 +--] [---+ CALL ANALOG +
 |        | (SUBROUTINE)|
 |        +-------------+

 | << RUNG 5  STEP #0003 >>
 |
 |ALW_ON
 |%S0007  +-------------+
 +--] [---+ CALL ALARMS +
 |        | (SUBROUTINE)|
 |        +-------------+
 |
 | << RUNG 6  STEP #0005 >>
 |
 |ALW_ON
 |%S0007  +-------------+
 +--] [---+ CALL OUTPUT +
 |        | (SUBROUTINE)|
 |        +-------------+
 |
 |[       END OF PROGRAM LOGIC          ]
 |
```

FIG. 3.10j
GE Fanuc PLC program structure.

TABLE 3.10k
GE Fanuc PLC Memory Identifiers and Address Format

Type of Memory	Memory Identifier	Address Format
Digital input	%I	%I0001
Digital output	%Q	%Q0001
Analog input	%AI	%AI001
Analog output	%AQ	%AQ001
Internal coil	%M	%M0001
Register	%R	%R0001
System	%S	%S0001
Temporary	%T	%T0001
Global	%G	%G0001

TABLE 3.10l
GE Fanuc PLC Variable Declaration Table

Reference	Nickname	Description
%I0001	PSL001	Tank TK-1 low pressure
%Q0001	SV001	Tank TK-1 air solenoid
%AI001	PT001	Tank TK-1 pressure TX
%AQ001	PCV001	Tank TK-1 pressure control valve
%M0001	TK1PLTD	Tank TK-1 low pressure for 10 s timer done
%R0001	TK1PLTMR	Tank TK-1 low pressure 10 s timer

Figure 3.10o shows a typical PLC logic for alarming with deadband.

Figure 3.10p shows a typical PLC logic for annunciator outputs.

Typical Allen-Bradley PLC

The program in an Allen-Bradley PLC is organized in a file structure. First a project with appropriate name is created. In the project, there are program files, as explained below.

File No.	Name	Description
0	System	PLC operating system—Unavailable to programmers
1	Undefined	PLC operating system—Unavailable to programmers
2	Main Program	Main program code and subroutine calls
3–999	Subroutines	Subroutine code—called by main program

In the project, there are data files, organized as explained below.

Data files O0 (output) and I1 (input) are physical I/O status files.

Data file S2 (status) is a general status file.

Data files B3 (binary), T4 (timers), C5 (counters), R6 (control word), N7 (integer), F8 (floating) are reserved for main program file.

Other data files can be designated for subroutines as needed.

Addressing is done in following format:

XY:WW:ZZ

where

X is file identifier such as I for input, O for output, S for status, B for binary, T for timers, C for counters, R for real, N for integer, and F for floating point.

Y is file number (in decimal) such as 0 for output, 1 for input, etc.

WW is word number (in decimal) in file such as 00 for word 1, etc.

ZZ is bit number (in octal) in word such as 00 for bit 1, etc.

Typical MODICON PLC 984

The program in MODICON PLC is organized in networks. First, a program file is created. In the program file, there are segments that are made up of logic networks. In PLC 984, there are 1 to 32 segments and logic networks can be written in each of these segments. Network numbers start from 1 and are continuous for the program.

Addressing is done in following format for data tables:

XYYYY

where

X is a data type identifier such as 0 for discrete o/p or coil (used to drive real outputs through the output module or to set internal coils), 1 for discrete input (used to drive contacts in the logic program, controlled by input module), 3 for input register (holds numeric input from external source like analog or high-speed counter input module), and 4 for output holding registers (used to store numeric data for internal use or analog output module).

YYYY is the reference number in decimal such as 1, 2, etc. with an upper limit as configured.

Total memory for all data types should not exceed processor user memory.

PLC ACCESS AND PLC PROGRAMMING MODES

Following are a few ways of connecting a programming device with the PLC.

1. Serial connection
2. Ethernet
3. Data Highway Plus (DH+)

```
|------+------+--- RACK 0 --+------+------|
|  PS  |  1   |  2   |  3   |  4   |  5   |
|======= PROGRAMMED  CONFIGURATION =======|
|      |      |      |      |      |      |
|PWR321|CPU351|MDL240|MDL340|ALG221|ALG391|
|      |      |      |      |      |      |
|      |      |I AC16|Q AC16|IALGI4|QALGI2|
|      |      |      |      |      |      |
|      |      |RefAdr|RefAdr|RefAdr|RefAdr|
|      |      |%I0001|%Q0001|AI0001|%AQ001|
|      |      |      |      |      |      |
|      |      |      |      |      |      |
|      |      |      |      |      |      |
|      |      |      |      |      |      |
+------+------+------+------+------+------+
```

```
SERIES 90-30 MODULE IN RACK 0 SLOT  1
+------+------------------ SOFTWARE  CONFIGURATION ----------------
| SLOT |  Catalog #: IC693CPU351       SERIES 90-30 CPU, MODEL 351
|    1 |
|      |
|CPU351|-----------------------------------------------------------
|      |  IOScan-Stop: NO        Baud Rate  : 19200
|      |  Pwr Up Mode: LAST      Parity     : ODD
|      |  Logic/Cfg  : RAM       Stop Bits  : 1
|      |  Registers  : RAM       Modem TT   : 0  1/100 Second /Count
|      |  Passwords  : ENABLED   Idle Time  : 10 Seconds
|      |  R/S Switch : DISABLED
|      |
|      |  Chksum Wrds:   8       Sweep Mode : NORMAL
|      |  Tmr Faults : DISABLED  Sweep Tmr  : N/A     msec
|      |
+------+
CPU MEMORY CONFIGURATION FOR Model 351 CPU:

      Discrete Input     (%I)  2048  Points
      Discrete Output    (%Q)  2048  Points
      Internal Discrete (%M)  4096  Points
      System Use         (%S)   128  Points
      Temporary Status  (%T)   256  Points
      GENIUS Global      (%G)  1280  Points
                               ----  -----
      TOTAL DISCRETE MEMORY:   9856  Points

      Analog Input      (%AI)  2048  Words
      Analog Output     (%AQ)   512  Words
      Register Memory   (%R)   9999  Words

      TOTAL LOGIC MEMORY      56802  Bytes

      CPU MEMORY TOTAL        81920  Bytes
```

FIG. 3.10m
GE Fanuc PLC rack and CPU configuration.

```
|
(****************************************************************************)
| (* LOGIC TO CONVERT RAW ANALOG INPUT INTO ENGINEERING UNITS FOR REACTOR  *)
| (* TEMPERATURES USING FOLLOWING FORMULA                                    *)
| (* TEMP (IN ENGG. UNIT) = TEMP (RAW ANALOG INPUT)*FULL SCALE RANGE/32000 *)
|
(****************************************************************************)
|
| << RUNG 4  STEP #0002 >>
|
|ALW_ON
|%S0007  +-----+                 +-----+
+--] [---+ DIV_+-----------------+ MUL_+-
|        | INT |                 | INT |
|        |     |                 |     |
|REACTOR |     | REACTOR REACTOR |     | REACTOR
|TEMP. 1 |     | TEMP. 1 TEMP. 1 |     | TEMP. 1
|ANALOG  |     | IN DEG  IN DEG  |     | IN DEG
|INPUT   |     | C       C       |     | C
|TI_1051 |     | TI1051  TI1051  |     | TI1051
|%AI0001-+I1  Q+-%R0001  %R0001 -+I1  Q+-%R0001
|        |     |                 |     |
| CONST -+I2   |         CONST -+I2    |
| +32000 +-----+         +00100 +-----+
|
| << RUNG 5  STEP #0005 >>
|
|ALW_ON
|%S0007  +-----+                 +-----+
+--] [---+ DIV_+-----------------+ MUL_+-
|        | INT |                 | INT |
|        |     |                 |     |
|REACTOR |     | REACTOR REACTOR |     | REACTOR
|TEMP. 2 |     | TEMP. 2 TEMP. 2 |     | TEMP. 2
|ANALOG  |     | IN DEG  IN DEG  |     | IN DEG
|INPUT   |     | C       C       |     | C
|TI_1052 |     | TI1052  TI1052  |     | TI1052
|%AI0002-+I1  Q+-%R0002  %R0002 -+I1  Q+-%R0002
|        |     |                 |     |
| CONST -+I2   |         CONST -+I2    |
| +32000 +-----+         +00200 +-----+
|
|
|[    END OF SUBROUTINE LOGIC     ]
|
```

FIG. 3.10n
GE Fanuc PLC logic for analog inputs.

Once we have connected the programmer to the PLC, we can place the CPU in one of the following modes.

1. Run—In this mode the CPU does all three tasks, i.e., read inputs, execute program logic, write to outputs.
2. I/O Scan Only—In this mode the CPU does only one task, i.e., read inputs.
3. Stop/Program—In this mode the CPU stops all three tasks, i.e., read inputs, execute program logic, write to outputs; the PLC can be programmed in this mode.

PLC programming and configuration can be done in following two ways.

1. Offline Programming—In this mode, a PLC can be programmed on a PC-based programmer without

```
|[    START OF SUBROUTINE LOGIC     ]
| (************************************************************************)
| (* HIGH ALARM WITH DEADBAND OF 2 DEG C                                  *)
| (************************************************************************)
| << RUNG 4  STEP #0002 >>
|                                                                REACTOR
|                                                                TEMP. 1
|                                                                > 70
|                                                                DEG C
|ALW_ON                                                          TI1051U
|%S0007  +-----+                                                 %M0001
+--] [---+ GT_ |+------------------------------------------------( )--
|        | INT ||
|        |     ||
|REACTOR |     ||
|TEMP. 1 |     ||
|IN DEG  |     ||
|C       |     ||
|TI1051  |     ||
|%R0001 -+I1  Q++
|        |     |
| CONST -+I2   |
| +00070 +-----+
| << RUNG 5  STEP #0005 >>
|                                                                REACTOR
|                                                                TEMP. 1
|                                                                > 68
|                                                                DEG C
|ALW_ON                                                          TI1051L
|%S0007  +-----+                                                 %M0002
+--] [---+ GT_ |+------------------------------------------------( )--
|        | INT ||
|        |     ||
|REACTOR |     ||
|TEMP. 1 |     ||
|IN DEG  |     ||
|C       |     ||
|TI1051  |     ||
|%R0001 -+I1  Q++
|        |     |
| CONST -+I2   |
| +00068 +-----+
| << RUNG 6  STEP #0008 >>
|
|REACTOR                                                         REACTOR
|TEMP. 1                                                         TEMP. 1
|> 70                                                            HIGH
|DEG C                                                           ALARM
|TI1051U                                                         TI1051A
|%M0001                                                          %M0003
+--] [---------+-------------------------------------------------( )--
|              |
|REACTOR REACTOR|
|TEMP. 1 TEMP. 1|
|HIGH    > 68   |
|ALARM   DEG C  |
|TI1051A TI1051L|
|%M0003  %M0002 |
+--] [-----] [--+
|
|[     END OF SUBROUTINE LOGIC     ]
```

FIG. 3.10o
GE Fanuc PLC logic for alarm with deadband.

```
|[   START OF SUBROUTINE LOGIC     ]
|
| << RUNG 3  STEP #0001 >>
|
|REACTOR                                                              REACTOR
|TEMP. 1                                                              TEMP. 1
|HIGH                                                                 HIGH
|ALARM                                                                ANN. OP
|TI1051A                                                              TAH1051
|%M0003                                                               %Q0001
+--] [----------------------------------------------------------------( )--
|
| << RUNG 4  STEP #0003 >>
|
|REACTOR                                                              REACTOR
|TEMP. 2                                                              TEMP. 2
|LOW                                                                  LOW
|ALARM                                                                ANN. OP
|TI1052A                                                              TAL1052
|%M0006                                                               %Q0002
+--] [----------------------------------------------------------------( )--
|
|[    END OF SUBROUTINE LOGIC     ]
|
```

FIG. 3.10p
GE Fanuc PLC logic for annunciator outputs.

connecting the programmer to the PLC. The program can be saved on computer hard disk or on a floppy disk for download to the PLC later.

2. Online Programming—In this mode, a PLC can be programmed on a PC-based programmer after connecting the programmer to the PLC on its programming port. The program can be saved on computer hard disk or on a floppy disk and at the same time downloaded to the PLC. Some PLCs allow a program to be downloaded into PLC only in stop/program mode. If the PLC was in run mode prior to program download, a message is given to the programmer that the PLC is in run mode and its mode needs to be changed to stop/program mode before download can be done.

DEVELOPING PLC PROGRAM LOGIC

There are no set rules or guidelines on how to develop PLC program logic. However, the following steps will help in developing PLC program logic:

1. Write a process description or obtain it from the process or operations engineer.
2. Make a sketch of the process or obtain a sketch from the process or operations engineer.

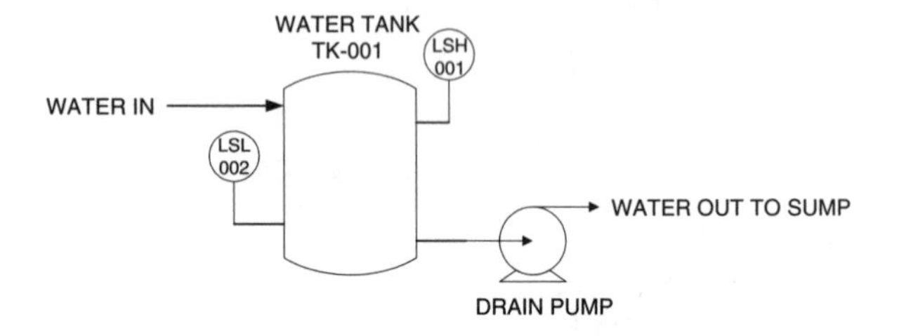

FIG. 3.10q
Tank level control.

3. Prepare control strategy narratives.
4. Make a list of field devices and their inputs and outputs that will be connected to the PLC.
5. Assign addresses to inputs/outputs.
6. Develop flowcharts (if required).
7. Develop Boolean logic diagrams.
8. Develop PLC hardware configuration based on number and type of I/Os.
9. Write PLC ladder logic, i.e., convert Boolean logic into actual ladder logic.
10. Test the PLC ladder logic.

A tank-level control application using a remote-operated pump is shown in Figure 3.10q. This section develops a ladder logic program for a GE Fanuc PLC using the steps described above.

1. Process description—Water tank TK-001 is being filled continuously and its level is to be controlled by a drain pump.
2. Make a sketch of the process—Make a sketch as shown in Figure 3.10q.
3. Prepare control strategy narratives—Pump has two modes of operation:

   - Remote mode—In this mode its operation is controlled automatically by the level in the tank. When the level goes high (say, 90%) in TK-001, the water pump will start to drain the water to a sump. The pump will continue to drain water until level goes low (say, 10%) in TK-001.
   - Local mode—In this mode its operation is controlled by an operator in the field and the automatic level control logic is bypassed.

4. Make a list of field devices (sensors, etc.) and their inputs and outputs that will be connected to PLC:

Field Device	*Input*	*Output*	*Remarks*
Local/remote switch	02	—	For pump operation control
Local start push button	01	—	To start pump in local mode
Local stop push button	01	—	To stop pump in local mode
Water pump	01	01	Input = Run status, Output = Run command
Motor fault alarm	—	01	For annunciation
LSH001	01	—	High-level switch
LSL002	01	—	Low-level switch

5. Assign addresses to inputs/outputs:

Field Device	*Input*	*Output*	*PLC Address*
Local/remote switch	02	—	%I0001 (Remote), %I0002 (Local)
Local start push button	01	—	%I0003
Local stop push button	01	—	%I0004
Water pump	01	01	%I0005 (Run status), %Q0001 (Run command)
Motor fault alarm	—	01	%Q0002
LSH001	01	—	%I0006
LSL002	01	—	%I0007

6. Develop Boolean logic diagram as shown in Figure 3.10r.
7. Develop PLC hardware configuration as shown in Figure 3.10m without the analog modules in slot 4 and 5.
8. Write PLC ladder logic, i.e., convert Boolean logic developed in previous step into ladder logic as shown in Figure 3.10s.
9. Download program in PLC and test the program as explained below.

PROGRAM TESTING AND DEBUGGING

Once the logic is entered in the PLC, the next important step is to debug the program thoroughly to make sure that the logic works the way it is supposed to work, and to correct any errors made while developing and writing the PLC program.

Some of the methods to check the PLC program are discussed below.

1. There are input modules available with switches to simulate the input signals. Install such a simulation module in place of the digital input module in the PLC rack. Put the PLC in run mode and simulate inputs by closing various switches. Check to see if the program turns various outputs on or off as per the logic. Use the PLC data table to see the status of various internal coils.
2. Instead of using simulation modules, it is also possible to use FORCE I/O function in PLC to enable or disable inputs to simulate field inputs. FORCE ON can be used to turn input on and FORCE OFF can be used to turn input off. Force inputs on or off and check to see if the program turns various outputs on or off as per the logic. Use the PLC data table to see the status of various internal coils.
3. When we have a large PLC system with a lot of I/Os, use of graphical process simulator packages available in the market or developed using MMI packages make it easier to check the PLC logic. Processes are displayed graphically on the screen, are fully animated, and will respond to signals of PLC in the same manner that actual process equipment and sensors would respond. Equipment can be operated from the screen and their status can be seen on the screen. There are interface cards available to connect the PLC to a computer running simulator programs.

OTHER DEVELOPMENTS/ADVANCES IN PLC PROGRAMMING

1. Until recently, most of the PLC programming software available on the market were DOS based for ladder logic programming. With PCs or laptops that use Windows

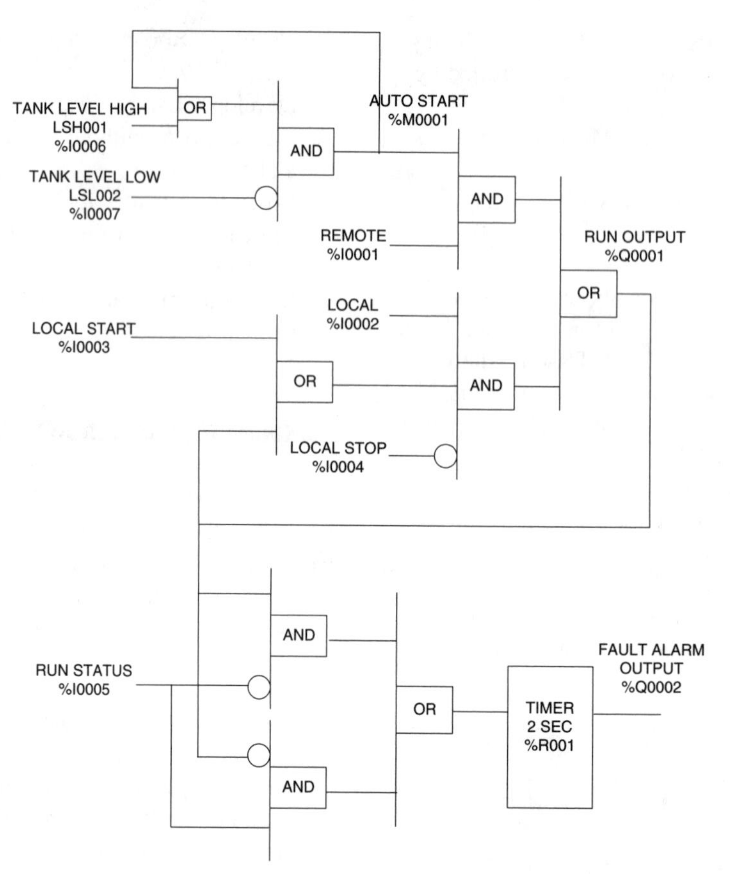

FIG. 3.10r
Boolean logic for tank level control example.

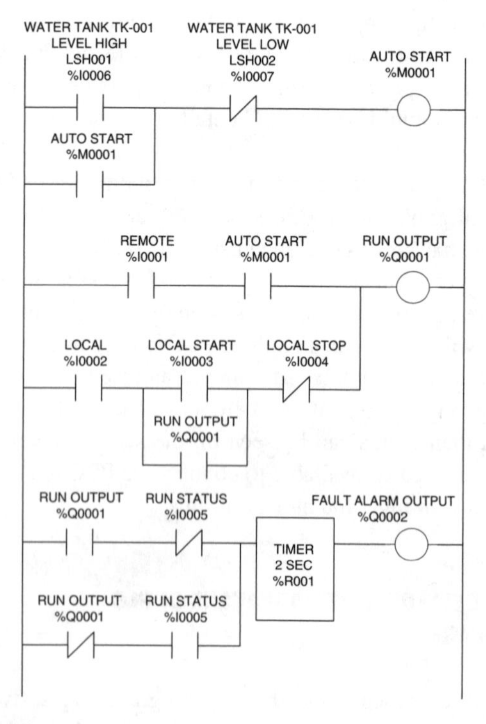

FIG. 3.10s
PLC logic for tank level control example.

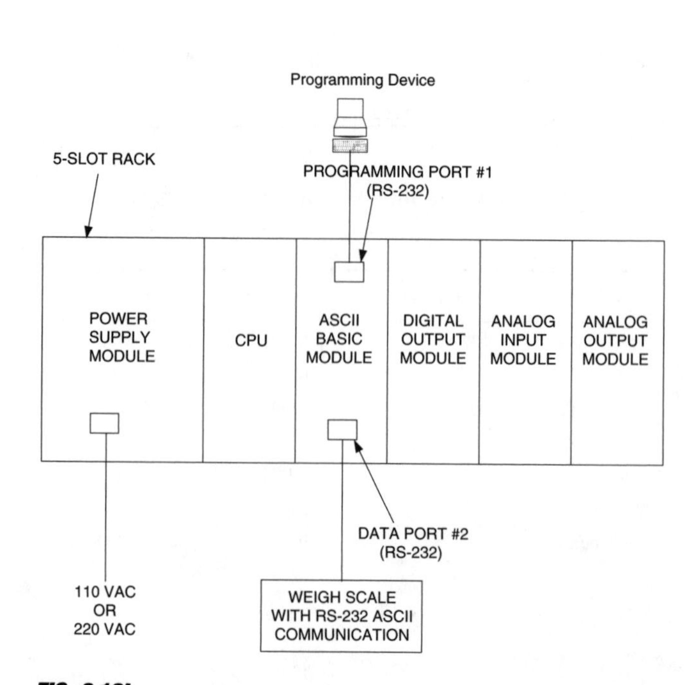

FIG. 3.10t
PLC system with ASCII basic module for weigh scale interface.

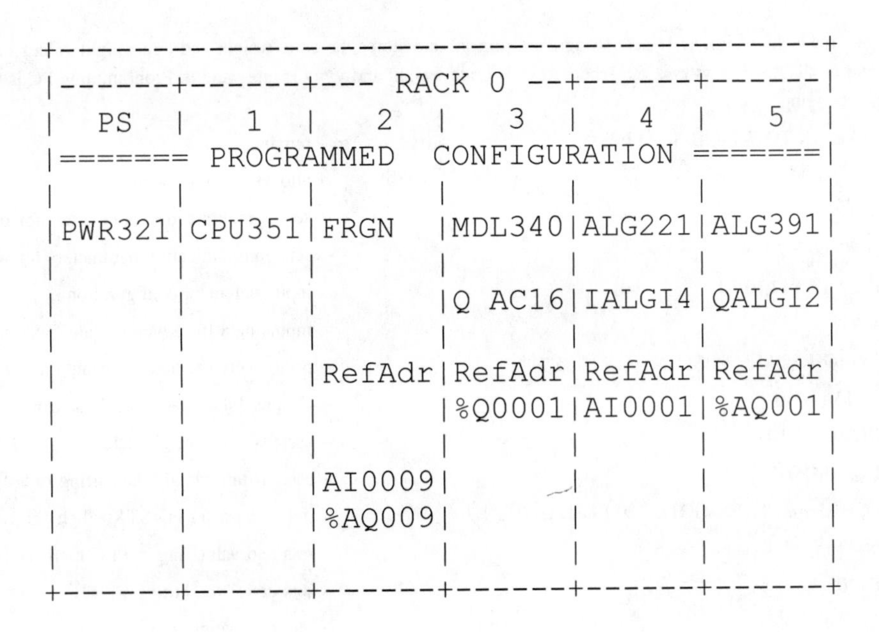

```
+-----------------------------------------+
|------+------+--- RACK 0 --+------+------|
|  PS  |   1  |   2  |   3  |   4  |   5  |
|======= PROGRAMMED  CONFIGURATION =======|
|      |      |      |      |      |      |
|PWR321|CPU351|FRGN  |MDL340|ALG221|ALG391|
|      |      |      |      |      |      |
|      |      |      |Q AC16|IALGI4|QALGI2|
|      |      |      |      |      |      |
|      |      |RefAdr|RefAdr|RefAdr|RefAdr|
|      |      |      |%Q0001|AI0001|%AQ001|
|      |      |      |      |      |      |
|      |      |AI0009|      |      |      |
|      |      |%AQ009|      |      |      |
|      |      |      |      |      |      |
+------+------+------+------+------+------+
```

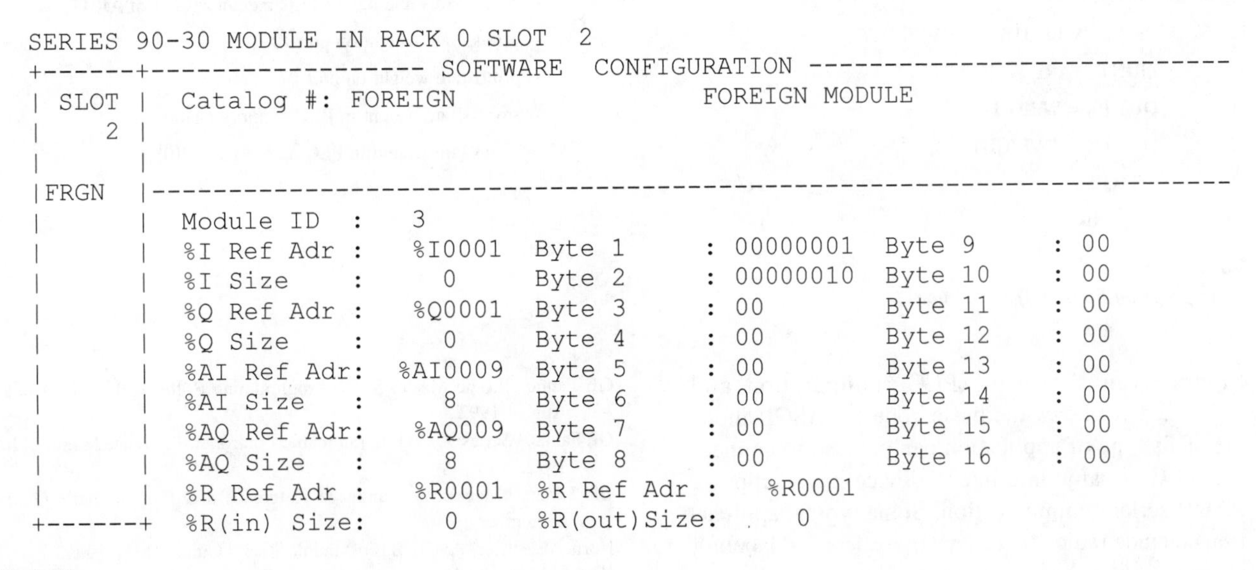

```
SERIES 90-30 MODULE IN RACK 0 SLOT  2
+------+----------------- SOFTWARE  CONFIGURATION ---------------------------
| SLOT |  Catalog #: FOREIGN                   FOREIGN MODULE
|    2 |
|      |
|FRGN  |---------------------------------------------------------------------
|      |  Module ID  :    3
|      |  %I Ref Adr :   %I0001  Byte 1     : 00000001  Byte 9      : 00
|      |  %I Size    :     0     Byte 2     : 00000010  Byte 10     : 00
|      |  %Q Ref Adr :   %Q0001  Byte 3     : 00        Byte 11     : 00
|      |  %Q Size    :     0     Byte 4     : 00        Byte 12     : 00
|      |  %AI Ref Adr:  %AI0009  Byte 5     : 00        Byte 13     : 00
|      |  %AI Size   :     8     Byte 6     : 00        Byte 14     : 00
|      |  %AQ Ref Adr:   %AQ009  Byte 7     : 00        Byte 15     : 00
|      |  %AQ Size   :     8     Byte 8     : 00        Byte 16     : 00
|      |  %R Ref Adr :   %R0001  %R Ref Adr :   %R0001
+------+  %R(in) Size:     0     %R(out)Size:     0
```

FIG. 3.10u
GE Fanuc PLC rack and ASCII basic module configuration.

operating system becoming very popular and less expensive as PLC programmers, PLC manufacturers have started coming out with programming software that are Windows based. Windows-based programming software has many advantages of Windows operating system, as given below.

- Ease of programming
- Use of multiple windows to do programming and troubleshooting
- Better online help for programming

2. Another significant advancement in PLC systems is the ability of the PLC to communicate with intelligent devices using an interface card that is programmable in BASIC language. This interface module is called the ASCII basic module and is physically installed in the PLC rack like any other module, and is programmed using the serial port on the module. One such system for a crate weight application is shown in Figure 3.10t. In this system, the ASCII basic module is programmed in basic language to read weight information from a weigh scale using RS-232 serial communication.

This basic module code reads the weigh scale data string from port #2 and stores the crate weight and tare weight values in PLC memory. Weigh scale data string format is <STX>xxxx yyyy<CR> where xxxx is crate weight and yyyy is tare weight. Programming PC is to be connected at port #1 for programming/debugging.

```
10    REM DATALINK TO WEIGH SCALE                         -- remark
20    STRING 1026,40                                      -- allocates string memory
30    SETCOM 9600                                         -- sets communication parameters for programming (port #1)
40    SETCOM #9600,E,7,1,N,0                              -- sets communication parameters for weigh scale (port #2)
50    SETINPUT 1,1,13,17,10,10                            -- input statement configuration
60    INPUT # $(1)                                        -- inputs data from weigh scale (port #2)
70    PRINT "WEIGH SCALE RESPONSE IS : ", $(1)            -- prints weigh scale data string on port #1
80    $(2) = LEFT$ ($(1),1)                               -- assigns 1st character of data string to $(2)
90    $(3) = MID$ ($(1),2,4)                              -- assigns crate weight data string to $(3)
100   $(4) = MID$ ($(1)7,4)                               -- assigns tare weight data string to $(4)
110   IF $(2) = CHR$ (02) THEN GOTO 160 ELSE GOTO 120     -- if 1st character is <STX> then valid data
120   OUT(2,0) = 1                                        -- sets bad value flag in PLC memory (bit 1 of AI011)
130   WEIGHT = 0                                          -- sets crate weight to 0
140   TWEIGHT = 0                                         -- sets tare weight to 0
150   GOTO 190
160   WEIGHT = VAL ($(3))                                 -- sets crate weight to value read from weigh scale
170   TWEIGHT = VAL ($(4))                                -- sets tare weight to value read from weigh scale
180   OUT(2,0) = 0                                        -- resets bad value flag in PLC memory (bit 1 of AI011)
190   PRINT "WEIGHT IS : ", WEIGHT                        -- prints crate weight on port #1
200   PRINT "TARE WEIGHT IS : ", TWEIGHT                  -- prints tare weight on port #1
210   OUT (0) = WEIGHT                                    -- stores crate weight in PLC memory (AI009)
220   OUT (1) = TWEIGHT                                   -- stores tare weight in PLC memory (AI010)
230   GOTO 50
240   RETURN
```

FIG. 3.10v
Basic program for ASCII basic module.

Figure 3.10u shows the PLC configuration and Figure 3.10v shows the basic code for ASCII module. This type of application can be used to connect the PLC to any intelligent device that supports ASCII serial communication. Some typical applications include radio-frequency tag readers and labeling stations.

Bibliography

Allen-Bradley, "PLC-5 Programming Software, Instruction Set Reference," Release 4.4.

Allen-Bradley, "PLC-5 Programming Software, Software Configuration and Maintenance."

Cox, R. D., *Technician's Guide to Programmable Controllers,* New York: Delmar Thomson Learning, 1995.

GE Fanuc, "Logic Master 90-70 Programming Software User's Manual," August 1992.

GE Fanuc, "Series 90-70 Programmable Controller Reference Manual," July 1992.

GE Fanuc, "MegaBasic Language Reference and Programmer's Guide," September 1994.

Horner Electric, "ASCII Basic Module User's Guide," May 1995.

IEC 1131-3, "International Standards for Programmable Controllers—Programming Languages," 1993.

MODICON, "Ladder Logic Block Library User Guide," 1994.

MODICON, "984 Controller Loadable Function Block Programming Guide," February 1992.

Petruzelka, F. D., *Programmable Logic Controllers,* New York: Glencoe McGraw-Hill, 1998.

Simpson, C., *Programmable Logic Controllers,* Englewood Cliffs, NJ: Prentice-Hall, 1994.

Webb, J. W. and Reis, R. A., *Programmable Logic Controllers: Principles and Applications,* Englewood Cliffs, NJ: Prentice-Hall, 1999.

3.11 Fault-Tolerant Programming and Real-Time Operating Systems

G. B. SINGH

Suppliers of Fault-Tolerant Control Systems:	General Electric; Woodward Controls; Triconex Systems; Foxboro Systems
Suppliers of Real-Time Operating Systems:	Accelerated Technology, Inc., 334-661-5788, www.atinucleus.com Embedded System Products, 281-561-9980, www.rtxc.com Eonic Systems, 301-572-5005, www.eonic.com Green Hills Software, 805-965-6343, www.ghs.com Integrated Systems, Inc., 408-542-1950, www.isi.com Lynx Real-Time Systems, 408-879-3920, www.lynx.com Microsoft Corp, 852-256-2217, www.microsoft.com Mentor Graphics, 503-685-8000, www.mentor.com/embedded QNX Software Systems, 613-591-3579, www.qnx.com Real-Time Innovations, 408-734-5009, www.rti.com VenturCom, Inc., 617-577-1607, www.vci.com Wind River Systems, 81-3-5467-5877, www.wrs.com Zentropic Computing, 703-471-2108, www.zentropix.com

FAULT TOLERANCE

Hardware and software faults in computer systems cause economic losses and endanger humans. To ensure trouble-free operation, monitoring, and control of the computerized systems the availability and robustness of computers in various areas of automation must be improved. Today, there is an increasing demand to make both the hardware and software of a computerized system more tolerant of failures or *faults*. Today, there are increasing demands to make application software systems more tolerant of failures. Fault-tolerant applications detect and recover from system failures that are not handled by underlying hardware or operating system of the application, thus increasing the availability of these applications. Such applications are being increasingly implemented on various operating systems.

What are the features that a user demands in a fault-tolerant system?

The fault must be detected, recorded, annunciated automatically.

There should be no interruption in the tasks being carried out by the computer system.

Fault of hardware and software must be eliminated through redundancy.

There should be an automatic recovery with minimum or no intervention from the operator.

The fault-tolerant system must be independent of hardware or software on which it is implemented.

Degree of redundancy must be freely selectable (dual redundant, triple redundant, quadruplets).

There should be strong cohesion between operating system and redundancy management software and hardware.

Characteristics of a Fault-Tolerant System

For a system to provide high availability or fault tolerance, the following characteristics must be present in the design of the system:

1. Design to withstand a single point defect
2. Identification of defect
3. Isolation of defected components
4. Repair of defected components
5. Restoration of defected components
6. Resurrection of complete system

A fault-tolerant system must have inherent design to withstand a single point failure. A single point failure should not lead to the stoppage of the complete system. Once the fault occurs, the system must be able to identify the defect and this defect must be well annunciated to alert the users to its faulty state. After identifying the defective component, the system must isolate the faulty components. Once the faulty component has been identified, the repair of the faulty component system should be possible online while the system is running. Once the faulty component has been repaired, it must be able to be restored back into the system online without disrupting the process/system activity. Once the faulty component has been replaced, it must be resurrected as a part of the complete system and must be synchronized with the existing application. It must start participating in the system. During all the above steps there should be no interruption in the quality of service provided by the system.

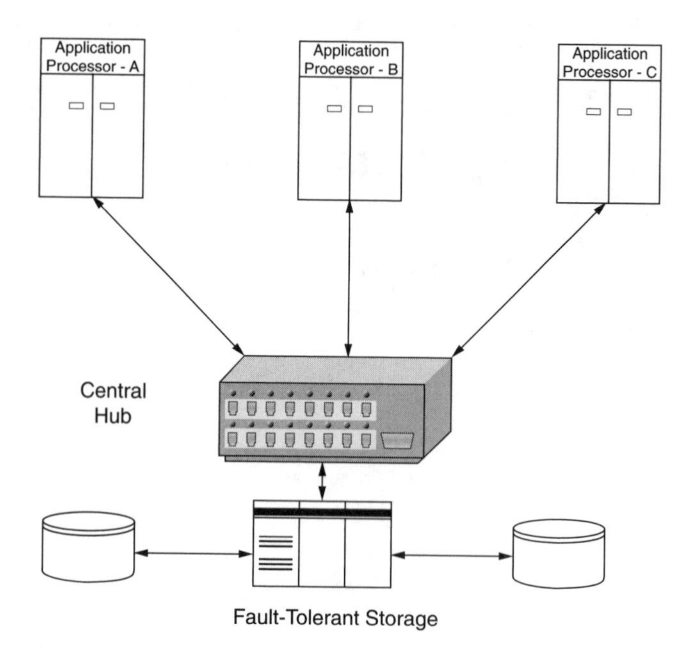

FIG. 3.11a
Cluster-based fault-tolerant system.

Classification of Fault-Tolerant Systems

There are many approaches to build a fault-tolerant system; however, in any fault-tolerant system the essence is the way redundancy is managed. There must exist an identical component that must perform the task identical to its predecessor.

Cold Standby Systems: In cold standby systems, the standby component has to be manually activated to take over from the normal component that has failed, and this may involve loss of data or temporary disruption in the service being performed by the system. In such systems, the current state of the application is interrupted and, on the standby system, a new application has to be restarted. These types of system are being used in many commercial or office applications (like file servers, clustered servers, etc.).

Standby Server/Cluster-Based Systems: Each server is connected through a redundant high-speed hub or switch to each of the other servers and to the fault-tolerant storage system. The fault-tolerant storage system provides high-speed access, disk caching, and mirroring, as well as handling backup to archival storage. Note that because the storage system is independent of the servers, the repair/recovery process does not require rebooting the whole system and the user is not interrupted. If the server crashes, the standby server (application processor) can be brought into service without affecting other systems. Also if one storage disk fails, there is a fault-tolerant disk available as backup. However, these changes have to be done manually or with minimal manual intervention required (Figure 3.11a).

Hot Standby Systems: In hot standby systems, the standby component is automatically activated and takes over the control of the application from its predecessor without any manual intervention. However, this change may or may not lead to loss of data or service for a brief period. Upon taking over from the faulty component, the application in the standby component is automatically restarted. Many control systems of the 1980 and 1990s have evolved around this principle. In a mission-critical application, any sort of interruption in the services is intolerable because it may lead to serious economic or human loss. Therefore, the modern-day automation systems are built with tolerance to any fault. These types of systems have a unique architecture and software design. The idea is to have at least two or more components looking at and manipulating the same data at the same time and keeping each other informed of their healthy status (heartbeat) all the time. The standby component executes control computations at all times in synchronization with the normal component. Thus, control continuity is maintained regardless of the component transfers. Whenever the heartbeat of one is lost, the other starts providing the control (Figure 3.11b).

All commercial, fault-tolerant systems use at least two processors and custom hardware in a fail-stop configuration as the basic building block. A typical fail-stop system runs two processors in microcycle lockstep and uses hardware comparison logic to detect a disagreement in the output of the two systems (Figure 3.11c). As long as the two processors agree, operation is allowed to continue. When the output disagrees, the system is stopped. These types of systems are also known as pair-and-spare fault-tolerant systems, configured so that each pair backs up the other. In each pair, special error-detection logic constantly monitors the two processors and then fail-stops if an error or failure is detected, leaving the other pair to continue execution. Each pair of processors is also connected to an input/output (I/O) subsystem and a common memory system, which uses error correction to mask memory failures. Thus, there are two processing CPUs (central processing unit), memory, and an I/O system in each half of the system. The operating system (OS), which is unique to the architecture and, hence, proprietary, is used to provide error handling, recovery, and resynchronization support after repair (Figure 3.11d).

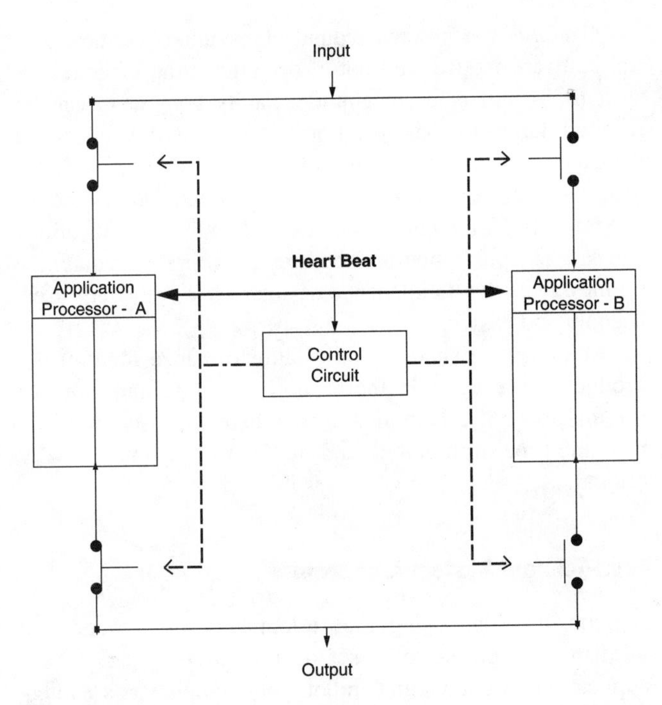

FIG. 3.11b
Example of hot stand by system.

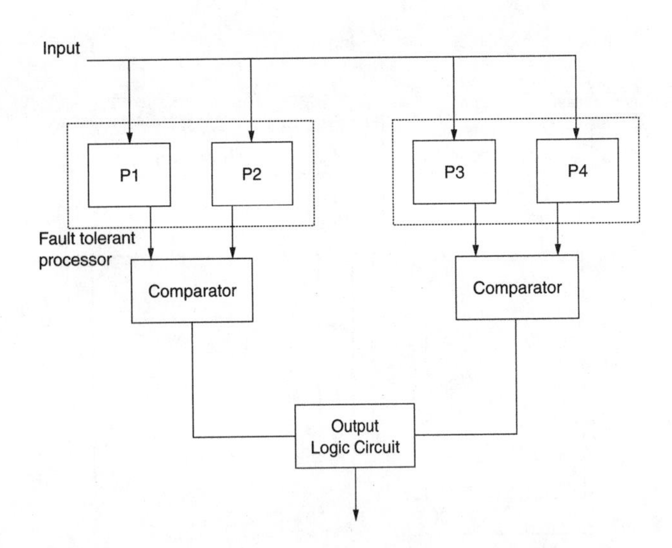

FIG. 3.11d
Example of pair-and-spare fault tolerance.

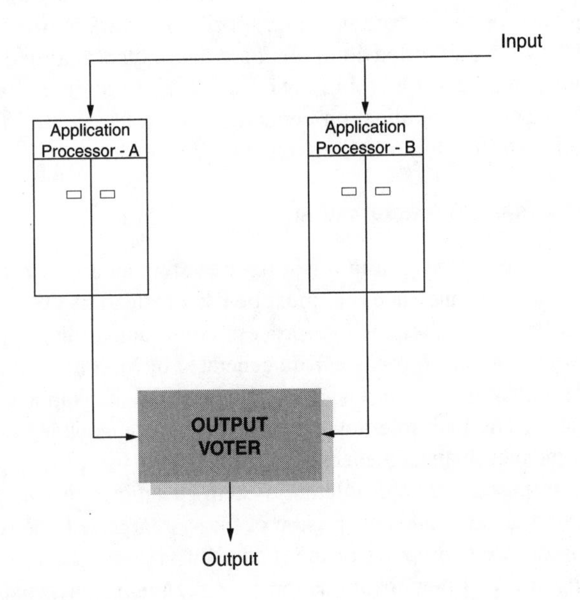

FIG. 3.11c
Active redundant system.

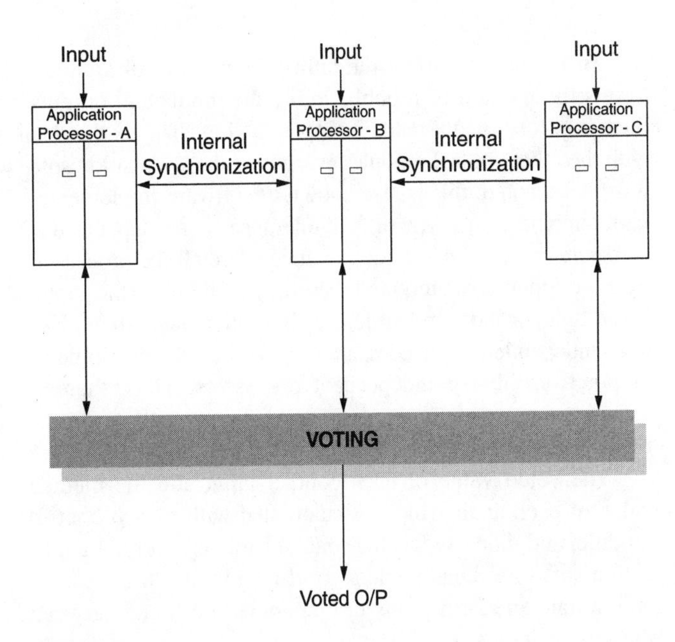

FIG. 3.11e
Software-implemented fault tolerance with software voter.

Another method for providing fault tolerance uses a concept called triple modular redundancy. This approach predates paired architectures and came at a time when processing errors were as prevalent as hard failures. In a triple modular redundant system, three CPUs simultaneously execute and pass through a voter; the majority result is the one the system uses. Obviously, the voter is the weak point in these systems, so one must either make the voter fast and extremely reliable or use multiple voters. Think of the voter as an extension of the output comparison logic in the pair-and-spare architecture with three, rather than two, inputs. With this method of providing fault tolerance, special problems exist with synchronization, and active participation of the OS is required for normal operation as well as recovery and resynchronization. These systems are characterized by custom proprietary hardware, a custom proprietary OS, and custom applications. Several shortcomings in all the current active redundant systems are due to the proprietary nature of the hardware, OS, and applications. This drives up both the initial acquisition cost and the life-cycle cost (Figure 3.11e).

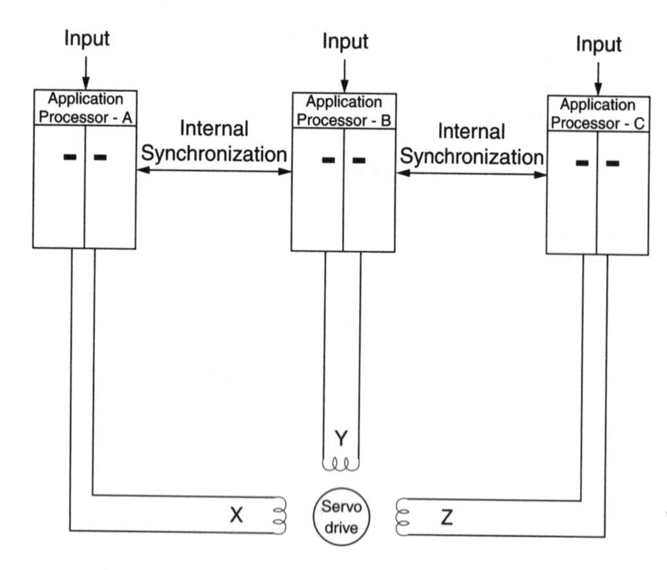

FIG. 3.11f
Software-implemented fault tolerance with hardware voter.

An important part of the fault-tolerant control architecture is the method of reliably *voting* the inputs and outputs. Each control module reads its inputs and exchanges the data with the other control modules every time the application software is executed; this is also known as software-implemented fault tolerance. This type of fault tolerance is one step ahead of hardware-implemented fault tolerance. Here all the processors read the input separately and then they manipulate the inputs based on their independent logic. Their outputs are then given to an independent logic comparator who votes the manipulated output from all the independent processors. Thus, there is software voting of all the outputs of the processor and the most-voted output goes as a final output of the system.

The voted value of each contact input and the median value of each analog input is calculated within each control module, and then used as the control parameter for the application software. Diagnostic algorithms monitor these inputs and initiate an alarm if any discrepancies are found between the sets of inputs.

An equally important part of the fault tolerance is the hardware voting of analog and contact outputs. For example, in a hardwired voting on a actuator, the three coil servos are separately driven from each control module; thus, the resultant position of actuator is the vector sum of the three signals (Figure 3.11f).

Advantages of Fault Tolerance

Safety: By applying the compare and selection mechanisms of fault tolerance, faulty control data are recognized and eliminated. Therefore, the safety of the control system is improved. In safety-critical control systems, fault-tolerant systems can be used in threefold or higher redundant systems.

Availability: The active redundant operation of critical control software reduces the out-of-operation time of computer systems by orders of magnitude. Faulty computers can be repaired during normal operation of the system. After restart, these are recovered by the fault-tolerant system. High availability can already be achieved through twofold redundant systems.

Ease of Maintenance: Because of online modifications and ease of troubleshooting and because of prevoted diagnostic messages, the maintenance of such systems is very user-friendly and easy.

Reliability: Because of the fault tolerance feature, the products have become increasingly reliable and suitable for mission-critical applications, where the user may not have the time to intervene and the control actions are very critical.

Fault-Tolerant System Architecture

A control system using fault-tolerant system redundancy mechanisms consists of a set of 1 to n computers where application programs run parallel. These computers are interconnected to exchange both fault-tolerant system internal and application data for the purpose of synchronization. The links between redundant computers may also be redundant. It is possible to choose the degree of redundancy of each application program according to its specific needs. Control data sent by the controlled equipment are distributed to all redundant computers. Outgoing data are selected (compared and filtered) by an application-dependent control logic and forwarded to the controlled equipment (Figure 3.11g).

Fault-Tolerant Programming

Fault-tolerant programming software offers an application–programmer interface that must be used instead of OS services (system calls) for interprocess communication. Application programs (processes) are generated by linking the fault-tolerant system library and OS libraries to the application objects. The fault-tolerant system library contains parts of the redundancy management mechanisms. Apart from the application processes, fault-tolerant system-specific programs for synchronization and supervision of the system have to be running on each computer of the redundant system. The control software is written in various forms, e.g., rung display, control statements, function blocks, etc., based on the choice of the vendor. A typical 2 of 3 voting and 2 of 4 voting logic in the form of a ladder diagram is shown in Figure 3.11h.

REAL-TIME OPERATING SYSTEM

Concepts

Real Time The simple meaning of *real time* is occurring immediately. Embedded designers are flooding our lives with dedicated, non-PC devices. The average person interacts with

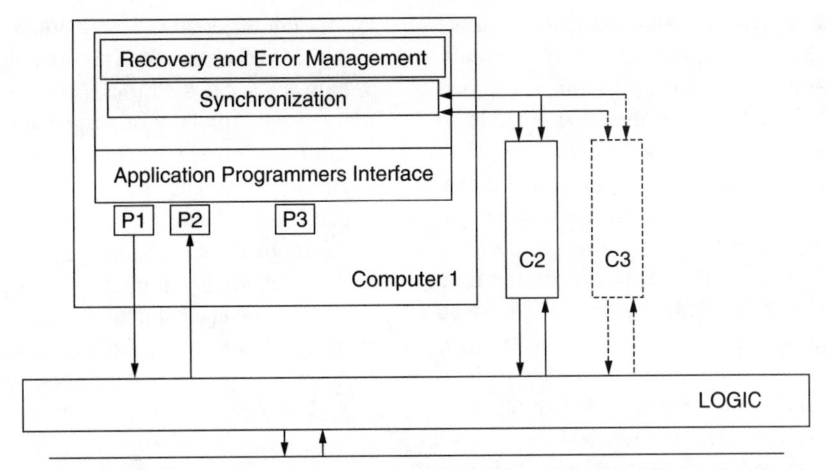

FIG. 3.11g
Fault-tolerant system architecture.

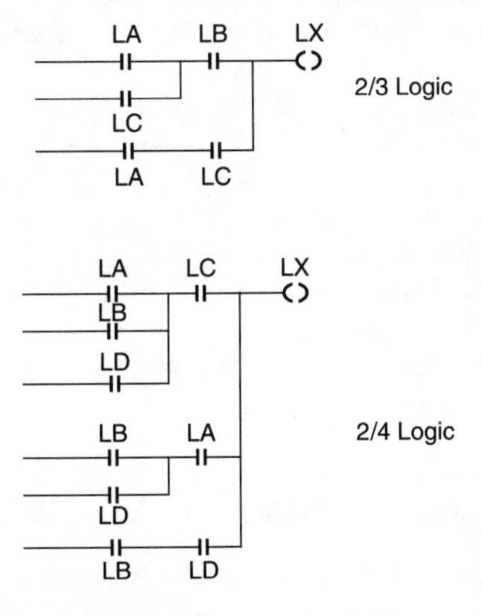

FIG. 3.11h
Example of fault-tolerant ladder logic.

hundreds of embedded processors every day in phones, automobiles, home appliances, toys, cash registers, entertainment electronics, security systems, environmental controls, and personal electronics. Now the second wave is under way, as many of these devices connect to the Internet or a local communications network. The common link between all of these products is their ability to react in real time to the user, external events, and the communications channel. Watching a teenager interact with a video game exemplifies the definition of real time. A real-time operating system (RTOS) is needed to ensure that routines execute on time in response to external events. The user sets up the priorities and data dependencies, and the RTOS manages the application software even with a flurry of external, real-time activity. A user can write each of the software routines independently without being bogged down with intertask timing problems. Messages that pass between routines give the user the perception that everything is running simultaneously.

There are several definitions of real time, most of them contradictory. Unfortunately, the topic is controversial, and there does not seem to be 100% agreement about the terminology. The canonical definition of a real-time system (from Donald Gillies, gillies@cs.ubc.ca) is the following: "A real-time system is one in which the correctness of the computations not only depends upon the logical correctness of the computation but also upon the time at which the result is produced. If the timing constraints of the system are not met, system failure is said to have occurred." Others have added: "Hence, it is essential that the timing constraints of the system are guaranteed to be met. Guaranteeing timing behavior requires that the system be predictable. It is also desirable that the system attain a high degree of utilization while satisfying the timing constraints of the system."

A good example is a robot that has to pick up something from a conveyor belt. The piece is moving, and the robot has a small window to pick up the object. If the robot is late, the piece will not be there anymore, and thus the job will have been done incorrectly, even though the robot went to the right place. If the robot is early, the piece will not be there yet, and the robot may block it. Another example is the servo loops in an airplane when on autopilot. The sensors of the plane must continuously supply the control computer with proper measurements. If a measurement is missed, the performance of the airplane can degrade, sometimes to unacceptable levels. POSIX Standard 1003.1 defines "real time" for operating systems as, "Real time in operating systems: the ability of the operating system to provide a required level of service in a bounded response time."

Hard or Soft? Real-time systems automatically execute software routines or tasks in response to external events. RTOS vendors have invented terms such as "hard" and "soft" real-time to describe the operation of their systems. Hard real-time systems are scheduled so tasks are guaranteed to start within a precise length of time from an external event. Hard real-time systems are deterministic. Soft real-time systems generally list the average length of time to start the routine, but a small probability exists that the maximum time will be much longer. Mission-critical applications must be deterministic. For example, an air bag controller, antilock brakes, and even arcade games must react in a known time. Soft real-time applications usually respond within a few seconds but an occasional slow response is not serious.

A classification can be made into hard and soft real-time systems based on their properties. The properties of a hard real-time system are:

- No lateness is accepted under any circumstances
- Useless results if late
- Catastrophic failure if deadline missed
- Cost of missing deadline is infinitely high

A good example of a hard real-time system is nuclear power plant control, industrial manufacturing control, medical monitoring, weapon delivery systems, space navigation and guidance, reconnaissance systems, laboratory experiments control, automobile engines control, robotics, telemetry control systems, printer controllers, antilock breaking, burglar alarms—the list is endless.

A soft real-time system is characterized by:

- Rising cost for lateness of results
- Acceptance of lower performance for lateness

Examples are a vending machine and a network interface subsystem. In the latter it is possible to recover from a missed packet by using one or another network protocol asking to resend the missed packet. Of course, by doing so, one accepts system performance degradation.

The difference between a hard and a soft real-time system depends on the system requirements: it is called hard if the requirement is "the system shall not miss a deadline" and soft if "the system should not miss a deadline."

David Sonnier adds a distinction: In the robot example, it would be hard real time if late arrival of the robot caused completely incorrect operation. It would be soft real time if late arrival of the robot meant a loss of throughput. Much of what is done in real-time programming is actually a soft real-time system.

Real-Time System A real-time system responds in a timely, predictable way to unpredictable external stimuli arrivals. To fulfill this, some basic requirements are needed:

1. Meet deadlines. After an event occurred an action has to be taken within a predetermined time limit. Missing a deadline is considered a (severe) software fault. On the contrary, it is not considered as a software fault when a text editor reacts slowly and so enervating the user. This lack of response is cataloged as a performance problem, which can probably be solved by putting in a faster processor. It can be demonstrated that using a faster processor will not necessarily solve the problem of missing deadlines.
2. Simultaneity or simultaneous processing. Even if more than one event happens simultaneously, all deadlines for all these events should be met. This means that a real-time system needs inherent parallelism. This is achieved by using more than one processor in the system or by adopting a multitask approach.

We define an RTOS as an OS that can be used to build a hard real-time system.

People often confuse the notion of real-time systems with real-time operating systems (RTOS). From time to time people even misuse hard and soft attributes. They say this RTOS is a hard RTOS or this one is a soft one. There is no hard RTOS or soft RTOS. A specific RTOS can only allow one to develop a hard real-time system. However, having such an RTOS will not prevent one from developing a system that does not meet deadlines.

If, for example, one decides to build a real-time system that should respond to an Ethernet TCP/IP connection, it will never be a hard real-time system as the Ethernet itself is never predictable. Of course, if one decides to build an application on top of an OS such as Windows 3.1, the system will never be a hard real-time system because the behavior of the "OS" software is by no means predictable.

Characteristics of an RTOS

An RTOS Has To Be Multithreaded and Preemptible As mentioned above, an RTOS should be predictable. This does not mean that an RTOS should only be fast, but that the maximum time to do something should be known in advance and should be compatible with the application requirements. Windows 3.1—even on a Pentium Pro 200 MHz—is useless for a real-time system, as one application can keep the control forever and block the rest of the system (Windows 3.1 is cooperative).

The first requirement is that the OS be multithreaded and preemptible. To achieve this the scheduler should be able to preempt any thread in the system and give the resource to the thread that needs it most. The OS (and the hardware architecture) should also allow multiple levels of interrupts to enable preemption at the interrupt level.

Thread Priority Has To Exist The problem is to find which thread needs a resource the most. In an ideal situation, an RTOS gives the resources to the thread or driver that has the closest deadline to meet (we call this a deadline-driven OS).

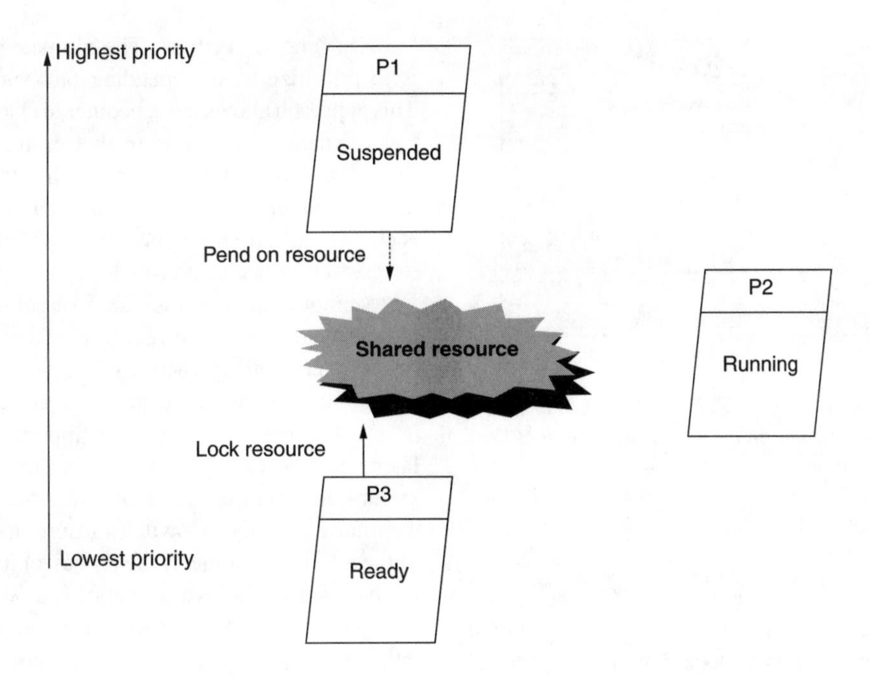

FIG. 3.11i
Priority inversion.

To do so, however, the OS has to know when a thread has to finish its job and how much time each thread needs to do so. At present there is no RTOS for which this is the case, as it is so far too difficult to implement.

Therefore, the OS developers took another point of view: they introduced the concept of priority levels for threads. The designer is responsible for converting deadline requirements in thread priorities. As this human activity is prone to error, a real-time system can easily go amiss. The designer can obtain help in this transformation process by using scheduling theories and some simulation software.

The OS Has To Support Predictable Thread Synchronization Mechanisms As threads share data (resources) and need to communicate, it is logical that locking and interthread communication mechanisms exist.

A System of Priority Inheritance Has To Exist Before having a priority inversion condition, at least three threads have to be involved. When the lowest-priority thread has locked a resource (shared with the highest one) while the middle priority thread is running, the highest-priority thread is then suspended until the locked resource is released and the middle-priority thread stops running. In such a case the time needed to complete a high-priority-level thread depends on a lower-priority-level "priority inversion." It is clear that in such a situation it is difficult to meet deadlines (Figure 3.11i).

To avoid this, an RTOS should allow priority inheritance to boost the lowest-priority thread above the middle one up to the requester. Priority inheritance means that the blocking thread inherits the priority of the thread it blocks (of course, this is only the case if the blocked thread has a higher priority).

OS Behavior Should Be Known Last, the timing requirements should be considered. In fact, the developer should know all the timing of system calls and the system behavior in all the designs. Therefore, the following figures should be clearly given by the RTOS manufacturer:

- The interrupt latency (i.e., time from interrupt to task run). This has to be compatible with application requirements and has to be predictable. This value depends on the number of simultaneous pending interrupts.
- For every system call, the maximum time it takes. It should be predictable and independent from the number of objects in the system.
- The maximum times the OS and drivers mask the interrupts.

The developer should also know the following points:

- System interrupt levels
- Device driver interrupt request queue (IRQ) levels, maximum time they take, etc.

When all the previous metrics are known, one can imagine development of a hard real-time system on top of this OS. Of course, the performance requirements of the system to develop should be compatible with the RTOS and the hardware chosen.

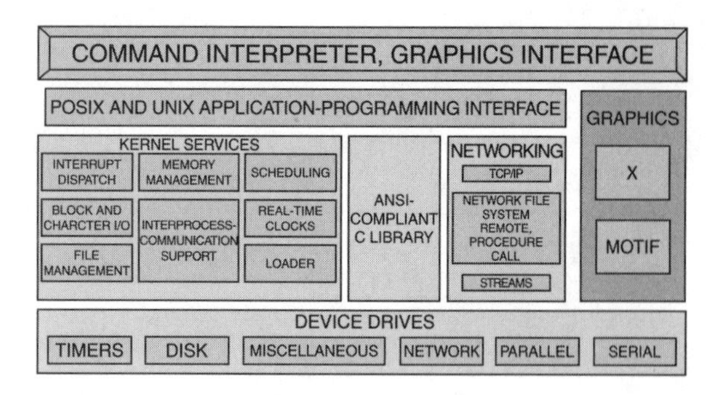

FIG. 3.11j
Architecture of RTOS. (From Webb, W. W., EDN Magazine, February 27, 2000. © Cahners Business Information 2001, a Division of Reed Elsevier, Inc. With permission.)

Architecture of RTOS*

The basic architecture of a multitasking RTOS includes a program interface, the kernel, device drivers, and optional service modules. Figure 3.11j is a block diagram of a typical full-featured RTOS. The kernel is the core of the operating system and provides an interrupt handler, a task scheduler, resource-sharing flags, and memory management. Calls to the application–programming interface of the kernel request the services of the kernel. The kernel is active continuously during real-time operation and must remain memory resident.

One of the primary functions of the kernel is to process interrupts that external or internal events cause. When an interrupt occurs, the processor transfers control to an interrupt service routine that logs in the interrupt, sends message to the scheduler, and returns to the active code. The kernel also includes the scheduler that sets up the order of execution of the application code. The software will consist of many individual tasks, each with an entry point, a priority, and stack space. The priority ensures that a higher-priority task can preempt a lower-priority task to maintain a deterministic response. RTOSs vary in their scheduling techniques. They can schedule threads of equal priority by a first-in, first-out, or round-robin, method. A simple modification of the round-robin scheduling technique adds time slicing, guaranteeing each task some processor time even if some tasks take a long time to complete. However, time slicing equally allocates processor time between tasks, and there is no way to ensure that the RTOS performs time-critical tasks first.

The most popular scheduling technique is preemptive, prioritized scheduling. Tasks can preempt a lower-priority task and keep the processor until they are finished or until a higher-priority task preempts them. This technique requires careful assignment of priorities to ensure that tasks are completed on time. Most RTOS vendors include tools to analyze priority scheduling and view the results.

On time, every time: The ultimate scheduling technique is to prioritize tasks depending on their execution deadline. This approach allows the scheduler to change priorities of other tasks dynamically to ensure that each task is completed on time. The drawback to this approach is that it is time-consuming to compute the deadline of each task before scheduling each new task. So far, dynamic scheduling has been restricted to research projects and robotics.

Memory allocation is also a kernel service, and it affects the performance of the real-time system. As each new task begins execution, the memory-allocation module searches for a free memory block adequate for the task.

The kernel design is also important in minimizing the latencies that can degrade the performance of a real-time system. Interrupt latency is the worst-case delay between an external event, like a switch closure, to the first instruction of the interrupt routine. If the processor interrupts are enabled, the hardware delay is very short, usually nanoseconds, but it can vary from processor to processor. The processor needs only to complete the current instruction before jumping to the requested interrupt location. If the processor only has a single interrupt line, the time required to poll inputs to determine which interrupt routine to call is part of the interrupt latency.

In addition to the kernel, RTOS vendors are adding a variety of optional modules to entice users to apply their product in their next application. For example, almost every RTOS includes Internet communications protocols such as transmission control protocol/Internet protocol (TCP/IP). Most of the major RTOS vendors also offer graphical user interface routines. The user can add or delete these modules as needed depending on the application.

If the application involves heavy data processing, RTOSs that can easily scale up to multiple processors should be investigated. One can spread tasks across several processors to gain a significant performance boost. Applications such as radar and sonar signal analysis use hundreds of processors. Most RTOSs are designed for 32-bit processors. The application complexity and multiple real-time inputs result in large software-development teams. European manufacturers have created a slimmed-down RTOS standard called OSEK to serve automotive applications. (The OSEK initials are from the German phrase for "Open systems and interfaces for in-car electronics.") Although 8-bit microprocessors handle most of the world's real-time applications, few 8-bit RTOSs exist. However, it is possible to find commercial RTOSs to support the 8051 and 6800 processors.

Bibliography

Catlin, B., "Design of a TCP/IP Router Using Windows NT," paper presented at the 1995 Digital Communications Design Conference, Catlin & Associates, Redondo Beach, CA, 1995.
Custer, H., *Inside Windows NT,* Seattle: Microsoft Press, 1992.
G&J Softwareentwicklungsges.m.b.H, Fault Tolerant System Design Manuals.
Glorioso, R., "Tolerant to a Fault," *Industrial Computing,* October 1999.

* As detailed by W. W. Webb. Reprinted from *EDN Magazine*, February 27, 2000. © Cahners Business Information 2001, a Division of Reed Elsevier, Inc. With permission.

Kortmann, P. and Mendal, G., "PERTS Proto-typing Environment for Real-Time Systems," *Real-Time Magazine,* Issue 1, 1996.

Microsoft, "Technology Brief: Real-Time Systems with Microsoft Windows NT," Microsoft Developer Network, Microsoft Corporation, 1995.

Microsoft® Win32® Software Development Kit Documentation.

Microsoft® Windows NT™ Version 3.51 Device Driver Kit Documentation.

"Mutex Wait Is FIFO but Can Be Interrupted," in *Win32 SDK Books Online,* Seattle: Microsoft Corporation, 1995.

Powell, J., *Multitask Windows NT,* Corte Madera, CA: Waite Group Press, 1993.

Richter, J., *Advanced Windows, The Developer's Guide to the Win32 API for Windows NT 3.5 and Windows 95,* Seattle: Microsoft Press, 1995.

Webb, W. W., "Real-Time Software Reigns in Post-PC Products," *EDN Magazine,* February 27, 2000.

Zalewski, J., "What Every Engineer Needs to Know about Rate-Monotonic Scheduling: A Tutorial," *Real-Time Magazine,* Issue 1, 1995.

3.12 Practical Logic Design

A. TUCK

An operator notices a demineralized water-forwarding pump rapidly cycling on and off during the regeneration cycle of a demineralized water system. The plant has been in service for years. "Why," the operator wonders, "after all this time, has the pump logic stopped working properly? Who changed the logic?"

The logic has not changed at all but contains an error that appears only under a certain combination of conditions—in this case, when the supply tank level is hovering around the low-low set point and the demineralized water system requires the pump to run.

Many simple logic schemes that were tested and thought to work properly have subtle errors that are discovered only after they are placed in service. It is easy to test for the things the logic *should* do, but it is much harder to test for the things it *should not* do.

To help avoid such problems when designing logic, we have provided standard logic diagrams and solutions for control in this section. We have also provided information, compiled from years of correcting logic in various plants, that will help avoid the most common problems found in logic. This information will be especially useful while redesigning logic during commissioning.

Refer to Figures 3.12a and 3.12b for an explanation of the logic elements used in this section.

DESIGN PHILOSOPHY

Make It Simple. Somehow, the simplest and most direct way to solve a problem usually works best. When correcting logic that does not work properly, the best way to solve the problem often involves removing extraneous logic rather than adding more. Logic can be made to do almost everything anyone can dream up, provided there is enough input/output (I/O), but such lengths usually are not necessary. Find a simple way of solving the problem and do that instead.

The purpose of the logic is to keep systems controlled at steady state and to protect the equipment—it works best when it does not attempt to make all of the decisions for the operator. It is better for the operator to understand what the logic is doing and to stay involved in the operation of the plant. If too much is controlled *for* the operators, they will not be clear about where the logic stops and their job begins, and they are more likely to miss something that needs their attention. The only safe alternative would be to have the logic control everything and replace the operator entirely, and that usually is not practical.

Be Consistent. Writing logic that is consistent throughout the plant is fundamental to good design, and is important for safety as well. All valves and pumps of the same type should work the same way so that the operator knows what to expect when a button is pushed. Keep in mind all of the operators who will be hired in the years to come—they will not have the advantage of learning all the quirks of the plant during its start-up.

OPEN/CLOSE VALVES

A valve opens automatically upon a steam turbine trip. While the turbine is down, the operator tries to close the valve; however, the steam turbine trip indication is still in place and locks the valve open. The operator then puts the valve in manual mode and closes it. The steam turbine is started again a few days later, and by this time the operator has forgotten about the valve and does not return it to auto mode. When the steam turbine trips again, the valve does not open.

Many would call this operator error, but it also is a design flaw that makes the plant hard to control. There are many different ways to design open/close valve logic, and it is not always clear how difficult it is to control until the plant is up and running.

Following are some tips for creating safe and easy-to-control logic, and some diagrams of standard logic for motor-operated valves (MOV) and solenoid valves.

Definitions

Input to the logic:

Open and close command: The commands from the graphic display, initiated by operator, to open and close the valve. Note that these are always momentary.

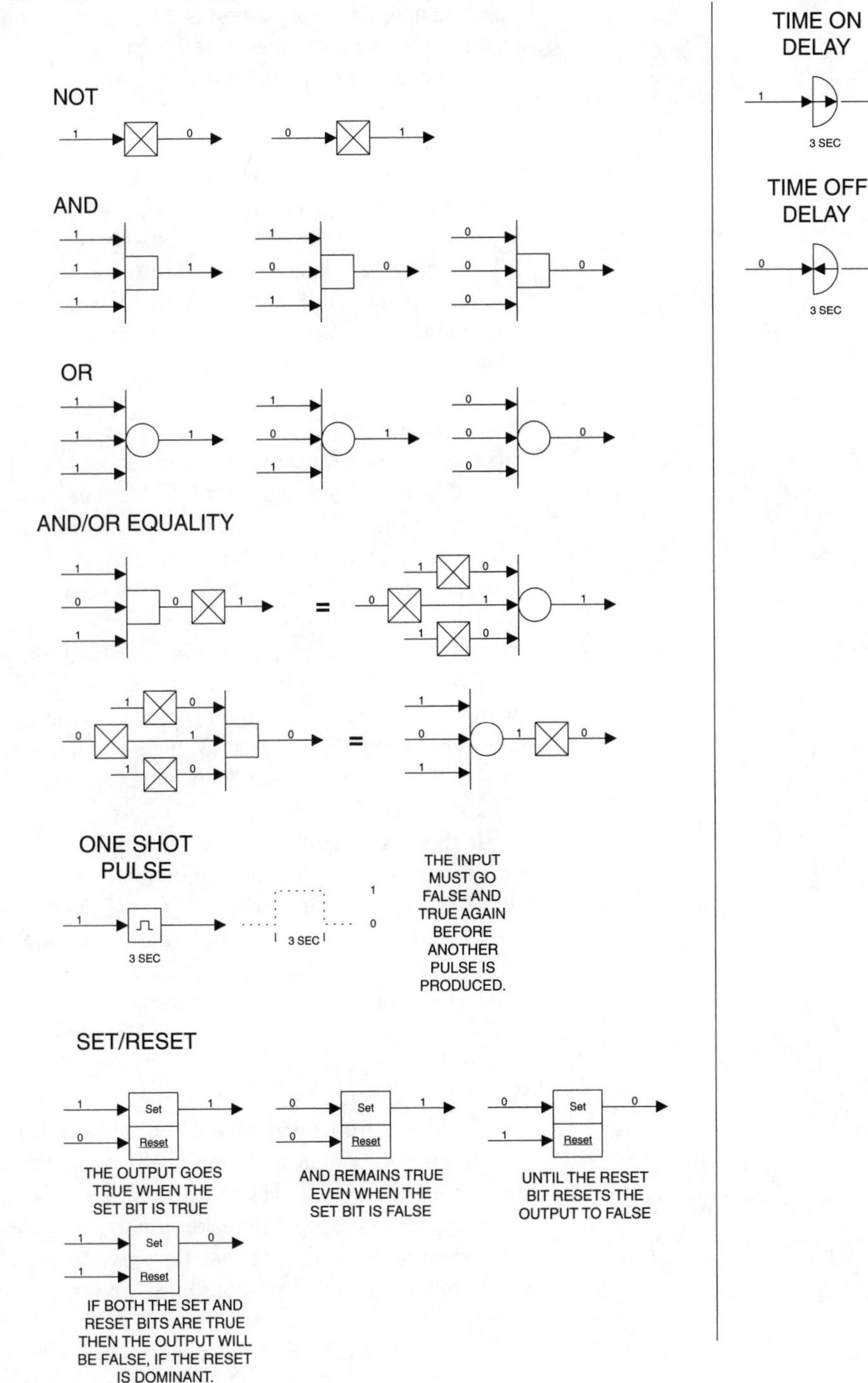

FIG. 3.12a
Logic diagram basics—digital.

Open and close limits: The indication from the valve limit switches that tells the operator and the logic when the valve is fully open or fully closed.

Open and close permissive: Process conditions that are required before the valve can be opened or closed.

Auto open or close: Conditions that open and close the valve automatically while the valve is in auto mode.

Output from the logic:

Open and close outputs: The output from the logic to the valve that strokes it open and closed.

Indications and alarms:

Auto mode: Shows that the valve is in auto mode and will be operated automatically by the **auto open** and **auto close** commands.

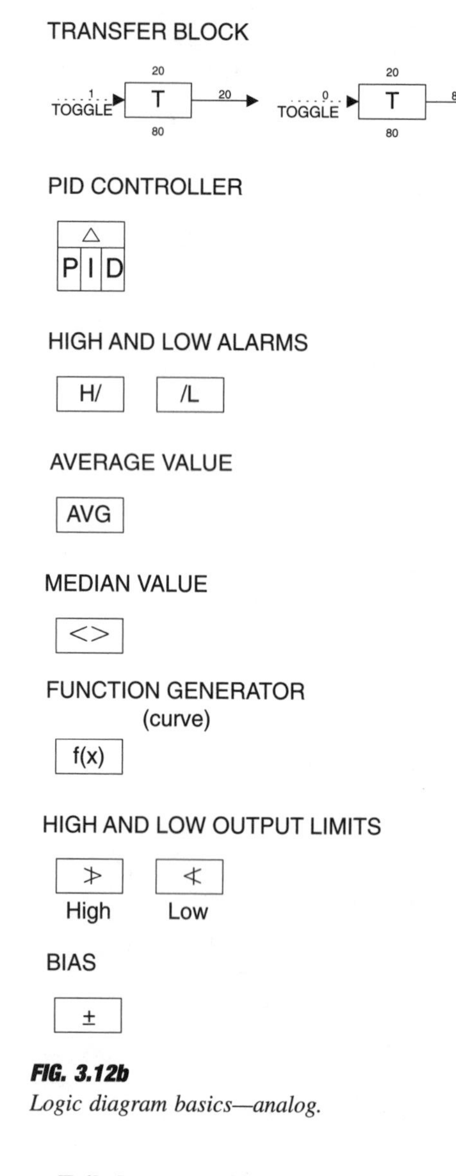

FIG. 3.12b
Logic diagram basics—analog.

Failed to open/failed to close: An alarm that shows that the valve was given an **open** or **close** output, but the fully open or fully closed limit was not reached in a specified amount of time.

Mismatch: An alarm that shows that both the open and close limit are true at the same time.

Auto Mode

One common approach to designing auto/manual logic is to design the logic to automatically switch the valve to manual mode upon an open or close command from the operator. Unfortunately, that makes it even easier for the operator to forget to place the valve in auto mode, because it can go unnoticed that the valve switched to manual mode. Logic designers also often elect not to use an auto mode at all to avoid the risk of the operator's forgetting to put the valve into auto mode; however, this makes the operator's job quite difficult since he or she will need to stroke the valve for maintenance when the system is not in service and will have to force the auto commands to do that.

A better way to handle this is not to require that the valve be in manual for the operator to give a command and to reserve manual mode for locking out the auto commands when the system is not in service—this can be done by simply making all auto open and auto close commands momentary as shown in the standard logic for valves in Figures 3.12f, 3.12g, and 3.12h, later in this section.

The operator in the example above would need only to give the valve a close command; then, when the steam turbine trips again, the valve will open as it should.

By making the auto command momentary, one need not worry about conflicting auto open and auto close commands that are True at the same time or about having a valve that strokes in the direction of the auto command as soon as it is placed into auto mode (the operator should always know what to expect when placing a valve into auto mode.) Also, if the auto commands are latched, then they will lock the reset of the fail alarm.

If an auto command should hold the valve in place while it is True, then the command should also be a permissive. For example, an auto open command that should hold a valve open should also be inverted and fed into the close permissive. This way, when the auto open condition is True, the valve will open and cannot be closed even when it is put into manual mode (Figure 3.12c).

If there is more than one auto open or auto close command, use a pulse for each individual auto command as shown in Figure 3.12d; otherwise, one of the commands may be ignored. For example, if there are two auto open commands, and one of them opens the valve and remains True, and then an auto close command shuts the valve, when the other auto open command becomes True, the valve will not reopen because the pulse will not have occurred since the first auto open command is still True.

With motor-operated valves in particular, it is important not to write logic that will continually cycle the valve open and closed as this will eventually destroy the motor. Take the time to analyze the logic to make sure the auto open and auto close commands will not cause the valve to cycle.

For example, it is best not to use the same input for both the auto open and (inverted) auto close commands, which can cause the valve to cycle open and closed. If it is necessary, because of hardware restrictions, be sure to add a hefty time delay on either auto mode input, as shown in Figure 3.12e.

Motor-Operated Valves

The standard logic for two of the most common types of MOV logic is shown in Figures 3.12f and 3.12g. The logic in Figure 3.12f has momentary open and close outputs that will command the MOV to move. The circuit for the MOV will then take over and continue to stroke the valve and will stop stroking when the limit is reached. The logic in Figure 3.12g sends latched outputs to the field that stay True until the limits are reached.

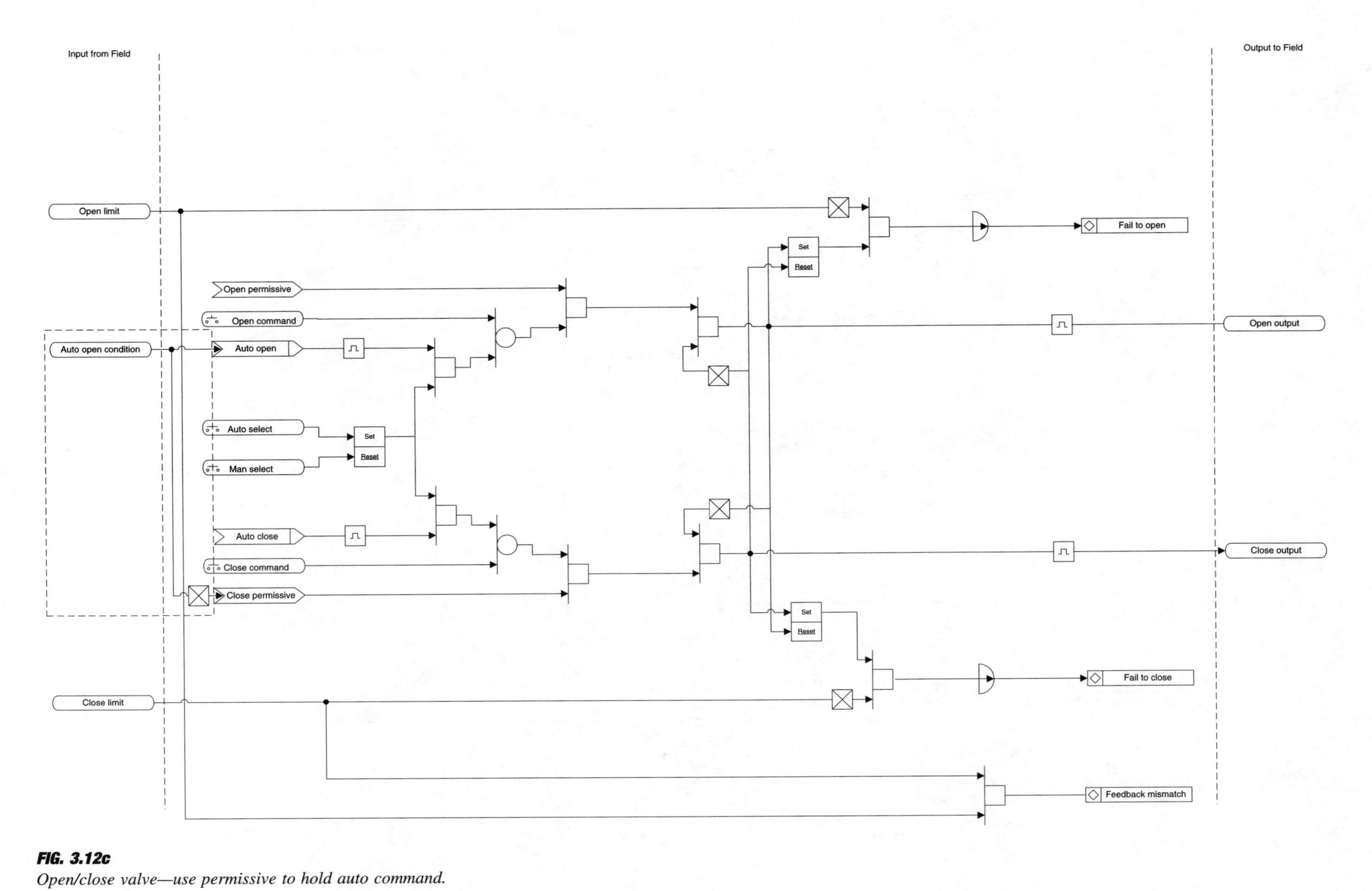

FIG. 3.12c
Open/close valve—use permissive to hold auto command.

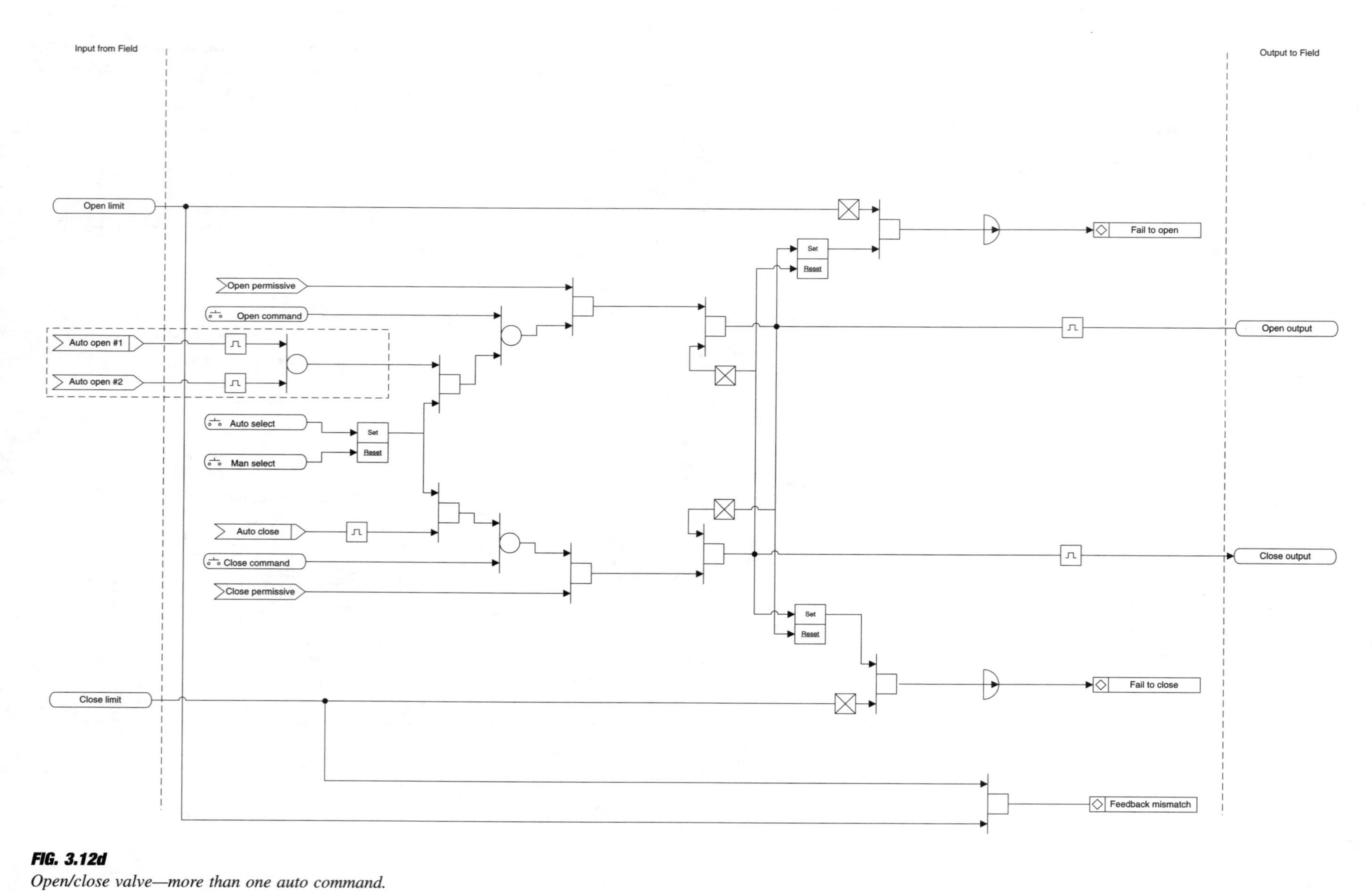

FIG. 3.12d
Open/close valve—more than one auto command.

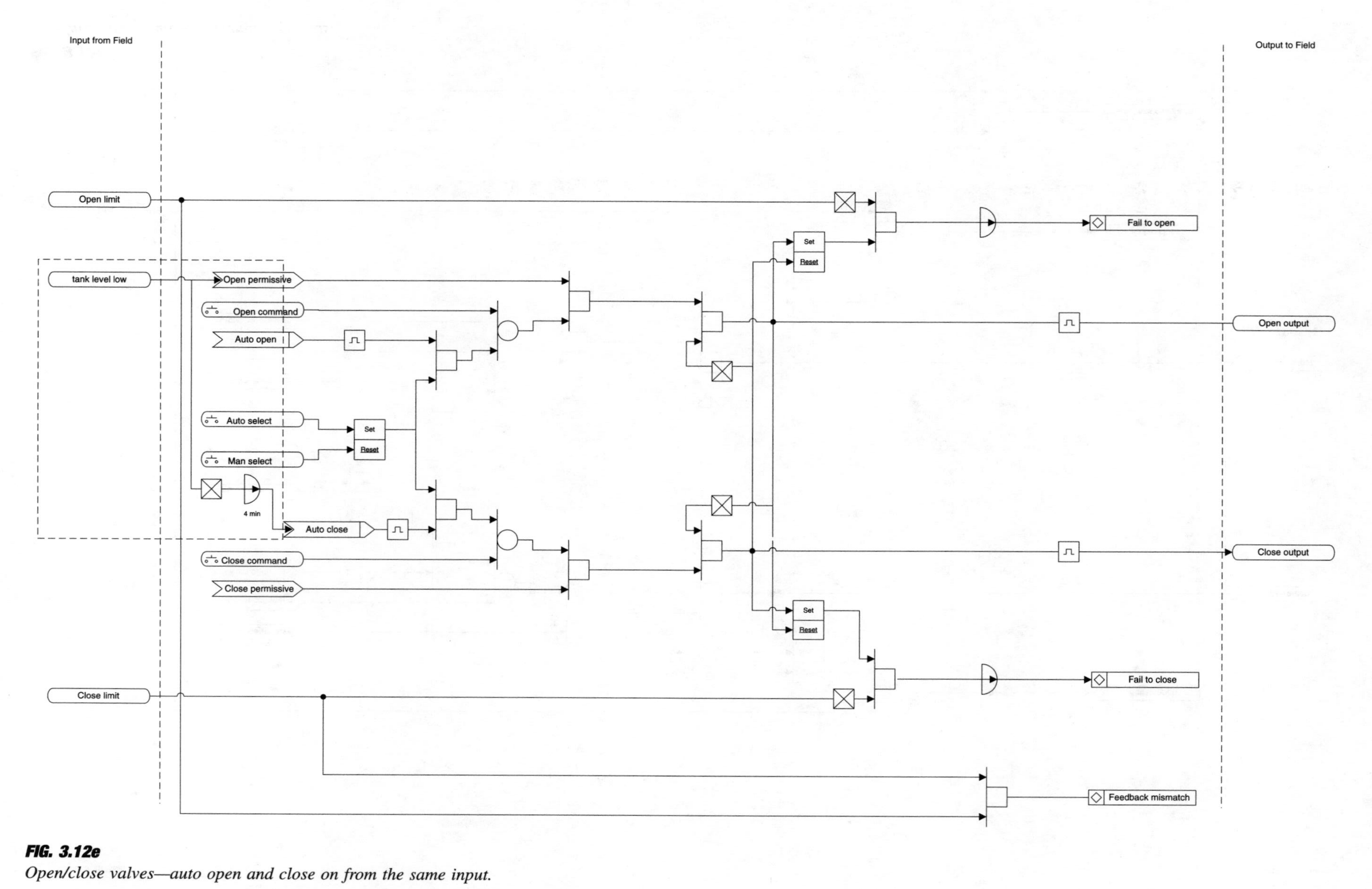

FIG. 3.12e
Open/close valves—auto open and close on from the same input.

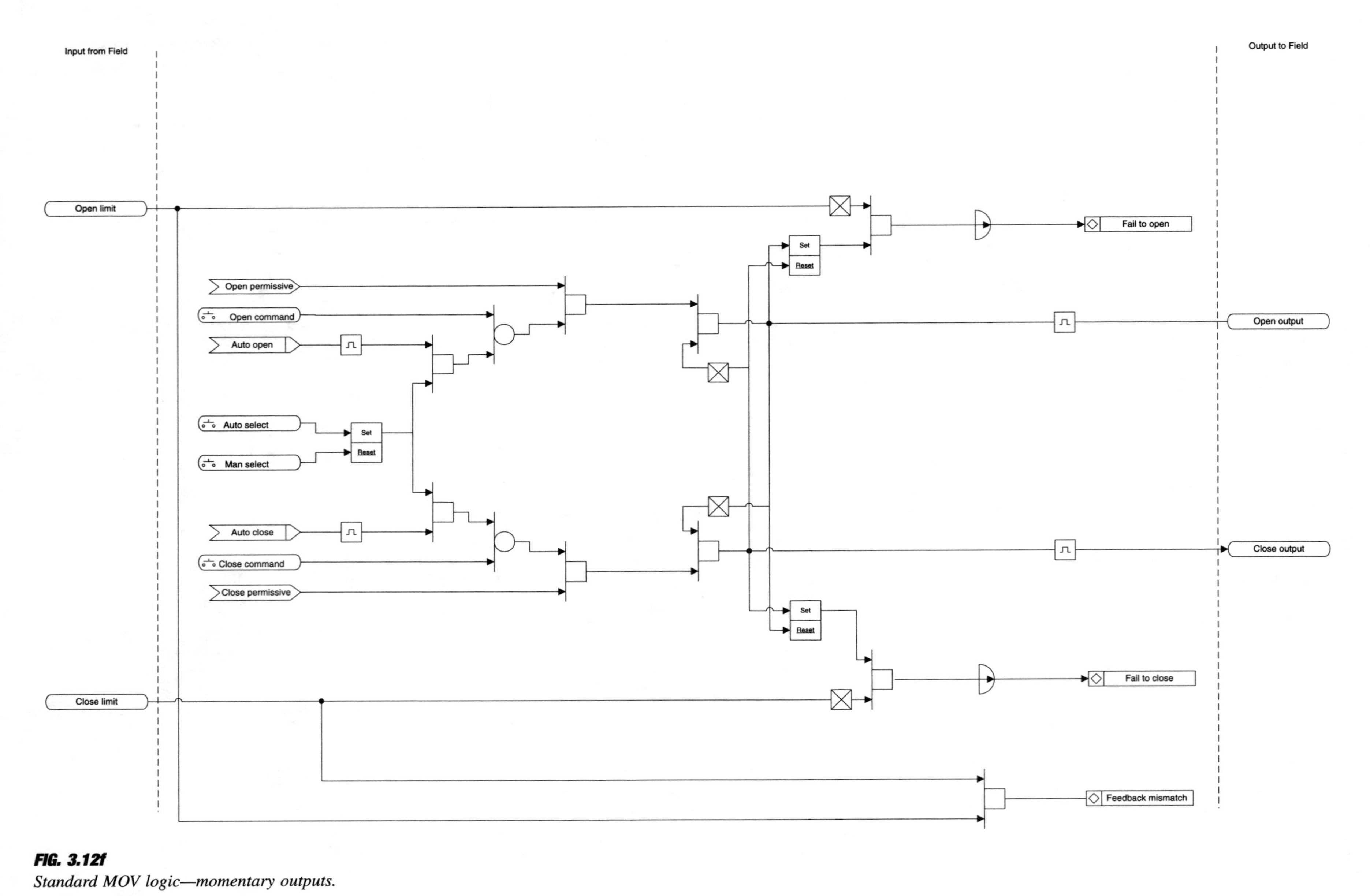

FIG. 3.12f
Standard MOV logic—momentary outputs.

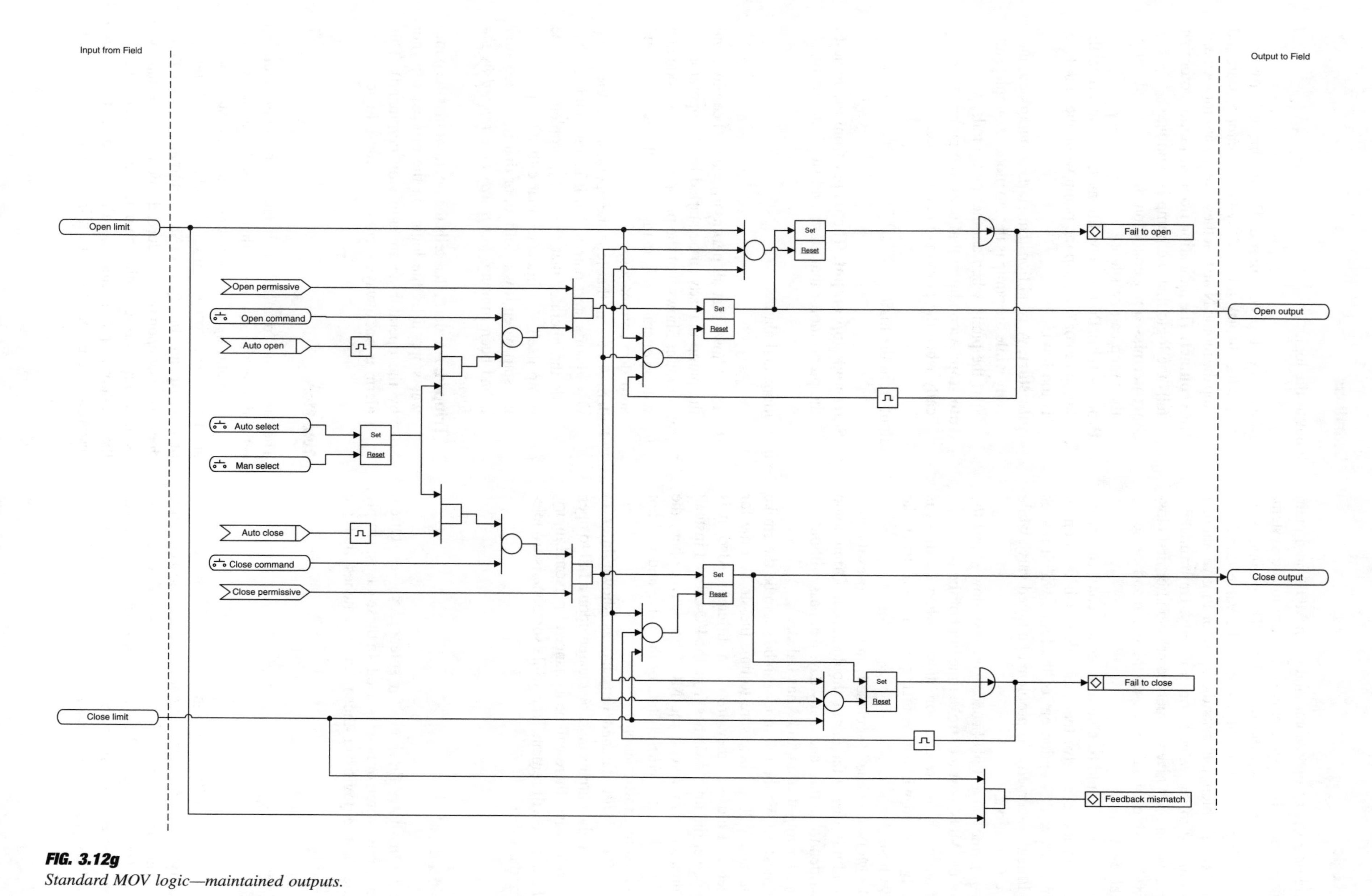

FIG. 3.12g
Standard MOV logic—maintained outputs.

Failure Logic

An operator gives an open command to an MOV on a main steam line. The valve opens, releasing steam to the system, but does not quite reach the open limit. The maintenance engineer who has been troubleshooting the valve thinks he has found the problem and asks the operator to give another open command to the valve to open it fully. Unfortunately, the operator can only give an open command after first closing the valve, which would bottle up the steam and cause the relief valves to open.

The operator should be able to issue a second command to attempt to move a valve that has failed. It is a common problem to leave this option out of the logic, which makes it very difficult to remedy the problem of a failed valve while a system is in service.

Also, if the valve fails because process flow slows the stroke of the MOV past the stroke time limit of the fail timer, the fail alarm should reset automatically when the limit is eventually reached without requiring the operator to stroke the valve first.

If the MOV is opened (or closed) and later loses its open (or close) limit, then a fail alarm should appear. Often, logic is not written to do this, and many days can pass without the operator knowing that an MOV has failed.

For most MOVs the fail timer can be set to be the stroke time plus about 20 s. Make sure to time the stroke time for both open and close as they are often different. Also, it is wise to power the limit switches from the I/O cabinet instead of the breaker. This way, if the MOV breaker is open, the logic will still see the correct state of the valve and will not react to incorrect information.

For MOVs with latched outputs, as shown later in Figure 3.12m, the outputs must be cleared upon a fail to open or close; otherwise, there will be a chance of personal injury, since the output will remain True while someone is troubleshooting the valve.

Solenoid Valves

Figure 3.12h shows the logic for solenoid valves. Often, only one limit switch or none is used. In these cases, simply omit the failure logic for those limit switches that are missing.

PUMPS

Figures 3.12i and 3.12j show standard logic for the two most common types of pump motor. Figure 3.12i, the most common type of pump motor, has one maintained output for starting the pump (True = start and False = stop). The pump motor with two momentary outputs shown in Figure 3.12j is often used for very large pump motors.

Definitions

Input to the logic:

Start and stop command: The commands from the graphic display initiated by the operator to start and stop the motor. Note that these are always momentary.

Run contact: The indication from the pump motor that tells the logic that the pump is running.

Start permissive: A condition that is required before the pump can be started.

Process trip: Process conditions that shut down the pump motor (i.e., trip the pump when the tank level is too low).

Auto start: A condition that starts a pump automatically while the pump is in auto mode (example: Auto start the pump when another pump fails).

Auto stop: A condition that stops the pump automatically when the pump is in auto mode.

Output from the logic:

Start and stop output: The output from the logic to the pump motor that starts and stops the motor.

Indications and alarms:

Ready: Shows that all permissives have been met and the pump is ready to be started by the operator.

Auto mode: Shows that the pump is in auto mode and will be operated automatically by the auto start and auto stop commands.

Failed: An alarm that tells the operator that the pump was given a start output, but it is not running. This tells the operator that there is a problem with the pump itself and indicates that either the pump did not start when it was told to, or that it stopped after it had been running, *but it was not stopped by the logic*.

Tripped: An alarm that tells the operator that the pump was stopped by the logic via the process trip (not by the operator or an auto stop command). This alarm is not always included in pump logic.

Auto Mode

The "auto" and "manual" push buttons shown in Figure 3.12k set and reset the auto mode that allows the pump to start and stop automatically when the auto start and auto stop commands are True. The auto mode reset should also include pump fail, process trip, and the operator stop command. If the auto mode does not reset upon pump failure, the pump will restart as soon as the failure is reset while the auto start is True. Similarly if the operator stop does not reset the auto mode the pump will restart after the operator stops it if the auto condition is True.

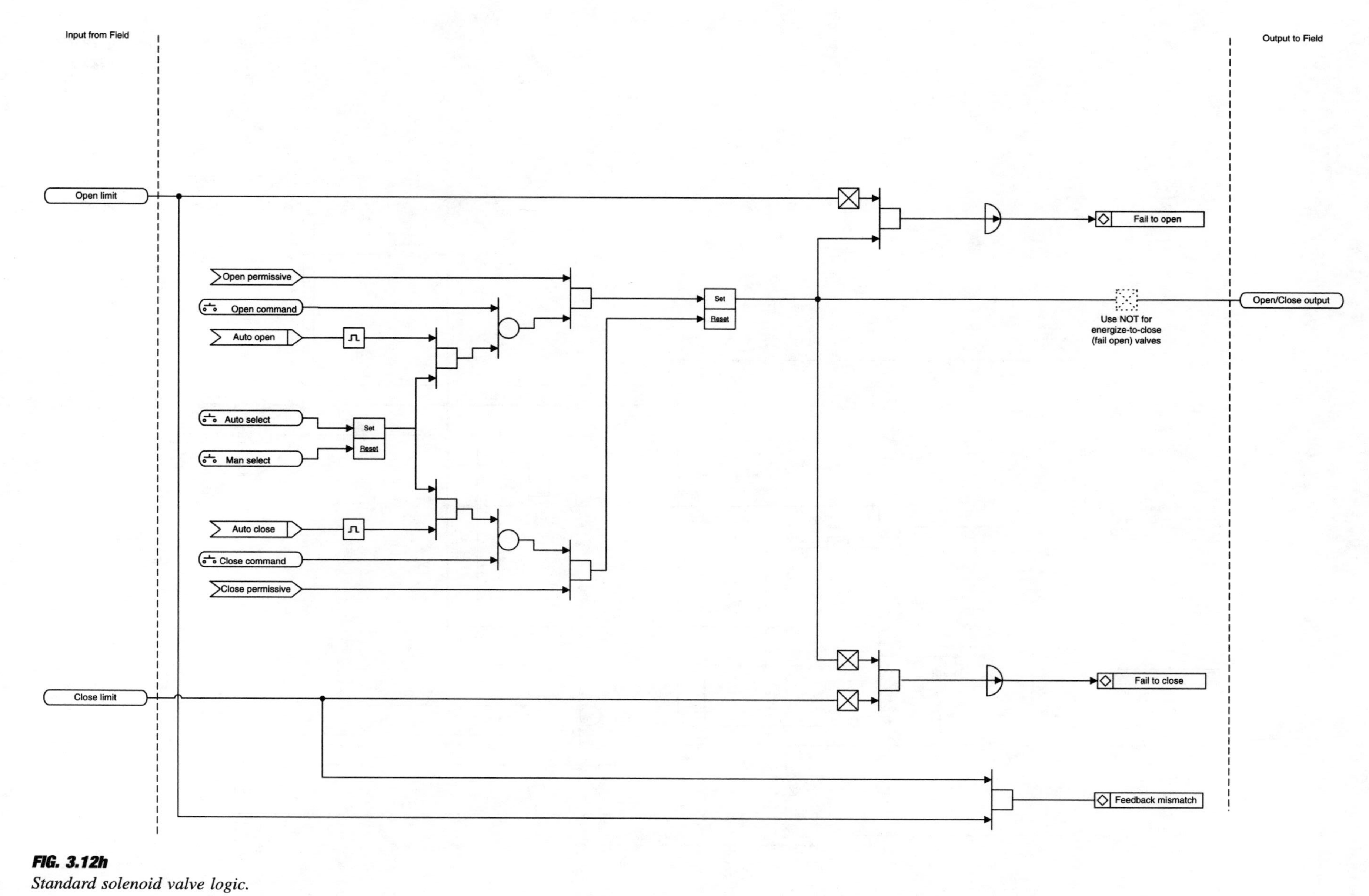

FIG. 3.12h
Standard solenoid valve logic.

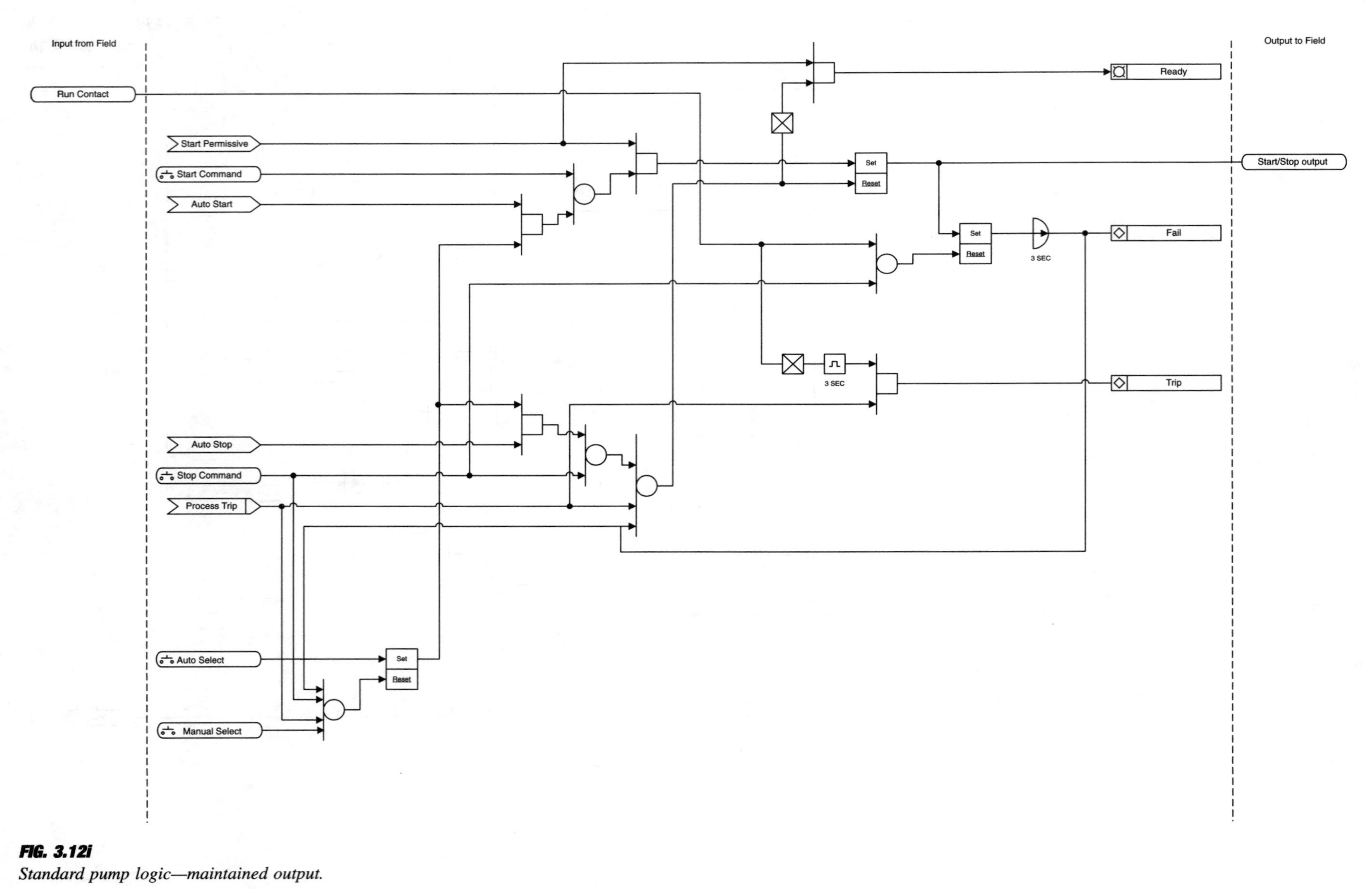

FIG. 3.12i
Standard pump logic—maintained output.

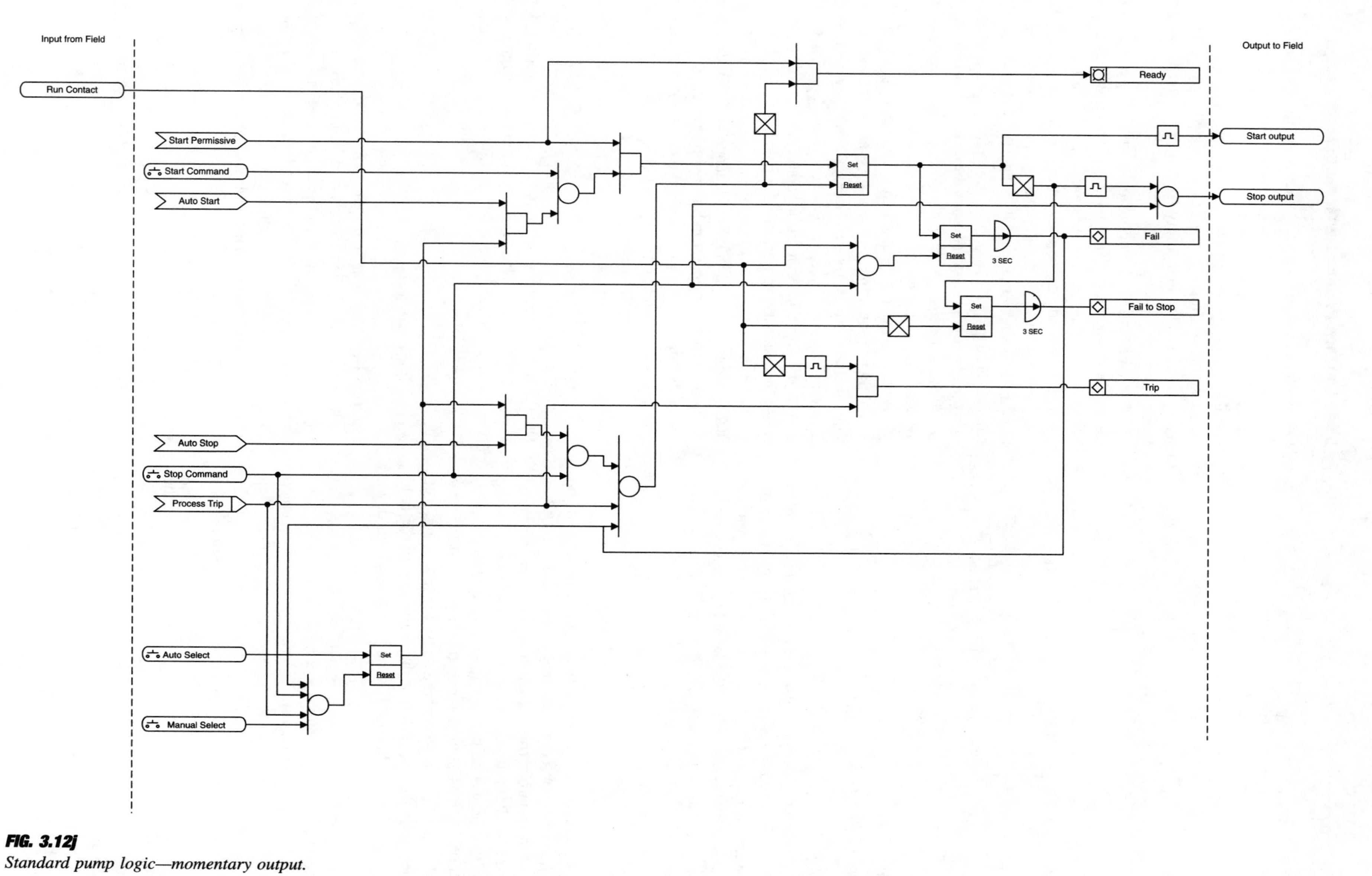

FIG. 3.12j
Standard pump logic—momentary output.

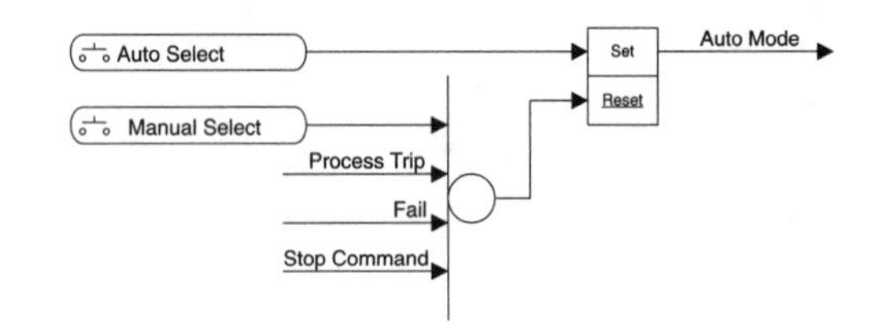

FIG. 3.12k
Typical pump auto mode.

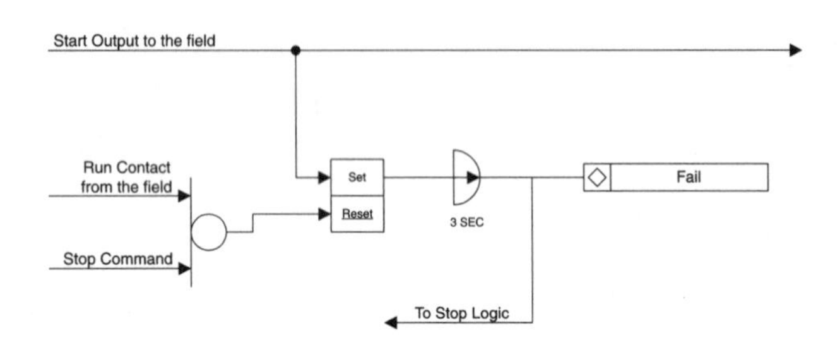

FIG. 3.12l
Typical pump failure logic.

If the process trip does not reset the auto mode, a problem like the demineralized water pump behavior described in the introduction will occur. If the trip condition cycles on and off (as a level trip may do while the level hovers around the set point) while the auto start condition is True, the pump will cycle on and off.

Pump Fail

A pump failure occurs when a start output is sent to the motor and no run contact is detected (Figure 3.12l). This is different from a pump trip, which indicates that the pump was tripped by the logic upon a process condition. Making the distinction between fail and trip eliminates much confusion about what has shut down a motor—something that happened in the field or something caused by the logic. The "fail" alarm tells the operator that the problem is in the field and the "trip" alarm indicates that the motor was shut down by the logic.

For safety purposes, when a motor fails, the start output must be reset so that a person inspecting the failed pump will not be surprised by a sudden start. Although this is accepted as standard design practice for safety purposes, it is often erroneously omitted.

The failure indication will remain True and will inhibit the start of the motor until the operator manually resets it by giving a stop command. The start inhibit is performed by feeding the fail bit into the stop logic (as shown in Figures 3.12i and 3.12j) and should not be pulsed before it is fed to the stop circuit or the pump will be allowed to start again without first clearing the fail indication, and will not clear the start command after it has failed a second time.

Other Common Problems with Pump Logic Design

Quite often, motor logic is written so that when the run contact is seen by the logic, a start output is given to the motor. If, for example, the run contact is forced True during simulation, or if the run contact is accidentally made True when wiring is altered in the I/O cabinet, the pump will start unexpectedly. This is a safety hazard that must be removed from the logic.

The start and stop commands from the operator should always be momentary. Otherwise, the pump will start unexpectedly when the permissives are later satisfied if an earlier attempt was made to start the pump when the permissives were not satisfied. A maintained command will also cause the pump to start unexpectedly when the trip condition that caused the trip is cleared. The momentary pulse is usually executed by the graphic and not the programmed logic. For all of the diagrams given in this section, it is assumed that the command push buttons are momentary.

Problems often appear when permissives are overridden by another input from the logic such as an auto start. Make sure that the permissives are always used as they are intended. If in rare cases it is necessary to override a permissive, then design the logic so that the operator will have to make a conscious decision to perform the override.

Often a "red tag" or logic lockout system is used to show that a pump is undergoing maintenance and to lock out start commands from the operator. Surprisingly often, it is mistaken for a substitute for proper locking and tagging out of the motor breaker, and when used alone it does not provide adequate prevention against starting the motor. It is acceptable to put the motor in manual mode while using proper maintenance procedures for locking out the breaker instead.

After a pump trips upon a process trip, the operator should not be required to initiate a stop command before restarting the pump. This is another common design flaw that is not a safety hazard, but it is a nuisance to the operator and should be fixed.

Pumps with Two Outputs

Figure 3.12j shows standard logic that can be used for pumps with two outputs. When motors of this type fail, it is important to send a stop output to the motor even though it may already have stopped. If the motor fails after losing the run contact but is still running, then the motor will not trip when a process trip occurs because the logic "thinks" that the motor has already stopped.

The stop output for this type of motor can be made fail-safe by inverting it so that it is always True unless a stop command is given. If the stop output is not fail-safe, then it is necessary to include a "fail to stop" alarm in the motor logic. The logic in Figure 3.12b has been designed so that the operator can make a second attempt to stop the motor upon a fail to stop without having to give a start command first.

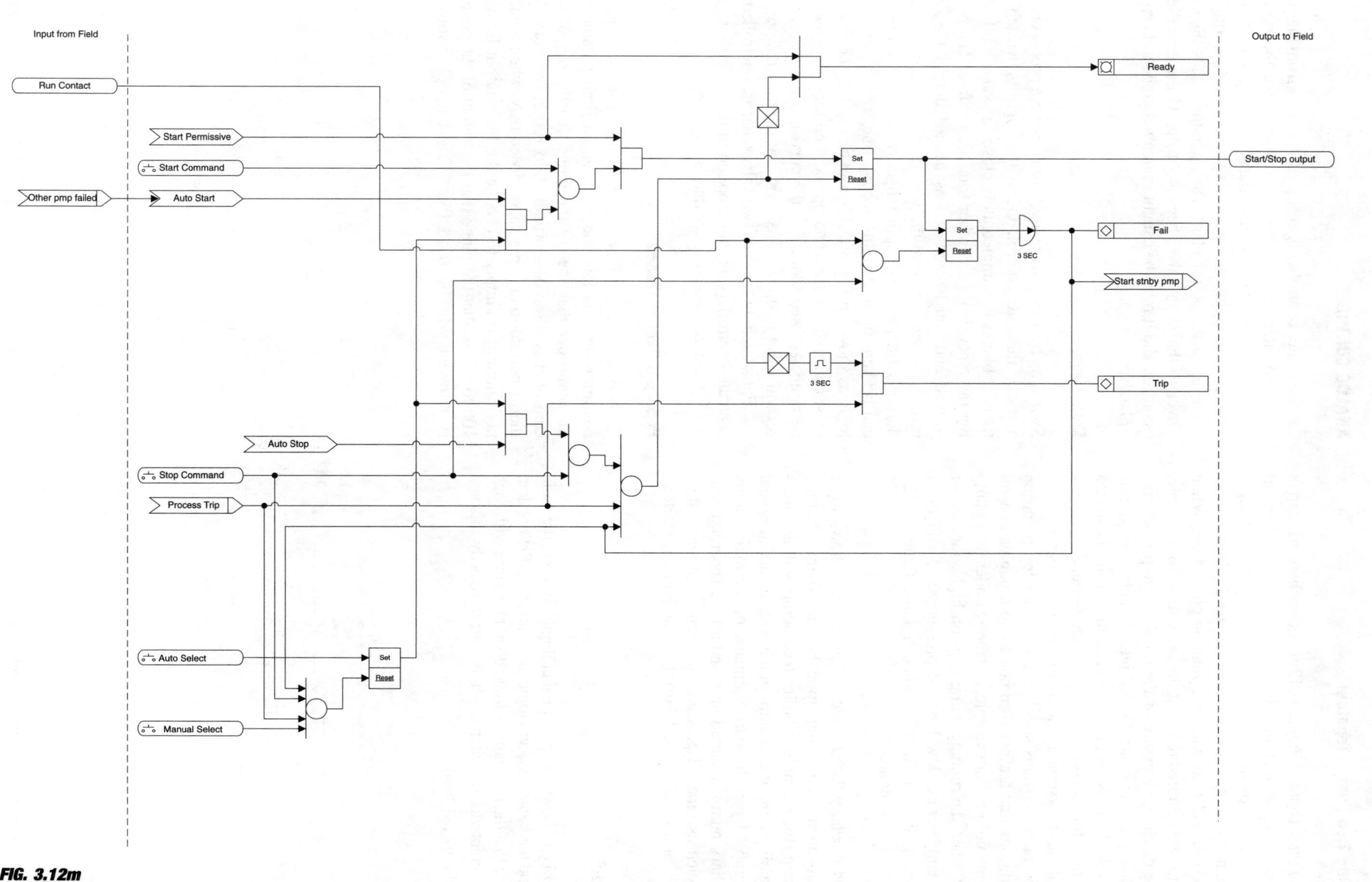

FIG. 3.12m
Pump—standby start.

Controlling Two Pumps Together

Standby Pump Logic Figure 3.12m shows how to start a standby pump upon the failure of the running pump. The fail indication of one pump is simply fed to the auto start input of the standby pump so that when the standby pump is in auto mode, it will start when the running pump fails. When two pumps are required to run at the same time, for example, when the discharge pressure of the running pump is insufficient, a low pressure contact is fed into the auto start. In this case, a time delay is required after the first pump is started to allow time for pressure to build up; otherwise, the standby pump will start as soon as the main pump has started.

Some pump systems should start the standby pump on trip conditions that are not common to both pumps as well as on pump failure. For example, a standby boiler feed pump can start when the running pump trips on high vibration or bearing temperature, low lube oil pressure, and any other noncommon trips, as well as when it fails. Figure 3.12n shows how this is done.

Main and Auxiliary Pump Logic Figure 3.12o shows logic for starting an auxiliary pump upon the start command of the main pump (for example, a boiler feed pump with an auxiliary lube oil pump, or a vacuum pump with an auxiliary seal water pump). Upon the start command of the main pump, the auxiliary pump is started first, and after a specified condition (such as lube oil pressure sufficient, seal water flow sufficient, or a simple time delay) the main pump is started.

BREAKERS

Figure 3.12p shows simple breaker logic. If both the field circuit and the control system logic are controlling the breaker, make sure that the two parts of logic do not interfere with each other. A remote/local switch is often used to remedy this and is fed into the permissive.

ANALOG CONTROLS

A feedwater controller is designed to switch from the start-up feedwater valve to the main valve when the start-up valve reaches 100%—immediately closing the start-up valve and popping the main valve open to 20% during the start-up of the power plant. The operator dreads the swap over to the main valve, because it upsets the drum level control, and it must be caught in manual mode to keep the plant from tripping.

Switchover

Switching over from one valve to another in a process is often a tricky thing to control. Often the logic is written with transfer blocks that immediately close one valve while opening the other. For a smoother transition, curves can be used instead and can be adjusted as needed during the tuning process to better control the way the valves open and close during switchover. For split ranged valves, if the first valve can be kept open after the second one starts to open, then logic as shown in Figure 3.12q can be used. If the first valve must close after the second opens, then the first valve must be ramped closed (not using the curve), while releasing the second valve to open on its own as it controls the process.

For switching between two valves that are controlled with separate controllers, the hold option in the controller can be used to freeze the controller that is not in use.

Pop Open/Clamp Closed

To pop a valve open or clamp it closed upon a specified condition, use the low and high output limit parameters of the controller that specify the range that the output of the controller is allowed—normally 0 to 100%. For example, all that is needed to clamp a valve closed (say, when a pump is not running) is simply to change the high output limit from 100 to 0% so that the range that the output of the controller is allowed will be 0 to 0%. This example, shown in

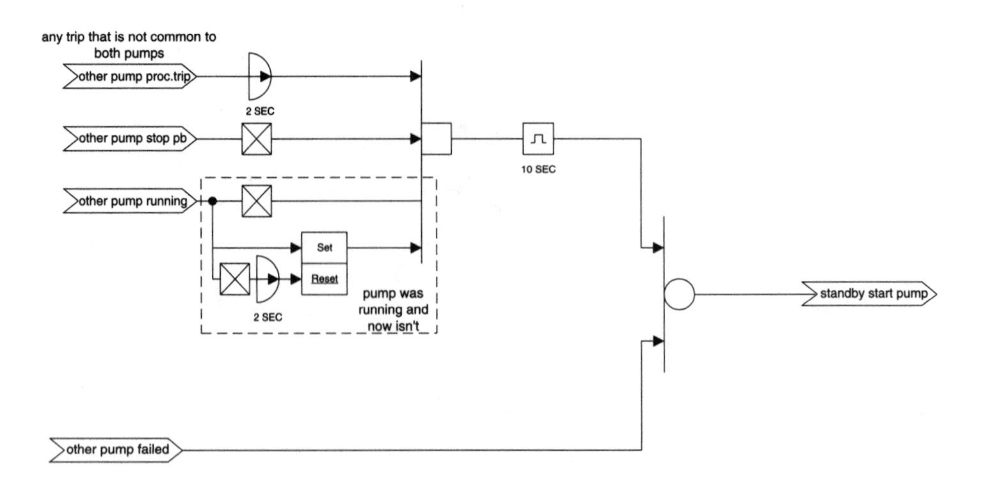

FIG. 3.12n
Standby start upon fail and noncommon trips.

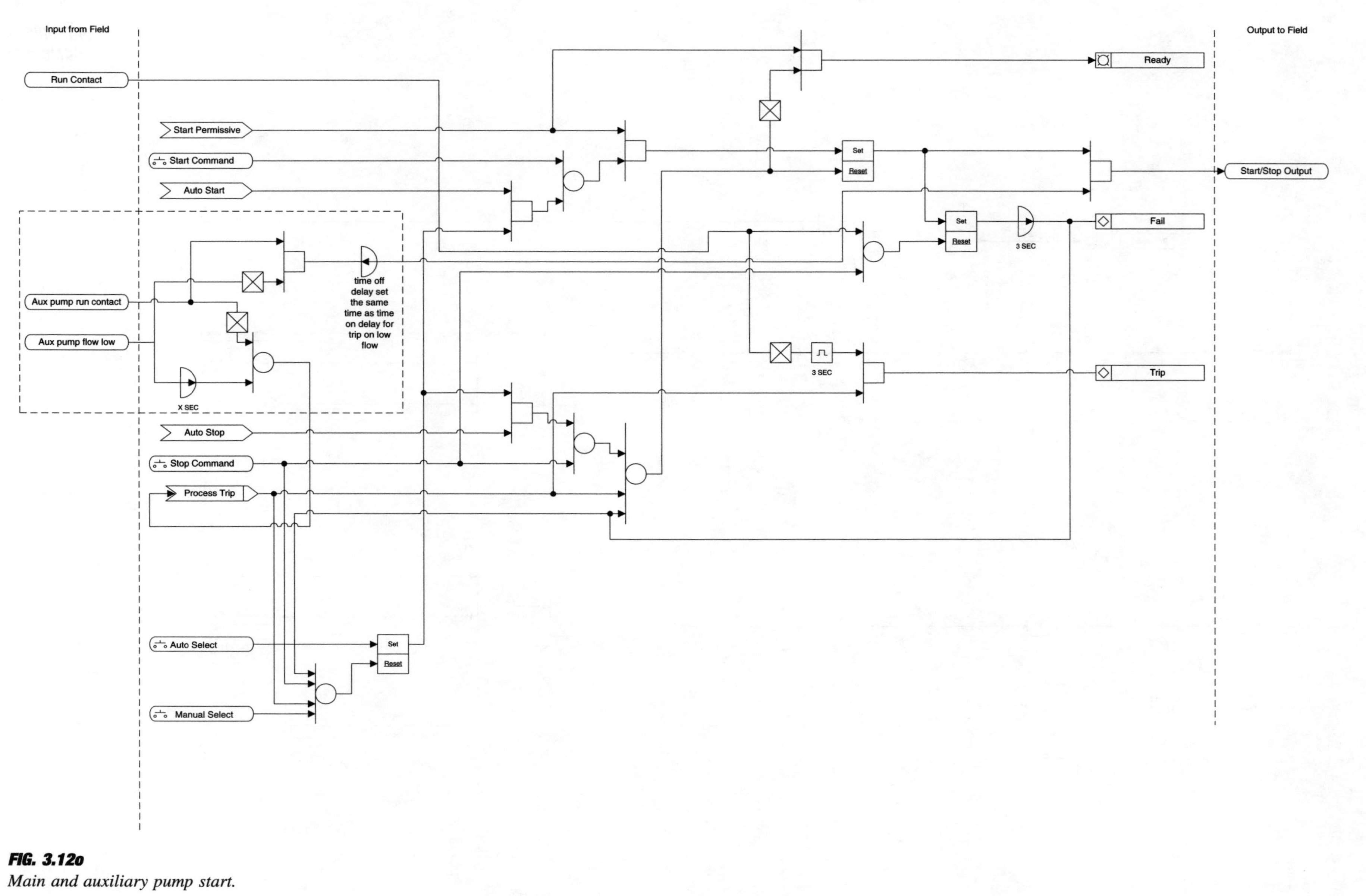

FIG. 3.12o
Main and auxiliary pump start.

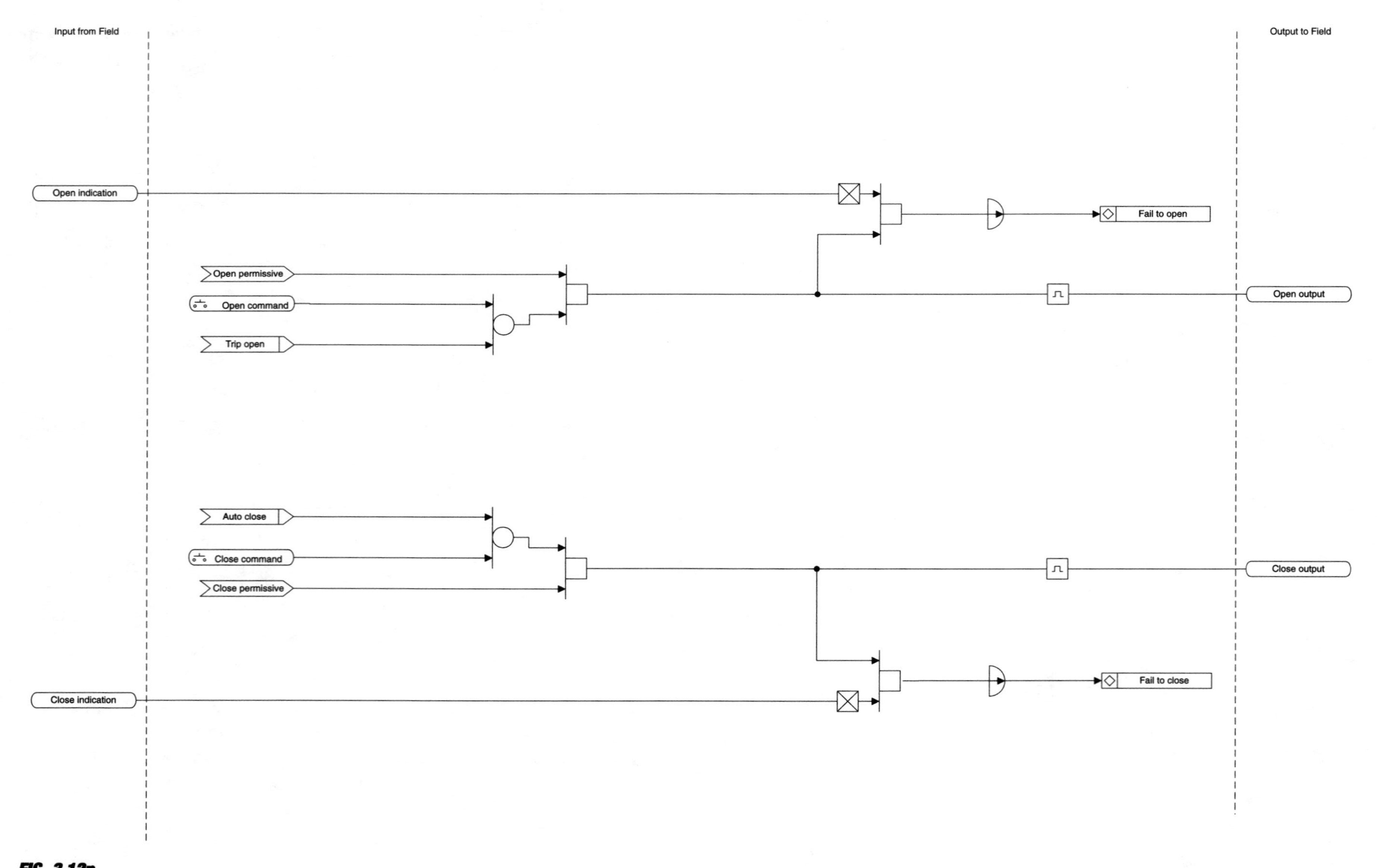

FIG. 3.12p
Standard breaker logic.

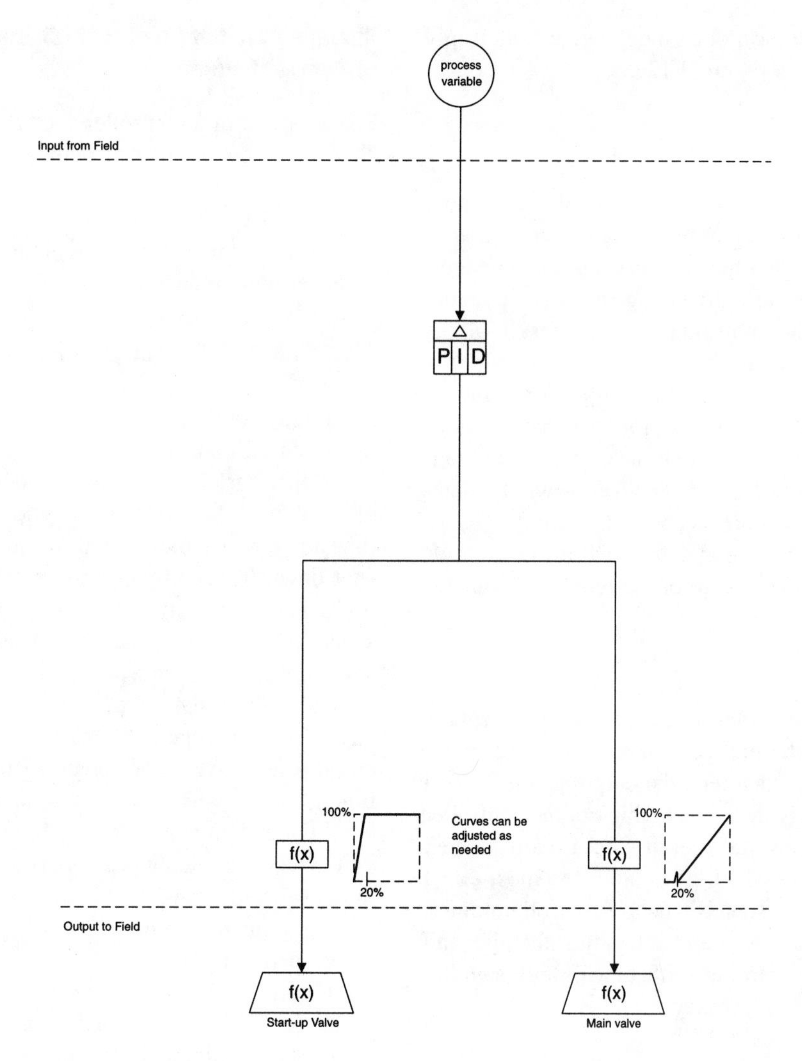

FIG. 3.12q
Analog—split range valves.

Figure 3.12r, shows how the change from 100 to 0% is handled with a transfer block. The valve is clamped closed the whole time that the pump is not running, so that there is no need also to switch the controller to manual mode, as is often done. When the pump runs again, the controller is automatically released and there is no need to have to remember to reset the controller to auto mode before starting the pump.

To pop open a controller, simply increase the low output limit. The controller will open to the specified position and start controlling from that point. In these cases, it is best to make the pop-open condition momentary so that the operator can manually close the valve if needed (Figure 3.12s). If a smoother transition is desired, a lagged transfer block can be used.

Override Open and Close

To override the controller to ensure that a limit on the process variable is maintained (for example, to push a level controller closed upon high water level in a drum), use a function block to generate a curve to feed the high output limit. As shown in Figure 3.12t, the input to the curve is the process variable to be controlled. When the limit is nearly reached, the output of the curve will start to decrease the high output limit from 100 to 0%, to close the valve until the process is normal. These curves can be adjusted during the tuning stage to best control the limit.

Similarly, one can use a controller to make sure that a limit is not reached on a process variable that one is not directly controlling and feed the controller output into the high or low limit of another related controller. This is different from cascade control (explained below), which controls two process variables at the same time, or feedforward control (explained below), which modifies the main controller as another process variable changes. In this case, one normally controls one process variable and uses the override to control only during the "out of spec" condition. For example, to control recirc flow in a system while making sure not to overpressurize the main discharge pipe, simply feed the overpressure controller output into the low output limit of the flow controller. This will push open the flow controller only when needed, as shown in Figure 3.12u.

Feedforward

If the change of one process affects another process under control, one can use the bias input of the controller to modify the controller output as the other process variable changes. This is especially handy for anticipating the need for more or less output before the controller sees the result of the change. The bias simply adds or subtracts a value from the output of a controller and is often added during the tuning process as needed (Figure 3.12v). The process variable should be passed through a multiplier so that the effect of the feedforward can be adjusted as needed. The feedforward adjustment can be made even more accurate by using a curve instead of a multiplier if it is known exactly how the controller should change with the change of the process variable.

Cascade

To control two process variables at once, use two controllers and feed the output of the first controller into the set point of the second controller. Then feed the output of the second controller to the valve (Figure 3.12w). Think of the controller that feeds the valve as the worker, because it is usually located at the point in the process where it can have the most effect or can see process changes sooner. The top controller adjusts that controller as necessary. Remember to adjust the high and low output limits of the controller so that the output is within range of the set point of the worker.

Three-Transmitter Select

When three redundant transmitters are used, a selection method as shown in Figure 3.12x is used to choose the median value, if the data from all three of the transmitters are of good quality, or to use the average of two if one transmitter has a faulty reading, or to select one transmitter. If the output of a transmitter has bad quality (usually specified as below 4 mA or above 20 mA), the logic automatically throws out its value. The operator can also choose to bypass a transmitter that has a reading that does not "look right." If all of the transmitters are bad or bypassed, any controller that is using the selected value as a process variable may be set to manual mode.

Switch to Manual Mode

If a controller is automatically switched to manual mode by the logic, there should always be an alarm to the operator to show what has happened. Switch to manual mode only if absolutely necessary (for example, where there is no available process variable). Otherwise, it is best to leave the decision to switch to manual mode to the operator's discretion.

Changing the Set Point with Changes in the Number of Pumps Running

The set point of a controller such as a flow controller can be changed as the number of pumps running changes by using the logic shown in Figure 3.12y. If the pump is tripped when the flow is too high or low, one must make sure to add a hefty time delay in the trip logic to allow for the process to meet set point after the set point has been changed.

START-UP AND SHUTDOWN SEQUENCES

Automatic start-up sequences will start an entire plant with the touch of a button. Because they take a lot of control out of the operator's hands, they should be reserved only for those plants that are particularly complicated and would take more time to start manually, or those plants that are started and shut down frequently. Operators often have a difficult time following these fully automatic sequences, especially if the sequence graphics are not sufficiently detailed.

Think of the sequence logic as acting in place of the operator during the start-up sequence. It should not include any safety logic that is normally included in the logic for the pumps and valves themselves so that the safety logic will function even when the auto sequence is not used or is on hold. For example, logic that automatically closes boiler vent valves when the steam pressure reaches 25 psig should reside in the valve logic, not in the sequence, because the automatic start-up sequence could be on hold while the pressure passes the 25 psig set point.

Figure 3.12z shows the basic structure of a start-up sequence. The logic should include a hold button so that, if there is an equipment malfunction during the sequence, it will wait while the problem is remedied. In this way, the operator will not have to shut down the sequence and start over. A sequence abort should be included in case the problem cannot be fixed quickly.

OPERATION AND CUSTOMIZATION

Now is the time to operate the equipment with the logic the designer has written. By now the designer should have simulated the logic, independently of the system that it controls, to work out any bugs. This is very important, as it will take years to find all of the bugs as they occur during operation, and *it will actually save time to do all the testing beforehand.* Even logic that the designer has personally programmed, and most certainly logic that has already been tested in a factory acceptance test, should be retested.

Now that the equipment is running, the logic designer will soon find that the systems do not always behave exactly as expected. This stage of the job is a humbling experience because the designer inevitably finds problems with logic that was thought to be well designed for the system. Designers will produce a much better product if they are willing to make

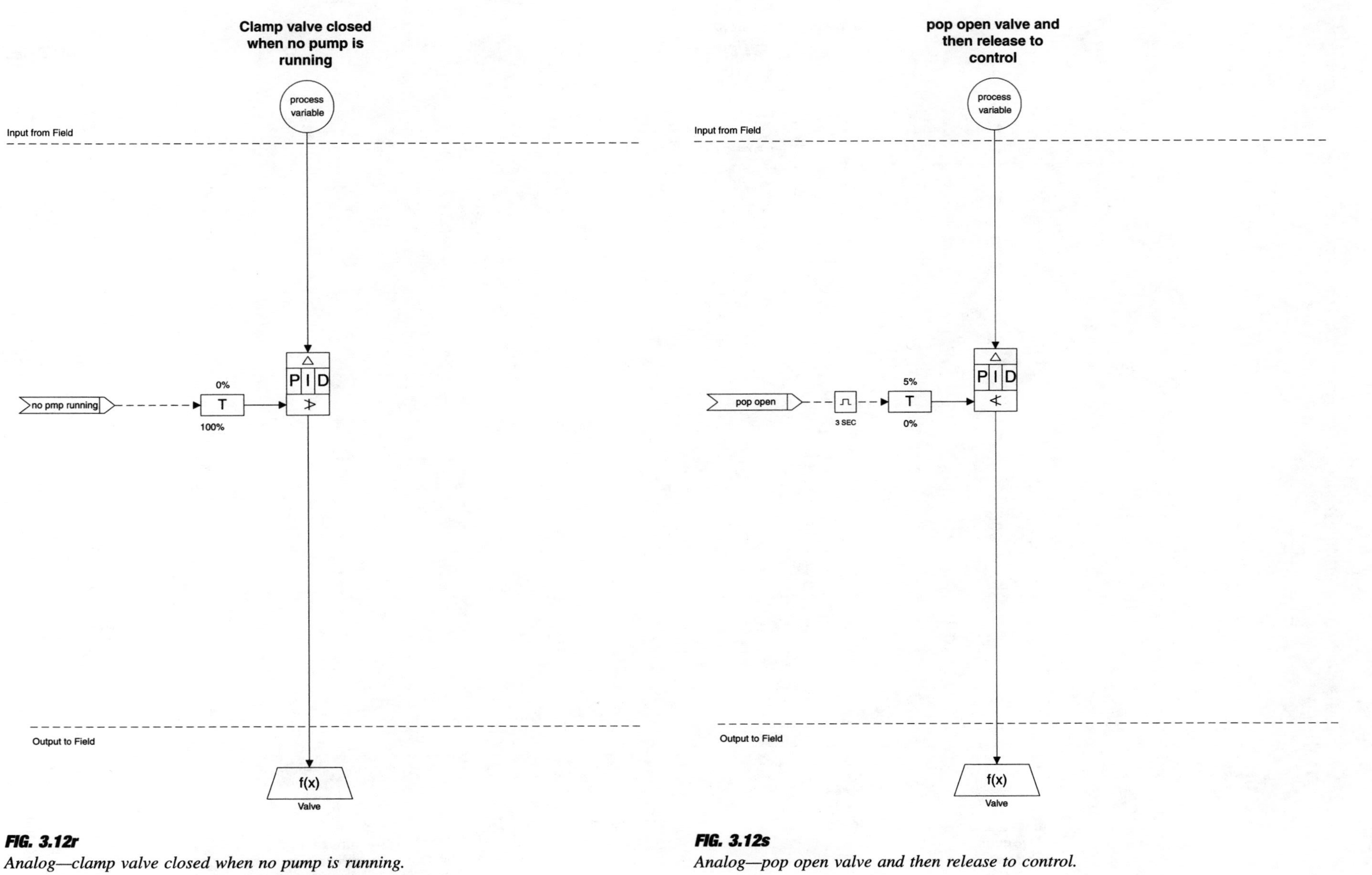

FIG. 3.12r
Analog—clamp valve closed when no pump is running.

FIG. 3.12s
Analog—pop open valve and then release to control.

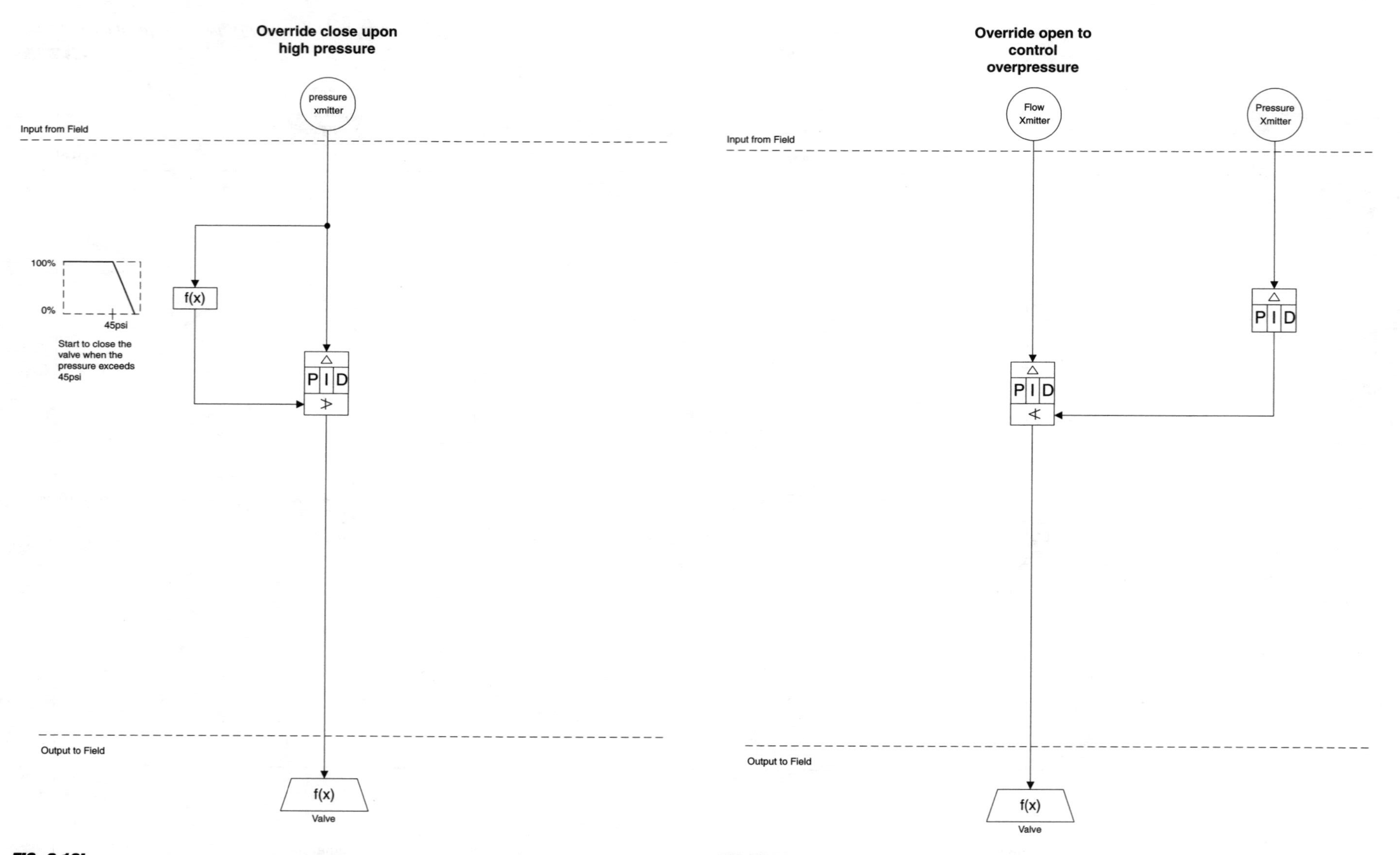

FIG. 3.12t
Analog—override close.

FIG. 3.12u
Analog—override open.

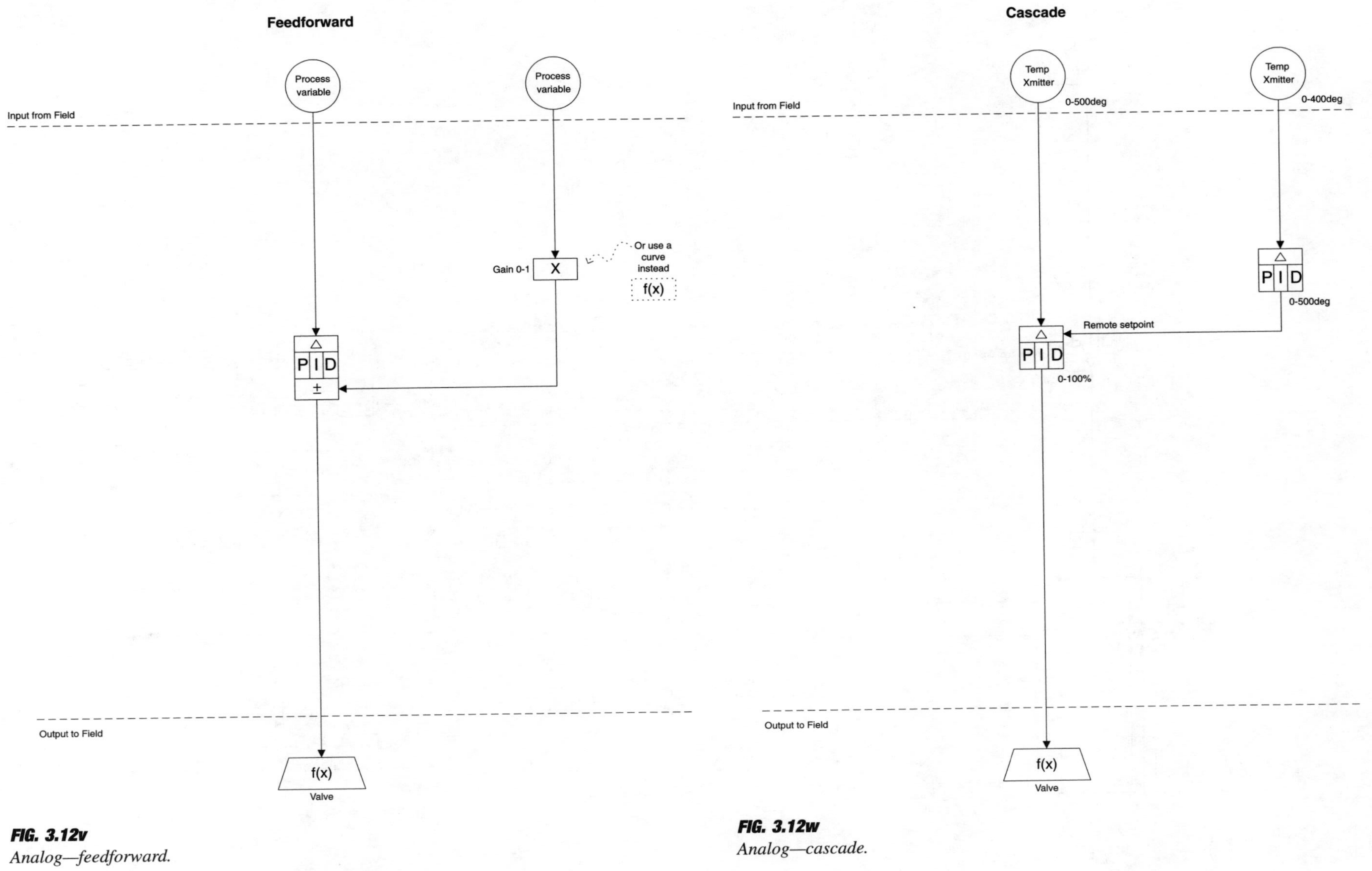

FIG. 3.12v
Analog—feedforward.

FIG. 3.12w
Analog—cascade.

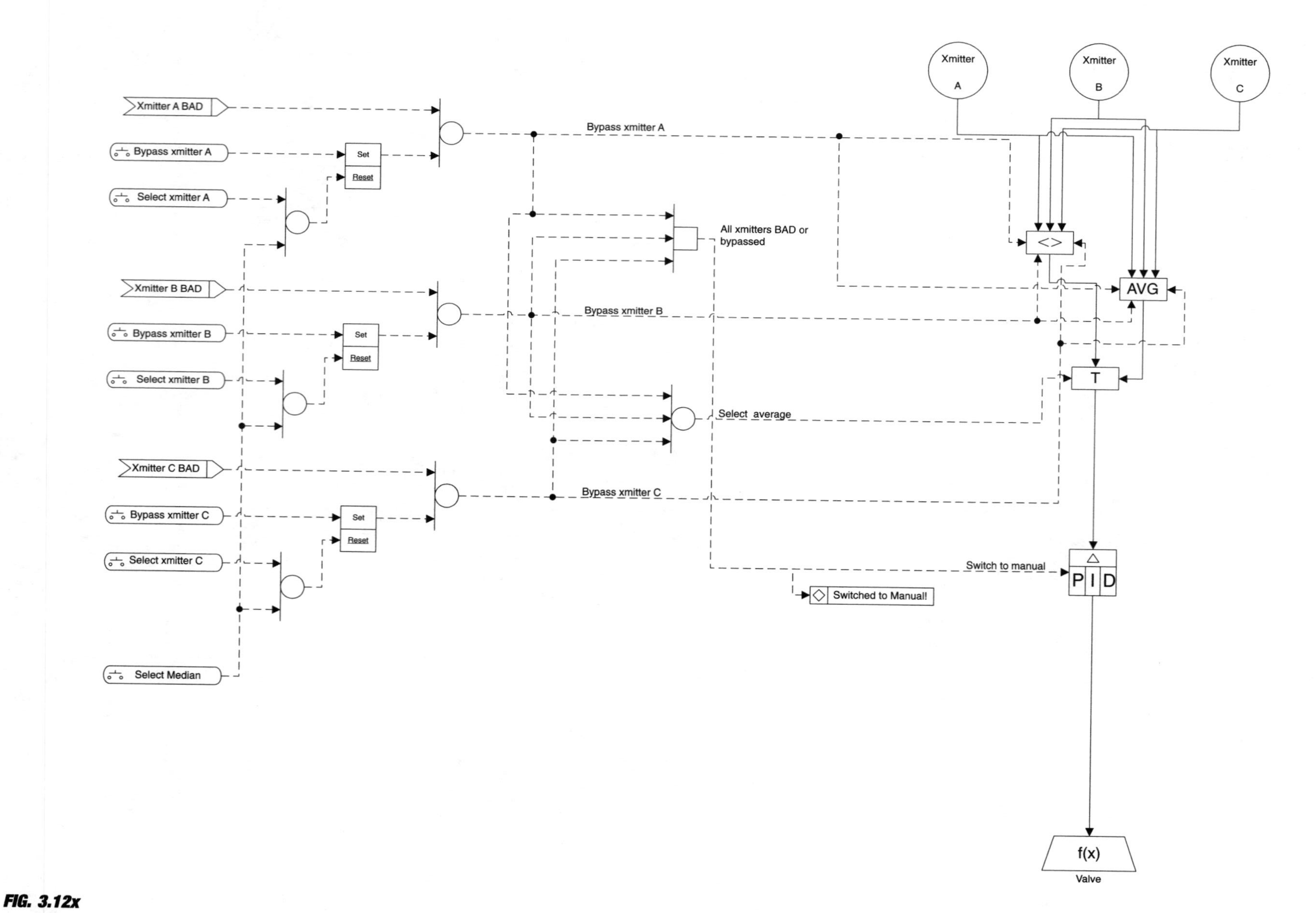

FIG. 3.12x
Analog—three transmitter selector.

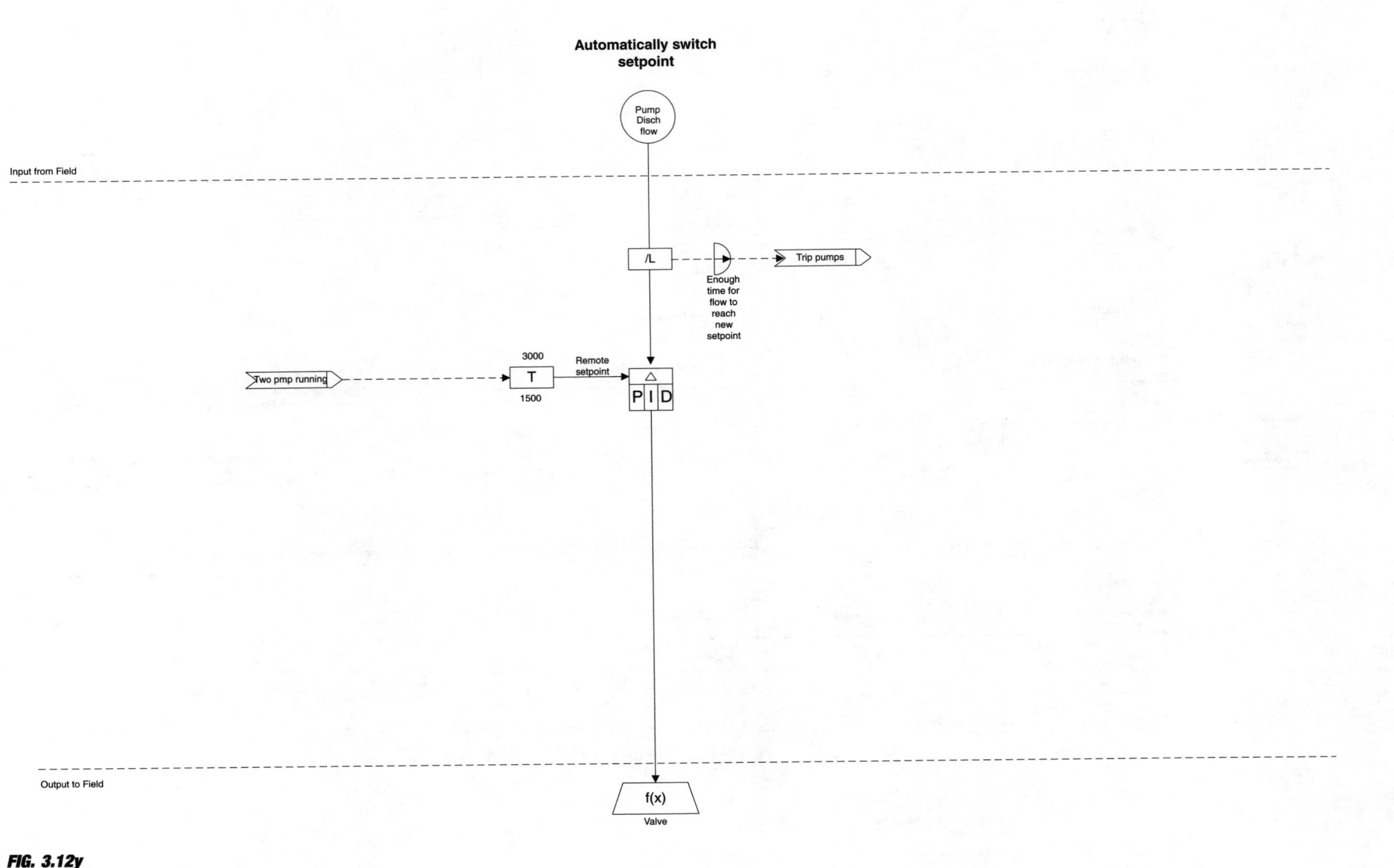

FIG. 3.12y
Analog—automatically switch set point.

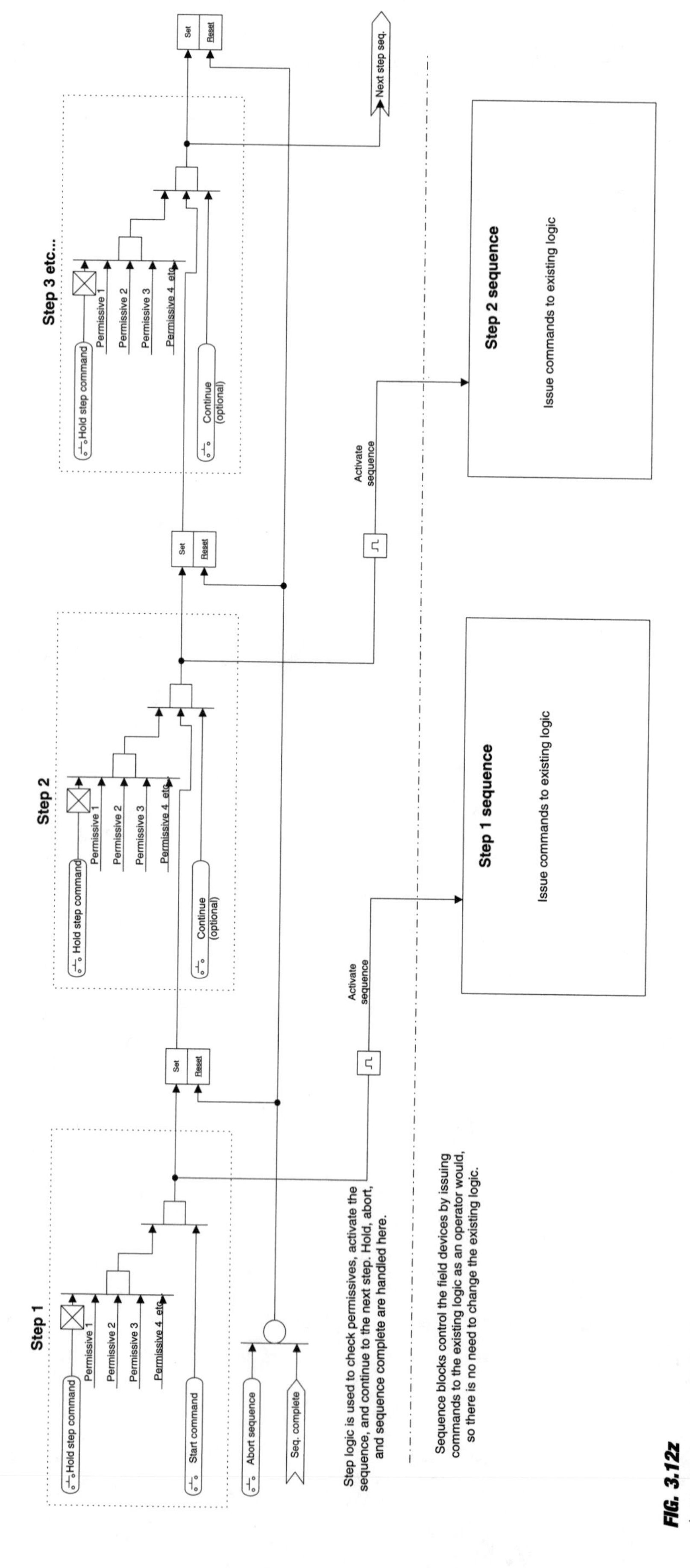

FIG. 3.12z
Auto start sequence logic structure.

changes to the design, customizing it to fit the equipment in the field.

Listen to the operators. The logic designer must rely on operators for information on errors or upsets that have occurred while the designer was away. Operators can often offer important insights on the behavior of the equipment, as well.

A good historian system is a useful tool. If possible, make sure to collect data on all of the process variables and the outputs and set points for all of the controllers—it will make troubleshooting and tuning much easier. Keep all the changes simple. Before adding new logic or feedforward to controllers, make sure the system is well tuned first—additions may not be necessary. Go easy on the fancy stuff—tuning parameters that change automatically and controllers that switch make the whole process less stable.

A NOTE ON SAFETY

Not enough has been said in this section about safety. Fortunately, safety goes hand in hand with good logic design. Safety should be the first priority. When safe systems are designed, well-designed logic will follow.

A few simple points to remember: Ensure that the equipment does not start, open, close, etc. unexpectedly. Remove any outputs to the field when in the failed state. Make sure the logic is consistent throughout the plant so that the operator will know what to expect when a button is pushed.

Document your system well so that others can understand how it works. Design user-friendly graphics that show permissives and trips so the operators can easily determine what is going on. Fewer accidents happen when there is less confusion in the control room.

Bibliography

Babcock and Wilcox Co., "Steam/Its Generation and Use," New York: Babcock and Wilcox, 1972.

Chesmond, C. J., *Basic Control System Technology,* New York: Van Nostrand Reinhold, 1990.

Dorf, R. C., *Modern Control Systems,* 4th ed., Reading, MA: Addison-Wesley, 1986.

Lipták, B. G., *Optimization of Industrial Unit Processes,* 2nd ed., Boca Raton, FL: CRC Press, 1999.

Polonyi, M. J. G., *Power & Process Control System,* New York: McGraw-Hill, 1991.

Stallings, W., *Data and Computer Communications,* 4th ed., Englewood Cliffs, NJ: Prentice-Hall, 1997.

Ulanski, W., *Valve and Actuator Technology,* New York: McGraw-Hill, 1991.

Woodruff, E. B., Lammers, H. B., and Lammer, T. F., *Steam Plant Operation,* 5th ed., New York: McGraw-Hill, 1984.

Buses and Networks

4

4.4
SORTING OUT THE PROTOCOLS 478

4.5
OVERALL FIELDBUS TRENDS 495

4.6
FIELDBUS ADVANTAGES AND DISADVANTAGES 505

4.7
FIELDBUS DESIGN, INSTALLATION, ECONOMICS, AND DOCUMENTATION 513

4.8
INSTRUMENTATION NETWORK DESIGN AND UPGRADE 522

4.9
GLOBAL SYSTEM ARCHITECTURES 534

4.10
ADVANTAGES AND LIMITATIONS OF OPEN NETWORKS 540

4.11 HART NETWORKS 547

4.12 FOUNDATION FIELDBUS NETWORK 564

4.13 PROFIBUS-PA 578

4.18 FIBER-OPTIC NETWORKS 638

4.19 SATELLITE, INFRARED, RADIO, AND WIRELESS LAN NETWORKS 649

4.1 An Introduction to Networks in Process Automation

P. G. BERRIE K.-P. LINDNER

When this section first appeared in the *IEH* in 1995, the 486 chip had just appeared on the market and personal computers ran at around 60 MHz. Today, we are three chip generations further on and PC speed has increased 200-fold and more. It is predicted that this trend will continue throughout the decade, until the physical limits of the present technology are reached.

This increase in computing speed and power has been accompanied by a breakthrough in network technology. Whereas in 1995 the company computer network was mainly a local affair, large corporations and computer manufacturers excepted, today many employees conduct their daily business electronically over the Internet. Strangely enough, this development was not business driven, but obtained most of its impetus through the home computer market. The Internet has developed from a collection of university and government servers into a hub of world communication. On the way it has spawned cheap network equipment and reliable and ever faster communication protocols. It is not surprising, therefore, that industry has been looking to integrate this and similar technologies, e.g., cellular telephone techniques, into manufacturing and process automation.

This section discusses the impact of these developments on process automation. It looks at the use of networks within a production company, the models used in digital communication, and the elements that form the basis of all networks. Finally, it takes a look at possible future developments. The section also includes an overview of the fieldbus and control systems of interest to process automation, most of which are covered in greater detail in the sections that follow.

INFORMATION FLOW REQUIREMENTS

Figure 4.1a shows a typical information flow within a company producing, e.g., chemicals, minerals, food, or beverages. As far as the production is concerned, a number of steps may be involved: batching, mixing and reaction, drying, conveying, storing, etc. Normally, each step will be individually controlled, requiring the measurement of process variables such as pressure, temperature, level, and flow. To optimize the performance of each step, and the process as a whole, data must be exchanged between the field devices, controllers, and a central host system.

From the administrative point of view, the scheduling, monitoring, and maintenance of production will be in the hands of a production planning department. It will rely on data from the sales department, which handles order entries for the finished product, arranges storage and dispatch, and bills the customer. Before production can start, the purchasing department must ensure that sufficient quantities of raw materials for a particular recipe are at hand. To assess the overall profitability and efficiency of the enterprise, financial management will also need additional information such as energy balances.

It can be seen, therefore, that the information provided by field devices is required not only for control, but also for administrative tasks. With it, raw materials can be ordered automatically when required, energy balances can flow directly into productivity calculations, output can be matched to order volume, and bills can be generated for the products shipped. The information exchange between the plant and enterprise management is handled by so-called management information or management executive systems (MES/MIS).

In recent years, asset management has also begun to play a role in the production process. Asset management is concerned with the optimal use of the plant, in the sense that the plant equipment, vessels, pipes, instrumentation, actuators, etc. are continually monitored regarding their state of health. The information required here is not whether a signal is good or bad or how a device is configured, but rather whether a device or plant unit requires maintenance or is about to fail. Such information is evaluated by maintenance programs and requires an independent data collection mechanism. With the development of e-business, there also is a call for automatic ordering of replacement equipment, an action that requires the presence of device identification data within the company administration.

The idea of computer integrated processing (CIP), as described above, is not new. Over the years, however, it has developed from the notion of controlling plant procedures with a mainframe computer to the vision of the complete exchange of data between all levels of the production and administrative process. Essential to its success is the reliable acquisition, presentation, and timely processing of data in all parts of the organization. This, in turn, requires open communication throughout the company network.

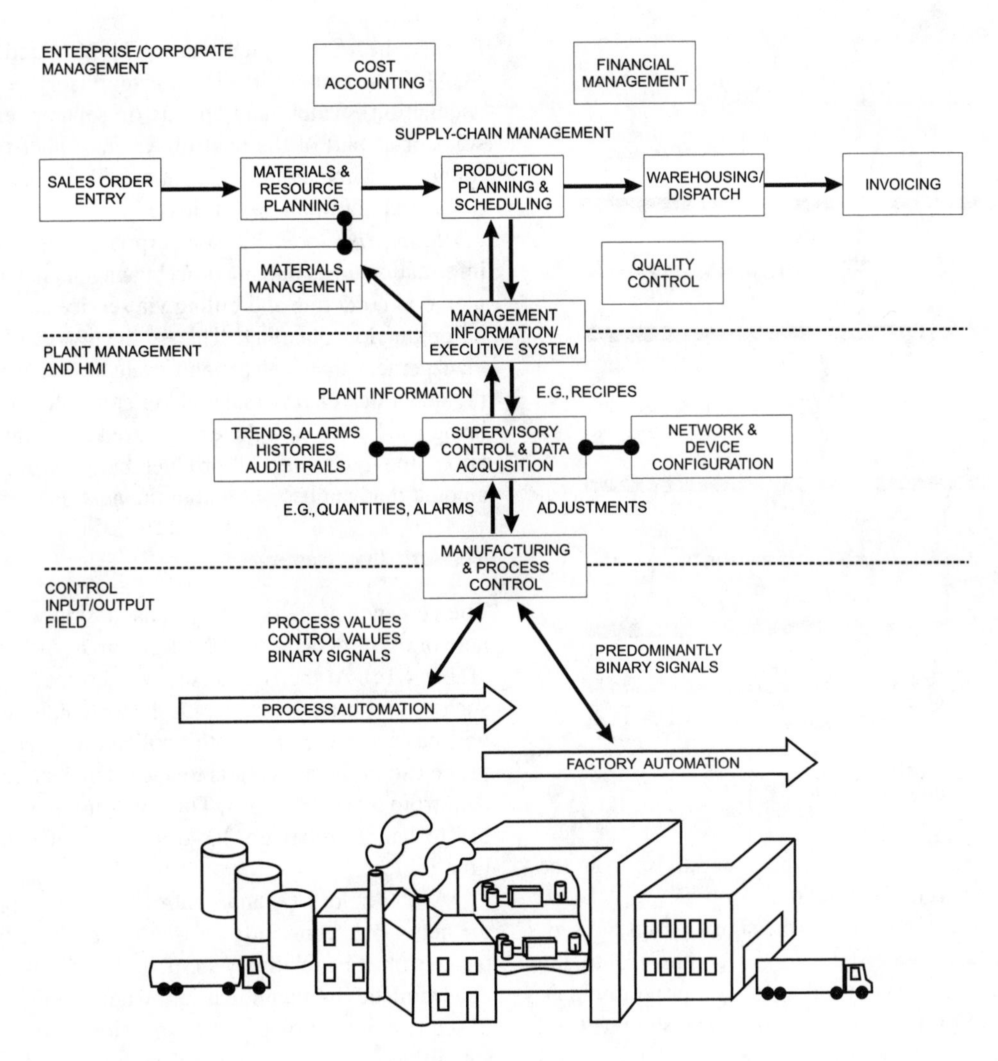

FIG. 4.1a
Information flow within a company.

HIERARCHICAL COMMUNICATIONS MODEL

Figure 4.1b shows a typical communications hierarchy by which CIP might be attained. This traditionally comprises five levels: field level, input/output (I/O) level, process control level, plant control or human–machine interface (HMI) level, enterprise (or corporate) level. The way in which this hierarchy is related to the information flow is also shown in Figure 4.1a. Please note that there are other ways of classifying the information flow and control, some of which are introduced in later sections of this chapter.

The five-level hierarchy model is, however, a convenient way to describe the equipment required to implement CIP, although it is somewhat out-of-date. For example, the latest generation of fieldbus devices are able to drive control loops, so that the difference between control, I/O, and field level is becoming blurred. In addition, the shape of the pyramid is not representative of the effort and cost at the various levels. The enterprise level is just as extensive as the field level, and the control levels require more programming effort. Nevertheless, the pyramid provides a simple overview of how data should be exchanged, even if the physical reality is somewhat different.

Hierarchical Levels

The function of each level within the model is as follows:

Field. The field level comprises sensors and actuators that are installed in the tanks, vessels, and pipelines that make up the processing plant. The sensors provide process variables, i.e., temperature, pressure, level, flow, etc., and pass the signals up to the I/O level. The actuators receive signals from the I/O level and perform an action, e.g., start a pump when the signal changes state or close a valve in proportion to the signal value.

Input/Output. The prime function of the I/O level is to marshal together input and output signals. The signals from

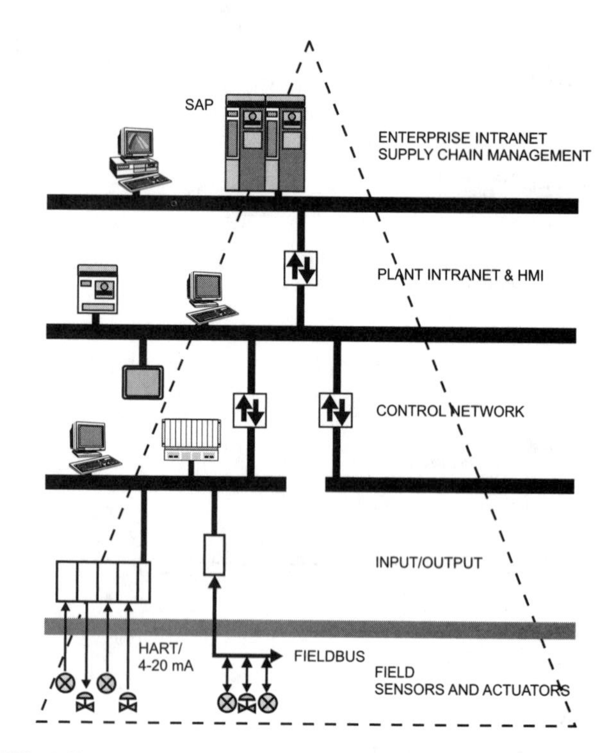

FIG. 4.1b
The factory communications hierarchy.

the sensors are directed to the controller, and those from the controller are directed to the actuators.

Control. At the control level, the signals from the sensors in the field are processed and the commands to the actuators are generated. The usual agent is a programmable logic controller (PLC) or process control system (PCS). Based on the signals received, valves are opened and closed or pumps and motors stopped and started.

Plant management. The plant management and HMI level is primarily concerned with the collection of plant data passed on from the control level. Network engineering tools, recipes, asset management, documention, and maintenance programs may also be resident here. Operators have a plant schematic on the monitor before them. They can display the process variables from any part of the plant, are warned when a value is out of limit, and can change the device configuration if need be. Such functionality is provided by so-called SCADA programs (SCADA = supervisory control and data acquisition), which may operate on separate engineer's consoles or as part of the control system. Other functions to be found are daily signal logs, audit trails of operator interventions, and alarm and event historians.

Enterprise level. The enterprise level is where process information flows into the office management world, for example, to aid ordering and billing via service access point (SAP) or production planning. This is the world of supply chain management that collates and evaluates such information as the quantities of raw materials in store, the amount of water being used every day, the energy used in heating and refrigeration, the quantities of product being processed, and the amount that can be sold within the next day, week, or month.

Network Requirements

The demands placed on data transmission within a company network will depend on the source and type of data handled (Table 4.1c). At the enterprise level, normal office programs, such as text processing, spreadsheets, databases, and e-mail, will be in use together with applications, such as SAP. The prime use of the network is to allow clients to call up or store data from a central server. Thus, it is not in constant use, but has to handle relatively large amounts of data that are not time critical.

At the field, I/O, and control level, the opposite is true: the network is constantly in use, the response times and data packets must be relatively short, and communication must be deterministic. In manufacturing, where there is a high degree of sequential control, high transmission speeds and very short signals are at a premium. In contrast, in processes where temperature, level, etc. are used for control, the speed is less critical, but the signal must contain more status information.

The plant management level must handle both types of data. Archiving, visualization, and maintenance programs are in regular contact with the control level, scanning, and mapping selected process data. There is also the need to change device configurations online. On the other hand, there are often exchanges with the enterprise level, e.g., computer-aided design (CAD) data, CNC programs, recipes, and spreadsheets

TABLE 4.1c
Fieldbus and Management-Level Data Characteristics

	Data Type					
Feature	Graphics	Data	NC Programs	Clock Signals	Process Values	Alarms
Response time	1–100 s	1–100 s	1–100 s	1–100 ms	20–100 ms	0.1–80 ms
String length	>10 kbit	1–10 kbit	>10 kbit	8–64 bits	<10 kbits	8–64 bits
Frequency	Seldom	Very seldom	Very seldom	Very frequent	Frequent	Seldom
Time	Uncritical			Critical		

Source: Electronics, February 1990.

with plant data. There is also the same demand for client/server services as on the enterprise level.

This variety of applications means that no single network protocol can operate efficiently at all levels. The protocol defines the principal network characteristics such as the speed of data transmission, length of data packets, degree of data security, the method by which data are distributed, etc. Thus, in a brewery, the beer production will use a protocol suitable for process control, whereas the bottling and packing facility will use one more suited to automated manufacturing. Information from both systems will be required at the plant management and enterprise levels.

To promote efficient data exchange, there is a need for standardization. This ensures that the devices on a particular network work together and that there is transparency between the various levels of the communications hierarchy. The central pillar of all network standards is the so-called ISO-OSI reference model.

OSI REFERENCE MODEL

The Open Systems Interconnection (OSI) Model[1] was first published in 1984 by the International Standards Organization (ISO) to bring uniformity and transparency into the development of communication network standards. It defines a communication model applicable to all network devices, from mainframe computers operating at the enterprise level to simple actuators in the field. It makes no attempt to make physical specifications. Instead, it provides a framework in which existing and future communications standards can be placed.

The OSI model splits the communication process into seven functional levels or "layers" (Table 4.1d), each of which performs a clearly defined task. By defining the function of each layer with respect to those above and below, the model ensures the compatibility of all devices using a particular network protocol, and between one network protocol and another. Above Layer 7 is the application process that requests or receives transmissions and which may be an office program, a controller, a database, a SCADA program, a workstation, a sensor, an actuator, and so forth. This so-called Layer 8 or "user layer" is not part of the standard.

The seven layers can be further classified into two functional groups. Layers 1 to 4 serve the network and are set the task of transporting data from one part of the network to another. Layers 5 to 7 serve the application and must present the transported data in an understandable form to the user of the network. The layers and the terms, protocols, and standards associated with them are discussed in more detail in Section 4.4, and for PLC networks in Section 4.2. Readers requiring more detailed information are also referred to Thompson.[2]

TABLE 4.1d
The Layers of the OSI Model

Layer	Function	Task	Examples: Standards/Realizations for Fieldbuses
Layer 7	Application	Provides the user with specific network commands and functions	Dynamic host configuration protocol (DHCP) Simple network time protocol (SNTP) Simple network management protocol (SNMP)
Layer 6	Presentation	Encodes the application layer data before forwarding it for transmission and decodes incoming data before forwarding it to the application layer	Not relevant to fieldbuses
Layer 5	Session	Synchronizes communication sessions between two applications	Not relevant to fieldbuses
Layer 4	Transport	Prepares the data string for transmission and ensures that it is reliably exchanged	Transmission control protocol (TCP) User datagram protocol (UDP)
Layer 3	Network	Selects the data route and ensures that the network is not overloaded	Internet protocol (IP)
Layer 2	Data link	Establishes and maintains connection between two participants	IEEE 802.2, Token passing IEEE 803.3, CMSA/CD IEC 61158/PROFIBUS-PA: Master/Slave IEC 61158/FF: Bus arbitration (LAS)
Layer 1	Physical	Puts the data on the physical medium and takes it off at its destination	EIA RS-232, EIA RS-422, EIA RS-485 IEC 61158-2 100BaseT (IEC 802.3u)

OSI Implementations

There are three OSI implementations that are of historical interest to the development of today's network standards: MAP, TOP, and ETA-MAP/Mini-MAP.

MAP. The first OSI implementation was the Manufacturing Automation Protocol (MAP), developed in 1983 by General Motors and standardized in IEEE 802. It supports data exchange on a wide-area network and defines all seven OSI layers. At the time of development, GM was experiencing problems with communications on its robot lines and required a protocol that allowed open data exchange between all automation equipment that the company used. Around the same time Boeing was developing its office protocol (see below), and the two were combined to produce MAP Version 3.0 at the beginning of the 1990s. As a result, many vendors of automated manufacturing equipment began to offer MAP interfaces. MAP allowed several alternatives at the various OSI layers: token passing and CSMA/CD for the data link layer.

TOP. The Technical Office Protocol (TOP) was defined by Boeing for use in technical scientific and office automation environments. It emphasizes the exchange of CAD/CAM documents, graphics, and text. The standard also supported a range of transmission media and access methods. It was combined with MAP in MAP Version 3.0.

EPA-MAP. Shortly after the publication of MAP, the need was recognized for a similar standard to cover the requirements of local networks. Here devices are equipped with identical or very similar operating systems. Data coding is either nonexistent or very simple, there is no requirement for a transport layer, and connections within the network do not have to be sought. These considerations led to the idea of enhanced performance architecture (EPA), which provides less powerful, but far more efficient communication than MAP. In addition, it is particularly suited to real-time applications. EPA describes a method, added later to the OSI model, in which only the application, data link, and physical layers are used for data communication. This method is also known today as Mini-MAP and forms the basis of many fieldbus protocols.

A number of Internet protocols are also of considerable interest to modern control systems. These include the TCP/IP (transmission control protocol/Internet protocol) and UDP/IP (universal data protocol/Internet protocol) data transport protocols and several protocols for file management and time distribution in the application layer. The Internet is particularly interesting because it is market driven. This means that developments are rapidly adopted and incorporated into the Internet official protocol standards, which are controlled by the Internet Activities Board. Because the Internet preceded the original OSI standard, which in any case did not address the possibility of connecting networks together, it does not follow it. Later editions of the OSI standard were adapted to include internetting.

Control Networks

It can be seen from Table 4.1d that quite a number of standards used in control networks are already in place within, or accepted as being part of, the OSI framework. A network protocol is characterized primarily by its physical and data link layers. The physical layer often defines the transmission medium, mode, transfer rate, extent of the network, etc., as can be seen from Table 4.1e. Among other things, the data

TABLE 4.1e
Attributes of Various Interface Standards

Attributes	*RS-232C*	*RS-422*	*RS-485*	*IEC 61158-2*	*100BaseTX*
Max devices	1 transmitter 1 receiver	1 transmitter 16 receivers	32 transmitters/ receivers	32 transmitters/ receivers	Limited only by address range[a]
Signal coding	±12 V pulses	Differential	Differential	Manchester II	8B6T
Topology	Star	Bus	Bus	Bus	Star
Max line length	15 m	1200 m	1200 m	1900 m	100 m
No. of lines	min 3	4	2	2	4
Max rate of transmission	19.2 kbit/s	10 Mbit/s	10 Mbit/s	37.5 kbit/s	100 Mbit/s
Communication mode	Duplex	Duplex	Half-duplex	Half-duplex	Duplex
Transmission type	Asynchronous	Asynchronous	Synchronous	Synchronous	Synchronous
Connectors	9- or 25-pin connectors	Not specified	Not specified	Not specified	RJ45 connectors

[a]There is a practical limit determined by the response time of the system.

link layer determines how the network devices communicate with each other. This is of particular interest to process control systems, because it also influences the time when refreshed information is available to the network participants and the determinism of the system. More general information on the physical and data link layers can be found in Section 4.2; specific information can be found in the sections that follow.

Until recently, standardization at the application layer was less advanced. For this reason there is less openness, and many control protocols are proprietary at this level. Table 4.1f lists those that are included in the IEC 61158 standard as well as other fieldbus protocols of interest to process automation. Even with the advance of so-called "industrial Ethernet" protocols, differences still remain at the application level and, for example, devices running ControlNet TCP/IP and Modbus TCP/IP will not understand each other if operated together on the same network. Of those protocols included in IEC 61158, only PROFIBUS-DP, Foundation Fieldbus, and ControlNet can really be seen to be of interest to process automation. They also enjoy a great deal of multivendor support. Using this criterion, Modbus must also be seen as an industrial standard that may run on a variety of physical layers.

It can be seen that both PROFIBUS and Foundation Fieldbus offer a second open protocol for the field level and, although strictly speaking proprietary, ControlNet and DeviceNet stand in a similar relationship. This is in accordance with the original concept for a fieldbus standard, which foresaw a high-speed (H2) network at control level and a low-speed (H1) network at field level. The change in speed necessitates an interface between H2 and H1 networks, which may also bridge a change in physical layer. Thus, Foundation Fieldbus changes from high-speed Ethernet (HSE) to IEC-61158-2 and PROFIBUS from RS-485 to IEC 61158-2. Although Foundation Fieldbus H1 and PROFIBUS-PA devices could be physically operated on the same segment, the data link layer differs, and the devices are unable to understand each other. The other protocols of interest at the field level are the Modbus, HART, and ASI (actuator/sensor interface), which are covered in more detail in the following sections of this book.

THE COMPANY NETWORK TODAY

In the previous sections, the theoretical basis for open communication within a company network has been discussed, but what is encountered in practice? In general, there is more standardization in protocols at the higher levels of the communications hierarchy than in the field, where legacy technologies with no communication interface must be integrated into a network. In addition, the requirements of the application, process, or manufacturing will have an influence on whether one or more protocols are implemented there.

Enterprise and Plant Management Level

The state of the art at enterprise and plant management level is Ethernet TCP/IP. Networks are constructed with standard Internet components such as hubs, switches, and routers. Within a company, there is free exchange of data between the two levels, based, for example, on SAP or Microsoft Office programs. For e-business, intercompany data exchange, or data-warehousing concepts, however, there is still a need for standard interchange formats. The STEP initiative is one possibility (see later); the other is a further development in the PDF format, which can already be regarded as an *ad hoc* standard. XML text and the associated CMG drawing formats also provide interesting possibilities for exchanging structured data.

As regards the transfer from plant management to control level, practically every soft controller or PLC now provides an Ethernet port for connection to a supervisory station. Workstations are office or industrial PCs, with Microsoft Windows NT, Windows 2000, and Windows XP. These run proprietary SCADA programs that may have interfaces to SAP and standard office applications.

Control Level

At the control level, the IEC 61158 standard has done nothing else but to standardize a number of competing, and in many cases, proprietary standards. As mentioned above, PROFIBUS-DP, Foundation Fieldbus HSE and ControlNet are the ones of most interest to process automation.

- ControlNet is coupled with DeviceNet, but also offers a linking device to the Foundation Fieldbus H1 bus. Foundation Fieldbus H2 devices are not be compatible. It is essentially a manufacturing protocol that can be adapted to process automation. It is supported by two major American manufacturers, and a good range of components, although not necessarily interchangeable, is available.
- Fieldbus Foundation HSE was designed for process automation and foresees the use of standard Internet components to build networks. In addition, HSE devices, frequency converters, remote I/Os, and gateways to other network systems are likely to come onto the market in the near future. Foundation Fieldbus has suffered from the late publication of the HSE specification (Spring, 2000), which resulted in a number of proprietary solutions for linking the H1 bus to a Foundation Fieldbus controller. This situation has now been remedied because the first HSE/FF H1 linking devices have arrived on the market.
- PROFIBUS-DP has been on the market since 1996, and as a result the user has a full range of control-level equipment to choose from. Version DPV1 and

TABLE 4.1f
Attributes of Various Control and Fieldbus Standards

Protocol	Standard	Industry	Special Features	Processing	Medium Access	Nodes[a]	Medium	Segment Length	Bus Safety Concept	Further Information
Ethernet	IEEE 802.3	Office, factory, process	Most widespread network protocol	Decentral	CSMA/CD	Max 30[b]	100BaseT4, copper 100BaseFL, fiber optics 100BaseTX, copper	Max 100 m Max 3000 m Max 100 m	None	Industrial Ethernet Association, www.industrial ethernet.com
Foundation Fieldbus FF HSE	IEC 61158	Factory, process	Function blocks for decentralized control	Decentral	CSMA/CD	Max 30	Copper or fiber optics	Max 100 m	None	Fieldbus Foundation, www.fieldbus.org
Foundation Fieldbus FF H1	IEC 61158	Process	Function blocks for decentralized control	Central/decentral	Token passing	Max 32	Copper	Max 1900 m	Exd or Exi	Fieldbus Foundation
PROFIBUS-DP	IEC 61158	Factory, process	Optimized for remote I/O	Central	Token passing	Max 126	Copper or fiber optics	Max 1200 m (copper) Several kilometers with optical fibers	None	PROFIBUS User Organization, www.profibus.com
PROFIBUS-PA	IEC 61158	Process	Standard certification process for hazardous areas	Central	Token passing but normally operated as master/slave	Max 32	Copper	Max 1900 m	Exi to FISCO model	PROFIBUS User Organization
World FIP	IEC 61158	Factory	Distributed real-time database	Decentral	Token passing	Max 256	Copper or fiber optics	Max 10 km	None	WorldFIP organization, www.worldfip.org
Interbus	IEC 61158	Factory	Optimized for remote I/O	Central	Single master with synchronized shift register	Max 256	Copper	Max 13 km	None	Interbus Club, www.interbusclub.com
ControlNet	IEC 61158	Factory	Optimized for factory applications	Central	TDMA	Max 99	Copper (coax)	Max 6 km (with repeaters)	None	ControlNet International, www.controlnet.org
SwiftNet	IEC 61158	Aircraft	Optimized for aircraft	Decentral	Token passing	Max 1024	Copper	Max 360 m	None	SwiftNet
P-Net[a]	IEC 61158	Factory, shipbuilding	Multinetting capability	Central	Token passing	Max 32 masters Max 125 slaves	Copper	Max 1200 m	None	International P-NET User Organization, www.p-net.com

MODBUS	Industrial standard	Process	Simple structure, widely used	Central	Master/slave	Max 247	Copper	Max 1200 m	None	MODBUS users Web site, www.modbus.org
DeviceNet	EN 50 325	Factory	High immunity to electromagnetic interference	Decentral	CSMA/CD	Max 200	Copper (4-wire)	Max 500 m	None	ODVA, www.devicenet.org
ASI	EN 50 295	Factory, process	Power over bus, for simple actuators and sensors	Central	Master/slave	Max 124	Copper	Max 100 m	None	ASI association, e.g., www.infoside.de/infida/asi/asi000.htm
HART	Industrial standard	Process	Integrates into existing 4–20 mA systems	Central	Token passing	Max 16	Copper	Max 3000 m, dependent on number of devices	Exi, Exd, or Exe	HART Communication Foundation, ww.hartcomm.org

[a]Number of nodes that can be physically connected per segment; in some cases the number of logical addressable nodes can far exceed this number.
[b]Practical guideline for control networks: the more nodes there are, the longer the refresh time of the system.

PROFIBUS-PA are used for process automation. Practically every major PLC manufacturer offers a PROFIBUS DP interface, and standard equipment such as frequency converters, remote I/Os, gateways, data recorders, and segment couplers are well supported. There are also a number of measurement devices, in particular flowmeters, on the market.

Interbus, although in the standard, is essentially a manufacturing protocol, and is found only in hybrid applications such as bottling and packaging machines. Modbus TCP/IP is not in the standard. It is Ethernet based, but again not compatible with ControlNet or Fieldbus Foundation HSE. There are also many proprietary protocols on the market, which enjoy considerable support because of their market position, but which are beyond the scope of this section.

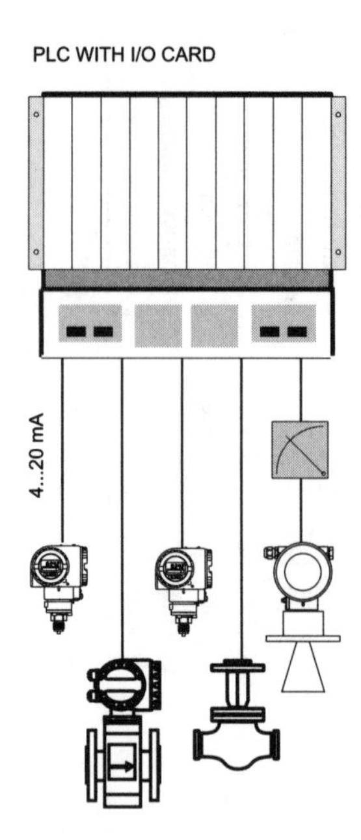

FIG. 4.1g
Connection of 4–20 mA and binary transmitters to a PLC.

I/O Level and Field Level

The fieldbus is not so far advanced as to be the dominant technology at the field level. For example, it was only in 2001, 5 years after the introduction of PROFIBUS-PA, that the German chemical industry gave its unqualified blessing to fieldbus technology. For some years to come, therefore, 4–20 mA/HART and fieldbus technologies will exist side by side in the field. Moreover, binary devices such as level limit switches, solenoids, and open/close valve positioners must also be integrated into the network. Modbus is still very strong in this area and ASI is finding increasing interest for binary devices. New standards, such as the NAMUR standard, have also appeared recently. There will be many cases, therefore, where more than one technology has to be integrated into a single network. The way in which this is done will depend on the control system, the I/O equipment available for this, and the protocols supported by the field devices.

Binary devices. The standard method of integrating binary devices is to connect them to the binary I/O card of a programmable logic controller via a point-to-point connection (Figure 4.1g). A fieldbus controller will require the use of a remote I/O to integrate these signals. When the field devices are equipped with the appropriate interface binary signals can also be integrated via a Modbus or ASI card in the control system.

4–20 mA. Where 4–20 mA devices exist at the field level, it is possible to connect them directly to the analog I/O card of a PLC via a point-to-point connection (Figure 4.1g). For fieldbus integration a remote I/O or intelligent I/O (see "legacy devices"), is required. If the 4–20 mA device is loop-powered, the PLC or I/O unit provides the power.

4–20 mA/HART. If the digital communication signal is not required, HART devices can be connected to a controller in the same way as a 4–20 mA device. If the HART digital signal is required, e.g., to configure the devices or monitor their readings, a multiplexer, remote I/O, or intelligent I/O may be used (Figure 4.1h). When integrated in this way, HART data are frequently accessed by a SCADA program running on a workstation in parallel with the controller. More information on HART can be found in Section 4.11.

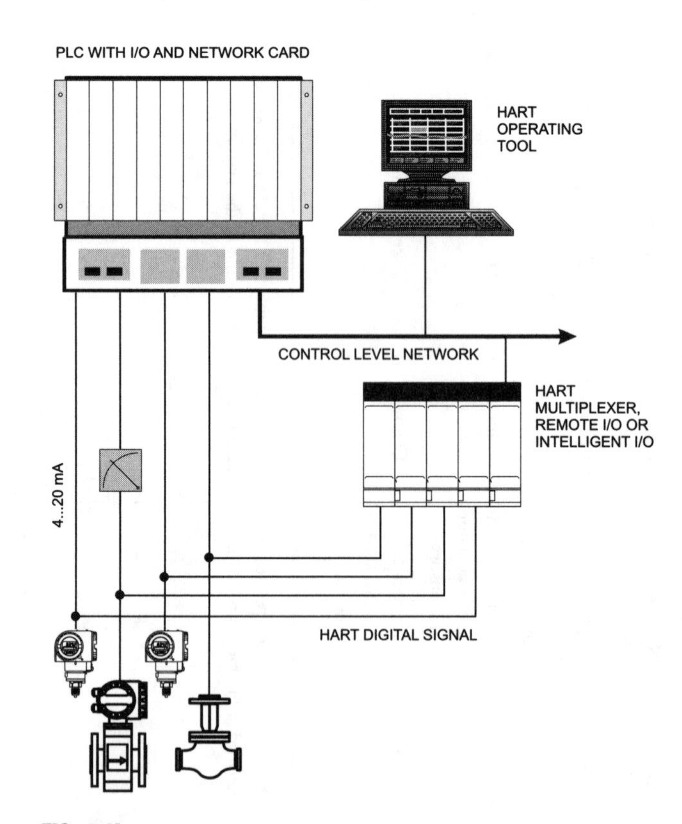

FIG. 4.1h
Connection of HART devices to a PLC with device monitoring.

Legacy devices. Legacy devices use older standards or proprietary bus or device signals. They can be integrated via proprietary gateways, remote I/Os, or intelligent I/Os. The latter comprise a number of input and output units that are

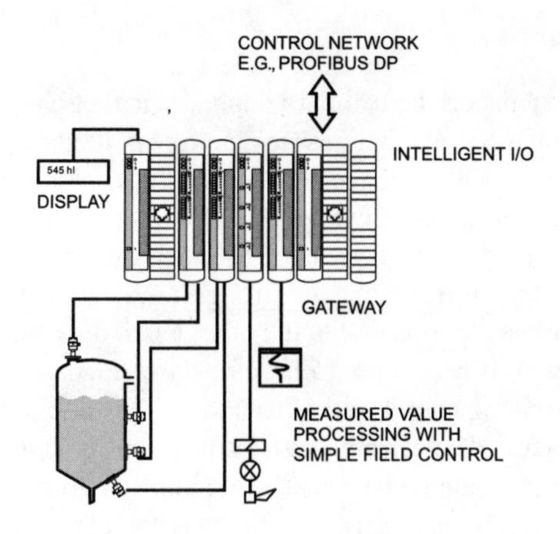

FIG. 4.1i
Connection of intelligent I/Os to a supervisory system.

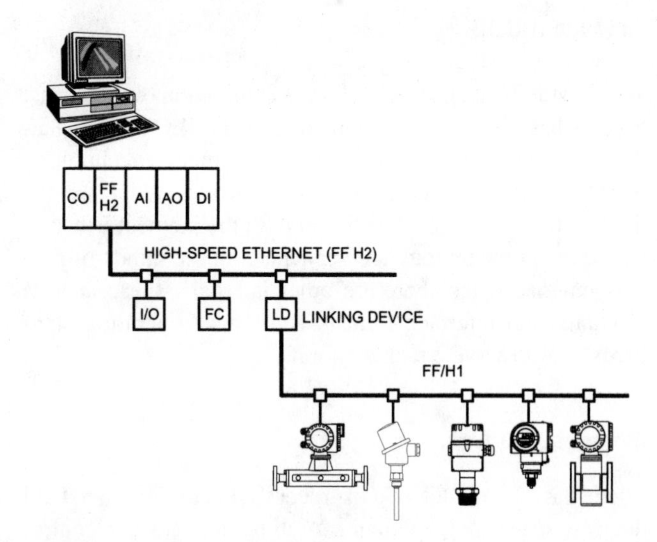

FIG. 4.1k
Connection of a Foundation Fieldbus segment to a supervisory system.

linked over an internal proprietary bus (Figure 4.1i) and often allow the connections of 4–20 mA, temperature and binary devices as well. The signals to and from the field are first processed before being passed on to the controller. Thus, it is possible to link signals from several sensors to produce a secondary process variable, e.g., density or mass. Alternatively, control loops can be built by linking sensor signals to actuators, e.g., to close a valve because the product in a tank has reached a preset level. The information from the system is passed on to the control level through an appropriate gateway, such as a PROFIBUS-DP or Modbus network.

PROFIBUS-PA. PROFIBUS-PA devices are connected to a bus segment that is integrated into a PROFIBUS-DP network via a segment coupler or link (Figure 4.1j); see Section 4.13. A coupler or link acts as an interface between the two networks and also supplies (intrinsically safe) power to the devices on the segment. The controller has access to all devices on the PROFIBUS-PA segment. Because most binary devices (limit switches, etc.) are conventional, they are integrated into the PA or DP network by remote or intelligent I/Os. This situation will change in the near future, as the first PA limit switches are about to be introduced to the market.

Foundation Fieldbus. The recommended method of connecting a Foundation Fieldbus H1 segment to a controller is through a linking device (Figure 4.1k). This performs only the protocol/baud rate conversion; bus power has to be supplied by a special power conditioner, where necessary through a barrier. Because the Foundation Fieldbus HSE was first published in 2000, HSE linking devices are only now coming onto the market. For this reason, some vendors offer an FF/H1 input card for their systems. Alternatively, ControlNet can be used with a ControlNet linking device. At present there are no possibilities for integrating binary devices into the H1 bus—this is foreseen via remote I/Os at the H2 level.

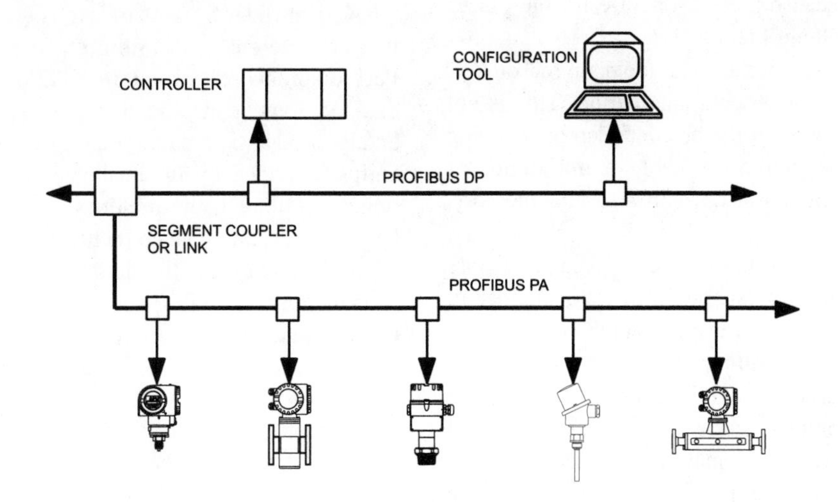

FIG. 4.1j
Connection of a PROFIBUS segment to a supervisory system.

THE WAY AHEAD

The most radical change in process automation over the past 5 years has not been the publishing of the IEC 61158 standard, but the increasing penetration of Internet technology into the instrumentation and control world. What are the likely influences in the next 5 years? At the moment Ethernet and wireless technology are very much in the minds of process engineers, but there are other internal issues, such as standard data interface formats and device operation standards, that deserve equal attention.

Ethernet

There has been a lot of talk recently about Ethernet field devices. After the penetration of Ethernet into the control level, is it not just a question of time before it arrives in the field? The question is complex and there are three factors that may have an influence:

1. The cost and availability of suitable chips
2. Information content and response times
3. (Intrinsic) powering of the devices

The first is probably a question of supply and demand; if there is a move toward Ethernet devices, the costs will decrease in proportion to the numbers produced. In the long-term there is no hindrance, but they will never be as cheap as the 8-bit processor used in today's field devices.

The standard TCP/IP frame has a minimum length of 72 bytes. Although the transmission of information across the network is fast, the protocol overhead increases the turn-around time and hence the response time of the system. At present, Ethernet is no match to a binary protocol such as ASI and, in this case, would in any case definitely be more expensive. The increase in transmission speed provides little advantage where process variables such as temperature and level are concerned, because the sensor refresh time places a lower limit on the polling interval.

Intrinsically safe power on the bus is the main reason IEC 61158-2 was developed for process automation. This is not just a case of getting power on the bus, but also of reducing the power consumption of the device to a minimum. At present, this places a limitation on chip size and memory. Ethernet chips are power-hungry and require more memory. Any substitute must solve this problem before it can penetrate at the field level. Because intrinsic power is not of interest to home and office applications, however, it is unlikely that there will be much pressure in this direction.

Summing up, Ethernet at the field level is far in the future, but devices operating in safe areas and having external power supplies, e.g., flowmeters, may well be provided with Ethernet connections for integration at control level in the not-too-distant future.

Wireless

Developments on the cellular telephone market have focused interest on the use of wireless technology in field devices; see also Section 4.19. All the devices within a plant would be integrated into a local wireless network, which would allow configuration and monitoring of parameters. This might have several advantages; among the strongest is the savings in cable and interface equipment. On the other hand, the devices must be powered, if necessary, to Ex e or Ex d standards.

Arguing against wireless technology at the moment are the problems of electromagnetic compatibility (interference from plant equipment) and the amount of iron and steel present, which may restrict communication. There is also no agreement on a standard. The leading candidate, Bluetooth, suffered a setback in Summer 2001, when a major supporter, Microsoft, decided not to include tooling in its new Windows XP operating system. One of the reasons was the inability of manufacturers to bring equipment onto the market. In this climate of uncertainty, it is questionable whether wireless will become available in the near future.

Interchange Formats

One of the reasons for the development of MAP and TOP was the exchange of office data such as documents, drawings, data bank information, etc. Most people working within a production company will be familiar with the problems: the drawing produced by the CAD system cannot be imported into the program used to produce the data sheet. The text in the data sheet cannot be imported into the plant documentation, etc. The result is that drawings and text are redrawn and retyped many times by manufacturer and customer alike.

The STEP initiative (Standard for the Exchange of Product Model Data, ISO 10303) aims to solve this problem. It is well advanced within the oil and gas industries and is finding increasing interest in the chemical industry. STEP aims to introduce standard exchange formats for all items that are interchanged within the life cycle of a plant, e.g., P&I diagrams, electrical installation diagrams, functional data, equipment specifications, etc. Among other things, this entails producing a data model of the entire plant, with each entity described by its attributes and relationships to other objects. Although the majority of items will be exchanged between the plant management and the enterprise level, some of the data concerns the field devices themselves.

One result of STEP could be that life-cycle data must be made available when a device is addressed in a SCADA or maintenance program. This requires standard structures and terminology: the same parameters must have the same names and appear at the same position within a data structure, irrespective of vendor. The proposed standard IEC 61987 and IEC 61380,[4,5] or indeed the Foundation Fieldbus and PROFIBUS-PA function blocks, may provide a basis. If the data content of field devices can be standardized, this opens the way for much simpler integration into software tools at both control and plant management levels, a step that will lead to more openness throughout the company network.

Device Operation Standards

One of the drawbacks of digital communication is that field devices have to be integrated into various control systems before they can be operated. To this end, the manufacturer supplies a device description with each device, which tells the operating software which parameters the device contains and how incoming data are to be interpreted. Unfortunately, device integration is seldom a matter of plug-and-play, and quite detailed knowledge of the system is required if the device is to work properly the first time.

At the moment, different device descriptions are required for HART, PROFIBUS-PA, and Foundation Fieldbus, and it is often the case that additional data or proprietary formats are required for particular systems. Moreover, it seldom occurs that all device functionality can be integrated into the third-party system, and, of course, there is always the danger that an update of the engineering software will cause devices to fail because the updated device descriptions are not immediately available. In addition, although the device appears in a uniform environment, it may be totally different from the one described in the device operating manual.

One initiative to solve this problem is the adoption of the FDT standard (field device tool) by a number of device and control companies. This also uses device descriptions, the DTMs (device type manager), which can be used in all programs and control systems supporting the FDT standard. The DTM is independent of the communication system, in as far as the driver for HART, PROFIBUS-DP, or Foundation Fieldbus is embedded in the FDT frame application, but because of the different parameter sets, separate device descriptions are still required for HART, PROFIBUS, and Foundation Fieldbus. The DTM also contains the complete data concerning how the configuration parameters are to be displayed within the host application, however, so that the vendor and not the control system environment is seen. The full functionality of the device is also ensured.

FDT provides plug-and-play integration of field devices into supporting systems. It still does not tackle the problem of a uniform configuration interface, but it is quite likely that there will be a general initiative among the user groups to standardize this as well.

CONCLUSION

It was stated at the beginning of this section that the five-layer communications model is out of date. The penetration of Ethernet technology to the control level and the introduction of control functions to the field level have caused a blurring of the edges. This trend will continue, especially because with PROFIBUS-PA and Foundation Fieldbus the possibility now exists of integrating field devices into an open network structure.

As far as process automation is concerned, HART will continue to play a significant role for a long time to come. It is the means by which many sensors and actuators are connected to the numerous control systems on the market today. Modbus and ASI continue to be strong in their own particular fields.

In the opinion of the authors, however, the future at the field level lies with the fieldbus. In the last 4 years, the number of PROFIBUS-PA devices sold has doubled from year to year, despite the fact that many companies have waited for the publication of the international standard. Now we can expect that fieldbus technology will really make inroads into traditional 4–20 mA business. The IEC 61158 standard was a bad compromise, but in standardizing the state of the art, it has removed the politics from the fieldbus debate. Now users can decide whether they prefer PROFIBUS-PA or Foundation Fieldbus (or even HART) and, in examining their relative merits, discover that they have much in common. Perhaps this is the true migration path to a common, open fieldbus standard.

References

1. International Standard Organization/ International Electrotechnical Commission, ISO/IEC 7498-1, "Information Technology—Open Systems Interconnection—Basic Reference Model: The Basic Model," 1994.
2. Thompson, L. H., *Industrial Data Communications, ISA Resources for Measurement and Control Series,* Research Triangle Park, NC: Instrument Society of America, 1991.
3. International Standard Organization, ISO 10303, "Product Data Representation and Exchange, Part 1: Overview and Fundamental Principles; Part 212: Electrotechnical Design and Installation; Part 221: Functional Design and 2D Representation; Part 231: Process Design/ Specs of Major Equipment." See also, e.g., www.ukceb.org/step/.
4. Committee draft to IEC 61987, "Industrial Measurement and Control: Data Structures and Elements in Process Equipment Catalogues, Part 1: Measuring Equipment with Analogue and Digital Output."
5. International Electrotechnical Commission, IEC 61360 (Series), "Standard Data Element Types with Associated Classification Scheme for Electric Components." See also, e.g., www.eclass-online.com.

4.2 PLC Proprietary and Open Networks

C. S. BARTON

—o—o—o—o—o—o—o—o—o—o— ISA Communication Line/Network Connection

or or ISA Computer Equipment or Function

Fiber-Optic Interface

or Switch

Router

Common Symbols

Parial List of Suppliers: 3Com; Allen-Bradley; Belden; Black Box; Cabletron; Cisco Systems; General Electric; Modicon; Siemens; Siecor

Typical Costs: Twisted Pair Cable: Belden 9841 (MB+), $0.24/ft; Belden 9463 (Blue Hose, DH+), $0.20/ft; Cat. 6, 4 Pair, UTP, General Rating, $0.35/ft; Cat. 5, 4 Pair, UTP, General Rating, $0.15/ft; Cat. 5, 4 Pair, STP, General Rating, $0.25/ft

Coax: RG 6, $0.13/ft; RG 11, $0.25/ft

Fiber: 12 Fiber, Multimode, Loose Tube, Exterior, $2.50/ft; 24 Fiber, Multimode, Loose Tube, Exterior, $4.00/ft
12 Port, 10BaseT Hub, $300
24 Port, 100BaseT Switch, $1500

Standards Organizations: ISA, IEEE, EIA-TIA, NEC

THE GENERAL PHILOSOPHY OF COMMUNICATION

Communication, even in its most basic form, is remarkably complex: something we take for granted on a day-to-day basis, but something that has gradually developed throughout history. Not until we tried to reproduce it artificially and to break it down into its various layers and components did we realize how intricate it can be.

Imagine two people, facing one another, wanting to say something. How, exactly does one go about expressing a complex idea to another?

We must first make sounds, which involves a medium on which to transmit (air) and a means by which to form and receive a signal (the mouth/ear set). There must be a common form of signaling (modulating the air into sound waves), and an accepted organization to those sounds (words). A language strings those words into meaningful sentences, conveying ideas, which must be interpreted into complex thoughts. All known forms of communication have similar a structure.

The OSI (Open Systems Interconnection) Reference Model (developed by the International Standards Organization, or ISO; the two should not be confused), shown in Figure 4.2a, has been developed to define this structure, and to help in the understanding of the process and nuances of communication. It contains seven layers, and is independent of technology:[1,2*]

Layer 1, *Physical*: Defines the mechanical and electrical characteristics necessary to establish, maintain, and terminate the link between physical devices. It includes voltage levels, modulation techniques, etc.

Layer 2, *Data link*: Concerned with physical addressing; network topology; establishing, maintaining and releasing links between nodes; assembling data

* Selected definitions within this section reprinted from Groth, D., *Network+ Study Guide*, 2nd ed., Alameda, CA: Sybex, Inc., © 2001. With permission.

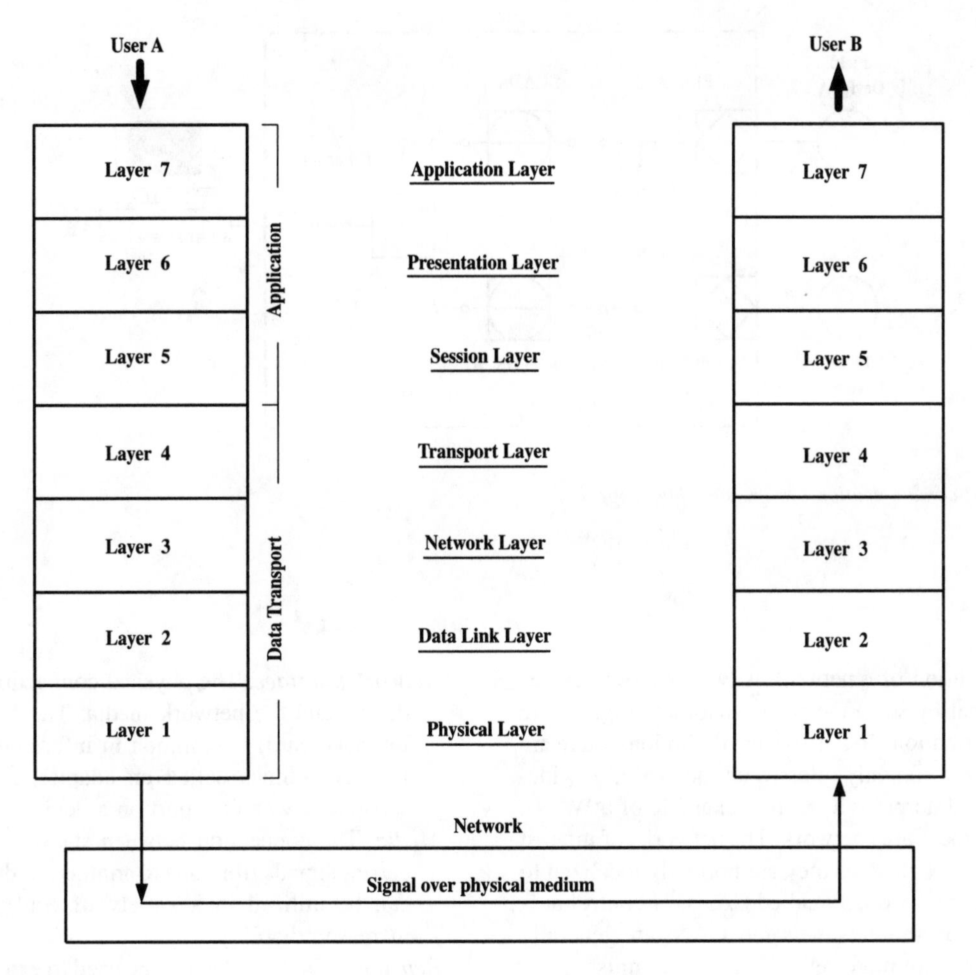

FIG. 4.2a
Diagram of OSI model.

into packets; as well as error detection and correction at the bit level.

Layer 3, *Network*: Provides connectivity and path selection between two end systems with logical addressing. It is the layer at which routing across subnets occurs.

Layer 4, *Transport*: Responsible for reliable network communication between end nodes, midlevel control of message delivery, including handshaking, message-level error detection, and retransmission.

Layer 5, *Session*: Establishes, manages, and terminates sessions between applications and manages data exchange between presentation layer entities, the higher level addressing of messages, as well as system control that controls communication sequencing and timing.

Layer 6, *Presentation*: Ensures that information sent by the application layer of one system will be readable by the application layer of another by translating the message into the proper format. It is concerned with the data structures used by programs.

Layer 7, *Application*: Provides services to application processes, the user programs that make a transmission request (such as e-mail, file transfer, and terminal emulation) that are outside the OSI model. The application layer identifies and establishes the availability of intended communication partners (and the resources required to connect with them), synchronizes cooperating applications, and establishes agreement on procedures for error recovery and control of data integrity.

This discussion will generally be limited to the lower layers of the model, which includes the physical medium, the modulated signal, message packet handling, and error checking. However, a basic understanding of the overall theory is pivotal to grasping the fundamental details of nuts-and-bolts operation, and all protocols, whether open or proprietary, are based on these basic concepts. The scope of consideration is shown in Figure 4.2b.

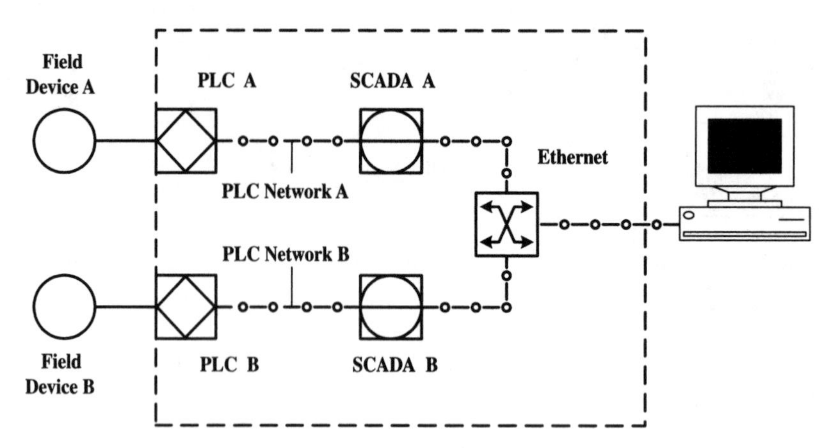

FIG. 4.2b
Diagram of typical communication system, showing scope.

Architectures

The general definition of a network is two or more devices connected together by some means, so that they might share resources or information. To narrow that definition, there are two types that are commonly referred to: the WAN, or wide-area network (the Internet is a familiar example of a WAN); and the LAN, or local-area network. The networks of interest here fit into the LAN class, as they are normally localized to one facility. They are normally called CANs, or control area networks, but the operation is the same. CANs are generally constructed so they are more reliable and deterministic.

The manner in which these devices are connected is called *topology*. Topology refers to the physical layout, and should not be confused with the transmission methods discussed later that can have similar names. Transmission methods are on a higher level of the OSI model, and are generally independent of physical topology. There are only three basic arrangements, along with a few variations (Figure 4.2c):

1. *The Bus*. A linear LAN architecture in which transmissions from network stations are broadcast over the entire length of a single continuous medium and are received by all other stations. A common physical signal path composed of wires or other media across which signals can be sent from one part of a computer to another. A bus is typically simpler and less expensive to install than the other topologies.
2. *The Star*. A LAN topology in which end points on a network are connected to a common central point by separate links. This topology is easiest to expand and troubleshoot.
3. *The Ring*. A network topology that consists of a series of nodes connected to one another by unidirectional transmission links to form a single closed loop. Each station on the network connects to the network at a repeater.

Building Blocks[2]

Network interface. The physical connection between a device and the network media. The NIC (network interface card) is common in informational LANs. It is sometimes called an adapter, and can be a separate device or a port on a device.

Media. The connection between stations that actually carries signals from one to another. Redundant paths can be utilized in a variety of configurations, to increase reliability.

Repeaters. Devices that can be used to extend the capabilities of a network by increasing the total length or number of units on a segment. Repeaters simply "repeat" everything they hear from one segment to the next.

Hubs. Hubs are basically multiport repeaters. They tie several devices together onto a common connection, in essentially a bus connection.

Bridges. Two-port devices used to divide a network into two logical segments (or collision domains) which are connected, but separate. Because the bridge operates at the lower levels of the OSI model, it does not know or care about higher-level protocols, so the two segments need not be the same type of network. Bridges simply pay attention to the address of a packet, so a message passes through to the other segment only if it needs to, and stays in its own segment if it does not. The bridge will ignore messages with addresses on the local segment.

Gateways. Similar to a bridge, except that they operate at the higher levels in the OSI model for the expressed purpose of connecting two dissimilar networks.

Switches. Basically, a multiported bridge that can open dedicated channels between two distinct stations. They operate by sorting data packets and sending them to a specific address.

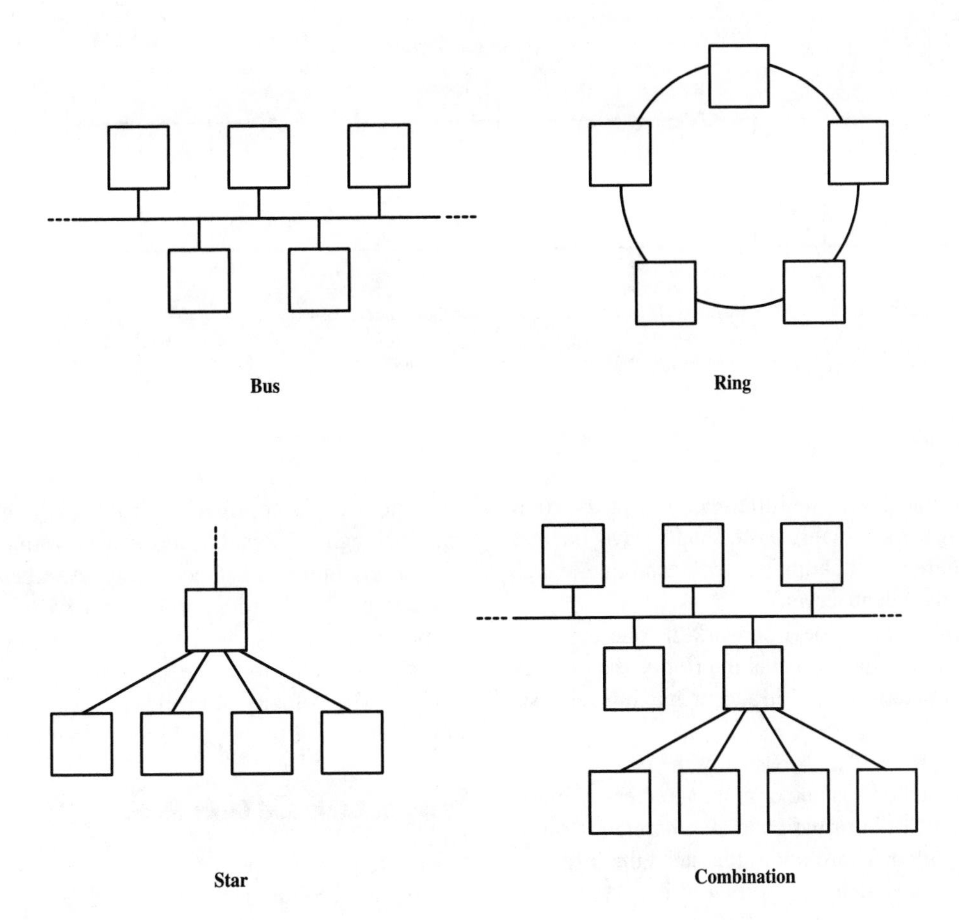

FIG. 4.2c
Diagram of topologies.

Routers. Similar to a switch, but with enhanced capabilities. Routers are usually used to connect split segments into subnets, and handle packets more efficiently. Routers operate on a higher level of the OSI model and can connect different types of networks.

DTE and DCE. Terms that describe the relationship between devices that communicate. The DTE (data terminal equipment) is the device ultimately attempting to communicate. The DCE (data communication equipment), typically a modem, is the interface that actually performs the business of communicating.

Physical Media Access and Arbitration Methods

Token Passing. A method of access control in which devices take turns using the network, by passing a special packet, known as a token. When a device has possession of the token, it may access the network for a specific period of time before passing the token on to the next device.

Multiple Random Access. MRA; a method of access control, in which all devices have the same priority and negotiate for control of the media. CSMA/CD (Carrier Sense, Multiple Access/Collision Detect) is the most common form of MRA methods. The "multiple access" portion refers to the many stations that share access to a common media. "Carrier sense" means that all devices listen to the media and will only claim control if no other devices are currently active. A device may only retain control for a limited amount of time before giving other stations a chance. The "collision detect" feature is a method of arbitrating the inevitable event where two stations attempt to claim control simultaneously. Upon hearing the resulting corrupted signals (a collision), both stations stop and wait for an arbitrary time before trying again.

Logical Media Access and Arbitration Methods

Client/Server and Polling (Master/Slave). In a client/server arrangement, a client makes a request of the server, which responds to the client. Servers may not speak until spoken to. There may be many clients and many servers; however, if more than one client is present, another arbitration method (usually physical) is required among them. Roles of devices may change, and a node that at one point is a client making a request of another may later be a server to that very node. Polling is a similar access method in which one device makes a request from another, but the connotation is that of a more regular, timed interval than that of client/server. The master/slave arrangement is similar in nature, in that the master

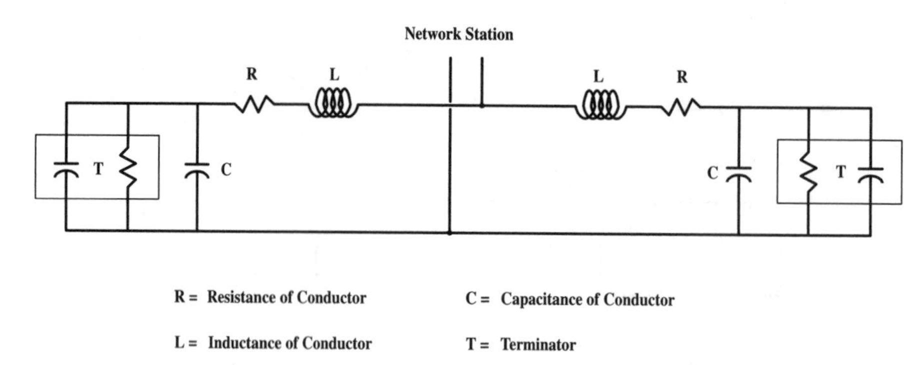

FIG. 4.2d
Diagram of equivalent circuit.

makes requests of the slave. The difference is that the roles do not change: masters remain masters and slave devices remain slaves. (There is a "floating master" technique, which is very similar to the client/server.)

Peer-to-Peer. In a peer-to-peer network, no one device is the controlling authority. Each is equal in priority, that is, they are peers. Devices take turns writing to or making requests of one another.

Publisher/Subscriber. This is basically the reverse of the client/server. This method has one or more suppliers of information, which broadcast information to all other nodes, without waiting to be polled. Those nodes that need the information receive it, and those that do not, ignore it.

PHYSICAL LAYER—OVERALL THEORY OF OPERATION

In its most basic form, network communication depends on the discrimination of one or more signal states or levels. The meaning of any discrete state or level varies widely, depending on the type of signal or protocol used. The definition of these signal types and meanings are the purpose of protocols.

Most signal types on most media operate by detecting a signal level with respect to some reference. For example, a voltage signal on an electrical conductor may be 0 or 5 V DC, with respect to a common signal ground reference wire (not necessarily earth ground). The job of the interface port is to discern whether the signal wire has 0 or 5 V on it at any given time by comparing it to the reference. The characteristics of the port must match those of the media and network as a whole to be able to make the measurement. Networks using electrical media have typical electrical qualities and, as such, can be represented by equivalent circuits that take into consideration conductor resistance, capacitance, frequency of operation, etc. Termination resistors and capacitors are attached to the end points of some electrical media types to attenuate the signal at the ends, in order to minimize signal reflections that can corrupt signal integrity and disrupt network operation. The value of the terminating resistor is determined by the characteristic impedance of the cable. It is intended to make the impedance of the equivalent circuit of the whole cable appear electrically as if the cable were of infinite length. A signal cannot reflect from the end of a cable that is an infinite distance away. Another quality that is affected by the electrical properties of a cable is the velocity of propagation. This is the ratio of the signal propagation speed in the cable to the speed of light, usually expressed as percent. The velocity of propagation is an important consideration in signal timing[3,4] (Figures 4.2d and 4.2e).

Types of Cable and Other Media

There are several types of media in widespread use in industry today: coax, twin axial, twisted pair, and optical fiber. There are several variations on each of these basic types:

Coaxial Cable (Coax). Coax is a type of cable characterized by a single solid conductor surrounded by a tubular shield, or outer conductor, separated by insulation and oriented about a common axis (hence the name). The shield reduces electrical interference (EMI, electromagnetic interference, and RFI, radio-frequency interference). The shield also acts as the signal reference conductor (Figure 4.2f). There are many different types and sizes of coaxial cable that are commonly used in industrial networks, designated by an "RG" number. "RG" is a U.S. Army descriptor (U.S. military standard MIL-C-17) that stands for "radio guide." Coax usually supports radio frequencies (RF signals) from 50 to 500 MHz, and is commonly used for broadband signals. In general, the larger-diameter coaxial cables offer greater performance because of their more rigorous construction; however, they have more stringent installation requirements and are more difficult to install in tight spaces, such as control cabinets. Table 4.2g outlines some of the more common coax cables.

Twin Axial Cable (Twinax). This cable is similar to coax, except with two inner conductors that are insulated from one another and twisted to reduce interference.

Triaxial Cable (Triax). This cable is also similar to coax, except with two outer conductors, one around the other, with dielectric insulation between.

Shielded Twisted Pair (STP). This cable has a number of individually insulated conductors, twisted into pairs, and surrounded by a shield to reduce EMI/RFI (Figure 4.2h). The pairs

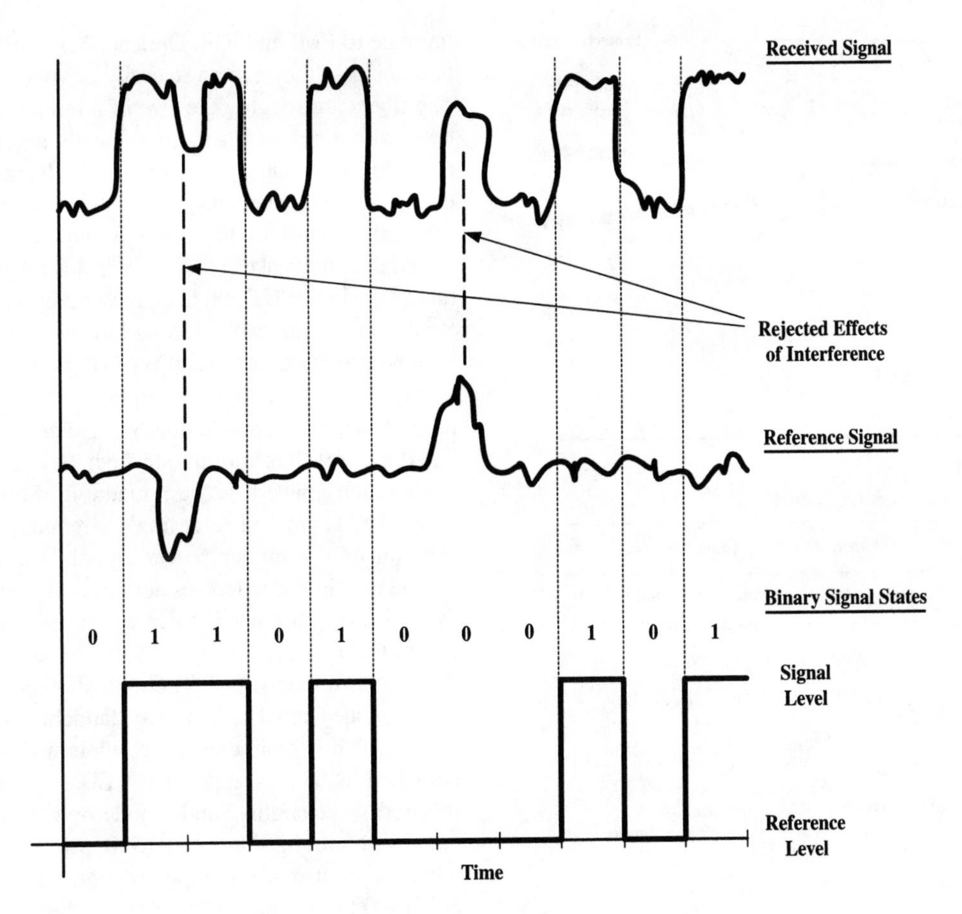

FIG. 4.2e
Diagram of signal states.

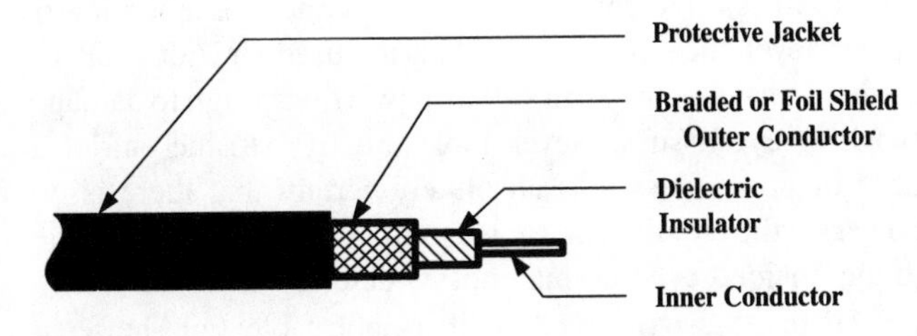

FIG. 4.2f
Diagram of coaxial cable.

TABLE 4.2g
Characteristics of Coaxial Cables

Cable Type	*Characteristic Impedance*	*Common Usage*
RG-6	75	Broadband, carrier band drop
RG-8	50	Thick Ethernet
RG-11	75	Broadband, carrier band trunk
RG-58	50	Thin Ethernet
RG-59	75	Broadband drop
RG-62	93	ARCnet

Note: Some references include the dash in RG-X; others do not.

may be individually shielded or have one overall shield, or both. A single-pair STP cable might appear similar to a twin axial cable, except that the outer shield is grounded, and not intended to be used as another conductor (signal reference or otherwise). STP is more expensive than UTP, but is more resistant to interference, making it the cable of choice for most industrial networks.

Unshielded Twisted Pair (UTP). This cable is very similar to STP, except without any shielding. It is very commonly used in informational Ethernet applications because of its less expensive construction and ease of use. A system has been devised by the EIA/TIA (EIA/TIA-568 standard) to define physical specifications and rate performance of cables for different applications, as shown in Table 4.2i. Category ratings are determined by physical factors such as the number of twists per foot, characteristic impedance, etc. Installation of cables should be made with cable construction in mind to maintain these qualities. More on installation practices will be covered later in this section. Bulk cable is typically constructed of solid conductors, which performs better over long distances, whereas patch cables are usually stranded conductors to offer more flexibility inside cabinets.

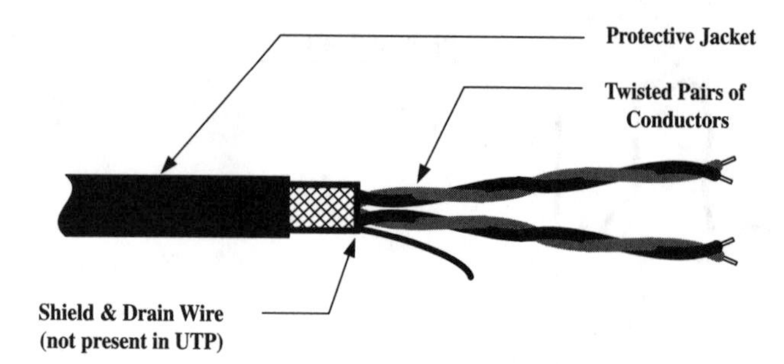

FIG. 4.2h
Diagram of STP/UTP cable.

TABLE 4.2i
Characteristics of UTP Cables

Cable Type	*Data Rate*	*Common Usage*
Category 1	N/A	Voice grade analog (not suitable for data)
Category 2	4 Mbps	Digital voice
Category 3	10 Mbps	10BaseT
Category 4	16 Mbps	Token ring
Category 5	100 Mbps	100BaseT
Category 6 (5e)	1000 Mbps	1000BaseT

Optical Fiber Cable

A medium of a totally different sort is optical fiber cable. Frequencies of light are launched from an LED (light emitting diode) or laser (light amplification from the stimulated emission of radiation), usually from a laser diode, into a shaft of plastic or glass (glass is more common) with a particular index of refraction. This shaft is surrounded by another, outer shaft of material with a different index. The difference in indices causes the light to reflect back toward the center of the inner shaft. Optical fiber has the ability to transmit higher data rates for longer distances than copper media; however, it costs much more and is usually more difficult to work with. Fiber-optic cables have the added important benefit of being completely immune to EMI and RFI. Optical fiber systems are nonspark-producing so they can more easily be used in explosive areas. The fibers are roughly the size of a human hair and are insulated with a buffer. There are typically several fibers, all surrounded by an outer jacket in a cable. Because of the size of the actual fibers, they make up only a small part of the overall cable size. Most of the cable is buffer, filler, and strength materials. This makes most fiber cables roughly the same size (about 0.5 in., or 1.27 cm in diameter), regardless of how many strands are in the cable; the size of the cable has more to do with construction and jacket type (Figure 4.2j).

There are two types of optical fiber: single mode and multimode. Single-mode fiber is smaller in diameter (usually 8.3 μm core) than multimode fiber (usually 62.5 μm core). Both usually have a 125 μm cladding. Single-mode fiber is meant to accommodate a single (frequency) signal, and so can transmit a higher power signal. Multimode fiber can handle multiple frequencies at once, but the signals can interfere with one another if their power level is too great. This means that a single-mode signal can be sent a greater distance than a multimode signal. There are, however, trade-offs: the single-mode signal is typically launched with laser equipment, which is more expensive, while multimode signals are launched with less expensive LEDs operating around the infrared wavelength. Single-mode optical applications typically operate in the 1300/1550 nm (nanometer) range, whereas multimode equipment normally operates around 850/1300 nm (Figure 4.2k). Some cables contain both types and are called "hybrid" cables.

***Important Safety Note*:** Even though the lasers and LEDs used in fiber-optic communications operate at a much lower power level than the ones ordinarily used in industrial or medical services, they can still be powerful enough to damage sensitive eye tissues. **Never** look directly into the end of a fiber or port unless it is absolutely certain that there is no power on the circuit! The wavelength of light used is invisible to the unaided eye and may not be detected until too late.

There are also two types of fiber-optic cable: tight-buffered and loose-tube cable. Loose-tube cables are filled with a gel, which insulates and protects the fibers but allows movement

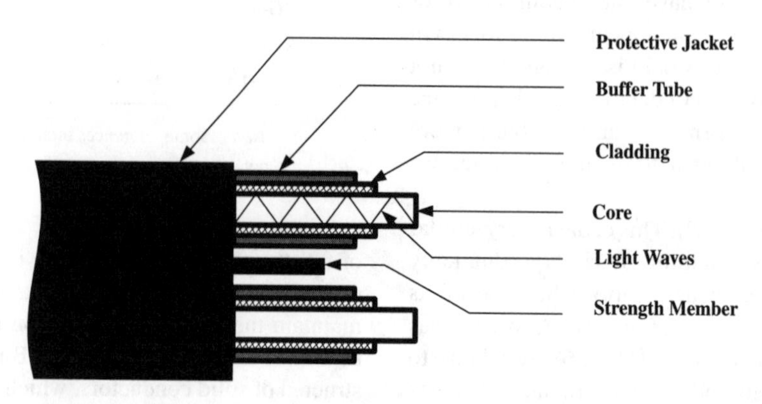

FIG. 4.2j
Diagram of fiber-optic cable.

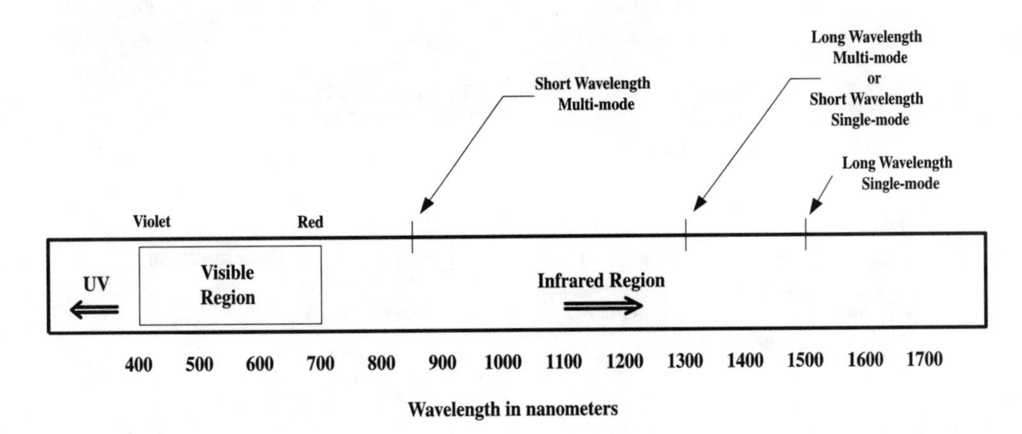

FIG. 4.2k
Diagram of wavelength spectrum.

within the jacket so that vibration, thermal expansion, and contraction, etc. do not inflict catastrophic damage on the delicate fibers. Tight-buffered cable is smaller in physical diameter, more flexible, easier to work with and terminate, and generally more suited to indoor applications than loose-tube cable. Fiber-optic cables are typically installed inside a sheath called "inner-duct." This is a specially constructed flexible tube, which can go inside of conduit to afford further protection from tangles, kinks, and abrasions.[5]

General Installation Guidelines

Special attention needs to be given to cable ratings and intended applications. In addition to the environmental considerations on the interior of the cable, the materials used in the jacket are chosen to suit certain situations. Outdoor cables can be designed for direct-burial (resistant to moisture and rot), open-air cable tray (ultraviolet and corrosion resistant), among others. Indoor cable can be rated for general office environments (less expensive, easier to install), or be riser or plenum rated with low flammability and toxic gas emission specifications. The NEC (National Electrical Code) Articles 700 and 800 have definitions of various applications and corresponding cable requirements. Local building codes and any other industry standards as well as area classifications should be checked before any installation.

Aside from design-type installation considerations, care needs to be given to physical installation of network systems, especially cabling. The cables used in modern communication systems have very particular construction specifications, which, if not maintained, will degrade the performance of the whole system. Factors such as cable stretch during installation and use, bending radius, and termination techniques can alter cable construction. In general, EIA/TIA 568A requires no more than 0.5 in. of UTP be untwisted at the termination point, because it may affect the performance at high bit rates. The standard specifies number of twists per foot, and any installation procedure that violates the specification (untwisting at termination, stretching of cable, etc.) may compromise the performance of the cable. The bending radius should always be greater than four times the outside diameter of the cable (for copper; 12 times for fiber), and a good general rule of thumb is that pulling tension should not exceed a force of the magnitude of 25 lb (110 N). These guidelines are general orders of magnitude; be sure to check the specifications of the cable to be installed. General care should be taken to avoid cutting or pinching the cable in any way. Cable should not be run near sources of heat or electromagnetic radiation. If a copper cable must be installed near power wiring, it should be oriented at a 90° angle to reduce induced voltages.

While shielding can reduce the effects of EMI/RFI, improper grounding, particularly grounding at both ends of a long run, can allow ground loops to form and actually introduce interference.

Types of Connectors[6]

DB or Sub-D connectors are shown in Figure 4.2l.

BNC (Bayonet-Neill-Concelman) and F connectors for coaxial cables are shown in Figure 4.2m.

RJ-XX (Registered Jack) for twisted pair cables are shown in Figure 4.2n.

Connectors for optical-fiber cables are shown in Figure 4.2o.

Transmission Techniques

At a slightly higher level than the physical circuitry, although still on Layer 1 of the OSI model, are the data representation and transmission techniques. The issue that bridges the two topics (i.e., physical circuitry and data transmission) is that of serial vs. parallel transmission. Serial transmissions send 1 bit at a time over a single transmission path, whereas parallel transmissions send many bits simultaneously over as many paths as there are bits. Parallel transmissions are much faster, but because of the number of conductors required, they are seldom used in networking applications. They are only found in short-distance settings, such as peripheral equipment (printers, storage media drives, etc.) (Figure 4.2p).

DB 9
DB 25
DB 9 Front View
DB 25 Front View
1 2 3 4 5
6 7 8 9
Male Connector Numbering
5 4 3 2 1
9 8 7 6
Female Connector Numbering
(looking at front)
Side View (of Male)

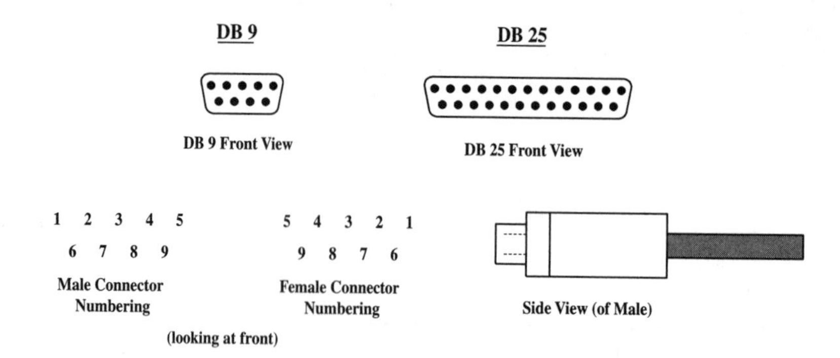

FIG. 4.2l
Diagram of DB connector.

BNC (Bayonet–Neill–Concelman)
F-Style
BNC Front View
F-Style Front View
Twist-Lock
Threaded
BNC Side View
F-Style Side View

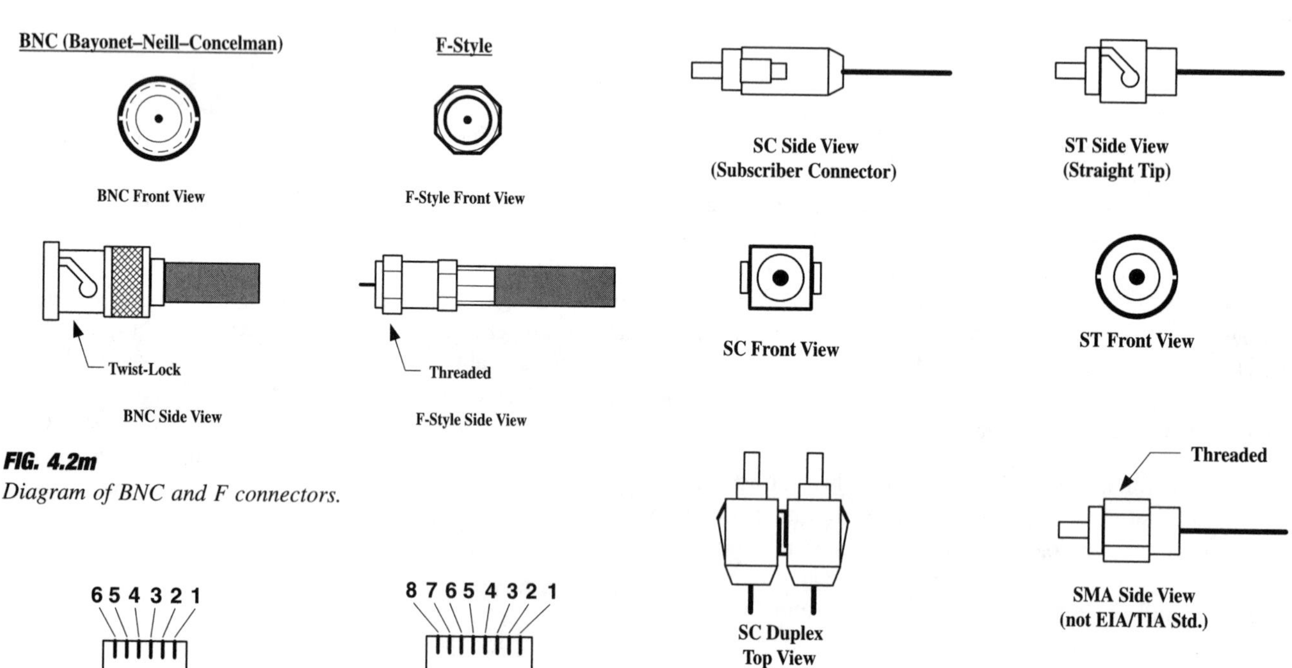

FIG. 4.2m
Diagram of BNC and F connectors.

SC Side View
(Subscriber Connector)
ST Side View
(Straight Tip)
SC Front View
ST Front View
Threaded
SC Duplex
Top View
SMA Side View
(not EIA/TIA Std.)

FIG. 4.2o
Diagram of optical-fiber connectors.

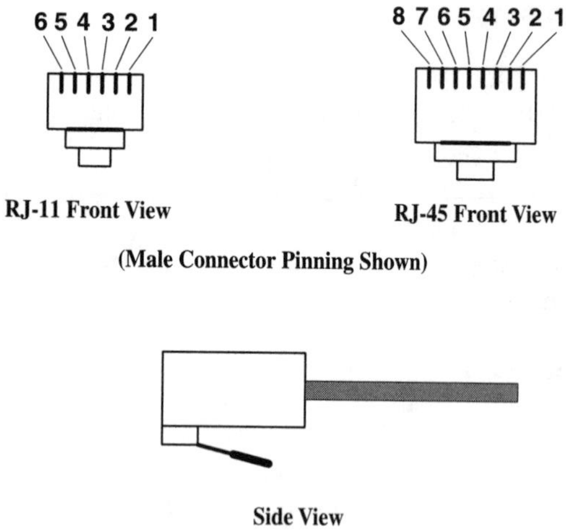

6 5 4 3 2 1
8 7 6 5 4 3 2 1
RJ-11 Front View
RJ-45 Front View
(Male Connector Pinning Shown)
Side View

FIG. 4.2n
Diagram of RJ-XX connector.

As previously mentioned, there is a requirement to distinguish information on the medium. There are several techniques used to carry information and enable the distinction of messages. Nearly all modern data network systems use binary digits (bits), a series of 1s (marks) and 0s (spaces) to send information, but there also must be a method of carrying the bits across the network.

Baseband involves the use of the entire bandwidth of a channel to transmit a single signal, using one carrier frequency. Broadband divides multiple analog signals into different RF frequencies and transmits them simultaneously (similar to multimode fiber-optic signals). Carrier band transmits a single analog signal. Baseband hardware is typically less expensive, as is the cable. Broadband requires high-quality coaxial cable, which has the disadvantage of being expensive, as well as heavy, stiff, and difficult to work with. Bandwidth is the range of frequencies that a given carrier is able to transmit effectively. These frequencies are modulated in one of several ways to encode data. Figure 4.2q depicts some of the more common methods. The rate at which data can be sent depends on the bandwidth of the cable, and is expressed in "baud." Baud is actually the rate of signaling events (changes in frequency, amplitude, etc.). Bits per second

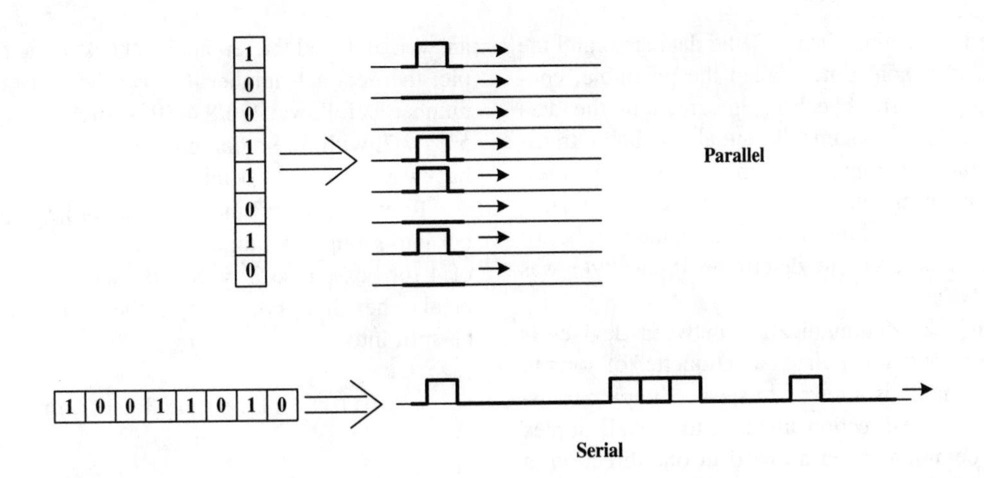

FIG. 4.2p
Diagram of transmission techniques.

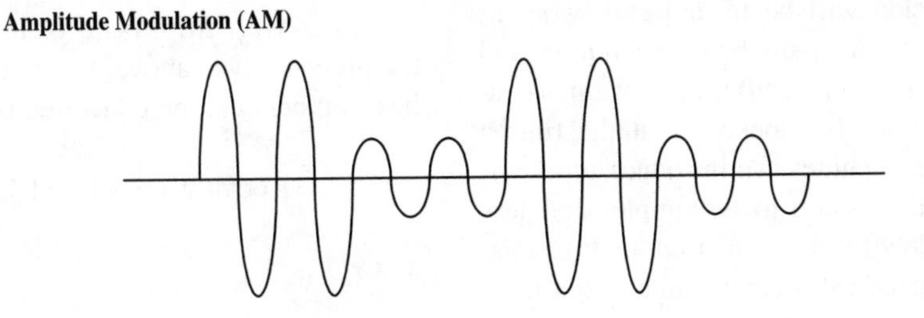

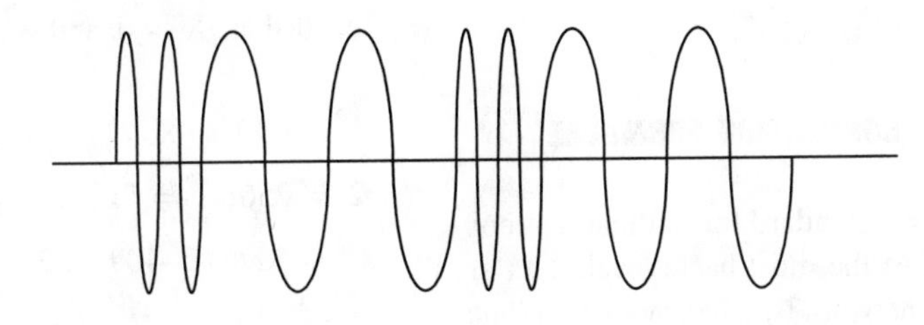

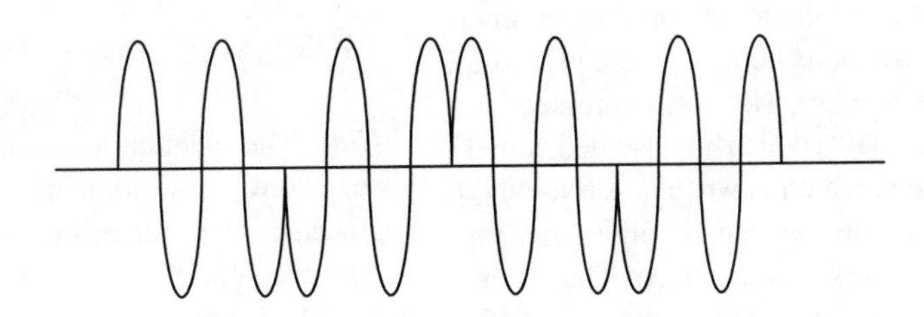

FIG. 4.2q
Diagram of modulation techniques.

is not necessarily the same as the baud rate and should not be confused, even though the terms are routinely used interchangeably. For example, if the modulation allows for four distinct states rather than the traditional two (1 and 0), two bits could be conveyed with every state. This would mean that the rate of bits-per-second would be twice as fast as the baud rate. The baud rate is the same as the bits-per-second rate if and only if each signal element is equal to 1 bit exactly. In a typical RS-232C arrangement, 11 bits are required to transfer 7 bits of data. This computes to 600 bits being sent over the line (baud) to transfer 382 bits of data (bps).

After solving the issue of placing bits on the network, they must be organized in such a manner to allow them to be translated into meaningful information. Messages are assembled into packets with formatting and addressing information, along with the data. The general form of a message packet or frame is a leading "header," the data area, and the "trailer." The header, sometimes called the preamble, contains addressing and error-checking information; the data area, called the payload, contains the actual data being transmitted; and the trailer contains more error-checking and message management information (e.g., parity and stop bits). Parity, a simple error-checking method, uses the number of 1s in a byte (odd or even) to determine if the byte was received correctly.[7]

The mechanics of communication between devices is managed by methods that stipulate an "etiquette" of sorts to determine when a node is allowed to speak. Simplex transmissions are only in one direction, all of the time. Half-duplex is bidirectional communication allowed in one direction at any given time, and full-duplex is bidirectional transmission in both directions simultaneously. In addition to this, synchronous transmissions are timed so that both devices know exactly when a transmission will begin and end, whereas asynchronous transmissions must mark the beginning and end of messages. Synchronous (clocked) transmission is usually faster than asynchronous (unclocked), but the timing issue between two remote machines can introduce problems causing asynchronous transmission to be simpler and less expensive, thus more widely used. Asynchronous transmission does, however, introduce extra control bits into a message, slowing the rate at which actual useful data can be transferred. (Refer to the previous discussion on baud rate vs. bits-per-second.)

NUMBERING SYSTEMS AND CONVERSION FORMULAE[8]

Data must be represented by a standard format that is recognized by the various users. At the most basic level, data are usually transmitted over the network by a binary system (ones and zeros), even when an analog method is used. Further up the OSI model, numbers in other base systems can be more meaningful. An understanding of the more common numbering conventions and the conversions between them is helpful. A common notation for a number, NNNN, expressed in a base system other than base 10 (which is assumed, unless otherwise specified) is base x, which is written as $(NNNN)_X$.

Numbers in a given base are written by digits that represent multipliers of the powers in that place. That is, the decimal (base 10) number 5732 is really 5 thousands (5×10^3), plus 7 hundreds (7×10^2), plus 3 tens (3×10^1), plus 2 ones (2×10^0). Similarly, the binary (base 2) number 1011 is 1×2^3, plus 0×2^2, plus 1×2^1, plus 1×2^0.

To convert from any base to base 10 (decimal), add up the powers of the original base. For example, the previous example, 1011, can be converted to decimal in the following manner: 1011 binary = $1 \times 2^3 = 8$, plus $0 \times 2^2 = 0$, plus $1 \times 2^1 = 2$, plus $1 \times 2^0 = 1$, which totals 11 decimal.

To convert from base 10 to any base, divide the base 10 number by the desired base. Record the remainder, and repeat until the remainder is less than the new base. Use the final answer, and read the remainders back to the front. For example, the decimal number 476 can be converted to an octal number as follows: 476/8 = 59, with 4 remaining (record 4). 59/8 = 7, with 3 remaining (record 3, answer is 7). Reading backward gives 734 octal.

To convert from binary to octal or hex, split binary number into groups of 3 digits for octal (since $2^3 = 8$), or groups of 4 for hex (since $2^4 = 16$), then assign the corresponding octal or hex digit. For example, the binary number 11111 can be split into

(0)11 and 111, giving 3 and 7, or 37 octal

or

(00)01 and 1111, giving 1 and F, or 1F hex

To convert from octal or hex to binary, simply reverse the process given above. For example, the previous octal and hex numbers can be converted back as follows:

37 octal gives 11 and 111, giving 11111

and

1F hex gives 1 and 1111, giving 11111

Recall that: $X^0 = 1$ and $X^1 = X$

$$2^2 = 4;\ 2^3 = 8;\ 2^4 = 16;\ 2^5 = 32;\ 2^6 = 64;\ 2^7 = 128;$$
$$2^8 = 256;\ 2^9 = 512;\ 2^{10} = 1024;\ 2^{11} = 2048;$$
$$2^{12} = 4096\ (0\text{–}4095);\ 2^{16} = 65{,}536\ (0\text{–}65{,}535)$$

$$8^2 = 64;\ 8^3 = 512;\ 8^4 = 4096;\ 8^5 = 32{,}768;\ 8^6 = 262{,}144$$

$$16^2 = 256;\ 16^3 = 4096;\ 16^4 = 65{,}536;\ 16^5 = 1{,}048{,}576$$

Hint: The calculator included in Microsoft Windows has a very convenient numbering system conversion feature when placed in the "scientific mode." Table 4.2r is a quick cross-reference chart.

Most industrial controllers when reading and converting an analog value to a digital representation (which is required to store it in a memory register) use a 0–4095 range, to which the analog value is scaled. This comes from the fact that binary registers, which hold 1s or 0s (the default method of storing data) in its bit locations, have historically been able to hold 12 bits of data (along with other bookkeeping bits, like sign, etc.). These 12 bits can be, at a minimum, all 0s (0), and at a maximum, all 1s (4095). Newer, more powerful devices have gone to 16 data bits, resulting in the 65,535 capability, but the 0–4095 convention has persisted. Integer numbers are typically stored in a single register, while IEEE

TABLE 4.2r
Number System Quick Cross-Reference Chart

Binary	*BCD*	*Octal*	*Decimal*	*Hex*
000000	0000 0000	0	0	0
000001	0000 0001	1	1	1
000010	0000 0010	2	2	2
000011	0000 0011	3	3	3
000100	0000 0100	4	4	4
000101	0000 0101	5	5	5
000110	0000 0110	6	6	6
000111	0000 0111	7	7	7
001000	0000 1000	10	8	8
001001	0000 1001	11	9	9
001010	0001 0000	12	10	A
001011	0001 0001	13	11	B
001100	0001 0010	14	12	C
001101	0001 0011	15	13	D
001110	0001 0100	16	14	E
001111	0001 0101	17	15	F
010000	0001 0110	20	16	10
⋮	⋮	⋮	⋮	⋮
111111111111 (12 bits)		7777	4095	0FFF
1000000000000 (12 bits + 1)		10000	4096	1000
1111111111111111 (16 bits)		177777	65,535	FFFF
10000000000000000 (16 bits + 1)		200000	65,536	10000

floating point numbers take two (for a total of 32 bits). A method of representing data, especially characters, is using the ASCII (American Standard Code for Information Interchange) code. ASCII uses 8 bits (7 data bits, plus 1 parity bit) to code letters, numerals, punctuation, and special control characters. Compare this with the Modbus RTU representation that is commonly used by many manufacturers (more on Modbus ASCII and RTU later).

OPEN PROTOCOLS

The term *open protocol* literally refers to a clearly defined communication standard that is available for use by the general public. The implication is that it is not developed by a specific company, but a cooperative industry consortium, but in reality many protocol standards have been developed by companies and later released, then adopted by many manufacturers. In absolute terms, there are few truly open systems; rather, there are mostly degrees of openness. No one standard is wholeheartedly embraced by every manufacturer. There are, however, some standards that are common and pervasive throughout the industry. MODBUS, for example, was developed by the Modicon company for proprietary use, but the standard is freely distributed and has become an industry *de facto* "open" standard. The standard is, however, still controlled by Modicon. Systems using open protocols usually cost less to install and modify than those using proprietary standards.

Industry Open Standards

Because parallel communication is not prevalent in industrial network arenas, this discussion focuses on some of the more common serial standards. The EIA (Electronic Industries Association) has developed several "Recommended Standards" (RS-XXX) to aid in the ease of connection.

RS-232C Probably the most versatile and widely used physical protocol, this standard was developed mainly for the interface between DCEs and DTEs in public telephone network services. Serial ports on PCs are the most common

TABLE 4.2s
RS-232C Standard

Pin Number	Abbreviation	Direction	EIA Circuit	Function
1	EG	—	AA	Equipment ground
2	TD	To DCE	BA	Transmitted data
3	RD	To DTE	BB	Received data
4	RTS	To DCE	CA	Request to send
5	CTS	To DTE	CB	Clear to send
6	DSR	To DTE	CC	Data set ready
7	SG	—	AB	Signal ground
8	DCD	To DTE	CF	Data carrier detect
9	—	To DTE	—	+ DC test
10	—	To DTE	—	– DC test
11	N/A	N/A	N/A	Unassigned
12	SDCD	To DTE	SCF	2nd data carrier detect
13	SCTS	To DTE	SCB	2nd clear to send
14	STD	To DCE	SBA	2nd transmitted data
15	TC	To DTE	DB	Transmitter clock
16	SRD	To DTE	SBB	2nd received data
17	RC	To DTE	DD	Receiver clock
18	N/A	N/A	N/A	Unassigned
19	SRTS	To DCE	SCA	2nd request to send
20	DTR	To DCE	CD	Data terminal ready
21	SQ	To DTE	CG	Signal quality detect
22	RI	To DTE	CE	Ring indicator
23	—	To DCE	CH	Data rate selector
24	TC	To DCE	DA	External transmitter clock
25	N/A	N/A	N/A	Unassigned

application of RS-232C, but many industrial applications use it as well. Because RS-232C is an unbalanced system, the distance between devices is limited to roughly 15 m (45 ft, at 19.2 Kbps), which is its largest downfall, but the inherent flexibility makes it a popular choice for many manufacturers (compare to RS-422 and 485). Unbalanced systems modulate a voltage signal on one data line, with respect to a signal ground (Figure 4.2u). Even though it is possible to transmit data successfully at a slower rate for longer distances (maximum of 100 ft, or 30.3 m), the standard recommends that the cable be no longer than 50 ft (15.2 m). The EIA standard allows both synchronous and asynchronous transmissions. It also permits a range of data rates up to 19.2 Kbps, varying data lengths and bit codes (parity, stop bits, etc.), which are normally selectable in the port configuration. The various modes (simplex, half/full duplex) are also available. This flexibility, while convenient, allows for a wide variety in interpretations, and can lead to some incompatibility between manufacturers (which is contrary to the concept of a standard). RS-232C was originally designed to use a DB-25 connector (Table 4.2s), but in practice most of the handshaking and timing capabilities are rarely used, so a DB-9 connector (Figure 4.2t) can be substituted at the user's discretion. The RS-232C interface is usually fashioned with a female connector on the DCE end, and with a male connector on the DTE end. The most common use involves six data lines, along with two ground connections:

Protective Ground—May be connected to an outer screening shield conductor, and tied to the equipment chassis (only on one end to prevent ground loop problems)
TD—Serial "Transmitted Data," an output
RD—Serial "Received Data," an input
RTS—"Request to Send," an output
CTS—"Clear to Send," an input

(RTS and CTS manage speaking and listening functions in a half-duplex arrangement.)

DSR—"Data Set Ready," an input
DTR—"Data Terminal Ready," an output

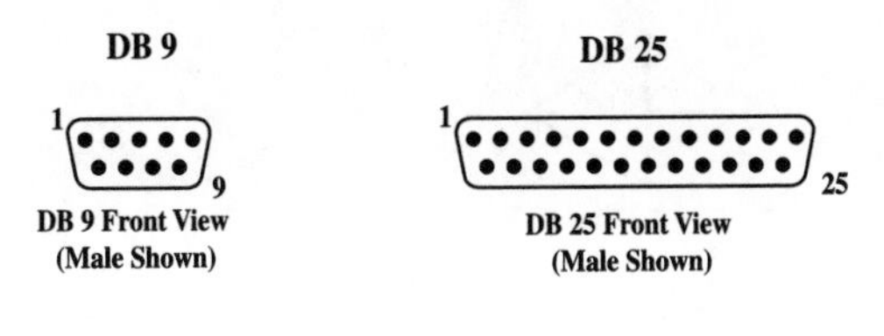

Common Usage

DB 9		DB 25	
1	Chassis Ground	1	Chassis Ground
2	Transmit Data	2	Transmit Data
3	Receive Data	3	Receive Data
4	Data Terminal Ready	4	Request to Send
5	Signal Ground	5	Clear to Send
6	Data Set Ready	6	Data Set Ready
7	Request to Send	7	Signal Ground
8	Clear to Send	8	Carrier Detect
9	Not Used	20	Data Terminal Ready

Typical Null Modem

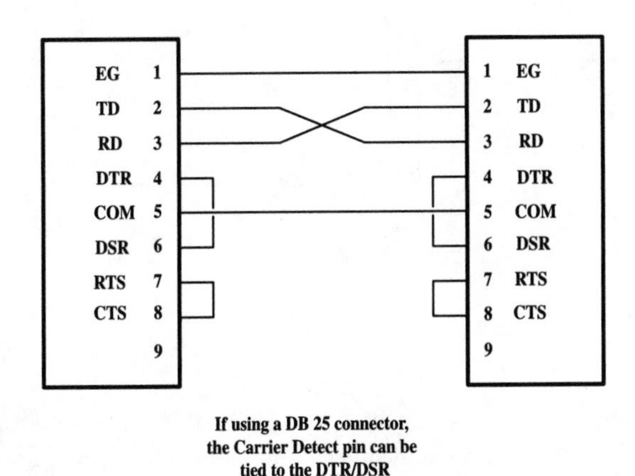

FIG. 4.2t
Diagram of RS-232C.

(DSR and DTR give an indication that the device is powered and ready to communicate.)

Signal Ground—Common signal reference; should not normally be connected to chassis ground

Signal levels that differentiate between a 1 or a 0 assume two, bipolar voltages. Standard circuitry uses a range of from +3 to +25 V DC to represent a 0, and a range of from −3 to −25 V DC to represent a 1. These circuits are normally powered by 12-V power supplies, so this is what is normally seen in field settings. TTL components typically use 0 V DC to designate a 0, and +5 V to designate a 1. Obviously, it is important not to intermingle the two, or serious damage to equipment ports could result. The relatively large voltage levels used help make the RS-232C link more resistant to EMI/RFI noise, but with a common signal ground there is no double-ended signaling (unbalanced), which allows common-mode noise to exist. Open circuit voltage (relative to the signal reference) should never exceed 25 V DC, and short-circuit current must not exceed 500 mA. The loading resistance should be between 3K and 7K Ω, with a shunt capacitance less than 2.5 nF.[3]

A null modem is used to connect two DTEs together, and the transmit and receive lines must be crossed to allow for proper function. (Refer to the crossover cable mentioned later in the Ethernet discussion.) If the RTS and CTS, and the DTR and DSR pins are jumped as shown, the interface operates as strictly data transfer, with no handshaking.

RS-422 A balanced system, RS-422 makes possible longer distances and faster transmission rates than RS-232C. (RS-423 is simply an unbalanced version of RS-422.) RS-422 can have multiple receivers, but only one line driver per twisted pair of wires. To facilitate full-duplex operation, two separate channels are used that would be two completely separate RS-422 links except for the fact that they reside on the same devices. Because it is a balanced, or differentiated system, it is very resistant to EMI/RFI. A balanced system uses two data lines and modulates a signal voltage between the two (compare to an unbalanced system; see Figure 4.2u). It is capable of 10 Mbps at distances of 4000 ft (1219 m). The RS-422 standard recommends a 24 AWG (American wire gauge) twisted pair cable with a 100 Ω characteristic impedance, and a 16 pF/ft shunt capacitance. Category 5 cables, used in Ethernet, meet these requirements and are widely available, making them a good choice for RS-422 installations. In keeping with the practice of terminating the cable with the equivalent of the characteristic impedance, a 100 Ω resister should be connected across the "A" and "B" lines.[9]

RS-485 This is a serial interface standard, which (like RS-422) is a balanced system, except that it uses a tri-state line driver (Figure 4.2v). The third, high-impedance off-state, allows an inactive device to sit quietly on the network, making multiple drops (up to 32 bidirectional line drivers) easier to manifest. When a device is not transmitting, its line driver must go into the high impedance state, so as not to interfere with any other transmission. It also permits even faster rates over longer distances—100 Kbps at up to 4000 ft (1219 m). It can operate on one pair of twisted pair in half-duplex mode, or on two pairs in full-duplex mode. The full-duplex mode is limited, with only a master/slave arrangement is allowed. Unlike 422, the "enable" function must be used. Exactly how this is accomplished varies between manufacturers. The terminals on which the transmission lines land are referred to as "A" and "B" (sometimes "−" and "+," respectively), and will be labeled "T(A)" and "T(B)" for the transmission port, and "R(A)" and "R(B)" for the receiving port.[9]

Notice that the two-wire/four-wire moniker is somewhat misleading, as a ground wire is also required. This would appear to blur the distinction between balanced and unbalanced

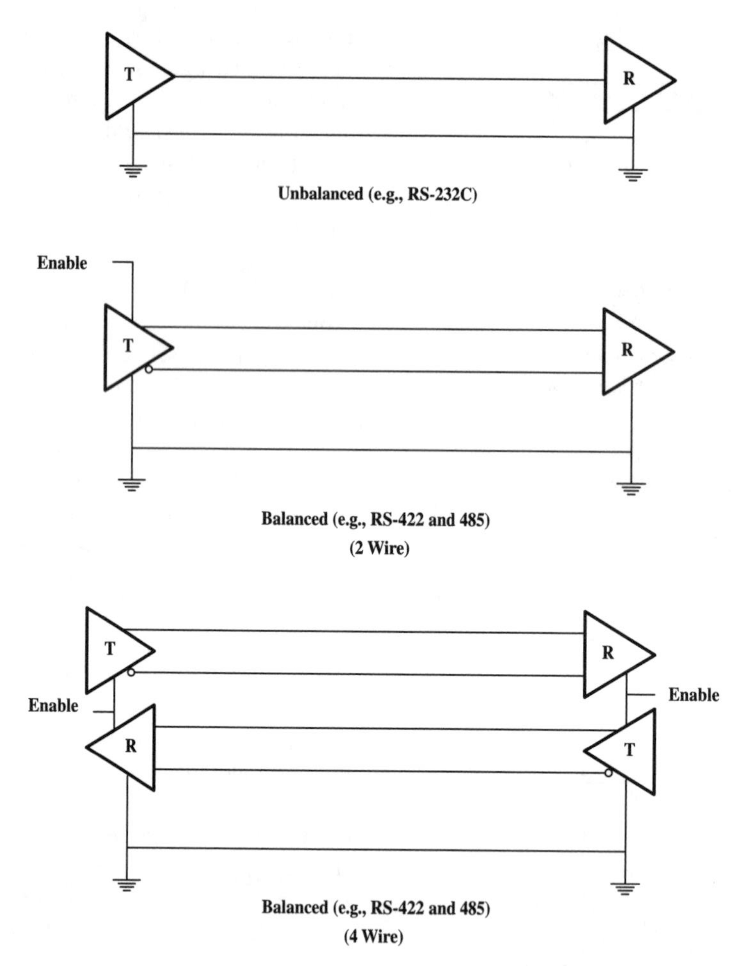

FIG. 4.2u
Diagram of balanced vs. unbalanced.

systems, except that the signals are not referenced to the ground in a balanced system. The two-wire system is less expensive to implement than the four-wire version. Even though the 485 standard does not specify a cable, the same twisted pair wiring used in 422 applications can be used to implement 485 as well. The two-wire system is slower because bidirectional transmissions occur over the same channel (typical of any full-duplex vs. half-duplex comparison), and the line driver must be put into the high impedance state when it is finished before any other driver can begin.

Ethernet

Ethernet is the most popular network in use today. It is a mature, highly developed open protocol mainly found in corporate settings, but is rapidly finding acceptance in the control arena. It was developed in the 1970s by Xerox, but hardware is widely available from a multitude of vendors. It is very similar to IEEE 802.3, but is technically not the same, although the term *Ethernet* is loosely used to describe both. Both are serial, broadcast-type networks, with the main difference that 802.3 specifies several different physical layers, whereas Ethernet defines only one.

Originally, Ethernet was meant for the office environment, so connectors and other equipment seldom met common industrial standards. This is changing as traditional manufacturers of industrial networking equipment develop hardened components that can withstand harsh conditions. Because it uses the CSMA/CD media access method, earlier implementations of Ethernet had the added disadvantage of being rather indeterminant, which was not a problem in most office settings, but is not tolerable in most control applications. This problem was often the result of poorly designed network arrangements. Office network traffic tends to be bursty, which minimizes problems with message collisions, coupled with the fact that these messages are not typically time sensitive. This is the exact opposite of most control strategies, which exchange data fairly constantly and require that values be refreshed frequently. Adding more devices only served to increase traffic, causing more collisions. For these reasons, Ethernet was deemed inappropriate for industrial control use.

The advent of switches and routers has alleviated most of these problems by separating collision domains, minimizing collisions and facilitating more timely message delivery. For all of these reasons, Ethernet is poised to become one of

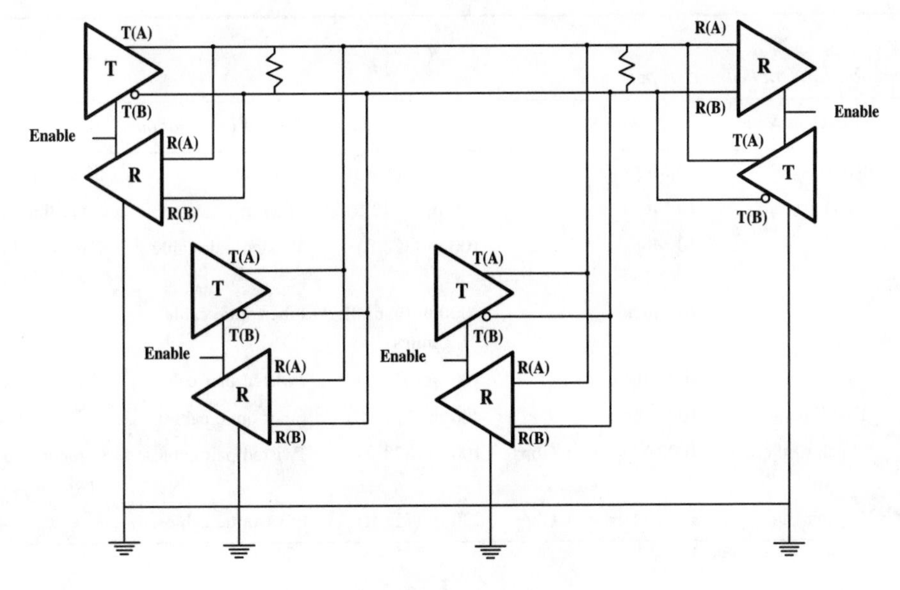

2-Wire Multidrop Network

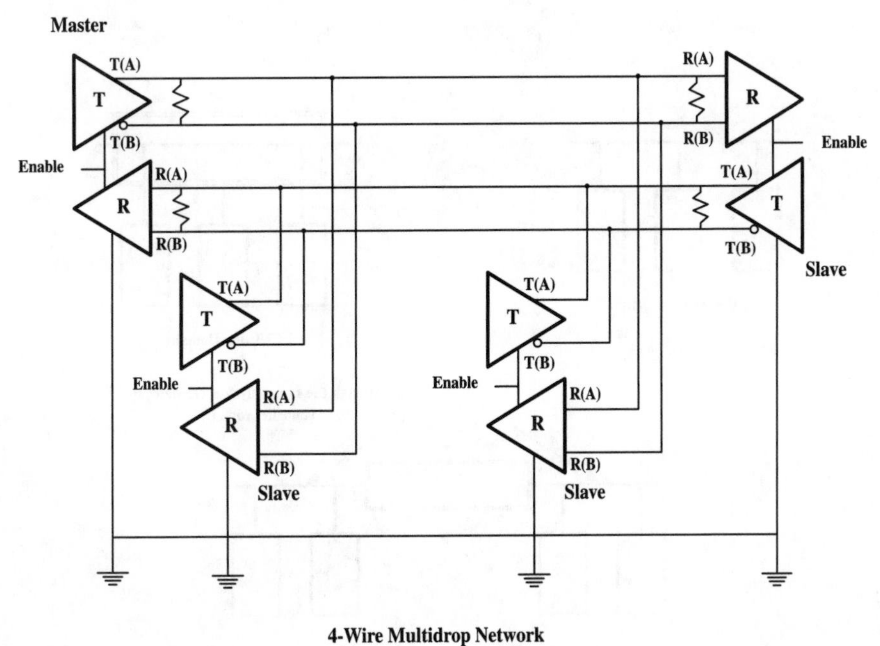

4-Wire Multidrop Network

FIG. 4.2v
Diagram of RS-485 networks.

the most popular industrial networking protocols, as most major manufacturers are introducing industrially hardened Ethernet interfaces, switches, and connectors, as well as industrial control message protocols (Modbus/TCP, Ethernet/IP, Profinet, Fieldbus HSE, to name a few).

Ethernet utilizes a baseband signal. The most common configuration is over twisted pair wires in either a bus or star topology, although optical fiber is becoming more prevalent. A notation for describing Ethernet signal types is N <Signal-Type> X, where N represents the signal rate in megabits per second (Mbps), the signal type is either baseband or broadband, and the X represents a special characteristic, such as "5" meaning the maximum distance is 500 m, or "T" meaning the signal is transmitted over twisted pair, or "F" meaning it is transmitted over optical fiber. A designation using "Base" as the signal type (e.g., 10Base2) does not refer to the base 2 numbering system that was described earlier[1,10] (Table 4.2w).

The most common protocol to ride on the Ethernet network is TCP/IP, and this is the route that the new industrially hardened Ethernet implementations are following. TCP (transmission control protocol) defines the source port and destination port numbers that allow data to be sent back and forth to the correct application running on each device. The TCP portion of the message also includes error checking and other housekeeping information. It operates on the transport layer of the OSI model. IP (Internet protocol) is responsible

TABLE 4.2w
Table of Common Ethernet Signal Types

Signal Name	*Common Name*	*Data Rate*	*Max. Distance*	*Special Designation*	*Media*
10Base2	Thinnet	10 Mbps	185 m (606.8 ft)	~200 m	50 Ω thin (RG-58) coax
10Base5	Thicknet	10 Mbps	500 m (1640 ft)	~500 m	50 Ω thick (RG-8) coax
10BaseT	—	10 Mbps	100 m (328 ft)	Twisted pair cable	Category 3 or 4 twisted pairs
10BaseF	—	10 Mbps	2000 m (6561 ft; 1.2 miles)	Fiber-optic cable	—
100BaseT	Fast Ethernet	100 Mbps	100 m (328 ft)	Twisted pair cable	Category 5 twisted pairs
100BaseF	Fast Ethernet	100 Mbps	400 m (1312 ft)	Fiber-optic cable	—
1000BaseT	Gigabit Ethernet	1000 Mbps or 1 Gbps	100 m (328 ft)	Twisted pair cable	Category 5e or 6 twisted pairs
1000BaseF	Gigabit Ethernet	1000 Mbps or 1 Gbps	220 m (722 ft)	Fiber-optic cable	—

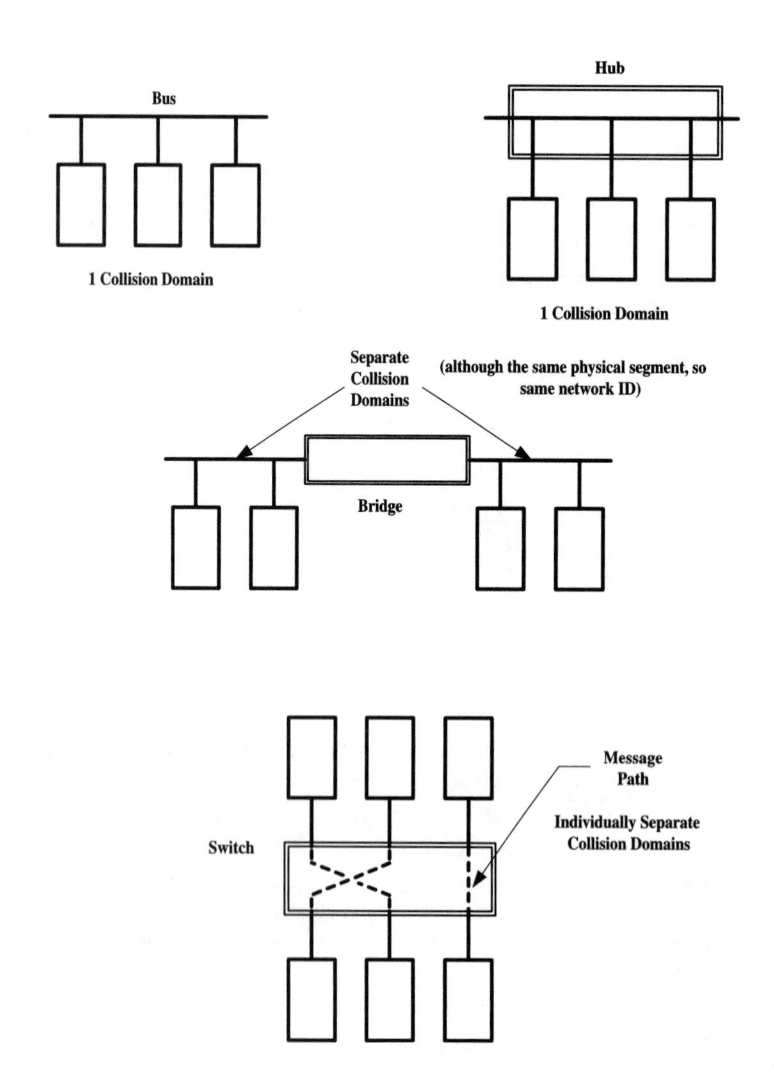

FIG. 4.2x
Collision domains.

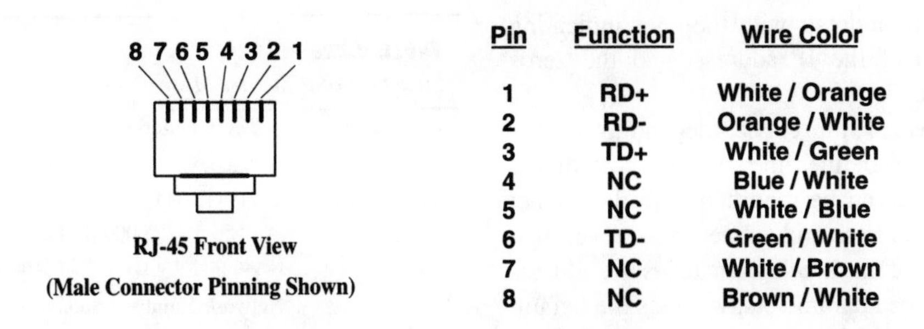

(Even though only two pairs are shown used, the standard requires all pairs be terminated.)

(The 568B standard is shown, which is common in data applications. The 568A standard is mostly used in telecom applications, and has the orange and green pairs swapped.)

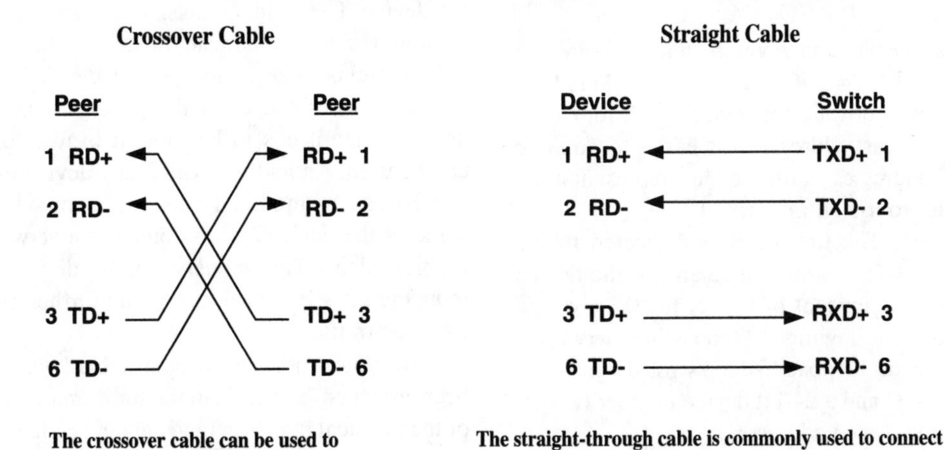

The crossover cable can be used to connect peer devices directly (useful in troubleshooting).

The straight-through cable is commonly used to connect devices to network infrastructure equipment (whose RD and TD ports are crossed internally).

FIG. 4.2y
Ethernet wiring and pinout.

TABLE 4.2z
TCP/IP Addressing

Network Number | Device Number (General Structure)
XXX . XXX . XXX . XXX (Decimal Digit Representation)
XXXXXXXX . XXXXXXXX . XXXXXXXX . XXXXXXXX (Binary Digit Representation)

for inserting routing information into the header of each message on the network layer. It is said to be "connectionless" because it does not concern itself with any handshaking practices. When IP does not know the path to a destination, it forwards the packet to the default gateway of the host. The default gateway is then responsible for routing the packet to its destination. That default gateway might, in turn, pass the packet to its own default gateway if it does not know the path.

The IP address of a device can be statically or dynamically assigned. A statically assigned address is one that is manually typed into the configuration of the device, and does not change. A dynamically assigned address is one that is assigned by a network master when the device boots up and joins the network environment, and is generally different every time. Dynamically assigned addresses make large corporate networks easier to manage, but are not common in control networks.[1,11]

An IP address (Table 4.2z) consists of 32 bits, and is split into four groups of 8 bits (called octets), separated by periods whenever it is written or typed in, to make it easier for people to deal with. The first part gives the network ID number, and the second part gives the particular device ID number, which must be unique within that network segment. All devices on a particular physical segment share a common network ID. Which bits correspond with which part depend on the class of address and the subnet mask. The subnet mask has a structure similar to the IP address, with the ones indicating the network ID portion of the IP address, and the zeros indicating the device number.

IP addresses are divided into classes depending on the type of service. The class defines the allowed size of the network. The defined size can be modified with the subnet mask. Generally, the IP address and subnet mask (even if it is simply the default subnet mask, which matches the address class definition) are required parameters, whereas the default gateway is optional. If a device will ever connect to the global Internet, the address must be assigned a public address by a governing body known as ARIN (American Registry of Internet Numbers). This will assure that the address is unique to all others on the Internet. If the device will remain local to a facility (most common for control networks), the address may be assigned a private address by a designated network administrator.[12]

Each octet contains 8 bits, and therefore, may be addressed from 0 to 255, decimal ($2^8 = 256$ or $0 \rightarrow 255$, $11111111 = 255$), although some are not valid for devices (255, for example, is a reserved broadcast address). It is easier to *examine* the properties of IP addresses in their binary representation, although much easier to *assign* in decimal.

Class A Addresses: The first octet is dedicated to the network address, and the remaining three are for the device ID. The HSB (highest significant bit) is set to "0," giving a 7-bit network number (allowing 127 possible networks, although 127 is a special loopback address for diagnostics, so 126 usable addresses) and a 24-bit device number (allowing 16,777,214 unique device addresses)

NW . DV . DV . DV

001.XXX.XXX.XXX → 126.XXX.XXX.XXX

Class B Addresses: The first two octets are dedicated to the network ID, and the remaining two to the device ID. The highest two significant bits are set to "10," giving a 14-bit network number and a 16-bit device number. This gives 16,382 networks, with 65,534 devices.

NW . NW . DV . DV

128.000.XXX.XXX → 191.255.XXX.XXX

Class C Addresses: The first three octets are dedicated to the network number, and the remaining one is for the device ID. The three highest significant bits are set to "110," giving a 21-bit network number, and an 8-bit device number. This gives 2,097,152 networks, with 254 devices.

NW . NW . NW . DV

192.000.001.XXX → 223.255.254.XXX

Classes D and E are defined but have special restricted uses, and therefore are not valid choices for control networks.

TABLE 4.2aa
Structure of a Subnet Mask

IP Address:	Network Number | Device Number (general structure)
Subnet Mask:	11111111.11111111.11111111.00000000 or 255.255.255.000 decimal (example of a class "C" default subnet mask)
	Network Number | Subnet Number | Device Number (structure of an I/P address with a custom subnet mask)

The subnet mask is used to interpret the IP address. The subnet mask works somewhat like a template that when superimposed on top of the IP address indicates which bits in the address identify the network and which bits identify the device. The 1s in the mask designate the network address portion, and the 0s indicate the device address portion. If the assigned class size is acceptable, the default subnet mask, which matches the class designation, can be used. If the network size needs to be modified, a custom subnet mask can be used. An address with many device bits (e.g., class A and B) can be modified with a custom subnet mask to use some of the device address bits as subnetwork designations (Table 4.2aa). The subnets can be divided with a router, reducing the size of a domain, and further reducing network message traffic.[12]

Every network interface has a hardware address that has been assigned by the manufacturer and stored in the ROM of the physical interface (and cannot be altered). This address is different from the IP address and is routinely called the MAC address because it is meaningful on the MAC sublayer of the OSI data link layer.

Some utilities are included in TCP/IP that can be used to troubleshoot problems on an Ethernet network.

The "*PING*" command (C:\ping XXX.XXX.XXX.XXX, or Computer Name) is an echo request that can test connectivity between points (physical as well as functional). It is executed from the DOS command line, and will return one of the following:

"Destination Host Unreachable" is an indication that an unreasonable request has been made, due to either the NIC not operating, or some addressing mistake.

"Request Timeout" means that an acceptable request has been made, but that the target device is not responding.

"Reply from XXX.XXX.XXX.XXX," along with timing information, is a successful response from the other device.

An interface device can "ping" itself. This is useful to determine if the interface is functioning properly.

The computer name and/or the IP address may be obtained from "Control Panel/Network" configuration information, or by running the "IPConfig" command at the DOS prompt.

TABLE 4.2bb
MODBUS Standard Register Designation

Data Type	Register Designation
Discrete Output	0xxxx(x)
Discrete Input	1xxxx(x)
Analog Input	3xxxx(x)
Holding Register	4xxxx(x)
Extended Memory (not universal)	6xxxx(x)

MODBUS[13]

Another type of open protocol standard, which was originally developed by a private company, is MODBUS. It was introduced in 1979 as one of the first industrial network standards. It has been so widely accepted that it is a *de facto* industry standard, even though the standard is still controlled by Modicon. It was developed by Modicon as a structure for data representation, and is independent of the physical layer. It can be implemented over any transmission medium, but is most commonly seen used with RS-422, RS-485, and most often with RS-232. It is a serial transmission technique that uses master/slave arbitration. A master can communicate half-duplex style with up to 247 slaves. It can make broadcasts to all slave devices, but in that case, slaves are not allowed to send a response.

MODBUS can operate in two modes: ASCII and RTU. The mode is selected along with the other serial port communication parameters (baud rate, parity, etc.). It simply defines how data are packed into the message frame (Table 4.2bb). In ASCII, 8 bits of data are sent as two ASCII characters. This allows for more reliable transmission, with gaps of up to one full second to elapse without causing errors. RTU mode packs two 4-bit hexadecimal characters into the message data area. RTU offers greater data density, and therefore higher throughput for a given baud rate, so it should be used whenever possible (which is nearly every case). The MODBUS packet is included in the data area of a standard message packet. A newer version, MODBUS/TCP, has been issued to ride on the Ethernet specification. The MODBUS message packet is inserted into the data area of the Ethernet message packet.

MODBUS can typically transfer 300 registers per second at 9.6 Kb, with a maximum speed of 19.2 Kb. Its speed is not as fast as most other industrial standards, but its simplicity and availability make it widely popular.

PROPRIETARY PROTOCOLS

Proprietary protocols are those developed and maintained by a particular company. Equipment is typically limited to a few manufacturers. The advantages of a proprietary protocol have mostly to do with strict compatibility between components, giving ease of use and increased performance. The disadvantages are that the user is restricted to the offerings of a few vendors, with limited flexibility and greater general cost.

The following discussion is intended to be representative of typical control networks, and is not intended to encompass all available varieties.

MODBUS Plus[14]

MODBUS Plus is an RS-485 based, peer-to-peer network protocol, which uses the MODBUS data structure. It transmits at a rate of 1 mbaud. It allows up to 32 nodes on a segment of up to 1500 ft (455 m). A repeater allows an additional segment to be attached for a maximum of 64 nodes, over 3000 ft (909 m). Another repeater may be used to increase the maximum distance to 6000 ft (1818.2 m), but the 64-node limit remains. This distance restriction is for the recommended Belden 9841, shielded twisted pair cable, but the maximum distance using optical fiber is several miles, depending on network and repeater configuration. MODBUS Plus uses the bus physical topology, with a token passing arbitration method. It is capable of transferring 20,000 registers per second. Devices on a MB+ network are given network addresses, and the device with the lowest address assumes the role of network master. After a node uses its allotted time, it releases the token and passes it to the next higher addressed device. Multiple network segments can be joined together, and MODBUS Plus allows routing up to 5 bytes. By using appropriate modules with dual ports, the network can use redundant cabling.

The old style of connectors went directly on the trunk cable in a daisy-chain fashion (the light gray casings were the terminators, and the dark gray casings were the in-line connectors). A new style has been introduced with taps and drop cables. These make for a neater installation, and the taps can be configured as in-line or terminating with jumpers. It is very important to adhere to the published guidelines that require a minimum of 10 ft (3 m) between taps to prevent reflected and corrupted signals. Both the old and new versions incorporate a 120 Ω terminating resister on both ends. The two network conductors in the twisted pair are connected to pins 2 and 3 of a connector, with pin 1 tied to the drain wire.

A properly assembled cable can be checked, and should have the following characteristics (with all connectors disconnected from the devices). There should be an open circuit between pin 2 or 3 and pin 1, which is the shield. If the impedance is checked between pins 2 and 3, there should be between 60 and 80 Ω. This accounts for the two terminating resistors (in parallel, so 120/2), plus cable impedance (typically 1 Ω/100 ft, or 30 m). With the terminators removed, there should be an open circuit between signal conductors.

Data Highway Plus[15]

The Allen-Bradley/Rockwell Automation company has developed a similar proprietary network communication system that has become so common and popular it is pseudo-open. As in previous examples, DH+ is proprietary, but many manufacturers produce compatible equipment that fills a variety of requirements. It is token passing, operates at 57.6 Kbps, and also uses a pair of wires (twin axial, in this case). DH+ specifies Belden 9463, which is commonly known as "Blue Hose." The network can be arranged in a daisy-chain or bus (trunk/drop) configuration, which uses taps with BNC connectors. There are 64 nodes allowed, and the trunk length depends on the configuration, but the maximum is 10,000 ft (3050 m). The trunk should be terminated with a 150 Ω resistor at each end. The cable may be checked as before, making certain that there is no short between either conductor and the drain wire, and there is appropriate impedance between the two signal wires. In this case, between 75 and 125 Ω (accounting for the two terminating resistors, in parallel, so 150/2), and the cable (about 1 Ω/100 ft, or 30 m). If the terminators are removed, there should be an open circuit between the signal conductors.

Like MODBUS Plus, Data Highway Plus uses a mirror image of the binary number to address a node. The difference is that "0" is a valid address in a DH+ network, so there is no offset of "1."

Other Control Networks

Of course, there are many other network protocols in use by the many systems that are available in the marketplace. Most of them are similar in nature to the ones discussed here, and are built on the same basic principles. Some PLC (programmable logic controllers) manufacturers have adopted some of the newer fieldbus standards that have emerged in recent years. These are discussed in detail in later sections of this chapter.

GENERAL TROUBLESHOOTING

As seen in the preceding paragraphs, communication network systems are amazingly complex, with rigorous specifications and exacting requirements. One can easily imagine problems occurring, but most of these systems have evolved over many years in many installations, and will perform satisfactorily if installed and maintained properly. Problems still do arise, both during initial installation and during operation. The following discussion offers advice on how to avoid or correct these problems.

Preventing Problems

During the discussion of the various standards, protocols, cables, and connectors, many specifications and parameters are mentioned. These qualities must be maintained to ensure continued performance. Careful attention to installation instructions and general guidelines will go a long way to minimizing troubleshooting later.

1. When installing cables, maintain structural integrity by not pinching, kinking, stretching, or abrading it.
2. When terminating conductors and connectors, manufacturers' guidelines offer criteria for preserving the ratings for speed and noise resistance. For example, untwisting too much of a twisted pair cable at the termination will degrade the performance of the whole system. A Category 5 cable must have no more than 0.5 in. (1.27 cm) untwisted at the end, or it will no longer meet Category 5 requirements, and may not support a 100 Mbps signal. Many topologies have segment length requirements (both maximum and minimum) that are meant to reduce signal reflections and corruption. (Most problems this author has encountered have come from a single strand of wire hanging out of its allotted terminal and touching another or shorting to ground.)
3. Be sure segments are terminated properly (with terminating resistors or capacitors), according to instructions, to prevent reflections and corrupted data.
4. Keep signal wires away from sources of EMI and RFI. If communication cables must be run near power wiring, be sure to cross at a 90° angle to reduce induced voltages.
5. When installing interface cards and other network equipment, be sure that all hardware and software configuration settings are correct and consistent (speed, parity, etc.). Be sure that settings for cards installed into PCs do not interfere with other devices. NICs require configuration, including hardware interrupt (IRQ) and memory locations, which if in conflict with another device will cause malfunctions. IRQs are requests from system hardware or software for attention from the CPU. Check with PC documentation to ensure correct settings. Conflicts will cause the card not to start, and may cause other abnormal behavior. Table 4.2cc gives typical values.[16] IRQ 3 (Com 2 may have to be disabled) or 5 is usually a good choice; IRQ 7, which is intended for the printer port, is not generally used for most printing applications, so it may be used without conflict.
6. Do not ask too much from the system. Attempting to transfer an excess amount of data will cause problems.
7. Be sure all components are powered with proper voltage, and grounded per manufacturers' specifications.

Diagnosing Problems

The first rule of troubleshooting is to check the obvious first:

1. Check to be sure that all connections are made securely.
2. Make sure all components are seeing the network by observing LEDs. Most interfaces will have status lights that will flash a particular code to indicate errors.

TABLE 4.2cc
Common Settings for ATPCs

Device	I/O Memory Address (within the 0–3FF memory range) (in hex)	Hardware Interrupt
Com 1	3F8–3FF	IRQ 4
Com 2	2F8–2FF	IRQ 3
Com 3	3E8–3EF	IRQ 4
Com 4	2E8–2EF	IRQ 3
LPT 1 (parallel port)	378–37F	IRQ 7
LPT 2	278–27F	IRQ 5
Floppy disk controller	3F0–3F7	IRQ 6
Hard disk controller	1F0–1F7	IRQ 14

3. Make sure communication is enabled or being initiated by some node.

If devices are still not able to establish a link, more aggressive methods are in order:

4. Completely disassemble the network, and rebuild it one piece at a time (segment, device, etc.). Try to establish communication between only two nodes and then add slowly.
5. Many manufacturers will supply diagnostic software with equipment or on Web sites (see the list following the bibliography to this section). These may be used to monitor COM port activity and diagnose potential errors. Some software works in conjunction with a loopback plug, which simply connects the transmit and receive pins (along with the control pins, like a null modem) to send and receive test messages to itself. More drastic measures would involve specialized test equipment that can put a network through extensive tests (protocol analyzers, cable testers, etc.). An inexpensive piece of equipment that is a "must have" in any technician's tool bag is a breakout box. This small device comes with a variety of standard connections on either side, and a miniature patch panel in the middle (inside), with jumpers that can be arranged to match any configuration. Some breakout boxes come with status LEDs that show activity on some or all of the lines. This, with a few common adapters, will prove invaluable.
6. Optical fiber needs to be tested by a trained and competent technician when installed, but the correct fiber can be verified by shining a flashlight into one end, and observing the tiny spot of light on the other.

Important Safety Note: Even though the lasers and LEDs used in fiber-optic communications operate at a much lower power level than the ones ordinarily used in industrial or medical services, they can be of sufficient power to damage sensitive eye tissues. **Never** look directly into the end of a fiber or port unless it is absolutely certain that there is no power on the circuit. The wavelength of light used is invisible to the unaided eye, and may not be detected until too late.

7. Check each connector for continuity (with all connectors unplugged from devices). Remember that the resistance read should be that of the equivalent circuit of the cable, including cable resistance (typical value is 1 Ω/100 ft, or 30 m), as well as that of any terminating resistors (in parallel, so R/2). With the terminators removed, there should be an open circuit between signal conductors.
8. Do not forget to use the resources available. Most manufacturers have very helpful technical departments. Again, much useful information can be obtained from the Web sites listed following the bibliography.

References

1. Groth, D., *Network + Study Guide,* 2nd ed., Alameda, CA: Sybex, 2001.
2. "Guide to Internetworking," Lawrence, PA: Black Box Corp., 1995.
3. Stone, H. S., *Microcomputer Interfacing,* Reading, MA: Addison-Wesley, 1983.
4. "Characteristic Impedance of Cables at High and Low Frequencies," Richmond, IN: Belden-Wire.com, 2001.
5. "Universal Transport System Design Guide," Hickory, NC: Siecor, 1995.
6. "Guide to Cable Properties," Lawrence, PA: Black Box Corp., 1998.
7. Miklovic, D. T., *Real-Time Control Networks,* Research Triangle Park, NC: Instrument Society of America, 1993.
8. Manno, M. M., *Digital Design,* Englewood Cliffs, NJ: Prentice-Hall, 1984.
9. "RS-422/485 Application Note," Ottawa, IL: B&B Electronics, 1997.
10. Byres, E., "Shoot-Out at the Ethernet Corral," *InTech,* February 2001.
11. "Catalyst 3500 Series Hardware Installation Guide," San Jose, CA: Cisco Systems, 2000.
12. Botsford, C., learntosubnet.com[online], Kirkland, WA.
13. "MODBUS Protocol Reference Guide," North Andover, MA: Modicon, June 1996.
14. "MODBUS Plus Network Planning and Installation Guide," North Andover, MA: Modicon, April 1996.

15. "Data Highway, Data Highway Plus Cable Installation Manual," Milwaukee, WI: Allen-Bradley, April 1994.
16. Glover & Young, *Pocket PCRef,* Littleton, CO: Sequoia, 1997.

Bibliography

Anixter, "Ethernet Switching White Paper," Skokie, IL, 1996.

Black Box Corp., "Black Box Pocket Glossary of Data Communication Terms," Lawrence, PA, 1994.

Black Box Corp., "Guide to Premise Cabling," Lawrence, PA, 1997.

Butt, F. M., "Open Systems Networks: Understand and Plan," *InTech,* July 1994.

Herb, S. M., "Networks for Factory Automation," in *Understanding Distributed Processor Systems for Control,* Research Triangle Park, NC: Instrument Society of America, 2000.

Palmer-Stevens, D., "Guide to Local Area Networking," Rochester, NH: Cabletron Systems, 1992.

Peterson, C. P., "Open Networking: From the Sensor to the Boardroom," Research Triangle Park, NC: Instrument Society of America, 1999.

Pieronek, D., "Open Protocol or Open System Architecture?" Research Triangle Park, NC: Instrument Society of America, 1997.

White, H. P., TCP/IP makes the "Factory Floor to Enterprise" Connection, *Control Solutions,* July 1999.

Web Sites for Reference

ab.com
anixter.com
bb-elec.com (B&B Electronics)
belden-wire.com
blackbox.com
cableu.net
isa.org
learntosubnet.com
modbus.org
modicon.com

4.3 Hardware Selection for Fieldbus Systems

I. VERHAPPEN

Just as a change in thinking was required during the conversion from pneumatic to analog, some of the thought processes associated with selection of sensors for fieldbus systems will also need to adapt. Fortunately, because of the processing power of today's field devices, these changes are minor.

This section discusses some of the changes and, more importantly, similarities between fieldbus systems and the technologies with which we are already familiar.

FIELDBUS DEVICES

At present, fieldbus devices are not significantly different from the technology that has been in use for the past decade. Fieldbus simply adds digital communications to existing sensors.

Similarities to Conventional Devices

Fieldbus represents a step change in process control communications technology. The basic sensor technology from which the process measurements are made remains unchanged. This is only logical because the physical properties being measured have not changed: vortex meters still measure "swirls"; Coriolis meters measure momentum; pressure meters, capacitance, or in the case of digital meters, frequency; and so on. What has changed is the interface between the sensor/transducer and the way this information is shared with the outside world.

Most manufacturers have in fact kept the form factors and dimensions of their devices *exactly* the same as previous "smart" transmitters. This makes sense, not only for their manufacturing processes, which they will not have to retool, but also for their customers who will have to stock only minimal additional spare parts. Dimensional similarities include the device housing as well as the process connection, again reaffirming how the changes between a fieldbus device and a conventional version of the same product are only in the communications technology.

The ISA has some work under way to standardize process connection dimensions as well, so devices will not only be interoperable but truly interchangeable as well.

Because the only change to a fieldbus device is its communications, it will "work" the same as any other similar device already installed. This means the technicians do not have to learn an entire new set of skills to maintain these new transmitters because the majority of their knowledge on how to install, maintain, and repair them will be the same as it is at present. Impulse lines are installed and "blown down" just like always, upstream and downstream lengths for flow measurement installations are unchanged, magnetic flow meters still need conductive process solutions, and analyzers still need sample systems.

Differences from Conventional Devices

Fieldbus changes the way we interface with field devices and the host control system. This is because with fieldbus the communications among the field, host, and human are now digital and network based. Being digital provides a number of key differences and enhancements to the capabilities and to the amount of information that can be shared between devices, the devices and the host, and of course with the human operator.

All digital communications as represented by fieldbus networks provide several advantages vs. both traditional analog loops as well as HART protocol devices. Because the entire system uses the same protocol, there is only one analog-to-digital conversion in the loop. This reduces the potential loop error in high accuracy systems that can result. It also prevents the computational delays associated with each conversion.

Digital networks as represented by fieldbus also mean that the additional information transmitted over the fieldbus network does not have to be processed in a separate system; it is integral with the host. Being fully digital also allows the system to transmit much more information than can be included in an analog or combination protocol.

The most important thing digital networks provide is the ability to connect multiple devices on a single wire, thus creating the network in the first place.

The biggest difference between Foundation Fieldbus and all other communications protocols is the device description and the device's three types of blocks. Each device contains a minimum of three blocks, which are shown in Figure 4.3a and briefly described below.

Function block—These blocks describe the "functions" or things that can be done by the device. Typical

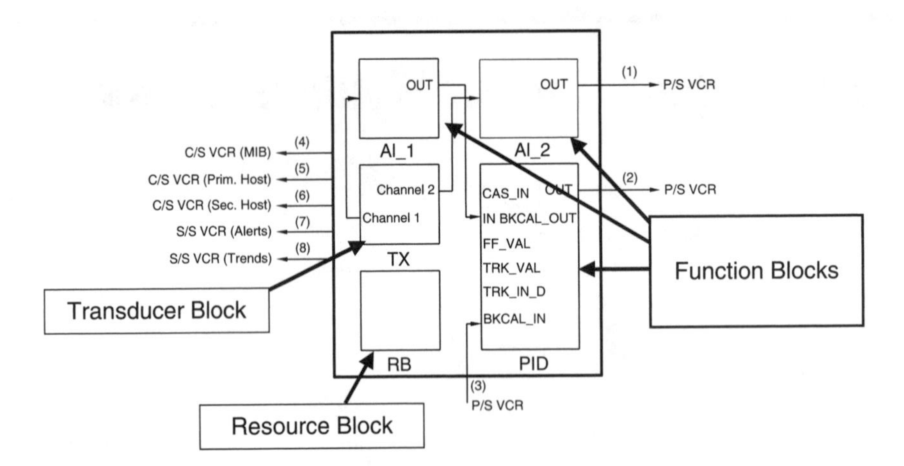

FIG. 4.3a
Function blocks in a typical device.

functions include the analog input and output blocks, PID block, and control algorithm blocks. At present, 29 different function block types have been identified with 21 defined.

Resource block—Each device has only one resource block, which provides the general information related to all the other blocks included in the device.

Transducer block—Converts the sensor signal to a form that is usable by the network.

Function blocks and their use will be described further later in this section.

Device description is the unique feature of Foundation Fieldbus *and* the key component that makes it possible to define how a device operates as well as what features it contains. Device description also enables the fieldbus "plug-and-play" and software configuration capabilities. The capabilities file is a text file that allows for offline configuration of fieldbus devices by "describing" the device description file for the host configuration computer so that the device can be configured offline prior to its installation on the network.

Fieldbus devices connect to the network in parallel and to be truly specification compatible are therefore polarity insensitive. This means that the signal terminals in fieldbus devices differ from those in traditional transmitters. This is often difficult to see from visual inspection as most manufacturers realize that because these devices will be connected to conventional cabling there must be enough "connections" for each wire in the cable. What does happen is that some of the terminals do not have anything connected to their reverse side and thus serve as wire holders only.

The other obvious but not yet mentioned difference is the device communications card itself. Once again, this will be difficult to observe visually because one piece of silicon looks much like another.

Figures 4.3b and 4.3c show how an all-digital communications standard makes it possible to see further into the enterprise while also providing more information critical to higher reliability of the unit operations.

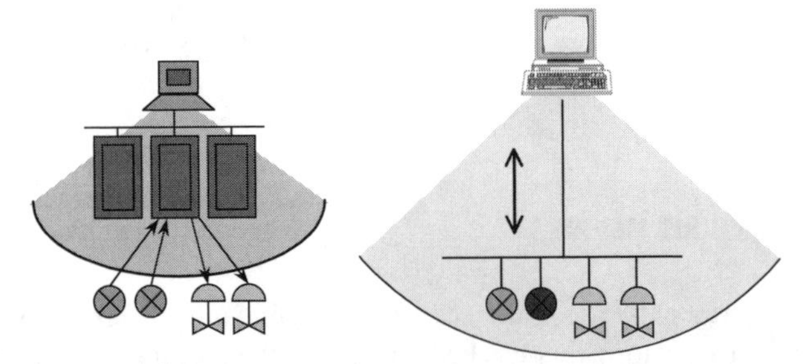

Analog: View does not include diagnostics and other information from field devices.

Fieldbus: Field devices are part of the system.

FIG. 4.3b
Expanded system view.

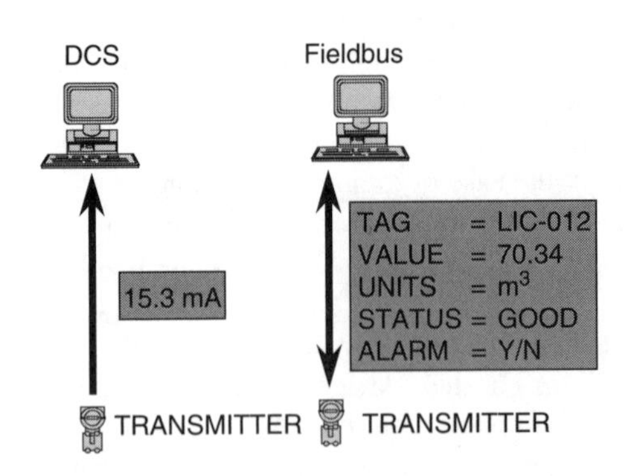

FIG. 4.3c
Increased data reliability.

Future Capabilities

As they are all digital, fieldbus devices will, just like the computers on our desktops, become "smarter." They will not only have more computing power with lower power demands but they will also have increased functionality.

Because fieldbus devices are now part of an integrated control system, in the future, devices and hosts will be even more closely linked.

Most major distributed control system (DCS) suppliers are migrating from a proprietary network backbone to one based on some form of Ethernet. This is consistent with the trends observed in the computing world, especially as this provides economies of scale and ready access to the various components needed to manufacture computer-based equipment. This trend has been evident for a few years now as most DCS (now called host) manufacturers have moved to a Windows environment, for their human–machine interface (HMI) as a minimum, and in some cases their advanced applications environment as well. As a result of this move to a common platform, host suppliers are now focusing more effort on how to gain maximum value from the information available in that environment and hence the increased shift to concentrate on software and services. This thirst for more information will be one of the key drivers to support the need for all-digital networks as represented by fieldbus and Ethernet.

The migration to all-digital communication networks for the host system–HMI of the future will eventually lead to the acronym DCS, meaning "digital control system," along with all the extra data flow and information associated with it.

The integration trend is clear and gaining momentum, control system engineers and process control practitioners in general are in a period of transition from the analog to the digital world. More and more, that digital world is based on the Microsoft operating system.

> Windows NT is suitable for 80% of manufacturing and process control applications. Windows CE's proposed real time capabilities will allow it to address 16% of the remainder, leaving a mere 4% for specialist embedded solutions.[1]

The above quote reveals one reason the Windows environment is the operating system of choice for the majority of host systems today. Add to this the fact that if the process control and corporate environments are similar, integration of data across the enterprise also becomes easier. The majority of corporate local-area networks (LANs) also operate on Ethernet; in fact, the Windows 2000 environment makes it possible to create one large domain from the desktop to the control system. Once again, this emphasizes the pressure on network designers to merge to as common a hardware platform as possible, and that platform will be Ethernet and Internet based. That Foundation Fieldbus HSE (high-speed Ethernet) uses a publish/subscribe environment to transfer information between devices and system components, while supporting control in the field, also makes it possible to deploy a control system without a host as part of the control loop.

With Ethernet-based control systems, it will be possible to implement a widely distributed fieldbus control loop without any of the signals being shown on the host. A virtual private network (VPN) can be created to connect two HSE gateways and their associated devices. The control is implemented in the field, and it is not necessary to have any intervention by the host. The host will connect to the network by another port on the switch and can receive the process variable updates as a passive subscriber whenever they are published.

Ethernet, TCP/IP (transmission control protocol/Internet protocol), and UDP (universal data protocol) represent only the lower layers of the OSI model. How is all the information on the network communicated to the other layers of both the OSI model and the computer networks in a facility? HSE and the other fieldbus networks all describe the necessary protocols to move data to the user layer, but it still remains to transfer this information between the various applications at the user layer and above.

One option that is receiving a lot of attention and appears viable at the process data management (PDM) layer used for control is OLE for process control (OPC). OPC is another open standard that links information from one database to another, in real time with minimal overhead. And because it is an open standard, it too is vendor neutral. OPC is supported by most host system suppliers of all the fieldbuses and therefore has the opportunity to be the "glue" linking disparate systems at higher levels in the organization/network.

At higher layers, where the Internet and corporate information transfer occur, eXtensible Mark-up Language (XML) is the present odds-on favorite. XML was recently approved by the World Wide Web Consortium as a standard language for use on the Internet and Microsoft has adopted this technology as the core for its new net operating system. This is certainly good news for acceptance of XML at multiple layers in the network.

The convergence of tools for the entire plant network is well on its way; we the engineering community must take action on this area as we plan for the future—not only at the data collection layer as represented by process control systems, but also by working with the corporate information people to ensure data can flow seamlessly throughout the organization. Some of the other future possibilities resulting from digital control systems include features such as:

Statistical quality control—The ability to use the internal diagnostic features of the device as inputs to control algorithms that will increase its ability to provide predictive maintenance information to the host and associated computer-based maintenance system.

Integration—With HSE it will be possible to connect across networks and if necessary over wide-area networks (WANs) to optimize operations across plant boundaries.

Function Blocks

As shown in Figure 4.3a, function blocks make it possible for devices to have unique characteristics and be "configured" to suit the application for which they are intended. There are a total of 29 different function block types have been identified

TABLE 4.3d
Types of Function Blocks

Basic Function Blocks	*Advanced Function Blocks*	*Flexible Function Blocks*
PID control	Analog alarm	8-channel analog input/output
Ratio control	Arithmetic	8-channel discrete input/output
Manual loader	Deadtime	Application specific (IEC 61131-3)
PD control	Device control	
Ratio	Input selector	
Control selector	Integrator	
Discrete input	Set point ramp generator	
Discrete output	Splitter lead/lag	
Analog input	Timer signal characterizer	
Analog output		
Bias		
Gain		

with 21 defined function blocks defined in the Fieldbus standard. These blocks are divided into three groups as shown in Table 4.3d.

For each block there is a set of parameters that, to a certain extent, defines the maximum functionality of a block. However, manufacturers may implement such blocks in their own way. For example, in the PID control block there must be a GAIN parameter and, therefore, the manufacturer may use this parameter as gain or proportional band.

Analog values are passed as floating points in engineering units, but are scaled to percentage (for example, in the PID block) to enable dimensionless tuning parameters, and thus implementation of a block is ensured to be supplier independent.

Resource Blocks

Each device has only one resource block, which provides the general information related to all the other blocks included in the device and is responsible for monitoring the operation of the whole device, thus providing it with its diagnostic capabilities. The resource block also contains device information such as final assembly numbers and materials. The resource block cannot be linked to other blocks and all its parameters are self-contained. The block does, however, contain global parameters, which may be used by any block in the device.

The fieldbus network does not schedule transducer block execution. The manufacturer may therefore control the execution of the resource block to suit the needs of the device.

Transducer Blocks

Transducer blocks are responsible for the interface between the function blocks and the device input/output (I/O) hardware. This block converts the sensor signal to a form that is usable by the network, thus making the device independent of the function blocks. There is one transducer block for each hardware point, such as a sensor, I/O terminal, or display. The standard itself does not specify any parameters for the transducer blocks, but the end user council has identified parameters for various device types.

Like the resource block, the fieldbus network does not schedule transducer block execution. The manufacturer may therefore control the execution of the transducer block to optimize it against the sensing technique of the device for which it is designed. The execution cycle is often faster than the execution of the remaining function blocks in the device.

Because these blocks are the difference between a conventional device and a fieldbus device, it is important that the following information be obtained for the device to allow the network to be properly designed:

- What is the execution time in milliseconds for each of the function blocks (e.g., PID, AI, AO, etc.)?
- LAS capability—Does the device support being the link active scheduler (LAS)?
- Device current draw (mA)—How much current does the device draw under normal operating conditions and at start-up?
- Device minimum voltage—What is the minimum voltage the device requires?
- Device capacitance—What is the capacitance of the device in nanofarads (nF)?
- Polarity sensitive—Is the device polarity sensitive?
- VCRs—How many virtual communication resource blocks are available in the device that are unassigned?
- DD revision—What is the device description revision number?
- CCF revision—What is the capabilities file revision number?

Converting an Existing Installation

Provided the conditions for the devices listed above are met and the device supplier supports an online conversion, moving to fieldbus can be done today. In fact, the author has done this with a number of devices in his facility over the last year and hopes to convert more devices this same way in the future.

If, on the other hand, the present device and host suppliers do not fully support fieldbus and device migration, there are alternatives. The most obvious alternative is to change out the device for one from another manufacturer. The only difficulty here is that not all devices are interchangeable. Fieldbus makes devices interoperable and interchangeable electronically, but not all devices have the same process connection dimensions. ISA (the Instrumentation, Systems, and Automation Society) has a standards committee in the process of addressing this need. ISA S-97 is currently out for committee review and approval of a draft standard for face-to-face dimensions of vortex meters is expected in early 2003. The IEC already has a standard for magnetic flowmeters, and fortunately, pressure transmitters already meet a *de facto* standard of 58 mm centerline for their tap connections.

The second method is to use a pair of devices available from Smar International. Smar has developed a current-to-fieldbus (I/F) and fieldbus-to-current (F/I) converter. Figure 4.3f shows how they might be used with a fieldbus host system, and Figure 4.3g shows how they could be used as a multiplexer/demultiplexer pair on a nonfieldbus system. The I/F converter is the model IF132 and the F/I device is model FI302.

There are a number of conditions associated with a successful fieldbus conversion:

- Not all suppliers have a path that provides an upgrade of field devices online, so the conversion may require planning to coincide with a plant outage opportunity, or a second device must be installed in parallel on an alternate process connection.
- If the installed host system does not support fieldbus, then the options shown in Figure 4.3f or 4.3g are possible.
- Use a multiplexer/demultiplexer combination in the field and at the host or central control system.
- Use a complete fieldbus system and then connect from it to the existing control system using some alternate protocol such as HSE to a server compatible with the control system and its Ethernet network ports.
- The infrastructure must meet the fieldbus design conditions for the physical layer.

Another feature that the host system should have to support the migration to fieldbus is incremental back build. This allows the addition or removal of a device from the network without having to reload the entire segment or network. Loading normally requires that all the devices on that portion of

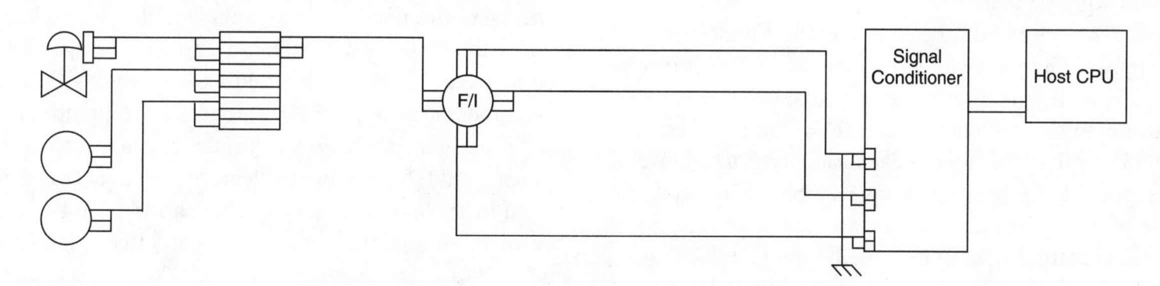

FIG. 4.3f
Traditional host system unable to support fieldbus.

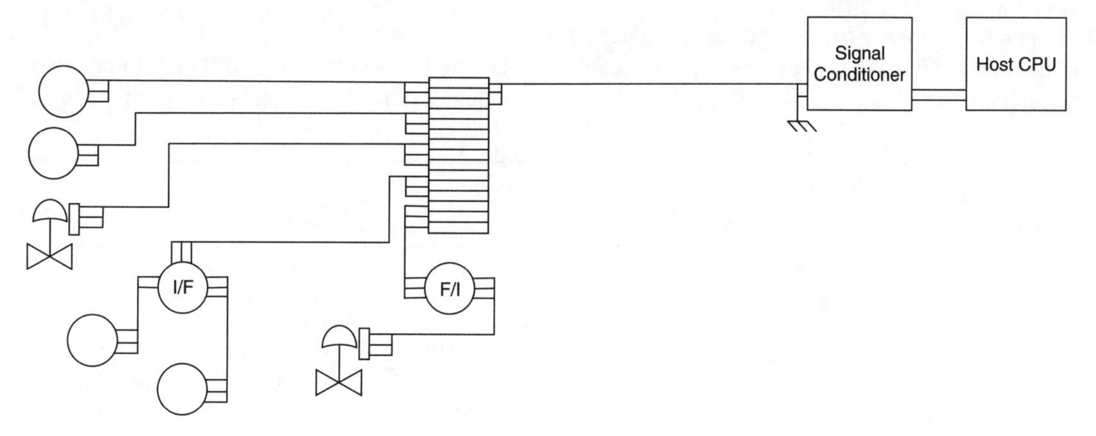

FIG. 4.3g
Fieldbus host with nonfieldbus devices.

the network be "off line" and hence on manual for, in some cases, up to 20 to 30 min, depending on the complexity of the network. Incremental back build is done during the acyclic times and only affects the device being changed.

CABLE

IEC 61158-2 specifies a variety of cable types to maximize the "backward compatibility" of the existing cable infrastructure. The standard supports optical fiber as well.

Fieldbus has designated four different cable types, each with a different *total combined length* limit. This means that if one uses the type of cable, A, B, C, or D, exclusively, the total length of the home run segment *and* all the spurs or branches must be less than this total. The maximum overall length of cable when mixing cable types is determined according to the formula

$$\frac{L_x}{L_{max_x}} + \frac{L_y}{L_{max_y}} \leq 1$$

where L_x = length of cable x, L_y = length of cable y, L_{max_x} = maximum length of cable type x alone, and L_{max_y} = maximum length of cable type y alone.

The same ratio calculation can be performed if more than two cable types are present in the system, as long as the sum of the ratios is less than 1.

Note: The system may reach an electrical limit in its ability to supply sufficient power to the devices, *before* it reaches the physical wire limitation as calculated above.

If using some combination of cables, the total length is calculated based on the ratio of each cable type to the overall total as per the following example. Table 4.3h shows an example of this calculation for a team that wishes to install a segment constructed of Type A and Type C cables with a combined length of 1900 m. The team members have available 300 m of Type C for one long spur from an existing installation that they wish to use and will use 1500 m of Type A for the remainder of the segment.

Because this ratio is *greater* than 1, the proposed design will not work and they will have to pull a new Type A cable to the existing device to bring the total under 1900 m.

TABLE 4.3h
Cumulative Ratio Calculation for Cable Combinations

Cable Type	*Max Length, m*	*Length Desired, m*	*Max Length Ratio*	*Cumulative Ratio*
Type A	1900	1500	0.79	0.79
Type C	400	300	0.75	1.54

Note: This is the first check only and they will still have to perform the other cable sizing calculations for power, resistive capacitance, and bandwidth.

The overall design is determined from the number of I/O points, the physical location of the field devices within the plant, and the maximum distances permitted for the fieldbus cable runs. A Foundation Fieldbus network is the total wire pair length, trunk length, and spur lengths. A trunk is the longest cable run between any two devices on a network. A spur can vary in length from 1 to 120 m. A spur that is less than 1 m is considered a splice.

Note: A spur that is less than 10 m is negligible as a transmission line and can accurately be modeled as an equivalent capacitor. Quarter-wavelength at H1 frequencies is in excess of 2 km.

Although unterminated spur lengths up to 120 m are allowed, any spur over 30 m requires careful review. The intent of the selected multidrop bus wiring method is to eliminate the need for long spur lengths and to keep spurs under the recommended length of 10 m. Longer spurs may be needed to keep the main bus out of high-risk areas such as immediately adjacent to furnaces or in pump alleys.

The total spur length is limited according to the number of spurs and the number of devices per spur. This is summarized in Table 4.3i. The spur table shown as Table 4.3i is not absolute. It merely serves as a guideline when designing networks.

The maximum spur length column based on Type A cable is determined by

$$\text{Devices} \times \text{Cable Length}$$

and can be used as a rule of thumb to determine the density of devices that can be installed on a network.

TABLE 4.3i
Guideline Table for Spur Network Design

Total Devices per Network	*One Device per Spur m(ft)*	*Two Devices per Spur m(ft)*	*Three Devices per Spur m(ft)*	*Four Devices per Spur m(ft)*	*Maximum Total Length m(ft)*
1–12	120 (394)	90 (295)	60 (197)	30 (98)	1440 (439)
13–14	90 (295)	60 (197)	30 (98)	1 (3)	1260 (384)
15–18	60 (197)	30 (98)	1 (3)	1 (3)	1080 (329)
19–24	30 (98)	1 (3)	1 (3)	1 (3)	720 (220)
25–32	1 (3)	1 (3)	1 (3)	1 (3)	32 (10)

TABLE 4.3j
Recommended Twisted-Pair Features for FF

Wire size	18 gauge (0.8 mm)
Shield	90% coverage
Attenuation	3 dB/km at 39 kHz
Wire size	18 gauge (0.8 mm)
Characteristic impedance	100 Ω ±20% at 31.25 kHz

The Foundation Fieldbus (FF) network uses twisted-pair wires. A twisted pair is used, rather than a pair of parallel wires, to reduce external noise from getting onto the wires. A shield over the twisted pairs reduces the noise further.

For new installations or to get maximum performance for an FF network, twisted-pair cable designed especially for FF should be used. The important twisted-pair cable characteristics are listed in Table 4.3j.

Various types of cables are usable for fieldbus. Table 4.3k contains the types of cables identified by the IEC/ISA Physical Layer Standard. Cables shall be labeled Type TC (16 gauge), PLTC (18 gauge), or ITC (18 gauge) and shall be installed in tray or conduit. All cables shall be single twisted pair with a shield. Multiconductor signal cable without shield has a maximum length of 400 m.

Network Layout

Fieldbus has a variety of ways in which its networks and segments can be installed. Below is a summary of each individual topology commonly employed. It is important to note that it is also possible to combine the topologies on a single segment or network to optimize cable use and installation costs.

Point-to-Point or One-to-One Topology This topology consists of a segment having only two devices. The segment could be entirely in the field (e.g., a transmitter and valve, with no connection beyond the two) or it could be a field device connected to a host (performing control or monitoring). This method would not be used often as it is only one measurement *or* control device per segment. This topology is illustrated in Figure 4.3l and is not normally used because it is not an economic design for systems other than safety systems.

Spur Topology This topology (Figure 4.3m) consists of fieldbus devices that are connected to the bus segment through a length of cable called a spur. Spurs shall be connected to current limiting (30 mA) connections to the bus to provide short-circuit protection and to provide the ability to work on field devices without a hot work permit. This current limiting connection should provide a nonincendive or intrinsically safe connection to the field device. The drops and current limiting can be provided by Relcom blocks in junction boxes or by InterlinkBT bricks that are field mounted. These bricks shall have the minifast (large) connectors if used. It is expected that each drop may have up to six devices connected on individual unterminated spurs of up to 10 m. This topology is illustrated below.

When using spurs, designers must remember that each individual tap into the network home run cable requires nine wire terminations (plus, shield, and minus times three; one to break, one for the spur, and one to remake); a fieldbus segment with eight spurs would have over 70 individual wire terminations. In contrast, using a chickenfoot topology would have fewer than 30 wire terminations. Simple math says that this system, the chickenfoot, is less than half as likely to suffer from a poor wire termination. On top of this, if there is a bad wire termination in a chickenfoot segment, it is possible to go to the *one* place (be it a junction box or "brick")

TABLE 4.3k
Cable Types Used in FF Installations

Type	*Description*	*AWG*	*Capacitance, pF/m*	*Attenuation, dB/km*	*Max Length, m*
A	FF, Shielded, twisted pair (preferred cable)	#18 AWG ($0.8\ mm^2$)	150	3	1900 (6232 ft)
B	FF, Multi-twisted-pair, with shield	#22 AWG ($0.32\ mm^2$)	150	5	1200 (3936 ft)
C	FF, Multi-twisted-pair, without shield	#26 AWG ($0.13\ mm^2$)	150	5	400 (1312 ft)
D	FF, Multicore, without twisted pairs and having one overall shield (not acceptable for new installations)	#16 AWG ($1.25\ mm^2$)			200 m (656 ft)
Xo	Multiconductor with overall shield	20	75	4	1200
Xi	Multiconductor with individual and overall shield	20	98	5	1900
Xs	Single pair	11	44	6	1900

AWG = American wire gauge.

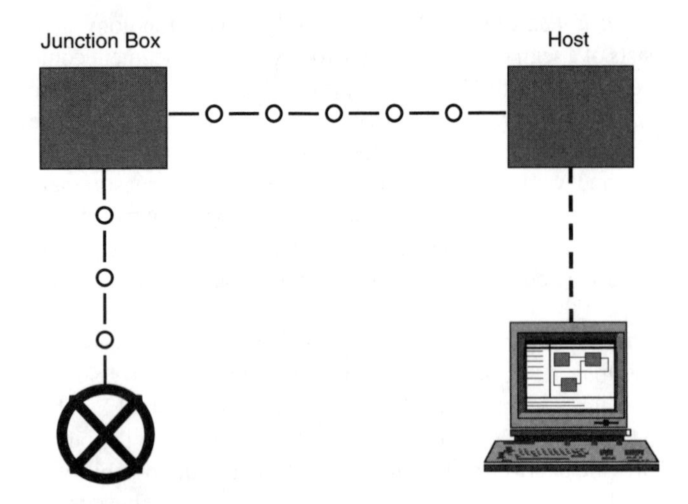

FIG. 4.3l
Simple point-to-point topology.

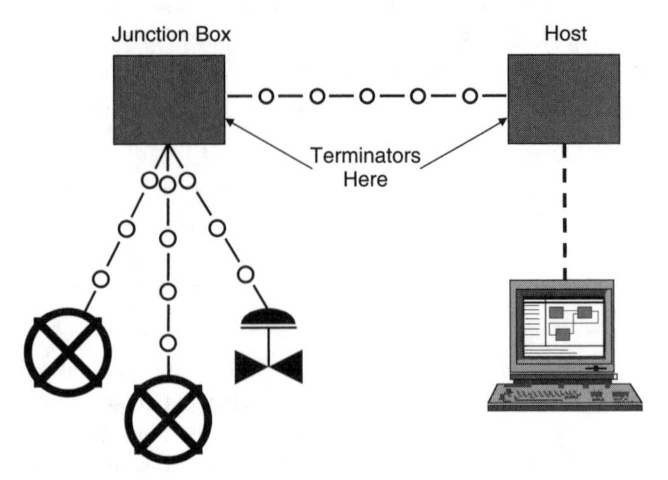

FIG. 4.3n
Tree or chickenfoot topology.

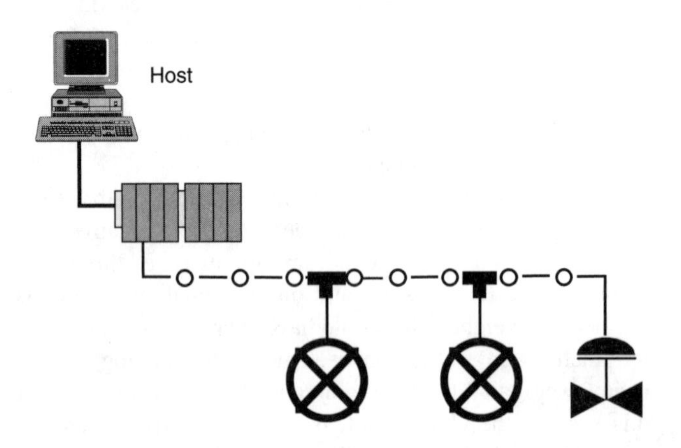

FIG. 4.3m
Bus with spurs topology.

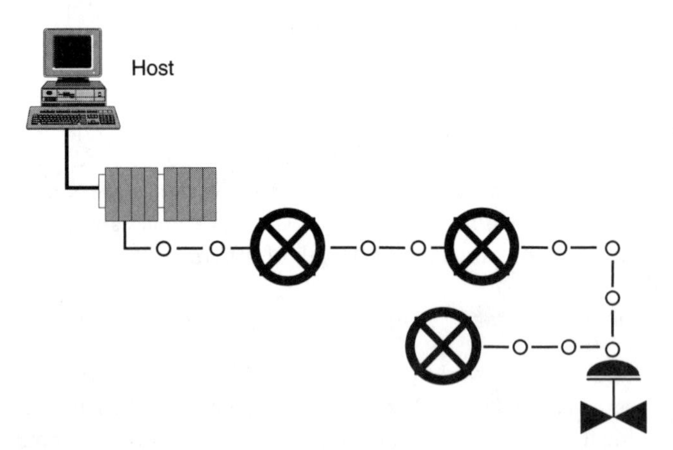

FIG. 4.3o
Daisy-chain topology.

where the wire terminations are and find the problem. With a bus topology with spurs, one is likely to spend a lot of time locating all of the places where the different spurs tap into the home run trying to find that one faulty termination.

Tree or Chickenfoot Topology This topology consists of a single fieldbus segment connected to a common junction box. This topology can be used at the end of a home run cable. It is practical if the devices on the same segment are well separated, but in the general area of the junction box. When using this topology, the maximum spur lengths must be considered. Maximum spur lengths were shown in Table 4.3i. This topology is illustrated in Figure 4.3n. Chickenfoot is the preferred topology to be used for fieldbus, as it is most similar to the traditional installation and will therefore provide the optimal use of existing infrastructure.

This topology also allows maximum flexibility when configuring and assigning devices to segments.

Daisy Chain Figure 4.3o represents daisy-chain technology in which each device is connected to the other in "series" much like older-style Christmas tree light strings. The cable is routed from device to device, where it is directly connected at the terminals of the fieldbus device. Installations using this topology should use connectors or wiring practices such that disconnection of a single device is possible without disrupting the continuity of the whole segment. This topology is not recommended, since it is unacceptable for maintenance purposes because, just like the tree lights of old, if one bulb or device fails, the entire segment is brought down as well.

POWER SUPPLY

Because fieldbus networks operate on AC signal, circuitry is needed between a regular power supply and the fieldbus network to prevent the active circuitry of the power supply from absorbing this AC fieldbus signal. It may be helpful to think of this circuitry as a "power supply impedance" because its main function is to act as an impedance.

Voltage regulation and noise reduction are of secondary importance for the power supply conditioner because the voltage out of most DC power supplies these days is pretty clean. The impedance is required because the fieldbus network carries both DC power and AC communications information. The communication signal is essentially a 31.25 kHz frequency alternating current. A DC power supply connected directly to the network "short-circuits" this signal and communication is lost. By putting an impedance between the power supply and the network one obtains a 3-kΩ impedance at 31.25 kHz, thus preventing the short. At the same time this impedance has near-zero resistance to DC so the current drawn by the devices causes little drop on the network.

It should also be noted that per IEC 1158-2, "The first 500uS is excluded (from the 'no more than 20 mA above quiescent current draw from 500 μs to 20 ms') to allow for charging of RFI filters and other device capacitance." The fact that some devices may draw more than 20 mA in the first 499.9999 μs is 100% allowable per the version of IEC 1158.2 even though it may present problems when adding or removing devices from the network. Fortunately, most manufacturers are not "abusing" this loophole and do not draw large currents on device start-up. It is expected that this loophole will be addressed in the next revision of the IEC standard to limit the current draw during this period to a certain value as it will make it possible then to design additional integrity and safeguards in intrinsically safe networks.

The 250-Ω resistor present in the HART network fills the same function as the signal conditioner and terminator in IEC 61158-2. The resistor prevents the power supply from short-circuiting the HART signal (1200/2200 Hz) and at the same time serves as the shunt that converts the modulated current from the transmitting device into a voltage that can be picked up by all the receiving devices.

Location on Segment

Subject to the conditions about the integrity of the power and its necessary impedance, power supplies can be installed anywhere on a fieldbus network. Some equipment suppliers now offer redundant power supplies with each of the power supplies at opposite ends of the network, i.e., one end is located in the central control room with the host and the other is in the field with the devices. In most cases, however, the power supply and conditioner are located near the host for the following reasons:

- This is normally an unclassified or safe area so no extra precautions are needed to prevent an explosion due to a hazardous atmosphere.
- Uninterruptible power supply is available as a source for the power conditioning equipment.
- This is similar to the way present systems are designed, and hence reuse of the existing infrastructure can be maximized.

It should also be noted that the power supply can be connected anywhere along the segment, including the middle if that makes economic sense and does not compromise the integrity of the network.

SEGMENT DESIGN RULES

Power

Fieldbus devices are specified to work with voltages of 9 to 32 V DC.

Note: The 9 V DC specified is a minimum: it is highly desirable that a margin of at least 1 V (i.e., a minimum of 10 V DC) be maintained. Some devices do not conform to FF standards and require 11 V to operate. Any segment designed to operate below 15 V normally should carry a warning about additional loads in the segment documentation. Minimum segment voltage should always be shown in the segment documentation.

If an ordinary power supply were to be used to power the fieldbus, it would absorb the signals on the cable because it would try to maintain a constant voltage level. Fieldbus power supplies are "conditioned" by putting an inductor between the power supply and the field cable and a resistor is added to the inductor to prevent "ringing." The power conditioner consists of the equivalent of a 50-Ω resistor and 5-mH inductor in series.

In addition to the physical limitations described earlier, which are generic guidelines based on voltage and capacitive limitations, the following equation can be used to calculate the maximum trunk length on a system with approximately equal spur lengths and devices with nearly equivalent current, voltage, and capacitance needs.

$$L_{Tmax} < \frac{((V_{PS} - V_{min}) \times 10^6 - I_D \times 2 \times R_S \times L_S)}{\sum I_D \times 2 \times R_T} \quad \mathbf{4.3(1)}$$

where

L_{Tmax} = maximum voltage of trunk cable, m
V_{PS} = power supply voltage, V
V_{min} = largest minimum voltage of all the field devices, V
I_D = DC current draw of the field device with the largest minimum voltage, mA
R_S = manufacturer-specified resistance of spur cable, Ω/km
L_S = length of spur cable, m
ΣI_D = total (sum) of DC current draw of *all* field devices, mA
R_T = manufacturer-specified resistance of trunk cable, Ω/km

If the installation is a chickenfoot arrangement, or if each of the field devices has very different minimum voltages (i.e., a temperature transmitter and a valve positioner) and current specifications, then the voltage available at each device on the segment should be calculated using the following formula:

$$V_D = V_{PS} - \left(\sum I_D \times 2 \times R_T \times L_T + I_D \times 2 \times R_S \times L_S\right) \times 10^{-6} > V_{min} \quad \textbf{4.3(2)}$$

where

V_{min} = minimum voltage of the field devices, V
V_D = DC voltage available at the field device, V
V_{PS} = power supply voltage, V
ΣI_D = total (sum) of DC current draw of *all* field devices, mA
R_T = manufacturer-specified resistance of trunk cable, Ω/km
L_T = length of trunk cable, m
I_D = DC current draw of the field device, mA
R_S = manufacturer-specified resistance of spur cable, Ω/km
L_S = length of spur cable, m

Capacitance constraints must also be considered because the effect of a spur less than 300 m long on the signal is very similar to that of a capacitor. In the absence of actual data from the manufacturer, a value of 0.15 nF/m can be used for fieldbus cables.

$$C_T = \sum (L_S \times C_S) + C_D \quad \textbf{4.3(3)}$$

where

C_T = total capacitance of network, nF
C_S = capacitance of wire for segment, nF/m (use 0.15 if no other number is available)
C_D = capacitance of device, nF

The attenuation associated with this capacitance is 0.035 dB/nF. The following formula provides a reasonable estimate of the attenuation associated with the installation. Total attenuation must be less than 14 dB, which is the minimum signal that can be detected between the lowest-level transmitter and the least-sensitive receiver.

$$A = C_T \times L_T \times 0.035 \frac{dB}{nF} < 14\,dB \quad \textbf{4.3(4)}$$

where

A = attenuation, dB

Bandwidth

As with any network, the more information transmitted in a given time frame, the faster the network and all its components must operate to keep pace with this information. Because Fieldbus H1 is constrained to a rate of 31.25 kb/s the only way to transmit additional information is to increase the cycle time. The cycle time must also take into account the actual process lag time since a PID loop should have at least three "samples" per process cycle. For example, if the process response time was 1 min (60 s) the cycle time for the loop must be less than 20 s.

A link can carry about 30 scheduled messages/s. This means the network could have 3 devices, each sending 10 messages/s or 120 devices, connected by repeaters, each sending one message every 4 s.

Some host systems are also constrained by their processor speed and the number of registers or data bytes it can process per second.

VCRs

The fieldbus access sublayer (FAS) uses the scheduled and unscheduled features of the data link layer to provide a service for the fieldbus message specification (FMS). The types of FAS services are described by virtual communication relationships (VCRs). There are different types of VCRs: client/server, sink/source, and publish/subscribe.

The client/server VCR type is used for queued, unscheduled, user-initiated, and one-to-one communication between devices on the fieldbus. Queued means that messages are sent and received in the order submitted for transmission, according to their priority, without overwriting previous messages. When a device receives a pass token (PT) from the LAS, it may send a request message to another device on the fieldbus. The requester is called the "client," and the device that received the request is called the "server." The server sends the response when it receives a PT from the LAS. The client/server VCR type is used for operator-initiated requests such as set point changes, tuning parameter access and change, alarm acknowledgment, and device upload and download.

The report distribution VCR type is used for queued, unscheduled, or user-initiated one-to-many communications. When a device with an event or a trend report receives a pass token (PT) from the LAS, it sends its message to a "group address" defined for its VCR. Devices that are configured to listen on that VCR will receive the report. Fieldbus devices to send alarm notifications to the operator consoles typically use the report distribution VCR type.

The publisher/subscriber VCR type is used for buffered, one-to-many communications. Buffered means that only the latest version of the data is maintained within the network. New data completely overwrite previous data. When a device receives the compel data (CD), the device will "publish" or broadcast its message to all devices on the fieldbus. Devices that wish to receive the published message are called "subscribers." The CD may be scheduled in the LAS, or subscribers may send it on an unscheduled basis. An attribute of the VCR indicates which method is used. The publisher/subscriber VCR type is used by the field devices for cyclic, scheduled publishing of user application function block input and outputs such as process variable (PV) and primary output (OUT) on the fieldbus.

TABLE 4.3p
VCR Types and Uses

Client/Server VCR Type	*Report Distribution VCR Type*	*Publisher/Subscriber VCR Type*
Used for operator messages	Used for event notification and trend reports	Used for publishing data
Set point changes Mode changes Tuning changes Upload/download Alarm management Access display views Remote diagnostics	Send process alarms to operator consoles Send trend reports to data historians	Send transmitter PV to PID control block and operator console

Table 4.3p summarizes the different VCR types and their uses. Each device requires a minimum number of VCRs to communicate with each other and a host. Below are suggested minimum guidelines to be used when selecting devices on a segment.

Each device requires:

One client/server for each MIB (management information base)
One client/server for the primary host
One client/server for the secondary host or maintenance tool
One sink/source for alerts
One sink/source for trends

Each function block requires:

One publisher/subscriber for each I/O

If the function block is used internally it does not need a VCR.

An example for a device with two analog inputs (AI), 1 PID, 1 transducer block, and 1 resource block:

2 AI blocks	2 publisher/subscriber VCRs
1 PID block	5 publisher/subscriber VCRs (In, Out, BckCal In, BckCal Out, Track)
Basic device blocks	5 blocks

Hence, a single dual AI device for which the trends and alerts are being used could require:

$$5 + 2 + 1 + 5 = 13 \text{ VCRs}$$

The following example illustrates a more typical scenario.

Assuming the two AI blocks and the PID block reside in the same device and assuming that one of the AI blocks is associated with the PID function and the other one sends its data to some other device, the VCRs required are as follows:

AI_1_OUT	1 publisher/subscriber VCR
AI_2_OUT	Internal communication
PID_IN	Internal communication
PID_OUT	1 publisher/subscriber VCR
PID_BKCAL_IN	1 publisher/subscriber VCR
Basic device blocks	5 VCRs

BKCAL_OUT is only required for cascade PID control. TRK_VAL is used when the output needs to track a certain signal, but this also requires a TRK_IN_D signal to turn tracking on and off. Therefore, the total number of VCRs needed is

$$1\ (\text{AI\_1}) + 2\ (\text{PID}) + 5\ (\text{Basic Device}) = 8$$

Moving the PID block from the input device to the control device reduces this requirement by 1 VCR.

TERMINATIONS

As mentioned above, fieldbus technologies, as they are digital networks, require that they be impedance matched to prevent the signal from being "short-circuited." The resistive/capacitive circuit devices to prevent this from becoming an issue are called terminators. Also, because fieldbus is a parallel wiring circuit, the terminal blocks also need to be installed differently. To simplify the installation of fieldbus networks, terminal suppliers have developed special terminal blocks for this purpose.

Fieldbus devices can be terminated on conventional terminal blocks although it will require a significant amount of "cross wiring" at each junction as well as the addition of capacitors and resistors to fabricate segment terminators. It is for this reason that manufacturers have developed fieldbus termination devices. These devices normally come in two forms; one is similar to the traditional terminal strip and the other uses a preassembled "brick" with quick-connect couplings.

Prewired Blocks

The quick-connect product is manufactured by TURCK under the name Interlink BT, commonly referred to as a "brick." These products have two different types of end connectors as well. One is larger and more rugged than the other.

Special cables are used to connect field devices to these bricks. The cables can be factory-assembled in predetermined lengths with military-style connectors at both ends. Alternatively, they can be obtained with the connector at one end only with the option of terminating the other end on regular terminals or with a special kit to the military connection form factor.

Terminals

The "conventional" terminals all have passive and in some cases active components associated or incorporated in them to accommodate the fact that a fieldbus network connects to all its devices in parallel. Manufacturers of a suite of products of this type are Relcom, Pepperl & Fuchs, and Weidmueller.

"Short-Circuit" Protection

All terminations to a network segment should use spur current protection devices to protect the segment from a short circuit in the spur. This will prevent loss of the entire segment should there be a short in one of the spurs or branches of the network.

These devices normally contain an active diode circuit to shunt any excess current beyond the rating of the device to ground, thus protecting the remainder of the segment from errors in an individual spur. When activated, these devices introduce an "additional" load to the segment. It is therefore important to include this "extra" load in segment calculations. Otherwise, if one of the spurs has a short causing a trip, it will introduce an extra load on the segment higher than the power supply can support and the entire segment will fail. This is the opposite of the reason the guards were installed in the first place and, hence, the need to design in the "extra" capacity to accommodate a short-circuit trip.

BARRIERS

Terminators

A terminator is an impedance-matching module used at or near each end of a transmission line. Only two terminators can be used on a single H1 segment. A matching network used at (or near) the end of a transmission line has the same characteristic impedance of the line. It is used to minimize signal distortion, which can cause data errors by converting current variations to voltage variations, and vice versa. As shown in Figure 4.3q, terminators in their simplest form are a resistive capacitance (RC) circuit at either end of a fieldbus segment made of a 100-Ω resistor and 1-μF capacitor. Most commercial terminators are more complex than this, as they include additional components to increase the reliability of the device.

There are two and only two terminators in a fieldbus segment. In many cases, the fieldbus specification is robust enough to operate with only one terminator; however, the signal may experience periodic "failures" that can be sporadic.

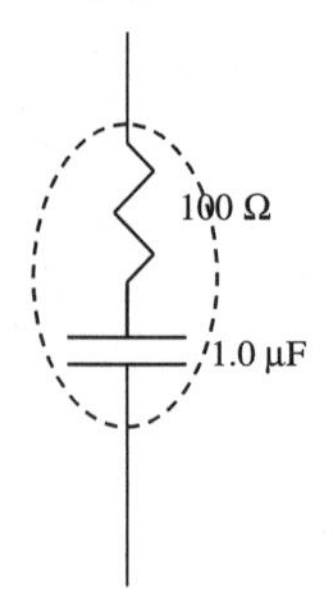

FIG. 4.3q
Fieldbus terminator circuit.

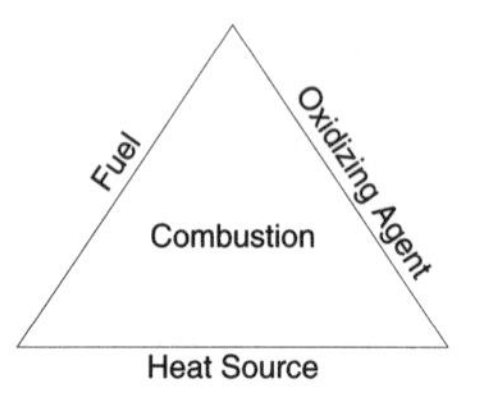

FIG. 4.3r
Fire triangle.

Similarly, a network may continue to operate with three or more terminators, although the signal will be significantly reduced in its peak-to-peak amplitude and therefore once again be susceptible to periodic sporadic failures. Remember, however, that each repeater on a network in effect creates a new segment so that a pair of terminators is required for the new segments created by the repeaters.

IS/NIS/FISCO

Fieldbus devices, like traditional installations, can be purchased and installed as either intrinsically safe (IS) or non-intrinsically safe (NIS), depending on the area classification in the field and facility practices.* IS circuits use either galvanic isolators or barriers to limit the amount of energy available in the hazardous area.

NIS fieldbus designs will most likely be constrained by some form of bandwidth constraint *before* they reach the limit of their ability to provide power to the segment. Fieldbus NIS design "allows" up to 32 devices on a single segment. Practical experience, however, has found that the maximum is reached at 8 to 10 devices.

IS operates on the principle of reducing the available energy to a system to a level below which ignition of a flammable mixture can occur. This removes the power side of the fire triangle shown in Figure 4.3r; the other two sides are fuel and oxygen.

* Some facilities prefer to use explosion-proof concepts in place of intrinsic safety. Intrinsic safety is more common in Europe, whereas many facilities in North America instead prefer to use explosion-proof designs to meet the area classification requirements.

IS designs by their nature reduce the power available on a segment to approximately 90 mA at 24 V. This obviously affects how many devices the segment can power. Galvanic isolators can be used in the field for transmission of signals from field-powered devices onto segments that are IS because of their installed location.

A number of manufacturers provide the necessary barriers and isolators for fieldbus networks. A relatively new concept has been introduced as a way to increase the power available to digital devices in the field. This new idea is called the Fieldbus Intrinsically Safe COncept (FISCO). As with the IS concept, FISCO is also based on experimental data to verify that it indeed maintains that even with a failure in the hazardous area, the energy level will remain below the explosive limits. There are several differences between the FISCO model and the regular entity concept. These are some of the highlights:

FISCO Barrier: FISCO has trapezoidal output instead of linear, therefore providing some 100 mA at 14 V instead of 60 mA at 9 V; therefore, one can power more devices. Because a typical fieldbus device consumes 12 mA one can safely put eight devices on a FISCO segment. FISCO barriers do not specify a capacitance and inductance limit.

Device: A FISCO device needs low inductance and capacitance and must handle nearly 2 W from the barrier compared to, say, 1.2 W for regular entity device. This is simply because the FISCO barrier emits more power.

Cable: The capacitance and inductance are considered distributed. Therefore, a capacitance and inductance "budget" need not be calculated for a FISCO network. As a result, FISCO barriers do not specify capacitance and inductance, and one can run cable as far as 1000 m compared to a few hundred meters for regular barriers, provided that the cable is within the fieldbus specifications.

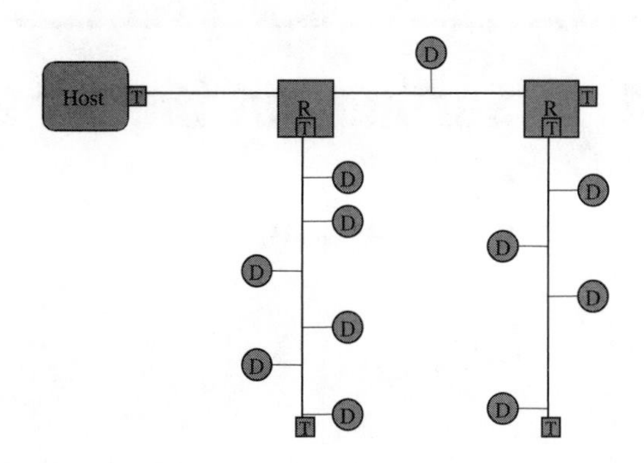

FIG. 4.3s
Fieldbus repeaters.

Additional information on these concepts and associated design criteria are presented in Section 4.14.

Repeaters

Repeaters are used in the event that it is necessary to extend the network beyond what is possible due to some physical constraint such as distance and resulting power restrictions. The specifications only allow a maximum of four repeaters on a single network. Figure 4.3s shows how a repeater may be installed on a single H1 link to increase the length or number of devices on that link. At 31.25 kHz, a single H1 link can carry about 30 scheduled messages/s. This means the network could have 3 devices, each sending 10 messages/s or 120 devices, connected by repeaters, each sending one message every 4 s.

Reference

1. "Distributed Communication Architecture, Industrial Networking Solutions with a Future," Hirschman Network Systems white paper, available at www.anixter.com, 2000.

4.4 Sorting out the Protocols

W. A. PRATT, JR.

Communications protocols have been around a very long time. Some of the earliest data transfer protocols used semaphores, and messages were transferred using flags and torches in Greek and Roman times. Semaphores and signaling flags are still in use. One of the earliest electronic communication protocols (starting in 1841, and still used today) was the Morse code.[4] Today, electronic communication is a rapidly growing field. Over 100 protocols can easily be identified.[11] Communication protocols will continue to evolve and the number of protocols will continue to increase.

Today, a large number of communications protocols compete for the attention of instrument engineers and for their business. Of course, not all of these protocols are "open" and "interoperable." This section attempts to sort out this large number of protocols and provide some insights to evaluate them.

Unfortunately, it is not possible to review all fieldbus contenders here. This was not intended to slight any particular protocol. Rather, the network examples highlighted were chosen to illustrate a diversity of solutions. This section should provide a basis for categorizing and evaluating any protocol the reader encounters.

THE PERFECT NETWORK

There is no single "perfect network." Furthermore, there will always be many different viable networks. Communications simply cannot be solved using a "one-size-fits-all" approach.

Communications protocols are tools just like saws (e.g., table saws, jig saws, and chain saws). Each saw has a function for which it was invented. Even though a chain saw can be used to build a china cabinet, who would want to? The industrial networks discussed in this section are simply tools. Understanding the origins and specialization of a network will help when selecting the tools.

PROTOCOL LAYERING

When evaluating industrial protocols, understanding the basic organization of a communications system can be very helpful. This subsection discusses two technical approaches to organizing and specifying a communications protocol. Understanding these organizational models assists in evaluation of an industrial protocol. Because communications protocols are very detail oriented, it is common practice to separate the problem into parts with clearly defined responsibilities.

These domains of responsibilities are called protocol layers or protocol hierarchies. These are standard models for describing any protocol. However, actual implementations may not completely adhere to these divisions of labor (e.g., because of throughput optimizations, or to simplify communications software design). Two different approaches to modeling communications systems are common: the Open System Interconnection (OSI) Reference Model (layered) and the model used by the Internet protocol (hierarchy).[1,2]

The layered approach divides a communications protocol into objects containing specific responsibilities. In addition, an object can only access services provided by an immediately adjacent layer. In theory, one data link layer could then be replaced with another without affecting the adjacent physical or network layers. In practice, this is not the case.

The hierarchical approach is less restrictive. It considers a communications system to be a collection of objects. An individual object on one computer communicates with a peer object on another. The hierarchical approach blurs the OSI layers.

OSI Model

The OSI model is the best-known scheme for organizing communications specifications and teaching communications technology (Figure 4.4a). The OSI model, released in 1984, describes seven communications layers[7] and details how a communications layer should be specified (see also Sections 4.2 and 4.3).

Note: Because the OSI model does not specify a communications system, it does not directly affect interoperability or how "open" a protocol may be.

In reality, no set of protocol specifications follows the OSI model 100% of the time. Even industrial protocols that claim to require only the physical data link and applications layer often mix in functions found in other layers. For example, many protocols have the ability to associate a tag name with a particular node. This is especially true with device networks. The protocols also generally have a mechanism to

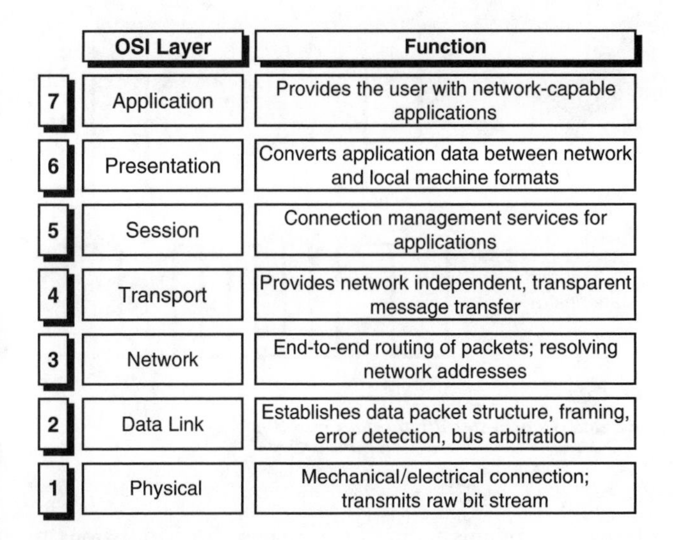

	OSI Layer	Function
7	Application	Provides the user with network-capable applications
6	Presentation	Converts application data between network and local machine formats
5	Session	Connection management services for applications
4	Transport	Provides network independent, transparent message transfer
3	Network	End-to-end routing of packets; resolving network addresses
2	Data Link	Establishes data packet structure, framing, error detection, bus arbitration
1	Physical	Mechanical/electrical connection; transmits raw bit stream

FIG. 4.4a
OSI reference model.

translate the tag into a physical network address. This function is, strictly speaking, a network layer responsibility.

Despite the lack of precision used in discussing the relation of a protocol to the OSI model, understanding its basic nature is useful. The lower, network-specific layers (physical, data link, and network) standardize interconnection of devices via a physical medium (e.g., twisted-pair wiring, radio transmission, fiber-optic cable). This is the most basic standardization a protocol can provide and some (e.g., Ethernet) provide little other standardization.

The upper layers (application, presentation, and session) generally contain a variety of protocol specifications. These levels enable the meaningful exchange of information. Although the transportation of data is important, the meaning of the data is crucial.[22] Communications address many problem domains each with its own needs, jargon (language semantics), and device types (e.g., limit switches, bar code readers, robot welders, weigh scales, pressure transmitters). This results in many different specifications tailored to specific problem domains.

This is also the subject of contentious debate. Users want the meaning standardized of as much data as possible. Manufacturers often feel threatened and are concerned about disclosing proprietary technology. They resist standardization of application layer data. Even when the data can be standardized, the standardization sometimes is vague and abstract, an optional protocol requirement, or only defines the absolute lowest common denominator set of data. In other cases, interoperability is claimed by wrapping the application layer in a presentation or integration tool (e.g., EDS, GSD, DDL, FDT/DTM, and others). These tools certainly make universal configuration tools possible and they are very valuable tools. However, accessing the data in the low-level controllers, programmable logic controllers (PLCs), and input/output (I/O) frequently requires software, processing power, and technology that is not available to these embedded, real-time products.

When evaluating industrial networks, separating the lower-level and upper-level functionality can be useful. Pursuing a clear understanding of what the upper levels provide is very beneficial in protocol evaluations. Be persistent. Anyone truly knowledgeable should be able to explain technology in a simple and easy-to-understand way.

TCP/IP Model

TCP/IP (transmission control protocol/Internet protocol) was developed in the 1970s and its basic structure was established between 1977 and 1979.[3] TCP/IP development was initially funded by the U.S. Department of Defense (DOD) and is actually a collection of protocols. The DOD had a wide assortment of communications equipment including radios, land lines, and satellite systems. As a result, the principal purpose behind TCP/IP is to link together different communications networks. This resulted in the hierarchy shown in Figure 4.4b.

The Internet level of the TCP/IP suite starts at about level 3 in the OSI model. In other words, TCP/IP does not specify the details of the communications across a single local-area network (LAN). It delegates responsibility for the LAN. One result is that TCP/IP operates on almost any communications network including packet radio, Ethernet, satellite systems, and dial-up telephone lines. This flexibility has led to widespread acceptance worldwide. In fact, TCP/IP was in widespread use long before Web browsers became popular.

As mentioned above, the purpose of the hierarchy is to provide a loose collection of building blocks to support network-capable applications. Unlike in the OSI model, the differing standards are not required to access services only in adjacent levels.[1] They can access whichever services best suit their objective.

The Internet Activities Board (www.iab.org and www.ietf.org) governs TCP/IP standards.[3] The standards documents are called RFCs (requests for comment) and the official set are listed in the latest RFC titled, "Internet Official Protocol Standards." The Internet set of standards is continuously growing (there are hundreds of RFCs in effect today). Consequently, this RFC is replaced several times per year with a new RFC with the same title. An RFC is never modified (i.e., there are no revision numbers); instead, it is replaced by a new RFC.

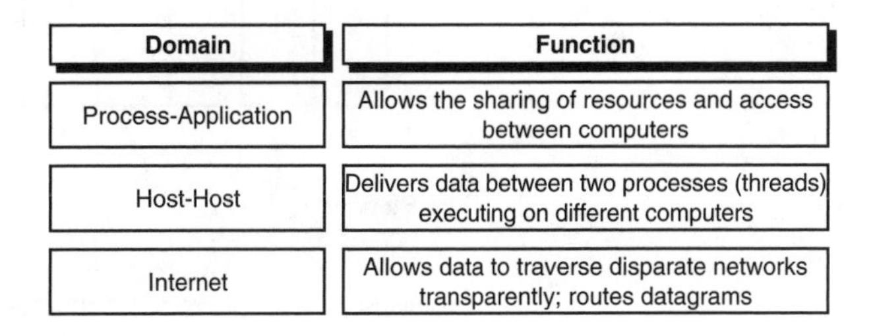

Domain	Function
Process-Application	Allows the sharing of resources and access between computers
Host-Host	Delivers data between two processes (threads) executing on different computers
Internet	Allows data to traverse disparate networks transparently; routes datagrams

FIG. 4.4b
TCP/IP communications hierarchy.

COMMUNICATION BASICS

This subsection will review some basic concepts that allow communications to be successful.

Data Transmission

There are four fundamental methods of data transmission:[1]

1. Digital data using digital signaling (e.g., LANs and fieldbus)
2. Digital data using analog signaling (e.g., HART and telephone modems)
3. Analog data using digital signaling (e.g., CD-ROMs and telephone networks)
4. Analog data using analog signaling (e.g., 4–20 mA current loop and AM/FM radio)

In some cases, these techniques are also mixed. For example, a cell phone digitizes voice (digital data representing analog signal). These digital data are then modulated (analog signal transmitting digital data) and transmitted. This subsection will briefly discuss the first two methods: digital communications and modulation (analog signaling) techniques.

Most computer networks use digital signaling for communications. Digital signaling transmits data one bit at a time using a series of pulses. The data bits must be encoded and transmitted synchronously or asynchronously.

Two encoding schemes are NRZ (nonreturn to zero) and Manchester (Figure 4.4c). NRZ encoded data are used by the serial port found in many PCs. The signal level (high or low) indicates a 1 or 0. NRZ encoding results in very simple transmitters and receivers. However, NRZ encoding does not allow the boundary between bits to be directly identified. Additional information is required. In other words, the bit rate at which the data are being transmitted is needed before the data can be received. With NRZ data the bit rate often must be manually set and this, at times, can result in confusion during the commissioning of a system. Such questions as "What direction was I supposed to set the dip switch?" can arise.

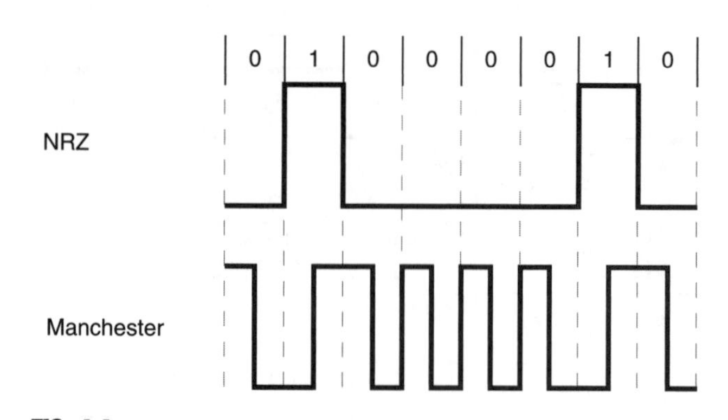

FIG. 4.4c
NRZ and Manchester digital data encoding.

Asynchronous Communications and Serial Ports Asynchronous communication is often used with NRZ encoding. Because the clock generators have finite tolerances, there will be slight frequency differences between the transmitter and the receiver. Over the course of a long data packet, the clock differences accumulate and can result in communications errors. Asynchronous communication compensates for the clock differences. For each byte of data, a start and one or more stop bits are added.

A start bit is a 1-to-0 transition that tells the receiver that the byte is about to start (Figure 4.4d). In effect, the start bit triggers the receiver's clock. The data are received (clocked in). The stop bit ensures that the signal line returns to a 1. The communication can remain idle for a while with the line remaining high. This is called marking. The next start bit resynchronizes the transmitter and receiver. The overall result is that any minor clock speed differences do not accumulate for more than 1 byte and cause communication errors.[8]

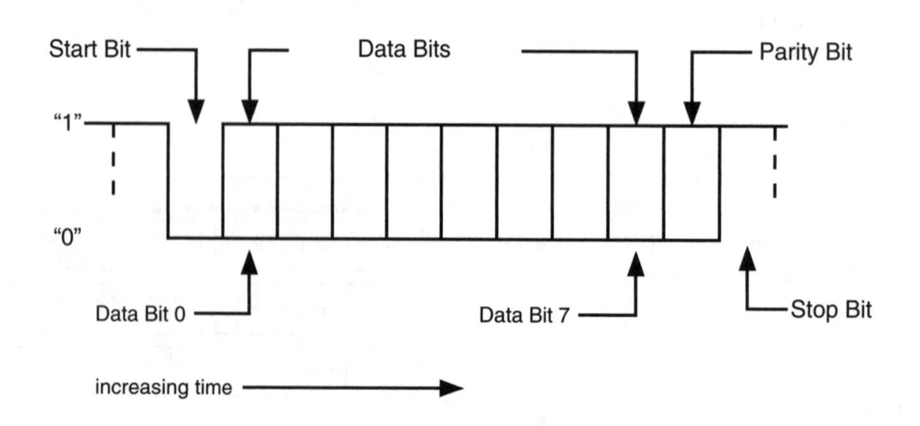

FIG. 4.4d
RS-232 asynchronous serial data stream.

PC serial ports typically communicate using RS-232 signal amplitudes, NRZ digital encoding, and asynchronous data transmission.

Note: RS-232 and RS-485 (see Section 4.2) specifies voltage levels for communication. They are simply physical layers. No assumptions should be made about effort required to communicate with a device supporting RS-232 or RS-485. The number of protocols based on these physical layers is huge.

Parity Parity (odd or even) is a simple error-checking scheme often used in asynchronous communications. Parity generators count the number of 1 bits in the byte and the parity bit is set accordingly. The receiver counts the total 1 bits (including the parity). If odd parity is specified and the receiver counts an even number of 1 bits, then the data have been corrupted. In some communications protocols, this concept is carried one step farther by adding a check byte to the message.[12] A check byte or block check character provides a parity check on the entire message. In other words, some protocols use vertical parity across an individual byte and horizontal parity (i.e., the check byte) across the entire message.

Manchester Encoding and Synchronous Communications
Manchester encoding includes the clock information in the data stream. In other words, the receiver does not need to be supplied the data rate. The receiver recovers the clock from the data stream itself, thus eliminating errors due to any clock speed differences with the transmitter. This has advantages, especially at higher data rates where clock tolerances become critical. Consequently, Manchester encoding is widely used in networks.

In Manchester-encoded signals there is always a signal transition in the center of the bit time.[1] This allows the clock to be recovered and ensures the value of the bit is always sampled by the receiver at the right time. The direction of the signal transition indicates the value of the bit. A low-to-high transition indicates a 1 and a high-to-low a 0.

Because the data transmission is synchronized by embedding the bit clock, synchronous communication is often used with Manchester-encoded digital signaling. In synchronous communications there are no start and stop bits and the bytes are transmitted back to back with no space between them. This (slightly) improves efficiency.

Cyclic Redundancy Codes

More sophisticated error detection coding is usually employed with synchronous transmissions. One of the more common is a cyclic redundancy code (CRC). CRCs are calculated by considering the data being transmitted a single large number. This large number is divided by a specified value.[8] The quotient is discarded and the remainder is sent with the message. The divisor is specified as a polynomial and the division uses modulo-2 arithmetic. This means that CRCs can be calculated in software or in hardware using a relatively simple circuit. CRCs are very good at detecting a burst of errors in data. CRCs were originally used with magnetic media (e.g., floppy disk drives) and are now widely used in communications protocols.

Today 16- and 32-bit CRCs are common, and the longer the CRC, the longer the error burst that can be detected. In addition to knowing the length of the CRC, the value of the polynomial used in calculating the CRC must be known.

Analog Signaling Digital data are also widely communicated using analog signaling. PC modems, cell phones, and wireless networks all modulate the digital data for communication. There are basically four modulation techniques[1] (Figure 4.4e):

1. Amplitude shift keying (ASK) varies the signal level to transmit the data.
2. Frequency shift keying (FSK) encodes the data by varying the frequency. HART and public telephone network caller ID services use FSK signaling.
3. Phase shift keying (PSK) varies the phase angle of the transmitted signal to encode the data. Cell phones and 802.11 wireless networks use PSK variants.
4. Quadrature amplitude modulation (QAM) varies both the phase and amplitude. QAM is primarily used in high-speed telephone modems.

Modulated communication can require a more complex interface chip than digital communications. However, analog

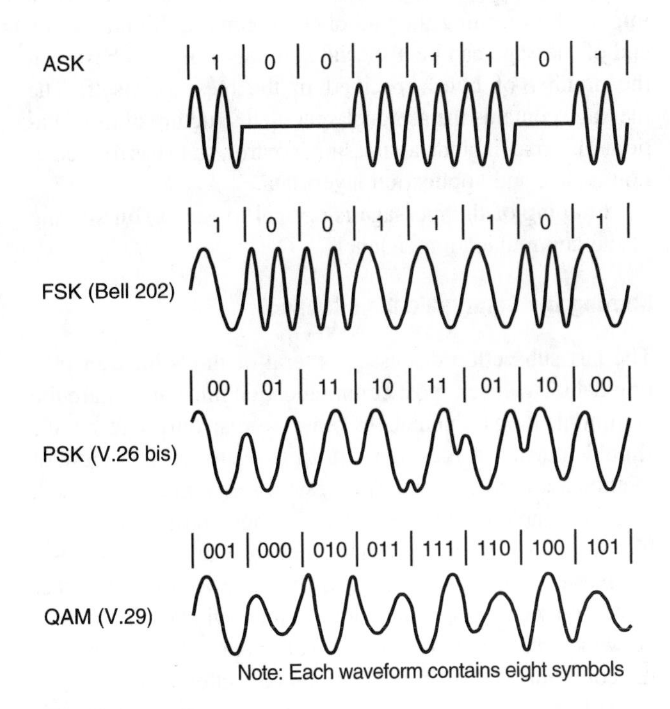

FIG. 4.4e
Data modulation techniques. Note: Each waveform contains eight symbols.

signaling requires less bandwidth than digital signaling and can operate at lower signal-to-noise levels.

With digital signaling, it is relatively easy to identify the data values in a picture of a waveform. With modulated communication this may not always be the case. For example, binary FSK uses only two signal frequencies and the data can be seen. They can be heard, as well. However, a V.34 modem uses 1024 phase–amplitude combinations and the "noise" it makes negotiating a connection is data.

Unlike digital communications, baud and bits per second are not equal to each other with modulated communications. Baud is symbols per second. For digital signaling there is normally 1 bit per baud. For modulated signals there can be many bits per baud. For example, a V.34 modem operates at (up to) 3429 baud with 9 bits transmitted per baud.[10]

Framing Messages

So far, basic techniques for transmitting data have been discussed. In addition, the start and end of a message and the message payload (i.e., the data) must be reliably extracted from the transmission. All messages have a preamble, start of message, data, and end of message.

The preamble allows the receiver to synchronize to the data stream. The preamble is a special sequence that ends with the start of message. Consider the Hays modem protocol. Every message starts with the characters "AT."[8] This allows the modem to change the bit rate and identify the start of a message. In some cases, the preamble is much more sophisticated. For example, some modulation standards send special phase–amplitude combinations as a preamble.

The start of message can be a special symbol or can be implied by reaching the end of the preamble. Similarly, the end of message can be a special symbol or derived based on the number of bytes specified in the message itself. The message contains the protocol-specified structure of information. This may include addressing, control, status, error detection codes, and application layer data.

Framing of the message is critical to successful sharing of the communication channel.

Sharing the Communication Channel

The last subsection discussed several methods for transmitting data. However, most communication must also share the communications channel. In other words, only one device should transmit a message at a time. Here, an overview of methods to arbitrate (share) access to a channel is provided.

The simplest networks contain a single master and (possibly) several slave devices. In these networks arbitration is easy: the master sends a message and the slave answers. In other words, only the master can initiate a messaging cycle. In most networks, the communication is half-duplex.[2] In other words, the communications flow in only one direction at a time.

When several nodes can initiate message cycles, bus arbitration (i.e., medium access control) becomes much more important. Three basic arbitration techniques are discussed here: carrier sense multiple access with collision detection (CSMA/CD); token passing; and time division multiple access (TDMA).

Note: The physical wiring of a network may not have anything to do with the bus arbitration technique in use. Ethernet is often wired in a star topology (e.g., 100BaseT) and some token-passing networks are wired as a bus.

CSMA/CD allows any node to initiate communications. The device listens to the network and, if no signal is present, begins transmitting a preamble (e.g., Ethernet sends 7 bytes of 10101010).[20] At the same time, the device listens to the bus and, if the device receives something different, it cancels its transmission. The device continues monitoring its transmission and aborts if the network signal is corrupted. This signifies a collision. In effect, everyone monitors the networks and speaks if the network is silent. If a collision is detected, then the devices back off a random amount of time and try again. Collisions reduce CSMA/CD network throughput. In the case of a busy Ethernet network with many nodes, throughput can be a fraction of the maximum rated capacity.

Token-passing networks are much more efficient and can easily operate at their maximum rated capacity. In a token-passing network, the device with the token has permission to transmit. The token may be a special message or implied by previous communications. The main drawback to token passing is the time it takes to recover a lost token. Algorithms are designed into the protocol to allow a new token to be generated when a communications error disrupts normal bus arbitration.

TDMA establishes a network cycle time. This is then subdivided into slots for messages. During that cycle, all devices are provided a time slot for transmitting data. This allows network access to all devices on a very predictable basis. Many of the emerging high-speed networks use this technique,[5] and it is suitable for time-critical applications like voice and video data transmission. Although there is a possibility that network capacity will be wasted, determinism is optimized.

Bus arbitration techniques do not end here and, in many cases, combinations or variations on the above techniques are employed in industrial network protocols.

NETWORK SOUP

There have been a number of efforts to compare networks.[15,16] Here, some common industrial communication protocols are classified by focusing on the native application domain. Understanding network classification should permit choice of the right tool for the job. This subsection segregates communications protocols into the following four basic categories[6] (Figure 4.4f):

1. Sensor networks that encompass protocols initially designed to support discrete I/O
2. Device networks that were originally focused on process instrumentation

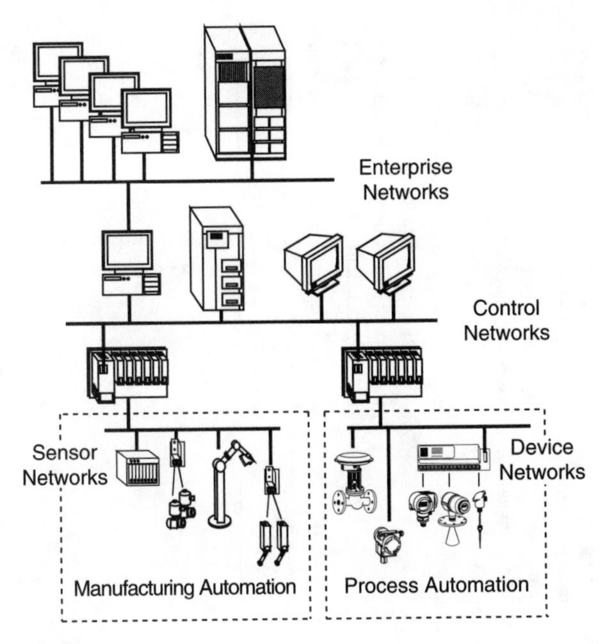

FIG. 4.4f
Network classifications.

3. Control networks that are typically used to connect controllers and I/O systems
4. Enterprise networks that focus on information technology applications

Sensor-level protocols are the simplest networks around today. These networks are principally focused on supporting discrete sensors and actuators (i.e., digital). These buses target manufacturing automation: sensors such as limit switches and push buttons, and actuators such as solenoid valves and motor starters. These protocols tend to have very fast cycle times. Because these buses are frequently promoted as an alternative to PLC discrete I/O, the cost of a network node must be relatively low.

Device-level protocols tend to support process automation and more complex transmitters and valve-actuators. In the process automation domain, the activities tend to be fundamentally continuous and analog. Typical transmitters include pressure, level, flow, and temperature. Actuators may include motorized valves, I-P controllers, and pneumatic positioners. Data tend to be floating point, and status information is usually available. Because the functions performed are more sophisticated than in a simple discrete device the cost per node tends to be higher.

Control-level protocols are typically used as backbones for communications between I/O systems, controllers, operator stations, and supervisory systems. Consequently, these networks often move chunks of heterogeneous data and operate at high data rates.

The large number of industrial networks signifies that control systems are evolving from simple inputs and outputs toward bidirectional communication with "smart" field devices. Although bidirectional communication moves from the field up the automation pyramid, enterprise networks are pushing down ever further into the historic domain of the control system. This results in a very blurred line between control networks and enterprise networks.

Enterprise networks are really a collection of LANs, wide-area networks (WANs), and a wide range of communication protocols. Frequently, several protocols are in simultaneous use on the same physical connection. The data carried on these networks are diverse, including process and production data, e-mail, music, images, and a wide range of business and financial transactions. The number of network-capable applications is growing rapidly and adding to the network traffic. Because there is a huge and growing number of nodes, the cost per node is very low.

Of course, these categories are a simplification because the separation between the categories is often blurred. In many cases, a protocol that best fits in one category can perform some or many of the functions from another. For example, AS-i (developed for discrete I/O) can communicate analog data (like temperature), and some process instruments support MODBUS (a control network).

Sensor Level

This subsection provides an overview of several sensor-level protocols.[14] In a given plant there may be 10 or 100 discrete points for every process transmitter and positioner. Consequently, sensor-level networks have the potential to encompass the largest number of nodes in the industry. Any successful sensor-level bus must meet the following requirements:

- *Simplicity.* Discrete I/O is simply an on/off contact, an indicator light, or a motor starter. Although there is often important high technology involved, the driving force is not generally electronics. Simple networks best serve simple transducers.
- *Fast cycle times.* High-speed manufacturing equipment like automatic welders, robots, and packaging equipment use a large amount of discrete I/O. Consequently, this is a prime application area for sensor networks and these applications require fast cycle times.
- *Very low cost.* Adding a sensor network to high-volume, discrete transducers adds cost to what is already a cost-sensitive product. The sensor network must be sufficiently low cost to compete against dumb transducers connected to discrete I/O cards. Because the machines using the network may be standard products selling on competitive bids, pricing is very sensitive.

Five networks are reviewed in this subsection: AS-i, CAN, DeviceNet, Interbus, and LON. All these networks have an established and substantial base of business and meet the above requirements. In addition, these examples encompass four very different approaches to sensor network design. The sensor-level networks are summarized in Table 4.4g.

TABLE 4.4g
Sensor Network Characteristics

	Masters	Signaling	Nodes	Data (bytes)	Pros	Cons	Installed Base
AS-i	Single	Alternating pulse modulation	31	4 bits	Extremely simple with innovative two-wire cabling; no connectors, instead wiring clamps to the AS-i node	Limited network size; not well suited for anything other than discrete transducers	Widespread use in manufacturing automation especially for replacing discrete I/O wiring in assembly and packaging equipment
CAN	Multiple CSMA	NRZ; bit rates programmable	64	8	Simple, fast, and reliable with unique, priority-based collision avoidance scheme; high volume results in low-cost interface chips	Only physical and link layer specified; limited message length complicates support for complex devices	CAN was invented by Bosch and is used in a wide variety of automobiles; huge number of nodes and many microcontrollers provide integrated CAN support; provides basis for several others (e.g., DeviceNet CANOpen, SDS)
DeviceNet	Multiple CSMA	NRZ; bit rates programmable	64	8	CAN-based with well-defined application layer; connectors, cabling, and power distribution standardized; devices can be "hot-swapped"	Wide range of messaging schemes can make trouble-shooting difficult; limited message length best supports simple devices	Recommended network for semiconductor industry; widespread use for discrete I/O, operator panels, simple devices, and assembly and packaging systems
InterBus	Single	Ring topology; virtual "shift register" model	256	64	High-speed deterministic cyclical communication; network address automatically assigned, simplifies commissioning	Bad node can break the ring	Large installed base in manufacturing automation connected to discrete transducers; widely used in material handling, assembly, and packaging
LON	Multiple CSMA/CD	Multiple physical layers; bit rates programmable	32K	228	Standardized data (SNVTs); Internetworking support (i.e., networks of networks) built into LON	Network collisions are possible and reduce determinism	Large installed base in building automation and HVAC controls; limited support in manufacturing and process automation

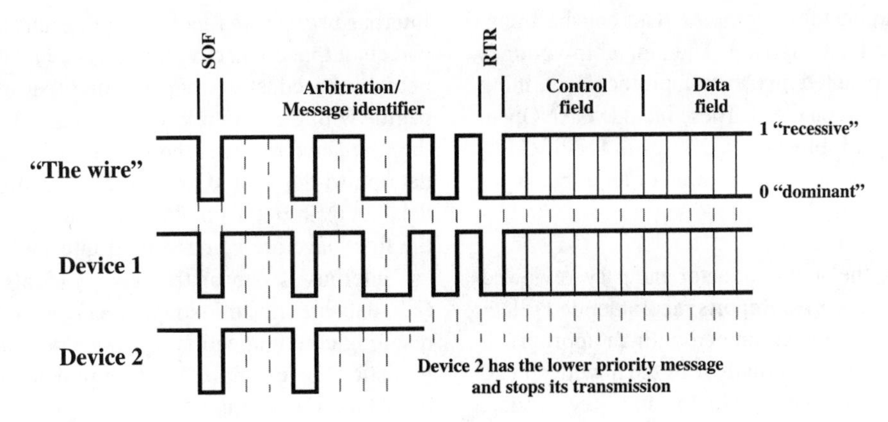

FIG. 4.4h
CAN medium access control.

AS-i AS-i can be thought of as a network-based replacement for a discrete I/O card. Therefore, AS-i offers a simple network consisting of up to 31 slave devices. The master polls each slave. The master message contains four output bits and the slave answers immediately with four input bits. Diagnostics is included in each message. In other words, the messages are fixed with four input and four output bits per message. A complete network of 31 slaves results in 248 I/O bits and a worst-case scan time of less than 5 ms.

AS-i is perhaps the simplest network in use. This simplicity carries over to the wiring as well. Standard AS-i wiring consists of a yellow, flat, mechanically coded, two-wire cable. This cable is clamped to the AS-i device and, in doing so, the cable is pierced, making an environmentally sealed electrical connection. The molded shape of the cable ensures the correct connection with the device. The two-wire connection provides both power and communications to the device; connectors or terminal blocks are optional. Data rates are fixed and, as a result, no confusion over baud rates and parity settings occur.

AS-i slave devices can be very simple as well. In most cases the AS-i interface chips (ASICs) support the entire protocol and the network electronics can be very small (often less then 2 cm^2). Because the ASICs provide most of the AS-i support needed, networking and software-related development expenses are minimized for manufacturers of discrete transducers (e.g., limit switches and solenoid valves).

If AS-i is a replacement for a discrete I/O card, then connectivity is important for success of the protocol. AS-i has good support and connectivity to a variety of PLC and control system vendors, single board computers, and other networks (e.g., PROFIBUS, MODBUS, DeviceNet, and Interbus).

CAN Bosch developed the controller area network (CAN) in the 1980s for automotive applications. Today it is used in a large number of vehicles and in a variety of other applications. As a result, a large number of different chips and vendors support CAN. The total chip volume is huge, and the parts cost is small (less than U.S. $1).

CAN was developed to provide simple, highly reliable, and prioritized communication between intelligent devices, sensors, and actuators. Because of its origins in target vehicle applications, reliability is paramount. Network errors that occur while on a highway or interstate are a "bad thing."

CAN defines only basic, low-level signaling, and medium-access specifications. These are both simple and unique. Although CAN medium access is technically CSMA/CD, this classification can be misleading (Figure 4.4h). All data are encoded as an NRZ bit stream, and when the network is idle (i.e., the signal is high) a device can begin a message by sending a start of frame. All devices on the network synchronize to the resulting high-to-low transition.

Note: Because CAN sends NRZ data, the bit rate must be configured. In addition, propagation delays limit the maximum bit rate suitable for a given cable run length.

The start of frame is followed by the arbitration sequence. CAN uses a "nondestructive bit-wise arbitration" technique. In effect, all devices wanting to access the bus place a message identifier on the bus during the arbitration interval. As the 1s and 0s are clocked onto the network, the device monitors the network. When it sees a 0 where it placed a 1 on the network, it loses its bid for the bus. The devices losing their bids for the bus listen to the communication and when the bus goes idle again they can attempt to access the network. This arbitration technique ensures there are no collisions on the bus and highest priority message receives first access.

Note: A CAN device generally supports more than one message type and the different messages can have different priority levels. CAN application layer protocols sometimes use a combination of a message group and the node address to form the message identifier.

The CAN data field can contain up to 8 bytes of data. This is consistent with CAN objectives for supporting discrete transducers. It also allows simple numeric information to be transferred. Larger, more complex data can be transported using segmented data transfer. The high chip volumes and low costs have resulted in several protocols defining upper layers that are based on CAN. These include CANOpen, J1939 (for trucks and buses), SDS, and DeviceNet. DeviceNet will be reviewed as an example of an upper-layer protocol for CAN.

DeviceNet In 1994, the semiconductor industry reviewed a variety of industrial networks. Reports recommended selection of a CAN-based solution combined with an appropriate upper-layer protocol.[19] This eventually resulted in widespread use of DeviceNet in the semiconductor industry. Today, DeviceNet is a well-established machine and manufacturing automation network supported by a substantial number of products and vendors.

DeviceNet specifies physical (wiring, power distribution, connectors, network terminators) and application layer operation based on the CAN standard. Because CAN medium arbitration is very flexible with no explicit master, DeviceNet provides configurable structure to the operation of the network. This allows (for example) the selection of polled, strobed, cyclic, or event-driven network operation. All the popular application layer hierarchies are supported (e.g., master/slave; peer-to-peer; client/server; and publisher/subscriber). However, in practice, most devices only support master/slave operation, resulting a significantly lower cost position.

Data can be accessed simply as blocks of I/O data or explicitly as part of an application layer object. Explicit data access allows application layer objects to be organized into profiles that support different kinds of transducers and applications. Its node address, object and instance identifiers, and attribute number identify a specific property (i.e., datum) within the object. This combination of four enumerations allows data to be identified and explicitly transferred across the DeviceNet network.

Note: Several protocols use the term *profile*. A profile generally refers to a standardized set of data and procedures that supports a particular transducer or application layer function. Example profiles might be defined for (1) a tangible device, such as a motor starter, variable speed drive, limit switch, or a position sensor, or (2) an abstract concept, such as a message router or a connection.

The combination of low-cost network chips, moderate message size (up to 8 bytes of data), flexible messaging, and standardized application objects allows DeviceNet to address a significant number of automation problems. Connectivity is good with support from many PLC and control system companies. In addition, connection of DeviceNet to other protocols (e.g., AS-i; MODBUS, PROFIBUS-DP, Ethernet) is available.

Interbus Interbus was developed in 1984 by Phoenix Contact and is a popular industrial network (especially in Europe). Interbus uses a ring topology with each slave having an input and an output connector. Interbus can be thought of as a large network-based shift register. In other words, a bus cycle begins with the network master transmitting a bit stream. As the first slave receives the bits, they are echoed, passing the data on to the next slave in the ring. Simultaneous with the data being shifted from the master to the first slave, data from the first slave are being shifted into the master.

Interbus is one of the few protocols that is full duplex (i.e., data are transmitted and received at the same time). The resulting communication is cyclical, efficient, fast, and deterministic. For example, 4096 digital inputs and outputs can be scanned in 14 ms.

One nice feature of Interbus is the simplicity of commissioning. Because of its ring topology, node addresses are not required. The master can automatically identify the nodes on the network. In addition, slaves provide identification information that allows the master to determine the quantity of the data provided by the slave. Using this identification data, the master explicitly maps the data to and from the slave into the bit stream as it shifts through the network. This also allows precise identification of the fault position in the ring to be provided, thus simplifying troubleshooting (if a malfunction should occur). Unlike other high-speed industrial networks, terminators are not required.

Connectivity is good with many PLCs and systems supporting Interbus. Interbus also supports connections to other networks (e.g., AS-i). Standardized profiles are defined for such equipment as robotic controllers, encoders, and variable-speed drives.

LON Echelon Corp. developed LON (local operating network) in the 1980s. The concept was to simplify sensor and control networks dramatically by embedding the entire communications stack in a microcontroller (i.e., the Neuron chip). Therefore, LON is focused on moving simple sensing and control information between LON devices.

To support product development Echelon produces some of the finest (albeit expensive) network development tools in use.[19] The combination of a network on a chip and the corresponding tools allows a LON device to be developed very quickly. As a result, one strong application area is using LON chips to create a proprietary network that is embedded in a variety of products (e.g., between sensors in a flow computer or within an office telephone system).

Of course, many open systems products exist and the LONMark user group supports conformance testing. In terms of open system products, the vast majority are for building automation. In this industry segment, LON may be the preferred sensor-level solution.

LON supports a wide range of physical layers including twisted-pair wiring and even via the mains power. The data communication bit rate depends on the physical layer and in some cases the device is powered via the network wiring. Medium access uses CSMA and collisions can degrade response times and determinism. However, LON includes a prioritization algorithm that minimizes the probability of a collision.

LON devices can be grouped into a domain consisting of subnets (communication channels) containing a number of nodes. This allows data to be routed across multiple subnets (i.e., internetworking is supported). This subnet scheme, similar to that used by TCP/IP, is discussed in Section 4.2.

LON has not made significant penetration into manufacturing and process automation. Connectivity to other industrial networks is limited.

Device Networks

Whereas sensor networks are mainly a product of manufacturing automation, device networks are a product of process automation. For manufacturing automation, functions like discrete I/O and high speed are main concerns. On the other hand, process automation grew from the need to accurately measure and control continuous physical processes. In effect, manufacturing automation is predominantly digital and process automation is analog.

The following characteristics are commonly found in device networks:

- The devices tend to be more complex than those found in sensor networks. The functions are basically analog and measure or control a continuous process. Because the devices are more complex, the status information becomes richer and more varied.
- Floating-point numbers with engineering units generally represent the data. Precision and accuracy are critical and the calibration of the device is fundamental. Message sizes tend to be larger and more complex.
- The devices tend to have a one-to-one connection to the process, as opposed to sensor buses where many discrete devices are often connected to a single node. Of course, multivariable devices are popular. However, all the measured or derived process data tend to relate to the process connection (e.g., differential pressure and calculated flow, level gauges calculating volume).
- Continuous processes tend to be much slower, and communication can be slower. However, determinism is still important and consistent sampling intervals tend to reduce aliasing of the data.

Three networks are reviewed here: Foundation Fieldbus H1, HART, and PROFIBUS-PA (Table 4.4i).

Foundation Fieldbus H1 The concept of "fieldbus" has been discussed in international technical committees for many years. Generating consensus has been difficult. One approach to solving these problems was the Interoperable Systems Project (ISP). This organization merged with FIP North America and formed the Fieldbus Foundation in 1994. This, in turn, led to the release of the Foundation Fieldbus (FF) H1 specifications in August 1996 (see Section 4.12 for more information).

Perhaps the most important characteristic is the FF focus on "the network is the control system."[21] This is fundamentally different from the controller and I/O approach used in traditional systems. In other words, FF specifies not only a communication network but control functions as well. As a result, the purpose of the communications is to pass data to facilitate proper operation of the distributed control application. The success of FF in this endeavor relies on synchronized cyclical communication and on a well-defined applications layer.

Communications occurs within framed intervals. These frames are of fixed time duration (i.e., a fixed repetition rate) and are divided into two phases. In the first phase, scheduled cyclical data exchange occurs and, in the second, acyclic (e.g., configuration and diagnostics) communication is performed. The "Link Active Scheduler" (LAS) controls the communication. The LAS polls the network, prompting the process data to be placed on the bus. This is done by passing a special token to grant the bus to the appropriate device. The net result is that the cyclic data are generated at regular intervals.

The application layer defines function blocks. These include analog in, analog out, transducer, and the blocks traditionally found in a distributed control system (DCS) (e.g., PID and ratio control). The data on the network are transferred from one function block to another in the network-based control system. The data are complex and include the digital value, engineering units, and status of the data (e.g., to indicate the PID is manual or the measured value is suspect).

Note: "Function blocks" are supported by several industrial protocols. The role of function blocks differs from protocol to protocol. Great care should be taken when comparing function block capabilities.

This sophistication comes at a cost. The FF network has significant overhead and even the simplest devices contain large amounts of memory and processing power. The protocol stack is sufficiently complex that a certified stack and function blocks are usually licensed by the product developer.

The FF installed base is relatively small: however, it is being promoted heavily by several major vendors. Because FF represents a major shift in thinking, plant procedures and engineering practices should be changed before its benefits can be maximized.[17]

HART HART was developed by Rosemount Measurement in the 1980s and today is owned by the independent HART Communication Foundation (see Section 4.11). HART is different from every other network discussed in this section because it is fundamentally an analog communications protocol. In other words, all the other protocols use digital signaling and HART uses modulated communications. This allows HART to support two communications channels simultaneously. The first is a one-way channel carrying a single process value (i.e., the 4–20 mA signal). The second is a bidirectional channel used to communicate digital process values, status, and diagnostics. "HART digital communications"

TABLE 4.4i
Device Network Characteristics

	Masters	*Signaling*	*Nodes*	*Data (bytes)*	*Pros*	*Cons*	*Installed Base*
Foundation Fieldbus (H1)	Multiple using token passing	Manchester 31.25 kbps (SP-50 physical layer)	30	128	The network is the control system; process-related standardized function blocks (e.g., AO, AI, PID, etc.)	Complex, expensive devices requiring large computer resources and high communications overhead	Process automation principal focus; small but growing installed base; sometimes considered the U.S. fieldbus
HART	Multiple using token passing	Modulated using FSK/PSK	64	255	Large range of devices and applications; simple for plant personnel to understand; lowest-cost process instruments	Lots of multivariable field devices; however, support for access to secondary process variables is poor in some hosts (i.e., DCSs, PLCs)	Supported by most "smart" 4–20 mA devices; diverse range of products including many niche applications; huge and growing installed base
PROFIBUS-PA	Single	Manchester 31.25 kbps (SP-50 physical layer)	30	244	Profiles for common instrument types simplify configuration	Requires PROFIBUS-DP backbone connection	Process automation principal focus; small but growing installed base; sometimes considered the European fieldbus

is actually a modulated analog signal centered in a frequency band separated from the 4–20 mA signaling.

HART focuses on enhancing smart 4–20 mA field devices by providing two-way communication that is backward-compatible with existing installations. In other words, HART devices can be used in traditional 4–20 mA applications. In addition, HART is compatible with current plant personnel and practices. Furthermore, utilization of HART technology can be gradual, avoiding the step change forced by purely digital protocols.

One of the biggest challenges facing a HART field device developer is power. The typical two-wire HART-compatible device must operate on 30 to 40 mW. Consequently, HART is a simple (not trivial) protocol designed with the field device designer in mind. Other facts about HART include (see Section 4.11):

- Two masters are supported using token passing to provide bus arbitration.
- HART also allows a field device to publish process data ("burst-mode").
- Cyclical process data include a floating-point digital value, engineering units, and status.
- Operating procedures are standardized (e.g., for loop test, current loop re-ranging, and transducer calibration).
- Standardized identification and diagnostics are also provided.

There are millions of HART field devices installed and its acceptance continues to grow. In many cases, HART field devices have become low-cost, commodity-priced instruments. The market share and cost position have resulted in widespread connectivity to PLCs and control systems. In addition, some industrial protocols have developed standards for embedding HART communication. This is resulting in the interesting trend of using HART field devices connected to remote I/O supported by a backbone protocol (e.g., PROFIBUS-DP). This technical approach reduces costs and wiring in new systems and in control system retrofits.

PROFIBUS-PA PROFIBUS-PA (process automation) was developed to extend PROFIBUS-DP to support process automation and was introduced in 1997. PROFIBUS-PA operates over the same H1 physical layer as Foundation Fieldbus H1. PROFIBUS-PA is essentially a LAN for communication with process instruments. PROFIBUS-PA meshes well with traditional control strategies using DCSs and PLCs. In other words, control-in-the-field is not normally employed via a PROFIBUS-PA network.

A PROFIBUS-PA network can be thought of as a PROFIBUS-DP spur extending into, for example, the hazardous areas of a process plant. The connection to PROFIBUS-PA network is always via a segment coupler to a PROFIBUS-DP network.[25] PROFIBUS-PA networks are fundamentally master/slave and sophisticated bus arbitration is not needed or employed.

PROFIBUS-DP slot-index addressing is used to support simple function blocks in PROFIBUS-PA devices, such as physical, transducer, and analog input (AI) and output (AO) blocks. The physical and transducer blocks contain some device-specific data. AO and AI blocks are used for scaling and to provide process data.

PROFIBUS-PA also defines profiles for common process instruments. These profiles include both mandatory and optional properties (data items). When a field device supports a profile, some configuration of the device should be possible without device-specific knowledge (such as provided by a Device Description (DD) or Device Type Manager (DTM)).

The PROFIBUS-PA installed base is relatively small; however, it is being promoted heavily by several major vendors.

Control Networks

Device and sensor networks are principally focused on communication with primary elements such as solenoid valves, limit switches, positioners, and pressure transmitters in a process unit or manufacturing cell. Control networks are more focused on providing a communication backbone that allows integration of controllers, I/O, and subnetworks.

Control networks are at the crossroads between the growing capabilities of industrial networks and the penetration of enterprise networks into the control system. Technology and networks at this level are in a state of flux and, in many cases, the technology is relatively new. Some characteristics of control networks include:

- Control networks can handle large amounts of complex data. Usually, a control-level protocol includes specifications for modeling the data.
- Control networks generally support multiple masters. Communications techniques like peer–peer, client/server, and publisher/subscriber are generally offered.
- In many cases, support for subnetworks and internetworking is included, allowing access to large numbers of nodes and devices.

Some control networks have evolved from manufacturing automation (e.g., ControlNet, PROFIBUS-DP, MODBUS) and other process automation (FF-HSE). In some cases, the networks are an adaptation of another network to TCP/IP (e.g., FF-HSE, Ethernet/IP).

Six networks are reviewed in this section: BACnet, ControlNet, Ethernet/IP, Foundation Fieldbus HSE, MODBUS, and PROFIBUS-DP (Table 4.4j).

BACnet BACnet (Building Automation and Control Network) was developed by a working group of the American Society of Heating, Refrigerating and Air-Conditioning Engineers (ASHRAE) between 1987 and 1995. As its name suggests, this protocol is focused on building and facility management.

TABLE 4.4j
Control Network Characteristics

	Masters	Signaling	Nodes	Data (bytes)	Pros	Cons	Installed Base
BACnet	Multiple	ARCNET, TCP/IP, LON, Ethernet	Huge		WAN for buildings; standardized objects and functions (e.g., lighting, energy management, air quality)	System-level focus rather than hard real-time determinism within a local network of controllers and instruments	Targets building automation and facility management
ControlNet	Multiple—TDMA	Fiber, coax (Manchester)	99	510	Good support for large sets of complex data; redundancy standard; very deterministic	Relatively expensive; does not support twisted pair wiring	Targets control system networks and communications between automation cells; small but growing installed base
Ethernet/IP	Multiple—CSMA/CD	Ethernet (Manchester)	Huge		Allows integration of DeviceNet and ControlNet object model and services into TCP/IP	Relatively new technology; released in 2000	Relatively new; targets DeviceNet and ControlNet upper layer over TCP/IP
Foundation Fieldbus (HSE)	Multiple—token passing	Ethernet (Manchester)	—	128	Process control function blocks can operate across Ethernet; allows redundant networks to be constructed from standard Ethernet components	Relatively new technology; released in 2000; requires special linking devices to FF H1; first approved products 2001	Relatively new high speed backbone for internetworking FF H1 subnets
MODBUS	Single	Asynchronous, NRZ, selectable up to 19.2 kbps	247	254	Simple-to-write drivers using standard PC serial port; large numbers of hosts and RTUs	Difficult to exchange complex sets of data due to weak application layer standardization	Widely used backbone bus for integrating controllers and I/O systems; supported by a wide variety of products; mature with a huge and growing installed base
PROFIBUS-DP	Multiple—token passing	9600 to 12 Mbps	127	244	Scalable in both communication speed and sophistication; operations can be synchronously triggered	A collection of several optional upper protocol layers (DP, DPV1, DPV2, and FMS) can be confusing	Mature with huge and growing installed base as a backbone bus for integrating controllers, I/O systems, and instruments; mostly used in manufacturing automation

RTU = remote terminal units.

The principal focus of BACnet is on the integration of such building systems as:

- Fire, safety, and security systems
- Lighting and energy management
- HVAC systems
- Integration with utility providers

In general, BACnet does not focus on hard real-time requirements. Instead, it focuses on data integration between building infrastructure systems. Consequently, its principal focus is on defining standard (albeit abstract) object definitions for common data, instrumentation, and control functions. Many of these objects have standard services like reading the properties of the object. This standardization of common functionality and objects (i.e., building blocks) is similar to the profiles found in other protocols.

An interesting fact about BACnet is found in its lower protocol layers. BACnet allows the selection of any combination of the lower protocol layers. The lower layers it supports include Arcnet, TCP/IP (i.e., BACnet/IP), LON, and several others. In effect, BACnet starts at the network layer and defers the physical and link layers to other protocols.

BACnet appears to have general acceptance in its industry segment and a substantial amount of manufacturer support. Many of the issues it addresses (e.g., integration of equipment from different manufacturers) are very similar to the problems addressed by other protocols in this section.

ControlNet Allen-Bradley developed ControlNet in 1996 as a very high performance network suitable for both manufacturing and process automation. ControlNet uses time division multiple access (TDMA) to control access to the network. This means a network cycle is assigned a fixed repetition rate. Within that bus cycle, data items are assigned a fixed time division for transmission by the corresponding device. Basically, data objects are placed on ControlNet within a designated time slot and at precise intervals. Once the data to be published are identified along with their time slot, any device on the network can be configured to use the data.

In the second half of a bus cycle, acyclic communications is allowed to occur. The ControlNet TDMA algorithm results in very low jitter and consistent sampling intervals. Not only is the ControlNet throughput more efficient than polled or token-passing protocols, data transmission is also very deterministic.

Data objects are defined and instantiated in basically the same fashion used by DeviceNet and Ethernet/IP. This common application layer data model is called control and information protocol (CIP). This strategy allows objects (data items and services) to be shared transparently across any of the three (DeviceNet, ControlNet, and Ethernet/IP) protocols.

ControlNet supports both coaxial and fiber-optic physical layers. Unlike Ethernet and RS-485, ControlNet can be connected within the hazardous areas of plants. Media and communications redundancy is mandatory for all ControlNet nodes.

Connectivity for ControlNet is good with support and linking devices for a number of other protocols (e.g., FF-H1). Although ControlNet is a standard part of the systems offerings of several vendors, the installed base seems relatively small.

Industrial Ethernet Ethernet and the use of commercial off-the-shelf networking products is generating a lot of interest.[23,24] In addition, many of the industrial communication protocols are specifying mechanisms that embed their protocols in Ethernet. However, Ethernet only addresses the lower layers of communications network and does not address the meaning of the data it transports. In fact, Ethernet is quite good at communicating many protocols simultaneously over the same wire and provides no guarantees that the data can be exchanged between the different protocols.

Furthermore, the protocol being adopted is TCP/IP and not Ethernet at all. TCP/IP is generally used to support the session, presentation, and application layers of the corresponding industrial protocol. Two approaches are used. In the first approach the industrial protocol is simply encapsulated in the TCP/IP. This allows the shortest development time for defining industrial protocol transportation over TCP/IP. The second approach actually maps the industrial protocol to TCP and UDP (universal data protocol) services. Although this strategy takes more time and effort to develop, it results in a more complete implementation of the industrial protocol on top of TCP/IP.

UDP is a connectionless, unreliable communication service. This works well for broadcasts to multiple recipients and fast, low-level signaling. Several industrial protocols uses UDP (e.g., for time synchronization). TCP is a connection-oriented data stream. This service can be mapped to the data and I/O functions found in some industrial protocols.

Note: Although the industrial protocols may be using TCP/IP services, they will not understand each other's communications. In fact, one protocol will not even hear the other's communications occurring on the same wire.

There has also been considerable discussion about all primary process instrumentation eventually supporting an Ethernet connection directly.[26] Although this may occur in some cases it will not, in general, be prevalent. Even the slowest 10-mbps link requires substantial memory and computing power. Despite continuing price decreases for memory and processing power, they still are not free. Some of the protocols discussed in this section do not require a microprocessor in an individual node and many of the protocols can be supported with low-cost, highly integrated 8-bit microprocessors. These small microprocessors will continue to be produced in larger volumes and at lower costs than microprocessors capable of Ethernet communications.[26]

Ethernet/IP Ethernet/IP is a companion protocol to ControlNet and DeviceNet.[24] A ControlNet working group developed Ethernet/IP between 2000 and 2001. Despite its name,

Ethernet/IP is a mapping of the the ControlNet and DeviceNet "control and information protocol" (CIP) to TCP/IP (not Ethernet). All of the basic functionality of ControlNet is supported. Of course, the hard real-time determinism that ControlNet offers is not present.

CIP is defined in Volume 1 of the Ethernet/IP standard and the Ethernet/IP standard is available as a free download. CIP is being promoted as a common, object-oriented mechanism for supporting manufacturing and process automation functions.

Ethernet/IP is a good contribution to the growing discussion of industrial Ethernet. However, like most industrial Ethernet offerings, Ethernet/IP is a recent development and basically the application of an existing industrial network application layer to the TCP/IP.

FF-HSE FF-HSE (high-speed Ethernet) was released by the Fieldbus Foundation in 2000,[21] and is a faithful extension of the FF "the network is the control system" concept to the TCP/IP environment. In fact, FF-HSE allows a function block application process (FBAP) to operate from one FF-H1 subnet across FF-HSE to another FF-H1 subnet transparently. Time synchronization is also included.

The key to FF-HSE is the linking device that acts as a gateway to an FF-H1 spur. Among other things, the linking device (using standard TCP/IP services) maps FF-H1 device addresses to IP addresses. This allows the routing of FF communications using normal IP techniques. It also allows FF-H1 devices to be "pinged" (see Section 4.2).

FF-HSE also supports redundancy across standard Ethernet equipment and networks. As a background activity, diagnostic messages are dispatched. The results are assembled into a table that provides a snapshot of network statistics. Because TCP/IP allows datagrams to traverse across multiple network paths, the table allows the FF-HSE datagrams to use the fastest available path. In addition, since multiple linking devices can reside on the same FF-H1 link, redundancy becomes available at the H1 level, as well.

FF-HSE is relatively new. Linking devices were first registered in 2001.

MODBUS MODBUS was developed by Modicon in 1979 and is one of the first open PLC and remote I/O protocols (see Section 4.2). MODBUS Plus and MODBUS/TCP are also available. MODBUS is a master/slave protocol. Communications can be binary (MODBUS-RTU) or a text stream (MODBUS-ASCII).

Note: MODBUS Plus is proprietary successor to MODBUS. It is not as widely supported as the original MODBUS.

Although there are standard function codes for some operations, the application layer is largely unspecified. Data transport generally resembles accessing a set of registers in a PLC. This allows the data of interest to be specified and accessed. It does not provide many clues about the meaning of the data. Even floating-point numbers are defined by convention rather than being an integral part of the protocol specification.

MODBUS is very simple, easy to implement, and widely supported. There are a large number of products including complex instrumentation that support MODBUS. MODBUS drivers are available for almost all PLCs, DCSs, and human–machine interface (HMI) software packages. However, specific knowledge of the device connected to MODBUS is generally required when integrating a system.

PROFIBUS-DP PROFIBUS-DP (decentralized peripherals) was originally developed in cooperation with the German government in 1989.[18] PROFIBUS-DP actually encompasses several protocols.

- The basic PROFIBUS-DP is a master/slave protocol primarily used to access remote I/O. Each node is polled cyclically updating the status and data associated with the node. Operation is relatively simple and fast.
- FMS (fieldbus message specification) is also supported by PROFIBUS. It is a more complex protocol for demanding applications that include support for multiple masters and peer-to-peer communication. Use of this protocol is expected to decline.[25]
- DPV1 is a multimaster enhancement to PROFIBUS-DP that supports multiple masters using token passing. Network access is divided into cyclic and acyclic. These occur at a fixed repetition rate. Cyclic access is similar to that used in the original PROFIBUS-DP and focuses on real-time continuous scanning of the nodes connected to a "Class 1" master. Acyclic access is via a "Class 2" master and provides for accessing configuration information or other, occasional communications needs.
- DPV2 is a PROFIBUS-DP variant, targeting motion control.

These variations allow PROFIBUS-DP to be configured to best serve the application including a simple master/slave remote I/O network or a complex multimaster token-passing network.

Slot and index address application layer data. A slot, for example, may correspond to a physical I/O card in a remote I/O chassis. The index refers to a block of data. Class 1 cyclic access is always to the same index and is limited to 244 bytes. However, Class 2 access can be to any random slot/index combination. This allows the data to be logically organized as an object instance residing at a specific index. Standard profiles are defined for common automation equipment. In addition, data definitions can be described in GSD files and DDL. PROFIBUS-DP also supports FDT/DTM for applications running on MS-Windows PCs.

PROFIBUS-DP is supported by most PLCs and control systems. The installed base is large and the majority of the products support manufacturing automation. As with MODBUS, there are process instruments that connect directly to PROFIBUS-DP.

Enterprise Networking

As industrial networks add communications starting at the grass roots of the plant, information technology departments push enterprise networks further into all the nooks and crannies of a plant. In some cases, this results in competition as IT moves down the automation pyramid and industrial networks endeavor to provide improved system integration.

Enterprise communications is characterized by a diversity of protocols, LANs, WANs, and communications technologies. In addition, enterprise communications is rapidly increasing in capacity and raw communications speed. The product volumes and the technology rate of change dwarf manufacturing and process automation industry efforts. This section briefly reviews:

- LANs
- Protocols
- The Internet

LANs Within LANs, the rapid increase in bandwidth and networked applications feed each other's growth. Two basic technologies can be found: Token Ring and Ethernet. Of the two, Token Ring is the most efficient and technically superior. Token passing prevents collisions on the LAN, and, consequently, the full available bandwidth can be used. Ethernet, using CSMA/CD, can rarely use more than 30% of the available bandwidth when several nodes are simultaneously active. However, Ethernet clearly is more widely used, surpassing Token Ring as a result of its lower cost and its ever-increasing bandwidth. This is a classic example that the best technology does not always win the battle.

Token-passing networks were the first LAN technology deployed and supported by Arcnet (DataPoint), and Token Ring (IBM). Both technologies are still in use. Arcnet has evolved to be a control network contender and Token Ring is still promoted by IBM. In fact, there is a substantial installed base of Token Ring (mostly in large organizations like government agencies).

Xerox conceived Ethernet in the 1970s and a consortium consisting of DEC, Intel, and Xerox released formal specifications in 1980.[20] This is called the DIX standard. Later, IEEE adopted Ethernet under the 802.3 series of publications. The format of an 802.3 packet is different from an Ethernet (DIX) packet.

Initially, Ethernet used a bus topology operating over coaxial cable (see Section 4.2). For technical reasons and to simplify troubleshooting, most Ethernet installations now use twisted-pair Category-3 or Category-5 cable and point-to-point wiring.[27] Initial commercial Ethernet data rates were 10 mbps. However, the twisted-pair point-to-point installations allow backward compatible upgrades to 100 mbps, 1000 mbps, and even 10 gbps. The higher data rates are made possible by using more than one pair of wires in the cable. At the higher data rates, collision avoidance is incorporated into some hubs.

Although not deterministic from a purist's view, real-time data are transported over Ethernet (e.g., voice over IP and video conferencing). If one is fast enough, uncertainty and jitter can be reduced to tolerable (and, perhaps, insignificant) levels.

Protocols Both Ethernet and 802.3 specify only the lowest-level communications (i.e., physical and data link layers). In fact, Ethernet was designed to support a variety of protocols from its inception. Today, there are a large number of protocols that operate over Ethernet. This allows the medium to be used to solve a wide variety of problems. Three protocols in particular are commonly used with Ethernet: IPX, NetBEUI, and TCP/IP.

- When MS DOS was originally introduced, it contained no support for networks. IPX was one of the original networks that supported MS DOS. IPX was developed by Novell and is still commonly used by many IT departments. IPX initially supported file and printer sharing.
- NetBEUI is a protocol developed by Microsoft for use with its Windows operating systems. NetBEUI supports common network activities like printer and file sharing. NetBEUI is expected to fade as Microsoft moves toward TCP/IP.
- TCP/IP was developed for the DOD and is now widely used by many applications in addition to the Internet. TCP/IP actually encompasses a large number of protocols. In addition, many protocols are able to be embedded in IP datagrams.

As can be seen, the terms *Ethernet* and *TCP/IP* have specific technical meanings; however, these terms are often used in a generic and imprecise way. Care should be used to verify the actual meaning and the extent of actual protocol support implied by product claims.

The Internet The word *Internet* is often used to refer to the World Wide Web. The Internet actually consists of a wide range of protocols[13] and networks. The Internet is fueled by the abundance of bandwidth. The availability of communications bandwidth and tools such as Web browsers are enabling solutions not imagined even a few years ago. Research on "Internet2" ensures this trend will continue.

Web browsers are commonplace and their use as an HMI is becoming a reality. In fact, some products are even beginning to embed Web servers. OPC (see Section 5.4) is based on DCOM, which is based (in turn), on remote procedure calls (RPC), one of the Internet protocols. OPC can allow data to be captured directly into a spreadsheet. SOAP and XML (more Internet protocols) are also being studied for application into automation systems. We are living in interesting times.

On the surface, the Internet appears to be a simple and low-cost alternative to offerings by traditional automation companies. However, within plants or businesses a lot of IT and network managers keep networks and servers running.

In addition to network management problems, security can be a major issue. Already computer viruses make news headlines. As Ethernet and Internet technologies invade the plant floor, network security will become an increasing concern. In too many cases networks are configured on a trial-and-error basis. As networks that encompass islands of automation are connected, security risks will become a serious concern.

Ethernet and the Internet are not silver bullets that solve all the world's problems. These technologies are simply additional tools to consider and exploit.

CHOOSING THE RIGHT NETWORK

This section has provided an overview of many protocols. As demonstrated, all have strengths and weaknesses. Each has grown out of solving a particular problem. Within their native problem domains, they are strong. Many have capabilities extending their usefulness into other application areas.

However, do not become seduced by the technology. Take a pragmatic approach to solving problems. When evaluating particular needs, possible communications alternatives, and vendor offerings:

- Focus on the application much more than on the technology. Focusing on the application must be paramount. The person involved understands the particular problems better than anyone else. The specific, measurable benefits being sought must drive the selection process.
- Consider the costs involved. All networks claim to save money and time. The competition for capital is intense and initial costs should be weighed carefully against the less tangible promised future savings.
- Assess the network connectivity. In the end, it is data that are wanted and connectivity is how the data are obtained. Once the data are obtained, the particular network used in one area or another becomes less important.
- Understand the hidden changes and impacts. Remember that people can make the worst technology successful and the best technology fail. All communications networks result in changes that affect the people and practices in the plant. The effect of these changes must be realistically assessed although the actual changes that result can be very difficult to foretell accurately.
- Evaluate the real interoperability. All networks claim to be interoperable. In fact, the word *interoperability* has been twisted to the point that it has no meaning. Insist on specific and tangible definitions and examples of interoperability. Take a "show me" attitude. If the interoperability is abstract and difficult to "touch and feel," it may not really exist.

Most of this advice is common sense, and applying normal evaluation practices can show the way through the networking maze. Remember that, in the end, networks are simply tools and there is no perfect network.

References

1. Stallings, W., *Data and Computer Communications,* New York: Macmillan, 1994.
2. Halsall, F., *Data Communications, Computer Networks and Open Systems,* Reading, MA: Addison-Wesley, 1992.
3. Comer, D. E., *Internetworking with TCP/IP,* Englewood Cliffs, NJ: Prentice-Hall, 1991.
4. Thompson, L. M., *Industrial Data Communications,* Research Triangle Park, NC: Instrument Society of America, 1997.
5. Internet2 Website, www.internet2.org.
6. Pinto, J., "Fieldbus—Conflicting Standards Emerge, but Interoperability Is Still Elusive," http://www.jimpinto.com/writings/fieldbus99.html.
7. ISO 7498-1. *Open Systems Interconnection—Basic Reference Model,* Geneva, Switzerland: International Standards Organization, 1984.
8. Campbell, J., *C Programmers Guide to Serial Communications,* Carmel, IN: Howard W. Sams and Company, 1987.
9. Kretzmer, E. R., "Evolution of Techniques for Data Communication over Voiceband Channels," *IEEE Communications Society Magazine,* January 1978.
10. Brown, R., "A Study of Local Loop Modem Limitations," http://www.agcs.com/supportv2/techpapers/modem_limit.htm.
11. Hulsebos, R. A., "The Fieldbus Reference List," http://ourworld-top.cs.com/rahulsebos/index.htm.
12. Leung, C., "Evaluation of Undetected Error Performance of Single Parity Check Product Codes," *IEEE Transactions on Communications,* Vol. COM-31, Number 2, 1983.
13. RFC 2800, *Internet Official Protocol Standards,* Internet Engineering Task Force, May 2001.
14. "Device-Level Fieldbus Facts," Manufacturing Automation, June 1999, http://www.automationmag.com/downloads/fieldbus.zip.
15. "Fieldbus Portal," Steinhoff Automation & Fieldbus Systems, 2000, http://www.steinhoff.de/fb_comp.htm.
16. "The Synergetic Fieldbus Comparison Chart," Synergetic Micro Systems, http://www.synergetic.com/compare.htm.
17. Sheble, N., "Foundation Fieldbus Disappointment," *InTech Online,* May 31, 2001.
18. Sink, P., "Eight Open Networks and Industrial Ethernet," Synergetic Microsystems, 2001.
19. Moyne, J. R., Najafi, N., Judd, D., and Stock, A., "Analysis of Sensor/Actuator Bus Interoperability Standard Alternatives for Semiconductor Manufacturing," *Sensors Expo Conference Proceedings,* September 1994.
20. Spurgeon, C., "Charles Spurgeon's Ethernet Web Site," http://www.ots.utexas.edu/ethernet/.
21. "The Foundation Fieldbus Primer," Fieldbus, Inc., www.fieldbusinc.com 2001.
22. Newman, H. M., "Control Networks and Interoperability," *HPAC Engineering,* May 2001.
23. Cleaveland, P. and Van Gompel, D., "Moving Popular Industrial Protocols to Ethernet TCP/IP," *Control Solutions,* July 2001.
24. "Realtime Control on Ethernet," ControlNet International White Papers, http://www.controlnet.org/cnet/ethernet.pdf.
25. "PROFIBUS Technical Description," PROFIBUS Nutzerorganisation e.V., September 1999.
26. Sink, P., "Industrial Ethernet: The Death Knell of Fieldbus?" *Manufacturing Automation Magazine,* April 1999.
27. Byres, E., "Are Buses Traveling Dead-End Roads?" *WorldBus Journal,* March 2001.

4.5 Overall Fieldbus Trends

S. VITTURI

FACTORY COMMUNICATION SYSTEMS

Figure 4.5a shows the structure of an automation system for industrial plants where communication networks are extensively used to connect all the components involved. As can be seen, three factory communication systems, based on different types of networks, are present.

At the *device level*, fieldbuses are mainly used to connect device controllers and sensors/actuators. The typical operation of the device controllers consists in the cyclic exchange of data with the sensors/actuators, performed at very high speed, which can be interrupted by alarms from the field. Usually, the amount of data exchanged by the device controller with each sensor/actuator is small (some tens of bytes, or even less), but the cycle times required are of a few milliseconds.

At the *cell level*, the operations of device controllers are coordinated and monitored. This implies the transfer of considerable quantities of data, for configuration, calibration, trending, etc. In this case, however, the times allowed for the data transfer can be on the order of hundreds of milliseconds or more. These requirements can be satisfied using suitable protocols available either for local-area networks (LANs) or fieldbuses.

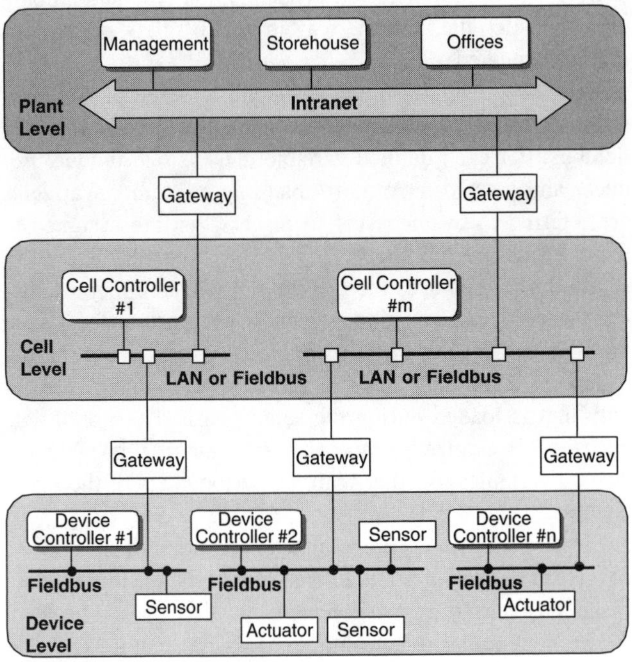

At the *plant level*, the overall production strategies are planned. There is, at this level, the necessity of interconnecting different entities such as enterprise resource planning (ERP), resource allocation, management, etc. The performance of a communication system operating at the plant level must guarantee the reliable transfer of large amounts of information in times that are not critical. Currently, the most widely used networks at this level are either LANs or intranets because their application protocols, such as FTP,[1] HTTP,[2] based on TCP/IP,[3,4] are particularly suitable to implement the required functions. Moreover, intranets use the well-known World Wide Web facilities.

FIG. 4.5a
The automation system of plants using communication networks.

USE OF FIELDBUSES IN INDUSTRIAL PLANTS

It has been shown that fieldbuses can be employed both at the device and at the cell levels of industrial automation systems.

At the device level, the use of fieldbuses allows for the replacement of the point-to-point (analog and/or digital) links between the device controller and sensors/actuators. This solution offers a considerable number of advantages:

- A substantial decrease in the cabling with the consequent reduction of the costs and of the system complexity
- The possibility of executing tests and calibrations remotely
- An increase of the signal-to-noise ratio deriving from the digital transmission techniques
- The possibility of adding/removing stations without additional cabling

At the cell level, a fieldbus may be used to realize the connection between device controllers and cell controllers. In the past, this type of communication, which can involve the transfer of considerable amounts of data, took place in several different ways. For example, data could be exchanged simply by means of a storage medium, such as a diskette, manually moved from one controller to another. Another possibility was represented by serial communication channels, often used with proprietary protocols.

FIELDBUS FUNCTIONS

The use of fieldbuses at different levels of factory automation systems is only possible if they are capable of satisfying the specific functions required by these levels:

- Cyclic data exchange
- Asynchronous traffic
- Messaging system

The implementation of these functions is realized at different layers of the typical communication profile of a fieldbus, as shown in Figure 4.5b. In particular, the cyclic data exchange and the asynchronous traffic functions, for efficiency reasons, are implemented by real-time protocols that use the services of the data link layer directly.

The messaging system, however, is realized by suitable application-layer protocols. In practice, as will be explained in the following, it is not usual to find a station connected to a fieldbus implementing the three functions contemporaneously. This is obvious because, for example, the messaging system function, while executing a bulk data transfer, could drastically slow the cyclic data exchange being performed on the same network.

Cyclic Data Exchange

With this function, a device controller is able to perform the polling of sensors/actuators connected to the fieldbus. Clearly, this operation takes place at the device level.

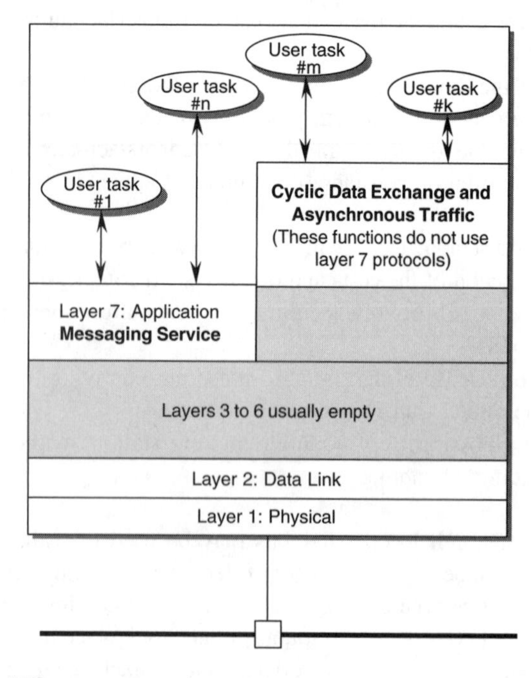

FIG. 4.5b
Typical communication profile of a fieldbus.

At every cycle a device controller sends the output signals to the actuators and reads the input signals from the sensors. The time employed to execute the complete polling is referred to as the cycle time. Normally this operation is required to be executed very rapidly, with the cycle time kept constant. Unfortunately, because of the protocols used to access the transmission medium (data link layer protocols), slight variations of the cycle time, known as *jitter*, can be introduced. This perturbation may influence, for example, control loops, or applications where a good synchronism is required, such as motion control systems. Consequently, it is very important to keep jitter as low as possible.

There are basically two possible techniques to implement the cyclic data exchange function for a fieldbus: message based and producer/consumer.

The message-based technique foresees the presence on the fieldbus of one master station (the device controller) and of some slaves (the sensors/actuators). Each station on the fieldbus has its own physical address by which it can be identified.

A possible implementation of this technique, shown schematically in Figure 4.5c, makes use of the data transfer facility of the underlying data link layer. In particular, the master, in order to poll a slave, issues a data link layer service request with which it sends a message carrying the output data for that slave; subsequently, the slave sends a message in response to the service request in which it encapsulates the input data.

The cyclic data exchange is formed by the repetitive calling of such a service by the master. Normally, this phase must be preceded by an initialization in which some indispensable information for correct operation is exchanged between master and slaves, such as the number of input/output (I/O) bytes to be transferred cyclically, the type and amount of diagnostic data, and the way in which safety procedures are implemented.

The producer/consumer technique is based on the definition of a set of variables that can be exchanged on the fieldbus. For each defined variable there is one unique producer and there can be more than one consumer. Variables are referred to by means of identifiers that are assigned to them during a setup phase.

In these fieldbuses there is normally a station coordinating the network operation, which is responsible for giving the right to transmit to the variable producers. This station, called the scheduler, works in accord with a table of periodicity that is loaded during the setup phase. The cyclic data exchange is realized simply by entering in this table the desired variables together with the periodicities of their production.

Figure 4.5d shows an example of operation of a producer/consumer fieldbus. Variable A has only one consumer, while variable B is used by two consumer stations. As can be seen, the producer/consumer technique does not require that stations present on the network have physical addresses assigned to them.

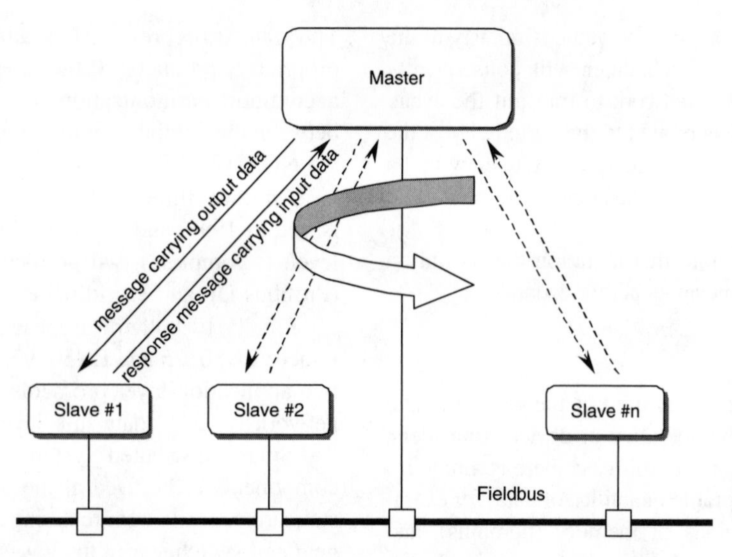

FIG. 4.5c
Message-based implementation of the cyclic data exchange function.

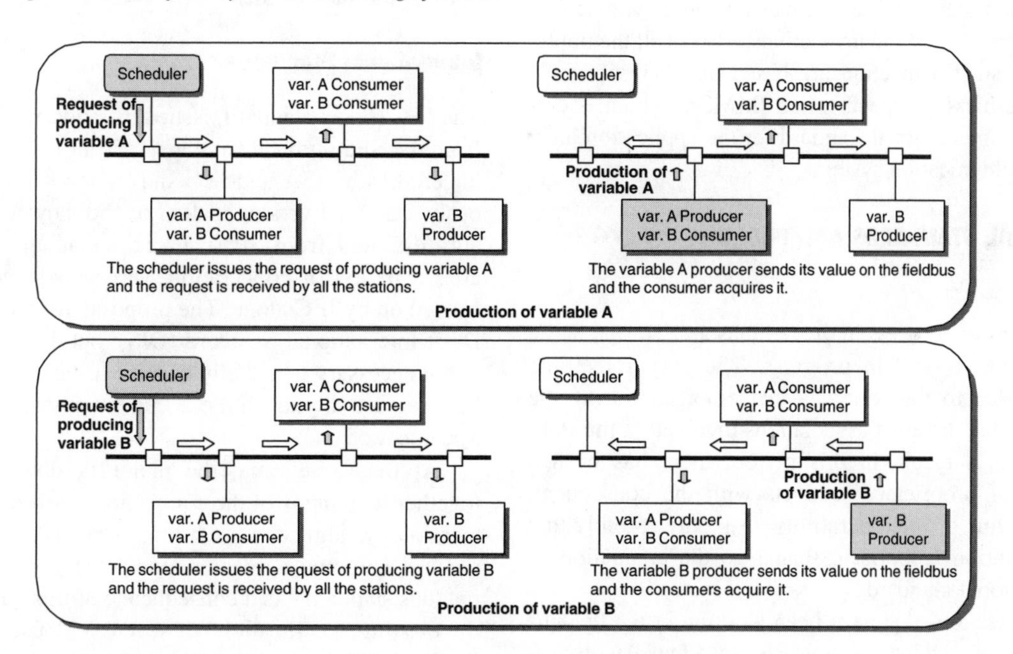

FIG. 4.5d
Producer/consumer fieldbus: example of operation.

Asynchronous Traffic

This device-level function is used to handle nondeterministic events, such as alarms. In particular, a sensor detecting a critical situation uses this function to notify the event to the device controller, which, in turn, will undertake the appropriate actions. Often, this function is also required to be able to indicate different levels of priority for the events. As can be imagined, it is of great importance to keep the reaction time to these events very short and bounded.

Normally, the asynchronous traffic function is implemented using the same protocol as that used for the cyclic data exchange. Consequently, the way in which this function is realized depends on the technique adopted for the cyclic data exchange. If the message-based technique is used, a suitable field in the slave response message can be reserved for notifying asynchronous events together with their priority.

It is worth noticing how, in this case, if an event occurs immediately after the polling of a slave, then it will be acquired by the master in the next cycle. That is, there can be a latency time in acquiring an alarm, which, in the worst case, can be equal to the cycle time.

With the producer/consumer technique a possible implementation of the asynchronous traffic exploits the frame used to produce variables. That is, a station wishing to notify an event waits until it is requested to produce a variable, whereupon it fills in a field in the frame carrying the value, similarly to the message-based technique. The scheduler will consequently grant that station access to the network to transmit the event. The latency time in this case is related to the periodicity of the produced variables. In general, the higher the frequency that a station produces variables, the lower the latency will be for that station to signal an event.

It is also interesting to note that a station that is only a consumer of variables can never generate events.

Messaging System

The messaging system function is used at the cell level and allows for the handling of the operation of device controllers. Unlike the above two functions, in this case there is a requirement for transferring considerable quantities of data, for example, configuration files, regions of memory (domains), and executable programs (program invocations). Consequently, the messaging system function has to make available services like read/write of complex variables, domain upload/download, and remote start/stop of program invocations. Almost all the implementations of such a function are based on a subset of the manufacturing message specification (MMS),[5] which is currently the only international standard for the application layer of factory communication systems.

INTERNATIONAL STANDARDS AND PROPRIETARY PRODUCTS

The standardization process in the fieldbus area started in the early 1990s and is still in progress. The very long time employed is due to the contrasts that emerged during the years between the different associations involved in the standardization. As a result, in this period, there has been a proliferation of proprietary products with the consequent, obvious problems of interoperability. It is only recently that a (partial) solution has arisen, thanks to the emanation of some international standards.

The organizations that have been working in the fieldbus standardization are the International Electrotechnical Commission (IEC), the International Standard Organisation (ISO), the European Committee for Electrotechnical Standardisation (CENELEC), the Instrumentation, Systems, and Automation Society (ISA), the American National Standard Institute (ANSI), and the Electronic Industries Alliance (EIA).

European Standards

The most important European Standard is EN50170, issued by CENELEC in 1996, which grouped together three fieldbuses previously recognized as national standards by three European countries: P-Net,[6] Denmark; Profibus,[7] Germany; WorldFIP,[8] France. Subsequently, two other fieldbuses have been added: Foundation Fieldbus[9] and ControlNet.[10]

The fieldbuses standardized by EN50170 are intended for a general-purpose use as they comprise protocols that make them suitable for use at either the device or the cell level. The standard represented an attempt to stop the increase of proprietary products. At the same time, it also tried to create a common harmonization to achieve software portability between the fieldbuses it incorporates. For this latter purpose, the NOAH ESPRIT project[11] was set up.

Another important European standard is EN50254, issued in 1998, and explicitly intended for use at the device level. It comprises two protocols extracted from EN50170 (Profibus DP and WorldFIP Profile 1) and Interbus.[12]

Finally, two other European standards have recently been issued: EN50325 and EN50295. EN50325 is concerned with two application-layer protocols based on the controller area network (CAN)[13] data link layer specification: DeviceNet[14] and Smart Distributed System.[15] This standard also refers to components to be used at the device level. EN50295[16] is a standard conceived for realizing the interface between control gear and switchgear in the low-voltage electrical distribution environment. It comprises a German fieldbus named Actuator Sensor interface (AS-i).

International Standards

The IEC 61158 standard, issued in January 2000, originated from the SP50 project for a general-purpose fieldbus, jointly undertaken by ISA and IEC and started several years previously. The final version of IEC 61158, however, is considerably different from SP50. In particular, this latter project, after a period of cooperation, was abandoned by ISA and carried on by IEC alone. The proposal arrived at the stage of Draft International Standard (DIS), but did not pass the last ballot necessary to actually become a standard. However, the IEC decided to publish the document anyway as a Technical Specification.[17,18]

At the same time, the major fieldbus organizations, together with many of the companies involved in the fieldbus technology, signed a Memorandum of Understanding to reach a common agreement for achieving an international fieldbus standard. As a consequence of this "understanding," the existing specifications of seven fieldbuses—ControlNet, Profibus, P-Net, Fieldbus Foundation HSE,[19,20] SwiftNet,[21] WorldFIP, and Interbus—were added to the IEC Technical Specification.

The resulting composite document was published as the IEC 61158 standard. The IEC 62026 standard, issued in 2000, encompasses the fieldbuses already included in the EN50295 and EN50325 documents. IEC 62026 explicitly restricts the use of all the specified fieldbuses to the low-voltage electrical distribution environment.

In the near future, a sixth section (Part 6) will be added to IEC 62026; it refers to the commercial product named Seriplex.[22] The standard is currently under evaluation by the IEC committees as a document approved for draft international standard circulation (ADIS).

The ISO 11898 standard, better known as CAN, was primarily conceived for the cabling of automotive applications inside vehicles. However, its features, together with the great availability of low-cost components, have made the adoption of CAN both possible and convenient in other environments, where fieldbuses traditionally are used. The ISO 11898 standard specifies only the first two layers of the communication profile reported in Figure 4.5b. This has caused the appearance on the market of several proprietary products implementing the application layer protocols and functions. Some of these, as mentioned above, have been included into both European and international standards.

The Lonworks fieldbus, which has widespread use in several application areas, has recently become an ANSI/EIA standard.[23] There was also the intention of including Lonworks in the IEC 62026 standard, as part 4. This possibility, however, has for the moment been abandoned by the IEC.

Proprietary Fieldbuses

In addition to all the above standards, there are some other proprietary fieldbuses, which, although not officially recognized by standards organizations, are well established. Two products, in particular, have to be mentioned: CANopen[24,25] and Modbus.[26] CANopen is based on the ISO 11898 data link layer specification and it supplies all the functions required of a fieldbus. Modbus is a master/slave protocol that can be implemented over several different types of physical layers: RS232, RS422, RS485. Moreover, recently a version of Modbus that uses Ethernet and the TCP/IP suite, also described in Reference 26, has been implemented.

Table 4.5e reports an overview of IEC and CENELEC fieldbus standards, together with their corresponding market names.

TABLE 4.5e
Fieldbus Standards and Market Names

IEC	*CENELEC*	*Market*
IS 61158—Type 1	EN50170 part 4	Foundation Fieldbus H1
IS 61158—Type 2	EN50170 part 5	ControlNet
IS 61158—Type 3	EN50170 part 2	Profibus
IS 61158—Type 4	EN50170 part 1	P-Net
IS 61158—Type 5	—	Foundation Fieldbus HSE
IS 61158—Type 6	—	SwiftNet
IS 61158—Type 7	EN50170 part 3	WorldFIP
IS 61158—Type 8	EN50254 part 2	Interbus
IS 61158—Type 3	EN50254 part 3	Profibus DP
IS 61158—Type 7	EN50254 part 4	WorldFIP profile 1
IS 62026—2	EN50295 part 2	Actuator–Sensor interface
IS 62026—3	EN50325 part 2	DeviceNet
IS 62026—5	EN50295 part 3	Smart Distributed System
ADIS 62026—6	—	Seriplex

IS = International Standard; ADIS = Approved for Draft International Standard circulation; EN = European Standard.

PERFORMANCE

The performance required of a fieldbus depends on the automation level at which it must operate. At the cell level, the services of the messaging system have to be executed employing times on the order of some hundreds of milliseconds. Considering that a typical communication service at this level can involve the transfer of some tens of kilobytes, then the data transfer rates have to be greater than 1 Mbit/s.

The most widely used physical medium is (shielded) twisted pair and the distances to cover are typically of some hundreds of meters. Many fieldbus specifications also foresee, as an alternative, the use of fiber-optic cables for very long distances or for environments characterized by high electrical noise.

The application-layer protocol of a fieldbus operating at the cell level usually introduces a model by means of which each station is considered as a container of communication objects. These communication objects are the only entities that can be exchanged on the fieldbus and they represent, for example, variables, domains, program invocations, events, etc. The data exchange can take place either with the connectionless or connection-oriented type of transmission, and for this purpose the client/server model is often used.

In Figure 4.5f a typical example of cell-level operation for a fieldbus is reported. As can be seen, two logical connections have been established, between stations A and B and between stations C and D, respectively. Over the first connection the variable identified by the object #1 is exchanged. In detail, an application task resident on station B, which has client functionality, uses the service *var_read* to issue a request to read object #1. Upon receiving this request, station A, which is a server, responds with the value of the variable. In the same way, an application task of station D uploads a domain resident on station C identified by the object #2. The service used in this case is *dom_upload*.

At the device level the correct operation of fieldbus has to ensure cycle times in the order of some tens of milliseconds with a very low jitter. However, as the amounts of data to be transmitted are low and the protocols used are very efficient, the data transfer speeds can be less than those required at cell level. In general, it may be sufficient that these speeds are greater than 500 kbit/s.

As for the cell level, the most common physical medium is (shielded) twisted pair, with fiber optic as a possible alternative. The distances to be covered at the device level are normally short, not greater than some tens of meters. A very important parameter to be taken into account when operating at this level is the efficiency of the protocols used to implement the fieldbus functions. There are actually three different definitions for the efficiency, related to the symbol coding, the data coding, and the medium access. These three efficiency definitions are not correlated among them, but should be separately evaluated when choosing a fieldbus to be used at device level.

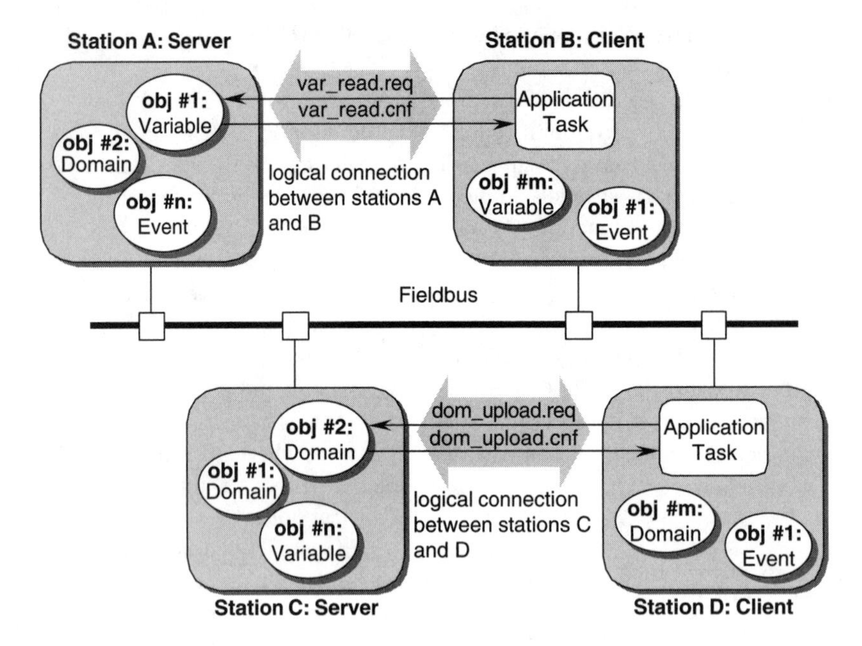

FIG. 4.5f
Example of fieldbus performance at the cell level.

The *symbol coding efficiency*, η_{SC}, is relevant to the transmission of a symbol on the network and it can be expressed as

$$\eta_{SC} = \frac{N_{USF}}{N_{TRN}} \qquad \textbf{4.5(1)}$$

where N_{USF} is the number of useful bits of the symbol, and N_{TRN} is the number of bits actually transmitted (that is, including all those necessary for the coding of the symbol).

The evaluation of η_{SC} is particularly simple when using character-oriented protocols; for example, PROFIBUS-DP has $\eta_{SC} = 8/11$, as for each 8-bit symbol it transmits 11 bits, adding start, stop, and parity bits.

However, for a bit-oriented protocol it is more difficult to determine η_{SC}, because, normally, control bits are added in a way that depends on the contents of the transmission. For example, the CAN data link layer protocol, in order to force a transition in the output signal, adds one "stuff" bit after every five consecutive transmitted bits of the same value. In this case η_{SC} is not a constant, but it can be reasonably estimated as not less than 4/5.

The *data coding efficiency*, η_{DC}, is concerned with the services of the data link layer. In this case, it is expressed as the ratio between the number of user data bits actually transmitted and the overall number of bits necessary for the execution of the service (considering also the delays introduced by the stations). Obviously, the data coding efficiency depends on the type of service and on the number of user data bits. In References 27 and 28 the behaviors of the data coding efficiency for some popular fieldbuses are reported.

The *medium access efficiency*, η_{MA}, in general allows for the evaluation of the overhead introduced by the medium-access technique used by the fieldbus. It is expressed by the ratio, within a cycle, between the time actually employed for data transfer and the total cycle time. To give an example, the data link layer protocol of PROFIBUS-DP is based on a token-passing access method. The medium-access efficiency for this fieldbus takes into account the time spent in circulating the token between the master stations. It can be expressed as

$$\eta_{MA} = \frac{T_C - NT_{TK}}{T_C} \qquad \textbf{4.5(2)}$$

where T_C is the cycle time; T_{TK} is the time necessary to pass the token from one master station to another; and N is the number of master stations connected to the fieldbus.

Reference 29 provides a detailed description of the medium-access efficiency for both PROFIBUS-DP and the IEC Technical Specification.

REDUNDANCY AND FAILURE CONSIDERATIONS

There is, in general, the necessity of protecting the operation of a fieldbus against several sources of failure. The concepts illustrated in this section are applicable to fieldbuses used either at device or cell level.

Physical Medium Failures

At the physical layer, the most serious fault that can occur is the interruption of the connection between the stations; this can be caused, for example, by a break in the network cable or, if active components are used (e.g., hubs), by an interruption of the power supply. In this case the only solution to guarantee the correct continuation of the fieldbus operation is the use of a redundant physical medium, realizing a configuration similar to that shown in Figure 4.5g. As can be seen, every station is connected to two physical mediums of

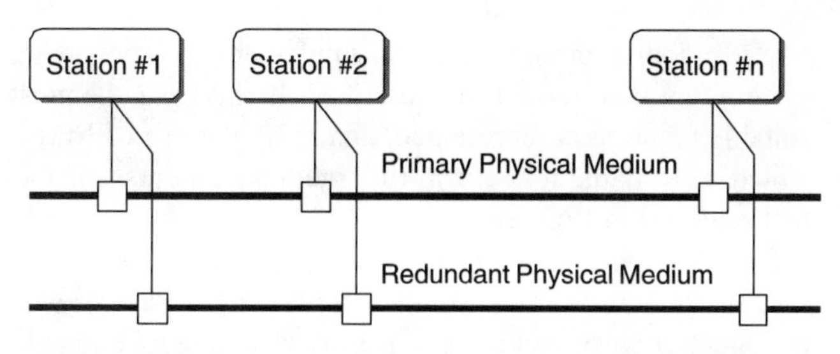

FIG. 4.5g
Physical layer redundancy.

which only one (the primary) is active. When the primary fails, communication is switched to the redundant physical medium. The way in which this switching takes place is particularly important for the features of the stations and for the overall performance of the fieldbus system.

If the switching is manual, then when a fault occurs, the system has to be stopped, the stations have to be commuted onto the redundant physical medium, and subsequently restarted. In this way, there is an interruption of the operation, but the stations are not required to supply additional features, with the (possible) exception of the connection to two separate physical mediums.

If the switching is required to be automatic, then each station has to be able to recognize a malfunction of the primary physical medium and switch over to the redundant physical medium. As can be imagined, this procedure is particularly complicated and its implementation can be considerably expensive. On the other hand, manual switching, if necessary, is easily implemented.

In practice it is infrequent to find redundant fieldbuses where automatic switching is realized.

Station Failures

These types of malfunctions occur when a station has a failure that compromises its participation in the fieldbus operation. It is obvious that the effects of the failure on the network depend on the role played by that station.

If the role is critical (typical examples are masters or schedulers for fieldbuses used at the device level), then there can be serious consequences. Generally, in these cases, the operation of the fieldbus is stopped and the other stations, which are no longer updated (slaves and producers/consumers), have to guarantee that the parts of the plant under their control are put in a safe state.

It is important to note that some fieldbus specifications foresee the possibility of having redundancy for critical stations. For example, WorldFIP and the IEC Technical Specification may have more than one station with scheduler capability connected; only one is working as scheduler at a time and one of the others can automatically replace it in case of fault.

When a problem occurs on a noncritical station (e.g., a slave), then the fieldbus operation can continue. Normally, the defective station is marked as inactive and this situation has to be signaled to the fieldbus administrator that, in turn, will initiate suitable recovery actions. More importantly, however, the defective station has to maintain the plant under its control in a safe state.

Transmission Errors

All communications systems are subject to transmission errors that can corrupt the contents of the information they carry. If, as in the case of fieldbuses, digital transmission techniques are used, the occurrence and the effects of such errors may be carefully bounded.

In particular, every fieldbus is able to detect, and in some cases correct, any possible loss or corruption of information during transmission. These objectives are achieved using suitable control bits or check fields, which are added to the useful data to transmit. It is worth noting, however, that these techniques have the disadvantage of reducing the transmission efficiency. In practice, two types of checks may be executed: on the characters and on the frames. The checks on the characters are performed by fieldbuses, which use character-oriented protocols to access the physical medium. This is the case, for example, of PROFIBUS, which uses a universal asynchronous receiver transmitter (UART) by means of which each octet is transmitted using 11 bits. On the receiver side, the control bits of each character are checked. If an error is detected, the frame to which the character belongs is discarded and, if possible, requested to be retransmitted.

A similar technique, known as block checksum, can also be applied to a block of characters to evaluate a global parity bit determined by the single parity bits of each character.

Frame checks are typically executed by all fieldbuses. They are based on the calculation of one or more octets obtained as result of a polynomial function applied to the frame to be transmitted. These octets are appended to the frame and then sent on the network. The receiver applies the same function to the received data and, if the results are different, it deduces that a transmission error has occurred. The octets obtained by the computation are referred to as frame check sequence (FCS) or cyclic redundancy check (CRC).

An important additional technique that is frequently used to detect and correct transmission errors is the use of a high Hamming distance for fields of a frame whose meanings are critical. Considering a field of a frame for which only a defined set of values (codewords) are possible, the Hamming distance is the minimum number of bits in which the contents of two codewords must differ. If an error occurs during transmission and a codeword is corrupted, then a sufficiently high Hamming distance allows the receiver to detect errors and, possibly, correct them by assigning the value of the "closest" codeword to the received field.

MARKET SITUATION AND TRENDS

The fieldbus market has suffered in recent years from the absence of a leading standard. This has made the integration between different installations difficult and has compelled the manufacturers of devices to develop and maintain equivalent products with different fieldbus interfaces.

Recently, the standards organizations have come to a partial solution of this age-old problem with the issuing of some definitive standards that amalgamate several existing products. However, these products remain incompatible with each other and, hence, it is necessary to undertake a massive action of harmonization as soon as possible.

However, notwithstanding the above problems, the number of fieldbus installations has grown considerably and this trend looks set to continue.

Use of Ethernet Networks

The impressive growth of Ethernet[30] is modifying the scenario of factory communication systems depicted in Figure 4.5a. In particular, there is a strong tendency to extend the deployment of such networks to industrial automation systems, both at the device and the cell levels where, traditionally, fieldbuses were used.

At the cell level, the replacement of a fieldbus with Ethernet is usually straightforward, because the performance of the latter equals, or often betters, that of the most popular fieldbuses. There is, however, at least for Ethernet, the necessity to adopt a protocol able to implement the messaging system functions previously described. In this case, the most suitable solution is represented, again, by the use of MMS, which is available for several Ethernet components, such as that reported in Reference 31.

At the device level, the use of Ethernet has to be considered very carefully because this network, as is well known, is nondeterministic and, consequently, it may not be able to

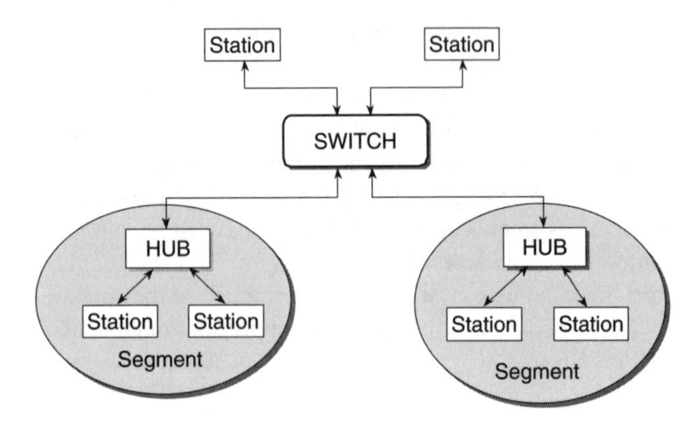

FIG. 4.5h
Example of an Ethernet switched configuration.

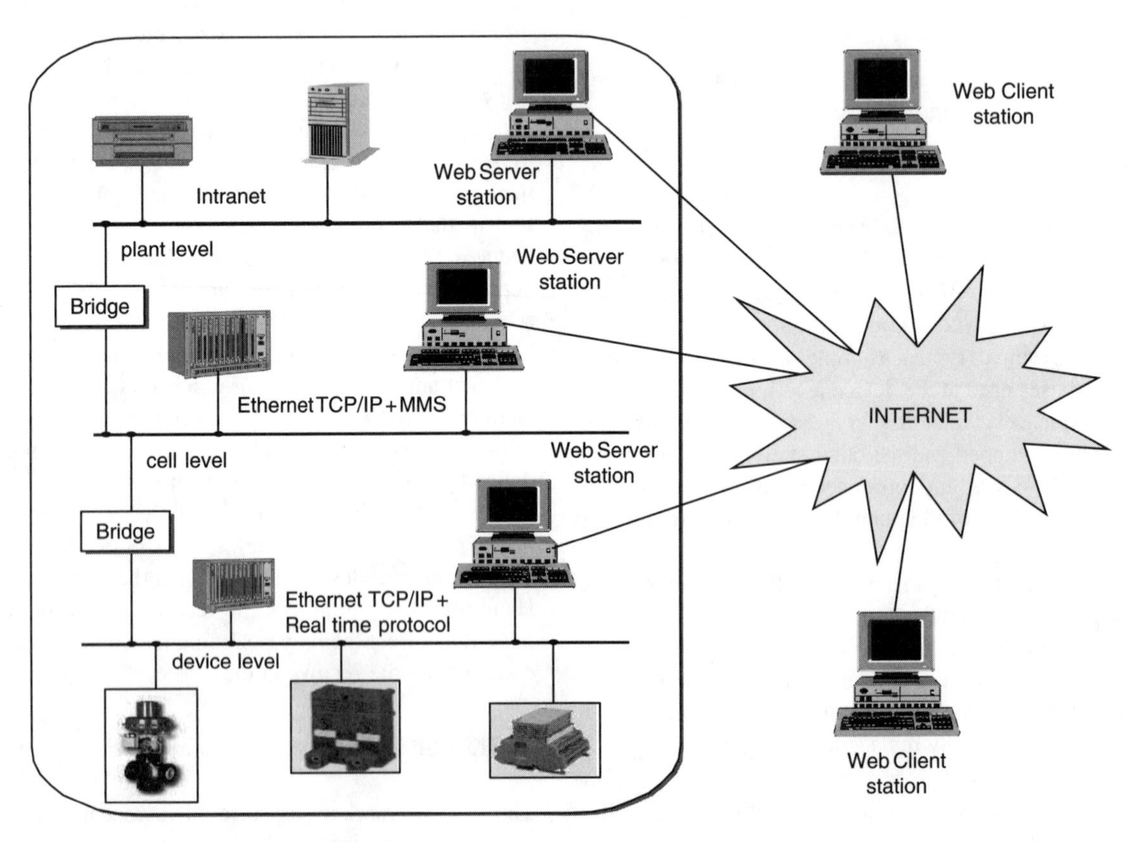

FIG. 4.5i
Remote access to industrial automation systems via Internet.

perform correctly the functions required at this level. However, in recent years the performance of Ethernet has been significantly enhanced by the introduction of new standards (IEEE 802.3u: Fast Ethernet, and IEEE 802.3z: Gigabit Ethernet, both described in Reference 30) and by the introduction of switches. The maximum data transfer speed has been elevated to 100 Mbit/s maintaining the same protocol for the data link layer; this new version is commercially known as Fast Ethernet, or by the acronym 100BASE-T. Another upgrade, in terms of speed, is represented by Gigabit Ethernet, which operates at 1000 Mbit/s, but this is not yet used in industrial automation.

Switches are "intelligent" devices that operate at the data link level, interconnecting stations and/or network segments. A network based on a switched architecture, an example of which is shown in Figure 4.5h, allows for a very efficient handling of the traffic because in this case frames are sent only to the addressed stations or segments. Ethernet networks implemented using switches are characterized by a low incidence of collisions and high overall throughputs. The above two important improvements have made possible the use of Ethernet also at the device level.

However, an as-yet-unsolved problem is the absence of a well-defined/established, worldwide standards-based, real-time protocol, to implement the device level functions. There are several benefits arising from such a new scenario: first of all, the widespread use of Ethernet has led to a great availability of low-cost components and know-how. Moreover, as intranet networks are increasingly adopted at the plant level, if Ethernet is used both at the device and the cell levels, the different factory communication systems can be easily integrated, as they may use the same protocols. Finally, remote access to all levels of industrial automation systems can be facilitated by using the World Wide Web functions, as depicted in Figure 4.5i; the presence of stations with Web server functionalities and connected to the Internet makes the automation system directly accessible from any Web client station.

It is worth noting that, in order to implement such a functionality, it is necessary to introduce the use of the TCP/IP protocols also at the device level. This solution has to be critically evaluated, because it could severely impact the performance of any real-time protocol used to implement the device-level functions.

References

1. Postel, J. and Reynolds, J., "RFC 959, File Transfer Protocol," Marina del Rey, CA: Information Sciences Institute, University of Southern California, October 1985.
2. Berners-Lee, T., Fielding, R., Frystyk, H., Gettys, J., Leach, P., Masinter, L., and Mogul, J., "RFC 2616, Hypertext Transfer Protocol Version 1.1," Geneva, Switzerland: The Internet Society, June 1999.
3. Postel, J., Ed., "RFC 793, Transmission Control Protocol," Marina del Rey, CA: Information Sciences Institute, University of Southern California, September 1981.
4. Postel, J., "RFC 791, Internet Protocol," Marina del Rey, CA: Information Sciences Institute, University of Southern California, January 1981.
5. International Standards Organization, "Manufacturing Message Specification Parts 1 and 2," ISO IS 9506, 1990.
6. European Committee for Electrotechnical Standardization, "General-Purpose Field Communication System—Volume 1, P-Net," EN50170-1, December 1996.
7. European Committee for Electrotechnical Standardization, "General Purpose Field Communication System—Volume 2, Profibus," EN50170-2, December 1996.
8. European Committee for Electrotechnical Standardization, "General Purpose Field Communication System—Volume 3, WorldFIP," EN50170-3, December 1996.
9. European Committee for Electrotechnical Standardization, "General Purpose Field Communication System—Volume 4, Foundation Fieldbus (H1)," EN50170-4, February 2000.
10. European Committee for Electrotechnical Standardization, "General Purpose Field Communication System—Volume 5, ControlNet," EN50170-5, July 2000.
11. Dietrich, D., Neumann, P., and Schweinzer, H., *Fieldbus Technology,* New York: Springer-Verlag, 1999.
12. European Committee for Electrotechnical Standardization, "High Efficiency Communication Subsystem for Small Data Packages—Volume 2, Interbus," EN50254-2, 1998.
13. International Standards Organization, "Road Vehicles—Interchange of Digital Information—Controller Area Network for High-Speed Communication," ISO IS 11898, November 1993.
14. European Committee for Electrotechnical Standardization, "Industrial Communication Subsystem Based on ISO 11898 (CAN) for Controller-Device Interface—Part 2: DeviceNet," EN50325-2, 2000.
15. European Committee for Electrotechnical Standardization, "Industrial Communication Subsystem Based on ISO 11898 (CAN) for Controller-Device Interface—Part 3: Smart Distributed System (SDS)," EN50325-3, 2000.
16. European Committee for Electrotechnical Standardization, "Low-Voltage Switchgear and Controlgear—Controller and Device Interface Systems—Actuator Sensor Interface (AS-i)," EN50295-3, 1999.
17. International Electrotechnical Commission, "Digital Data Communications for Measurement and Control—Fieldbus for Use in Industrial Control Systems—Part 4: Data Link Protocol Specifications," project number 61158-4 FDIS, 1998.
18. International Electrotechnical Commission, "Digital Data Communications for Measurement and Control—Fieldbus for Use in Industrial Control Systems—Part 3: Data Link Service Definitions," project number 61158-3, FDIS, 1998.
19. International Electrotechnical Commission, IEC 61158-6: "Digital Data Communications for Measurement and Control—Fieldbus for Use in Industrial Control Systems—Part 5: Application Layer Service Definition, Type 4," January 2000.
20. International Electrotechnical Commission, IEC 61158-6, "Digital Data Communications for Measurement and Control—Fieldbus for Use in Industrial Control Systems—Part 6: Application Layer Protocol Specification, Type 4," January 2000.
21. Ship Star Associates, "SwiftNet Fieldbus," 36 Woodhill Drive, Suite 19, Newark, DE 19711-7017, USA.
22. Seriplex Technology Organisation, Inc., "Seriplex Control Bus Version 2, Standard Specification," August 2000.
23. American National Standard Institute (ANSI), Electronic Industries Alliance (EIA), "Control Network Protocol Specification, ANSI/EIA 709.1," 1999.
24. CAN in Automation International Users and Manufacturers Group e.V: "CAL-Based Communication Profile for Industrial Systems," CiA/DS301, 1996.
25. CAN in Automation International Users and Manufacturers Group e.V, "CAN Application Layer for Industrial Applications," Germany, CiA/DS201-207, 1996.
26. Schneider Automation, "Modbus Protocol," 1978.
27. Cena, G., Demartini, C., and Valenzano, A.: "On the Performances of Two Popular Fieldbuses," in *Proceedings of the WFCS'97 IEEE*

Workshop on Factory Communication Systems, Barcelona, Spain, October 1997, pp. 177–186.
28. Vitturi, S., "On the Use of Ethernet at Low Level of Factory Communication Systems," *Computer Standard & Interfaces,* Volume 23, pp. 267–277, 2001.
29. Vitturi, S., "Some Features of Two Fieldbuses of the IEC 61158 Standard," *Computer Standard & Interfaces,* Volume 22, Issue 3, pp. 203–215, 2000.
30. Institute of Electrical and Electronics Engineers, "IEEE 802-3: 2000 Standard, Part 3: Carrier Sense Multiple Access with Collision Detection (CSMA/CD) and Physical Layer Specifications," 2000.
31. Siemens AG, Automation and Drives Division, Technical Support, "Simatic Net IE CP1613 Communication Processor with Technological Functions Protocol," 1998.

4.6 Fieldbus Advantages and Disadvantages

I. VERHAPPEN

As with any technology, the fieldbus has both advantages and disadvantages. In addition, engineers are inherently a risk-adverse portion of the population and need an incentive to try something new. Project managers are also risk adverse, although plain old economics can convince this group and management to try a new technology. The hope is that this section will provide readers with a few ideas to consider before installing an initial fieldbus system. Experience has shown that once these benefits have been recognized, there is no going back.

ADVANTAGES

Because of their all-digital design, fieldbus systems are capable of providing an abundance of information between the field and central computer system. Some of the information that is of most importance includes device diagnostics, device status, and remote calibration.

Device diagnostics. All fieldbus devices contain diagnostic information. The various "flavors" of fieldbus do this to varying degrees. But all, by virtue of their bidirectional communications, have some form of diagnostics. Foundation Fieldbus, which uses the Device Description (DD) language similar to that used by HART,* is much richer in its information content. It can therefore assist the user in diagnosing and, in many cases, remedying the problem remotely over the network. HART has two levels of condition-based maintenance (CBM) support. Generic devices allow users to read universal commands (such as manufacturer, model, and revision) and read and write common-practice commands (such as range and units). The generic DD does not allow access to device-specific commands, such as construction material. To obtain full HART support a "CBM–aware" device is necessary.[1]

Device status. Again, because of their bidirectional communication, fieldbus systems provide information to the host system about their operational condition. Foundation Fieldbus integrates this seamlessly into the host system and provides status information, typically good—MAN (manual) and AUTO (automatic)—or bad—O/S (out of service). This alone is justification for use of a fieldbus system because it is now possible to inform panel operators, in real time, whether the signal being used for control is actually valid or requires work by an instrument technician. This same status information can be used to ensure that any applications controlling the process only use valid data to make changes to the operation.

Remote calibration. Fieldbus has the ability to change the range and calibration of a device remotely, without requiring a maintenance technician to remove it from service and bring it into the shop. Experience has in fact shown that, in most cases, removing a digital transmitter and calibrating it on a shop bench actually decreases its accuracy vs. the original factory settings.

In addition, Foundation Fieldbus devices have the following capabilities:

Field-based control. Foundation Fieldbus is capable of migrating regulatory control from the host system to the field devices themselves. There are some "restrictions" necessary to implement this feature, but they are fairly simple and not much different from those the industry has been accustomed to applying from its beginnings in the days of pneumatic loops. This means Foundation Fieldbus provides a way to achieve control and diagnostics at the lowest level, which represents a return to single-loop integrity.

"Plug-and-Play" configuration. The device description and capabilities files that come with every Foundation Fieldbus–certified device ensure that when a device is added to a fieldbus network, it will be "automatically" recognized and visible on the host/configuration system. There is no need to set dipswitches or other mechanical settings to install a new device on the network, while the network is operating. In many cases, all that will be required is to rename the device with its actual system tag name so the operators can associate it with the service to which it is applied. Foundation Fieldbus through its certification and testing programs ensures truly interoperable devices from diverse manufacturers.

Function blocks. These blocks of memory in the device determine its characteristics and, because they are memory based, they can be configured and updated as additional features are defined for the device.

High-speed Ethernet (HSE) connection. Foundation Fieldbus introduced devices in 2001 that will enable integration of the H1 31.25 kbit/s network into an Ethernet-based high-speed network. This network, called HSE, operates at

* Highway Addressable Remote Transducer.

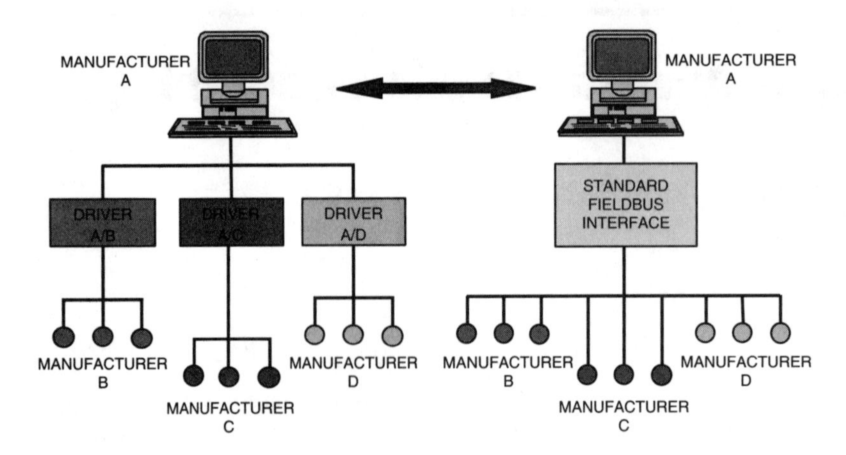

FIG. 4.6a
Open technology vs. drivers.

100 Mbit/s using commercial off-the-shelf technology to connect "best in class" devices for each layer of the computer environments in a corporation.

As one can see, fieldbus networks offer numerous advantages and challenges vs. the analog technology installed in today's facilities. The remainder of this section provides more detail on these subjects.

Bidirectionality

Fieldbus represents digital bidirectional open communications technology, which makes it possible to share more information between network components more accurately and easily. Openness also means that it is vendor independent. Figure 4.6a shows how this can make a difference. It is no longer necessary to develop and maintain a special driver for every protocol-to-protocol interface.

Condition-Based Maintenance

The previously mentioned bidirectional digital communications enables transferring more than simply process variable (PV) information, and some of this "extra" information is device diagnostics—not only diagnostic information such as status for the PV, but also information on how the various components and subcomponents of the device itself are performing. Figure 4.6b represents the anticipated increase in information that can be transmitted over a digital networking protocol. Note there is little change in the level of "control" information, but the associated management information content, according to some sources, is almost 200 times greater than that available with the hybrid systems typically installed today.

Figure 4.6c shows how a CBM system would be installed using today's "traditional" technology. Note that this technology, which is typically HART based, requires the addition of a parallel system to recover the superimposed digital signal and to process it for use by the CBM application. Compare this with Figure 4.6d using all digital fieldbus technology. Note that the reduction in the number of required components

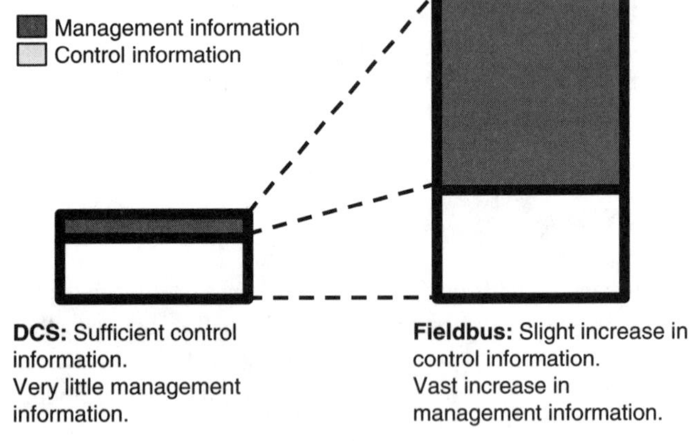

FIG. 4.6b
Increased information with fieldbus.

is due to the following:

- Diagnostic data are within the fieldbus data packets.
- With Foundation Fieldbus technology, control has migrated to the field, thus minimizing the I/O system to such a level that it can be incorporated on a single card within the host.
- CBM can be connected to the same network as the balance of the control system, although the processing itself will be done in parallel.

The subject of CBM will be covered in greater detail later in this section.

DISADVANTAGES

New technologies are not always the panacea for all concerns and in fact often introduce their own sets of challenges or disadvantages. This subsection identifies some of the disadvantages associated with installing fieldbus technology, especially if it is planned for an existing operation.

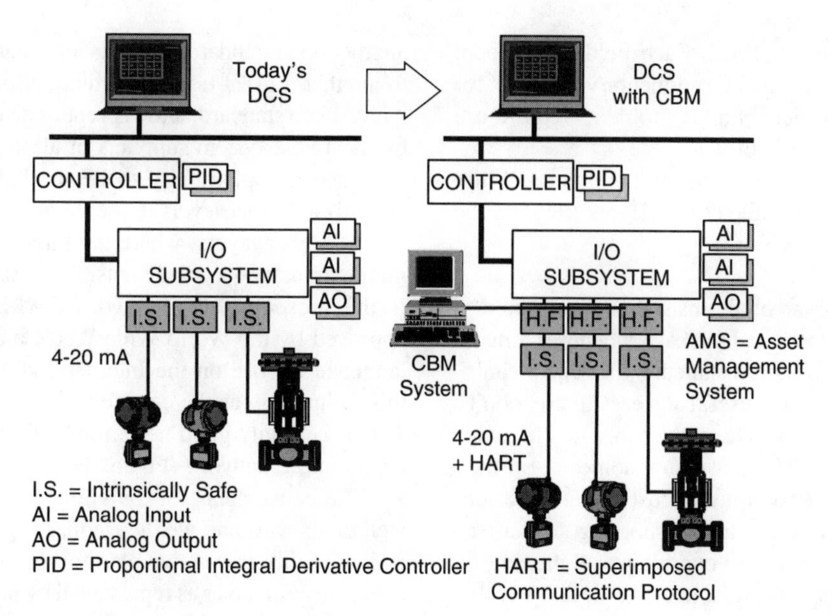

FIG. 4.6c
Traditional CBM savings vs. analog.

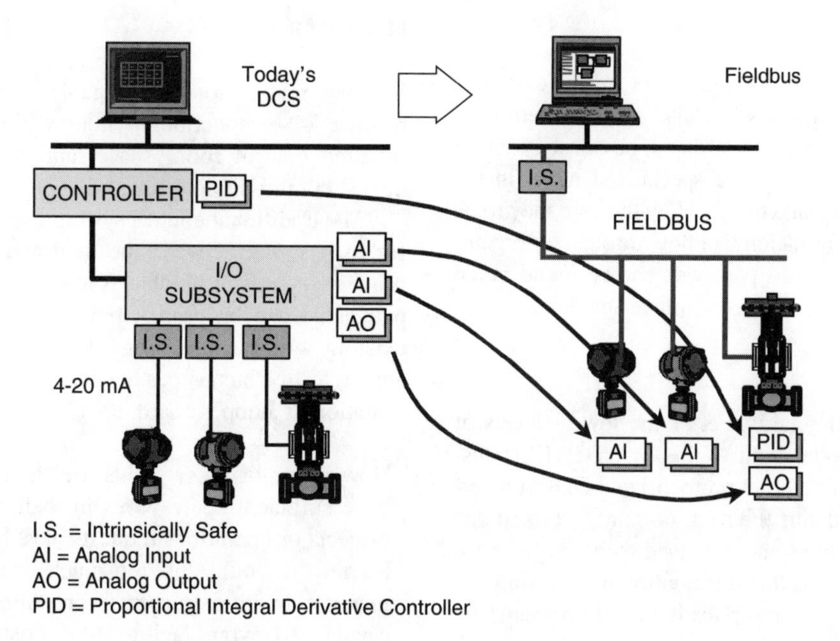

FIG. 4.6d
Fieldbus system cost savings.

"Mixed" Signals

The most "obvious" disadvantage of installing fieldbus technology in an existing facility is that technicians, operators, and engineers will have to be familiar with and support mixed technologies, analog and digital, at the same time.

Architecture

As seen in Figure 4.6d, fieldbus control systems have fewer components. This is intuitive as they do not have to convert between analog and digital as the signals "travel" from one system to another. The implications of these architectural changes are more completely described elsewhere in this volume, and in fact are the basis for most of this work, and hence will not be discussed further here.

Maintenance Tools

As digital networks gain prominence, technicians will no longer be able to get by with a multimeter and signal generator. The tool for today's digital networks is an industrial laptop or notebook with appropriate network diagnostic software. The buzz phrase now is "multimedia instead of multimeter," i.e., diagnose with PC instead of DMM. Each fieldbus protocol has varying degrees of handheld rudimentary devices for initial problem diagnosis, but complex problems require use of a computer and associated software.

What else needs to be done to ensure the high reliability of networks once they are installed? Eric Byres suggests the following.

> What should users do? First of all, insist that good maintenance tools are part of the initial package whenever a new fieldbus or industrial network is installed. If the vendor can't demonstrate tools that are both useful and easy to use, don't buy the network. Second, purchase the tools for the networks you currently own and make sure someone on the maintenance staff really knows how to use them. It's a lot more expensive to be learning troubleshooting skill after the network crashes (or to rely on consultants all the time). It also helps to encourage manufacturers to develop tools if they see that users are actually buying them.[2]

INTERCONNECTIVITY

There are in excess of 30 different digital protocols in existence that all call themselves "fieldbus." "Only" eight of them have been incorporated in the IEC standard and, obviously, each of these protocols serves at least one specialized niche in the controls and automation marketplace. How then are they to be all connected so that information can flow through an organization? The answer, or at least part of it, can be found in the Open System Interconnect (OSI) seven-layer model.

Ethernet Connectivity

Ethernet defines only the hardware, or the lowest levels of the OSI seven-layer model. Add to this that TCP/IP (transmission control protocol/Internet protocol) and UDP (universal data protocol) then define the next, or data, layer and one can quickly see that some standards need to be in place for the user to be able to access this information in a meaningful way. This is where the user or application layer protocols are required. In the case of the Internet, this interface is normally Hypertext Markup Text Language (HTML) but in many ways, HTML is not rich enough to describe all the information required for the control environment.

Connecting to the Corporation

How is all the information on the network communicated to the other layers of both the OSI model and the computer networks in a facility? HSE and the other fieldbus networks' Ethernet-based protocols all describe the necessary "mechanics" to move data to the user layer, but it still remains to transfer this information between the various applications at the user layer and above. One option that is receiving a lot of attention and appears viable at the process data management (PDM) layer used for control is OLE for Process Control (OPC). OPC is another open standard that links information from one database to another, in real time with minimal overhead. And because it is an open standard, it too is vendor neutral. OPC is supported by most host system suppliers of all the fieldbuses and therefore has the opportunity to be the "glue" linking disparate systems at higher levels in the organization/network.

At higher layers, where the Internet and corporate information transfer occurs, eXtensible Mark-up Language (XML) is the present odds-on favorite. XML has recently been approved by the World Wide Web Consortium as a standard language for use on the Internet and Microsoft has adopted this technology as the core for its new .net operating system. This is certainly good news for XML acceptance at multiple layers in the network (Figure 4.6e).

The convergence of tools for the entire plant network is well on its way, and we the engineering community must take action on this area as we plan for the future—not only at the data collection layer as represented by process control systems, but also by working with the corporate information people to ensure data can flow seamlessly throughout the organization.

ECONOMICS

Fieldbus installations are capable of delivering formidable results: 74% reduction in field wiring, 93% reduction in required control room space, and 80% reduction in field device commissioning time.

The traditional control system uses point-to-point wiring for each signal between the field and the host. One of the primary focuses of fieldbus is to ensure that existing twisted-pair cable can be used. Obviously, one would always use existing wiring on an upgrade project or to migrate from analog to fieldbus as per the following example from a presentation at Jump Aboard 2001.[3]

> How many of older plants are finding that they do not have sufficient spare wires in their facility? How many projects or operations requests have had to be "canceled" because it would cost too much to install the necessary increase in infrastructure represented by lack of cable capacity? Is your facility, like most plants, discovering that the first place to run out of these spares is in the most remote part of the plant? Fieldbus is the answer to these questions. Fieldbus is the only standard multiplexer available on the market. By converting from analog to Fieldbus, a person can "automatically" at least triple the number of signals on existing cables in a facility, and if the facility does not use Intrinsic Safety (I.S.) circuits, this factor is even better, closer to a sixfold increase in signal capacity. Never mind the "old" wiring savings stories associated with Fieldbus, this represents real savings, since for a grassroots facility, the project will be installing cables anyway. Everyone knows that at least 75–80 percent of cable installation costs are related to labor and cable tray, the cable itself is only a fraction of the cost. Therefore, if you have an existing facility that needs new cable and this

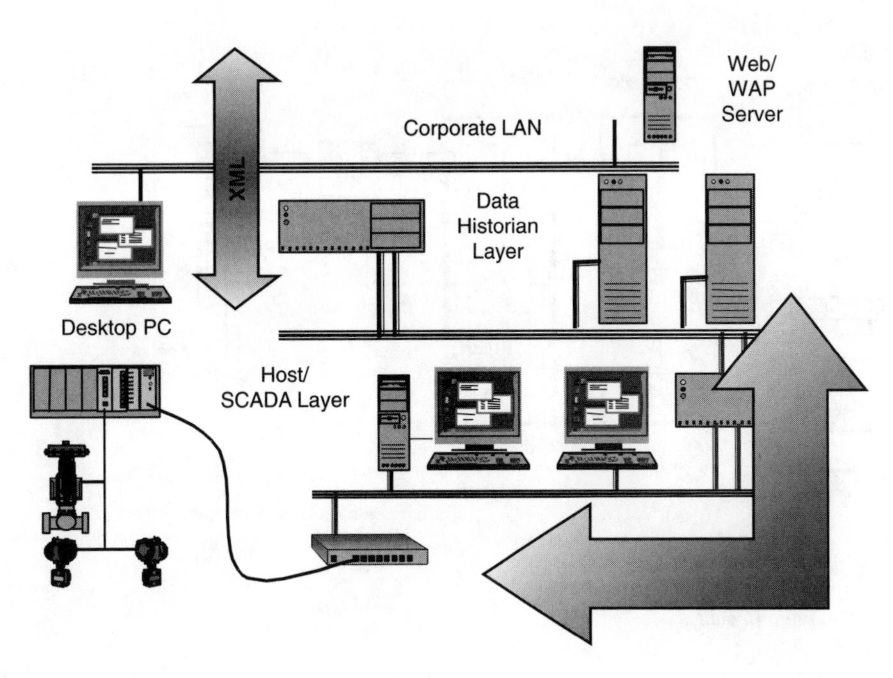

FIG. 4.6e
Data transfer mechanisms.

"new" cable can be found without the associated labor costs, management will quickly see the economics.

As shown in this example, wiring savings are very dependent on the particular case involved. So why focus on them when other benefits of digital instrumentation can be enjoyed by all? The cost savings can go a long way to convincing project management that Foundation Fieldbus is a good choice. It can help offset the higher costs of the Foundation Fieldbus–capable devices. It can help offset the cost to a project of an engineering procurement construction (EPC) firm doing Foundation Fieldbus for the first time.

Initial Capital

This is such a complex issue that it may be quite impossible to make blanket statements. However, there is big potential for savings in a new plant, although a case-by-case analysis is clearly warranted.

If one takes advantage of the multidrop capability of Foundation Fieldbus and uses armored cable instead of conduit (as is commonly done in Europe), the savings in time and material for cabling infrastructure can be substantial. In addition, regarding spare capacity for future expansion, if one keeps initial device loading per segment low there is built-in spare capacity.

Figure 4.6f shows how fieldbus networks can result in a significant reduction in the number of terminations on a project. The number to the left of each "terminal strip" shows how many terminations are required if traditional analog signals were used, whereas the number on the right shows how the same number of devices could be connected using fieldbus technology. The figure represents the number of terminations for four devices on an intrinsically safe installation, and the reduction is from 76 terminations to 34 terminations.

How does one calculate this, for example, for 24 devices as part of a nonintrinsically safe (NIS) installation? Very simply, 24 conventional devices require a pair per device from the junction box to the device, which is $24 \times 5 = 120$ terminations (no shield termination at the device). Then each end of the 24 pairs needs to be terminated. That is another 6 per pair times $24 = 144$. Then, there is the marshaling cable between the home run and the host input/output (I/O). That is another 120 terminations. So a conventional installation requires a total of 384 terminations.

For fieldbuses with six devices per segment, the field to junction box is the same at 120 terminations. But now there are just four pairs going to the host, so instead of 144 terminations, only 24 are required. Fieldbus has no marshaling cable, just another pair (perhaps) between the fieldbus power conditioner and the H1 interface, or another 24 terminations at the most. Based on the above it has been possible to save at least 216 terminations on this single multiconductor pair and field junction box.

Both of these examples show that fieldbus terminations for a single loop are over 50% less than those required to install a conventional system. This also adds to the reliability of the system, as the fewer the number of terminations and components, the fewer potential points there are for failure.

A study by NAMUR in Europe determined that fieldbus systems have design cost savings of 28% vs. a conventional system.[4]

Engineering for terminations, Contractor for termination costs = \$30/terminal

Based on project estimating "rules of thumb," automation is typically 15% of the cost of a project, and based on some

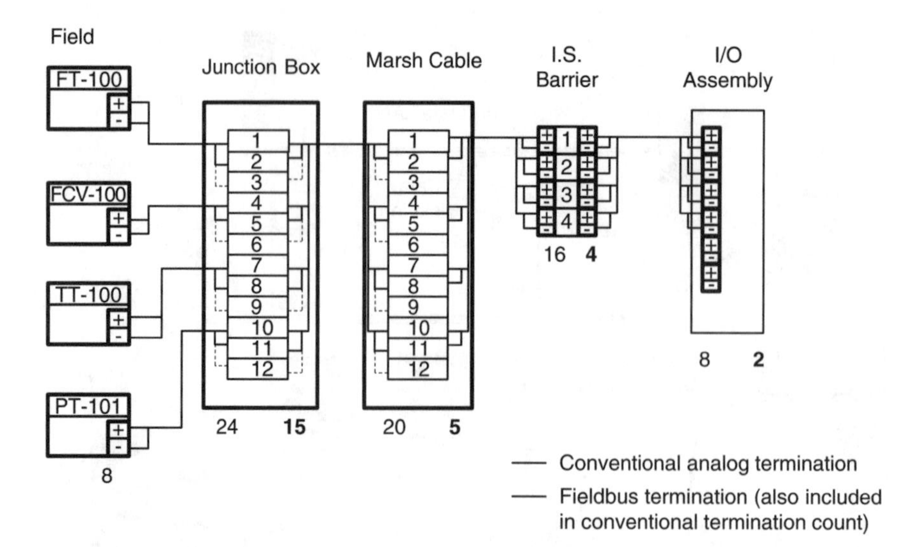

FIG. 4.6f
Conventional vs. fieldbus termination count.

FIG. 4.6g
Fieldbus cable reductions. Fieldbus will significantly impact the field installation.

work done several years ago, which is likely to be even more favorable now, Foundation Fieldbus saves at least 4% of the automation budget.[5] Combining these figures results in a 0.6% decrease in total project costs (Figure 4.6g).

One other avenue that is site specific is the degree to which a site is willing to use multivariable transmitters. Fieldbus devices are capable of transmitting multiple process values from multiple sensors over the same wire and with the same electronics. The more obvious choices are multielement temperature transmitters, or valves that also transmit the associated auxiliary "limit switch" positions. However, the near future will also see devices that use internal calculation blocks along with an input from another sensor on the network or remote sensing element, for example, temperature correction for a flow, to determine mass flow or flow at standard conditions.

Commissioning

As discussed above, actual savings are application specific, but it should not be difficult to confirm that some significant savings exist for nearly every application, even if it is just commissioning time. During commissioning, loop checking is no longer on the critical path for project completion, without augmenting the original loop check crew.

The capabilities file makes it possible to configure fieldbus networks off line so that all that is required during commissioning is to "plug" the device in. The fieldbus "plug-and-play" capabilities will recognize the device and add it to the network, fully configured and ready to operate. If the device is not the one expected, the engineer has only to move it to the correct loop or alternatively download the proper configuration to the device.

Commissioning time with fieldbus systems is typically at least an order of magnitude less than with traditional systems, decreasing from 2 h to a maximum of 20 min/loop. In addition, being able to commission and start up a facility "early" means "feed-in" that much sooner as well. What is an extra day's production worth at a facility?

Life Cycle

Because of their digital nature, fieldbus systems are less susceptible to ground faults, which inherently means that the system will be more reliable.

Once the fieldbus infrastructure in place, adding new devices is much easier than with a conventional system. In a recent case, shared on a Web discussion group, it was necessary to add four Foundation Fieldbus valves (three quarter-turn shutoff valves and one control valve) and a vortex shedding flow meter on top of a 115-ft reactor. This was done by installing 5 m of conduit. The facility just added a junction box at an existing device at that elevation and relocated the segment terminator from the box at grade to the new little box at the top the reactor.

As mentioned earlier in this section, the diagnostics features of fieldbus systems, which include online process parameter status summary to the HMI, data quality, and alarm conditions, are available to the process operator at all times.

This information can also be used by the control system to "instruct" the associated applications that they should not use a process value in the event the data points are tagged as "bad" or out of service.

Detail diagnostics is used with the engineering tool to determine such items as failures (sensor, output, and memory), configuration errors, communication errors, and access to parameters allowing external analysis. An example is the use of valve position and valve travel time to detect stick–slip action of a final control element.

This same information can be integrated with a predictive maintenance system to schedule maintenance based on device diagnostics and to schedule calibration based on device calibration data. The result is that it will be possible to perform maintenance and calibration of several devices at the same time, with shorter shutdowns, and no time wasted on maintenance of equipment that does not need it (true preventive maintenance).

Because fieldbus systems are open, sensors are field-upgradable and interoperable. This means that, provided the process connection dimensions are the same, it is possible to use one manufacturer's device in place of another, should one fail. For example, if a pressure transmitter with a range of 0 to 100″ WC from Manufacturer A fails and the warehouse has only a pressure transmitter with a similar range from Manufacturer B, the device from Manufacturer B can be installed. The process involves ensuring the device description for the new device is on the host, then "swapping out" the two devices, and downloading the configuration to the new device after it has been recognized by the system.

Benchmark studies[6] from the North Sea indicate that maintenance represents 40% of a facility's operating costs, with control system at 45% of maintenance; this means that control system maintenance represents 18% of a facility's operating costs. The remote calibration that is possible with all versions of fieldbus conservatively results in a 15% per field device maintenance savings.

Another study[7] has determined that the typical annual cost for a single control loop in a refinery is $4,700.00 for a conventional analog loop and $1,200.00 for a fieldbus loop, utilizing field-based control.

Other savings possible as a result of the embedded diagnostic capabilities of fieldbus systems include elimination of a significant portion of the control system maintenance expense. The previously referenced NAMUR study indicated that digital fieldbus technology makes a 20% maintenance support reduction and a 40% inventory reduction possible. For example, considering that 80% of operations maintenance requests are unnecessary, if embedded diagnostics can identify 75% of these as false requests,* the result is a 60% savings in this portion of the maintenance budget.

* With fieldbus and its inherent status indication for every data point, there is no reason that these requests should even be initiated; thus, it is reasonable to assume that this 75% target reduction can be obtained.

Similarly, experience indicates that a technician only spends about 33% of the scheduled maintenance time on the actual repair. The remainder of the time is spent obtaining the necessary permits, parts from the warehouse, and diagnosing what the possible problem needing resolution really is. Assuming the embedded diagnostics capability of the fieldbus system captures half of diagnostic time, an additional 17% operational maintenance expense savings is possible.

To conclude this section, it is instructive to look at an example from a typical application perspective. At the end of the day, what counts is safe and effective process control. Assume there is a pH transmitter on a batch application and the probe is fouled, but is still giving a reading. The problem is that the reading is not correct. If control action is based on this value, the result will be off-specification product. Maybe it can be blended, maybe it is waste. If the operator does not know, and this off-specification product is added to a holding tank, several batches of good product have been compromised. Worse yet, the product may be shipped to a customer. What is needed is to diagnose the condition, communicate it to the host and the operator, and take corrective action so the result is good product, not waste.

HART and Foundation Fieldbus can both diagnose the condition, and set a status bit identifying the problem; however, HART must be polled by a host to communicate. If the host does not poll the device for that specific piece of information, the error is never detected.

Foundation Fieldbus can set the PV status to "bad." This goes with the PV to both the control strategy and the operator. Remember that status is a part of *every* fieldbus PV, and is provided automatically. Once the status enters the host, how is it used? With HART the status is totally independent of the PV. Communicating the bad PV status to the control strategy, the operator, and maintenance is not automatic. It must be engineered and frequently manually monitored. With fieldbus, the PV status is immediately displayed on the operator screen, and alarmed. The operator knows immediately that there is a problem. The control strategy can be configured to go to a safe operating condition. It may be anything from hold last value, through shutdown, to switching production to another unit. The key is the bad product is not produced, nor is the good product corrupted.

Now maintenance needs to fix the problem. Both HART and fieldbus can provide the same information to the maintenance team so the device can be fixed just as quickly and effectively. To use HART diagnostics effectively a CBM system is required as well as a continuous alert monitor. Offline tools, database managers, and handhelds are effective for configuration or data management, but do nothing for diagnostics.

The net result is that fieldbus notifies the affected people automatically if there is a problem; HART does not. This is one of the big differences between the diagnostic capabilities of the two technologies at present.

Most importantly, the engineering and maintenance tool is permanently connected to all the H1 networks and constantly monitors the devices. Host systems can provide several levels of troubleshooting diagnostics, for operators, technicians, and engineers. The process visualization software can be configured to show not only the value but also the associated quality (status: good, bad, or uncertain) so that a pretty good picture of the problem can be obtained even before bringing out the other diagnostic tools.

In case the status is bad, e.g., due to no communication, sensor failure, configuration error, etc., the value should not "freeze up" showing a valid number but instead show ***** or equivalent to further make clear it is invalid.

The maintenance tool has a "live list" to show which devices are connected and operating properly. The device configuration tree indicates with a red cross or equivalent any devices that are not communicating. To see overall device status, there is no need to get down to the block level. If they are not communicating, they are most likely not properly connected or are not getting sufficient power. Power is still checked with the good old multimeter so do not throw it away just yet.

If the devices are communicating, but something else is the matter, it will be necessary to view information at the block level. The maintenance tool will "burrow" down the devices to display block by block and parameter by parameter what is wrong. The transducer blocks and resource block are compared against the device configuration done with handheld before, to find sensor problems. Investigate the function blocks as the control strategy just as in a controller before, to determine if there are any application-related problems. As mentioned above, many dynamic parameters have an associated status that is useful, including a number of standard diagnostic parameters, which are available in every block.

Additional device specific parameters for sensor and actuator diagnostics and operational statistics are found in the transducer blocks.

References

1. Masterson, J., "Open Communications Essential for Successful Asset Management," *The HART Book*, Vol. 8, www.thehartbook.com/articles/h8ams.asp.
2. Byres, E., "Working in the Dark," *Manufacturing Automation Magazine,* March/April 2001.
3. Verhappen, I., "Converting from Analogue to Fieldbus—The Latest Upgrade," *Jump Aboard* conference, Perth, Australia, June 2001.
4. "How Fieldbus May Impact Your Next Project," based on work by NAMUR Study Group 3.5—Communication on the Field Level. Order information available at www.namur.de/en/publi%5FENlist.html.
5. Verhappen, I., "Foundation Fieldbus Economics Comparison," *ISA Tech'99 Proceedings,* also available at http://www.isa.org/journals/intech/brief/1,1161,215,00.html?userid=&first=101&last=275.
6. Schellekens, P. L., "A Vision for Intelligent Instrumentation in Process Control," *Control Engineering,* October, pp. 89–94, 1996.
7. Payne, J. T., "Field-Based Control Changes the Cost of Ownership of Control Systems," *Hydrocarbon Online,* August 18, 1998, www.news.hydrocarbononline.com/featurearticles/19980818-1557.html.

4.7 Fieldbus Design, Installation, Economics, and Documentation

S. C. CLARK

PROCESS AUTOMATION BUSES

Fieldbuses for the process industry have seen a slow but steady growth in the past 5 years, and for good reason. Many of the reliability, integration, redundancy, and end user acceptance issues have been resolved. International standards are evolving, operability testing is common, and users have commissioned and enjoyed the economics of more and more fieldbus systems.

This section presents a project plan for the implementation of a fieldbus project in a process automation setting from the viewpoint of an end user and discusses how to approach the design, installation, economics, and documentation aspects of a fieldbus system. This section explores not only what the technology causes us to rethink, but also the cultural adjustments that it brings.

BUS BASICS

In general, a fieldbus system is an automation communications network. Automation buses can be divided into three categories based on their level of complexity and whether they provide logic control or process control functionality. In the process industry, the word *fieldbus* is sometimes used loosely to encompass the three categories listed in Table 4.7a.

Refer to Figure 4.7b for a diagram of the basic components of a typical H1 fieldbus segment, which includes the following:

- Physical media—wire/fiber optics
- Power supply/conditioner
- Terminators to suppress communication reflections
- Host/master or link/coupler, to coordinate communications
- Bus-compliant field device(s)
- Field device/host management software

Teams

Implementing fieldbuses on a sizable automation project will require participation from the teams or organizations shown in Table 4.7c. These groups are similar to a project installing conventional input/output (I/O) except for the introduction of a host/master device firm and a field device/host management software firm.

TABLE 4.7a
Bus Types

Bus Name/Type	*Description*	*Examples*
Sensorbus	Simple devices with bit-level communications such as switches or on/off valves	AS-i
Devicebus	Medium complexity with byte-level communications where the functionality of the device can be mapped to other devices and logic control can be embedded within the devices	DeviceNet, PROFIBUS-DP
Fieldbus	Complex devices with block-level communications that provide control in the form of function blocks; this bus has a highly developed application layer	PROFIBUS-PA, Foundation Fieldbus

Fieldbus Resources

Refer to Table 4.7d for a list of technical resources that should be made available to the project teams during the execution of a fieldbus project. There are many excellent references and installation guidelines to help in the design and installation of the chosen fieldbuses, including Sections 4.3 and 4.14. Additional resources can be found in the bibliography of this section.

DESIGN

Strategies

One of the first steps in the design phase is to decide which buses will be used. Currently, this choice may be largely dependent upon the compatibility of the field devices with a chosen host/master, or vice versa. If a more internationally standardized fieldbus becomes a reality, this decision will be based on the best field device to perform the task and less on how successfully the device integrates with the host.

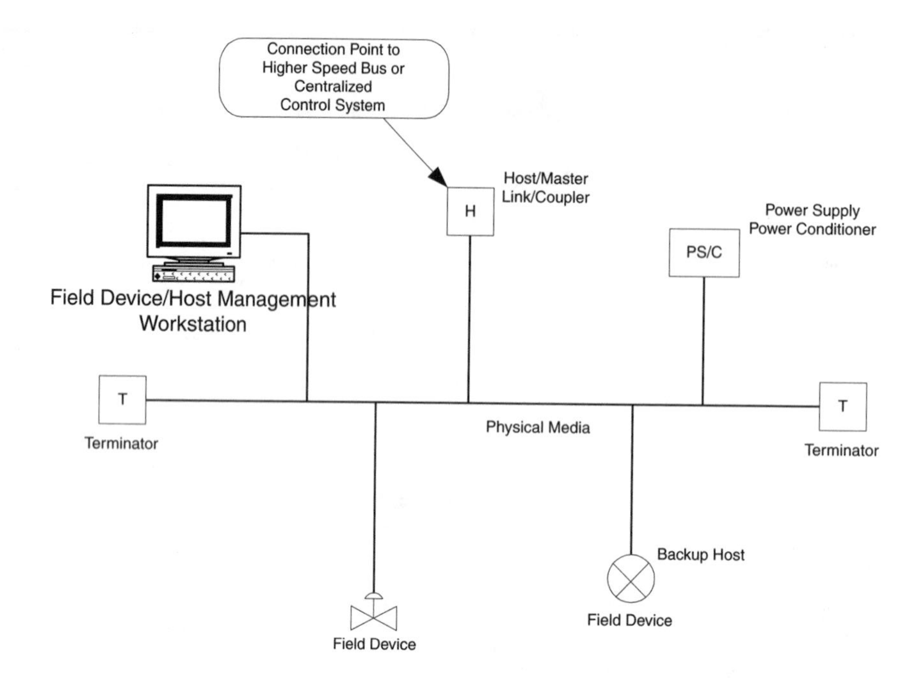

FIG. 4.7b
Typical H1 Segment.

TABLE 4.7c
Project Team Responsibility Matrix

Team/Organization	*Responsibility*
Corporate engineering standards team	Defining project/company standards to be used during the project
Project engineering team	Designing the facility and equipment for the specified process
Design engineering team	Designing the fieldbus networks
Control system configuration team	Designing, coding, and testing the application software for the project
Field device firms	Supplying the field devices
Control system firm	Supplying the centralized control system
Host/master device firms	Supplying the bus host/master device(s)
Field device/host management software firm	Supplying the software configuration tools to configure and/or manage the field devices/host
Electrical and instrumentation contractor	Installing the fieldbus networks
Commissioning team	Starting up the facility
Plant maintenance personnel	Maintaining the facility after it is turned over by the commissioning team
End user	Operating the fieldbus system

During the design phase, the development of a fieldbus general design document may prove beneficial. This document may address segment assignment strategies, redundancy strategies, electrical safety strategies, control loop execution speeds, and future expandability, and may serve as a technology bridging document to existing corporate guidelines that might not otherwise address fieldbus. In writing the fieldbus general design document, consult with the resources listed in Table 4.7d.

Component manufacturers have developed excellent design guidelines and segment design software. These can be used to verify the working designs against fieldbus specifications.

Integration

Integration is key to a successful fieldbus installation as there are components from various manufacturers that must work together successfully. Integration can encompass the following component relationships:

- Field devices with the bus protocol
- Field devices with other devices on the bus
- Field devices with a centralized control system
- Field devices with data servers[1]
- Field devices with the software that manages and configures them
- The process automation bus with higher-speed buses

Field device firms, control system firms, and bus organizations may advertise compatibility, but the integration burden is ultimately laid upon the end user.[3] It is important that a project strategy exist to ensure that the required level of integration can be achieved. This strategy might include the following:

- Independent, qualified evaluations of the chosen fieldbus

TABLE 4.7d
Fieldbus Resources

Resource Category	*Resource Name*	*Description*
Specifications	Fieldbus protocol specifications	To be used for reference in all design-related activities; this documentation can often be provided by the organization that supports the bus technology
Project document	Economic study	Supporting documentation to justify the use of fieldbus and for the development of a strategy to maximize the economic benefits of fieldbus
Design	Corporate standards	Company-specific wiring and cabinet design guidelines
	Bus-specific wiring design standards and guidelines	Wiring guidelines including intrinsically safe applications; this documentation is often published by bus component manufacturers
	Fieldbus general design	General design document defining which type of buses will be used and addressing the strategies for segment assignments, redundancy, electrical safety, execution speeds, and expandability
	Documentation index	Document that calls out the information to be contained on each type of document
Design tools	Segment design software	Tool to validate and revalidate that the physical network design remains within the bus specifications as the project progresses
Testing	Interoperability testing results	Qualification of devices by an independent testing firm to provide evidence that each device type is compliant with the bus specification
	Prototype testing results	Host to device testing on a loaded segment; this testing provides the ability to confirm fieldbus application layer functionality with the chosen host system and guide the control system application software interface to the field devices

- Discussions with experienced end users and system integrators
- Field trips to validate marketing claims
- Detailed device specifications that include integration requirements
- Appropriate training for each of the design engineering team members
- Prototype testing of segments

Application Software

If sufficient prototype testing is not already available, then consider prototyping application software to test with each field device type at the control system configuration team's site. With this testing, the control system configuration team will be able to take full advantage of the features of each device and better understand their limitations as an integrated system. The author has found this testing to be very helpful.

Consider how the application software, whether it resides in a centralized control system or decentralized field devices, is to be tested during an application software factory acceptance test and to what extent. The control system configuration team must understand the goals for the centralized and decentralized application software testing so that it can be properly executed.

INSTALLATION

Strategies

General strategies for a successful fieldbus installation would include well-thought-out and well-documented designs, appropriate electrical and instrumentation (E&I) contractor and commissioning team training, strong E&I supervision, good communication between the design and construction teams, and the appropriate level of testing.

Training

It is possible that the E&I contractor may have no specific experience in installing digital communication cabling or that the commissioning team has never been involved in troubleshooting a fieldbus system. Consider the cultural issues as people familiar with conventional I/O may be learning new definitions, drawings, and practices to comply with fieldbus specifications. Training for the E&I and commissioning team personnel is important so that they can provide strong leadership to make decisions that avoid rework in the cases where the design documentation is not specific.

The E&I contractor will be addressing more stringent designs in the areas of cable segregation as well as interpreting new specifications, and the contractor needs to feel comfortable doing so. The commissioning team will require training on device check procedures and how to approach troubleshooting a communication network. The commissioning team, while executing device checks, will be using the device configuration software and may be called on first to fix any design errors.

Commissioning

Commissioning provides the opportunity to perform fully staged bus, device, and control qualification testing. In general, the commissioning process will progress much more quickly than with conventional I/O as there are fewer terminations, considerably less wire, fewer opportunities for human error, automatic device recognition, and enhanced diagnostic information available from the field devices. The commissioning process should consist of segment, field device, and operational testing.

Segment Testing Prior to any field device checks, a segment test should be executed. Bus segment testing will confirm the physical installation including power supplies, terminators, wire lengths, and other compulsory electrical characteristics. This testing may require specialized equipment and troubleshooting skills so it important to find the appropriate technical resource. Clear as-built documentation is necessary for successful segment testing and future expandability, so confirm the as-built status of segment drawings during the segment test.

When the segment testing is complete, issues identified in the following tests will likely be due to unique integration issues. Unique integration issues may require the expertise of the field device firms, so identify this contact for each manufacturer prior to commissioning. If prototype testing was previously conducted, it will reduce the possibility of such integration issues affecting the project schedule.

Field Device Testing Field device testing will confirm that the subject device is correctly wired, properly addressed, communicating on the bus, and providing basic I/O functionality. If the field device executes control logic using standardized function blocks in the application software, then field device testing should verify the device is configured to perform this functionality.

Some devices provide functionality employed in the control strategy through the use of a data map. For example, an AC drive may provide current, on/off status, fault information, and diagnostic information using a data map, which in effect must be "unmapped" by the other devices on the bus if they are to use this information. In this case, data-map testing could be included in the device test procedure.

Operational Testing After field device testing is complete, an operational test is warranted. This operational test would include specific test cases to confirm that the bus and connected devices are capable, as a working unit, of executing the process control logic embedded within the devices. This stress testing would include verifying cycle/execution times and reviewing diagnostic information within the field devices and host. It would also include testing the functionality of any backup host/masters, data server connections, higher-speed network connections, and bus redundancy to confirm the overall robustness of the bus configuration.

DOCUMENTATION

Strategies

During a fieldbus project it is worthwhile to evaluate the need and use of certain conventional I/O project documentation and to decide how these new or modified documents will be efficiently managed. A documentation index describing what information will be placed on each deliverable document or drawing would be helpful. The documentation index will help the design engineering team and control system configuration team manage project information associated with the fieldbus. It can also serve as a supplement to the training required for the E&I contractor, commissioning team, and plant maintenance personnel later in the project.

P&ID Representation

Whereas most other documentation associated with a fieldbus project is the responsibility of the instrument and electrical engineers on the design engineering team, this drawing is often codeveloped by many disciplines on the project engineering team. In this case, any changes made to the P&IDs with respect to fieldbus devices impact the project engineering team. If changes are to be made on the P&IDs to represent fieldbus devices, include the project engineering team in this decision to ensure that the representations will be properly implemented during the entire project.

ISA standard S5.1 was developed, in part, to ensure a consistent, system-independent means of communicating instrumentation, control, and automation intent.[4] The standard was designed to depict physical locations of devices and the control functionality that they provide. The challenge to the team implementing a fieldbus system is how far to extend the current international standards, or company standards, to accommodate fieldbus implementations. As field devices become more standardized in their offering of functionality, then representing them on P&IDs can also become more standardized.

Whichever approach is chosen to document fieldbus devices on the P&IDs, it should be easy to maintain, and duplicate the least amount of design information. Be sure the P&IDs depict the control intent while being cautious not to provide information that is better documented elsewhere, such as on P&ID lead sheets and application software configuration designs.

Segment List

A segment list can replace the conventional I/O index and detail the strategy for assigning devices to segments. This list should provide all the design data necessary for checking the

TABLE 4.7e
Example Segment List

Tag	Description	Controller	Card	Port	Origin	Destination	Length (ft)	Current Draw (mA)	Backup Master
TT-1	TA-1 Vessel Temperature	5	1	2	TT-1	JBOX-05-01-02A	28	17	N
TY-1	TA-1 Vessel Temperature Control Valve	5	1	2	TV-1	JBOX-05-01-02A	32	26	N
PT-2	TA-2 Vessel Pressure	5	1	2	PT-2	JBOX-05-01-02A	15	17.5	Y
PY-2	TA-2 Vessel Pressure Control Valve	5	1	2	PV-2	JBOX-05-01-02A	10	26	N
JBOX-05-01-02A	Junction Box	5	1	2	JBOX-05-01-02A	IO-05	115	N.A.	N.A.
FT-5	TA-5 Outlet Flowrate	5	1	1	FT-5	JBOX-05-01-01A	10	18	Y
JBOX-05-01-01A	Junction Box	5	1	1	JBOX-05-01-01A	JBOX-05-01-01B	35	N.A.	N.A.
TT-5	TA-5 Vessel Temperature	5	1	1	TT-5	JBOX-05-01-01B	19	17	N
JBOX-05-01-01B	Junction Box	5	1	1	JBOX-05-01-01B	IO-05	142	N.A.	N.A.

segments against the fieldbus specification constraints and to develop the segment drawings. Refer to Table 4.7e for an example segment list. Include what information is to reside on the segment list in the documentation index.

Segment Drawings

Fieldbus segment drawings can be designed to replace conventional I/O loop drawings, marshaling cabinet maps, and terminal box maps. Refer to Figure 4.7f for an example drawing. Segment drawings can provide the following information:

- Tagging information for all major components
- Segment identification and device addressing
- Device interconnection details inside terminal boxes and marshaling cabinets
- Segment wire lengths
- Power supplies and conditioners
- Device and electrical component physical locations including junction boxes and cabinets
- Bus communication components including host/master, terminators, and coupler/link devices
- Control system interfaces
- Device DIP switch settings
- Backup host/master identification

Because a fieldbus system is a communications network, make clear distinctions between fieldbus cabling and conventional I/O cabling to eliminate installation, troubleshooting, and maintenance errors. Include what information is to reside on the segment drawings in the documentation index.

Specifications

Specifications not found on conventional I/O projects are required for the host/master and the field devices in a fieldbus installation.

Host Specifications The host/master specification should provide detailed information on the features of the host/master. This might include the configuration of field devices, configuration of function block linkages, redundancy, communication speeds, cyclic and acyclic communications, and interoperability testing. Documentation is available through the fieldbus organizations to help develop the requirements for the host/master.[5,6]

When the host is an integral part of a centralized control system, the host compliance specifications should be included in the control system specification and/or functional requirements.

Field Device Specifications Field device specifications should include compliance with fieldbus standards, device current draw, voltage range, polarity sensitivity, backup host capability, device addressing, data mapping requirements, function block availability, interoperability testing requirements, and host compatibility.

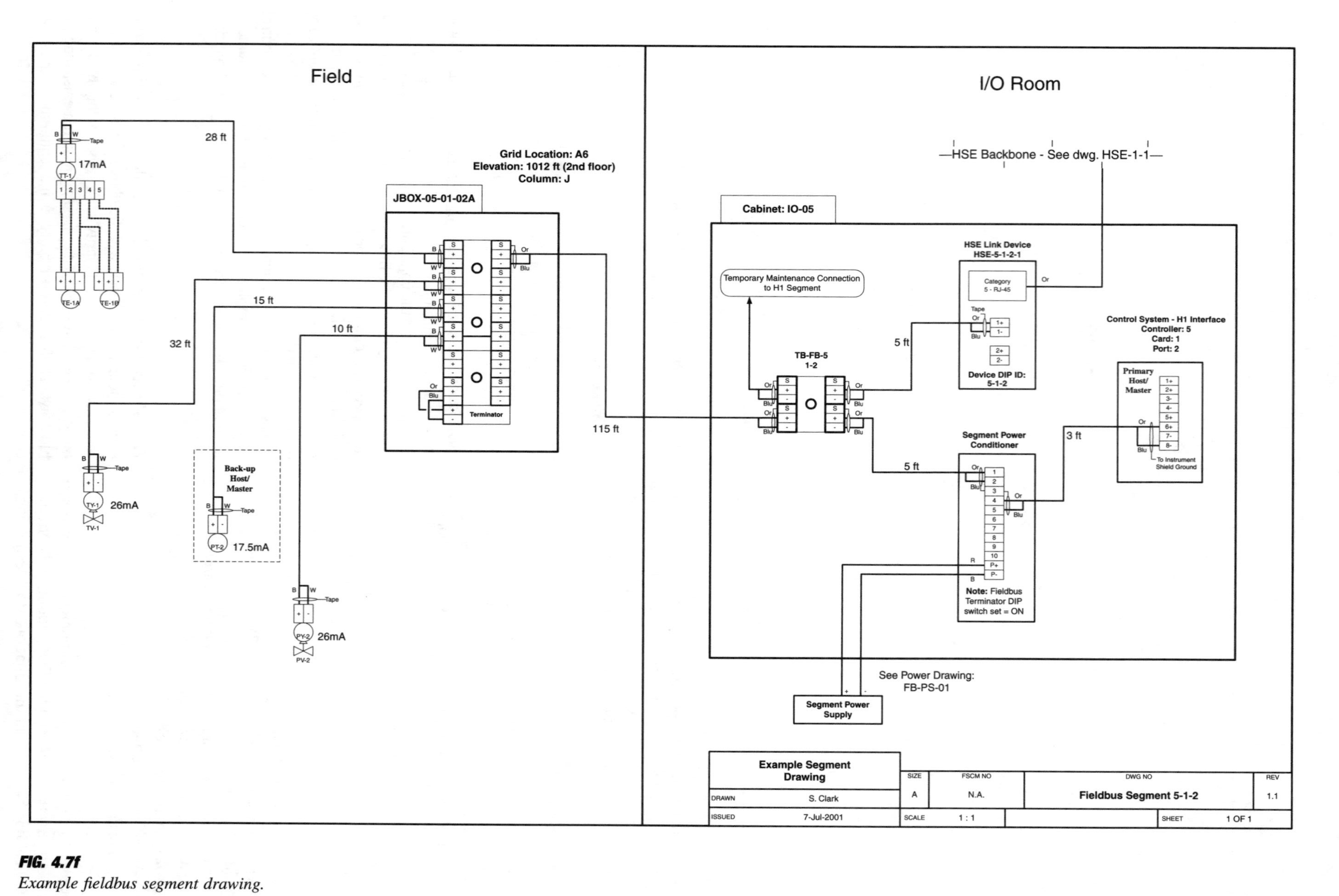

FIG. 4.7f
Example fieldbus segment drawing.

Procedures

The following procedures may require review and possible updates when implementing a fieldbus system:

- Commissioning procedures for the new tools and methods involved with commissioning field devices that are auto recognized by the control system and contain no digital-to-analog circuitry.
- Calibration procedures to describe how to verify that devices are accurate to within their specified ranges.
- Preventative maintenance procedures to help guide proactive troubleshooting using the enhanced diagnostic information available from the field devices and setting intervals at which devices need to be calibrated/rechecked. These procedures may include the use of the device management software to obtain and document changes made to devices.

Maintenance

As-built segment drawings and instrument specifications are important documents for long-term maintenance. This documentation can prove invaluable during troubleshooting, when purchasing spares, and during device replacement.

Software design tools are available to document how the devices are configured and the control strategies with which the field devices participate. As-built documentation denoting where control resides will need to be kept current for maintenance purposes, and these documentation tools can simplify the task of keeping an installation maintained in an as-built state. When the field devices are part of a validated system, these tools can be used to provide audit trails and maintain device history and calibration information. If these software tools are to be used on validated systems, they will need to be validated as well.

ECONOMICS

Strategies

It has been shown that the more digitally integrated the system, the better the economics.[2] General strategies for obtaining the greatest economic returns are to consult with applicable economic studies when developing the fieldbus general design document, reducing the number of possible integration issues by not using multiple bus solutions, and researching opportunities to reduce wiring, terminations, and panel real estate. Provide the appropriate level of training to the project engineering team members involved with cost estimating so that they understand the areas where cost reductions can take place. Refer to Section 4.6 for additional information on fieldbus economics.

Design

Additional expenses will be incurred during the design phase of a fieldbus project as compared with conventional I/Os. Fieldbus designs require more detail than conventional I/O cabling such as accurate physical device locations, wire lengths, and communication speed restrictions. This is especially true as facility changes are made during the course of a project, and the fieldbus designs must be reverified against bus specifications.

Field devices can be configured, in many cases, to provide multiple process values. This feature should be exploited to help reduce overall instrumentation costs. The cost for fieldbus devices compared with conventional devices has been shown to be only marginally higher.[2] According to case studies provided by Profibus International, the PROFIBUS device ASICs can cost from $12 for an intelligent slave, to $30 for a master ASIC. These ASIC costs are offset by the removal of expensive digital-to-analog circuitry in devices.[6]

Home run wiring from the field to the I/O rooms can be reduced by at least 75% for fieldbus devices when the segments are designed with four devices per segment. Wiring systems and connector components have been shown to speed terminations and reduce wiring errors. Current limiting, plug-type terminal blocks can safeguard against segment failures during maintenance procedures that require devices to be disconnected from the bus.

Currently, fieldbus interoperability must be approached with caution by the end user during the design phase. This caution can be mitigated with prototype testing, which does introduce additional testing costs. It has been the author's experience that this prototype testing has always paid for itself when it comes time to commission the fieldbus.

Documentation

Documentation costs can be reduced with the use of segment drawings in place of conventional I/O loop drawings, marshaling panel maps, and terminal box drawings. Including the appropriate information on segment drawings can help minimize the need for additional drawings. Use fieldbus as an opportunity to reevaluate the documentation requirements and to streamline where possible.

To reduce costs, seek out electronic documentation tools that can record and document the information available from the field devices regarding configuration, calibration, maintenance, and troubleshooting. The value of this as-built documentation becomes immediately apparent when adding new devices to existing segments. In a validated process these software tools themselves must be validated and this cost should be estimated.

Installation and Commissioning

In case studies conducted by Profibus International and Foundation Fieldbus end users, installation and commissioning cost savings have been well documented in the 20 to 50%+ range based on the designs and strategies employed. The fact that fieldbus devices automatically identify themselves, require fewer terminations, and are capable of being managed remotely has contributed to the reduced commissioning times.

Maintenance

There will be initial costs associated with maintenance such as training the plant maintenance personnel on the diagnosis of field devices, replacing field devices (as they may be more integral to the centralized control system), using the field device management software, and updating maintenance procedures. In fieldbus devices, the digital-to-analog circuitry no longer exists and any procedures related to this circuitry must be modified. The calibration procedure becomes more of a device check to confirm that the device is within its specified range.

To capitalize on the reduced maintenance costs associated with fieldbus devices, the project should provide the appropriate software, procedures, and any test rigs necessary to obtain and interpret the additional information available from the field devices.

CULTURAL ASPECTS

New Technology

With new technology comes the decision when and where it should be implemented. New technology is risky. There needs to be acceptance from the corporate engineering standards team, project engineering team, and plant maintenance personnel. Each group will be impacted by the decision to use the technology. Bringing all the teams into alignment is important to the overall acceptance and benefit of the technology to the company.

The groups responsible for maintaining the fieldbus may include control engineers, information technology staff, electrical support staff, and instrumentation support staff. These groups may need to rethink how they perform certain activities when dealing with a communication protocol in a process automation setting. Fieldbus may bring shifts in responsibility for these groups and in some cases these groups will need to work more closely with each other. Some of the responsibilities to consider include:

- Field device/host management software that maintains the revision levels of field devices
- Fieldbus devices that connect lower-speed buses with higher-speed buses
- Field device diagnostics software
- Control logic that is executing inside the field devices
- The fieldbus power supply
- The communication bus wiring
- Higher-speed bus networking, data servers, and firewalls

Integration

As fieldbus devices provide more functionality, they also inherit more complexity and a greater potential for integration problems. It is important to remember that the fieldbus integration responsibility ultimately lies with the end user.[3]

Integration and compatibility issues will require the cooperation of end users, field device firms, control system firms, and fieldbus component manufacturers. This may affect the strategy of a company for long-term support as the responsibility lines shift. Being able to support the process automation system from field devices up to the boardroom may require a larger pool of support organizations working together.

Training

Training is important for the plant maintenance personnel and the end users so that the technology transfer to these groups is well received. The training should be developed to ensure that the maintenance personnel and end users can sustain the economic benefits realized during the project through the entire facility life cycle. Training is important for the management staff of the end user as well so that they feel the same level of comfort that they have with conventional I/O installations.

Teamwork

Along with new technology comes the need for all the teams involved to keep an open mind and place more effort into learning a new approach to achieve the same results. The teams will have to resist the temptation to make judgments too quickly as they are faced with new ideas, new problems, and new solutions. New team interdependencies will need to be developed to satisfy the design and maintenance requirements of a fieldbus.

CONCLUSIONS

This section has explored the design, installation, economics, documentation, and cultural aspects of installing a fieldbus system from an end user's point of view. Fieldbus device functionality and compatibility have significantly increased over the past 5 years, and this technology has demonstrated proven benefits and economic savings once the end user effectively addresses each of the technical and cultural issues associated with fieldbuses. Knowledge is one of the most effective tools in a successful implementation, and it has been the intent of this section to provide a starting point, plan, and the appropriate references to ensure a successful fieldbus implementation.

References

1. Glanzer, D. A., "An Open Architecture for Information Integration," *Sensors,* May 2001.
2. Verhappen, I., "Fieldbus Cost Comparisons Show Going Digital Can Save Big Dollars," *InTech,* August 1999.
3. Studebaker, P., "Be Careful What You Ask For," *Control,* January 1998.
4. Harrold, D., "Back to Basics—How to Read P&IDs," *Control Engineering,* August 2000.
5. Official Fieldbus Foundation Web site, http://www.fieldbus.org.
6. Official Profibus International Web site, http://www.profibus.org.

Bibliography

"Fieldbus Wiring Design and Installation Guide by Relcom (H1 Foundation Fieldbus)," http://www.relcominc.com/fieldbus/tutorial.htm.

Official ISA Fieldbus Portal, http://www.isa.org/portals/fieldbus/.

"The Synergetic Fieldbus Comparison Chart," http://www.synergetic.com/ompare.htm, Synergetic Micro Systems, Inc.

"Whitepaper—Cost and Analysis Cable Study Using Foundation Fieldbus Instruments," Emerson Process Management, April 2001.

4.8 Instrumentation Network Design and Upgrade

M. J. BAGAJEWICZ

This section presents techniques for the optimal selection and location of instruments in process grassroots designs as well as in retrofit scenarios to perform data gathering for control and monitoring, including fault detection and online optimization. The main goal is usually to minimize the instrumentation investment cost subject to performance constraints. Some of these constraints are as follows:

1. The network should be able to determine estimates of measured and unmeasured variables with certain precision and with accuracy.
2. Malfunctioning instrumentation should be able to be detected and its readings filtered through an algorithm that makes use of all measurements.
3. Unsafe or faulty operating conditions and their origin should be able to be detected.
4. Conditions and events that can impact efficiency and quality of products should be identified.

When data acquisition is supported solely by instrument measurements, each particular value of a variable is directly associated with the instrument that measures it. Very precise and failure-proof instrumentation, albeit also costly, would suffice to obtain precise and reliable estimates. However, because the probability of a sensor failure is finite, redundancy is usually used as means of guaranteeing data availability. With redundancy comes discrepancy. If the noise associated with the signal is small enough, these discrepancies can be ignored, but normally these readings have to be reconciled.

Reconciliation also brings improved precision of the estimates. In turn, redundancy is classified as:

- *Hardware Redundancy*: When two or more sensors are used to measure the same variable
- *Software or Analytical Redundancy*: When a set of values of different variables is supposed to satisfy a mathematical model

A simple example of analytical redundancy is a unit with several input and output streams. If one measures the flow rate of all streams with a single instrument per flow measured, there is no hardware redundancy. However, because the sum of the input flow rates has to be equal to the sum of the outputs, redundancy is analytical. In other words, there are two estimates available for each flow rate, one from its direct measurement and the other obtained using the material balance equations, which constitutes a conflict that needs to be resolved.

Data reconciliation and gross error detection are techniques that help obtain precise estimates and at the same time identify instrument malfunction (Tasks 1 and 2). While data reconciliation relies on analytical constraints and mostly on least-square estimation, other techniques used to obtain estimates, like Kalman filtering, pattern recognition, neural networks, principal component analysis (PCA), partial least squares, among others, rely on different concepts. There are several books that outline many aspects of data reconciliation.[1–5] In addition, hundreds of articles have been devoted to the problem. Kramer and Mah[6] discussed how good estimates of data can be obtained by means of models and it is for this reason that these activities have been named model-based monitoring. Thus, the field of sensor network design and upgrading has traditionally relied on model-based concepts simply because it is the reverse engineering problem of monitoring.

Instrument malfunction is a term that relates to situations that range from miscalibration or bias to total failure. In the absence of redundancy, miscalibration or bias cannot be detected (unless the deviation is so large that it becomes obvious). Redundancy, and especially software or analytical redundancy, is the only way to contrast data and determine possible malfunctions of this sort. This approach was named "gross error detection" by one group of researchers[7–12] and lately "sensor/data validation" or "signal reconstruction" by another group.[13–16] Data reconciliation and gross error detection are also discussed in Section 5.2.

MEASURED AND KEY VARIABLES

Traditionally, an *a priori* selection of measured variables has been the starting point of the design of control and monitoring strategies. Usually, the variable whose estimate is desired is the one that is measured. Even the emerging field of parameter estimation for online optimization relies in practice on the information obtained through traditional monitoring. Analytical redundancy can provide better and more reliable estimates. *We therefore call key variables those for which estimated values are of interest for control, monitoring, and parameter estimation. Thus, in general, measured variables are not necessarily the same as the key variables.*

The concept of key variable is already used in inferential control[17,18] and procedures for the selection of key variables in control are reviewed elsewhere (see, for example, Seborg et al.,[19] or Ogunnaike and Ray[20]). Several techniques have been developed to establish the existence of a fault (observability), determine its nature (diagnosis), and take appropriate action (alarm and/or shutdown), while assuming that the instrumentation is free of biases and is well calibrated. A few books cover all this in detail.[21–23] These techniques range from fault trees, cause–effect digraphs, neural networks, and knowledge-based systems to model-based statistical approaches like PCA and partial least squares. The field is intertwined with the data reconciliation/rectification field to the extent that the same techniques are used sometimes in both fields under different names.

INSTRUMENTATION DESIGN GOALS

Several goals exist for every designer of a network of instruments in a plant:

- *Cost*, including maintenance cost.
- *Estimability*, a term coined to designate the ability to estimate a variable using hardware or software.[24] The term is a generalization of the concepts of both observability and redundancy (Section 5.2 provides more specific definitions of these terms).
- *Precision.*
- *Reliability* and *availability* of the estimates of the key variables, which are functions of the reliability/availability of the measurements.
- *Gross-error robustness.*[25] This encompasses three properties:
  —*Gross-error detectability*, which is the ability to detect a certain number of gross errors larger than a specified size.
  —*Residual precision*, which is the precision of estimates left when a certain number of instruments fail.
  —*Gross-error resilience*, which is defined as control over the smearing effect of undetected gross errors.
- *Ability to perform process fault diagnosis.*
- *Ability to distinguish sensor failure from process failure.*

Although all these goals seem equally important, cost has been the traditional objective function used in design, although other goals, such as precision, have also been used as objective functions. The cost-based mathematical programming design procedure is the following:

Minimize {*Total Cost*}
s.t.
- *Estimability of Key Variables*
- *Precision of Key Variables*
- *Reliability of Key Variables*
- *Gross-Error Robustness*

The total cost includes the maintenance cost, which regulates the availability of variables, a concept that substitutes reliability when the system is repairable. The above design procedure does take into account the ability of a sensor network to detect process faults and a logic for alarm systems. Even though important attempts are being made to address this issue,[26] a model based on cost-efficient alarm design is yet to be produced. Similarly, the direct incorporation of control performance measures as additional constraints of a cost-optimal representation has not yet been fully investigated. Finally, methods for cost-optimal instrumentation design corresponding to the implementation of several other monitoring procedures, like PCA, projection to latent structures (PLS), wavelet analysis, and neural networks, among others, have not been yet proposed.

From the exclusive point of view of fault detection, the problem of the design of instrumentation is:

Minimize {*Total Cost*}
s.t.
- *Desired Observability of Faults*
- *Desired Level of Resolution of Faults*
- *Desired Level of Reliability of Fault Observation*
- *Desired Level of Gross-Error Robustness in the Sensor Network*

The combination of both goals, that is, the design of a sensor network capable of performing estimation of key variables for production accounting, parameter estimation for online optimization, as well as fault detection, diagnosis, and alarm (all of this in the context of detecting and assessing instrument biases and process leaks) is perhaps the long-term goal of this field.

In the rest of this section all existing methods to design and upgrade instrumentation for monitoring purposes, mainly focusing on the minimization of cost, are presented. Methods to design sensor networks for fault detection and observation also will be briefly reviewed.

COST-OPTIMAL AND ACCURATE SENSOR NETWORKS

Let c_i be the cost of the candidate instrument to measure variable i, and let q be a vector of binary variables to denote whether a variable is measured ($q_i = 1$) or not ($q_i = 0$). Then cost-optimal design of an accurate sensor network is obtained solving the following problem:[25]

$$\left.\begin{array}{ll} \text{Min} \sum_{i \in M_1} c_i q_i & \\ \text{s.t.} & \\ \sigma_j(q) \le \sigma_j^* & \forall j \in M_P \\ q_i \in \{0, 1\} & \forall i \in M \end{array}\right\} \quad \mathbf{4.8(1)}$$

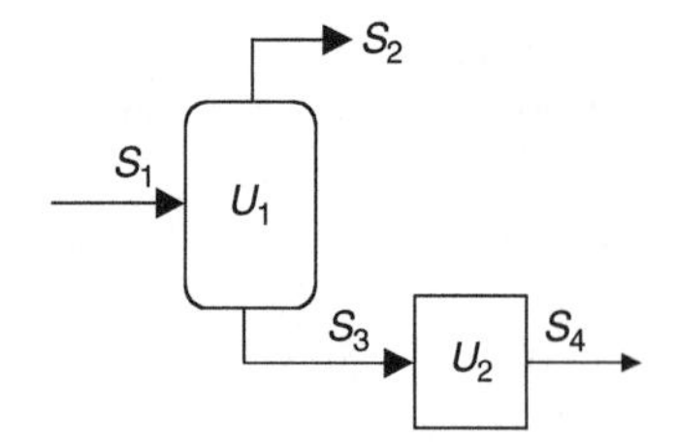

FIG. 4.8a
Two units-four streams example.

where M_P is the set of all variables for which precision constraints are imposed and M the set of variables where measurements can be placed. In the formulation of the objective function, it is assumed that there is only one potential measuring device with associated cost c_i for each variable (that is, no hardware redundancy), but this condition can be relaxed.[25,27] Furthermore, $\sigma_j(q)$ represents the standard deviation of the estimated value of variable j obtained after data reconciliation and σ^*_j its threshold value. The following small example, taken from Bagajewicz,[25,27] will be used to illustrate the procedure.

Example Consider the process flow diagram of Figure 4.8a. The flow rates are F = (150.1,52.3,97.8,97.8). Assume that for each rate, flowmeters of precision 3, 2, and 1% are available at costs 800, 1500, and 2500, respectively, regardless of size. Precision is only required for variables F_1 and F_4, that is, $M_P = \{S_1,S_4\}$, with $\sigma^*_1 = 1.5\%$ and $\sigma^*_4 = 2.0\%$. Two solutions are obtained featuring a cost of C = 3000. The corresponding meters are shown in Table 4.8b. Assume now that the cost of the 3% sensors drops to $700. Then, the optimal solution is no longer the one shown in Table 4.8b. In this case two solutions of equal cost (C = 2900) are shown in Table 4.8c. This solution is redundant and cheaper.

Although precision is achieved using nonredundant networks, gross errors are impossible to detect. Therefore, if at least one degree of redundancy is requested, that is, at least two ways of estimating each key variable, then there are two solutions with a cost of C = 3100 (Table 4.8d). This can be formally achieved adding a constraint requiring the estimability degree of each key variable to be two. The concept of degree of estimability is a generalization of the concepts of degrees of observability and redundancy,[24] and it basically relates to the number of ways a variable can be estimated, regardless of whether it is measured or not.

TABLE 4.8b
Solutions of the Precision Constrained Problem

Solution	S_1	S_2	S_3	S_4
A	—	2%	2%	—
B	—	2%	—	2%

TABLE 4.8c
Solutions of the Precision Constrained Problem (new cost of instruments)

Solution	S_1	S_2	S_3	S_4
C	3%	3%	2%	—
D	3%	3%	—	2%

TABLE 4.8d
Solutions of the Precision Constrained Problem (redundancy required)

Solution	S_1	S_2	S_3	S_4
E	3%	3%	2%	—
F	3%	3%	—	2%

Bagajewicz[25] proposed a tree enumeration procedure as a means to solve the problem. In a recent work, it was shown that the problem can in principle be reduced to a convex MINLP formulation.[28,29] Finally, Bagajewicz and Cabrera[30] showed that the model can be written as an MILP problem.

A series of techniques, which produce cost-optimal networks subject to constraints on estimability, which were introduced by Bagajewicz and Sánchez,[24] were extended to bilinear systems (typically those for which component and energy balances are added to overall mass balances) by Bagajewicz.[27]

DESIGN FOR MAXIMUM PRECISION

Madron and Veverka[31] proposed to design sensor networks minimizing the mean square error of the required quantities:

$$\text{Min } \frac{1}{n_M}\sum_{i=1}^{n_M}\sigma_i^2 \qquad \textbf{4.8(2)}$$

where σ_i is the standard deviation of the i-th required quantity and n_M is the total number of required measurements. This problem was efficiently solved by Madron[2] using the concept of minimum spanning tree of a graph. A problem maximizing the precision of only one variable was proposed by Alhéritière et al.,[32–34] who unfortunately did not use integer variables. The generalized maximum precision problem[27] considers the minimization of a weighted sum of the precision of the parameters.

$$\left.\begin{aligned} &\text{Min} \sum_{j\in M_P} a_j\sigma_j^2(q) \\ &\text{s.t.} \\ &\sum_{i\in M_1} c_i q_i \le c_T \\ &q_i \in \{0,1\} \quad \forall i \in M \end{aligned}\right\} \qquad \textbf{4.8(3)}$$

where c_T is the total resource allocated to all sensors, σ is the vector of measurement standard deviations, and a_j the weights.

The result of this generalized problem is a design for multiple parameter estimation; it is more realistic due to the discrete variables, and takes into account redundancy as well as all possible forms of obtaining the parameters. The only difficulty associated is the determination of meaningful values of weights. Finally, the minimum overall variance problem given by the expression in Equation 4.8(2) is a particular case where c_T is chosen to be a large value, i.e., the constraint on total cost is dropped. A mathematical connection between the maximum precision and the minimum cost representations of the problem exists.[35] More precisely, *the solution of one problem is one solution of the other, and vice versa.*

PARAMETER ESTIMATION

Considerable attention is placed today on the issue of parameter estimation, especially in the context of the increasing popularity of online optimization. One of the early papers[36] proposed the use of maximum likelihood principle to obtain parameters in implicit models. Reilly and Patino-Leal[37] initiated a line of work that bases parameter estimation on linearization but Kim et al.[38] proposed the use of nonlinear programming. The effect of data reconciliation and gross error detection in parameter estimation has been analyzed by MacDonald and Howat[39] as well as by Serth et al.[40] and Pages et al.[41] Among the approaches based on linear algebra, Kretsovalis and Mah[42] proposed a combinatorial search based on the effect of the variance of measurements on the precision of reconciled values. Tjoa and Biegler[43] explored methods for the estimation of parameters in differential-algebraic equation systems. To select measurements for such goals, Krishnan et al.[44,45] presented a three-step strategy and Loeblein and Perkins[46] discussed the economics.

The method proposed by Krishnan et al.[44] relies on a screening procedure that involves three steps:

1. A first step performs a structural analysis (singular value decomposition), which disregards measurements with little or no effect on the parameters.
2. A second step disregards measurements that have insignificant effect on the axis length of the confidence region of the parameter estimates.
3. The last step determines the interaction between the parameter estimates by means of calculating a covariance matrix.[47]

If the off-diagonal elements are too large, then the parameters are highly interactive and, therefore, any problem with the set of measurements that affects one parameter will also affect the other. The "best" set of measurements will have a small confidence region and lead to low interaction between the parameters. Unfortunately, this method does not take into account cost and does not offer a systematic procedure to make a final selection of the "best" set. In contrast, Bagajewicz[27] discusses linearization techniques in the context of cost-based minimization design procedures.

PRECISION UPGRADE

As pressure mounts to obtain more reliable and accurate estimates of variables and parameters, the use of data reconciliation techniques is the first response, but in most cases, the existing redundancy level is not sufficient to guarantee the level of accuracy required. Thus, one is faced with an upgrade problem.

There are three possible ways of performing the upgrade of a sensor network: (1) by the addition of new instruments, (2) by the substitution of existing instruments by new ones, and (3) by relocation of existing instruments.

Typically, addition of new instruments has been the response first considered. Kretsovalis and Mah[42] proposed a combinatorial strategy to incorporate measurements one at a time, to an observable system, but no constraints were considered. Substitution and relocation are options that are sometimes cheaper. One example is the substitution of thermocouples by thermoresistances or their relocation. However, it is in the case of laboratory analysis where the options of substitution and relocation should be strongly considered. Because upgrading requires capital expenditure, it must be done on the basis of a cost–benefit analysis. The costs of the instrumentation are straightforward to obtain. However, the benefits need to be somehow quantified. In the case of data accuracy needs for accounting purposes, the benefit can be quantified as the decrease of lost revenue due to imprecise data. Describing it in simple terms, the larger the uncertainty in the assessment of the amount of raw materials purchased or of the products sold, the larger the probability of lost revenue. Thus, every percent accuracy of these flows can be assigned a monetary value.

In the case of monitoring and parameter estimation for online optimization, an economic measure can also be developed. First, in the case of monitoring, one can associate a revenue loss for product that is not in specification. Finally, in the case of online optimization, Loeblein and Perkins[46] discuss the use of measures of loss of economic performance due to offset from the correct optimum due to plant–model mismatches. In other cases, industry is looking less at the benefit in monetary terms and simply plots the increased precision as a function of investment.[34] Although this approach is intuitive, its importance relies on the possibility of visualizing the effect of instrumentation cost.

The upgrading of a sensor network by the simple addition of instrumentation has to be done at minimum cost while reaching the goals of precision in key variables. Hence the following minimum cost problem:[48]

$$\left.\begin{array}{lr} \text{Min} \sum_{i\in M_1} \sum_{k\in K_i} c_{ik} q_{ik}^N & \\ \text{s.t.} & \\ \sigma_j(q) \le \sigma_j^* & \forall j \in M_P \\ \sum_{k\in K_i} q_{ik}^N + N_i \le N_i^* & \forall i \in M \\ q_{ik}^N \in \{0,1\} & \forall i \in M,\ \forall k \in K_i \end{array}\right\} \quad \mathbf{4.8(4)}$$

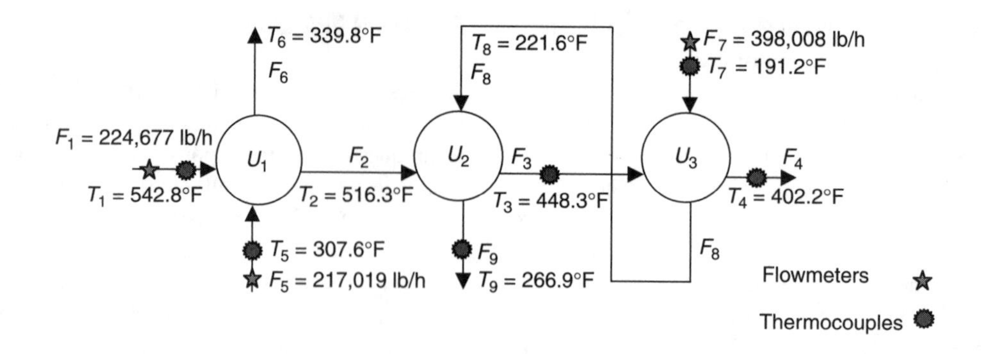

Heat Exchanger	Area (ft²)	Fr	Cp_h (BTU/lb °F)	Cp_c (BTU/lb °F)
U_1	500	0.997	0.6656	0.5689
U_2	1100	0.991	0.6380	0.5415
U_3	700	0.995	0.6095	0.52

FIG. 4.8e
Industrial heat exchanger network.

where K_i is the set of new sensors available to measure variable i, N_i is the number of existing sensors measuring variable i, c_{ik} is the cost of the k-new sensor for measuring variable i, N_i^* is the maximum number of sensors that are allowed to be used to measure variable i, and q^N is the vector of binary variables for new sensors.

The first constraint establishes a bound on precision, which is a function of the set of fixed existing sensors and new instrumentation. The second constraint establishes an upper bound on the number of sensors used to measure each variable. This number is usually 1 for the case of flow rates, but it can be larger in the case of laboratory measurements of concentrations. In this case, these constraints can be lumped in one constraint for all concentration measurements to express the overall limitation of the laboratory on a daily basis, or fraction thereof.

When maximum precision is requested, and cost is a constraint, then a maximum precision problem can be employed, using a bound on the capital expenditure. The two problems are equivalent.

Example Consider the industrial heat exchanger network of Figure 4.8e. It corresponds to a set of heat exchangers where crude is heated using hot gas–oil coming from a column. In this case, the heat transfer coefficients for the heat exchangers are estimated in terms of temperature and flow rate measurements. The existing instrumentation is indicated in the figure (flowmeters are of 3% precision and thermocouples have a precision of 2°F).

The standard deviations of heat transfer coefficients of the heat exchangers calculated using the installed set of instruments are 12.27, 2.96, 3.06 BTU/h ft^2°F. To obtain these values, all redundant measurements have been used. To enhance the precision of the parameter, new instruments should be added. In this example, hardware redundancy is considered. Furthermore, different types of new instruments are available to measure some temperatures. Data for new instrumentation are presented in Table 4.8f, where cost and standard deviation are shown. The maximum number of allowed instruments for measuring each variable is given in Table 4.8g, where a zero value indicates there is a restriction for measuring the corresponding variable.

TABLE 4.8f
Availability of New Instrumentation

	Flowmeters		Temperature Sensors	
Stream	Standard Deviation, %	Cost	Standard Deviation, °F	Cost
F_1	3	2250	2/0.2	500/1500
F_2	3	2250	2	500
F_3	3	2250	2	500
F_4	3	2250	2/0.2	500/1500
F_5	3	2250	2	500
F_6	3	2250	2	500
F_7	3	2250	2	500
F_8	3	2250	2	500
F_9	3	2250	2/0.2	500/1500

Table 4.8h presents the results for the upgrade problem using the minimum cost problem. When there are two possible instruments to measure a variable, the type of instrument is indicated between parentheses in the optimal solution set. Thus, for example, $T_4(1)$ indicates that the first instrument available to measure the temperature in stream S_4 is selected; in this case a thermocouple with a precision of 2°F is selected. The weights for the maximum precision problem are assumed equal to 1. One can notice that as precision requirements are

TABLE 4.8g
Maximum Number of Instruments for the Heat Exchanger Network

Variable	N_i^*	Variable	N_i^*
F_1	1	T_1	2
F_2	1	T_2	1
F_3	1	T_3	1
F_4	1	T_4	2
F_5	1	T_5	1
F_6	0	T_6	1
F_7	1	T_7	1
F_8	0	T_8	0
F_9	1	T_9	2

made more and more stringent, the cost and the number of instruments added increase. Case 3 has three alternative solutions.

RESOURCE REALLOCATION

As mentioned, in many cases measurements can be easily transferred at no cost from one stream to another. This is the case of concentration measurements that are performed in the laboratory. Pressure gauges and thermocouples can also be transferred from one place to another at a relatively small cost. However, flowmeters are probably an exception. Because one can consider that the resource reallocation does not involve cost, or eventually neglects it, the minimum cost problem would reduce to a set of mixed integer nonlinear algebraic inequalities from which one solution is to be extracted. However, even in the case where cost is not considered, one would like to minimize the number of changes. In addition, these reallocation costs may overcome the simple addition of new instrumentation. Therefore, any reallocation and upgrade program should consider the trade-off between all these decisions. A mathematical programming representation, which is an extension of Equation 4.8(4) has been developed.[48] We illustrate it next.

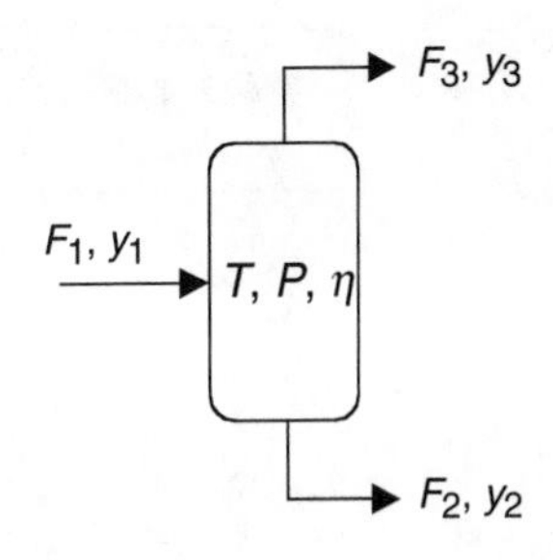

FIG. 4.8i
A flash unit.

Example Consider the flash tank of Figure 4.8i. The number of existing instruments to measure each variable (N_i) and the corresponding standard deviations $\sigma_{i,k}$ for different types of installed instruments are included in Table 4.8j. In this example the mass fractions of all the components of a stream are measured on line, but a laboratory analysis may be done as a second alternative to know their values.

The standard deviation of the vaporization efficiency coefficient estimated using the existing instrumentation is 0.00434. As this value is not satisfactory, a reallocation and possibly an incorporation of new flowmeters and laboratory composition analysis are proposed. The sets of streams from which measurements can be transferred to other streams is $M_T = \{1, 2, 3\}$ and the set of new locations is $M_R = \{1, 3, 4, 6\}$. The costs of feasible reallocations between sets M_T and M_R are given in Table 4.8k. For example, the cost of relocating a sensor from F_1 to F_2 is 80. Infinite costs are used for forbidden reallocations. Typically, this relocation pattern is constructed based on engineering judgment.

Table 4.8l shows the results. The first row represents the case for the existing instrumentation. A reduction of the standard deviation from 0.00438 to 0.00347 results if the laboratory analysis for the feed stream is relocated to the liquid stream and a pressure sensor is added. The cost of this case is 100. Higher precision is obtained by means of the reallocation and addition of instruments. For $\sigma^* = 0.0031$ no reallocation and instrument addition can achieve this goal.

TABLE 4.8h
Results for the Minimum Cost Problem

Case	$\sigma^*_{U_1}$	$\sigma^*_{U_2}$	$\sigma^*_{U_3}$	σ_{U_1}	σ_{U_2}	σ_{U_3}	Cost	Optimal Set
1	4.0	4.0	4.0	3.6160	1.9681	2.7112	500	$T_6(1)$
2	3.5	2.0	2.5	2.7746	1.6892	2.3833	1500	$T_2(1)\ T_4(1)\ T_6(1)$
3	3.0	1.5	2.5	2.7230	1.4972	2.2844	6500	$F_2\ F_3\ T_2(1)\ T_4(1)\ T_6(1)\ T_9(1)$
								$F_2\ F_4\ T_2(1)\ T_4(1)\ T_6(1)\ T_9(1)$
								$F_3\ F_4\ T_2(1)\ T_4(1)\ T_6(1)\ T_9(1)$
4	3.5	2.0	2.0	—	—	—	—	—

TABLE 4.8j
Installed and New Instrumentation Data

Variable Index	Variable	N_i	Installed $\sigma_{i,k}$		New Available $c_{i,k}$	New Available $\sigma_{i,k}$
1	F_1	2	2.5	2.5	350	2.
2	y_1	2	0.015	0.01	2700	0.01
3	F_2	1	1.515	—	350	1.48
4	y_2	1	0.01	—	2700	0.01
5	F_3	1	1.418	—	400	1.38
6	y_3	1	0.01	—	2700	0.01
7	P	1	14.	—	100	14.

TABLE 4.8k
Flash Drum: Costs of Relocation

	F_1	F_2	y_2	y_3
F_1	∞	80	∞	∞
y_1	∞	∞	0	50
F_2	80	∞	∞	∞

TABLE 4.8l
Flash Drum: Results for Sensor Reallocation and Upgrade

Case	σ^*	σ	Cost	Reallocations	New Instruments
1	∞	0.00438	—	—	—
2	0.0038	0.00352	100	—	P
		0.00347	100	y_1 to y_2	P
3	0.0033	0.00329	2800	y_1 to y_2	y_3 P

RELIABLE SENSOR NETWORKS

Bagajewicz[27] discusses the concepts of availability and reliability for the estimates of key variables and distinguishes them from the same concepts applied to instruments. The former are called estimation reliability/availability, whereas the latter are called service reliability/availability. Ali and Narasimhan[49] proposed to use the system reliability given by the minimum estimation reliability throughout the whole network as the objective function. In other words, the reliability of the system is defined by its weakest element. To address the fact that their representation does not control cost, they proposed to limit the number of sensors to the minimum possible that will still guarantee observability. Although their procedure does not guarantee global optimality, it produces good results. Bilinear networks are also discussed by Ali and Narasimhan[50] in detail. Finally, genetic algorithms are successfully used by Sen et al.[51] not only for reliable networks, but also for a variety of other objective functions.

The cost-based representation for the design of the sensor network subject to reliability constraints is[52]

$$\left.\begin{array}{ll} \text{Min} \sum_{\forall i \in M_1} c_i q_i & \\ \text{s.t.} & \\ R_k^v(q) \geq R_k^* & \forall k \in M_R \\ q_i \in \{0, 1\} & \forall i \in M \end{array}\right\} \qquad 4.8(5)$$

where $R_k^v(q)$ is the reliability of variable k, R_k^* the threshold, and M_R the set of variables whose reliability is to be constrained. The reliability of each variable is calculated using the failure probabilities of all the sensors participating in the corresponding cutsets (material balances). If all sensors have the same cost c and N is the number of sensors, one obtains a problem where the number of sensors is minimized. Finally, the representation due to Ali and Narasimhan[49] can be put in the form of a minimum cost problem. The details of such equivalency can be found in the article by Bagajewicz and Sánchez,[52] where examples are shown. The models in Equations 4.8(1) and 4.8(5) have the same objective function. Therefore, a single model containing precision and reliability constraints can be constructed.

REPAIRABLE SENSOR NETWORKS

When repairs are not present, the service availability of a sensor is equal to its service reliability. In addition, the failure rate has been considered in a simplified way as a constant. However, in the presence of repairs, failure is no longer an event that depends on how many hours the sensor has survived from the time it was placed in service. It is also conditioned by the fact that due to preventive or corrective maintenance, the sensor has been repaired at a certain time after being put in service. These events condition the failure rate. We thus distinguish unconditional from conditional events in failure and repair. These concepts are important because sensor

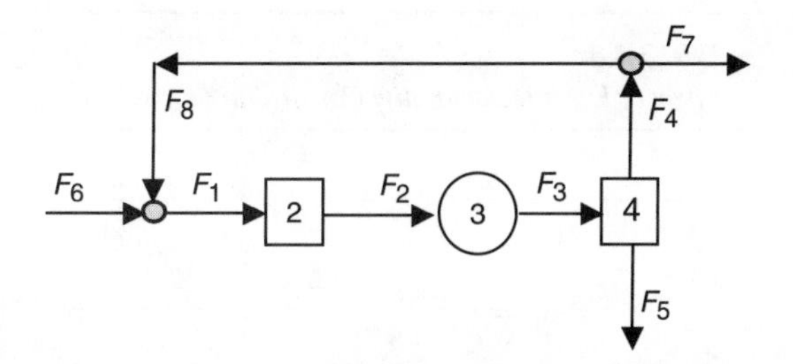

FIG. 4.8m
Simplified ammonia plant network. (Adapted from Bagajewicz.[27])

TABLE 4.8n
Instrumentation Data

	No. 1	No. 2	No. 3
Purchase cost	350	250	200
Precision	1.5%	2.5%	3%
Failure rate (failure/yr)	0.3	0.6	0.7

maintenance cost accounts for nearly 20% of all maintenance cost.[53] Its reduction or containment is therefore essential. The connection between failure rate, repair rate, and the expected number of repairs as well as illustrations of the impact of maintenance on sensor network design are described by Bagajewicz.[27]

Example We now show a design of a sensor network for the simplified ammonia network (Figure 4.8m). In this example, sensors for each stream may be selected from a set of three instruments with different precision, purchase cost, and failure rate. These data are included in Table 4.8n.

Maintenance corrective costs are evaluated considering spare part cost and labor cost of 10 and 40, respectively, a life cycle of 5 years, and an annual interest of 6%. Constraints of precision, residual precision, and availability are included for only two flow rates. The limits on these variables are presented in Table 4.8o. The repair rate of instruments, a parameter that is a characteristic of the plant in consideration, has been varied between 1 and 20. The results of the optimization problem are presented for each case in Table 4.8p.

In the first case, the repair rate is comparatively low. Consequently, the availability of instruments in the life cycle is also relatively low. To satisfy the availability of key variables, the optimal solution includes a set of six instruments. Three of these instruments are of type 1, which are sensors of low failure rate, high precision, and high cost. For this reason precision and residual precision constraints are not binding.

When the repair rate is 2, an optimal solution exists that consists of five instruments. Two of these instruments are of type 1 and the rest are of type 3. Consequently, the total instrumentation cost decreases.

TABLE 4.8o
Constraints of the Optimization Problem

Stream	Precision Requirements	Residual Precision Requirements	Availability Requirements
F_1	—	—	0.9
F_2	1.5%	2.0%	
F_5	2.5%	3.0%	
F_7	—	—	0.9

TABLE 4.8p
Optimization Results for the Simplified Ammonia Process Flowsheet

Repair Rate	Measured Variables	Instrument Precision (%)	Cost	Precision (%) (F_2) (F_5)	Availability (F_1) (F_7)
1	F_1 F_4 F_5 F_6 F_7 F_8	3 1 1 1 3 2	2040.2	0.8067 1.2893	0.9021 0.9021
2	F_4 F_5 F_6 F_7 F_8	3 3 1 3 1	1699.8	0.9283 1.9928	0.9222 0.9062
4	F_4 F_5 F_6 F_7 F_8	3 3 1 3 3	1683.7	1.2313 1.9963	0.9636 0.9511
20	F_4 F_5 F_6 F_7 F_8	3 3 1 3 3	1775.2	1.2313 1.9963	0.9983 0.9969

A lower instrumentation cost is obtained for a repair rate equal to 4. Even though sensors are located on the same streams as in the previous case, one sensor of higher failure rate is installed on stream S_8. This occurs because the repair rate is now higher, making the constraint on availability of variable S_7 not binding.

The results of the last case show that the influence of availability constraints decreases for high repair rates. The cost increases because of the effect of increasing the repair rate μ (from 4 to 20) in the maintenance cost model.

As a conclusion, the repair rate has a direct influence on the availability of a variable. If the repair rate is high, the design follows the requirements of precision and residual precision constraints. Thus, the availability of a variable may be a binding constraint for lower repair rates. In this situation, cost may increase because it is necessary to incorporate more instruments to calculate the variable alternative ways.

ROBUST SENSOR NETWORKS

When data reconciliation is performed, systematic errors (biases) produce a smearing effect in all the data so that the whole data set has to be discarded. Without analytical redundancy, no smearing exists, but then there is no way other than direct inspection to detect sensor failure.

A robust sensor network provides meaningful values of precision, residual precision, variable availability, error detectability, and resilience. These five properties encompass all the most desired features of a network. Indeed, precision and residual precision guarantee that data are always of the desired quality. Residual precision controls redundancy and reliability and allows the data to be most of the time at hand. Finally, error detectability and resilience ensure that the probability of data being free of gross errors is large. While the two first properties are deterministic and depend directly on the quality of the instrument selected, the other three are statistical in nature. It is expected that more properties will be added to define robustness in the future. For example, for neural networks, wavelet analysis, PCA, partial least squares (PLS), and other techniques for process monitoring, very little analysis regarding goals that a sensor network design or upgrade procedure can undertake was investigated. It is expected that in the next few years, such methods will emerge. Robustness can be incorporated explicitly in the form of constraints to the minimum cost problem 4.8(1).[27] We illustrate them next.

Example We now add residual precision capabilities to the example of Figure 4.8a. We formally define residual precision of order k for stream i, $\psi_i^*(k)$, the precision that the estimate of the flow rate of stream i will have if any k measurements of the network are eliminated. Consider now that residual precision of order k = 1 is added to variables S_1 and S_4 as follows: $\psi_1^*(1) = 1.5\%$ and $\psi_4^*(1) = 3\%$. The solution is to put sensors of precision 2%, 3%, 3%, 3% in S_1 through S_4, respectively. The cost is C = 3900. Assume now that residual precision is requested to the same level as precision. Then two alternative solutions with cost C = 5500 are obtained (Table 4.8q).

Not only is the cost higher, but also there is one more degree of redundancy. For larger problems, the number of alternatives will increase, requiring new criteria to screen alternatives further.

We now turn to adding error detectability to this example. As the error detectability increases, so does the precision of the sensor network. However, it can also make the design infeasible. Consider adding an error detectability level of $\kappa_D = 3.9$ to the problem. Requiring such a value of κ_D means that the network should be able to detect gross errors of size 3.9 times the standard deviation of the measurement or larger in any measured variable. A statistical power of 50% is assumed in the algorithm used to detect the gross errors. Two solutions from a set of only four feasible solutions are found with cost C = 4800 (Table 4.8r).

TABLE 4.8q
Solutions of the Residual Precision Constrained Problem

S_1	S_2	S_3	S_4
1%	2%	2%	—
1%	2%	—	2%

TABLE 4.8r
Effect of Error Detectability Constraints ($\kappa_D = 3.9$)

S_1	S_2	S_3	S_4
1%	3%	—	2%
1%	3%	2%	—

If an error detectability of $\kappa_D = 3.4$ is requested, the problem has only one solution, namely, sensors with precision of 1%, 3%, 1%, 1% in S_1 through S_4, respectively, with a cost C = 8300.

A resilience requirement means that the network will be able to limit the smearing effect of gross errors of a certain size without affecting the other properties. If resilience at a level of three times the standard deviation for all measurements is added, then the solution is to put sensors with precision 1%, 3%, 1%, 1% in S_1 through S_4, respectively, with a cost of C = 8300. Relaxing (increasing) the resilience levels maintaining the error detectability at the same level may actually lead to solutions of higher cost, even to infeasibility.

DESIGN OF SENSOR NETWORKS FOR PROCESS FAULT DIAGNOSIS

Whereas the basic goal of monitoring systems is to provide a good estimate of the state of the system, alarm systems are designed to alert personnel of process malfunction. In turn, process faults, which typically are rooted in some unit, propagate throughout the process, altering the readings of instruments (pressures, temperatures, flow rates, etc.). Thus, these sensors should be able to determine departures from normal operation. In this sense, this task is different from that of gross error detection, which concentrates on instrument malfunction. As a consequence, the discrimination between instrument malfunction and process fault is an additional task of the alarm system. As a consequence, the problem of designing an alarm system consists of determining the cost-optimal position of sensors, such that all process faults, single or multiple and simultaneous, can be detected and distinguished from instrument malfunction (biases).

A process fault is a departure from an acceptable range of operation or "degradation from normal operation conditions, and includes symptoms of a physical change (such as deviations in measured temperature or pressure) as well as the physical changes themselves (scaling, tube plugging, etc.) and deviations in parameters (such as a heat transfer coefficient)."[54] Faults originate in a process and propagate to a set of sensors. These sensors are also subject to faults themselves. These faults on the sensors are either biases or catastrophic failures. In the latter case, the fault detection is compromised, whereas in the former case the process fault either can go undetected or false alarms may be induced.

Therefore, a good alarm system should be able to filter the disturbances affecting the sensors as well as the gross errors induced by their faults. The next step is the process fault detection itself, although these two steps can be performed simultaneously. This procedure also needs to have some capabilities of distinguishing process disturbances from real faults. Once a process fault has been identified, the final step consists of taking corrective actions, or determining a shutdown of the process. This is performed by implementing an alarm logic. Fault detection and diagnosis have been addressed in several books.[21–23] Bagajewicz[27] offers a review of the different approaches.

The first attempt to present a technique to locate sensors was done by Lambert,[55] where fault trees are used based on failure probabilities. Failure probabilities are difficult to assess and fault trees cannot handle cycles and the construction of the tree is cumbersome for large-scale systems. Because of these limitations, the technique has not been developed further. Raghuraj et al.[56] proposed an algorithm for one and multiple fault observability and resolution. Their design procedure is not based on cost and does not include other features, like filtering of sensor failures. Other aspects that have not been considered yet are failure probabilities, sensor failure probability, and severity of particular faults and the cost of the sensors.

Raghuraj et al.[56] used directed graphs (DG), that is, graphs without signs. The arcs of the DG represent a "will cause" relationship, that is, an arc from node A to node B implies that A is a sufficient condition for B, which in general is not true for a signed DG, where an arc represents a "can cause" relationship. The strategy used to solve the problem is based on identifying directed paths from root nodes where faults can occur to nodes where effects can be measured, called the observability set. Of all these paths, the objective is to choose the minimal subset of sensors from the observability set that would have at least one directed path from every root node. Raghuraj et al.[56] proposed a greedy search, which is later modified to remove redundant members of the observability set by a "backtracking algorithm." They do not guarantee optimality, that is, a minimum set may not be found by this algorithm. It is likely that a formal mathematical programming version of the problem will accomplish it.

Consider the DG of Figure 4.8s. All the nodes of the observability set are in the top row and the root nodes are in the bottom row. The problem of choosing the minimum number of sensors (nodes in the top row) that would cover all the root nodes is the well-known "minimum set covering problem."[57] All the root nodes are said to be "covered" if a directed path exists from every root node to at least one of the nodes of the observability set.

The solution to this problem obtained by Raghuraj et al.[56] is that nodes C_7 and C_8 constitute the minimal set. Indeed, R_1 and R_2 are observable from C_7 and the rest from C_8.

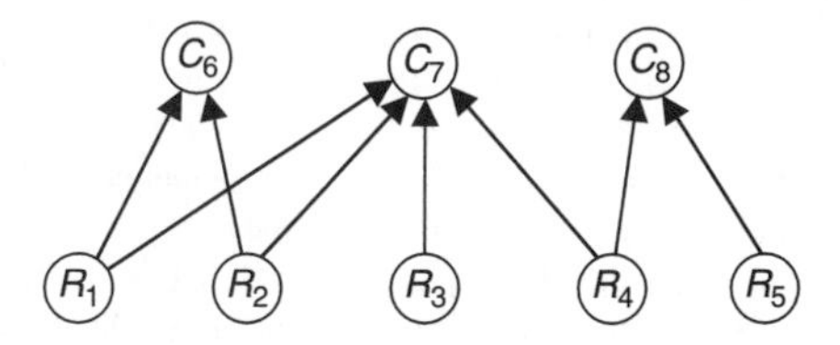

FIG. 4.8s
Bipartite graph.

DESIGN OF SENSOR NETWORKS FOR PROCESS FAULT RESOLUTION

For a fault-monitoring system to be useful in practice, it should not only be able to observe all the faults but also identify their location and number to the maximum extent possible. This is called maximum fault resolution. Therefore, *a sensor network for maximum fault resolution is such that each fault has one and only one set of nodes from which it is observable.*

Consider three fault nodes R_1, R_2, and R_3 (Figure 4.8t). Clearly, if only one fault is expected to occur at a time, then the set $[C_1, C_3]$ would be adequate to distinguish between the three faults. Indeed, a fault in R_1 is reflected in C_1, but not in C_3. Similarly, a fault in R_2 is reflected in C_3, but not in C_1. Finally, a fault in R_3 is reflected in both C_1 and C_3 simultaneously.

Raghuraj et al.[56] proposed a procedure to obtain such sets of nodes for maximum resolution. They also show that the sensor-location problem for multiple faults and maximum resolution can be solved as an extension of the single-fault assumption problem. The application of these algorithms to a continuous stirred tank reactor (CSTR) and a fluidized catalytic cracker (FCC) unit are presented by Raghuraj et al.[56]

Bhushan and Rengaswami[58,59] made considerable progress in the study of several important aspects of the problem. Very recently, Bagajewicz and Fuxman[60] presented a methodology to obtain cost-optimal networks for fault observation and multiple fault resolution. Moreover, Bhushan and Rengaswami[61] presented a methodology for fault detection that minimizes cost and takes into account the reliability of the network.

CONCLUSIONS

In this section, the cost-optimal formulation of the instrumentation network design and upgrade problem was presented. The design of networks with precision objectives in

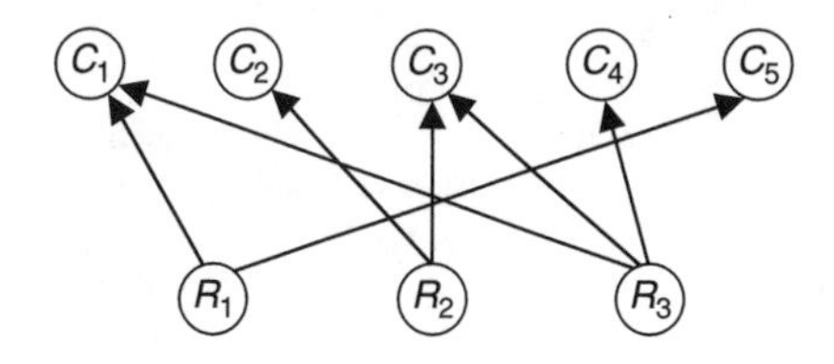

FIG. 4.8t
Maximum fault resolution.

key variables, including its upgrade, was illustrated. Additional features as reliability and the ability to handle instrument malfunction were also discussed. The design of instrumentation networks for fault detection and resolution was also briefly discussed. One emerging cost-optimal paradigm is the sensor network design for simultaneous fault detection and variable estimation.

NOMENCLATURE

a_j	weight factor
c	vector of sensor costs
C	total cost
c_T	bound on total cost
F	vector of flow rates
K_i	set of new sensors available for variable i
M	set of variables where sensors can be placed
M_P	key variables for precision
M_R	key variables for reliability
n_M	number of measurements
N_i	number of existing sensors for variable i
N^*	number of fixed sensors
P	pressure
q	binary vector indicating whether sensors are located (1) or not (0) in variables
q^N	binary vector for new sensors
S_i	stream i
T	temperature
U_i	unit i
y_i	mole fraction
σ	precision of estimates
σ^*	precision threshold

References

1. Mah, R. S. H., *Chemical Process Structures and Information Flows,* Stoneham, U.K.: Butterworths, 1990.
2. Madron, F., *Process Plant Performance, Measurement Data Processing for Optimization and Retrofits,* West Sussex, England: Ellis Horwood, 1992.
3. Veverka, V. V. and Madron, F., *Material and Energy Balances in the Process Industries,* Amsterdam: Elsevier, 1997.
4. Romagnoli, J. and Sánchez, M., *Data Processing and Reconciliation for Chemical Processes Operations,* San Diego, CA: Academic Press, 1999.
5. Narasimhan, S. and Jordache, C., *Data Reconciliation & Gross Error Detection. An Intelligent Use of Process Data,* Houston: Gulf Publishing Company, 2000.
6. Kramer, M. A. and Mah, R. S. H., "Model-Based Monitoring," *FOCAPO Proceedings,* Crested Butte, 1993.
7. Romagnoli, J. and Stephanopoulos, G., "On the Rectification of Measurement Errors for Complex Chemical Plants," *Chemical Engineering Science,* Vol. 35, Number 5, pp. 1067–1081, 1980.
8. Mah, R. S. H. and Tamhane, A. C., "Detection of Gross Errors in Process Data," *AIChE Journal,* Vol. 28, p. 828, 1982.
9. Rollins, D. K. and Davis, J. F., "Unbiased Estimation of Gross Errors in Process Measurements," *AIChE Journal,* Vol. 38, p. 563, 1992.
10. Crowe, C. M., Garcia Campos, Y. A., and Hrymak, A., "Reconciliation of Process Flow Rates by Matrix Projection. I. The Linear Case," *AIChE Journal,* Vol. 29, p. 818, 1983.
11. Crowe, C. M., "Data Reconciliation. Progress and Challenges," *Journal of Process Control,* Vol. 6, pp. 89–98, 1996.
12. Bagajewicz, M. and Jiang, Q., "Gross Error Modeling and Detection in Plant Linear Dynamic Reconciliation," *Computers and Chemical Engineering,* Vol. 22, Number 12, pp. 1789–1810, 1998.
13. Dunia, R., Qin, J., Edgar, T. F., and McAvoy, T. J., "Identification of Faulty Sensors Using Principal Component Analysis," *AIChE Journal,* Vol. 42, pp. 2797–2812, 1996.
14. Qin, S. J., Yue, H., and Dunia, R., "Self-Validating Inferential Sensors with Application to Air Emission Monitoring," *Industrial Engineering and Chemical Research,* Vol. 36, Number 5, pp. 1675–1685, 1997.
15. Tay, M. E., "Keeping Tabs on Plant Energy and Mass Flows," Chemical Engineering, September 1996.
16. Tham, M. T. and Parr, A., *"Succeed at On-Line Validation and Reconstruction of Data,"* in G. F. Nalven, Ed., *Plant Operation and Optimization,* New York: AIChE, 1996.
17. Moore, C., "Application of Singular Value Decomposition to the Design, Analysis and Control of Industrial Processes," *Proceedings American Control Conference,* p. 643, 1986.
18. Moore, C., "Determining Analyzer Location and Type for Distillation Column Control," 14th Annual Meeting of the Federation of Analytical Chemistry and Spectroscopy Societies, Detroit, 1987.
19. Seborg, D. E., Edgar, T. F., and Mellichamp, D. A., *Process Dynamics and Control,* New York: John Wiley & Sons, 1989.
20. Ogunnaike, B. A. and Ray, W. H., *Process Dynamics, Modeling and Control,* New York: Oxford University Press, 1994.
21. Himmelblau, D., *Fault Detection and Diagnosis in Chemical and Petrochemical Processes,* Amsterdam: Elsevier, 1978.
22. Pau, L. F., *Failure Diagnosis and Performance Monitoring,* New York: Marcel Dekker, 1981.
23. Gertler, J. J., *Fault Detection and Diagnosis in Engineering Systems,* New York: Marcel Dekker, 1998.
24. Bagajewicz, M. and Sánchez, M., "Design and Upgrade of Non-Redundant and Redundant Linear Sensor Networks," *AIChE Journal,* Vol. 45, Number 9, pp. 1927–1939, 1999.
25. Bagajewicz, M., "Design and Retrofit of Sensor Networks in Process Plants," *AIChE Journal,* Vol. 43, Number 9, pp. 2300–2306, 1997.
26. Tsai, C. S. and Chang, C. T., "Optimal Alarm Logic Design for Mass-Flow Networks," *AIChE Journal,* Vol. 43, Number 11, p. 3021, 1997.
27. Bagajewicz, M., *Design and Upgrade of Process Plant Instrumentation,* Lancaster, PA: Technomic, 2000.
28. Chmielewski, D., Palmer T. E., and Manousiouthakis, V., "Cost Optimal Retrofit of Sensor Networks with Loss Estimation Accuracy," AIChE Annual Meeting, Dallas, 1999.
29. Chmielewski, D., "Convex Methods in Sensor Placement," *Proceedings of the 4th IFAC Workshop on On-Line Fault Detection & Supervision in the Chemical Process Industries,* June 8–9, Seoul, Korea, 2001.
30. Bagajewicz, M. and Cabrera, E., "A New MILP Formulation for Instrumentation Network Design and Upgrade," *Proceedings of the 4th IFAC Workshop on On-Line Fault Detection & Supervision in the Chemical Process Industries,* June 8–9, Seoul, Korea, 2001.
31. Madron, F. and Veverka, V., "Optimal Selection of Measuring Points in Complex Plants by Linear Models," *AIChE Journal,* Vol. 38, Number 2, p. 227, 1992.
32. Alhéritière, C., Thornhill, N., Fraser, S., and Knight, M., "Evaluation of the Contribution of Refinery Process Data to Performance Measures," AIChE Annual Meeting, Los Angeles, 1997.
33. Alhéritière, C., Thornhill, N., Fraser, S., and Knight, M., "Cost Benefit Analysis of Refinery Process Data: Case Study," *Computers and Chemical Engineering,* Vol. 22, Suppl., pp. S1031–S1034, 1998.
34. Alhéritière, C., Thornhill, N., Fraser, S., and Knight, M., "Cost Benefit Analysis of Process Data in Plant Performance Analysis," AIChE Annual Meeting, Miami, 1998.

35. Bagajewicz, M. and Sánchez, M., "Duality of Sensor Network Design Models for Parameter Estimation, *AIChE Journal,* Vol. 45, Number 3, pp. 661–664, 1999.
36. Britt, H. I. and Luecke, R. H., "The Estimation of Parameters in Nonlinear Implicit Models," *Technometrics,* Vol. 15, Number 2, pp. 233–247, 1973.
37. Reilly, P. M. and Patino-Leal, H., "A Bayesian Study of the Error-in-Variables Model," *Technometrics,* Vol. 23, Number 3, p. 221, 1981.
38. Kim, I., Kang, M. S., Park, S., and Edgar, T. F., "Robust Data Reconciliation and Gross Error Detection: The Modified MIMT Using NLP," *Computer and Chemical Engineering,* Vol. 21, Number 7, pp. 775–782, 1997.
39. MacDonald, R. and Howat, C., "Data Reconciliation and Parameter Estimation in Plant Performance Analysis," *AIChE Journal,* Vol. 34, Number 1, 1988.
40. Serth, R., Srinkanth, B., and Maronga, S., "Gross Error Detection and Stage Efficiency Estimation in a Separation Process," *AIChE Journal,* Vol. 39, p. 1726, 1993.
41. Pages, A., Pingaud, H., Meyer, M., and Joulia, X., "A Strategy for Simultaneous Data Reconciliation and Parameter Estimation on Process Flowsheets, *Computer and Chemical Engineering,* Vol. 18, Suppl., pp. S223–S227, 1994.
42. Kretsovalis, A. and Mah, R. S. H., "Observability and Redundancy Classification in Multicomponent Process Networks," *AIChE Journal,* Vol. 33, pp. 70–82, 1987.
43. Tjoa, I. B. and Biegler, L. T., "Simultaneous Solution and Optimization Strategies for Parameter Estimation of Differential-Algebraic Equation Systems," *Industrial Engineering and Chemical Research,* Vol. 30, pp. 376–385, 1991.
44. Krishnan, S., Barton, G., and J., Perkins, "Robust Parameter Estimation in On-line Optimization—Part I. Methodology and Simulated Case Study," *Computer and Chemical Engineering,* Vol. 16, pp. 545–562, 1992a.
45. Krishnan, S., Barton, G., and Perkins, J., "Robust Parameter Estimation in On-line Optimization—Part II. Application to an Industrial Process," *Computer and Chemical Engineering,* Vol. 17, pp. 663–669, 1992b.
46. Loeblein, C. and Perkins, J. D., "Economic Analysis of Different Structures of On-Line Process Optimization Systems," *Computer and Chemical Engineering,* Vol. 22, Number 9, pp. 1257–1269, 1998.
47. Beck, J. V. and Arnold, K. J., *Parameter Estimation for Engineering and Science,* New York: Wiley, 1977.
48. Bagajewicz, M. and Sánchez, M., "Reallocation and Upgrade of Instrumentation in Process Plants," *Computers and Chemical Engineering,* Vol. 24, Number 8, pp. 1961–1980, 2000.
49. Ali, Y. and Narasimhan, S., "Sensor Network Design for Maximizing Reliability of Linear Processes," *AIChE Journal,* Vol. 39, Number 5, pp. 2237–2249, 1993.
50. Ali, Y. and Narasimhan, S., "Sensor Network Design for Maximizing Reliability of Bilinear Processes," *AIChE Journal,* Vol. 42, Number 9, pp. 2563–2575, 1996.
51. Sen, S., Narasimhan, S., and Deb, K., "Sensor Network Design of Linear Processes Using Genetic Algorithms," *Computer and Chemical Engineering,* Vol. 22, Number 3, pp. 385–390, 1998.
52. Bagajewicz, M. and Sánchez, M., "Cost-Optimal Design of Reliable Sensor Networks," *Computers and Chemical Engineering,* Vol. 23, Number 11/12, pp. 1757–1762, 2000.
53. Masterson, J. S., "Reduce Maintenance Costs with Smart Field Devices," *Hydrocarbon Processing,* January 1999.
54. Wilcox, N. A. and Himmelblau, D. M., "The Possible Cause and Effect Graphs (PCEG) Model for Fault Diagnosis—I. Methodology," *Computer and Chemical Engineering,* Vol. 18, Number 2, pp. 103–116, 1994.
55. Lambert, H. E., "Fault Trees for Locating Sensors in Process Systems," *Chemical Engineering Progress,* August, pp. 81–85, 1977.
56. Raghuraj, R., Bhushan, M., and Rengaswamy, R., "Locating Sensors in Complex Chemical Plants Based on Fault Diagnostic Observability Criteria," *AIChE Journal,* Vol. 45, Number 2, pp. 310–322, 1999.
57. Parker, R. G. and Rardin, R. L., *Discrete Optimization,* San Diego, CA: Academic Press, 1988.
58. Bhushan, M. and Rengaswamy, R., "Design of Sensor Location Based on Various Fault Diagnosis Observability and Reliability Criteria," *Computers and Chemical Engineering,* Vol. 24, p. 735, 2000.
59. Bhushan, M. and Rengaswamy, R., "Design of Sensor Network Based on the Signed Directed Graph of the Process for Efficient Fault Diagnosis," *Industrial Engineering and Chemical Research,* Vol. 39, pp. 999–1019, 2000.
60. Bagajewicz, M. and Fuxman, A., "An MILP Model for Cost Optimal Instrumentation Network Design and Upgrade for Fault Detection," *Proceedings of the 4th IFAC Workshop on On-Line Fault Detection & Supervision in the Chemical Process Industries,* June 8–9, Seoul, Korea, 2001.
61. Bhushan, M. and Rengaswamy, R., "A Framework for Sensor Network Design for Efficient and Reliable Fault Diagnosis," *Proceedings of the 4th IFAC Workshop on On-Line Fault Detection & Supervision in the Chemical Process Industries,* June 8–9, Seoul, Korea, 2001.

4.9 Global System Architectures

R. H. CARO

The architecture of a control system is defined by the network serving to join all elements of the structure into a unified operating entity. In 1975, Leeds & Northrup (L&N) defined its MAX 1 control system as a collection of microprocessor-powered controllers and color CRT-based operator stations all linked by a fiber-optic network. In 1976, Honeywell introduced its TDC 2000 with a similar architecture calling it a "distributed control system" introducing the term DCS. By 1981, all the process control system suppliers had introduced their own DCS, and DCS was established as the dominant architecture for continuous process control systems.

During this same period, General Motors stimulated the discrete manufacturing industry with its requirements to unite "islands of automation" with a common network protocol—MAP (Manufacturing Automation Protocol). With partners from the automotive industry, suppliers, and all of the major PLC (programmable logic controller) vendors, MAP was fully specified and each of the protocols well on its way to international standardization.

The IEEE had formed its 802 committee to standardize on office local-area networks (LANs), thinking that it only needed to endorse Ethernet as IEEE 802.3, which had been submitted by the original consortium of Digital Equipment Corp., Intel, and Xerox. Industrial automation interests convinced the committee that a more deterministic protocol using token passing was necessary, which gave rise to token bus, IEEE 802.4. Defensively, IBM offered its fully developed deterministic protocol called Token Ring for committee consideration as well. Unable to achieve consensus for any one of these protocols, the committee established all three as standards. All were accepted by the ISO as well, creating the ISO/IEC 8802 series of protocols with corresponding numbers to IEEE.

In the early 1980s, there was no formal architecture for a wide-area network (WAN) for data, but the voice telephone network was adapted to handle data flows. This is not very strange because the voice telephone network was at the time in the process of change from analog to digital technology in any case. The Internet was being deployed by the U.S. Department of Defense as its ARPAnet, and it used a totally separate data network to link research centers all across North America and extending eventually all around the world. These were the beginnings of the WAN. Corporations and ARPAnet leased copper wires and, later, optical fibers from telephone carriers to form their connections.

PURPOSE OF EACH NETWORK

Data communications networks are formed to transport data reliably from one location to another. Typically, the volume and frequency of data exchanges increases as we progress down from the WAN, to the LAN, to the control-level network, and finally down to the device- and input/output (I/O)-level networks. Inversely, the length of the transaction increases as we progress upward through the series of networks from the device level to the WAN. Although all network users will argue that their transactions are always critical, the only ones with a valid claim to being "time-critical" are those networks involved in actual manipulation of manufacturing process conditions, which we refer to as control-level, device-level, and I/O-level networks.

WANs and LANs

LANs are designed to move data on private networks constrained by law from "crossing a public way." In fact, we often bend that requirement by using some wireless link such as radio or infrared when plant buildings have a public road traversing the property. The requirement for a LAN is that it is totally in the control of one authority, with any delegation of it to a common carrier, such as a local exchange carrier (LEC), the official term for the local telephone company. Although LANs are designed to pass data, the most common functions of LANs are shown in Table 4.9a.

The WAN is the domain of the LEC or the IXC (IntereXchange Carrier), which we know to be the long-distance suppliers, or the ISPs (Internet service providers). We can lease bandwidth from them as either dedicated or on-demand time slots using their TDM (time division multiplexing), or we can lease copper (private lines) from them if we wish to dedicate a circuit to our private use, perhaps to link company locations.

TABLE 4.9a
Common Functions of LANs

Allow centralized storage of data (database) with distributed access
Share printers
Access the Internet using a shared service line
Provide backup data repositories for workstations
Serve as a conduit for the exchange of electronic mail

The Internet is a special kind of WAN defined as a "network of networks" and is dedicated to the transport of data, not voice. Since 1996, the Internet has been released from control by the U.S. government, and placed under commercial management. Although most of the origins of the Internet were to allow remote access to remote computers, that use is almost invisible since the domination of the Internet by the World Wide Web. The fact that beautiful graphic Web pages are merely cosmetic overlays for the access of remote computers is almost unknown to the casual user. The Web also is supported by powerful search engines stemming from the early research uses of the Internet to access databases at many locations.

Control-Level Network

Control-level networks are really LANs but have the property of being time-critical, a more exact expression than the loose phrase "real-time." Although the time property is important, the need for delivery of data without retransmission (a common error recovery solution) is equally important. These properties have been blended together in a requirement for determinism that does not exist for LANs in general.

Determinism is the quality of delivering network data transmissions with a maximum and predictable worst-case delay. All control systems will adequately perform with some variability in the time of delivery of data across a network, but process control needs a uniform time for the sampling interval for data delivered across the network. If the worst-case delay cannot be predicted, instability of the control may result. Nondeterminism is a property of any protocol that uses a random waiting period after the detection of a collision. Many network protocols have the potential for collision, but use procedures to avoid collision. Only IEEE 802.3 (also ISO/IEC 8802-3), commonly referred to as Ethernet, uses a random waiting period after collision detection, and is generally considered a nondeterministic network as a result, because nothing is done to remove collisions.

Field Network

When a network is especially designed for installation on the shop floor in an industrial plant, it is known as a field network. Although Ethernet dominates office networks, there are many field networks typically designed to solve a particular problem or meet a specific need. Table 4.9b tabulates the most common field networks for both process control and factory automation.

NETWORK SEGMENTATION

Networks provide a common medium for sharing information among nodes. The nodes of a network are interconnected in a variety of ways called a topology. The list below provides some of the existing network topologies and networks that use them.

1. Star—nodes are wired to a common hub (10/100BaseT Ethernet), as shown in Figure 4.9c.
2. Multidrop—also called a bus; nodes connected to a common trunk wire (PROFIBUS), as shown in Figure 4.9d.
3. Daisy chain—nodes connected to each other requiring forwarding of messages in both directions (IEEE 1394, Firewire), as shown in Figure 4.9e.
4. Ring—a type of daisy chain network in which the ends are closed allowing single-direction communications (IEEE 802.5 Token Ring), as shown in the upper part of Figure 4.9f.

TABLE 4.9b
Common Field Automation Networks

Network Name	Type of Protocol	Speed	Type of Wiring	Bus Power
PROFIBUS-DP	Master/slave and token passing	12 Mbps	UTP	No
PROFIBUS-PA	Master/slave	31,250 bps	STP	Yes
Foundation Fieldbus H1	Arbitrated	31,250 bps	STP	Yes
Foundation Fieldbus HSE	Arbitrated	100 Mbps	UTP	No[a]
DeviceNet	Master/slave	500 Kbps	UUP(2) or flat	Yes
AS-interface	Master/slave	500 Kbps	Flat	Yes
ControlNet	Timed slot	5 Mbps	CATV coaxial	No
Interbus	Slot-ring	500 Kbps	UTP	No

[a]IEEE802.3af will define power from switch.

UTP = unshielded twisted pair, STP = shielded twisted pair, UUP = unshielded untwisted pair, FSK = frequency shift keying, HSE = high-speed Ethernet, CATV = community antenna television (cable).

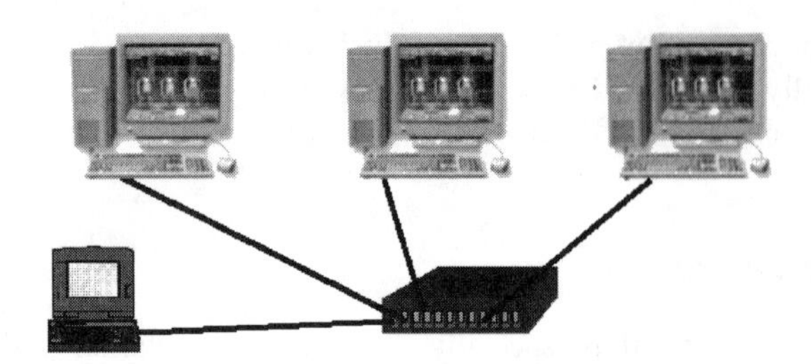

FIG. 4.9c
Star network.

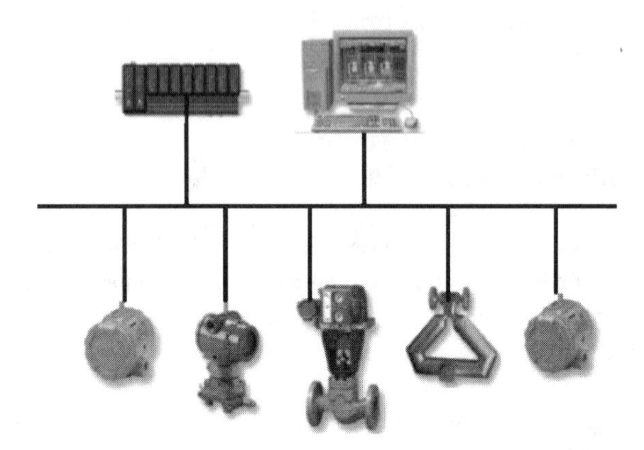

FIG. 4.9d
Multidrop (bus).

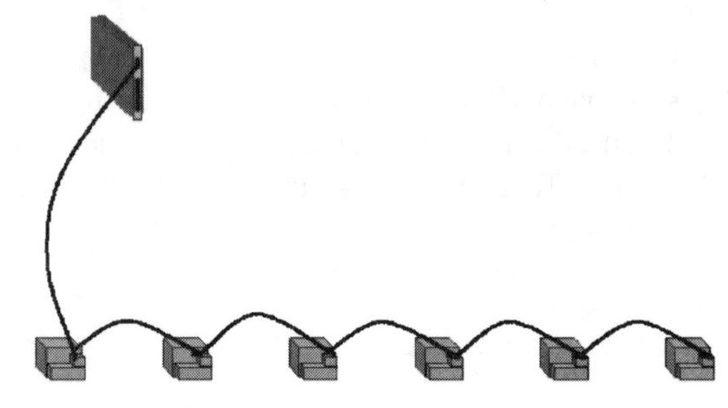

FIG. 4.9e
Daisy chain.

5. Mesh—cross-connection in which each node is connected to all of its closest neighbors (Internet), as illustrated in Figure 4.9g.

Eliminate Unnecessary Traffic

Each of these network topologies can grow from small to large by adding nodes. When all network traffic reaches all nodes, this is called a "flat" network. However, when typical network traffic is mapped, we usually see many clusters of local data traffic such as the PCs in a work group accessing a common server and printer. When we install network switches or routers to exclude local traffic from the entire network, it is called "segmenting." Segmenting a network

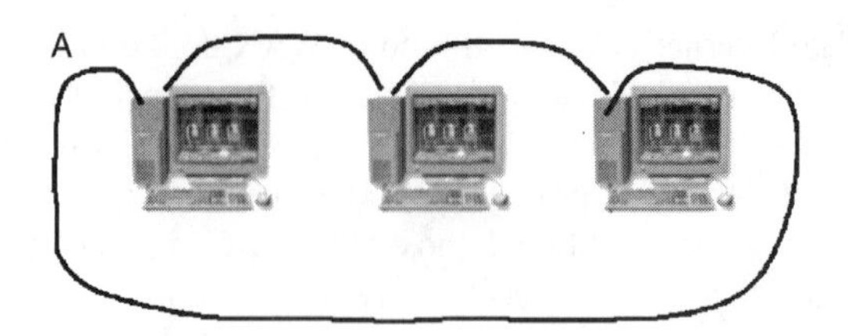

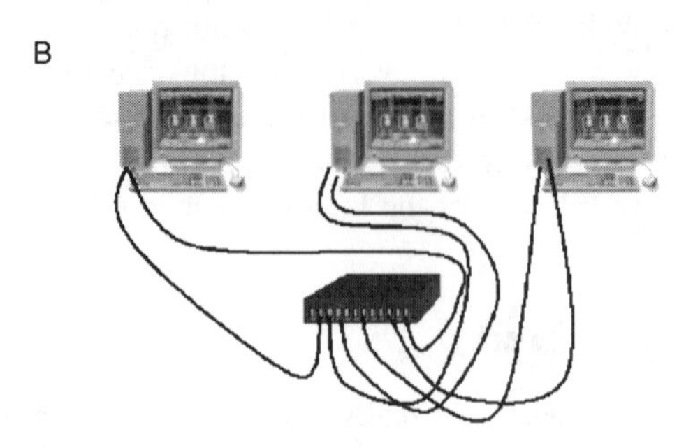

FIG. 4.9f
Real ring (A) and star-connected ring (B).

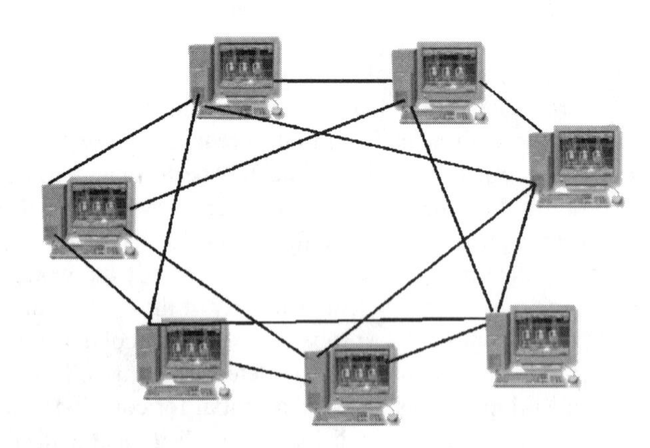

FIG. 4.9g
Mesh/Internet.

increases the overall capacity of the network as each segment operates simultaneously not loading up the entire network with redundant data.

Achieve Deterministic Behavior

Switches segment networks to remove redundant messages or traffic not intended for the segment. It is also possible to create networks with only one node per segment to prevent collisions that cause loss of determinism. Ethernet switches provide both network segmentation and deterministic behavior when all possible collisions are removed.

To remove the last potential source of collisions, networks must operate in full duplex, meaning that nodes must be able to transmit and receive at the same time. This is most important for nodes executing the Ethernet protocol. Half-duplex nodes can only receive or transmit at any one time. If a node is transmitting, and a new message appears on the receive line, Ethernet protocol will cause the node to stop transmitting and apply the Ethernet protocol as it interprets the event as a collision. All modern Ethernet nodes and Ethernet switches are configured for full-duplex operation by default, but some legacy devices still may be configured for half-duplex. Note that coaxial cable versions of Ethernet naturally operate in half-duplex because there is only one bidirectional data path, unlike the more modern 10/100BaseT versions of Ethernet using two pairs of Category 5 network cable.

NETWORK SECURITY AND PRIVACY

Automation networks of the past were isolated from the corporate LAN and other outside contact, which made them free from intrusion and other security concerns. Modern automation networks are connected not only to the corporate LAN, but the LAN gives open access to the Internet. Therefore, it becomes necessary to make sure that the automation network is protected from intrusion from external sources via the Internet, and is made inaccessible to unauthorized individuals.

The Function of Firewalls

A firewall is a network device designed to eliminate access to a network from outside for transactions not meeting access criteria. Firewalls exclude access to networks using the assigned port number. Each of the types of Internet applications has its own port number assigned by the IETF (Internet Engineering Task Force). Firewalls are set to inspect only the port number designation of each transaction or service. For example, Web pages are sent via port 80, while Foundation Fieldbus HSE transactions are at ports 1089 to 1091. Other ports are reserved in the master list of port assignments that can be seen at http://www.iana.org/assignments/port-numbers. Unless added to the list by the firewall administrator, Foundation Fieldbus transactions cannot pass a typical corporate firewall because their port assignments are filtered out.

Functions Required for Automation Systems

Automation systems need the same firewall protection as the corporate LAN to keep the data inaccessible to unauthorized individuals. Usually, the connection to the corporate LAN is the path to the Internet, and the corporate firewall is generally adequate. One of the problems, however, is that maintenance of corporate networks often requires access to network nodes from inside the firewall. For this reason, it is advised to install a unique firewall at the interface to the corporate LAN, and not to allow any access to ports other than for the supported network, and for Voice over IP. For example, most of the IT network-broadcast pop-up messages must be excluded from automation networks.

Automation Gateways

An automation gateway is more than a firewall and more than a network translation device. It should provide a network filtering function, normally called a layer 4 router, to prevent improper access to the network data. Once access to data is allowed, the automation gateway may provide a proxy server for that data, rather than to forward the request to the data source. As always, proxy servers provide local access to data to eliminate unnecessary network traffic.

The automation gateway is but one of many possible solutions to guarantee access security and privacy to the automation network. The technology has not yet been fully developed or explored, so there is no firm recommendation. The problem remains that automation networks must be protected from unauthorized access and the random local traffic of the corporate LAN. Yet, the security and privacy solutions must not hinder or slow the necessary access of authorized personnel and applications to the rich manufacturing data present only on the automation network.

NETWORK TECHNOLOGY

Automation networks have many unique requirements not shared with the corporate LAN. Most of the differences are in the physical layer of the ISO seven-layer stack. Automation environments are often harsh compared with the office. Harsh means some combination of chemically active environment, high vibration, high electrical or electrostatic noise, or extremes of temperature. These are conditions not normally considered in the design of a LAN.

Cabled Networks

Many types of cables have been used for networks, but most low-speed connections tend to be based on the use of copper wire. High-speed and long-distance connections also use copper wire, but the clear trend is for the use of fiber-optic cables. A “cable” consists of the conductor, strength members, and environmental and electrical protection. Table 4.9h gives some standard designations for communications cables.

Optical Free-Air Networks

Although infrared energy is often used in fiber optics, infrared can also be used in free-air for data communications. In fact, infrared is one of the most commonly available data communications methods available because it is standard on almost every laptop computer and personal digital assistant (PDA) ever sold, and on most mobile telephones. More than 70 million infrared ports were shipped in 2001 into an installed base several times as large. Yet, it is one of the least-used connection methods in spite of conformance to the IrDA (Infrared Data Association) standard. IrDA conforming infrared is very inexpensive.

Although properly aligned IrDA ports will usually exchange data on command, there is the software application problem of compatibility. Making infrared communications actually work has been highly problematic within the PC community even when running the same operating system,

TABLE 4.9h
Common Communications Cable Types

Designation	*Meaning*	*Application*	*Example*
		Copper wire	
UUP	Unshielded untwisted pair	Low-speed communications	DeviceNet
UTP	Unshielded twisted pair	Higher-speed communications	Ethernet Category 5
STP	Shielded twisted pair	Analog instrumentation	4–20 mA DC
SUP	Shielded untwisted pair	Not used	—
Multicore	Many conductors in a common cable	Instrumentation home run cable	4–20 mA DC home run cable
		Fiber optics	
Single mode	Few optical paths inside conductor	High-speed, long-distance data communications	Undersea laser-powered fiber-optic cables
Multimode	Allow multiple paths inside conductor	Short-distance communications	Industrial LED-powered fiber-optic cables

but making it work between a PC and a PDA or cell phone is very problematic. Every one of these problems can be corrected with sufficient work and correct software, but the simple alternative is communications via the wired network or by exchange of portable media such as floppy disk.

IrDA uses low-power LEDs (light emitting diode) for communications. An emerging technology uses lasers for free-air optical networking. Unlike the natural scattering and reflecting properties of IrDA communications, free-air optics uses the property of lasers, which are highly collimated light beams suitable only for point-to-point communications in free space. There are a few niche products in this market, but no standards. The potential low cost and high performance of an installed network will be very attractive, however.

Wireless Networks

In this section, the term *wireless* means communications via radio. Infrared and other optical communications have been covered previously. The generally accepted position is that wireless will replace all forms of wired communications—if all the problems can be resolved. That is a very big IF. Will there be enough bandwidth to allow wireless communications without interference? Will it be reliable enough? What unknown problems will emerge?

Usable bandwidth for wireless data communications keeps expanding, probably faster than usage of the bandwidth—so far. Expansion of bandwidth has come through development of low-cost circuitry capable of exploiting higher frequencies, which have always been available to those applications willing to expend resources on exotic components. Military applications have been among those using bandwidth above 2.5 GHz, to date only practical using expensive GaAs (gallium arsenide) semiconductors. However, each year sees advances in making commercial complementary metal oxide semiconductor (CMOS) technology smaller, denser, and capable of processing higher frequencies. The year 2001 marked the leap to the 5 GHz band for CMOS integrated circuits. Moore's law applied to this dimension should see the next leap to 10 to 11 GHz by 2003. Each frequency doubling opens many more data communications channels.

The wireless spectrum is a scarce resource allocated by governments and treaties between governments. Low frequencies have all been carefully allocated to avoid interference as much as possible, because the lowest frequencies are capable of circumnavigating the globe. Higher frequencies are the new frontier for wireless as they tend to be line-of-sight. Unfortunately, they also are subject to blockage by massive objects—they cannot penetrate walls or buildings. Sometimes making wireless communications work is like magic.

One of the spectacular technologies to be invented for wireless is spread-spectrum radio. The original spread-spectrum idea was conceived by movie actress Hedy Lamar (real name: Hedwig Maria Eva Kiesler), and her piano composer friend, George Antheil, who patented the method and donated it to the U.S. government in 1943. The idea for frequency hopping was conceived as a method to allow secure radio communications, without the fear of interception, and one that would be able to function in the presence of high-power jamming signals in use by the Nazis to disrupt battlefield communications. The very same jamming avoidance provides spread-spectrum radio with a high degree of noise rejection. Two different forms of spread spectrum have emerged: frequency hopping and direct sequence. The market in 2001 was dominated by Wi-Fi or "wireless Ethernet"—names assigned to a protocol standardized by IEEE as 802.11b, which is also an ISO/IEC standard 8802-11b. This is a direct sequence spread-spectrum radio signal transmitting 11 Mbps of digital data over distances in the order of 100 m at a carrier frequency of 2.4 GHz. All we know says that this will change by 2003 to faster speeds, and perhaps at higher carrier frequencies.

We cannot ignore the wireless telephony market, also to be dominated by direct sequence spread-spectrum radio. In 2001, this market underwent transformation from second-generation digital technology (analog radio was first generation) to the worldwide standard third-generation called 3G wireless. This transformation allows high-speed wireless Internet access as well as end-to-end digital voice. The technology is all founded on the use of CDMA (code division multiple access), a form of direct sequence spread spectrum. The uncertainties are in the actual frequency assignments in the various countries of the world. 3G wireless will be fully functional in most of the developed world by 2003, and in North America by 2006.

Will wireless be used for automation networks? Most certainly the answer is *yes*, but it is too early to predict the exact form. For many applications, cabled networks will continue to be cost-effective and heavily used. Wireless will be attractive for portions of the network that are too costly to connect by cables, such as when it becomes necessary to cross public roadways. The uncertainty is between these extremes. Will wireless reach down to the field transmitter and remote I/O multiplexer? Sometimes, but cabled networks often deliver power to the remote units that wireless cannot. However, the cost of running new cabling in an existing factory is so high that wireless remains an attractive solution for retrofitting new devices in old plants. The technology problem is not the wireless itself, but rather the power technology. Batteries have never been a good answer in automation, but there continues to be considerable development of alternative power sources in many fields suitable to industrial automation. Once the alternative power source becomes economical, then wireless to the field device will follow.

4.10 Advantages and Limitations of Open Networks

R. H. CARO

The word *open* in relation to networks and software has been used and abused over recent years. What does "open" actually mean in these contexts?

The classical meaning of open is for a network (or software interfaces) to be standardized such that access follows some established rules documented by a recognized standards body, not by just a single individual or corporation. Traditional standards bodies are those that are established by international treaty and the organizations within nations to administer and develop national standards under international rules. International standards exist to foster competition and international commerce by opening the borders of nations to products of other countries unfettered by protectionist local standards. Within nations, standards also exist to foster competition and commerce by establishment of rules not favoring the products of a single company. Standards have traditionally benefited the end user by fostering international competition and open markets, hence the use of the word open.

Standards for networks have always deviated from this classic definition because formal standards have not been able to keep pace with the rapidly evolving technology. Network standards have traditionally been established by research laboratories such as Bell Labs for telecommunications or by government research funds such as DARPA (Defense Advanced Research Projects Agency.) Bell Labs, now a part of Lucent Technologies, and a one-time affiliate, BellCore, now Telcordia Technologies, Inc., an SAIC Company, wrote most of the rules (standards) used for the world's telecommunications networks. These former Bell Telephone research laboratories developed the specifications and tested telephone network equipment for conformance. They still play an important role in the development of new telecommunications technology, but eventually assign most of their specifications to international standards committees.

Research money from the U.S. government was used to develop ARPAnet for the U.S. Advanced Research Project Agency in the 1970s. The objective was to interconnect university, defense contractors, and defense agency networks so that the surviving network nodes would continue to communicate with each other in the event of nuclear disaster. Eventually, ARPAnet became the Internet when the U.S. government *opened* it to commercial use. The standards developed for the Internet to allow open access were administered by a standardization group not established by international treaty, but established under a formal set of rules and primarily controlled by academic interests. Hence, the naming of the group, IETF (Internet Engineering Task Force).

Today, the classification of a network specification as open is used to mean that the specification for access to the network is not controlled by a single supplier. The opposite of open is *proprietary*, meaning that a single supplier controls the specification for access and implementation of the network. Most of the open networks were developed originally by a single supplier, and donated to a nonprofit agency for administration by a committee populated by individuals interested in the propagation of that network. In fact, the original network developer typically provides most of the ongoing technical staffing, does most of the development work, and carries a much greater influence on the committees than other members. The golden rule of open network standards applies: He who has the gold—rules.

Even among the standardized networks, development of standards belongs to the larger and best-financed companies, which can afford to send their technology leaders to standards meetings and either help forge the standards in the image they desire or prevent the standards from being issued if they are moving in adverse directions. Fortunately, the market ultimately determines the survivors and establishes the true standards for a particular communications or industrial market. The history of communications is spotted with many obsolete standards that did not gain sufficient market volume to achieve full acceptance. The industrial data communications industry is especially rich with failed network standards, many of which experienced popularity for a while, but eventually fell to more widely used networks that may have actually been inferior at one time, but have been adapted for acceptable performance.

ISO/IEC 7498 OPEN SYSTEM INTERCONNECTION (OSI)

All networks, standard, open, and proprietary, are defined in terms of a landmark international standard ISO/IEC 7498-1:1994 "Information Technology—Open Systems Interconnection—Basic Reference Model: The Basic Model." Originally defined in 1986, this is the reference model for all networks—some conforming to the model, and others not in conformance. The model defines data transport starting at the top of a layer of

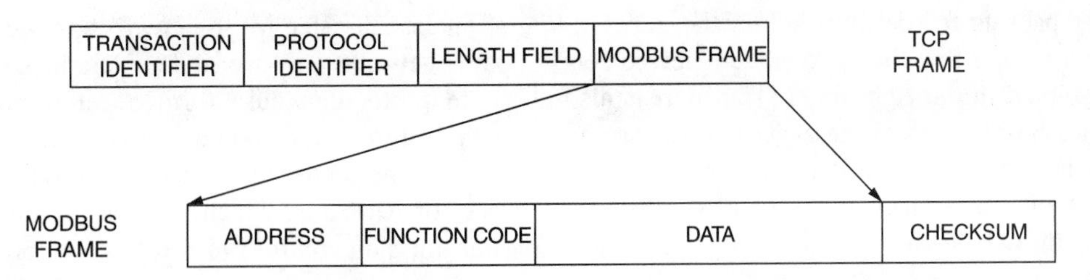

FIG. 4.10a
MODBUS TCP encapsulation of a MODBUS message.

TABLE 4.10b
ISO 7498 Open System Interconnection Seven-Layer Model

Number	*Layer Name*	*Meaning*
7	Application	Software interface to the network
6	Presentation	Data conversion or formatting
5	Session	Connecting and disconnecting circuits
4	Transport	Delivery of complete messages
3	Network	Basic underlying protocol for message delivery
2	Data link	Protocol for message frames
1	Physical	Media or cable and connectors and the electrical interfaces

functions by some application program. Each layer processes the message transforming it as specified until the final layer at which the message is transformed into an electrical signal with a specified modulation designed to be transported across the network and detected at the designated receiver. The receiver demodulates the electrical impulses, converting them into a binary bit stream. The message then passes through each layer in the receiver until it is passed to the intended application. This is conceptually illustrated in Figure 4.10a.

Seven-Layer Standard

ISO 7498 defines seven layers for all communications networks as shown in Table 4.10b. This does not mean that every network must have all seven layers, but it does divide the network functionality into these seven functions. Early experience in using networks with all seven layers has shown that inefficiencies may be created when no valid reason exists for a layer. Similarly, by assigning the functionality of all seven layers to a hardware, firmware, or software entity can make networks operate more efficiently.

Functions of the Layers

Application The interface between software running on the computer and the network. Notice that this is not the actual "application," but is the system software supporting application programs. Many times the application layer is given the name "API" for application–program interface, and in some networks it may be that. In other networks, it is just the lowest level software allowing invocation of network functions.

Presentation Often this layer is omitted, but its original intent is to provide services to convert from one data organization to another. ASCII (ISO 646:1991—7-bit coded character set for information interchange) to EBCDIC (Extended Binary Code for Information Interchange) character conversion. Many times the presentation layer is used to invert bit order when computers sending and receiving messages use different bit orders, sometimes called the Big/Little-endian problem. This problem still exists as not all microprocessors use the same bit order for representation. Fortunately, many operating systems and communications software packages shield us from this problem, and we do not often see it.

One problem that is with us and is still unresolved is called Unicode, effectively the standard for encoding characters requiring more than 1-byte length. ISO 10646 is the international standard that is supposed to create the standard for 2-byte characters, and Unicode is a working organization responsible for development of the character sets and all the rules for their use. Unfortunately, international standards take a long time to ratify, and tend to remain static between revisions. Unicode is backed by an organization found at www.unicode.org, which is constantly revising these encodings for all of the major languages of the world, and attempts to make sure that the next revisions of ISO 10646 have the latest revisions.

Session Establishes and breaks connected services between the transmitting system and the receiving system. The session layer was originally conceived as the protocol necessary to dial the other computer on dial-up networks, but it is now also used to establish virtual connections across systems that are already electrically connected. Early use of the Internet with Telnet protocol required a "connection" to begin use, also requiring a session layer. Modern use of the World Wide Web is all encoded in TCP (transmission control protocol) and performs its own connections and disconnections without use of a session layer.

Transport Protocol to guarantee delivery of messages between sender and receiver; sometimes called end-to-end delivery mechanism. Assumes that messages are long and segments them into shorter packets for delivery across the network. Transport protocol assures that the packets are received correctly and reassembled in the right order. Transport is also responsible for network routing of messages and for pacing message delivery to allow the network to learn about optimal routing between sender and receiver. An end-to-end check code is added to the whole message to validate complete delivery.

Network The basic protocol of network packets. Packets are sequence numbered for correct reassembly at the other end by the transport layer. Network packets contain the address information for both the source and destination network addresses.

Data Link The basic protocol of the original unsegmented network message. Data link layer provides for orderly sharing of the network so that only one node may transmit a message at a time. Data link specifies the frame format of the message, resolves all contention problems, and assures valid delivery of error-free data between two nodes using a redundancy check field.

Physical Defines the cable types, topology, and connector choices for the network. In addition, the electrical interface to the cable is defined, including the signal levels, modulation technology, and encoding of network symbols. Although modulation techniques, signal levels, and symbol encoding technology are well beyond the interests of the end user, it is important to know that these electrical or optical properties must be tightly defined if interoperability is to be achieved. The user is very interested in cable choices, connector types, and cable topology, however, as these factors relate to the cost of a network and its suitability to a particular application.

Missing Layers

As indicated previously, not all layers need to be implemented in any one protocol. Many times, it is far more efficient to combine layer functions into a single layer. For example, some networks combine data link, network, and transport layers. The purpose of the network may also allow skipping of some layers. For example, if the nodes of the network are all local, there is no need for a session layer that exists for dial-up networks. Finally, there may be no need for a layer. For example, if all devices use ISO 646 (ASCII) encoded characters, then no data conversion is necessary and the presentation layer need not be used. Many networks also do not have a real application layer, but allow programmers to access the data link layer directly.

TCP/IP NETWORK OF THE INTERNET

While the ISO 7498 seven-layer stack was being designed, ARPAnet was also being designed. ARPAnet evolved the mesh network with lots of interconnected nodes and alternative paths designed for "resiliency" or survivability in event of nuclear disaster. Mesh networks provide many choices of paths between any two nodes including some that, without some protocol, would form never-ending loops. Clearly, branched tree, star, and bus networks did not have this problem and therefore did not need a protocol solution. The protocol of ARPAnet, which became the Internet, was TCP/IP (transmission control protocol/Internet protocol).

Common Network Protocols Included in TCP/IP

TCP certainly includes OSI layers 4 and 5, transport and session. There is no attempt to include layer 6 presentation, but there is a well-defined API corresponding to application layer 7. The most important part of TCP, however, is routing, not included at all in the OSI stack. In fact, routing is generally included as part of the data link layer in simple networks like Ethernet, where routing only means "finding the network addresses reporting to a switch." In fact, a routing protocol is included in the IEEE 802.1d specification and is called the "spanning tree bridge." Switches use spanning tree logic to learn the network addresses of their lower-level connections (children) so that messages passed between children need not appear on the "parent" or backbone network. Similarly, messages addressed to a child can be recognized and passed to them. Spanning tree is very effective at segmenting network traffic, a very minor routing function.

Before we discuss more about TCP, first the functions of IP must be defined. IP, or Internet protocol, was initially designed to allow interconnection of many different computer networks. At the time of design, the concept of more than 32,000 different computer networks was unknown. Therefore, IP through its many revisions (Version 4 is currently used on the Internet) has been designed for a maximum number of network addresses (32-bit address) which is 2^{32} or 4,294,967,296, about half of which are already assigned (2001). With the explosion of Internet addresses, since the mid-1990s, it became obvious that we would soon run out of valid network IP addresses; therefore, IP version 6 IPV6 was created to extend network addressing to 128 bits. IPV6 is not yet in common use, mostly because ways were found to extend the life of 32-bit addresses through the use of private local network addresses using NAT (network address translation) and local proxy servers. The proxy server recognizes the local network address and substitutes its own address for the Internet access, then passes the data on to the local node. Every data packet on the Internet is encoded using IP.

TCP is designed for "heavy-duty" routing. IP addresses are defined by the assignment of a 32-bit address as part of the Internet domain name assignment, an administrative function of the Internet. Most Internet service providers (ISPs) are assigned a high-level Internet address, and allocate lower-level addresses to their subscribers. The mechanism is detailed and is covered in other sections. There is no location information found in IP address assignment! At first when there were only a few networks, it did not seem to matter.

Now, with thousands of networks and millions of nodes, it would seem to be a good idea to contain geographic information in the IP address, but even IPV6 contains no geographic content. IP address resolution is contained in the network routers, a necessary part of the Internet. As a side benefit, network nodes may move around the world as well because location is a dynamic network property.

To understand how IP addressing can work, it is useful to follow the data flow about the Internet for a message. A message is received at the Internet router closest to the origin of the message, containing both the IP address of the source and of the destination. The router looks up the destination high-level domain address in its routing tables. It will find this address if it has recently sent a message to the same high-level host, or has received this information as part of a router exchange of information. If the high-level IP address exists, the message will be sent to the next router in the router's tables in the direction of that address. If the address is not found, then the message is forwarded to another Internet router at random. Most Internet routers are actually part of a hierarchy of routers on the private network of an ISP so that messages are not really routed at random, but are routed toward a focal point of that service provider closest to an actual Internet NAP (Network Access Point) for that network.

Once the message is at the NAP, after perhaps two to three hops on the ISP network, it is routed to the ISP having the destination's high-level domain address. NAPs are themselves routers for one of the Internet backbone providers, a small list. The message is generally routed within one or two additional hops to the NAP for the destination network, and so on down to the ISP for the destination, and finally to the destination site. It is highly unusual for any message to take more than 15 hops, and most are delivered in fewer than seven hops. With the rapid growth of the Internet, router tables have grown to many megabytes, but router processors have also become very fast.

TCP helps in the routing process. TCP contains the pacing logic to help the Internet adapt to new IP addresses with short, efficient messages. The message at the start of a transmission is made smaller than those messages toward the end. When TCP begins to send a message, it first prepares a short message containing perhaps only a few bytes of the actual message and sends these as a "probe" on the network. The short message acts to prepare the receiving node that a longer message will be arriving, thereby establishing a session. Since TCP requires acknowledgment for each transmission, successful delivery of the short message serves to cause the routers along the way to update their routing tables with the destination address. Once the probe is acknowledged, then longer messages can be sent until the route is found to be secure and acknowledgments are received for all message segments. Once the acknowledgment is received for the last segment, TCP follows it with a specific message to terminate the connection for that message.

Another protocol, UDP (universal datagram protocol), should also be mentioned. UDP sends IP messages without segmentation, pacing, or error checking. It is the protocol underlying SNMP (simple network management protocol) and other messaging methods not needing extra support. If any of these missing services are necessary, they must be defined by the user or other authority to solve the application purpose.

When the term "TCP/IP" is used, often the meaning is that there is a full Internet protocol stack provided. This includes TCP/IP, UDP, and SNMP, and also several applications such as FTP (file transfer protocol), Telnet (teletype networking), SMTP (simple mail transfer protocol), and NTP (network time protocol). Details of each of these is outside the scope of this section, but they are used to varying degrees in automation systems.

OPEN APPLICATION LAYERS AND INDUSTRIAL NETWORKS

Industrial networks have well-defined protocols as well, but often they are not as widely used as the Internet protocols. Industrial networks are usually developed as proprietary protocols, but because the objective is usually to connect plant equipment, the protocol often must be shared beyond the original developer by "opening" the protocol. Open protocols are given to independent agencies to maintain. Sometimes this may be a standards committee of a recognized standards body, but most often it is a nonprofit membership organization formed to publish the protocol.

DeviceNet, ControlNet, and Ethernet/IP

Rockwell Automation developed DeviceNet, ControlNet, and EtherNet/IP protocols. DeviceNet was given to ODVA (Open DeviceNet Vendors Association) to maintain. ControlNet was given to ControlNet International (CNI) to maintain. When EtherNet/IP was created it was assigned to a joint committee of ODVA and CNI to maintain, but the final disposition appears to be in the hands of ODVA. An application layer called CIP (control and information protocol) was established first by ODVA, and later made common among ODVA, CNI, and EtherNet/IP. CIP contains common functions needed for discrete automation and is defined as a set of electronic data sheets (EDS) defined in object form. Table 4.10c lists the CIP functions.

DeviceNet is a unique network designed about the use of the CAN (control and automation network) chip used widely in automobiles and trucks to reduce wiring harnesses. DeviceNet is a simple network with a frame length of 8 bytes. Collisions are allowed and resolved with a bitwise arbitration protocol that is part of the chip logic. However, most of the DeviceNet implementations control scanning from the host interface and do not allow collisions. Cabling for DeviceNet allows power distribution, and one of the cable types supports power on a separate pair from data. Maximum data rate is 500 kbps, but long cable runs at slower data rates are allowed. DeviceNet appears mostly as a protocol for remote multiplexing, but it can actually be used to connect end devices such as limit switches and push buttons.

TABLE 4.10c
CIP Functions, Electronic Data Sheets

Identity	*Message Router*	*Assembly*
Connection Manager	Register	Discrete Input Point
Discrete Output Point	Analog Input Point	Analog Output Point
Presence Sensing	Parameter	Parameter Group
Group	Discrete Input Group	Discrete Output Group
Discrete Group	Analog Input Group	Analog Output Group
Analog Group Object	Position Sensor Object	Position Control Supervisor Object
Position Controller Object	Block Sequencer Object	Command Block Object
Motor Data Object	Control Supervisor Object	AC/DC Drive Object
Overload Object	Softstart Object	Selection Object
ControlNet Object	Keeper Object	Scheduling Object
Connection Configuration Object	Port Object	TCP/IP Object
Ethernet Link Object		

ControlNet is designed to be used at the control network level, but it is also used to connect remote I/O (input/output) multiplexers. ControlNet protocol divides time into a critical time segment and free time. Critical time is used to schedule updates from critical control devices on a fixed time schedule, while free time allows open sharing of the network. Network speed is fixed at 5 Mbps. The cabling is commercial CATV (community antenna television) hard-shell rigid foam dielectric 75-Ω coaxial cable with flexible RG-58 drops. A wide variety of low-cost cable connectors, amplifiers, and terminations are available for this cable.

EtherNet/IP is a logical extension from both ControlNet and DeviceNet. The protocol is simpler than either as it is based on conventional 10/100BaseT Ethernet for the physical and data link layers. The upper layers are simply the functions of CIP mapped to either TCP or UDP over IP messages. Whereas DeviceNet and ControlNet require special cabling, EtherNet/IP uses common Category 5 Ethernet cable. Recent EIA/TIA cable and connector standards for industrial Ethernet are making it possible to keep the economic advantages of using off-the-shelf cable and connectors where they can be used, and still to use more rugged cable and connectors where necessary. Category 5E cabling is preferred for EtherNet/IP applications. Ruggedized RJ45 and pin-and-socket M12-style connectors for bulkhead use and in environmentally harsh conditions are also available for use with EtherNet/IP and Foundation Fieldbus HSE as well.

Power delivered on EtherNet/IP and Foundation Fieldbus HSE is now available using conventional Category 5 or 5E cable and Ethernet switches supporting the IEEE 802.3af standard for DTE (data terminal equipments) power. DeviceNet and Foundation Fieldbus H1 have both supplied DC power to field devices through the data cable, and can now have power delivered via these Ethernet-based networks.

MODBUS and MODBUS/TCP

The original open specification for connection of a Modicon PLC (programmable logic controller) to a computer was MODBUS, a master/slave protocol developed around 1976. Since then, MODBUS has been implemented on a variety of communications media, at many different speeds and for a wide variety of PLCs, SCADA remote stations, instruments, and many other devices.

MODBUS protocol was originally designed to overcome some of the system problems of early minicomputers and for application programs written in a high-level language such as FORTRAN. Although PLC data is represented as a binary register, presentation of that kind of data to a FORTRAN program was not easily handled, so MODBUS protocol encodes each 16-bit register into ASCII characters that can be used by the program to represent register data.

MODBUS clients initiate all requests by command, which is interpreted by the PLC (server) and one or more data values returned. MODBUS commands may also be used to set the value of (force) outputs directly, or to write to a register (preset). The MODBUS commands reflect the type of data available on a PLC, binary bit values, or full registers of data. Many of the PLCs other than those of Schneider/Modicon that have copied the MODBUS commands have a difference hidden from the user: Modicon has always numbered its registers starting at one (1), while others have numbered registers more in the "computer" tradition by starting the numbers with zero (0). The meaning and use of registers on different PLCs is also quite different. This means that a given command presented to a Modicon PLC will have quite different effects on the process or machine than the same MODBUS command sent to a non-Modicon PLC. Table 4.10d lists the most common MODBUS commands.

TABLE 4.10d
Modbus Commands

Function Code	Command
1	Read Coil Status
2	Read Input Status
3	Read Holding Registers
4	Read Input Registers
5	Force Single Coil
6	Preset Single Register
7	Read Exception Status
8	Fetch Comm Event Counter
9	Fetch Comm Event Log
10	Force Multiple Coils
11	Preset Multiple Registers
12	Report Slave ID
13	Read General Reference
14	Write General Reference
15	Mask Write 4x Registers
16	Read / Write 4x Registers
17	Read FIFO Queue

MODBUS TCP is a pure encapsulation of the MODBUS commands into a TCP/IP transaction. The pure information part of the original MODBUS command is inserted as a character string into a conventional TCP/IP message appearing eventually as an Ethernet frame on an Ethernet network. Figure 4.10a illustrates this encapsulation.

Not all the MODBUS commands (function codes) have been implemented in MODBUS TCP. Only MODBUS function codes 1, 2, 4, 5, 6, and 7 are universally defined for MODBUS TCP. The details of MODBUS TCP are completely defined on the www.modbus.org Web site. Follow the links to the complete specification for MODBUS TCP including all of the details of mapping each of the MODBUS function codes to the TCP/IP message.

The most interesting side effect of MODBUS TCP is the immediate high-speed Internet access to a wide variety of devices already implementing MODBUS protocol by adding a low-cost Ethernet interface. The opposite effect is to provide a low-cost interface from a PC with software already recognizing MODBUS to a wide variety of controllers using the Ethernet interface.

Foundation Fieldbus and Foundation Fieldbus/HSE

In 1985, the ISA standards committee SP50 met to consider the future standardization of the digital link necessary to replace the old analog 4–20 mA signal. The committee did this by defining two very different sets of network requirements: H1, for installation on the plant floor as a digital replacement for 4–20 mA DC transmission; and H2, a higher-speed, "backbone" network to link the H1 segments to each

FIG. 4.10e
Facsimile of the Fieldbus Foundation Registration Mark.

other and to host systems. The committee could not think of a proper name for this kind of application classification so the "H" in H1 and H2 stands for "Hunk."

H1 was defined narrowly for process control: 4–20 mA replacement including the desirable feature of installation of new digital instruments directly replacing existing 4–20 mA analog instruments retaining the existing wiring. Limits of intrinsic safety and power delivery forced H1 segments to have few nodes and a fairly low speed. The speed was also the result of compromise with the functional requirements for process control.

Fieldbus Foundation was formed in 1994 from two previous competing organizations, ISP (Interoperable System Project) and WorldFIP North America. Fieldbus Foundation set about writing an implementation specification around the H1 version of S50.02 to satisfy its process control users. This specification, called Foundation Fieldbus, was completed in 1996 and released to all member companies. It was strongly oriented to solving the 4–20 mA DC replacement problem for process control instrumentation and control systems. In 1998, the Fieldbus Foundation completed work on conformance testing for H1 and announced its "Registration" program. Instrumentation suppliers submit their instruments for a suite of interoperability testing, and if they pass these rigorous tests, they receive a Foundation Registration Mark similar to that shown in Figure 4.10e.

Early in 1999, the Foundation Steering Committee made the decision to create an implementation specification for H2 based on use of commercial off-the-shelf high-speed (100-Mbps) Ethernet because it was unlimited, inexpensive, and more than fast enough. The project was completed in 2000 with the release of an implementation specification for Foundation Fieldbus HSE (high-speed Ethernet).

With the completion of HSE and products beginning to appear on the market, the full two-level architecture of S50.02 has been completed, although with a physical/data link combination not contemplated in 1993. HSE demands the use of full-duplex switched Ethernet, which itself accomplishes a very subtle correction to the "old" Ethernet—elimination of collisions through the use of active buffers in the switches. Elimination of collisions itself makes Ethernet fully deterministic, and removes the annoying minimum message length requirement, which was necessary to allow time to detect collisions on long cables. Now the H2 can be implemented with any variety of Ethernet apparatus including repeaters and especially with any of the fiber-optic Ethernet cable extensions found in communications catalogs. This is very important as one of the primary applications for H2 is the replacement of the "home run" cable connecting field instrumentation to the control room. A single fiber cable with HSE may be as much as 6 km in length to connect many thousands of H1 instruments in the field. The two-level architecture of Foundation Fieldbus is designed to support control in the field device such as the control valve positioner and field instrumentation.

Finally, the interface between Foundation Fieldbus and control systems is defined very loosely in the S50.02 standard, and with only the functions at the application layer defined by the Fieldbus Foundation specifications. Interoperability of host systems is now defined by a test called HIST (Host Interoperability Support Test), a specification of the Fieldbus Foundation. Systems passing HIST are registered with the foundation and are most likely to interoperate well with the full two-level architecture of Foundation Fieldbus.

PROFIBUS and PROFInet

PROFIBUS protocol began as an early response to the fieldbus requirements dating from 1988. PROFIBUS was one of the final candidates for protocol to be selected by SP50 during the early development of the data link layer, and is functionally included in S50.02 and the International Fieldbus Standard, IEC 61158. The protocol of PROFIBUS calls for a token bus architecture with token passing and lost token recovery fully specified. However, in practical applications, token passing is less efficient and the alternative master/slave protocol is usually used. Token passing still exists within a PROFIBUS network, but usually as the methodology to switch between redundant (fault-tolerant) master systems.

The original PROFIBUS is called FMS (Fieldbus Messaging System) as it is typically used at the control bus level where its native speed of 9600 bps is not an obstacle. PROFIBUS became a German national standard in 1989 as DIN 19245, and in July 1996 it became a European Standard as EN 50 170. In 1999, PROFIBUS became Type 3 of the International Fieldbus Standard IEC 61158. FMS is a proper subset of ISO 9506—MMS (Manufacturing Messaging System). FMS is also fully contained in the Foundation Fieldbus specifications, reflecting the close relationship of PROFIBUS and Foundation Fieldbus in their origins. Today, Profibus FMS is no longer a hardware specification but is just one of the PROFIBUS protocols supported on a PROFIBUS system.

PROFIBUS is also a two-level architecture with PROFIBUS-DP providing both the backbone communications structure and the connection with discrete manufacturing remote I/O multiplexing. For process control applications, PROFIBUS-PA has been defined using identically the same physical layer as Foundation Fieldbus H1, but the PROFIBUS data link layer. PROFIBUS-PA provides intrinsic safety and DC power delivery to field instruments. PROFIBUS-PA does not support control in the field device, but it does support a wide variety of signal conditioning computations, which are specified by the PROFIBUS application profiles for process automation.

FUTURE OF INDUSTRIAL NETWORKS

The entire industrial automation industry is too small to support even one proprietary network. There are no reasons remaining why adaptation of an existing protocol cannot be used. There are two such protocols: CAN and Ethernet. Both are incomplete. CAN needs a physical layer and an application layer to be defined, which is what SDS, DeviceNet, CAN Kingdom, and CAN in Automation do. Unfortunately, these are all different and noninteroperable. Ethernet needs a few fittings and cable defined for the industrial automation community, and that has been defined by the EIA/TIA. Ethernet also needs an application layer closely suited to its end application, and these are being supplied in the form of complete suites of IP-based protocols that are not interoperable, but may peacefully coexist on the same wire when each is assigned to its own port number.

AS-interface is narrowly defined to reduce the cost of cabling for low-cost binary (two-state) sensors and actuators. It does not provide for device intelligence. The trend is for all devices to have some intelligence using very low cost microprocessors. If this is the case, there is no future for AS-interface as any network providing device power provides the low-cost installation. The survival network of the future will most likely be a mix of IP-based protocols operating on industrial Ethernet. Because of the conservative nature of the industrial automation market, and the large installed base of specialty vendor–supported proprietary networks, replacement will take many years. However, the scalability and low cost of IP-based protocols on top of Ethernet with all of its fiber-optic and wireless options is too compelling to resist.

4.11 HART Networks

W. A. PRATT, JR.

HART enhances the 4–20 mA standard by providing two-way communication with smart field devices. The two-way communication is modulated to be 4–20 mA signal at a higher frequency than normally observed by process control equipment. This allows communication via HART to occur simultaneously with 4–20 mA signaling. Because HART field devices are fundamentally 4–20 mA devices, backward compatibility with existing plant systems and personnel is maintained. Consequently, plant personnel can utilize HART-compatible field devices with little knowledge of the protocol and gradually adopt HART protocol features at their own pace. In some cases plant personnel may even unknowingly utilize the HART protocol (e.g., via a handheld communicator) during the course of their normal workday.

HART has been well established and accepted for many years. Device types are available supporting both mainstream and niche process instrumentation needs. In addition, many multivariable instruments are available to simplify system design and reduce costs.

In addition to supporting the 4–20 mA standard, HART-compatible field devices provide (for example):

- Digital process variables (in IEEE 754 floating-point) with standardized engineering unit codes and status
- Device status in every response from the field device, allowing system integrity to be continuously monitored
- Extensive calibration support and diagnostic information

Significant benefits can be realized by utilizing capabilities that are found in all 4–20 mA HART-compatible field devices. This section will provide:

- An overview of basic HART capabilities
- Review of the architecture of smart field devices
- Evaluation of HART-compatible equipment and software
- Summary of the evolution of HART

This section will not focus on bits and bytes, provide guidance on the development of a HART field device, or provide all the details a system integrator may desire. Other resources are available to fill this need at no cost or for a nominal fee. Instead, this section will focus on the capabilities provided by the HART application layer and the benefits they provide.

Note: Unless otherwise noted, all features and capabilities discussed in this section are standardized in the HART specifications and the vast majority of the features are mandatory (i.e., universal). Consequently, any host can support the capabilities discussed in this section with moderate development effort and without requiring any device-specific drivers.

EVOLUTION OF THE "SMART" FIELD DEVICE

A smart field device is a microprocessor-based process transmitter or actuator that supports two-way communications with a host; digitizes the transducer signals; and digitally corrects its process variable values to improve system performance. Many field devices contain sophisticated signal processing algorithms to perform the measurements or control action required. The value of a smart field device lies in the quality of the data it provides.

HART and the capabilities of the smart field device were developed in tandem. When HART was invented, smart field devices were just emerging. Most field devices were analog and most of the engineers developing smart field devices had little experience with microprocessors. Consequently, engineers did not have a clear image of the benefits a microprocessor could provide to their product.

Because 4–20 mA transmitters provide a single process variable to the control system via the loop current, early HART development focused on the primary process variable (PV). In HART, PV includes all of the properties and data items necessary to support the loop current (Figure 4.11a). HART universal commands provide access to all loop current data and properties. Process values associated with PV include the digital value, percent range, and loop current. Properties of the PV include upper and lower range values, upper and lower transducer limits, transducer serial number, and minimum span. Status information includes "loop current fixed," "loop current saturated," "PV out of limits," and whether or not the device has malfunctioned.

Soon developers began finding many uses for the microprocessor. Most development efforts have focused on improving accuracy and reliability. For example, adding an ambient

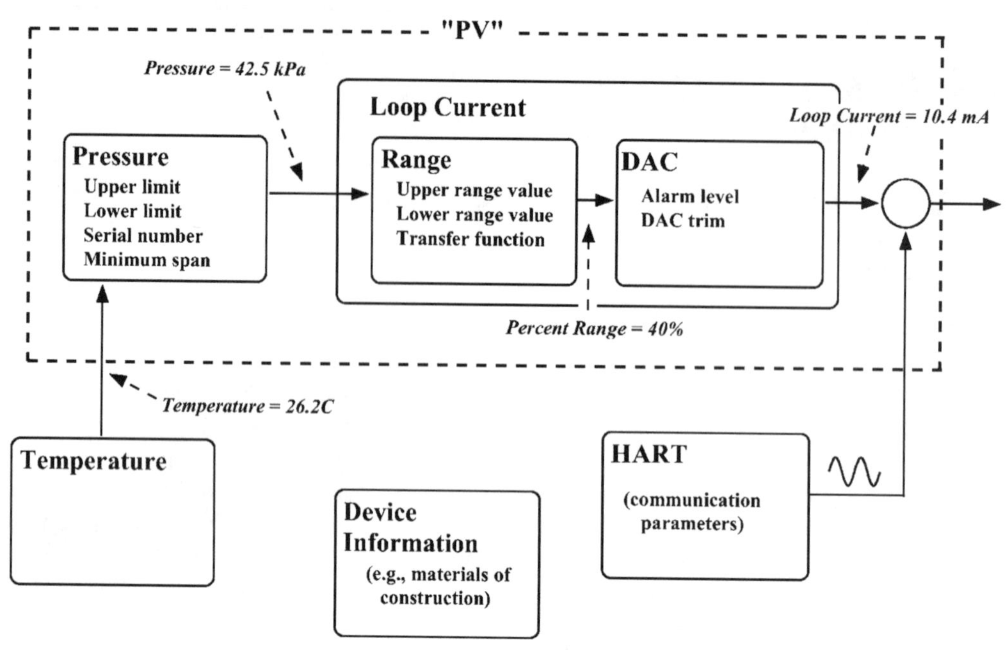

FIG. 4.11a
Smart pressure transmitter.

sensor (e.g., temperature) along with linearization and compensation software dramatically improved accuracy. Even simple pressure and temperature transmitters now have an embedded temperature sensor.

Today, virtually all HART field devices are multivariable, and ever more sophisticated multivariable field devices are becoming available. For example, there are pressure transmitters that provide the sensor temperature, level transmitters that can calculate volume, and flowmeters that contain totalizers and perform data logging. The growth in multivariable field devices continues to accelerate as microprocessor capabilities increase and power requirements decrease. At the same time, HART communications has evolved to provide access to secondary process variables in multivariable field devices.

HART was initially developed by transmitter companies and later adopted by intelligent valve controllers. Unfortunately, systems manufacturers have been slow to accept and fully support HART. Consequently, the huge installed base of HART-compatible field devices offers a wealth of largely untapped capabilities. However, system suppliers are beginning to recognize this opportunity and the continuing growth of multivariable field devices provides them with additional encouragement. Today, many system suppliers are adding support for HART, while others are improving their existing support.

EVOLUTION OF THE HART PROTOCOL

HART was invented by Rosemount in the 1980s as an enhancement to its process transmitters. Initial HART development focused on supporting the 4–20 mA loop current and providing digital process variables. Much later HART gained a reputation for its diagnostics and troubleshooting capabilities. While these are important strengths, digital process variables and competition motivated development of the HART protocol. HART revision 4 was the first version of the protocol to be included in a significant number of (mostly Rosemount) field devices. Although several hundred thousand HART 4 devices were produced, the vast majority of installed HART devices (more than 10 million) supports revision 5 or later of the protocol.

In 1989, Rosemount released HART revision 5 and made the protocol available to any manufacturer desiring to incorporate it into its product. HART became (and still is) the only open protocol supporting 4–20 mA devices. This led to the formation of the HART User Group in 1990. Soon a large number of manufacturers were supporting HART and the technical support needs resulted in the formation of the HART Communication Foundation (HCF) in 1993. When the HCF was established, Rosemount transferred ownership of all HART technology to the HCF including patents, trademarks, and other intellectual property.

The HART Communication Foundation

Today, all members of the HCF have an equal voice in enhancing or modifying the HART protocol. The HCF is supported by a professional and technical staff responsible to the entire membership with the mission to support the application of HART technology worldwide as the standards-setting body for HART communications. The HCF:

- Educates both end users and manufacturers
- Maintains the HART protocol specifications and develops enhancements to the HART technology

TABLE 4.11b
HART Field Communications Protocol Specifications

Document Title	*Rev.*	*Doc. No.*
HART Field Communications Protocol Specification	6.0	HCF_SPEC-12
FSK Physical Layer Specification	8.1	HCF_SPEC-54
C8PSK Physical Layer Specification	1.0	HCF_SPEC-60
Data Link Layer Specification	8.0	HCF_SPEC-81
Command Summary Specification	8.0	HCF_SPEC-99
Universal Command Specification	6.0	HCF_SPEC-127
Common Practice Command Specification	8.0	HCF_SPEC-151
Device Families Command Specification	1.0	HCF_SPEC-160
Common Tables Specification	13.0	HCF_SPEC-183
Block Data Transfer Specification	1.0	HCF_SPEC-190
Command Response Code Specification	5.0	HCF_SPEC-307

- Directs the HCF quality assurance (QA) program to ensure adherence to protocol requirements

As an educator, the HCF provides many materials at no charge, including white papers, application notes, and technical overviews. The HCF makes presentations at conferences worldwide. In addition, developer and end-user workshops are offered at a nominal charge. Assistance is provided to member companies and nonmember companies alike.

When this section was written, the HART specifications[1] included 11 documents, each with its own revision number (Table 4.11b). The revision level of the protocol is indicated by the document titled "HART Field Communications Protocol Specification." Within this document, the revision level of all other specifications documents is specified. All HART products must meet all requirements of a specific protocol revision. All HART devices must support all universal commands exactly as specified. The HART specifications are revised every 6 to 9 months, principally to add additional manufacturer codes and engineering units. All versions of the protocol, including the early Rosemount versions, are available from the HCF.

The HART specifications are available (upon request) to anyone, anywhere.

To ensure adherence to protocol requirements the HCF operates a QA program. All products claiming HART compatibility must pass all applicable tests included in the QA program. When evaluating a product, users should insist the supplier complete these tests.

There are logos and trademarks frequently associated with the HART protocol and restrictions on the use of these logos and trademarks. For example, the round "HARTability" logo may be used only by members of the HCF. HART is an open protocol and a manufacturer is not required to be a member of the HCF. However, any product describing itself as HART compatible must meet all the requirements of a specific protocol revision. HART is a registered trademark of the HCF. Consequently, products that do not meet protocol requirements are not allowed to claim "HART" compatibility.

Early Adopters of HART

HART began in the grassroots of process automation with transmitter and valve companies and, in process plants, first became popular with instrument technicians. Time-saving tools such as handheld configurators and instrument-specific software facilitated a rapid growth and acceptance in the instrument shop. Purchasing agents began specifying HART at the insistence of the instrument technician. However, purchasing agents also began requiring HART compatibility to shorten the bidder list and simplify the bid evaluations. Demands by instrument technicians and purchasing agents prompted even more field device manufacturers to support HART.

As instrument manufacturers began supporting HART, many began using it in their manufacturing processes. HART can reduce field device production costs. For example, sensor characterization can be performed by the field device itself rather than using expensive specialized production test equipment. Furthermore, by providing increased field device software capabilities, overall parts costs and the number of optional assemblies can be reduced. Utilization of the protocol during field device production is so common that HART allocates a range of command numbers for device-specific factory-only use. In some cases, cost savings produced by using HART in field device production even help justify the manufacturer's investment in the protocol.

Another early adopter of HART communications was in SCADA applications. Pipeline monitoring, custody transfer, and water/waste water applications all have significant communications requirements. HART communications simplified SCADA systems and reduced costs. In SCADA applications, analog input/output (I/O) cards were dropped in favor of direct acquisition of digital process data using HART communications. Multidropping HART field devices reduced power consumption and allowed smaller power supplies, batteries, and solar panels to be used. The use of HART in SCADA systems also led to *ad hoc* embedding of HART messages in other protocols such as packet radio, satellite communication, TCP/IP (transmission control protocol/Internet protocol), MODBUS, and others.

Climbing the Automation Pyramid

Somewhere along the way, the myth that HART was invented as a maintenance protocol became ingrained. Some system suppliers subscribed to this myth and used it to justify token support for HART. For example, many systems were supplied only with painfully slow, multiplexed I/O. A few other system suppliers even justified not supporting HART at all. Starting around 1997 and after years of acceptance by instrument technicians, end users began demanding better HART support from systems suppliers. End-user demands are prompting many systems suppliers to either improve their support for HART or to develop totally new interfaces.

HART is relatively simple (albeit not trivial). The protocol specifications include many mandatory requirements for all HART-compliant equipment. Consequently, considerable functionality can be included in any system without resorting to special drivers or supporting any device-specific commands. Table 4.11c lists the data and simple functions that any host system should be able to provide. Knowledgeable end users should insist on this minimum level of support when evaluating system offerings.

In addition to improving system support, the embedding of HART messages within other protocols is growing as well. This allows backbone control networks to support and exploit the HART installed base. Both standardized and *ad hoc* support are available using several protocols including PROFIBUS-DP, TCP/IP, and MODBUS. In addition, HART messages are often passed through control system proprietary networks and I/O.

HART FIELD DEVICES

All HART field devices support two communication channels: the traditional current loop and HART communication. The current loop occupies the 0 to 25 Hz band used by all 4–20 mA devices to transmit a single process value continuously. HART resides in a higher 500 to 10,000 Hz band. Although frequently referred to as HART digital communication it is actually analog. Specifically, HART communicates digital data using analog signals (see Section 4.4, Sorting out the Protocols).

There are minimum requirements that all HART field devices must meet:

- Adherence to physical and data link layer requirements
- Supporting minimum application layer requirements (e.g., device status information, engineering units, IEEE floating-point)
- All universal commands must be implemented in all devices exactly as specified

In addition, most HART field devices support a variety of common practice commands. Whenever a field device supports a common practice command, it must implement that command exactly as specified. The minimum data content of any field devices is summarized in Table 4.11c. The minimum command set is listed in Table 4.11d.

Note: Table 4.11d includes several common practice commands. Field device support for these commands is not required. However, surveying the hundreds of device types registered in the HCF DD (device description) library shows that a preponderance of field devices support the listed common practice commands. Of course, some devices support many additional common practice commands.

In even the earliest revisions of HART, support for digital process variables was mandatory. These were accessed using universal commands and called "dynamic variables." Four dynamic variables may be supported HART field devices: the primary (PV), secondary (SV), tertiary (TV), and quaternary (QV) variables. All devices must support the dynamic variable commands. However, a device may not return (for example) the TV and QV if the device is not multivariable.

The 4–20 mA current loop communication channel connects the HART field device to the control system (see Figure 4.11e). The current loop is always connected to the PV (see the calibration subsection below). Furthermore, some devices support more than one current loop connection. When this occurs, the second current loop is connected to the SV and so on. In other words, the dynamic variables connect a HART field device to the control system.

Note: For transmitters, PV and the loop current are outputs. For valves, they are inputs.

However, as multivariable field devices began to proliferate, the dynamic variable concept proved too restrictive. Revision 7.0 of the Common Practice Command Specification added support for "device variables" and provided additional standardized capabilities to developers of multivariable field devices.

A **device variable** directly or indirectly characterizes the connected process. For example, in Figure 4.11e, device variables 0, 1, and 2 are connected to the process and device variable 4 is calculated based on device variables 0 and 1. Device variable 3 is within the field device itself (e.g., it may be an onboard temperature sensor).

Sophisticated multivariable field devices allow the device variable mapped to PV (i.e., the current loop) to be configured based on application requirements. For example, a level gauge may measure level and calculate volume. Sophisticated level gauges will allow the user to choose whether the current loop reports level or volume. This allows the data that are most important to the user to be transmitted continuously via the current loop. Simultaneously on the same wire, the other process variables can be accessed using HART communications. Because devices supporting multiple current loops are rarely used, HART is usually the only way to access secondary process variables.

As many control systems support only one measurement (i.e., process variable) per I/O connection, multivariable devices can present them with serious problems. As we will see later, this problem can be compounded when a poor I/O system design is chosen. Older systems with poor or no HART support can still indirectly support multivariable field devices. Some products can convert digital HART secondary process variables into multiple current loop signals. These devices make

TABLE 4.11c
Overview of Standardized HART Field Device Data and Capabilities

Process Variable Values

Process Variable Values
Primary Process Variable (analog PV) The 4–20 mA current signal continuously transmitted to the host
Primary Process Variable (digital PV)
Percent Range Primary process variable expressed as percent of calibrated range
Loop Current Digital loop current value in milliamps
Secondary Process Variable 1 (“SV”)
Secondary Process Variable 2 (“TV”)
Secondary Process Variable 3 (“QV”)

Notes on Process Variables:

1. All digital values are IEEE floating-point with engineering units and (HART6) data quality status.
2. Secondary process variables are available in multivariable devices.

Status and Diagnostics Bits

Status and Diagnostics Bits
Device Malfunction Device self-diagnostics has detected a problem in device operation
Device Needs Maintenance (HART6) Device self-diagnostics indicates that preventive maintenance is required
Primary Process Variable out of Limits Primary process variable value is outside the transducer limits (i.e., the digital value is bad)
Secondary Process Variable out of Limits A secondary variable value is outside the transducer limits (i.e., the digital value is bad)
Loop Current Saturated The process value is out of range and the 4–20 mA signal is saturated
Loop Current Fixed The 4–20 mA signal is fixed and does not represent the process
Cold Start Device has gone through a power cycle
Configuration Changed Indicates device configuration (e.g., range values, units code, etc.) has been changed
More Status Available Indicates additional devices status data available using Command 48

Primary Process Variable Properties and 4–20 mA Current Loop Support

Loop Current Transfer Function (Code) Relationship between Primary Variable digital value and 4–20 mA current signal	**Upper Transducer Limit** (Read-only) Largest process variable value that is accurate
Loop Current Alarm Action Loop current action on device failure (upscale/downscale/hold last value)	**Lower Transducer Limit** (Read-only) Smallest process variable value that is accurate
Upper Range Value Primary Variable Value in engineering units for 20 mA output point	**Transducer Minimum Span** (Read-only) Indicates transducer precision; smallest delta between upper and lower range values that produces device-specified accuracy
Lower Range Value Primary Variable Value in engineering units for 4 mA output point	**Transducer Serial Number** Allows replaceable process transducers to be tracked
Date 3-byte code; date of last calibration	**Upper Transducer Trim Point** (HART6) The point last used to calibrate the upper end of the digital process variable
PV Damping Primary Process Variable Damping Factor (time constant in seconds)	**Lower Transducer Trim Point** (HART6) The point last used to calibrate the lower end of the digital process variable

Notes on Properties and Calibration:

1. Unless noted, all values are IEEE floating-point with engineering units.
2. Upper and lower transducer trim points were supported by device-specific commands in HART5.
3. Transducer limits, serial number, and minimum span mandatory for PV and optional for other process variables.

Device Identification

Instrument Tag User-defined abbreviated name of the device (8 characters)	**Manufacturer Name (Code)** (Read-only) Code defined by HCF that indicates the device manufacturer
Long Instrument Tag (HART6) User-defined name of the device (32 international characters)	**Device Type** (Read-only) A code defined by the manufacturer indicating the model of the device
Message User defined (32 characters), often used to record asset info (e.g., replacement ordering information)	**Device Revision** (Read-only) Indicates the revision of the device command set (i.e., the set of ALL device data available via HART)
Descriptor User defined, often used to describe instrument location	**Firmware Revision** (Read-only) Incremented every time the device firmware is modified
Write Protect Status Device Write Protect indicator	**Hardware Revision** (Read-only) Incremented every time the device hardware design is changed
	Device Serial Number (Read-only) Unique for every manufactured device; allows device to be tracked

Standardized Host Functions

Identify Device. A host function used to establish a connection to a field device.
Perform Loop Test. Forces the loop current to different values to allow the field device milliamp value to be verified.
Calibrate Loop Current. Forces the field device loop current to the specified value and adjusts its value to match a milliamp reference meter. Does not affect device range values or the digital PV value.
Re-Range Loop Current. Change the 4 and 20 mA operating points to different set of high and low process variable values. Does not affect loop current calibration or the digital PV value.
Set PV Damping Value. Change the time constant used to dampen the primary process variable value response rate.
Set Tag / Long Tag
Set Message
Set Date
Set Descriptor
Select PV Units. Change the engineering units returned with the primary process variable. The PV digital value and all of the PV properties are now calculated in the selected engineering units.
Initiate Self-Test

TABLE 4.11d
Minimum Set of Commands Supported by HART Field Devices

Cmd No.	*Description*
	Universal Commands
0	**Read Unique Identifier** Uses the "polling address" to establish a connection with the field device
1	**Read Primary Variable**
2	**Read Loop Current and Percent of Range**
3	**Read Dynamic Variables and Loop Current** Reads the loop current, PV, and (if device is multivariable) SV, TV, QV
6	**Write Polling Address**
7	**Read Loop Configuration** Reads the polling address and whether the loop current is active or not
8	**Read Dynamic Variable Classifications** (HART6) Reads the type of each process variable (pressure, temperature, mag-flow, etc.)
9	**Read Device Variables with Status** (HART6) Reads up to four process variables with data quality status
11	**Read Unique Identifier Associated with Tag** Uses the 8-character tag to establish a connection with the field device
12	**Read Message**
13	**Read Tag, Descriptor, Date**
14	**Read Primary Variable Transducer Information**
15	**Read Device Information** Reads upper and lower range values and other device-related information
16	**Read Final Assembly Number**
17	**Write Message**
18	**Write Tag, Descriptor, Date**
19	**Write Final Assembly Number**
20	**Read Long Tag** (HART6)
21	**Read Unique Identifier Associated with Long Tag** (HART6) Uses the 32-character tag to establish a connection with the field device
22	**Write Long Tag** (HART6)
	Common-Practice Commands
34	**Write Primary Variable Damping Value**
35	**Write Primary Variable Range Values** Changes primary process variable values used as the 4 and 20 mA operating points
38	**Reset Configuration Changed Flag**
40	**Enter/Exit Fixed Current Mode** Allows the loop current value to be forced; used in loop test and calibrating the loop current (see commands 45 and 46)
41	**Perform Self-Test**
42	**Perform Device Reset** The device performs a hard reset (i.e., the same effect as cycling the power on and off)
44	**Write Primary Variable Unit**
45	**Trim Loop Current Zero** Adjusts the loop current to 4 mA; does not affect the range values or the digital primary process value
46	**Trim Loop Current Gain** Adjusts the loop current to 20 mA; does not affect the range values or the digital primary process value

Notes:

1. All devices must support all universal commands.
2. There are many common practice commands. However, the ones listed are commonly used in the vast majority of devices including even the simplest ones.
3. Commands 0, 11, 21 are three of the identity commands. All return the same identity data (see Table 4.11f).

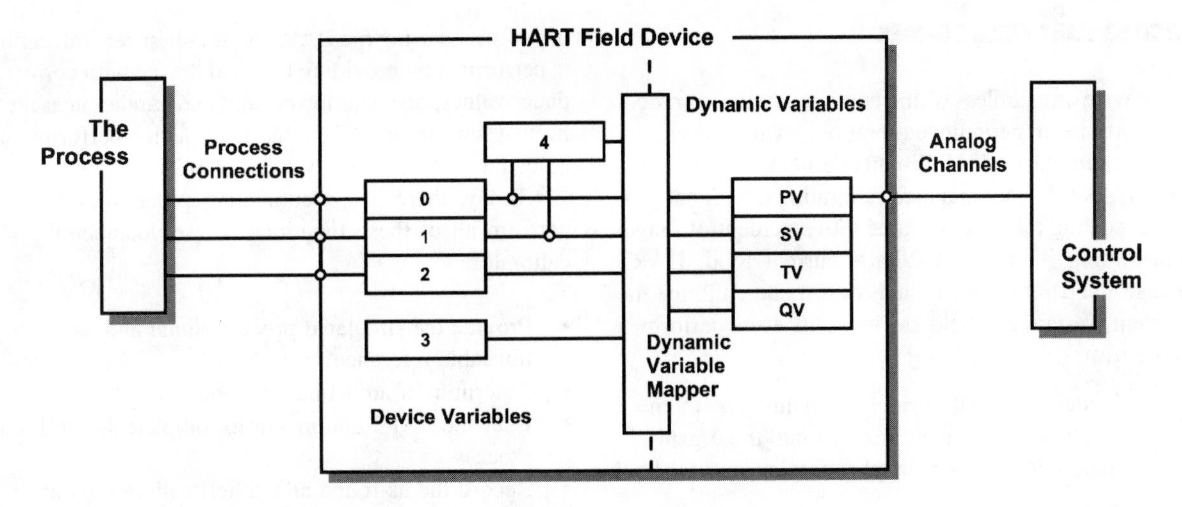

Fig. 4.11e
HART device variables vs. dynamic variables.

TABLE 4.11f
Application of HART Field Device Identifying Data

Data Items	*Purpose*
Tag; Long Tag; Manufacturer ID; Device Type	User identification of the field device
Manufacturer ID; Device Type; Device ID (i.e., serial number)	These three data items, when combined, identify a single, unique field device. Frequently, these data items are used to track the field device through an installation, calibration, plant operation, and refurbishment cycle. These three data items also provide the basis for the 5-byte address of the device.
Manufacturer ID; Device Type; Device Revision	Field device command and data item set identification. The combination of these three numbers always indicates a unique command set. For example, if a new device-specific command is added, the device revision number is changed. If a device-specific command is deleted, a new device type number must be used (to maintain backward compatibility).
Device Revision; Software Revision; Hardware Revision; Universal Command Revision	Field device revision information. The hardware and software revisions track the configuration of the field device and its software. The device revision reflects the set of commands supported by the device. The universal command revision ties the field device to a major revision of the HART protocol. For example, a software revision can be made to correct a defect or enhance device operation without necessarily affecting the device revision level (i.e., the HART commands it supports).

a field device look like it generates four current loops instead of one. Of course, the control system needs four I/O points instead of one, as well.

Backward Compatibility

Clearly, the HART protocol is backward compatible with 4–20 mA equipment. However, the backward compatibility of HART extends far beyond that. HART protocol specifications include specific backward compatibility requirements. These requirements apply to both modifications to an existing HART product and to modifications of the HART specifications themselves.

Backward compatibility rules, in general, allow data and commands to be added. However, the meaning of the data already present cannot be changed. Furthermore, commands and data may not be deleted under any circumstances. For example, a field device adding a data byte to device-specific command 136 is only required to increment the device revision (Table 4.11f). However, if command 136 was deleted, the developer must change the device type number and, in effect, create a new type of field device.

Backward compatibility is critical to ensuring continued operation of installed plant systems. A host using any feature in a field device can be assured that the same feature will always be available and operate the same way with all future revisions of that field device. This allows any field device to be safely (i.e., without breaking or reconfiguring the control system) replaced by a newer revision device even if the system is using device-specific commands or data.

CALIBRATING HART FIELD DEVICES

All field devices, regardless of the communications protocol they support, need periodic calibration.[2] This subsection focuses on calibration of HART-capable field devices. Calibration of HART devices includes calibrating the digital process value, scaling the process value into a percent of range, and transmitting it on the 4–20 mA current loop. HART devices use standardized commands to test and calibrate the loop current, re-range a field device, and even perform a transducer trim.

Note: Understanding the principles in this subsection will provide valuable insights into the organization of the HART application layer.

Figure 4.11g is a block diagram showing the processing steps associated with generating PV and will be used to illustrate the calibration process. Three blocks are used for data acquisition and communication of PV via the current loop. From left to right, these blocks include the transducer block, range block, and DAQ block.

Calibrating the Process Variable

The transducer block is responsible for generating a precise digital value representing the connected process. In HART, the transducer block contains properties like the upper and lower transducer limits, transducer trim points, as well as device-specific transducer characterization data. Access to characterization data is usually not available or adjustable in the field. All smart field devices (no matter the communication protocol supported) contain something equivalent to the HART transducer block. Furthermore, all devices require periodic calibration of the digital value produced by its transducers.

The calibration of the transducer blocks (i.e., transducer trim) consists of supplying a simulated transducer value and comparing the transducer value provided by the field device with the value measured by a traceable reference to determine whether calibration is required. If calibration is required, it is performed using the HART protocol. In general, calibration is performed by providing the field device with correct transducer values, one near lower limit and another near the upper limit. Using these values, the field device performs internal adjustments to correct its calculations.

Today, there are documenting process calibrators that perform all of these functions. These documenting process calibrators:

- Provide the simulated process signal and an accurate, traceable reference value
- Determine if adjustment is required
- Issue the HART commands to complete the calibration process
- Record the as-found and as-left calibration data

Some instrument management systems are capable of automatically uploading the calibration records for trending and later analysis. Standardized Microsoft Windows-based protocols exist to facilitate communications with documenting process calibrators.[3]

The latest release of the HART specifications provides standardized commands to support transducer calibration.

Note: Earlier versions of the protocol did not include standardized transducer trim commands. Despite this limitation, most HART calibrators support a wide range of HART-compatible field devices. In many cases, the developers analyzed the contents of the HCF DD Library to learn the device-specific commands required for each individual field device.

Scaling the Process Variable

The next two blocks in Figure 4.11g translate the transducer value into a 4–20 mA signal. Traditionally, this has been referred to as the "zero and span" of the device. HART has always supported standardized commands to re-range a field device and to calibrate the loop current.

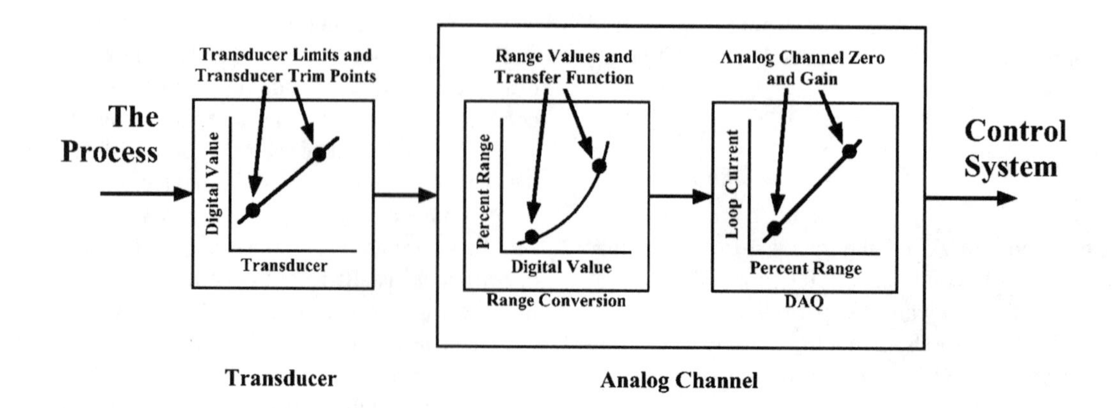

FIG. 4.11g
PV blocks that can be calibrated.

In the range block, HART uses the upper and lower range values to scale the transducer value into "percent range." The lower range value sets the transducer value that produces a 4-mA signal. The upper range value specifies the transducer value that will produce a 20-mA signal. In many field devices, the upper range value can be smaller than the lower range value. For example, a level gauge could be configured to provide 4 mA when the tank is full and a 20 mA value when it is empty. This can be useful, for example, when the level gauge 4–20 mA signal is connected to a proportioning valve.

Note: Understanding Figure 4.11g and the differences between ranging and calibrating a device is critical. Considerable confusion can result if the difference between calibrating the digital value of the PV and re-ranging the instrument is not understood.[2] Many hours can be spent attempting to get an "accurate" 4–20 mA signal if the functions of the transducer and range blocks are ignored.

The range block can also apply a transfer function (e.g., linear, square root, cubic spline, etc.) to the PV signal. Square root functions are frequently used to approximate a flow from a differential pressure measurement.

Producing the Loop Current

The DAQ block is responsible for the actual production of the 4–20 mA signal. HART has always provided standardized commands to calibrate the loop current. Because the range block already scales the process value, the responsibility of the DAQ block is to ensure 0% produces 4 mA and 100% produces 20 mA. The objective of calibrating this block is to ensure that when the device thinks the loop current is, for example, 7.956 mA, there are exactly 7.956 mA on the wire. The meaning of 7.956 mA, in terms of the process value it represents, has been already established in the previous two blocks.

From a system viewpoint, the whole purpose of the 4–20 mA current loop is to communicate with the control system. Consequently, the most critical requirement is that the field device and the control system both agree on the value of the current loop. Agreement between the field device and the control system on the milliamp value is much more critical than the absolute accuracy in most applications.

HOST APPLICATIONS

Although the minimum requirements for field devices are quite strict, host applications are provided some leeway in their implementations. This is due to the wide range of applications supporting HART. Applications vary from simple ones like an embedded application closing relay contacts based on HART status information to complete process control systems. Of course, all host applications are required to adhere to all physical and data link layer requirements.

HART hosts identify field devices using standardized procedures and "identity commands." There are several identity commands and each provides a different procedure to initiate communications. Among other things, identification procedures allow the 5-byte communication address of the field device to be ascertained. Hosts are required to support all field device identification procedures (e.g., connect to the device by poll address, tag, or by using the "find device" command).

Note: HART requires all field devices to provide all identity data exactly as specified.

All identity commands return the same data and hosts must analyze these data to support communications properly. Identity commands return such information as the device type, who made the device, what is the device, and software and hardware revisions. The data are used to learn specific properties of the field device (see Table 4.11f).

Host Conformance Classes

All hosts utilize standardized identification procedures to initiate field device communications. Beyond that, however, the services provided by a host become application specific. To simplify assessment and ranking of host systems the HART specifications require hosts to disclose their "conformance class" (Table 4.11h). The classes range from zero (the simplest, least-capable host) up to five (a universal host). As the conformance class number increases, so does the capability of the host. For each conformance class, support for specific HART commands and procedures is required.

- Class 0 and 1 hosts tend to be simple embedded or special-purpose applications.
- Class 3 is called a "generic host." Even at this level, no special device-specific drivers are needed for compliance. Any reasonably capable host, for example, a programmable logic controller (PLC) or control system,

TABLE 4.11h
Host Conformance Classes

Class	*Description*
0	The host does not meet the minimum requirements of Conformance Class 1
1	The host can utilize cyclical process data from any field device
2	The host can supply the user with basic identification and configuration data about any field device
3	The host can perform basic configuration of any field device; minimum level required to be classified as a "generic host"
4	The host provides basic commissioning and calibration support for any device
5	The host is capable of accessing all field device data items and all device-specific commands for any field device

should be able to adhere to generic host requirements. In addition, most day-to-day operations can be supported by any generic host.
- Levels 4 and 5 frequently need device-specific knowledge supplied (e.g., by a DD or a DTM). Level 4 and 5 hosts, although very important, may not be necessary for daily communications with every field device in the plant.

Understanding I/O System Trade-Offs

All communications systems can be characterized by their throughput and latency (Table 4.11i):

- **Throughput** indicates the maximum number of transactions per second that can be communicated by the system.
- **Latency** measures the worst-case maximum time between the start of a transaction and the completion of that transaction.

Throughput and latency are affected by the communication protocol requirements and by implementation decisions. Knowledgeable end users will insist on receiving these two statistics when evaluating system components. The HART Application Guide includes a host capability checklist.[4]

The performance (throughput and latency) of I/O systems can vary dramatically. This subsection reviews design decisions and architecture that affect HART I/O system performance.

In general, HART can provide 2.65 PV updates per second via command 1. This is a typical value and actual throughput can be less. For example, slave devices may begin their response later in the allotted transmit time window or the message might be a HART command containing a longer data field. In addition, hosts have a small time window to begin their message transmission. If that window is missed, the HART network becomes "unsynchronized" and the host must back off an extra time interval. The resulting extra time lowers performance to 0.88 updates per second (or about three times slower) or less.

Note: All calculations in this discussion assume the HART FSK physical layer is in use. If the HART PSK physical layer is used, then all values are increased about fivefold. For example, the FSK physical layer allows 2 to 3 tps and the PSK physical layer provides 10 to 12 tps.

Missed host transmit windows are seen in several HART hosts (e.g., those running on Microsoft Windows systems have very poor latency). Consequently, many point-to-point HART networks are running at significantly reduced speeds. Missed transmit windows can be avoided by including message FIFOs in the HART interface. I/O systems containing the small amount of extra memory for FIFOs provide significantly better performance.

TABLE 4.11i
Summary of Latency and Throughput for I/O Systems

Number Channels	*Point–Point*	*Point–Point (unbuffered)*	*Multidrop*	*Multiplexed*
		Latency (seconds)		
1	0.38	1.14	0.38	0.38
4	0.38	1.14	1.51	2.73
8	0.38	1.14	3.02	5.46
16	0.38	1.14	6.04	10.92
32	0.38	1.14	12.08	21.84
		Throughput (transactions per second)		
1	2.65	1.14	2.65	1.47
4	10.60	4.56	2.65	1.47
8	21.19	9.12	2.65	1.47
16	42.38	18.24	2.65	1.47
32	84.77	36.48	2.65	1.47

Multidrop Connections All HART-compatible host and slave devices must support multidrop. In multidrop, several HART field devices are connected via the same pair of wires. Latency is increased because several commands may be enqueued and waiting for transmission.

Note: In multidrop applications, all field devices are typically parked at a fixed current and only HART communication is available. There are installations, however, where the field devices may use current loop signaling although several devices are on the same wire. Two proportioning valves wired in series for split-range operation is a good example of these applications.

In multidrop networks, throughput remains the same (approximately 2.65 tps). However, latency increases and is proportional to the number of devices on the loop. Even with the degraded latency, multidrop is suitable for many applications (e.g., temperature monitoring or tank inventory management).

Multichannel I/O Systems For optimum performance, I/O systems should include one (hardware or software) modem for each channel. With such an I/O system, utilization of all HART capabilities is practical including continuous status and diagnostics, remote configuration of instruments, acquisition of secondary process variables from multivariable field device, and the deployment of multidrop networks.

Lower-cost I/O systems multiplex HART communications by using one HART modem chip to support several I/O channels. With multiplexed I/O, all transactions must be serialized and the host must resynchronize its communications each time the HART channel is changed. Consequently, the execution of the last requested command may be delayed by the time required to process all commands already enqueued plus the resynchronization delay time.

Table 4.11i summarizes system performance-based vs. I/O architecture. Although actual performance can vary, this table provides a basis for evaluating I/O system designs. Multiplexed I/O is clearly the worst performer and is even worse than multidropped I/O. For 8-channel analog I/O (a common format in many systems), multiplexed I/O throughput is 14.5 times worse than when a modem is used for each channel. Clearly, the use of multiplexed systems dramatically degrades system performance. In fact, real-world performance may be even worse (e.g., due to delays introduced by other system overhead).

Note: Multivariable HART field devices are very common and secondary process variables can only be accessed via HART communication. Multiplexed I/O severely limits utilization of secondary variables found in multivariable field devices.

Impacts of Burst Mode on I/O Performance

Burst mode allows a field device to be configured to transmit digital process values continuously. All HART hosts must support burst mode; however, utilization of burst mode will depend on the application. If common multivariable field devices are employed, burst mode maximizes the update rate of secondary process variables. For example, secondary process variables can be normally updated (using command 3) 1.8 times per second. With burst mode the secondary process variables can be updated 2.4 times per second (about a 30% improvement).

However, configuration of a field device can take somewhat longer. When burst mode is enabled, the burst mode device is responsible for token recovery. Consequently, hosts only have network access after a burst message from the field device. This results in a throughput of 0.80 configuration read commands per second or about 2.5 times slower than with burst mode not in use. This is about the same throughput as if the host missed its transmit window (discussed above).

In summary, burst mode improves system support for multivariable field devices while slowing somewhat uploads and downloads of instrument configurations. Of course, uploads and downloads are infrequent compared with cyclic updates of secondary process variables. For multiplexed I/O systems, performance is difficult to estimate. In some cases, burst mode may actually provide an overall benefit to multiplexed systems. This is certainly the case for multiplexed systems supporting cyclic access to secondary process variables in multivariable field devices.

INSTALLING HART NETWORKS

HART communication is designed to be compatible with existing 4–20 mA systems. In fact, the HART signal is a modulated analog signal. Consequently, no special requirements are needed for most installations. That being said, no discussion of HART can be complete without providing an overview of basic HART signaling and the mechanics behind installing a HART network.

HART communication is based on communication techniques long used in telephone and telecommunications and, consequently, HART is very forgiving. For example, there are several ways to calculate HART cable lengths. No matter the technique used, the result is very conservative (i.e., HART will normally be successful on cables 25 to 50% longer than calculated).

Note: HART cable run lengths are a function of the quality of the cable used. The lower the capacitance, the longer is the possible cable run. In many cases, the cable runs are determined by intrinsic safety capacitance limitations before HART limits are reached. Because HART is based on modems designed for analog telephone networks, in most cases, HART communication will be successful over some of the worst imaginable wiring to be found.

Understanding the HART Physical Layers

All HART devices support two communications channels: (1) the traditional 4–20 mA signaling and (2) the modulated HART communications. These exist in separate communications bands (see Figure 4.11j) and the signals do not interfere with each other. This allows both communications to occur simultaneously. In fact, a significant portion of the physical layer specification is dedicated to defining the filtering and response times to ensure the separation of these bands.

Process control equipment normally contains a low-pass filter to block out noise. Consequently, HART looks like noise to most existing systems and is filtered out. In most cases, this feature allows HART to be used with existing systems with no special modifications.

All HART devices are required to support the HART FSK physical layer (the PSK physical layer is optional). The FSK physical layer is based on the Bell 202 modem standard. FSK modulates the digital data into frequencies ("1" is 1200 Hz and "0" is 2200 Hz). FSK is quite robust, can be communicated

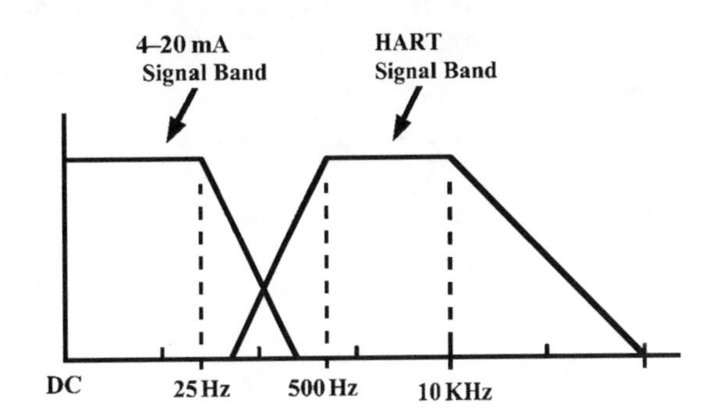

FIG. 4.11j
HART frequency bands.

long distances, and has good noise immunity. A single, simple HART modem chip is used by a device to send and receive the HART modulated signals.

Figure 4.11k provides an overview of different physical connections. Depending on the device, the HART signal is modulated by varying the loop current or by directly modulating a voltage onto the loop. Process transmitters vary the loop current ±0.5 mA. On a typical loop the current sense resistor is 250 Ω, which produces a 250 mV peak-to-peak HART signal. Other equipment (e.g., handhelds, DCSs, valves) modulate a voltage on the loop. For each type, the specification establishes signaling levels, device impedances, and internal filter requirements (Table 4.11l). All these requirements are fulfilled in the device design and, consequently, no special terminations or installation practices are required.

HART Network Guidelines

As previously discussed, HART was designed to work in typical 4–20 mA loops. Consequently, guidelines for HART networks are simple (Table 4.11m) and virtually identical to normal 4–20 mA instrumentation practices.

The loop with a two-wire transmitter in Figure 4.11k is an example of a typical HART loop. The loop is wired using standard 4–20 mA practices and by default meets all of the guidelines in Table 4.11n. It has a current sensor (i.e., one low impedance device), a transmitter controlling the 4–20 mA signal (i.e., only one device varying the analog signal), and a single HART secondary host (when a handheld is connected to the loop). Virtually all control systems use a 250 Ω current sense resistor and a linear power supply.

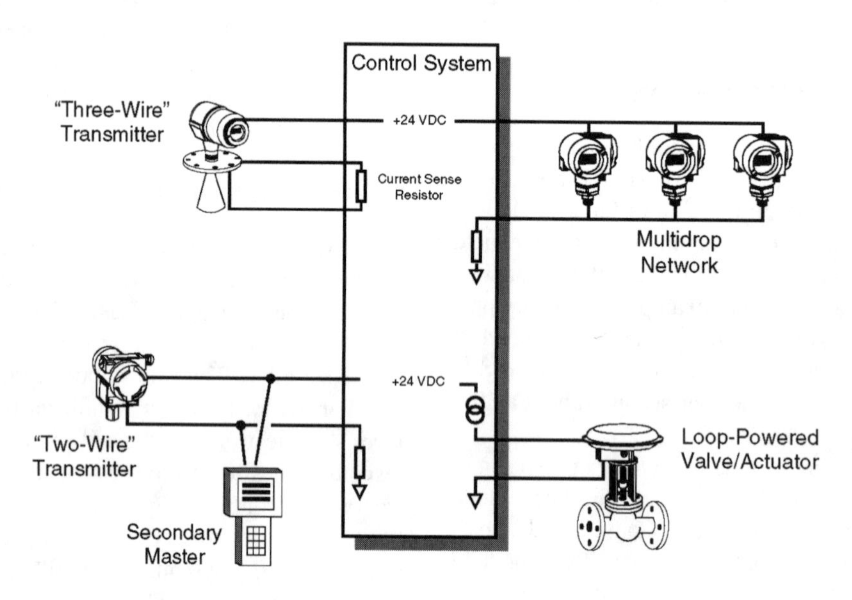

FIG. 4.11k
HART devices and networks.

TABLE 4.11l
Characteristics of the HART Devices Shown in Figure 4.11k

	Two-Wire Transmitter	*Three-Wire Transmitter*	*Multidrop Network*	*Control System Current Input*	*Control System Current Output*	*Secondary Master*	*Valve-Actuator*
Loop-powered	Yes	No	Yes	No	No	No	Yes
4–20 mA signaling	Current sink	Current source	None	Current receiver	Current source	None	Current receiver
4–20 mA impedance	High	High	High	Low	High	Very high	Low
HART signaling	Current	Current	Current	Voltage	Current	Voltage	Voltage
HART impedance	High	High	High	Low	High	Medium	Low
Isolated from ground	Yes	No	Yes	No	No	Yes	Yes

TABLE 4.11m
HART Network Guidelines

A network must have at least one, typically only one, low impedance device. Total loop resistance must be between 170 and 600 Ω.
A network must have no more than one device varying the 4–20 mA signal.
Only one secondary device is allowed.
Cable run lengths to 3000 m for single pair cable and 1500 m for multiconductor cables are typical. Actual length depends on the number of multidropped field devices and the quality of the cable used (see Table 4.11n).
Low capacitance shielded twisted pair cable is recommended. However, HART has been successfully used over poor quality, unshielded wiring. Do not replace wiring until HART communication has been attempted.
Linear power supplies should be used.
HART is compatible with intrinsic safety (IS) rules and HART communicates across most IS barriers. In general, zener diode barriers are acceptable as they do not prevent two-way communication. However, isolating barriers must be HART compatible (i.e., some isolating barriers support one-way communication only).

TABLE 4.11n
Allowable Cable Lengths for 1.02 mm (#18 AWG) Shielded Twisted-Pair Cable

	Cable Capacitance, pf/m			
No. Devices	*65*	*95*	*160*	*225*
1	2769 m	2000 m	1292 m	985 m
5	2462 m	1815 m	1138 m	892 m
10	2154 m	1600 m	1015 m	769 m
15	1846 m	1415 m	892 m	708 m

As seen in Figure 4.11k, HART also supports multidrop networks. In a multidrop network the current loop is used to power two-wire devices and all devices park at a fixed current (e.g., 4 mA). Although supporting more field devices is possible, a network typically only has up to 16 multidropped field devices. Of course, the number of devices affects the process variable scan rate and fewer are allowed when intrinsic safety rules are in effect.

In summary, the HART physical layers are simple with most of the details managed by a low-cost modem chip. HART signaling is in a separate frequency band from the 4–20 mA analog signal and, in most cases, does not interfere with existing 4–20 mA systems. The impedances, filtering, and signaling of HART devices are specified to ensure reliable HART communication when using standard 4–20 mA installation practices.

DEVICE DESCRIPTIONS

As this section demonstrates, considerable functionality is available using the standard HART commands that all field devices support. While these commands address routine, continuous systems requirements like acquisition of cyclical process data and online status and diagnostic monitoring, from time to time device-specific features and configuration properties must be accessed. Device descriptions (DDs) facilitate access to the occasionally needed device-specific features.

The Device Description Language (DDL) is a very powerful technology that provides significant benefits for the configuration and setup of device-specific features (Table 4.11o). Although DDs are an optional part of the HART protocol, most HART-compatible devices have a corresponding DD registered with the HCF.

The DDL is an object-oriented modeling language tailored to describing field devices, and DDs do this very well.

TABLE 4.11o
Summary of Device Description Strengths and Weaknesses

Strengths	*Weaknesses*
DDs allow the development of very beneficial universal hosts supporting both commonly used standardized HART features and infrequently accessed device-specific features.	DDs do not enhance interoperability.
When problems arise, a universal host using DDs greatly simplifies the diagnosis and troubleshooting of field device problems.	Although DDs model a field device, they do not tell a host what to do with the field device. Interpretation of the DD and all the possible combinations of DD constructs requires significant host software development effort.
DDs allow the capture of 100% of a field device configuration. This allows device configurations to be engineered in advance of plant commissioning, thus reducing start-up times and costs.	DDs allow modeling of field device features that do not comply with protocol requirements.
DDs are technology independent and can be used in Microsoft Windows platforms, embedded real-time devices, and other computing environments.	DDs have limited HMI support and no support for graphical user interfaces. Additional development by the device manufacturer may be needed to support some DD-based hosts.
DDs allows the schema of the real-time database of the field device to be defined and exported.	DDs allow device-specific commands and data to be used in place of standardized (e.g., common practice) commands.

The HART DDL has been in heavy use for over 7 years and there are DDs for hundreds of different field device types registered with the HCF. Elements of the HART DDL can also be found in the DDL associated with Foundation Fieldbus and PROFIBUS. DDL consists of three major functional areas:

1. Modeling the field device application layer data
2. Describing the commands used to transport that data
3. Specifying the standard operating procedures (SOPs) used for periodic maintenance

Data modeling is the central role of DDL and the majority of both DDL constructs and most of the lines of code in a DD are used to describe the real-time database found in the field device.

A system consisting of controllers, human–machine interface (HMI), I/O, and smart field devices can be considered a distributed database system. The control system and HMI caches copies of the database of the device to control the process and keep the plant operator informed. DDs describe that database and HART is used to keep the system copy of the database up to date. This is a radical notion to many system and field device developers. Furthermore, control systems that do not view the smart field device as part of "The System" ignore the valuable process-related data found in HART-compatible field devices. The system should not stop at the 4–20 mA input card.

DDL provides significant benefits and allows any field device to be fully modeled. The principal benefit to device manufacturers and end users is that DDL allows universal hosts to support 100% of the features of a field device, including features rarely used. Device developers do not need to develop a custom application or a handheld, and end users do not need to acquire and track hundreds of applications and special device drivers.

APPLICATIONS

Thus far, the basic concepts underlying the HART protocol have been covered. This subsection looks at a cross section of typical HART applications. Each application is briefly discussed, highlighting the HART capabilities utilized. As noted in the rest of this section, with few exceptions, the applications in this section are relatively simple to implement using commands and data common to all HART field devices.

Using HART in Control Systems

Many plants already have large numbers of HART-compatible field devices and consequently a significant investment in HART technology. This subsection discusses emerging system architectures and how to exploit the untapped capabilities found in all HART field devices. This section outlines methods for maximizing the value of existing installed HART field devices.

Historically, 4–20 mA field wiring is frequently brought to a junction box and then a multiconductor "home run" cable connects the junction box and the I/O located in a marshaling room. I/O system data is then communicated using a proprietary backbone bus to the controller. The controller in turn provides data to operator consoles, data historians, and engineering workstations. In this traditional architecture, the only information available from the field device is the 4–20 mA signal.

This architecture is now frequently replaced by a distributed system consisting of hardened remote I/O, controllers, operator consoles, and engineering stations all connected using "open" communications protocols. The remote I/O is mounted close to the process units with short 4–20 mA analog cable connecting to HART field devices. The expensive multiconductor cable from the junction box is replaced by a network cable connecting the remote I/O to the balance of the control system. In this emerging architecture, continuous communication of the primary variable is provided via the 4–20 mA signal with HART communication used to enhance the capabilities and integrity of the system.

One of the simplest and most valuable assets HART communication provides is access to the information on device status. Device status is provided with every HART response message. Using HART communication, detailed knowledge of the health of the field device can permeate the control system. Monitoring device status can predict device failures and unscheduled plant outages can be reduced. The integrity and availability of the system are improved and the effectiveness of predictive maintenance programs can be significantly enhanced using HART status data.

Furthermore, the latest release of the HART specifications provide standardized data quality status that supports development of self-validating field devices. As these devices emerge, the value provided by utilizing HART status information will be enhanced even further.

Control system can provide further value by using HART communications to tap the wealth of secondary process variables found in the many multivariable HART field devices. Multivariable process data can be updated two to three times a second (10 to 12 times with PSK). Although not particularly fast, this update rate is much faster than the time constants associated with many processes. For example, the temperature sensor bonded to the transducer in most HART pressure transmitters can be monitored. This can be used to provide temperature indication of the process or provide early warning of potentially damaging freezing conditions during winter. Utilizing this "free" data may eliminate the need for additional transmitters and I/O in some applications.

Maximizing the Value of HART Current Loop Support

The current loop is fundamentally a communication channel. The properties of the current loop are universally available from HART field devices. Any system designed to utilize the power of HART should be able to:

- Automatically configure the I/O channel. The range values can be read by the control system to configure the calculation of the process value from the received current loop.
- Confirm the integrity of the low-level current loop signal. The transmitted current loop value can be read from the field device and compared to that received by the system. The control system can alert the plant operator to any discrepancy.
- Perform an end-to-end validation of process variable communication. The process value calculated by the control system from the current loop can be verified against the digital process value originating in the transducer block of the field device.

A smart transmitter is the primary indicator of process conditions. The current loop signal and HART communications allow the process values in the control system to be accurately slaved to the process values originating in the HART field device.

As the current loop is a communication channel, agreement between the current loop reading in the control system and the field device is critical. HART communication allows the value of the current loop perceived by the field device to be continuously verified against that measured by control system (see universal command 2). In addition, the control system can test current loop integrity by forcing the HART field device to output any milliamp value desired (see common practice command 40). Any discrepancies between the control system and the field device can then be brought to the operator's attention.

If desired, a control system supporting HART should be able to calibrate automatically the loop current value generated by the field device (e.g., using common practice commands 45 and 46). The I/O system knows the milliamp value it is measuring and as long as the value from the HART field device agrees, the 4–20 mA analog communication channel is secure.

The control system has a fundamental interest in the process value the 4–20 mA signal represents. When HART communication is not supported, the equation used by a system to convert the milliamp value into a usable process value must be manually configured. For example, the controller must be told that 20 mA = 100 kPa. Furthermore, if a technician should re-range the field device, the process value calculated by the control system would no longer be accurate. All HART devices contain the information necessary to configure the I/O channel automatically (see universal command 15). Significant savings in system engineering and commissioning can result. While in operation, the system should be able to periodically monitor the range values and detect any inadvertent changes in the system configuration. If any changes occur, the operator should be alerted.

Although the primary process value is continuously updated using the current loop, it can also be read using HART communications (see universal commands 1, 3, and 9). Continuous HART communication allows end-to-end validation of primary process value used by the system. All of the intermediate calculations and the current loop signal itself are tested together to ensure the system functions correctly as a whole. Any deviations detected should cause the operator to be alerted.

SCADA Applications

SCADA applications usually focus on acquiring real-time data. The speeds of operations are frequently measured in minutes as opposed to a control system requiring fractions of a second response times. HART communication is very complementary to the objectives of SCADA systems. Some of the SCADA applications using HART communications include pipeline monitoring, custody transfer, inventory and tank farm management, automated meter reading, and the monitoring of remote petroleum production platforms.

The key to the successful use of HART in SCADA applications is the digital process values supported in all HART field devices. In addition, the status information provided via HART can be used to schedule field trips to perform preventive maintenance and used to reduce unscheduled downtime. HART provides other benefits:

- HART communication can be transmitted directly over leased telephone lines (HART FSK physical layer is based on Bell 202). The leased lines can create a virtual multidrop network of instruments, for example, distributed along a pipeline. Dedicated RTUs (remote terminal units) may not be required.
- Power consumption can be reduced, minimizing battery and solar panel sizes. A system with, for example, multidropped devices requires significantly less power (4 mA times 8 devices = 32 mA total) and I/O (1 point) when compared with a point–point SCADA implementation (160 mA and 8 I/O points). With the multidropped approach, process values can be read once every 4 s, which is much faster than required for most SCADA applications.
- Some spread-spectrum radio equipment supports HART communications. With this equipment, field devices scattered within a 100-km radius can be linked together into virtual multidrop networks, thus greatly reducing costs.

Control in the Field

HART devices have been used to provide a simple control in the field almost since their inception. The early uses may have been in tank-level control. In this application, the HART standardized support for re-ranging is utilized (see common practice command 35). The level gauge can be configured with the lower range value set to the target tank level and the upper range value used to provide proportional action. The loop current from a level gauge is directly connected to a control valve filling the tank. The upper range value (i.e., the 20-mA point) sets the valve full open tank level.

HART requires that process variables be correct any time the process is between the lower and upper transducer limits. In other words, when the process is outside the upper and lower range values and the current loop is saturated, the HART digital process variable is still available and accurate. Consequently, the tank level can be monitored using HART even while the current loop is used to control the tank level.

PID controllers in HART field devices also utilize the unique hybrid nature of the protocol. Some HART transmitters have supported PID control functionality since the early 1990s and, more recently, HART intelligent valve controllers have begun including PID controllers. These controllers allow a transmitter and valve to be connected in series to create a stand-alone control loop in the field. When the PID is in the transmitter, the current loop continuously updates control signal and, when in the valve, the process signal is carried on the current loop. In these applications, the current loop provides the basis for continuous closed-loop control. This is similar to the single-loop controller topology used for many years. HART communication is used to monitor control loop operation (e.g., read process variables and monitor status) and change set points.

Device-specific commands were used to interact with PID controllers in early device implementations. However, HART specifications have now standardized PID support. This allows devices that support a PID controller to use standardized commands and data, thus allowing even simple hosts to utilize this functionality.

Instrument Management

Instrument management systems are specialized database applications. Their database allows instrument configurations to be tracked and maintenance activities planned and recorded. When combined with documenting process calibrators, instrument calibration histories can be automatically captured.

Simple systems utilizing common HART capabilities can provide valuable assistance in monitoring basic instrument configurations. Standard HART communications allows devices needing maintenance to be identified and allows any changes to device configurations to be detected. Range values and device status can be continuously monitored by the management system. When plant calibration procedures reset the date field in devices as they are calibrated even a simple system should be able to automatically list devices requiring recalibration.

HART provides two standard data items that are specifically directed to tracking configuration. The first is the "configuration changed" flag (see Table 4.11c). The second is a "configuration change counter" (see universal commands 0, 11, and 21) that is incremented every time a command is received by the field device that changes the instruments configuration. Both of these data items are easily accessed and can be tracked by even the simplest host. This allows any host to detect a configuration change.

High-end instrument management packages have sophisticated databases and many automated reports can be generated (e.g., an instrument technician work order for devices needing maintenance). Frequently, these systems are universal hosts that can capture 100% of the configuration data of each field device and many can even access infrequently required device-specific features. To support 100% of the device data items, universal hosts usually support DDL and utilize the hundreds of DDs registered in the HCF DD library. By utilizing the HART device identity information (see Table 4.11f), these packages can also document the entire life of the plant field devices. Analyses of these histories allow predictive maintenance activities to be optimized and maintenance costs to be reduced.

For either a simple or a complex instrument management system HART communication can automate many activities, simplify record keeping, and reduce manual data entry.

Valve Diagnostics

Valve diagnostic software packages are specialized tools using HART two-way communication to support intelligent valves and actuators. The protocol standardizes some valve properties, but many of the diagnostics are device-specific and use proprietary algorithms. Any HART-capable host should be able to access some basic valve information:

- "Device needs maintenance," "valve in manual," and "valve upper or lower limited" status is available (HART6). These provide basic information about the operation of the valve.
- Actual valve position is generally returned as a secondary process variable. This information can be monitored and trended to confirm basic valve operation.
- For pneumatic positioners, actuator pressure is generally available as a secondary process variable.

In addition, many other device-specific, smart valve capabilities can be accessed remotely using HART:

- Manually stroking the valve or manually controlling valve position
- Linearization of the flow characteristics of the valve (i.e., ensuring a change in the loop current causes a linear proportional change in the flow)
- Tracking of valve position vs. set point to assess the mechanical condition of the valve
- Accessing predictive maintenance indicators like total valve travel, number of reversals, and stroke count data stored in the smart valve
- Characterization of the valve (e.g., testing the valve travel speed)

Many of these diagnostics are available while the valve is online and additional diagnostic features continue to be added.

OPC

OPC (OLE for process control) provides high-level communication between software modules. OPC servers comply with specific software interfaces, which encapsulate the underlying details of the hardware and communications supported by the server. This allows OPC clients to connect to any compliant OPC server to create a whole system.

Several OPC servers exist that support HART. These allow, for example, standard HMI and historical trending packages to access HART data. These OPC servers support direct connection to HART networks and communication via intermediate I/O systems. While HART inherently supports two hosts, OPC allows many clients to access HART data via Microsoft Windows networks and share connections to HART field devices. HART OPC servers broaden and simplify access to HART data.

OPC allows clients to access "named data items." These named data items are defined by an OPC-compatible server. The OPC client "browses" the named data items provided by the server and subscribes to the data item of interest. The data items can be grouped and the group turned off and on by the client. When a group is turned on, the server will publish the data items based on the criteria set by the server. These criteria include update rate and dead band. This allows the client, for example, to obtain the data item only when it changes.

As a minimum, a HART-compatible OPC server should provide access to the application layer data shown in Table 4.11c. In addition, some universal hosts support an OPC server interface that allows access to device-specific data as well.

However, one source of potential mischief should be recognized. Standard HART data items should all be referred to by the same name. If the naming scheme varies by device, then the client application risks becoming linked to one and only one device (with the primary variable named "PRESSURE"). If that device is replaced with a similar product from another manufacturer and the primary variable is named "DRUCK," the client application will fail. Utilization of standardized OPC naming schemes is important to ensure interoperability and the openness of the protocol.

CONCLUSION

The simplicity of the HART protocol contributes to its popularity and to the low cost of HART-compatible equipment. This section demonstrates the power, value, and the wide range of HART capabilities and applications. Growth of HART capabilities and applications are certain to continue for many more years.

Readers should use this information to understand the capabilities of HART and to evaluate the field devices and host systems they encounter.

References

1. "HART Field Communications Protocol Specification," HART Communication Foundation, HCF_SPEC-12 (see http://www.hartcomm.org), 2001.
2. Holladay, K. L., "Calibrating HART Transmitters," ISA Field Calibration Technology Committee, http://www.isa.org/~pmcd/FCTC/FCTC_Home.htm, 1996.
3. Holladay, K. L. and Lents, D., "Specification for Field Calibrator Interface," ISA Field Calibration Technology Committee, http://www.isa.org/~pmcd/FCTC/FCTC_Home.htm, 1999.
4. "HART Application Guide," HART Communication Foundation, HCF_LIT-34 (see http://www.hartcomm.org), 1999.

Bibliography

"The Complete HART Guide," HART Communication Foundation (see http://www.hartcomm.org), 2001.

"HART Technical Overview," HART Communication Foundation, HCF_LIT-20 (see http://www.hartcomm.org), 1994.

"Unleash the Power," *Control Magazine,* August 2001.

4.12 Foundation Fieldbus Network

R. H. CARO

The Fieldbus Foundation is the nonprofit consortium responsible for the support and the technical specifications for Foundation Fieldbus. The specifications of the Fieldbus Foundation are based on the ISA/ANSI standard S50.02 and Type 1 of IEC 61158, the fieldbus standards. It is important to note that standards define communications requirements and methods, but do not define an implementation. Foundation Fieldbus is an implementation of these standards for a particular application, and is only the specification for that implementation. Manufacturers will still need to decide on interface designs, and other items to turn Foundation Fieldbus specifications into real products. Finally, since Foundation Fieldbus is intended to be implemented by many suppliers, conformance testing is required to assure interoperability.

HISTORY

In 1985, the ISA standards committee SP50 met to consider the future standardization of the digital link necessary to replace the old analog 4–20 mA signal. At the time, these committee members were aware that every instrumentation company was in the process of considering such a communications link. It just seemed that developing a common link as an international standard made more economic sense than for each supplier to develop a unique communications link. At the time, it seemed like a relatively simple task that should take only a couple of years. Little did they know what would happen.

The first effort in the SP50 standardization effort was to create a functional specification. While producing this document, it became obvious that there were physical cabling problems such as intrinsic safety and power delivery on the data wire for process control applications that had to be separated from the more conventional protocol functions. The committee did this by defining two very different sets of network requirements: H1, for installation on the plant floor as a digital replacement for 4–20 mA DC transmission, and H2, a higher-speed, "backbone" network to link the H1 segments to each other and to host systems. The committee could not think of a proper name for this kind of application classification so the "H" in H1 and H2 stands for "Hunk." H1 was defined narrowly for process control: 4–20 mA replacement including the desirable feature of installation of new digital instruments directly replacing existing 4–20 mA analog instruments retaining the existing wiring. Limits of intrinsic safety and power delivery forced H1 segments to have few nodes and a fairly low speed. The speed was also the result of compromise with the functional requirements for process control.

One of the early debates in discussion of the physical layer was: Shall Fieldbus carry the traditional 4–20 mA signal specified by ISA S50.1, the original standard for analog signal transmission? The final analysis was that to carry an analog signal would require that the digital transmission speed be slower. The committee decided that reduction in the speed of the digital data transmission was unacceptable just to preserve backward compatibility with analog transmission. The digital fieldbus represents a clean break with prior technology except for the wiring plan.

H2 had proposals from several groups, each of which had constructed a similar network at about the same speeds we thought H2 needed. In addition to the need for H2 as the backbone for H1, several PLC (programmable logic controller) suppliers also added to the H2 functional requirements. At the time, it was considered that a properly configured H2 would satisfy PLC suppliers for their common network needs for remote input/output (I/O) and the data network needs as well. The ISA S50.02 standards were completed by 1993 and the 7-year cycle for international standards had begun.

Fieldbus Foundation was formed in 1994 from two previous competing organizations when major U.S.-based process control end users made it very clear to their suppliers that they did not want two different competing networks for process control. One organization, ISP (Interoperable Systems Project), had already made some progress basing its work on the S50 physical layer, but PROFIBUS upper layers. The other group was WorldFIP North America, which favored the use of the full fieldbus standard and used both its physical and data link layers. Users also made it known that they had invested time and effort in the ISA S50.02 Fieldbus standard, and this is what they wanted and expected their suppliers to deliver.

The newly constituted Fieldbus Foundation set about to write an implementation specification around the H1 version of S50.02 to satisfy the process control users. This specification, called Foundation Fieldbus, was completed in 1996 and released to all member companies. It was strongly oriented to solving the 4–20 mA DC replacement problem for process control instrumentation and control systems. In 1998, the Fieldbus Foundation completed work on conformance testing for H1 and announced its "Registration" program. Instrumentation suppliers submit their instruments for a suite of interoperability testing, and if they pass these rigorous tests, they will receive a Foundation Registration Mark:

Fieldbus Foundation had not resolved the H2 network issues. None of the member companies wanted to invest the time and effort to prepare implementation specifications for any of the versions of H2 recommended by the S50.02 standard. They felt them to be too limited, too slow, and too expensive. Early in 1999, the Foundation Steering Committee made the decision to create an implementation specification for H2 based on use of commercial off-the-shelf high-speed (100 Mbps) Ethernet because it was unlimited, inexpensive, and more than fast enough. The project was completed in 2000 with the release of an implementation specification for Foundation Fieldbus HSE (high-speed Ethernet).

With the completion of HSE and products beginning to appear on the market, the full two-level architecture of S50.02 has been completed, although with a physical/data link combination not contemplated in 1993. HSE demands the use of full-duplex switched Ethernet, which itself accomplishes a very subtle correction to the "old" Ethernet—elimination of collisions through the use of active buffers in the switches. Elimination of collisions itself makes Ethernet fully deterministic, and removes the annoying minimum message length requirement that was necessary to allow time to detect collisions on long cables. Now the H2 can be implemented with any variety of Ethernet apparatus including repeaters and especially with any of the fiber-optic Ethernet cable extensions found in communications catalogs. This is very important as one of the primary applications for H2 is the replacement of the "home run" cable connecting field instrumentation to the control room. A single fiber cable with HSE may be as much as 6 km in length to connect many thousands of H1 instruments in the field. Figure 4.12a shows the architecture for the full two-level Foundation Fieldbus conforming to the S50.02 specification, but with HSE replacing one of the standard H2 physical layers.

The two-level architecture of Foundation Fieldbus is designed to support control in the field device such as the control valve positioner and field instrumentation. The architecture even supports cascade control links between field devices and control loops in dedicated controllers, wherever they are located. Finally, the architecture supports control only in field devices with no control room equipment should that be required.

Finally, the interface between Foundation Fieldbus and control systems is defined very loosely in the S50.02 standard,

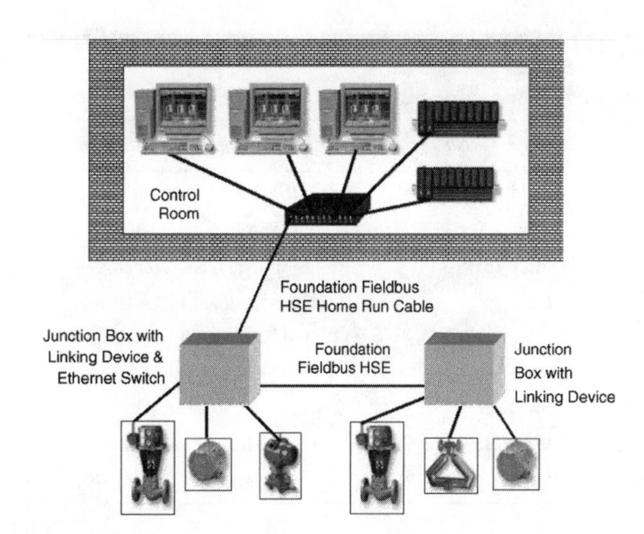

FIG. 4.12a
Foundation Fieldbus two-level architecture.

with only the functions at the application layer defined by the Fieldbus Foundation specifications. Interoperability of host systems is now defined by a test called HIST (Host Interoperability Support Test), a specification of the Fieldbus Foundation. Systems passing HIST are registered with the foundation and are most likely to interoperate well with the full two-level architecture of Foundation Fieldbus.

BASED ON ISA/ANSI S50.02 AND IEC 61158 STANDARDS

The ISA (Instrumentation, Systems, and Automation Society) is a standards-making body operating under ANSI rules, which in turn follow the rules for international standards established by the United Nations. The S50.02 standard was completed in 1993 and held for coordination with the international (IEC) standard for several years. Eventually, the international standard IEC 61158 was completed at the end of 1999 and it contains the total S50.02 standard as one of eight types.

As mentioned above, a standard is not a description or specification of an implementation. The purpose of a standard is to give a set of requirements and a methodology in enough detail that conformance to the methodology can be measured. S50.02 and the corresponding Type 1 of IEC 61158 are such requirements and methodology documents. The physical layer, data link layer, and application layer of the ISO 7498 seven-layer stack are fully specified in this format. When the IEC standard in this form was balloted, it was defeated because strong and influential supplier interests were greatly concerned that approval of such a document would make their existing digital communications buses obsolete. The compromise was to allow a few suppliers actively participating to include specification documents closely following the existing digital communications buses in popular use at the time. With the compromise, the fieldbus standard passed at

TABLE 4.12b
Source of Specifications for Types of IEC Fieldbuses

Type	*Name*	*Controlling Source of Specification*
1	Fieldbus	ISA SP50 standards committee
2	ControlNet	ControlNet International
3	PROFIBUS	Profibus International
4	P-Net	International P-Net User Organization
5	Foundation Fieldbus HSE	Fieldbus Foundation
6	SwiftNet	Boeing and Shipstar
7	WorldFIP	WorldFIP Organization
8	Interbus	Interbus Club

the international level. This now gives each type the recognition as an international standard, but also makes it more difficult to change, as it is a standard. Table 4.12b gives the origin of each of the "types" of IEC 61158 fieldbus standards.

Notice that there is a difference between Type 1 and Type 5 in terms of its support and the very nature of this standard. Type 1 reflects the original standards development, and the work of the Fieldbus Foundation in developing H1 specifications, as well as the pilot programs of various suppliers that have influenced the standard, but the contents is a pure standard. Type 5 is somewhat different in that it reflects almost completely the work of the Fieldbus Foundation HSE specification cast in the language of the standard. In this respect, Type 5 is much more like the specifications for PROFIBUS, ControlNet, P-Net, WorldFIP, and Interbus that reflect the implementation specifications for those networks.

FIELDBUS ARCHITECTURE

Foundation Fieldbus is a two-level bus architecture:

1. H1 is the low-level bus interconnecting field instruments.
2. H2 is the high-level bus interconnecting H1 segments and linking to host-level services.

Both bus levels run the same protocol and perform the same services. It was only the electrical reality of being a 4–20 mA DC replacement that has required this two-level bus architecture. The basic differences are found in the physical layer, although the data link layers are also quite different. Bus architecture is probably the biggest difference between Foundation Fieldbus and the S50.02 standard on which it is based.

In S50.02, the H2 bus was different from the H1 bus only in the physical layer. The data link layer was identical for both H1 and H2. The arbitrary division of bus functionality along the lines of differences in the physical layer is also found in other automation bus structures, such as the division between PROFIBUS-PA and PROFIBUS-DP. However, the underlying S50.02 standard does not require this dual bus structure, it only allows it.

When the Fieldbus Foundation decided to base its H2 bus on high-speed Ethernet, several compromises were necessary. Use of Ethernet within the H2 bus architecture meant that a unique physical and data link layer was no longer possible as those of Ethernet would be used. The decision was made, however, to capture all the data link layer functionality by implementation of a complete protocol suite using the TCP/IP (transmission control protocol/Internet protocol) stack commands.

The scope of S50.02 and Foundation Fieldbus is far more than simply to replace the analog 4–20 mA DC transmission signal with a digital data path. The mission was to fully enable control in intelligent field instrumentation and smart actuators. One of the problems fieldbus had to solve was the cascade control problem—how to link control functions between different stations or nodes of the fieldbus network. It is not enough simply to link data between the elements of a cascade control scheme, the data must be made available with the same kind of synchronization that the control loop would have if it were implemented in the controller of a DCS (distributed control system). In other words, Foundation Fieldbus is designed for control in field devices—the very same kinds of control that have been implemented in pneumatics, analog electronics, DDC (direct digital control) computer control, and in DCS. The complexity of the fieldbus data link layer is all designed to implement critical final loop control as though it were not distributed. Furthermore, the complexity is designed to be hidden from the end user and left to the suppliers building these complex cascade control loops with portions in field devices and other portions in dedicated controllers. The complexity of Foundation Fieldbus field control is achieved with a layer above the application layer called "Layer 8" or the user layer. The primary building block of the user layer is the predefined "function block," a common term in process control systems since the concept of "block diagrammatic control"[1] was invented by George Woodley of the Foxboro Company in 1966 and first implemented on the PDP-7 computer.

Block diagrammatic control is based on the control diagram familiar to textbook courses in process control and servomechanism theory. The assumption is that the process itself is controlled by a set of manipulated variables and also responds to inputs from uncontrolled or environmental factors that may, or may not, be measured. We make another assumption that steady-state control of the process can be accomplished by the process of regulation of these manipulated variables using a conventional three-term PID (proportional/integral/derivative) controller with feedback from the controlled variable. When the uncontrolled variables have a significant effect on the stability of the process, and they can be measured, feedforward control can also be used assuming that the process can be modeled with relatively simple digital simulations of a dynamic lead/lag function with perhaps a dead-time function. Control schemes, called strategies, can be built by graphically constructing the control loop diagrams with a format specified by ANSI/ISA-5.4-1991—"Instrument Loop Diagrams." Vendors of Foundation Fieldbus instruments and systems supply strategy building software based on these graphical diagrams.

With Foundation Fieldbus instrumentation, smart control valve positioners, and intelligent variable-speed motor drives containing function blocks, both signal conditioning and final loop controls are moved out of the DCS controller. With H1 terminating in a controller, each H1 segment terminates from 2 to as many as 32 instruments in a digital interface replacing from 2 to 32 analog terminations. When Foundation Fieldbus HSE is used to bridge H1 segments, even the H1 terminations in controllers are no longer necessary. Additionally, there are many simple control systems not needing a DCS controller at all. However, for many complex feedforward control loops and for more conservative control engineers, considerable control will continue to be assigned to DCS controllers for many years. The economic effect is inescapable, however; movement of the final loop controls and the analog interfaces out of the DCS controller reduces the cost of the total control system and increases its performance.

Physical Layer

The ANSI/ISA 50.02 and IEC 61158 Part 2 standards are followed by both Foundation Fieldbus and Profibus-PA in order to replace 4–20 mA DC connections with a moderate-speed bidirectional digital data link that can support intrinsic safety and supply power to field instruments using the typical instrumentation cable previously used for analog instruments. This data link was explicitly designed to operate at the highest speed thought possible in 1990 with the constraints of power, intrinsic safety, distance, and for the type of wire specified. The compromise speed specified is 31,250 bps. If an analog signal had been specified, the data rate would have been greatly reduced, so the analog option was not included.

One of the design criteria was to use conventional analog instrument cable, twisted shielded pair (STP) cable of indeterminate capacitance. Cable capacitance, resistance, and dielectric properties all affect the maximum length, speed, and signal-to-noise ratio. It would have been possible to allow the speed to vary with cable characteristics, but the standards committee chose to fix the speed and work against a maximum length specification and to establish rules for cable termination and topology. In fact, considerable experimentation has proved the very conservative nature of the standards and the Foundation Fieldbus specifications based on these standards. It has usually been possible to exceed the maximum cable length by more than 100% without affecting detectable errors by using low-resistance (larger-diameter) wire with low capacitance. Such wire is made by several cable suppliers and marketed as "Fieldbus" wire. The original specifications in both the standards and in the Fieldbus Foundation specify a very conservative number for the intrinsically safe field instruments that can be terminated on a single Foundation Fieldbus H1 segment. These specifications are based on worst-case current draw by every unit. In fact, every unit will have a true worst-case current draw specified that is typically lower than the assumption upon which the standard was based. A very recent change called the FISCO specifications has been accepted by the standards committees and is being written into the standards and into the Foundation Fieldbus specifications. It has been in use by Profibus-PA for several years. FISCO changes allow the actual current draw for each node to be used to calculate the total current draw for each H1 segment, and the total is not to exceed the previous assumed current draw. The net effect is that the maximum number of stations for an intrinsically safe H1 segment can safely be increased from about 4 to 6 nodes to a more realistic 10 to 14 nodes without adversely affecting intrinsic safety.

The "bus" concept for Foundation Fieldbus is followed, and indeed it can be wired as a multidropped bus with spurs as illustrated in Figure 4.12c. In practical plant designs, however, bus wiring is often impractical for maintenance reasons. Therefore, the Foundation Fieldbus wiring specifications provide for a hub-and-spoke form of wiring, which the standards committee referred to as a "chickenfoot," illustrated in Figure 4.12d.

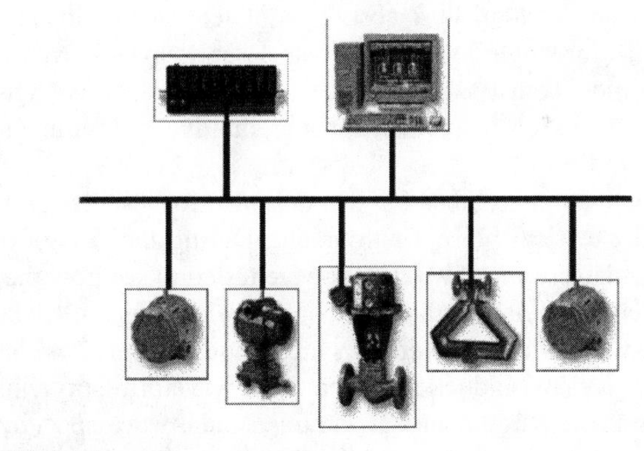

FIG. 4.12c
Multidropped bus.

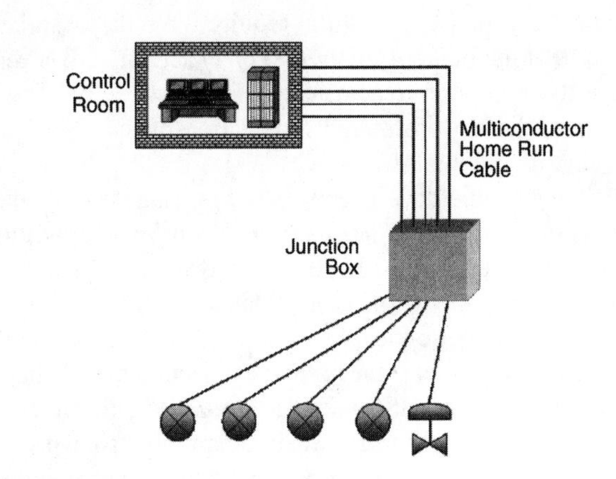

FIG. 4.12d
Hub and spoke or chickenfoot wiring of H1.

The hub of the chickenfoot is a junction box where bus power is usually supplied.

How long can an H1 fieldbus segment be? The original specifications called for maximum cable length of 1600 m, which combined with the capacitance problems previously noted, the speed selected, the wire resistance, and the expected signal-to-noise ratio of the modulation waveform, was a complete design. Obviously, use of lower-capacitance and lower-resistance wire has a desirable effect of increasing maximum cable length. Several experiments have shown that use of fieldbus cable from several suppliers allows maximum cable lengths two or three times the specified length.

It is also possible to insert fieldbus repeaters to increase bus length. A fieldbus repeater may be a simple analog signal amplifier that can be expected to increase cable length by 50 to 100%, but can only be repeated a maximum of two times before the signal is distorted too much for error-free performance. A more expensive digital repeater can also be used to double the cable maximum length each time the signal is repeated. Eventually, the delays introduced by the buffering and digital repeating will cause timeout conditions on that bus segment. Finally, various fiber-optic cable extenders can be used both to increase bus length and to provide galvanic isolation between plant grounding planes. Conventional 50/62.5/125 μm multimode fiber cables using light-emitting diode (LED) sources will allow several kilometers for each fiber segment at modest cost. Use of telecommunications grade 5 μm single-mode fiber with laser sources currently allows more than 100 km for each segment, and passive light amplifiers for long distances, but at prohibitive cost for industrial automation.

Foundation Fieldbus H1 was designed for installation in a high electrical noise environment. During the design of the standard, several waveforms were tested to see how they survived the noise in the heart of a steel rolling mill with thousands of horsepower of synchronous AC, DC brush motors, and AC induction motors. The waveform surviving best in this environment was the trapezoidal wave specified by the fieldbus standards and the Foundation Fieldbus specifications. Additionally, Manchester biphase encoding is specified so that common impulse noise can be rejected. The Manchester biphase encoding required by the standard requires that the binary symbol (a 0 or 1) actually cross zero voltage twice in order to be detected. Figure 4.12e illustrates the trapezoidal waveform and modulation of the signal for Foundation Fieldbus H1.

Foundation Fieldbus HSE is based on standard, off-the-shelf 10/100BaseT Ethernet. Entire books have been written on the construction of 10/100BaseT networks, the standard of the local-area network (LAN) industry. This too can be connected with fiber optics, where it is called 10/100BaseFx. 10/100BaseT Ethernet is the normal connection to the Internet as well, allowing HSE messages to traverse the Internet as easily as the local wiring. Again, many books have been written on making such networks both robust and secure.

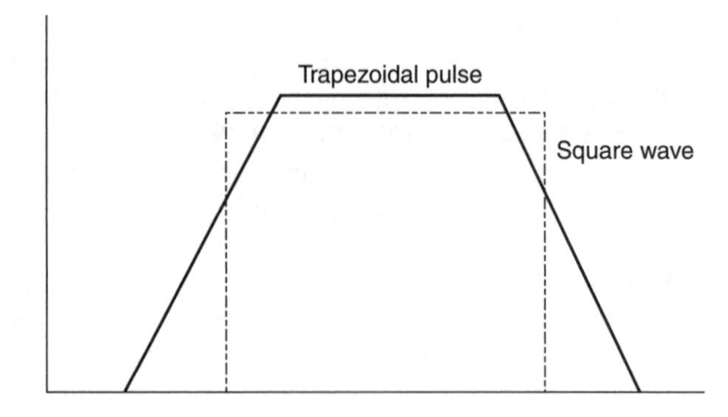

FIG. 4.12e
Comparison of trapezoidal pulse vs. square wave.

Type 5 of the IEC 61158 standard and the Foundation Fieldbus HSE specifications do not specify anything at all about the HSE physical layer. Since these documents were written, there has been a body of work to specify modest improvements to the Ethernet media to qualify it for use in the factory. This work appears as EIA/TIA 42.9 standards. The EIA/TIA specifies Category 5 cable for use in 10/10/100BaseT Ethernet cabling. A new cable specification called Category 5E has been defined for use in factory environments and is recommended for all industrial automation applications. Additionally, EIA/TIA 42.9 specifies a slightly ruggedized form of the RJ-45 plug/jack combination for cable bulkhead and environmentally sensitive applications. It also specifies an M12-style round plug/jack for high-vibration applications also requiring environmental protection.

Wireless connections for Foundation Fieldbus are not yet being offered, but appear to be interesting to a significant number of users. Most of the discussion of wireless currently relates to the use of wireless Ethernet, or IEEE 802.3b, to be used in applications where Ethernet cable is needed but may be difficult or impractical to run, for example, crossing a public roadway. It is not beyond comprehension for forms of wireless to be used to connect field instrumentation otherwise using Foundation Fieldbus H1, but finding a source of electric power for the field instrument would be necessary. All of these wireless applications would be expected to use Foundation Fieldbus HSE.

An emerging technology suitable for field instrument connection is Bluetooth, soon to be standardized as IEEE 802.15a. The appeal of Bluetooth is that it is designed for low power consumption, is used in the commercial market in thousands of units, uses a fast error-correcting interface, and is designed to operate in the presence of high electrical noise. The application to Foundation Fieldbus connections as a new physical layer is not currently offered by any supplier. If the problem of electrical power in the field instrument is solved inexpensively, then Bluetooth technology may be used to connect field instruments without wires.

Data Link Layer

The S50.02 standard provides methods supporting three different architectures for sharing data traffic on the bus:

1. Master/slave—Also called polling. A master station continually queries each slave station for new values.
2. Delegated or free token passing with lost token recovery—Stations on the bus receive bus mastership for a dedicated time period (token) and pass the token to the next station in a declared order so that all may be bus masters for a brief time.
3. Bus arbitration—A bus master determines which station shall be bus master next and for some time period according to some dedicated set of rules.

It should be noted that the data link layer for Foundation Fieldbus H1 provides many functions normally associated with ISO protocol layers 3 to 7. This was done on purpose to avoid the inefficiencies of separating these functions into many software layers. When Foundation Fieldbus HSE was created, it was still required to perform the identical functions of H1 but not implemented in the data link layer. The following discussion of link active scheduler (LAS) and publish/subscribe are just as appropriate for HSE as they are for H1, but are implemented differently through the functions of the TCP/IP communications suite. The Foundation Fieldbus application layer, however, is identical for both H1 and HSE so that the user need not be concerned about the technology used at layers 2 to 6.

Link Active Scheduler Foundation Fieldbus uses a combination of bus arbitration and delegated token for sharing the fieldbus segments. Each Foundation Fieldbus segment requires at least one LAS, which is the bus arbitrator. Each station needing to transmit a message notifies the LAS that it needs the bus, and the LAS will grant that request according to the set of rules established for that bus segment. Often, there will be more than one LAS in each bus segment, where the first LAS to begin communicating will be the master. All other LASs monitor the behavior of the LAS master, and if it falters, the second LAS to begin communicating will take over as the bus master. An LAS may also transfer mastership to another LAS using the token passing method.

Use of the LAS is similar in Foundation Fieldbus and WorldFIP, but no other common communications bus. Using bus arbitration overcomes one of the weaknesses of ordinary token passing, recovery from lost tokens that just do not occur in Foundation Fieldbus. Bus arbitration allows Foundation Fieldbus to schedule the token holders for exactly the time they need and at the time they need it. This may be called "smart tokens" as opposed to the blind token passing of other buses, which pass tokens to stations whether they need it or not. As the LAS schedule is determined by the configuration program, the bandwidth of each H1 segment is predetermined, providing a highly focused quality of service (QoS) not possible with other networks in which the QoS is determined dynamically using a priority mechanism.

The LAS provides deadline scheduling for all scheduled activities. This is a scheduling mechanism in which the completion of data transfers are scheduled, rather than their beginning. It is only practical in networks such as Foundation Fieldbus where the size of data packets are known, and allows the events that are dependent upon a data transfer to be scheduled when the required data are received. Control block execution is one of those events that are dependent upon the exact timing of the receipt of data. The deadline schedule accounts for the network anomalies such as transportation delay in case the data are not on a local segment, but must be received from another segment through a bridge arrangement. This is only possible in fieldbus networks where the bridge protocol is fully specified and participates in LAS scheduling.

Publish/Subscribe Data Distribution Foundation Fieldbus uses the basic data distribution mechanism of ANSI/ISA S50.02 and Type 1 of IEC 61158 fieldbus standards data link layer, which is called Publish/Subscribe. When the fieldbus network first starts, there are no scheduled data transfers. Each function block or another processes operating on the distributed nodes of the network needing data for its calculations "subscribes" to the data by tag name and specifies the frequency at which it is required. This creates or adapts the publishing schedule at the node owning the data. The same data may be published at different intervals depending upon the demands of each subscription. Data published by the data owner may be any data structure of the data at that network node. Data are typically organized into either buffers or queues. A queue is a time-ordered data set with the most recent item at the top of the queue with each element time-stamped. Reading a queue empties the queue. If the queue should become full, the oldest data are lost to make room for the most recent data. Queues are usually used for the latest alarms. Buffers are sets of data organized into a data structure all sampled at the same time and also includes a time stamp. The buffer may have multiple data sets included, but publishing a buffer only sends the most recent data set and clears it from the buffer. Polling the buffer sends all data, but does not empty the buffer. When buffers become full, the newest data are overwritten so that the oldest unpublished data are not lost to aid in process diagnostics.

Processes subscribing to a data set have an expectation of receiving that data on time, and are awaiting its arrival. If the data are not published, an exception is created. In the case of function blocks, the exception may be used to open active control loops or take any other programmed action appropriate for that control. Because the arrival of data via publishing is highly synchronous, it is the signal used to complete the computation of a time-dependent operation such as a PID control loop.

Application Layer

The Foundation Fieldbus application layer is a very specific implementation of the generic application layer contained in ANSI/ISA 50.02 and Type 1 of IEC 61158 parts 4. Typically, the end user does not see the application layer at all, but it is used by the user layer function blocks for their communications to the data within the nodes. For completeness, all the functions of FMS (fieldbus messaging services), the application layer of PROFIBUS, is included and is available if any custom applications are necessary. Additionally, all the interfaces necessary to enable the data transfer functions of the H1 data link layer are supported.

The simplest of the application layer functions is the READ and WRITE data features that can be used to poll buffers and queues established in the remote devices. Commands are available to ESTABLISH and DISESTABLISH buffers and queues to support the read/write service. The establish command specifies the structure and source of the data to be buffered or queued, the name of the buffer or queue, the buffer or queue selection, and the update frequency. Since all fieldbus data refer to function blocks by tag name, the variables are identified by the name of the database entry for that variable. For example, if the function block has been set up for control, then a buffer may contain data for the set point, process variable, output, and status of the PID function block. If the block name is FC101a, then the set point will be referred to as FC101a.SP, and the process variable as FC101a.PV. Succeeding READ statements would then specify the buffer name and would be data access by polling. The same data item may be established in as many different buffers as desired so that it may be part of several different published data sets.

Publish/subscribe is a connected service, in the framework of the application layer. Once the buffer or queue is established, it becomes available for publishing. Applications such as an HMI (human–machine interface) may SUBSCRIBE to a buffer by name. If there have been no prior subscribers, then the data owner will begin to publish it according to the parameters established when the buffer was created. This may be a cyclic update at a declared frequency, an exception update when the value has a significant change, or the combination that will update periodically, even when there is no significant change.

Function blocks are actually remote procedures to the application layer. Utilities of supplier systems download the block database, initialize the procedure, and make the function block ready to run, but do not start it with the functions of the application layer. Typically, HMI software uses the WRITE function to change the MODE variable of the function block to change its operating state, after it has been loaded and initialized.

Mapping to Data Link Layer for H1 The Foundation Fieldbus application layer is implemented by mapping each of its commands to lower-layer functions. In the case of H1, that lower layer is the data link layer with all of its functionality. Many times the application layer command is simply translated to the same command of the data link layer, whereas, at other times, a series of data link layer commands must be used. This is called "mapping" of the commands between layers. The details of this mapping are not of any real importance to the user, except to know that it is accomplished by the system software. The fact that mapping of application functionality to the lower layers is done by the application layer is important to the creator of applications since Foundation Fieldbus has more than one lower layer. It is the application layer software itself that makes the determination of which lower layer is mapped, and it is not the responsibility of the application. This allows applications to run exactly the same independent of which lower layers are actually used in any system.

Mapping to TCP/IP and UDP/IP over Ethernet for HSE When the layer below the application layer is Foundation Fieldbus HSE, the application layer maps all its commands to a complex set of messages passed across the Ethernet network using Internet protocols. The most efficient protocol in the TCP/IP suite is called UDP (universal datagram protocol), which also has the least functionality. UDP is especially suited for fieldbus functions as it does not provide, nor is it burdened with, the hidden features of TCP that was designed to operate across the World Wide Web implemented on the Internet. The Internet is a "mesh network," as described in Section 4.9, in which there are multiple paths between any two nodes of the network. The linking of the Internet uses adaptive routing technology to "find" the best path through the mesh connections between source and destination. Errors in routing are frequent and recovery is planned, but is hardly real time.

When HSE was being designed, it was recognized that the following truths allowed a simpler method of message encoding to be used when Ethernet data link layer and physical layer were to be used under the application layer:

- There was almost always a single path between source and destination.
- Error detection was provided by a strong algorithm, the 32-bit cyclic redundancy code of Ethernet.
- Most messages were short and no message would require segmentation.
- Distances in most cases were in-plant and short in terms of time delay.

Encoding all the application layer commands into UDP messages provides efficient data transfer without the adaptive routing of TCP. UDP message frames were created to provide the ID field that identifies the contents of published messages sent using the general IP multicast protocol. In this way, stations receiving the IP multicast message can quickly identify if there is a need for that published data for an application in that node. The specific encoding used within the UDP messages is of little importance except to the programmers of the systems software. Even programmers of application software for Foundation Fieldbus devices will code programs to the Foundation Fieldbus application layer commands and will not need to know which type of encoding or physical layer is being used. The application layer shields the software from the knowledge of the lower layers, allowing the same application to be used on any Foundation Fieldbus system.

User Layer

The ISA SP50 standards committee began by clearly identifying the process control applications that would be implemented in field devices. This became known as "Layer 8," the process control user layer, but was never completed as a standard. The process control user layer has been issued by the ISA as a Technical Report,[2] to capture the work and serve as a seed for future standards development. Many details of the implementation plan of this report were not used by the Fieldbus committee, but the embedded function block model has served as the basis for the Foundation Fieldbus function blocks and as the basis for the international standard still in progress by the IEC in the SC65C/WG7 committee eventually to be published as IEC 61804, "Function Blocks for Process Control." The corresponding function blocks for discrete automation was proposed, but was never drafted.

The most notable difference between the ISA TR50.02 and Foundation Fieldbus is in the implementation of the schedule for function blocks. The underlying LAS of the network itself provides an exact scheduling mechanism for the distributed function blocks, removing the necessity for the user layer to provide any block scheduling. Small differences also exist in the algorithms of the function blocks, but the basic mechanisms of block initialization, status, and cascade structure are identical.

Field Control In the early 20th century, feedback loop control was distributed to individual panelboard-mounted instruments, and occasionally to field-mounted equipment. Any cascade linkages were directly piped (pneumatic) or wired (analog) between the controllers. Control was simple, but installation was expensive, hard to tune, and operators often found better process stability on the manual setting—effectively disabling control. Operators generally could supervise about 30 to 50 control loops, typically by scanning the panel every few minutes looking for excessive deviation between set point and the process variable being controlled.

The improvement made to dedicated loop control was to substitute a digital computer to provide several loops of control, and this was called DDC (direct digital control). Early DDC systems were used mostly for economic reasons—the digital controller cost less than the loop controllers it replaced. However, few companies would risk operations on this computer control without 100% analog backup, effectively removing most of the cost justification and greatly adding to the cost of the system. Eventually, DDC provided a platform for development of more complex controllers not really practical or even possible with analog control. Computer control did not become too popular, but was used in the steel, aluminum, chemical, and petroleum industries.

The DCS (distributed control system) was introduced in 1976 and by 1980 had become the preferred control strategy for practically all process control applications. DCS took advantage of the microprocessor essentially to do DDC for a small number of loops in a self-contained controller, with many controllers in each system. It also introduced networks into control systems, primitive networks at first, but eventually high-performance networks to link controllers to each other and to the HMI units serving as the operator consoles. The problem was that early DCS could not depend upon data access timing between controllers, and did not allow cascades links to be built between controllers.

The work of the SP50 process control user layer was directed to take the technology of the DCS and relocate it into intelligent (contains a microprocessor) field instruments (also called sensors) and control valve positioners (also called actuators.) It was always understood that there would be the fieldbus to link instruments and control valves together. The problems associated with synchronously transporting a process variable from a sensor to the location of the controller at exactly the right time were solved by LAS logic.

For many years, field instruments have made process variable measurements and converted them to a dimensionless fraction of scale represented by either a 3 to 15 psig pneumatic signal or a 4–20 mA DC analog electronic signal. With the availability of a microprocessor in field sensors, transmitters, and actuators, S50.02 and Foundation Fieldbus specifications call for the transmission of digital data in scaled engineering units. Furthermore, the raw data from the sensor are processed in the same way as DCS performs signal processing. The essentials of raw signal processing are as follows:

- Smoothing, usually using an exponential smoothing or a moving average algorithm
- Linearization; square root extraction for orifice flows
- Conversion to engineering units
- Alarm limit checking for various limits

One important concept in Foundation Fieldbus is that the scan time for sensor devices is independent of the processing time for control loops, which is usually not the case for DCS. Often sensors/transmitters scan their inputs at least twice as fast and often ten times faster than the signal is needed for control loop processing. This allows quality smoothing to occur without delaying the process variable. Because sensor inputs are processed by an AI (analog input) function block (so called AI even if the actual device is purely digital), each function block has its own scan processing time cycle.

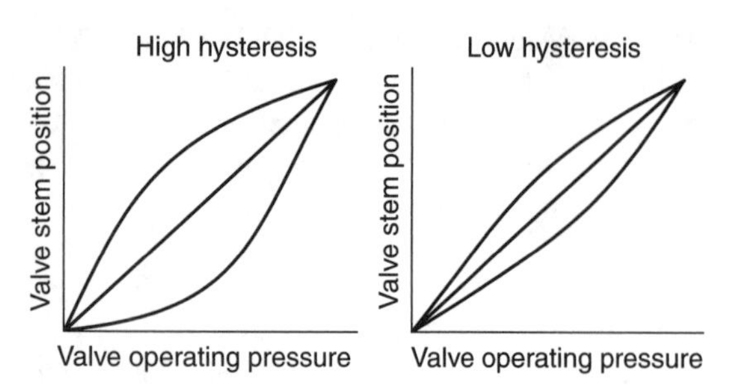

FIG. 4.12f
Illustration of the effect of hysteresis.

The logical field device for the location of the PID control loop function block is the control valve positioner. Process control valves are highly nonlinear devices in which there is an electronic or pneumatic servomechanism to move the valve plug to a desired position determined by the output of a PID controller. Although a retransmitting slide wire for the valve stem position has long been available for panel-mounted controllers and DCS, it has rarely been used within a control loop. Control valve positioners are classically proportional-only controllers, or at least close to that. After a period of use, control valves exhibit hysteresis, as illustrated by Figure 4.12f. Hysteresis, often called valve stickiness or sticktion, eventually becomes so bad that the valve appears unresponsive to the controller signals, causing control loop sluggishness. If the controller is tuned for higher-speed response, the control loop may become unstable. The problem is the unresponsiveness of the control valve due to hysteresis.

Using the same microprocessor that is used for Foundation Fieldbus function blocks, modern control valve positioners can use the full PID algorithm for their local servopositioning loop and can also include compensation for hysteresis, which can only be measured at the valve itself. Once the control valve becomes responsive to the position output of the primary loop controller, overall control performance is improved. This is one primary example of the potential for increased performance of control in the field using Foundation Fieldbus.

Configuration of process control function blocks is done without necessarily knowing where each function block will be executed. As long as the function blocks are selected from the standard set of 29 Foundation Fieldbus function blocks, they can be executed on any of the devices of the network where it makes sense to run that function block. What does not make sense is to run the AI block at a node that does not have that primary input. It also does not make sense to run the AO (analog output) function block on a node without a control valve or a variable-speed drive. On the other hand, the PID function block may be run on any node of the network, even if it has no inputs or outputs.

Function Block Diagram Programming Control strategies are drawn graphically to describe how they are to work. Process control theory is drawn directly from servomechanism theory, and servo loops have always been drawn graphically as well. There are even ISA standard paper forms for engineers to draw control loop diagrams. The engineering step used to construct control strategies for all process control systems is called "configuration," although in modern language it can also be called "programming." Configuration of process controls is done by selection of the proper function blocks, linking AI blocks to supply the process variable values, the latest set point value perhaps from another loop's output, and the valve/actuator output to an AO block.

Function block diagramming (FBD) is one of the standard languages of IEC 61131-3, "Programming Languages for Programmable Logic Controllers." While the title cites PLCs, the language is perfectly general and is included mostly because it is the graphical programming language used to build control strategies. Early DCSs knew nothing about this standard, but had proprietary ways to draw and define control strategies. Most of the modern DCS and PAS (process automation system) are now using standards conforming FBD languages for building process control strategies.

To build a strategy with most of the FBD programming software, it is necessary to select a function block from a template library, drag it to a screen position, then click on the block to get a drop-down menu for configuration of the block details—each block in the linked chain of the control strategy, starting with the AI block to process the process data, any calculation blocks necessary, the PID control block, and finally the AO block to deliver the desired position to the control valve. Arrows are drawn from the data source to the inputs of each function block, and from each output to the input of the next function block. Most graphic software packages allow the user, who is not a graphic artist, to locate function blocks in any pleasing way and to route arrows automatically for best effect.

One of the problems with the FBD form of programming is that there is no universally accepted "source language," as a source code is not specified in the standard. Fortunately, control block diagrams are themselves familiar to control engineers, and are easily copied between systems. Foundation Fieldbus specifies that the field instrumentation hardware in a fieldbus system itself have a transducer block. The name "block" is unfortunate as transducer blocks do not themselves behave like function blocks as they are bound to the actual hardware locations. There are also "resource blocks" associated with the field instrumentation hardware devices. Both transducer and resource blocks do have a static data structure defining the source of process variable data and the valve output for control. Note that AI function blocks might ordinarily be considered always to operate in the corresponding measurement transducer, but this often is not the case.

Function Block Library One of the most important utilities of the Foundation Fieldbus user layer is found in its rich library of function blocks. There are ten basic function blocks required to be supported on all systems, and on the appropriate devices. For example, the AO function block is intended

TABLE 4.12g
Foundation Fieldbus Basic Function Blocks

AI	Analog input
AO	Analog output
B	Bias
CS	Control selector
DI	Discrete input
RA	Ratio
DO	Discrete output
ML	Manual loader
PD	Proportional/derivative
PID	Proportional/integral/derivative

TABLE 4.12h
Foundation Fieldbus Extended Function Blocks

ISEL	Signal selector
SPLT	Splitter
DT	Dead time
LL	Lead lag
CAO	Complex analog output
DC	Device control
AALM	Analog alarm
DA	Discrete alarm
ARTH	Arithmetic
PUL	Pulse input
CDO	Complex digital output
CALC	Calculation block
AHI	Analog human interface
DHI	Discrete human interface
SG	Set point generator
INTG	Integrator
CHAR	Characterizer
STEP	Step control

to be supported on actuator devices, but need not be supported on sensor devices. The intent is that suppliers support the basic standard function blocks and the extended set as well. Suppliers are certainly welcome to add new function blocks to augment the basic and extended set as well as to provide many new functions never before supported by standard devices. Table 4.12g lists the basic function blocks, and Table 4.12h lists the extended function blocks.

Because basic function blocks must be supported on all systems, the parameters or attributes must be known to all systems. Foundation Fieldbus provides a method for the complete definition of a function block in terms of its attributes: the data type, name, and the maximum and minimum values if appropriate. This is contained in a DD, a data definition block, one for every type of Foundation Fieldbus function block, basic, extended, or custom. In this way, suppliers can develop new function blocks that anyone can use without knowing anything about the internals and implementation details of that function block. The only other item required to use a function block is the description of the contained algorithm or transfer function of the block, and each of the block attributes or parameters in a descriptive document. The basic and extended function blocks are documented by the Fieldbus Foundation in the function block specifications, and by the block supplier in its function block specifications.

Analog Input The basic input block of process control. The data source for the AI block is the data input value of the transducer block of the sensor device. The AI block is usually located in the same device as the transducer block, but this need not be the case. The purpose of the AI block is to perform the signal processing on the basic process variable signal received by the transducer block. The first step is to determine from the transducer block if there is any active fault with the process variable. If so, then the status is set to BAD indicating one of the eight possible causes of failure identified by the transducer hardware, and no further processing occurs. The next step is to perform the smoothing or filtering operation using the static parameters of the exponential smoothing algorithm, or the moving average algorithm, if selected. The next step is to convert the filtered process variable value to engineering units using the conversion equation specified at configuration and the static variable values of the conversion equation. Temperature variables are converted using a table lookup method with linear interpolation between points of the table. Finally, a whole series of limit tests occur using both dynamic and static limits as shown in Table 4.12i. A deadband value is applied below high limits and above low limits and also to the rate of change limit, so that values close to these limits do not cause the

TABLE 4.12i
Alarm Limit Testing in the AI Block

Rate of change limits	Current value is more than a specified number of units more or less than last reading (*note:* applied after smoothing)
High range limit	Reading exceeds the maximum value of the transmitter
Low range limit	Reading is less than or equal to the minimum value of the transmitter
High-high limit	Value is higher than the high-high limit value
Low-low limit	Value is lower than the low-low limit value
High limit	Value is above the high limit value, but less than the high-high limit
Low limit	Value is below the low limit value, but above the low-low limit
Alarm deadband	A specified amount used to prevent alarms from "chattering" as the PV varies close to an alarm limit

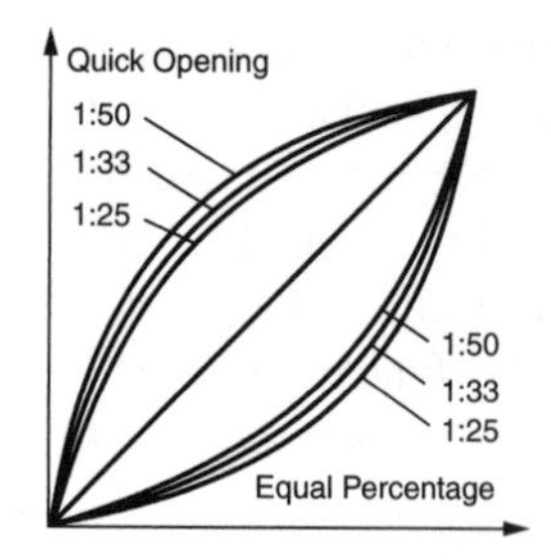

FIG. 4.12j
Control valve configuration options.

status to show limit violations with small changes of the actual value.

Analog Output The basic signal processing for actuator outputs. Control blocks such as PID generate a positional signal for the output in dimensionless terms, between 0 and 100% valve opening. Inherently, control valves are nonlinear devices in which hardware cams used to be inserted to make them work with a linear opening signal. The AO block provides a series of "electronic CAMS" for control valves of different types, as illustrated in Figure 4.12j. The input for the AO block is usually the output of a control block. The output of the AO block is to a transducer block of the actuator device, most often a control valve positioner.

Several alarms can be raised by an AO block as well. When the actual feedback from the control valve positioner indicates that the valve stem position is either full open or full closed, then the control valve is said to be in "saturation," meaning that the controller is really not doing any good by expecting additional corrective action by the valve. Although it is possible with the Foundation Fieldbus PID function blocks to configure them not to wind up, the better solution is to correct the condition causing saturation.

Process control valves have another characteristic called hysteresis or valve stickiness, which means that the valve will have a different stem position while it is opening, than it will have while it is closing. This is illustrated in Figure 4.12f. New valves usually exhibit small amounts of hysteresis, but older valves can have a considerable amount of hysteresis. The effect upon the PID controller is to continue to try to put out higher and higher values as the integral term causes wind-up. Although it is possible to limit reset wind-up with some PID options, the instability caused by hysteresis is one of the primary causes of control loop instability. Most of the Foundation Fieldbus control valve positioners today report on the hysteresis of the attached control valve as one of the functions of the AO block.

Bias A simple arithmetic addition or subtraction function.

Control Selector Used when two different control strategies must be configured for the same control function, but based on different conditions. The selection of the control strategy to be used may be from the process operator, a batch recipe, or based on a process condition. One situation in which the control selector would be used is for heating and cooling of exothermic reactions. During the early part of such a process, it is necessary to heat the contents of a reactor requiring a strategy to control steam heat flow to a reactor jacket. When a certain point in the reaction is reached, the reaction generates heat and now requires water cooling in the reactor jacket. These two strategies are similar, and are based on the same temperature sensor, but demand very different control tuning and output valves. The control selector is located after the AO for the reactor temperature and routes the PV to the correct controller for the selected PID. In this case, the unselected PID is placed in Manual.

Discrete Input A simple function block to allow up to 16 binary bit values to be read. When the 16 bit values are configured, each may be given a tag name and treated as an individual True or False logical state value. It is also possible to configure any group of these bits as a single scalar value that might come from a thumbwheel or from an instrument. Data conversions for these bit string values are provided for bit strings reading in binary, BCD, or Grey code. It is also possible to link the resulting scalar value to an AI function block so that conventional signal processing can be performed with the value.

Ratio A relatively simple function block to provide the multiply function. Used in blending and other processes to establish one operator or recipe set value as a master, and to slave many different set point values based on the master. In blending, a master flow is set as a constant with other flows proportioned to the master flow. A ratio block would be configured in front of each set point of the flow control loops for each blended ingredient. As the master flow is changed, each ingredient flow will then be changed in constant ratio to the master flow.

Discrete Output A simple function block to allow up to 16 binary bit values to be written. Each bit output is configured with a tag name and can be logically processed as either a True or False logical value. Any string of bits can be treated as a scalar value as well with the contents expressed in binary, BCD, or Grey code.

Manual Loader A simple scalar output function block to provide a place for an operator manual set function to be integrated into the system. The manual loader is a control block capable of all of the functions of a control block except computation.

Proportional/Derivative (PD) A variation of the traditional feedback control function block with the Integral term missing so that it cannot wind up. Certain processes have found that the PD block is very responsive and accurate enough for the process scaling. Note that for all gain values of the PD other than unity, there will be some offset from the desired set point. In the PD form of control, the proportional gain term is *not* multiplied by the derivative gain as in the standard PID with integral gain set to zero.

Proportional/Integral/Derivative The classic and traditional process control form of the control algorithm that simulates the action of a pneumatic controller. The proportional gain term is multiplied by both the integral and derivative gain terms. This form of the PID equation has been used in all DCS and digital single-loop controllers. One of the primary reasons to continue the use of this form of the PID equation is that the body of tuning procedures, both manual and programmatic, is based on the Ziegler Nichols methods developed in the 1930s on pneumatic controllers. No modern automatic loop tuning has yet replaced the work of Ziegler and Nichols, leading to the conclusion that the so-called noninteractive form of the equation developed for rocket thrust vector control and motion control should be confined to these nonprocess applications. The PID algorithm provides for several options necessary for various control strategies. There is the external feedback to prevent reset wind-up when the output is either deselected by a control select block, or in some other alternative control strategy. The derivative term may be based on error or based only on PV change to eliminate the effects of operator set point change on derivative action.

The following are the extended function blocks not required for all Foundation Fieldbus devices. The descriptions and titles of these function blocks may vary between systems, but blocks similar to these have existed for years on DCS. Although the defined source for these blocks are the Fieldbus Foundation Specifications,[3] other more public sources exist.[4]

Output Signal Selector A switch used to select one of two different control strategies as part of an overall control strategy. Also called "Auto Select" in some DCSs, this has often been recommended in the control of boilers and some other processes. In boilers, most of the time the air damper flow control loop is based on excess oxygen in the flue gas, but during boiler transitions, the air damper position should be based on the fuel flow since excess oxygen has too much deadtime. The output signal selector is positioned typically just before the air damper AO block and takes the valve position signals from both the excess oxygen PID control block and the fuel flow control block. One of these blocks will be selected, in this case the highest of the two (configuration is high select), and the AO is set accordingly. It is strongly recommended that the external feedback term of both PIDs be linked to the output of the output signal selector block to allow automatic correction of the unselected controller to prevent reset wind-up.

Splitter Used for split-range control with a single three-way control valve or with two control valves. Another option for the control of a single process for two different control mechanisms can use the Splitter function block. For example, control of a single reactor requiring both heating and cooling to the reactor jacket can use one PID control block and have its output go through a splitter to select either the heating control valve (steam) or cooling (water).

Dead Time (DT) Used in process simulation, dynamic simulation, and in feedforward loop control, the dead-time function block delays the input by the designated time period. DT is not just a time delay function such as exists in PLCs. DT is a full "bucket brigade" time delay stack. Every datum entered into the stack at the sampling interval is returned after the time delay in the same order as entered. This is necessary in modeling true dead-time processes.

Lead Lag The dynamic modeling tool for real processes. Provides an easy-to-use, second-order time lag model of the process. Combined with the DT function block, it is possible to create a complete process model for a first- or second-order dynamic process.

Complex Analog Output Provides a buffer for the simultaneous output of up to eight analog signal values to AO blocks or as cascade set points.

Device Control Provides for construction of relay ladder logic for discrete interlock control.

Analog Alarm Provides for alarm limit testing similar to the AI block for any scalar value.

Discrete Alarm Provides for alarm testing for up to eight discrete values.

Arithmetic Provides several preprogrammed arithmetic functions (equations) for scalar inputs.

Pulse Input Creates a scalar value from a digital pulse count input.

Complex Digital Output Provides a buffer for the simultaneous output of up to eight discrete signal values to digital output blocks, or for use in any other logic.

Calculation Block Provides a calculator-like capability for scalar inputs to be combined.

Analog Human Interface Provides the capability to drive a local display for a single scalar value.

Discrete Human Interface Provides the capability to drive a local display for a single discrete value.

Set Point Generator Provides a time-based ramp-and-soak cycle needed in heat treatment and batch process control. Set point ramping is a timed function capable of being held at constant values with a discrete input. Soak is a timed hold at constant value.

Integrator At the sample interval, adds the current value of a scalar input creating a time-based summation.

Characterizer Converts a scalar input to a scalar output using table look-up and interpolation. This is similar to the signal characterizer used in valve control in the AO block except that any table of input and output values can be used.

Step Control Functions the same as an AO block except that the output device is driven as either a pulse motor or as a timed output function as for electric motor-driven actuators. The output change per pulse or per time period must be defined.

Cascade Control Concepts Building a control strategy involves the linking of function blocks. There are two kinds of links for function blocks: cascade and control. Cascade links are fully bidirectional and are intended to pass data along with status information. A control linkage is simply a one-way command link using discrete data and used typically to change the state of a function block.

The concept of cascade control is as old as control itself. Pneumatic controllers would be piped to allow the output of one, the upstream controller, to change the set point of the downstream controller. Why would anyone want to do this? The most typical cascade involves use of a motive force that has high speed, but somewhat noisy or wild characteristics. A controller is needed to control this stream, but the actual controlled value is of little interest. The upstream control loop is the variable that must be controlled, and the downstream loop is only a means by which control can be effected. This is best illustrated with an example. The controlled variable may be temperature so a control loop is built from the PID function block to perform the control action by the manipulation of a steam flow control valve. Unfortunately, the steam flow itself varies due to disturbances in the steam header as other processes use steam, which causes the temperature to vary even though there is no movement in the control valve. A secondary or downstream control loop is then built to measure steam flow and control it with a PID function block using the output of the temperature control loop as its set point. This is the very common temperature cascaded to flow cascade control loop.

One of the things that enables cascade control is that the output of the PID function block is a dimensionless number indicating a “desired valve position,” but can also be interpreted as “desired set point” but in dimensionless terms. With analog control, everything was dimensionless, but in digital fieldbus field control, only controller outputs are dimensionless, and all else has engineering units dimensions. Therefore, it is usually necessary to provide a dimension conversion constant to cascade controls.

Controllers exist in several states, but only one state at a time. Foundation Fieldbus and the TR50.02 define a very complex structure called MODE. The MODE of a controller describes two things: (1) the on/off state of the control calculation and (2) where the control computation gets its set point. The recognized states of all Foundation Fieldbus function blocks are given in Table 4.12k. There are only eight states or modes, but clearly more could have been defined. Only these eight are used, however.

TABLE 4.12k
Foundation Fieldbus Cascade MODES

MODE	*State of Computation*	*Source of Set Point*
Manual	Off	Operator
Automatic	On	Operator
Cascade	On	Output of another block
Initialization	Initialization	Output
Remote cascade	On	A remote source
Remote output	Off	Operator but not effective
Local override	Off	Tracking override input
Set point tracking	Off	Tracking a process variable

When control loops are defined, they are initialized in the manual mode. This allows the operator to change the process control valve manually. There are two options for going to automatic from manual that are configured into each PID controller: bumpless and nonbumpless. Bumpless transfer from manual to automatic means that the current value of the PV (process variable) is installed as the set point. Then, on transfer to automatic, nothing happens. For nonbumpless configurations, the set point remains where it is as set by the operator and the process control action moves the process to that set point according to the computations of the PID controller. Normally, on transfer of control, all derivative action is suppressed because there has been no sharp change in the process variable. Another option of the PID suppresses all control action due to set point change, which also occurs on this transfer. This means that a properly configured nonbumpless transfer will only gradually move to the fixed set point under integral-only control, a desired state.

From the automatic mode, the operator may advance the control by closing the cascade link that has already been established by changing the mode to cascade. At that time, the mode changes to cascade, in order to achieve bumpless transfer, using the same rules as for bumpless transfer for any controller that is configured in this way. Bumpless transfer for the cascade controller means that the set point of the upstream controller must be backcalculated so that its output does not change. The output of the controller is set to the current set point value of the downstream loop, and its own set point must be calculated—a process called cascade initialization.

All the methods and options for cascade initialization, as well as the methods used for function blocks to communicate upstream and downstream, are fully defined in the Foundation Fieldbus function block specifications. These are too detailed for these concepts, but are very complete and are based on the TR50.02 work. It is important to understand that provision is made in these cascade rules for a "supervisory" or "DDC" computer, meaning a DCS, to be operating with its own set of function blocks that are not Foundation Fieldbus compatible. This is the reason for the remote cascade and remote output modes. It becomes very much easier when the DCS runs control function blocks that are completely compatible with Foundation Fieldbus function blocks. Then, remote cascade and remote output are not needed at all.

References

1. Woodley, G., "Block Diagrammatic Control," Foxboro Company internal white paper, 1966.
2. SP50.8 Committee, *ISA TR50.02 Process Control User Layer,* Research Triangle Park, NC: Instrument Society of America, 2000.
3. Foundation Fieldbus Specifications, Vol. 3, Austin, TX: Fieldbus Foundation.
4. SMAR Web site: http://www.smar.com/Products/Pdf/FBook00/FunctionBlocks.pdf.

4.13 PROFIBUS-PA

P. G. BERRIE L. FÜCHTLER

PROFIBUS is an open fieldbus standard. It was developed by a German consortium that quickly and pragmatically produced German Standard DIN 19 245 after attempts to produce an international fieldbus failed in 1992. The European Standard EN 50 170 followed roughly a year later. The application profiles PROFIBUS-DP and PROFIBUS-PA, which are the protocols used in process automation, are incorporated into the IEC 61158 standard published in 2000. PROFIBUS is supported by an international network, the PROFIBUS User Organization (PNO), and all activities are coordinated by PROFIBUS International.[1]

Figure 4.13a shows a typical control network for a processing facility with both process automation (measurement, actuation, control) and factory automation tasks (conveyance, filling, packing).

- The process is controlled by a process control system or a programmable logic controller (PLC). The controller uses cyclic services to acquire measurements and to output control commands. The configuration tool allows bus participants to be configured during installation and start-up and device parameters to be changed when the production process is running. It uses acyclic services.
- PROFIBUS-DP is used to handle the fast communication applications (2 to 10 ms), e.g., drives, remote input/outputs (I/Os), etc., met in factory automation. In a PROFIBUS-PA system, PROFIBUS-DP also provides the high-speed H2 link between controller and fieldbus segment.
- PROFIBUS-PA is designed for use in process automation applications. The segment coupler (or link) serves both as interface to the PROFIBUS-DP system and as power supply for the PROFIBUS-PA field devices. Depending upon the type of coupler, the PROFIBUS-PA segment can be installed in safe or hazardous areas.

PROFIBUS-DP BASICS

It can see from the above that PROFIBUS-DP (DP = decentralized periphery) is an essential part of any PROFIBUS-PA application. Version DPV1 is used in process automation: the principal technical data are listed in Table 4.13b and a more detailed description of the operation is to be found in Section 4.16.

When used in conjunction with PROFIBUS-PA, an architecture is often used that comprises a PLC and a personal or laptop computer equipped with a PROFIBUS-DP card, on which the engineering tools are installed. All other devices, whether PROFIBUS-DP or PROFIBUS-PA, are slaves.

- The PLC acts as a Class 1 master and communicates cyclically or acyclically with the slaves. A slave may be assigned to one Class 1 master only, but this rule is not relevant in a mono-master architecture.
- The engineering tool acts as a Class 2 master and communicates acyclically with any device, i.e., on demand.

A hybrid method of centralized master/slave and decentralized token passing regulates the media access (Figure 4.13c).

- The masters build a logical token ring.
- When a master possesses the token, it has the right to transmit. It can now talk with its slaves in a master/slave relationship for a defined period of time.
- At the end of this time, the token must be passed on to the next active device in the token ring, thus ensuring deterministic communication.

The network topology is normally linear, with connection to PROFIBUS PA segments made via one or more segment couplers or links; see below. The maximum permissible cable length depends upon the transmission rate; see Table 4.16m in Section 4.16. In addition to copper cable, optical fiber networks can also be used. All devices operate at the same baud rate: this may range from 45.45 kbit/s to 12 Mbit/s, depending on the components used.

PROFIBUS-PA BASICS

PROFIBUS-PA (PA = process automation) has been designed to satisfy the requirements of process engineering. There are three major differences from a PROFIBUS-DP system:

1. PROFIBUS-PA supports the use of devices in explosion-hazardous areas.

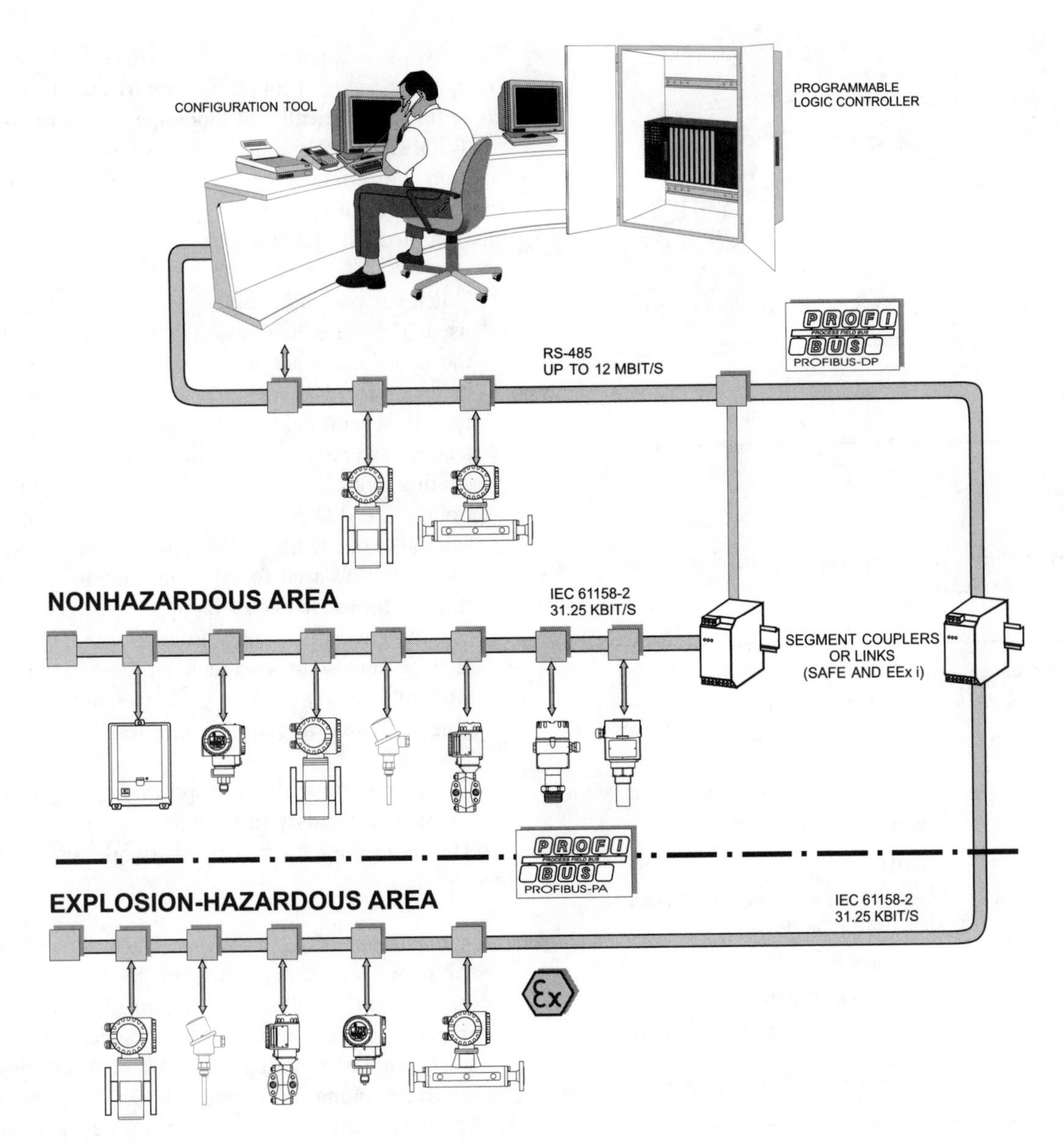

FIG. 4.13a
Typical PROFIBUS control network in a processing facility.

TABLE 4.13b
Main Technical Data for PROFIBUS-DP Version DPV1

Property	*Description*
Standard	DIN 19245 Parts 1 to 3, Version DPV1; EN 50179; IEC 61158
Protocol	PROFIBUS
Physical layer	RS-485 and/or fiber optics
Length	Up to 1200 m (3900 ft), depending upon transmission rate
Transmission rate	9600 bit/s to 12 Mbit/s
Bus access method	Master/slave, with token passing
Participants	Max. 126 (using repeaters), including max. 32 as masters Per segment: max. 32

2. The devices can be powered over the bus cable.
3. The data are transferred via the IEC 61158-2 physical layer, which allows greater freedom in the selection of the bus topology and longer bus segments.

The most important technical data are listed in Table 4.13d.

System Architecture

PROFIBUS-PA is always used in conjunction with PROFIBUS-DP. The field devices, i.e., sensors and actuators, on the PROFIBUS-PA segment communicate with a PROFIBUS-DP master. A segment coupler or link interfaces the RS-485 (PROFIBUS-DP) and IEC 61158-2 (PROFIBUS-PA) physical layers: the same protocol is used throughout the network; see also Tables 4.13b and 4.13d.

TOKEN WITH MASTER CLASS 1: CYCLIC COMMUNICATION

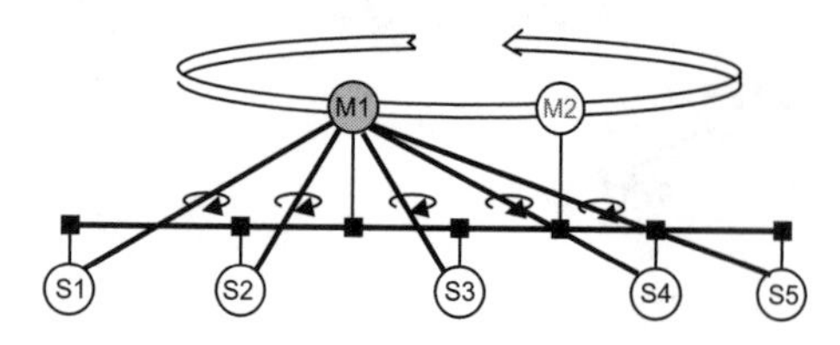

TOKEN WITH MASTER CLASS 2: ACYCLIC COMMUNICATION

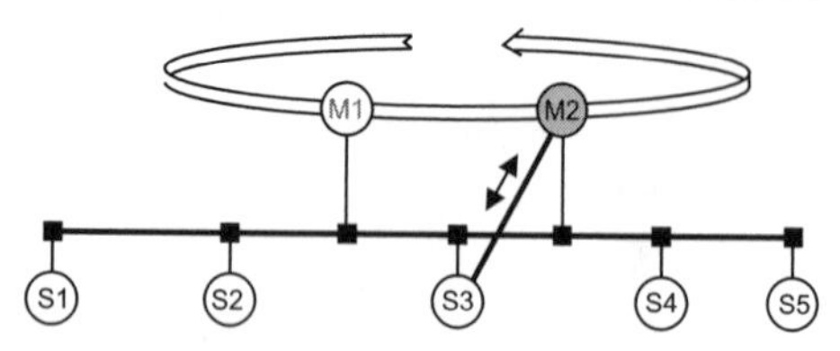

FIG. 4.13c
PROFIBUS-DP bus access method of centralized master/slave and decentralized token passing.

TABLE 4.13d
Main Technical Data of PROFIBUS-PA

Property	*Description*
Standard	DIN 19245 Parts 4; EN 50179; IEC 61158
Protocol	PROFIBUS
Physical layer	IEC 61158-2
Length	Max. 1900 m for safe and EEx ib areas Max. 1000 m for EEx ia
Transmission rate	31.25 kbit/s
Bus access method	Handled from PROFIBUS-DP side
Participants	32 in safe areas, ~24 in EEx ib and ~10 in EEx ia

TABLE 4.13e
Technical Data of Standard PROFIBUS-PA Segment Couplers

Coupler	*Type A*	*Type B*	*Type C*
Type of protection	EEx [ia/ib] IIC	EEx [ib] IIB	None
Supply voltage Us	13.5 V	13.5 V	24 V
Max. power	1.8 W	3.9 W	9.1 W
Max. supply current Is	110 mA	280 mA	400 mA
No. of devices	~10	~24	Max. 32

- A segment coupler comprises a signal coupler and bus power unit. Three types of segment couplers have been specified according to the type of protection required (Table 4.13e).
- A link comprises an intelligent gateway and one or more segment couplers, whereby the couplers may exhibit different types of protection.

The network is designed according to the rules for PROFIBUS-DP up to the segment coupler or link. Within the PROFIBUS-PA segment, practically all topologies are permissible.

PROFIBUS-PA stipulates a two-core cable as transmission medium. An informative annex to IEC 61158-2 lists the characteristics of four cable types that can be used as transmission medium (Table 4.13f).

- Wherever possible, cable types A and B should be used. These are the ones offered as fieldbus cables by the cable manufacturers. They offer the best EMC behavior for data transmission. In the case of cable type B, several fieldbuses (with the same type of protection) can be operated with one cable; other current-bearing circuits in the same cable are not permitted.
- Cables C and D are included in the table to check the suitability of existing cable runs for fieldbus use. They should not be used for new installations and are not suitable for use in explosion hazardous areas designed according to FISCO rules. Problems with the communication are also to be expected if the cables are routed through plants with heavy electromagnetic interference, e.g., near frequency converters.

Cable for intrinsically safe applications as per the FISCO model must also satisfy the requirements in Table 4.13g.

The basic rules for physical network design are as follows:

- A maximum of 32 participants are allowed in safe applications and, depending on the power rating of the segment coupler, a maximum of 10 participants are allowed in explosion hazardous areas (EEx ia). The actual number of participants must be determined during the planning of the segment; see Section 4.14.
- The maximum permissible length is dependent on the type of cable used; see Table 4.13f and Section 4.14, Table 4.14a.
- For systems that are to be realized according to the FISCO model in type of protection EEx ia, the maximum bus length is 1000 m.
- A terminator is required at each end of the segment.
- The bus length can be increased by using repeaters. A maximum three repeaters are allowable between a participant and the master.

The following rules apply to spurs:

- Spurs longer than 1 m are counted in the total cable length.
- The length of the individual spurs in safe areas is dependent on the number of participants; see Section 4.14, Table 4.14b.
- For segments designed according to the FISCO model, the spurs in intrinsically safe applications may not exceed 30 m in length.

TABLE 4.13f
Specifications of Cable Types for PROFIBUS-PA

Property	*Type A*	*Type B*	*Type C*	*Type D*
Cable construction	Twisted pairs, shielded	One or more twisted pairs, common shield	Several twisted pairs, unshielded	Several untwisted pairs, unshielded
Core cross section	0.8 mm^2 AWG 18	0.32 mm^2 AWG 22	0.13 mm^2 AWG 26	1.23 mm^2 AWG 16
Loop resistance (DC)	44 Ω/km	112 Ω/km	254 Ω/km	40 Ω/km
Characteristic impedance at 31.25 Hz	100 Ω ± 20%	100 Ω ± 30%	—	—
Attenuation constant at 39 kHz	3 dB/km	5 dB/km	8 dB/km	8 dB/km
Capacitive unsymmetry	2 nF/km	2 nF/km	—	—
Envelope delay distortion (7.9–39 kHz)	1.7 μs/km	—	—	—
Degree of coverage of shielding	90%	—	—	—
Max. bus length including spurs	1900 m/6175 ft	1200 m/3900 ft	400 m/1300 ft	200 m/650 ft

TABLE 4.13g
Specifications for FISCO Cables

Type of Protection	*EEx ia/ib IIC*	*EEx ib IIB*
Loop resistance (DC)	15–150 Ω/km	15–150 Ω/km
Specific inductance	0.4–1 mH/km	0.4–1 mH/km
Specific conductance	80–200 nF/km	80–200 nF/km
Max. spur length	≤30 m/100 ft	≤30 m/100 ft
Max. bus length	≤1000 m/3250 ft	≤1900 m/6175 ft

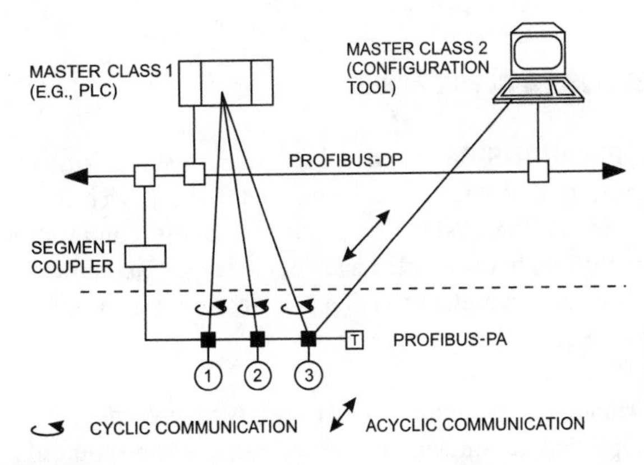

FIG. 4.13h
PROFIBUS-PA bus access via a segment coupler.

Bus Access Method

PROFIBUS-PA uses the central master/slave method to regulate bus access. The masters, e.g., a PLC as Class 1 master controlling the process and engineering/configuration tool as Class 2 master, are installed in the PROFIBUS-DP system. The field devices on the PROFIBUS-PA segment are the slaves. They are accessed via a segment coupler or a link.

Segment couplers are transparent as far as the PROFIBUS-DP master is concerned, so that they are not engineered in the PLC. They simply route all signals directly to the devices (Figure 4.13h).

- Couplers are not assigned a PROFIBUS-DP address. The field devices in the PROFIBUS-PA segment are each assigned a PROFIBUS-DP address and behave as PROFIBUS-DP slaves.
- A Class 1 master, e.g., the PLC, uses the cyclic polling services to fetch the process value data provided by the field devices for the process automation application.
- A Class 2 master, e.g., a configuration tool, transmits and receives configuration data by using the acyclic services to address the field device directly.

A link is recognized by the PROFIBUS-DP master and must be engineered in the PLC. As the link is opaque, the PROFIBUS-PA devices connected to it are not seen by the PROFIBUS-DP master (Figure 4.13i).

- On the PROFIBUS-PA side, the link acts as the bus master. It polls the field device data cyclically and stores them in a buffer. Every field device is assigned a PROFIBUS-PA address that is unique for the link, but not for other PROFIBUS-PA segments. The PROFIBUS-DP and PROFIBUS-PA addresses are numbered independently of each other.
- On the PROFIBUS-DP side, the link acts as a slave. It is assigned a PROFIBUS-DP address. A Class 1 master, e.g., the PLC, uses the cyclic polling services to fetch the process value data provided by the field devices, which the link has packed together in a telegram.
- When the link is accessed by a Class 2 master with the acyclic services it is quasi-transparent. The desired

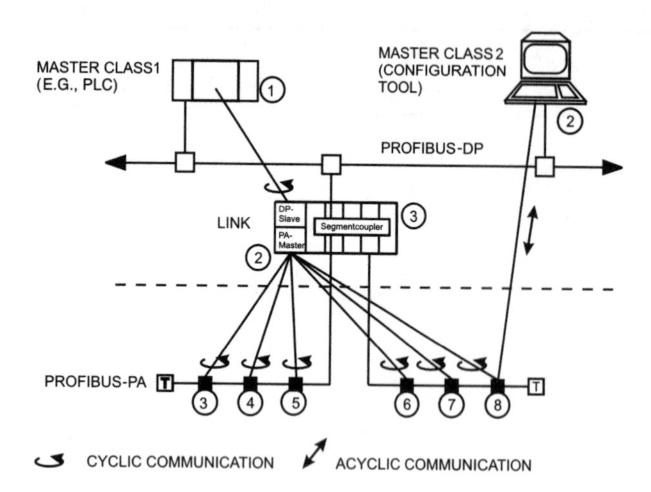

FIG. 4.13i
PROFIBUS-PA bus access via a link.

field device can be accessed directly by specifying the link address (DP address) and the device address (PA address).

PROFIBUS-PA BLOCK MODEL

The PROFIBUS-PA standard specifies a set of universal parameters that must be implemented if the device is to be certified by the PNO. Certification ensures configuration compatibility between identical "device types" manufactured by different manufacturers, e.g., pressure transmitters or flowmeters.

- Mandatory parameters must always be present.
- Optional parameters are only present when required, e.g., for a particular transmitter type such as a Coriolis mass flowmeter.
- Manufacturer-specific parameters are used to realize device functions that are not in the standard profile. A manufacturer's configuration tool or a device description file (GSD file) is required for their operation.

In the case of PROFIBUS-PA devices that conform with the standard, these parameters are managed in block objects. Within the blocks, the individual parameters are managed using relative indices.

Figure 4.13j shows the block model of a simple sensor. It comprises four blocks: device management, physical block, transducer block, and function block. The sensor signal is converted to a measured value by the transducer block and transmitted to the function block. Here the measured value can be scaled or limits can be set before it is made available as the output value to the cyclic services of the PLC. For an actuator, the processing is in the reverse order. The PLC outputs a set point value that serves as the input value to the actuator. After any scaling, the set point value is transmitted to the transducer block as the output value of the function block. It processes the value and outputs a signal that drives, e.g., the valve, to the desired position.

The parameters assigned to the individual blocks use the data structures and data formats that are specified in the PROFIBUS standard. The structures are designed such that the data are stored and transmitted in an ordered and interpretable manner. All parameters in the PROFIBUS-PA profile, whether mandatory or optional, are assigned an address (slot/index). The address structure must be maintained, even if optional parameters are not implemented in a device. This ensures that the relative indices in the profile are also to be found in the devices.

With the exception of the device management, standard parameters are to be found at the beginning of every block. They are used to identify and manage the block: an almost identical set is to be found in Foundation Fieldbus. The user

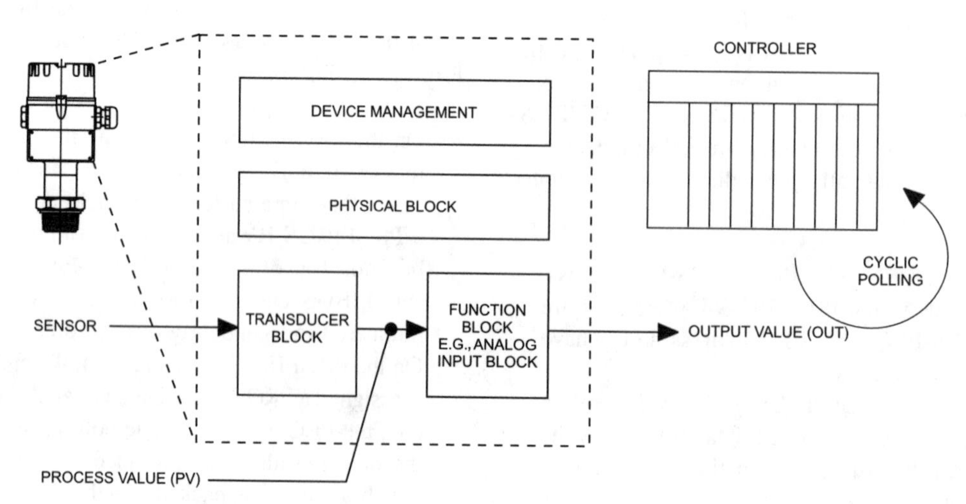

FIG. 4.13j
Block model of a simple PROFIBUS-PA sensor.

can access these parameters using the acyclic services, e.g., by means of a configuration tool.

Device Management

Device management comprises the directory for the block and object structure of the device. It gives information about:

- Which blocks are present in the device
- Where the start addresses are located (slot/index)
- How many objects each block holds

By using this information, the application program of the master can find and transmit the mandatory and optional parameters of a profile block.

Physical Block

The physical block contains the properties of the field device. These are device parameters and functions that are not dependent on the measurement method. This ensures that the function and transducer blocks are independent of the hardware. The physical block contains the following information:

- Device identification data
- Device installation data
- Device operational data
- Diagnostic messages, both standard and manufacturer specific

Transducer Blocks

Transducer blocks stand as separating elements between the sensor (or actuator) and the function block. They process the signal from the sensor (or actuator) and output a value that is transmitted via a device-independent interface to the function block. The transducer blocks reflect the measurement (or actuator) principles. They also contain generic parameters that allow every device to be configured by a basic operation mode, even when the corresponding GSD file is not available to the configuration tool. The following blocks are currently available:

- Pressure, temperature, level, flow, and analysis for "analog" sensors
- Discrete input/output for switches
- Electromagnetic and pneumatic transducer blocks for actuators

Function Blocks

Function blocks contain the basic automation functions. Because the application program demands that a cyclic value always behaves in the same manner, the blocks are designed to be as independent as possible from the actuator/sensor and the fieldbus. There are currently five standardized function blocks:

1. Analog input blocks, fed by a sensor transducer block, allowing simulation, characterizing, and scaling of the output value. The output value is collected by the Class 1 master.
2. Totalizor block, used when a process variable, must be summed over a period of time, e.g., for flowmeters, and fed by a sensor transducer block. The output value is collected by the Class 1 master.
3. Discrete input block, fed by a limit switch transducer block, allowing signal inversion. The output value is collected by the Class 1 master.
4. Discrete output block, feeding the transducer block of a binary actuator. The input value is provided by the Class 1 master.
5. Analog output block, feeding the transducer block of an "analog" actuator. The input value is provided by the Class 1 master.

Details of the structure and parameters of the blocks can be found in the PROFIBUS standard and manufacturers' literature.

APPLICATIONS IN HAZARDOUS AREAS

The explosion protection concept for PROFIBUS-PA is based on the type of protection "intrinsic safety i." In contrast to other types of explosion protection, intrinsic safety is not confined to the individual unit, but extends over the entire electrical circuit. Not only must all circuits running into hazardous areas exhibit this type of protection, but all devices and terminators as well as all associated electrical apparatus (e.g., PA links or segment couplers) must also be approved for the corresponding atmospheres.

To reduce the proof of intrinsic safety of a fieldbus system comprising different devices from different vendors to a justifiable level, the German PTB and various equipment manufacturers developed the FISCO model (**F**ieldbus **I**ntrinsically **S**afe **CO**ncept), which is described in more detail in Section 2.2. The basic idea is that only one device supplies power to a particular segment and the devices are limited in their capacitance and inductance values. The field devices are divided into those that draw their power from the bus itself, and those that must be powered locally. In addition to the type of protection intrinsic safety, the latter devices, which require more energy, must also exhibit a further type of protection for their power circuit. The auxiliary energy required by the segment coupler and the locally powered devices must be galvanically isolated from the intrinsically safe circuits.

As is the case for all intrinsic circuits, special precautions must be observed when installing the bus. The aim is to maintain the separation between the intrinsically safe and all other circuits.

Grounding. The intrinsically safe fieldbus circuit is operated earth-free, which does not preclude that individual sensor circuits can be connected to ground. If an overvoltage protector is installed before the device, it must be integrated into the chosen grounding system in accordance with the instructions in the certificate or device manual. Particular attention must be paid to the grounding of the conducting cable screening because if it is to be earthed at several positions, a high-integrity plant grounding system must be present.

Category. The category of the bus segment is determined by the device circuit with the worst rating, i.e., if one device on the segment has the type of protection EEx ib, then the whole segment falls into Category ib. All the devices connected to a Category ia circuit must have an EEx ia rating (requirements as per certificate). This applies only to device circuits that are connected directly to the segment, not to external power circuits, which may have other types of protection.

Explosion group. Devices that are approved for different explosion groups (IIC, IIB or IIA) can be operated on the same segment. The type of protection and explosion group of the device must, however, be permitted for the explosive atmosphere with which it is in contact.

FISCO approval. All devices, terminators, and cables that are installed in hazardous areas as well as all associated electrical apparatus (e.g., PROFIBUS-PA links or segment couplers) must have FISCO approval for the corresponding atmospheres.

Thanks to the FISCO model, the design of an intrinsically safe circuit is greatly simplified, as can be seen in the example in Section 4.14. In addition, the replacement of a device by one from a different manufacturer becomes easier because there is no need to calculate the capacitance and impedance of the segment.

NETWORK DESIGN

When a PROFIBUS-PA segment is designed, two factors must be considered:

1. The physical limitations of the segment dictated by the available power and voltage, the cable used, and the type of protection that must be upheld, etc.
2. The data transfer limitations dictated by the permissible telegram length and cycle times

The physical limitations are discussed in more detail in Section 4.14, where several calculation examples that apply to both PROFIBUS-PA and Foundation Fieldbus H1 segments are shown. In this respect both fieldbuses are basically the same. Where data transfer is concerned, however, PROFIBUS-PA differs.

Telegram Length

Data are exchanged over the PROFIBUS-PA segment by means of standard telegrams. An "analog" value requires 5 bytes (4 bytes measured value in IEEE 754 floating-point format, 1 byte status) and a discrete value 2 bytes (1 byte value, 1 byte status). A device may output more than one measured value; for example, a Coriolis flowmeter may offer up to ten different outputs.

When the segment is operated through a segment coupler, i.e., each device is polled individually, data exchange presents no restriction to segment size. Even when all ten measured values of our Coriolis flowmeter are activated, only 50 bytes are required.

If, however, the segment is operated through a link, the PROFIBUS-DP telegram length comes into play because the segment data are collected before being forwarded in one telegram. The permissible telegram length is limited:

1. By the buffer size of the link, e.g., 244 bytes
2. By the maximum telegram length of the PLC
3. By the PROFIBUS-PA specification = 244 bytes

At the time of writing, masters are available that transmit 122 or 244 bytes (Table 4.13k). This imposes an upper limit on the number of bytes that can be generated within the PROFIBUS-PA segment, and hence on the number of devices it can support. If the master supports only 122 bytes, then only 24 "analog" devices offering 5 bytes each can be accommodated on the segment. This number is reduced when, like our flowmeter, the devices have been configured to deliver more than one measured value.

Cycle Times

In addition to the amount of data, the cycle times must also be considered when the PROFIBUS-PA segment is planned. Data exchange between a PLC (a Class 1 master) and the field devices occurs automatically in a fixed, repetitive order. The cycle times determine how much time is required until the data of all the devices in the network are updated. The more complex a device, the greater the amount of data to be exchanged and the longer the response time for the exchange between the PLC and the device.

The total cycle time for the updating of network data can be estimated as follows:

Total cycle time
= Sum of the cycle times of the field devices
+ internal PLC cycle time
+ PROFIBUS-DP system reaction time

The cycle time of the PLC is typically 100 ms, and that of the devices typically 10 ms for one measured value and 1.5 ms

TABLE 4.13k
Properties of Selected PROFIBUS-DP Controllers

PLC/Interface	*DP/PA Coupler*	*No. of Slaves per DP Interface*	*DP Telegram Length*
Siemens S7-300 315-2 DP	Pepperl+Fuchs	64	244 bytes
	Siemens	64	244 bytes
	Siemens DP/PA link	Max. 64 links with max. 24 slaves[a]	122 bytes read 122 bytes write
Siemens S7-400 414-2 DP	Pepperl+Fuchs	96	244 bytes
	Siemens	96	244 bytes
	Siemens DP/PA link	Max. 96 links with max. 24 slaves[a]	122 bytes read 122 bytes write
Siemens S5-135U IM 308C	Pepperl+Fuchs	122	244 bytes read 244 bytes write
	Siemens	122	244 bytes read 244 bytes write
Siemens S5-155U IM 308C	Pepperl+Fuchs	122	244 bytes read 244 bytes write
	Siemens	122	244 bytes read 244 bytes write
	Siemens DP/PA link	Max. 20 links with max. 24 slaves[a]	122 bytes read 122 bytes write
Allen-Bradley PLC-5 SST-PFB-PLC5	Pepperl+Fuchs	125	244 bytes read 244 bytes write
	Siemens DP/PA link	Max. 125 links with max. 48 slaves[a]	244 bytes read 244 bytes write
Allen-Bradley SLC 500 SST-PFB-SLC	Pepperl+Fuchs	96	244 bytes read 244 bytes write
Mitsubishi Melsec AnS A1S-J71PB92D	Pepperl+Fuchs	60	244 bytes read 244 bytes write
Schneider TSX Quantum + 140 CRP 81100	Pepperl+Fuchs	125	244 bytes read 244 bytes write
Schneider Premium + TSX PBY 100	Pepperl+Fuchs	125	Max. 244 bytes
HIMA H41 (MODBUS) + PKV 20-DPM	Pepperl+Fuchs	125	Max. 244 bytes
Klöckner–Moller PS 416 + PS416-NET-440	Pepperl+Fuchs	30, 126 with repeaters	244 bytes read 244 bytes write
ABB Freelance 2000 + Fieldcontroller	Pepperl+Fuchs	125	244 bytes read 244 bytes write
Bosch CL 250 P + BM DP12	Pepperl+Fuchs	125	244 bytes read 244 bytes write
	Siemens DP/PA link	Max. 125 links with max. 48 slaves[a]	244 bytes read 244 bytes write

Note: The table is for information only; it does not claim to list all systems on the market. The data were correct at the time of writing.

[a] Dependent on the telegram length of the slaves.

for each additional measured value. The PROFIBUS-DP system reaction time is normally given by the token rotation time. In a mono-master system, however, it is dependent on the number of slaves and the retry setting. Both are dependent on the baud rate, which may range from 45.45 kbit/s to 12 Mbit/s depending upon the components used. The total cycle time of a large system can be reduced considerably by the use of links, because the controller does not have to poll all the slaves in the network individually. As the internal cycle of the link runs independently of the controller, the total refresh time is then primarily dependent on the slowest link.

SYSTEM CONFIGURATION

In general, a PROFIBUS-DP/PA system is configured as follows:

- The network participants are stipulated in a PROFIBUS-DP network design program. Normally, the configuration software will scan the network for "live" devices, i.e., those that are operational and that have a unique PROFIBUS-DP address. Where devices have DIP switches for address setting, the network can be preconfigured prior to start-up. Devices with software addresses can be configured individually before commissioning or must be introduced one by one to the network, and then assigned an address by the configuration tool.
- The network is configured off line with the planning software. To this end, the GSD files are first loaded into the specified directory of the program. A device database file (GSD file) contains a description of the properties of the PROFIBUS-DP/PA device, e.g., the values that it outputs, formats, transmission rates supported, etc. Bit map files also belong to the GSD files; these allow the measuring point to be represented by an icon in the system tree. To distinguish among GSD files, every device is allocated an identity code (IDENT_number) by the PNO that appears in the file name.
- The PLC application program must now be written. This is done using the manufacturer's software. The application program controls the input and output of data and determines where the data are to be stored. If necessary, an additional conversion module must be used for PLCs that do not support the IEEE 754 floating-point format. Depending on the way the data are stored in the PLC (LSB or MSB), a byte-swapping module may also be required.
- After the network has been designed and configured, the result is loaded into the PLC as a binary file.

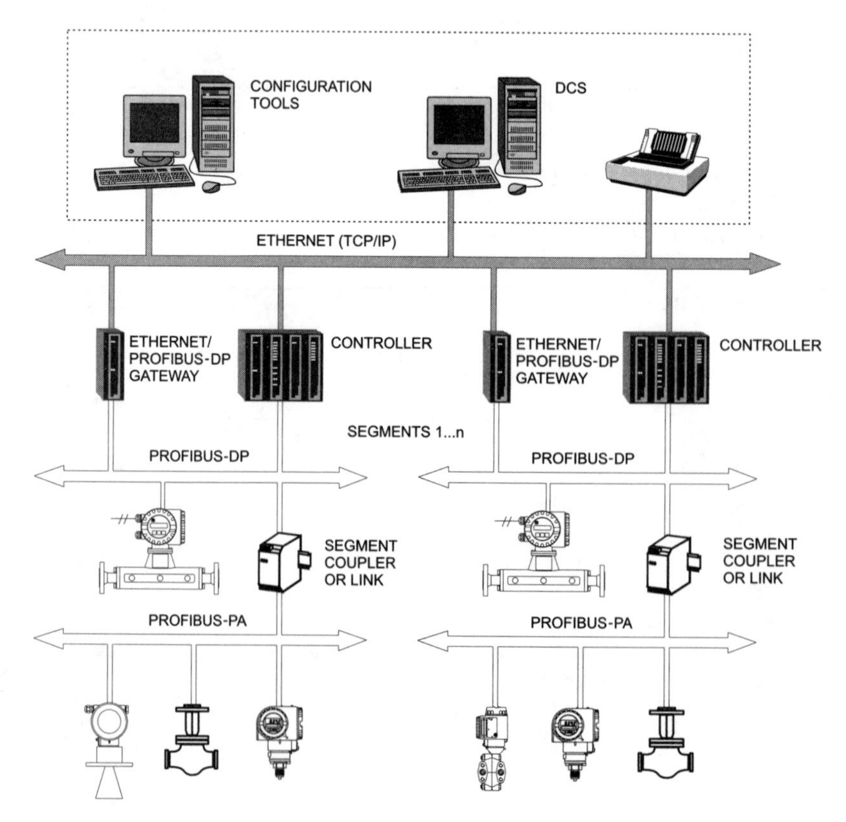

FIG. 4.13l
Network structure using Ethernet/PROFIBUS interface.

- When the PLC configuration is complete, the system can be started up. The master opens a connection to each individual device. By using a device configuration tool as a Class 2 master, the devices parameters can now be set.

The most frequent network fault is a mismatch of the bus parameters set in the various devices, particularly on the PROFIBUS-DP side, where all devices must be set to operate at the same baud rate. The component manufacturers give recommended settings here. If a device does not appear in the system tree, the most frequent cause is an address conflict. If the network has been designed according to the criteria in Section 4.14, and PNO certified devices have been used throughout, no further problems should be encountered.

FUTURE DEVELOPMENTS

Since PROFIBUS-PA devices were first introduced to the market in 1997, the number of devices sold has doubled each year. The standard has penetrated into most processing industries; particular strongholds are food and beverages and water/wastewater applications. Its introduction into chemical and pharmaceutical plants has been slowed by the conservative stance of the industry to its use in hazardous applications; however, the recent decision of the influential NAMUR chemical manufacturer's organization to approve its use also opens this field. Contrary to popular belief, PROFIBUS-PA is not a purely German activity, but has been installed in working plants all over the world, including the United States.

PROFInet. The year 2001 saw the introduction of PROFInet to the market. As shown in Figure 4.13l, this allows distributed control using centralized engineering in an IT network (Ethernet TCP/IP). PROFInet completes the PROFIBUS concept of a plantwide automation.

FDT/DTM. A second interesting development is the adoption by the PNO of the FDT/DTM standard for device operation. The field device tool (FDT) is a standard for operating instruments, which seeks to separate the configuration and operation of a device from the communication protocol it uses. More information can be found in Section 4.1.

CONCLUSION

These and other developments show that PROFIBUS-PA will continue to play an important part in fieldbus systems of the future. Whether PROFIBUS-PA and Foundation Fieldbus will migrate further toward each other will depend on how the market develops in the next few years. Much of the acrimony in the fieldbus standardization process was caused by the realization that an international fieldbus standard would bring more control functionality to the field level and more flexibility to networks in general, resulting in a loss of traditional markets for control system manufacturers. In the past 5 years there has been a noticeable realignment of the market, most companies either offering a complete solution from field to management level or entering into strategic alliances to support the same. At this point, the demands of the customer for open systems begin to tell again, demands that PROFIBUS-PA ably fulfills.

Reference

1. See Web site www.profibus.com or www.profibus.org.

4.14 Designing PROFIBUS-PA and Foundation Fieldbus Segments

P. G. BERRIE L. FÜCHTLER K. H. KORSTEN

Both Foundation Fieldbus H1 and PROFIBUS-PA use the IEC 61158-2 standard as the physical layer. This means that as far as the physical implementation of a segment is concerned, both are identical. Only the protocol and the interface to the higher-level H2 network differ, whereby, even in the case of the latter, several basic principles regarding powering of the bus remain the same.

This section examines three aspects of designing an IEC 61158-2 bus segment:

1. General considerations
2. How to set about designing the segment
3. How to ground the devices

A fourth topic, the optimization of cycle times, is dependent on the bus protocol. For a Foundation Fieldbus network operating with control in the field, the location of the various control function blocks within the devices can have a significant effect on the total time that is required to refresh the system. For PROFIBUS-PA, the use of a link can under certain circumstances significantly speed the communication process. For more details, see Section 4.13.

GENERAL CONSIDERATIONS

Before a bus segment is designed, a few general issues must be considered regarding the topology required, the cable type and length, and the number of devices, as well as the way the components are to be connected.

Although there is a great amount of freedom in the choice of fieldbus topology, the basic rule is to keep it as simple as possible. The introduction of elaborate branches produces additional signal distortion, and it is often not clear where the bus terminators, which are required at both ends of the segment, must be placed. Daisy chains are not recommended because a single bad connection kills the segment. In contrast, a trunk line with spurs to single devices or junction boxes to several devices provides a clean solution with a defined beginning and end.

Cable Length

The maximum cable length is dependent upon the type of cable used and the type of protection required for the segment. Table 4.14a summarizes the dependencies. Details of the cable specifications for standard and FISCO applications can be found in Section 4.13 in Tables 4.13f and 4.13g, respectively. It should be noted that the permissible length in a particular network configuration may be less than the figures quoted. The segment length can be increased by using repeaters, whereby a maximum of three repeaters is allowed between a participant and the start of the segment. All extensions must be terminated at both ends.

Another restriction to topology is the permissible length of any spurs. A spur is a length of cable connecting the device to the T-piece/junction box on the trunk cable. The length is dependent on the number of devices on the segment and the type of protection, and is summarized in Table 4.14b. In addition, according to the standard, only four devices may be connected to a spur. In practice, it is better to avoid such

TABLE 4.14a
Maximum Cable Length for an IEC 61158-2 Segment as a Function of Cable Type and Type of Protection

	Max. Length as a Function of Cable Type			
Type of Protection	*Type A*	*Type B*	*Type C*	*Type D*
Safe area/Ex i/Ex d	1900 m/6200 ft	1200 m/3900 ft	400 m/1300 ft	200 m/650 ft
FISCO EEx ia IIB/IIC	1000 m/3250 ft	1000 m/3250 ft	Not suitable	Not suitable
FISCO EEx ib IIB	1900 m/6200 ft	1200 m/3900 ft	Not suitable	Not suitable

TABLE 4.14b
Maximum Spur Length of an IEC 61158-2 Segment as a Function of Number of Devices and Type of Protection

	No. of Devices				
	1 to 12	*13 to 14*	*15 to 18*	*19 to 24*	*25 to 32*
Max. spur length	130 m/420 ft	90 m/300 ft	60 m/200 ft	30 m/100 ft	1 m/3 ft
FISCO EEx ia/ib	30 m/100 ft	30 m/100 ft[a]	30 m/100 ft[a]	30 m/100 ft[a]	Not applicable

[a]Applicable to EEx ib only, as the physical limit for EEx ia is approximately 10 devices.

TABLE 4.14c
Examples of Foundation Fieldbus and PROFIBUS-PA Power Components

Type	U_s	I_s	R_Q	*Remarks*
MTL 5053	18.4 V	80 mA	105 Ω	IS power supply with power conditioner and switchable terminator
MTL 5995	19.0 V	≤350 mA	<2 Ω	Non-IS power supply with power conditioner and switchable terminator
Relcom FCS-PC	$V_{input} - 5$ V	≤330 mA	—	Power conditioner
Relcom FCS-PCT	$V_{input} - 5$ V	≤330 mA	—	Power conditioner with terminator
Siemens 6ES7-157-0 AD00 0XA0	12.5 V	100 mA[a]	—	PROFIBUS segment coupler EEx [ia] IIC
Siemens 6ES7-157-0 AC00 0XA0	19.0 V	400 mA[a]	—	PROFIBUS segment coupler for safe areas
Pepperl+Fuchs KFD2-BR-EX1.2PA.93	13.0 V	110 mA[a]	—	PROFIBUS segment coupler EEx [ia] IIC
Pepperl+Fuchs KFD2-BR-1PA.93	25.0 V	380 mA[a]	—	PROFIBUS segment coupler for safe areas

Note: Inclusion in this table does not imply a recommendation; other suppliers exist.

[a]The bus current has already been subtracted from these specifications.

configurations. If they are absolutely necessary, then devices should be connected either directly to T-pieces, or if a junction box is used, the connecting cable should have a maximum length of 1 m only.

Fieldbus Devices

A maximum of 32 devices is allowed in safe applications. The maximum number of devices allowed on a segment running into an explosion hazardous area depends on the explosion protection concept chosen as well as the power available on the bus. In all cases, the actual number of devices must be determined during the planning of the bus.

Loop-powered devices operate with a direct voltage in the range 9 to 32 V and draw their power from the bus. Externally powered devices can also be connected to the segment. If operating in hazardous areas, the external power supply and device connection compartment must offer an appropriate type of protection.

Power Supply

Direct connection of a power supply to the segment would disturb communication. For this reason, power is routed through a fieldbus power conditioning unit. This is normally installed at the start of the fieldbus segment and often contains a bus terminator. If the segment runs into an explosion hazardous area, a barrier is also required. As far as Foundation Fieldbus is concerned, the power supply and conditioner are separate components. They must be chosen according to the application; Table 4.14c gives some examples.

In contrast, PROFIBUS-PA uses a link or segment coupler, which supplies the power while acting as interface to the PROFIBUS-DP control network. The FISCO concept, which is part of the PROFIBUS-PA standard, specifies three types of couplers, which are classified according to type of protection; see Table 4.13e in Section 4.13. The Type C coupler (for nonhazardous areas) could also be used with an appropriate barrier and, for example, Ex d/Ex i multidrop barriers, to build up non-FISCO-conforming segments; see below.

TABLE 4.14d
Typical Specifications of FISCO-Conforming Field Instrumentation as Used in the Examples

Device	U_{Basic}	I_{Basic}	I_{FDE}	$I_{Startup}$
Flowmeter	9–32 V	12 mA	0 mA	0 (<IBasic)
Pressure transmitter	9–32 V	10.5 mA	0 mA	0 (<IBasic)
Radar level transmitter	9–32 V	11 mA	0 mA	0 (<IBasic)
Positioner	9–32 V	13 mA	4 mA	0 (<IBasic)
	Max. fault current		4 mA	

The choice of instruments is left to the discretion of the user. There are currently many types available for both Foundation Fieldbus and PROFIBUS-PA. Those originally designed for PROFIBUS-PA FISCO conformance, however, usually draw less current than standard fieldbus instruments. This has the advantage that more devices can be connected to a given segment. Table 4.14d shows typical values for FISCO devices from the authors' company that are used in the examples which follow. The values apply to both PROFIBUS-PA and Foundation Fieldbus. When designing a Foundation Fieldbus Ex-segment, the safety values of each device are also required. This situation will change in the near future, as Fieldbus Foundation is in the process of adopting the FISCO model. The issuing of an Ex-certificate then implies that the inductance and capacitance of a FISCO device is less than that of the FISCO-approved power supply.

Finally, the way the connections between the various bus components are made should be considered. Two types of connection are recommended:

1. Cordsets with 7/8 in. (Foundation Fieldbus) or M12 (PROFIBUS-PA) cable connectors. These can be used for standard and Ex i applications.
2. Wireable cables and cable glands, with the cable connected to the screw terminals in the device housing. This method can be used in standard, Ex i, and Ex d applications.

Before a decision on connection mode is made, however, attention should be paid to the grounding scheme; see below for more details.

NETWORK DESIGN

There are two aspects to designing an IEC 61158-2 segment:

1. *Function*: The physical network structure must be checked against the technical data for the transmission technology laid down in IEC 61158-2. It must always be ensured that the total basic current of all bus participants does not exceed the maximum permissible feed current of the bus power supply and that the field devices are always supplied with a minimum bus voltage of 9 V.
2. *Technical safety*: Where an IEC 61158-2 runs into an explosion hazardous area, proof of intrinsic safety must be provided for the entire bus segment by checking its technical safety. The safety considerations correspond to those in a conventional 4–20 mA measuring circuit. The only difference is that more than one field device is supplied by the power supply. The safety considerations must result in the knowledge that the fieldbus power supply does not exceed the safety parameters (P, U, and I values) of the field devices and that the inductance and capacitance are within the permissible limits.

The functional considerations apply to all IEC-61158-2 segments and are described in the following on the basis of examples for a segment for a nonhazardous area, an intrinsically safe circuit to IEC 61158-2, and a FISCO intrinsically safe circuit. The technical considerations apply to standard intrinsically safe segments only. For segments conforming to the FISCO model, it is not necessary to supply separate proof of technical safety of the segment regarding inductance and capacitance.

Fieldbus Segment for a Nonhazardous Area

Figure 4.14e shows an example of segment for a nonhazardous area, which is to be checked for functionality. It is constructed as follows:

- For Foundation Fieldbus (IEC 61158-2), the segment starts at an input/output (I/O) card (1) or linking device, and a power supply with conditioner (2) adds power to the bus. For PROFIBUS-PA these functions are combined in the link or segment coupler, e.g., of Type C.
- The bus is terminated at its start and at the furthest junction box.
- Both loop-powered and four-wire devices are connected to the bus, maximum 32 per segment.
- Cable Type A with a loop-resistance of 44 Ω/km is used.

The functional check is made on the basis of the values in Tables 4.14a to 4.14d (power supply MTL 5995) as follows:

Step	*Procedure*	*Calculation/Condition*	*Result*
1.1	Calculate the cable length by adding together the length of the trunk and all spurs that are longer than 1 m (3 ft) $l_{SEG} = l_{trunk} + \Sigma l_{spurs}$ where l_{trunk} = length of trunk cable Σl_{spurs} = sum of all spur lengths	$l_{SEG} = l_{trunk} + \Sigma l_{spurs}$ $= 570$ m + 100 m **= 670 m**	
1.2	Check that it lies within specifications for Type A cable.	The segment length l_{SEG} is less than 1900 m. There is no spur longer than 120 m. There is no spur with more than four devices connected to it.	**True** **True** **True**
1.3	When all the conditions are met, the network can be structured as required. If one of the conditions is not met, then the network structure must be revised.	All conditions met?	**Yes**
2.1	Calculate the current I_{SEG} drawn by the segment $I_{SEG} = \Sigma I_B + \max I_{FDE} + I_{MOD} + \Sigma I_{startup}$ where I_B = the basic current drawn by a device I_{FDE} = the largest fault disconnect electronics current in the segment I_{MOD} = the modulation current of the segment (= 9 mA)—not required for PROFIBUS-PA $I_{startup}$ = the extra current a device may draw above the basic current on start-up	$I_{SEG} = \Sigma I_B + \max I_{FDE} + I_{MOD} + \Sigma I_{startup}$ $= 2 \times 12$ mA + 4×10.5 mA + 2×11 mA + 2×13 mA + 4 mA + 9 mA = 24 mA + 42 mA + 22 mA + 26 mA + 4 mA + 9 mA **= 127 mA**	
3.1	Check that the current drawn by the segment I_{SEG} is within the current limits of the fieldbus power supply.[a]		
3.2	Calculate the resistance of the cable R_{cable}: $R_{cable} = R_{loop} \times l_{SEG}$ where R_{loop} = loop resistance of the cable l_{SEG} = length of the segment	$R_{cable} = R_{loop} \times l_{SEG}$ = 44 Ω/km × 0.670 km **= 29.48 Ω**	
3.3	Calculate the supply current I_{Smax}, assuming a minimum device supply voltage of 9 V[b]: $I_{Smax} = (U_S - 9\text{ V})/(R_Q + R_{cable})$ where U_s = supply voltage R_Q = internal resistance of the power supply R_{cable} = resistance of the cable	$I_{Smax'} = (U_S - 9\text{ V})/(R_Q + R_{cable})$ = (19 V − 9 V)/(2 Ω + 29.48 Ω) **= 317.7 mA**	
3.4	Is the maximum supply current $I_{Smax} \geq$ segment current I_{SEG} calculated in Step 2.1?	$I_{Smax} \geq I_{SEG}$ 317.7 mA ≥ 127 mA	**True**
4.1	Calculate the voltage U_{FG} at the last field device: $U_{FG} = U_S - I_{SEG} \times (R_Q + R_{cable})$	$U_{FG} = U_S - I_{SEG} \times (R_Q + R_{cable})$ = 19 V − 127 mA × (2 Ω + 29.48 Ω) = 19 V − 4 V **= 15 V**	
4.2	Is the voltage at the last device $U_{FG} \geq 9$ V?	$U_{FG} \geq 9$ V 15 V ≥ 9 V	**True**
5.1	If all conditions are met, the segment will function.	All conditions met?	**Yes**

[a] In the case of a PROFIBUS-PA segment coupler, this calculation is not required and I_{SEG} can be compared directly to the I_S quoted in its specification; see FISCO example.

[b] Some devices may require a minimum supply voltage of more than 9 V. This calculation assumes the worst case of all devices being at the end of the bus, however, so that there is no problem in using 9 V unless the offending device is the last one on the bus.

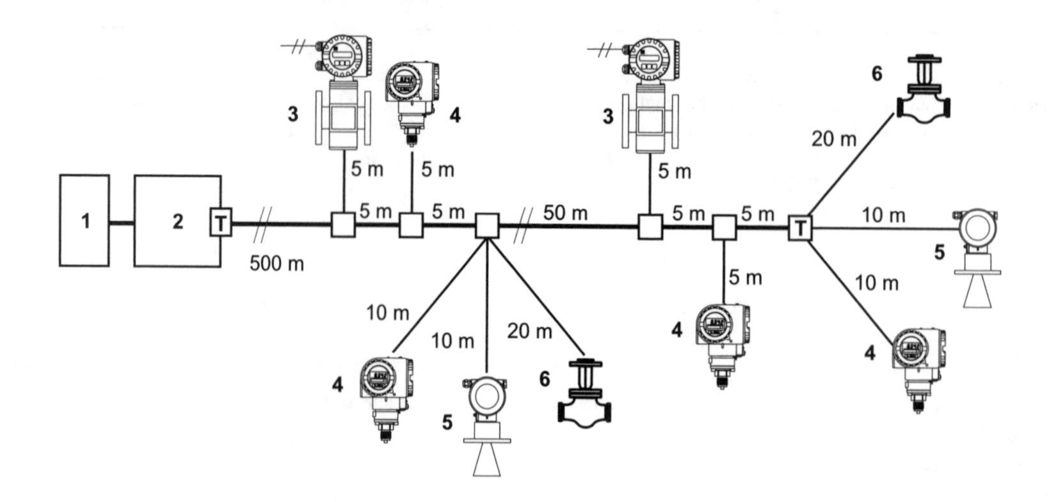

FIG. 4.14e
Example of a bus segment in a nonhazardous area. 1 = I/O card or linking device; 2 = power supply and conditioner; 3 = flow device with external power supply; 4 = pressure device; 5 = level device; 6 = valve positioner; T = terminator. For PROFIBUS-PA, elements 1 and 2 are replaced by a segment coupler.

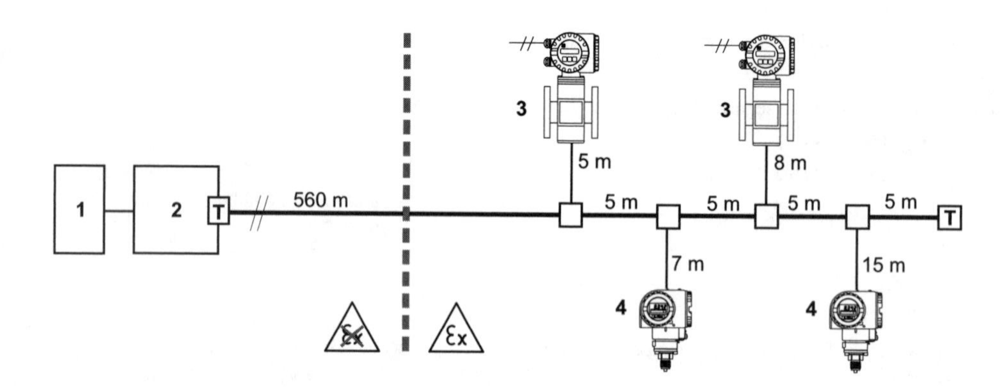

FIG. 4.14f
Example of an IEC 61158-2 bus segment in an explosion-hazardous area: 1 = I/O card or linking device; 2 = intrinsically safe power supply and conditioner or power supply, conditioner, and safety barrier; 3 = intrinsically safe flow device with EEx de external power supply; 4 = intrinsically safe pressure device; T = intrinsically safe terminator.

Foundation Fieldbus (IEC 61158-2) Segment to Ex i

Up to four fieldbus devices can be connected to a standard Ex i segment. Figure 4.14f shows the typical structure and the associated components. Replacement devices from different manufacturers can be used without problem in the event of a failure, provided the data relevant to safety are observed. These data comprise the electrical safety parameters, device group and category, and temperature class. The segment is constructed as follows:

- The segment starts at an I/O card (1) or linking device.
- An intrinisically safe power supply with conditioner (2) adds power to the bus. Alternatively, a standard power supply and safety barrier can be used.
- The bus is terminated at the power supply or barrier and at a *separate* terminator at the end of the bus.
- Both loop-powered and four-wire devices are connected to the bus.
- Cable Type A with a loop-resistance of 44 Ω/km is used.

All components in the explosion-hazardous area must have Ex i certification.

The functional check is made on the basis of the values in Tables 4.14a to 4.14d (power supply MTL 5053) as follows. The key to the parameters can be taken from the previous subsection.

Step	*Procedure*	*Calculation/Condition*	*Result*
1.1	Calculate the cable length by adding together the length of the trunk and all spurs that are longer than 1 m (3 ft): $l_{SEG} = l_{trunk} + \Sigma l_{spurs}$	$l_{SEG} = l_{trunk} + \Sigma l_{spurs}$ = 580 m + 35 m = **615 m**	
1.2	Check that it lies within specifications for Type A cable.	The segment length l_{SEG} is less than 1900 m. There is no spur longer than 120 m.	**True** **True**
1.3	When all the conditions are met, the network can be structured as required. If one of the conditions is not met, then the network structure must be revised.	All conditions met?	**Yes**
2.1	Calculate the current I_{SEG} drawn by the segment $I_{SEG} = \Sigma I_B + \max I_{FDE} + I_{MOD} + \Sigma I_{startup}$	$I_{SEG} = \Sigma I_B + \max I_{FDE} + I_{MOD} + \Sigma I_{startup}$ = 2 × 12 mA + 2 × 10.5 mA + 0 mA + 9 mA + 0 mA = 24 mA + 21 mA + 9 mA = **54 mA**	
3.1	Check that the current drawn by the segment I_{SEG} is within the current limits of the fieldbus power supply.		
3.2	Calculate the resistance of the cable R_{cable}: $R_{cable} = R_{loop} \times l_{SEG}$	$R_{cable} = R_{loop} \times l_{SEG}$ = 44 Ω/km × 0.615 km = **27.06 Ω**	
3.3	Calculate the supply current I_{Smax}, assuming a minimum device supply voltage of 9 V [a]: $I_{Smax} = (U_S - 9\text{ V})/(R_Q + R_{cable})$	$I_{Smax} = (U_S - 9\text{ V})/(R_Q + R_{cable})$ = (18.4 V − 9 V)/(105 Ω + 27.06 Ω) = **70.9 mA**	
3.4	Is the maximum supply current $I_{Smax} \geq$ segment current I_{SEG} calculated in Step 2?	$I_{Smax} \geq I_{SEG}$ 70.9 mA ≥ 54 mA	**Yes**
4.1	Calculate the voltage U_{FG} at the last field device: $U_{FG} = U_S - I_{SEG} \times (R_Q + R_{cable})$	$U_{FG} = U_S - I_{SEG}$ x $(R_Q + R_{cable})$ = 18.4 V − 54 mA × (105 Ω + 27b Ω) = 18.4 V − 7.1 V = **11.2 V**	
4.2	Is the voltage at the last device $U_{FG} \geq 9$ V?[a]	$U_{FG} \geq 9$ V 11.2 V ≥ 9 V	**True**
5.1	If all conditions are met, the segment will function.	All conditions met?	**Yes**
6.1	Using the technical safety values recorded in the equipment certificates, check the technical safety of the segment.		
6.2	First check that the individual inductance and capacitance values of the devices and terminator (see Table 4.14g) fulfill the conditions in Table 4.14h.	For every device: Voltage $[U_i] \geq [U_o]$ Current $[I_i] \geq [I_o]$ Power $[P_i] \geq [P_o]$ Capacitance $[C_i] \leq [C_o]$ Inductance $[L_i] \leq [L_o]$	**Yes**
6.3	Calculate the cable capacitance C'_{SEG} of the bus segment as follows: $C'_{SEG} = (C' + 0.5 \times C'_{LS}) \times l_{SEG}$ where C' = capacitance per unit length between the two wires or "wire against shielding" for shielded wires C'_{LS} = capacitance "shielding against wire" for shielded wires. l_{SEG} = total length of the segment	$C'_{SEG} = (C' + 0.5 \times C'_{LS}) \times l_{SEG}$ = (82 nF/km + 0.5 × 147 nF/km) × 0.615 km = **95.6 nF**	
6.4	Calculate the cable inductance L'_{SEG} of the bus segment as follows: $L'_{SEG} = L' \times l_{SEG}$ where L' = inductance per unit length	$L'_{SEG} = L' \times l_{SEG}$ = 623 μH/km × 0.615 km = **383.1 μH**	
6.5	Now calculate the total inductance and capacitance of the segment: $C_{tot} = \Sigma C_{i/devices} + C'_{SEG} + C_{i/termination}$ $L_{tot} = \Sigma L_{i/devices} + L'_{SEG} + L_{i/termination}$	$C_{tot} = \Sigma C_{i/devices} + C'_{SEG} + C_{i/termination}$ = (5 nF × 4) + 95.6 nF = **115.6 nF** $L_{tot} = \Sigma L_{i/devices} + L'_{SEG} + L_{i/termination}$ = (10 μH × 4) + 383.1 μH = **423.1 mH**	
6.6	Check that the technical safety conditions for the segment in Table 4.14h are met.	For the segment: Capacitance $[C_{tot}] < [C_o]$ for IIC Inductance $[L_{tot}] < [L_o]$ for IIC	**Yes** **No!!**

[a]Some devices may require a minimum supply voltage of more than 9 V. This calculation assumes the worst case of all devices being at the end of the bus, however, so that there is no problem in using 9 V unless the offending device is the last one on the bus.

TABLE 4.14g
Technical Safety Characteristics of the Components Used in the Examples

Safety Element	U_o	I_o	P_o	C_o	L_o
MTL 5083	22 V	216 mA	1.2 W	165 nF	320 μH
Pepperl+Fuchs KFD2-BR-EX1.2PA.93	15 V	207 mA	1.93 W	FISCO	FISCO
Devices	U_i	I_i	P_i	C_i	L_i
Flowmeter	30 V	500 mA	5.5 W	5 nF	10 μH
Pressure transmitter	17.5 V	500 mA	5.5 W	5 nF	10 μH
Radar-level transmitter	17.5 V	280 mA	4.9 W	FISCO	FISCO
Valve positioner	20 V	220 mA	3.2 W	FISCO	FISCO
Terminator	30 V	—	1.2 W	—	—

TABLE 4.14h
Technical Safety Conditions for the Individual Devices and for the Segment

Hazardous Area (Device)	*Condition*	*Safe Area (Barrier)*
Maximum voltage [U_i]	≥	Idle voltage of safety element [U_o]
Maximum current [I_i]	≥	Short-circuit current [I_o]
Maximum power [P_i]	≥	Maximum output power [P_o]
Maximum internal capacitance [C_i]	≤	Maximum permitted external capacitance [C_o]
Maximum internal inductance [L_i]	≤	Maximum permitted external inductance [L_o]
	Hazardous Area (Segment)	
Total capacitance [C_{tot}]	<	Permitted capacitance for IIC [C_o]
Total inductance [L_{tot}]	<	Permitted inductance for IIC [L_o]

In our example the condition $L_{tot} < L_o$ was not met since the cable is too long. Possible solutions to this problem are to:

- Mount the power supply/barrier in a separate field housing at least 175 m down the trunk line in the immediate vicinity of the hazardous area. This effectively reduces the trunk length from 580 to 405 m, fulfilling the inductance condition.
- Check whether and under what conditions the manufacturer of the safety barrier or segment connector allows the use of the L/R ratio for technical safety considerations.

FISCO EEx ia IIC Segment

Figure 4.14i shows a FISCO intrinisically safe segment, which is to be checked for functionality. It is constructed as follows:

- The segment starts with a link or segment coupler that contains the barrier and adds intrinsically safe power to the bus.
- The bus is terminated at its start and with a separate terminator at its end.
- Both loop-powered and four-wire devices are connected to the bus. The externally powered devices must have an EEx e or EEx d power supply, depending on design. The circuit connected to the bus must be EEx i.
- Cable Type A with a loop-resistance of 44 Ω/km is used.
- All segment components are certified as FISCO Ex i conforming.

The functional check is made on the basis of the values in Tables 4.14a to 4.14d (bus coupler Pepperl+Fuchs) as follows.

Step	Procedure	Calculation/Condition	Result
1.1	Calculate the cable length by adding together the length of the trunk and all spurs that are longer than 1 m (3 ft): $l_{SEG} = l_{trunk} + \Sigma l_{spurs}$	$l_{SEG} = l_{trunk} + \Sigma l_{spurs}$ $= 550.5$ m $+ 90$ m **= 640.5 m**	
1.2	Check that it lies within FISCO specifications for Type A cable.	The segment length l_{SEG} is less than 1000 m.	**True**
		There is no spur longer than 30 m.	**True**
		Separate terminator.	**True**
1.3	When all the conditions are met, the network can be structured as required. If one of the conditions is not met, then the network structure must be revised.	All conditions met?	**Yes**
2.1	Calculate the current I_{SEG} drawn by the segment $I_{SEG} = \Sigma I_B + \max I_{FDE} + I_{MOD}{}^{*} + \Sigma I_{startup}$ $^{*}I_{MOD}$ is accounted for in the specification of the P + F coupler.	$I_{SEG} = \Sigma I_B + \max I_{FDE} + \Sigma I_{startup}$ $= 2 \times 12$ mA $+ 2 \times 10.5$ mA $+ 2 \times 11$ mA $+ 2 \times 13$ mA $+ 4$ mA $+ 0$ mA **= 97 mA**	
3.1	Check that the current drawn by the segment I_{SEG} is within the current limits of the fieldbus power supply I_S	$I_S \geq I_{SEG}$ 110 mA ≥ 97 mA	**Yes**
4.1	Check that there is sufficient voltage at the last device.		
4.2	Calculate the resistance of the cable R_{cable}: $R_{cable} = R_{loop} \times l_{SEG}$	$R_{cable} = R_{loop} \times l_{SEG}$ $= 44$ Ω/km × 0.6405 km **= 28.12 Ω**	
4.3	Calculate the voltage U_{FG} at the last field device: $U_{FG} = U_S - I_{SEG} \times R_{cable}$	$U_{FG} = U_S - I_{SEG} \times R_{cable}$ $= 13.0$ V − 97 mA × 28.1 Ω $= 13.0$ V − 2.73 V **= 10.27 V**	
4.2	Is the voltage at the last device $U_{FG} \geq 9$ V?[a]	$U_{FG} \geq 9$ V 10.27 V ≥ 9 V	**True**
5.1	If all conditions are met, the segment will function.	All conditions met?	**Yes**
6.1	Check the technical safety of the segment.		
6.2	The individual inductance and capacitance values of the devices and terminator must fulfill the conditions in Table 4.14h. This is implied when they have FISCO certification.	Are all components on the segment certified according to FISCO rules?	**Yes**
6.3	If the answer to Item 6 is yes, the segment is technically safe. If one of the components is not FISCO certified, it must be replaced by one that is.		

[a]Some devices may require a minimum supply voltage of more than 9 V. This calculation assumes the worst case of all devices being at the end of the bus, however, so that there is no problem in using 9 V unless the offending device is the last one on the bus.

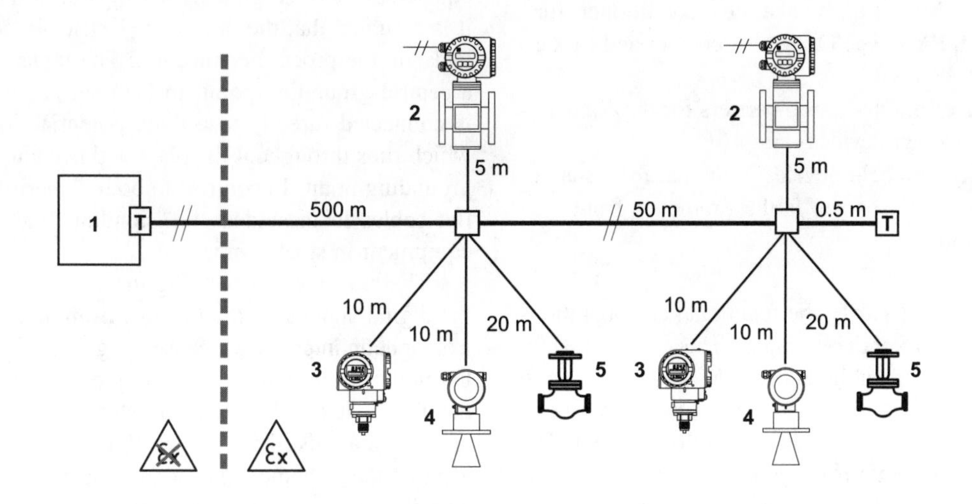

FIG. 4.14i
Example of a FISCO bus segment in an explosion-hazardous area: 1 = intrinisically safe segment coupler EEx [ia] IIC; 2 = intrinsically safe flow device with EEx de external power supply; 3 = intrinsically safe level device; 4 = intrinsically safe pressure device; 5 = intrinsically safe valve positioner; T = separate EEx i terminator.

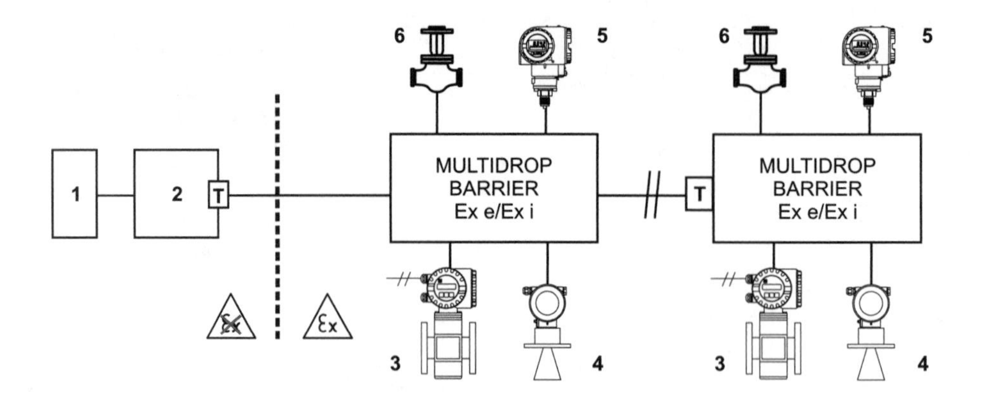

FIG. 4.14j
Example of an EEx e/EEx i bus segment in an explosion-hazardous area using multidrop barriers: 1 = I/O card or linking device; 2 = power supply and conditioner; 3 = intrinsically safe flow device with EEx de external power supply; 4 = intrinsically safe level device; 5 = intrinsically safe pressure device; 6 = intrinsically safe valve positioner; T = EEx i terminator. For PROFIBUS-PA elements 1 and 2 can be replaced by a safe segment coupler and an EEx e barrier.

Other Possibilities

There are two other possibilities for designing segments for hazardous areas, which are not described in detail here:

1. *Ex d concept*: In this case an appropriate power supply must be selected (Ex e) and the entire bus run in Ex d approved conduits. The devices must have a corresponding type of protection. 7/8 in. connectors are not allowed in this concept, i.e., the bus connections must be made with screw terminals. The disadvantage of this method is that the devices cannot be connected or disconnected when the system is running.
2. *Ex e/Ex i concept*: Up to 32 field devices can be connected to a mixed Ex e/Ex i segment. The concept requires the following components; see also Figure 4.14j:
   - An Ex e power supply and bus conditioner for PROFIBUS-PA a Type C segment coupler and EEx e barrier can be used.
   - Special Ex e/Ex i multidrop barriers for the connection of up to four devices
   - Intrinsically safe field devices in the hazardous area
   - Termination at the barrier farthest removed from the power supply

The complete bus must satisfy the functional considerations described in Steps 1 to 5 of the examples. The technical safety considerations are based on the entity concept. In contrast to the standard Ex i concept, however, the technical safety calculation is made for each individual multidrop barrier. The devices are considered safe if the junction box output characteristics (U_o, I_o, P_o, C_o, L_o) do not exceed the input parameters of the field devices (U_i, I_i, P_i, C_i, L_i) and the inductance and capacitance are within the permissible limits.

GROUNDING SCHEMES

This section describes three possible grounding schemes for an IEC 61158-2 segment.

1. Single-point grounding
2. Multipoint grounding
3. Multipoint grounding with capacitive isolation

The difference between the schemes lies in the grounding of the bus cable shield. In one case it is connected to a separate ground and in the others it is integrated into the general plant grounding system.

Plant grounding schemes also differ according to national and or local practice. For example, British practice is neutral star earth bonding, German practice is the use of a potential equalization line (Figure 4.14k). In neutral star earth bonding it is assumed that the device is electrically connected to the tank via the process connection. The tanks are grounded at a central grounding point. In German practice, each device is connected directly to a thick potential equalization line, which runs throughout the plant and is connected to a single grounding point. In contrast to both, American practice is to run cables in grounded steel conduits and enclose control equipment in steel cabinets.

The reason for grounding the cable shield is to protect the digital signals on the fieldbus from high-frequency electromagnetic interference caused, e.g., by cellular telephones, harmonics from frequency converters, heavy electrical equipment, etc. It can be seen, therefore, that local installation practice may also have some influence on the suitability of a particular scheme. The conditions for transmission may also be poorer in extensive networks or in networks with many branches. Finally, Ex considerations may rule out some schemes in some countries.

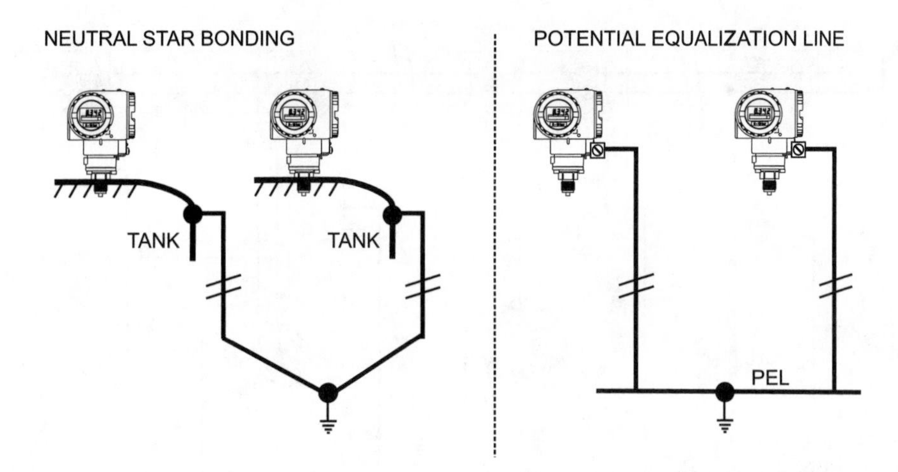

FIG. 4.14k
Device grounding schemes: British neutral star bonding and German potential equalization line.

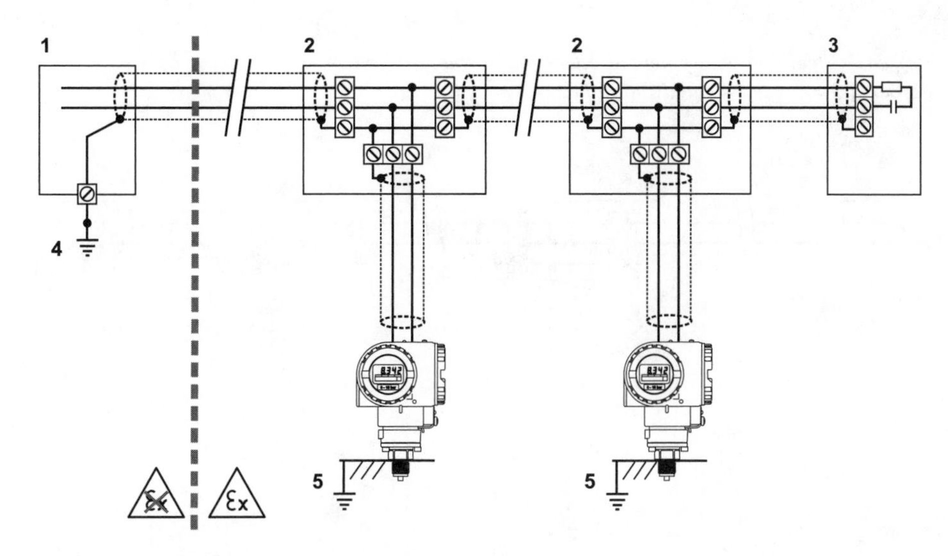

FIG. 4.14l
Single-point grounding of cable shielding with optional grounding of devices: 1 = power supply; 2 = T-box or junction box; 3 = bus terminator; 4 = grounding point for bus cable shield; 5 = optional grounding of devices, e.g., by neutral star bonding or conduit.

Experience has shown that a great deal of care and attention must be paid to the wiring of a fieldbus segment. Two factors may influence overall bus performance:

1. *Choice of cable*: Cable Type A undoubtedly gives the best results, no matter the total length of the bus. If one has the choice, use Type A cable.
2. *Continuity in the cable shielding*: The bus cable shielding must exhibit electrical continuity throughout the length of the bus.

The grounding schemes shown below assume that the T-boxes/junction boxes are serving plant sections that are some distance from each other and the control room, making it likely that each is connected to a different earth potential. They address the problem of preventing current loops developing within the cable shield, should it be grounded at both ends. For simplicity, the wiring is shown with screw terminals. If a connector is used, the device connection compartment should be opened to check whether the shield wire is already connected to the internal device grounding terminal. If this is the case, it must be detached and isolated if an isolated scheme is required. For T-boxes and junction boxes, the manufacturer's instructions must be followed.

Single-Point Grounding

Single-point grounding of the cable shielding (Figure 4.14l), is the scheme described in IEC 61158-2, and is the favored method in Britain and the United States. In this case, the

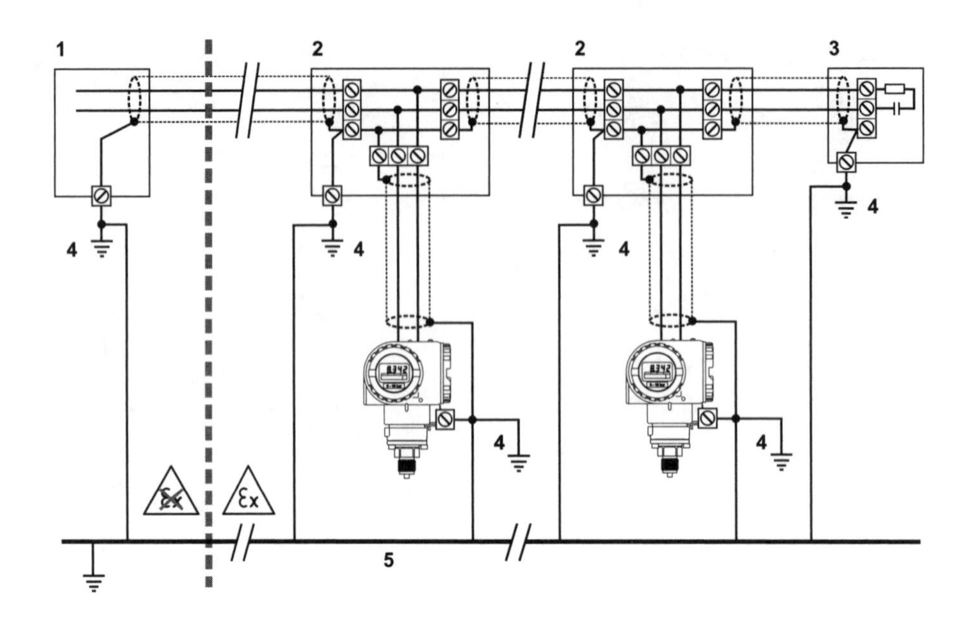

FIG. 4.14m
Multipoint grounding of cable shielding with potential equalization line: 1= power supply/linking device; 2 = T-box, junction box; 3 = terminator; 4 = local ground; 5 = potential equalization line.

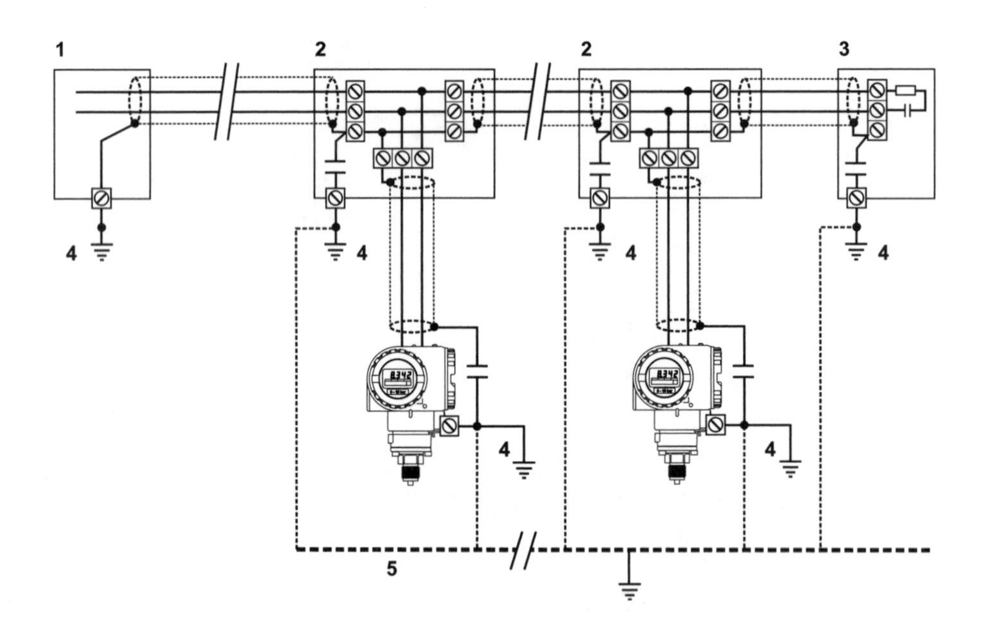

FIG. 4.14n
Multipoint grounding with capacitive isolation in a nonhazardous area: 1 = power supply/linking device; 2 = T-box, junction box with capacitive shield grounding; 3 = terminator with capacitive shield grounding; 4 = local grounds; 5 = potential equalization line (optional).

cable shield ground is fully isolated from the device grounds. The shield is grounded at the power supply or safety barrier only. In German practice the device and component grounds would lead to a common potential equalization line.

EMC tests have shown that with this scheme the bus signal is not optimally protected from high-frequency interference. Just how much this disturbs communication depends on the length of the bus, its topology, and the sources of interference. If care is taken to avoid parallel runs of bus and unshielded power cables or the bus runs in grounded metal conduits, as, e.g., in American installations, then it will probably work quite well.

Multipoint Grounding

Multipoint grounding (Figure 4.14m) provides enhanced protection against electromagnetic interference in noisy environments. It is the favored method in Germany. All devices and cable shields are grounded locally. Each local ground is connected to a thick potential equalization line,

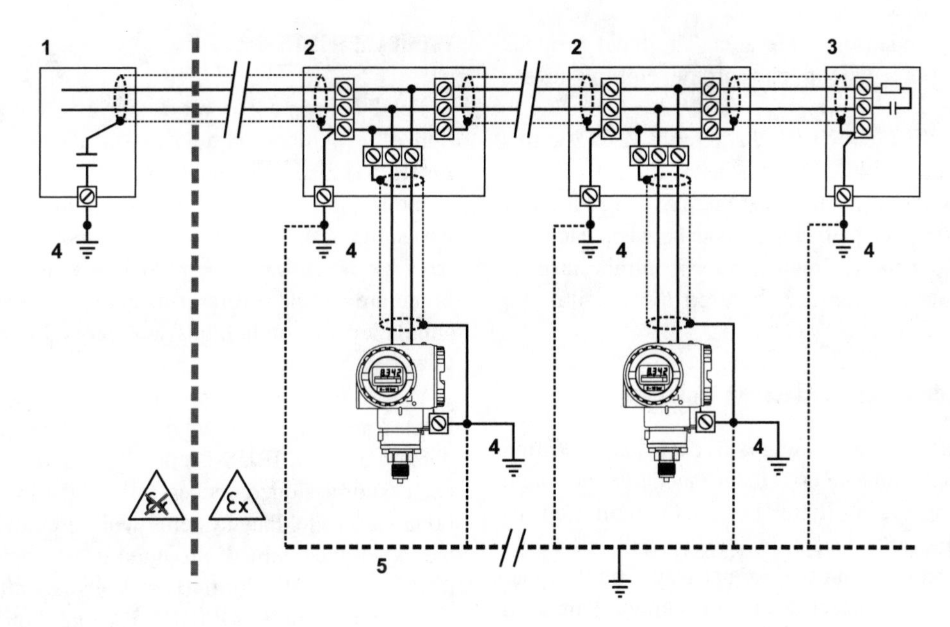

FIG. 4.14o
Multipoint grounding with capacitive isolation in an explosion-hazardous area: 1 = power supply/linking device with capacitive grounding; 2 = T-box, junction box; 3 = terminator; 4 = local grounds; 5 = potential equalization line (optional).

TABLE 4.14p
Troubleshooting Table

Symptom	*Cause*	*Action/Remedy*
No communication on bus	Bus interrupted	Check the integrity of the bus line
	Last device has less than 9 V/too many devices on the bus	Check with voltmeter; check all connections; if necessary remove device from bus
	Terminator not optimally placed	Check that the terminator is at farthest point from the power conditioner/coupler
	Additional terminators switched on	Check all components between the start and end of the bus with switchable terminators
	Bus parameters not set correctly (PROFIBUS)	Check that baud rate set at all DP components corresponds to the baud rate of the coupler or link
Device not found	Device not connected	Check the bus connections; if they are OK, perhaps the device is defective or not receiving enough power (see above)
	Device not correctly integrated into system	Check that the device description files are correct for the device version at hand
	Double addressing (PROFIBUS)	Either reset the address at the device or take all of the offending devices from the bus and progressively introduce and readdress each one using the configuration tool
Poor communication	Bus has too many branches	Check the signal distortion; maybe better shielding or a stricter grounding scheme will help
	No continuity of shielding	Check the integrity of the shield at all points, including prefabricated cord sets
Sporadic communication breaks	Faulty contact	Check all contacts including screw connectors; maybe one has come loose due to vibration
	One particular device is disturbing the communcations	Monitor bus; check which device is talking and remove it from the bus; contact the device manufacturer regarding a solution
	A machine is interfering with communication	Check the shield gounding; maybe better protection is required
	Cellular telephone is interfering with communication	Check the shield gounding; maybe better protection is required

which itself is grounded in a safe area. The local grounds are necessary to prevent loop currents developing in the cable shielding.

According to IEC 60079-13, Paragraph 12.2.2.3, this method can be used provided that the installation is effected and maintained in such a manner that there is a high level of assurance that potential equalization exists between each end of the circuit. Under these circumstances it fulfills intrinsic safety requirements and can also be used in a neutral star bonding scheme.

Multipoint Grounding with Capacitive Isolation

In the case of multipoint grounding with capacitive isolation (Figure 4.14n), capacitors are used to DC decouple the shield from the grounding system (here shown as potential equalization line) at all points except the power supply or safety barrier, which is grounded in the normal way. Small capacitors (e.g., 1 nF, 1500 V, dielectric strength, ceramic) are used and the total capacitance connected to the shielding may not exceed 10 nF. The capacitors are normally built into the T-box and junction box connectors, and offered as such by the component manufacturers. This scheme may be used as an alternative to multipoint grounding where this is not permitted or where the ground is different in different plant areas.

If the bus runs into a hazardous area the T-boxes and junction boxes must be wired in the normal way and the power supply or barrier grounded via a capacitor (Figure 4.14o).

TROUBLESHOOTING

A fieldbus segment is more than just the physical connection of a number of devices on a single line; its correct function is also dependent on the correct operation and integration of the devices within the control system. Table 4.14p lists some of the most common faults that have been encountered by the authors' company in 4 years of installing buses. Some are simple errors in installation; others have required simulation in our fieldbus laboratory before they could be localized.

Most faults are easy to remedy once they have been located and the cause diagnosed. In the time since the introduction of PROFIBUS-PA in 1997, hundreds of systems have been commissioned successfully and run without problem. Foundation Fieldbus systems will undoubtably benefit from the experience gained, provided its champions are prepared to use it. It is not helpful to users when conflicting statements are made about PROFIBUS-PA and Foundation Fieldbus installation, when we should really be talking about the IEC 61158-2 physical layer.

Bibliography

Foundation Fieldbus, see website www.fieldbus.org.
PROFIBUS-PA, see website www.profibus.com or www.profibus.org.

4.15 Ethernet and TCP/IP-Based Systems

E. J. BYRES

Partial List of Suppliers: 3Com Corporation; Cisco Systems; Hirschmann Electronics; Lucent Technologies; Schneider Electric; Sixnet; Woodhead Connectivity

There is little doubt that Ethernet and TCP/IP (transport control protocol/Internet protocol) networking are among the hottest technologies to hit the plant floor in the past decade. Look at any process controls magazine or attend any industrial conference and one will quickly notice that Ethernet and TCP/IP are topics that generate a lot of interest. At the Instrument Systems and Automation Society ISA Expo 2000, for example, presentations with the term *Ethernet* in the title experienced four times the average attendance rate.

Few experts are surprised Ethernet is doing so well in the industrial marketplace. It has completely won over the commercial sector, relegating former contenders like ARCnet and Token Ring to the same fate as hula hoops and Beta videotapes. Furthermore, due to its huge market share, it has become a very inexpensive technology and one well understood by network specialists, who find it one of the simplest network technologies to implement. Similarly, TCP/IP has emerged the victor over its main competitor, IPX/SPX, which has been reduced to a historical footnote.

Clearly, both users and vendors unanimously agree Ethernet and TCP/IP-based technologies have an important place in the design of the modern control system. Together, they offer the possibility of a truly open connectivity standard. PLC (programmable logic controller) or DCS (distributed control system) vendors may often only support their favorite field device network such as PROFIBUS or Foundation Fieldbus, whereas every vendor supports Ethernet–TCP/IP connectivity of some form. These protocols have become the *lingua franca* of process control, the new RS-232 of the 21st century.

Despite this overwhelming popularity, Ethernet and TCP/IP are poorly understood by most controls professionals. Few understand the limitations of Ethernet technology as it applies to the industrial environment because its underlying technologies and design strategies are shrouded in myth and misinformation. This section attempts to provide the reader with a working knowledge of the facts behind industrial Ethernet and TCP/IP, including the core elements of Ethernet, the use of repeaters, switches, and routers, and the strengths and weaknesses of Ethernet on the plant floor.

THE CORE ELEMENTS OF ETHERNET

Ethernet technology was originally developed by Xerox in the early 1980s and was adopted as the IEEE 802.3 standard by the Institute of Electrical and Electronics Engineers in 1988. Since then, it has become the dominant standard for local-area networks (LANs) throughout the world, with over 200 million nodes installed at the close of the year 2000.[1]

Ethernet is what we call a *physical layer* and *data link layer* protocol.[2] The physical layer part of the standard defines the cable types, connectors, and electrical characteristics. The data link layer defines the format for an Ethernet frame, the error-checking method, and the physical addressing method. It also defines the protocol Ethernet uses to determine when nodes can transmit on the network. The scheme is known as carrier sense, multiple access with collision detection (CSMA/CD) and is the defining technology for Ethernet. (For more details on the OSI Reference Model for protocol layers see Section 4.4—Sorting out the Protocols).

Although the physical and data link protocols are critical to the operation of a network, it is important to understand that they are not enough to make a system operational. For example, Ethernet cannot help a message find its way through a complex network like the Internet. Nor can it define how to carry out specific tasks on a network, such as transferring a file, sending e-mail, or reading a block of registers in a PLC. To accomplish these goals, we need to add additional protocols on top of Ethernet to create a *protocol suite*.

The TCP/IP protocols are almost universally used on top of Ethernet to provide the network and transport layers in the OSI protocol model and to solve issues of routing and end-to-end data integrity (Figure 4.15a). In fact, Ethernet and TCP/IP are used together so often that many people mistakenly believe they are synonymous rather than very separate standards. That said, even the Ethernet–TCP/IP combination is not enough. Every network needs a top layer of protocols called *application layer protocols*. These protocols provide for the task and command definitions to be executed over the network, such as a procedure to request that ten words of data be transferred between two PLCs.

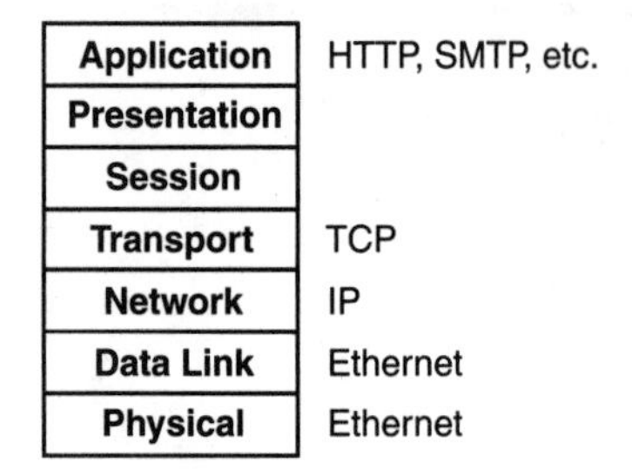

FIG. 4.15a
Ethernet and TCP/IP in the OSI Reference Model.

Understanding CSMA/CD

As noted earlier, one of the core technologies of Ethernet is its CSMA/CD technology. CSMA/CD provides the media access control (MAC) method used to determine which device on a network has the right to transmit at any given moment. Despite its complicated name, CSMA/CD is a very simple concept. To the Ethernet device "carrier sense" simply means listen before sending, while "multiple access" means any device can send data if the cable is quiet. In many ways, CSMA is like a telephone party line. Before calling, one picks up the telephone to make sure a neighbor is not using the line. If someone is already on the line, one can only hang up and try again later.

The "collision detection" part of the term deals with the situation when two nodes both come onto the network at the same time and start to transmit simultaneously. This collision is detected electronically and both nodes will send a jam signal to let other devices know a collision has occurred. The nodes will then *back off* (i.e., stop transmitting) and wait a random amount of time before trying to resend their message. The randomness of the backoff time helps prevent the two nodes from colliding repeatedly.

Collisions are not bad unless there are lots of them. In fact, Ethernet networks have been known to continue to operate with collision rates as high as 40%. However, this collision detection scheme does have important ramifications to the operation and design of Ethernet networks. Traditional CSMA/CD Ethernet is a nondeterministic network, as there is no guarantee when a node will get on the network. If the network is busy, then a node with a message, no matter how important, has to wait. One can guess how long the wait would be, but the guess is based on probabilities.

Ethernet opponents have long used the nondeterministic nature the Ethernets as a reason to avoid it on the plant floor. To understand why nondeterminism might be a problem, imagine a PID (proportional/integral/derivative) controller collecting data from a field transmitter over an Ethernet segment. To tune the PID loop, one needs very consistent sample times. In theory, traditional Ethernet could not guarantee consistent timing for message delivery. Moreover, people widely believed that any Ethernet network loaded above 37% would experience an exponential growth in transmission delay times and fail.

In reality, the determinism and loading issue has been a bit of a red herring. Studies conducted in the late 1980s showed that, in practice, Ethernet delays are linear and can be consistently maintained less than 2 ms for a lightly loaded network and 30 ms for a heavily loaded network.[3] These delays are inconsequential for most process control applications. Other studies have shown that factors such as network card buffer handling will impact the message delivery time long before the collisions have any impact.[4]

The main concern is to keep the traffic level on an Ethernet control network reasonably low so that collisions occur infrequently. For most systems, this is done by restricting the number of devices and the amount of traffic each device generates so the total segment traffic never rises above 10 to 20%. If more devices are needed, they are simply put on separate segments and then connected through a router or switch.

Plenty of evidence is available that demonstrates this concept can work well in the field. For the past decade a number of DCS vendors such as ABB and Foxboro have been relabeling Ethernet and using it for their intercontroller communications (MasterBus 300 and Nodebus, respectively). More recently Modicon, Opto-22, and other PLC vendors have been releasing Ethernet I/O (input/output). All are based on the strategy of keeping traffic levels low enough that the probability of significant delays from collisions is in the same range as delays from noise, an issue all networks face. In addition, new developments in Ethernet switch equipment have made many of the questions of Ethernet collisions and nondeterminism irrelevant and will be discussed later in this section.

Ethernet Frame Format

Another Ethernet constant is that every message has the same structure known as the *Ethernet frame*. Each frame can be no smaller than 64 octets and no larger than 1514 octets.

Note: An octet is simply 8 bits and is officially used in Ethernet literature instead of the more common term *byte*.

As shown in Figure 4.15b, the frame structure includes:

- *Preamble* that allows the receiver to synchronize with the transmitter (7 octets)
- *Start of Frame Delimiter (SFD)* that tells the receiver that the preamble is over (1 octet)
- *Destination Address* field indicating the intended receiver of the frame (6 octets)
- *Source Address* field indicating who generated the frame (6 octets)
- *Length/Type* field determining length or type of frame; if the value is less than 1500 [Decimal] or 05D0 [Hex] it indicates the frame length, otherwise it indicates the frame type (2 octets)
- *Data* field containing the headers for the layers above Ethernet and the actual data

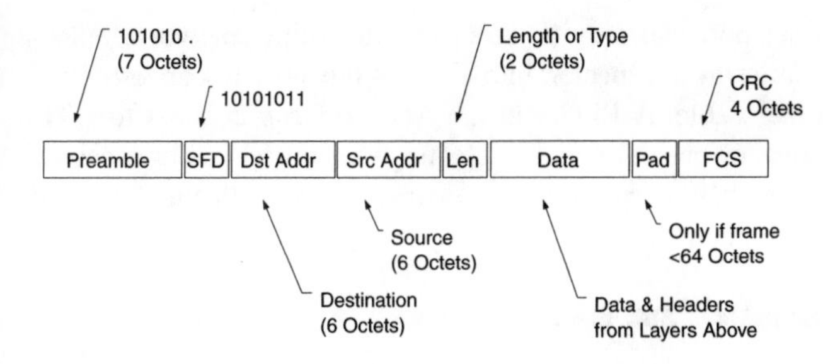

FIG. 4.15b
The Ethernet frame.

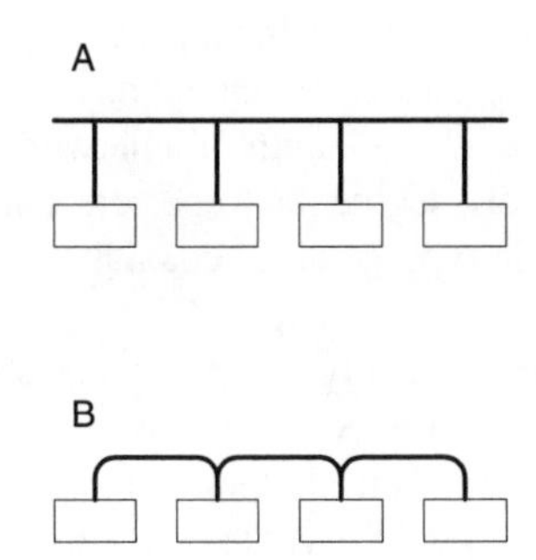

FIG. 4.15c
Trunk–drop vs. daisy-chain bus topologies.

- *Pad* field, extra meaningless octets that are inserted to ensure a minimum of 64 octets if the frame is smaller than 64 octets
- *Frame Check Sequence (FCS)*, an error check field (4 octets)

There are actually two variations of Ethernet frame depending whether IEEE 802.3 or traditional Digital/Intel/Xerox (DIX) Ethernet is being used. The difference is largely defined by how each uses the length/type field. Regardless, every modern Ethernet device can handle both formats transparently and is of no concern to most network designers.

Note: The use of the 802.1p/Q protocols causes a 4-octet field to be inserted into the traditional Ethernet frame. This will make the Ethernet frame look slightly different from the above description.

Topology Overview

A network topology is the general strategy for connecting the devices on the network. When the Ethernet standard was initially defined, it was based on bus topology with a main trunk cable that connects all devices in linear fashion. The first version of Ethernet, known as 10Base5 or Thicknet Ethernet, had devices connected to a coaxial trunk using a drop lines, a topology known as a trunk-and-drop bus. Later 10Base2 or Thinnet Ethernet was created with devices that connected directly on to the trunk with no drop lines, a topology described as a daisy-chain bus (Figure 4.15c).

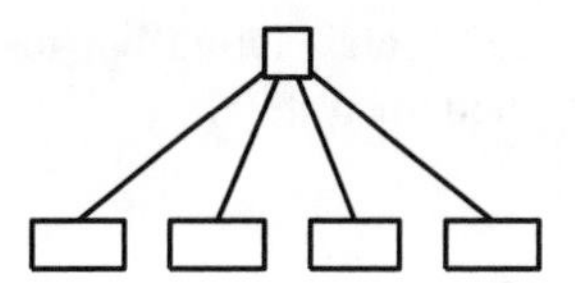

FIG. 4.15d
Star topology.

Anyone who has worked with buses quickly learns that trunk lines must be terminated at both ends with a termination resistor. The high-frequency signals used in Ethernet travel down the trunk until they encounter a change in the *characteristic impedance* of the cable, at which point the signals reflect back, much like a wave hitting a swimming pool wall. Because the end of a cable causes a change in its characteristic impedance, the signals will reflect back down the trunk, interfering with the rest of the data signal. The termination resistor prevents that reflection by absorbing the signal when it reaches the end of the trunk cable.

Crushed cables, shorted cables, cut cables, mismatched cables, and poor connectors also cause changes in impedance and resulting reflections. As a result, buses are very susceptible to cable damage anywhere along the trunk cable and it is notoriously difficult to locate the source of reflections on a cable. For example, a computer with a damaged connection to a trunk may operate perfectly well; yet reflections from that connection may knock out devices on the far side of the plant. In effect, every splice, connector, or inch of trunk cable is a potential point of failure for the entire network system.

An alternative to the bus is the star topology (Figure 4.15d), where all cables radiate from distribution centers. Typically, these central points utilize intelligent electronic hubs and the failure of a cable only impacts the one device connected to it. This way, cable damage does not affect the entire network and can be isolated quickly.

By the early 1980s network designers began to realize that a star topology is an inherently more stable solution than the bus topology. In 1988, the influential Electrical Industries Alliance/Telecommunication Industries Association EIA/TIA-568 standard for commercial communications cabling was published and it recognized star topology as the only acceptable design. The impact on Ethernet was that all new versions were based on the star, including the modern unshielded twisted pair (UTP) and fiber-optic systems such as Fast and Gigabit Ethernet.

The use of star topologies in industrial settings it is not without cost. There are many situations that using a bus simply makes good cabling sense. For example, in November 2000, the Industrial Ethernet Association newsgroup had an interesting posting about the problems faced by an engineer who wanted to use Ethernet to connect 21 devices that were scattered along a lengthy (300 m/1000 ft) conveyor. A Thick Ethernet (10Base5) bus system would have made the cabling task simpler, but would have forced the client to use outdated technology. In the end, the engineer selected a star-based solution that required more cabling but resulted in a more reliable and more modern system.

ETHERNET HARDWARE

In any industrial facility there is likely to be a number of Ethernet networks, ranging from business LANs to PLC control highways. Sometimes these networks are identical except for location—for example, the LAN in the accounting department is likely to be functionally identical to the LAN in engineering department. Other times, the networks are very different, such as Foundation Fieldbus high-speed Ethernet (HSE) network vs. a business network. Ethernet designers will use one of the two types of network attachment devices to connect devices to networks, namely, the network interface card (NIC) and the transceiver. To interconnect the various networks, the designer will use one of the five different types of network interconnection devices, known as repeaters, bridges, routers, gateways, and switches. The difference between an Ethernet repeater, bridge, router, or gateway is found in the protocol layer the device is designed to work with. Repeaters work at only the lowest layer, whereas bridges, switches, routers and gateways can interpret protocols at progressively higher layers. Each of these is addressed in some detail below.

Network Interface Card

The most basic Ethernet device is the NIC. This is the adapter that is installed in the computer or PLC to allow it to communicate over the Ethernet network. NICs in computers are usually card based and can be installed or removed, whereas the NICs in many PLCs are built in. Regardless, they have the same functions; they convert the Ethernet electrical signals to a signal the device can understand. They assemble the Ethernet frame for transmission and check incoming messages to see if they are intended for their device and if the messages have been corrupted in transit. Most NICs also carry out the CSMA/CD procedure so their device only transmits when it is approporiate and backs off if there is a collision. However, this feature can be turned off if a device also uses a transceiver.

Transceivers

When Ethernet was first implemented in the form of 10Base5 thick coax networks, the collision management features were handled in a separate device instead of in the NIC. This device was called a transceiver and it attached to the NIC via a special 15-pin connector called an attachment unit interface (AUI).

Today, 10Base5 is rarely used, but the AUI connectors and transceivers are still popular on many industrial systems. The reason is that they allow devices some independence from the cable type. For example, a PLC with an Ethernet fiber port can only be connected to a fiber network, unless some sort of repeater, hub, or switch is used to convert to the other cable. A PLC with an AUI port can connect to UTP, coax, or fiber, simply by selecting the appropriate transceiver. Note that transceivers only can be used with the 10-Mbps versions of Ethernet.

Repeaters and Hubs

As noted above, repeaters work at only the lowest layer, i.e., the physical layer. When a signal travels along a network cable, it tends to lose strength. An Ethernet repeater can boost that signal by electrically reshaping the signal received on one port and then rebroadcasting it out its other ports. Repeaters can simply extend the length of a network by connecting two or more network segments. Repeaters also allow conversion between cable types such as coaxial cable to fiber-optic cable.

Most hubs are typically a type of repeater, and cannot tell if the signal they are receiving is part of a valid Ethernet message or just noise. Thus, hubs will forward bad messages just as effectively as they forward good messages. In addition, hubs do not understand who the message is intended for, so they do not make any attempt to direct messages—they just send them to everyone. Think of hubs as parrots—they repeat everything to everyone in hearing range, and, like parrots, that sometimes can be very inappropriate.

Bridges

A bridge is used to connect separate but related networks, or divide a larger network into two or more small networks. Working at the second protocol layer (the data link layer), bridges open and check packets as they are received. Most can learn addresses of the devices on each port, forwarding only the necessary traffic through.

This feature makes bridges very useful for controlling network loading. If the traffic between two computers is overloading a network, a bridge will prevent the traffic from spreading over other networks and overloading them as well. Furthermore, a bridge will check the physical integrity of each message, preventing any invalid packets from passing onto other networks. For example, a bridge isolating the administration network from the process control network would prevent heavy e-mail traffic in accounting from tying up the process network. It would also stop noise-corrupted packets from engineering from getting onto the process network.

Routers

Operating at the third, or network, layer of the OSI model, routers interconnect complex networks, such as the Internet or a corporate wide-area network (WAN). Communicating with other routers, they select the best possible route for a message, based on criteria such as availability, cost, loading, and speed.

Routers are intelligent devices used to divide networks logically rather than physically. For example, an IP router can divide a network into various subnets so that only traffic destined for particular IP addresses can pass between segments.

Another router feature is filtering. Protocols such as TCP contain information about the purpose of packets (e.g., e-mail, file transfer, video, etc.), and IP contains information about their source and destination. This can be useful for network security because it allows use of the router to filter out certain types of messages and prevent them from ever entering a network. This filtering of network traffic through a router is known as firewalling a network.

Gateways

Gateways provide support for all seven layers of protocol and thus can connect completely different systems. For example, gateways are often used to connect a DCS highway to Ethernet or a Windows/Ethernet LAN to an IBM mainframe. However, the fact that they have to interpret all seven layers of protocols makes most gateways relatively slow and expensive.

Switches

The newest Ethernet device attracting lots of attention is the network switch. A switch is basically a multiport bridge (a layer-two switch) or multiport router (a layer-three switch) with a very high speed backplane. Each port connects to an independent network and the high-speed backplane transfers the messages between ports.

Like bridges and routers, switches open and check every packet they receive, only forwarding packets that are error-free. Most also learn addresses of the devices attached to each port, directing only the necessary traffic through to each port, a feature that makes switches very useful for controlling network loading. Think of switches as intelligent traffic cops who try to make sure traffic only goes where it should and that bad traffic is prevented from spreading.

Choosing the Right Hardware

Switching technology has emerged as the evolutionary heir to bridges and hubs in most modern networks. Where bridging technology was once used, switches now dominate, due in part to their superior performance, lower per-port cost, and greater flexibility. Most network engineers consider it good design strategy to only use switches for a plant floor network. A few of the reasons include:

1. *Prevent Spread of Bad Packets*: Switches prevent the propagation of ill-formed packets throughout the network while hubs do not. There are numerous reports of entire hub-based networks being brought to their knees by one jabbering network interface card. A switch would prevent this from happening.
2. *Superior Traffic Control*: Switches provide superior traffic control to hubs, making collisions a thing of the past and solving the Ethernet determinism question.
3. *Prevent Broadcast Storms*: Broadcasts are special packets that are addressed to every device on the network. In small numbers, they are important to proper operations but, when the network is accidentally flooded with broadcast packets, a broadcast storm results. These storms can make a mess of the best-planned network and hubs do nothing to control them. Layer 2 switches can be configured to block broadcasts when they go over a set threshold, and Layer 3 switches can prevent the propagation of broadcast storms altogether.
4. *Seamless Connection of 10-Mbps and 100-Mbps Devices*: Most PLC systems utilize 10Base-T Ethernet, while most PC systems are 100Base-T based. Unless it is acceptable to slow everything to 10 Mbps, a switch must be used to allow blended data rates. Similarly, most uplinks in industrial sites are 100Base-FX. If a switch is not used, these cannot be connected to the slower PLC Ethernet networks.
5. *Simplified Ethernet Design*: As discussed later in this section, hub/repeater count rules for Ethernet are so complex that few people can use them properly. Using switches removes this very serious design issue because the switches are not subject to these rules.

There are several arguments against using switches on the plant floor; the most prevalent is that switches are too expensive. But this is false economy—switches are typically only a few hundred dollars more than a hub of similar quality. In addition, the cost of network electronics on most projects is small compared with the cost of cabling, engineering, and maintenance. The bottom line is that only switches should be used on the plant floor.

OVERVIEW OF THE ETHERNET VERSIONS

One of the reasons for the outstanding success of Ethernet in capturing the LAN market is that it has not been a static standard like EIA-232, but rather has evolved over the years as technology evolved. When Dr. Robert Metcalfe of Xerox first introduced it in the mid-1970s, Ethernet ran at 10 Mbps over a thick coaxial cable. In the early 1980s, the IEEE adopted Ethernet as the 802.3 standard and eventually extended it to operate over other media such as UTP and fiber-optic cable. The 1990s saw Ethernet speeds increase to 100 Mbps and then 1 Gbps, with 10 Gbps due to be released by time of publication of this volume.

Network specialists use a nomenclature to identify the various flavors of Ethernet that includes the data rate, the signaling method, and the cable length for coax or cable type for fiber and UTP. For example, the designation 10Base5

TABLE 4.15e
Common Ethernet Variations

	10Base-5	*10Base-2*	*10Base-T*	*10Base-FL*	*100Base-T*	*100Base-FX*	*1000Base-TX*
Common name	Thicknet	Thinnet	10Base-T	10Base-FL	Fast Ethernet	Fast Ethernet	Gigabit Ethernet
Media access	CSMA/CD	CSMA/CD	CSMA/CD	CSMA/CD	CSMA/CD	CSMA/CD	CSMA/CD
Topology	Bus	Bus	Star	Star	Star	Star	Star
Cabling	RG-8 coax	RG-58 coax	UTP	Fiber	UTP	Fiber	UTP
Data rate	10 Mbps	10 Mbps	10 Mbps	10 Mbps	100 Mbps	100 Mbps	1000 Mbps
Segment length	500 m	185 m	100 m	2000 m	100 m	412 m	100 m

indicates a data rate of **10** Mbps, **base**band signaling and **5**00 m maximum cable length without repeaters. Similarly, the 100Base-T indicates a data rate of **100** Mbps, **base**band signaling and unshielded **t**wisted pair cable. Table 4.15e shows the most common Ethernet variations, which are describe in more detail below.

10 Mbps Ethernet Standards

The original Ethernet, 10Base5, is commonly known as "Thicknet" because the network trunk consists of thick RG-8U or RG-11U coaxial cable. Each Thicknet device uses a media access unit (MAU) or transceiver to attach to the coax cable through "vampire taps." These taps must be situated at marked intervals of 2.5 m on the trunk cable. A maximum of 100 taps per segment is allowed and the maximum segment length (without repeaters) is 500 m. Drop cables from transceivers to computers are called AUI cables and these have a maximum length of 50 m.

10Base-5 was soon relegated to backbone runs only, replaced by a new Ethernet variation called 10Base2 or "Thinnet." It used thinner RG-58 A/U coaxial cable at the same data rate of 10 Mbps. The thinner cable resulted in less expensive electronics at the cost of fewer stations and poor noise immunity. It also meant that segment lengths were reduced to a maximum of 185 m.

10Base2 uses a daisy-chain topology, rather than a trunk/drop topology. BNC style T-connectors attach each NIC directly to trunk cable with no drop line or external transceiver. If one should happen to install 10Base-2, it is crucial to remember that the T-connectors must be connected directly to the NIC; do not allow for any drop. As well, use only RG-58A/U cable—RG-58 U, RG-58B/U, and RG-58C/U either have the wrong dimensions or characteristic impedance.

Both 10Base5 and 10Base2 are now considered dead technologies. With the advent of UTP and fiber optics, neither of the two coaxial types was specified for new installations. By the early 1990s 10Base-T dominated the connection to the device, while 10Base-FL ruled the backbone.

10Base-T networks use a star topology where each node is connected directly to a switch or hub. The cable is a four-pair UTP cable and the connector is a small plastic jack called an RJ-45. Interestingly, only two of the four twisted pairs are used—one pair for transmitting and one pair for receiving. The maximum length for a segment is 100 m.

Fiber-based Ethernet also uses a star topology with a single pair cable. There are three 10 Mbps versions for fiber:

1. 10Base-FOIRL: An obsolete version for extending copper segments. It has a maximum segment length of 1 km.
2. 10Base-FB: An obsolete version for creating Ethernet backbones. It is not compatible with the other versions.
3. 10Base-FL: The current version for 10 Mbps point-to-point links between switches or repeaters. It has a maximum segment length of 2 km.

Regardless of media type, any two Ethernet segments can be joined through the use of repeaters. However, this cannot be done haphazardly. The CSMA/CD algorithm requires that a station involved in a collision still be in the process of transmitting its frame when the collision occurs. Because the Ethernet signals take time to travel through repeaters and along cables, extending a network carelessly can result in propagation delays that are excessive and collisions that are missed. The IEEE provides two models to determine the maximum cable lengths and repeater counts in a network. The simplest is Model 1, which describes a set of multisegment configuration rules for combining various 10 Mbps Ethernet segments. It is often referred to as the "5-4-3" rule: a maximum of 5 segments can be connected in series, with 4 repeaters, but only 3 segments may be populated with nodes. This turns out to be a gross oversimplification and should be used with care.

Model 2 details how to determine repeater/distance maximums based on calculation aids that verify the round-trip signal delay time and interframe gap shrinkage. Neither Model 1 nor Model 2 is trivial to use correctly, particularly for mixed-media Ethernet. A better solution is not to use repeaters at all and instead use switches, a device not subject to the repeater rule issues.

Fast Ethernet Standards

The most widespread type of Ethernet being installed today is 100Base-T and 100Base-FX, commonly called *Fast Ethernet*. These two versions operate at speeds of 100 Mbps over UTP or fiber cable and use a star topology. The 100Base-T version preserves the 100-m maximum UTP segment cable length defined under the 10Base-T standard. However, because of the increased speed, there are greater restrictions on the maximum distance between two nodes when using repeaters and the calculations are even more complex. 10Base-FX is limited to 412 m (due to the CSMA/CD technology) unless it is run in full-duplex mode. As a result, switches are used almost exclusively for all Fast Ethernet.

Gigabit Ethernet

Gigabit Ethernet (GbE) runs at 1000 Mbps over a variety of twisted pair and fiber media. However, its segment distances can be very limited due to current cable technology. Most issues exist around use of existing fiber-optic cabling for Gigabit Ethernet as the most commonly installed fiber cable (62.5/125–200 MHz) will only allow distances of 220 m. The rules for cable lengths are complex, but a rough guide to the variations is:

- 1000Base-T: UTP cable with 100-m maximum
- 1000Base-CX: Shielded cable with 25-m maximum
- 1000Base-SX: Short wavelength light source over fiber with 220- or 550-m maximum, depending on cable type
- 1000Base-LX: Long wavelength light source over fiber with 550- or 5000-m maximum, depending on cable type

Gigabit Ethernet can only be used with switches in full-duplex mode.

Future Standards

The future of Ethernet is clearly in the direction of 10 Gbps and a standard is expected by 2002. However, it is likely that Ethernet at these speeds will remain restricted to corporate and telecommunication company backbone runs and is unlikely to be seen on the plant floor for some time. The cost of both the Ethernet components and the more sophisticated fiber infrastructures make it uneconomic for short-distance applications.

TCP/IP OVERVIEW

What Is TCP/IP?

TCP/IP is a family of protocols that grew out of the early Defense Advanced Research Projects Agency (DARPA) to connect various Defense Department computer networks. The intent was to create a system that would be resilient in the face of nuclear missile strikes. As various universities and the public joined in, the network became what we know as the Internet, and TCP/IP evolved into the dominant set of protocols for routing and transport control in today's LANs and WANs. TCP/IP is designed to allow very complex networks to evolve (such as the Internet) and is constantly changing and growing as new demands are placed on the Internet. For most process control applications, TCP/IP is probably overly complex, but, like Ethernet, its universal adoption in the business world has advanced its use into the industrial world as well.

If one were to define the function of TCP and IP in a sentence, it would be fair to say that IP helps messages find a route through a complex network like the Internet and TCP makes sure the messages arrive at the end destination intact and in the correct order. However, the actual execution of these tasks is surprisingly complex. Both TCP and IP are protocols in their own right but are always in the company of a number of other protocols grouped under the TCP/IP banner. At a minimum, there are several network layer protocols that IP requires to function properly:

- *IP—Internet Protocol*: The actual routing datagram protocol that defines the frame format, the addressing scheme, and the routing control information of all messages
- *ARP—Address Resolution Protocol*: A protocol that translates IP addresses to physical Ethernet addresses
- *ICMP—Internet Control Message Protocol*: A protocol needed to pass IP error and status information between devices

At the transport layer, there are actually two protocols to choose from: TCP and the user datagram protocol (UDP). TCP sets up a connection between the two end computers and checks to see if every packet gets through. It also makes sure the packets get through in the correct order. In contrast, UDP sends packets and then forgets about them.

To a software application, using TCP is like using registered mail; one expects an acknowledgment to indicate the letter was received. On the other hand, UDP is like using regular mail, as there is no acknowledgment of the delivery. Although it might seem that TCP is superior to UDP, the TCP acknowledgments can be superfluous if an application layer protocol already provides acknowledgments. In fact, although TCP is much more famous, UDP is more commonly used for industrial control networks because it involves much less overhead.

On top of these five protocols are myriad other protocols such as the simple mail transfer protocol (SMTP), which provides a standard method of sending e-mails across a network, and the hyper text transfer protocol (HTTP), which allows a Web browser to read the layout of a Web page so it can display it on a user's computer. There are also protocols like open shortest path first (OSFP) and routing information protocol (RIP) that allow the network routers to pass route information between them. In all, there are over 100 protocols in the TCP/IP family and more are being added every month.

Whether one sees TCP/IP as simply a half dozen core protocols or a huge and complex suite of protocols depends on the application. Web developers will focus on all the upper layer protocols like HTTP and SMTP, whereas router designers consider the routing protocols like OSPF and RIP the key to TCP/IP. For the purposes of most industrial network designs and for the remainder of this section, the focus is solely on IP itself.

Basics of TCP/IP Addressing

Every device in a TCP/IP network needs a unique IP address to identify it from all the other devices on the network. This address is a 32-bit address written in the form:

192.168.24.16

where each number is the decimal coding for 8 bits. For example, the above *dotted decimal* address actually represents the bit pattern:

1100 0000 1010 1000 0001 1000 0001 0000

where the 192 represents the first 8 bits, 168 represents the second 8 bits, and so on.

Every IP address is divided into two parts. The first portion is known as the *network-prefix*, and the second part is known as the *host-number*.

Note: The term *host* in TCP/IP jargon simply refers to any IP-capable computer or device.

The network-prefix indicates the network on which a given host resides, while the *host-number* identifies the particular host on the given network.

What makes IP addressing particularly tricky is that the boundary point between network-prefix and host-number can vary from network to network. For example, in the IP address 10.1.2.3 the "10" is the network-prefix and the "1.2.3" is the host-number. Conversely, in the IP address 192.1.2.3, the "192.1.2" is the network-prefix and the "3" is the host-number.

When the Internet first evolved, the first few most significant bits in the IP address indicated the boundary point between network-prefix and host-number. For example, if the most significant bit was a zero, then only the first 8 bits indicated the network-prefix and the remaining 24 bits indicated the host-number. In the 1980s this would be called a "Class A" address, but today it would be referred to as /8 ("slash-eight") address. Table 4.15f shows the common IP classes.

Obviously, in the /8 address case there are only a very limited number of networks (126), but each can contain a huge number of hosts (16,777,214), whereas in the /24 case the situation is reversed. The result is that a few early applicants secured most of the available IP addresses in the form of /8 addresses and the rest of the world corporations have been forced to fight over the leftovers. As pointed out in "Understanding IP Addressing: Everything You Ever Wanted to Know":[5]

> The classful A, B, and C octet boundaries were easy to understand and implement, but they did not foster the efficient allocation of a finite address space. Problems resulted from the lack of a network class that was designed to support medium-sized organizations. A /24, which supports 254 hosts, is too small while a /16, which supports 65,534 hosts, is too large. In the past, the Internet has assigned sites with several hundred hosts a single /16 address instead of a couple of /24s addresses. Unfortunately, this has resulted in a premature depletion of the /16 network address space. The only readily available addresses for medium-size organizations are /24s which have the potentially negative impact of increasing the size of the global Internet's routing table.

Companies that were lucky enough to obtain a /8 or /16 address space quickly realized that their network would be more efficient if they subdivided their large address space into smaller "subnets" based on regional or functional divisions. For example, a company that had the /16 address 135.123.X.X might assign 135.123.1.X to the accounting department, 135.123.2.X to the engineering department, and so on. To let the various network devices know that the network-prefix/host-number boundary has shifted from the /16 position to the /24 position, the company would implement a *subnet mask* in every device. This mask could be written as simply /24 but is most commonly indicated as 255.255.255.0, where the three 255s designate 24 ones (255 decimal equals 1111 1111 binary).

IP Routing Basics

One of the special features of an IP address is that it is a routable address. This means there is location information in the address to assist in routing messages to a device anywhere in the system. To understand this concept, consider another

TABLE 4.15f
Common IP Address Classes

Class	*Notation*	*First Network in Class*	*Last Network in Class*	*No. of Networks*	*No. of Hosts/Network*
A	/8	1.X.X.X	126.X.X.X	126	16,777,214
B	/16	128.1.X.X	191.254.X.X	16,384	65,534
C	/24	192.1.1.X	223.255.254.X	2,097,152	254

network with routable addresses—the telephone system. Every North American telephone number contains three digits to indicate the area code, three digits to indicate the exchange, and four digits to indicate the phone. For example, if the number 919-990-2376 is dialed from the 713 area code in Houston, the phone system knows immediately not to search the local or state databases for the number. Instead, the call is immediately transferred to North Carolina (919). Once in North Carolina, the system searches the database of exchanges, discovers that 990 is in the Research Triangle Park area, and directs the call there. Only after the call is in the correct exchange is a database searched for the last four digits. The advantage of this system is that there is no need to maintain and search enormous telephone number databases. Instead, much smaller local databases can be used.

IP addresses operate in a similar fashion, with the area code replaced by a network number, the exchange replaced by the subnet number, and the local phone number replaced by the host number. If a host wants to send an IP message, it first checks the network/subnet portion of destination IP address to see if it is a "local call." If the destination network/subnet portion is the same as source network/subnet portion, then the destination host is assumed to be on the local network and the message is sent directly. However, if the destination network/subnet portion is different in any way, then the sending host assumes that the destination host must be on a remote network and it will hand off the message to its default router. The router then starts looking for a path to a router that accepts the particular destination network address, using a protocol such as OSPF. Once it knows the path, it can forward the message onward.

When configuring IP addresses, it is important to remember that a host always compares its network/subnet portion IP address with the address of the host it is trying to send to. If the network/subnet portions are not identical, it will always pass the message to its "default gateway" router. It will do this even if the two computers are attached directly to each other. If no default router has been configured, the message will never get through.

SHOULD ETHERNET AND TCP/IP BE USED ON THE PLANT FLOOR?

Ask any controls vendor about Ethernet in industrial environments and, nine times out of ten, a tale will be told about determinism and how it makes Ethernet either suitable or unsuitable for the plant floor. Although the determinism question is important in a small number of situations, it tends to obscure far more important questions regarding industrial Ethernet. As discussed earlier, the use of switches has largely removed the determinism problem from most applications.

In the following, some of the more critical issues regarding the implementation of Ethernet on the plant floor are discussed.

Tough Enough for the Plant Floor?

As Gary Workman of General Motors Controls Robotics and Welding has pointed out, one of the problems with defining an industrial Ethernet standard is there is no single industrial environment. For example, the environmental requirements of an industrial control room and a wet bottling line are very different.

At its simplest, industrial Ethernet can be separated into three categories; business interface Ethernet, a control-level industrial Ethernet, and a field-level industrial Ethernet. The control-level Ethernet is typically used in place of a traditional data highway, connecting PLCs, DCSs, and human–machine interfaces (HMI). The field-level industrial Ethernet takes the place of field I/O or fieldbuses, and connects directly to field devices, such as pressure transmitters or motor starters.

Note: These levels correspond to the plant, cell, and device levels described in Section 4.5 and the corporate LAN, control level, device/sensor levels described in Section 4.9.

Understanding this difference means an engineer can start using Ethernet on the plant floor without having to solve all of the issues surrounding its use. All that is necessary is to keep in mind what Ethernet needs to accomplish in a particular application. For example, the nodes on a control-level industrial Ethernet tend to be controllers that are independently powered and are housed in enclosures. As a result, a user of control-level industrial Ethernet can be far less concerned with issues such as waterproof connectors and power over the wire.

Ethernet Cabling Ethernet in the commercial world is based heavily on UTP copper-based cabling. Unfortunately, this cable, which is commonly known as Category 5E or Category 6 cable, may not be suitable for the plant floor. It has very low pull strength (only 25 lb) and is not crush resistant. According to many cable manufacturers, even over by tightened tie-wraps UTP can become damaged and it is not resistant to many chemicals. In the control room, these restrictions can be tolerated, but in an actual machinery area, unprotected UTP cable is likely to be destroyed.

At the present time, the designer who wants to employ Ethernet in a machinery area is faced with two possible solutions: either use fiber-optic cabling or protect the UTP in metal conduit or armor. Unfortunately, preliminary research presented by Bob Lounsbury of Rockwell Automation at the May 2000 ISA Industrial Ethernet Conference indicates that the conduit/armor solution may have its own problems. Placing a single Category 5 cable in 1-in. metal conduit resulted in slight degradation of return loss, attenuation, and near end cross talk (NEXT), three key indicators of cable performance. The cable in the tests was still within specification, but was worse in all cases when placed in conduit. Capacitive coupling between the conduit and the cable was believed to be responsible.

Another question is whether the lack of shield on UTP will make the network too susceptible to electromagnetic interference (EMI) from typical industrial machinery such as variable frequency drives and welding machines. Although some proponents have recommended that industry use a shielded cable, the preliminary reports in this area indicate that Ethernet can operate effectively without a shield. For example, Lounsbury also provided results from a test where a variety of shielded and unshielded Ethernet cables were run adjacent to the 480 V cables between a drive and 20 horsepower motor. No errors were detected for any of the cables. Shielded cables definitely performed better than unshielded cables, but the problems that shield terminations can introduce into industrial networks will likely make them undesirable in the long run.[6]

The future direction of copper-based Ethernet cabling is likely a UTP cable with a thicker, chemical-resistant jacket. The thicker jacket will both improve the cable strength and reduce capacitive coupling, by keeping the copper pairs farther away from the metal conduits or armor. As of summer 2001 no product had been released to market, but a number of cable manufacturers, such as Belden and Commscope, have produced test products.

At the present time fiber-optic cabling is the surest solution for field Ethernet cabling. Not only are fiber cables immune to EMI, arcing, short circuits, ground loops, and loss of ground, but they also are considerably more rugged than most coaxial and twisted-pair cables. Some unarmored fiber is even designed to have army tanks drive over it—even the worst forklift driver cannot beat that. Fiber also has a very large bandwidth so it can carry a great deal of data over very long distances without repeaters (see Section 4.18 for more details on designing fiber-optic networks).

Ethernet Connectors Another critical problem is the need for standard field-hardened connectors, particularly for field-level industrial Ethernet. Nearly all the commercial Ethernet equipment uses a RJ-45 jack, which is neither rugged nor waterproof. The ideal industrial Ethernet connector needs to be rugged enough to withstand the shocks and loads typically found on the plant floor, preferably meeting the IEC 68-2-27 standards. It must also withstand damage from vibration, which has been shown to destroy the gold contacts on some commercial Ethernet connectors. It should have the option to be waterproof and dustproof, meeting the IP67 and IP65 specifications. Ideally, it should be compatible with commercial off-the-shelf (COTS) hardware; otherwise, industry loses the enormous cost benefits from using low-cost commercial Ethernet electronics. Finally, it must meet the Ethernet electrical specifications for data rates of at least 100 Mbps (Gigabit Ethernet compatibility would be ideal, but is not vital).

To date, a number of possible solutions have been proposed for an industrial Ethernet connector and these fall into three general classes. The first is the use of the DB-9 style connector, the same connector that most PCs use for their EIA-232 port. For vibration and ruggedness, this is an improvement over the RJ-45, but it will be very difficult to waterproof. Also it will not connect to any COTS equipment.

In August 2000, InterlinkBT proposed a solution to the TIA based on an M12 connector, the same circular connector that is commonly used for both Foundation Fieldbus and DeviceNet. The difference is this new Ethernet connector would have 8 pins, rather than the typical 4 or 5. The beauty of this design is that it is easy to seal for IP65/67 dust- and waterproofness. It has also proved to be a very robust connector in the field. Unfortunately, there is some question whether it can be made to operate at 100 Mbps, and it is certainly not compatible with existing COTS equipment.

The most promising solutions to this dilemma is combining an Ethernet RJ-45 connector in an M-Style connector, which is sealed to give it an IP-67 industrial waterproof rating. As of 2001, Woodhead Connectivity and a Rockwell/Belden/Panduit/Commscope/Anixter consortium independently proposed solutions based on this design. The Woodhead product is already in production. The hope is that one of these designs will become the standard for industrial Ethernet in the near future.

Power over Ethernet Cabling Over the past 20 years, the industrial control world has come to expect that device power be delivered over the same wire as the signal, leading to two-wire devices. Fieldbuses, such as Foundation Fieldbus and DeviceNet, have continued this tradition, allowing the end user to reduce wiring costs. One of the criticisms of Ethernet is that it has not been able to supply power to field devices. Obviously, this is not an issue for a control-level industrial Ethernet, but it could be for a field-level industrial Ethernet.

Surprisingly, Ethernet may be able to offer this feature very soon. Driven by the need for a standard to supply power to Ethernet-based telephones, the IEEE formed a working group in 1999 to create standard IEEE 802.3af. At the present time, it appears it plans to supply power by superimposing it over the data signal.

There is still work to do in this area, such as answering the question of how to prevent the power system from inducing noise into data lines. Furthermore, the industrial sector has been very poorly represented at the meetings, so the plant floor issues are not being addressed. However, a standard for power over Ethernet is anticipated in 2002 that it is hoped will meet the needs of the automation market as well.

Interface Performance Issues

In the past 20 years, considerable attention has been focused on the question of whether the Ethernet infrastructure (i.e., hubs and switches) can provide deterministic traffic delivery. Unfortunately, the performance and stability of the NIC drivers and TCP/IP stack in the end device has largely been ignored. Just because an Ethernet NIC is rated at 100 Mbps does not mean it can sustain that rate for more than one packet—once the first packet leaves at 10 Mbps, a card might take 100 ms to get the next packet ready, resulting in an effective data rate of 1 Mbps. All the best switches in the world cannot help if the packet is waiting in the PLC to be handled for sending.

Industry needs a specification of meaningful performance metrics for describing the performance of an Ethernet–TCP/IP interface and a specification for measuring those metrics. Such a specification would provide control engineers with the performance capabilities of the products they are intending to interconnect so they can be certain that they will be capable of supporting the control applications. Otherwise, people are going to be burned by bad implementations of drivers and blame Ethernet for it.

Application Layer Protocols

The use of Ethernet and TCP/IP is certainly bringing some standardization to the plant floor. For example, as the industrial world adopts an "all-Ethernet" stance, we will be much closer to having a unified cabling and interface card standard. Users will be able to install a single cable type for communications in their plant and know that it will work, independent of the controls vendor they select. This is certainly not the situation now—even buses from the same vendor, such as ControlNet and DeviceNet, can require different cabling schemes.

Unfortunately, as noted earlier in this section, the Ethernet standard only defines the physical layer and data link layer protocols, while TCP/IP only defines the network and transport layer protocols. Every network needs a top layer of protocols, called application layer protocols, to provide the definition for the tasks or commands that can be executed over the network.

Exactly which protocol should be used for the application layer is shaping up to be one of the new battlegrounds in industrial control. We are seeing war break out between Foundation Fieldbus, PROFIBUS, and DeviceNet/ControlNet (plus a few others like IDA and MODBUS) for the dominance of the application layer.

It would be easy to blame the proliferation of proposed application layer standards on the greed and desire for market dominance by the major controls equipment vendors. However, this is only partly the cause. The fact is the entire industrial market is very divided and different industries have very different needs in a network. For example, an oil pipeline will expect the network to deliver different services from those an automotive assembly plant network will expect. Vendors who dominate in a particular industry sector will quite reasonably want to champion an application layer solution that best meets their particular customers' needs.

Regardless, the impact of a standards war at the application layer is significant to end users. The number of different Ethernet I/O systems for sale at the ISA Expo 2000 show in New Orleans was staggering. One of the main sales themes was that, with Ethernet, the user will receive instant interoperability. This is simply not true. Most of these Ethernet I/O systems will not talk to each other because they do not have the same application layer protocol. If users think that the Ethernet will give them some sort of wide-open I/O interoperability, they will be sorely disappointed. It is critical that design engineers understand exactly what they are implementing at the application layer and how it will interface with other systems.

CONCLUSIONS

There is no doubt that Ethernet and TCP/IP will play a major role in the industrial controls environment over the next decade. They are dominant technologies at the corporate, business interface level and control levels. The only question now is whether Ethernet will dominate at the field level, replacing other bus technologies such as Profibus or Foundation Fieldbus.

Certainly, Ethernet and TCP/IP standards have the flexibility to address most of the concerns industry faces. Many of these industrial issues, like determinism and power over Ethernet, correspond to similar issues in the commercial market, where thousands of times more research dollars are being poured into Ethernet development. Industrial users are likely to be able to take advantage of commercial Ethernet developments. However, as we have seen, issues such as ruggedness and industrial application layer protocols are unique to the industrial world and it is important that the automation industry begin to become involved in the creation of new Ethernet standards. As this happens, Ethernet and TCP/IP will allow controls engineers to interface everything from the sensor to the boardroom with a revolutionary level of convenience and flexibility.

References

1. Holley, D., "Ethernet Provides Open Connectivity," *AutomationView,* Vol. 4, Number 1, p. 4, Spring 1999.
2. ANSI/IEEE Std. 802.3-1985, ISO DIS 8802/3, May 1988.
3. Boggs, D. R., Mogul, J. C., and Kent, C. A., "Measured Capacity of an Ethernet: Myth and Reality," *Proceedings of the SIGCOMM '88 Symposium on Communications Architectures and Protocols, ACM SIGCOMM,* Stanford, CA, August 1988.
4. Falk, H., "Test Methodologies, Setup, and Result Documentation for EPRI Sponsored Benchmark of Ethernet for Protection Control," Systems Integration Specialists Company, Inc. (SISCO), May 1997.
5. Semeria, C., "Understanding IP Addressing: Everything You Ever Wanted to Know," 3COM Corporation, 1996.
6. Lounsbury, B. and Westerman, J., "Ethernet: Surviving the Manufacturing and Industrial Enviroment," Rockwell Automation white paper, Rockwell Automation/Anixter, Inc., May 2001.

Bibliography

Spurgeon, C. E., "Ethernet: The Definitive Guide," O'Reilly and Associates, 2000.

Stevens, W. R., *TCP/IP Illustrated,* Vol. 1, *The Protocols,* Reading, MA: Addison-Wesley, 1994.

www.iaona.com.

4.16 Fieldbus Networks Catering to Specific Niches of Industry

S. VITTURI

This section describes the features of some very popular fieldbuses that are employed in several applications related to different types of industrial plants. For each fieldbus, the communication profile is reported and compared with that, typical for fieldbuses, shown in Figure 4.5b. Considerable attention is dedicated to the data link layer characteristics (in particular to the protocol data units that, at this layer, are also called *frames*) as they influence the overall operation of a communication network. The data link layer, actually, is responsible for the access to the physical medium and for the correct transfer of the data.

In addition, this section, with regard to Figure 4.5a, indicates the factory automation levels at which the fieldbus under consideration is employed.

DEVICENET

DeviceNet is a fieldbus included in both the EN 50325[1] and IEC 62026[2] standards. It has been realized for use at the device level of industrial automation systems, where it implements the communication between the device controller and sensors/actuators.

DeviceNet is widely used and has been adopted in various applications. Some of the most important are in electrical energy distribution plants, the food and beverage industry, chemical plants, oil and gas production, and distribution systems.

The communication profile for DeviceNet is shown in Figure 4.16a. As can be seen, the name DeviceNet refers to an application protocol that uses the controller area network (CAN)[3] protocol to access the physical medium.

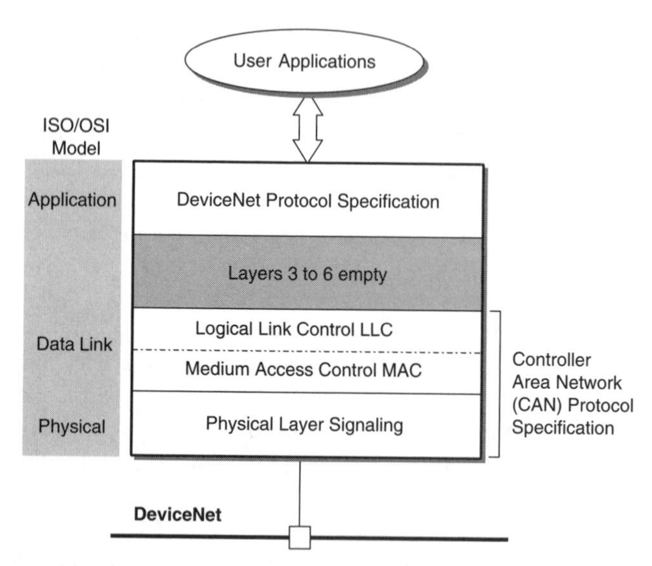

FIG. 4.16a
DeviceNet communication profile.

DeviceNet Physical Layer

DeviceNet uses the CAN physical layer signaling. One of the major advantages deriving from this choice is the great availability of chips implementing the CAN protocol from many suppliers.

The typical network configuration is a bus constituted by a trunk line from which several drop lines are derived to connect stations. There is also the possibility of arranging the drop lines with daisy chains and branches especially for use inside control panels. A maximum of 64 nodes can be connected to the network.

DeviceNet offers the possibility of powering the sensors directly from the network. For this reason the trunk cable comprises two twisted pairs: one is used for the data transfer, the other for the power distribution.

Three different data transfer speeds are selectable: 125, 250, and 500 kbit/s. The correspondent maximum distances coverable by the trunk line are, respectively, 500, 250, and 100 m.

DeviceNet Data Link Layer

The DeviceNet data link layer uses the CAN medium access control (MAC) and logical link control (LLC) sublayers.

CAN is based on a producer/consumer model whose technique to access the physical medium is known as carrier sense multiple access with collision detection (CSMA/CD). It is similar to that used by Ethernet networks,[4] but in this case a deterministic procedure, based on the priority of the transmitted frames, is used to resolve the contentions on the bus deriving from collisions. This procedure, which is non-disruptive, avoids wasting of bandwidth.

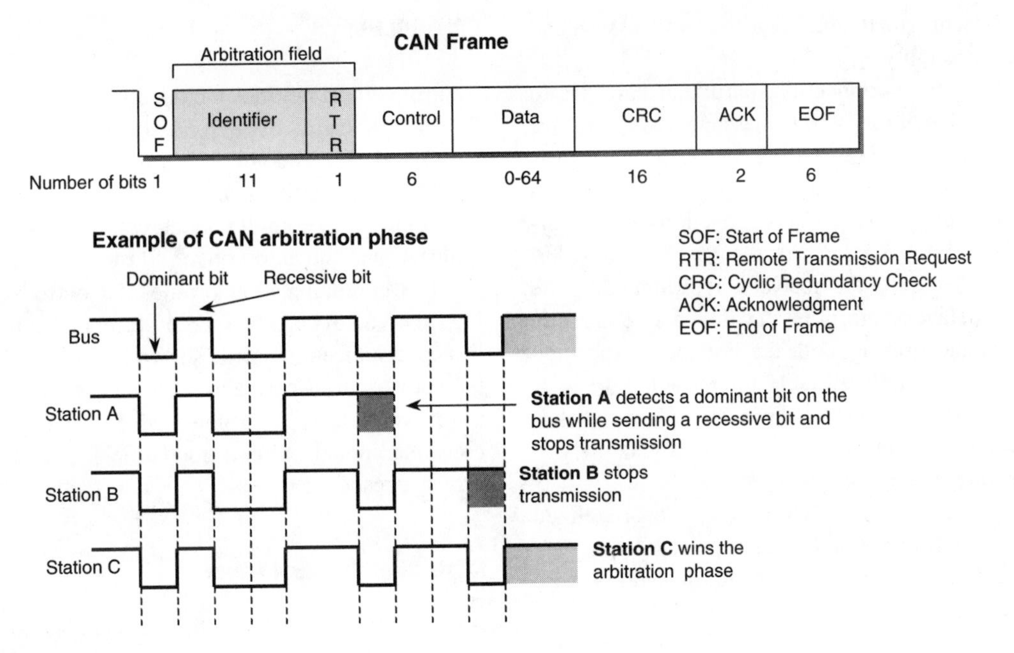

FIG. 4.16b
CAN frame and example of arbitration phase.

In detail, as shown in Figure 4.16b, each data frame transmitted on the network has an arbitration field composed of an 11-bit identifier field plus another bit called remote transmission request (RTR). The arbitration field determines the priority of the frame. A 0 logical value is called dominant, and a 1 logical value is called recessive. As CAN stations are connected to the network via an open collector stage, when two or more of them transmit simultaneously, dominant bits overrule recessive bits. Consequently, if a station while transmitting a recessive bit detects on the network a dominant bit, it realizes that at least another station is sending a frame of higher priority and hence stops its transmission. The stations that lost the arbitration will attempt to resend the frames at the end of the current transmission.

The structure of a CAN frame is shown in Figure 4.16b, which illustrates that a maximum of 64 bits of user data can be transmitted with each identifier. Figure 4.16b also shows an example of the CAN arbitration phase.

A station wishing to send on the network the data associated with an identifier simply transmits the relevant frame. If instead a station needs to know the data associated with an identifier that it did not produce, then it has to send a frame, known as a remote frame, specifying the desired identifier, with the RTR bit set to the recessive value, and with the data field empty.

It should be pointed out, however, that DeviceNet does not make use of the CAN remote frames.

DeviceNet Application Layer

The DeviceNet application layer protocol is based on the definition of *objects* that represents an abstract description of the components of a station. In practice, a station connected to DeviceNet is considered a collection of objects.

A *class* is defined as a group of objects representing the same type of component. An *instance* is the real occurrence of an object and the *attributes* describe its characteristics. The *behavior* specifies how an object performs, and a *service* identifies a function supported by an object. The protocol assigns to each station connected to the network an address, which is called the media access control identifier (MAC ID).

DeviceNet specifies that any data exchange between stations can only take place over logical connections, which are communication paths established between applications resident on different stations. Two types of connections are defined: input/output (I/O) connections and explicit messaging connections. I/O connections are intended for time-critical data transfer; they provide for the data exchange between a producing application and one or more consuming applications. Explicit messaging connections are used for general-purpose point-to-point communication between devices. They are particularly suitable for low-priority, asynchronous data exchange. DeviceNet uses CAN frames to implement the functions specified by its application layer protocol. These frames have been divided into four groups, from Group 1 to Group 4, with decreasing priority.

The 11-bit CAN identifier field contains the information necessary to classify the frames: in particular, it includes a message ID and (except for Group 4) the MAC ID which, depending on the message group, can be either the source or destination MAC ID.

The 64-bit CAN data field is used to transmit the information related to the connection between devices. For example, it can be used either to specify a service request or to carry process data. If the available 64 bits are not sufficient to complete a transmission, then the protocol makes use of a fragmentation technique to split data for transmission and its subsequent reassembly.

When establishing a connection, particular care has to be dedicated to the choice of the message group: if, for example, a Group 1 message (which has high priority) is used to set an explicit messaging connection, then it could slow the data exchange related to I/O connections also occurring in Group 1.

An important feature of DeviceNet is represented by the predefined master/slave connection set, which defines a group of customized connections of both the previously specified types which are particularly suitable for master/slave relationships. The predefined master/slave connection set has been designed to considerably reduce the steps necessary to set up a connection between applications.

In practice a master device to activate the predefined master/slave connection set must first send a service request (allocate master/slave connection set) to the slaves with which it intends to exchange data, specified by a scan list loaded in precedence on the master. A subsequent request issued by the master (set attribute single, relevant to the expected_packet_rate attribute of the connection) has the effect of establishing the connection.

With the predefined master/slave connection set the data exchange may take place using one of four different methods. The master selects which method to use in the above-mentioned allocate master/slave connection set service request. These methods are the bit strobe command/response messages, intended for the exchange of small amounts of data; the poll command/response messages, to be used for the transfer of any amount of data; the change of state/cyclic messages, used for triggered data exchange; and the explicit response/request messages.

The bit strobe command is an I/O message sent by the master to all the slaves defined in its scan list. This message carries 64 bits of output data, one bit for each slave. Obviously, if a slave is not present, or not included in the scan list, the relevant bit has no meaning. The bit strobe responses are I/O messages sent from each slave to the master in response to the command message. They can contain up to 64 bits of input data (and hence each of them is transmitted with just one CAN frame).

The poll command is an I/O message, containing any amount of data, sent by the master to a single slave. The latter, with the poll response I/O message, may also return data.

The change of state/cyclic are I/O messages that can be transmitted either by the master or by a slave. They are normally triggered by specific events such as, for example, an alarm or the expiring of a timer. These messages are used to implement the cyclic data exchange and the asynchronous traffic functions mentioned in Section 4.5.

The explicit response/requests are general-purpose messages used, for example, to transfer the values of some attributes.

CONTROLNET

ControlNet is a fieldbus included in both the EN 50170[5] and IEC 61158 standards.[6–9] It has been designed to be used at both the device and cell levels of industrial automation systems. Typical applications of ControlNet can be encountered, for example, in complex batch control systems, the process industry, and the automotive industry.

The communication profile of ControlNet is shown in Figure 4.16c. As can be seen, it differs from the typical fieldbus communication profile shown in Figure 4.5b; in this case, the network, transport, and presentation layers are not empty.

Nevertheless, ControlNet, which is based on a producer/consumer model, exhibits good efficiency and provides deterministic response times.

ControlNet Physical Layer

As shown in Figure 4.16c, the physical layer is split into two sublayers: the physical medium attachment (PMA) and the physical layer signaling (PLS). The PMA is responsible for the transmission/reception of signals to/from the bus and comprises all the necessary circuitry that is not part of the ControlNet stations. The PLS implements the bit representation and timing functions required for the correct transmission of the signals on the network. Moreover, the PLS realizes the interface to the data link layer.

The ControlNet standard basically recommends two types of physical media: coaxial cable and fiber optic. A third media type, network access port (NAP), allows for a

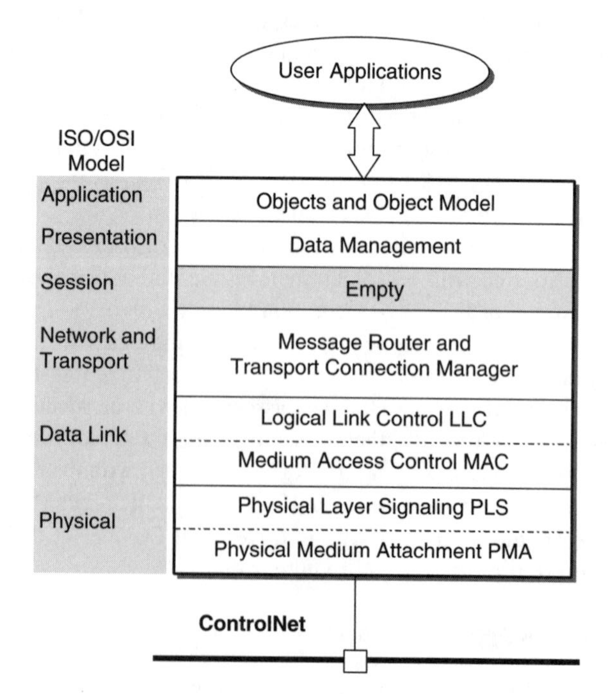

FIG. 4.16c
ControlNet communication profile.

point-to-point temporary connection between two nodes. A NAP is often used, for example, to realize a link between a programming unit and a station already connected to ControlNet. The NAP cable has eight conductors and an overall shield.

The specified coaxial cable is a RG-6 quad shield, of the same type as that used for the cable television industry. Several topologies (e.g., bus, tree, and star networks) may be realized with ControlNet. Repeaters can be used to link segments. Each station is connected to a segment by means of a tap to which it is linked via a drop cable. The maximum segment length is related to the number of stations connected: 1000 m with two stations, decreasing to 250 m with 48 stations.

The fiber-optic physical medium makes use of active star or active hub topology and foresees two variants depending on the distances to be covered. The first variant specifies a "short range" system for distances up to 300 m, and the second variant refers to "medium range" systems with distances of up to 7000 m.

Considerable distances (up to 30 km) can be covered by complex ControlNet installations using either type of physical media. ControlNet has a fixed transmission speed of 5 Mbit/s and may connect a maximum of 99 stations. Finally, redundant physical media are available as an option for ControlNet. In this case, each station is connected to two lines and transmits/receives contemporaneously on both.

ControlNet Data Link Layer

Figure 4.16c shows that the ControlNet data link layer is split into two sublayers: medium access control (MAC) and the logical link control (LLC).

The MAC is based on a time-slice algorithm called concurrent time domain multiple access (CTDMA), whose operation is shown in Figure 4.16d. As can be seen, a fixed period of time, the network update time (NUT) to which all the stations are strictly synchronized, is repeated indefinitely on the network. The NUT, as can be noted in Figure 4.16d, is divided in three intervals of time: scheduled, unscheduled, and guardband. The scheduled interval is typically reserved for the transmission of real-time data. For example, both the cyclic data exchange and the asynchronous traffic described in Section 4.5, if present, have to be performed in this interval. The MAC grants each station the possibility of transmitting one and only one frame during the scheduled interval. This is accomplished by means of an implicit token-passing procedure, which works as follows.

Each station is identified by an address called the MAC ID. At the beginning of the scheduled interval, the station with the lowest MAC ID is granted permission to transmit a MAC frame. Every station connected to the network maintains an implicit token register that contains the MAC ID of the transmitting station. This value is incremented by one at

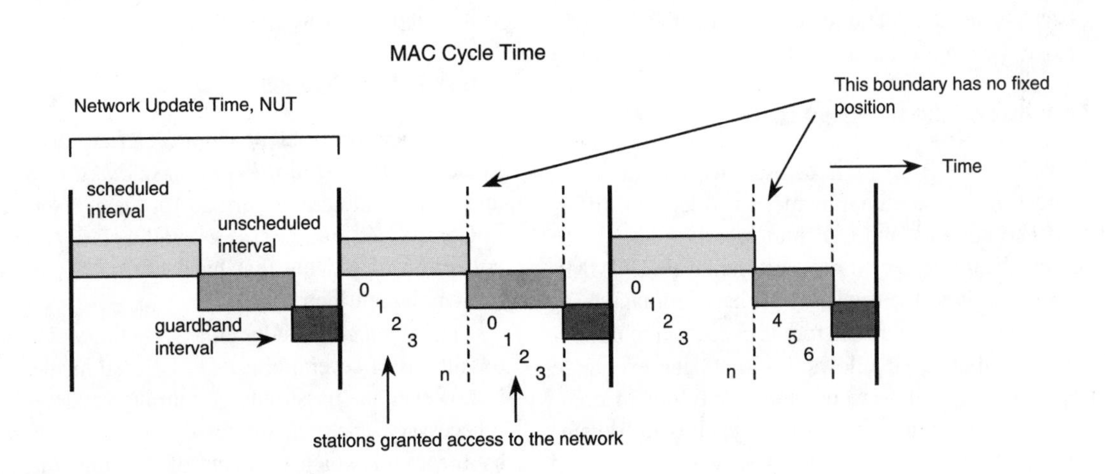

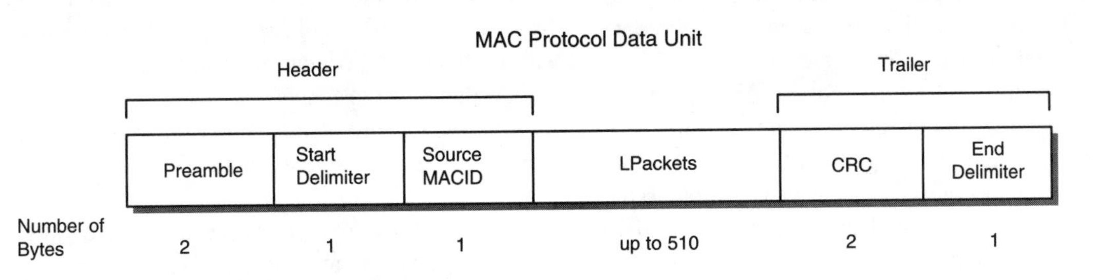

FIG. 4.16d
ControlNet MAC cycle time and protocol data unit.

the end of the frame transmission. Immediately after, each station compares its MAC ID with that contained in the implicit token register and, if they match, that station has the right to transmit. If a station is not present, in the fraction of the scheduled interval reserved to that station there will not be activity on the network. In this case, after a time-out, the MAC ID is newly incremented by all the stations and the implicit token is passed on. The scheduled interval ends when the maximum value for the MAC ID is reached.

Subsequently, the unscheduled interval is started. In this period the implicit token is still circulated among stations according to a round-robin scheme. The duration of such an interval, however, is variable (because it is related to the utilization of the scheduled interval), as shown in Figure 4.16d, and a station can be granted zero, one, or more times to access the network. As a consequence, the unscheduled interval does not allow for a deterministic access to the network and hence is used for transmission of noncritical data.

When the guardband interval is reached, the stations stop transmitting and access to the network is granted to the station with the lowest MAC ID, which is called the moderator. This station transmits a special message (the moderator frame) with which synchronizes all the stations for the beginning of a new cycle and dispatches a set of parameters necessary for correct network operation.

The structure of the MAC protocol data unit is shown in Figure 4.16d. As can be seen, it comprises a header, a trailer, and a field with the data from the upper layers. This last comprises zero or more structured sets of data called Lpackets, which may contain a maximum of 510 bytes in total.

ControlNet Network and Transport Layers

The major task of these layers is to establish and maintain logical connections between applications resident on different stations. The network and transport layers contain three basic modules: the unconnected message manager (UCMM), the message router, and the transport connection manager.

The UCMM provides a means for the execution of unconnected services; that is, it allows for the data exchange between applications that were not previously put in communication by a connection. In this case, each data transfer is independent from the others and the messages have to carry the full details of the destination and source applications. The UCMM is mainly used for handling noncritical traffic and one of its most important applications is given by the data exchange services necessary to set up a connection.

The message router realizes the correct dispatching of services, inside a station, to the addressed applications.

The transport connection manager is actually responsible for the handling of connections. It has to make available the services necessary to create and delete connections and to exchange data over an established connection. As ControlNet is based on a producer/consumer model, the transport connection manager must allow setting up of either point-to-point or multicast connections. This latter type actually allows for the data produced by an application (producer) to be shared among all the other connected applications (consumers).

The transport connection manager defines seven classes of connections (from class 0 to class 6), which have different features and complexity, and consequently can be used for different purposes. Class 0, for example, is the simplest, and is called null or base. It can be either multicast or point to point and is commonly used in applications such as cyclic block transfers, acquisition of diagnostic events, etc. Class 3 is more complex; it uses a sequence number to identify the frames to avoid possible duplication of the transmitted data, and, again, such a connection can be either point to point or multicast. Moreover, class 3 adds a return point-to-point connection used to notify a source application that the destinations read the data previously transmitted. Typical examples of class 3 uses include communication between controllers and operator interface systems, uploads and downloads of structured sets of data, and asynchronous block transfer.

ControlNet Presentation Layer

This layer is concerned with the data management. In particular, it defines a common means for specifying the format of the data handled by the application layer. The ControlNet presentation layer is mainly based on the IEC 1131-3 standard,[10] which defines elementary and derived data types. Examples of elementary data types handled by the presentation layer are Boolean, integer, floating point, character string, and bit string.

ControlNet Application Layer

The application layer of ControlNet is based on the object modeling. The definitions of *object, class, instance, behavior,* and *service*, already described for DeviceNet, are applicable to ControlNet. In this case as well a station can be seen as a collection of objects that have attributes, provide services, and implement behaviors.

The standard includes an object library, which contains the definitions of several objects to be used inside stations. There is, however, the possibility of defining vendor-specific objects, if necessary. The standard also describes a set of *device profiles* by means of which products of different manufactures that adhere to the same profile can be interchanged with no effect on the network. Products belonging to the same device profile have the same behavior, produce and/or consume the same group of data, and contain the same set of attributes.

To illustrate the concepts relevant to object modeling used by ControlNet, two standard objects are briefly described in the following: the *discrete output point object* and the *position sensor object*.

The discrete output point object is used to model an output bit in a hardware component. The *class attributes* of such an object can be accessed by means of the Get_Attributes_All and Get_Attributes_Single services: they supply general information about the object such as the current revision, the maximum instance number of an object currently created in this class, etc. The *instance attributes*, depending on their characteristics, can be either accessed or set. To access these attributes the same services described above are used; to set the attributes, the services Set_Attributes_All and Set_ Attributes_Single are available. The most important attribute that can be set is the *value*, which allows for the discrete output point object to determine the value of the output bit.

Another important attribute is the *object state*, which reports the operational state of the object: it is defined as an unsigned integer number that can assume values defined in a specified range. The behavior of the discrete output point object is described in the standard by means of a state event matrix where, for each operational state, all the possible incoming events and the consequent state transitions or actions are specified.

The position sensor object models an absolute position sensor hardware component, for example, an absolute encoder. This object is actually more complete than a simple position sensor, as it introduces a zero offset and the limit position check.

The class attributes are the same already described for the discrete output point object.

The instance attributes are related to the actual use of the object. The most important of them are the value (an unsigned double integer that reports the absolute position), the low and high limits for the position, the zero offset, and the resolution. The services available in this case are Get_Attributes_Single and Set_Attributes_Single.

As for the other objects, the behavior of the position sensor object is described by a state event matrix.

LONWORKS

Lonworks is a communication system, designed to operate in several application areas, which may be configured to work as a fieldbus. This type of design is the reason, as shown in the following, Lonworks does not have the typical communication profile of a fieldbus. Moreover, the functions specified in Section 4.5 required of a fieldbus are not directly implemented, but have to be realized using the available communication services. Lonworks has recently become an ANSI/EIA standard.[11]

The most important application area of Lonworks is in the building automation industry, but it is also employed in the transportation area and in home automation (in which there is currently very considerable growth). Lonworks also has a widespread use in industrial automation, where, thanks to its features, it can be adopted as the unique communication system for the three levels of automation shown in Figure 4.5a.

The communication profile of Lonworks is completely based on the OSI Reference Model[12] and implements all the protocols and services specified from layer 1 to layer 7. The most significant of them are described in the following.

Lonworks Physical Layer

There are several different physical mediums that can be used to implement a Lonworks network. The stations are connected to segments, called *channels*, by means of transceivers and a great variety of configurations may be realized. Channel length, data transmission speeds, and maximum number of stations depend on the chosen channel and are completely described in the standard. For example, the channel type named TP/XF-1250 is based on a twisted-pair cable and has a bus topology; its transmission speed is 1.25 Mbit/s and it can connect 64 stations with a maximum channel length of 125 m. Some typical channel configurations are shown in Figure 4.16e. The use of repeaters and routers allows for the implementation of very complex Lonworks configurations with considerable geographic extensions.

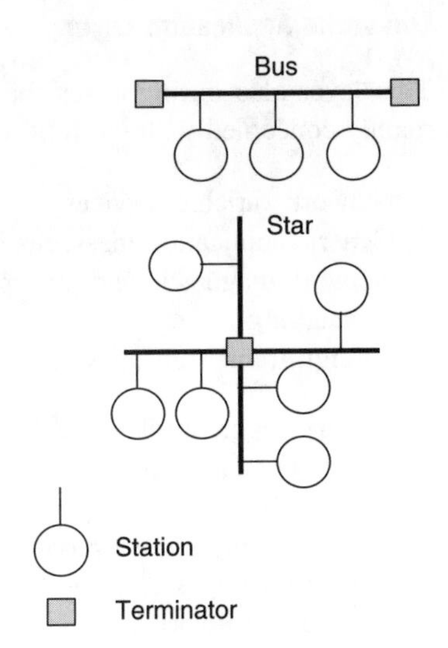

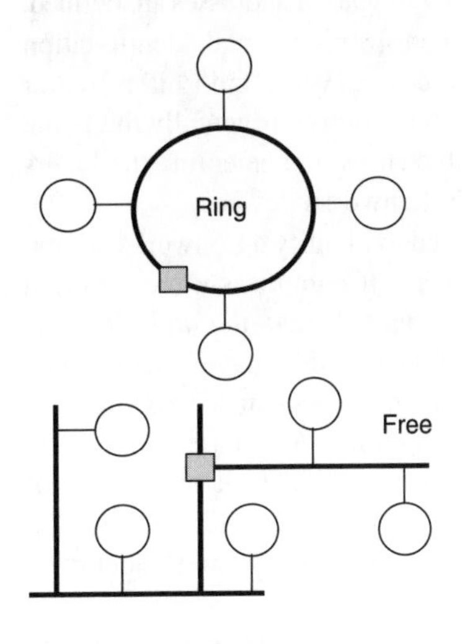

FIG. 4.16e
Typical Lonworks channel configurations.

Lonworks configurations can also be realized using traditional IP networks. In this case, the datagrams specified by the IP protocol[13] are used to encapsulate the Lonworks protocol data units.

Lonworks Data Link Layer

The data link layer of Lonworks is composed of two parts: the medium access control (MAC) sublayer and the link layer.

The MAC uses a carrier sense multiple access (CSMA) protocol similar to that adopted by Ethernet networks but with some modifications necessary to limit the occurrence of collisions. This protocol is called *predictive p-persistent CSMA*, and it works as follows.

A station wishing to transmit senses the network and, if it is idle, computes a random time to be waited before the transmission takes place. When the time expires, if the network is still idle, then the station transmits; otherwise, it receives the incoming frame and then restarts the delay calculation procedure.

The delay is computed as a multiple of a time interval, called *Beta2 slot*, whose definition is given in Reference 11. A station has a minimum of 16 Beta2 slots available; in this situation, the average delay calculated for each frame to transmit is 8 Beta2 slots. Every station, however, basing on a prediction of the network load, has the possibility of dynamically increasing the number of Beta2 slots up to 1008. With this technique it has been demonstrated that the collision rate can be kept constant and independent of the network load.

The link layer realizes the interface with the upper layer to which it supplies a connectionless type of service necessary to transmit and receive data to/from the MAC. The link layer performs framing, data encoding, and error checking.

Lonworks Network Layer

The concept of *addressing* is of particular relevance at the network layer. In Lonworks, several types of addresses are defined.

The *physical address* is a 48-bit internal identification number assigned to every device when built; it is better known by the term Neuron ID. (Neuron is actually the name of the commercially available chips implementing the layers from 1 to 6, used to realize Lonworks devices).

The *device address* is used to identify a Lonworks station uniquely on a defined network. It comprises three different fields: the *domain ID*, the *subnet ID*, and the *node ID*. The domain ID specifies a set of devices that may exchange data with one another. A domain may contain a maximum of 32,385 devices. The subnet ID refers to a maximum of 127 devices connected either to a single channel or to a set of channels linked together by means of repeaters. Up to 255 subnets can be included in a domain. The node ID identifies a single node of a subnet.

Moreover, a *group address* can be used to specify a group of devices belonging to the same domain, but connected to different subnets.

Finally, a *broadcast address* may be used to identify all the devices of either a subnet or a domain.

Lonworks Transport Layer

The transport layer allows for both sending and receiving messages across the Lonworks network. It makes available three different services to the upper layer: Send_Message, Trans_Completed, and Rcv_Message.

With the Send_Message service a message may be sent to a single device or to a group of devices. This service, when invoked, returns an identifier, named TID, which is used by the Trans_Completed service to notify the correct execution of the sending procedure. With the Rcv_Message, the transport layer notifies the upper layer of the arrival of a message received from another station of the network.

There are basically three different ways of exchanging messages:

1. The acknowledged messaging specifies that a source station, after sending a message, receives acknowledgment of the correct reception. A maximum of 64 destination stations can be addressed as destinations and acknowledgments are expected from each. In case of time-out in receiving the acknowledgment, the message is retransmitted up to a number of times specified by a network configurable parameter.
2. The repeated messaging allows for a station to repeatedly send a message to any number of destinations. The messages do not require any acknowledgment, and consequently, the whole transmission is highly efficient. The repetition interval is determined by an internal timer handled by the transport layer and set by the user requiring the service.
3. The unacknowledged messaging is similar to the repeated messaging with the difference that, in this case, the periodic transmission is not allowed.

Lonworks Application Layer

This layer also encompasses the presentation layer and is mainly concerned with the following functions:

Network variable services
Generic application messages handling
Network diagnostic and management messages handling
Foreign frame transmission

These services are implemented by the transfer of application layer protocol data units (APDUs), whose structure is shown in Figure 4.16f.

The network variable services are dedicated to the propagation of messages carrying a special type of data simply referred to as network variables.

Generic application messages are exchanged between application processes resident on different stations of a

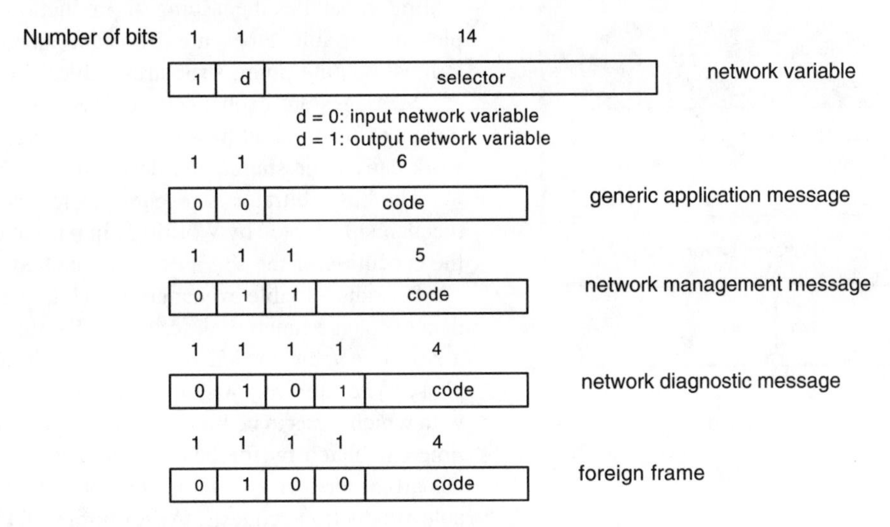

FIG. 4.16f
Lonworks application layer protocol data unit.

Lonworks network; they typically transport data directly related to the application processes involved.

The diagnostic and management messages allow for the execution of a set of services, such as router tables maintenance and the querying of stations to obtain their status.

A foreign frame is defined as a frame belonging to networks different from Lonworks (for example, IP networks). Lonworks explicitly offers the possibility of being used as a gateway between stations connected to different networks. To this purpose the Lonworks application layer offers a set of services necessary to handle the foreign frames.

The network variable services represent the most important module of the application layer. A network variable is a set of data, normally related to a physical process, uniquely identified within a Lonworks system. Network variables can be either inputs or outputs. Examples are temperature and pressure set points (as outputs), switch positions, valve status (as inputs), etc.

The Lonworks application layer specifies that every network variable must be of a defined type and, for this reason, a set of standard network variable types has been defined. Moreover, the device manufacturers also have the possibility of specifying a set of user-defined types.

A procedure defined by the application layer, called *binding*, allows creation of a relationship (connection) between input and output network variables resident on different stations. In practice, an output network variable can have some readers, located on different stations, and defined in an offline phase by means of the binding procedure. When this happens, an address table specifying all the readers is associated with that output network variable. Similarly, these readers must have some input network variables defined, which have a field internal to their description, called a selector (Figure 4.16f), identifying that of the output network variable.

The propagation of network variables takes place as follows: when an output network variable is updated, the Lonworks application layer checks if there is any reader for such a variable (i.e., it checks for an address table). If so, the new value is automatically sent to all the stations specified in the address table. The application layers of the destination stations compare the selector field of the received variable with that of the connected input network variables. If they match, the value is accepted and passed to the application programs through that variable.

Finally, there is also the possibility that an output network variable is connected to an input network variable defined on the same station; this is called turnaround connection.

WORLDFIP

WorldFIP is a general-purpose fieldbus designed for use at both the device and cell levels of industrial automation systems. It is included in both the EN 50170[14] and IEC 61158 standards.[15–18]

The features of WorldFIP allow for several applications: for example, in the automotive industry, building automation, process control, and the energy production industry. The communication profile for WorldFIP is shown in Figure 4.16g.

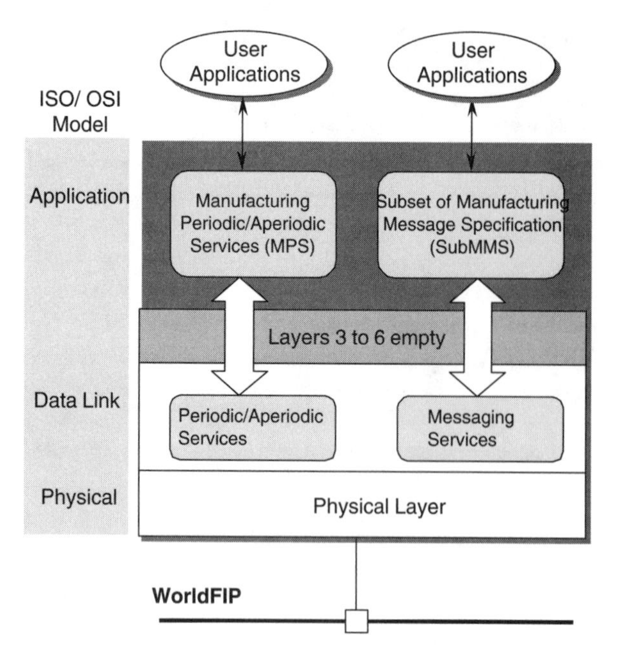

FIG. 4.16g
WorldFIP communication profile.

WorldFIP Physical Layer

The standard allows for the use of either twisted-pair or fiber-optic cables to implement the physical layer. Possible data transmission speeds are 31.25 kbit/s, 1 Mbit/s (specified as the standard speed), 2.5 Mbit/s, and 5 Mbit/s. This last speed is only available for fiber-optic cables. If a twisted-pair cable is used, several topologies can be realized. In particular, the network is divided into segments where passive components called junction boxes may be used to implement derivations. The connection of devices to the network can take place using junction boxes, or taps, or active star components called diffusion boxes.

A maximum of 32 devices can be connected to a segment. The maximum length of a segment is related to the transmission speed: 1900 m at 31.25 kbit/s, 750 m at 1 Mbit/s, and 500 m at 2.5 Mbit/s.

The covered distances can be incremented using repeaters. Up to four repeaters can be inserted between two stations; in this way the network maximum distance at 31.25 kbit/s is 9500 m and the maximum number of stations is 128.

If a fiber-optic cable is used, the network has to be realized by means of active star couplers to which the stations are directly connected; star couplers can also be connected to each other.

WorldFIP Data Link Layer

As shown in Figure 4.16g, the data link layer provides two types of services: periodic/aperiodic services for the exchange of identified variables and message services for the handling of messages.

WorldFIP uses two different addressing schemes dependent on the type of services to be performed.

Variables, which are distributed across the stations connected to the network, can be exchanged simply by referring to their identifiers, which are unique on the network. For each variable one and only one producer exists, while there can be more than one consumer. A special station, called the bus arbitrator, handles the issuing of production requests. Examples of variable types are Boolean, integer, bit and byte strings, floating point, structures, tables, etc.

Messages are exchanged between stations identified by means of physical addresses. Each message sent on the network carries the source and destination addresses.

The bus arbitrator is responsible for the medium access mechanism adopted by WorldFIP. In particular, it has to grant the execution of the services mentioned above.

Periodic variable exchange (which implements the cyclic data exchange function described in Section 4.5) is based on a scanning table loaded on the bus arbitrator in an offline phase. The table reports, for each variable, the periodicity with which it has to be produced, the type, and the production time calculated by the bus arbitrator. During operation, the bus arbitrator scans the table and generates the suitable variable production requests. An example of this procedure is reported in Figure 4.16h. As can be seen, variable A, which has the shortest periodicity, has to be produced every 10 ms and determines the duration of an elementary cycle. Every elementary cycle is characterized by the production of one or more variables. A sequence of these cycles repeated periodically identifies a macrocycle.

Scanning Table

Variable	Periodicity (ms)	Type	Time (us)
A	10	INT_16	178
B	20	UNS_32	194
C	30	INT_8	170

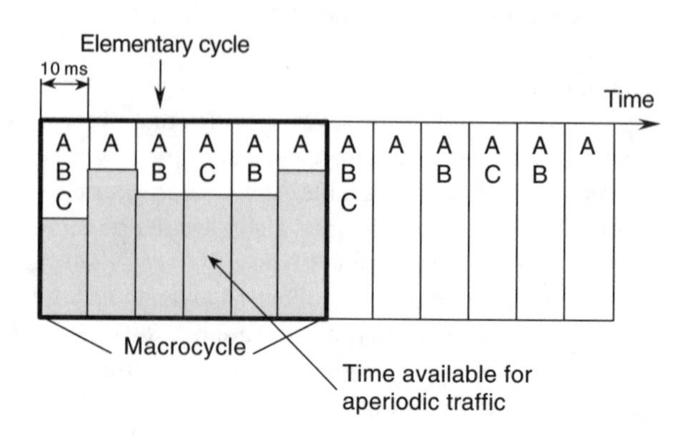

FIG. 4.16h
WorldFIP examples of scanning table and variable production cycles.

Figure 4.16h also shows how the time interval between the end of an elementary cycle and the beginning of the following is available for aperiodic traffic, also handled by the bus arbitrator. In detail, a station wishing to issue a request for aperiodic activities (either referring to the production of variables or to the sending of messages) sets a special bit in any frame used to produce a variable during the cyclic operation. Consequently, if the request concerns the production of variables, the bus arbitrator, in the aperiodic window, will request from that station the list of the variables that have to be produced. Depending on the time availability, the bus arbitrator will then broadcast the production requests on the network.

If, instead, the aperiodic request is for the sending of messages, the bus arbitrator, when possible, will grant the requesting station the right to access the network. Subsequently, this latter party will send the message(s) to the addressed station(s). The message transfer services can be either acknowledged or unacknowledged. It is worth noting that requests for aperiodic activities can only be issued by stations that are producers of variables exchanged cyclically.

WorldFIP Application Layer

Figure 4.16g shows that the WorldFIP application layer comprises two different protocols: the subset of manufacturing message specification (subMMS)[19] and the manufacturing periodic/aperiodic services (MPS).

The subMMS relies on the messaging services offered by the data link layer. It allows for the execution of some activities, such as setup and configuration of network entities, download of programs, monitoring of the variables, and alarm management.

The MPS is concerned with the identified variables that can be exchanged on the network. In particular, MPS is responsible for the synchronization of produced and consumed variables, for the consistency of the data exchanged, and for the reading/writing of either local or remote variables.

ACTUATOR SENSOR INTERFACE

The actuator sensor interface (AS-i) is a fieldbus conceived for use at the device level of industrial automation systems. It has been designed for highly efficient exchange of very reduced sets of data between a master device and binary sensors/actuators referred to as slaves. The AS-i is included in both the EN 50295[20] and IEC 62026[21] standards. Although these documents specify the use of such a fieldbus for realizing the interface between control gear and switchgear in the low-voltage electrical distribution environment, the features of AS-i and the low cost of its components make its use very convenient also in other fields of application. As will be shown, AS-i is particularly suitable to connect any type of binary field devices to a controller, replacing the traditional cabling.

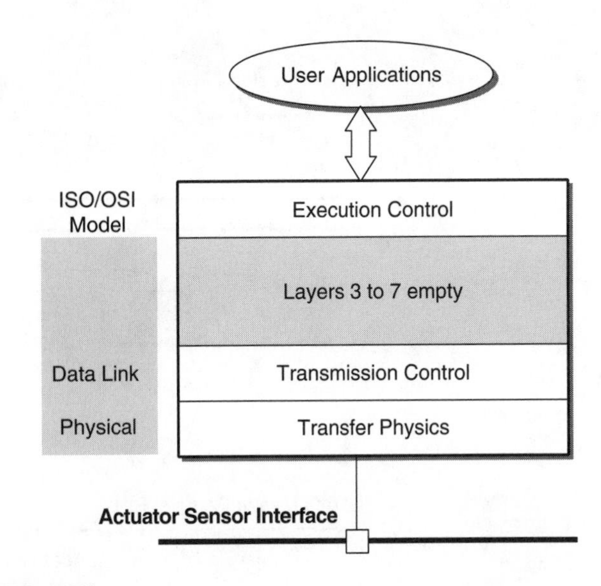

FIG. 4.16i
AS-i communication profile.

The communication profile of an AS-i master station is reported in Figure 4.16i.

AS-i Physical Layer

This layer, called transfer physics, is responsible for the physical connection between master and slaves. The standard specifies the use of an unshielded and untwisted cable, which is available in two versions: a general-purpose cable to which stations are connected by means of traditional screw terminals, and a special AS-i cable of equivalent electrical characteristics, which allows for a direct connection of stations by means of contacts that penetrate the cable isolation. In an AS-i installation, the same cable is used to realize the data transfer and to power the connected stations, as shown in Figure 4.16j. Moreover, several topologies, such as bus and star, can be adopted.

The standard specifies a maximum of 31 slaves for AS-i fieldbuses. The longest allowed distance between master and slave in an AS-i installation is limited to 100 m. Greater distances can be covered using repeaters.

The data transmission technique used by AS-i is alternating pulse modulation (APM), which produces a baseband signal, superimposed on the power supply DC voltage. Starting from the bit sequence to be transmitted, the resulting modulated signal is shaped as $\sin^2$ pulses which, by virtue of a low content of frequencies, limit the radiation emissions.

The above technique is used with a transmission speed of 167 kbit/s, which is equivalent to a bit time (the time necessary to transmit a bit) of 6 μs.

AS-i Data Link Layer

The AS-i data link layer, named transmission control, is responsible for the access of stations to the physical medium and for the correct transfer of information. The AS-i transmission

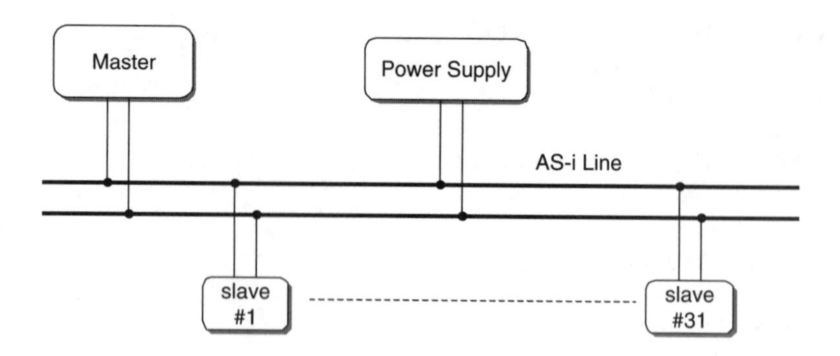

FIG. 4.16j
AS-i configuration.

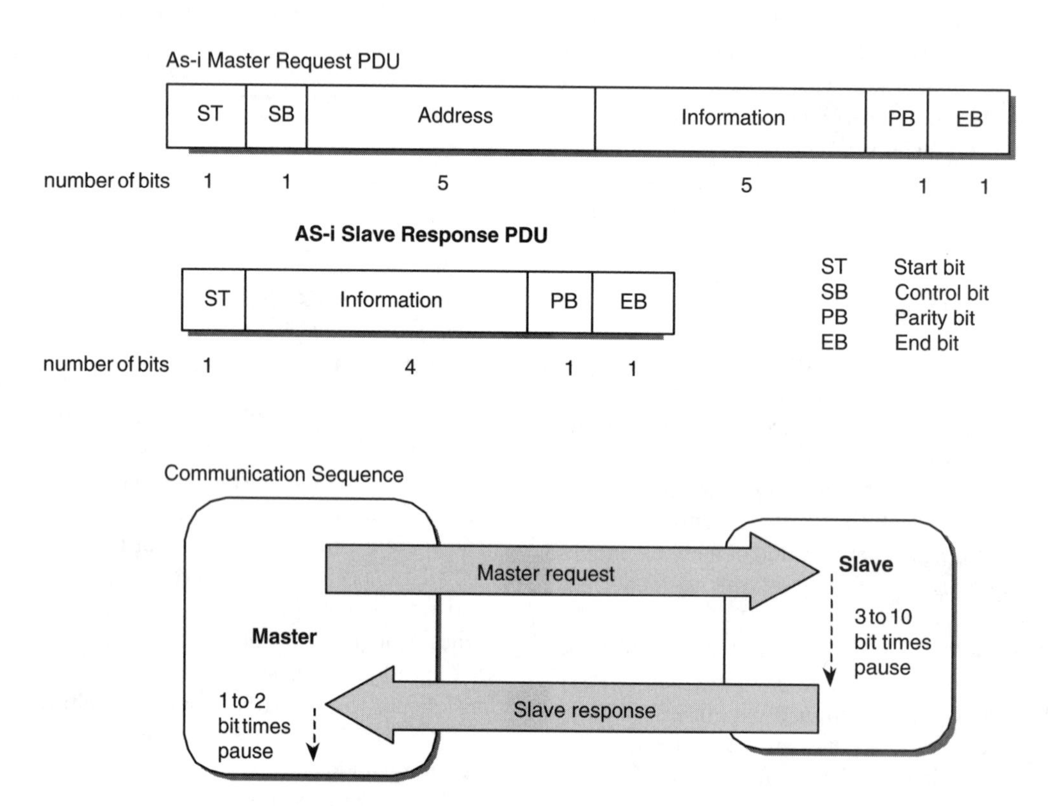

FIG. 4.16k
AS-i protocol data units and master/slave communication sequence.

control is a bit-oriented protocol whose protocol data units (PDU) are shown in Figure 4.16k. As shown, only two PDUs are defined: the master request and the slave response. In the master request PDU, the *address* field is used to identify the polled slave, while the *information* field contains the data transferred between master and slave (i.e., process data or parameters).

The communication sequence specified by the protocol, also shown in Figure 4.16k, imposes a pause of 3 to 10 bit times between the two PDUs and a subsequent additional pause of 1 to 2 bit times after the slave response.

The master request specifies, by means of suitable combinations of the information field bits, the operation that has to be executed on the slaves. The most common of them are Data_Exchange, Write_Parameter, and Address_Assignment. With the Data_Exchange master request, four output bits are sent to the slave. The slave, on answering, encloses four input bits in the slave response. The Write_Parameter master request allows for the remote setting of some slave options. With the Address_Assignment the master is able to set the address of a slave. This operation is possible only if the slave has a default address that does not allow for data exchange.

AS-i Execution Control

The execution control layer manages the overall operation of AS-i. It receives the function requests from the user application and generates the necessary master calls to the transmission control. Three different operational phases are implemented by the execution control: initialization, start-up, and normal operation.

The initialization phase comprises a set of actions that have to be performed off line. First, some operational parameters of the master have to be set to their initial values; then the AS-i power supply has to be tested to verify the presence of sufficient electrical power for the operation of the slaves. After that, the next phase can be entered.

In the start-up phase the master detects and activates the slaves. This operation can be performed in two different ways:

1. In the *protected operation mode*, the master activates, among the detected slaves, only those recorded in a "list of projected slaves" table, previously prepared by the user.
2. Conversely, in the *configuration mode*, all the slaves detected by the master on the network are activated.

The normal operation phase realizes the cyclic data transfer between master and slaves, (possibly) some acyclic management tasks, and the inclusion function. With the cyclic data transfer, the master updates each slave with the output bits and acquires from that slave the input bits. This operation is performed automatically without the intervention of any user request. If a slave does not answer to three consecutive master requests, it is marked as faulty and excluded from the polling list.

The management tasks are triggered by user requests. At the end of every data transfer cycle, the execution control analyzes the occurrence of such requests and, if they are present, it forces the transmission control to generate the correspondent master calls. Typical operations that can be carried out are, for example, reading the slave status, setting slave parameters, resetting a slave. The inclusion function allows for the addition of slaves to the cyclic data transfer. In particular, with this function the master first detects the presence of new slaves on the network and then activates them. This operation may take several master cycles.

PROFIBUS-DP

PROFIBUS-DP is a fieldbus designed to be used at the device level of industrial automation systems. It is included in both the EN 50254[22] and IEC 61158 standards.[23–26] PROFIBUS-DP is widely used in several application areas, such as the

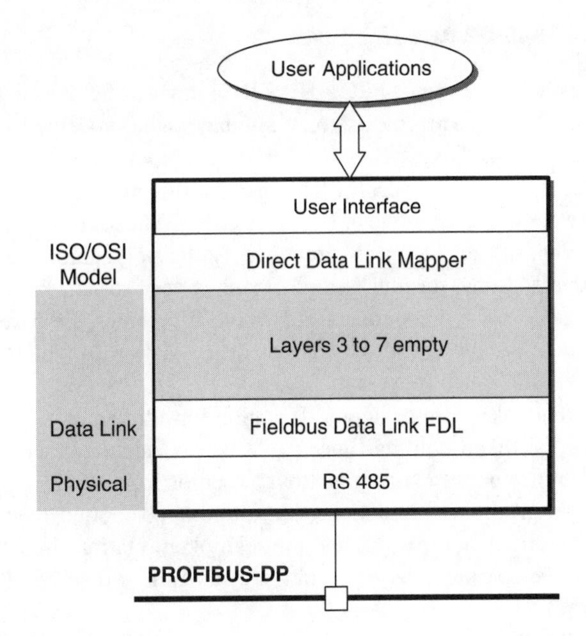

FIG. 4.16l
PROFIBUS-DP communication profile.

food and beverage industry, research laboratories, the automotive industry, and paper production plants. The communication profile of PROFIBUS-DP is shown in Figure 4.16l. As can be seen, the application layer is empty, and the typical functions required of a fieldbus are implemented by the two upper layers: the direct data link mapper (DDLM) and user interface.

PROFIBUS-DP Physical Layer

The physical layer of PROFIBUS-DP is specified by the well-known RS-485 standard.[27] The basic network configuration is a bus comprising a maximum of four segments connected by means of repeaters.

Up to 126 stations can be connected to PROFIBUS-DP and the most commonly used physical medium is a twisted-pair cable, which, if necessary, can be shielded. However, several PROFIBUS-DP implementations use different physical mediums (and possibly different configurations), such as fiber optics, infrared, and other wireless transmission systems.

Many transmission speeds are available, ranging from 9.6 kbit/s to 12 Mbit/s. The maximum length of a segment is related to the transmission speed, as shown in Table 4.16m.

TABLE 4.16m
PROFIBUS-DP Transmission Speeds vs. Maximum Segment Length

Transmission speed (kbit/s)	9.6	19.2	93.75	187.5	500	1500	6000	12000
Maximum segment length (m)	1200	1200	1200	1000	400	200	100	100

PROFIBUS-DP Data Link Layer

The data link layer of PROFIBUS-DP, known as fieldbus data link (FDL), is a protocol very similar to that specified by IEEE 802.4, "Token Bus."[28]

As with IEEE 802.4, FDL specifies the presence on the network of either active or passive stations. The token (which is a particular PDU) is circulated exclusively among the active stations, forming a logical ring. Every station knows the address of the station from which it receives the token (previous station) and the address of the station to which it passes the token (next station).

The token can be kept by a station for a time not greater than the token holding time, T_{th}, which is calculated as the difference between the target token rotation time, T_{tt} (a value set on all the active stations that is the upper bound of the token circulation period) and the real token rotation time T_{tr} (a value computed by each station every time it receives the token).

The FDL protocol specification is slightly different from IEEE 802.4. In particular, it defines only two levels of priority for messages (high and low) and it always grants a station the possibility of sending a high-priority message even if T_{th} is less than or equal to zero.

The services available to an FDL user are only of the connectionless type and allow for the data transfer between users of different stations. Two of these services—send and request data with reply (SRD) and send data with no acknowledge (SDN) described later on—are of particular relevance, as they are explicitly used to implement the PROFIBUS-DP functions.

With the confirmed service SRD, a source station can send a maximum of 246 bytes to a selected destination. The latter in its response message (necessary to confirm the arrival of the data) can include up to 246 bytes. In this way, SRD realizes a confirmed bilateral data exchange between the two stations.

SDN is an unconfirmed service by means of which a source station can send a maximum of 246 bytes. The destination in this case can be one or more of the stations present on the fieldbus. This service is mainly used either for broadcast or multicast transmissions, as it does not generate confirmation messages.

User Interface and Direct Data Link Mapper

The user interface represents the core of the PROFIBUS-DP protocol and is responsible for the correct execution of all operations specified by the standard. The direct data link mapper has the task of mapping the requests coming from the user interface onto FDL services.

PROFIBUS-DP is a master/slave fieldbus in which every master station cyclically polls its slaves. As will be shown in the following, this exchange is realized using the SRD service. Consequently, a maximum of 246 output bytes and 246 input bytes can be transferred between master and slave for each polling cycle.

The PROFIBUS-DP standard allows for the presence on the network of two classes of masters.

- Class 1 masters, which are devices actually used to implement control tasks
- Class 2 masters, which are devices used for network administration purposes

The standard also specifies two types of network configuration: monomaster and multimaster.

- A monomaster network comprises one class 1 master, up to 125 slaves, and, optionally, one class 2 master.
- A multimaster network can connect more than one class 1 master, one class 2 master, and some slaves.

Both the configurations allow for a maximum of 126 devices connected to the network.

As PROFIBUS-DP is based on the circulation of a token, the presence of several master stations has negative effects on the efficiency of the network because part of the transmission bandwidth has to be used for token transmission. Consequently, to achieve the best performances from PROFIBUS-DP, it is important to limit the number of master stations.

During the operation of a PROFIBUS-DP network, a master station can execute several functions on its slaves. The most important are:

- Reading diagnostic data
- Setting parameter data
- Checking configuration data
- Exchanging cyclic data
- Emission of global control commands

These functions are implemented by means of messages exchanged between master and slaves.

With the *diagnostic message*, a slave, if requested, reports the general information about its status to a master. The message comprises 6 bytes of data explicitly specified by the standard and up to 238 bytes of extended diagnostic data, which can be configured by the user.

The *parameter message*, sent by a master to a slave, specifies some important features that will regulate the operation of that slave. For example, the master can set the value of a watchdog timer used by the slave to force itself into a safe state if the timer expires without being polled by the master. The parameter data message is composed of 7 bytes specified by the standard and up to 237 user-defined bytes.

The *checking configuration message* is used by the master, which has already parameterized a slave to communicate to it the number of input/output bytes that will be exchanged cyclically. If the slave accepts, then the cyclic data exchange phase is entered.

The *global control commands* are sent from a master to one or more of its slaves and are principally used to

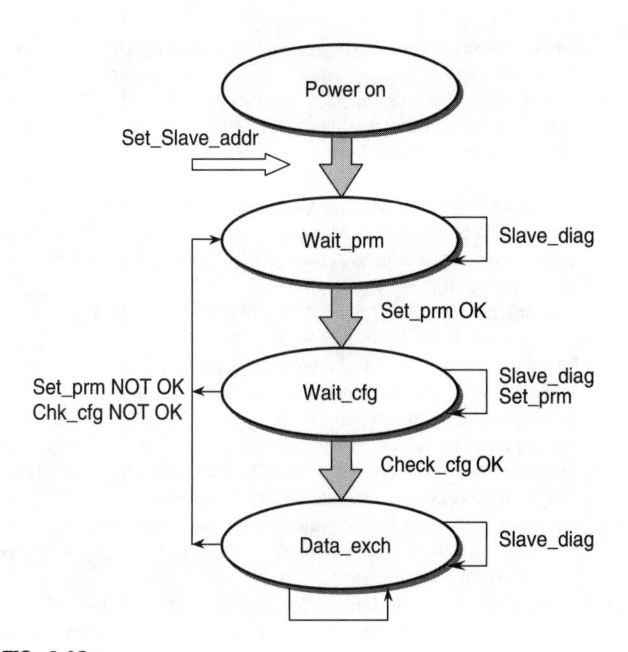

FIG. 4.16n
PROFIBUS-DP slave state diagram.

synchronize input and outputs (freeze and sync commands). With the global control commands there is also the possibility of forcing all the outputs of the addressed slaves to a safe state (clear command).

The state diagram of a PROFIBUS-DP slave, shown in Figure 4.16n, illustrates the use of the above functions. As can be seen, after power on, a slave looks for its address: if it is set (usually via hardware), then it immediately enters the Wait_prm state. Otherwise, the slave waits for the arrival of a Set_slave_address message from a master (the slave, while waiting, assumes the default address 255).

After the address has been set, in all the states of the diagram, a master can request the slave to supply its diagnostic data.

In the Wait_prm state the device awaits the arrival of the parameter message; on receipt, the slave can decide whether or not to accept the proposed parameters. If it accepts, then the Wait_cfg state is entered; otherwise, the slave remains in Wait_prm.

In the Wait_cfg state, when the checking configuration message arrives, the slave compares its own configuration with that proposed by the master and, if they match, it enters the Data_exch state.

In the Data_exch state, the master cyclically polls the slave, exchanging exactly the number of bytes specified by the checking configuration message. Moreover, in the Data_exch state, global control commands can be issued by the master.

It is interesting to note how the PROFIBUS-DP functions are implemented, and in particular the way in which they are mapped onto the data link layer services.

In practice, PROFIBUS-DP makes a "proprietary" use of the data link layer service access points (SAPs), normally used for addressing purposes, to identify the functions. In particular,

TABLE 4.16o
Correspondence between PROFIBUS-DP Functions and SAPs

Source SAP	*Destination SAP*	*PROFIBUS-DP Function*
62	54	Master/master functions
62	55	Set slave address
62	56	Read inputs
62	57	Read outputs
62	58	Global control commands
62	59	Read configuration
62	60	Read diagnostic
62	61	Set parameters
62	62	Check configuration
255	255	Cyclic data exchange

all the functions, except for the global control commands, are mapped onto SRD services. For example, an SRD request with the source SAP set to the value 62 and the destination SAP set to 60 specifies a request of reading the diagnostic data.

The global control commands also use the SAPs, but in this case the chosen service is SDN, as this is the only service that allows addressing more than one slave contemporaneously. Table 4.16o shows the correspondence between PROFIBUS-DP functions and SAPs.

PROFIBUS-DP Extended Functions

PROFIBUS-DP has recently been enhanced with the introduction of some extended functions, which, however, are specified to be optional. These new functions are the acyclic read/write (from/to a slave) and the alarm acknowledgment.

The acyclic read/write functions can be required by a user application on the master and are performed at the end of the cyclic polling of the slaves. In the request, the master has to specify the location, inside the slave, of the data to be read or written.

The extended PROFIBUS-DP functions also define a list of alarms that a slave may signal by means of the diagnostic message. When one or more of these alarms are communicated, the master has to answer that slave by means of the alarm acknowledgment function.

References

1. European Committee for Electrotechnical Standardization, "Industrial Communication Subsystem Based on ISO 11898 (CAN) for Controller-Device Interface—Part 2: DeviceNet," EN 50325-2, 2000.
2. International Electrotechnical Commission, IEC 62026-3: "Low-Voltage Switchgear and Controlgear—Controller-Device Interfaces (CDIs)—Part 3: DeviceNet," July 2000.
3. International Standard Organization, "Road Vehicles—Interchange of Digital Information—Controller Area Network for High-Speed Communication," ISO IS 11898, November 1993.

4. Institute of Electrical and Electronics Engineers, IEEE 802-3: 2000 standard, Part 3: "Carrier Sense Multiple Access with Collision Detection (CSMA/CD) and Physical Layer Specifications," October 2000.
5. European Committee for Electrotechnical Standardization, "General Purpose Field Communication System—Volume 5, ControlNet," EN 50170/5, July 2000.
6. International Electrotechnical Commission, IEC 61158-6: "Digital Data Communications for Measurement and Control—Fieldbus for Use in Industrial Control Systems—Part 3: Data Link Service Definitions, Type 2," January 2000.
7. International Electrotechnical Commission, IEC 61158-6: "Digital Data Communications for Measurement and Control—Fieldbus for Use in Industrial Control Systems—Part 4: Data Link Protocol Specifications, Type 2," January 2000.
8. International Electrotechnical Commission, IEC 61158-6: "Digital Data Communications for Measurement and Control—Fieldbus for Use in Industrial Control Systems—Part 5: Application Layer Service Definition, Type 2," January 2000.
9. International Electrotechnical Commission, IEC 61158-6: "Digital Data Communications for Measurement and Control—Fieldbus for Use in Industrial Control Systems—Part 6: Application Layer Protocol Specification, Type 2," January 2000.
10. International Electrotechnical Commission, IEC 1131-3: "Programmable Controllers—Part 3: Programming Languages," March 1993.
11. American National Standard Institute (ANSI), Electronic Industries Alliance (EIA), "Control Network Protocol Specification, ANSI/EIA 709.1," April 1999.
12. International Standard Organization/International Electrotechnical Commission, ISO/IEC 7498-1: "Information Technology—Open Systems Interconnection—Basic Reference Model: The Basic Model," 1994.
13. J. Postel, Ed., "RFC 791, Internet Protocol," Marina del Rey, CA: Information Sciences Institute, University of Southern California, January 1981.
14. European Committee for Electrotechnical Standardization, "General Purpose Field Communication System—Volume 3, WorldFIP," EN 50170/3, December 1996.
15. International Electrotechnical Commission, IEC 61158-6: "Digital Data Communications for Measurement and Control—Fieldbus for Use in Industrial Control Systems—Part 3: Data Link Service Definitions, Type 7," January 2000.
16. International Electrotechnical Commission, IEC 61158-6: "Digital Data Communications for Measurement and Control—Fieldbus for Use in Industrial Control Systems—Part 4: Data Link Protocol Specifications, Type 7," January 2000.
17. International Electrotechnical Commission, IEC 61158-6: "Digital Data Communications for Measurement and Control—Fieldbus for Use in Industrial Control Systems—Part 5: Application Layer Service Definition, Type 7," January 2000.
18. International Electrotechnical Commission, IEC 61158-6: "Digital Data Communications for Measurement and Control—Fieldbus for Use in Industrial Control Systems—Part 6: Application Layer Protocol Specification, Type 7," January 2000.
19. International Standard Organization, "Manufacturing Message Specification Parts 1 and 2," ISO IS 9506, 1990.
20. European Committee for Electrotechnical Standardization, "Low-Voltage Switchgear and Controlgear—Controller and Device Interface Systems—Actuator Sensor Interface (AS-i)," EN 50295-3, 1999.
21. International Electrotechnical Commission, IEC 62026-2: "Low-Voltage Switchgear and Controlgear—Controller-Device Interfaces (CDIs)—Part 2: Actuator Sensor Interface (AS-i)," July 2000.
22. European Committee for Electrotechnical Standardization, "High Efficiency Communication Subsystem for Small Data Packages—Volume 3, Profibus DP," EN 50254–3, 1998.
23. International Electrotechnical Commission, IEC 61158-6: "Digital Data Communications for Measurement and Control—Fieldbus for Use in Industrial Control Systems—Part 3: Data Link Service Definitions, Type 3," January 2000.
24. International Electrotechnical Commission, IEC 61158-6: "Digital Data Communications for Measurement and Control—Fieldbus for Use in Industrial Control Systems—Part 4: Data Link Protocol Specifications, Type 3," January 2000.
25. International Electrotechnical Commission, IEC 61158-6: "Digital Data Communications for Measurement and Control—Fieldbus for Use in Industrial Control Systems—Part 5: Application Layer Service Definition, Type 3," January 2000.
26. International Electrotechnical Commission, IEC 61158-6: "Digital Data Communications for Measurement and Control—Fieldbus for Use in Industrial Control Systems—Part 6: Application Layer Protocol Specification, Type 3," January 2000.
27. Telecommunication Industry Association, Electronic Industry Alliance, TIA/EIA-485: "Electrical Characteristics of Generators and Receivers for Use in Balanced Digital Multipoint Systems," 1998.
28. Institute of Electrical and Electronics Engineers, IEEE 802-4 standard: "Information Processing Systems—Local Area Networks—Part 4: Token-Passing Bus Access Method and Physical Layer Specifications," 1990.

4.17 Proprietary Networks

D. E. CAPANO

In the networking environment that exists today, two types of networks are broadly defined: open networks and proprietary networks. The former are familiar networks such as Ethernet and Token Ring. The latter are networks as defined by a specific vendor or for a specific purpose. Of these myriad specialized networks, three will be discussed as a representative sample. MODBUS is the first system discussed. Although MODBUS was developed for use on the Modicon Company proprietary network of the same name, the underlying concept defined by the protocol documents has been extended for use on other, "open" networks, such as the aforementioned Ethernet. The command set defined by the protocol is used as command and data information in the messaging and signaling formats of the native network protocol.

Data Highway (DH) is another proprietary network that will be discussed in this section. Data Highway, and Data Highway plus (DH+), is a system developed by the Allen-Bradley Company to support its line of network-enabled hardware, such as PLCs (programmable logic controllers), industrial terminals, and remote input/output (I/O) hardware. The DH protocol is specific to this type of network, and the command structure used is defined by company protocol documents.

The last system to be considered is the Genius Bus, defined by the General Electric Company. Genius Bus is similar in implementation to the DH system described above. The protocol and physical layout are specific to this type of network and were developed to support the GE line of automation products.

It should be noted that, at this writing, the above-named networks are being largely supplanted by standard networks such as Ethernet. Command structures are preserved, however, to allow the relationships defined in the protocols to be implemented. Put another way, regardless of the physical messaging or cabling scheme used, proprietary protocol-specific control and monitoring commands are used and are embedded in the native protocol framing or messaging structure.

PRIMARY USES OF A PROPRIETARY BUS

Proprietary buses are used in environments where an installed base of a particular manufacturer's devices is found. This may be as few as two devices on a serial link, or hundreds of devices. Determining factors in deciding on an open standard or a proprietary one are expandability, flexibility, and convenience. Should a user decide that a processing line is to be expanded, for example, the user would be concerned about adding more control and monitoring devices to the process control network. If the user decides to use the same manufacturer, then the simplest implementation is to add devices that are capable of communicating on a proprietary protocol. Should the expansion involve the addition of devices that adhere to a different protocol or to an open protocol, an open protocol would probably be the best solution, allowing the user to communicate with all devices over one common link, instead of trying to meld different networks through gateways and other devices. Using one protocol across the network eliminates the extraordinary complexity involved in translation between protocols. Using an open protocol also lends flexibility to the network, allowing many different network-capable devices to communicate over the network. A proprietary network is typically specialized for use with a particular manufacturer's line of equipment. This equipment may be specialized to the point that only a few devices are offered, limiting choices and opportunities for innovation. Expansion within an existing proprietary network is relatively simple as all that is required is to cable a location and connect a network-ready device. This is a very convenient feature in that a minimum of design overhead is used, including programming time. Open networks offer the same type of convenience, but mixed networks pose several problems in implementation and expansion because of the nature of cross protocol communication and translation.

DISADVANTAGES OF A PROPRIETARY BUS

One of the chief disadvantages of a proprietary bus is the lack of flexibility in equipment selection and command capabilities. Although most manufacturers have gone to great lengths to offer their clients every available option in process control and data acquisition, the user is still limited to those devices that are compatible with that particular network. This is not a negative, however, but merely a limiting factor for the designer of the network or control system. When a proprietary network must be integrated into a larger network, such as a distributed control system (DCS), then the vagaries of message and protocol translation come into play. This added step incurs more cost in the design and implementation phases of the project. An open network, on the other hand, allows a broader range of equipment to be implemented on a given process, allowing the designer a good deal more freedom. Open network protocols, such as Ethernet, are mature, robust technologies with most compatible components available "off-the-shelf." This translates into lower costs in both the design and implementation phases. Open protocols also come without licenses and fees, further driving down implementation costs. Both types of networks have their optimal application. Proprietary networks, however, are best relegated to small, closed process control applications that employ the use of one line of device.

MODBUS NETWORKS

The MODBUS protocol defines a method of access and control of one device by another, regardless of the type of physical network on which the protocol is used. The protocol describes the process by which one device requests access to another device, how information is received, and how queries are responded to. The protocol also describes how errors are handled and what messages are developed and sent to the user. Communications may take place on a MODBUS network, or commands defined by the MODBUS protocol may be embedded into data packets defined by other network protocols, such as Ethernet. Communication on a MODBUS network involves device address recognition, message recognition, and message interpretation and response. On other types of networks, where the MODBUS message is embedded in the particular frame or packet defined for that network, addressing and device address resolution are handled by the native protocol. When the packet is delivered to the intended device successfully, the MODBUS information is removed from the frame and interpreted by the receiving device. At this point the appropriate action is taken based upon the contents of the message. In either case, the receiving device either performs a control function or responds to the sending device with an acknowledgment or with information.

MODBUS-capable controllers can communicate with each other over a variety of networks. Among these are standard networks such as MAP (manufacturing access protocol), defined as a token-bus network and Ethernet. The protocol is the standard method of communication between devices manufactured by Modicon for use over its MODBUS network. At the simplest level, two devices may communicate over a serial link, such as RS-232C, or RS-485, whether single link or multidrop. Controllers may be remote and connected via modem. The following section deals with the setup of communication and data exchange between devices.

Communication Transactions

Devices communicate on a MODBUS network using a master/slave relationship. The device initiating the transaction is the master and the accessed device is the slave. On any network, for any transaction, there is only one master device.

On other networks, devices communicate on a peer-to-peer relationship, that is, the devices communicate with one another on an as-required basis and without any master/slave hierarchy. A master may communicate with one slave or may broadcast inquiries, called *queries* to all slaves. Slaves respond by returning information to the master or by performing the control action requested by the master. Slaves respond to queries specifically addressed to them; broadcast messages are not responded to by any slaves on the network. The query structure used is defined by the protocol as the destination device address, a function code for the action requested, any data, which may need to be sent, and an error-checking field. The response structure is defined as information confirming the action taken, any data to be returned, and an error-checking field.

At the message level, the master/slave relationship is reestablished between the initiator (master) and the responder (slave). The master device still expects a response from the slave device. The exchange of data and instructions still occurs much as it would in a MODBUS network. In a peer-to-peer network, a device may be a master or a slave simultaneously in different transactions.

The query contains four parts, as illustrated in Figure 4.17a:

1. Device address: This is the address of the slave to which the query is being sent.
2. Function code: The function code tells the slave what action to perform, such as operating a relay coil.
3. Data: This field provides additional information on the action to be taken, such as the relay tag that is to be operated.
4. Error Checking: This performs one of several error-checking methods to determine the validity of the message sent.

The response structure is identical to the query structure. The device address is the address of the master to which the slave is responding. The function code, in a normal response, is the same function code sent in the query. The data bytes return information about the function, i.e., whether the function was carried out successfully. The data field will also return any information requested by the master, such as the contents of a specific register. If an error has occurred, either with the query by the master, or with the action performed by the slave, the function code is modified to indicate that the response is an error message. The data field will then contain

Device Address
Function Code
Data Bytes
Error-Checking

FIG. 4.17a
Query and response structure.

an error code indicating what error occurred. The error-checking field allows the master to determine if the message is valid.

Transmission Modes

Two types of transmission modes are possible on a serial MODBUS network: ASCII and RTU (remote terminal unit) modes. Either mode is selected as part of the device configuration parameters, along with other serial parameters such as baud rate and parity. Once a transmission mode is selected, all devices on the network must conform to the same parameters.

ASCII Mode In ASCII (American Standard Code for Information Interchange) mode, each 8-bit byte is sent as two ASCII characters. This allows for a slower transmission rate without generating errors. The format for each byte in ASCII mode is as follows:

- Coding System: Hexadecimal, ASCII characters 0 to 9, A to F. One hexadecimal character is contained in each ASCII character of the message.
- Bits per Byte: 1 start bit; 7 data bits, LSB sent first; 1 bit for odd/even parity, no bit for no parity; 1 stop bit if parity is used, 2 bits if no parity.
- Error Check Field: Longitudinal redundancy check (LRC).

RTU Mode In RTU mode, each byte in a message contains two 4-bit hexadecimal characters. The greater character density allows for better throughput than ASCII mode at the same baud rates. Each message is transmitted in a continuous stream. The format for each byte in RTU mode is as follows:

- Coding System: 8 bit binary, hexadecimal 0 to 9, A to F. Two hexadecimal characters are contained in each 8-bit field of the message.
- Bits per Byte: 1 start bit; 8 data bits, LSB sent first; 1 bit for odd/even parity, no bit for no parity; 1 stop bit if parity is used, 2 bits if no parity.
- Error Check Field: Cyclical redundancy check (CRC).

MODBUS Message Framing

In either transmission mode, a frame is constructed by the transmitting device, which allows determination of the beginning and end of the message and the destination device address of the device, or devices, in the case of a broadcast message. On standard networks the framing is done by the native protocol, with addressing converted from that in the MODBUS message by the network adapter.

ASCII Framing In ASCII mode, messages start with a colon (:) character, which is ASCII character 3Ah (h signifies that the code is in hexadecimal and will be used throughout); messages end with a carriage return-line feed (CRLF) which is the ASCII character pair 0Dh and 0Ah. Allowable characters for the remaining fields are hexadecimal 0 to 9 and A to F.

All devices continuously monitor the network for the colon character signifying the start of a message. Each device then decodes the following field, the address field, to determine if the message is for that device. If it is not the addressed device, it ignores the remainder of the transmission and resumes monitoring of the network for the next colon character. If the address field indicates that the message is intended for that device, then the following field, the function code, is decoded. A typical ASCII message frame is shown in Figure 4.17b.

In ASCII mode, intervals in transmission up to 1 s are acceptable. If the transmission lapses for more than 1 s, the receiving device assumes an error has occurred and generates an error response.

RTU Framing In RTU mode, messages start with a silent interval of at least 3.5 times character length. The first field any monitoring device hears is the address field, which it decodes to determine if the message is intended for that device. The message is ended with the same silent interval. The entire frame must be transmitted in one continuous stream or an error will be generated. If there is a lapse in transmission over 1.5 character times, then the receiving device assumes an error has occurred and deletes any received information. The receiving device assumes that the next byte received will be the address field of a new message. Similarly, if a message starts within the 3.5 character times ending the message, the receiving device assumes that the information is part of the previous message. Either condition will cause a CRC error. A typical RTU frame is shown in Figure 4.17c.

Field Descriptions Following is a description of the particulars of each field as shown in Figures 4.17b and 4.17c. Refer to these illustrations during the following discussion.

Address Field The address field contains two characters (ASCII) or 8 bits (RTU). Valid slave addresses are 0 to 247. Slaves are addressed in the range 1 to 247, with 0 reserved for broadcast to all slaves; all slaves recognize this address.

Start (:) 1 character: 3Ah	Address 2 Characters	Function Code 2 Characters	Data *N* Characters	LRC Check 2 Characters	End **(CRLF)** 2 characters: 0Dh, 0Ah

FIG. 4.17b
A typical ASCII frame.

Start 3.5 character times	Address 8 bits	Function Code 8 bits	Data $N \times 8$ bits	CRC Check 16 bits	End 3.5 character times

FIG. 4.17c
A typical RTU frame.

When a master addresses a slave, it places the slave's address in the address field. When a slave responds, it places its address in the address field of the response message to inform the master which slave is responding.

Function Code The function code field contains two characters (ASCII) or 8 bits (RTU). Valid codes are 1 to 255 *decimal.* Function codes are used either to request information from a slave or to request that particular task be performed. In either case, the slave responds informing the master device of the status of the operation. If the task is completed successfully, then the slave responds with the same function code that was sent by the master. If an error occurred, then the slave responds with an *exception response*, which is the function code with the most significant digit (MSD) or most significant bit (MSB) set to 1. In addition to setting the MSB to 1, the slave also sends a unique error code in the data field. The application program of the master device has the responsibility of dealing with the error code. Typical responses are message retries and diagnostic routines between the master and slave. The master device also notifies operators of the error. Not all MODBUS commands are supported by all devices. Some function codes are shown below:

- 01 Read Coil Status
- 02 Read Input Status
- 03 Read Holding Registers
- 04 Read Input Registers
- 05 Force Single Coil
- 06 Preset Single Register
- 07 Read Exception Status
- 08 Diagnostics
- 09 Program
- 10 Poll
- 11 Fetch Communication Event Counter
- 12 Fetch Communication Event Log
- 13 Program Controller
- 14 Poll Controller
- 15 Force Multiple Coils
- 16 Preset Multiple Registers
- 17 Report Slave ID
- 18 Program
- 19 Reset Communication Link
- 20 Read General Reference
- 21 Write General Reference

Data The data field is constructed of two hexadecimal numbers, in the range 00 to FF. These are made up of two ASCII characters or one RTU character, depending on which mode is in use. The data field contains information required by the slave to properly perform the function defined in the function code field. This information may be a range of inputs or registers to be read, for example, or discrete device addresses to be acted upon. In a normal response, the slave will return the data or status requested by the master. If an error occurs, an exception response will be generated. A data field may also be zero length if the function specified requires no additional data.

Error Checking MODBUS serial networks utilize two types of error checking: parity and frame checking. Parity checking is applied to each character and can be even or odd. Frame checking is accomplished in one of two ways and is dependent on the mode of transmission used. In ASCII mode, longitudinal redundancy checking (LRC) is used; in RTU mode, cyclical redundancy checking (CRC) is used. Both character and frame checking information is generated at the master device and applied to the message prior to transmission. The slave checks each character and the entire frame while decoding the information in the message.

DATA HIGHWAY

The DH protocol was developed by the Allen-Bradley Company to effect communication over its proprietary network among PLCs, other PLCs, and computer terminals (operator workstations). The DH is a peer-to-peer network, allowing any station on the network to communicate with any other station on the network. The DH protocol may also be implemented over an asynchronous link, such as RS-232C or RS-422A. Two types of communication protocol are utilized on an asynchronous link:

1. Half Duplex: Only one station may communicate at one time, in one direction. After one station is finished transmitting, the other station may reply. In this mode, the message transaction method becomes one based upon a master/slave relationship. Each station on the link is polled by the master, to which it replies status information or data.
2. Full Duplex: Both stations can communicate in both directions at the same time. Communication is again peer to peer and stations are unpolled. Full duplex is used where the highest maximum throughput is desired through a given transport medium.

Physical Implementation

The physical link between nodes, called stations on a DH, consists of any available media commonly used for computer networking. This may be shielded cable, fiber-optic cable, or wireless. A station is defined as any intelligent device that is capable of being attached to and communicating on the network while adhering to the protocol. The DH is a local-area network (LAN) that is capable of supporting up to 64 stations on a trunk up to 10,000 ft long. Stations may be connected to the trunk by drops up to 100 ft. The media used is a twin-axial shielded cable, which is "daisy-chained" between devices using a simple compression-type terminal strip.

To communicate on the DH network, a modified token-passing scheme is used, called a "floating master." In this method of access control, each station "bids" for the opportunity to control the network and place a message on the medium for transmission to another station. Each station has an equal opportunity to become the master. The master token is awarded to a station based on message urgency or priority and on the relative amount of access the station has attained during a preset time period. For example, a mission-critical message has higher priority than a status request; the former will be granted transmission privileges (master token) before the latter. Similarly, a station that has just finished a transmission will not be awarded the token if another station is bidding for it. If no priority exists for the first station, then the second station will become the master and be allowed to transmit its information. No polling is involved and this allows higher data throughput without errors generated by collisions. A DH has a data transmission rate of 57,600 bps and uses a half-duplex transmission protocol. A simple DH is illustrated in Figure 4.17d.

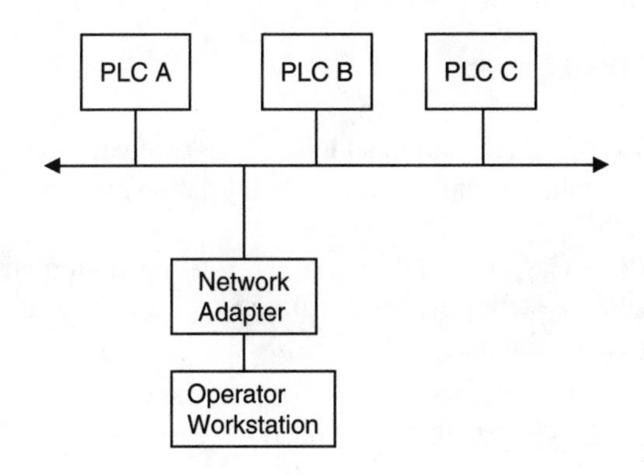

FIG. 4.17d
A simple DH network.

Another less common implementation of the DH protocol is over a LAN defined by Allen-Bradley as a peer communication link (PCL). Communication is controlled on a PCL by a token-passing scheme, which rotates mastership among the connected stations. The token is passed from station to station; if the station receiving the token has data or command information to transmit, the information is input into the data field of the token. The token is then passed along to the next station and continues until it reaches the intended station. The receiving station decodes the information and sends an acknowledgment. This method of token passing is similar to that seen in the MAP protocol, which is described as a token-passing bus. Various time-out settings prevent the network from jamming should a station become disabled for any reason. A typical PCL implementation is shown in Figure 4.17e.

PCLs are typically used in small stand-alone applications with several PLCs or other devices connected together, usually on one machine or on a localized process. A PCL can also be connected, through the appropriate adapter, to a DH network.

Software Layers

Each of the links just described requires three layers of software to communicate:

1. The Applications Layer: Controls communication between stations and executes commands sent between stations.
2. The Network Management Layer: Handles the sequencing, scheduling, and error-handling functions on the LAN.
3. The Data Link Layer: Controls the communication link by establishing, maintaining, and releasing the link between stations.

Application Layer

This layer is concerned with the actual command sent over the link. There are two types of application programs used on a DH network: command initiators and command executors. Command initiators specify which command is to execute and when the execution should occur. The command initiator sends messages called command messages. The command executor is the receiving station that decodes the

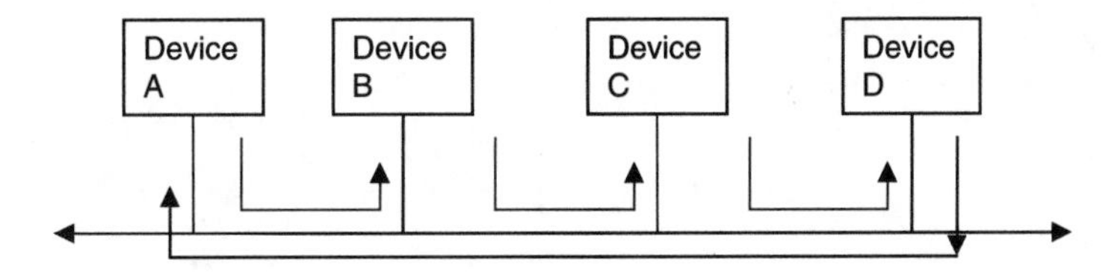

FIG. 4.17e
A typical PCL network showing token rotation.

command information and executes the command specified. The executor replies to every message it receives. If there is a malfunction either in receiving or executing the message, then the appropriate error message is generated by the command executor.

A network transaction consists of two messages, a command and a reply. This ensures that the command message is always acknowledged, either with an error message or with data or status form a successful message and command execution. The command initiator is referred to as the *local* station and the command executor is referred to as the *remote* station. Messages are also assigned priorities. A message can have either normal or high priority. The level of priority of a message of a station determines the order in which stations are awarded mastership of the network. If a particular station has a high-priority message to send, while a neighboring machine has a message with normal priority, the higher-priority message will be sent first.

Command Structures

There are four basic types of command structure on a DH network:

1. *Read*: Any read command can request up to 122 16-bit words from the specified memory location. There are two types of read command:
   - Physical reads: Allow a physical read of any part of the memory of a remote machine. PLCs cannot initiate a physical read command; only a workstation or other computer can initiate this type of read.
   - Unprotected reads: Allow access to only the data tables of a PLC or other device. Both workstations and PLCs can initiate this command.
2. *Write*: Write commands are classified by their application and their level of memory access. As an application classification, there are two types of write commands:
   - Bit writes: Bit writes allow a local station to manipulate bits in the data table of a remote station.
   - Word writes: Word writes allow a local station to write up to 121 contiguous words into the remote station memory, depending upon the level of memory access involved. Memory access levels are described below.

   Memory access is also used to classify write commands. There are two levels:
   - Physical: Can access all memory at a remote station, whether that station is a PLC or a workstation.
   - Nonphysical: Can access only the data table at a remote station. Nonphysical writes can be further divided into two types:
     —Protected: Can access only specific areas of the remote station data table as defined by the memory protection rungs (program code) in the remote station program.
     —Unprotected: Allows unrestricted access to the entire data table of the remote station.
3. *Diagnostic*: Used to perform status checking and to alter parameters at a remote station. Used during start-up or debugging.
4. *Mode Select*: Used to load a new program into a remote compute. This command can only be initiated by a workstation.

Message Format

Command and reply messages travel on the network in the form of message packets. These packets are defined by the protocol and the standard packet contains fields, which delineate the sending and receiving addresses of the local and remote stations, respectively, status and message control data, and payload data. The command and reply messages have different formats (Figures 4.17f and 4.17g).

Field Descriptions

- DST: The destination address of the receiving station, also called a station number. Valid station numbers are in the range of 0 to 254 decimal.
- SRC: The source address of the sending station, also called a station number. Valid station numbers are 0 to 254 decimal.

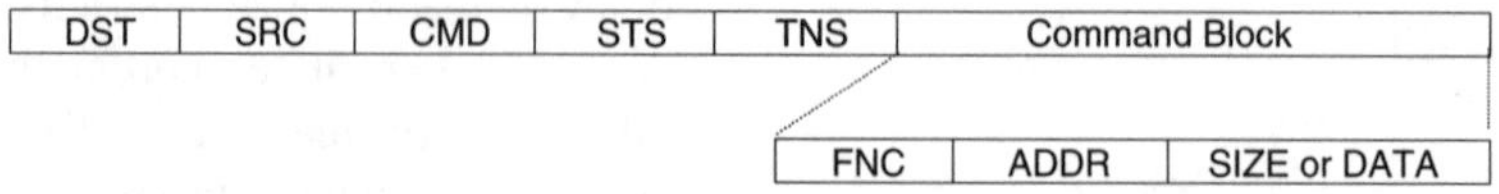

DST	SRC	CMD	STS	TNS	Command Block

FNC	ADDR	SIZE or DATA

FIG. 4.17f
The DH command frame format.

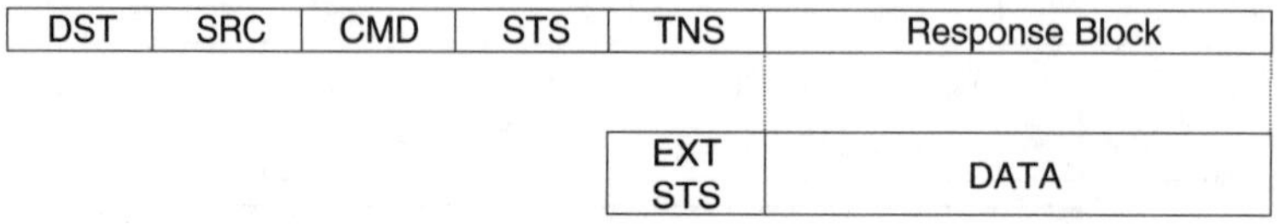

DST	SRC	CMD	STS	TNS	Response Block

EXT STS	DATA

FIG. 4.17g
The DH reply frame format.

7	6	5	4	3	2	1	0
0	R	P	0	Command			

FIG. 4.17h
The CMD byte format.

TABLE 4.17i
The Basic DH Command Set

Command	*CMD Code*	*FNC Code*
Diagnostic Counters Reset	06	07
Diagnostic Loop	06	00
Diagnostic Read	06	01
Diagnostic Status	06	03
Protected Bit Write	02	N/A
Protected Write	00	N/A
Set ENQs	06	06
Set NAKs	06	05
Set Timeout	06	04
Set Variables	06	02
Unprotected Bit Write	05	N/A
Unprotected Read	01	N/A
Unprotected Write	08	N/A

- CMD: The command field defines the command type. Together with the function byte (FNC below) the CMD byte defines the activity or task to be performed at the remote or receiving station. The command is a single hexadecimal character and occupies a specific position in the command byte as shown in Figure 4.17h. The command itself occupies bit positions 0 through 3. Bits 4 and 7 are always set to zero. Bit 5 is the priority flag: it is set to 0 for normal priority and 1 for high priority. Bit 6 is the command/reply indicator: it is set to 0 in a command message and 1 for a reply message. The basic command set is shown in Table 4.17i.
- STS: The status field indicates the status of the transmission. In command messages, this field is always set to zero. In reply messages, a zero indicates a successful transmission. Otherwise, an error code is returned in the reply message in this field.
- TNS: The transaction field assigns each transaction a unique 16-bit (2-byte) identifier. The first byte, RNG, indicates the rung number in the initiator PLC program that generated the command. The second byte, SQN, indicates the transmission sequence number. The RNG and SQN bytes are incremented by the command initiator for each successive message. These bytes also use dot track replies to commands. If the RNG and SQN numbers of a reply match those of a command message, then that reply is the appropriate reply for that command.
- FNC: The function field, together with the CMD (command) field, determines the action to be taken by the receiving station. A list of function codes is shown in Table 4.17i.
- EXT STS: The external status field contains additional status information.
- ADDR: This field contains the address of the location in the memory of the command executor where the command is to begin executing. This is a specific byte address, not a word address.
- SIZE: This field indicates the number of data bytes to be transferred by the message. This field is sent with read commands, and tells the executor how many bytes to be sent back with its reply.
- DATA: This field contains binary data from the application programs. abcd edhg sgstgts lkjk xc jhjx cx

Table 4.17i lists the basic DH command set. Commands are sent by the initiator and are defined by the combination of the CMD field and the FNC field. To request that a command in the first column be executed, the following two hexadecimal values must be entered into the CMD and the FNC fields, respectively.

Asynchronous Link Protocols

The last implementation of the DH protocol is by utilizing serial links conforming to the RS-232C and RS-422A standards. A link protocol is defined as a method for moving information over the link. The protocol does not decode the message in any way and is not concerned with the contents of the message. The protocol simply moves information from the sending station to the receiving station. The upper end for data throughput on an asynchronous link is 9600 bps. Two types of link protocols are defined:

1. Full-duplex transmission: Allows transmission in both directions simultaneously; this allows faster throughput, but is more difficult to implement. Used for point-to-point communication. Also referred to as DF1 protocol.
2. Half-duplex transmission: Allows transmission in only one direction at a time. Slower throughput, but easier to implement. Used in a master/slave communication scheme. Also called polled-mode protocol.

Each of these protocols will be discussed in detail below.

Full-Duplex Transmission Transmission codes are grouped into two categories: message codes, which are sent by the

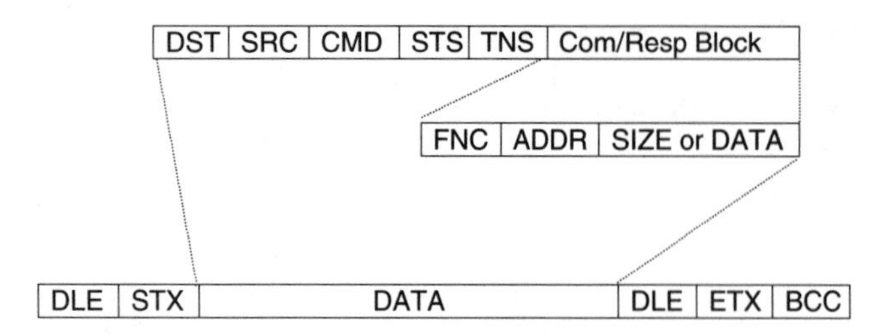

FIG. 4.17j
The full-duplex frame format.

DLE	ENQ	STN	BCC

FIG. 4.17k
A polling packet.

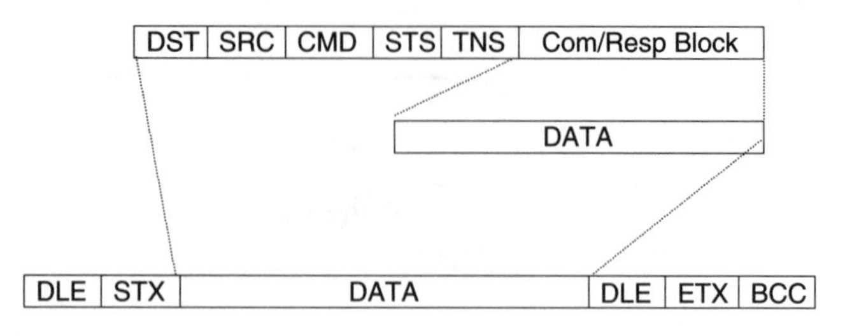

FIG. 4.17l
Slave message link packet.

initiator or transmitting station, and response codes, which are sent from the executor or receiving station. The following transmission codes are used in full duplex transmissions:

- DLE STX (Data Link Escape Start of Text) is a message code and indicates the start of a message.
- DLE ETX BCC/CRC (Data Link Escape End of Text Block Check Character/Cyclical Redundancy Check) is a message code and is used to terminate a message.
- Data 00-0F and 11-FF hexadecimal are message codes that are used to encode values of the codes into the message. The data code DLE DLE is used to encode the value 10h in the message.
- DLE ENQ (Data Link Escape Enquiry) is a message code that requests retransmission of the last received code.
- DLE ACK (Data Link Escape Acknowledge) is a response code that signals that a message has been successfully received.
- DLE NAK (Data Link Escape Negative Acknowledgment) is a response code that signals that an attempt to transfer data was unsuccessful.

The format for the full-duplex packet frame is shown in Figure 4.17j.

Half-Duplex Transmission Half-duplex transmission is used in a master/slave communication environment. It is defined as a multidrop protocol for use with one master and one or more slaves. There may be as many as 256 stations connected to the link simultaneously. One station is designated as the master, and the rest of the stations are designated as slaves. Slaves cannot initiate transmission and must wait for permission from the master before transmitting. Each slave has a unique identification number ranging from 0 to 256.

The master can send and receive messages to and from any slave on the network. This can be done on a single or dual circuit system. In the former, master and slaves transmit and receive over the same circuit; in the latter, the master sends and slaves receive on one circuit and slaves send and the master receives on the other. Multiple masters are allowed only if the second master acts as a backup to the primary. Two masters cannot be active on the network.

Transmission codes for half-duplex transmission are described below.

- DLE SOH (Data Link Escape Start of Header) indicates the start of the message.
- DLE STX (Data Link Escape Start of Text) separates the link level header from the data field of the message.
- DLE ETX BCC/CRC (Data Link Escape Block Check Character/Cyclical Redundancy Check) is used to terminate a message.
- DEL ACK (Data Link Escape Acknowledge) is used to signal that a message has been successfully transmitted.
- DLE NAK (Data Link Escape Negative Acknowledgment) is used as a global link reset command and is sent only by the master to all slaves on the network. This command causes all slaves to delete any messages scheduled for transmission to the master.
- DLE ENQ (Data Link Escape Enquiry) is used to start a poll command.
- DLE EOT (Data Link Escape End of Transmission) is used by a slave station as a response to a polling command when it has no information to send.

Figures 4.17k, 4.17l, and 4.17m are illustrations of the various packet frames used in this system. Three types of packet frames are used:

- Polling Packets: Used to poll slaves for status.
- Slave Message Packets: Used to respond to master message packets with status or information.
- Master Message Packets: Used to initiate commands to slave stations.

GENIUS BUS

Overview

The Genius Bus is a proprietary network designed for use with equipment manufactured by the General Electric Company. This LAN is very similar to the others described in this section.

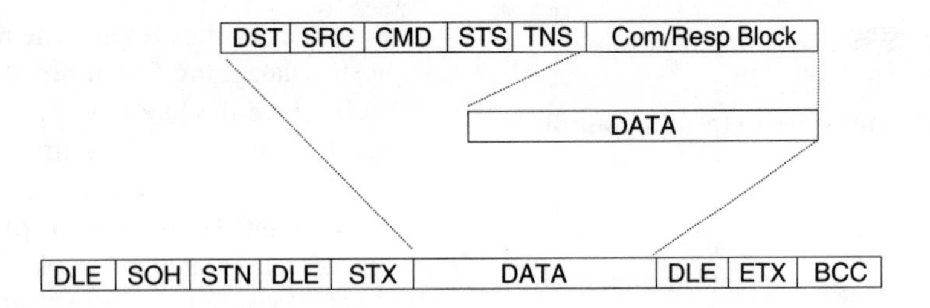

FIG. 4.17m
Master message link packet.

The LAN utilizes a wide variety of media to connect network devices, ranging from twisted shielded pairs to wireless. The protocol is not media specific, however, except at the physical layer. Devices are "daisy-chained" together in a network (bus) that can support up to 32 devices. A system can support multiple buses, with each bus utilizing a separate bus controller. Maximum bus length is 7500 ft, with a maximum baud rate of 153.6 kbps. The method of bus control is a token-passing scheme.

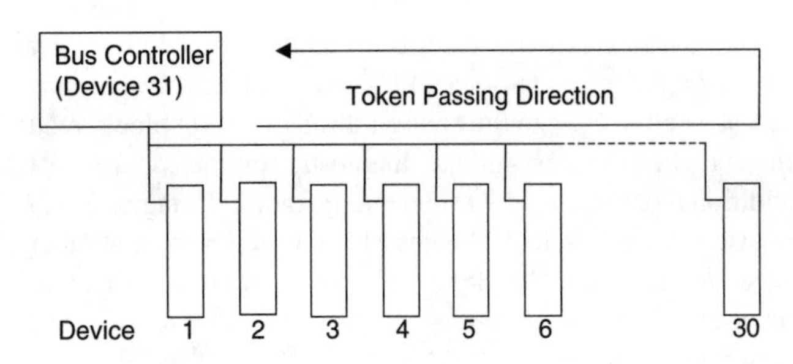

FIG. 4.17n
Token passing on a Genius Bus.

Physical Implementation

A Genius Bus is a collection of up to 32 devices, including the bus controller, on a LAN referred to as a bus. Individual nodes are allowed network access by acquiring bus mastership through possession of a token, possession of which allows that node to transmit and receive data. The node is not allowed to transmit any information unless it has the permission to do so by possessing the token. Two types of token-passing schemes are used on a Genius Bus:

1. Implicit Token: This method ensures that any devices that are coming online or going offline do not affect the operation of the rest of the network.
2. Fast Token Recovery: Restores device access after any system transients, which may disturb the operation of the network.

A bus has a potential of 32 devices, numbered 0 through 31. The bus controller is always device 31 on any bus. The bus controller manages the transfer of information from the bus to the PLC or the PC. Figure 4.17n illustrates the scheme.

All devices automatically log on to the network by transmitting information such as data length, I/O configuration, reference address, and device number.

Communication services that take place on a Genius Bus are grouped into three categories, which are described in the following paragraphs.

I/O Service Each time a network device, referred to as a *block*, receives the token, it broadcasts all its inputs. The bus controller reads all information broadcast by the devices on the bus and passes this information to the host CPU (central processing unit); this CPU may reside in either a PLC or in a workstation. Any fault or diagnostic information is also

Start of Block
Function Code
Source Address
(Omitted if broadcast)
Subfunction Code
Datagram Content
End of Block

FIG. 4.17o
The Genius Bus datagram format.

decoded and passed to the host computer. When the bus controller acquires the token, it transmits whatever current output and control data it has queued. These data are not broadcast; rather, the data are addressed to the individual block for which the data were intended. When all transmissions are completed, the token is passed to the next device.

Datagrams A datagram is a message sent from one device on a bus to one or more other devices. Regardless of who sends the datagram, the content remains the same. All datagrams follow the same format, as shown in Figure 4.17o. The format is used to send any type of datagram. The LSB is sent first in this system. The sending device adds all information, except for the datagram content, automatically.

To send a datagram, a device or CPU sends information to the bus controller describing the content of the datagram it wants to send, including information about how to send the datagram and to whom. Datagrams may be sent with normal or high priority. High-priority messages take precedence over all other transmissions, including fault reporting by other blocks. In general, the procedure for sending a datagram is similar for all devices on the bus:

1. The CPU or block sends a message to the bus controller describing the action to be taken.
2. The bus controller automatically performs the requested action:
   a. The bus controller sends the supplied datagram to the device specified.
   b. The bus controller supplies a datagram it has received to that block.

Datagrams can be sent from the PLC or CPU to any device on the bus or to other CPUs on other buses. Datagrams can be used to change the configuration of an I/O block, read the status of devices, initiate diagnostic routines, or request additional or detailed data about an operation. Datagrams can be used to read up to 128 bytes of information from another CPU or write up to 128 bytes of information to one or more devices. The same bus may be used for I/O service and communication.

The types of datagrams used on the Genius Bus are as follows:

- Read Identification: Sent at start-up by the bus controller to discover the identities of the other devices on the bus.
- Read ID Reply: This datagram is the reply to the Read Identification datagram. It contains all device configuration data, including device address and I/O status.
- Read Configuration: Used to read up to 12 bytes of configuration data from a device on the bus.
- Read Configuration Reply: The reply datagram to the Read Configuration datagram. It returns all pertinent device configuration data.
- Write Configuration: Used to write up to 16 bytes of configuration data to any I/O block on the bus.
- Assign Monitor: This datagram directs a block to send a copy of the Configuration Change and Report Fault datagrams (explained below) to the device number specified in the Assign Monitor datagram.
- Begin Packet Sequence: Forces a receiving block to accept an entire transmission without decoding or otherwise analyzing the received data. The use of this datagram allows a rapid download of information to the block.
- End Packet Sequence: Ends the download sequence started by the Begin Packet Sequence datagram. The receiving block will take no action on the received data until it receives the End Packet Sequence datagram.
- Read Diagnostics: Used to query a block for its current diagnostics.
- Read Diagnostics Reply: The reply to the Read Diagnostics datagram. The information returned may be all faults since the block was powered up or only faults that have occurred since the last Clear Circuit Faults or Clear All Faults datagrams were received.
- Write Point: Used to set or reset up to 16 individual bits of data in another block.
- Read Block I/O: Used to read input and output data from specific Genius analog and thermocouple blocks.
- Read Block I/O Reply: This datagram is the reply to the Read Block I/O query and contains the requested information.
- Report Fault: This datagram is sent automatically if a block experiences any type of fault. The datagram can be sent to a maximum of two bus controllers and the optional assigned monitor of the block (see Assign Monitor above).
- Pulse Test: Sent to a specific block directing it to pulse-test all of its output circuits. Any circuit faults located are reported through the normal Report Fault datagram.
- Pulse Test Complete: Sent by a block that has completed its pulse testing to the device that initiated the Pulse Test datagram.
- Clear Circuit Fault: Sent to a specific block directing it to clear a specific circuit number on that block. If the fault is not cleared, then a new Report Fault datagram will be generated.
- Clear All Circuit Faults: Directs a specific block to attempt to clear all circuits on the block. If the faults are not cleared then a new Report Fault datagram will be sent.
- Switch BSM: Sent to the bus switching module, if the system is a dual bus system. The datagram instructs the BSM to switch to the specified bus in the dual system.
- Read Device: Used to read data from another device on the bus. The memory map of the other device must be known to process this datagram.
- Read Device Reply: The reply datagram to the Read Device datagram returns the requested data.
- Write Device: This datagram allows one device to write to the memory of another device on the bus.
- Configuration Change: Sent to the bus controller is a block that determines one of its critical parameters has been changed.
- Read Data: Used with a specialized high-speed counter block. Allows another device to read specific data from the block memory.
- Read Data Reply: The reply datagram to the Read Data datagram containing the requested information.
- Write Data: Sends temporary data to a high-speed counter block memory.
- Read Map: Used in specialized PLC operations.
- Read Map Reply: The reply to the Read Map datagram containing the requested information.
- Write Map: Used in specialized PLC operations.

Global Data Any device connected to the Genius Bus can send and receive global data. Global data are data that are automatically and regularly broadcast by a bus controller. There are two basic differences between global data and datagrams:

1. Global data are sent repeatedly and require very little, if any, programming effort. Datagrams require a programming input when sending or receiving and the status of the datagram must be monitored.
2. Some devices respond only to global data and do not recognize datagrams.

Bibliography

The following items (all available on the Web) are recommended reading for those who wish to pursue further study on the subjects covered in this section:

"Data Highway and Data Highway Plus Asynchronous (RS-232-C and RS-422-A) Interface User Manual," Cat. No. 1770-KF2, Allen-Bradley Corporation.

IEEE 802.4, "Token Passing Bus Networks," Piscataway, NJ: IEEE, 1997.

"Genius I/O System and Communication User's Manual," Cat. No. GEK90486F-1, GE Fanuc Automation.

"MODBUS Protocol Reference Guide," North Andover, MA: Modicon Corporation.

4.18 Fiber-Optic Networks

E. J. BYRES

Partial List of Suppliers: Siecor; 3M; Lucent Technologies; Amp; Cabletec; Belden; Berktec; Corning; Optical Cable Corporation; Tyco Electronics; ADC Broadband Connectivity; Fotec; Alcatel

PRINCIPLES OF FIBER-OPTIC NETWORKS

Fiber-optic technology was originally developed to allow high-speed data and voice communications over long distances. Its initial market was the commercial communications industry, particularly the telephone companies and television or radio broadcast companies. Within a decade, the excellent immunity of fiber to electromagnetic interference (EMI) quickly attracted the attention of heavy industry, which led to its use in a few high-noise areas. The acceptance in these areas and the growing need for more sophisticated communications systems have resulted in the widespread use of fiber optics in industrial communications.

However, fiber-optic technology is borrowed from the telephone industry, and the needs of industrial communications can be very different from those of the commercial market. The harsh environment, relatively short cabling distances, and the need for maintenance by electrical and instrumentation tradespeople not exclusively dedicated or trained to do fiber installation call for different, more functional products and installation methods than those practiced by a telephone company.

This section explains how fiber-optic networks work and then discusses the design of fiber-based communications systems in industrial settings.

Why Use Fiber Optics?

When engineers initially consider using fiber optics in an industrial facility, the initial reasons are usually related to some of the outstanding physical properties of fiber. Certainly, fiber cabling offers an electrical robustness that copper-based systems cannot match. It is immune to EMI, arcing, short circuits, ground loops, and loss of ground. It also has a very large bandwidth, so it can carry a great deal of data over very long distances without repeaters (up to 120 km for single-mode fiber).

However, there are other less obvious but important reasons for using fiber optics in industrial settings. First of all, fiber-optic cable is considerably more rugged than most coaxial and twisted-pair cables. For example, the pull strength of a fiber pair is typically 200 lb, as compared to only 25 lb for Category 5 twisted-pair cable and 45 lb for RG-58 coax cable.[1] The crush strength of fiber is equally high when compared with copper communications cables. In fact, some military-specified fiber cables are robust enough to allow tanks to be driven over unarmored fiber.

Fiber optics offers another advantage: it can be cheaper than copper because it offers multiple pathways and requires no special protection from EMI. As a result, the total cost of ownership for a fiber-based industrial network is usually much less than a copper solution, even though the per foot price of fiber cable is roughly twice that of coaxial cable and eight times that of twisted-pair cable.

Finally, fiber optics offers an industrywide standard that no other cable can match. Every vendor has different specifications for copper communications cable. However, a single correctly chosen fiber meets the communications needs of every major programmable logic controller (PLC), distributed control system (DCS), drive, CATV (community antenna television), and fire alarm vendor in the world. This ability of fiber to provide such a wide variety of services can have a significant cost impact in wiring an industrial site.

The bottom line is that fiber optics networks are making significant inroads in the industrial setting at the expense of copper solutions and the reasons are numerous. In 1999, one of North America's major paper companies started a major expansion project with the goal of minimizing copper-based communications outside the control room. On completion, the senior electrical engineer stated, "If we were to do this project again, we would get rid of all the copper communications cabling. Copper accounted for 20% of our communications cable, yet it caused 80% of our commissioning delays."[2]

How Fiber Optics Works

Regardless of the application or industry where fiber optics is being used, all fiber cable follows the same basic principles of construction and operation (Figure 4.18a). The typical fiber has a cylindrical glass or plastic *core* surrounded by an outer glass or plastic layer called the *cladding*. A protective cover called the *coating,* in turn, then covers the cladding.

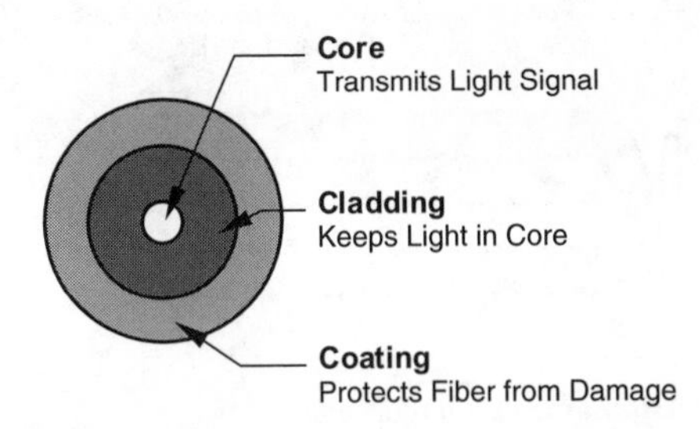

FIG. 4.18a
Basic fiber construction.

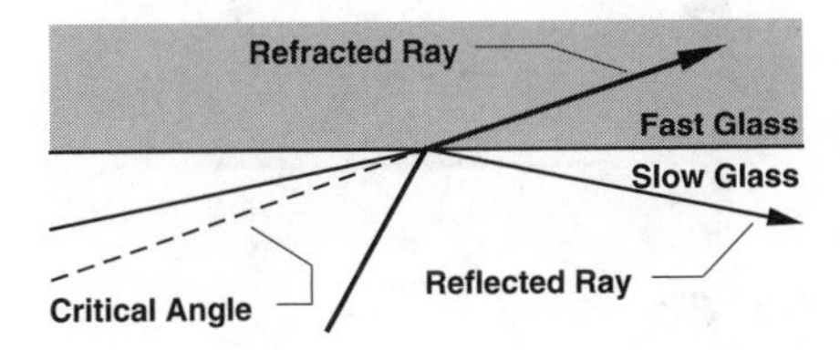

FIG. 4.18b
Reflected vs. refracted light rays.

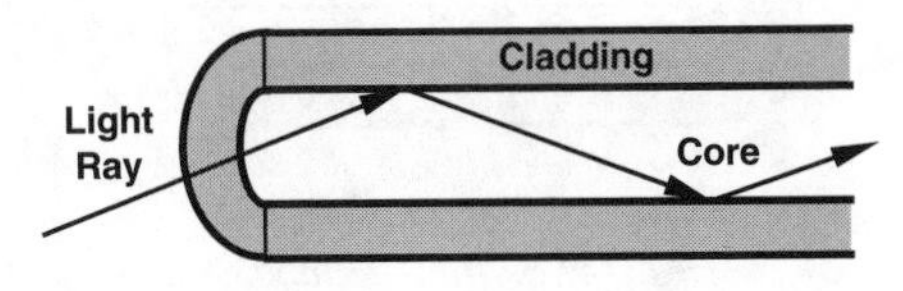

FIG. 4.18c
A ray of light traveling down an optical fiber.

Both the core and the cladding are made of similar materials that can pass light. However, each has a slightly different density and it is this difference that makes the fiber cable transmit light. The reason is that when a light ray passes from a higher-density medium to a lower-density medium, it changes speed. This, in turn, causes a change in direction called *refraction*. The change in direction depends on two factors: the difference in velocity of light (*refractive index*) in each medium, and the angle at which the light ray approaches (*angle of incidence*) from one medium to the other. As the angle of incidence increases, the angle of refraction increases until it reaches the *critical angle* when the light is reflected back into the originating medium, rather than entering the new medium (Figure 4.18b).

In a fiber, the core and the cladding each has a different refractive index. Light enters the fiber core and hits the core–cladding boundary at an angle greater than the critical angle. This causes the ray of light to be reflected back into the core until it bounces off the opposite core–cladding boundary. Because the angle of incidence and reflection are equal, the light ray continues down the core in a zigzag pattern. When this is achieved, the light signal stays within the core of the fiber and travels down the cable to the end where a receiver converts the light signal into an electrical signal (Figure 4.18c).

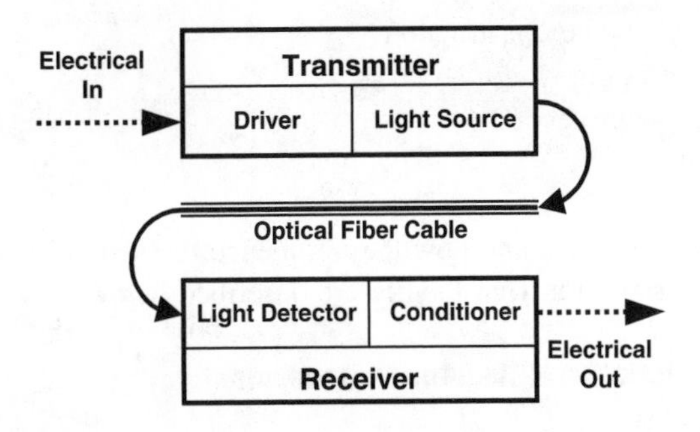

FIG. 4.18d
The major components of a fiber-optic system.

Components

The basic components of an industrial fiber-optic system are very simple: a transmitter, a receiver, and the fiber cable (Figure 4.18d). Except in very high-speed, long-distance applications, fiber is used as a digital signaling medium. That is, the light is simply turned on and off to signal 1s and 0s in the data. The fact that a cable could contain many different frequencies of light is rarely capitalized on in industry—it is simply less expensive to install more fiber than it is to vary the frequency of the light.

The typical fiber-optic transmitter receives electrical input signals and converts them into light signals. It contains a driver and a light source. The driver circuit converts a standard electrical signal to a digital pulse that modulates the light. The light source is typically a light-emitting diode (LED) or laser that turns ON when it receives an "ON" electrical signal and OFF when it receives an "OFF" signal.

The receiver is the interface to the computer or controller that converts the light signals back to electrical signals. It contains a light detector and a conditioner circuit. The light detector converts light signals back into electrical signals through the use of an optical-to-current converter (usually a photodiode). The conditioner circuit amplifies the signals produced by the light detector and determines whether each bit of information is a binary 0 or 1. The binary signal is then converted to a form that can be used by the computer or controller hardware.

Because most communication is two way, most fiber cable is used in pairs, with one fiber for the signal coming into the device (receive) and one fiber for the signal being sent out from the device (transmit).

TYPES OF FIBER-OPTIC CABLES

Fiber-optic cables can be classified by a large variety of properties that affect both their light-carrying characteristics and their suitability for industrial environments. However, the most important considerations for selecting a fiber in industrial applications are the following:

- Core/cladding dimensions
- Single vs. multimode

- Buffer/tube construction
- Fire rating
- Cable protection
- Fiber count

These properties and how they influence the suitability of a fiber for an industrial facility are described below.

Fiber-Optic Core/Cladding Dimensions

As discussed earlier, the center of an optical fiber is a cylindrical core that is surrounded by an outer layer called the cladding. The core is constructed of glass while the cladding is made of either plastic or a different type of glass. Slight differences in refractive indexes between the two materials keep the light within the core region as it travels down the fiber.

Fiber size is determined by the diameter of the core and cladding, expressed in microns. For example, a fiber with a core dimension of 62.5 μm and cladding dimension of 125 μm is designated 62.5/125 cable.

There are a wide variety of sizes available, ranging from 8 to 230 μm. Not surprisingly, the larger sizes are easier to handle and less fragile, but this comes at a price—the larger the core diameter, the lower the data-carrying capacity or *bandwidth*.

For industrial purposes, the most common standard core/cladding diameter is either 50/125 μm or 62.5/125 μm. These sizes give an excellent mix of high data rate and relative ease of handling.

Most other sizes of fiber cabling are slowly being phased out and cannot be used with all types of equipment. For example, many PLC manufacturers offer fiber-optic repeaters with the choice of 62.5/125, 100/140, and 100/200 fiber sizes. Networking equipment, however, typically uses either 50/125 or 62.5/125 fiber. Only the 62.5/125 fiber is common to both.

Single Mode vs. Multimode

When a ray of light travels down a fiber, it can follow many paths or *modes*. For example, it can either pass straight down the axis of the core or it can reflect off the core–cladding interface. A ray that is traveling straight has a shorter path than one that is reflected and, as a result, it will arrive at the receiver sooner. These different modes will cause a square-edged light pulse to spread out, much like capacitance in a copper cable will spread a square wave electrical pulse. The spreading of this pulse is called *dispersion* and it can severely reduce the bandwidth (data-carrying capacity) of a cable over long distances (Figures 4.18e and 4.18f).

Fiber-optic cables are divided into two categories, single mode and multimode, based on how they are designed to address the issue of modal dispersion. Single-mode fiber is an optical glass fiber with a core diameter so small it has only one path (or mode) for light to travel. It usually has a core size between 8 and 9 μm and, because of this straight path, dispersion is negligible.

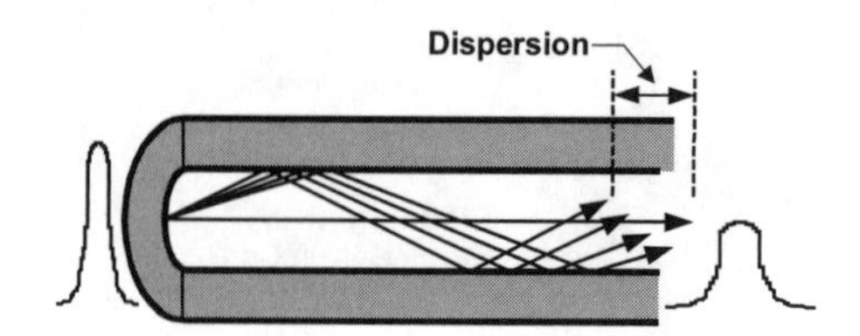

FIG. 4.18e
Modal dispersion in a multimode fiber.

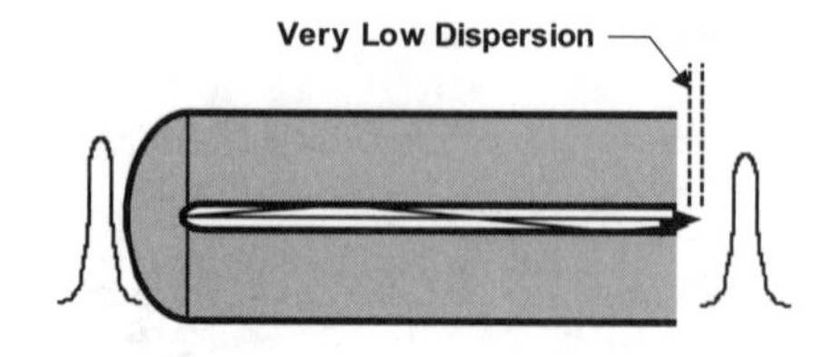

FIG. 4.18f
Light rays in a single-mode fiber.

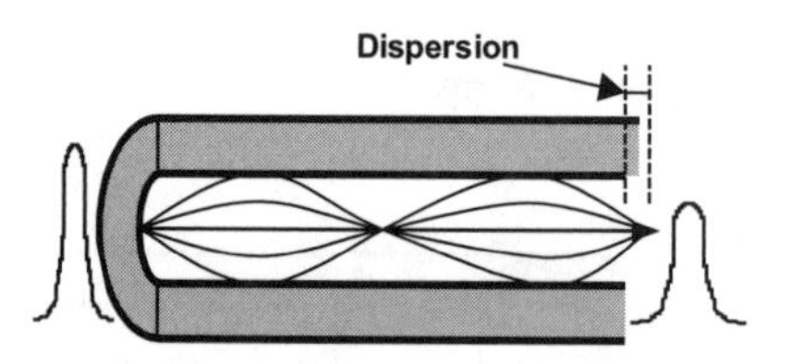

FIG. 4.18g
Light rays in a graded-index multimode fiber.

Single-mode fiber is used in long-distance (up to 120 km) and extremely high bandwidth applications (10 Gbps), but the small diameter makes terminating and connecting difficult. It is rarely required in industrial applications at this time. However, some engineers include two or three pairs of single-mode fiber in their multimode cable bundles to future-proof their design.

Multimode fiber has a core with a relatively large diameter (50 to 200 μm), and light rays can travel along many modes or paths. This causes dispersion and limits the distances and bandwidth of multimode cable. The dispersion can be reduced considerably by the use of graded-index multimode (Figure 4.18g), where the core–cladding boundary is gradual rather than a sharp transition. Regardless, the large diameter of multimode fiber facilitates termination and requires less expensive light sources, making it very popular for most indoor applications. Multimode fibers are typically used for industrial applications requiring distances not exceeding 2 km and data rates not exceeding 200 Mbps. As noted earlier, the most common standard core–cladding diameter for multimode is 62.5/125 μm, although 50/125 μm is gaining considerable acceptance for gigabit Ethernet applications.

Cable Construction

Another important consideration is the cable construction that protects the fibers. Typically, the cladding is surrounded by a buffer that protects the glass from abrasion and blocks out

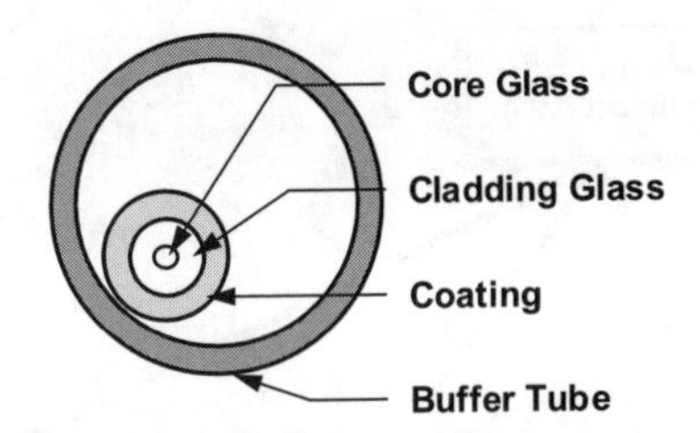

FIG. 4.18h
Loose tube cable.

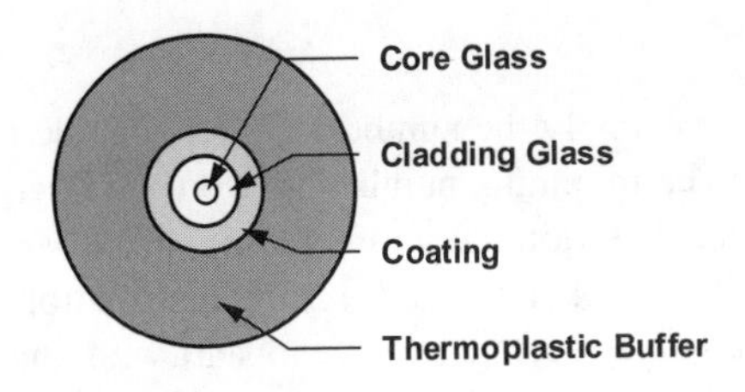

FIG. 4.18i
Tight-buffered cable.

any light that might enter the cladding layer from the outside. Next, an inner jacket surrounds the buffer and protects the fiber from being crushed or bent. A layer of strength members surrounds the inner jacket and limits how much the cable can be stretched. Finally, an outer jacket made of PVC, PE, or PVDF protects the cable from bending, crushing, and environmental hazards. For industrial applications, the cable is often wrapped in an interlocked aluminum or steel tape armor. The major difference in fiber construction is the way the fiber is designed to be protected by the buffer and jacket.

Loose Tube Cables

Loose tube cable (Figure 4.18h) was originally developed for long-haul telephony applications that required rugged, low-cost, high-fiber-count cables. These cables had to be able to withstand large temperature fluctuations and direct in-ground burials.

In loose tube construction, the coated fiber floats within a stiff abrasion-resistant tube filled with waterproof gel. The outer tube protects the fiber from external pressures and mechanical stress caused by temperature change. The gel prevents any stress caused by expansion or contraction of cable materials and keeps water from entering the tube. Often, more than one fiber will be in each tube.

This cable type is designed for outdoor applications such as long runs between cities. Unfortunately, the fact that the fibers are loose means they lack protection or mechanical support once the tubing is stripped back for termination. This makes handling for terminations more sensitive and much more time-consuming. Often *fanout* kits are installed at the cable ends to protect the bared fibers and prevent the gel from leaking out of the cut cable.

One particular problem with loose tube cable is that it is often designed with only outdoor use in mind and is not flame-rated for indoor use. If a loose tube cable will be run more than 50 ft inside a building it must have an FT-4 flame rating, which means switching to an indoor cable. This can be a severe disadvantage in an industrial environment, because a cable will often transit many buildings as it crosses a site.

Another potential problem with loose tube cable occurs on long vertical runs. Because the fiber is not supported by the tube, it can slip downward in the tube, causing excessive stress at connectors. Manufacturers deal with this issue by designing special "riser-rated" loose tube cable that is acceptable for specified vertical distances. Cable installed in long risers must be rated for the vertical distance.

Tight-Buffered Cables

In tight-buffered cables, a thermoplastic material wraps directly over the fiber to add strength, protection, and mechanical support for vertical runs (Figure 4.18i). It also increases the effective size of the individual fibers, making handling easier and giving connectors more to hold on to. Other advantages are ease of termination, color coding for identification, and better standards for flammability and flexibility. This cable type is often rugged enough for interbuilding runs, but this must be confirmed with the cable manufacturer.

Cable Protection

Two main types of armor protection for fiber are available on the market. One is "steel tape armor," which is the most common armor on the market. The other is "aluminum interlocked armor," which is the most suitable for industrial environments.

Steel tape armor is primarily used by telephone companies whose work is often damaged by rodents chewing up the cable for breakfast. The steel tape is not "rodent friendly" and withstands their attempts at gnawing. It also offers superior protection against water intrusion, again a more important issue for telephone companies with long outdoor runs.

Aluminum interlock armor, commonly known as TEC armor, is recommended for use in industrial settings. Electricians are familiar with it and find it much easier to "strip back." Thus, it is considered "craft friendly." Aluminum interlock armor is also stronger and more flexible than steel tape armor.

A third alternative that is gaining increasing acceptance is the use of "blown" fiber. In this system plastic tubing is installed throughout the facility and then bare fibers are blown through the tubing with compressed air. The plastic tubing provides both protection to the fiber and a guide for installation. Additional fibers can be installed without interrupting the existing fiber network, so this method allows considerable room for expansion. The biggest disadvantage is the plastic tubing is often unable to withstand subzero temperatures, and thus cannot be used in northern climates.

Fiber Count

Fiber count is simply the number of fibers bundled together in a cable. The minimum number is two (one fiber for each transmission direction). A more reasonable fiber count is 24 for facility backbones and 12 for distribution cabling. Although this may seem excessive to start with, once a fiber cable is installed, its ability to be used for many different communications systems will quickly use up much of the spare fiber.

FIBER-OPTIC NETWORK DESIGN

Distribution and Layout Planning

For many years the design of communications cabling in industrial plants has been based on an "as-needed" strategy. Each control system would have a unique cable type and cables would be run on a point-to-point distribution model. The result was that maintenance staff was faced with installing and maintaining many different cable systems, often with very inefficient cable layouts. With the explosive increase in networks on the plant floor, coupled with the use of fiber-optic cabling, it has become both necessary and possible to design standardized cabling infrastructures. These infrastructures treat communications as a sitewide utility, minimizing the overall cabling cost and complexity.

The basis for these design strategies is a standard known as the Electrical Industries Alliance/Telecommunications Industries Association "Commercial Building Telecommunications Cabling Standard 568B" (EIA/TIA-568B). This standard recommends a design strategy for cabling commercial structures that is independent of what will be passed over the cable. For example, the cable plant of a building might be used simultaneously for voice, video, computer networking, control, and alarm systems. Furthermore, the type of network equipment should not influence the type of cable being installed.[3]

The architecture of the EIA/TIA standard is based on a tree/star topology as shown in Figure 4.18j. The top of the hierarchy is a main communications center or main cross-connect (MC) feeding a maximum of two sublayers of telecommunications closets. These are known as intermediate cross-connects (IC) and horizontal cross-connects (HC). Equipment such as computers or controllers can be connected at any layer. This is an extremely flexible design because it provides a clear path between any two points in a facility. At the same time, as the EIA/TIA points out in its standard, it can mimic other topologies such as the ring or bus.

It is important to understand that the EIA/TIA does not suggest that two different communications functions should share the same wire. It simply intends that if a designer is installing one wire into a location, then the designer might as well install two wires, as the increased cost would be minimal. In addition, it suggests to equipment manufacturers that all communications equipment, regardless of brand or function, should use the same type of cabling. This cable is either twisted-pair cable or fiber-optic cable.

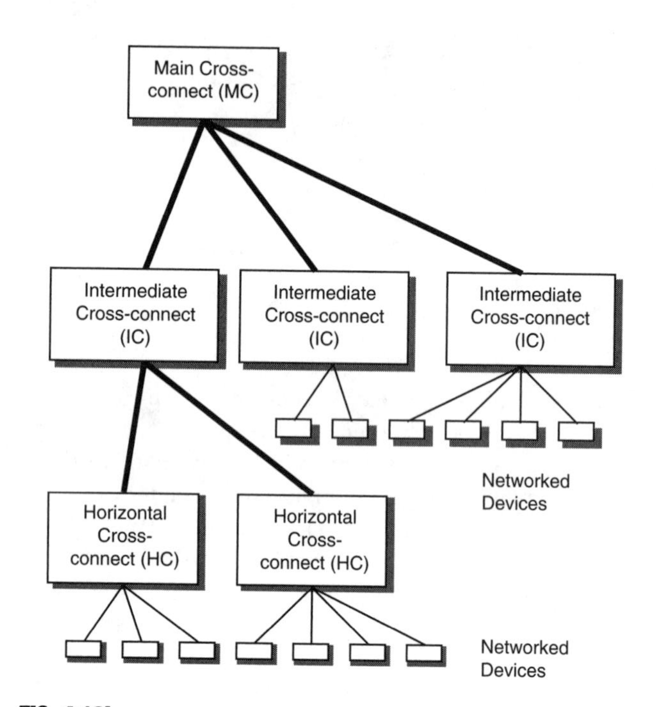

FIG. 4.18j
The EIA/TIA topology model for commercial building communications.

The impact of this EIA/TIA standard in the commercial communications industry has been significant. By following the specification, users found that they could migrate between technologies by changing only the electronics; the cabling was unaffected. For example, the move from 10 to 100 Mbps Ethernet has been relatively painless for companies that followed the standard. In addition, the cost of installing communications infrastructures in commercial buildings has dropped significantly in the past 10 years.

Unfortunately, as of January 2002 none of the standards bodies has been able to develop a similar cabling standard for industrial facilities, although the EIA/TIA has created the TR42.9 subcommittee to look into the issue. That said, the EIA/TIA-568B is not perfectly suited for all industrial applications but it is an excellent starting point.

The first step in designing a communications infrastructure is to catalog all the communications requirements in a facility. From this, select the main communications center and the backbone cable runs radiating from it. Some of the things to consider include:

- Locations of existing telecommunications closets
- Availability of cabling pathways
- Central server locations
- Areas with a high density of devices

The key is to pick the MC and work from there. It is important to be able to get from an MC to any closet in 2 hops and to any device in 3.

Another central guideline in the EIA/TIA-568B standard is the cable distance rules. Basically, fiber runs must be kept under 2000 m (6500 ft) from an MC to an HC. If any ICs exist in the design, then the distance from IC to MC is 1700 m and from IC to HC is 300 m. In industrial sites, the latter rule is often ignored, but the overall 2000-m rule should not be.

Cable Selection

Once the distribution plan is developed, the next stage is the selection of the correct fiber cable. Remember the bulk of the fiber-optic cable installed today is for either telephony or office applications. Industrial sites offer challenges that are not addressed by many of the common fiber installation designs. Just because the local telephone company prefers a particular type of fiber cable does not mean it is the correct one for a facility.

The general rule of thumb for selecting cable in industrial settings is to use 62.5/125 μm or 50/125 μm multimode fiber. The cable should be rated for both indoor and outdoor use and must have an FT-4 flame rating if it is to be installed indoors. Aluminum interlock armor is preferred over steel tape for all but long, buried runs. Fiber counts should be a minimum of 12, with 24 fibers as the standard for main backbones. More specific selection details depend on the area where the cable will be installed.[4]

Indoor Office Installations Fiber selection for office applications is relatively simple. The fiber must be flame-rated for either FT-4 for general use or FT-6 for plenum use. Typically, tight-buffered cable with kevlar strength members and a light jacket is used. There is little reason to use loose tube as it is more difficult to install and usually does not meet the flame-rating standards. As well, cables in these environments do not require armor as the chance of crush or pull damage is relatively low. Because jacketed fiber cable is more rugged than most coaxial and twisted-pair cables, plan to armor fiber only in the places where coaxial cables would be armored.

Indoor Industrial Installations If fiber-optic cable is being installed in plant-floor conditions, it is likely to be installed in existing cable trays and be subject to more stresses than office cable systems. Thus, some form of armor is recommended, usually aluminum interlocked (TEC style) armor. This armor must be electrically bonded to ground at all distribution cabinets.

Interbuilding Installations Industrial sites often require a combination of indoor and outdoor fiber routing. Telecommunication industry guidelines recommend switching between indoor and outdoor cable types at each transition, a solution that is not practical for most industrial sites. On a typical site, this would require numerous patch boxes or splices and is unworkable for both cost and *attenuation* (signal loss) reasons. Instead, FT-4 flame-rated, outdoor, tight-buffered cable should be used so that the cable can transit both indoor and outdoor environments.

Long-Run Outdoor Installations Outdoor fiber cable generally falls into three categories: direct burial, underground conduit, and aerial. These cables are manufactured specifically for outdoor applications and are recommended for any long outdoor cable runs, especially in regions subject to cold weather. Most are loose tube designs with high tensile strength, to withstand environmental conditions, and gel filling, to prevent water migration. The jacket materials are specially selected to be abrasion and ultraviolet resistant. If a facility is planning to install long outdoor runs it will need to work closely with the manufacturer to determine the correct cable for its application.

Power Loss Budgeting

Power loss or attenuation budgeting is the process of calculating the signal losses that are expected in a cabling system and making sure that the attached network devices can function with this level of loss. Loss budgeting used to be standard practice 10 years ago, but today is less common. If a network will be used for standard network protocols such as Ethernet and follows the EIA/TIA-568B guidelines, then the signal loss is unlikely to be an issue. However, if the fiber network will be used for more proprietary DCS and PLC protocols, then a loss budget should be carried out. Regardless, some loss budget calculations need to be carried out to act as a reference point for later attenuation testing once the cable is installed.

Cable attenuation is measured in decibels (dB), which are the log of the ratio of power in vs. power out. The use of logarithms allows the loss values to be added rather than multiplied, making calculations simpler. Each component in calculation has a standard value that is either specified by the manufacturer or by EIA/TIA standards. For example, connectors are allowed a maximum loss of 0.75 (per the EIA/TIA 568B standard), but multimode connectors will typically have losses of 0.2 to 0.5 dB. For splices the EIA/TIA maximum is 0.3 dB but one can expect values of 0.1 to 0.5 dB for multimode splices; 0.2 is a good average for an experienced installer. Loss in cable is specified in dB/km, and 3.5 dB/km is the EIA/TIA maximum for multimode fiber. So if a plant has a fiber run 2 km long with four connections (two connectors at each end and two connections at patch panels in the link) and one splice in the middle, the calculations would be:

End Connector Loss:	2×0.5 dB = 1.0 dB
Panel Connector Loss:	2×0.5 dB = 1.0 dB
Splice Loss:	1×0.2 dB = 0.2 dB
Cable Loss:	2 km $\times$ 3.5 dB/km = 7.0 dB
Total Expected Loss:	9.2 dB

This value would be compared to the maximum loss acceptable to the equipment in use and later to the measured values after installation.

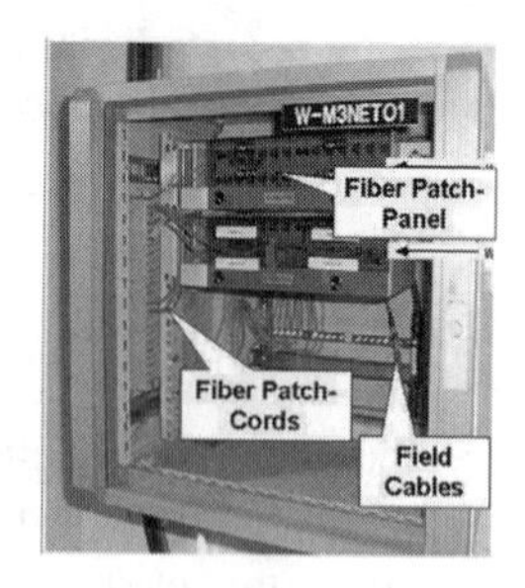

FIG. 4.18k
Wall-mounted cabinet with patch panel and termination tray.

Cabinets and Cable Management

Proper protective cabinetry and cable management is critical to a successful network, and creating a standard design for these cabinets will make engineering and maintenance much easier. The key is to create communications cabinets where all field cabling is terminated and any splices and connectors are protected from abuse. Typically, most industrial facilities create two designs; a full-sized freestanding cabinet and a smaller wall-mounted field cabinet (Figure 4.18k).

Fiber cables from the field are brought into the rear of the cabinets and terminated on the back of fiber patch panels. The connections from field cables to electronics are then made using patch cords. Every cabinet and every fiber should be designed with at least 30% spare capacity. This will allow future expansion by simply adding patch cords and electronics and will reduce the need to install new cables.

Never connect the fiber cables from the field directly to any electronic devices. Most of the wear and tear on fiber-optic networks occurs at the end points, and the risk is that the field cables will be cut back for repairs until they are too short and need to be reinstalled.

FIBER CABLE INSTALLATION AND FINISHING

Fiber Installation Concepts

As noted earlier, fiber-optic cable is surprisingly rugged. However, it is not immune to damage. There are three measures that help cable installation without harming it: the pull strength, crush strength, and minimum bend radius of the cable.

Pull Strength One of the goals in any fiber cable installation is to apply as little stress to the fibers as possible. For this reason, all cables have a tensile load rating that should not be exceeded. The tensile or pull strength is a value that represents the highest load that can be placed on a cable before any damage occurs to the fiber. This value is not the breaking strength, but a realistic allowable limit and will depend on cable construction and the application.

Most manufacturers will specify two maximum pull strength values: installation and long-term load. The installation or short-term load is the load a cable can withstand during the actual installation process. It accounts for the additional stresses caused by pulling cable through and around stationary objects such as ducts, corners, and conduits. After the cable has been installed, it will be subject to lower loads. Manufacturers refer to this value as installed, long-term static, or operating load.

Bend Radius The minimum bend radius is the value representing the smallest bend a cable can withstand. Bending fiber beyond the recommended limits can cause an increase in the fiber attenuation at those points. As with tensile strength, there are two values associated with bend radius: installation and final training. These values will again depend on size of cable, construction, and application. Although it is tempting to bend cables tightly over corners or around itself to stuff into walls or cabinets, bending the cable can seriously affect its performance.

Crush and Impact Strength Cable crush and impact are often listed but rarely understood details of optical fiber cables. They provide some guidelines for cable installation, as they follow EIA standards for crush resistance. These standards were developed with the intention of determining the ability of fiber cable to withstand repeated impact loads such as might occur during installation. In real-life terms, a fiber cable must be able to withstand a determined amount of pressure with a minimal amount of increased attenuation. Because optical fiber cables can be run in the same cable trays as much heavier power cables, consideration must be given to avoid placing the crushing forces of the heavy cables on the fiber cables, particularly if they are unarmored.

Before one begins installing fiber cable it is necessary to determine the pull strength and bend radius values for the cable being installed and then not to exceed these values. Installing fiber-optic cable is not difficult as fiber manufacturers take great pains to make sure the cable design protects the fibers during installation. The key is to pull on the strength members only and not to pull on the fibers. For unarmored fiber, do not pull on the jacket unless the cable manufacturer specifically approves it and an approved cable grip is used. On long runs, use proper lubricants and make sure they are compatible with the cable jacket. On really long runs, pull from the middle out to both ends. If possible, use an automated puller with tension control.[5]

The other important point to remember is not to exceed the cable bend radius. Fiber is stronger than steel when pulled straight, but it breaks easily when bent too tightly. If a kink is put in the cable, it will harm the fibers, maybe immediately, maybe not for a few years, but the cable will become damaged and must be replaced.

Finally, the cable should not be twisted as this can also stress the fibers. Spinning the cable off the spool end will put a twist in the cable for every turn on the spool. Instead, roll the cable off the spool or, if laying cable out for a long pull, use a "figure-8" on the ground to prevent twisting.

Always use a swivel pulling-eye, because pulling tension will cause twisting forces on the cable.

Terminating the Fiber

Once a fiber cable is installed, some means of terminating the ends of the cable is required. These fall into two categories:

1. **Splices:** Splices are permanent connections between two fibers made by welding them in an electric arc (fusion splicing) or aligning them in a fixture and gluing them together (mechanical splicing).
2. **Connectors:** Connectors are metal or plastic fasteners that are used to mate two fibers together or attach fibers to equipment.

Fiber termination is the most critical part of the installation process because terminations are where the greatest losses are likely to occur in the cable system. It is also the most common point of failure.

An Overview of Splices

Ideally, cable runs will use one continuous length of cable from one device to another. This is the most economical and convenient solution with the least amount of *attenuation* (signal loss). However, because of site layout, length, or existing cable runs, this is not always possible and sometimes the fiber cable has to be spliced.

In most industrial sites the cable will run from area cabinet to area cabinet, so that the cable does not require any splicing between patch panels. Fibers that need to pass on to another area are patched through the patch panel to other patch panels. Terminating all fibers at each patch panel will also make it easy to manage expansion and topological changes, as well as making cable numbering easier to follow.

Splicing falls into two categories, fusion and mechanical. Both types are field proven, have excellent long-term reliability, and can be used for termination of fiber-optic cables in indoor and outdoor environments. Splices are usually protected by a splice closure that contains the splices in splice trays or organizers.

1. **Fusion Splicing:** Fusion splicing employs the use of an electric fusion splicer. The cables to be connected are fused together with a high-voltage electrical arc. Contrary to common belief, fusion splicers are very simple to operate. The advantage of a fusion splice is a slightly lower loss connection, but this is minor in most multimode installations. For single-mode fiber, fusion splicing is preferred. The disadvantage with fusion splicing is that it requires an expensive ($20,000 to $40,000) fusion splice machine.
2. **Mechanical Splicing:** This type of splicing aligns the two fiber ends to a common centerline, thereby aligning the cores. The ends are butted together and permanently secured with adhesive, friction, or strain relief. For multimode fiber this system is very effective. It is easy to perform, requires only a few special tools, and is ideal for emergency field splices. However, it can be expensive if many splices are required.

An Overview of Connectors

Connectors are used to attach the end of a fiber into the transmitter or receiver electronics or into a fiber-optic patch panel. There are many different types of connectors on the market, but four connector types dominate in modern industrial fiber installations (Figure 4.18l):

1. **ST:** A steel bayonet-style connector that is the most widely used in industrial plants
2. **SC:** A plastic rectangular connector that is officially recommended to replace the ST style
3. **MT-RJ:** A duplex-style connector that contains both transmit and receive fibers in single, small form-factor connector
4. **VF-45:** A competing duplex-style connector that contains both transmit and receive fibers in single connector similar in size to an RJ-45 phone jack

Although it is preferable to use only one type of connector on an industrial site, it is rarely possible. To solve this problem, most manufacturers make patch cables that will convert from one connector type to another. Modern designs recommend selecting either the MT-RJ or VF-45 connectors for the patch panels and then purchasing converting patch cables to attach any equipment with the other styles of connectors into the fiber network.

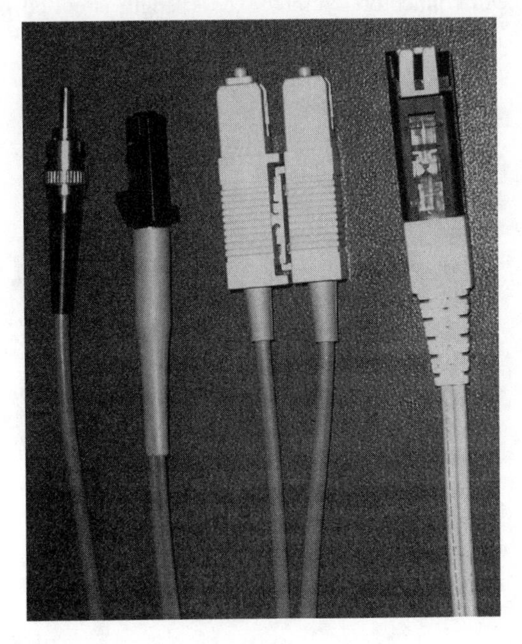

FIG. 4.18l
ST, MT-RJ, SC, and VF-45 fiber-optic connectors.

Once a connector type is selected, the connectors must be attached to the end of each fiber. There are two basic methods of doing this:

1. **Splicing Factory-Made Pigtails:** Optical fiber pigtails are short cable segments that have factory-finished fiber-optic connectors on one end and are unterminated at the other end. The unterminated end is field-spliced to a cable needing termination. Pigtail splicing is the fastest way to terminate a cable, particularly if loose tube fiber is being used.
2. **Field Connectorization:** With this technique the connectors are directly attached to the optical fiber, so there is no splice, and therefore, no need for a splice tray. This method is more economical than buying pre-connectorized pigtails but requires that the tradesperson learn how to cleave, polish, and fasten connectors. There are three main techniques of field connectorization that are suitable for industrial use: crimp, epoxy, and hot-melt. All of these methods are quick and can be easily learned by electrical tradespeople.

Cable Management and Labeling

Poor cable management is the single biggest cause of cable failure in industrial applications. No matter how good the connector finish is, it will not help if the cable is caught in a cabinet door. Figure 4.18m shows exactly this type of problem—the two lower fibers in the fiber wall-mounted cabinet are likely to be pinched when the cabinet door is closed. Problems of this nature are extremely common on industrial sites and likely due to inexperience and a lack of clear standards for fiber installations.

To prevent problems, make sure that once fiber cables have been connectorized, any excess length is coiled into a termination tray and that all the fibers are attached to the back of a patch panel. The tray can be closed and, from then on, all connections to equipment can be effected using a patch cable from the front of the patch panel. The worst problem most industrial sites experience is connectors broken at the back end inadvertently by people working in communications closets. Once it is installed, fiber does not need maintenance or inspection. Lock the back of the patch panel and only unlock it when something needs to be moved.

It is also important that all fibers be clearly labeled so that physically tracing the cable is not unnecessarily difficult. Both ends of a fiber run should be labeled with the same information. This information will usually show (1) the source ports at the local patch panel, (2) the local cabinet number, (3) the local equipment room number, (4) the remote equipment room number, (5) the remote cabinet number, and (6) the remote patch panel destination ports.

All patch panels should be labeled, and should show summary information giving details of the remote patch panel to which the cable is connected. Fibers should be attached to the patch panel from left to right as viewed from the front, with port numbers from 1 to X. This means that they will seem to be backward when viewed from the rear of the panel.

FIG. 4.18m
Poor fiber cabling management.

INSPECTION AND TESTING

After a cable system is installed, testing is crucial for assuring the overall integrity and performance of the optical fiber system. There are a number of different tests that can be carried out both during installation and later as an acceptance test. This section examines three possible tests.

Inspecting Connectors with a Microscope

Visual inspection of the end surface of a connector is one of the best ways to determine the quality of the termination procedure and to diagnose problems during installation. A well-made connector will have a smooth, polished, scratch-free finish, and the fiber will not show any signs of cracks or pistoning (where the fiber is either protruding from the end of the ferrule or pulling back into it).

The proper magnification for viewing connectors is generally accepted to be 30 to 100 power. Lower magnification, typically with a jeweler's loupe or pocket magnifier, will not provide adequate resolution for judging the finish on the connector. Too high a magnification tends to make small, insignificant faults look worse than they really are. A better solution is to use medium magnification, but inspect the connector three ways: viewing directly at the end of the polished surface with side lighting, viewing directly with side lighting and light transmitted through the core, and viewing at an angle with lighting from the opposite angle.

Viewing directly with side lighting allows one to determine if the ferrule hole is of the proper size, whether the fiber is centered in the hole, and if a proper amount of adhesive

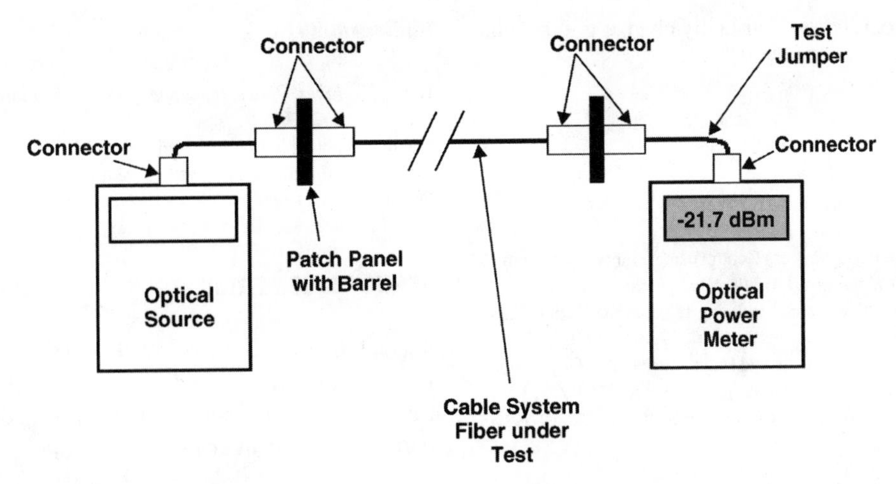

FIG. 4.18n
Testing attenuation using a light source and optical power meter.

has been applied. However, only the largest scratches will be visible this way. Adding light transmitted through the core will make cracks in the end of the fiber, caused by pressure or heat during the polish process, visible.

Viewing the end of the connector at an angle, while lighting it from the opposite side at approximately the same angle, will allow the best inspection for the quality of polish and possible scratches. The shadowing effect of angular viewing enhances the contrast of scratches against the mirror-smooth polished surface of the glass.

That said, it is important to be careful in inspecting connectors. The tendency is to be overly critical, especially at high magnification. Only defects over the fiber core are a problem. Chipping of the glass around the outside of the cladding is not unusual and will have no effect on the ability of the connector to couple light in the core. Similarly, scratches only on the cladding will not cause any loss problems.

Attenuation Testing

Attenuation or loss testing is a simple test where a light source of known strength is injected into a cable and the resulting light strength is measured at the other end. Attenuation is optical power loss measured in decibels (dB) and is a good indication of the physical quality of fiber splices, connectors, and cable. Badly finished connectors, tight bends, or excessive force placed on the cable during installation will typically show up as excessive loss during the tests. Because this test is both simple and quick, it is a good rule never to accept an installation without at least a basic attenuation test in both directions.

Attenuation is measured by using an optical light source, short launch cables, and an optical power meter. The light source provides input light at a set wavelength. The launch cables are short lengths of fiber-optic cable used as a standard for comparison in determining attenuation. The optical power meter is set to the same wavelength as the light source and measures the light transmitted through the cables. It contains a photodiode detector and a meter that indicates the amount of signal loss in the cable system.

The single most important test is the end-to-end attenuation test, which measures the optical power loss between the cable and termination points. The end-to-end fiber path loss should be checked and compared with the budgeted design loss, and unacceptable deviations checked and rectified. Figure 4.18n shows a typical test equipment setup for measuring attenuation.

Optical Time Domain Reflectometers

Optical time domain reflectometers (OTDR) are a more sophisticated (and more expensive) testing technique. The OTDR works by injecting a light pulse into one end of the fiber. Any imperfections in the cable or connectors will cause small amounts of light to be reflected back to the OTDR. The OTDR measures these reflections and their time delay and determines the type of flaw and its distance along the cable.

Although OTDRs give superior information about a cable as compared to attenuation meters, their cost restricts their use. OTDRs are most commonly used while installing long-distance runs where the cost of cable repair can be very high. Their other use is for determining the location of faults in a cable that is known to be defective.

CONCLUSIONS

The use of fiber-optic networks in industrial settings is well proven. Fiber provides superior ruggedness and reliability, while initial costs are only marginally higher, if at all. In the long term, a well-designed fiber-optic network can result in considerable cost savings in communications cabling yet provide significant maintenance and flexibility benefits. Properly planned out, it will provide a stable standard for communications while PLC and DCS equipment constantly change in our mills and factories.

References

1. Belden Wire and Cable Specification Sheets for RG-58 Coax, CAT-5 UTP and Premises 1-Pair Fibre Cable.
2. Byres, E. J. and Mitchell, C., "Process Control using a Fiber Optic Unified Cabling System," *IEEE Industrial Applications Magazine,* in press.
3. ANSI/EIA/TIA-568B, "Commercial Building Telecommunications Cabling Standard," Telecommunications Industries Association, 2001.
4. Byres, E. J., "What Superintendents and Engineers Need to Know about Fiber Optics," *Pulp and Paper Canada,* Volume 8, pp. T270-273, 1999.
5. "Lennie Lightwave's Guide to Fibre Optic—Fibre Optic Cable," Fotec Inc., 1997.

Bibliography

Hayes, J., *Fiber Optics Technician's Manual,* Florence, KY: Delmar Publishers, 2000.

"Lennie Lightwave's Guide to Fibre Optics," Fotec Inc., 1998, http://www.fotec.com/fiberu-online/fuonline.htm.

Pearson, E. R., *Complete Guide to Fiber Optic Cable Systems Installation,* Florence, KY: Delmar Publishers, 1996.

ACKNOWLEDGMENTS

Special thanks to Dave Wall, Gord Gillespie, and Alden Hagerty of Artemis Industrial Networking in Vancouver, Canada for their technical assistance and permission to use portions of their fiber installation guide.

4.19 Satellite, Infrared, Radio, and Wireless LAN Networks

D. E. CAPANO

The wireless local-area network (WLAN) is a network that communicates without a physical connection such as wires or cables. This allows networks to exist without the limitation of physical boundaries. The possibilities offered by this type of arrangement are truly a realization of the promise of a fully integrated computer network. Utilizing modern wireless networking techniques, virtually every object of utility, including computers, PDAs (personal digital assistants), automobiles, production machinery, home appliances, power plants—every accoutrement of modern life—can and will be networked together to allow seamless interoperability.

Wireless sensors have also become very popular. Wireless sensor networks offer flexibility in installation and freedom from the vagaries of signal wiring in industrial or hostile environments. Enhancement and simplification of instrumentation and control systems are possible by allowing the designer or end user unprecedented flexibility in the application of instrumentation to a system. Utilizing radio-frequency (RF) modulation techniques such as spread spectrum or media such as diffused infrared (DIR), the effects of noise on the transfer of commands and information are virtually eliminated. Reliable, hardened, and self-contained sensor and control packages can be used to safely monitor and control remote processes or equipment.

The cost of these devices is only now becoming competitive with wired networks. Factoring out labor costs associated with cable and conduit installation, and including the reduction, if not complete elimination of power wiring, could make a wireless control system cost-effective over the long haul. In hazardous locations, the elimination of rated conduit and equipment makes the WLAN an attractive option. In less-demanding applications, available data transfer speeds are more than adequate for monitoring or controlling a process parameter in real time.

The chief advantage of WLANs is the elimination of cables and wires, which suffer from the tendency for mechanical failure. Unprotected wires can become broken or become disconnected. Physical damage because of accidents or electromagnetic interference from other equipment can render a cable system unusable. These adverse characteristics require the installation of wire and cable into rigid conduit systems for protection. A conduit system effectively triples the cost of a network wiring system. In any installation, wire and cable must be protected and supported to ensure continued and reliable operation, adding additional expense. Use of fiber-optic (FO) cable also requires a conduit system and specialized equipment and labor for termination and commissioning aside from the network configuration itself. FO cable can be more susceptible to physical damage by the very nature of the materials used to fabricate it. Recent advantages in jacketing materials allow FO to withstand a moderate amount of abuse without degradation. FO always requires support, however, to avoid any tensile stress, which will cause serious degradation of the cable with a requisite degradation of performance. In all cases, wire and cable require additional support systems and the labor to install and maintain them. WLANs effectively eliminate these major costs associated with information networks.

WLANs eliminate the redundancy normally required for networks. A handheld wireless device may be used to communicate with any node on the network at any physical location within range of the network. There is no need for remote workstations with cabling and support systems, and duplicate or triplicate cable runs, with support systems. A wireless LAN will eliminate or reduce the need for field or factory floor terminals, data interfaces, and even desktop computers. All the functions required to access and control a network will be consolidated in powerful mobile network access devices or "access points" (AP), such as PDAs or like enabled laptop computers. Users will enjoy freedom of movement and the ability to access any node or process directly from anywhere in the physical plant. Cost savings from the elimination of redundant wired equipment add to the cost savings associated with WLANs.

In this section, the major types of WLANs being implemented at the time of this writing are discussed. Standards being developed will allow the successful implementation of the technology, avoiding the confusion of interface and configuration difficulties associated with early, wired networking. Standardization at this early stage will allow the orderly development of the most promising technologies without the influence of disparate interests. The Institute of Electrical and Electronic Engineers (IEEE) has defined wireless standards for use by the industry. We explore these standards as part of our discussion. An emerging technology, satellite networking, is also explored as an alternative for those applications involving communication with or monitoring of remote or inaccessible locations.

RADIO SYSTEMS

Standards have evolved that define the form and function of a WLAN. IEEE has and is developing WLAN standards to deal with the burgeoning market for wireless devices. Without these standards, many different technologies would arise, preventing an orderly progression to a feasible and preeminent technology in the shortest possible time. In the case of wireless networking, the IEEE has taken the lead in developing workable standards to avoid the confusion and wasted time that was witnessed in the development of wired LANs. IEEE 802.11 is the first WLAN standard developed by the IEEE. It treats the physical (PHY) and medium access control (MAC) layers of the ISO model. IEEE 802.15 defines the configuration of a personal area network (PAN) and uses the Bluetooth standard as its starting point. IEEE 802.16 defines broadband wireless networks. IEEE 802.11 is being implemented at this writing. IEEE 802.15 and IEEE 802.16 are still in development.

IEEE 802.11 (CSMA/CA)

Based upon the standard, wireless networks can be set up in two ways: as *ad hoc* and infrastructure networks. In an *ad hoc* network, computers, PDAs, or other wireless devices are brought together informally, "on the fly," to communicate and exchange information. There is no physical structure to the network and no boundaries save the ranges of the devices themselves. Each node can communicate with each other node. This, as can be imagined, could cause confusion and diminish the effectiveness of the network. There is no order to the information exchange, in other words. To offset this problem, two techniques are used to establish an orderly flow and exchange of data. In the first method, a spokesman election algorithm (SEA) is implemented to "elect" one machine to be the "master" while the other machines are slaves. Another method is simply to broadcast information to every node. Identification of each wireless node is transmitted along with the data sent. An *ad hoc* network is extremely useful where it is difficult, if not impossible, to set up a wired network and an exchange of information is essential to a project or situation. A group of engineers, for example, could quickly and easily set up a network by simply booting up each wireless-capable device or laptop computer while sitting down for lunch or while gathering at a conference table. This type of functionality sets this technology apart from other wireless technologies in that it has the greatest potential for facilitating the exchange of information and ideas without the excessive physical overhead required for wired networks. The time savings alone make this technology extremely competitive from the standpoint of lost production time due to a wired network setup.

The second type of wireless network is called an "infrastructure" wireless network. This type of network is a hybrid of wired and wireless computers or devices. Wireless or mobile nodes communicate with fixed network access points by which a link is made between the two mediums. In this type of network, a mobile node will stay in communication with the infrastructure network as long as it stays within range of a wired network access point (NAP). As the mobile unit passes out of the range of one NAP to another, a handoff will take place that is very similar to those that occur on cellular telephone systems when a mobile unit passes from one cell to the next. This type of network has the greatest potential in an environment where there may be a large installation of fixed nodes, which may or may not be in close proximity to one another. An example of this type of environment would be a factory or process facility where a network of wireless sensors is installed. As a mobile unit moved within range of the wireless node, a recognition sequence would occur between devices, followed by a download of data from that sensor. This could also be handled on a continuous basis utilizing other nodes to repeat the information over large geographical distances. An infrastructure network could use several different types of communication technologies to effect continuous communication. A remote node might communicate with another node using cellular or satellite technology, with the rest of the network either being wired or using short-range wireless such as that defined by the standard.

Table 4.19a is a summary of the standard.

CSMA/CA

Carrier sense multiple access/collision avoidance (CSMA/CA) allows communication between nodes while reducing the occurrence of lost or corrupted data due to "collisions" between transmissions from different wireless nodes. A collision occurs when two nodes attempt to transmit information at the same time. In wired networks conforming to the IEEE 802.3 standard (Ethernet), CSMA/CD is employed to deal with the same problem. The CD, in this case, stands for collision detection. In the first method, avoidance, each node listens to the carrier frequency to determine if any other node is transmitting. Hearing no other transmission, the node transmits a "ready to send" packet (RTS) to the intended destination node. The destination node then transmits a "clear to send" packet back to the sending node. The data are then transmitted. If another node happens to be transmitting when the sending node wishes to transmit, the sending node will back off for a random time period and then attempt a retransmission. Upon a successful transfer of data, an acknowledgment signal is sent from the receiving node. In the second method, a collision is detected by data being corrupted or rendered unusable by a competing signal. The same back-off

TABLE 4.19a
802.11 Wireless LAN Standard

Frequency	2.4 GHz		2.4 GHz		Infra-red (850–950 mm)	
Spread spectrum	DSSS		FHSS			
Baseband modulation	DBPSK	DQPSK	2GFSK	1GFSK	16-PPM	4 PPM
Max bandwidth	1 Mbps	2 Mbps	1 Mbps	2 Mbps	1 Mbps	2 Mbps
Access control	CSMA/CA, RTS/CTS					
Area (open orifice)	100–300 m/20–100 m				20–30 m/about 5 m	

DBPSK: Differential Binary Phase Shift Keying; DQPSK: Differential Quadrature Phase Shift Keying; 2GFSK: 2Level Gaussian-filtered Frequency Shift Keying; 4GFSK: 4Level Gaussian-filtered Frequency Shift Keying; 16-PPM: 16 Pulse Position Modulation; 4-PPM: 4 Pulse Position Modulation; RTS/CTS: Request to Send/Clear to Send.

rules apply, whereupon the transmitting node tries again. Collision detection is unsuitable for wireless networks, however, as the signal of the transmitting node would tend to overpower any other and the node would not hear the interference.

802.11 specifies the physical parameters for media that are suitable for transmission of wireless data. These specifications define the physical (PHY) layer of the standard, and describe the physical characteristics of the physical apparatus of the network. In dealing with wireless networks, which have no actual physical apparatus, the standard specifies method of the manipulation of the available media, i.e., radio and infrared frequencies. In manipulation of these, which is referred to as modulation, the medium is used in much the same way as its wired counterpart. Radio frequencies are made to change in a way that is directly related to the impressed information. The same is true for the manipulation of light waves in the infrared spectrum. Direct sequence and frequency-hopping spread spectrum and infrared are currently defined as suitable media for wireless transmission.

Bluetooth (IEEE 802.15/Wireless Personal Area Networks/WPAN)

The Bluetooth/IEEE 802.15 standard is a short-range wireless PAN protocol being developed by the Bluetooth Special Interest Group (SIG) and the IEEE. The name, Bluetooth, incidentally, has nothing to do with any operating characteristics of the system. The name derives from that of a Danish king, Harald Bluetooth, who ruled in the 10th century; the use of the name underscores the heavy involvement of the Nordic communications industry in the development and implementation of the system. The IEEE is leading standardization efforts in the United States in the body of the 802.15 working group. At this writing, the specification is still a work in progress.

General Physical Characteristics The system operates in the industrial, scientific, and medical (ISM) RF band between 2400 and 2483.5 MHz, with 79 1-MHz channels. This band of frequencies is unregulated and users do not need a license to operate a transceiver in this band. The system utilizes a frequency-hopping scheme to coexist in the ISM band with native devices. The frequency hops at a rate of 1600 hops per second; the nominal time slot duration is 625 μs.

The system range is 10 m and supports data rates at up to 723 kbps in one direction while maintaining a return channel of 57.6 kbps in the reverse direction. The system is capable of maintaining full duplex communication at speeds up to 432.6 kbps if requirements dictate. One asynchronous data channel can be supported, or three synchronous voice channels, or one asynchronous data channel combined with one synchronous voice channel. Each voice channel can support a 64-kbps synchronous channel in either direction.

Bluetooth supports communication on a point-to-point basis, or on a point-to-multipoint basis. In a point-to-point system, only two units are involved in the network transaction. In point-to-multipoint systems, one unit communicates with many others (up to seven other units). Two or more units form a *piconet*, which is controlled by one unit, referred to as the *master*. Up to seven units, called *slaves* can be connected to one master, with many more being synchronized to the master, but not actively transmitting. Multiple piconets, which may overlap coverage areas, form what is referred to as a *scatternet*. Each piconet in a scatternet has its own

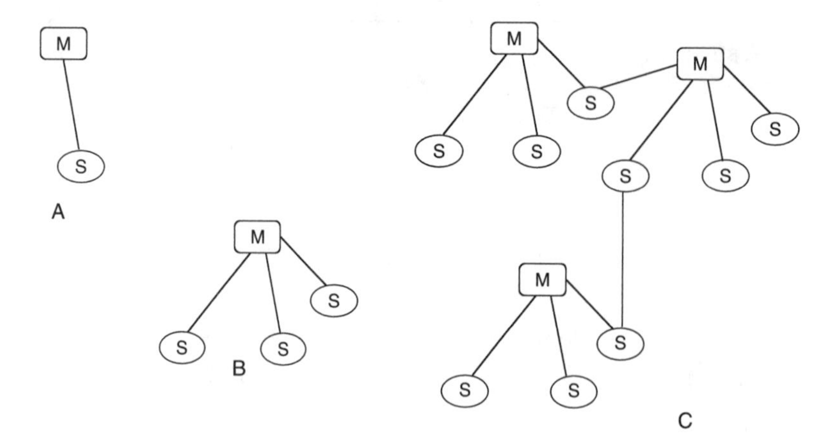

FIG. 4.19b
Piconets with (A) one slave; (B) multiple slaves; (C) scatternet operation.

hopping frequency. Each piconet can only have one master, but that master can be a slave in another piconet. The relationships are shown in Figure 4.19b.

Time Slots Each of the 79 available channels is divided into time slots of 625-μs each. The time slots are numbered and controlled by the clock of the piconet master. Both master and slave can transmit in these time slots, the master transmitting in even-numbered slots, the slave transmitting in the odd-numbered slots. The start of the packet transmission by either master or slave starts at the beginning of the time slot and can extend to five additional time slots. The RF hop frequency, which ordinarily changes with each successive hop, is held constant for the duration of any single packet that extends over a single time slot.

Physical Links Two types of links can be established between master and slave(s):

- Synchronous connection-oriented (SCO)
- Asynchronous connection-less (ACL)

An SCO link is used between a master and a single slave and is a symmetrical, point-to-point link. The SCO link reserves time slots between master and slave and could be considered a circuit-switched connection. The SCO link is well suited for the transmission of voice information and can support up to three SCO links to either one slave or three. A slave can support up to three SCO channels from one master, or two links from two different masters. Some time slots are reserved solely for master-to-slave or slave-to-master SCO packet transmission. The slave-to-master slots immediately follow the master-to-slave slots. The master and slave maintain regular communication through these reserved time slots to maintain synchrony.

ACL links can only occur in time slots not reserved for SCO transmission. ACL communication can occur between a master and several slaves on a slot-by-slot basis. ACL provides the equivalent of a packet-switched network between the master and all slaves participating on the piconet. A slave is permitted to communicate with a master only if the slave was addressed in the previous master-to-slave time slot. ACL packets not addressed to a specific slave are considered to be broadcast and are received by every slave. If no data are available to be sent, then no transmission occurs.

Access Code	Header	Payload

LSB MSB

FIG 4.19c
A standard Bluetooth packet.

Packets On a Bluetooth piconet, data are transmitted in packets. Each packet is made up of three fields: The access code, the header, and the payload. The general arrangement of a Bluetooth packet is shown in Figure 4.19c. The access code field is fixed at 72 bits and the header is fixed at 54 bits. The payload can contain a maximum of 2745 bits.

Access Code Field Each packet starts with an access code. A standard access code that is followed by a header is 72 bits wide; if no header follows, it is only 68 bits wide. The access code is used for synchronization and identification. The access code identifies all packets sent on a particular piconet; each packet is preceded by the channel access code of that piconet. The access code field is divided into three parts.

1. Preamble: This field consists of a fixed pattern of 4 bits in the sequence 0101 or 1010 depending on the starting bit of the following sync word. If the sync word begins with a one, then the preamble will *end* with a zero, and vice versa.
2. Synchronization word: This is a 64-bit field that provides the necessary synchronization between the master and slave prior to the delivery of data.
3. Trailer: This field is present only if a packet header follows the access code. This field is a 4-bit pattern similar to that in the preamble except that the pattern is determined by the value of the last bit of the sync word.

Access Code Types Three different types of access codes are used in a Bluetooth packet and identify the packet as performing a specific function. These access codes are described below:

1. Channel Access Code (CAC): The CAC identifies a piconet. This code is included in all packets on the piconet channel.
2. Device Access Code (DAC): The DAC is used for special signaling procedures such as aging and paging response.
3. Inquiry Access Code (IAC): The IAC does not contain a trailer and is 68 bits wide. There are two variations on the IAC:
   a. General Inquiry Access Code (GIAC): This is used to determine which Bluetooth devices are within range.
   b. Dedicated Inquiry Access Code (DIAC): This is used for all Bluetooth units within range, which share a common characteristic, such as voice-only service.

Packet Header The packet header contains the link control (LC) information and consists of six fields. It is 18 bits wide, which, after encoding, becomes a header 54 bits wide. The components of the packet header field, from the LSB to the MSB, are as follows:

- ADM_ADDR: This field represents the address of the piconet member to which the packet is being sent. Each slave is assigned a temporary 3-bit address to differentiate between them. Packets exchanged between the master and slave carry this address. An all-zero address denotes a broadcast packet, which goes to all slaves.
- TYPE: There are several different types of packets available on the Bluetooth system. These types depend on the physical link used, i.e., a SCO or ACL link. The 4-bit code identifies the link type, packet type, and the number of time slots the packet will occupy, allowing nonaddressed receivers to ignore the packet efficiently. The packet types are briefly described:
  —ID packet: Contains the DAC or the IAC and is fixed at 68 bits.
  —NULL packet: Contains the CAC and no payload and is fixed at 126 bits.
  —POLL packet: Similar to the NULL packet in that it has no payload. A POLL packet requires a response from the recipient.
  —FHS packet: Contains control, address, and clock information and is used for control of frequency hopping and channel synchronization.

The following packets are used on SCO links only. SCO packets do not contain error checking and are never retransmitted. There are four types of SCO packet:

1. HV1: Contains 1.25 ms of speech at 64 kbps
2. HV2: Contains 2.5 ms of speech at 64 kbps
3. HV3: Contains 3.75 ms of speech at 64 kbps
4. DV: Contains combined voice and speech data and is divided into a voice field of 80 bits and a data field of 150 bits

The following packets are used on ACL links only. The information carried is either user or control data.

- DM1 (Data-Medium rate): Carries data up to 18 bytes plus error correction.
- DH1 (Data-High rate): Carries up to 28 bytes.
- DM3: An extension of the DM1 packet; carries up to 123 bytes and occupies up to three time slots.
- DH3: An extension of the DH1 packet, carrying up to 185 bytes over three time slots.
- DM5: A DM3 packet with an extended payload; can carry up to 226 bytes of information over five time slots.
- DH5: Can carry up to 342 bytes over five time slots
- AUX1: Similar to the DH1 packet; can carry up to 30 bytes and is restricted to one time slot.
- FLOW: When the recipient's receive buffer is full, this bit is set to zero to stop any further transmission; conversely, if the bit is set to one, transmission continues.
- ARQN: If this bit is set to one, the transmission was successful (ACK = 1); if the bit is set to zero, the transmission was not received or errors were encountered (NAK = 0).
- SEQN: This bit provides a sequential numbering scheme to transmitted packets to track retransmissions and allow discard of retransmitted packets not required by the recipient.
- HEC (Header Error Check): This is used to check header integrity.

Channel Control This section describes how a piconet channel is set up and controlled. Several states of operation are described as well as the operation of a scatternet. In every piconet, there is one master and one or more slaves. The device address of the master is the determining parameter for setting the frequency hopping (FH) sequence and the CAC. The clock of the master controls the phase of the FH sequence and sets the timing for the piconet. The master also controls the flow of information utilizing a polling routine. A master, by definition, is the unit that initiates transmission; the slave is defined as the unit that responds.

The master clock is responsible for all timing and synchronization of the piconet. Every unit has an internal clock that is free running and has no relation to the actual time of day; the clock can be initialized at any value. When a piconet is established, the master clock is transmitted to the slave units, which then synchronize to the master. The clock determines critical periods such as time slots, which are predetermined to allow transmission by the master or slave. Master-to-slave transmissions can occur only in even-numbered time slots, while slave-to-master transmission can occur only in odd-numbered slots. In different modes and states, the master clock has different functions:

- CLKN: This is the free-running native clock and is the reference to all other clocks.
- CLKE: This is the estimated clock that is the native clock plus an offset.
- CLK: This is the master clock for the piconet and controls all timing and scheduling of activity on the piconet.

Operating Procedures and States There are two operating states in the Bluetooth system: standby and connection. The standby state is the default-operating mode of the unit. The unit is in a low-power mode, with only the native clock running. The unit will leave the standby state to enter the connection state to scan for page or inquiry messages from another unit, or to initiate page or inquiries to determine if any other units are within range. When responding to a page message, the unit will enter the connection mode as a slave; when initiating a page or inquiry, it will enter connection mode as a master.

To establish a connection with another unit, certain procedures are defined for the orderly setup of the link. These procedures are the inquiry and paging procedures. Inquiry procedures are used to discover what units are in range, what their device addresses are, and to obtain clock information. Paging procedures accomplish the same task, but only the device address is shared between devices, enabling a connection to be set up immediately. In these procedures, the DAC and IAC are used to establish the link between the devices. Two mandatory paging schemes are used when two units "meet" for the first time. These schemes are described below:

1. Page Scan: In the page scan substate, the unit listens for its own DAC. If the unit hears its DAC, it will enter the slave response substate.
2. Page: In the page substate, the master (source) activates a link to the slave (destination), which periodically scans for its DAC in the page scan substate. The master sends out repeated transmissions of the slave DAC on different hop frequencies in order to "capture" the slave. The master then enters the master response substate.

When a slave responds to a page from a master, their clocks are roughly synchronized to provide a preliminary "lock" between the two units. The units exchange vital information such as CAC and FH sequence and clock synchronization. This exchange of information proceeds in five steps. These steps are described below:

Step 1: The master is in the page substate and the slave is in the page scan substate.
Step 2: The slave receives its DAC and enters the slave response substate. The slave responds to the master.
Step 3: The master receives a response from the slave and enters the master response substate.
Step 4: If the slave receives data without error, the slave sends a response to the master. Final synchronization of the master and slave occurs in this step.
Step 5: The slave enters the connection state. Hereafter, the slave will use the master clock to determine FH sequence and CAC. The connection process is started by the master transmitting a POLL packet to the slave, which will respond with any type of packet.

The unit that initiated the paging is the master unit, a distinction that is valid only for the duration of the existence of the piconet. At the beginning of the connection, all communication will be from master to slave. After successful connection setup, information packets are exchanged.

Inquiry procedures are used when the destination address is unknown to the source. The inquiry procedure is also used to discover which other compatible units are within range. Upon entering the inquiry substate, the discovering unit collects clock and address information about compatible units within range. Units within range respond to the inquiry with this basic information. The inquiring unit can then connect to any unit if required. An inquiry packet is broadcast from the source and contains no information about the source. The source may, however, request response from only those units with some unique feature or characteristic. Two types of packet are broadcast, determined by either the GIAC, which is used to inquire about any device, or the DIAC, which is used to inquire about only specific types or classes of device.

A unit wishing to discover and possibly connect to other compatible units will enter the inquiry substate. The unit will then continuously broadcast an ID packet containing either a GIAC or a DIAC. A unit that allows itself to be discovered will enter the inquiry scan substate and respond to the inquiry. The master transmits the inquiry and then listens for a response. For an inquiry response, only a slave response is defined. The slave response to an inquiry is quite different from that of a page response. Only the recipient's device address must be returned; this is accomplished using a conventional FHS packet. This packet also contains other vital information about the responding unit. If several units respond simultaneously, a contention procedure is implemented. Each slave is directed to back off for a random period of time and then respond again. This allows each respondent to respond without interference and achieve synchronization with the master.

The Connection State The connection state starts with the transmission of a POLL packet from the master, which verifies the switch to the master clock and FH sequence. The slave can respond with any type of packet. If the slave does not receive the POLL packet, or does not respond to the master, both units enter the page/page scan substates. The first information packets sent contain information that characterizes the link between the units. The procedure of transferring user information is started by alternately transmitting and receiving packets. Using one of the two following commands can end the connection. The *detach* command is used to effect a normal termination of the connection, leaving all configuration data valid in the controller. The *reset* command is used to perform a hard reset on the controller, requiring the controller to be reconfigured through a complete link setup procedure. Bluetooth units can operate in four different modes while in the connection state:

1. *Active Mode*: In the active mode, both master and slave actively participate on the channel. The master schedules and controls traffic to and from slave units according to demand. The master transmits regular synchronization information to the slaves to keep all units in synchrony.
2. *Sniff Mode*: In sniff mode, the time required for slaves to monitor traffic is reduced. Slaves are required to monitor only specifically determined time slots instead of every time slot, as in the case of an active slave. A slave or a master may issue a sniff command to place the slave in *sniff* mode.
3. *Hold Mode*: In hold mode, the slave temporarily suspends ACL activity on the link. In some circumstances, SCO traffic will be supported. In hold mode, a slave is freed to perform other activities such as scanning, paging, inquiries, or responding to a master or slaves in another piconet. The slave can also enter a low-power sleep mode. During the hold mode, the slave retains its piconet address. Prior to entering hold mode, the master and slave agree upon the length of time the slave is to remain in hold. At the end of this time, the slave wakes up, synchronizes to the master, and awaits further instructions.
4. *Park Mode*: If a slave unit is not required to participate in the traffic on the piconet, then the slave can enter park mode. In park mode, the unit releases its piconet address and is assigned two new addresses, which classify it as a parked unit. The unit does not remain synchronized, but does wake up periodically to synchronize and listen for broadcast messages. This function is supported by a periodic "beacon," which is regularly broadcast by the master. Both master and slave can initiate the unpark procedure. This mode is used to conserve power and to connect more than seven units to one master. The master can selectively park and unpark slaves to connect to the maximum number of slaves allowed (255). There is no limit on the number of parked slaves.

Scatternets Scatternets are overlapping multiple piconets that cover the same area and may contain slaves and masters from several different piconets. Because each piconet has a different master, they each hop independently, with the FH sequence and phase determined by each master. In addition, the packets carried on each piconet carry different CACs. As more piconets occupy the same area, the probability of collisions increases. A master or slave can become a slave by being paged by a unit in another piconet.

Time division multiplexing is used in interpiconet communication. In addition, a unit may request to be placed on hold or in park mode, during which time it can participate in the traffic of another piconet. Units in sniff mode may also participate between sniff time slots. Another consideration the concept of a master–slave (MS) switch. An MS switch is desirable when the master of a piconet is paging the master of another piconet. Another possibility is when a slave in one piconet is involved in setting up another piconet as the master. These situations involve the synchronization of any new piconet members to the new master clock and FH sequence.

CELLULAR TELEPHONY

We know wireless as "cell phones" because of the technology that makes their use possible. The theory behind cellular telephones is simple; the actual implementation is more difficult. The problems of the efficient use of available channels, power and size of mobile units, and connection to the wired telephone network had to be addressed and dealt with before any attempt could be made to implement mobile telephone service for the masses.

Early in 1981, a new mobile telephone standard was introduced. Called "the advanced mobile phone service (AMPS)," it was the first system to use the concepts of cells and frequency reuse. Cells are defined as the area of coverage of a given base station and can cover from 2 to 10 miles depending on location. Cells are represented by hexagons for planning and statistical purposes. The true shape of the covered area may be completely different depending on the physical terrain of the cell. The optimum shape for a cell is a circular pattern of transmission coverage. These comparisons are illustrated in Figure 4.19d.

Without the technique of frequency reuse, cellular telephones would not be possible. Each cell is assigned a specific

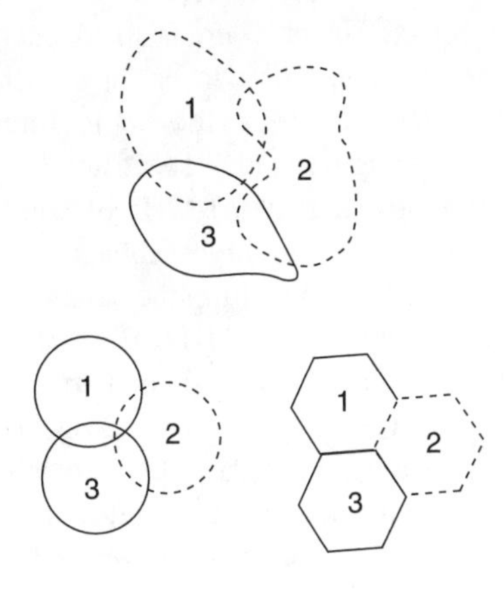

FIG 4.19d
Idealized vs. actual cell coverage.

number of channels (frequencies) to cover mobile units in that cell area. As a user travels from cell to cell, the user is switched into another channel, freeing up the previously used channel. This is a departure from the older IMTS system where a user remained on the assigned channel until the call was terminated or moved out of range of the transmitter. Sophisticated switches called mobile telephone switching offices (MTSO) connect the cell site to the wired telephone network and to other cell sites. The terms *cell site* and *base station* will be used interchangeably. Both terms describe the radio transmitter that serves the cell. In addition to frequency reuse, multiplexing techniques are used to allow a higher concentration of calls over a given channel. The multiplexing techniques used are similar to those used in wired telephone technology. One variant of the AMPS system is NAMPS, which narrows the channel bandwidth to enable more voice channels. Fiber optics has also allowed a larger volume of calls to be completed reliably. Most cell sites are connected to the MTSOs using fiber optics. All connections between MTSOs and the switches of the wired system are T1 and run over a fiber-optic trunk.

Channels and Clusters

To truly exploit the available bandwidth, a new system called DAMPS (Digital Advanced Mobile Phone Service) was introduced. DAMPS is the digital implementation of the original AMPS system. DAMPS uses the same frequency band, but utilizes multiplexing techniques to triple the number of channels available. DAMPS is rapidly replacing the older analog network.

The cellular frequency band in the AMPS system starts at 824.04 MHz and ends at 893.7 MHz. When the cellular network was originally set up, the FCC assigned each area one half of the frequency band. These are called the "A" band and the "B" band. This allows two providers in each coverage area. Coverage areas are referred to as metropolitan statistical areas (MSAs), or rural service areas (RSAs). The A band automatically went to the local exchange carrier (LEC), and the B band went to any group or individual who could meet the rigorous financial and professional requirements for ownership of the frequency band. There is no advantage to having either the A or B band. One carrier may, with the proper application of technology, provide up to three times the capacity with the same frequency spectrum.

Within the approximately 70 MHz of bandwidth available for cellular telephones, three subbands exist. Cellular (mobile) telephones transmit within the band between 824.04 and 848.97 MHz, roughly 25 MHz. Base stations, or cell sites, operate within the frequency band between 869.04 and 893.97 MHz, again roughly 25 MHz. A 20-MHz guard band separates the transmit and receive frequency subbands. In addition, 45 MHz separates each transmit frequency from the corresponding receive frequency to prevent interference. For example, 824.04 MHz is 45 MHz below 869.04 MHz. Remember that transmit and receive frequencies are "paired" and fixed in order to maintain the 45 MHz separations. These frequency pairs constitute single full-duplex channels. Each channel in each band is 30 kHz wide. This allows a total of 832 channels (832×30 kHz = 24.96 MHz).

Of 832 channels, 790 are assigned to carry voice traffic. The remaining 42 channels are used for control signals between the mobile unit and the base station. The control channels are also full duplex. Control channels carry information, which controls the operation of the telephone, and are integral to the efficient use of available frequencies. Each cell has two control channels, one for each system. The mobile unit monitors one of the two frequencies for information needed to initiate or complete a call. Control channels are used to set up the call; the actual conversation takes place on a voice channel. The control channel in each cell is usually the first channel assigned to the cell.

Frequency reuse is the underlying principle of cellular communications. The same bands of frequencies can be reused in different cells, allowing the most efficient use of the frequency spectrum. Each cell may support up to 45 voice channels. Each cell has one control channel assigned to it. Channels are assigned to cells to arrange the maximum geographic distance and assure channel separation. As a mobile travels from one cell to another, the circuit is automatically switched to another channel in the adjacent cell. Continuing through several cells, a mobile telephone may be switched to many different voice channels without any perceptible interruption in service.

Cells are generally grouped into clusters. In less congested areas, a 12-cell cluster, or frequency reuse pattern, is used to handle all traffic. A typical cluster is seven cells. Figure 4.19e illustrates an idealized seven-cell cluster.

A cluster utilizes 42 channel sets, 21 for each system. All data will pertain to the total system channel capacity of 832 channel. Each cell is assigned 6 channel sets, each channel separated from another by 7 channels. No frequency pair is reused within the cluster. In an adjoining cluster, channels are similarly assigned. This method of channel assignment and physical separation of cell sites is employed to avoid any occurrence of cross-channel interference.

Cells can be further broken into sectors. In the example above, an ideal cell was illustrated. An omnidirectional

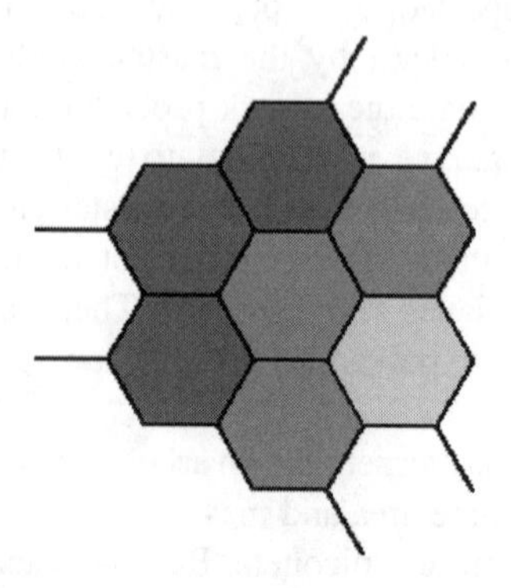

FIG. 4.19e
A seven-cell cluster.

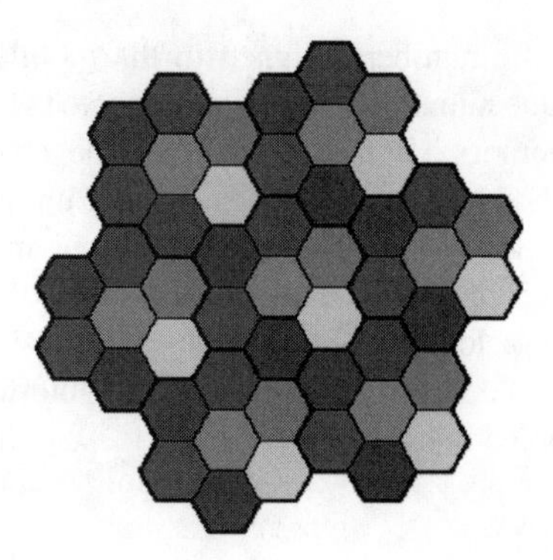

FIG. 4.19f
Frequency reuse patterns.

antenna is used to transmit in all directions outward from the base station. Within the cell, a mobile unit uses the assigned channel until it is handed off to another cell and assigned another channel. Using directional antennas, cells may be separated into three or six sectors depending on the number of antennas used. The effect of sectorization is to provide much better coverage in the cell. Theoretically, the amount of traffic handled is also increased. A disadvantage is the greater chance of cross-channel interference.

Frequency reuse is only possible with the careful assignment of frequencies and channels to cells that are separated by adequate distances. Figure 4.19f illustrates typical cell frequency reuse patterns based upon the seven-cell cluster. Ideally, this is the smallest possible distance.

The larger the number of cell sites in a given geographic area, the greater the chance for interference. The minimum practical size for a cell is a coverage area of about a mile. Tunnels, bridges, and highly congested urban areas are often served by "microcells," which may be up to 300 m in diameter. Picocells are even smaller, covering an area of only 30 m. Picocells can be used in offices or college campuses in wireless networking applications. At the high end, a cell could be as much as 35 miles in diameter, the major constraint being the transmitting strength of the mobile unit. Other factors affecting propagation are trees, hills and valleys, mountain ranges, and buildings. At present, cellular telephones are limited to a maximum radiated power of 7 W. This power is referred to as ERP, or the effective radiated power. Larger cells do not depend as much on frequency reuse as do smaller cells. This is a direct consequence of the amount of traffic in the coverage area of a cell. As traffic increases in a given area, cells are split and clustered to take advantage of frequency reuse.

Call Switching

Before we discuss the mechanics of making and receiving a call, call switching must be understood. The central control for activity in a cellular system is the MTSO. The MTSO monitors every call in the cells to which it is connected. The MTSO tracks every mobile unit within the range of its controlled transmitters. The MTSO measures the signal strength of each mobile and controls the handoff of that mobile to an adjoining cell while dynamically reassigning channels. The MTSO handles an incredible amount of wireless traffic while also connecting the wireless network to the PSTN (public switched telephone network).

The MTSO supervises and administers the wireless network. This is a fantastically complicated task. Many cells and cell clusters must be monitored and controlled. The same cells are connected to the MTSO for connection to the public network and each other. The MTSO is the brain of the system. All customer data travel through the switch and must be validated prior to handling a call. Customer data include the status of the customer account, the class of service requested, the electronic serial number (ESN) of the mobile telephone, and the signal strength. The signal strength is used to determine the relative position of the mobile. The MTSO must also constantly monitor channel usage, assigning channels as they become available.

Call Processing

Call processing is possible through the use of the 42 control channels on the system. Each cell has two control channels, one for each carrier. It is on these control channels that the mobile and MTSO negotiate for the system resources. Each mobile telephone monitors the control channel for the system to which it is subscribed. This can lead to delays when several thousand subscribers are trying to access the system.

Of the two frequency pairs required for a voice channel, one frequency is dedicated to transmission from the mobile unit to the cell site. This is the lower of the two frequencies and is called the reverse channel. Communications from the cell site to the mobile unit is the higher frequency and is referred to as the forward channel. Each call uses both forward and reverse channels. Both voice and control channels are divided into forward and reverse channels. Calls are set up on control channels and then assigned a voice channel by the MTSO.

The first step in the process is registration. Registration begins when the cell phone is turned on. The unit immediately begins a sweep of the available frequency band, looking for the strongest forward control signal. The nearest base station and cell site typically supply this signal. The phone then locks on to the signal and begins transmission of data on the corresponding reverse control channel. The data are sent to tell the system that the unit is operating and active and ready to take calls. These data include the phone ESN, the telephone number assigned to the phone, and the home system ID. The cell site then transmits the information to the MTSO, which then tells the world. The entire process takes only a few hundred milliseconds. Registration occurs every 15 min or so and whenever the phone travels into another cell. The MTSO can also request registration data from the mobile by transmitting the request on the forward control channel.

The phone responds by transmitting registration data to the cell site on the reverse control channel.

After the phone is successfully registered, it is ready to accept calls. The unit idles while monitoring a forward control channel for a paging signal. A paging signal is sent whenever a call is made to the mobile's number.

The mobile unit monitors the forward control channel and recognizes its telephone number. The mobile responds by sending registration data and an acknowledgment of the page on the reverse channel of the cell site. The MTSO sends a supervisory audio tone (SAT) over the assigned forward voice channel. This does two things. First, it completes the loop between the cell site and the mobile telephone, verifying that the phone is on the correct frequency. If the SAT is not returned, the connection is not made and the call is not completed. Second, it gives the MTSO a better idea of where the mobile is, owing to the time it took for the SAT to be returned from the phone. The MTSO could then find a cell site closer, if possible, to improve reception.

Making a call from the mobile unit is slightly more complicated. After the telephone number is punched into the phone and the send button is pushed, the mobile sends the number and a request for service over the strongest reverse control channel available. After the MTSO validates this information, it assigns a voice channel and transmits the assignment to the mobile, which tunes to that channel. The MTSO transmits an SAT over the forward voice channel assigned. The mobile, if it detects the SAT at the assigned channel, transmits the SAT back to the cell site on the reverse voice channel (RVC). The MTSO receives the SAT and switches the call through from the mobile.

Although it seems that this process is just the reverse of the process of receiving a call, the similarity ends there. Before the MTSO even assigns a voice channel, it must determine if the caller requesting service is, in fact, a valid user. This "pre-call validation" is necessary because of fraud and the enormous loss to cellular carriers, which are responsible for the cost of most fraudulent calls.

The mobile requests service when the send button is pressed on the mobile telephone. This request is sent over the reverse control channel (RCC) to the cell site along with information about the telephone. This information allows the MTSO to check this information against the databases maintained by GTE and others. These databases are regularly updated and allow carriers to keep track of subscribers. This information is sent repeatedly to ensure that it is received. Four separate identification numbers identify the telephone:

1. The home system identification number (SID)
2. The mobile identification number (MIN)
3. The electronic serial number (ESN)
4. The station class mark (SCM)

The SID identifies the cellular telephone carrier to which the user is subscribed. This number tells the MTSO if the mobile is in its home area. If it is not, the roaming indicator is activated. The MIN is the number assigned to the mobile phone. The ESN is a unique number, which is burned into the mobile unit's read-only memory (ROM) and cannot be changed. Every mobile telephone in existence possesses a unique ESN. The SCM informs the MTSO at what power level and frequencies the telephone is operating. The cell site can instruct the mobile to turn down or lower its transmitting power level if a lower power level can do the job. This avoids any interference within the cell coverage area.

Multiplexing

A brief description of three common multiplexing techniques closes out this discussion of cellular telephony. Three methods are currently used:

1. FDMA—Frequency division multiple access
2. TDMA—Time division multiple access
3. CDMA—Code division multiple access

The first, FDMA, is widely used in analog systems, and is being replaced by the latter two techniques, which are more suitable for use in digital systems. Aside from providing advanced features such as caller ID, digital telephony does not provide any great advances in the technology. Coupled with advanced multiplexing techniques, however, it allows marked increases in cellular frequency use efficiencies. TDMA is used with analog systems as well as the newer digital systems. CDMA is, by definition, a purely digital system.

FDMA Frequency division multiple access was the first technique used for increasing the capacity of available cellular channels. It was the basis for the NAMPS system, the "N" designating narrowband. Each 30-kHz cellular frequency was broken into three 10-kHz subchannels over which a conversation was transmitted. This allowed three times as many conversations on a given channel.

TDMA TDMA is another multiplexing technique used to increase frequency use efficiency. In TDMA, each narrowband subfrequency is divided into six repeating time slots, or frames. Two slots in each frame are assigned to each call. TDMA suffers from "multipath fading" and distortion, phenomena that are caused by reflections of the signal from any number of physical interferences. These reflections cause interference and distortion because of time differences in the arrival of the signal from different paths. Fading is caused by the same physical interferences. ETDMA, or enhanced TDMA makes even more efficient use of the frequency spectra. ETDMA actively monitors a conversation, sensing pauses or dead spots in the communication. The system utilizes these holes in the transmission and stuffs them with bits from other conversations. This system is not universally used; however, this technique makes the most effective use of the available time slots.

CDMA Code division multiple access is the latest method for improving spectral efficiency. CDMA works by assigning a unique code to each message and then transmitting the encoded message over single or multiple frequencies. By using this method, channel efficiency can be increased by a factor of 20. This is a vast improvement over FDMA and TDMA, which allow up to a maximum of 6-to-1 improvement in channel efficiency. The principles of CDMA are well known, as they have been used in military communication for the last 35 years. Chief among the advantages of CDMA are its excellent immunity to noise and improved security from interception (improved privacy).

Spread Spectrum

Spread spectrum is a modulation technique, which spreads the transmitted signal over a much larger frequency bandwidth. Fundamentally, spreading a signal over many narrowband (read, "noisy") channels increases the probability that the signal will be received with little or no interference. By spreading the signal, the signal-to-noise ratio (SNR) is decreased. Lower power may be used over a larger range of frequencies as opposed to a high-power signal over a narrow range of frequencies. This has a direct effect on noise. Noise is an integral component of the signal in the form of other frequencies, which are multiples of the base, or carrier frequency. By using lower power, the effect of these multiples is also decreased. The effects of other noise components due to atmospheric factors are also similarly decreased. Also, the information is spread over a large number of frequencies, allowing the system to ignore noise specific to each narrowband frequency channel. The effect of this technology is to produce secure and reliable wireless communication while dramatically increasing system capacity over previous systems. Two main types of spread spectrum are in use: direct sequence spread spectrum (DSSS) and frequency hopped spread spectrum (FHSS).

Direct Sequence Spread Spectrum DSSS is the most widely used form of spread spectrum. It is an all-digital system using "pseudo-random" or "pseudo-noise" codes to control frequency spreading. These codes are called pseudo because they are not actually random or noise. Information to be transmitted is modulated with the PN code. Modulation techniques in use are BPSK (binary phase shift keying) and QPSK (quadrature phase shift keying).

Frequency Hopped Spread Spectrum FHSS also utilizes PN codes to spread the transmitted data. FHSS is the easier of the two main spread spectrum methods to implement. The technique employs precision frequency synthesizers, which continuously shift the center carrier frequency based upon the impressed PN code. As the signal is "hopped" from frequency to frequency, care must be taken to avoid interference with other users on the same channel who may still be using older technology. The period of time a signal remains on a specific channel is called the dwell time. The dwell time is typically very brief, less than 10 ms. The manipulation of dwell time is very effective in eliminating interference to other users. There are two methods within this technique: slow frequency hopping (SFH) and fast frequency hopping (FFH). SFH transmits very few (<2) data bits. This allows improved data detection at the receiving end. If a frequency is blocked or jammed, however, data may be lost. This requires the use of error-correcting codes. FFH divides the same bit over several frequencies. Error-correcting codes are not needed, but the method requires additional processing overhead at the receiving end to improve data detection.

WIRELESS DATA

Cellular telephone systems also support wireless data communications. This technology lends itself to such applications as:

- Remote information retrieval
- Remote inventory checking
- Warehouse management systems
- Wireless credit card authorization and point of sale
- Remote utility meter reading
- Wireless fax and e-mail

Three systems are described here, showing the evolution of this technology: Specialized mobile radio (SMR), cellular digital packet data (CDPD), and personal communication services (PCS).

SMR

SMR is a half-duplex method of communication designed for the trucking industry to aid in the dispatching of vehicles. SMR is capable of sending small amounts of data over existing analog channels. Transmission rates of only 4.8 kbps are possible on this system, and interface software is usually custom and therefore costly. SMR has been rapidly expanding into the digital realm in recent years, offering paging, telephone, and data services to its subscribers. Many large trucking and package-handling firms still use these cell-type networks to track packages and dispatch vehicles.

CDPD

CDPD is a method of transmitting data over wireless networks. Originally developed as a technique for transmitting data over existing analog cellular networks, CDPD has made the transition to modern CDMA-based cellular networks. CDPD supports many different data transfer protocols, allowing interoperability with many different applications and networks. Also, CDPD supports a wide range of equipment options and standard interface applications. There are two types of CDPD networks:

1. Dedicated channel
2. Channel hopping

In a dedicated channel network, certain channels or a range of channels are set aside exclusively for the use of the CDPD network. This is similar to the channel/frequency allocation to cellular networks. These channels are outside the range of channels specified for cellular telephone service. In a channel-hopping network, cellular telephone and CDPD devices share all channels within the cellular band. Data transfers are sent over vacant channels until the CDPD software senses voice traffic, at which time the data transfer "hops" to another vacant channel and continues the transfer. The cellular call has preemptive priority over the CDPD transaction, and if there are no vacant channels available for the data transfer, the transfer will be dropped. The CDPD standard specifies that the transfer must occur within 40 ms of the detection of voice traffic; the cellular caller will never know that a CDPD transfer is taking place. Each channel in the network can handle up to 30 calls at a time. CDPD transfers operate, as would many users, on an Ethernet computer network. Each device listens to the channel to make sure that no other device is transmitting; it then places data on the air. If two devices transmit at the same time, a "collision" occurs, and each device resets and retransmits after a random delay. A CDPD network consists of five components, as shown in Figure 4.19g:

1. The mobile end station (MES)
2. The mobile data base station (MDBS)
3. The mobile data intermediate system (MDIS)
4. The intermediate system (IS)
5. The fixed end system (FES)

The MES is the mobile device and wireless modem. This can be a laptop computer, a point-of-sale terminal, or a remote measurement device; each of these devices is typically connected to a wireless (cellular) data modem. The MES sends data to the network via the MDBS at a maximum rate of 19.2 kbps. Finally, the MES packetizes the data and may encrypt it for security before sending it out over the airwaves. The MDBS is installed at the cell site, or base station. It uses the same antenna as the cellular voice network and controls the activities of the CDPD network. The MDBS actively monitors all voice channels in the cell whenever a CDPD transfer is taking place. The MDBS controls the "hopping" of transfers whenever it is necessary and is responsible for monitoring all unused voice channels for use by the CDPD network. The MDIS is the interface between the wireless and wired portions of the network.

Registration is similar to that performed for voice calls and is handled by the MDIS. The MDIS keeps track of all MES in its coverage area. Two types of MDIS are defined: home and serving. A home MDIS is the primary MDIS for a given MES, or the one to which it normally registers. A serving MDIS is located in another area to which the given MES has traveled. The serving MDIS communicates with the home MDIS to transfer packetized data between the home MDIS and its MES.

An IS can be any of the various systems that comprise the fixed network. Routers, switches, and the wiring on the landlines are considered the IS. The FES is any of the destinations possible at the other end of the landline. This could be an office network or just another computer. FES are typically dedicated to serving wireless clients, allowing the additional system overhead to be absorbed by an isolated machine rather than the entire enterprise.

CDPD is very cost-effective because of its use of the existing cellular infrastructure. CDPD does not overly impact

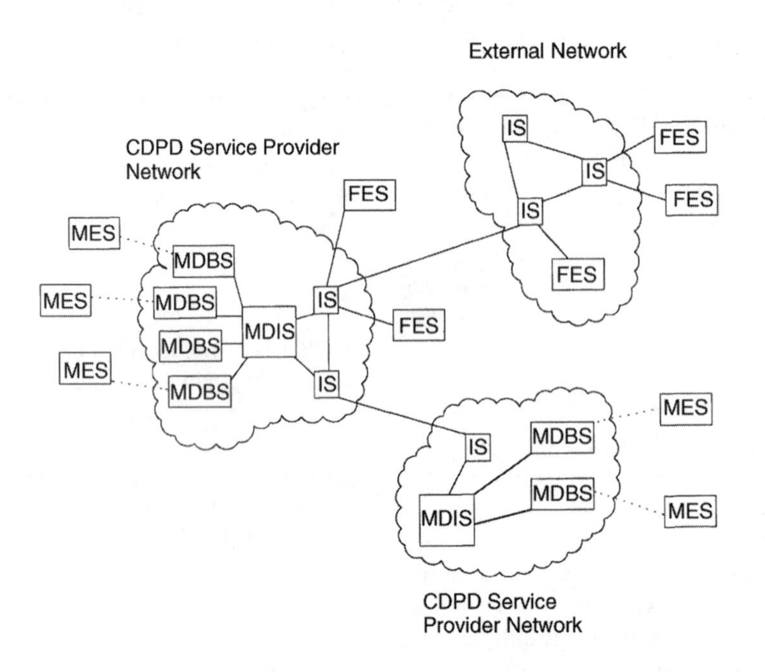

FIG 4.19g
A CDPD network.

the existing systems and can be a genuine enhancement to current modes of customer service. CDPD allows users to engage in multimode forms of communication, allowing voice, fax, and data from the same equipment.

PCS

Personal communication services is the newest wireless data technology to hit the market. PCS offers enhanced messaging capability, allowing services such as paging, voice, and data transmission, and Internet access to subscribers through the use of a single, integrated mobile unit. Some enhanced features of PCS include:

- Voice and data privacy with increased protection against eavesdropping
- Sleep mode allowing extended battery performance
- Enhanced data services and accessibility
- Increased use of private and residential cellular systems
- Improved radio spectrum efficiency

Although present units offer simple text messaging and e-mail retrieval in addition to voice, palmtop computers with integrated digital communication capability have been developed and will be deployed. PCS utilizes both cellular frequency allocations and has also been assigned to a new frequency band for use of PCS technology exclusively. PCS has been assigned the frequency band between 1850 and 1990 MHz, allowing approximately 1840 channels for each of the transmit and receive bands. A significant enhancement of digital communication brought by PCS is the ability of a mobile to exhibit different characteristics and offer expanded features depending upon the system it registers to. Systems may be public, private, or residential.

SATELLITE LANS

A satellite is any natural or artificial body that describes an orbit around another body. In our discussion, satellites will be defined as manmade objects that orbit the Earth in the space surrounding it. These satellites are capable of a variety of tasks, from military missions to direct broadcast television. Until recently, the concept of satellite networking was just that—a concept. With the advent of affordable cellular technology, the need for a mobile method of communication and networking had been satisfied, at least as long as one was within range of a cell. Satellite networking has become a new means of mobile connectivity to expand networking capability further, particularly to those areas where it would be difficult or impractical to provide wired or cellular service, such as at sea or in the air. Future wireless networks will rely heavily on satellites for coverage of areas that were heretofore unable to obtain coverage. Advances in this technology will enhance existing cellular networks. Remote monitoring and control of facilities will not only become more feasible and reliable, but also much less costly to operate and maintain.

Basic Satellite Technology

Artificial satellites are unlike aircraft in that there is very little control over the route on which they fly, or orbit. Satellites typically move about the Earth in elliptical or circular orbits. The altitude of orbit can be adjusted by the amount of force used to launch the object from Earth; this altitude has a profound effect on the time it takes for the satellite to orbit the Earth, called the *period*. Modern communication satellites are found in many different orbital positions, each with a characteristic period.

Satellites stay in orbit because of the interaction between several natural forces. Gravity is chief among them, exerting a constant attractive force upon the orbiting object. Counteracting this gravitational pull is centrifugal force. Centrifugal force acts upon any object moving in a circular motion, forcing the object to move away from the center of its rotational plane. For any object, this force is proportional to the square of its velocity. In actuality, the orbiting object is continuously falling into the Earth, the object's *angular velocity*, or sideways motion, acting to "move it out of the way" of the Earth before it collides. Centrifugal forces acting on the satellite compensate for the gravitational force also acting on the object, allowing it to remain in a predetermined orbit above the Earth as long as the angular velocity remains constant.

The laws describing planetary motion were explained by the work of Johannes Kepler, whose three laws of planetary motion are the basis for the understanding of satellite orbital mechanics (Figure 4.19h). Kepler's first law states that the planets orbit the sun in an elliptical path, with the sun occupying one foci of the ellipse. Applying this to an artificial satellite, the object's path can be described as elliptical, with the Earth occupying one foci of the ellipse. This means that the distance between the sun and the planet is constantly changing.

Kepler's second law states that the area swept by a straight line from the center of the sun to the center of the planet sweeps equal areas in equal time periods. The angular velocity of the planet is constantly changing as it traces its

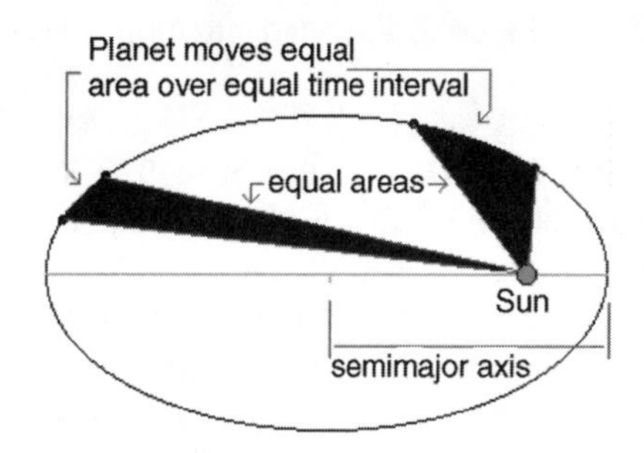

FIG 4.19h
Kepler's laws illustrated.

orbit. In other words, the orbit of a body is moving more slowly at the apogee of its orbit, the point furthest from the Earth, than at its perigee, the point closest to the Earth.

Kepler's third law states that the squares of the planets' orbital periods are proportional to the cubes of the semimajor axes of the orbits. Simply put, if an object's orbit (height above the Earth) is known, then the object's orbital period can be determined; the converse is also true, allowing one to determine orbit if the period is known.

The major thrust of Kepler's laws is to establish that planets or satellites travel at different speeds during their orbit; these laws allow the accurate measurement or prediction of these speeds. This information is especially useful in planning and implementing a satellite system. By manipulating the orbit of a satellite, the period can also be manipulated, allowing a large degree of control over coverage areas.

Satellite Orbits

Low Earth Orbit Satellites Low Earth orbit satellites (LEOS) orbit between 500 and 4000 km (300 and 2500 miles) above the Earth. LEOS complement the terrestrial cellular systems and are based on small lightweight satellites, which are less costly to build and place in orbit. Because these satellites are close to the Earth, low power transceivers can easily communicate with the nearest satellite. Coverage area is small for a particular satellite and comprehensive coverage requires multiple satellites with overlapping "cells" similar to the method of coverage used by terrestrial systems. Because of the much shorter distances to be covered by the signals involved, users can communicate using inexpensive handheld devices, such as what is commonly referred to as a "cell phone."

Different systems have been proposed, each using a different number of satellites, referred to as a *constellation*, to effectively provide coverage to the entire surface of the Earth. Proposals range from a system of 840 satellites in 21 orbital planes, with 40 satellites to a plane, to a system of 288 satellites in 12 planes, with 24 satellites to a plane. The satellites would orbit the Earth at speeds between 16,000 and 17,000 miles per hour, with a period of approximately 90 min. The distributed nature of the system offers superior reliability over a fixed satellite. If a fixed or single satellite is damaged or malfunctions, the entire network can be rendered inoperative (redundancy makes this unlikely); LEOS offers multiple satellites and therefore redundant equipment and signal paths.

Little LEOS and Big LEOS A little LEOS is a small, low-cost satellite weighing between 50 and 100 kg. Little LEOS systems have been allocated the bands of 137 to 138 MHz and 400.15 to 401 MHz for downlinks, and 148 to 149.9 MHz for uplinks. These bands have been used for meteorological satellites, general research, and fixed and mobile communication systems. Proposed systems will use these bands for slow data communication, paging, and messaging systems. Voice services will be excluded from little LEOS systems.

Big LEOS, with satellites weighing between 350 and 500 kg, are aimed at data communications and real-time voice into handheld units. A new allocation of spectrum between 1610 and 1626.5 MHz currently used by aeronautical radio navigation services will be used to implement this system. Big LEOS can carry voice and high-speed data services, unlike little LEOS. Also, big LEOS will use new technologies such as onboard processing and intersatellite backbone linking at millimetric (30 GHz) frequencies.

Iridium, a little LEOS system built and implemented by Motorola, placed 66 satellites into orbit before declaring bankruptcy. Big LEOS holds the most promise for universal connectivity, touting itself as the "internet in the sky." For comparison, signal latency, the time it takes for a signal to make the roundtrip from Earth to the LEOS and back, is only 20 to 40 ms as opposed to upward of 0.5 s for a geosynchronous Earth orbit satellite (GEOS). Demand for LEOS services is expected to be primarily in rural telephone service, mobile, and broadband data communication.

Medium Earth Orbit Satellites MEOS operate in the range of 8000 to 16,000 km (5000 to 10,000 miles). While offering greater area coverage than LEOS systems, the distances involved require larger terrestrial transceivers. Also signal latency for MEOS is in the range of 50 to 150 ms. MEOS are often used as space-based relays, operating in conjunction with GEOS systems.

Geosynchronous Earth Orbit Satellites GEOS systems offer the distinct advantage of being able to be "parked" at a specific location over the Earth's surface. GEOS operate at an orbit of 35,860 km (22,282 miles) above the Earth. Theoretically, three satellites in geosynchronous orbit above the equator could cover the Earth's entire surface, with the exception of the polar regions. Allowing for the agreed-upon 2° separation between satellites, a possible "constellation" of 180 satellites could be parked above the equator. Given the distances involved, however, transceivers capable of communicating with GEOS systems are land based and usually not mobile. Signal latency approaches 0.5 s on a round-trip between Earth and the satellite.

Figure 4.19i illustrates the orbits of LEOS, MEOS, and GEOS.

Communication Bands

Satellite communication occurs within four allocated bands of frequencies. These are the C, L, Ku, and Ka bands. One of the

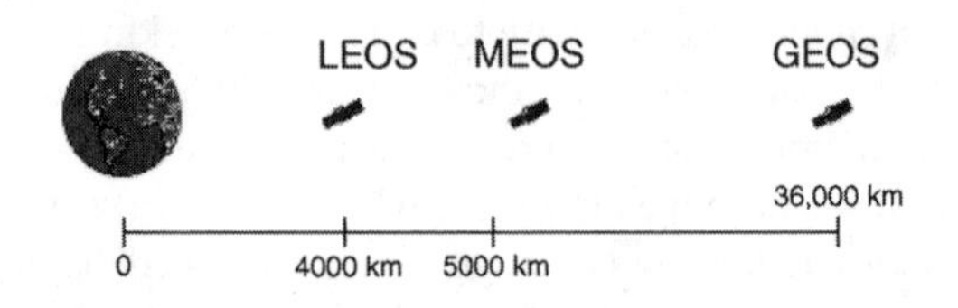

FIG. 4.19i
Comparison of satellite orbits.

problems facing the industry is finding space within the allocated frequency spectra. Most of the available band is already allocated to other services, which have to contend with a large installed base of equipment designed to operate in the allotted band. This is, and will continue to be, a problem as the technology is implemented.

L Band L band communication occurs within the spectrum at frequencies between 1.53 and 2.7 GHz. This range of frequencies is a "long" wave and is capable of penetrating many structures and materials. Also, less powerful transmitters may be used. This band is allocated to existing services, leaving very little space for new satellite communication services.

C Band The C Band has traditionally been used for the direct broadcast television market. Although this is still the primary use for this frequency band, the market is being supplanted by DSS systems. C band systems utilize large (7.5 ft) diameter antennas to capture the relatively weak signals from the geosynchronous satellites used.

Ku Band The Ku band operates in the range of frequencies between 11.7 and 12.7 GHz (downlink) and 14.0 and 17.8 GHz (uplink). This medium wave transmission also has good penetrating power and, owing to the large bandwidth per channel, can carry quite a bit of data. This band is also largely allocated to existing services.

Ka Band The Ka band operates between 18 and 31 GHz and wavelengths are referred to as "millimetric." Although there is an abundance of available spectrum, the signals do not have the same penetrating power as the longer wavelengths and are subject to fading, particularly from atmospheric conditions such as rain and snow. Also, much more powerful transceivers are required to effect reliable communication. The Ka band will be used by most of the new satellite services being offered.

Satellite Networks

Satellite networks will operate much the same way as any other network, with the major difference in the type of equipment used to convey the signals. Practical wireless networks will use a combination of terrestrial systems such as existing cellular and wired networks, which will connect through Earth stations to an available satellite. Direct satellite communication will also be common, but limited bandwidth may make this type of satellite networking more costly and possibly out of the range of the average user. One unique facet of satellite networking is the methods of signal routing developed to deal with the requirements of a space-based system.

Routing Two types of routing are in common use: bent-pipe routing and satellite-to-satellite routing. In bent-pipe routing, a user sends a signal to the nearest satellite (assuming a LEOS system), which reflects the signal back to Earth to either the intended receiver or to an Earth station. The signal is then routed on conventional terrestrial wireless (cellular) or landlines to another Earth station. The signal is then uplinked to another satellite, which then finds the intended receiver, and downlinks the signal, completing the network transaction.

Satellite-to-satellite routing involves transmission of an uplinked signal to several other satellites, each of which attempts to find the intended receiver. The closest satellite then downlinks the signal. Within the satellite constellation, however, a method of intersatellite linking (ISL) must be established to effect the orderly transfer of information between satellites. These methods of ISL are accomplished in two ways: interorbit and intersatellite. Interorbit linking involves the transfer of information between satellites in different orbital planes. Interorbit linking has the capability, when fully implemented, to eliminate the need for a terrestrial link, as in bent-pipe routing. Intersatellite linking is simply communication between satellites in the same orbital plane.

Bent-pipe routing introduces the most signal latency, while the higher level of sophistication of the onboard electronics required for signal processing and routing will drive up the cost of satellite systems utilizing satellite-to-satellite routing.

Signal Handoff Each satellite, as it orbits the Earth, will cover a "footprint," which is divided into multiple cells. This technique allows greater transmission power per cell, a big consideration. Signals must therefore be passed not only between satellites, but also between cells within the footprint of a satellite. This process is called a *handoff* and occurs much the same way as in cellular technology. As the signal moves between cells, signal strength will vary. Each cell has upper and lower signal strength thresholds. As the signal strength falls outside of the allowable limits, the signal is transferred to the next cell.

Protocols Satellite communication and networking will make extensive use of existing multiple-access protocols. Among these protocols are TDMA, CDMA, and FDMA. These protocols are covered in detail in the subsection "Cellular Telephony" (see p. 655).

INFRARED SYSTEMS

Infrared energy is the range of electromagnetic energy between visible light and radio frequencies. Infrared means "below red," referring to the range of frequencies beyond the lowest frequency of visible light, red (Figure 4.19j). This is in contrast to the range of frequencies above the highest visible light frequency, violet. This range of frequencies is referred to as ultraviolet, or "above violet." Early experimenters found that infrared has the same properties and characteristics as

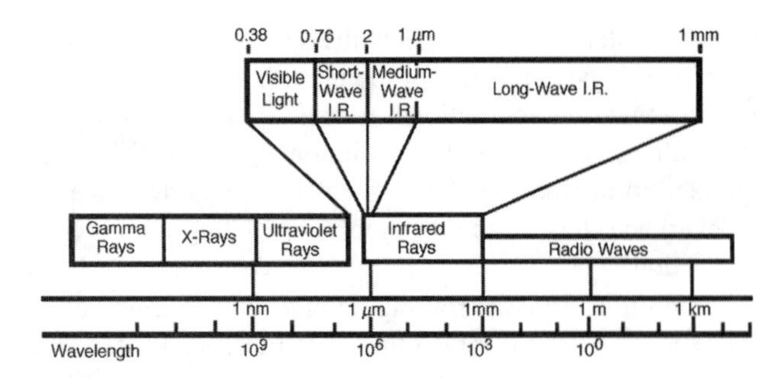

FIG. 4.19j
The infrared spectrum.

TABLE 4.19k
Regions within the Infrared Spectrum

Spectral Region	*Wavelength Range, μm*
Near-infrared	(0.7–1) to 5
Mid-infrared	5 to (25–40)
Far-infrared	(25–40) to (200–350)

visible light. Infrared is divided into three spectral types: near infrared, mid-infrared, and far infrared. In reality, the spectral regions shown have a significant overlap. The near-infrared region is the closest to the visible spectra, explaining the naming of the regions (Table 4.19k).

The near-infrared region is further subdivided into two regions: the very near-infrared, in the region between 0.7 and 1 μm, and the shortwave infrared, between 1 and 5 μm. Infrared networking uses electromagnetic energy with wavelengths of 0.8 μm, corresponding to a frequency of 350,000 GHz.

Infrared LANs

Infrared energy cannot penetrate opaque objects such as walls and floors. It is therefore limited to line-of-sight applications that require the communicating devices to be within a short distance of each other. Infrared LANs are also susceptible to disruption due to sunlight or other bright lights. Two types of infrared propagation systems are in common use. The first is known as directed infrared and requires each communicating device to be within sight of each other. The range of this type of device is between 3 and 19 ft. The second type is known as a diffused infrared (DFIR) LAN and uses the walls, floors, and ceilings to bounce IR around a small area in order to communicate. Both of these LANs are unsuitable for any application requiring large areas and serve well for very short and fast links between PDAs or between closely located devices such as a desktop and its peripherals, or several computers in a small room, constituting a subnetwork of either the *ad hoc* or infrastructure variety.

Infrared LANs are currently defined by two standards, both developed and endorsed by the Infrared Data Association (IRDA). The first standard is referred to as the IrLAP standard, which stands for serial link access protocol. The serial infrared physical layer standard is also discussed. The second standard, referred to as very fast infrared (VFIR), defines a system that is capable of data transfers up to 16 Mbps. The IEEE 802.11 standard defines DFIR (diffused infrared) as an acceptable physical medium for WLANs. The projected data throughput is 16 Mbps.

Infrared Serial Link Access Protocol

The IrLAP standard defines the methods for utilizing the infrared medium to effectively establish reliable communication between two infrared-enabled devices. The standard deals mainly with the operating characteristics of systems of a mobile or *ad hoc* nature. This information is available to manufacturers in order to facilitate communication among electronic devices such as computers, peripherals, or appliances using directed half-duplex infrared links in free space. This standard describes the features, functions, protocols, and services at the data-link layer (as defined in the OSI model). Because of the mobile nature of the devices involved, several concepts are presented that are unique to this medium. Dynamic discovery of devices, called stations, recovery of data, and medium contention are among the concepts defined.

Physical Layer Characteristics There are two different sets of transmitter/receiver specifications. The first is the core standard for links operating between 0 and 1 m. The second standard is called the low power option and has a shorter operating range. Each standard can operate three possible links. Low power transmission to low power transmission, standard transmission to low power transmission, standard transmission to standard transmission. These links are illustrated in Table 4.19l.

The minimum signaling rate for all infrared links is 9.6 kb/s. Pulse durations vary with data speed. Table 4.19m lists available signaling rates and associated pulse durations.

Data Encoding At fractional megabit speeds, a system called 4PPM is used. This designation is translated as 4 position pulse position modulation. Each frame is divided into four "chips" of fixed duration. A 4PPM frame is illustrated in Figure 4.19n.

TABLE 4.19l
Infrared Transmitter/Receiver Specifications

	Low Power to Low Power	*Standard to Low Power*	*Standard to Standard*
Link distance, lower limit, m	0	0	0
Minimum link distance, upper limit, m	0.2	0.3	1.0

TABLE 4.19m
Infrared Signaling Rates and Pulse Durations

Signaling Rate	*Pulse Duration*
2.4 kb/s	1.41 μs
9.6 kb/s	1.41 μs
19.2 kb/s	1.41 μs
38.4 kb/s	1.41 μs
57.6 kb/s	1.41 μs
115.2 kb/s	1.41 μs
0.576 Mb/s	295.2 ns
1.152 Mb/s	147.6 ns
4 Mb/s	115–240 ns

CHIP 1	CHIP 2	CHIP 3	CHIP 4

FIG. 4.19n
A complete 4PPM symbol.

PA	STA	DD (Information payload and error checking)	STO

FIG 4.19o
PPM packet format.

A complete data packet is made up of four distinct data fields (Figure 4.19o). The first field is the preamble (PA), which is broadcast to establish phase lock with the other station. The PA field consists of exactly 16 transmissions of a particular sequence of symbols (four chips, one symbol). The next symbol is the start (STA) flag, which, if received and interpreted correctly, enables symbol synchronization and demodulation of the data payload information. The STA field consists of one transmission of eight chips (two symbols) containing a particular bit sequence. Data delivery is accomplished in the DD field, which contains all of the transmitted data or commands to be transferred. The receiver continues to receive and interpret data from the DD field until the stop flag (STO) is received. The STO flag consists of one transmission of eight chips (two symbols) containing a particular bit sequence.

Configurations and Operating Characteristics Data links can exist in several different states and modes corresponding to different stages of connection and disconnection of the links. A link can exist in a *connection state*, in which two devices, or stations, are connected and exchanging information. A link can exist in a *contention state* in which there is no connection. Any time a device or station is not connected, it is said to be in the contention state.

All connections require two or more communicating stations. For purposes of control and connection setup, one of the communicating stations must be considered the *primary station*. The primary station assumes responsibility for connection setup and control; it transmits frames known as *command frames* containing the information necessary for the proper flow of data between stations. All communication is based upon the flow of data to and from the primary station. All other stations on the link are called *secondary stations*. Secondary stations receive instructions and data from primary stations and transmit *response frames* to the primary station. Only one station on a link can be the primary station. Although some stations are not designed to perform as primary stations, the ideal situation is for all devices to have both primary and secondary capability.

Two modes of operation characterize connections between devices. These are referred to as normal response mode (NRM) and normal disconnect mode (NDM), which correspond to device connection state and contention state, respectively. When the system is in NRM, two or more stations are connected and communicating information; when the system is in NDM, the stations are disconnected and ready for link setup. When in NRM, each station is set to the proper relationship, i.e., as a primary or secondary station. A typical link setup occurs when a primary station communicates with a secondary station; the secondary station receives explicit permission to respond, which it does by transmitting one or more response frames. The end of transmission will be indicated by the secondary station, which will then cease transmission until given further permission. The mechanics of connection setup and termination will be discussed further below, following a description of the various types of frames used by the system.

Frame Format An IrLAP frame is the basic unit of data communication used to transfer control information and data between stations. A typical frame consists of three separate fields:

1. An address (A) field that contains the address of the secondary station(s) the primary station wishes to communicate with
2. A control (C) field that specifies the function of the particular frame
3. An information (I) field that contains any information that may be transmitted; the I field is optional based on the function of the frame

The frame structure is similar to frames found in other protocols and is based on a sequence of bytes (8 bits). The general structure of the frame is shown in Figure 4.19p. Each frame is "wrapped" by information required for information management and for the establishment of the beginning and end of the frame. This information consists of, at a minimum:

ADDRESS	CONTROL	INFORMATION
8 BITS	8 BITS	8 × BITS

FIG 4.19p
Frame format.

- A start flag (STA) marking the beginning of the frame
- An error-checking algorithm to check the accuracy of the transmitted data
- A stop flag (STO) marking the end of the frame

Address Field Address field contents change depending on whether a primary or secondary station is transmitting the frame. If a primary station is transmitting the data, then the A field contains the address of the secondary station for which it is intended. If a secondary station is transmitting the frame, then the A field contains the address of the station from which it was transmitted. The field also contains a bit that indicates whether it is a command frame, sent by a primary station, or a response frame, sent by a secondary station; In addition to station addresses, two addresses are defined by the standards that are used only by primary stations:

- "00000000" is the NULL connection address and is not assigned to any secondary station.
- "11111111" is the global, or broadcast, address and is used as the address of all receiving stations. Only primary or stand-alone stations can use this address.

Control Field The control field in an IrLAP frame utilizes different formats to accomplish the different tasks required by the system for connection setup, link management, and information transfer. The possible formats (and corresponding frame types) are unnumbered (U), supervisory (S), and information (I). These format types are described below.

Unnumbered Format (U). Unnumbered frames are used for establishing and managing the link. This includes the discovery of secondary stations and subsequent activation and initialization of that station while managing the response mode of the secondary station. These frames are also used to log and report any errors, such as unrecoverable data loss. Information may also be transferred in an unnumbered frame if the information to be transferred does not require error checking. This information is usually in addition to the normal I field of the frame.

Supervisory Format (S). Supervisory frames are used to control the flow of information between stations, although these frames do not carry information themselves. Control fields of the S type report information such as busy or ready status of a station; frame sequencing errors resulting from incomplete or unreliable links; or for acknowledging the receipt of information. S frames may be used by primary stations to poll secondary stations and, conversely, by secondary stations to respond to a request for status or confirmation from a primary station. As previously stated, these frames do not contain an information field.

Information Format (I). Information frames contain information. This information is what must be moved from station to station. The information is unrestricted in size or content. Aside from indicating the format of the frame, the control field also tracks the send and receive counts to determine if frames are sent and received in the proper order. The send count, Ns, is used to ensure that the frames are sent in the proper order; the receive count, Nr, is used to ensure that the frames were received. Both I and S frames track sequence counts; unnumbered frames do not. The Ns count indicates the numbered position of the transmitted frame within the sequence of frames transmitted. The Nr count is the number of the frame that the transmitting station expects to receive next. If the incoming Ns does not agree with the Nr, then the frame is out of sequence.

Poll/Final (P/F) Bit The last piece of information required for the efficient transfer of information is the P/F bit. This bit occupies the 4th bit position in C, S, or I fields and, depending on station function, has different meanings. If the frame is transmitted from a primary station operating in normal response mode (NRM), then the bit is a polling bit and is used to solicit a response or a series of responses from a secondary station. If the frame is transmitted from a secondary station, then the bit is the final bit, used to indicate the final frame of the transmission.

Operating Procedures for an Infrared Link

To establish and maintain a reliable and error-free link between stations, certain procedures are followed according to the rules set down in the IrLAP protocol. The following section details the procedures for the orderly establishment and termination of an infrared link. The flowchart shown in Figure 4.19q illustrates the sequence of procedures in the proper order.

Procedure Descriptions *Link Start-up*: This procedure is defined as the event by which the station goes from an off-line condition, such as being powered off or having infrared capabilities disabled, to the online or enabled state. The station can transmit and receive frames at this point. The link is shut down when the station is disabled, shut down, or if the physical layer is down. This condition can exist if the atmospheric conditions are preventing the link from operating (such as bright sunshine) or if there is physical damage to the equipment.

Address Discovery: This procedure is used to determine the device addresses of all stations within communication range. The station performing the discovery procedure is called the *initiator*, and the stations that reply are called *responders*. Four pieces of information are returned to the initiator from discovered machines:

1. Device address, the 32-bit address of the discovered device
2. Discovery information, information about the key attributes and features of the device
3. Solicited/unsolicited, which indicates whether the information received was found by the initiator of discovery (solicited), or sent by the responder, which

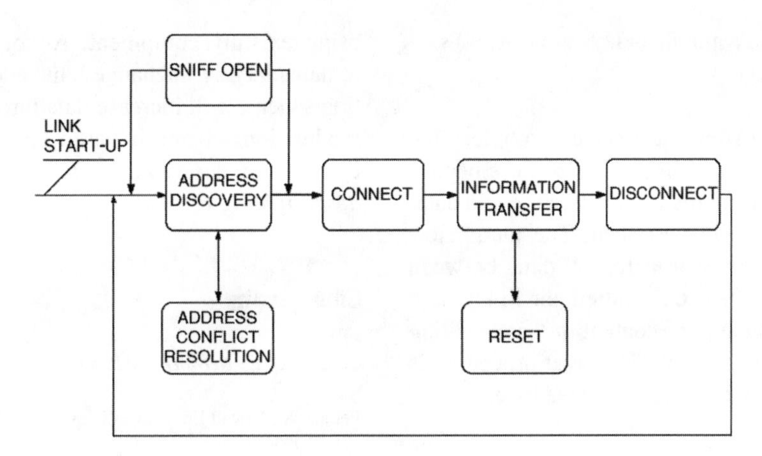

FIG. 4.19q
Infrared link operational procedures.

can discover the initiator from information in received frames
4. Sniffer/nonsniffer, which indicates whether a discovered device is a sniffer or nonsniffer (refer to the explanation below for further information)

Sniff Open: This procedure allows a device to broadcast its availability for connection in such a way as to conserve power. A sniffing device "wakes up" and listens for traffic for a short period of time. If no traffic is heard, the device transmits its availability to be connected as a secondary station. The information transmitted also identifies it as a sniffer. It then waits for a response from a primary station. If no traffic is transmitted back to the device, the device will "go back to sleep" for 2 to 3 s and then repeat the procedure. If frames are transmitted in response to its transmission, the discovered device then transmits the information described previously. If the device hears traffic not intended for it, it also goes to sleep.

Address Conflict Resolution: In the event that two or more devices respond to discovery with the same device addresses, this procedure serves to allow the initiator to resolve the conflict and guide the responders in selecting new device addresses. Address conflicts can also arise from a situation where a primary station is attempting to communicate with multiple devices and one or more share an address. The initiator transmits information to the shared address of the devices instructing them to select new device addresses and delineating a time slot for reinitiation of the links using the new addresses. Each responder then selects a new address and waits for the initiator to "rediscover" the device during the predetermined time slot. The responders then send a confirmation to the initiator. The new addresses become effective at the completion of the conflict resolution procedure.

Connection: Once the discovery process is completed and any conflicts are resolved, the devices establish a link. If the station accepts the connection, then there are seven basic parameters that must be agreed upon to establish the connection. This is accomplished through a *negotiation* process, which is defined in the protocol. These parameters are:

1. Baud Rate: The baud rate is the speed at which information is transferred between devices. Both devices must agree to the same baud rate. Available baud rates are between 2400 bps and 4 Mbps.
2. Maximum Turnaround Time: This is the maximum time that a station can hold the P/F bit. Along with baud rate, this parameter determines the maximum amount of information a station can transmit before relinquishing the connection to another station. This parameter is used by one station to indicate to another station how long it can transmit before turning the link around.
3. Data Size: This is the maximum number of data bytes allowed in each frame during a given connection. Data size is defined as all bytes in the I field of a frame. Data sizes range from 64 to 2048 bytes.
4. Window Size: This parameter determines the maximum number of unacknowledged I frames that can be received before an acknowledgment must be sent. Available window sizes vary between a minimum of one frame window to a maximum of seven frame windows.
5. Additional BOFs (Beginning of Frame): This parameter determines how many additional BOF flags are added to the beginning of the frame. Up to 48 additional BOFs can be added. Note that one BOF contains two STA flags.
6. Minimum Turnaround Time: This parameter deals with the time a receiver requires to recover from saturation from signals sent from a given transmitter. This time delay prevents the reception of the next frame before the last frame is fully received and acknowledged. Available time delays are between 0 and 10 ms.
7. Link Disconnect/Threshold Time: This is the time a station will wait without receiving valid frames before

disconnecting the link. Available delays are 3 to 40 s before device disconnect.

Information Transfer: After the devices complete the negotiation process, the information transfer process begins. Information is transferred using the frame structure and data-encoding techniques previously described. The parameters negotiated govern the orderly transfer of data between machines. When the last frame is transmitted, the link is shut down and the devices are put into the contention state, waiting for the establishment of another link. The reset procedure is used to reset a connection and can be requested by either the primary or secondary station.

Very Fast Infrared

A proposed standard defines a new modulation scheme that will enable infrared links to operate at speeds of 16 Mbps using existing equipment. Although the physical standard remains largely unchanged, the addition of the new modulation scheme will increase data throughput fourfold. The new modulation scheme is based on a scrambling/descrambling mechanism described as *frame synchronized scrambling/descrambling* (FSS).

Bibliography

Capano. D. E., *Network Cabling for Contractors,* New York: McGraw-Hill, 2000.

Feibel, W., Novell Encyclopedia of Networking, Alameda, CA: Sybex, Inc., 1995.

IEEE 802.11, Piscataway, NJ: IEEE, 1997.

Nemzow, M., *Implementing Wireless Networks,* New York: McGraw-Hill, 1995.

Westman, H. P., Ed., Reference Data for Radio Engineers, Indianapolis, IN: Howard Sams, Inc., 2001.

Software Packages

5

5.1 CONTROL LOOP OPTIMIZATION 672

5.2 DATA RECONCILIATION 687

5.3 SEQUENCE OF EVENT RECORDERS AND POST-TRIP REVIEWS 703

5.4 OPC SOFTWARE ARCHITECTURE 708

5.5 BATCH CONTROL STATE OF THE ART 714

5.6 PLANTWIDE CONTROL LOOP OPTIMIZATION 728

5.7 PLANTWIDE CONTROLLER PERFORMANCE MONITORING 749

5.8 THE "VIRTUAL PLANT," A TOOL FOR BETTER UNDERSTANDING 761

5.1 Control Loop Optimization

J. P. GERRY

DESIGN CONSIDERATIONS

How much does choice of a controller affect the end goal of good process control, at a reasonable cost? Surprisingly, in almost all cases, a well-designed PID (proportional/integral/derivative) controller is still the best. The key is a well-designed, properly operating, and well-tuned PID loop. Well-designed means that:

- The loop has been linearized.
- Interactions with other loops have been eliminated.
- Valves are properly sized and working properly.

Fuzzy Logic

Fuzzy logic is a buzzword that sounds very high-tech. In the process industries, it is rarely used and should only rarely be used. Perhaps one loop in 10,000 justifies the large amount of data collection and programming required to set up a fuzzy controller. The question is when to use it. The answer is only after all other attempts at controlling the process using a well-designed and optimized PID loop fail, *and* where there is a human operator who can somehow control the process. This human operator supplies the exhaustive amount of data for the fuzzy controller.

Model Predictive

Model predictive control is also a fashionable buzzword that takes on such other names as internal model control (IMC), pole-cancellation, Dahlin, or direct synthesis controller. But, is it any better than PID? The short answer is that model predictive is about the same as PID on loops that are dominated by dead time. On loops dominated by lags, PID is *much* better when load upsets occur. Model predictive is better on set point changes, assuming the process model is known exactly. However, in the process industries, it is usually the response to load upsets that is the more important.

Loop Dominated by Dead Time

This is the case where model predictive control has slightly higher performance than PID. The extreme case of a loop dominated by dead time is a pure dead time process. Figures 5.1a and 5.1b compare a model predictive controller (dashed line) to a PI controller (solid line). Figure 5.1a is the response of the loops to a step load injected at the output of the controller. Figure 5.1b is the robustness of the loops. Robustness plots are explained in connection with Figures 5.1u and 5.1v. Refer to that subsection for a detailed discussion of robustness. The response of a loop must always be compared along with its robustness, as these two trade off.

The PI controller has been tuned for optimal response:

$$100/PB = \text{controller gain} = 0.34/(\text{process gain})$$

$$I = 0.45\ (\text{Dead Time})$$

The tuning constant of the model predictive controller (closed loop response time) has been tuned to give robustness as good as a PI controller in most regions.

$$\text{Relatively robust closed loop time constant} = \text{Dead Time}/3.14$$

As one can see from Figure 5.1a, the two time responses are very similar. However, note in the robustness plot for the model predictive controller that the process dead time can either increase or decrease and the loop will be unstable. This type of tuning is probably unacceptable in most plants.

Loop Dominated by Lag

On loops dominated by lag, the PID controller is far superior to the model predictive. Figures 5.1c and 5.1d show load time response and robustness plots for PID (solid line) and model predictive (dashed line) controller on a process with gain = 1, dead time = 1, and lag time = 50. Figure 5.1c shows the PID controller performing much better than the model predictive. The integral of absolute error (IAE) for the PID controller is nine times better than the model predictive controller. (Refer to this section for a detailed discussion on IAE.) Both loops were tuned for as similar robustness as possible, as Figure 5.1d shows. However, note that, again, the robustness for the model predictive controller suffers. Again, if the process dead time were to decrease, the model predictive controller would become unstable.

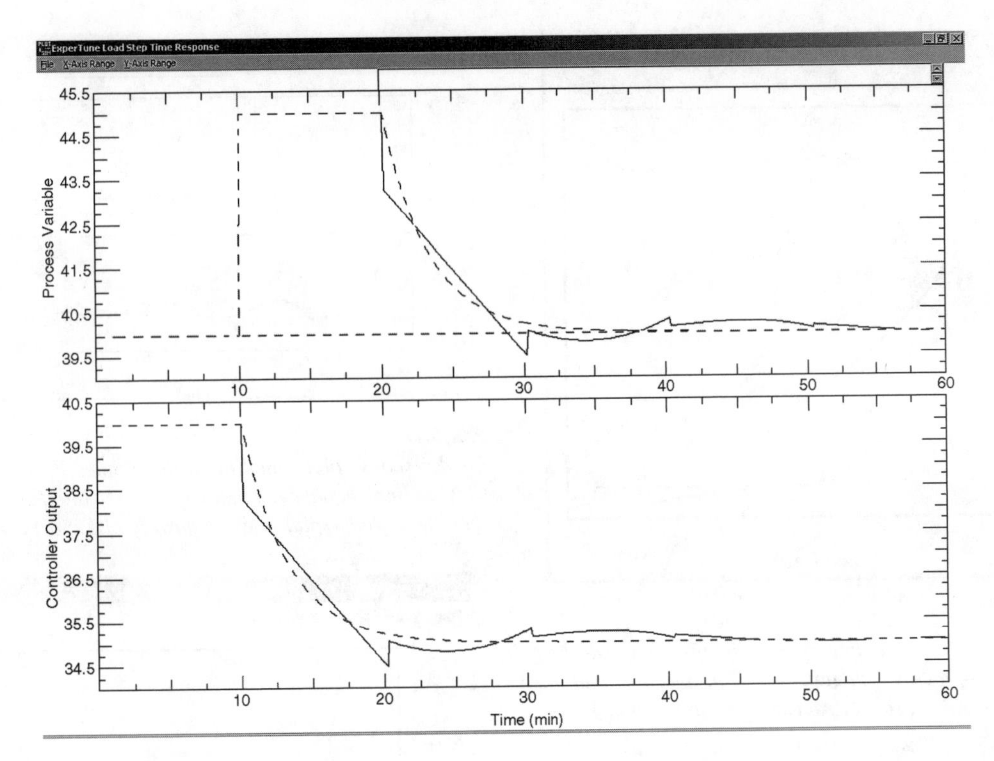

FIG. 5.1a
Load response of a PI controller (solid line) compared with the load response of a model predictive controller (dashed line) on a pure dead time process. (Courtesy of ExperTune, Inc.)

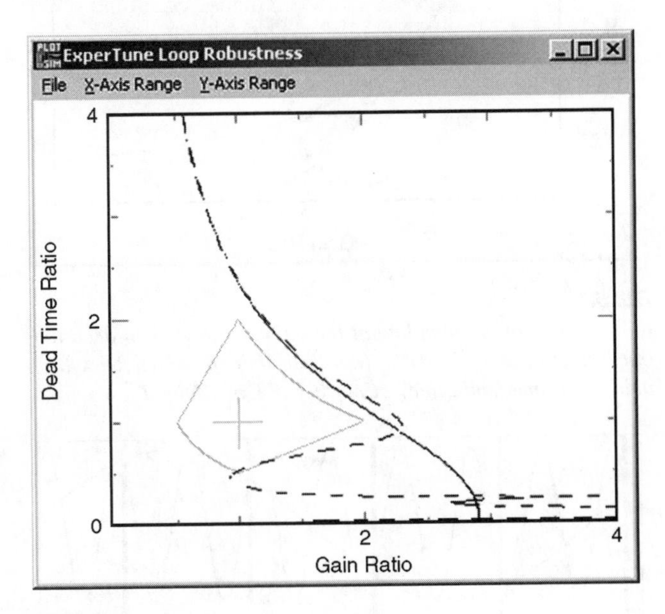

FIG. 5.1b
Robustness of a PI controller (solid line) compared with that of a model predictive controller (dashed line) on a pure dead time process. (Courtesy of ExperTune, Inc.)

Design Goal of Model Predictive Controllers

The cause for the poor response of the model predictive controller is rooted in its fundamental design. Model predictive controllers are designed to give good response to set point changes. This is achieved by canceling the lag with

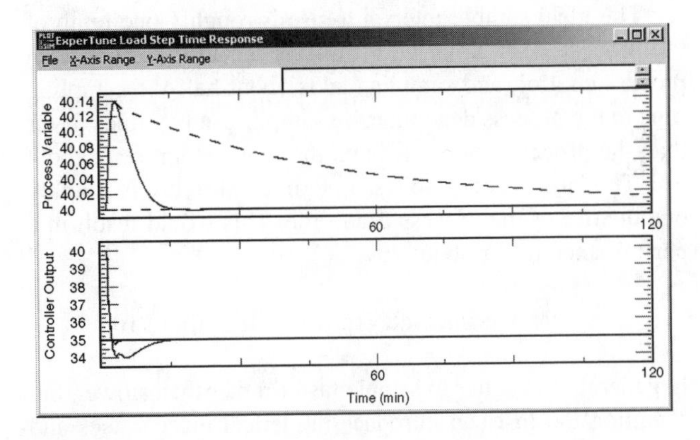

FIG. 5.1c
Load response plots for PID (solid line) and model predictive (dashed line) controller on a process with gain = 1, dead time = 1, and lag time = 50. Loops tuned for similar robustness. (Courtesy of ExperTune, Inc.)

the controller. The longer the lag time, the longer the effective integral action in the controller to cancel that lag. The result is little integral action in loops with large lags, resulting in poor response to load upsets compared with PID controllers, even though the robustness is the same.

Sample Interval

For optimal control, the sample interval of the controller must be properly set. This is independent of the controller algorithm chosen. If the controller is unaware of an upset until

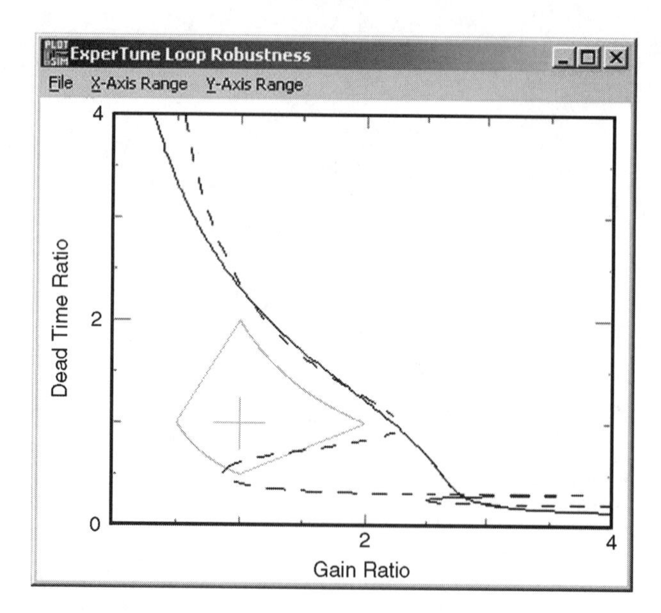

FIG. 5.1d
Robustness plots for PID (solid line) and model predictive (dashed line) controller on a process with gain = 1, dead time = 1, and lag time = 50. Loops tuned for similar robustness. (Courtesy of ExperTune, Inc.)

the sample interval and must wait for the sample interval to pass before it can act, the performance of the loop suffers.

The ideal sample interval to use is roughly one tenth of the process dead time. Dead time is the limiting factor in process control, and sampling adds about half the sampling time to the process dead time. So sampling at ten times faster than the process dead time hurts the loop response by about 5%. The largest acceptable sample interval to use is roughly one quarter of the process dead time. This would result in a performance hit of about 20%.

ideal sample interval = (process dead time)/10

In general, it is better to sample faster rather than slower, but sampling too fast can introduce numerical inaccuracies and can place unnecessary load on the computer.

Sampling Too Slowly

The effects of sampling too slowly can be seen graphically in Figure 5.1e, which shows the response of a control loop with three different sample intervals. With the dashed line, the sample interval is three times longer than the process dead time. This slow sampling hurts the response compared with the dot-dashed line, where the sampling is the same as the dead time. Using a sample interval equal to one tenth of the dead time results in the best response, as shown by the solid line in Figure 5.1e. The controller tuning in the first cases is the same. In the solid line where the controller was at the proper sample interval the tuning was adjusted to roughly match the robustness in the other two loops. Figure 5.1f shows the robustness plot for these cases.

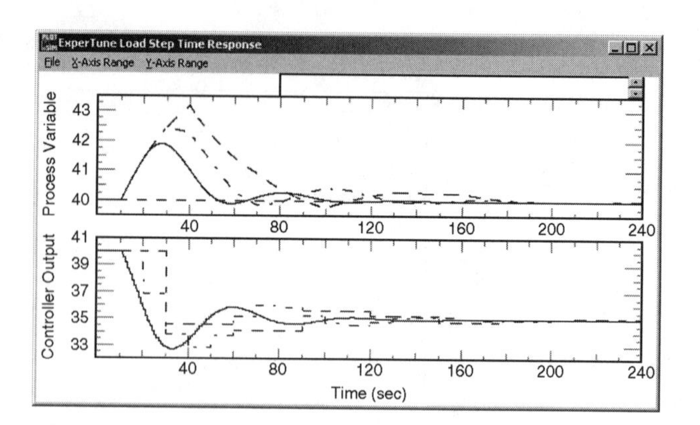

FIG. 5.1e
Load response plot sampling at three times slower than dead time (dashed line), equal dead time (dot-dashed line), and ten times faster than dead time (solid line). (Courtesy of ExperTune, Inc.)

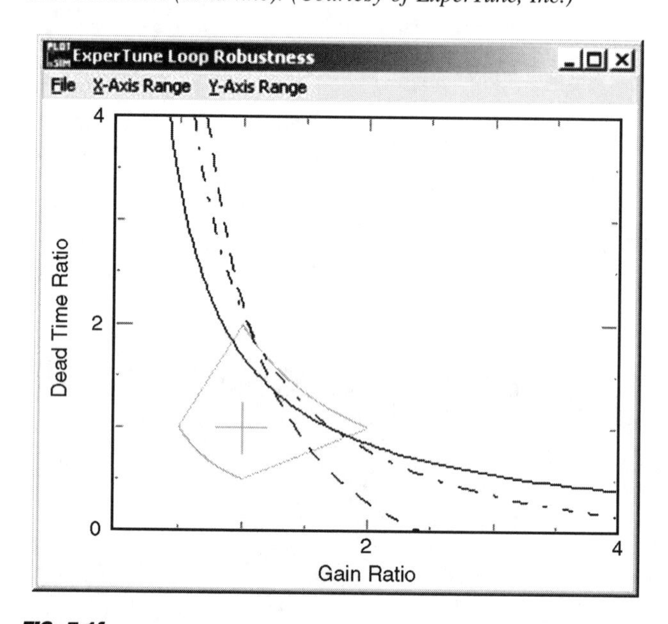

FIG. 5.1f
Robustness plot for sampling at three times slower than dead time (dashed line), equal dead time (dot-dashed line), and ten times faster than dead time (solid line). (Courtesy of ExperTune, Inc.)

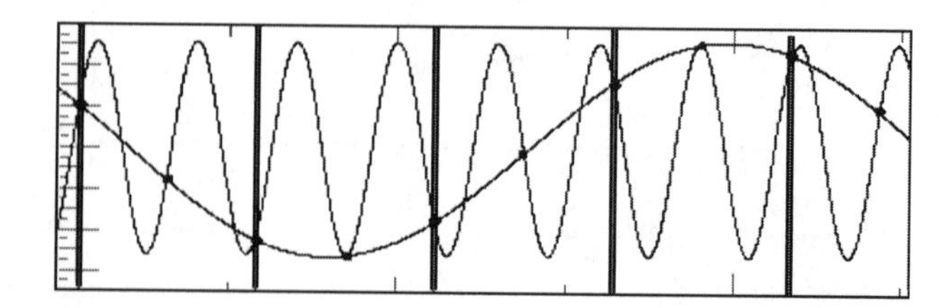

FIG. 5.1g
Aliasing caused by sampling. (Courtesy of ExperTune, Inc.)

Aliasing and Hypersampling

Whenever a signal is sampled, there is a risk of obtaining false aliasing signals. In Figure 5.1g, the signal is sampled at the vertical bars. The slow sine wave is an alias signal occurring as a result of sampling. It is not in the original signal. This phenomenon is called aliasing.

One approach to eliminate aliasing is to sample very fast, at frequencies 1000 times faster than the dead time. The digital signal is then filtered. This approach works for frequencies slower than the hyperfast sampling frequency but not for those faster than this. This approach also is fairly costly as more data must be processed in a short time.

A better approach to reduce aliasing is to use an analog filter on the signal before it is digitized. This is the approach used by most distributed control systems (DCS) and most industrial I/O (input/output). Some I/O has special filters to decimate noise in the 50 to 60 Hz range. This analog filtering solution is less expensive and more effective than the hypersampling approach. The analog filter almost completely decimates higher frequency signals that the hypersampling approach leaks through.

Digital Input Resolution

The resolution of input sensors is often overlooked, especially in analog-to-digital converters. Analog-to-digital converters affect the resolution by

$$1/2^n * 100\%$$

where n is the number of bits in the analog-to-digital converter. For example, an 8-bit analog-to-digital converter restricts your resolution to 1/2^n * 100 or 0.39%. An 8-bit analog-to-digital converter is usually not considered to have enough resolution in industrial control applications; 12- or 16-bit analog-to-digital converters are more common and give much better resolution. Loops with small input resolution can end up cycling between the bits of resolution.

Valve Sizing and Output Resolution

Valves also have a resolution or a smallest increment by which they can change, usually called the turndown ratio. This can also affect how the loop responds. A loop operating in a small output range can end up cycling between areas at which the valve can actually operate. A valve that is oversized will exacerbate any turndown problems.

PID Controller Algorithm

Every industrial PID controller is unique. Because descriptions and nomenclature differ among manufacturers—there are no adhered-to standards—there is no sure way to tell how a controller will respond and how to tune it without looking in detail at the equation for the algorithm. ExperTune, Inc. has a suggested standard for PID algorithms available at http://www.expertune.com/PIDspec.htm.

It should also be noted that such control loop optimization software as the ExperTune PID Tuner/Analyzer is aware of most of the nuances of each algorithm and takes these nuances into consideration in its analysis. The user simply chooses the industrial controller from a list.

Some industrial controllers use gain, whereas others use proportional band. Some use what they call proportional gain that is really proportional band.

$$\text{Gain} = 100/(\text{Proportional Band})$$

Integral action is often called reset. It can have the following units:

Minutes/repeat
Repeats/minute
Seconds/repeat
Repeats/second

Derivative action is in either minutes or seconds.

Generally, there are three types of algorithms used. In simplified form these are

$$\text{Ideal:}\quad \text{Output} = Kc\left[e(t) + \frac{1}{I}\int e(t)d(t) + D\frac{de(t)}{dt}\right]$$

$$\text{Parallel:}\quad \text{Output} = Kp[e(t)] + \frac{1}{I}\int e(t)d(t) + D\frac{de(t)}{dt}$$

$$\text{Series:}\quad \text{Output} = Kc\left[e(t) + \frac{1}{I}\int e(t)d(t)\right]\left[1 + D\frac{d}{dt}\right]$$

Again, manufacturers do not adhere to these terms. For example, some call "interacting" an algorithm that is the same as the series one above, whereas others call "interacting" an algorithm that is the same as the ideal one above.

Using Measurement or Error

Another difference in control algorithms is whether the set point is applied to the derivative or whether the set point is applied to the gain. If the derivative or gain operates on the error signal, then the set point is applied to it.

There is no difference in how the loop will respond to upsets, but the set point response will be very different. It is almost always good practice not to have the set point applied to the derivative. Use derivative on measurement or derivative on process variable (PV) based algorithms instead of derivative on error.

For smooth set point response, one may want to have the inner loop of a cascade use proportional band on measurement instead of error.

Derivative Filter or Limit

Because the derivative uses a change or slope in the PV signal, it amplifies noise. Without filtering or limiting, derivative action can be punishing for the final control element. If the anti-reset windup circuitry of the controller is not properly designed and the derivative is not limited, using it can have a catastrophic effect on the loop.

Some industrial controllers do not have a derivative gain limit. Do not use derivative action in these controllers. Many others limit derivative gain to between 8 and 10.

Derivative action can have a stabilizing effect on the control of a loop, *if* it is set properly. On every loop there is only a small range of D values that are appropriate. Too little D and the output becomes noisy with no benefit. Too much D and the loop is unstable.

Single Value Performance Measures

Integrated absolute error to a step load upset at the controller output is a very good measure of the performance of regulatory control loops. The IAE can be tied to economic benefits.

A conservative performance measure of the robustness of a loop is the closest distance from the cross in the robustness plot to the marginal stability line. This measure assumes that gain and dead time are equally likely to change. The distance represents the least amount of process change to drive the system unstable.

Percentage Improvement Performance Indices

A quick numerical measure of the robustness or performance compares the current condition to a proposed one. In this case, the percent increase (or decrease) in performance is defined by

$$\text{(Current} - \text{Proposed)/Smallest} * 100 \qquad \mathbf{5.1(1)}$$

where

Current = current performance value
Proposed = proposed performance value
Smallest = the smaller of Current or Proposed

This method yields a useful percentage value. For example, a good measure of performance is the IAE; the current tuning values result in a current IAE and proposed tuning values may result in a proposed IAE.

Here is an example of performance indices:

Current IAE	*Proposed IAE*	*% Improvement in Performance*
100	50	100
100	200	−100
100	20	400
100	500	−400
100	80	25
100	125	−25

In the first row, the performance would increase by a factor of two; hence, the improvement is 100%. In the second row, the performance decreased by a factor of two; hence, a worsening by 100%. Note that the percent increase and decrease were the same even though the difference between current and proposed was not the same. This same principle holds for the other examples. In each case, the differences between the current and proposed IAEs are unequal, yet the percent increase or decrease is the same.

Equation 5.1(1) can be applied to the robustness, yielding the same kind of performance measure. In this way, robustness and performance can be compared quickly by looking at two numbers. These are not a substitute for viewing the complete robustness plot and time response plot, but they can help to arrive more quickly at tuning or optimization values worth additional consideration.

General Robustness and Performance Guidelines

It is important when optimizing and tuning a control loop to consider robustness as a trade-off with performance. Knowing the trade-offs allows the designer to exercise good engineering judgment in the design of the system.

Robustness plots are useful for examining the effects of optimization on sensitivity to process changes. Percentage improvement performance indices can provide a quick, rough estimate of the trade-off between performance and robustness.

STEPS TO OPTIMIZE A CONTROL LOOP

The first step in optimizing a loop is to make sure the loop is sampled fast enough, and has enough input and output resolution as discussed earlier.

Of the steps discussed below, not all may be necessary to perform on the loop. This is meant to be a thorough testing and optimization and not all the steps may be required. One can use good engineering judgment to determine which ones to perform. Notice that finding optimal PID tuning is the last of these steps.

Performance Objectives

Before optimizing a loop, one needs to know what optimized means on this particular loop. Most of the time in the process industries the loop is used for regulatory control and optimized means good recovery from upsets. However, on some loops valve movement or controller output movement is a consideration. On still others, such as level loops used for surge control, good regulatory response is not an optimal condition.

Check with operation personnel to glean what is known and to let them know of the desire to improve the loop response. If possible, go in to the field and look at the valve and sensor.

Interaction with Other Loops

Is the loop cycling? If so, is this because of interactions with other loops? To test this, put the loop in manual. If the cycling continues, the problem is most likely from an external source. Sometimes, there may still be a cycle present, but it is hidden by noise.

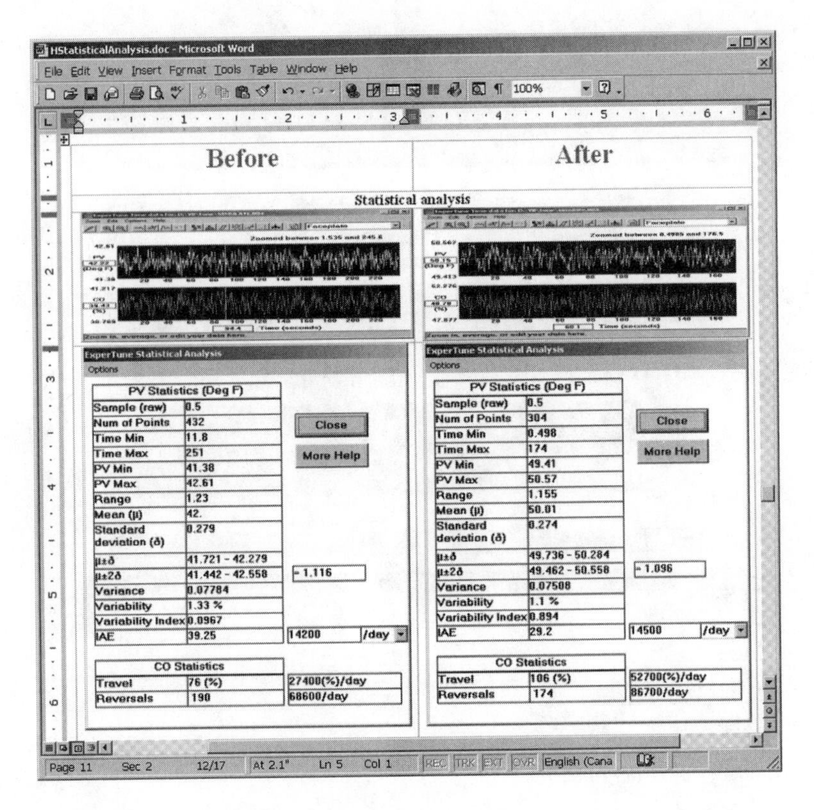

FIG. 5.1h
Statistical analysis of loop performance. (Courtesy of ExperTune, Inc.)

To uncover a hidden cycle, collect data with the loop in manual; then, using analysis software, perform a power spectral density analysis on the data. Hidden cycles will show as peaks.

The power spectral density separates out all the sine wave frequencies present in the signal.

Statistical Analysis

For a before-and-after comparison, collect some normal operating data with the controller in automatic. The statistical analysis gives many valuable indications for how the loop performance improves, including variance, variability, variability index, valve travel and reversals over time, current IAE, etc. Figure 5.1h shows a typical statistical analysis on a loop.

Valve Stiction Tests

Quite often when a loop is causing problems, the trouble is with the valve, valve actuator, or valve positioner. Testing the valve should almost always be done. In many cases, when the stiction and hysteresis in the valve are eliminated, the cycling in the loop disappears.

Stiction is a combination of stiction and friction. Consider attempting to move a large object. One constantly applies more force until suddenly the object breaks free. This sticking and slipping is called stiction. Stiction can be an insidious problem in control loops that will almost certainly result in cycling and variability in the end product. In most loops, stiction over 0.5% is too much.

Here is how to test for stiction:

1. Put the loop in manual.
2. Make a large (5% or so) controller output change.
3. Wait for the loop to settle.
4. Make a small output change of 0.1 or 0.2% in the same direction as step 2.
5. Repeat steps 3 and 4 until the PV moves as a result of the change that was made in the controller output (CO).

The stiction is the total amount of CO changes from step 4, minus roughly one change. Figure 5.1i shows a report section of a stiction check on a flow loop. This report is from Microsoft Word and was created with ExperTune analysis software.

Hysteresis Check

Hysteresis in valves is also a cause of cycling in control loops. Checking the hysteresis is recommended, along with the stiction check. Hysteresis is sloppiness or looseness in the valve. With hysteresis, the air signal to the valve will not move the valve until the air overcomes the hysteresis. Once overcome, the valve will continue to move in that direction if the air signal stays in that direction. Once the air signal switches to the opposite direction, the valve again will not move until the hysteresis is overcome.

If the hysteresis is more than 1% for valves with positioners or 3% for valves without positioners, then reducing the hysteresis by repairing the valve will improve control.

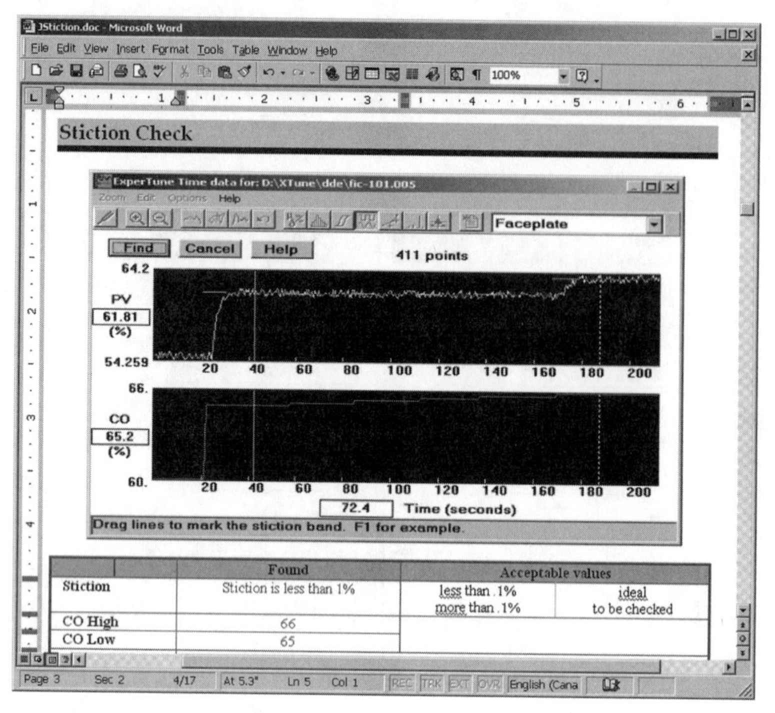

FIG. 5.1i
Example of a stiction check. (Courtesy of ExperTune, Inc.)

Here are the steps to check for hysteresis:

1. Put the loop in manual. Wait for the loop to settle.
2. Make a large change (5% or so) in the CO. Wait for the loop to settle.
3. Make a large change (5% or so) in the CO in the same direction as step 3. Wait for the loop to settle.
4. Make a large change (5% or so) in the CO in the opposite direction as the previous steps. Wait for the loop to settle.

The hysteresis is calculated with the following formula:

$$\% \text{ hysteresis} = C3 - P3 * C2/P2$$

Figure 5.1j shows a report created in Microsoft Word with ExperTune analysis software. Figure 5.1j diagrams where to obtain the data for the above formula.

Linearity and Characterization

If the loop cycles sometimes while others do not, the loop could be nonlinear. The loop in Figure 5.1k is nonlinear. At the low end of the range, the loop is stable; toward the upper end of the range, the loop cycles. If the loop is detuned, then it will be sluggish toward the lower end of the range.

In a nonlinear loop, there are three choices for control:

1. Characterize the nonlinearity and apply a characterizer to the loop. This effectively linearizes the control loop and is the preferred solution.
2. Detune the loop so it is stable over the entire range. This solution will not result in optimal control because the loop will be sluggish in the low-gain area of the loop.
3. Apply an adaptive controller to the loop. This is more complicated and can result in an unstable solution.

Of the three solutions above, the first is the best choice. The second results in sluggish response in the low-gain area. The third, applying an adaptive controller, is risky because:

1. A cycle may cause the adapter to detune the loop unnecessarily.
2. If the loop moves to a different gain area quickly, the adaptive controller will not be able to keep up, causing either sluggish or unstable control.
3. Adapters are usually difficult to set up and for the above two reasons must be monitored closely.

By applying a characterizer to the loop, one is assured of having the proper characterizer gain in the loop all the time.

The nonlinearity in the loop can be dependent on the output or it could be dependent on the measurement. These two different nonlinearities require two different solutions. Most loops are nonlinear with the output. In these cases, when the output is in a different region, the gain of the loop will change. Most flow loops are somewhat nonlinear this way. If the nonlinearity is in the measurement, then the gain of the loop will be different depending on where the measurement is. pH loops are a classic example of a loop whose gain depends on the measurement.

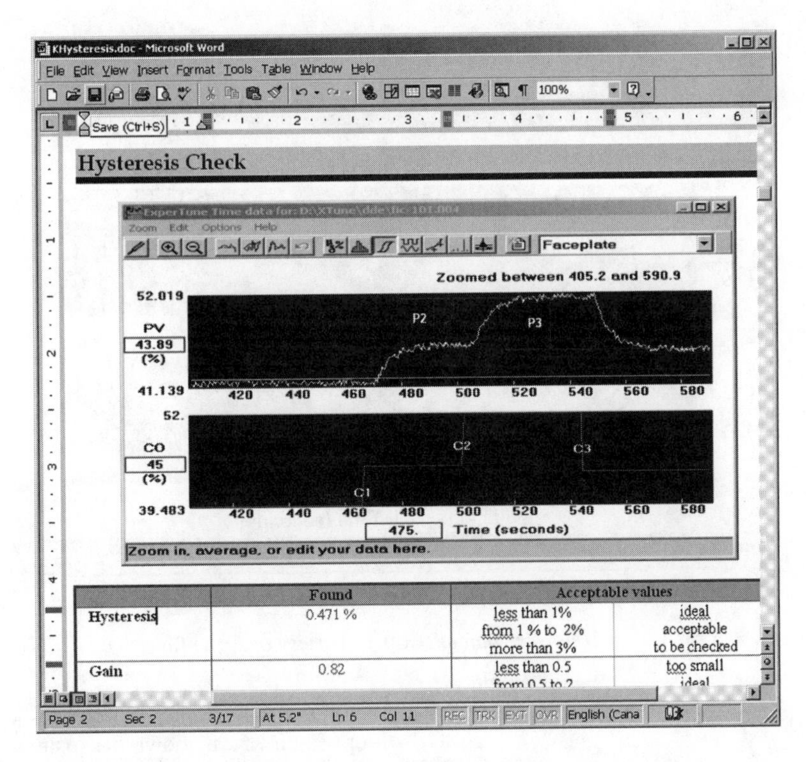

FIG. 5.1j
The test required for calculating the hysteresis in a control valve. C2 = second CO change for hysteresis check; C3 = third CO change for hysteresis check; P2 = distance between second and third steady-state areas for hysteresis check; P3 = distance between third and fourth steady-state areas for hysteresis check. (Courtesy of ExperTune, Inc.)

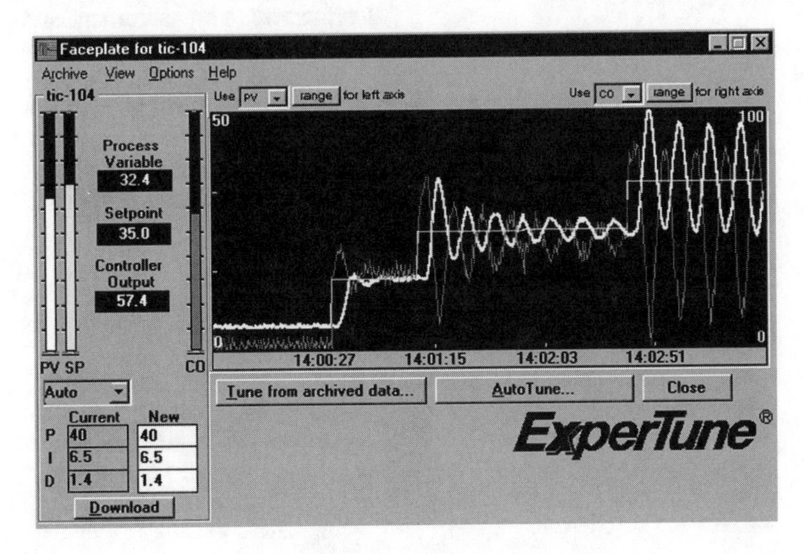

FIG. 5.1k
Set point responses in a nonlinear flow loop, whose gain varies with flow. (Courtesy of ExperTune, Inc.)

Nonlinearity in the Output and Characterization

With a loop that is nonlinear dependent on output, one needs to collect settled data at different operating locations. For example, Figure 5.1l has data collected from a flow loop while the loop is in manual mode. Figure 5.1m shows a plot of the PV vs. CO for these settled areas. Here one can again see the nonlinearity: a change in CO at the low end of the range causes a small change in PV, but the same change in CO at the high end of the range causes a large change in the PV. Also, the points are not in a line; hence, they are nonlinear.

Finding an appropriate characterizer to linearize the loop using these data is not a straightforward process using pencil and paper. However, using analysis software, it is simple. For example, ExperTune analysis software finds the characterizer in BASIC, FORTRAN, C, XY pairs, or structured text. XY pairs are often used by DCS systems. The analysis software

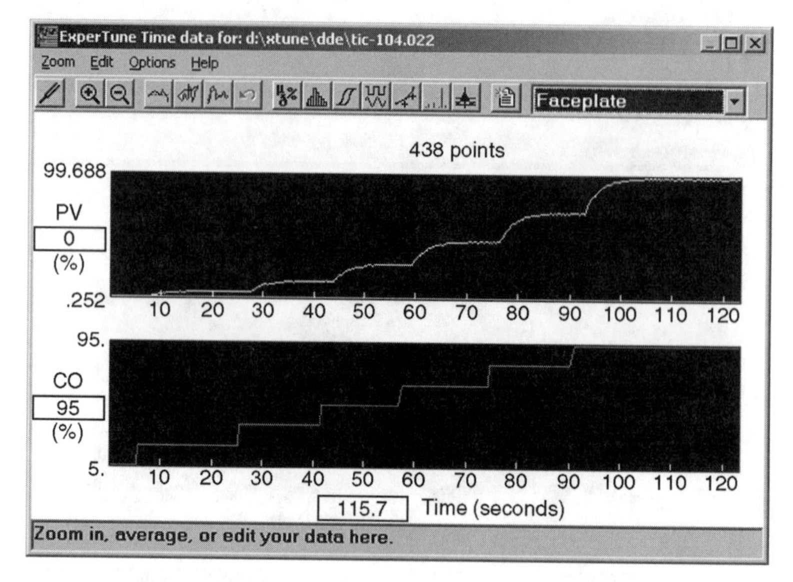

FIG.5.1l

Nonlinear flow loop response to step changes in controller output (CO). (Courtesy of ExperTune, Inc.)

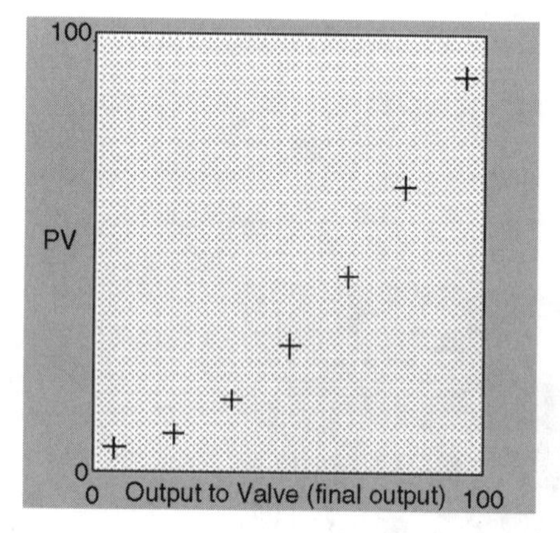

FIG. 5.1m

Response of a flow loop. PV vs. controller output (CO). (Courtesy of ExperTune, Inc.)

will optionally find a hyperbolic equation to linearize the loop. Figure 5.1n shows the resultant characterizer.

To linearize the loop, the output of the controller is passed through the characterizer and then out to the valve or final control element as shown in Figure 5.1o.

Nonlinearity in Measurement and Characterization—pH

To linearize a loop that is nonlinear with the measurement, the first step is to characterize the nonlinearity. A curve of gain vs. measurement is needed. With the most common form of these loops, pH, a titration curve is needed. This is usually readily available from a laboratory. The titration curve defines the nonlinearity.

The curve is entered into the ExperTune analysis software that designs the characterizer. Because the nonlinearity is

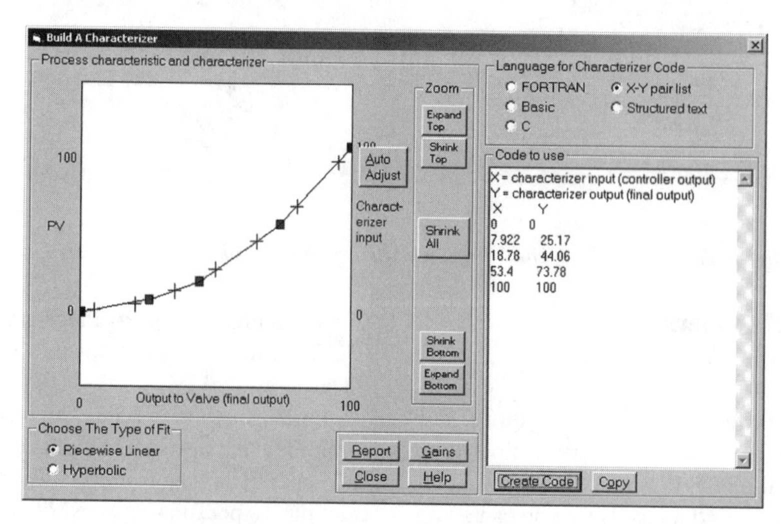

FIG. 5.1n

Characterizer for linearizing a nonlinear flow loop. (Courtesy of ExperTune, Inc.)

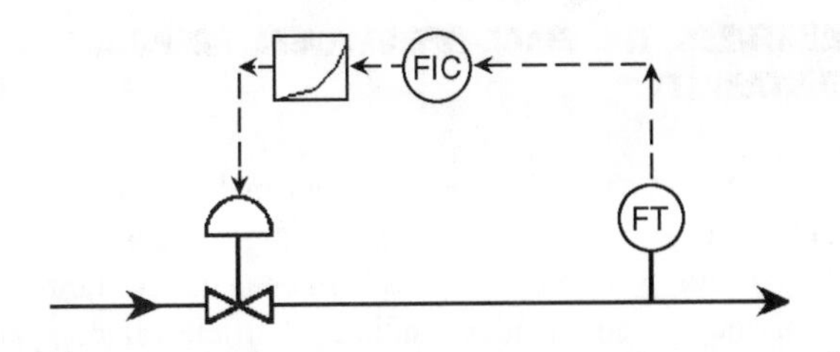

FIG. 5.1o
Output characterizer in a flow control loop. (Courtesy of ExperTune, Inc.)

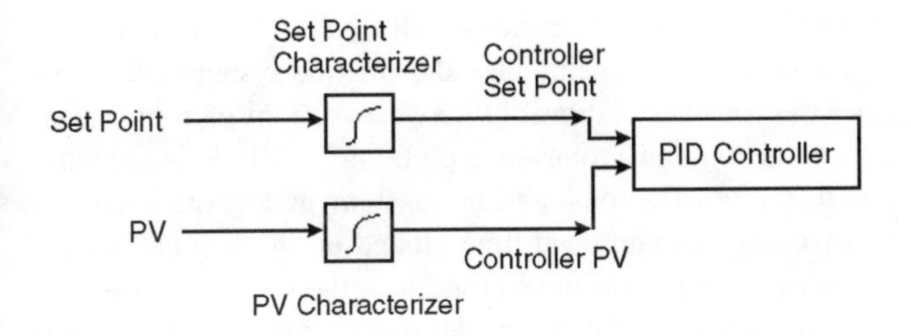

FIG. 5.1p
When the measurement is nonlinear, the characterizer must be applied to both the set point and the PV signals. (Courtesy of ExperTune, Inc.)

with the measurement, the characterizer must be applied to the measurement signal. So, the characterizer must be applied to both the measurement and the set point, as shown in Figure 5.1p. The set point for display and for the operation personnel should be the value before the characterizer. After characterization, the signal will have no physical meaning to operation personnel.

Figure 5.1q shows an example of the characterized pH signal.

PID Tuning and Filter Selection

The last step in optimizing a control loop should be finding the optimal PI or PID tuning parameters and best PV filter for the control loop.

The controller can be tuned for good rejection of upsets or for good set point response. By using a set point filter, these two goals are not mutually exclusive. However, if a typical PID loop is tuned for great set point response, the resultant rejection of load upsets may suffer.

PID Tuning for Load Rejection This is the most typical case in the process industries. Most of the time the aim is simply to keep the loop running smoothly. The most useful index to optimize toward is the IAE to a step load change in the controller output. Step loads at the output of the controller are the extreme case of load upsets in a loop. IAE is a measure that can be most closely related to economic benefit in the control loop.

A poorly tuned loop results in product that is off specification, which requires rework and increased cost. To stay on specification with a poorly tuned loop, the product is run richer than it needs to be. This unnecessarily richer product costs profits. With tuning optimized for IAE one gives away less while staying on spec.

PID Tuning for Set Point A simple technique for PID tuning based on pole-cancellation or the Dahlin controller is set point or lambda tuning. The goal of lambda tuning is to obtain smooth set point response with the closed-loop shape of response of a first-order lag with time constant equal to lambda. Lambda is often set equal to the lag time in the loop. Table 5.1r lists the equations for working out the lambda tunings on a common first-order plus dead time process.

This "lambda" tuning or pole-cancellation technique suffers from the same design problem and hence the same response problem as the model predictive controllers: it has poor response to load upsets. Figure 5.1s shows the load upset response of a PID controller tuned for load (solid line) and one tuned for set point response (dashed line). Figure 5.1t shows the corresponding robustness plot. As can be seen, the tuning optimized for load rejection was just as robust as the sluggish tuning optimized for set point response.

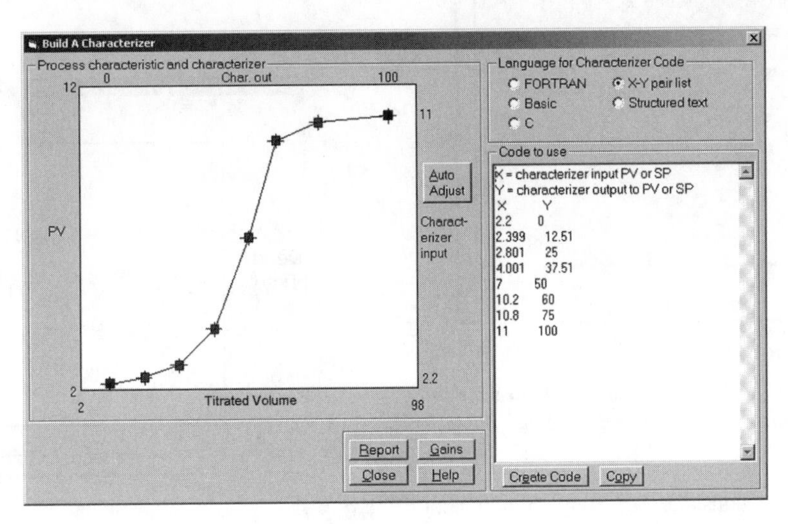

FIG. 5.1q
An example of a characterized pH signal. (Courtesy of ExperTune, Inc.)

TABLE 5.1r
Settings for a PID Controller If Smooth Set Point Response Is Desired. (Courtesy of ExperTune, Inc.)

Controller Type	Controller Gain (no units)	Integral Time (seconds)	Derivative Time (seconds)
PI control	$\frac{\tau}{K(\lambda+\theta)}$	γ	Not applicable
PID control	$\frac{\tau}{K(\lambda+\theta/2)}$	γ	$\theta/2$

τ = process time constant (seconds); θ = process dead time (seconds); K = process gain (dimensionless); λ = desired closed loop time constant (often set to loop lag time).

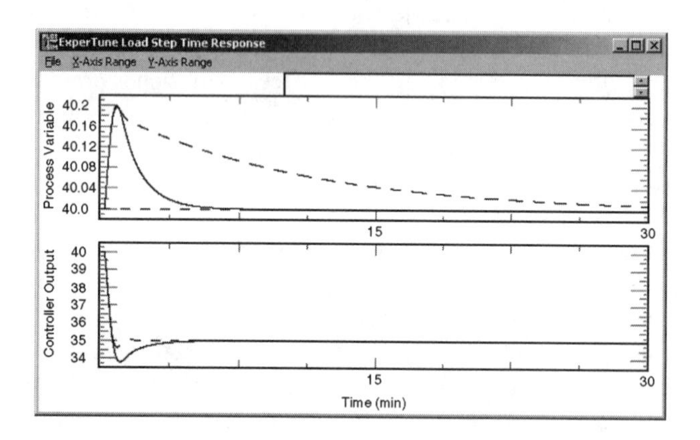

FIG. 5.1s
Load upset response of loop optimized for load (solid line) and set point response (dashed line). (Courtesy of ExperTune, Inc.)

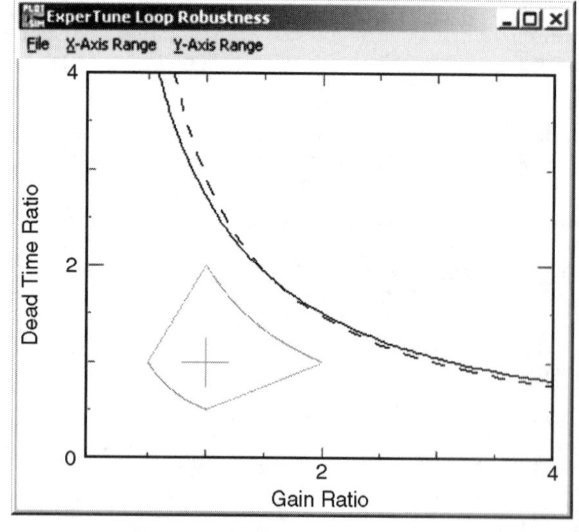

FIG. 5.1t
Robustness for loop optimized for load response (solid line) and for set point response (dashed line). (Courtesy of ExperTune, Inc.)

The PV filter is sometimes considered the fourth tuning parameter in a control loop. Proper setting of the PV filter can reduce valve wear while not compromising the response of the loop.

ROBUSTNESS, THE TRADE-OFF BETWEEN RESPONSE AND STABILITY

There is always a trade-off between optimal performance and sensitivity to process or controller changes. This sensitivity of the control system to system changes is called the robustness of the system. Understanding the trade-off of system robustness and optimal controller performance is paramount to achieving good system design.

The system robustness is quantifiable and can be shown in a graphical format. A graphical representation makes it easier to examine the trade-off with optimal performance.

One graphical method for showing the system robustness is based on the relationship between system gain and dead time and is called a robustness plot. Figure 5.1u is an example of this plot. The plot has the system gain plotted on the horizontal axis and dead time plotted on the vertical axis.

A universal robustness criterion is the limit when the system becomes marginally stable. A marginally stable system would have a time response showing sustained continual sine wave–shaped oscillation. At various values of system gain and dead time, the system will be marginally stable. These values can be calculated and plotted as a line or curve on the robustness plot. This marginal stability curve is shown in Figure 5.1v.

The lower right area of the plot has the smallest system gain and dead time. Hence, the lower right area of the plot represents a stable area. Conversely, the upper left area of the plot represents a less stable area. Areas to the above right of the marginal stability curve represent an unstable space. Areas below and to the left of the marginal stability curve represent a stable system space.

The current system gain and dead time can be pinpointed on the robustness plot. This is shown as a cross on the plot in Figure 5.1v. To be able to compare different plots easily, the gain axis and dead time axis of the plot are adjusted so that the current system gain and dead time (the cross) occur on the same location on every plot.

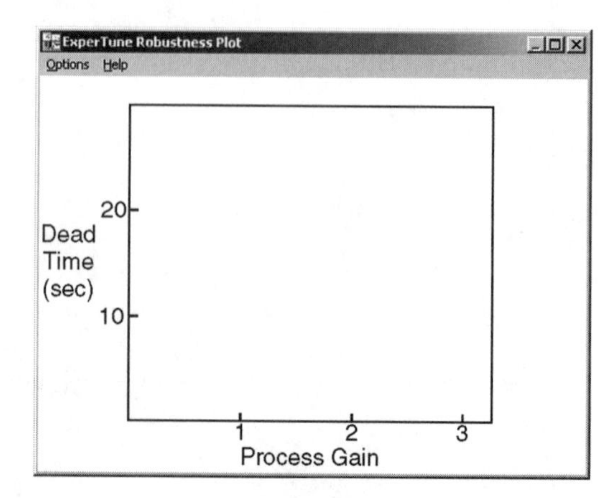

FIG. 5.1u
The robustness plot shows the values of system dead time as a function of process gain. (Courtesy of ExperTune, Inc.)

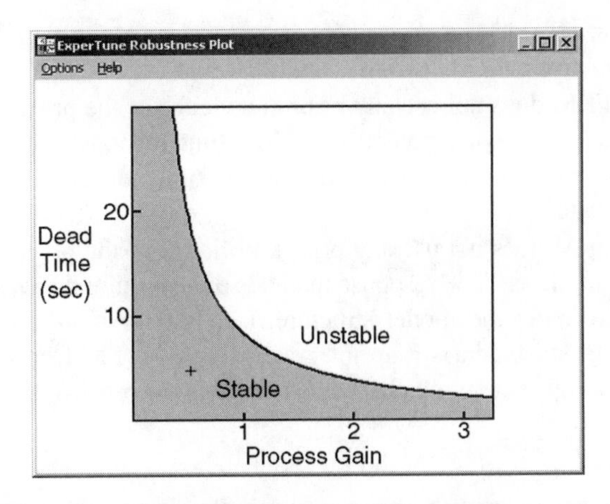

FIG. 5.1v
The marginal stability line. (Courtesy of ExperTune, Inc.)

If the marginal stability curve were to go through the cross, the system would be marginally stable.

Dead Time, Lags, and Controller Parameters

Dead time is the limiting element in speed of performance and robustness in control loops. Almost all process industry control loops contain dead time or enough small lags that combine to form characteristics similar to dead time in the control loop. Hence, it makes sense to use gain as one axis of the robustness plot.

System gain is a factor of the process gain and controller gain. Changing either the process gain or controller gain will shift the marginal stability curve either to the left or right. The shape of the curve remains the same with changes in system gain. An increase in system gain will shift the curve toward the left, moving it closer to the cross, taking the current system closer to instability.

The frequency of sustained oscillation on the marginal stability curve increases as one moves lower and to the right on the curve. Lags in the system will change the shape of the marginal stability curve. Large lags and integration times affect low-frequency stability, and hence the upper left portion of the curve will change the most. Changing high-frequency components of the controller like derivative action will change the shape of the lower right portion of the curve the most.

Guidelines

A factor of two is often considered a reasonable safety margin in process control systems. The four-sided object in Figure 5.1w has corners as a factor and divisor of two in both the gain and dead time axis. The top corner of the object represents twice the system dead time with the gain unchanged and the bottom corner of the object represents half the system dead time. The right corner of the object represents twice the system gain with the dead time unchanged. The left corner represents half the system gain. The corners of the object are connected with lines that on a log–log plot would appear to be straight.

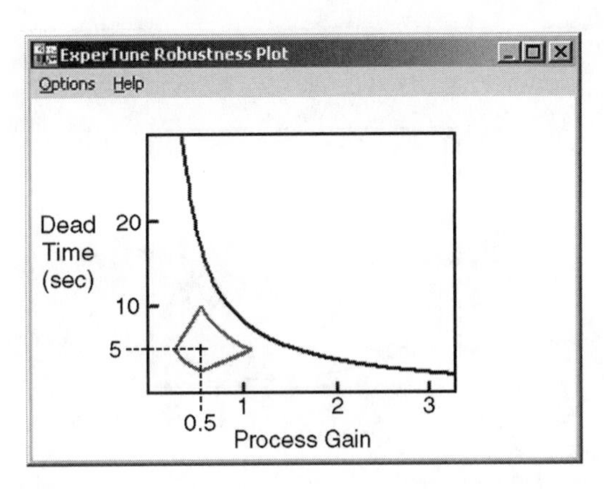

FIG. 5.1w
A full robustness plot includes the area within which the stability margin is two in any direction. (Courtesy of ExperTune, Inc.)

With a design goal of a stability margin of two in any direction, the inside area of the four-sided object represents a region to keep out of. If the line were inside the object, the stability margin would be less than two. The idea behind the four-sided object is as a design aid only and is not absolute for any system.

An Example

An example helps to understand how to use a robustness plot. In Figure 5.1w, the current system has a dead time of 5 s and an overall system gain of 0.5.

If the dead time of the system remained constant while the gain of the system increased, this would be represented by moving horizontally from the cross toward the right. If the system gain were to increase by a factor of three (to about 1.5), the system would become marginally stable when it encountered the marginal stability line.

If the gain of this system remained constant while the dead time of this system increased, this would be represented by moving vertically up from the cross. If the system dead time were to increase by a factor of three (to about 15), the system would become marginally stable when it encountered the marginal stability line.

If both the gain and dead time of the system increased, this would be represented by moving diagonally toward the upper right. With enough of an increase, the system would become unstable when the marginal stability line would be encountered.

Figures 5.1x and 5.1y show robustness plots and time response plots of a system to an increase of the controller gain to the point of instability.

RELATIVE RESPONSE TIME

The relative response time (RRT) is a useful parameter for decoupling interacting loops and for setting the speed of loops operating together. Use of the RRT is discussed in more detail in Section 5.6.

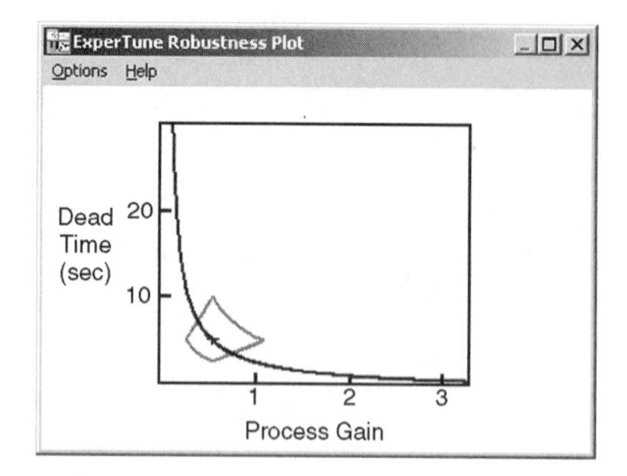

FIG. 5.1x
Robustness plot showing loop at the point of instability. (Courtesy of ExperTune, Inc.)

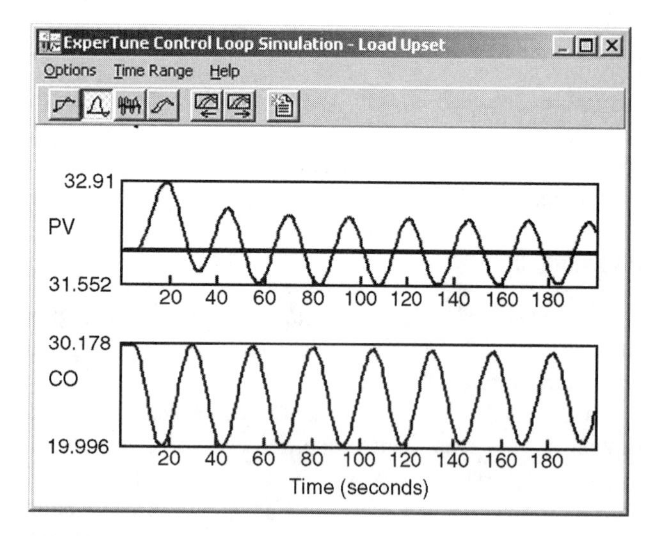

FIG. 5.1y
Time response plot showing loop at the point of instability. (Courtesy of ExperTune, Inc.)

The RRT is the period at the peak amplitude ratio in the closed-loop frequency response to loads. For example, Figure 5.1z shows the closed-loop frequency response to a load upset at the controller output. From this frequency response, the frequency at the peak amplitude ratio is roughly 0.0677 cycles/s. Convert this frequency to a period by taking its inverse or $1/0.0677 = 15$ s. Thus, the RRT for this loop is 15 s.

ADVANTAGES OF FREQUENCY RESPONSE METHODS

Some analysis software uses frequency response methods (FRM) to obtain the best PID tuning and process model. FRM provides several advantages over other methods:

1. FRM requires only one process bump to identify the process. The bump can be a change in manual, either a pulse, step, or other bump, or the bump can be made in closed loop. A set point change provides excellent data for FRM.
2. FRM does not use any prior knowledge of the process dead time or time constant. With time response methods one often needs dead time and time constant estimates.
3. FRM does not use any prior knowledge of the process structure. Time response methods often require the user to know the model structure, i.e., is it first order or second order, is it an integrator, etc. With FRM, none of this is required; all that is needed is the process data.

EXPERTUNE TUNING TOOLS WITH INTEGRATED, COHESIVE WINDOWS

Advanced software tools, such as the one from ExperTune, Inc., neatly combine the above optimization steps into one cohesive integrated package.

PID Tuning and Filter

ExperTune provides optimal tuning in five categories—three categories of tuning for load rejection and two for set point tuning. Figure 5.1aa shows a window from ExperTune with all the settings. The load rejection categories include:

1. Optimal or minimum IAE to a load upset
2. 25% overshoot to a step load upset
3. 10% overshoot to a step load upset

A safety factor can be set to make the tuning more conservative or aggressive. A safety factor of one is optimal control, five is backed off by factor of five, and the default safety factor is 2.5. The optimal or minimum IAE is the most useful category. The second and third categories exist for historical reasons only. The second category, or 25% overshoot, is closest to Ziegler–Nichols values, but will outperform Ziegler–Nichols on loops dominated by dead time or second-order processes. The first category or minimum IAE will outperform the second and third on all loops.

The two categories of set point tuning both use the pole-cancellation or lambda method:

1. Set point tuning with closed-loop time constant optimized for fast response
2. Set point tuning with the closed-loop time constant set equal to the lambda time

It should be noted that by using a set point filter, both good response to set point and good load rejection are achievable in the same control loop.

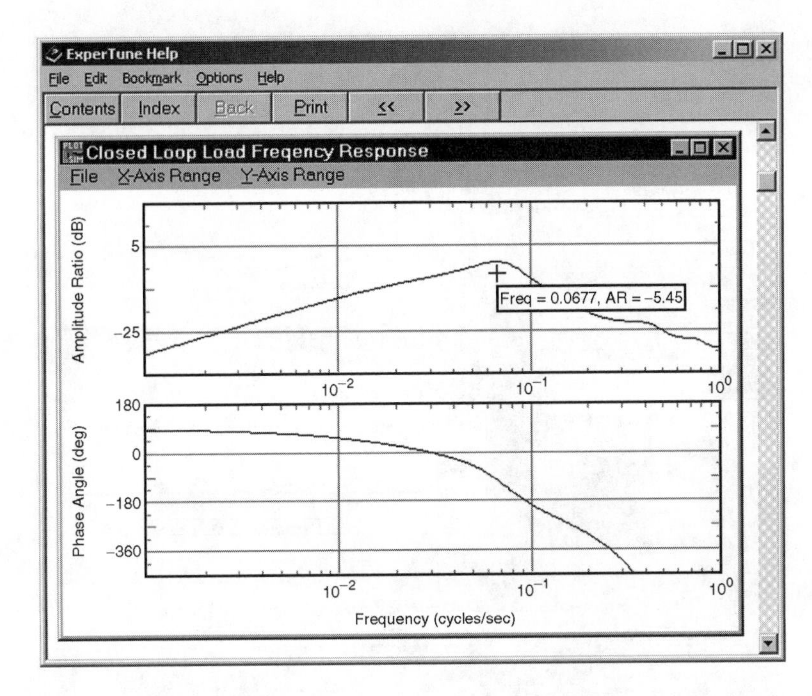

FIG. 5.1z
Closed-loop frequency response to a load upset, diagramming the source of the RRT. (Courtesy of ExperTune, Inc.)

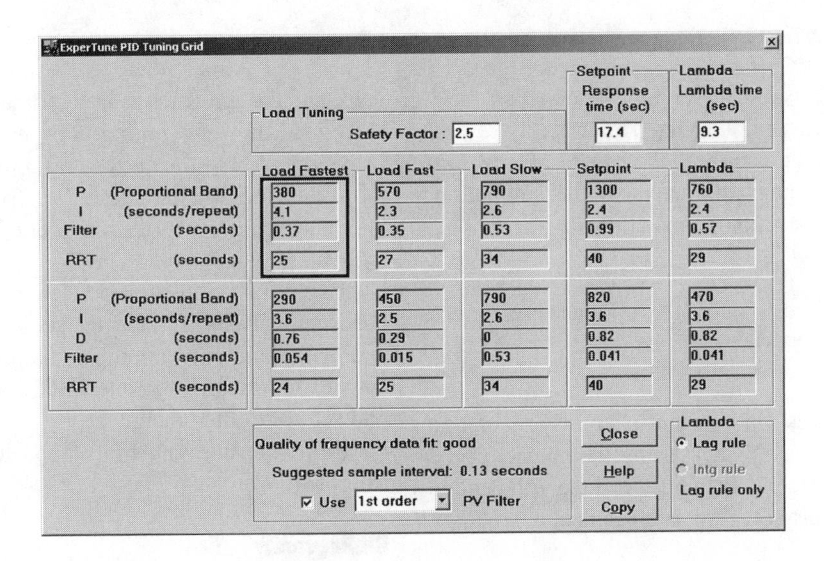

FIG. 5.1aa
Window showing three sets of tuning recommendations when tuning for load upsets and two sets when tuning for set point changes. (Courtesy of ExperTune, Inc.)

Optimal Filter

ExperTune software automatically finds the best PV filter to use. This filter is the largest possible that does not significantly affect the response or robustness of the control loop. The PV filter increases the life of control valves by making the output of the loop less jittery.

ExperTune software allows one to choose among:

1. First order
2. Second order
3. Butterworth
4. Averaging filter

Modeling and Simulation

ExperTune provides a model of the process of up to second order with dead time. The model can also contain an integrator. One does not need to specify the model structure; the software optimizes all models and automatically provides the structure that most closely matches the process.

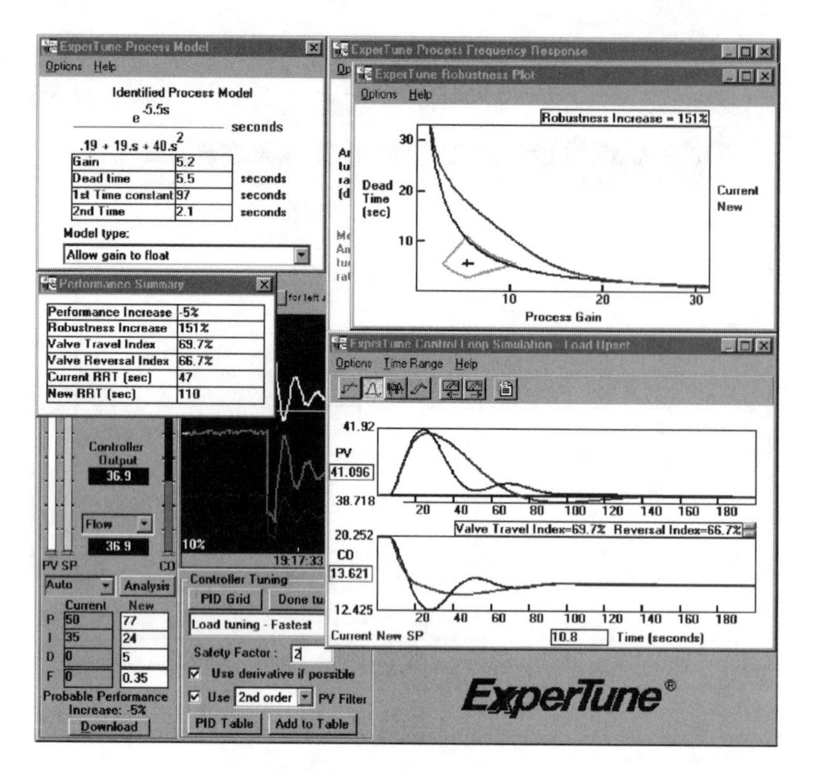

FIG. 5.1bb
Analysis windows in ExperTune PID Tuner software. (Courtesy of ExperTune, Inc.)

ExperTune lets users simulate the time response of their process with their industrial controller. This way, users can experiment offline without disturbing the process. Users can compare two sets of tuning parameters in the simulation window. They can choose among simulating:

1. Set point response
2. Load response
3. Model compared to actual
4. Response to noise

ExperTune robustness plots permit examination of the trade-offs of the tuning parameters. Using the mouse, users can drag the robustness line to adjust for the robustness wanted. Compare two sets of tuning parameters in the robustness window.

Cohesive Windows

The analysis option in ExperTune brings up five windows, as shown in Figure 5.1bb.

1. PID tuning
2. Time simulation
3. Robustness plot interchangeable with process frequency response
4. Model
5. Performance summary

Changing a value in one of the windows instantly updates the other windows. For example, change a PID or the filter value and the simulation and robustness plots instantly update to reflect the new tuning. For ease of comparison, the responses of both the current (PID parameters in the control loop) and the new (suggested or trial PID parameters that one can change) tunings are plotted in the simulation and robustness plots.

ExperTune software also includes hysteresis and stiction checkers, valve wear analysis, statistical analysis, variability and histogram, output and pH characterizers, power spectral density, loop summary table, RRT, inverse response analysis, and set point filter design.

Full reporting capability for each analysis is done in Microsoft Word.

Bibliography

Gerry, J. P., "Tune Loops for Load Upsets vs. Setpoint Changes," *Control Magazine,* September 1991.

Gerry, J. P., "How to Control Processes with Large Dead Times," *Control Engineering,* March 1998.

Gerry, J. P., "Tuning Process Controllers Starts in Manual," *InTech,* May 1999.

Gerry, J. P., "Control Fundamentals: PID Settings Sometimes Fail to Make the Leap," *InTech,* November 1999.

Ruel, M., "Stiction: The Hidden Menace," *Control Magazine,* November 2000.

Ruel, M. and Gerry, J. P., "Quebec Quandry Solved by Fourier Transform," *InTech,* August 1998.

Shinskey, F. G., *Process Control Systems,* New York: McGraw-Hill, 1996.

Shinskey, F. G., "Characterizers for Control Loops (White Paper)," May 1999, http://www.expertune.com/artCharact.html.

Shinskey, F. G. and Gerry, J. P., "PID Controller Specification (White Paper)," 1999, http://www.expertune.com/PIDspec.htm.

5.2 Data Reconciliation

M. J. BAGAJEWICZ D. K. ROLLINS

Data reconciliation is a term used to refer to the problem of adjusting plant measurements containing errors (e.g., flow rates, temperatures, pressures, concentrations, etc.) so that they conform to a certain chosen model, usually conservation laws. This model typically consists of a set of differential algebraic equations (DAE). For the simplest case of material balances, the model is the following:

$$\frac{d\mathbf{w}}{dt} = \mathbf{Af} \qquad 5.2(1)$$

$$\mathbf{Cf} = 0 \qquad 5.2(2)$$

where **f** is the vector of flows of the different streams connecting the units and **w** the vector of holdups of the respective units. Of all these state variables $\mathbf{x} = (\mathbf{w}, \mathbf{f})$, only a subset $\mathbf{x}_M = (\mathbf{w}_M, \mathbf{f}_M)$, is commonly measured. Thus, the general data reconciliation problem is stated as follows:

> Given a set of measurements of a subset of state variables $\mathbf{z}_{M,k} = (\mathbf{w}^+_{M,k}, \mathbf{f}^+_{M,k})$ at different instances of time t_k ($k = 1,\ldots,n_M$), it is desired to obtain the best estimate of as many variables as possible (measured and/or unmeasured), denoted by $\tilde{\mathbf{w}}_k$ and $\tilde{\mathbf{f}}_k$. Such estimates need to satisfy the model equations 5.2(1) and 5.2(2).

The vector of measurements ($\mathbf{z}_{M,k}$) and the corresponding vector of true values ($\mathbf{x}_k$) is given by $\mathbf{z}_{M,k} = \mathbf{x}_{M,k} + \varepsilon_{M,k}$, where $\varepsilon_{M,k}$ is the vector of errors. The errors are assumed to be random and normally distributed. In the absence of systematic errors (biased instrumentation), the problem of reconciliation typically consists of minimizing the weighted square of the difference between the measurements $\mathbf{z}_M$ and the estimates $\tilde{\mathbf{x}}_M$ at the $\mathbf{n}_M$ instances of time at which the measurements were made using as weights the variance–covariance matrix of measurements, **Q**. That is, it consists of solving the following optimization problem:

$$\left.\begin{aligned} &\min \sum_{k=0}^{n_M} [\tilde{\mathbf{x}}_{M,k} - \mathbf{z}_{M,k}]^T \mathbf{Q}^{-1} [\tilde{\mathbf{x}}_{M,k} - \mathbf{z}_{M,k}] \\ &\text{s.t.} \\ &\left(\frac{d\mathbf{w}}{dt}\right)_k = \mathbf{Af}_k \\ &\mathbf{Cf}_k = 0 \end{aligned}\right\} \qquad 5.2(3)$$

When steady state is assumed, averages of several measurements of each variable are usually used to construct one "measurement vector," denoted by $\mathbf{f}^+_M$. We assume momentarily that no accumulation terms exist. The problem becomes

$$\left.\begin{aligned} &\min [\mathbf{f}_M - \mathbf{f}^+_M]^T \mathbf{Q}^{-1} [\mathbf{f}_M - \mathbf{f}^+_M] \\ &\text{s.t.} \\ &\mathbf{Cf} = 0 \end{aligned}\right\} \qquad 5.2(4)$$

However, because accumulation terms are fairly common in practice, these accumulation terms have been added to this model by defining pseudo-streams. This issue will be discussed later in more detail.

This least-squares problem can be derived from different sources using the assumption that the distribution of errors follows a joint normal distribution, that is $\varepsilon_M \sim N_n(0, \mathbf{Q})$. Johnston and Kramer[1] presented a maximum likelihood derivation of the steady-state linear reconciliation model and Crowe[2] showed that the same result can be derived using information theory. These are recent references, but there are a number of earlier ones.

The models are usually based on first principles. However, there are some dangerous exceptions that have permeated into practice. Notoriously, the petroleum refining industry uses volumetric units (barrels) to measure production, which has prompted practitioners to believe that a "volumetric balance" is the proper model. The pressure has been so intense that some data reconciliation software vendors agreed to introduce such models.

The above linear and steady-state data reconciliation problem was extended to the nonlinear case of estimating other than mass flows, and to the dynamic case of estimating variables throughout time. All these extensions are discussed later.

Finally, the use of bounds can also be added, especially to prevent negative reconciled values when flows are small.

VARIABLE CLASSIFICATION

We now describe a variable classification for linear systems, represented by $\mathbf{Cf} = 0$, which is needed to differentiate which set of variables can be estimated.

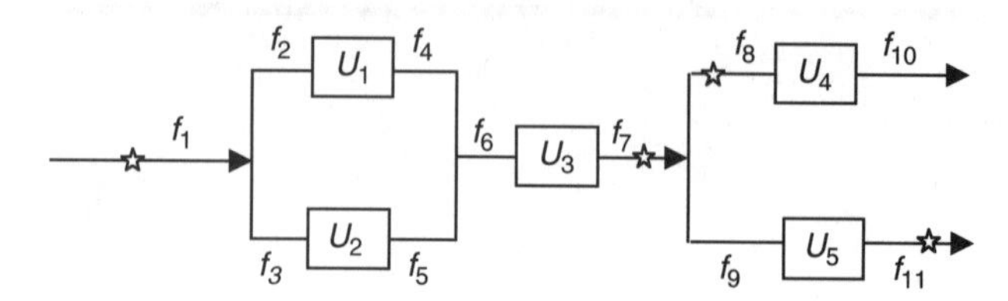

FIG. 5.2a
A system of 5 units and 11 streams.

Consider the system shown in Figure 5.2a. It consists of 5 units and 11 streams. Measured streams are indicated by a small star (☆). Consider that the holdups of all units are negligible.

The material balances on each unit are

$$\left.\begin{aligned} f_1 - f_2 - f_3 &= 0 \\ f_2 - f_4 &= 0 \\ f_3 - f_5 &= 0 \\ f_4 + f_5 - f_6 &= 0 \\ f_6 - f_7 &= 0 \\ f_7 - f_8 - f_9 &= 0 \\ f_8 - f_{10} &= 0 \\ f_9 - f_{11} &= 0 \end{aligned}\right\} \qquad 5.2(5)$$

The matrix **C** corresponding to these balances is

$$\mathbf{C} = \begin{array}{c} \begin{matrix} f_1 & f_2 & f_3 & f_4 & f_5 & f_6 & f_7 & f_8 & f_9 & f_{10} & f_{11} \end{matrix} \\ \begin{bmatrix} 1 & -1 & -1 & & & & & & & & \\ & 1 & & -1 & & & & & & & \\ & & 1 & & -1 & & & & & & \\ & & & 1 & 1 & -1 & & & & & \\ & & & & & 1 & -1 & & & & \\ & & & & & & 1 & -1 & -1 & & \\ & & & & & & & 1 & & -1 & \\ & & & & & & & & 1 & & -1 \end{bmatrix} \end{array} \qquad 5.2(6)$$

where the columns correspond to the variables indicated on top of the matrix and the blank spaces are zeros.

For convenience, the set of state variables is divided into measured variables ($\mathbf{f}_M$) and unmeasured variables ($\mathbf{f}_U$). In our example,

$$\mathbf{f}_M = \begin{bmatrix} \mathbf{f}_1 \\ \mathbf{f}_7 \\ \mathbf{f}_8 \\ \mathbf{f}_{11} \end{bmatrix} \qquad \mathbf{f}_U = \begin{bmatrix} \mathbf{f}_2 \\ \mathbf{f}_3 \\ \mathbf{f}_4 \\ \mathbf{f}_5 \\ \mathbf{f}_6 \\ \mathbf{f}_9 \\ \mathbf{f}_{10} \end{bmatrix} \qquad 5.2(7)$$

The objective is to determine which of the variables are observable, that is, for which variable is an estimate possible. Because all measured variables are estimable, this objective is really only applied to the set of unmeasured variables. In other words, because the unmeasured variables can be obtained using the material balance equations, one needs to determine exactly which ones can be obtained.

To motivate the concepts involved, we first notice that all the flow rates after the first split (f_2, f_3, f_4, and f_5) cannot be calculated using any of the material balances. Indeed, for example, only the sum of flows f_2 and f_3 is known from the measurements. These variables are called unobservable variables. The rest of the unmeasured variables (f_6, f_9, and f_{10}) can be obtained from material balances using the measured values. We call these variables, observable variables. This leads to our first classification:

$$\text{Variables} \begin{cases} \text{Measured (M)} \\ \text{Unmeasured (UM)} \begin{cases} \text{Observable (O)} \\ \text{Unobservable (UO)} \end{cases} \end{cases}$$

Let us turn our attention to the measured streams in Figure 5.2a. If flow rate f_1 is not measured, it can be estimated by the measurement of f_7. However, another way is by adding f_8 and f_{11}. It is then said that the set $\{f_1, f_7, f_8, \text{and } f_{11}\}$ is redundant.

Suppose now that the flow rates f_7 and f_8 are not measured. Then, because f_1 cannot be estimated by any balance equation using other measurements, we call it nonredundant. In other words, removing its measurements makes it unobservable. Thus, we have the following definition:

> A measured variable is nonredundant if, after removing its measurement, the variable is unobservable, i.e., it cannot be calculated using a balance equation involving the other variables of the system. Otherwise, the measured variable is called redundant.

Such redundancy is called "software or analytical redundancy." It is a desirable property because in the case when an instrument fails, its variable can be estimated through balances. Moreover, if the number of different balances that can be used increases, there will be additional ways to calculate the variable.

We therefore complete our classification of variables as follows:

$$\text{Variables} \begin{cases} \text{Measured (M)} \begin{cases} \text{Redundant (R)} \\ \text{Nonredundant (NR)} \end{cases} \\ \text{Unmeasured (UM)} \begin{cases} \text{Observable (O)} \\ \text{Unobservable (UO)} \end{cases} \end{cases}$$

Redundancy has been traditionally understood as the use of more than one instrument to measure the same variable and in such case it is called hardware redundancy. It has an effect on the accuracy and the reliability with which the estimates of the variables are obtained, but it has no effect on observability of other variables.

Kalman[3] introduced the concept of observability for linear dynamic systems. Stanley and Mah[4] discussed this issue of observability of systems described by steady-state models in depth. In particular, they discuss conditions under which observability can be attained. Definitions of degree of observability and redundancy were first introduced by Maquin et al.[5,6] Bagajewicz and Sánchez[7] unified both concepts by introducing the concept of estimability.

CANONICAL REPRESENTATION

By using simple rearrangement of columns in matrix **C**, the system can be rewritten in the following way:

$$\mathbf{Cf} = \begin{bmatrix} \mathbf{C}_U & \mathbf{C}_M \end{bmatrix} \begin{bmatrix} \mathbf{f}_U \\ \mathbf{f}_M \end{bmatrix} = 0 \qquad 5.2(8)$$

In the case of Figure 5.2a, matrices $\mathbf{C}_U$ and $\mathbf{C}_M$ are

$$\mathbf{C}_U = \begin{array}{c} \begin{array}{ccccccc} f_2 & f_3 & f_4 & f_5 & f_6 & f_9 & f_{10} \end{array} \\ \begin{bmatrix} -1 & -1 & & & & & \\ 1 & & -1 & & & & \\ & 1 & & -1 & & & \\ & & 1 & 1 & -1 & & \\ & & & & 1 & & \\ & & & & & -1 & \\ & & & & & & -1 \\ & & & & & 1 & \end{bmatrix} \end{array}$$

$$\mathbf{C}_M = \begin{array}{c} \begin{array}{cccc} f_2 & f_7 & f_8 & f_{11} \end{array} \\ \begin{bmatrix} 1 & & & \\ & & & \\ & & & \\ & & & \\ & -1 & & \\ & 1 & -1 & \\ & & 1 & \\ & & & -1 \end{bmatrix} \end{array} \qquad 5.2(9)$$

We now perform a Gauss–Jordan factorization of matrix **C**, concentrating on the unmeasured part only. The basic elementary steps are

1. The multiplication of two rows of matrix **C** by a number different from zero
2. The interchange of position of two rows (or columns) of the matrix
3. The addition of a row to another row

The resulting matrix is called the canonical form. In this canonical form, one can identify the unit matrix in the left upper corner, while the other rows, if present, are equal to zero in the same columns. This procedure was first proposed for data reconciliation by Madron.[8] For our example, this matrix is

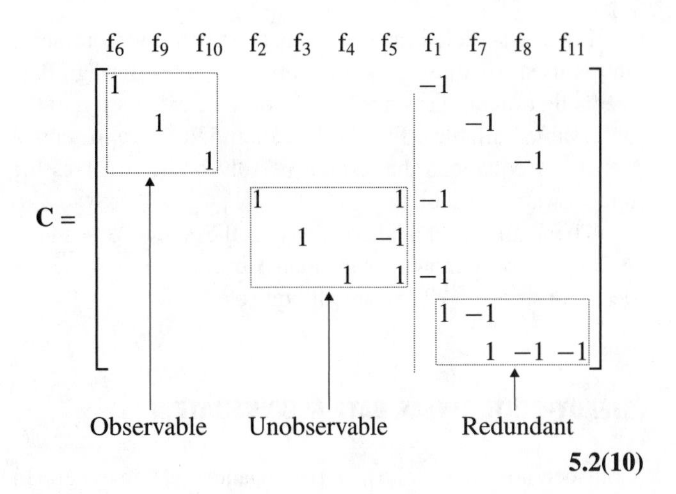

5.2(10)

If, in addition, variable S_3 is measured, then the result is

$$\mathbf{C} = \begin{array}{c} \begin{array}{ccccccccccc} f_6 & f_9 & f_{10} & f_2 & f_4 & f_5 & f_1 & f_7 & f_8 & f_{11} & f_3 \end{array} \\ \begin{bmatrix} 1 & & & & & & -1 & & & & \\ & 1 & & & & & & -1 & 1 & & \\ & & 1 & & & & & & -1 & & \\ & & & 1 & & & -1 & & & & 1 \\ & & & & 1 & & -1 & & & & 1 \\ & & & & & 1 & & & & & -1 \\ & & & & & & 1 & -1 & & & \\ & & & & & & & 1 & -1 & -1 & \end{bmatrix} \end{array} \qquad 5.2(11)$$

Observable — Redundant — Nonredundant

In general, we obtain a matrix with the following structure:

$$\left[\begin{array}{cc|cc} \mathbf{I} & 0 & -\mathbf{E}_{RO} & -\mathbf{E}_{NRO} \\ 0 & \mathbf{E}_{UO} & -\mathbf{E}_{RUO} & -\mathbf{E}_{NRUO} \\ 0 & 0 & \mathbf{E}_R & 0 \end{array}\right] \begin{bmatrix} \mathbf{f}_O \\ \mathbf{f}_{UO} \\ \mathbf{f}_R \\ \mathbf{f}_{NR} \end{bmatrix} = 0 \qquad 5.2(12)$$

(Labels: Observable, Unobservable, Redundant, Nonredundant; Unmeasured, Measured) where $\mathbf{f}_O$, $\mathbf{f}_{UO}$, $\mathbf{f}_R$, and $\mathbf{f}_{NR}$ are the vectors of observable, unobservable, redundant, and nonredundant flows. Rewriting the system $\mathbf{Cf} = 0$ using these vectors, one obtains:

$$\mathbf{f}_O = \mathbf{E}_{RO}\mathbf{f}_R + \mathbf{E}_{NRO}\mathbf{f}_{NR} \qquad 5.2(13)$$

$$\mathbf{E}_{UO}\mathbf{f}_{UO} = \mathbf{E}_{RUO}\mathbf{f}_R + \mathbf{E}_{NRUO}\mathbf{f}_{NR} \qquad 5.2(14)$$

$$\mathbf{E}_R\mathbf{f}_R = 0 \qquad 5.2(15)$$

Equation 5.2(15) cannot be satisfied by the measurements and is used to adjust these measured variables through data reconciliation. Equation 5.2(13) allows the calculation of the observable variables. Finally, Equation 5.2(14) represents a system of equations that cannot be solved and involves the unobservable variables.

There are several alternatives to the above procedure. Among the best known are the matrix projection[9] and the QR decomposition.[10,11] They are all equivalent.

STEADY-STATE LINEAR DATA RECONCILIATION

Consider now that only material balances are involved. In such a case, only flow rates are estimated. The data reconciliation problem is the one shown in Equation 5.2(4), which after variable classification becomes

$$\left.\begin{aligned} \min\ [\tilde{\mathbf{f}}_R - \mathbf{f}_R^+]^T \mathbf{Q}_R^{-1} [\tilde{\mathbf{f}}_R - \mathbf{f}_R^+] \\ \text{s.t.} \\ \mathbf{E}_R\tilde{\mathbf{f}}_R = 0 \end{aligned}\right\} \qquad 5.2(16)$$

Notice that the nonredundant variables have been eliminated from consideration and a variance–covariance matrix $\mathbf{Q}_R$ corresponding only to redundant variables is used. This is correct only when the matrix is diagonal, which is the usual assumption made in industrial practice.

This problem and its solution was the object of the 1965 seminal paper of this field,[12] but it has older roots in statistics. It is formally a quadratic programming problem with linear equality constraints. The solution is analytical:

$$\tilde{\mathbf{f}}_R = \left[\mathbf{I} - \mathbf{Q}_R\mathbf{E}_R^T(\mathbf{E}_R\mathbf{Q}_R\mathbf{E}_R^T)^{-1}\mathbf{E}_R\right]\mathbf{f}_R^+ \qquad 5.2(17)$$

After reconciliation is performed, the following estimators are obtained:

$$\tilde{\mathbf{f}}_{NR} = \mathbf{f}_{NR}^+ \qquad 5.2(18)$$

$$\tilde{\mathbf{f}}_O = \mathbf{E}_{RO}\tilde{\mathbf{f}}_R + \mathbf{E}_{NRO}\tilde{\mathbf{f}}_{NR} \qquad 5.2(19)$$

Precision of Estimators

Once the steady-state data reconciliation problem is solved, it is desired to estimate the standard error or mean square error of the estimator. In general, if $\mathbf{y} = \Gamma\mathbf{x}$, then the variance of $\mathbf{y}$ is given by

$$\tilde{\mathbf{Q}}_R = \Gamma\mathbf{Q}_R\Gamma^T \qquad 5.2(20)$$

and, consequently, applying this general formula to this case, the variance-covariance matrices of the flow rate estimators are given by the following expressions:

$$\tilde{\mathbf{Q}}_R = \mathbf{Q}_R - \mathbf{Q}_R\mathbf{E}_R^T(\mathbf{E}_R\mathbf{Q}_R\mathbf{E}_R^T)^{-1}\mathbf{E}_R\mathbf{Q}_R \qquad 5.2(21)$$

$$\tilde{\mathbf{Q}}_O = [\mathbf{E}_{RO} \quad \mathbf{E}_{NRO}]\begin{bmatrix}\tilde{\mathbf{Q}}_R \\ \mathbf{Q}_{NR}\end{bmatrix}[\mathbf{E}_{RO} \quad \mathbf{E}_{NRO}]^T \qquad 5.2(22)$$

Example

Consider first the system of Figure 5.2b and the corresponding set of measured data given in Table 5.2c. The reader can verify that the canonical form of this matrix is

$$\mathbf{C} = \begin{array}{c} \begin{matrix} f_7 & f_2 & f_3 & f_4 & f_5 & f_1 & f_6 \end{matrix} \\ \begin{bmatrix} 1 & & & & & & -1 \\ & 1 & & & 1 & -1 & \\ & & 1 & & -1 & & \\ & & & 1 & 1 & -1 & \\ & & & & & 1 & -1 \end{bmatrix} \end{array} \qquad 5.2(23)$$

From this canonical form, one can conclude that all measured streams are redundant, that nonredundant variables are not present, and that S_2, S_3, S_4, and S_5 are unobservable variables, whereas S_7 is observable. After data reconciliation is performed, the results shown in Table 5.2d are obtained.

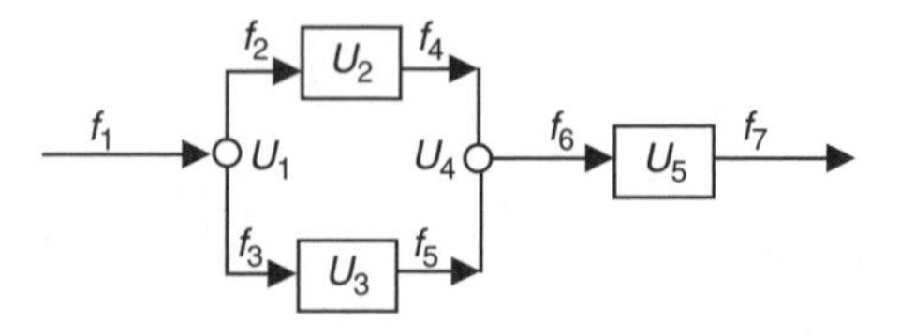

FIG. 5.2b
Example of a system.

TABLE 5.2c
Measurements for Figure 5.2b

Stream	Measurement	Variance
S_1	101.3	2.1
S_6	102.7	1.9

TABLE 5.2d
Data Reconciliation Results for Figure 5.2b

Stream	Measurement	Variance	Reconciled/ Estimated Value	Variance of Estimator
f_1	101.3	2.1	102.07	0.9975
f_6	102.7	1.9	102.07	0.9975
f_7	—	—	102.07	0.9975
f_5	33.8	0.3	33.8	0.3
f_3	—	—	33.8	0.3
f_2	—	—	68.27	1.2975
f_4	—	—	68.27	1.2975

Suppose now that flow rate f_5 is also measured with a variance of 0.5. Then, the new canonical form of the system matrix is

$$\mathbf{C} = \begin{array}{c} \begin{matrix} f_7 & f_2 & f_3 & f_4 & f_1 & f_6 & f_5 \end{matrix} \\ \left[\begin{array}{cccc|cc|c} 1 & & & & & -1 & \\ & 1 & & & -1 & & 1 \\ & & 1 & & & & -1 \\ & & & 1 & -1 & & 1 \\ \hline & & & & 1 & -1 & \end{array}\right] \end{array} \qquad 5.2(24)$$

As a result of the measurement of f_5, which is nonredundant, all the unmeasured variables become observable. Reconciled values are the same, because the redundant system has not changed. The previously observable variable S_7 is not calculated using the new nonredundant measurement. Thus, its variance does not change. The rest of the variances are shown in Table 5.2d.

PRACTICAL CHALLENGES OF THE STEADY-STATE MODEL

The following problems and challenges have been identified:

1. Tank holdup measurements cannot be directly used in steady-state data reconciliation models. In practice, to add redundancy, holdup changes are modeled as pseudo-streams. In addition, tanks are very often an important part of a chemical plant, especially in the case of refineries. It has therefore been of great interest for practitioners to be able to include all the transfers of raw material between tanks, from tanks to processes, or vice versa (called transactions or custody transfer when done at battery limits), as part of data reconciliation. However, tank holdup changes do not fit in the description of a steady state. Thus, when steady state is assumed over a period of time, the changes in holdup are usually divided by the time elapsed and considered as a pseudo-stream leaving (or entering) the system (Figure 5.2e). If the level is measured, then the pseudo-stream is considered a measured stream. Real streams entering or leaving the tank are considered separately. Because transactions between tanks are reported, this conversion has been very useful in adding a substantial number of redundant streams to refinery installations and has proved to be valuable to perform refinerywide oil accounting and oil-loss assessment.

2. Plants are never truly at a steady state. As practitioners have resorted to perform averages of several measurements to obtain one single number per stream to use in the objective function of Equation 5.2(19), they are averaging process variations as well. Process variations, due to all kinds of disturbances, are often ignored and lumped into one so-called measurement for each instrument. The assumption that gives some validity to such a procedure is that the plant is at this so-called pseudo steady state where, loosely speaking, state variables fluctuate and/or drift, "but not that much." From the purely theoretical point of view, this practice has been considered questionable. In principle, accumulation terms are not considered in these models. They are only included in integral form using the idea of considering accumulation as an input or output stream of a unit, a procedure that cannot capture the changes due to the dynamics of the system. An average, no matter how weighted or transformed, cannot possibly be validated as a "pseudo-steady-state" value, the criticism goes.

In spite of all these criticisms, steady-state data reconciliation often produces results that are acceptable for the practitioner, who does not seem to see the averaging of slowly drifting systems as a problem.

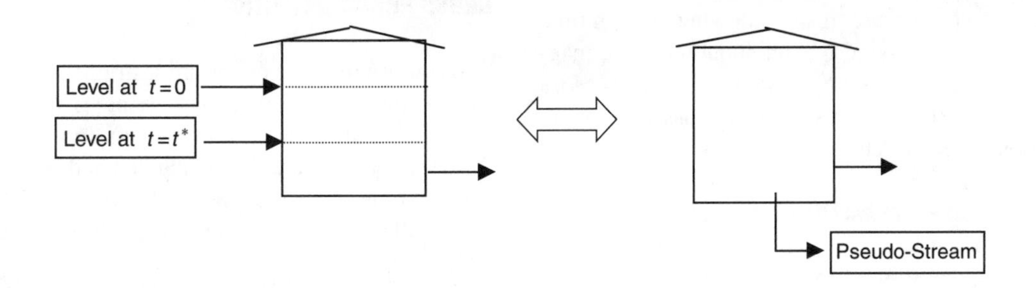

FIG. 5.2e
Use of pseudo-streams in tanks.

However, users of data reconciliation software wonder how most effectively to deal with this averaging: "Should one average in shorter periods of time?" "How small should the time interval be?" etc. Practitioners are also increasingly aware of this deficiency, and it will probably become an issue when pressure mounts to get more accurate and realistic figures for accounting, control, and monitoring purposes. This issue is later covered in more detail when the detection of biases and leaks is discussed.

The answer to this could be to apply dynamic data reconciliation methods, which will be briefly reviewed later. However, quite recently, a connection between dynamic and steady-state data reconciliation was found[13] for the case of linear systems (material balances only). Specifically, it is proved that, provided the variance–covariance matrix is diagonal, then *the results of the steady-state data reconciliation of averaged values are equal to the average of the dynamic data reconciliation values* (which are obtained for all time intervals). The aforementioned article also explores the effect of holdup and concludes that the steady state still constitutes a reasonable approximation. In a recent study,[14] another dynamic data reconciliation model[15] was applied to non-smooth signals, proving that the approximation is also good in these cases. The above findings provide a reasonable argument for continuing to use the technology available in commercial software.

For processes with high sampling rate for all variables, one can collect data over a short duration before the process significantly deviates from some steady state. However, for processes with slow sampling rates relative to process changes, it will not likely be possible to collect multiple samples at a single steady state. What one wants to do is to take enough samples so that process variations are somehow averaged. Consider the following measurement model for pseudo steady state:

$$\mathbf{f}_R^+ = \mathbf{f}_R + \lambda_R + \varepsilon_R + \delta_R \qquad 5.2(25)$$

where $\mathbf{f}_R$ is the "steady-state" value, λ_R is the process variation around $\mathbf{f}_R$, ε_R is the random error and δ_R is the bias. This model is valid for each instance of time a measurement is made. Although λ_R is assumed to change independently over time, in real processes it is likely to be serially correlated (i.e., changing according to some pattern). This assumption will not likely be too critical as long as the mean of λ_R is close to zero. One can ensure this by allowing enough time for the process to cycle around $\mathbf{f}_R$ with about the same magnitude of positive deviations as negative deviations. More guidelines regarding the sample size are discussed below in connection to gross error detection.

3. It has been a problem to pick the values of the variance–covariance matrix Q. Typically, covariances are ignored and a diagonal form of **Q** is used. In the absence of hard variance data, as questionable as this practice may seem, these variances are chosen in industrial practice using vendor information or, in some cases, the standard deviation of the measurements.

If measurements are independent, then they are not correlated (which is nonetheless not likely in the case of plant data), the variance–covariance matrix **Q** is diagonal and the elements of the diagonal are the variances of the individual measurements. In the case of steady-state data reconciliation, estimates of these values can be obtained by calculating the variance of the distribution of data around the mean value. This is called the *direct method*.[16] This is a correct procedure if the two aforementioned assumptions, independence and steady state, hold. In addition, outliers must not be present or they should be removed.

Because the system is never at a true steady state, the above formulas of the direct method incorporate the process variations, that is, the variance of the natural process oscillations or changes as part of the measurement variance. In a simple case like a ramp function, for example, the variance will be a composite of one half the change in true value of the measured value during the sampling interval and the true variance.

To ameliorate this problem and to assess the existence of variable interdependence (nondiagonal variance matrix) the indirect method was proposed.[16] In this method, the sum of the squares of the off-diagonal elements of the variance–covariance matrix of the constraint residuals ($\mathbf{r} = \mathbf{E}_R\mathbf{f}^+$) is minimized. The method was later slightly modified using an iterative procedure based on the solution of a nonlinear optimization resulting from using a maximum likelihood estimator.[17] Finally, Keller et al.[18] extended this work to nondiagonal covariance matrices. All these approaches still suffer from the problem that they do not consider the possible presence of outliers. Chen et al.[19] proposed the application of an M-estimator that applies a weight to each data based on its distance to the mean. They called this the *robust indirect method*. The discussion of the estimation using dynamic data is omitted.

While all these methods for variance estimation have been tested using computer simulations and have shown their power in these controlled experiments, there is no assessment of how they behave in practice. In particular, data reconciliation software notoriously lacks any module to perform such estimation, and there are no published results regarding the efficiency of these methods in practice.

GROSS ERROR DETECTION

The sources of gross errors are instrument biases, leaks, and true outliers (occasional measurements that depart significantly from all other measurements). In the past, departures from steady state have been included in this list. As discussed above, some of this has been clarified for linear systems.[13] The challenging task is to accomplish the following:

- Identify the existence of gross errors (biases, leaks, outliers)
- Identify the gross errors location

- Identify the gross error type
- Determine the size of the gross error and eliminate its influence on the final estimates

There exists at least one method that allows the detection of the existence of gross errors. Some of the methods for gross error identification are to a certain extent capable of discerning the location and type. After the gross errors are identified, two responses are possible and/or desired:

1. Eliminate the measurement containing the gross error, or
2. Correct the measurements or the model and run the reconciliation again.

The first alternative is the one implemented in commercial software, which in general only considers biases and is incapable of detecting leaks. This leaves the system with a smaller degree of redundancy and, as we saw, the quality of the reconciliation is deteriorated. If one is able to identify the gross errors and obtain a reliable estimate of their values, then the second alternative becomes appealing because redundancy is not lost. Some methods perform such tasks with reasonable success.

Identification Using Closure

One popular way of identifying measured variables with significant bias and nodes with significant leaks is testing for material and energy balance closure in nodes (interconnecting point or units). When closure is achieved, the streams entering and leaving the node are concluded to be unbiased and leaks at this node are concluded to be negligible. This strategy is illustrated using the flowsheet of Figure 5.2f, which is assumed to be in a steady state and where, for simplicity, random errors are zero.

Consider that there are no leaks and that measurement bias exists only in the measured mass flow rate for Stream 2. Then the measurements $\mathbf{f}^+$, are given by

$$\left.\begin{aligned} f_i^+ &= f_i \qquad \forall i \neq 2 \\ f_2^+ &= f_2 + \delta_2 \end{aligned}\right\} \qquad 5.2(26)$$

where f_i are the true values and δ_i are the biases. Overall mass

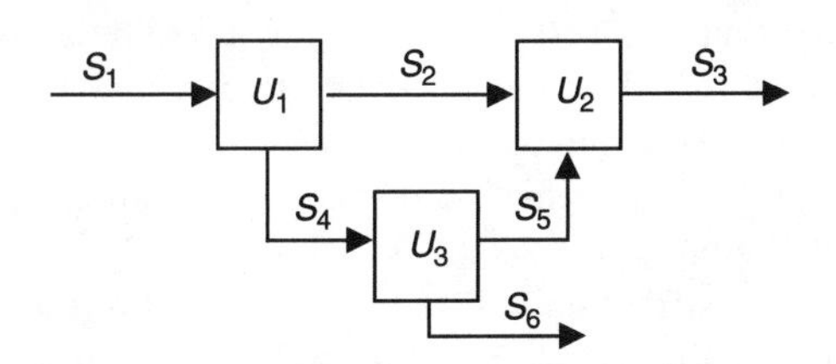

FIG. 5.2f
Process network used to illustrate a closure strategy.

balances on the nodes give

$$\left.\begin{aligned} r_1 &= f_1^+ - f_2^+ - f_4^+ \\ r_2 &= f_2^+ - f_3^+ + f_5^+ \\ r_3 &= f_4^+ - f_5^+ - f_6^+ \end{aligned}\right\} \qquad 5.2(27)$$

Combining Equations 5.2(26) and 5.2(27), and recognizing that at steady state the true flow rate values provide node closure, one obtains

$$\left.\begin{aligned} r_1 &= f_1 - f_2 - \delta_2 - f_4 = -\delta_2 \\ r_2 &= f_2 + \delta_2 - f_3 + f_5 = \delta_2 \\ r_3 &= f_4 - f_5 - f_6 = 0 \end{aligned}\right\} \qquad 5.2(28)$$

If closure is concluded upon testing, then all measured variables (i.e., stream flow rates) entering and leaving the node or combination of nodes are concluded to be unbiased (i.e., $\delta = 0$) and the nodes are concluded not to have significant leaks. Error cancellation, that is, cases where biases or leaks exist but add to zero, is typically not likely and is assumed not to occur.[20] Nevertheless, in the strict sense, this test makes conclusions only regarding zero biases and leaks, but it does not make conclusions regarding them as nonzero.

Selection of Nodes

For the process in Figure 5.2f there are seven balances one could test for closure. They are balances around the individual units, U_1, U_2, and U_3 and balances around combination of nodes, that is: U_1U_2, U_1U_3, U_2U_3, and $U_1U_2U_3$. Assume one decides to test closure for the individual units U_1, U_2, and U_3, then the nodal combination of $U_1U_2U_3$. With measurement random errors removed, the conclusion can be known for each test and there is no need to rely on statistical hypothesis testing. Thus, Equation 5.2(28) gives the results for the balances around the individual nodes and $r_{123} = r_1 + r_2 + r_3 = 0$. Therefore, no conclusions are made regarding closure for nodes U_1 and U_2. In contrast, because $r_3 = 0$, measured streams S_4, S_5, and S_6 are concluded to be unbiased (i.e., $\delta_4 = \delta_5 = \delta_6 = 0$). Next, the test closure from r_{123} concludes that $\delta_1 = \delta_3 = \delta_6 = 0$. Thus, from this strategy of testing, the overall conclusion is that $\delta_1 = \delta_3 = \delta_4 = \delta_5 = \delta_6 = 0$. As stated above, it is not proper for nodal testing methods to conclude that $\delta_2 \neq 0$. One could just simply state that it may not be zero. Therefore, a statistical test is needed to determine if such an overall conclusion can be made in the presence of measurement error. In order for this approach to have practical merit, tests and strategies are needed to control error rates for false conclusions. This discussion is restricted, for simplicity, to the case of pseudo steady state.

Hypothesis testing has been used to detect and identify gross errors. We now discuss a few of these methods. We also assume that averages of a certain number of measurements (N_M) have been taken. Thus, an average of all residual vectors (denoted by $\mathbf{r}$) is also defined.

Global Test

The null hypothesis (H_0) is that no bias is present. Thus, if **r** is the vector or residuals of the expected value of **r** ($= \mathbf{E_R f^+}$) is zero, that is E[**r**] = 0 and N_M is the number of measurements, then, in the absence of gross errors and under the assumption of multivariate normality for $\mathbf{f}^+$, the statistics

$$\chi_m^2 = N_M \mathbf{r}^T \mathbf{J}^{-1} \mathbf{r} \qquad 5.2(29)$$

follows a chi-squared distribution with m degrees of freedom, where m is the number of rows of $\mathbf{E_R}$. Matrix **J** is the variance–covariance matrix of **r** and is given by

$$\mathbf{J} = \text{cov}(\mathbf{r}) = \mathbf{E}_R \mathbf{Q}_R \mathbf{E}_R^T \qquad 5.2(30)$$

where $\mathbf{Q}_R$ is the variance–covariance matrix of $\mathbf{f}^+$.

The test therefore consists of rejecting H_0 if $N_M \mathbf{r}^T \mathbf{J}^{-1} \mathbf{r}$ is larger than a threshold value $\chi^2_{m,\alpha}$, the upper αth percentile of the χ^2 distribution with m degrees of freedom.

Notice that this test can be performed without actually implementing data reconciliation. If this number falls within the nonrejection region, then the null hypothesis is not rejected, that is, no gross error is suspected. On the other hand, if $N_M \mathbf{r}^T \mathbf{J}^{-1} \mathbf{r}$ is larger than the critical value, a gross error cannot be ruled out, although in practice it is usually assumed it has been detected. Its location, however, cannot be determined using this test.

Nodal Tests

In the absence of gross errors the constraint residuals **r** follow an m-variate normal distribution (m is the rank of $\mathbf{E_R}$), with zero mean. Therefore, the following test statistic

$$\mathbf{Z}_i^N = N_M^{1/2} \frac{\mathbf{r}_i}{\sqrt{\mathbf{J}_{ii}}} \qquad 5.2(31)$$

follows a standard normal distribution, N(0,1), under H_0.

Therefore, if $\mathbf{Z}_i^N$ is larger than the critical value ($Z_{\alpha/2}$) based on the level of significance α, then one concludes that there is at least one gross error in the set of measurement that participates in the corresponding nodal balance. Notice that $\alpha/2$ is used instead of α because the test is a two-tail test, that is, it tests for a positive or a negative deviation.

Assume that there are k tests that one makes on a single set of data. Multiple tests are then made using the same critical value, and for this reason, the likelihood of a Type I error (wrongly reject H_0, that is, declare the existence of a gross error when there is none), increases. Mah and Tamhane[21] proposed the use of a new smaller level of significance (β) derived using the Sidák inequality.[22] This new level of significance is given by

$$\beta = 1 - (1 - \alpha)^{1/k} \qquad 5.2(32)$$

that is, to use $Z_{\beta/2}$ instead of $Z_{\alpha/2}$ as a threshold value. The Bonferroni correction for multiple testing is

$$\beta = \alpha/k \qquad 5.2(33)$$

which is another alternative.[21] Note that when k is large, Equation 5.2(32) approaches Equation 5.2(33), but even for small values of k they are very similar. Because Sidák's inequality assumes independent tests and Bonferroni does not, the use of the Bonferroni correction is recommended.

As the square of a variable that follows a normal distribution follows a chi-squared one, then the statistics

$$\mathbf{Z}_i^C = N_M \frac{\mathbf{r}_i^2}{\mathbf{J}_{ii}} \qquad 5.2(34)$$

can also be used to test the presence of gross errors. Thus, H_0 is rejected if $\mathbf{Z}_i^C \geq \chi^2_{m,\alpha}$.

Tests Using Linear Combination of Nodes

It was discussed above that different combinations of nodes lead to different conclusions that one can draw. A particular combination of nodal residuals can be represented by $\mathbf{l}_k^T \mathbf{r}$. For example, for the network of Figure 5.2f, the nodal residual combination $r_1 + r_2$ is obtained using $\mathbf{l}_k^T = (110)$. There are $2^m - 1$ such vectors. Thus, the nodal test statistic based on the normal distribution for these combinations becomes

$$\mathbf{Z}_k^{Nl} = N_M^{1/2} \frac{|\mathbf{l}_k^T \mathbf{r}|}{\sqrt{\mathbf{l}_k^T \mathbf{J} \mathbf{l}_k}} \qquad 5.2(35)$$

while the nodal test statistic based on the chi-squared distribution becomes

$$\mathbf{Z}_k^{Cl} = N_M \frac{(\mathbf{l}_k^T \mathbf{r})^2}{\mathbf{l}_k^T \mathbf{J} \mathbf{l}_k} \qquad 5.2(36)$$

In principle, one could perform this test on all combinations of nodes, but because the number of such tests can be very high, Rollins et al.[23] proposed guidelines to disregard certain nodes in an effort to maximize power. Tests that are suitable for the case where variances and covariances are not known were also developed.[24]

Note that the test based on the normal distribution [Equation 5.2(35)] is independent of the number on tests contrasting to the one based on the chi-squared distribution [Equation 5.2(36)], which depends on the number of tests. When the number of tests k is small, the nodal test based on the normal distribution will be a more powerful test than the one based on the chi-squared distribution. In other words, the Bonferroni test will be more powerful if $(Z_{a/2k})^2$ is less than $\chi^2_{m,\alpha}$. For $k \leq m$ the test based on the normal distribution is usually more powerful than the chi-squared test. However, when k > m, the test based on the chi-squared distribution can be more powerful. This is true in particular when $k \gg m$.

Finally, Crowe[25] proposed a linear transformation for which the nodal test has maximum power, that is, minimum of failures to identify existing gross errors, which are known as Type II errors. Such test is equivalent to the nodal test given by Equation 5.2(35).

Measurement Test

The measurement test (MT) is based on the vector of measurement adjustments (or corrections) denoted by **a** and defined by

$$\mathbf{a} = \mathbf{f}_R^+ - \tilde{\mathbf{f}}_R \qquad 5.2(37)$$

The test is based on the assumption that the random errors for measurements are independently and normally distributed. Under the null hypothesis (H_0), the expected value of **a** is zero. That is, the following statistic

$$Z_i^{MT} = \frac{a_i}{\sqrt{(\tilde{Q}_R)_{ii}}} \qquad 5.2(38)$$

follows a normal distribution with zero mean under H_0.

Because multiple tests are involved, the test should be based on a level of significance given by β obtained from Equation 5.2(32) or 5.2(33).

Finally, Mah and Tamhane[21] also proposed a linear transformation for which the measurement test has maximum power under some rather restrictive assumptions. The linear transformation consists of using the new transformed adjustments as follows:

$$\mathbf{d} = \mathbf{Q}_R^{-1}\mathbf{a} \qquad 5.2(39)$$

and therefore the maximum power test statistics, which is used the same way as the regular measurement test, is given by

$$Z_i^{MMP} = \frac{|d_i|}{\sqrt{(\tilde{Q}_R)_{ii}}} \qquad 5.2(40)$$

The measurement test is "statistically inadmissible," and therefore unacceptable, because it has a property called "inconsistency." This means that it does not give the correct answer with infinite sampling. In short, with the measurement error removed (i.e., using the deterministic solution) and one biased variable, the test may point to the wrong variable, which is incorrect. In order for a method to be admissible, it has to give the correct deterministic solution. Nevertheless, the method has been included here because of its widespread use in the field for the identification of biases.

Generalized Likelihood Ratio Test

The alternative hypothesis in this test consists of a particular bias in a stream associated to a node or a leak,[26] that is, H_1: $\mu_r = b\mathbf{h}_i$, where μ_r is the expected value of the node residual, g_i a vector in the direction of a bias ($\mathbf{h}_i = \mathbf{E}_R\mathbf{e}_i$), or in the direction of a leak ($\mathbf{h}_i = \mathbf{m}_i$), and b is the size of this gross error. This is an unorthodox way of posing an alternative hypothesis because it contains an unknown number (b) and an unknown vector $\mathbf{h}_i$. Nevertheless, this is equivalent to H_0: $\mu_r = 0$ and H_1: $\mu_r \neq 0$. The test is based on finding the supremum of the likelihoods of each hypothesis, that is:

$$\lambda = \sup \frac{\Pr\{\mathbf{r}|H_1\}}{\Pr\{\mathbf{r}|H_0\}} \qquad 5.2(41)$$

where the supremum is computed over all possible bias and leaks. When the probability distributions used are normal, the test consists of computing:

$$\mathbf{T} = 2\ln\lambda = \sup_{\forall i} \frac{(\mathbf{h}_i^T\mathbf{Q}_R^{-1}\mathbf{r})^2}{\mathbf{h}_i^T\mathbf{Q}_R^{-1}\mathbf{h}_i} \qquad 5.2(42)$$

which is compared with the corresponding threshold.

Unlike the measurement test, this test appears to be consistent when one gross error is present. However, when multiple gross errors are present, the test procedure can fail under deterministic conditions, and is therefore inconsistent.

Principal Component Test

Tong and Crowe[27] proposed the use of principal components, that is, eigenvalue decomposition of the variance of the residuals ($\mathbf{E}_R\mathbf{Q}_R\mathbf{E}_R^T$), given by

$$\mathbf{\Lambda}_r = \mathbf{U}_r^T(\mathbf{E}_R\mathbf{Q}_R\mathbf{E}_R^T)\mathbf{U}_r \qquad 5.2(43)$$

where $\mathbf{U}_r$ is the matrix of orthonormalized eigenvectors of $\mathbf{E}_R\mathbf{Q}_R\mathbf{E}_R^T$. Accordingly, the principal components are

$$\mathbf{p}_r = (\mathbf{\Lambda}_r^{-1/2}\mathbf{U}_r)^T\mathbf{r} \qquad 5.2(44)$$

which are tested against the normal distribution. Once a suspect has been identified, the error is traced back to the corresponding node by looking at the largest contributions to this component. A similar measurement test using principal components was developed.[27]

Sample Size

Issues related to process variations were discussed above. However, no matter which strategy one chooses for data collection, it is important to recognize that sampling should be limited. If the number of samples N_M is too large, the power (i.e., probability of detecting a nonzero bias δ or leak λ) will be too large.[28]

Because all measured streams are likely to have some degree of bias, very high power will declare that all streams bias, which will not be a helpful conclusion. Thus, one should attempt to control power by selecting the sample size *a priori* based on control of the Type II error (i.e., false detection). This issue is directly related to the power of these tests, a discussion that is entirely omitted in this review for space reasons, but can be consulted in the books cited in the bibliography as the end of the section.

MULTIPLE GROSS ERROR IDENTIFICATION

In the case where multiple gross errors (biases, leaks, and outliers) are present, a strategy to identify them is needed. We briefly discuss two strategies; the first, serial elimination, is widely used in commercial software.

Serial Elimination

In this strategy, a selected test (measurement test or principal component analysis) is coupled with an elimination strategy. If the test flags the existence of gross errors, then a strategy is proposed to identify one or more variables, which are the "most suspected ones." The measurements of these variables are eliminated and the test is run again, even if that implies performing the data reconciliation again. Commercial software has used the measurement test in this strategy with relatively good success, but with some troubling counter-examples. In the presence of multiple gross errors, the measurement test usually (but not always) gives the largest value for the variable where the gross error exists, and this variable is singled out for elimination. The reader is reminded of the limitations regarding the measurement test stated above.

Unbiased Estimation (UBET)[20]

This method renders unbiased estimators. Candidate gross errors are identified, and a simple formula is used to estimate their size to obtain unbiased estimates for process variables and leaks.

UBET is developed from the balance residuals **r** and its expected value

$$\mu_r = \mathbf{A}\delta + \mathbf{M}\gamma \qquad 5.2(45)$$

where γ represents the possible leaks. By partitioning **A**, **M**, δ, γ, one obtains

$$\mu_r = \begin{bmatrix} \mathbf{A}_{11} & 0 \\ \mathbf{A}_{21} & \mathbf{M}_{22} \end{bmatrix} \begin{bmatrix} \delta_1 \\ \gamma_2 \end{bmatrix} = \mathbf{C}_1 \theta_1 \qquad 5.2(46)$$

Finally, by introducing the vector $\mathbf{l}_i$ by means of $\mathbf{l}_i^T = \mathbf{e}_i^T \mathbf{C}_1^{-1}$, one obtains

$$\mathbf{l}_i^T \mu_r = \mathbf{e}_i^T \theta_1 \qquad 5.2(47)$$

Thus, $\mathbf{l}_i^T \mathbf{r}$ (for all i) are unbiased estimators of the components of δ and γ contained in θ_1. Originally, the method was proposed to be preceded by an identification step, at which point the procedure would stop. Some difficulties of this method have been studied and measures to overcome them have been suggested.[29] They are connected to the issue of equivalency of gross errors, which is discussed below. In particular, Bagajewicz and Jiang[30] showed that UBET can be used by assuming m locations that correspond to a spanning tree of the graph of the flow sheet. Once the errors are estimated, hypothesis testing can proceed. These errors are then equivalent to several alternative sets.

Uncertainties

We now illustrate some uncertainties that arise in practice, especially when the measurement test is used. We reiterate the limitations of this test made above, but we include this discussion in view of its widespread use in commercial software.

In many cases (actually very often) the measurement test gives the same value for many variables. Consider the system of Figure 5.2a and assume that only S_1, S_8, and S_{11} are measured. They are redundant. Assume the values given by Table 5.2g.

If one adds a bias of size 5 to the measurement of the flow rate f_1, making the measurement 15.03, the reconciliation renders

$$\tilde{\mathbf{f}}_R = \begin{bmatrix} 13.29 \\ 6.51 \\ 6.78 \end{bmatrix} \qquad 5.2(48)$$

The values of the Z-statistics for the measurement test are all the same. These value are Z = 9.37. This illustrates the uncertainty regarding which variable should be eliminated in the case where the strategy is based on the elimination of the variable with the largest Z-statistics. This poses a problem for the serial elimination strategy because there is no criterion to determine which one of these three measurements should be eliminated. This phenomenon was first pointed out by Iordache et al.[31] and is now well understood.[30] In the case of two gross errors, it happens because the corresponding columns of E_R are proportional. Indeed, the last two columns (corresponding to S_8 and S_{11}) are equal, and the proportionality constant between the first and the two last columns is (–1).

TABLE 5.2g
Values for Figure 5.2a

Stream	Measured Flow Rate	Variance
f_1	10.03	0.1
f_8	5.99	0.03
f_{11}	3.99	0.16

In the next section, cases involving more streams are explained. As will be shown, if one understands the nature of these uncertainties, the serial elimination strategies are not flawed (in this sense) after all.

EQUIVALENCY OF GROSS ERRORS

Two sets of multiple biases are equivalent when they have the same effect in data reconciliation, that is, when eliminating either one leads to the same value of the objective function. Therefore, the equivalent sets of gross errors are theoretically indistinguishable. In other words, when a set of gross errors is identified, there exists an equal possibility that the true locations of gross errors are in one of its equivalent sets.

Consider the case of Figure 5.2h and the measurements given in Table 5.2i. Random errors have been eliminated in this table so that the phenomenon is clearly visible.

We now analyze the set $\{f_2, f_4, f_5\}$. As shown in Table 5.2i, three different sets of biases and reconciled values can explain these measurements (Cases 1, 2, and 3). Therefore, without *a priori* additional knowledge of where the biases might be, or some additional information about the reconciled values, the three cases are impossible to detect.

We noticed that every case has two biases. This is no coincidence. These are called *equivalent sets*. Every equivalent set has a certain *minimum* number of biases that can represent all situations. Indeed, there are infinite numbers of sets of three gross errors that can be represented by two biases. Equivalent sets can be identified using the method developed by Bagajewicz and Jiang.[30] Several intricacies of the gross error equivalency are discussed in their article.

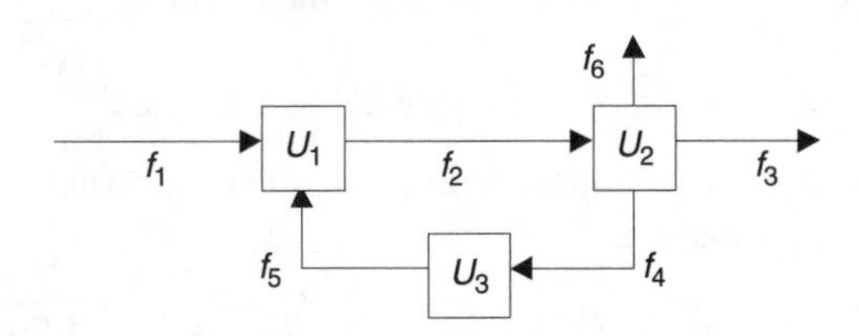

FIG. 5.2h
Flow sheet to illustrate gross error equivalencies.

TABLE 5.2i
Illustration of Equivalent Sets in $\{f_2, f_4, f_5\}$ of Figure 5.2h

		f_1	f_2	f_3	f_4	f_5	f_6
Measurement		12	18	10	4	7	2
Case 1	Reconciled data	12	18	10	6	6	2
(Bias in f_4, f_5)	Estimated biases				−2	1	
Case 2	Reconciled data	12	19	10	7	7	2
(Bias in f_2, f_4)	Estimated biases		−1		−3		
Case 3	Reconciled data	12	16	10	4	4	2
(Bias in f_2, f_5)	Estimated biases		2			3	

In addition, in the presence of random errors, one can define quasi-equivalency as an approximation to true equivalency to detect leaks by searching for an equivalent set of biases.[32]

This has very important consequences in practice. For example, if the biases are in reality in $\{f_2, f_4\}$, *any* strategy for gross error detection can identify *with equal probability* any of the three cases. The impact of this is not only in the location and size of the biases, but also in the reconciled values. The importance of this in production accounting is paramount.

Once all the equivalent sets have been identified, one may want to determine the sizes of the gross errors in one equivalent set, when the sizes in another equivalent set have been calculated with some gross error detection and estimation method. In addition, one wants to know the new values of the reconciled streams.

GROSS ERROR SIZE ESTIMATION

Gross errors can be estimated using several methods. The oldest method is the compensation model. It consists of assuming a set of instrument biases and introducing the new "measurement":

$$\mathbf{f}_R^+ = \tilde{\mathbf{f}}_R + \boldsymbol{\delta} \qquad 5.2(49)$$

where $\boldsymbol{\delta}$ is a vector that contains biases for bias candidates in a limited number of positions. The rest of the elements of $\boldsymbol{\delta}$ that are not candidates are zero. Conducting a data reconciliation using these measurements renders estimates that, in turn, can be used to obtain the values of the nonzero elements of $\boldsymbol{\delta}$. To avoid running into singularities, one equivalent set needs to be picked for the nonzero elements of $\boldsymbol{\delta}$. The complete procedure is illustrated by Bagajewicz and Jiang.[30]

Jiang and Bagajewicz[32] used this model in a serial strategy where the measurement test is used at each step to identify a candidate gross error, which is added to a list of candidates. In contrast to other serial procedures where the sizes of the gross errors identified are determined at each step, this procedure evaluates the size of all the gross errors in the candidate list at each step. Finally, the UBET method presented earlier provides the size of gross errors.

USE OF ALTERNATIVE OBJECTIVE FUNCTIONS

Some alternative objective functions that are capable of handling the gross errors in the data simultaneously with data reconciliation have been proposed. We cite only those that have reached publicly offered software. Tjoa and Biegler[33] proposed a mixture distribution as the objective (likelihood) function:

$$\min\{a\mathbf{P}(\varepsilon, \sigma) + (1 - a)\mathbf{P}(\varepsilon, \sigma/b)\} \qquad 5.2(50)$$

where a is the probability associated with the absence of gross errors, σ^2 is the variance of such errors, and $b^2\sigma^2$ is the variance of the biases. A test follows the reconciliation to determine the gross error presence. The shortcomings of this approach[34] were addressed by Albuquerque and Biegler,[35] who proposed the use of the Fair function:

$$\rho(\varepsilon) = c^2\left[\frac{|\varepsilon|}{c} - \log\left(1 + \frac{|\varepsilon|}{c}\right)\right] \qquad 5.2(51)$$

where c is a parameter, and which has the advantage of being convex with continuous first and second derivatives.

In turn, Johnston and Kramer[1] proposed use of the Lorentzian distribution:

$$\rho(\varepsilon) = \frac{1}{1 + \frac{\varepsilon^2}{2}} \qquad 5.2(52)$$

which is a robust estimator because it has the ability to filter large gross errors.

All these methods are centered on outlier detection due to a large variance and not on bias detection.

NONLINEAR STEADY-STATE DATA RECONCILIATION

Steady-state nonlinear data reconciliation is represented by

$$\left.\begin{aligned} &\min[\tilde{x}_M - z_M]^T Q^{-1}[\tilde{x}_M - z_M] \\ &\text{s.t.} \\ &g(\tilde{x}) = 0 \end{aligned}\right\} \qquad 5.2(53)$$

For process plants z_M includes the typical state variables, flow rates, concentrations, temperatures, pressures, and the model $g(\tilde{x})$ can include any type of unit operations and equipment. In addition, $\tilde{x}$ usually contains parameters that are not measured directly.

Several methods have been proposed to solve this problem, especially when $g(\tilde{x})$ is bilinear. Because the solution is supposed to be close to the measurement (unless gross errors are present), then one can linearize $g(\tilde{x})$, perform a classification of variables, and extract the redundant system of equations. Once this is done, one can solve successively, updating the Jacobian in each iteration until convergence is achieved. Other approaches using nonlinear programming, such as the popular sequential quadratic programming codes (SQP), can also be used.

Commercially available software performs nonlinear steady-state data reconciliation to a good extent. Some have sophisticated property estimation routines such that they are capable of performing material, energy, and component balances, all simultaneously, reconciling temperature, pressure, concentration, and flow rates, and of performing parameter estimation, for example, the heat transfer coefficient of heat exchangers. In the absence of systematic errors and leaks in the system, there is no reason these models cannot be made as sophisticated and complex as the optimization techniques used to solve them can handle. However, software vendors are reluctant to introduce such models. One of the reasons is certain conviction that undetected gross errors may still be largely amplified by the reconciliation, especially when nonlinearities correspond to very non-ideal systems. Bagajewicz and Mullick[36] discuss this issue in more detail.

DYNAMIC DATA RECONCILIATION

Early work in dynamic data reconciliation is rooted in the problem of process state estimation using the concept of filtering. Lately, the problem has been solved using the concept of model-based data smoothing. Consider the three types of state estimation problems that are illustrated in Figure 5.2j. Assume an estimation of the state of the system is desired at time t.

When only measurement values prior to the time of prediction t are used, including the measurement at time t, the estimation is called filtering. When time t is not included, the estimation is called prediction, and when data for times larger than t are used, the estimation process is called smoothing. Finally, when discrete measurements are used, the estimators are called discrete estimators.

Consider now a discrete system whose behavior is given by the following model:

$$\mathbf{x}_k = \Phi_{k-1}x_{k-1} + w_k \qquad 5.2(54)$$

where $\mathbf{x}_k$ is a vector of system states and w_k is called process noise and has zero mean and variance R_k. For example, the discrete model for a constant is $x_k = x_{k-1}$. Consider also the following measurement model:

$$z_k = H_k x_k + v_k \qquad 5.2(55)$$

where v_k is a zero mean random error with variance S_k, and z_k is the vector of measurements at time t_k, namely, $z_k = [z_{k,1}\ z_{k,2} \ldots z_{k,m}]^T$.

A linear estimator is constructed as follows:

- The *a priori* estimate at time t_k is obtained using the system equation:

$$\hat{x}_k^{(-)} = \Phi_{k-1}\hat{x}_{k-1}^{(+)} \qquad 5.2(56)$$

where we have ignored the process noise.

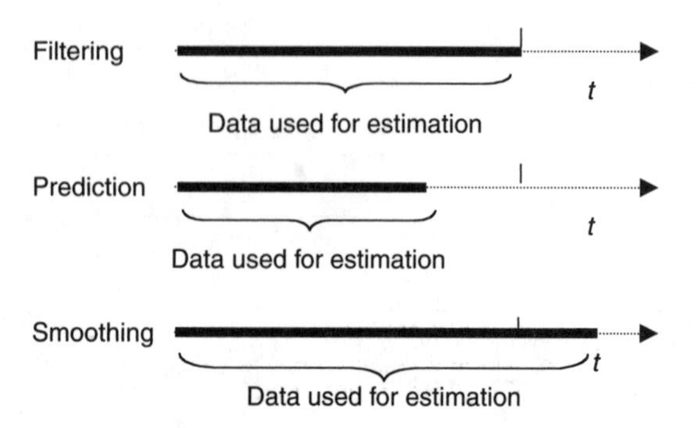

FIG. 5.2j
Different types of estimation.

- Given the estimate at time t_k denoted $\hat{x}_k^{(-)}$, we seek an update estimate $\hat{x}_k^{(+)}$ based on z_k using the following linear, recursive form:

$$\hat{x}_k^{(+)} = K_k^{(-)}\hat{x}_k^{(-)} + K_k^{(z)}z_k \qquad \textbf{5.2(57)}$$

One such model, the well-known Kalman filter,[3] provides both the estimates and an estimate of their variance matrix. If one assumes that the system is at steady state, then the model is $x_k = x_{k-1}$ and a simplified Kalman filter can be obtained. If, in addition, one also assumes that the model satisfies the balance equations ($Cx_k = 0$), a quasi-steady-state Kalman filter is obtained.[37] Other variants of this model have been proposed. For example, one could use neural networks in conjunction with the extended Kalman filter.[38]

A discrete estimator, called a singular or generalized dynamic estimator, is now presented. Consider the following balance equations:

$$B_R\frac{dw_R}{dt} = A_R f_R \qquad \textbf{5.2(58)}$$

$$C_R f_R = 0 \qquad \textbf{5.2(59)}$$

The discrete version of Equations 5.2(58) and 5.2(59) is

$$B_R(w_{R,i+1} - w_{R,i}) = A_R f_{R,i+1} \qquad \textbf{5.2(60)}$$

$$C_R f_{R,i+1} = 0 \qquad \textbf{5.2(61)}$$

which can be expressed as follows:

$$E_R z_{R,i+1} + G_R z_{R,i} = 0 \qquad \textbf{5.2(62)}$$

where

$$G_R = \begin{bmatrix} 0 & B_R \\ 0 & 0 \end{bmatrix} \qquad \textbf{5.2(63)}$$

Unfortunately, E_R is usually not square and, therefore, singular. Therefore, the discrete Kalman filter cannot be applied. Instead, all estimates for all times up to t_N are reconciled at the same time using Equation 5.2(62) as constraints, that is:

$$\left.\begin{array}{c} \min\sum_{k=0}^{N}[z_{R,k} - \overset{+}{z}_{R,k}]^T Q_R^{-1}[z_{R,k} - \overset{+}{z}_{R,k}] \\ \text{s.t.} \\ E_R z_{R,i+1} + G_R z_{R,k} = 0 \quad \forall k = 0, N-1 \end{array}\right\} \qquad \textbf{5.2(64)}$$

Darouach and Zasadzinski[39] developed recursive formulas that would allow the solution to these models. Rollins and Devanathan[15] presented an alternative methodology that is not as computationally intensive and allows one to trade computational speed for estimation accuracy.

Finally, a smoothing approach based on polynomial representations and integration of the differential equations can be used.[40] Consider the following s-order polynomial representation of f_R and w_R:

$$f_R \approx \sum_{k=0}^{s}\alpha_k^R t^k \qquad \textbf{5.2(65)}$$

$$w_R \approx w_{R0} + \sum_{k=0}^{s}\omega_{k+1}^R t^{k+1} \qquad \textbf{5.2(66)}$$

Therefore,

$$\int_0^t f_R(\xi)d\xi \approx \sum_{k=0}^{s}\frac{\alpha_k^R}{k+1}t^{k+1} \qquad \textbf{5.2(67)}$$

Thus, Equation 5.2(58) is equivalent to

$$B_R[w_R - w_{R0}] = B_R\sum_{k=0}^{s}\omega_{k+1}^R t^{k+1} = A_R\sum_{k=0}^{s}\frac{\alpha_k^R}{k+1}t^{k+1} \qquad \textbf{5.2(68)}$$

As this equation is valid for all t, then

$$B_R\omega_{k+1}^R = \frac{A_R\alpha_k^R}{k+1} \qquad k = 0,\ldots,s \qquad \textbf{5.2(69)}$$

In turn, Equation 5.2(59) is equivalent to the following set of equations:

$$C_R\alpha_k^R = 0 \qquad k = 0,\ldots,s \qquad \textbf{5.2(70)}$$

For k + 1 measurements the dynamic data reconciliation problem assumes the following form:

$$\left.\begin{array}{ll} \min \sum_{k=1}^{N}[z_{R,k} - \overset{+}{z}_{R,k}]^T Q_R^{-1}[z_{R,k} - \overset{+}{z}_{R,k}] & \\ \text{s.t.} & \\ B_R(w_{R,k} - w_{R,0}) = A_R\int_0^t f_R(\xi)d\xi & (k = 0,\ldots,N) \\ C_R f_{R,k} = 0 & (k = 0,\ldots,N) \end{array}\right\} \qquad \textbf{5.2(71)}$$

Typical results are shown in Figure 5.2k.

Although a great deal of work has been performed in the field of dynamic data, its implementation in industrial sites is rare. Several approaches that are of practical value have not been described in this section. One good example is the use of orthogonal collocation[41] and the use of differential algebraic solvers.[35,42]

Software vendors are still considering dynamic data reconciliation as too computationally intensive and logistically an effort in programming that is not worthwhile. Ultimately, it will be the pressure from the practitioners, who will be not satisfied with the steady-state-based approach, that will force the commercial implementation of these techniques.

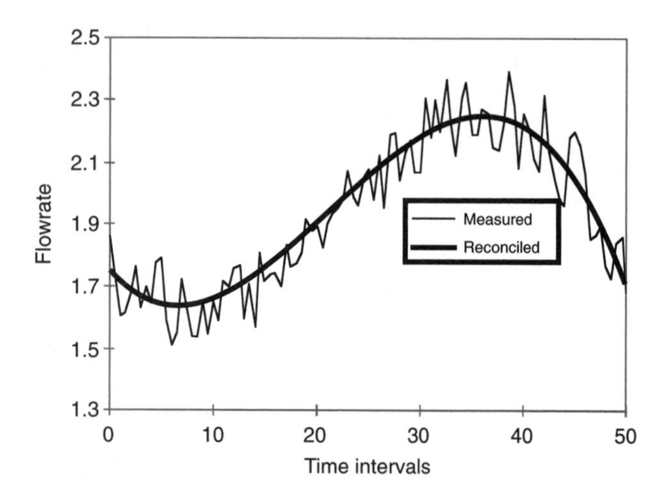

FIG. 5.2k
Typical results obtained using integral dynamic data reconciliation.

AVAILABLE SOFTWARE

Several types of software exist that are either devoted directly to data reconciliation or have data reconciliation embedded as a functionality within other programs, typically yield accounting or online optimization. Table 5.2l depicts some of these programs available to the authors by August 2001. Most of the information given in this section was obtained directly from vendors, and in many cases no independent verification was performed.

Almost all packages make use of the weighted least squares and a preprocessing that allows the determination of redundant measurements. However, some use alternative objectives. IOO (Louisiana State University) allows the use of the combined probability distribution and the Lorentzian objective function. In turn, Production Balance (Honeywell) uses ridge regression, which is a technique that avoids classification by treating unmeasured variables as measured with a high variance. This package has also some alternative classification methods.

Many packages report the precision of the estimators (reconciled values), and some perform studies of interaction between measurements and percent contributions of each measurement to the estimator (Datacon, Vali). A variety of packages also allow consideration of bounds on system variables.

Table 5.2m depicts the gross error detection method capabilities. The incorporation and enhancement of multiple gross error handling seem to be the next step for all commercial codes, and many already have it. Indeed, Datacon (Simsci) and Sigmafine (OSI) perform serial elimination based on the measurement test, but Datacon also allows the user to use principal components. Massbal (Hyprotech) performs serial compensation using generalized likelihood ratio (GLR), and Vali (Belsim) will shortly add serial elimination. Finally, departing from pure statistical approaches, Aspen Advisor relies on an expert-based system to identify gross errors and a simultaneous least-squares reconciliation for random errors.

Table 5.2n summarizes some usability features. Several packages are capable of performing material, component, and energy data reconciliation. Among those that perform energy balance reconciliation, some are equipped with very powerful property prediction methods (even liquid vapor equilibrium calculations and handling of pseudo-components), whereas others rely on limited correlation methods. This feature should be of concern at usage time because the errors included in the model due to inaccurate correlations propagate to the estimators obtained. Finally, some are able to handle pseudo-components (Datacon, Vali) and even reconcile them (Vali).

Many of these packages function together with production accounting software. The Simsci Datacon works with Open Yield, and the OSI Sigmafine reconciliation capability is just one of many other accounting features. Reconciler is available as a stand-alone DLL/ActiveX component, which

TABLE 5.2l
Data Reconciliation Packages (in alphabetical order)

Package	*Nature*	*Offered by*
Aspen Advisor	Commercial	Aspentech (USA)
Datacon	Commercial	Simulation Sciences (USA)
IOO (Interactive On-Line Optimization)	Academic	Louisiana State University (USA)
Massbal	Commercial	Hyprotech (Canada)
Production Balance	Commercial	Honeywell (USA)
Recon	Commercial	Chemplant Technologies (Czech Republic)
Reconciler	Commercial	Resolution Integration Solutions (USA)
Sigmafine	Commercial	OSI (USA)
Vali	Commercial	Belsim (Belgium)

TABLE 5.2m
Gross Error Handling in Data Reconciliation Packages

Package	*Gross Error Detection/Identification Options*
Aspen Advisor	Expert system
Datacon	Global test; serial elimination using the measurement test and principal components
IOO (Interactive On-Line Optimization)	Joint distributions and robust Lorentzian on function
Massbal	GLR with serial compensation
Production Balance	Global test; measurement test using Student t distribution
Recon	Measurement test
Reconciler	GLR and simplified measurement and nodal tests
Sigmafine	Serial elimination using *ad hoc* nodal and measurement tests
Vali	Global test; serial elimination to be added

TABLE 5.2n
Functionality Features in Data Reconciliation Packages

Package	*Models*	*Built-In Thermodynamic Property Prediction*	*Process Units Models*
Aspen Advisor	Material balances Component balances Utility balances	No	No
Datacon	Material balances Component balances Energy balances Reactors	Yes (all)	Yes
IOO (Interactive On-Line Optimization)	Defined through user-defined equations	No	No (equation oriented)
Massbal	Material balances Component balances Energy balances	Yes	Yes
Production Balance	Material balances	No	No-only nodes
Recon	Material balances Component balances	No	Yes
Reconciler	Material balances	No	No (equation oriented)
Sigmafine	Material balances Component balances Energy balances	Yes (enthalpy correlations)	No-only nodes
Vali	Material balances Component balances Energy balances Reactors	Yes (all)	Yes

can be embedded into any program, and Production Balance, which is the name used for transfers between tanks, tanks and units, or battery limits, has specific gross error detection in movements.

References

1. Johnston, L. P. M. and Kramer, M. A., "Maximum Likelihood Data Rectification. Steady State Systems," *AIChE Journal,* Vol. 41, p. 11, 1995.
2. Crowe, C. M., "Formulation of Linear Data Reconciliation Using Information Theory," *Computers and Chemical Engineering,* Vol. 51, Number 12, pp. 3359–3366, 1996.
3. Kalman, R. E., "New Approach to Linear Filtering and Prediction Problems," *Journal of Basic Engineering,* ASME, Vol. 82D, p. 35, 1960.
4. Stanley, G. M. and Mah, R. H. S., "Observability and Redundancy in Process Data Estimation," *Chemical Engineering Science,* Vol. 36, pp. 259–272, 1981.
5. Maquin, D., Luong, M., and Ragot, J., "Observability Analysis and Sensor Placement," Presented at Safe Process '94 IFAC/IMACS Symposium on Fault Detection, Supervision and Safety for Technical Process, June 13–15, Espoo, Finland, 1994.
6. Maquin, D., Luong, M., and Paris, J., "Dependability and Analytical Redundancy," Presented at IFAC Symposium on On-Line Fault Detection in the Chemical Process Industries, Newcastle, U.K., 1995.
7. Bagajewicz, M. and Sánchez, M., "Design and Upgrade of Non-Redundant and Redundant Linear Sensor Networks," *AIChE Journal,* Vol. 45, Number 9, pp. 1927–1939, 1999.
8. Madron, F., *Process Plant Performance,* Chichester, U.K.: Ellis Horwood, 1992.
9. Crowe, C. M., García Campos, Y. A., and Hrymak, A., "Reconciliation of Process Flow Rates by Matrix Projection. Part I: Linear Case," *AIChE Journal,* Vol. 29, pp. 881–888, 1983.
10. Swartz, C. L. E., "Data Reconciliation for Generalized Flowsheet Applications," Presented at American Chemical Society, National Meeting, Dallas, TX, 1989.
11. Sánchez, M. and Romagnoli, J., "Use of Orthogonal Transformations in Data Classification—Reconciliation," *Computer and Chemical Engineering,* Vol. 20, pp. 483–493, 1996.
12. Kuehn, D. R. and Davidson, H., "Computer Control. II. Mathematics of Control," *Chemical Engineering Progress,* Vol. 57, p. 44, 1961.
13. Bagajewicz, M. and Jiang, Q., "Comparison of Steady State and Integral Dynamic Data Reconciliation," *Computers and Chemical Engineering,* Vol. 24, Number 11, pp. 2367–2518, 2000.
14. Bagajewicz, M. and Gonzales, M., "Is the Practice of Using Unsteady Data to Perform Steady State Reconciliation Correct?" Presented at AIChE Spring Meeting, Houston, TX, April 2001.
15. Rollins, D. K. and Devanathan, S., "Unbiased Estimation in Dynamic Data Reconciliation," *AIChE Journal,* Vol. 39, p. 8, 1993.
16. Almasy, G. A. and Mah, R. S. H., "Estimation of Measurement Error Variances from Process Data," *Industrial and Engineering Chemistry Research. Process Design and Development,* Vol. 23, p. 779, 1984.
17. Darouach, M., Ragot, R., Zasadzinski, M., and Krzakala, G., "Maximum Likelihood Estimator of Measurement Error Variances in Data Reconciliation," *IFAC AIPAC Symp.,* Vol. 2, pp. 135–139, 1989.
18. Keller, J. Y., Zasadzinski, M., and Darouach, M., "Analytical Estimator of Measurement Error Variances in Data Reconciliation," *Computers and Chemical Engineering,* Vol. 16, p. 185, 1992.

19. Chen, J., Bandoni, A., and Romagnoli, J. A., "Robust Estimation of Measurement Error Variance/Covariance from Process Sampling Data," *Computers and Chemical Engineering,* Vol. 21, Number 6, pp. 593–600, 1997.
20. Rollins, D. K. and Davis, J. F., "Unbiased Estimations of Gross Errors in Process Measurements," *AIChE Journal,* Vol. 38, pp. 563–572, 1992.
21. Mah, R. S. H. and Tamhane, A. C., "Detection of Gross Errors in Process Data," *AIChE Journal,* Vol. 28, p. 828, 1982.
22. Sidák, Z., "Rectangular Confidence Regions for the Means of Multivariate Normal Distributions," *American Statistical Association,* Vol. 62, pp. 626–633, 1967.
23. Rollins, D. K., Cheng, Y., and Devanathan, S., "Intelligent Selection of Hypothesis Tests to Enhance Gross Error Identification," *Computers and Chemical Engineering,* Vol. 20, Number 5, pp. 517–530, 1996.
24. Rollins, D. K. and Davis, J. F., "Gross Error Detection When Variance–Covariance Matrices Are Unknown," *AIChE Journal,* Vol. 39, Number 8, pp. 1335–1341, 1993.
25. Crowe, C. M., "Observability and Redundancy of Process Data for Steady State Reconciliation," *Chemical Engineering Science,* Vol. 44, pp. 2909–2917, 1989.
26. Narasimhan, S. and Mah, R. S. H., "Generalized Likelihood Ratio Method for Gross Error Identification," *AIChE Journal,* Vol. 33, pp. 1514–1521, 1987.
27. Tong, H. and Crowe, C. M., "Detection of Gross Errors in Data Reconciliation Using Principal Component Analysis," *AIChE Journal,* Vol. 41, pp. 1712–1722, 1995.
28. Chen, V. C. P., Melendez, M., and Rollins, D. K., "The Problem of Too Much Power in Detecting Biases in Real Chemical Processes," *ISA Transactions,* Vol. 37, pp. 329–336, 1998.
29. Bagajewicz, M., Jiang, Q., and Sánchez, M., "Removing Singularities and Assessing Uncertainties in Two Efficient Gross Error Collective Compensation Methods," *Chemical Engineering Communications,* Vol. 178, pp. 1–20, 2000.
30. Bagajewicz, M. and Jiang, Q., "Gross Error Modeling and Detection in Plant Linear Dynamic Reconciliation," *Computers and Chemical Engineering,* Vol. 22, Number 12, pp. 1789–1810, 1998.
31. Iordache, C., Mah, R., and Tamhane, A., "Performance Studies of the Measurement Test for Detection of Gross Errors in Process Data," *AIChE Journal,* Vol. 31, p. 1187, 1985.
32. Jiang, Q. and Bagajewicz, M., "On a Strategy of Serial Identification with Collective Compensation for Multiple Gross Error Estimation in Linear Data Reconciliation," *Industrial and Engineering Chemical Research,* Vol. 38, Number 5, pp. 2119–2128, 1999.
33. Tjoa, I. B. and Biegler, L. T., "Simultaneous Strategies for Data Reconciliation and Gross Error Detection of Nonlinear Systems," *Computers and Chemical Engineering,* Vol. 15, Number 10, pp. 679–690, 1991.
34. Mah, R. S. H., "Letter to the Editor," *Computers and Chemical Engineering,* Vol. 21, Number 9, p. 1069, 1997.
35. Albuquerque, J. S. and Biegler, L. T., "Data Reconciliation and Gross-Error Detection for Dynamic Systems," *AIChE Journal,* Vol. 42, Number 10, p. 2841, 1996.
36. Bagajewicz, M. and Mullick, S., "Reconciliation of Plant Data. Applications and Future Trends," Presented at AIChE Spring Meeting and First International Plant Operations and Design Conference, Houston, TX, March 1995.
37. Stanley, G. M. and Mah, R. H. S., "Estimation of Flows and Temperatures in Process Networks," *AIChE Journal,* Vol. 23, Number 5, p. 642, 1977.
38. Karjala, T. W. and Himmelblau, D. M., "Dynamic Rectification of Data via Recurrent Neural Nets and the Extended Kalman Filter," *AIChE Journal,* Vol. 42, p. 2225, 1996.
39. Darouach, M. and Zasadzinski, M., "Data Reconciliation in Generalized Linear Dynamic Systems," *AIChE Journal,* Vol. 37, Number 2, p. 193, 1991.
40. Bagajewicz, M. and Jiang, Q., "An Integral Approach to Dynamic Data Reconciliation," *AIChE Journal,* Vol. 43, p. 2546, 1997.
41. Liebman, M. J., Edgar, T. F., and Lasdon, L. S., "Efficient Data Reconciliation and Estimation for Dynamic Process Using Nonlinear Programming Techniques," *Chemical Engineering Science,* Vol. 16, Number 10/11, p. 963, 1992.
42. Albuquerque, J. S. and Biegler, L. T., "Decomposition Algorithms for On-line Estimation with Nonlinear DAE Models," *Computers and Chemical Engineering,* Vol. 19, p. 1031, 1995.

Bibliography

Madron, F., *Process Plant Performance,* Chichester, U.K.: Ellis Horwood, 1992.

Mah, R. S. H., *Chemical Process Structures and Information Flows,* Boston: Butterworths, 1990.

Narasimhan, S. and Jordache, C., *Data Reconciliation and Gross Error Detection. An Intelligent Use of Process Data,* Houston, TX: Gulf Publishing Company, 2000.

Sánchez, M. and Romagnoli, J., *Data Processing and Reconciliation for Chemical Process Operations,* San Diego, CA: Academic Press, 2000.

5.3 Sequence of Event Recorders and Post-Trip Reviews

A. ROHR

Sequence of event recorders and post-trip review packages are typically used in the power-generation industry with the aim of giving operators and plant management the tools to examine:

1. Reasons a plant upset has brought fatal consequences such as a plant shutdown or operation to island mode (operation disconnected from the grid, with the generator feeding the auxiliaries only)
2. Consequences of a sudden equipment stop

Even though these two functions are distinct, quite often they are considered together because they are complementary in the study of a shutdown (causes and effects, respectively).

SEQUENCE OF EVENT RECORDERS

The sequence of event recorder (SOE or SER) can be considered as having evolved from the first-out sequence of alarms in use for many years in process plants, where a limited number of initiating causes are involved in tripping the equipment, and only the initial cause is of real interest. In these cases, the process is normally relatively slow, and therefore does not require highly sophisticated equipment for determining the first cause of a trip.

In the power industry, the process is constituted not only by boilers and turbines, but also by electrical equipment (generators, transformers, and the grid) that have a faster response and are sensitive to many independent causal variables that can lead to a trip. Some of these causes, it should be noted, can be imported from the grid rather than originating in the plant itself. The number of possible causes involved normally ranges in the hundreds for a large power plant down to 50 in a simple, small power plant connected to the national grid. Contrary to what happens in the process industry, the first initiating cause is not the only one of interest, because sometimes the various events are linked together and other times they are independent and superimposed. This originates the need to know in detail the correct sequence of switching of the primary devices under consideration, in order for the study of the initiation and evolution of a trip to be carried out successfully.

Because of the speed of electrical phenomena, the SOE should have a time resolution in the millisecond range; i.e., it should be able to show in the correct sequence two events that have taken place with only 1 ms difference between them.

The time resolution of 1 ms is not strictly necessary in all instances, because the switching time of many electrical devices being monitored is normally much longer (several milliseconds). A time resolution up to 4 to 5 ms is normally adequate for the purpose, and is acceptable for the National Grid Dispatching Authority. However, because of the current availability of systems capable of a time resolution of 1 ms or less, the standard specifications of many companies call for such time resolution.

Architectures

Up until the early 1980s, the only solution was to use dedicated equipment, capable of recognizing and time-stamping the switching conditions of several devices with a time resolution of 1 to 5 ms. Dedicated equipment is still currently in use, and is required by the electric companies that operate the national grid. As compared with the earlier equipment, they now have much faster processors, incorporate a CRT (cathode ray tube) in the cabinet with the electronics, and drive a printer, thus providing a complete interface to the operator. Additionally, they are capable of being interfaced with slave SOE via serial link, covering large and geographically wide applications. In this case, the time correlation of the events managed by the SOE and of the other events managed by the control system is not so easy and reliable, both because of the synchronization of the various apparatuses and because of the independent presentations to the operators. Figures 5.3a and 5.3b show, respectively, a diagram and a picture of a self-standing SOE.

The architecture of a self-standing SOE can be distributed, in that several satellite processors read every 2 to 5 ms all the input cards connected to them and time-stamp the detected switching of the field contacts. A main processor keeps the satellite processors synchronous, merges all data coming from the satellite processors, and makes them available in a time-ordered manner. The input cards do not have a processor on board, because the cycle time of the satellite processors is kept within the required time resolution by limiting the number of the connected input cards.

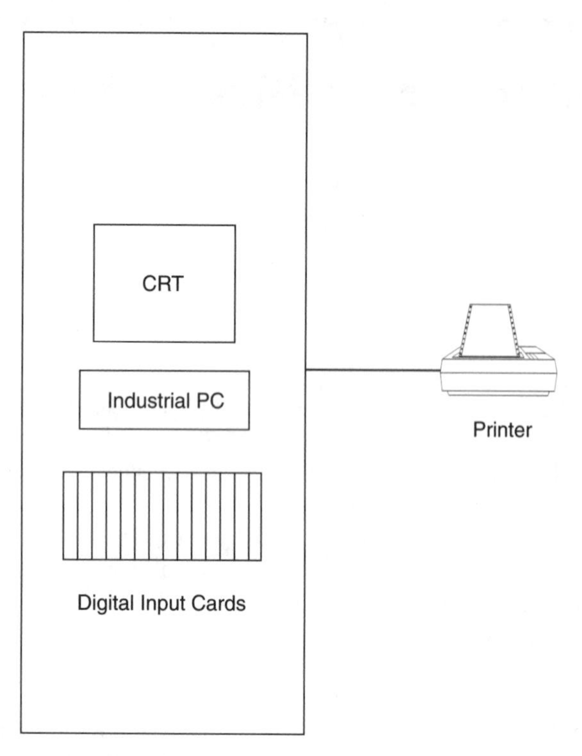

FIG. 5.3a
Self-standing SOE.

FIG. 5.3b
Self-standing SOE. (Courtesy of TELEGYR Systems, Italy.)

At the time of introduction of the distributed control systems (DCS), the idea of integrating the SOE functionality inside the DCS was not broadly implemented because of limited capacity of the DCS and because the first DCS were targeted to the process (oil and gas) industry, where the need for SOE was not especially relevant. Only companies with long histories in the power-generation industry built the possibility of SOE functionality into their DCS, even though with some time resolution limitations (i.e., resolution of 2 to 4 ms, instead of the 1 ms normally used and accepted today).

Current high-level DCS make available specific digital input cards dedicated to SOE function, whereas entry-level DCS normally do not have this facility.

The specific DCS hardware needed to perform sequence-of-event recording consists of a set of digital input cards with a microprocessor on board capable of time-stamping the switching events in the process, because the cycle time of the multiloop controller is much longer than the required time resolution. The events, with associated time of occurrence, are read by multiloop controllers and then made available to the HMI (human–machine interface) for display on-screen and in printout form. Figure 5.3c shows the architecture of a fully integrated SOE into a DCS. It is to be noted that the printing can be delayed by several seconds with respect to the event occurrence, because of the time needed for formatting the information and for printing the full set of events, but the important fact is that the correct sequence is maintained.

If the DCS hardware does not include special digital input cards as mentioned before, an external, independent SOE can

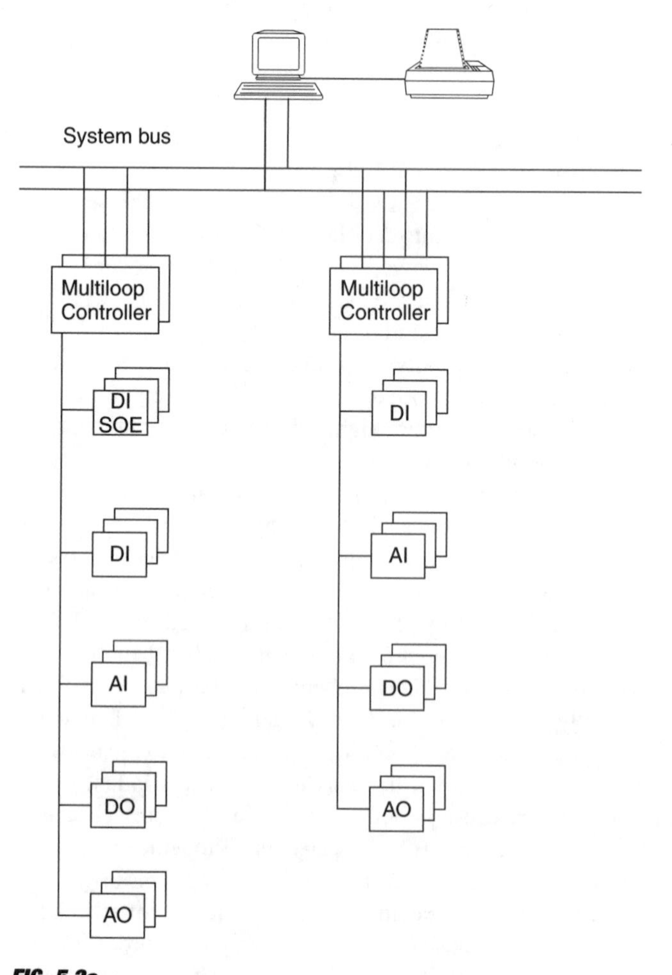

FIG. 5.3c
DCS with fully integrated SOE.

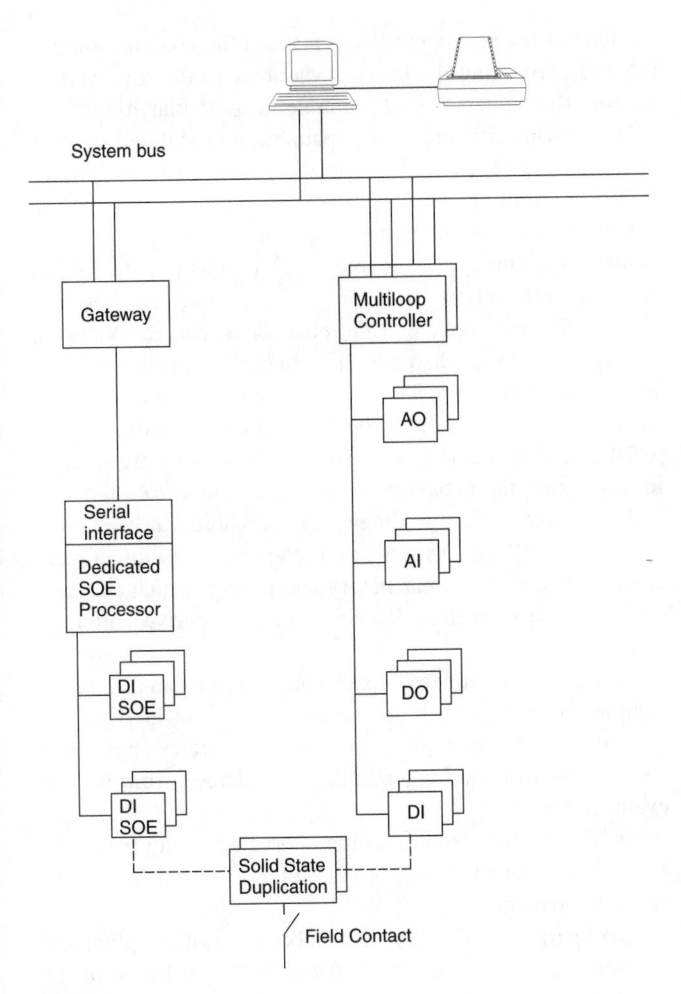

FIG. 5.3d
Independent SOE with serial link to a DCS and solid-state duplication of digital input.

be provided and serially linked to the DCS. The time stamping occurs at the level of the external SOE and is transmitted over the serial link. The serial link is normally a MODBUS, whose broad application makes it easily implementable. Figure 5.3d depicts this architecture, which is often used with entry-level DCS.

Time Consistency and Integrity

The major issue in an SOE application is the consistency of time, both with respect to all points involved in the SOE and with respect to all other alarm points in the plant.

Consistency among all SOE points of a DCS is easily obtained if all SOE digital input cards are connected to the same multiloop controller that sets the time to all the digital input card clocks, to keep them synchronous. If, instead, SOE digital input cards are connected to different multiloop controllers, it is possible for the relevant clocks to slip by more than 1 ms, with variable timing, thereby introducing different time references for the cards connected to different multiloop controllers, and therefore resulting in an inconsistent time stamping.

The consistency in time between SOE points and other points of the plant is essential for correlating the behavior of the electric plant with the process, in order to understand the causes of a malfunction. This is easily accomplished if the SOE function originates inside the same DCS, but it may become difficult if an independent SOE system is used. To overcome this problem, one must synchronize the DCS with the independent SOE, for example, by using the same synchronizing clock, a so-called plant master clock, to give the same reference time to the different plant systems.

The same problem occurs if the SOE functions of different power stations or electrical substations have to be correlated and merged to understand what went wrong in the electrical grid. Nowadays, this issue—a difficult problem in the past—is solved by synchronizing each of the systems with their respective system clocks, and all the system clocks are synchronized with the same reference time at millisecond, by radio or satellite signals, for example.

There is another potential problem related to the time consistency of the points pertaining to the SOE. In some instances there is a need to use the same initiating contact for both SOE and for other purposes, e.g., in a logic circuit. If the latter is independent from the SOE, a duplication of the initiating contact is the standard way to proceed. If duplication is obtained via an electromechanical relay, a time delay will be introduced, ranging from few milliseconds to dozens of milliseconds, possibly making the order of event occurrences unascertainable. This situation is more frequent when the SOE is independent from the DCS, because an independent SOE has no logic capabilities, whereas the DCS with integral SOE also has logic capabilities using the digital inputs from SOE cards. This problem can be overcome by using solid-state duplicating modules, which do not introduce a delay of a magnitude that would impair their use for this purpose, in that the delay ranges in the microseconds, if the equipment has been properly selected. Figure 5.3d shows this solid-state duplicating module, feeding the signal both to the DCS and to the SOE.

The dedicated SOE have the possibility of associating to each point an independently adjustable lead time, to take into consideration the possible inertia or the switching time of the primary sensing element. This possibility is not so common in the DCS, or it can be cumbersome, because the SOE functionality is not the main task of a DCS.

Printout

The presentation to the operator of an SOE is very similar to the alarm presentation, with the difference that the time is expressed to the millisecond.

An example is given in Table 5.3e.

Correlating the printout of the SOE with the printout from the plant alarms facilitates a review of the events leading to a major plant upset. Such a review defines the reasons for a shutdown and the sequences of other actions on the overall

TABLE 5.3e
A Sequence of Events Recording

Date	*Time*	*Description*	*Status*
03-05-01	12:23:34:653	Ground fault line A1 detected	Alarm
03-05-01	12:23:34:655	Circuit breaker G-1 open	Alarm
03-05-01	12:23:34:660	Bus-tie 11 kV close	Alarm
03-05-01	12:23:34:914	Steam turbine overspeed trip	Alarm
03-05-01	12:23:35:213	Steam turbine bypass open	Alarm
03-05-01	12:23:38:715	Ground fault line A1 recovered	Normal
03-05-01	12:23:38:717	Circuit breaker G-1 available	Normal

plant, to verify if the behavior is in line with expectations and has not led to a dangerous situation.

POST-TRIP REVIEW

In processes characterized by high heat flux, a sudden shutdown of the plant can lead to abnormal conditions, mainly for temperature and pressure, that can exceed the design conditions of equipment and piping.

This situation is particularly true when solid fuel is burned, because it cannot be suddenly extinguished, and keeps generating heat, even when the heat cannot be transferred or removed with the same efficiency and effectiveness as before the trip. This is what happens in waste-to-energy plants. The fuel typically burned in the furnace consists of solid waste materials whose high heat flux cannot be extinguished in a reasonable amount of time with the consequential risk of serious damage to the structures of the equipment directly involved in the combustion process even if precautions are adopted.

Reference is made in particular to the following equipment:

- Combustion section (rotary or grate furnace)
- Heat recovery steam generator
- Gas treatment section
- Slag and ash inertization section
- Thermal cycle
- Turbogeneration unit
- BOP systems

For the above-mentioned systems, it is absolutely mandatory to monitor all the parameters (typically no more than ten analog signals per system), which can reveal if allowable stresses have been exceeded and for how long.

Even when burning liquid or gaseous fuel, the heat stored in the refractories can be high enough to overstress the construction materials.

To verify if allowable stresses have been exceeded, software packages have been designed to verify the values of some significant variables (e.g., skin point temperature, pressures, etc.) to be able to monitor their behavior both some time before and some time after the trip.

One of the possible ways to achieve the goal of monitoring and correlating the process variables to the trip event is to store the values of each variable in a circular file, of the FIFO (first in, first out) type, spanning a predefined number of minutes, selectable by the user (typically 15 or 20 min). If, for example, the chosen time span is 15 min, this means that at any given moment the history of the last 15 min of the points under survey is available, and after this time the oldest data are overwritten.

In this case, on the occurrence of a trip, the software package is configured so that the data collection continues for 10 min and then stops. The file would thus contain the values pertaining to 5 min before the trip and the values pertaining to 10 min after the trip. Data from the first 5 min are suitable for analyzing the behavior of the equipment before the trip, and hence for helping to diagnose its possible causes.

The last 10 min, subsequent to the trip, show the process values reached in the machinery and piping, which can then be compared with the allowable values to derive abnormal stress conditions.

With some commercial packages, the user can also select sample intervals, which can be either constant in period or variable, in the latter instance with a shorter sample time close to the trip and longer as the time elapsed from the trip event increases.

With variable sampling, time intervals ranging from 10 s to 1 min can be used, depending on the time constant of the variable involved.

To facilitate understanding the behavior of the plant and evaluate the stresses incurred, it can be helpful to split the data into different groups, each one referenced to a predefined plant area (boiler, turbine, etc.). Each group will generate its own report on a trip event.

These files, related to each shutdown event, have to be saved and stored to document the operational conditions of the plant during emergency shutdown.

With current DCS, the post-trip review function can be integrated with the historical data archiving and retrieval package. The sampling time can be adjusted so that the stored information is actually representative of the behavior of the plant. To preserve such a history, however, it is necessary to copy the set of values into a permanent file before the data are squeezed or reduced because of their aging.

In fact, some DCS archive data in the following way: when the data become "old," they are squeezed or reduced to obtain for a certain time duration the average, minimum, and maximum values. Although this information is sufficient in steady-state (or quasi-steady-state) operation, after a trip, this information can no longer provide enough detail to determine whether the equipment has been overstressed and, if so, to what extent.

Other software packages have been developed to control closely the behavior of the main machinery of a power station, such as the stress evaluators for turbines and boilers. These packages can be usefully adopted in conjunction with a post-trip review package, to go deeper into the evaluation of the consequences of a sudden shutdown. The information obtained as a synthesis from the set of the above-described software packages is suitable to lead operators and management to a complete knowledge of plant behavior, to take the proper correcting actions necessary to avoid the repetition of improper shutdown, and to evaluate if plant safety has been impaired.

PRICE CONSIDERATIONS

The following considerations are based on the implementation of a system that has previously been fully studied and engineered, so that the configuration time does not include any search for information, nor discussions on the best solution, etc.

SOE System

A self-standing SOE, with 700 points capability, has a price of approximately $40,000; such a system includes an industrial PC complete with CRT and printer. It does not include the serial link or the synchronization with an external clock, whose price ranges from $3000 to $5000.

An SOE fully integrated in a DCS has a price defined by the cost of the license (about $5000), plus the price of DI-SOE cards. The price of the DCS multiloop controller has to be considered separately because different levels of redundancy can be implemented and it can be shared with other control functions.

A simple DI card with 32 points has a price in the range of $800, while a DI-SOE card with 16 points costs about $1000.

A solid-state duplicating input module with one input channel and two output channels has a price of about $25.

The configuration price can be based on the cost of one manhour per 15 points.

Post-Trip Review

In this case the price is almost irrelevant, in that no additional licenses are normally needed and the configuration requires only to fill some additional fields in the existing blocks, requiring a minimal time consumption.

5.4 OPC Software Architecture

J. BERGE

- Pay attention to the system software architecture when choosing a system.
- Use hardware with native auto-configured OPC servers.
- Use software with direct-access OPC clients.
- Use redundant OPC servers where high availability is required.
- OPC servers such as the one for Smar DFI302 Foundation™ Fieldbus linking device cost from U.S.$280 up.
- OPC clients such as Iconics Genesis32 process visualization software cost from U.S. $1000 up.

A modern control system needs more than just configuration and monitoring software, more than inflexible, "closed" applications and inaccessible data associated with proprietary distributed control systems (DCS). A DCS has fixed proprietary operation software that cannot be changed, and hardware and databases, that cannot easily be accessed. If users have a software application they prefer based on previous experience, there is no way to change the DCS to use different process visualization software. Similarly, adding new applications for statistics, advanced control, or other data processing is difficult or even impossible. A system based on OPC (OLE* for Process Control) software architecture is interoperable and much more flexible.

OPC is a standard interface between software applications, which allows individual software components to exchange data appearing to the user as a single homogeneous application. Because OPC is supported by many companies, it is possible to interconnect software applications from different manufacturers into an integrated system. OPC is based on Microsoft Windows COM (Component Object Model) architecture, using DCOM (Distributed COM) among different computers over a network.

In the past, device drivers to make software talk to hardware and databases from different manufacturers were developed based on unique and complex APIs (application programming interfaces). Such driver development was time-consuming because documentation was often lacking and one driver for every software and hardware combination had to be developed, especially if several different kinds of hardware and software were used in the same system. The API solution was costly to the end user because manufacturers had to spend a lot of effort on every driver amortized over few copies sold of each. API requires programming knowledge; hence, users could not perform the integration themselves. Frequently, a desired combination of hardware and software was not possible because of lack of drivers. Microsoft Windows DDE (Dynamic Data Exchange) was a definite improvement, as it is a simpler interface between applications for exchanging data, but it was not well accepted because various different and incompatible implementations for mapping data onto DDE exist. That is, true interoperability was not ensured. DDE performance was also slow, lacked network and multiple client support, and was not user-friendly as a great deal of manual configuration had to be done by the user. DDE was replaced by OLE, which still was an inefficient and unreliable mechanism for passing data between applications, with performance and reliability limitations. Because of driver conflict, usually only one piece of software at a time could be used; tedious switching between configuration tool and process visualization software was often necessary. Because software manufacturers or third parties had to develop one driver for every piece of hardware, they were unable to keep the drivers up to date with the latest device versions or even to utilize the latest features in their own software. No independent testing of the driver existed to ensure the quality of the interoperability.

To sum up device driver problems:

- Driver not available for the desired hardware, database, and software combination
- Drivers not fully supporting hardware or database capabilities
- Drivers not fully supporting software capabilities
- Drivers not supporting latest version of hardware or database
- Drivers not supporting latest version of software
- No network support
- Driver conflicts limiting access to one application at a time
- No interoperability testing
- Expensive

Other past problems include that the client application cannot see what data are available from the driver. Data therefore

*OLE = Object Linking and Embedding.

had to be keyed in by the user to enter the database. Easy integration between applications from different manufacturers could not be achieved in the past.

OPC INTERFACES

Numerous proprietary interfaces existed. Clearly, the situation was unacceptable to both users and manufacturers of hardware as well as software. OPC was developed by the OPC Foundation to provide a standard software interface to hardware and databases. That is, OPC is useful not only for access of data in hardware, but also for databases such as alarm and event logging and historical trending. COM is the Microsoft Windows software infrastructure that is the basis for OPC. COM and DCOM are major improvements over DDE and OLE, because they allow software functions to be distributed across computers on a network and ensure the efficient exchange of data.

Today, OPC is a widely accepted industry standard client/server technology used to interchange data between applications tailored for automation. The server is a piece of software associated with the data source, such as hardware or a database. The client is the application that interfaces to the user for display and data entry. The OPC client collects data from OPC servers that are running on any computer on the network, and the OPC client application can also write data through the OPC server (Figure 5.4a). There are three flavors of OPC: data access (DA), alarms and events (A&E), and historical data access (HDA). A DA server is used for access of data in hardware, an HDA server is used for access of data in a historian database engine, and an A&E server notifies clients of occurrences of an alarm condition or event. There are two types of interfaces between clients and servers: custom and automation. Custom interfaces are the most common and are used between most clients and servers purchased from software and hardware suppliers. Automation interfaces allow users to develop in-house scripting and applications, using, e.g., Visual Basic, which can access the server.

In a system, one server for each device type or protocol is usually required. For example, there may be one server for MODBUS devices and another for Foundation Fieldbus devices (Figure 5.4b). Multiple clients can simultaneously access one or more servers located in the same computer or remotely over a network. All workstations access the same information, eliminating inconsistencies.

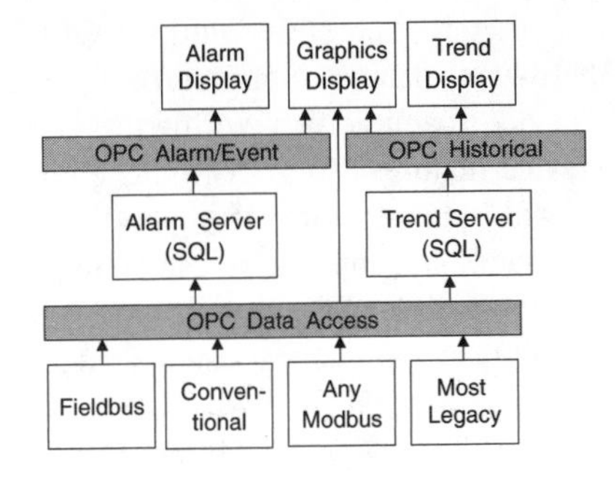

FIG. 5.4a
Three flavors of OPC.

FIG. 5.4b
Hardware such as a Foundation HSE Fieldbus linking device comes with an OPC server. (Courtesy of Smar.)

OPC is a software interface for exchange of data between software applications, i.e., an OPC client is one application and an OPC server is another. OPC does not take the place of fieldbus networks. Rather, OPC takes the place of the "device driver" software previously required for software applications to communicate with hardware. Thus, OPC can be likened to a "software backplane" or a "software bus" where various applications simply plug in.

The OPC clients automatically start the OPC servers. There is no need to initialize manually. Launch and access permission is controlled by the DCOM security in Windows.

OPC does not standardize semantics of data. Care must therefore be taken when OPC is used to integrate different device types and protocols. For example, in one device, manual mode may be achieved by writing the value 16 to a parameter called MODE_BLK, while in another device manual mode is achieved by setting a bit to one in register 00041. For this reason data manipulation is usually required in the client end.

Modern networked system architecture carries so much information that mapping device address, memory register, data type, scaling, etc. into a database the traditional way is impractical. In most modern systems the OPC server is automatically configured based on the control strategy, making all data available. To simplify client application configuration, all the server data are visible and available simply by pointing and clicking in a universal tag browser in the OPC client (Figure 5.4c). For example, if a function block is included in the control strategy using the Foundation Fieldbus programming language, all the parameters in that block become available from the OPC server automatically. When building a screen in the OPC client, users simply select the parameters to be displayed without having to key in any tag or location.

Data in OPC servers are named, often with tags defined by the user. Typically, the tags are defined in the control strategy tool and exported to the OPC server together with

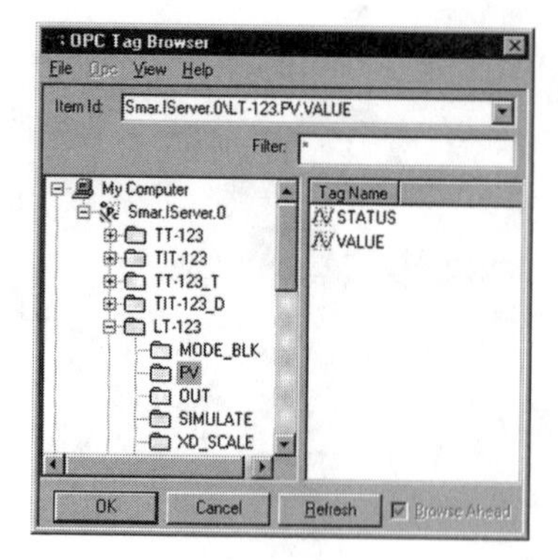

FIG. 5.4c
Point and click in tag browser to access data.

all location information required for access. Users simply click on the desired parameter in the OPC client and the server will access the data. This eliminates mapping, cross-referencing, and all worries about addresses. The browser interface is optional but is supported by a vast majority of servers and clients. Data in the same or networked computer are accessed the same way. The data are arranged into a hierarchical namespace, a treelike structure organized in some logical way.

A DA OPC server communicates with the hardware and therefore is designed to handle communication protocol and other specific characteristics of the hardware, i.e., it essentially contains an invisible device driver but serves data in a common, standard way. OPC encapsulates and hides the complexity of hardware and communication from the user. Users need not know the intricacies of how, e.g., fieldbus communication works. All one need do is specify the information needed and the OPC server will obtain it and the OPC client will display it. OPC clients create groups in OPC servers with the subset of information in which they are interested. Items are added and deleted without shutting down. OPC clients set the sampling rate for each group. The OPC server can start and stop scan of groups depending on whether or not one displayed the data in it or used it otherwise.

Until now historians in systems have been using proprietary interfaces, making it difficult or impossible to access the data live from a DCS without transfers and conversion. A historical database engine stores the raw data, typically in an SQL server database, and an OPC HDA server is used for client applications to retrieve data from the historical archive to display process variable trends.

BENEFITING FROM OPC

OPC is tailored for the needs of industrial software. One of the major benefits of OPC is that it makes information from the field available to a vast array of software for supervisory control, historical trending, advanced control, and auto-tuning, as well as inferential sensing, modeling, advanced control, optimization, and historians, without the need for any drivers. There are even OPC bridges that pass data from one OPC server to another such as from one system or device to another. All kinds of software from different suppliers can therefore be added to a system. New types of software from many suppliers become available every day. Because OPC servers have been developed for new as well as old field equipment such as legacy DCS, etc., OPC is a convenient way to connect existing equipment in a plant retrofit scenario, or to integrate other subsystems. OPC makes things work together.

OPC is not only an interesting technology, it also translates into real savings. Savings resulting from OPC can be divided into two categories: savings thanks to simpler integration and savings as a result of leveraging increased information access. In addition to these user benefits, OPC brings many benefits to manufacturers as well, which is a major reason for the success of OPC.

OPC is plug-'n'-play and therefore extremely easy and convenient to use. Servers throughout the system are automatically listed and can easily be selected. Engineers do not have to learn intricacies of communication protocols, etc. Because OPC is completely tag based using the universal tag browser interface, one receives a list of data available from the server, and simply points and clicks to obtain the desired data. Mapping of addresses, data types, scaling, bits and bytes, etc. as was common with old PLC (programmable logic controller) and HMI (human–machine interface) combinations is not required. Thus, the beauty of the universal tag browser is that using OPC it is possible not only to link to data, but also to find the piece of information through simple pointing and clicking. Typically, once a tag has been created in the engineering tool, it becomes available for access by all other client applications with the same name throughout the system. It is not necessary to retype any tags, eliminating problems due to typographical errors. OPC instead gives an unprecedented level of integration and ease of use, yet complete openness and security. Ultimately, ease of use results in lower training costs.

The interoperability of OPC makes it possible to select any software and any hardware from thousands of devices and applications available. In the past, even for "open systems," device drivers for many combinations of hardware and software were not available. This limited options. OPC interoperability dramatically increases the number of available choices and therefore possible solutions. One can choose the desired hardware, database, and software combination. Because the device manufacturer writes the OPC server it supports all device features. Similarly, OPC clients are written by the software manufacturer to support all OPC capabilities. Thus, devices and application can be improved while still remaining compatible. Project risk is lowered because OPC interoperability is compliance-tested, resulting in a minimum of surprises. Thus, OPC standardization ensures that applications from different manufacturers can interoperate. In the past users were limited to a narrow selection of software that supported the desired hardware, or a limited choice of hardware supported by the software. With OPC, users are not restricted to a single vendor.

Many different types of software quickly became available for OPC because standardization provided stability, making it worthwhile for suppliers to develop software based on OPC. Simple systems could at an early stage be fitted with statistical process control (SPC), auto tuning, and a tie-in with advanced control, enterprise resource planning (ERP), and more functions than proprietary DCS costing several times as much.

For example, operation, historical trending, and alarm log can be software applications from different vendors. For example, a "standard" off-the-shelf software package may include trending and alarm, but more powerful third-party historian and alarm analysis software can be added later if required. Users are no longer forced to choose particular software for a system.

Because a software or hardware supplier need only develop a single OPC client or server instead of multiple drivers, the product cost is reduced. A single OPC server from which all clients access data represents the data source. OPC thus gives the system a single integrated database even if software from several third-party suppliers is used, eliminating inconsistencies and duplication of work. Because OPC is a standard, a minimum of custom configuration is required, reducing integration costs. Drivers are eliminated and projects can also be executed faster, which results in additional savings.

In systems using device drivers and database mapping, future expansions after the original project team had been disbanded could be very time-consuming, difficult, and costly. In contrast, the standard nature and simplicity of OPC make the system easier and cheaper to maintain and to lower long-term software maintenance costs. As compared with proprietary DCS the dependency on costly after-sale support is reduced since well-publicized standard technologies empower the user to solve many problems.

Modern systems based on a networked architecture in general, but Foundation Fieldbus in particular, have access to far more information from the field than was available in the past. Process information and detailed diagnostics can now be accessed and disseminated through the enterprise without having to rely on manual data collection. Because OPC is easy to use engineers can spend less time on system integration and more time on value-added functions based on the improved access to information. Important data are no longer restricted to the plant floor; the data can propagate throughout the enterprise at every level. Disseminating and using information in various software applications can improve the plant bottom line. OPC should be used for digital integration of process information with the business environment, which is much faster than manual data collection and entry and has many fewer errors. Up-to-date information can be used in production planning, etc. OPC clients are available to integrate live data in real time into advanced ERP systems or Excel. Production-related information becomes available for use not only by operators and engineers, but also in the business domain.

In the past, upgrading software or a piece of hardware was very difficult because a single change would require one or more new drivers and might result in incompatibilities. The standard OPC interface eliminates these problems. Any software or hardware can be changed without affecting other parts of the system because the OPC interface remains the same. Any hardware differences are handled in the new OPC server so the client side is unaffected.

In the past, this was not possible and thus product development was hampered by lack of standards. In a system with OPC software architecture, newer and better products can be integrated easily. Therefore, manufacturers can introduce new features without causing incompatibilities.

SOFTWARE ARCHITECTURE

The software architecture of the control system has to be designed based on the information flow requirements in the plant. Because many important functions in a plant are performed in software, an interoperable software architecture based on OPC is the best solution. The DCS used in the past had a proprietary database requiring custom device drivers to communicate with third-party hardware. Additional drivers were usually required to interface to plant information systems and historians. Advanced control may connect using a driver to the DCS database or often through the historian. A software architecture based on OPC eliminates the need for all these drivers (Figure 5.4d).

Although OPC is a standard technology, the extent to which it can be of benefit depends on the implementation in the system. Some system architectures supporting OPC remain completely proprietary using the manufacturer's own network protocols and with fixed operator software and databases that cannot be changed. OPC is implemented as a gateway, typically in a dedicated computer. Parameters from the DCS database are mapped to an OPC server that in turn can be accessed by third-party clients. The gateway machine may also run an OPC bridge application allowing data to be

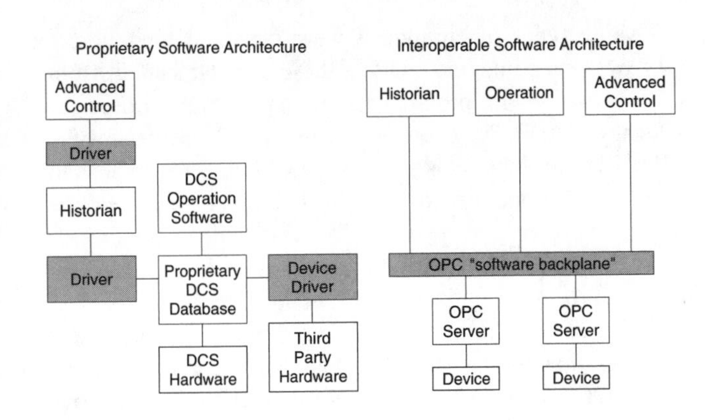

FIG. 5.4d
To integrate with other applications a DCS may require multiple drivers, whereas an interoperable system relies on OPC.

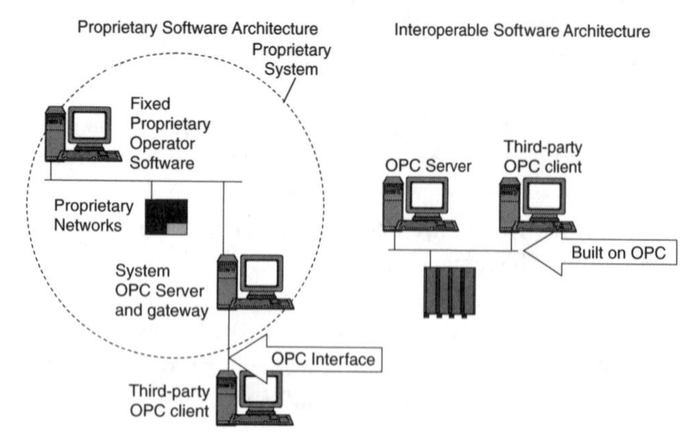

FIG. 5.4e
The OPC implementation is important to ensure interoperability among software.

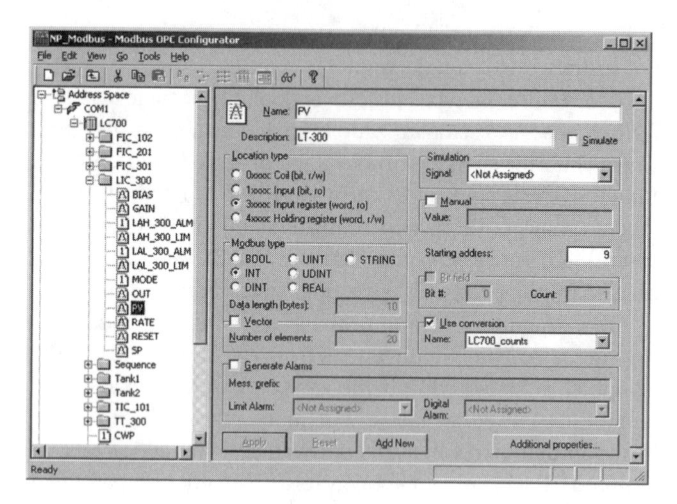

FIG. 5.4f
Configuration of generic OPC server for MODBUS protocol.

exchanged with other servers (Figure 5.4e). With this solution not all data in the system may be accessible and users are still not able to select process visualization software of their choice.

In contrast, in interoperable system software architecture, all software including third-party applications can access all data freely. Users can select process visualization and other functions of their choice.

Modern devices come with "native" OPC servers that are specifically designed for a device type and that are tightly integrated with their configuration tool. This means that when the control strategy is built all the tags and parameters are automatically exported to the OPC server and made available to all clients by the same name without any configuration effort from the user. Not only is this convenient, it also eliminates the likelihood of mistakes. A typical example of this is a control strategy using the Foundation Fieldbus programming language where the function blocks and parameters are exported to an OPC server. The list of parameters in the function blocks is based on the device description of the instruments connected to the linking device. The tag browsers in all clients all see the same tags as configured in the engineering tool.

Many types of equipment were designed long before OPC was even imagined. Often, these can also benefit from OPC using generic protocol servers or a server tailored for a legacy system. For example, a generic OPC server for the MODBUS protocol exists that allows MODBUS devices to interface to modern software. Third parties have developed OPC servers for most legacy DCS, which allows the proprietary DCS consoles with dull graphics to be replaced by modern process visualization OPC clients. Furthermore, OPC can also be used to integrate the existing DCS with a new system, allowing data to be exchanged. For a generic protocol OPC server such as for MODBUS all communications settings have to be done and, if there is no integration with the configuration tool, all parameters have to be mapped. That is, it is necessary to configure port, baud rate, parity bits, stop bit, flow control, device address, time-outs, and to map every parameter for register number, data type, scaling, etc. (Figure 5.4f).

Although such a generic server is not as convenient as a native OPC server, it offers many benefits over device drivers, for example, interoperability, multiple client access, network access, browser interface, etc. The communication settings and parameter mapping are configured only once, in the server, instead of again and again in every client. A MODBUS OPC server is an ideal way to integrate shutdown systems, weighing scales, and other subsystems.

Not all OPC clients implement OPC in the same way. For example, whereas some OPC-enabled process visualization software retain a proprietary internal database, other software, based entirely on the OPC software architecture, have completely abandoned proprietary databases. The most common type of OPC client is process visualization software from which operators can monitor the process and initiate control actions. Many process visualization software applications implement OPC just as an old device driver, i.e., they use it to map data into a proprietary configuration database from which it is then used for display, trending, alarming, etc. In such an implementation some parameter mapping and renaming is required, i.e., one piece of information may be known by different names in different applications. Modern OPC clients based on OPC use the value directly without intermediate conversions or naming. This reduces work and errors, and ensures that values are named and displayed consistently throughout the system (Figure 5.4g).

Some applications may offer both options. In this case it may make sense to take OPC via the internal database only for process variables for alarm and trending, but to bring other values directly onto the screen where alarm, trending, scaling, etc. do not make sense.

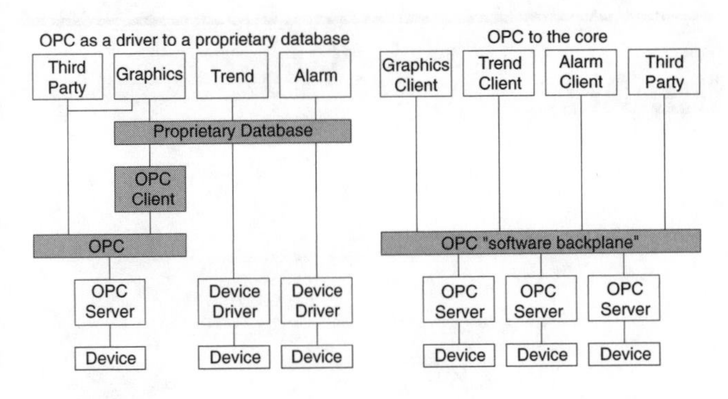

FIG. 5.4g
For complete interoperability, OPC should be used to eliminate proprietary databases.

AVAILABILITY, SAFETY, AND SECURITY

The OPC server is a critical component in the control system because, if it fails, the operators go blind and any bridging between systems is severed. If the OPC server fails the system must shut down for safety reasons.

Where a shutdown must be avoided, implementing fault-tolerant measures such as redundant OPC servers can increase the system availability. Redundant OPC servers are executed in separate computers connected to primary and secondary host-level network and device, respectively (Figure 5.4h).

A status capability is built into OPC that indicates for each variable if it is accessed "good" or cannot be accessed "bad." This allows for detection of a failed server and automatic and bumpless switching to a working one. OPC redundancy manager software is executed in each client workstation. This software looks at the primary and secondary OPC server and switches to the secondary should the primary fail. The OPC client in turn looks at the redundancy manager. The switchover is completely bumpless and automatic, and there is no need to reconfigure the client. Diagnostics is available to see if the primary is being used, which allows notification to technicians in case of server failure. Once the primary recovers, there is an option to switch back to the primary, or not.

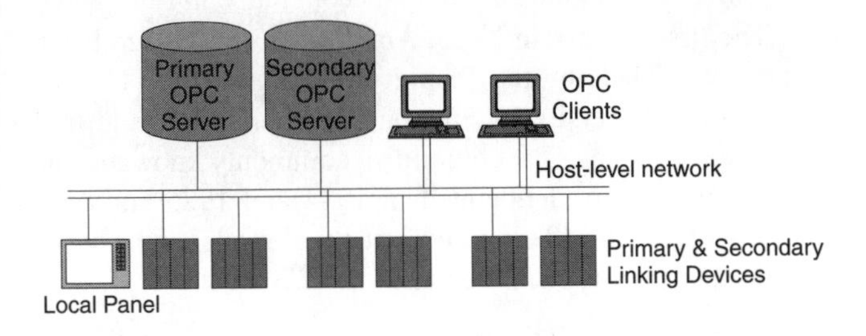

FIG. 5.4h
OPC redundancy for greater availability.

OPC is an open technology used to disseminate information across the enterprise. Thus, there is a need for security. OPC is built on DCOM and therefore inherits all the security features built into the Windows NT security framework. The security can be based on Windows NT domain or Windows NT workgroup requiring log-on with user name and password. The Windows NT server domain-based scheme has a single security database and is therefore easy to manage. DCOM provides security for distributed applications even though these applications are not specifically designed to be secure. The security for any application component, such as the OPC servers, can be customized. Windows security uses a concept of "users" and "user groups" that are assigned different access rights to files and software components. There are three security functions:

1. Authentication
2. Launch
3. Access

Before an application like an OPC server is launched or accessed, DCOM checks the rights of the client user. If the user does not have the rights, access is rejected.

Status is also used to increase safety. An OPC client must indicate "bad" OPC communication as "*" or other nonvalid symbols to the operator. To show a frozen or stale last value, which may not be valid, is dangerous. Similarly, OPC bridges must propagate "bad" status to allow for fail-safe shutdown in the receiving end. The status gives operators more confidence because they know if the value were not valid it would be displayed as "bad." A numerical value can be trusted.

INTEGRATING INCOMPATIBLE SYSTEMS AND NETWORKS

In the past, tight integration between a DCS and other systems such as emergency shutdown, paper scanner, gas chromatograph, etc. could only be achieved if the same manufacturer made them. OPC removes this limitation. Other systems in the plant, such as existing legacy controls, critical control systems for emergency shutdown, or paper machine quality controls, with large amounts of data can be tied into an OPC-based system with relative ease. A dedicated server station acting as a gateway may be fitted with the appropriate interface and run an OPC server. An OPC bridge application can be used to mirror data in one system onto the other, and vice versa. This way information can be exchanged for supervisory and control. By using OPC, more things can be integrated, things that could not be integrated before. For example, a system overview screen can show the status of the host-level network switches, etc.

5.5 Batch Control State of the Art

A. GHOSH

Partial List of Suppliers:

Batch Control Software: ABB; Emerson Process Management (Fisher-Rosemount & Intellution); GSE Systems; Honeywell; Invensys (Foxboro & Wonderware); Rockwell; Siemens; Yamatake; Yokogawa

General Recipe and Formulation Software: Agri-Data Systems; Arrow Scientific; DTW Associates; Formation Systems; Format International; Owl Software; Sequencia

Software That Help Maintaining Electronic Batch Records and Signatures: AspenTech; Documentum; Doxis

Batch Control Optimization Software: AEA Tech (Hyprotech); AspenTech; Batch Process Technologies; GSE Systems

Control and Enterprise Integration Software: AspenTech; Hewlett Packard; IBM; OSI Soft

Enterprise & Supply Chain Optimization Software: ABB (Skyva); AspenTech; Camstar; EXE Technologies; i2 Technologies; Manhattan Associates; Manugistics; SAP

The worldwide market for batch control systems (BCSs) is growing at a faster pace than the overall process control market. The need for traceability and increased EPA and FDA requirements are fueling the growth of BCSs in North America, Europe, and the rest of the world. With the approval of electronic batch record and electronic signature by the FDA in the United States, there is now a greater incentive for regulated industries, such as pharmaceutical and biotech, to move to paperless systems. Many companies in these regulated industries are already using electronic data collection to produce paper records. These electronic-paper systems must be validated and made compliant or be replaced by a system capable of compliance. This is leading to increased investments in automating and upgrading their control systems.

Growth of E-business and the need to optimize the supply chain are increasing the need for real-time plant and production information, which is fueling the growth of manufacturing automation and its integration to business systems. Market globalization and increased competition are fueling the need for multifunction plants with improved operational efficiencies. In turn, these are increasing the need for more efficient and flexible manufacturing.

Increasing competition between different control system suppliers and continued consolidation in the entire automation supplier market is significantly changing the depth and breadth of products and services. Control systems suppliers continue to acquire other control systems companies and companies that offer automation products and services in the enterprise level. Major benefits of acquisitions for suppliers include an increased distribution network, an increased installed base, broader market presence, and expansion of product and services offerings.

BATCH CONTROL STANDARDS

Until 1995, there was no universal batch control standard. The lack of batch standardization raised the degree of difficulty for successful batch automation. The ISA/SP88 committee was formed in 1988 to rectify this situation, and in 1995, it published the first part of the standard.[1] The committee took into account the earlier recommendations of the TC-4 committee of Purdue University and the NAMUR subcommittee in Europe.

The first attempt in North America to standardize batch control was made by the Technical Committee 4 (TC-4) of the Purdue University International Workshop on Industrial Computer Systems. The committee, commonly known as the Purdue Workshop, deliberated in 1984 and 1985 and published its recommendation on a structural model and a set of standard terms for batch control.

NAMUR

In 1949, several large chemical companies from West Germany, the Netherlands, and Switzerland established a committee for the standardization of industrial practices in process measurement and control, called NAMUR (Normenarbeitsgemeinschaft für Meß- und Regelungstechnik in der chemischen Industrie). In 1966, NAMUR formed a working group because it felt there were problems in batch automation that are common to chemical and pharmaceutical industries. Two years after its formation, the committee published its first recommendations: that sequence programs be interchangeable and be reusable for different products, and that sequential programs be able to be changed without going outside of control.

Since then, the NAMUR subcommittee has developed and updated its recommendations a number of times. NAMUR is finding a great deal of acceptance throughout the European chemical industry and has made an impact in the United States. The ISA/SP88 committee took many recommendations from NAMUR into consideration while developing the ISA-S88.01 batch standard.

NAMUR looks at batch automation from a process activity viewpoint. It states that certain process activities, such as charge, mix, heat, and cool, are repeatable and, therefore, can be shared across many processes and industries. Beginning with a batch automation hierarchy, NAMUR includes structures, terminology, and definitions for classification of production facilities, processing equipment, recipe structure, batch operational sequences, and phases.

To create working batch programs for a process plant under NAMUR, a library of basic process operations written in a control-oriented language would be made available from a control system supplier. With this library, a user would develop a master recipe. This master recipe uses only those process operations necessary to construct the recipe for the particular plant. When a specific product is to be manufactured, the master recipe is appended with specific process units. This becomes the "control" recipe, which is used for the actual control of the batch process.

Batch Control Standard Part 1

ISA/SP88 committee generated Part 1 of the batch standard in 1995. Soon after its publication, it was approved by ANSI (American National Standards Institution), making it a U.S. standard. The full designation of this standard is ANSI/ISA-88.00.01-1995. For brevity, it will be termed ISA-88 Part 1 in the rest of this section. In 1997, it became the international batch control standard when the International Electrotechnical Commission (IEC) approved this with minor modifications. The IEC standard is designated IEC 61512-01.

Part 1 of the standard deals with models and terminology and fills a long-standing need in the batch control community. In generating this part, the ISA/SP88 committee drew heavily on the earlier works by TC-4 of the Purdue Workshop and NAMUR. Soon after the completion of Part 1 of the standard, the ISA/SP88 committee started work on Part 2, which deals with the data structures and language guidelines. The main benefits of the standard are as follows:

- Improved communications between suppliers and users of batch control
- Easier end-user needs identification
- Straightforward recipe development
- Reduced cost of automating batch processes
- Reduced life-cycle engineering effort

Part 1 defines standard terminology and a number of models for batch control. The key models are physical, control activity, recipe, and unit. The terminology, structures, and concepts used in the standard will affect everyone in the batch control business. In the past, batch control packages were monolithic, and suppliers used their own terminology. Most suppliers are now adopting standard terminology and are designing BCSs with a modular set of functions and hierarchy based on the control activity model. This modularity allows for easier integration of third-party packages to do functions such as production planning, scheduling, and production information management. It also makes a control system easier to integrate with production management and business planning systems. As the user community becomes more knowledgeable about batch control, it is demanding that system suppliers comply with the standards.

Although the ISA-88 Part 1 standard is meant for batch process control, its broad-based models have wider applications. These models and their interrelationship concepts can be applied to other applications to gain a significant advantage.

Batch Control Standard Part 2

Work is now complete on Part 2 (ANSI/ISA-88.00.02-2001) of the batch control standard.[2] Part 2 defines data models and data structures to facilitate communications within and between batch control implementations. The standard also defines language guidelines for representing recipe procedures. The IEC has circulated a draft standard (IEC 61512-2), which is almost identical to the ISA standard, to solicit comments from standards organizations around the world. The IEC standard, which is in English and French, was published in 2001.

The ISA-88 Part 2 standard is in three parts: data models, information exchange tables, and procedure function charts. The data model section provides formal representation of entities specified in Part 1 of the standard, such as recipe, equipment, planning and scheduling, and information management using Universal Modeling Language (UML) notation. The information exchange section uses SQL relational tables to specify exchange requirements among recipes, process equipment, schedules, and batch production (Figure 5.5a).

The final part of the standard deals with a graphical representation of procedures, such as master and control recipes, using Procedure Function Chart (PFC) notation (Figure 5.5b). PFC is somewhat similar to Sequential Function Chart (SFC) notation as defined in IEC 61131-3 standard. PFC notation

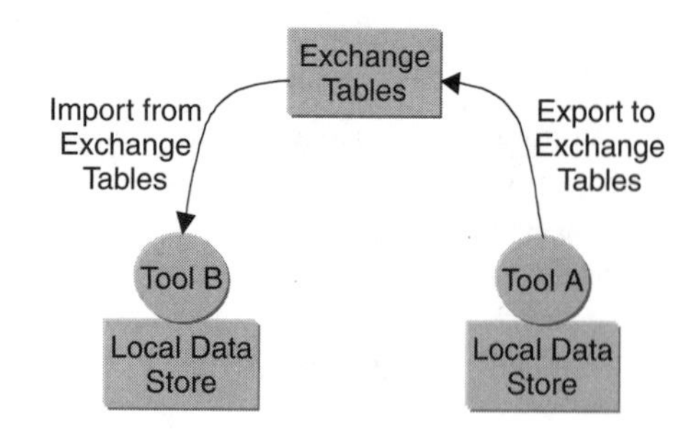

FIG. 5.5a
Data transfer using exchange tables. (© Copyright 2001 ISA—The Instrumentation, Systems, and Automation Society. All rights reserved. Used with permission. From ANSI/ISA-88.00.02-2001, "Batch Control Part 2: Data Structures and Guidelines for Languages." For information, contact ISA Standards, www.isa.org.)

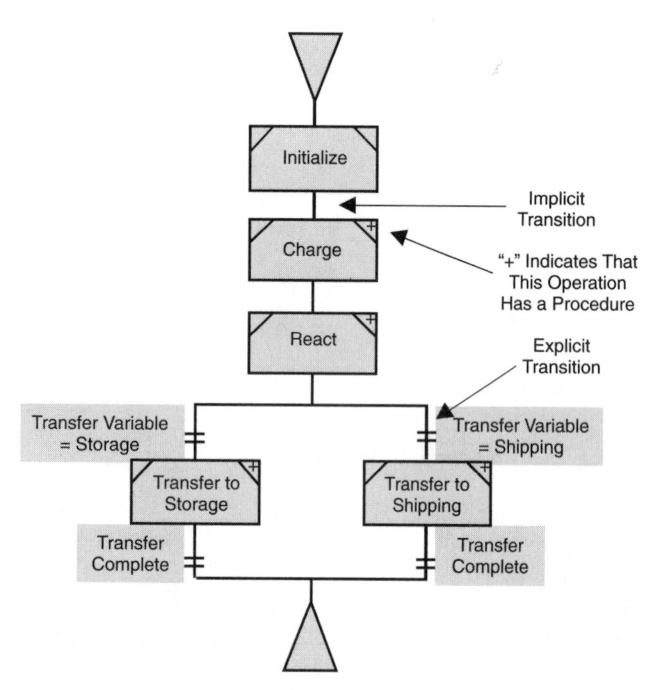

FIG. 5.5b
A unit procedure in PFC.

addresses procedural control and execution, whereas SFC notation was developed primarily for state machines. PFC notation meets the requirements of recipe procedures better than SFC.

The Part 2 standard does not address general and site recipes and phase level interfaces, which are expected to be considered in Part 3 or beyond. The scope and timelines for future parts have yet to be determined. The standard does not deal with exception handling procedures in any great length. That is a serious shortcoming in the proposed standard, which will significantly hinder recipe transformation between dissimilar systems. The compliance requirements are not specified explicitly in the standard, but are implicit throughout the document. Rigorous data models and information exchange tables will help make BCSs more modular. However, more work is needed to define data transfer between higher-level control activities and equipment phases and to develop exception handling to make master recipes more transportable.

The benefits of the standard go beyond batch process control. The data models and information exchange tables provide a significant advantage in the automation and modularization of any manufacturing process. The shift from a monolithic centralized control architecture to a modular architecture allows a process to be more flexible in manufacturing different products and allows for easier maintenance and upgrade of control systems. It will also facilitate the exchange of information among different systems.

PFC notation is intuitive and easy to follow. Users familiar with SFC notation will find many similarities between the two. However, explicit specification of functions such as process equipment allocation and synchronization of phases makes PFC easier to follow than SFC notation.

Providing tools for information exchange, as specified in the standard, will allow suppliers to design more modular BCSs. This will reduce the high cost of maintenance and version upgrades. It is felt that PFC notation will soon become the common way for representing procedural elements in a recipe. To gain competitive advantage, a supplier should conform to PFC notation as soon as possible.

Expanding Interests in Batch Control Standard The interest generated since the publication of ISA-88 Parts 1 and 2 of the standard is expected to continue. A number of independent nonprofit organizations have given their support. These include World Batch Forum (WBF), which covers North America and Europe, and Japan Batch Forum (JBF). In addition, various countries have formed local organizations such as the French Batch Forum. These organizations provide facilities for information exchange related to batch process control and the standard and to explain the standard to the control community (Figure 5.5c).

IEC 61131-3

One shortcoming of Part 1 of the ISA standard is that it does not specify any language or documentation standard, but this has been partly rectified in Part 2. However, in the meantime, the programming languages specified in IEC 61131-3 have been widely accepted for specifying sequential steps in recipes and phases.[3]

The intent of the IEC 61131-3 is to provide general programming compatibility between different brands of PLCs. With IEC 61131, it should be possible, for example, to transfer code written for an Allen-Bradley PLC (programmable logic controller) to a Siemens device. However, the standard can also be applied in the DCS environment to similar advantage. It defines four languages, two textual and two graphical.

Batch Standard	US Standard	International Standard	Scope
Part 1	ANSI/ISA-88.00.01	IEC 61512-01	Models & Terminology
Part 2	ANSI/ISA-88.00.02	IEC 61512-02	Data Structures & Language Guidelines

FIG. 5.5c
Batch control standard.

The Textual Languages

- Instruction List (IL) is similar to assembly language with only one operation per line. It is useful only for relatively simple logic control or safety interlocks.
- Structured Text (ST) is a high-level block structure language similar to Pascal that can be used to express complex statements. Structured text can handle a wide range of different data types including analog and digital values. The language also has support for iteration loops and conditional tests.

The Graphical Languages

- Ladder Diagram (LD) has been in use since PLCs, where it was first introduced, and is a favorite with electricians and electrical maintenance organizations.
- Function Block Diagram (FBD) is a graphic language that allows program elements that appear as blocks to be connected together. Function blocks are standard blocks that execute PID, Boolean, and other functions.

Additionally, SFC, while not a stand-alone language, has been approved. SFCs are a way of organizing, rather than writing, programs. Using the other four languages as a base, SFCs partition a program organization into a set of interconnected steps and transitions.

In North America, LD has been the language of choice for PLC configuration. This is changing, with more suppliers and third-party software houses offering other standard languages for PLC configuration. The DCS (distributed control system) suppliers are also becoming part of the act by providing one or more of these standard languages for the configuration of their controllers.

Control and Enterprise Integration Standard

Although both business and control systems have been evolving independently, they have now reached a point where they can meet and exchange data. Open control systems are making business system integration easier. This area has been in dire need of a set of interface definitions and nomenclature standards. To solve these difficulties, several major manufacturing companies, process automation system suppliers, and ERP software suppliers are working with the ISA-SP95 Committee to develop the enterprise-control system integration standard (Figure 5.5d).

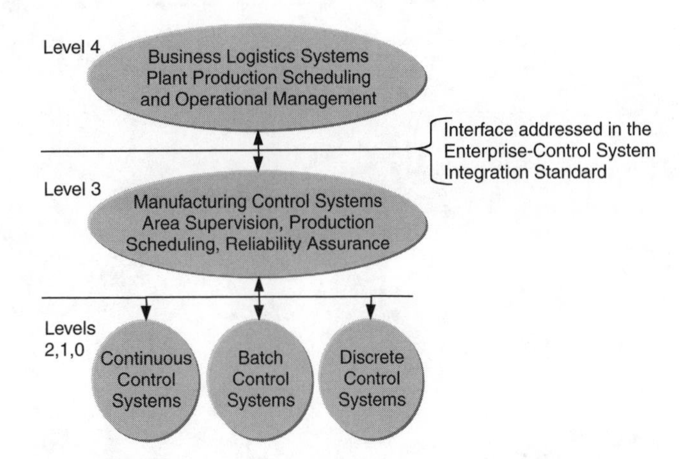

FIG. 5.5d
Enterprise-control domain. (© Copyright 2001 ISA—The Instrumentation, Systems, and Automation Society. All rights reserved. Used with permission. From ANSI/ISA-95.00.01-2000, "Enterprise-Control System Integration Part 1: Models and Terminology." For information, contact ISA Standards, www.isa.org.)

The scope of work for the SP95 committee includes:

- Defining an abstract model of the control domain and identifying target domains to which it will interface
- Establishing common terminology for the description and understanding of the control domain and its information exchange
- Defining electronic information exchange between the control domain and target domain including data models and exchange definitions

In 2000 the committee published ANSI/ISA-95.00.01-2000, which is the first part of the standard.[4] It provides standard terminology and a consistent set of concepts and models for integrating control systems with enterprise systems. The models and terminology emphasize good integration practices of control systems with enterprise systems during the entire life cycle of the systems (Figure 5.5e).

The second part of the standard was published in October 2001.[5] Part 2 does not add any significant new concepts to the integration model. Rather, it contains additional details and examples to help explain and illustrate the objects in Part 1.

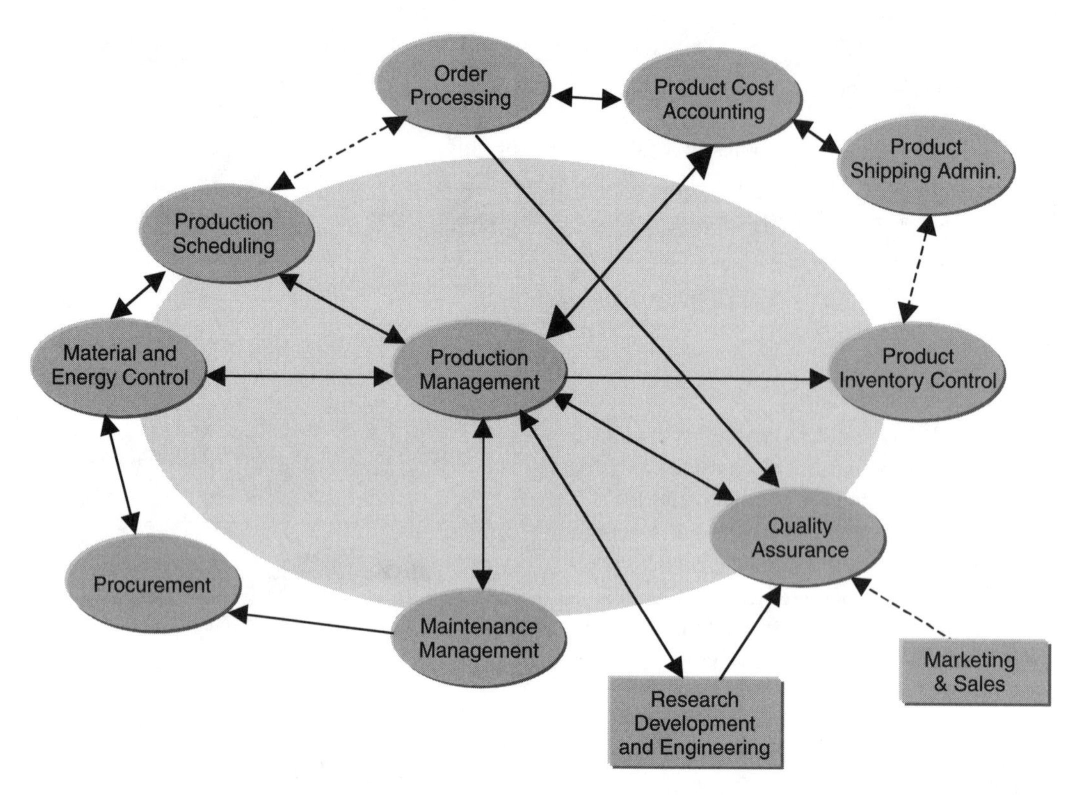

FIG. 5.5e
Enterprise-control functional model. (© Copyright 2001 ISA—The Instrumentation, Systems, and Automation Society. All rights reserved. Used with permission. From ANSI/ISA-95.00.01-2000, "Enterprise-Control System Integration Part 1: Models and Terminology." For information, contact ISA Standards, www.isa.org.)

The standard defines electronic exchange formats for shared information, and emphasizes good integration practices between enterprise and control systems during design and operation. The goal of the standard is to improve existing integration capabilities of manufacturing control systems and enterprise systems.

The SP95 work complements the ISA-88 batch control standard by establishing a standard data definition for the interface between batch automation systems and business systems. SP95 will also encompass other control aspects of a manufacturing company seeking to meet the general goal of enterprisewide automation.

The ISA-95 standard will go a long way toward satisfying a vital need for common terminology and models for enterprise and plant systems integration. Manufacturers should feel positive that integration is no longer amorphous, but will follow a consistent methodology similar to the ISA-88 batch control standard. It will enable users to better identify their needs, thus reducing life-cycle engineering efforts. It will also make it easier for suppliers to provide appropriate tools to integrate plant operations with the enterprise system.

Open control systems have moved from proprietary hardware to standard hardware, and are now moving from proprietary networks to standard networks, while proprietary operating systems are evolving to standard operating systems. The work of the SP95 committee is the next logical step in that direction. It will help open control systems move from proprietary data sets to standard data sets.

GENERAL AND SITE RECIPES

A common challenge in process manufacturing today is the pace of change. To remain competitive, manufacturers must be increasingly responsive, reacting quickly to changes in fluctuating customer demands, emerging technologies, new product development, mergers and acquisitions, and global competition. The ability to respond in a cost-effective manner can mean the difference between profit and loss in the highly competitive process industries.

General and site recipes can accelerate the time-to-market and time-to-volume production. They provide an equipment-independent way to describe the specific manufacturing processing actions required to make a product. They are also the source of information for ERP bills of material and routing. Companies that have implemented the general recipe model have considerably reduced their time-to-volume production. The reduction in time-to-volume production will make manufacturers more responsive to customer needs, more competitive with new products, and will obtain faster returns on assets. The general recipe model, as defined in the batch control standards, can provide significant corporate benefits to all manufacturing companies (Figure 5.5f).

A general recipe contains generic information for the manufacture of a product and does not include equipment or site-specific information. This is the first recipe that may be generated when a new product is developed at the pilot plant level. The general recipe is created by people with knowledge of both the chemistry and processing requirements of the product. It defines manufacturing-specific information. A site recipe contains site-specific information that may include peculiarities of raw material available, types of units that are used, and other site-specific constraints. A site recipe is also generally derived from a general recipe (Table 5.5g).

In the increasing complexity of global manufacturing, many companies find it a challenge to maintain a single definition for a manufactured product. The general recipe meets this challenge as the central repository of product manufacturing information. A general recipe lets one unambiguously communicate processing requirements to multiple manufacturing locations.

Today, most manufacturers who generate and maintain their general and site recipes do so in text or tabular format. The transformations of these recipes to master and control recipes are done manually.

Soon after the completion of Part 2 of the batch control standard the ISA/SP88 committee started working on general and site recipes. The committee is expected to complete its work in 2002, and it will be published as Part 3 of the standard.

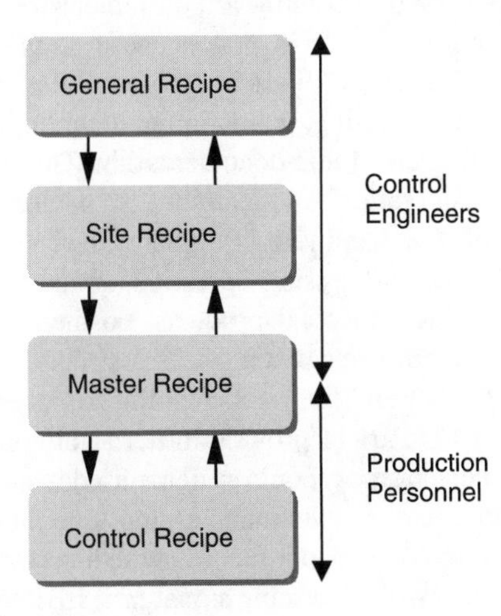

FIG. 5.5f
Recipe development and execution.

TABLE 5.5g
Procedural Element Relationships

General/Site Recipe	*Master/Control Recipe*
Process	Procedure
Process stage	Unit procedure
Process operation	Operation
Process action	Phase

Most BCS suppliers are yet to provide general recipe function. The Sequencia gRecipe is a notable exception. It allows a user to specify general and site recipes in graphical format and allows a user to significantly automate the transformation to master recipes.

General recipes can significantly reduce the corporate information management requirements of an end user. When changes to recipes and new product recipes are performed manually, it can take a significant amount of time and effort to get them written, approved, and sent to production. Once there, they still have to be implemented on the local control system or converted to manual instructions. Use of the general recipe automates these tasks, cutting the time to full volume production by months. Production recipes specific to the equipment capabilities and layout can be automatically generated with much less effort, leading to faster production start-up. It is expected that BCS suppliers will include general recipe function in their future product offerings.

COMPUTER-AIDED FORMULATION

For most process manufacturers, variability in raw materials and changing customer specifications require frequent adjustments in formulas and batches. Computer-aided formulation tools can effect major productivity improvements by streamlining the development and application of formulas and automatically integrating product development information and activities with manufacturing (Table 5.5h).

Clear visibility into the product development process is a key to the success of a company in research and development (R&D) activity, manufacturing, and in supply-chain management. The product development process, by its nature, also involves a wide cross section of people from various parts of the organization. Marketing, for example, will need to understand and clearly define the marketing claims associated with newly developed products. R&D will set technical parameters, specifications, and testing requirements. Quality control, legal, and purchasing departments

TABLE 5.5h
Benefits of Computer-Aided Formulation

- Faster time to market for new products
- Reduced material costs
- Lower regulatory-driven costs
- Increased productivity for chemists, food technicians, and schedulers
- Lower losses from work off
- Fewer unexpected scheduling changes from batch noncompliance
- Fewer adjustments per batch, resulting in lower labor and equipment cost
- The possibility of using less skilled employees to adjust batches

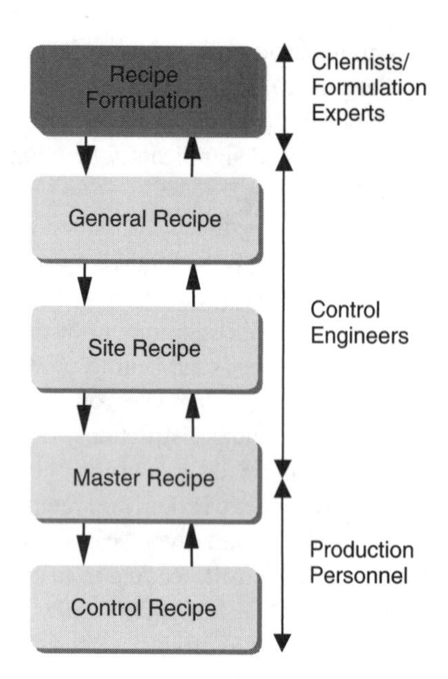

FIG. 5.5i
Recipe formulation development.

will ensure local regulatory compliance as newly acquired, unfamiliar formulas are introduced into the manufacturing and supply-chain processes. The challenge is to integrate these separate R&D-driven activities into a cohesive, enterprisewide whole that enables the manufacturing company to rationalize its product line and supply chain quickly, while driving the creation of those competitively priced products that customers want (Figure 5.5i).

Traditionally, it has been difficult for product developers to share information not just with manufacturing, but also even among themselves. Formulas and other product development data are scattered across the enterprise in various R&D labs and isolated in spreadsheets, legacy systems, disorganized file cabinets, or lab notebooks.

Because of the lack of a centralized product development information system linked to manufacturing execution system/enterprise resource planning (MES/ERP), formulas have usually been manually keyed, inevitably creating batch errors. A new breed of product development software has emerged that centralizes all product development data, automates the product development process, and creates a seamless link between R&D and MES/ERP, ensuring accurate and timely communication of new product formulas and modifications to existing formulas.

One technology tool helping manufacturing companies address the challenge of meeting their post-integration product development information management needs is enterprisewide product development software. It provides a common product data repository, tools to help identify rationalization opportunities, global access to information, stronger links to marketing and manufacturing, standardization of work processes and approvals, and improved collaboration. The combination of these benefits results in a more effective business process.

A formulation package considers a list of constraints such as the cost of raw materials and energy usage to calculate the minimum-cost formula that meets the product specification. It recommends adjustments to any or all of the ingredients in a batch to bring it within specification. Advances in computing capabilities over the last few years have made computer-aided formulation within the reach of most process manufacturers.

In the past, major manufacturers of chemical, pharmaceutical, food and beverage, and animal and pet food relied on the acquired knowledge of chemists and formulators to develop and maintain their product formulas. With the increased use of personal computers, some of them have developed custom packages to meet their needs. That is now changing as a number of suppliers are offering software packages to address these issues.

A computer-aided formulation system cannot replace the human judgment of a skilled product development and manufacturing management team. However, it can perform all the tedious calculations necessary to convert a specification and a set of assumptions into a new product. It can also optimize formulas based on various constraints far more quickly than is humanly possible. BCS suppliers should integrate one or more available formulation software package with their BCS.

ELECTRONIC BATCH RECORDS AND SIGNATURES

An electronic batch records (EBR) system with integral electronic signatures is becoming a requirement for most batch control systems.[6] This is in part because of the 1997 U.S. FDA ruling, U.S. 21 CFR Part 11. These regulations stipulate requirements for creating, maintaining, archiving, retrieving, and transmitting electronic documentation. They include criteria for consideration of electronic signatures to be the equivalent of full handwritten signatures. The regulations affect many batch processing operations in the pharmaceutical, biotechnology, medical products, cosmetics, food, and fine chemicals industries under the U.S. Federal Food, Drug, and Cosmetic Act and the Public Health Services Act.

U.S. 21 CFR Part 11 provides criteria for compliance of all regulated electronic records systems used in an enterprise. It applies to existing and future systems. It includes EBR systems used to create paper records with handwritten signatures or systems with electronic signatures. EBR systems and the regulated processes vary from the simple and well-defined low-acid canned foods process to the more complex one-of-a-kind biotechnology-based hormone production process.

As with any new law, compliance requires interpretation and adaptation to individual product-specific manufacturing processes. This is, by its nature, a slow process that is further complicated by the diversity of industries and processes involved plus the limited resources of users, suppliers, and regulators. User industries are continuing to evaluate and define the impact on existing and future manufacturing operations. Recognizing that installed and currently available technology lacks some of the required compliance enabling

TABLE 5.5j
Benefits of Electronic Batch Records and Signatures

- Lower regulatory-driven costs
- Shorter time to validation
- Reduced record-keeping costs
- Reduced human errors
- Reduced vulnerability to abuse
- Lower losses from work off
- Increased speed of information exchange
- Improved data integration and trending
- Improved manufacturing and business efficiency
- Improved product safety

functionality, the FDA has required the pharmaceutical industry to complete a gap analysis, develop a plan to achieve compliance based on sound scientific assessment of risk, and begin execution of the plan.

The benefit of EBR over paper records is well documented (Table 5.5j). It can be even more significant in the regulated industries because of the cost and time associated with validation and compliance. Site-specific application engineering services and documentation of application software are a significant part of the cost of achieving validation and compliance. In fact, the EBR system extends beyond EBR and security functionality to include the administrative functionality of EBR systems that automate procedures necessary for compliance, such as automated documentation and automated validation methodologies. Suppliers of EBR systems are continuing to increase their product functionality and services capabilities to help reduce implementation and validation time and cost. This is most pressing in the pharmaceutical industry where the reduction in the drug development cycle and being the first-to-market with a new drug are vital to economic success.

Impact of 21 CFR Part 11

As a result of increasing FDA enforcement activities, the immediate impact of 21 CFR Part 11 is on the regulated pharmaceutical and medical products industries. It is also having an impact on their automation suppliers, who are upgrading their products with functionality designed to minimize the costs of achieving site-specific and product-specific compliance, which includes the predicate rules for good manufacturing practices. However, the impact of 21 CFR Part 11 is more far-reaching (Table 5.5k)

Product Safety and Traceability

Product safety is based on close monitoring of every aspect of the supply chain and most importantly the actual manufacturing process. With increased global ingredient procurement and productization, safety has become a great concern, because contaminated food and beverage ingredients and

TABLE 5.5k
Impact of 21 CFR Part 11

- Standardizes a global approach to accelerate the utilization of automation, with clearly demonstrated benefits
- Provides global economic benefits to manufacturers who use electronic records and signatures
- Accelerates adoption of additional global product safety standards for consumer products

products, as well as disease-bearing organisms, can no longer be controlled through traditional geographic isolation. Ingredients are procured from many parts of the world. Packaging protects the quality of the material in the package, but may have external contamination. Packaged ingredients and finished products are shipped all over the globe.

In most enterprises, product safety and traceability extends from development through manufacturing and through the distribution channel to the customer. This is especially critical in industries where being the first to market determines the success of the enterprise. For example, it is estimated that in pharmaceutical industries enterprise-centric electronic record systems can reduce Phase 3 clinical trial times by 75%. Product safety and traceability are very complex issues. They will require substantial automation of the entire manufacturing process supply chain to address this issue, but 21 CFR Part 11 provides the first definitive set of guidelines for records security and traceability for both regulated and nonregulated industries across the globe.

More Granular Consumer Information

Most enterprises with hybrid manufacturing processes deliver product through third-party sales and distribution channels. Consumers then purchase from these channels. In the alcoholic beverages segment, there are two channel layers prior to the consumer. Manufacturers are unable to obtain enough granular information about consumer buying habits and preferences because of this channel insulation. Stock keeping units (SKUs) have been used for many years to provide valuable information for close cooperation between manufacturers and mass merchandisers. The information is useful, but still not granular enough to forecast changing trends in consumer demand.

Manufacturers, pharmacy, grocery, and hardware chains, as well as mass merchandisers, will be required or find it economically compelling to implement electronic records with products designed to achieve compliance with 21 CFR Part 11. This has already begun to occur in some industries not governed by this regulation.

Enterprise-Centric Approach

The scope of batch control systems with electronic batch records systems will continue to expand both horizontally and vertically throughout the enterprise. Horizontally, it extends from ingredient specification and receiving through manufacturing, packaging, and warehousing. Vertically, it extends from sensor validation procedures to data integration in the enterprise business systems. It is likely to include more systemwide applications software rather than the currently segmented pieces of software.

Users in the regulated pharmaceutical, biotechnology, food and beverage, cosmetics, and fine chemicals industries have begun to recognize the need for enterprise-centric electronic records systems that will enable compliance with 21 CFR Part 11 over the life cycle of the system. BCS suppliers have begun to recognize and respond to the needs of users for traceability across the entire enterprise, but increasing the functionality of current products and developing new products takes time. As more functional, reliable, and common off-the-shelf technology becomes available, it can be expected to become part of compliance requirements, further raising the requirements bar.

BATCH CONTROL OPTIMIZATION IN THE SUPPLY CHAIN

Supply-chain management is a set of coordinated activities that include procurement, manufacturing, storing, and marketing. Even though every factory and distribution center at a large manufacturer can be operating at peak efficiency, the organization as a whole can still be operating suboptimally. It is analogous to a sports team with individual players who have great statistics, but cannot play as a team and, therefore, cannot win. Supply-chain management addresses this problem by optimizing the performance of the system as a whole (Figure 5.5l).

Understanding the type of supply chain that a company has is an important step in its optimization. For some industries, the critical constraints are likely to be found in procurement; for others it is manufacturing or distribution. Manufacturers in similar industries tend to be in the same supply-chain segment.

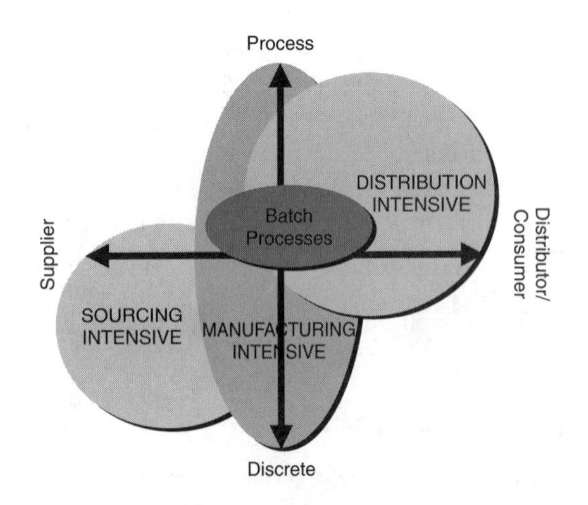

FIG. 5.5l
Supply-chain positioning.

Distribution-intensive supply-chain manufacturers include consumer packaged goods producers who must meet the demands of large retailers or lose business. In recent years, there has been a fundamental shift in market power from manufacturers to retailers. Historically, manufacturers dictated the terms of trade with retailers and organized their businesses primarily to increase manufacturing efficiency and output. Today, large chain retailers increasingly are choosing suppliers based on their ability to match product flow to actual customer demand.

Manufacturing-intensive industries include industrial equipment, aerospace and defense, heavy metals, and semiconductors. In this segment, it is not unusual to find manufacturing facilities that cost over a billion dollars. Labor and material acquisition costs are relatively insignificant. The key is to keep those expensive machines up and running, with minimum transfer times and minimum queues. For this reason, real-time scheduling is more important for such companies.

Companies that compete in industries with short product life cycles tend to be in the sourcing-intensive sector. The two primary industries in this segment are consumer electronics and apparel.

Specialty chemicals, pharmaceuticals, food and beverage, and consumer products constitute a majority of the batch processes. Specialty chemicals and pharmaceutical industries fall between the manufacturing-intensive and distribution-intensive areas, whereas food and beverage and consumer products are mostly distribution intensive. Therefore, batch-manufacturing enterprises generally require optimization in the manufacturing and distribution areas.

The Three Levels of Optimization

Until recently, process plant optimization was the main focus for control engineers. We were satisfied when the control loops were properly tuned and the recipes and phases ran in proper order producing products that were within the required quality specifications. With globalization of the market, increased competition, and the need for custom products, batch manufacturers are rapidly moving toward flexible just-in-time manufacturing. In this environment, supply-chain and enterprisewide optimizations are becoming increasingly important. It is no longer sufficient for an enterprise to have islands of automation along with the isolated supply chains. To maximize the potential of batch process control, manufacturing plants, enterprises, and the supply chains need to work together closely in an optimized fashion (Figure 5.5m).

In this new environment, with integrated enterprises and supply chains, manufacturing will continue to maintain its central role. In this environment, it is not enough to leave the total responsibility of supply-chain and enterprisewide optimization to information technology personnel, who have little understanding of the manufacturing processes. Today, to maximize the potential of batch process control, engineers need to broaden their focus considerably in the industry and supply-chain optimization areas.

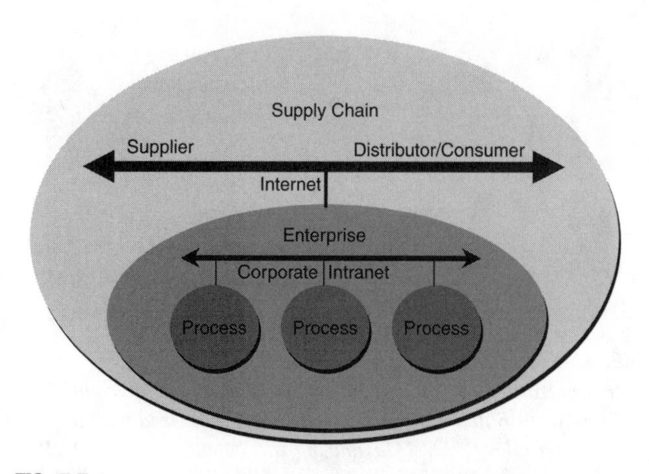

FIG. 5.5m
The three levels of optimization. (© Copyright 2000 WBF—World Batch Forum. All rights reserved. Used with permission. From Ghosh, A., "Maximizing the Potential of Batch Processes," presented at WBF Conference, Brussels, October 2000. For information, contact WBF, www.wbf.org.)

Process Plant Optimization

Today's engineering design, training, and model-based control are converging, but primarily for the benefit of continuous processes. Batch processes bring at least one added complexity: they rarely reach the steady state. Batch-oriented industries such as pharmaceuticals, fine chemicals, and food and beverage have not used simulation-based tools for process optimization and operator training to the same extent as their counterparts in the continuous process industries.

Because the current optimization technology used in refining and other continuous processes must sense steady state before they can calculate an optimum, they are somewhat limited in their applicability to batch processes. Many manufacturers, however, have found approaches to optimize their batch processes. For example, if the best practices of a master brewer can be captured as a series of rules, then expert system applications have exhibited the ability to optimize the beer production process.

Other types of simulation and analysis tools also become useful when approaching hybrid and batch processes with the goal of optimization. Discrete event simulation software can optimize the process by which the work is done. In this context, the word *process* applies to the procedure or methods used from start to finish during an operation, whether it is a physical process or a human one. For example, raw materials flowing through a plant that manufactures finished goods will encounter many processes or procedures. In question are issues such as how many reactors are needed to handle a number of batches, how fast a packaging line can move, and how many of them are needed to meet the production demand.

Efforts are now under way to integrate planning and scheduling tools with online batch control products (Figure 5.5n). It is not difficult to envision how integrated simulation and design tools can allow a product introduction cycle of 18 months to be shortened to as little as 6 months. A pharmaceutical company that is trying to beat the competition to market with the latest miracle drug could find it is worth millions of dollars.

Enterprisewide Optimization

The primary objective of a manufacturing facility is to add value. The function of a business process is to convert this added value into profit. That is pivotal for the success of an

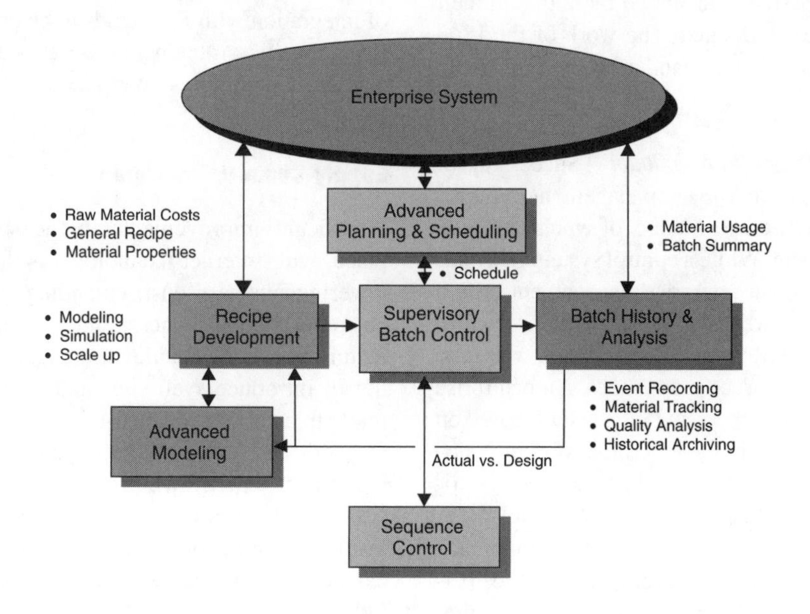

FIG. 5.5n
Integrated batch process optimization. (© Copyright 2000 WBF—World Batch Forum. All rights reserved. Used with permission. From Ghosh, A., "Maximizing the Potential of Batch Processes," presented at WBF Conference, Brussels, October 2000. For information, contact WBF, www.wbf.org.)

enterprise and it is leading the batch process industry to shift its emphasis from process optimization to business optimization. Access to manufacturing data is critical for business optimization because continuous improvement is based on patterns and relationships contained in these data. Manufacturing data originate in plant systems. Enterprise integration minimizes or eliminates duplication of data.

An enterprise needs to understand its manufacturing requirements, its business needs, and their relationships before it can come up with a viable integration strategy. A company may have a corporate business information system or may be in the process of setting up one. It needs to determine the type and quantity of information the business system will need and provide it to the automation system. The information exchange between business and automation systems can range from simple production schedules to elaborate quality, inventory, and process management instructions. The amount and type of information that a business system will need from the control system can also vary widely. They may only be averaged process values or could include information on alarm, quality, operator action, production record, and materials movement. The amount of information exchange will determine the level of integration required between the batch automation and business systems.

There are a number of production management functions such as quality, documentation, maintenance, production scheduling, and quality management. One needs to decide whether these functions are to be performed by the business system or by the automation system, or partially by both. It is important to decide which system will be the main repository for process information. Inventory management and the tracking of material movements in production facilities may be carried out either by the business system or by the process automation system. Often, they are shared by both and their interfaces need to be clearly defined. The work of the ISA-95 Enterprise-Control Integration standard is helping us to make these decisions.

The Issue Is Not Real-Time vs. Transactional Successful integration of a BCS with enterprise management requires proper synchronization. There are a number of synchronization gaps. Time is an obvious gap. While a control system responds in seconds and milliseconds, an enterprise management system measures its time in months, weeks, days, and hours. A control system gathers real-time or event-based information with time frames consistent with plant process dynamics. An enterprise system maintains transaction-processing information based on business dynamics. Execution in the enterprise system is oriented toward planning and scheduling, while execution is oriented toward engineering and control at the plant level.

However, the enterprise of the future will need to approach closer to the essence of optimization. From a front-office perspective, this involves optimized planning for business requirements. Optimization from a back office perspective involves manufacturing responsiveness to business requirements. For the enterprise of the future to satisfy its performance

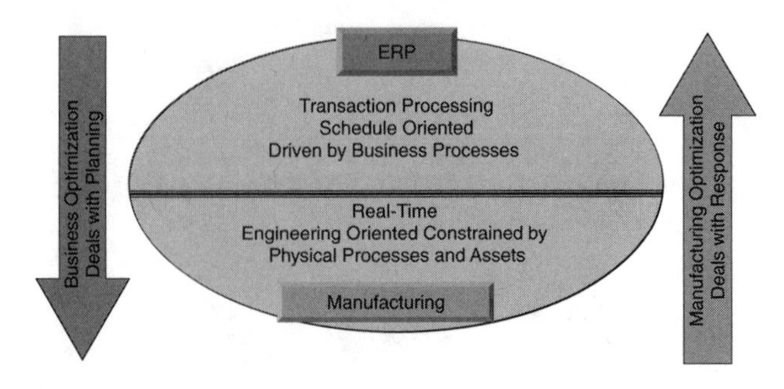

FIG. 5.5o
Synchronization of information as needed. (© Copyright 2000 WBF—World Batch Forum. All rights reserved. Used with permission. From Ghosh, A., "Maximizing the Potential of Batch Processes," presented at WBF Conference, Brussels, October 2000. For information, contact WBF, www.wbf.org.)

requirements, the front office and the back office need to be synchronized, not in a time sense, but in an information sense. Both need to be mutually supportive, driving toward the same goals, and operating on the same information. This level of optimization requires both business and manufacturing processes to interoperate in a single environment and information domain. This single environment will require a new generation of open plant systems based on a common component model (Figure 5.5o).

Cultural gaps between control and business management are a major problem in many organizations. Effective integration of these two worlds requires significant technical, educational, and planning efforts.

The general and site recipe concepts in the batch control standards and the ISA-95 standard are conceptually helping the integration, while eXtensible Markup Language (XML) and component technology are the main enablers. Today, a majority of integration with ERP tends to be done on a custom basis to solve specific problems. However, standard integration products are emerging as enterprise software companies develop these interfaces.

Supply-Chain Optimization

Significant improvements in the supply chain are taking place, with Internet technology as the primary enabler. It is lowering business cost, extending global reach, increasing customer responsiveness, and making made-to-order manufacturing easier. Manufacturers and suppliers alike are rushing to introduce Web sites and business-to-business (B2B) marketplaces that can be used to market and procure all types of goods and services. Most manufacturers currently offer or plan to offer their own products over the Web via their own sites. The next move, which has already commenced among leading manufacturers in the automotive, aerospace, chemical, and other industries, is to align with one of the emerging industry-specific buy-side Internet portals to purchase raw materials, finished parts, and subassemblies. The end result is that the Internet is emerging as the great equalizer, leveling the playing field of competing suppliers and giving the

TABLE 5.5p
Supply-Chain Optimization Benefits

- Lowering business cost
- Extending global reach
- Increasing customer responsiveness
- Allowing made-to-order manufacturing

TABLE 5.5q
Supply-Chain Optimization Challenges

- Flexible production facilities
- Smaller production lots
- Faster shipments
- Effective information management
- Transaction security
- Change in corporate culture

manufacturing customer greater visibility and control over a procurement process that should ultimately result in improved procurement processes and significantly lower prices for purchased goods. Projections are that B2B business will grow to $1 trillion in 2003 (Tables 5.5p and 5.5q).

The Promise of the Internet However, the promise of the Internet is much greater than just lower prices. Strategic use of Web-based procurement capabilities can drive significant reductions in the costs associated with the procurement process itself, including request for quote, supplier selection, the bid process, and supplier management, among numerous other tasks. This promise of lower prices coupled with drastically reduced procurement process costs is what makes the Internet such a powerful force behind the sea of change currently under way in procurement strategies.

A key element fueling the proliferation of the Internet is its tremendous ease of use and resulting ease of doing business. While some manufacturers may not have moved to actually purchasing products on the Web, a vast majority relies on it to expedite the information-gathering phase due to the zero cost and minimal time associated with Web research. The Web offers a speedier and more efficient search process, particularly relative to the typical loop of asking the distributor a question and then waiting for a response after the supplier responds to the distributor.

Internally, the greater visibility offered by a Web-based system allows tremendous control over the purchasing process and improved supplier management. Web-based procurement introduces lower transaction costs, particularly if the Web is used to internally standardize, control, and monitor the procurement process. Improved control and visibility over the procurement process throughout the entire enterprise will also result in lower inventory-carrying costs because the now-agile procurement system can monitor inventory levels, time elapsed since last purchase, and calculate the optimum time to place a new order.

In the automation space, the Internet has been widely used to provide self-service support. This is true at the level of both suppliers to user and within manufacturing companies themselves, where second- and third-shift operators can access application, machine, or product-specific information via the intranet. Often, no human contact is required. Furthermore, the information required for service or support is available around the clock.

Availability of Web-based support and information has been a tremendous boon to small and mid-sized manufacturers. With the Internet leveling the playing field, these Tier 2 and Tier 3 manufacturers are now able to access technical databases and other information that were previously available only to Tier 1 manufacturers. Some suppliers are using this market reach to their strategic advantage.

Both single supplier sites and multivendor B2B marketplaces can offer transaction efficiencies through Web-based selection, procurement, and post-sales support. Because they contain the offerings of only one supplier and perhaps their partners, however, single supplier sites cannot deliver the added advantage of competitive pricing that can ultimately drive down end costs. Suppliers use customer extranets to present customer-specific negotiated pricing on their Web site, but as recent experience shows, manufacturers can achieve tremendous savings by conducting a reverse auction with project request for quotes (RFQs). Moving from single supplier sites to a B2B marketplace also shifts the focus resoundingly to the buyer's application requirements rather than the sellers and their products. Thus, the Internet technology is helping the optimization of supply chain in a very positive way.

BATCH CONTROL SYSTEM SELECTION CRITERIA

The selection of the best control system for an application is largely dependent on the type and complexity of the application. The complexity of an application is dependent on many factors such as the topology of the plant, the number of different products and grades, the criticality of the application, and the size of the plant.

The important supplier selection criteria are the following:

- Batch control experience
- Industry application experience
- Project management capabilities
- Business system integration experience
- Knowledge of regulatory requirements

Batch processes are generally considered highly critical because a control failure may result in a hazardous situation, equipment damage, or loss of very high value product. The most common hazardous processes involve very fast exothermic reactions in which all corrective actions need to be taken automatically to ensure a safe operation. The manufacture of polyvinyl chloride or rocket propellants are examples of this

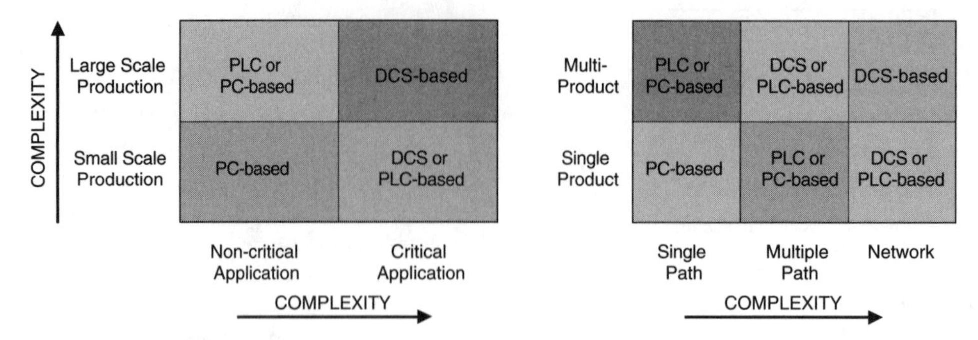

FIG. 5.5r
Recommended type of BCSs based on process complexity.

type of process. Many pharmaceutical and biotechnology manufacturing processes involve very high value product. Critical applications generally require a high level of fault tolerance, a high level of safety interlocking, and extensive exception-handling logic to take care of abnormal conditions. This combination of requirements significantly increases the complexity of the control solution.

PC-, PLC-, or DCS-based BCSs perform similar functions, but some of their characteristics, such as capacity, fault tolerance, and human interface, vary significantly. PC-based systems provide a lower-cost alternative to either a PLC- or DCS-based system, but offer little or no fault tolerance, and thus are not generally well suited for critical applications. PLCs are well suited for logic control, permissives, and fast interlocking. PLCs also perform sequence control quite well and some continuous control. Some PLCs also provide good fault tolerance making them suitable for critical applications. However, PLCs are generally not as good as PC- and DCS-based systems in the areas of human interfacing, expandability, and flexibility. Modular programming has been difficult in PLCs, in which only ladder logic is used for configuration. PLC-based systems are well suited for small- to medium-sized batch control applications and are generally not used for complex batch applications or those applications that require a high degree of flexibility (Figure 5.5r).

DCSs are well suited for continuous, sequential, and some basic logic control. They are generally not as good as PLCs at safety interlocking and other high-speed logic functions. Most DCS controllers can be configured in a redundant manner with automatic switchover for a high degree of fault tolerance. DCSs also offer more flexible programming facilities than PLCs and easier-to-use human interfaces. However, the key strengths of many DCSs are their scalability, expandability, and the ability to interface seamlessly with PLCs and other types of controllers and business systems. These attributes make DCS-based batch control systems technically suitable for a wide range of batch applications. However, they are generally more expensive than PC or PLC systems, which makes them less attractive for small- and medium-sized applications. Thus, DCS-based systems are most ideally suited for flexible, large, complex applications.

The lines that separate the DCS- and PLC-based systems are starting to blur. A new type of system, called a hybrid system, is now being targeted for batch and discrete control applications. A typical hybrid system has DCS architecture, but uses controllers that are similar to PLCs. The functionality and scalability of hybrid systems are similar to that of the DCSs. Examples of these new hybrid control systems include the DeltaV from Fisher-Rosemount, PlantScape from Honeywell, and ProcessLogix from Rockwell. It is also important to note that PC-based systems are beginning to acquire many of the DCS attributes.

End users should move to a supplier model that entails reducing the number of direct suppliers from as many as five down to two key automation partners, one for the business systems and the other for the control systems. This drastic reduction in the number of suppliers is possible and enabled by the emergence of both open control systems and open business systems. The scope of coverage of these new open systems is significantly expanded over traditional systems. The solution sets of the open batch control system partners will typically be a combination of their own batch software and third-party software running in an integrated environment.

To meet users' requirements in batch process automation, suppliers must exhibit many important characteristics. These include batch process control experience, industry application experience, enterprisewide integration experience, project management capabilities, and knowledge of appropriate regulatory requirements.

Successful batch process control solutions have been elusive, and it is important to work with suppliers with proven track records. Experience in one's industry and specific application area is also very important in the selection process. Knowledge of specific industry segment characteristics, such as regulatory requirements, can be critical to the success of the project.

Management of batch projects is a difficult task because of the extended scope and integration typically involved. Batch control projects involve personnel from diverse disciplines and organizations. One of the toughest challenges in implementing batch control projects is convincing project teams to work together effectively. No generally accepted methodology has been developed to address this issue. Clearly, however, there is no substitute for solution partners who offer high-quality batch project managing experience.

Successful automation of batch processes has traditionally been very difficult. The scope of batch projects often grows as the project progresses, causing schedule delays and taxing budgets. One key reason is that users typically do not define their batch automation requirements to a detailed enough level prior to project execution. Manufacturers need to develop a good requirement definition to ensure successful batch control projects that truly meet their needs. The development of a detailed definition of the automation requirements through the generation of a project application specification is strongly recommended. With the acceptance of the standard-based models, these specifications can utilize standard structures that will be general across all control systems. Once a project application specification has been completed, intelligent decisions on the suitability of a system and supplier for the application can be made.

ACKNOWLEDGMENTS

The author acknowledges the help and support provided by John Blanchard and Steve Banker of ARC Advisory Group. Special thanks to Edward Bassett of ARC Advisory Group for proofreading and suggesting improvements.

References

1. ANSI/ISA-88.01-1995, "Batch Control Part 1: Models and Terminology," Research Triangle Park, NC: Instrument Society of America, 1995.
2. ANSI/ISA-88.00.02-2001, "Batch Control Part 2: Data Structures and Guidelines for Languages," Research Triangle Park, NC: Instrument Society of America, 2001.
3. IEC 61131-3, "International Standard, Programmable Controllers—Programming Languages," Geneva, Switzerland: International Electrotechnical Commission, 1993.
4. ANSI/ISA-95.00.01-2000, "Enterprise-Control System Integration Part 1: Models and Terminology," Research Triangle Park, NC: Instrument Society of America, 2000.
5. ANSI/ISA-95.00.02-2001, "Enterprise-Control System Integration Part 2: Object Model Attributes," Research Triangle Park, NC: Instrument Society of America, 2001.
6. "Code of Federal Regulations: 21 CFR Part 11," Food and Drug Administration, March 1997.

Bibliography

ARC Report, "Batch Process Automation Strategies," ARC Advisory Group, October 1999.

ARC Report, "The Real Deal Present & Future of Hybrid Manufacturing Integration," ARC Advisory Group, June 2001.

Dayton, N. A., "A Practical Approach to Compliance for 21 CFR Part 11," Institute of Validation Technology, May 2001.

Ghosh, A., "Batch Processes and Their Automation," *Instrument Engineers' Handbook,* 3rd ed., *Process Control,* Lipták, B. G., Ed., Radnor, PA: Chilton, 1995, Sec. 8.3.

Ghosh, A., "Maximizing the Potential of Batch Process Control," presented at WBF Conference, Brussels, October 2000.

5.6 Plantwide Control Loop Optimization

M. RUEL

Software Suppliers: ABB; Emerson; ExperTune; Invensys; Matrikon; Honeywell; Techmation

PROCESS OPTIMIZATION

Plantwide process optimization involves the systems and strategies required to control an entire plant according to control objectives. A plant consists of many unit operations, and control objectives could be conflicting.

Rationale

One of the most challenging tasks a process control engineer faces is how to optimize the processes. The following questions must be answered:

- How do we tune the control loops and the systems?
- Given control objectives, restrictions, and operation procedures, how do we proceed to optimize each unit?
- How do we optimize the entire plant?
- What issues do we need to consider?
- Do we need to modelize the system?
- Are the strategies appropriate?
- Do we need advanced control strategies?
- Do we need to tune all loops? If so, in what order?
- When optimizing a unit, do we leave the loops in manual mode? In automatic mode?
- Which steps should be done first?

This section will guide the reader through the optimization process. Process performance depends on effective utilization of process control. Included in process control are transmitters, control systems (logic, algorithm, interlocks, etc.), but also pumps, valves, pipes, and other process equipment.

The reader will discover that optimizing a process is similar to a conductor directing an orchestra: coordinate, accelerate here, slow down there, synchronize other parts, etc. The "conductor" must know about control, but also about the process itself. To optimize a control system, one must understand the process perfectly.

Complex processes are best controlled by simple control systems. Many academic researchers promote multivariable control systems. The reality is that most of these complex systems are turned off. We should move to multivariable systems when simple techniques fail, even if properly tuned. A common mistake is to use advanced multivariable control with bad equipment or with loops not tuned to obtain the expected performance.

When optimizing, it is important to consider design, equipment performance, control strategies, operation procedures, performance monitoring, logic, and special control strategies for start-up, for shutdown, for grade changes, for abnormal conditions.

There are many reasons to use process control:

- Quality
- Safety
- Efficiency
- Improved uptime
- Reliability
- Labor cost
- Operability
- Environmental regulations

Control objectives depend on the goal for the unit.

Plant Organization Impacts on Tuning

A plant consists of units connected together, mainly in series. When optimizing an entire plant, it is divided in units, and each unit optimized before looking at the big picture. Each unit is then divided in loops and systems.

A Plant, a Series of Units If a surge tank is used between the units, they become almost independent of one another. Each unit should have its own control objective and performance criteria.

Units in Parallel If units are in parallel, it could be useful to synchronize them to ensure the reaction time is similar in each. Again, the use of surge tanks will reduce interaction.

Recirculation If a part of the output from a unit is recirculated, then the dynamic is different. The plantwide control problem becomes more complex and its optimization less obvious. When recirculation varies too much, one should consider using different tuning parameters.

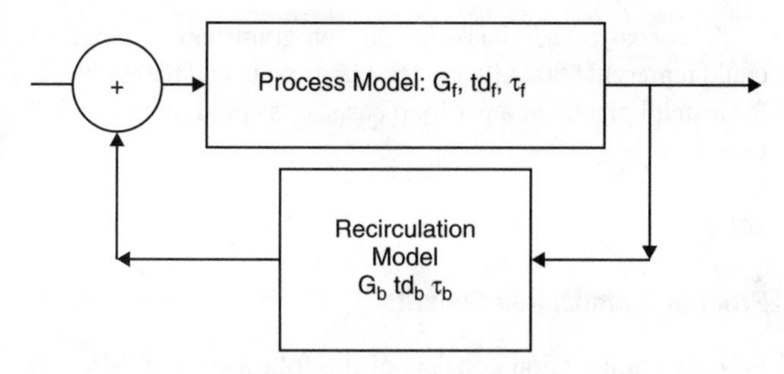

FIG. 5.6a
A simple process with recirculation.

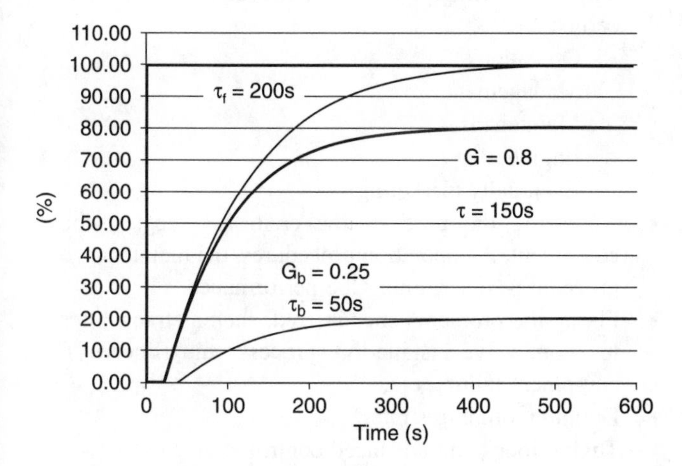

FIG. 5.6b
Equivalent model: G = 0.8, τ = 150 s.

Recirculation is used to:

- Modify process time constant
- Improve yields
- Improve thermal dissipation
- Improve control reaction
- Operate the equipment at maximum efficiency

The impacts of increased recirculation are as follows:

- The equivalent process gain is reduced.
- The equivalent time constant is reduced.
- The equivalent dead time is almost unaffected.

Figure 5.6a is an example of a simple process with recirculation. The process model is process gain G_f is 1.0, dead time td_f is 10 s, and time constant τ_f is 200 s. The recirculation model is process gain G_b varies, the dead time td_b is 25 s, time constant τ_b is 50 s. The equivalent overall process gain is $G = Gf/(1 + G_f \cdot G_b)$. The equivalent time constant also varies, as Figures 5.6b and 5.6c illustrate.

Global markets have forced many factories to rationalize their operations. World overproduction has enabled consumers to become very demanding on quality—defining desired characteristics of finished product with precision. The good old days when the only measure of success was quantity are over and done. Large-scale changes are occurring.

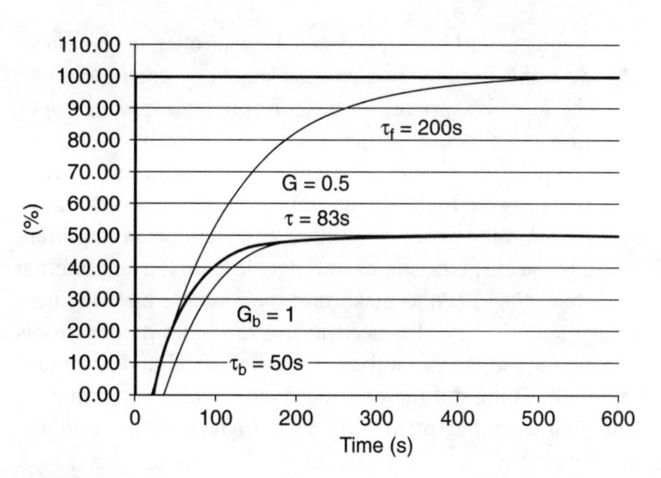

FIG. 5.6c
Equivalent model: G = 0.5, τ = 83 s.

PROCESS OPTIMIZATION PLANS

A good optimization plan requires a methodical approach. Traditional methods can be used; however, modern tools, with computers and software, should be favored. More important than the tools, the approach needs to make sure constant efforts are exerted to maintain the performance improvement. The performance of process control systems will decline if the system is not optimized on a regular basis.

In most plants, the decline will result in a performance improvement decrease of 50% every 6 months (Figure 5.6d).

Often during the start-up, when the loop is placed in automatic mode, that loop will be forgotten if it appears to work (not perform), as there are more important issues. And there is no time later to refine the tuning. When a manager decides to optimize this unit, then finally the design will be verified,

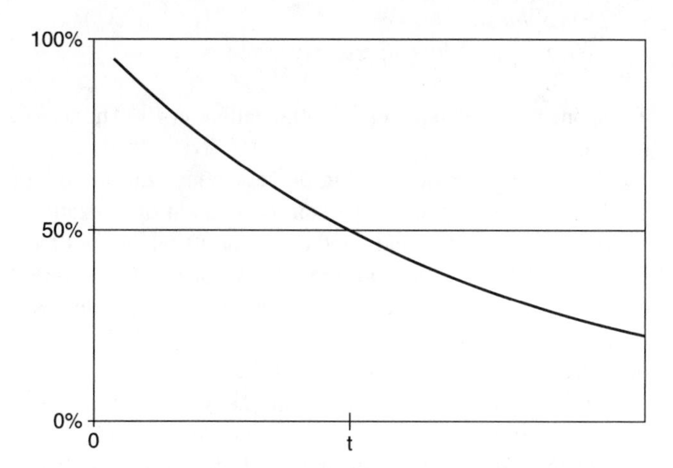

FIG. 5.6d
Performance decreases by 50% after 6 months if no effort is made.

the equipment will be checked, and the controller will be tuned. However, this optimization process should be continuous.

The level of support required to maintain process optimization is a function of the process complexity.

The plant first needs to decide if it has all the expertise and the time to maintain optimization internally. In most cases, even if the expertise is there, time will be an important issue. In the event the senior staff decide to do it using internal resources, they need to make sure their people have the necessary skills. They also need to use the most modern tools to achieve results efficiently. For example, trial-and-error tuning methods are not the best means to achieve performance. The first step the plant will need to take is to train the employees.

The staff must be experienced and motivated. External resources combined with key people is always essential when consultants work in an open manner and share their know-how with plant employees.

Budget and Workforce

For a typical plant, the workforce should consist of *trained people* using *proper tools*. This does not include the workforce needed to calibrate, repair, program, etc.

In a typical plant, 20% of loops have a direct impact on product quality. These loops should be optimized twice a year.

- (8 loops/day)/person
- 2 times/year
- 240 day/year
- Hence, ~(1000 critical loops/person)/year

Other loops should be verified every 2 years (or each time a process change occurs: modification, piece of equipment replaced, resizing, etc.).

- (8 loops/day)/person
- 0.5 time/year
- 240 day/year
- Hence, ~(4000 loops/person)/year

Thus, one person can maintain optimization on 4000 noncritical loops.

The workforce needed for process optimization for a plant with 5000 loops: one person to maintain optimization on 1000 critical loops and another person to maintain 4000 noncritical loops. These numbers vary according to the type of industry, process control system, control system strategies, etc. At $75,000/year, one person can maintain 2500 loops optimized; the price per loop is $30/year.

The price for software, hardware, networking, and configuration varies greatly; $100,000 is considered a good average. If amortized over a 5-year period, this corresponds to $20,000/year. For a plant of 5000 loops, this is $4/year and for a plant of 500 loops, this represents $40/year.

The cost of administering the program and for training could represent $50,000/year for a large plant and $20,000/year for a small plant. On a per loop basis, this could represent $10 to $40.

Total Cost The total cost is ~$100/year per loop.

Process Optimization Benefits

Process optimization consists of the following:

- Defining performance objectives. This has to be done with the participation of all the departments concerned:
  - Operations
  - Management
  - Maintenance
  - Engineering
  - Eventually marketing
- Analyzing the process, the control strategies, the equipment, the operation procedures, the maintenance procedures, determining the performance
- Fixing the problems encountered, such as transmitter location, valve installation, process gain (linearity), equipment failure, etc.
- Defining a tuning strategy
- Tuning loops and advanced control strategies
- Measuring performance
- Producing a report
- Planning the follow-up

Process optimization assures several benefits:

- Fewer process upsets—With controls properly tuned, there are fewer process upsets. The control loops can do what they were designed to do—maintain consistent process results.
- Smoother process start-ups—Process start-ups will be smoother and more consistent. Operations can quickly move into full production. With the process under control, there are fewer alarms.
- Faster grade changes—Product or grade changes can be accomplished smoothly and quickly, with a minimum of scrap.
- Better operation.

Each of the benefits reduces the amount of required operator attention for routine operation. This frees the operator to work on other tasks.

Process Optimization Methods

Loop tuning is different from process optimization. To optimize a process, it is essential to work on the global picture. To do so, one must organize the work, and tune each loop according to interacting loops.

Process and control system audits assess the performance capabilities of operating units to determine where improvements can be made and compare this performance against industry standards. An audit is done by observing many variables at the same time to determine if the performance criterion for each loop is reached. From an audit, it is possible to observe cycling problems, instability problems, and operation problems.

Software looks at the data from the process and computes performance indices at regular intervals. If a threshold is reached, an alarm is triggered. Usually, the process is first optimized, then the performance monitor will watch the performance over time. The program identifies the areas of the plant where the greatest economic gains are possible. The program is designed to help make the greatest impact on the plant. It pinpoints areas that will yield the greatest economic return.

Optimization Plan

A good optimization plan should include and define:

- Budget
- Plan
- Training
- Software, tools
- Data acquisition strategy
- Schedule
- Work plan: auditing, tuning, fixing problems, etc.
- Reporting
- Follow-up
- Justifications
- Return on investment

Process optimization concerns include:

- Quality improvement
- Security
- Cost of material
- Grade change
- Reduction in maintenance costs

Process optimization consists of adjusting each loop and system to match its control objective and at the same time reaching the main goal for this unit.

PERFORMANCE INDICES

Many performance indices can be used to determine if the goal has been reached:

- Variability (the relative value of twice the standard deviation expressed as a percentage of the mean)
- Response time, settling time
- Overshoot
- Phase and gain margins
- IAE (integral of absolute error)
- Valve travel and valve reversal, variance of controller output
- Others: ISE, ITAE, ITSE, Harris index, comparison to minimum variance control, % time not in normal mode, % time at saturation, area in autocorrelation, flat power spectral density, numbers of alarms per day, % of power at a specific frequency (or a cluster of frequencies) on a power spectral density graphic, etc.

TABLE 5.6e
Acceptable Equipment Performance

Process gain	>0.5 and <2
Hysteresis	<2%
Stiction	<0.5%
Noise	<2%
Positioner overshoot	<20%
Linearity (G max/G min)	<2

TABLE 5.6f
North American Reality

Typical Control Loop Problems and Performances	*Percentage of Loops*
Control valves in poor quality or in poor condition	30
Poor controller tuning (unacceptable values)	30
Poor controller tuning (not selected according to the performance goal)	85
Poor loop design	15
Controller in manual mode	30
Control loop not performing according to control objectives	85
Loops that perform better in automatic than in manual mode	25

Table 5.6e gives some criteria of what kind of control loop performance can be considered good or acceptable.

Different studies confirm that the performance of control loops operating in North America are as listed in Table 5.6f (these have remained the same for the last 10 years).

Even if 25% of loops perform better in automatic than manual mode, they do not necessarily perform to maximum capacity. In three of four cases, not only does the controller not improve the performance of the finished product, it worsens it.

Expected Performance

The order of magnitude values for common performance indices and the average values for a loop are given in Table 5.6g.

To analyze the process and the control systems, various techniques can be used: simple observations and determining control parameter values. For example, knowing the type of

TABLE 5.6g
Expected Values for Performance Indices

Performance Measurement	*Units*	*Usual*	*Ideal*	*Realistic Target*
Variability	%	2–3	0	<1
IAE	%·day	4000	0	<1000
RRT and settling time	s	$20 \cdot t_d$	<<	$<10 \cdot t_d$
Valve travel	%/day	5000	<<	<1000
Valve reversal	N/day	7500	<<	<2000
Power spectral density, power in a single frequency component (or cluster or harmonics)	% of total power	4	<<	<2
Robustness	Absolute	2	>	>2
% time in normal mode	%	85	~100	99

RRT = relative response time, time to remove 90% of a disturbance.

TABLE 5.6h
Usual Tuning Parameter Values for Common Loops

	P, K_p	I, T_i	D, T_d	F, τ_F	*RRT and Settling Time*
Flow, pressure (liquid)	<1	seconds	0	seconds	seconds
Level	>1	minutes	0	seconds	minutes
Temperature		minutes	seconds	seconds	minutes
Analysis		minutes	seconds	seconds	minutes

process, the actual values can be compared to usual values; example filter time constant, integral time, and derivative time should make sense (Table 5.6h).

The most common problems found are as follows:

- Tuning too aggressive—Process variable (PV) and controller output (CO) trends have a sinusoidal waveform in the time domain and we observe a single peak in power spectral density.
- Stiction problem—CO trend has a sawtooth waveform in the time domain and we observe peaks multiple times during the cycling period (fundamental + harmonics) in power spectral density. The data are not distributed normally (bell curve) on a distribution graphic.
- Backlash or hysteresis problem—PV and CO trends have an oscillation but the period increases as the PV reaches set point (SP).
- Loops interacting—Cycling at the same period in many trends and we observe similar power spectral density on many signals; cross-correlation analysis will be strong.

Process Optimization Tools

- Power spectral density: Should be flat.
- Cumulative power spectral density: Should be continuous.
- Statistical analysis: Data distribution should be a bell curve, variability should be small, valve movement should be minimized.
- Process modeling: To validate the process and to find tuning parameters.
- Robustness analysis: To validate tuning parameters.
- Process analysis: Hysteresis and backlash, stiction, noise, process model, hidden cycling, etc.
- Autocorrelation: Emphasizes the oscillation; the area under the curve is a good indicator of good control.
- Cross correlation and multivariate techniques: To measure interaction between signals and loops; also to determine how helpful multivariable control could be.
- Performance index monitoring: Variability, IAE, Harris index, etc.

Figures 5.6i through 5.6n represent a flow loop in a refinery before (left, in each figure) and after (right) optimization. Because stiction was detected, the valve was repaired and the loop was retuned.

When problems have been identified, they should be fixed:

- Process: Mechanical, electrical, process, operation
- Nonlinear and changing process: A more elaborate control strategy
- Interacting loops: A good tuning strategy is generally enough; if not, multivariable control could help

Main Goals for a Control Loop

- Remove disturbances
- Follow set point
- Reduce variability

The goal of a control system is to maintain the process variable equal to the set point in the presence of load changes, disturbances, or set point changes.Tuning a controller means adjusting the parameters (P, I, D) to achieve "good control." Good control is a performance criterion that depends on the application.

Is the trace from the flow loop in Figure 5.6o an example of good control? The measurement is being maintained at set point, but is the PID controller properly tuned? The truth is that we do not know. We must induce a change in the set point or the load to see the dynamic loop behavior. A loop that looks good in steady state may not behave as desired

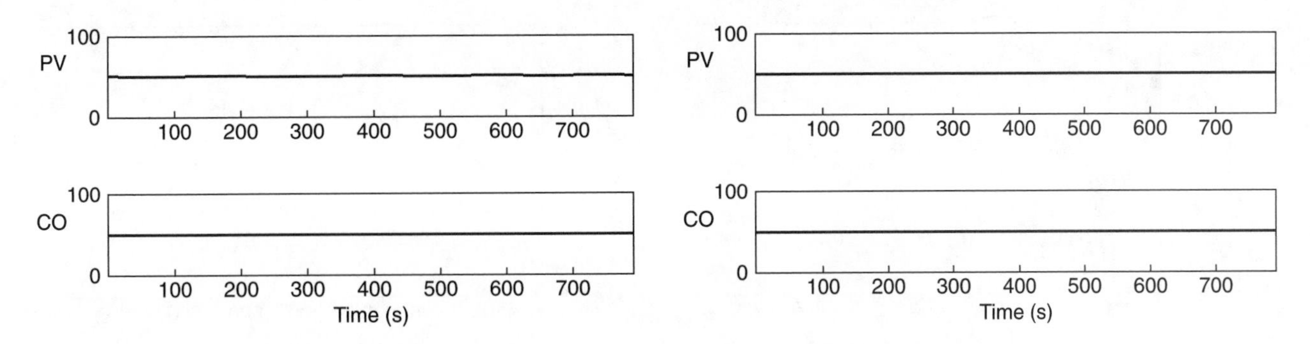

FIG. 5.6i
PV and CO trends as observed on the control system screen for the flow loop.

PV FIC141 Statistics		
Sample (raw)	0.33	
Num of Points	1819	
Time Min	0	
Time Max	801	
PV Min	46.47	
PV Max	49.56	
Range	3.088	
Mean (μ)	47.99	
Standard deviation (δ)	0.544	
μ±δ	47.446 – 48.534	
μ±2δ	46.902 – 49.078	= 2.176
Variance	0.2959	
Variability	2.27%	
Variability	39	

CO FIC141 Statistics		
Travel	8.13	878/day
Reversals	14	1510/day

PV FIC141 Statistics		
Sample (raw)	0.33	
Num of Points	1817	
Time Min	0	
Time Max	800	
PV Min	46.82	
PV Max	48.97	
Range	2.156	
Mean (μ)	47.65	
Standard deviation (δ)	0.258	
μ±δ	47.392 – 47.908	
μ±2δ	47.134 – 48.166	= 1.032
Variance	0.06656	
Variability	1.08%	
Variability	0.475	

CO FIC141 Statistics		
Travel	0	0/day
Reversals	1	108/day

FIG. 5.6j
Statistical analysis for the flow loop.

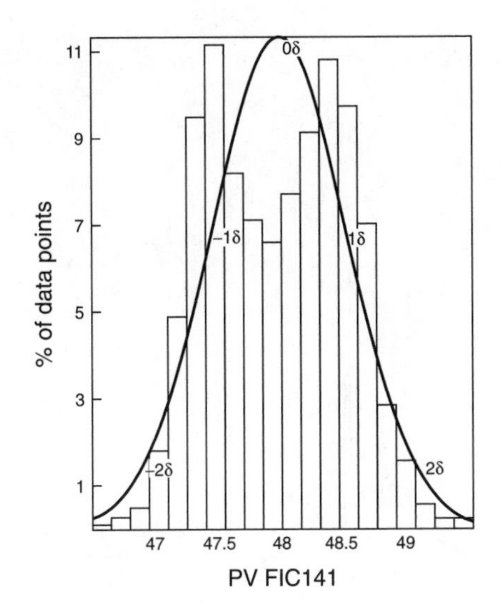

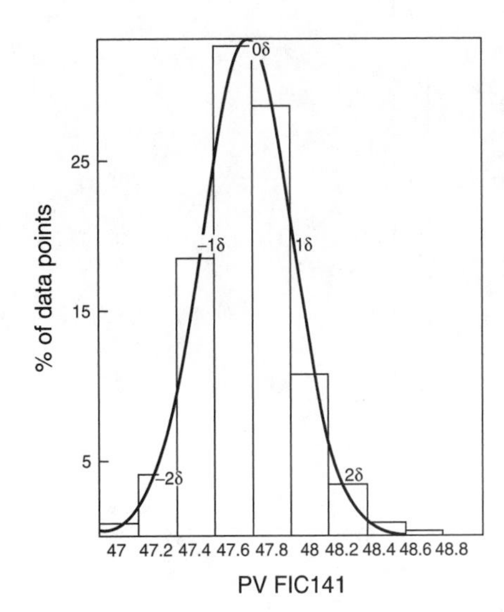

FIG. 5.6k
Data distribution for the flow loop.

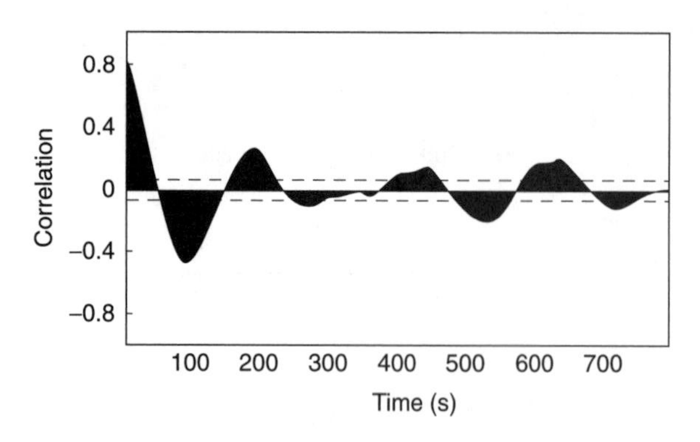

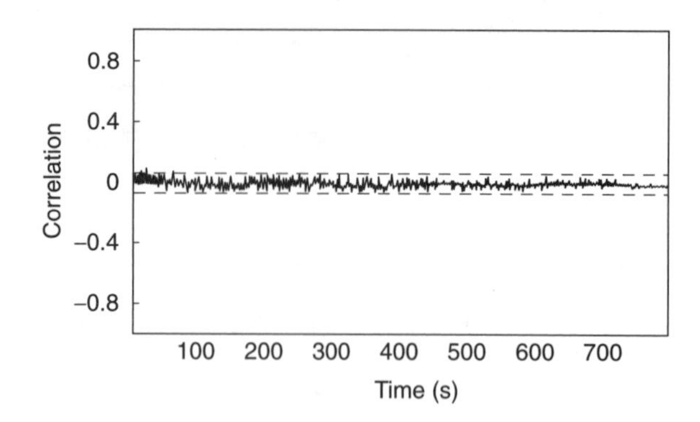

FIG. 5.6l
Autocorrelation for the flow loop.

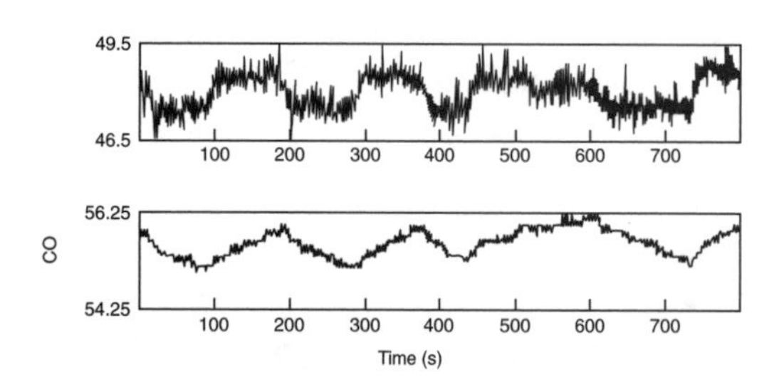

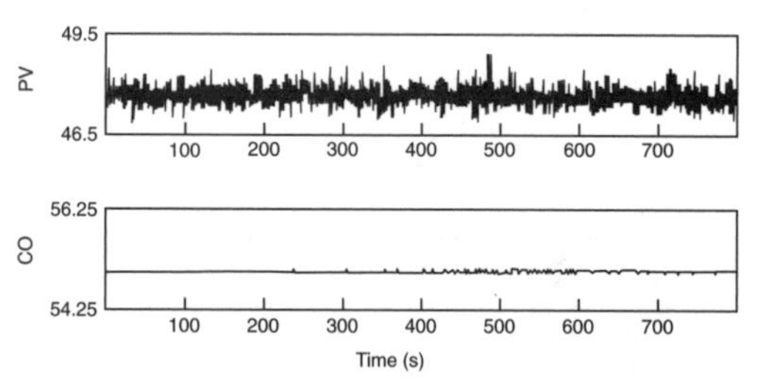

FIG. 5.6m
Zoomed data for the flow loop.

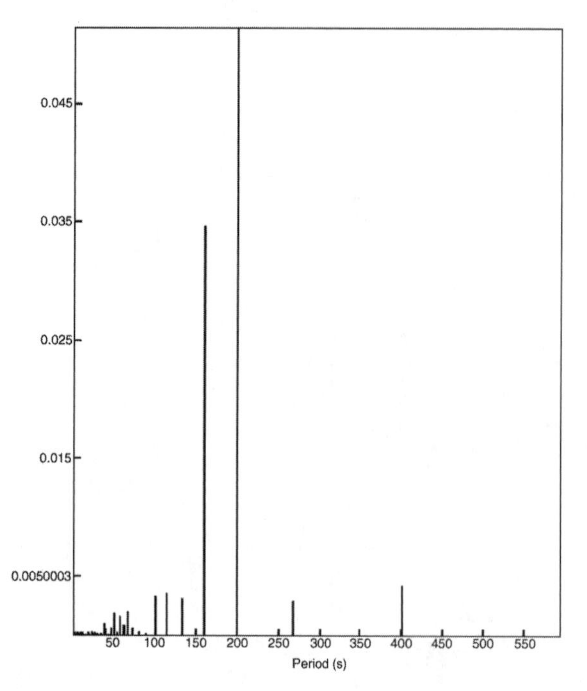

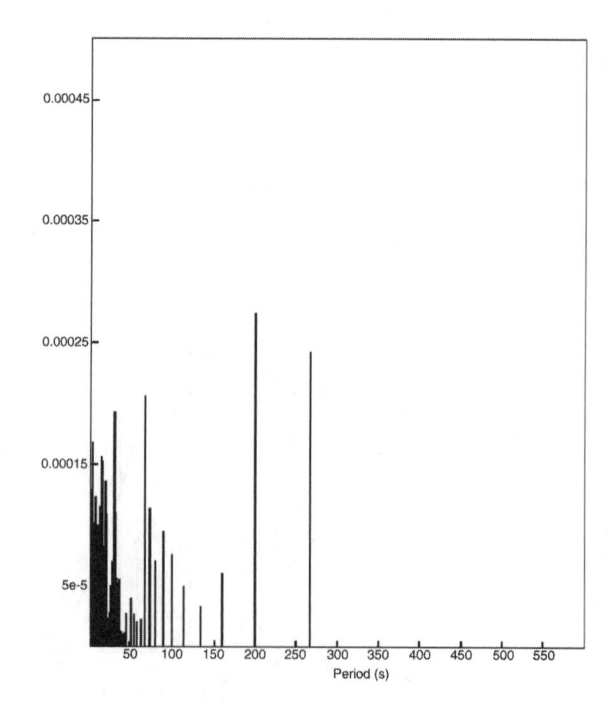

FIG. 5.6n
Power spectral density for the flow loop.

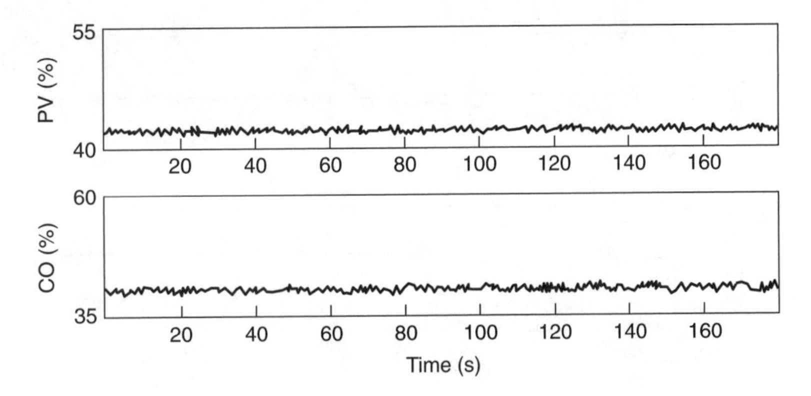

FIG. 5.6o
A loop at steady state without disturbances.

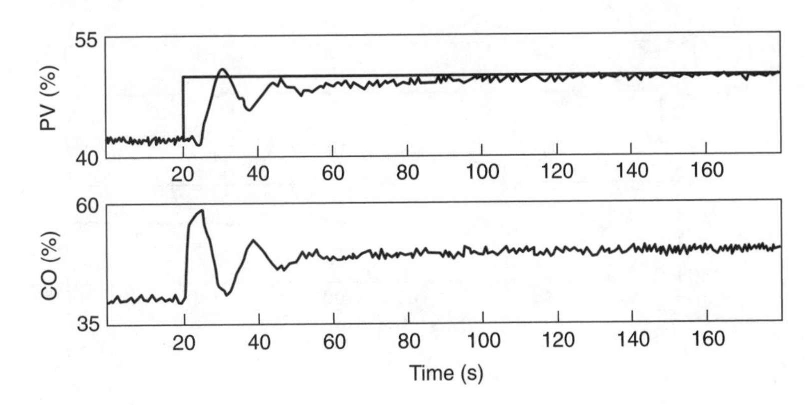

FIG. 5.6p
Loop response to a set point change.

when changes occur. Figure 5.6p shows a different picture. The response probably does not meet our objectives.

- To determine if a loop is properly tuned, one must induce a change. This is normally done by creating a set point change or a load change.
- A set point change is more drastic than a load change, because with a set point change, the error changes instantly. A load change must go through the process and is more gradual.
- Just looking at steady-state data will determine whether or not the loop is stable. To know if the performance objective is reached, one must observe the response following a set point change or a disturbance.

What is good control? One must ask

- Is it fast?
- Is it without error?
- Is it without overshoot?
- Does it minimize variability?
- Is it without oscillations?
- Does it minimize errors, tight control?

Tuning objectives will normally vary from loop to loop. Very fast control with slight continuous oscillation may be fine for one loop but dangerous for another loop. For example, the controller of a robot arm moving a bottle filled with nitroglycerin will be tuned without overshoot and without oscillation. The response will be slow and sluggish (Figure 5.6q). On the other hand, a flow controller in a loop filling milk bottles will be tuned tightly and aggressively to ensure the maximum speed of the production line with minimum error (Figure 5.6r). In such a system, the process tolerates oscillations but must minimize the error.

Tuning a controller not only makes the process operational, it is also makes it meet the performance criterion.

Common Performance Criteria

Figures 5.6s through 5.6v illustrate different performance criteria.

The least aggressive tuning is called *critical damping* (Figure 5.6s). It gives the quickest possible response without overshoot. Trying to go even a slight bit faster than critical damping will create overshoot. For this reason, processes that can tolerate no overshoot will normally use even less aggressive tuning, just to be safe. A method developed for this purpose is called *lambda tuning* which is extensively used in

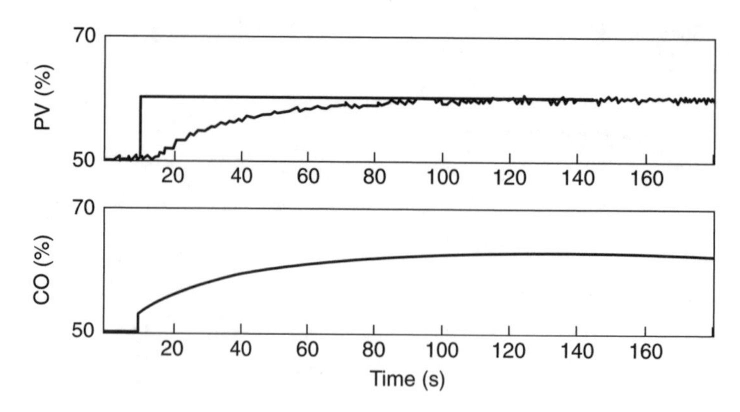

FIG. 5.6q
Loop turned for critical damping.

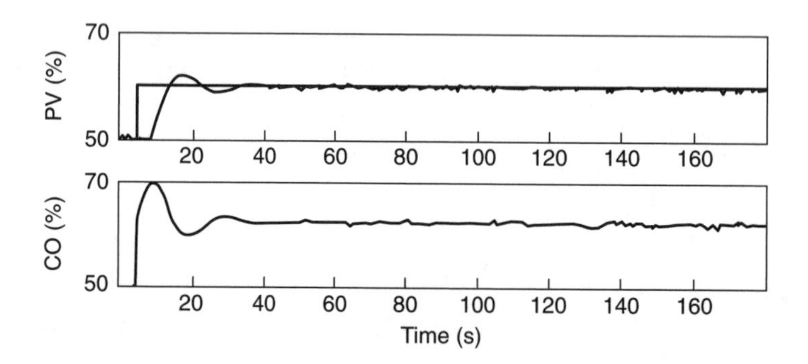

FIG. 5.6r
Loop tuned aggressively.

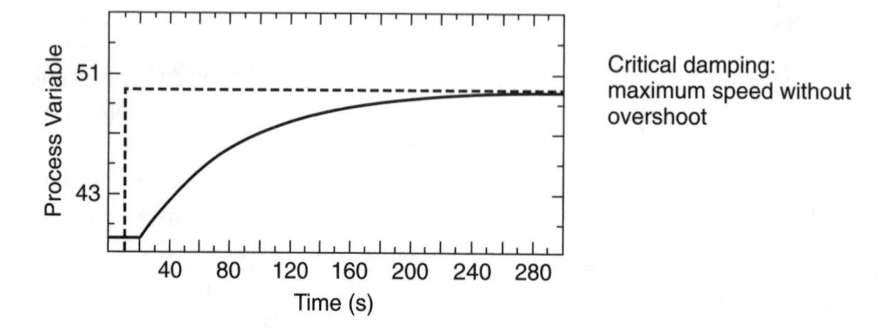

FIG. 5.6s
Critical damping.

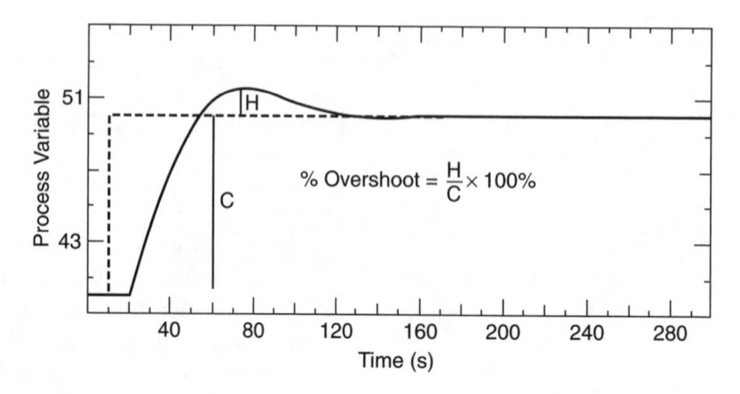

FIG. 5.6t
Overshoot less than a specified value.

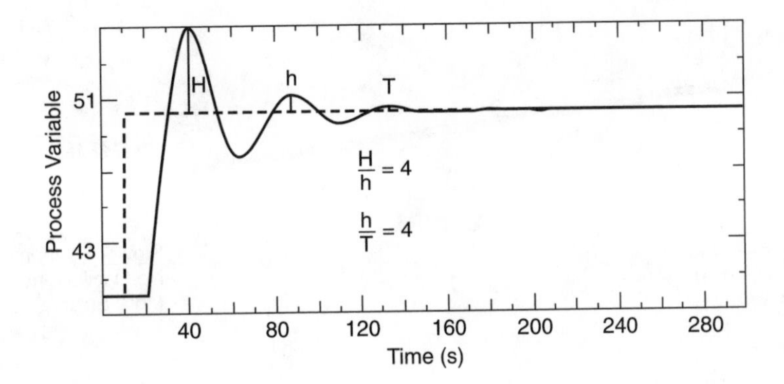

FIG. 5.6u
Quarter amplitude decay.

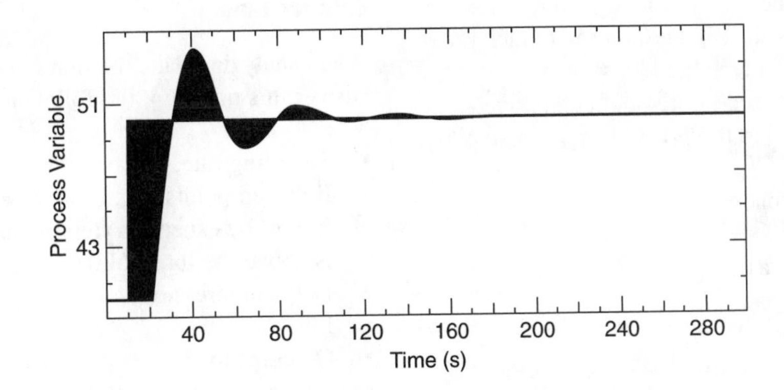

FIG. 5.6v
Integral of absolute error.

the paper industry. The closed-loop time constant is usually chosen three times longer than the process time constant. When the criterion is variability reduction, the goal is to remove a disturbance without overshoot or cycling, trying to minimize variability. For example, when producing paper, the variability is embedded into the sheet of paper so it must be minimized everywhere. In contrast, when producing fuel gas, because the product goes in a large storage mixed tank, the variability is averaged by mixing.

The second response has an overshoot H (Figure 5.6t). It is normally expressed as a percentage of how big it is in reference to the steady-state value (Percent Overshoot = H/C × 100%). Common tuning goals are 5 or 10% overshoot. This tuning gives a fast response, but some oscillations are to be expected.

The third and fourth responses are *quarter amplitude decay* (Figure 5.6u) and *minimum integral of absolute error* (IAE) (Figure 5.6v). These two have a very similar shape and give very quick response with obvious oscillations.

Quarter amplitude decay specifies a dampened oscillation in which each successive positive peak is one fourth of the magnitude of the preceding positive peak. This criterion is popular because it is easy to understand and easy to apply in the field. It differs from IAE in that IAE will be slightly quicker and will have slightly less error (process variable compared to set point). A faster response than IAE is not practical, as it will entail a greater total error, more overshoot, and will take longer to stabilize. Loops tuned to have quarter amplitude decay or IAE response should be retuned more frequently, as any error in tuning or changes in the process will give a highly oscillatory response that could become unstable.

When tuning for tight control, tuning is aggressive, with the goal of minimizing IAE. A pressure loop where set point is close to the relief valve set point pressure adjustment is a good example. A good compromise is tuning with 10% overshoot. This response is quick with very little oscillation. This tuning also provides peace of mind. If the process dynamic changes slightly, the response may become more aggressive, but it will not become unstable, as may occur with IAE or quarter amplitude decay.

There is a relationship between these responses that is very useful when tuning loops on the spot. Reducing the controller gain K_p by a factor of two will take the response from aggressive to moderate. Reducing it by half again takes the response from moderate to slow. It is important to note that this is only the case with ideal and series-type controllers. If a controller with a parallel structure controls the loop, the integral and derivative gains must also be cut.

TUNING: A COMPROMISE

Tuning a controller consists of making compromises between speed and stability (Figure 5.6w). The stabler the loop, the more it reacts slowly, and the larger are the errors after a disturbance. In industry, control loops are nonlinear (the process

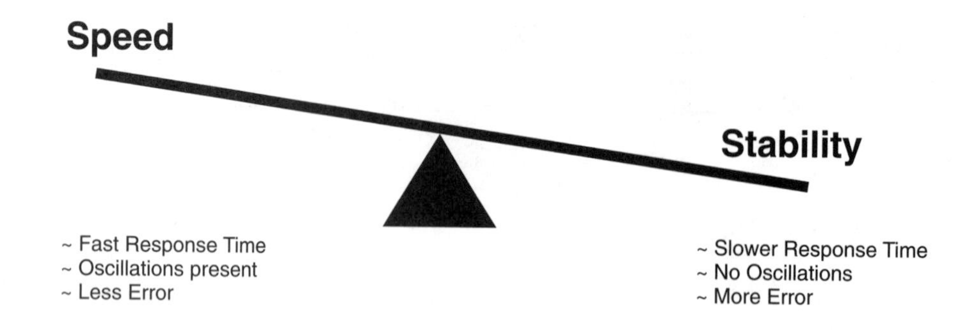

FIG. 5.6w
Tuning, a compromise between stability and speed.

model is not constant), and it is essential to tune them for the worst case. When possible, if the process model varies too much, the control algorithm should also vary.

The more the loop is tuned tightly and aggressively, the more it will react quickly, but the price to pay is instability, cycling, and overshooting.

Before tuning a loop, one must ask:

- Is the process slow or fast?
- Is the process self-regulating?
- Is the process symmetrical?
- Is the process linear, is the right valve characteristic used?
- Is the process gain near one, is the valve properly sized?
- Is the transmitter properly installed and is the transmitter working well?
- Does the valve have hysteresis?
- Does the valve have stiction?
- Is dead time dominant?
- Is noise excessive?
- Will the tests represent the worst cases?

After tuning, the tuning parameters will be valid until a change is made to the process or to a component in the control loop.

The role of a control loop is usually to maintain the process variable at set point. To achieve this, the controller output modulates a final control element to reject disturbances. For some loops, it is also essential to obtain a good set point response; for example, the inner loop of a cascade control system. Finally, in a batch control system, the process variable must follow the profile set point, generally without overshoot.

Hence, when optimizing a process control system, the process control engineer must consider the process control objective for the loop, the unit, and the entire plant. How well each controller accomplishes its task depends on the selection of the controller parameters, according to the process and the control objectives; this is loop tuning.

Single-loop tuning is covered in other sections. Most of the time, a well-designed PID controller is still the best algorithm to handle the job. Various algorithms are available for difficult processes.

Data for Tuning

When analyzing data, the first question is how good is it? To answer this question, the following should be considered:

- Sampling rate, fast enough?
- Between points, snap shot or averaging?
- Report by exception (report change only if the change is above the threshold)?
- Data compression?
- Filtering?
- Quantization?
- Integer or real (floating) number?
- Other functions inserted?
  - Scaling
  - Filtering
  - Averaging
  - Characterizer
  - Limiters (amplitude, rate)
  - Interlocks
- How good is the configuration and the programming?
  - Scan rate
  - Program sequence
  - PID function and selection
  - Scaling in PID function
  - Algorithm
  - Bumpless transfer
  - Filtering, averaging, moving average
  - Derivative on process variable or on error
  - Anti-reset wind-up active
  - Dead band
  - Rate limiters, set point filter
  - Integer
  - Limits on I, D, unit multipliers, quantization

Types of Controller Algorithms

For most loops, a simple PID controller is used. In fact, a filter should usually be used to remove the noise. The noise corresponds to fast disturbances; the PID controller cannot handle fast disturbances. The filter will not remove the noise, but will hide it from the controller. Hence, the controller output will be noise-free, and this will reduce valve movement.

Sometimes, the simple PID controller needs help to ensure performance will be achieved. If the process gain varies, performance will vary as well. To maintain good control in all conditions, the controller must behave the same way at all rates and in all conditions.

To obtain these conditions, it could be useful to make the following changes:

- Add a characterizer at the controller output, to change the valve characteristic, for example
- Add a characterizer at the input (process variable and CO), pH loops, for example
- Modify the tuning parameters based on process conditions or on a variable, for example, for grade changes
- Modify the control algorithm, as explained in the next paragraph

A standard PID controller is normally implemented using one of the three most common structures—ideal, parallel, or series. Although the structures differ, the behavior of all three is similar; the input to the controller is always the error (PV-SP or SP-PV) and the controller output is composed of a proportional, an integral, and a derivative contribution. This standard algorithm is a universal solution. It works on all processes, but it does not always provide the desired performance. To increase performance, especially on nonlinear processes, the standard algorithm can be modified. Special algorithms change a standard PID controller and make it nonlinear when some facet of control needs improving.

A nonlinear controller is usually used to compensate for process nonlinearities. For a linear process, standard PID control is a better choice than a modified algorithm. However, when the situation warrants a modified algorithm and if it is implemented properly, the result is improved performance. Some controller manufacturers make these algorithms available as a choice in the controller setup. If they are not available, it may be possible to program their behavior into a programmable controller or a DCS.

- Standard algorithm: The implementation shown in Figure 5.6x uses an ideal (also called ISA structure) controller structure.

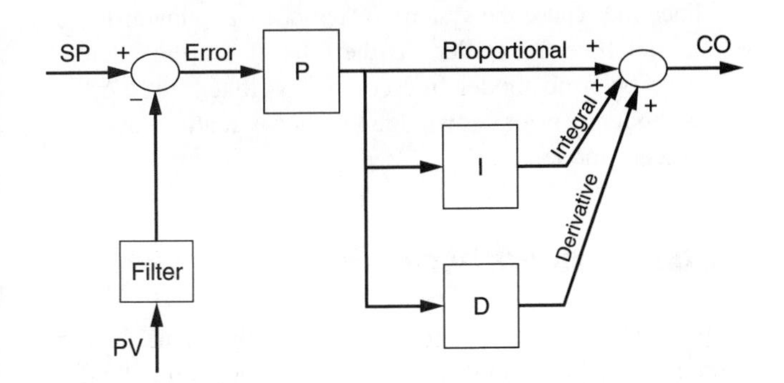

FIG. 5.6x
Ideal or ISA controller.

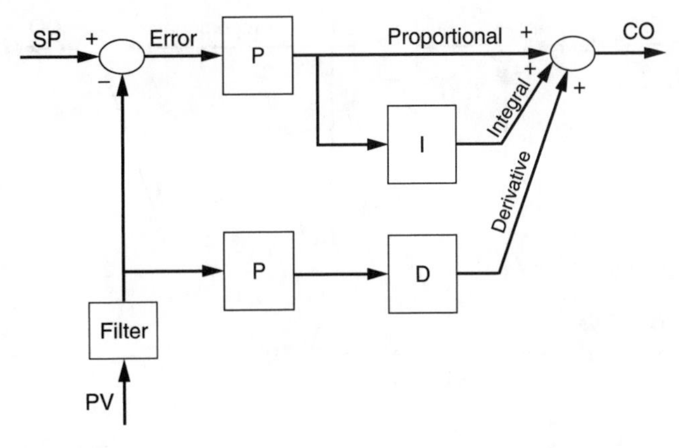

FIG. 5.6y
Derivative on PV.

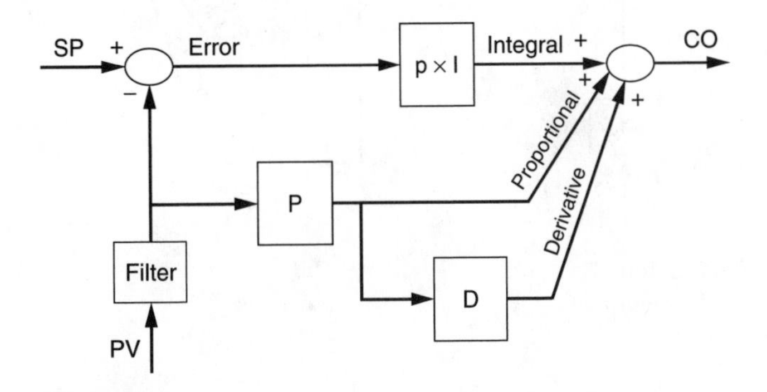

FIG. 5.6z
Derivative and proportional on PV.

- D on PV: A set point change will be handled by the proportional and the integral components only. The derivative is only active when the process variable moves. This algorithm is recommended any time the derivative is used, and the algorithm can be implemented (Figure 5.6y).
- D and P on PV: A set point change will be handled by the integral component only. As a result, there should be no overshoot on a set point change. The proportional and derivative are only active when the process variable moves. This algorithm allows for more aggressive tuning (Figure 5.6z).
- Set Point Filtering: A set point change will not provoke as aggressive a response (Figure 5.6aa). Overshoot on a set point change will be less and the proportional gain can be increased as a result. This algorithm allows for more aggressive tuning and is recommended on loops that have frequent set point changes and disturbances. A filter is preferred over a ramp limiter.
- P^2ID (error squared): An error-squared controller will act quite aggressively when the error is large and quite sluggishly when the error is less than 1%. At 10% error, the input to the PID is 100%. At 0.5% error, the input to the PID is 0.25%. When properly implemented this algorithm can give aggressiveness and steady-state stability.

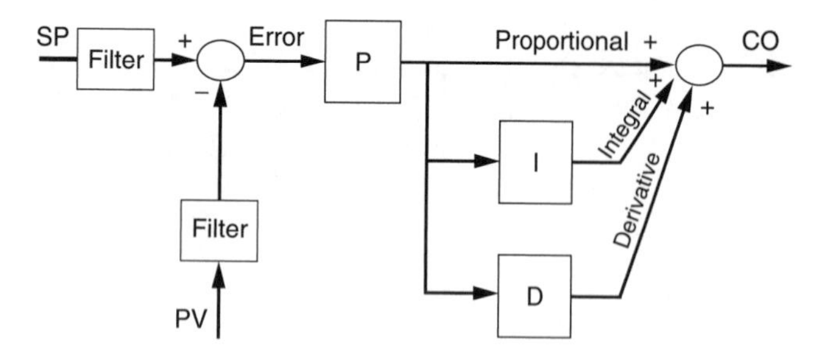

FIG. 5.6aa
Set point filtering.

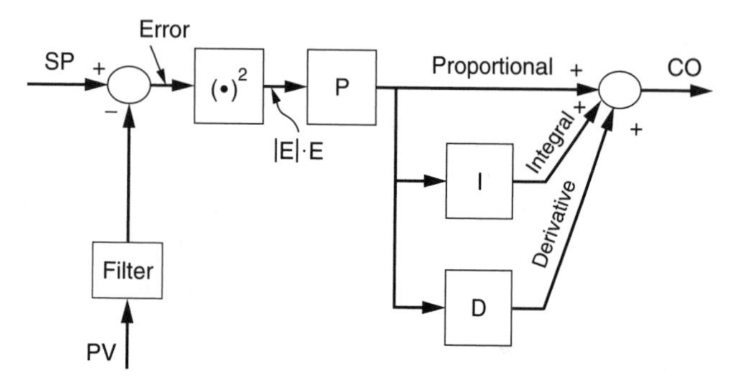

FIG. 5.6bb
P^2ID algorithm.

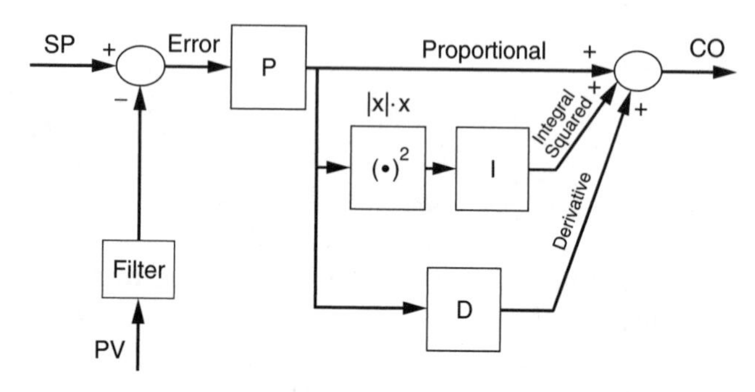

FIG. 5.6cc
PI^2D algorithm.

The most common uses for this algorithm are for level control (averaging and surge) (Figure 5.6bb).

- PI^2D (integral squared): A less commonly used algorithm is the PI^2D (Figure 5.6cc). Because integral action is normally linked with dead time, it makes sense that integral-squared action works best to compensate for processes with a variable dead time. Integral action causes overshoot in level loops—with integral squared as the level approaches set point, the integral contribution goes to zero.
- PID+gap: When the error is within the gap, it is considered 0. Outside the gap, the normal tuning parameters apply. Using this algorithm, the controller output changes are reduced, increasing the terminal element life expectancy (Figure 5.6dd).

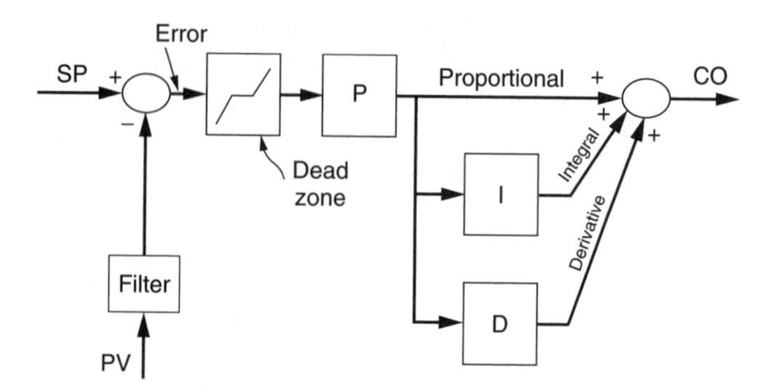

FIG. 5.6dd
PID-gap algorithm.

- PID+gap+adaptive tuning: To fight stiction when inside the gap, integral action is removed; to fight hysteresis when inside the gap, the proportional action is reduced; to fight nonlinearity when inside the gap, proportional action is reduced.

Many other algorithms have been suggested to replace the PID controller.

- Smith Predictor: This controller receives feedback from a model of the process without the dead time rather than from the real process. The real process variable is compared with models to correct errors and handle disturbances. It is good for processes with large dead time that have set point changes; no overshoot on set point changes; and it is as fast or only slightly faster than normal PID control on load changes.
- Dahlin Controller/Model Predictive Controllers: The controller behaves as one over the process [process × (1/process) = 1], giving a closed-loop response that has the same form as the open-loop response. This is similar to lambda tuning: beneficial for loops that can tolerate no overshoot on a set point change. There is no overshoot on set point changes, and it is very sluggish on load changes.
- Adaptive Model Controller: Same as model predictive controllers but the model is continuously updated. It is very good in mechanical systems, but does not handle process problems very well. Hysteresis, stiction, and nonlinearities cause the system to "remodel" continuously.
- Model Free Controller: At the time of writing, there have been no model free controllers that truly work without any prior knowledge (or approximation) of the process model.

MULTILOOP AND INTERACTING SYSTEMS

Unfortunately, when many loops are in the same system, there is no magic formula that tells one if one loop will affect another. This information will only come through knowledge of the process. If one loop directly feeds another, oscillation

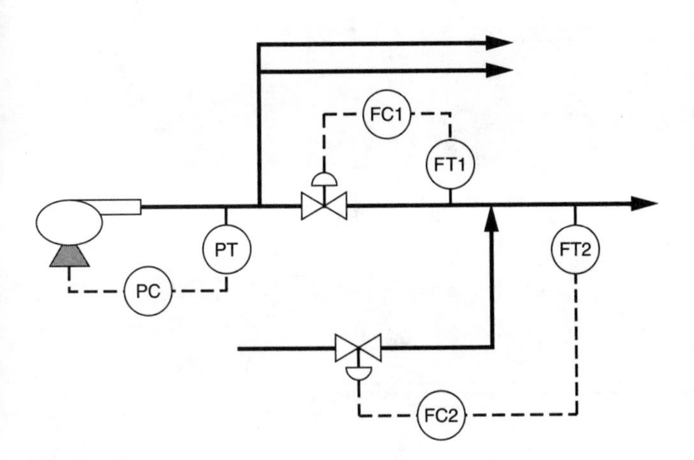

FIG. 5.6ee
Interacting loops.

in the first loop will cause oscillation in the second and possibly any other downstream loops. If two flow loops are fed from the same pump, oscillation in one loop could cause oscillation in the second. Other times, loops do not interact, but it is imperative that they respond with the same speed.

Figure 5.6ee shows a solution flow loop. Two liquids are brought together. The top flow is an expensive product and is maintained at 100 USGPM. The lower flow is water; it is adjusted to give the total flow required (between 200 and 400 USGPM).

All three of these loops have the potential to be fast. A response time of less than 30 s is attainable on all three loops, but which loop *needs* to be fastest?

In this case, the pressure loop must be faster than flow loop 1; its job is to maintain pressure for the flow loop, regardless of the flow. If the pressure loop and flow loop 1 are tuned at the same speed, they may work for a while, but, eventually, a disturbance will occur that will cause the two loops to oscillate. For example:

1. A disturbance causes the line pressure to increase.
2. If the pump is too slow, the flow will increase.
3. The valve closes to compensate, causing the pressure to increase.
4. The pump eventually slows, causing the flow to decrease.
5. The valve opens to compensate, causing the pressure to decrease.
6. The pump speeds up, causing the flow to increase.

Steps 3, 4, 5, and 6 repeat continuously—the two loops will oscillate and potentially resonate. Similarly, because flow loop 1 feeds flow loop 2, flow loop 1 must be faster than 2. Otherwise, a disturbance in flow 1 could cause both flow controllers to react—oscillation would result. When loops interact together, we need to select response speeds that differ. The exact same speed could cause oscillation. Speeds that differ but are close have the potential to oscillate. To be safe, we should choose response speeds that differ by a factor of 3 to 5. If a speed difference of less than 3 is selected, the loops may one day start to oscillate. To decouple loops when they are highly interactive, a speed difference of up to 10 may be required.

For the system pictured in Figure 5.6ee, this would mean that response time of the pressure loop determines the response time of flow loop 1. The response time of flow loop 1 in turn determines the response time of flow loop 2. The procedure for tuning this system would be to put the downstream loops in manual and tune the first loop, the pressure loop. It should be tuned fairly aggressively because its response time determines the system response time. Once tuned, put the pressure loop in auto; its behavior is now part of the process. Flow loop 1 is tuned next, leaving flow loop 2 in manual. This loop must be tuned for a response time that is at least three times slower than the pressure loop response time, ideally five to ten times slower. Once tuned, leave the pressure and flow loop 1 in auto and tune flow loop 2. It too must be tuned at least three times slower than flow loop 1 (ideally five to ten times slower).

Step 1. Tune fastest loop PC for quick response; other loops are in manual mode (Figure 5.6ff).
Step 2. Tune moderate loop FC1 for moderate response; previous loop PC remains in auto while the other loop is in manual mode (Figure 5.6gg).
Step 3. Tune slow loop FC2 for slow response; previous loops PC and FC1 remain in auto (Figure 5.6hh).

Synchronizing Loops

Sometimes loops do not interact, but they do work together. When loops need to have the same response time, they are called synchronizing loops. It is important to note that synchronizing loops should be designed so that there is no physical link between them that could cause interaction.

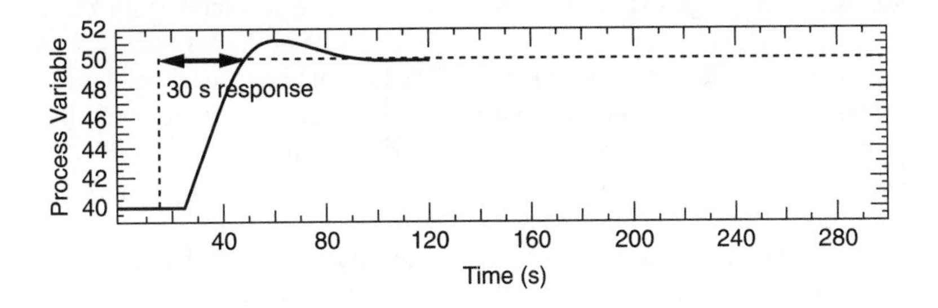

FIG. 5.6ff
Fastest loop settling time is 30 s.

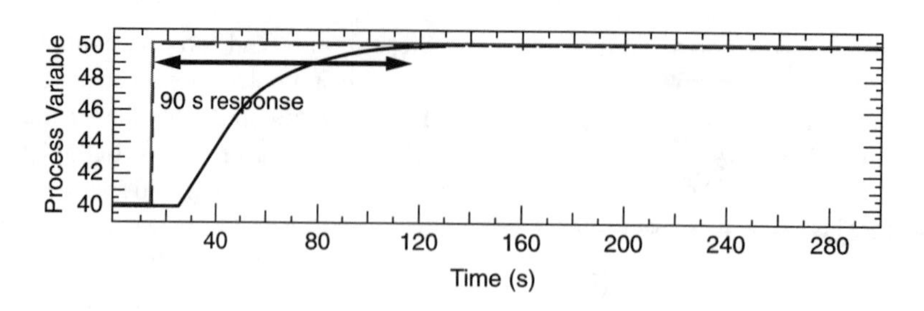

FIG. 5.6gg
Fast loop settling time is 90 s.

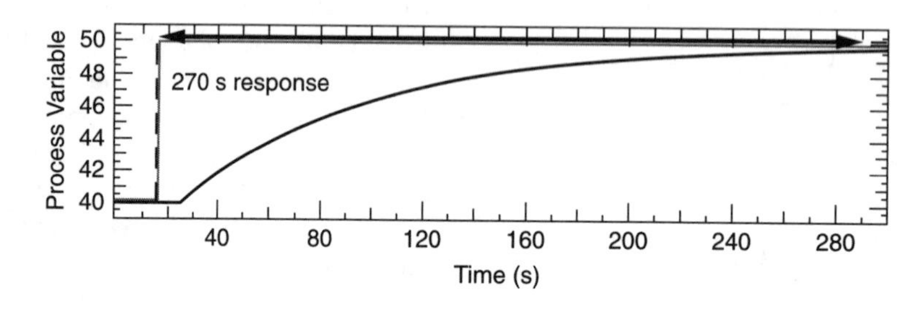

FIG. 5.6hh
Fast loop settling time is 270 s.

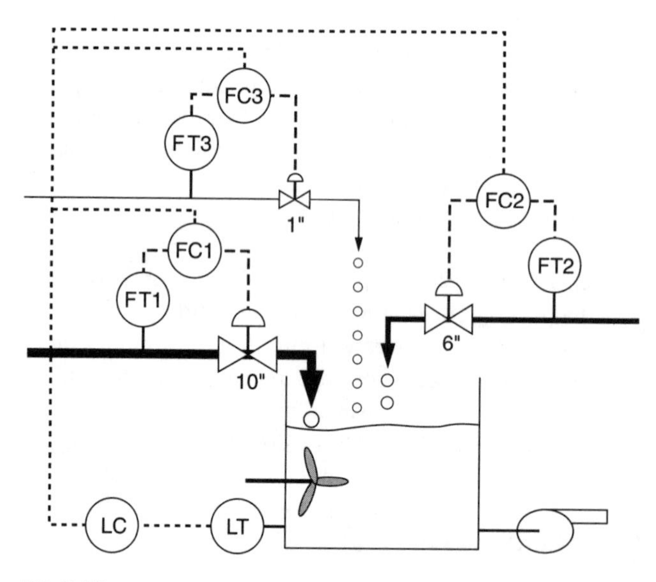

FIG. 5.6ii
Synchronizing loops.

Batch mixing is one example of a system that needs to be synchronized. Figure 5.6ii shows three ingredients that are sent into a mix tank. The proper ratio of ingredients must always be maintained—even when the system starts, speeds up, or shuts down.

If placed in manual and bumped, all three flow loops would probably have different process gains, dead times, and time constants, especially with such a large difference in valve size. As such, if all three loops were tuned for 10% overshoot, the response times would not be the same. When the level loop calls for increased flows, the recipe will be out of balance until all three flows are stable and at set point.

To ensure that all three loops work in harmony, determine which loop is the slowest and match the response time of the others to it. Normally, the slowest loop is the one with the largest dead time.

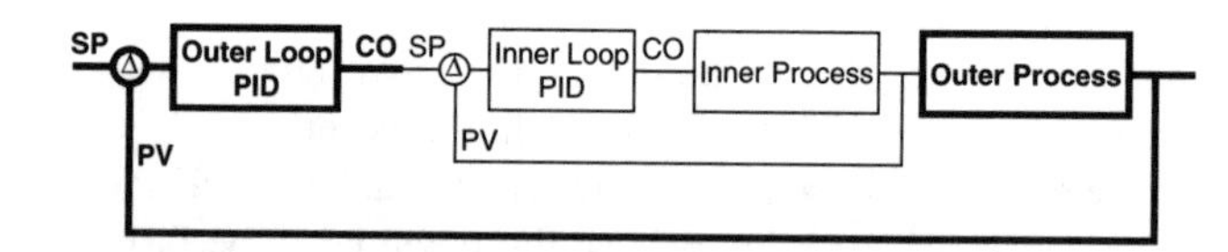

FIG. 5.6jj
Cascade loop setup.

Procedure for tuning synchronizing loops:

- Do a bump test on each loop.
- Determine which is slowest.
- Tune slowest aggressively and measure speed of response when done.
- Determine tuning parameters for other loops that will give approximately the same response time.

When tuning loops that need to work in harmony, select tuning parameters that give similar response times.

Cascade Loops

A system is in cascade if the output of one controller becomes the set point for another (or others), as in Figure 5.6jj. The loop receiving the set point is called the inner or slave loop. The loop sending the set point is called the outer or master loop.

The outer process is usually the primary concern and could be controlled by the outer PID loop only. However, adding the inner loop can greatly increase performance if it has disturbances that can be quickly eliminated by it, but not the outer. For example, the level controller in Figure 5.6kk can control the level on its own. However, its response to a disturbance in the incoming flow is poor. By adding a flow controller in cascade, the performance is increased by the ratio of the

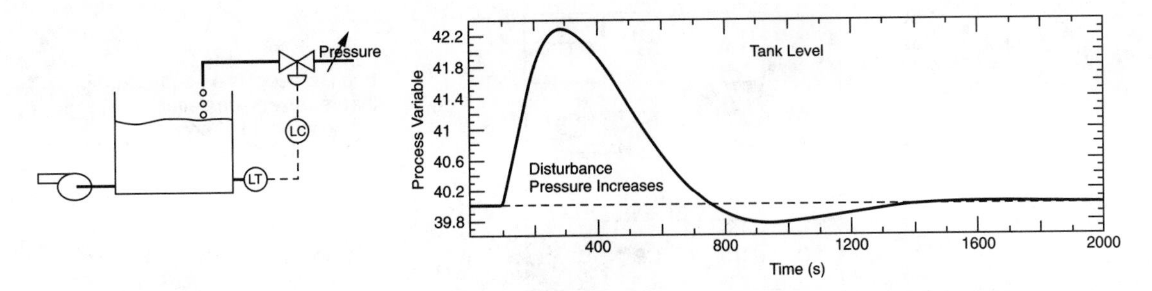

FIG. 5.6kk
Simple level loop when a pressure disturbance occurs.

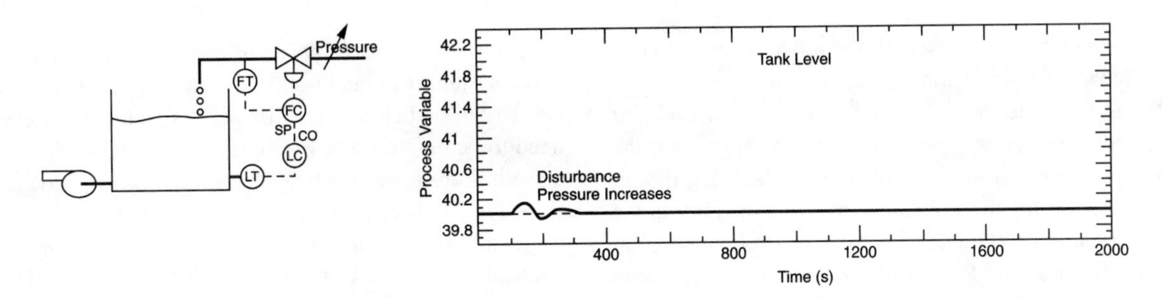

FIG. 5.6ll
Cascade level loop when a pressure disturbance occurs.

outer loop response time over the inner loop response time (normally a factor of 10 or greater). Figure 5.6kk shows a simple level loop's response, when a pressure disturbance occurs, while Figure 5.6ll illustrates the response of a cascade level loop when a pressure disturbance occurs.

$$\text{Performance Increase} = \frac{\text{Outer Loop Response Time}}{\text{Inner Loop Response Time}}$$

As cascade loops are interacting, the inner loop needs to be three to five times faster than the outer loop.

Tuning cascade loops:

Step 1. Tune the inner loop.
Step 2. Tune the outer loop, leaving the inner loop in cascade mode (auto and remote set point); select tuning to obtain a response three to five times slower than the inner loop.

Cascade systems work best when the inner loop can eliminate disturbances quickly.

Feedforward Loops

Feedforward is a simple but often misunderstood concept. Feedforward is added to "help" a properly tuned PID loop. Without the "help," the loop will act as any other PID loop, deviating from set point when disturbances occur. When feedforward from a specific disturbance is added, the process should not deviate from set point when that particular disturbance occurs.

Feedforward is added when a disturbance that causes an error can be measured and its effect on the controller output is repeatable. By predicting the effect of the disturbance and reacting to it before it affects the process, there is a huge gain in performance. For example, if driving a car on a cold day and a passenger opens a window, one could act like a PID controller and wait for the temperature to drop before reacting (turning up the heat). If one reacts in this manner, after a few minutes the temperature will be back in the comfort zone. The better choice would be to turn up the heat as soon as the window is opened, before the temperature drops. This thinking ahead, similar to feedforward, will ensure that the cabin temperature hardly even changes.

Feedforward may be added, subtracted, or in some cases multiplied. If the speed needs to be matched, a dynamic compensator (mostly, lead/lag functions) is added. To add feedforward:

- First, make sure that the loop is properly tuned.
- In automatic, observe the effect of a disturbance on the controller output several times (make sure there are no other disturbances or changes that occur during the test).
- Determine the feedforward factor by finding the relationship between Δdisturbance and ΔCO.
- Test the loop performance with feedforward.

Fuzzy Logic

Fuzzy logic is a very interesting topic as its approach for solving problems differs from conventional methods. Whether fuzzy logic is used for process control or to fly a helicopter, one does not need to know anything about the mathematical dynamics of the element being controlled. For example, when

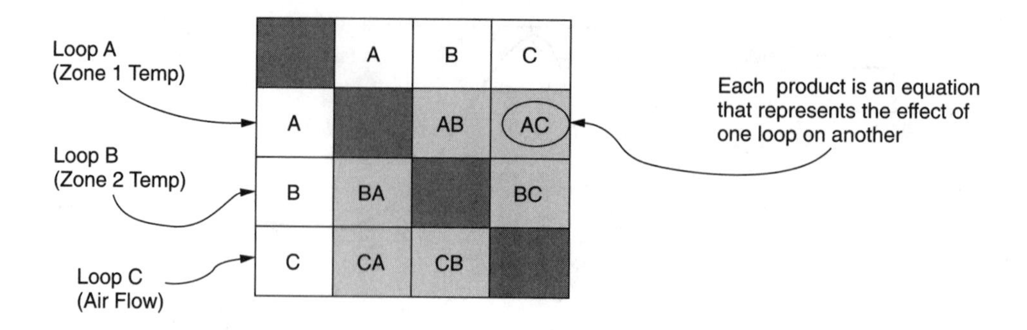

FIG. 5.6mm
A 3 × 3 multivariable system.

a baseball player chases a baseball, the player does not need to know Newton's laws of motion to catch it.

Fuzzy logic controllers are used to replace an operator when this operator is doing better than a control system; the implementation varies from one process to another. The design consists of transferring the knowledge from the operator to the fuzzy logic controller. By adjusting the rules and the strength of each rule, the fuzzy logic controller will mimic the operator.

Fuzzy logic can also be used with PID controllers. For example, a PID controller manufacturer has developed an autotuner for its PID controller that uses fuzzy logic to determine the optimal PID tuning parameters.

Examples of fuzzy logic in process control:

- As a nonlinear controller (PID replacement): By adjusting rules and "fuzzification," the controller becomes nonlinear.
- As a multivariable controller: The fuzzy logic controller can control multiple loops to cut cost or maximize production. It can maintain tight control of loops that would normally interact with one another.

Neural Networks

Neural networks are similar to fuzzy logic in that the mathematical model relating inputs to outputs does not need to be known; the behavior response is known instead. One major difference between fuzzy logic and neural networks is that we cannot modify the gains or functions in the neural network; we can train the system with data, but cannot train it with human reasoning. Fuzzy logic allows for modification of each controller parameter.

Multivariable Controllers

In a multivariable control system, the control algorithm is designed for the specific system. For example, a model is computed that links each variable. This is similar to adding feedforward between all loops and potential disturbances in a system. In reality, the controller is an equation set that allows each component to know the impacts of each move it makes in advance. Hence, the controller will send a new controller output to each valve (or a new set point to each loop) each time it makes a move. When properly set up, this allows the entire system to maintain stability when applying a correction.

To build such a controller, a lot of models or equations need to be defined. For example, a 3 Input/3 Output system, 6 models are needed (Figure 5.6 mm); with N variables, $N * (N - 1)$ models are needed; with 10 variables, 90 models are necessary. If the number of variables is high, a very complex controller is needed as well as a lot of data to obtain the models.

Because of the number of relationships that need to be determined, the multivariable controller is relatively expensive to set up. Once it is up and running, it will do a good job until something changes. For example, if the process is redesigned or equipment starts to wear (hysteresis, stiction, noise, etc.), the controller performance will decrease or become unstable. In many cases, a multivariable controller controls entire production rates and loop set points rather than individual loops. If any of the individual PID loops are retuned, system performance will change.

The controller links all of the control loops and disturbances together; it compensates where required to maintain stability in all linked loops when manipulating a variable. Such a controller is most difficult to implement and to test compared with fuzzy logic and neural network systems, *unless* the system is small (3 × 3 to 5 × 5).

The most important problem is that the system must be retuned when the process changes.

OPTIMIZATION TOOLS

Many software packages are available to tune or to assist in the tuning of industrial controllers. The criteria listed below are not met by all of them. Most packages are single-loop tuners and are insufficient for process optimization.

The software should as a minimum:

- Run under Windows
- Accept ASCII data and/or provide a data acquisition system and/or allow for direct connection to the DCS or PLC and/or communicate through DDE or OPC
- Include tools for multiloop tuning and analysis (cross correlation, cascade, feedforward, etc.)

- Assure quality of the data
  - Missing points
  - Data validation
  - Scan time
  - Data filtering
- Compute the tuning parameters from data obtained in manual or automatic mode
- Compute the tuning parameters based on user-selectable performance criteria
- Analyze the data to determine and compute:
  - Process model (automatically), including inverse response, complex models
  - Hysteresis, backlash, stiction
  - Noise, cycling
  - Power spectrum
  - Autocorrelation, cross correlation
  - Statistical analysis
  - Bode plot
  - Robustness plot
  - Quality of the tests
  - Quality of data
  - Tuning according to different criteria
  - And so forth
- Know the structure and characteristics of the industrial controller used (the user specifies the model and manufacturer only), including derivative and integral specific methods
- Compute the filter time constant for process variable and set point
- Inform the user of any disadvantages of the specific controller
- Have an excellent support and help file

Many software packages on the market do not meet all these criteria. Some suggest strange values and use noncoherent methods.

Most software packages use the frequency domain to analyze and find the tuning parameters.

Autotune, Pretune, Self-Tune

Manufacturers of industrial controllers (stand-alone, DCS, etc.) claim that their controllers are intelligent, are self-tuning, and are able to adapt themselves to the process. Reality tells a different story.

There are two different approaches to controller automatic tuning:

1. Pretuning
2. Adaptive tuning

A program makes a test on the process and computes parameter tunings by analyzing the process response. The most common tests that the programs perform are either a bump test in open loop or ultimate cycling test in closed loop. The last is called the relay method. A person can do the same tests and analysis in less time and the results will be better.

The program frequently needs the intervention of the user and generates error messages.

A program analyzes the behavior of the process in automatic mode and computes parameter tuning. The program is based on heuristic rules or on complex mathematical models and formulas. Programs that use heuristic rules are sometimes able to adapt correctly to the process, but they need the supervision of experienced and skilled people. Frequently, the programs compute bad tunings, but if properly supervised (limits, pretuning, etc.), these bad tunings will not differ too much from initial tunings. The second method is based on complex formulas and elaborate mathematical models. These programs are rarely able to adapt correctly to the process and they need supervision of experienced and skilled people.

Both pretuning and adaptive tuning work well when the process model is constant, as in mechanical systems. If the process contains hysteresis, backlash, stiction, or is nonlinear, the algorithm behaves strangely.

Performance Monitor

A performance monitor program identifies the areas in a plant where the greatest economic gains are possible. The software is designed to help make the biggest impact on the plant. It pinpoints areas that will yield the greatest economic return.

This determines which controllers in a plant are not performing well and which would benefit from retuning or require maintenance.

The software automatically surveys any (or all) controllers in the plant and highlights candidates for performance improvement. A simple performance index compares the performance of each controller. The software digests data coming from the plant and generates a list of loops outside predetermined performance limits.

Additionally, the software can be used to determine the economic benefits for advanced control implementation, which can include the following:

- Reduces process variance, hence improving plant profitability
- Maximizes throughput
- Minimizes requirements for raw material and additives
- Reduces engineering time
- Improves plant safety
- Maximizes profit subject to various steady-state production and process restrictions
- Reduces equipment downtime
- Helps to determine if a unit is a good candidate for multivariable control and other advanced techniques; provides numbers on quality and performance

EXAMPLE OF OPTIMIZATION STEPS

Step 1. Understand the process and the performance objectives, work in cooperation with the operators, explain the work.

Step 2. A loop is rarely alone, define a tuning strategy.
Step 3. Check the process before tuning, check equipment.
Step 4. Tuning loops, simple PID control should be the first choice.
Step 5. Use modern tools to analyze the process and select the performance criterion.
Step 6. Validate, check, then write a report.

An example is a paper machine optimization. A mill (in Canada) produces newsprint paper. For many years, one of the older machines at the mill had a low efficiency compared with the others; the uptime for this machine was 83% and below the budget.

The variability for this machine was high; 1.4% for the basis weight and 6% for the moisture content. The quality needed improvement and the machine broke down too often.

Workers verified the process, the mechanical parts, the electrical components, and the operation procedures. The improvements were minimal. Then, members of the process control department suggested using a global approach to optimize the paper machine. This subsection describes how they did it and their results.

The average uptimes for the previous years were:

1997	83.3%
1998	84.3%
1999	83.2%
Average	**83.6%**

The average variability remained the same month after month for the last 3 years:

Basis weight variability	1.4%
Humidity variability	6.1%

They trained the staff to analyze and optimize the process. They went through the entire paper machine, and each part of the process was analyzed, repaired, or modified, if necessary. They hired a process control consulting company to train plant staff on the methods. This company also provided support for the more complex problems. Technicians, engineers, and supervisors were trained in process control basics, troubleshooting analysis, and efficient use of optimization software. They used loop optimization software to help analyze and troubleshoot the process.

The most important loops were properly tuned once the problems in operation and equipment were fixed. The loops were tuned to work in harmony, to remove interactions, and to reduce variability. For example, they synchronized the flow loops in the mixing tank and reduced interaction between level and pressure in the head box. At the same time, the PID parameters and the PV filter were selected to reduce valve effort, therefore ensuring valve maintenance would be minimized in the future.

The operation and production people fully collaborated with the instrumentation. They worked hard to improve performance, for example, by installing fast cameras to analyze the origin of breaks on the sheet of paper. They also installed a system to detect holes in the sheet.

Hence, the results not only originated from the instrumentation effort, but also from the operation, production, and management staff.

Changes and Resulting Improvements

As a result of their analysis work, they made the following interventions:

- Repaired four valve positioners, six valves
- Replaced four valves
- Relocated two valves
- Repaired one transmitter
- Configured two transmitters
- Modified two control strategies
- Modified the PID tuning parameters on 42 loops
    - Mostly reduced proportional gain, and reduced integral time
    - Added derivative on four loops
    - Added filtering on 21 loops
- Modified the process operation
    - Pressure in the rejection tank
    - Valve opening for cleaner cyclones
    - Vacuum control in rejection tank
- Modified the process
    - Removed a buffer tank
    - Modified piping arrangement
- Replaced pieces of equipment
    - A mixer
    - Manual valves

To obtain a solid assessment of their efforts, they initially collected data on all the important loops of the paper machine. Each variable was analyzed to detect hidden oscillations (using correlation and power spectral density analysis), to determine variability, to verify if tuning parameters were appropriate. Interaction was also analyzed using cross-correlation analysis. Even the amount of valve movement or valve effort was recorded.

The following data were analyzed:

- Variability
- Power spectral density
- Valve performance (hysteresis, stiction, and process gain)

The overall performance of the paper machine was computed. To reduce the time and the cost, only the most important loops (42 out of 85 loops) were analyzed.

Once the main problems were fixed and the loops were properly tuned, they again collected data on the same loops.

The same variables were analyzed. They compared variability, cycling, valve wear, robustness, and performance of the control loops, as well as process stability, product quality, downtime, and efficiency.

The following average results were found on the most important loops:

- The oscillations of the loops were reduced by a factor of over 200.
- The variability of loops was reduced by a factor of 2, on average.
- Valve movement was reduced by a factor of at least 5 (average for 42 loops).
- The overall variability (basis weight, humidity, and dry weight) was reduced by a factor of 2.
- The uptime (efficiency) was increased from 83 to 87%.
- The time to reach steady state after a grade change was greatly reduced.

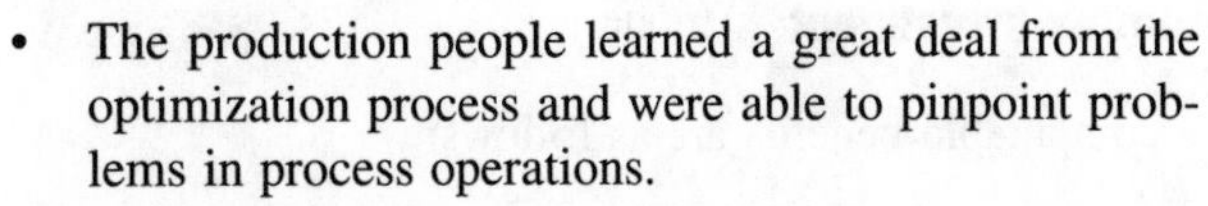

- The production people learned a great deal from the optimization process and were able to pinpoint problems in process operations.
- Valve maintenance was reduced since valve movement was greatly reduced.

The basis weight (weight of an area of a sheet of paper, expressed in g/m^2) was then analyzed using the software, as shown in Figures 5.6nn through 5.6pp. The left side of each figure is before; the right side is after.

As shown by the power spectral density graphics, the hidden oscillations were greatly reduced. The same results were observed for other loops. Looking at the basis weight informs us about all cycling on the paper machine as this measurement is done at the end of the process. The hidden oscillations disappeared after the optimization; for example, the cycling at 820 s was from the machine chest level loop. The results have been the same for months.

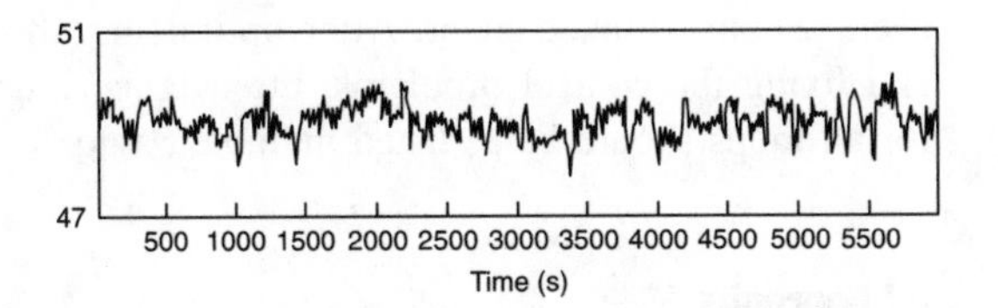

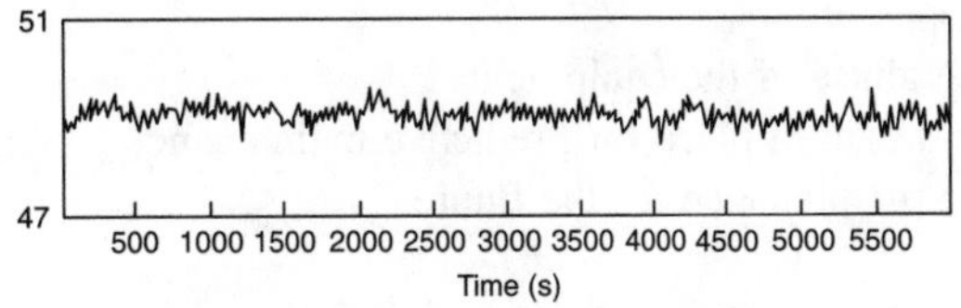

FIG. 5.6nn
Time data from the quality control system.

Time Min	0
Time Max	6000
PV Min	47.88
PV Max	50.1
Range	2.213
Mean (μ)	49.07
Standard deviation (δ)	0.347
μ±δ	48.723 – 49.417
μ±2δ	48.376 – 49.764
Variance	0.1204
Variability	1.41%

= 1.388

Time Min	0
Time Max	5500
PV Min	48.51
PV Max	49.66
Range	1.151
Mean (μ)	49.07
Standard deviation (δ)	0.167
μ±δ	48.903 – 49.237
μ±2δ	48.736 – 49.404
Variance	0.02789
Variability	0.681%

= 0.668

FIG. 5.6oo
Statistical analysis for the data from the quality control system.

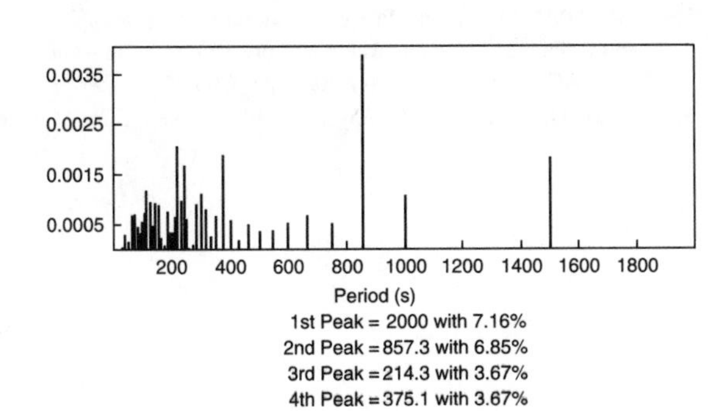

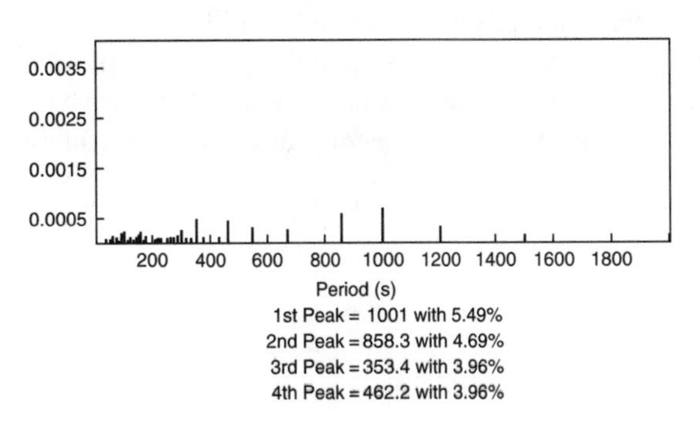

FIG. 5.6pp
Power spectral density for the data from the quality control system.

Return on Investment

The computable benefits are as follows:

- Better performance of the paper machine: it starts easily after a grade change and breaks down less often.
- Better efficiency of the paper machine.
- Increased production rate.
- The value of the above benefits can be estimated to be $1,800,000 per year.

The noncomputable gains are as follows:

- Better knowledge of the paper machine
- Better operation of the paper machine
- People better trained to troubleshoot a problem on a paper machine
- Better product quality
- Replicable expertise for other paper machines
- Smooth operation
- No more abuse of the equipment
- Tools and data in place for predictive maintenance
- Reduced maintenance for the future

The costs for the entire process are as follows:

- Training and support by a consultant
- Purchase of analysis and data acquisition software
- Labor time for optimization process
- Maintenance by mill staff

Equipment replacement was not included in the total cost of optimization, as the equipment would have been replaced eventually. The total cost was $68,000.

If equipment cost is excluded, this gives a return on investment of less than 1 month. Process optimization therefore can provide one of the highest returns on investment in a plant today.

Preventive maintenance instead of reactive maintenance is another major benefit, because the maintenance people now have the tools and the skills to maintain their gains. They are able to detect and to fix problems before they have an impact on production. They now use the same techniques on other paper machines and other sections in the mill.

Because in process optimization, the plan is to use the equipment at its best, one does not need new installations with all their procurement costs, engineering, installation, and future maintenance cost. In fact, the goal is to achieve the maximum performance possible with the existing equipment.

TABLE 5.6qq
Expected Results from Process Optimization

Performance	*Impact from Optimization*	*Improvement, %*
Variability	Reduced by a factor of 2	50
IAE	Reduced by a factor of 2	50
Settling time	Reduced by a factor of 2	50
Valve travel	Reduced by a factor of 5	80
Valve reversal	Reduced by a factor of 5	80
% time not in normal mode	Reduced by a factor of 2	50
Cycling, oscillations	Removed	—
Robustness	Increased by a factor of 2	100

Table 5.6qq lists the likely benefits for optimizing an average, unoptimized plant. After optimizing all the loops and fixing the control problems, the average improvement for all loops should be as listed in Table 5.6qq.

Bibliography

Astrom, K. J., *PID Controllers: Theory, Design, and Tuning,* 2nd ed., Research Triangle Park, NC: Instrument Society of America, 1995.

Gerry, J. P., "Tune Loops for Load Upsets vs. Setpoint Changes," *Control Magazine,* September 1991.

Gerry, J. P., "Tuning Process Controllers Starts in Manual," *InTech Magazine,* May 1999.

Luyben, W. L., Luyben, M. L., and Bjorn, T. D., *Plantwide Process Control,* New York: McGraw-Hill, 1998.

Ruel, M. and Gerry, J. P., "*Quebec Quandry Solved by Fourier Transform,*" *InTech,* August 1998.

Ruel, M., "Loop Optimization: Before You Tune," *Control Magazine,* Vol. 12, Number 3, pp. 63–67, 1999.

Ruel, M., "Loop Optimization: Troubleshooting," *Control Magazine,* Vol. 12, Number 4, pp. 64–69, 1999.

Ruel, M., "Loop Optimization: How to Tune a Loop," *Control Magazine,* Vol. 12, Number 5, pp. 83–86, 1999.

Ruel, M., "How Valve Performance Affects the Control Loop," *Chemical Engineering,* Vol. 107, Number 10, pp. C13–C18, 2000.

Ruel, M., "Stiction: The Hidden Menace," *Control Magazine,* Vol. 13, Number 11, pp. 69–76, 2000.

Ruel, M., "PID Tuning and Process Optimization Increased Performance and Efficiency of a Paper Machine," presented at 87th Annual Meeting, PAPTAC, Book C, February 2001, pp. C63–C66.

Shinskey, F. G., *Process Control Systems,* 4th ed., New York: McGraw-Hill, 1996.

5.7 Plantwide Controller Performance Monitoring

K. A. HOO   M. J. PIOVOSO

With the advent of automation, most modern industrial plants have hundreds and even thousands of automatic control loops. These loops can be simple proportional/integral/derivative (PID) or more sophisticated model-based linear and nonlinear control loops. It has been reported that as many as 60% of all industrial controllers have performance problems.[1] However, assessing their performance is not a trivial task. Because they play a vital role in product quality, safety, and ultimately economics, it is prudent to have an automated means of *detecting* when a loop is not performing well and then *diagnosing* the root cause. A primary obstacle that prevents this automatic assessment from being a part of a routine maintenance program is a lack of easy-to-use, readily understandable, and reliable tools. In addition, education of the operations staff is essential to make full use of some of the currently available time and frequency methodologies. The papers by Bialkowski,[2] and Kozub[3] provide an industrial perspective on controller performance challenges.

Some of the common causes of poor performing loops are as follows:[4,5]

1. Incorrect tuning
2. Changing process dynamics (transitions, unmeasured disturbances)
3. Limited controller output range
4. Large dead times or inaccurate determination of the dead time
5. Incorrect sampling interval
6. Incorrect controlled and manipulated variable pairings
7. Poor hardware (sensors, actuators) maintenance

Performance assessment is concerned with the analysis of available process data against some benchmark. Many researchers have presented different benchmarks to assess controller performance. Most notable are the works by Aström,[6] Shinskey,[7] Harris,[8] Stanfelj et al.,[5] and Huang and Shah[9] for univariate feedforward and feedback controllers, and Harris et al.,[10] Huang et al.,[11] and Kozub and Garcia[12] in the case of multiple loop controllers.

The most used benchmark is the *minimum variance controller* (MVC) first introduced by Harris[8] for single-loop feedback controllers (see Figures 5.7a and 5.7k). It provides a theoretical lower bound on the closed-loop process output variance. To determine the MVC involves the use of open-loop or closed-loop routine output data and knowledge of the process dead time. By fitting an autoregressive integrated moving average (ARIMA) time series model to the output data, this lower bound on the performance can be calculated. Implicit in the calculation of this benchmark and others is that the process can be represented adequately by a linear time-invariant (LTI) transfer function model with additive disturbances [see Equation 5.7(4)].

Aström[6] assessed achievable performance in terms of bandwidth, rise time, and peak errors for servo and regulatory PID controllers. Shinskey[7] used the integral of the absolute error (IAE) to assess the performance of PID controllers and introduced a performance index called the *absolute performance index* (API) to select the structure and tuning parameters of the controller. Desborough and Harris[13] introduced a *normalized performance index* [see Equation 5.7(13)], which gives the ratio of the variance *in excess* of that which can be theoretically achieved *if* the controller was a minimum variance controller. They also provide a useful discussion on the relationship of this measure to that of the squared correlation coefficient, r^2, that is usually calculated in multiple regression analysis. Huang et al.[12] developed an efficient, stable filtering and correlation method (FCOR) to estimate the minimum variance benchmark of Harris[8] and the normalized performance index of Desborough and Harris.[13]

Stanfelj et al.[5] and Desborough and Harris[14] addressed the issue of feedback and feedforward control loops (see Figure 5.7p). The idea is that if the feedback controller is already providing satisfactory performance, but this still exceeds the overall performance, then alternatives such as feedforward control can be used to compensate for the disturbances that are measurable. In the former contribution, a cross-correlation test was developed to assess the feedforward and feedback control loop performance. In the latter study, a procedure was provided to calculate the minimum variance performance when both types of control strategies are present. The procedure is an extension of the methods in the single-loop feedback case.

DeVries and Wu[15] and Kozub and Garcia[12] proposed a *closed-loop potential* (CLP) factor [see Equation 5.7(17)], which is similar to the normalized performance index of Desborough and Harris.[13] Both are based on a normalization of the closed-loop output error variance and both indicate the performance of the controller relative to the minimum variance benchmark.[8] When a model-based feedback control strategy (see Figure 5.7k) is used, Harris et al.[16] and Kozub[3] propose an *extended horizon performance* index. Unlike the normalized performance index, this one does not require exact knowledge of the process dead time and permits an evaluation of long settling times. An interpretation of the extended horizon performance index is that it corresponds to the square of the correlation between the current error and the least-squares estimate of the prediction made *in the past*. If there is no correlation, then the performance of the feedback controller is deemed satisfactory because the disturbance was rejected after the dead time elapsed.

In the realm of multivariate controllers of multiple-input multiple-output (MIMO) processes, Harris et al.[16] employed a multivariate spectral factorization and the solution of a multivariate Diophantine identity, which is an extension of the univariate case. Huang and Shah[9,11] also extended the single-input single-output (SISO) performance measures and added the notion of an *interactor* matrix or time-delay matrix.[17] In both of these contributions, knowledge of all process time delays is a requirement.

The minimum variance controller is not the only acceptable benchmark with which the univariate assessment methods are compared. Others include settling time, rise time, overshoot, and damping.[6] In the multivariate case, Shah et al.[18] have also suggested the use of linear quadratic Gaussian (LQG) performance benchmarks and a design performance benchmark.

It has been pointed out by many researchers that the implementation of a minimum variance controller may not be desirable because (1) it may require excessive controller action, (2) it may not be robust to modeling errors, and (3) the process dynamics may not be invertible.[5,8,14] Harris et al.[10] and Shah et al.[18] have proposed spectral factorization to address the issue of non-invertible zeros. Tyler and Morari[19] and Sternad and Söderström[20] also address this issue and the issue of unstable systems.

If the goal is to develop an automated online technology then features, such as computational simplicity, reliability, easy interpretability, prioritization in the case of multivariable process systems, and assessment of controller redesign vs. controller retuning, ought to be a part of the methodology.

There is commercial software available to provide automated process loop monitoring, tuning, and diagnostics, for example, the INTUNE™ software tools by Control Soft (Cleveland, OH). The literature on this software states that it can detect and rectify poor control loop performance. Additional tools in the set can also provide optimal PID tuning parameters for univariate and multivariate situations.

UNIVARIATE SYSTEMS: THE FEEDBACK CONTROLLER

The Minimum Variance Controller

Consider the block diagram of a feedback control system shown in Figure 5.7a. Assume that this single-loop system can be represented adequately by a linear time-invariant, discrete transfer function model and additive disturbance.

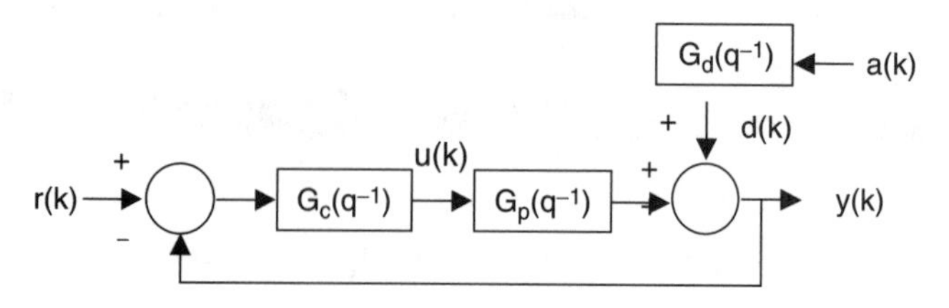

FIG. 5.7a
Block diagram of a feedback controller. r(k): set point, y(k): output variable, d(k): unmeasured disturbance, u(k): controller output.

The process transfer function given by G_p, and G_c and G_d denote the controller and disturbance transfer functions, respectively,

$$y(k) = G_p(q^{-1})u(k-b) + d(k) = \frac{\omega(q^{-1})}{\delta(q^{-1})}u(k-b) + d(k) \qquad 5.7(1)$$

where q^{-1} is the backward shift operator such that

$$q^{-j}(y(k)) = y(k-j)$$

$y_p(k)$ and u(k) are deviations of the measured process output and controller outputs, respectively, from their nominal operating values; d(k) is a bounded disturbance; $\omega(q^{-1})$ and $\delta(q^{-1})$ are polynomials in the backward shift operator; and $b \geq 1$ is the number of whole periods of delay in the process. The term d(k) is assumed to represent all unmeasured disturbances acting on $y_p(k)$ and it may be deterministic or stochastic.

Let d(k) be given as a linear function of past values of a statistically independent random sequence of variables, $\{a_j\}$,

$$d(k) = G_d(q^{-1})a(k) = \frac{\theta(q^{-1})}{\phi(q^{-1})}a(k) \qquad 5.7(2)$$

The terms $\theta(q^{-1})$ and $\phi(q^{-1})$ are assumed to be stable polynomials. Let the feedback controller that regulates y(k) be a single degree of freedom controller that can be represented by the following linear time-invariant model:

$$u(k) = G_c(q^{-1})(r(k) - y(k)) \qquad 5.7(3)$$

where r(k) is the deviation of the set point from its reference value. Substitution of Equations 5.7(2) and 5.7(3) into Equation 5.7(1) gives

$$y(k) = G_p(q^{-1})G_c(q^{-1})[r(k) - y(k)] + G_d(q^{-1})a(k) \qquad 5.7(4)$$

If the set point is constant, it then follows that

$$y(k) = \frac{G_d(q^{-1})}{1+G_c(q^{-1})G_p(q^{-1})q^{-(b-1)}}a(k)$$
$$y(k) = \frac{\theta(q^{-1})/\phi(q^{-1})}{[1+\omega(q^{-1})/\delta(q^{-1})G_c(q^{-1})q^{-(b-1)}]}a(k) \qquad 5.7(5)$$
$$y(k) = \frac{\alpha(q^{-1})}{\beta(q^{-1})}a(k) \equiv \Psi(q^{-1})a(k)$$
$$y(k) = [\psi_0 + \psi_1 q^{-1} + \cdots + \psi_b q^{-b} + \cdots +]a(k)$$

where ψ_j are impulse weights.[19] The series in Equation 5.7(5) is convergent if the closed loop between y(k) and d(k) is stable. Because of the delay term, $q^{-(b-1)}$, the first b terms in Equation 5.7(5) are identical to those computed from the disturbance transfer function $G_d(q^{-1})$ and can be interpreted as *system invariant.* Thus, only terms at lag b and beyond are affected by the current controller action. The reason for this is seen as follows. Once a disturbance appears at the output, it is fed back to the controller and the controller makes a correction. However, because of the delay, that corrective action has no effect on the output for b time intervals into the future. No disturbance compensation can occur at the output until the dead time of the system has expired. Desborough and Harris[14] offer the following interpretation of Equation 5.7(5), the first b terms form the b-step ahead forecast errors and the remaining ones are the b-step ahead forecast.

The variance of the controlled variable can be calculated by squaring Equation 5.7(5) and then applying the expectation operator E{˙},

$$\sigma_y^2 = E\left\{\sum y_i(k)^2\right\} = [\psi_0^2 + \psi_1^2 + \cdots + \psi_{b-1}^2 + \cdots +]\sigma_a^2 \qquad 5.7(6)$$

where $\sigma_a^2 = E\{a(k)a(k)\}$. Terms such as $E\{a(k)a(k+h)\}$ evaluate to zero because the sequence, $\{a_j\}$, is assumed zero-mean and independent. Equations 5.7(5) and 5.7(6) verify that the variance of the controlled variable is related to the feedback process dynamics, the disturbance model, the controller, and the variance of $\{a_j\}$. If the feedback controller is a minimum variance controller, then the b-step ahead forecast (terms at and beyond b) equals zero and the output variance is given by

$$\sigma_{mv}^2 = \lfloor \psi_0^2 + \psi_1^2 + \cdots + \psi_{b-1}^2 \rfloor \sigma_a^2 \qquad 5.7(7)$$

This controller then rejects the prediction of the disturbance after the dead time has elapsed leaving only the prediction errors. Thus, the controlled variable under minimum variance control will depend on only the most recent b past disturbances,

$$y_{mv}(k) = \lfloor \psi_0^2 + \psi_1^2 + \cdots + \psi_{b-1}^2 \rfloor a(k) \qquad 5.7(8)$$

The finite stochastic process in the above equation is called a *moving average* process of order b. It then follows that any controller that is not minimum variance must inflate the variance, that is,

$$\sigma_y^2 = \sigma_{mv}^2 + \sigma_{\hat{y}}^2 \qquad 5.7(9)$$

where the term $\sigma_{\hat{y}}^2$ means the variance of the b-step ahead forecast. Harris[21] has shown that this minimum variance controller also minimizes a wide class of symmetric and asymmetric objective functions.

Performance Measures

An important consideration in the calculation of the theoretical minimum variance, as presented by Harris,[8] is that routine process data, with or without feedback control, are used rather than specially designed tests. The only requirement is that the number of observations is large and representative of the process. Kozub[3] and others have commented that the data sequence ought to contain information on important disturbance upsets to avoid using data corrupted by unusual process operations. Furthermore, too short a data record may result in statistical estimates whose variability renders them useless while too lengthy a data set may lead to erroneous results because different response characteristics are contained in the data set. How to select the appropriate data window size remains an open research issue. Thornhill et al.[22] discuss the related issues of sampling interval, model order, and data compression on data selection and analysis.

The theoretical autocorrelations of the moving average process described by Equation 5.7(5) at and beyond lag b are zero when the controller is minimum variance. For real, representative data of size n, the sample autocorrelation value must be compared with some statistical confidence interval because, in practice, the autocorrelations values are never truly zero. If there exist many large autocorrelations beyond lag b, then it can be concluded that the performance of the controller deviates substantially from the minimum variance performance bounds. On the other hand, if only a few values at and beyond lag b are significant, then the performance is *close to* the lower bound. See Figure 5.7b, which shows the autocorrelation function for an example process. The confidence bands are the 95% or two standard deviation probability limits. Methods to calculate the autocorrelation function can be found in Box and Jenkins.[23]

The sampled variance of the output error can be calculated by

$$s_y^2 = \frac{\sum_{k=1}^{n}(y(k)-\bar{y})^2}{n-1} \qquad 5.7(10)$$

where $\bar{y}$ is the mean value of the deviations of sampled data from nominal value. The ratio of the current variance to the

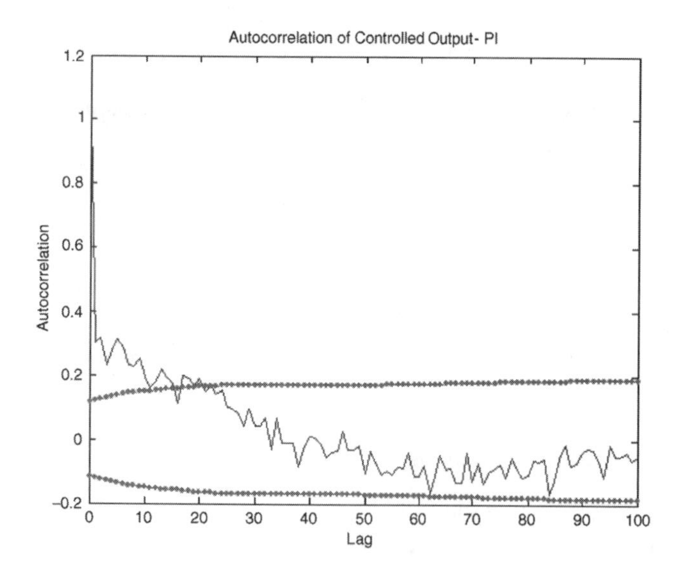

FIG. 5.7b
Autocorrelation values of the two-state CSTR example with 95% confidence bands.

theoretical minimum variance is given by

$$\frac{s_y^2}{\sigma_{mv}^2} \approx \frac{\sigma_y^2}{\sigma_{mv}^2} \geq 1 \qquad 5.7(11)$$

which indicates the maximum improvement possible by modification of the controller. This ratio is related to the performance index defined by Desborough and Harris[13]

$$\xi(b) = \frac{s_y^2 + \bar{y}^2}{\sigma_{mv}^2} \qquad 5.7(12)$$

When $\bar{y}^2$ term is zero, Equations 5.7(11) and 5.7(12) have the same meaning and it is clear from Equation 5.7(12) and the meaning of σ_y^2 that $\xi(b) \geq 1$ but has no upper bound.

It is very possible that the controller is providing minimum variance but its performance exceeds the overall performance (product or process specification). This means that reductions in the output variance can only be realized by modifying the process. In other words, no other modification of the present controller structure will reduce the output variance. The only recourse is to change the system structure. For example,

1. Reduce the dead time
2. Reduce the variance of the disturbance
3. Eliminate disturbances through process modification
4. Introduce feedforward control on measured disturbances
5. Change manipulated variables

When the performance is significantly greater than the lower bound, further analysis must be done to ascertain the cause for the variance inflation. This is the diagnosis or third level in the monitoring flowchart given in Stanfelj et al.[5] Some common causes are

1. De-tuning or poor tuning
2. Errors in the process model
3. Wrong sampling interval
4. Changing process dynamics
5. Saturation of manipulated variables
6. Errors in input/output models
7. Faulty sensors and actuators

The normalized performance index of Desborough and Harris[13,14] is defined as follows:

$$\eta(b) = 1 - \frac{\sigma_{mv}^2}{\sigma_y^2} = 1 - \frac{\psi_0^2 + \psi_1^2 + \psi_2^2 + \cdots + \psi_{b-1}^2}{\psi_0^2 + \psi_1^2 + \psi_2^2 + \cdots + \psi_b^2 + \cdots} \qquad 5.7(13)$$

This index represents the fractional increase in the variance of the output that arises from not implementing a minimum variance controller. Further, unlike $\xi(b)$, $\eta(b)$ is bounded between [0, 1]. When $\eta(b) = 0$, the controller is a minimum variance controller. At the other extreme, $\eta(b)$ approaches a value of 1. The closer that $\eta(b)$ gets to 1, the larger the variance of the process output, y, relative to its best possible performance, σ_{mv}^2. Thus, although theoretically the controller is capable of eliminating some of the effect of disturbances on the output, it fails to perform as might be expected.

This index was shown to be analogous to the multiple coefficient of determination, r^2, encountered in multiple regression analysis.[14] Whereas r^2 indicates the predictability of a times series given past information ($r^2 = 1$), $\eta(b)$ provides a measure of the predictable component in the process output, b steps into the future.

Calculation of $\eta(b)$ can be done using linear regression, an autocorrelation approach, or recursive least- squares.[24] For example,

$$r_k = \sum_{j=1}^{m} r_{b+j+k-1}\hat{a}_j \qquad k = 0, 1, \ldots \qquad 5.7(14)$$

These methods generate parameter estimates, $\{\hat{a}_j\}$, that are biased (order 1/n) and for systems that are or nearly unstable, the estimates themselves may be unstable. Huang et al.[11] have also introduced a stable filtering and correlation procedure (FCOR) to calculate σ_{mv}^2 and ultimately $\eta(b)$. These estimates are also biased. Biased estimates mean that if an experiment is repeated an infinite number of times and the results are averaged, the computed average will not be the correct answer.

Alternatively, $\eta(b)$ might be estimated from the data themselves by solving a least-squares problem.[4] First, a set of parameters, $\{\alpha_j\}$, are found using a least-squares method.

From these parameters, an estimate of $\eta(b)$ can be generated:

$$\tilde{y} = \tilde{X}\alpha$$

$$\tilde{y} = \begin{bmatrix} \tilde{y}_n \\ \tilde{y}_{n-1} \\ \vdots \\ \tilde{y}_{b+m} \end{bmatrix}; \quad \tilde{X} = \begin{bmatrix} \tilde{y}_{n-b} & \tilde{y}_{n-b-s} & \cdots & \tilde{y}_{n-b-m+1} \\ \tilde{y}_{n-b-1} & \tilde{y}_{n-b-2} & \cdots & \tilde{y}_{n-b-m} \\ \vdots & \vdots & \vdots & \vdots \\ \tilde{y}_m & \tilde{y}_{m-1} & \cdots & \tilde{y}_1 \end{bmatrix}$$

where $\tilde{y}$ is the mean-centered controlled output, b is the time delay, and m is an estimate of the autoregressive order of the process. The set $\{\alpha_j\}$ can be readily found. Having $\{\alpha_j\}$ allows $\eta(b)$ to be calculated using the following equation:

$$\eta(b) = 1 - \frac{n-b-m+1}{n-b-2m+1}\frac{(\tilde{y}-\tilde{X}\alpha)^T(\tilde{y}-\tilde{X}\alpha)}{\tilde{y}^T\tilde{y}+\bar{y}^2} \quad \textbf{5.7(15)}$$

where $\bar{y}$ is the mean value of the controlled variable.

The *extended horizon performance index*[3,10] is defined as

$$\eta(b+h) = 1 - \frac{\psi_0^2+\psi_1^2+\psi_2^2+\cdots+\psi_{b+h-1}^2}{\psi_0^2+\psi_1^2+\psi_2^2+\cdots+\psi_b^2+\cdots} \quad \textbf{5.7(16)}$$

which gives the proportion of the variance arising from nonzero impulse coefficients ψ_j, $j > (b + h)$. The extended horizon performance index is said to correspond to the square of the correlation between the current error and the least-squares estimate of the prediction made at (b + h) control periods in the past. Similar to $\eta(b)$, $\eta(b+h) \sim 0$ means that the controller is performing satisfactorily since, at current sample k, y(k) is not well predicted by the samples older than (b + h) in the past. Thus, knowledge about the output at times older than (b + h) gives no indication to discern what is occurring at the present time. A controller regulating a process should be able to eliminate disturbances satisfactorily so that at some point, not too far in the past, the process behavior is unrelated to the events at the present time.

The use of $\eta(b + h)$ also implies that minimum variance control may not be feasible. The minimum variance controller will often generate large control moves to eliminate the disturbance effect at a point in the future after the dead time. Such large moves are undesirable because of their detrimental effect on actuators. Furthermore, assumptions of approximate linearity that are made in the design of minimum variance controllers are more likely to be violated significantly by large control efforts.

Unlike $\eta(b)$, $\eta(b+h)$ does not require precise knowledge of the process delay. Ericksson and Isaksson[25] examined the use of $\eta(b)$ and $\eta(b+h)$ for PI and PID controllers. They concluded that $\eta(b)$ and $\eta(b+h)$ indicate performance deterioration but that $\eta(b + h)$ also permits diagnosis of long settling times, which is not always possible with $\eta(b)$.

For such systems, $\eta(b)$ may be large, and the system is well controlled. Because of constraints around safety and limitations on actuator moves, the closed loop might be far from minimum variance. By using $\eta(b + h)$, the controller may be monitored because h can be chosen according to the desired settling. When the controller performance deteriorates, a greater change will be observed in the value of $\eta(b + h)$ as compared to that of $\eta(b)$.

The *closed-loop potential factor* given by[3,15]

$$CLP = \frac{\sigma_{mv}^2}{\sigma_y^2} \quad \textbf{5.7(17)}$$

is another normalization of the closed-loop output error variance. It is not difficult to show that the CLP is related to $\eta(b)$ by

$$\eta(b) = 1 - CLP \quad \textbf{5.7(18)}$$

Thus, CLP $\in$ [0, 1]. When CLP is 0, the existing controller can be improved. A value near 1 means that improvements can only be obtained by process modifications such as the addition of feedforward control, if possible. See the earlier discussion.

Example: Two-State CSTR

A reduced-order, dimensionless model of a first-order, exothermic reaction

$$A \rightarrow B$$

occurring in a jacketed continuous-stirred tank reactor (CSTR) is given by[26]

$$\dot{x}_1 = \phi x_1 \kappa(x_2) + q(x_{1f} - x_1)$$
$$\dot{x}_2 = \beta\phi x_1 \kappa(x_2) - (q+\delta)x_2(x_{1f} - x_1) + \delta x_3 + q x_{2f}$$
$$\kappa = \exp\left(\frac{x_2}{1 + x_2/\delta}\right)$$

with x_1 the dimensionless conversion, x_2 the dimensionless reactor temperature, and x_3 the jacket temperature. The controlled variable is x_2, and x_3 is the indirect manipulated variable. The steady operating conditions are $x_1^* = 0.1039$, $x_2^* = 0.4875$, $x_3^* = -0.6567$. From a practical point of view, the jacket temperature is cascaded to the flow rate of the cooling fluid, and with the assumption of fast jacket dynamics this permits the two-state model representation. In the next subsection, a modified version of this system will be used to demonstrate model-based feedback control. Table 5.7c

TABLE 5.7c
Dimensionless Parameters in the CSTR Example

$\phi = 0.072$	Damköhler number
$\beta = 8$	Heat of reaction
$\delta = 0.3$	Heat transfer coefficient
$\gamma = 20$	Activation energy
q = 1	Flow rate
$x_{1f} = 1$	Feed A concentration
$x_{2f} = 0$	Feed temperature

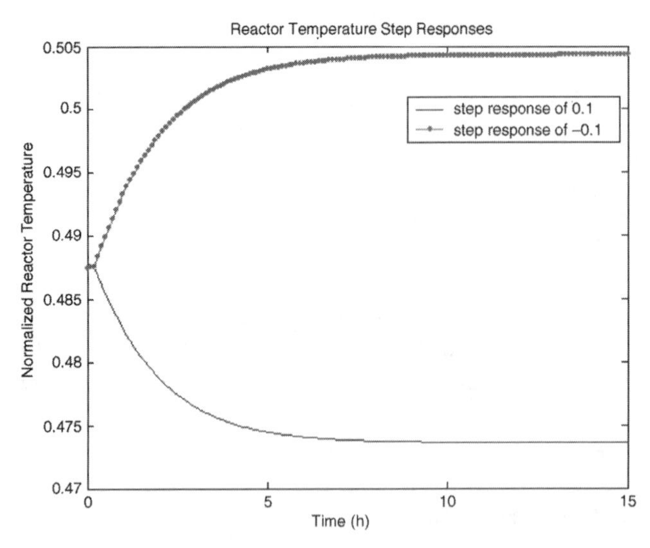

FIG. 5.7d
Performance of the two-state CSTR reactor temperature when subjected to step inputs of magnitude ±0.1.

provides the definition of the parameters and their nominal values.

A step test of size ±0.1 was imposed on the nonlinear system. The results are shown in Figure 5.7d. The positive and negative step responses were first normalized by the magnitude of the change. From the two normalized responses, an average response was computed as 0.5 * ([positive step response – original steady state] – [negative step response – original steady state]). This was then used to compute a model for the process. The identified process transfer function model is found to be

$$G_p(q) = \frac{-0.003739z - 0.003739}{z - 0.9512}$$

Figure 5.7e shows the fit of the response of this model (labeled by —•—) to an average of the step response data (solid line). From these data, a dead time of three samples was found. The response appears to be first order. A PI controller is designed to provide closed-loop poles at –0.5, –0.5. This

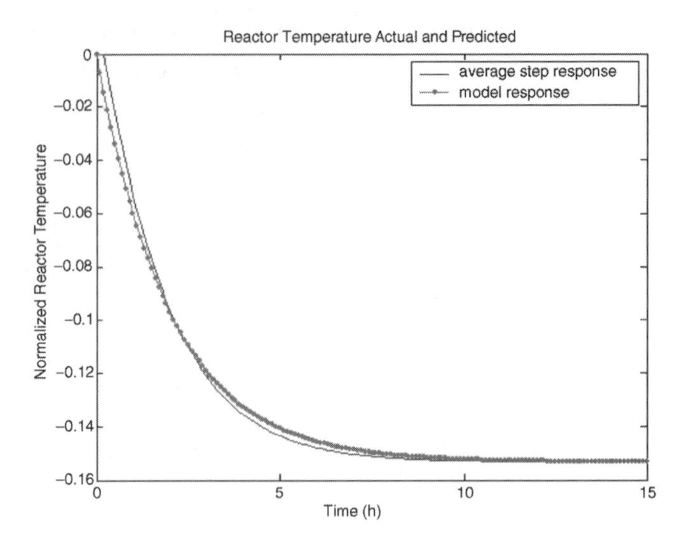

FIG. 5.7e
Fit of the temperature model to the step response data of the two-state CSTR.

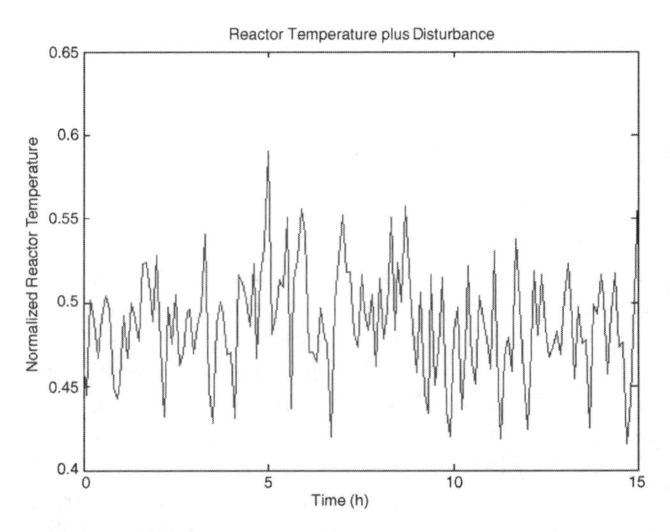

FIG. 5.7f
Reactor temperature of the two-state CSTR with added white noise.

should yield a closed-loop performance with a settling time of 90 min, a damping coefficient of 1 (critically damped), and a closed-loop time constant of 1/2 h. The controller has a gain of 6.5232 and integral time constant of 2 h. Note the PI controller was not optimally tuned. Figure 5.7f shows the reactor temperature with a filtered white noise disturbance. The disturbance is described by

$$d(k) = \frac{1 - 0.9q^{-1}}{1 - 0.99q^{-1}} a(k)$$

where a(k) is normally distributed white noise with variance of 0.001. The estimated value of $\eta(b = 3)$ is calculated using

TABLE 5.7g
Model Orders vs. Estimate of the Normalized Performance Index

m	*1*	*2*	*3*	*4*
$\hat{\eta}(b)$	0.5894	0.5930	0.5937	0.5962

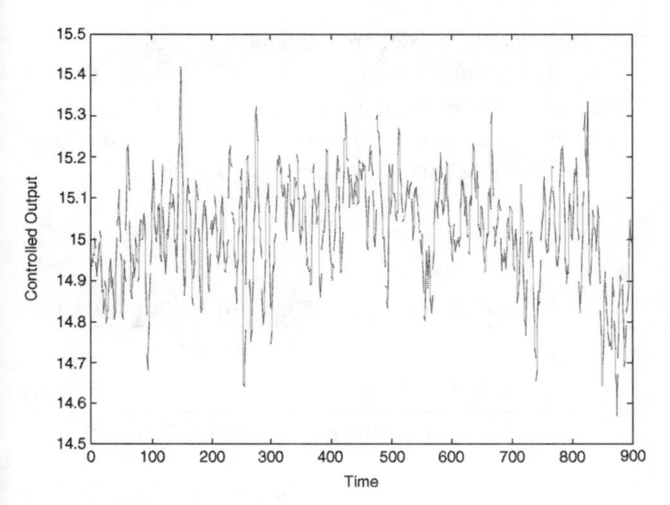

FIG. 5.7h
Closed-loop composition data of an industrial polymer reactor.

a linear regression approach. The estimated value essentially stops declining with a model order of m = 1. This concurs with the model assumption of first order. The estimate of σ_y^2 is given as $(\tilde{y} - \tilde{x}\hat{\alpha})^T(\tilde{y} - \tilde{x}\hat{\alpha})/(n - b - 2m + 1)$ in Desborough and Harris.[14] The value estimated from the data is 0.0011, which is consistent with the variance of the white noise of 0.001. The autocorrelations are shown in Figure 5.7b.

The values of $\hat{\eta}(b)$ for model orders that varied over the range [1, 4] are shown in Table 5.7g.

Industrial Example

Figure 5.7h illustrates data taken from a polymer reactor. It constitutes 15 h of composition data sampled once per minute. This period constitutes normal operation without any major process upset. The process is controlled by a set of PID controllers. For this process, b = 5 and m = 2 provides the minimum $\eta(b)$. For this situation, $\eta(b) = 0.1007$. Figure 5.7i is the autocorrelation of the output.

MODEL-BASED FEEDBACK CONTROL

In the preceding discussion, Equation 5.7(5) was obtained assuming a constant set point, uncorrelated disturbances, and a feedback control structure such as a PID (see Figure 5.7a). The reality is that the disturbances may have significant autocorrelation and the set point may be varying. Stanfelj et al.[5] considered this problem using the model-based feedback concept, shown in Figure 5.7j, to design a feedback controller between the controller output and the model prediction error. They showed that in this situation, it is not possible to attribute poor control performance *definitively* to modeling errors or to poor tuning using *only* routine operating data. As such, they suggest the generation of additional data whereby perturbations in the set point are made to facilitate model development. These perturbations must be *rich* enough to provide sufficient excitation.[24]

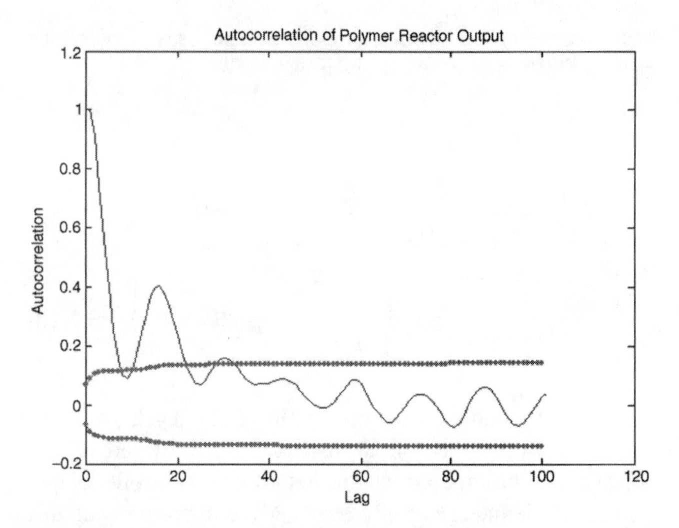

FIG. 5.7i
Autocorrelation plot of the industrial composition data. The data are within the 95% confidence limit after 33 lags.

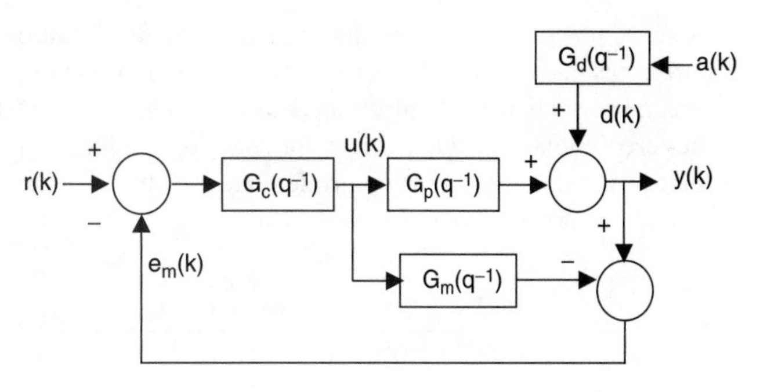

FIG. 5.7j
Block diagram of a model-based feedback controller. e(k): feedback model prediction error.

To observe this, consider the block diagram in Figure 5.7j, where the model prediction error, $e_m(k)$, between the process, $G_p(q^{-1})$, and an approximate model of the process, $G_m(q^{-1})$, is used to design the feedback controller, $G_c(q^{-1})$,

$$e_m(k) = \frac{G_d(q^{-1})}{1 + G_c(q^{-1})[G_p(q^{-1}) - G_m(q^{-1})]} a(k) \qquad 5.7(19)$$

with the other definitions as before. The cross-covariance between u(k) and $e_m(k)$ is given as

$$E\{u(k)e_m(k+h)\}$$

$$= E\left\{\frac{G_c(q^{-1})G_d(q^{-1})}{1+G_c(q^{-1})[G_p(q^{-1})-G_m(q^{-1})]}a(k)\right.$$

$$\left.\times\frac{G_d(q^{-1})}{1+G_c(q^{-1})[G_p(q^{-1})-G_m(q^{-1})]}a(k+h)\right\} \quad 5.7(20)$$

When the model is perfect, that is, $G_m(q^{-1}) = G_p(q^{-1})$, the modeling error is just the disturbance signal d(k). If d(k) is white noise, the cross-correlation between the controller output and the modeling error will be zero. More importantly, in this case, any nonzero cross-correlation values can be attributed to model error. In a more realistic scenario, d(k) is not white noise. In this case, it is impossible to determine whether unacceptable controller performance is due to modeling errors or to poor controller settings.

Consider set point changes that are independent of future disturbances as a way of deciphering between modeling errors and poor choice of controller parameters. The feedback model prediction error due to set point changes is a function of the disturbance model, the controller, and the approximate model of the process, $G_m(q^{-1})$,

$$e_m(k) = \frac{G_d(q^{-1})a(k)+G_c(q^{-1})[G_p(q^{-1})-G_m(q^{-1})]r(k)}{1+G_c(q^{-1})[G_p(q^{-1})-G_m(q^{-1})]} \quad 5.7(21)$$

When there is plant/model mismatch, there will be nonzero correlation values between the prediction error and the set point. When there is no significant correlation present, then other reasons for unsatisfactory performance may be poor controller tuning and even poor pairing between the controlled and manipulated variables.

Example: Three-State CSTR

The example CSTR, with parameters defined in Table 5.7c, was modified to include a third state, the jacket temperature, which can be described by

$$\dot{x}_3 = \delta_1 U(x_{3f}-x_3)+\delta\delta_1\delta_2(x_2-x_3)$$

where U is the flow rate ratio, $\delta_1 = 1$ is the volume ratio of the reactor to the jacket, $\delta_2 = 1$ is the density-heat capacity ratio, and q = 1. A model was identified in a similar fashion as in the two-state CSTR case. Here, the model identified is a third-order transfer function:

$$G_m(s) = \frac{-(0.1030s+0.1149)}{s^3+2.9245s^2+2.6810s+0.7566}$$

A controller was developed from this model to have all the closed-loop poles at −1. The transfer function of this controller is given by

$$G_c(s) = \frac{-s^3-2.295s^2-2.681s-0.7566}{0.103s^3+0.3209s^2+0.3328s+0.1149}$$

The controller and model are implemented digitally. This is done by simulation of the analog systems. There are a number of techniques for doing this, and in this case it was implemented using trapezoidal integration. Thus, the transfer functions for the model and controller are given by

$$G_p(q^{-1}) = \frac{10^{-3} * (-0.2424q^{-3}-0.2680q^{-2}+0.1912q^{-1}+0.2168)}{q^{-3}-2.7710q^{-2}+2.5669q^{-1}-0.7952}$$

$$G_c(q^{-1}) = \frac{-9.3545q^{-3}+25.9213q^{-2}-24.0122q^{-1}+7.4391}{q^{-3}-2.7039q^{-2}+2.4369q^{-1}-0.7321}$$

The same disturbance used in the two-state CSTR case is also applied here. Using the structure in Figure 5.7j, the closed-loop response for this three-state CSTR, with a model given by $G_p(q^{-1})$ and controller by $G_c(q^{-1})$, is shown in Figure 5.7k. Observe that the response reflects the disturbance but appears to remain about the nominal value (0.4875).

The auto- and cross-correlations for this example are shown in Figures 5.7l and 5.7m. Note that the autocorrelations for the reactor temperature after lag 14 are within the 95%

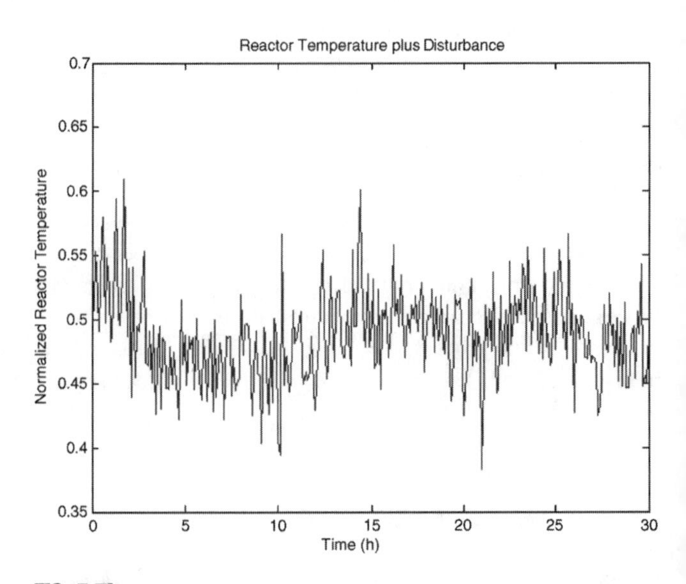

FIG. 5.7k
Closed-loop performance of the three-state CSTR reactor temperature when model-based feedback control is applied in the presence of an unmeasured disturbance.

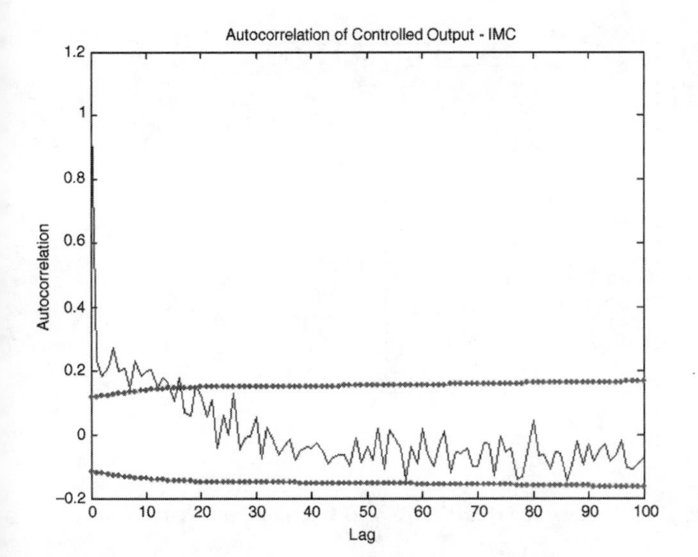

FIG. 5.7l
Autocorrelation of the three-state CSTR reactor temperature with the 95% confidence bands.

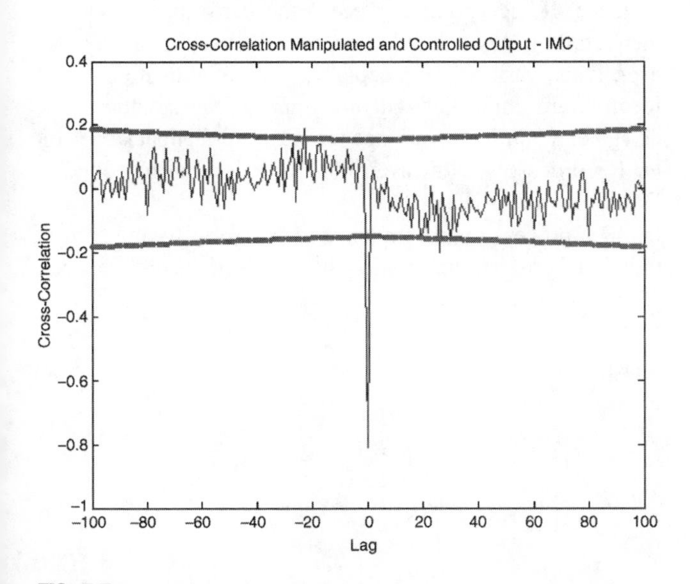

FIG. 5.7m
The cross-correlations of the three-state CSTR between the reactor temperature and the cooling fluid flow rate.

TABLE 5.7n
Model Orders vs. Estimate of the Normalized Performance Index

m	1	2	3	4
$\hat{\eta}(b)$	0.0362	0.0928	0.1052	0.0868

confidence bands. The estimated values of $\eta(b)$ using the same procedure and for the same range of the model order yield values shown in the Table 5.7n. Note the improvement in the values of $\eta(b)$. The estimate of the minimum variance is essentially unchanged from before. It is 0.0012. The spike at 0 lag is due to the fact that the controller has direct feed through.

UNIVARIATE SYSTEMS: FEEDFORWARD/FEEDBACK

Harris et al.[10] and Eriksson and Isaksson[25] partition the disturbance signals into measured and unmeasured ones. The former are said to be *deterministic* and can be represented by step or pulse functions, whereas the latter are *stochastic*. This separation allows a comparison of the deterministic component to a desired reference trajectory. In fact, Kozub[3] provides an example in which the desired reference trajectory is a random walk.

The closed-loop error can be represented by

$$e_m(k) = \frac{\alpha(q^{-1})}{\beta(q^{-1})}a(k) + \sum_{j=1}^{J}\frac{\alpha_j(q^{-1})}{\beta_j(q^{-1})}f_j(k) \qquad 5.7(22)$$

where $f_j(k)$ is the deterministic portion of the disturbance. It is not difficult to compute the performance bounds and the autocorrelation function. Furthermore, the desired trajectory can be used to fit the parameters of the model.

When the feedback controller is operating within its performance bound, further reductions in process variability may be achieved by implementing feedforward control.[5,14] Consider Figure 5.7o where the controller output consists of two components:

$$\begin{aligned} u(k) &= u_{fb}(k) + \sum_{j=1}^{J} u_{ff,j}(k) \\ u_{ff,j}(k) &= G_{ff,j}(q^{-1})D_j(k) \end{aligned} \qquad 5.7(23)$$

a feedback and a feedforward part. $D_j(k)$ represents the jth deviation of the measured feedforward disturbance from some average value, and $G_{j,ff}(q^{-1})$ is the model for the jth feedforward controller. The output is given by

$$\begin{aligned} y(k) = {} & [G_d(q^{-1}) + G_{ff}(q^{-1})G_p(q^{-1})]D(k) \\ & + d(k) + G_p(q^{-1})u_{fb}(k) \end{aligned} \qquad 5.7(24)$$

which is driven by the unmeasured and measured disturbances. When the feedforward controller compensates for the measured disturbance, then Equation 5.7(23) becomes

$$y(k) = d(k) + G_p(q^{-1})u_{fb}(k) \qquad 5.7(25)$$

By using the cross-correlation test of Stanfelj et al.,[5] the expectation is that there is no significant correlation between the future values of the output and the measured disturbances,

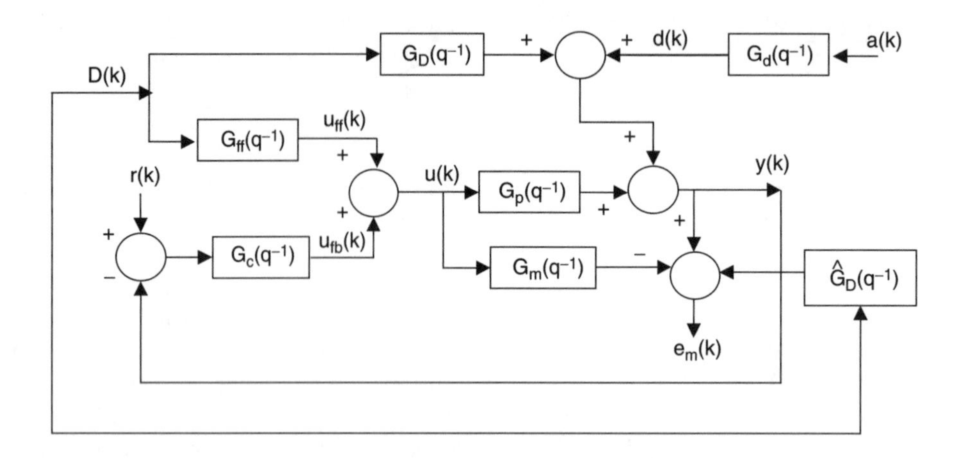

FIG. 5.7o
Block diagram of a model-based feedforward/feedback control strategy. D(k): measured disturbance, $u_{ff}(k)$: feedforward controller output, $u_{fb}(k)$: feedback controller output.

$D_j(k)$, whenever the measured and unmeasured disturbances themselves are independent of each other. Thus, the feedforward controller is well within acceptable performance bounds.

Similar analysis (use of the cross-correlation test) of the prediction error, $e_m(k)$ indicates that it is a function of D(k), d(k), the controllers, $G_c(q^{-1})$ and $G_{ff}(q^{-1})$, and the models $G_m(q^{-1})$ and $\hat{G}_D(q^{-1})$. If the feedforward controller is performing well, it can be determined that deviation from minimum variance feedback performance is due to controller tuning and modeling errors. On the other hand, when the feedforward controller is not performing well but the feedback controller is adequate, the deviation from minimum variance is due to poor tuning of the feedforward controller. When both are operational and the performance exceeds product or process specifications the cross-correlation test must be supplanted by other methods to arrive at an unequivocal diagnostics.

To address this, Desborough and Harris[14] provide a means of estimating the minimum variance performance when both feedforward and feedback control loops are present. Analogous to the earlier discussion, the closed-loop output can be represented by a superposition of ARIMA time series models, that is,

$$y(k) = \Psi_0(q^{-1})a(k) + \sum_{j=1}^{J} \Psi_{fj}(q^{-1})a_{fj}(k) \qquad 5.7(26)$$

where $\Psi_0(q^{-1})$ and $\Psi_{fj}(q^{-1})$ are the closed-loop transfer functions between the unmeasured and the jth measured disturbances, respectively, and $\{a(k)\}$ and $\{a_{fj}(k)\}$ are statistically independent random variates with variance σ_a^2 and σ_{fj}^2. Consider only the feedforward forcing function in Equation 5.7(26). The variance of this is found by taking the expectation, $E(\cdot)$, of the square of Equation 5.7(26),

$$\sigma_y^2 = E\left(\sum_{j=0}^{J} \Psi_{fj} a_{fj}\right)^2 = \sum_{j=0}^{J}\sum_{i=o}^{\infty} \Psi_{fj,i}^2 \sigma_{fj}^2 \qquad 5.7(27)$$

In practice only a finite sum is used. This equation provides an analysis of the process variance, which in turn permits an identification of the significant components of the variance. If the variance exceeds the performance bounds, then retuning may be used to improve the performance. However, if the variance is acceptable, then modification of the feedforward controller will not improve the situation. The design of a minimum variance feedforward/feedback controller for processes with invertible dynamics can be found in Desborough and Harris.[14]

The process output under MVC is given by the sum of individual errors in forecasting the effects of the disturbances,

$$y(k) = \sum_{j}^{J} e_j(k)$$

$$e_j(k) = \begin{cases} 0 & \tau_j > b \\ \left(\psi_{j,k_j} q^{k_j} + \psi_{j,k_j} q^{k_j} + \cdots + \psi_{j,b-1} q^{b-1}\right) a_j(k) & \tau_j \le b \end{cases} \qquad 5.7(28)$$

where τ_j is the feedforward delay, $e_j(k), j > 0$ is the contribution of the jth measured disturbance to the total forecast error, and $e_0(k)$ is the effect of the forecast error for the unmeasured disturbance on the closed-loop process output, which is assumed independent of $e_j(k)$, $j > 0$. The first b terms are identical in terms of the expansion given in Equation 5.7(2) since $k_0 = 0$.

The process variance under MVC is given by

$$\sigma_{mv}^2 = \sum_{j=0}^{J} \mathrm{var}(e_j(k)) \qquad 5.7(29)$$

which is similar to Equation 5.7(27) but truncated to $h = b - 1$. The term, σ_{mv}^2, is a lower bound on the variance of the process output. If σ_{mv}^2 is greater than the maximum allowed process variability, then retuning or redesigning the controllers will not decrease the process variance to an acceptable level. Rather, process modifications must be considered. When σ_y^2 exceeds the lower bound, improvements are possible by retuning or implementing alternative control strategies. There are no guarantees and, especially in the case of non-invertible unstable dynamics, alternative methods such as those recommended by Tyler and Morari[19] and Sternad and Soderström[20] should be investigated.

Kozub and Garcia[12] and Thornhill et al.[22] provide information on practical experiences in implementing large-scale performance assessment of feedforward/feedback controllers. Their main message is that both formal mathematical analysis and engineering insights are necessary for diagnosis leading to root cause identification. Other examples include Lynch and Dumont,[27] Huang et al.,[11] Miller and Huang,[28] and Vishnubhotla et al.[29]

As an example of the application of feedforward control, consider the three-state CSTR discussed above. In this case, the same PI controller as in the earlier example is applied. The measured disturbance is the cooling fluid flow rate. The response between a step in the disturbance and the normalized controlled vaiable is shown in Figure 5.7p. The transfer function approximation, $G_d(s)$, to the above system is given by

$$G_d(s) = \frac{0.452}{2.8s + 1}$$

This, together with the earlier-defined transfer function between the manipulated variable and the controlled variable, $G_p(s)$, defines the feedforward compensator as

$$G_{comp}(s) = \frac{5.901s + 2.95}{2.8s + 1}$$

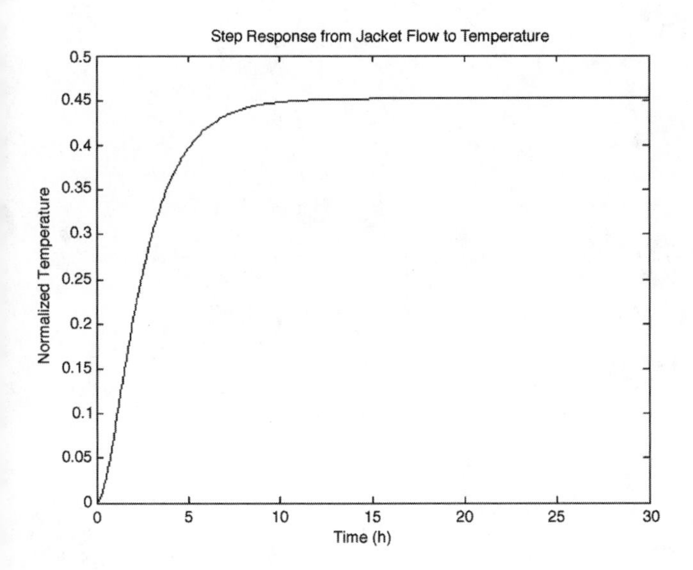

FIG. 5.7p
Step response from cooling water flow rate to reactor temperature for the three-state CSTR.

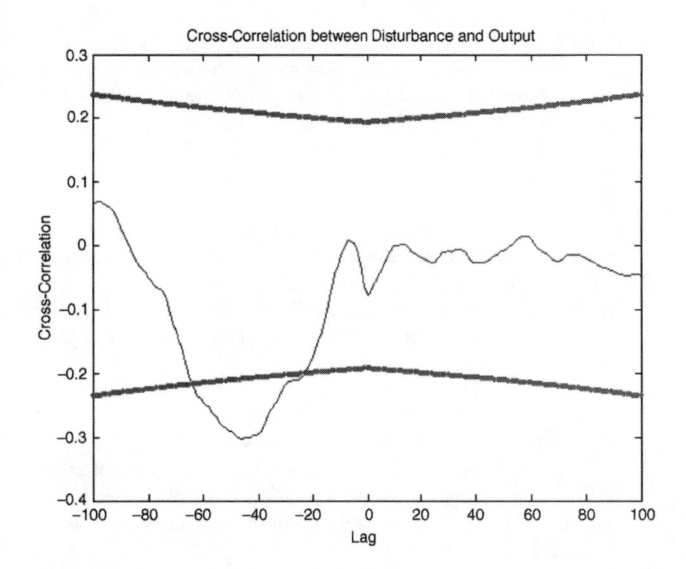

FIG. 5.7q
Cross-correlation between the measured disturbance and the reactor temperature under feedforward control.

Using trapezoidal integration for simulation, the discretized compensator is given by

$$G_{comp}(q^{-1}) = \frac{2.122 - 2.019q^{-1}}{1 - 0.9649q^{-1}}$$

Figure 5.7q shows the cross-correlation between the temperature output and the measured disturbance together with the 95% confidence limits. According to Stanfelj et al.,[5] cross-correlation should have no statistically significant values under perfect feedforward. In this case, the control is almost entirely within the confidence limits.

References

1. Ender, D., "Process Control Performance: Not as Good as You Think," *Control Engineering*, p. 180, September 1993.
2. Bialkowski, W. L., "Dreams versus Reality: A View from Both Sides of the Gap," *Pulp & Paper Canada*, 94, p. 19, 1993.
3. Kozub, D. J., "Controller Performance Monitoring and Diagnosis: Experiences and Challenges," *AIChE Symposium Series*, Vol. 93, pp. 83–96, 1997.
4. Desborough, L. D., "Performance Assessment Measures for Univariate Control," thesis, Queen's University, Kingston, Ontario, Canada, 1992.
5. Stanfelj, N., Marlin, T., and MacGregor, J. F., "Monitoring and Diagnosing Process Control Performance: The Single-loop Case," *Industrial Engineering and Chemical Research*, Vol. 32, pp. 301–314, 1993.
6. Aström, K.-J., "Assessment of Achievable Performance of Simple Feedback Loops," *International Journal of Adapative Control & Signal Processing*, Vol. 3, pp. 3–19, 1991.
7. Shinskey, F. G., "Putting Controllers to the Test," *Chemical Engineering*, pp. 96–106, December 1990.
8. Harris, T., "Assessment of Control Loop Performance," *Canadian Journal of Chemical Engineering*, Vol. 67, pp. 856–861, 1989.
9. Huang, N. and Shah, S. L., *Performance Assessment of Control Loops*, London: Springer-Verlag, 1999.

10. Harris, T., Seppala, C. T., and Desborough, L. D., "A Review of Performance Monitoring and Assessment Techniques for Univariate and Multivariate Control Systems," *Journal of Process Control,* Vol. 9, pp. 1–18, 1999.
11. Huang, B., Shah, S. L., and Kwok, E. K.,"Good, Bad or Optimal? Performance Assessment of Multivariable Processes," *Automatica,* Vol. 33, pp. 1175–1183, 1997.
12. Kozub, D. J. and Garcia, C. E., "Monitoring and Diagnosis of Automated Controllers in the Chemical Process Industries," presented at AIChE Annual Meeting, St. Louis, MO, 1993.
13. Desborough, L. D. and Harris, T. J., "Performance Assessment Measure for Univariate Feedback Control," *Canadian Journal of Chemical Engineering,* Vol. 70, pp. 1186–1197, 1992.
14. Desborough, L. D. and Harris, T. J., "Performance Assessment Measure for Univariate Feedforward/Feedback Control," *Canadian Journal of Chemical Engineering,* Vol. 71, pp. 605–616, 1993.
15. DeVries, W. R. and Wu, S. M., "Evaluation of Process Control Effectiveness and Diagnosis of Variation in Paper Basis Weight via Multivariate Time Series Analysis," *IEEE Transactions on Automatic Control,* Vol. AC-23, pp. 702–708, 1978.
16. Harris, T., Boudreau, F., and MacGregor, J., "Performance Assessment of Multivariable Feedback Controllers," *Automatica,* Vol. 32, pp. 1505–1518, 1996.
17. Peng, Y. and Kinnaert, M., "Explicit Solution to the Singular LQ Regulation Problem," *IEEE Transactions on Automatic Control,* Vol. AC-37, pp. 633–636, 1992.
18. Shah, S., Patwardhan, R., and Huang, B., "Multivariate Controller Performance Analysis: Methods, Applications and Challenges," in Preprints of Chemical Process Control-VI, *Assessments and New Directors for Research,* Rawlings, J. B. and Ogunnaike, B. A., Eds., Tucson, AZ, pp. 187–219, January 7–12, 2001.
19. Tyler, M. and Morari, M., "Performance Monitoring of Control Systems Using Likelihood Methods," *Automatica,* Vol. 12, pp. 1145–1162, 1996.
20. Sternad, M. and Söderström, T., "LQG-optimal Feedforward Regulators," *Automatica,* Vol. 24, pp. 557–561, 1988.
21. Harris, T., "One-Step Optimal Controllers for Nonsymmetric and Non-quadratic Loss Functions," *Technometrics,* Vol. 34, pp. 298–306, 1992.
22. Thornhill, N. F., Oettinger, M., and Fedenczuk, P., "Refinery-wide Control Loop Performance Assessment," *Journal of Process Control,* Vol. 9, pp. 109–124, 1999.
23. Box, G. E. P. and Jenkins, G. M., *Time Series Analysis: Forecasting and Control,* Englewood Cliffs, NJ: Holden-Day, 1976.
24. Aström, K.-J. and Wittenmark, B., *Adaptive Control,* Reading, MA: Addison-Wesley, 1989, chapter 3.3.
25. Eriksson, P.-G. and Isaksson, A. J., "Some Aspects of Control Loop Performance Monitoring," 3rd *IEEE Conference on Control Applications,* Glasgow, Scotland, 1994, pp. 1029–1034.
26. Kosanovich, K. A. (Hoo), Charboneau, J. G., and Piovoso, M. J., "Operating Regime-based Controller Strategy for Multi-product Processes," *Journal of Process Control,* Vol. 7, pp. 43–56, 1997.
27. Lynch, C. B. and Dumont, G. A., "Control Loop Performance Monitoring," *IEEE Transactions on Control System Technology,* Vol. 4, p. 185, 1996.
28. Miller, R. M. and Huang, B., "Perspectives on Multivariate Feedforward/Feedback Controller Performance Measures for Process Diagnosis," *Proceedings IFAC ADCHEM 97,* p. 435, Branff, Alberta, 1997.
29. Vishnubhotla, A., Shah, S., and Huang, B., "Feedback and Feedforward Performance Analysis of the Shell Industrial Closed-Loop Data Set," *Proceedings IFAC ADCHEM 97,* p. 295, Branff, Alberta, 1997.

5.8 The "Virtual Plant," A Tool for Better Understanding

G. K. MCMILLAN

There is a long history in instrumentation and process control of learning the hard way by being thrown into an application. In general, the more mistakes that a person made, the more that was learned. This "trial by fire" is exciting but is hard on the people and the processes. This section proposes an alternative to the school of "hard knocks." A virtual plant is described that allows the user to work on processes that are as difficult as the real ones. An opportunity is depicted to demonstrate key concepts and best practices with a "hands-on" approach using new industrial advanced control tools, all residing on a standard desktop or laptop computer.

The control engineer can use the virtual plant to develop an understanding and confidence in the application of advanced process control. The technologies include advanced regulatory control techniques, online performance indicators, auto tuners, dynamic online property estimators, power spectrum analyzers, constrained multivariable predictive controllers, and real-time optimization. The process industry will reach and probably exceed at a small fraction of the cost what has been accomplished by the aerospace and nuclear power industry in the use of high-fidelity dynamic simulators for failure analysis, studies, prototyping, testing, and training.

The heart of the virtual plant is a high-fidelity dynamic model of the plant that includes everything that is in the field. The foremost requirement is a dynamic model of the process that easily evolves from the steady-state model used for process design and process flow diagram development by the addition of volumes and characteristic curves. Equipment and piping flow coefficients are automatically sized to provide the pressure drops in the system. The dynamics of field instrumentation and control valves, often overlooked, but important for control studies, testing, and training, is included. The location of the sensor and control valve and the associated process transport and mixing delays, the resolution, repeatability, and span of the measurement, the resolution (stick–slip), trim characteristic, size of the control valve, and noise are added where important. For compositions, the sample transport delay and cycle time of the analyzer are critical. For temperatures, the thermal lag associated with the air gap and the design of the thermowell assembly are significant. Transfer functions are used to provide these instrument dynamics and noise. Once constructed, the dynamic model becomes the warehouse of the plant knowledge. The configuration of the basic and advanced process control and batch systems, safety interlock system, data historian, and tools for time series and neural network modeling and advanced data analysis is loaded to complete the virtual plant and enable it to become the warehouse of control and interlock system knowledge. Figure 5.8a summarizes the detail merged with a steady-state model to make it a dynamic plant model.

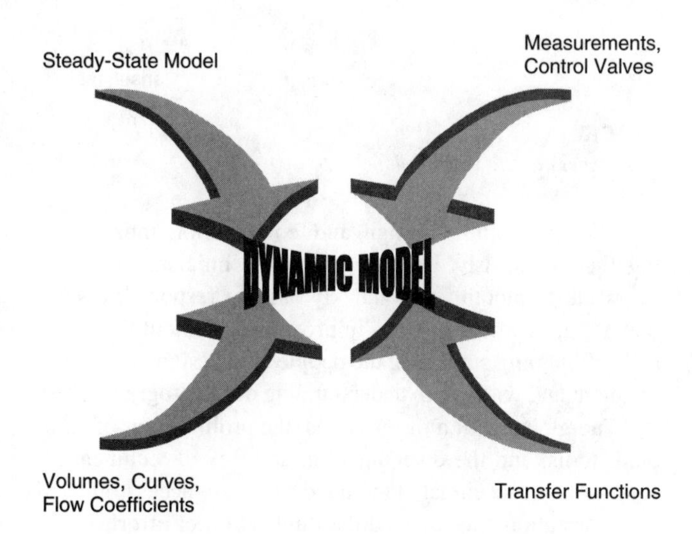

FIG. 5.8a
The evolution of the dynamic plant model.

KNOWLEDGE BUILDING AND STORAGE

The decentralization and more fluid nature of technical resources and the proliferation of new computer-aided design tools have created the need for a more extensive and efficient sharing of plant knowledge. The scope, front-end design, specifications, drawings, configuration, and construction may each be done in different parts of the world with different cultures, languages, and software packages. Furthermore, companies and engineers have their own individual favorite computer programs for sizing and specifying equipment, piping, instrumentation, and valves, and for generating the process flow and piping and instrument diagrams and control system functional diagrams. Often these programs come and go with the individual or the firm. This, coupled with a higher turnover rate of personnel, poses a challenge to the efficient execution of a project.

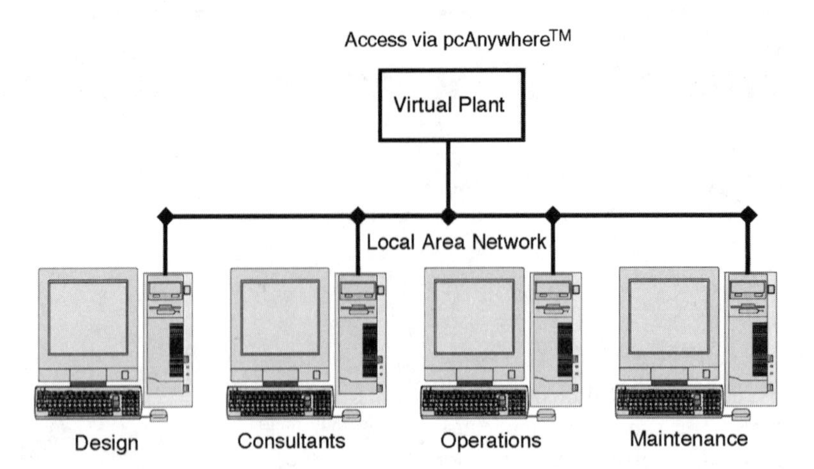

FIG. 5.8b
LAN access to the virtual plant.

Yet, the whole design and construction must come together seamlessly. Ultimately, local technicians and engineers must smoothly and quickly assume responsibility for operation, maintenance, and improvement without the assurance of ongoing support of the people who designed and built the plant and without an understanding of the programs used.

The globalization of resources, the proliferation of computer tools, and the diversification and flux of technical resources have accentuated the need to ensure accessibility of all information and to avoid the duplication of effort.

The virtual plant naturally evolves from the design and becomes the common source of information for operating, maintaining, and improving the plant. The new work needed to create this capability is less than 1% of a medium-size project. The improved efficiency from sharing information and avoiding duplication and inconsistency is greater than 1%. To summarize, the virtual plant is a real-time representation of the actual plant and the control, interlock, and historian systems running on a standard desktop or laptop computer. No special hardware is required. Design, consulting, operation, and maintenance engineers have access to the virtual plant as depicted in Figure 5.8b from desktop and laptop computers wherever there is a connection provided to the local-area network (LAN) through inexpensive software packages, such as pcAnywhere™, that allow a user to take over control of a remote computer.

The virtual plant starts out as a process flow diagram (PFD) created from a graphical high-fidelity steady-state model. However, additional detail is included that is usually associated more with a piping and instrument diagram (P&ID) than a PFD. Finally, an interface table to a programmable electronic system (PES) is completed. Measurements are exported to the PES and final control elements (valve positions and pump or compressor speeds) are initialized by the model and imported from the PES. A link enable button is used to set up and check the communication between the dynamic model and the PES via object link embedding for process control (OPC). The integrator is turned on to start the model and initialize the final control elements. The dynamic model provides all of the signals to the basic process control system, advanced control tools (online performance indices, auto tuners, model predictive controllers, neural networks), sequential functions, batch operations, recipe management, and safety interlock systems.

There may be some reluctance to invest in a dynamic simulation mostly due to the high initial cost and maintenance of previous implementations where the dynamic model had to be written from scratch and the control system and displays had to be emulated by programmers. State-of-the-art modeling and PES software eliminate these efforts. A properly set up steady-state model can be readily made dynamic and interfaced to the actual configuration and displays of the PES. Also, the graphical Windows environment of the modeling program and PES means that the average engineer can run or modify the virtual plant. The ease of the new approach means that the virtual plant is not restricted to large or complex processes but can be employed for much smaller applications.

Models are sometimes thought to be artificial or impractical until one conceptually understands the fundamental role that properly constructed models play in building knowledge. We sometimes forget that all advances in engineering and physics use mathematical models to represent reality. In the virtual plant, high-fidelity process models use drag and drop blocks of the mathematical relationships for common unit operations in plant processes. While all technical knowledge is based on a model, not all models are based on technical knowledge. In particular, tieback models that connect the PES inputs to the PES outputs with a simple delay, gain, and/or lag make the displays functional for basic checkout and operator training, but do not reflect or convey process knowledge.

Some key points for the efficient building and effective use of models for the storage of plant knowledge and the application of advance control are listed below.

1. Steady-state models are used for process design and the real-time optimization (RTO) of continuous processes.

2. Dynamic models are at present used for control system design, testing, and training. They will be increasingly used to prototype and tune advanced control systems and to provide dynamic online property estimators. Batch process RTO must be done with a dynamic model because there is no steady state. There is also a possibility of adapting a dynamic model faster than converging a steady-state model for continuous processes.
3. Models must be reconciled online to the actual process to be used for real-time optimization or to supplement or replace the plant testing needed to develop models for dynamic online property estimators or model predictive controllers.
4. The detail required for steady-state RTO is also needed for dynamic models.
5. Only knowledge that can be derived from actual online and lab measurements is worth putting into the model for RTO and advanced process control.
6. The art of simulation involves the knowledge of how to add and match empirical relationships and what level of detail is actually needed to meet the objectives.

VIRTUAL PLANT SETUP

The control and analysis software must also be included without any modification or translation or hardware that would require extra time and expense or compromise the transportability of the results. Fortunately, Windows-based tools and PES all can be run on a standard PC and communicate via the OPC or the open database communication (ODBC) interface.

Virtual plants consist of real-time dynamic plant models, multivariate principal component analysis tools, real-time data historians, multivariable advanced control systems, Visual Basic displays and programs, and a suite of advanced control tools that are integrated into a field-based PES. All of the software runs on standard desktop or laptop computers with dual Ethernet cards so that the computers can communicate over both the LAN and the PES control network. The speed of data access over the local hub of the LAN between the local computers is high and faster than the control network because communication occurs preferentially between the local computers. The software and configuration used in the virtual plant is exactly the same as that used in the actual plant installations. No special configurations or PES hardware is required. Figure 5.8c shows a virtual plant setup that consists of three desktop computers interconnected by local area and control networks.

PRESENT PRACTICES

A traditional steady-state modeling practice is to omit utility streams, flow resistances, elevations, and prime movers. Because pressure drops do not determine flow in these steady-state models, very little attention is paid to specifying the proper pressure profile. Often, there are negative pressure gradients where flow is shown coming from a source at a lower pressure than the destination. Heaters and coolers with heat duties computed from a temperature specification are used rather than heat exchangers with heat transfer coefficients and utility streams. The steady-state model starts and ends in the office of the simulation specialist.

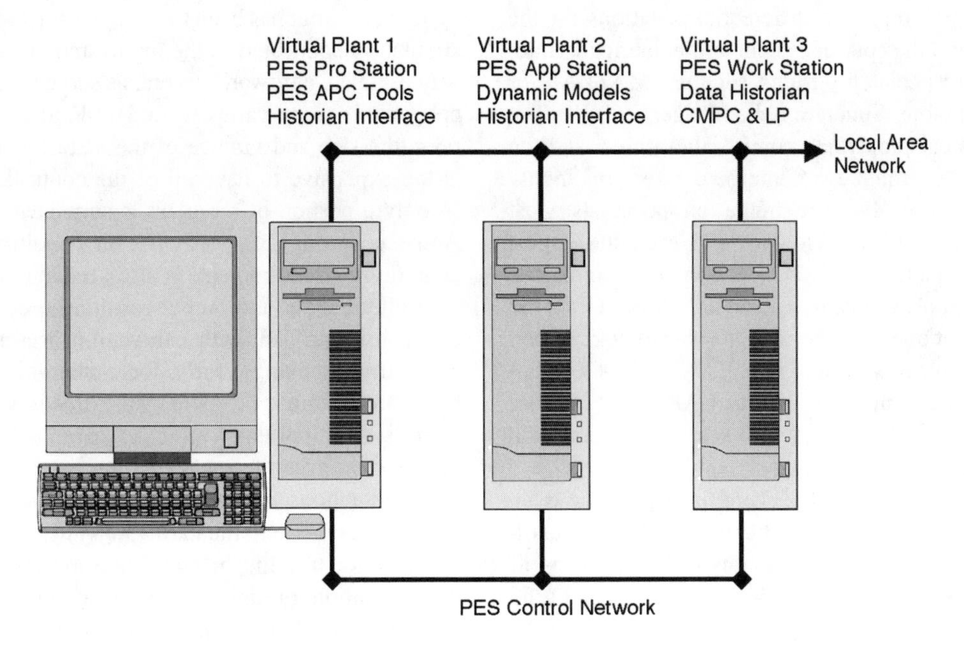

FIG. 5.8c
Virtual plant setup.

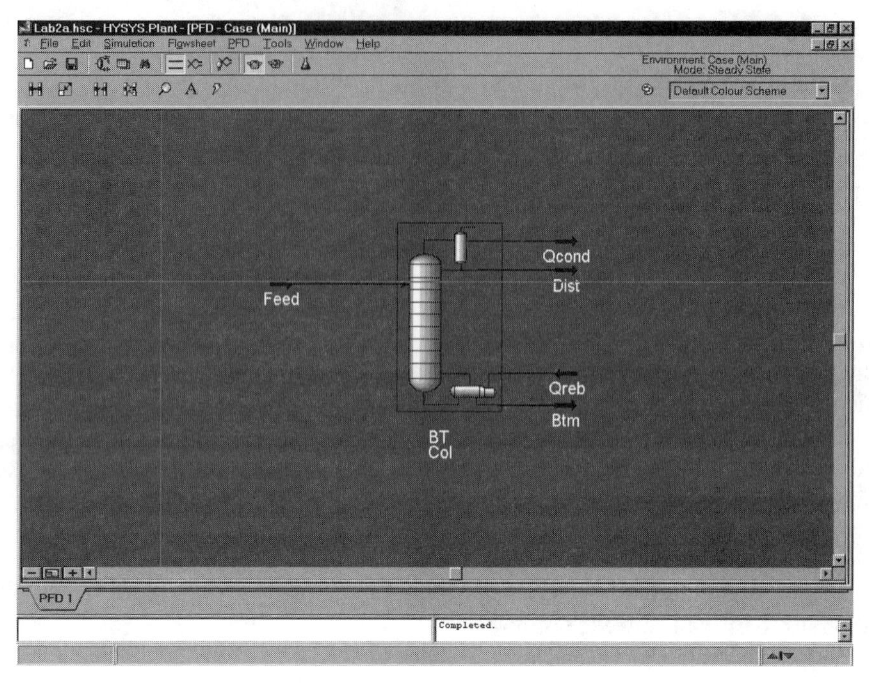

FIG. 5.8d
Typical steady-state PFD type model of a column.

Figure 5.8d shows a typical steady-state model for a distillation column. Note that the sump, reboiler, condenser, and overhead receiver are built into the column sub flow sheet. It is not possible to insert valves, nozzle locations, pumps, and pressure drops for the reflux or reboiler circulation flow or to simulate the utility streams.

Custom dynamic models for control system studies have not been able to build upon the investment in the steady-state model or its physical property package. These dynamic models required programming the differential equations for the unit operations and the control system and numerically integrating them in special programs such as MATLAB or Advanced Continuous Simulation Language (ACSL). The addition of physical properties, control algorithms, and the initial steady state required an intensive effort by highly trained and experienced dynamic simulation specialists. Most companies had no internal resources and even the largest companies had just a few. These custom dynamic simulations are a luxury that most companies could not afford. Often the simulation used for control system studies had to be rewritten in a different language such as FORTRAN or C++ or configured in a graphical simulator such as LABVIEW in order to be connected to a PES for control system checkout and operator training.

Tieback models came on the scene in the 1980s as an inexpensive and efficient method of creating simple dynamics. For discrete control devices such as pumps and on–off valves, the tieback would read the discrete outputs and send back the appropriate motor run contact or valve limit switch contact action for the discrete inputs after a suitable delay. For control loops, the tieback would read the analog output and respond with an analog input in the right direction with an adjustable gain, lag, and delay. For levels, ramps were initially devised and later integrators were set up to simulate the change in inventory. Tieback models can be automatically generated from the configuration. These tieback models make the displays and color conditionals operational, but do not provide an independent test and do not directly show flaws in logic or mistakes in assignments.

The practice to date for operator training and control system checkout has been to use a tieback or custom dynamic simulation connected to the inputs and outputs or data highway of a PES hardware system, as shown in Figure 5.8e. The cost of this setup varies from \$100K to \$1000K depending upon the size and vintage of the PES. For large systems, it is too expensive to have all of the controllers and consoles so only a portion of the plant is simulated at any one time. A special room with particular air conditioning and power conditioning requirements is also needed to handle the actual controllers, input and output card racks, network, and operator consoles. Also, with hardware processing and memory capability changing yearly for a state-of-the-art PES, the hardware system is obsolete by the time it is running. Finally, there is no opportunity either to speed up or to slow down the PES execution to run scenarios faster or to analyze confusing situations.

An alternative to the hardware system that has been used for operator training in larger and more complex plants is the emulation of the PES in a computer. However, this requires a translation of proprietary configuration blocks and displays that is expensive and vulnerable to mismatch. Often there is a different interface because the actual consoles are

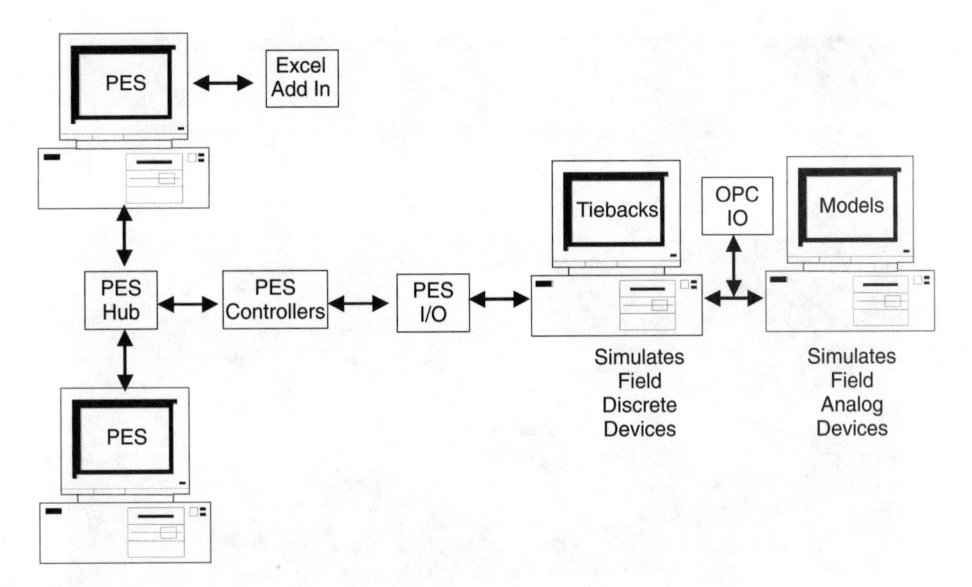

FIG. 5.8e
Hardware-based system for configuration check out and operator training.

not used so that learning the navigation of displays for start-up, shutdown, or abnormal conditions, which is one of the major objectives of operator training, is compromised. The cost of the custom dynamic simulation and PES emulation can easily exceed \$1000K. Typically, only nuclear power plants or high-capacity petrochemical plants justify such a large expenditure.

PROPOSED PRACTICES

The expected pressure drops for valves, exchangers, filters, and piping should be added to a steady-state model and built-in sizing routines used to compute the flow coefficients from these pressure drops, flows, and the physical property data (densities and viscosities) that are in the fluid basis of the steady-state model. Nozzle locations should be specified or pseudo pumps inserted to show the effect of liquid head. The individual split ranged and on–off (isolation) valves should be depicted to enable the detailing of separate characteristics and responses. The efficiencies and characteristic curves of pumps and compressors should be added to provide the proper pressure–flow relationships and individual series or parallel pumps or compressors inserted to show the interaction or transition from one unit to another.

Actual heat exchangers should be used with utility streams rather than heaters and coolers with heat duties, to simulate the effect of pressure and temperature changes, fouling, frosting, piping configuration, and control valves in the brine, steam, Therminol™, Dowtherm™, and water systems. The omission of this extra effort has historically created a fertile ground for performance problems. The unexpected periodic upsets from these utility streams due to interactions and control valve stick and slip are a major source of variability. It is usually too expensive to fix these utility systems after the fact. Data analysis techniques such power spectrum analysis are used to track the source of the oscillations, and special tuning, responsive control valves, and control techniques such as feed forward are used to reduce the variability. The real solution is to use modeling to alert one early enough to prevent the installation of the problem.

Figure 5.8f shows the same column but with the sump and overhead receiver modeled as separators and the reboiler and condenser modeled as heat exchangers. Pumps, valves, measurements, volumes, steam, and cooling tower water streams have been added. A transfer function has been used to simulate the thermowell dynamics for the temperature measurement and a separator has been placed on the condensate stream from the reboiler to simulate a steam trap. Valves are inserted and sized to provide the expected piping pressure drops. This structure also facilitates the modeling of a thermosyphon reboiler by a spreadsheet calculation where the bottoms pump pressure rise and hence reboiler circulation flow is a function of sump level. The pump pressure rise is used because nozzle locations do not affect pressure in the steady-state mode. The increase in time for the steady-state model to converge is a matter of minutes. This more detailed model is saved out as a template that can be brought into other models as a starting point.

Figure 5.8g shows the interface table that has been added to connect the virtual process to a virtual PES via OPC. Any operating condition in the model (composition, density, level, pressure, and temperature) can be inserted into the process variable (PV) export table to be sent from the process to the control system. Field switches such as level switches can be readily set to activate via spreadsheets in the process model and placed into the same PV export table. Control valve analog signals and on–off valve and pump discrete command

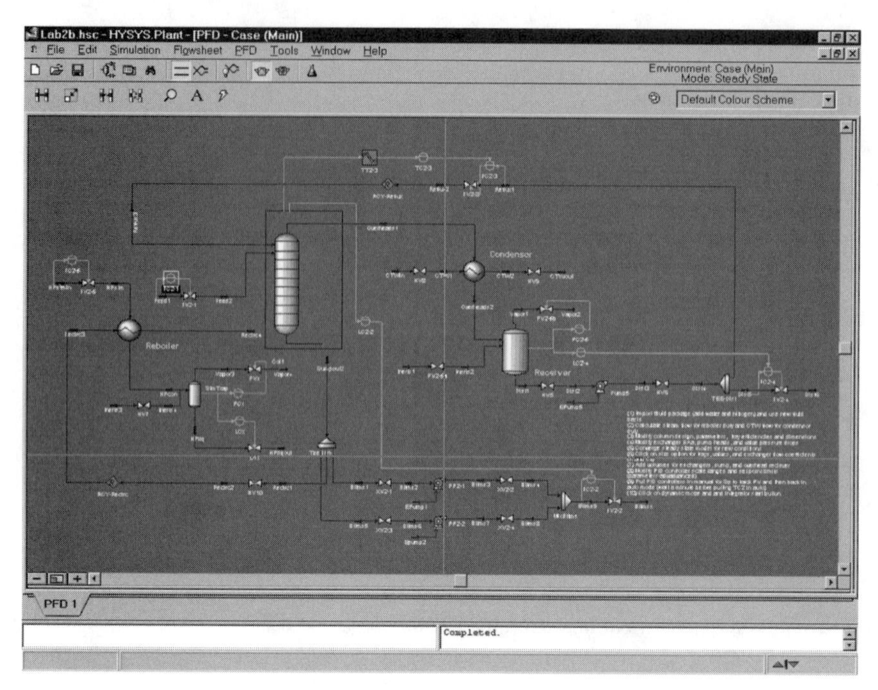

FIG. 5.8f
Column model with P&ID type details added.

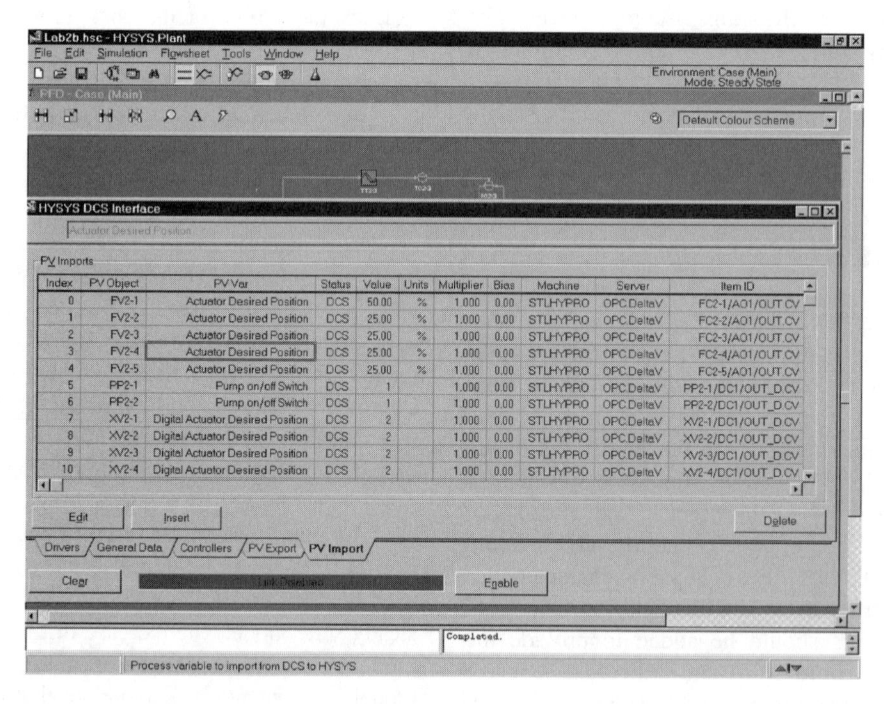

Index	PV Object	PV Var	Status	Value	Units	Multiplier	Bias	Machine	Server	Item ID
0	FV2-1	Actuator Desired Position	DCS	50.00	%	1.000	0.00	STLHYPRO	OPC.DeltaV	FC2-1/AO1/OUT.CV
1	FV2-2	Actuator Desired Position	DCS	25.00	%	1.000	0.00	STLHYPRO	OPC.DeltaV	FC2-2/AO1/OUT.CV
2	FV2-3	Actuator Desired Position	DCS	25.00	%	1.000	0.00	STLHYPRO	OPC.DeltaV	FC2-3/AO1/OUT.CV
3	FV2-4	Actuator Desired Position	DCS	25.00	%	1.000	0.00	STLHYPRO	OPC.DeltaV	FC2-4/AO1/OUT.CV
4	FV2-5	Actuator Desired Position	DCS	25.00	%	1.000	0.00	STLHYPRO	OPC.DeltaV	FC2-5/AO1/OUT.CV
5	PP2-1	Pump on/off Switch	DCS	1		1.000	0.00	STLHYPRO	OPC.DeltaV	PP2-1/DC1/OUT_D.CV
6	PP2-2	Pump on/off Switch	DCS	1		1.000	0.00	STLHYPRO	OPC.DeltaV	PP2-2/DC1/OUT_D.CV
7	XV2-1	Digital Actuator Desired Position	DCS	2		1.000	0.00	STLHYPRO	OPC.DeltaV	XV2-1/DC1/OUT_D.CV
8	XV2-2	Digital Actuator Desired Position	DCS	2		1.000	0.00	STLHYPRO	OPC.DeltaV	XV2-2/DC1/OUT_D.CV
9	XV2-3	Digital Actuator Desired Position	DCS	2		1.000	0.00	STLHYPRO	OPC.DeltaV	XV2-3/DC1/OUT_D.CV
10	XV2-4	Digital Actuator Desired Position	DCS	2		1.000	0.00	STLHYPRO	OPC.DeltaV	XV2-4/DC1/OUT_D.CV

FIG. 5.8g
Interface table for the OPC connection of the model to the PES.

signals are read from the PES via the PV import table. They are also automatically initialized in the PES by the model, momentarily putting the analog output block or discrete control block in auto and writing a set point to match the position or status in the process model.

When the PES link button is enabled, the model verifies each of the OPC connections and pinpoints any missing ones. When the integrator start button is selected, each controller output, on–off valve, and pump in the PES is automatically initialized to match the current valve position and pump status in the process. No programs or scripts are needed. This initialization is critical to provide a fast and bumpless transition to the desired operating conditions. For slow processes such as columns it is essential because it can take hours to

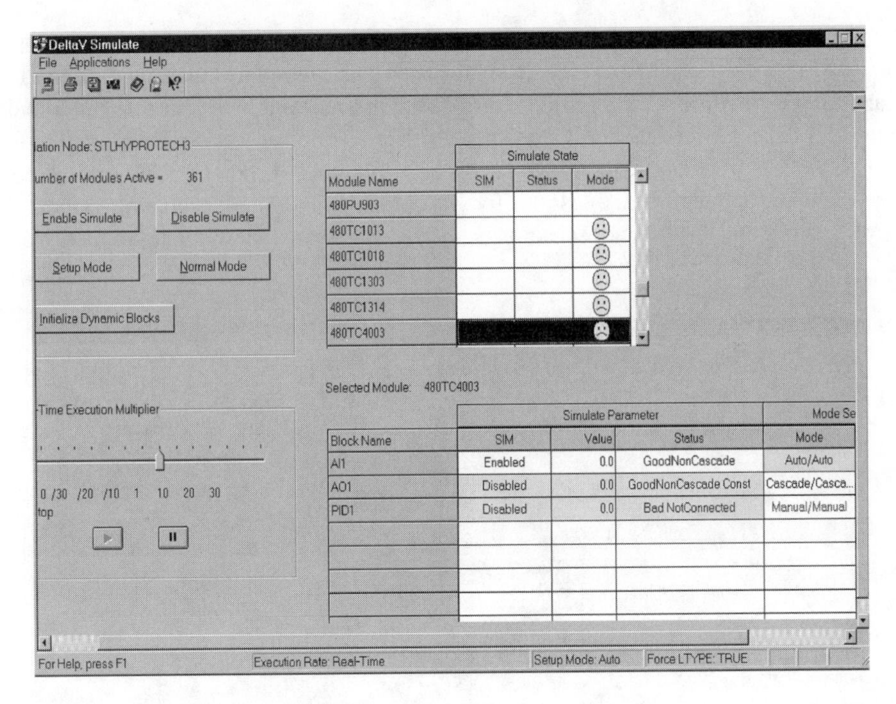

FIG. 5.8h
Interface for monitoring and controlling the speed of the virtual PES.

recover from a valve that is in the wrong position or a pump that is not running.

A key concept is that there is no duplication, translation, or modification of the actual control system. This is of great significance in terms of reduced cost and increased confidence. The duplication of a complex or advanced control strategy in the process modeling program can easily add 50% more time to the modeling effort, especially because the model software lacks a complete set of function blocks used in a PES. The investment in these blocks is incredible and cannot be expected to be a feature of a process simulation program. There is no assurance that the simulated control system will behave the same as the real control system. Consequently, changes would need to be made in both places. The process model should be used to represent the field equipment, measurements, and final elements and connected to the actual PES configuration and operator displays to create a virtual plant.

The virtual plant resides in a single desktop computer or, for large processes, in a network of desktops. These desktop computers can be anywhere and connected to another desktop or laptop anywhere on the LAN, and can use inexpensive software such as pcAnywhere. Access and navigation via analog telephone connection is slower than through a digital cable or digital subscriber line (DSL) connection but is sufficient for demonstration purposes. Thus, the virtual plant can be accessed while at home, traveling, or visiting another site. The virtual plant can be enhanced. Other Windows-based software systems for loop performance monitoring, multivariate principal component analysis, neural networks, and real-time optimization can be loaded and interfaced via OPC. The only limits are an individual's imagination and the power of the desktop.

A tool built into the virtual PES changes the necessary parameters of the desired functional blocks to enable simulation without making any permanent changes to the actual configuration. The interface for this tool as shown in Figure 5.8h displays the status of each functional block and allows the user to run the PES system either 30 times slower or faster than real time to match the acceleration factor set in the process model.

There is, of course, some reluctance to add these details. The size and convergence time for a steady-state model will increase. However, as the processing speeds of PCs increase, this becomes less of an issue. Steady-state modeling techniques have been around a long time compared to those for dynamic flow sheet modeling. The use of templates for common unit operations and sub flow sheets will eventually make the transition from a steady-state PFD to a dynamic P&ID model faster than the custom approach at present used by providing a common starting point and method. The result is a much more accurate representation of the plant plus a standardization of techniques that expedites the effort and expands the potential number of people who can create or interrogate the model. An analogous benefit to the 25% or more effort savings in developing and maintaining control system configurations gained from the standardization on the control system type and configuration module library is expected. Of greater significance is the wider audience of contributors and users of the knowledge base in the virtual plant.

PROCESS MODELS

Figure 5.8i shows Lab100, which consists of a static mixer on the feed to a neutralizer. The process model shows the interaction and nonlinearity of the pH loops, the thermal lags for cascade control of temperature, the process dynamics for vessel pressure and level control, and the effect of sensor lags and noise set in the transfer functions. This model can be saved as a template that can be inserted multiple times as sub flow sheets and interconnected in a main flow sheet to simulate a series of static mixers and neutralizers.

Figure 5.8j shows Lab200, which consists of a distillation column with a thermosyphon reboiler. The rigorous model of the column includes a material, energy, and component balance

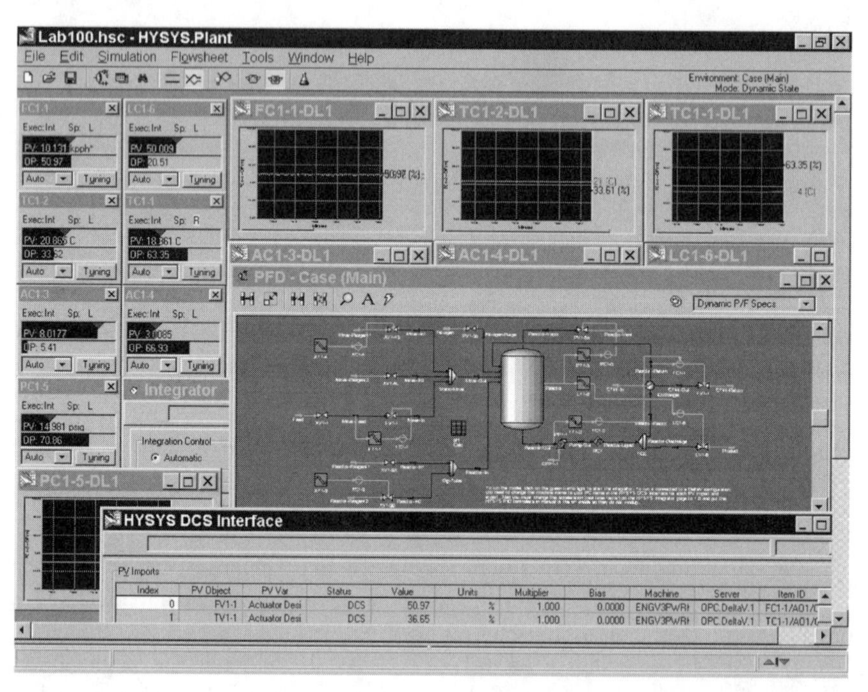

FIG. 5.8i
Static mixer and neutralizer.

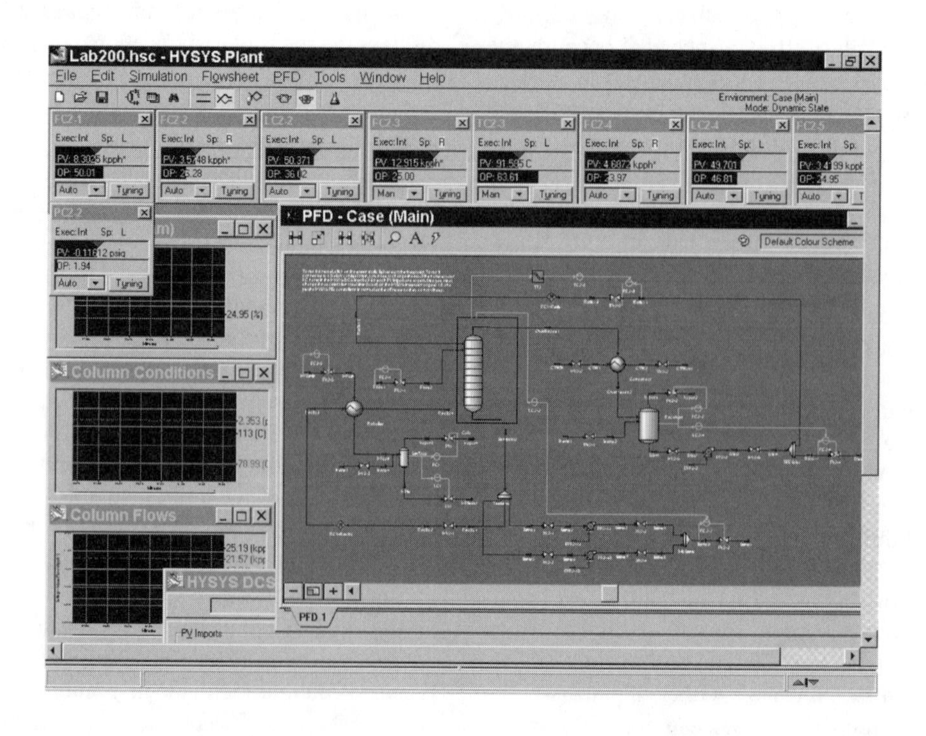

FIG. 5.8j
Distillation column with a thermosyphon reboiler.

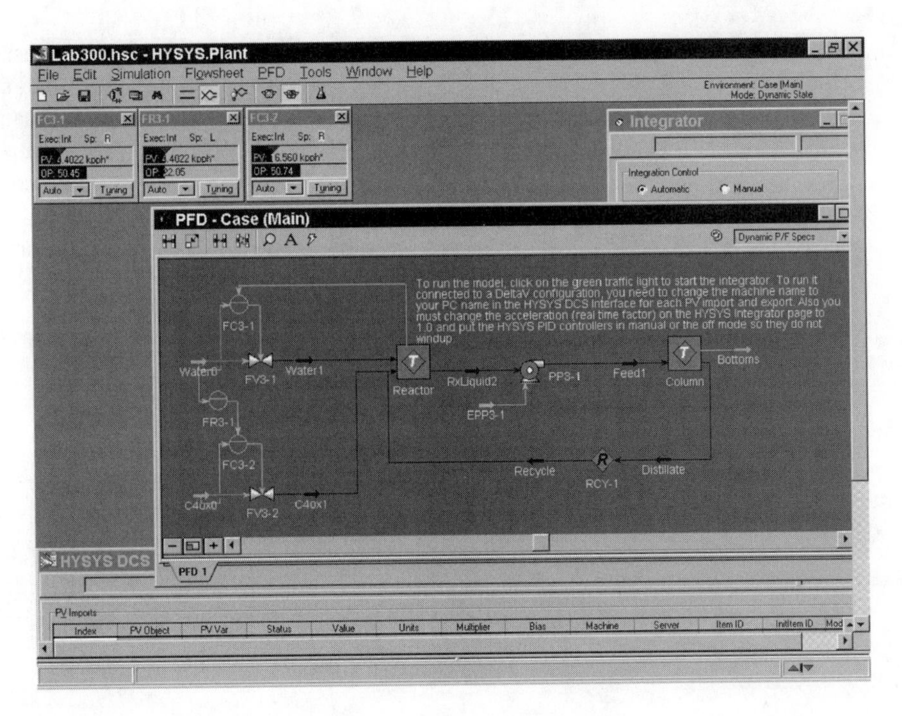

FIG. 5.8k
Reactor and column with a recycle stream.

and vapor–liquid equilibrium for each tray and volume. The effect of interacting lags from the trays and the separate lag of the overhead receiver volume on the composition and temperature response can be studied. The use of time series models, neural networks, principal component analysis, and first principal relationships for dynamic online property estimators can be explored to provide a faster and more reliable measurement of distillate or bottoms composition than online or lab analyzers. The effect of sump level on the reboiler circulation and column vapor flow can also be investigated.

Figure 5.8k shows Lab300, which consists of a feed tank, exothermic reactor, and waste heat boiler in a reactor sub flow sheet whose product is purified in the column sub flow sheet. The column distillate is recycled to the feed tank. The model can be used to show the effect of kinetics and the interaction between the reactor and the boiler on the temperature and composition control of the product. As with the column model, various methods for the development of virtual analyzers can be explored. The dynamics and control of recycle systems detailed in References 3 and 4 can be illustrated along with the applicability of constrained multivariable predictive controllers. The use of strategies to maximize a feed by the use of an internal optimization variable or external pusher algorithm can be demonstrated.

Figure 5.8l shows Lab404, which consists of a boiler with an economizer, superheater, and desuperheater. The model can be used to show the interactions as well as the inverse response of furnace temperature, pressure, and boiler drum level. The model can be saved as a template and inserted as a sub flow sheet multiple times and connected into a steam header system with a turbine generator to illustrate the effect of various continuous and batch steam users on pressure control. Online energy balances and real-time optimization of boiler load allocation can be developed and demonstrated.

Not shown is the process model for a plant that consists of complex batch reaction and continuous recovery areas. This virtual plant is used to develop a multivariate analysis of batch trajectories using principal components and a dynamic model for real-time optimization of batch conditions. Because the data historian can poll actual plant conditions via the LAN as well as access and set any value in the virtual plant, parameter estimation techniques are planned for the dynamic model so that it better matches the plant performance. This opens the door for the use of the dynamic model for extensive process testing, process and estimator identification, and advanced process control prototyping without upsetting the actual plant.

Previous attempts at reconciliation and parameter estimation of dynamic models have used complex custom techniques. A conceptual breakthrough has led to a general method for an easy adaptation of dynamic models to match the actual process using standard advanced control tools that exist in the virtual plant. Although the proprietary nature of this discovery precludes its disclosure here, similar inventions are expected to unfold at a rapid pace that will revolutionize the utility of dynamic models.

CONTROL STUDIES

The virtual plant offers the opportunity to conduct the following list of process control studies to provide an understanding that leads to a series of key concepts and best practices for the

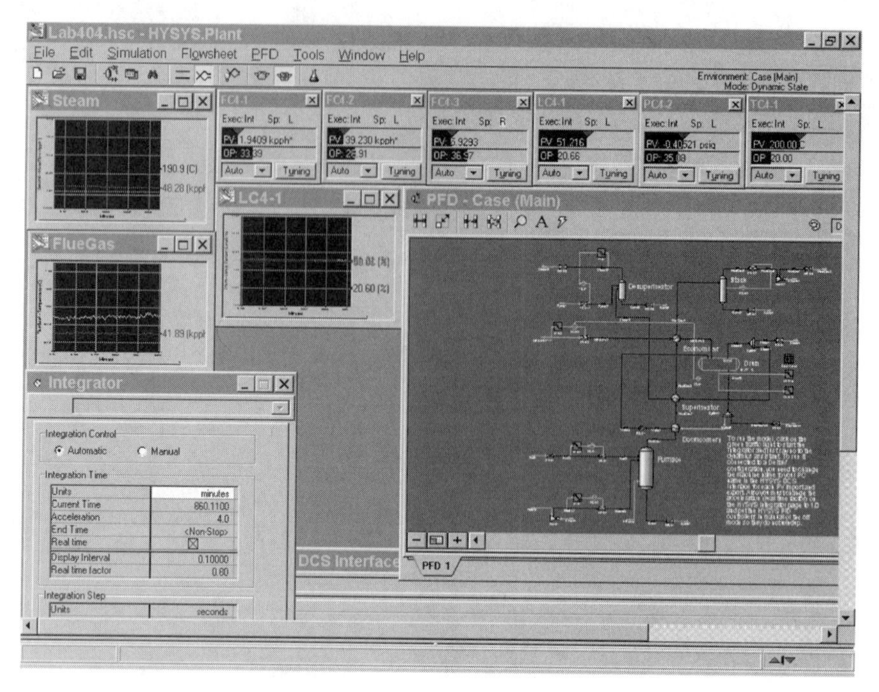

FIG. 5.8I
Utility boiler with an economizer, superheater, and desuperheater.

application of wide range of basic and advanced data analysis and process control.

1. Correction of the ramp rate to gradually return a surge tank level to target using an adapted velocity limited feedforward to minimize flow variability for a batch-to-continuous transition. Comparison of the performance of the adapted feedforward to an error-squared controller and investigation of whether the cycling, which is caused by reset action in a controller with a gain below the window of allowable gains for an integrating process, can be avoided.
2. Signal characterization of rotary valves for feed maximization and for compensation of the process gain from the manipulation of coolant and bypass flow to control a heat exchanger temperature for various ratios of valve to system pressure drop.
3. Effect of controller tuning on the switchover point of the approach to set point and overshoot for override loops.
4. Bang-bang, pulse width modulation, error-squared, fuzzy logic control comparisons.
5. Conditions that create positive feedback for pressure correction of column temperature set point and moisture correction of dryer inferential moisture control.
6. Performance of valve position control, split range control, and dual P&PI loops for big-small valves and various types of upsets.
7. The minimum ratio of reset to gain settings to maintain the Wade relationship to suppress level oscillations for various amounts of dead time and dead band.
8. The minimum ratio of reset to gain contribution that slows a restart of a control loop (faltering of the approach to set point).
9. Relative effectiveness of signal filters, scan times, P-only control, and lambda tuning to break the interaction between loops.
10. Tuning rules for feedforward, model predictive, and fuzzy logic control.
11. Performance of different signal filter designs to attenuate noise and minimize lag for load upsets.
12. Trade-off of an increase in a filter and gain vs. an addition of a feedforward of set point as a function of valve dead band and nonlinearity and the controller gain.
13. Robustness and performance of an adaptive controller that uses knowledge of the dead time and the relative contribution of the controller modes to adjust the tuning.
14. Robustness and performance of constrained multivariable predictive control (CMPC) vs. a detuned PID for changes in process delay, lag, and gain.
15. Performance of feed ratio control corrected to maintain mass totals in a vessel during start-up and rate changes.
16. Pattern recognition of a CMPC prediction array to decipher model mismatch and degree of unaccounted load upsets. For example, if the actual process gain is higher or the load upset is a ramp where the unmeasured disturbance is greater or faster than the move, will the prediction array be in the opposite direction of the true trajectory?

17. Exploration of CMPC model mismatch or tuning as the cause of a staircase or falter in the approach to set point.
18. Investigation into rate action as the primary beneficial mode for highly nonlinear processes (pH and organic-water distillation temperature).
19. Performance of a CMPC compared to a PID with signal characterization for a nonlinear process and valve and frequent severe unmeasured load upsets.
20. Performance of a CMPC compared to a PID for noise and A/D chatter (measurement resolution limits).
21. Performance of a CMPC compared to a PID for valve dead band and stick–slip.
22. Ability of various data reconciliation and parameter estimation techniques to improve the fidelity of a real-time model running in sync with a real process.
23. The value in the use of first principles to update the delay and gain in a CMPC.
24. The CMPC matrix and tuning weights needed for constraint and ratio control for maximization of feed.
25. The robustness, performance, and ease of using principal component analysis, neural networks, and first principle dynamic models as dynamic online property estimators.
26. The sensitivity and capability of online performance indicators to diagnose noise, tuning, interaction, and valve problems.

BENEFITS

The virtual plant facilitates the simulation of a wide range of operating conditions, process upsets, control schemes, and equipment conditions such as condenser and reboiler fouling, control valve stick–slip, and pump failures.[1] Pressures, temperatures, and concentrations can be viewed and changed by a simple double click on the stream and either the condition or composition option. The user can also easily navigate to process and ambient heat transfer coefficients, volumes, and flow coefficients. A pump can be started or stopped and a valve opened or closed from the operator display.

The greatest, most immediate benefit is the ability to tune feedback controller gain, reset, and rate settings and feedforward signal gain, delay, and lead-lag settings for realistic process responses and interactions. Auto tuning tools can be used and the simulation can be run faster than real time to speed the tuning process.

Trials of improvements in batch sequences and regulatory controls can be conducted and the actual implementation in the configuration checked out. Expert systems can be developed for advisory control and neural networks can be trained as intelligent sensors to predict compositions. Model predictive control and optimization strategies can be quickly prototyped. The virtual plant can be used to uncover a flaw common in these advanced control techniques: the ability to handle a wide range of normal and abnormal operating conditions and plant dynamics. A virtual plant can go where one does not want the real plant to go or does not think the real plant can go. Thus, it can show how to recover from undesirable situations and reveal unforeseen or doubted capabilities. This last advantage of being able to demonstrate set points closer to constraints builds operator confidence.

There is also the classical benefit from operator training using the virtual plant. Often overlooked are the additional training opportunities of other people who support the on-stream time of the advanced control system, such as process and instrument engineers.

The portability of the virtual plant means that training is not relegated to a training room or a particular audience or purpose. Operator training systems have traditionally been used to gain familiarity with the interface and process, and to learn how to start up, shut down, and to deal with failures.[2] It is particularly important to provide continual access to these systems to refresh skills in units with frequent transitions, trips, campaigns, and high personnel turnover rates. Less obvious is the need to replace operating skills lost from the lack of interaction with the process and loops caused by extensive automation.

There are huge additional benefits if one considers that the operators are potentially the biggest constraint to pushing plant capacity, especially when it is beyond the original nameplate. The gap between the operating point and the constraint is often set based on a perceived relationship that may have only occurred one time to one operator. The gap based on a war story is typically an order of magnitude larger than the gap dictated by process variability. Consider how much plant production has increased when an experienced process or control system engineer is in the control room or an advanced control system stays online for all of the shifts. High-fidelity process simulations and actual control system configurations can be used to increase the knowledge base and skill level of the operator to push limits safely and reliably to achieve this incremental capacity. This can be accomplished by the operator just choosing better set points or by keeping control systems in their highest mode despite rough or confusing conditions.

Access to the virtual plant should be opened to nontraditional audiences such as maintenance, mechanical, and control system engineers so that they can test, calibrate, troubleshoot, and adjust set points, equipment, instrumentation, and control schemes for various scenarios and improvements. The time to prototype an idea can be a matter of a few hours or days. The effort and cost to explore and demonstrate a new idea is no longer an obstacle. Investments in more advanced control tools can be analyzed, optimized, and justified. PES configurations with new operating points, equipment, or control capability can be checked out, tuned, exported, and downloaded without modification.

The virtual plant and simulator will also decrease the cost, time, and the disruption, and increase the performance

TABLE 5.8m
A Cost and Benefit Comparison of Simulation Methods

	Initial Cost	*Life Cost*	*Model Scope*	*Model Fidelity*	*Model Access*	*Control Studies*	*Ease of Idea Prototyping*	*Operator Training*
Custom	☹	☹	😐	😐	☹	😐	☹	😐
Tieback	☺	☺	☺	☹	☺	☹	😐	😐
Virtual	😐	☺	☺	☺	☺	☺	☺	☺

☺ = good; 😐 = fair; ☹ = poor.

and on-stream time of APC implementation. Dynamic models in the virtual plant that are adapted to match the process in the actual plant will accomplish the following:

1. Reduce the time from concept to implementation by offering studies and rapid prototyping of APC ideas. In other words, virtual trials are used instead of debates and meetings to establish the relative merits of various approaches.
2. Decrease the disruption to the process by experimentation, response testing, and parameter tuning with the virtual plant rather than with the actual process.
3. Minimize the time to commission an APC application by configuration testing and parameter tuning.
4. Eliminate misleading measurements and identify inaccurate or improperly calibrated transmitters from the mass and energy balances in the model.
5. Increase the on-stream time for the APC from hands-on training of engineers, technicians, and operators. The biggest constraint is often the degree of understanding of the objectives, function, and trajectory of the APC.
6. Improve the performance of the online APC from off-line parameter tuning.

Table 5.8m summarizes the cost and benefits of the present practices that use hardware or emulated systems with tieback or custom models and the proposed practice of creating a virtual plant.

The virtual plant offers the opportunity of a warehouse of plant understanding where each person develops a stake in using, updating, and upgrading the knowledge base.

References

1. Lo, P., Chen, D. H., Bean, W. C., and Thompson, W., "Distillation Tower Pump Failure," *Control,* pp. 71–82, October 2000.
2. Chin, K., "Learning in a Virtual World," *Chemical Engineering,* pp. 107–110, December 2000.
3. Tyreus, B. D. and Luyben, W. L., "Dynamics and Control of Recycle Systems. Part 4. Ternary Systems with One or Two Recycle Streams," *Industrial Engineering and Chemical Research,* Vol. 32, pp. 1154–1162, 1993.
4. Luyben, W. L. and Luyben, M. L., *Essentials of Process Control,* Chemical Engineering Series, New York: McGraw-Hill, 1997, pp. 184–222.

Appendix

A.1 International System of Units

The decimal system of units was conceived in the 16th century when there was a great confusion and jumble of units of weights and measures. It was not until 1790, however, that the French National Assembly requested the French Academy of Sciences to work out a system of units suitable for adoption by the entire world. This system, based on the meter (meter) as a unit of length and the gram as a unit of mass, was adopted as a practical measure to benefit industry and commerce. Physicists soon realized its advantages and it was adopted also in scientific and technical circles. The importance of the regulation of weights and measures was recognized in Article 1, Section 8, when the U.S. Constitution was written in 1787, but the metric system was not legalized in this country until 1866. In 1893, the international meter and kilogram became the fundamental standards of length and mass in the United States, both for metric and customary weights and measures. The tables of conversion factors presented here are intended to serve two purposes:

1. To express the definitions of miscellaneous units of measure as exact numeral multiples of coherent "metric" units. Relationships that are exact in terms of the base unit are followed by an asterisk. Relationships that are not followed by an asterisk are either the results of physical measurements or are only approximate.
2. To provide multiplying factors for converting expressions of measurements given by numbers and miscellaneous units to corresponding new numbers and metric units.

Conversion factors are presented for ready adaptation to computer readout and electronic data transmission. The factors are written as a number equal to or greater than 1 and less than 10 with six or fewer decimal places. This number is followed by the letter E (for exponent), a plus or minus symbol, and two digits that indicate the power of 10 by which the number must be multiplied to obtain the correct value. For example:

$$3.523\,907 \text{ E–02 is } 3.523\,907 \times 10^{-2}$$

or

$$0.035\,239\,07$$

Similarly,

$$3.386\,389 \text{ E+03 is } 3.386\,389 \times 10^{3}$$

or

$$3\,386.389$$

An asterisk (*) after the sixth decimal place indicates that the conversion factor is exact and that all subsequent digits are zero.

When a figure is to be rounded to fewer digits than the total number available, the procedure should be as follows:

1. When the first digit discarded is less than 5, the last digit retained should not be changed. For example, 3.463 25, if rounded to four digits, would be 3.463; if rounded to three digits, 3.46.
2. When the first digit discarded is greater than or if it is a 5, followed by at least one digit other than 0, the last figure retained should be increased by one unit. For example 8.376 52, if rounded to four digits, would be 8.377; if rounded to three digits, 8.38.
3. When the first digit discarded is exactly 5, followed only by zeros, the last digit retained should be rounded upward if it is an odd number, but no adjustment made if it is an even number. For example, 4.365, when rounded to three digits, becomes 4.36. The number 4.355 would also round to the same value, 4.36, if rounded to three digits.

Where fewer than six decimal places is shown, more precision is not warranted.

TABLE A.1a
International System of Units

Quantity	*Unit*	*SI Symbol*	*Formula*
	Base Units		
length	meter	m	—
mass	kilogram	kg	—
time	second	s	—
electric current	ampere	A	—
thermodynamic temperature	kelvin	K	—
amount of substance	mole	mol	—
luminous intensity	candela	cd	—
	Supplementary Units		
plane angle	radian	rad	—
solid angle	steradian	sr	—
	Derived Units		
acceleration	meter per second squared	—	m/s^2
activity (of a radioactive source)	disintegration per second	—	(disintegration)/s
angular acceleration	radian per second squared	—	rad/s^2
angular velocity	radian per second	—	rad/s
area	square meter	—	m^2
density	kilogram per cubic meter	—	kg/m^3
electric capacitance	farad	F	A·s/V
electrical conductance	siemens	S	A/V
electric field strength	volt per meter	—	V/m
electric inductance	henry	H	V·s/A
electric potential difference	volt	V	W/A
electric resistance	ohm	Ω	V/A
electromotive force	volt	V	W/A
energy	joule	J	N·m
entropy	joule per kelvin	—	J/K
force	newton	N	kg·m/s^2
frequency	hertz	Hz	(cycle)/s
illuminance	lux	lx	lm/m^2
luminance	candela per square meter	—	cd/m^2
luminous flux	lumen	lm	cd·sr
magnetic field strength	ampere per meter	—	A/m
magnetic flux	weber	Wb	V·s
magnetic flux density	telsa	T	Wb/m^2
magnetomotive force	ampere	A	—
power	watt	W	J/s
pressure	pascal	Pa	N/m^2
quantity of electricity	coulomb	C	A·s
quantity of heat	joule	J	N·m
radiant intensity	watt per steradian	—	W/sr
specific heat	joule per kilogram-kelvin	—	J/kg·K
stress	pascal	Pa	N/m^2
thermal conductivity	watt per meter-kelvin	—	W/m·K
velocity	meter per second	—	m/s
viscosity, dynamic	pascal-second	—	Pa·s
viscosity, kinematic	square meter per second	—	m^2/s
voltage	volt	V	W/A
volume	cubic meter	—	m^3
wavenumber	reciprocal meter	—	(wave)/m
work	joule	J	N·m

TABLE A.1b
Alphabetical List of Units (Symbols of SI units given in parentheses)

To Convert from	*To*	*Multiply by*
	A	
abampere	ampere (A)	1.000 000*E+01
abcoulomb	coulomb (C)	1.000 000*E+01
abfarad	farad (F)	1.000 000*E+09
abhenry	henry (H)	1.000 000*E+09
abmho	siemens (S)	1.000 000*E+09
abohm	ohm (Ω)	1.000 000*E+09
abvolt	volt (V)	1.000 000*E+08
acre foot (U.S. survey)[a]	meter^3 (m^3)	1.233 489 E+03
acre (U.S. survey)[a]	meter^2 (m^2)	4.046 873 E+03
ampere hour	coulomb (C)	3.600 000*E+03
are	meter^2 (m^2)	1.000 000*E+02
angstrom	meter (m)	1.000 000*E–10
astronomical unit	meter (m)	1.495 979 E+11
atmosphere (standard)	pascal (Pa)	1.013 250*E+05
atmosphere (technical = 1 kgf/cm^2)	pascal (Pa)	9.806 650*E+04

[a] Since 1893, the U.S. basis of length measurement has been derived from metric standards. In 1959, a small refinement was made in the definition of the yard to resolve discrepancies both in this country and abroad, which changed its length from 3600/3937 m to 0.9144 m exactly. This resulted in the new value being shorter by two parts in a million.

At the same time it was decided that any data in feet drived from and published as a result of geodetic surveys within the United States would remain with the old standard (1 ft = 1200/3937 m) until further decision. This foot is named the U.S. survey foot.

As a result all U.S. land measurements in U.S. customary units will relate to the meter by the old standard. All the conversion factors in these tables for units referenced to this footnote are based on the U.S. survey foot, rather than the international foot.

Conversion factors for the land measures given below may be determined from the following relationships:

1 league = 3 miles (exactly)
1 rod = $16^1/_2$ feet (exactly)
1 section = 1 square mile (exactly)
1 township = 36 square miles (exactly)
1 chain = 66 feet (exactly)

To Convert from	*To*	*Multiply by*
	B	
bar	pascal (Pa)	1.000 000*E+05
barn	meter^2 (m^2)	1.000 000*E–28
barrel (for petroleum, 42 gal)	meter^3(m^3)	1.589 873 E–01
board foot	meter^3(m^3)	2.359 737 E–03
British thermal unit (International Table)[b]	joule (J)	1.055 056 E+03
British thermal unit (mean)	joule (J)	1.055 87 E+03
British thermal unit (thermochemical)	joule (J)	1.054 350 E+03
British thermal unit (39°F)	joule (J)	1.059 67 E+03
British thermal unit (59°F)	joule (J)	1.054 80 E+03
British thermal unit (60°F)	joule (J)	1.054 68 E+03
Btu (International Table) · ft/h · ft^2 · °F (*k*, thermal conductivity)	watt per meter-kelvin (W/m·K)	1.730 735 E+00
Btu (thermochemical) · ft/h · ft^2 · °F (*k*, thermal conductivity)	watt per meter-kelvin (W/m·K)	1.729 577 E+00
Btu (International Table) · in./h · ft^2 · °F (*k*, thermal conductivity)	watt per meter-kelvin (W/m·K)	1.442 279 E–01
Btu (thermochemical) · in./h · ft^2 · °F (*k*, thermal conductivity)	watt per meter-kelvin (W/m·K)	1.441 314 E–01
Btu (International Table) · in./s · ft^2 · °F (*k*, thermal conductivity)	watt per meter-kelvin (W/m·K)	5.192 204 E+02
Btu (thermochemical) · in./s · ft^2 · °F (*k*, thermal conductivity)	watt per meter-kelvin (W/m·K)	5.188 732 E+02
Btu (International Table)/h	watt (W)	2.930 711 E–01
Btu (International Table)/s	watt (W)	1.055 056 E+03
Btu (thermochemical)/h	watt (W)	2.928 751 E+01
Btu (thermochemical)/min	watt (W)	1.757 250 E+01
Btu (thermochemical)/s	watt (W)	1.054 350 E+03
Btu (International Table)/ft^2	joule per meter^2 (J/m^2)	1.135 653 E+04

[b] This value was adopted in 1956. Some of the older International Tables use the value 1.055 04 E+03. The exact conversion factor is 1.055 055 852 62*E+03.

TABLE A.1b Continued

Alphabetical List of Units (Symbols of SI units given in parentheses)

To Convert from	*To*	*Multiply by*
Btu (thermochemical)/ft^2	joule per meter^2 (J/m^2)	1.134 893 E+04
Btu (thermochemical)/ft^2 · h	watt per meter^2 (W/m^2)	3.152 481 E+00
Btu (thermochemical)/ft^2 · min	watt per meter^2 (W/m^2)	1.891 489 E+02
Btu (thermochemical)/ft^2 · s	watt per meter^2 (W/m^2)	1.134 893 E+04
Btu (thermochemical)/in.^{2} · s	watt per meter^2 (W/m)^2	1.634 246 E+06
Btu (International Table)/h · ft^2 · °F (*C*, thermal conductance)	watt per meter^2-kelvin (W/m^2 · K)	5.678 263 E+00
Btu (thermochemical)/h · ft^2 · °F (*C*, thermal conductance)	watt per meter^2-kelvin (W/m^2 · K)	5.674 466 E+00
Btu (International Table)/s · ft^2 · °F	watt per meter^2-kelvin (W/m^2 · K)	2.044 175 E+04
Btu (thermochemical)/ s · ft^2 · °F	watt per meter^2-kelvin (W/m^2 · K)	2.042 808 E+04
Btu (International Table)/lb	joule per kilogram (J/kg)	2.326 000*E+03
Btu (thermochemical)/lb	joule per kilogram (J/kg)	2.324 444 E+03
Btu (International Table)/lb · °F (*c*, heat capacity)	joule per kilogram-kelvin (J/kg · K)	4.186 800*E+03
Btu (thermochemical Table)/lb · °F (*c*, heat capacity)	joule per kilogram-kelvin (J/kg · K)	4.184 000 E+03
bushel (U.S.)	meter^3 (m^3)	3.523 907 E–02
	C	
caliber (inch)	meter (m)	2.540 000*E–02
calorie (International Table)	joule (J)	4.186 800*E+00
calorie (mean)	joule (J)	4.190 02 E+00
calorie (thermochemical)	joule (J)	4.184 000*E+00
calorie (15°C)	joule (J)	4.185 80 E+00
calorie (20°C)	joule (J)	4.181 90 E+00
calorie (kilogram, International Table)	joule (J)	4.186 800*E+03
calorie (kilogram, mean)	joule (J)	4.190 02 E+03
calorie (kilogram, thermochemical)	joule (J)	4.184 000*E+03
cal (thermochemical)/cm^2	joule per meter^2 (J/m^2)	4.184 000*E+04
cal (International Table)/g	joule per kilogram (J/kg)	4.186 800*E+03
cal (thermochemical)/g	joule per kilogram (J/kg)	4.184 000*E+03
cal (International Table)/g · °C	joule per kilogram-kelvin (J/kg · K)	4.186 800*E+03
cal (thermochemical)/g · °C	joule per kilogram-kelvin (J/kg · K)	4.184 000*E+08
cal (thermochemical)/min	watt (W)	6.973 333 E–02
cal (thermochemical)/s	watt (W)	4.184 000*E+00
cal (thermochemical)/cm^2 · min	watt per meter^2 (W/m^2)	6.973 333 E+02
cal (thermochemical)/cm^2 · s	watt per meter^2 (W/m^2)	4.184 000*E+04
cal (thermochemical)/cm · s · °C	watt per meter-kelvin (W/m · K)	4.184 000*E+02
carat (metric)	kilogram (kg)	2.000 000*E–04
centimeter of mercury (0°C)	pascal (Pa)	1.333 22 E+03
centimeter of water (4°C)	pascal (Pa)	9.806 38 E+01
centipoise	pascal second (Pa · s)	1.000 000*E–03
centistokes	meter^2 per second (m^2/s)	1.000 000*E–06
circular mil	meter^2 (m^2)	5.067 075 E–10
clo	kelvin meter^2 per watt (K · m^2/W)	2.003 712 E–01
cup	meter^3 (m^3)	2.365 882 E–04
curie	becquerel (Bq)	3.700 000*E+10
	D	
day (mean solar)	second (s)	8.640 000 E+04
day (sidereal)	second (s)	8.616 409 E+04
degree (angle)	radian (rad)	1.745 329 E–02
degree Celsius	Kelvin (K)	

TABLE A.1b Continued

Alphabetical List of Units (Symbols of SI units given in parentheses)

To Convert from	To	Multiply by
degree centigrade	[see footnote[c]]	$t_K = t_{°C}+273.15$
degree Fahrenheit	degree Celsius	$t_{°C} = (t_{°F}-32)/1.8$
degree Fahrenheit	kelvin (K)	$t_K = (t_{°F}+459.67)/1.8$
degree Rankine	kelvin (K)	$t_K = t_{°R}/1.8$
°F·h·ft^2/Btu (International Table) (*R*, thermal resistance)	kelvin meter^2 per watt (K·m^2/W)	1.761 102 E–01
°F·h·ft^2/Btu (thermochemical (*R*, thermal resistance)	kelvin meter^2 per watt (K·m^2/W)	1.762 280 E–01
denier	kilogram per meter (kg/m)	1.111 111 E–07
dyne	newton (N)	1.000 000*E–05
dyne/cm	newton meter (N·m)	1.000 000*E–07
dyne/cm^2	pascal (Pa)	1.000 000*E–01
	E	
electronvolt	joule (J)	1.602 19 E–19
EMU of capacitance	farad (F)	1.000 000*E+09
EMU of current	ampere (A)	1.000 000*E+01
EMU of electric potential	volt (V)	1.000 000*E–08
EMU of inductance	henry (H)	1.000 000*E–09
EMU of resistance	ohm (Ω)	1.000 000*E–09
ESU of capacitance	farad (F)	1.112 650 E–12
ESU of current	ampere (A)	3.335 6 E–10
ESU of electric potential	volt (V)	2.997 9 E+02
ESU of inductance	henry (H)	8.987 554 E+11
ESU of resistance	ohm (Ω)	8.987 554 E+11
erg	joule (J)	1.000 000*E–07
erg/(cm^2·s)	watt per meter^2 (W/m^2)	1.000 000*E–03
erg/s	watt (W)	1.000 000*E–07
	F	
faraday (based on carbon-12)	coulomb (C)	9.648 70 E+04
faraday (chemical)	coulomb (C)	9.649 57 E+04
faraday (physical)	coulomb (C)	9.652 19 E+04
fathom	meter (m)	1.828 8 E+00
fermi (femtometer)	meter (m)	1.000 000*E–15
fluid ounce (U.S.)	meter^3 (m^3)	2.957 353 E–05
foot	meter (m)	3.048 000*E–01
foot (U.S. survey)[a]	meter (m)	3.048 006 E–01
foot of water (39.2 °F)	pascal (Pa)	2.988 98 E+03
ft^2	meter^2 (m^2)	9.290 304*E–02
ft^2/h (thermal diffusivity)	meter^2 per second (m^2/s)	2.580 640*E–05
ft^2/s	meter^2 per second (m^2/s)	9.290 304*E–02
ft^3 (volume; section modulus)	meter^3 (m^3)	2.831 685 E–02
ft^3/min	meter^3 per second (m^3/s)	4.719 474 E–04
ft^3/s	meter^3 per second (m^3/s)	2.831 685 E–02
ft^4 (moment of section)[d]	meter^4 (m^4)	8.630 975 E–03
ft/h	meter per second (m/s)	8.466 667 E–05
ft/min	meter per second (m/s)	5.080 000*E–03
ft/s	meter per second (m/s)	3.048 000*E–01
ft/s^2	meter per second^2 (m/s^2)	3.048 000*E–01

[c]The SI unit of thermodynamic temperature is the kelvin (K), and this unit is properly used for expressing thermodynamic temperature and temperature intervals. Wide use is also made of the degree Celsius (°C), which is the SI unit for expressing Celsius temperature and temperature intervals. The Celsius scale (formerly called centigrade) is related directly to thermodynamic temperature (kelvins) as follows:

1. The temperature interval one degree Celsius equals one kelvin exactly.
2. Celsius temperature (t) is related to thermodynamic temperature (T) by the equation $t = T - T_0$, where $T_0 = 273.15$ K by definition.

[d]This is sometimes called the moment of inertia of a plane section about a specified axis.

TABLE A.1b Continued

Alphabetical List of Units (Symbols of SI units given in parentheses)

To Convert from	To	Multiply by
footcandle	lux (lx)	1.076 391 E+01
footlambert	candela per meter^2 (cd/m^2)	3.426 259 E+00
ft·lbf	joule (J)	1.355 818 E+00
ft·lbf/h	watt (W)	3.766 161 E–04
ft·lbf/min	watt (W)	2.259 697 E–02
ft·lbf/s	watt (W)	1.355 818 E+00
ft·poundal	joule (J)	4.214 011 E–02
free fall, standard (g)	meter per second^2 (m/s^2)	9.806 650*E+00
	G	
gal	meter per second^2 (m/s^2)	1.000 000*E–02
gallon (Canadian liquid)	meter^3 (m^3)	4.546 090 E–03
gallon (U.K. liquid)	meter^3 (m^3)	4.546 092 E–03
gallon (U.S. dry)	meter^3 (m^3)	4.404 884 E–03
gallon (U.S. liquid)	meter^3 (m^3)	3.785 412 E–03
gallon (U.S. liquid) per day	meter^3 per second (m^3/s)	4.381 264 E–08
gallon (U.S. liquid) per minute	meter^3 per second (m^3/s)	6.309 020 E–05
gallon (U.S. liquid) per hp·h (SFC, specific fuel consumption)	meter^3 per joule (m^3/J)	1.410 089 E–09
gamma	tesla (T)	1.000 000*E–09
gauss	tesla (T)	1.000 000*E–04
gilbert	ampere (A)	7.957 747 E–01
gill (U.K.)	meter^3 (m^3)	1.420 654 E–04
gill (U.S.)	meter^3 (m^3)	1.182 941 E–04
grad	degree (angular)	9.000 000*E–01
grad	radian (rad)	1.570 796 E–02
grain (1/7000 lb avoirdupois)	kilogram (kg)	6.479 891*E–05
grain (lb avoirdupois/7000)/gal (U.S. liquid)	kilogram per meter^3 (kg/m^3)	1.711 806 E–02
gram	kilogram (kg)	1.000 000*E–03
g/cm^3	kilogram per meter^3 (kg/m^3)	1.000 000*E+03
gram-force/cm^2	pascal (Pa)	9.806 650*E+01

To Convert from	To	Multiply by
	H	
hectare	meter^2 (m^2)	1.000 000*E+04
horsepower (550 ft·lbf/s)	watt (W)	7.456 999 E+02
horsepower (boiler)	watt (W)	9.809 50 E+03
horsepower (electric)	watt (W)	7.460 000*E+02
horsepower (metric)	watt (W)	7.354 99 E+02
horsepower (water)	watt (W)	7.460 43 E+02
horsepower (U.K.)	watt (W)	7.457 0 E+02
hour (mean solar)	second (s)	3.600 000 E+03
hour (sidereal)	second (s)	3.590 170 E+03
hundredweight (long)	kilogram (kg)	5.080 235 E+01
hundredweight (short)	kilogram (kg)	4.535 924 E+01
	I	
inch	meter (m)	2.540 000*E–02
inch of mercury (32°F)	pascal (Pa)	3.386 38 E+03
inch of mercury (60°F)	pascal (Pa)	3.376 85 E+03
inch of water (39.2°F)	pascal (Pa)	2.490 82 E+03
inch of water (60°F)	pascal (Pa)	2.488 4 E+02
in.^{2}	meter^2 (m^2)	6.451 600*E–04
in.^{3} (volume; section modulus)[e]	meter^3 (m^3)	1.638 706 E–05
in.^{3}/min	meter^3 per second (m^3/s)	2.731 177 E–07
in.^{4} (moment of secion)[4]	meter^4 (m^4)	4.162 314 E–07
in./s	meter per second (m/s)	2.540 000*E–02
in./s^2	meter per second^2 (m/s^2)	2.540 000*E–02
	K	
kayser	1 per meter (1/m)	1.000 000*E+02
kelvin	degree Celsius	$t_{°C} = t_K - 273.15$
kilocalorie (International Table)	joule (J)	4.186 800*E+03

[e]The exact conversion factor is 1.638 706 4*E–05.

TABLE A.1b Continued

Alphabetical List of Units (Symbols of SI units given in parentheses)

To Convert from	To	Multiply by
kilocalorie (mean)	joule (J)	4.190 02 E+03
kilocalorie (thermochemical)	joule (J)	4.184 000*E+03
kilocalorie (thermochemical)/min	watt (W)	6.973 333 E+01
kilocalorie (thermochemical)/s	watt (W)	4.184 000*E+03
kilogram-force (kgf)	newton (N)	9.806 650*E+00
kgf·m	newton meter (N·m)	9.806 650*E+00
kgf·s^2/m (mass)	kilogram (kg)	9.806 650*E+00
kgf/cm^2	pascal (Pa)	9.806 650*E+04
kgf/m^2	pascal (Pa)	9.806 650*E+00
kgf/mm^2	pascal (Pa)	9.806 650*E+06
km/h	meter per second (m/s)	2.777 778 E–01
kilopond	newton (N)	9.806 650*E+00
kW·h	joule (J)	3.600 000*E+06
kip (1000 lbf)	newton (N)	4.448 222 E+03
kip/in^2 (ksi)	pascal (Pa)	6.894 757 E+06
knot (international)	meter per second (m/s)	5.144 444 E–01
	L	
lambert	candela per meter^2 (cd/m^2)	$1/\pi$ *E+04
lambert	candela per meter^2 (cd/m^2)	3.183 099 E+03
langley	joule per meter^2 (J/m^2)	4.184 000*E+04
league	meter (m)	[see footnote a]
light year	meter (m)	9.460 55 E+15
liter[f]	meter^3 (m^3)	1.000 000*E–03
	M	
maxwell	weber (Wb)	1.000 000*E–08
mho	siemens (S)	1.000 000*E+00
microinch	meter (m)	2.540 000*E–08

To Convert from	To	Multiply by
micron	meter (m)	1.000 000*E–06
mil	meter (m)	2.540 000*E–05
mile (international)	meter (m)	1.609 344*E+03
mile (statute)	meter (m)	1.609 3 E+03
mile (U.S. survey)[a]	meter (m)	1.609 347 E+03
mile (international nautical)	meter (m)	1.852 000*E+03
mile (U.K. nautical)	meter (m)	1.853 184*E+03
mile (U.S. nautical)	meter (m)	1.852 000*E+03
mi^2 (international)	meter^2 (m^2)	2.589 988 E+06
mi^2 (U.S. survey)[a]	meter^2 (m^2)	2.589 998 E+06
mi/h (international)	meter per second (m/s)	4.470 400*E–01
mi/h (international)	kilometer per hour (km/h)	1.609 344*E+01
mi/min (international)	meter per second (m/s)	2.682 240*E+01
mi/s (international)	meter per second (m/s)	1.609 344*E+03
millibar	pascal (Pa)	1.000 000*E+02
millimeter of mercury (0°C)	pascal (Pa)	1.333 22 E+02
minute (angle)	radian (rad)	2.908 882 E–04
minute (mean solar)	second (s)	6.000 000 E+01
minute (sidereal)	second (s)	5.983 617 E+01
month (mean calendar)	second (s)	2.628 000 E+06
	O	
oersted	ampere per meter (A/m)	7.957 747 E+01
ohm centimeter	ohm meter (Ω·m)	1.000 000*E–02
ohm circular-mill per foot	ohm millimeter^2 per meter (Ω·mm^2/m)	1.662 426 E–03
ounce (avoirdupois)	kilogram (kg)	2.834 952 E–02
ounce (troy or apothecary)	kilogram (kg)	3.110 348 E–02
ounce (U.K. fluid)	meter^3 (m^3)	2.841 307 E–05
ounce (U.S. fluid)	meter^3 (m^3)	2.957 353 E–05

[f]In 1964 the General Conference on Weights and Measures adopted the name liter as a special name for decimeter. Prior to this decision the liter differed slightly (previous value, 1.000028 dm^3) and in expression of precision volume measurement this fact must be kept in mind.

TABLE A.1b Continued

Alphabetical List of Units (Symbols of SI units given in parentheses)

To Convert from	To	Multiply by
ounce-force	newton (N)	2.780 139 E–01
ozf·in.	newton meter (N·m)	7.061 552 E–03
oz (avoirdupois)/gal (U.K. liquid)	kilogram per meter^3 (kg/m^3)	6.236 021 E+00
oz (avoirdupois)/gal (U.S. liquid)	kilogram per meter^3 (kg/m^3)	7.489 152 E+00
oz (avoirdupois)/in^3	kilogram per meter^3 (kg/m^3)	1.729 994 E+03
oz (avoirdupois)/ft^2	kilogram per meter^3 (kg/m^3)	3.051 517 E–01
oz (avoirdupois)/yd^2	kilogram per meter^2 (kg/m^2)	3.390 575 E–02
	P	
parsec	meter (m)	3.085 678 E+16
peck (U.S.)	meter^3 (m^3)	8.809 768 E–03
pennyweight	kilogram (kg)	1.555 174 E–03
perm (0°C)	kilogram per pascal second meter^2 (kg/Pa·s·m^2)	5.721 35 E–11
perm (23°C)	kilogram per pascal second meter^2 (kg/Pa·s·m^2)	5.745 25 E–11
	kilogram per pascal second meter (kg/Pa·s·m)	1.453 22 E–12
perm·in. (0°C)	kilogram per pascal second meter (kg/Pa·s·m)	1.453 22 E–12
perm·in. (23°C)	kilogram per pascal second meter (kg/Pa·s·m)	1.459 29 E–12

To Convert from	To	Multiply by
phot	lumen per meter^2 (lm/m^2)	1.000 000*E+04
pica (printer's)	meter (m)	4.217 518 E–03
pint (U.S. dry)	meter^3 (m^3)	5.506 105 E–04
pint (U.S. liquid)	meter^3 (m^3)	4.731 765 E–04
point (printer's)	meter (m)	3.514 598*E–04
poise (absolute viscosity)	pascal second (Pa·s)	1.000 000*E–01
pound (lb avoidrupois)[g]	kilogram (kg)	4.535 924 E–01
pound (troy or apothecary)	kilogram (kg)	3.732 417 E–01
lb·ft^2 (moment of inertia)	kilogram meter^2 (kg·m^2)	4.214 011 E–02
lb·in.^{2} (moment of inertia)	kilogram meter^2 (kg·m^2)	2.926 397 E–04
lb/ft·h	pascal second (Pa·s)	4.133 789 E–04
lb/ft·s	pascal second (Pa·s)	1.488 164 E+00
lb/ft^2	kilogram per meter^2 (kg/m^2)	4.882 428 E+00
lb/ft^3	kilogram per meter^3 (kg/m^3)	1.601 846 E+01
lb/gal (U.K. liquid)	kilogram per meter^3 (kg/m^3)	9.977 633 E+01
lb/gal (U.S. liquid)	kilogram per meter^3 (kg/m^3)	1.198 264 E+02
lb/h	kilogram per second (kg/s)	1.259 979 E–04
lb/hp·h (SFC, specific fuel consumption)	kilogram per joule (kg/J)	1.689 659 E–07
lb/in.^{3}	kilogram per meter^3 (kg/m^3)	2.767 990 E+04
lb/min	kilogram per second (kg/s)	7.559 873 E–03
lb/s	kilogram per second (kg/s)	4.535 924 E–01
lb/yd^3	kilogram per meter^3 (kg/m^3)	5.932 764 E–01
poundal	newton (N)	1.382 550 E–01
poundal/ft^2	pascal (Pa)	1.488 164 E+00
poundal·s/ft^2	pascal second (Pa·s)	1.488 164 E+00

[g] The exact conversion factor is 4.535 923 7*E–01

TABLE A.1b Continued

Alphabetical List of Units (Symbols of SI units given in parentheses)

To Convert from	To	Multiply by
pound-force (inf)[h]	newton (N)	4.448 222 E+00
lbf/ft	newton meter (N·m)	1.355 818 E+00
lbf·s/ft^2	pascal second (Pa·s)	4.788 026 E+01
lbf·s/in^2	pascal second (Pa·s)	6.894 757 E+03
lbf/ft	newton per meter (N/m)	1.459 390 E+01
lbf/ft^2	pascal (Pa)	4.788 026 E+01
lbf/in.	newton per meter (N/m)	1.751 268 E+02
lbf/in.^{2} (psi)	pascal (Pa)	6.894 757 E+03
lbf/lb (thrust/weight [mass] ratio)	newton per kilogram (N/kg)	9.806 650 E+00
	Q	
quart (U.S. dry)	meter^3 (m^3)	1.101 221 E–03
quart (U.S. liquid)	meter^3 (m^3)	9.463 529 E–04
	R	
rad (radiation dose absorbed)	gray (Gy)	1.000 000*E–02
rhe	1 per pascal second (1/Pa·s)	1.000 000*E+01
rod	meter (m)	[see footnote a]
roentgen	coulomb per kilogram (C/kg)	2.58 E–04
	S	
second (angle)	radian (rad)	4.848 137 E–06
second (sidereal)	second (s)	9.972 696 E–01
section	meter^2 (m^2)	[see footnote a]
shake	second (s)	1.000 000*E–08
slug	kilogram (kg)	1.459 390 E+01
slug/ft·s	pascal second (Pa·s)	4.788 026 E+01

To Convert from	To	Multiply by
slug/ft^3	kilogram per meter^3 (kg/m^3)	5.155 788 E+02
statampere	ampere (A)	3.335 640 E–10
statcoulomb	coulomb (C)	3.335 640 E–10
statfarad	farad (F)	1.112 650 E–12
stathenry	henry (H)	8.987 554 E+11
statmho	siemens (S)	1.112 650 E–12
statohm	ohm (Ω)	8.987 665 E+11
statvolt	volt (V)	2.997 925 E+02
stere	meter^3 (m^3)	1.000 000*E–04
stilb	candela per meter^2 (cd/m^2)	1.000 000*E+04
stokes (kinematic viscosity)	meter^2 per second (m^2/s)	1.000 000*E–04
	T	
tablespoon	meter^3 (m^3)	1.478 676 E–05
teaspoon	meter^3 (m^3)	4.928 922 E–06
tex	kilogram per meter (kg/m)	1.000 000*E–06
therm	joule (J)	1.055 056 E+08
ton (assay)	kilogram (kg)	2.916 667 E–02
ton (long, 2240 lb)	kilogram (kg)	1.016 047 E+03
ton (metric)	kilogram (kg)	1.000 000*E+03
ton (nuclear equivalent of TNT)	joule (J)	4.184 E+09[i]
ton (refrigeration)	watt (W)	3.516 800 E+03
ton (register)	meter^3 (m^3)	2.831 685 E+00
ton (short, 2000 lb)	kilogram (kg)	9.071 847 E+02
ton (long)/yd^3	kilogram per meter^3 (kg/m^3)	1.328 939 E+03
ton (short)/yd^3	kilogram per meter^3 (kg/m^3)	1.186 553 E+03

[h] The exact conversion factor is 4.448 221 615 260 5*E+00.

[i] Defined (not measured) value.

Table A.1b Continued

Alphabetical List of Units (Symbols of SI units given in parentheses)

To Convert from	To	Multiply by
ton (short)/h	kilogram per second (kg/s)	2.519 958 E–01
ton-force (2000 lbf)	newton (N)	8.896 444 E+03
tonne	kilogram (kg)	1.000 000*E+03
torr (mm Hg, 0°C)	pascal (Pa)	1.333 22 E+02
township	meter^2 (m^2)	[see footnote a]
	U	
unit pole	weber (Wb)	1.256 637 E–07
	W	
W·h	joule (J)	3.600 000*E+03
W·s	joule (J)	1.000 000*E+00
W/cm^2	watt per meter^2 (W/m^2)	1.000 000*E+04
W/in^2	watt per meter^2 (W/m^2)	1.550 003 E+03
	Y	
yard	meter (m)	9.144 000*E–01
yd^2	meter^2 (m^2)	8.361 274 E–01
yd^3	meter^3 (m^3)	7.645 549 E–01
yd^3/min	meter^3 per second (m^3/s)	1.274 258 E–02
year (365 days)	second (s)	3.153 600 E+07
year (sidereal)	second (s)	3.155 815 E+07
year (tropical)	second (s)	3.155 693 E+07

A.2 Engineering Conversion Factors

TABLE A.2a
Conversion Factors

To Convert	Into	Multiply by
	A	
abcoulomb	statcoulombs	2.998×10^{10}
acre	sq. chain (Gunters)	10
acre	rods	160
acre	sq. links (Gunters)	1×10^{5}
acre	hectare or sq. hectometer	.4047
acres	sq. ft	43,560.0
acres	sq. meters	4,047.
acres	sq. miles	1.562×10^{-3}
acres	sq. yards	4,840.
acre-feet	cu. ft	43,560.0
acre-feet	gallons	3.259×10^{5}
amperes/sq. cm	amps/sq. in.	6.452
amperes/sq. cm	amps/sq. meter	10^{4}
amperes/sq. in.	amps/sq. cm	0.1550
amperes/sq. in.	amps/sq. meter	1,550.0
amperes/sq. meter	amps/sq. cm	10^{-4}
amperes/sq. meter	amps/sq. in.	6.452×10^{-4}
ampere-hours	coulombs	3,600.0
ampere-hours	faradays	0.03731
ampere-turns	gilberts	1.257
ampere-turns/cm	amp-turns/in.	2.540
ampere-turns/cm	amp-turns/meter	100.0
ampere-turns/cm	gilberts/cm	1.257
ampere-turns/in.	amp-turns/cm	0.3937
ampere-turns/in.	amp-turns/meter	39.37
ampere-turns/in.	gilberts/cm	0.4950
ampere-turns/meter	amp/turns/cm	0.01
ampere-turns/meter	amp-turns/in.	0.0254
ampere-turns/meter	gilberts/cm	0.01257
Angstrom unit	in.	$3{,}937 \times 10^{-9}$
Angstrom unit	meter	1×10^{-10}
Angstrom unit	micron or μm	1×10^{-4}
are	acre (U.S.)	.02471
ares	sq. yards	119.60
ares	acres	0.02471
ares	sq. meters	100.0
Astronomical Unit	kilometers	1.495×10^{8}
atmospheres	ton/sq. in.	.007348
atmospheres	cms of mercury	76.0
atmospheres	ft of water (at 4°C)	33.90
atmospheres	in. of mercury (at 0°C)	29.92
atmospheres	kgs/sq. cm	1.0333
atmospheres	kgs/sq. meter	10,332.
atmospheres	pounds/sq. in.	14.70
atmospheres	tons/sq. ft	1.058
	B	
barrels (U.S., dry)	cu. in.	7056.
barrels (U.S., dry)	quarts (dry)	105.0
barrels (U.S., liquid)	gallons	31.5
barrels (oil)	gallons (oil)	42.0
bars	atmospheres	0.9869
bars	dynes/sq. cm	10^{6}
bars	kgs/sq. meter	1.020×10^{4}
bars	pounds/sq. ft	2,089.
bars	pounds/sq. in.	14.50
baryl	dyne/sq. cm	1.000
bolt (U.S. cloth)	meters	36.576
Btu	liter-atmosphere	10.409
Btu	ergs	1.0550×10^{10}
Btu	foot-lbs	778.3
Btu	gram-calories	252.0
Btu	horsepower-hrs	3.931×10^{-4}
Btu	joules	1,054.8
Btu	kilogram-calories	0.2520
Btu	kilogram-meters	107.5
Btu	kilowatt-hrs	2.928×10^{-4}
Btu/hr	foot-pounds/s	0.2162

TABLE A.2a Continued
Conversion Factors

To Convert	Into	Multiply by
Btu/hr	gram-cal/s	0.0700
Btu/hr	horsepower-hrs	3.929×10^{-4}
Btu/hr	watts	0.2931
Btu/min	foot-1bs/s	12.96
Btu/min	horsepower	0.02356
Btu/min	kilowatts	0.01757
Btu/min	watts	17.57
Btu/sq. ft/min	watts/sq. in.	0.1221
bucket (U.K. dry)	cu. cm	1.818×10^{4}
bushels	cu. ft	1.2445
bushels	cu. in.	2,150.4
bushels	cu. meters	0.03524
bushels	liters	35.24
bushels	pecks	4.0
bushels	pints (dry)	64.0
bushels	quarts (dry)	32.0
	C	
calories, gram (mean)	Btu (mean)	3.9685×10^{-3}
candle/sq. cm	lamberts	3.142
candle/sq. in.	lamberts	.4870
centares (centiares)	sq. meters	1.0
Centigrade	Fahrenheit	$(C° \times 9/5) + 32$
centigrams	grams	0.01
centiliter	ounce fluid (U.S.)	.3382
centiliter	cu. in.	.6103
centiliter	drams	2.705
centiliters	liters	0.01
centimeters	feet	3.281×10^{-2}
centimeters	inches	0.3937
centimeters	kilometers	10^{-5}
centimeters	meters	0.01
centimeters	miles	6.214×10^{-6}
centimeters	millimeters	10.0
centimeters	mils	393.7
centimeters	yards	1.094×10^{-2}
centimeter-dynes	cm-grams	1.020×10^{-3}
centimeter-dynes	meter-kgs	1.020×10^{-8}
centimeter-dynes	pound-feet	7.376×10^{-8}
centimeter-grams	cm-dynes	980.7
centimeter-grams	meter-kgs	10^{-5}
centimeter-grams	pound-feet	7.233×10^{-5}
centimeters of mercury	atmospheres	0.01316
centimeters of mercury	feet of water	0.4461
centimeters of mercury	kgs/sq. meter	136.0
centimeters of mercury	pounds/sq. ft	27.85
centimeters of mercury	pounds/sq. in.	0.1934
centimeters/s	feet/min	1.1969
centimeters/s	feet/sec	0.03281
centimeters/s	kilometers/hr	0.036
centimeters/s	knots	0.1943
centimeters/s	meters/min	0.6
centimeters/s	miles/hr	0.02237
centimeters/s	miles/min	3.728×10^{-4}
centimeters/s/s	feet/s/s	0.03281
centimeters/s/s	kms/hr/s	0.036
centimeters/s/s	meters/s/s	0.01
centimeters/s/s	miles/hr/s	0.02237
chain	inches	792.00
chain	meters	20.12
chains (surveyors' or Gunter's)	yards	22.00
circular mils	sq. cms	5.067×10^{-6}
circular mils	sq. mils	0.7854
circumference	radians	6.283
circular mils	sq. inches	7.854×10^{-7}
cords	cord feet	8
cord feet	cu. ft	16
coulomb	statcoulombs	2.998×10^{9}
coulombs	faradays	1.036×10^{-5}
coulombs/sq. cm	coulombs/sq. in.	64.52
coulombs/sq. cm	coulombs/sq. meter	10^{4}
coulombs/sq. in.	coulombs/sq. cm	0.1550
coulombs/sq. in.	coulombs/sq. meter	1,550.
coulombs/sq. meter	coulombs/sq. cm	10^{-4}
coulombs/sq. meter	coulombs/sq. in.	6.452×10^{-4}
cubic centimeters	cu. ft	3.531×10^{-5}
cubic centimeters	cu. in.	0.06102
cubic centimeters	cu. meters	10^{-6}
cubic centimeters	cu. yards	1.308×10^{-6}
cubic centimeters	gallons (U.S. liq.)	2.642×10^{-4}
cubic centimeters	liters	0.001
cubic centimeters	pints (U.S. liq.)	2.113×10^{-3}
cubic centimeters	quarts (U.S. liq.)	1.057×10^{-3}
cubic feet	bushels (dry)	0.8036
cubic feet	cu. cms	28,320.0
cubic feet	cu. in.	1,728.0

TABLE A.2a Continued
Conversion Factors

To Convert	*Into*	*Multiply by*
cubic feet	cu. meters	0.02832
cubic feet	cu. yards	0.03704
cubic feet	gallons (U.S. liq.)	7.48052
cubic feet	liters	28.32
cubic feet	pints (U.S. liq.)	59.84
cubic feet	quarts (U.S. liq.)	29.92
cubic feet/min	cu. cms/s	472.0
cubic feet/min	gallons/s	0.1247
cubic feet/min	liters/s	0.4720
cubic feet/min	pounds of water/min	62.43
cubic feet/sec	million gals/day	0.646317
cubic feet/sec	gallons/min	448.831
cubic inches	cu. cms	16.39
cubic inches	cu. feet	5.787×10^{-4}
cubic inches	cu. meters	1.639×10^{-5}
cubic inches	cu. yards	2.143×10^{-5}
cubic inches	gallons	4.329×10^{-3}
cubic inches	liters	0.01639
cubic inches	mil-feet	1.061×10^{5}
cubic inches	pints (U.S. liq.)	0.03463
cubic inches	quarts (U.S. liq.)	0.01732
cubic meters	bushels (dry)	28.38
cubic meters	cu. cms	10^{6}
cubic meters	cu. ft	35.31
cubic meters	cu. in.	61,023.0
cubic meters	cu. yards	1.308
cubic meters	gallons (U.S. liq.)	264.2
cubic meters	liters	1,000.0
cubic meters	pints (U.S. liq.)	2,113.0
cubic meters	quarts (U.S. liq.)	1,057.
cubic yards	cu. cms	7.646×10^{5}
cubic yards	cu. ft	27.0
cubic yards	cu. in.	46,656.0
cubic yards	cu. meters	0.7646
cubic yards	gallons (U.S. liq.)	202.0
cubic yards	liters	764.6
cubic yards	pints (U.S. liq.)	1,615.9
cubic yards	quarts (U.S. liq.)	807.9
cubic yards/min	cu. ft/s	0.45
cubic yards/min	gallons/s	3.367
cubic yards/min	liters/s	12.74
	D	
dalton	gram	1.650×10^{-24}
days	seconds	86,400.0

To Convert	*Into*	*Multiply by*
decigrams	grams	0.1
deciliters	liters	0.1
decimeters	meters	0.1
degrees (angle)	quadrants	0.01111
degrees (angle)	radians	0.01745
degrees (angle)	seconds	3,600.0
degrees/s	radians/sec	0.01745
degrees/s	revolutions/min	0.1667
degrees/s	revolutions/sec	2.778×10^{-3}
dekagrams	grams	10.0
dekaliters	liters	10.0
dekameters	meters	10.0
drams (apothecaries' or troy)	ounces (avoidupois)	0.1371429
drams (apothecaries' or troy)	ounces (troy)	0.125
drams (U.S., fluid or apothecaries)	cu. cm	3.697
drams	grams	1.7718
drams	grains	27.3437
drams	ounces	0.0625
dyne/cm	erg/sq. millimeter	.01
dyne/sq. cm	atmospheres	9.869×10^{-7}
dyne/sq. cm	inch of mercury at 0°C	2.953×10^{-5}
dyne/sq. cm	inch of water at 4°C	4.015×10^{-4}
dynes	grams	1.020×10^{-3}
dynes	joules/cm	10^{-7}
dynes	joules/meter (newtons)	10^{-5}
dynes	kilograms	1.020×10^{-6}
dynes	poundals	7.233×10^{-5}
dynes	pounds	2.248×10^{-6}
dynes/sq. cm	bars	10^{-6}
	E	
ell	cm	114.30
ell	in.	45
em, pica	in.	.167
em, pica	cm	.4233
erg/s	dyne-cm/s	1.000
ergs	Btu	9.480×10^{-11}
ergs	dyne-centimeters	1.0
ergs	foot-pounds	7.367×10^{-8}
ergs	gram-calories	0.2389×10^{-7}
ergs	gram-cms	1.020×10^{-3}
ergs	horsepower-hrs	3.7250×10^{-14}

TABLE A.2a Continued
Conversion Factors

To Convert	Into	Multiply by
ergs	joules	10^{-7}
ergs	kg-calories	2.389×10^{-11}
ergs	kg-meters	1.020×10^{-8}
ergs	kilowatt-hrs	0.2778×10^{-13}
ergs	watt-hours	0.2778×10^{-10}
ergs/s	Btu/min	$5{,}688 \times 10^{-9}$
ergs/s	ft-lbs/min	4.427×10^{-6}
ergs/s	ft-lbs/s	7.3756×10^{-8}
ergs/s	horsepower	1.341×10^{-10}
ergs/s	kg-calories/min	1.433×10^{-9}
ergs/s	kilowatts	10^{-10}
	F	
farads	microfarads	10^{6}
faraday/s	ampere (absolute)	9.6500×10^{4}
faradays	ampere-hours	26.80
faradays	coulombs	9.649×10^{4}
fathom	meter	1.828804
fathoms	feet	6.0
feet	centimeters	30.48
feet	kilometers	3.048×10^{-4}
feet	meters	0.3048
feet	miles (naut.)	1.645×10^{-4}
feet	miles (stat.)	1.894×10^{-4}
feet	millimeters	304.8
feet	mils	1.2×10^{4}
feet of water	atmospheres	0.02950
feet of water	in. of mercury	0.8826
feet of water	kgs/sq. cm	0.03048
feet of water	kgs/sq. meter	304.8
feet of water	pounds/sq. ft	62.43
feet of water	pounds/sq. in.	0.4335
feet/min	cms/s	0.5080
feet/min	ft/s	0.01667
feet/min	kms/hr	0.01829
feet/min	meters/min	0.3048
feet/min	miles/hr	0.01136
feet/s	cms/s	30.48
feet/s	kms/hr	1.097
feet/s	knots	0.5921
feet/s	meters/min	18.29
feet/s	miles/hr	0.6818
feet/s	miles/min	0.01136
feet/s/s	cms/s/s	30.48
feet/sec/s	kms/hr/s	1.097
feet/sec/s	meters/s/s	0.3048
feet/sec/s	miles/hr/s	0.6818
feet/100 feet	percent grade	1.0
foot-candle	lumen/sq. meter	10.764
foot-pounds	Btu	1.286×10^{-3}
foot-pounds	ergs	1.356×10^{7}
foot-pounds	gram-calories	0.3238
foot-pounds	hp-hrs	5.050×10^{-7}
foot-pounds	joules	1.356
foot-pounds	kg-calories	3.24×10^{-4}
foot-pounds	kg-meters	0.1383
foot-pounds	kilowatt-hrs	3.766×10^{-7}
foot-pounds/min	Btu/min	1.286×10^{-3}
foot-pounds/min	foot-pounds/s	0.01667
foot-pounds/min	horsepower	3.030×10^{-5}
foot-pounds/min	kg-calories/min	3.24×10^{-4}
foot-pounds/min	kilowatt	2.260×10^{-5}
foot-pounds/s	Btu/hr	4.6263
foot/pounds/s	Btu/min	0.07717
foot-pounds/s	horsepower	1.818×10^{-3}
foot-pounds/s	kg-calories/min	0.01945
foot-pounds/s	kilowatts	1.356×10^{-3}
furlongs	miles (U.S.)	0.125
furlongs	rods	40.0
furlongs	feet	660.0
	G	
gallons	cu. cms	3,785.0
gallons	cu. ft	0.1337
gallons	cu. in.	231.0
gallons	cu. meters	3.785×10^{-3}
gallons	cu. yards	4.951×10^{-3}
gallons	liters	3.785
gallons (liq. Br. Imp.)	gallons (U.S. liq.)	1.20095
gallons (U.S.)	gallons (Imp.)	0.83267
gallons of water	pounds of water	8.3453
gallons/min	cu. ft/s	2.228×10^{-3}
gallons/min	liters/s	0.06308
gallons/min	cu. ft/hr	8.0208
gausses	lines/sq. in.	6.452
gausses	webers/sq. cm.	10^{-8}
gausses	webers/sq. in.	6.452×10^{-8}
gausses	webers/sq. meter	10^{-4}
gilberts	ampere-turns	0.7958
gilberts/cm	amp-turns/cm	0.7958

TABLE A.2a Continued
Conversion Factors

To Convert	Into	Multiply by
gilberts/cm	amp-turns/in.	2.021
gilberts/cm	amp-turns/meter	79.58
gills (British)	cu. cm	142.07
gills	liters	0.1183
gills	pints (liq.)	0.25
grade	radian	.01571
grains	drams (avoirdupois)	0.03657143
grains (troy)	grains (avdp)	1.0
grains (troy)	grams	0.06480
grains (troy)	ounces (avdp)	2.0833×10^{-3}
grains (troy)	pennyweight (troy)	0.04167
grains/U.S. gal	parts/million	17.118
grains/U.S. gal	pounds/million gal	142.86
grains/Imp. gal	parts/million	14.286
grams	dynes	980.7
grams	grains	15.43
grams	joules/cm	9.807×10^{-5}
grams	joules/meter (newtons)	9.807×10^{-3}
grams	kilograms	0.001
grams	milligrams	1,000.
grams	ounces (advp)	0.03527
grams	ounces (troy)	0.03215
grams	poundals	0.07093
grams	pounds	2.205×10^{-3}
grams/cm	pounds/inch	5.600×10^{-3}
grams/cu. cm	pounds/ cu. ft	62.43
grams/cu. cm	pounds/cu. in.	0.03613
grams/cu. cm	pounds/mil-foot	3.405×10^{-7}
grams/liter	grains/gal	58.417
grams/liter	pounds/1,000 gal	8.345
grams/liter	pounds/cu. ft	0.062427
grams/liter	parts/million	1,000.0
grams/sq. cm	pounds/sq. ft	2.0481
gram-calories	Btu	3.9683×10^{-3}
gram-calories	ergs	4.1868×10^{7}
gram-calories	foot-pounds	3.0880
gram-calories	horsepower-hrs	1.5596×10^{-6}
gram-calories	kilowatt-hrs	1.1630×10^{-6}
gram-calories	watt-hrs	1.1630×10^{-3}
gram-calories/s	Btu/hr	14.286
gram-centimeters	Btu	9.297×10^{-8}
gram-centimeters	ergs	980.7
gram-centimeters	joules	9.807×10^{-5}
gram-centimeters	kg-cal	2.343×10^{-8}
gram-centimeters	kg-meters	10^{-5}
	H	
hand	cm	10.16
hectares	acres	2.471
hectares	sq. ft	1.076×10^{5}
hectograms	grams	100.0
hectoliters	liters	100.0
hectometers	meters	100.0
hectowatts	watts	100.0
henries	millihenries	1,000.0
hogsheads (U.K.)	cu. ft	10.114
hogsheads (U.S.)	cu. ft	8.42184
hogsheads (U.S.)	gallons (U.S.)	63
horsepower	Btu/min	42.44
horsepower	ft-lbs/min	33,000.
horsepower	ft-lbs/sec	550.0
horsepower (metric) (542.5 ft lb/s)	horsepower (550 ft-lb/s)	0.9863
horsepower (550 ft lb/s)	horsepower (metric) (542.5 ft-lb/s)	1.014
horsepower	kg-calories/min	10.68
horsepower	kilowatts	0.7457
horsepower	watts	745.7
horsepower (boiler)	Btu/hr	33.479
horsepower (boiler)	kilowatts	9.803
horsepower-hrs	Btu	2,547.
horsepower-hrs	ergs	2.6845×10^{13}
horsepower-hrs	ft-lbs	1.98×10^{6}
horsepower-hrs	gram-calories	641,190.
horsepower-hrs	joules	2.684×10^{6}
horsepower-hrs	kg-calories	641.1
horsepower-hrs	kg-meters	2.737×10^{5}
horsepower-hrs	kilowatt-hrs	0.7457
hours	days	4.167×10^{-2}
hours	weeks	5.952×10^{-3}
hundredweights (long)	pounds	112
hundredweights (long)	tons (long)	0.05
hundredweights (short)	ounces (avoirdupois)	1,600
hundredweights (short)	pounds	100
hundredweights (short)	tons (metric)	0.0453592
hundredweights (short)	tons (long)	0.0446429

TABLE A.2a Continued
Conversion Factors

To Convert	Into	Multiply by	To Convert	Into	Multiply by
	I		kilograms/cu. meter	pounds/cu. ft	0.06243
inches	centimeters	2.540	kilograms/cu. meter	pounds/cu. in.	3.613×10^{-5}
inches	meters	2.540×10^{-2}	kilograms/cu. meter	pounds/mil-foot	3.405×10^{-10}
inches	miles	1.578×10^{-5}	kilograms/meter	pounds/ft	0.6720
inches	millimeters	25.40	kilograms/sq. cm	dynes	980,665
inches	mils	1,000.0	kilograms/sq. cm	atmospheres	0.9678
inches	yards	2.778×10^{-2}	kilograms/sq. cm	feet of water	32.81
inches of mercury	atmospheres	0.03342	kilograms/sq. cm	inches of mercury	28.96
inches of mercury	feet of water	1.133	kilograms/sq. cm	pounds/sq. ft	2,048.
inches of mercury	kgs/sq. cm	0.03453	kilograms/sq. cm	pounds/sq. in.	14.22
inches of mercury	kgs/sq. meter	345.3	kilograms/sq. meter	atmospheres	9.678×10^{-5}
inches of mercury	pounds/sq. ft	70.73	kilograms/sq. meter	bars	98.07×10^{-6}
inches of mercury	pounds/sq. in.	0.4912	kilograms/sq. meter	feet of water	3.281×10^{-3}
inches of water (at 4°C)	atmospheres	2.458×10^{-3}	kilograms/sq. meter	inches of mercury	2.896×10^{-3}
inches of water (at 4°C)	inches of mercury	0.07355	kilograms/sq. meter	pounds/sq. ft	0.2048
inches of water (at 4°C)	kgs/sq. cm	2.540×10^{-3}	kilograms/sq. meter	pounds/sq. in.	1.422×10^{-3}
inches of water (at 4°C)	ounces/sq. in.	0.5781	kilograms/sq. mm	kgs/sq. meter	10^{6}
inches of water (at 4°C)	pounds/sq. ft	5.204	kilogram-calories	Btu	3.968
inches of water (at 4°C)	pounds/sq. in.	0.03613	kilogram-calories	foot-pounds	3,088.
International ampere	ampere (absolute)	.9998	kilogram-calories	hp-hrs	1.560×10^{-3}
International volt	volts (absolute)	1.0003	kilogram-calories	joules	4,186.
	J		kilogram-calories	kg-meters	426.9
joules	Btu	9.480×10^{-4}	kilogram-calories	kilojoules	4.186
joules	ergs	10^{7}	kilogram-calories	kilowatt-hrs	1.163×10^{-3}
joules	foot-pounds	0.7376	kilogram meters	Btu	9.294×10^{-3}
joules	kg-calories	2.389×10^{-4}	kilogram meters	ergs	9.804×10^{7}
joules	kg-meters	0.1020	kilogram meters	foot-pounds	7.233
joules	watt-hrs	2.778×10^{-4}	kilogram meters	joules	9.804
joules/cm	grams	1.020×10^{4}	kilogram meters	kg-calories	2.342×10^{-3}
joules/cm	dynes	10^{7}	kilogram meters	kilowatt-hrs	2.723×10^{-6}
joules/cm	joules/meter (newtons)	100.0	kilolines	maxwells	1,000.0
joules/cm	poundals	723.3	kiloliters	liters	1,000.0
joules/cm	pounds	22.48	kilometers	centimeters	10^{5}
			kilometers	feet	3,281.
	K		kilometers	inches	3.937×10^{4}
kilograms	dynes	980,665.	kilometers	meters	1,000.0
kilograms	grams	1,000.0	kilometers	miles	0.6214
kilograms	joules/cm	0.09807	kilometers	millimeters	10^{6}
kilograms	joules/meter (newtons)	9.807	kilometers	yards	1,094.
kilograms	poundals	70.93	kilometers/hr	cms/s	27.78
kilograms	pounds	2.205	kilometers/hr	ft/min	54.68
kilograms	tons (long)	9.842×10^{-4}	kilometers/hr	ft/s	0.9113
kilograms	tons (short)	1.102×10^{-3}	kilometers/hr	knots	0.5396
kilograms/cu. meter	grams/cu. cm	0.001	kilometers/hr	meters/min	16.67

TABLE A.2a Continued
Conversion Factors

To Convert	Into	Multiply by
kilometers/hr	miles/hr	0.6214
kilometers/hr/s	cms/sec/s	27.78
kilometers/hr/s	ft/s/s	0.9113
kilometers/hr/s	meters/s/s	0.2778
kilometers/hr/s	miles/hr/s	0.6214
kilowatts	Btu/min	56.92
kilowatts	ft-lbs/min	4.426×10^4
kilowatts	ft-lbs/s	737.6
kilowatts	horsepower	1.341
kilowatts	kg-calories/min	14.34
kilowatts	watts	1,000.0
kilowatt-hrs	Btu	3,413.
kilowatt-hrs	ergs	3.600×10^{13}
kilowatt-hrs	ft-lbs	2.655×10^6
kilowatt-hrs	gram-calories	859,850.
kilowatt-hrs	horsepower-hrs	1,341
kilowatt-hrs	joules	3.6×10^6
kilowatt-hrs	kg-calories	860.5
kilowatt-hrs	kg-meters	3.671×10^5
kilowatt-hrs	pounds of water evaporated from and at 212°F	3.53
kilowatt-hrs	pounds of water raised from 62° to 212°F	22.75
knots	ft/hr	6,080.
knots	kilometers/hr	1.8532
knots	nautical miles/hr	1.0
knots	statute miles/hr	1.151
knots	yards/hr	2,027.
knots	ft/s	1.689
	L	
league	miles (approx.)	3.0
light year	miles	5.9×10^{12}
light year	kilometers	9.46091×10^{12}
lines/sq. cm	gausses	1.0
lines/sq. in.	gausses	0.1550
lines/sq. in.	webers/sq. cm	1.550×10^{-9}
lines/sq. in.	webers/sq. in.	10^{-8}
lines/sq. in.	webers/sq. meter	1.550×10^{-5}
links (engineer's)	inches	12.0
links (surveyor's)	inches	7.92
liters	bushels (U.S. dry)	0.02833
liters	cu. cm	1,000.0
liters	cu. ft	0.03531
liters	cu. in.	61.02
liters	cu. meters	0.001
liters	cu. yards	1.308×10^{-3}
liters	gallons (U.S. liq.)	0.2642
liters	pints (U.S. liq.)	2.113
liters	quarts (U.S. liq.)	1.057
liters/min	cu. ft/s	5.886×10^{-4}
liters/min	gals/s	4.403×10^{-3}
lumens/sq. ft	foot-candles	1.0
lumen	spherical candle power	.07958
lumen	watt	.001496
lumen/sq. ft	lumen/sq. meter	10.76
lux	foot-candles	0.0929
	M	
maxwells	kilolines	0.001
maxwells	webers	10^{-8}
megalines	maxwells	10^6
megohms	microhms	10^{12}
megohms	ohms	10^6
meters	centimeters	100.0
meters	feet	3.281
meters	inches	39.37
meters	kilometers	0.001
meters	miles (naut.)	5.396×10^{-4}
meters	miles (stat.)	6.214×10^{-4}
meters	millimeters	1,000.0
meters	yards	1.094
meters	varas	1.179
meters/min	cms/s	1.667
meters/min	ft/min	3.281
meters/min	ft/s	0.05468
meters/min	kms/hr	0.06
meters/min	knots	0.03238
meters/min	miles/hr	0.03728
meters/s	ft/min	196.8
meters/s	ft/s	3.281
meters/s	kilometers/hr	3.6
meters/s	kilometers/min	0.06
meters/s	miles/hr	2.237
meters/s	miles/min	0.03728
meters/s/s	cms/s/s	100.0
meters/s/s	ft/s/s	3.281

TABLE A.2a Continued
Conversion Factors

To Convert	Into	Multiply by	To Convert	Into	Multiply by
meters/s/s	kms/hr/s	3.6	milliliters	liters	0.001
meters/s/s	miles/hr/s	2.237	millimeters	centimeters	0.1
meter-kilograms	cm-dynes	9.807×10^{7}	millimeters	feet	3.281×10^{-3}
meter-kilograms	cm-grams	10^{5}	millimeters	inches	0.03937
meter-kilograms	pound-feet	7.233	millimeters	kilometers	10^{-6}
microfarad	farads	10^{-6}	millimeters	meters	0.001
micrograms	grams	10^{-6}	millimeters	miles	6.214×10^{-7}
microhms	megohms	10^{-12}	millimeters	mils	39.37
microhms	ohms	10^{-6}	millimeters	yards	1.094×10^{-3}
microliters	liters	10^{-6}	million gals/day	cu. ft/sec	1.54723
microns	meters	1×10^{-6}	mils	centimeters	2.540×10^{-3}
miles (naut.)	miles (statute)	1.1516	mils	feet	8.333×10^{-5}
miles (naut.)	yards	2,027.	mils	inches	0.001
miles (statute)	centimeters	1.609×10^{5}	mils	kilometers	2.540×10^{-8}
miles (statute)	feet	5,280.	mils	yards	2.778×10^{-5}
miles (statute)	inches	6.336×10^{4}	miner's inches	cu. ft/min	1.5
miles (statute)	kilometers	1.609	minims (U.K.)	cu. cm	0.059192
miles (statute)	meters	1,609.	minims (U.S., fluid)	cu. cm	0.061612
miles (statute)	miles (naut.)	0.8684	minutes (angles)	degrees	0.01667
miles (statute)	yards	1,760.	miles (naut.)	feet	6,080.27
miles/hr	cms/sec	44.70	miles (naut.)	kilometers	1.853
miles/hr	ft/min	88.	miles (naut.)	meters	1,853.
miles/hr	ft/sec	1.467	minutes (angles)	quadrants	1.852×10^{-4}
miles/hr	kms/hr	1.609	minutes (angles)	radians	2.909×10^{-4}
miles/hr	kms/min	0.02682	minutes (angles)	seconds	60.0
miles/hr	knots	0.8684	myriagrams	kilograms	10.0
miles/hr	meters/min	26.82	myriameters	kilometers	10.0
miles/hr	miles/min	0.1667	myriawatts	kilowatts	10.0
miles/hr/s	cm/sec/sec	44.70			
miles/hr/s	ft/sec/sec	1.467		*N*	
miles/hr/s	kms/hr/sec	1.609	nepers	decibles	8.686
miles/hr/s	meters/sec/sec	0.4470	Newton	dynes	1 × 105
miles/min	cms/sec	2,682.		*O*	
miles/min	ft/sec	88.	ohm (International)	ohm (absolute)	1.0005
miles/min	kms/min	1.609	ohms	megohms	10^{-6}
miles/min	knots/min	0.8684	ohms	microhms	10^{6}
miles/min	miles/hr	60.0	ounces	drams	16.0
mil-feet	cu. in.	9.425×10^{-6}	ounces	grains	437.5
milliers	kilograms	1,000.	ounces	grams	28.349527
millimicrons	meters	1×10^{-9}	ounces	pounds	0.0625
milligrams	grains	0.01543236	ounces	ounces (troy)	0.9115
milligrams	grams	0.001	ounces	tons (long)	2.790×10^{-5}
milligrams/liter	part/million	1.0	ounces	tons (metric)	2.835×10^{-5}
millihenries	henries	0.001	ounces (fluid)	cu. in.	1.805

TABLE A.2a Continued
Conversion Factors

To Convert	Into	Multiply by	To Convert	Into	Multiply by
ounces (fluid)	liters	0.02957	pounds	dynes	44.4823×10^4
ounces (troy)	grains	480.0	pounds	grains	7,000.
ounces (troy)	grams	31.103481	pounds	grams	453.5924
ounces (troy)	ounces (avdp.)	1.09714	pounds	joules/cm	0.04448
ounces (troy)	pennyweights (troy)	20.0	pounds	joules/meter (newtons)	4.448
ounces (troy)	pounds (troy)	0.08333	pounds	kilograms	0.4536
ounce/sq. in.	dynes/sq. cm	4309	pounds	ounces	16.0
ounces/sq. in.	pounds/sq. in.	0.0625	pounds	ounces (troy)	14.5833
			pounds	poundals	32.17
	P		pounds	pounds (troy)	1.21528
parsec	miles	19×10^{12}	pounds	tons (short)	0.0005
parsec	kilometers	3.084×10^{13}	pounds (troy)	grains	5,760.
parts/million	grains/U.S. gal	0.0584	pounds (troy)	grams	373.24177
parts/million	grains/Imp. gal	0.07016	pounds (troy)	ounces (avdp.)	13.1657
parts/million	pounds/million gal	8.345	pounds (troy)	ounces (troy)	12.0
pecks (U.K.)	cu. in.	554.6	pounds (troy)	pennyweights (troy)	240.0
pecks (U.K.)	liters	9.091901	pounds (troy)	pounds (avdp.)	0.822857
pecks (U.S.)	bushels	0.25	pounds (troy)	tons (long)	3.6735×10^{-4}
pecks (U.S.)	cu. in.	537.605	pounds (troy)	tons (metric)	3.7324×10^{-4}
pecks (U.S.)	liters	8.809582	pounds (troy)	tons (short)	4.1143×10^{-4}
pecks (U.S.)	quarts (dry)	8	pounds of water	cu. ft	0.01602
pennyweights (troy)	grains	24.0	pounds of water	cu. in.	27.68
pennyweights (troy)	ounces (troy)	0.05	pounds of water	gallons	0.1198
pennyweights (troy)	grams	1.55517	pounds of water/min	cu. ft/s	2.670×10^{-4}
pennyweights (troy)	pounds (troy)	4.1667×10^{-3}	pound-feet	cm-dynes	1.356×10^7
pints (dry)	cu. in.	33.60	pound-feet	cm-grams	13,825.
pints (liq.)	cu. cms	473.2	pound-feet	meter-kgs	0.1383
pints (liq.)	cu. ft	0.01671	pounds/cu. ft	grams/cu. cm	0.01602
pints (liq.)	cu. in.	28.87	pounds/cu. ft	kgs/cu. meter	16.02
pints (liq.)	cu. meters	4.732×10^{-4}	pounds/cu. ft	pounds/cu. in.	5.787×10^{-4}
pints (liq.)	cu. yards	6.189×10^{-4}	pounds/cu. ft	pounds/mil-foot	5.456×10^{-9}
pints (liq.)	gallons	0.125	pounds/cu. in.	gms/cu. cm	27.68
pints (liq.)	liters	0.4732	pounds/cu. in.	kgs/cu. meter	2.768×10^4
pints (liq.)	quarts (liq.)	0.5	pounds/cu. in.	pounds/cu. ft	1,728.
Planck's quantum	erg/s	6.624×10^{-27}	pounds/cu. in.	pounds/mil-foot	9.425×10^{-6}
poise	gram/cm s	1.00	pounds/ft	kgs/meter	1.488
pounds (avoirdupois)	ounces (troy)	14.5833	pounds/in.	gms/cm	178.6
poundals	dynes	13,826.	pounds/mil-foot	gms/cu. cm	2.306×10^6
poundals	grams	14.10	pounds/sq. ft	atmospheres	4.725×10^{-4}
poundals	joules/cm	1.383×10^{-3}	pounds/sq. ft	feet of water	0.01602
poundals	joules/meter (newtons)	0.1383	pounds/sq. ft	inches of mercury	0.01414
poundals	kilograms	0.01410	pounds/sq. ft	kgs/sq. meter	4.882
poundals	pounds	0.03108	pounds/sq. ft	pounds/sq. in.	6.944×10^{-3}
pounds	drams	256.00	pounds/sq. in	atmospheres	0.06804

TABLE A.2a Continued
Conversion Factors

To Convert	Into	Multiply by
pounds/sq. in	feet of water	2.307
pounds/sq. in	inches of mercury	2.036
pounds/sq. in	kgs/sq. meter	703.1
pounds/sq. in	pounds/sq. ft	144.0
	Q	
quadrants (angle)	degrees	90.0
quadrants (angle)	minutes	5,400.0
quadrants (angle)	radians	1.571
quadrants (angle)	seconds	3.24×10^{5}
quarts (dry)	cu. in.	67.20
quarts (liq.)	cu. cms	946.4
quarts (liq.)	cu. ft	0.03342
quarts (liq.)	cu. in.	57.75
quarts (liq.)	cu. meters	9.464×10^{-4}
quarts (liq.)	cu. yards	1.238×10^{-3}
quarts (liq.)	gallons	0.25
quarts (liq.)	liters	0.9463
	R	
radians	degrees	57.30
radians	minutes	3,438.
radians	quadrants	0.6366
radians	seconds	2.063×10^{5}
radians/s	degrees/s	57.30
radians/s	revolutions/min	9.549
radians/s	revolutions/s	0.1592
radians/s/s	revs/min/min	573.0
radians/s/s	revs/min/s	9.549
radians/s/s	revs/s/s	0.1592
revolutions	degrees	360.0
revolutions	quadrants	4.0
revolutions	radians	6.283
revolutions/min	degrees/s	6.0
revolutions/min	radians/s	0.1047
revolutions/min	revs/s	0.01667
revolutions/min/min	radians/s/s	1.745×10^{-3}
revolutions/min/min	revs/min/s	0.01667
revolutions/min/min	revs/s/s	2.778×10^{-4}
revolutions/s	degrees/s	360.0
revolutions/s	radians/s	6.283
revolutions/s	revs/min	60.0
revolutions/s/s	radians/s/s	6.283
revolutions/s/s	revs/min/min	3,600.0
revolutions/s/s	revs/min/s	60.0
rod	chain (Gunters)	.25
rod	meters	5.029
rods (surveyors' meas.)	yards	5.5
rods	feet	16.5
	S	
scruples	grains	20
seconds (angle)	degrees	2.778×10^{-4}
seconds (angle)	minutes	0.01667
seconds (angle)	quadrants	3.087×10^{-6}
seconds (angle)	radians	4.848×10^{-6}
slug	kilogram	14.50
slug	pounds	32.17
sphere	steradians	12.57
square centimeters	circular mils	1.973×10^{5}
square centimeters	sq. ft	1.076×10^{-3}
square centimeters	sq. in.	0.1550
square centimeters	sq. meters	0.0001
square centimeters	sq. miles	3.861×10^{-11}
square centimeters	sq. millimeters	100.0
square centimeters	sq. yards	1.196×10^{-4}
square feet	acres	2.296×10^{-5}
square feet	circular mils	1.833×10^{8}
square feet	sq. cms	929.0
square feet	sq. in.	144.0
square feet	sq. meters	0.09290
square feet	sq. miles	3.587×10^{-8}
square feet	sq. millimeters	9.290×10^{4}
square feet	sq. yards	0.1111
square inches	circular mils	1.273×10^{6}
square inches	sq. cms	6.452
square inches	sq. ft	6.944×10^{-4}
square inches	sq. millimeters	645.2
square inches	sq. mils	10^{6}
square inches	sq. yards	7.716×10^{-4}
square kilometers	acres	247.1
square kilometers	sq. cms	10^{10}
square kilometers	sq. ft	10.76×10^{6}
square kilometers	sq. in.	1.550×10^{9}
square kilometers	sq. meters	10^{6}
square kilometers	sq. miles	0.3861
square kilometers	sq. yards	1.196×10^{6}
square meters	acres	2.471×10^{-4}
square meters	sq. cms	10^{4}

TABLE A.2a Continued
Conversion Factors

To Convert	*Into*	*Multiply by*
square meters	sq. cms	10^4
square meters	sq. in.	1,550.
square meters	sq. miles	3.861×10^{-7}
square meters	sq. millimeters	10^6
square meters	sq. yards	1.196
square miles	acres	640.0
square miles	sq. ft	27.88×10^6
square miles	sq. kms	2.590
square miles	sq. meters	2.590×10^6
square miles	sq. yards	3.098×10^6
square millimeters	circular mils	1,973.
square millimeters	sq. cms	0.01
square millimeters	sq. ft	1.076×10^{-5}
square millimeters	sq. in.	1.550×10^{-3}
square mils	circular mils	1.273
square mils	sq. cms	6.452×10^{-6}
square mils	sq. in.	10^{-6}
square yards	acres	2.066×10^{-4}
square yards	sq. cms	8,361.
square yards	sq. ft	9.0
square yards	sq. in.	1,296.
square yards	sq. meters	0.8361
square yards	sq. miles	3.228×10^{-7}
square yards	sq. millimeters	8.361×10^5
	T	
temperature (°C) + 273	absolute temperature (K)	1.0
temperature (°C) + 17.78	temperature (°F)	1.8
temperature (°F) + 460	absolute temperature (°R)	1.0
temperature (°F) − 32	temperature (°C)	5/9
tons (long)	kilograms	1,016.
tons (long)	pounds	2,240.
tons (long)	tons (short)	1.120
tons (metric)	kilograms	1,000.
tons (metric)	pounds	2,205
tons (short)	kilograms	907.1848
tons (short)	ounces	32,000.
tons (short)	ounces (troy)	29,166.66
tons (short)	pounds	2,000.
tons (short)	pounds (troy)	2,430.56
tons (short)	tons (long)	0.89287
tons (short)	tons (metric)	0.9078
tons (short)/sq. ft	kgs/sq. meter	9,765.
tons (short)/sq. ft	pounds/sq. in.	2,000.
tons of water/24 hrs	pounds of water/hr	83.333
tons of water/24 hrs	gallons/min	0.16643
tons of water/24 hrs	cu. ft/hr	1.3349
	V	
volt/inch	volt/cm	.39370
volt (absolute)	statvolts	.003336
	W	
watts	Btu/hr	3.4129
watts	Btu/min	0.05688
watts	ergs/s	107.
watts	ft-lbs/min	44.27
watts	ft-lbs/s	0.7378
watts	horsepower	1.341×10^{-3}
watts	horsepower (metric)	1.360×10^{-3}
watts	kg-calories/min	0.01433
watts	kilowatts	0.001
watts (abs.)	Btu (mean)/min	0.056884
watts (abs.)	joules/s	1.
watt-hours	Btu	3.413
watt-hours	ergs	3.60×10^{10}
watt-hours	foot-pounds	2,656.
watt-hours	gram-calories	859.85
watt-hours	horsepower-hrs	1.341×10^{-3}
watt-hours	kilogram-calories	0.8605
watt-hours	kilogram-meters	367.2
watt-hours	kilowatt-hrs	0.001
watt (International)	watt (absolute)	1.0002
webers	maxwells	10^8
webers	kilolines	10^5
webers/sq. in.	gausses	1.550×10^7
webers/sq. in.	lines/sq. in.	10^8
webers/sq. in.	webers/sq. cm	0.1550
webers/sq. in.	webers/sq. meter	1,550.
webers/sq. meter	gausses	10^4
webers/sq. meter	lines/sq. in.	6.452×10^4
webers/sq. meter	webers/sq. cm	10^{-4}
webers/sq. meter	webers/sq. in.	6.452×10^{-4}
	Y	
yards	centimeters	91.44
yards	kilometers	9.144×10^{-4}
yards	meters	0.9144
yards	miles (naut.)	4.934×10^{-4}
yards	miles (stat.)	5.682×10^{-4}
yards	millimeters	914.4

TABLE A.2b
Units of Area

1 cir mil[a]
= .000000785 sq. in.

1 sq. in.
= 1,273,200 cir mils
= .00694 sq. ft
= .000772 sq. yd

1 sq. ft.
= 144 sq. in.
= .01111 sq. yd
= .00002296 acres

1 sq. yd
= 1296 sq. in.
= 9 sq. ft
= .0002066 acres

1 acre
= 43,560 sq. ft
= 4840 sq. yd
= 4096 sq. m

[a]A cir (circular) mil is the area of a circle of 1/1000 in. dia. Thus, a round rod of 1-in. dia. has an area of 1,000,000 cir mils.

TABLE A.2c
Units of Density

1 lb/cu. in.
= 1728 lb/cu. ft
= 0.864 tons[a]/cu. ft
= 23.3 tons/cu. yd
= 231 lb/gal

1 lb/cu. ft
= .000579 lb/cu. in.
= .000500 tons/cu. ft
= .0135 tons/cu. yd
= .1337 lb/gal

1 ton/cu. ft
= 1.157 lb/cu. in.
= 2000 lb/cu. ft
= 27 tons/cu. yd
= 267 lb/gal

1 ton/cu. yd
= .0429 lb/cu. in.
= 74.1 lb/cu. ft
= .0370 tons/cu. ft
= 9.90 lb/gal

1 lb/gal
= .00433 lb/cu. in.
= 7.48 lb/cu. ft
= .00374 tons/cu. ft
= .1010 tons/cu. yd

[a]Tons are short = 2000 lb.

TABLE A.2d
Units of Work, Energy, Heat

1 Btu
= 9340 in. lb
= 778.3 ft-lb
= .0002938 kwhr[a]
= .0003931 hphr

1 in. lb
= .0001070 Btu
= .0833 ft-lb
= .00000003137 kwhr
= .0000000421 hphr

1 ft-lb
= .001284 Btu
= 12 in. lb
= .000000376 kwhr
= .000000505 hphr

1 kwhr
= 3413 Btu
= 31,873,000 in. lb
= 2,656,100 ft-lb
= 1.342 hphr

1 hphr
= 2544 Btu
= 23,760,000 in. lb
= 1,980,000 ft-lb
= 0.7455 kwhr

[a]1 kilowatthour = 3413 Btu and 1 Btu = 778.3 ft-lb.

TABLE A.2e
Units of Mass Flow

1 lb/s
= 60 lb/min
= 3600 lb/hr
= 86,400 lb/day
= 2,628,000 lb/mo[a]
= 31,536,000 lb/yr

1 lb/min
= .01667 lb/s
= 60 lb/hr
= 1440 lb/day
= 43,800 lb/mo
= 525,600 lb/yr

1 lb/hr
= .0002778 lb/s
= .01667 lb/min
= 24 lb/day
= 730 lb/mo
= 8760 lb/yr

1 lb/day
= .00001157 lb/s
= .000694 lb/min
= .0417 lb/hr
= 30.4 lb/mo
= 365 lb/yr

1 lb/mo
= .000000381 lb/s
= .0000228 lb/min
= .001370 lb/hr
= .0329 lb/day
= 12 lb/yr

1 lb/yr
= .0000000317 lb/s
= .000001903 lb/min
= .0001142 lb/hr
= .002740 lb/day
= .0833 lb/mo

[a]Month used is exactly 1/12 year = 30.4 days.

TABLE A.2f
Units of Volume Flow

1 cu. ft/s
= 60 cu. ft/min
= 3600 cu. ft/hr
= 7.48 gal/s
= 448.8 gal/min
= 26,930 gal/hr

1 cu. ft/min
= .01667 cu. ft/s
= 60 cu. ft/hr
= .1247 gal/s
= 7.48 gal/min
= 448.8 gal/hr

1 cu. ft/hr
= .0002778 cu. ft/s
= .01667 cu. ft/min
= .002078 gal/s
= .1247 gal/min
= 7.48 gal/hr

1 gal/s
= .1337 cu. ft/s
= 8.02 cu. ft/min
= 481 cu. ft/hr
= 60 gal/min
= 3600 gal/hr

1 gal/min
= .002228 cu. ft/s
= .1337 cu. ft/min
= 8.02 cu. ft/hr
= .01667 gal/s
= 60 gal/hr

1 gal/hr
= .0000371 cu. ft/s
= .002228 cu. ft/min
= .1337 cu. ft/hr
= .0002778 gal/s
= .01667 gal/min

TABLE A.2g
Units of Length

1 in.
= .0833 ft
= .0277 yd
= .0000158 miles

1 ft
= 12 in.
= .333 yd
= .000189 miles

1 yd
= 36 in.
= 3 ft
= .000568 miles

1 mile
= 63360 in.
= 5280 ft
= 1760 yd

TABLE A.2h
Linear Conversion

Inches to Millimeters 1 inch = 25.4 millimeters						*Millimeters to Inches* 1 millimeter = .03937 inches							
Inches	*Millimeters*	*Inches*	*Millimeters*	*Inches*	*Millimeters*	*Millimeters*	*Inches*	*Millimeters*	*Inches*	*Millimeters*	*Inches*	*Millimeters*	*Inches*
1/16	1.6	2.0	50.8	4.5	114.3	1.0	.0394	3.5	.1378	6.0	.2362	8.5	.3346
1/8	3.2	2.1	53.3	4.6	116.8	1.1	.0433	3.6	.1417	6.1	.2402	8.6	.3386
3/16	4.8	2.2	55.9	4.7	119.4	1.2	.0472	3.7	.1457	6.2	.2441	8.7	.3425
1/4	6.4	2.3	58.4	4.8	121.9	1.3	.0512	3.8	.1496	6.3	.2480	8.8	.3465
5/16	7.9	2.4	61.0	4.9	124.5	1.4	.0551	3.9	.1535	6.4	.2520	8.9	.3504
3/8	9.5	2.5	63.5	5.0	127.0	1.5	.0591	4.0	.1575	6.5	.2559	9.0	.3543
7/16	11.1	2.6	66.0	5.5	139.7	1.6	.0630	4.1	.1614	6.6	.2598	9.1	.3583
1/2	12.7	2.7	68.6	6.0	152.4	1.7	.0669	4.2	.1654	6.7	.2638	9.2	.3622
9/16	14.3	2.8	71.1	6.5	165.1	1.8	.0709	4.3	.1693	6.8	.2677	9.3	.3661
5/8	15.9	2.9	73.7	7.0	177.8	1.9	.0748	4.4	.1732	6.9	.2717	9.4	.3701
11/16	17.5	3.0	76.2	7.5	190.5	2.0	.0787	4.5	.1772	7.0	.2756	9.5	.3740
3/4	19.1	3.1	78.7	8.0	203.2	2.1	.0827	4.6	.1811	7.1	.2795	9.6	.3780
13/16	20.6	3.2	81.3	8.5	215.9	2.2	.0866	4.7	.1850	7.2	.2835	9.7	.3819
7/8	22.2	3.3	83.8	9.0	228.6	2.3	.0906	4.8	.1890	7.3	.2874	9.8	.3858
15/16	23.8	3.4	86.4	9.5	241.3	2.4	.0945	4.9	.1929	7.4	.2913	9.9	.3898
				10.0	254.0							10.0	.3937
1.0	25.4	3.5	88.9			2.5	.0984	5.0	.1969	7.5	.2953		
1.1	27.9	3.6	91.4			2.6	.1024	5.1	.2008	7.6	.2992		
1.2	30.5	3.7	94.0			2.7	.1063	5.2	.2047	7.7	.3031		
1.3	33.0	3.8	96.5			2.8	.1102	5.3	.2087	7.8	.3071		
1.4	35.6	3.9	99.1			2.9	.1142	5.4	.2126	7.9	.3110		
1.5	38.1	4.0	101.6			3.0	.1181	5.5	.2165	8.0	.3150		
1.6	40.6	4.1	104.1			3.1	.1220	5.6	.2205	8.1	.3189		
1.7	43.2	4.2	106.7			3.2	.1260	5.7	.2244	8.2	.3228		
1.8	45.7	4.3	109.2			3.3	.1299	5.8	.2283	8.3	.3268		
1.9	48.3	4.4	111.8			3.4	.1339	5.9	.2323	8.4	.3307		

By moving the decimal place, conversions for figures larger than 10 may be obtained.

TABLE A.2i
Units of Power

1 kw	1 ft-lb/hr
= 1.3415 hp	= .000000376 kw
= 738 ft-lb[a]/s	= .000000505 hp
= 44,268 ft-lb/min	= .000278 ft-lb/s
= 2,656,100 ft-lb/hr	=. 01667 ft-lb/min
= .948 Btu/s	= .000000357 Btu/s
= 56.9 Btu/min	= .00002141 Btu/min
= 3413 Btu/hr	= .001284 Btu/hr
1 hp	1 Btu/s
=.7455 kw	= 1.055 kw
= 550 ft-lb/s	= 1.416 hp
= 33,000 ft-lb/min	= 778 ft-lb/s
= 1,980,000 ft-lb/hr	= 46,700 ft-lb/min
= .707 Btu/s	= 2,802,000 ft-lb/hr
= .424 Btu/min	= 60 Btu/min
= 2544 Btu/hr	= 3600 Btu/hr
1 ft-lb/s	1 Btu/min
= .001355 kw	= .01759 kw
= .001818 hp	= .02359 hp
= 60 ft-lb/min	= 12.98 ft-lb/s
= 3600 ft-lb/hr	= 778 ft-lb/min
= .001284 Btu/s	= 46,700 ft-lb/hr
= .0771 Btu/min	= .01667 Btu/s
= 4.62 Btu/hr	= 60 Btu/hr
1 ft-lb/min	1 Btu/hr
= .00002259 kw	= .0002931 kw
= .0000303 hp	= .0003932 hp
= .01667 ft-lb/s	= .2163 ft-lb/s
= 60 ft-lb hr	= 12.98 ft-lb/min
= .00002141 Btu/s	= 778 ft-lb/hr
= .001284 Btu/min	= .0002778 Btu/s
= .0771 Btu/hr	= .01667 Btu/min

[a]Ft-lb means foot pound, the work done in moving against one pound force a distance of one foot.

Table A.2j
Units of Pressure

1 in. water[a]	1 oz/sq. ft
= .0833 ft water	= .01203 in. water
= .0735 in. Hg	= .001002 ft water
= .577 oz/sq. in.	= .000886 in. Hg
= .83.1 oz/sq. ft	= .00694 oz/sq. in.
= .0361 lb/sq. in.	= .000434 lb/sq. in.
= 5.20 lb/sq. ft	= .0625 lb/sq. ft
1 ft water	1 lb/sq. in.
= 12 in. water	= 27.71 in. water
= .882 in. Hg	= 2.31 ft water
= 6.93 oz/sq. in.	= 2.04 in. Hg
= 998 oz/sq. ft	= 16 oz/sq. in.
= .433 lb/sq. in.	= 2304 oz/sq. ft
= 62.4 lb/sq. ft	= 144 lb/sq. ft
1 in. Hg	1 lb/sq. ft
= 13.61 in. water	= .1924 in. water
= 1.131 ft water	= .01604 ft water
= 7.84 oz/sq. in.	= .01418 in. Hg
= 1129 oz/sq. ft	= .1111 oz/sq. in.
= .491 lb/sq. in.	= 16 oz/sq. ft
= 70.5 lb/sq. ft	= .00694 lb/sq. in.
1 oz/sq. in.	
= 1.732 in. water	
= .1443 ft water	
= .1276 in. Hg	
= 144 oz/sq. ft	
= .0625 lb/sq. in.	
= 9 lb/sq. ft	

[a]in. water means inches of water at 60°F.
in. Hg means inches head of mercury at 32°F.

TABLE A.2k
Pressure Conversion

1 pound per square inch = .0703 kilograms per square centimeter

lbs. per sq. in.	kgs per sq. cm	lbs. per sq. in.	kgs per sq. cm	lbs. per sq. in.	kgs per sq. cm	lbs. per sq. in.	kgs per sq. cm
1.00	.0703	2.25	.1582	4.50	.3164	8.25	.5800
1.10	.0773	2.30	.1617	4.75	.3339	8.50	.5976
1.20	.0844	2.40	.1687	5.00	.3515	8.75	.6151
1.25	.0879	2.50	.1758	5.25	.3691	9.00	.6327
1.30	.0914	2.60	.1828	5.50	.3867	9.25	.6503
1.40	.0984	2.70	.1898	5.75	.4042	9.50	.6679
1.50	.1055	2.75	.1933	6.00	.4218	9.75	.6854
1.60	.1125	2.80	.1969	6.25	.4394	10.00	.7030
1.70	.1195	2.90	.2039	6.50	.4570		
1.75	.1230	3.00	.2109	6.75	.4746		
1.80	.1265	3.25	.2285	7.00	.4921		
1.90	.1336	3.50	.2461	7.25	.5097		
2.00	.1406	3.75	.2636	7.50	.5273		
2.10	.1476	4.00	.2812	7.75	.5448		
2.20	.1547	4.25	.2988	8.00	.5624		

1 kilogram per square centimeter = 14.223 pounds per square inch

kgs per sq. cm	lbs. per sq. in.	kgs per sq. cm	lbs. per sq. in.	kgs per sq. cm	lbs. per sq. in.	kgs per sq. cm	lbs. per sq. in.
1.0	14.22	2.5	35.56	4.0	56.89	7.5	106.67
1.1	15.65	2.6	36.98	4.1	58.31	8.0	113.78
1.2	17.07	2.7	38.40	4.2	59.74	8.5	120.90
1.3	18.49	2.8	39.82	4.3	61.16	9.0	128.01
1.4	19.91	2.9	41.25	4.4	62.58	9.5	135.12
1.5	21.33	3.0	42.67	4.5	64.00	10.0	142.23
1.6	22.76	3.1	44.09	4.6	65.43		
1.7	24.18	3.2	45.51	4.7	66.85		
1.8	25.60	3.3	46.94	4.8	68.27		
1.9	27.02	3.4	48.36	4.9	69.69		
2.0	28.45	3.5	49.78	5.0	71.12		
2.1	29.87	3.6	51.20	5.5	78.23		
2.2	31.29	3.7	52.63	6.0	85.34		
2.3	32.71	3.8	54.05	6.5	92.45		
2.4	34.14	3.9	55.47	7.0	99.56		

TABLE A.2I

Pressure Head Conversion

The center column, marked "Known," may be used either for head in feet or for pressure in pounds per square inch—when used as head, the corresponding pressure is found in the column at the left designated as "Pressure Wanted"; when used as pressure, the corresponding head is found in the column at the right designated as "Head Wanted." For example: a 10-foot head has a pressure of 4.33 pounds per square inch, and a 10-pound pressure has a head of 23.094 feet. By moving the decimal place, quantities larger than 100 may be used.

Pressure Wanted lb. per sq. in.	Known Pressure or Head	Head Wanted ft. H_2O	Pressure Wanted lb. per sq. in.	Known Pressure or Head	Head Wanted ft. H_2O	Pressure Wanted lb. per sq. in.	Known Pressure or Head	Head Wanted ft. H_2O	Pressure Wanted lb. per sq. in.	Known Pressure or Head	Head Wanted ft. H_2O
0.433	1	2.309	11.259	26	60.044	22.084	51	117.779	32.909	76	175.514
0.866	2	4.619	11.692	27	62.354	22.517	52	120.089	33.342	77	177.824
1.299	3	6.928	12.125	28	64.663	22.950	53	122.398	33.775	78	180.133
1.732	4	9.238	12.558	29	66.973	23.383	54	124.708	34.208	79	182.433
2.165	5	11.547	12.991	30	69.282	23.816	55	127.017	34.642	80	184.752
2.598	6	13.856	13.424	31	71.591	24.249	56	129.326	35.075	81	187.061
3.031	7	16.166	13.857	32	73.901	24.682	57	131.636	35.508	82	189.371
3.464	8	18.475	14.290	33	76.210	25.115	58	133.945	35.941	83	191.680
3.897	9	20.785	14.723	34	78.520	25.548	59	136.255	36.374	84	193.990
4.330	10	23.094	15.156	35	80.829	25.981	60	138.564	36.807	85	196.299
4.763	11	25.403	15.589	36	83.138	26.414	61	140.873	37.240	86	198.608
5.196	12	27.713	16.022	37	85.448	26.847	62	143.183	37.673	87	200.918
5.629	13	30.022	16.455	38	87.757	27.280	63	145.492	38.106	88	203.227
6.062	14	32.332	16.888	39	90.067	27.713	64	147.803	38.539	89	205.537
6.495	15	34.641	17.321	40	92.376	28.146	65	150.111	38.972	90	207.846
6.928	16	36.950	17.754	41	94.685	28.579	66	152.420	39.405	91	210.155
7.361	17	39.260	18.187	42	96.995	29.012	67	154.730	39.838	92	212.465
7.794	18	41.569	18.620	43	99.304	29.445	68	157.039	40.271	93	214.774
8.227	19	43.879	19.053	44	101.614	29.878	69	159.349	40.704	94	217.084
8.660	20	46.188	19.486	45	103.923	30.311	70	161.658	41.137	95	219.393
9.093	21	48.497	19.919	46	106.232	30.744	71	163.967	41.570	96	221.702
9.526	22	50.807	20.352	47	108.542	31.177	72	166.277	42.003	97	224.012
9.959	23	53.116	20.785	48	110.851	31.610	73	168.586	42.436	98	226.321
10.392	24	55.426	21.218	49	113.161	32.043	74	170.896	42.869	99	228.631
10.825	25	57.735	21.651	50	115.470	32.476	75	173.205	43.302	100	230.940

TABLE A.2m
Temperature Conversion

Degrees–Fahrenheit to Centigrade
−°C = 5/9 (°F + 32)
+°C = 5/9 (°F − 32)

°F	0	25	50	75
−200	−128.9	−142.8	−156.7	−170.6
−100	−73.3	−87.2	−101.1	−115.0
−0	−17.8	−31.7	−45.6	−59.4
+0	−17.8	−3.9	+10.0	+23.9
100	+37.8	+51.7	65.6	79.4
200	93.3	107.2	121.1	135.0
300	148.9	162.8	176.7	190.6
400	204.4	218.3	232.2	246.1
500	260.0	273.9	287.8	301.7
600	315.6	329.4	343.3	357.2
700	371.1	385.0	398.9	412.8
800	426.7	440.6	454.4	468.3
900	482.2	496.1	510.0	523.9
1000	538.0	551.7	565.6	579.4
1100	593.2	607.2	621.1	635.0

Degrees–Centigrade to Fahrenheit
−°F = (9/5 × °C) − 32
+°F = (9/5 × °C) + 32

°C	0	25	50	75
−200	−328	−373	−418	
−100	−148	−193	−238	−283
−0	+32	−13	−58	−103
+0	+32	+77	+122	+167
100	212	257	302	347
200	392	437	482	527
300	572	617	662	707
400	752	797	842	887
500	932	977	1022	1067
600	1112	1157	1202	1247
700	1292	1337	1382	1427

TABLE A.2n
Units of Time

1 s
= .01667 min
= .0002778 hr
= .00001157 days
= .0000003805 mo[a]
= .0000000317 yr

1 min
= 60 s
= .01667 hr
= .000694 days
= .0000228 mo
= .000001903 yr

1 hr
= 3600 s
= 60 min
= .0417 days
= .001370 mo
= .0001142 yr

1 day
= 86,400 s
= 1440 min
= 24 hr
= .0329 mo
= .002740 yr

1 mo
= 2,628,000 s
= 43,800 min
= 730 hr
= 30.4 days
= .0833 yr

1 yr
= 31,536,000 s
= 525,600 min
= 8760 hr
= 365 days
= 12 mo

[a]Month used is exactly 1/12 year.

TABLE A.2o
Units of Velocity

fps
= 60 fpm
= 3600 fph
= .01136 mpm
= .682 mph

1 fpm
= .01667 fps
= 60 fph
= .0001894 mpm
= .01136 mph

1 fph
= .002778 fps
= .01667 fpm
= .00000316 mpm
= .0001894 mph

1 mpm
= 88 fps
= 5280 fpm
= 316,800 fph
= 60 mph

1 mph
= 1.467 fps
= 88 fpm
= 5280 fph
= .01667 mpm

f = feet; h = hours; m = miles or minutes; p = per; s = seconds.

TABLE A.2p
Viscosity Conversion

Kinematic Viscosity Centisokes = K	*Seconds Saybolt Universal*	*Seconds Saybolt Furol*	*Seconds Redwood*	*Seconds Redwood Admiralty*	*Degrees Engler*	*Degrees Bardey*
1.00	31	—	29.1	—	1.00	6200
2.56	35	—	32.1	—	1.16	2420
4.30	40	—	36.2	5.10	1.31	1440
5.90	45	—	40.3	5.52	1.46	1050
7.40	50	—	44.3	5.83	1.58	838
8.83	55	—	48.5	6.35	1.73	702
10.20	60	—	52.3	6.77	1.88	618
11.53	65	—	56.7	7.17	2.03	538
12.83	70	12.95	60.9	7.60	2.17	483
14.10	75	13.33	65.0	8.00	2.31	440
15.35	80	13.70	69.2	8.44	2.45	404
16.58	85	14.10	73.3	8.86	2.59	374
17.80	90	14.44	77.6	9.30	2.73	348
19.00	95	14.85	81.5	9.70	2.88	326
20.20	100	15.24	85.6	10.12	3.02	307
31.80	150	19.3	128	14.48	4.48	195
43.10	200	23.5	170	18.90	5.92	144
54.30	250	28.0	212	23.45	7.35	114
65.40	300	32.5	254	28.0	8.79	95
76.50	350	35.1	296	32.5	10.25	81
87.60	400	41.9	338	37.1	11.70	70.8
98.60	450	46.8	381	41.7	13.15	62.9
110.	500	51.6	423	46.2	14.60	56.4
121.	550	56.6	465	50.8	16.05	51.3
132.	600	61.4	508	55.4	17.50	47.0
143.	650	66.2	550	60.1	19.00	43.4
154.	700	71.1	592	64.6	20.45	40.3
165.	750	76.0	635	69.2	21.90	37.6
176.	800	81.0	677	73.8	23.35	35.2
187.	850	86.0	719	78.4	24.80	33.2
198	900	91.0	762	83.0	26.30	31.3
209	950	95.8	804	87.6	27.70	29.7
220	1000	100.7	846	92.2	29.20	28.2
330	1500	150	1270	138.2	43.80	18.7
440	2000	200	1690	184.2	58.40	14.1

TABLE A.2p Continued
Viscosity Conversion

Kinematic Viscosity Centisokes = K	*Seconds Saybolt Universal*	*Seconds Saybolt Furol*	*Seconds Redwood*	*Seconds Redwood Admiralty*	*Degrees Engler*	*Degrees Bardey*
550	2500	250	2120	230	73.00	11.3
660	3000	300	2540	276	87.60	9.4
770	3500	350	2960	322	100.20	8.05
880	4000	400	3380	368	117.00	7.05
990	4500	450	3810	414	131.50	6.26
1100	5000	500	4230	461	146.00	5.64
1210	5500	550	4650	507	160.50	5.13
1320	6000	600	5080	553	175.00	4.70
1430	6500	650	5500	559	190.00	4.34
1540	7000	700	5920	645	204.50	4.03
1650	7500	750	6350	691	219.00	3.76
1760	8000	800	6770	737	233.50	3.52
1870	8500	850	7190	783	248.00	3.32
1980	9000	900	7620	829	263.00	3.13
2090	9500	950	8040	875	277.00	2.97
2200	10000	1000	8460	921	292.00	2.82

The viscosity is often expressed in terms of viscosimeters other than the Saybolt Universal. The formulas for the various viscosimeters are given opposite.

If viscosity is given at any two temperatures, the viscosity at any other temperature can be obtained by plotting the viscosity against temperature in degrees Fahrenheit on special Log paper. The points for a given oil lie in a straight line.

$$\text{Kinematic viscosity} = \frac{\text{absolute viscosity}}{\text{specific gravity}}$$

$$\text{Redwood K} = 26t - \frac{180}{t}\text{(British)}$$

$$\text{Redwood Admiralty K} = 2.7t - \frac{20}{t}\text{(British)}$$

$$\text{Saybolt Universal K} = .22t - \frac{195}{t}\text{(American)}$$

$$\text{Saybolt Furol K} = 2.2t - \frac{184}{t}\text{(American)}$$

$$\text{Engler K} = .147t - \frac{374}{t}\text{(German)}$$

TABLE A.2q
Viscosity Conversion Chart

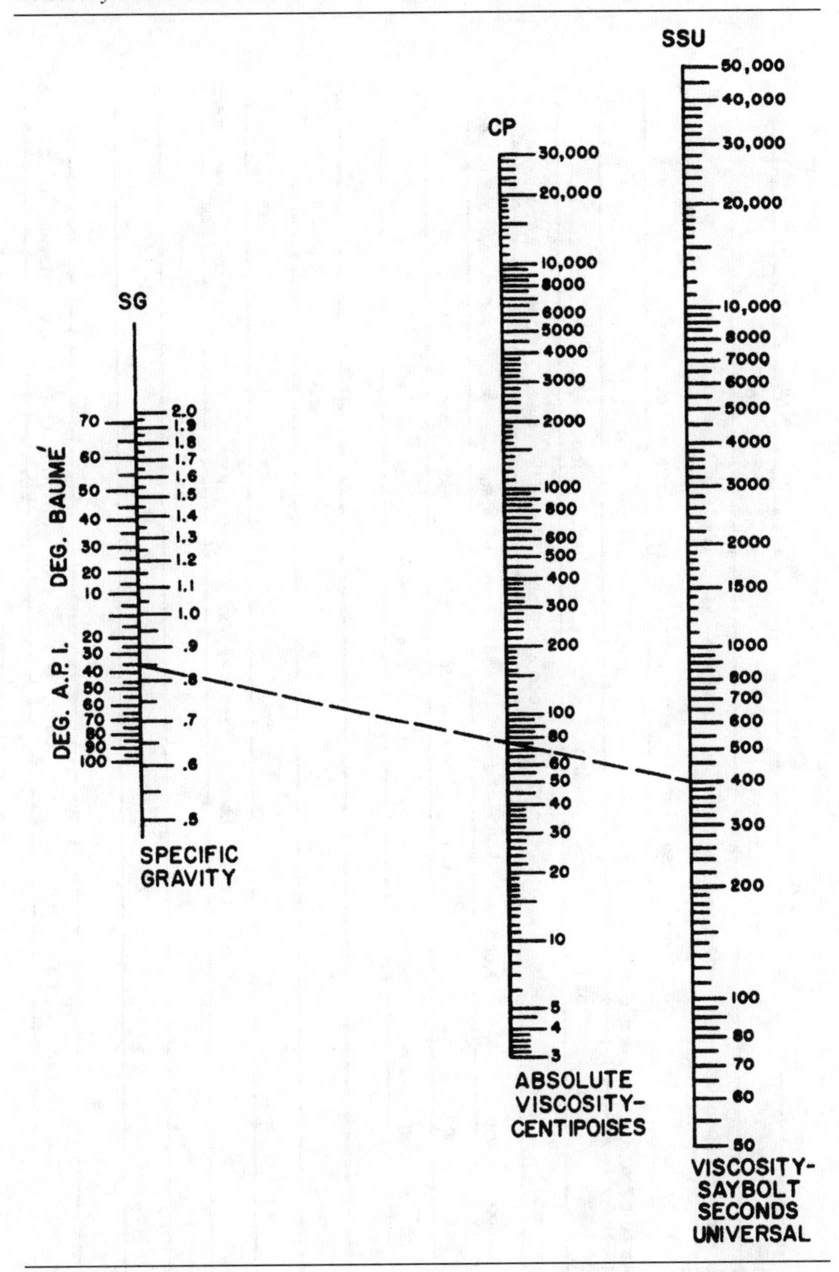

This chart enables the direct conversion of a viscosity in centipoises to a SSU viscosity. As an example, suppose the liquid under consideration has a specific gravity of .85 and a viscosity of 75 centipoises. To determine the viscosity in SSU, lay a straight edge between 75 on the CP scale and .85 on the G scale. The viscosity in SSU can be read directly on the SSU scale. In this instance, the SSU viscosity is 400 (see dotted line).

If the viscosity value is given in centistokes (kinematic viscosity), it can be used directly on the viscosity correction nomograph. The relationship between the absolute viscosity and the kinematic viscosity is expressed by the following formula:

$$\text{Centipoises} = \text{Centistokes} \times \text{Specific Gravity}$$

TABLE A.2r
Approximate Viscosity Conversion Chart

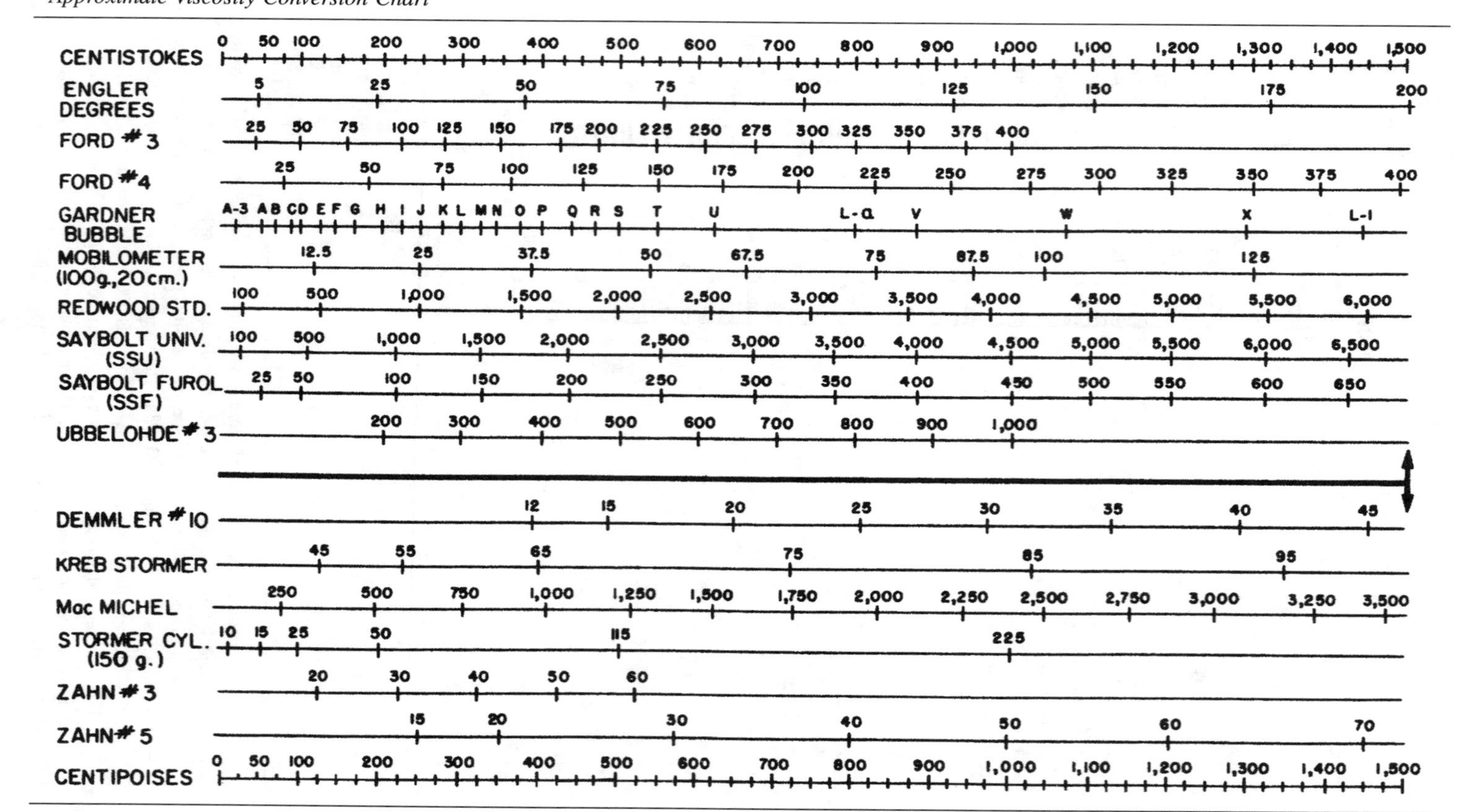

Note: Scales in lower section read directly in centipoises. Those above read in centistokes (to convert into centipoises, multiply by liquid specific gravity).

TABLE A.2s

Units of Volume

1 cu. in.
- = .00433 gal
- = .000579 cu. ft
- = .0000214 cu. yd

1 gal
- = 231 cu. in.
- = .1337 cu. ft.
- = .00495 cu. yd
- = .00000307 acre ft[a]

1 cu. ft
- = 1728 cu. in.
- = 7.48 gal
- = .0370 cu. yd
- = .0000230 acre ft

1 cu. yd
- = 46,656 cu. in.
- = 202.0 gal
- = 27 cu. ft.
- = .000620 acre ft

1 acre ft
- = 325,800 gal
- = 43,560 cu. ft
- = 1613 cu. yd

[a] Acre ft of water is the volume in 1 ft of depth covering 1 acre.

TABLE A.2t

Units of Weight

1 gr
- = .00229 oz[a]
- = .0001429 lb
- = .0000000714 tons

1 oz
- = 438 gr
- = .0625 lb
- = .00003125 tons

1 lb
- = 7000 gr
- = 16 oz
- = .000500 tons

1 ton
- = 14,000,000 gr
- = 32,000 oz
- = 2000 lb

[a] Avoirdupois oz and lb and short ton of 2000 lb.

TABLE A.2u

Weight Conversion

1 kilogram = 2.2046 pounds   1 pound = 0.4536 kilograms

lbs	*kgs/ lbs*	*kgs*	*lbs*	*kgs/ lbs*	*kgs*	*lbs*	*kgs/ lbs*	*kgs*	*lbs*	*kgs/ lbs*	*kgs*	*lbs*	*kgs/ lbs*	*kgs*
2.205	1.0	.454	6.614	3.0	1.361	11.023	5.0	2.268	15.432	7.0	3.175	19.841	9.0	4.082
2.425	1.1	.499	6.834	3.1	1.406	11.243	5.1	2.313	15.653	7.1	3.221	20.062	9.1	4.128
2.646	1.2	.544	7.055	3.2	1.452	11.464	5.2	2.359	15.873	7.2	3.266	20.282	9.2	4.173
2.866	1.3	.590	7.275	3.3	1.497	11.684	5.3	2.404	16.094	7.3	3.311	20.503	9.3	4.218
3.086	1.4	.635	7.496	3.4	1.542	11.905	5.4	2.449	16.314	7.4	3.357	20.723	9.4	4.264
3.307	1.5	.680	7.716	3.5	1.588	12.125	5.5	2.495	16.535	7.5	3.402	20.944	9.5	4.309
3.527	1.6	.726	7.937	3.6	1.633	12.346	5.6	2.540	16.755	7.6	3.447	21.164	9.6	4.355
3.748	1.7	.771	8.157	3.7	1.678	12.566	5.7	2.586	16.975	7.7	3.493	21.385	9.7	4.400
3.968	1.8	.816	8.377	3.8	1.724	12.787	5.8	2.631	17.196	7.8	3.538	21.605	9.8	4.445
4.189	1.9	.862	8.598	3.9	1.769	13.007	5.9	2.676	17.416	7.9	3.583	21.826	9.9	4.491
												22.046	10.0	4.536
4.409	2.0	.907	8.818	4.0	1.814	13.228	6.0	2.722	17.637	8.0	3.629			
4.630	2.1	.953	9.039	4.1	1.860	13.448	6.1	2.767	17.857	8.1	3.674			
4.850	2.2	.998	9.259	4.2	1.905	13.669	6.2	2.812	18.078	8.2	3.720			
5.071	2.3	1.043	9.480	4.3	1.950	13.889	6.3	2.858	18.298	8.3	3.765			
5.291	2.4	1.089	9.700	4.4	1.996	14.109	6.4	2.903	18.519	8.4	3.810			
5.512	2.5	1.134	9.921	4.5	2.041	14.330	6.5	2.948	18.739	8.5	3.856			
5.732	2.6	1.179	10.141	4.6	2.087	14.550	6.6	2.994	18.960	8.6	3.901			
5.952	2.7	1.225	10.362	4.7	2.132	14.771	6.7	3.039	19.180	8.7	3.946			
6.173	2.8	1.270	10.582	4.8	2.177	14.991	6.8	3.084	19.400	8.8	3.992			
6.393	2.9	1.315	10.803	4.9	2.223	15.212	6.9	3.130	19.621	8.9	4.037			

A.3 Chemical Resistance of Materials

TABLE A.3

Chemical Resistance of Materials

MATERIALS
× — *Very Good Service*
+ — *Moderate Service*
– — *Limited or Variable Service*
O — *Unsatisfactory*
Blank — *No Information*

CHEMICALS
Solids Assumed in Solution
Room Temperatures Assumed
Unless Otherwise Stated

METALS

Chemicals	Carbon Steel; Fe	Cast Iron & Ductile Iron; Fe	304 Stainless Steel; Fe, 18Cr, 8Ni	316 Stainless Steel; Fe 16Cr, 10Ni, 2Mo	347 Stainless Steel; Fe, 17Cr, 9Ni; (Cx10)Cb	Ni-Resist Iron; Fe 14Ni, 2Cr, 2Si	Durimet 20; Carpenter 20; Fe; 4Cu, 20Cr, 29Ni, 2Mo, 1Si	Worthite; 3Mo, 2Cu, Fe, 20Cr, 24Ni, 3Si	Durorpm; Fe, 14Si; Durichlor, Fe, 14Si, 3Mo*	Copper; Brass; Bronzes; Everdur	Aluminum; Al (and Alloys)	Lead; Pb	Monel; 67Ni, 30Cu, 1.4Fe	Nickel; Ni	Inconel; 76Ni, 15Cr, 8Fe	Hastelloy B; Ni, 26Mo, 4Fe	Hastelloy C; Ni, 16Mo, 4Fe, 14Cr, 4W	Hastelloy D; Ni, 8Si, 3Cu	Chlorimet 3; 3Fe, 1Si, 60Ni, 18Mo, 18Cr	Chlorimet 2; 63Ni, 32Mo, 3Fe, 1Si	Stellite; Co, 28Cr, 4W	Zirconium; Z	Tantalum; Ta	Silver; Ag	Platinum; Pt	Dowmetal; (Mg alloys)	Titanium; Ti	Molybdenum; Mo
Acetic Acid, 100% CH_3COOH	O	–	+	×	–	O	×	×	×	O	×	O	–	+	–	×	×	×	×	×	×	×	×	×	×		×	
Acetic Acid, Dilute	O	O	×	×	–	O	×	×	×	×	–	–	–	O	+	×	×	+	×	×	×	×	×	×	×		×	×
Acetic Anhydride, $(CH_3CO)_2O$	–	+	×	×	×	–	×	×	×	–	–	O	–	–	+	+	×	+	×	×	×		×	×	×		×	
Acetone, CH_3COCH_3	×		×	×	×	×	×	×	×	×	×	×	×	×	×	×	×	×	×	×		×		×			×	
Acetyl Chloride, CH_3COCl	O		–	×			×		×	O			+						×	×								
Aluminum Chloride, $AlCl_3$	O	O	O	O	O	–	×	O	×	–	O	O	O	–	–	×	+	×	×	×	–	×	×	×	×		×	
Aluminum Hydroxide, $AL(OH)_3$			×	×			×	×	+	×	×		+	+					×	×			×					
Aluminum Sulfate, $Al_2(SO_4)_3$	O		–	–	×	–	×	×	×	–	×	×	–	–	–	×	×	×	×	×	×	×	×	×	×		×	
Alums, Conc., $Al_2(SO_4)_3K_2SO_4$, etc.	O		O	–	–		+		×	–	–	×							×	×			×		×			
Alums, Dilute	O	O	×	–	–	–	×	×	×	–	×	+	+	+		+	+		×	×		×	×	×	×			
Amines, various	×	×	×	×	×		×	×	×	–	–					×	×	×	×	×	×		–	×	×			
Ammonia (Gas), Moist, NH_3	×		×	×	×	×	×		×	O	O	–	–	–	–	×	×	×	×	×	×		–	×	×	×	–	
Ammonium Carbonate, $(NH_4)_2CO_3$	+	+	+	+	×		×		×	–	–		+	×	+	+	+	+	×	×			×					
Ammonium Chloride, NH_4Cl	O	O	O	–	–	×	×	×	×	O	O	–	+	+	–	+	+	×	×	×	–	×	×	×	×		×	+
Ammonium Hydroxide, NH_4OH	×	×	×	×	×	×	×		+	O	–	+	O	O	×	+	+	+	×	×	×	×	–	×	×	×	×	–
Ammonium Nitrate, NH_4NO_3	–		×	×	×	–	×	×	×	O	O		×		–	O	×	O	×	×			×					
Ammonium Persulfate, $(NH_4)_2S_2O_8$	O		×	×	×		×	×	×	–	–		O	O	–	O	×	O	×	×			×					
Ammonium Phosphate, $(NH_4)_1H_2PO_4$	O		×	×		–	×		×	O									×	×								
Ammonium Phosphate, $(NH_4)_2HPO_4$	–		×	×		×	×		×	–	–								×	×								
Ammonium Phosphate, $(NH_4)_3PO_4$	×	–	×	×	×	×	×	×	×	O	–	×	×	×	×	×	×	×	×	×	×		×					
Ammonium Sulfate, $(NH_4)_2SO_4$	×		–	–	+	×	×	×	×	–	–	–	+	+	×	×	×	×	×	×	×	×	×	×	×		×	
Amyl Acetate, $C_5H_{11}COOCH_3$	–		×	×	×	×	×	×	×	–	×		+	+	+	×	×	×	×	×				×				
Amyl Alcohol, $C_5H_{11}OH$	×		×	×			×		×	×	×								×	×								

CARBONS & CERAMICS / RUBBERS

Chemicals	Carbon & Graphite	Glass, "Pyrex" brand	Silicaware	Silicate Cements	Chemical Stoneware	Transite (asbestos & cement)	Chemical Porcelain	Concrete—Unbonded	Concrete—Motor Bonded	Hard Rubber (Natural)	Soft Rubber (Natural)	Neoprene	Butadiene Derivatives	Nitrile Rubber (Chemigum)
Acetic Acid, 100% CH_3COOH	×	×	×	×	×		×			×	–	O	O	O
Acetic Acid, Dilute	×	×	×		×		×			×	–	O	×	O
Acetic Anhydride, $(CH_3CO)_2O$	×	×	×	×	×		×			×	–	O	–	O
Acetone, CH_3COCH_3	×	×	×		×	×	×			O	O	O	×	–
Acetyl Chloride, CH_3COCl		×					×							
Aluminum Chloride, $AlCl_3$	×	×	×	×	×		×			×	×	×	×	×
Aluminum Hydroxide, $AL(OH)_3$		×					×							
Aluminum Sulfate, $Al_2(SO_4)_3$	×	×	×		×	×	×			×	×	×	×	–
Alums, Conc., $Al_2(SO_4)_3K_2SO_4$, etc.	×	×					×					×		
Alums, Dilute	×	×	×	×	×	×	×			×	×	×	×	×
Amines, various	×	×			×		×			–				
Ammonia (Gas), Moist, NH_3	×	×	×	O	×	×	×			–	–	×	×	
Ammonium Carbonate, $(NH_4)_2CO_3$		×		×			×			×				O
Ammonium Chloride, NH_4Cl	×	×	×	×	×	×	×	×		×	×	×	×	–
Ammonium Hydroxide, NH_4OH	×	–			×		×			–	×	–		×
Ammonium Nitrate, NH_4NO_3	×	×	×		×		×	×		–	–	–	×	
Ammonium Persulfate, $(NH_4)_2S_2O_8$		×												
Ammonium Phosphate, $(NH_4)_1H_2PO_4$		×					×							
Ammonium Phosphate, $(NH_4)_2HPO_4$		×					×							
Ammonium Phosphate, $(NH_4)_3PO_4$	×	×			×		×			×				
Ammonium Sulfate, $(NH_4)_2SO_4$	×	×	×		×		×			×	×	×	×	
Amyl Acetate, $C_5H_{11}COOCH_3$	×	×	×		×		×	×		O	O	O	O	
Amyl Alcohol, $C_5H_{11}OH$	×	×	×		×	×	×			–	–	×	×	

THERMOPLASTICS

Chemicals	Viton	Asphaltic, Bitumastic	Cellulose Acetate	Cellulose Acetatebutyrate	Ethyl Cellulose (Ethocel)	Cellulose Nitrate	Acrylic (Lucite, Plexiglas)	Coumarone Resins	Polyethylene	Polyvinyl Chloride, Rigid or Unplasticized	Tygon (PVC & Copolymers)	Saran (Vinyl Chloride, Vinylidene Chloride)	Kel-F (Polytrifluorochloroethylene)	Teflon (Polytetrafluoroethylene)	Uscolite CP (Styrene–Acrylonitrile–Butadiene)
Acetic Acid, 100% CH_3COOH	O	O	O		×	O	O		O	O	O	–	×	×	O
Acetic Acid, Dilute		×	×	×	×	×	×	×	×	–	–	×	×	×	–
Acetic Anhydride, $(CH_3CO)_2O$	O	O	O		O	O	O			O	O	–		×	
Acetone, CH_3COCH_3	O	O	O	O	O	O	O	O	O	O		O		×	O
Acetyl Chloride, CH_3COCl													×	×	
Aluminum Chloride, $AlCl_3$	×	×				×			×	×	×	×	×	×	×
Aluminum Hydroxide, $AL(OH)_3$	×								×	×			×	×	
Aluminum Sulfate, $Al_2(SO_4)_3$	×	×				×	×	×	×	×	×	×	×	×	×
Alums, Conc., $Al_2(SO_4)_3K_2SO_4$, etc.									×	×	×	×	×	×	×
Alums, Dilute		×				×	×	×	×	×	×	×	×	×	×
Amines, various	O									O			×	×	
Ammonia (Gas), Moist, NH_3	O		–	×	×		×		×	–		O	×	×	–
Ammonium Carbonate, $(NH_4)_2CO_3$		×							×	×		×	×	×	×
Ammonium Chloride, NH_4Cl	×	×	×	×		×	×	×	×	×	×	×	×	×	×
Ammonium Hydroxide, NH_4OH	+	–					×		×	×	–	O	×	×	×
Ammonium Nitrate, NH_4NO_3	+	×	×			×	×		×	×		×	×	×	×
Ammonium Persulfate, $(NH_4)_2S_2O_8$		×								×	×		×	×	
Ammonium Phosphate, $(NH_4)_1H_2PO_4$										×			×	×	×
Ammonium Phosphate, $(NH_4)_2HPO_4$										×			×	×	×
Ammonium Phosphate, $(NH_4)_3PO_4$									×	×	×		×	×	×
Ammonium Sulfate, $(NH_4)_2SO_4$		×				×	×	×		×	×	×	×	×	×
Amyl Acetate, $C_5H_{11}COOCH_3$	O	O	O	O	O	O	O	O	–	O		O	×	×	
Amyl Alcohol, $C_5H_{11}OH$	×	O	×	O	O	O	×	–	–	×		×	×	×	×

THERMOSETTING PLASTICS / WOODS

Chemicals	Penton (Chlorinated Polyether)	Shellac Compounds	Organic Polysulfides	Polystyrene (Styron)	Vinylidene Chlorides	Vinyl Chloride Acetates	Cast Phenol Formaldehyde	Haveg 41 (Phenolic w. Asbestos)	Heresite (Phenol Formaldehyde)	Molded Phenolformald (Durez)	Phenol Furfural Plastics	Urea Formaldehyde	Casein Plastics	Epoxy Resins	Furane Resins (Haveg 61, Duralon)	Silicone Resins	Permanite (Furan, Glass Fiber)	Nylon (Adipic Acid–Hexameth, Diamine)	Durcon 6 (Modified Epoxy)	Cypress	Fir	Maple	Oak	Pine	Redwood
Acetic Acid, 100% CH_3COOH	+		×	O	O	O	–	–	×	×	–			–	×	O					–	×	×	×	–
Acetic Acid, Dilute	×	×	×	O	–	–	×	×	×	×	×		×	×	×	×	×	O	×	×	×	×	×	×	×
Acetic Anhydride, $(CH_3CO)_2O$	+			O			–	O	×	×	×				O	–	×								
Acetone, CH_3COCH_3	+	O	×	–	O	O	–	O		–	–	×	×	×		×	×								
Acetyl Chloride, CH_3COCl																×									
Aluminum Chloride, $AlCl_3$	×	×		×	×	×	×	×	×	×	×			×	×	×	×	×							
Aluminum Hydroxide, $AL(OH)_3$	×															×	×								
Aluminum Sulfate, $Al_2(SO_4)_3$	×	×	×	×	×	×	–	×	×	×			×		×	×	×	×	×		×	×	–	×	×
Alums, Conc., $Al_2(SO_4)_3K_2SO_4$, etc.	×							×	×					×		×									
Alums, Dilute	×	×	×	×	×	×	–	×	×	×			×	×	×	×	×				×	×	–	×	×
Amines, various								×	×							×	×								
Ammonia (Gas), Moist, NH_3	×	×		O	×	×	×			×	–			×	×	–	×								
Ammonium Carbonate, $(NH_4)_2CO_3$	×			×						×	×			×	×	×	×								
Ammonium Chloride, NH_4Cl	×	×		×	×	×	×	×	×	×	×		×	×	×	×	O		+**						
Ammonium Hydroxide, NH_4OH	×			×	O			×	×	O	O			×	×	–	×	×	×						
Ammonium Nitrate, NH_4NO_3	×	–		×	×	×	×			×	×		×	×		×	×								
Ammonium Persulfate, $(NH_4)_2S_2O_8$	×													×	×	×									
Ammonium Phosphate, $(NH_4)_1H_2PO_4$	×															×									
Ammonium Phosphate, $(NH_4)_2HPO_4$	×															×									
Ammonium Phosphate, $(NH_4)_3PO_4$	×							×	×							×									
Ammonium Sulfate, $(NH_4)_2SO_4$	×	×	×		×	×	×	×	×	×	×		×	×		×	O								
Amyl Acetate, $C_5H_{11}COOCH_3$	×	O		O	O	×	–			–	–		×	×		×	×	×							
Amyl Alcohol, $C_5H_{11}OH$	×	O		×	–	–	×			×	×		×	×		×	×								

*Note: Duriron is as shown. Durichlor is also satisfactory on chlorides and HCl.

**Durcon 5 would be the preferred formula.

TABLE A.3 Continued
Chemical Resistance of Materials

MATERIALS
× — *Very Good Service*
+ — *Moderate Service*
– — *Limited or Variable Service*
O — *Unsatisfactory*
Blank — No Information

CHEMICALS
Solids Assumed in Solution
Room Temperatures Assumed
Unless Otherwise Stated

METALS

Chemical	Carbon Steel; Fe	Cast Iron & Ductile Iron; Fe	304 Stainless Steel; Fe, 18Cr, 8Ni	316 Stainless Steel; Fe 16Cr, 10Ni, 2Mo	347 Stainless Steel; Fe, 17Cr, 9Ni; (Cx10)Cb	Ni-Resist Iron; Fe 14Ni, 2Cr, 2Si	Durimet 20; Carpenter 20; Fe; 4Cu, 20Cr, 29Ni, 2Mo, 1Si	Worthite; 3Mo, 2Cu, Fe, 20Cr, 24Ni, 3Si	Durorpm; Fe, 14Si; Durichlor, Fe, 14Si, 3Mo*	Copper; Brass; Bronzes; Everdur	Aluminum; Al (and Alloys)	Lead; Pb	Monel; 67Ni, 30Cu, 1.4Fe	Nickel; Ni	Inconel; 76Ni, 15Cr, 8Fe	Hastelloy B; Ni, 26Mo, 4Fe	Hastelloy C; Ni, 16Mo, 4Fe,14Cr, 4W	Hastelloy D; Ni, 8Si, 3Cu	Chlorimet 3; 3Fe, 1Si, 60Ni,18Mo, 18Cr	Chlorimet 2; 63Ni, 32Mo, 3Fe, 1Si	Stellite; Co, 28Cr, 4W	Zirconium; Z	Tantalum; Ta	Silver; Ag	Platinum; Pt	Dowmetal; (Mg alloys)	Titanium; Ti	Molybdenum; Mo
Amyl Chloride, $C_5H_{11}Cl$	×		–	–	–		×	–	×	×	O		–	+	–	×	×	×	×	×			×					
Antimony Trichloride, $SbCl_3$	O		O	O	O				×	–	O	×	×	×	×	×	×		×	×			×					
Arsenic Acid, $HAsO_8$	O		–	–		×	×	×	×	–		–			–				×	×			×	×				
Barium Carbonate, $BaCO_3$			×	×	×		×	×	×	×	O		+	×	×	×	×	×	×		×	×					×	
Barium Hydroxide, $Ba(OH)_2$	×			×		×	×	×	+	O	O		×	×		+	+	+	×	×		×	×				+	
Barium Sulfide, BaS	O		×	×		×	×	×	×	O	O	–							×	×				O				
Benzaldehyde, C_6H_5CHO	×		×	×			×		×		–	O							×	×								
Benzene, C_6H_6	+	+	×	×	×	×	×	×	×	×	×	+	+	+	×	+	+	+	×	×	×		×	×	×	×		
Benzoic Acid, C_6H_5COOH	O		×	×	+		×	×	×	+	×	O	+	+	+	×	×	×	×	×			×	×				
Borax, $Na_2B_4O_7$	×	–	×	×	×	×	×		×	–	–	×	×	×					×	×				O				
Boric Acid, H_3HO_3	O	O	+	+	+	–	×	×	×	+	+	+	+	+	×	×	×	×	×	×	×		×	×	×		×	
Bromine, Wet, Br_2	O		O	O	O	O	O	O	O	O	O	–	O	O	O	O	×	O	×	O	–		×	O	–		O	
Butanol, C_4H_2OH	×		×	×			×		×	×	×		×	×	×	×	×	×	×	×								
Butyl Acetate, $C_4H_9COOCH_3$	–			×			×	×	×	×	×	×	+			×	×	×	×	×			×			–		
Butyric Acid, C_3H_7COOH	O	O	O	×	O		×	×	×	–	–	O	+	+	+	×	×	×	×	×			×				×	
Calcium Bisulfate, $CaHSO_4$	O		O	×		O	×			O	O		O	O					×	×								
Calcium Bisulfate, $CaHSO_3$	×		O	×		O	×	×	O	–	–	×	O						×	×							×	
Calcium Carbonate, $CaCO_3$	×		×	×		×	×	×	×	×	O	×	×	×	×	×	×	×	×	×			×				×	
Calcium Chlorate, $CaClO_3$			–	×	O		+	–	+	–			+	+	+		+	+	+	+			×					
Calcium Chloride, $CaCl_2$	+	+	–	–	–	–	+	×	×	+	–	O	+	+	+	+	×	+	×	×	–	×	×				×	
Calcium Hydroxide, $Ca(OH)_2$	×		×	×	×	×	×	×	+	+	–		×	×	×	×	×	×	×	×			×				×	
Calcium Hypochlorite, $Ca(OCl)_2$	–	O	O	–	–	–	–	–	×	O	O	O	O	O	O	O	–	O	–	O	O	×	×		×		×	
Calcium Sulfate, $CaSO_4$	–		×	×			×	×	×	×	–		+			+	+	+	×	×			×					
Carbon Dioxide (Dry), CO_2	×		×	×	×	×	×		×	×	×	×	×	×	×	×	×		×	×				×				×
Carbon Dioxide (Wet or H_2CO_3)	–	–	×	×	×	–	×	×	×	–	–	–	–	×	×	×	×	×	×	×	×		×	×				
Carbon Disulfide, CS_2	×	+	+	+	×	×	×	×	×	–	–	+	+	+	+				×	×			×	–				

CARBONS & CERAMICS / RUBBERS

Chemical	Carbon & Graphite	Glass, "Pyrex" brand	Silicaware	Silicate Cements	Chemical Stoneware	Transite (asbestos & cement)	Chemical Porcelain	Concrete—Unbonded	Concrete—Motor Bonded	Hard Rubber (Natural)	Soft Rubber (Natural)	Neoprene	Butadiene Derivatives	Nitrile Rubber (Chemigum)
Amyl Chloride, $C_5H_{11}Cl$	×	×	×		×		×			–	O	–	×	
Antimony Trichloride, $SbCl_3$		×	×		×		×						×	
Arsenic Acid, $HAsO_8$	×	×					×					×		
Barium Carbonate, $BaCO_3$		×					×							
Barium Hydroxide, $Ba(OH)_2$		×					×							
Barium Sulfide, BaS		–					×			×				
Benzaldehyde, C_6H_5CHO	×	×	×		×	×	×	×		O	O	O	×	
Benzene, C_6H_6	×	×		×	×		×			O	O	O		O
Benzoic Acid, C_6H_5COOH		×					×							×
Borax, $Na_2B_4O_7$	×	×	×		×	×	×	×		×	×	×	×	O ×
Boric Acid, H_3HO_3	×	×		×	×		×			×	×	×		× ×
Bromine, Wet, Br_2	O	×	×		×		×	×		–	O		–	× ×
Butanol, C_4H_2OH		×					×			–	–	–		× ×
Butyl Acetate, $C_4H_9COOCH_3$	×	×	×		×	×	×	×	×	O	O	O	O	
Butyric Acid, C_3H_7COOH		×					×			×	–	O		
Calcium Bisulfate, $CaHSO_4$		×					×							
Calcium Bisulfate, $CaHSO_3$		×	×		×		×			×	×	×	×	
Calcium Carbonate, $CaCO_3$		×					×			×	×	–		
Calcium Chlorate, $CaClO_3$	×	×					×							
Calcium Chloride, $CaCl_2$	×	×	×	×	×	×	×	×		×	×	×	×	×
Calcium Hydroxide, $Ca(OH)_2$		×					×			×	×	×		
Calcium Hypochlorite, $Ca(OCl)_2$	×	×	×	O	×		×	×		×	×	O	×	–
Calcium Sulfate, $CaSO_4$		×					×			×	×	×		
Carbon Dioxide (Dry), CO_2		×					×			×	×	×		
Carbon Dioxide (Wet or H_2CO_3)	×	×	×		×	×	×	×	×	×	×	×	×	
Carbon Disulfide, CS_2	×	×	×	×	×		×			O	O	O	×	

THERMOPLASTICS

Chemical	Viton	Asphaltic, Bitumastic	Cellulose Acetate	Cellulose Acetatebutyrate	Ethyl Cellulose (Ethocel)	Cellulose Nitrate	Acrylic (Lucite, Plexiglas)	Coumarone Resins	Polyethylene	Polyvinyl Chloride, Rigid or Unplasticized	Tygon (PVC & Copolymers)	Saran (Vinyl Chloride, Vinyldene Chloride)	Kel-F (Polytrifluorochloroethylene)	Teflon (Polytetrafluoroethylene)	Uscolite CP (Styrene–Acrylonitrile–Butadiene)
Amyl Chloride, $C_5H_{11}Cl$	–	O								O		×	×	×	
Antimony Trichloride, $SbCl_3$	×				–					×		×	×	×	×
Arsenic Acid, $HAsO_8$	×								–	×			×	×	
Barium Carbonate, $BaCO_3$	×									×			×	×	×
Barium Hydroxide, $Ba(OH)_2$	×	×							×	×			×	×	×
Barium Sulfide, BaS	×								×	–	×	×	×	×	×
Benzaldehyde, C_6H_5CHO		O							O	O		–	×	×	
Benzene, C_6H_6	–	O	–	O	O		O		–	O	O	–	×	×	
Benzoic Acid, C_6H_5COOH		×								×	×	×	×	×	
Borax, $Na_2B_4O_7$	×	×	×	×	×	×	×	×		×			×	×	×
Boric Acid, H_3HO_3	×									–	×	×	×	×	×
Bromine, Wet, Br_2									O	O	–	O	×	×	
Butanol, C_4H_2OH	–	O					×			×		×	×	×	×
Butyl Acetate, $C_4H_9COOCH_3$	O	O	O	O	O	O	O			O		O	×	×	
Butyric Acid, C_3H_7COOH	O	O								O		–	×	×	
Calcium Bisulfate, $CaHSO_4$													×	×	
Calcium Bisulfate, $CaHSO_3$										×		×	×	×	×
Calcium Carbonate, $CaCO_3$	×									×		×	×	×	
Calcium Chlorate, $CaClO_3$		×								×			×	×	×
Calcium Chloride, $CaCl_2$	×	×	×	×	×	×	×	×	×	×	×	×	×	×	×
Calcium Hydroxide, $Ca(OH)_2$	×	×								×		×	×	×	×
Calcium Hypochlorite, $Ca(OCl)_2$		–							×	×	×	×	×	×	×
Calcium Sulfate, $CaSO_4$	×	×								×		×	×	×	×
Carbon Dioxide (Dry), CO_2		×								×		×		×	×
Carbon Dioxide (Wet or H_2CO_3)		×	×	×	×	×	×	×	×	×	×	×		×	×
Carbon Disulfide, CS_2	×	O	O	O	O	×	×		O	O		–	×	×	

THERMOSETTING PLASTICS / WOODS

Chemical	Penton (Chlorinated Polyether)	Shellac Compounds	Organic Polysulfides	Polystyrene (Styron)	Vinylidene Chlorides	Vinyl Chloride Acetates	Cast Phenol Formaldehyde	Haveg 41 (Phenolic w. Asbestos)	Heresite (Phenol Formaldehyde)	Molded Phenolformald (Durez)	Phenol Furfural Plastics	Urea Formaldehyde	Casein Plastics	Epoxy Resins	Furane Resins (Haveg 61, Duralon)	Silicone Resins	Permanite (Furan, Glass Fiber)	Nylon (Adipic Acid–Hexameth, Diamine)	Durcon 6 (Modified Epoxy)	Cypress	Fir	Maple	Oak	Pine	Redwood
Amyl Chloride, $C_5H_{11}Cl$	×				×	×	–			–	–		×	×		×									
Antimony Trichloride, $SbCl_3$	×													×		O	×								
Arsenic Acid, $HAsO_8$	×															×	×								
Barium Carbonate, $BaCO_3$	×													×		×	×								
Barium Hydroxide, $Ba(OH)_2$	×													×		×	×								
Barium Sulfide, BaS	×															×	×								
Benzaldehyde, C_6H_5CHO	–	O			×		×			×	×			–		×	×								
Benzene, C_6H_6	+			O		O		–	×	×	×			×	×	×	×	×**							
Benzoic Acid, C_6H_5COOH	×													×		O	+**								
Borax, $Na_2B_4O_7$	×	×		×	×	×	×			×	×					×	×	×			–	–	–	×	–
Boric Acid, H_3HO_3	×							×	×	×	×				×	×	×	×							
Bromine, Wet, Br_2	–			–				O	×	–	–	–		O	O	O	O	×**							
Butanol, C_4H_2OH	×													–		×	×								
Butyl Acetate, $C_4H_9COOCH_3$	+			O	O	O	–			–	–	O	×	×		×									
Butyric Acid, C_3H_7COOH	×													–		O	×								
Calcium Bisulfate, $CaHSO_4$																×									
Calcium Bisulfate, $CaHSO_3$	×						×			×	×					×									
Calcium Carbonate, $CaCO_3$	×															×	×								
Calcium Chlorate, $CaClO_3$	×													×		×									
Calcium Chloride, $CaCl_2$	×	×	×	×	×	×	–	×	×	–	–			×	×	×	O	×		×	×	×	×	×	–
Calcium Hydroxide, $Ca(OH)_2$	×													×		×	×	×							
Calcium Hypochlorite, $Ca(OCl)_2$	×		×	×		×	O	–	–	O	O			×	×	O	×	O			O	O	O	–	O
Calcium Sulfate, $CaSO_4$	×													×		×	×								
Carbon Dioxide (Dry), CO_2	×													×		O									
Carbon Dioxide (Wet or H_2CO_3)	×	×	×	×	×	×	×	×	×	×	×	×	×			×	×				×	×	×	×	×
Carbon Disulfide, CS_2	–	–		O	O		×			×	×			O	×	×	×	×							

**Note*: Duriron is as shown. Durichlor is also satisfactory on chlorides and HCl.
**Durcon 5 would be the preferred formula.

TABLE A.3 Continued

Chemical Resistance of Materials

MATERIALS
× — *Very Good Service*
+ — *Moderate Service*
– — *Limited or Variable Service*
O — *Unsatisfactory*
Blank — No Information

CHEMICALS
Solids Assumed in Solution
Room Temperatures Assumed
Unless Otherwise Stated

METALS

CHEMICALS	Carbon Steel; Fe	Cast Iron & Ductile Iron; Fe	304 Stainless Steel; Fe, 18Cr, 8Ni	316 Stainless Steel; Fe 16Cr, 10Ni, 2Mo	347 Stainless Steel; Fe, 17Cr, 9Ni; (Cx10)Cb	Ni-Resist Iron; Fe 14Ni, 2Cr, 2Si	Durimet 20; Carpenter 20; Fe; 4Cu, 20Cr, 29Ni, 2Mo, 1Si	Worthite; 3Mo, 2Cu, Fe, 20Cr, 24Ni, 3Si	Durorpm; Fe, 14Si; Durichlor, Fe, 14Si, 3Mo*	Copper; Brass; Bronzes; Everdur	Aluminum; Al (and Alloys)	Lead; Pb	Monel; 67Ni, 30Cu, 1.4Fe	Nickel; Ni	Inconel; 76Ni, 15Cr, 8Fe
Magnesium Chloride, $MgCl_2$	–	–	–	+	–	×	×	×	×	+	–	O	+	+	+
Magnesium Hydroxide, $Mg(OH)_2$	×		×	×	+		×	×	+	–	O		×	×	×
Magnesium Sulfate, $MgSO_4$	+	O	+	+	+	×	×	×	×	+	×	+	+	+	+
Maleic Acid; $CO_2HC_2H_2CO_2H$	O		×	×			×	×	×						
Malic Acid, $CO_2HCH_2CHOHCO_2H$	O		×	×	×		+		×	O	–		+	+	+
Mercuric Chloride, $HgCl_2$	O		O	O	O		–	–	+	O	O	–	–		+
Mercury, Hg	×		×	×	×		×		×	O	O	O	+	+	×
Methanol, (Conc.), CH_3OH	+	+	+	+	+	×	×	×	×	+	–	+	+	+	+
Methanol, (Dilute)	–						×		×	–	–				
Methyl Chloride, CH_3Cl	O		–	–	+		×	×	×	×		×	×	×	×
Naphtha, Petroleum	×		×	×	×	×	×		×	×	×		+	+	+
Nickel Chloride, $NiCl_2$	O		–	–			+	O	+	–	O		–	×	
Nickel Sulfate, $NiSO_4$	O	O	–	–			×	×	×	+	–	+	+	–	–
Nitrating Acid (>15% H_2SO_4)	×	×	–	–	–		×	×	×	O	O		O	O	
Nitrating Acid (<15% H_2SO_4)	O		–	–	–		×	×	×	O	O		O	O	
Nitrating Acid (<15% HNO_3)	O		–	O	–		×		×	O	O		O	O	
Nitrating Acid (>1% Acid)	O		–	×	–		×		×	O	O		O	O	
Nitric Acid (Conc.), HNO_3	O	O	+	+	×		×	×	×	O	–	–	O	O	–
Nitric Acid, Dilute	O	O	×	×	×	O	×	×	×	O	O	O	O	O	O
Nitrobenzene, $C_6N_5NO_2$	×		×	×			×	×	×			O			
Nitrous Acid, HNO_2	O		×	×	–		×	×	×	×			×	O	
Oleic Acid, $C_8H_{17}CH{:}CH(CH_2)_7CO_2H$	–	–	–	×	×	–	×	×	×	–	×	O	×	×	×
Oxalic Acid, CO_2HCO_2H	O	O	–	–	–	–	×	×	×	+	–	O	+	–	×
Phenol (Conc.), C_6H_5OH	+	+	–	×	–	×	×	×	×	O	–	–	+	+	×
Phenol (Dilute)	+	+	×	×	×		×		×	–	×	–	+	+	–
Phosphoric Acid (100%), H_3PO_4	O	O	–	–	×		×		×		O	+	–	O	–
Phosphoric Acid (>45% Hot)	O	O	O	O	×	O	×	×	×	O	O	+	O	O	O

METALS (continued)

CHEMICALS	Hastelloy B; Ni, 26Mo, 4Fe	Hastelloy C; Ni, 16Mo, 4Fe, 14Cr, 4W	Hastelloy D; Ni, 8Si, 3Cu	Chlorimet 3; 3Fe, 1Si, 60Ni, 18Mo, 18Cr	Chlorimet 2; 63Ni, 32Mo, 3Fe, 1Si	Stellite; Co, 28Cr, 4W	Zirconium; Z	Tantalum; Ta	Silver; Ag	Platinum; Pt	Dowmetal; (Mg alloys)	Titanium; Ti	Molybdenum; Mo
Magnesium Chloride, $MgCl_2$	×	×	×	×	×	–		×	×			×	
Magnesium Hydroxide, $Mg(OH)_2$	×	×	×	×	×			×	×			×	
Magnesium Sulfate, $MgSO_4$	×	+	×	×	×		×	×	×			×	
Maleic Acid; $CO_2HC_2H_2CO_2H$	×			×	×				×				
Malic Acid, $CO_2HCH_2CHOHCO_2H$	×	×	×	×				×					
Mercuric Chloride, $HgCl_2$	O	×	O	+	O		×	×	O			×	
Mercury, Hg	×	×	×	×	×				O			–	
Methanol, (Conc.), CH_3OH	×	×	+	×	×		×	×	×				
Methanol, (Dilute)	×	×	–	×	×			×					
Methyl Chloride, CH_3Cl				×	×			×					
Naphtha, Petroleum	×	×	×	×	×								
Nickel Chloride, $NiCl_2$	O	×	O	×	×		×	×				×	
Nickel Sulfate, $NiSO_4$	O	+	O	×	×		×	×					
Nitrating Acid (>15% H_2SO_4)	O	×	O	×	×	×		–	×	×		–	
Nitrating Acid (<15% H_2SO_4)	O	×	O	×	×	×		–	O	×		×	
Nitrating Acid (<15% HNO_3)	O	×	O	×	O	×	O	–	O	×		–	
Nitrating Acid (>1% Acid)	O	×	O	×		×	O	–	O	×			
Nitric Acid (Conc.), HNO_3	O	+	O	×	O	×	×	×	O	×		×	+
Nitric Acid, Dilute	O	×	O	×	O	×	×	×	O	×		×	O
Nitrobenzene, $C_6N_5NO_2$				×	×			×					
Nitrous Acid, HNO_2				×				×					
Oleic Acid, $C_8H_{17}CH{:}CH(CH_2)_7CO_2H$	×	×	×	×	×	×		×	×	×			
Oxalic Acid, CO_2HCO_2H	+	+	+	×	×	×	×	×	×	×		–	+
Phenol (Conc.), C_6H_5OH	+	+	+	×	×	×	×	×	×	×	×	–	
Phenol (Dilute)	+	+	+	×	×		×						
Phosphoric Acid (100%), H_3PO_4	+	×	×	×	×	×	×	×	+	×			×
Phosphoric Acid (>45% Hot)	–	–	–	×	×	×	–	×		×			

CARBONS & CERAMICS / RUBBERS

CHEMICALS	Carbon & Graphite	Glass, "Pyrex" brand	Silicaware	Silicate Cements	Chemical Stoneware	Transite (asbestos & cement)	Chemical Porcelain	Concrete—Unbonded	Concrete—Motor Bonded	Hard Rubber (Natural)	Soft Rubber (Natural)	Neoprene	Butadiene Derivatives	Nitrile Rubber (Chemigum)
Magnesium Chloride, $MgCl_2$		×	×	×	×	×	×	×	×	×	×	×	×	
Magnesium Hydroxide, $Mg(OH)_2$		×					×			×			×	
Magnesium Sulfate, $MgSO_4$		×	×	×	×	×	×	×	×	×	×	×	×	
Maleic Acid; $CO_2HC_2H_2CO_2H$		×					×					×		
Malic Acid, $CO_2HCH_2CHOHCO_2H$		×	×		×		×			×	×	×	×	
Mercuric Chloride, $HgCl_2$		×					×			×	–			
Mercury, Hg		×					×							
Methanol, (Conc.), CH_3OH		×		×			×			–	×	–		×
Methanol, (Dilute)		×					×			×	×			
Methyl Chloride, CH_3Cl		×					×			O	O			
Naphtha, Petroleum	×	×	×		×		×			–	O	×	×	
Nickel Chloride, $NiCl_2$	×	×					×			×	×			
Nickel Sulfate, $NiSO_4$	×	×					×			×	×	×		
Nitrating Acid (>15% H_2SO_4)	O	×			×		×			–	–			
Nitrating Acid (<15% H_2SO_4)	O	×			×		×			–	–			
Nitrating Acid (<15% HNO_3)	O	×			×		×			–	–			
Nitrating Acid (>1% Acid)	O	×			×		×			–	–			
Nitric Acid (Conc.), HNO_3	O	×	×	×	×		×			O	O	O	O	O
Nitric Acid, Dilute	–	×	×		×		×			–	O	O	×	O
Nitrobenzene, $C_6N_5NO_2$		×					×			O	O	O		
Nitrous Acid, HNO_2		×	×		×		×			–	O	O	×	
Oleic Acid, $C_8H_{17}CH{:}CH(CH_2)_7CO_2H$	×	×	×		×	×	×	×	×	O	O	O	×	
Oxalic Acid, CO_2HCO_2H	×	×	×	×	×	×	×	×		×	×	–	×	
Phenol (Conc.), C_6H_5OH	×	×	×		×	×	×	×		–	O	–	×	
Phenol (Dilute)		×					×			–		–		
Phosphoric Acid (100%), H_3PO_4	+	–	×		×	×	×	×	×	×	×	×	×	
Phosphoric Acid (>45% Hot)	×	–		×			O			–	–	–		×

THERMOPLASTICS

CHEMICALS	Viton	Asphaltic, Bitumastic	Cellulose Acetate	Cellulose Acetatebutyrate	Ethyl Cellulose (Ethocel)	Cellulose Nitrate	Acrylic (Lucite, Plexiglas)	Coumarone Resins	Polyethylene	Polyvinyl Chloride, Rigid or Unplasticized	Tygon (PVC & Copolymers)	Saran (Vinyl Chloride, Vinyldene Chloride)	Kel-F (Polytrifluorochloroethylene)	Teflon (Polytetrafluoroethylene)	Uscolite CP (Styrene–Acrylonitrile–Butadiene)
Magnesium Chloride, $MgCl_2$	×	×	×	×		×		×	×	×	×	×	×	×	×
Magnesium Hydroxide, $Mg(OH)_2$	×									×		×	×	×	×
Magnesium Sulfate, $MgSO_4$	×	×	×	×		×		×	×	×		×	×	×	×
Maleic Acid; $CO_2HC_2H_2CO_2H$	+	O								–		–	×	×	×
Malic Acid, $CO_2HCH_2CHOHCO_2H$	–									×			×	×	
Mercuric Chloride, $HgCl_2$	×	×	×							–	×	×	×	×	×
Mercury, Hg	×	×								–		×		×	
Methanol, (Conc.), CH_3OH	–	×					O		–	×		×	×	×	×
Methanol, (Dilute)										×			×	×	×
Methyl Chloride, CH_3Cl	–	O								O		–	×	×	
Naphtha, Petroleum	+	O				×	×			×			×	×	
Nickel Chloride, $NiCl_2$	×	×							×	×	×	×	×	×	×
Nickel Sulfate, $NiSO_4$	×	×							×	×	×	×	×	×	×
Nitrating Acid (>15% H_2SO_4)		O									O		×	×	
Nitrating Acid (<15% H_2SO_4)		O									–		×	×	
Nitrating Acid (<15% HNO_3)		O									–		×	×	
Nitrating Acid (>1% Acid)		O									–		×	×	
Nitric Acid (Conc.), HNO_3	O				O				O	O	O	–	×	×	O
Nitric Acid, Dilute	–	–			–		×		–	×	–	×	×	×	×
Nitrobenzene, $C_6N_5NO_2$	O	O							–	O		–	×	×	
Nitrous Acid, HNO_2		×											×	×	
Oleic Acid, $C_8H_{17}CH{:}CH(CH_2)_7CO_2H$	+	O	×	×		×			O	×	×	×	×	×	×
Oxalic Acid, CO_2HCO_2H	–	×							×	×	×	×	×	×	×
Phenol (Conc.), C_6H_5OH	×	O	O	O	–		×			–	–	–	×	×	O
Phenol (Dilute)		–								×	–	–	×	×	O
Phosphoric Acid (100%), H_3PO_4		×	×		O				×	×	×		×	×	×
Phosphoric Acid (>45% Hot)	+								–	×	×		×	×	×

THERMOSETTING PLASTICS

CHEMICALS	Penton (Chlorinated Polyether)	Shellac Compounds	Organic Polysulfides	Polystyrene (Styron)	Vinylidene Chlorides	Vinyl Chloride Acetates	Cast Phenol Formaldehyde	Haveg 41 (Phenolic w. Asbestos)	Heresite (Phenol Formaldehyde)	Molded Phenolformald (Durez)	Phenol Furfural Plastics	Urea Formaldehyde	Casein Plastics	Epoxy Resins	Furane Resins (Haveg 61, Duralon)
Magnesium Chloride, $MgCl_2$	×	×	×	×	×	×	×	×	×	×	×	×		×	×
Magnesium Hydroxide, $Mg(OH)_2$	×														
Magnesium Sulfate, $MgSO_4$	×	×	×	×	×	×	×			×	×	×		×	×
Maleic Acid; $CO_2HC_2H_2CO_2H$	×													×	
Malic Acid, $CO_2HCH_2CHOHCO_2H$	×			×											
Mercuric Chloride, $HgCl_2$	×													×	
Mercury, Hg	×													×	
Methanol, (Conc.), CH_3OH	×			×				–		×	×			×	×
Methanol, (Dilute)	×			×						×	×			×	
Methyl Chloride, CH_3Cl	×													×	
Naphtha, Petroleum	×			–	–	–	–			–	–		×	×	
Nickel Chloride, $NiCl_2$	×													×	×
Nickel Sulfate, $NiSO_4$	×			–						×	×			×	×
Nitrating Acid (>15% H_2SO_4)								O	×					O	
Nitrating Acid (<15% H_2SO_4)								–	×						
Nitrating Acid (<15% HNO_3)								–	×						
Nitrating Acid (>1% Acid)								–	×						
Nitric Acid (Conc.), HNO_3	O			O	O	O	O	O	×	O	O			O	O
Nitric Acid, Dilute	+			×	–	–	–	–	×	–	–			O	
Nitrobenzene, $C_6N_5NO_2$	+													–	
Nitrous Acid, HNO_2	×		×												
Oleic Acid, $C_8H_{17}CH{:}CH(CH_2)_7CO_2H$	×			×	×	×	–	×	×	×	–	×		×	
Oxalic Acid, CO_2HCO_2H	×			×			×	×	×	×	×			×	×
Phenol (Conc.), C_6H_5OH	+			O			–	–	×	–	–				×
Phenol (Dilute)				O						×	×			–	×
Phosphoric Acid (100%), H_3PO_4					×	×		×	×						–
Phosphoric Acid (>45% Hot)								×	×	O	O			×	×

THERMOSETTING PLASTICS (continued) / WOODS

CHEMICALS	Silicone Resins	Permanite (Furan, Glass Fiber)	Nylon (Adipic Acid–Hexameth, Diamine)	Durcon 6 (Modified Epoxy)	Cypress	Fir	Maple	Oak	Pine	Redwood
Magnesium Chloride, $MgCl_2$	×	×								
Magnesium Hydroxide, $Mg(OH)_2$	×	×								
Magnesium Sulfate, $MgSO_4$	×	×								
Maleic Acid; $CO_2HC_2H_2CO_2H$	×	+**								
Malic Acid, $CO_2HCH_2CHOHCO_2H$	×	×								
Mercuric Chloride, $HgCl_2$	×									
Mercury, Hg	×	×								
Methanol, (Conc.), CH_3OH	×	×								
Methanol, (Dilute)	×	×								
Methyl Chloride, CH_3Cl	×	×								
Naphtha, Petroleum	×	×								
Nickel Chloride, $NiCl_2$	×									
Nickel Sulfate, $NiSO_4$	×	×								
Nitrating Acid (>15% H_2SO_4)	×**									
Nitrating Acid (<15% H_2SO_4)	×									
Nitrating Acid (<15% HNO_3)	×									
Nitrating Acid (>1% Acid)	×**									
Nitric Acid (Conc.), HNO_3	–	O	O	×						
Nitric Acid, Dilute	–	O	×							
Nitrobenzene, $C_6N_5NO_2$	×									
Nitrous Acid, HNO_2	×									
Oleic Acid, $C_8H_{17}CH{:}CH(CH_2)_7CO_2H$	×	×	×			×	×	×		–
Oxalic Acid, CO_2HCO_2H	×	O	×							
Phenol (Conc.), C_6H_5OH	×	O	×							
Phenol (Dilute)	×									
Phosphoric Acid (100%), H_3PO_4	–	–	–	×						
Phosphoric Acid (>45% Hot)	O	×								

**Note*: Duriron is as shown. Durichlor is also satisfactory on chlorides and HCl.
**Durcon 5 would be the preferred formula.

TABLE A.3 Continued

Chemical Resistance of Materials

MATERIALS
× — *Very Good Service*
\+ — *Moderate Service*
– — *Limited or Variable Service*
O — *Unsatisfactory*
Blank — No Information

CHEMICALS
Solids Assumed in Solution
Room Temperatures Assumed
Unless Otherwise Stated

METALS

CHEMICALS	Carbon Steel; Fe / Cast Iron & Ductile Iron; Fe / 304 Stainless Steel; Fe, 18Cr, 8Ni / 316 Stainless Steel; Fe 16Cr, 10Ni, 2Mo / 347 Stainless Steel; Fe, 17Cr, 9Ni; (Cx10)Cb	Ni-Resist Iron; Fe 14Ni, 2Cr, 2Si / Durimet 20; Carpenter 20; Fe; 4Cu, 20Cr, 29Ni, 2Mo, 1Si / Worthite; 3Mo, 2Cu, Fe, 20Cr, 24Ni, 3Si / Durorpm; Fe, 14Si; Durichlor, Fe, 14Si, 3Mo* / Copper; Brass; Bronzes; Everdur	Aluminum; Al (and Alloys) / Lead; Pb / Monel; 67Ni, 30Cu, 1.4Fe / Nickel; Ni / Inconel; 76Ni, 15Cr, 8Fe	Hastelloy B; Ni, 26Mo, 4Fe / Hastelloy C; Ni, 16Mo, 4Fe, 14Cr, 4W / Hastelloy D; Ni, 8Si, 3Cu / Chlorimet 3; 3Fe, 1Si, 60Ni, 18Mo, 18Cr / Chlorimet 2; 63Ni, 32Mo, 3Fe, 1Si	Stellite; Co, 28Cr, 4W / Zirconium; Z / Tantalum; Ta / Silver; Ag / Platinum; Pt	Dowmetal; (Mg alloys) / Titanium; Ti / Molybdenum; Mo
Phosphoric Acid (>45% Cold)	O O – × ×	– × – × O	O + – – –	+ × + × ×	× × × × ×	–
Phosphoric Acid (<45% Cold)	O O – × ×	– × – × O	– + – – –	+ × + × ×	× × × × ×	×
Phosphoric Anhydride, Dry or Moist	O × ×	× ×		× ×		
Phosphoric Anhydride Molten, P_2O_5	× × ×	× × O	O O O	× ×		
Phthalic Anhydride, $C_6H_4(CO)_2O$	× ×	× × ×		× × × ×	×	
Picric Acid, Solution, $HOC_6H_2(NO_3)_3$	× × +	× × × O	– – O O O	× × × × ×		
Potassium Bromide, KBr	O – –	× ×	– × + ×	× × × × ×	× O	×
Potassium Carbonate, K_2CO_3	– × ×	– × × × –	O × × × +	× × × × ×	–	
Potassium Chlorate, $KClO_3$	× × ×	× – × –	+ + + +	O × O × O	×	
Potassium Chloride, KCl	× – –	× × × × –	– + +	× × × × ×	×	×
Potassium Cyanide, KCN	× × ×	× × × O	O ×	× × × × ×	× O ×	
Potassium Dichromate, $K_2Cr_2O_7$	× × –	× × × ×	× – – +	O × O × ×	× × O	×
Potassium Ferrocyanide, $K_4Fe(CN)_6$	O × ×	× × × ×	– × × × ×	× × × × ×	× O	
Potassium Hydroxide, KOH	+ – + + –	– × × + O	O O × × ×	+ + + × ×	× – ×	– +
Potassium Nitrate, KNO_3	× × × ×	× × × ×	× – + + +	O × O × ×	× O	
Potassium Permanganate, $KMnO_4$	+ + + + +	× × × ×	× O – + +	O × O × O	×	×
Potassium Sulfate, K_2SO_4	× + + +	× × × × +	× × + + +	+ + + × ×	×	×
Potassium Sulfide, K_2S	– × ×	– × ×		× ×	O	
Pyrogallol, $C_6H_3(OH)_3$	× ×	× × ×	+	× × × × ×		
Silver Nitrate, $AgNO_3$	× ×	× × × O	O O +	+ + + × ×	× ×	
Sodium, Molten 210 –400°F	× × × × ×	× × × ×	× × × ×	× × × ×	× ×	– ×
Sodium Acetate, $NaCH_3COO$	O – ×	× × × ×	– × +	× × × × ×		×
Sodium Bicarbonate, $NaHCO_3$	– × × ×	× × × × –	× × × × ×	× × × × ×	× × ×	
Sodium Bisulfate, $NaHSO_4$	O O – × ×	× × × –	O × +	× × × × ×	× – ×	×
Sodium Bisulfite, $NaHSO_3$	O × ×	× × –	× ×	× ×		

CARBONS & CERAMICS, RUBBERS, THERMOPLASTICS

CHEMICALS	Carbon & Graphite / Glass, "Pyrex" brand / Silicaware / Silicate Cements	Chemical Stoneware / Transite (asbestos & cement) / Chemical Porcelain / Concrete—Unbonded / Concrete—Motor Bonded	Hard Rubber (Natural) / Soft Rubber (Natural) / Neoprene / Butadiene Derivatives / Nitrile Rubber (Chemigum)	Viton / Asphaltic, Bitumastic / Cellulose Acetate / Cellulose Acetatebutyrate / Ethyl Cellulose (Ethocel)	Cellulose Nitrate / Acrylic (Lucite, Plexiglas) / Coumarone Resins / Polyethylene / Polyvinyl Chloride, Rigid or Unplasticized	Tygon (PVC & Copolymers) / Saran (Vinyl Chloride, Vinylidene Chloride) / Kel-F (Polytrifluorochloroethylene) / Teflon (Polytetrafluoroethylene) / Uscolite CP (Styrene–Acrylonitrile–Butadiene)
Phosphoric Acid (>45% Cold)	× ×	– –	– – –	×	× ×	× × × ×
Phosphoric Acid (<45% Cold)	× ×	– –	× × –		× ×	× × × ×
Phosphoric Anhydride, Dry or Moist	× ×				×	×
Phosphoric Anhydride Molten, P_2O_5	O					×
Phthalic Anhydride, $C_6H_4(CO)_2O$	×	×	× ×	×		× ×
Picric Acid, Solution, $HOC_6H_2(NO_3)_3$	×	×	–	O	O	× O ×
Potassium Bromide, KBr	×	×	× ×		×	× × ×
Potassium Carbonate, K_2CO_3	–	×	× ×	× ×	× –	× × ×
Potassium Chlorate, $KClO_3$	×	×		×	×	× × ×
Potassium Chloride, KCl	×	×	× ×	× ×	× ×	× × ×
Potassium Cyanide, KCN	–	×	× ×	×	×	× × × ×
Potassium Dichromate, $K_2Cr_2O_7$	×	×	– × ×	×	× ×	× × × ×
Potassium Ferrocyanide, $K_4Fe(CN)_6$	×	×		×	×	× ×
Potassium Hydroxide, KOH	– O	–	× × × O	× ×	× ×	– × ×
Potassium Nitrate, KNO_3	×	×	× ×	×	× ×	× × ×
Potassium Permanganate, $KMnO_4$	×	×	– O O	× O	× ×	– × ×
Potassium Sulfate, K_2SO_4	× ×	×	× × ×	× ×	×	× × ×
Potassium Sulfide, K_2S	–	×			×	× ×
Pyrogallol, $C_6H_3(OH)_3$	×	×			×	× ×
Silver Nitrate, $AgNO_3$	×	×	× ×	×	× × ×	× × ×
Sodium, Molten 210 –400°F	–	O		O	O O	O O –
Sodium Acetate, $NaCH_3COO$	×	×	×	O ×	×	× ×
Sodium Bicarbonate, $NaHCO_3$	× × ×	× × × × ×	× × × ×	× × ×	× × ×	× × ×
Sodium Bisulfate, $NaHSO_4$	× × ×	× ×	× × × ×		×	× × ×
Sodium Bisulfite, $NaHSO_3$	× × ×	× ×	× × – ×	×	× ×	× × × ×

THERMOSETTING PLASTICS, WOODS

CHEMICALS	Penton (Chlorinated Polyether) / Shellac Compounds / Organic Polysulfides / Polystyrene (Styron) / Vinylidene Chlorides	Vinyl Chloride Acetates / Cast Phenol Formaldehyde / Haveg 41 (Phenolic w. Asbestos) / Heresite (Phenol Formaldehyde) / Molded Phenolformald (Durez)	Phenol Furfural Plastics / Urea Formaldehyde / Casein Plastics / Epoxy Resins / Furane Resins (Haveg 61, Duralon)	Silicone Resins / Permanite (Furan, Glass Fiber) / Nylon (Adipic Acid–Hexameth, Diamine) / Durcon 6 (Modified Epoxy) / Cypress	Fir / Maple / Oak / Pine / Redwood
Phosphoric Acid (>45% Cold)	× ×	× ×	×	– ×	
Phosphoric Acid (<45% Cold)	×	× ×		– × O ×	
Phosphoric Anhydride, Dry or Moist				×	
Phosphoric Anhydride Molten, P_2O_5				×	
Phthalic Anhydride, $C_6H_4(CO)_2O$			×	×	
Picric Acid, Solution, $HOC_6H_2(NO_3)_3$	×			×	
Potassium Bromide, KBr	×			×	
Potassium Carbonate, K_2CO_3	×		×	× ×	
Potassium Chlorate, $KClO_3$	×			O ×	
Potassium Chloride, KCl	×	×	× ×	× × ×	
Potassium Cyanide, KCN	×			× ×	
Potassium Dichromate, $K_2Cr_2O_7$	×		×	×**	
Potassium Ferrocyanide, $K_4Fe(CN)_6$	×		×	× ×	
Potassium Hydroxide, KOH	× ×	O O	O × ×	× × ×	
Potassium Nitrate, KNO_3	×		×	×	
Potassium Permanganate, $KMnO_4$	× ×	×	× × ×	O ×	
Potassium Sulfate, K_2SO_4	× ×	× ×	× × ×	× × ×	
Potassium Sulfide, K_2S				× ×	
Pyrogallol, $C_6H_3(OH)_3$				×**	
Silver Nitrate, $AgNO_3$	×	– –	×	× ×	
Sodium, Molten 210 –400°F	O O O	O O		×	
Sodium Acetate, $NaCH_3COO$	×		×	– ×	
Sodium Bicarbonate, $NaHCO_3$	× × × ×	× × × ×	× × ×	× × ×	
Sodium Bisulfate, $NaHSO_4$	× × ×	× – × × ×	– × × ×	× ×	
Sodium Bisulfite, $NaHSO_3$	×	×	× ×	O	× × × ×

**Note*: Duriron is as shown. Durichlor is also satisfactory on chlorides and HCl.
**Durcon 5 would be the preferred formula.

TABLE A.3 Continued

Chemical Resistance of Materials

MATERIALS
× — *Very Good Service*
+ — *Moderate Service*
– — *Limited or Variable Service*
O — *Unsatisfactory*
Blank — *No Information*

CHEMICALS
Solids Assumed in Solution
Room Temperatures Assumed
Unless Otherwise Stated

METALS

Chemical	Carbon Steel; Fe	Cast Iron & Ductile Iron; Fe	304 Stainless Steel; Fe, 18Cr, 8Ni	316 Stainless Steel; Fe 16Cr, 10Ni, 2Mo	347 Stainless Steel; Fe, 17Cr, 9Ni; (Cx10)Cb	Ni-Resist Iron; Fe 14Ni, 2Cr, 2Si	Durimet 20; Carpenter 20; Fe; 4Cu, 20Cr, 29Ni, 2Mo. 1Si	Worthite; 3Mo, 2Cu, Fe, 20Cr, 24Ni, 3Si	Duriron; Fe, 14Si; Durichlor, Fe, 14Si, 3Mo*	Copper; Brass; Bronzes; Everdur	Aluminum; Al (and Alloys)	Lead; Pb	Monel; 67Ni, 30Cu, 1.4Fe	Nickel; Ni	Inconel; 76Ni, 15Cr, 8Fe	Hastelloy B; Ni, 26Mo, 4Fe	Hastelloy C; Ni, 16Mo, 4Fe, 14Cr, 4W	Hastelloy D; Ni, 8Si, 3Cu	Chlorimet 3; 3Fe, 1Si, 60Ni, 18Mo, 18Cr	Chlorimet 2; 63Ni, 32Mo, 3Fe, 1Si	Stellite; Co, 28Cr, 4W	Zirconium; Z	Tantalum; Ta	Silver; Ag	Platinum; Pt	Dowmetal; (Mg alloys)	Titanium; Ti	Molybdenum; Mo
Sodium Borate $NaBO_2$			×	×			×		×	×	–		+			×	×	×	×	×			×					
Sodium Carbonate, Na_2CO_3	+	+	+	+	×	×	×	×	×	+	O	–	+	+	–	+	×	×	×	×	×		–	×	×	×	×	
Sodium Chlorate, $NaClO_3$	–		×	×			–	–	×	+	+				+	O	×	O	×	O			×				×	
Sodium Chloride, NaCl	+	+	–	–	–	–	×	×	+	+	–	+	+	+	–	+	+	–	×	×	–	×	×	×			×	
Sodium Cyanide, NaCN	×		×	×			×	×	×		O	×	×						×	×				O			×	
Sodium Fluoride, NaF	O		–	–			+	–	O	–	–		+			×	×	×	×	×		×	O				×	
Sodium Hydroxide, (Conc.), NaOH	–	–	–	–	×	×	×	×	–	–	O	O	×	×	×	×	+	×	×	×	×	×	O	×	×		×	
Sodium Hydroxide (Dilute)	×	+	+	+	×		×	×	+	×	O	O	–	–	–	×	+	×	×	×	×	×	–		×	–	×	+
Sodium Hydrosulfite							×		×	–	×		O	O	O	–	×	–	×	×								
Sodium Hypochlorite, NaOCl	O	O	–	–	–	–	–	–	×	O	O	O	O	O	O	O	–	O	+	O	–		–	×	×		×	
Sodium Hyposulfate	–		×	×			×		×	O	O		×						×	×				O				
Sodium Nitrate, $NaNO_3$	×	+	×	×	×	–	×	×	×	+	×	O	+	+	×	O	+	O	×	O	×		–	–	×			
Sodium Peroxide, Na_2O_2	–		×	×	+		×		×	–	–		×	+	+	+	×	+	×	×			O	O				
Sodium Phosphate, (Tri) Na_3PO_4	×		×	×	+		×		×	–	O		+	+		×	×	×	×	×			×	×			×	
Sodium Silicate, Na_2SiO_3	×		×	×	+	×	×		×	–	–		+	×	+	×	×	×	×	×			×					
Sodium Sulfate, Na_2SO_4	×	×	–	–	×	×	×	×	×	×	×	–	+	×	×	×	×	×	×	×	–		–	×			×	
Sodium Sulfide, Na_2S	+	×	–	×	×	×	×	×	×	O	O	–	+	+	+	×	×	×	×	×	×		–	O			×	
Sodium Sulfite, Na_2SO_3	+	×	–	–	–	×	×	×	O	–	–	×	–	–	–	O	×	O	×	×	×		–				×	
Stannic Chloride, $SnCl_4$	O		O	O	O		×	–	×	O	O	×	O	O	O	O	×	O	–	O			×				×	
Stannous Chloride, $SnCl_2$	O		O	–			×	–	×	O	O		O	O	O	×	×	×	×	×		×	×				×	
Stearic Acid $CH_3(CH_2)_{16}COOH$	–	–	×	×	×	–	×	×	×	–	–	–	×	×	×	×	×	×	×	×	×		×	×			×	
Sulfur, Molten, S	–	×	–	–	–	–	×	×	×	O	×	O	–	–	×	O	O	O	×	–	O		×	O			×	+
Sulfur Chloride, (Wet), S_2Cl_2	–		O	–		×	–	–	×	O	O		–	–					×					O				
Sulfur Dioxide (Dry), SO_2	×	×	×	×	–	×	×	×	×	×	×	×	O	O	O	O	×	O	×	–	×		×	O		×		
Sulfur Dioxide (Wet)	O		–	×			×		–	O		×	O	O	O	O	×	O	×				×				+	
Sulfur Trioxide, SO_3	×		×	×		×	×	×	–	–	×		O	O	O				×				O	O				
Sulfuric Acid (Fuming to 98%)	–	–	O	O		–	×	×	O	O	–	O	O	O	O	×	O	×	×	×		O	O	O			O	
Sulfuric Acid (Hot Conc.) H_2SO_4	O		O	O			+		×	O	O	–	O	O	O	O	O	×	+	+	O	O	–				O	

CARBONS & CERAMICS; RUBBERS

Chemical	Carbon & Graphite	Glass, "Pyrex" brand	Silicaware	Silicate Cements	Chemical Stoneware	Transite (asbestos & cement)	Chemical Porcelain	Concrete—Unbonded	Concrete—Motor Bonded	Hard Rubber (Natural)	Soft Rubber (Natural)	Neoprene	Butadiene Derivatives	Nitrile Rubber (Chemigum)
Sodium Borate $NaBO_2$		×	×			×	×	×		×	×	×	×	
Sodium Carbonate, Na_2CO_3	×	–	–	O	×	×	×	×		×	×	×	×	×
Sodium Chlorate, $NaClO_3$		×	×		×	×	×	×	×	×	×	×	×	
Sodium Chloride, NaCl	×	×		×	×		×			×	×	×		×
Sodium Cyanide, NaCN		–					×			×	×	×		
Sodium Fluoride, NaF		–					×			×	×			
Sodium Hydroxide, (Conc.), NaOH	×	–	O	O	O		O			×	×	×	×	×
Sodium Hydroxide (Dilute)	×	–	O	O	–		–			×	×	×	×	×
Sodium Hydrosulfite		×	×		×		×			×	–		×	
Sodium Hypochlorite, NaOCl	O	×			×		×			×	–	O	×	
Sodium Hyposulfate		×					×							
Sodium Nitrate, $NaNO_3$	×	×	×	×	×		×			×	–	×	×	×
Sodium Peroxide, Na_2O_2		–					×							
Sodium Phosphate, (Tri) Na_3PO_4		–	–		×	×	×			×	×	–	×	
Sodium Silicate, Na_2SiO_3		–					×	×		×	×	×	×	
Sodium Sulfate, Na_2SO_4	×	×			×		×			×	×	×		
Sodium Sulfide, Na_2S	×	–	–		×	×	×			×	–	×	×	
Sodium Sulfite, Na_2SO_3	×	×	×		×		×			×	×	O	×	
Stannic Chloride, $SnCl_4$		×					×							
Stannous Chloride, $SnCl_2$		×					×			×	×	–		
Stearic Acid $CH_3(CH_2)_{16}COOH$	×	×			×		×			–		O		
Sulfur, Molten, S	–	×				–		–						
Sulfur Chloride, (Wet), S_2Cl_2		×					×							
Sulfur Dioxide (Dry), SO_2	×	×	×		×		×			–	–	O	–	
Sulfur Dioxide (Wet)	×	×	×				×			–		O		
Sulfur Trioxide, SO_3		×			×		×			×	×	O	×	
Sulfuric Acid (Fuming to 98%)	O	×	×	×								O		O
Sulfuric Acid (Hot Conc.) H_2SO_4		×								O	O	O		

THERMOPLASTICS

Chemical	Viton	Asphaltic, Bitumastic	Cellulose Acetate	Cellulose Acetatebutyrate	Ethyl Cellulose (Ethocel)	Cellulose Nitrate	Acrylic (Lucite, Plexiglas)	Coumarone Resins	Polyethylene	Polyvinyl Chloride, Rigid or Unplasticized	Tygon (PVC & Copolymers)	Saran (Vinyl Chloride, Vinyldene Chloride)	Kel-F (Polytrifluorochloroethylene)	Teflon (Polytetrafluoroethylene)	Uscolite CP (Styrene–Acrylonitrile–Butadiene)	Penton (Chlorinated Polyether)	Shellac Compounds	Organic Polysulfides	Polystyrene (Styron)	Vinylidene Chlorides
Sodium Borate $NaBO_2$	×	×	×	×		×	×			×	×			×	×		×		×	×
Sodium Carbonate, Na_2CO_3	×	×	–	×		–	O		×	×	×	×	×	×	×	×			×	
Sodium Chlorate, $NaClO_3$	×	×	×	×		×	×		×	×		×		×	×	×	×	×	×	×
Sodium Chloride, NaCl	×	×	×	×			×		×	×	×	×	×	×	×	×			×	
Sodium Cyanide, NaCN	×	×								×	×	–		×	×	×				
Sodium Fluoride, NaF	×									×		×		×	×	×				
Sodium Hydroxide, (Conc.), NaOH	–	×	O	–	×	O	×		×	×	–	O	×	×	×	×			×	
Sodium Hydroxide (Dilute)		×	–	×	×	–	×		×	×	×	–	×	×	×	×		×	×	×
Sodium Hydrosulfite														×						
Sodium Hypochlorite, NaOCl	–	–	–			×				×	–	×		×	×	×			×	×
Sodium Hyposulfate														×						
Sodium Nitrate, $NaNO_3$	×	×	×	×		×				×	×	×		×	×	×			×	
Sodium Peroxide, Na_2O_2										×				×	×					
Sodium Phosphate, (Tri) Na_3PO_4		×			×		×			×	×			×	×				–	
Sodium Silicate, Na_2SiO_3	×	×								×				×	×				×	
Sodium Sulfate, Na_2SO_4	×	×							×	×	×	×		×	×	×				
Sodium Sulfide, Na_2S	×	×							×	×	×	×		×	×				×	
Sodium Sulfite, Na_2SO_3	×	×								×	×	×		×	×	×				
Stannic Chloride, $SnCl_4$	×	O								×	×		×	×	×	×				
Stannous Chloride, $SnCl_2$		×							×	–		×		×	×	×				
Stearic Acid $CH_3(CH_2)_{16}COOH$	×	O							×	×	×	×		×	×	×				
Sulfur, Molten, S										O	O			×		O				
Sulfur Chloride, (Wet), S_2Cl_2														×						
Sulfur Dioxide (Dry), SO_2		×			–					×	–	–		×		×				×
Sulfur Dioxide (Wet)	×									–		–		×		×				
Sulfur Trioxide, SO_3	×	O			O					×				×						×
Sulfuric Acid (Fuming to 98%)		×	O	O	O	O	O			O		–		×	O	O			O	
Sulfuric Acid (Hot Conc.) H_2SO_4	O													×	O	–				

THERMOSETTING PLASTICS; WOODS

Chemical	Vinyl Chloride Acetates	Cast Phenol Formaldehyde	Haveg 41 (Phenolic w. Asbestos)	Heresite (Phenol Formaldehyde)	Molded Phenolformald (Durez)	Phenol Furfural Plastics	Urea Formaldehyde	Casein Plastics	Epoxy Resins	Furane Resins (Haveg 61, Duralon)	Silicone Resins	Permanite (Furan, Glass Fiber)	Nylon (Adipic Acid–Hexameth, Diamine)	Durcon 6 (Modified Epoxy)	Cypress	Fir	Maple	Oak	Pine	Redwood
Sodium Borate $NaBO_2$	×	–			–	–					×	×**								
Sodium Carbonate, Na_2CO_3		×	×	×	×	×		O	×	×	×	×	×**		+	+	+	×	+	+
Sodium Chlorate, $NaClO_3$	×	×			×	×	×	×			O	×	×			×	×	×	×	
Sodium Chloride, NaCl			×	×	×	×			×	×	×	×	×	×	×	×	×	×	+	+
Sodium Cyanide, NaCN									×		–	×								
Sodium Fluoride, NaF											×	×								
Sodium Hydroxide, (Conc.), NaOH		O	O	O	O	O	O		×	×	–	×	×	×	–	–	O	O	O	
Sodium Hydroxide (Dilute)	×	–	O	–	O	O	×		×	×	–	×	×	–		–	O	O	–	–
Sodium Hydrosulfite		×			×	×					×									
Sodium Hypochlorite, NaOCl	×	–	O	×	–	–			O	O	O	O								
Sodium Hyposulfate																				
Sodium Nitrate, $NaNO_3$		×	×	×	×	×			×	×	O	×								
Sodium Peroxide, Na_2O_2											×									
Sodium Phosphate, (Tri) Na_3PO_4		–	–		×	–			×	×	×	×	×							
Sodium Silicate, Na_2SiO_3		×			×	×					×	×								
Sodium Sulfate, Na_2SO_4			×	×					×	×	×	×	×							
Sodium Sulfide, Na_2S			×	×				–	×		×	+**	–			–	O	O	–	–
Sodium Sulfite, Na_2SO_3		–	×	×	–	–			×	×	×	×	×							
Stannic Chloride, $SnCl_4$			×								×	O	×							
Stannous Chloride, $SnCl_2$									×		×									
Stearic Acid $CH_3(CH_2)_{16}COOH$			×	×					×		×	×								
Sulfur, Molten, S			×	O							×									
Sulfur Chloride, (Wet), S_2Cl_2											×									
Sulfur Dioxide (Dry), SO_2	×		×						×		–	×								
Sulfur Dioxide (Wet)											O	×								
Sulfur Trioxide, SO_3	×								×		×	×								
Sulfuric Acid (Fuming to 98%)		O			O	O			O	O	O	×	O			O	O	O	O	O
Sulfuric Acid (Hot Conc.) H_2SO_4			O						O	O	×	O				O	O	O	O	O

**Note*: Duriron is as shown. Durichlor is also satisfactory on chlorides and HCl.

**Durcon 5 would be the preferred formula.

TABLE A.3 Continued

Chemical Resistance of Materials

MATERIALS
× — *Very Good Service*
+ — *Moderate Service*
− — *Limited or Variable Service*
O — *Unsatisfactory*
Blank — No Information

CHEMICALS
Solids Assumed in Solution
Room Temperatures Assumed
Unless Otherwise Stated

METALS

Chemical	Carbon Steel; Fe	Cast Iron & Ductile Iron; Fe	304 Stainless Steel; Fe, 18Cr, 8Ni	316 Stainless Steel; Fe 16Cr, 10Ni, 2Mo	347 Stainless Steel; Fe, 17Cr, 9Ni; (Cx10)Cb	Ni-Resist Iron; Fe 14Ni, 2Cr, 2Si	Durimet 20; Carpenter 20; Fe; 4Cu, 20Cr, 29Ni, 2Mo, 1Si	Worthite; 3Mo, 2Cu, Fe, 20Cr, 24Ni, 3Si	Durorpm; Fe, 14Si; Durichlor, Fe, 14Si, 3Mo*	Copper; Brass; Bronzes; Everdur	Aluminum; Al (and Alloys)	Lead; Pb	Monel; 67Ni, 30Cu, 1.4Fe	Nickel; Ni	Inconel; 76Ni, 15Cr, 8Fe
Sulfuric Acid (Cold Conc.)	+	+	×	×	−		×	×	×	O	−	×	+		
Sulfuric Acid (75%–95%)	+	−	O	O	−	−	×	×	×	O	O	×	O	O	−
Sulfuric Acid (10%–75%)	O	O	O	O	−	−	×	×	×	O	O	×	−	O	−
Sulfuric Acid (<10%)	O	O	O	−	−	−	−	×	×	O	−	×	−	−	−
Sulfurous Acid, H_2SO_3	O	O	−	×	−	O	×	×	O	O	−	−	O	O	O
Sulfuryl Chloride, SO_2Cl_2							+		×						
Tannic Acid	−		×	×			×	×	×	−	×		+	+	+
Tartaric Acid, $(CHOH COOH)_2$	−		×	×			×	×	×	−	O	×	+	+	
Toluene, $CH_3C_6H_5$	×		×				×		×	×	×	×	×		
Trichlorethylene, Dry, Cl_2CCHCl	−	−	−	×	−	−	×	×	×	−	−		×	×	×
Water, Fresh, H_2O	×		×	×			×		×	×	×		×	×	
Water, Distilled Lab.	O		×	×		O	×		×	×	O		O	−	
Zinc Chloride, $ZnCl_2$	O	O	O	−	O	−	×	×	×	O	O	−	+	+	
Zinc Sulfate, $ZnSO_4$	O	O	+	+	+	×	×	×	×	+	−	+	+	+	+

Chemical	Hastelloy B; Ni, 26Mo, 4Fe	Hastelloy C; Ni, 16Mo, 4Fe, 14Cr, 4W	Hastelloy D; Ni, 8Si, 3Cu	Chlorimet 3; 3Fe, 1Si, 60Ni, 18Mo, 18Cr	Chlorimet 2; 63Ni, 32Mo, 3Fe, 1Si	Stellite; Co, 28Cr, 4W	Zirconium; Z	Tantalum; Ta	Silver; Ag	Platinum; Pt	Dowmetal; (Mg alloys)	Titanium; Ti	Molybdenum; Mo
Sulfuric Acid (Cold Conc.)	×	×	×	×	×	−	O	−		×		O	−
Sulfuric Acid (75%–95%)	×	×	−	×	×	−	−	−	O	×		O	
Sulfuric Acid (10%–75%)	×	×	×	×	×	−	×	−	×	×		−	×
Sulfuric Acid (<10%)	×	×	×	×	×	−	×	×	×	×		×	
Sulfurous Acid, H_2SO_3	O	+	+	×	×	×	×	×	×			×	
Sulfuryl Chloride, SO_2Cl_2				+	+			×					
Tannic Acid	×	×	×	×	×		×	×	×			×	
Tartaric Acid, $(CHOH COOH)_2$	×	×	×	×	×		×	×	×			×	+
Toluene, $CH_3C_6H_5$				×	×		×						
Trichlorethylene, Dry, Cl_2CCHCl	×	×	×	×	×	×		×			×	×	
Water, Fresh, H_2O				×	×			×	×				×
Water, Distilled Lab.				×	×				×				
Zinc Chloride, $ZnCl_2$	+	+	+	×	×		×	×	×			×	
Zinc Sulfate, $ZnSO_4$	+	+	+	×	×			×	×			×	

CARBONS & CERAMICS, RUBBERS

Chemical	Carbon & Graphite	Glass, "Pyrex" brand	Silicaware	Silicate Cements	Chemical Stoneware	Transite (asbestos & cement)	Chemical Porcelain	Concrete—Unbonded	Concrete—Motor Bonded	Hard Rubber (Natural)	Soft Rubber (Natural)	Neoprene	Butadiene Derivatives	Nitrile Rubber (Chemigum)
Sulfuric Acid (Cold Conc.)	O	×					×			O	O	O		
Sulfuric Acid (75%–95%)	×	×			×		×					O		O
Sulfuric Acid (10%–75%)	×	×	×		×		×			×	×	−	×	−
Sulfuric Acid (<10%)	×	×	×		×		×			×	×	×	×	×
Sulfurous Acid, H_2SO_3	×	×	×	×	×		×			−	−	×		
Sulfuryl Chloride, SO_2Cl_2		×	×		×		×							
Tannic Acid	×	×			×		×	×		×	−	×		
Tartaric Acid, $(CHOH COOH)_2$	×	×	×		×		×			×	×	×		
Toluene, $CH_3C_6H_5$	×	×	×		×		×			O	O	O	O	
Trichlorethylene, Dry, Cl_2CCHCl	×	×			×		×			O	O	O		
Water, Fresh, H_2O	×	×	×		×	×	×	×	×	×	×	×	×	
Water, Distilled Lab.		+					×			×		×		
Zinc Chloride, $ZnCl_2$	×	×		×			×	×		×	×	×	×	×
Zinc Sulfate, $ZnSO_4$	×	×					×			×	×	×		

THERMOPLASTICS

Chemical	Viton	Asphaltic, Bitumastic	Cellulose Acetate	Cellulose Acetatebutyrate	Ethyl Cellulose (Ethocel)	Cellulose Nitrate	Acrylic (Lucite, Plexiglas)	Coumarone Resins	Polyethylene	Polyvinyl Chloride, Rigid or Unplasticized	Tygon (PVC & Copolymers)	Saran (Vinyl Chloride, Vinyldene Chloride)	Kel-F (Polytrifluorochloroethylene)	Teflon (Polytetrafluoroethylene)	Uscolite CP (Styrene–Acrylonitrile–Butadiene)
Sulfuric Acid (Cold Conc.)	+						−			×	O	−		×	O
Sulfuric Acid (75%–95%)		×							×	×	O	−		×	O
Sulfuric Acid (10%–75%)	×	×	×	×	O	O	×		×	×	−	×	×	×	−
Sulfuric Acid (<10%)	×	×	×	×	−	−	×		×	×	×	×	−	×	×
Sulfurous Acid, H_2SO_3		×							O	×	−	−		×	×
Sulfuryl Chloride, SO_2Cl_2									O					×	
Tannic Acid		×			−				×	×		×		×	×
Tartaric Acid, $(CHOH COOH)_2$	×	×	×	×						×		×		×	×
Toluene, $CH_3C_6H_5$	−		−	−	O		O		−	O	O	−	−	×	
Trichlorethylene, Dry, Cl_2CCHCl	−	O					O		O	O	O	−	O	×	
Water, Fresh, H_2O	×	×	×	×	×	×	×	×		×		×		×	×
Water, Distilled Lab.							×			×				×	×
Zinc Chloride, $ZnCl_2$	×	×			×				×	×		×		×	×
Zinc Sulfate, $ZnSO_4$	×	×							×	×	×	×		×	×

THERMOSETTING PLASTICS

Chemical	Penton (Chlorinated Polyether)	Shellac Compounds	Organic Polysulfides	Polystyrene (Styron)	Vinylidene Chlorides	Vinyl Chloride Acetates	Cast Phenol Formaldehyde	Haveg 41 (Phenolic w. Asbestos)	Heresite (Phenol Formaldehyde)	Molded Phenolformald (Durez)
Sulfuric Acid (Cold Conc.)	−							−		
Sulfuric Acid (75%–95%)	+			−				−		−
Sulfuric Acid (10%–75%)	×	×	×	×	×	×	−	×		−
Sulfuric Acid (<10%)	×	×	×	×	×	×	−	×		×
Sulfurous Acid, H_2SO_3	×			×			×	×		×
Sulfuryl Chloride, SO_2Cl_2										×
Tannic Acid	×			×			×	×		×
Tartaric Acid, $(CHOH COOH)_2$	×						−	×		−
Toluene, $CH_3C_6H_5$	+	O	O	O	−	−	−	−		−
Trichlorethylene, Dry, Cl_2CCHCl	+							×		
Water, Fresh, H_2O	×	×	×	×	×	×	×			×
Water, Distilled Lab.	×									
Zinc Chloride, $ZnCl_2$	×			×			−	×		×
Zinc Sulfate, $ZnSO_4$	×			×				×		×

Chemical	Phenol Furfural Plastics	Urea Formaldehyde	Casein Plastics	Epoxy Resins	Furane Resins (Haveg 61, Duralon)	Silicone Resins	Permanite (Furan, Glass Fiber)	Nylon (Adipic Acid–Hexameth, Diamine)	Durcon 6 (Modified Epoxy)	Cypress
Sulfuric Acid (Cold Conc.)				O		O	O	×	O	
Sulfuric Acid (75%–95%)	−			−	−	−	×			
Sulfuric Acid (10%–75%)	−		O	×	×	−	−	×	−	
Sulfuric Acid (<10%)	×	×	−		×	−	×	−	×	−
Sulfurous Acid, H_2SO_3	×		×	×	×	−	×	O	+**	
Sulfuryl Chloride, SO_2Cl_2						×				
Tannic Acid	×	×	×	×	×	×	×			
Tartaric Acid, $(CHOH COOH)_2$	−	×		×	×	×	×	×		
Toluene, $CH_3C_6H_5$	−		×		×	×	×	×	−	
Trichlorethylene, Dry, Cl_2CCHCl				×		×	×			
Water, Fresh, H_2O	×	×	O			×	×			
Water, Distilled Lab.						×				
Zinc Chloride, $ZnCl_2$	−			×	×	×	−	×		
Zinc Sulfate, $ZnSO_4$	×			×	×	×	×	×		

WOODS

Chemical	Fir	Maple	Oak	Pine	Redwood
Sulfuric Acid (Cold Conc.)	O	O	O	O	O
Sulfuric Acid (75%–95%)					
Sulfuric Acid (10%–75%)	−	−	−	−	−
Sulfuric Acid (<10%)	−	×	−	−	−
Sulfurous Acid, H_2SO_3	×	×			
Sulfuryl Chloride, SO_2Cl_2					
Tannic Acid	×	×			
Tartaric Acid, $(CHOH COOH)_2$					
Toluene, $CH_3C_6H_5$	−	×	×	×	
Trichlorethylene, Dry, Cl_2CCHCl					
Water, Fresh, H_2O	×	×	×	×	×
Water, Distilled Lab.					
Zinc Chloride, $ZnCl_2$					
Zinc Sulfate, $ZnSO_4$					

**Note*: Duriron is as shown. Durichlor is also satisfactory on chlorides and HCl.

**Durcon 5 would be the preferred formula.

A.4 Composition of Metallic and Other Materials

TABLE A.4
Composition of Metallic and Other Materials

No.	Material	Manufacturer	Composition or Description
		Metals	
17	Aluminum		
19	Aloyco-20	Alloy Steel Products Co.	Fe; 19–21 Cr; 28–30 Ni; 4.0–4.5 Cu; 2.5–3.0 Mo; 1.5 max. Si; 0.65–0.85 Mn; 0.07 max. C
19a	720 Alloy	General Plate	20 Mn; 20 Ni; Cu
54–60	Brass		Various commercial grades ranging 60–65 Cu; 35–40 Zn; 0.5–3.0 Pb
63	Brass, red		85 Cu; 15 Zn
66	Bronze, comm.		90 Cu; 10 Zn
73	Bronze, phosphor, 5% A		94.8–95.5 Cu; 4.3–5 Sn; P
74	Bronze, phosphor, 8% C		Cu; 7–9 Sn; 0.03–0.25 P
75	Bronze, phosphor 10% D		89.5–90 Cu; 10–10.5 Sn; P
76	Bronze, phosphor, spec. free cutting		88 Cu; 4 Zn; 4 Sn; 4 Pb
81	CA-FA20	Cooper Alloy	Fe; 19–21 Cr; 28–30 Ni; 3.5 Mo; 4–4.5 Cu; 0.07 max. C
82	CA-MM	Cooper Alloy	67 Ni; 30 Cu; 1.4 Fe; 0.1 Si; 0.15 C
86	Cast iron		Ordinary unalloyed cast iron
88	Chlorimet 2	Duriron Co.	63 Ni; 32 Mo; 3 max. Fe; 0.15 max. C; 1 Si; 1 Mn
89	Chlorimet 3	Duriron Co.	60 Ni; 18 Mo; 18 Cr; 2 Fe; 0.07 max. C; 1 Si; 1 Mn
111	Copper		99.9+ Cu
112	Copper, Be		97.5 Cu; 2.15 Be; 0.35 Ni
119	Corrosiron	Pacific Fdry.	Fe; 14.5 Si
140	Durichlor	Duriron Co.	Fe; 0.85 C; 14.5 Si; 3 Mo; 0.35 Mn
141	Durimet 20	Duriron Co.	Fe; 20 Cr; 29 Ni; 0.07 max. C; 2 Mo; 4 Cu; 1 Si
142	Durimet T	Duriron Co.	Fe; 19 Cr; 22 Ni; 0.07 max. C; 2 Mo; 1 Cu; 1 Si
143	Duriron	Duriron Co.	Fe; 0.80 C; 14.5 Si; 0.35 Mn
148	Everdur 1000	Amer. Brass	94.9 Cu; 4 Si; 1.1 Mn
149	Everdur 1010	Amer. Brass	95.8 Cu; 3.1 Si; 1.1 Mn
150	Everdur 1015	Amer. Brass	98.25 Cu; 1.5 Si; 0.25 Mn
156	Gold		99.99 Au
156a	Green gold		75% Au; 25% Ag
159	Hastelloy A	Haynes Stellite	Ni; 17–21 Mo; 17–21 Fe
160	Hastelloy B	Haynes Stellite	Ni; 24–32 Mo; 3–7 Fe; 0.02–0.12 C
161	Hastelloy C	Haynes Stellite	Ni; 14–19 Mo; 4–8 Fe; 0.04–0.15 C; 12–16 Cr; 3–5.5 W
162	Hastelloy D	Haynes Stellite	Ni; 8–11 Si; 2–5 Cu; 1 max. Al
163	Stellite 1	Haynes Stellite	Co; 28–34 Cr; 11–15 W

TABLE A.4 Continued
Composition of Metallic and Other Materials

No.	Material	Manufacturer	Composition or Description
165	Stellite 6	Haynes Stellite	Co; 25–31 Cr; 3–6 W
184	Inconel	Int'l Nickel	79.5 Ni; 13 Cr; 6.5 Fe; 0.08 C; 0.2 Cu; 0.25 Mn
191	Lead		99.9 + Pb
192	Lead, antimonial		94 Pb; 6 Sb
193	Lead, antimonial		Pb; 4–12 Sb
196	Lead, chemical		99.93 Pb; 0.06 Cu
200	Lead, Te		99.88 Pb; 0.045 Te; 0.06 Cu
216	Monel	Int'l Nickel	67 Ni; 30 Cu; 1.4 Fe; 0.1 Si; 0.15 C
219	Muntz Metal		60 Cu; 40 Zn
224	Nickel	Int'l Nickel	99.4 Ni; 0.2 Mn; 0.1 Cu; 0.15 Fe; 0.05 Si
224a	Z-Nickel	Int'l Nickel	95 + Ni
226	Nickel–Silver 18% A		65 Cu; 18 Ni; 17 Zn
227	Nickel–Silver 18% B		55 Cu; 18 Ni; 27 Zn
227a	Ni-Span	Int'l Nickel	Ni, Ti, Cr, C, Mn, Si, Al
229	Ni-Hard	Int'l Nickel	Fe; 3.4 C; 1.5 Cr; 4.5 Ni; 0.6 Si
231	Ni-Resist	Int'l Nickel	Fe; 2.8 C; 14 or 20 Ni; 6 Cu (optional); 2 Cr; 2 Si
240	Platinum		99.99 Pt
268	Silver		99.9+ Ag
275	S.S. 301		Fe; 16–18 Cr; 6–8 Ni; 0.08–0.15 C
276	S.S. 302		Fe; 17–19 Cr; 8–10 Ni; 0.08–0.15 C
278	S.S. 303		Fe; 17–19 Cr; 8–10 Ni; 0.15 max. C; 0.07 min. P, S, Se; 0.6
279	S.S. 304		Fe; 18–20 Cr; 8–11 Ni; 0.08 max. C; 2 max. Mn
282	S.S. 310		Fe; 24–26 Cr; 19–22 Ni; 0.25 max. C
283	S.S. 316		Fe; 16–18 Cr; 10–14 Ni; 0.1 max. C; 1.75–2.75 Mo
284	S.S. 317		Fe; 17.5–20 Cr; 10–14 Ni; 0.1 max. C; 3–4 Mo
285	S.S. 321		Fe; 17–19 Cr; 8–11 Ni; Ti, 5xC min.
286	S.S. 347		Fe; 17–19 Cr; 9–12 Ni; Cb, 10xC min.
287	S.S. 403		Fe; 11.5–13 Cr; 0.15 max. C
290	S.S. 410		Fe; 11.5–13.5 Cr; 0.15 max. C
292	S.S. 416		Fe; 12–14 Cr; 0.15 max. C; 0.07 min. P, S, Se; 0.6 max. Zr, Mo
295	S.S. 430		Fe; 14–18 Cr; 0.12 max. C
303	S.S. 446		Fe; 23–27 Cr; 0.35 max. C; 0.25 max. N
360a	Steel		Plain carbon steel
368	Tantalum	Fansteel	99.9+ Ta
390	Worthite	Worthington Pump	Fe; 20 Cr; 24 Ni; 0.07 max. C; 3.25 Si; 3 Mo; 1.75 Cu; 0.5 Mn
		Carbon and Graphite	
401	Karbate (carbon)	National Carbon	Impervious carbon
402	Karbate (graphite)	National Carbon	Impervious graphite
		Ceramics	
611	Lapp Porcelain	Lapp Insulator Co.	Chemical porcelain
614	Pfaudler Glass Lining	Pfaudler Co.	Glass-lined steel equipment
615	Plate Glass		Polished plate glass, flat or bent
616	Pyrex	Corning Glass Wks.	Glass

TABLE A.4 Continued
Composition of Metallic and Other Materials

No.	Material	Manufacturer	Composition or Description
		Plastics	
700	Ace Saran	American Hard Rubber	Vinylidene chloride
710	Geon	B. F. Goodrich	Polyvinyl chloride
711	Haveg 41	Haveg Corp.	Phenolic-asbestos
712	Haveg 43	Haveg Corp.	Phenolic-graphite
713	Haveg 60	Haveg Corp.	Furan-asbestos
714	Haveg 63	Haveg Corp.	Furan-graphite
715	Heresite M 66	Heresite & Chem. Co.	Transparent molding powder
716	Heresite MF 66	Heresite & Chem. Co.	Black molding powder
717a	Kel-F	M. W. Kellogg	Polymerized trifluoroethylene
718	Koroseal	B. F. Goodrich	Plasticized polyvinyl chloride
731	Nylon FM-101	E. I. du Pont	Injection, compression and extrusion moldings (tubing, sheeting, wire covering, gasketing)
731a	Plastisol		Polyvinyl chloride
735	Polythene	E.I. du Pont	Polyethylene
740	Saran	Dow Chemical	Vinyl chloride-vinylidene chloride copolymer
740a	Sirvene	Chicago Rawhide	Synthetic rubber
742	Teflon	E.I. du Pont	Polymerized tetrafluoroethylene
746	Tygon	U.S. Stoneware	Synthetic compounds
		Rubber	
800	Ace Hard Rubber	American Hard Rubber	Vulcanized rubber
805	Butyl (GR-I)	Stanco Distributors	Solid copolymer of isobutylene and isoprene
820	Hycar (GR-A)	B. F. Goodrich	Nitrile type synthetic rubber
829	Neoprene	E. I. du Pont	Polymer of chloroprene
836	Natural (soft)		
837	Natural (hard)		
838	GR-S (soft)		
839	GR-S (hard)		
853	Thiokol (GR-P)	Thiokol Corp.	

A.5 Steam and Water Tables

TABLE A.5a

Dry Saturated Steam: Temperature Table

Temp., °F/°C	Abs. Press., PSIA P^a	Specific Volume, ft^3/lbm[a] Sat. Liquid v_f	Evap. v_{fg}	Sat. Vapor v_g	Enthalpy, Btu/lbm[a] Sat. Liquid h_f	Evap. h_{fg}	Sat. Vapor h_g	Entropy, Btu/lbm R[a] Sat. Liquid s_f	Evap. s_{fg}	Sat. Vapor s_g
32/0	0.08854	0.01602	3306	3306	0.00	1075.8	1075.8	0.0000	2.1877	2.1877
35/1.7	0.09995	0.01602	2947	2947	3.02	1074.1	1077.1	0.0061	2.1709	2.1770
40/4.4	0.12170	0.01602	2444	2444	8.05	1071.3	1079.3	0.0162	2.1435	2.1597
45/7.2	0.14752	0.01602	2036.4	2036.4	13.06	1068.4	1081.5	0.0262	2.1167	2.1429
50/10	0.17811	0.01603	1703.2	1703.2	18.07	1065.6	1083.7	0.0361	2.0903	2.1264
60/15.6	0.2563	0.01604	1206.6	1206.7	28.06	1059.9	1088.0	0.0555	2.0393	2.0948
70/21.1	0.3631	0.01606	867.8	867.9	38.04	1054.3	1092.3	0.0745	1.9902	2.0647
80/26.7	0.5069	0.01608	633.1	633.1	48.02	1048.6	1096.6	0.0932	1.9428	2.0360
90/32.2	0.6982	0.01610	468.0	468.0	57.99	1042.9	1100.9	0.1115	1.8972	2.0087
100/37.8	0.9492	0.01613	350.3	350.4	67.97	1037.2	1105.2	0.1295	1.8531	1.9826
110/43	1.2748	0.01617	265.3	265.4	77.94	1031.6	1109.5	0.1471	1.8106	1.9577
120/49	1.6924	0.01620	203.25	203.27	87.92	1025.8	1113.7	0.1645	1.7694	1.9339
130/54	2.2225	0.01625	157.32	157.34	97.90	1020.0	1117.9	0.1816	1.7296	1.9112
140/60	2.8886	0.01629	122.99	123.01	107.89	1014.1	1122.0	0.1984	1.6910	1.8894
150/66	3.718	0.01634	97.06	97.07	117.89	1008.2	1126.1	0.2149	1.6537	1.8685
160/71	4.741	0.01639	77.27	77.29	127.89	1002.3	1130.2	0.2311	1.6174	1.8485
170/77	5.992	0.01645	62.04	62.06	137.90	996.3	1134.2	0.2472	1.5822	1.8293
180/82	7.510	0.01651	50.21	50.23	147.92	990.2	1138.1	0.2630	1.5480	1.8109
190/88	9.339	0.01657	40.94	40.96	157.95	984.1	1142.0	0.2785	1.5147	1.7932
200/93	11.526	0.01663	33.62	33.64	167.99	977.9	1145.9	0.2938	1.4824	1.7762
210/90	14.123	0.01670	27.80	27.82	178.05	971.6	1149.7	0.3090	1.4508	1.7598
212/100	14.696	0.01672	26.78	26.80	180.07	970.3	1150.4	0.3120	1.4446	1.7566
220/104	17.186	0.01677	23.13	23.15	188.13	965.2	1153.4	0.3239	1.4201	1.7440
230/110	20.780	0.01684	19.365	19.382	198.23	958.8	1157.0	0.3387	1.3901	1.7288
240/116	24.969	0.01692	16.306	16.323	208.34	952.2	1160.5	0.3531	1.3609	1.7140
250/121	29.825	0.01700	13.804	13.821	218.48	945.5	1164.0	0.3675	1.3323	1.6998
260/127	35.429	0.01709	11.746	11.763	228.64	938.7	1167.3	0.3817	1.3043	1.6860
270/132	41.858	0.01717	10.044	10.061	238.84	931.8	1170.6	0.3958	1.2769	1.6727
280/138	49.203	0.01726	8.628	8.645	249.06	924.7	1173.8	0.4096	1.2501	1.6597
290/143	57.556	0.01735	7.444	7.461	259.31	917.5	1176.8	0.4234	1.2238	1.6472
300/149	67.013	0.01745	6.449	6.466	269.59	910.1	1179.7	0.4369	1.1980	1.6350
310/154	77.68	0.01755	5.609	5.626	279.92	902.6	1182.5	0.4504	1.1727	1.6231
320/160	89.66	0.01765	4.896	4.914	290.28	894.9	1185.2	0.4637	1.1478	1.6115
330/166	103.06	0.01776	4.289	4.307	300.68	887.0	1187.7	0.4769	1.1233	1.6002
340/171	118.01	0.01787	3.770	3.788	311.13	879.0	1190.1	0.4900	1.0992	1.5891

TABLE A.5a Continued
Dry Saturated Steam: Temperature Table

Temp., °F/°C	Abs. Press., PSIA P[a]	Specific Volume, ft³/lbm[a] Sat. Liquid v_f	Evap. v_{fg}	Sat. Vapor v_g	Enthalpy, Btu/lbm[a] Sat. Liquid h_f	Evap. h_{fg}	Sat. Vapor h_g	Entropy, Btu/lbm R[a] Sat. Liquid s_f	Evap. s_{fg}	Sat. Vapor s_g
350/177	134.63	0.01799	3.324	3.342	321.63	870.7	1192.3	0.5029	1.0754	1.5783
360/182	153.04	0.01811	2.939	2.957	332.18	862.2	1194.4	0.5158	1.0519	1.5677
370/188	173.37	0.01823	2.606	2.625	342.79	853.5	1196.3	0.5286	1.0287	1.5573
380/193	195.77	0.01836	2.317	2.335	353.45	844.6	1198.1	0.5413	1.0059	1.5471
390/199	220.37	0.01850	2.0651	2.0836	364.17	835.4	1199.6	0.5539	0.9832	1.5371
400/204	247.31	0.01864	1.8447	1.8633	374.97	826.0	1201.0	0.5664	0.9608	1.5272
410/210	276.75	0.01878	1.6512	1.6700	385.83	816.3	1202.1	0.5788	0.9386	1.5174
420/216	308.83	0.01894	1.4811	1.5000	396.77	806.3	1203.1	0.5912	0.9166	1.5078
430/221	343.72	0.01910	1.3308	1.3499	407.79	796.0	1203.8	0.6035	0.8947	1.4982
440/227	381.59	0.01926	1.1979	1.2171	418.90	785.4	1204.3	0.6158	0.8730	1.4887
450/232	422.6	0.0194	1.0799	1.0993	430.1	774.5	1204.6	0.6280	0.8513	1.4793
460/238	466.9	0.0196	0.9748	0.9944	441.4	763.2	1204.6	0.6402	0.8298	1.4700
470/243	514.7	0.0198	0.8811	0.9009	452.8	751.5	1204.3	0.6523	0.8083	1.4606
480/249	566.1	0.0200	0.7972	0.8172	464.4	739.4	1203.7	0.6645	0.7868	1.4513
490/254	621.4	0.0202	0.7221	0.7423	476.0	726.8	1202.8	0.6766	0.7653	1.4419
500/260	680.8	0.0204	0.6545	0.6749	487.8	713.9	1201.7	0.6887	0.7438	1.4325
520/271	812.4	0.0209	0.5385	0.5594	511.9	686.4	1198.2	0.7130	0.7006	1.4136
540/282	962.5	0.0215	0.4434	0.4649	536.6	656.6	1193.2	0.7374	0.6568	1.3942
560/293	1133.1	0.0221	0.3647	0.3868	562.2	624.2	1186.4	0.7621	0.6121	1.3742
580/304	1325.8	0.0228	0.2989	0.3217	588.9	588.4	1177.3	0.7872	0.5659	1.3532
600/316	1542.9	0.0236	0.2432	0.2668	617.0	548.5	1165.5	0.8131	0.5176	1.3307
620/327	1786.6	0.0247	0.1955	0.2201	646.7	503.6	1150.3	0.8398	0.4664	1.3062
640/338	2059.7	0.0260	0.1538	0.1798	678.6	452.0	1130.5	0.8679	0.4110	1.2789
660/349	2365.4	0.0278	0.1165	0.1442	714.2	390.2	1104.4	0.8987	0.3485	1.2472
680/360	2708.1	0.0305	0.0810	0.1115	757.3	309.9	1067.2	0.9351	0.2719	1.2071
700/371	3093.7	0.0369	0.0392	0.0761	823.3	172.1	995.4	0.9905	0.1484	1.1389
705.4/374.1	3206.2	0.0503	0	0.0503	902.7	0	902.7	1.0580	0	1.0580

Source: Abridged from *Thermodynamic Properties of Steam*, by Joseph H. Keenan and Frederick G. Keyes. © 1936, by Joseph H. Keenan and Frederick G. Keyes. Published by John Wiley & Sons, Inc., New York.

[a]PSIA = 0.069 bar (abs); ft³/lbm = 62.4 l/kg; Btu/lbm = 0.556 Kcal/kg

TABLE A.5b
Properties of Superheated Steam

Abs. Press., PSIA[a] (Sat. Temp. °F)	Temperature, °F/°C												
	200/93	220/104	300/149	350/177	400/204	450/232	500/260	550/288	600/316	700/371	800/427	900/482	1000/538
υ	392.6	404.5	452.3	482.2	512.0	541.8	571.6	601.4	631.2	690.8	750.4	809.9	869.5
1 h	1150.4	1159.5	1195.8	1218.7	1241.7	1264.9	1288.3	1312.0	1335.7	1383.8	1432.8	1482.7	1533.5
(101.74) s	2.0512	2.0647	2.1153	2.1444	2.1720	2.1983	2.2233	2.2468	2.2702	2.3137	2.3542	2.3923	2.4283
υ	78.16	80.59	90.25	96.26	102.26	108.24	114.22	120.19	126.16	138.10	150.03	161.95	173.87
5 h	1148.8	1158.1	1195.0	1218.1	1241.2	1264.5	1288.0	1311.7	1335.4	1383.6	1432.7	1482.6	1533.4
(162.24) s	1.8718	1.8857	1.9370	1.9664	1.9942	2.0205	2.0456	2.0692	2.0927	2.1361	2.1767	2.2148	2.2509
υ	38.85	40.09	45.00	48.03	51.04	54.05	57.05	60.04	63.03	69.01	74.98	80.95	86.92
10 h	1146.6	1156.2	1193.9	1217.2	1240.6	1264.0	1287.5	1311.3	1335.1	1383.4	1432.5	1482.4	1533.1
(193.21) s	1.7927	1.8071	1.8595	1.8892	1.9172	1.9436	1.9689	1.9924	2.0160	2.0596	2.1002	2.1383	2.1744
υ		27.15	30.53	32.62	34.68	36.73	38.78	40.82	42.86	46.94	51.00	55.07	59.13
14.696 h		1154.4	1192.8	1216.4	1239.9	1263.5	1287.1	1310.9	1335.8	1383.2	1432.3	1482.3	1533.1
(212.00) s		1.7624	1.8160	1.8460	1.8743	1.9008	1.9261	1.9498	1.9734	2.0170	2.0576	2.0958	2.1319
υ			22.36	23.91	25.43	26.95	28.46	29.97	31.47	34.47	37.46	40.45	43.44
20 h			1191.6	1215.6	1239.2	1262.9	1286.6	1310.5	1334.4	1382.9	1432.1	1482.1	1533.0
(227.96) s			1.7808	1.8112	1.8396	1.8664	1.8918	1.9160	1.9392	1.9829	2.0235	2.0618	2.0978
υ			11.040	11.843	12.628	13.401	14.168	14.93	15.688	17.198	18.702	20.20	21.70
40 h			1186.8	1211.9	1236.5	1260.7	1284.8	1308.9	1333.1	1381.9	1431.3	1481.4	1532.4
(267.25) s			1.6994	1.7314	1.7608	1.7881	1.8140	1.8384	1.8619	1.9058	1.9467	1.9850	2.0214
υ			7.259	7.818	8.357	8.884	9.403	9.916	10.427	11.441	12.449	13.452	14.454
60 h			1181.6	1208.2	1233.6	1258.5	1283.0	1307.4	1331.8	1380.9	1430.5	1480.8	1531.9
(292.71) s			1.6492	1.6830	1.7135	1.7416	1.7678	1.7926	1.8162	1.8605	1.9015	1.9400	1.9762
υ				5.803	6.220	6.624	7.020	7.410	7.797	8.562	9.322	10.077	10.830
80 h				1204.3	1230.7	1256.1	1281.1	1305.8	1330.5	1379.9	1429.7	1480.1	1531.3
(312.03) s				1.6475	1.6791	1.7078	1.7346	1.7598	1.7836	1.8281	1.8694	1.9079	1.9442
υ				4.592	4.937	5.268	5.589	5.905	6.218	6.835	7.446	8.052	8.656
100 h				1200.1	1227.6	1253.7	1279.1	1304.2	1329.1	1378.9	1428.9	1479.5	1530.8
(327.81) s				1.6188	1.6518	1.6813	1.7085	1.7339	1.7581	1.8029	1.8443	1.8829	1.9193
υ				3.783	4.081	4.363	4.636	4.902	5.165	5.683	6.195	6.702	7.207
120 h				1195.7	1224.4	1251.3	1277.2	1302.5	1327.7	1377.8	1428.1	1478.8	1530.2
(341.25) s				1.5944	1.6287	1.6591	1.6869	1.7127	1.7370	1.7822	1.8237	1.8625	1.8990
υ					3.468	3.715	3.954	4.186	4.413	4.861	5.301	5.738	6.172
140 h					1221.1	1248.7	1275.2	1300.9	1326.4	1376.8	1427.3	1478.2	1529.7
(353.02) s					1.6087	1.6399	1.6683	1.6945	1.7190	1.7645	1.8063	1.8451	1.8817

TABLE A.5b Continued
Properties of Superheated Steam

Abs. Press., PSIA[a] (Sat. Temp. °F)	Temperature, °F/°C												
	200/93	220/104	300/149	350/177	400/204	450/232	500/260	550/288	600/316	700/371	800/427	900/482	1000/538
v					3.008	3.230	3.443	3.648	3.849	4.244	4.631	5.015	5.396
160 h					1217.6	1246.1	1273.1	1299.3	1325.0	1375.7	1426.4	1477.5	1529.1
(363.53) s					1.5908	1.6230	1.6519	1.6785	1.7033	1.7491	1.7911	1.8301	1.8667
v					2.649	2.852	3.044	3.229	3.411	3.764	4.110	4.452	4.792
180 h					1214.0	1243 5	1271.0	1297.6	1323.5	1374.7	1425.6	1476.8	1528.6
(373.06) s					1.5745	1.6077	1.6373	1.6642	1.6894	1.7355	1.7776	1.8167	1.8534
v					2.361	2.549	2.726	2.895	3.060	3.380	3.693	4.002	4.309
200 h					1210.3	1240.7	1268.9	1295.8	1322.1	1373.6	1424.8	1476.2	1528.0
(381.79) s					1.5594	1.5937	1.6240	1.6513	1.6767	1.7232	1.7655	1.8048	1.8415
v					2.125	2.301	2.465	2.621	2.772	3.066	3.352	3.634	3.913
220 h					1206.5	1237.9	1266.7	1294.1	1320.7	1372.6	1424.0	1475.5	1527.5
(389.86) s					1.5453	1.5808	1.6117	1.6395	1.6652	1.7120	1.7545	1.7939	1.8308
v					1.9276	2.094	2.247	2.393	2.533	2.804	3.068	3.327	3.584
240 h					1202.5	1234.9	1264.5	1292.4	1319.2	1371.5	1423.2	1474.8	1526.9
(397.37) s					1.5319	1.5686	1.6003	1.6286	1.6546	1.7017	1.7444	1.7839	1.8209
v						1.9183	2.063	2.199	2.330	2.582	2.827	3.067	3.305
260 h						1232.0	1262.3	1290.5	1317.7	1370.4	1422.3	1474.2	1526.3
(404.42) s						1.5573	1.5897	1.6184	1.6447	1.6922	1.7352	1.7748	1.8118
v						1.7674	1.9047	2.033	2.156	2.392	2.621	2.845	3.066
280 h						1228.9	1260.0	1288.7	1316.2	1369.4	1421.5	1473.5	1525.8
(411.05) s						1.5464	1.5796	1.6087	1.6354	1.6834	1.7265	1.7662	1.8033
v						1.6364	1.7675	1.8891	2.005	2.227	2.442	2.652	2.859
300 h						1225.8	1257.6	1286.8	1314.7	1368.3	1420.6	1427.8	1525.2
(417.33) s						1.5360	1.5701	1.5998	1.6268	1.6751	1.7184	1.7582	1.7954
v						1.3734	1.4923	1.6010	1.7036	1.8980	2.084	2.266	2.445
350 h						1217.7	1251.5	1282.1	1310.9	1365.5	1418.5	1471.1	1523.8
(431.72) s						1.5119	1.5481	1.5792	1.6070	1.6563	1.7002	1.7403	1.7777
v						1.1744	1.2851	1.3843	1.4770	1.6508	1.8161	1.9767	2.134
400 h						1208.8	1245.1	1277.2	1306.9	1362.7	1416.4	1.469.4	1522.4
(444.59) s						1.4892	1.5281	1.5607	1.5894	1.6398	1.6842	1.7247	1.7623

TABLE A.5b Continued

Properties of Superheated Steam

Abs. Press., PSIA[a] (Sat. Temp. °F)	*500/260*	*550/288*	*600/316*	*620/327*	*640/338*	*660/349*	*680/360*	*700/371*	*800/427*	*900/482*	*1000/538*	*1200/649*	*1400/760*	*1600/871*
v	1.1231	1.2155	1.3005	1.3332	1.3652	1.3967	1.4278	1.4584	1.6074	1.7516	1.8928	2.170	2.443	2.714
450 h	1238.4	1272.0	1302.8	1314.6	1326.2	1337.5	1348.8	1359.9	1414.3	1467.7	1521.0	1628.6	1738.7	1851.9
(456.28) s	1.5095	1.5437	1.5735	1.5845	1.5951	1.6054	1.6153	1.6250	1.6699	1.7108	1.7486	1.8177	1.8803	1.9381
v	0.9927	1.0800	1.1591	1.1893	1.2188	1.2478	1.2763	1.3044	1.4405	1.5715	1.6996	1.9504	2.197	2.442
500 h	1231.3	1266.8	1298.6	1310.7	1322.6	1334.2	1345.7	1357.0	1412.1	1466.0	1519.6	1627.6	1737.9	1851.3
(467.01) s	1.4919	1.5280	1.5588	1.5701	1.5810	1.5915	1.6016	1.6115	1.6571	1.6982	1.7363	1.8056	1.8683	1.9262
v	0.8852	0.9686	1.0431	1.0714	1.0989	1.1259	1.1523	1.1783	1.3038	1.4241	1.5414	1.7706	1.9957	2.219
550 h	1223.7	1261.2	1294.3	1306.8	1318.9	1330.8	1342.5	1354.0	1409.9	1464.3	1518.2	1626.6	1737.1	1850.6
(476.94) s	1.4751	1.5131	1.5451	1.5568	1.5680	1.5787	1.5890	1.5991	1.6452	1.6868	1.7250	1.7946	1.8575	1.9155
v	0.7947	0.8753	0.9463	0.9729	0.9988	1.0241	1.0489	1.0732	1.1899	1.3013	1.4096	1.6208	1.8279	2.033
600 h	1215.7	1255.5	1289.9	1302.7	1315.2	1327.4	1339.3	1351.1	1407.7	1462.5	1516.7	1625.5	1736.3	1850.0
(486.21) s	1.4586	1.4990	1.5323	1.5443	1.5558	1.5667	1.5773	1.5875	1.6343	1.6762	1.7147	1.7846	1.8476	1.9056
v		0.7277	0.7934	0.8177	0.8411	0.8639	0.8860	0.9077	1.0108	1.1082	1.2024	1.3853	1.5641	1.7405
700 h		1243.2	1280.6	1294.3	1307.5	1320.3	1332.8	1345.0	1403.2	1459.0	1513.9	1623.5	1734.8	1848.8
(503.10) s		1.4722	1.5084	1.5212	1.5333	1.5449	1.5559	1.5665	1.6147	1.6573	1.6963	1.7666	1.8299	1.8881
v		0.6154	0.6779	0.7006	0.7223	0.7433	0.7635	0.7833	0.8763	0.9633	1.0470	1.2088	1.3662	1.5214
800 h		1229.8	1270.7	1285.4	1299.4	1312.9	1325.9	1338.6	1398.6	1455.4	1511.0	1621.4	1733.2	1847.5
(518.23) s		1.4467	1.4863	1.5000	1.5129	1.5250	1.5366	1.5476	1.5972	1.6407	1.6801	1.7510	1.8146	1.8729
v		0.5264	0.5873	0.6089	0.6294	0.6491	0.6680	0.6863	0.7716	0.8506	0.9262	1.0714	1.2124	1.3509
900 h		1215.0	1260.1	1275.9	1290.9	1305.1	1318.8	1332.1	1393.9	1451.8	1508.1	1619.3	1731.6	1846.3
(531.98) s		1.4216	1.4653	1.4800	1.4938	1.5066	1.5187	1.5303	1.5814	1.6257	1.6656	1.7371	1.8009	1.8595
v		0.4533	0.5140	0.5350	0.5546	0.5733	0.5912	0.6084	0.6878	0.7604	0.8294	0.9615	1.0893	1.2146
1000 h		1198.3	1248.8	1265.9	1281.9	1297.0	1311.4	1325.3	1389.2	1448.2	1505.1	1617.3	1730.0	1845.0
(544.61) s		1.3961	1.4450	1.4610	1.4757	1.4893	1.5021	1.5141	1.5670	1.6121	1.6525	1.7245	1.7886	1.8474
v			0.4532	0.4738	0.4929	0.5110	0.5281	0.5445	0.6191	0.6866	0.7503	0.8716	0.9885	1.1031
1100 h			1236.7	1255.3	1272.4	1288.5	1303.7	1318.3	1384.3	1444.5	1502.2	1615.2	1728.4	1843.8
(556.31) s			1.4251	1.4425	1.4583	1.4728	1.4862	1.4989	1.5535	1.5995	1.6405	1.7130	1.7775	1.8363
v			0.4016	0.4222	0.4410	0.4586	0.4752	0.4909	0.5617	0.6250	0.6843	0.7967	0.9046	1.0101
1200 h			1223.5	1243.9	1262.4	1279.6	1295.7	1311.0	1379.3	1440.7	1499.2	1613.1	1726.9	1842.5
(567.22) s			1.4052	1.4243	1.4413	1.4568	1.4710	1.4843	1.5409	1.5879	1.6293	1.7025	1.7672	1.8263
v			0.3174	0.3390	0.3580	0.3753	0.3912	0.4062	0.4714	0.5281	0.5805	0.6789	0.7727	0.8640
1400 h			1193.0	1218.4	1240.4	1260.3	1278.5	1295.5	1369.1	1433.1	1493.2	1608.9	1723.7	1840.0
(587.10) s			1.3639	1.3877	1.4079	1.4258	1.4419	1.4567	1.5177	1.5666	1.6093	1.6836	1.7489	1.8083

TABLE A.5b Continued
Properties of Superheated Steam

Abs. Press., PSIA[a] (Sat. Temp. °F)	Temperature, °F/°C													
	500/260	550/288	600/316	620/327	640/338	660/349	680/360	700/371	800/427	900/482	1000/538	1200/649	1400/760	1600/871
v				0.2733	0.2936	0.3112	0.3271	0.3417	0.4034	0.4553	0.5027	0.5906	0.6738	0.7545
1600 h				1187.8	1215.2	1238.7	1259.6	1278.7	1358.4	1425.3	1487.0	1604.6	1720.5	1837.5
(604.90) s				1.3489	1.3741	1.3952	1.4137	1.4303	1.4964	1.5476	1.5914	1.6669	1.7328	1.7926
v					0.2407	0.2597	0.2760	0.2907	0.3502	0.3986	0.4421	0.5218	0.5968	0.6693
1800 h					1185.1	1214.0	1238.5	1260.3	1347.2	1417.4	1480.8	1600.4	1717.3	1835.0
(621.03) s					1.3377	1.3638	1.3855	1.4044	1.4765	1.5301	1.5752	1.6520	1.7185	1.7786
v					0.1936	0.2161	0.2337	0.2489	0.3074	0.3532	0.3935	0.4668	0.5352	0.6011
2000 h					1145.6	1184.9	1214.8	1240.0	1335.5	1409.2	1474.5	1596.1	1714.1	1832.5
(635.82) s					1.2945	1.3300	1.3564	1.3783	1.4576	1.5139	1.5603	1.6384	1.7055	1.7660
v							0.1484	0.1686	0.2294	0.2710	0.3061	0.3678	0.4244	0.4784
2500 h							1132.3	1176.8	1303.6	1387.8	1458.4	1585.3	1706.1	1826.2
(668.13) s							1.2687	1.3073	1.4127	1.4772	1.5273	1.6088	1.6775	1.7389
v								0.0984	0.1760	0.2159	0.2476	0.3018	0.3505	0.3966
3000 h								1060.7	1267.2	1365.0	1441.8	1574.3	1698.0	1819.9
(695.36) s								1.1966	1.3690	1.4439	1.4984	1.5837	1.6540	1.7163
v									0.1583	0.1981	0.2288	0.2806	0.3267	0.3703
3206.2 h									1250.5	1355.2	1434.7	1569.8	1694.6	1817.2
(705.40) s									1.3508	1.4309	1.4874	1.5742	1.6452	1.7080
v								0.0306	0.1364	0.1762	0.2058	0.2546	0.2977	0.3381
3500 h								780.5	1224.9	1340.7	1424.5	1563.3	1689.8	1813.6
s								0.9515	1.3241	1.4127	1.4723	1.5615	1.6336	1.6968
v								0.0287	0.1052	0.1462	0.1743	0.2192	0.2581	0.2943
4000 h								763.8	1174.8	1314.4	1406.8	1552.1	1681.7	1807.2
s								0.9347	1.2757	1.3827	1.4482	1.5417	1.6154	1.6795
v								0.0276	0.0798	0.1226	0.1500	0.1917	0.2273	0.2602
4500 h								753.5	1113.9	1286.5	1388.4	1540.8	1673.5	1800.9
s								0.9235	1.2204	1.3529	1.4253	1.5235	1.5990	1.6640
v								0.0268	0.0593	0.1036	0.1303	0.1696	0.2027	0.2329
5000 h								746.4	1047.1	1256.5	1369.5	1529.5	1665.3	1794.5
s								0.9152	1.1622	1.3231	1.4034	1.5066	1.5839	1.6499
v								0.0262	0.0463	0.0880	0.1143	0.1516	0.1825	0.2106
5500 h								741.3	985.0	1224.1	1349.3	1518.2	1657.0	1788.1
s								0.9090	1.1093	1.2930	1.3821	1.4908	1.5699	1.6369

Source: Abridged from *Thermodynamic Properties of Steam*, by Joseph H. Keenan and Frederick G. Keyes. © 1936, by Joseph Keenan and Frederick G. Keyes. Published by John Wiley & Sons, Inc., New York.

[a]For SI units see Section A.1.

TABLE A.5c

Properties of Water at Various Temperatures from 40 to 540°F (4.4 to 282.2°C)

Temp., °F	Temp., °C	Specific Volume[a] ft^3/lb	Specific Gravity	Weight[a] (lb/ft^3)	Vapor Pressure[a] PSIA
40	4.4	.01602	1.0013	62.42	0.1217
50	10.0	.01603	1.0006	62.38	0.1781
60	15.6	.01604	1.0000	62.34	0.2563
70	21.1	.01606	0.9987	62.27	0.3631
80	26.7	.01608	0.9975	62.19	0.5069
90	32.2	.01610	0.9963	62.11	0.6982
100	37.8	.01613	0.9944	62.00	0.9492
120	48.9	.01620	0.9901	61.73	1.692
140	60.0	.01629	0.9846	61.39	2.889
160	71.1	.01639	0.9786	61.01	4.741
180	82.2	.01651	0.9715	60.57	7.510
200	93.3	.01663	0.9645	60.13	11.526
212	100.0	.01672	0.9593	59.81	14.696
220	104.4	.01677	0.9565	59.63	17.186
240	115.6	.01692	0.9480	59.10	24.97
260	126.7	.01709	0.9386	58.51	35.43
280	137.8	.01726	0.9293	58.00	49.20
300	148.9	.01745	0.9192	57.31	67.01
320	160.0	.01765	0.9088	56.66	89.66
340	171.1	.01787	0.8976	55.96	118.01
360	182.2	.01811	0.8857	55.22	153.04
380	193.3	.01836	0.8736	54.47	195.77
400	204.4	.01864	0.8605	53.65	247.31
420	215.6	.01894	0.8469	52.80	308.83
440	226.7	.01926	0.8328	51.92	381.59
460	237.8	.0196	0.8183	51.02	466.9
480	248.9	.0200	0.8020	50.00	566.1
500	260.0	.0204	0.7863	49.02	680.8
520	271.1	.0209	0.7674	47.85	812.4
540	282.2	.0215	0.7460	46.51	962.5

[a] ft^3/lb = 62.4 l/Kg; lb/ft^3 = 0.016 Kg/l; PSIA = 0.069 bar (abs.).

Computed from Keenan & Keyes Steam Table.

A.6 Friction Loss in Pipes

Friction loss moduli for laminar flow are shown by the 45° lines in the upper left-hand portion of each chart. Moduli for turbulent flow are shown by the steeper curves in the lower right-hand portion. Both of these regions represent stable states of flow. A diagonal line separates the regions of laminar and turbulent flow and represents the critical zone, a region in which it is difficult to predict the state of flow and, hence, the friction loss. The critical zone usually represents a region of unstable flow. The critical zone line gives approximate moduli on the high side for this region of unstable flow.

The bottom scale of each chart represents flow in gallons per minute, GPM. An auxiliary top scale shows the average velocity in the pipe in feet per second. Read vertically from the GPM scale to find the corresponding velocity in feet per second. The vertical scales, labeled "Friction Loss Modulus for 100 Feet of Pipe," represent values of the ratio:

$$M = \frac{\Delta p}{SG}$$

where

M = friction loss modulus for 100 feet (30 m) of pipe
Δp = pressure loss in pounds per square inch per 100 feet of pipe
SG = specific gravity of fluid at 60°F (15.6°C)

The loss due to pipe friction may be obtained as follows:

$$\Delta p = M \times SG$$

To use the charts, proceed as follows:

1. Select the chart for the size of pipe in question.
2. Follow the vertical line representing the flow in GPM to its intersection with the desired viscosity curve, and read the modulus at the left.
3. If the vertical line representing the flow in GPM does not intersect the viscosity line in either turbulent or laminar flow, use the intersection with the critical zone line.
4. Compute the friction loss in pressure drop from the equation above.

1" STEEL PIPE (1.049" I.D.)
PIPE VELOCITY FT. PER SEC.

FRICTION LOSS MODULUS FOR 100 FEET OF PIPE
LOSS – PSI = MODULUS X SPECIFIC GRAVITY LOSS—FEET OF LIQUID = MODULUS X 2.31

GALLONS PER MINUTE

Loss—Lbs. per sq. in. = modulus × Specific Gravity
Loss—Feet of liquid = modulus × 2.31

*For SI units see Section A.1.

FIG. A.6a
*Friction loss modulus for 100 ft of 1" steel pipe.**

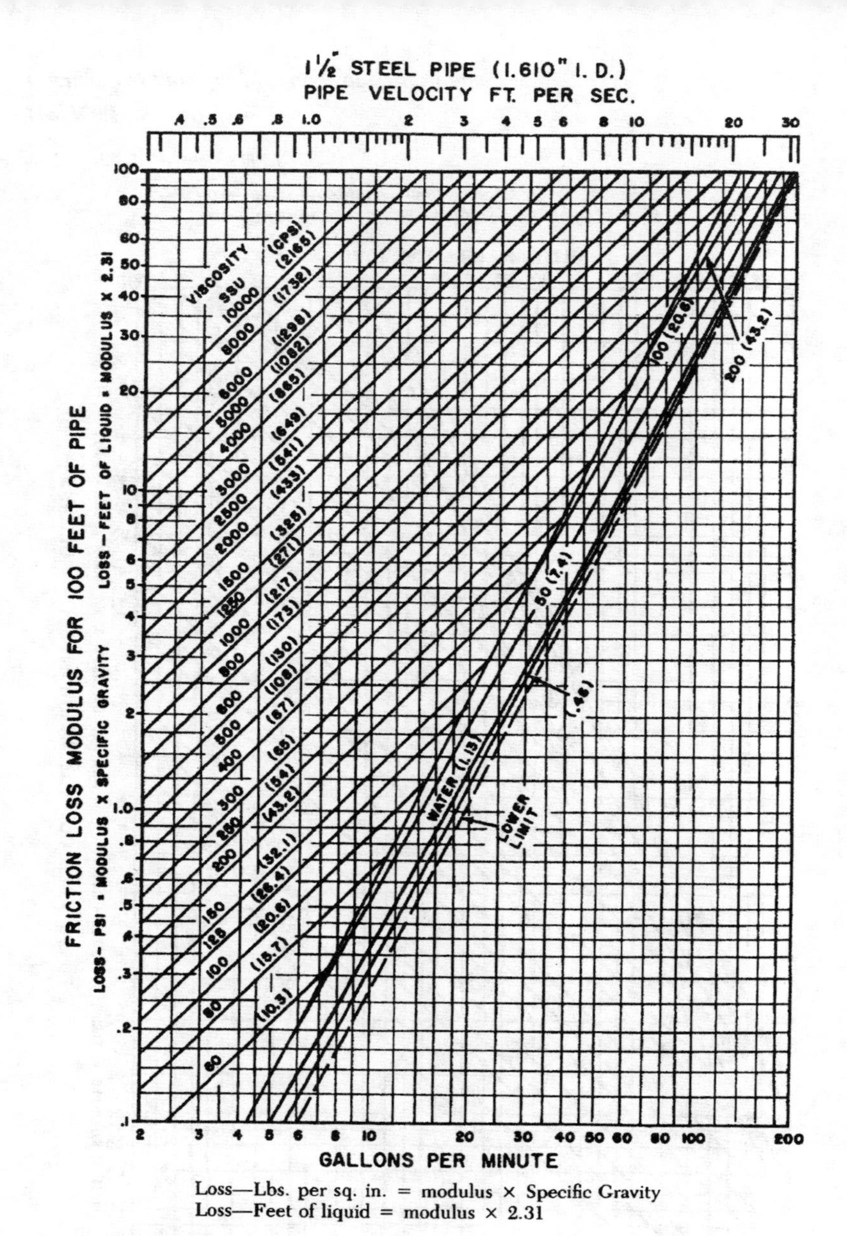

Loss—Lbs. per sq. in. = modulus × Specific Gravity
Loss—Feet of liquid = modulus × 2.31

*For SI units see Section A.1.

FIG. A.6b
*Friction loss modulus for 100 ft of 1½" steel pipe.**

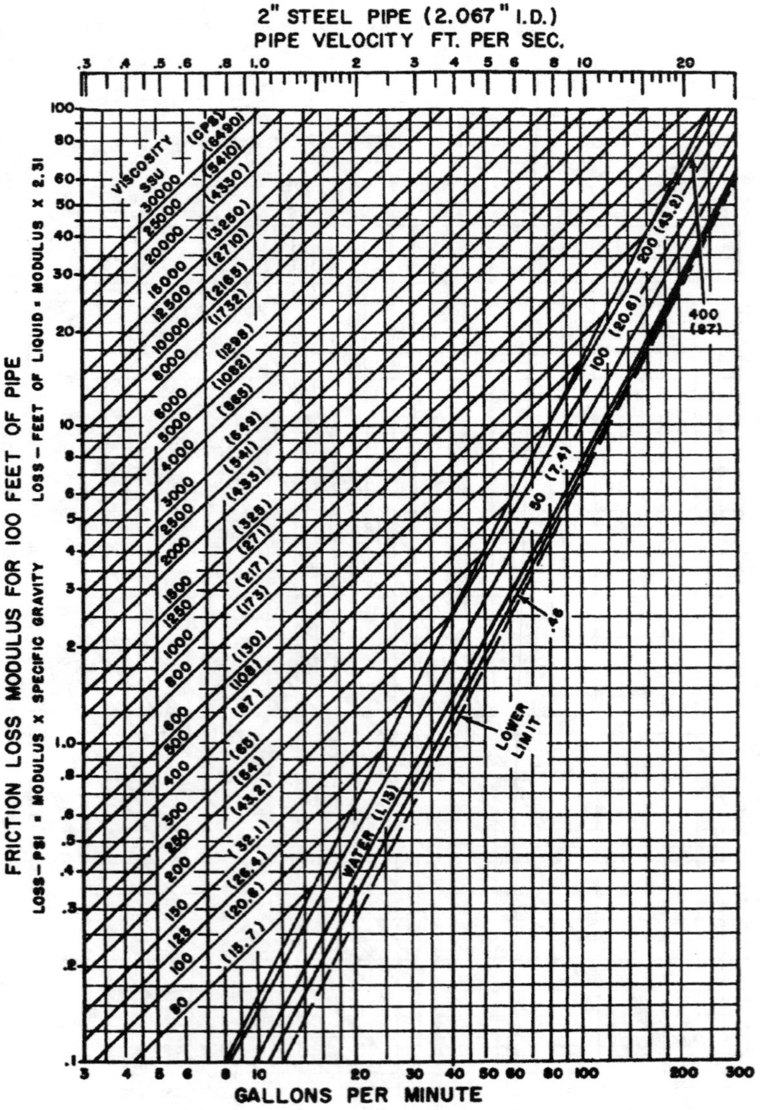

2" STEEL PIPE (2.067" I.D.)
PIPE VELOCITY FT. PER SEC.

FRICTION LOSS MODULUS FOR 100 FEET OF PIPE

LOSS–PSI = MODULUS × SPECIFIC GRAVITY LOSS – FEET OF LIQUID = MODULUS × 2.31

GALLONS PER MINUTE

Loss—Lbs. per sq. in. = modulus × Specific Gravity
Loss—Feet of liquid = modulus × 2.31

*For SI units see Section A.1.

FIG. A.6c
*Friction loss modulus for 100 ft of 2″ steel pipe.**

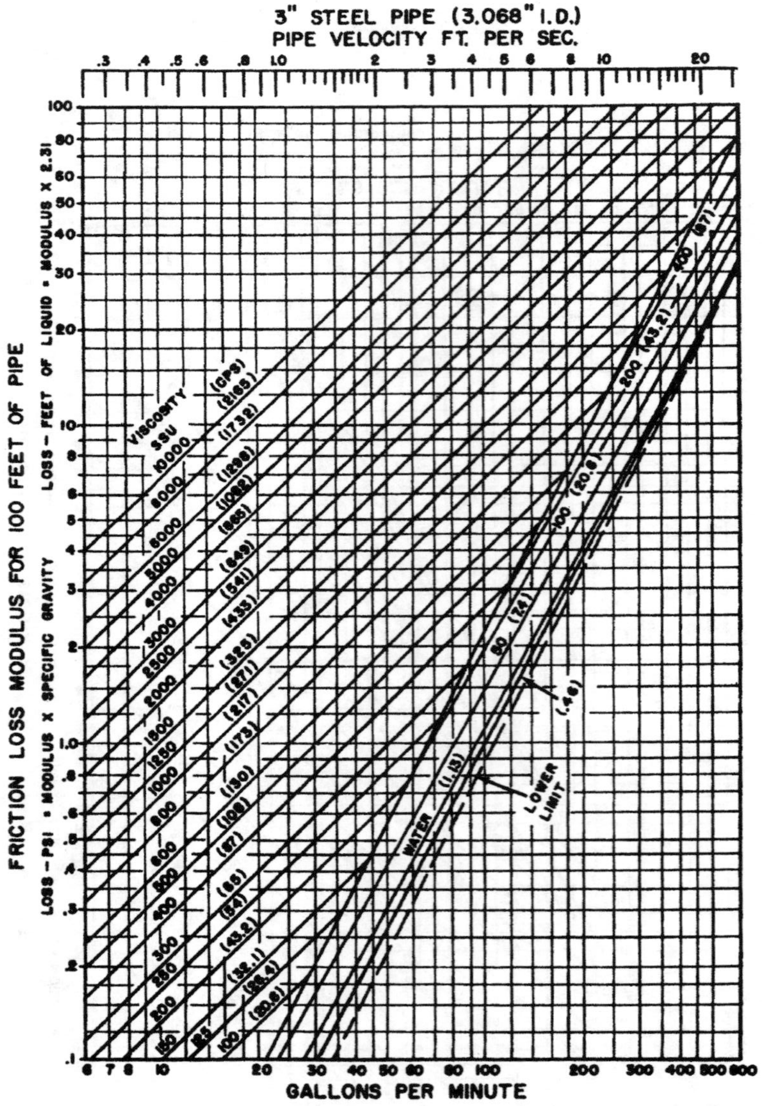

3" STEEL PIPE (3.068" I.D.)
PIPE VELOCITY FT. PER SEC.

FRICTION LOSS MODULUS FOR 100 FEET OF PIPE

LOSS–PSI = MODULUS × SPECIFIC GRAVITY LOSS – FEET OF LIQUID = MODULUS × 2.31

GALLONS PER MINUTE

Loss—Lbs. per sq. in. = modulus × Specific Gravity
Loss—Feet of liquid = modulus × 2.31

*For SI units see Section A.1.

FIG. A.6d
*Friction loss modulus for 100 ft of 3″ steel pipe.**

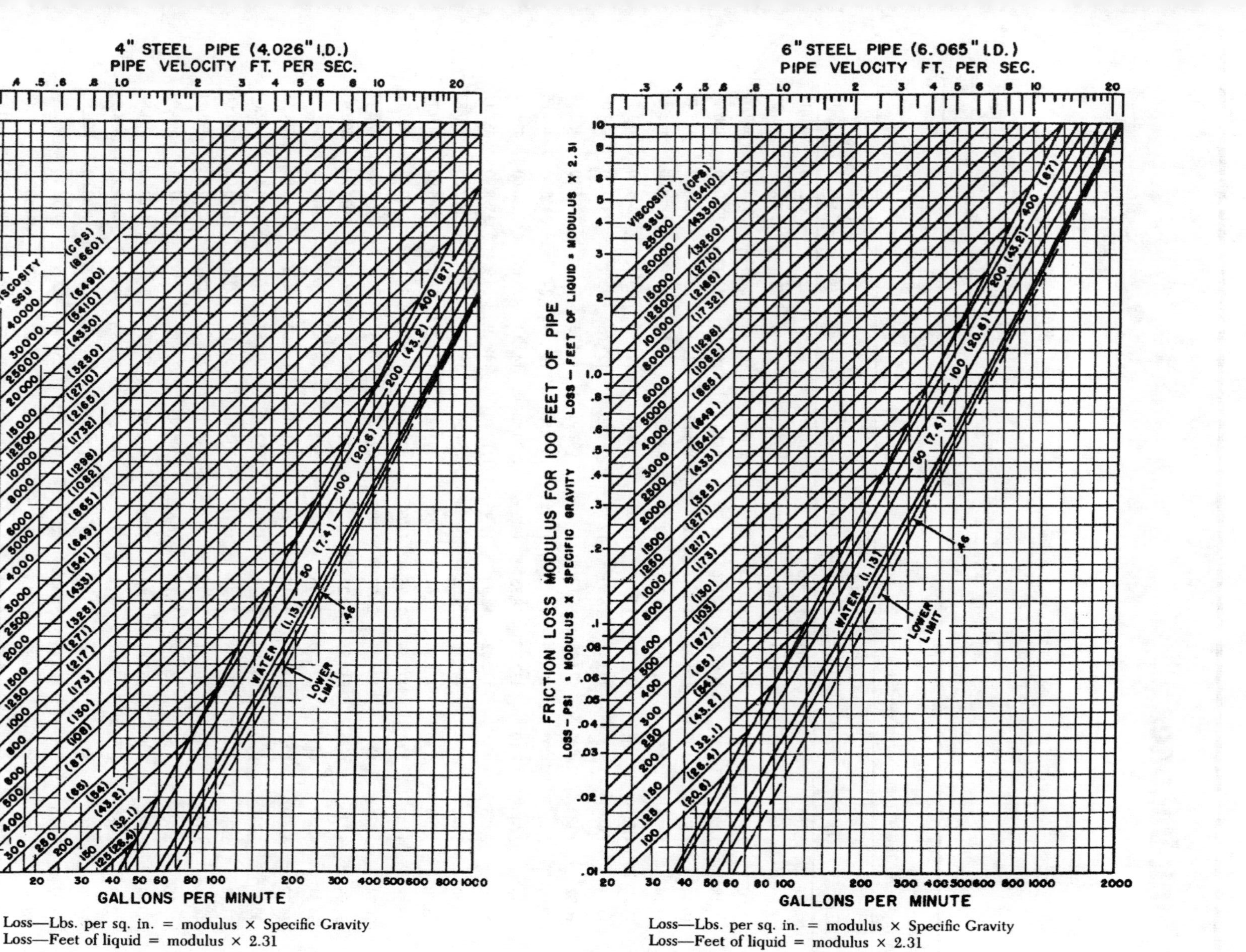

Loss—Lbs. per sq. in. = modulus × Specific Gravity
Loss—Feet of liquid = modulus × 2.31

*For SI units see Section A.1.

FIG. A.6e
*Friction loss modulus for 100 ft of 4" steel pipe.**

Loss—Lbs. per sq. in. = modulus × Specific Gravity
Loss—Feet of liquid = modulus × 2.31

*For SI units see Section A.1.

FIG. A.6f
*Friction loss modulus for 100 ft of 6" steel pipe.**

A.7 Tank Volumes

TABLE A.7a
Capacity of Round Tanks[a] *(per foot of depth)*

Diam.	*Gals.*	*Area sq. ft*	*Diam.*	*Gals.*	*Area sq. ft*	*Diam.*	*Gals.*	*Area sq. ft*	*Diam.*	*Gals.*	*Area sq. ft*
1′	5.87	.785	4′	94.00	12.566	11′	710.90	95.03	22′	2843.60	380.13
1′1″	6.89	.922	4′1″	97.96	13.095	11′3″	743.58	99.40	22′3″	2908.60	388.82
1′2″	8.00	1.069	4′2″	102.00	13.635	11′6″	776.99	103.87	22′6″	2974.30	397.61
1′3″	9.18	1.227	4′3″	106.12	14.186	11′9″	811.14	108.43	22′9″	3040.80	406.49
1′4″	10.44	1.396	4′4″	110.32	14.748	12′	846.03	113.10	23′	3108.00	415.48
1′5″	11.79	1.576	4′5″	114.61	15.321	12′3″	881.65	117.86	23′3″	3175.90	424.56
1′6″	13.22	1.767	4′6″	118.97	15.90	12′6″	918.00	122.72	23′6″	3244.60	433.74
1′7″	14.73	1.969	4′7″	123.42	16.50	12′9″	955.09	127.68	23′9″	3314.00	443.01
1′8″	16.32	2.182	4′8″	127.95	17.10	13′	992.91	132.73	24′	3384.10	452.39
1′9″	17.99	2.405	4′9″	132.56	17.72	13′3″	1031.50	137.89	24′3″	3455.00	461.86
1′10″	19.75	2.640	4′10″	137.25	18.35	13′6″	1070.80	142.14	24′6″	3526.60	471.44
1′11″	21.58	2.885	4′11″	142.02	18.99	13′9″	1110.80	148.49	24′9″	3598.90	481.11
2′	23.50	3.142	5′	146.76	19.62	14′	1151.50	153.94	25′	3672.00	490.87
2′1″	25.50	3.409	5′3″	161.86	21.64	14′3″	1193.00	159.48	25′3″	3745.80	500.74
2′2″	27.58	3.687	5′6″	177.66	23.75	14′6″	1235.30	165.13	25′6″	3820.30	510.71
2′3″	29.74	3.976	5′9″	194.27	25.97	14′9″	1278.20	170.87	25′9″	3895.60	520.77
2′4″	31.99	4.276	6′	211.51	28.27	15′	1321.90	176.71	26′	3971.60	530.93
2′5″	34.31	4.587	6′3″	229.50	30.68	15′3″	1366.40	182.65	26′3″	4048.40	541.19
2′6″	36.72	4.909	6′6″	248.23	35.18	15′6″	1411.50	188.69	26′6″	4125.90	551.55
2′7″	39.21	5.241	6′9″	267.69	35.78	15′9″	1457.40	194.83	26′9″	4204.10	562.00
2′8″	41.78	5.585	7′	287.88	38.48	16′	1504.10	201.06	27′	4283.00	572.66
2′9″	44.43	5.940	7′3″	308.81	41.28	16′3″	1551.40	207.39	27′3″	4362.70	583.21
2′10″	47.16	6.305	7′6″	330.48	44.18	16′6″	1599.50	213.82	27′6″	4443.10	593.96
2′11″	49.98	6.681	7′9″	352.88	47.17	16′9″	1648.40	220.35	27′9″	4524.30	604.81
3′	52.88	7.069	8′	376.01	50.27	17′	1697.21	226.87	28′	4606.20	615.75
3′1″	55.86	7.467	8′3″	399.80	53.46	17′6″	1798.51	240.41	28′3″	4688.80	626.80
3′2″	58.92	7.876	8′6″	424.48	56.75	18′	1902.72	254.34	28′6″	4772.10	637.94
3′3″	62.06	8.296	8′9″	449.82	60.13	18′6″	2009.92	268.67	28′9″	4856.20	649.18
3′4″	65.28	8.727	9′	475.89	63.62	19′	2120.90	283.53	29′	4941.00	660.52
3′5″	68.58	9.168	9′3″	502.70	67.20	19′6″	2234.00	298.65	29′3″	5026.60	671.96
3′6″	71.97	9.621	9′6″	530.24	70.88	20′	2350.10	314.16	29′6″	5112.90	683.49
3′7″	75.44	10.085	9′9″	558.51	74.66	20′6″	2469.10	330.06	29′9″	5199.90	695.13
3′8″	78.99	10.559	10′	587.52	78.54	21′	2591.00	346.36	30′	5287.70	706.86

TABLE A.7a Continued

Capacity of Round Tanks[a] (per foot of depth)

Diam.	Gals.	Area sq. ft	Diam.	Gals.	Area sq. ft	Diam.	Gals.	Area sq. ft	Diam.	Gals.	Area sq. ft
3′9″	82.62	11.045	10′3″	617.26	82.52	21′3″	2653.00	354.66	30′3″	5376.20	718.69
3′10″	86.33	11.541	10′6″	640.74	86.59	21′6″	2715.80	363.05	30′6″	5465.40	730.62
3′11″	90.13	12.048	10′9″	678.95	90.76	21′9″	2779.30	371.54	30′9″	5555.40	742.64

To find the capacity of tanks greater than shown above, find a tank of one-half the size desired, and multiply its capacity by four, or find one one-third the size desired, and multiply its capacity by 9.

[a]foot = 0.3048 m; gals = 3.785 l; sq. ft = 0.0929 m^2; inch = 25.4 mm

TABLE A.7b

Capacity of Partially Filled Horizontal Tanks[a]

Diam. (in ft)	Gallons per ft of Length When Tank Is Filled 1/10	1/5	3/10	2/5	1/2	3/5	7/10	4/5	9/10
1	.3	.8	1.4	2.1	2.9	3.6	4.3	4.9	5.5
2	1.2	3.3	5.9	8.8	11.7	14.7	17.5	20.6	22.2
3	2.7	7.5	13.6	19.8	26.4	33.0	39.4	45.2	50.1
4	4.9	13.4	23.8	35.0	47.0	59.0	70.2	80.5	89.0
5	7.6	20.0	37.0	55.0	73.0	92.0	110.0	126.0	139.0
6	11.0	30.0	53.0	78.0	106.0	133.0	158.0	182.0	201.0
7	15.0	41.0	73.0	107.0	144.0	181.0	215.0	247.0	272.0
8	19.0	52.0	96.0	140.0	188.0	235.0	281.0	322.0	356.0
9	25.0	67.0	112.0	178.0	238.0	298.0	352.0	408.0	450.0
10	30.0	83.0	149.0	219.0	294.0	368.0	440.0	504.0	556.0
11	37.0	101.0	179.0	265.0	356.0	445.0	531.0	610.0	672.0
12	44.0	120.0	214.0	315.0	423.0	530.0	632.0	741.0	800.0
13	51.0	141.0	250.0	370.0	496.0	621.0	740.0	850.0	940.0
14	60.0	164.0	291.0	430.0	576.0	722.0	862.0	989.0	1084.0
15	68.0	188.0	334.0	494.0	661.0	829.0	988.0	1134.0	1253.0

[a]foot = 0.3048 m; gallon = 3.785 l.

TABLE A.7c

Capacities of Various Cylinders in U.S. Gallons[a]

Diam. (in inches)	Length of Cylinder																	
	1″	1′	5′	6′	7′	8′	9′	10′	11′	12′	13′	14′	15′	16′	17′	18′	20′	22′	24′
1		0.04	0.20	0.24	0.28	0.32	0.36	0.40	0.44	0.48	0.52	0.56	0.60	0.64	0.68	0.72	0.80	0.88	0.96
2	0.01	0.16	0.80	0.96	1.12	1.28	1.44	1.60	1.76	1.92	2.08	2.24	2.40	2.56	2.72	2.88	3.20	3.52	3.84
3	0.03	0.37	1.84	2.20	2.56	2.92	3.30	3.68	4.04	4.40	4.76	5.12	5.48	5.84	6.22	6.60	7.36	8.08	8.80
4	0.05	0.65	3.26	3.92	4.58	5.24	5.88	6.52	7.18	7.84	8.50	9.16	9.82	10.5	11.1	11.8	13.0	14.4	15.7
5	0.08	1.02	5.10	6.12	7.14	8.16	9.18	10.2	11.2	12.2	13.3	14.3	15.3	16.3	17.3	18.4	20.4	22.4	24.4
6	0.12	1.47	7.34	8.80	10.3	11.8	13.2	14.7	16.1	17.6	19.1	20.6	22.0	23.6	25.0	26.4	29.4	32.2	35.2
7	0.17	2.00	10.0	12.0	14.0	16.0	18.0	20.0	22.0	24.0	26.0	28.0	30.0	32.0	34.0	36.0	40.0	44.0	48.0
8	0.22	2.61	13.0	15.6	18.2	20.8	23.4	26.0	28.6	31.2	33.8	36.4	39.0	41.6	44.2	46.8	52.0	57.2	62.4
9	0.28	3.31	16.5	19.8	23.1	26.4	29.8	33.0	36.4	39.6	43.0	46.2	49.6	52.8	56.2	60.0	66.0	72.4	79.2
10	0.34	4.08	20.4	24.4	28.4	32.6	36.8	40.8	44.8	48.8	52.8	56.8	61.0	65.2	69.4	73.6	81.6	89.6	97.6
11	0.41	4.94	24.6	29.6	34.6	39.4	44.4	49.2	54.2	59.2	64.2	69.2	74.0	78.8	83.8	88.8	98.4	104.	118.
12	0.49	5.88	29.4	35.2	41.0	46.8	52.8	58.8	64.6	70.4	76.2	82.0	87.8	93.6	99.6	106.	118.	129.	141.
13	0.57	6.90	34.6	41.6	48.6	55.2	62.2	69.2	76.2	83.2	90.2	97.2	104.	110.	117.	124.	138.	152.	166.
14	0.67	8.00	40.0	48.0	56.0	64.0	72.0	80.0	88.0	96.0	104.	112.	120.	128.	136.	144.	160.	176.	192.
15	0.77	9.18	46.0	55.2	64.4	73.6	82.8	92.0	101.	110.	120.	129.	138.	147.	156.	166.	184.	202.	220.
16	0.87	10.4	52.0	62.4	72.8	83.2	93.6	104.	114.	125.	135.	146.	156.	166.	177.	187.	208.	229.	250.
17	0.98	11.8	59.0	70.8	81.6	94.4	106.	118.	130.	142.	153.	163.	177.	189.	201.	212.	236.	260.	283.
18	1.10	13.2	66.0	79.2	92.4	106.	119.	132.	145.	158.	172.	185.	198.	211.	224.	240.	264.	290.	317.
19	1.23	14.7	73.6	88.4	103.	118.	132.	147.	162.	177.	192.	206.	221.	235.	250.	265.	294.	324.	354.
20	1.36	16.8	81.6	98.0	114.	130.	147.	163.	180.	196.	212.	229.	245.	261.	277.	294.	326.	359.	392.
21	1.50	18.0	90.0	108.	126.	144.	162.	180.	198.	216.	238.	252.	270.	288.	306.	324.	360.	396.	432.
22	1.65	19.8	99.0	119.	139.	158.	178.	198.	218.	238.	257.	277.	297.	317.	337.	356.	396.	436.	476.
23	1.80	21.6	108.	130.	151.	173.	194.	216.	238.	259.	281.	302.	324.	346.	367.	389.	432.	476.	518.
24	1.96	23.5	118.	141.	165.	188.	212.	235.	259.	282.	306.	330.	353.	376.	400.	424.	470.	518.	564.
25	2.12	25.5	128.	153.	179.	204.	230.	255.	281.	306.	332.	358.	383.	408.	434.	460.	510.	562.	612.
26	2.30	27.6	138.	166.	193.	221.	248.	276.	304.	331.	359.	386.	414.	442.	470.	496.	552.	608.	662.
27	2.48	29.7	148.	178.	208.	238.	267.	297.	326.	356.	386.	416.	426.	476.	504.	534.	594.	652.	712.
28	2.67	32.0	160.	192.	224.	256.	288.	320.	352.	384.	416.	448.	480.	512.	544.	576.	640.	704.	768.
29	2.86	34.3	171.	206.	240.	274.	309.	343.	377.	412.	446.	480.	514.	548.	584.	618.	686.	754.	824.
30	3.06	36.7	183.	220.	257.	294.	330.	367.	404	440.	476.	514.	550.	588.	624.	660.	734.	808.	880.
32	3.48	41.8	209.	251.	293.	334.	376.	418.	460.	502.	544.	586.	668.	668.	710.	752.	836.	920.	1004.
34	3.93	47.2	236.	283.	330.	378.	424.	472.	520.	566.	614.	660.	708.	756.	802.	848.	944.	1040.	1132.
36	4.41	52.9	264.	317.	370.	422.	476.	528.	582.	634.	688.	740.	793.	844.	898.	952.	1056.	1164.	1268.

[a] inch = 25.4 mm; foot = 0.3048 m; gallon = 3.785 1.

A.8 Partial List of Suppliers*

3M TOUCH SYSTEMS, 800 Carleton Ct., Annarcis Island, New Westminster, BC, V3L-6L3 Canada, 604/521-3962, Fax: 604/521-4629, www.dynapro.com*

4B COMPONENTS LTD, 729 Sabrina Dr., East Peoria, IL 61611, 309/698-5611, Fax: 309/698-5615, www.go4b.com

A

AALBORG INSTRUMENTS & CONTROLS, 20 Corporate Dr., Orangeburg, NY 10962, 800/866-3837, Fax: 845/398-3165, www.aalborg.com

AALIANT, 150 Venture Blvd., Spartanburg, SC 29306, 864/574-8060, Fax: 864/574-8063, www.bindicator.com

ABB AUTOMATION—ANALYTICAL DIV., 843 N. Jefferson St., Lewisburg, WV 24901, 304/647-1761, Fax: 304/645-4236, www.abb.com/usa

ABB AUTOMATIONS, INC., DRIVES & POWER PRODUCTS, 16250 W. Glendale Dr., New Berlin, WI 53151, 800/752-0696, Fax: 262/785-0397, www.abb-drives.com

ABB, INC., 29801 Euclid Ave., Wickliffe, OH 44092, 800/626-4999, Fax: 716/273-7014, www.abb.com

ABB INSTRUMENTATION, 125 E. County Line Rd., Warminster, PA 18974, 215/674-6000, Fax: 215/674-7183, www.abb.com/us/instrumentation

ABB WATER METERS, INC., 1100 SW 38th Ave., Ocala, FL 34474, 800/874-0890, Fax: 352/368-1950, www.abbwatermeters.com

ABSOLUTE PROCESS INSTRUMENTS, INC., 1029 Butterfield Rd., Vernon Hills, IL 60061, 800/942-0315, Fax: 800/949-7502, www.api-usa.com

AC DATA SYSTEMS, 806 Clearwater Loop, Suite C, Post Falls, ID 83854, 800/890-2569, Fax: 208/777-4466, www.surgeblox.com

ACCES I/O PRODUCTS, INC., 10623 Roselle St., San Diego, CA 92121, 858/550-9559, Fax: 858/550-7322, www.acces-usa.com

ACCUTECH, 577 Main St., Hudson, MA 01749, 800/879-6576, Fax: 978/568-9085, www.savewithaccutech.com

ACME ELECTRIC CORP., 4815 W. 5th St., Lumberton, NC 28358, 910/738-1121, Fax: 910/739-0024, www.acmepowerdist.com

* From *CONTROL Magazine Buyer's Guide*, May 2001. With permission.

ACOPIAN, PO Box 638, Easton, PA 18044, 610/258-5441, Fax: 610/258-2842, www.acopian.com

ACP, 6865 Shiloh Rd. E., Alpharetta, GA 30005, 770/205-2475, Fax: 770/888-5362, www.acpthinclient.com

ACROMAG, INC., 30765 South Wixom Rd., Wixom, MI 48393, 248/624-1541, Fax: 248/624-9234, www.acromag.com

ACTION INSTRUMENTS, 8601 Aero Dr., San Diego, CA 92123, 585/279-5726, Fax: 858/279-6290, Millard Schewe, millards@actionio.com, www.actionio.com

AD PRODUCTS CO., 4799 W. 150th St., Cleveland, OH 44135, 800/325-4935, Fax: 216/267-5392, www.adproductsco.com

ADALET, 4801 W. 150th St., Cleveland, OH 44135, 216/267-9000, Fax: 216/267-1681, www.adalet.com

ADAPTIVE MICRO SYSTEMS, 7840 N. 86th St., Milwaukee, WI 53224, 414/357-2020, Fax: 414/357-2029, www.adaptivedisplays.com

ADAPTIVE RESOURCES, 2 Park Dr., Lawrence, PA 15055, 724/746-4969, Fax: 724/746-9260, www.adaptiveresources.com

ADTECH, 3750 Monroe Ave., Pittsford, NY 14534, 716/383-8280, Fax: 716/383-8386, www.adtech-inst.com

ADVANCED CONTROL TECHNOLOGY, INC., 7050 E. Hwy. 101, Shakopee, MN 55379, 952/882-0000, Fax: 952/890-3644

ADVANCED MOTION CONTROLS, 3805 Calle Tecate, Camarillo, CA 93012, 805/389-1935, Fax: 805/389-1165, www.a-m-c.com

ADVANCED SEPARATIONS AND PROCESS SYSTEMS, INC., 6111 Pepsi Way, Windsor, WI 53716, 608/846-1130, Fax: 608/846-1144, www.asapsys.com

ADVANCED SYSTEMS & DESIGNS, INC., 1100 Owendale #J, Troy, MI 48083, 248/689-4800, Fax: 248/689-8811, www.asdspc1.com

ADVANTECH AUTOMATION CORP., 1320 Kemper Meadow Dr. #500, Cincinnati, OH 45240, 513/742-8895, Fax: 513/742-8892, www.advantech.com

ADVANTECH TECHNOLOGIES, INC., 15375 Barranca Pkwy., Irvine, CA 92618, 949/789-7178, Fax: 949/789-7179, www.advantech.com/epc

AEROTECH, INC., 101 Zeta Dr., Pittsburgh, PA 15238, 412/963-7470, Fax: 412/963-7459, www.aerotech.com

AFCON CONTROL & AUTOMATION, INC., 1014 E. Algonquin Rd. #102, Schaumburg, IL 60173, 847/397-6900, Fax: 847/397-6987, www.afcon-inc.com

AGILE SYSTEMS, 575 Kumpf Dr., Waterloo, Ontario, N2V 1K3 Canada, 519/886-2000, Fax: 519/886-2075, www.agile-systems.com

AIR DIMENSIONS, INC., 1015 W. Newport Center Dr., Deerfield Beach, FL 33442, 954/428-7333, Fax: 954/360-0987, www.airdimensions.com

AIR INSTRUMENTS & MEASUREMENTS, INC., 13300 Brooks Dr. Suite A, Baldwin Park, CA 91706, 626/813-1460, Fax: 626/338-2565, www.aimanalysis.com

AIRPAX CORP., 807 Woods Rd., Cambridge, MD 21613, 410/228-1500, Fax: 410/228-3456, www.airpax.net

ALLEN-BRADLEY, 1201 S. 2nd St., Milwaukee, WI 53204, 414/382-2000, Fax: 414/382-4444, www.ab.com

ALLEN-BRADLEY CO. LLC, Two Executive Dr., Chelmsford, MA 01824, 978/446-3476, Fax: 978/446-3322, www.ab.com/sensors

ALLIED MOULDED PRODUCTS, INC., 222 N. Union St., Bryan, OH 43506, 419/636-4217, Fax: 419/636-2450, Valerie Hageman, www.enclosures.alliedmoulded.com

ALNOR INSTRUMENT CO., 7555 N. Linder Ave., Skokie, IL 60077, 847/677-3249, Fax: 847/677-3514, www.alnor.com

ALTECH CORP., 35 Royal Rd., Flemington, NJ 08822, 908/806-9400, Fax: 908/806-9490, www.altechcorp.com

ALTEK INDUSTRIES CORP., 35 Vantage Point Dr., Rochester, NY 14624, 716/349-3520, Fax: 716/349-3510, www.altekcalibrators.com

ALTERSYS, INC., 555 d'Auvergne St., Longueuil, Quebec, J4H 4A3 Canada, 405/674-7774, Fax: 405/674-7344, www.altersys.com

AMERICAN LED-GIBLE, INC., 1776 Lone Eagle St., Columbus, OH 43228, 614/851-1100, Fax: 614/851-1121, www.ledgible.com

AMERIPACK, 70 S. Main St., Cranbury, NJ 08512, 609/395-6969, Fax: 609/395-8753, www.ameripack.com

AMETEK DIXSON, 207 27 road, Grand Junction, CO 81503, 970/242-8863, Fax: 970/245-6267, www.dixson.com

AMETEK DREXELBROOK, 205 Keith Valley Rd., Horsham, PA 19044, 215/674-1234, Fax: 215/674-2731, www.drexelbrook.com

AMETEK NATIONAL CONTROLS, 1725 Western Dr., W. Chicago, IL 60185, 630/231-5900, Fax: 630/231-1377, www.nationalcontrols.com

AMETEK PATRIOT SENSORS, 1080 N. Crooks Rd., Clawson, MI 48017, 248/435-0700, Fax: 248/435-8120, www.patriotsensors.com

AMETEK POWER INSTRUMENTS, 255 N. Union St., Rochester, NY 14605, 716/238-4054, Fax: 716/454-7805, www.rochester.com

AMETEK PROCESS INSTRUMENTS, 455 Corporate Blvd., Newark, DE 19702, 302/456-4400, Fax: 302/456-4444, www.ametekpi.com

AMETEK TEST + CALIBRATION INSTRUMENTS, 8600 Somerset Dr., Largo, FL 33773, 727/536-7831, Fax: 727/539-6882, www.ametek.com/tci

AMETEK THERMOX, 150 Freeport Rd., Pittsburgh, PA 15238, 412/828-9040, Fax: 412/826-0399, www.thermox.com

AMETEK U.S. GAUGE, 900 Clymer Ave., Sellersville, PA 18960, 215/257-6531, Fax: 215/257-3058, www.ametekusg.com

AMETEK U.S. GAUGE/PMT, 820 Penna Blvd., Feasterville, PA 19053, 215/355-6900, Fax: 215/355-2937, www.ametekusg.com

AMOT CONTROLS, 401 N. First St., Richmond, CA 94801, 510/307-8315, Fax: 510/234-9950, www.amotusa.com

AMPRO COMPUTERS, INC., 4757 Hellyer Ave., San Jose, CA 95138, 408/360-0200, Fax: 408/360-0222, www.ampro.com

ANALAB LLC, PO Box 34, Sterling, PA 18463, 570/689-3919, Fax: 570/689-9360, www.analab1.com

ANALOG DEVICES, INC., 3 Technology Way, Norwood, MA 02062, 781/987-1428, Fax: 781/937-1021, www.analog.com

ANALOGIC CORP., 8 Centennial Dr., Peabody, MA 01915, 978/977-3000, Fax: 978/977-6809, www.analogic.com

ANDERSON INSTRUMENT CO., 3950 Greenbriar, Fultonville, NY 12072, 518/922-5315, Fax: 518/922-8997, www.andinst.com

ANN ARBOR TECHNOLOGIES, PO Box 1247, Ann Arbor, MI 48106, 734/995-1360, www.a2t.com

ANORAD CORP., 110 Oser Ave., Hauppauge, NY 11788, 631/231-1990, Fax: 631/435-1612, www.anorad.com

ANSIMAG, INC., 1090 Fox Ct., Bloomingdale, IL 60108, 303/940-3111, Fax: 303/940-3141, www.sundyne.com

ANT COMPUTER, INC., 20760 E. Carrey Rd., Walnut, CA 91789, 909/598-3315, Fax: 909/598-3215, www.antcomputer.com

ANTEK INDUSTRIAL INSTRUMENTS, 700 Gateway Parkway, Marble Falls, TX 78654, 830/693-5671, Fax: 830/798-8208, www.antekhou.com

ANTON PAAR USA, 10201 Maple Leaf Court, Ashland, VA 23005, 804/550-1051, Fax: 804/550-1057, www.anton-paar.com

API MOTION, INC., 45 Hazelwood Dr., Amherst, NY 14228, 716/691-9100, Fax: 716/691-9181, www.apimotion.com

APPLICOM INTERNATIONAL, INC., 4340 Redwood Hwy, Suite D-309, San Rafael, CA 94903, 415/472-1595, Fax: 415/472-1596, www.applicom-int.com

APPLIED CHEMOMETRICS, INC., PO Box 100, Sharon, MA 02067, 781/784-7700, www.chemometrics.com

APPLIKON ANALYZERS, INC., 1701 North Park Drive, #20, Kingwood, TX 77339, 281/354-2211, Fax: 281/354-0050, www.applikin.com

APV, 9525 W. Bryn Mawr Ave., Rosemont, IL 60018, 847/678-4300, Fax: 847/678-4313, www.apv.com

APW ZERO CASES, 500 W. 200 N., North Salt Lake City, UT 84054, 801/298-5900, Fax: 801/292-9450, www.zerocases.com

AQUA MEASURE INSTRUMENTS CO., 1712 Earhard Ct., La Verne, CA 91750, 909/392-5833, Fax: 909/392-5838, www.moisturerregisterproducts.com

ARI INDUSTRIES, 381 ARI Ct., Addison, IL 60101, 630/953-9100, Fax: 630/953-0590

ARIES ELECTRONICS INC., PO Box 130, Frenchtown, NJ 08825, 908/996-6841, Fax: 908/996-3891, www.arieselec.com

ARIZONA INSTRUMENT CO., 1912 W. 4th St., Tempe, AZ 85281, 602/470-1414, Fax: 480/804-0656, www.azic.com

AROMAT CORP., 629 Central Ave., New Providence, NJ 07974, 800/228-2350, Fax: 908/464-4128, www.aromat.com/acsd

ASAHI/AMERICA, INC., PO Box 653, Malden, MA 02148-6834, 781/321-5409, Fax: 781/321-4421, www.asahi-america.com

ASCO, 50 Hanover Rd., Florham Park, NJ 07932, 973/966-2372, Fax: 973/966-2448, www.ascovalve.com

ASL, INC., 100 Brickstone Sq., Andover, MA 01810, 978/658-0000, Fax: 978/658-5444, www.aslinc.com

ASPEN TECHNOLOGY, 10 Canal Park, Cambridge, MA 02141, 617/949-1000, Fax: 617/949-1030, www.aspentech.com

ASTEC, 6339 Paseo Del Lago, Carlsbad, CA 92009, 760/930-4745, Fax: 460/930-4700, www.astec.com

ASTRO-MED., INC., 600 E. Greenwich Ave., W. Warwick, RI 02893, 401/828-4000, Fax: 401/822-2430, www.astro-med.com

ASTRODYNE CORP., 300 Myles Standish Blvd., Taunton, MA 02780, 508/823-8080, Fax: 508/823-8181, www.astrodyne.com

ATA SENSORS, 4500 Anaheim Ave., B-6, Albuquerque, NM 87113, 505/823-1320, Fax: 505/823-1560, www.atasensors.com

ATHENA CONTROLS, 5145 Campus Dr., Plymouth Meeting, PA 19462, 610/828-2490, Fax: 610/828-7084, www.athenacontrols.com

ATS RHEOSYSTEMS, 52 Georgetown Rd., Bordentown, NJ 08505, 609/298-2522, Fax: 609/298-2795, www.atsrheosystems.com

AUMA ACTUATORS, INC., 4 Zesta Dr., Pittsburgh, PA 15205, 412/787-1340, Fax: 412/787-1223, www.auma-usa.com

AUTOMATA, INC., 104 New Mohawk Rd., Suite A, Nevada City, CA 95959, 530/478-5882, Fax: 530/478-5881, www.automata-inc.com

AUTOMATED CONTROL CONCEPTS, INC., 3535 Route 66, Neptune, NJ 07753, 732/922-6611, Fax: 732/922-9611, www.automated-control.com

AUTOMATED SOLUTIONS, INC., 1415 Fulton Rd., Suite 205-A12, Santa Rose, CA 95403, 707/578-5882, Fax: 707/579-5756, www.automatedsolutions.com

AUTOMATIC TIMING & CONTROLS, 1827 Freedom Rd., Lancaster, PA 17557, 717/295-0500, Fax: 717/481-7240, www.automatictiming.com

AUTOMATION PRODUCTS, INC., 3030 Maxmy St., Houston, TX 77008, 800/231-2062, Fax: 713/869-7332, www.dynatrolusa.com

AUTOMATION SYSTEMS INTERCONNECT, INC., PO Box 1230, Carlisle, PA 17013, 877/650-5160, Fax: 717/249-5542, www.asi-ez.com

AUTOMATIONDIRECT.COM, 3505 Hutchinson Rd., Cumming, GA 30040, 800/633-0405, Fax: 770/889-7876, www.automationdirect.com

AUTOMATIONTECHIES.COM, PO Box 44759, Eden Prairie, MN 55344, 877/300-6792, Fax: 877/593-6792, www.automationtechies.com

AVG PRESS AUTOMATION SYSTEMS, 343 St. Paul Blvd., Carol Stream, IL 60188, 800/TEC-ENGR, Fax: 630/668-4676, www.avg.net

AVO INTL., PO Box 9007, Valley Forge, PA 19485, 610/676-8500, Fax: 610/676-8610, www.avointl.com

AVX CORP., 801 17th Ave. S., Myrtle Beach, SC 29578, 843/946-0601, Fax: 843/626-5814, www.avxcorp.com

AXIOM TECHNOLOGY, INC., 18138 Rowland St., City of Industry, CA 91748, 626/581-3232, Fax: 626/581-3552, www.axiomtek.com

AYDIN DISPLAYS, INC., 700 Dresher Rd., Horsham, PA 19044, 215/784-5335, Fax: 215/830-9545, www.aydindisplays.com

AZONIX CORP., 900 Middlesex Turnpike, Billerica, MA 01821, 978/670-6300, Fax: 978/670-8855, www.azonix.com

B

BABBITT INTERNATIONAL, INC., PO Box 70094, Houston, TX 77270, 800/835-8012, Fax: 713/467-8736, www.babbittlevel.com

BACHARACH, INC., 625 Alpha Dr., Pittsburgh, PA 15238, 412/963-2161, Fax: 412/963-2091, www.bacharach-inc.com

BADGER METER, INC., PO Box 245036, Milwaukee, WI 53224, 800/876-3837, Fax: 414/371-5932, www.badgermeter. com

BALDOR ELECTRIC CO., 5711 R.S. Boreham, Jr. St., Fort Smith, AR 72903, 501/646-4711, Fax: 501/648-5792, www.baldor.com

BALLUFF, 8125 Holton Dr., Florence, KY 41042, 606/727-2200, Fax: 606/727-4823, www.balluff.com

BANNER ENGINEERING CORP., 9714 10th Ave. N., Minneapolis, MN 55441, 763/544-3164, Fax: 763/544-3213, www.baneng.com

BARBER COLMAN INDUSTRIAL INSTRUMENTS, 741-F Miller Dr., Leesburg, VA 20175, 703/443-0000, Fax: 703/669-1300, www.barber-colman.com

BARKSDALE, 3211 Fruitland, Los Angeles, CA 90058, 323/583-6218, Fax: 323/586-3060, www.barksdale.com

BARNANT CO., 28W092 Commercial Ave., Barrington, IL 60010, 847/381-7050, Fax: 847/381-7053, www.barnant.com

BARNETT ENGINEERING LTD, 215 7710 5th St. S.E., Calgary, Alberta, T2H 2L9 Canada, 403/255-9544, Fax: 403/259-2343, www.barnett-engg.com

BARTEC U.S. CORP., 9902 E. East 43rd St., Tulsa, OK 74146, 918/627-1889, Fax: 918/627-1890, www.bartecus.com

BARTON INSTRUMENT SYSTEMS, 900 S. Turnbull Canyon Rd., City of Industry, CA 91745, 626/961-2547, Fax: 626/961-4452, www.barton-instruments.com

BAUMER ELECTRIC LTD, 122 Spring St., Southington, CT 06489, 860/621-2121, Fax: 860/628-6280, www.baumerelectric.com

BEAMEX, INC., 2270 Northwest Pkwy., Marietta, GA 30067, 800/888-9892, Fax: 770/951-1927, www.beamex.com

BEBCO INDUSTRIES, 600 Gulf Frwy., Texas City, TX 77591, 800/OK-BEBCO, Fax: 409/938-4189, www.okbebco.com

BECKHOFF AUTOMATION LLC, 12204 Nicollet Ave. S., Minneapolis, MN 55337, 952/890-0000, Fax: 952/890-2888, www.beckhoff.com

BEI—INDUSTRIAL ENCODER DIV., 7230 Hollister Ave., Goleta, CA 93117, 800/350-2727, Fax: 805/968-3154, www.beiied.com

BELDEN ELECTRONICS, 2200 US Hwy 27 S., Richmond, IN 47374, 765/983-5200, Fax: 765/983-5294, www.belden.com

BELL TECHNOLOGY, INC., 6120 Hanging Moss Rd., Orlando, FL 32807, 407/678-6900, Fax: 407/678-0578, www.belltechinc.com

BENTLY NEVADA, 1631 Bently Pkwy S., Minden, NV 89423, 775/782-3611, Fax: 775/782-9337, www.bently.com

BERGOTECH, INC., 32 Clarissa Dr., PH 22, Richmond Hill, Ontario, L4C 9R7 Canada, 416/456-7533, Fax: 908/883-6202

BETTIS/EMERSON VALVE AUTOMATION, 18703 GH Circle, Waller, TX 77484, 281/463-5100, Fax: 281/463-5103, www.emersonvalveautomation.com

BI TECHNOLOGIES, 4200 Bonita Pl., Fullerton, CA 92835, 714/447-2666, Fax: 714/447-2400, www.bitechnologies.com

BI-LOK TUBE FITTINGS, 18254 Technology Dr., Meadville, PA 16335, 814/337-0380, Fax: 814/337-8468, www.bilok.com

BINDICATOR, 150 Venture Blvd., Spartanburg, SC 29306, 864/574-8060, Fax: 864/574-8063, www.bindicator.com

BINMASTER, 4200 N. 48th St., Lincoln, NE 68504, 402/434-9100, Fax: 402/434-9133, www.binmaster.com

BLACK & VEATCH, PO Box 8405, Kansas City, MO 64114, 913/458-2000, Fax: 913/458-2934, www.bv.com

BLANCETT, 100 E. Felix St., Suite 190, Fort Worth, TX 76115, 817/920-9998, Fax: 817/921-5282, www.blancett.com

BLUE BOX VIDEO, 3101 Bee Caves #290, Austin, TX 78746, 512/330-9990, Fax: 512/330-9996, www.bluevideo.com

BLUE MOUNTAIN QUALITY RESOURCES, INC., 208 W. Hamilton Ave., State College, PA 16801, 814/234-2417, Fax: 814/234-7077, www.coolblue.com

BLUE-WHITE INDUSTRIES, 14931 Chestnut St., Westminster, CA 92683, 714/893-8529, Fax: 714/894-9492, www.bluwhite.com

BOSCH AUTOMATION TECHNOLOGY, 7505 Durand Ave., Racine, WI 53406, 262/554-7100, Fax: 262/554-8103, www.boschat.com

BRAY CONTROLS, 13333 Westland East Blvd., Houston, TX 77041, 281/894-5454, Fax: 281/894-0022, www.bray.com

BRINKMANN INSTRUMENTS, INC., One Cantiague Rd., Westbury, NY 11590-0207, 516/334-7500, Fax: 516/334-7506, www. brinkmann.com

BRISTOL BABCOCK, 1100 Buckingham St., Watertown, CT 06795, 860/945-2295, Fax: 860/945-2278, www.bristolbabcock.com

BRISTOL EQUIPMENT CO., PO Box 696, Yorkville, IL 60560, 630/553-7161, Fax: 630/553-5981, www.bristolequipment.com

BROOKFIELD ENGINEERING LABORATORIES, 11 Commerce Blvd., Middleboro, MA 02346, 508/946-6200, Fax: 508/946-6262, www.brookfieldengineering.com

BROOKS AUTOMATION, 15 Elizabeth Dr., Chelmsford, MA 01824, 978/262-2400, Fax: 978/262-2500, www.brooks.com

BROOKS INSTRUMENT, 407 W. Vine St., Hatfield, PA 19440, 215/362-3700, Fax: 215/362-3745, www.brooksinstrument.com

BRUEL & KJAER, 2815 Colonnades Ct., Norcross, GA 30071, 800/332-2040, Fax: 770/447-8440, www.bkhome.com

BRUKER DALTRONICS, INC., 15 Fortune Drive, Billerica, MA 01821, 978-667-9580, Fax: 978-667-5993, www.daltronics.bruker.com

BRUKER OPTICS, INC., 19 Fortune Drive, Billerica, MA 01821, 978-667-9580, Fax: 978-663-9177, www.bruker.com/optics

BURKERT, 2602 McGaw Ave., Irvine, CA 92614, 949/223-3139, Fax: 949/223-3198, www.burkert-usa.com

BURNS ENGINEERING, INC., 10201 Bren Rd. E., Minnetonka, MN 55343, 952/935-4400, Fax: 952/935-8782, www.burnsengineering.com

BW TECHNOLOGIES, 242, 3030-3rd Ave. NE, Calgary, Alberta, T2A 6T7 Canada, 403/248-9226, Fax: 403/273-3708, www.gasmonitors.com

C

CAIG LABORATORIES, INC., 12200 Thatcher Ct., Poway, CA 92064, 858/486-8388, Fax: 858/486-8398, www.caig.com

CAL CONTROLS, INC., 1580 S. Milwaukee Ave., Libertyville, IL 60048, 847/680-7080, Fax: 847/816-6852, www.cal-controls.com

CALEX MFG. CO., INC., 2401 Stanwell Dr., Concord, CA 94520, 925/687-4411, Fax: 925/687-3333, www.calex.com

CALIBRON SYSTEMS, INC., 7861 E. Gray Rd., Scottsdale, AZ 85260, 480/991-3550, Fax: 780/998-5589, www.calibron.com

CAMSOFT CORP., 32295-8 Mission Trail #299, Lake Elsinore, CA 92530, 909/674-8100, Fax: 909/674-3110, www.camsoftcorp.com

CANARY LABS, Brownstone Bldg., Martinsburg, PA 16662, 814/793-3770, Fax: 814/793-3145, www.canarylabs.com

CAPE SOFTWARE, INC., 650 N. Sam Houston Pkwy. #313, Houston, TX 77060, 713/661-7100, Fax: 281/448-2607, www.capesoftware.com

CAPITAL CONTROLS CO., 3000 Advance Lane, Colmar, PA 18915, 215/997-4000, Fax: 215/997-4062, www.capitalcontrols.com

CARDINAL SCALE MFG. CO., 203 E. Daugherty, Webb City, MO 64870, 417/673-4631, Fax: 417/673-5001, www.ardinalscale.com

CARLO GAVAZZI AUTOMATION COMPONENTS, 750 Hastings Ln., Buffalo Grove, IL 60089, 847/465-6100, Fax: 847/465-7373, www.gavazzionline.com

CARLON, 25701 Science Park Dr., Cleveland, OH 44122, 800/3-CARLON, Fax: 216/831-5579, www.carlon.com

CAROLIN CORP., PO Box 2649, Woburn, MA 01888, 781/935-0146, Fax: 781/937-3499

CASHCO, 607 W. 15th St., Ellsworth, KS 67439, 785/472-4461, Fax: 785/472-3539, www.cashco.com

CCI, 22591 Avenida Empresa, Rancho Santa Margarita, CA 92688, 949/858-1877, Fax: 949/858-1878, www.ccivalve.com

CEA INSTRUMENTS, INC., 16 Chestnut St., Emerson, NJ 07630, 201/967-5660, Fax: 201/967-8450, www.ceainstr.com

CHROMALOX—PRECISION HEAT & CONTROL, 701 Alpha Dr., 3rd Fl., Pittsburgh, PA 15238, 412/967-3800, Fax: 412/967-5148, www.chromalox.com

CI TECHNOLOGIES, 4828 Parkway Plaza Blvd. #140, Charlotte, NC 28217, 704/329-3838, Fax: 704/329-3839, www.citect.com

CICOIL CORP., 24960 Avenue Tibbitts, Valencia, CA 91350, 661/295-1295, Fax: 661/295-0813, www.cicoil.com

CIMLOGIC, INC., 402 Amherst St., Suite 203, Nashua, NH 03063, 603/881-9918, Fax: 603/595-0381, www.cimlogic.com

CIRCLE SEAL CONTROLS, 2301 Wardlow Cir., Corona, CA 92880, 909/270-6280, Fax: 909/270-6201, www.circle-seal.com

CIRCUIT COMPONENTS, INC., 2400 S. Roosevelt, Tempe, AZ 85282, 480/967-0624, Fax: 480/967-9385, www.surgecontrol.com

CIRRONET, INC., 5375 Oakbrook Pkwy., Norcross, GA 30093, 678/684-2000, Fax: 678/684-2001, www.cirronet.com

CITADEL COMPUTER CORP., 29 Armory Rd., Milford, NH 03055, 603/672-5500, Fax: 603/672-5590, www.citadelcomputer.com

CITEL, INC., 1111 Parkcentre Blvd, Suite 340, Miami, FL 33169, 305/621-0022, Fax: 305/621-0766, www.citelprotection.com

CJM CONSULTING, 107 Oak Newel, Peachtree City, GA 30269, 678/637-9062, Fax: 770/487-1712

CLAROSTAT SENSORS & CONTROLS, INC., 12055 Rojas Dr., Suite K, El Paso, TX 79936, 915/858-2632, Fax: 915/872-3333, www.clarostat.com

CLEVELAND MOTION CONTROLS, 7550 Hub Pkwy., Cleveland, OH 44125, 216/642-2178, Fax: 216/642-2100, www.cmccontrols.com

CLIFFORD OF VERMONT, INC., Rte. 107, PO Box 51, Bethel, VT 05032, 802/234-9921, Fax: 802/234-5006, www.cliffordvt.com

CLIPPARD INSTRUMENTS LAB., INC., 7390 Colerain Ave., Cincinnati, OH 45235, 513/521-4261, www.clippard.com

COGNEX CORP., One Vision Dr., Natick, MA 01760, 800/677-2646, Fax: 508/650-3344, www.cognex.com

COLDER PRODUCTS CO., 1001 Westgate Dr., St. Paul, MN 55114, 800/444-2474, Fax: 651/645-5404, www.colder.com

COLE-PARMER INSTRUMENT CO., 625 E. Bunker Ct., Vernon Hills, IL 60061, 800/323-4340, Fax: 847/247-2929, www.coleparmer.com

COLLINS INSTRUMENT CO., Inc., PO Box 938, Analeton, TX 77516, 979/849-8266, Fax: 979/848-0783, www.collinsinst.com

COMARK CORP., 93 West St., Medfield, MA 02052, 508/359-8161, Fax: 508/359-2267, www.comarkcorp.com

COMPUTER DYNAMICS, 7640 Pelham Rd., Greenville, SC 29615, 864/627-8800, Fax: 864/675-0106, www.cdynamics

COMPUTER INSTRUMENTS CORP., 1000 Shames Dr., Westbury, NY 11590, 516/876-8400, Fax: 516/876-9153, www.computerinstruments.com

CONAX BUFFALO TECHNOLOGIES, 2300 Walden Ave., Buffalo, NY 14225, 716/684-4500, Fax: 716/684-7433, www.conaxbuffalo.com

CONBRACO INDUSTRIES, INC., 701 Matthews-Mint Hill Rd., Matthews, NC 28105, 704/841-6000, Fax: 704/ 841-6020, www.conbraco.com

CONCOA, 1501 Harpers Rd., Virginia Beach, VA 23454, 757/422-8330, Fax: 757/422-3125, www.concoa.com

CONDEC, 3 Simm Lane, Newtown, CT 06470, 888/295-8475, Fax: 203/364-1556, www.4condec.com

CONSILIUM US, INC., 59 Porter Rd., Littleton, MA 01460, 978/486-9800, Fax: 978/486-0170, www.consiliumus.com

CONSTANT POWER MFG., INC., 600 Century Plaza, Bldg. 140, Houston, TX 77073-6033, 281/821-3211, Fax: 281/821-6093, www.constantpowermfg.com

CONTEC MICROELECTRONICS, 744 S. Hillview Dr., Milpitas, CA 95035, 800/888-8884, Fax: 408/719-6750, www.contecusa.com

CONTEMPORARY CONTROLS, 2431 Curtiss St., Downers Grove, IL 60515, 630/963-7070, Fax: 630/963-0109, www.ccontrols.com

CONTREX, INC., Box 9000, Maple Grove, MN 55311-9000, 612/424-7800, Fax: 612/424-8734, www.contrexinc.com

CONTROL CHIEF, 200 Williams St., Bradford, PA 16701, 814/362-6811, Fax: 814/368-4133, www.controlchief.com

CONTROL FOR LESS, 101 Copperwood Way, Suite L, Oceanside, CA 92054, 760/433-7633, Fax: 760/433-6859, www.controlforless.com

CONTROL & MEASUREMENT INTL., INC., 421 Homewood Blvd., Delray Beach, FL 33445, 561/330-8144, Fax: 561/330-8134, www.cmi-temp.com

CONTROL INSTRUMENTS CORP., 25 Law Dr., Fairfield, NJ 07004, 973/575-9114, Fax: 973/575-0013, www.controlinstruments.com

CONTROL MICROSYSTEMS, 28 Steacie Dr., Kanata, Ontario, K2K 2A9 Canada, 613/591-1943, Fax: 613/591-1022, www.controlmicrosystems.com

CONTROL SYSTEMS INTERNATIONAL, 8040 Nieman Rd., Lenexa, KS 66214, 913/599-5010, Fax: 913/599-5013, www.csiks.com

CONTROL TECHNOLOGY CORP., 25 South St., Hopkinton, MA 01748, 508/435-9595, Fax: 508/435-2373, www.ctc-control. com

CONTROLAIR, INC., 8 Columbia Dr., Amherst, NH 03031, 800/216-3636, Fax: 603/889-1844, www.controlair.com

CONTROL.COM, INC., 134 Flanders Rd., Westborough, MA 01581, 508/898-9111, Fax: 508/621-3614, www.control.com

CONTROLOTRON, 155 Plant Ave., Hauppauge, NY 11788, 631/231-3600, Fax: 631/231-3334, www.controlotron.com

CONTROLSOFT, INC., 14077 Cedar Rd., Suite 200, Cleveland, OH 44118, 216/397-3900, Fax: 216/381-5001, www.controlsoftinc.com

CONVEYOR COMPONENTS CO., 130 Seltzer Rd., Croswell, MI 48422, 800/233-3233, Fax: 810/679-4510, www.conveyorcomponents.com

COOPER BUSSMANN, 114 Old State Rd., Ellisville, MO 63021, 636/527-1642, Fax: 636/527-1340, www.bussmann.com

COSA INSTRUMENT CORP., 55 Oak St., Norwood, NJ 07648, 201/767-6600, Fax: 201/767-8604

COSENSE, INC., 155 Rice Field Ln., Hauppauge, NY 11788, 631/231-0735, Fax: 631/231-0838, www.cosence.com

CROMPTON INSTRUMENT, INC., 1640 Airport Rd., Suite 109, Kennesaw, GA 30144, 770/425-8903, Fax: 770/423-7194, www.crompton-instruments.com

CROUSE-HINDS, PO Box 4999, Syracuse, NY 13221, 315/477-5110, Fax: 315/477-5118, www.crouse-hinds.com

CROUSE-HINDS MOLDED PRODUCTS, 4758 Washington St., LaGrange, NC 28551, 252/566-3014, Fax: 252/556-9337

CRYSTAL ENGINEERING, 1450 Madonna Rd., San Luis Obispo, CA 93405, 800/444-1850, Fax: 805/595-5466, www.crystalengineering.net

CTC PARKER AUTOMATION, 50 W. TechneCenter Dr., Milford, OH 45150, 513/831-2340, Fax: 513/831-5042, www.ctcusa.com

CTI ELECTRONICS CORP., 110 Old South Ave., Stratford, CT 06615, 203/386-9779, Fax: 203/378-4986, www.ctielectronics.com

CURTISS-WRIGHT FLOW CONTROL CORP., 1966 E. Broadhollow Rd., E. Farmingdale, NY 11735, 631/293-3800, Fax: 631/293-4949, www.cwfc.com

CUTLER-HAMMER, 4201 N. 27th St., Milwaukee, WI 53216, 414/449-6000, Fax: 414/449-7319, www.cutlerhammer.eatoncom

CYBOSOFT, 2868 Prospect Park Dr., Suite 300, Rancho Cordova, CA 95670, 916/631-6313, Fax: 916/631-6312, www.cybosoft.com

D

DAISY DATA, INC., 2850 Lewisberry Rd., York Haven, PA 17370, 717/932-9999, Fax: 717/932-8000, www.daisydata.com

DANAHER CONTROLS, 1675 Delany Rd., Gurnee, IL 60031, 800/873-8731, Fax: 847/662-6633, www.dancon.com

DANIEL MEASUREMENT AND CONTROL, PO Box 19097, Houston, TX 77224, 713/827-5184, Fax: 713/827-4360, www.danielind.com

DATA INDUSTRIAL CORP., 11 Industrial Dr., Mattapoisett, MA 02739, 508/758-6390, Fax: 508/758-4057, www.dataindustrial.com

DATAFORTH CORP., 3331 E. Hemisphere Loop, Tucson, AZ 85706, 520/741-1404, Fax: 520/741-0762, www.dataforth.com

DATALUX CORP., 155 Aviation Dr., Winchester, VA 22602, 540/662-1500, Fax: 540/662-7385, www.datalux.com

DATANET QUALITY SYSTEMS, 24567 Northwestern Hwy., 4th Fl., Southfield, MI 48075, 248/357-2200, Fax: 248/357-4933, www.winspc.com

DATA TRANSLATION, 100 Loche Dr., Marlboro, MA 01752, 800/525-8528, Fax: 508/481-8620, www.datatranslation.com

DATATRONICS, INC., 28151 Hwy 74, Romoland, CA 92585, 909/928-7700, Fax: 909/928-7701, www.datatronics.com

DATAVIEWS CORP., 47 Pleasant St., Northhampton, MA 01060, 413/586-4144, Fax: 413/586-3805, www.dvcorp.com

DATEL, INC., 11 Cabot Blvd., Mansfield, MA 02048, 508/339-3000, Fax: 508/339-6356, www.datel.com

DAYTRONIC CORP., 2211 Arbor Blvd., Dayton, OH 45439, 937/293-2566, Fax: 937/293-2586, www.daytronic.com

DELTRONIC, INC., 290 Wissahickon Ave., North Wales, PA 19454, 215/699-2310, www.deltronic.com

DENSITRON CORP., 10430-2 Pioneer Blvd., Santa Fe Springs, CA 90670, 562/941-5000, Fax: 562/941-5757, www.densitron.com

DESERT MICROSYSTEMS, INC., 3381 Chicago Ave., Riverside, CA 92507, 800/633-0448, Fax: 909/982-8506, www.desertmicrosys.com

DETCON, INC., 3200 A1 Research Forest, The Woodlands, TX 77381, 281/367-4100, Fax: 281/292-2860, www.detcon.com

DEVAR, INC., 706 Bostwick Ave., Bridgeport, CT 06605-2396, 203/368-6751, Fax: 203/368-3747, www.devarinc.com

DEZURIK/COPES—VULCAN, A Unit of SPX Corp., 250 Riverside Ave. N., Sartell, MN 56377, 320/259-2000, Fax: 320/259-2227, www.dezurikcopesvulcan.com

DGH CORP., PO Box 5638, Manchester, NH 03108, 603/622-0452, Fax: 603/622-0487, www.dghcorp.com

DH INSTRUMENTS, INC., 1905 W. 3rd St., Tempe, AZ 85281, 480/967-1555, Fax: 480/968-3574, www. dhinstruments.com

DIAMOND POWER INTL., INC., 2600 E. Main St., Lancaster, OH 43130, 740/687-4277, Fax: 740/687-4304, www.diamondpower.com

DIETRICH STANDARD, INC., 5601 N. 71st, Boulder, CO 80301, 720/622-2626, Fax: 303/530-7064, www.annubar.com

DIGI INTERNATIONAL, 11001 Bren Rd. E., Minnetonka, MN 55343, 952/912-3361, Fax: 952/912-4953, www.digi.com

DIGITAL SYSTEMS ENGINEERING, 4325 S. 34th St., Phoenix, AZ 85040, 602/426-8588, Fax: 602/426-8688, www.digitalsys.com

DIGITAL WIRELESS CORP., 1 Mega Way, Norcross, GA 30093, 770/564-5540, Fax: 770/564-5541, www.digital-wireless.com

DIMENSION TECHNOLOGIES, INC., 315 Mt. Read Blvd., Rochester, NY 14611, 716/436-3530, Fax: 716/436-3280, www.dti3d.com

DIONEX CORP., 1228 Titan Way, Sunnyvale, CA 94086, 408/737-0700, Fax: 408/739-4398, www.dionex.com

DIRECT MEASUREMENT CORP., 4040 Coriolis Way, Longmont, CO 80026, 303/702-7400, Fax: 303/702-7488, www.directmeasurement.com

DIVELBISS CORP., 9778 Mt. Golend Rd., Fredericktown, OH 43019, 740/694-9015, Fax: 740/694-9035, www.divelbiss.com

DJSCIENTIFIC, 5200 Dickey-John Rd., Auburn, IL 62563, 217/438-3371, Fax: 217/438-2609, www.djscientific.com

DORIC INSTRUMENTS, 4750 Viewridge Ave., San Diego, CA 92123, 888/423-6742, Fax: 858/569-8474, www.doric-vas.com

DPL SYSTEMS ENGINEERING, 1216 Sand Cove Rd., Saint John, New Brunswick, E2M 5V8 Canada, 506/635-1055, Fax: 506/ 635-1057, www.dpl.ca

DRAEGER SAFETY, INC., 101 Technology Dr., Pittsburgh, PA 15275, 412/787-8383, Fax: 412/787-2207, www.draeger.net

DRANETZ-BMI, 1000 New Durham Rd., Edison, NJ 08818, 800/372-6832, Fax: 732/248-1834, www.dranetz-BMI.com

DRESSER, INC., Masoneilan Operations, Dresser Flow Control, 85 Bodwell St., Avon, MA 02322, 508/586-4600, Fax: 508/427-8971, www.masoneilan.com

DRESSER INSTRUMENT DIV., 250 E. Main St., Stratford, CT 06614, 203/378-8281, www.dresserinstruments.com

DRIVE CONTROL SYSTEMS, 6111 Blue Circle Dr., Minnetonka, MN 55343, 952/930-0196, Fax: 952/930-0180, www.drivecontrolsystems

DRUCK, INC., 4 Dunham Dr., New Fairfield, CT 06812, 203/746-0400, Fax: 203/746-2494, www.pressure.com

DUTECH, PO Box 964, Jackson, MI 49204, 800/248-1632, Fax: 517/750-4740, www.dutec.net

DWYER INSTRUMENTS, 102 Indiana Hwy., Suite 212, Michigan City, IN 46360, 219/879-8000, Fax: 219/872-9057, www.dwyer-inst.com

DYNAMIC DISPLAYS, INC., 1625 Westgate Rd., Eau Claire, WI 54703, 715/835-9440, Fax: 715/835-2436, www.dynamicdisplay.com

DYNAPRO, 800 Carleton Ct., Annacis Island, New Westminster, BC, U3M 6L3, Canada, 604/521-3962, Fax: 604/521-4629, www.dynapro.com

E

E-MON CORP., One Oxford Valley, Suite 418, Langhorne, PA 19047, 800/334-3666, Fax: 215/752-3094, www.emon.com

E-T-A CIRCUIT BREAKERS, 1551 Bishop Ct., Mt. Prospect, IL 60056, 847/827-7600, Fax: 847/827-7655, www.etacbc.com

EASON TECHNOLOGY, 241 Center St., Heardsburg, CA 95448, 707/433-2854, Fax: 707/433-3706, www.eason.com

EATON CORP., 15 Durant Ave., Bethel, CT 06801, 800/736-1557, Fax: 203/796-6196, www.eaton.com

ECHELON CORP., 4015 Miranda Ave., Palo Alto, CA 95050, 650/855-7456, Fax: 650/843-1242, www.echelon.com

ECHIP, INC., 724 Yorklyn Rd., Hockessin, DE 19702, 302/239-5429, Fax: 302/239-6227, www.echip.com

ECKARDT—INVENSYS FLOW CONTROL, 1790 Satelite Blvd., Suite 100-04, Duluth, GA 30097, 678/474-1500, Fax: 678/474-1515

ECOM INSTRUMENTS, INC., 2000 Dairy Ashford, Suite 3061, Houston, TX 77077, 281/496-5930, Fax: 281/496-2321, www.ecom-ex.com

ECT INTERNATIONAL, INC., 4100 N. Calhoun Rd., Brookfield, WI 53005, 262/781-1511, Fax: 262/781-8411, www.ecti.com

EDGETECH, 455 Fortune Blvd., Milford, MA 01757, 800/276-3729, Fax: 508/634-3010, www.edgetech.com

EDO ELECTRO-CERAMIC PRODUCTS, 2645 South 300 West, Salt Lake City, UT 84115, 801/461-9259, Fax: 801/484-3301, www.edocorp.com

EDWARD VOGT VALVE CO-INVENSYS FLOW CONTROL, 1900 S. Saunders St., Raleigh, NC 27603, 800/225-6989, Fax: 919/831-3254

EL-O-MATIC, 135 English St., Hackensack, NJ 07601, 201/489-5550, Fax: 201/489-9171, www.elomaticusa.com

ELCON INSTRUMENTS, 2700 Garden Rd., Swuanee, GA 30024, 770/271-5519, Fax: 770/271-5049, www.elconinst.com

ELDRIDGE PRODUCTS, INC., 2700 Garden Rd., Monterey, CA 93940, 800/321-3569, Fax: 831/648-7780, www.epiflow.com

ELECTRIC POWER & HEAT CO., 65 White Oak Dr., Smithfield, NC 27577, 919/934-9448, Fax: 919/934-0345

ELECTRO CAM CORP., 13647 Metric Rd., Roscoe, IL 61115, 815/389-2620, Fax: 815/389-3304, www.electrocam.com

ELECTRO SENSORS, INC., 6111 Blue Circle Dr., Minnetonka, MN 55343, 952/930-0100, Fax: 952/930-0130, www.electrosensors.com

ELECTRO-MECH COMPONENTS, INC., 1826 Floradale Ave., South El Monte, CA 91733, 626/442-7180, Fax: 626/ 350-8070, www.electromechcomp.com

ELECTRONIC SENSORS, INC., 1611 W. Harry, Wichita, KS 67213, 800/886-2511, Fax: 316/267-2819, www.leveldevil.com

ELECTROSWITCH, 180 King Ave., Weymouth, MA 02188, 781/607-3303, Fax: 781/335-4253, www.electroswitch.com

ELECTROSWITCH ELECTRONIC PRODUCTS, 2010 Yonkers Rd., Raleigh, NC 27604, 888/768-2797, Fax: 800/909-9171, www.electro-nc.com

ELGAR, 9250 Brown Deer Rd., San Diego, CA 92121, 858/450-0085, Fax: 858/458-0267, www.elgar.com

ELIPSE SOFTWARE, 40190 Jarvis Gray Ln., Avon, NC 27915, 252/995-6885, Fax: 252/995-5686, www.elipse-software.com

ELO TOUCHSYSTEMS, INC., 6500 Kaiser Dr., Fremont, CA 94555, 800/ELO-TOUCH, Fax: 510/790-0627, www.elotouch.com

EMATION, 89 Forbes Blvd., Mansfield, MA 02048, 508/337-9200, Fax: 508/337-9201, www.emation.com

EMCO FLOW SYSTEMS, 600 Diagonal Hwy., Longmont, CO 80501, 303/651-0550, Fax: 303/678-7152, www.emcoflow.com

ENDEVCO, 30700 Rancho Viejo Rd., San Juan Capistrano, CA 92675, 949/493-8181, Fax: 949/661-7231, www.endevco.com

ENDRESS + HAUSER, INC., 2350 Endress Place, Greenwood, IN 46143, 317/535-1391, Fax: 317/535-2171, www.us.endress.com

ENDRESS + HAUSER SYSTEMS & GAUGING, INC., 5834 Peachtree East, Norcross, GA 30096, 770/447-9202, Fax: 770/622-8939, www.systems.endress.com

ENEA OSE SYSTEMS, 5949 Sherry Ln., Suite 625, Dallas, TX 75225, 214/346-9339, Fax: 214/346-9344, www.enea.com

ENERPRO, INC., 5780 Thornwood Dr., Goleta, CA 93117, 805/683-2114, Fax: 805/964-0798, www.enerpro.thomasregister.com

ENRAF, INC., 4333 W. Sam Houston Pkwy N., Houston, TX 77043, 832/467-3422, Fax: 832/467-3441, www.enrafinc.com

ENTERTRON INDUSTRIES, INC., 3857 Orangeport Rd., Gasport, NY 14067, 716/772-7216, Fax: 716/772-2604, www.entertron.com

ENTIVITY, 935 Technology Dr., Suite 200, Ann Arbor, MI 48108, 734/205-5000, Fax: 734/205-5100, www.entivity.com

ENTRAN DEVICES, INC., 10 Washington Ave., Fairfield, NJ 07004, 973/227-1002, Fax: 973/227-6865, www.entran.com

ENTRELEC, INC., 1950 Hurd Dr., Irving, TX 75038, 800/431-2308, Fax: 800/862-5066, www.entrelec.com

ENTRON COMPUTER CORP., 9001 Airport Blvd., Suite 409, Houston, TX 77061, 713/941-7007, Fax: 713/941-3852, www.entron.com

EPLAN, 16650 Bluemound Rd., Suite 600, Brookfield, WI 53005, 262/789-0428, Fax: 262/789-0428, www.eplan.org

ERGOTRON, 1181 Trapp Rd., St. Paul, MN 55121, 800/888-8458, www.ergotron.com

ESSENTIALS CONTROL, INC., 128 Elmore Dr., Acton, Ontario, L7J 1T2 Canada, 519/853-3830, Fax: 519/853-5073, www.essentials-control.com

ETI SYSTEMS, INC., 2251 Las Palmas Dr., Carlsbad, CA 92009, 760/929-0749, Fax: 760/929-0748, www.etisystems.com

EUROTHERM CHESSELL, 8601 Aero Dr., San Diego, CA 92123, 800/801-5099, Fax: 858/514-0426, www.chessell.com

EUROTHERM CONTROLS, INC., 741-F Miller Dr., Leesburg, VA 20175, 703/443-0000, Fax: 703/669-1300, www.eurotherm.com

EVANS CONSOLES, INC., 1616 27th Ave. N.E., Calgary, Alberta, T2E 8W4 Canada, 403/717-3009, Fax: 403/717-3320, www.evansonline.com

EVEREST INTERSCIENCE, 1891 N. Oracle Rd., Tucson, AZ 85705, 520/797-0927, Fax: 520/797-0938, www.everestinterscience.thomasregister.com

EXERGEN CORP., 51 Water St., Watertown, MA 02472, 617/923-9900, Fax: 617/923-9911, www.exergen.com

EXOR ELECTRONIC R&D, INC., 3420 Fairlane Farms Rd., Wellington, FL 33414, 561/753-2250, Fax: 561/753-2291, www.exor-rd.com

EXPERTUNE, INC., 4734 Sonseeahray Dr., Hubertus, WI 53033, 262/628-0088, Fax: 262/628-0087, www.expertune.com

EXPO-TELEKTRON SAFETY SYSTEMS, INC., 67 E. Washington St., Chagrin Falls, OH 44022, 440/247-5314, Fax: 440/247-5409, www.expotelektron.com

EXTECH INSTRUMENTS, 285 Bear Hill Rd., Waltham, MA 02451, 781/890-7440, Fax: 781/890-7864, www.extech.com

F

FACTORYSOFT, INC., 89 Forbes Blvd., Mansfield, MA 02048, 508/337-9200, Fax: 508/337-9201, www.emation.com

FAIRBANKS SCALES, 821 Locust, Kansas City, MO 64106, 816/471-0231, Fax: 816/471-5951, www.fairbanks.com

FAIRCHILD INDUSTRIAL PRODUCTS CO., 3920 West Point Blvd., Winston-Salem, NC 27103, 336/659-3400, Fax: 336/659-9323, www.fairchilproducts.com

FASTECH, A SUBSIDARY OF BROOKS AUTOMATION, 15 Elizabeth Dr., Chelmsford, MA 01824, 978/262-2400, Fax: 978/262- 2500, www.fastech.com

FASTRAK SOFTWORKS, INC., 6659 W. Mill Road, Milwaukee, WI 53218, 414/358-8088, Fax: 414/358-8066, www.fast-soft.com

FLUID COMPONENTS INTL., 1755 La Costa Meadows Dr., San Marcos, CA 92069, 760/744-6950, Fax: 760/736-6250, www.fluidcomponents.com

FEDERAL PRODUCTS CO., 1144 Eddy St.; PO Box 9400, Providence, RI 02940, 401/784-3100, Fax: 401/784-3246, www.fedgage.com

FESTO CORP., 395 Moreland Rd., Hauppauge, NY 11788, 631/435-0800, Fax: 631/435-8026, www.festo-usa.com

FIBOX ENCLOSURES, 6675 Santa Barbara, Elkridge, MD 21075, 888/FIBOXUS, Fax: 410/496-3298, www.fiboxusa.com

FIELDBUS FOUNDATION, 9390 Research Blvd. Suite 11-250, Austin, TX 78759, 512/794-8890, Fax: 512/794-8893, www.fieldbus.org

FIELDSERVER TECHNOLOGIES, 1991 Tarob Ct., Milpitas, CA 95035, 408/262-2299, Fax: 408/262-9042, www.fieldserver.com

FIKE CORP., 704 South 10th St., Blue Springs, MO 64015, 816/229-3405, Fax: 816/228-9277, www.fike.com

FISHER CONTROLS INTERNATIONAL, INC., 205 S. Center St., Marshalltown, IA 50158, 641/754-3011, Fax: 641/754-2830

FISHER-ROSEMOUNT, 12001 Technology Dr., Eden Prairie, MN 55344, 952/828-3180, Fax: 952/828-3033, www.assetweb.com

FISHER-ROSEMOUNT SYSTEMS, 8301 Cameron Rd., Austin, TX 78754, 512/835-2190, Fax: 512/418-7505, www.frco.com

FLEX-CORE, 6625 McVey Blvd., Columbus, OH 43235, 614/889-6152, Fax: 614/876-8538, www.flex-core.com

FLO-TORK, INC., 1701 N. Main St., Orrville, OH 44667, 330/682-0010, Fax: 330/683-6857, T.M. Leaver, executive@flo-tork.com, www.flo-tork.com

FLOW AUTOMATION, 9303 W. Sam Houston Pkwy. S., Houston, TX 77099-5298, 713/272-0404, Fax: 713/272-2272, www.flowautomation.com

FLOW RESEARCH, 27 Water St., Wakefield, MA 01880, 781/245-3200, Fax: 781/224-7552, www.flowresearch.com

FLOW TECHNOLOGY, INC., 4250 E. Broadway, Phoenix, AZ 85040, 602/437-1315, Fax: 602/437-4459, www.ftimeters.com

FLOW-TEK, INC., a subsidiary of Bray Intl., Inc., 7404 Fairfield, Columbia, SC 29203, 803/754-8201, Fax: 803/754-2501

FLOWLINE, 10500 Humbolt St., Los Alamitos, CA 90720, 562/598-3015, Fax: 562/431-8507, www.flowline.com

FLOWSERVE, FLOW CONTROL DIV., PO Box 2200, Springville, UT 84663, 801/489-8611, Fax: 801/489-3719, www.flowserve.com

FLUID METERING, INC., 5 Aerial Way, Suite 500, Syosset, NY 11791, 800/223-3388, Fax: 516/624-8261, www.fmipump.com

FLUKE CORP., PO Box 9090, Everett, WA 98206, 800/44-Fluke, Fax: 425/446-5116, Sales & Applications, fluke-info@fluke.com, www.fluke.com

FMC BLENDING & TRANSFER, 20 N. Wacker Dr., Suite 1300, Chicago, IL 60606, 805/495-7111, Fax: 805/379-3365, www.fmcblending-transfer.com

FORMOSA USA, INC., 21540 Prairie St., Unit A, Chatsworth, CA 91311, 818/407-4965, Fax: 818/407-4966, www.goformosa.com

FOSS NIRSYSTEMS, 12101 Tech Rd., Silver Spring, MD 20904, 301/680-9600, Fax: 301/236-0157, www.foss-nirsystems.com

FOX INDUSTRIES, 505 Mayock Rd., Suite A4, Gilroy, CA 95020, 408/847-2090, Fax: 408/847-1806

FOXBORO CO., The, 33 Commercial St., Foxboro, MA 02035, 508/549-6240, Fax: 508/549-4834, www.foxboro.com

FRANEK TECHNOLOGIES, INC., 13821 Newport Ave., Suite 100, Tustin, CA 92780, 714/734-6957, Fax: 714/544-6957, www. franek-tech.com

FRANK W. MURPHY MFR., PO Box 470248, Tulsa, OK 74147, 918/627-3550, Fax: 918/664-6146, www.fwmurphy.com

FUJI ELECTRIC CORP. OF AMERICA, Park 80 West, Plaza II, Saddle Brook, NJ 07663, 201/712-0555, Fax: 201/368-8258, www.fujielectric.com

FUJIKIN OF AMERICA, INC., 4 Alsan Way, Little Ferry, NJ 07643, 201/641-1119, Fax: 201/641-1137, www.fujikin.com

FURNESS CONTROLS, INC., 3801-A Beam Rd., Charlotte, NC 28217, 800/898-5325, Fax: 704/357-1103, www.furnesscontrols.com

G

GALIL MOTION CONTROL, 203 Ravendale Dr., Mountain View, CA 94043, 650/967-1700, Fax: 650/967-1751, www.galilmc.com

GAUMER PROCESS, 13616 Hempstead Rd., Houston, TX 77040, 800/460-5200, Fax: 800/460-1444, www.gaumer.com

GE CONTINENTAL CONTROLS, 10,000 Richmond, Houston, TX 77042, 713/978-4401, Fax: 713/978-4444, www.geindustrial.com

GE FANUC AUTOMATION, Rte. 29N & Rte. 606, Charlottesville, VA 22911, 804/978-5508, Fax: 804/978-5620, www.gefanuc.com

GEMS SENSORS, INC., 1 Cowles Rd., Plainville, CT 06062, 800/378-1600, Fax: 860/793-4500, www.gemssensors.com

GENERAL EASTERN, 20 Commerce Way, Woburn, MA 01801, 781/938-7070, Fax: 781/938-1071, www.geinet.com

GENERAL MICRO SYSTEMS, PO Box 2689, Rancho Cucamonga, CA 91729, 800/307-4863, Fax: 909/987-4863, www.gms4ume.com

GENERAL MONITORS, 26776 Simpatica Circle, Lake Forest, CA 92630, 949/581-4464, Fax: 949/581-1151, www.generalmonitors.com

GENSYM CORP., 125 Cambridge Park Dr., Cambridge, MA 02140, 617/547-2500, Fax: 617/547-1962, www.gensym.com

GEORGE FISCHER, INC., 2882 Dow Ave., Tustin, CA 92780, 800/854-4090, Fax: 714/731-6923, www.us.piping.georgefischer.com

GEOSPHERE EMERGENCY, RESPONSE SYSTEM, 100 S. Main St., Doylestown, PA 18901, 215/340-2204, Fax: 215/340-2205, www.plantsafe.com

GESTRA STEAM PRODUCTS—INVENSYS FLOW CONTROL, 5171 Maritime Rd., Jeffersonville, IN 47130, 800/225-6989, Fax: 812/218-7777, www.edwardvogt.com

GIDDINGS & LEWIS CONTROLS, MEASUREMENT & SENSING, 660 S. Military Rd., Fond du Lac, WI 54935, 920/921-7100, Fax: 920/906-7669, www.giddings. com

GLI INTERNATIONAL, 9020 W. Dean Dr., Milwaukee, WI 53224, 414/355-3601, Fax: 414/355-8346, www.gliint.com

GLOBAL WEIGHING, 5110 Old Ellis Pointe, Suite 200, Roswell, GA 30076, 678/393-9960, Fax: 678/393-9961, www.global-weighing.com

GMC INSTRUMENTS, INC., 250 Telser Rd., Unit F, Lake Zurich, IL 60047, 800/462-4040, Fax: 847/540-7242, www.gmcinc.com

GO REGULATOR, 2301 Wardlow Circle, Corona, CA 92880, 909/270-6280, Fax: 909/270-6201, www.circle-seal.com

GORDON PRODUCTS, 67 Del Mar Dr., Brookfield, CT 06804, 800/315-9233, Fax: 203/775-1162, www.gordonproducts.com

GP:50, 2770 Long Rd., Grand Island, NY 14072, 716/773-9300, Fax: 716/773-5019, www.gp50.com

GRACE ENGINEERED PRODUCTS, 5000 Tremont, #203, Davenport, IA 52807, 319/386-9596, Fax: 319/386-9639, www.grace-eng.com

GRAYHILL, INC., 561 Hillgrove Ave., LaGrange, IL 60525, 708/354-1040, Fax: 708/354-2820, www.grayhill.com

GREAT PLAINS INDUSTRIES, INC., 5252 E. 36th St. N., Wichita, KS 67220, 316/686-7361, Fax: 316/686-6746, www.greatplainsindustries.com

GRECO SYSTEMS, 372 Coogan Way, El Cajon, CA 92020, 619/442-0205, Fax: 619/447-8982, www.grecosystems.com

GSE SCALE SYSTEMS, 23900 Haggerty Rd., Farmington Hills, MI 48335, 248/471-5880, Fax: 248/471-5844, www.gse-inc.com

H

HACH CO., PO Box 389, Loveland, CO 80539, 970/669-3050, Fax: 970/669-2932, www.hach.com

HALLIBURTON ENERGY SERVICES, 410 17th Street, Suite 900, Denver, CO 80202, 800-654-3760, Fax: 303-573-7856

HALLIBURTON ENERGY SERVICES, 2600 S. 2nd St., Duncan, OK 73536, 303/899-4715, Fax: 303/573-7856, www.halliburton.com

HAM-LET VALVES & FILTERS, 2153 O'Toole Ave., Suite B, San Jose, CA 95131, 408/473-8880, Fax: 408/473-8882, www.ham-let.com

HAMMOND ENCLOSURES, 394 Edinburgh Rd. N., Guelph, Ontario, N1H 1E5 Canada, 519/822-2960, Fax: 519/822-2301, www.hammfg.com

HARDY INSTRUMENTS INC., 3860 Calle Fortunada, San Diego, CA 92123, 800/821-5831, Fax: 858/278-6700, www.hardyinstruments.com

HART COMMUNICATIONS FOUNDATION, 9390 Research Blvd., Suite 1-350, Austin, TX 78759, 512/794-0369, Fax: 512/ 794-3904, www.hartcomm.org

HATHAWAY PROCESS INSTRUMENTATION BETA CALIBRATORS DIV., 2309 Springlake Rd., Suite 600, Farmers Branch, TX 75234, 972/241-2200, Fax: 972/241-6752, www.hathawayprocess.com

HAYS CLEVELAND, 1111 Brookpark Rd., Cleveland, OH 44109, 216/398-4414, Fax: 216/398-8553, www.hayscleveland.com

HBM, 2815 Colonnades Ct., Norcross, GA 30071, 888/816-9006, Fax: 770/447-8440, www.hbmhome.com

HELM INSTRUMENT CO., 361 W. Dussel Dr., Maumee, OH 43537, 419/893-4356, Fax: 419/893-1371, www.helminstrument.com

HEM DATA CORP., 17336 W. 12 Mile Rd. #200, Southfield, MI 48076, 248/559-5607, Fax: 248/559-8008, www.hemdata.com

HERTZLER SYSTEMS, INC., 2312 Eisenhower Dr. N., Goshen, IN 46526, 219/533-0571, Fax: 219/533-3885, www.hertzler.com

HEWLETT-PACKARD/AGILENT, 100, 3030-3rd Ave. N.E., Calgary, Alberta, T2A 6T7 Canada, 403/299-4756, Fax: 403/272-2299, www.agilent.com/find/ais

HI-SPEED CHECKWEIGHER, 5 Barr Rd., Ithaca, NY 14850, 607/257-6000, Fax: 607/257-6396, www.hispeedcheckweigher.com

HIMA PAUL HILDEBRANDT GMBH & CO., Albert-Bassermann-Str. 28, Gruehl, 68782 Germany, 49/6202-709-145, Fax: 49/6202-709-123, www.hima.com

HIRSCHMANN ELECTRONICS, INC., 30 Hook Mountain Rd., Pine Brook, NJ 07058, 973/830-2000, Fax: 973/830-1470, www.hirschmann-usa.com

HMW ENTERPRISES, INC., 207 N. Franklin St., Waynesboro, PA 17268, 717/765-4690, Fax: 717/765-4660, www.hmwent.com

HNU SYSTEMS, INC., 160 Charlemont St., Newton Highlands, MA 02461, 617/964-6690, Fax: 617/558-0056, www.hnu.com

HOFFER FLOW CONTROLS, INC., 107 Kitty Hawk Ln., Elizabeth City, NC 27909, 252/331-1997, Fax: 252/331-2886, www.hofferflow.com

HOFFMAN, 2100 Hoffman Way, Anoka, MN 55303, 800/355-3560, Fax: 612/942-6940, www.hoffmanonline.com

HOKE, INC., 2301 Wardlow Circle, Corona, CA 92880, 909/270-6280, Fax: 909/270-6201, www.circle-seal.com

HONEYWELL INDUSTRIAL CONTROL, 16404 N. Black Canyon Hwy., Phoenix, AZ 85053, 800/288-7491, Fax: 319/294-2968, www.iac.honeywell.com

HONEYWELL SENSING AND CONTROL, 11 W. Spring St., Freeport, IL 61032, 800/537-6945, Fax: 815/235-6545, www.honeywell.com/sensing

HORIBA INSTRUMENTS, INC., 17671 Armstrong Ave., Irvine, CA 92614, 949/250-4811, Fax: 949/250-0924, www.horiba.com

HORNER APG, LLC, 7821 Carinthian Dr., Indianapolis, IN 46201, 317/916-4274, Fax: 317/916-4280, www.heapg.com

HTM ELECTRONICS INDUSTRIES, 8651 Buffalo Ave., Niagara Falls, NY 14304, 800/644-1756, Fax: 888/283-2127, www.htm-sensors.com

HYDE PARK ELECTRONICS, INC., 1875 Founders Dr., Dayton, OH 45420-4017, 937/252-2121, Fax: 937/258-5830, www. hpsensors.com

HYPROTECH LTD, 707/Eighth Ave. SW, Suite 800, Calgary, Alberta, T2P 1H5 Canada, 403/870-6193, Fax: 403/520-6060, www.hyprotech.com

I

IAI AMERICA, INC., 2360 W. 205th St., Torrance, CA 90501, 310/320-3978, Fax: 310/320-4553, www.iaiamerica.com

IBOCO CORP., 10 Alvin Ct., Suite 100, East Brunswick, NJ 08816, 732/238-0200, Fax: 732/238-0304, www.iboco.com

IC FLUID POWER, 63 Drive Hwy., Rossford, OH 43460, 419/661-8811, Fax: 419/661-8844, www.icfluid.com

ICONICS, 100 Foxborough Blvd., Foxborough, MA 02035, 508/543-8600, Fax: 508/543-1503, www.iconics.com

ICS ADVENT, 6260 Sequence Dr., San Diego, CA 92121, 800/523-2320, Fax: 858/677-0898, www.icsadvent.com

ICS TRIPLEX, 16400-A Park Row, Houston, TX 77084, 281/492-0604, Fax: 281/578-3268, www.icstriplex.com

IDEC CORP., 1175 Elko Dr., Sunnyvale, CA 94089, 800/262-4332, Fax: 800/635-6246, www.idec.com

IDS ENGINEERING, INC., 600 Century Plaza Dr., Houston, TX 77073-6033, 281/821-7100, Fax: 281/821-3230, www.idsengr.com

IFM EFECTOR, INC., 805 Springdale Dr., Exton, PA 19341, 610/524-2004, Fax: 610/524-2745, www.ifmefector.com

IMAGEVISION, INC., PO Box F, La Grange, TX 78945, 979/247-4068, Fax: 979/247-4062, www.imagevisioninc.com

IMAGING & SENSING TECHNOLOGY, 14737 NE 87th St., Redmond, WA 98052, 425/881-0778, Fax: 425/869-0667, www.istimaging.com

INDUSOFT, PO Box 164073, Austin, TX 78716, 877/INDUSOFT, Fax: 512/527-0792, www.indusoft.com

INDUSTRIAL DATA SYSTEMS, INC., 15031 Woodham, Suite 360, Houston, TX 77073-6026, 281/821-3200, Fax: 281/821-5488, www.inddata.com

INDUSTRIAL INDEXING SYSTEMS, 626 Fishers Rd., Victor, NY 14564, 716/924-9181, Fax: 716/924-2169, www.iis-servo.com

INDUSTRIAL NETWORKING SOLUTIONS, PO Box 540, Addison, TX 75001, 972/248-7466, Fax: 972/248-9533, www.industrialnetworking.com

INDUSTRIAL SCIENTIFIC CORP., 1001 Oakdale Rd., Oakdale, PA 15071, 412/788-4353, Fax: 412/788-8353, www.indsci.com

INNOVA ELECTRONICS, 9207 Emmott Rd., Suite 105, Houston, TX 77423, 713/690-9909, Fax: 713/466-0210, www.innovaelec.com

INNOVATIVE SENSORS, INC., 4745 E. Bryson St., Anaheim, CA 92807, 800/835-5474, Fax: 714/779-9315, www.innovativesensors.com

INSPIRED SOLUTIONS LLC, 2451 Cumberland Pkwy, Suite 3455, Atlanta, GA 30339, 770/803-0555, Fax: 770/803-0580, www.inspiredsolutions.net

INTECOLOR, 2150 Boggs Rd., Duluth, GA 30096, 770/622-6242, Fax: 770/622-6370, www.intecolor.com

INTEK, INC., 751 Intek Way, Westerville, OH 43082, 614/895-0301, Fax: 614/895-0319, www.intekflow.com

INTELLIGENT INSTRUMENTATION, 3000 S. Valencia Rd., Suite 100, Tucson, AZ 85706, 800/685-9911, Fax: 520/573-0522, www.instrument.com

INTELLIGENT MOTION SYSTEMS, 370 N. Main St., Marlborough, CT 06447, 860/295-6102, Fax: 860/295-6107, www.imshome.com

INTELLUTION, 325 Foxborough Blvd., Foxborough, MA 02035, 508/698-3322, Fax: 508/698-6973, www.intellution.com

INTERBUS CLUB, THE, PO Box 25141, Philadelphia, PA 19147, 888/281-2871, www.interbusclub.com

INTERFACE, INC., 7401 E. Butherus Dr., Scottsdale, AZ 85260, 480/948-5555, Fax: 480/948-1924, www.interfaceforce. com

INTERGRAPH PROCESS & BUILDING SOLUTIONS, 300 Intergraph Way, Madison, AL 35758, 800/260-0246, Fax: 256/730-3028, www.intergraph.com/pbs

INTERLINKBT, LLC, 3000 Campus, Dr., Plymouth, MN 55441, 763/694-2332, Fax: 763/694-2399, www.interlinkbt.com

INTERNATIONAL RECTIFIER CORP., 233 Kansas St., El Segundo, CA 90245, 310/252-7019, Fax: 310/252-7171, www.irf.com

INTERSTATE ELECTRONICS CORP., 602 East Vermont St., Anaheim, CA 92803, 714/758-0500, Fax: 714/758-4148, www.iechome.com

INTRINSYC, 700 W. Pender St., 10th Fl., Vancouver, BC V6C 1E8, 604/801-6461, Fax: 604/801-6417, www.intrinsyc.com

INVENSYS FLOW CONTROL, 1900 S. Saunders St., Raleigh, NC 27603, 800/225-6989, Fax: 919/831-3254, www.invensysflowcontrol.com

INVENSYS SENSOR SYSTEMS/CLAROSTAT, 12055 Rojas Dr., Suite K, El Paso, TX 79936, 915/858-2632, Fax: 915/872-3333, www.clarostat.com

IOTECH, INC., 25971 Cannon Rd., Cleveland, OH 44146, 440/439-4091, Fax: 440/439-4093, www.iotech.com

IRCON, INC., 7300 N. Natchez Ave., Niles, IL 60714, 847/967-5151, Fax: 847/647-0948, www.ircon.com

ISA—THE INSTRUMENTATION, SYSTEMS & AUTOMATION SOCIETY, 67 Alexander Dr., Research Triangle Park, NC 27709, 919/990-9215, Fax: 919/549-8411, www.isa.org

IST-QUADTEK, 14737 NE 87th St., Redmond, WA 98052, 425/881-0778, Fax: 425/869-0667, www.istimaging.com

ITAC SYSTEMS, INC., 3113 Benton St., Garland, TX 75042, 972/494-3073, Fax: 972/494-4159, www.itacsystems.com

ITS ENCLOSURES, 271 Westech Dr., Mt. Pleasant, PA 15666, 800/423-9911, Fax: 724/696-3333, www.itsenclsures.com

ITT CONOFLOW, 5154 Highway 78, St. George, SC 29477, 843/563-9281, Fax: 843/563-2131, www.ittconoflow.com

ITW SWITCHES, 7301 W. Ainslie St., Harwood Heights, IL 60706, 708/667-3370, Fax: 708/667-3440, www.itwswitches.com

J

J-TEC ASSOCIATES, INC., 5255 Blairs Forest Ln. N.E., Suite L, Cedar Rapids, IA 52402, 319/393-5200, Fax: 319/393-5211, www.j-tecassociates.com

J. H. INSTRUMENTS, 9800 N. Southern Pine Blvd., Charlotte, NC 28273, 704/527-6920, Fax: 704/527-8406, www.jhinstrument.com

J. R. MERRITT CONTROLS, 55 Sperry Ave., Stratford, CT 06615, 203/296-2272, Fax: 203/381-0400, www.jrmerritt.com

JAVLYN, INC., 3136 Winton Rd. South, Rochester, NY 14623, 716/697-5212, Fax: 716/424-3110, www.javlyn.com

JC SYSTEMS, INC., 11535 Sorrento Valley Blvd. #400, San Diego, CA 92121, 858/793-7117, Fax: 858/793-1931, www.jcsystemsinc.com

JENSEN TOOLS, INC., 7815 S. 46th St., Phoenix, AZ 85044, 800/426-1194, Fax: 800/366-9662, www.jensentools.com

JH TECHNOLOGY, INC., 4233 Clark Rd., Unit 9, Sarasota, FL 34233, 800/808-0300, Fax: 941/925-8774, www.jhtechnology.com

JOGLER, INC., 9715 Derrington Rd., Houston, TX 77064, 281/469-6969, Fax: 281/469-0422, www.jogler.com

JUMO PROCESS CONTROL, INC., 885 Fox Chase, Coatesville, PA 19320, 610/380-8002, Fax: 610/380-8009, www.jumousa.com

K

K-PATENTS, INC., 1804 Centre Point Cir., Suite 106, Naperville, IL 60563, 630/955-1545, Fax: 630/955-1585, www.kpatents.com

K-TEK, 18321 Swamp Rd., Prairieville, LA 70769, 225/673-6100, Fax: 225/673-2525, www.ktekcorp.com

KAHN INSTRUMENTS, INC., 885 Wells Rd., Wethersfield, CT 06109, 860/529-8643, Fax: 860/529-1895, www.kahn.com

KALGLO ELECTRONICS CO., 5911 Colony Dr., Bethlehem, PA 18017-9348, 610/837-0700, Fax: 610/837-7978, www.kalglo.com

KAMAN INSTRUMENTATION OPERATIONS, 3450 N. Nevada Ave., Colorado Springs, CO 80907, 719/635-6979, Fax: 719/634-8093, www.kamaninstrumentation.com

KAVLICO CORP., 14501 Los Angeles Ave., Moorpark, CA 93021, 805/523-2000, Fax: 805/523-7125, www.kavlico.com

KAY-RAY/SENSALL, 1400 Business Center Dr., Suite 100, Mt. Prospect, IL 60056, 800/323-7594, Fax: 847/803-5466, www.kayray-sensall.com

KEITHLEY INSTRUMENTS, INC., 28775 Aurora Rd., Cleveland, OH 44139, 440/248-0400, Fax: 440/248-6168, www.keithley.com

KEPWARE, 81 Bridge St., Yarmouth, ME 04096, 207/846-5881, Fax: 207/846-5947, www.opcsource.com

KESSLER ELLIS PROD., 10 Industrial Way E., Eatontown, NJ 07724, 732/935-1320, Fax: 732/935-0184, www.kep.com

KEY INSTRUMENTS, 250 Andrews Rd., Trevose, PA 19053, 215/357-0893, Fax: 215/357-9239, www.keyinstruments.com

KEYENCE CORP. OF AMERICA, 50 Tice Blvd., Woodcliff Lake, NJ 07675, 201/930-0100, Fax: 201/930-1883, www.keyence.com

KIN-TEK, 504 Laurel, LaMarque, TX 77568, 409/938-3627, Fax: 409/938-3710

KINETROL USA, INC., 4557 Westgrove Dr., Addison, TX 75001, 972/447-9443, Fax: 972/447-9720, www.kinetrol.com

KING ENGINEERING CORP., 3201 S. State St., Ann Arbor, MI 48108, 734/662-5691, Fax: 734/662-6652, www.king-gage.com

KING INSTRUMENT CO., 12700 Pala Dr., Garden Grove, CA 92841, 714/891-0008, Fax: 714/891-0023

KISTLER-MORSE CORP., 19021 120th Ave. NE, Bothell, WA 98011, 425/486-6600, Fax: 425/402-1500, www.kistler-morse.com

KNF NEUBERBER, INC., 2 Black Forest Rd., Trenton, NJ 08691, 609/890-8600, Fax: 609/890-8323, www.knf.com

KNOWLEDGESCAPE SYSTEMS, 669 W. 200 S., Salt Lake City, UT 84106, 801/526-2351, Fax: 801/526-2701, www.kscape.com

KOBOLD INSTRUMENTS, INC., 1801 Parkway View Dr., Pittsburgh, PA 15205, 800/998-1020, Fax: 412/788-4890, www.koboldusa.com

KOLLMORGEN, 201 Rock Road, Radford, VA 24141, 540/633-4124, Fax: 540/731-0847, www.kollmorgen.com

KRAISSL CO., INC., 299 Williams Ave., Hackensack, NJ 07601-5225, 201/342-0008, Fax: 201/342-0025, www.strainers.com

KROHNE, INC., 7 Dearborn Rd., Peabody, MA 01960, 978/535-6060, Fax: 978/535-1720, www.krohne.com

KSR KUEBLER, 7516 Precision Dr., Raleigh, NC 27613, 919/596-3800, Fax: 919/598-8128, www.ksr-usa.com

KURT MANUFACTURING CO., Electronics Div., 2130 107th Ln. NE, Minneapolis, MN 55449, 763/572-4597, Fax: 763/784-6055, www.kurt.com

KURZ INSTRUMENTS, INC., 2411 Garden Rd., Monterey, CA 93940, 800/424-7356, Fax: 831/646-0427, www.kurz-instruments.com

KW SOFTWARE, 3536 Edwards Rd., Cincinnati, OH 45208, 513/321-9385, Fax: 513/321-6992

L

L & J TECHNOLOGIES, 5911 Butterfield Rd., Hillside, IL 60162, 708/236-6000, Fax: 708/236-6006, www.ljtechnologies.com

LABTECH, 2 Dundee Park, Suite B09, Andover, MA 01810, 978/470-0099, Fax: 978/470-3338, www.labtech.com

LAKE SHORE CRYOTRONICS, INC., 575 McCorkie Blvd., Westerville, OH 43082, 614/891-2243, Fax: 614/818-1600, www.lakeshore.com

LAMBDA, 3055 Del Sol Blvd., San Diego, CA 92154, 619/628-2871, Fax: 619/429-1011, www.lambdapower.com/control

LAND INFRARED DIV. OF LAND INSTRUMENTS INTL., INC., 10 Friends Ln., Newtown, PA 18940, 800/523-8989, Fax: 215/781-0723, www.landinst.com

LEVITON MFG. CO., 59-25 Little Neck Pkwy., Little Neck, NY 11362, 718/281-6031, Fax: 718/631-6508, www.leviton.com

LIEBERT CORP., 1050 Dearborn Dr., Columbus, OH 43229, 614/841-5798, Fax: 614/841-6022, www.liebert.com

LIGHTHAMMER SOFTWARE, 690 Stockton Dr., Exton, PA 19341, 610/903-8000, Fax: 610/903-8006, www.lighthammer.com

LIMITORQUE, PO Box 11318, Lynchburg, VA 24506, 800/366-4401, Fax: 804/522-9858, www.limitorque.com

LINSEIS, INC., PO Box 666, Princeton Junction, NJ 08550, 609/799-6282, Fax: 609/799-7739, www.linseis.com

LION PRECISION, 563 Shoreview, St. Paul, MN 55126, 651/484-6544, Fax: 651/484-6824, www.liorprecision.com

LIQUID CONTROLS, INC., A UNIT OF IDEX, 105 Albrecht Dr., Lake Bluff, IL 60044, 847/295-1050, Fax: 847/295-1057, www.lcmeter.com

LOAD CONTROLS, INC., 10 Picker Rd., Storbrodge, MA 01566, 508/347-2606, Fax: 508/347-2064, www. loadcontrols.com

LOCKWOOD GREENE, PO Box 491, Spartanburg, SC 29304, 864/578-2000, Fax: 864/599-4117, www.lg.com

LOCON SENSOR SYSTEMS, INC., PO Box 789, Holland, OH 43528, 419/865-7651, Fax: 419/865-7756, www.locon.net

LOGIC BEACH, INC., 8363-6F Center Dr., La Mesa, CA 91942, 619/698-3300, Fax: 619/469-8604, www.logicbeach.com

LUCENT SPECIALTY FIBERTECHNOLOGIES, 55 Darling Dr., Avon, CT 06001, 860/678-0371, Fax: 860/674-8818, www.lucent.com/ofs/specialtyfiber

LUMBERG, INC., 14121 Justice Rd., Midlothian, VA 23113, 804/379-2010, Fax: 804/379-3232, www.lumbergusa.com

LUMENITE CONTROL TECH., 2331 N. 17th Ave., Franklin Park, IL 60131, 847/455-1450, Fax: 847/455-0127

LUTZE, INC., 13330 S. Ridge Dr., Charlotte, NC 28273, 704/504-0222, Fax: 704/504-0223, www.lutze.com

LYNX REAL-TIME SYSTEMS, INC., 2239 Samaritan Dr., San Jose, CA 95124, 408/879-3900, Fax: 408/879-3920, www.lynx.com

M

M-SYSTEM, 15028 Beltway Dr., Addison, TX 75001, 800/544-3181, Fax: 972/385-2277, www.m-system.com

MAGNETEK, 16555 W. Ryerson Rd., New Berlin, WI 53151, 414/782-0200, Fax: 414/782-3418, www.magnetekdrives.com

MAGNETROL INTL., INC., 5300 Belmont Rd., Downers Grove, IL 60516, 630/969-4000, Fax: 630/969-9489, www.magnetrol.com

MALUERN/INSITEC, 2110 Omega Rd., Suite D, San Ramar, CA 94583, 925/837-1330, Fax: 925/837-3864, www.insitec.com

MANNESMANN REXROTH CORP., REXROTH MECMAN DIV., 1953 Mercer Rd., Lexington, KY 40511, 606/254-8031, Fax: 606/281-3491, www.rexrothmecman.com

MAPLE SYSTEMS, INC., 808 134th St. SW, Suite 120, Everett, WA 98204, 425/745-3229, Fax: 425/745-3429, www.maple-systems.com

MARKLAND SPECIALTY ENG. LTD, 48 Shaft Rd., Toronto, Ontario M9W 4M2 Canada, 416/244-4980, Fax: 416/244-2287, www.sludgecontrols.com

MARSH BELLOFRAM, State Route 2; Box 305, Newell, WV 26050, 304/387-1200, Fax: 304/387-4417, www.marshbellofram.com

MARSH-MCBIRNEY, INC., 4539 Metropolitan Ct., Frederick, MD 21704, 301/874-5599, Fax: 301/874-2172, www.marsh-mcbirney.com

MARTEL ELECTRONICS, PO Box 897, Windham, NH 03087, 603/893-0886, Fax: 603/898-6820, www.martelcorp.com

MASSA PRODUCTS CORP., 280 Lincoln St., Hingham, MA 02043, 781/749-4800, Fax: 781/740-2045, www.massa.com

MATHESON INSTRUMENTS, 166 Keystone Dr., Montgomeryville, PA 18936, 215/648-4026, Fax: 215/641-0656, www.mathesoninstruments.com

MATRIC, RDI, Box 421A, Seneca, PA 16346, 800/462-8742, Fax: 814/678-1301, www.matric.com

MATRIKON CONSULTING, INC., Suite 1800, 10405 Jasper Ave., Edmonton, Alberta, T5J 3A4 Canada, 780/448-1010, Fax: 780/448-9191, www.matrikon.com

MAX CONTROL SYSTEMS, INC., 1180 Church Rd., Lansdale, PA 19446, 215/393-3900, Fax: 215/393-3921, www.maxcontrols.com

MAX MACHINERY, INC., 1420 Healdsburg Ave., Healdsburg, CA 95448, 707/433-7281, Fax: 707/433-0571, www.maxmachinery.com

MAXITROL CO., 23555 Telegraph Rd., Southfield, MI 48037-2230, 248/356-1400, Fax: 248/356-0829, www.maxitrol.com

MCCROMETER, 3255 W. Stetson Ave., Hemet, CA 92545, 909/652-6811, Fax: 909/652-3078, www.mccrometer.com

MCG SURGE PROTECTION, 12 Burt Drive, Deer Park, NY 11729, 800/851-1508, Fax: 516/586-5120, www.mcgsurge.com

MCLEAN MIDWEST, 11611 Business Park Blvd. N., Champlin, MN 55316, 612/323-8200, Fax: 612/576-3200, www.2010corp.com

MCMILLAN CO., PO Box 1340, Gerogetown, TX 78627, 800/861-0231, Fax: 512/863-0671, www.mcmillancompany.com

MDSI, 220 E. Huron, Suite 600, Ann Arbor, MI 48104, 734/769-9900, Fax: 734/769-9112, www.mdsi2.com

MDT SOFTWARE, 2520 NorthWind Pkwy., Suite 100, Alpharetta, GA 30004, 678/297-1000, Fax: 678/297-1003, www.mdtsoft.com

MEE INDUSTRIES, INC., 204 W. Pomona, Monrovia, CA 91016, 626/359-4550, Fax: 626/359-4660, www.meefog.com

MEECO, INC., 250 Titus Ave., Warrington, PA 18976, 215/343-6600, Fax: 215/343-4194, www.meeco.com

MELCOR, 1040 Spruce St., Trenton, NJ 08648, 609/393-4178, Fax: 609/393-9461, www.melcor.com

MEN MICRO, INC., 1940 Camden Way, Suite 100, Carrollton, TX 75007, 972/939-2675, Fax: 972/466-5986, www.menmicro.com

MERIAM INSTRUMENT, 10920 Madison Ave., Cleveland, OH 44102, 216/281-1100, Fax: 216/281-0228, www.meriam.com

METERSANDINSTRUMENTS.COM, 10 Vantage Point Dr., Unit 5, Rochester, NY 14624, 800/773-0370, Fax: 800/773-0371, www.metersandinstruments.com

METRIX INSTRUMENT CO., 1711 Townhurst Dr., Houston, TX 77043, 713/461-2131, Fax: 713/461-8223, www.metrix1.com

METTLER-TOLEDO, 1900 Polaris Pkwy., Columbus, OH 43240, 800/523-5123, Fax: 614/438-4544, www.mt.com

METTLER-TOLEDO PROCESS/INGOLD, 299 Washington St., Woburn, MA 01801, 800/352-8763, Fax: 781/939-6392, www.mt.com/pro

MGE UPS SYSTEMS, 1660 Scenic Ave., Costa Mesa, CA 92626-1410, 714/557-1636 Fax: 714/434-7652, www.mgeups.com

MGR INDUSTRIES, INC., 3013 E. Mulberry St., Ft. Collins, CO 80524, 970/221-2201, Fax: 970/484-4078, www.mgrind.com

MICON SYSTEMS, 4955 Gulf Frwy., Houston, TX 77023, 713/921-1899, Fax: 713/921-1882, www.miconsystems.com

MICRO DECISIONS CORP., 3206-A Cascade Dr., Valparaiso, IN 46383, 219/477-2002, Fax: 219/477-3910, www.micro-decisions.com

MICRO MOTION, INC., 7070 Winchester Circle, Boulder, CO 80301, 800/760-8119, Fax: 303/530-8459, www.micromotion.com

MICROGROUP, INC., 7 Industrial Park Rd., Medway, MA 02053, 508/533-4925, Fax: 508/533-5691, www.microgroup.com

MICROTOUCH SYSTEMS, 300 Griffin Brook Pk., Methuen, MA 01844, 978/659-9000, Fax: 978/659-9051, www.microtouch.com

MID-WEST INSTRUMENT, 6500 Dobry Dr., Sterling Heights, MI 48314, 810/254-6500, Fax: 810/254-6509, www.midwestinstrument.com

MIDAC CORP., 17911 Fitch Ave., Irvine, CA 92614, 949/660-8558, Fax: 949/660-9334, www.midac.com

MIE, INC., 7 Oak Park, Redford, MA 01730, 781/275-1919, Fax: 781/275-2121, www.mieinc.com

MIKRON INSTRUMENT CO., Inc., 16 Thornton Rd., Oakland, NJ 07436, 201/405-0900, Fax: 201/405-0090, www.mikroninst.com

MILLTRONICS, INC., 709 Stadium Dr., Arlington, TX 76011, 817/277-3543, Fax: 817/277-3894, www.milltronics.com

MITAC INDUSTRIAL CORP., 42001 Christy St., Fremont, CA 94538, 510/656-5288, Fax: 510/656-2669, www.mitacinds.com

MITSUBISHI ELECTRIC AUTOMATION, INC., 500 Corporate Woods Pkwy., Vernon Hills, IL 60061, 847/478-2419, Fax: 847/478-2396, www.meau.com

MITSUBISHI ELECTRIC & ELECTRONICS USA, 1050 E. Arques Ave., Sunnyvale, CA 94085, 408/730-5900, Fax: 408/245-2690, www.angleview.com

MKS INSTRUMENTS, INC., 6 Shattuck Rd., Andover, MA 01810, 978/975-2350, Fax: 978/975-0093, www.mksinst.com

MODCOMP, INC., 1650 W. McNab Rd., Ft. Lauderdale, FL 33309, 954/977-1380, Fax: 954/977-1900, www.modcomp.com

MODULAR INDUSTRIAL COMPUTERS, INC., 6025 Lee Hwy., Suite 340, Chattanooga, TN 37421, 423/499-0700, Fax: 423/892-0000, www.mic.com

MOELLER ELECTRIC CORP., 25 Forge Pkwy., Franklin, MA 02038, 508/520-7080, Fax: 508/520-7084, www.moellerusa.net

MOISTURE REGISTER PROD., 1712 Earhart Ct., La Verne, CA 91750, 909/392-5833, Fax: 909/392-5838, www.moistureregisterproducts.com

MONARCH INSTRUMENT, 15 Columbia Dr., Amherst, NH 03031, 603/883-3390, Fax: 603/886-3300, www.monarchinstrument.com

MONITOR TECHNOLOGIES, LLC, 44W320 Keslinger Rd., Elburn, IL 60119, 630/365-9403, Fax: 630/365-5646, www.monitortech.com

MONITROL MFG. CO., INC., PO Box 6296, Tyler, TX 75711, 903/561-0742, Fax: 903/561-3559, www.monitrolmfg.com

MOOG, INC., Seneca & Jaminson Rd., E. Aurora, NY 14052, 716/687-4785, Fax: 716/687-4467, www.moog.com/imc/product

MOORE INDUSTRIES-INTERNATIONAL, INC., 16650 Schoenborn St., Sepulveda, CA 91343, 818/894-7111, Fax: 818/891-2816, www.miinet.com

MOORE PROCESS AUTOMATION SOLUTIONS, 1201 Sumneytown Pike, Spring House, PA 19477, 215/646-7400, Fax: 215/ 283-2802

MOREHOUSE INSTRUMENT CO., 1742 South Ave., York, PA 17403, 717/843-0081, Fax: 717/846-4193

MOSAIC INDUSTRIES, 5437 Central Avenue, Suite 1, Newark, CA 94560, 510/790-1255, Fax: 510/790-0925, www.mosaic-industries.com

MOTORTRONICS, 13214-38th St. N., Clearwater, FL 33762, 727/573-1819, Fax: 727/573-1803, www.motortronics.com

MOYNO RKL CONTROLS, 1895 W. Jefferson Ct., Springfield, OH 45506, 937/327-3540, Fax: 937/327-3619, www.moyno.com

MSA INSTRUMENT DIV., PO Box 427, Pittsburgh, PA 15230, 800/MSA-4678, Fax: 724/776-3280, www.msanet.com

MSDI, 220 E. Huron St., Ann Arbor, MI 48104, 734/327-8246, Fax: 734/769-9112, www.mdsi2.com

MTL, INC., 9 Merrill Industrial Dr., Hampton, NH 03842, 603/926-0090, Fax: 603/926-1899, www.mtl-inst.com

MTS SYSTEMS CORP., 3001 Sheldon Dr., Cary, NC 27513, 919/677-0100, Fax: 919/677-0200, www.levelplus.com

MUSTANG ENGINEERING, INC., 16001 Park Ten Pl., Houston, TX 77084, 713/215-8000, Fax: 713/215-8590, www.mustangeng.com

MYPLANT.COM, a Honeywell business, 7047 E. Greenway Parkway, Suite 400, Scottsdale, AZ 85254, 877/848-8831, Fax: 480/850-6301, www.myplant.com

N

NAF—INVENSYS FLOW CONTROL, 1790 Satelite Blvd., Ste 100-04, Duluth, GA 30097, 678/474-1500, Fax: 678/474-1515, www.nafcontrols.com

NATIONAL INSTRUMENTS, 11500 N. Mopac Expwy., Austin, TX 78759, 512/683-6863, Fax: 512/683-5759, www.ni.com

NATIONWIDE PERSONNEL RECRUITING & CONSULTATION, 20834 SW Martinazzi Ave., Tualatin, OR 97062, 503/692-4925

NELES AUTOMATION, 7000 Hollister, Fl. 3, Houston, TX 77040, 713/346-0600, Fax: 713/346-0602, www.nelesautomation.com

NEMATRON, 5840 Interface Dr., Ann Arbor, MI 48103, 734/214-2000, Fax: 734/994-8074, www.nematron.com

NETSILICON, INC., 411 Waverly Oaks Rd., Bldg. 26, Waltham, MA 02452, 781/893-1234, Fax: 781/893-1338, www.netsilicon.com

NEURALWARE, 230 E. Main St., Suite 200, Carnegie, PA 15106, 412/278-6280, Fax: 412/278-6289, www.neuralware.com

NEUTRONICS, INC., 456 Creamery Way, Exton, PA 19341, 610/524-8800, Fax: 610/524-8807, www.neutronicsinc.com

NEWPORT ELECTRONICS, INC., 2229 S. Yale St., Santa Ana, CA 92704, 714/540-4914, Fax: 714/546-3022, www.newportinc.com

NEXUS ENGINEERING, 900 Rockmead, Suite 250, Kingwood, TX 77339, 281/359-5190, Fax: 281/359-5278, www.nexusengineering.com

NIBCO, INC., 1516 Middlebury St., Elkhart, IN 46516, 219/295-3000, Fax: 219/295-3307, www.nibco.com

NKK SWITCHES, 7850 E. Gelding Dr., Scottsdale, AZ 85260, 480/991-0942, Fax: 480/998-1435, www.nkkswitches.com

NORDSTROM/AUDCO—INVENSYS FLOW CONTROL, 1511 Jefferson St., Sulphur Springs, TX 75482, 903/885-4691, Fax: 903/439-3411, www.nordstromaudco.com

NORRISEAL, 11122 W. Little York, Houston, TX 77041, 713/466-3552, Fax: 713/896-7386, www.norriseal.com

NORTECH FIBRONIC, INC., 240-500 Avenue St. Jean-Baptiste, Quebec, QC, G2E 5R9 Canada, 418/872-4686, Fax: 418/872-2894, www.nortech.ca

NORTH EAST ELECTRONICS CONTROLS, 1545 Holland Rd., Maumee, OH 43537, 419/893-4158, Fax: 419/893-4171, www.nee-controls.com

NORTHSTAR TECHNOLOGIES, INC., 575 McCorkle Blvd., Westerville, OH 43082, 614/818-1150, Fax: 614/891-6909, www.northstarencoders.com

NORTHWEST ANALYTICAL, INC., 519 SW Park Ave., Portland, OR 97205, 503/224-7727, Fax: 503/224-5236, www.nwasoft.com

NOSHOK, INC., 1010 W. Bagley Rd., Berea, OH 44133, 440/243-0888, Fax: 440/243-3472, www.noshok.com

N-TRON, 578 Azalea Rd., Suite 105, Mobile, AL 36609, 334/666-9878, Fax: 334/666-9833, www.n-tron.com

NUMATICS, INC., 1450 N. Milford Rd., Highland, MI 48357, 248/889-6227, Fax: 248/887-4768, www.numatics.com

O

OCEANA SENSOR TECHNOLOGIES, 1632 Corporate Landing Pkwy., Virginia Beach, VA 23454, 757/426-3678, Fax: 757/426-3633, www.oceanasensor.com

OGDEN MANUFACTURING CO., 64 W. Seegers Rd., Arlington Heights, IL 60005, 847/593-8050, Fax: 847/593-8062, www.ogdenmfg.com

OHIO SEMITRONICS, INC., 4242 Reynolds Dr., Hilliard, OH 43026, 614/777-1005, Fax: 614/777-4511, www.ohiosemitronics.com

OHMART/VEGA CORP., 4241 Allendorf Dr., Cincinnati, OH 45209, 513/272-0131, Fax: 513/272-0133, www.ohmartvega.com

OI DIRECT, 343 St. Paul Blvd., Carol Stream, IL 60188, 888/OI-DIRCT, Fax: 888/OI-FAX-US, www.oidirect.com

OLFLEX WIRE & CABLE, INC., 30 Plymouth St., Fairfield, NJ 07004, 973/575-1101, Fax: 973/575-1267, www.olflex.com

OMEGA ENGINEERING, INC., One Omega Dr., Stamford, CT 06907, 203/359-1660, Fax: 203/359-7700, www.omega.com

OMEGA VANZETTI, INC., 6 Merchant St., Sharon, MA 02067, 781/784-4733, Fax: 781/784-2447, www.vanzetti.com

OMNI FLOW COMPUTERS, 10701 Corporate Dr, #300, Stafford, TX 77477, 281/240-6161, Fax: 281/240-6162, www.omniflow.com

OMRON ELECTRONICS INC., One E. Commerce Dr., Schaumburg, IL 60173, 800/55-OMRON, Fax: 847/843-7787, www.omron.com/oei

ONCUITY, INC., 1410 Blalcok Rd., Houston, TX 77055, 888/271-6726, Fax: 713/682-8066, www.oncuity.com

ONTRAK CONTROL SYSTEMS, INC., 422 Arnley St., Sudbury, Ontario, P3C 1E7 Canada, 705/671-2652, Fax: 705/671-6127, www.ontrak.net

OPTEK-DANULAT, INC., 279 S. 17th Ave., Suite 10, West Bend, WI 53095, 800/371-4288, Fax: 262/335-4299, www.optek.com

OPTIMATION, INC., PO Box 4107, Huntsville, AL 35815, 256/883-3050, Fax: 256/883-3070, www.optimate.com

OPTO 22, 43044 Business Park Dr., Temecula, CA 92590, 909/695-3000, Fax: 909/695-3095, www.opto22.com

OP/STATION, A DIV. OF AUTOMATED CONTROL CONCEPTS, INC., 3535 Route 66, Neptune, NJ 07753, 732/922-6611, Fax: 732/922-9611, www.opstation.com

ORBITAL SCIENCES CORP., 2771 N. Garey Ave., Pomona, CA 91767, 909/593-3581, Fax: 909/392-3207, www.orbital-ait.com

ORENA CONTRONET, INC., 8320 NW Hawkins Blvd., Portland, OR 97229, 503/297-1854, Fax: 503/297-1914, www.orena.com

ORSI-GROUP, 502 Earth City Expwy., Suite 139, St. Louis, MO 63045, 314/298-9100, Fax: 314/298-1729, www.orsigroup.com

OSI SOFTWARE, 777 Davis St., Suite 250, San Leandro, CA 94577, 510/297-5836, Fax: 510/357-8136, www.osisoft.com

OSMONICS, INC., 5951 Clearwater Dr., Minnetonka, MN 55343-8995, 719/264-3937, Fax: 719/536-3301, www.osmonics.com

OTEK CORP., 4016 E. Tennessee St., Tucson, AZ 85714, 520/748-7900, Fax: 520/790-2808, www.otekcorp.com

OTTO ENGINEERING, 2 E. Main St., Carpenterville, IL 60110, 847/428-7171, Fax: 847/428-1956, www.ottoeng.com

P

PACIFIC SCIENTIFIC, 4301 Kishwaukee St., Rockford, IL 61105-0106, 815/226-3100, Fax: 815/226-3080, www.pacsci.com

PAI PARTNERS, 135 Fort Lee Rd., Leonia, NJ 07605, 201/585-2050, Fax: 201/585-1968

PANAMETRICS, INC., 221 Crescent St., Waltham, MA 02453, 781/899-2746, Fax: 781/894-8582, www.panametrics.com

PARKER COMPUMOTOR, 5500 Business Park Dr., Rohnert Park, CA 94928, 800/358-9068, Fax: 707/584-8015, www.compumotor.com

PARKER FILTRATION, 100 Ames Pond Dr., Tewksbury, MA 01876, 800/343-4048, Fax: 978/858-0625, www.parker.com/balston

PARKER HANNIFIN, 6035 Parkland Blvd., Cleveland, OH 44124, 256/435-2130, Fax: 256/435-7710, www.parker.com

PARKER HANNIFIN CORP., INSTRUMENTATION CONNECTORS DIV., 9400 S. Memorial Pkwy., Huntsville, AL 35803-2197, 256/881-2040, Fax: 256/881-5730, www.parker.com/icd/

PARKER HANNIFIN CORP., INSTRUMENTATION VALVE DIV., 2651 Alabama Hwy. 21 North, Jacksonville, AL 36265-9681, 256/435-2130, Fax: 256/435-7718, www.parker.com/ivd

PARKER HANNIFIN CORP., SKINNER VALVE DIV., 250 Canal Blvd., New Britain, CT 06051, 860/827-2300, Fax: 860/827-2384, www.parker.com/skinner

PARKER HANNIFIN CORP., VERIFLO DIVISION, 250 Canal Blvd., Richmond, CA 94947, 510/412-1166, Fax: 510/232-7396, www.veriflo.com

PATLITE CORP., 3860 Del Amo Blvd., Suite 404, Torrance, CA 90503, 310/214-5286, Fax: 310/214-5288, www.patlite.com

PAVE TECHNOLOGY CO., 2751 Thunderhawk Ct., Dayton, OH 45414-3445, 937/890-1100, Fax: 937/890-5165, www.pavetechnologies.com

PAVILION TECHNOLOGIES, INC., 11100 Metric Blvd., Austin, TX 78758, 800/886-8432, Fax: 512/438-1401, www.pavtech.com

PAYNE ENGINEERING CO., PO Box 70, Scott Depot, WV 25302, 304/757-7353, Fax: 304/757-7305, www.payneg.com

PC SOFT INTERNATIONAL, 89 Forbes Blvd., Mansfield, MA 02048, 508/337-9200, Fax: 508/337-9201, www.emation.com

PCD, INC., 2 Technology Dr., Peabody, MA 01960, 978/532-8800, Fax: 978/532-6800, www.pcdinc.com

PCME LTD, Clearview Building, Edison Rd., St. Ives, Cambs PE27 3GH UK, 01480 468200, Fax: 01480 463400, www.pcme.co.uk

PELICAN PRODUCTS, INC., 23215 Early Ave., Torrance, CA 90505, 310/326-4700, Fax: 310/326-3311, www.pelican.com

PENBERTHY, 320 Locust St., Prophetstown, IL 61277, 815/537-2311, Fax: 815/537-5764, www.pcc-penberthy.com

PENTAIR ELECTRONIC PACKAGING, 170 Commerce Dr., Warwick, RI 02886, 800/451-8755, Fax: 401/738-2904, www.pentair-ep.com

PENTEK, One Park Way, Upper Saddle River, NJ 07458, 201/818-5900, Fax: 201/818-5904, www.pentek.com

PEP MODULAR COMPUTERS, 750 Holiday Dr., Pittsburgh, PA 15220, 412/921-3322, Fax: 412/921-3356, www.pepusa.com

PEPPERL+FUCHS, INC., 1600 Enterprise Pkwy., Twinsburg, OH 44087, 330/425-3555, Fax: 330/425-4607, www.am.pepperl-fuchs.com

PERMA PURE, INC., PO Box 2105, Toms River, NJ 08754, 732/244-0010, Fax: 732/244-8140, www.permapure.com

PGI INTERNATIONAL, 16101 Vallen Dr., Houston, TX 77041, 713/466-0056, Fax: 713/744-9899, www.pgiint.com

PHOENIX CONTACT, INC., PO Box 4100, Harrisburg, PA 17111, 800/322-3225, Fax: 717/944-1625, www.phoenixcon.com

PHOENIX DIGITAL, 7650 E. Evans Rd., Bldg. A, Scottsdale, AZ 85260, 480/483-7393, Fax: 480/483-7391, www.phoenixdigitalcorp.com

PHONETICS, INC., 701 Tryens Rd., Aston, PA 19014, 610/558-2700, Fax: 610/558-0222, www.sensaphone.com

PILZ AUTOMATION SAFETY L.P., 24850 Drake Rd., Farmington Hills, MI 48335, 248/473-1133, Fax: 248/473-3997, www.pilzusa.com

PLANAR SYSTEMS, 1400 NW Compton, Beaverton, OR 97006, 503/690-1100, Fax: 503/690-1493, www.planar. com

PLAST-O-MATIC VALVES, INC., 1384 Pompton Ave., Cedar Grove, NJ 07009, 973/256-3000, Fax: 973/256-4745, www.plastomatic.com

PMV-USA, INC., 1440 Lakefront Circle #160, The Woodlands, TX 77380, 281/292-7500, Fax: 281/292-7760, www.pmvusa.com

POMONA ELECTRONICS, 1500 E. Ninth St., Pomona, CA 91766-3835, 909/469-2900, Fax: 909/469-3317, www.pomonaelectronics.com

POWERCUBE, 9340 Owensmouth Ave., Chatsworth, CA 91311, 818/734-6500, Fax: 818/734-6540, www.powercube.com

POWERS PROCESS CONTROLS, 3400 Oakton St., Skokie, IL 60076, 847/568-6256, Fax: 847/673-9044, www.powerscontrols.com

PQ SYSTEMS, INC., 10468 Miamisburg-Springbordo Rd., Miamisburg, OH 45342, 800/777-3020, Fax: 937/885-2252, www.pqsystems.com

PRECISION DIGITAL CORP., 19 Strathmore Rd., Natick, MA 01760, 508/655-7300, Fax: 508/655-8990, www.predig.com

PRECISION SOLUTIONS, INC., 3101 Bee Caves Rd. #290, Austin, TX 78746, 512/330-9990, Fax: 512/330-9996, www.psivideo.com

PRESSURE SYSTEMS, INC., 34 Research Dr., Hampton, VA 23666, 757/865-1243, Fax: 757/865-8744, www.pressuresystems.com

PRESYS INSTRUMENTS, INC., 3000 SW 77 Pl., Miami, FL 33155, 305/262-8488, Fax: 305/262-7225, www.presys.com.br.

PRINCO INSTRUMENTS, INC., 1020 Industrial Blvd., Southampton, PA 18966, 215/355-1500, Fax: 215/355-7766, www.princoinstruments.com

PRO-TECH, 3600A Swiftwater Park Dr., Suwanee, GA 30024, 770/271-0048, Fax: 770/271-2796, www.protech1.com

PROCONSUL, INC., PO Box 1823, Georgetown, TX 78627, 512/863-8000, Fax: 512/869-4999, www.proconsul.net

PROFIBUS TRADE ORGANIZATION (PTO), 16101 N. 82nd St., Suite 3B, Scottsdale, AZ 85260, 480/483-2456, Fax: 480/483-7202, www.profibus.com

PROSOFT TECHNOLOGY, INC., 9801 Camino Media, Suite 105, Bakersfield, CA 93311, 661/664-7208, Fax: 661/664-7233, www.prosoft-technology.com

PROSYS, INC., 11814 Coursey Blvd., Suite 408, Baton Rouge, LA 70816, 225/291-9591, Fax: 225/291-9594, www.prosysinc.com

PSI-TRONIX TECHNOLOGIES, INC., 3950 South K St., Tulare, CA 93274, 559/686-0558, Fax: 559/686-0609, www.psi-tronix.com

PULSE, A TECHNITROL CO., 12220 World Trade Dr., San Diego, CA 92128, 619/385-8031, Fax: 619/674-8262, www.pulseeng.com

PYROMATION, INC., 209 Industrial Pkwy., Fort Wayne, IN 46825, 219/484-2580, Fax: 218/482-6805, www.pyromation.com

PYROMETER INSTRUMENT CO., 209 Industrial Pkwy., Northvale, NJ 07647, 201/768-2000, Fax: 201-768-2570, www.pyrometer.com

Q

QSI CORP., 2212 South W. Temple, #50, Salt Lake City, UT 84115, 801/466-8770, Fax: 801/466-8792, www.qsicorp.com

QUATECH INC., 662 Wolf Ledges Pkwy., Akron, OH 44311, 800/553-1170, Fax: 330/434-1409, www.quatech.com

QUEST INTERNATIONAL, INC., 65 Parker, Irvine, CA 92618, 800/231-6777, Fax: 949/581-4011, www.questinc.com

R

R. STAHL, INC., 45 Northwestern Dr., Salem, NH 03079, 603/870-9500, Fax: 603/870-9290, www.rstahl.com

RACO MANUFACTURING ENGINEERING, 1400 62nd St., Emeryville, CA 94608, 510/658-6713, Fax: 510/658-3153, www.racoman.com

RADSTONE TECHNOLOGY CORP., 50 Craig Rd., Montvale, NJ 07645, 201/391-2700, Fax: 201/391-2899, www.radstone.com

RAECO, INC., 9324 Gulfstream Rd., Frankfort, IL 60423, 815/464-6200, Fax: 815/464-8720, www.raeco.com

RAYTEK CORP., 1201 Shaffer Rd., Bldg. 2, Santa Cruz, CA 95061, 831/458-1110, Fax: 831/458-1239, www.raytek.com

RC ELECTRONICS, INC., 6464 Hollister Ave., Santa Barbara, CA 93117, 805/685-7770, Fax: 805/685-5853, www.rcelectronics.com

RC SYSTEMS CO., 2513 Hwy. 646, Santa Fe, TX 77510, 409/925-7808, Fax: 409/925-1078, www.members.aol.com/rcsci/res.htm

RCM INDUSTRIES, INC., 110 Mason Cir., Suite D, Concord, CA 94520-1238, 925/687-8363, Fax: 925/671-9636, www.flo-gage.com

RDF CORP., 23 Elm Ave., Hudson, NH 03051, 603/882-5195, Fax: 603/882-6925, www.rdfcorp.com

REAL-TIME INNOVATIONS, 155A Moffett Park Dr., Sunnyvale, CA 94089, 408/734-4200, Fax: 408/734-5009, www.rti.com

REBIS, 1600 Riviera Ave., Suite 300, Walnut Creek, CA 94596, 925/933-2525, Fax: 925/933-1920, www.rebis.com

RED LION CONTROLS, 20 Willow Springs Circle, York, PA 17402, 717/767-6511, Fax: 717/764-0839, www.redlion-controls.com

RED VALVE CO., 700 N. Bell Ave., Carnegie, PA 15106, 412/279-0044, Fax: 412/279-7878, www.redvalve.com

RELIABLE POWER METERS, 400 Blossom Hill Rd., Los Gatos, CA 95032, 408/358-5100, Fax: 408/358-4420, www.reliablemeters.com

REMOTE CONTROL, INC., 386 Dry Bridge Rd., North Kingstown, RI 02852, 401/294-1400, Fax: 401/294-3388, www.rciactuators.com

REXROTH MECMAN PNEUMATICS, 1953 Mercer Rd., Lexington, KY 40511, 859/254-8031, Fax: 859/281-3491, www.us.rexroth.com

RICE LAKE WEIGHING SYSTEMS, 230 W. Coleman St., Rice Lake, WI 54868, 715/234-9171, Fax: 715/234-6967, www.rlws.com

RITEPRO, INC., A SUBSIDIARY OF BARY INTL., INC., One Rittal Place, Montreal-North, Quebec, H1G 3K7 Canada, 514/324-8900, Fax: 514/324-9525

RITTAL CORP., One Rittal Place, Springfield, OH 45504, 937/390-0500, Fax: 937/390-8392, www.rittal-corp.com

ROBERTSHAW IPD, AN INVENSYS CO., 1602 Mustang Dr., Maryville, TN 37801, 865/981-3100, Fax: 865/981-3168

ROBICON, 500 Hunt Valley Dr., New Kensington, PA 15068, 724/339-9500, Fax: 724/339-9505, www.robicon.com

ROCHESTER INSTRUMENT SYSTEMS, 255 N. Union St., Rochester, NY 14605, 716/238-4078, Fax: 716/454-7805, www.rochester.com

ROCKWELL AUTOMATION, 1201 S. 2nd St., Milwaukee, WI 53204, 414/382-2000, Fax: 414/382-4444, www.automation.rockwell.com

ROCKWELL AUTOMATION, 1 Allen-Bradley Dr., Mayfield Heights, OH 44124, 440/646-5000, Fax: 414/382-4444, www.ab.com

ROCKWELL AUTOMATION PRESENCE SENSING BUSINESS, Two Executive Dr., Chelmsford, MA 01824, 978/446-3476, Fax: 978/446-3322, www.ab.com/sensors

ROCKWELL SOFTWARE, 2424 S. 102nd St., West Allis, WI 53227, 414/321-8000, Fax: 414/321-9647, www.software.rockwell.com

RONAN ENGINEERING CO., PO Box 1275, Woodland Hills, CA 91367, 818/883-5211, Fax: 818/992-6435, www.ronan.com

ROSE & BOPLA, 7330 Executive Way, Frederick, MD 21704, 301/696-9800, Fax: 301/696-9494, www.rose-bopia.com

ROSEMOUNT ANALYTICAL PROCESS ANALYTIC DIV., 1201 N. Main St., Orrville, OH 44667, 330/682-9010, Fax: 330/684-4434, www.processanalytic.com

ROSEMOUNT ANALYTICAL, INC., Uniloc Div., 2400 Barrance Pkwy., Irvine, CA 92606, 949/863-1181, Fax: 949/474-7250, www.rauniloc.com

ROSEMOUNT, INC., 8500 Market Blvd., Chanhassen, MN 55317, 952/949-5165, Fax: 952/949-5114, www.rosemount.com

ROTEK INSTRUMENT CORP., 390 Main St., Waltham, MA 02452, 781/899-4611, Fax: 781/894-7273, www.rotek.com

ROTORK CONTROLS, INC., 19 Jetview Dr., Rochester, NY 14624, 716/328-1550, Fax: 716/328-5848, www.rotork.com

ROTRONIC INSTRUMENT CORP., 160 E. Main St., Huntington, NY 11743, 631/427-3898, Fax: 631/427-3902, www.rotronic-usa.com

ROYCE INSTRUMENT CORP., 13555 Gentilly Rd., New Orleans, LA 70129, 800/347-3505, Fax: 504/254-6855

RTP CORP., 2705 Gateway Dr., Pompano Beach, FL 33069, 954/974-7210, Fax: 954/975-9815, www.rtpcorp.com

RVSI ACUITY CI MATRIX, 5 Shawmur Rd., Canton, MA 02021, 781/821-0830, Fax: 781/828-8942, www.rvsi.com

S

S. HIMMELSTEIN & CO., 2990 Pembroke Ave., Hoffman Estates, IL 60195, 847/843-3300, Fax: 847/843-8938, www.himmelstein.com

SAAB TANK CONTROL, INC., 10700 Hammerly Blvd., Suite 115, Houston, TX 77043, 713/722-9199, Fax: 713/722-9115, www.saabradar.com

SAMSON CONTROLS, INC., 4111 Cedar Blvd., Baytown, TX 77520, 281/383-3677, Fax: 281/383-3690, www.samson-usa.com

SANDELIUS INSTRUMENTS, INC., PO Box 30098, Houston, TX 77249, 713/861-1100, Fax: 713/861-9136, www.sandelius.com

SARTORIUS CORP., 131 Heartland Blvd., Edgewood, NY 11717, 631/254-4249, Fax: 631/254-4253, www.sartorius.com

SBS TECHNOLOGIES, INC., 8371C Central Ave., Newark, CA 94560, 510/742-2500, Fax: 510/742-2501, www.sbs.com

SCANIVALVE CORP., 1722 N. Madson St., Liberty Lake, WA 99019, 800/935-5151, Fax: 509/891-9481, www.scanivalve.com

SCHLUMBERGER MEASUREMENT DIV., 1000 Lucas Way, Greenwood, SC 29646-8800, 800/833-3357, Fax: 864/223-0341, www.slb.com/rms/measurement

SCHROFF NORTH AMERICAN, 170 Commerce Dr., Warwick, RI 02886, 800/451-8755, Fax: 401/738-7988, www.schroffus.com

SCHNEIDER ELECTRIC, 1415 S. Roselle Rd., Palatine, IL 60067, 800/392-8781, Fax: 800/824-7151, www.schneiderautomation.com

SDRG CONTROLS, 8234 Braniff, Houston, TX 77061, 713/242-0822, Fax: 713/644-8294, www.sdrg.com

SEALEVEL SYSTEMS, INC., 155 Technology Place, Liberty, SC 29657, 864/843-4343, Fax: 864/843-3067, www.sealevel.com

SEIKO INSTRUMENTS U.S.A., INC., 2990 West Lomita Blvd., Torrance, CA 90505, 909/975-5637, Fax: 909/975-5699, www.seiko-usa-ecd.com

SELCO PRODUCTS CO., 709 N. Poplar St., Orange, CA 92868, 714/712-6200, Fax: 714/712-6222, www.selcoproducts.com

SENSIDYNE, 16333 Bay Vista Dr., Clearwater, FL 33760, 800/451-9444, Fax: 727/530-3602, www.sensidyne.com

SENSO-METRICS, INC., 4584 Runway St., Simi Valley, CA 93063, 805/527-3640, Fax: 805/584-2960, www.senso-metrics.com

SENSOR ELECTRONICS CORP., 5500 Lincoln Dr., Minneapolis, MN 55436, 952/938-9486, Fax: 952/938-9617, www.sensorelectronic.com

SENSOR PRODUCTS, INC., 188 Rte. 10 W., Suite 307, E. Hanover, NJ 07936, 973/884-1755, Fax: 973/884-1699, www.sensorprod.com

SENSOTEC, INC., 2080 Arlingate Ln., Columbus, OH 43204, 614/850-5000, Fax: 614/850-1111, www.sensotec.com

SEQUENCIA CORP., 15458-B N. 28th Ave., Phoenix, AZ 85053, 602/896-3700, Fax: 602/896-3896, www.sequencia.com

SERVOMEX, 90 Kerry Place, Norwood, MA 02062, 781/769-7710, Fax: 781/769-2834, www.servomex.com

SETRA SYSTEMS, INC., 159 Swanson Rd., Boxborough, MA 01719, 978/266-3629, Fax: 978/264-0292, www.setra.com

SEW EURODRIVE, INC., 1295 Old Spartanburg Hwy., Lyman, SC 29365, 864/439-7537, Fax: 864/661-1276, www.seweurodrive.com

SICK, INC., 6900 W. 110th St., Bloomington, MN 55438, 952/941-6780, Fax: 952/941-9287, www.sickoptic.com

SIEMENS APPLIED AUTOMATION, 500 W. Highway 60, Bartlesville, OK 74003, 918/622-7000, Fax: 918/662-7052, www.aai-us.com

SIEMENS ENERGY & AUTOMATION, INC., 3333 Old Milton Pkwy., Alpharetta, GA 30005, 800/964-4114, Fax: 678/475-5840, www.sea.siemens.com

SIERRA INSTRUMENTS, INC., 5 Harris Court, Bldg. L, Monterey, CA 93940, 800/866-0200, Fax: 831/373-4402, www.sierrainstruments.com

SIERRA MONITOR CORP., 1991 Tarob Court, Milpitas, CA 95035, 408/262-6611, Fax: 408/262-9042, www.sierramonitor.com

SIMPSON ELECTRIC CO., 853 Dundee Ave., Elgin, IL 60120, 847/697-2260, Fax: 847/697-2272, www.simpsonelectric.com

SIMULATION SCIENCES, INC., 601 Valencia Ave., Brea, CA 92823, 714/579-0412, Fax: 714/579-7927, www.simsci.com

SIXNET, PO Box 767, Clifton Park, NY 12065, 518/877-5173, Fax: 518/877-8346, www.sixnetio.com

SL CORP., 240 Tamal Vista Blvd., Corte Madera, CA 94925, 415/927-8400, Fax: 415/927-8401, www.sl.com

SMAR INTERNATIONAL CORP., 7240 Brittmoore #118, Houston, TX 77041, 713/849-2021, Fax: 713/849-2022, www.smar.com

SMITH METER, INC., an FMC Energy Systems Business, 1602 Wagner Ave., Erie, PA 16514, 814/898-5264, Fax: 814/899-8927, www.smithsystems-inc.com

SMOOT CO., 1250 Seminary, Kansas City, KS 66103, 913/362-1710, Fax: 913/362-7863, www.smootco.com

SNELL INFRARED, PO Box 6, Montpelier, VA 05601, 800/636-9820, Fax: 802/223-0460, www.snellinfrared.com

SOFTPLC CORP., 25603 Red Brangus Dr., Spicewood, TX 78669, 512/264-8390, Fax: 512/264-8399, www.softplc.com

SOFTWARE HORIZONS, INC., 100 Treble Cove Rd., N. Billerica, MA 01862, 978/670-8700, Fax: 978/670-8787, www.instanthmi.com

SOFTWARE TOOLBOX, INC., 148A E. Charles St., Matthews, NC 28105, 704/849-2773, Fax: 704/849-6388, www.softwaretoolbox.com

SOLA/HEVI-DUTY, 7770 N. Frontage Rd., Skokie, IL 60077, 800/377-4384, Fax: 800/367-4384, www.sola-hevi-duty.com

SOLAREX, 630 Solarex Ct., Frederick, MD 21703, 301/698-4200, Fax: 301/698-4201, www.solarex.com

SOLARTRON, INC., 19408 Park Row, Suite 320, Houston, TX 77084, 281/398-7890, Fax: 281/398-7891, www.solartron.com

SOLBERG MFG., INC., 1151 W. Ardmore Ave., Itasca, IL 60143, 630/773-1363, Fax: 630/773-0727, www.solbergmfg.com

SONY PRECISION TECHNOLOGY AMERICA, 20381 Hermana Circle, Lake Forest, CA 92630, 949/770-8400, Fax: 888/910-7669, www.sonypt.com

SORENSEN, 9250 Brown Deer Rd., San Diego, CA 92121-2294, 858/450-0085, Fax: 858/458-0267, www.sorensen.com

SPARLING INSTRUMENTS, INC., 4097 N. Temple City Blvd., El Monte, CA 91731, 626/444-0571, Fax: 626/452-0723, www.sparlinginstruments.com

SPC PRESS/STATISTICAL PROCESS CONTROLS, INC., 5908 Toole Dr., Suite C, Knoxville, TN 37919, 423/584-5005, Fax: 423/588-9440, www.spcpress.com

SPENCE ENGINEERING CO., INC., 150 Coldenham Rd., Walden, NY 12586, 800/398-2493, Fax: 914/778-1072, www.spenceengineering.com

SPIRAX SARCO, INC., 1150 Northpoint Blvd., Blythewood, SC 29016, 803/714-2071, Fax: 803/714-2224, www.spiraxsarco-usa.com

SPONSLER CO., INC., 2363 Sandifer Blvd., Westminster, SC 29693, 864/647-2065, Fax: 864/647-1255, www.sponsler.com

SST, A PART OF WOODHEAD CONNECTIVITY, 50 Northland Rd., Waterloo, Ontario, NV2 IN3 Canada, 519/725-5136, Fax: 519/725-1515, www.sstech.on.ca

STACOSWITCH, 1139 Baker St., Costa Mesa, CA 92626, 714/549-3041, Fax: 714/549-0930, www.stacoswitch.com

STAHLIN, 500 Maple St., Belding, MI 48809, 616/794-0700, Fax: 616/794-7564, www.stahlin.com

STANLEY ELECTRIC, 2660 Barranca Pkwy., Irvine, CA 92606, 949/222-0777, Fax: 949/222-0555, www.stanleyelec.com

STAT-EASE, INC., 2021 E. Hennepin Ave, #191, Minneapolis, MN 55413, 612/378-4449, Fax: 612/378-2152, www.statease.com

STATISTICAL PROCESS CONTROLS, INC., 5908 Toole Dr., Suite C, Knoxville, TN 37919, 865/584-5005, Fax: 865/588-9440, www.spcpress.com

STOCHOS, INC., 14 N. College St., Schenectady, NY 12305, 518/372-5426, Fax: 518/372-4789, www.stochos.com

STONEL, One StoneL Dr., Fergus Falls, MN 56537, 218/739-5774, Fax: 218/739-5776, www.stonel.com

SUN ELECTRONIC SYSTEMS, INC., 1900 Shepard Dr., Titusville, FL 32780, 321/383-9400, Fax: 321/383-9412, www.sunelectronics.com

SUNX SENSORS, 1207 Maple St., West Des Moines, IA 50265, 800/280-6933, Fax: 515/225-0063, www.sunx-ramco.com

SUPERIOR ELECTRIC, 383 Middle St., Bristol, CT 06010, 860/585-4500, Fax: 860/584-1483, www.superiorelectric.com

SVF FLOW CONTROLS, INC., 13560 Larwin Cir., Santa Fe Springs, CA 90670, 562/802-2255, Fax: 562/802-3114, www.svfflowcontrols.com

SWAGELOK, 31400 Aurora Rd., Solon, OH 44139, 440/349-5934, Fax: 440/349-5843, www.swagelok.com

SYMCOM, INC., 2880 N. Plaza Dr., Rapid City, SD 57702, 605/348-5580, Fax: 605/348-5685, www.symcominc.com

SYNERGETIC, 2506 Wisconsin Ave., Downers Grove, IL 60515, 630/434-1770, Fax: 630/434-1987, www.synergetic.com

T

TA ENGINEERING CO., INC., 1150 Moraga Way, Moraga, CA 94556, 925/376-8500, Fax: 925/376-4977, www.aimax.com

TAIYO YUDEN (USA), INC., Power Systems Group, 1770 La Costa Meadows Dr., San Marcos, CA 95131, 760/510-3200, Fax: 760/471-4021, www.t-yuden.com

TEAC AMERICA, INC., 7733 Telegraph Rd., Montebello, CA 90640, 323/726-0303, Fax: 323/727-7674, www.teac.com

TECHNE, INC., 743 Alexander Rd., Princeton, NJ 08540, 609/452-9275, Fax: 609/987-8177, www.techneusa.com

TECHNICAL MARINE SERVICE, INC., 6040 N. Cutter Cir., Suite 302, Portland, OR 97217, 503/285-8947, Fax: 503/285-1379, www.tms-usa.com

TECHNOLAND, INC., 1050 Stewart Dr., Sunnyvale, CA 94086, 408/992-0888, Fax: 408/992-0808, www.technoland.com

TECNOMATIX TECHNOLOGIES, 21500 Haggerty Rd., Suite 300, Noriville, MI 48167, 248/471-6140, Fax: 248/471-6147, www.tecnomatix.com

TEL-TRU MANUFACTURING CO., 408 St. Paul St., Rochester, NY 14605, 716/232-1440, Fax: 716/232-3857, www.teltru.com

TELEDYNE ANALYTICAL INSTRUMENTS, 16830 Chestnut St., City of Industry, CA 91748, 626/934-1507, Fax: 626/961-2538, www.teledyne-ai.com

TELETROL SYSTEMS, INC., 286 Commercial St., Manchester, NH 03101, 603/645-6061, Fax: 603/645-6174, www.teletrol.com

TEMPROX, 2915 Parkway St., Lakeland, FL 33811, 863/619-5999, Fax: 863/619-5274, www.temprox.com

TESCOM CORP., 12616 Industrial Blvd., Elk River, MN 55330, 763/441-6330, Fax: 763/241-3224, www.tescom.com

TESTO, INC., 35 Ironia Rd., Flanders, NJ 07836, 800/227-0724, Fax: 973/252-1724, www.testo.com

TEXAS INDUSTRIAL PERIPHERALS, 2621 Ridgepoint Dr. Suite 235, Austin, TX 78754, 800/866-6506, Fax: 512/837-0207, www.ikey.com

TEXAS MICRO, INC., 5959 Corporate Dr., Houston, TX 77036, 713/541-8200, Fax: 713/541-8226, www. texasmicro.com

THERMACAL, INC., 30275 Bainbridge Rd., Solon, OH 44139, 440/498-1005, Fax: 440/498-1062, www.thermacal.com

THERMAL INSTRUMENTS CO., 217 Sterner Mill Rd., Trevose, PA 19053, 215/355-8400, Fax: 215/355-1789, www.thermalinstrument.com

THERMO ANDERSEN, 500 Technology Court, Smyrna, GA 30082, 770/319-9999, Fax: 770/319-0336

THERMO BLH, 75 Shawmut Rd., Canton, MA 02021, 781/821-2000, Fax: 781/828-1451, www.thermoblh.com

THERMO BRANDT INSTRUMENTS, 3333 Air Dark Rd., Fuquay, NC 27526, 919/552-9011, Fax: 919/552-9717, www.brandtinstruments.com

THERMO ELECTRIC CO., INC., 109 N. Fifth Ave., Saddle Brook, NJ 07663, 201/843-5800, Fax: 201/843-4568, www.thermo-electric-direct.com

THERMO GASTECH, 8407 Central Ave., Newark, CA 94560, 510/745-8700, Fax: 510/794-6201, www.thermogastech.com

THERMO MEASURETECH, 2555 North IH-35, Round Rock, TX 78664, 800/736-0801, Fax: 512/388-9200, www.thermomt.com

THERMO NICOLET CORP., 5225 Verona Rd., Madison, WI 53711, 608/276-6100, Fax: 608/273-5046, www.thermonicolet.com

THERMO ONIX, 1201 N. Velasco, Angleton, TX 77515, 979/849-2344, Fax: 979/849-2166, www.thermoonix.com

THERMO POLYSONICS, 10335 Landsbury Dr. #300, Houston, TX 77099, 281/879-3700, Fax: 281/498-7721, www.thermopolysonics.com

THERMO RAMSEY, 501-90th Ave. NW, Minneapolis, MN 55433, 763/783-2500, Fax: 763/780-2315, www.thermoramsey.com

THERMO WESTRONICS, 22001 North Park Dr., Suite 100, Kingwood, TX 77339, 281/348-1800, Fax: 281/348-1288, www.thermowestronics.com

THIELSCH ENGINEERING, 195 Frances Ave., Cranston, RI 02910, 401/467-6454, Fax: 401/467-6454, www.thielsch.com

TIPS, INC., 2402 Williams Dr., Georgetown, TX 78628, 512/863-3653, Fax: 512/863-5392, www.tipswweb.com

TISCOR, 12250 Parkway Centre Dr., Poway, CA 92064, 800/227-6379, Fax: 858/513-8497, www.tiscor.com

TOL-O-MATIC, 3800 CR 116, Hamel, MN 55340, 612/478-4322, Fax: 612/478-8080, www.tolomatic.com

TOPWORX, 3300 Fern Valley Rd., Louisville, KY 40213, 502/969-8000, Fax: 502/964-5911, www.topworx.com

TOSHIBA INTERNATIONAL CORP., 13131 West Little York Rd., Houston, TX 77041, 713/466-0277, Fax: 713/896-5225, www.tic.toshiba.com

TOTAL CONTROL PRODUCTS, INC., 2001 N. Janice Ave., Melrose Park, IL 60160, 708/345-5500, Fax: 708/345-5670, www.total-control.com

TOUCH CONTROLS, INC., 520 Industrial Way, Fallbrook, CA 92028, 800/848-4385, Fax: 760/723-7910, www.touchcontrols.com

TRACEWELL SYSTEMS, INC., 567 Enterprise Dr., Columbus, OH 43081, 800/848-4525, Fax: 614/846-4450, www.tracewellsystems.com

TRANSCAT, 10 Vantage Pt. Dr., Rochester, NY 14624, 800/828-1470, Fax: 800/395-0543, www.transcat.com

TRANSDUCER TECHNIQUES, 43178 Business Park Dr., Temecula, CA 92590, 909/676-3965, Fax: 909/676-1200, www.ttloadcells.com

TRANSICOIL, 43178 Business Park Dr., Norristown, PA 19403, 800/323-7115, Fax: 616/539-3400, www.transicoil.com

TRANSMATION, 35 Vantage Point Dr., Rochester, NY 14624, 716/349-3520, Fax: 716/349-3510, www.transmation.com

TRANSTECTOR SYSTEMS, 10701 Airport Dr., Hayden Lake, ID 83835, 208/762-6055, Fax: 208/762-6080, www.transtector.com

TRANSYSOFT, INC., 11 Merrill Dr., Suite C5, Hampton, NH 03842, 603/929-6330, Fax: 603/929-6331, www.transyssoft.com

TRENDVIEW RECORDERS, PO Box 141489, Austin, TX 78714-1489, 512/927-7800, Fax: 512/834-4333, www.trendview.com

TRENTON TECHNOLOGY, INC., 2350 Centennial Dr., Gainesville, GA 30504, 800/875-6031, Fax: 770/287-3150, www.trentonprocessors.com

TRICONEX CORP., 15345 Barranca Pkwy., Irvine, CA 92618, 949/885-0714, Fax: 949/753-9101, www.triconex.com

TRIHEDRAL ENGINEERING LTD, 1160 Bedford Hwy., Suite 400, Bedford, Nova Scotia, B4A ICI Canada, 902/835-1575, Fax: 902/835-0369, www.trihedral.com

TRIPLETT CORP., One Triplett Dr., Bluffton, OH 45817, 800/TRIPLETT, Fax: 419/358-7956, www.triplett.com

TRW SENSORS & COMPONENTS SCHAEVITZ SENSORS, 1000 Lucas Way, Hampton, VA 23666, 757/766-1500, Fax: 757/766-4297, www.schaevitz.com

TSI, INC., 500 Cardigan Rd., Shoreview, MN 55126, 651/490-2711, Fax: 651/490-2874, www.tsi.com

TTI, 8 Leroy Rd., Williston, VT 05495, 800/235-8367, Fax: 802/863-1193, www.ttiglobal.com

TURCK, INC., 3000 Campus Dr., Minneapolis, MN 55441, 800/544-7769, Fax: 612/553-0708, www.turck.com

TUTHILL TRANSFER SYSTEMS, 8825 Aviation Dr., Fort Wayne, IN 46809, 919/460-6000, Fax: 919/460-7595, www.tuthill.com

TWO TECHNOLOGIES, INC., 419 Sargon Way, Horsham, PA 19044, 215/441-5305, Fax: 215/441-0423, www.2t.com

TYCO VALVES & CONTROLS LP, 9700 W. Gulf Rd., Houston, TX 77040, 713/466-1176, www.tycovalves.com

U

UE SYSTEMS, INC., 14 Hayes St., Elmsford, NY 10523, 800/223-1325, Fax: 914/347-2181, www.uesystems.com

ULTRAFLO CORP., a Subsidiary of Bray Intl., Inc., PO Box 423, St. Genevieve, MO 63670, 573/883-8881, Fax: 573/883-8882

ULTRAMAX CORP., 110 Boggs Ln., Suite 255, Cincinnati, OH 45246, 513/771-8629, Fax: 513/771-7185, umaxcorp.com

UNITED ELECTRIC CONTROLS, 180 Dexter Ave., Watertown, MA 02471, 617/926-1000, Fax: 617/926-4354, www.ueonline.com

UNIVERSAL DYNAMICS TECHNOLOGIES, INC., 100-13700 International Place, Richmond, British Columbia, V6V 2X8 Canada, 604/214-3456, Fax: 604/214-3457, www.brainwave.com

USDATA, 2435 N. Central Expwy., Richardson, TX 75080, 972/497-0233, Fax: 972/669-9556, www.usdata.com

USFILTER/WALLACE & TIERNAN PRODUCTS, 1901 W. Garden Rd., Vineland, NJ 08360, 856/507-9000, Fax: 856/507-4125, www.wallaceandtiernan.usfilter.com

V

VAISALA, INC., 100 Commerce Way, Woburn, MA 01801, 781/933-4500, Fax: 781/933-8029, www.vaisala-usa.com

VALCOR SCIENTIFIC, 2 Lawrence Rd., Springfield, NJ 07081, 973/467-8400, Fax: 973/467-9592, www.valcor.com

VALIDYNE ENGINEERING, 8626 Wilbur Ave., Northridge, CA 91324, 818/886-2057, Fax: 818/886-6512, www.validyne.com

VAS ENGINEERING, 4750 Viewridge Ave., San Diego, CA 92123, 619/569-1601, Fax: 619/569-8474, www.doric-vas.com

VERANO, 310 E. Caribbean Dr., Sunnyvale, CA 94089, 408/541-7658, Fax: 408/541-7601, www.verano.com

VERIS, INC., 6315 Monarch Park Place, Niwot, CO 80503, 303/652-8550, Fax: 303/652-8552, www.veris-inc.com

VERSALOGIC CORP., 3888 Stewart Rd., Eugene, OR 97402, 541/485-8575, Fax: 541/485-5712, www.versalogic.com

VERTACROSS, INC., PO Box 14526, Research Triangle Park, NC 27709, 866/248-4088, Fax: 919/248-4099, www.vertacross.com

VIA DEVELOPMENT CORP., PO Box 3268, Marion, IN 46953, 765/677-3232, Fax: 765/674-3964, www.viadevelopment.com

VIATRAN CORP., 300 Industrial Dr., Grand Island, NY 14072, 716/773-1700, Fax: 716/773-2488, www.viatran.com

VICOR CORP., 25 Frontage Rd., Andover, MA 01810, 978/470/2900, Fax: 978/475-6715, www.vicr.com

VIKING PUMP, INC., 406 State St., Cedar Falls, IA 50613, 319/266-1741, Fax: 319/273-8157, www.vikingpump.com

VISIONEX, 430 Tenth St. NW, #N205, Atlanta, GA 30318, 404/873-9775 x.14, Fax: 404/873-0535, www.mindspring.com/~visionex

VISTA CONTROLS SYSTEMS, INC., 176 Central Park Square, Los Alamos, NM 87544, 505/662-2484, Fax: 505/662-3956, www.vista-control.com

VISUAL SOLUTIONS, INC., 487 Groton Rd., Westford, MA 01886, 978/392-0100, Fax: 978/692-3102, www.vissim.com

VMIC, 12090 S. Memorial Pkwy., Huntsville, AL 35803, 256/880-0444, Fax: 256/882-0859, www.vmic.com

VMR SOFTWARE, PO Box 1463, Edmonds, WA 98020, 425/774-2483, Fax: 206/727-8650, www.vmrsoftware.com

VORNE INDUSTRIES, INC., 1445 Industrial Dr., Itasca, IL 60143, 888/DISPLAYS, Fax: 630/875-3609, www.vorne.com

VYNCKIER ENCLOSURE SYSTEMS LTD, 249 McCarty Dr., Houston, TX 77029, 713/374-7850, Fax: 713/672-8632, www.enclosuresonline.com

W

WAGO CORP., N120 W19129 Freistadt Rd., Germantown, WI 53022, 800/DINRAIL, Fax: 262/255-3232, www.wago.com

WARNER ELECTRIC MOTORS & CONTROLS, 383 Middle St., Bristol, CT 06010, 860/585-4500, Fax: 860/582-3784, www.warnernet.com/sev_main.html

WATLOW, 12001 Lackland Rd., St. Louis, MO 63146, 800/4-WATLOW, Fax: 314/878-6814, www.watlow.com

WEED INSTRUMENT, 707 Jeffrey Way, Round Rock, TX 78664, 512/434-2844, Fax: 512/434-2851, www.weedinstrument.com

WEG MOTORS AND DRIVES, 1327 Northbrook Pkwy., Suwanee, GA 30024, 770/338-5656, Fax: 770/338-1632, www.webelectric.com

WEIDMULLER, INC., 821 Southlake Blvd., Richmond, VA 23236, 804/379-6027, Fax: 804/379-2593, www.weidmuller.com

WEIGH-TRONIX, 1000 Armstrong Dr., Fairmont, MN 56031, 507/238-8253, Fax: 507/238-8258, www.weigh-tronix.com

WESCO DISTRIBUTION, INC., 1600 N. Sixth St., Milwaukee, WI 53212, 414/264-6400, www.wescodist.com

WESTERN RESERVE CONTROLS, 1485 Exeter Dr., Akron, OH 44306, 330/733-6662, Fax: 330/733-6663, www.wrcakron.com

WESTINGHOUSE PROCESS CONTROL, 200 Beta Dr., Pittsburgh, PA 15238, 412/963-2727, Fax: 412/963-3644, www.westinghousepc.com

WESTLOCK CONTROLS CORP., 280 Midland Ave., Saddle Brook, NJ 07663, 201/794-7650, Fax: 201/794-0913

WHESSOE VAREC, 10800 Valley View St., Cypress, CA 90630, 714/761-1300, Fax: 714/952-2701, www.whessoevarec.com

WIELAND ELECTRIC, 49 International Rd., Burgaw, NC 28425, 910/259-5050, Fax: 910/259-3691, www.wielandinc.com

WIKA INSTRUMENT CORP., 1000 Wiegand Blvd., Lawrenceville, GA 30043, 770/338-5229, Fax: 770/277-2668, www.wika.com

WILKERSON INSTRUMENT CO., INC., 2915 Parkway St., Lakeland, FL 33811, 800/234-1343, Fax: 863/644-5318, www.wici.com

WINTERS INSTRUMENTS, 600 Ensminger Rd., Buffalo, NY 14150, 716/874-8700, Fax: 716/874-8700, www.winters.ca

WOLFRAM RESEARCH, INC., 100 Trade Center Dr., Champaign, IL 61820, 217/398-0700, www.wolfram.com

WONDERWARE, 100 Technology Dr., Irvine, CA 92618, 949/453-6568, Fax: 949/450-5098, www.wonderware.com

WOODHEAD CONNECTIVITY, 3411 Woodhead Dr., Northbrook, IL 60062, 847/272-7990, Fax: 847/272-8133, www.connector.com

WORCESTER CONTROLS CORP., 33 Lock Drive, Marlborough, MA 01752, 508/481-4800, Fax: 508/481-4454, www.worcestercc.com

X

XENTEK POWER SYSTEMS, 1770 La Costa Meadows Dr., San Marcos, CA 92069, 760/471-4001, Fax: 760/471-4021, www.xentek.com

XYCOM AUTOMATION, INC., 750 N. Maple Rd., Saline, MI 48176, 734/429-4971, Fax: 734/429-3087, www.xycom.com

XYMOX TECHNOLOGIES, INC., 9099 W. Dean Rd., Milwaukee, WI 53224, 414/362-9000, Fax: 414/362-9090, www.xymoxtech.com

XYNTEK, INC., 301 Oxford Valley Rd., Bldg. 1402A, Yardley, PA 19067, 215/493-7091, Fax: 215/493-7094, www.xyntekinc.com

Y

Y2 SYSTEMS, 3101 Pollok Dr., Conroe, TX 77303, 936/788-5526, Fax: 936/788-5698, www.y2systems.com

YASKAWA ELECTRIC AMERICA, INC., 2121 Norman Dr. S., Waukegan, IL 60085, 800/YASKAWA, Fax: 847/887-7310, www.yaskawa.com

YCV, A YAMATAKE CO., 11225 North 28th Dr., Suite A106, Phoenix, AZ 85029, 602/548-1800, Fax: 602/548-1127, www.ycv.com

YOKOGAWA CORP. OF AMERICA, 2 Dart Road, Newnan, GA 30265, 770/258-2552, www.yca.com

Z

Z-WORLD, 2900 Spafford St., Davis, CA 95616, 530/757-3737, www.zworld.com

ZELLWEGER ANALYTICS, INC., 405 Barclay Blvd., Lincolnshire, IL 60069, 847/955-8200, Fax: 847/955-8208, www.zelana.com

ZENITH PRODUCTS DIV., PARKER HANNIFIN CORP., 5910 Elww Buchanan Dr., Sanford, NC 27330, 919/775-4600, Fax: 919/774-5952, www.zenithpumps.com

ZONEWORX, 40925 County Center Dr., Suite 200, Temecula, CA 92591, 909/296-1226, Fax: 909/506-9309, www.zoneworx.com

ZONTEC INC., 1389 Kemper Meadow Dr., Cincinnati, OH 45240, 513/648-0088, Fax: 513/648-9007, www.zontec-spc.com

INDEX

A

B

C

D

E

F

I

P

Q

R

S

T

X

Z